XX International Grassland Congr
Offered papers

XX International Grassland Congress: Offered papers

edited by:
F.P. O'Mara
R.J. Wilkins
L. 't Mannetje
D.K. Lovett
P.A.M. Rogers
T.M. Boland

Wageningen Academic
P u b l i s h e r s

Subject headings:
Agriculture
Environment
Land use

ISBN 9076998817

First published, 2005

Wageningen Academic Publishers
The Netherlands, 2005

The XX International Grassland Congress took place in Ireland and the UK in June-July 2005. The main congress took place in Dublin from 26 June to 2 July and was followed by post congress satellite workshops in Aberystwyth, Belfast, Cork, Glasgow and Oxford. The meeting was hosted by the Irish Grassland Association and the British Grassland Society and was organised by the bodies represented on the Local Organising Committee.

Organising Committee of the XX International Grassland Congress

Mr J. Flanagan, President, Teagasc
Dr. F.P. O'Mara, Secretary, University College Dublin
Prof. R.J. Wilkins, British Grassland Society
Dr. P McFeely, Irish Grassland Association
Mr. N.P. McGill, Department of Agriculture
    and Food, Ireland
Prof. M.P Boland, University College Dublin
Prof. J.H. Roche, University College Dublin
Dr. M. Camlin, Department of Agriculture and Rural
    Development of Northern Ireland
Mrs. J. Crichton (deceased), British Grassland Society
Dr. T.F. Nolan, Teagasc

Scientific Committee of the XX International Grassland Congress

Prof R.J. Wilkins, Chairperson
Mr. D.A. Davies
Dr. P. French
Prof. M.B. Jones
Dr. A.S. Laidlaw
Mrs. C.A. Marriott
Dr. D. McGilloway
Dr. P. O'Kiely

Session moderators (in addition to Scientific Committee members)

R. Cook	J. Frame	M.H. Humphreys	D.C. Patterson
C.P. Ferris	J.A. Milne	S.C. Jarvis	

Reviewers

M.T. Abberton	C.J. Duller	F.W. Kirkham	L. Philipps
R.E. Agnew	B. Evans	G.P.F. Lane	G. Purvis
J.C. Alliston	R.J. Fallon	A.H. Marshall	M. Rath
P.D. Barrett	J. Feehan	J. McAdam	D.J. Roberts
A.D. Black	C.P. Ferris	M. McGee	S.M. Rutter
A.J. Brereton	P.D. Forristal	J.A. Milne	R. Schulte
C. Campbell	M. Fothergill	J.M. Moorby	F. Taube
D.R. Chadwick	M.D. Fraser	F.J. Mulligan	G.E.D.Tiley
R.O. Clements	P. French	J.J. Murphy	C.R.S. Topp
R.P. Collins	T.J. Gilliland	P.J. Murray	H. Tunney
J. Connolly	D.I. Givens	T.F. Nolan	M. Wallace
T.F. Crosby	D.J. Hatch	B.J. O'Brien	C.A. Watson
D.R. Davies	M.C. Hickey	M.A. O'Donovan	J.R. Weddell
O. Davies	D.H. Hides	N.W. Offer	J. Whelan
L. Dawson	A. Hopkins	P. O'Kiely	R.J. Wilkins
L. Delaby	J. Humphreys	R.J. Orr	P.W. Wilkins
R.J. Dewhurst	S.C. Jarvis	R.J. Parkinson	J.D. Wood
P. Dillon	D.T. Johnston	S. Peel	D. Younie
M.J. Drennan	A.H. Kingston-Smith	J.L. Peyraud	

Members of the Continuing Committee of the International Grassland Congress (2001–2005)

Region 1 (Canada, USA)	Dr. Vivien Gore Allen (Chair)	Texas Tech University, USA
Region 2 (Central America)	Dr. Luis Ramirez Aviles	Universidad Autonoma de Yucatan, Mexico
Region 3 (South America)	Dr. Raul R. Vera	Pontificia Universida del Chile
Region 4 (South East Asia)	Dr. Chaisang Phaikaew	Department of Livestock Development, Thailand
Region 5 (Australia, New Zealand)	Dr. Gavin W. Sheath	AgResearch Limited, New Zealand
Region 6 (East Asia)	Dr. Masakazu Goto	Mie University, Japan
Region 7 (Middle East)	Dr. Hossein Arzani	University of Tehran, Iran
Region 8 (Mediterranean, Near East)	Dr. Maria Ermelinda Vaz Lourenço	Universidade de Évora, Portugal
Region 9 (Europe excluding Regions 8 and 10)	Professeur Alain Peeters	Université Catholique de Louvain, Belgium
Region 10 (Northern Eurasia)	Dr. Geza Nagy	Debrecen University, Hungary
Region 11 (Africa excluding Regions 7 and 8)	Dr. Apollo Bwonya Orodho	Kenya Agricultural Research Institute
Representative of previous host country	Dr. Sila Carneiro da Silva	University of Sao Paulo, Brazil

Supporters of the XX International Grassland Congress, Dublin

Department of Agriculture and Food, Ireland
Teagasc, Ireland
Bord Bia, Ireland
Allied Irish Banks, plc
FBD Trust, Ireland
Environmental Protection Agency, Ireland
University College Dublin, Ireland
USDA-Agricultural Research Service
Food and Agriculture Organization of the United Nations
Department for Environment Food and Rural Affairs, UK
USDA-National Resources Conservation Service
USDA-Cooperative State Research, Education, and Extension Service
US Bureau of Land Management
US Environmental Protection Agency

Alltech (Ireland) Ltd
ACCBank, Ireland
Irish Farmers Journal
Greenvale Animal Feeds
Biotechnology and Biological Sciences Research Council, UK
Irish Farmers Association
British Grassland Society
Irish Grassland Association
American Society of Agronomy
Crop Science Society of America
Soil Science Society of America
Glanbia Plc
K & S Kali, Germany
Stapledon Memorial Trust, UK
Grassland Fertilizers Ltd, Ireland
Grazing Lands Conservation Initiative, USA

Organisations that assisted the attendance of delegates from developing countries

Development Cooperation Ireland
Swiss Agency for Development and Cooperation
CTA Technical Centre for Agricultural and Rural Cooperation ACP-EU
Joint FAO/IAEA Division of Nuclear Techniques in Food and Agriculture
Australian Centre for International Agricultural Research
Syngenta Foundation for Sustainable Agriculture, Switzerland
Japanese Society of Grassland Science
Ministry for Foreign Affairs of Finland

An appreciation of Jan Crichton

Many delegates will have had contact with Jan Crichton, a member of the IGC Organising Committee. Jan died on 1 May 2005 after a short illness. Anyone who met Jan will recall her enthusiasm, vitality and commitment to anything she was involved with. Her untimely death is a grievous loss to her husband Charlie, her family and the whole grassland community.

Jan worked on the scientific staff at the Grassland Research Institute, Hurley, from 1977-1982, before joining the British Grassland Society to provide support for Jim Corrall and then Mike Helps in their roles as Secretary of BGS. On Mike's retirement in 2001, it was an easy decision for the Society to offer the newly-created post of Chief Executive Officer to Jan, a post she filled with distinction.

Jan's organisational ability and inter-personal skills made her an ace in making conferences and events a success. One of her ambitions was to be involved in organising an International Grassland Congress. Mike Helps suggested that Jan took the lead from the BGS office in preparations for this Congress. She was central to the first meeting of the UK Planning Group in 1998 and contributed much to evolving the UK contribution to the Congress. Her infectious enthusiasm was very evident when she was one of our ambassadors at the XIX Congress in Brazil and she subsequently joined the Organising Committee for this Congress. Amongst her specific responsibilities were the planning of the Pre-Congress tour in England and Wales and the organisation of the Producer Programme on behalf of BGS, the Irish Grassland Association and the Ulster Grassland Society. As a tribute to Jan, the Societies have dedicated this as 'The Jan Crichton Producer Programme'. Jan has always striven to improve the effectiveness of communication between farmers and scientists, which is exactly the object of this Programme.

The grassland world has lost a wonderful person totally dedicated to the objectives of Grassland Societies and IGC. She played a major part in the evolution of this Congress and would have wanted it to be a great success.

Foreword

This book contains a compilation of offered papers presented at the main congress of the XX International Grassland Congress held in University College Dublin, Ireland from 26 June to 1 July, 2005. It is complemented by six other books arising from the XX IGC as listed on the back cover: the book of invited papers from the main congress and five books containing the proceedings of five satellite workshops held immediately after the main congress at locations in the UK and Ireland (Aberystwyth, Belfast, Cork, Glasgow and Oxford). The workshops were designed to facilitate more in-depth presentations and discussions on more specialised topics of worldwide significance.

The main congress brought together scientists from many disciplines, policy makers, consultants and producers involved directly in grass production and utilisation, as well as people in associated industries. They discussed issues around the theme of the congress, *Grasslands – a Global Resource*. The congress programme was drawn up by the Scientific Committee with substantial international consultation, and it was organised around three main thematic areas:
- Efficient Production from Grassland
- Grassland and the Environment
- Delivering the Benefits from Grassland

Within each thematic area, there were from seven to thirteen sessions, with Efficient Production from Grassland having the most sessions, and attracting the most papers. There were particularly strong sessions on grass and forage plant improvement, forage quality for animal nutrition, overcoming seasonality of production, animal-plant relations, grass and forage agronomy and animal production. A landmark session was held on advances in sown tropical legumes, which outlined many success stories from different regions.

As has been the trend in recent congresses, there were a large number of papers offered on topics concerned with the environmental issues related to grassland. They included issues related to climate change, greenhouse gases and carbon sequestration, biodiversity, water resources, soil quality and nutrients, and the multifunctional use of grasslands. Overall these papers reflected the many diverse roles and functions of grassland throughout the world.

In the third thematic area of the congress, tools for grassland management and decision support systems for grassland had large numbers of offered papers. Other strong sessions were concerned with adoption of new technology, participatory and on-farm research, and improved livelihoods from grassland. A unique session examined the role of the IGC and grassland societies in technology interaction and influencing policy, with papers contributed by over 20 societies/ organisations involved in grasslands with producer, professional and scientific members.

Over 1000 delegates from 80 countries attended the XX IGC in Dublin. The 829 papers in this book came through a thorough review and editorial process.

There have now been twenty International Grassland Congresses, with the first taking place in Leipzig in 1927. We felt it fitting to document the progress of the meeting over the years. We are much indebted to Professor Ross Humphreys for agreeing to write a brief history of the IGC. This very interesting and useful article is included in this book and charts the evolution of the congress, and themes discussed over the years. It will also serve as a reference source for information about the IGC.

The XIX IGC was held in São Pedro, São Paulo, Brazil, in 2001. The proceedings were published at the time of the meeting and, thus there was no opportunity to include the minutes of the Opening Business Meeting and the Closing Business Session. They are, therefore, included in this book.

Frank O'Mara   Roger Wilkins

Table of contents

Offered papers **11**

Offered papers 13

Section 3: Improving quality of products from grassland   **177**

Section 5: Forage quality for animal nutrition 217

Section 8: Grassland management 349

Section 9: Integrated production systems 375

Section 10: Industrial products from grassland 401

Section 11: Grass and forage agronomy

Section 12: Overcoming seasonality of production   **445**

Section 13: Animal-plant relations 485

Section 16: Carbon sequestration   585

A brief history of the International Grassland Congress

L.R. Humphreys

School of Land and Food Sciences, University of Queensland, Brisbane, 4072, Australia, Email: l.humphreys@uq.edu.au

Key points

1. Nineteen International Grassland Congresses met over the period 1927-2001 in every continent except Africa. Scientists from North America, Western Europe and Australia and New Zealand dominated proceedings.
2. Analysis of 6 representative Congresses indicates a considerable homeostasis of disciplinary content. The plant genetic base for grassland improvement, plant physiology, plant ecology and soil science contributed 46 to 57 per cent of papers, which were mainly complemented by studies of grazing management and animal production from forage.
3. Environmental science, systems theory, socio-economic perspectives and technology transfer emerged with more force in recent Congresses.

Keywords: International Grassland Congress, history, scientific disciplines, plant and animal production, environment

Locations and attendance

The dates and locations of the Congresses are listed in Appendix Table 1. The International Grassland Congress first met in Germany from 20-31 May 1927. The principal participants were 16 scientists from Austria, Denmark, Finland, Germany, Norway, Sweden and Switzerland, who assembled in Bremen and made a study tour through north-west Germany, visiting Emden, Berlin and Dortmund before taking the train to Leipzig. Here there were two days of scientific discussion at the Zoo, revisited subsequently as the site of the 50th Anniversary XIII Congress in 1977. The Congress under the presidency of Prof. A. Falke of Leipzig had a further study tour through grassland production sites in Saxony before dissolving at Dresden.

The second Congress, which met in 1930 in Sweden and Denmark under the presidency of Dr A. Elofson of Uppsala was enlarged to 58 participants from 13 countries (including Canada). The third Congress in Switzerland in 1933 with Prof. A. Volkart of Zurich as President had scientists from Turkey and South Africa present, but it was not until the IV Congress in 1937 at Aberystywyth, United Kingdom, that the meeting could claim a global constituency. There were some 365 participants from 37 countries; all 11 regions of the world as defined by the 1977 International Grassland Congress Constitution were represented with the exception of the Middle East. The leadership of R.G. Stapledon of the Welsh Plant Breeding Station was pre-eminent. At this meeting it was agreed that the funds of the International Grassland Congress Association be banked in Germany and that the next Congress be held in the Netherlands in 1940. The intervention of the Second World War delayed the V Congress until 1949 and the funds of the Association were not recovered.

The VI International Grassland Congress, held at State College, Pennsylvania, USA in 1952, built on the European foundations of the movement to enlarge its scientific content and global representation, and accorded a new maturity. The world regions with an established history of grassland research (North America, Western Europe, Australia and New Zealand) accounted for 75% of the 271 scientific papers presented, and the participation of other regions increased to 25%.

The location of subsequent Congresses usually alternated between continents: America (4), Oceania (3), Asia (1), Europe (5) but no Congress has been held in Africa. The VII Congress in 1956 at Palmerston North, New Zealand, had a restricted representation but the VIII Congress at Reading, UK (591 participants from 53 countries), indicated the continued strength of grassland science. At this Congress an Executive Committee representative of eight regions of the world was elected with a rotating membership so that members would serve for a period covering the two intervals between three Congresses. This Committee was charged with providing a continuing organisation which would advise future host country committees. The full membership of the Congress voted for the IX Congress venue of Brazil, and this was held in 1965 at São Paulo, the first Congress to be located in a tropical country. In Brazil it was decided that the venue of the XI Congress would be Australia (118 votes, Canada 63 votes, USSR 63 votes). The X Congress moved closer to the Arctic Circle in 1966 at Helsinki, Finland, where USSR (128 votes) was selected over Canada (108 votes) for the XII Congress site. The designation of Executive Committee was altered to that of a "Continuing Committee", which was *inter alia* given the responsibility "to select the host country for the forthcoming Congress and to announce the name of that host country at the intermediate Congress". The XI Congress was mounted in 1970 at Surfers' Paradise, Queensland, Australia.

The question of the venue of the XIII Congress aroused controversy at the XII Congress in 1974 in Moscow. The Continuing Committee, empowered by the Constitution adopted in 1966 at Helsinki, determined the Republic of Ireland as the venue. This decision was challenged by the Host Committee in Moscow who put the question to a free vote of full Congress members, of whom 64% were from the northern Eurasia region. This resulted in a decision for the XIII Congress to be held in 1977 at Leipzig, German Democratic Republic. (It is reported that at this meeting a USSR official on the platform turned to R.J. Bula, the North American proxy delegate on the Continuing Committee, when the vote was announced and asked "So how do you enjoy democracy?"). A further resolution led to the promulgation of a new constitution which was adopted at the Leipzig Congress and which reaffirmed the power of the Continuing Committee to determine future venues, subject to one country-one vote procedure at the Congress in the event of a disagreement in the Continuing Committee. The Continuing Committee was enlarged to representatives of 11 regions and an additional representative from the previous host country.

S.C. Pandeya, the outgoing chairman of the Continuing Committee, had expected to invite the XIV Congress to India, but the defeat of the Gandhi government by Mr Desai put paid to this proposition and no invitation from other countries was forthcoming. Canada had previously sought to host congresses but 1977 was not a propitious time to find support. The American Forage and Grassland Council, led by R.F. Barnes and J.E. Baylor, ventured in faith and the XIV Congress at Lexington, Kentucky, USA, resulted in 1981. The XV Congress in 1985 was the first Congress to be held in Asia and at Kyoto, Japan, a large delegation of scientists from China attended for the first time.

Previous Congresses in Europe had been held in cold northern latitudes and the XVI Congress in 1989 at Nice, France, was the first Mediterranean location and attracted a higher proportion of participants (13%) from the designated Mediterranean region countries, whilst France provided 24% of the attendance. The XVII Congress in 1993 was unusual in that it arose from the joint invitation of New Zealand and Australia, and its locations in Palmerston North, Hamilton, and Christchurch, New Zealand, and Rockhampton, Queensland, provided a range of ecological conditions including both temperate and tropical pastures. This was the largest and most representative Congress with 1200 delegates from 82 countries. The scientific contribution and leadership of indigenous participants from the developing countries increased substantially at the XVI and XVII Congresses; in the early Congresses their rather meagre representation often arose from expatriate scientists from developed countries. The invitation of Canada to host the XVIII Congress in 1997 was accepted by the 1993 Continuing Committee, and this led to a similarly large and representative Congress. A resolution was adopted at this Congress to continue to explore the possibilities for closer collaboration with the International Rangeland Congress (IRC). The XIX Congress took place in 2001 at São Pedro, São Paulo, Brazil. An up-dated and revised Constitution was adopted at that Congress. The XX Congress is being held in 2005 in the Republic of Ireland and the United Kingdom.

Table 1 Regional participation (per cent) in International Grassland Congresses

Region	Period					
	1927-1937	1949-1952	1956-1966	1970-1981	1985-1993	1997-2001
North America	4	27	19	21	15	19
Central America	<1	3	2	2	1	3
South America	<1	2	20	2	4	25
Southern Asia	<1	2	<1	1	2	4
Oceania	1	6	15	21	22	15
East Asia	<1	<1	<1	3	23	9
Middle East	<1	<1	<1	<1	1	3
Mediterranean	1	3	1	2	6	5
Western Europe	87	54	36	27	20	11
Northern Eurasia	3	0	4	18	2	3
Africa	2	3	2	1	3	5

The regional participation (Table 1) is estimated for non-orthogonal periods designated to coincide with the Congresses chosen for later discussion of the evolution of thematic content. The naming of the regions in Table 1 has been modified to reflect current understanding. Changes in regional representation partly reflect the location of Congresses in each period but there has never been strong participation from the countries of Central

America, Middle East and Africa. More detail is available in Humphreys (1997). Office bearers of the Congresses are listed in Appendix Table 1 and members of the Congress committees are listed in Appendix Tables 2 and 3.

The International Rangeland Congress

The management of rangelands, focused on natural pastures in the arid and semiarid zones, has always been a topic at International Grassland Congresses and has received varying attention. However, some scientists working in this general area considered there was a need for a separate international meeting directed to developing a better science of the manipulation, improvement and utilisation of rangelands. This was exacerbated in the USA by the dichotomy of effort between members of the Society of Range Management and of the American Forage and Grassland Council, whose primary interests were in sown grasslands. The decision to form a separate organisation which would mount International Rangeland Congresses was further stimulated by the decision at the XII International Grassland Congress in 1974 to reject the Continuing Committee's acceptance of the Republic of Ireland as the venue for the XIII Congress and to retain the Congress in what was perceived as the Eastern Bloc venue of the German Democratic Republic.

The first International Rangeland Congress was held in 1978 at Denver, Colorado, USA, and was succeeded in 1984 by the second Congress at Adelaide, Australia. This was attended by 499 participants from 42 countries; of these 79% came from Oceania, North America and Western Europe. The third Congress in 1988 met in New Delhi, India, and the fourth Congress was held in 1991 at Montpellier, France, whilst the fifth Congress returned to the USA in Utah. Further Congresses were held in Townsville, Australia, and Durban, South Africa. Reciprocal representation on the two Congress Continuing Committees was arranged from 1981, and plans are being made to hold a joint IGC/IRC Congress at Hohhot, Inner Mongolia, China, in 2008, which would integrate the thrust of the two movements.

Changes in the balance of themes

Overview

The changing themes which have occupied scientists at International Grassland Congresses were analysed by identifying 110 topics grouped within 10 main themes, and additionally including four miscellaneous themes: synoptic papers, biometrics, agricultural engineering and animal production not specifically related to grassland improvement (Table 2). Papers presented at Congresses were allocated to each sub-theme according to its major content; this was not necessarily the theme of the Congress session to which it may have been allocated for convenience.

The content of six Congresses that were held at a mean interval of 11 years from 1937 to 2001 was studied; 1937 was chosen as the first Congress that could claim a good international status. All six Congresses accepted voluntary papers and were held in regions with a history of research in grassland science.

This analysis revealed a considerable homeostasis of disciplinary content over the 64 years. The science of grassland improvement has relied first on an interest in its plant genetic base, and plant genetics, plant physiology, plant ecology and soil science contributed 46 to 57 per cent of the subject matter at all six Congresses. Animal nutrition and systems of animal production arising from study of the animal-plant-soil interface were the other key preoccupations of grassland scientists, whilst environmental science, systems theory and socio-economic perspectives emerged with more force in recent Congresses.

The 1937-1952 period

The general theme of the first subject in Table 2, which was designated as styles of grassland development, included the papers with general or integrative themes which were insufficiently specific to be allocated elsewhere and whose main interest was regional or local. These constituted 19% of papers in 1937 and mainly dealt with humid or sub-humid temperate grasslands; in 1952 this category decreased to 12% with a predominance of non-specific tropical papers. The balance of content focused on intensity of land use, tree crops with pastures, leys and turf.

The plant genetic basis for grassland development in 1937 was oriented to evaluation of and selection within improved species; in 1952 there were more papers on hybridisation, induced polyploidy, disease resistance and certification of seed for varietal purity. Edaphic constraints on grassland development in 1952 were defined less in terms of general fertiliser needs and responses and more in terms of specific nutrients, including sulphur, and soil toxicities; soil conservation and watershed management became significant emphases. More interest in the physiology of flowering and seed production emerged, whilst in plant succession, the control of weed and shrub encroachment and the production of inventories of grassland resources were of significant interest in grassland ecology.

In the 1952 Proceedings studies of selective grazing and foraging strategy, stocking rate and forage allowance, and the methodology of grazing experiments appeared. More sophisticated approaches to nutritive value of forage were evident in the attention to energy value, digestibility and intake, mineral content, and anti-quality factors. Continuity of forage supply was addressed through irrigation and techniques of crop processing, which were especially dependent upon innovations in agricultural engineering. Characterisation of climate emerged as a topic, as did the transfer of technology to farmers.

The 1966 Congress

The trend to fewer general papers of regional interest continued, especially at this Finnish venue with respect to tropical grasslands. Papers dealing with specialist techniques of plant breeding such as induced polyploidy were again presented. The intensive use of fertiliser N was a new emphasis and there were 19 papers on this topic. Plant physiology was accorded greater importance through papers on growth analysis, tillering, plant response to defoliation, and the role of carbohydrate "reserves", but there were fewer papers on plant ecology.

Stocking method, stocking rate and forage allowance were further addressed, together with the spatial transfer of nutrients under grazing. Mixed grazing and the innovative choice of animal species were canvassed. Nutritive value received increased attention relative to 1952, especially in relation to forage intake, digestibility and anti-quality factors. Animal responses to systems of forage conservation were described and systems modelling in grassland research appeared as a topic (Table 2).

The 1981 Congress

This Congress was marked by considerable advances in tropical pasture science, and 133 of the 480 papers presented bore directly on grassland development in the tropics and subtropics, mainly in specialist areas. Styles of grassland development embraced interest in the intensity of land use, integration of land classes, deforestation and woodland management, long-term trends in production, the use of shrub legumes and intercropping.

Wide approaches to the improvement of the plant genetic base were enunciated which displayed increased emphasis on species evaluation, the conservation of germplasm, and the identification of elite material, whilst *in vitro* embryo culture signalled the nascency of molecular biology.

Table 2 Themes represented at International Grassland Congresses (per cent papers with main theme)

Year of Congress	1937	1952	1966	1981	1993	2001	Mean
Congress number	IV	VI	X	XIV	XVII	XIX	
Subject theme							
Styles of grassland improvement; regional themes	23	18	10	9	16	8	14
Plant genetic base	23	20	19	21	25	17	21
Edaphic constraints	19	13	14	14	11	12	14
Perspectives from plant physiology	9	9	17	15	14	11	13
Ecology of grasslands	6	10	4	6	7	6	7
Grazing systems	4	4	8	8	8	16	8
Nutritive value	4	7	15	10	7	13	9
Continuity of forage supply	7	11	8	10	5	11	9
Systems approach	-	0.4	1	3	3	3	2
Socio-economic perspectives	1	3	3	4	4	3	2
Miscellaneous	3	6	3	1	2	1	3
Number entries	69	256	220	480	943	499	

Scientists at all six Congresses emphasised the role of legumes and of biological N fixation in grassland production; associative mechanisms of N fixation were mentioned at the 1981 Congress, and soil N, together with nutrient cycling, stream pollution, soil toxicities and salinity received increased attention. Perspectives from plant physiology incorporated more interest in pathways of photosynthesis, efficiency of conversion of radiation, moisture use, stress resistance, growth regulators and the understanding of constraints to pasture establishment. The dynamics of change in plant communities, the role of fire and the control of shrub encroachment figured in grassland ecology, and some 91 papers were directed to the conservation and improvement of natural grasslands.

The influence of grazing on the balance of legumes and grasses and studies of foraging strategies figured in the 1981 Congress. The effects of endophytes, the potential of growth regulators and of chemical processing of crop materials were canvassed. Modelling of grassland systems and the development of decision support systems emerged as strong emphases, whilst technology transfer and the development of the human skills base in grassland science were accorded more significance.

The 1993 Congress

A wider series of topics was structured in depth at the 1993 Congress than had occurred previously. Environmental science was a strong feature of the Congress and the fashionable term `sustainable development' was explored in its various facets: the properties of systems of land use of varying intensity, tree crops with pastures, alley farming, the role of leys, relict areas, deforestation and woodland management. Atmospheric pollution and global warming, stream pollution, nutrient leaching and nutrient cycling were components of the agenda, whilst a recurrence of interest in organic matter and soil biological activity reinforced these trends.

Studies of the genetic basis of grassland improvement included more attention to the definition of criteria of merit and of disease resistance, and the rise of genetic engineering and of molecular biology in the allocation of research resources was evident. Many of the themes previously attacked in plant physiology continued from 1981, with more attention to the control of flowering and the processes of seed production. In grassland ecology the dynamics of change in plant communities, the utility of state and transition models and the use of remote sensing in producing inventories and current assessments of grassland resources figured strongly.

The perennial themes within the concepts of nutritive value, the devising of grazing systems and the maintenance of continuity of forage supply were elaborated further but in a new context of this description within systems theory.

The socio-economic perspectives which emerged at the 1981 Congress were enlarged by reference to social equity in grassland development, the participation of farmers in grassland research, and to the larger canvases of institutional policies with respect to resource transfer and international trade.

The 2001 Congress

A return to Brazil, 36 years after the IX Congress, revealed a much increased investment in grassland research in the countries of South America. The expense of travel was one factor limiting attendance, and c.700 scientists from 67 countries met in congenial social and intellectual circumstances at São Pedro, São Paulo.

The traditional IGC themes of plant improvement, ecophysiology, soil fertility and plant nutrition were complemented by studies of grazing ecology and management, forage nutritive value, continuity of forage supply and fodder conservation. However the trend at recent Congresses to reduce the emphasis on maximising efficient animal production from grassland and to pay greater attention to the sustainable use of grassland as an environmental resource continued. There were fewer general papers on regional themes (Table 2), reflecting an increasing sophistication and specialisation of grassland research and perhaps the growth of regional meetings elsewhere, sometimes stimulated under the aegis of the International Grassland Congress.

At this Congress many topics concerned with the wider aspects of land use were canvassed: de-intensification with grasslands, especially in relation to the policies of the European Community, deforestation, grassland degradation, the maintenance of biodiversity and the role of agro-silvipastoral systems. Increases in grassland growth and legume nitrogen fixation due to atmospheric carbon dioxide enrichment were quantified, together with speculation about the associated changes in climate. Socio-economics of pastoral development and the constraining effects of trade policies on grassland production were examined. The development of pragmatic information and analytical systems were central both to the efficient use of research resources and to the

adoption of grassland improvement, whilst the dynamics of technology transfer and its basis in interactive education were recognised.

These trends will be intensified at the XX Congress in 2005 when only about a third of invited papers will deal with themes of grassland production and the overall title of 'Grasslands – a Global Resource' will embrace many environmental topics such as biodiversity, carbon sequestration and the best uses of water. The basic targets of food security, reduction of rural poverty and better livelihoods arising from improved grassland management and altered socio-economic policies will be discussed.

A central experience of grassland scientists over the decades under review is that the International Grassland Congresses have helped people working in specialist areas to conceptualise their work in wider contexts. The great world movement of International Grassland Congresses has delivered better managed ecosystems, greater equanimity in rural communities and more efficient production of food and fibre.

Acknowledgements

I am indebted to Roger Wilkins, Vivien Allen and Tom Nolan for assistance. Permission from Cambridge University Press to reprint the main body of this paper from Humphreys (1997) is gratefully acknowledged; the statistics in the paper are drawn from the Proceedings of the 19 International Grassland Congresses. The support of the School of Land and Food Sciences, University of Queensland, is acknowledged.

Reference

Humphreys, L.R. (1997). The Evolving Science of Grassland Improvement. Cambridge University Press, Cambridge, U.K., 202-209.

Appendix

Office bearers of the International Grassland Congresses

Appendix Table 1 Presidents and Chairpersons of Continuing Committee of IGC*

Congress		President	Chair of Continuing Committee
I	Leipzig, Germany 1927	A Falke	
II	Sweden/Denmark 1930	A Elofson	
III	Switzerland 1933	A Volkart	
IV	Aberystwyth, UK 1937	RG Stapledon	
V	Netherlands 1949	DS Huizinga	
VI	Pennsylvania, USA 1952	PV Carden	
VII	New Zealand 1956	B Levy	
VIII	Reading, UK 1960	HG Sanders	
IX	São Paulo, Brazil 1965	AJR Filho	
X	Finland 1966	P Saarinen	HA Steppler, Canada
XI	Australia 1970	EM Hutton	RM Moore, Australia
XII	Moscow, USSR 1974	PI Morosov	DE McCloud, USA
XIII	Leipzig, GDR 1977	R Lemke	SC Pandeya, India
XIV	Kentucky, USA 1981	RF Barnes	WR Childers, Canada
XV	Kyoto, Japan 1985	I Nikki	LR Humphreys, Australia
XVI	Nice, France 1989	J Picard	Y Maki, Japan
XVII	Australia/New Zealand 1993	RW Brougham, NZ	D Crespo Portugal/FAO, Italy
XVIII	Canada 1997	BR Christie	T Nolan, Ireland
XIX	São Pedro, Brazil 2001	SC da Silva	RJ Clements, Australia
XX	Ireland/UK 2005	J Flanagan, Ireland	V Allen, USA

*Presidents were designated by Host Country; Chairpersons of Continuing Committee were elected at the end of the Congress preceding the listed Congress

Appendix Table 2 Executive (1960-1965) and Continuing (1966-1974) Committee membership

Region	VIII Reading 1960	IX São Paulo 1965	X Helsinki 1966	XI Australia 1970	XII Moscow 1974
North America	K Rasmussen Canada	K Rasmussen Canada	DE McCloud USA	DE McCloud USA	WR Childers Canada
Central America	WF Kugler Argentina	GI da Rocha Brazil	GI da Rocha Brazil	F Perez- Infante Cuba	F Perez-Infante Cuba
Oceania	JG Davies Australia	RM Moore Australia	RM Moore Australia	RHM Langer NZ	RHM Langer NZ
Southern Asia	S Emasiri Thailand	S Emasiri Thailand	-	SC Pandeya India	SC Pandeya India
East Asia	T Yamada Japan	T Yamada Japan	S Nishimura Japan	S Nishimura Japan	Y Maki Japan
Mediterranean, Near East	G Haussmann Italy	JVC Malato- Beliz Portugal	JVC Malato- Beliz Portugal	J Cizek Yugoslavia	J Cizek Yugoslavia
Europe	TA Robotnov USSR	TA Robotnov USSR	DFR Bommer W Germany	DFR Bommer W Germany	RJ Wilkins UK
Africa	CEM Tidmarsh South Africa	L Mukendi Congo	L Mukendi Congo	JA Agyare Ghana	JA Agyare Ghana
Host Country Representative	-	AR Filho Brazil	P Saarinen Finland	EM Hutton Australia	MA Smurygin USSR

Appendix Table 3 Continuing Committee membership from 1977

Region	XIII Leipzig 1977	XIV Lexington 1981	XV Kyoto 1985	XVI Nice 1989	XVII NZ/Aust 1993	XVIII Canada 1997	XIX São Pedro 2001
North America	WR Childers Canada	RF Barnes USA	RF Barnes USA	R Michaud Canada	R Michaud Canada	V Allen USA	V Allen USA
Central America	JJ Paretas Cuba	RA Martinez Mexico	RA Martinez Mexico	F Funes Cuba	F Funes Cuba	L Ramirez Mexico	L Ramirez Mexico
South America	A Gallardo Venezuela	A Gallardo Venezuela	C Lascano Colombia	C Lascano Colombia	EA Serrao Brazil	EA Serrao Brazil	R Vera Chile
Southern Asia	Vacant	IM Nitis Indonesia	IM Nitis Indonesia	P Singh India	P Singh India	C Phaikaew Thailand	C Phaikaew Thailand
Oceania	LR Humphreys Australia	LR Humphreys Australia	RW Brougham NZ	RW Brougham NZ	RJ Clements Australia	RJ Clements Australia	G Sheath NZ
East Asia	Y Maki Japan	Y Maki Japan	Z Tingchen China	Z Tingchen China	Dong Am Kim S Korea	Dong Am Kim S Korea	M Goto Japan
Middle East	F Tosun Turkey	F Tosun Turkey	AET Osman Syria	AET Osman Syria	M Munzur Turkey	M Munzur Turkey	H Arzani Iran
Mediterranean	A Corleto Italy	A Corleto Italy	D Crespo FAO Italy	D Crespo FAO Italy	E Piano Italy	E Piano Italy	M Vaz Lourenco Portugal
Western Europe	RJ Wilkins UK	A Hentgen France	A Hentgen France	T Nolan Ireland	T Nolan Ireland	A Peeters Belgium	A Peeters Belgium
Northern Eurasia	VG Iglovikov USSR	VG Iglovikov USSR	I Vinczeffy Hungary	I Vinczeffy Hungary	R Dapkus Lithuania	R Dapkus Lithuania	G Nagy Hungary
Africa	VA Oyenuga Nigeria	VA Oyenuga Nigeria	EA Asare Ghana	EA Asare Ghana	B Dzowela Zimbabwe	B Dzowela Zimbabwe	AB Orodho Kenya
Host Country Representative	E Wohjahn GDR	JE Baylor USA	Y Maki Japan	A Hentgen France	J Hodgson NZ	R Michaud Canada	SC da Silva Brazil

Minutes of the business sessions at the XIX International Grassland Congress
São pedro, São Paulo, Brazil, 11-21 February 2001

Opening business meeting

The opening business session of the XIX International Grassland Congress was held on Sunday 11 February 2001. Dr Sila Carneiro da Silva (Chair, Organising Committee, XIX IGC) introduced Dr Bob Clements (Chair, IGC Continuing Committee), who formally opened the XIX IGC and presented the following report on behalf of the members of the IGC Continuing Committee, namely: Dr Vivien Gore Allen (USA); Dr Luis Ramirez-Aviles (Mexico); Dr Adilson Serrão (Brazil); Ms Chaisang Phaikaew (Thailand); Professor Dong Am Kim (South Korea); Dr Mehmet Munzur (Turkey); Dr Efisio Piano (Italy); Dr Alain Peeters (Belgium); Dr Rimantas Dapkus (Lithuania); Dr Ben Dzowela (Zimbabwe); Dr Réal Michaud (Canada);and Dr Robert Clements (Australia).

Report on the activities of the Continuing Committee: R J Clements

Distinguished guests; members of the International Grassland Congress; ladies and gentlemen:

One of my lasting memories of this Congress will be the sea of golden shirts that were worn by the team of grassland researchers that welcomed us at São Paulo airport. After travelling 15,000 kilometres it was indeed a cheerful sight, and in my mind this will always remain the Golden Congress. On behalf of the Continuing Committee of the International Grassland Congress, and on behalf of the international community of grassland scientists, I have great pleasure in declaring the Golden Congress – the XIX International Grassland Congress – open. Eu declaro aberto o congresso de ouro!

It is my privilege on behalf of the Continuing Committee of the IGC to present this report on the activities of the Continuing Committee since the Canadian Congress in 1997.

The work of the Continuing Committee during this four-year period was dominated by four major issues. These were the selection of the host country for the XX IGC (2005); consideration of opportunities to enhance collaboration between the International Grassland Congress and the International Rangeland Congress; the re-writing of the constitution of the International Grassland Congress; and the provision of advice and assistance to the Organising Committee of the XIX International Grassland Congress – the Congress we are attending now. I will deal with each of these issues in turn. I will also make a few comments about future challenges for the International Grassland Congress.

Venue for the XX International Grassland Congress

Two bids were received to host the XX International Grassland Congress. One was from the Irish Grassland Association and the British Grassland Society, in combination. The other was from the Chinese Grassland Society. Both bids were of good quality. In accordance with rule 6(d)(ii) of the constitution, the bids were considered by the members of the Continuing Committee. The Committee voted by a clear majority to support the bid from Ireland and the United Kingdom. I therefore declare that the XX International Grassland Congress will be held in Dublin, Ireland, from 26 June to 1 July 2005, with five subsequent satellite meetings at Aberystyth, Belfast, Cork, Glasgow and Reading. Please join me in congratulating the successful bidders.

The theme of the XX International Grassland Congress will be "Grasslands: A Global Resource". A feature of the Congress will be consideration of the role of grasslands not only in providing feed resources for livestock and in generating income for farmers, but also as a global resource for wildlife, biodiversity, soil stabilisation, and water catchment and quality. The environmental aspects of grasslands are receiving greater attention every year, and this trend seems certain to continue. Grasslands also have significant social and amenity values, and these have become increasingly important during the last twenty years and seem likely to increase even further.

I am conscious that the Peoples Republic of China has now twice invited the International Grassland Congress to meet in China. Naturally they are very disappointed not to have been selected. I have personally visited the proposed venue for a Congress in China at Hohhot, Inner Mongolia, and I am satisfied that it would have been suitable and that our Chinese colleagues are more than capable of running a successful Congress. I hope that the Chinese Grassland Society will continue with its efforts to attract the Congress to China.

Offered papers **53**

Collaboration between the International Grassland Congress and the International Rangeland Congress

At the XVIII International Grassland Congress in Canada in 1997, delegates instructed the Continuing Committee to make representations to the Continuing Committee of the International Rangelands Congress (IRC) on the possibility of a joint meeting of the two Congresses, and on possible eventual amalgamation of the two Congresses.

In accordance with this instruction, I met with the then Chairperson of the IRC Continuing Committee, Dr Margaret Friedel. Together we developed a discussion paper that was circulated to all members of the Continuing Committees of both Congresses, and to a number of senior members of the international grassland research community. Feedback from members of both Continuing Committees showed a considerable diversity of opinions, with no groundswell of support for a shared Congress at that time. On your behalf, I attended the VI International Rangeland Congress in July 1999 and spoke to the delegates about the possibilities for greater collaboration between the two Congresses. In close consultation with Dr Vivien Allen and representatives of a number of rangeland societies in North America, three resolutions were drafted for consideration by the delegates at the VI International Rangeland Congress. These were:

1. To promote a more efficient and effective interchange of information on all aspects of range and grassland science, and to meet common goals and objectives, the IRC endorses the concept of closer cooperation with the IGC.
2. The Chair of the IRC Continuing Committee should explore mechanisms for meeting common goals and objectives with the Chair of the IGC Continuing Committee.
3. The IRC endorses the concept of a shared conference with the IGC by the year 2007 and requests the Continuing Committee of the IRC to develop in collaboration with the Continuing Committee of the IGC the framework for a shared conference program and procedures for selection of a host country.

The first two resolutions were supported by a considerable majority, but the third was lost by a vote of 46 votes to 71.

Despite this disappointing outcome, a groundswell of support for closer collaboration is now emerging. For the last two years, the grassland and rangeland societies of North America have consistently supported a shared Congress. These include the American Society of Agronomy, the Soil Science Society of America, the Crop Science Society of America, the American Forage and Grassland Council, the American Society of Animal Science, the Society for Range Management, and the Canadian Society of Animal Science. I expect that this momentum for change could be maintained. However, I believe that if the negotiations with the IRC are to be continued, the new IGC Continuing Committee will need a clear indication of support from the delegates at this Congress. I expect that the Resolutions Committee will be giving this matter its close consideration during the next few days, and will be consulting widely with the delegates present.

On a positive note, one example of strong positive collaboration between the two Congresses is worth mentioning. Many delegates will recall that, in 1991, the Forage and Grazing Lands Committee published a book entitled "Terminology for Grazing Lands and Grazing Animals". The Committee was chaired by Dr Vivien Allen. This publication was the result of the combined efforts of six scientific societies in North America, numerous research organisations, and representatives from other countries. Recognising that a revision of this book would be timely, the IGC and IRC are collaborating in a revision that will be published under the auspices of both Congresses. The team of writers is led by Dr Mort Kothmann from Texas A&M University, and contains representatives from both Congresses and five countries. A second team of reviewers will be chaired by Vivien Allen, and will contain representatives from many more countries to ensure that the terminology has wide support. This shared venture between the two Congresses is a good example of the benefits that could be achieved from greater collaboration. On your behalf, I thank Vivien Allen for her considerable efforts to champion and support this initiative.

Our re-written constitution

At the XVIII International Grassland Congress in Canada in 1997, delegates instructed the Continuing Committee to review the IGC Constitution, to incorporate a number of suggestions for change, and to present the constitution to the XIX Congress. Acting on these instructions, a small team led by Professor Roger Wilkins undertook the task of re-writing the existing constitution. Early drafts of the re-written constitution were widely circulated, and the completed document was published on the IGC web-site in February 2000. No suggestions for change have been received since that time, so clearly the constitution has the approval of IGC members. The re-written constitution does not contain any changes that were not proposed and adopted in Canada or earlier, so there is no need for a formal vote of endorsement, and I commend the writing team for a sterling effort. Copies of the constitution are available at the Congress registration desk.

The Brazilian Congress

It is already clear that the Brazilian Congress is destined to be stimulating and memorable. The Organising Committee has put together a solid program of subject areas and invited presentations, and the smorgasbord of mid-Congress tours looks truly tempting. The Continuing Committee provided a number of suggestions in relation to invited speakers and topics, and the Organising Committee has picked up the best of these suggestions. I was privileged to visit the Organising Committee during 1998, and I was impressed by the vigour and commitment that was displayed by the small but dynamic team. I'm sure that by the end of the Congress we will be unanimous in congratulating our Brazilian hosts.

Other matters

The publication of the Proceedings of the XVIII Congress (Canada, 1997) was complicated by a challenging financial situation. In the circumstances, our Canadian colleagues did well to publish the entire Proceedings electronically, in CD-ROM format, and on your behalf I thank them for this. Electronic publication is growing rapidly both in frequency and in acceptability, and is surely a sign of things to come. However, I must say that it is pleasing to see that our Brazilian colleagues have already published the Proceedings of the present Congress both in hard copy and in CD-ROM format.

The IGC constitution requires me to establish two committees. One, called the **Nominating Committee**, has the task of identifying the replacements on the Continuing Committee for the next four years. We are fortunate that Dr Tom Nolan (Ireland), who was the previous Chair of the Continuing Committee, has agreed to chair the Nominating Committee during this Congress. Other members of the Nominating Committee are Dr Carlos Lascano (CIAT, Colombia), Dr Efisio Piano (Italy), and Professor Dong Am Kim (South Korea). This committee has a significant task, with seven members of the current Continuing Committee to be replaced. The Committee will be consulting carefully with delegates during the next few days.

The second committee I am required to establish is the **Resolutions Committee**. This committee has the task of receiving suggestions from delegates concerning resolutions that are to be considered during the closing business session. I am delighted that Dr Réal Michaud (Canada) has agreed to chair this committee. Other members of the Resolutions Committee are Professor Rainer Schultze-Kraft (Germany), Dr Vivien Allen (USA), Dr Luis Ramirez-Aviles (Mexico) and Professor Alain Peeters (Belgium). Please consult one of these people if you wish to propose a resolution for consideration by the delegates during the final business session.

Thoughts on the next ten years

It has become a tradition for the Chair of the Continuing Committee to make a few comments about the future, and I would like to close by doing this. In doing so, I am aware that I am following in the footsteps of some illustrious predecessors. In 1993, Dr David Crespo championed the use of legumes and noted the distortions in resource use that are sometimes caused by inadequate or ill-conceived government policies. At the same Congress in 1993, Dr Ray Brougham urged us to become involved in the fight to raise public awareness of the benefits of grasslands research. He especially urged us to get involved in influencing policy-makers. In 1997, at the Canadian Congress, Dr Tom Nolan spoke about the complexity of the modern research environment, and again mentioned the need to influence policy formulation at the local, regional and global levels.

I speak to you from a different perspective. I speak as the Director of a research funding body. My organisation, ACIAR (the Australian Centre for International Agricultural Research) is a facilitator and funder of collaborative agricultural research, with a firm eye on delivering benefits to developing countries. I think it may be the first time that the Chair of the Continuing Committee has come from such a position, and it certainly does give one a different view of the world. As I look at the bulk of the current research on grasslands, I find a good deal of it simply irrelevant to the needs of developing countries in the Asia-Pacific region. This is not entirely a new observation. In 1993, at the XVII International Grassland Congress in Rockhampton, Australia, the participants at the session on feeding animals in subtropical and tropical forage systems concluded that much of the grassland research being conducted in the tropics was of little relevance to end-users in developing countries, and was not likely to be adopted because it paid inadequate attention to the economic, social, biological and farming systems constraints to adoption. Therefore, not surprisingly, my organisation funds very little grassland research in the

Asia-Pacific region. We are very proud of the grassland research we **do** support, and I am delighted that several of the project teams we are supporting are present at this Congress. Of course, there are many sources of research funds, and not every funder has such a strong focus on delivering benefits from research. But I urge every delegate here to think hard about the relevance of your research. Who is going to use the technologies you develop? How will they access those technologies? How can you make your research more relevant to end-users? If your research is not relevant, not only will the funds eventually dry up, but you may lead other researchers into irrelevance, to the detriment of our profession.

My second comment is about research innovation. All of us admire the truly innovative scientist – the person who moves us in new directions, who applies new science to attack old problems, who shatters myths, who forces us to re-think our comfortable paradigms, who sees room for progress where the rest of us can only see complexity. We need to recognise that many aspects of grassland science are now mature. This means beneficial changes in many fields of grassland science will be modest and incremental unless we take positive steps to seek truly novel approaches. We need to redefine grassland science. We need to bring it into the 21st century. We need to apply to it the most modern adaptations of information technology, biotechnology and modern social sciences. If we don't do this, we again run the risk of irrelevance and, in this case, the associated risk that by delaying the application of new science to grasslands we slow down the rate of progress in managing and improving our grasslands for the benefit of mankind.

Australians have a reputation for speaking their minds plainly, and perhaps I should apologise for being so blunt. However, these are sobering thoughts, and they deserve careful consideration. Whether the targeted end-user of our research is the farmer, the conservationist, the policy-maker, our fellow scientists, or even our students, we have an obligation to be as relevant to their needs as we can possibly be. The importance of grasslands in the world demands nothing less. As we share our thoughts and results during the next few days, let us make a conscious effort to seek and provide evidence that our research is relevant and innovative.

Concluding comments

Thank you for the privilege of chairing the Continuing Committee of the International Grassland Congress. I commend to you the strong contribution of my colleagues on the Continuing Committee, and assure you that any deficiencies in our performance are not their fault, but mine. I will have one more opportunity to speak to you again briefly at the conclusion of this Congress. In the meantime, like you, I look forward with enormous anticipation and excitement to the next few days. It is great to be with you once again, and to be at what promises to be a memorable Congress. On your behalf, I extend congratulations to the Organising Committee.

Closing business session

The closing business session of the 19th International Grassland Congress was held on 21 February 2001. Dr Bob Clements, Chair of the Continuing Committee, and Dr Sila Carneiro da Silva, Chair of the Organising Committee of the XIX IGC, jointly chaired this business session.

Resolutions

The Resolutions Committee brought forward eight resolutions for the consideration of the delegates. These were as follows:

Resolution 1 (supported by acclamation)

The members of the XIX International Grassland Congress express their sincere thanks to the government of Brazil and to all sponsoring organisations who enabled this congress to be held ;

The Congress congratulates the Organising Committee that has been outstandingly successful in putting together an excellent program and in handling the logistics of the Congress ;

Special thanks and deep appreciation are extended to Professor Sila Carneiro da Silva, President of the XIX International Grassland Congress, to Professor Wilson Mattos, Executive Secretary, and to all those involved in making this Congress a success.

Resolution 2 (supported by acclamation)

The members of the XIX International Grassland Congress would like to recognise the very good contribution of the working group, chaired by Professor Roger Wilkins, that edited the existing Constitution to comply with the various suggestions that were agreed upon in Canada in 1997 and at previous Congresses.

Resolution 3 (supported unanimously)

The members of the XIX IGC ask that the Brazilian Organising Committee makes available a supplement to the Proceedings which contains
1. a list of participants and their addresses
2. an account of the two business meetings

Resolution 4 (supported unanimously)

To promote a more efficient and effective interchange of information on all aspects of grassland and range science, and to meet common goals and objectives, the International Grassland Congress (IGC) endorses the concept of closer cooperation with the International Rangeland Congress (IRC).

Resolution 5 (supported unanimously)

The chair of the IGC Continuing Committee should explore mechanisms for meeting common goals and objectives with the chair of the IRC Continuing Committee.

Resolution 6 (supported unanimously)

The members of the XIX IGC request that the Chair of the IGC Continuing Committee meets with the Chair of the IRC Continuing Committee within the next 12 months to jointly identify and promote shared activities for meeting common goals and objectives.

Resolution 7 (supported unanimously)

The members of the XIX IGC ask that the Organising Committee of the XX IGC, as far as possible, considers incorporating in the main Congress program specific topics dealing with the contribution of grasslands to a more sustainable agriculture. Thus the program should include environmental, socio-economical and political issues across diverse ecosystems, as well as strategies to maximise the impact of research and extension.

Resolution 8 (supported unanimously)

As it was resolved at the XVII International Grassland Congress, it is recommended that the practices evident at the1993 IGC in New Zealand/Australia, which stimulated a strong attendance of grassland scientists from developing countries, be continued and explored in future Congresses.

Membership of the Continuing Committee, 2001-05

The chair of the Nominating Committee, Dr Tom Nolan, announced the nomination of the following members to represent the various regions defined in the IGC constitution:

Region 1 (North America): Dr Vivien Gore Allen (USA)
Region 2 (Central America): Dr Luis Ramirez-Aviles (Mexico)
Region 3 (South America): Dr Raul Vera (Chile)
Region 4 (South and South-East Asia): Ms Chaisang Phaikaew (Thailand)
Region 5 (Oceania): Dr Gavin Sheath (New Zealand)
Region 6 (East Asia): Dr Masakazu Goto (Japan)
Region 7 (Middle East): Dr Hossein Arzani (Iran)
Region 8 (Mediterranean): Dr Maria E V Lourenço (Portugal)
Region 9 (Europe): Dr Alain Peeters (Belgium)
Region 10 (North Eurasia): Dr Geza Nagy (Hungary)
Region 11 (Africa other than Regions 7 and 8): Dr Apollo Bwonya Orodho (Kenya)
Immediate past host country: Dr Sila Carneiro da Silva (Brazil).

The nominations were endorsed by acclamation. The members of the new Continuing Committee then retired briefly to elect a new Chair, Dr Vivien Allen. Dr Allen subsequently addressed the Congress delegates, emphasising the opportunities and challenges that would confront the Continuing Committee during the next four years.

Presentation by the host countries of the next Congress

Dr John Walsh (Ireland) and Dr Roger Wilkins (United Kingdom) presented a brief overview of the next (XX) International Grassland Congress and the proposed venue, and formally invited the delegates present in Brazil to attend.

Closure

Dr Sila Carneiro da Silva thanked delegates for attending the XIX Congress. He thanked the staff of the Hotel Fazenda Fonte Colina Verde (the Congress venue), and the people who had organised the social functions. He mentioned particularly the work of his colleagues on the Organising Committee, and the "golden team" that had worked so hard during the Congress to satisfy the requirements of the delegates.

Theme A: Efficient production from grassland

Section 1

Grass and forage plant improvement

Participatory collection of forage species in Uruguay

M. Rebuffo, F. Condon and M.J. Cuitiño

INIA, National Institute of Agricultural Research, La Estanzuela, Colonia, Uruguay, Email: rebuffo@inia.org.uy

Keywords: germplasm collection, *Lotus*, lucerne, clover

Introduction Local landraces are potential valuable gene sources with benefits for local agriculture. However, their *in situ* conservation depends on the personal motivation of farmers and the permanence of traditional farming methods (Negri, 2003). There is the latent risk of loosing landraces and site-specific naturalized germplasm due to rapid socio-economic changes in Uruguayan agriculture. Participatory conservation can provide the necessary synergy between farmers, extension services, geneticists and genebank to achieve the goal of conserving these genetic resources. Although farmers' involvement in germplasm collection is common, particularly for vegetable crops, most forage collections had been done from fields or sideroad areas, therefore lacking passport information in relation to cultural practices. This work outlines the farmers' presence in *ex situ* germplasm conservation and explores current and past uses of crops.

Materials and methods Having defined the collection scope, we delineated the participatory procedures, using historic information on seed marketing. The objective was to ensure a representative sampling for each forage species, yet remain within manageable sample sizes. Collection of exhaustive passport data, including site, cultivar origin, period of on-farm multiplication, grazing management and cultural practices, was a main priority. Collection began in 1999, with the support of PROCISUR and the participation of 21 farmers' associations.

Results The geographical range included virtually all cultivated regions. We identified 132 farmers that multiplied their own seed through the cooperation of Extension Services, representing over a 70% of the total contacted farmers. Of these, 25% provided at least 2 species with 73% of the accessions being perennial legumes, reflecting their importance in the crop-pasture rotation use in Uruguay. The prevalent species were birdsfoot trefoil (*Lotus corniculatus* L.), red clover (*Trifolium pratense* L.) and lucerne (*Medicago sativa* L.), with 99, 33 and 22 accessions, respectively. *Lotus* (27%) and clover (17%) had been multiplied for more than 10 years, this increased to 43% for lucerne, reflecting the prevalence of traditional farming for the latter (Table 1a). Accessions of unknown genetic origin represented 20, 22 and 28% for lucerne, *Lotus* and clover, respectively. The collection stands out in its representation of old public varieties. 73% of *Lotus* accessions were originally cv. San Gabriel; Estanzuela 116 and Estanzuela Chaná represented 72% and 65% of clover and lucerne, respectively. The collection reflected the relevance of seed production within each farming system. Traditional farming and on-farm seed multiplication was more frequent for farmers involved in beef production (Table 1b). Daily grazing is routinely used with lucerne and clover, whereas non-bloating *Lotus* is grazed in a more relaxed schedule (Table 1c). Forage oats characterization has already shown the collection value in traits related to adaptation under grazing (Vilaró *et al*, in press).

Table 1 Accessions categorized by:

a) on-farm multiplication (% within each species)

Years	Lotus	Red Clover	Lucerne
Less than 3	2	14	5
3 to 10	71	69	52
10 to 20	20	17	43
over 20	7	0	0

b) farm production (% within each species)

System	Lotus	Red Clover	Lucerne
Dairy	19	36	45
Beef	55	49	52
Sheep	15	18	14
Agriculture	33	23	15

c) grazing management (% within each species)

System	Lotus	Red Clover	Lucerne
Daily	37	61	73
Weekly	39	15	18
Monthly	9	12	5
Continuous	14	11	4

Conclusions The collection accomplished the task of reaching a broad spectrum of farming production systems and management, integrating farmers, extension services and researchers to elaborate detailed information of each accession. *In situ* on-farm conservation sustains the evolutionary avenue through which genetic variability is generated, but ex situ conservation protects landraces vulnerability from extreme climatic changes, as with the drought of 2000. Detailed passport information on cultural practices will be a key feature for future utilization of the collection, as well as the analysis of current and future threats to genetic erosion.

References

Negri V. (2003). Landraces in Central Italy: where and why they are conserved and perspectives for heir on-farm conservation. *Genetic Resources and Crop Evolution,* 50, 871-885.

Vilaró M, C. Miranda, C. Pritsch, M. Rebuffo & T. Abadie. Morphological characterization of *Avena sativa* L. collection in Uruguay. *Plant Genetic Resources Newsletter* (in press)

Effects of a recurrent selection scheme, applied to an interspecific hybrid *Pennisetum purpureum* Schum. (elephantgrass) *x Pennisetum glaucum* (L.) R. Br. Stuntz (pearl millet), on several seed quality parameters

R. Usberti[1], J.A. Usberti Jr.[2], R.H. Aguiar[3], L.M.T.A. Carneiro[3], J.B. Fantinati[3], F.G. Francisco[3]
[1]*Plant Protection Agency, P.O. Box 960, CEP 13073-001, Campinas, Brazil, Email: usberti@cati.sp.gov.br*
[2]*Agronomic Institute, Campinas, Brazil, [3]Faculty of Agricultural Engineering, Campinas, Brazil*

Keywords: recurrent selection, interspecific hybrid, seed quality traits

Introduction Elephantgrass cultivars and introductions show practically no viable pure seeds and their uses in cultivated grasslands are exclusively dependent on vegetative propagation. Therefore, in large areas, sowing operation costs make unfeasible setting up new pastures. On the other hand, pearl millet is a high seed producer species though presenting some forage constraints (poor forage production, low regrowth potential after cutting or grazing and low field persistence). Recently, an hexaploid interespecific hybrid between the two species was developed (Schank & Diz, 1996), which is able to produce viable pure seeds, in variable amounts according to the genotype considered (Diz & Schank, 1995). This research aimed to check several seed quality parameters in two selected populations, derived from the original F_2 interespecific hybrid population.

Materials and methods Sample seeds of selected individuals of cutting (CT, tall plants with low-tillering potential, long and broad leaves, thick stems with long internodes and erect growth habit populations) and grazing types (GT, small and high-tillered individuals with short and fine leaves, thin stems with short internodes and prostrated growth habit) were collected during two selection cycles. The following seed traits were recorded for both selected populations as well as for control original F_2 population: 1,000 seed weight, number of seeds/g, physical purity, standard germination and vigour percentages (accelerated ageing test after 24, 48 and 72h) (ISTA, 1985) as well as mean germination times (T_{50}) (Alvarado & Bradford, 1988).

Results CT and GT populations presented marked increases in physical purity percentages mostly due to increases of phenotypic uniformity, after 2 recurrent selection cycles (47.3 and 45.1%, as compared to F_2 population - 13.6%)., CT population showed the highest value for number of seed/g (375.7) and the smallest for 1,000 seed weight (2.878g), clearly indicating that the selection scheme favoured, in this case, the occurrence of a higher amount of smaller seeds (Table 1). Seeds of both populations revealed higher germination percentages than those of the control though mean germination times (T_{50}) were similar for all populations. Vigour results revealed the superiority of selected populations, when compared to the control (24 and 48h); after 72h CT population significantly outperformed the others. Few differences among populations were detected as (T_{50}); however GT population presented a delayed germination at 72h as compared to the others.

Table 1 Mean physical seed quality traits results for two selected populations derived from an interspecific hybrid *P. purpureum* x *P. glaucum*, as compared to those of the original F2 population, after 2 selection cycles

Population	Purity	1,000 seed	Number	Germination		Vigour (Accelerated Ageing Test)					
	(%)	weight (g)	seeds / g	%	T_{50} (d)	%	T_{50}	%	T_{50}	%	T_{50}
Cutting type	47.3a	2.878b	375.7a	59.6a	4.9a	60.5a	4.4a	51.6a	4.1a	54.5a	3.9b
Grazing type	45.1a	3.163a	328.2b	51.3a	4.5a	54.9a	3.9a	52.6a	4.3a	33.8b	5.4a
Original F_2	13.6b	3.269a	308.1b	40.4b	4.5a	41.7b	3.8a	41.5b	3.9a	30.5b	4.0b
Mean	35.3	3.103	337.3	50.4	4.7	52.4	4.0	48.5	4.1	39.6	4.4
C.V (%)	7.5	3.4	3.3	7.5	3.5	7.8	4.1	8.2	3.9	8.7	3.4

C.V. (%) = coefficient of variation; b) Means followed by different small letters, in the same column, are statistically different according to the Duncan test at p<0.05.

Conclusions The recurrent selection scheme used was effective in improving physical purity, germination and vigour percentage results for both selected populations.

References
Alvarado, N.A.& K.J. Bradford (1988). Priming and storage of tomato (*Lycopersicon lycopersicum*) seeds. 1. Effects of storage temperature on germination rate and viability. *Seed Science & Technology*, 16, 601-612.
Diz, D.A., S.C. Schank, D.S. Wofford. (1995). Defoliation effects and seed yield components in pearl millet x elephantgrass hybrids. *Agronomy Journal*, 87, 56-62.
ISTA, International Seed Testing Association. (1985). International rules for seed testing. RULES 1985. *Seed Science and Technology*, 13, 299-355; 356-513.
Schank, S.C., D.A. Diz, P.J. Hoghe, C.V. Vann. (1996). Evaluation of pearl millet x elephantgrass hybrids for use as high quality forage for livestock. *Soil and Crop Society of Florida Proceedings*, 55, 120-121.

Thoughts on breeding for increased forage yield

E.C. Brummer
Raymond F. Baker Center for Plant Breeding, Department of Agronomy, Iowa State University, Ames, IA 50011 USA, Email: brummer@iastate.edu

Keywords: Forage breeding, biomass, recurrent selection, population genetics, genetic mapping

Introduction Most forage crops have not experienced yield gains as impressive as those observed in annual grains crops such as maize (*Zea mays* L); in fact, yield improvement in lucerne appears to have stopped in the Midwestern USA (Riday and Brummer, 2002). I contend that much of this disparity can be explained by a failure of breeders to pursue long term recurrent selection programs within populations to capitalize on small, incremental improvements in yield over time. Many selection programs last only two or three cycles, resulting in a germplasm or cultivar release. Either no further selection is attempted or the new population is mixed with a larger germplasm pool in the belief that genetic variation is running low, a belief with little empirical support.

Context Yield is "the 'bottom line' in most plant breeding programs" (Burton, 1982), yet virtually no selection experiments have been conducted in any forage crop expressly for yield. The one striking exception is in Pensacola bahiagrass, in which linear yield improvements have been realized both in a broad based population over the course of 22 cycles and in extremely narrow based populations over the course of 14 cycles (Burton and Mullinix, 1998; Burton, 1982). The selection program was solely phenotypic mass selection; no progeny testing was used. No comparable selection experiments are available in any other forage crop. The success of Burton (1982) is based on a set of "restrictions" that he devised to streamline traditional mass selection, making it faster and more effective at concentrating desirable alleles and improving yield.

Discussion Whether phenotypic selection is the most effective method to improve yield in other forage species is unknown as experiments comparing breeding methods are scarce. Coors (1999) identified 133 maize studies that compared at least two selection methods over at least four cycles of selection. The number of comparable studies in lucerne is zero. Further, an understanding of the optimum population size and structure needed for long term, continual yield improvement is utterly lacking. Narrow populations would facilitate the concentrating of desirable alleles, yet broad based synthetics are the rule in most forage crops today. One of the populations used by Burton (1982) was derived from a single hybrid individual; through six cycles of selection, it showed linear yield gains even larger than those realized in a broad based population (Burton and Mullinex, 1998).

Suggestions Improving yield of forage crops will require a concerted effort to address several fundamental deficiencies in current breeding programs. I suggest the following: (1) Improving yield must be based on evaluation of yield *per se*; few current programs are explicitly selecting for yield *per se*. (2) Careful consideration of selection methodology, revisiting the "restrictions" proposed by Burton (1982) should be undertaken to improve efficiency. (3) Long term recurrent selection programs, continually turning cycles, are needed to evaluate gain and to result in the continual development and release of superior cultivars. (4) Selection should be conducted within populations to avoid disrupting favorable gene complexes. (5) Alternative breeding methods should be evaluated in both narrow and broad based populations to understand the effects of genetic structure on selection response. (6) Multiple improved populations, maintaining genetic diversity within the crop metapopulation, will enhance the development of hybrid or semihybrid cultivars to capture heterosis (Brummer, 1999). (7) Within the context of the foregoing research, molecular markers and genomics tools can be assessed for their utility in breeding quantitative traits. Breeding is often considered slow and inefficient, and biotechnological methods are proposed as a means of making them more effective (e.g., Cook, 1998). Given that a considerable amount of knowledge is needed about appropriate breeding methods, biotechnology does not appear to be the most likely avenue by which to improve yield.

References

Brummer, E.C. (1999). Capturing heterosis in forage crop cultivar development. *Crop Science,* 39, 943-945.
Burton, G.W. (1982). Improved recurrent restricted phenotypic selection increases bahiagrass forage yields. *Crop Science,* 22, 1058-1061.
Burton, G.W. & B.G. Mullinix (1998). Yield distributions of spaced plants within Pensacola bahiagrass populations developed by recurrent restricted phenotypic selection. *Crop Science,* 38, 333-336.
Coors, J.G. (1999). Selection methodology and heterosis. In: J.G. Coors and S. Pandey (eds.) The Genetics and Exploitation of Heterosis in Crops. ASA, CSSA, CIMMYT, Madison, WI, 225-245.
Cook, R.J. (1998). Toward a successful multinational crop plant genome initiative. *Proceedings of National Academy of Science, USA*, 95, 1993-1995.
Riday, H. & E.C. Brummer (2002). Forage yield heterosis in alfalfa. *Crop Science,* 42, 716-723.

Inheritance of yield, morphological and quality characteristics in cocksfoot (*Dactylis glomerata*)

A. Jafari[1], H.M. Arefi[1] and H. Nasri[2]
[1]Research Institute of Forests and Rangelands, Tehran, Email: aajafari@rifr-ac.ir, [2]Islamic Azad University, Brojerd, Iran

Introduction This study estimated genetic variation in cocksfoot (*Dactylis glomerata*) for yield, tiller number, heading date, dry matter digestibility (DMD), water soluble carbohydrate (WSC) and crude protein (CP).

Material and methods A ploycross nursery containing 29 genotypes was established in 2000 in Alborz Research Center, Karaj, Iran. Prior to planting, each parent was vegetatively propagated to give 5 clones. At harvest, seed from clonal replicates of each genotype was combined. Seed of half sib (HS) families and their clonally propagated parents were grown as spaced plants using a randomised complete block design with 2 replicates. Spaced plants were established in rows 40×40 cm spacing. The plants were harvested three times per year during 2001-02 with each plant cut, weighed and dried at 70°C for 24 h. The DOMD, WSC and CP were measured using near infrared spectroscopy following the methodology and calibrations of Jafari *et al.* (2003b). The data were analysed for individual years and combined over 2 years. Variance components were estimated from mean squares linear functions. Broad sense heritability (h^2_b) was estimated based on components of variance from analysis of parents, narrow sense heritability (h^2_n) was calculated from analysis of HS families and narrow sense heritability (h^2_{op}) based on linear regression of offspring on one parent was also estimated. The reference unit of these estimates was based on mean of clonal parents and HS families. Heritability estimates were obtained assuming a diploid inheritance model without epistasis and non-inbred parents chosen randomly.

Results and discussion Tables 1 and 2, respectively, show estimates of variance component S^2 from combined analysis of parents and HS families over 2 years for yield, morphological and quality. All traits had significant variances between clonal parents except CP and DMD. In the HS generation, between family variances were significant for WSC, tiller number and heading date. Clone×year (S^2_{GY}) and Family×year interactions (S^2_{FY}) were significant for all traits except for forage yield and tiller number in HS families. The broad sense heritability (h^2_b) estimated from parental analysis had moderate to high values for heading data, tiller number and dry matter yield (Table 3). Narrow sense heritability (h^2_n) estimates from analysis of progenies and narrow sense heritability (h^2_{op}) estimates from regression of parents on progenies had relatively the same values as (h^2_b) for heading date and tiller number but lower for forage yield (Table 3). Additive genetic variance was the main component controlling these 2 traits but non-additive genetic variance was important for forage yield. The results for WSC over 2 years gave high values for both h^2_b and h^2_{op}, indicating the importance of additive genetic variance in controlling this trait (Table 3). For DMD and CP moderate values of h^2_b were obtained for individual years (data not shown), but heritability values were decreased in combined analysis over years. The h^2_n and h^2_{op} estimates were low in progenies for both traits. Jafari *et al.* (2003a) in ryegrass found the same h^2_b values for yield and WSC, but, their h^2_b values were higher for CP and DMD.

Table 1 Estimates of variance components derived from combined analysis of parents over 2 years for dry matter yield, morphological and quality traits

Traits	S^2_G	S^2_{GR}	S^2_{GY}	S^2_e
Yield (t/ha)	1.25*	1.77**	0.21*	2.84
Heading date	53.8**	27.9**	15.5**	55.2
Tiller number	0.33**	0.23**	0.04**	0.31
CP%	0.27	0.89**	1.01**	1.41
Carbohyd. %	0.93**	0.54**	1.11**	1.57
Digestibility%	0.05	1.05**	1.84**	1.64

*, ** Significant at 5 and 1% level, respectively.

Table 2 Estimates of variance components derived from combined analysis of HS families over 2 years for dry matter yield and quality traits

Traits	S^2_F	S^2_{FB}	S^2_{FY}	S^2_w
Yield (t/ha)	0.01	1.51**	0.08	3.78
Heading date	17.3*	16.9**	22.4**	78.1
Tiller number	0.29**	0.21**	0.003	0.41
CP%	0.01	0.81**	0.84**	1.19
Carbohyd. .%	0.36*	0.39**	0.36**	2.63
Digestibility%	0.01	1.66	1.7**	2.32

*, ** Significant at 5 and 1% level respectively.

Table 3 Summary of three estimate of heritability (h^2 ±s.e.) from combined analysis of parents and HS families over 2 years

Traits	h^2_b	h^2_n	h^2_{op}
Yield (t/ha)	.43 ±.08	.01 ±.21	.20 ±.12
Heading date	.60 ±.01	.31 ±.02	.40 ±.08
Tiller number	.61 ±.14	.58 ±.15	.89 ±.11
CP%	.17 ±.19	.01 ±.29	.24 ±.13
Carbohyd. %	.81 ±.16	.28 ±.14	.68 ±.11
Digestibility%	.03 ±.19	.01 ±.19	.19 ±.14

References

Jafari, A., V. Connolly & E. J. Walsh (2003a). Genetic analysis of yield and quality in full sib families of perennial ryegrass (*Lolium perenne* L) under two cutting management. *Irish Journal of Agricultural and Food Research*, 42, 275-292.

Jafari, A., V. Connolly, A. Frolich & E.J. Walsh (2003b). A note on estimation of quality parameters in perennial ryegrass by Near infrared reflectance spectroscopy. *Irish Journal of Agricultural and Food Research*, 42, 293-299.

Leaves of high yielding perennial ryegrass contain less aggregated Rubisco than S23

A. Kingston-Smith and P.W. Wilkins

Institute of Grassland and Environmental Research, Plas Gogerddan, Aberystwyth, Ceredigion SY23 3EB, UK, Email: pete.wilkins@bbsrc.ac.uk

Keywords: dry matter yield, nitrogen use efficiency, plant breeding

Introduction   Breeding diploid perennial ryegrass for improved dry matter yield under nitrogen-limiting conditions has reduced the nitrogen (N) concentration of the herbage (Wilkins *et al.*, 2003). Reduced N concentration in the ruminant diet is one potential way to reduce losses of N to the environment by reducing the amount of N that animals excrete. The underlying physiological basis of this increased N-use efficiency in ryegrass was investigated.

Materials and methods   Leaf samples were taken from the third harvest year (2004) of a field plot trial with 4 replicate fully randomised blocks containing two perennial ryegrass varieties that had varied consistently in N concentration throughout the first two harvest years (Wilkins *et al.*, 2003): Ba13582, which had the lowest mean N concentration over all harvests, and S23, which had the highest. Ba13582 produced significantly more dry matter than S23 in both these harvest years (2002 and 2003). Ten fully expanded leaves from each plot were frozen in liquid N_2 and stored at -80^0C. Samples were ground to a fine powder and protein was extracted by grinding in a neutral buffer (0.1 M HEPES, pH 7.5, 1 mM EDTA, 2 mM DTT, 0.1% Triton X-100, 1 mM PMSF, 1 μM E64) at a ratio of 25 ml per g dry weight. After centrifugation (5 min at 10,000g_{av}), protein contents of the supernatants were determined (Bradford, 1977) while protein separation was achieved by denaturing gel electrophoresis (Laemmli, 1970). Gels were loaded with 10 μg protein per sample track plus molecular weight standards. They were stained with Coomassie blue and analysed by densitometry (BioRad GS710 equipped with Qantity One software, BioRad UK, Hemel Hempstead). Analyses of variance were carried out using GENSTAT.

Results   Densitometric analysis of the major leaf protein bands of Ba13582 and S23 did not reveal significant differences between the varieties in concentration of the large and small subunits of Rubisco (Table 1). However, Ba13582 contained less than half the amount of high molecular weight polypeptide (~205 kDa) that was present in S23. This 205 kDa polypeptide is typical of non-heat dissociable, aggregated Rubisco subunits.

Table 1 Densitometric analysis (OD x mm^2) of Rubisco protein bands in leaf protein extracts from Ba13582 and S23 resolved by denaturing electrophoresis

Variety	Large subunit	Small subunit	Aggregated
Ba13582	31.3	10.4	2.9
S23	24.8	8.0	6.9
s.e.d.	2.23	1.18	0.56
p	NS	NS	0.006

NS, not significant at p=0.05

Conclusions   Since aggregated Rubisco is unlikely to function in capturing CO_2 from the atmosphere, this result suggests a possible mechanism for the superior N-use efficiency of Ba13582. The *in vivo* significance of aggregated rubisco is unclear. It may represent an N storage pool, to which Ba13582 partitions less assimilated N than S23. Alternatively, it may indicate protein damage. In either case, Ba13582 would be predicted to achieve efficient photosynthesis with less protein N than S23. Families derived from Ba13582 are currently being used at IGER for genetic mapping. If our hypothesis is correct, it should be possible to identify quantitative trait loci that control both N-use efficiency and the amount of aggregated Rubisco.

References

Bradford, M.M (1977). A rapid and sensitive method for the quantitation of microgram quantities of protein utilizing the principle of protein-dye binding. *Analytical Biochemistry*, 72, 248-254.

Laemmli, U.K. (1970). Cleavage of structural proteins during the assembly of the head of bacteriophage T4. *Nature*, 227, 680-685.

Wilkins, P.W., J.A. Lovatt & M.L. Jones (2003). Improving annual yield of sugars and crude protein by recurrent selection within diploid ryegrass breeding populations, followed by chromosome doubling and hybridisation. *Czech Journal of Genetics and Plant Breeding*, 39 (Special Issue), 95-99.

Molecular clone of the Na$^+$/H$^+$ antiporter gene *AtNHX1* and study of transgenic salt tolerant lucerne

Q. Yang[1], H. Zhou[2] and Y.N. Tongquan[1]
[1]*Department of Biotechnology, Beijing Agricultural College, Beijing, 102206, Email: yangqijian@hotmail.com,*
[2]*Institute of Grassland Science, China Agricultural University, Beijing 100094, China*

Keywords: Na$^+$/H$^+$ antiporter, lucerne, tissue culture, salt tolerance

Introduction Lucerne (*Medicago sativa*) with its good quality and ease of cultivation occupies an important position in animal feeding. Salinity is a major constraint of crop productivity, because it reduces yield and limits expansion of agriculture. Na$^+$/H$^+$ antiporter catalyses the counter transport of Na$^+$ and H$^+$ across membranes. Vacuolar Na$^+$/H$^+$ antiporter plays an important role in developing salt-tolerance of plants. Therefore, we could use the gene involved in this mechanism to modify salt tolerance of lucerne.

Materials and methods This experiment in molecular biology is based on Gene Bank data. We designed a pair of primers and used reverse-transcription PCR to isolate a DNA fragment of 1.6kb in length from *Arabidopsis thaliana* (Columbia type). The target fragment was ligated into pGEM-T easy vector and subject to DNA sequencing. It was introduced into lucerne by *Agrobacterium*-mediated transformation.

Results DNA sequencing with the target fragment shows that it contains a whole open reading frame. We selected a fragment of *AtNHX1* encoding peptide rich in hydrophilic residues and designed primers. We induced callus of seedling cotyledon and hypocotyls of lucerne. A plant expression vector harbouring the *AtNHX1* gene was constructed and introduced into lucerne by *Agrobacterium*-mediated transformation. The promoter used in our experiment was CaMV 35S. Three cultivars of lucerne were used for the transgenic experiment. Regenerated plants were subjected to genome-PCR and Southern blotting to attest the stability of *McNHX1* in acquired salt tolerant lucerne plants. Based on the data from EST of Crystalline Iceplant (*Mesembryanthemum crystallinum*) to obtain *McNHX1* from vacuole membrane by RACE and RT-PCR, we will introduce it into lucerne to get a more salt resistant cultivar and apply a heterologous Na$^+$/H$^+$ antiporter gene to improve salt tolerance of other agricultural crops.

Acknowledgements Supported by Foundation of Beijing key laboratory (KF 2003-03) and Beijing Science Foundation (2002KJ074)

The genetic characteristic of salt tolerance at the stage of seed germination in lucerne

H. Zhou[1] and Q. Yang[2]
[1]Institute of Grassland Science, China Agricultural University, Beijing 100094, China, [2]Department of Agronomy, Beijing Agricultural College, Beijing 102206, China

Keywords: Lucerne, genetic characteristic, Superweak luminescence, Salt resistance,

Introduction Superweak luminescence exists in all animal and plants; it shows important genetic information. We report the relationship between superweak luminescence and genetic characteristics to salt resistant in lucerne (*Medicago sativa*). The method can quickly and exactly evaluate the salt-resistance of plants.

Materials and methods Superweak luminescence of germinating lucerne seed was measured in distilled water and in NaCl using BACKMAN-5801 in constant temperature, avoiding light.

Results The physiological action of the salt-stressed germinating lucerne seeds was restricted to some extent. The luminescence value of lucerne seed in NaCl solution was lower than that in distilled water. Its luminescence characteristic peak value appeared 24h later than it did in the others treatments. There was a concordant luminescence tendency but large differences in luminescence intensity among the different cultivars of lucerne. The tolerant cultivar Duoye had a little change both in germinating metabolism and growth rate under the salt stress. The luminescence value of germinating seed in 0.5% and 1% NaCl solution was similar to that in distilled water. But for a sensitive cultivar such as Yongji, 0.5% and 1% NaCl solution had serious effects. The peak value at 96h had respectively decreased 26% for 0.5% NaCl solution and 65% for 1% NaCl solution. Thus, there was a significant difference between two cultivars Yongji and Duoye.

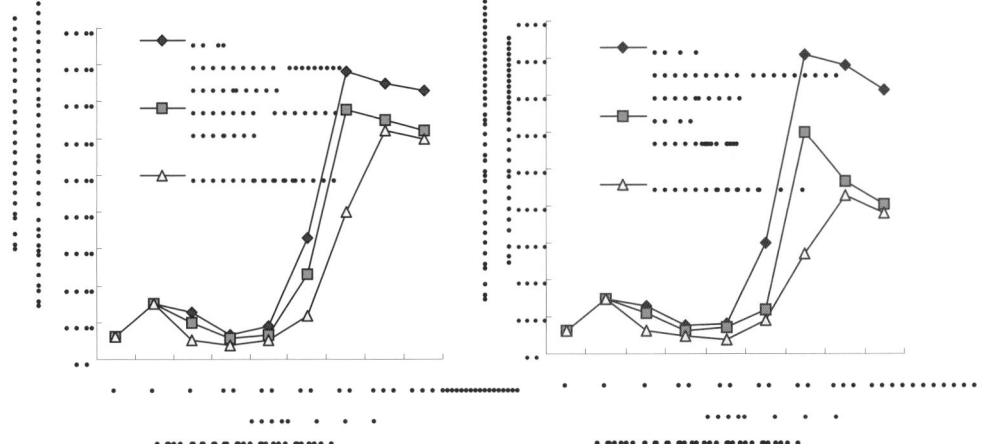

Figure 1 Comparison of luminescence among two varieties of alfalfa in the distilled water and in NaCl solution

Conclusions There was a significantly greater lag in the luminescence curve of germinating seed in 1% NaCl solution than that in distilled water and 0.5% NaCl solution. Under the same salt stress, the different cultivars of lucerne emitted different levels of superweak luminescence. On the basis of this difference we can determine the salt tolerance degree of lucerne.

Persistence and yield of ten lucerne cultivars under dryland and heavy continuous grazing in the Western Cape Province of South Africa

J.M. van Heerden
ARC Range & Forage Institute, PO Box 3320, Matieland, 7602, South Africa, Email: jmvh@sun.ac.za

Keywords: plant cover, agronomic performance, Merino sheep, winter activity

Introduction Pastures in the Rûens region of the Western Cape are mainly legume based with lucerne being the most important and productive legume used (van Heerden & Tainton, 1987). Most pastures are sown to the land race cultivar SA Standard (SAS). Grazing resistance of SAS is high, but relative to other imported cultivars resistance to endemic insect and other pests is poor. This study therefore involves the evaluation for yield and persistence of nine new cultivars compared to SAS under local grazing conditions.

Methods The cultivars WL320 (320), WL414 (414), PAN4546 (546), PAN4764 (764), Meteor (MET), Alfagraze (ALF), Aurora (AUR), Genesis (GEN) and Aquarius (AQU) were compared with SAS under dryland conditions and heavy (10 sheep/ha) continuous grazing with Merino sheep in the Caledon district with an average winter rainfall of 470 mm, for three seasons (2001/02 to 2003/04). The cultivars varied in winter activity (see Table 1). The soil of the trial site was fertilised with P and lime and well cultivated. Individual plots were 5m x 50m and the respective cultivars were randomly allocated to plots in three blocks and sown in rows at 20 kg/ha during May 2001. All seed was inoculated with standard commercial Rhizobium. Grazing started during October 2001. An index of lucerne plant cover (number of grids with lucerne plant material in a 60 grid 0.25 sq m quadrate) was derived for each cultivar during September 2002 and August 2004 and expressed relative to SAS. Yield was determined, using round 0.70 sq m exclosure cages constructed of wire mesh, taking 0.17 sq m samples six-weekly in- and outside each cage and moving the cages to new positions in the plots after sampling. Cut samples were washed, dried and weighed. The data was analysed over three seasons, using analysis of variance.

Results Plant cover and average annual yield of the respective cultivars, are shown in Table 1. The relative cover of the cultivars varied ($p<0.05$) between sampling dates. Cultivar PAN4546 had the highest ($p<0.05$) cover in September 2002, but did not differ from WL320, Alfagraze and WL414 in August 2004. Plant cover for SAS and Genesis was lower ($p<0.05$) than WL320, Alfagraze, PAN4546, WL414, Meteor and Aurora in September 2002 and than WL320, Alfagraze and PAN4546 in August 2004. The cover of WL320, Alfagraze, Meteor, Aquarius and Genesis relative to SAS did not decline ($p<0.05$) between sampling dates. The relative yield of the cultivars did not vary ($p<0.05$) between seasons. Yield for SAS was lower ($p<0.05$) than WL320, WL414 and PAN4546.

Table 1 Index of plant cover and average annual yield of 10 grazed lucerne cultivars

	Cultivar (winter activity)									
	546 (5)	320 (5)	ALF (2)	414 (7)	MET (2)	AUR (7)	AQU (8)	764 (7)	SAS (6)	GEN (7)
Date	Cover (relative to SAS)									
2002	1.65 a*	1.58 b	1.54 bc	1.37 cde	1.28 defg	1.22 defg	1.18 efgh	1.14 gh	1.00 hi	0.86 ij
2004	1.35 cdef	1.40 bcd	1.38 cd	1.16 fgh	1.10 gh	1.04 hi	1.12 gh	0.60 kl	1.00 hi	0.87 ij
	Average yield (kg DM/ha)									
	3047ab	3258 a	2695 abc	3103ab	2792 abc	2588 bc	2677 bc	2561 bc	2235 c	2251 c

* Data followed by the same letters do not differ significantly ($p < 0.05$)

Conclusions With the possible exception of PAN4764, all the cultivars evaluated in the trial are better than or at least equal to SAS in suitability for use in pastures in the area where the trial was conducted. The three highest yielding and most persistent cultivars were also the least winter active types. These cultivars therefore are recommended for use within the area.

References

van Heerden, J.M., & N.M. Tainton (1987). Potential of medic and lucerne pastures in the Rûens area of the Southern Cape. *Journal of the Grassland Society of Southern Africa*, 4, 95-99.

Aluminum tolerance in the model legume *Medicago truncatula*

M.K. Sledge, B. Narasimhamoorthy, P. Pechter and J.H. Bouton
The Samuel Roberts Noble Foundation, 2510 Sam Noble Parkway, Ardmore, Oklahoma, 73401, USA, Email: mksledge@noble.org

Keywords: lucerne, plant breeding, aluminum toxicity, microarray

Introduction Aluminum (Al) is the most abundant metal found in the earth's crust, comprising up to 7% of its mass. At low pH, Al becomes soluble and available to plants, resulting in inhibition of root elongation and reduced plant growth. Aluminum toxicity associated with acid soils has been a major obstacle in alfalfa (*Medicago sativa*) production. The objective of this study is to identify genes that are differentially expressed under normal and Al stress conditions in the model legume *M. truncatula*, with the long term goal of using these genes to improve cultivated alfalfa.

Materials and methods A hydroponics procedure was used to screen 272 accessions of *M. truncatula* obtained from the USDA National Plant Germplasm System for aluminum tolerance. Sterilized and sprouted *M. truncatula* seeds (Cohn *et al.*, 2001), placed in plastic mesh-bottomed cups with the roots threaded through the mesh, were floated in a hydroponics culture tank containing 25 L of a modified Blaydes media (Blyades, 1966), with 25% macronutrients (0.5mM CaCl2), pH 4.3, both with and without 50M Al. After five days, seedling root lengths were measured, and relative root length was calculated. Root tips from tolerant and sensitive genotypes were stained with the Al-sensitive stain lumogallion (3-[2,4 dihydroxyphenylazo]-2-hydroxy-5-chlorobenzene sulfonic acid), and visualized using confocal microscopy. Microarray analysis was used to investigate gene expression in response to Al toxicity in an Al sensitive genotype and an Al tolerant genotype. Seedlings were grown without aluminum for 5 days in modified Blaydes medium with 25% macronutrients, 0.5mM $CaCl_2$, and pH 4.3. Seedlings were then transferred to fresh medium with 50 μM Al, and control and Al stressed root tips are harvested at 0 h, 1 h, 6 h, and 24 h. Root tip mRNA was isolated at each time point, and cDNA was prepared from the isolated RNA. Labelled cDNA was hybridized to 16K *M. truncatula* 70mer oligonucleotide microarrays.

Results Tolerance of Al was normally distributed, and sensitive and tolerant genotypes were identified. Images of root tips stained with lumogallion indicate that root tip cells of Al tolerant genotypes accumulate less Al than those of Al sensitive genotypes. Microarray results indicate differences in gene expression between Al sensitive and Al tolerant genotypes.

Conclusions Variability for Al tolerance exists within the USDA collection of *M. truncatula*. Therefore, it is feasible to construct a population segregating for Al tolerance, with the objective of mapping quantitative trait locus for Al tolerance in *M. truncatula*, and this work is currently underway. Microarray results could aid in identifying candidate genes for Al tolerance.

References

Blaydes, D.F. (1966) Interaction of kinetin and various inhibitors in the growth of soybean tissue. *Physiologia Planttarum*, 19, 748-753.
Cohn, J.R., T. Uhm, S. Ramu, Y.-W. Nam, D.-J. Kim, R.V. Penmetsa, T.C. Wood, R.L. Denny, N.D. Young, D.R. Cook, & G. Stacey (2001) Differential regulation of a family of apyrase genes from Medicago truncatula. *Plant Physiology*, 125, 2104-2119.

Genetic variability in different lucerne (*Medicago sativa*) genotypes

V. Miki••; J. Radovi•; Z. Lugi•• and D. Lazarevi••
Institute Serbia,Centre for Forage Crops, Trg Rasinskih partizana 50, 37000 Kruševac, Serbia and Montenegro, Email: vajap@hotmail.com

Keywords: plant height, tillering, lateral branching, leaf length, leaf width

Introduction One of the basic goals of modern lucerne breeding programmes is creation of new cultivars with a great potential for high quality and stable yields of both forage and hay (Riday & Brummer, 2002). Such cultivars meet increased needs of animal husbandry and must contribute to diverse farming systems (Luki••; 2000). Our trial was aimed at determining genetic variability of yield components in 7 lucerne genotypes, as well as at evaluation of their breeding potential as gene donors to new lucerne cultivars.

Materials and methods A trial was conducted between 2001 and 2003 at the Forage Crops Centre Experiment Field in Ma•kovac. It included 7 lucerne genotypes selected from hybrid populations. Since morphological traits often represent yield components, plant height (cm), number of tillers per plant, number of lateral branches per plant, leaf length (mm), leaf width (mm) and plant mass (g) were monitored.

Results The greatest plant height was found in genotype 7 and the smallest in genotype 6 (Table 1), which was in accordance with Urbano & Davilla (2003). There was a positive correlation between number of tillers per plant and plant mass and a negative correlation between these two traits and number of lateral branches. Genotype 1 had the largest number of tillers per plant and the largest plant mass, but the smallest number of lateral branches per plant as well. On the other hand, genotype 7 had the smallest number of tillers per plant and the smallest plant mass, but also the largest number of lateral branches per plant. The greatest leaf length was measured in the genotype 6 and the smallest in genotype 4. Genotype 5 had the widest leaves and genotype 1 the narrowest.

Table 1 Genetic variability of seven lucerne genotypes during 2002-2004

Genotype	Plant height (cm)	No. of tillers per plant	No. of lateral branches per plant	Leaf length (mm)	Leaf width (mm)	Plant mass (g)
1	68.1	57.7	11.3	24.3	7.8	294.60
2	63.3	45.5	11.7	23.3	8.6	214.41
3	71.0	49.4	11.7	24.9	9.0	276.05
4	69.0	53.7	12.3	22.8	7.9	232.69
5	64.6	51.3	12.0	25.0	9.8	257.08
6	61.5	41.3	12.0	26.1	8.9	199.97
7	72.8	35.1	13.6	21.7	9.5	194.21
LSD 0.05	3.2	3.8	0.7	1.6	1.0	32.30
0.01	4.2	5.0	0.9	2.1	1.4	42.66

Conclusions Although the trial will be continued in the years to come, enriched by introducing new genotypes and monitoring more traits, it is obvious that a certain genetic variability of 7 examined lucerne genotypes can be successfully used in future breeding work. This should lead to the creation of new cultivars with an optimal relationship between yield components, resulting in high adaptability to the prevailing environmental conditions and high yield and quality of forage and hay.

References

Luki••; D. (2000). Lucerka [Lucerne]. Institute of Field and Vegetable Crops, Novi Sad, 458 pp.
Riday, H. & E.C. Brummer (2002). Forage yield heterosis in alfalfa. *Crop Science*, 42 (3), 716-723.
Urbano, D. & C.Davilla (2003). Yield and chemical composition evaluation of eleven alfalfa (*Medicago sativa*) varieties under cutting in high land region at Merida state, Venezuela. *Revista de la Facultad de Agronomia*, Universidad del Zulia, 20 (1), 97-107.

Factor analysis of components of yield and quality traits in lucerne (*Medicago sativa*)

A. Jafari and A. Ghamari Zare
Scientific boards of Research Institute of Forests and Rangelands, Tehran, Iran, Email: aajafari@rifr-ac.ir

Introduction Lucerne (*Medicago sativa*) originated from Iran and is one of the most important forage species in this country. The improvement in yield and quality traits are important objectives in breeding programmes. Smith *et al.* (1997) indicated that in the context of dairy production ranked digestibility was the most important criterion whilst high crude protein and low fibre content were of moderate priority. This study sought to determine the dependence relationships between yield, morphological and quality traits in 64 genotypes.

Materials and methods Sixty four lucerne genotypes were planted at Sept. 2000 at Alborz research center, Karaj, Iran, using a simple lattice (8•8) with two replicates. In each plot, six individual plants of each genotype were allocated in 40•40 cm spacing. Data were collected and analysed for 3 harvests of each of 2 years for plant height, tiller number, growth score, dry matter yield (TDM), plant stand, flowering date, plant diameters and leaf stem ratio (LSR). Crude protein (CP), dry matter digestibility (DMD), water soluble carbohydrates (WSC), crude fibre (CF) and acid detergent fiber (ADF) were measured using near infrared spectroscopy. Phenotypic correlation was determined between traits. Estimation of factor loading was based on data averaged over cuts and replications on 13 characteristics of 64 genotypes. The number of factors was estimated using principal components extraction and varimax rotation method.

Results and discussion Table 1 illustrates the changes in mean expression of average values for yield, morphological and quality traits. There was significant variation among genotypes for all traits. For TDM, mean values were relatively small, but standard deviation was large. There were strong positive correlations between TDM and tiller number, growth score, plant stand, plant height and plant diameter (data not shown), in agreement with Jafari *et al.* (2003). There were negative and significant relationships between DMD and plant growth, stand, plant height, plant diameter, CF and ADF. The relationships between DMD with CP and LSR were positively significant. Some 64% of total variance was accounted for in the first three factors. The loading of factors indicate the contribution of each variable (Table 2). Factor 1, which accounted for 35% of variation, was strongly associated with quality traits and considered as the quality factor with CP and DMD having negative and CF and ADF positive loading. Factor 2, which accounted for about 20% of variation, was named as the productivity factor, since it included several traits which are components of TDM, all with positive loading. The third factor was named as phenology factor, since it contained flowering date. The negative relationships between flowering date and LSR indicate that late maturity genotypes had lower LSR. These results indicated that selection of variables for the productivity factor could enable breeders to increase forage yield of lucerne.

Table 1 Mean, standard deviation, maximum and minimum of evaluated traits

Trait	Mean	Std	Min	Max
Tiller number	26.1	6.5	10.3	40.7
Plant growth	34.6	7.5	15.4	50.9
Plant stand	3.24	0.48	1.82	4.09
Plant height	50.3	4.9	34.8	61.4
TDM (t/ha)	9.6	1.6	4.4	13.3
50% flowering	68.0	3.0	62.4	74.3
Plant diameter	32.3	5.4	20.0	41.3
CP %	18.9	1.4	15.7	22.1
DMD %	65.4	3.4	59.2	73.4
LS	1.04	0.11	0.75	1.34
WSC %	12.9	0.8	10.8	14.6
CF %	37.7	2.1	31.3	41.5
ADF %	43.6	2.8	37.1	49.2

Table 2 Factor matrix after varimax rotation and total variance explained for each factor on 64 alfalfa genotypes

Trait	Factors			Commu.
	F1	F2	F3	
CP %	**-0.81**	-0.26	-0.07	0.73
DMD %	**-0.96**	-0.11	-0.04	0.93
CF %	**0.95**	0.07	0.03	0.90
ADF %	**0.92**	0.06	-0.01	0.85
Tiller number	0.04	**0.69**	-0.34	0.60
Plant growth	0.32	**0.54**	0.31	0.49
Plant stand	0.21	**0.65**	-0.16	0.49
Plant height	0.13	**0.86**	0.12	0.77
TDM (t/ha)	-0.06	**0.77**	0.31	0.69
Plant diameter	0.14	**0.58**	-0.28	0.44
50% flowering	0.12	-0.31	**0.51**	0.36
LSR	-0.28	-0.15	**-0.65**	0.52
WSC %	-0.30	-0.04	**0.54**	0.41
% of variance	35	20	10	

References

Jafari, A., M. Nosrati Nigjeh & H. Haidari Sharifabad (2003). Comparison of yield, morphological and quality traits in 18 ecotypes and varieties of alfalfa (*Medicago sativa*) grown under irrigated and non-irrigated conditions. *Proceedings of the Seventh International Rangelands Congress*, 1403-1405.

Smith, K. F., K.F.M. Reed & J.Z. Foot (1997). An assessment of relative importance of specific traits for the genetic improvement of nutritive value in dairy pasture. *Grass and Forage Science*, 52, 167-175.

Evaluation of lucerne cultivars for dry matter and seed production in Latvia

S. Rancane and M. Sparnina
Skriveri Research Centre of LUA, Skriveri-1, LV 5126, Aizkraukles reg., Latvia, Email: rancanei@inbox.lv

Keywords: lucerne, cultivars, winterhardiness, seed production, yield

Introduction Lucerne is highly productivity and has a high quality. Nevertheless, lucerne growing in Latvia is limited, because of a large area of acid soils (pH_{KCl}<5.8) as well as low and unstable seed yields (Berzins, 2002). In recent years lucerne has taken a more important place in forage production in Latvia, because it is a fast growing crop and in good years gives 3-4 harvests per season. It is important to determine which cultivars of lucerne are more productive and give good seed yields every year in Latvia (Luksa, 2002).

Materials and methods For the estimation of green mass and dry matter (DM) yield of lucerne cultivars, a field trial was conducted using a split plot design with 4 replicates. Plot size was 1.5 x 7.0 m. Seed productivity was estimated at the 5.0 m long rows from the centre to the outside of a circle, where space between rows was 0.1 m in the centre of the circle and 1.20 m on the outside. The purpose of that arrangement was to evaluate the optimal space between rows for successful seed production. The field trial was conducted on a podzolic loamy sand , with a 24-25 cm top layer, a soil organic carbon content of 20 g/kg, soil pH_{KCl} of 5.8- 6.2, a content of plant available P_2O_5 of 88-124 mg/kg and of K_2O 126-150 mg/kg. The experiment lasted for 5 years.

Results High green mass and DM yields were harvested for 4 – 6 years depending on the cultivar (Figure 1). In order to obtain high lucerne seed yields, it is necessary, during the vegetation period that the sum of temperatures above +10 ^{0}C is > 2000 ^{0}C. During the first year, when summer was hot and the sum of temperatures above +10 ^{0}C was 2413^{0}C, seed production was extremely high (53-90 g/m^2). During the fifth year, however, the yields obtained went down markedly as the summer was cooler; and subsequently the production diminished. The productivity in the case of the U.S. seeds fell faster than that of local and East European cultivars (Table 1). Highest yields of green mass were obtained during the 2nd year (84.5–98.0 t/ha). The sum of temperatures above +10 ^{0}C in that year was 2255 ^{0}C.

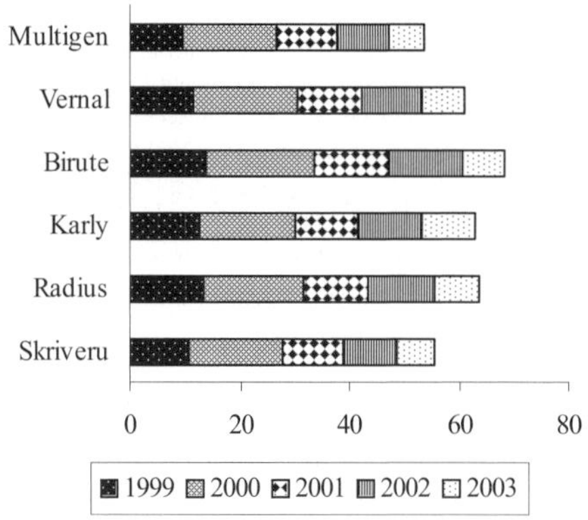

Table 1 Evaluation of lucerne cultivars

Figure 1 DM yield (t/ha) of lucerne cultivars

Legend: ■ 1999 ▨ 2000 ☒ 2001 ■ 2002 ☐ 2003

Cultivar (country)	Winter-hardiness 1 - 10 points	Re-growth at spring 1 - 10 p.	Yield of seeds g/m^2 1st year	Yield of seeds g/m^2 5th year
Skriveru (Latvia)	9.0	7.0	90.1	21.0
Radius (Poland)	9.0	9.5	90.7	6.6
Karly (Estonia)	9.5	7.5	66.3	21.6
Birute (Lithuania)	9.0	9.0	80.6	20.3
Vernal (USA)	6.0	8.0	62.3	5.2
Multigen (USA)	7.0	8.0	53.2	12.2

Conclusions It was concluded that the cultivar 'Birute' (Lithuania) stood out with high and durable green mass and seed yields. The cultivar ' Skrīveru' (Latvia) stood out with durable and high yields also in more acid soils.

References
Berzins P., B. Jansone, S. Bumane, M. Sparnina. & S.Luksa (2002) Results obtained in breeding of forage grasses and legumes. *Proceedings in Agronomy-* (Latvian) Nr.4, 181- 185.
Luksa S., M Sparnina & S. Bumane. (2002) The forage quality of legume and perennial ryegrass varieties in SRC, Latvian Multi-function grasslands. Quality Forages, Animal products and Landscapes. *Proceedings of the 19th General Meeting of the European Grassland Federation*, La Rochelle, France 27-30 May 2002, 440-441.

Relationships between nitrogenase activity, dry forage and water soluble carbohydrates content of selected red clover genotypes

T. Simon[1] and H. Jakesova[2]

[1]Research Institute of Crop Production, Drnovska 507, 161 06 Prague 6 – Ruzyne, CZ, Email: simont@vurv.cz,
[2]Plant Breeding Station, Fulnecka 100, 742 47 Hladke Zivotice, CZ

Keywords: Trifolium pratense

Introduction Improving N_2 fixation of perennial forage legumes through selection and breeding reduce reliance on soil N and N fertiliser, whilst enhancing residual benefits to subsequent crops and increasing legume forage dry matter (DM) yield. Simultaneously with the selection for enhanced N fixation, new legume genotypes should be evaluated on forage quality, especially nutrient concentration and dry matter digestibility.

Materials and methods Hydroponic perlite experiments were used for screening 1200 individual plants of 38 tetraploid red clover cv 'Dolina' genotypes. The selection was based on two previous cycles of recurrent breeding for enhanced nitrogenase activity. Standard cultivar 'Tempus' was included as a criterion for the selection. Effective *Rhizobium leguminosarum* bv. *trifolii* strain HZ6 was used for inoculation. Plants were cultivated in plastic pots in a greenhouse (Jakesova *et al.*, 1995). Total nitrogenase activity (TNA) (μmol C_2H_4/plant/hour) was measured according to Hardy *et al.* (1973) at the onset of anthesis. Simultaneously with TNA, root volume and forage DM were determined. Water soluble carbohydrates content (WSC) was measured in forage DM using NIRS spectrometry.

Results Mean values of measured characteristics are given in Table 1. Hundred and fifty one plants that exceeded TNA of cv. Tempus by 10 % were selected. TNA significantly correlated (p<0.01) with forage DM yield (r=0.4075) and root volume (r=0.5023). Some selected plants had extremely high TNA. WSC content was positively but weakly related to TNA (Fig. 1).

Table 1 Mean values of TNA, dry forage, root volume and WSC of selected red clover plants

Plants	TNA (μmol C_2H_4/plant/hour)	Dry forage (g/plant)	Root volume (cm³)	WSC (mg/g dry forage)
all plants selected	1.57	3.29	15.50	17.26
above the average plants	1.89	3.68	18.14	18.91
poor plants	0.04	1.42	3.06	9.29

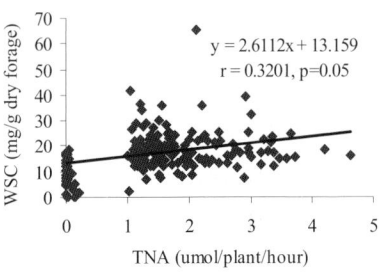

Figure 1 Relationship between TNA and WSC

Conclusions New genotypes of tetraploid red clover were selected on the basis of TNA. Positive significant correlations of TNA and other measured characteristics provided promising plant material for future breeding.

References
Hardy, R.W.F., R.C. Burns & R.D. Holsten (1973). Application of the acetylene-ethylene assay for measurement of nitrogen fixation. *Soil Biology and Biochemistry*, 5, 47-81.
Jakesova, H., T. Simon & M. Machova (1995). Plant selection for a high nitrogenase activity and its using in red clover breeding (In Czech). *Rostlinna Vyroba*, 41, 571-575.

Evaluation of a white clover variety with increased resistance to stem nematode (*Ditylenchus dipsaci*) under sheep grazing and cutting

T.A. Williams, M.T. Abberton, K.A. Mizen, P. Olyott and R. Cook
Institute of Grassland and Environmental Research, Plas Gogerddan, Aberystwyth, Ceredigion, SY 23 3EB, UK,
Email: michael.abberton@bbsrc.ac.uk

Keywords: yield, persistence, agronomic features, management regime

Introduction Stem nematode (*Ditylenchus dipsaci* (Kühn) Filipjev) is a major pest of white clover (*Trifolium repens* L.) in UK pastures (Cook *et al.*, 1992a) and in other parts of the world. In a previous trial, resistant and susceptible selections yielded the same in three years in the absence of the nematode but, on infested plots, the susceptible yielded the same as the resistant selection in year 1 but only 68 and 58% in years 2 and 3, as the nematode infestation increased (Cook *et al.*, 1992b). We have now developed varieties with enhanced resistance to this pest by screening under controlled conditions (Plowright *et al.*, 2002). We describe an experiment to test the hypothesis that stress imposed on the plant by grazing as opposed to cutting management would exacerbate the effects of nematode infestation and accentuate the advantages of resistance through longer survival of clover plants.

Materials and methods Two morphologically similar white clover lines of small leaf size, selected from the same parental material for stem nematode resistance (AC63) and susceptibility (Ac4586) and a 50:50 mixture were sown (3kg/ha) in 2000 in 5 x 4m plots with perennial ryegrass (25kg/ha cv. Fennema). Two adjacent experiments were sown, each in a randomised complete block design with four replicates, one as the stem nematode-free (control) and the other infested with nematodes at the end of the establishment year. Plots were divided into areas subjected to either cutting or continuous sheep grazing. Cutting and sampling of all areas was carried out 6 times during the growing season at 5 week intervals to determine dry matter yield and clover content. Statistical analyses (ANOVA) were conducted separately on each experiment.

Results On the infested trial, stem nematode symptoms were present during the first harvest year but severe infestations did not develop in the second year. Ac4586 yielded least in 2002 (*P*<0.001) under both managements (Figure 1). Grazing reduced yields of all three clovers in the infested trial in 2002 (*P*<0.05) and in both trials in 2003 (*P*<0.001). To compare the trials, yields were expressed as 100*(AC63-Ac4586) /Ac4586.

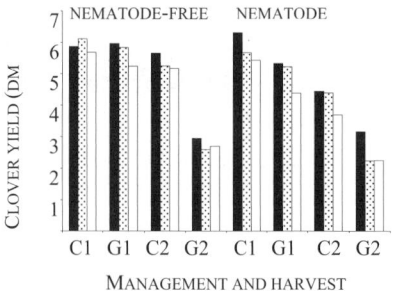

Resistant AC63 yielded 6% and 11% more than Ac4586 under cutting and grazing, respectively, on the nematode-free trial and 20% and 41% on the infested trial.

Conclusions In the absence of a severe infestation of stem nematode, grazing had the major impact on clover yields and persistence. The resistant clover yielded more than the susceptible and the differences appeared to be greater under grazing than cutting. To date, the results lend support to the previous trial indicating an advantage of resistance. The resistant AC63 has good yields and agronomic features.

Figure 1 Yields in two harvest years (1: 2002; 2: 2003) of three clover varieties (black column: AC63, resistant; white: Ac4586, susceptible; stipple: 50:50 mixture) in trials managed by cutting (C) or grazing (G) and infested or not by stem nematode

Acknowledgements This project was funded by the Department for Environment, Food and Rural Affairs. IGER is grant aided by the Biotechnology and Biological Sciences Research Council.

References
Cook R., D.R., Evans, T.A., Williams, & K.A. Mizen (1992a). The effect of stem nematode on establishment and early yields of white clover. *Annals of Applied Biology,* 120, 83-94.
Cook R., D.R. Evans, T.A. Williams, & K.A. Mizen (1992b). Resistance to stem nematode and persistence of white clover. Proceedings of 21st International Nematology Symposium, Albufeira, Portugal, April 1992. *Nematologica,* 38, 405.
Plowright, R A, G. Caubel, & K.A. Mizen (2002). *Ditylenchus* species. In: J.L. Starr, J. Bridge & R. Cook (eds) Evaluating plants for resistance and tolerance to nematodes. CABI International, Wallingford, 107-139.

Breeding white clover with improved tolerance of nitrogen fertiliser

T.A. Williams, M.T. Abberton and P. Olyott
Legume Breeding and Genetics Team, Institute of Grassland and Environmental Research, Plas Gogerddan, Aberystwyth, Ceredigion SY 23 3EB, Wales, U.K., Email: michael.abberton@bbsrc.ac.uk

Keywords: persistence, yield, agronomic performance, mixed sward

Introduction White clover (*Trifolium repens* L.) is often considered a forage legume with a primary use in 'low input/ low output' systems. One facet of this is the perception that the persistency of this species is poor when Nitrogen (N) fertiliser is applied. However, new varieties of white clover are able to play a significant role in highly productive systems (Williams *et al.*, 2000) and show consistent yields over ten years at a range of applied N levels (Williams *et al.*, 2003). Germplasm improvement for nitrogen tolerance has been carried out with the aim of not only allowing white clover to perform well under applied N but also to dampen the oscillations in clover yield that may be a consequence of the build up of N fixed by the clover itself. The former aspect is illustrated in this paper with respect to the variety AberConcord.

Materials and methods Four medium leaf size varieties (AberConcord, Menna, Merwi and Grasslands Huia all at 3 kg/ha) were sown in 2 x 1.4m plots with perennial ryegrass (variety Fennema at 25 kg/ha). Plots were sown in 1998 in a randomised complete block design with four replicates. Four different treatments were imposed: No Nitrogen (N 0), 150kg N/ha per year (N1), 300kg N/ha per year (N2) and 450kg N/ha per year (N3). The experiment was sampled six times in each of 1999, 2000 and 2001 to obtain clover and grass dry matter.

Table 1 Clover content (as proportion dry matter of total sward) of four varieties at four N levels over three years

N0 Treatment	Year 1	2	3
AberConcord	75.6	79.8	58.5
Menna	73.7	73.8	50.8
Merwi	74.7	78.4	59.0
Huia	77.5	73.0	58.8
N1 Treatment			
AberConcord	50.0	63.8	42.2
Menna	46.3	43.5	33.0
Merwi	43.1	54.4	32.1
Huia	46.3	48.6	38.1
N2 Treatment			
AberConcord	29.8	55.9	37.5
Menna	21.1	39.9	14.9
Merwi	20.4	46.8	28.8
Huia	20.3	37.7	24.9
N3 Treatment			
AberConcord	15.5	47.0	42.8
Menna	12.6	27.4	14.4
Merwi	20.6	40.7	28.4
Huia	10.9	31.6	30.2

Results AberConcord showed consistently higher yields than the other varieties at all N levels (Table 1). However, differences were more pronounced at higher N levels. Thus, at N 2, AberConcord gave a clover dry matter yield of 4.3 t/ha in the third year compared to a mean of 3.1t/ha across all varieties. At the highest N application, AberConcord had yields of 5.8 and 4.9 t/ha in years 2 and 3, significantly higher than the other varieties.

Indeed, AberConcord had a higher yield at N3 than at N1 or N2 in the third year. Clover contents of 47% and 42% in the 2nd and 3rd years were achieved by AberConcord at the highest N level with total dry matter yield of 12.4 and 11.6 t/ha respectively.

Conclusions The results suggest that breeding for improved tolerance of N allows a considerable white clover contribution to be maintained even at levels of N fertiliser application of 450 kg/ha per year. Although N fixation is likely to be reduced, gains in animal production, derived from the forage quality and intake characteristics of white clover are still likely to be realised. The competitiveness of the white clover allows a balanced sward to be maintained along with high total dry matter content.

Acknowledgements The authors thank the Department of Environment Food and Rural Affairs for financial support. IGER is grant-aided by the Biotechnological and Biological Sciences Research Council.

References
Williams, T.A. M.T. Abberton, D.R. Evans, W. Thornley, & I. Rhodes (2000). Contribution of white clover varieties in high-productivity systems under grazing and cutting. *Journal of Agronomy and Crop Science,* 185, 121-128.
Williams, T.A. D.R. Evans, I. Rhodes, & M.T. Abberton (2003). Long-term performance of white clover varieties grown with perennial ryegrass under rotational grazing by sheep with different nitrogen applications. *Journal of Agricultural Science,* 140, 151-159.

Forage quality of white clover (*Trifolium repens* L.) X Caucasian clover (*T.ambiguum* Bieb.) hybrids over three harvest years

T.P.T. Michaelson-Yeates, A.H. Marshall, M.T. Abberton, T.A. Williams, P. Olyott and H.G. Powell
Institute of Grassland and Environmental Research, Plas Gogerddan, Aberystwyth, Ceredigion, SY23 3EB, UK, Email: terry.michaelson-yeates@bbsrc.ac.uk

Keywords: interspecific hybrids, introgression, rhizomatous trait, chemical composition

Introduction Interspecific hybrids have been produced from crosses of white clover, a stoloniferous species with Caucasian clover, a rhizomatous species. Using white clover as the recurrent parent first and second generation backcross (BC_1 and BC_2) plants have been produced that have both rhizomes and stolons and are more drought tolerant than white clover (Marshall *et al.*, 2001). Forage quality of these interspecific hybrids was investigated to determine whether introgression of the rhizomatous trait has any impact on forage quality.

Material and methods In September 1998 16 plots of 2.7m x 1.5m were established, with 32 plants of *T.repens, T. ambiguum*, BC_1 or BC_2 in each plot. After transplanting each plot was oversown with the intermediate perennial ryegrass, Fennema, at a seed rate of 25 kg ha^{-1}. Once established, plots were cut five times in 1999 and 2000 and six times in 2001, from May with the final cut in October. Clover and grass fractions were milled separately through a 1 mm sieve. In vitro dry matter digestibility (DMD), nitrogen (N) and water soluble carbohydrates (WSC) concentrations were predicted by near infrared reflectance spectrophotometer. Crude protein (CP) concentration was then calculated as N concentration x 6.25.

Results After the establishment year parental material of *T.ambiguum* was absent, and forage quality of the clover fraction was analysed without the *T.ambiguum* plots (Table 1). Legume fractions from the BC_1 and BC_2 hybrid plots had a higher WSC, a lower CP concentration but an *in vitro* DMD value comparable with white clover, (supporting the results of a previous glasshouse study by Abberton *et al.*, 2002). These differences were observed in all harvest years, and, generally, the backcross hybrids and *T.repens* followed a similar trend throughout the growing season.

Table 1 In vitro DMD, WSC and CP (g kg^{-1}DM), of the clover fraction of plots of *Trifolium repens,* and the backcross hybrids (BC_1 and BC_2). Data is a mean of three harvest years

	DMD	WSC	CP
T. repens	0.809	58.8	236.5
BC_1	0.810	66.7	210.3
BC_2	0.810	64.8	222.3
s.e.d.	0.0019 NS	1.50***	3.02***

s.e.d., standard error of difference of means; NS, not significant; *$P< 0.05$; **$P< 0.01$; ***$P< 0.001$

Conclusions Compared to white clover the backcrosses had a lower CP concentration and this may contribute to a more efficient utilization of protein in the rumen, thus reducing the release of nitrogenous waste. The significant differences between the backcross hybrids and *T.repens* were largely attributed to differences in forage quality at the beginning of the growing season and further work will determine whether these differences are beneficial in livestock feeding experiments. The fertility of these hybrids is poor and more cytological studies need to be made using fluorescence *in situ* hybridisation at meiosis to select plants with stable integration of introgressed genes.

Acknowledgements We gratefully acknowledge the financial support of Defra and BBSRC.

References
Abberton M.T., A.H. Marshall, T.P.T. Michaelson-Yeates, T.A. Williams & I. Rhodes (2002). Quality characteristics of backcross hybrids between *Trifolium repens* and *Trifolium ambiguum. Euphytica*, 127, 75-80.
Marshall A.H., C. Rascle, M.T. Abberton, T.P.T. Michaelson-Yeates & I. Rhodes (2001). Introgression as a route to improved drought tolerance in white clover (*Trifolium repens* L.). *Journal of Agronomy and Crop Science*, 187, 11-18.

Forage quality of white clover (*Trifolium repens* L.) X Ball clover (*T.nigrescens Viv.*) hybrids over three harvest years

T.P.T. Michaelson-Yeates, M.T. Abberton, A.H. Marshall, T.A. Williams, H.G. Powell and P. Olyott
Institute of Grassland and Environmental Research, Plas Gogerddan, Aberystwyth, Ceredigion, SY23 3EB, UK,
Email: terry.michaelson-yeates@bbsrc.ac.uk

Keywords: backcrosses, reproductive characters, chemical composition

Introduction Introgression of reproductive traits from the annual, profusely flowering species ball clover into white clover is one route to improve seed yields in *T.repens*. The interspecific cross produced F_1, backcross 1 (BC$_1$), backcross 2 (BC$_2$) and backcross 3 (BC$_3$) plants with white clover as recurrent parent (Marshall *et al.*, 2002). These hybrids were found to be comparable with white clover for yield and persistency but produced 30% more flowers and their forage quality, relative to white clover was investigated.

Material and methods In August 1997 24 plots of 3.0m x 1.5m were established, with 40 plants of *T.repens, T. nigrescens*, F_1, BC$_1$, BC$_2$ or BC$_3$ in each plot. After transplanting each plot was oversown with the intermediate perennial ryegrass, Fennema, at a seed rate of 25 kg ha^{-1}. Plots were cut seven times in 1998, six times in 1999 and four times in 2000, from May with the final cut in October. Clover and grass fractions were milled separately and *in vitro* dry matter digestibility (DMD), nitrogen (N) and water soluble carbohydrates (WSC) concentrations were predicted by near infrared reflectance spectrophotometer. The CP concentration was then calculated as N concentration x 6.25.

Results After the establishment year the legume component of the *T.nigrescens* and F_1 plot was absent and there were no significant differences between white clover and the hybrids for DMD in any of the harvest years. The WSC concentration of the backcross hybrids was less than that of white clover, significantly so in 1998 and 1999, and the CP concentration greater but only significantly so in 1999 and 2000 (Table 1). There was some evidence that the differences between white clover and the hybrids were less in the later backcross generations.

Table 1 WSC and CP concentration (g kg^{-1}DM) of the legume component of plots of *Trifolium repens*, BC$_1$, BC$_2$ and BC$_3$ hybrids in three harvest years

	WSC			CP		
	1998	1999	2000	1998	1999	2000
T.repens	47.0	56.1	76.4	247.6	254.6	242.8
BC$_1$	37.7	43.9	60.5	258.6	271.6	261.4
BC$_2$	38.2	45.3	67.9	270.3	266.3	255.3
BC$_3$	37.5	47.5	64.1	266.4	263.9	253.4
s.e.d.	1.91	2.49	4.87	10.39	3.12	6.22
Significance	**	***	n.s.	n.s.	***	*

s.e.d., standard error of difference of means; NS, not significant; *P< 0.05; **P< 0.01; ***P< 0.001.

Conclusions The difference in CP concentration between white clover and hybrids generally reduced with each backcross generation. This suggests that introgression of reproductive traits from *T.nigrescens* can be achieved without sacrificing forage quality and may improve some aspects of forage quality, although these improvements may be removed by further backcrossing. Currently a range of BC3 hybrids, placed into three leaf size groups is being evaluated for agronomic traits and this material will be the basis for the first hybrid varieties.

Acknowledgements We gratefully acknowledge the financial support of Defra and BBSRC.

References
Marshall A.H., T.P.T. Michaelson-Yeates, M.T. Abberton, A. Williams & H.G. Powell (2002). Variation for reproductive and agronomic traits among *T.repens* x *T. nigrescens* third generation backcross hybrids in the field. *Euphytica*, 126, 195-201.

New approaches to clover breeding

M.T. Abberton, T.A. Williams, T.P.T. Michaelson-Yeates, A.H. Marshall, C. Jones, E. Sizer-Coverdale and R.P. Collins
Institute of Grassland and Environmental Research, Plas Gogerddan, Aberystwyth, Ceredigion, SY 23 3EB, UK, Email: michael.abberton@bbsrc.ac.uk

Keywords: white clover, red clover, environment, quality, breeding programmes

Introduction White clover (*Trifolium repens* L.) and red clover (*T. pratense*) are the major forage legumes of temperate pastures. Breeding efforts have focused on overcoming the constraints to productivity and reliability in this species and thereby optimising their contribution to mixed swards. In recent years there has been an increased emphasis on livestock production and the efficient utilisation of forage material in the rumen. In this paper we report on a shift in the aims of forage legume breeding at IGER, building on a strong agronomic platform but giving greater consideration to the environmental footprint of our varieties and the contribution that they can make to the quality of meat and milk.

White clover In white clover we are focusing on reducing phosphorus and nitrogen leaching. We have shown variation in the field for ability to yield well at reduced P levels and are relating genotypic variation to leaching propensity. We are also approaching the problem of nitrogenous pollution by improving the efficiency of protein utilisation in the rumen. Protein levels in agronomically superior individual plants show considerable, hitherto unsuspected, variation implying scope for selection to promote a balanced supply of energy (primarily from companion grasses) and protein. In collaboration with the Centre for Ecology and Hydrology at Bangor, UK we are also evaluating ozone tolerance in relation to white clover agronomic performance.

Red Clover The red clover breeding programme at IGER recommenced in 1998 after an interval of more than twenty years; there are considerable opportunities for improvements in yield, persistency, grazing tolerance and pest and disease resistance. Nonetheless, work on this species too is bringing into focus the needs of a more environmentally friendly livestock sector. Thus, we are developing lines with enhanced protein protection, due to increased levels of the enzyme polyphenol oxidase (PPO), which have the potential to reduce nitrogenous pollution from the silo. Ongoing work in collaboration with animal scientists at IGER is exploring variation in polyunsaturated fatty acid profiles in red clover and their consequences in terms of meat and milk quality.

Breeding strategies and methods The evaluation of breeding lines through the ruminant animal or analysis of impacts on the environment poses considerable logistical challenges to classical approaches to improvement, necessitating novel strategies and methods. Molecular marker approaches are being developed consequent to the creation of the first molecular genetic map in white clover (Jones *et al.* 2003) in a collaboration between IGER and the Plant Biotechnology Centre in Victoria, Australia. Isobe *et al.* (2003) recently reported the development of the first map in red clover. The use of marker assisted selection (MAS) offers the possibility of making gains in some of the traits described above and the transfer of genetic and genomic information, such as sequence information and transcriptome arrays, from the 'model legumes' *Medicago truncatula* and *Lotus japonicus* will greatly facilitate this process. At IGER, we are using simple sequence repeat markers derived from *M. truncatula* expressed sequences to assess their utility in white and red clover and determine the degree of synteny between these three species. While the evaluation of clover content and quality of mixed swards is now carried out by near infrared reflectance spectrophotometry (NIRS) and the combined application of NIRS and MAS through multivariate statistical methods offers considerable potential for the future.

The scale at which environmental impacts such as reduced pollution of waterways are manifest is beyond that of the small plots traditionally used to assess clovers under grazing or cutting regimes. Evaluation and improvement for these traits and on this scale will require not only the use of MAS but also a combination of high resolution environmental monitoring and modelling approaches.

Acknowledgements The authors thank the Department of Environment Food and Rural Affairs for financial support. IGER is grant-aided by the Biotechnological and Biological Sciences Research Council.

References
Isobe, S., I. Klimenko, S. Ivashuta, M. Gau, & N.N. Kozlov (2003). First RFLP linkage map of red clover (*Trifolium pratense* L.) based on cDNA probes and its transferability to other red clover germplasm. *Theoretical and Applied Genetics,* 108, 105-112.
Jones, E.S, L.J. Hughes, M.C. Drayton, M.T. Abberton, T.P.T. Michaelson-Yeates, C. Bowen & J.W. Forster (2003). An SSR and AFLP molecular marker-based genetic map of white clover (*Trifolium repens* L.) *Plant Science,* 165, 447-479.

Quantifying the variation in protein content in white clover (*Trifolium repens* L.)

A.H. Marshall, E. Sizer, A. Kingston-Smith, A. Williams and M.T. Abberton
Institute of Grassland and Environmental Research, Plas Gogerddan, Aberystwyth, Ceredigion, SY23 3EB, UK,
Email: athole.marshall@bbsrc.ac.uk

Keywords: white clover, protein content, Bradford assay

Introduction White clover (*Trifolium repens* L.) is the main legume in temperate pastures. It has relatively low levels of water-soluble carbohydrate but produces forage of high quality with a high crude protein (CP) content and dry-matter digestibility (Beever, 1993). Some studies have suggested that the forage quality of white clover can be problematic because its high CP content may contribute to inefficient use of nitrogen in the rumen and exacerbate diffuse pollution via excreta (Waghorn & Caradus, 1994). The development of white clover germplasm with lower CP content would potentially benefit forage production and grassland management. A study was carried out to quantify the variation in CP content within an existing gene pool and develop high throughput techniques for protein determination appropriate to a plant breeding programme.

Materials and methods A medium and large leaf gene pool comprising 1000 genotypes (50 plants each of 10 medium and 10 large leaved varieties) grown as spaced plants in the field were sampled for protein content in summer 2003. The youngest fully expanded trifoliate leaf with petiole on three primary stolons per plant was sampled from each plant at peak flowering. Chlorophyll content of the middle trifoliate leaf was also measured using a SPAD meter. Protein content was measured using a modified version of the assay described by Bradford (1976), used by Kingston-Smith et al., 2003 and chlorophyll content by spectrophotometery. A sub-sample of 48 plants, across the range of measured protein content, was identified for further study. A clone of these plants was produced by removing a stolon core which was planted into pots in an unheated glasshouse. In summer 2004 protein content was measured on this subset of plants in the glasshouse and field as previously described.

Results Protein contents (mg/g fresh weight) in the medium leaf and large leaf gene pools were 0.48-8.09 and 0.29-4.99, respectively. There was no significant correlation between protein content and the chlorophyll content derived by the SPAD meter, or by assay, for either the medium or large leaf gene pools. The mean protein content of the sub-sample plants in the field did not differ significantly from that of the plants grown in the glasshouse (2.81 (range 0.92-8.53) vs. 2.97 (range 0.16-6.46) respectively). The correlation between the protein content of plants in the field and the cloned plants grown in the glasshouse was low and not significant. Similarly the correlation between the field and the glasshouse of the ranking of genotypes on the basis of their protein content was not significant (r=0.06; ns).

Conclusions Foliar protein content was lower than previously reported for white clover (Kingston-Smith et al., 2003), perhaps due to the fact that petioles were included in the assay. Although protein content varied widely among the medium and large leaf gene pools in the field, the non-significant correlation between protein and chlorophyll content confirmed the SPAD to be an inappropriate way to measure protein content. Although the wide range of variation in protein content between genotypes in the field was also observed in the glasshouse, there was no significant correlation between the protein content in the field and in the glasshouse. Similarly the correlation between the ranking of the genotypes in terms of their protein content in the glasshouse and field was not significant. This suggests that individual genotypes differed in their response to environment in terms of plant protein content, making identification of plants with consistently high or low protein content difficult to select. It also suggests that the environment where screening takes place is important and should be selected to represent the environments where white clover will be utilised if meaningful results are to be obtained. Further work is necessary, in terms of sampling plant material and in replication, if this technique is to provide reliable information on plant protein content that can be used in a plant breeding programme.

Acknowledgements The financial support of Defra and the BBSRC is acknowledged gratefully.

References

Beever, D. E. (1993). Ruminant animal production from forages: present position and future opportunities. In. Baker, M. J. (ed.) *Grasslands for our World*, pp.158-164. Wellington, New Zealand: SIR Publishing.
Bradford, M. M. (1976). A rapid and sensitive method for the quantitation of microgram quantities of protein utilizing the principle of protein-dye binding. *Analytical Biochemistry*. 72, 248-254.
Kingston-Smith, A. H., A. L.Bollard, I. P. Armstead, B. J. Thomas & M. K. Theodorou. (2003). Proteolysis and cell death in clover leaves is induced by grazing. *Protplasma*. 220, 119-129.
Waghorn, G. C. & J. R. Caradus (1994). Screening white clover cultivars for improved nutritive value-development of a method. *Proceedings of the New Zealand Grassland Association*, 56, 49-53.

Quantitative trait locus analysis of morphogenetic and developmental traits in an ssr- and aflp-based genetic map of white clover (*Trifolium repens* L.)

M.T Abberton[1], N.O.I. Cogan[2], K.F. Smith[3], G. Kearney[3], A.H. Marshall[1], A. Williams[1], T.P.T. Michaelson-Yeates[1], C. Bowen[1], E.S. Jones[4], A.C. Vecchies[2] and J.W. Forster[2]

[1]*Legume Breeding and Genetics, Institute of Grassland and Environmental Research, SY 23 3EB, UK, Email: michael.abberton@bbsrc.ac.uk* [2]*Primary Industries Research Victoria, Plant Biotechnology Centre, La Trobe University, Bundoora, Victoria 3086, and Molecular Plant Breeding Cooperative Research Centre, Australia,* [3]*Primary Industries Research Victoria, Hamilton Centre, Mount Napier Road, Hamilton, Victoria 3300, and Molecular Plant Breeding Cooperative Research Centre, Australia,* [4]*Crop Genetics, Pioneer Hi-Bred International, 7300 NW 62nd Avenue, Johnston, Iowa 50131-1004, USA*

Overview Molecular marker-assisted plant breeding is a key target for the temperate legume pasture crop white clover (*Trifolium repens* L.). The first genetic linkage map of white clover has been constructed using self-fertile mutants to derive an intercross based fourth and fifth generation inbred parental genotypes (F_2[I.4R x I.5J]). The framework map was constructed using simple sequence repeat (TRSSR) and amplified fragment length polymorphism (AFLP) markers. Eighteen linkage groups (LG) corresponding to the anticipated 16 chromosomes of white clover (2n = 4x = 32), with a total map length of 825 cM were derived from a total of 135 markers (78 TRSSR loci and 57 AFLP loci). The F_2(I.4R x I.5J) family has been subjected to intensive phenotypic analysis for a range of morphogenetic and developmental traits over several years at IGER, Aberystwyth, Wales and East Craigs, near Edinburgh, Scotland. The resulting phenotypic data were analysed independently to identify QTL (quantitative trait loci) for the various traits, using single marker regression (SMR), interval mapping (IM) and composite interval mapping (CIM) techniques. Multiple coincident QTL regions were identified from the different years and different sites for the same or related traits. The data were reanalysed using a meta-analysis across years and sites and Best Linear Unbiased Estimates (BLUEs) were derived for the plant spread, petiole length, leaf width, leaf length, leaf area, internode length, plant height and flowering date traits. A total of 24 QTLs were identified on 10 of the linkage groups. Three regions on LGs 2, 7 and 12 all demonstrated overlapping QTLs for multiple traits (Figure 1). A meta-analysis approach can quickly identify regions of the genome that control the trait in a robust predictable manner across multiple spatial and temporal replication for rapid targeted genetic enhancement via marker-assisted breeding. This first genetic dissection of agronomic traits in white clover provides the basis for comparative trait-mapping studies and the enhanced development and implementation of marker-assisted breeding strategies.

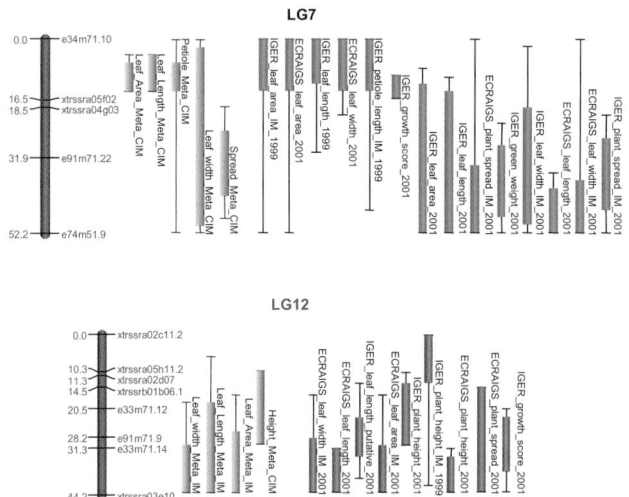

Figure 1 Linkage groups 7 and 12 from genetic map of the F_2(I.4R x I.5J) mapping population, with QTL regions identified. The QTL identified using the meta-analysis are indicated, along with comparison to QTLs from the separate datasets.

Acknowledgements The authors are grateful for financial support from (UK) Defra and BBSRC and (Australia) Victorian DPI, DA, MLA, GGDF and MPB CRC

Evaluation of white clover breeding lines in the Australasian region

M.Z.Z. Jahufer[1], D.R. Woodfield[1], J.L. Ford[1], K.H. Widdup[2], J.F. Ayres[3] and L.A. Lane[3]
[1]AgResearch Limited, Tennent Drive, Private Bag 11008, Palmerston North, New Zealand, Email: zulfi.jahufer@agresearch.co.nz, [2]AgResearch Limited, PO Box 60, Lincoln, New Zealand, [3] New South Wales Department of Primary Industries, PBM, Glen Innes, NSW 2370, Australia

Keywords: genotype-by-environment interaction, pattern analysis, multi-site, persistency

Introduction The accuracy of predicting breeding line performance across target environments is a significant criterion in the development of cultivars with broad or specific adaptation. This paper characterises the type and magnitude of genotype-by-environment (GE) interactions estimated from a multi-site white clover (*Trifolium repens* L.) breeding line evaluation trial conducted across sites in New Zealand and Australia.

Materials and methods A total of 17 experimental breeding lines and 2 control cultivars, Nu Siral and Sustain, were evaluated across 4 sites; Palmerston North and Lincoln in New Zealand, and Armidale and Glen Innes in New South Wales, Australia. Seasonal clover content during years 2 and 3 were expressed as a percentage relative to cultivar Sustain and analysed. Variance component analysis was conducted across years 2 and 3 within sites, and across years and sites, using a random effects linear model. The data were log transformed prior to analysis. The mean values generated from the individual site analysis were used in pattern analyses (Jahufer *et al.*, 2002). Lack of genetic correlation among sites (L_c) and heterogeneity of genotypic variance (V) were calculated from the line-by-site interaction component (Cooper and DeLacy, 1994).

Results There were significant ($P<0.05$) genotypic differences within lines for mean clover content across years 2 and 3 at all sites. There was a significant ($P<0.05$) line-by-site interaction. The large L_c relative to V indicates changes in relative performance of the lines across the 4 sites (Table 1). The biplot indicates the presence of 3 breeding line groups and two individuals that are distinct from the rest (Figure 1). One of these lines was developed into the new cultivar Grasslands Trophy based on broad adaptation. The directional vectors in figure 1 indicate that Armidale, Lincoln and Palmerston North sites are closely associated in comparison to Glen Innes. Reanalysis of the data without Glen Innes, still indicated significant GE interaction (results not presented).

Table 1 Genotypic variance components and standard errors from within site analysis of seasonal clover content across years 2 and 3. (P, Palmerston North; L, Lincoln; G, Glen Innes; A, Armidale)

	P	L	G	A	GE
σ^2	0.055	0.268	0.085	0.017	0.086
±SE	0.024	0.094	0.04	0.007	0.019
$L_C = 0.079$		V = 0.007			

GE, genotype by environment interaction variance component from analysis across sites and years,
L_C, lack of genetic correlation,
V, heterogeneity of genotypic variance

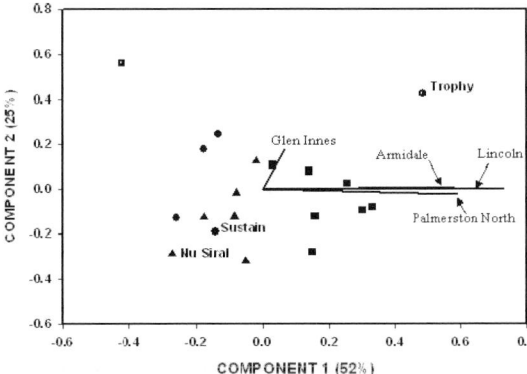

Figure 1 the vectors in the biplot represent the sites. The symbols represent line groups. Positions of the two controls and the breeding line developed into the new cultivar, Grasslands Trophy, are indicated

Conclusion This study indicates the importance of conducting multi-site trials to evaluate breeding lines for selection of superior performers for broad and specific adaptation in the Australasian region.

References

Cooper, M., & I. H. DeLacy (1994). Relationships among analytical methods used to study genotypic variation and genotype-by-environment interaction in plant breeding multi-environment trials. *Theoretical and Applied Genetics*, 88, 561-72.

Jahufer, M.Z.Z., M. Cooper, J.F. Ayres, & R.A. Bray (2002). Identification of future research to improve the efficiency of conventional white clover breeding strategies in Australia – A review. *Australian Journal of Agricultural Research*, 53, 239-57.

Improving the utilisation of germplasm of *Trifolium spumosum* L. by the development of a core collection using ecogeographical and molecular techniques

K. Ghamkhar, R. Snowball and S.J. Bennett
Centre for Legumes in Mediterranean Agriculture, University of Western Australia, 35 Stirling Highway, Crawley, WA 6009, Email: kioumars@cylenne.uwa.edu.au

Keywords: Amplified Fragment Length Polymorphisms, Arcview, MStrat, passport data, pasture

Introduction A core collection is a sub-set encompassing more than 70% of the variability of all accessions held in a collection (Brown 1995), the development of one for *Trifolium spumosum* (bladder clover) could assist in future development of the cultivar within southern Australia. The aim of this work is to develop a core collection of *Trifolium spumosum* as a model for other pasture legume species using molecular and ecogeographical data.

Materials and methods Accessions with near complete ecogeographical data were selected from the Australian *ex situ* collection of *Trifolium spumosum*. This collection of 317 accessions was grouped into 5 geographical regions. MStrat Software (Gouesnard *et al.*, 2001)was used to select the preliminary core of 30% of the collection. Fluorescent Amplified Fragment length polymorphism (FAFLP) will be used to screen the diversity within the species. The primers producing the highest number of bands will be used to screen the preliminary core collection. Mstrat will be used to develop final core collections containing 30% of the preliminary core.

Results A preliminary core collection of 95 accessions was selected. In the randomly selected cores the scores (based on Nei index) were different for each repeat, however, scores were constant for cores selected using the maximising strategy (OPT in Table 1). A final core of 32 accessions will be selected using AFLP and ecogeographical data. The AFLP markers with the green fluorescent labelled EcoR-I primer (TET) showed the greatest amount of data with the highest diversity (Figure 1). The genetic profiles of the preliminary core will be scored and recorded in a database with ecogeographical data.

Table 1 Active scores generated from the optimisation (OPT) and random (RAN) sampling methods using MStrat for core sizes of 76 and 109

Core size	Method	Final score#								
Repeat		1	2	3	4	5	6	7	8	9
76	OPT	101	101	101	101	101	101	101	101	101
	RAN	84	85	84	83	82	84	83	80	81
109	OPT	101	101	101	101	101	101	101	101	101
	RAN	89	91	92	82	88	90	91	92	86

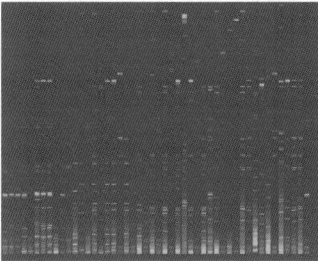

Figure 1 AFLP bands obtained from 48 samples in green fluorescent labelled EcoRI primers

Conclusions The present study hopes to demonstrate that a combination of AFLP marker and ecogeographical data can be used to develop an effective core collection that maintains the majority of the genetic diversity. This model should be used to develop core collections of other pasture legume species that are too large for efficient utilisation.

References

Brown, A.H.D. (1995). The core collection at the crossroads. In: T. Hodgkin, A.H.D. Brown, Th. J.L. van Hintum, & E.A.V. Morales (eds.) Core collections of plant genetic resources John Wiley & Sons, 3-20.
Gouesnard, B., T.M. Bataillon, G. Decoux, C. Rozale, D.J. Schoen & J.L. David (2001). MSTRAT: an algorithm for building germplasm core collections by maximising allelic or phenotypic richness. *Journal of Heredity*, 92, 93-94.

Concurrent selection for low coumarin and multi-stemmed crowns in annual sweetclover

G.R. Smith and G.W. Evers

Texas Agricultural Experiment Station, Texas A&M University System, PO Box 200, Overton, TX 75684 USA, Email: g- smith@tamu.edu

Keywords: forage legume, coumarin, dicoumarol, plant breeding

Introduction Annual sweetclover (*Melilotus alba* Desr) is a reseeding, drought tolerant forage legume that is well adapted to the alkaline and neutral soils of the US Southern Great Plains. Coumarin is a phytochemical found in sweetclover that is a precursor to dicoumarol, which causes a toxic bleeding disease in cattle. The inheritance of low coumarin in sweetclover is determined by a single gene (Goplen *et al.*, 1957), but current low coumarin cultivars are all biennial, northern types. The ability of annual sweetclover to develop a multi-stemmed crown, similar to alfalfa (*Medicago sativa* L.), is also under genetic control (Hartwig, 1942). The objective of this breeding program is to combine the traits of annual flowering, low coumarin and multi-stemmed crowns into improved sweetclover germplasm.

Materials and methods Hand pollinations and bee cage crosses were made between "Denta" (low coumarin biennial) and "Emerald" (high coumarin, multi-stemmed annual) sweetclover with the high coumarin allele used as a genetic marker to identify true hybrids. The Emerald parents were grouped by maturity with a 60 day range of flowering date. The Denta parents were grown under a 20 hr daylength in a growth chamber to force flowering. The low coumarin status of each Denta plant was confirmed before the crossing program began and all seed were harvested from the Denta parents. Hybrids were identified based on the presence of coumarin using a rapid assay (Gorz and Haskins, 1958). All hybrids were self-pollinated in the greenhouse and F_2 seed produced in the spring and summer of 2002. Fifteen hundred F_2 seedlings from each of seven parental maturity groups, for a total of 10,500 plants, were planted for evaluation and screening in the fall of 2002. Emerald and Denta seedlings were also planted for use as checks. A simultaneous seedling screen was initiated for low coumarin (*cu cu*) and the multi-stemmed trait.

Results From 338 hand crosses, 36 hybrids were identified. Forty-seven hybrids were identified from bee cage crosses. About 240,000 F_2 seed were produced from hybrids between Denta and seven different maturity groupings of the Emerald parent. Based on a preliminary study, we developed a screening technique to identify the multiple stem trait in young (6 to 8 weeks of age) sweetclover seedlings. The growth of secondary stems from the axillary bud of the unifolioate leaf and from the axillary buds of the cotyledons was used as a positive signal for the potential development of the multi stem trait. About 2700 (25.7%) sweetclover plants were identified with the potential to express the multiple stem trait. Each seedling was then evaluated for coumarin content and those testing positive were discarded. Four hundred and ninety-one (4.6%) sweetclover seedlings were identified with the trait combination of multiple stem crowns and low coumarin. Artificial lighting was used beginning in early November, 2002 to extend the daylength and force early flowering. By mid December 2002 about 25% of the F_2 selections were flowering. Evaluation, selection and seed production continued in the greenhouse on the low coumarin + multi-stemmed selections. This group of plants represents a selection intensity of about 4% on the original F_2 population of 10,500 plants. In February 2003 another 150 plants were discarded due to severe powdery mildew infection and/or general low vigor. Flowers on each plant were hand tripped to insure pollination. Seed was produced on 193 plants from late spring to mid summer with seed yields ranging from 0.5g to 5.0g per plant. Plants with annual (*AA* and *Aa*) growth habit were identified based on flowering date, relative to biennial (*aa*) plants. About 25 biennial plants were noted that did not flower until spring 2004. Progeny testing will be necessary to identify F_3 families that will breed true for the annual trait.

Conclusions A concurrent seedling screen was successful in the intermediate state of an annual sweetclover breeding program to develop F_3 families with both low coumarin and multi-stemmed crowns.

References

Glopen, B.P., J.E.R. Greenshields, & H. Baenziger (1957). The inheritance of coumarin in sweetclover. *Canadian Journal of Botany*, 35, 583-593.

Gorz, H.J. & F.A. Haskins (1958). Rapid tests for free and bound coumarin in sweetclover. *Agronomy Journal*, 50, 211-214.

Hartwig, E.E. (1942). Inheritance of growth habit, cotyledon color, and cup-leaf in *Melilotus alba*. *Agronomy Journal*, 34, 160-166.

Analysis of genetic diversity in white clover (*Trifolium repens*) breeding populations using agro-morphological and RAPD markers

T.R. Sharma, S. Singh, R. Rathour and S.K. Sharma
Advanced Centre of Hill Bioresources & Biotechnology, CSK H.P. Agricultural University, Palampur-176 062 India, Email: sharmat88@yahoo.com

Keywords: white clover, RAPD, descriptors, genetic diversity

Introduction White clover is an important forage legume for temperate regions, but very little is known about the genetic organisation of its breeding populations. The low amount of variability in the Indian collections of white clover for genetic improvement warrants the introduction of new germplasm and collecting local ecotypes for characterisation, utilisation and conservation. Several molecular techniques have been used for germplasm characterisation, variety identification, marker development and identification, molecular diagnostics, phylogenetic studies and diversity analysis. Because of its simplicity, rapidity and reliability, the RAPD technique has been used extensively for diversity analysis. The present study aims at characterising white clover genotypes of distinct geographical origin using standard descriptors and RAPD markers.

Materials and methods Twenty-eight white clover accessions acquired from Western Regional Plant Station, WSU, Pullman, USA, including some locally adapted populations, were grown in a randomised complete block design. The data on 30 descriptors and descriptor states developed by the IPGRI for white clover were recorded. Cluster analysis was done following Beale (1969). Genomic DNA was also extracted from ten randomly taken plants. DNA was isolated, using the CTAB method of Murray & Thompson (1980). The Numerical Taxonomy System of Multivariate Statistical Program (NTSYS) software package was used for data analysis (Rohlf, 1993). Jaccard's similarity coefficient was used for construction of a dendrogram by the Unweighted Paired Group Method of Arithmetic Averages (UPGMA).

Results The Non Hierarchial Euclidean Cluster analysis grouped accessions into 7 broad clusters, showing a high level of genetic divergence. Clusters 7 and 2 were found to be distantly placed. The accessions clustered arbitrarily with genotypes of the same region were found distributed in more than one cluster, while the genotypes of a heterogeneous region were grouped in the same cluster. The clustering pattern depicted the presence of sufficient genetic diversity and showed the lack of correspondence with geographical affinities of accessions. The morphological characterisation proved useful for cultivar discrimination and diversity analysis and gene bank management for white clover. A local collection RRCP-L-42 was found highly resistant for powdery mildew and clover rot, so that this can be used in resistance breeding. A dendrogram constructed by using molecular data, illustrated that genetic diversity and geographical origin of accessions did not show much correlation. The genotypes of the same regions were distributed in more than one cluster and also the genotypes of heterogeneous regions were grouped in the same clusters. However, one group had all the accessions of local origin. Molecular variation of higher order was resolved in the exotic accessions compared to indigenous germplasm. This suggests the need to introduce exotic germplasm in the Indian white clover gene pool to increase the genetic diversity.

Conclusion The amount and patterns of diversity observed in the white clover accessions can be of value in identifying the populations that parents of synthetic cultivars are derived from and to exploit the variation available in the populations analysed. The characterisation based upon standard descriptors can be of great value in the management of germplasm collections in the gene bank.

References

Beale, E.M.L. (1969). Euclidean cluster analysis. 37th session of the International Statistical Institute, U.K.
Murray, M.G. & W.F. Thompson (1980). Rapid isolation of high molecular weight plant DNA. *Nucleic Acids Research,* 8, 432-4325.
Rohlf, F.J. 1993. NTSYS-pc : Numerical Taxonomy and Multivariate Analysis System, Version 1.80. Exeter Software: Setauket, New York.
Zohary, M. 1970. Trifolium. In Davis, P.H. (Ed.) Flora of Turkey, Vol. 3. Edinburgh, UK; Edinburgh University Press, 384-448.

Breeding *Lotus australis* Andrews for low cyanide content

D. Real[1], G.A. Sandral[1,2], J. Warden[1], L. Nutt[1], R. Bennett[1] and D. Kidd[1]

[1]*Cooperative Research Center for Plant-Based Management of Dryland Salinity, The University of Western Australia, University Field Station, 1 Underwood Avenue, Shenton Park, WA 6009, Australia Email: dreal@cyllene.uwa.edu.au,* [2]*NSW Department of Primary Industries, Wagga Wagga Agricultural Institute, PMB, Pine Gully Road, Wagga Wagga, NSW 2605, Australia*

Keywords: *Lotus australis*, cyanide

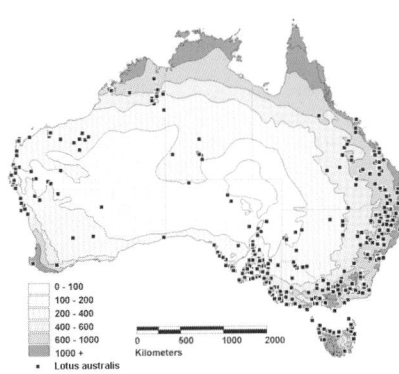

Figure 1 Annual rainfall and *L. australis* distribution in Australia

Introduction *Lotus australis* Andrews is a native perennial tetraploid legume (2n=4x=28) widely distributed throughout Australia (Figure 1). It is highly variable with 14 botanical varieties reported in the Australian Plant Name Index (http://www.anbg.gov.au). Despite broad adaptation within *L. australis* no cultivars have been developed for cultivation. One of the main barriers to cultivation is the reported cyanogenic nature of the species (Foulds, 1982), which makes it potentially toxic when plant cyanogenic glycosides are fully hydrolysed to form hydrogen cyanide (HCN). Foulds (1982) also reported that the cyanophoric trait was polymorphic at seedling and adult stages with 12% of plants acyanogenic in some populations. . The Cooperative Research Center for Plant-Based Management of Dryland Salinity, financially supported by Australian Wool Innovation has commenced a breeding program to develop a non-toxic cultivar of *L. australis*. The selection criterion of the first phase of the breeding programme was for low HCN production. Once this trait is stabilised, forage production and seed yield as well as general plant health will be the main breeding objectives.

Materials and methods A preliminary test was conducted to compare the semi-quantitative Feigl-Anger paper (FA) test (Feigl and Anger, 1966) with a wet-chemistry method (APHA, 1988) to estimate HCN content in 4 *Lotus* species including *L. australis*. Subsequently a test was performed on 66 plants corresponding to 10 accessions of *L. australis*, using the FA test. Time was recorded for three distinct stages observed in the FA test. The first stage is observed when the white filter paper starts to change colour, the second when the filter paper is light blue and finally when the filter paper is dark blue. The more rapid the colour change, the more HCN present. Each test was completed within 1 h and 3 h were required to complete the test on all 66 samples.

Results The ranking obtained from the FA test and the wet-chemistry test were the same. Subsequently the former test was chosen for its speed, low cost and sample size efficiencies (eg. 5 to 10 leaves). From the 66 samples tested with the FA test, 6 samples turned dark blue within 6 min of exposure, whilst 10 were light blue or had not changed colour after 1 h. There was only 1 sample that did not record a colour change. Thirty-nine of the lowest HCN content plants have been selected from the paper test and 60 seeds per plant were scarified, pre-germinated in Petri-plates and subsequently planted to re-select the low HCN lines.

Conclusions The FA test is an accurate, rapid and low cost method for screening cyanogenesis in *L. australis*. The low-HCN material identified in this research will form the basis of a new breeding programme for *L. australis*.

References

APHA (1988). Standard methods for the determination of water and wastewater 20th Edn., American Public Health Association, American Water Works Association and Water Environment Federation. Washington DC American Public Health Association.

Feigl, F. & V. Anger (1966). Replacement of benzidine by copper ethylacetoacetate and Tetrabase as spot-test reagent for hydrogen cyanide and cyanogens. *Analyst*, 91, 282-284.

Foulds, W. (1982). Polymorphism for Cyanogenesis in *Lotus australis* Andr. Populations at Greenough front flats, Western Australia. *Australian Journal of Botany,* 30, 211-217.

Development of high-yielding and early-maturing Korean hairy vetch

C.N. Shin and K.H. Ko
Depatment of Animal production, Keimyung College, Daegu, Korea, Email: scn225@km-c.ac.kr

Keywords: Korean hairy vetch, early maturing, dry matter yield

Introduction There are a number of forage cultivars recommended by the government, but only a few winter-spring legumes are available for the farmers in Korea. Hairy vetch has superior winter hardiness and produces high dry matter (DM) yield compared with crimson clover, Persian clover, rose clover, common vetch and Chinese milk vetch in Korea. However, under a double cropping forage production system an early maturing vetch cultivar was more productive than a late maturing one (Shin *et al.*, 2000). The objective of this research was to develop a high yielding and early flowering new hairy vetch cultivar derived from an accession in Korea.

Materials and methods Most vigorous early flowering hairy vetch cultivars were selected and crossed by open pollination and their seeds were bulked and plants were reselected (Seedco, 1999). A performance trial was conducted to evaluate agronomic characteristics, forage quality and dry matter (DM) yield of a new Korean hairy vetch (KHV) (*Vicia villosa*), an early maturing introduced cultivar, (Haymaker Plus) (HP) (*Vicia villosa ssp. dasycarpa*) and a late maturing hairy vetch cultivar (Balosa) at Seongju in the Keongbuk inland region and Sacheon in the southern coastal region for one year.

Results Seedling vigour and cold tolerance of HP were a little lower than those of Barlosa and KHV at Seongju, but there were no differences at Sacheon (Table 1). Fifty percent-flowering dates of KHV and HP were similar, but Barlosa did not flower until harvest. At harvest, KHV was taller than Balosa and HP. DM yield of KHV was higher than that of Balosa and HP at Seongju, but that of HP was higher than that of Balosa and KHV at Sacheon. Vetch cultivars were high in CP and low in ADF.

Table 1 Agronomic characteristics, forage quality and forage yield of vetch at Seongju and Sacheon, 2003-2004

Cultivar	Seedling vigour	Cold tolerance	Flowering 9day/month	Plant height (cm)	DM yield (t/ha)	CP %	ADF %	NDF %
Barlosa	8(8)	9(9)	-(-)	21(50)	2.5(9.2)	24(22)	23(30)	32(44)
KHV	8(8)	9(9)	9/4(10/4)	41(69)	2.9(8.8)	22(20)	26(34)	35(47)
HP	7(8)	8(9)	11/4(12/4)	30(59)	2.7(10.4)	23(20)	24(33)	33(46)
Mean	8(8)	9(9)	10/4(11/4)	31(59)	2.7(9.4)	23(21)	24(32)	33(46)
LSD (P<0.05)					0.3(0.8)			

( ) = Sacheon results. Rating (Seedling vigour, Cold tolerance): 9=outstanding, 1=poor.

Conclusions DM yield of early maturing KHV was the highest (p<0.05) of the cultivars at Seongju, but that of HP was the highest (p<0.05) at Sacheon. Growth of HP at Seongju was restricted by low temperatures during winter, more so than hairy vetch cultivars were.

References
Seedco (1999). Haymaker Plus. Forbes Seeds & Grain Inc., USA.
Shin, C. N., D. A. Kim, K. H. Ko & Y. W. Kim (2000). Forage performance of introduced vetch cultivars and Korean native vetch. *Journal of Korean Grassland Science* 20, 251-258.

Seasonal variation of tannin content on pigeon pea (*Cajanus cajan* (L.) Millsp) plants

R. Godoy[1], P.M. Santos[1] and L.A.R. Batista[1]
[1]*Embrapa Pecuária Sudeste, Caixa Postal 339-13560-970 São Carlos, SP Brazil, Email: godoy@cppse.embrapa.br*

Keywords: forage digestibility, palatability, leguminous crop

Introduction Tannins reduce forage digestibility and palatability by reacting with proteins (Makkar, 1989) and due to that, tannin content is considered to be an important characteristic of leguminous crops and it has been used as a selection criterion on pigeon-pea improvement programs (Godoy et al., 1994). In Southeastern Brazil, pigeon pea is less consumed by bovines in the rainy season (October through March) than in the dry season (from April on), when flowering occurs. Alencar et al. (1991) found very little consumption by Canchim cows during the rainy season, and Lourenço et al. (1984) recommend pigeon-pea for feeding bovines in the dry season.

Materials and methods In 5 locations (São Carlos, Pirassununga, Jaboticabal, Itapuí and Pratânia) in São Paulo State, Brazil, trials with 17 pure pigeon pea lines and 3 cultivars were planted in December 1998. In all locations 4 cuttings were performed, in April and August 1999, January and May 2000. In the first 2 locations, 2 other cuttings were also performed, in December 2000 and June 2001. Plants were cut at 40 cm height and the material was dried, ground and their tannin contents, expressed in equivalent percents of tannic acids, were determined, according to the Folin-Denis method, described by Burns (1963). To make possible the performance of cluster analysis, as described in the SAS cluster procedure (SAS, 1999-2001), data of the 3 first cuttings of 18 genotypes were submitted to an analysis of variance for each location. This analysis revealed no significant interaction of cutting date and genotypes for Itapuí and Pratânia but that interaction was significant in the other 3 locations. Two cluster analysis were then performed, one with the data from Jaboticabal, Pirassununga and São Carlos and the other with the data from Itapuí and Pratânia. Variance analysis was performed within the clusters and a genotype group was considered to be made when no significant genotype interaction was found. Average tannin content, expressed as dry matter percentage, was then calculated for each time of cutting and these means were compared with each other by the Tukey test, as described in the SAS GLM procedure (SAS, 1999-2001). Assuming these groupings to be correct, means for all 3 of cuttings of these 18 genotypes were compared.

Results Three groups of genotypes were found within the data from Jaboticabal, Pirassununga and São Carlos. Group 1 had with 9 genotypes, Fava Larga, g3-94, g6-95, g18-95, g19b-94, g27-94, g29b-94, g47-94 and g66-95. Group 2 had 6 genotypes, Caqui, g124-95, g146-97, g154-95, g167-97 and g184-97; Group 3 had only three, g17c-94, g101-97 and g127-97. In all 3 groups, the August 1999 means (3.31, 2,70 and 2,58%) were significantly higher than the January 2000 means (2.41, 2.05 and 2.25%), that were significantly higher than the April 1999 means (2.00, 1.75 and 2.02%). One single group was formed with the Itapuí and Pratânia data and also in this case, the August 1999 mean (2.20%) was significantly higher than the January 2000 mean (2.00%) which was significantly higher than the April 1999 mean (1.63%). The same trend occurred when means for the other cutting dates were calculated. Then, when animal consumption is higher (Alencar *et al.*, 1991 and Lourenço *et al.*, 1994), during the dry season, after flowering, tannin contents were also higher.

Conclusions Tannin contents increase in the period when bovine consumption of pigeon pea is higher. This indicates that tannins are not responsible for its low palatability in summer. Tannin contents should not be used as a selection criterion in pigeon pea selection programmes.

References

Alencar, M. M., R. Godoy, L. A. Corrêa, R. R. Tullio & M. Bugner (1991) Desempenho de vacas da raça Canchim em pastagens de braquiária e guandu. *Pesquisa Agropecuária Brasileira*, 26, 1717-1723.

Burns, R. E. (1963). Methods of tannin analysis for forage crop evaluation. Georgia: Georgia Agriculture Experimental Station. 11p. (Technical Bulletin N.S., 32).

Godoy, R., L. A. R. Batista & G. F. Negreiros (1994). Avaliação agronômica e seleção de guandu forrageiro (*Cajanus cajan* (L.) Millsp). *Revista Brasileira de Zootecnia*, 23, 730-742.

Lourenço, A. J. & E. Matsui (1994). Composição botânica da forragem disponível e da dieta selecionada por bovinos em pastos de capim-colonião consorciado com centrosema e ou galactia com acesso ao banco de proteína de guandu. *Revista Brasileira de Zootecnia*, 23, 100-109.

Makkar, H. P. S. (1989). Protein precipitation methods for quantitation of tannins: a review. *Journal of Agricultural Food Chemistry*, 37, 1197-1202.

SAS Institute. SAS/STAT 1999-2001. User's guide: statistics, version 8, v.2, SAS Cary, NC, USA.

New pigeon pea (*Cajanus cajan*) hybrids with desirable forage traits

P.B. Alcantara[1], J.A. Usberti Jr.[2], C.A. Colombo[2], V.B.G. Alcantara[1], R. Usberti[3], M.A.C. Lucena[1] and M. Harris[2]

[1]*Zootechnic Institute, Nova Odessa, Brazil P.O. Box 28, CEP 13073-001, Email: bardauil@iz.sp.gov.br,* [2]*Agronomic Institute, Campinas, Brazil.* [3] *Plant Protection Agency,Campinas. Brazil*

Keywords: pigeon pea, hybrids, genetic variation, forage traits

Introduction Pigeon pea is a tropical forage legume usually sown in mixed pastures with tropical forage grasses. Most of the available cultivars shows erect and tall plants with poor tillering potential, breakable thick stems, low leaf/stem ratios (fresh/dry matter) and low persistence under animal grazing. It shows a high dry matter production, due to low leaf/stem ratios (Barnes & Addo, 1997). Pigeon pea shows good crude protein levels/dry matter (ranging from 14-23%) and regular *in vitro* digestibility indexes (52-58%) (Karachi & Matata, 1996); animal consumption is affected by high tannin levels of young leaves. Being a self-pollinated species, the variability for forage traits occurs **among** cultivars available at germplasm banks. No significant variation is observed for any forage character **within** a given population. Effective selection and releasing of new genetic materials bearing desirable morpho-agronomic and forage traits is mostly dependent on increases of genetic variation, which may be accomplished through artificial crossings between selected parentals. This research work was aimed at the synthesis of new pigeon pea hybrids, hopefully bearing new desirable forage characters.

Materials and methods Sixteen pigeon pea accessions, available at germplasm banks, were genetically characterized through the RAPD methodology. The 6 most divergent ones were chosen as parentals, aiming to get a broadest genetic variability in F_2 and subsequent generations, so making effective the selection of individual genotypes carrying desirable forage traits (Sidhu et al., 1996). Hand-made artificial crossings were carried out under greenhouse-controlled conditions during the 2002-growing season. The F_1 hybrid seeds of each crossing were sown to get F_1 hybrid plants, from which we picked up F_2 hybrid seeds, during the 2003-growing season. Eight segregant F_2 populations were established at 3 different experimental sites, during the 2004-growing season. Superior genotypes were chosen **within** each F_2 population. The following traits were scored: plant height and growth habit, number of basal and lateral branches, stem thickness and flowering cycles.

Results Evaluation results are presented in Table 1. Significant variation for plant height was detected, ranging from <1.0m up to >3.0m. However, most of the genotypes were concentrated in the interval 1.6-3.0m. Most of the individuals showed only one basal branch, though 2-3 basal branches were not rare events. On the other hand, remarkable variability as to the number of lateral branches was recorded, ranging from <15 to >30; as this parameter is closely related to fresh and dry matter yielding potential, indirect selection might be feasible. A broad variation was also observed for stem diameter, with a significant frequency of individuals with thin stems. Considerable variation on growth habit and flowering cycle was observed but there was the occurrence of new and desirable types never detected before in the available pigeon pea cultivars, like individuals showing prostrated growth habit and/or very late flowering cycle.

Table 1 Frequency distributions of 6 morpho-agronomic traits scored in 8 segregating F populations of pigeon pea (*Cajanus cajan*)

SP	NI	PH (m)						NBB			NLB					GH			CD (mm)					FC			
		<1	1-1.5	1.6-2	2.1-2.5	2.6-3	>3	1	2	3	<15	15-20	21-25	26-30	>30	E	SE	P	2-4	5-6	7-8	9-10	>10	E	M	L	VL
1	38	-	-	10	9	19	-	29	5	4	3	14	11	7	3	13	22	3	14	12	8	4	-	4	13	12	9
2	30	1	1	6	4	16	2	18	12	-	6	13	2	5	5	8	18	4	10	7	7	5	1	2	15	4	9
3	17	-	-	7	8	2	-	10	6	1	4	10	1	-	2	7	9	1	4	5	5	3	-	3	5	9	-
4	12	-	-	4	6	2	-	5	4	3	3	4	2	-	3	5	6	1	3	5	2	2	-	4	5	3	-
5	17	-	-	5	3	6	-	12	3	2	2	1	2	5	7	6	5	6	-	-	2	2	16	9	2	3	3
6	20	8	5	2	4	1	-	17	3	-	5	3	3	6	3	5	8	7	4	2	3	1	7	7	4	6	3
7	20	1	-	-	7	8	2	-	2	3	3	3	3	9	10	8	2	12	3	2	1	2	11	4	2	3	
8	8	-	-	1	6	1	-	8	-	-	-	4	3	1	1	7	-	-	-	-	2	6	2	4	-	2	

Observations: SP = segregating F2 population; NI = number of individuals; PH = plant height; NBB = number of basal branches; NLB = number of lateral branches; GH = growth habit (E = erect; SE = semi-erect; P = prostrated); CD = stem diameter; FC = flowering cycle (E = early; M = medium; L = late; VL = very late).

Conclusions Effective selection of superior individuals is now feasible due to broad genetic variation detected.

References

Barnes, P. & Addo, K. A. (1997). Evaluation of introduced forage accessions for fodder production at a subhumid site in southern Ghana. *Tropical Grasslands,* 30(4), 422-425.

Karachi, M. & Matata, Z. (1996). Preliminary evaluation of pigeon-pea (*Cajanus cajan* (L.) Millsp.) accessions for forage production and nutritive value and seed yield in western Tanzania. *Tropical Agriculture,* 73(4), 253-258.

Sidhu, P. S., M. M. Verma, R. S. Sarlach, R. S. Sekhon and D. Sandhu, D. (1996). Identification of superior parents and hybrids for improving pigeon pea. *Crop Improvement,* 23(1), 66-70.

Agronomic characteristics of Novi Sad winter vetch cultivars

V. Mihailović, A. Mikić, B. Ćupina, S. Vasiljević and D. Milić
Institute of Field and Vegetable Crops, Forage Crops Department, Maksima Gorkog 30, 21000 Novi Sad, Serbia and Montenegro, Email: vojamih@ifvcns.ns.ac.yu

Keywords: crude protein, Vicia spp.

Introduction Genus *Vicia*. contains important annual food and forage species such as field bean, *V. faba.*, narbon vetch, *V. narbonensis*, and common vetch, *V. sativa* (Maxted, 1995), while Hungarian *V. pannonica* Crantz and hairy vetch *V. villosa* also play an important role in the Balkans. Winter vetches are excellent forage catch crops useful for sustainable agriculture and organic farming (Ćupina *et al.*, 2004). Our study was aimed at determining the agronomic characteristics of the winter vetch cultivars developed in Novi Sad, assessing thus their ability for successful growing in the prevailing conditions of Serbia and Montenegro.

Materials and methods A three-year small-plot trial (2002-2004) was established at the Experiment Field Institute at Rimski Šančevi. It included all Novi Sad winter vetch cultivars, NS Sirmium (*V. sativa*), NS Violeta (*V. villosa*), NS Panonika (*V. pannonica*) and an unreleased cultivar L-92 (*V. sativa*). They were sown during early October, at a crop density of 150 viable seeds/m^2, and were cut when the first pods began to appear (Mihailović *et al.*, 2004). Yield of fresh weight and hay crude protein (CP) in the dry matter of the plants that survived until cutting, were determined.

Results The cultivar NS Violeta had the greatest plant height as well as the greatest number of internodes, the greatest number of stems per plant and the largest mass/plant. The smallest plant height and the smallest plant mass were found in NS Panonika. The smallest number of internodes was found in NS Sirmium and L-92, while NS Sirmium also had the smallest number of stems per plant. On the basis of number of plants that survived until cutting, NS Sirmium and L-92 proved the most winter hardy, with 123 and 125 plants per m^2. Apart from an excellent winter hardiness, NS Violeta had the smallest number of plants per m^2, mainly due to severe competition within the stand. Thanks to an optimal relationship between yield components, L-92 had the highest yield of both fresh weight and hay. CP varied from 1.96 g/kg DM in NS Panonika to 2.14 g/kg DM in NS Violeta, which was in accord with Mišković (1986).

Table 1 Agronomic characteristics of Novi Sad winter vetches cultivars during 2002-2004

Cultivar name	No. of plants per m^2 before cutting	Plant height (cm)	No. of stems per plant	No. of inter-nodes	Plant mass (g/plant)	Yield of fresh weight (t/ha)	Yield of hay (t/ha)	Level of CP (g/kg DM)
NS Sirmium	123	89	2.4	17.0	31.61	32.0	6.29	2.03
NS Violeta	88	169	3.3	30.8	42.59	30.7	6.16	2.14
NS Panonika	113	71	2.8	20.9	25.41	30.0	5.17	1.96
L-92	125	89	2.8	17.3	36.70	32.7	6.53	2.07
LSD 0.05	11	35	1.7	4.5	9.78	2.5	0.6	0.11
0.01	15	47	2.4	5.9	14.22	4.1	1.0	0.17

Conclusions All winter vetch cultivars that were developed at the Institute of Field and Vegetable Crops in Novi Sad are winter hardy and well adapted to the environmental conditions of Serbia and Montenegro. Being able to take a significant part in diverse farming systems, winter vetches can supply animal husbandry with high yields and high quality feed in one of the easiest and least expensive possible ways.

References
Ćupina, B., P. Erić, V. Mihailović & A. Mikić (2004). The importance and role of cover crops in sustainable agriculture. *Zbornik radova (a periodical of scientific research of field and vegetable crops)*. 40, 419-430.
Maxted, N. (1995). An ecogeographical study of *Vicia* subgenus *Vicia*. IPGRI, Rome, Rome, 1.
Mihailović, V., T. Prentovik, S. Vasiljević, S. Katić & A. Mikić (2004). Forage yield and forage yield components of spring vetch genotypes (*Vicia sativa* L.). *Acta Agriculturae Serbica*. IX:17, 407-411.
Mišković, B. (1986). Krmno bilje. Naučna knjiga, Beograd, 36-48.

Genetic variability of yield and its components in winter forage pea cultivars

V. Mihailovi•; A. Miki•; Đ. Karagi•; S. Kati• and I. Pataki
Institute of Field and Vegetable Crops, Forage Crops Department, Maksima Gorkog 30, 21000 Novi Sad, Serbia and Montenegro, Email: vojamih@ifvcns.ns.ac.yu

Keywords: Pisum spp., plant height

Introduction   The genus *Pisum* (peas) is rich in variability of morphological traits. It provides an excellent basis for breeding, but is also one of the reasons for the still undefined status of *Pisum* taxons (Mihailovi•*et al.*, 2004a). The majority of forage pea cultivars used belongs to subspecies *sativum* and variety *arvense* (Maxted & Ambrose, 2000). The objective of our study was to determine genetic variability of yield and its components in six winter forage pea cultivars of different origin and to evaluate their breeding potential.

Materials and methods   A trial was conducted between 2002 and 2004 at the Experimental Field Institute in Rimski Šan•evi. It included six winter forage pea cultivars, i.e. accessions from the Annual Forage Legumes Genetic Collection (AFLGC) of the Forage Crops Department. All of them were sown in early October, at a density of 120 viable seeds/m^2, and were cut in May, when the first pods began to appear (Đuki•; 2002). A study on genetic variability of these cultivars was carried out, on the basis of yield of fresh weight and hay and its main components.

Results   The greatest plant height was found in the Serbian NS Pionir, while the smallest was in the Bulgarian Mir (Table 1). Thanks to the smallest number of plants at the time of cutting, caused by low winter hardiness, the Ukrainian Odesskiy 58 had the greatest number of stems per plant and the largest plant mass, as well as the lowest yield of both fresh weight and hay. Novi Sad cultivars NS Pionir and NS Dunav had the greatest number of plants per m^2 before cutting and of internodes and the smallest number of stems per plant. The smallest values of number of internodes and plant mass were found in Mir. NS Pionir and Bulgarian Pleven 10 had the highest yield of fresh weight and of hay.

Table 1 Agronomic characteristics of winter forage pea cultivars during 2002-2004

Accession (cultivar) number in AFLGC	Accession (cultivar) name	No. of plants per m² before cutting	Plant height (cm)	No. of stems per plant	No. of inter-nodes	Plant mass (g/plant)	Yield of fresh weight (t/ha)	Yield of hay (t/ha)
PIS 001	NS Pionir	110	204	1.0	24.5	65.72	42.7	8.5
PIS 005	Odesskiy 58	30	160	2.6	21.8	116.51	16.3	3.5
PIS 015	Pleven 10	103	134	1.1	18.9	56.47	42.0	8.3
PIS 016	Champagne	92	137	1.6	18.9	58.23	38.7	7.8
PIS 018	Mir	83	129	1.3	14.9	39.36	31.7	7.3
PIS 020	NS Dunav	105	178	1.0	23.7	59.52	37.7	7.4
LSD	0.05	24	27	1.8	3.8	11.20	3.1	1.1
	0.01	29	39	2.6	4.6	14.67	6.2	2.4

Conclusions   There were significant differences in most yield components between six winter forage pea cultivars, as well as between yields of fresh weight and hay. The winter hardy domestic cultivars, such as NS Pionir, and the introduced ones with good relationship between yield components and small yield variation, like the French Champagne or the Bulgarian Pleven 10, can be considered good gene donors for new winter forage pea cultivars.

References
Đuki•; D (2002). Biljke za proizvodnju sto•ne hrane. Faculty of Agriculture, Novi Sad, 283-294.
Maxted, N. & M. Ambrose (2000). Peas (*Pisum* L.). In: N. Maxted N & S. Bennet (eds.) Plant genetic resources of legumes in the Mediterranean. Kluwer.
Mihailovi•; V., A. Miki••& B. • npina (2004a). Botanical and agronomic classification of fodder pea (*Pisum sativum* L.). *Acta Agriculturae Serbica*, IX:17, 61-65.
Mihailovi•; V., P. Eri••& A. Miki••(2004b). Growing pea and vetches for forage in Serbia and Montenegro. *Grassland Science in Europe*, 9, 457-459.

Physical and quality seed traits observed in new pigeon pea (*Cajanus cajan*) hybrids

R. Usberti[1], J.A. Usberti Jr.[2], R.H. Aguiar[3], F.C. Simoes[3], F.N. Cavalcante, Jr.[3], M. Harris[3], M.T. Colozza[4] and J.C.T. Freitas[4]

[1]*Plant Protection Agency, P.O. Box 960, CEP 13073-001, Campinas, Brazil, Email: usberti@cati.sp.gov.br,* [2]*Agronomic Institute, Campinas, Brazil,* [3]*Faculty of Agricultural Engineering, Campinas, Brazil,* [4]*Zootechnic Institute, Nova Odessa, Brazil*

Keywords: hardseededness, germination, purity

Introduction Pigeon pea seed production may be affected by factors such as % of pure seeds, mean seed weight, incidence of pests and diseases and environmental stresses. Harvested seeds from different cultivars may also vary in germination %, hardseededness and germination speed. Hardseededness (seed coat impermeability to water) commonly occurs in forage legume species (Hopkinson, 1993). There is considerable variation among different entries for seed characters but this is not considered within genetic materials. This research analysed harvested seeds of selected individuals of two segregating F_2 pigeon pea populations for the above cited traits and assessed the range of variation for them resulting from the hybridisation process.

Materials and methods Selected individuals of two segregating F_2 pigeon pea populations, arisen from previous artificial crossings, had their seeds carefully harvested in special cloth bags and transferred to the laboratory. After a manual cleaning for separating seeds damaged by insects and fungi (Kashyap & Punia, 1996), the following physical seed traits were scored: physical purity (percentage of pure seeds/sample) and number of seeds/10g (ISTA, 1985). Seed standard germination tests were then performed using 4x50 replicates and the number of hard seeds scored at the first count (4th day), when their coats were partially removed by forceps; mean germination time test (T_{50}) was also carried out (Alvarado & Bradford, 1988).

Results In both segregating F_2 populations, most of the genotypes analysed showed low incidences of damaged seeds by insects and fungi (0-10%). For physical purity, there was a marked concentration of individuals in the ranges of 61-90 and >90% (population 1 and 2, respectively). There was a clear difference in the number of seeds/10g between the populations, the first showing high concentration of individuals in the range 91-100 and >100 while the second had a more uniform distribution (Table 1). Most of the selected individuals of both populations (75 and 60%, respectively) showed hardseededness, varying from <20 to 40%, while population 1 showed a higher frequency of individuals in the germination percentage interval of 61-100%. Both segregating populations presented similar distributions of their individuals as to mean germination times.

Table 1 Frequency distribution of physical and physiological seed traits in two pigeon pea (*Cajanus cajan*) segregating populations

SP	Damaged seeds (%)					Physical purity (%)			Number of seeds/10g			
	Insects			Fungi								
	0-10	11-20	>20	0-10	11-20	30-60	61-90	>90	71-80	81-90	91-100	>100
1	16	5	3	18	6	4	13	7	-	2	9	13
2	15	3	2	17	3	1	6	13	2	8	7	3

SP	Hardseededness (%)				Germination (%)				Mean germination time (T_{50}, days)		
	<20	21-40	41-60	>60	<40	41-60	61-80	>81	4-5	5-6.3	>6.3
1	4	11	6	3	1	1	16	6	9	7	8
2	6	6	4	4	2	9	7	2	6	6	8

SP, segregating population with SP1 and SP2 having 24 and 20 individuals respectively; physical purity is pure seed weight/working seed sample weight

Conclusions The hybridisation process used was effective as to significantly increasing the available genetic variability for several seed quality traits, making feasible effective selection for them in the near future.

References

Alvarado, N.A.& K.J. Bradford, (1988). Priming and storage of tomato (*Lycopersicon lycopersicum*) seeds. I. Effects of storage temperature on germination rate and viability. *Seed Science & Technology*, 16, 601-612.

Hopkinson, J.M. (1993). Tropical pasture establishment .2. Seed characteristics and field establishment. *Tropical Grasslands*, 27, 276-290.

ISTA, International Seed Testing Association. (1985). International rules for seed testing. RULES 1985. *Seed Science and Technology*, 13, 299-355; 356-513.

Kashyap, R.K. & R.C. Punia, (1996). Seed quality as influence by pod infesting insect pests of pigeon pea *Cajanus cajan* L. Millsp. *Seed Science and Technology*, 23, 873-876.

Breeding a psyllid-resistant interspecific hybrid *Leucaena* for beef cattle production in northern Australia

S.A. Dalzell, F. Lemos de Moraes and G. Solis Pasos
School of Land and Food Sciences, The University of Queensland, Queensland 4072, Australia, Email: s.dalzell@uq.edu.au

Keywords: psyllid-resistance, recurrent selection, *Leucaena*

Introduction Production of the valuable fodder tree legume *Leucaena leucocephala* (leucaena) is limited to the subhumid (600-800 mm annual rainfall) areas of northern Australia by the psyllid insect pest *Heteropsylla cubana*. Defoliation caused by severe psyllid infestations can suppress forage yields of commercial leucaena varieties by 50-80%. Susceptibility to psyllid damage is a major impediment to grazier adoption of leucaena pastures in the more humid tropical areas of Australia. A comprehensive international agronomic evaluation of the entire *Leucaena* genus (Mullen et al., 2003) revealed that the artificial interspecific F_1 hybrid of *L. pallida* × *L. leucocephala* ssp. *glabrata* (called KX2) had a high degree of psyllid resistance, excellent vigour and broad environmental adaptation. The KX2 F_1 hybrid also had superior forage quality compared to other psyllid-resistant taxa, such as *L. pallida*, *L. trichandra* and *L. diversifolia*. Commercial utilization of the KX2 F_1 hybrid by Australian graziers has been prevented by a lack of planting material. To date, seed production of the F_1 hybrid has only been possible by laborious hand pollination. The KX2 F_1 hybrid has been successfully vegetatively propagated for smallholders in SE Asia, however cloned cuttings are expensive to produce and are not suited to broad acre leucaena planting in Australia. A recurrent selection breeding program was initiated to produce a genetically stable, advanced generation KX2 hybrid that breeds true-to-type and is suitable for commercial release. We anticipate that 4 cycles of selection will be required to achieve this objective. This paper reports the agronomic evaluation of the KX2 F_2 generation.

Materials and methods The base population of the breeding program comprised 5 superior KX2 F_1 hybrids originally bred by The University of Hawaii: K806 × cv. Tarramba; K748 × cv. Tarramba; K748 × K584; K748 × K658; and K748 × K481. These parental lines were open-pollinated (bees) to produce F_2 seed. Field evaluation of 4,900 KX2 F_2 seedlings was conducted under high-psyllid pressure at Brisbane (27° 37'S, 153° 19'E), Queensland, Australia. Ten seedlings (2 from each half-sib family) were randomly planted at 0.5 m spacings between reference trees in contiguous rows 2 m apart in December 2001. Alternating reference trees of cv. Tarramba and the KX2 F_1 hybrid were used to benchmark psyllid damage (susceptibility) and biomass yield respectively, and were arranged in an ∝-lattice design. Psyllid damage rating (PDR) (Wheeler, 1988), tree form rating (degree of branching), leaf production rating and self-compatibility were recorded in May 2002. Dry matter (DM) yield was estimated from a non-destructive yield index (plant height × basal stem diameter^2) and analysed using a linear mixed model by residual maximum likelihood (REML) (ASReml, New South Wales Agriculture, Australia) to produce best linear unbiased predictors for the DM yield for each plant. Superior trees were selected by interrogating the data with Microsoft Access® and subsequent field validation. Later selection cycles will also evaluate forage quality, i.e. condensed tannin activity, to ensure the bred KX2 hybrid variety is of high nutritive value.

Results The F_2 population segregated widely for all traits observed, with individuals exhibiting the full range of traits characteristic of both parental species. Elite F_2 individuals (n = 240) that were highly psyllid resistant, vigorous, of good form and self-incompatible were selected. Their average DM yield was 190% that of cv. Tarramba and 70% that of the KX2 F_1, indicating that a significant amount of heterosis was retained. Median psyllid resistance of the elite trees (PDR = 1, no damage) was greater than that of the KX2 F_1 (PDR = 2, minor damage) and much higher than that of psyllid-susceptible cv. Tarramba (PDR = 6, loss of 50% of young leaves). Elite trees had excellent form for forage shrubs, with greater basal and secondary branching than cv. Tarramba.

Conclusion Elite F_2 individuals were identified that were highly psyllid resistant, vigorous, of good form and self-incompatible. These trees were retained for F_3 seed production. Heritability of these traits will be estimated from the F_3 progeny that are currently under evaluation. Each selection cycle takes approximately 18 months to complete. We anticipate F_6 seed of a psyllid-resistant KX2 hybrid will be commercially available in 2009.

Acknowledgements Meat and Livestock Australia Limited funded this research program (NBP.307).

References
Wheeler, R. A. (1988). Leucaena psyllid trial at Waimanalo, Hawaii. *Leucaena Research Reports*, 9, 25-29.
Mullen, B. F., H. M. Shelton, R. C. Gutteridge & K. E. Basford (2003). Agronomic evaluation of Leucaena. Part 1. Adaptation to environmental challenges in multi-environment trials. *Agroforestry Systems*, 58, 77-92.

Assessment of inter-specific diversity of the *Hedysarum* genus in Tunisia

S. Marghali, H. Chennaoui, M. Marrakchi and N. Trifi-Farah
Laboratoire de Génétique Moléculaire, Immunologie & Biotechnologie. Faculté des Sciences de Tunis, Tunisia,
Email: sonia.marghali@fst.rnu.tn

Keywords: pasture, phylogenetic relationships, molecular markers, rDNA, intergenic spacer polymorphism (IGS)

Introduction In Tunisia, many grassland and pasture species were menaced by genetic erosion. Thus, we were interested in the *Hedysarum* species which constitute a very important phytogenetic patrimony able to produce forage and restore destroyed pasture land especially in arid and semi-arid areas. In order to facilitate fodder improvement, we investigated the phenetic relationships among *Hedysarum* species using rDNA intergenic spacer (IGS) polymorphism.

Materials and methods Appropriate conserved oligonucleotides flanking the rDNA IGS were used to amplify total cellular DNA extracted from eight *Hedysarum* species. Polymorphic bands were scored for their presence/absence and employed to produce genetic relationships among species based on Nei and Li's distance matrix (Nei & Li, 1979).

Results Our data show that this IGS amplification constituted an efficient tool to examine the genetic diversity between species in this genus and showed a high level of polymorphism in the *Hedysarum* genus.

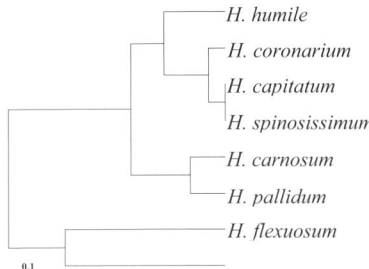

Figure 1 Phylogenetic relationships of *Hedysarum* species based on IGS sequence

The clustering supports two main groups (Figure 1). The first one is represented by two species, *H. flexuosum* and *H. aculeolatum.* All the remaining species are ranged in the second cluster. It seems that the two sub-species *H. capitatum* and *H. spinosissimum* are closely related to each other and characterised by a very similar sequence of rDNA intergenic spacer.

Conclusions There is a high degree of polymorphism at the inter-specific levels and this permitted the establishment of the genetic relationships among species. Our data provide evidence of a nuclear lineage between included species. In spite of their classification as sub-species of *H. spinosissimum* and their distinctiveness by agronomic characters, mating systems and geographical distribution, *H. capitatum* and *H. spinosissimum* are characterised by great similarities of the IGS sequence. Both sub-species, which are closely related to *H. coronarium,* can assist in the selection of genetic materials for improvement programmes.

References
Nei, M., & W.S. Li (1979). Mathematical model for studying genetic variation in terms of restriction
    endonuclease. *Proceedings of the National Academy of Sciences*, USA, 76, 5269-5273.

Small leaf mid-rib xylem related to leaf freeze tolerance trait in Bahia (*Paspalum notatum* Flugge) grass lines

J.W. Breman[1], A.R. Blount[2] and T.R. Sinclair[2]

[1]Union County Extension Office 25 NE 1st Street, Lake Butler, Florida 32054 USA, Email: union@ifas.ufl.edu, [2]North Florida AREC-Marianna, 3925 Highway 71, Marianna, Florida 32446, USA, [3]Agronomy Physiology Building 350, PO Box 110965, Gainesville, Florida 32611-0965, USA

Keywords: cold tolerance, Bahia grass, *Paspalum notatum* Flugge, xylem diameter, xylem area

Introduction Controlled freeze (-60° C) trials of 31 bahiagrass selections from a breeding program for cold tolerance by Blount *et al.* (2001) showed diverse genotype Leaf Tissue Cold Damage (LTCD). Breman *et al.* (2003) defined LTCD on a rating scale (1 = no damage to 9 = 100% leaf damage). Unique midrib damage was observed as part of LTCD in bahiagrass under transpiration stress after a freeze trial. Small xylem conduit diameter and area has been strongly correlated with reduced cavitation caused by freeze thaw cycles which maintain leaf tissue in evergreen temperate woody plants; further shown in twelve woody species by Davis *et al.* (1999). Air bubbles in vessel ice columns prevent normal refill and function upon thawing. The purpose of this study was to test whether genotype xylem diameter and area could be used to predict LTCD.

Materials and methods Permanent stained (crystal violet) slides of four cold sensitive and four cold tolerant genotype lamina cross sections (I.E.C. CTF Microtome Cryostat) taken at the first, second and third emerged leaf 2 cm above the leaf collar from four different plants per genotype, five reps per plant in February. Olympus BH2 microscope at 400X with an ocular micrometer calibrated with a stage micrometer was used to measure midrib abaxial vessel diameters from which vessel area was calculated.

Results Xylem area produced the most consistent determination of whether a genotype was cold tolerant. Simple effects of xylem diameter and xylem area were significant (P<0.001) for LTCD regardless of leaf position and predicted whether a genotype was freeze sensitive or tolerant.

Table 1 Mean Midrib Abaxial Xylem Diameters and Areas

Cold Tolerance Category	LTCD Rating	Genotype	Mean Xylem Diameter (μ)	Mean Xylem Area ($\mu2$)
Sensitive	**9.0**	**1-30-4**	**242 a**	**46260 a**
Sensitive	**9.0**	**1-30-3**	**223 b**	**39364 b**
Sensitive	**6.2**	**FL9**	**209 c**	**35911 c**
Sensitive	**9.0**	**2-22-1**	**209 c**	**34735 c**
Tolerant	2.6	OK1	187 d	27570 d
Tolerant	2.2	OK2	170 e	22971 e
Tolerant	2.0	FL67	157 f	22024 e
Tolerant	2.0	CO6	158 f	20109 e

Conclusions Critical values for xylem diameter must lie between 187 and 209 μ and for xylem area between 27,570 and 34,735 μ^2, determining whether a genotype is susceptible or tolerant to freezing. This method has the potential for screening bahiagrass breeding lines for low LTCD (leaf freeze tolerance) without cold chamber trials or natural freeze events in the field with small amounts of plant leaf tissue. This data also suggests a mode of action by which some bahiagrass genotypes tolerate freeze events. Davis et al. (1999) postulated smaller ice volumes in frozen vessels with small diameters maintain dissolved air in the ice column. No gas bubbles form to disrupt vessel transport function upon thawing. This might explain why bahiagrass lines with small leaf xylem diameters or areas maintain a green and functioning lamina under transpiration demand after a freeze event.

References

Blount, A.R., K.H., Quesenberry, P. Mislevy, R.N. Gates, & T.R. Sinclair (2001). Bahiagrass and other Paspalum species: An overview of the plant breeding efforts in the Southern Coastal Plain. *Proceedings, 56th Southern Pasture and Forage Crop Improvement Conference*, Springdale, AR, April 21-22. http://www.agr.okstate.edu/spfcic/procedures/2001/breeders/blount.htm.

Breman, J.W., A.R. Blount, K.H. Quesenberry, T.R. Sinclair, R.N. Gates, R.D. Barnett & S.W. Coleman (2003). Discrimination of Cold Damage in Bahiagrass. *Soil and Crop Science Society of Florida*, 62:94.

Davis, S.D., J.S. Sperry & W.G. Hacke (1999). The relationship between xylem conduit diameter and cavitation caused by freezing. *American Journal of Botany*, 86, 1367-1372.

Development and evaluation of *Pennisetum purpureum* mutants through irradiation with ^{60}Co

R.S. Herrera, M. García, R. Tuero, A.M. Cruz, A. Romero and N. Fraga
Instituto de Ciencia Animal, Apdo 24, San José de las Lajas, La Habana, Cuba, Email: rherrera@ica.co.cu

Keywords: dry matter yield, drought tolerance

Introduction   The development of plant mutants is a commonly used procedure nowadays. One way of obtaining mutant plants has been seed irradiation with gamma rays. As these studies are not common in tropical pasture species (Micke *et al.*, 1993, Herrera, 2001), the objective of this study was the development and evaluation of *Pennisetum purpureum* mutants through irradiation with ^{60}Co.

Materials and methods   Agamic seeds of *Pennisetum purpureum* cv. King grass were irradiated in a source of ^{60}Co (MPX-gamma-25M) with a potency of 0.21 Gy/s and an acute dosage of 0-100 Gy. The methodology of Herrera *et al.* (1996) was used for the identification and the initial evaluation of the mutants. The promising mutants were evaluated for five years in a field experiment under dry conditions (without irrigation in the dry season and with fertilisation of 180 kg N/ha in the rainy season).

Results   The highest (P < 0.01) yield was obtained in the mutants (Table 1), specially in the dry season, when tropical pastures have considerably reduced yields. This was an advantage for the mutants because larger amounts of forage for the animals are produced with the same inputs and reduced production costs. At the same time, this could indicate drought tolerance. Besides, the lowest plant loss (10-16 %) at the end of the experimental period was recorded in the mutants and they had the highest (P < 0.01) content of leaves (52-55 %) in the dry season. There were no differences for crude protein and the highest (P < 0.01) in vitro organic matter digestibility was obtained in the stems of one of the mutants (57 and 63 % for the rainy and dry season, respectively). There was no incidence of pest and diseases during the five years of the exploitation.

Table 1 Yield (t DM/ha) accumulated during five years

Varieties	Rainy season	Dry season	Total	Difference
King grass	58.09[b]	39.44[b]	97.53[b]	
CUBA MF-1	61.28[ab]	46.22[a]	107.50[a]	+ 9.97
CUBA MF-2	63.16[a]	44.99[a]	108.15[a]	+ 10.62
SE ±	1.21**	1.53**	2.60**	

[ab] Values with different superscripts within each column differ at P < 0.05 (Duncan, 1955)
** P < 0.01

Conclusions   The mutants were superior to King grass and they indicate possible drought tolerance.

References

Duncan, D.B. (1955). Multiple range and multiple F test. *Biometric,* 11, 1.
Herrera, R.S. (2001). Mejoramiento de pastos. Vías no clásicas. I *Simposium de biotecnología para la producción de alimento animal.* La Habana, CD-ROM
Herrera, R.S. Cruz, R. Martínez, R.O. García, M. Tuero, R. Cruz, A.M. Romero, A. & Fraga, N. (1996). Metodología para la obtención de mutantes de Pennisetum mediante técnicas nucleares y evaluación de las mismas. *XI Forum de Ciencia y Técnica,* La Habana
Micke, A. Domini, B, & Maluszynski, M. (1993). Les mutations induites en amelioration des plantes. *Mutation Breeding Review* No. 19 Vienna.

Yield and quality parameters of an interspecific hybrid *Pennisetum purpureum* Schum. (elephant-grass) *x Pennisetum glaucum* (L.) R. Br. Stuntz (pearl millet)

J.A. Usberti, Jr.[1], R. Usberti[2], P.B. Alcantara[3], V.B.G. Alcantara[3], R.A. Possenti[3] and M.A.C. Lucena[3]
[1]*Agronomic Institute, Campinas, Brazil. P.O. Box 28, CEP 13073-001, Email: usberti@iac.sp.gov.br,* [2]*Plant Protection Agency, Campinas. Brazil,* [3] *Zootechnic Institute, Nova Odessa, Brazil*

Keywords: recurrent selection, interspecific hybrid, forage yielding and quality traits

Introduction Elephant-grass is a tropical forage grass used either as a supplement fodder or for direct grazing. It usually shows regular nutritive value (6-13% crude protein, CP, and 55-60% forage digestibility) (Alcantara *et al.*, 1981). Most of the available cultivars produce no viable seeds. On the other hand, pearl millet has high seed yielding potential along with high quality forage (>15% CP and 70% forage digestibility). However, it shows poor forage production, low field persistence under grazing and low regrowth potential after cutting or grazing. During the 90's, an interspecific hybrid between the two species was developed, trying to combine the elephant-grass adaaptability and forage yielding potential with the pearl millet forage quality and seed yielding potential (Schank *et al.*, 1993; Schank, 1996). The new genetic material was able to produce viable seeds in variable amounts (Diz *et al.*, 1995). The main aim of this research was to produce selected populations with high phenotypic uniformities, showing high average forage production and quality.

Materials and methods Cutting and grazing types populations (CT and GT) were established, through the selection of 250 individual plants within the original F_2 population. The CT was made up of tall plants with low-tillering potential, long and broad leaves, thick stems with long internodes and erect growth habit while GT presented small and high-tillered plants, short and narrow leaves, thin stems with short internodes and prostrated growth habit. During two consecutive selection cycles, both were allowed free intercrossing in isolated fields. Plant height, tiller number, leaf length and width, stem diameter, internode length, inflorescence axis length, inflorescence length and leaf/stem ratios were scored, as well as dry matter (DM) production; leaf and stem CP contents; leaf+stem CP production and leaf and stem digestibility.

Results The CT and GT populations showed marked gains in phenotypic uniformities, while maintaining their original differential traits. Similar results were showed for internode length, inflorescence axis length and inflorescence length (data not shown here). However, there were differences among them for leaf/stem ratios (Table 1). Both selected populations showed DM production higher than the control with GT outstanding. The highest leaf and stem CP contents were also in GT, but there were no significant differences in leaf and stem digestibility. Therefore, GT and CT gave higher overall CP production than the original F_2 population .

Table 1 Mean results for selected morpho-agronomic, yield and quality traits in two selected populations compared with original F_2 population for 2004 selection cycle

Population	Plant height	Tiller number	Stem diameter	Leaf stem ratio	DM yield (t/ha)	Leaf CP (%)	Stem CP (%)	Leaf digestibility (%)	Stem digestibility (%)	CP yield (kg/ha)
Cutting type	3.41a	18.0b	1.80a	32.3b	15.2a	8.1b	4.3b	71.9a	56.3a	779b
Grazing type	3.01b	34.4a	1.21b	38.2a	14.1a	9.4a	5.8a	70.9a	59.0a	965a
Original F_2	3.32a	19.5b	1.73a	26.5a	11.8b	7.8b	4.6b	69.0a	57.4a	620c
Mean	3.25	24.0	1.58	32.3	13.6	8.4	4.9	70.6	57.6	788
C.V (%)	7.09	10.4	14.1	15.2	17.2	12.0	12.8	10.1	10.3	15.4

C.V. (%) = coefficient of variation; b) Means followed by different small letters, in the same column, are statistically different according to the Duncan test at p<0.05

Conclusions The phenotypic uniformities of both selected populations were improved; forage yielding and quality parameters were enhanced, making feasible the use of those populations in cultivated pastures.

References
Alcantara, P.B., V.B.G. Alcantara & J.E. Almeida (1981). *Proceedings of the Fourteenth International Grassland Congress*, 91.
Diz, D.A., S. C. Schank & D. S. Wofford (1995). Defoliation effects and seed yield components in pearl millet X elephantgrass hybrids. *Agronomy Journal*, 87, 56-62.
Schank, S. C. & D. P. Chynoweth (1993). The value of triploid, tetraploid, and hexaploid Napier grass derivatives as biomass and/or forage. *Tropical Agriculture*, 70, 83-87.
Schank, S. C., D. A. Diz, P. J. Hogue, C. V. Vann. (1996). Evaluation of pearl millet X elephantgrass hybrids for use as high quality forage for livestock. *Soil and Crop Science Society of Florida Proceedings*, 55, 120-121.

Growth characteristics of ecotype superior line of bermudagrass and development of its rDNA markers

Y.W. Rim, K.Y. Kim, M.J. Kim, B.R. Sung, Y.C. Lim and E.S. Chung
The National Livestock Research Institute, RDA, Suwon, Korea, E-mail:ywrim58@rda.go.kr

Keywords: bermudagrass, growth characteristics, DNA marker

Introduction Interest in turfgrass has steadily increased in Korea since the 2002 Korea-Japan World Cup . Use of zoysiagrass (*Zoysia japonica* L.) has been limited due to its slow recovery, low shoot density, short green period and low wear tolerance during dormancy (Lee *et. al.*, 1999). Bermudagrass has high quality and fast recovery, but has low cold tolerance (Richardson *et. al.*, 1978). This research compared the growth characteristics of a superior line of bermudagrass, named as Joyspy with other standard cultivars and to develop its rDNA markers.

Materials and methods Joyspy had been collected in Korea and growth characteristics were compared with the standard domestic cultivar Konwoo, and the standard imported cultivar Tifway-419 during 2003 and 2004. Plant height, leaf length, leaf width, leaf hair, covering speed, fourth internode thickness, fourth internode length, density (quality), disease resistance, green period and cold tolerance were examined. To find the rDNA markers of Joyspy, internal transcribed space (ITS) 1 primer detecting ITS 1 region of nuclear rDNA genes of fungi was used for PCR amplification and a new primer was constructed to detect the rDNA markers by PCR amplification.

Results Growth characteristics of Joyspy are summarised in Table 1. Joyspy had good growth characteristics such as high covering speed, narrow internode length, high density (quality) and cold tolerance compared to Konwoo and Tifway-419. From nucleotide sequencing of band amplified by ITS 1 primer, we found that eight nucleotides (CGGGAGTT) were missing in Joyspy. A new unique primer (front : 5'- GGC ATA ACA TGA CGT CAG GA – 3' : rear: 5'- GCG GAA GGA TCA TTG TCA – 3') was constructed to detect the missing region of nucleotides and a new rDNA marker was found for Joyspy by PCR amplification using this primer (Figure 1).

Table 1 Growth characteristics of the bermudagrass superior line ecotype Joyspy

Cultivar (line)	Plant height (cm)	Leaf length (cm)	Leaf width (mm)	Leaf hair (HML)	Covering speed (HML)	Internode thickness (mm)	Internode length (cm)	Density (Quality) (1~9)	Disease resistance (1~9)	Green period (HML)	Cold tolerance (1~9)
Konwoo	16	3.3	1.9	Low	High	0.9	2.5	3	1	Medium	5
Tifway-419	17	4.0	1.2	Low	Medium	1.0	4.5	5	1	Medium	7
Joyspy	10	3.5	1.5	Low	High	1.0	2.1	1	1	Medium	3

[#] 1: Strong (High), 9: Weak (Low)

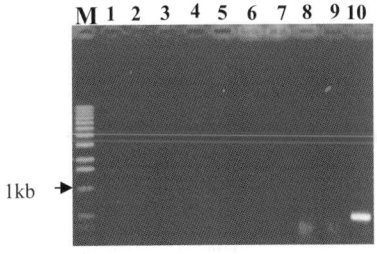

Figure 1 PCR amplification using unique primer
[#] Unique primer F : 5'- GGC ATA ACA TGA CGT CAG GA – 3', R : 5'- GCG GAA GGA TCA TTG TCA – 3'

[##] Lane 1-7: Zoysiagrass cultivars and lines, lane 1: Konhee, lane 2: Anyang-jungji, lane 3: Meyer, lane 4: S-94, lane 5: J01106, lane 6: J01067, lane 7: J01122, Lane 8-10: Bermudagrass, lane 8: Konwoo, lane 9: Tifway-419, lane 10: Joyspy (unique band indicated by arrow)

Conclusions Joyspy had good growth characteristics compared to other common cultivars of bermudagrass. Use of primers for amplification of fungal ribosomal RNA genes, specially ITS 1 primer detecting ITS 1 region of nuclear rDNA genes, was effective in developing the rDNA marker by PCR amplification.

References

White, T.J., T. Brunes, S. Lee & J. Taylor. (1990). Amplification and direct sequencing of fungal ribosomal RNA genes for phylogenetics. PCR Protocols: A Guide to Methods and Applications, 314-322.
Lee, J.P., J.B. Kim, J.Y. Kim & D.H. Kim. (1999). Development of cultivar 'Konwoo' in bermudagrasss. *Korean Turfgrass Science*, 13, 153-158.
Richardson, W.L., C.M. Taliaferro & R.M. Ahring. (1978). Fertility of eight bermudagrass clones and open-pollinated progeny from them. *Crop Science*, 16, 247-250.

Growth characteristics of superior lines of Zoysia grass (*Zoysia japonica*) and development of its DNA markers

Y.W. Rim, K.Y. Kim, M.J. Kim, B.R. Sung, Y.C. Lim and E.S. Chung
The National Livestock Research Institute, RDA, Suwon, Korea, Email: ywrim58@rda.go.kr

Keywords: Zoysia grass, growth characteristics, DNA marker

Introduction Demand for turf grass has steadily increased for recreation and sport fields after the 2002 Korea-Japan World Cup in Korea. Zoysia grass has the advantage of easy management, including low water and fertiliser requirement, but has limitations such as low recovery, low shoot density and short green period (Kim *et al.*, 1999). Objectives of this research were to select superior lines in the collected clones, compare the superior lines of zoysia grass with other standard cultivars for growth characteristics and to develop the DNA markers of superior lines.

Materials and methods Zoysia grass clones were collected in Korea and their growth characteristics (Table 1) were compared to standard cultivars "Konhee", "Anyang-Jungji" and imported cultivars "Meyer" and "S-94" during 2003~2004. To find the unique band of superior lines of zoysia grass, SRILS UniPrimer (URP primer) originally developed from rice repetitive sequences (Kang *et al.*, 1997) was used.

Results Growth characteristics of superior lines of zoysia grass are summarised in Table 1. A superior line named "Joydent" had good growth characteristics such as narrow leaf width, high covering speed, high density (quality) and disease resistance compared to "Konhee" and "Anyang-Jungji", "Meyer" and "S-94". PCR amplification using primer no. 6 from 12 URP primers tested appeared to have the unique band of superior line in lane 6 (Figure 1). New primer (front, 5F : 5'- AGC CTT GAA GTG TTC GAG TG – 3'; rear, 5R : 5'- ACC ATG CAT AGC GTA GCT TC – 3') from the nucleotide sequencing of this unique band was constructed for PCR amplification. "Joydent" had negative band using this primer (Figure 2).

Table 1 Growth characteristics of superior lines of Zoysia grass

tivar (line)	Plant height (cm)	Leaf length Cul (cm)	Leaf width (mm)	Leaf hair (HML)	Covering speed (HML)	Internode thickness (mm)	Internode length (cm)	Density (Quality) (1~9)	Disease resistance (1~9)	Green period (HML)
Konhee	9	5.2	1.8	Low	High	1.0	2.5	1	1	High
Anyang-Jungji	29	12.0	3.8	Low	High	1.9	5.4	3	5	Medium
Meyer	26	9.1	4.2	Medium	High	1.8	4.9	3	5	Medium
S-94	27	12.8	5.2	Medium	High	2.1	4.5	3	5	Medium
Joydent	14	5.2	1.5	Low	High	1.3	2.7	1	1	High
J01067	24	9.3	3.8	Medium	Medium	2.1	5.1	3	3	Very high
J01122	23	8.7	4.6	High	High	1.8	3.6	1	1	Medium

[#] 1: Strong (High), 9: Weak (Low)

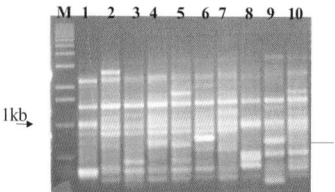

Figure 1 PCR amplification using URP primer
[#] URP primer no. 6(GC50%)

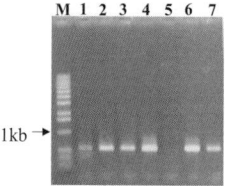

Figure 2 PCR amplification using unique primer
[#] Primer 5F : 5'- AGC CTT GAA GTG TTC GAG TG – 3'
5R : 5'- ACC ATG CAT AGC GTA GCT TC – 3'
[##] Lane 1-7: Zoysiagrass cultivars and lines, lane 1: Konhee,
lane 2: Anyang-jungji, lane 3: Meyer, lane 4: S-94, lane 5: Joydent,
lane 6: J01067, lane 7: J01122

Conclusions Superior lines of zoysia grass had good growth characteristics compared to other common cultivars. URP primers were primarily used to detect unique bands of zoysia grass lines by PCR amplification. New primer constructed from the result of nucleotide sequencing was found to have a negative unique band by PCR amplification.

References

Kang, H.W., Y.G. Cho & M.Y. Eun. (1997). DNA fingerprint of rice varieties (*Oryza sativar* L.) using primers designed from repetitive sequence of Korean rice and its application on other organisms. *5th International Conference on Plant and Animal Genome.* San Diego, CA. USA, 81.

Kim, D.H., J.P. Lee, J.B. Kim & S.Y. Mo,. (1999). Development of narrow leaf type cultivar 'Konhee' in zoysiagrasss. *Korean Turfgrass Science*, 13(3), 147-152.

The economic benefit of increased yield and digestibility in a perennial C_4 grass

R. Mitchell, K.P. Vogel and G. Sarath
*USDA-ARS, 344 Keim Hall, University of Nebraska, PO Box 830937, Lincoln, Nebraska 68583-0937, USA,
Email: rmitchell4@.unl.edu*

Keywords: big bluestem, breeding, digestibility, economics

Introduction Big bluestem (*Andropogon gerardii* Vitman) is a perennial C_4 grass native to the North American tallgrass prairie (Weaver, 1954). It provides productive, high quality forage during late spring and summer in the Great Plains, USA (Mitchell *et al.*, 1994). Increasing forage yield and digestibility can increase livestock performance and grassland profitability (Casler & Vogel, 1999). This study aimed to compare the economic value of 2 big bluestem strains developed by 3 generations of breeding for increased forage yield and digestibility with the base populations from which they were derived.

Materials and methods Bonanza and Goldmine are big bluestem cultivars developed for high yield and digestibility by 3 cycles of selection from the Pawnee and Kaw base populations, respectively. Pawnee and Kaw cultivars are based on germplasm from USDA Plant Hardiness Zones (HZ) 5 and 6, respectively. Plots (n=12, each 0.4 ha) near Mead, Nebraska were seeded in 1998 with 3 replicates/cultivar x 4 cultivars. Ammonium nitrate was applied at 112 kg N/ha each spring. Crossbred yearling steers (350-400 kg; 3 steers/plot; stocking rate 7.5 steers/ha) grazed the plots in 2000, 2001, and 2002. All plots were grazed for the same number of days/year. The value of steers at the start and end of grazing was calculated at 1.87 and 1.74 US$/kg live weight, respectively. Steer values were average Nebraskan market prices for the representative weight classes across years.

Results The big bluestem pastures gave 38-62 days of continuous grazing during each of the 3 years. Beef production and average daily gain (ADG) of steers grazing the Bonanza strain were 14-16% more than in steers grazing the Pawnee base population (Table 1). Economic returns to the producer improved by 32%, or 109 US$/ha. Cattle performance on the Goldmine strain was 5-7% more than for the Kaw base population. Economic returns were 10% more for Goldmine compared to Kaw. Cattle performance and economic returns did not differ significantly between the 2 selected strains of Bonanza and Goldmine.

Table 1 Beef production, ADG, gross return, and the economic value of improvement for 4 big bluestem cultivars grazed with 7.5 steers/ha in 2000, 2001, and 2002. The standard error is in parentheses

Cultivar	Beef production (kg/ha)	ADG (kg/hd/d)	Gross return (US$/ha)	Value of improvement (US$/ha)
Pawnee	398 (24)	1.12 (0.04)	340	
Bonanza	455 (19)	1.30 (0.03)	449	**109**
Kaw	424 (29)	1.19 (0.05)	392	
Goldmine	444 (18)	1.27 (0.04)	431	**39**

Conclusions Cultivars bred for increased yield and digestibility can increase significantly beef production and the profitability of grazing operations for producers in the tallgrass prairie region of the Great Plains, USA. These improved cultivars can be used to convert marginal cropland or degraded grazinglands to highly productive perennial grasslands. These grasslands can give co-benefits of ecosystem services to the public in the forms of perennial vegetative cover, carbon sequestration and storage, and reduced soil erosion. Bonanza and Goldmine were released officially in 2004. They are recommended in USA for use in HZ 5 and HZ 6, respectively.

References

Casler, M. D., & K. P. Vogel (1999). Accomplishments and impact from breeding for increased forage nutritional value. *Crop Science*, 39, 12-20.
Mitchell, R. B., R. A. Masters, S. S. Waller, K. J. Moore, & L. E. Moser (1994). Big bluestem production and forage quality responses to burning date and fertilizer in tallgrass prairies. *Journal of Production Agriculture*, 7, 355-359.
Weaver, J. E. (1954). North American Prairie. Johnsen Publishing Company, Lincoln, NE.

Analysis of genomic affinity between *Brachiaria ruziziensis* and *B. brizantha* through meiotic behaviour

C.B. do Valle[1], M.S. Pagliarini[2], A.B. Mendes-Bonato[2] and C. Risso-Pascotto[2]
[1]*Embrapa Beef Cattle Center, C.P. 154, Campo Grande MS 79002-970, Brazil, Email: cacilda@cnpgc.embrapa.br,* [2]*Department of Cell Biology and Genetics, State University of Maringá, 87020-900 Maringá PR Brazil*

Keywords: genome analysis, interspecific hybridization

Introduction Genetic divergence between polyploid hybrids is displayed in chromosome pairing and in the rate of chromosome elimination due to differences in cell cycle between the two combined genomes (Sundberg *et al.* 1991). In *Brachiaria*, a genus of African grasses reaching continental proportions as a tropical pasture in Latin America, genome analysis has never been performed. The majority of accessions in this genus is polyploid and apomictic, which restricts breeding. The relative ease of obtaining fertile interspecific hybrids once ploidy barriers are overcome (Pereira *et al.* 2001) confirms the phylogenetic proximity among *B. ruziziensis, B. decumbens* and *B. brizantha.* Hybrids were synthesised using sexual artificial 4x as the female genitor and natural apomictic 4x as the pollen donors. Genome affinity is a pre-requisite for chosen genitors to produce fertile hybrids and plenty of viable seed to assure adoption of the new cultivar. Microsporogenesis of a hybrid between *B. ruziziensis* and *B. brizantha* is described in this paper, focusing on the behaviour of both genomes.

Materials and methods Cytogenetic studies were done on an interspecific hybrid, where the male genitor was *B. brizantha* (B genome) and the female an artificially tetraploidised sexual accession of *B. ruziziensis* (R genome). Inflorescences for meiotic studies were fixed in a mixture of ethanol 95%, chloroform, and propionic acid (6:3:2 v/v) during 24 hrs. Microsporocytes (PMCs) were squashed and stained with 0.5% propionic carmine. Over 1800 microsporocytes were analysed. Chromosome associations were evaluated at diakinesis. Images were photographed with Kodak Imagelink – HQ, ISO 25 black and white film.

Results Chromosomes associated predominantly as bivalents, equally distributed in two metaphase plates. In 70% of PMCs, one genome did not divide synchronically, with chromosomes lagging behind or not segregating at all. The second division was very irregular, resulting in polyads. Based on previous results from analysis of a triploid hybrid between these species where the R genome was eliminated by asynchrony during meiosis (Risso-Pascotto *et al.*, 2004), it is suggested that the laggard genome in this hybrid also belongs to *B. ruziziensis*.

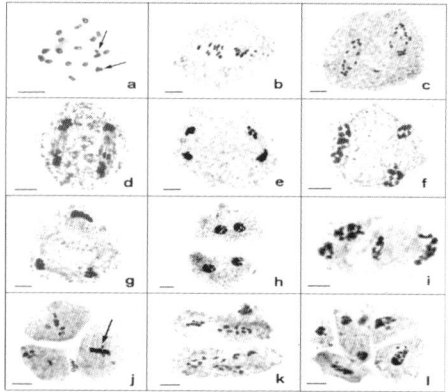

Figure 1 Chromosome behaviour in microsporogenesis: (a) DI with 17II and 2I (arrows) (b) MI with B and R genomes arranged in two metaphase plates (c, d) AI with 2 distinct spindles. In c, the 9 chromosomes are migrating to the poles (e) TI with both genomes properly segregated. (f, g) Trinucleate TI with only one segregated genome; the other remained non-segregated. (h, i) Early and late PII with normal genome segregation. (j) MII where only one genome underwent chromosome segregation at first division. Arrow indicates the cell with the non-segregated genome. Cytokinesis in the other cell yielded a triad in the second division. (k) Irregular chromosome distribution in AII. (l) Polyad with differently sized nuclei and microspores. (Bars = 1 µm).

Conclusions Abnormalities detected in this interspecific hybrid compromise pollen fertility. The eliminated genome is probably of the sexual parent *B. ruziziensis*. Cytological analyses are essential as a screening tool in this breeding program, if viable interspecific hybrids are to be selected and advanced to cultivar status.

References
Sundberg E., U. Lagercrantz & K. Glimelius, (1991) Effects of cell type for fusion on chromosome elimination and chloroplast segregation in *Brassica oleracea* (+) *Brassica napus* hybrids. *Plant Science*, 78, 89-98.
Pereira A.V., C.B. do Valle, R.P. Ferreira & J.W. Miles (2001) Melhoramento de forrageiras tropicais. In: Nass, L.L. (ed.) Recursos Genéticos & Melhoramento de Plantas, Fundação MT, Rondonópolis, pp. 549-601.
Risso-Pascotto C., M.S Pagliarini., C.B. Valle & L. Jank, (2004) Asynchronous meiotic rhythm as the cause of selective chromosome elimination in an interspecific *Brachiaria* hybrid. *Plant Cell Reports*, 22, 945-950.

Meiotic arrest compromises pollen fertility in an interspecific hybrid between *Brachiaria ruziziensis* x *Brachiaria decumbens* (Gramineae)

M.S. Pagliarini[1], C.B. do Valle[2], A.B. Mendes-Bonato[1] and C. Risso-Pascotto[1]
[1]*Department of Cell Biology and Genetics, State University of Maringá, 87020-900 Maringá PR Brazil, Email: mspagliarini@uem.br,* [2]*Embrapa Beef Cattle, C.P. 154, Campo Grande MS 79002-970, Brazil*

Keywords: cell fusion, meiosis arrest, pollen sterility, syncytes

Introduction Disruptions in meiosis, development of the free microspores, microspore mitosis, pollen differentiation or anthesis can result in male-sterile plants (Glover *et al.,* 1998). An understanding of the meiotic process is pivotal to work on reproduction, fertility, genetics and breeding in plants, with serious implications in crop production (Armstrong & Jones, 2003). Some African species of *Brachiaria* are the most important for pastures in the American tropics due to good adaptation and production. Artificial hybridization is underway in Embrapa to improve production, quality and insect resistance (Valle & Miles, 2001). For a cultivar to be successfully adopted good seed production and pollen viability are required. This paper reports on meiotic abnormalities impairing pollen fertility in a hybrid between *B. ruziziensis* x *B. decumbens*.

Materials and methods Cytological studies were carried out on an interspecific hybrid between an artificial 4x sexual accession of *B. ruziziensis* (2n=4x=36) and the most common cultivar of *B. decumbens*, a natural apomict (2n=4x=36). Inflorescences for microsporogenesis studies were collected and fixed in a mixture of ethanol 95%, chloroform and propionic acid (6:3:2 v/v) over 24 hours. Microsporocytes were prepared by squashing and staining with 0.5% propionic carmine. Pollen fertility was estimated using Alexander reactive (Alexander, 1969).

Results Syncytes involving a large number of cells were recorded in 15.40% of meiocytes. Meiosis was arrested in metaphase I and pycnotic nuclei and micronuclei were formed. Abnormal cytokinesis fractionated the syncyte into abnormal meiotic products that were covered by the pollen wall. Meiocytes in leptotene were recorded during both meiotic divisions and abnormal "pollen grains" with well-developed pollen walls but containing leptotene nuclei were recorded in 9.18% of grains analyzed. These findings suggest that the meiocytes received the signal to enter meiosis but lacked the signal to proceed beyond leptotene. Despite the absence of the meiotic process, such cells were covered by pollen grain walls. Total pollen sterility resulted from these abnormalities.

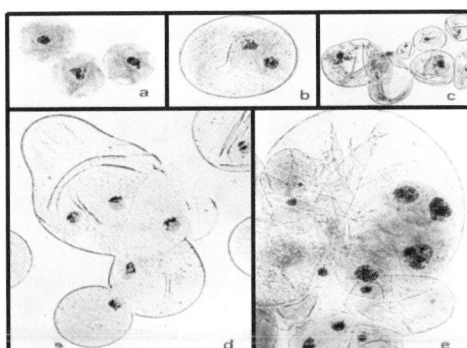

Figure 1 Abnormal "pollen grains". a) "Pollen grains" with leptotene nucleus. b) Binucleate "pollen grain" with leptotene nuclei. c) Abnormal pollen grains of different sizes and content of chromatin. d) Anomalous "pollen grain" with five leptotene nuclei. e) Anomalous pollen grain with pycnotic nuclei.

Conclusions Pollen sterility as a result of disrupted chromosome behavior was observed and is the single most important hindrance to the interspecific approach to *Brachiaria* breeding. Hybrids such as this, with these types of problems need to be identified and discarded early to avoid contributing defective genes to the breeding population.

References
Alexander M.P. (1969). Differential stain of aborted and non aborted pollen. *Stain Technology*, 44, 117-122.
Armstrong S.J. & G.H. Jones (2003). Meiotic cytology and chromosome behavior in wild-type *Arabidopsis thaliana. Journal of Experimental Botany*, 54, 1-10.
Glover J., M. Grelon, S. Craig, A. Chaudhury, & E. Dennis (1998). Cloning and characterization of *MS5* from *Arabidopsis*: a gene critical in male meiosis. *Plant Journal,* 15, 345-356.
Valle, C.B., & J.W. Miles (2001). Breeding of apomictic species. In: Savidan, Y.H., Carman, J.C. & Dresselhaus, T. (eds.) The flowering of apomixis: from mechanisms to genetic engineering. CYMMIT, IRD, European Commission, Mexico City, 137-152.

Shearing strength and chemical composition in the selection for quality in Brachiaria brizantha

C.B. Valle[1], F.E. Torres[2] and B. Lempp[3]

[1]Embrapa Beef Catttle, PO Box 154, Campo Grande MS Brazil, Email: cacilda@cnpgc.embrapa.br, [2]State University of Mato Grosso do Sul, Aquidauana, MS Brazil, [3]Federal University of Mato Grosso do Sul, Dourados, MS Brazil

Keywords: correlation, nutritive value, physical x chemical parameters, structural components

Introduction   Selection of quality pasture forages for productivity, nutritive value and animal performance require long-term, expensive trials. Simpler and accurate techniques to detect quality differences among genotypes have been proposed (Mackinnon et al., 1988; Hughes et al., 2000). This paper discusses the use of shearing strength in *Brachiaria brizantha* ecotypes to correlate physical traits with chemical composition: the objective being the identification of cultivars of improved quality forage suitable to the savannas of Brazil.

Materials and methods   One-thousand-square-meter paddocks of nine ecotypes were evaluated, in a randomized block design with two replicates. Forage was sampled after 42 days regrowth in two seasons: rainy (summer) and dry (winter). Total forage produced was sampled using a 0.25 m frame but only leaves were used for chemical analysis performed by near infrared reflectance spectroscopy. The second fully expanded leaf was used to evaluate shearing strength (F) with a Warner-Brazler equipment, after weight, area, width and length had been measured to standardize the F value to account for different leaf sizes.

Results   Shearing strength (F) and F standardized by leaf weight and area (F/DA) were the two best parameters to discriminate among ecotypes. Smaller differences were observed in the winter due to slower growth and less fiber accumulation. Higher significant differences were found in the summer (Table 1) and B2, B4, B5 e B6 were the best when compared to the check, particularly when total dry matter production was considered (5.7; 5.2; 3.5; 6.8 t/ha respectively vs. 2.5 of the check). Tropical forages tend to accumulate cell walls thus lower F or F/AD values signify less structural fibers which according to Jung & Allen (1995) should indicate better digestibility and consumption. The correlations of these parameters with chemical composition were highly significant (Table 2) and can be used as selection criteria for quality among ecotypes.

Table 1 Shearing strength and chemical composition on leaves of *B. brizantha* in the wet season

B. brizantha Ecotypes	F (kg)	F/AD (kg/g/cm^2)	ADF %	Cel. %	IVOMD %
check	2,9^c	132.2^{abc}	36^c	26^{de}	62^{ab}
B1	3,9^{ab}	159.2^{ab}	38^{bc}	29^b	59^{bc}
B2	2,7^c	122.9^{cd}	38^c	27^{cd}	62^{ab}
B3	4,1^{ab}	166.6^a	40^a	30^a	55^c
B4	3,2^{bc}	118.4^{cd}	37^c	28^{bc}	63^{ab}
B5	1,7^d	92.6^d	33^d	25^e	66^a
B6	2,8^c	123.8^{bcd}	38^c	27^{cd}	57^{bc}
B8	4,3^a	160.2^a	40^{ab}	29^{ab}	58^{bc}
B9	3,0^c	133.2^{abc}	37^c	27^d	62^{ab}

Cel. = cellulose
Columns containing the same superscript do not differ significantly (P < 0.05)

Table 2 Correlation coefficients between physical and chemical parameters in the wet season

	F (kg)	F/AD (kg/g/cm^2)
ADF[#]	0.88**	0.82**
Cellulose	0.90**	0.82**
IVOMD	-0.74**	-0.75**

[#]ADF= acid detergent fiber
IVOMD= in vitro organic matter digestibility
** highly significant (P<0.001)

Conclusions   Shearing strength (F) and F/Area density were the best predictors of quality in this study with ecotype being better discriminated in the rainy season. Both F and F/AD are highly correlated to fiber content and digestibility, proving to be a reliable, inexpensive and accurate selection technique.

References

Hughes, N.R.G., C.B. Valle, V. Sabatel, J. Boock, N.S. Jessop & M. Herrero (2000). Shearing strength as additional selection criterion for quality in *Brachiaria* pasture ecotypes. *Journal of Agricultural Science,* 135, 123-130.

Mackinnon, B.W., H.S. Easton, T.N. Barry, & J.R. Sedcole (1988). The effect of reduced leaf shear strength on the nutritive value of perennial ryegrass. *Journal of Agricultural Science*, 111, 469-474.

Jung. H.G. & M.S. Allen (1995). Characteristics of plant cell walls affecting intake and digestibility of forages by ruminants. *Journal of Animal Science*, 73, 2774-2790.

Hymenachne amplexicaluis [(Rudge) Nees] genetic resources collection in México, a suitable grass for flood plains in tropical areas

J.F. Enriquez-Quiroz, A.R. Quero-Carrillo, J. Perez-Perez, A. Hernandez-Garay and E. Garcia-Moya
INIFAP-*Veracruz; CP Iturbide 73, Salinas, San Luis Potosi. 78600, México, Email: queroadrian@colpos.mx*

Keywords: *Hymenachne amplexicaulis*, tropical grasses, flooded savannas, grass genetic resources

Introduction *Hymenachne amplexicaluis* [(Rudge) Nees; 2n= 2x= 24; Azuche, West Indian marsh grass] is a native Central and South America C_3 grass that grows well under intermittent flooding conditions. It produces good seed set and stolons to thrive on new areas assuring its survival, combined with an efficient N metabolism to promote vigorous new growing leaves and tillers (Antel *et al.*, 1998). Azuche is a dual attribute species when introduced to new areas; it has valuable forage attributes but also is a potential weed (Hill, 2000). As Azuche is a native species, one must deal with in the best possible way within Tropical Latin America areas (Enríquez *et al.*, 2004). No report has been found to date on living genetic resources collection and evaluation for this species.

Materials and methods In 2002 and 2003, an expedition by staff of **CP** and **INIFAP** collected Azuche from 90 sites in its natural diversity in tropical México from Puerto Vallarta to Tapachula on the Pacific and from Palizada, Campeche to Veracruz on the Gulf of México, and also on sections through the Tehuantepec isthmus and through the Sierra Madre de Chiapas (Figure 1). On each collection site, 25 tillers were tagged and stored in a commercial soil mix until establishment in 6 l pots; no fertiliser was applied. Pots were watered regularly to mimic flooded conditions and plant material was cut every 30 days. Pots were clipped 3 times and a 90-days growth was evaluated for morphology: (1) number of shoots, (2) central leaf length, and (3) central leaf width.

Results There were more collection sites on the Gulf of México than on the Pacific shore and the sierra routes; 88/90 sites were <150m; 2/90 were at 619 and 853m above the sea level. Azuche was not found on lagoons or areas of salt concentrations. Small patches of Azuche may indicate disappearance of its ecological niche and a drastic reduction in the surface it permeates. Azuche is a good alternative for wildlife. Farmers in flooding areas preferred Azuche to other grasses, mainly C_4, for grazing. These morphological attributes had high variation (Table 1) for future selection and hybridization studies. There was a low negative correlation (-0.5) between shoot number and leaf length, and a positive correlation between length and width of leaf (0.7; p <0.05).

Figure 1 Hymenachne amplexicaulis [(Rudge) Nees] collection route

Table 1 Basic statistics for three measured attributes on *Hymenachne amplexicaulis* [(Rudge) Nees] accessions collected within tropical Mexico

		Shoot Number	Central leaf width (cm)	Central leaf length (cm)
Coef.		49.2	25	21.5
Variance		21.2	0.2	53.4
Mean		9.4	1.8	34.1
Highest	1	26	3.6	52.1
Lowest	2	2.4	1.1	17.1

1. Five highest registered values
2. Five lowest registered values

Conclusions Most Azuche populations were <150 m above the sea level, and occurred in small isolated patches throughout tropical México. Farmers recognized Azuche as an important grass for flooded areas. There is a valuable morphological variation among the collected materials.

References
Antel, N. P. R., M. J. A. Werger & E. Medina (1998). Nitrogen distribution and leaf area indices in relation to photosynthetic nitrogen use efficiency in savanna grasses. *Plant Ecology, 138*: 63-75.
Enríquez-Quiroz, J. F., A. R. Quero-Carrillo & A. Hernandez-Garay (2004). Pastos para zonas tropicales inundables. *In*: INIFAP-PRODUCE-Veracruz. (eds.) Memorias del día del ganadero 2004. C. Experimental Playa Vicente SAGARPA-INIFAP (in press). México.
Hill, K. U. (2000). *Hymenachne amplexicaulis*: A review of literature and summary of work in Florida. http://www.naples.net/~kuh/hymen.htm

Critical analysis of tropical forage breeding in Brazil

J.A. Usberti, Jr.[1] and P.B. Alcantara[2]
[1]Agronomic Institute, Campinas, Brazil, P.O. Box 28, CEP 13073-001, Email: usberti@iac.sp.gov.br; [2]Zootechnic Institute, Nova Odessa, Brazil

Keywords: plant breeding, forage grasses and legumes, cultivar releasings, research needs

Introduction Forage grasses account for 90% of the Brazilian forage seed market while the genera *Brachiaria*, *Panicum* and others are responsible for 85, 10 and 5% of the traded grass seeds, respectively. Most of the forage grass and legume cultivars available for sowing in Brazil were selected in germplasm banks during the last 20 years, while few of them were derived from artificial crossings, followed by selection for desirable forage traits. The selection of new genetic materials in germplasm banks (exploitation of naturally-occurring genetic variability) is still feasible but the chances of success are decreasing through time. From now on, a clear trend is becoming quite evident: the exploitation of new genetic variation, to be accomplished through artificial crossings between selected parentals, in each forage species, aiming at the synthesis, selection and releasing of new hybrids showing high field performance.

Materials and methods A comprehensive search was carried out about the forage grass and legume cultivars released in Brazil since the 80's, including their origins (germplasm banks or plant breeding programs).

Results Out of the 19 released forage grass cultivars, only 6 (31,5%) were selected in plant breeding programs; however, only 5 of them are *Brachiaria* cultivars, which are unable to attend the seed market needs (Table 1). Recent releasings of **interspecific** *Brachiaria* hybrids (*B. ruziziensis* x *B. brizantha*, *B. ruziziensis* x *B. decumbens*) have been unable to succeed, mostly because of their extremely poor seed yielding potentials. Dealing with the forage legumes, only 7 cultivars were released, belonging to several species (6, selected in germplasm banks and only one as a result of a breeding program).

Table 1 Some forage cultivars released in Brazil since 1982 and their origins

Forage grass species	Cultivar	Institution[2] Releasing Year	Origin[1] GB	PBP
	Tobiata	IAC 1982	X	
	Aruana	IZ 1989	X	
Panicum maximum	Tanzania	EMBRAPA 1993	X	
	Atlas	MSCo. 2003		X
	Aries	MSCo. 2003		X
Brachiaria brizantha	Marandu	EMBRAPA 1983	X	
	MG-5	MSCo. 2000	X	
Brachiaria dictyoneura	Llanero	MSCo. 2000	X	
B. ruziziensis x *B. brizantha*	Mulato	PSCo 2003		X
P. purpureum x *P. glaucum*	Paraiso	MSCo. 1997		X
Macrotyloma axillare	Java	MSCo. 2003		X
Stylosanthes guianensis	Mineirao	EMBRAPA 1998	X	
S. capitata + *S. macrocephala*	Campo Grande	EMBRAPA 2000	X	
Arachis pintoi	Amarillo	MSCo. 1996	X	

[1] GB = germplasm bank; PBP = plant breeding program; [2] IAC = Campinas Agronomic Institute; IZ = Zootechnic Institute; EMBRAPA = Brazilian Agricultural Research Corporation; MSCo. = Matsuda Seeds Co.; PSCo. = Papalotla Seeds Co.

Conclusions Research effort and resources are needed for the synthesis and selection of new **intraspecific** *Brachiaria* hybrids, showing high forage quality, resistance/tolerance to acidic soils, pest and diseases and better ruminant acceptability. The new forage legumes hybrids should present high persistences under grazing, highest seed settings, low levels of tannins and phenols and abilities to persist in mixed pastures with forage grasses.

References

Tmanntje, L. (1997). Potential and prospects of legume-based pastures in the tropics. *Tropical Grasslands*, 31(2), 81-94.

Usberti Jr., J.A. (1993). Capim-colonião (Guineagrass). In: Viegas, G.P.; Furlani, A.M.C. (ed.). Melhoramento de plantas no Instituto Agronômico. (*Plant breeding at Agronomic Institute*). p.95-109.

Miles, J.W.; C.B. Valle. (1996). Manipulation of apomixis in *Brachiaria* breeding. In: *Brachiaria*: biology, agronomy and improvement. CIAT Publication, 259, 165-177.

Differential behaviour of guineagrass (*Panicum maximum Jacq.*) hybrids, with different Al^{+3} reactions, as to major nutrient translocations to the leaves

R.S. Paterniani[1], J.A. Usberti, Jr.[2] and J.C. Werner[3]
[1]Esalq, Piracicaba, Brazil. Sao Paulo State University; [2]Agronomic Institute, Campinas, Brazil. P.O. Box 28, CEP 13073-001. Email: usberti@iac.sp.gov.br, [3]Zootechnic Institute, Nova Odessa, Brazil

Keywords: guineagrass, hybrids, aluminium reaction, nutrient translocation

Introduction Most of the Brazilian cultivated pasture fields presents soils with high Al^{+3} levels and liming is economically unfeasible. So, there is an urgent need for grasses with good forage yielding potentials that can withstand Al^{+3} deleterious effects (abnormal root development: short, thick and poorly branched roots, which are unable to effectively translocate water and essential nutrients to the leaves) (Foy, 1984); as a consequence, susceptible genetic materials have their field persistences greatly affected, mainly during drought periods. Researches on Al^{+3} reaction are usually compare supposedly resistant/tolerant genotypes with a resistant control check; doing so, the genotypic effect is not isolated, making unreliable the comparisons made (Thomas & Lapointe, 1989). Guineagrass hybrids were tested as to nutrient translocations to the leaves, through comparisons of results obtained in treatments with and without N, P and K applications to the soil, for each genotype.

Materials and methods Greenhouse pot trials were carried out with 6 guineagrass hybrids showing different Al^{+3} reactions (Susceptible, Tolerant, Resistant: S, T, R) and flowering cycles (Early, Intermediate, Late: E, I, L) (Oliveira et al., 2000). A complete fertiliser formula was developed with the proper levels of N, P, K, Ca, Mg and S (control check); the treatments were additionaly supplied with N, P and K. During the vegetative stage, full developed leaves were picked up and properly analysed (AOAC, 1995). Hybrid differences as to N, P, K, Ca, Mg and S leaf levels were calculated through comparisons between treatments with additional N, P and K and the control checks. The effects of Al^{+3} resistance/tolerance were analysed, by observing the hybrids that showed the two highest percent increases or the two smallest percent decreases in the leaf nutrient levels.

Results Hybrid differences as to nutrient translocations to the leaves were detected, after N, P and K applications to the soil (Table 1). After N addition, no clear trend was observed in leaf nutrient levels which could be related to Al^{+3} reaction; however, after P and K addictions, resistant and tolerant genotypes outperformed the susceptible ones (83.3 and 70.0% of the cases, respectively), suggesting that both are highly correlated to Al^{+3} reaction.

Table 1 Effects of N, K and P application to the soil on the N, P, K leaf contents of six guineagrass (*Panicum maximum* Jacq.) F$_1$ apomictic hybrids, variable as to flowering cycles and Al^{+3} reactions.

| Hybrid | Percent leaf content increases / reductions[1] | | | | | | | | |
| | N | | | P | | | K | | |
	N[2]	P[2]	K[2]	N[2]	P[2]	K[2]	N[2]	P[2]	K[2]
H31 (E, S)	+ 23.8c	- 15.2 b	- 32.6b	- 7.1a	+ 42.8bc	- 36.8b	- 41.6bc	+ 17.7b	+ 82.3b
H33 (E, T)	+ 40.5bc	+ 3.1^a	- 17.7a	- 31.2bc	+ 28.6c	- 31.8b	- 52.8cd	+ 8.9bc	+ 125.0a
H55 (I, S)	+ 67.5a	- 29.5c	- 25.5ab	- 27.2bc	+ 25.0c	- 30.4b	- 34.4ab	- 38.2d	+ 43.8c
H54 (I, R)	+ 45.0b	- 5.2ab	- 24.5ab	- 34.6c	+ 57.1ab	- 17.4a	- 24.4a	- 4.7cd	+ 86.9b
H64 (L, S)	+ 48.8b	- 34.2c	- 28.1ab	- 30.4bc	+ 37.5c	- 3.8a	- 59.0d	- 12.8d	+ 38.4b
H79 (L, R)	+ 50.0b	- 14.3b	- 34.6b	- 21.1b	+ 66.7a	- 36.0b	- 50.2bcd	+ 32.2a	+ 78.3c
lsd p<0.05	16.8	13.5	14.8	12.6	17.8	12.8	15.8	14.2	18.2
CV %	12.4	11.2	16.4	10.5	13.4	15.4	13.4	10.3	13.2

[1] Calculated through comparisons between treatments with and without N, for each genotype; b) [2] N, P, K application; c) Means followed by different letters, in the same column, are different according to least significant difference test at p<0.05; d) CV % = coefficient of variation

Conclusions Al^{+3} resistant/tolerant genotypes are able to translocate nutrients to the leaves in a more effective way than the susceptible ones, mainly after P and K applications to the soil.

References
AOAC, Association of Official Analytical Chemistry. (1995). Official methods for analysis. Arlington, 430p.
Foy, C.D. (1984). Physiological effects of hydrogen, aluminium and manganese toxicities in acid soils. **In**: Soil acidity and liming. *American Society of Agronomy*, Madison, p.57-97.
Oliveira, A.C.; J.A. Usberti Jr, W.J. Siqueira. (2000). New methodology of aluminium resistance evaluation in guineagrass. *Brazilian Journal of Agricultural Research*, 35(11), 2261-2268.
Thomas, D.; S. Lapointe. (1989). Testing news accessions of guineagrass (*Panicum maximum* Jacq.) for acid soils and resistance to spittlebug (*Aeneolamia reducta*). *Tropical Grasslands*, 23(4), 232-239.

Animal performance and productivity of new ecotypes of Brachiaria brizantha in Brazil

V.P.B. Euclides, M.C.M. Macedo, C.B. do Valle, R. Flores and M.P. Oliveira
Embrapa Gado de Corte, Caixa Postal 154, 79002-970, Campo Grande, MS, Brazil, E-mail: val@cnpgc.embrapa.br

Keywords: forage availability, NPK fertilisation, nutritive value, oxisol, savannas

Introduction Brazil has the competitive advantage of a very dynamic and cost effective animal production system on pastures over other countries. The pursuit for more productive forages that will result in higher quality beef at a lower cost is then justified. *Brachiaria* is the most important forage genus utilised in Brazil, thus an intense search for new cultivars amongst collected and introduced ecotypes from Africa is underway. Following agronomic evaluation of this material in plots, 8 pre-selected *Brachiaria* ecotypes were tested under intermittent grazing in paddocks (Euclides *et al.*, 2001). Continuing on the process of cultivar development, two out of the eight, selected for superior agronomic characteristics were compared to the standard cultivar Marandu, under grazing and the results are presented in this paper.

Materials and methods The experiment was carried out at the National Beef Cattle Research Centre, Campo Grande, MS, Brazil, from March 2001 to February 2004. The *B. brizantha* selected ecotypes were Xaraés, Piatã, and the commercial cv Marandu was used as control. The experiment had a randomised block design with three treatments and two replicates. Six paddocks measuring 2 ha were divided in half and each was submitted to alternated grazing with a 28-day grazing and resting cycle. Three steers (testers) stayed in each paddock for a whole year, additional steers were allocated and removed according to forage availability, to assure the planned residues (3 t/ha of DM). All treatments received lime (2.25 t/ha) and 400 kg/ha of 0-20-20 NPK fertiliser at establishment. Maintenance fertiliser was 80, 60 and 60 kg/ha/year of N, P_2O_5 and K_2O, respectively. Forage samples, before and after grazing, were taken and liveweight gain was measured at 28-day intervals.

Results Steers grazing Piatã and Marandu pastures performed better than those grazing Xaraés; however, Xaraés pastures sustained a higher stocking rate than the others grasses (Table 1), which resulted in greater productivity (795, 715 and 670 LW kg/ha/year, respectively for Xaraés, Piatã and Marandu). The amount of total dry matter and green dry matter leaf percentage availabilities can explain these differences since there was no difference in nutritive value among the cultivars (Table 2). Differences ($P < 0.05$) between rainy and dry periods were also observed for all variables (Table 1 and 2). Dry matter availability after grazing, was always greater than 3 t/ha, indicating that this was not limiting animal performance, independently of the season of the year.

Table 1 Means for average daily gain (ADG, kg/steer per day) stocking rate (SR, steers/ha),over a period of 3 years

Cultivars	ADG	SR
Rainy		
Xaraés	0.718[b]	6.85[a]
Piatã	0.782[a]	5.19[b]
Marandu	0.770[a]	5.07[b]
Dry		
Xaraés	0.286[b]	2.25[a]
Piatã	0.349[a]	1.82[b]
Marandu	0.312[b]	1.97[b]

Means n the same column, within year period, bearing different superscript letters are different (P<.05), by Tukey.

Table 2 Means for herbage dry matter (DM, kg/ha) green dry matter (GDM, kg/ha) and percentages of leaf, crude protein (CP), *in vitro* organic matter digestibility (IVOMD), neutral detergent fiber (NDF) and lignin (Lig)),over a period of 3 years

Cultivars	Rainy Xaraés	Piatã	Marandu	Dry Xaraés	Piatã	Marandu
DM	4550[a]	4050[b]	4056[b]	3830[a]	3740[a]	3640[a]
GDM	3532[a]	3355[a]	2970[b]	2120[a]	1915[ab]	1655[b]
Leaf	51.5[a]	51.5[a]	48.4[a]	25.1[a]	24.8[a]	19.6[b]
CP	10,4[a]	9.5[a]	10.4[a]	8.1[a]	7.3[a]	7.9[a]
IVOMD	59.3[a]	59.9[a]	61.0[a]	53.0[a]	51.9[a]	53.5[a]
NDF	72.2[a]	73.8[a]	70.7[b]	74.5[a]	75.8[a]	73.4[a]
Lignin	2.52[a]	2.79[a]	2.46[b]	2.88[b]	3.10[a]	3.00[ab]

Means in the same row, within year period, bearing different superscript Letters are different (P< 0.05), by Tukey.

Conclusions Cultivar Xaraés was released by EMBRAPA Beef Cattle based on these results as a contribution to pasture diversification. Although this new cultivar had an inferior animal performance than cv. Marandu, it presented higher forage production, consequently sustaining higher stocking rate and greater productivity. These traits suggest it as a new alternative to be used under different production systems.

References

Euclides, V.P.B., C.B do Valle, M.C.M. Macedo & M. P. Oliveira (2001). Evaluation of *Brachiaria brizantha* ecotypes under grazing in small plot In: *Proceedings of the 19th International Grassland Congress*, Piracicaba, Brazil, 535-536.

Development of a bahiagrass *Paspalum notatum* Flugge with increased short-day biomass

P. Mislevy[1], A.R. Blount[2] and T.R. Sinclair[3]
[1]University of Florida, Agricultural Research and Education Centre, Ona, Florida 33865, USA, Email: pmislevy@ifas.ufl.edu, [2]UF AREC, Marianna, Florida 32446, [3]USDA-ARS Gainesville, Florida 32611, USA

Keywords: daylength, cool season forage, photoperiod, physiological dormancy

Introduction Low herbage productivity of subtropical grasses during the short-day winter months of October through to March can place a severe burden on livestock producers in Southeastern U.S. Researchers at the University of Florida (Sinclair *et al.*, 2001) hypothesised that the decrease in forage production might result from physiological dormancy induced by short day length. A study using artificial lights to extend the day length demonstrated that maintaining the day length at 15 hr during the short-day length period increased 'Pensacola' bahiagrass *P. notatum* Flugge *saure* Parodi forage yield 122% when compared with normal photoperiod (Mislevy *et al.*, 2001). A Pensacola-derived bahiagrass population was selected for increased vegetative growth under short-day length using restricted recurrent phenotypic selection for three cycles (UF Cycle 3) to increase forage yield. Plants that comprise this population were less sensitive to short photoperiod and produced increased forage mass during the short days. The objective of this clipping study was to evaluate forage production and forage nutritive value of UF Cycle 3 compared with selected standard entries during short and long day length periods.

Materials and methods The experiment was conducted at University of Florida, Ona, FL (82° 55' W and 27° 26' N) over 2 years. The study consisted of eight entries (Table 1) in a randomised complete block with ten replications. Plots were clipped (to 7.5 cm) every 5 wk during short days and every 4 wk during long days.

Table 1 Dry biomass yield (Mg/ha) during short and long days and crude protein (CP), in vitro organic matter digestion (IVOMD g/kg), of Paspalum entries grown during 2002-2004

Entry	Total biomass yield		CP			IVOMD		
	Short days	Long days	Summer	Autumn	Winter	Summer	Autumn	Winter
Atra paspalum (Suerte)	7.0 a†	17.2 a	156 a	140 d	136 c	659 a	636 a	683 a
UF Cycle 3	6.9 a	15.7 b	153 a	149 c	178 b	608 b	539 c	679 ab
UF Turf	6.1 b	14.4 b	151 a	155 c	179 b	594 bc	506 d	670 ab
Tifton 7	5.0 c	14.7 b	157 a	162 b	179 b	593 bc	584 b	645 cd
Tifton 9	5.4 c	14.5 b	151 a	154 c	180 b	591 c	500 d	670 ab
Sand Mountain	3.7 d	11.0 c	154 a	164 b	181 b	595 bc	477 e	645 cd
Pensacola	3.4 d	11.7 c	155 a	173 a	190 a	573 d	456 f	660 bc
Argentine	2.4 e	11.7 c	154 a	173 a	183 ab	575 d	535 c	633 d

†Means within the column followed by the same letter (s) are not different (*P*>0.05)

Results Dry biomass yield of UF Cycle 3 during the short and long day photoperiod was 22 and 8% greater than 'Tifton 9' and 51 and 25% greater than Pensacola, respectively. Digestibility of UF Cycle 3 was 40 and 80 g/kg and 10 and 20 g/kg higher than Tifton 9 and Pensacola during October and January, respectively. Data indicated plants less sensitive to day length will produce increased above ground biomass during short days.

Conclusions Data demonstrate that UF Cycle 3 out yielded Tifton 9 and Pensacola 1.5 and 1.2 and 3.5 and 4.0 Mg/ha during the short and long day length,, respectively. Forage nutritive value was generally equal or higher for UF Cycle 3 compared with standard cultivars of Tifton 9 and Pensacola.

References

Mislevy, P. T.R. Sinclair, & J.D. Ray (2001). Extended daylength to increase fall/winter yields of warm-season perennial grasses. p 256-257. *Proceedings of the Nineteenth International Grassland Congress* San Pedro, San Paulo,Brazil. Brazilian Society of Animal Husbandry.

Sinclair, T.R., P. Mislevy, & J.D. Ray (2001). Short photoperiod inhibits winter growth of subtropical grasses. *Planta*, 213, 488-491.

Genetic and molecular characterization of temperate and tropical forage maize inbred lines

B. Alarcón-Zúñiga, E. Valadez-Moctezuma, T. Cervantes-Martinez, T. Cervantes-Santana and M. Mendoza
Animal and Crop Science Depts. Universidad Autónoma Chapingo, Mexico, 56230, Email: camilaa@iastate.edu

Keywords: genetic similarity, maize races, forage quality, PIC, SSR

Introduction The livestock feeding in the Central highland of Mexico is based on harvest, grazing and annual forage conservation, being forage maize the most important silage crop (Alarcón, 1995). Even though forage maize is extensively bred in Europe, USA and Asia since 1900's, this started in Mexico in the 1960's, and little is known about the genetic diversity in both agronomic and nutritive value traits. Our breeding program goals are to analyze combining ability of biomass and quality predictors and to study the genetic relationship of inbred lines between lowland tropical and temperate races from Mesa Central, by genetic and molecular approaches.

Material and methods Fourteen inbred lines (IL) highly selected for forage biomass and quality value were used. 6 S_5 IL were collected from Mesa Central, Mexico, and 8 S_9 IL were originally single crosses from the tropical races: Tuxpeno, Vandeno, Olotillo, Naltel, Blandito, Reventador, Comiteco, Tepexintle, Celaya and Oloton. The single crossbred tropical races were recombined up to F_{13}, and selfpollinated up to S_9 (Cervantes et al., 1978). In early 2004, the ILs and temperate x tropical crosses were field established in two locations, and agronomic and quality traits evaluated when the kernel 2/3 milklined: total and per component dry matter, plant height (PH), days at flowering, soluble (SP) & insoluble protein (ISP), NDF, ADF & ADL, IVDMD, FAME, volatile fatty acids, sucrose and starch, were assayed. 27 out of 40 SSR markers were used to estimate genetic similarities among ILs, and to compute a discriminatory analysis by PCGA. The genetic components of variance, additive genotypic correlations and narrow sense heritabilities were estimated by MANOVA and standard errors computed by the delta method (Lynch and Walsh, 1998).

Results Genotypic effects were highly significant for all investigated traits (P<0.001), and much higher than genotype x environment interaction effects in all tropical ILs, but only higher in two of four temperate ILs. Transgressive segregations were observed in both tropical and temperate ILs for traits related to total DM, plant height, LSR, fiber predictors, free and volatile fatty acids; however, for each of the investigated digestibility traits, sucrose and starch, transgressive segregations were observed only on temperate Cacahuasintle and tropical ILs Tuxpeno y Vandeno. Narrow sense heritabilities on an entry basis ranged from low (~0.15; total and per component DM, IVDMD, FAME, ADL), medium (~0.3, PH, ISP, NDF, ADF, sucrose) and high (~0.6, days at flowering, volatile FA, starch). The average dry matter ear content was 51%, ranged 2-3% among ILs, and showed high heritability (0.5), with a high DM ear weight on ILs Cacahuasintle, Vandeno and Tuxpeno x Naltel. Positive additive genetic correlations between ear size, sucrose, starch or ADF content and IVDMD had similar absolute values, 0.55, so each of these two traits was an important but not the unique determinant of silage maize quality. A low genetic correlation was found between ADL/NDF and IVDMD, suggesting that digestibility can be improved in both tropical and temperate ILs independent on lignin content. The 27 SSR marker primers detected 86 alleles, with a range per locus from 2 to 7 (avg=3.19). The PIC values ranged from 0.09 to 0.75, with an average of 0.48. The marker analysis leads to the 14 inbred lines were classified into three distinct groups: temperate inbreds were included in group 1, with two distinct subgroups: Cacahuasintle and Chalqueno. The tropical inbreds were clustered into two groups: group 2 derived from dent grain germplasm included Tuxpeno, Blandito, Comiteco, and Tepexintle; and group 3 (flint germplasm) included Vandeno, Naltel and Oloton.

Conclusions These results suggest that tropical ILs Tuxpeno and Vandeno can be used as top parental ILs for forage maize in the highland Mexico; same was observed for Cacahuasintle as temperate IL. The primer loci of SSR markers did not cover all genomes completely, leading just to determine genetic similarities among ILs, and more SSR primers and ILs are ongoing to associate with loci that positively determine heterotic groups, supported by a field diallel analysis in 2005.

References

Alarcon Z.B. 1995. Tropical and temperate forage crops for the Mexican Livestock. XII Livestock Nat. Cong. Mexico.

Cervantes S.T., M.M. Goodman, E.D. Casas and J.O. Rawlings. 1978. Use of genetic effects and genotype by environmental interactions for the classification of Mexican races of Maize. *Genetics*, 90, 339-348

Lynch M. and B. Walsh. 1998. Genetics and analysis of quantitative traits. Sinauer Assoc. Pub. USA.

Preliminary performance of *Panicum maximum* accessions and hybrids in Brazil

L. Jank, R.M.S. Resende, S. Calixto, M.M. Gontijo Neto, V.A. Laura, M.C.M. Macedo and C.B. do Valle
Embrapa Gado de Corte, CxP. 154, 79002-970, Campo Grande, MS, Brazil, Email: liana@cnpgc.embrapa.br

Keywords: apomixis, breeding, selection, network trial

Introduction Due to the lack of high quality forages adapted to the distinct ecosystems in Brazil, a national network to evaluate 14 accessions, 4 hybrids and 5 commercial standards of *Panicum maximum* was established in 2002, in 5 regions of Brazil (states of Mato Grosso do Sul, Acre, Rondônia, Minas Gerais and in the Federal District). Staff of the Embrapa Beef Cattle group co-ordinated the network and pre-selected the accessions and hybrids (Jank, 1995; Jank *et al.*, 2001; Resende *et al.*, 2004). We present their performance as to leaf dry matter yield (LDMY) and leaf percentage (LP) obtained from 7 harvests in 2003-2004 in Mato Grosso do Sul.

Materials and methods The experiment was established at Embrapa Beef Cattle station in 24m^2 plots in a randomised complete blocks design, with 3 replicates. The commercial cultivars were Tanzania-1, Mombaça, Massai, Aruana and Milênio. The first harvest (dry season, 192 days growth) was on 16 Oct 2003 and 6 subsequent harvests were at 35-day intervals until 11 May 2004. The characters LDMY and LP were analysed using Selegen REML-BLUP (Resende, 2002), and a selection index was calculated using the weights 4.7 and 2 (rainy season) and 2.3 and 1 (dry season), for LDMY and LP, respectively.

Results Annual broad sense heritabilities for LDMY and LP were high, 0.55 and 0.75, respectively, with a high accuracy, indicating precision in the selection of the best accessions. The production in the dry season was low and varied from 4.9-11% of the year-round production, a characteristic of the species and already observed previously (Jank, 1995). Five accessions (PM35, PM40, PM36, PM30 and PM39) had better indexes than cv. Mombaça. The above accessions and the hybrids PM46 and PM47, had better indexes than cv. Tanzânia-1 (Table 1). The accession with the best index also had the best annual LDMY, and its selection should result in 23 and 12% gains, respectively, over the overall mean and the best commercial cultivar for this character. The selection index allowed characters associated to production in the rainy and dry seasons to be considered simultaneously as a selection criterion.

Table 1 Ranking of *Panicum maximum* accessions and hybrids based on selection index and genotypic values for the characters leaf dry matter yield (LDMY, t/ha) and leaf percentage (LP), annual data

Rank	Index	LDMY	LP	Rank	Index	LDMY	LP
1	PM35	14.57	78.02	13	PM37	10.66	78.85
2	PM40	13.86	83.76	14	PM34	10.56	81.68
3	PM36	13.43	74.74	15	Massai	11.55	76.27
4	PM30	13.80	73.98	16	PM33	10.58	79.11
5	PM39	12.66	83.14	17	PM45	10.19	79.41
6	Mombaça	13.03	79.92	18	PM44	10.80	76.09
7	PM47	12.04	76.22	19	PM31	10.72	76.97
8	PM46	11.84	77.23	20	PM41	10.52	75.45
9	Tanzânia	11.26	76.39	21	PM38	9.46	74.21
10	Milênio	12.50	76.09	22	PM43	9.19	65.63
11	PM32	11.50	75.72	23	Aruana	8.29	62.29
12	PM42	11.12	77.69	**Mean**	-	**11.84**	**76.25**

Conclusions The tested accessions had variable LDMY (8.3-14.6 t/ha) and LP (62.3-83.8%). Therefore superior genotypes may be selected, and the hybrids have potential for increased *P. maximum* productivity.

References
Jank, L. (1995). [Selection and breeding *Panicum maximum*] (Melhoramento e seleção de variedades de Panicum maximum). *Simpósio sobre Manejo da Pastagem*, FEALQ, Piracicaba, 12, 21-58.
Jank, L., C.B. do Valle, J. de Carvalho & S. Calixto (2001). Evaluation of guineagrass (*Panicum maximum* Jacq) hybrids in Brazil. *Proccedings of the XIX International Grassland Congress*, São Pedro, 498-499.
Resende, R.M.S., L. Jank, C.B. do Valle & A.L.V. Bonato (2004). Biometrical analysis and selection of tetraploid progenies of Panicum maximum using mixed model method. *Pesquisa Agropecuária Brasileira*, 39, 335-341.
Resende, M.D.V. (2002). Software SELEGEN REML/BLUP. Embrapa Florestas. Documentos, 77. 67p.

Agronomic performance and genetic variability of *Panicum maximum* accessions in the Cerrado of Federal District, Brazil

F.D. Fernandes[1], G.B. Martha, Jr[1], F.G. Faleiro[1], A.K.B. Ramos[1], R.P. Andrade[1], C.T. Karia[1], L. Vilela[1] and L. Jank[2]

[1]Embrapa Cerrados, C.P. 08223, 73.310-970, Planaltina-DF, Brazil, Email: duarte@cpac.embrapa.br, [2]Embrapa Gado de Corte, C.P. 154, 79.002-970, Campo Grande-MS, Brazil

Keywords: germplasm, selection, tropical forages

Introduction In the last three decades, the Brazilian Savanna (locally called "Cerrado") became the most important beef cattle production region in Brazil. Around 90% of all beef produced in the region comes from pasture-based systems. Intensively-managed and fertilised *Panicum maximum* pastures can be highly productive and economic. As a result, farmers are demanding new *P. maximum* cultivars for using in well-fertilised pastures or in crop-pasture rotation systems. This study aimed to evaluate the agronomic performance and the genetic variability of *P. maximum* accessions in the Brazilian Cerrado.

Material and methods Twenty-four *P. maximum* genotypes, being six cultivars (Mombaça, Tanzânia, Massai, Vencedor, Milênio and Aruana) and 18 previously selected accessions were studied. The experiment was established on 21 Nov. 02 at Embrapa Cerrados (15°35'30" S, 47°42'30" W, altitude 1007 m), on a highly fertilised clayey Dark Red Latosol, in a randomised complete block design with three replicates. Each plot was 12.5 m². Six cuts at a 20-cm stubble height were made in 2003 on 5 Feb., 12 March, 16 April, 25 June, 27 Oct. and 1 Dec. The leaf lamina (LDMY) and stem (SDMY) dry matter yields (kg/ha) in each cut and the crude protein (CP), neutral detergent fibre (NDF) and *in vitro* organic matter digestibility (IVOMD) contents in cuts 1 to 4 were evaluated. Means comparisons were carried out at the *P*<0.05 significance level. Random Amplified Polymorphic DNA (RAPD) molecular markers were used to estimate the genetic variability.

Results The overall LDMY and SDMY means were 11,266 kg/ha (LSD=3,683 kg/ha) and 3,763 kg/ha (LSD=2,091 kg/ha), respectively (Table 1). The accessions PM31, PM33 and PM34 were the most promising, because of highest leaf production (mean + 1 standard deviation) and lowest stem production (mean - 1 standard deviation). The CP, IVOMD and NDF contents varied similarly throughout the cuts. Crude protein and IVOMD decreased (P<0.05) in the fourth cut while NDF remained fairly constant during the experiment. The genetic distances between the genotypes ranged from 0.054 to 0.415 with the lowest distances occurring between PM39 and PM40 (0.054), PM31 and Massai (0.110), and PM42 and Tanzânia (0.132). Cultivars Mombaça, Milênio, Vencedor and Aruana are genetically distinct and are not related to the remaining collection.

Table 1 Agronomic performance of *Panicum maximum* genotypes in the Brazilian Cerrado

	Accumulated LDMY in six cuts						Accumulated SDMY in six cuts					
Overall mean	11,266						3,763					
Cultivars	10,392						4,158					
Accessions	12,598						3,632					
LSD (0.05)	3,683						2,091					
	CP (g/kg)				IVOMD (g/kg)				NDF (g/kg)			
	1	2	3	4	1	2	3	4	1	2	3	4
Overall mean	178	180	169	102	687	686	703	610	722	746	755	725
Cultivars	183	188	173	105	699	698	704	618	706	745	756	744
Accessions	176	178	168	100	683	682	702	607	727	746	755	719
LSD (0.05)												
genotype (cut)	39				102				76			
cut (genotype)	22				60				54			

Conclusions There were no differences in LDMY and SDMY between the means of cultivars and accessions. However, there were differences among the accessions, thus permitting the identification of promising accessions on the basis of forage production components. Molecular characterisation was an efficient tool to show the variability among the accessions and cultivars.

Forage yield and nutritive value of 30 cultivars of maize for silage in the Highland Valleys of Central Mexico

G. Tovar[1], J.L Arellano V[1], M.E. Sosa[2], C. Sánchez[2], H.P. Pérez[1], E. Vera[1], U.M Vera[2] and J.I. Vázquez[2]
[1]INIFAP-CEVAMEX, [2]Universidad Autónoma de Chapingo, México, Email: tovar.rosario@inifap.gob.mx

Keywords: Zea maïs, chemical composition, in vitro digestibility, days to silk

Introduction In Mexico, the selection of maize cultivars for forage has mainly been based on dry matter (DM) yields, not considering nutritional quality as an important evaluation parameter. The objective of this study was to assess forage yield and nutritive value of Highland and Subtropical maize cultivars for silage in the Highland Valleys of Central México.

Materials and methods Thirty genotypes of maize were studied during two years (spring 1999 and 2000) in Texcoco State, México. The experiment was made using a randomised complete block design with four replications. The harvest was performed during the silage stage (30-35 % DM of the whole plant). The registered variables were: Days to silk (DS), plant height (PH), green matter yield (GMY), dry matter yield (DMY) and digestible DM yield (DDMY), crude protein (CP), cellular walls (NDF), lingo-celluloses (ADF), lignin (ADL) and in vitro DM digestibility (IVDMD). Cell wall constituents (NDF-ADF-ADL) and IVDMD were determined according to Goering & Van Soest (1970).

Results and discussion The average DM content of the evaluated maize genotypes at harvest time was of 32.2 ± 3.4 %. There were differences among the genotypes of all the evaluated variables. The interaction cultivar*year was significant in all variables. Genotypes 3, 11, 12 y 23 of the Highland Valleys and the 14 subtropical genotypes had the highest GMY and DMY production (Table 1).

Table 1 Comparisons of means for forage yield and nutritional value of Highland Valleys and subtropical maize cultivars, 1999-2000

Regions	Cultivar/Hybrid	DS	PH (m)	GMY (t /ha)	DMY (t /ha)	DDMY (t /ha)	% CP	% IVDMD	% NDF	% ADF	% ADL
	Ganador (11)	97	3.1	83.8	26.8	18.6	8.1	69.7	64.0	37.8	6.0
	HS2 (23)	91	2.9	89.0	28.2	19.5	7.9	69.5	60.3	36.0	5.7
	H-135 (3)	101	2.9	96.7	31.8	22.0	8.3	69.4	60.3	36.1	6.4
Highlands	VS-22 (28)	87	2.7	77.6	24.6	16.8	8.6	68.9	58.7	34.5	5.6
	ZINA-1 (10)	92	2.9	77.9	24.9	17.0	8.0	68.6	56.6	35.7	5.6
	Triunfo (12)	95	3.0	96.9	28.7	19.3	8.1	68.0	60.2	34.9	6.4
	Trueno (2)	102	2.5	70.7	22.8	14.2	7.4	62.5	64.8	38.5	6.4
	A-791 (14)	104	2.6	87.9	28.5	17.3	7.8	60.7	66.8	37.9	5.6
Subtropical	Pantera (16)	105	2.3	77.7	24.2	14.8	7.9	60.8	62.1	37.0	6.3
	Tromba (17)	99	2.5	74.4	24.6	15.0	7.5	61.2	63.5	37.7	6.8
	Tukey 0.05	4.2	0.30	16.4	5.0	3.6	1.3	7.1	7.6	4.3	2.3

The genotypes with significantly lower IVDMD were the hybrids 2, 14, 16 y 17 integrated with subtropical germplasm; they also were among the ones with a higher content of NDF and ADF. Cultivars of significantly high digestibility were 10, 11, 12, 23 y 28 with adaptability to the Highland Valleys of Mexico and those that were integrated with germoplasm of "Chalqueño." Subtropical and Highland Valleys cultivars had equal CP content. However, among the cultivars of higher IVDMD, the 3, 11 and 23 showed a significantly higher content of FDN; this can suggest that the fibre quality of these genotypes might be better, due to composition changes and cell wall digestibility with higher hemicelluloses concentration and reduced contribution of lignified material and increased cell contents. In an in vitro study, Doane *et al.* (1997) demonstrated that a higher digestibility of NDF in maize stover was related to a higher hemicelluloses concentration, which was available for fermentation.

Conclusion The material with the best productive and nutritive behaviour were the hybrids H-135 and HS-2 as well as the cultivars Ganador and Triunfo of the late cycle and among the ones of the intermediate cycle the cultivar VS-22; however, the subtropical genotypes showed a better plant structure and root and stalk lodging resistance.

References
Doane, P.H., P. Schofield & N. Pell (1997). Neutral detergent fiber disappearance and gas volatile fatty acid production during the in vitro fermentation of six forages. *Journal of Animal Science,* 75, 3342-3352.
Goering, H.K. & P.J. Van Soest (1970). Forage fiber analysis. Apparatus, reagent, procedures and some applications. Agricultural Research Service. (Agriculture Handbook No. 379). Washington, D.

Grazing effects on the seed pool of *Stipa krylovii* and its genetic diversity in relationship to the plant population on a typical Steppe community in Inner Mongolia

B. Han[1], M. Zhao[1] and W.D. Willms[2]
[1]*College of Ecology and Environmental Science, Inner Mongolia Agricultural University, Huhhot, Inner Mongolia 010018, Peoples Republic of China, Email: mengli.zhao@yahoo.com, [2]Agriculture and AgriFood, PO Box 3000, Lethbridge, Alberta, Canada, T1J 4B1*

Keywords: coefficient of gene differentiation

Introduction *Stipa krylovii* is an important tufted forage species on the typical steppe in Inner Mongolia and is sensitive to heavy grazing pressure. Vegetative recovery of plant density is dependent on the seed bank, which is a genetic reservoir that supports the vegetative expression of the species thus enhancing its resilience (McCue and Holtsford 1998). The ability of the seed bank to support the *Stipa krylovii* population is dependent on its size and genetic diversity. Therefore, we conducted a study to determine the effects of heavy grazing pressure on its seed reserves and examine its genetic diversity in relation to surviving plants.

Materials and methods Thirty plants (spaced at least 10 m apart) of *Stipa krylovii* were randomly sampled from a site (41° 07' N, 115° 42' E, average annual precipitation = 368 mm, elevation = 1300 asl, soil = Typical Chestnut) that had been protected from grazing since 1984 and from a nearby community that had been heavily grazed for over 50 years. The plant cover of *Stipa krylovii* varied from 60 to 90% on the protected site and from 5 to 20% on the heavily grazed site. Nevertheless, plant composition was similar on each. The seed bank was sampled from 1, 50 x 50 x 5 cm plot associated with each sampled plant after seed rain. Seed numbers were determined using two methods: One was to screen the soil and count the seeds and the second was to germinate the seeds in trays and count the seedlings. Leaves were collected from each plant and from one seedling per plot (some plots produced no seedlings). The collected leaves were prepared to extract genomic DNA and RAPD markers that were used to detect the genetic diversity. Thirteen arbitrary primers were used and their markers analyzed using POPGENE 1.31 (Yeh *et al.*, 1997).

Results Grazing severely depleted the number of *S. krylovii* seeds and frequency of occurrence in the seedbank (Table1). Of the total number of seeds founding in grazed plot, only 28% produced seedlings, while 60% did so in the protected area. The seedlings contained greater genetic diversity than the plants (Table 2). Of the total variation, 33% was found between the populations.

Table 1 Number and frequency of seeds and seedlings (germinated seeds) from grazed and protected sites in Inner Mongolia

Material	Grazed		Protected	
	No.	Fr. (%)	No.	Fr. (%)
Seeds	7	32	35	85
Seedling	2	7	21	50

Table 2 Genetic diversity (H_o) of *S. krylovii* plants and seed pool, as measured from germinated seeds, from protected sites in Inner Mongolia

Material	Sample (no.)	Loci Total	Loci Poly.	(H_o)	H_t	H_s	H_s/H_t (%)	G_{st}* (%)
Plants	30	111	85	0.194	0.260	0.173	66.73	33.27
Seeds	18	92	71	0.153				

*Coefficient of gene differentiation

Conclusion The seed pool is essential for imparting resilience to the plant population. Grazing affected the *S. krylovii* population not only by reducing its ground cover but also by nearly eliminating the seed pool. However, 27 years after removing grazing pressure, the ground cover and seed pool of the *S. krylovii* population had recovered. The seed pool contained a greater amount of genetic diversity and thus contributes to the recovery of the population following disturbances. Grasslands need to be managed to allow seed production to maintain the seed pool and the health of the grassland.

Acknowledgements This work was funded by the National Natural Science Foundation of China (30060015)

References
McCue, K. A. & T. P. Holtsford. (1998). Seed bank influences on genetic diversity in the rare annual *Clarkia springvillensis* (Onagraceae). *American Journal of Botany*, 85, 30-36.
Yeh, F.C., R-C. Yang, T. B.J. Boyle, Z-H. Ye, & J.X. Mao (1997). POPGENE, the user-friendly shareware for population genetic analysis. Molecular Biology and Biotechnology Center, University of Alberta, Canada.

Grazing effects on genetic diversity of *Festuca campestris* Rydb. and *Stipa grandis* L. on the native grasslands in Canada and China, respectively

M. Zhao[1], W.D. Willms[2], B. Han[1] and G. Han[1]
[1]College of Ecology and Environmental Science, Inner Mongolia Agricultural University, Huhhot, Inner Mongolia 010018 Peoples Republic of China, Email: mengli.zhao@yahoo.com, [2]Agriculture and AgriFood, PO Box 3000, Lethbridge, Alberta, Canada T1J 4B1

Keywords: genetic identity, genetic distance, gene flow

Introduction Genetic drift or selectively neutral mutation in finite populations may result in genetic diversity within a natural population (Kimura, 1986). Genetic diversity influences the resilience of a species to survive perturbations or adapt to changes in its environment. Grazing by livestock may affect genetic diversity by exerting selection pressure on grazing sensitive species. In this study, we examine the effects of heavy sustained grazing pressure on the genetic diversity of *Festuca campestris* Rydb. and *Stipa grandis* L. These species are found on the Canadian Plains and the steppes of Inner Mongolia, respectively. Each is an important forage species that dominates their respective grasslands but decline readily when subjected to heavy grazing pressure.

Materials and methods Single natural grassland sites were subjected to heavy grazing pressure (~80% of ANPP) annually for more than 50 years in Canada (Fescue site, 50° 12' N, 113° 54' W) and Inner Mongolia (Stipa site, 43° 33' N, 116° 42' E) by cattle and sheep, respectively. Each site included contiguous grazing exclosures (> 1 ha) erected 1949 on the Fescue site and 1979 on the Stipa site to protect areas from livestock grazing. The Fescue and Stipa sites had thin Black Chernozemic and Typical Chestnut soils, respectively, and average annual precipitation of about 500 and 350 mm, respectively. Thirty four and 43 plants were collected from the protected and grazed areas, respectively, in the Fescue site and 30 and 30, respectively, in the Stipa sites. The collected leaves were prepared to extract genomic DNA and RAPD markers were used to detect the genetic diversity. welve and 18 arbitrary primers were used for the samples from the Fescue and Stipa sites, respectively. The data were analyzed using POPGENE 1.31 (Yeh *et al.*, 1997).

Results Average genetic diversity (H_o) of both populations at each site was similar (Table 1) with most diversity within populations (H_s/H_t). Overall, the coefficient of gene differentiation (G_{st}) was relatively low. The number of migrants per generation (Nm) was large, suggesting high gene flow (Table 1). Nei's genetic identity was high for both sites suggesting that the grazed and protected populations were genetically similar. The gene flow and genetic identity were somewhat less in the Stipa population vs the Fescue population.

Table 1 Mean estimates of genetic diversity (H_0) produced by 12 or 18 primers for plants of two species that were either protected from grazing or heavily grazed for an extended period

Site	Years of protection	Primers	Population (H_o)		H_t	H_s	H_s/H_t (%)	G_{st} (%)	N_m*	I**
			Heavy	Zero						
Fescue	50	12	0.33	0.34	0.35	0.33	96.52	3.20	16.25	0.969
Stipa	26	18	0.32	0.31	0.38	0.35	92.10	4.50	11.61	0.924

*Gene flow: $N_m=0.5(1-G_{st})/G_{st}$, **Nei's genetic identity

Conclusion The 3.5 and 7.9 % inter-population variation observed for *F. campestris* and *S. grandis* is not clear evidence that grazing affected their genetic diversity. Reduced diversity might be expected if sensitive plants were killed and sensitivity had a genetic control. The lack of effect suggests either no genetic link between plant vulnerability to grazing or the masking of its expression with the replacement of killed genotypes. However, with over 30% of genetic diversity (H_o, Table 1), the populations of the two species contain a large amount of genetic potential to respond to selection pressure produced by grazing and other perturbations. The relatively small effect that grazing may have on genetic diversity and the apparent lack of genetic drift caused by grazing indicates that any loss of genetic diversity would recover if grazing pressure were relieved.

Acknowledgements This work was funded by the National Natural Science Foundation of China (30060015).

References
Kimura, M. (1986). DNA and the neutral theory. Philosophical Transactions of the Royal Society of London, Series B, *Biological Sciences*, 312, 343-354.
Yeh, F.C., R-C, Yang,, T.B.J. Boyle, Z-H, Ye & J.X. Mao (1997). POPGENE, the user-friendly shareware for population genetic analysis. Molecular Biology and Biotechnology Center, University of Alberta, Canada.

Hybrids between meadow and smooth bromegrass: a new forage crop for Canada

B.E. Coulman and Y.S.N. Ferdinandez
Agriculture and Agri-Food Canada, Saskatoon Research Centre, 107 Science Place, Saskatoon, SK S7N 0X2 Canada, Email: coulmanb@agr.gc.ca

Keywords: hybrid bromegrass, forage quality, animal performance, yield

Introduction Smooth bromegrass (*Bromus inermis* Leyss.) has been an important hay grass in the agricultural regions of western Canada for approximately 50 years. Meadow bromegrass (*B. riparius* Rehm.) has become the most important pasture species in this region over the last 15 years. It is possible to produce hybrids between these species, which could lead to the development of a type of bromegrass which would be useful for both hay and pasture purposes.

Materials and methods In 1976 and 1977, three hybrid bromegrass populations were created by the late R.P. Knowles by crossing plants of meadow bromegrass (cv Fleet and Paddock) and smooth bromegrass (cv Signal). By bringing the two species into flower at the same time under a controlled environment and enclosing panicles of the two species in a crossing bag, approximately 20% of the florets of the meadow brome parent developed viable seed. The hybrid plants grown had variable floret fertility, with a mean slightly lower than the parents. Two lines were developed, S9197 and S9073, which were selected over several cycles over a 15 year period for increased vigor, improved floret fertility, improved seed type (no awns or pubescence), reduced creeping habit and fast regrowth. In 1981-82, hybrid plants were backcrossed to smooth bromegrass (a reduced creeping line) and seed harvested from the smooth brome parent. This led to the development of S9183 which was selected for uniform plant type, floret fertility and vigour. A line designated S9356, which had increased height, large seed size and reduced creeping habit, was selected out of S9183. These four hybrid lines were extensively evaluated in western Canadian trials from 1995-2002.

Results and Discussion In an analysis of the morphology of hybrid plants relative to those of the parental species (Ferdinandez and Coulman, 2000), the hybrids were found to be intermediate (leaf to stem ratio) or did not differ (tiller height). In other characteristics, hybrid plants more closely resembled meadow brome (leaf area index, leaf pubescence) or smooth brome (tiller density, panicle density, hay dry matter yield and leaf disease incidence). S9356 and S9183 were more "smooth-brome" like in appearance than S9197 and S9073 with wider leaves and fewer basal leaves.

In forage nutritive value, the hybrid populations had consistently lower NDF and ADF than the parental species at a vegetative stage of growth. This suggests that the hybrid has potential as a high quality forage grass for grazing or hay prior to, or at, heading. At anthesis, the hybrids had lower concentrations of crude protein than either of the parental species.

In grazing experiments conducted in two eco-zones of western Canada, beef steers gained more per unit area on S9197 hybrid brome, than on either smooth or meadow bromegrass. In simulated grazing studies which measure regrowth after clipping, the hybrids produced more dry matter than smooth bromegrass and a similar amount to meadow bromegrass.

Hay yield trials including one or more of the hybrid lines were conducted at from six to ten locations in western Canada. The hybrid lines produced more dry matter than meadow brome in all eco-zones. The mean hybrid yields were from 2-6% less than smooth bromegrass, but in the drier eco-zones, the yields of the hybrid were similar to, or higher than, smooth bromegrass. In the sub-humid climate of eastern Canada, hay yields of the hybrid lines were inferior to both smooth and meadow bromegrass.

S9197 hybrid bromegrass was released to the Canadian seed industry in 2000 under the name of 'AC Knowles', with certified seed available for planting in spring, 2004. S9356 was released as 'AC Success' in 2003.

Conclusions Hybrid bromegrass is a new dual purpose (hay and pasture) type of grass available to producers in Canada, with two cultivars having been recently released. Hybrid brome produces hay yield similar to smooth bromegrass, with regrowth similar to meadow brome. Grazing experiments have shown higher beef gain ha^{-1} for hybrid bromegrass than for either of the parental species.

Reference

Yasas F.S.N. & B.E. Coulman (2000). Characterization of meadow X smooth bromegrass hybrid populations using morphological characteristics. *Canadian Journal of Plant Science*, 80, 551-557.

Stem anatomy of switchgrass plants developed by divergent breeding cycles for tiller digestibility

G. Sarath[1], K.P. Vogel[1], R. Mitchell[1] and L.M. Baird[2]
[1]USDA-ARS, 344 Keim Hall, University of Nebraska, PO Box 830937, Lincoln, NE 68583-0937, Email: gsarath1@unl.edu, [2]University of San Diego, San Diego, CA 92110, USA

Keywords: breeding strategy, anatomy, herbage digestibility

Introduction Switchgrass (*Panicum virgatum*. L.) is an important perennial forage and biomass crop that is native to the temperate prairies of the North America east of the Rocky Mountains. Breeding for improved forage *in vitro* dry matter digestibility (IVDMD) has been conducted using post-heading, whole-tiller IVDMD as the selection criterion (Hopkins *et al.*, 1993; Vogel *et al.*, 2002). One breeding cycle (C-1) for low IVDMD and three cycles for high IVDMD (C1, C2, C3) were completed in a switchgrass population adapted to the USA mid-latitudes. Sward trials demonstrated that whole plant IVDMD had been improved (Hopkins *et al.*, 1993). This study reports on changes in plant anatomy of plants from populations divergently bred for whole tiller IVDMD.

Materials and methods Seedlings of each population were transplanted on 1.1 m centres into a replicated field nursery in eastern Nebraska, USA in 2002. After heading, ten tillers of each plant from the C-1 and C3 populations were sampled in 2002 and 2003 and dissected into leaf, stem, and sheath. Based on tissue IVDMD, lignin, cell wall, cellulose, and hemicellulose concentrations, ramets of specific plants were transplanted into greenhouse pots where the cloned plants were grown to maturity. Plants from the C-1 and C3 populations were sampled for anatomical analysis. The anatomy of the second internode below the peduncle was evaluated by light microscopy using conventional procedures.

Results Four representative micrographs of internode sections stained for lignin (Maule's stain) are shown in Figure 1. Sections from C-1 plants (low IVDMD population) showed a large amount of lignified fibres cells just below the epidermis (A & B). Additional zones of lignification were seen around vascular bundles. The cell walls of the cortical parenchyma in these plants exhibited greater secondary cell wall deposition. Plants from the C3 (high IVDMD) population had fewer cortical fibres below the epidermis and in the lignified sheaths surrounding the vascular bundles (C). There also was substantially less secondary cell wall thickening. Some plants of the C3 population had high stem lignin concentration but still had altered plant anatomy (D).

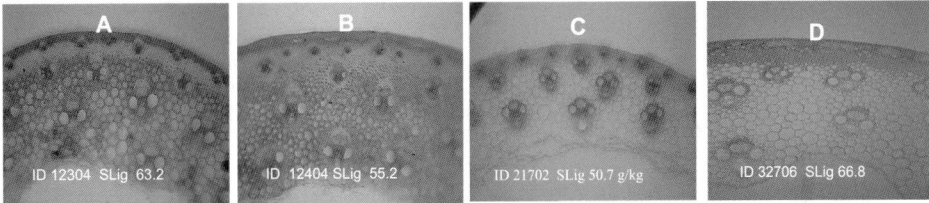

Figure 1 Internode sections stained for cell wall composition. ID = plant ID number; SLig = stem lignin

Conclusions Multiple cycles of breeding for IVDMD produced plant populations which differ significantly in digestibility and lignin concentration (data not shown). In the stems of switchgrass, changes in lignin appear to involve two primary mechanisms: (1) the loss or decrease in cortical fibres and (2) changes in secondary cell wall deposition. These results indicate that plant anatomy can be significantly impacted through a simple selection tool, strongly suggesting that the genes that control these functions will be attractive targets for future manipulation.

References
Hopkins, A.A., K.P. Vogel, & K.J. Moore (1993). Predicted and realized gains from selection for in vitro dry matter digestibility and forage yield in switchgrass. *Crop Science*, 33, 253-258.
Vogel, K.P., A.A. Hopkins, K.J. Moore, K.D. Johnson, & I.T. Carlson (2002). Winter survival in switchgrass populations bred for high IVDMD. *Crop Science,* 42, 1857-1862.

Divergent breeding for tiller digestibility modified leaf, sheath, and stem composition of switchgrass (*Panicum virgatum* L.)

K.P. Vogel, G. Sarath and R. Mitchell
USDA-ARS, 344 Keim Hall, University of Nebraska, PO Box 830937, Lincoln, NE 68583-0937, USA, Email: kpv@unlserve.unl.edu

Keywords: breeding strategy, digestibility, lignin, plant morphology

Introduction Switchgrass (*Panicum virgatum.* L.) is a cross-pollinated, C_4 species that is native to the prairies of temperate North America. Breeding to improve its forage quality has been conducted using post-heading, whole-tiller *in vitro* dry matter digestibility (IVDMD) as the selection criterion. One breeding cycle (C-1) for low IVDMD and three cycles for high IVDMD (C1, C2, C3) were completed in a switchgrass population adapted to the USA mid-latitudes. Sward trials demonstrated that whole plant IVDMD had been improved (Hopkins *et al.*, 1993). The objective of this study was to determine the effect of breeding for tiller IVDMD on leaf, sheath, and stem digestibility and composition of plants of the derived populations.

Materials and methods Seedlings of each population were transplanted into a replicated (r=4) field nursery in eastern Nebraska, USA in 2002. A plot consisted of 50 plants on 1.1m centres. After heading, ten tillers of each plant from the C-1 and C3 populations were sampled in 2002 and 2003. Tillers were dissected into leaf, sheath, and stem. Panicles were excised. Neutral detergent fibre (NDF), acid detergent fibre (ADF), acid detergent lignin (ADL), ash, N and IVDMD concentrations were determined using near infrared reflectance spectroscopy (NIRS) and wet laboratory procedures as described by Casler *et al.* (2004). Cell wall (CW) concentrations were calculated assuming NDF g/kg = cell wall g/kg.

Results Breeding for increased tiller IVDMD had a greater effect on sheath and stem IVDMD, ADL, cell wall, hemicellulose, cellulose, lignin, and cell wall digestibility than on the same leaf traits (Table 1). Leaf cell wall, hemicellulose, and cellulose concentrations of the C-1 and C3 populations did not differ. Stem hemicellulose and cellulose concentrations did not differ at $p<0.05$. The C3 plants with improved IVDMD differed from plants of the C-1 population by having lower stem and sheath cell wall and lignin concentrations.

Table 1 Mean digestibility (DMD), cell wall (CW), lignin (L), cellulose (C), hemicellulose (HC), and nitrogen (N) of leaf, sheath, and stem of switchgrass plants from a low (C-1) and high (C3) IVDMD population

Strain	Tissue	IVDMD (g/kg)	ADL (g/kg)	CW (g/kg)	HC(CW) (g/kg)	C(CW) (g/kg)	L(CW) (g/kg)	DMD(CW) (g/kg)	N (g/kg)
C-1	Leaf	672*	25.9*	589	500	448	43.5*	440*	16.4**
C3		682	23.8	586	500	450	39.9	453	16.8
SE		21	3.4	23	5	9	4.1	17	0.4
C-1	Sheath	581**	37.7**	704*	448**	489	53.5**	427**	6.4
C3		623	32.4	697	457	487	46.6	482	6.3
SE		21	3.3	7	8	10	4.0	29	1.0
C-1	Stem	464**	69.5**	756*	416	485[+]	91.4**	291**	3.7**
C3		502	60.3	747	419	493	80.4	333	4.2
SE		16	4.8	15	15	13	6.5	16	0.3

+, *, ** Indicate populations differ at the 0.10, 0.05, and 0.01 levels of probability; SE = standard error. Two-year means are expressed on a dry weight or cell wall (CW) basis

Conclusions Leaf, sheath, and stem IVDMD, cell wall, cellulose, hemicellulose, and lignin concentration of switchgrass responded differently to breeding for tiller IVDMD indicating tissue specific regulation of plant composition. Breeding switchgrass for use in grazing systems should emphasize leaf and sheath composition. Stem composition can be modified to develop switchgrass cultivars for biomass energy.

References
Casler, M.D., K.P. Vogel, C.M. Taliaferro, & R.E. Wynia (2004). Latitudinal adaptation of switchgrass populations. *Crop Science*, 44, 293-403.
Hopkins, A.A., K.P. Vogel, & K.J. Moore (1993). Predicted and realized gains from selection for *in vitro* dry matter digestibility and forage yield in switchgrass. *Crop Science*, 253-258.

Geographic patterns in the genetic diversity of Elymus species from Qinghai-Tibetan and Inner Mongolian Plateau

Y. Xuebing[1,2], G. Yuxia[3], Z. He[1] and W. Kun[1]
[1]China Agricultural University, Beijing 100094, PRC, Email: yxbbjzz@163.com, [2]Zhengzhou College of Animal Husbandry, zhengzhou 450002, PRC, [3]Henan Agricultural University Zhengzhou 450002, China,

Keywords: *Elymus*, genetic diversity, allozyme, SSR, geographic variation

Introduction The genus *Elymus* is the largest genus in the tribe Triticeae with about 150 species distributed in most temperate regions of the world (Dewey, 1984). The genetic diversity of *Elymus* spp. from alpine regions is very important for improving resistance to adverse condition. The goals of this study were to investigate microsatellite and enzyme polymorphism and population structure of different regions and *Elymus* spp. in China.

Materials and methods Forty eight natural populations of 10 *Elymus* spp. from different elevations of the Qinghai-Tibetan Plateau except for 8 populations of 3 spp. from the Inner Mongolian Plateau were examined for morphological, allozyme, micro-satellite (SSRs) markers.. Five enzyme systems, alcohol dehydrogenase (ADH), aspartate amino transferase (AAT), Aryl esterase (EST), peroxidase (PRX) and malic dehydrogenase (MDH), were analysed against five plants of each population. SSR variations were observed using six primer pairs specific for *Elymus* spp., ie. ECGA22, ECGA210, ECGA125, EAGA13, EAGA53, EAGA101, described by Sun *et al.* (1998a,1998b). Diversity analyses were performed against 48 populations using SPSS ,the software package Popgene and NTSYS-PC program for morphology, allozyme and micro-satellite, respectively. To graphically display the relationship of the spp. and populations, a dendrogram was generated from a pair-wise distance matrix using an unweighted pair group method with arithmetic average (UPGMA) (Rohlf, 1993).

Results The five enzyme systems and six SSR markers assayed in this study were found to be encoded by 14 loci and 28 alleles, respectively. Data of allozyme showed that the average total genetic diversity (HT) was 0.37, ranging from 0.23 to 0.48. The mean within population component of diversity (HS) was 0.03, ranging from 0.0 to 0.07. The mean gene diversity among populations (DST) and species were 0.34 and 0.47, ranging from 0.21 to 0.41 and 0.27 to 0.62. The proportion of total genetic variation among populations (GST) was, on average, 0.92. For micro-satellites, a total of 28 alleles were found at the six loci analyzed. Among all spp. and populations, the percentage of polymorphic loci was 100% expected heterozygosis and Shannon's Information index were 0.25 and 0.38. Between populations of each species, the percentage, heterozygosis and Shannon's Information index ranged from 16.67 to 83.33, 0.07 to 0.37 and 0.10 to 0.53, respectively (Table 1). Primer pair ECGA22 amplified six fragments in many species. The others produced one to four polymorphic markers. All amplified products were within the range of 100-500bp.

Conclusion The majority of the total genetic diversity of selfing *Elymus* spp. reside among populations based on allozyme analyses. This also indicates a high degree of differentiation among the populations. Micro-satellite is useful for differentiating very close species.

Table 1 *Micro-satellite diversity of Elymus spp. at different levels*

		%	h*	I*
Elymus level		100.00	0.25	0.38
	E.nutans	50.00	0.13	0.20
	E.sibiricus	66.67	0.21	0.30
Species	E.breviaristatus	50.00	0.17	0.25
	E.purpuraristatus	66.67	0.21	0.30
level	E.cylindricus	83.33	0.37	0.53
	E. glaucus	50.00	0.19	0.27
	E.geminatus	50.00	0.19	0.27
	E.tangutorum	16.67	0.07	0.10
	E.atratus	33.33	0.14	0.20
	E.submuticus	33.33	0.18	0.21

h*=Nei's (1973) gene diversity, I* = Shannon's Information index

References

Dewey D. R. (1984). The genome system of classification as a guide to intergeneric hybridization with the perennial Triticeae. *In:* Gustafson J P (Ed.) Gene manipulation in plant improvement. New York: Plenum Press, 209-280.

Wang ZH-R.(1996). Plant allozyme analyses. Beijing: Science Publishing Press.

Sun, G.-L., B Salomon & R. von Bothmer, (1998a). Characterization of microsatellite loci from Elymus alaskanus and length polymorphism in several Elymus species (Triticeae: Poaceae). *Genome*, 41, 455-463.

Sun, G.-L., Salomon, B., & R. von Bothmer (1998b) Characterization and analysis of microsatellite loci in Elymus caninus (Triticeae: Poaceae). *Theoretical and. Applied Genetics*, 96, 676-682.

Rohlf F.J. (1993). Numerical taxonomy and multivariate analysis system. Version 1.80. Exeter Software, Setauket, NY.

Genetic structure of Mongolian Wheatgrass (*Agroypron mongolicum* Keng) in Inner Mongolia of China

Y. Jinfeng, Z. Mengli and X. Xinmin
College of Ecol. and Env. Sci., Inner Mongolia Agricultural University, Huhhot, Inner Mongolia 010018 P.R..China, Email: csgrass@public.hh.nm.cn

Keywords: genetic diversity, RAPD

Introduction Mongolia wheatgrass (*Agroypron mongolicum*) is a cross-pollinated, long-lived, cool-season and drought-resistant perennial bunchgrass, which plays an important role in arid and semi-arid grasslands of Inner Mongolia. Collections of *A. mongolicum* from different areas of Inner Mongolia are valuable sources of useful genes for its breeding. The genetic diversity of 8 accessions of *A. mongolicum* were examined in this study. A dendrogram was constructed to obtain information on the relationship between cultivated and wild *A. mongolicum* genotypes, which is basic information to explore the possibility of its use in intra- and inter-specific breeding programs.

Materials and methods A total of 8 accessions (6 wild and 2 cultivated) of *A. mongolicum* were collected from 7 areas in arid and semi-arid grasslands of Inner Mongolia. Fifteen plants (spaced at least 10 m apart) were randomly sampled from each site. Seeds were collected from each plant and planted in a greenhouse. Leaves were collected from one seedling from each plant. The collected leaves were prepared to extract genomic DNA and RAPD markers that were used to detect the genetic diversity. Seventeen arbitrary primers were used and their markers were analysed by using SPSS 8.0 to generate Jaccard's similarity coefficient (S_{ja}), genetic distance ($D=1-S_{ja}$), and diversity coefficient (DC). UPGMA dendrogram was constructed using MEGA software.

Results The diversity coefficients (DC) ranged from 0.147 (Zhenlan) to 0.273 (Qingshuihe), with an average of 0.237 (Table 1) and the six wild populations showed a higher diversity than the cultivated populations. The diversity coefficient (DC) among the 8 populations was 0.222, among the six wild populations 0.250, while among the two cultivated varieties it was 0.181. This evidence was supported with a relatively small genetic distance between the two cultivated populations (average 0.290 among wild populations and 0.213 among cultivated populations). The UPGMA dendrogram showed that the eight populations could be divided into 3 groups, based on their geographic origin and on the soil type of their distribution areas.

Table 1 Diversity coefficients (DC) within populations

Site	Xiwu	Baiyinxi	Qingshuihe	Zhenlan	Yijihuole	Sunitezu	Var.1	Var.2	Average
DC	0.243	0.269	0.272	0.226	0.258	0.273	0.210	0.147	0.237

Conclusion The average diversity coefficient of 0.237 indicated considerable genetic diversity among populations of *A. mongolicum*. The open pollination and out-crossing system, as well as the strong gene flow led to a great diversity of *A. mongolicum* being retained in recessive genes in the heterozygotic state. As a geographically widely distributed species, *A. mongolicum* has a lot of ecotypes, each of them with their own physiological traits and adaptive capacity. It is noteworthy that there was a clear tendency for *A. mongolicum* that originated from the same location to be clustered together in the dendrogram.

References
P.Virk, B.,M. Ford-lloyd, Jackson & H. John, 1994, Use of RAPD for the study of diversity within plant germplasm collections, *Heredity*, 74:170-179
G. Yan, S. Zhang, 1985, Karyotype analysis of Mongolia wheat grass, *Grassland of China*, 2: 38-42

Genetic engineering for breeding for drought resistance and salt tolerance in Agropyron spp. (wheatgrass)

Mi Fugui, Y. Jinfeng and H. Xiuwen
Inner Mongolia Agricultural University, 010018, Huhhot, China, Email: mfgui@ yahoo.com.cn

Keywords: genetic transformation, drought resistance, salt tolerance

Introduction Genetic engineering for breeding for drought resistance and salt tolerance in wheatgrass, lucerne and tall fescue is one of the main projects in major national programs of 10th five-year national plan: "Research of gene transfer in plants and its industrialisation". The project is a large one that has the financial support for forage crops in China and many research institutes and universities take part in it. The Inner Mongolia Agricultural University is in charge of the project on wheatgrass. The research was started in Nov. 2002. The general situation and the primary results are introduced and summarised in this paper.

Material and method Four species of wheatgrass (*Agropyron mongolicum, A .cristatum* cv. 'Fairway', *A. desertorum* cv. 'Nordan' and *A .cristatum* ×*A. mongolicum* cv. 'Hycrest-Mengnong') were used as plant materials. Based on the established regeneration system, the p5CS gene, which regulates the last step of proline synthesis, was transformed into the species, with phosphinothricin acetyltransferase (*bar*) conferring herbicide resistance as selecting gene. The transformation was conducted through microprojectile bombardment of callus derived from immature inflorescence and the transgenic plants were examined by PCR Southern and RT-PCR analysis.

Results The results showed that callus initiation from immature inflorescence and plant regeneration could occur under induction in all four species. The medium for callus induction was the improved culture MS+2,4-D (2.0mg/L), with induction frequency up to 83.5%. The differentiation medium was hormone free MS+KT (0.2mg/L) and the differentiation ratio about 74.5%. One hundred procent of roots could be induced under the culture medium as 1/2 MS.

The transgenic plants were obtained by microprojectile bombardment of callus induced from immature inflorescence, the results of PCR and Southern analysis of those plants displayed that the exogenous p5CS gene had integrated into the genome of the plants and the assay of RT-PCR showed the transgenic p5CS had expressed at transcript level.

Conclusions Wheatgrass is one of the grasses which are suitable to conduct the genetic transformation. The culture system of callus and plant regeneration was optimized. The integration of the exogenous p5CS in the research materials was displayed by the molecular examinations, and the transgenic frequency of p5CS gene was about 0.1%. The obtained transgenic plants provide the new genetic resources, in which the further works of breeding should be improved.

Germplasm collection and dry matter production of Mongolian forage plants

D. Tsogoo and Sh. Batsukh
Animal Husbandry Research Institute, Ulaanbaatar, Mongolia, Email: tsogoo@gobi.initiative.org.mn

Keywords: germplasm collection, forage plants, plantation

Introduction Mongolian rangelands harbour 564 genera, 128 families and 2,823 species of plants (Gubanov, 1996). Approximately 600 species can be used as forage plants (Yunatov, 1968). There are 5 to 6 species of perennial forage plants that are unique to Mongolia and the country has a domestic cultivar *"Burgaltai"* of lucerne. This paper provides the results of a study on vegetative and seed yield of germplasm collected from naturally occurring forage plants during 1976-2003. To date, seeds have been collected and are involved in the experimental work from approximately 70 percent of forage plants available at the forage plant seed bank of the Animal Husbandry Research Institute where there are over 2000 samples available.

Materials and methods One thousand one hundred and fifty samples of 230 species from 11 natural vegetation zones of Mongolia were collected during several expeditions. The experiments involved planting the seeds under irrigated and non-irrigated conditions in the steppe of Mongolia.

Results From the Mongol-Daurian Mountains forest steppe, 376 samples of 200 species, 120 genera and 40 families were collected. This zone has the richest collection of forage plant germplasm within Mongolia and forage plants such as *Medicago falcata, Elymus dauhricus* and *Melilotus dentatus* were found within the natural rangeland of this zone. From the botanical-geographical region passing through the Depression of the Great Lakes, Trans-Altai Gobi desert and Dzungarian Gobi, plant samples with high nutritional value such as *Reaumuria zoongorica, Zygophyllum xanthoxylon, Tamarix gracilis* and also precious and rare plants such as *Rosa laxa, Elaeagnus Moorcroftii and Halimodendron halodendron* were collected. Forage plants were found to be unique in their germination period, seed viability, winterhardiness, seed and dry matter yield. The method of planting was also found to influence the production results. If the plant species were first grown in sheds and then transplanted as seedlings into the field, 8.3 to 50.0 % of the legume plants germinated in 19-29 days. If the plant species were planted as seeds directly into the experimental fields then the germination rate dropped to 5.3 to 33.3 percent in 58-71 days (Tsogoo & Batsukh (2003). Forage plants such as *Medicago falcata, , Astragalus adsurgens, Trifolium lupinaster, Agropyron cristatum , Bromus inermis, Elymus dahuricus* under irrigation will grow in 44-133 days and one plant can produce between 3.9 and 156.2 grams of forage and 0.1 to 4.4 grams of seeds (Table 1) (Tsogoo & Turtgtoh, 2003).

Table 1 Yield of some forage plants (2001-2002)

Plant names	Nr. Mongolian gene bank catalogue	Plant yield, g/pl		Seed yield, g/pl		Conclusions
		Transplanted	Seed	Transplanted	Seed	
1,Elymus dahuricus	1118	62.0	45.0	4.4	4.3	
2. E. Gmelinii	1446	11.5	21.3	1.2	2.0	
3.E.sibiricus	895	66.0	20.8	8.0	2.7	
4.Poa attnuata	1192	17.5	44.0	0.3	0.3	
5.Agropyron cristatum	1311	42.7	13.6	3.7	2.5	
6.A. michnoi	1068	26.4	9.35	1.4	1.1	
7.Bromus inermis	420	52.7	7.2	0.4	0.1	
8.Stipa sibirica	920	36.7	37.0	0.6	2.4	
9.Medicago falcata	1131	20.0	30.0	0.1	0.5	
10.Astragalius adsurgens	1390	156.2	29.2	2.2	4.2	
11.Trifolium lupinaster	1371	5.8	3.9	0.6	0.3	

Conclusions As expected, most of the forage species did not grow under dry or non-irrigated conditions. *Hedysarum , Festuca, Poa, Cleistogenes, Ajania, Salsola* could be used under these conditions.

References
A. Gubanov (1996). *Conspectus of flora of outer Mongolia*, Moscow, 136 p.
D. Tsogoo &, H. Batsukh (2003). *Extract from the results of Genetic resources survey on Medicago ruthenica*, "Scientific papers of livestock science", Journal, 30, Ulaanbaatar, 68-71.
D.Tsogoo & B. Turtogtoh, (2003). *Germplasm research of some legumes*, Livestock Science Journal, UB, 182-188.
A.A.Yunatov (1968). *Forage plants in the pasture of Mongolia*, Ulaanbaatar, 43.

Morphological characteristic to discriminate *Festulolium* hybrids (*Festuca pratensis* × *Lolium perenne*)

M. Kulik[1], Z. Zwierzykowski[2] and W. Jokś[3]
[1]*Agricultural University of Lublin, 20-950 Lublin, Akademicka 15, Poland Email: qliusz@agros.ar.lublin.pl,*
[2]*Institute of Plant Genetics, Polish Academy of Sciences, Strzeszyńska 34, 60-479 Poznań, Poland,* [3]Szelejewo Plant Breeding Ltd., 63-820 Piaski, Poland

Keywords: *Festulolium loliaceum,* inner glume, hybrid, plant breeding, inflorescences

Introduction Environmental change and uncertainty is likely to pose new challenges in plant breeders. Recently attention has focused on the crossing of *Lolium* and *Festuca* species to obtain hybrids exhibiting many desirable traits of both parents. Key objectives of such programs are to combine the persistency, winter hardiness and drought tolerance of fescues with the high herbage yields and quality of ryegrasses (Zwierzykowski, Naganowska, 1994). One of the hybrids with great practical significance is a *F. pratensis* × *L. perenne* hybrid [*Festulolium loliaceum* (Huds.) P.V. Fourn]. Many morphological traits of *Festulolium* hybrids demonstrate intermediate character, however, in relation to inflorescence type they are similar to *L. perenne*; the hybrids and perennial ryegrass have spike-like inflorescences, though they may be rarely a little-branched. Occurrence of a reduced inner glume in hybrid spikelets is a trait, which enables discrimination between *Festulolium* and *L. perenne* plants. The aim of this work was to analyse the morphological trait of inflorescences to aid the identification of the hybrids *Festulolium* in relation to *L. perenne*.

Materials and methods The initial hybrids between tetraploid forms of *Festuca pratensis* and *Lolium perenne* were obtained at the Institute of Plant Genetics PAS in Poznań. Breeding materials were developed at Szelejewo Plant Breeding. Plants from two *Festulolium* strains were used in preliminary research in Sosnowica (south-east part of Poland). Monocultures of both strains were sown in 2003 on mineral light soil. Plants for analysis were selected at random in 2004 (50 individuals per strain) at the stage of promotion or at the beginning of maturation. Basic morphological analyses were qualified with special regard of spikelet number on inflorescence as well as occurrence of inner glume.

Results *Festulolium loliaceum* is a loosely tufted grass. It has folding in the bud leaves, short ligule and characteristic auricles. Spikes are 20 to 30 cm long with 12 to 20 spikelets alternately arranged directly to the axis of inflorescence. Spikelets are shorter (1,7 to 2,5 cm long) and contain 7 to 14 florets. All these traits are similar to perennial ryegrass (Falkowski, 1982), however one characteristic trait was a reduced inner glume, which is absent from the spikelets (Figure 1). In both strains, frequency of occurrence initially decreased before increasing within the higher situated spikelets (numbers 11 to 12) thereafter frequency again declined. Mean numbers of inner glumes were 6,98 for strain I and 4,7 for strain II. Significant correlation were not observed between spikelets numbers per spike and the number of inner glumes per spikelet. Furthermore, this trait was characterized by high variability (Table 1).

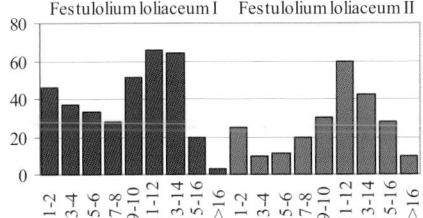

Figure 1 Frequency of reduced inner glume per spikelet in particular ranges

Table 1 Relation between number of spikelets per inflorescence (a) and number of inner glumes per spikelet (b) in *Festulolium loliaceum* I and II strains

Strain	Trait	Mean	Variability factor	Correlation factor
I	a	15.04	9.49	- 0.17
	b	6.98	53.87	
II	a	15.76	11.99	- 0.05
	b	4.7	64.08	

Conclusions These findings confirm that visual distinction between *Festulolium loliaceum* and *L. perenne*, is possible due to the occurrence of an inner glume within the spikelet. Visual identification should focus on spikelet number 11 and 12 as it is here where the trait is frequently exhibited.

References

Falkowski M., (1982). Trawy polskie. Państwowe Wydawnictwo Rolnicze i Leśne, Warszawa.
Zwierzykowski Z., & B. Naganowska (1994). The use of *Lolium-Festuca* hybrids in breeding. Genetica Polonica, 35A, 11-17.

Introgression breeding for improvement of winter hardiness in *Lolium /Festuca* complex using androgenenesis

T. Yamada[1], Y.D. Guo[1,2] and Y. Mizukami[3]
[1]*National Agricultural Research Center for Hokkaido Region, Sapporo, 062-8555, Japan, Email: Toshihiko.Yamada@affrc.go.jp,* [2]*China Agricultural University, Beijing,100094, China,* [3]*Aichi Agricultural Research Center, Nagakute, Aichi 480-1193, Japan*

Keywords: androgenesis, *Festulolium*, freezing tolerance, genomic *in situ* hybridisation, pollen fertility

Introduction Intergeneric hybrids between closely related *Lolium* and *Festuca* species are used to broaden the gene pool and provide plant breeders with options to combine complementary traits to develop robust but high quality grass varieties. Androgenesis was found to be an effective procedure for selecting *Lolium-Festuca* genotypes comprising gene combinations rarely or never recovered by conventional backcross breeding programs. Here we describe the optimisation of androgenesis in *Lolium perenne* x *Festuca pratensis*. The male fertility and freezing tolerance of the *Festulolium* microspore-derived progenies were analysed and these progenies were also analysed by using genomic *in situ* hybridisation (GISH). The object of this study is to initiate introgression breeding for the improvement of winter hardiness in *Lolium /Festuca* complex.

Materials and methods Genotypes of *Lolium perenne* x *Festuca pratensis* (*Festulolium* hybrid), 'Prior', 'Bx350' and 'Bx351'were investigated in this study. PG-96 (Guo *et al.,* 1999) with 2 mgl⁻¹ 2,4-D, 0.5 mgl⁻¹ kinetin was used as embryo (calli) induction media. Calli were transferred to the solid medium 190-2 (Wang & Hu, 1984) supplemented with 0.1 mgl⁻¹ 2,4-D, 1.5 mgl⁻¹ kinetin for green plants regeneration. The ploidy level of androgenic progenies was analysed by Partec CAII flow cytometry (Münster, Germany) with DAPI staining. GISH was carried out according to the method described by Mizukami *et al.* (1998) with some modification. Androgenic-derived plants were grown outdoor for natural hardening during autumn. Crown tissues each genotype were analysed for freezing tolerance cooled to –17°C. Male fertility was measured by staining pollen with 1% acetocarmine and counting the frequency of stainable pollen grains.

Results The calli and green plants were obtained from all three accessions, but the genotype responses differed; accessions 'Bx350' and 'Bx351' were more active than 'Prior' in androgenesis. Among microspore-derived progenies the diploids were dominant (68.2%). High levels of chromosome pairing and recombination were observed by GISH due to close homology between genomes of *L. perenne* and *F. pratensis*. These androgenic-derived *Festulolium* progenies showed a wide range of variation in freezing tolerance, 19 progenies (6.5%) exceeding that in the *F. pratensis* cv. 'Tomosakae'. More than 60% of flowing progenies produced dehiscing anthers with pollen stainability ranging from 5% to 85% in all three accessions (Table 1). The diploid progenies with both freezing tolerance and fertility potential have been crossed with *L. perenne*.

Table 1 Pollen fertility in amphidiploid *Festulolium* anther-derived progenies. F & PF, fertile and partial fertile, with pollen stainability ranging from 5% to 85%. MS, male sterile, no pollen or very few pollen (<5%) stained

B x 350				B x 351				Prior			
2n=2x=14		2n=4x=28		2n=2x=14		2n=4x=28		2n=2x=14		2n=4x=28	
F+PF	MS	F+PF	MS	F+PF	MS	F+PF	MS	F+PF	MS	F+PF	MS
25	12	8	9	43	19	11	3	26	11	4	4
46.3%	22.2%	14.8%	16.7%	56.6%	25.0%	14.5%	3.9%	57.8%	24.4%	8.9%	8.9%

Conclusions High frequency androgenesis in *L. perenne* x *F. pratensis* was established. The diploid microspore-derived progenies with both freezing tolerance and fertility potential are promising to introduce winter hardiness of *F. pratensis* to *L. perenne* by backcrossing with *L. perenne* as introgression breeding.

References
Guo, Y.D., P. Sewón & S. Pulli (1999). Improved embryogenesis from anther culture and plant regeneration in timothy. *Plant Cell, Tissue and Organ Culture*, 57, 85-93.
Mizukami, Y., S. Sugita, N. Ohmido & K. Fukui (1998). Agronomic and cytological characterization of F₁ hybrids between *Lolium multiflorum* Lam. and *Festuca arundinacea* var. *glaucescens* Boiss. *Grassland Science*, 44, 14-21.
Wang, X. & H. Hu (1984). The effect of potato II medium for Triticale anther culture. *Plant Science Letter*, 36, 237-239.

The feasibility of autoclave-assisted water soluble carbohydrate extraction to distinguish annual ryegrass genotypes at the seedling stage

L.P. Passos, F.B. de Sousa, A. Mittelmann, M.C. Vidigal, I.G. Perry, L.O. Cruz and J.R. Magalhães
Embrapa Dairy Cattle, The National Dairy Cattle Research Centre, Juiz de Fora, 36016-210 MG, Brazil, Email: lpassos@cnpgl.embrapa.br

Keywords: carbohydrate, forage, *Lolium multiflorum*, selection

Introduction Annual ryegrass (*Lolium multiflorum*) swards are being used increasingly in Southern Brazil as animal forage (Carvalho *et al.*, 2001). As observed elsewhere, reduced forage availability and quality during dry or cold seasons often limits the efficacy of pasture utilisation. As reported by Meissner *et al.* (1992), this demands breeding efforts to improve forage nutritive value. However, lack of concurrent examination of large number of samples for important nutritional parameters, like water-soluble carbohydrate (WSC) levels, which could lead to more precise selection strategies, is a common constraint of those programs. This study aimed to verify the suitability of using autoclave-assisted WSC extraction for concurrent screening of seedlings of annual ryegrass genotypes, aiming to establish early selection criteria.

Materials and methods Field-grown seedlings of 24 annual ryegrass genotypes (Table 1) were sampled at cutting 1, following 60 days after germination, and evaluated for dry matter (DM), crude protein (CP), neutral detergent fiber (NDF), in vitro digestible dry matter (IVDDM), and in vitro digestibility of organic matter (IVDOM). Also, WSC contents of stem samples were determined by the autoclave-based extraction procedure of Passos *et al.* (2003). The trial was conducted as completely randomized block design with 2 replicates. The data were analyzed statistically through ANOVA and WSC mean contrasts among treatments compared by the Tukey test.

Results WSC levels were correlated positively to DM (r=0.45, p<0.001), and negatively to CP (r=-0.60, p<0.001) and IVDOM (r=-0.31, p<0.01). WSC did not relate with NDF or IVDDM. There were WSC genotypic differences (Table 1), with cv. Zorro yielding higher contents than cv. ETB AZ 096. No other significant contrast was observed among the studied genotypes, despite a consistent tendency of decreasing values apparently related to DM production. A possible association of that trend with physiological or morphological attributes needs research. Further comparisons are to be conducted during forage production.

Table 1 Mean WSC levels (mg/g DM) of annual ryegrass seedlings following 60 days after germination [#]

Genotype	WSC	Genotype	WSC	Genotype	WSC	Genotype	WSC
Zorro	353.77 a	ETB AZ 003	336.92 ab	ETB AZ 022	328.12 ab	ETB AZ 055	320.28 ab
Jeanne	343.43 ab	Riga	335.64 ab	Tetragold	327.76 ab	ETB AZ 085	318.24 ab
ETB AZ 007	339.59 ab	INIA Titan	331.87 ab	ETB AZ 080	322.41 ab	CPPSUL1	317.44 ab
INIA Cetus	338.40 ab	ETB AZ 089	330.97 ab	ETB AZ 011	321.06 ab	ETB AZ 078	317.13 ab
Kemal	338.32 ab	ETB AZ 071	330.19 ab	ETB AZ 079	320.97 ab	ETB AZ 077	309.44 ab
ETB AZ 049	337.08 ab	Hercules	328.57 ab	ETB AZ 097	320.49 ab	ETB AZ 096	301.38 b

[#] Means followed by the same letter are not significantly different (p<0.001) by the Tukey test

Conclusions Autoclave-assisted WSC extraction is suitable for concurrent evaluation of large numbers of annual ryegrass genotypes. It may be possible to use WSC content to predict annual ryegrass behaviour and forage production. However, confirmation of this applicability needs verification, by comparing the reported results with profiles of plants undergoing simple forage production.

References

Carvalho, P. C. F., L. S. Pontes, E. O. Silveira, C. H. E .C. Poli, C. Nabinger & O. A. Pereira Neto (2001). Sheep performance in Italian ryegrass swards at contrasting sward height In: International Grassland Congress, XIX. Proceedings. São Pedro, USP-ESALQ, 845-846.

Meissner, H. H., M .M. Du Preez, A. D. Enslin & E. B. Spreeth (1992). Utilization of *Lolium multiflorum* by sheep. 1. Influence of dry matter content and correlated factors on voluntary intake. *Journal of the Grassland Society of Southern Afric*a, 9, 11-17.

Passos, L. P., M. C. Vidigal, F. B. de Sousa, H. S. Barud, A. F. C. Paiva & A. R. Santos (2003). Comparative efficacy of autoclave-based extraction of soluble carbohydrates in various forage grasses. *Proceedings of the World Conference in Animal Production, 9*, Porto Alegre, World Association of Animal Production, CD-ROM.

A quantitative trait locus analysis of root distribution in perennial ryegrass (*Lolium perenne* L.)

M.J. Faville[1], J.R. Crush[2] and H.S. Easton[1]

[1]*AgResearch Ltd., Grasslands Research Centre, Private Bag 11008, Palmerston North, New Zealand, Email: marty.faville@agresearch.co.nz,* [2]*AgResearch Ltd., Ruakura Research Centre, Private Bag 3123, Hamilton, New Zealand*

Keywords: simple sequence repeats, expressed sequence tag, map, quantitative trait locus

Introduction   Root system architecture impacts perennial ryegrass performance, with deeper roots potentially contributing to drought tolerance, nutrient interception, and anchoring of plants. Root mass in a perennial ryegrass sward is typically shallow, concentrated in the top 10 cm of soil (Troughton 1957). Phenotypic selection for deeper root systems in breeding programmes is limited by the inaccessibility of underground plant components. We aim to use quantitative trait locus (QTL) analysis to discover genetic factors influencing root architecture traits, including vertical root distribution, in perennial ryegrass. Ultimately, markers linked to root architecture QTL may be used in a marker-assisted selection strategy that would alleviate the limitations of conventional selection, and lead to ryegrass cultivars with improved production and environmental performance.

Materials and methods   A bi-parental genetic linkage map was constructed (JoinMap 3.0) using 165 EST-SSRs (simple sequence repeat markers derived from expressed sequence tags) in population IxS, a full-sib F_1 (n=198) developed by pair-crossing individual heterozygous genotypes from cv. 'Grasslands Impact' and cv. 'Grasslands Samson'. Tillers from three clonal replicates of each F_1 and parental genotype were planted in sand-filled 1m deep x 0.09m wide rigid plastic tubes. Tubes were irrigated daily with low-ionic strength nutrient solution, in a temperature-controlled glasshouse. Root dry weight (RDW) (g) in 11 depth increments was recorded after 60 days of treatment. Data analysed were the ratio of RDW in the 10-20 cm depth to RDW in the 0-10 cm depth. ANOVA indicated significant differences amongst genotypes for the RDW ratio (F value = 2.86). Mean RDW ratio for each genotype was used for QTL analysis by MQM mapping implemented in MapQTL 4.0 software. Permutation testing (n=1000) established a logarithm-of-odds (LOD) threshold of 3.60 to declare a QTL at a genome-wide significance of α=0.05.

Results   Mean RDW ratio varied from 0.138 to 1.010 across 198 genotypes (overall mean = 0.470). Values of 0.245 and 0.313 were measured in the Impact and Samson parents, respectively. Linkage analysis located 160 EST-SSR loci on seven linkage groups (LG1 – 7) with assignments consistent with other ryegrass maps (Faville *et al.*, 2004). The IxS map is 462 Kosambi cM long, with mean density of 2.9 cM/marker and only 8 intervals > 10 cm. Five genomic regions affecting RDW ratio were identified by MQM QTL analysis, with two exceeding the significance threshold of LOD 3.60 - one on LG3 (LOD 6.07, peak 49 cM, 11% variance explained) and another on LG6 (LOD 3.61, peak 18 cM, 6% variance explained). *In silico* analysis suggests the QTL occur in genomic regions syntenic with regions of the rice genome (chromosomes 1 and 2) where QTL for root morphological traits have been identified (Yadav *et al.*, 1997; Hemamalini *et al.*, 2000).

Conclusions   We have detected two QTL affecting vertical root distribution in perennial ryegrass, the first reported for root architecture in this species. The results demonstrate that the experimental system is suitable for detecting genetic variation in root architecture of perennial ryegrass, and for investigation of QTL underlying that variation. This system will be used for validation and expansion of the current experimental results, and will serve as a vehicle for further investigation of the genetic basis of root architecture in perennial ryegrass.

References

Faville M.J., A.C. Vecchies,, M. Schreiber, M.C. Drayton, L.J. Hughes, E.S. Jones, K.M. Guthridge, K.F. Smith, T. Sawbridge, G.C. Spangenberg, G.T. Bryan & J.W. Forster (2004).   Functionally-associated molecular genetic marker map construction in perennial ryegrass (*Lolium perenne* L.).   *Theoretical and Applied Genetics,* 110, 12-32.

Hemamalini, G.S., H.E. Shashidhar & S. Hittalmani (2000).   Molecular marker assisted tagging of morphological and physiological traits under two contrasting moisture regimes at peak vegetative stage in rice (*Oryza sativa* L.).   *Euphytica*, 112, 69-78.

Troughton, A. (1957). The underground organs of herbage grasses. *Commonwealth Bureau of Pastures and Field Crops, Bulletin Number*. 44, 41-42.

Yadav, R., B. Courtois, N. Huang & G. McLaren (1997).   Mapping genes controlling root morphology and root distribution in a doubled-haploid population of rice. *Theoretical and Applied Genetics,* 94, 619-632.

Variability and correlations of some investigated traits of perennial ryegrass populations

D. Sokolovic[1], S. Ignjatovic[1] and Z. Tomic[2]
[1]Institute Serbia, Center for forage crops Krusevac, Trg rasinskih partizana 50, 37000 Krusevac, Serbia, Email: vojasoko@ptt.yu, [2]Institute for animal husbandry, 11080, Beograd-Zemun

Keywords: *Lolium perenne*, breeding, tillering, dry matter yield

Introduction Perennial ryegrass (*Lolium perenne*) is one of the most important perennial forage grasses for temperate climates. It is a highly productive grass with the highest nutritive value (Sokolović *et al.*, 2002). In Serbia, breeders have developed perennial ryegrass cultivars with high stabile yield and quality with different times of maturity and resistance to drought and frost. The initial breeding material were usually wild populations (Charmet *et. al.*, 1996) with high variability and adaptability. These characteristics lend themselves for selection of superior genotypes. But breeding for some important agronomic traits may influence others. This relationship between traits and breeding population variability is the objective of this article.

Materials and methods Perennial ryegrass populations originating from the Serbian flora were investigated in a second breeding cycle. The trial was designed in a space - plant design (60x60cm). Time of tillering (days after April 01.), plant height, top internode length and dry matter yield (DMY) were assessed over three years. Traits are shown as three year-mean values. Variability (coefficients of variations (%)) and coefficients of correlation between traits were determined.

Results The earliest population is Kopaonik, but it has lowest DMY (Table 1). The population with the best breeding characteristics is Jastrebac. It has highest yield with medium tillering date and height. Coefficients of variations are between 6 and 20%, except for DMY (29-46%), but they are lower in comparison with the first breeding cycle (over 50%) (Sokolović *et al.*, 2003). Correlations between time of tillering and plant height and internode length are significant and negative (Table 2), whilst DMY was not significantly affected by tillering date. Humphreys (1989) reported that time of tillering greatly affected DMY per plant (0.86). Plant height was positively correlated with internode length and DMY. Plants with longer internodes had lower DMY (r = -0.24).

Table 1 Means and variability of perennial ryegrass populations properties

Trait Population	Tillering date $\overline{X}$	CV	Plant height (cm) $\overline{X}$	CV	Internode length (cm) $\overline{X}$	CV	DMY (g per plant) $\overline{X}$	CV
Jastrebac	49.6	8.7	74.5	10.4	20.3	19.9	137	29.7
Kopaonik	44.9	6.9	73.7	10.9	24.7	15.3	109	41.6
Divci	51.2	7.7	75.1	12.3	19.5	18.5	131	45.8
Lomnica	51.4	9.5	70.0	14.8	20.3	19.5	112	39.7
Goč	50.0	9.2	72.4	14.1	22.4	16.9	129	42.3
Vlasina	47.8	5.9	74.1	10.1	23.5	18.5	129	40.6
Javor	47.1	11.3	75.3	12.9	26.2	17.8	121	46.4
Radočelo	53.0	7.9	71.0	13.4	20.9	18.0	123	39.6

Table 2 Correlations between traits

	Tillering date	Plant height	Internode length	DMY
Tillering date				
Plant height	-0.65			
Internode length	-0.86	0.30		
DMY	0.35	0.49	-0.24	

Conclusions Lower coefficients of variations were estimated, but variability necessary for breeding was maintained within populations. The Jastrebac population had the highest average DMY per plant and contains the most promising genotypes. Late genotypes had lower heights and internode lengths, whilst DMY was not influenced significantly by date of tillering (time of maturity). Breeding for early genotypes may cause inferior plant height and internode length. Increasing internode length had a negative impact on DMY, whilst taller plants showed higher DMY.

References

Charmet, G., F. Balfourier, C. Ravel, D. Leconte, B. Debote, J.C. Vezine, C. Astier, & G. Leau, (1996): Study of a French collection of natural populations of perennial ryegrass. *Fourrages*, 146, 107-121.

Humphreys, M.O. (1989): Assessment of perennial ryegrass (*Lolium perenne* L.) for breeding. II. Components of winter hardiness. *Euphytica*, 41, 99-106.

Sokolović, D., Z. Tomić, S. Ignjatović, G. Šurlan-Momirović, & T. Živanović, (2002): Genetic variability of perennial ryegrass (*Lolium perenne* L.) autochthonous populations. II. Dry mater yield and chemical composition. *Grasslands Science in Europe*, 7, 92-93.

Sokolović, D., Z. Tomić, and Z. Lugić, (2003): Dry matter yield components of perennial ryegrass (*Lolium perenne* L.) populations. *Grasslands Science in Europe*, 8, 126-130.

Variability in quantity and composition of water soluble carbohydrates among Irish accessions and European varieties of perennial ryegrass

S. McGrath[1,2], S. Barth[1], A. Frohlich[1], M. Francioso[1], S.A. Lamorte[1] and T.R. Hodkinson[2]
[1]Teagasc Crops Research Centre, Oak Park, Carlow, Ireland, Email: smcgrath@oakpark.teagasc.ie,
[2]Department of Botany, University of Dublin, Trinity College, Ireland

Keywords: quality, fructose, glucose, high pressure liquid chromatography, plant breeding

Introduction The objective of this study was to identify perennial ryegrass accessions displaying high fructose and glucose contents and an improved ratio between fructose and glucose fractions across different time points throughout the year. Fructose and glucose are the main constituents of the water soluble carbohydrate (WSC) fraction in perennial ryegrass. For animal nutrition the amount of WSC is crucial as it is the primary energy source available to metabolise the intake of plant protein. The ratio between fructose and glucose fractions is important since fructosan chains, which are an excellent energy source for ruminants, are built from fructose. Furthermore the seasonal variability of WSC content in feed reflects the changing balance between protein and carbohydrates.

Materials and methods In the summer of 2003, 33 perennial ryegrass entries were grown from true seed and planted as spaced plants in the field. Forty plants per accession were divided into 4 pools for analysis. The plant material was selected from a collection of historic indigenous Irish accessions held at Oak Park (23 entries) and current commercially grown varieties (10 entries). In 2004, at three time points during the growing season, samples were taken and processed for WSC analysis via HPLC as described by Jafari et al. (2003). Means and standard deviations were calculated and entries were assigned to one of four classes (1 = very good to 4 = poor) based on percentage of dryweight attributed to carbohydrates.

Results At the three time points across ecotypes and varieties a high variability was found for both the WSC content and the ratio of fructose/glucose (Table 1), e.g. contents of fructose ranging between 1.65 and 18.99%. Generally the material displayed wide genetic variation across the traits investigated. Among the ecotypes, several entries were superior to the commercial varieties at the third cutting time point (Table 2).

Table 1 Means (x), standard deviations (SD), minimum and maximum percentage of fructose, glucose and total water soluble carbohydrate (WSC), and ratio of fructose/glucose (ratio) across three cuts

	% fructose				% glucose				% WSC				ratio			
	X	SD	min	max	X	SD	min	max	X	SD	min	max	X	SD	min	max
cut 1	8.89	2.90	2.36	15.77	3.99	1.08	1.57	6.89	12.90	3.85	4.11	21.49	2.22	0.40	1.35	3.51
cut 2	6.34	1.96	2.47	11.29	4.39	1.24	2.04	6.68	10.73	3.13	4.54	17.37	1.44	0.19	1.04	2.29
cut 3	8.93	3.74	1.65	18.99	4.40	1.81	1.07	12.34	13.33	5.19	4.18	27.19	2.11	0.64	0.65	5.10

Table 2 Number of varieties and ecotypes within the four classes of carbohydrate content

index	fructose 1	2	3	4	glucose 1	2	3	4	WSC 1	2	3	4	ratio 1	2	3	4
							varieties									
cut 1	2	4	4	-	1	5	4	-	2	5	3	-	-	6	2	2
cut 2	1	6	3	-	1	6	3	-	1	6	3	-	-	7	3	-
cut 3	1	7	2	-	1	6	3	-	1	8	1	-	-	7	3	-
							ecotypes									
cut 1	1	17	5	-	2	16	5	-	1	17	5	-	4	13	4	2
cut 2	1	17	5	-	1	16	6	-	2	16	5	-	-	20	3	-
cut 3	2	15	6	-	2	2	4	-	5	15	6	-	3	15	4	1

Conclusions Perennial ryegrass genetic resource collections such as that held at Teagasc Oak Park, hold a great potential for improving the quality of ryegrass varieties. Further ryegrass traits should also be examined, e.g. digestibility and fatty acids. The high quality ecotypes identified in this study will be investigated further in 2005-2006.

Acknowledgements We are grateful to the grass breeding group in Oak Park for their support and to the Irish Department of Agriculture for partial funding of this study.

Reference
Jafari, A., V. Connolly, A. Frohlich, & E. J. Walsh (2003). A note on estimation of quality parameters in perennial ryegrass by near infrared reflectance spectroscopy. Irish Journal of Agricultural & Food Research, 42, 293-300.

Breeding of CMS-F1-Hybrids in Lolium perenne with improved nitrogen use efficiency

W. Luesink[1], B. Ingwersen[1], L. Wolters[2] and J. de Riek[3]
[1]Norddeutsche Pflanzenzucht Hans-Georg Lemke KG, Hohenlieth, 24363 Holtsee, Germany, Email:
W.Luesink@npz.de, [2]Zelder B.V., P.O. Box 26, 6590 AA Gennep, The Netherlands, [3]Department of Plant
Breeding, CLO-Gent, Caritasstraat 21, 9090 Melle, Belgium

Keywords: nitrogen use efficiency, low input varieties, genetic distance, plant breeding, CMS-F1-Hybrids

Introduction The environmental pollution by nitrogen losses from dairy farms can be reduced by improving the nitrogen use efficiency (NUE) of grass varieties. The main goal is to develop varieties with a better nitrogen utilisation. These "low input varieties" can produce acceptable yields at a low level of N-fertilisation. High input varieties express their high yield potential only at high N-supply. These varieties are less preferable, because N-losses increase at higher levels of nitrogen application. The breeding of CMS-F1-Hybrids can be a successful strategy for developing varieties with a higher NUE. In F1-Hybrid varieties higher heterosis effects can be achieved than in populations or synthetic varieties.

Material and methods In a field trial at 2 locations the responses of 35 F1-Hybrid varieties and 2 high yielding control varieties, 'Respect' and 'Fennema', at 3 different levels of N-fertilisation (200, 270 and 360 kg N/ha) were investigated. The time of cutting was the same for all N-levels at a location. In total 4 cuts were made. By using amplified fragment-length polymorphism (AFLP)-markers, genetic distances between the F1-Hybrid combinations were calculated. Between 12 F1-Hybrids the heterosis could be calculated with the formula: F1-value – mid-parent value. This experiment was part of the EU-project NIMGRASS.

Results Table 1 shows that low input varieties can be found with a 9 – 12% higher dry matter yield (DM-yield) at the low N-level than the control varieties. At the high N-level the production of these varieties was not significantly higher. One variety, Hybrid 2, had a 10% yield elevation at both the low and high N-level.

Table 1 Relative DM-yield (%) of F1-Hybrid varieties and control varieties at 3 N-levels (kg N/ha)

N-level	200	270	360
Fennema	100	101	98
Respect	100	99	102
Hybrid 1	112	104	99
Hybrid 2	110	105	110
Hybrid 3	109	111	109

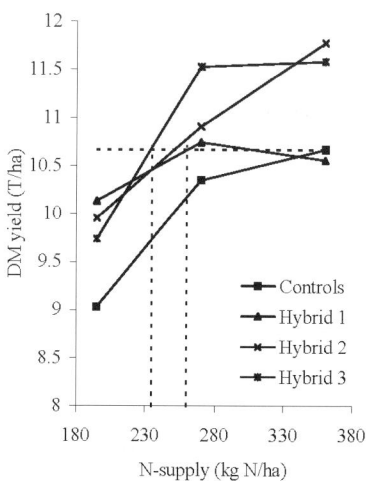

Figure 1 DM yields of F1-Hybrids

An increase of 10% in the NUE allows a reduction by 25-35% N-input for the same dry matter production: The F1-Hybrids reach the DM-yield of the controls at the high level of 360 kg N/ha with 100-125 kg N/ha less Nitrogen (figure 1).
The relative high correlation between heterosis and the genetic distance (GD) (figure 2) shows that crossing of unrelated parent lines probably will be worthwhile, as in this trial the combination with the highest genetic distance had the highest heterosis, but there were also exceptions.

Conclusions The results of the experiment show, that F1-Hybrids can be selected with a higher nitrogen use efficiency (low input varieties). The higher NUE allows an N-input reduction of 25-35% to achieve the same yield as the control at the high N-level. Estimating genetic distance by using AFLP can be a useful tool for selecting good combining parent lines.

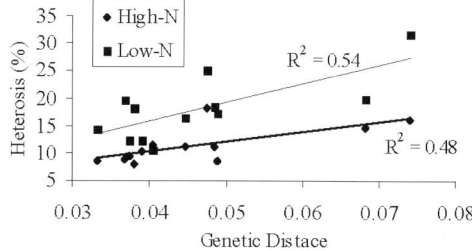

Figure 2 Relation between heterosis and GD

Effects of perennial ryegrass cultivars on traits for improved animal performance

H.J. Smit[1], B.M. Tas[2], H.Z. Taweel[1,2], J. Dijkstra[2], A. Elgersma[1,3] and S. Tamminga[2]
[1]Plant Sciences Group and [2]Animal Sciences Group, Wageningen University, the Netherlands and [3]Ghent University, Belgium,. Email: harm.smit@wur.nl

Keywords: *Lolium perenne*, dairy cows, selection

Introduction The use of quality parameters in grass breeding is limited. There may be options to improve grass cultivars (cvs) for improved animal performance.

Material and methods Grass breeders' options to select perennial ryegrass cvs on traits associated with improved animal performance were examined in a 4-year project. To examine the effects on dry matter intake and milk production, 3 experiments with 6 commercial perennial ryegrass were conducted. In 2000 and 2001, 12 dairy cows were stall-fed 6 diploid perennial ryegrass as fresh herbage in two 3x3 Latin squares (Tas *et al.*, 2005). In 2002 and 2003, 12 dairy cows grazed 4 of these in a 4x4 Latin square (Tas *et al.*, 2005; Smit *et al.*, 2005). In 2003 and 2004, 3 grazing experiments were conducted with 3 groups of dairy cows that could choose freely among 6 perennial ryegrass (Smit *et al.*, 2005).

Results Cvs differed for earliness: 1, 2 and 6 were intermediate- and 3, 4 and 5 late-heading. 1 and 4 had higher water soluble carbohydrates and lower neutral detergent fibre concentrations than the other. Cvs 1 and 4 had a higher resistance to crown rust (*Puccinia coronata* f.sp. *lolii*). Table 1 shows the effects of cv on herbage intake (DMI; kg DM/cow/d) and milk production (MP; kg FPCM[*]/cow/d) during stall-feeding and grazing. When offered one cv for a period of 2 weeks, cows showed no significant differences in DMI and MP.

Table 1 Herbage DMI (kg/cow/d) and milk production (MP; kg FPCM[*]/cow/d) of dairy cows fed on perennial ryegrass cultivars in 2 experiments

Experiment	Trait	Cultivar						Sig.	Mean
		1	2	3	4	5	6		
Stall feeding	DMI	15.5	16.2	15.9	14.9	15.4	15.7	NS[**]	15.6
	MP	25.4	25.8	24.8	24.2	25.2	25.5	NS	25.1
Grazing	DMI	17.3	17.4	16.6	18.2			NS	17.4
	MP	25.9	24.5	25.1	25.8			NS	25.3

[*]FPCM: Fat and Protein Corrected Milk = ((0.337+ 0.116 Fat (%) + 0.06 Protein (%)) x MP (kg/cow/d)
[**]NS: Not Significant

Cows given a free choice to graze all 6 at high allowance showed clear selective grazing behaviour (Table 2). As a significantly higher proportion of the diet consisted of cv 1, it was concluded that dairy cows preferred cv 1. Cv 3 was least preferred. DMI in mixed swards was higher than in pure stands.

Table 2 Herbage intake (kg DM/cow/d) on 6 perennial ryegrass cultivars in 3 free choice experiments

Exp.	Trait	Cultivar						Sign.	Total
		1	2	3	4	5	6		
Free Choice	DMI	4.32[c]	3.70[bc]	2.99[a]	3.55[ab]	3.51[ab]	3.22[ab]	**	21.3

[a,b,c,d] : means with the same subscript are not significantly different (P>0.05). **: P<0.01

Conclusions Options to improve cow performance through breeding perennial ryegrass for forage quality are limited. DMI and MP variation was limited among the perennial ryegrass; differences during grazing or stall feeding were not significant. However, dairy cows offered free choice preferred certain perennial ryegrass cvs and ingested more total herbage.

References

Smit, H. J. & A. Elgersma. Cattle grazing preference among six cultivars of perennial ryegrass (2005). (Submitted to *Journal of Animal Science*)

Smit, H. J., B. M Tas, H. Z. Taweel, S. Tamminga & A. Elgersma (2005). Cultivar effects of perennial ryegrass on grass production, nutritive quality and herbage intake under grazing. (Subm to *Grass and Forage Science*)

Tas, B. M., H. Z. Taweel, H. J. Smit, A. Elgersma, J. Dijkstra & S. Tamminga (2005). Nitrogen utilisation of perennial ryegrass cultivars of stall-fed lactating dairy cows. (Submitted to *Journal of Animal Science*)

Tas, B. M., H. Z. Taweel, H. J. Smit, A. Elgersma, J. Dijkstra & S. Tamminga (2005). Intake and milk production by dairy cows grazing four perennial ryegrass cultivars. (Submitted to *Journal of Dairy Science*).

Selection for tillering in *Lolium multiflorum* L. in Texas USA

L.R. Nelson, G.W. Evers and M.J. Parsons
*Texas A&M University Agricultural Research and Extension Center, P. O. Box 200, Overton, Texas 75684 USA,
Email: lr-nelson@tamu.edu*

Keywords: annual, ryegrass, plant breeding

Introduction Annual ryegrass is an important cool season forage in Texas and across the southern USA with approximately 2 million ha planted annually. Early tillering will enhance the leaf area index and the amount of photosynthesis which will improve seedling vigor and early forage production. The objective of this study was to determine if it was possible to select for early tiller (shoot) production in a diploid and tetraploid annual ryegrass breeding population. We expect early tiller production would be correlated with improved early season forage production with annual ryegrass.

Materials and methods Two ryegrass breeding lines TXR2000-2 (2n) and TXR2000-T2 (4n) were used as base populations (designated p populations). Seed were planted in growing mix in plastic pot packs in the greenhouse. Number of plants in the TXR2000-2 and TXR2000-T2 populations were 700 and 500 plants, respectively. Seedlings were monitored daily and were marked by placing a toothpick with a flag indicating the date the first tiller appeared until all plants had tillered. Approximately 10% of the each population was selected as early or late plants resulting in four populations. Each population consisting of about 50 plants were transplanted into isolation blocks in the field and were allowed to cross-pollinate and produce seed. Seed was harvested, cleaned and stored until the following season when the 2nd cycle of selection occurred. In the 2nd cycle, from the early tiller population, only early tillering plants were selected, and from the late tillering population, only late tillering plants were selected. Therefore we had two parental populations, 4 S_1 populations and 4 S_2 populations. These 10 populations were grown in growth chambers and date of tillering was recorded. The test had 10 treatments and four replications, and each treatment was a mean of 24 plants.

Results The figures below show the mean time required for plants from each population to have tillered. Note that on individual days after planting, the early selection group had a higher % of plants which had tillered, while the late selection group had a lower % plants tillered. Selection for early tillering or late tillering resulted in the S_1 and S_2 populations moving toward the desired directions. On day 15 with the 4 n genotype, the percent plants which had tillered was 62 and 66% (early selection) and 52 and 33% (late selection) for the S_1 and S_2 populations respectively. Progress in the 2 n populations was equally successful. A different data point where significant differences were noted was with days required for 98% of plants to have tillered. With the 4 n populations, in the S_2 populations, the early selection group required 20 days, while the late selection group required 24 days.

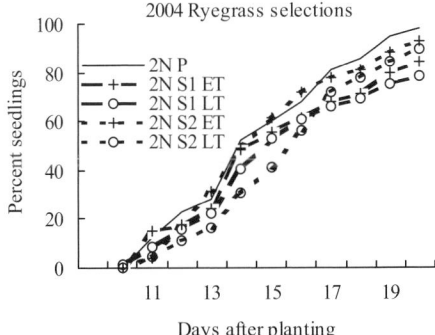

Figure 1 Cumulative percentage of diploid annual ryegrass seedling with first tiller

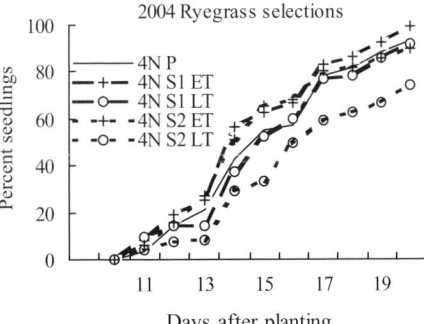

Figure 2 Cumulative percentage of tetraploid annual reygrass seedlings with first tiller

Conclusions After two cycles of selection, we were able to select for early and late tillering in both diploid and tetraploid annual ryegrass populations.

Varietal differences in perennial ryegrass for fructan metabolism and their relationship to grazing tolerance

B. Lasseur[1], J. Lothier[1], A. Morvan-Bertrand[1], M.O Humphreys[2], M.P. Prud'homme[1]
[1]UMR INRA-UCBN EVA, IRBA, Université de Caen, Esplanade de la Paix, 14032 Caen cedex, France, Email: prudhomme@ibfa.unicaen.fr, [2]IGER, Plas Gogerddan, Aberystwyth, SY23 3EB, UK

Keywords: defoliation tolerance, fructan exohydrolase (FEH), 1-sucrose:sucrose fructosyltransferase (1-SST)

Introduction Perennial ryegrass (*Lolium perenne* L.) is the most important grass in Europe; it often is defoliated. The link between fructan metabolism and defoliation tolerance has been studied in 2 *Lolium perenne* varieties, Aurora (high sugar perennial) and Perma (low-to-normal sugar perennial) (Turner *et al.*, 2002).

Materials and methods The 2 *Lolium perenne* varieties were sown separately and were grown hydroponically in controlled environment for 11wk before being cut 4cm above ground. They were defoliated again 4, 8, 13 and 18d later in order to deplete the carbohydrate reserves. Leaf growth, fructan content, fructan synthesizing and degrading enzyme activities (1-SST and FEH, respectively) were measured (Morvan-Bertrand *et al.* ,2001).

Results Aurora fructan level was higher than that of Perma in leaf sheaths and elongating leaf bases. It fell after defoliation (data not shown). Defoliations every 4-5d markedly decreased regrowth for both varieties, but regrowth was always higher for Aurora than for Perma (Figure 1). Activity of 1-SST fell strongly in leaf tissues of both varieties after the first defoliation (Figure 2). It fell thereafter or remained constant at a low level. FEH activity rose in both leaf tissues of Aurora only after the first defoliation but it kept rising during the following regrowth in Perma, so that FEH activity in Perma was higher than in Aurora.

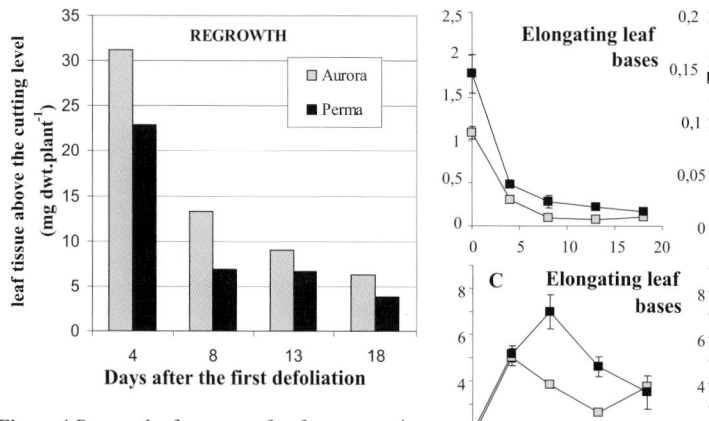

Figure 1 Regrowth of ryegrass after four successive defoliations at day 0, 4, 8 and 13

Figure 2 1-SST (A, B) and FEH (C, D) activities (nkat.g^{-1} FW) in elongating leaf bases and in leaf sheaths of ryegrass after four successive defoliations at day 0, 4, 8 and 13.

Conclusions Based on leaf dry matter production after repeated defoliations, Aurora is more tolerant to defoliation than Perma. This was attributed to the initial reserve carbohydrate, namely fructans, which was higher in Aurora than in Perma. However, the use of fructans is not related to the FEH activity measured *in vitro* which is often higher in Perma than in Aurora and which increased more in Perma than in Aurora in response to repeated defoliations.

References

Turner, L.B., Humphreys, M.O., Cairns, A.J.& C.J. Pollock (2002). Carbon assimilation and partitioning into non structural carbohydrate in contrasting varieties of Lolium perenne. *Journal of Plant Physiology*, 159, 257-263.

Morvan-Bertrand, A., Boucaud, J., Le Saos, J. & M.P. Prud'homme (2001). Roles of the fructans from leaf sheaths and from elongating leaf bases in the regrowth following defoliation of *Lolium perenne* L. *Planta*, 213, 109-120.

Water soluble carbohydrate content of two cultivars of perennial ryegrass *(Lolium perenne)* at eight European sites

M.A. Halling[1], A.C. Longland[2], S. Martens[3], L. Nesheim[4] and P. O'Kiely[5]
[1]*Swedish University of agricultural Sciences P.O. Box 7043,SE-050 07, Uppsala, Sweden, Email: magnus.halling@evp.slu.se,* [2]*Institute of Grassland and Environmental Research, (IGER) Plas Gogerddan, Aberystwyth, SY23 3EB, UK,* [3]*Institute of Crop and Grassland Science, Federal Agricultural Research Centre (FAL), Bundesalle 50, D-38116, Braunschweig, Germany,* [4]*The Norwegian Crop Research Institute, Kvithamar Research Centre, N-7500 Stjordal, Norway,* [5]*Teagasc, Grange research Centre, Dunsany, Co. Meath, Ireland*

Keywords: perennial ryegrass, plant quality traits, water soluble carbohydrates

Introduction Grasses with high levels of WSC have been shown to enhance livestock production (Miller *et al.*, 1999). This has led to the development of perennial ryegrass (PRG) cultivars that can accumulate high levels of WSC. The aim of this experiment was to determine if the genetic potential to accumulate high levels of WSC was expressed under varying conditions.

Materials and methods The PRG cultivar Aberdart, bred for high WSC accumulation and the control Fennema were established in small-plot field trials at 8 European sites (Table 1). During 2001-2003 grass was harvested when Fennema had reached the early boot stage at each site at 1400h and analysed for WSC content.

Results Aberdart usually contained more WSC in the first cut than Fennema, but this effect was not always significant. There was a strong G x E interaction, such that Aberdart accumulated significantly more WSC at all of the Swedish sites and in Ireland in 2001, and at Ultuna and Tvååker in all years harvested. On no occasion were significant differences in WSC content found between the two cultivars at the Norwegian or Welsh sites (Table 1).

Table 1 Concentration of WSC (g kg/DM) in first cut at nine sites during the years 2001 to 2003

Variety	Kvithamar Norway	Fureneset Norway	Saerheim Norway	Ultuna Sweden	Rådde Sweden	Tvååker Sweden	Traws-goed Wales	Grange Eire
				2001				
Aberdart (2n)	297	288	329	232	281	263		101
Fennema (2n)	277	286	315	195	227	228		89
Aberdart (2n)#	107	101	104	119***	124**	115***		113**
CV %	8.9	5.2	4.9	6.2	9.7	5.3		6.3
				2002				
Aberdart (2n)		215	217	166	182	186	242	174
Fennema (2n)		245	215	125	172	172	183	172
Aberdart (2n)#		88	101	133***	106	108*	132	101
CV %		13.6	13.0	14.9	8.1	4.1	21.3	18.0
				2003				
Aberdart (2n)		206	235	215			227	231
Fennema (2n)		204	218	174			309	211
Aberdart (2n)#		101	108	124***			74	109
CV %		10.8	10.4	3.4			13.1	12.2

#=relative Fennema=100, 2n = diploid variety, * *P*<0.05 **p<0.01; ***p<0.001

Conclusions Aberdart did not always accumulate significantly higher levels of WSC than Fennema at all sites. There was a strong G x E interaction, which was consistently expressed in Sweden and Norway. These results serve to illustrate the importance of trialling cultivars in different environments, to determine the robustness of desired characteristics. Further studies are needed with a mechanistic model to explain the G x E interaction.

Acknowledgements This work was as funded by the EU Framework V project (QLK-CT-2001-0498)

References
Miller L.A., Neville M.A., Baker D.H., Evans R.T., Theodorou M.K., MacRae J.C., Humphreys M.O. and Moorby J.M. (1999) Milk production from dairy cows offered perennial ryegrass selected for high water soluble carbohydrate concentrations. *Proceedings of the British Society of Animal Science, Scarborough*, pp 208.

A survey of European regional adaptation in Italian ryegrass varieties

T.J. Gilliland[1] and A.J.P. van Wijk[2]
[1]*Department of Agriculture for Northern Ireland, Plant Testing Station, Crossnacreevy, Belfast BT6 9SH, UK, Email: trevor.gilliland@dardni.gov.uk*, [2]*Centre for Genetic Resources the Netherlands, P.P. Box 16, 6700 AA Wageningen, The Netherlands*

Keywords: plant testing, registration, agro-climatic zones, environmental response

Introduction Ryegrass is widely adapted to cool temperate eco-zones and breeders often submit individual varieties for testing in a number of EU countries. National testing programmes often combine data from several trial sites that may differ climatically, but not from sites in other member states, despite the possibility of high ecological similarity. Given increasing interest in 'animal value' characters (soluble sugars, lipids, sward geometry), additional testing for these would be valuable but is prohibited by capped or declining funding. Data sharing between EU national authorities could be advantageous but is inhibited by the lack of statistically valid data on the sensitivity of each performance parameter to agro-climatic conditions across the EU. This paper, reports the preliminary stages of the 'EuroVCU' (herbage) desktop study of ryegrass variety performances across an extensive range of EU national test centres. Analysis of the resulting data sets quantifies the genotype by environment responses of current varieties and could provide a validated protocol for future data sharing.

Materials and methods Due to its wide use and similarity in management, Italian ryegrass (*Lolium multiflorum* LAM.) was chosen as the model grass species for this pilot study. In 2003, the scientists responsible for conducting official grass variety tests in each of the current member states of the EU were contacted. They were invited to join the EuroVCU (herbage) consortium and asked to compile information on the testing history, test decisions and availability of data for each of the 169 varieties listed on the 22nd EU Common Catalogue. This constituted Phase I of the study and the data were summarised and processes with the aim of deriving a core set of varieties that would provide 10 pair-wise variety-links between each of the contributing member states. Phase II of the study involved gathering performance, management and metrological data for the core varieties from each site. This constituted the full data set from which analyses of the consistency/inconsistency of individual varieties across management and climatic variation could be measured, of stability/instability of individual performance characteristics across varieties and protocols and climatic variants, and finally the similarity/dissimilarity of test centres for common variety performance results on a character-by-character basis.

Results Twenty EU regions responded to the Phase I call (Austria, Belgium, Croatia, England & Wales, Estonia, Finland, France, Germany, Hungry, Italy, Lativa, Netherlands, N. Ireland, Norway, Republic Ireland, Scotland, Slovenia, Slovakia, Sweden, Switzerland), which generated a total of 679 variety x country reports. The pattern of variety testing appeared disjointed and impossible to interpret prior to Phase II data being gathered. For example, of the 169 EU registered varieties, only three were tested in half or more of the countries surveyed (Ajax, Danergo and Mondora). Conversely, 22 varieties were tested in only a single country and a further 18 were not tested by any contributors (AM1, Bella Bionda, Califa, Classic, EF486 Dasas, Kitil, Locobelo, Marvel, Menichetti, Multisolc AX9, Primadonna, Ralino, Rouky, Sultano, Tauro, Teanna, 110DE, 111DE). The average number of varieties tested per country was 35, but there were big differences, with some having tested more than 70 varieties, while others having only tested a few. This probably indicates the different importance of Italian ryegrass in each region, which may be linked to climatic differences. In total, Phase I generated over 2,600 data entries on variety test history, current status and results availability, from which 44 varieties were selected to provide the pair-wise comparisons in Phase II (Abercomo, Adin, Ajax, Atalja, Atos, Baresi, Barextra, Barmultra, Bartolini, Bartoluchi, Bofur, Danergo, Exalta, Fabio, Fenil, Fredrik, Gemini, Gisel, Gordo, Lemtal, Ligrande, Lolita, Lubina, Malmi, Meroa, Minaret, Mondora, Montblanc, Multimo, Rio, Sabalan, Sikem, Sultan, Taurus, Tetraflorum, Tonic, Total, Tribune, Tur, Turgo Pajbjerg, Urbana, Zarastro, Zenith, Zorro). Phase II was sent to all contributors plus the authorities in the remaining countries (Bulgaria, Cyprus, Czech Republic, Denmark, Greece, Luxembourg, Poland, Romania and Spain). It is anticipated that the participant numbers in Phase II may increase by the October 2004 deadline, which has been set to allow for final report completion by Spring 2005.

Conclusions The concept of this study is to use common varieties as replicates of each national site. This strategy makes it possible to compare how each performance characteristic varies between different locations and to identify common agri-climatic zones on a character-by-character basis. The EuroVCU (herbage) study, therefore, provides the necessary knowledge to validate future data sharing between variety testing authorities.

Acknowledgements The members of the EuroVCU consortium are thanked for their essential contribution to this study.

Theme A: Efficient production from grassland

Section 2

Animal production

Milk production potential of different dairy pasture types in southern Australia

J. Tharmaraj[1], D.F. Chapman[1], Z.N. Nie[2] and A.P. Lane[1]
[1]The University of Melbourne, Parkville, Victoria, Australia 3010, Email: jthar@unimelb.edu.au, [2]Department of Primary Industries, Hamilton, Victoria, Australia 3300

Keywords: pasture types, pasture quality, milk production, dairy systems

Introduction The growth rate of traditional perennial ryegrass-based pastures commonly fails to meet herd feed requirements through winter and summer in non-irrigated dairy systems in southern Australia. Alternative pasture species can improve seasonal feed supply in this region (Tharmaraj & Chapman, 2005). However, the feeding value and milk production of these pastures must at least match perennial ryegrass if they are to be adopted successfully on dairy farms. This paper reports results of a comparison of the milk production potential of pasture types similar to those investigated by Tharmaraj & Chapman (2005) for their agronomic performance.

Materials and methods The experiment was conducted at Glenormiston College, southwest Victoria ($38°09'S$, $142°58'E$). Four pasture types were established in April 2002 in two replicate blocks in paddocks of 1.2-1.9 ha: 1) short-term winter active (STW), based on Italian ryegrass; 2) long-term winter active (LTW), based on Mediterranean tall fescue; 3) long-term summer active (LTS), based on Continental tall fescue plus chicory and red clover; and 4) perennial ryegrass (Control). Pastures were grazed in spring 2002 and in summer, autumn and winter 2003 by small herds (n=3) of cows balanced for age, liveweight and stage of lactation. The LTW treatment was not included in spring because of slow pasture establishment, and STW was not included in autumn because the pasture contained only dead matter. A chicory and white clover treatment, established at the same time as the other pastures, was included in summer only. Each experimental run lasted 10 d and was preceded by a 7-d uniformity period when animals grazed on a common ryegrass pasture. Pasture allowance was 45-60 kg DM/cow per d, and animals received no supplements. Pre-grazing pasture mass was in the range 2200-2600 kg DM/ha. Cows were on average 124, 227, 234 and 64 d into lactation for the spring, summer, autumn and winter runs, respectively. Milk yield (kg/cow) was measured daily, and milk fat and protein content was measured on days 6, 8 and 10. Data for milk solids production per cow (mean of 3 cows/ group) were analysed in three ways. Firstly, treatments (pasture types) 1-4 and all seasons were included in a split-plot ANOVA with treatment as main plot and season as sub-plot using missing values for LTW in spring and STW in autumn. Secondly, treatments 1-4 were included in a balanced split-plot ANOVA design using only summer and winter data. Finally treatments 1-4 plus the chicory treatment were analysed for summer only, using one-way ANOVA.

Results and discussion When all treatments and seasons were included in the ANOVA, the main effect of treatment and the season x treatment interaction were not significant. When the milk yield of cows grazing treatments 1-4 was analysed for summer and winter only, there was a significant treatment x season interaction, due to a greater difference between seasons for the control treatment compared to all other treatments (Table 1). This was a result of factors such as limited ability of animals to select a diet high in energy and protein in summer on ryegrass and possible sub-clinical effects of endophyte alkaloids on animals grazing ryegrass in summer. When all five treatments were analysed in summer only, the main effect of treatment was significant ($P<0.05$); the chicory pasture resulted in substantially greater milk solids (1.40 kg/cow per d) production than all other treatments (Table 1) due to its much higher digestibility and protein content.

Table 1 Effects of pasture type on milk fat plus protein yield (kg/cow per d) in summer and winter

	Pasture type			
	STW	LTW	LTS	Control
Summer	0.92	0.93	1.04	0.80
Winter	1.89	1.87	1.93	2.00

Season x pasture interaction: $P<0.05$; s.e.d =0.011

Conclusions When managed so that green leaf content is maximised within the limits of environmental constraints, pastures based on alternative species such as tall fescue can result in milk yields that are at least similar to perennial ryegrass-based pastures and less variable between summer and winter – the seasons when pasture feed deficits commonly occur. The feeding value of chicory is very high and this plant should be able to play a role in improving milk production in southern Australia dairy systems, especially in summer.

Reference

Tharmaraj, J., D.F. Chapman & Z.N. Nie (2005). Seasonal herbage accumulation of different dairy pasture types in southern Australia. *XX International Grassland Congress - offered papers* (in press).

Effect of strain of Holstein-Friesian cow and feed system on reproductive performance in seasonal-calving milk production systems over four years

B. Horan[1,2], J.F. Mee[1], M. Rath[2], P. O'Connor[1] and P. Dillon[1]

[1]Dairy Production Department, Teagasc, Moorepark Production Research Centre, Fermoy, Co. Cork, Ireland, Email: bhoran@moorepark.teagasc.ie, [2]Department of Animal Science, Faculty of Agriculture, University College Dublin, Belfield, Dublin 4, Ireland

Keywords: Holstein-Friesian, cow strain, feed system, reproductive performance

Introduction In Ireland most dairy farms operate seasonal calving grass-based milk production systems. Feed demand and supply are matched by having calving highly concentrated in spring. This requires high pregnancy rates within a short time following the start of mating in late April or early May, but has become increasingly difficult to achieve due to declining fertility in Irish dairy herds (Mee, 2004). In New Zealand, cows of North American Holstein-Friesian origin have poorer fertility than New Zealand Holstein-Friesians on pasture-based seasonal calving systems (Harris & Kolver, 2001). The present study sought to determine the effect of strain of Holstein-Friesian (HF) cow and feed system on reproductive performance within Irish milk production systems.

Materials and methods Three strains of Holstein-Friesian (HF) cows: high production North American (HP), high durability North American (HD) and New Zealand (NZ) were assigned, within strain, to one of three pasture-based feed systems: Moorepark (MP; 350 kg concentrate/cow, stocking rate of 2.47 cows/ha), high concentrate (HC; 1,500 kg concentrate/cow, stocking rate of 2.47 cows/ha), and high stocking rate (HS; 350 kg concentrate/cow, stocking rate of 2.74 cows/ha). The total number of animals in each of the four years ranged from 99 to 126. Cows were bred by artificial insemination (AI) over a 13-week period, starting in late April. Pregnancy detection was performed by ultrasound imaging at 30 to 37 d and again 60 to 67 d after AI. A final manual pregnancy examination was carried out 150 d after mating start date. Data for number of services were analysed using a non-parametric model (PROC NPAR1WAY) (SAS, 2002). Differences in 24-d submission rate and pregnancy rates were investigated taking account of year, parity, strain of HF and feed system using the PROC GENMOD (binomial distribution and logit link function) procedure of SAS (SAS 2002).

Results There was no significant interaction between strain of HF and feed system for any of the reproductive variables measured and therefore only the main effects are shown (Table 1). The HP strain received a greater number of services per cow and had a lower submission rate than the other strains. The NZ strain had a higher pregnancy rate to first service and six-week pregnancy rate compared to the HP strain, with the HD strain intermediate. The HP strain had the lowest overall pregnancy rate and the NZ strain had the highest. Feed system had no significant effect on reproductive performance.

Table 1 Effect of strain of Holstein-Friesian and feed system on reproductive performance (2001-2004)

	Strain of HF			Sig.[†]	Feed system			Sig.[†]
	HP	HD	NZ		MP	HS	HC	
24-day submission rate (%)	78[a]	90[b]	88[b]	**	90	79	87	NS
Conception rate to first service (%)	45[a]	54[ab]	62[b]	**	56	51	54	NS
Services per cow (no.)	2.07[a]	1.79[b]	1.61[b]	**	1.86	1.77	1.83	NS
6-week pregnancy rate (%)	54[a]	65[b]	74[b]	***	66	65	61	NS
Overall pregnancy rate (%)	74[a]	86[b]	93[c]	***	84	82	86	NS

HP = High production, HD = High durability, NZ = New Zealand, MP = Moorepark feed system, HS = High stocking rate feed system, HC = High concentrate feed system, [a b] Means with different superscripts within the same row are significantly different (P<0.05). Significance: ***= P<0.001, **= P<0.01, NS = Not significant

Conclusions Both the NZ and HD strains, selected for lower milk production and better reproductive traits, had better reproductive performance than a North American HF strain selected for high milk production. The results indicate that offering higher levels of concentrate supplementation will not alleviate the reduced reproductive performance of the HP strain in a pasture-based feed system.

References

Harris, B.L. & E.S. Kolver (2001). Review of Holsteinization of intensive pastoral dairy farming in New Zealand. *Journal of Dairy Science*, 84 (E. Supplement), E56-E61.

Mee, J.F. (2004). Temporal trends in reproductive performance in Irish dairy herds and associated risk factors. *Irish Veterinary Journal*, 57, 158-166.

Phenotype x herbage allowance interactions in reproduction of first calf heifers grazing semiarid rangeland

J.W. Holloway, B.G. Warrington and M.K. Owens
Texas Agricultural Experiment Station, 1619 Garner Field Road, Uvalde, TX 78801, USA, Email: jw-holloway@tamu.edu

Keywords: herbage allowance, growth curve, milk production, environmental interactions

Introduction Cattle are differentially adapted to nutritional environments. The most sensitive measure of adaptation is reproduction of first-calf heifers. We studied the role of maturation rate and milk production on reproductive performance of first-calf heifers allowed different levels of herbage in semiarid rangeland.

Materials and methods Data was collected over 4 y for 197 primiparous Brahman x Hereford F1 females grazing a semiarid Acacia savannah (Study 1, 29^0 lat. 99^0 52' long. 52 cm rain/y). Females were the residual from 345 heifers allotted at weaning each year to 4 pastures (1,950 to 6,100 ha) with 4 stocking rates (17 to 146 kg body weight [BW]/ha) to give a total of 16 herbage allowances (HA) over the four years (400 to 2,800 kg DM/100 kg BW). Calving (Jan.-Mar) was as 2- and then as 3-y olds. The HA was the mean of about 55 monthly hand clipped and visually estimated $0.5m^2$ quadrats/pasture. Maturation rate (k [-kg/d per d; this is the rate of decline in weight accretion and is negatively correlated with frame size] from $Wt=B+A \, loge^{-kt}$ of Brody, 1945) was computed (Beltran *et al.*, 1992). First-lactation milk production (MP) was the mean of three estimates based on weighing calves before and after suckling. Distance traveled/d was estimated in Study 2 for 72 GPS-collared cows varying in frame size and allowed 1,200 ha pastures.

Results Figures 1 and 2 show the probability of a heifer being bred (PB = 1.28 -(log (-30.7149 + 0.0171771*HA + 398.689*k + 0.0143144*MP - 0.005403*HA*MP - 0.237244*HA*k -120.042*MP*k + 0.071795*HA*k*MP)-0.08)) (Chi Squares, P<.05). For the mean MP, earlier maturing females (EMF, k=0.080) had higher PB at low HA. At high HA, the reverse was true. For late maturing females (LMF, k=0.075), HA influenced PB more than for EMF. In Study 2, for frame sizes 4, 5, 6, 7 and 8 the distances traveled were 4,837, 5,255, 5,541, 5,531, and 5,227 m/d respectively, indicating an asymptote at 6. The frame size in Study 1 was

from 2 to 6. The data suggest that in Study 1 the LMF, with higher frame size, would have traveled further in search of feed. At high HA, the search was successful, resulting in increased PB. At low HA, the search was less successful, resulting in lower PB. In Figure 2, at the mean HA, MP influenced PB much more than k.

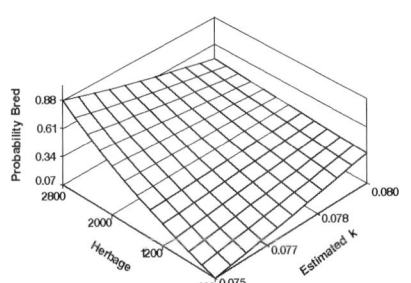 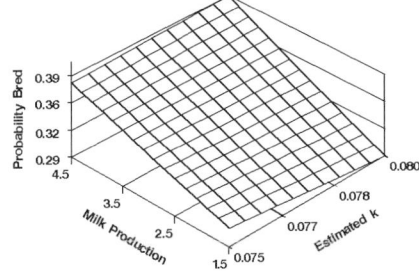

Figure 1 Effect of HA, kg DM/100 kg BW and estimated k, kg/kg/d on probability bred (milk=5 kg/d)

Figure 2 Effect of milk, kg/d and k kg/kg/d on probability bred (HA = 1600 kg DM/100 kg BW)

Conclusions The cow type that gives the best reproductive performance in semiarid rangeland depends on HA. Late maturing cows have ability to travel further from water and thus acquire more pasture at high HA. At limited HA, this is a detriment because LMF have larger working maintenance requirements and derive little extra feed. Adapted females gave more milk and had greater PB than less adapted females.

References

Brody, S. (1945). Bioenergetics and Growth. Reinhold Publishing Corp., New York.
Beltran, J.J.. W.T. Butts, Jr., T.A. Olson & M. Koger (1992) Growth patterns of two lines of Angus cattle selected using predicted growth parameters. *Journal of Animal Science*, 70, 734-741.

The impact of the level of feed-on-offer available to Merino ewes during winter-spring on the wool production of their progeny as adults

C. Oldham[1], M. Ferguson[2], B. Paganoni[1], A. Thompson[2], G. Kearney[2] and M.A. van Burgel[1]
[1]Department of Agriculture, Locked Bag No. 4, Bentley Delivery Centre, Western Australia, Australia 6983, Email: coldham@agric.wa.gov.au, [2]epartment of Primary Industries, Hamilton, Private Bag 105, Hamilton, Victoria 3300, Australia

Keywords: maternal nutrition, feed-on-offer, progeny, wool, dose-response

Introduction New opportunities for developing optimum ewe management systems, based on achieving liveweight and body condition score (CS) targets at critical stages of the reproductive cycle, have emerged from the acceptance that nutrition during pregnancy can have substantial impacts on the lifetime wool performance of the progeny (Kelly *et al,*. 1996). However, most studies of the impacts of nutrition on foetal growth and development tended to focus on late pregnancy and have also only considered extreme nutritional regimes often outside the boundaries of commercial reality. Hence, the 'Lifetime Wool' team (Thompson & Oldham, 2004) conducted dose-response experiments to determine the levels of feed-on-offer (FOO; kg dry matter/ha; Hyder *et al.*, 2004) needed at different stages of the reproductive cycle to optimise both wool and meat production per ha in the short term and the lifetime performance of the progeny in the long term. This paper reports the response in the first two years of the experiment of clean fleece weight (CFW) and fibre diameter (FD) of the progeny as adults to the level of FOO available to their mother in late pregnancy and lactation.

Materials and methods Research sites were located at Coleraine, Victoria ($36^0$58'S, $141^0$17'E) and Kendenup, Western Australia ($34^0$27'S, $117^0$35'E). Following artificial insemination (day 0), ewes at each site were fed to maintain or lose weight in early and mid-pregnancy (CS 2 or 3 by day 90). Ewes were then grazed at five different levels of FOO (from a low of 900 to a high target of 3000 kg dry matter (DM)/ha; Hyder *et al.*, 2004) in late pregnancy and lactation (design = 2 CS x 5 FOO x 2 or 3 replicates of 20-30 pregnant ewes in each of 2 years). After weaning, all progeny grazed in common and were shorn as lambs and then again after 12 m . Linear mixed models (REML; Genstat 7, VSN International Limited) were fitted to the CFW and FD data with fixed effects and significant two-way interactions of site, year, CS, FOO, sex, rear type of lamb and ram source.

Results

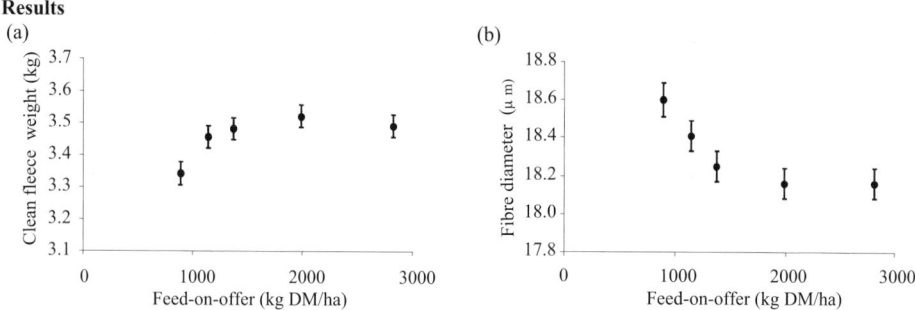

(a) Clean fleece weight (kg) vs Feed-on-offer (kg DM/ha)
(b) Fibre diameter (µ m) vs Feed-on-offer (kg DM/ha)

Figure 1 Effect of the average feed-on-offer (kg DM/ha) available to ewes during late pregnancy and lactation on (a) clean fleece weight (p = 0.001) and (b) mean fibre diameter (p < 0.001) of their progeny at first adult shearing (REML predicted means ± sem at each level of FOO with significance tested by chi-sq)

Conclusions The difference between the extremes is similar for CFW but much greater for FD than previously reported (Kelly *et al.*, 1996). In practical terms the 'Lifetime Wool' project aims to produce nutritional guidelines for the optimum management of ewes by coupling these results with response curves for wool production from the ewes, survival of the lambs and carry over effects on the CS of the ewes at their next mating.

References

Hyder, M.W., D.J. Gordon & K. Tanaka (2004). Lifetime wool. Pasture growth, utilisation and ewe stocking rates. *Animal Production in Australia* 25, 265, www,publish.csiro.au.

Kelly, R.W., I. Macleod P. Hynd & J. Greeff (1996). Nutrition during fetal life alters annual wool production and quality in young merino sheep. *Australian Journal of Experimental Agriculture*, 36, 259-267.

Thompson, A.N. & C.M. Oldham (2004). Lifetime wool. Project overview. *Animal Production in Australia.* 25, 326. www,publish.csiro.au

Effects of forage chicory (*Cichorium intybus*) on farmed deer growth and internal parasitism

S.O. Hoskin, PR. Wilson, WE. Pomroy and T.N. Barry

Institute of Veterinary, Animal and Biomedical Sciences, Massey University, PN 452, Private Bag 11222, Palmerston North, New Zealand, Email: S.O.Hoskin@massey.ac.nz

Keywords: deer, chicory, growth, lungworm, gastrointestinal parasites

Introduction Internal parasitism (particularly lungworm - *Dictyocaulus* sp) significantly limits post-weaning growth of deer. Endoparasite control using anthelmintics may be unsustainable, due to the increasing risk of anthelmintic resistance and the risk or perception of chemical residues in animal products. Chicory has a high feeding value and contains sesquiterpene lactones and low levels of condensed tannins, both with anti-parasitic activity (Molan *et al.*, 2003). Grazing chicory during autumn may reduce the requirement for anthelmintic treatment of young deer compared with grazing ryegrass-based pasture (Hoskin *et al.*, 1999). The objective of this study was to investigate the effect of withholding anthelmintic treatment of young deer grazing grass-based pasture or chicory on autumn growth and internal parasitism.

Materials and methods Newly weaned red and wapiti hybrid deer (*Cervus elaphus*) calves (n=68) were allocated to a 2x2 factorial design, rotationally grazing either perennial ryegrass (*Lolium perenne* cv Grasslands Nui, 90%) -based pasture or chicory (*Cichorium intybus* cv Grasslands Puna, 75%) *ad libitum* from March to May 2002 and receiving either the topical anthelmintic Moxidectin ("treated") monthly or remaining untreated ("control"), unless criteria for individual deer treatment were triggered on the basis of fortnightly monitoring (Hoskin *et al.*, 2003). During May, when clinical parasitism became evident in the pasture-grazed control group, five sentinel deer were randomly selected for slaughter from each group. Adult lungworm recovered from the lungs and gastrointestinal (GI) nematodes recovered from the abomasum, small and large intestines were counted. Data were compared by ANOVA (GLM, SAS Institute Inc, USA) with forage type, anthelmintic treatment, animal sex and genotype and interactions as factors.

Results Growth of deer grazing chicory exceeded that of deer grazing pasture (P<0.001) and growth of treated deer exceeded growth of control deer on pasture (P<0.001), but not on chicory (Table 1). A significant type of forage by anthelmintic treatment interaction was found for deer growth (P<0.05). Clinical parasitism was not observed in deer grazing chicory, resulting in zero anthelmintic usage in control deer, whereas 35% of control deer on pasture required anthelmintic. Table 1 also shows that the control deer grazing chicory had half the lungworm population of deer grazing pasture and also harboured 18% fewer GI parasites (both P<0.10).

Table 1 Mean liveweight gain, percentage of control animals requiring anthelmintic and total lungworm and gastrointestinal (GI) nematode populations per animal of deer grazing either ryegrass-based pasture or chicory

Forage	Pasture		Chicory		SEM
Anthelmintic treatment	Treated	Control	Treated	Control	
Liveweight gain (g/d)	134[a]	60[b]	208[c]	175[c]	11.8[1]
Clinical parasitism (%)	-	35	-	0	[1]
Lungworm (no.)	1	643	0	311	115.2[2]
GI nematodes (no.)	0	2642	52	2240	373.7[2]

[abc] letters designate significant differences between treatments (P<0.05 or better). [1]n=17, DF=32, [2]n=5, DF=8.

Conclusions There is an important potential role for alternative forages in sustainable deer production systems. Further research is required on the relative contributions to endoparasitism of secondary-compound-mediated, direct anthelmintic effects, compared with indirect nutritional, plant morphological and sward structure effects.

References

Hoskin, S.O., T.N. Barry, P.R. Wilson, W.A.G. Charleston & J. Hodgson (1999) Effect of reducing anthelmintic input upon growth and faecal egg and larvae counts in young farmed deer grazing chicory and perennial ryegrass/ white clover pasture. *Journal of Agricultural Science, Cambridge*, 132, 335-345.

Hoskin, S.O. W.R. Pomroy, I. Reijrink, P.R. Wilson & T.N. Barry (2003). Effect of withholding anthelmintic treatment on autumn growth and internal parasitism of weaner deer grazing perennial ryegrass-based pasture or chicory. *Proceedings of the New Zealand Society of Animal Production*, 63, 269-273.

Molan, A.L., A. Duncan, T.N. Barry & W.C. McNabb (2003). Effect of condensed tannins and crude sesquiterpene lactones extracted from chicory on the motility of deer lungworm and gastrointestinal nematodes. *Parasitology International*, 52, 209-218.

Cattle production from native pastures in the semi-humid grasslands of Uganda

S. Okello[1] and E.N. Sabiiti[2]
[1]Department of Veterinary Physiological Sciences, Makerere University, P. O. Box, 7062, Kampala, Uganda, Email: sokello@vetmed.mak.ac.ug, [2]Department of Crop Science, Faculty of Agriculture, Makerere University, P. O. Box, 7062, Kampala, Uganda

Keywords: Ankole cattle, body condition score, herbage yield, *in situ* digestibility, predictive model

Introduction The cattle population of Uganda, estimated at 6 million, consists of more than 95% indigenous stock, raised on the semi-arid and semi-humid grasslands that make up 48% of the total land area and which supply over 85% of the marketed milk and meat. Native grassland pastures, which vary seasonal in quantity and quality due to rainfall and temperature variations, are the sole feed resource for cattle (Mbuza *et al.*, 1992). This study examined the effects of seasonal herbage mass (HM), dietary crude protein (CP_d), detergent lignin (ADL_d) and digestibility (DiG) on body condition scores of milking cows grazed only on natural grassland pastures.

Materials and methods Ten milking cows of the Ankole breed, of similar body size early in their second or third lactation were selected into a cohort to record daily milk yield and body condition scores at two week intervals over a twelve-month period. Body condition scores were determined on the nine point scale (Nicholson & Butterworth, 1986). Herbage mass was estimated from two transects on protected paddocks using 1 m² quadrats placed randomly on 10 m sampling segments. Grazing cows were carefully observed to record plant and dietary preference every 10 minutes during peak grazing. Herbage eaten was plucked to obtain representative dietary samples, which were analysed for crude protein and acid detergent lignin. Some of the samples were used for *in situ* dry matter disappearance in two fistulated steers of the same breed. Results were analysed by multiple regression, with body condition score as dependent variable, using the STATISTICA programme.

Results and discussion Seasonal variations in herbage, milk yield and body condition of cows are shown in Figure 1. Cows gained condition as milk yield also increased with each herbage growth wave and lost condition

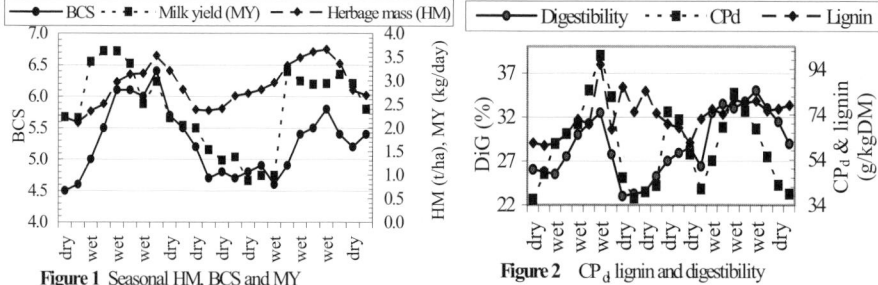

Figure 1 Seasonal HM, BCS and MY   **Figure 2** CP_d lignin and digestibility

with declining milk yield as herbage mass waned. Figure 2 shows that dietary crude protein was below the critical 60-80 g/kg DM required for optimal ruminal digestion and feed intake (Minson, 1981), except at the peaks herbage growth. Lignin was less affected by season. Digestibility increased with crude protein, but was little affected by lignin. Ankole cows selected higher quality diets to maintained body condition in spite of the seasonal variation in herbage quantity and quality. Regression analysis showed that HM was the most reliable predictor (p = 0.00087) of BCS according to the model: BCS = 3.12+0.75HM, which explained 40% of the variation (R^2 = 0.4). High herbage mass allows maximum dietary selection and digestibility, enabling cattle to easily meet the daily feed intake requirement.

Conclusions Body condition of Ankole cattle remained fairly stable across seasons due to increased selection of better quality from seasonally varying herbage, leading to higher digestibility. Body condition scores were most reliably predictable from herbage mass.

References
Mbuza, F. M.B., J. Holmes, R. Beilharz, & G. Rimmington (1992). Stocking rates and herd structure for sustainable grassland utilisation in Uganda. *Nomadic Peoples*, 31, 97-106.
Minson, D.J. (1981). Effects of physical and chemical composition of herbage eaten upon intake. In: J.B. Hacker (ed.). Nutritional Limits to Animal Production from Pastures, CAB, 167-182.
Nicholson, M. J., and M.H. Butterworth (1986). A guide to body condition scoring of zebu cattle. ILCA, Addis Ababa.

The seasonal dry matter production and carrying capacity of kikuyu oversown with ryegrass and clover

P.R. Botha[1], R. Meeske[1] and H.A. Snyman[2]

[1]*Agriculture Western Cape, Outeniqua Experimental Farm, P.O. Box 249, George, 6530, South Africa, Email: philipb@elsenburg.com,* [2]*Department of Animal, Wildlife and Grassland Sciences, University of the Free State, P.O. Box 339, Bloemfontein, 9300, South Africa*

Keywords: kikuyu, over sow, dry matter production, carrying capacity

Introduction In the main milk producing areas of the Southern Cape, kikuyu (*Pennisetum clandestinum*) is considered an important summer and autumn pasture which is climatologically well adapted. The main problem experienced with kikuyu is that winter and spring production is low and the exclusion of legumes make it dependent of nitrogen fertiliser and that increases the input cost. The aim of the study was to quantify the seasonal dry matter (DM) yield and carrying capacity of kikuyu (K) through treatments involving kikuyu oversown with annual ryegrass (*Lolium multiflorum* spp.) (KR), kikuyu oversown with a mixture of perennial ryegrass (*L. perenne*) and perennial white (*Trifolium repens*) and red (*Trifolium pratense*) clovers (KRC) and kikuyu oversown with a mixture of only perennial white and red clovers (KC).

Materials and methods The trial was carried out under irrigation using Jersey cows in a put-and-take grazing system. Annual ryegrass (KR) was oversown (25 kg/ha) into kikuyu using a mulcher (1.6 meter Nobili with 32 blades) and perennial clover (red clover 6 kg/ha plus white clover 5 kg/ha) (KC) and perennial ryegrass-clover (perennial ryegrass 10 kg/ha, red clover 4 kg/ha, white clover 4 kg/ha) (KRC) were oversown with a rotavator (1.55 meter Celli with 36 blades) during May. Fertiliser was applied to raise phosphorus level to 35 mg/kg, potash level to 80 mg/kg and the pH (KCl) to 5.5. No nitrogen was applied to the KC and KRC pastures. The K pasture was fertilised at a rate of 420 kg N/ha in seven applications of 60 kg N/ha and the KR pasture at a rate of 600 kg N/ha in ten applications of 60 kg N/ha. Dry matter production, growth rate and grazing capacity were determined. Cows were fed 2 kg of dairy concentrate after each milking. The number of animals per paddock was adjusted daily to insure a forage availability of 10 kg DM/cow per day.

Results The growth rates during spring were highest for Kikuyu-clover (KC) (59 kg DM/ha per d), KR (58-66 kg DM/ha per d) and KRC (57 kg DM/ha per d) pastures ($p<0.05$). During summer and autumn K (67 and 72 kg/ha per d) and KR (66-82 and 70-76 kg DM/ha per d) were more productive ($p<0.05$) than KC (55-58 and 38-43 kg DM/ha per d) and KRC (52 and 47 kg DM/ha per d). Kikuyu clover (KC) and KRC produced 27 and 26 kg DM/ha per d during winter. The second year autumn growth rate of KC (49-57 kg DM/ha per d) was higher ($p<0.05$) than that of the first year (38-43 kg DM/ha per d).

Both KR (15953-19292 kg DM/ha) and KRC produced higher ($p<0.05$) annual yields than K (13786 kg DM/ha). The lowest annual yield was produced by KC (12609-12954 kg DM/ha) during the second year of growth ($p<0.05$) but this was not significantly different ($p>0.05$) from K (13786 kg DM/ha).

Kikuyu-ryegrass (KR) had a higher ($p<0.05$) grazing capacity (7.2-7.6 cows/ha) during spring than K (3.9 cows/ha), KRC (5.8 cows/ha) and KC (6.5-6.8 cows/ha) during the first year of growth and was not statistically different ($p>0.05$) from KC (6.81 cows/ha) during the second year of growth. The grazing capacity of K during summer and autumn (8 and 8.5 cows/ha) and KR (8.1-9.1 and 8.4-10.8 cows/ha) was higher ($p<0.05$) than KRC (5.3 and 5.0 cows/ha). It was also higher ($p<0.05$) than the grazing capacity of KC during the summer of the first year (6.4 cows/ha) but with no statistical difference ($p>0.05$) during the summer of the second year (7.6 cows/ha) and higher ($p<0.05$) during the autumn of year 1 and year 2 (4.9-5.5 cows/ha).

Conclusions The incorporation of annual ryegrass, perennial clover or perennial ryegrass-clover into kikuyu changed the fodder flow and increased the spring dry matter production and grazing capacity. The incorporation of clovers into kikuyu pasture resulted in a lower grazing capacity. The oversowing of kikuyu with a annual ryegrass during May had no effect on the dry matter production of kikuyu during the summer and autumn. Kikuyu and kikuyu-ryegrass, fertilised with nitrogen, have a high dry matter production rate resulting in a high grazing capacity.

The seasonal botanical composition, calcium and phosphorus content of kikuyu oversown with ryegrass and clover

P.R. Botha[1], R. Meeske[1] and H.A. Snyman[2]
[1]Agriculture Western Cape, Outeniqua Experimental Farm, P.O. Box 249, George, 6530, South Africa, Email: philipb@elsenburg.com, [2]Department of Animal, Wildlife and Grassland Sciences, University of the Free State, P.O. Box 339, Bloemfontein, 9300, South Africa

Keywords: kikuyu, over sow, ryegrass, clover, calcium content

Introduction Kikuyu (*Pennisetum clandestinum*) is one of the major grasses used for summer and autumn grazing in the Southern Cape coast area of South Africa. Annual ryegrass (*Lolium multiflorum* spp.), perennial white (*Trifolium repens*) and red clover (*Trifolium pratense*) can be incorporated into an existing kikuyu stand to improve pasture quality and spring production. The aim of this study was to determine the persistence of these species and effects on the calcium and phosphorus content of kikuyu (K), kikuyu oversown with annual ryegrass (KR), kikuyu oversown with a mixture of perennial ryegrass (*L. perenne*) and perennial white and red clovers (KRC) and kikuyu oversown with a mixture of perennial white and red clovers (KC).

Materials and methods The trial was carried out under irrigation using Jersey cows in a put-and-take grazing system. Fertiliser was applied to raise phosphorus level to 35 mg/kg, potash level to 80 mg/kg and the pH (KCl) to 5.5. No nitrogen fertiliser was applied to the KC and KRC pastures. The K pasture was fertilised at a rate of 420 kg N/ha in seven applications of 60 kg N/ha and the KR pasture at a rate of 600 kg N/ha in ten applications of 60 kg N/ha. Botanical composition, calcium and phosphorus content of each pasture was monitored at monthly intervals during the year. The number of animals per paddock was adjusted daily to insure a forage availability of 10 kg DM/cow per day.

Results The clover content of the KC pasture was respectively 86%, 85%, 79% and 70% during the spring, summer, autumn and winter of the first year and 66%, 64% and 48% during the following spring, summer and autumn. The clover content of this pasture declined during spring (41%) and summer (15%) of the third year after it was oversown with annual ryegrass during the previous autumn and received a monthly application of 60 kg N/ha. The clover content of the KRC pasture was respectively 48%, 52%, 49% and 30% during the spring, summer, autumn and winter. The grass content of the KR pastures consisted mainly of annual ryegrasses during spring.

Table 1 The Ca and P content and Ca:P ratio of kikuyu (K), kikuyu oversown with annual ryegrass (KR), kikuyu oversown with a mixture of perennial ryegrass and perennial white and red clover (KRC) and kikuyu oversown with a mixture of perennial white and red clover (KC)

Year		Pasture	%	Pasture	%	Pasture	%	LSD 0.05
1	Ca	KC	0.87	KR	0.30	K	0.34	0.081
	P		0.40		0.41		0.54	0.049
	Ca:P		2.18:1		0.73:1		0.63:1	-
2	Ca	KC	0.85	KC	1.18	KR	0.43	0.081
	P	(Second	0.43		0.46		0.49	0.049
	Ca:P	year growth)	1.98:1		2.57:1		0.88:1	-
3	Ca	KR	0.46	KC	0.60	KRC	0.66	0.081
	P		0.52	(Second year growth)	0.51		0.53	0.049
	Ca:P		0.89:1		1.18:1		1.24:1	-

Discussion and conclusions The clover content of the KC pasture was maintained at levels higher than 30% for more than two years. The clover content of the KRC pastures decreased to 30% within a year. The Ca content of the KC pasture was higher than the nutritional requirement for dairy cattle (0.67%), but decreased as the grass content increased (KRC). The Ca content of the grass pastures (K and KR) was low and cows should receive Ca supplementation. The P content of both the legume and grass pastures exceeded the requirement for dairy production (0.38%). The low Ca content in the grass pastures resulted in a Ca:P imbalance that was lower than the 1.6:1 ratio needed by dairy cows.

The evaluation of kikuyu oversown with ryegrass and clover in terms of milk production

R. Meeske[1], P.R. Botha[1] and H.A. Snyman[2]

[1]*Agriculture Western Cape, Outeniqua Experimental Farm, P.O. Box 249, George, 6530, South Africa, Email: robinm@elsenburg.com,* [2]*Department of Animal, Wildlife and Grassland Sciences, University of the Free State, P.O. Box 339, Bloemfontein, 9300, South Africa*

Keywords: kikuyu, oversow, ryegrass, clover, milk production

Introduction Kikuyu (*Pennisetum clandestinum*) comprises the greater part of irrigated summer and autumn pasturage for milk production in the Southern Cape. Milk production per cow is limited by low forage quality. The aim of the study was to determine the milk production from kikuyu (K), kikuyu oversown with annual ryegrass (*Lolium multiflorum* spp. cv Energa) (KR), kikuyu oversown with a mixture of perennial ryegrass (*Lolium perenne* cv Yatsyn, Dobson) and perennial white clover (*Trifolium repens* cv Haifa, Waverley) and red clover (*Trifolium pratense* cv Kenland, Cherokee) (KRC) and kikuyu oversown with a mixture of perennial white and red clover (KC). The trial was carried out under irrigation using Jersey cows in a put-and-take grazing system. Fertiliser was applied to raise phosphorus level to 35 mg/kg, potash level to 80 mg/kg and the pH (KCl) to 5.5. No nitrogen fertiliser was applied to the KC and KRC pastures.

Materials and methods The study was carried out on 9 ha kikuyu pasture divided into seven blocks. Each block was divided into three experimental paddocks and pasture treatments were randomly allocated to paddocks. Cows strip grazed four days on each paddock resulting in a 28-day grazing cycle. The K pasture was fertilised at a rate of 420 kg N/ha in seven applications of 60 kg N/ha and the KR pasture at a rate of 600 kg N/ha in ten applications of 60 kg N/ha. Dry matter production, growth rate and grazing capacity were determined. Thirty-six mid-lactation cows were randomly allocated to three different pasture treatments (12 cows per treatment) at the start of spring, summer, autumn and winter. The groups were balanced for milk production (four weeks prior to experimental period), days in milk and lactation number. The number of cows per paddock was adjusted daily to ensure a forage availability of 10kg DM/cow per day. Cows were fed 4 kg of dairy concentrate per day during milking and were milked twice daily. Milk production and number of cows on each paddock was recorded daily. Milk composition was determined monthly.

Results The results are presented on a yearly basis in Table 1. The KR carried more cows/ha than KC during the three years of the study. During year 1, milk production per cow was higher (P <0.05) on KC than on KR and K. Milk production/ ha did not differ (P > 0.05) between KR and KC during its first year of growth in years 1 and 2 of the study.

Table 1 The carrying capacity (cows/ha), average milk production per cow and milk production per hectare of kikuyu (K), kikuyu oversown with annual ryegrass (KR), kikuyu oversown with a mixture of perennial ryegrass and perennial white and red clover (KRC) and kikuyu oversown with a mixture of perennial white and red clover (KC)

Year	Parameter	KC first year of growth	KR	K
1	Cows/ha	5.27[de]	8.03[b]	6.72[c]
	Milk/cow per d (kg)	15.7[b]	14.0[c]	13.8[c]
	Milk/ha (kg)	25940[bcd]	25953[bcd]	21377[d]
		KC second year of growth	KC first year of growth	KR
2	Cows/ha	5.37[de]	5.78[d]	9.03[a]
	Milk/cow per d (kg)	16.8[ab]	17.4[a]	17.0[ab]
	Milk/ha (kg)	22761[cd]	34615[a]	38406[a]
		KR	KC second year of growth	KRC
3	Cows/ha	6.76[c]	5.77[d]	4.80[e]
	Milk/cow per d (kg)	16.8[ab]	17.2[a]	18.1[a]
	Milk/ha (kg)	27109[bc]	24148[cd]	29298[b]

[a, b, c, d] Means with no common superscript differ (P< 0.05)

Conclusions The KC supported higher milk production per cow than KR and K during the first year. During year two of the study, milk production per hectare of KC and KR was higher than that of KC in its second year of growth. The oversowing of kikuyu with clover and/or ryegrass increased milk production per cow and milk production per hectare. Milk produced per hectare was very high on KR and KC pastures. Carrying capacity was higher on KR pasture than on KC and KRC pastures.

The seasonal nutritional value of kikuyu oversown with ryegrass and clover

R. Meeske[1], P.R. Botha[1] and H.A. Snyman[2]
[1]Agriculture Western Cape, Outeniqua Experimental Farm, P.O. Box 249, George, 6530, South Africa, Email: robinm@elsenburg.com, [2]Department of Animal, Wildlife and Grassland Sciences, University of the Free State, P.O. Box 339, Bloemfontein, 9300, South Africa

Keywords: kikuyu, over sow, metabolisable energy, crude protein, neutral detergent fibre

Introduction   To overcome the seasonality and relatively low forage quality of kikuyu (*Pennisetum clandestinum*), annual ryegrass (*Lolium multiflorum* spp.), perennial white (*Trifolium repens*) and red clover (*Trifolium pratense*) can be incorporated into an existing kikuyu stand to improve pasture quality and spring production. The aim of this study was to determine the quality of kikuyu (K), kikuyu oversown with annual ryegrass (KR), kikuyu oversown with a mixture of perennial ryegrass (*Lolium perenne*) and perennial white and red clover (KRC) and kikuyu oversown with a mixture of white and red clover (KC), in swards grazed by dairy cows.

Materials and methods   The trial was carried out under irrigation using Jersey cows in a put-and-take grazing system. Fertiliser was applied to raise phosphorus level to 35 mg/kg, potash level to 80 mg/kg and the pH (KCl) to 5.5. No nitrogen fertiliser was applied to the KC and KRC pastures. The K pasture was fertilised at a rate of 420 kg N/ha in seven applications of 60 kg N/ha and the KR pasture at a rate of 600 kg N/ha in ten applications of 60 kg N/ha. Dry matter production, growth rate and grazing capacity were determined. Cows were fed 2 kg of dairy concentrate after each milking. During each grazing cycle, two 0.09 m^2 samples were cut at a height of 50 mm before grazing on three paddocks for each pasture. Samples were dried at 60°C for 72h and *in vitro* organic matter digestibility (IVOMD), crude protein (CP) and neutral detergent fibre (NDF) was determined. Metabolisable energy (ME) (MJ/kg DM) was calculated from IVOMD values (ME = 18.4 X IVOMD% / 100 X 0.81).

Results   The DM % of K, KR and KC was respectively 16.8, 12.0 and 10.0 during spring, 14.2, 11.7 and 11.7 during summer and 16.6, 11.6 and 11.6 during autumn of year 1 (LSD = 0.97 at P = 0.05). During year 2 the DM % of KR, KC (first year of growth), KC (second year of growth) was 12.1, 9.6 and 13.2 during spring, 14.8, 12.4 and 15.7 during summer and 13.3, 12.4 and 15.8 during autumn respectively (LSD =0.67). In year 3 the DM% of KR, KC (second year of growth) and KRC was 12.8, 13.9 and 11.7 during spring, 16.6, 16.2 and 14.6 during summer and 15.9, 16.5 and 16.8 during autumn respectively (LSD = 0.93).

The ME (MJ/kg DM) of K, KR and KC was respectively 8.9, 10.2 and 11.4 during summer and 8.1, 8.2 and 11.1 during autumn of year 1 (LSD = 0.91). During year 2 the ME (MJ/kg DM) of KR, KC (first year of growth), KC (second year of growth) was 11.5, 11.3 and 11.1 during spring, 9.1, 10.4 and 10.2 during summer and 8.0, 10.1 and 8.8 during autumn respectively (LSD = 0.87). In year 3 the ME (MJ/kg DM) of KR, KC (second year of growth) and KRC was 11.3, 11.0 and 11.5 during spring, 9.3, 9.5 and 9.3 during summer and 7.9, 8.1 and 8.7 during autumn respectively (LSD = 0.73).

The CP% of K, KR and KC was respectively 23.7, 23.7 and 28.8 during summer and 23.1, 23.5 and 26.9 during autumn of year 1 (LSD = 3.43). During year 2 the CP % of KR, KC (first year of growth), KC (second year of growth) was 21.8, 27.9 and 27.2 during spring, 18.9, 25.6 and 21.8 during summer and 23.1, 25.3 and 19.5 during autumn respectively (LSD = 2.94). In year 3 the CP % of KR, KC (second year of growth) and KRC was 20.8, 24.2 and 22.8 during spring, 16.1, 18.4 and 18.5 during summer and 17.3, 15.8 and 18.4 during autumn respectively (LSD = 2.46).

The NDF % of K, KR and KC was respectively 64.7, 56.8 and 37.4 during summer and 62.6, 65.6 and 40.9 during autumn of year 1 (LSD = 8.38). During year 2 the NDF % of KR, KC (first year of growth), KC (second year of growth) was 50.1, 36.4 and 37.0 during spring, 66.9, 42.2 and 48.7 during summer and 67.4, 50.9 and 58.8 during autumn respectively (LSD = 6.20). In year 3 the NDF % of KR, KC (second year of growth) and KRC was 46.0, 44.6 and 43.2 during spring, 64.4, 59.6 and 54.6 during summer and 70.1, 69.9 and 60.2 during autumn respectively (LSD = 4.89).

Conclusions   Oversowing of kikuyu with clover resulted in lower (P < 0.05) DM and NDF values and higher (P<0.05) CP and ME values. The ME value of KC and KR pasture was very high during spring and therefore a lower milk response to concentrate feeding can be expected during spring compared to summer and autumn. The lowest CP % in KR pasture was found summer and autumn. The CP content of the concentrate supplement fed to cows should be increased during summer and autumn when cows graze kikuyu dominant pasture.

Liveweight gain of lambs grazing six short-term ryegrass cultivars

W.W. Nichol and M.G. Norriss
Wrightson Research, P.O. Box 939, Christchurch, New Zealand, Email: waynenichol@wrightson.co.nz

Keywords: liveweight gain, ryegrass, lambs

Introduction   Increasing dry matter (DM) production per ha is a key goal in ryegrass (*Lolium*) breeding programmes (Easton *et al.*, 2002), based on the assumption that increased DM yield will in turn increase profitability per ha, through an increase in animal productivity. However, the performance of animals grazing pasture can also be modified by the quality of the forage on offer, and the presence of toxins within the pasture. The objective of this study was to compare the liveweight gain per hectare of lambs grazing six short-term ryegrasses, which had been selected for various combinations of improved DM production and nutritive value.

Materials and methods   Thirty pure swards of six commercially available short-term ryegrass cultivars were sown in five replicates across three sites during Feb. 2000 in Canterbury, New Zealand. Each site consisted of 10 plots, with plots measuring 0.325, 0.425 and 0.45 ha, for sites 1, 2 and 3 respectively. Sowing rates were 18 kg/ha and 25 kg/ha for diploid and tetraploid cultivars respectively.   Treatments were: tetraploid annual ryegrasses, cv. Winter Star and cv. Tetila (*Lolium multiflorum var westerwoldicum*); diploid Italian ryegrass cv. Flanker; tetraploid Italian ryegrasses cv. Moata and cv. Feast II (*L. multiflorum*); and the diploid hybrid ryegrass cv. Maverick Gold (*L. hybridum*). Treatments were stocked with Coopworth ewe lambs. Lambs were weighed before and after each grazing period. There were five grazing periods during the trial: winter 14 June 00 to 15 Aug. 00; spring 6 Oct. 00 to 13 Nov. 00 and 15 Nov. 00 to 11 Dec. 00; summer 14 Dec. 00 to 25 Jan 01 and 25 Jan. 01 to 20 Feb. 01. Base stocking rate for each grazing period was determined by the cultivar with the lowest pasture mass (assessed visually). A common post-grazing pasture mass was achieved across all cultivars by adding extra lambs to higher DM yielding plots. These lambs were recorded as extra grazing days, and added to weight gain per ha. The trial was irrigated, independently managed by AgResearch LTD, and all cultivars were coded to avoid bias.

Results   Liveweight gain (LWG) per ha during each season is presented for each cultivar in Figure 1. Total LWG (kg/ha) for lambs grazing ryegrass cultivars Winter Star (974), Maverick Gold (971) and Feast II (997), was significantly greater ($LSD_{0.05} = 94.8$) than those grazing Moata (805), Flanker (867) and Tetila (766). However, significant seasonal differences between cultivars only occurred during summer (P < 0.05) and between Winter Star (326) and Moata (292) in winter (P < 0.05).

Discussion and conclusions Annual ryegrasses: We suggest that improved LWG on Winter Star compared with Tetila during summer was due to a combination of improved forage quality (higher leaf/stem ratio) and improved persistence (more DM on offer). Italian/hybrid ryegrasses: Moata had the lowest DM production and persistence of this group (Easton *et al.* 2002), and these traits offer an explanation for the low overall LWG of this cultivar. Maverick Gold, Feast II and Flanker produce similar amounts of DM in Canterbury (M. Norriss pers. comm), and this is reflected in the winter and spring LWG results. Feast II and Maverick Gold have both been selected for improved leaf to stem ratio during summer and this was likely to contribute to the significantly higher LWG relative to Flanker during this period. Results support the hypothesis that both forage quantity and forage quality are important predictors of animal performance on grazed pastures.

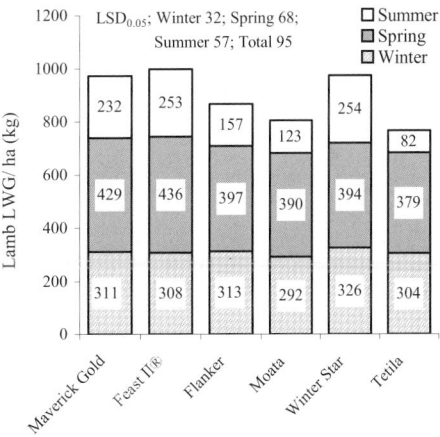

Figure 1 Liveweight gain of lambs grazing short-term ryegrasses

Reference

Easton, H.S., J.M. Amyes, N.E. Cameron, R.B. Green, G.A. Kerr, M.G. Norriss & A.V. Stewart (2002). Pasture plant breeding in New Zealand: where to from here? *New Zealand Grassland Association,* 64, 173-179.

Year-round grazing of beef cows on pangolagrass (Digitaria decumbens cv.Transvala) pasture in southern area of Japan

Y. Nakanishi, K. Hirano and A. Shouji
National Agricultural Research Center for Kyushu Okinawa Region, Kumamoto, 861-1102, Japan, Email: yuji@affrc.go.jp

Keywords: year-round grazing, transvala, pangolagrass, beef cows

Introduction The southern area of Japan (Okinawa) has a sub-tropical climate. In this area beef calf production is now based on year-round grazing on giant stargrass (Cynodon aethiopicus Clayton & Haylan). However, the numbers of beef cows in this area are increasing rapidly and a grass with higher productivity than giant stargrass is required. The objective of this experiment was to examine the possibility of using pangolagrass (*Digitaria decumbens* cv. Transvala) pasture in this area.

Materials and methods Japanese black cows were grazed on pangolagrass (Transvala) pasture all year round. The period from May to Nov. involved whole-day grazing and the period from Dec. to Feb. gazing only in daytime. Rotational grazing was based on 2-week rotations with a stocking rate of 6 cows/ha. Live weight and blood biochemical levels of the beef cows were measured every 30 d. Grass samples for chemical composition and stem density were collected every 30 d. The pasture was fertilised with 200 kg N, 60 kg P_2O_5, 84 kg K_2O/ha per year.

Results The crude protein content of pangolagrass varied from 13.5% to 19.5% and was higher than that normally found in giant stargrass. The percentages of crude fibre and the organic b fraction in the cell wall (Ob) of pangolagrass varied from 29% to 41% and from 63% to 69% respectively (Table 1). These values were lower than those generally found in giant stargrass. Mean stem density of pangolagrass was 6818 (5325-8320)/m^2 and dry matter yield was 34.6 t/year, higher values than would be expected with giant star grass. Mean live weight gain of the beef cows was 66.3 kg (55.0-75.0) during pregnancy and blood biochemical levels changed normally through the year (Table 2).

Table 1 Chemical composition (%) and stem density (number/m^2) of pangolagrass (Transvala)

	Spring	Summer	Winter
Crude protein	17.9	13.5	19.5
Crude fibre	28.9	41.1	30.1
Ob	62.8	68.6	62.0
NFE	45.1	34.7	36.0
Stem density	5825	8320	5325

Ob:organic b fraction in cell wall

Table 2 Blood biochemical levels of beef cows

	Summer	Winter
Total protein (g/dl)	7.4	7.5
Urea nitrogen (mg/dl)	12.3	18.9
Cholesterol (mg/dl)	98.3	115.3
3-OHBA (μmol/L)	337.5	292.5
GPT (IU/L)	21.8	19.1

3-OHBA:3-htdroxybutyric acid
GPT:glutamic pyruvic transaminase

Conclusions These results demonstrate that pangolagrass (Transvala) maintained high nutritive value and high productivity through the year in this area with the result that beef cows kept in good body condition and had good blood biochemical levels all year round. Pangolagrass (Transvala) is, therefore, a suitable grass for year-round grazing in this area.

Economic comparison of pasture based lamb production systems in southern Australia

A.J. Kennedy and A.N. Thompson
Primary Industries Research Victoria, Department of Primary Industries, Hamilton, Victoria 3300, Australia,
Email: Andrew.Kennedy@dpi.vic.gov.au

Keywords: lamb production, economics, pasture

Introduction Lamb production enterprises in southern Australia utilise a pasture base as their primary nutrition source due to its low cost. Holmes Sackett & Associates (2003) identified that increasing total lamb weight per hectare by increasing ewe stocking rate, animal genetic potential and weaning percentage can increase enterprise profitability. The limitation of these enterprises is the seasonal and geographic variations impeding pasture production and quality. The 'MoreLamb Quality Pastures' project is demonstrating the benefits of mixing high performing grass, legume and herb species to extend the pasture-growing season and increase pasture quality. Key economic indicators of three pasture systems and commercial lamb enterprises are reported.

Materials and methods On five commercial lamb enterprises situated around Hamilton in southeastern Australia, the following pasture systems (10-25 ha/system per site) were established in June 2002: (i) System A - perennial ryegrass, subterranean clover and white clover; (ii) System B - tall fescue, subterranean clover, white clover and red clover; and (iii) 'System C' - perennial ryegrass, tetraploid Italian ryegrass, subterranean clover, balansa clover, red clover, white clover and chicory. Rainfall is between 650-750 mm/yr and winter/spring dominant. Systems are rotational grazed from Dec. to June and are set-stocked from July to Nov. Key economic indicators for systems A, B, C and the five commercial lamb enterprises (LE) were formulated for 2003 data. The LE is used as a control unit and is representative of lamb production systems for the region. General analysis of variance was undertaken on the key economic indicators using GenStat 7.1 (Genstat Committee, 2003).

Results System C produced more lamb/ha than System B (84 kg, P<0.05), and systems A, B and C produced more lamb per hectare than LE (61-145 kg, P<0.05) (Table 1). There were no significant differences in weaning percentage between systems A, B and C, but values for these systems were 16-26% higher (P<0.05) than for LE. There were no significant differences for ewe stocking rate between systems. Weaning percentage had a greater impact than ewe stocking rate on lamb output/ha when systems A, B and C are compared to the LE. The higher weaning percentages of systems A, B and C compared to LE were attributed to higher reproductive rate, as lamb mortality to weaning was constant at 25-30% for all systems.

Gross income/ha was higher for systems A and C ($110 and $180 respectively, P<0.05) compared to LE, but no significant differences were found for enterprise costs between any systems. The gross margin/ha was greater for systems A and C than LE ($98 and $112 respectively, P<0.05). Hassell & Associates (2003) reported that a 20% increase in weaning percentage could return a 22% increase in gross margin/ha.

Table 1 Key economic indicators for three pasture systems and lamb enterprises for 2003

System	Lamb (kg live/ha)	Weaning (%)	Stocking rate (ewes/ha)	Gross income ($/ha)	Enterprise costs ($/ha)	Gross Margin ($/ha)
A	472 (6.16)	140 (4.94)	10.5	1064 (6.97)	421 (20.5)	642 (25.3)
B	433 (6.07)	143 (4.96)	9.7	990 (6.90)	404 (20.1)	588 (24.2)
C	517 (6.25)	150 (5.01)	10.6	1136 (7.04)	481 (21.9)	656 (25.6)
LE	372 (5.92)	124 (4.82)	9.6	954 (6.86)	406 (20.2)	544 (23.3)
l.s.d (P=0.05)	0.12[#]	0.08[#]	1.4	0.10[#]	2.48[*]	1.72[*]

[#] l.s.d in reference to log values in parenthesis. [*] l.s.d in reference to square root values in parenthesis.

Conclusions Lamb enterprises that use systems to increase the supply of high quality pasture can increase ewe reproductive rate and weaning percentages. The increases in lamb output can return a positive economic outcome that offset any increase in the costs of these systems. At this level, pursuing reproductive and lamb survival efficiencies should be considered before increasing stocking rates.

References
Hassell & Associates (2003). Economic Analysis of Sheep Production Systems. Hassall & Associates Pty Ltd, NSW, Australia.
Holmes Sackett & Associates (2003). Economic and Situation Analysis of the Australian Sheep Industry, Holmes Sackett & Associates Pty Ltd, NSW, Australia.
GenStat Committee (2003). GenStat for Windows.Release 7.1, Seventh ed., VSN International Ltd, Oxford, UK.

The impact of tillage system for small-grain pasture establishment on the performance of growing beef calves in Arkansas

P. Beck[1], S. Gunter[1], M. Anders[2], K. Lusby[3] and D. Hubbell[4]
*University of Arkansas, [1]Southwest Research & Extension Center, 362 Hwy 174 N, Hope, AR 71801, USA,
Email: pbeck@uaex.edu, [2]Rice Research & Extension Center, 2900 Hwy 130E, Stuttgart, AR 72160, USA,
[3]Department of Animal Science, AFLS 114, Fayetteville, AR 72701, USA, [4]Livestock & Forestry Branch Station,
70 Experiment Station Drive, Batesville, AR 72501, USA*

Keywords: small grain, pasture, beef cattle, tillage, no-till

Introduction In the United States, governmental regulations mandate the improvement of farming practices to improve environmental quality. There is a requirement to reduce the siltation of waterways, soil carbon losses, and nutrient runoff along the Mississippi River Delta. The use of small-grain forages by grazing cattle offers real opportunities to produce high-quality forage for cattle production during the winter and spring months. No-till and reduced tillage practices developed primarily for grain production may offer environmental and economic solutions for both grain farmers and cattle producers. Producers are slow to adopt conservation tillage practices because of a perceived risk of reduced production. The objective of this project was to compare conventional tillage to reduced tillage and no-till systems for production of small-grain forage for grazing livestock.

Materials and methods Four hundred eighty-two weaned calves were used in a 2-year study evaluating the effects of tillage method on small-grain forage production and animal performance. Three tillage methods were evaluated: 1) conventional tillage, consisting of chisel ploughing, heavy discing, and light discing, 2) reduced tillage with a target of 50% soil surface residue, and 3) no-till seeding. Wheat and rye were planted in the first week of Sept. annually at a rate of 68 kg/ha of each. In the first year, grazing was managed using put-and-take stocking, while set stocking rates were used in the second year. In the first year, 90 calves (213 kg) were stocked in Nov. when forage height reached 20 cm in each pasture and removed when forage became limiting in late Jan. A second group of 167 calves (270 kg) was stocked beginning on 31 Jan. and removed by 13 May. In the second year, 90 calves (208 kg) were stocked on 28 Oct. and removed on 23 Jan. On 2 March, the pastures were restocked with 135 calves (233 kg), and removed on 27 April. In the second year, forage availability of each pasture was measured using a calibrated disk meter. Data pooled across the 2-year study were analysed using the mixed procedure of SAS (SAS Inst., Inc.; Cary, NC); least-square means were separated using contrast statements.

Results Autumn average daily gains (ADG) by the steers in no-till pastures were 0.14 kg/d higher ($p<0.05$) than steers in conventional tillage pastures. When data were pooled across years, there were no differences ($p>0.05$) in ADG during the spring-grazing period, grazing days/ha, or gain/ha. Forage production was higher ($p<0.05$) in no-till pastures than conventional-tillage pastures at the initiation of autumn grazing (1,879 vs. 1,525 kg), the end of autumn grazing (1,254 vs. 1,015 kg), and the initiation of spring grazing (1,170 vs. 856 kg). No-till pastures contained more ($p<0.05$) forage at the initiation of spring grazing than reduced-tillage pastures (1,170 vs. 913 kg).

Table 1 Effect of tillage system on average daily gains, grazing-day/ha, and gain/ha of calves grazing wheat-rye pasture

	Conventional	Reduced	No-Till	s. e. m
Autumn ADG, kg	0.65[a]	0.75[ab]	0.79[b]	0.21
Spring ADG, kg	1.04	1.11	1.06	0.004
Grazing days/ha	664	578	627	53.1
Gain/ha, kg	576	550	593	68.7

[a, b] LS means in rows with differing superscripts differ ($p<0.05$).

Conclusion Establishment of small-grain pastures using no-till methods appears to be superior to conventional tillage in autumn and winter forage production. This increase in forage production may be the mechanism for improved performance during the late autumn and early winter observed in the 2-year study. These results indicate no-till production systems are a viable alternative for establishment of small-grain pastures for livestock grazing, with no change in animal gain/ha and increased forage production in autumn compared to conventional farming methods.

Urea applied to puccinellia-based pastures increases pasture and sheep production

M.L. Hebart[1], N.J. Edwards[1], A.D. Craig[1], EA. Abraham[1], JD. McFarlane[2] and JE. Hocking Edwards[1]
[1]South Australian Research and Development Institute, Struan Agricultural Centre, P.O. Box 618, Naracoorte, SA, 5271, Australia, Email: hebart.michelle@saugov.sa.gov.au, [2]Rural Solutions SA, Struan Agricultural Centre, P.O. Box 618, Naracoorte, SA, 5271, Australia

Keywords: dryland salinity, puccinellia, nitrogen, pasture production, liveweight gain

Introduction In the 1950's large areas of native vegetation in the upper south east of South Australia (SA) were replaced with highly productive Hunter River lucerne. This maintained groundwater recharge at near pre-clearing levels. The area of lucerne was reduced dramatically in the late 1970's by a combination of lucerne aphids, wingless grasshoppers and drought. In 1981 severe flooding inundated large areas of the region, causing the saline groundwater to rise to the soil surface. Since that time, dryland salinity has been a feature of the local farming system and salt-tolerant pastures based on puccinellia (*Puccinellia ciliata*) were widely established. Despite this, few agronomic studies have been conducted on puccinellia to enable management guidelines to be determined. The aim of this experiment was to compare animal and pasture production on volunteer saline pasture and improved saline pasture with and without fertiliser inputs.

Materials and methods One-hundred and twenty 15-month-old Merino wethers were allocated to one of four treatment groups: 1) unimproved saltland pasture (predominantly sea barley grass (*Hordeum marinum*), samphire (*Halosarcia* spp.) and salt scalds); 2) improved saltland pasture (puccinellia-based) with no fertiliser inputs; 3) improved saltland pasture with 75 kg/ha superphosphate (SP); and 4) improved saltland pasture with 75kg/ha SP and 100 kg/ha urea (U). Each treatment was replicated three times, giving a total of 12 plots. The nine improved plots were 2 ha each and stocked at 5 Dry Sheep Equivalents (DSE)/ha. The unimproved plots were stocked at 2 DSE/ha but were 5 ha each to maintain 10 animals per plot. Grazing began at the end of April 2003 and fertiliser was applied in July after the season-opening rains had fallen. Liveweight, condition score, soil salinity and pH and pasture composition and mass were recorded monthly.

Results The three puccinellia-based pastures produced more dry matter than the unimproved pastures from late-autumn (May) to mid-spring (Oct; Fig 1a). Phosphorus did not limit pasture growth in the improved treatment, as there was no difference in pasture growth between this and the improved + SP treatment. This is explained by the fertiliser history of the paddock, which resulted in Colwell P levels of 25mg/kg (0.5M sodium bicarbonate extract) at the start of the project. Consequently, the body weight and condition score of animals grazing the improved + SP pasture was not different from those grazing the improved pasture with no fertiliser (Fig 1b). The addition of SP and U increased pasture mass (Fig. 1a) and resulted in sheep being 9% heavier than those grazing the two other improved pastures (Fig. 1b). Furthermore, during late spring/early summer, when feed availability was declining in other improved pastures, the pasture mass in the SP + U treatment was maintained (Fig 1a).

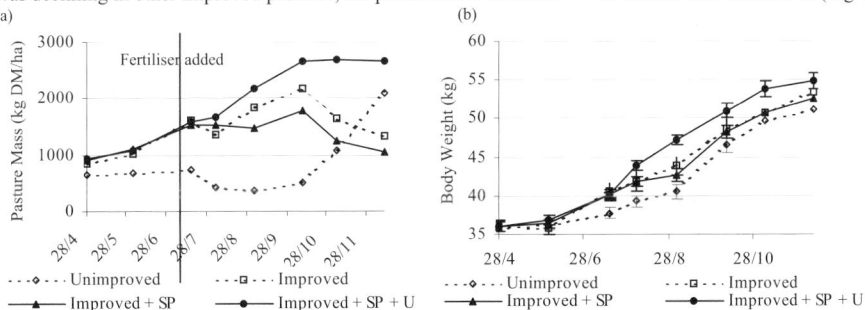

Figure 1 a) Pasture mass, and b) animal liveweight (mean ± sem) during the first year of grazing

Conclusions Saltland pastures based on puccinellia are more productive from both an animal and pasture perspective than if the pastures are left in an unimproved state. GrazFeed modelling indicates that the excess feed grown in the SP + U treatment could have supported approximately 13 extra DSE/ha during 2003. Increased stocking rates are therefore planned for these treatments in spring 2004 to control pasture accumulation and capture some of the benefits in animal production (i.e. liveweight, wool growth and wool quality).

Acknowledgement This work is part of the Sustainable Grazing on Saline Land (SGSL) research into profitable use of saline land that is an initiative of Australian Wool Innovation Pty Ltd, with Land and Water Australia and support from the Cooperative Research Centre for Plant-based Management of Dryland Salinity and Meat and Livestock Australia.

Sustainability of beef, dairy and goat production with Batiki grass (*Ischaemum aristatum var. Indicum*) in the dry season in Samoa

E.M. Aregheore
The University of the South Pacific, School of Agriculture, Animal Science Department, Alafua Campua, Apia, Samoa, Email: aregheore_m@samoa.usp.ac.fj

Keywords: *Ischaemum aristatum*, *Leucaena leucocephala*, supplements, nutritive value, animal performance

Introduction Batiki grass (*Ischaemum aristatum* var. *indicum*) was introduced to Samoa from Fiji in the early 1970s to complement other existing natural or unimproved grasses and is now the most common propagated pasture grass in Samoa. The contrast between the wet and the dry seasons has a great impact on the nutritive value of batiki grass. Aregheore (2002) observed reduced growth rate, poor body condition score and low performance in steer calves and goats offered batiki as the sole diet during the dry season. This paper reports on the sustainability of beef, dairy and goat production with batiki grass in the dry season in Samoa.

Results The mean nutrient composition of batiki grass for four years (2000-3) is given in Table 1, together with the results of a feeding experiment with goats. Decline in crude protein (CP) and high fibre contents in the dry season are responsible for low voluntary dry matter (DM) intake and growth rate in steer calves (Aregheore, 2003) and goats (Aregheore, 2002), and milk production of dairy cows (personal communication, Lome, 2002). Crude protein content in the rainy season was well above 7 %, the level below which DM intake becomes depressed, but was below that value in the dry season. Poor performance by goats and steers when no supplements are fed was confirmed in Table 2. However supplementation with *Leucaena leucocephala* increased DM intake and growth rates significantly (P<0.05) (Aregheore, 2003). Apparent digestibility of DM and CP also significantly (P<0.05) improved with supplementation resulting in increases in digestible energy intake and nutritive value index.

Conclusion The nutrient content of batiki grass declines rapidly in the dry season. Lactating cows on batiki grass alone have poor body condition scores, but beef cattle, goats and sheep utilised it to meet their requirements for maintenance of body conformation without depleting their body reserve, as is usually the situation in most tropical countries during the dry season. Supplementation with *Leucaena leucocephala* improved intake and animal performance and is one way to sustain beef, dairy and goat production in Samoa. In other experiments we have shown similar responses to supplementation with foliage of *Moringa oleifera* and *Erythrina* spp.

Table 1 Nutrient composition (% of dry matter) of batiki grass, voluntary dry matter intake, growth rate and nutrient digestibility in the dry and rainy seasons (10 goats per treatment)

Nutrients (%)	Rainy	Dry
Crude protein	11.1	5.3
Crude fibre	32.8	52.2
Animal response	10	10
DM intake (g/kg$^{0.75}$per d)	71.2a	50.6b
Daily live-weight gain (g)	71a	36b
DM digestibility (%)	76.6a	54.8b
CP protein digestibility (%)	76.6	58.2
Digestible energy intake (MJ/kg $^{0.75}$per d)	305.0a	415.2b

[a,b] Means within row with different superscript differ (P<0.05)

Table 2 Effects of supplementation with 50 % of *Leucaena leucocephala* (fresh weight) on DMI and growth and nutrient digestibility of goats and steer calves in the dry season

Animal species	Goats		Steer calves	
Supplementation	No	Yes	No	Yes
DM intake (g/kg$^{0.75}$per d)	41.2	50.8	4.9	6.2
Daily live weight gain (g)	57a	90b	172a	375b
DM digestibility (%)	58.3a	70.2b	50.1a	65.1b
CP digestibility (%)	45.4a	66.7b	48.3a	70.6b
DE (MJ/kg $^{0.75}$per d)	11.6	13.3	10.0	12.5
NVI (Kj/kg $^{0.75}$ per d)	226a	438b	211a	408b

[a,b] Means within row with different superscript differ (P<0.05)

References

Aregheore, E.M. (2002). Intake and digestibility of *Moringa oleifera*-batiki grass mixtures by growing goats. *Small Ruminant Research*, 46, 23-28.

Aregheore, E. M. (2003). Nutritional characterization and evaluation of batiki grass (*Ischaemum aristatum* var. *indicum*) and batiki grass-forage legume mixtures with steer calves. *Journal of Animal and Veterinary Advances*, 2, 404-412.

Goose meat production responses to grass based diets

P. Gyüre[1], G. Nagy[1] and S. Mihók[2]
[1]Department of Rural Development, University of Debrecen, Hungary, Email: gyurep@helios.date.hu,
[2]Department of Animal Breeding, University of Debrecen, Hungary

Keywords: goose, chopped grass, feeding, live weight

Introduction Goose meat and feather production are important elements of farming in Hungary. There were no available data on grass intake by geese or the production potential of grass in goose farming, therefore, we conducted a series of experiments between 2000 and 2003 on goose production responses to grass-based diets.

Materials and methods Between 2000 and 2003 four feeding trials were conducted at the Animal Research Station of the Debrecen University Agricultural Sciences Centre using four-week-old growing geese. Previously, goslings had been intensively reared with complex grain pellets. We investigated goose production responses to different grass proportions in the diet (Treatment 1: 75%, Treatment 2: 50%, Treatment 3: 25%, Treatment 4: 0% chopped grass). The remainder of the diet comprised complex grain pellets for geese. Each group consisted of 25 birds. Geese were kept indoors with free exit to an open yard. Food was offered twice a day. Grass for feeding was cut every day from a mixed sward. Data were recorded daily for offered and rejected grass and on weekly for live weight gain. Live weight gains were statistically analysed by analysis of variance.

Results and discussion Throughout the experiment in 2003, the average live weight of geese was significantly lower in Treatments 1 and 2 (75% and 50% chopped grass in the diet) compared to other treatments. As a result, the final live weights for Treatments 3 and 4 were significantly higher than those for Treatments 1 and 2 (Table 1.). These results were probably due to the higher energy concentration of the diet. These results confirmed those from previous goose feeding trials with the same treatments in 2001 and 2002.

Table 1 Average live weight of geese in the weeks of the experiment in 2003 (kg per goose)

Treatments proportion of chopped grass			Weeks								
			1		2		3		4		5
75%	mean value	s.e	2.41	0.16	2.47	0.16	2.46	0.20	2.68	0.25	2.82	0.33
50%	mean value	s.e.	2.41	0.24	2.70	0.41	2.77	0.34	2.97	0.31	3.23	0.33
25%	mean value	s.e.	2.47	0.23	2.81	0.28	3.21	0.31	3.60	0.33	3.86	0.41
0	mean value	s.e.	2.38	0.38	2.79	0.28	3.25	0.38	3.52	0.32	3.85	0.33
$LSD_{0.05}$			0.08		0.12		0.13		0.12		0.14	

s.e. - standard error of mean, LSD - Least Significant Difference

Geese are typically grazing waterfowl that can utilise grass quite effectively, because microbial degradation of fibre during digestion provides energy for maintenance and production. The digestive system of geese not only utilises fibre, but requires 4-10% fibre in the diet, depending on age (Anrique *et al.*, 1982). For reasonable meat and feather production grazing meat geese need at least 15% grain supplementation during the grazing season (Nagy & Mihók, 1992). Our results indicate that farmers can include grass (chopped or grazing) in goose feeding up to c. 25% with positive effects on feeding and production costs. A higher proportion of grass in the diet would result in both poorer live weight gains and final live weights.

References

Anrique, G.R., C.J. Gajardo, S.S. Voullieme, B.E. Cuevas & C.D. Alomar (1982). Nutritive value of pasture for geese. *Agro Sur*, Chile, 2, 65-69.
Nagy, G. & S. Mihók (1992). Grazing geese on permanent and seeded pastures. *Proceedings of the Fourteenth[h] General Meeting of the European Grassland Federation*, 303-306.
Nagy, G., P. Gyüre & S. Mihók (2001). Reaction of geese to grassland based diets. *Grassland Science in Europe*, 6, 170-172.

Effect of urea-treated *Pennisetum pedicellatum* and supplementation of concentrates with urea on milk production of "Mossi" ewes

V.MC Bougouma-Yameogo and A.J. Nianogo

Institut of Rural Development (IDR) Polytechnic University of Bobo-Dioulasso, 01 BP 1091 Bobo-Dioulasso Burkina Faso, Email:Bouval2000@yahoo.fr

Keywords: *Pennisetum pedicellatum*, urea treatment, urea supplementation, milk production, milk composition, "Mossi" ewes

Introduction The "Mossi" sheep is a near parent of "Djallonke" sheep that live in sudano-sahelian area of Burkina Faso. However, there are few available results on dairy production from this breed. The treatment of straw with urea is a technique used in several developing countries to improve the nutritional value of gramineous forages (Sourabié *et al.*, 1995). The aim of this study was to test the influence on the performance of "Mossi" ewes and on milk composition of treatment of *Pennisetum pedicellatum* (Pp) with urea in comparison with addition of urea to the concentrate feed.

Materials and methods Twenty-three "Mossi" ewes in early lactation were used. Three dietary rations were tested: (1) untreated Pp + 22% concentrate (treatment NoU); (2) untreated Pp + 2.8% urea + 20% concentrate (UCo); (3) Pp treated with 6% urea + 22% concentrate (UPp) (Table 1). The Pp was harvested at the straw stage at the beginning of the dry season. The composition of the concentrate was: 25% whole cottonseed, 25% cottonseed cake and 50% of ground corn grain. The diet dry matter was offered at 4.4-6 % of bodyweight (BW). Measurements were made of milk yield, by the oxytocin method, milk composition (AOAC, 1984), fat (Babcock) and body condition (Russel *et al*,. 1969).

Results The effects on milk yield, body weight changes, feed intake and milk composition are given in Table 2. The ADY was significantly higher for UPp than for UCo, but not higher than for the NoU treatment, which had no added urea. The differences in average daily milk yield were reflected in differences between treatments in the yields of milk solids, milk protein and milk fat. There were no significant differences in milk protein %. Body weight (BW), however, showed a clear advantage for UPp over the other two treatments.

Table 1 Chemical composition of straw and experimental diets

	Pp-U	Pp+U	UPp	NoU	UCo
CP	3.7	13.9	8.0	15.9	15.9
EFUL	0.58	0.66	0.69	0.68	0.75

EFUL=French Energy Feed Unit for lactation/DM; Pp-U, *P. pedicellatum* without urea; Pp+U, *P. pedicellatum* treated with urea

Conclusion Whilst milk production in local ewes cannot be sustained by poor quality straws alone, natural grasses such as Pp may be significantly improved by treatment with urea. This treatment gave better results than addition of urea at feeding.

Table 2 Effect of urea treatment on milk yield, body weight and condition changes, and on feed intake and on milk composition

	NoU	UCo	UPp	Prob
ADY (g)	257[ab]	197[b]	316[a]	0.0001
BW	-3.3	-1.3	+0.2	-
TFI/MW	93	88	96	NS
Milk composition				
ES (g/d)	44[b]	35[b]	56.7[a]	0.0001
Ash (g/d)	2.8[ab]	2.3[b]	3.1[a]	0.01
CP (g/d)	9.5[b]	7.5[ab]	12.9[a]	0.0001
Fat (g/d)	17.2	13.8	21.5	NS
CP (%)	3.8[b]	4.0[a]	4.1[a]	0.05

a, b : means in a row with the same superscript are not significantly different (p>0.05). ADY, average daily milk yield; BW, bodyweight; TFI/MW, total feed intake g/kgBW$^{0.75}$; ES, milk solids

References

AOAC (1993). Official Methods of Analysis, Association of Official Analytical Chemists. Washington DC, 114 pp.

Russel, A.J.F., J.M. Doney & R.G. Gunn (1969). Subjective assessment of body fat in live sheep. *Journal of Agricultural Science, Cambridge*, 72, 451-454.

Sourabié, K.M., C. Kayouli & C. Dalibard (1995). Le traitement des fourrages grossiers à l'urée : une technique très promoteuse au Niger. *World. Animal Review*, 82, 3-13.

Supplementing dairy cows in late lactation with high quality silages

T.A. White[1], T.L. Knight[1], M.G. Hyslop[2] and T.J. Fraser[1]

[1]AgResearch Limited, PO Box 60, Lincoln 8152, New Zealand, Email: todd.white@agresearch.co.nz, [2]Heinz-Wattie's Ltd, PO Box 16083, Christchurch 8004, New Zealand

Keywords: dairy cattle, legumes, cereals, Canterbury, New Zealand

Introduction Agriculture on the Canterbury Plains of New Zealand is a mixture of integrated cropping and pastoral enterprises. Cropping farmers often provide supplementary feed for dairy farmers by growing forages for high quality silage. Such silages can improve milk production by increasing dry matter (DM) intake and/or by alleviating deficiencies of either soluble carbohydrate or protein in pasture (Woodward *et al.,* 2002). Legumes and/or cereals have potential to make large quantities of high quality silage (de Ruiter *et al.,* 2002). This trial aimed to determine milk production and composition differences between three silages fed during late lactation.

Materials and methods In April 2003, three groups of 20 cows (balanced for production, calving date and age) were selected from a 220-cow dairy farm on the Canterbury Plains, New Zealand. The cows consumed 12 kg DM/hd per d of perennial ryegrass/white clover (RGC) pasture. In addition, each group was allocated 5 kg DM/hd per d of silage made from either RGC pasture, whole-crop (WC) triticale (cv. *DoubleTake*) or triticale (cv. *DoubleTake*)/forage pea (cv. *Magnus*). Silages were fed in troughs (with nil wastage) for 34 d. The silages were made as large plastic-wrapped square bales. The triticale silage was cut at the "cheesy dough" stage and the triticale/pea silage was cut when the pods were beginning to fill. Cows were milked twice daily and weighed at the start and end of the trial. Milk yield and composition were measured weekly. Individual cows were treated as replicates. Several forage quality traits were determined for the silages and green pasture (FeedTECH, AgResearch, New Zealand).

Results There were no significant differences in milk yield, composition or live weight gain of cows fed either RGC pasture, WC triticale or triticale/pea silage (Table 1). Triticale/pea silage and green RGC pasture were significantly higher in crude protein (CP) than WC triticale and RGC pasture silage (Table 2). WC triticale silage had higher organic matter digestibility (OMD), soluble sugars/starch (SSS) and metabolisable energy (ME) than the other silages and green RGC pasture.

Table 1 Daily milk production, composition and live weight gain of cows supplemented with three different silages

	Vol. (l)	Solids (kg)	Fat (%)	Prot. (%)	LWG (kg)
RGC pas.	20.1	1.58	4.40	3.59	0.70
WC Trit.	20.0	1.58	4.48	3.59	0.54
Trit./pea	20.0	1.60	4.55	3.62	0.78
LSD(5%)	1.20	0.09	0.21	0.10	0.36

NDF: neutral detergent fibre

Table 2 Composition of three silages and fresh green RGC pasture fed to cows in the trial

	CP (%)	OMD (%)	NDF (%)	SSS (%)	ME (MJ/kg)
RGC pas.	11.3	63.9	59.7	9.1	10.2
WC Trit.	11.8	68.5	48.5	27.1	11.0
Trit./pea	17.2	64.8	54.2	2.3	10.4
Fr. green	18.2	65.7	46.4	8.9	9.9
LSD(5%)	1.7	1.5	3.4	3.0	0.3

Conclusions In this trial highly nutritious cereal and cereal/legume silages did not improve milk production during late lactation. The most probable reason for this was that the non-improved forage (RGC pasture and pasture silage) was of high enough quality to meet cow requirements for energy and protein. Greater productivity gains from feeding the alternative silages examined here are more likely to be achieved in situations where pasture quality deteriorates over summer and autumn. In Canterbury this is most likely to occur on farms without irrigation or those that have irrigation restrictions during late summer/autumn.

Acknowledgements The authors thank John Greenslade for use of his farm, Chris McLeod for technical assistance and Dave Saville for statistical analysis.

References

de Ruiter, J.M., R. Hanson, A.S. Hay, K.W. Armstrong & R.D. Harrison Kirk (2002). Whole-crop cereals for grazing and silage: balancing quality and quantity. *Proceedings of the New Zealand Grassland Association,* 64, 181-189.

Woodward, S.L., A.V. Chaves, G.C. Waghorn & P.G. Laboyrie (2002). Supplementing pasture-fed dairy cows with pasture silage, maize silage, Lotus silage or sulla silage in summer - does it increase production? *Proceedings of the New Zealand Grassland Association,* 64, 85-89.

Whole crop cereal silage in dairy production

J. Wallsten, L. Ericson and K. Martinsson
Dept. of Agricultural Research for Northern Sweden, Box 4097, 904 03 Umeå, Sweden. Email: Johanna.Wallsten@njv.slu.se

Keywords: barley, silage, dairy cows, intake, milk production

Introduction Whole-crop cereal silages (WCCS) are used to some extent in Sweden, but knowledge about the use of this feed for high yielding dairy cows is scarce. The crop is often harvested at different stages of maturity, from heading to yellow ripeness, which gives forages that differ in chemical composition. The purpose of this trial was to compare intake and milk production of dairy cows fed a WCCS based on barley harvested at three different stages of maturity.

Materials and methods Three cuts of whole-crop barley were taken at heading (BSH), milk (BSM) and dough stage (BSD) and ensiled with the additive Kofasil Ultra (4-6 l/t fresh matter) as round big bales. The grass silage (GS) consisted of timothy and red clover. The experiment was conducted as a change-over design over three periods of four weeks each, using fifteen multiparous Swedish red and white dairy cows. Average milk production at the start was 31.6 kg energy corrected milk (ECM)/d. All diets consisted of concentrate, GS, and WCCS of one of the cuts (Table 1). Two of the diets were given as a mixed ration of either 70% GS and 30% BSH (Mix 1) or 30% GS and 70% BSH (Mix 2) on a dry matter (DM) basis. Refusals were collected every morning. Milk yield and milk composition were analysed every week and the animals were weighed on two consecutive days in each period.

Table 1 Composition of the five diets (kg DM/day)

Diet	Concentrate	GS	BSH	BSM	BSD	Mix 1	Mix 2
K1	10.6	4	*Ad lib*				
K2	10.6	4		*Ad lib*			
K3	10.6	4			*Ad lib*		
M1	10.6					*Ad lib*	
M2	10.6						*Ad lib*

Results Total intake was 3.25-3.36 kg DM/100 kg of live weight, with no significant difference in intake of different diets. Milk yield was lower for animals fed diet K2 ($p<0.05$) and K3 ($p<0.001$) compared with M1, M2 and K1 (Table 2). Diet K1 gave higher lactose content than K2 ($p<0.001$) and K3 ($p<0.01$), and higher protein content than K3 ($p<0.05$). Fat content was lowest for animals fed diet K3 compared to M1 ($p<0.001$), M2, K1 and K2 ($p<0.01$). Diet M1 gave higher fat content than the other diets containing BSH (M2 and K1; $p<0.05$).

Table 2 Milk production (kg ECM/d) and milk concentration of lactose, protein and fat (g/kg milk)

Diet	ECM	Lactose	Protein	Fat
K1	30.9	47.1	37.1	49.3
K2	29.5	46.3	36.4	49.5
K3	28.2	46.5	36.0	46.5
M1	31.2	46.8	37.5	51.2
M2	31.2	46.7	37.0	49.2
SED[a]	0.6	0.2	0.4	0.8

[a] SED=standard error of the difference

Conclusions Whole-crop cereal silage harvested at the dough stage gave lower milk yield and lower concentrations of fat, protein and lactose in the milk compared to animals fed WCCS harvested at heading. Higher intake of grass silage (M1) tended to increase fat content in the milk, but otherwise there were no differences in milk yield or composition between the three BSH treatments.

Effect of potato pulp silage supplementation on milk production in cows grazing temperate pasture

M. Hanada, Y. Aibibula, D. Okumura and M. Okamoto

Obihiro University of Agriculture and Veterinary Medicine, Obihiro Hokkaido, 080-8555, Japan, Email: hanada@obihiro.ac.jp

Keywords: potato pulp silage, dairy cows, grazing

Introduction In a dairy farming system based on pasture in Japan, maize grain is generally used as an energy source for milking cows, with almost all grain been imported. Potato-pulp is one of the agricultural by-products derived from the starch industry in the northern island of Japan. In our previous study (Aibibula *et al.*, 2004), it was demonstrated that potato pulp could be preserved for a long time by ensiling without additives, and that the digestible energy value of potato pulp silage (13 MJ/kg DM) was almost the same as beet pulp. From these results, it is possible that some part of the maize grain fed to grazing cows could be substituted with potato pulp silage (PPS). The objective of this study was to compare PPS with rolled-maize as an energy source for cows grazing on temperate pasture.

Materials and methods Potato pulp silage ensiled in Oct. 2003 was used. From 14 May 2004, twelve primiparous Holstein cows were rotationally grazed on temperate pasture for 20 h a day. During first 34 d (preliminary period), all cows were supplemented with a concentrate, rolled-maize and maize silage. After the preliminary period, the cows were divided into two groups (CG and PP) according to milk yield. In the experimental period, the cows in CG were supplemented with concentrate, rolled-maize and maize silage and the cows in PP were supplemented with concentrate, PPS and maize silage for 28 d. The supplementation level of rolled-maize and PPS was equivalent to 17.5% of the digestible energy requirement of the cows. Throughout the trial, the supplements were offered before and after milking in restricted quantities., Herbage, supplements, milk, blood and rumen fluid samples were collected in the last seven days of the preliminary and experimental periods.

Results Herbage mass before grazing in the preliminary and experimental periods were 222 and 136 g DM/m^2 respectively. Sward heights before grazing in the preliminary and experimental periods were 49.8 and 21.5 cm respectively. The crude protein (CP) content in herbage was higher in the experimental period (21%) than in the preliminary period (11%). Dry matter (DM) intake of the supplement was higher in PP than in CG (Table 1), because the feeding level of supplement in PP was slightly higher compared with CG to meet the digestible energy supply from the supplement. Milk yield did not differ between CG and PP (Table 1). Milk composition was not influenced by PPS supplementation, but milk urea nitrogen concentration (MUN) in the experimental period was lower in PP than in CG (P<0.05). Serum urea nitrogen concentration (SUN) in CG was higher than in PP (P<0.05). This result suggested that the cost for urea excretion was lower in PP than in CG. Although MUN and SUN in the experimental period were higher in CG than in PP, there was no significant difference in ruminal ammonium nitrogen concentration between CG and PP. Total cholesterol concentration in the serum did not differ between CG and PP, though Hanada *et al.* (2004) showed that total cholesterol in serum was depressed when maize grain was replaced with PPS in the diet of growing steers.

Conclusions These results suggest that maize grain fed to grazing cows as an energy source can be substituted with PPS without decrease in milk production, and that PPS may be an effective energy source for decreasing milk and serum urea levels in cows grazing temperate pasture.

Table 1 Milk yield, urea nitrogen in milk and serum, ruminal ammonium nitrogen and total cholesterol in serum

	Preliminary period		Experimental period	
	CG	PP	CG	PP
DM intake of supplement, kg/d	10.2	10.2	9.7[b]	11.7[a]
Milk yield, kg/d	31.3	31.7	29.0	29.2
Milk composition, %				
Milk fat	3.08	3.16	3.33	3.38
Milk protein	3.21	3.28	3.15	3.26
Lactose	4.84	4.72	4.74	4.61
Milk urea nitrogen, mgN/dl	13.3	13.9	18.4[a]	13.7[b]
Serum urea nitrogen, mgN/dl	12.8	12.5	20.5[a]	15.7[b]
Ruminal ammonium nitrogen, mg/dl	9.7	9.9	10.0	8.2
Total cholesterol in serum, mg/dl	218	227	214	198

[a,b]:P<0.05

References

Aibibula, Y., M. Hanada, O. Abdulrazak, S. Murata & M. Okamoto (2004). Effect of potato pulp silage supplementation on nitrogen utilization in ruminants. *Proceedings of 11th Animal Science Congress Volume 3*, The Asian-Australasian Association of Animal Production Societies, 331-333.

Hanada, M., Y. Aibibula, O. Abdulrazak, S. Murata, K. Ikehata & M. Okamoto (2004) Comparison of potato pulp silage with rolled corn as an energy source of growing Holstein steers. *Proceedings of 11th Animal Science Congress Volume 2*, The Asian-Australasian Association of Animal Production Societies, 367-369.

Weight gain of Nellore x Red Angus, Holstein x Zebu, and Nellore steers supplemented on *Brachiaria brizantha* cv. Marandú pasture

R.A. Reis, D. Freitas, F.L. Fregadolli, L.M.A. Bertipaglia, T.T. Berchielli, K.T. de Resende, D.S. Ferreira and A.G. Caselli
Faculdade de Ciências Agrárias e Veterinárias – UNESP, Jaboticabal, Via de acesso Professor Paulo Donato Castelane, km 5 Jaboticabal – São Paulo – Brazil 14884 100, Email: rareis@fcav.unesp.br

Keywords: genetic groups, supplementation, tropical grass

Introduction Supplementation is a very efficient approach for improving animal performance and gain/ha in grazing systems in tropical conditions. However, care needs to be taken to avoid substitution effects (Moore *et al.*, 1999). Results may also depend on animal genetics and the availability and nutritional value of forage. This trial aimed to evaluate different levels of supplementation using steers from different genetic groups on palisade grass (*Brachiaria brizantha* cv. Marandú) pasture during the rainy season.

Materials and methods The trial was carried out at "Julio de Mesquita Filho" University (UNESP), São Paulo State, Brazil. Fifty-four steers from three genetic groups, Nellore x Red Angus (n=18), Holstein x Zebu (n=18) and Nellore (n=18) were evaluated on palisade grass pasture during the rainy season (from 10 Dec. 2002 to 30 March 2003) in a rotational system with 36 d of resting and 6 d of grazing. A put and take technique was used to maintain forage availability around 6.4% of dry matter in relation to body weight (BW) and stocking rate around 5.74 AU/ha. An oesophageal-fistulated animal was used for pasture sampling, and extrusa samples were analysed to determine crude protein (CP), neutral detergent fibre (NDF) and *in vitro* dry matter digestibility (IVDMD). Steers received a supplement (77.86% citrus pulp, 12.32% corn gluten meal, and 9.82% of cotton meal) with low protein degradability and containing 19.7% CP and 76.4% TDN, in individual pens at 08.00h. The treatments involved supplement offered at 0.2, 0.6 and 1.0% of BW. Starved weights were determined at 28-d intervals. Data were analysed according to a split plot design, considering treatments (genetic groups x supplement level)as the main plot,, grazing time as the sub-plot and the three replicates (pastures).

Results and discussion During the experimental period, forage had 10.7% CP, 66.1% NDF, and 56.6% IVDMD. Supplementation increased (P < 0.05) weight gain (WG), however Nellore and Holstein x Zebu steers showed lower WG than Red Angus x Nellore (Table 1). Nellore x Red Angus and Holstein x Zebu steers maintained the same supplement intake during the whole grazing period, but Nellore reduced supplement intake during the second and third grazing days, particularly when offered at the highest level (Table 1) suggesting preferential consumption of the fresh grass.

Table 1 Weight gain of three genetic groups offered 0.2%, 0.6% and 1.0 % BW of supplement and intake of the highest level of supplement in relation to the period of occupation of a paddock

Genetic groups	Weight gain (kg/d)			
	Supplementation level			
	0.2 % BW	0.6 % BW	1.0% BW	Mean
Nellore x Red Angus	0.81	0.98	1.03	0.94 A
Holstein x Zebu	0.65	0.85	0.94	0.81 B
Nellore	0.64	0.80	0.97	0.80 B
Mean	0.70 c	0.88 b	0.98 a	

| | Supplement offered at 1.0 % BW | | | | | | |
| | Days from entering a paddock | | | | | | |
Genetic group	1	2	3	4	5	6	Mean
Nelore x Red Angus	0.90 Aa	0.90 ABa	0.92 Aa	0.92 Aa	0.92 Aa	0.93 Aa	0.92
Holstein x Zebu	1.00 Aa	0.99 Aa	0.99 Aa	1.00 Aa	0.99 Aa	0.99 Aa	0.99
Nelore	0.95 Aa	0.87 Bc	0.89 B bc	0.92 Aab	0.92 Aa	0.92 Aa	0.91
Mean	0.95	0.92	0.93	0.95	0.94	0.95	

Means followed by different capital letters within column and lowercase within line are different (P < 0.05) by Tukey test.

Conclusions Supplementation of tropical pasture during the rainy season increased steer live weight gains.

References
Moore, J.E., M.H. Brant, W.E. Kunkle & D.I. Hopkins (1999). Effects of supplementation on voluntary forage intake, diet digestibility, and animal performance. *Journal of Animal Science*, Supplement.2, 77, 122-135.

Suitability of selenate containing silage additives for the supply of beef cattle

H. Laser
Justus-Liebig-UniversityGiessen, Department of Grassland Management and Forage Growing, Ludwigstr. 23, 35390 Giessen, Germany, Email: Harald.Laser@agrar.uni-giessen.de

Keywords: selenium deficiency, Na_2SeO_4, additives, silage quality, aerobic stability

Introduction Selenium concentrations in herbage are frequently insufficient (< 100 µg Se/kg dry matter (DM)) to meet the requirements of ruminants. Whereas increasing Se supply by feeding mineral mixtures is a reliable method to prevent Se deficiencies for dairy cows, adequate Se supplementation is more difficult to achieve in extensive systems (McDowell, 1996). A suitable measure could be the addition of Se to herbage before ensiling.

Materials and methods A laboratory ensiling experiment with four replicates was carried out with the following factors: addition of Na_2SeO_4 (= 0, 75, 150, 300, 1200 µg Se/kg DM), pre-wilting (= 30, 40% DM) and source material (= primary growth of *Festuca arundinacea,* secondary growth of *Lolium perenne,* no further additives). The Se concentration in silage after a storage period of 120 d was determined by hydride generation atomic absorption spectroscopy following microwave digestion. Silage quality was assessed by pH (potentiometric determination), lactic acid (colorimetric determination) and volatile fatty acids and ethanol (by gas chromatography). Ammonia-N was measured with an ion sensitive electrode. Aerobic stability was assessed as the time needed to increase temperature by 1 °C above the ambient temperature.

Results Measured Se concentrations are in accord with the Se amounts added to fresh herbage plus the initial Se in herbage (Table 1). The addition of 75µg Se/kg DM was sufficient to give total concentrations > 100 µg Se/kg DM, but even a 16-fold higher dosage did not affect silage quality as reflected in values for pH, lactic and acetic acid. Concentrations of other volatile acids, including butyric acid, were negligible and differences between treatments in NH_3-N and ethanol were not significant (data not shown). Aerobic stability was not affected.

Table 1 Effect of the addition of selenate to herbage on silage properties

Herbage	DM content	Selenate µg Se/kg DM	Se in silage µg/kg DM	pH	Lactic acid g/kg DM	Acetic acid g/kg DM	Aerobic stability days
Primary	30 %	0	25	4.7	59	18	4.1
growth of		75	110	4.7	61	20	4.4
Festuca		150	181	4.7	58	19	3.9
arundinacea		300	356	4.6	57	17	4.1
		1200	1170	4.7	57	18	4.1
	40 %	0	20	5.0	35	13	4.6
		75	109	5.0	36	14	4.6
		150	178	5.0	36	14	4.3
		300	350	5.0	35	14	5.1
		1200	1163	5.0	37	13	5.3
Secondary	30 %	0	44	4.4	98	16	4.6
growth of		75	130	4.4	95	17	5.0
Lolium		150	212	4.4	97	18	5.4
perenne		300	368	4.4	95	16	6.2
		1200	1187	4.4	99	17	6.3
	40 %	0	43	4.6	77	14	7.3
		75	136	4.7	79	14	6.2
		150	214	4.7	77	15	6.6
		300	359	4.7	78	14	7.4
		1200	1120	4.7	81	13	6.7
$LSD_{0.05}$			33.4	0.1	4.9	1.9	0.58

Conclusions Because silage additives containing nitrate or nitrite are usually necessary anyway to produce well-fermented silage from herbage of extensively managed grassland, the addition of selenate would be a reliable and cheap method to improve the supply of selenium deficient cattle in low-input systems.

Reference

McDowell, L.R. (1996). Feeding minerals to cattle on pasture. *Animal Feed Science Technology,* 60, 247-271.

Effect of supplementation on performance and faecal particle size distribution for yearling horses and weaned calves grazing coastal bermudagrass

F.M. Rouquette, Jr., K.N. Grigsby, D.K. Hansen, G.D. Potter and W.C. Ellis.
Texas A & M University Agricultural and Extension Center at Overton, USA, Email: m-rouquette@tamu.edu

Keywords: bermudagrass, faecal particle, horse, calf, supplementation

Introduction Supplements are often fed, especially to horses, without realistic expectations of the magnitude of performance response nor knowledge of biological or economic efficiencies of supplementation. The objectives of this experiment were to compare performance of weaned calves vs yearling horses grazing 'Coastal' bermudagrass [Cynodon dactylon (L.) Pers] (COS) pastures with and without a protein-energy supplement, and to assess faecal particle size distribution for both calves and horses to document digesta dynamics.

Materials and methods Weaned, 10-m-old calves (n=24) at 300 kg liveweight and 16-m-old horses (n=18) at 350 kg grazed in separate COS pastures from late May to Sept. (140 d). Two replicate pastures each were used for : calves grazing COS only (CPAS); CPAS plus self-limiting 34% CP supplement (CSUP); horses grazing COS only (HPAS); HPAS plus daily, hand-fed ration supplying 25% of energy requirements (H25SUP); and HPAS plus daily, hand-fed ration supplying 50% of energy requirement (H5SUP). At mid-trial, faecal samples were collected *per rectum* from all calves and from horses on HPAS and H5SUP (n=12). The samples were frozen until wet-sieved in duplicates using a Fritsch apparatus with mesh openings of 1.0, 0.40, 0.160, 0.100, 0.071, and 0.032 mm.

Results Herbage allowance averaged 150 kg DM/100 kg BW for all replicate COS pastures (low stocking rate). Forage CP was about 17% and NDF about 72%. Both CPAS and HPAS had identical ADG of 0.47 kg/d (Table 1). Calves on CSUP had increased (P<.05) ADG of 0.69 kg/d with a partial feed conversion ratio (SUP fed:extra gain) of 4:1. Horses on H25SUP showed no extra gain over HPAS; whereas H5SUP had improved (P<.05) ADG of 0.58 kg/d with a partial feed conversion ratio of 21:1. Horses had a higher percent of large faecal particles compared to calves (19% vs. 10%) on grass only, and calves had more small-sized faecal particles retained on sieves than horses (8% vs. 3%) (Table 2). Supplementation did not affect faecal particle size for calves, nor for horses, except at the 0.160 seive size.

Table 1 Performance of weaned calves and yearling horses grazing Coastal bermudagrass (PAS) only or with supplement (SUP)

Treatment	ADG[1]	ADC[2]	Extra gain	SUP:Extra Gain
	----------- kg -----------			
Calves-PAS	0.47 a[3]			
Calves-SUP	0.69 b	0 .87	0.22	4:1
Horses-PAS	0.47 a			
Horses-25 SUP	0.40 a	1.14	---	---
Horses-5 SUP	0.58 b	2.3	0.11	21:1

[1,2]Average daily gain (ADG) and average daily consumption (ADC).
[3]Means followed by a different letter, differ (P<.05).

Table 2 Distribution of faecal particles from calves and horses grazing bermudagrass pastures (PAS) only or with supplement (SUP)

Sieve Size (mm)	Calves PAS	Calves PAS + SUP	Horses PAS	Horses PAS + 5 SUP
	-------------- % retained --------------			
>1.00	10.4 b	14.1 b	18.7 a	21.3 a
0.400	17.1 b	17.2 b	21.8 a	23.8 a
0.160	30.2 b	26.1 b	34.7 a	29.7 b
0.100	20.0 b	19.2 b	15.6 a	14.6 a
0.071	14.7 b	14.8 b	6.4 a	7.5 a
0.032	7.5 b	8.6 b	3.2 a	2.7 a
Groupings				
0.160 + .100	50.2	42.3	50.3	44.3
.071 + .032	22.2	23.4	9.6	10.2

[1]Numbers in the same row and followed by a different letter, differ (P<.01).

Conclusions Weaned calves and yearling horses can be expected to perform similarly during the summer on bermudagrass pastures. Supplementation of calves showed an additive effect; whereas, supplementation of horses showed a substitution effect. Supplementation was economically positive for calves, but not profitable for horses from a weight gain perspective. Smaller faecal particles with calves indicated differences due to rumination processes and possibly diet selection. We consider that documentation of faecal particle size between livestock classes has major implications for ration formulation and potential performance from forages.

West African Dwarf goat response to supplementary feeding in Cameroon

E. Tedonkeng Pamo[1], B. Boukila[2], F. Tendonkeng[1] and J.R. Kana[1]

[1]Dept. of Animal Prod., FASA, University of Dschang. P.O.Box.: 222 Dschang, Cameroon, E-mail: pamo_te@yahoo.fr, [2] I.N.S.A.B., Université des Sciences et Techniques de Masuku, B.P. 941. Gabon

Keywords: *Calliandra calothyrsus, Leucaena leucocephala,* supplementation, West African Dwarf Goat

Introduction The production of the West African Dwarf goat (WADG) in Cameroon is very low, because of lack of proper nutrition. Nitrogen content is generally very low and fibre content is high, both in the grass and the crop residues which form the basis of their diet particularly during the dry season. Supplementation of these roughages is a promising way of alleviating nutrient deficiencies. Different types of supplementary feeding have been advocated to boost goat production (Leng, 2003), of which supplementary feeding with leguminous tree leaves is of high merit. The present study was undertaken to evaluate the effects of supplementary feeding of *Calliandra calothyrsus* and *Leucaena leucocephala* leaves on growth and reproduction of WADG.

Materials and methods The study was conducted on 24 WADG in both the dry season (Nov. 2001 to April 2002) and in the rainy season (March to Sept. 2003). The WADG grazed on mixed pasture comprised of *Brachiaria ruziziensis* and *Pennisetum purpureum* between 09.00 and 17.00 h daily. After about a month, two bucks were introduced into the herd for two months and breeding allowed. The males were removed thereafter. Twelve goats received supplementary feeding with *C. calothyrsus* and *L. leucocephala* leaves mixed in equal quantities by weight. The mixture was left in the pens in the afternoon (16.00h) at the rate of 800 g/goat for eating at night. The remaining 12 goats served as unsupplemented controls. Observations were made of: (i) consumption of the supplement, (ii) goat growth by weighing on the day of the start of supplementation, just before and after kidding and every two weeks thereafter up to three months, (iii) number and sex of kids born. The data were analysed statistically (Steel & Torrie, 1980).

Results On average the goats consumed between 700 and 800 g of the foliage supplement per head per day. Neither the stage of reproduction of WADG or the season had any significant effect on the quantity of supplement consumed. Similar proportions of goats in the control and the supplemented groups became pregnant (Table 1), in line with conception taking place before commencement of supplementation. A substantially higher proportion of goats (91.7%) became pregnant during the rainy season as compared to 87.5% during the dry season. Abortions were recorded in 33% of the control goats compared to only 5% of the supplemented animals. While 24 supplemented goats gave 24 kids, only 20 kids were obtained from the unsupplemented goats. The body weight of and the profile of the dam's body weight losses after parturition during the dry and rainy season are shown in Figure 1. On average, the control and the supplemented goats lost similar weight (7.5 vs 7.4%) at kidding. During the three month post-kidding period, the supplemented goats continued to have 11-15% higher body weight than their respective controls in the dry season.

Table 1 Reproductive performance of goats in different seasons and on different feeding regimes

	Season		Pregnant	Abortion	Number of kids	Males	Females
Control	Dry	n=12	11	4	8	5	3
	Rainy	n=12	11	2	12	6	6
Supplemented	Dry	n=12	10	1	10	6	4
	Rainy	n=12	11	0	14	7	7

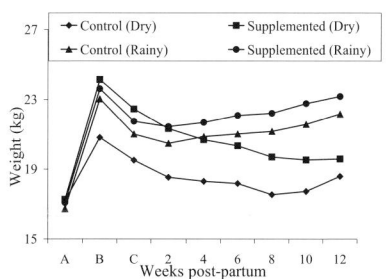

Figure 1 Change in body weight in goats in the dry and rainy seasons with different feeding regimes (A, at beginning; B, before kidding; C, at kidding)

Conclusion Supplementary feeding with multipurpose leguminous tree browse in Cameroon proved to be highly beneficial for goat production. It helped to substantially reduce the incidence of abortion and increased the overall yield of kids per animal. Pregnancy rate was higher during the rainy season than in the dry season.

References

Leng, R.A. (2003). Drought and dry season feeding strategies for cattle, sheep and goats. Penambul Books, Queensland, Australia. 271pp.

Steel, R.G. & J.H. Torrie (1980). Principles and procedures of statistics. McGraw Hill Book C, New York, 633pp.

Urea molasses multi-nutrient block (UMMB) as a feed supplement: effect on reducing liveweight losses in yaks during the cold season of Qinghai-Tibetan Plateau, China

R.J. Long, S.K. Dong, X.H. Wei and J.P. Wu
Faculty of Grassland Science, Gansu Agricultural University, Lanzhou, 730070 China, Email: longrj@gsau.edu.cn

Keywords: livewight loss, yak, urea molasses multi-nutrient block

Introduction Urea molasses multi-nutrient block (UMMB) has been widely used as a feed supplement to balance nutrition of ruminants fed low-quality forage-based diets in tropical and sub-tropical areas. This study was carried out to determine the effects of UMMB in reducing liveweight loss of both yak cows and calves grazing on low-quality pastures during winter on the Qinghai-Tibetan Plateau, China.

Materials and methods The UMMB contains molasses, urea, bentonite, rape-seed meal, sesame-seed meal, dry hay meal, salt, wheat flour, wheat bran and mineral mixture in the proportions of 200, 100, 300, 100, 100, 90, 50, 30, 20 and 10 g/kg (crude protein, 36.1%). Three ages of yak - 1-year calves; 2-year calves and yak cows - were used. Each age had a control (C) group of 10 and a supplement (S) group of 20. Control yaks had grazing only, whilst S-yaks were given UMMB to lick *ad libitum* at night This gave average intakes of 150, 250 and 500 g/d for 1-, 2-year calves and cows respectively, from 15 Jan. to 9 May. All yaks grazed on the natural pasture during the day. Animals were weighed every 30 d before grazing in the morning. Data were analysed with SYATAT.

Results In January all animals gained in bodyweight, but rates of gain were higher for S than for C (Figures 1, 2 and Table1) (P<*0.01*). Thereafter the monthly body weight gain for all animals was negative, even when UMMB was offered, except for the 1-year calves of S-group in April. In all cases weight loss was lower for S than for C.

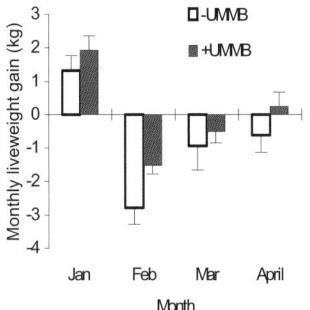

Figure 1 Monthly gain in body weight of 1-year calves with or without UMMB(±SE)

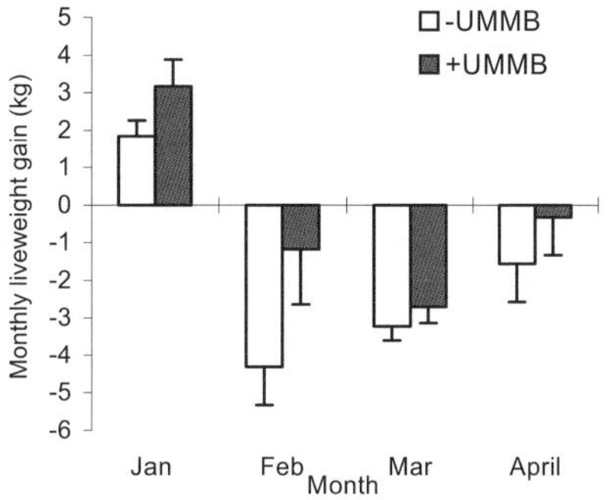

Figure 2 Monthly gain in body weight of 2-year calves with or without UMMB (±SE)

Table1 Monthly gain in body weight of cows supplementing with or without UMMB (kg±SE)

Month	Jan	Feb	Mar	Apr
-UMMB	0.74±0.51	-7.46±-2.13	-5.83±-0.61	-5.76±-1.55
+UMMB	1.81±0.37	-3.72±-1.25	-2.82±-1.14	-2.07±-0.54

Conclusions The effects of UMMB supplement in reducing body weight loss of yaks during the period with inadequate forage are important. Further research should focus on investigating optimum levels of UMMB intake for yaks and its effects on milk production and reproductive performance.

Acknowledgements The authors gratefully acknowledge financial assistance from ACIAR.

Reference
Long, R.J., D.G. Zhang, X. Wang, Z.Z. Hu & S.K. Dong (1999). Effect of strategic feed supplementation on productive and reproductive performance in yak cows. *Preventive Veterinary Medicine*, 38, 195-206.

New advance in forage production and dairy industry in China

Z.Q. Li and J.G. Han
Department of Grassland Science, College of Animal Science and Technology, China Agricultural UniversityNo 2, Yuan Ming Yuan Western Road, Hai Dian District, Beijing city, P. R. China 100094, Email: lizhiqiang8888@163.com

Keywords: forage, dairy cow

Introduction There were 6.9 million dairy cows in China in 2002 with milk yield of 13 million t. These values were 21.4 and 26.7% higher respectively than in 2001. Milk yield per cow was 1891 kg, 4.4 % higher than in the previous year, but still one third lower than the world average. Milk fat and protein contents were about 3.2 and 3.0 % respectively, a little below the world average. Milk consumption was 10.9 kg per person, much lower than the world average (97.6 kg) So, although great changes have taken place in dairy breeding, forage production and disease control, further improvements in milk yield and milk quality are badly needed.

New advances in forage production and the dairy industry

(1) Grassland improvement and forage-based diets: More and more attention has been paid to grassland improvement. Large quantities of lucerne hay and maize silage have been used since 2001. Before this year, maize straw and native chinese wildrye hay were the common forages fed to dairy cows, but they are much inferior to lucerne hay and maize silage in nutritive value and resulted in the low milk yield and poor quality. In 2002, the total area of improved grassland and artificial grassland is 24 million ha with lucerne occupying 2 million ha. A recent study (Li et al., 2003) has shown the optimal level of lucerne hay in the diet of high producing dairy cows in China to be 9 kg (Table 1).

(2) Total mixed rations (TMR): During 2001-2003, more than 200 TMR mixers were used in state owned dairy farms. Although still expensive, the use of TMR can produce more milk, improve milk quality, maintain the health of dairy cow, save labour and make the Chinese dairy industry more competitive.

Table 1 Effect of lucerne hay addition on milk production

	3kg group (control)	6kg group	9kg group
Concentrate (% of DM)	58.0	46.6	38.1
Maize silage (%DM)	29.4	18.2	8.1
Lucerne hay (%DM)	12.6	25.2	37.8
DDGS (%DM)	0	10.0	16.0
NEL (MJ/kgDM)	6.7	6.7	6.7
CP (%DM)	16.4	17.9	19.4
NDF (%DM)	37.8	37.8	37.8
Milk yield (kg/d)	29.06Bc	31.35ABb	33.55Aa
DMI (kg/d)	19.3b	20.0ab	20.4a
Fat (%)	3.7a	3.7a	3.6a
4% FCM (kg/d)	27.8B	29.9A	31.5A
Fat yield (kg/d)	1075b	1160a	1208a
Protein (%)	3.3a	3.3a	3.2a
Protein yield (g/d)	959C	1035B	1074A
Lactose (%)	5.00a	4.95a	5.03a
Dry matter (%)	12.7a	12.6a	12.4a
SCC (×103/ml)	162a	149a	86b

Note: The means in the row with different capital letter (small letter) indicate significant difference at 0.01 (0.05) level

References

Li, Z.Q., J G. Han, S.L. Li, X.S. Li, Y.Q. Huang, C.X. Zhang & Z Q. Wang (2003). [Study on the optimal amount of alfalfa hay addition in the diet of high producing dairy cow] (in Chinese). *China Agriculture Science*, 36, 950-954.

Productivity of Sahiwal and Friesian –Sahiwal crossbreds in marginal grasslands of Kenya

W.B. Muhuyi, F.B. Lukibisi and S.N. ole Sinkeet
National Animal Husbandry Research Centre, P.O. Box 25 Naivasha, Kenya, Email: karinaiv@kenyaweb.com

Keywords: Sahiwal, Friesian-Sahiwal crossbreds, grasslands, productivity

Introduction Dual-purpose cattle can be used to exploit the production potential of semi-arid grasslands of Kenya for milk and meat production. Although the Sahiwal is adapted to these grasslands, its productivity is low. In order to increase milk and meat productivity, the Sahiwal has been crossed with the Friesian to produce Friesian-Sahiwal crossbreds (McDowell *et al.*, 1996) adapted to the tropical environment. The objective of this study was to evaluate the productivity of Sahiwal and Friesian-Sahiwal crossbreds.

Materials and methods Data were obtained from Sahiwal and Friesian-Sahiwal crossbred cattle at Naivasha Research Centre, situated in agro-ecological zone IV. The average rainfall is 680mm per annum. Cattle grazed on natural pastures. The herd consisted of young and breeding cattle. Data on performance traits were analysed using a fixed effect model (Harvey, 1990) to obtain mean estimates. Measures of productivity were production efficiency (milk yield in kg/d of calving interval) and pry productivity index (total output value in Kenya Shillings per kg dry matter intake), derived using the Pry productivity model (Baptist, 1988).

Results Age at first calving was significantly different ($P < 0.01$) between the two genotypes, with mean values (± standard deviation) of 35.0±4.05 and 41.2±5.42 months for the Friesian-Sahiwal crossbreds and Sahiwal, respectively. The decrease in this parameter is similar to that reported in India (Bhat *et al.*, 1978). Similarly the calving interval was significantly lower ($P < 0.01$) among the crossbreds (400±61 days) compared to the Sahiwal cattle (446±105 days. Mean milk yield was 2,210±808.9 kg and 1,500±551.7 kg for the crossbred and Sahiwal cattle, respectively ($P < 0.01$) and the values for the crossbred are similar to those reported by McDowell *et al.* (1996). Productivity indices are different with the Friesian-Sahiwal crossbreds superior to the Sahiwal (Table 1).

Table 1 Productivity indices of the Sahiwal and Friesian-Sahiwal crossbreds

Productivity Index	Sahiwal	Friesian-Sahiwal
Production efficiency (kg/dci[1])	3.46	5.54
PRY Productivity index (Ksh*./kg DMI[2])	5.71	6.47

[1] Milk yield in kg/d of calving interval, [2] Total output value in Kenya shillings per kg of dry matter intake, * 1 US$ =75 Kenya Shillings.

Conclusion The crossbred genotype gave improved productivity for all the measured parameters. The Friesian - Sahiwal crossbreds, which combine the high production of the Friesian and hardiness of the Sahiwal, have high overall productivity. The utilisation of crossbreds is of economic importance for smallholder farmers, since the Friesian - Sahiwal is well suited to multiple uses in marginal areas.

Acknowledgement I am grateful to the Kenya Agricultural Research Institute for sponsorship.

References
Baptist, R. (1988). Herd and flock productivity assessment using the standard offtake and the demogram. *Agricultural Systems*, 28, 67-78.
Bhat, P.N., V.K. Taneja & R.C. Garg (1978). Effects of crossbreeding on reproduction and production traits. *Indian Journal of Animal Science*, 48, 71-78.
Harvey, W.R. (1990). Mixed model least squares and maximum likelihood computer program. Ohio State University, Columbus.
McDowell, R.E., J.C. Wilk & C.W Talbott (1996). Economic viability of crosses of *Bos taurus* and *Bos indicus* for dairying in warm climates. *Journal of Dairy Science*, 79, 1292-1303.

Comparison of a pasture-based system of milk production on a high rainfall, heavy-clay soil with that on a lower rainfall, free-draining soil

L. Shalloo[1,2,3], P. Dillon[1], J. O'Loughlin[1], M. Rath[3] and M. Wallace[2]
[1]Dairy Production Department, Teagasc, Moorepark Production Research Centre, Fermoy, Co. Cork, Ireland, Email: lshalloo@moorepark.teagasc.ie, [2]Department of Agribusiness, Extension and Rural Development, [3]Department of Animal Science, Faculty of Agriculture, University College Dublin, Belfield, Dublin 4, Ireland

Keywords: soil type, rainfall, grazed grass, costs, farm profit

Introduction Dairy farming in Ireland depends to a large extent on the efficient conversion of grass to milk (Dillon et al., 1995) and grass is the cheapest feed available on most dairy farms (O'Kiely, 1994). In Ireland, the two factors that have the largest influence on production from grassland farming are soil type and climatic conditions. The objective of this study was to compare the biological and economic efficiencies of milk production on a well-drained (infiltration 10mm/h) acid brown earth soil of a loam texture with medium rainfall of approximately 1000mm per annum (Moorepark site (MPN) 51^0N 8^0W) to a poorly-drained (infiltration 0.5mm/h), fine texture soil with a compact plastic sub-soil, with high rainfall approximately 1600mm per annum (Kilmaley site (KMY) 52^ON 8^OW).

Materials and methods The feeding systems at both sites were based on current best practice for seasonal spring-calving pasture based systems (O' Donovan 2000; O'Loughlin et al., 2001) which aim to maximise the amount of grazed grass in the diet. The biological information (milk yield and composition, liveweight, calving pattern, feed quality, and concentrate feeding level) were based on the averages for 1998, 1999 and 2000 for the two sites from cows of similar genetic potential for milk production. The Moorepark Dairy Systems Model (MDSM) (Shalloo et al., 2004), which is a stochastic budgetary simulation model, was used to compare the two sites. The farm size and milk quota were assumed to be 40 ha and 468,000 kg respectively with all costs and price based on Teagasc (2003).

Results Table 1 shows the milk production, feed budget and financial results for KMY and MPN for the three years of the study. Milk production was on average 640 kg higher at MPN, with a lower fat content of 1.0 g/kg and a higher protein content 0.3 g/kg. The total feed budget (kg/cow) was 115kg DM higher at the KMY site than at the MPN site. On average, 0.70 of the diet was grazed grass at the MPN site, while it was only 0.40 at the KMY site, with grass silage constituting 0.23 of the diet in MPN while the corresponding figure for KMY was 0.45. Total receipts and total farm costs were €1276 and €29,733 higher in KMY while farm net profit was €28,417 lower in KMY.

Table 1 Milk production, feed budget and financial results of the KMY - MPN comparison

	Milk yield kg	Fat g/kg	Protein g/kg	Total feed budget kg DM	Grass kg DM	Silage kg DM	Conc kg DM	Total receipt (€)	Total cost (€)	Net profit (€)
MPN	6421	38.6	33.8	5255	3590	1200	350	170,369	124,483	45,583
KMY	5781	39.6	33.5	5140	2121	2375	759	171,645	154,216	17,469

Conclusions The results indicate that the profitability of a spring-calving and grass based milk production system on a site with a lower rainfall and free draining soil is much greater than on a high rainfall and heavy clay soil. The higher profitability obtained on the site with low rainfall and free draining soil type was a result of lower cost of milk production and higher milk output per cow. The lower costs of milk production and the higher output per cow were directly related to the number of grazing days at both sites.

References

Dillon, P., S. Crosse, G. Stakulem & F. Flynn (1995). The effect of calving date and stocking rate on the performance of spring-calving dairy cows. Grass and Forage Science, 50, 286-299.
O'Donovan, M. (2000). The relationship between the performance of dairy cows and grassland management on intensive dairy farms in Ireland. PhD Thesis, University College, Dublin.
O'Kiely P. (1994). The cost of feedstuffs for cattle. R & H Hall Technical Bulletin No. 6.
O'Loughlin, J., F. Kelly & L. Shalloo (2001). Development of grassland management systems for milk production in wet land situations. Teagasc IE Project Report 4585.
Shalloo L., P. Dillon, M. Rath & M. Wallace (2004). Description and validation of the Moorepark Dairy Systems Model. Journal of Dairy Science. 87, 1945-1959.
Teagasc (2003) Management Data for Farm Planning.

The effects of a high grass input feeding system compared to high concentrate input feeding system offered to spring calving dairy cows in early lactation

E. Kennedy[1,2], M. O'Donovan[1], J.P. Murphy[1], F.P. O'Mara[2] and L. Delaby[3]
[1]Teagasc, Dairy Production Research Centre, Moorepark, Fermoy, Co. Cork, Ireland, Email: ekennedy@moorepark.teagasc.ie, [2]Faculty of Agri-Food and Environment, NUI Dublin, Belfield, Dublin 4, Ireland, [3]INRA, UMR Production du Lait 35590 St. Gilles, France

Keywords: dairy cows, early-turnout, spring grazing, total mixed rations

Introduction Grazed grass is the cheapest feed available on Irish dairy farms. The inclusion of grass in the diet of the spring-calving dairy cow in early lactation is recommended. Previous studies focused on introducing grazed herbage into the cow's diet in early spring in conjunction with grass silage and concentrate, and compared this to cows fed indoors. The objective of this study was to compare the milk production and feed budget of two contrasting early lactation feeding regimes. One regime was based on a high herbage inclusion with a low concentrate level (HG), while the other was based on a high concentrate inclusion with grass silage (HC).

Materials and methods Sixty-four spring calving dairy cows, (mean calving date - 2 Feb.) were randomised on lactation number, milk yield and composition, days in milk, bodyweight and body condition score. They were assigned to one of two feeding regimes. Each regime (n=32) consisted of 16 primiparous and 16 multiparous animals. The HC herd remained indoors and was offered a total mixed ration (TMR) consisting of grass silage in combination with a high concentrate level. Mean feed allowance was 19.6kg DM/cow per day. The HG animals were offered a high daily herbage allowance (DHA) of 15.1kg DM/cow (s.d. 3.7) above a height of 4cm with a mean concentrate allowance of 3.0kg DM/cow per day (s.d. 1.0). The study began in mid Feb. and finished in early April, lasting seven weeks. The stocking rate for the HG herd was 2.4 cows/ha. A fresh TMR mix was offered daily to the HC herd, refused feed was weighed and removed before each feeding. Both fresh and refused feed was sampled at each feeding. Pre- and post-grazing sward heights were measured daily while herbage mass and sward density were measured twice weekly. Milk yield was recorded daily and milk composition was determined weekly from samples collected over two consecutive morning and evening milkings. Cow dry matter intake (DMI) was measured once during the study using the n-alkane technique. Live weight was recorded weekly and body condition score was measured every three weeks. Milk production carryover effects of each of the feeding regimes were measured in a subsequent grazing study.

Results The TMR offered to the HC herd had a concentrate inclusion of 11.1kg DM/cow per day (s.d. 2.3) or 0.56 of the diet; grass silage constituted the remaining 0.44 of the diet, or 8.5kg DM/cow per day (s.d. 1.6). Table 1 summarises the effect of feeding regime on milk production: there was no significant effect of feeding regime on milk yield. The HG herd produced milk with significantly ($P<0.001$) higher protein concentration than the HC herd, however milk from the HC herd had a significantly ($P<0.001$) higher fat concentration. There was no significant difference in total DMI between treatments, the HC herd had a DMI of 15.3 kg DM/cow compared to an intake of 15.7 kg DM/cow for the HG herd. The HC herd had a higher live weight ($P<0.006$) than the HG herd, even though the HG herd had a numerically higher live weight gain. The HG herd maintained a significant difference ($P<0.01$) in milk protein concentration for the 12-week carryover period, there was no significant difference in any other production parameter.

Table 1 Effect of feeding regime on milk yield, milk composition, bodyweight and body condition score

	HC	HG	SED	Sig
Milk yield (kg/day)	27.3	28.3	0.72	NS
Fat (g/kg)	41.6	38.6	0.86	***
Protein (g/kg)	30.7	33.6	0.35	***
Lactose (g/kg)	48.7	49.0	0.28	NS
SCM yield	25.9	26.6	0.73	NS
Live weight (kg)	517.2	498.9	4.78	***
Live weight gain (kg/day)	+0.03	+0.20	+0.12	NS
BCS	2.92	2.87	0.035	NS

SCM=Solids Corrected Milk yield. NS=Non-significant, ***=$P\leq0.001$.

Conclusion Increased milk production performance can be achieved by offering a high DHA to spring-calving dairy cows in early lactation. Economic benefits are achieved by the production of milk with a high value combined with the efficient utilisation of a low-cost feed. The results of this study question the role of grass silage in the cow's diet in early lactation.

A comparison of a full time grazing and a partial storage feeding system, for dairy cows

C.P. Ferris and D.C. Patterson
The Agricultural Research Institute of Northern Ireland, Hillsborough, Co. Down BT26 6DR, UK, Email: conrad.ferris@dardni.gov.uk

Keywords: dairy cows, grazing systems, partial storage feeding

Introduction Partial storage feeding has been adopted by a number of Northern Ireland dairy farmers in recent years. This is due in part to increasing cow numbers, and as such, insufficient pasture close to the milking parlour to permit full time grazing. Partial storage feeding may also have environmental benefits, as well as reducing labour requirements associated with 'droving' and pasture management. In view of this, a study was undertaken to examine animal performance with either a full-time grazing, or a partial storage feeding regime.

Materials and methods Seventy-six Holstein-Friesian dairy cows, including 22 primiparous animals, were allocated to one of two treatments - full time grazing (FG) or partial storage feeding (PSF). Animals were 171 d calved when the study started on 5 May. With treatment FG, animals were given access to grazing both by day (morning through to evening milking) and by night (evening through to morning milking). With treatment PSF, animals grazed by day, and were housed by night. During the day the two treatment groups grazed perennial ryegrass swards separately within fixed paddock grazing systems, with FG paddocks 0.3 ha in size, and PSF paddocks 0.18 ha in size. Six grazing cycles were completed during the study, with these being 24, 21, 24, 24, 27 and 27 d in length respectively. Animals were given access to a fresh paddock following morning milking, while concentrates were offered in the parlour during milking at a flat rate of 3.0 kg/d. Animals on treatment PSF were offered grass silage during the period of night-time housing. The silage offered was of medium feed value, and had dry matter (DM), crude protein, and ammonia N concentrations of 257 g/kg, 117 g/kg DM, 114 g/kg total N respectively, and a DM digestibility (determined *in vivo*) of 706 g/kg. Silage was offered twice weekly alongside a feed barrier mounted on wheels. This allowed the cows to push the barrier out, while eating their way through the blocks of silage. Animals remained on the study for a mean of 127 (s.d., 22.8) d, being removed either eight weeks pre-calving, or on completion of the study on 29 Sept.

Results Mean stocking rates and residual sward heights were 5.2 and 8.6 cows/ha, and 6.3 and 6.7 cm, with treatments FG and PSF respectively. Throughout the season grazing conditions were good, and grass quality remained high. Total milk output and milk protein content was significantly higher with treatment FG than with treatment PSF (P<0.001), while milk fat content and somatic cell counts were unaffected by treatment (P>0.05). Although the differences were small, animals on treatment FG had significantly higher condition scores and live-weights at the end of the study, compared to animals on treatment PSF. The lower milk yield with treatment PSF conflicts with the findings of Ferris *et al.* (2003) who observed the opposite effect. This is probably due to differences in silage quality and grazing conditions between the two studies, with grazing conditions during the current study being ideal, while silage quality was medium, while the reverse was true in the study described by Ferris *et al.* (2003).

Table 1 Animal performance with a full time grazing system and a partial storage feeding system

	FG	PSF	s.e.m.	Significance
Daily milk yield (kg)	20.0	18.4	0.341	**
Milk fat (g/kg)	42.4	41.0	1.14	NS
Milk protein (g/kg)	35.8	33.4	0.20	***
Milk fat + protein yield (kg/day)	1.538	1.358	0.0413	***
Somatic cell count (log10)	2.22	2.16	0.083	NS
Condition score at end of study	2.6	2.5	0.02	**
Live weight at end of study (kg)	594	580	4.8	*

Conclusions Milk output and milk protein content were reduced with partial storage feeding in this study. The response to partial storage feeding appears to be largely influenced by the quality of the forage supplement on offer and the grazing conditions encountered.

Acknowledgements Funded by DARDNI, AgriSearch, John Thompsons and Son Ltd and Devenish Nutrition

References
Ferris, C.P.,J.P. Frost, R.C. Binnie & D.C. Patterson (2003). A comparison of full-time grazing and a part-grazing part-housing management regime for dairy cows. *Seventh Research Conference of the British Grassland Society*, 31-32.

Effect of a summer period at pasture on the performance of young bulls offered concentrates *ad libitum*

R.J. Fallon and M.G. Keane
Teagasc, Grange Research Centre, Dunsany, Co. Meath, Ireland, Email: rfallon@grange.teagasc.ie

Keywords: bulls, pasture, concentrates *ad libitum*, carcass

Introduction Holstein bull calves can produce a 250 kg carcass at 11 to 12 months of age on indoor feeding of *ad libitum* concentrates. An outdoor period at pasture would reduce housing costs. Two experiments were undertaken to determine the effect of an outdoor period at pasture on concentrate intake, liveweight gain and selected carcass traits of Holstein young bulls.

Materials and methods In year 1, thirty-six 14-week-old spring-born Holstein-Friesian bull calves (initial weight of 126 kg) were allocated on a liveweight basis to indoors on concrete slats for 250 d (In) or outdoors (Out) at pasture for 112 d and then indoors on slats for 138 d. Throughout the experimental period both treatment groups had *ad libitum* access to a rolled barley/soyabean meal ration (150 g/kg crude protein). At pasture the animals were stocked heavily to minimise grass consumption to that required to maintain rumen health and to maximise concentrate consumption. Thus, the pasture area was divided into three paddocks of 1500 m^2. Within each paddock, the space allowance was 83 m^2/animal. While indoors the animals had a daily barley straw allowance of 50 g/kg total dry matter intake and each animal had a pen space allowance of 2.5 m^2 in a naturally-ventilated house (6 animals/pen according to treatment). Water was available to all animals at all times. In year 2, thirty-six 12-week-old spring-born Holstein-Friesian bull calves with an initial weight of 114 kg were assigned to either indoors on concrete slats for 266 d (In) or to outdoors at pasture for 140 d (Out), followed by a period indoors of 126 d. Feeding and management of animals were similar to year 1.

Results Concentrate intake, liveweight gain and selected carcass traits for each of the years are presented in Table 1 There were no differences between the treatments in any of the measured traits.

Table 1 Effect of an outdoor period at pasture on the performance of young bulls offered *ad libitum* concentrates

	Year 1			Year 2		
	In	Out	sed	In	Out	sed
Initial weight (kg)	125	126	1.7	114	114	1.6
Outdoors (days)	0	112		0	126	
Concentrate intake (kg DM/d)						
Pasture period	5.0	4.6		5.1	4.6	
Indoor period	7.6	7.6		6.9	7.3	
Total period	6.3	6.2		6.0	5.9	
Liveweight gain (g/d)						
Pasture period	1390	1420	23	1260	1340	54
Indoor period	1180	1220	32	1240	1300	39
Total period	1270	1310	28	1250	1320	47
Final liveweight (kg)	442	452	7.5	447	464	6.4
Carcass weight (kg)	237	237	0.21	237	244	4.2
Kill-out (g/kg)	536	524	8.4	531	525	4.2
Kidney and channel fat (kg)	8.47	8.49	0.016	6.88	6.76	0.087
Conformation score[1]	1.89	1.97	0.059	1.85	1.90	0.042
Fat score[2]	3.08	3.47	0.27	3.18	2.94	0.036

[1]Based on E = 5, U = 4, R = 3 and 0 = 2; [2]Based on 1 (leanest) to 5 fattest.

Conclusion It was concluded that a period of 4 to 5 months at pasture did not have any negative effects on performance. Such an outdoor period reduces slurry storage capacity and handling costs and may improve animal welfare.

Spring calving suckler beef systems: influence of grassland management system on herbage availability, utilisation, quality and cow and calf performance to weaning

M.J. Drennan, M. McGee, S. Kyne and B. O'Neill
Teagasc, Grange Research Centre, Dunsany, Co. Meath, Ireland, Email: mdrennan@grange.teagasc.ie

Keywords: grassland management systems, suckler cow

Introduction Suckler beef systems in Ireland are primarily based on grass. Suckler systems vary in intensity but many operate low input systems and participate in REPS (Rural Environmental Protection Scheme). As there is a considerable cost associated with second-cut silage this research compared a two-cut system with a simplified low input one-cut system.

Materials and methods Data were collected over three consecutive years from two, rotationally grazed (mid-April to Oct./Nov.) systems using a total of 188 spring-calving Limousin × Friesian and Simmental × (Limousin × Friesian) cows and their progeny to weaning. The systems were (i) High (H); stocking rate (SR) of 0.77 ha/cow unit, 206 kg/ha nitrogen (N), two silage cuts and (ii) Low (L); SR of 0.95 ha/cow unit, 102 kg/ha N and one silage cut. Pre- and post-grazing sward heights and mass were measured using a rising plate meter and cutting (4 cm stubble height) and weighing strips (0.54 m x 4.5 to 5 m) of grass, respectively. Herbage yield and grass crude protein (CP) and dry matter digestibility (DMD) were determined in years 1 and 3.

Results There was no significant effect of grazing system on cow liveweight or body condition score changes or calf liveweight gains at pasture over the entire grazing season in any of the three years (Table 1). Pre-grazing heights were similar for both systems in the three years, but post-grazing heights (and yield) were lower (P<0.05) for H than L in year 1. There was no significant difference between systems in herbage DMD either pre- or post-grazing. In year 1 herbage CP was lower pre-grazing (n.s.) and post-grazing (P<0.01) and, in year 3 lower (P<0.001) both pre- and post-grazing for L than H (Figure 1).

Table 1 Cow liveweight and body condition score changes, calf liveweight gains and, herbage availability and *in vitro* digestibility for the High (H) and Low (L) grazing systems over three years

System		Year 1			Year 2			Year 3		
		H	L	s.e.	H	L	s.e.	H	L	s.e.
Cow weight gain (kg)										
	Turnout – June	26.8	33.4	9.6	58[a]	45[b]	4.3	46	58	5.4
	June – housing	31.0	31.5	8.7	26[a]	41[b]	4.2	37	36	3.4
	Turnout – housing	57.8	65.4	11.4	84	86	6.0	83	94	5.4
Cow body condition score change										
	Turnout - housing (units)	-0.02	-0.19	0.14	0.44	0.47	0.12	0.41	0.59	0.09
Calf weight gain (kg)										
	Turnout - housing	252	256	6.8	237	234	3.6	238	241	3.8
Grazing heights (cm)	Pre	12.1	12.6	0.48	11.4	11.4	0.22	11.6	10.9	0.24
	Post	5.7[a]	6.3[b]	0.17	5.6	5.8	0.10	6.3	6.2	0.11
Grazing mass (kg)	Pre	2022	2369	163.0	-	-	-	2325	2541	140.0
	Post	424[a]	555[b]	33.0	-	-	-	1005	1003	72.7
In vitro dry matter digestibility (g/kg)										
	Pre	750	764	10.4	-	-	-	761	747	5.7
	Post	674	655	14.0	-	-	-	640	641	7.3

* Columns, within year with different superscripts are significantly different, P<0.05

Conclusions Cow and calf performance at pasture was similar between the management systems. Grass DMD did not differ between the systems but CP levels were lower for L than H reflecting the lower N fertiliser application.

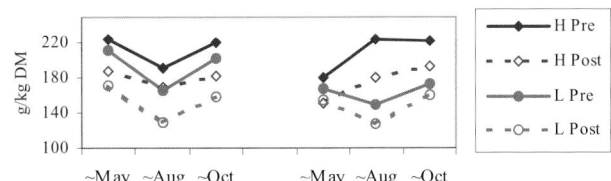

Figure 1 Pre- and post-grazing herbage crude protein for systems H and L in years 1 and 3

Effect of time spent on pasture on milk production and quality in Latxa dairy sheep

A. Garcia-Rodriguez, N. Mandaluniz and L.M. Oregui
NEIKER, A.B. - Granja Modelo de Arkaute, Apartado 46, 01080 Vitoria-Gasteiz, Spain, Email: loregi@neiker.net

Keywords: dairy ewes, milk yield, milk quality, grazing time

Introduction Dairy sheep production systems in the Basque country are pasture based. Forage resources are managed by shepherds who match herbage, forage and supplement availability with production requirements. This is achieved through a part-time grazing system, where ewes spend 3 to 7 hours grazing. The aim of this study was to evaluate the effect of restricted access to pasture on milk production and quality.

Materials and methods The experiment was conducted during the spring of four consecutive grazing seasons (2000 - 2003). Each year 48 multiparous Latxa dairy ewes were blocked into homogeneous groups of 12, on the basis of lactation number, day of lactation (DIM), milk yield (MY) and body weight (BW). Initial mean values for MY for the different years were respectively 1333, 1307, 1583 and 1818 ml d^{-1}, for BW 62.5, 68.4, 62.2 and 58.3 kg, and for DIM 28, 40, 42 and 44 days. Each year blocks were randomly assigned to one of the following experimental treatments: i) 4 hour-access to pasture (4H) or ii) 7 hour-access to pasture (7H). In 2001 all ewes were offered the same concentrate, while in 2003 ewes were offered concentrates with a different protein content (130 *vs.* 190 g kg^{-1} CP), and in 2000 and 2002, concentrates had a different degradability. Concentrate intake was measured for each ewe. Sward heights were maintained between 6 and 8 cm. After the evening milking each ewe was offered 250 g DM of lucerne hay. Grazing behaviour was recorded once weekly, with ewe activity assessed every 5 minutes. Milk yield for each ewe was recorded and collected once a week at morning and evening milking. Milk samples were analysed for protein and fat contents. Standardised milk yield (SMY) was calculated as described by Bocquier *et al.,* (1993). Data were analysed using the GLM procedures with year, concentrate, time on pasture, week and all possible interactions as fixed factors, and initial values as covariates.

Results With the exception of 2003**,** the interaction between supplement type and time spent on pasture was not significant. There were no significant differences between any of the variables in 2001. However, for each of the other years, time spent on pasture significantly increased MY (Figure 1), decreased milk fat content, and resulted in a significant increase in SMY (Figure 2). Although significant differences in total grazing time were observed (246 *vs.* 196 min/day, P<0.001), the impact of this on production was limited (<10%). This is probably due to non-significant differences on herbage intake (Perojo *et al.,* 2003).

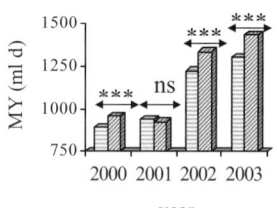

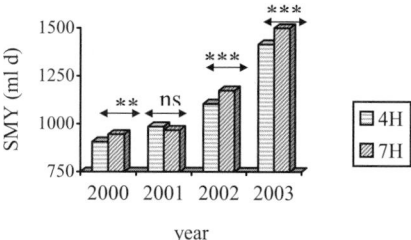

Figure 1 Effect of time spent on pasture (4H vs 7H) on milk production (ml d^{-1})
  *: P<0.05; **: P<0.01; ***:P<0.001

Figure 2 Effect of time spent on pasture (4H vs 7H) on standardised milk yield
  *: P<0.05; **: P<0.01; ***:P<0.001

Conclusion Increasing the time spent on pasture from 4 to 7 hours per day had a limited effect on milk yield and milk quality (10%), probably a consequence of a similar herbage intake.

References

Bocquier, F., T. Barillet, P. Guillouet, P. & M.. Jacquin. M. (1993). Prévision de l'énergie du lait de brebis à partir de différents résultats d'analyses: proposition de lait standard pour les brebis laitières. *Annales de Zootechnie,* 42, 57.

Perojo, A., A. Garcia-Rodriguez, J. Arranz, J. & L.M. Oregui (2003). Effects of time spent on pasture on milk yield, body reserves, herbage intake and grazing behaviour. In: *First joint seminar of the FAO-CIHEAM sheep and goat nutrition and mountain and Mediterranean pastures sub-networks. "sustainable grazing, nutritional utilization and quality of sheep and goat products".* 65 Granada, Spain.

Animal unit of sheep (Zel breed) grazing in Mazandaran grasslands in Iran

H. Arzani[1], R. Erfanzadeh[2] and S. Farazmand[2]
[1]College of Natural Resources, University of Tehran, Karaj- Iran, Email: harzani@ut.ac.ir, [2]College of Natural Resources, University of Tarbiat Modares, Iran

Keywords: animal unit, forage quality, phenological stage, grazing animal, Zel breed

Introduction Animal unit equivalents (AUEs), provide a means of summarising grazing capacity, calculating stocking rates and other stocking variables (Scarnecchia, 1990). Animal units (AU) are often defined in relation to the dominant animal type in an area In Iran there are 27 sheep breeds which are classified into three main body size classes, namely: small, medium and large. The objective of this experiment was to define animal units, animal and forage requirements for sheep grazing in Mozandaran grasslands.

Material and methods The experiment was conducted in a grassland area in Mazandaran province, north Iran, where the Zel breed of sheep is common. Five herds were randomly selected and in each herd five sheep from each of five age classes of 1, 2, 3, 4, and 5 years old were weighed at the beginning and the end of the grazing season. Three and six month old lambs and rams were also weighed. To determine forage quality, three samples from each edible species were collected at three phenological stages (rapid growth, flowering and maturity) and crude protein (CP), Acid detergent fibre (ADF), dry mater digestibility (DMD) and metabolizable energy (ME) were assessed according to Oddy et al. (1983). For determination of animal unit day requirement, and in view of topographic conditions, distances to watering points and villages and vegetation density, values calculated by the equation of ME= 1.8+0.1W (MAFF,1984), where W is live weight, were increased by 50%.

Results The average live weight of the five classes of sheep was 30.8 kg (Table 1). This was considered as 1 animal unit, with a requirement at grazing of 7.33.MJ ME per day. Based on forage quality (Table 2) and vegetation composition, the forage requirement to meet the requirement of 1 animal unit were 0.8, 0.91 and 0.96 kg DM at the phenological stages of vegetative (rapid growth), flowering and maturity respectively.

Table 1 Average live weight (kg) of different sheep classes of the Zel breed

End of Grazing	Beginning of grazing	Age (year or month)	Type of animal
24.83±5.77	27.16±2.47	1	Sheep
32.05±5.81	30.60±2.88	2	Sheep
32.97±3.67	31.46±3.46	3	Sheep
33.96±5.40	33.37±3.50	4	Sheep
31.50±5.98	30.26±2.50	5	Sheep
19.75±1.83		3	Lamb
23.53±3.60		6	Lamb
47.67±3.27		3&4	Ram

Table 2 The results of the Dankan test showing effects of species on quality

Species	ADF%	CP%	ME (MJ/Kg DM)
Stipa barbata	46a	7.7j	6.1f
Poterium sangoisorba	32ihg	8.9i	8.3b
Atriplex sp.	34fe	16.9c	8.4b
Festuca ovina	30jih	8.5i	8.4b
Bromus briziformis	31ba	8.7i	8.3b
Anthemis altissima	44d	8.5i	6.5e
Thymus kotschyanus	37gfe	9.8h	7.6c
Astragalus microcephalus	34hgf	13.9d	8.3b
Taraxcum officinali	32ji	14d	8.6b
Poa bulbosa	29j	8.2ji	8.6b
Medicago sativa	29j	23.3a	9.6a
Medicago coronata	26k	22.4b	9.9a
Artemisa aucheri	36ed	12.1f	7.9c
Stachys inflata	43cb	10.7g	6.9d
Agropyron tauri	38d	13.1e	7.7c
Achilea millefolium	41c	11.1g	7.1d

Discussion and conclusions A clear concept of an animal unit is basic to consideration of range capacity and utilisation (Scarnecchia, 1990). These results indicate that the Zel is a small breed of sheep with the adult ewe (1 animal unit) having average weight of 32 kg with three and six month lambs and rams being 0.65, 0.8 and 1.6 animal unit respectively. The stage of growth greatly affects nutritive levels in the forage plants. This influences the quantity of forage required to supply animal requirements, but, more importantly, low levels of intake mean that animal requirements are unlikely to be able to be met with mature forages.

References
MAFF (Ministry of Agriculture Fisheries and Food) (1984). Energy Allowance and System for Ruminants, Her Majesty's Stationery Office, London, Reference Book No. 43.
Oddy, V.H. G.E. Robards & S.G. Low (1983). Prediction of in vivo digestibility from the fibre nitrogen content of a feed. In: G.E. Robards & R.G. Pakham (eds.). Feed Information and Animal Production, Commonwealth Agricultural Bureaux, Australia, 395-398.
Scarnecchia, D.L. (1990). Concept of carrying capacity and substitution ratios: systems viewpoint, Journal of Range Management, 43, 553-555.

Mixed fattening of steers and lambs on improved grasslands in Uruguay: I. pasture performance

D.F. Risso, F. Montossi, E.J. Berretta, R. Cuadro, I. De Barbieri, R. San Julián, A. Dighiero and A. Zarza.
Instituto Nacional de Investigación Agropecuaria (INIA), Ruta 5 Km 386, Tacuarembó, Uruguay, E-mail: drisso@tb.inia.org.uy

Keywords: improved grasslands, mixed grazing, availability

Introduction The use of P fertilisers together with legume broadcasting is a low cost and high impact technology for improving native grassland (Risso *et al.*, 2001). Its use is increasing in Uruguay, although not for mixed grazing, even though this management is a common practice on native grasslands. Good pasture response may occur under mixed grazing when it is adequately managed (Nolan & Connolly, 1989). The following trials characterise pasture response with such management, in Uruguayan conditions.

Materials and methods The trials encompassed an average of 265 d per year (March-Nov.) for three years (2000-2002), in the granitic soils region of Uruguay. Type of pasture improvement (IT) and lamb/steer ratio (LSR) were evaluated. Two 9 to 11-year-old pastures were studied. These were oversown with: a) a mixture of *Trifolium repens* cv. LE Zapicán plus *Lotus corniculatus* cv. San Gabriel (WCL) or b) *Lotus subbiflorus* cv. El Rincón (LR). The LSRs evaluated were 1.5:1 (L) and 4:1 (H) in 2000 with these values increased to 4:1 (L) and 7:1 (H) in 2001 and 2002. A rotational paddock grazing system was applied, with 10 d of grazing and 20 d of resting. Average stocking rate was 470 kg LW/ha. The experimental design was a complete randomised block, with a factorial arrangement and two replicates. Pasture estimates were: sward mass at ground level (SM; kg DM/ha) before (BG) and after (AG) grazing, sward height (SH; cm) measured by ruler and spring legume percentage (LP).

Results No general treatment effects were detected, although high LSR resulted in significantly higher BG and AG, more frequently than did low LSR, possibly as a response to a differential increase in stocking rate, based on the higher proportion of steers gaining weight in both cases (Table 1). There was a slight trend for higher values of BG and AG in favour of WCL rather than LR. The tendency to lower LP (23% vs. 33%) when managing the high LSR, was probably due to lamb selectivity. Legume percent was also influenced by IT, with LR resulting in higher values than WCL (33% vs. 25%). Sward mass and SH were highly correlated (Table 2). The lower BG values per cm height during the autumn-winter period were the result of reduced fresh sward growth after the late summer grazing. The increase in BG per cm in spring was associated with a higher DM content in the swards.

Table 1 Sward mass (kg DM/ha) before (BG) and after (AG) grazing according to treatment

	Year	2000	2001	2002
BG	WCL H	1797a	1868	1326b
	WCL L	1502b	1931	1755a
	LR H	1675ab	1863	1378b
	LR L	1554b	1873	1236b
AG	WCL H	1147	1478	1315a
	WCL L	1046	1496	997b
	LRH	1144	1418	1121ab
	LR L	1028	1300	1047b

* Values with different letters within columns, are significantly different (P<0.05)

Table 2 Regression equations for predicting sward mass (kg DM/ha) before (BG) and after (AG) grazing using sward height (cm) (SH)

IT	Fall-Winter	R²	RSD	Spring	R²	RSD
WCL	BG=73.3SH+772	0.852	228	BG=119.4SH+461	0.846	366
	AG=136.5SH+370	0.895	148	AG=132.8SH+105	0.725	491
LR	BG=76.5SH+744	0.776	306	BG=142.8SH+467	0.868	368
	AG=134.9SH+328	0.928	109	AG=118.0SH+573	0.752	534

Conclusions Throughout the study period, both IT maintained their good condition. No consistent effect on pasture attributes was observed due to LSR at the ratios used. The high associations between SM and SH both before and after grazing indicated that SH provided a reasonable estimate of the seasonal pasture on offer in both IT.

References

Nolan, T. & J. Connolly (1989). Animal/vegetation relationships in mixed and mono grazing systems. *Proceedings General Meeting of the European Association for Animal Production*, 40, Dublin, Ireland.
Risso, D. F., Berretta, E.J. & G. Carracelas (2001). Productivity and composition of two improved native pastures under different grazing managements in Uruguay. *Proceedings Nineteenth International Grassland Congress*, 860-861.

Mixed fattening of steers and lambs on improved grasslands in Uruguay: II. animal performance and productivity

D.F. Risso, F. Montossi, E.J. Berretta, R. Cuadro, I. De Barbieri, R. San Julián, A. Dighiero and A. Zarza.
Instituto Nacional de Investigación Agropecuaria (INIA), Ruta 5 Km 386, Tacuarembó, Uruguay, E-mail: drisso@tb.inia.org.uy

Keywords: mixed fattening, steer, lamb, performance

Introduction In cow-calf operations in Uruguay, mixed cattle and sheep grazing on rangelands is predominant, while fattening is a specialised process. Within certain limits of the lamb/steer ratio and stocking rate, a complementary grazing effect occurs under mixed grazing, improving net results (Nolan & Connolly, 1977; Risso *et al.*, 2002). These trials characterise animal performance under such management.

Materials and methods Two types of pasture improvement (IT), *Trifolium repens* cv. Zapicán (WCL) and *Lotus subbliflorus* cv. El Rincón (LR) were studied, together with three lamb/steer ratios (LSR, 1.5:1, 4:1 and 7:1), for 265 d for three years (see Part 1). Average stocking rate was 470 kg live weight (LW)/ha although this varied between years as a result of different LSR (524, 563 and 356 kg LW/ha.). Twenty-eight steers (initial LW 326 kg) and 72 lambs (during 2000) or 148 lambs (2001 and 2002) were used. There were two cycles of lamb fattening (autumn-winter, C1, and early-late spring, C2) each year except that in 2002, only the low LSR was used in C2. Animals were weighed every 28 d and regression analyses were performed to estimate daily gains (DG). Live weight and wool/ha were calculated using DG, stocking rate and period length. The experimental design was a complete randomised block (two replicates), with a factorial arrangement combining IT and LSR.

Results Steer DG was high in all cases (Table 1). There was no consistent significant effect of LSR on DG, with the exception of year 2001. The highest DG in 2000 derived from the lower LSR in that year as well as from an intermediate stocking rate. The small difference in DG between the last two years, was caused by lower forage availabilities (see Part I). Lambs in C1 had a significantly (P<0.05) higher DG (89 g/d) on WCL, than on LR (74 g/d). In addition, low LSR resulted in significantly (P<0.05) better DG than high LSR (88 vs. 76 g/d), probably as a result of intraspecific competition with the higher number of lambs. Lambs of C2, had a significantly (P<0.05) lower DG on WCL. The DG averages for the three years were 164 and 173 g/d for WCL and LR respectively, probably resulting from the higher legume percentage with LR (see Part I). As in C1, low LSR resulted in better DG than high LSR (178 vs. 165 g/d), for the two years. A consistent tendency for higher production in WCL at the lower LSR

Table 1 Daily gain (g/d) of steers according to treatment and year

Treatment	2000	2001	2002
WCL-H	979	763a	772
WCL-L	938	684b	773
LR-H	915	736ab	735
LR-L	896	695ab	723

*Different letters within columns, are significantly different (P<0.05)
Note: L (Low LSR) and H (High LSR)

occurred, probably due to the higher steer density (Figure 1). The decrease in animal production with a higher stocking rate was associated with lower DG.

Conclusions Both IT resulted in good production in mixed fattening production systems, with no consistent effect of the ration of lambs to steers over the range studied.

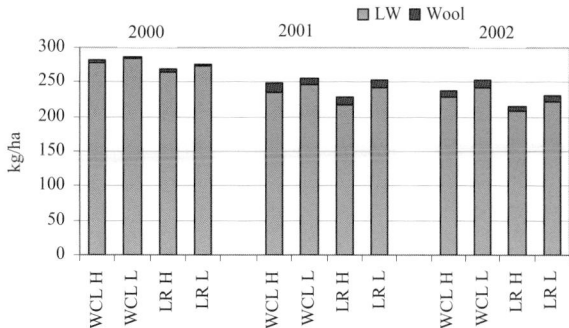

Figure 1. Live weight wool production (kg/ha) due to IT and LSR per year

References

Nolan, T. & J. Connolly. (1977). Mixed stocking by sheep and steers: a review. *Herbage Abstracts*, 47, 367-374.

Risso, D. F., F. Montossi, E. J. Berretta, R. Cuadro, I. De Barbieri, R. San Julián, A. Dighiero & A. Zarza (2002). Comportamiento de novillos en engorde y persistencia productiva de las pasturas. In: D.F. Risso & F. Montossi (eds.). Mejoramientos de campo en la región de Cristalino: Fertilización, Producción de carne de calidad y persistencia productiva. INIA Tacuarembó, Serie Técnica 129, Ed. Hemisferio Sur, Montevideo, 45-58.

The Rostock energetic feed evaluation on the base of net energy

W. Jentsch, A. Chudy, M. Beyer and S. Kuhla
Research Institute for the Biology of Farm Animals, Research Unit Nutritional Physiology "OSKAR KELLNER", 18196 Dummerstorf, Germany, Email: kuhla@fbn-dummerstorf.de

Keywords: feed evaluation, metabolisable energy, net energy

Introduction Feed evaluation was an emphasis of research from the foundation of the "Oskar-Kellner-Institute of Animal Nutrition" in 1953 at Rostock by Prof. Kurt Nehring. The aim of the research work was the elaboration of a feed evaluation system containing reference numbers of feed values and requirements of farm animals. The approach and the present system are outlined in this paper.

Materials and methods By means of indirect calorimetry using respiration chambers for cattle (4), sheep (2), pigs (4) and fowl (6) many metabolic experiments were carried out at different levels of nutrition measuring digestibility of energy (DE) and nutrients, metabolisable energy (ME) and net energy (NE) of 110 and 92 cattle and pig rations respectively (Schiemann *et al.*, 1971). On the basis of the experimental results, equations for the estimation of energy reference numbers of feedstuffs were calculated by multiple regression analysis (Table 1).

Results Net Energy Retention (NER) was taken as a uniform measure for energetic feed evaluation and characterisation of energy requirement of farm animals. The NER is an experimentally measurable and universal measure of the biologically utilisable feed energy and correlates closely with ATP potential of nutrients and feedstuffs as well as with energetic potential for all energy consuming reactions of the animal body and reflects the energy requirement for different levels of performance (Schiemann *et al.*, 1971).

Table 1 Equations for estimation of Net Energy Retention (NER) in kJ for cattle (NERc) (1) and pig (NERp) (2) (Collective of authors, 2003)

(1) $NERc = (7.2dCP + 20.0dCF + 10.1dST + 8.3dSU + 8.2dNFR)(-0.5574 + 0.0405DE - 0.0002633DE^2)$	$\pm\ 4.5\ \%$
(2) $NERp = 11.0dCP + 27.0dCF + 12.7dST + 11.6dSU + (12.0 - 0.14(80 - DE))\ dNFR$	$\pm\ 4.0\ \%$

(d=digestible, C=crude, P=protein, F=fat, ST=starch, SU=sugar, NFR=N-free residue, DE=energy digestibility)

Net energy retention is dependent on animal species (Table 1), but results in sheep were similar to those in cattle, so that energetic values of feedstuffs are subdivided into three tables: cattle in NERc, pigs in NERp and fowl in NERf (equation not shown). The energy requirements are tabulated for different species groups (Collective of authors, 2003): Cattle, sheep, goat and horse in NERc, Pig, rabbit, and fur-bearing animals in NERp, Fowl, all species of poultry and fish in NERf.

Energetic feed values at the maintenance level for examples of feedstuffs in NERc, ME, and Net Energy Lactation (NEL) and relative feed values are given in Table 2.

Table 2 Comparison of feed values for cattle, relations and utilisation of ME between different systems

Feedstuff	Feed value			Relative feed value			Utilisation of ME	
	NERc MJ	ME MJ	NEL MJ	NERc Barley (= 100)	ME Barley (= 100)	NEL Barley (= 100)	NERc/ ME %	NEL/ME %
Barley, corn	7.84	12.54	8.44	100	100	100	63	67
Maize silage, milk-wax ripeness	5.89	10.25	6.62	75	82	78	58	65
Meadow hay, full bloom	3.77	7.87	4.79	48	63	57	48	61
Oats straw	2.42	6.01	3.61	31	48	43	40	60

Conclusions Feed evaluation on the basis of NER is superior to NEL and ME because it provides greater differentiation between feedstuffs in their energetic values. The system was tested in farms specialised in milk production, cattle breeding, sheep breeding and fattening. The system has been used with high success in feed production, feeding, economic evaluation and feed planning in many farms.

References

Collective of authors (2003). Rostock Feed Evaluation System. Plexus Verlag, Miltenberg-Frankfurt, 392 pp.
Schiemann, R., K. Nehring, L. Hoffmann, W. Jentsch & A. Chudy (1971). Energetische Futterbewertung und Energienormen. VEB Deutscher Landwirtschaftsverlag Berlin, 340 pp.

Constraints on dairy cattle production from locally available forages in Bangladesh

M.A.S. Khan
Department of Dairy Science, Bangladesh Agricultural University, Mymensingh 2202, Bangladesh, Email: s-khan@royalten.net

Keywords: constraints, protein, parasitism, rice straw, grass

Introduction The productivity of milk producing animals in Bangladesh is low because of low individual yield and poor fertility. The reasons for the low productivity are complex but, in order of priority, appear to be (a) the imbalanced nature of the nutrients that arise from the digestion of the forage resources, (b) the incidence of disease/parasitism, and (c) the often harsh climatic circumstances. Thus, the purpose of this study was to find out the practical constraints on dairy cattle production from locally available forages under small holding village conditions of Bangladesh.

Materials and methods One typical village – Boira, which is about 2 km from Bangladesh Agricultural University, Mymensingh, was chosen for this study. The villagers were mostly resource-poor farmers. The cows of the villagers were used for multipurposes such as draught, dairy and meat. Rice straw was the main roughage source for the animals. Limited seasonal cut and carry grasses were also used. Animals were mainly stall fed. Sixty seven post-partum cows were taken from 65 smallholder farms. Nutrient metabolites were measured by FAO/IAEA Nutritional Metabolite kits according to Kaneko (1989) and concentration of milk progesterone (P4) for calving to first ovulation were measured using the solid-phase Radioimmunoassay (RIA) kits supplied by FAO/IAEA according to Plaizier (1993).

Results Among the blood metabolites studied, a considerable change in plasma urea values were noticed as shown in Table 1. It represents an important nutritional constraint to productivity. The spring value was low enough to suggest a shortage of RDP in the rations at that time. The calving to first service was higher than that of calving to first ovulation as shown in Table 2. It means that farmers were unable to detect heat of their cows at the proper time. The calving to first service interval was lower in autumn than in the other three seasons. It would be tempting to relate this to the urea levels which were highest in the autumn. Average parasitic egg counts were 54/g and were mainly of *Fasciola gigantica.*

Table 1 Group metabolite means within each season

Seasons	BHB (mmol/L)	SEM	Globulin (g/L)	SEM	Albumin (g/L)	SEM	Urea (mmol/L)	SEM	Pi (mmol/L)	SEM
Summer	0.342[a]	0.05*	37.00[a]	7.40**	38.00[c]	2.34**	4.52[b]	0.98**	1.33[c]	0.12*
Autumn	-------	------	36.67[ab]	7.80**	33.67[c]	2.78**	7.20[a]	0.09*	1.61[a]	0.12*
Winter	0.362[a]	0.06*	--------	-----	35.00[b]	2.74**	5.46[b]	0.93**	1.42[b]	0.09*
Spring	0.292[b]	0.04*	35.00[a]	5.00**	38.18[a]	3.85**	3.25[c]	1.08**	1.46[b]	0.13*

*p<0.5; **p<0.01, [abc]Figures with dissimilar superscript in the same column differ significantly (p<0.05)

Table 2 Reproductive intervals by seasons (days)

Seasons	Calving to 1st ovulation	SEM		Calving to 1st service	SEM		Calving to conception	SEM		Calving interval	SEM	
Summer	66	42	NS	272	147	NS	283	148	NS	544	162	NS
Autumn	67	25	NS	120	60	NS	136	67	NS	419	72	NS
Winter	187	105	NS	191	81	NS	197	92	NS	489	84	NS
Spring	51	14	NS	216	38	NS	223	38	NS	501	39	NS

NS p>0.05

Conclusions Feed protein deficiency, improper heat detection and parasitic infestation were the constraints on dairy cattle production from locally-available forages in Bangladesh.

References
Kaneko, J. (1989). Biochemistry of Domestic Animals. Third ed., Academic Press, New York.
Plaizier, J.C.B. (1993). Validation of the FAO/IAEA RIA kits for the measurement of progesterone in skim milk and blood plasma. In: Improving the Productivity of Indigenous African Livestock, IAEA-TCDOC-708, IAEA, Vienna. 151-156.

Effects of supplementary concentrate level and separate or mixed feeding of grass silage and concentrates on rumen fluid composition in steers

J. Caplis[1], M.G. Keane[1] and F.P. O'Mara[2]

[1]Teagasc, Grange Research Centre, Dunsany, Co. Meath, Ireland, Email: gkeane@grange.teagasc.ie,
[2]Department of Animal Science and Production, University College, Belfield, Dublin 4, Ireland

Keywords: concentrates, diet mixing, rumen fluid, steers

Introduction The effects of dietary concentrate level and method of feeding (separate or mixed) on performance and carcass traits of steers may be mediated through changes in rumen fluid composition. The objectives of this study were to determine the effects on rumen fluid composition of (1) supplementary concentrate level with grass silage, and (2) separate or mixed feeding of silage and concentrates,

Materials and methods In a 5 x 5 latin square design experiment, five rumen cannulated Friesian steers were offered five feeding treatments for five periods of 28 d each. The feeding treatments were:
1. Silage only (SO)
2. Silage + 3kg concentrate dry matter (DM) per day fed separately (LS)
3. Silage + 3kg concentrate DM per day fed mixed by feeder wagon (LM)
4. Silage + 6kg concentrate DM per day fed separately (HS)
5. Silage + 6kg concentrate DM per day fed mixed by feeder wagon (HM)

Rumen fluid samples were collected on day 28 of each period in the morning immediately before feeding and at 1, 2, 4, 8, 14 and 24 h after feeding. The pH, ammonia and volatile fatty acid (VFA) concentrations were measured.

Results There was no significant feeding treatment by sampling time interaction. Mean rumen pH was significantly lower for the high concentrate level than for silage only (Table 1). Differences between treatments in ammonia concentration were not significant but the silage only treatment had the lowest value. Total VFA was significantly higher for the high concentrates fed separately than for silage only but other differences between treatments were not significant. The acetate to propionate ratio tended to be lower for the silage only than for the concentrate supplemented groups. There was no effect of mixing but the acetate to propionate ratio tended to be lower for mixed compared with separate feeding.

Table 1 The effect of concentrate level and separate or mixed feeding on rumen fermentation variables

	Treatment					s.e.d.	Sig
	SO	LS	LM	HS	HM		
pH	6.81[b]	6.64[ab]	6.55[ab]	6.38[a]	6.48[a]	0.121	*
Ammonia[1]	12.60	15.10	14.30	15.79	13.19	1.685	NS
Total VFA[2]	85.6[a]	91.3[ab]	98.8[ab]	104.5[b]	94.5[ab]	8.53	*
Acet:prop. ratio[3]	3.58	4.12	3.82	4.14	3.99	0.191	P<0.07

[1]mg/1; [2]mmol/1; [3]acetate:propionate ratio

The effects of sampling time are shown in Table 2. There was a decrease in pH after feeding for 8 h, and then an increase to 24 h. Ammonia and VFA increased for 2-4 h after feeding and then decreased to 24 h. Acetate to propionate ratio decreased up to 8 h and then increased to 24 h.

Table 2 The effect of sampling time on rumen fermentation variables

	Time(h)							s.e.d	Sig
	0	1	2	4	8	14	24		
pH	6.80[a]	6.66[ab]	6.44[b]	6.28[bc]	6.24[bc]	6.45[b]	7.14[d]	0.143	***
Ammonia[1]	7.67[a]	14.34[b]	20.02[c]	18.85[c]	17.85[bc]	13.03[b]	7.59[a]	1.994	***
Total VFA[2]	75.8[a]	94.7[ab]	107.7[b]	108.9[b]	105.5[b]	101.8[b]	70.1[f]	10.10	***
Acet:prop. ratio[3]	4.60[a]	3.92[b]	3.63[bc]	3.67[bc]	3.37[c]	3.68[bc]	4.62[a]	0.226	***

[1]mg/1; [2]mmol/1; [3]acetate:propionate ratio

Conclusions Rumen pH decreased as concentrate level increased but ammonia concentration was not affected. Mixing had no effect on rumen pH or total VFA concentration. All rumen fermentation variables varied with time of sampling.

Body temperature in free-roaming beef cattle

T.L. Mader and S.L. Colgan
*University of Nebraska, Haskell Agricultural Laboratory, 57905 866 Road, Concord, Nebraska, USA 68728,
Email: tmader1@unl.edu*

Keywords: body temperature, metabolisable energy, cattle

Introduction Body temperature (BT) measurements are traditionally used in diagnosing sick animals, but may also be used as an indicator of stress or activity. Based on results of metabolism studies, Mader *et al.* (1999) reported that BT can vary as much as 0.9°C and can depend on metabolisable energy (ME) of the diet consumed. Acceptable measures of BT can be obtained from the rectum, vagina, or ear canal. Technologies are also being developed for continuously monitoring BT via radio-telemetry. The objectives of this study were to determine the effect of high concentrate (low fibre) versus high fibre diets on BT, assess the capabilities of obtaining BT in free-roaming cattle, and compare temperatures taken in the rumen with vaginal and tympanic temperatures.

Materials and methods Tympanic, vaginal, and ruminal temperatures were obtained from four, 400 kg crossbred heifers over two, four-day periods while being provided either a 10 MJ/kg high fibre (forage based) or a 13 MJ/kg, low fibre (non-forage based) diet. Heifers had a 400 m^2 roaming area and were allowed a ten-day acclimation period between periods before temperatures were monitored. Tympanic temperatures were recorded at 15-minute intervals using a Stowaway XTI® data logger (Onset Corporation, Pocasset, MA) and thermistor. The thermistor was inserted into the ear canal until the tip was near the tympanic membrane. Loggers, with thermistors attached, were secured to the inside of the ear. Vaginal and ruminal temperatures were recorded at hourly intervals using an ETD Bolus™ (CowTek, Inc., Santa Clarita, CA). Boluses were inserted into the rumen using a balling gun. Boluses were hand-placed inside the vagina, immediately behind the cervix. Boluses were activated via wireless signal.

Results To negate heat stress effects, the study was conducted during the winter in which temperatures averaged near 0°C. Over the two study periods, ruminal temperatures averaged 0.7°C greater (P < 0.05) than tympanic and vaginal temperatures (Table 1). Cattle consuming the high fibre diet had mean temperatures of 0.4°C (38.87 vs 39.27°C) lower (P < 0.05) than cattle consuming the low fibre diet. No diet by temperature measuring location interactions were found. However, diet by time of measure interactions were evident (P < 0.10) with mean temperature of cattle consuming the high fibre diet being 0.4°C lower (38.7 vs 39.1°C; P < 0.05) in late morning (09.00 h) versus late evening (21.00 h) while cattle consuming the low fibre diet had only a 0.1° lower (39.2 vs 39.3°C) BT in the morning. Although feed intakes are not known in the current study, the BT data agree with data previously reported by Mader *et al.* (1999) which were obtained from cattle fed in metabolism stalls. In these studies, cattle fed low fibre diets had 0.4°C greater BT during the 4-day hot period than cattle fed diets containing intermediate fibre levels, even though ME intakes for both groups, prior to the hot period, were equivalent at 82 MJ/head per day.

Table 1 Location of measurement and fibre level effects on mean daily body temperature[*]

| Diet: | Low Fibre | | | High Fibre | | |
Location:	Tympanic	Vaginal	Ruminal	Tympanic	Vaginal	Ruminal
Temperature, °C	39.0	39.0	39.8	38.6	38.7	39.3

[*]Fibre levels differ (P < 0.05); tympanic and vaginal measures differ from ruminal measures (P < 0.05)

Conclusion Tympanic, vaginal, and ruminal temperatures were all found to be acceptable measures of BT in free-roaming cattle, however ruminal temperatures would overestimate core BT. The greater ruminal temperatures can most likely be attributed to fermentation effects. Cattle BT can be influenced by diet ME content. Cattle consuming diets with higher ME are likely to have greater core BT, which will probably influence their susceptibility to health and climatic heat stress challenges.

References

Mader, T.L., J.B. Gaughan & B.A. Young (1999). Feedlot diet roughage level for Hereford cattle exposed to excessive heat load. *The Professional Animal Scientist*, 15 (1), 53-62.

Effect of applied biosolids to bahiagrass pastures on copper status of cattle

M.E. Tiffany, L.R. McDowell, G.A. O'Connor and N.S. Wilkinson
University of Florida, 125 Animal Science Bldg., P O Box 110910, Gainesville, FL 32611, USA, Email: McDowell@animal.ufl.edu

Keywords: biosolids, bahiagrass, copper deficiency, cattle

Introduction When grazing ruminants consume forages high in Mo but adequate in S, there is a risk of molybdenosis (a Mo-induced Cu deficiency). This occurs when Mo, S, and Cu join to form Cu-thiomolybdate complexes in the rumen that are not readily absorbed (Suttle, 1991). High dietary S reduces Cu absorption, possibly due to unabsorbable Cu sulphide formation, independent from its part in thiomolybdate complexes. The use of municipal sludge (biosolids) as a pasture fertiliser is of interest since some contain high Mo which may induce Cu deficiency. The objective of this study was to evaluate the performance and Cu status of cattle grazing pastures fertilized with biosolids.

Materials and methods Angus yearling steers (n = 96) were randomly assigned to bahiagrass (*Paspalum notatum* flugge) pasture (n = 32) treated with three high Mo-containing biosolids varying in mineral content for 151 d and evaluated for Cu status. Soils were well drained and acid (pH 5.0 to 5.8). The biosolids are classified as high quality and originated from Tampa and Largo, Florida and Baltimore, MD. Copper concentrations of biosolids varied from 431 to 989 pm and Mo from 12 to 60 ppm. Biosolids and NH_4NO_3 (control) fertiliser were applied to 0.81-ha pastures at a rate of either 179 kg N/ha (X) or 2, 3 and 6 X, for a total of six treatments. The treatments were control, L1x, L2x, B3x, B6x, T3x and T6x. There were three animals per pasture with from four to six replicates of treatments. One of three steers of each plot received a 3-ml subcutaneous injection of Cu glycinate (60 mg Cu/ml).

Results Weight gains were independent of treatments, and did not reflect potential Cu deficiencies. Forage Cu was low (5.0 to 8.0 ppm), Mo was low (<1.0 ppm) and S was high (0.4 to 0.45%). Forage Cu was less than NRC requirements of 10 ppm (McDowell, 2003) at termination. Mean liver Cu concentration for control animals was 110 ppm, compared to mean range of 29 to 83.4 ppm for biosolids treatments (Table 1). Plasma Cu was likewise less than control animals. Liver Cu was dramatically higher (<0.001) for Cu glycinate treatments.

Table 1 Plasma and liver concentrations as affected by biosolids treatments

Treatments[a]	Liver Cu (ppm, DM)		Plasma Cu (μg/ml)	
	Day 1	Day 151	Day 1	Day 151
Control 1X	84.2 ± 37.9	110.6 ± 33.6[cde]	1.21 ± 0.08[bc]	0.68 ± 0.09[c]
Largo 1X	86.0 ± 43.8	63.2 ± 38.8[cd]	1.13 ± 0.09[b]	0.55 ± 0.11[bc]
Largo 2X	80.9 ± 43.8	83.4 ± 38.8[cd]	1.17 ± 0.09[bc]	—
Baltimore 3X	69.5 ± 33.9	61.8 ± 30.1[cd]	1.37 ± 0.07[bcd]	0.47 ± 0.08[bc]
Baltimore 6X	84.5 ± 33.9	40.3 ± 30.1[bc]	1.29 ± 0.07[bcd]	0.50 ± 0.08[bc]
Tampa 3X	105.3 ± 31.0	29.0 ± 27.5[b]	1.32 ± 0.07[bcd]	0.37 ± 0.08[b]
Tampa 6X	87.5 ± 31.0	41.0 ± 27.5[bc]	1.29 ± 0.07[bcd]	0.51 ± 0.08[bc]

[a]Data represent treatment means and standard errors. Means with different superscripts are significantly different at P<0.05

Conclusions Pastures treated with high Mo containing biosolids resulted in a decline of Cu status of steers. The decline in Cu status was not due to Mo, but most likely from low forage Cu (<6 ppm) and high forage S (> 0.4%).

References

McDowell, L.R. (2003). Minerals in Animal and Human Nutrition, 2nd ed., Elsevier, Amsterdam.
Suttle, N.F. (1991). The interactions between copper, molybdenum and sulfur in ruminant nutrition. *Annual Review of Nutrition*, 11, 121-140.

Theme A: Efficient production from grassland

Section 3

Improving quality of products from grassland

An *in vitro* investigation of forage factors which affect the production of conjugated linoleic acid and *trans* vaccenic acid in the rumen. I. Grass species

M.R.F. Lee[1], C. Hodgkins[2], J.K.S. Tweed[1], N.D. Scollan[1] and R.J. Dewhurst[1]
[1]*Institute of Grassland and Environmental Research, Plas Gogerddan, Aberystwyth, Ceredigion SY23 3EH, UK,
Email: michael.lee@bbsrc.ac.uk, *[2]*Institute of Rural Studies, University of Wales Aberystwyth, Llanbadarn
Campus, Aberystwyth SY23 3AL, UK*

Keywords: conjugated linoleic acid, grass species, *trans* vaccenic acid, biohydrogenation

Introduction Extensive, pasture-based systems appear to offer the most cost-effective and natural means of *cis* 9 *trans* 11 conjugated linoleic acid (CLA) levels of milk fat. However, since most regions in Europe rely increasingly on conserved forages for winter feeding of lactating animals, it is necessary to develop feeding systems for both fresh and conserved forage diets. An understanding of the mechanisms that cause the differences in CLA response to conservation is an essential pre-requisite to this task. This study investigated whether grass species affected CLA and *trans* vaccenic acid (TVA) production in an *in vitro* system.

Materials and methods Four week regrowths of eight different grass species were evaluated on two separate occasions: Diploid PRG (DPR); Tetraploid PRG (TPR); Italian ryegrass (IRG); Hybrid ryegrass (HRG); Timothy (TIM); Cocksfoot (COC); Meadow fescue (MF) and Tall fescue (TF). Fresh grass was cut from experimental plots (*circa.* 100grams FW) 3cm above soil level. The tissue was then crushed and cut into 5mm strips. Two and a half grams of fresh material was then loaded under CO_2 into incubation bottles containing 10ml of strained rumen liquor and 10ml of Van Soest medium, the head space was gassed and the bottles sealed. Three bottles were used for each treatment in the first period and two in the second period. The bottles were incubated for 6 h in the dark at 39°C. At the end of the incubation the bottles were removed and 25ml of isopropanol: chloroform (1:1 v/v) was added along with 1ml of internal standard (2.5mg C19:0 / ml chloroform). This was then blended and the lipid extracted as described by Lee *et al.* (2004) and bimethylated as described by Kramer & Zhou (2001). Biohydrogenation of C18:2 and C18:3 were calculated as proportional loss of these fatty acids during the incubation. Genotypic differences in the concentrations of CLA, TVA and biohydrogenation were calculated using an unbalanced ANOVA with genotype × period as the treatment.

Results The production of CLA and TVA (as a proportion of C18 polyunsaturated fatty acid (PUFA) supply) and the extent of C18:2 and C18:3 PUFA biohydrogenation across the seven genotypes are shown in Table 1. Both Cocksfoot and Timothy produced significantly lower levels of CLA and TVA than the other genotypes, whilst Italian ryegrass resulted in the greatest formation of these biohydrogenation intermediates. Biohydrogenation of both C18:2 and C18:3 were lowest for Cocksfoot and highest for meadow fescue. The relatively large S.e.d.s, particularly in regards to CLA, were due to large variation between periods.

Table 1 CLA and TVA production (as a proportion of C18 PUFA supply; g/g) and the extent of C18 PUFA biohydrogenation (g/g)

	COC	DPR	HRG	IRG	MF	TF	TIM	TPR	S.e.d	Sig.
CLA x 10^3	0.44^a	0.75^{bc}	0.89^{bc}	1.39^d	0.63^b	0.95^c	0.52^{ab}	0.95^c	0.135	***
TVA	0.14^a	0.19^b	0.21^c	0.22^c	0.19^b	0.21^c	0.12^a	0.17^b	0.011	***
Biohydrogenation										
C18:2	0.43^a	0.53^{abc}	0.57^{bc}	0.52^{bc}	0.59^c	0.47^{ab}	0.54^{abc}	0.52^{abc}	0.053	*
C18:3	0.52^a	0.60^b	0.67^{cd}	0.64^{cd}	0.68^d	0.59^{bc}	0.55^{ab}	0.63^{bc}	0.039	*

Conclusions The lower concentration of CLA and TVA with Cocksfoot may be due to the lower level of C18 PUFA biohydrogenation, although this does not explain the lower concentration of these fatty acids in Timothy. Species (genotype) appears to have an effect on the production of CLA and TVA, although numerically the differences were small and period/stage of growth had a dramatic effect on levels of production. The experiment also shows the low levels of *cis* 9 *trans* 11 CLA which are produced in the rumen and hence the importance of TVA in the formation of CLA in milk.

References

Kramer, J.K.G. & J. Zhou (2001). Conjugated linoleic acid and octadecenoic acids: extraction and isolation of lipids. *European Journal of Lipid Science and Technology*, 103, 594-632.
Lee, M.R.F., A.L. Winters, N.D. Scollan, R.J. Dewhurst, M.K. Theodorou & F.R. Minchin (2004). Plant-mediated lipolysis in red clover with different polyphenol oxidase activities. *Journal of the Science of Food and Agriculture*, 84, 1639-1645.

An *in vitro* investigation of forage factors which affect the production of conjugated linoleic acid and *trans* vaccenic acid in the rumen. II. Wilting & cell damage

M.R.F. Lee, J.K.S. Tweed, N.D. Scollan and R.J. Dewhurst
Institute of Grassland and Environmental Research, Plas Gogerddan, Aberystwyth, Ceredigion SY23 3EH, UK,
Email: michael.lee@bbsrc.ac.uk

Keywords: conjugated linoleic acid, *trans* vaccenic acid, biohydrogenation, cell damage, wilt

Introduction Previous studies have shown that *cis* 9 *trans* 11 conjugated linoleic acid (CLA) concentrations are higher for summer milk produced from cows grazing fresh pastures than for winter milk when conserved forages are fed (Jahreis *et al.*, 1997). Furthermore, Offer (2003) showed that a similar depression in milk fat CLA occurred if grass was simply cut and fed after a short wilt. This experiment investigated the effect of wilting on the production of CLA and *trans* vaccenic acid (TVA) in an *in vitro* rumen simulation, and whether any differences could be related to changes in plant structure. It was hypothesised that fresh turgid grass may have a greater propensity for cell damage during mastication than the much more flaccid wilted grass, which in turn may increase microbial metabolism of the grass lipid.

Materials and methods Fresh grass, eight different grass species, were cut from experimental plots on day 0 and day 1 of the trial (*circa*. 200grams FW) 3cm above soil level. The grass cut on day 0 was left to wilt for 24 hr on a laboratory bench. Half of the fresh and wilted tissue was then crushed and cut (C&C) into 5mm strips and 2 ½ g loaded into incubation bottles under CO_2. The remaining grass was loaded into the incubation bottles and homogenised (H) also under CO_2. Two bottles contained 10ml of strained rumen liquor and 10ml of Van Soest medium were used for each treatment. After gassing the head space and sealing, the bottles were incubated for 6 h in the dark at 39°C. At the end of the incubation the bottles were removed and the lipid extracted and analysed as described by Lee *et al.* (2005). Biohydrogenation of C18:2 and C18:3 were calculated as proportional loss of these fatty acids during the incubation. The effect of wilting and cell damage was assessed using an unbalanced ANOVA model with wilt × cell damage as the treatment and blocking according to grass species. The effect of grass species has previously been reported (Lee *et al.* 2005).

Results The concentration of CLA and TVA in the incubation vessels and the extent of C18:2 and C18:3 biohydrogenation are shown in Table 1. Wilting had no significant effect on the concentration of CLA in the vessels but significantly reduced the concentration of TVA and the biohydrogenation of C18:2 and C18:3. The extent of cell damage appeared to have no effect on the concentration of either CLA or TVA but significantly increased the concentration of C18:0 (data not shown) and the biohydrogenation of C18:2 and C18:3.

Table 1 CLA and TVA concentration (g/kg dry matter input) and C18:2 and C18:3 biohydrogenation

	Unwilted		Wilted			Significance		
	C&C	H	C&C	H	S.e.d	Cell damage	Wilt	Interaction
CLA	0.011	0.011	0.010	0.011	0.0007	NS	NS	NS
TVA	3.79	3.36	2.62	2.81	0.162	NS	***	**
Biohydrogenation								
C18:2	0.37	0.52	0.33	0.43	0.037	***	*	NS
C18:3	0.48	0.65	0.46	0.58	0.027	***	*	NS

Conclusions The low level of CLA production in the batch cultures highlights the important role of TVA as a precursor of milk CLA. Increasing the extent of cell damage significantly increased biohydrogenation, but surprisingly did not increase the concentration of the intermediates CLA and TVA. It, did, however increase the concentration of C18:0, which is the end point of biohydrogenation, which suggests that the reaction went further to completion. The fact that H wilted grass was not similar to H unwilted grass in terms of TVA concentration may suggest that structural differences alone between wilted and unwilted grass are unlikely to explain the large differences in milk CLA produced from fresh pasture and cut/wilted grass (Offer, 2003).

References
Jahreis, G., J. Fritsche & H. Steinhart (1997). Conjugated linoleic acid in milk: high variation depending on production system. *Nutrition Research*, 17, 1479-1484.
Lee, M.R.F., C. Hodgkins, J.K.S. Tweed, N.D. Scollan & R.J. Dewhurst (2005). An *in vitro* investigation into forage factors which may influence the production of conjugated linoleic acid and *trans* vaccenic acid in the rumen. I. Genotype. *XX International Grassland Congress – offered papers* (in press).
Offer, N.W. (2002). Effect of cutting and ensiling grass on levels of CLA in bovine milk. XIII International silage conference, September 11th-13th, Auchincruive, Scotland, pp. 16-17.

Effects of grassland management on herbage lipid composition and consequences for fatty acids in milk

A. Elgersma[1,2], P. Maudet[1], I. Witkowska[1] and A.C. Wever[1]
[1]Crop and Weed Ecology, Plant Sciences Groups, Haarweg 333, 6709 RZ Wageningen, Wageningen University, The Netherlands, Email: anjo.elgersma@wur.nl, [2]University of Ghent, Belgium

Keywords: fatty acids, conjugated linoleic acid, protein, fertiliser, regrowth

Introduction Herbage provides bulk feed and is the basis for ruminant nutrition. Herbage lipids, especially C18:3, are a major source of beneficial fatty acids (FA) in milk. These desired FA are unsaturated FA such as CLA (conjugated linoleic acid), especially the isomer rumenic acid, and also vaccenic acid, both trans omega-7 FA (Ellen & Elgersma, 2004). As information on lipids in forages is scarce, effects were studied of N application level and regrowth period on the lipid concentration and FA composition of perennial ryegrass (*Lolium perenne* L.), the most important forage in temperate climate zones. A linear relation had previously been found between C18:3 intake of cows stall-fed with fresh grass and the amount of omega-7 FA in milk (Elgersma *et al.*, 2003).

Materials and methods N was applied at 100, 45 and 0 kg/ha, and swards were cut after various regrowth periods (20, 27 and 32 days after N application), resulting in six treatments abbreviated as 100N-20d, etc. Treatments were designed as randomised blocks with three replicates. For lipid analyses, herbage was frozen immediately after cutting and concentrations of individual FA were determined by gas chromatography. Canopy characteristics and herbage quality were also analysed.

Results The N applications and different harvest dates resulted in canopies with contrasting dry matter (DM) yields. The leaf blade proportion of DM declined from 0.70 to 0.55 to 0.45 in the various harvests, due to booting as time progressed. Five FA (C16:0, C16:1, C18:1, C18:2 and C18:3), representing 0.98 of total FA, were studied in detail. The mean concentration of FA in fresh grass was 15.5 g/kg DM and on average 0.68 of the FA consisted of C18:3. Regrowth period did not significantly affect the total FA concentration, but relatively less C18:3 and C16:1 and more C16:0, C18:0 and C18:2 were found after a longer period of regrowth. N application resulted in higher concentrations of total FA. There was a tendency for a relative increase in C18:3 and a decrease in C18:2 with higher N application. A strong positive overall linear relation was found between the concentrations of total FA and C18:3 and the protein content in the fresh herbage (Figure 1). This confirms findings of Boufaïed *et al.* (2003) with timothy.

The linear relation was due to both the increase in protein and C18:3 concentrations with N application and the declines in concentrations with longer regrowth period.

Conclusions These studies demonstrate opportunities to change the FA concentration and composition of FA in herbage through management strategies, which could favour an improved milk FA composition.

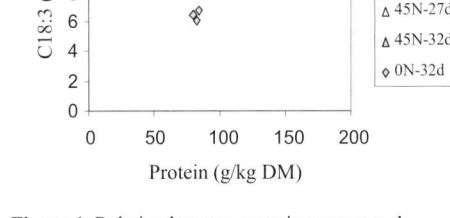

Figure 1 with legend: ■ 100N-20d, □ 100N-27d, ▲ 45N-20d, △ 45N-27d, △ 45N-32d, ◇ 0N-32d; axes C18:3 (g/kg DM) versus Protein (g/kg DM).

Figure 1 Relation between protein content and C18:3 concentration of perennial ryegrass

References

Boufaïed, H., P.Y. Chouinard, G.F. Tremblay, H.V. Petit, R. Michaud & G.Bélanger (2003). Fatty acids in forages. I. Factors affecting concentrations. *Canadian Journal of Animal Science*, 83, 501-511.

Elgersma, A., G. Ellen, P.R. Dekker, H. van der Horst, H. Boer & S. Tamminga (2003). Effects of perennial ryegrass (*Lolium perenne* L.) cultivars with different linoleic acids contents on milk fatty acid composition. *Aspects of Applied Biology*, 70, 107-114.

Ellen, G. & A. Elgersma (2004). Letter to the Editor: Plea for using the term n-7 fatty acids in place of C18:2 *cis*-9, *trans*-11, and C18:1 *trans*-11 or their trivial names rumenic acid and vaccenic acid rather than the generic term conjugated linoleic acids. *Journal of Dairy Science*, 87, 1131.

Effect of feeding L-carnitine and sunflower seeds on CLA content of pasture-fed beef

S.L. Scott, W.P. McCaughey and K.E. Buckley
Agriculture and Agri-Food Canada, Brandon Research Centre, P.O.Box 1000A, RR#3, Brandon, MB, R7A 5Y3, Canada, Email: sscott@agr.gc.ca

Keywords: CLA, beef, fatty acids

Introduction Pasture finishing enhances levels of conjugated linoleic acids (CLA) in beef lipids (Shanta et al. 1997). CLA (e.g., C18:2 c9, t11), formed during biohydrogenation of polyunsaturated fatty acids (PUFA) in the rumen, can reduce the incidence of heart disease, cancer and obesity in humans. However, pasture finishing cattle can reduce carcass grade. Feeding pasture-fed cattle a high-grain diet for a short finishing period (~60 d) improves grades but may reduce lipid CLA levels. A feeding regime is required that maintains the positive nutritional attributes of pasture-fed beef and improves the meat grade. The objective of this study was to investigate the effects of adding sunflower seeds (SFS), a good source of PUFA (Mir et al. 2000), or carnitine, a vitamin-like compound shown to increase fat deposition and marbling in cattle, to finishing diets of pasture-fed cattle on lipid fatty acid profiles (FAP).

Materials and methods Sixty-four steers grazed 11 paddocks (94.4% grass, 4.6% alfalfa) 13 May to 02 Sep 2003. Sixteen steers (503.1 ± 34.9 kg) were slaughtered off pasture (Time 0); carcass data and ribeye steaks were obtained. The remaining 48 steers (469.3± 34.5kg) received a basal diet of 80% barley grain and 20% hay (15.0 MJ DE/kg, 17.4% ADF, 28.2% NDF, 5.3% ash, 7.0 mg L-carnitine/kg, DM basis) and were allocated to one of the following diets: 1) 200 mg added carnitine (Carnipass®, Lonza, Inc., USA)/kg DM + 14% DM SFS (substituted for barley), 2) 0 mg added carnitine/kg DM + 14% DM SFS, 3) 200 mg added carnitine/kg DM + 0% DM SFS, and 4) 0 mg added carnitine/kg DM + 0% DM SFS (control). Blood samples were obtained every 14 d for analysis of plasma carnitine. Feed intake was measured every 7 d and feed sampled for analysis of nutrient and carnitine content. Body weights were measured every 14 d to calculate average daily gain (ADG). Every 28 d (Times 1, 2 & 3), 16 steers were slaughtered and carcass data and ribeye steaks were obtained.

Results Addition of carnitine to diets increased plasma levels of carnitine (μmol/L) and feeding SFS resulted in greater DM intakes (DMI, kg/d) and ADG (kg/d), with no effect on feed:gain (Table 1). FAP analysis of the lean portion of ribeye steaks showed that grain-finished animals had similar levels of total saturated fatty acids (ΣSFA) and total mono-unsaturated fatty acids (ΣMUFA), but lower levels of total PUFA (ΣPUFA) than steaks from pasture-fed animals. Feeding SFS increased total trans fatty acids (ΣTFA), mostly due to higher levels of C18:1 t11 (Table 1). With respect to pasture-fed steers, CLA content of beef (mg C18:2 c9, t11/100 mg lipid) increased by 50% in SFS-fed steers and decreased by 50% in barley-fed steers (Figure 1).

Table 1 Effect of diet on beef performance & FAP

Diet	Pasture	+SFS +CAR	+SFS -CAR	-SFS +CAR	-SFS -CAR
Carnitine	18.9	28.5	17.2	25.6	16.9
DMI	ND	13.2	13.1	11.7	12.8
ADG	ND	1.55	1.56	1.44	1.42
F : G	ND	8.48	8.40	8.12	9.02
ΣSFA	48.6	47.8	48.3	46.7	47.8
ΣMUFA	37.9	39.8	41.4	43.9	42.9
ΣPUFA	10.6	8.17	6.07	7.29	7.45
ΣTFA	2.21	3.60	3.56	1.75	1.59

ND=not determined

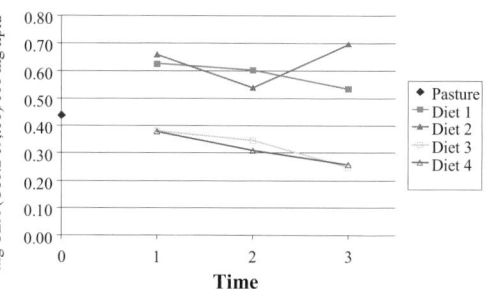

Figure 1 Effect of diet on CLA concentration (mg CLA/100 mg lipid) in meat from pasture-fed and finished steers

Conclusions The results demonstrate that adding SFS to diets of pasture-fed steers during a 28-, 56-, or 84-d finishing period can increase lipid levels of the C18:2 c9, t11 CLA isomer.

References

Mir, Z., M.L. Rushfeldt, P.S. Mir, L.J. Paterson & R.J. Weselake (2000). Effect of dietary supplementation with either CLA or linoleic acid rich oil on the CLA content of lamb tissues. *Small Ruminant Research* 36:25-31.

Shanta, N. C., W.G. Moody, Z. Tabeidi (1997). Conjugated linoleic acid concentration in semimembranous muscle of grass-and grain –fed and zeranol-implanted beef cattle. *Journal of Muscle Foods*. 8:105-110.

Content, pattern and esterification of fatty acids in fresh grass in relation to extraction solvents and sample storage conditions

M. Lourenço and V. Fievez
Laboratory for Animal Nutrition and Animal Product Quality, Ghent University, Proefhoevestraat 10, 9090 Melle, Belgium, Email: marta.lourenco@ugent.be

Keywords: fresh grass, storage, extraction solvents, fatty acid (FA) extraction

Introduction The relatively high content of linolenic acid (C18:3 n-3) in forages has lead to research on the role of forage based feeding strategies in the production of healthier milk. In these studies, forage samples are often stored, although analysis immediately after harvest has been considered to avoid oxidative deterioration and transformation of unsaturated fatty acids (FA) (Christie, 1993; Frankel, 1998). In the current study, we evaluated the effect of different storage conditions (fresh grass samples under liquid nitrogen (liq.N$_2$) or in a cool box during 3h or at -20°C, -80°C or in the extraction solvent at -20°C, during 24h) on the total amount of FA extracted, the FA pattern and the extent of FA esterification. The effectiveness of *iso*propanol to inhibit plant enzyme activity (Hawke, 1973), which has been reported to be particularly high in plant tissues (Christie, 1993), was considered. Measures to avoid thawing losses during sample handling were also compared.

Materials and methods *Experiment 1* Fresh grass samples of *Lolium multiflorum* were harvested. Samples (5g) of the control group were immediately analysed, while the rest were stored either under liq.N$_2$ or in a cool box for 3h. Samples were cut into 1 cm strips prior to FA extraction or storage during 24h at -20°C, -80°C or in the extraction solvent at -20°C. Alternatively, grass was stored whole for 24h and then cut into 1 cm strips. Half of the samples were extracted with chloroform/methanol (C/M, 2/1, v/v), while the other half were extracted with *iso*propanol/chloroform (I/C, 1/1, v/v). Separation of the lipid classes in the samples conserved in liq.N$_2$ was performed by thin layer chromatography. *Experiment 2* Fresh grass samples of *Lolium perenne* were harvested. Samples (5g) of the control group were analysed immediately. The rest, stored for 3h in liq.N$_2$, were freeze-dried (FD) and extracted (2.5g) with C/M (2/1, v/v). All samples were homogenised with an ultra-turrax prior to extraction and water content was adjusted to C/M/water and I/C/water ratios of 8/4/3 and 2/2/1, respectively. For FD samples 30 ml of C/M (2/1, v/v) and 20 ml of distilled water were used.

Results *Experiment 1* Higher amounts of total FA and proportions of C18:3 n-3 were obtained in samples stored for 3h either in liq.N$_2$ or in a cool box compared to the 24h storage (Table 1). Increased proportions of free FA (FFA) to total FA (%) are indicative of ongoing lipolysis during or after sample storage (1.3, 2.7, 6.0, 7.6, for direct analysis, 3h in liq.N$_2$, 24h storage at -20°C and at -80°C, respectively). Although the use of *iso*propanol has been recommended for FA extraction from plant tissues, significantly lower amounts of FA were extracted compared to extractions with C/M and separation in lipid classes showed proportionally more FFA to total FA (6.4 *vs.* 4.4 %). *Experiment 2* Reduced amounts of FA and lower C18:3 n-3 proportions were observed when performing FA analysis on FD samples (Table 2).

Table 1 Total FA content and C18:3 n-3 proportions in *Lolium multiflorum* samples stored under different conditions and extracted with either C/M (2/1, v/v) or I/C (1/1, v/v)

Fatty acids	Direct analysis (n=12)	3h liq.N$_2$ or cool box (n=4)	24h (n=8) -20°C	24h (n=8) -80°C	24h (n=8) Solvent	SEM	Mean[1]	
Total FA (mg/g fresh grass)	7.28^a	6.99^a	5.52^b	5.88^b	5.57^b	0.243	C/M	7.20**
							I/C	4.85
C18:3 n-3 (% of total FAME)	64.6^b	65.8^a	63.3^b	63.3^b	66.0^a	0.34	C/M	66.6***
							I/C	62.3

[1]Mean of extractions with C/M or I/C; ^{a,b,c} Values within a row lacking a common superscript differ significantly (p<0.05)

Table 2 Effect of freeze-drying grass samples before FA extraction on total FA content and proportion of C18:3 n-3 (n=3)

Fatty acids	Direct analysis	3h liq.N$_2$ and freeze-drying	SEM	Sign.
Total FA (mg/g DM)	38.7	19.2	1.36	***
C18:3 n-3 (% of total FAME)	65.4	60.5	0.42	***

Conclusions FA should be extracted from grass immediately after harvest or after a short storage in liquid N$_2$. *Iso*propanol/chloroform does not inhibit lipolysis and less FA is extracted compared to chloroform/methanol.

References
Christie, W.W. (1993). Advances in lipid methodology –Two. The Oily Press, Dundee, UK, Chapter 3, 195-213.
Frankel, E.N. (1998). Lipid Oxidation. The Oily Press, Dundee, UK, Chapter 10, 187-225.
Hawke, J.C. (1973). Lipids. In: Butler, G.W., Bailey, R.W. Chemistry and biochemistry of herbage, Vol.I, AP-Academic Press, London, UK, Chapter 5, 213-263.

Traditional cattle feeding stuffs: fatty acid profile

A.M. Peres[1,2], L. Dias[2], J. Sá Morais[2], F. Sousa[2] and J.M. Pires[2]
[1]LSRE, Escola Superior Agrária, 5301-855 Bragança, Portugal, Email: peres@ipb.pt, [2]CIMO, Escola Superior Agrária, 5301-855 Bragança, Portugal

Keywords: fatty acids, hay, meadow, forage, feeding stuffs

Introduction Dietary polyunsaturated fatty acids (PUFA) are perceived to be healthier than saturated fatty acids. Therefore, in order to be able to manipulate the fatty acid profile of meat and/or milk, to respond to the consumer demands, knowledge of the fatty acid profile of feeding stuffs for cattle is of major importance (LeDoux et al., 2002; Petit, 2002). In this work a preliminary study was made of the fatty acid profile of the cow's diet in a traditional farm production system.

Materials and methods A farm that produces beef from the "Barrosã" breed, located near Montalegre (north of Portugal) was monitored for a year (autumn 2002 – 2003). Two samples, of each feeding stuff given to the cattle, were collected and analysed for the fatty acid profile: wheat straw (ws), high-quality hay meadow (hhm), forage rye (fr), forage wheat (fw), regional white corn (rwc), concentrate (c), low-quality hay meadow (lhm), highland grassland (hg), regional yellow corn (ryc), meadow (m) and low-quality meadow (lm). The extraction was performed by soxhlet with hexane followed by an acid derivatisation using sulphuric acid:methanol reagent. The methylated fatty acids were then extracted by diethylic ether followed by water and chlorophyll removal. The extracts were then analysed in duplicate by GC with FID detector and a SUPELCO column (SP-2560). The fatty acid quantification was carried out using internal standard calibration with undecanoic acid (C11:0) as internal standard. For GC calibration, a SUPELCO 37 commercial FAME mix solution was used.

Results Fatty acids were only considered if the amount found in at least one of the samples analysed was higher than 5% of the total fatty acids (TFA) content: C12:0; C14:0; C16:0; C18:0; C20:0; C21:0; C22:0; C24:0; C18:1n9c; C18:2n6c; C18:3n3. The uncertainty of the results obtained in this study was evaluated taking into account the repeatability and reproducibility. The relative average deviation for the TFA was less than 7% and 15%, respectively. The data were analysed using principal components analysis by SPSS v11.0 program (Figure 1). The fatty acid profiles show three different groups: forage (rwc, ryc, c, fr, fw), meadow (hg, m, lm) and hay (hhm, lhm, ws). Analysis of the TFA data (Table 1) explains these three groups: the forage group with TFA>10000 µg/g; the meadow group with a 2000<TFA<7500 µg/g and the hay group with a TFA≈1000 µg/g. Moreover, this last group had a MUFA (monounsaturated fatty acids) content higher than the PUFA content, in contrast to the other two groups.

Conclusions From a nutritional point of view, the results obtained are in accordance with expectations. The TFA and PUFA contents were highest in the most nutritive feeding stuffs.

References

LeDoux, M., A. Rouzeau., P. Bas & D. Sauvant (2002). Occurrence of trans-C$_{18:1}$ fatty acid isomers in goat milk: effect of two dietary regimens. Journal of Dairy Science, 85, 190-197.
Petit, H.V. (2002). Digestion, milk production, milk composition and blood composition of dairy cows fed whole flaxseed. Journal of Dairy Science, 85, 1482-1490.

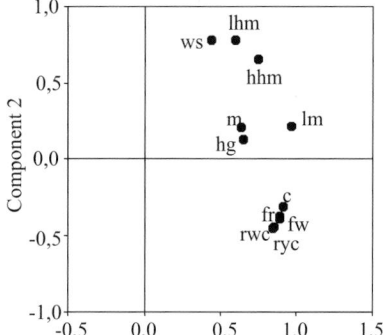

Figure 1 Component plot

Table 1 Total fatty acid content and MUFA percentage for the feed stuffs

Feed stuff	Total (µg/g)	MUFA (%)
Regional white corn	43208	16.6
Regional yellow corn	31681	16.7
Concentrate	30611	34.0
Forage rye	10566	17.9
Forage wheat	15057	18.6
Highland grassland	7429	24.1
Meadow	5552	27.8
Low-quality meadow	2190	48.0
High-quality hay meadow	1348	62.5
Low-quality hay meadow	1084	71.8
Wheat straw	1070	85.6

Finishing effect on fatty acid profile of intramuscular fat in extensively reared steers

N. Aldai[1], B.E. Murray[2], A.I. Nájera[3] and K. Osoro[1]

[1]S.E.R.I.D.A. Apdo. 13 – 33300 Villaviciosa, Asturias, Spain, Email: naldai@serida.org, [2]Teagasc, The National Food Centre, Dunsinea, Ashtown, Dublin 15, Ireland; [3]Tecnología de los Alimentos. Universidad del País Vasco. Paseo de la Universidad 7, 01006 Vitoria, Spain

Keywords: fatty acid composition, concentrate finishing, pasture

Introduction Both the amount and the composition of fat depots in beef may be influenced by several factors i.e. feeding system. Related to this factor, extensively reared cattle may produce beef with a more desirable fatty acid (FA) composition in terms of beneficial effect on human health, especially in relation to the content of *n-3* type FAs. However, concentrate finishing improves some carcass traits and meat quality. In this sense, the objective of this work was to study the effect of concentrate finishing on intramuscular (IM) FA profile of *Longissimus thoracis* (LT) muscle in pasture fed steers.

Materials and methods Eight yearling steers from "Asturiana de los Valles" were reared under extensive conditions on pasture. Four of them received a finishing diet (84% barley meal, 10% soya meal, 3% fat, 3% minerals, vitamins and oligoelements) during the last 60 days before slaughter. Average slaughter weight was 441 kg for animals fed with grass, and 504 kg for animals supplemented with concentrate. After 24h *post mortem* meat sample was vacuum packed and frozen at –80°C for subsequent FA analysis by GC using internal standard ($C_{23:0}$). The FAs were extracted and methylated by the modified method of Elmore *et al.* (1999). IM fat content (%) was determined by near infrared spectroscopy.

Results In comparison to animals fed with pasture, meat from animals finished with concentrate showed (Figure1): 1) significantly higher quantities of $C_{18:1}c11$, $C_{18:3}n-6$ and $C_{20:2}n-6$; 2) significantly lower quantities of $C_{18:3}n-3$ and $C_{20:5}n-3$, while some *n-6* type long chain FAs showed the same tendency ($C_{20:3}n-6$ & $C_{20:4}n-6$); 3) significantly higher *n-6/n-3* and a tendency to a lower P/S ratio. The first two principal components explained 79% of the variation observed on FA composition (Figure 2). PC1 is positively related to SFAs, MUFAs, BFAs (individuals & groups), *c9t11*CLA, $C_{18:3}n-6$, $C_{20:2}n-6$ and total FAs, and negatively related to P/S ratio and in a lower degree with *n-3* type of FAs. PC2 is positively related to $C_{18:3}n-3$, $C_{20:5}n-3$ and $C_{22:1}c13$, and negatively related to *n-6/n-3* ratio.

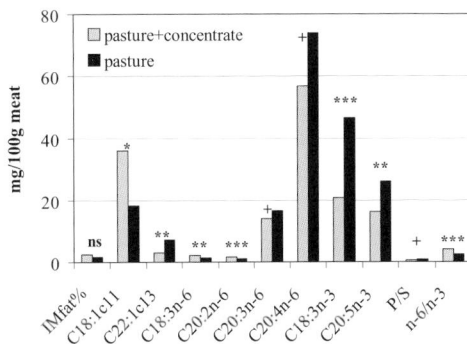 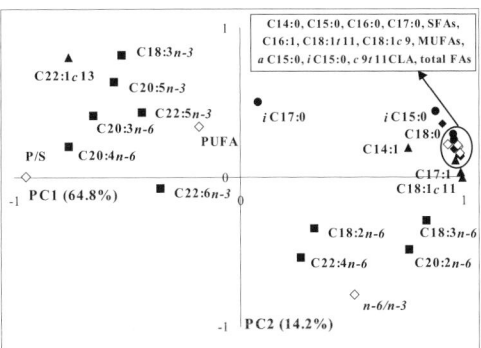

Figure 1 Finishing effect on IM fat%, FA quantities (only significantly different ones are represented, $p \le 0.1$) and ratios

Figure 2 Biplot representation of principal components (PC1 & PC2) of different variables studied

Conclusions According to the FA profile, meat obtained from animals finished with concentrate had lower quantities of *n-3* and higher quantities of *n-6* type of FAs in comparison to animals fed only with pasture. However, in general, meat obtained from both production systems were well adapted to human nutritional requirements taking into account P/S and *n-6/n-3* ratios.

References

Elmore, JS; Mottram, DS; Enser, M; Wood, JD. (1999). Effect of the polyunsaturated fatty acid composition of beef muscle on the profile of aroma volatiles. *Journal of Agricultural Food Chemistry*, 47, 1619-1625.

Effect of perennial ryegrass cultivars on the fatty acid composition in milk of stall-fed cows

A. Elgersma[1,4], H.J. Smit[1], G. Ellen[2] and S. Tamminga[3]
[1]Plant Sciences Group, Crop and Weed Ecology, Wageningen University, Haarweg 333, 6709 RZ Wageningen, The Netherlands, Email: anjo.elgersma@wur.nl, [2]NIZO food research, PO Box 20, 6710 BA Ede, The Netherlands, [3]Animal Sciences Group, Animal Nutrition, Wageningen University, Marijkeweg 40, 6709 PG Wageningen, The Netherlands, [4]University of Ghent, Belgium

Keywords: fatty acids, conjugated linoleic acid, water-soluble carbohydrates, volatile fatty acids, herbage

Introduction Herbage provides bulk feed for ruminants and plant lipids, especially C18:3, are a major source of benefical fatty acids (FA) in milk. There are very few direct comparisons allowing a precise evaluation of the effects of the basal forage diet on milk FA composition. Grass quality differences can affect rumen metabolism and there could be opportunities to change the composition of ruminant products through choice of grass cultivar. To test this hypothesis, six cultivars were fed to dairy cows in a stall-feeding trial with fresh grass to evaluate the effect of grass cultivar on rumen VFA and milk FA composition during the growing season.

Materials and methods Twelve Holstein Friesian dairy cows were used in a stall-feeding trial with fresh grass to evaluate the effect of grass cultivar on milk fatty acid (FA) composition during the growing season. Six diploid perennial ryegrass (*Lolium perenne* L.) cultivars were used: Abergold, Respect and Agri (intermediate heading) and Herbie, Barezane and Barnhem (late heading). They were cut daily during three 14-d periods between July and August at the same target yield. The experiments consisted of two 3x3 Latin square trials, in each of which three cultivars were fed to two groups of three cows. Half of the cows had a rumen fistula. Dry matter intake (DMI), milk production (MP) and milk composition (MC) were recorded daily in individual cows. Rumen liquid samples were taken from the fistulated cows and analysed for volatile fatty acid (VFA) composition. Levels of individual FA in grass and milk were determined by gas chromatography.

Results The dry matter (DM) yield during the three harvest periods was on average 2433 kg/ha in early July and 2090 kg/ha thereafter. The leaf blade proportion of DM increased from 0.67 to 0.87 to 0.91 during the season. The biggest range among cultivars was found in early June (Table 1); in later harvests leaf blade proportions varied from 0.84 to 0.90 and from 0.90 to 0.93, respectively. The six cultivars were rather variable in their chemical characteristics. Barnhem and Abergold had the highest (P < 0.001) WSC and the lowest (P < 0.01) NDF concentrations. Barezane and Respect had a lower WSC concentration (P < 0.001) than the other cultivars. Barnhem had the lowest and Barezane the highest CP concentration (P < 0.001). However, there were no significant differences among cultivars in FA concentration (22.1 g/kg DM) or proportions of FA. Average proportions of the major FA C18:3, C16:0 and C18:2 were 0.74, 0.13 and 0.10, respectively.

Despite the variation in quality parameters among the cultivars, their DMI (16.6 kg DM/d) did not differ, and MP (27.4 kg/d) and MC (34 g/kg protein, 40 g/kg fat and 45 g/kg lactose) were similar (Smit *et al.*, 2005). Rumen VFA concentrations did not differ among cultivars. No variation in milk FA composition was found. The mean proportions of individual FA C16:0, C14:0, C18:0, C18:1 *cis*-9, vaccenic acid and rumenic acid were 263, 121, 108, 178, 33 and 14 g/kg FA, respectively.

Conclusions Despite variation in morphological and chemical characteristics of the herbage, no variation in DMI, VFA concentrations, milk production or milk FA composition was found. The latter may be due to the lack of variation in grass FA concentration and composition in the cultivars studied.

References

Elgersma A., G. Ellen, H. van der Horst, B.G. Muuse, H. Boer & S. Tamminga (2003). Influence of cultivar and cutting date on the fatty acid composition of perennial ryegrass. *Grass and Forage Science,* 58, 323-331 and Erratum in ibid. 59, 104.

Smit H.J., B.M. Tas, H.Z. Taweel, J. Dijkstra, A. Elgersma & S. Tamminga (2005). Effects of perennial ryegrass cultivars on traits for improved animal performance. *XX International Grassland Congress* (in press).

Table 1 Chemical characteristics (g/kg DM) of six perennial ryegrass cultivars, averaged over 3 cuts taken early and late July and late August, and leaf blade proportion of DM in early July

Cv	CP	NDF	WSC	Leaf
Abergold	160	399	192	0.66
Respect	159	429	152	0.59
Agri	157	423	170	0.57
Herbie	156	412	172	0.64
Barezane	166	414	158	0.73
Barnhem	150	400	195	0.79
Mean	158	413	173	0.67
s.e.d.	2	6	5	0.03
Sign.	***	**	***	***

Sign.: ** : P < 0.01; *** : P < 0.001

Quantitative and qualitative characteristics of the loin of grazing lambs from different production systems

A.L.G. Monteiro[1], C.H.E.C. Poli[2], G.J. Bosquetto[1], T.M.D. Ribeiro[1], O.R. Prado[1], M.A.M. Fernandes[1] and C. Gasperin[1]
[1]Universidade Federal do Paraná, Curitiba, PR, Brazil, Email: alda.lgm@ufpr.br, [2]Embrapa Pecuária Sul/Paraná, P.O. box 242, CEP 83411-000, Colombo, PR, Brazil

Keywords: commercial cuts, *M. longissimus*, subcutaneous fat

Introduction The international sheep market is supplied mainly from New Zealand and Australia, where there are advanced production systems and marketing organisations, and where the product is exported mainly in the form of carcasses and cuts. In Brazil, the farmers should aim to produce a younger animal, with an adequate fat level at an optimum stage of muscle development, in order to gain market share The proportions of muscle, bone and fat largely determine the value of the carcass, and the breed and age of animal, in addition to other factors such as feeding systems, cause variation in the proportions (Purchas *et al.*, 1991). Grazing systems may be considered in Brazil due to reduced production costs. Measuring *M. longissimus* (the loin muscle) traits is a way of evaluating carcass quality because this muscle is one of the most important commercial cuts and represents total carcass characteristics. The objective of this study was to determine qualitative and quantitative characteristics of the loin of lambs on different production systems.

Materials and methods The experiment was carried out at the Experimental Research Station of UFPR, Pinhais, PR, Brazil, in a randomised block design with three replications. Four Suffolk lambs/plot grazed bermuda grass (*Cynodon dactylon* hybrid Tifton 85) in three production systems during the summer season: (1) lambs weaned at 60 d of age and grazed until slaughter; (2) lambs with their mothers until slaughter; (3) the same as Treatment 2, but the lambs were supplemented by a creep feed each day at 1% of live weight with a concentrate (18% crude protein and 80% TDN). Male lambs were weighed each 14 d and slaughtered at 33-34 kg live weight. After slaughter, carcasses were cleaned and eviscerated. They were then cooled at 5ºC for 24 h and the left side was sectioned into: shoulder, neck, ribs, loin (*M. longissimus)* and leg. *M. longissimus* weight was recorded and its dressing-out (%) calculated. Linear loin measurements were made: A or loin maximum width (cm); B or loin maximum depth (cm); C, loin subcutaneous fat thickness (mm) and J, loin maximum fat thickness (mm). Statistical analysis was conducted using a generalised linear model procedure.

Results and discussion Genetics may be the main factor that affects carcass and cut characteristics (Purchas *et al.,* 1991). In this experiment, all animals were from the Suffolk breed and were slaughtered at similar live weights (33-34 kg LW), but with different ages. Treatment (1) lambs were slaughtered at 134 d of age; treatment (2) and (3) lambs were slaughtered at 105 d and 100 d, respectively. Lambs weaned and kept on pasture until slaughter had lighter loin (0.552 kg; P<0.05) compared to lambs receiving creep feed (0.719 kg) and unweaned lambs (0.680 kg); the latter two treatments did not differ significantly. All treatments had similar loin dressing-out percentages. Loin maximum width (A) and depth (B) showed similar results (P>0.05). The fat thickness assessments (C and J- mm) represent the depth of subcutaneous fat. Weaned lambs gave the lowest average for C and J (Table 1) compared to lambs from Treatment (2) and (3), that did not differ significantly from one another. The production systems did not give the subcutaneous fat depth and slaughter ages required by the Brazilian market, although the lambs kept with their dams gave more acceptable fat thickness (J=1.78 mm).

Conclusions Sheep production systems on summer pastures with lambs slaughtered at around 100 to 134 days of age and slaughter weight 33-34 kg did not reach the subcutaneous fat depth required by the Brazilian market.

Acknowledgement Study sponsored by Brazilian Government CNPq.

Reference

Purchas, R.W., A.S. Davies & A.Y. Abdullah (1991). An objective measure of muscularity: changes with animal growth and differences between genetic lines of Southdown sheep. *Meat Science*, 30, 81-94.

Table 1 Average values of loin characters of lambs produced in different systems

Treatments	1	2	3
Loin weight(kg)	0.547a*	0.720b	0.680b
Loin dressing-out (%)	9.24a	9.63a	9.82a
Loin linear measurements			
A (cm)	4.90a	4.15a	5.25a
B (cm)	2.30a	3.26a	2.78a
C (cm)	0.44a	0.77b	0.98b
J (cm)	0.56a	1.21b	1.78b

Different letters in rows represent statistically significant differences (P<.,05) by Tukey test.

Processing quality of organic and conventional milks from Irish pasture based systems

B. O'Brien[1], P. Murphy[2], N. Culleton[3] and P. O'Connor[2]
[1]Teagasc, Dairy Production Department, Moorepark Research Centre, Fermoy, Co. Cork, Ireland, Email: bobrien@moorepark.teagasc.ie, [2]Teagasc, Dairy Products Research Centre, Fermoy, Co. Cork, Ireland; [3]Teagasc, Johnstown Castle Research Centre, Co. Wexford, Ireland

Keywords: milk, organic, processing quality

Introduction The maintenance of white clover in the pasture sward is essential to viable organic farming in Ireland. Thus, the diet of the organically and conventionally managed cow is different. It is well documented that milk composition is affected by cow diet (Kefford *et al.*, 1995). This study addresses the issue of technological quality differences between conventionally and organically produced milks.

Materials and methods Two spring-calved Friesian herds under conventional (97 cows) and organic (21 cows) management were established. The conventional and organic herds were stocked at 2.5 cows/ha and 1.4 cows/ha, respectively. Both herds received 570 kg concentrate/cow. Samples from conventional and organic bulk herd milks were collected at 2 weekly intervals during the lactation of 2001 (May to October). Milks were analysed for gross composition (Milkoscan), rennet coagulation characteristics (rennet coagulation time in min and curd firmness at 60 min in mm of amplitude) (McMahon & Brown, 1982) and N-fractions of total protein (IDF, 1993). Both organic and conventional milks were also investigated with respect to somatic cell count (SCC) (Somacount 300), thiocyanate levels (Partanen *et al.*, 1998) and a shelf-life study.

Results Mean milk yields of organic and conventional herds were 4,469 l/cow and 5,228 l/cow, respectively. Mean fat and protein concentrations of organic and conventional milks were 4.07 ($\pm$0.272) and 3.42 ($\pm$0.239), and 4.11 ($\pm$0.380) and 3.51 ($\pm$0.253) g/100g, respectively (Figure 1). Mean casein number (indicator of cheese yield) of organic and conventional milk was 79.3 ($\pm$1.22) and 77.7 ($\pm$0.67), respectively. Mean rennet coagulation time and curd firmness for these milks were 19.9 ($\pm$4.21) min and 40.7 ($\pm$7.59) mm, and 20.0 ($\pm$4.02) min and 41.8 ($\pm$7.27) mm, respectively. Somatic cell count of organic and conventional milks was 230 ($\pm$39.3) x10^3 and 317 ($\pm$71.2) x10^3 cells/ml, respectively. Average thiocyanate values of 7.94 mg/l ($\pm$1.90) and 5.64 mg/l ($\pm$1.91) were recorded for organic and conventional milks, respectively (Figure 2).

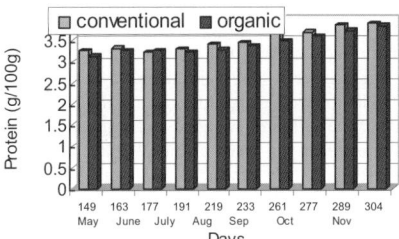

Figure 1 Protein content of organic and conventional milks

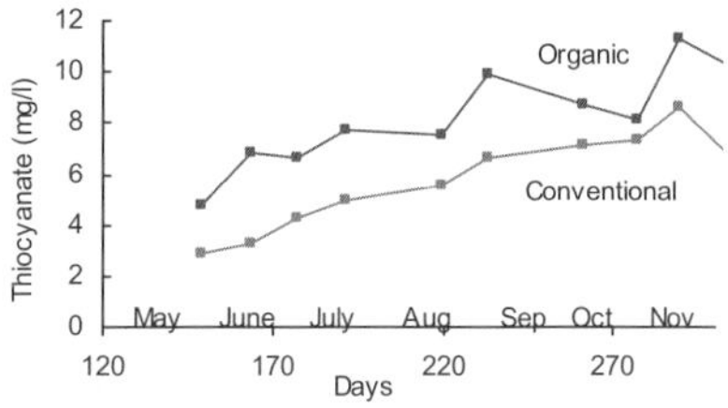

Figure 2 Thiocyanate content of organic and conventional milks

Conclusions Mean protein content and casein number of conventional and organic milks were higher by 0.1% and 1.6%, respectively. Fat content and rennet coagulation properties of the milks were similar. Both organic and conventional milks had a similar shelf-life with no advantage evident from the higher thiocyanate levels in the organic product. The absence of overall superior processing qualities of organic compared to conventional milk means the market for organic milk is very much dependent on consumer perception and emphasises the importance of strict monitoring of organic production rules in an environment of premium payments.

References

McMahon, D.J. & R.J. Brown (1982). Evaluation of Formagraph for comparing rennet solutions. *Journal of Dairy Science,* 65, 1639-1642.

Kefford, B., M. Christian, B. J. Sutherland, J. J. Mayes & C. Grainger (1995). Seasonal influences on Cheddar cheese manufacture: influence of diet quality and stage of lactation. *Journal of Dairy Research,* 62, 529-537 IDF (1993). *Milk: Determination of Nitrogen Content.* Brussels: IDF (*FIL-IDF Standard* no. 20B).

Partanen, L., K. Valkonen & T. Alatossava (1998). Determination of thiocyanate in milk. *Milchwissenschaft,* 53, (3), 132-135.

Utilizing forages to program steer growth patterns to achieve consistent quality beef

B.G. Warrington[1], J.W. Holloway[1], R.K. Miller[2] and H. Lippke[1]
[1]*Texas Agriculture Experiment Station, 1619 Garnerfield Road, Uvalde, TX 78801, Email: b-warrington@tamu.edu, [2]Texas A &M University, Kleberg Center, College Station, TX 77843*

Keywords: grazing, small grains, native range, carcass, shear force

Introduction Many options are available for programming stocker cattle growth patterns through forage selection. In semi-arid south Texas rapid growth rates can be achieved by grazing irrigated small grains (oats, wheat and ryegrass) and slow growth rates are possible grazing native range pastures. Ryegrass (RG) nutrient quality indicates potential gains greater than 1.0 kg/d for steers, while typical winter native range (NR) pasture indicates gains of 0.45 kg/d or less. The purpose of this experiment was to quantify the impact of different programmed growth patterns on beef retail product especially size, marbling and tenderness.

Materials and methods Over 3 years, 153 steers were utilized to compare rapid versus limited post weaning growth systems. Calves comprised known tropically adapted*temperate genotypes (Angus, Bonsmara, Braunvieh, Hereford, Senepol, and Tuli breeds) and were stratified by genotype to grazing treatment. Steers were weaned on October 12 and held in a preconditioning lot until approximately December 10 of each year, and then allotted to their respective grazing regime. Ultrasound measurements of backfat and rib eye area (REA), weight, visual body condition score (BCS-1=thin, 9=fat), and frame score (1=short, 9 =tall) were taken at initiation and termination. Grazing terminated May 10 of each year, after termination steers were fed a high concentrate ration in a commercial feedlot. Steers were harvested at a targeted backfat thickness of 10 mm. Standard USDA carcass traits were collected and one 50 mm thick steak (*longissimus dorsi muscle*) was excised from 12th/13th rib for Warner-Bratzler shear force (WBS) determination (mean of 6, 12.5 mm cores/steak).

Results Limited growth rate steers on NR had more variation in growth rate, marbling score and WBS (Table 1) than those programmed to grow faster. For each year, steers grazing NR had lower (p<0.05) end weight, BCS, grazing ADG, ultrasound backfat and REA, but their gain in the feedlot was similar (p>0.05) and, as planned, their carcass backfat was similar (p>0.05) to steers grazing RG (Table 2). Only in Yr 3 was WBS affected (p<0.05) by grazing treatment with steers programmed for rapid growth being 12 % tougher.

Table 1 Variation in grazing ADG, marbling score and WBS across 3 years[a]

	Rapid			Limit			% Increase
	High	Low	Range	High	Low	Range	
Grazing ADG, kg/d	0.86	0.68	0.18	0.59	0.34	0.25	28
Marbling score	462	424	38	462	386	76	50
WBS, kg	6.25	5.26	0.99	6.95	4.73	2.22	66

[a] Least squares means from model ŷ = calf sire

Table 2 Steer characteristics during grazing phase and their effect on carcass traits and tenderness[a]

	Yr 1			Yr 2			Yr 3		
Forage	Rapid	Limit	s.e.	Rapid	Limit	s.e.	Rapid	Limit	s.e.
N	32	25		28	20		24	24	
Start wt, kg	259	258	8.4	279	267	9.2	280	274	4.6
End wt, kg	338*	295	9.7	371*	316	6.0	395*	364	5.9
End BCS, 1-9	5.3*	4.5	0.1	5.8*	3.8	0.2	5.8*	5.2	0.1
End backfat, mm	7.0*	4.9	0.31	8.8*	2.9	0.2	4.3*	3.5	0.01
End REA, cm²	59.3*	51.5	1.4	78.7*	59.4	2.3	64.2*	53.8	1.2
GrazingADG,kg/d	0.86*	0.40	0.04	0.68*	0.34	0.02	0.76*	0.59	0.02
Slaughter wt, kg	505	522	10.0	542*	503	11.1	563*	531	9.7
Hot carcass wt, kg	304	319	9.4	343*	310	6.9	350*	330	5.9
Backfat, mm	9.7	10.9	0.72	14.5	13.4	0.6	12.5	12.5	0.04
Marbling score	433*	462	10.3	462	441	23.6	424	386	22.7
FeedlotADG, kg/d	1.59	1.62	0.06	1.35	1.32	0.06	1.44	1.42	0.05
Days on feed, d	111*	141	6.5	126*	143	3.7	121	117	3.0
WBS, kg	6.25	6.95	0.37	5.26	5.17	0.36	5.29*	4.73	0.22

[a] Least squares means from model ŷ = calf sire
* Least squared means within yr and row differ at p<0.05

Conclusions Steers programmed for a slow rate of post-weaning growth were older at harvest having been fed a high concentrate ration longer in the feedlot, and more variable in carcass quality having about twice the variation in marbling and 66% more variation in WBS. Production systems designed for rapid, controlled growth may produce more consistent quality beef than those dependent upon more extensive, less controlled, discontinuous growth rate systems.

The effect of post-weaning management on the physico-chemical and textural quality of beef from bulls and steers

M. Oliván, P. García, M.J. Martínez, M. Mocha, A. Martínez, P. Castro and K. Osoro
S.E.R.I.D.A. Apdo 13,33300 Villaviciosa, Asturias, Spain, Email: mcolivan@serida.org

Keywords: extensive system, castration, grain-finishing, beef quality

Introduction There is an increasing interest for extensification in Europe due to environmental and animal welfare concerns. Furthermore, forage-fed beef may present benefits for human health. However, animals fed at pasture produce in some cases darker and tougher meat. It has been shown that castration or a grain-finishing period before slaughter could improve some sensory traits of beef from pasture. The objective of this work was to study the impact of castration and four feeding systems (grazing, grazing + 70 days concentrate, grazing + 100 days concentrate, 200 days concentrate) on the quality of beef from yearling bulls and steers.

Materials and methods Ninety four bulls of "Asturiana de los Valles" breed were managed and slaughtered around 500 kg live weight (15 to 19 months age). Fifty five animals were castrated at 10 months age and 39 remained entire. For each physiological state, there were four feeding treatments: 1) grazing on ryegrass and clover pastures, 2) grazing + concentrate for 70 d before slaughter, 3) grazing + concentrate for 100 d, 4) intensive feeding with concentrate for 200 d. Animals were slaughtered in a commercial abattoir following approved EU procedures. At 24 h *post mortem* cold carcass weight (kg) was recorded and pH was measured on the *Longissimus* muscle of the left carcass. The loin from 6th to 8th ribs was extracted, sliced and aged for 7 d for subsequent analysis. Water-holding capacity was estimated as expressible juice (EJ) and intramuscular fat (IMF) content by Soxhlet extraction. The loin colour was measured after 6 d oxygenation in the CIE L* a* b* space. Toughness of cooked meat was determined by Warner-Bratzler (WB) shearing force.

Table 1 Effect of castration (C) and grain-feeding (F) on meat characteristics

Grain-feeding (days)	Bull				Steer				Effect	
	0	70	100	200	0	70	100	200	C	F
Carcass (kg)	288	295	349	324	253	266	293	244	***	***
pH24	5.7	5.4	5.4	5.4	5.5	5.4	5.4	5.4	**	***
EJ (%)	23.9	22.7	22.8	20.6	24.0	20.9	22.0	18.2	*	***
IMF (%)	1.1	1.9	2.3	2.9	2.3	2.9	3.8	4.1	***	***
WB (kg)	6.0	5.2	5.7	4.7	4.7	4.5	4.4	4.0	***	*
L*	35.8	41.7	41.8	42.7	38.0	41.5	41.3	42.8	NS	***
a*	18.3	21.0	21.6	18.6	21.3	21.7	23.0	21.6	***	***
b*	4.2	9.9	9.8	10.7	6.3	9.8	10.4	10.3	NS	***

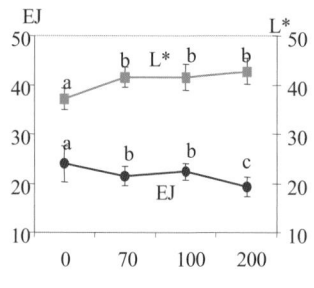

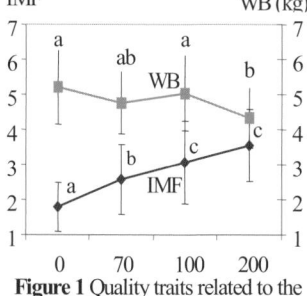

Figure 1 Quality traits related to the concentrate feeding period

Results Steers produced lighter carcasses (p<0.001), although they were slaughtered at the same age as bulls. Castration decreased the ultimate pH of carcass (p<0.01), but pH values were normal in all treatments. Meat from steers also showed lower juice losses (p<0.05), higher IMF (p<0.001), lower toughness (p<0.001) and higher redness values (p<0.001). Concentrate feeding affected significantly carcass weight and age, due to differences in fattening level. Ultimate pH was higher (p<0.001) in animals from pasture. Grain-feeding also decreased significantly juice losses (EJ, p<0.001), increased IMF (p<0.001) and decreased meat toughness (p<0.05). Meat produced from pasture was significantly darker than meat from any other treatment.

Conclusions Castration and the inclusion of a grain-finishing period improved the quality of meat produced from pasture. Meat from steers had higher intramuscular fat, lower juice losses and lower toughness than meat from bulls. Concentrate feeding reduced meat toughness and juice losses and increased the intramuscular fat content. However, this increase was not significant when comparing the intensive feeding with the treatment based on pasture + 100 days concentrate. Whilst the quality of meat from pasture improved when using a grain-finishing period, the increase of concentrate feeding period from 70 to 100 d did not produce significant changes of quality traits such as lightness, water-holding capacity or toughness.

Effects of supplementary concentrate level and separate or mixed feeding of grass silage and concentrates on carcass tissue colour traits in steers

J. Caplis[1], M.G. Keane[1], A.P. Moloney[1] and F.P. O'Mara[2]
[1]Teagasc, Grange Research Centre, Dunsany, Co. Meath, Ireland, Email: gkeane@grange.teagasc.ie,
[2]Department of Animal Science and Production, University College, Belfield, Dublin 4, Ireland

Keywords: concentrates, grass silage, fat colour, muscle colour, steers

Introduction The level of supplementary concentrates fed with grass silage and the method of feeding (separate or mixed) may affect carcass tissue colour in steers. The objectives were to determine the effects of (1) supplementary concentrate level with grass silage, and (2) separate or mixed feeding of silage and concentrates, on muscle and fat colour.

Materials and methods The experiment had 6 feeding treatments with 14 steers per treatment as follows:
1. Silage only (SO)
2. Silage + 3kg concentrate dry matter (DM) per day fed separately (LS)
3. Silage + 3kg concentrate DM per day fed mixed by feeder wagon (LM)
4. Silage + 6kg concentrate DM per day fed separately (HS)
5. Silage + 6kg concentrate DM per day fed mixed by feeder wagon (HM)
6. Concentrates *ad libitum* + 1kg silage DM per day (AL)

The animals were individually fed for a mean period of 132 days and the concentrate allowance was fed once daily to the animals fed separately. After slaughter the 6th -10th ribs joint was separated into its component tissues and fat and muscle colour values were measured using a Minolta chromometer. In the statistical analysis, the 5 degrees of freedom for treatment were partitioned into 5 orthogonal contrasts, one for the effect of mixing, one for the concentrate level x mixing interaction, and one each for the linear, quadratic and cubic effects of concentrate level.

Results There was a significant linear effect of concentrate level on muscle brightness (L value) which increased with increasing concentrate level (Table 1). There were significant linear and quadratic effects of concentrate level on muscle redness (a value) and yellowness (b value). Both of these values increased up to the high concentrate level but not beyond. There was no significant effect of concentrate level on fat brightness but there were significant quadratic effects on fat redness and yellowness. Fat a and b values were highest for the low and high concentrate levels and lower for the silage only and concentrates *ad libitum* treatments. There was no significant effect of mixing and no concentrate level by mixing interaction for any of the muscle or fat colour variables.

Table 1 Effects of concentrate level and separate or mixed feeding on muscle and fat colour measurements

| | Treatment | | | | | | | Significance | |
	SO	LS	LM	HS	HM	AL	s.e.	L[1]	Q[2]
Colour measurements									
Muscle "L"	34.2	36.0	35.6	36.5	35.7	36.2	0.50	**	NS
Muscle "a"	11.1	13.6	13.1	14.1	13.6	13.5	0.48	***	**
Muscle "b"	6.7	8.2	8.0	8.7	8.2	8.3	0.29	***	**
Fat "L"	66.9	64.3	65.3	65.8	64.5	66.0	1.04	NS	NS
Fat "a"	8.1	11.1	9.3	9.9	10.6	9.2	0.67	NS	**
Fat "b"	18.2	18.7	18.7	18.5	18.8	17.5	0.42	NS	*

[1]Linear effect of concentrate level; [2]Quadratic effect of concentrate level.
There was no significant effect of mixing and no significant concentrate level by mixing interaction.

Conclusions The main difference in muscle colour was between the silage only group and the concentrate supplemented groups. Differences amongst the latter were small. Similarly, fat a value was lowest for silage only and differed little between the concentrate supplemented groups. Fat yellowness was least for the concentrates *ad libitum* group. The relatively low yellowness of the silage only group may be a reflection of the low fat level of this group. There was no effect of mixing and no concentrate level by mixing interaction for any of the colour measurements.

Effects of supplementary concentrate level and separate or mixed feeding of grass silage and concentrates on carcass tissue composition in steers

J. Caplis[1], M.G. Keane[1] and F.P. O'Mara[2]
[1]Teagasc, Grange Research Centre, Dunsany, Co. Meath, Ireland, Email: gkeane@grange.teagasc.ie,
[2]Department of Animal Science and Production, University College, Belfield, Dublin 4, Ireland

Keywords: concentrates, grass silage, diet mixing, steers

Introduction Supplementary concentrate level may affect carcass composition in steers. Feeder wagons facilitate feeding and management. The objectives were to determine the effects of (1) supplementary concentrate level with grass silage, and (2) separate or mixed feeding of silage and concentrates, on ribs joint composition.

Materials and methods The experiment had 6 feeding treatments of 14 animals per treatment:
1. Silage only (SO)
2. Silage + 3kg concentrate dry matter (DM) per day fed separately (LS)
3. Silage + 3kg concentrate DM per day fed mixed by feeder wagon (LM)
4. Silage + 6kg concentrate DM per day fed separately (HS)
5. Silage + 6kg concentrate DM per day fed mixed by feeder wagon (HM)
6. Concentrates *ad libitum* + 1kg silage DM per day (AL)

The animals were individually fed for a mean period of 132 days. The concentrate allowance was fed once daily to the separate groups. After slaughter the 6-10th ribs joint was separated into its component tissues of fat, bone and muscle. In the statistical analysis the 5 degrees of freedom for treatment were partitioned into 5 orthogonal contrasts, one for the effect of mixing, one for the concentrate level x mixing interaction, and one each for the linear, quadratic and cubic effects of concentrate level.

Results Growth and slaughter data were reported previously (Caplis *et al.*, 2003). Carcass weight increased with increasing concentrate level with both the linear and quadratic components significant (Table 1). Subcutaneous fat, intermuscular fat and total fat proportions increased with increasing concentrate level with both the linear and quadratic components significant. *M. longissimus,* other muscle and total muscle proportions were unaffected by concentrate level. Bone proportion decreased with increasing concentrate level with both the linear and quadratic components significant. There were no significant effects of mixing and no significant concentrate level x mixing interactions.

Table 1 Effects of concentrate level and separate or mixed feeding on ribs-joint composition

	Treatment							Significance	
	SO	LS	LM	HS	HM	AL	s.e.	L[1]	Q[2]
Carcass weight (kg)	308	352	351	369	364	382	5.4	***	**
Ribs joint (g/kg)									
Subcutaneous fat	33	57	58	55	53	53	4.3	**	***
Intermuscular fat	115	142	154	151	140	142	9.3	P<0.06	*
Total fat	148	199	211	206	194	195	12.2	*	**
M. longissimus	225	215	208	217	219	224	7.0	NS	NS
Other muscle	416	399	397	403	408	403	9.0	NS	NS
Total muscle	640	614	604	620	627	627	11.7	NS	NS
Total bone	211	187	188	175	180	178	4.4	***	**

[1]Linear effect of concentrate level; [2]Quadratic effect of concentrate level
There was no significant effect of mixing and no significant concentrate level by mixing interactions.

Conclusions Carcass weight and total fat proportion were lowest and total muscle and bone proportions were highest on the silage only diet. The first concentrate increment resulted in big increases in carcass weight and fat proportion and big reductions in muscle and bone proportions. Above the low concentrate level further increases in concentrates increased carcass weight but did not increase fat or decrease muscle proportions. There was no effect of mixing or no concentrate level by mixing interaction for any variable.

References

Caplis, J., M.G. Keane & F.P. O'Mara (2003). Comparison of separate and mixed feeding of silage and concentrates for finishing cattle. In: *Proceedings of the Agricultural Research Forum* (2003) p38, ISBN 184174016.

Lipid oxidation and sensory characteristics of grass-fed beef: effect of duration of grazing prior to slaughter

A.P. Moloney[1], F. Noci[1,2], F.J. Monahan[2], G.E. Nute[3] and R.I. Richardson[3]
[1]Teagasc, Grange Research Centre, Co. Meath, Ireland, Email: amoloney@grange.teagasc.ie, [2]University College Dublin, Ireland, [3]University of Bristol, Langford, Bristol BS40 5DU, UK

Keywords: grass, beef, sensory characteristics, lipids

Introduction Beef from cattle produced from grass has a higher concentration of fatty acids considered to be beneficial to human health than beef produced from more intensive production systems and this increase in fatty acid concentration is dependant on the duration at pasture prior to slaughter (Noci et al., 2003). Improvements in the fatty acid composition of beef must not impair other quality characteristics of beef. Little information is available on the pattern of change of quality characteristics in grazing animals. The objective of this study was to determine the shelf-life and eating quality of beef from cattle produced from a standard Irish grass silage/concentrates finishing system but allowed to graze grass for different periods prior to slaughter.

Materials and methods Sixty Charolais crossbred heifers (BW = 338 kg) were used. One group was offered a silage/concentrates based diet indoors for 158 days (0 days at grass). One group grazed a predominantly perennial ryegrass pasture for 158 days. Two groups were initially offered a silage/concentrates diet, but grazed the above pasture for 40 and 99 days prior to slaughter after 158 days, respectively. Concentrate and grass allowances were adjusted periodically to achieve a similar mean carcass weight for all treatments. Carcasses were chilled for 48h at 4°C. A sample of longissimus muscle was aged for 14 days and stored frozen prior to lipid oxidation analysis (thiobarbituric acid reactive substances (TBARS)). A similarly-treated sample was used for sensory analysis by a 10 member trained taste panel. Panellists rated cooked steak using 0-100 line scales where low values are low ratings and high values higher ratings for a particular trait. Data were analysed according to a randomised block design.

Results Extending the grazing period increased the concentration of individual long chain n-3 polyunsaturated fatty acids (PUFA), total PUFA and conjugated linoleic acid (CLA) in muscle but did not affect lipid oxidation after 5 or 10 days of retail display. This was likely due to vitamin E supplied by the grass. Beef from 99-day grass fed animals was tougher than beef from animals fed silage/concentrates or grass for 158 days. Extending the grazing period increased the scores for "greasy" and "fishy". Beef from animals fed silage/concentrates was preferred to beef from 40 or 99-day grass fed animals. There were no differences for juiciness, or for beef, abnormal, bloody, livery, metallic, bitter, sweet, rancid, acidic, cardboard, vegetable/grassy or dairy flavours.

Table 1 Fatty acid (mg/100g tissue), TBARS (mg malonaldehyde/kg) and sensory perception of beef

	Days at grass				SED	P
	0	40	99	158		
C18:3	19.6 [a]	25.4 [b]	30.9 [c]	34.4 [c]	1.86	***
C20:5	5.6 [a]	5.5 [a]	6.4 [a]	7.7 [b]	0.50	***
C22:5	10.1 [a]	9.4 [a]	10.6 [a]	12.7 [b]	0.74	***
CLA	12.3 [a]	12.1 [a]	15.2 [a]	18.4 [b]	1.79	***
PUFA	129.4 [a]	136.7 [a,c]	145.6 [b,c]	158.4 [d]	6.93	***
Total	2641	2329	2754	2525	177.5	NS
TBARS-10 d	4.06	3.05	4.04	4.04	0.678	NS
Toughness	44.9 [ab]	48.8 [bc]	52.7 [c]	42.7 [a]	2.16	***
Greasy	10.4 [a]	12.7 [ab]	13.1 [ab]	14.4 [b]	1.44	*
Fishy	0.7 [a]	3.5 [b]	2.3 [ab]	3.3 [b]	1.07	*
Overall liking	22.5 [b]	18.1 [a]	18.8 [a]	19.9 [ab]	1.48	*

Figures with different superscripts differ significantly (P<0.05)

Conclusions Increasing the duration of grazing prior to slaughter improved the fatty acid composition, did not affect lipid oxidative stability, had minor effects on flavour but inconsistently influenced the toughness of beef.

Acknowledgements This study was supported by the European Commission (QLRT-2000-31423).

References

Noci, F., A.P. Moloney, P. French & F.J. Monahan (2003). Influence of duration of grazing on the fatty acid profile of M. Longissimus dorsi from beef. British Society of Animal Science, 23.

Antioxidative activities of alfalfa and timothy varieties

J.K. Lee[1], H.S. Park[1], J.G. Kim[1], B.H. Paek[1] and J.H. Fike[2]
[1]Hanwoo Experiment Station, National Livestock Research Institute, RDA, Chahang-ri, Doam-myeon, Pyeongchang-gun, Kangwon-do, South Korea, 232-952, Email: leejk58@rda.go.kr, [2]Crop and Soil Environmental Sciences Department, Virginia Tech. Blacksburg, VA 24061, USA

Keywords: antioxidative activity, DPPH, alfalfa, timothy

Introduction The term "functional foods" is often used as a generic description for the beneficial effects of ingested foods that go beyond their traditional nutritive value (Bauman et al., 2001). Milk and dairy products are important dietary sources of nutrients, providing energy, high quality protein, and a variety of vitamins and minerals. Recent research has focused on altering the fat and protein content of milk and other dairy products in order to improve their nutrient content to more aptly reflect current dietary recommendations and trends. As a result, additional focus is being given to designing foods that have beneficial effects on human health. This study was carried out to investigate the antioxidative activities of forages grown in Korea.

Materials and methods Nine alfalfa cultivars and eight timothy cultivars grown in an alpine area (altitude 800 m) were used in this study. Methanol extracts from each were tested for radical scavenging ability using the DPPH (1,1-diphenyl-2-picrylhydrazyl) method as determined by UV spectrophotometry (Uchiyama et al., 1968). Effect of drying method (oven drying vs. drying in shade) on antioxidative activities of alfalfa was also compared. Tests with timothy were conducted only on oven dried forage.

Results Examination of methanol extracts from alfalfa and timothy for radical scavenging effects revealed large differences among forage varieties. Extracts from 'Horizon' and 'DK 125' had the strongest antioxidant activity among alfalfas and 'Argus' and 'Itasca' extracts had the greatest antioxidant activity of the timothy varieties (Figures 1, 2). Antioxidative activities of shade-dried alfalfas (av. 0.101mg) were greater than those of oven dried forage (av. 0.056mg).

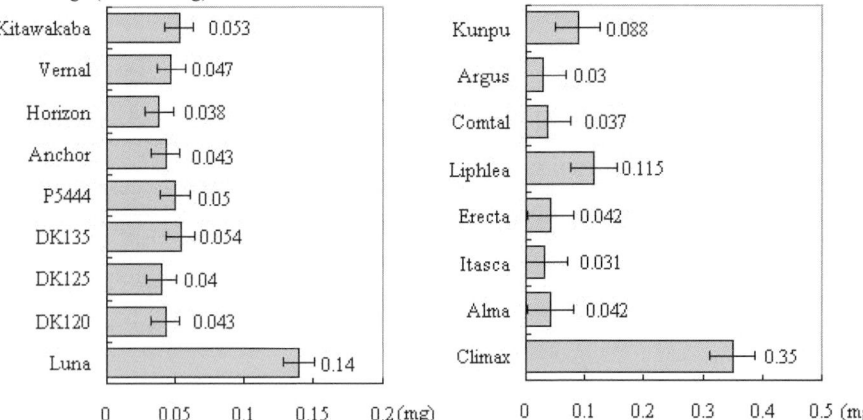

Figure 1 Antioxidative effect* of MeOH extracts from oven-dried alfalfa cultivars

Figure 2 Antioxidative effect* of MeOH extracts from oven-dried timothy cultivars

*Amount required for 50% reduction of DPPH(2X10[-7]ml, 0.079mg) solution

Conclusions Results of this study demonstrate that there can be large differences in antioxidant activities among forage cultivars, especially within timothy cultivars. These data suggest great potential for manipulating such potentially beneficial characteristics of forages. Further research on cultivar differences and management practices, impacts to animal production and health, and healthfulness of the product for human consumers is needed.

References

Bauman, D. E., B. A. Corl, L. H. Baumgard & J. M. Griinari (2001). Conjugated linoleic acid (CLA) and the dairy cow. Pages 221-250 in Recent Advances in Animal Nutrition-2001, P.C.Garnsworthy and J.Wiseman, eds. Nottingham University Press, Nottingham, UK.
Uchiyama, M., Y. Suzuki & K. Fukuzawa (1968). Biochemical studies of physiological function of tocopheronolactone. Yakugaku Zasshi, 88, 678-683.

Effect of three legumes containing different condensed tannin concentrations on the in vitro formation of the pastoral flavour compound; skatole

N.M. Schreurs[1,2], M.H. Tavendale[1], G.A. Lane[1], T.N. Barry[2] and W.C. McNabb[1]
[1]AgResearch Ltd, Grasslands Research Centre, Private Bag 11008, Palmerston North, New Zealand, Email: nicola.schreurs@agresearch.co.nz, [2]Institute of Veterinary, Animal and Biomedical Sciences, Massey University, Private Bag 11222, Palmerston North, New Zealand

Keywords: legumes, skatole, pastoral flavour, condensed tannins

Introduction Feeding legumes, such as white clover (*Trifolium repens*), results in higher intakes and increased animal production compared to grasses (Ulyatt, 1981). Skatole is produced in the rumen from plant protein fermentation and is associated with undesirable pastoral flavours in meat (Young *et al.* 2002). Feeding white clover causes a greater skatole concentration in the rumen compared to perennial ryegrass (*Lolium perenne*) or *Lotus corniculatus*, as the protein in white clover is highly soluble and rapidly degraded (Schreurs *et al.*, 2004). The condensed tannins (CT) in *Lotus* species slow protein degradation in the rumen (Aerts *et al.*, 1999). The aim of this study was to determine the effect of legumes with different concentrations of CT on skatole formation.

Materials and methods Fresh, minced white clover (WC), *Lotus corniculatus* (LC) and *Lotus pedunculatus* (LP) were incubated using the *in vitro* method of Barrell *et al.* (2000). Samples of the *in vitro* media were taken every hour for ten hours. Skatole concentration in the samples was determined by high performance liquid chromatography. Chemical composition and digestibility of the minced forages was measured by near infrared reflectance spectrophotometry.

Results Chemical composition of the three legumes is given in Table 1. Crude protein content (CP) and organic matter digestibility (OMD) was similar for WC and LC and lower with LP. Neutral detergent fibre concentration (NDF) was similar for all forages. The CT concentration was highest for LP, negligible for WC and intermediate for LC. Figure 1 shows that the skatole concentration at the end of the incubation, adjusted for CP added to the incubations, was greatest with WC and lowest with LP while LC was intermediate (P<0.001).

Table 1 Composition of forages in incubations

	WC	LC	LP
CP (g/kgDM)	276	267	194
NDF (g/kgDM)	210	186	221
OMD (%)	>87	86.4	80.9
CT (g/kgDM)	1.4	35.4	98.5

Conclusions This study shows that forages with a high CT concentration are associated with reduced formation of skatole in the rumen. The response is most likely due to the higher CT effectively slowing protein degradation.

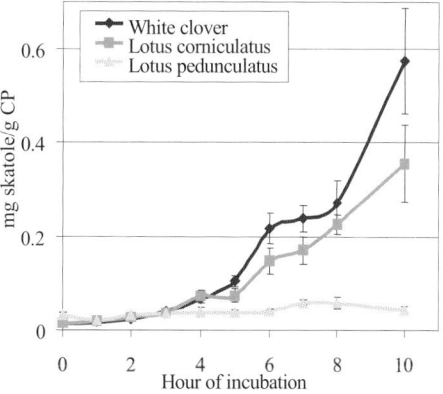

Figure 1 Skatole formation when incubating three legumes (4 replicates; error bars are SEM)

References

Aerts, R.J., T.N. Barry & W.C. McNabb (1999). Polyphenols and agriculture: beneficial effects of proanthocyanidins in forages. *Agriculture, Ecosystems and Environment, 75*, 1-12.

Barrell, L.G., J.L. Burke, G.C. Waghorn, G.T. Attwood & I.M. Brookes (2000). Preparation of fresh forages for incubation and prediction of nutritive value. *Proceedings of the New Zealand Society of Animal Production, 60*, 5-8.

Schreurs, N.M., M.H. Tavendale, G.A. Lane, T.N. Barry & W.C. McNabb (2004). Effect of white clover (*Trifolium repens*), perennial ryegrass (*Lolium perenne*) and *Lotus corniculatus* on *in vitro* skatole and indole formation. *Animal Production in Australia, 25*, 164-167.

Ulyatt, M.J. (1981). The feeding value of temperate pastures. In: F.H.W. Morely (ed.) World Animal Science. Elsevier Scientific Co., New York, 125-141.

Young, O.A., G.A. Lane, A. Priolo & K. Fraser (2002). Pastoral and species flavour in lambs raised on pasture, lucerne or maize. *Journal of the Science of Food and Agriculture, 83*, 93-104.

Theme A: Efficient production from grassland

Section 4

Grass and forage physiology

Prediction of canopy photosynthesis for cocksfoot pastures grown under different light regimes

P.L. Peri[1], D.J. Moot[2] and D.L. McNeil[3]
[1]INTA-Universidad Nacional de la Patagonia Austral, Casilla de Correo 332, CP 9400, Rio Gallegos, Santa Cruz, Argentina, Email: pperi@correo.inta.gov.ar, [2]Soil, Plant and Ecological Science Division, PO BOX 84, Lincoln University, Canterbury, New Zealand, [3]Department of Primary Industries, Victoria, Australia

Keywords: canopy photosynthesis, leaf area index, light regime, shade

Introduction Plants in field environments can experience frequent fluctuations in irradiance from full sun to shade caused by cloud cover, overstory shading (e.g. silvopastoral systems) and within canopy shading. Research with widely spaced radiata pine (*Pinus radiata* D. Don) has suggested that due to its shade tolerance cocksfoot (*Dactylis glomerata* L.) is a suitable grass for silvopastoral systems. However, there is limited explanation of the physiological basis for the responses, and consequently no predictive capacity. This limits the application of results to environments, sites and seasons outside of those in which they were measured. The objectives of this study were to simulate net daily canopy photosynthesis rates incorporating the leaf photosynthesis models into a canopy photosynthesis model when only shade was limiting, and to determine the optimum net canopy photosynthesis and LAI for each light regime.

Materials and methods The mathematical model of canopy photosynthesis developed for cocksfoot (Peri *et al.*, 2003) consists of four steps: (1) calculation of leaf light distribution and interception at different canopy depths; (2) calculation of gross canopy photosynthesis incorporating variations in photosynthetic capacity of individual cocksfoot leaves; (3) calculation of total respiration; (4) calculation of net canopy photosynthesis. Daily net canopy photosynthesis (*Pn*) was predicted for cocksfoot under different light regimes by integration of leaf photosynthesis models developed for the light-saturated rate (*Pmax*), the photosynthetic efficiency (α) and the degree of curvature (θ) of the leaf light-response curve. Values of the incident intensity of photosynthetic photon flux density (PPFD) on an area of leaf at the level Z in the canopy (*Iz*) were calculated under the following light regimes and intensities: (i) full sunlight (100% transmissivity), (ii) continuous moderate shade (50% transmissivity), (iii) a fluctuating light regime with alternating periods of full sunlight and severe shade (5% of the open PPFD) at intensities of: 10, 20, 30, 40, 50, 60, 70, 80 and 90% transmissivity. This range was used to represent overstorey canopies of different density or size.

Results *Pn* was parabolic against LAI in all simulations, but as light intensity decreased, the maximum *Pn*, optimum LAI and values of *Pn* after the maximum also decreased. As photosynthetic photon flux density (PPFD) fell from full sunlight to 10% of open PPFD in a fluctuating light regime, maximum *Pn* (*Pn*$_{max}$) decreased approximately linearly from 33.4 g $CO_2/m^2/d$ to zero (Figure 1). Also, it was predicted that for a continuous light regime (50% transmissivity) *Pn*$_{max}$ was higher than for a fluctuating light regime with the same intensity (10.4 vs 8.4 g $CO_2/m^2/d$).

Conclusions The light regimes of cocksfoot plants modified the utilisation of solar energy for net canopy photosynthesis. Continuous light regime overestimated *Pn* by 20% compared with fluctuating light regimes of the same light intensity over a day. Artificial slatted structures are more suitable for simulating the response of understorey pasture species in silvopastoral systems.

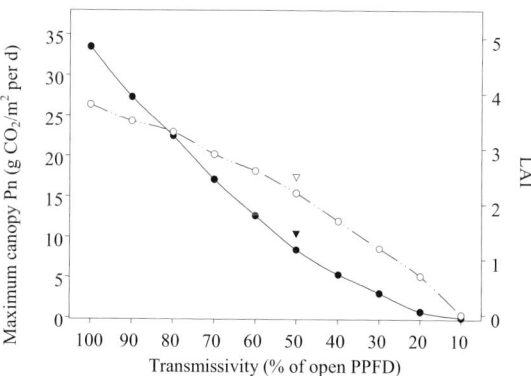

Figure 1 Predicted maximum daily net canopy photosynthesis (*Pn*$_{max}$) (•) and optimum leaf area index (LAI) (○) against different intensities of fluctuating light regime for a cocksfoot pasture. *Pn*$_{max}$ (▼) and optimum LAI (▽) values for a continuous 50% transmissivity light regime are also indicated.

References

Peri, P.L., D. J. Moot & D. L. McNeil (2003). A canopy photosynthesis model to predict the dry matter production of cocksfoot pastures under varying temperature, nitrogen and water regimes. *Grass and Forage Science*, 58, 416-430.

Net photosynthesis rate and chlorophyll content of Caucasian and white clover leaves under different temperature regimes

A.D. Black[1], R.J. Lucas[2] and D.J. Moot[2]
[1]Teagasc, Grange Research Centre, Dunsany, Co. Meath, Ireland, E-mail: ablack@grange.teagasc.ie,
[2]Agriculture and Life Sciences Division, Lincoln University, Canterbury, New Zealand.

Keywords: Caucasian clover, chlorophyll, photosynthesis, temperature, *Trifolium ambiguum*, *T. repens*

Introduction   In spring and summer in intensive temperate pastures, Caucasian clover (Cc) (*Trifolium ambiguum*) has higher dry matter (DM) production rates than white clover (wc) (*Trifolium repens*) (Black *et al.*, 2003). An examination of the physiological basis for these differences can provide a greater insight into the suitability of Cc for inclusion in temperate pastures. Specifically, leaf photosynthesis rate is a major driver of seasonal growth and is strongly regulated by temperature and chlorophyll content. This study aimed to compare the net photosynthesis rate (Pn) and chlorophyll content of Cc and wc leaves under different temperature regimes.

Materials and methods 'Endura' Cc and 'Demand' wc were grown under field irrigation at Lincoln University, New Zealand (Black *et al.*, 2003). Pn was measured on a random sample of 10 youngest fully expanded intact leaves using a photosynthesis system (LI-6400 LiCor, USA) at 7 light intensities, 0, 100, 250, 500, 750, 1000 and 2000μmol/m²/s photosynthetic photon flux density (PPFD) following Peri *et al.* (2002). Leaves were measured for chlorophyll using a chlorophyll meter (SPAD-502 Minolta, Japan) and estimated for chlorophylls *a* and *b* and total chlorophyll, after Salisbury & Ross (1985). Pn and chlorophyll were recorded when measured air temperatures were either limiting (12°C, T_{lim}) or non-limiting (23°C, T_{opt}) for the growth of both species (Black *et al.*, 2003).

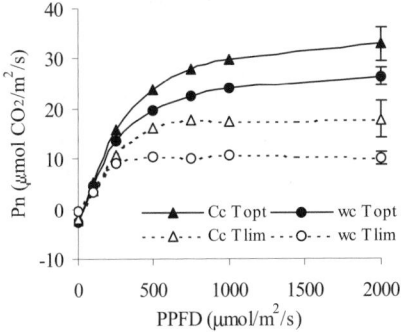

Figure 1 Pn against light intensity for Cc and wc with air temperatures either limiting (12°C, T_{lim}) or non-limiting (23°C, T_{opt}). Bars represent standard errors for Pn_{max}.

Results The response of Pn to light intensity followed a non-rectangular hyperbola for both species at 12°C and 23°C (Figure 1). The maximum rate of Pn (Pn_{max}) for Cc was 32μmol CO_2/m²/s at 23°C, but decreased to 17μmol CO_2/m²/s at 12°C. These Pn_{max} rates were about 6μmol CO_2/m²/s higher than for wc at both temperatures. The concentrations of chlorophylls *a* and *b* and total chlorophyll were higher for Cc than for wc irrespective of air temperature (Table 1).

Table 1 Chlorophyll content (mg/g) of Cc and wc leaves grown with air temperatures either limiting (12°C, T_{lim}) or non-limiting (23°C, T_{opt})

	T_{lim}		T_{opt}		
	Cc	wc	Cc	wc	s.e.d.
Chlorophyll *a*	2.01	1.79	1.90	1.74	0.033
Chlorophyll *b*	0.32	0.25	0.29	0.24	0.010
Total chlorophyll	2.33	2.05	2.20	1.99	0.044

Conclusions The higher Pn for Cc leaves can be explained by their higher chlorophyll content than wc leaves at the air temperatures experienced. Thus, for any given canopy leaf area index, the canopy photosynthesis rate of Cc can be expected to exceed that for wc and give more assimilate/unit leaf area. Confirmation of this would provide an explanation for the DM production advantage of Cc over wc observed in temperate pastures.

References
Black, A.D., D.J. Moot & R.J. Lucas (2003). Seasonal growth and development of Caucasian and white clovers under irrigated and dryland conditions. In: D.J. Moot (ed.) Legumes for dryland pastures. *Proceedings of a New Zealand Grassland Association symposium*, Lincoln University. Grassland Research and Practice Series. no. 11: 81–89.
Peri, P.L., D.J. Moot, D.L. McNeil, A.C. Varella & R.J. Lucas (2002). Modelling net photosynthetic rate of field-grown cocksfoot leaves under different nitrogen, water and temperature regimes. *Grass and Forage Science*, 57, 61–71.
Salisbury, F.B. & C.W. Ross (1985). Plant Physiology. Wadsworth Publishing Company, Belmont, California, 540pp.

Shading effect on production and protein concentration of *Dactylis glomerata* and *Agrostis tenuis*

A. Rigueiro-Rodríguez, S. Rodríguez-Barreira and M.R. Mosquera-Losada
Crop Production Department, High Politechnic School, University of Santiago de Compostela, Spain, Email: romos@lugo.usc.es

Keywords: shading, silvopasture, herbage growth, *Dactylis glomerata*, *Agrostis tenuis*

Introduction Silvopastoral systems make compatible livestock and timber production and provide important advantages from economic and ecological points of view (Sibbald, 1996). Around one million ha of new afforested areas promoted by the EU Common Agricultural Policy have been established in the last decade, that can be used as potential silvopastoral system areas. Pasture production is usually reduced in dense stands as trees grow up due to the light interception by the tree crown, but the radiation reaching the soil will depend on the tree type and this will affect herbaceous species composition and development. The aim of this work was to evaluate the shading effect (0 and 50 % of light interception) on pasture production and composition of monocultures of cocksfoot (*Dactylis glomerata* L. var. Artabro) and bent grass (*Agrostis tenuis* Sibth. cv Highland) in simulated conditions.

Materials and methods The experiment was conducted in Lugo (NW Spain) in a sandy soil. The experimental design was split-plot with three replicates. Shading was the main plot and herbaceous species the subplots. Monocultures of cocksfoot and bent grass were initially established in autumn 1996 in each sub-plot of 1 square meter with seed rates of 30 kg/ha and 8 kg/ha, respectively. Treatments consisted of two shading intensities (0 and 50% light interception) simulated with a black plastic mesh located at 0.5 m above ground and imposed from the time of sowing. Pasture production was estimated from sampling two sub-samples of 0.3x0.3 m prior to harvesting each sub-plot. Five harvests were taken in 1997 at a height of 0.05 m. Protein concentration was determined after microkjheldal digestion (Castro *et al*, 1990). Analysis of variance was used for statistical analyses and LSD for separation of means.

Results The year 1997 had an unusually rainy summer. Pasture production (Table 1) was similar in shaded and unshaded plots, with the exception of the middle spring cut which gave higher yields in unshaded conditions, when light was the limiting factor, and after summer (October) when shading allowed better growth when water supply was the main ecologic constraint for pasture production. Protein content was similar between treatments with the exception of post-summer harvest, when protein concentration was higher under shaded conditions and in the autumn harvest for bent grass when the reverse was found.

Table 1 Pasture production (t/ha) in 1997 (Unsh: unshaded, and Sh: shaded)

	Cocksfoot			Bent grass		
	Sh	Unsh	Sig	Sh	Unsh	Sig
April	3.56	3.30		3.72	3.20	
June	5.16	8.17	*	5.75	6.09	
July	4.05	3.57		4.26	3.83	
October	3.19	1.73	*	2.25	1.29	*
November	1.94	1.32		1.66	1.09	

Table 2 Protein concentration (%) in 1997 (Unsh: unshaded, and Sh: shaded)

	Cocksfoot			Bent grass		
	Sh	Unsh	Sig	Sh	Unsh	Sig
April	10.49	10.27		18.65	19.22	
June	10.35	10.44		14.94	12.52	
July	12.57	15.16		14.77	12.33	
October	19.11	15.68	*	14.38	16.92	
November	23.56	22.38		8.33	13.48	*

Conclusions In our conditions, shading positively affects pasture production in the summer and allows extension of the grazing season in areas with summer drought, but negatively affects pasture production when moisture conditions are adequate and light input can reduce pasture production. Quality of cocksfoot was higher late in the season when grown with shade, as indicated by increased content of protein.

References
Castro, P., A. Gonzalez & D. Prada (1990). Determinación simultánea de nitrógeno y fósforo en muestras de pradera. *XXX Reunión Científica de la Sociedad Española para el Estudio de los Pastos*, 200-207.
Mosquera, M.R. & A. Gonzalez (1999). Pasture production in Northern Spain dairy systems. *New Zealand Journal of Agricultural Research*, 42, 125-132.
Sibbald, A., 1996. Silvopastoral systems on temperate sown pastures: a personal perspective. In: Étienne (ed.) Western European Silvopastoral Systems. INRA editions, 23-37
Whitehead, D. C. (1995). Grassland Nitrogen. CAB International, Wallingford.

The effect of blue light on leaf growth and plant development in two morphologically contrasted perennial ryegrass genotypes: cellular basis and ecological implications

F. Gastal[1], A. Verdenal[1] and P. Barre[2]
[1]INRA-UEPF, F 86600 Lusignan, France. [2]INRA-UGAPF, F 86600 Lusignan, France, Email: gastal@lusignan.inra.fr

Keywords: grasses, leaf growth, tillering, blue light, cellular dynamics

Introduction Several major plant responses to competition for light are determined by responses to light signals, in particular to red/far-red ratio (R/FR) and blue light, besides responses mediated through photosynthesis and carbon assimilation (Gautier *et al.*, 1999). These responses to light signals allow plants to react to the presence of neighbours and to anticipate the impact of light competition on photosynthesis. The objective of the study was to evaluate the impact of blue light on leaf growth and its cellular basis, on two short and long leaved populations (FC and FL respectively), which were shown to have different competitive ability (Hazard *et al.*, 1996).

Material and methods Seeds of FL and FC population were germinated and set into two twin growth cabinets. In one of the cabinets, blue light (400-550 nm) was totally suppressed with plastic filter (Blue⁻ treatment), and supplemental lamps allowed to get similar PAR (320 μM m^{-2} s^{-1}). Leaf elongation rate (LER), leaf elongation duration (LED) and tiller number were recorded. Epidermal cell elongation rate (CER) and number of elongating cells (NEC) in the leaf growth zone were determined on leaf 6 according to MacAdam *et al.* (1989).

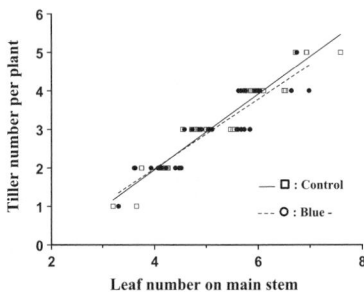

Figure 1 Plant tillers number in relation to leaf number (main stem)

Results Leaf appearance rate was not affected by blue light (not shown). In addition, the relationship between tiller and leaf number on the main stem was similar in control and Blue⁻ treatment, indicating no significant effect of blue on site usage (Fig. 1). Therefore, blue light did not alter the pattern of plant development (rhythm and co-ordination of organ initiation), in both ryegrass populations. In contrast, blue light suppression significantly altered leaf growth (Table 1). Lamina and sheath length were increased by +32% and + 52% respectively. LER was increased by +38%. The increase in LER was mostly due to increased cell elongation rate (FC) or increased division rate (FL). FL had a greater leaf length response to blue light suppression than FC.

Table 1 Effect of suppressing blue light on leaf growth and cellular dynamics

		Mature leaf 6				Growing leaf 6			
		Lamina length (mm)	Sheath length (mm)	Cell number (cel /file)	Cell length (μm)	LER (mm /d)	LED (d)	CER (μm /cel /h)	NEC (cel /file)
FC	Blue+	175 a	38 a	451 a	475 a	1.2 a	140 a	11.0 a	95 a
	Blue-	219 b	51 b	514 ab	535 ab	1.6 b	136 a	13.0 b	113 ab
FL	Blue+	251 b	46 bc	531 b	548 b	1.5 b	162 b	12.0 ab	115 b
	Blue-	346 c	77 c	625 c	676 c	2.0 c	167 b	12.7 ab	139 c

Conclusion Blue light does not alter the timing and the co-ordination of plant development, in contrast to R/FR (Gautier *et al.*, 1999). However, blue light has a major impact on leaf growth and leaf size, due to cell elongation or cell division depending on genotypes. From a grassland ecology point of view, the increase in leaf size under low blue light may i) lead to an increased light capture under situation of competition for light; ii) play a role in the tiller size – tiller density compensations observed according to grazing management.

References

Gautier H., C. Varlet-Grancher, and L. Hazard. 1999. Tillering responses to the light environment and to defoliation in populations of perennial ryegrass selected for contrasting leaf length. *Annals of Botany*, 83, 423-429.

Hazard L., Ghesquière M. and C. Barraux, 1996. Genetic variability for leaf development in perennial ryegrass populations. *Canadian Journal of Plant Science,* 76, 113-118

MacAdam, J.W., J.J. Volenec, C.J. Nelson. 1989. Effects of nitrogen on mesophyll cell division and epidermal cell elongation in tall fescue leaf blades. *Plant Physiology*, 89, 549-556.

Long term tiller population dynamics in swards of grasses with contrasting persistence strategy

F. Gastal[1] and C. Matthew[2]

[1]INRA- UEPF, F 86600 Lusignan, France, Email: gastal@lusignan.inra.fr, [2]Institute of Natural Resources, Massey University, Palmerston North, New Zealand

Keywords: grass sward, persistence, tiller dynamics, nitrogen fertilisation

Introduction The lifespan of individual grass tillers usually does not exceed 12-15 months, because of death of tillers after floral induction and development, or randomly from disease or other factors. Persistence of the tiller population over several years, and associated long term maintenance of the sward, thus depends on the rate of turnover of individual tillers. This study aimed to characterise seasonal and management conditions critical for tiller turnover and its components, tiller birth and tiller death. Two grasses were investigated: *Festuca arundinacea* and *Lolium multiflorum*, having high and low persistence, respectively.

Material and methods Swards of *Festuca arundinacea* (Fa) cv Florine and *Lolium multiflorum* (Lm) cv Fastyl were sown in spring 2000 and grown for 4 years under a cutting regime and under two N fertilisation treatments (not shown). Tiller density was evaluated every 6-8 weeks in 0.0375 m² frames. Within additional 0.0139 m² frames, successive cohorts of tillers were marked with coloured rings and counted every 6-8 weeks, according to the methodology followed by Matthew (1992). Relative tiller birth and death rates were calculated.

Results Tiller density for both species was higher during the winter and lower during the summer (Figure 1A). Tiller density of Lm declined progressively over the 3 years, in contrast to tiller density of Fa, reflecting a low persistence of Lm compared to Fa. Seasonal pattern of the 2 species also differed; the decline of Lm tiller density was large during the summer but recovered partially during the winter.

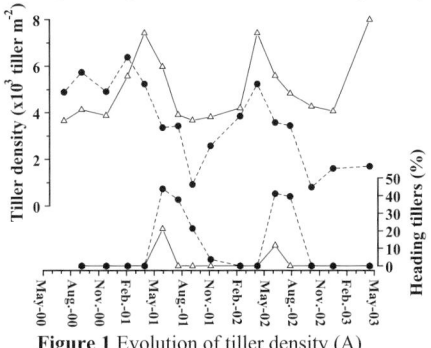

Figure 1 Evolution of tiller density (A) and proportion of heading tillers. (B) (Δ) : *Festuca a.* ; (●) : *Lolium m.*

Heading occurred in May for Fa and from May to Aug. for Lm (Figure 1B). The relative tiller birth rate of Fa was higher during winter (Dec-Mar) than the rest of the year for (Figure 2). For Lm, the winter peak occurred earlier (Oct.-Dec.) than for Fa and a second peak was observed during late spring (June). Thus for both species, tillering was active during periods of low LAI (winter and post flowering recovery period for Lm), in agreement with Simon *et al.*, (1987).

The relative tiller death rate was low during the winter period (Oct.-April) and high from the end of spring (May) to late summer (Sept.) for both species. Tiller death was partly associated with heading for both species, but clearly also partly occurred independently of heading (significant death rate during summer for Fa despite no heading; significant death rate for both species in year 2000 despite no flowering).

Conclusions Winter, which could be seen as a period of rest for the sward, is in fact a period of active tillering and recovery. In contrast, loss of tillers occurs during the growing seasons, not only due to the flowering strategy of the species but also due to other factors. Presumably, summer drought is detrimental for the young tillers that have not yet developed deep roots. A major conclusion is winter tillering is an important period of recovery for grass swards in this environment, and management should be considered accordingly.

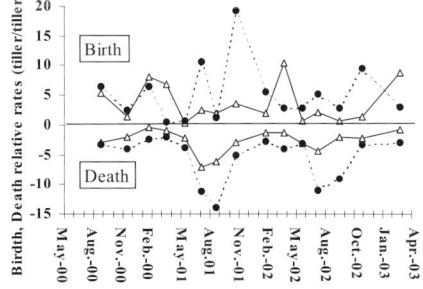

Figure 2 Relative tiller birth and death rates (Δ) : *Festuca a.* ; (●) : *Lolium m.*

References

Matthew, C. (1992). A study of seasonal root and tiller dynamics in swards of perennial ryegrass. PhD Thesis. Massey University, New Zealand.

Simon, J. C. & G. Lemaire (1987). Tillering and leaf area index in grasses in vegetative phase. *Grass and Forage Science*, 42:373-380.

Morphogenetic characteristics of *Panicum maximum* cv. Aruana subjected to five defoliation stubble heights and two frequencies

D.D. Carvalho and A.A. Giacomini
Instituto de Zootecnia, Nova Odessa, Rua Heitor Penteado, 56, Brazil, Email: dcarvalho@iz.sp.gov.br

Keywords: aerial tiller, basal tiller, leaf area, tiller number, volume

Introduction Tillers in grass swards are subject to size density compensation and this mechanism has been observed to follow the –3/2 self thinning rule. This theory assumes that tiller components (leaf lamina and stems) have a constant geometry as the sward is taller or shorter. In a re examination of this rule (SackvilleHamilton *et al.*, 1995) observed that in grass swards the slope can be different from -3/2 depending on the extremes of defoliation (Hernandez-Garay *et al.*, 1999). Therefore, the dimensionless measure, R (ratio tiller leaf area : volume) was proposed to isolate the tiller geometry component from the tiller size. Guinea grass is the second most sown grass in Brazilian pastures and cv. Aruana has been used successfully in the past ten years in sheep grazing systems. Since this bunch grass presents numerous aerial tillers that contribute to yield, the objective of this paper was to evaluate changes in tiller leaf area and volume on plants subjected to high frequency (simulating a continuous grazing), and low frequency defoliation (simulating an intermittent grazing), under different defoliation stubble heights.

Material and methods Treatments corresponded to five defoliation stubble heights (70, 140, 210, 280 and 350 mm above ground level) and two frequencies of defoliation: high (defoliation every 10 days in Spring/Summer and 15 days in Autumn/Winter) and low (defoliation at monthly intervals) in potted plants in a glasshouse during one year. At the end of the experimental period plants were destructively harvested and from each plant all basal and aerial tillers were counted and volume (Sbrissia et al., 2004), leaf area and dry weight of each tiller category calculated. Each frequency was considered a complete and independent experiment on a complete block design with five treatments (cutting heights) and four replicates.

Results A decrease in basal tillers associated to an increase in aerial tillers was observed in both frequencies with an increase in defoliation height (Table 1). Basal tillers dry weight and volume was greater compared to aerial tillers due to the great amount of stem component in both defoliation regimes resulting in smaller ratio leaf area: volume in basal tillers. New studies are being conducted to evaluate the contribution of each tiller group (basal and aerial) to dry matter production of this cultivar.

Table 1 Morphogenetic variables of basal and aerial tillers of Guinea grass cv. Aruana subjected to five defoliation stubble height and two harvest frequencies: High = 31 harvests/year (on the left) and Low = 13 harvests/year (on the right)

Variables	Tiller type	Stubble height (mm) 70	140	210	280	350	Mean	sem	Tiller type	Stubble height (mm) 70	140	210	280	350	Mean	sem
Tiller number	Bas	92.0	42.3	10.5	8.5	10.3	32.6	16.11	Bas	44.5	15.0	4.0	3.2	4.2	14.2	4.2
(plant/pot)	Aer	0	42.5	54.3	65.0	58.8	55.1	4.75	Aer	13.9	16.5	26.0	31.7	29.5	23.5	3.54
Dry weight/tiller	Bas	79	130	231	300	424	232	61.4	Bas	366	434	902	645	697	609	96.0
(mg)	Aer	---	55	92	130	181	114	27.0	Aer	212	321	315	402	345	319	30.9
Leaf Area/tiller	Bas	5.2	6.0	11.5	13.0	12.7	9.7	1.69	Bas	42.4	53.7	56.4	42.4	50.8	49.1	2.95
(cm2)	Aer	---	4.2	6.9	9.1	13.9	8.5	2.05	Aer	34.3	38.5	37.7	37.4	42.6	38.1	1.33
Volume/tiller	Bas	0.6	0.8	1.5	1.4	2.5	1.4	0.34	Bas	1.9	2,4	4,6	3,3	3.0	3,0	0,45
(cm3)	Aer	---	0.3	0.7	0.9	1.2	0.7	0.18	Aer	1.2	1.7	1.6	2.0	1.7	1.6	0.12
R	Bas	22.4	19.5	37.2	35.0	30.4	2894	3.46	Bas	143.8	172.7	100.0	92.7	139.7	129.8	14.82
(ratio leaf : area:volume)	Aer	---	28.6	29.5	32.4	46.6	34.3	4.18	Aer	157	151.8	149.3	159.1	165.1	156.4	2.78

Conclusion These results suggest a great phenotypic plasticity in this cv. which provides great flexibility of response to grazing management

References
Hernandez Garay, A., Matthew, C.& Hodgson, J. (1999) Tiller size-density compensation in ryegrass miniature swards subject to differing defoliation heights and a proposed productivity index. *Grass and Forage Science*, 54 (4), p.347-356,
Sackville Hamilton, N.R., Matthew, C. & Lemaire, G. (1995). In defence of the –3/2 boundary rule: a re-evaluation of self-thinning concepts and status. *Annals of Botany*, 76, 569-577.
Sbrissia, A.F., Da Silva, S.C., Molan, L.K., Sarmento, D.O.L., Andrade, F.M.E., Lupinacci, A.V.& Gonçalves, A.C. (2004) A simple methodology for measuring tiller volume. *Grass and Forage Science*, 59 (4), 406-410.

Effect of defoliation interval on regrowth of leaves and roots, and tiller number of cocksfoot plants

L.R. Turner[1], D.J. Donaghy[1] and P.A. Lane[2]
[1]Tasmanian Institute of Agricultural Research (TIAR), P.O. Box 3523, Burnie TAS 7320 Australia, [2]TIAR, University of Tasmania, Private Bag 98, Hobart TAS 7001 Australia Email: wilsonlr@utas.edu.au

Keywords: defoliation interval, regrowth, cocksfoot

Introduction The key to defoliation management for optimal production and persistence of pasture lies in the use of a physiological basis for defoliation interval (specific to plant type), as opposed to a regime based on time or the height of herbage. The full expansion of a particular number of leaves/tiller is a useful plant-related indicator of optimal defoliation timing. Leaf regrowth stage ('leaf stage') reflects the stage of plant recovery from defoliation as regards plant energy levels (Fulkerson & Donaghy, 2001). The level of water-soluble carbohydrate (WSC) reserves in grass tiller bases influences the rate of regrowth after defoliation, affecting the rate of shoot growth, root growth and tillering. This study was aimed to determine the influence of leaf stage-based defoliation interval on regrowth of leaves and root, and on tiller number of 'Kara' cocksfoot plants up to 24 days after defoliation.

Methods Each leaf stage was defined as the time required to produce the appropriate number of fully expanded leaves/tiller. Treatments were based on defoliation intervals of 1, 2 and 4-leaf stages of regrowth, with treatments terminated when the 1-leaf defoliation interval had been completed 4 times, the 2-leaf interval 2 times and the 4-leaf interval once. Selected plants were harvested destructively immediately after cessation of treatments (H1) and at 5 days, 10 days and 24 days after H1. Leaf, root and tiller dry matter (DM) yield were determined at each harvest event, and tiller number/plant was determined at 24 days after H1.

Results Leaf, root and tiller DM yields related closely to defoliation interval, with a general trend of increasing regrowth with increasing defoliation interval. A significantly lower (P<0.05) DM yield was associated with defoliation at 1 leaf/tiller compared with 4 leaves/tiller, for all plant components, lasting until 24 days after H1 (Table 1). A significant difference (P<0.05) in tiller number between all treatments had developed by 24 days after H1. Repetitive defoliation at the 1-leaf stage resulted in a mean of 42±5 tillers/plant, while defoliation at the 2-leaf and 4-leaf stages resulted in a mean 63±5 and 83±5 tillers/plant, respectively.

Conclusions Less frequent defoliation of cocksfoot plants leads to greater leaf, root and tiller DM accumulation during subsequent recovery. The critical

Table 1 Effect of defoliation interval on DM accumulation from H1 to 24 days after H1

	Leaf stage	H1	5d	10d	24d
Leaf	1L	0,40	0,20	0,52	2,24
(g/plant)	2L	3,06	0,37	1,14	6,24
	4L	9,22	0,91	2,68	7,39
LSD (P = 0.05)		1,52	0,29	0,79	3,43
Root	1L	1,72	1,27	1,37	1,49
(g/plant)	2L	2,53	1,68	2,23	2,84
	4L	6,14	4,31	5,36	5,56
LSD (P = 0.05)		1,15	1,44	1,42	1,42
Tiller	1L	37,0	35,0	39,0	30,0
(mg/tiller)	2L	52,0	43,0	38,0	37,0
	4L	75,0	62,0	52,0	57,0
LSD (P = 0.05)		10,0	20,0	6,0	12,0

defoliation interval in terms of influencing regrowth is at 2 leaves/tiller. Defoliation at 1 leaf/tiller is unsatisfactory for regrowth, most likely through depletion of WSC reserves. Defoliation at 2 leaves/tiller is adequate for recovery but does not maximise regrowth. This study suggests that a rotation based on the 4-leaf stage will maximise leaf, root and tiller DM accumulation. Rawnsley et al. (2002) found that WSC levels increase significantly in cocksfoot stubble and roots at the 4-leaf stage of regrowth, indicating that this is the point at which WSC supply from photosynthesis exceeds that needed for growth and respiration. Tiller number also reflects the effect of defoliation on energy reserves and provides further support for this defoliation management recommendation.

References
Fulkerson, W. J. & D. J. Donaghy (2001). Plant soluble carbohydrate reserves and senescence - key criteria for developing an effective grazing management system for ryegrass-based pastures: a review. *Australian Journal of Experimental Agriculture*, 41: 261-275.
Rawnsley, R. P., D. J. Donaghy, W. J. Fulkerson & P. A. Lane (2002). Changes in the physiology and feed quality of cocksfoot (*Dactylis glomerata* L.) during regrowth. *Grass and Forage Science*, 57: 203-211.

Microcalorimeter as a biologic activity monitor for the study of *Brachiaria brizantha* seed germination process

M.A. Barboza[1], P.L.O. Volpe[1], R. Usberti[2], J.F.G. Faigle[1] and R.H. Aguiar[3]

[1]*Chemical Institute, Campinas State University, P.O. Box 6154, CEP 13084-971 Brazil, Email: barboza@iqm.unicamp.br,* [2]*Plant Protection Agency, Campinas, Brazil* [3]*Faculty of Agricultural Engineering, Campinas, Brazil*

Keywords: heat flow germination, microcalorimetric, energetic cycle, *B. brizantha* seeds

Introduction Calorimetry helps better understanding of biological processes (Calvet & Prat, 1963). Very sensitive thermal sensors and microcalorimeters allow real time investigation and monitoring heat production of seed germination but few experiments have been performed in this area (Sigstad & Prado, 1999). Moreover, experimental procedures correlating germination phenomena and chemical thermodynamics are exceptional (Barboza, 2002). One can detect calorimetrically the heat flow produced during seed germination and compare the results with data recorded using standard germination methodology (ISTA, 1985). Seed germination and the biomass increase respiration and determination of the energy involved aids understanding of the energetic cycle involved. This work analysed the germination of *Brachiaria brizantha* seeds, including the water uptake phase.

Materials and methods Experiment used intact and chemically scarified (sulphuric acid, 96%, 36N, 15') seeds of *B. brizantha* cultivar Marandu. Germination rates were 28 and 78%, respectively, evidencing seed dormancy. An isothermic conduction microcalorimeter TAM, Thermometrics 2277 was used, with a twin system of heat detection i.e. of sample and reference vessels. A homemade calorimetric vessel was developed to permit the water addition and gas flow control through the vessel and to retain the water vapour. The reference vessel was identical, without the gas flow attachment. Heat effect due to gas flow was considered in the standard deviation.

Results Figure 1 shows germination heat flow of 10 *B. brizantha* scarified seeds. Heat flow rose during 14d of germination. Water uptake (physical chemical part) occurred until d2; a latency period (low biological activity) occurred from d2-5; an exponential growth of heat production + radicle protrusion (visible germination) occurred from d5. Figure 2 shows the mass increase of seed water uptake. Figure 3 shows the correlation of the inverse tendency between mass increase and the heat flow. The exothermic tendency indicates that the system releases energy during anabolic and catabolic processes. The trend (Figure 2) was unexpected (intact seeds released most); release probably was caused by the seed coats (highly hygroscopic), which were removed partially during acid scarification. Figure 4 shows that in phase 1 the water probably penetrates through seed coats, with a significant difference between scarified and intact seeds. In phase 2, we assume that water penetrates the endosperm, showing similar behaviour for both kinds of seeds, as the thermograms show.

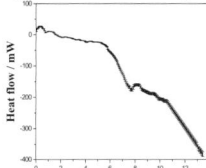 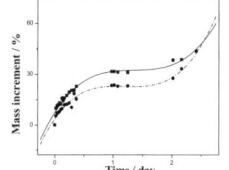 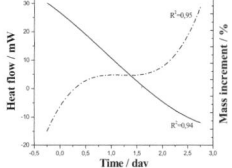 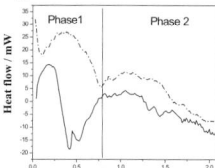

Figure 1 Typical thermogram of *B. brizantha* 10-seed germination; air flow 15 mL/h, error bar equivalent to 2 % of measured energy (232000 measurements)

Figure 2 *B. brizantha* seed water uptake fitted by a 3rd degree polynomial curve for intact (dotted line) and scarified (straight line)

Figure 3 Typical trends observed on heat flow and mass increment in *B. brizantha* seeds, both fitted by a 3rd degree function and showing typical fisiological phases

Figure 4 Thermograms of water uptake on intact (dotted line) and scarified (straight line) *B. brizantha* seeds, showing typical physiological phases

Conclusions Microcalorimetry showed important details of seed germination. Its use for biological studies on forage grasses is promising.

References

Barboza, M.A. (2002). Magnetic field action in some chemical and biological systems. MS Thesis. Unicamp, Campinas State University, Brazil. p 87

Calvet, E. & H. Prat (1963). *Recent progress in microcalorimetry.* Pergamon. London.

ISTA, International Seed Testing Association. (1985). International rules for seed testing. RULES 1985. *Seed Science and Technology,* 13, 299-355; 356-513.

Sigstad, E.E. & F.E. Prado (1999). A microcalorimetric study of *Chenopodium quinoa* willd. seed germination. *Thermochimica acta,* 326, 156-164.

Temperature response comparison of controlled and field environments for four tropical grasses

S. Fukagawa[1], Y. Ishii[2], K. Sato[3], R. Kobayashi[3] and I. Hattori[3]
[1]Nagasaki Prefectural Livestock Experiment Station, Email: s.fukagawa-123@pref.nagasaki.lg.jp, [2]Faculty of Agriculture, University of Miyazaki, [3]National Agricultural Research Center for Kyushu Okinawa Region, Japan

Keywords: temperature response, tropical grasses, controlled environment

Introduction Tropical grasses are cultivated mostly as annuals in the warm region of SW Japan. They have a long-term sowing time after harvesting temperate Italian ryegrass. We compared the early growth of tropical grasses in a controlled environment vs. field data at 2 sowing times to determine their temperate response.

Materials and methods From 19 December, 2001 to 28 January, 2002, a controlled environment facility (CEF) experiment was conducted at the National Agricultural Research Center, Kyushu Okinawa Region (32° 53' N, 130° 44' E). Rhodesgrass (*Chloris gayana* Kunth cv. Asatsuyu: Rg), guineagrass (*Panicum maximum* Jacq. cv. Natsukomaki: Gg), colored guineagrass (*Panicum coloratum* L. cv. Tamidori: Cg), and sudangrass (*Sorghum sudanense* (Piper) Stapf cv. Sugarslim: Sg) were sown with 3 seedlings per 1/5000 a Wagner pot. Temperature regimes were 20/15°C (LT), 25/20°C (MT), and 30/25°C (HT). Day length and relative humidity were 14h and 80%, respectively. The plants were sampled 40 days after sowing to investigate the tiller number, top dry matter weight (TDW), mean tiller weight (MTW) and growth rate (GR). From 14 May to 18 July, 2002 a field data (FD) experiment was conducted on an experimental field in the Nagasaki Livestock Experimental Station (32° 14' N, 130° 20' E). Sowing dates of the same 4 species were 14 May (MS) and 7 June (JS). Sowing rates of Sg and the other 3 grasses were 3 and 1g/m[2], respectively, at a 40-cm row distance. Plants were sampled 44–46 days after sowing to investigate growth characteristics and to calculate the crop growth rate (CGR).

Results Daily mean temperature increased with the increase in temperature regime in the CEF and was higher at JS in FD. As regards daily mean temperature, MS was plotted between LT and MT; JS was between MT and HT. Figure 1 shows TDW and MTW. TDW and MTW increased concomitant with the increase in daily mean temperature in both FD and CEF. The changes in MTW correlated almost exactly with those in TDW. Figure 2 shows relations of daily mean temperature with GR and CGR. Differences of GR in CEF among grasses were greater than those of CGR in FD. The GRs of Gg and Sg increased constantly from LT to HT, whereas those of Rg and Cg increased slowly from MT to HT. The high GRs of Sg and Rg at LT in CEF were related to high seed weight (Fukagawa *et al.*, 2003) and a high germination rate.

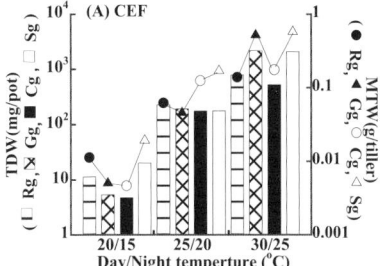

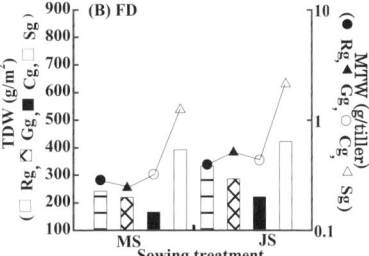

Figure 1 Effect of temperature on TDW and MTW

Conclusions Air temperature affected early growth for grasses sown from May to July in FD analogously to that in CEF. For that reason, optimal grass species for May-sowing would be Sg and Rg. Optimal species for late sowing in July would be Sg and Gg because of their prominent early growth.

References
Fukagawa S., K. Sato, R. Kobayashi & I. Hattori (2003). Effect of temperature treatment on the germination and early growth in five tropical grasses. *Kyushu Agricultural Research*, 65, 132.

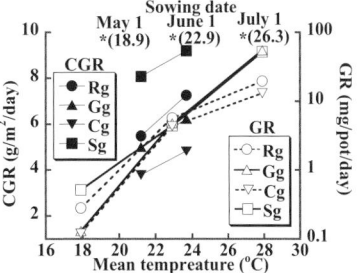

Figure 2 Relations of mean temperature with GR and CGR

Gas exchange and stomatal density of Brachiaria brizantha in a silvopastoral system during one dry period

M. Rakocevic, F.C. de Oliveira and J. Ribaski
Embrapa Florestas, Estrada da Ribeira Km 111, 83411-000 Colombo, Paraná, Brazil,
Email: mima@cnpf.embrapa.br

Keywords: net photosynthesis, stomata, stomatal conductance, transpiration

Introduction The integration of grasses and trees strips (silvopastoral systems, SPS) planted on the contours can be economically beneficial and at the same time reduce erosion (Schaller *et al..*, 2003). The production of the grass component varies depending on the orientation of the tree strips and distance between trees (Rakocevic & Ribaski, 2003). The aim of this study was to determine the leaf anatomical and functional properties of *Brachiaria brizantha* Hochst. ex A. Rich. (Bb), in a SPS with *Corymbia citriodora* Hook. (Cc), in relation to tree orientation and proximity, during one dry period.

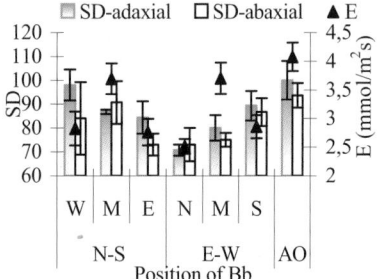

Figure 1 Stomatal density (SD) and transpiration (E) of Bb n SPS

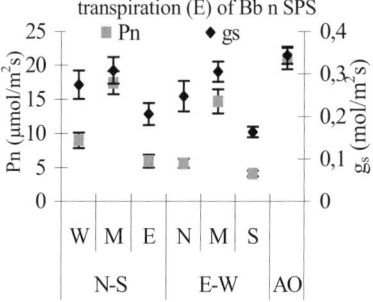

Position of Bb

Figure 2 Net photosynthesis (Pn) and stomatal conductance (g_s) of Bb

Material and methods The experiment was conducted on one eleven-year-old SPS established on 70 ha, with Cc and Bb components planted on the contours (30 m distance between tree strips), in Tamboara, north-west of Paraná State, southern Brazil (23° 10′ latitude and 52° 27′ longitude), average altitude of 480m. The climate is Cfa (Köppen´s classification) with droughts relatively frequent, but no defined drought season. The soil was classified as Rhodic Haplodux (depth to 40 m). Two distinct Cc orientations were defined, North-South (N-S) and East-West (E-W). *B. brizantha* was measured in the mid-distance (M) between tree strips (15 m) and at 3 m distance from tree lines, corresponding to W and E side of N-S orientation, and N and S side of E-W tree orientation. The open area (AO) of pasture was used as a control. *B. brizantha* shows C_4 photosynthetic pathway and was found to be amphistomatous. Ecophysiological parameters: net photosynthesis (Pn - μmolCO$_2$/m^{2}s), stomatal conductance (g_s - molH$_2$O/m^{2}s), and transpiration (E - mmolH$_2$O/m^{2}s) were measured on ten plants per treatment in two repetitions using a LICOR 6200 gas analyser, at the end of a 26 d-period without any precipitation in Nov. 2001. Stomatal density (SD per mm^2) on adaxial and abaxial surfaces was determined on three replicates per treatment. All parameters were measured on the youngest fully expanded leaf.

Results and discussion In SPS (Figures 1, 2) functional parameters and SD decreased compared with control (AO), as a result of Bb sensitivity to light reduction (Dias-Filho, 2002). Values for E and Pn were lower in relation to tree proximity (W, E, N, S), than in the middle (M), as a consequence of inter-specific competition for water in order to reduce water loss through stomatal regulation (SD in Figure 1 and g_s in Figure 2). This phenomenon was recently described by Earl (2002) for soybean under drought stress and enlarged by non-stomatal regulation (internal CO$_2$ concentration).

Conclusion The heterogeneity of Bb functional responses in the SPS is due to plasticity and high sensitivity to light reduction and water stress. The reduction of CO$_2$ assimilation (Pn) of Bb is regulated through stomatal restriction (stomatal density and stomatal conductance) under drought and light competition.

References

Dias-Filho, M. B. (2002). Photosynthetic light response of the C_4 grasses *Brachiaria brizantha* and *B. humidicola* under shade. *Scientia Agricola*, 59, 65-68.

Earl, H. J. (2002). Stomatal and non-stomatal restrictions to carbon assimilation in soybean (*Glycine max*) lines differing in water use efficiency. *Environmental and Experimental Botany*, 48, 237-246.

Rakocevic, M. & J. Ribaski (2003). The efficiency of *Brachiaria brizantha* Hochst. ex A. Rich., in a silvipastoral system in Southern Brazil. In: M. Zlatic, M.S. Kostadinov & N. Dragovic (eds.) Proceedings of Int. Conf. Natural and Socio-Economic Effects of Erosion Control in Mountainous Regions, Faculty of Forestry, Belgrade, 323-332.

Schaller, M., G. Schroth, J. Beer & F. Jimeénez (2003). Root interactions between young *Eucalyptus deglupta* trees and competitive grass species in contour strips. *Forest Ecology and Management*, 179, 429-440.

Are leaf traits stable enough to rank native grasses in contrasting growth conditions?

V. Poozesh, R. Al Haj Khaled, P. Ansquer, J.P. Theau, M. Duru, G. Bertoni and P. Cruz
Research unit UMR 1248 Managing grassland farming systems, BP 27, F - 31326 Castanet -Tolosan - France,
Email: vahid.poozesh@ensat.fr

Keywords: leaf dry matter content, specific leaf area, natural grasslands, functional traits

Introduction   The growing interest in classifying species in response groups relating to variations in environmental factors has triggered the search for functional traits that express differences in ecological behaviour among plant species (Lavorel & Garnier, 2002). Specific leaf area (SLA) and leaf dry matter content (LDMC) reflect a fundamental trade-off in plant functioning between a fast growth rate (high SLA, low LDMC species) and nutrient conservation (low SLA, high LDMC species). This study aimed to analyse the stability of ranking native grasses by SLA and LDMC values under different plant growing conditions.

Materials and methods   Twelve wild grass species, *Anthoxanthum odoratum* (Ao), *Agrostis capillaris* (Ac), *Arrhenatherum elatius* (Ae), *Avenula marginata* (Am), *Brachypodium pinnatum* (Bp), *Briza media* (Bm), *Dactylis glomerata* (Dg), *Danthonia decumbens* (Dd), *Holcus lanatus* (Hl), *Lolium perenne* (Lp), *Festuca rubra* (Fr), and *Molinia cœrulea* (Mc), harvested in natural Pyrenean meadows, were cultivated in a growth chamber with a complete nutrient solution (GC treatment: 260μ E/m^2 per s , 25-18°C day-night temperature) and in heavily fertilised, irrigated plots at Toulouse (T). LDMC and SLA were measured in the two treatments following the standard protocol described by Garnier *et al.*, (2001). Mean of 3 replicates of 3 plants in GC or 12 plants in T were compared to a database of traits measured in field conditions (F) in the Central Pyrenees. As Dd was not cultivated in T, correlations were calculated for 11 or 12 pairs of species.

Results   Pearson's correlations and Spearman's rank correlations between GC values and F values or T values are generally higher for LDMC than for SLA (Table 1). Growth chamber and field LDMC correlated fairly well for all the grasses excepted Bm. This species showed lower field LDMC than expected (Figure 1).

Table 1 Correlation between LDMC and SLA measured in contrasted environments [#]

Treatments	Correlation coefficients	Traits	
		LDMC	SLA
GC - T	r Pearson	0.913 ***	0.658 *
	r Spearman	0.884 ***	0.747 **
GC - F	r Pearson	0.828 ***	0.794 **
	r Spearman	0.886 ***	0.725 **
T - F	r Pearson	0.874 ***	0.753 **
	r Spearman	0.736 **	0.825 **

[#] *** P < 0.001; ** P< 0.01; * P < 0.05

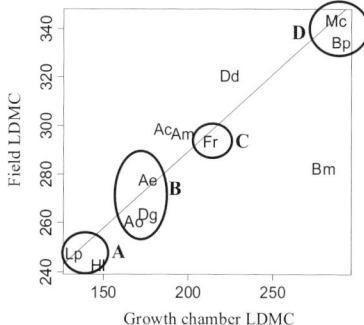

Figure 1 Relation between field and growth chamber LDMC of grasses (n = 11, Bm excluded; Pearson's r = 0.945 ***, Spearman's r = 0.943 ***). A, B, C and D are functional types of grasses (Ansquer *et al.*, 2004)

Conclusions   In spite of variation in growth factors (large fertility gradient between F and T or GC and lower photon fluxes in GC than in F and T), the LDMC was robust enough to rank grass species. Species ranking, based on correlation of agronomic characteristics and functional traits (Ansquer *et al.*, 2004), agree with the functional grass typology, as used to assess the utilisation value of natural grasslands according to their dominant grasses.

References

Ansquer P., J. P. Theau, P. Cruz, J. Viegas, R. Al Haj Khaled & M. Duru (2004). Caractérisation de la diversité fonctionnelle des prairies naturelles. Une étape vers la construction d'outils pour gérer les milieux à flore complexe. *Fourrages*, 179, 353-368.
Garnier, E., B. Shipley, C. Roumet & G. Laurent (2001). A standardized protocol for the determination of specific leaf area and leaf dry matter content. *Functional ecology*, 15, 688-695.
Lavorel, S., & E. Garnier (2002). Predicting changes in community composition and ecosystem functioning from plant traits: revisiting the "Holy Grail". *Functional Ecology*, 16, 545-556.

Persistence strategy of *Panicum maximum* cv. Tanzania in grazed pastures

P.M. Santos[1], L.G.H. de Camargo[2], M.M. Santoni[2], N.A. de Souza[2], C.H.B.A. Prado[2], C.M.B. Nussio[3]

[1]*Embrapa Southeast – Cattle Research Centre, Rod. Washington Luiz, km 234, Caixa Postal 339, CEP 13560-970, São Carlos – SP, Brazil, Email: patricia@cppse.embrapa.br,* [2]*Universidade Federal de São Carlos, Rod. Washington Luiz, CEP 13560-970, São Carlos – SP, Brazil,* [3]*University of Sao Paulo, Av. Pádua Dias, 11, CEP 13400-000, Piracicaba-SP, Brazil*

Keywords: flowering, *Panicum maximum*, tiller survival, rotational grazing

Introduction In many cases, tiller age cohorts survival diagrams show seasonal increases or decreases in rates of tiller birth and death, which may be regarded as persistence strategy (Matthew *et al.*, 2000). The aim of this work was to analyse tiller demographic information of *P. maximum* cv. Tanzania to determine its persistence strategy.

Materials and methods The experiment was conducted on a *Panicum maximum* cv. Tanzania pasture, in São Carlos, Brazil (21°57'42" S, 47°50'28" W). The pasture was rotationally grazed in a 30-day cycle (1-d grazing, 29-d rest). Tiller cohorts of five tussocks were identified by coloured plastic coated wire (Santos, 2001). Each tiller cohort represented a group of tillers marked during the same grazing cycle; Cohort 1 corresponded to the initial population marked in November 2002 and cohorts 2, 3, 4, 5, 6 and 7 to tillers that appeared in January, March, April, June, August and October 2003, respectively. A survival diagram for marked tiller age cohorts was made with means of five tussocks.

Results Initial mortality rates of tillers cohorts 1 and 2 were higher than for the others tiller cohorts (only 28 and 46% of tillers from cohorts 1 and 2, respectively, survived until the grazing cycle following its appearance). Tiller cohort 3 represented 25 and 13% of total tiller population in August and October, respectively; thus it seems to be important for sward persistence. Reproductive tillers represented 27, 20 and 10% of total tiller number in April, June and August, respectively. Sward renewal occurred mainly during spring and summer (from September till March). It is necessary to categorise new tillers according to the age cohort of the tillers that produced them to better characterise the persistence strategy of *Panicum maximum* cv. Tanzania.

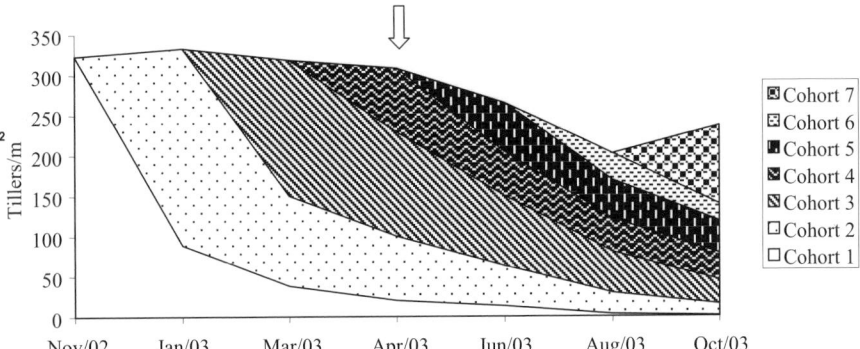

Figure 1 Survival diagram for marked tiller age cohorts of *Panicum maximum* cv. Tanzania. Arrow indicates the beginning of flowering

Conclusions Sward renewal in *Panicum maximum* cv. Tanzania occurs mainly during spring and summer. It is necessary to categorise new tillers according to the age cohort of the tillers that produced them to better characterise the persistence strategy of *Panicum maximum* cv. Tanzania.

References

Matthew C., S.G Assuero, C.K. Black & N.R. Sackville Hamilton (2000). Tiller dynamics of grazed swards. In: Lemaire, G, Hodgson, J., Moraes, A., Carvalho, P.C.F. & Nabinger, C., (eds.) *Grassland Ecophysiology and Grazing Ecology*. Wallington: CAB International, 127-150.

Santos, P.M. (2002). *Controle do desenvolvimento das hastes no capim-tanzânia: um desafio*. PhD. Thesis, Escola Superior de Agricultura "Luiz de Queiroz", Universidade de São Paulo.

Utilisation of photosynthetic active radiation by grasslands in time and in space

V.V. Kolomeychenko
Orel State Agrarian University, General Rodin St., 54, 10, Orel, 302019, Russia, Email: kolom_alla@inbox.ru, audit-kontakt@tula.net

Keywords: production process, energetic coefficients, landscape, cropping, forestry

Introduction Natural grasslands occupy about 3b ha worldwide This is twice the area of arable lands (McCloud, 1974), making it particularly important to understand the energetic relationships of grasslands. This paper contributes to such an understanding for the steppe and forest-steppe zones of Russia.

Materials and methods The theoretical and experimental research was carried out from 1963 to 2004 in the steppe and forest-steppe zones (Volgograd, Tula, Orel, Kursk). The experiments involved natural grass stands and field crops. Calculations were made of the following indices: leaf area index (LAI) (Kolomeychenko, 1982), leaf area duration (LAD), net assimilation rate (NAR) and coefficients of the utilisation of photosynthetic active radiation (PhAR) in time (Kt) and in space (Ks).

Results and discussion There are two important energetic coefficients of PhAR utilisation - in time (Kt) and in space (Ks). Our results from long-term research (1963-2004) indicated that maximum PhAR utilisation in time (Kt=100%) is possible only in highly-productive natural and cultivated plant communities (meadows, forests, bogs, perennial legumes and grasses) with optimum moisture conditions. The daily mean air temperature in spring and autumn must be about +5°C. This defines the potential possible growing season (PPGS). The average Kt for annual crops in the forest-steppe zone is in the range of 46 to 54% - relatively low values, because much of the PhAR is received by soil without green plants. This cropping pattern, typical of the 20th century, with only one crop per year may be considered to be very primitive. A challenge for the 21st century is to increase Kt by extending the PPGS. This may be achieved by the use of catch crops, perennial legumes and grasses on arable lands increasing Kt up to 80-100%.

The Central Chernozem Biosphere State Natural Reservation (Kursk) may be considered a model of ideal cultural landscape with the harmonious combination of different land uses: meadow steppes and forests occupy the whole territory and plough lands are totally absent. The good rainfall (500-600 mm) and fertile chernozem soils provide effective utilisation of PhAR from early spring to late autumn, with the large biodiversity of plant and production processes in the landscape being important features (Kt=100%). In the surrounding regions of forest-steppe and steppe zones some 75-80% of the area is in ploughed fields and Kt is 54-67%. In order to increase energetic efficiency in the forest-steppe and steppe zones there needs to be considerable reduction in therisks of erosion of plough lands and their conversion to more sustainable use. In areas of undulating relief, afforestation and establishment of grasses are the main approaches that can be followed to protect soil from erosion.

Global changes of climate will give the possibility of compensating for this reduction of arable lands in the southern regions by developing new lands in the north. According to our calculations, Ks of natural grasslands in the forest-steppe zone is only 0.1-0.3% (Kolomeychenko, 1985, 2003). Effective methods for grassland improvement can result in a 5- to 10-fold increase in Ks and in production. For instance, the annual top dressing of N180P90K90 increased LAI from 0.94 to 4.30; LAD from 229 to 1047 th.m^2 x day/ha and Kt from 0.33 to 1.46%. The improvement of mineral fertilisation allowed the optimisation of photosynthetic activity with the result that the above ground dry matter (AGDM) increased from 0.86 to 3.75 t/ha and protein yield (PY) from 100 to 540 kg/ha. Two and three cuts of a grass and legume mixture were compared with annual application of mineral fertilizers of N180P60K60. For two and three cuts respectively, LAD was 2054 and 838 th.m^2 x day/ha, average NAR was 3.9 and 6.0 g/m^2 x day and Ks was 1.54 and 1.06%. As a result, AGDM was 8.01 and 5.03 t/ha and PY was 1060 and 1050 kg/ha for two and three cuts respectively.

Conclusion Long-term investigations allowed elucidation of the main principles of the energetic efficiency of grassland, cropping and different landscape organizations for the Russian forest-steppe and steppe zones. This approach could be applied to other countries with similar relief, soils and climatic conditions.

References
Kolomeychenko, V. (1982). Gravimetric determination of leaf area in grasses. *Photosynthetica*, 16, 251-252.
Kolomeychenko, V. (1985). Optimization of photosynthetic activity and increase in productivity of natural grasslands on the slopes. *Proceedings XV International Grassland Congress,* 5-6.
Kolomeychenko, V. (2003). Ecological role of grasslands on slopes. *Grassland Science in Europe,* 8, 596-598.
Mc Cloud, D. (1974). Man's impact on world grassland. *Proceedings XII International Grassland Congress*, 51-64.

Modelling of nitrogen allocation and partitioning within lucerne (*Medicago sativa*) shoot tissues during recovery from defoliation: an approach to estimate forage production and nitrogen composition

F. Meuriot[1], A. Escobar-Gutiérrez[2], J-C, Avice[1], J-C. Simon[1], F. Lesuffleur[1] and F. Gastal[2]
[1]*UMR INRA EVA - Nutritions NCS, Esplanade de la Paix, 14032 CAEN Cedex, France, Email: avice@ibfa.unicaen.fr, [2]UEPF INRA, Domaine du Chêne, 86600 LUSIGNAN, France*

Keywords: forage, modelling, nitrogen, amino acids

Introduction Lucerne has been grown over centuries for forage. Its forage production is strongly correlated to the initial taproot and stubble N reserves (Avice *et al.,* 1996; Meuriot *et al.,* 2004). However, the influence of cutting management on the level of N storage and the contribution of these N reserves to forage production still remain unclear and need to be studied at the whole plant level. For this purpose, a deterministic model of N allocation within the different organs and partitioning within different biochemical N pools was developed for lucerne with high and low initial N status and cutting heights of 6 or 15 cm.

Materials and methods The model is based on the simple hypothesis that N allocation and partitioning are determined by local N supplies of nitrate and amino acids, and by the ability of the different organs or physiological functions to use these resources (Escobar–Gutiérrez *et al.,* 1998). In a sink organ, the demand for N is the sum of three elementary demands: (i) structural associated N growth, (ii) metabolic associated N compounds, and (iii) N storage. The supply of N readily accessible for the organs only comes from the local stores of nitrate and amino acids. N flows in the network are determined by the source / sink activities of the different plant parts and by the local demand and supplies.

Results After optimisation of the parameter values, the model simulated satisfactorily the partitioning and allocation of N during regrowth after cutting (Figs 1 and 2). The model also showed that N supply to regrowing shoots is primarily due to nitrate and amino acid fluxes from stubble and taproot tissues. Moreover, the amino acids were fully depleted in the remaining tissues after cutting, but mobilisation of soluble proteins allowed maintaining an adequate shoot N supply. This mobilisation was more or less severe, depending of the plant initial N status and cutting height.

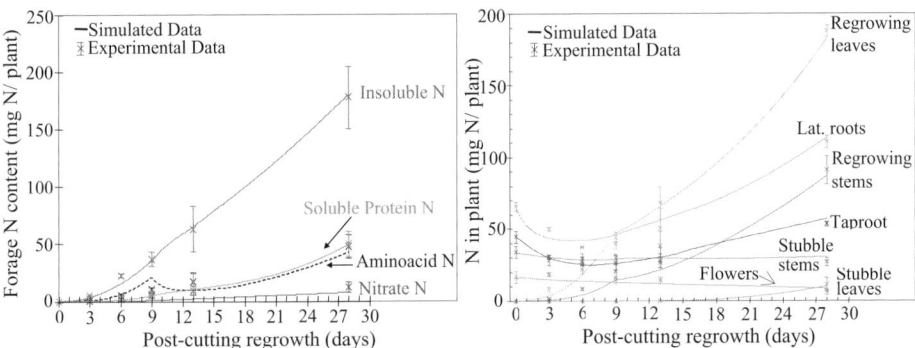

Figure 1 Forage N content during regrowth   **Figure 2** Total N in plant during regrowth

Conclusions The model allowed an accurate description of the N metabolism within the different plant tissues and biochemical N pools during post-cutting regrowth

References

Avice J-C., A. Ourry, J.J. Volenec & G. Lemaire (1996) Nitrogen and carbon flows estimated by ^{15}N and ^{13}C pulse-chase labelling during regrowth of alfalfa. *Plant Physio*logy, 112, 281-290.

Escobar-Gutiérrez A.J., F.A. Daudet, J-P. Gaudillère, P. Maillard & J.S. Frossard (1998) Modelling of allocation and balance carbon in walnut (*Juglans regia* L.) seedlings during heterotrophy-autotrophy transition. *Journal of Theoretical Biology,* 194, 29-47.

Meuriot F., J-C Avice, J-C. Simon, P. Lainé, M-L. Decau & A. Ourry (2004) Influence of the initial organic N reserves and residual leaf area on growth, N uptake, N partitioning and N storage in alfalfa (*Medicago sativa* L.) during post-cutting regrowth. *Annals of Botany,* 94, 311-321.

Root senescence in red clover (*Trifolium pratense* L.)

K.J. Webb, E. Tuck, S. Heywood and M.T. Abberton
*Institute of Grassland and Environmental Research, Plas Gogerddan, Aberystwyth, Ceredigion SY23 3EB, UK,
Email: judith.webb@bbsrc.ac.uk*

Keywords: red clover, roots, senescence

Introduction Legume root systems form a mosaic of living, ageing and dead roots and nodules. The balance between these stages alters during plant development. Stressful events (drought, temperature change, reduced carbon supply, etc.) disturb the balance (Butler *et al.*, 1959). Effects of root and nodule death on soil structure, composition and leaching and on plant persistency are understood poorly. Plants with differing senescence patterns are useful tools to study these effects. Molecular studies of root senescence need detailed knowledge of the process and timing of root senescence and death. Biochemical and histochemical markers of senescence were used to generate preliminary results of the effects of reduced carbon input, temporary (by defoliation, D) or permanent (by defoliation and shading, DS) on red clover shoot survival and root death.

Materials and methods In a controlled environment (20°C/15°C light/dark; 16h photoperiod; 400uM/m/sec; nitrogen-free nutrients), nodulated clover seedlings cv. Milvus were grown for 7 weeks in Agsorb. At day 0, plants were either defoliated (D), defoliated and heavily shaded with black polythene (DS), or left intact (control). Shoot re-growth, root death index (RDI, based on quantitative Evan's Blue staining (Baker & Mock, 1994)) and root catalase activities (Doulis *et al.*, 1997) were measured (3 replicates/treatment) at 0, 7, 14 and 21 days. After 21 days, DS plants were grown for a further 14 days in the light before assessment of RDI. Statistical analysis was by ANOVA (Genstat).

Results Red clover plants recovered from temporary (D) but not permanent (DS) reduction in carbon supply (Figure 1A). Quantitative differences in RDI were not significant between any of the treatments for the first 14 days, indicating that similar proportions of cells of these roots were still alive (Figure 1B). By 21 days, DS roots were significantly more strongly stained ($p<0.05$) indicating a higher level of cell death. Root catalase activity showed a similar pattern (Figure 1C); there were no significant differences in catalase activity between control, D and DS plants at 7 or 14 days. Root catalase activity of DS plants increased about 8-fold by day 21 ($p<0.05$). After a further 14 days growing in the light, all roots from DS plants stained strongly with Evan's Blue, resulting in OD similar to dead, control roots (Figure 1B).

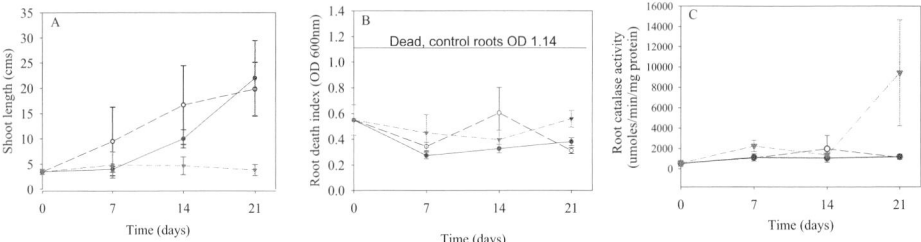

Figure 1 Effects of temporary or permanent reduction in carbon supply in red clover cv. Milvus over time. **A** Shoot growth. **B** Root death index. **C** Root catalase activity. $-\mathbf{O}-$ Control; $-\!\!\bullet\!\!-$ Defoliated (D); $-\!\!\blacktriangledown\!\!-$ Defoliated and shaded (DS).

Conclusions Red clover roots survived temporary reduction in carbon supply (D) and had no change in root senescence. Permanent reduction in carbon supply (DS) caused plant death by 21 days; shoots failed to re-grow when returned to light. These data show that defoliation alone is not enough to trigger root death and provide a temporal framework for studies on differential gene expression during root senescence.

References
Baker, C. J. & N. M. Mock (1994). An Improved Method for Monitoring Cell-Death in Cell-Suspension and Leaf Disc Assays Using Evans Blue. *Plant Cell Tissue and Organ Culture,* 39, 7-12.
Butler, G. W., R. M. Greenwood & K. Soper (1959). Effects of shading and defoliation on the turnover of root and nodule tissue of plants of Trifolium repens, Trifolium pratense and Lotus uliginosus. *New Zealand Journal of Agricultural Research,* 2, 415-426.
Doulis, A. G., N. Debian, A. H. Kingston Smith & C. H. Foyer (1997). Differential localization of antioxidants in maize leaves. *Plant Physiology,* 114, 1031-1037.

Seed production and resource allocation in three cultivars of *Achnatherum hymenoides*, Nevada, USA

B.L. Perryman[1], C.A. Busso[2] and H.A. Glimp[1]
[1]*Department of Animal Biotechnology, University of Nevada-Reno, MS 202, Reno, NV USA 89557, Email: bperryman@cabnr.unr.edu,* [2]*Departamento de Agronomia, Universidad Nacional del Sur, Altos del Palihue, 8000 Bahia Blanca, Argentina*

Keywords: seed production, dry matter, partitioning

Introduction Plant production is partially determined by resource allocation among various organs (Monsi & Murata, 1970), however, studies on dry matter partitioning among different plant organs are scarce in general (Marceli, 1996), and lacking in *Achnatherum hymenoides*. This study compared dry matter production and partitioning among three commercial cultivars (Paloma, Nezpar and Rimrock) of *A. hymenoides* and identified growth and developmental characteristics that could indicate potential seed production. In addition, the relationship between an organ weight as a percent of total aerial plant biomass was assessed.

Materials and methods All three cultivars were seeded at a rate of 3.9 kg pure live seeds/ha on 1m row spacings, under irrigation in central Nevada, USA. Eight plants of each cultivar were randomly selected on each of two internal rows and seeds were harvested every week as they ripened (12 July-12 Oct.). Sixteen plants of each cultivar were randomly selected for tiller and leaf length measurements, and harvested to ground level at the end of the 2003 growing season. Total standing crop was divided into blades, sheaths, stems, seeds and reproductive structures (glumes, rachis, etc.). All parts (except seeds) were oven dried at 70° C for 48 h and weighed. Seeds were dried at 35° C for 48 h so they could be used later in germination studies. Each category was then expressed as a percent of total aerial plant dry weight. The ratio between blade and sheath dry weights was also calculated. Seeds were individually separated from the lemma and palea. In addition to dry matter production, on each of two tillers per plant, total green leaf length, total (green + dead) leaf length, and tiller length were determined. Number of green, total (green + dead) and reproductive tillers was also obtained. One-way ANOVA was used in all analyses and significance was determined at P < 0.05. Seed production and biomass relationships were analysed with simple linear regression.

Results and discussion Paloma had the lowest blade production and the greatest production of stems, seeds and reproductive structures. Seed production was greater for Paloma (5 kg/ha) than Nezpar (1.9 kg/ha) and Rimrock (0.3 kg/ha). Dependent seed variables correlated best with Paloma (number of reproductive tillers P<0.001, R^2=0.84; leaf length P=0.006, R^2=0.62; total above ground biomass P<0.001,R^2=0.80). The relationship between weight of an organ (sink size) and the percentage of total aerial biomass assigned to this organ was significant for blades and seed weight in all three cultivars. This could indicate a correlation between size and resource needs of the sink. However, need of the sink for resources is probably not causally related to size, because it can also depend on age, potential growth rate, carbohydrate supply, presence of other sinks, and resistance of the transport pathway to the flow of assimilates (Marcelis, 1996; Minchin & Thorpe, 1996). These relationships were not significant for reproductive structures in any cultivar.

Conclusions Based on one year of observation, Paloma would be better choice for land managers if seed production is a priority, and Rimrock a better choice if leaf and stem biomass is important. Nezpar would be a good combination choice. A better understanding of dry matter partitioning in the aerial portion of the 3 commercially available cultivars of this grass species will provide managers with the information needed to select a variety based on its intended use.

References
Marcelis, L.F.M. (1996). Sink strength as a determinant of dry matter partitioning in the whole plant. *Journal of Experimental Botany*, 47, 1281-1291.
Minchin, P.E. & M.R. Thorpe (1996). What determines carbon partitioning between competing sinks? *Journal of Experimental Botany*, 47, 1293-1296.
Monsi, N. & Y. Murata (1970). Development of photosynthetic systems as influenced by distribution of matter. In: I. Setlik (ed.). Prediction and Measurement of Photosynthetic Productivity. Pudoc, Wageningen, 115-129.

Effect of pre-planting seed treatment options on dormancy breaking and germination of *Ziziphus mucronata*

A. Hassen[1], N.F.G. Rethman[1] and W.A. van Niekerk[2]

[1]*Department of Plant Production and Soil Science, University of Pretoria, Pretoria 0002, Republic of South Africa, Email: hassenabubeker@yahoo.com,* [2]*Department of Animal and Wildlife Science, University of Pretoria, Pretoria 002, Republic of South Africa *Permanent Address: Adami Tulu Research Centre, Zeway, Ethiopia*

Keywords: *Ziphus mucronata*, dormancy, hard seed coat, germination, seed mortality

Introduction *Ziziphus mucronata* (Buffalo thorn) is a multipurpose tree, widely adapted to a range of ecological conditions and tolerant of extreme climatic conditions, including frost and drought (Venter & Venter, 1996). It is a valuable fodder tree for livestock and game animals, especially in the drier parts of Africa (Rothauge *et al.* 2003). Similar to many other leguminous species, establishment is constrained by low and erratic germination of the seed, which has been attributed mainly to the physical barrier of the stony endocarp and dormancy associated with seed coat impermeability . This experiment aimed to compare the suitability of various seed treatment options as practical methods to break seed dormancy and enhance germination.

Material and methods *Z. mucronata* seeds were isolated from the dried fruit by light hammering on a concrete floor covered with a cloth sheet. Seeds were subjected to the following pre-planting treatment options: untreated seed (control); scarification using sandpaper; immersion in boiling water for 1 or 5 minutes; and immersion in 94% sulphuric acid for 10, 20, 30 or 45 minutes. Treated seeds were placed in petri dishes on moist filter paper and kept in a germination compartment adjusted to day/night temperature of 30/20°C with 12h of light. Germination of seeds was counted every day while non- germinating seeds were categorized into hard and dead after 15 days. The data were subjected, after arcsine transformation, to analysis of variance and means were compared using Tukey's test at the threshold of P<0.05. Arcsine transformed means were back transformed for presentation.

Results Hard seed percentage was significantly (P<0.05) lower than the control in seeds subjected to sandpaper scarification or treated with sulphuric acid for >/=10 minutes (Table 1). However, sulphuric acid treatment for >/=20 minutes was the most effective way to break hard seed dormancy. Regardless of its duration, boiling water treatment was not effective to break hard seed dormancy for *Z. mucronata*. Scarification with sandpaper or sulphuric acid treatment for 20 minutes significantly (P<0.05) increased the percentage germination compared to the control but the latter also resulted in significantly (P<0.05) higher seed death (Table 1).

Table 1 Effect of pre-planting seed treatment options on germination rate and proportions of hard and dead seed remaining of *Z. mucronata* after 2 weeks' incubation

Treatment options	Seed (%)		
	Hard	Germinated	Dead
Control	77.7a	10.7b	11.6d
Sandpaper scarification	23.9c	65.4a	10.7d
Immersion in boiling water for 1min	52.2abc	26.8ab	21.1cd
Immersion in boiling water for 5min	57.4ab	11.0b	31.7cd
Immersion in 94% H_2SO_4 for 10min	49.5bc	12.0b	38.5bcd
Immersion in 94% H_2SO_4 for 20min	2.2d	57.5a	40.4abc
Immersion in 94% H_2SO_4 for 30min	0d	37.2ab	62.9ab
Immersion in 94% H_2SO_4 for 45min	0d	31.5ab	68.5a

Means followed by the same letter within a column differ significantly at P<0.05

Conclusions Scarification using sandpaper was the best way to maximize germination without increasing the risk of seed death significantly. The duration of immersion was critical in the case of sulphuric acid treatment; 20 minutes was the optimum time of immersion.

References

Rothauge, A., G. Kaendji & M. L. Nghikembua (2003). Forage preference of Boer goats in the highland savanna during the rainy season II: Nutritive value of the diet. Agricola, pp 43-48

Venter, F. & J. A. Venter (eds.) (1996). Making the most of indigenous trees. Briza publications, Arcadia 0007. Pretoria, South Africa. pp. 304

Pollen viability and seed setting in Egyptian clover under open and caged conditions

B.L. Bhardwaj and A. Kumar
Department of Plant Breeding, Genetics and Biotechnology, PAU, Ludhiana, Punjab 141004 India, Email: drblbhardwaj@yahoo.com

Keywords: Egyptian clover, *Trifolium alexandrinum* L., pollen viability, seed setting

Introduction Grown on an area of about 200Kha annually, Egyptian clover (berseem) is the most important winter forage crop of Punjab. Its seed production is erratic due to several factors, including high temperature during flowering, which may affect pollen viability and the role of insect pollinators. Hence, berseem seed production is an entomological problem as much as a botanical one. We studied pollen viability under open and caged conditions, and the role of honeybees in seed setting.

Materials and methods Twenty diverse berseem genotypes were sown in a replicated randomized block design for 2 consecutive years. To restrict the movement if insect pollinators in a portion of flower heads, cages of 25 x 50 x 85 cm were built with the help of muslin cloth in each plot. Pollen viability was studied in each genotype in the morning as well as evening and in open as well as caged condition. At the time of harvesting 100 flower heads were collected/genotype from within and outside the cages. Seed yield was assessed for each cage by counting and weighing.

Results The seed count (Table 1) was much higher in open condition (5106) than that under caged condition (305). Seed weight also was higher in open condition (13.34g) than that under caged condition (0.563g).

Table 1 Seed count and weight under open and caged conditions in different genotypes of berseem (Mean of 2 years)

Genotype	Seed count (No.) Open	Cage	Mean	Seed wt (g) Open	Cage	Mean	Genotype	Seed count (No.) Open	Cage	Mean	Seed wt (g) Open	Cage	Mean
BL 1	5220	364	2792	15.12	0.843	7.98	BL 88	5386	280	2833	14.00	0.618	7.30
BL 2	6027	447	3237	16.29	0.988	8.64	BL 90	5446	276	2861	13.70	0.496	7.10
BL 10	4413	244	2328	9.89	0.499	5.19	PBL 145	4737	417	2577	12.81	0.941	6.87
BL 22	4677	233	2455	10.65	0.674	5.66	Wardan	3749	288	2018	11.59	0.787	6.18
BL 42	4804	310	2557	11.49	0.639	6.06	BL 112	5176	257	2716	15.02	0.515	7.76
BL 50	5833	376	3104	14.79	0.447	7.61	BL 114	5882	336	3109	14.27	0.392	7.33
BL 52	5696	215	2955	13.42	0.235	6.83	BL 130	5891	305	3098	16.54	0.246	8.39
BL 67	5036	210	2623	12.59	0.294	6.44	BL 167-1	4815	310	2562	13.39	0.370	6.88
BL 73	4520	341	2430	13.20	0.633	6.91	BL 189-1	5000	298	2649	12.73	0.590	6.66
BL 87	4627	197	2412	11.51	0.201	5.85	Mescavi	5189	392	2790	13.82	0.845	7.73
Mean								**5106**	**305**		**13.34**	**0.563**	

CD (P=0.05)
Seed count (A) : 95.6
Genotype (B) : 302.5
A x B : 427.8

Seed Weight (A) : 0.302
Genotype (B) : 0.957
A x B : 1.354

Pollen viability (means of 82.9 and 84.0%, respectively, under open and caged conditions) showed non-significant differences. This may be attributed to the absence of insect pollinators, which are reported to increase seed set by tripping (Dixit *et al*, 1989). Iannucci (2001) also reported that *Trifolium alexandrinum* was predominantly self-pollinated but requires tripping to improve seed setting. As pollen and stigma hide within the keel, when honeybees forage nectar from the base of the gynoecium, the keel is pressed sufficiently to give access for the bee's tongue and the central column of style and stamen, which was kept under tension, is released explosively. Hence, it was suggested that for better seed setting beehives should be kept within 1km of Egyptian clover field at the time of flowering (Dhaliwal & Atwal, 1974).

Conclusion Seed setting under open and caged condition showed that there was no difference in pollen viability under both these conditions but seed setting increased in the presence of honeybees, which were observed to increase the seed set by tripping.

References

Dhaliwal, J. S. & A. S. Atwal (1974). Foraging range of *Apis mellifera* L. on berseem (*Trifolium alexandrinum* L.) at Ludhiana. *Indian Bee Journal,* 96, 16-17.
Dixit, O. P., U. P. Singh & J. N. Gupta (1989). Significance of pollination in seed setting efficiency of berseem (*Trifolium alexandrinum* L.). *Journal of Agronomy and Crop Science,* 162, 93-96.
Iannucci, A. (2001). Effects of inbreeding and pollination modes on self fertility and seed weight in berseem populations. *Journal of Genetics and Breeding,* 55, 165-171.

Theme A: Efficient production from grassland

Section 5

Forage quality for animal nutrition

The death of plants in animals

M.K. Theodorou, A.H. Kingston-Smith and P. Morris
Institute of Grassland and Environmental Research, Plas Gogerddan, Aberystwyth, SY23 3EB, UK, Email: mike.theodorou@bbsrc.ac.uk

Keywords: plant cell death, plant quality traits, rumen function

Introduction It is necessary first to understand some of the basic concepts associated with the digestion of the plant biomass within the rumen when considering mechanisms for altering/enhancing N-conversion efficiency in the forage-fed ruminant. Although it is generally assumed that breakdown of plant proteins in the rumen is mediated by microbial enzymes, there is increasing evidence to suggest that both plant and microbial proteases are active during degradation of ingested fresh forage (Beha *et al.*, 2002; Kingston-Smith & Theodorou, 2000; Kingston-Smith *et al.*, 2003, 2004). After fresh plant biomass enters the rumen and prior to extensive plant cell-wall degradation, there is often a phase of rapid proteolysis in excess of that needed to maintain the rumen microbial population and we now believe that plant enzymes largely mediate this initial proteolysis. Recent evidence also suggests a role for plant lipases in the rumen (Lee *et al.*, 2003). An understanding of the mechanisms that underlie these processes is essential if we are to devise plant-based strategies to manipulate them. This paper presents a new rumen model which, by taking account of the plants biological attributes, provides us with a novel framework for describing the plant contribution to rumen function in grazing livestock.

Materials and methods Data from *in vitro* and *in vivo* experiments was used in provision of evidence upon which to base the model. The experiments involved freshly harvested (living) plant biomass which was (a) incubated *in vitro* with and without populations of rumen micro-organisms (b) incubated in the rumen in bags of varying pore size or where (c) ingested boli were retrieved prior to entering the rumen of recipient animals, placed in Dacron bags and incubated for timed intervals in donor animals. The methodologies associated with these experiments can be found in publications by Beha *et al.* (2002) and Kingston-Smith *et al,* (2003, 2004).

Results Taken collectively, the experiments referred to have enabled us to a construct rumen model which provides a conceptual framework of how plant status (on entering the rumen) can contribute to rumen function. According to the model, upon mastication and ingestion, plant cells are either intact (IC), partially damaged (PD) or entirely destroyed (ED). In each case the processes that occur subsequent to ingestion has a major impact on rumen function and the efficiency by which ruminants degrade and digest plant tissues. In the case of IC, for example, evidence suggests that plant cells respond to ruminal stresses by entering autolytic processes and begin to degrade their own proteins and membranes (plant-mediated proteolysis and lipolysis). Plant enzymes liberated from PD and ED cells may contribute to rumen function via the digestion processes in rumen fluid. Furthermore, evidence suggests that certain plant defence mechanisms active in herbivory, such as protein protection via the polyphenol oxidase reaction, are invoked when damaged cells undergo decompartmentation in the rumen. In light of this model and in terms of PD and ED cells, the relationship between tannins in tanniniferous forages and the possible inactivation of plant and microbial enzymes is worthy of further consideration.

Conclusions The model proposed in this paper suggests that on entering the rumen the biological status of plant cells can contribute significantly to rumen function. We anticipate that the concepts underlying the model will assist in elaborating new criteria to breed forage plants that are pre-disposed to behave in particular ways, or cause particular behaviour(s) during their ingestion, digestion and passage through the ruminant digestive tract.

References

Beha, E.M., M.K. Theodorou, B.J. Thomas & A.H. Kingston-Smith (2002). Grass cells ingested by ruminants undergo autolysis which differs from senescence: implications for grass breeding targets and livestock production. *Plant Cell & Environment,* 25, 1299-1312.

Kingston-Smith, A.H., & M.K. Theodorou (2000). Post ingestion metabolism of fresh forage. *New Phytologist (Tansley Review 118) 148 37-55.*

Kingston-Smith, A.H., A.L. Bollard, B.J. Thomas, A.E. Brooks & M.K. Theodorou (2003). Nutrient availability during the early stages of digestion and colonisation of fresh forage by rumen micro-organisms. *New Phytologist* 158, 119-130

Kingston-Smith, A.H., R.J. Merry, D.K. Leemans, H. Thomas, & M.K. Theodorou (2004). Evidence in support of a role for plant-mediated proteolysis in the rumens of grazing animals. *British Journal of Nutrition* (in press).

Lee, M.R.F., E.M. Martinez & N.D. Scollan (2003). Plant enzyme mediated lipolysis of *Lolium perenne* and *Trifolium pratense* in an *in vitro* simulated rumen environment. In: M.T. Abberton, M. Andrews, L Scot & M.K. Theodorou (eds) Crop Quality: Its Role in Sustainable Livestock Production. *Aspects of Applied Biology* 70 115-120.

Polyphenol oxidase activity and *in vitro* proteolytic inhibition in grasses

J.M. Marita, R.D. Hatfield and G.E. Brink
USDA-Agricultural Research Service, U.S. Dairy Forage Research Center, 1925 Linden Drive West, Madison, Wisconsin 53706 USA, Email: jmarita@wisc.edu

Keywords: Polyphenol oxidase (PPO), *o*-diphenols, proteolysis, grasses, caffeic acid

Introduction Harvesting and storing high quality forage in the cool humid regions remains a challenge due to the potential for protein degradation during ensiling. Red clover is an exception as high protein levels are maintained during ensiling. Decreased proteolytic activity in red clover is due to polyphenol oxidase (PPO) activity and appropriate *o*-diphenol substrates (Jones *et al.*, 1995, Sullivan *et al.*, 2004). This project was undertaken to determine if PPO activity is present in a range of grasses and the potential role in proteolytic inhibition in the presence of the *o*-diphenol caffeic acid.

Methods Fifteen grass species were established in the greenhouse under a 14/10 h (day/night) lighting regime. Leaf blades were harvested from plants at the vegetative stage, frozen in liquid nitrogen and stored at -80° C. Individual samples were processed and analyzed following modified methods of Sullivan *et al.* (2004). The PPO activity was determined in duplicate using *o*-diphenol substrates (caffeic acid, chlorogenic acid and catechol; 2mM final concentration). To determine the potential impact of PPO and *o*-diphenols on proteolytic activity leaf blades were prepared similar to above. Two samples were prepared with one processed through a G-25 sephadex spin column to remove low molecular weight materials and the other clarified by centrifugation and removal of the supernatant. At time zero, caffeic acid (3 mM final concentration) was added to the eluted protein (spin column). Aliquots were removed from each sample at specific time intervals (t_0, t_1, t_2, t_4 and t_{24} h) and soluble amino acids and small peptides were quantified to assess the degree of proteolysis.

Results and discussion Both species and the specific type of *o*-diphenol significantly altered PPO activity (Table 1). Orchardgrass, meadow fescue, ryegrass, and smooth bromegrass exhibited the highest PPO activities. Chlorogenic acid and/or caffeic acid were the preferred substrates, although there were differences among the most active grasses as to which was the best utilized. This suggests potential differences among the individual PPO enzymes. Generally, the addition of caffeic acid to isolated grass extracts resulted in proteolytic inhibition in grasses with substantial PPO activity. Such results suggest that several important grass species contain PPO activity, but may lack the appropriate *o*-diphenol substrates to effectively inhibit proteolysis. Initial results suggest that proteolytic inhibition can be achieved with the addition of caffeic acid.

Table 1 PPO activity in 15 grass species in the presence of three representative *o*-diphenol substrates and the percent reduction in *in vitro* proteolysis with the addition of caffeic acid after 24 h

Grass Species	Polyphenol Oxidase Activity (µmoles/µg/min)			Reduction in proteolysis
	Caffeic Acid	Chlorogenic Acid	Catechol	
Tall fescue (soft)	3.2 E-4	2.9 E-4	8.9 E-5	5%
Meadow fescue	3.5 E-3	4.1 E-3	9.0 E-4	90%
Timothy grass	1.5 E-4	2.2 E-4	1.1 E-4	6%
Smooth bromegrass	8.2 E-4	4.1 E-3	8.7 E-4	99%
Quackgrass	1.0 E-4	1.6 E-4	2.1 E-4	20%
Tall fescue	1.1 E-4	5.6 E-5	4.5 E-5	55%
Reed canary grass	5.1 E-5	3.4 E-5	3.1 E-5	49%
Ryegrass	2.6 E-3	1.3 E-2	7.9 E-4	78%
Spring wheat	2.1 E-5	1.0 E-5	2.6 E-5	8%
Winter wheat 1	2.5 E-5	1.8 E-5	2.5 E-5	16%
Winter wheat 2	1.1 E-5	1.0 E-5	2.1 E-5	15%
Rye	2.6 E-5	1.7 E-5	4.2 E-5	0%
Oat	2.6 E-5	1.7 E-5	1.9 E-5	0%
Orchardgrass	1.1 E-2	2.6 E-2	1.7 E-3	60%
Kentucky bluegrass	6.0 E-5	3.3 E-5	7.0 E-5	24%

References

Jones, B.A., R.E. Muck & R.D. Hatfield (1995). Red clover extracts inhibit legume proteolysis, *Journal of the Science of Food and Agriculture*, 67, 329-333.

Sullivan, M.L., R.D. Hatfield, S.L. Thoma & D.A. Samac (2004). Cloning and characterization of red clover Polyphenol oxidase cDNAs and expression of active protein in Escherichia coli and transgenic alfalfa, *Plant Physiology*, (accepted July 2004).

Ruminal proteolysis in forages with distinct endopeptidases activities

G. Pichard, C. Tapia and R. Larraín

Department of Animal Sciences, School of Agriculture, Pontificia Universidad Católica de Chile, Email: gpichard@puc.cl

Keywords: forages, proteolysis, rumen, plant peptidases

Introduction Improving livestock efficiency in utilization of nitrogen resources continues to be a major environmental and economic objective. Zhu *et al.* (1999) have shown that plant endopeptidases are activated as a response to cutting stress. Previous work in our laboratory explored over 300 entries of forage genotypes and found a broad diversity in enzymatic activity by means of hydrolysis in gelatine and direct autolysis assays in forage tissues. The objective of this work was to assess if the species previously identified as having high or low endopeptidase activity, would behave consistently when exposed to ruminal microbial proteolysis.

Materials and methods Two groups of forages were selected according with their level of peptidase activity (Table 1). They were grown in a greenhouse, fresh leaf samples were collected in vegetative stages, submitted to a molar-like pressing device and further chopped to 1cm. The rumen fluid inocula was subjected to 3 hours pre-incubation with sugar for depletion of free N and further 2 hours with hydrazine-cloramphenicol inhibitors (Broderick, 1987). Fresh forage samples (2 g) were incubated *in vitro* during 6 hours (T_0 to T_6) with inhibited rumen fluid (IRF). The residue of neutral detergent insoluble nitrogen (NDIN) (Licitra *et al.*, 1996) and non protein soluble nitrogen (NPSN) were determined.

Table 1 Extent and rates of solubility of N compounds incubated in inhibited rumen fluid

Species		EPA*	Total N (g/Kg)	NDIN (% of TN)			NPSN (% of TN)		
				T_0	T_6	k_{0-6}	T_0	T_6	k_{0-6}
Avena strigosa cv. Negra		High	32	46	20	4.4	19	32	2.
Festuca arundinacea cv. Conway		High	23	62	17	7.5	22	31	1.5
Medicago sativa cv. Innovator		High	35	56	18	6.3	21	35	2.4
Trifolium repens cv. Blanca		High	37	51	19	5.4	22	34	1.9
	Mean			54	19	5.9	22	33	1.9
Bromus unioloides cv. M. Fierro		Low	25	62	31	5.3	20	27	1.2
Lolium hybridum cv. Galaxy		Low	25	54	29	4.3	20	28	1.4
Trifolium pratense cv. Resistenta		Low	37	48	32	2.7	19	27	1.3
Trifolium repens cv. Kopu		Low	33	50	38	1.9	17	25	1.2
	Mean			53	32	3.6	19	27	1.3
Effect of endopeptidases (p value)				NS	0.003	0.06	NS	0.012	0.03

* Endopeptidases activity

Results and Discussion The extents and rates of NDIN disappearance and NPSN accumulation were statistically different between the groups of high and low enzymatic activity (Table 1). The extent of nitrogen solubilised from the ND residue was not accounted for by the fraction of NPSN, thus suggesting that it remains as a soluble true protein. Also, within the group with high endopeptidase activity, no major differences were observed between legumes and grasses.

Conclusion The activity of plant endopeptidases varies among different germplasms and affects ruminal proteolysis. Such plant diversity supports the idea that one way of improving nitrogen utilization in pasture based animal production systems is the genetic improvement of germplasms that have the potential for delayed protein hydrolysis. The laboratory enzymatic assays with gelatine showed to be consistent with the *in vitro* rumen microbial proteases. The large fraction of soluble proteins released in rumen soon after eating suggests that the potential for rumen by pass with the liquid phase may be high and should be reassessed.

Acknowledgements Funded by Fondecyt Grant 1030918

References

Broderick, G.A. (1987) Determination of protein degradation rates using a rumen in vitro system containing inhibitors of microbial nitrogen metabolism. *British Journal of Nutrition*, 58, 463-475.

Zhu W-Y., A.H. Kingston-Smith, D. Troncoso, R.J. Merry, D.R. Davies, G. Pichard, H. Thomas & M.K. Theodorou (1999). Evidence of a Role for plant proteases in the degradation of herbage proteins in the rumen of grazing cattle. *Journal of Dairy Science*, 82, 2651-2658.

Licitra, G., T.M. Hernández, & P.J. Van Soest (1996) Standardization of procedures for nitrogen fractionation of ruminant feeds. *Animal Feed Science and Technology*, 57, 347-358.

Mating ewes on condensed tannin-containing forages increases ewe reproductive rate and reduces lamb mortality

T.N. Barry, C.A. Ramirez-Restrepo, E.L. McWilliam, N. Lopez-Villalobos and P.D. Kemp
Massey University, Palmerston North, New Zealand. Email: T.N.Barry@massey.ac.nz

Keywords: condensed tannin, reproduction, lamb mortality

Introduction Action of condensed tannin (CT) reduces forage protein degradation in the rumen and increases the absorption of amino acids from the small intestine (Barry & McNabb 1999). This paper reports the effects of grazing ewes on two CT-containing forages during mating upon ewe reproductive rate and lamb mortality.

Materials and methods Mixed age Romney ewes grazed *L. corniculatus* (18-29 g CT/kg DM) or perennial ryegrass / white clover during mating (Expt 1), whilst in Expts 2, 3 and 4 ewes grazing low quality drought pasture during mating were fed 1.4 kg supplements of fresh willow (*Salix* sp 27-52 g CT/kg DM) or poplar (*Populus* sp 7-19 g CT/kg DM per day). Treatments were applied over a 70 day period to groups of 100 ewes, including two cycles of mating; groups were then joined and grazed on pasture until weaning.

Results Mating on *L. corniculatus* increased lambing percentage (P<0.05) and reduced lamb mortality (P<0.05; Table 1). Supplementation with willow and poplar during mating increased lambing percentage in two experiments (P<0.05) and reduced overall lamb mortality from 17.8 to 11.7% (P<0.05), with no interaction between treatment and years (Table 2).

Table 1 Mating on *Lotus corniculatus* and reproductive performance (Ramirez *et al.* 2005)

Experiment 1	Pasture	Lotus	P
Liveweight change (g/d)	-5	67	***
Lambing (%)	159	175	*
Lamb mortality (%)	22.9	11.7	*

Table 2 Willow and poplar supplementation during mating on drought pastures (McWilliam *et al.*, 2005)

Experiment	Control	Supplement	SEM
(2; poplar supplementation)			
Liveweight change (g/d)	-82	-67	5.2
Lambing (%)	121	155	5.8
Lamb mortality (%)[1]	20.3	16.3	
(3; willow supplementation)			
Liveweight change (g/d)	-103	-86	4.3
Lambing (%)	131	148	6.9
Lamb mortality (%)[1]	17.3	12.1	
(4; willow supplementation)			
Liveweight change (g/d)	-147	-96	4.5
Lambing (%)	124	127	5.6
Lamb mortality (%)[1]	16.0	8.0	

[1] Corrected for birth rank and sex.

Conclusions: Feeding CT-containing forages during mating and early pregnancy increased ewe reproductive rate and reduced lamb mortality. Further research is needed to confirm the effects on lamb mortality, using larger group sizes of ewes.

References

Barry, T.N., & W.C. McNabb (1999). The Implications of condensed tannins on the Nutritive value of temperate forages fed to ruminants. *British Journal of Nutrition* 81, 263-272.

McWilliam, E.L., T.N. Barry, N. Lopez-Villalobos, P.N. Cameron & P.D. Kemp (2005). The effects of willow (*Salix*) supplementation for 31 and 63 days on the reproductive performance of ewes grazing low quality drought pasture during mating. *Animal Feed Science and Technology,* 119, 87-106.

Ramirez-Restrepo, C.A., Barry, T.N., Lopez-Villalobos, N., Kemp, P.D., Harvey, T.G. (2005). Use of *Lotus corniculatus* containing condensed tannins to increase reproductive efficiency in ewes. *Animal Feed Science and Technology* (In Press).

Anthelmintic effects of sericea lespedeza hay fed to goats infected with *Haemonchus contortus*

S.A. Shaik[1], T.H. Terrill[1#], J.E. Miller[2], B. Kouakou[1], G. Kannan[1], R.M. Kaplan[3], J.M. Burke[4] and J.A. Mosjidis[5]
[1]*Fort Valley State University, Fort Valley, GA 31030, USA,* [2]*Louisiana State University, Baton Rouge, LA 70803, USA,*[3]*The University of Georgia, Athens, GA 30602, USA,* [4]*USDA, ARS-DBSFRC, Booneville, AR 72927, USA,* [5]*Auburn University, Auburn, AL 36849, USA,* [#]*Corresponding Author; Email: terrillt@fvsu.edu*

Keywords: *Haemonchus contortus*, sericea lespedeza, goats

Introduction Infection with gastrointestinal nematodes (GIN), particularly *Haemonchus contortus*, is the major hindrance to economic goat production in the southern USA. Grazing forages high in condensed tannins (CT) or adding purified CT to the diet has been shown to reduce numbers of parasite eggs in sheep and goat faeces (Min & Hart, 2003). An alternative to grazing is feeding hay from CT-containing forages to livestock to reduce the effects of GIN (Shaik *et al.*, 2004). The purpose of the current study was to test potential anthelmintic effects of feeding sericea lespedeza [(*Lespedeza cuneata* (Dum.-Cours.) G. Don] hay to goats.

Materials and methods Twenty 4-month-old Boer male goats were randomly assigned to two groups of 10 each based on faecal egg count (FEC). One group was fed long stem sericea lespedeza hay and the other bermudagrass hay. The diets were balanced for protein and energy with a small amount of supplement (ground maize, soybean meal poultry fat, trace mineral salt and vitamin premix). The diets comprised 80 % hay and 20 % supplement by weight. Hay was fed at 3.5% of body weight. During the pretrial and trial periods, a trickle infection was given three times a week (500 *H. contortus* larvae/animal). All the goats were fed bermudagrass hay for 5 weeks (pre-trial period), after which two pens were switched to the sericea hay ration for 6 weeks (trial period). Throughout the experiment, FEC and blood packed cell volume (PCV) were monitored on a weekly basis. Data were analysed by repeated measures analysis with the pre-trial and trial periods analysed separately.

Results Egg counts were similar between the two groups during the 5-week pre-trial period (Table 1). During the 6-week trial period, treatment, time, and treatment x time effects were all significant (P < 0.05). Egg counts dropped by 79.7 % the week after sericea feeding was started and were lower (P < 0.05) in the sericea-fed group during each week of the trial period. During the pre-trial period, PCV was higher (P < 0.05) in control animals (27.2) than in the sericea-fed goats (24.3), while PCV was higher (P < 0.05) in the treatment group than in the control animals during the trial period (19.7 versus 23.1, respectively).

Table 1 Worm egg counts in goats infected with *H. contortus* larva and fed sericea lespedeza or bermudagrass hay

Diet	Week of sampling[1]										
	1	2	3	4	5	6	7	8	9	10	11
	Parasite eggs per gram of faeces										
Bermudagrass hay + concentrates	275	325	825	2994	538	2106[a]	1872[a]	1350[a]	2083[a]	2622[a]	2467[a]
Sericea lespedeza hay + concentrates	250	306	756	1919	1856	428[b]	294[b]	700[b]	483[b]	294[b]	333[b]
Standard error	81	39	153	712	251	360	144	191	144	224	196

[1]Pre-trial weeks 1-5, trial weeks 6-11; [a,b]Column means with unlike superscripts differ significantly (P < 0.05)

Conclusions Feeding sericea lespedeza hay to goats reduced FEC by approximately 80% compared to bermudagrass hay and may be an effective means of reducing egg shedding on pasture. Although the main effect may be on worm fecundity, higher PCV in goats fed sericea hay compared with bermudagrass hay suggests a direct anthelmintic effect.

References
Min, B.R. & S.P. Hart (2003). Tannins for suppression of internal parasites. *Journal of Animal Science*, 81(E. Suppl. 2), E102-E109.
Shaik, S.A., T.H. Terrill, J.E. Miller, B. Kouakou, G. Kannan, R.K. Kallu, & J.A. Mosjidis (2004). Effects of feeding sericea lespedeza hay to goats infected with *Haemonchus contortus*. *South African Journal of Animal Science (Suppl. 1)*, 234-236.

Influence of internode length on degradability of lucerne stems

H.G. Jung[1], F.M. Engels[2] and J.F.S. Lamb[1]
[1]USDA-ARS, 411 Borlaug Hall, University of Minnesota, 1991 Upper Buford Circle, St. Paul, Minnesota, USA email: jungx002@umn.edu, [2]Wageningen University, Haarweg 333, NL-6709 RZ, Wageningen, The Netherlands

Keywords: lucerne, stem, internodes, cell wall, degradability

Introduction Lignification of plant tissues restricts degradation of forages by ruminants. The undegradable lignified middle lamella/primary cell wall prevents rumen microbes from accessing undamaged adjacent plant cells. In contrast, walls of non-lignified tissues are completely degradable. Preliminary observations of how deeply rumen microbes could degrade different tissues within 20- to 30-mm long lucerne (*Medicago sativa* L.) stem pieces indicated a two-fold range in depth of degradation among stem samples. Studies were undertaken to determine if extent of cell wall degradation in lucerne was influenced by length of stem internodes.

Materials and methods The seventh internode from the stem base was collected after four wk of regrowth from a single lucerne clone. Five short (20 ± 1 mm) and six long (53 ± 4 mm) internodes were selected, and 18- to 20-mm pieces were excised from each internode. Stem pieces were coated with bee's wax and a microtome was used to remove the wax at one end. After incubation with rumen fluid for 24 h, serial sections were made from the exposed end of the stem pieces and evaluated by light microscopy to determine maximal depth of degradation for various tissues. To validate the conclusions of the first study, 20 lucerne clones selected for low or high 16- or 96-h in vitro neutral detergent fibre degradability (IVNDFD) were grown in three field environments, sampled twice at flowering for two years, and evaluated for IVNDFD and internode length.

Results Use of a bee's wax coating effectively limited access of rumen microbes to the cut end of lucerne stem pieces. Maximal depth of degradation of non-lignified tissues (collenchyma and primary xylem parenchyma) was greater for stem pieces derived from longer internodes (Fig. 1). Of the lignified tissues, only phloem fibre primary walls were degraded to greater depth when internodes were longer. Based on measurements of cell length, long internodes had both longer cells for all tissues and more cells per internode. Lucerne clones selected for high 16-h IVNDFD had shorter average internode length than did low 16-h IVNDFD selections (Fig. 2), whereas the opposite result was found for high vs. low 96-h IVNDFD selections. The low IVNDFD lucerne selection groups were lower ($p<0.05$) in degradability than the corresponding high groups (16-h: 21.2 vs. 22.2%; 96-h: 39.4 vs. 43.5%). Tissues degraded within 16 h were primarily limited to those that were non-lignified.

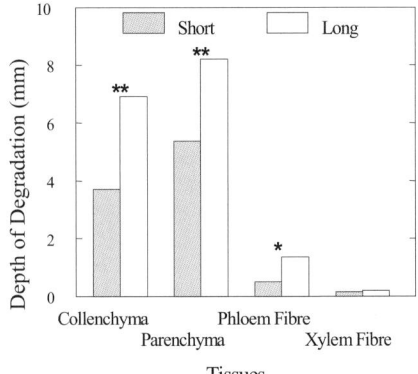

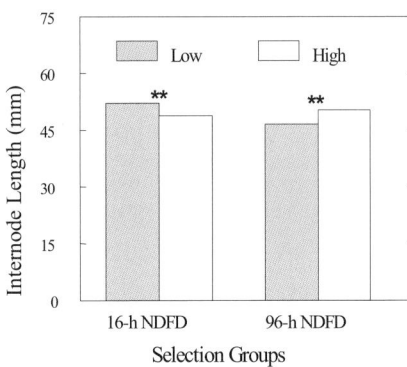

Figure 1 Depth of degradation observed in short and long lucerne stem internodes (* $p<0.10$, ** $p<0.05$)

Figure 2 Mean stem internode length of lucerne clones selected for low or high IVNDFD (** $p<0.05$)

Conclusions If the observed positive correlation between internode and cell length is generally true for lucerne, then several hypotheses can be proposed. Degradation of cell wall material from non-lignified lucerne tissues will be greater for short internodes because shorter cell lengths allow quicker movement of microbes between potentially degradable cell wall surfaces. In contrast, degradation of potentially available walls in lignified tissues requires mechanical rupture of cells for microbial access. Therefore, lucerne with longer internodes and associated longer cells should be more degradable because the probability of longer cells being ruptured by grinding or mastication will be higher.

Effect of ferulic acid esterase enzyme application on the in vitro digestibility and in situ rumen degradability of tropical grasses

A.T. Adesogan, N. Krueger, C.R. Staples, D.B. Dean and W. Krueger
Department of Animal Sciences, IFAS, University of Florida, PO Box 110910, Gainesville, Florida 32611, USA
Email: adesogan@animal.ufl.edu

Keywords: ferulic acid, esterase, enzyme, tropical grass

Introduction Tropical grasses are the primary staple diet of most of the domesticated ruminants in the tropical and subtropical regions of the world. However, dry matter (DM) digestibility (DMD) and intake levels of these C_4 grasses are considerably low, partly because of high lignin contents. Ferulic acid also impedes fibre digestion in such forages due to formation of cross linkages with digestible xylans. Certain esterase enzymes have been shown to cleave ferulic acid cross linkages in wheat bran but no studies have examined whether such enzymes can be used to increase the digestion of tropical grasses. Therefore, the objective of this work was to evaluate the effect of a ferulic acid esterase preparation applied at different rates on the digestibility of C_4 grasses.

Materials and methods An enzyme complex (Depol 740L, BioCatalyst UK) containing high esterase activity (32 U/ml) was sprayed on 12-week regrowths of ground (1 mm), Pensacola bahiagrass, Coastal bermudagrass, and Tifton 85 bermudagrass hays at 0, 0.5, 1, 2, 3 % DM. The samples were digested in rumen fluid within Ankom® Daisy II incubators, Ankom Technologies USA for 6, 24, and 48 hours. In vitro 96 h rumen fluid-pepsin DMD was measured using the two stage, rumen fluid-pepsin technique. Ground (4 mm) samples were also weighed in duplicate into porous, polyester bags and placed in the rumen of two cannulated, cows for 0, 3, 6, 9, 12, 24, 48, 72, 96, and 120 h to estimate the rate and extent of digestion. A 2 (cows) x 3 (hays) x 5 (enzyme rates) factorial arrangement of treatments was used for this study. The cows used for both parts of the study were fed Pensacola bahiagrass *ad lib.* and 900g of soybean meal. Ruminal degradation parameters were estimated using an exponential model. The results described are those for means across the three forages.

Results and discussion Forage neutral detergent fibre (NDF), acid detergent fibre (ADF) and lignin concentrations ranged from 814 to 889, 450 to 538 and 90 to 130 g/kg DM. The *in vitro* dry DMD of the bermudagrasses (440 g/kg) were greater than that of the bahiagrass (398 g/kg). As the enzyme application rate increased, NDF and hemicellulose concentrations decreased linearly (P=0.001) while there were linear and cubic increases in ADF concentration (P=0.033) and water-soluble carbohydrate concentration (P= 0.04), respectively. Increasing the rate of enzyme application also linearly increased the in vitro rumen fluid DMD of the hays incubated for 6 h (P= 0.001) or 24 h (P= 0.03). This suggests that enzyme application increased the rate of digestion of the grasses and could potentially increase their intake in cattle. Increasing the enzyme application rate also resulted in a cubic increase (P=0.001) in the 96 h in vitro rumen fluid-pepsin DMD of the hays. The 24 h NDF digestibility of the hays increased linearly (P=0.002) with increasing enzyme application but there were no increases in 6 or 48 h NDF digestibility. Enzyme application also resulted in a cubic increase in the wash value (P=0.001, Table 1), quartic increases in the total degradability (P=0.005) and the rate of degradation (P=0.005), and a cubic decrease in the lag phase (P= 0.03) which reflects how long it takes for the rumen microbes to begin to digest the hays.

Table 1 Effect of esterase enzyme application on the in situ rumen degradability of tropical hays. (g/kg DM)

	Enzyme application rate					Mean	SEM	P
	0x	0.5x	1x	2x	3x			value [d]
Wash loss [a]	90	93	100	124	110	103	4.3	0.001 C
Slowly degradable fraction	553	559	522	539	552	555	17.6	0.070
Total degradability	643	651	623	663	661	648	15.9	0.005 Q
Degradation rate[b]	0.028	0.026	0.033	0.026	0.026	0.028	0.003	0.005Q
Lag phase[c]	6	5	7	6	4	6	1.1	0.030 C

[a] Wash loss or immediately soluble fraction; [b] Expressed per hour; [c] L refers to the lag phase before the start of digestion (h). [d] Letters represent significant (P<0.05) polynomial contrasts: C = cubic effect, Q = quartic effect.

Conclusions Application of the esterase enzyme increased the *in vitro* DMD and NDF digestibility of the hays and also increased the rate and extent of degradation.

The nutritional value of cocksfoot (Dactylis glomerata L.) and perennial ryegrass (Lolium perenne L.) under leaf-stage based defoliation management

L.R. Turner[1], D.J. Donaghy[1] and P.A. Lane[2]
[1]Tasmanian Institute of Agricultural Research (TIAR), P.O. Box 3523, Burnie TAS 7320 Australia [2]TIAR, University of Tasmania, Private Bag 98, Hobart TAS 7001 Australia Email: wilsonlr@utas.edu.au

Keywords: cocksfoot, perennial ryegrass, nutritional value, productivity

Introduction The perception that cocksfoot is of lower nutritional value and less productive than perennial ryegrass is largely the result of studies in which management was based on perennial ryegrass (to the detriment of cocksfoot) or involved defoliation of both species at the same time (Greenhalgh & Reid, 1969; Johnson & Thomson, 1996). Maintaining plants in a vegetative state through species-specific management is essential to retain the high quality and productivity of cocksfoot. Defoliation at or before 4-5 fully expanded live leaves per tiller is recommended as the ideal physiological regrowth stage for defoliation of cocksfoot, resulting in metabolisable energy (ME) levels in excess of 11MJ/kg dry matter (DM) (Rawnsley et al., 2002). The aim of the current study was to provide an objective comparison of the nutritional value of cocksfoot and perennial ryegrass grown in dryland conditions under a management regime based on the physiological status of each grass species.

Methods Newly established cocksfoot (cv. Kara) and perennial ryegrass (cv. Impact) swards were defoliated repeatedly over an 8-month period at species-specific intervals (between 2 and 4 leaves/tiller). At each defoliation event DM yield tonnes/hectare (t/ha) was calculated and herbage samples were collected for analysis of neutral detergent fibre (NDF) and acid detergent fibre (ADF) using an Ankom$^{200/220}$ ® fibre analyser, while total nitrogen (N) was determined by the Kjeldahl method. Crude protein (CP), digestibility and ME were subsequently calculated. Means for herbage quality measurements from defoliations during spring, summer and autumn were compared by ANOVA using the statistical package SPSS and least significant difference (LSD), as defined by Steele and Torrie (1960).

Results There was no significant difference (P > 0.05) in total DM yield between cocksfoot and perennial ryegrass at the conclusion of defoliation treatments; both yielded a mean of 4.2t DM/ha. The slower growth rate of ryegrass resulted in less frequent defoliation than for cocksfoot (i.e. 7 defoliations at 2L for cocksfoot vs. 6 defoliations for ryegrass). The ME levels of both species increased significantly (P < 0.001) between summer and autumn, by at least 0.4 MJ/kg DM (which equates to approximately 2.4% digestibility). There was no difference in ME between species throughout the experimental period (see Figure 1). Cocksfoot had significantly higher levels of ADF (22.9 vs. 20.4%; P < 0.05) and CP (17.9 vs. 12.6%; P < 0.001) than perennial ryegrass.

Conclusions Under the imposed defoliation management regime, the nutritional value and productivity of cocksfoot is comparable to perennial ryegrass. The ME range of cocksfoot throughout the experimental period was between 10.8 and 11.2 MJ/kg DM, indicating the energy content of this species is acceptable in terms of meeting the high energy requirements of a lactating dairy cow. Although cocksfoot is relatively slow to establish, this study showed its productivity can be equivalent to that of perennial ryegrass within one year from sowing. Maintaining cocksfoot plants in a vegetative state through defoliation at or before 4-5 leaves/tiller was successful in maintaining the nutritional value and productivity of this species.

Figure 1 Metabolisable energy content of cocksfoot (▲) and perennial ryegrass (■) (MJ/kg DM). Vertical bars represent LSD at $P = 0.05$

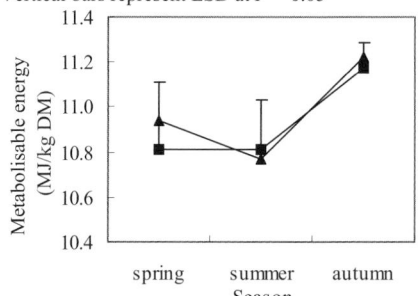

References

Greenhalgh, J.F.D., & G.W. Reid (1969). The herbage consumption and milk production of cows grazing S24 ryegrass and S37 cocksfoot. *Journal of the British Grassland Society*, 24, 98-103.

Johnson, R.J. & N.A. Thomson (1996). Effect of pasture species on milk yield and milk composition. *Proceedings of the New Zealand Grasslands Association*, 57, 151-156.

Rawnsley, R.P., D.J. Donaghy, W.J. Fulkerson & P.A. Lane (2002). Changes in the physiology and feed quality of cocksfoot (Dactylis glomerata L.) during regrowth. *Grass and Forage Science* 57, 203-211.

Steele, R.G.D. & J.H. Torrie (1960). Principles and Procedures of Statistics, McGraw-Hill, U.S.A

Perennial ryegrass variety differences in nutritive value characteristics

T.J. Gilliland[1], R.E. Agnew[2], A.M. Fearon[3] and F.E.A. Wilson[1]
[1]Department of Agriculture and Rural Development for Northern Ireland, [1]Plant Testing Station, Crossnacreevy Belfast BT6 9SH, Email: trevor.gilliland@dardni.gov.uk, [2]ARINI, Hillsborough, BT26 6DR, [3]Agriculture and Food Science Centre, Newforge Lane, Belfast BT9 5PX, UK.

Keywords: perennial ryegrass, varieties, herbage quality

Introduction Animal grazing performance at grass is predominately determined by herbage intake rates, with high yielding dairy cows requiring up to 20 kg/d DM within a limited grazing time (Gibb, 1998). Grass nutritional factors such as seasonal patterns in digestibility and water-soluble carbohydrate levels have been linked to animal productivity (Davies et al., 1991), while sward surface height, herbage mass, bulk density and green leaf mass have been shown to promote high grazing intake (Barrett et al., 2001). Furthermore, fatty acid profiles have been shown to improve the unsaturated fatty acid composition of milk, with potential human health benefits (Parodi, 1997). Recent CAP funding changes are expected to intensify the drive to optimise margin over costs. Given that grazed grass is the cheapest ruminant feed, it is expected that nutritive value characteristics of grass varieties will become increasingly important relative to total productivity, both as a breeding objective and as an evaluation criteria by variety testers and by farmers. This study examined the genetic diversity in such parameters among a wide range of perennial ryegrass (Lolium perenne L.) varieties, as an indicator of the heterogeneity among current varieties and the prospects for improvement by selective breeding.

Materials and methods A range of UK recommended perennial ryegrass diploid and tetraploid varieties of widely differing maturities were managed over four growing seasons, under a nine-cut simulated grazing system. In addition to yield parameters, at each cut, water-soluble carbohydrate (WSC) concentration was assessed on herbage dried within one hour of cutting (60°C). Further 350g dry samples were subjected to modified acid detergent fibre (MADF) analysis and 350g fresh herbage samples were stored at -20°C and then tested for lipid composition, including the linoleic acid (C18:2, cis-9, cis-12) and • •linolenic acid (C18:3, cis-9, cis-12, cis-15) fractions of total fatty acid content. Sward surface height (SSH), extended tiller height (ETH), bulk density (BD$_{>6}$) and herbage mass (HM$_{>6}$) characteristics of sward geometry were measured on a subset of varieties.

Results In overview, this study showed that existing registered perennial ryegrass varieties differ significantly in a range of nutritive value and sward structural parameters and would, therefore, differ in the output performances they could support in a grazing herd. Water-soluble carbohydrate concentration differed significantly ($P<0.001$) between varieties with tetraploids higher than all diploid varieties, except those selectively bred for this trait. Significant differences were also recorded between varieties of different maturity type and stage of physiological development ($P<0.001$). Similarly, differences in herbage digestibility were observed between varieties ($P<0.001$) and maturities ($P<0.05$), but only between ploidies when the very high digestibility, high WSC diploids were excluded. Significant varietal differences were also recorded in the proportion of linoleic acid ($P<0.05$) and of • •linolenic acid ($P<0.05$). These differences were not associated with either ploidy or maturity classes and were generally low in varieties with highest digestibility/WSC contents. Comparison of canopy structure characteristics among the intermediate varieties revealed significant differences ($P<0.001$), overlaid by temporal patterns of variation associated with maturity and season.

Conclusion The presented studies showed that overall, differences in animal value parameters were poorly associated with variety yield potential, but were strongly influenced by the physiological stage of development of the grass when harvested. It was concluded that the observed genetic diversity, indicated a good potential for achieving selective breeding improvements from within existing genepools. However, achieving such changes without a linked yield penalty may be difficult in certain cases and precise sward management to maintain grass in its optimum physiological condition, may be vital in fully exploiting 'animal value' factors on-farm.

References
Gibb, M.J. (1998). Animal grazing/intake terminology and definitions. In: M.G. Keane & E.G. O'Riordan (eds.) Pasture Ecology and Animal Intake. Proceedings of a Workshop held in Dublin, September 1996. Occasional Publication No. 3, 21-37. Dublin: Teagasc.
Barrett, P.D., A.S. Laidlaw, C.S. Mayne & H. Christie (2001). Pattern of herbage intake rate and bite dimensions of rotationally grazed dairy cows as sward height declines. Grass and Forage Science, 56, 362-373.
Davies, D.A., M. Fothergill & D. Jones (1991). Assessment of contrasting perennial ryegrasses with and without white clover, under continuous sheep stocking in the uplands. 3.Herbage production, quality and intake. Grass and Forage Science, 46, 39-49.
Parodi, P.W. (1997). Cow's milk fat components as potential anticarcinogenic agents. Journal of Nutrition, 127, 1055-1060.

Fibre degradation rate of perennial ryegrass varieties measured using three techniques: *in situ* nylon bag, *in vivo* rumen evacuation and *in vitro* gas production

H.Z. Taweel[1], B.M. Tas[1], B.A. Williams[1], A. Elgersma[2,3], J. Dijkstra[1] and S. Tamminga[1]
[1]*Animal Sciences Group, Animal Nutrition, P. O. Box 338, 6700 AH, Wageningen, The Netherlands, Email: Hassan.taweel@wur.nl,* [2]*Plant Sciences Group, Crop and Weed Ecology, Wageningen University, P. O. Box 430, 6700 AK, Wageningen, The Netherlands,* [3]*University of Ghent, Belgium*

Introduction In Western Europe, perennial ryegrass is the most widely used grass species for grazing cattle, because of its high productivity, palatability and nutritive value. However, the low dry matter intake (DMI) of perennial ryegrass pasture has been identified as a major factor limiting milk production of high producing dairy cows. Altering the chemical, physical and mechanical characteristics that contribute to its low DMI through grass breeding and the choice of variety may be a way forward in trying to maximise its DMI. This study aimed to examine whether perennial ryegrass varieties differ in their NDF degradation rates (kd_{NDF}).

Materials and methods The samples for the nylon bag and gas production measurements originated from an indoor feeding experiment, in which six diploid varieties of perennial ryegrass were fed to six high producing rumen-cannulated dairy cows using a double 3 × 3 Latin square design. In this experiment, NDF clearance rate (Kcl_{NDF}) and kd_{NDF} of each variety were estimated using two rumen evacuations separated by a 12-h period of feed deprivation. For the *in situ* measurements, freeze-dried grass samples were chopped to a 1-cm length, weighed into nylon bags and incubated in the rumen for 2, 4, 8, 12, 24, 48 and 336 h. For the *in vitro* gas measurements, freeze-dried grass samples were ground to pass a 1-mm sieve and incubated with inoculum from grass-fed cows for 72 hours using an automated gas production system. The resulting gas curves were fitted using the two-phase logistic model.

Results The *in situ* NDF degradation kinetics parameters (undegradable fraction (U), potentially degradable fraction (D) and rate of degradation (kd)) did not differ significantly among varieties, which was in agreement with the *in vivo* and *in vitro* data (Table 1). On average, the U-fraction of NDF was around 14 % and the D-fraction around 86 % with a degradation rate of around 2.5 %/h for all varieties. The three techniques showed that the different varieties used in this study did not differ significantly in their NDF degradation rates and characteristics. Moreover, the three techniques showed that there was a small range of less than 1 %/h in the NDF degradation rate. On average, the *in situ* and the *in vivo* estimates of kd_{NDF} were similar at around 2.5 %/h, whereas the *in vitro* gas production estimates were much higher at around 6.0 %/h. The coefficient of variation (SD/mean) of estimating kd_{NDF} was similar for the *in situ* and the *in vitro* techniques (8 %) and was lower in comparison to that of the *in vivo* technique (32 %).

Table 1 NDF degradation characteristics of six varieties of ryegrass measured *in situ*, *in vivo* and *in vitro*

Variable	cv1	cv2	cv3	cv4	cv5	cv6	Mean	SD[1]	P
			In situ nylon bag[2]						
U_{NDF}, %	14.2	13.8	12.5	13.7	14.5	14.5	13.9	1.4	0.3
D_{NDF}, %	85.8	86.2	87.5	86.3	85.5	85.5	86.1	1.4	0.3
kd_{NDF}	2.5	2.4	2.5	2.5	2.7	2.4	2.5	0.2	0.2
			In vivo rumen evacuations[3]						
kcl_{NDF}	5.4	5.9	5.7	5.3	4.9	5.1	5.4	0.9	0.2
kd_{NDF}	2.7	2.4	2.7	2.4	2.5	2.0	2.5	0.8	0.4
			In vitro gas production[3]						
Gas, ml	57	57	53	55	60	59	57	3.9	0.2
kd_2	6.5	5.9	5.9	5.7	6.6	6.0	6.1	0.5	0.3
Lag, h	4.6	5.4	4.9	5.3	4.2	4.6	4.8	0.6	0.2

kd_{NDF} is the degradation rate and kcl_{NDF} is the clearance rate, both expressed as %/h; kd_2 is the degradation rate of the second phase calculated from the gas production curves in %/h

Conclusions The different varieties used in this study did not differ significantly in their kd_{NDF}. Moreover, the difference between the fastest degrading variety and the slowest one was less than 1.0 %/h. This suggests that perspectives for choosing varieties with fast degrading fibre as a mean to improve pasture DMI is very limited.

The effect of cow-diet on the digestion kinetics of forages

J.L. Burke[1,2], G.C. Waghorn[3], S.L. Woodward[3] and I.M. Brookes[1]
[1] *Massey University, Private Bag 11 222, Palmerston North, New Zealand, Email: J.L.Burke@massey.ac.nz,*
[2]*Nutrition and Behaviour Group, AgResearch Grasslands, Private Bag 11 008, Palmerston North, New Zealand,*
[3] *Dexcel Limited, Private Bag 3221, Hamilton, New Zealand*

Keywords: forage degradation, *in sacco* rates, cow-diet effect

Introduction Recent research in New Zealand has used *in sacco* techniques to define the kinetics of digestion and fermentation of fresh forages minced to resemble chewed material. That work (Burke *et al.*, 2000) was undertaken in one cow fed a lucerne hay diet but Weimer *et al.* (1999) demonstrated significant effects of both cow and diet on rumen cellulolytic bacterial populations which influenced *in vitro* digestion kinetics (Mertens *et al.*, 1998). The objective of this study was to measure cow-diet effects *on in sacco* digestion of perennial ryegrass (P; *Lolium perenne*), Sulla (S; *Hedysarum coronarium*), Maize (M; *Zea maize*) silage and mixtures.

Materials and methods Four rumen fistulated cows were held in metabolism stalls and fed either ryegrass pasture (P), P:S (66:34 on a DM basis), P:M (66:34) or P:M:S (66:17:17). Digestion kinetics were measured from *in sacco* bags containing about 30 g (6 g DM) of freshly minced P, S, M as well as P:M, P:S and P:M:S mixtures (substrate). Ten bags of each forage and forage mixture were placed in the rumen in each cow and duplicate bags removed at 2, 6, 12, 24 and 72 hours. All bags were rinsed with cold water after removal from the rumen (including the 0 h bags) until no further colour appeared. Bags were dried at 60°C for 48 hours, weighed and residues removed for NIRS analysis to estimate protein (CP) and neutral detergent (NDF) contents. Rates of DM, CP and NDF degradation were determined by using a non-linear regression model.

Results and discussion Degradation kinetics of forages and mixtures (Table 1) were significantly slower when incubated in the cow fed P:M compared to cows fed P, P:S and P:M:S. The P:M diet lowered the digestion rate of DM by 16% relative to the mean of all diets and NDF by 21%, suggesting a significant impact upon rumen digestion in cattle fed diets containing maize silage. There was no cow-diet x substrate interaction. These results correspond with a lower *in vitro* gas production from substrates incubated using inocula from cows fed a maize based TMR diet compared to a lucerne based TMR (Mertens *et al.* 1998). Differences in rumen pH of the four cows used in this study were minor and unlikely to account for the slower degradability in the cow fed the P:M diet (Grant & Mertens, 1992), so the cow-diet appeared to effect a change in the rumen

Table 1 Composition (% of dry matter (DM)) and degradation rates (k, %/h) DM, protein (CP) and neutral detergent fibre (NDF) cows fed four pasture-based diets ([a, b] LS means with common superscripts within columns do not differ significantly; $P < 0.05$: SC = soluble carbohydrate)

Cow-diet	Chemical Composition			Degradation rate (%/h) of		
	SC	CP	NDF	DM	CP	NDF
P	7.4	18.7	52.1	7.7[c]	12.2[b]	5.7[b]
P:M	17.0	15.0	48.9	5.8[a]	8.8[a]	4.1[a]
P:S	10.7	18.9	28.9	7.2[b]	11.2[b]	5.6[b]
P:M:S	14.4	16.8	28.9	7.2[b]	11.3[b]	5.4[b]
SEM	-	-	-	0.21	0.38	0.14
$P <$	-	-	-	0.01	0.05	0.05

microflora. These results have implications for the New Zealand dairy cow grazing ryegrass-white clover-based pastures when maize silage is used to overcome feed deficits

Conclusions The results of this study illustrate the slower rate of digestion in the cow fed maize silage with ryegrass based pastures. Slower digestion will lower feed intakes and limit performance and further research is needed to separate cow and diet effects to define the cause of reduced degradation rates.

References
Burke, J.L., G.C. Waghorn, I.M. Brooks, G.T. Attwood & E.S. Kolver (2000). Formulating total mixed rations from forages – defining the digestion kinetics of contrasting species. *Proceedings of the New Zealand Society of Animal Production,* 60, 9-14.
Grant, R.J. & D.R. Mertens (1992). Influence of buffer pH and raw corn starch addition on *in vitro* fibre digestion kinetics. *Journal of Dairy Science,* 75, 2762-2768
Mertens, D.R., P.J. Weimer & G.C. Waghorn (1998). Inocula differences affect *in vitro* gas production kinetics. In: E.R. Deadville, E. Owen, A.T. Adesogan, C. Rymer, J.A. Huntington & T.L.J Lawerence (eds.) *In vitro* techniques for measuring nutrient supply to ruminants. Occasional Publication No. 22, British Society of Animal Science, 209-211.
Weimer, P.J., G.C. Waghorn, C.L. Odt & D.R. Mertens (1999). Effect of diet on populations of three species of ruminal cellulolytic bacteria in lactating dairy cows. *Journal of Dairy Science,* 82, 122-134.

The effect of cow-diet on the fermentation of forages

J.L. Burke[1,2], G.C. Waghorn[3], S.L. Woodward[3] and I.M. Brookes[1]
[1]Massey University, Private Bag 11 222, Palmerston North, New Zealand, Email: J.L.Burke@massey.ac.nz,
[2]Nutrition and Behaviour Group, AgResearch Grasslands, Private Bag 11 008, Palmerston North, New Zealand,
[3]Dexcel Limited, Private Bag 3221, Hamilton, New Zealand

Keywords: proteolysis, volatile fatty acids, *in vitro* fermentation, cow-diet effect

Introduction *In vitro* fermentation of fresh forages minced to resemble chewed material have enabled net proteolysis and volatile fatty acid (VFA) production to be measured using rumen inocula from a cow fed lucerne hay (Burke *et al.*, 2000). However both cow and diet affect the rumen cellulolytic bacterial populations (Weimer *et al.*, 1999) and are able to influence *in vitro* digestion kinetics (Mertens *et al.*, 1998). The objective of this study was to measure cow-diet effects on *in vitro* digestion and fermentation of perennial ryegrass (P; *Lolium perenne*), sulla (S; *Hedysarum coronarium*), maize (M; *Zea maize*) silage and mixtures.

Materials and methods Four rumen fistulated cows were held in metabolism stalls and fed either ryegrass pasture (P), P:S (66:34 on a DM basis), P:M (66:34) or P:M:S (66:17:17) to provide inocula for fermentation of freshly minced P, S, M and P:M, P:S and P:M:S mixtures. About 2.5 g of freshly minced wet material (0.5 g DM) was incubated in 50 mL vented bottles with buffer, reducing agent and strained rumen liquor (Burke *et al.* 2000) from each of the four cows. Duplicate samples of forages (and mixtures) were removed for sampling after 0, 2, 4, 6, 8, 12 and 24 h. *In vitro* pH was measured as well as net ammonia (NH_3) and VFA production (following correction for contribution from rumen contents) according to Burke *et al.* (2000). Effects of cow-diet were determined by analysis of variance.

Table 1 *In vitro* pH, net NH_3 (% forage N recovered as NH_3) and total VFA concentration (mmol/gDM) and the acetate:propionate (A:P) ratio of forages and mixtures incubated in inocula from cows fed four pasture-based diets

Cow-diet	pH	Net NH_3		VFA	
				Total	A:P
Hours	12	2	12	12	12
P	6.40[a]	8.6[b]	-0.5[b]	2.8[a]	2.2
P:M	6.09[b]	6.1[a]	-1.3[b]	3.3[b]	2.2
P:S	6.29[c]	8.4[b]	0.3[a]	2.7[a]	2.1
P:M:S	6.14[b]	9.0[b]	0.2[a]	2.8[a]	2.2
SEM	0.03	0.5	0.5	0.2	0.2
Pr <	0.01	0.01	0.01	0.01	NS

[a,b] LS means within columns with common superscripts do not differ significantly (Pr < 0.05)

Results and discussion Fermentation of forages and mixtures was faster when incubated *in vitro* with rumen inoculum from the cow fed P:M (Table 1) indicated by pH, net NH_3 concentrations and net VFA production. This rapid fermentation contrasts with a slow *in sacco* digestion in the same cow (Burke *et al.* 2005), but *in vitro* incubations include soluble and rapidly fermented components which are washed out of *in sacco* bags. These results indicate that the rumen inoculum is influenced by the diet fed to cows and, together with results of Burke *et al.* (2005), suggest that the P:M cow-diet had higher amylolytic and proteolytic, but lower fibrolytic activity than other cow-diets. Inclusion of M in ryegrass-white clover-based diets which form the basis of New Zealand dairy systems may lower fibre digestion rates.

Conclusions The inclusion of maize silage with ryegrass-white clover-based pastures may increase rumen degradation rate of soluble components at the expense of fibre digestion. Further research is needed to separate cow and diet effects to better define the cause of the more rapid *in vitro* fermentation.

References
Burke, J.L., G.C. Waghorn, I.M. Brooks, G.T. Attwood & E.S. Kolver (2000). Formulating total mixed rations from forages – defining the digestion kinetics of contrasting species. *Proceedings of the New Zealand Society of Animal Production*, 60, 9-14.
Burke, J.L., G.C. Waghorn, S.L. Woodward & I.M. Brookes (2005). The effect of cow-diet on the digestion kinetics of forages. *XX International Grasslands Conference – offered paper* (in press).
Mertens, D.R., P.J. Weimer & G.C. Waghorn (1998). Inocula differences affect *in vitro* gas production kinetics. In: E.R. Deadville, E. Owen, A.T. Adesogan, C. Rymer, J.A. Huntington & T.L.J Lawerence (eds.) *In vitro* techniques for measuring nutrient supply to ruminants. Occasional Publication No. 22, British Society of Animal Science, 209-211.
Weimer, P.J., G.C. Waghorn, C.L. Odt & D.R. Mertens (1999). Effect of diet on populations of three species of ruminal cellulolytic bacteria in lactating dairy cows. *Journal of Dairy Science*, 82, 122-134.

The effect of nitrogen fertiliser and season on the *in situ* degradability of Irish perennial ryegrass in cattle

V. Olsson[1,4], J.J Murphy[1], F.P O'Mara[3], K. O'Connell[2], J. Humphreys[1] and F.J. Mulligan[4]
[1]Teagasc, Dairy Production Research Centre, Moorepark, Fermoy, Co Cork, Ireland, Email: volsson@moorepark.teagasc.ie, [2]Teagasc, Johnstown Castle, Co Wexford, Ireland, [3]Department of Animal Science and Production, Faculty of Agri-Food and the Environment, [4]Department of Animal Husbandry and Production, Faculty of Veterinary Medicine, University College Dublin, D4, Ireland

Keywords: perennial ryegrass, rumen degradability, nitrogen, season

Introduction In light of increasing environmental and economic pressure on agriculture to utilise resources more efficiently, protein feeding and its effects are fundamentally important. As grazed grass is the predominant feed in Irish dairy and beef cattle production systems, it is necessary to establish protein values for different grass varieties and cultivars fed. It is also important to investigate the extent of ruminal nitrogen (N) degradability for these grasses since this characteristic greatly influences environmentally damaging urinary N excretion.

Materials and methods Experimental plots of perennial ryegrass (*Lolium perenne*) with different fertiliser application rates (0, 90 or 350 kg N/ha/yr) were grazed to 5 cm throughout the season of 2001, at Moorepark Research Centre. Fertiliser and grazing patterns were: 350 kg N/ha and a 3 week (wk) grazing rotation, 90 kg N/ha and a 4 wk grazing rotation and 0 kg N/ha and a 5 wk grazing rotation. Grass samples were harvested weekly, oven dried at 40°C for 48 hours (h), milled through a 1 mm screen and pooled by month (April to October). Twelve 2 g samples from each treatment (*n*=24) were incubated in nylon bags (5•10 cm; 50μm pore size) in each of four Holstein Friesian steers fitted with a ruminal cannula. All samples were incubated together and subsequently two bags per sample were removed at 0, 2, 4, 8, 12, 24 and 48 h. Immediately after removal the samples were immersed in cold water, then frozen and later treated with an *in vitro* buffer (4g NH_3HCO_3 and 35g $NaHCO_3$ per L distilled water) using 5 ml buffer per nylon bag in a Seward Lab Blender. The samples were then washed in a domestic washing machine (3•10 min rinse cycle) and oven dried at 40°C for 48 h. Nitrogen analysis was carried out using a Leco FP-328 analyser. The animals were offered a diet of 75% grass silage and 25% concentrate fed twice daily.

Results Effective degradability (ED) was calculated according to Ørskov and McDonald (1979) assuming a rumen outflow rate of 6% per hour. Data was analysed by repeated measures analysis using the PROC GLM statement of SAS. The crude protein concentration of the grass averaged 127, 157 and 247 g/kg for 0, 90 and 350 kg of N/ha. The in vitro DMD of the grass samples were 850, 840 and 855 g/kg for 0, 90 and 350 kg of N/ha respectively. There was a significant overall effect ($p<0.05$) of fertiliser on ED for dry matter (DM) but not for N ($p<0.14$). Season had a significant effect both on ED for DM and for N. The ED of DM was significantly reduced after each increase in fertiliser application in June (Table 2). In April ED of DM was significantly lower for the 350 kg of N/ha treatment (Table 2).

Table 1 Average effective degradability for N

Fertiliser	0	90	350	SEM
April	69.2	67.0	67.7	0.97
May	64.1	63.4	66.2	1.07
June	67.0	68.4	70.3	1.67
July	62.5	64.7	63.7	1.85
August	61.5	63.5	60.8	1.58
September	63.7	63.2	64.0	1.78
October	62.1	64.0	64.4	1.09

Table 2 Average effective degradability for DM

Fertiliser	0	90	350	SEM
April	76.0[a]	74.7[a]	73.2[b]	0.85
May	75.3	72.6	72.3	1.25
June	75.7[a]	68.1[b]	71.5[c]	1.05
July	68.1	70.7	67.6	1.43
August	64.5	65.8	64.4	1.12
September	69.9	69.0	66.9	1.28
October	67.2	66.3	65.9	1.43

[a, b, c] *Means within rows not sharing a superscript differ significantly (p<0.05)*

Conclusion Effective degradability of DM was reduced as fertiliser application rate increased and the grazing rotation length was reduced. The effective degradability of both DM and N was decreased as the grazing season advanced from April to October.

References
Ørskov, E.I., & I. McDonald (1979). The estimation of protein degradability in the rumen from incubation measurements weighted according to rate of passage *Journal of Agricultural Sciences (Cambridge)* 92, 499-503.

Effects of forage species and stage of maturity on *in situ* disappearance of organic matter and fibre fractions

H.S. Hussein

Department of Animal Biotechnology, University of Nevada-Reno, Mail Stop 202, Reno, NV 89557, USA, Email: hhussein@cabnr.unr.edu

Keywords: forages, grasses, *in situ*, nutritive value, maturity

Introduction Nutrient utilisation by ruminants is altered by the forage species and its maturity. Maturity is the major factor affecting forage morphology and quality. Forage quality is reduced with maturity due to a decrease in the leaf:stem ratio and an increase in fibre components (Ugherughe, 1986). Improving forage utilisation by ruminants depends on accurate measurements of their nutritive value by using *in vitro* and *in situ* methods. The objective of this study was to assess the nutritive value (i.e., extent of in situ disappearance of organic matter [OM], neutral detergent fibre [NDF], and acid detergent fibre [ADF]) of four grass species that were grown under the same conditions and were harvested at two stages of maturity.

Materials and methods Two ruminally-cannulated steers were used and had *ad libitum* access to lucerne/grass hay for 28 d before and during the experiment. The substrates were four grass hays (bromegrass [BG; *Bromus inermis*], cocksfoot [CK; *Dactylis glomerata*], ryegrass [RG; *Lolium perenne*], and tall fescue [TF; *Festuca arundinacea*]) that were harvested at the vegetative (pre-head) and mature (early head) stages. A total of 200 Dacron polyester bags (6 cm × 12 cm; 20- to 70-μm pore size) containing ground (1 mm) substrates (4 g dry matter [DM]) or blanks (empty bags to correct for weight change due to ruminal incubation) were used with 6 bags being assigned to each substrate, steer, and incubation time (24 or 48 h). Methods of incubation, residue preparation, and analysis of substrates and residues were according to Hussein et al. (1995). The data were analysed as a completely randomised design experiment by the general linear models procedure of SAS (2001).

Results With regard to the chemical analysis (NDF, ADF, and crude protein, respectively) of substrates (on DM basis), vegetative BG had 57.5, 33.8, and 19.8%, mature BG had 60.8, 36.1, and 13.5%, vegetative CK had 57.2, 33.5, and 15.7%, mature CK had 60.4, 35.6, and 12.2%, vegetative RG had 57.5, 32.6, and 18.1%, mature RG had 66.1, 35.9, and 13.8%, vegetative TF had 57.8, 35.6, and 15.5%, and finally mature TF had 61.4, 36.5, and 12.1%. The steer effect was not significant (p>0.05). Because the interactions also were not significant, effects of the main factors are presented in Table 1. Forage species did not affect disappearance of OM or ADF. Extent of NDF disappearance, however, was lower for TF than for the remaining forages. Maturity decreased extent of disappearance by 8.4, 9.1, and 6.1 percentage units for OM, NDF, and ADF, respectively.

Table 1 Extent of *in situ* disappearance (%) of substrate OM, NDF and ADF

Item	Forage species					Stage of maturity		
	BG	CK	RG	TF	s.e.m.	Vegetative	Mature	s.e.m.
OM	68.5	69.5	67.7	63.8	1.8	71.6[a]	63.2[b]	1.3
NDF	56.3[a]	57.5[a]	55.9[a]	48.9[b]	2.1	59.2[a]	50.1[b]	1.5
ADF	51.0	52.2	50.3	47.2	2.3	53.2[a]	47.1[b]	1.6

[a,b] Means in the same row and under the same factor with different superscripts differ (p<0.05)

Conclusions The results support the widely found decline in nutritive value with increasing grass maturity, but in the condition of this experiment there appeared to be little differences among the four species examined.

References

Hussein, H.S., M.R. Cameron, G.C. Fahey, Jr., N.R. Merchen & J.H. Clark (1995). Influence of altering ruminal degradation of soybean meal protein on *in situ* ruminal fiber disappearance of forages and fibrous by-products. *Journal of Animal Science*, 73, 2428-2437.

SAS (2001). SAS User's Guide: Statistics, Version 8.2, SAS Inst. Inc., Cary, NC, U.S.A.

Ugherughe, P.O. (1986). Relationship between digestibility of *Bromus inermis* plant parts. *Journal of Agronomy and Crop Science*, 157, 136-143.

Physical impediment towards digestive breakdown in leaf blades of Brachiaria brizantha

B. Lempp[1], C.B. do Valle[2], M. das G. Morais[1], R.A. Borges[1], E. Detmann[3]
[1]Federal University of Mato Grosso do Sul, Dourados MS Brazil, Email:blempp@ceud.ufms.br, [2]Embrapa Beef Cattle, Campo Grande MS Brazil, [3]Federal Universit y of Viçosa, Viçosa MG Brazil

Keywords: epidermal stegmata, girder structure, leaf anatomy, physical structure

Introduction Consumption of grasses is influenced by the physical properties of forages which confer resistance to digestive breakdown. Such barriers may be the proportion of indigestible tissues, girder structure and epidermal cell arrangements. Anatomical factors, if identified early are invaluable tools in breeding and selection programmes for forages of high quality. The objective of this study was to verify which anatomical attributes might be interfering in the physical resistance to rumen breakdown in *Brachiaria brizantha* ecotypes.

Materials and methods Four ecotypes of *Brachiaria brizantha*, B1, B4, B8 and B9, were grazed for two years in 1,000 m² paddocks, in two replicates, in Campo Grande, Brazil. After leveling, transect sampling of the second fully expanded leaf and of leaf fragments in oesophageal fistula samples were taken. The leaf fragments of herbage and extrusa were incubated *in vitro* for 24 h. *In situ* dry matter (DM) degradability (6, 24, 36, 48 and 72 h) was done using leaves, analysed according to Van Milgen *et al.* (1991). The residues of *in vitro* digestibility were then fixed, histologically cut (10 μm) and stained for optical microscopy observation (MO). The residues of *in vitro* digestibility from the extrusa were analysed with the samples *in natura*. The epidermis was isolated from leaf fragments using Jefrey solution followed by MO.

Results All genotypes displayed a high frequency (average of 95.6%) of girder I structure along the cross section. This did not interfere in digestive breakdown of B9, however (Figure 1A). The effect of girder I in B4 (Figure 1B), B8 and B1 genotypes (in decreasing order) was attributed to epidermal stegmata (Figure 1C); epidermal silica cells (ESC) lying over sclerenchyma fibers associated with vascular bundles, as caps or girds (Prychid *et al.*, 2003). In B9, ESCs were less associated with the sclerenchyma (Figure 1 D). The disappearance of parenchyma bundle sheaths (PBS) also interfered with digestive breakdown: in B4 there was greater liberation of tissues to the incubation media due to greater PBS degradation when compared to B8 and B1. The results of *in situ* degradability agree with the anatomical study. The soluble fraction was 14.9%; 28.7%; 9.7% and 21.3%; potentially degradable fraction was 52.6% ± 2.4; 36.0% ± 4.0; 24.4% ± 1.6 and 57.3% ± 9.4; undegradable fraction was 32.4% ± 2.4; 35.0% ± 3.6; 24.4% ± 1.6 and 66.4% ± 5.3; and rate of DM disappearance was 7.13; 5.05; 7.40 and 5.75%/h for B1; B4; B8 and B9 respectively. Differences among these genotypes could not be detected by chemical analysis including silica content or by IVDMD percentages (Torres, 2002).

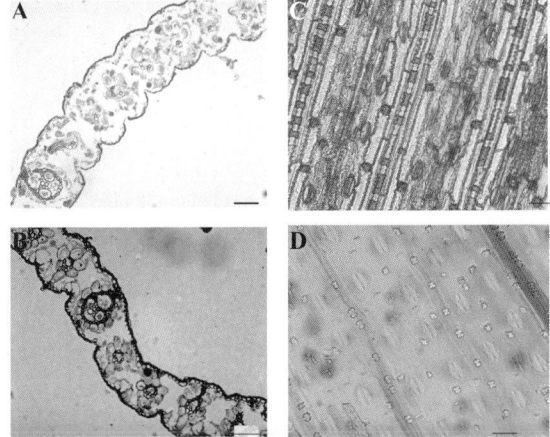

Figure 1 A and **B** Cross-section of leaf blades of *Brachiaria brizantha,* incubated in buffer for 24 hours (**A**.B9, **B**.B4). **C** and **D**. Abaxial epidermal cells (**C**. B4 and **D**. B9) (— 20 μm)

Conclusions The physical impediment to digestive breakdown of leaf blades of *Brachiaria brizantha* is attributed to epidermal stegmata and to parenchyma bundle sheath cells disappearance.

References
Prychid, C.J.,P.J. Rudall & M. Gregory (2003). Systematics and biology of silica bodies in monocotyledons. *The Botanical Review*, 69, 377-440.
Torres, F.E. (2002). Avaliação de produção e valor nutricional de ecotipos de Brachiaria brizantha. Master of Agonomy thesis, Universidade Federal de Mato Grosso do Sul, 56 pp.
Van Milgen, J., M.R. Murphy & L.L. Berger (1991). A compartmental model to analyze ruminal digestion. *Journal of Dairy Science*, 74, 2515-2529.

Herbage mass and *in situ* dry matter ruminal degradation kinetics of *Brachiaria* spp

E. Valencia-Chin[1], A.A. Rodríguez-Carías[2], R. Ramos-Santana[1] and S. Pagán-Riestra[2]
Agronomy Department[1] and Department of Animal Industry[2], University of Puerto Rico, Mayagüez Campus, Mayagüez, PR 00681, Puerto Rico, Email: Carias_Abner@hotmail.com

Keywords: herbage mass, *Brachiaria spp, in situ*, dry matter degradation

Introduction In Puerto Rico, *Brachiaria decumbens* cv. Basilisk has been promoted as a potential forage for acid soils and humid areas, but with limited success. Recently, *B. brizantha* cv. Marandú and a hybrid (*B. brizantha x B. Ruziziensis*) cv. Mulato were introduced for evaluation on acid soils and as a potential replacement for cv. Basilisk, but little information is available on yield performance under grazing or nutritive value. The objective of this study was to assess herbage mass and nutritive value of grazed pastures consisting of Basilisk, Marandú, and Mulato and determine the rate of *in situ* dry matter degradation.

Materials and methods This study was conducted at the Corozal substation, Agriculture Experiment Station of the University of Puerto Rico, Puerto Rico (18.0° and 18.5° N and 65.6° and 66.3° W). Fully established stands (three replicates of 0.5 ha each) of Basilisk, Marandú and Mulato were stocked every 28 d with yearling steers (mobgrazed to 15-cm height). Prior to grazing, a disc meter (Santillan *et al.*, 1979) and a double sample technique (Ortega-S. *et al.*, 1992) were used to determine herbage mass (all grass above soil level) of pastures. Double samples (plant height in cm, followed by clipping a 0.25-m^2 quadrat to ground level) were taken at four sites in each pasture every 28 d. In addition, disc heights (20 randomly selected sites) were taken from each pasture. Herbage mass of actual values were regressed on plant height to generate prediction equations. Means of the 20 plant height from each pasture and the prediction equations were used to predict herbage mass. Chemical composition was determined in forage samples using standard procedures (AOAC, 1990; Van Soest, *et al.*, 1991). The *in situ* dry matter (DM) degradation study was conducted using the suspended nylon bag technique in two fistulated cows maintained on a grass diet. Triplicate samples of each cultivar were incubated for 0, 6, 12, 24, 30 and 48 h and analysed for rate of degradation and DM disappearance. Data was analysed using the non-linear model; Degradation = a+b*(1 – exp (-c*t)), where; a = soluble fraction, b = degradable fraction, c = rate of degradation and t = incubation time (SAS, 1990).

Results Mean herbage masses of Basilisk (3.6 t/ha), Marandú (2.6 t/ha), and Mulato (3.0 t/ha) were not different (P>0.05), but there were differences in CP and NDF. Crude protein concentration of Basilisk (4.4%) was 1.6 units lower than Marandú (6%) and 2.6 units lower than Mulato 7%). Neutral detergent fiber values were also higher for Basilisk (72%), than for Marandú (67%) and Mulato (66%). *In situ* DM disappearance was higher in Marandú after 48 h incubation compared to the other two forage species (Figure 1). Ruminal degradation rate (Kd's) was faster (P<.05) for Mulato (.07) than Basilisk and Marandú (.09).

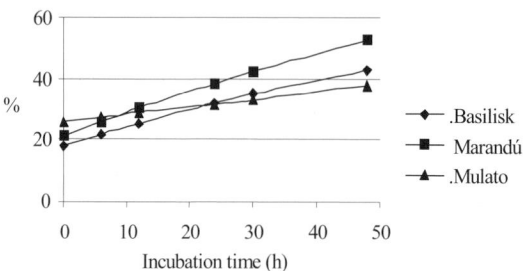

Figure 1   DM disappearance of forages species evaluated (point values according to the non linear regression)

Conclusions Mulato and Marandú exhibited comparable yields, higher CP and lower NDF values than Basilisk. *In situ* DM disappearance was higher in Marandú than either of the other forage species. Both Mulato and Marandú represent a potential replacement for Basilisk in the animal industry in Puerto Rico.

References
AOAC (1990). Official Methods of Analysis. 15th ed. Association of Official Analytical Chemists. Arlington, VA.
Santillan, R. A., W.R. Ocumpaugh & G.O. Mott (1979). Estimating forage yield with a disk meter. *Agronomy Journal*, 1, 71-75.
Ortega-S., J.A., L.E. Sollenberger, K.H. Quesenberry, J.H. Cornell & C.S. Jones. Jr. (1992). Productivity and persistence of rhizome peanut pastures under different grazing management. *Agronomy Journal*, 84, 930-934.
SAS Institute (1990). SAS/STAT® User´s Guide (Release 6.12). SAS Inst. Inc., Cary, N.C.
Van Soest, P.J, J.B. Robertson & B.A. Lewis (1991). Methods for dietary fiber, neutral detergent fiber, and non starch polysaccharides in relation to animal nutrition. *Journal of Dairy Science*, 74, 3583.

The effects of urease source and moisture content on the nutritive value of Brachiaria brizantha hay treated with urea

R.A. Reis, L.M.A. Bertipaglia, G.M.P. Melo, A.P. Oliveira and S. Luca
FCAV/ UNESP, Campus de Jaboticabal-SP, Brazil, Email: rareis@fcav.unesp.br

Keywords: *Brachiaria decumbens,* chemical treatment, elephant grass, leucaena, urease

Introduction Ammoniation has a high potential to increase the nutritive value (NV) of mature tropical grasses hays. Urea utilisation, like ammonia, seems to be a promising alternative for on-farm treatment of low quality forage in Brazil. Urea has the advantages over anhydrous ammonia of being widely available, easier to handle and, in some countries, less expensive. The aim of this work was to study the effects on the efficiency of urea treatment of palisade grass (*Brachiaria brizantha* cv. Marandu) hay of the moisture level, and the urease source.

Materials and methods The experiment was conducted at UNESP Jaboticabal-Brazil with palisade grass hay. The treatments were hays with 15 or 30% moisture either untreated, treated with urea (5% of dry matter) or treated with the combination of urea (5% of dry matter) and three urease sources. The urease sources were: signal grass (*Brachiaria decumbens)* hay, elephant grass (*Pennisetum purpureum*e), *and* leucaena (*Leucaena leucocephala*) green forage. The palisade grass hays were analysed for crude protein (CP), protein fractions (A, B1, B2, B3, and C), neutral detergent fiber (NDF), acid detergent fiber (ADF) (AOAC, 1984), and *in vitro* dry matter digestibility (IVDMD) (Tilley & Terry, 1963). The data were analysed as a completely randomised design with three replications using the Tukey test.

Results Urea application increased CP, and A fractions (P< .05), but did not affect the B1 and B2 fractions (Table 1). However, the B3 and C fractions decreased in urea treated hays (P < .05). Urea treatments decreased NDF value (P< .05. The treatments did not affect the ADF values. The urease source in combination with the highest moisture content increased the IVDMD of the hays.

Table 1 Neutral detergent fiber, acid detergent fiber, IVDMD, crude protein (in dry matter %), and protein fractions (in CP%) in palisade grass hay

Treatments	NDF	ADF	IVDMD	CP	A	B1	B2	B3	C
Hay 15% moisture + signal grass hay + urea	$73,3^{CDE}$	$40,6^{AB}$	$52,28^{B}$	$15,1^{AB}$	$86,2^{A}$	$0,08^{C}$	$6,4^{E}$	$2,5^{B}$	$4,7^{B}$
Hay 15% moisture + elephant grass + urea	$75,0^{BCD}$	$40,9^{AB}$	$59,27^{AB}$	$12,6^{ABC}$	$82,3^{AB}$	$0,19^{BC}$	$8,7^{DE}$	$2,7^{B}$	$6,1^{B}$
Hay 15% moisture + leuacena + urea	$76,6^{ABC}$	$41,1^{A}$	$59,62^{AB}$	$12,4^{ABC}$	$76,2^{BC}$	$0,16^{BC}$	$15,4^{ABC}$	$1,5^{B}$	$6,7^{B}$
Hay 15% moisture + urea	$69,8^{F}$	$37,7^{B}$	$65,08^{A}$	$16,3^{A}$	$85,7^{AB}$	$0,05^{B}$	$8,7^{CDE}$	$0,9^{B}$	$4,4^{B}$
Hay 15% moisture without urea	$77,5^{AB}$	$42,2^{A}$	$53,55^{B}$	$3,1^{D}$	$49,1^{E}$	$0,15^{BC}$	$19,2^{A}$	$5,8^{A}$	$25,7^{A}$
Hay 30% moisture + signal grass hay + urea	$74,9^{BCD}$	$41,7^{A}$	$66,44^{A}$	$10,6^{BC}$	$76,2^{BC}$	$0,14^{BC}$	$14,1^{ABCD}$	$1,0^{B}$	$8,5^{B}$
Hay 30% moisture + elephant grass + urea	$71,1^{EF}$	$42,3^{A}$	$64,47^{A}$	$8,7^{C}$	$71,2^{CD}$	$0,35^{A}$	$15,7^{AB}$	$2,4^{B}$	$10,3^{B}$
Hay 30% moisture + leucaena + urea	$70,7^{EF}$	$40,4^{AB}$	$66,99^{A}$	$11,0^{BC}$	$77,8^{ABC}$	$0,37^{B}$	$12,1^{BCDE}$	$1,1^{B}$	$8,5^{B}$
Hay 30% moisture + urea	$72,0^{DEF}$	$40,0^{AB}$	$66,52^{A}$	$12,5^{ABC}$	$80,6^{ABC}$	$0,29^{BC}$	$10,3^{BCDE}$	$1,0^{B}$	$7,7^{B}$
Hay 30% moisture without urea	$78,7^{A}$	$42,6^{A}$	$56,33^{A}$	$3,4^{D}$	$65,1^{D}$	$0,63^{A}$	$6,8^{E}$	$3,0^{AB}$	$26,9^{A}$
CV (%)	1,6	2,8	11,2	14,8	4,4	35,9	19,3	46,7	18,1

Means with different superscripts are significantly different (P< .05)

Conclusion Urea application, in association with a urease source, increased the nutritive value of palisade grass hay.

References

Association of Official Analytical Chemists (AOAC) (1984). Official Methods of Analysis, 14th ed. Arlington, Virginia, 1141pp.

Tilley, J.M A. & R.A. Terry (1963). Two-stage technique for the *in vitro* digestion of forage crops. *Journal of the British Grassland Society*, 18, 104-111.

Rumen degradation characteristics of four species of native pastures from central Mexico in three growing periods

A.A. Rayas[1], A. Espinoza[1], J. Estrada[1], C. Arriagal[1], F. Mould[2] and O.A. Castelán[1]
[1]CICA-UAEMex, Instituto Literario No. 100, col. Centro CP 50000 Toluca, México, Email: oaco@uaemex.mx, [2] Department of Agriculture, The University of Reading RG6 6AR Berkshire, UK

Keywords: native grasses, rumen degradation, Mexico

Introduction The smallholder cattle systems of the Toluca valley in central Mexico are based on the use of maize and native grasses. Research has been devoted to nutritional characterisation of improved pastures, but native species of grasses have not been studied, despite their importance. The aim of this experiment was to evaluate the *in vitro* rumen degradation kinetics of four species of native pastures.

Material and methods *Pennisetum clandestinum* (PC), *Sporobolus indicus* (SI), *Trifolium amabile* (TA), *Eleocharis dombeyana* (ED) and pooled samples (PS), which included all the species, were collected. Sample selection was based on the preference shown by grazing livestock and their abundance within four communities in Toluca valley. Four plots of native pastures were selected by community - 16 plots in total. Samples were harvested in 2003 in July (P1), September (P2) and November (P3). All samples were analysed for *in vitro* gas production (GP) as described by Theodorou *et al.* (1994); digestibility of the NDF (dNDF) and organic matter digestibility (OMD) were estimated from the GP analyses. The cumulative GP was fitted to the model of Jessop & Herrero (1996). A split plot design was used, where the main plots were the species and the split plots were growing periods and communities were blocks. Data was analysed by analysis of variance, Minitab v13.

Results and discussion As expected ,there was a significant decline from P1 to P3 for all degradation parameters and for gas production. Significant differences were observed between species for *a* (P<0.05), *b*, *c2* and *lag* time (P<0.001) (Table 1). There was a significant interaction between species and period for *b* and *lag* phase (P<0.05). The *a* fraction ranged from 38.9 ml/g dry matter in SI to 63.9 in TA. No significant differences (P>0.05) were observed for *c1*. The species with the highest *b* fraction (202.5 ml) was SI, while ED had the lowest value (139.1 ml). Highly significant differences (P<0.001) were observed for *c2*; TA having the highest rate (0.046/h) and SI the lowest (0.026(/h). As a consequence of its low NDF degradation rate, SI and ED had the highest lag times, whilst TA showed the shortest *lag* phase.

Table 1 Rumen degradation parameters

Species	a (ml)	c1 (/h)	b (ml)	c2 (/h)	lag (h)
PS	59.6	0.08	165.9	0.032	9
PS	55.9	0.076	180.7	0.038	8.7
SI	38.9	0.14	202.5	0.026	10.3
TA	63.9	0.089	164.3	0.046	5.1
ED	55.1	0.083	139.1	0.028	10.7
s.e.m	14.7	0.041	36.1	0.013	3.4
Period	*	NS	***	***	***
P1	61.9	0.113	192.1	0.039	6.4
P2	52	0.073	157.1	0.034	10.1
P3	50.1	0.095	162.37	0.029	9.8
s.e.m	7.1	0.022	21.1	0.005	2.2
	NS	NS	***	***	***

The opposite tendency was observed for the *lag* phase, as in PS it increased from 6.4 h in P1 to 9.8 h in P3, probably as a response to the increase in NDF content from P1 to P3. TA had the highest gas production rate at 8 h (9.4 ml gas/h), released probably from its pectin content. It also showed the highest GP at 24 h, which is probably through fermentation of the NDF. These data for TA suggest that most of the insoluble fraction was degraded before 48 h and at a faster rate than for the other species. Degradation data suggest that TA will permit high intakes for the animal because of its high degradation rate of the insoluble fraction and short *lag* time.

Conclusion The results obtained here suggest that nutritive value was highest from early July to late September (P1 and P2). TA showed the best degradation parameters.

References

Jessop, N.S. & M. Herrero (1996). Influence of soluble components on parameter estimation using the *in vitro* gas production technique. *Animal Science*, 62, 626-627.
Theodorou, M.K., B.A. Williams, M.S. Dhanoa, A.B. McAllan and J. France. (1994). A simple gas production method using a pressure transducer to determine the fermentation kinetics of ruminant feeds. *Animal Feed Science and Technology*, 48, 185-197.

Ruminal dry matter, neutral detergent fibre and acid detergent fiber degradation kinetics of dominant pasture forages in Kurdestan province of Iran

A.A. Sadeghi[1], P. Shawrang[2] and A. Nikkhah[2]
[1]Department Of Animal Science, Faculty of Agriculture, Science & Research campus, Islamic Azad University, Ponak, Hesarak, Tehran, Iran, Email: draasadeghi@yahoo.co.uk, [2]Department Of Animal Science, Faculty of Agriculture, Tehran University, Tehran, Iran

Introduction Neutral detergent fiber (NDF) is a major chemical component of forages and its degradability (dNDF) is an essential parameter in predicting their energetic value. Moreover, dNDF has been used in models to estimate the physical fill of fibrous feeds in the rumen and, therefore, the intake capacity of animals. As the available information on the nutritive value of pastoral forages is limited a study was undertaken to measure chemical composition and cell wall degradation kinetics of eight pasture forages in the rumen.

Materials and methods Forage samples (Vicia villosa, Bromus tomentellus, Hordeum bolbusum, Festuca ovina, Agropyron tauri, Agropyron trichophorum, Prangus ferulacea and Ferula orientalis) were collected at the pre-flowering stage (30th April) from pastures within the Kurdestan province of Iran (altitude 1480 m, latitude 35° 19' N and longitude 47° 00' E). Forage collected for *in situ* procedure was freeze dried and ground to pass a 3-mm screen. The samples were incubated for 0, 12, 24, 48 and 72-h within the rumen of three Varamini rams (BW = 46.4 ±4.2 kg) n . Zero time washing losses were determined by soaking 3 bags in warm water (39°C) for 1 h. All bags were then washed within a washing machine on a cold water cycle prior to freeze drying. The NDF and acid detergent fiber (ADF) concentrations were determined (Van Soest *et al.*, 1991). The kinetics of degradation for dry matter (DM), NDF and ADF were determined by way of the Ørskov and McDonald (1979) model. Effective rumen degradability (ERD) of DM, NDF and ADF were analyzed by a variance analysis GLM procedure of SAS (SAS, 1996) in a completely randomized design according to this model: Y= μ + Ti + Eij, where μ is overall average, Ti is the feed effect and Eij is the residual error.

Results and discussion The NDF and ADF contents of V. villosa, B. tomentellus, H. bolbusum, F. ovina, A. tauri, A. trichophorum, P. ferulacea and F. orientalis were 39.8 and 35.7, 45.1 and 27.3, 61.1 and 19.2, 62.6 and 19.0, 60.4 and 34.8, 63.8 and 37.8, 25.1 and 24.5, 23.9 and 23.3 % (DM basis), respectively. *In situ* results showed that the effective DM, NDF and ADF degradability of eight pasture forages differed significantly (Table 1). Effective DM degradability at rumen out flow rate 0.02/h was highest for P. ferulacea (78.0%) and lowest for F. ovina (37.1%). Effective NDF and ADF degradability was greatest for P. ferulacea (61.8% and 59.6%) and lowest one for H. bolbusum (32.0and F. ovina (31.4%) respectively.

Table 1 Ruminal DM, NDF and ADF degradation parameters of forage samples

Pasture forages	DM degradation traits (%)			ERD (%)	NDF degradation traits (%)			ERD (%)	ADF degradation traits (%)			ERD (%)
	a	b	c		a	b	c		a	b	c	
Vicia villosa	31.0	43.7	5.5	63.0d	8.1	54.6	3.6	43.3 e	6.5	52.5	3.8	41.1d
Bromus tomentellus	29.1	56.3	4.1	67.2c	10.6	72.3	1.7	52.8 c	7.3	76.1	2.7	51.0b
Hordeum bolbusum	16.5	36.2	4.9	39.4g	2.8	44.5	3.8	32.0 g	0.1	46.1	1.2	35.1g
Festuca ovina	11.3	33.0	7.3	37.1h	10.8	45.6	2.1	34.0 f	7.3	74.6	6.2	31.4f
Agropyron tauri	19.2	51.5	6.0	57.9e	2.7	60.5	5.8	47.7 b	1.2	61.1	5.7	46.6c
Agropyron trichophorum	19.9	45.0	3.7	49.4f	4.9	52.5	3.5	38.3 d	3.6	48.9	3.5	34.8e
Prangus ferulacea	41.6	48.3	6.0	78.0a	11.6	81.5	3.2	61.8 a	9.9	82.5	3.0	59.6a
Ferula orientalis	49.0	37.7	4.4	75.0b	9.4	53.3	2.5	39.2 c	1.0	79.5	3.0	49.2e

a: immediately soluble fraction, b: potentially degradable fraction, c: degradation rate; significant level (p<0.05).

Conclusion Differences between pasture species in the rate and extent of fiber degradation could affect intake, under such conditions they should be considered as main parameters in ration formulation of ruminants.

References
Ørskov, E.R., & I. McDonald (1979). The estimation of protein disappearance in the rumen from incubation measurements weighted according to rate of passage. *Journal of Agriculture Science (Cambridge)*, 92, 499-503.
SAS Institute Inc., 1996. Statistical Analysis System (SAS) User's Guide, SAS Institute, Cary, NC, USA.
Van Soest, P.J., J.B. Robertson & B.A. Lewis (1991). Methods for dietary fiber, neutral detergent fiber and non starch- polysaccharides in relation to animal production. *Journal of Dairy Science*, 74, 3583-3597.

Ruminal disappearance and passage rates in fresh Nezasa dwarf bamboo growing in Japanese native pasture

M. Yayota, J. Karashima, T. Kokestu, M. Nakano and S. Ohtani
Faculty of Applied Biological Science, Gifu University, 1-1 Yanagido, Gifu 501-1193, Japan, Email: yayo@cc.gifu-u.ac.jp

Keywords: native forage, disappearance rate, passage rate, rumen

Introduction Nezasa dwarf bamboo (*Pleioblastus chino* makino) is one of major native forages for grazing in Japan. However its nutritional utilisation in the rumen has been little studied. The object of this research was to measure ruminal disappearance and passage rates in fresh Nezasa dwarf bamboo compared with improved grass.

Materials and methods Six Suffolk ewes surgically fitted with ruminal cannulae were used in this study. Three ewes were assigned to receive fresh Nezasa dwarf bamboo (*Pleioblastus chino* makino) as native forage and the remaining three received fresh bahiagrass (*Paspalum notatum* Flluge) as improved forage. The diets were fed twice daily at 09.00 and 16.00 in maintenance quantities. The experiment was conducted in June, July - Aug. and Oct. in 2003. Each experimental period consisted of 17 d. The first 7 d of each period were used for adjustment to the diet. The following 5 d were used for measurement of ruminal disappearance rate using the *in sacco* technique. The diet samples were incubated in the rumen of ewes for 0, 4, 8, 16, 24, 48, 96 and 120 h. The last 5 d were used for measurement of ruminal passage rate using the forage labelling technique. At 09.00 of the first day (d 13), the ewes were fed a 15g single dose of each forage that had been labelled with ytterbium (Yb) by immersing fresh forages in a 0.5% w/v aqueous solution of $YbCl_3 \times H_2O$ for 24 h, rinsing in running water for 1h and drying for 48 h at 60°C. The faecal samples were collected every 4 h for 48 h and every 8 h for the next 72 h. The passage rate constant of digesta was calculated as described by Grovum & Williams (1973); $Y = Ae^{-K1(t-TT)} - Ae^{-K2(t-TT)}$ where Y = marker concentration in the faeces, K1 = ruminal passage rate, K2 = post-ruminal passage rate, t = time of sampling, TT = time of first appearance of marker in faeces. Total mean retention time (TMRT) in the whole tract was calculated as $1/K1 + 1/K2 + TT$. Data were analysed using ANOVA for a two-way factorial design with forage and season as the factor.

Results Extent of disappearance in the rumen at 120 h of Nezasa bamboo was significantly lower than that of bahiagrass for all seasons (Table 1). Also the extent of disappearance in the rumen of Nezasa bamboo decreased with the progress of season. Ruminal passage rate was slower and TMRT was longer in Nezasa bamboo than in bahiagrass during July - Aug. and Oct., but the differences were not significant (Table 1).

Table 1 Chemical composition, ruminal disappearance and passage rates in Nezasa bamboo and bahiagrass

	June		July - Aug.		Oct.		Significance		
	Nezsa	Bahiagrass	Nezsa	Bahiagrass	Nezsa	Bahiagrass [1]	F	S	F×S
Chemical composition (% of dry matter)									
CP	12.7	16.1	14.3	13.4	17.0	11.5			
NDF	72.5	53.6	70.4	64.0	62.4	66.6			
Extent of disappearance in the rumen at 120 h (%)									
DM	52.3	89.6	47.3	81.9	40.0	70.6	***	***	n.s.
	72.5	96.1	67.5	94.6	65.9	86.8	***	*	n.s.
NDF	58.9	83.7	52.9	81.6	45.0	68.9	***	*	n.s.
Ruminal passage rate (%/hr)									
	2.1	1.9	2.1	3.0	2.0	3.8	n.s.	n.s.	n.s.
TMRT (h)	90.1	97.9	108.5	61.3	87.0	65.9	n.s.	n.s.	n.s.

1) Because of health disorder, one ewe was removed from the treatment (n=2)
F = effect of forage, S = effect of season, F×S = interaction between F and S, * = P < 0.05, *** = P < 0.001

Conclusions This study suggests that Nezasa dwarf bamboo was nutritionally poorer than the improved grass. Therefore note must be taken of the chemical composition and digestion characteristics of native forages such as dwarf bamboo when considering the role that they can play in feeding grazing cattle.

Reference

Grovum, W.L. & V J.Williams (1973). Rate of passage of digesta in sheep. 4. Passage of marker through the alimentary tract and the biological relevance of rate constants derived from the changes in concentration of marker in faeces. *British Journal of Nutrition*, 30, 313-329.

Effects of particle size in forage samples for protein breakdown studies

G. Pichard and C. Tapia
Department of Animal Sciences, School of Agriculture, Pontificia Universidad Católica de Chile, Email: gpichard@puc.cl

Keywords: forages, particle size, proteolysis, rumen, nitrogen solubility

Introduction Coupling ruminal processes of hydrolysis and synthesis continues to be a research issue where more progress is needed. This requires the development of good protein assessment methods, particularly when representing the breakdown processes that occur in fresh pastures eaten by herbivores. Laboratory analyses need to deal with small and homogeneous samples, but the mechanical reduction of particle size may not reflect the actual digestion kinetics occurring when the original fresh forage is consumed. Such physical traits may alter the release of non-structural compounds and the penetration of microbial enzymes (Boudon *et al.,* 2002). The objective of this work was to assess in fresh samples the effect of reducing particle size upon the *in vitro* breakdown of proteins during the early rumen fermentation period.

Materials and methods Eight fresh forage samples with contrasting endopeptidase activities were subjected to different strategies for particle size reduction. Protein hydrolysis was assessed by measuring the residual neutral detergent insoluble nitrogen (NDIN) (Licitra *et al.*, 1996) and the accumulation of non-protein soluble nitrogen (NPSN) after 6 h *in vitro* rumen fermentation (IIV, Broderick, 1987). In fresh samples mastication-like damage was obtained with a device in which forage samples were pressed bewteen two stony surfaces that simulated the animal molar surfaces. During the development of this method, microscopic observation was used in order to obtain a similar damage to that observed in samples obtained from the cardias of a fistulated adult cow fed the same type of fresh long forage. Three chopping sizes and two macerations were tested. Chopping was preceded by laboratory-mastication and further cutting to 3cm, 1cm or 0.25cm; maceration was thoroughly done in a mortar with dry ice (CO_2) or liquid nitrogen. Treatments means were compared by Tukey-Kramer test at $p < 0.05$.

Results and discussion Sample size significantly affected ($p < 0.05$) the fractions of NDIN and NPSN (Figure 1), but the two macerates were essentially identical. As expected, the smaller the chopping the greater the solubility, with this effect being more pronounced in cultivars with lower endopeptidase activity. Mechanical particle comminution may facilitate access of external enzymes and activate the endogenous enzymatic system.

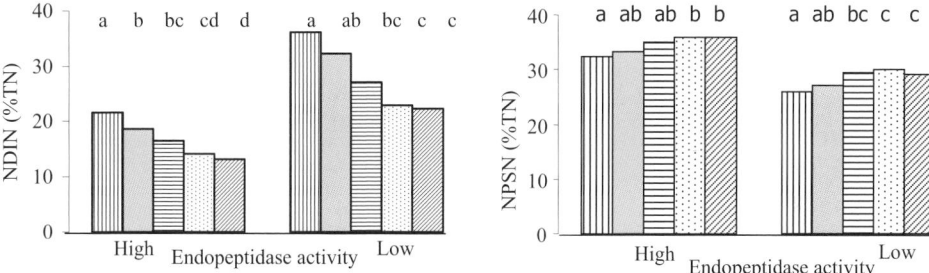

Figure 1 Effect of forage particle size on protein solubilisation during 6 h *in vitro* rumen fermentation.
▥ : 3cm.; ▨ : 1cm.; ▤ : 0.25cm.; ▧ : CO_2; ▨ : N $_{Liq}$

Conclusion Our results show that particle size is a major source of variation when studying kinetics of protein breakdown. The mechanical damage affects the release of fermentable substrates as well as the accessibility of bacteria or their enzymes into the plant cells and the activity of plant endopeptidases. Larger particles may be more representative of actual animal behaviour, but they present practical problems for analytical purposes.

Achnowledgement This research was funded by Fondecyt Grant No. 1030918

References

Boudon, A., S. Mayne, J-L Peyraud & A.S. Laidlaw (2002). Effects of stage of maturity and chop length on the release of cell contents of fresh ryegrass (*Lolium perennne* L.) during ingestive mastication in steers fed indoors. *Animal* Research, 5, 349-365.

Broderick, G.A. (1987). Determination of protein degradation rates using a rumen *in vitro* system containing inhibitors of microbial nitrogen metabolism. *British Journal of Nutrition*, 58, 463-475.

Licitra, G., T.M. Hernández, & P.J. Van Soest (1996). Standardisation of procedures for nitrogen fractionation of ruminant feeds. *Animal Feed Science and Technology*, 57,347-358.

Variation in protein quality of forage legumes during spring growth

M. Gierus, J. Kleen and F. Taube

Grass and Forage Science/Organic Agriculture, Institute of Crop Science and Plant Breeding, Christian Albrechts University of Kiel, 24118 Kiel, Germany, Email: mgierus@email.uni-kiel.de

Keywords: protein quality, forage legumes

Introduction The utilisation of forage legumes in combination with reduced N inputs as fertilisers is an alternative for the production of high quality forage. Although white clover (WC) is widely used in grassland and has a high content of crude protein (CP), a combination of this high CP content with a high proportion of rapidly-available proteins and fast degradation rate in the rumen may reduce the efficiency of N use by ruminants in comparison with other forage legumes. The objective of the present study was to investigate the variation in content of CP fractions A, B and C in different forage legume during the spring growth in comparison with WC.

Table 1 Content and fractions of N in different forage legumes

Date	25/4	12/5	22/5	3/6	12/6	23/6	1/7
Total-N, g/kg DM							
WC	45.3	44.6	37.8	34.5	29.5	28.8	26.9
RC	44.8	41.1	33.6*	26.2*	20.8*	21.2	17.3*
KU	42.6	40.1	39.1	28.6*	24.6*	25.5	22.0*
LU	38.7	36.0*	36.8	27.2*	26.3	27.4	24.5
BT	NA	37.8*	38.2	30.1*	25.5	27.3	25.6
SE	-	0.16	0.10	0.13	0.13	0.30	0.11
Fraction A, % total-N							
WC	8.8	14.4	22.2	28.5	22.1	28.0	27.4
RC	7.2	16.2	19.5	18.4*	16.6*	18.3*	20.9*
KU	6.7	13.4	16.7	30.1	20.5	27.8	29.4
LU	13.7	18.8	21.9	27.7	20.4	25.9	23.3
BT	NA	11.2	12.2*	16.7*	16.0*	18.0*	18.4*
SE	-	1.44	2.17	1.55	1.47	1.48	1.55
Fraction B, % total-N							
WC	86.8	80.6	76.9	67.7	81.1	62.9	66.4
RC	87.5	75.8*	72.8*	72.6	76.7	67.4	65.4
KU	87.4	81.4	78.5	67.2	79.9	61.3	61.7
LU	86.7	83.5	78.1	68.0	79.9	61.6	58.6*
BT	NA	75.4*	73.8*	65.9	73.2*	62.2	59.4*
SE	-	0.99	0.60	2.50	1.94	2.35	1.46
Fraction C, % total-N							
WC	3.0	4.3	5.6	6.1	3.9	7.1	9.1
RC	5.6	6.6	7.9	9.6	5.6	9.9	16.3
KU	3.4	4.8	3.2	5.7	3.5	7.3	9.3
LU	2.0	3.8	6.2	5.8	4.8	7.9	7.9
BT	NA	8.1	9.0	10.8	10.3	13.3	13.9
SE	-	0.87	1.52	1.11	1.13	1.72	1.13

* differ significantly (*P*<0.05) from white clover (WC); NA: not available

Material and methods The experiment was carried out at the experimental station Lindhof, which had an average temperature of 9.0°C and a precipitation of 528 mm in 2003. The swards were established in the late summer of 2002 as a complete randomised block design. Red clover (RC), birdsfoot trefoil (BT), lucerne (LU), kura clover (KU) and white clover (WC) were sown as binary mixtures of legume and perennial ryegrass in three replicates. Sampling occurred from the end of April to the beginning of July 2003. Samples were separated into legume, grass and weeds, dried at 60°C and milled in a Cyclotech mill to pass a 1-mm sieve. For each sampling time, quality analyses were performed on legumes. Samples were analysed for total N content and for fractions A, B and C as described by Licitra *et al.* (1996). The N content in B fraction was calculated as: (B = Total-N – A – C). Data was submitted to analysis of variance and means were compared to WC using LSD. Probabilities were adjusted using the Bonferroni-Holm test.

Results For total-N and fractions A and B, the interaction sampling dates x legume species was significant. For fraction C only the main effects were significant. For the fraction A, which contains mostly non-protein-N, lower contents were found for RC and BT (Table 1). The proportion of N in fraction A was low at the beginning of the vegetation period and increased subsequently. The fraction B (potentially degradable protein) followed a similar trend as observed for total-N, with RC and BT showing lower N content when compared to WC for the second and third sampling dates. For the fifth and sixth sampling dates, BT and later LU had lower N-content compared to WC. For fraction C, which corresponds to the N content in the ADF residue (ADIN), only the main factors were significant. Compared to WC, only RC and BT showed higher ADIN-content over all sampling dates. ADIN content increased through the experiment period.

Conclusions The results demonstrate that the proportions of N in the different fractions of CP in legumes differ seasonally and between legume species variation. Legumes like BT and RC, which contains tannins and polyphenol oxidase activity, respectively, may positively influence N use efficiency by ruminants.

Reference

Licitra, G., T.M. Hernandez & P.J. Van Soest (1996). Standardization of procedures for nitrogen fractionation in ruminant feed. *Animal Feed Science and Technology*, 57, 347-358.

Substrate-dependent activation of polyphenol oxidase in red clover

A.L. Winters, F.R. Minchin and P. Morris
Institute of Grassland and Environmental Research, Plas Gogerddan, Aberystwyth, SY23 3EB, UK, Email: ana.winters@bbsrc.ac.uk

Keywords: polyphenol oxidase, red clover, latent enzyme, protein protection

Introduction Polyphenol oxidases (PPO) are copper metaloproteins that catalyse the oxidation of *o*-diphenols to quinones, highly reactive molecules which readily bind to nucleophilic sites on cellular components and proteins. Red clover protein, due to this enzyme is resistant to protease degradation during. Theses enzymes (*circa*. 60-65 kDa) are located in the thylakoid lumen and can be converted to a 40-45 kDa form by proteolysis both *in vitro* and *in vivo* (Gelder *et al.*, 1997). Conversion to the smaller form has been demonstrated to confer activity at neutral pH. Other treatments, such as the presence of lipids or detergents *eg*. SDS, can also confer activity at pH7 (Gelder *et al.*, 1997). Here we describe studies on treatments that affect red clover PPO activity at neutral pH, which is equivalent to the physiological pH of macerated/homogenised leaf extracts.

Materials and methods Red clover leaf material (0.5g approx.) was extracted in McIlvaine buffer, pH7 with (+asc) or without (-asc) 50mM ascorbic acid. Extracts were desalted to remove ascorbic acid and endogenous substrates as descibed by Winters *et al.* (2003). After pre-treatment with the *o*-diphenols, methylcatechol, caffeic acid and chlorogenic acid, PPO activity was analysed in extracts and visualised in gels following separation of proteins by SDS-PAGE according to Winters *et al.* (2003).

Results and discussion Table 1 shows red clover PPO activity, following extraction with ascorbic acid (+asc) after 0 and 20h following treatment of extracts with *o*-diphenol compounds. Additional results are for leaves extracted in the absence of ascorbic acid (0h –asc), so that PPO was active during the extraction process. Results are presented as a percentage of maximum PPO activity of 0h +asc extracts, measured in the presence of SDS (491 ΔOD/min per g FW). Incubation of +asc extracts for 20h increased activity by approximately 2.5 fold. Extraction in the absence of ascorbic acid and treatment of +asc extracts with caffeic acid, chlorogenic acid and methylcatechol also increased PPO activity by between 1.8 and 3.6 fold. These findings suggest that the presence of endogenous or applied phenolic substrates can activate the PPO enzyme in the absence of ascorbic acid.

Table 1 PPO activity of red clover leaf extracts, prepared with and without ascorbic acid and following treatment with a range of *o*-diphenols. Results are presented as % of maximum potential activity

			Treatments*		
0h +asc	20h +asc	0h -asc	Caffeic acid	Chlorogenic acid	Methylcatechol
28	70	100	93	44	48

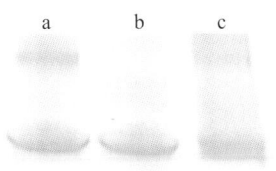

Figure 1 PPO activity following separation of proteins by SDS-PAGE. a) 0h +asc extract, b) 20h extract +asc and c) 0h –asc extract. Relative PPO activities were 28, 70 and 100% for a, b and c respectively.

Figure 1 shows that two size classes of PPO (*circa*. 63k Da and 45 kDa) occur in red clover and are evident in 0h +asc and 0h –asc extracts. The upper form had disappeared after incubation of +asc extracts for 20h. This was concomitant with an increase from 28 to 70% of maximal activity and is consistent with proteolytic cleavage of an inactive form to an active one. The two forms were still present in the –asc extracts, suggesting that the increased activity observed with this treatment is not due to cleavage of an inactive form. We propose that activation following incubation with *o*-diphenols is due to the interaction of quinones with an inactive, larger form of PPO. This results in altered conformation of the PPO protein, conferring activity in a similar manner to SDS treatment.

References

Van Gelder, C.W.G., W.H. Flurkey & H.J. Wichers (1997). Sequence and structural features of plant and fungal tyrosinases. *Phytochemistry*, 45, 1309-1323.

Winters, A.L., F.R. Minchin & P. Morris (2003). Comparison of polyphenol oxidase activity in red clover and perennial ryegrass. In: Abberton, M.T., Andrews, M., Skøt, L & Theodorou, M.K. (eds.) Crop Quality: Its role in livestock production. A*spects of Applied Biology*, 70, 121-128.

High floral tannin white clover reduces rumen ammonia concentrations in dairy cows

V.T. Burggraaf[1], G.C. Waghorn[2], S.L. Woodward[2], D.R. Woodfield[3], E.R. Thom[2] and P.D. Kemp[4]
[1]AgResearch Ltd, Private Bag 3123, Hamilton, New Zealand, Email: vicki.burggraaf@agresearch.co.nz, [2]Dexcel, Private Bag 3221, Hamilton, New Zealand, [3]AgResearch Ltd, Private Bag 11008, Palmerston North, New Zealand, [4]Massey University, Private Bag 11222, Palmerston North, New Zealand

Keywords: ammonia, condensed tannins, flower, rumen, white clover

Introduction White clover produces high quality forage for ruminant production, but it results in high rumen ammonia concentrations, indicating extensive protein degradation. The excess ammonia is absorbed through the rumen and excreted as urea in urine, at a cost to the animal and the environment. Condensed tannins (CT) contained in some forages reduce proteolysis in the rumen, which can lead to increased amino acid absorption and therefore improved animal performance. White clover produces CT in its flower heads, but concentrations are normally too low to benefit animals. This paper reports on comparisons of rumen ammonia concentrations in dairy cows grazing HT (high tannin) white clover (an experimental line of white clover with increased flowering) or Grasslands Huia white clover.

Materials and methods Ten rumen-fistulated lactating Friesian dairy cows grazed either Huia or HT white clover for 18 days on an unrestricted allowance. Treatments were balanced for cow age and milk production, and experiments were repeated in December 2001, February 2002, April 2002 and December 2002. Herbage CT (butanol-HCl method; Terrill et al., 1992) and crude protein concentrations (near infra-red spectroscopy; NIRS) were measured on samples cut to grazing height. Rumen fluid was collected from each cow three times per day on three of the last 6 days of each experimental period and was analysed for rumen ammonia concentrations. Treatment differences were determined by analysis of variance using Genstat 5.

Results Pre-grazing herbage mass was 3.8 to 3.9 t DM/ha for Huia and HT in both Decembers, 3.5 in February, and 2.5 for Huia and 2.4 for HT in April. Post-grazing herbage masses were 2.8 to 3.0 t DM/ha in December experiments, 2.2 for Huia and 2.1 for HT in February, and 1.2 for Huia and 1.0 for HT in April. Average rumen ammonia concentrations were 6 to 22% lower in cows grazing HT white clover than those grazing Huia. This was most likely due to the 0.3 to 3 times higher CT concentrations in HT, as the clovers had similar crude protein concentrations (Table 1). Rumen ammonia was consistently lower in cows fed HT than Huia 6.5 hours after the start of morning grazing, and was lower at other sampling times on some occasions (Table 1).

Table 1 Condensed tannin (CT) and crude protein (CP) concentrations as a % of herbage DM and average rumen ammonia concentrations (mM/L) of cows before the morning grazing and at 2 and 6.5 hours after the start of grazing Huia or HT white clover

| | CT | | CP | | Rumen ammonia | | | | | | | | |
| | | | | | Pre-grazing | | | 2 hours | | | 6.5 hours | | |
	Huia	HT	Huia	HT	Huia	HT	se	Huia	HT	se	Huia	HT	se
Dec 2001	0.18***	0.74	28*	27	29.5*	22.1	2.8	30.6***	22.0	1.6	45.1**	37.7	1.7
Feb 2002	0.48*	0.74	26	26	16.1*	11.6	2.3	23.1	21.8	2.5	32.1†	30.2	1.9
Apr 2002	0.42*	0.70	28*	26	11.7	12.3	1.3	22.8	24.9	1.3	35.7*	28.7	2.3
Dec 2002	0.50***	1.20	27	24	21.7	18.9	1.4	26.3**	22.8	1.0	28.0**	22.3	1.3

† = p<0.1, * = p<0.05, *** = p<0.001, se = standard error of the difference for differences between treatments within experiments and sampling times

Conclusions The CT in HT white clover appeared to reduce protein degradation in the rumen, resulting in lower rumen ammonia concentrations than in cows grazing Huia white clover. This may lower urinary N excretion, providing minor benefits to the environment and the animal. However, the herbage CT concentrations and reduction in rumen ammonia concentrations were low compared to reports from other legumes where animal production has been improved by CT (Waghorn et al., 1999).

References

Terrill, T.H., A.M. Rowan, G.B. Douglas & T.N. Barry (1992). Determination of extractable and bound condensed tannin concentrations in forage plants, protein concentrate meals and cereal grains. *Journal of the Science of Food and Agriculture,* 58, 321-329.

Waghorn, G.C., J.D. Reed & L.R. Ndlovu (1999). Condensed tannins and herbivore nutrition. Proceedings of the Eighteenth International Grassland Congress III, 153-166.

Condensed tannin concentration and herbage accumulation of a white clover bred for increased floral condensed tannin

V.T. Burggraaf[1], S.L. Woodward[2], D.R. Woodfield[3], E.R. Thom[2], G.C. Waghorn[2] and P.D. Kemp[4]

[1]AgResearch Ltd, Private Bag 3123, Hamilton, New Zealand, Email: vicki.burggraaf@agresearch.co.nz, [2]Dexcel, Private Bag 3221, Hamilton, New Zealand, [3]AgResearch Ltd, Private Bag 11008, Palmerston North, New Zealand, [4]Massey University, Private Bag 11222, Palmerston North, New Zealand

Keywords: condensed tannins, flower, white clover

Introduction White clover is a high quality feed for ruminants, however, its high protein content results in excessive urea excretion in urine and can cause bloat, reducing its potential value for animal production. The condensed tannins (CT) in some forages can reduce these problems, but plants may have poor agronomic performance. White clover produces CT in its flower heads, but herbage CT concentrations are normally too low to benefit animals. This paper reports CT concentrations and herbage accumulation over 2 years of an experimental line of white clover (HT) selected for increased flowering and floral CT concentrations.

Materials and methods Three 1 hectare (ha) replicates of each treatment (HT white clover and Grasslands Huia white clover) were sown in monoculture in April 2001, in the Waikato region, New Zealand, and grazed by cows. Clover herbage dry matter accumulation was measured monthly in 12 exclosure cages per treatment. Total CT concentrations in clover flower heads were measured every 1-2 months (when flowers were present) using the butanol-HCl method (Terrill et al., 1992). The CT concentration in the clover herbage (leaf plus flower) was calculated from the CT concentration in the flower heads multiplied by the proportion of herbage as flower head. Data were analysed by analysis of variance.

Results Huia swards grew approximately 500 kg/ha more dry matter (DM) than HT swards in each spring and summer ($p<0.05$), with annual herbage accumulations of 10.3 and 9.5 t DM/ha in year 1 ($p<0.05$, sed = 0.16), and 11.7 and 10.6 t DM/ha in year 2 ($p = 0.07$, sed = 0.31) for Huia and HT, respectively. Floral CT concentrations varied markedly over time, but were similar between treatments (Table 1). The higher flower head content of HT led to higher clover herbage CT concentrations than for Huia (Table 1).

Table 1 Flower head content, and total CT concentration in flower heads and clover herbage (leaves plus flowers) for Grasslands Huia and HT white clover, from November 2001 to October 2003

Month/year:	11/01	1/02	3/02	5/02	10/02	11/02	12/02	1/03	2/03	3/03	4/03	5/03	9/03	10/03
Flower head (% of DM)														
Huia	3.6*	4.9**	13.2*	1.2	0.0**	3.2*	4.7*	12.8*	12.6	11.5	5.2*	0.7**	0.1	0.5
HT	9.9	15.3	21.2	2.9	0.7	10.0	14.1	22.0	14.2	14.4	7.7	1.4	0.3	2.6
sed	0.69	0.59	1.35	0.42	0.01	0.27	2.01	0.96	0.45	0.95	0.45	0.04	0.11	0.68
Flower head CT (g/kg DM)														
Huia	65	70	39	45	-	75	71	33	47	28	17	53*	68[1]	75[1]
HT	70	79	40	53	46[1]	79	75	30	47	30	13	54	80[1]	60[1]
sed	2.3	2.6	6.8	5.7		2.1	2.8	2.2	1.5	2.1	1.2	0.2		
Clover herbage CT (g/kg DM)														
Huia	2.3*	3.4**	5.0	0.6**	-	2.4**	3.3*	4.3*	5.7	3.2	0.9	0.4**	0.1[1]	0.4[1]
HT	7.0	12.1	8.7	1.5	0.3	7.9	10.7	6.5	6.5	4.2	1.0	0.8	0.3[1]	1.6[1]
sed	0.98	0.72	1.39	0.07		0.35	1.53	0.48	0.31	0.27	0.08	0.07		

* = $p<0.05$, ** = $p<0.01$, - = no data, [1] = one sample per treatment bulked across replicates

Conclusions The HT white clover often had herbage CT concentrations above that thought to prevent bloat (5 g/kg DM; Li et al., 1996), but herbage accumulation was slightly lower than for Huia. In order to attain herbage CT concentrations that may improve ruminant production (20-40 g/kg DM; Aerts et al., 1999) an understanding of the mechanisms influencing white clover floral CT concentrations is required.

References

Aerts, A.U., T.N. Barry & W.C. McNabb (1999). Polyphenols and agriculture: beneficial effects of proanthocyanidins in forages. Agriculture, Ecosystems and Environment, 75, 1-12.

Li, Y–G., G. Tanner & P. Larkin (1996). The DMACA-HCl protocol and the threshold proanthocyanidin content for bloat safety in forage legumes. Journal of the Science of Food and Agriculture, 70, 89-101.

Terrill, T.H., A.M. Rowan, G.B. Douglas & T.N. Barry (1992). Determination of extractable and bound condensed tannin concentrations in forage plants, protein concentrate meals and cereal grains. Journal of the Science of Food and Agriculture, 58, 321-329.

Light intensity is positively correlated with the synthesis of condensed tannins in Lotus corniculatus

S. Arcioni, T. Bovone, F. Damiani and F. Paolocci

Plant Genetic Institute – Research Division of Perugia-CNR, via Madonna Alta 130, 06128 Perugia, Italy, Email: sergio.arcioni @igv.cnr.it

Keywords: condensed tannins, DFR, gene expression, light, real time PCR

Introduction The importance of Condensed Tannins (CT) in forage legumes has been well documented in several studies. The role of plant genetics in this field is the acquisition of competences in order to be able to modulate CT synthesis in leaves of these species. The role of light has been investigated in this work on the increase of condensed tannin levels in leaves of two contrasting genotypes of birdsfoot trefoil (*Lotus corniculatus*).

Materials and methods Well developed cuttings of two plants S41 and S50, with high and low CT, were grown under two contrasting light regimes: high light ($>1000~\mu E~m^{-2}~s^{-1}$), and low light ($150~\mu E~m^{-2}~s^{-1}$). After 4 weeks leaf CT, using DMACA protocol (Li *et al.*, 1996) and gene expression of DFR, through Real Time RT-PCR, were quantified. A dissociation protocol of amplified products was performed and information about the presence of polymorphic cDNA fragments was obtained.

Results Light affected CT accumulation (Figure 1) with high light intensity increasing plant staining for both S50 and S41 than at low light intensity, S41 produced more CT than S50 at both treatments. Only at a high light intensity did DFR expression differ significantly between plants (Table 1). The qualitative expression of DFR monitored through the dissociation protocol showed that the two plants mainly differed for the allelic expression at both light intensity. In fact from previous experiments the profile of amplification of different DFR alleles was established (Paolocci *et al.*, in press) so it was possible to determine which alleles were expressed in plants at different light intensity. Figure 2 show the DFR profiles: at high light S41 expressed mainly allele 3 and trace amounts of allele2 while S50 expressed allele 2 and 3 at similar rate; at low light S41 expressed only allele 3 and 16 mean while in S50 all the three alleles were present. It seems that allele 16 is less responsive to light than the two others and that allele 3 is the most relevant for CT synthesis.

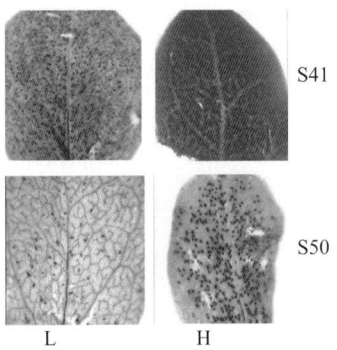

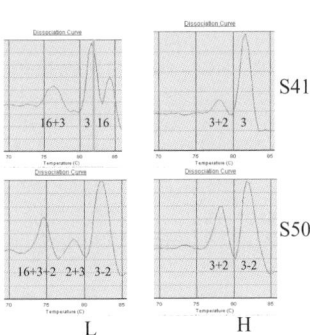

Table 1 DFR level measured through real time RT-PCR and expressed in relative standard unit

	L	H
S41	1.2a	1.6b
S50	0.9a	1.1a

Means with the same letter do not differ significantly

Figure 1 CT stain in S41 and S50 leaves at low (L) and high (H) and low (L) light

Figure 2 Dissociation profiles of S41 and S50 plants at high (H) and low (L) light

Conclusions Light positively affects CT accumulation. DFR, a key enzyme of the pathway, is quantitatively and qualitatively regulated by light and some alleles play a determinant role on the rate of CT accumulation while some others probably have a competitive negative effect.

References

Li, Y.G., G.J. Tanner & P.J. Larkin (1996) The DMACA-HCl protocol and the threshold proanthocyanidin content for bloat safety in forage legumes. *Journal of the Science of Food and Agriculture*, 70, 89-101.

Paolocci, F., T. Bovone, N. Tosti, S. Arcioni & F. Damiani (in press) Light and an exogenous transcription factor qualitatively and quantitatively affect the biosynthetic pathway of condensed tannins in *Lotus corniculatus* leaves. *Journal of Experimental Botany*.

Variation in tannin content and morphological traits in Lotus corniculatus L. (bird's-foot trefoil)

A.H. Marshall, F. Ribaimont, R.P. Collins, D. Bryant and M.T. Abberton
Institute of Grassland and Environmental Research, Plas Gogerddan, Aberystwyth, Ceredigion, SY23 3EB, UK.
Email: athole.marshall@bbsrc.ac.uk

Keywords: *Lotus corniculatus*, agronomic traits, tannin content

Introduction   *Lotus corniculatus* L. (bird's-foot trefoil) is a potentially valuable species for UK grassland agriculture. The herbage contains proanthocyanidins, or condensed tannins (CT's), which help to reduce bloat, have anthelmintic properties, can protect protein in the rumen, and thus potentially reduce N losses to the environment. It is currently a relatively minor species within UK grassland, as available varieties lack persistence in mixed swards (Hopkins *et al.*, 1996). The seed used is of foreign bred varieties with no varieties bred specifically for the UK environment. To explore the feasibility of breeding *L. corniculatus* for the UK, variation in morphological traits, dry matter yield and tannin content within existing varieties was measured.

Materials and methods   Forty plants of seven cultivars of *Lotus corniculatus* L. were grown from seed in an unheated glasshouse. At the onset of flowering, 3 stems were removed from each plant, separated into leaf and stem and tannin content of the leaf determined by the high throughput tannin assay developed at IGER (Bryant, pers.comm). Length and diameter of the longest stem on each plant was measured. Dry matter (DM) yield per plant was determined at harvest and after 45 days of recovery growth.

Results   There was a significant difference between and within varieties in yield per plant and in length and diameter of the longest stem. DM yield was highest in Gran San Gabrielle (GSG) and lowest in ARS2620. At the 2nd cut, DM yield was greatest in ARS2620 and lowest in AL531. GSG had the longest stem and Highgrove the shortest. Stem diameter was greatest in Leo and least in Highgrove and ARS2620. Leaf tannin content was greatest in AL531 and lowest in Leo and the range in tannin content between individual plants of each variety was greatest in AL531 and least in Steadfast.

Table 1 Dry matter yield (g/plant), length and width of longest stem and tannin content (mg/g dry weight) of 7 varieties of *Lotus corniculatus* L. (mean ± s.e.m.)

Variety and country of origin	Dry matter yield (g/plant)		Longest Stem		Tannin content (mg CT/g dry weight)	
	1st cut	2nd cut	Length (cm)	Diameter (cm)	Mean	Range
Inia Draco (Uruguay)	62.0±2.79	24.2±2.37	58.3±3.34	2.0±0.11	18.0	5.4-41.6
AL531 (UK)	47.9±3.32	20.9±2.47	63.4±4.11	2.2±0.13	58.8	42.8-103.5
Leo (Canada)	73.4±3.45	29.7±2.22	50.1±2.80	2.4±0.21	8.5	4.5-24.5
GSG (Uruguay)	81.4±3.89	27.6±4.06	70.5±29.7	1.9±0.12	22.6	11.7-35.1
Highgrove (UK)	56.9±6.93	28.4±2.26	35.9±3.55	1.2±0.11	30.4	9.9-66.1
Steadfast (USA)	63.1±3.55	30.3±3.38	58.8±2.46	1.3±0.11	14.2	3.6-23.7
ARS2620 (USA)	47.4±5.31	40.8±2.43	53.0±3.27	1.2±0.09	16.9	9.6-36.7

Conclusions   The variety that produced the greatest yield at the 1st cut produced only average yields at the 2nd cut, while the variety that yielded the least at the 1st cut produced the greatest yield at the 2nd cut. Selection of plants with appropriate yields at the 1st and 2nd cut will be important in future breeding programmes. Stem length and width exhibited significant variation and stem length was correlated (r=0.45) with DM yield suggesting it could have potential as a selection tool. Tannin content differed significantly between and within varieties and was generally below the 5% of DM reported as optimal for good animal health and nutrition (Waghorn *et al.*, 1990). Further work will be carried out to investigate the stability of tannin levels over the growing season.

Acknowledgements The financial support of Defra and the BBSRC is gratefully acknowledged.

References
Hopkins, A., T.M. Martyn, R.H. Johnson, R.D. Sheldrick & R.L. Lavender (1996). Forage production by two Lotus species as influenced by companion grass species. *Grass and Forage Science*, 51, 343-349.
Waghorn, G.C., W.T. Jones, I.D. Sheldon & W.C. McNabb (1990).Condensed tannins and the nutritive value of herbage. *Proceedings of the New Zealand Grassland Society*, 11, 171-176.

Condensed tannins in different varieties of *Lotus corniculatus*

C.L. Marley, R. Fychan and R. Jones
Institute of Grassland and Environmental Research, Plas Gogerddan, Aberystwyth, SY23 3EB, UK, Email: christina.marley@bbsrc.ac.uk

Keywords: *Lotus* varieties, *Lotus* species, condensed tannins, birdsfoot trefoil

Introduction   Research has shown that *Lotus corniculatus* may provide benefits as a grazing forage for ruminants, because of improved efficiency of protein utilisation in part due to the presence of condensed tannins (CT) in the forage. *Lotus* species have been found to contain different CT (Aerts *et al.*, 1999). The aim of this experiment was to determine if differences exist in CT concentration among *L. corniculatus* varieties.

Materials and methods   Replicate 2.4 x 6 m plots of pure stands of 13 varieties of *L. corniculatus* or one variety of *L. uliginosus* were sown at 12 kg/ha in a randomised block design. Plots were maintained by cutting to 100 mm using a Haldrup 1500 plot harvester, on 2, 3 and 3 occasions in the establishment and 2 subsequent harvest years, respectively. At Cut 2 of the second harvest year, sub-samples of forage were taken and separated according to *Lotus* and unsown species. Some 200g of *Lotus* was ground to <1 mm, freeze-dried and stored at -20°C prior to CT analysis by the HCl/butanol method (Terrill *et al.*, 1992). To determine the concentration of CT in each variety would require a standard CT isolated from each variety (Stewart *et al.*, 2000). Therefore, a bulked sample comprising of 1.2g of each *Lotus* variety (300 mg per replicate plot) was used as a relative standard in the assay. The HCl/butanol results were calculated as the difference between the standard and the sample, expressed as absorbance units (AU) at 550nm per 50 mg dry matter (I. Mueller-Harvey, pers. comm.).

Results   There were significant differences in the HCl/Butanol test absorbance units among varieties and species of *Lotus* (Table 1). The AU of Canadian varieties was found to differ from others, except Dawn and Norcen.

Table 1 Varieties of *Lotus*, their country of origin and differences in CT

Variety	Country of origin	AU at 550 nm
Oberhaunstaedter	Germany	0.139^{bcd}
Lotara	Czech Rep.	0.043^{cde}
Emlyn	Hungary	0.111^{bcd}
Leo	Canada	-0.533^{g}
Upstart	Canada	-0.209^{f}
Steadfast	USA	0.020^{de}
Georgia-1	USA	0.037^{cde}
Dawn	USA	-0.358^{fg}
Norcen	USA	-0.145^{ef}
AU-Dewey	USA	0.118^{bcd}
Inia Draco	Uruguay	0.255^{bc}
San Gabriel	Uruguay	0.314^{b}
Grasslands Goldie	New Zealand	0.226^{bcd}
Grasslands Maku[†]	New Zealand	1.194^{a}

[†] *L. uliginosus*; Means with different superscripts were significant ($p<0.001$)

Conclusions   Differences in AU suggest that differences in CT concentration exist among *L. corniculatus* varieties. Research is now needed to isolate and characterise these CT and determine their effects on protein metabolism in ruminants.

References

Aerts, R. J., W.C. McNabb, A. Molan, A. Brand, T.N. Barry & J.S. Peters (1999). Condensed tannins from *Lotus corniculatus* and *Lotus pedunculatus* exert different effects on the *in vitro* rumen degradation of ribulose-1,5-bisphosphate carboxylase/oxygenase (Rubisco) protein. *Journal of the Science of Food and Agriculture*, 79, 79-85.

Stewart, J.L., F. Mould & I. Mueller-Harvey (2000). The effect of drying treatment on the fodder quality and tannin content of two provenances of *Calliandra* calothysus Meissner. *Journal of the Science of Food and Agriculture*, 80, 1461-1468.

Terrill, T.H., A.M. Rowan, G.B. Douglas & T.N. Barry (1992). Determination of extractable and bound condensed tannin concentrations in forage plants, protein concentrate meals and cereal grains. *Journal of the Science of Food and Agriculture*, 58, 321-329.

Herbage production, nitrogen fixation and condensed tannin concentrations in *Lotus glaber* Mill. germplasm

H. Acuña, M. Figueroa, P. Hellman and A. Concha
Instituto de Investigaciones Agropecuarias, INIA. Centro Regional de Investigación Quilamapu, Casilla 426, Chillán, Chile, Email: hacuna@quilamapu.inia.cl

Keywords: *Lotus glaber*, N-fixation, extractable condensed tannins, bound condensed tannins

Introduction *Lotus glaber* (Lg) grows on clay, sandy and medium textured soils in central Chile (32° to 38° S). The diversity of environments where the species grows naturally supports the hypothesis that genetic variability would be found. The objectives of the experiment were to characterise accessions of Lg collected in the region for dry matter (DM) production, comparative capacity to fix atmospheric N and condensed tannins (CT) in Lg grown on clay soils used for cropping rice.

Materials and methods Eleven accessions of 20 plants of Lg were collected from different sites in the region. The plants were replicated vegetatively to establish an experiment in a clay soil with a randomised design with three blocks of 1 x 2 m plots comprising two 10-plant rows 40 cm apart. The same 20 original plants were present in each block. Two blocks were used to measure DM production (four cuts in the season 1999-2000) and CT concentration. Samples of whole plants from the late December cutting (five cm stubble height) were used for CT analysis. This cut was taken 39 days after the previous cut,. The samples were field frozen –20° C, freeze dried and ground to pass a 1 mm screen. The butanol-HCl procedure for extractable and bound CT determinations (Terrill et al., 1992) was used. On the third block N^{15} labelled fertiliser was applied to a 1 m^2 subplot after the first cut of the season and the percentage of N derived from the atmosphere (Ndfa) was estimated in samples from the following cut using the formula described by Marriott & Haystead (1993).

Table 1 Dry matter (DM) production (g/plant), N-fixation (% Ndfa) and concentration of extractable (Ext CT), protein bound (Prt. b. CT), fibre bound (Fib.b. CT) and total condensed tannins

Lg accessions	Lg1	Lg3	Lg4	Lg5	Lg6	Lg7	Lg8	Lg 11	Lg 12	Lg 14	Lg 15	s.e.	Signif.
DM	101	68	93	102	76	90	66	87	104	87	89	11.1	*
N-fixation	82.7	79.4	80.0	83.5	78.9	73.4	72.1	81.4	88.8	79.4	77.6	-	-
Ext. CT	2.3	2.5	2.7	2.3	2.4	2.5	2.4	2.1	2.1	2.4	2.5	0.18	*
Prt. b. CT	0.7	1.1	0.6	0.9	0.9	1.1	0.9	0.8	0.7	0.7	0.8	0.12	*
Fib. b. CT	1.2	1.3	0.7	1.3	1.2	1.5	1.4	1.1	1.0	1.6	0.8	0.27	*
Total CT	4.2	4.9	4.1	4.5	4.5	5.1	4.7	3.9	3.9	4.7	4.1	0.50	n.s.

Results There were significant differences (P< 0.05) among accessions for DM production. The N-fixation on average was 80%, with a difference of 18 points between those with the highest proportion of N coming from the atmosphere and those with least. The values for the CT concentration were, in general, higher than the data reported in the literature (Table 1). Some authors reported absence of CT in Lg leaves, presence in stems and abundance in roots, using vanillin-HCl reaction. The differences found in this experiment are small, but significant (P<0.05), for the three fractions.

Conclusions The concentrations of CT in Lg were higher than the data reported in the literature. The differences found among the Lg accessions in the studied variables, and in other attributes studied separately, show that the collected germplasm has enough variability to support a programme for genetic improvement.

References

Marriott, C.A. & A. Haystead (1993). Nitrogen fixation and transfer, In: A. Davies, R.D. Baker, S. Grant & A.S. Laidlaw (eds.). Sward Measurement Handbook (2nd ed.). *British Grassland Society*, 245-264.

Terrill, T.H., A. M. Rowan, G.B. Douglas & T. N. Barry (1992). Determination of extractable and bound condensed tannin concentrations in forage plants, protein concentrate meals and cereal grains. *Journal of the Science of Food and Agriculture*, 58,321-329.

Effect of condensed tannins in sainfoin on *in vitro* protein solubility of lucerne

J. Aufrère, M. Dudilieu, C. Poncet and R. Baumont
INRA, Unité de Recherches sur les Herbivores, Centre de Clermont-Theix-Lyon F-63122 St Genes Champanelle, France, Email: aufrere@clermont.inra.fr

Keywords: tannin, soluble nitrogen, protein, sainfoin, lucerne

Introduction Proteins of fresh legume forages such as lucerne are highly degraded in the rumen, resulting in their inefficient use by the animal. The condensed tannins (CT) present in some forages can improve the nutritional value of these forages and of associated feeds in the diet. Previous *in vitro* work (Waghorn & Shelton, 1997) showed that CT from *Lotus corniculatus* are able to bind with and precipitate protein from a ryegrass/clover pasture, but when these forages were fed to sheep, the CT effects on digestion and animal performance were weak. This revealed a need for a better understanding of the mechanism of CT interaction between feeds. The present work was designed to measure, *in vitro*, the effects of CT in sainfoin when mixed with fresh lucerne.

Materials and methods Fresh finely chopped lucerne (L) and sainfoin (S) were mixed in the proportions (L-S): 100-0, 75-25, 50-50, 25-75, 0-100. Six 8 g samples of each mixture were individually ground in a Waring blender with 100 ml of artificial saliva (Verité & Demarquilly, 1978); PEG 4000 (600 mg) was added to 3 samples to inhibit the effect of CT in sainfoin. These slurries were continuously stirred for 60 min at 20°C, and then centrifuged (27,000 g, 20 min). The supernatant was analysed for total N (tN) before and after protein precipitation with TCA 10% (v/v) to measure N solubility (Nsol) and protein N content in soluble N.

Results The Nsol and the protein N content in soluble N were much lower for sainfoin than for lucerne (Table 1). Increasing the proportion of sainfoin in the mixture strongly decreased its Nsol and the proportion of protein N in the soluble N (Table 1). As showed by results with PEG, this effect mainly arose from CT in sainfoin. The CT had a larger effect in reducing the solubility of the protein fraction than the non-protein fraction. Measured Nsol values in the mixtures were lower than theoretical values calculated from Nsol of each plant and the proportion of the plant in the mixture. This showed that CT in sainfoin are able to decrease the solubility of the lucerne protein.

Table 1 *In vitro* nitrogen solubility and protein N in soluble N for mixtures of lucerne (L) and sainfoin (S) in different proportions, measured with and without PEG 4000. Measured values are expressed as the mean and standard deviation (SD) of 3 replicates. Theoretical Nsol values of mixtures are calculated from Nsol values measured on L and S alone (100-0 and 0-100 treatments respectively)

	Proportion of lucerne (L) and sainfoin (S) tested				
L-S on dry matter basis	100-0	75-25	50-50	25-75	0-100
Nitrogen solubility (% tN)					
without PEG					
measured	57.1 (2.5)	42.1 (1.7)	22.7 (2.1)	12.0 (0.2)	10.1 (0.5)
theoretical values of mixtures		*45.1*	*33.6*	*22.0*	
with PEG					
measured	54.6 (1.5)	52.8 (1.0)	48.3 (1.8)	46.9 (1.8)	40.6 (2.6)
theoretical values of mixtures		*51.8*	*48.5*	*45.0*	
Protein N in soluble N (%)					
measured without PEG	69.2 (3.1)	58.4 (1.9)	44.6 (3.4)	25.9 (0.7)	5.9 (8.2)
measured with PEG	61.5 (0.4)	65.1 (1.1)	62.6 (0.4)	66.9 (2.7)	62.7 (2.5)

Conclusion CT in excess in sainfoin can efficiently reduce in vitro nitrogen solubility of other forages, here lucerne, when intimately mixed. The right conditions must now be found to reproduce this effect in animals.

References

Vérité, R. & C. Demarquilly (1978). Qualité des matières azotées pour ruminants. In: La Vache Laitière, INRA-CNRA, Versailles, 143-147.
Waghorn, G.C. & I.D. Shelton (1997). Effect of condensed tannins in *Lotus corniculatus* on the nutritive value of pasture for sheep. *Journal of Agricultural Science*, 128, 365-372.

Leaf, pod and whole plant tannin contents in pigeon pea (*Cajanus cajan* (L.) Millsp)

R. Godoy, P.M. Santos[1] and F.H.D. Souza
Embrapa Pecuária Sudeste - Caixa Postal 339 13560-970 São Carlos, SP Brazil, Email: godoy@cppse.embrapa.br

Keywords: forage digestibility, palatability, tannin evaluation, leguminous crop

Introduction Tannin content is an important characteristic of leguminous crops and it has been used as a selection criterion in pigeon-pea improvement programmes (Godoy *et al.*, 1994). In south-eastern Brazil, pigeon pea is often consumed by bovines in the dry season (from April though October), after flowering occurs, and is recommended in some cases, specifically for that time of the year (Lourenço *et al.*, 1994). Since tannin content is being used as a selection criterion and the animals in the dry season preferentially eat pods and leaves, an experiment was conducted to compare whole plant, leaf and pod tannin content.

Materials and methods Tannin contents of the whole plant, leaf and pod were determined according to the Folin-Denis method (Burns, 1963). The trials were planted in December 1998 at five locations in the State of São Paulo, Brazil, with seventeen pure pigeon pea lines and three cultivars. Cuts were taken in the months shown in Table 1. Plants were cut at 40 cm and the material was dried and ground. Data for two lines, which survived less than a year, were not considered. Principal components and Pearson correlation analysis were performed as described, respectively, in the SAS princomp and corr procedures (SAS, 1999-2001), using the three tannin contents as variables. Average whole plant and leaf data from each location and time of harvest were plotted.

Results The principal components analysis showed that the tannin contents of the whole plant and leaf were responsible for 91 per cent of the observed variance and the Pearson correlation coefficient between those variables was found to be highly significant, indicating that all those data follow similar patterns. Besides, as shown in Table 1, in all locations, leaf and whole plant tannin contents follow very similar patterns, indicating that, if tannin content is used as a selection criterion, there is no need to determine it in all plant parts.

Table 1 Content of tannins in leaf, pod and whole plants when grown at five locations and cut at different dates

Locations		Itapui		Jaboticabal		Pirassununga		Pratânia		São Carlos	
Cutting time		Mean	SE	Mean	SE	Mean	SE	Mean	SE	Mean	SE
April 1999	Plant	1.46	0.18	1.88	0.34	1.64	0.23	1.80	0.27	2.32	0.31
	Leaf										
	Pod										
August 1999	Plant	2.31	0.47	2.56	0.55	3.40	0.67	2.10	0.43	2.85	0.68
	Leaf	3.36	1.07	3.46	0.97	4.36	1.27	3.09	0.62	5.78	1.23
	Pod	2.48	0.54	2.15	0.54	3.69	1.18	2.13	0.44	2.31	0.54
Jan. 2000	Plant	1.81	0.39	2.49	0.38	2.32	0.33	2.20	0.31	1.90	0.31
	Leaf	2.83	0.94	3.01	0.45	2.71	0.41	2.62	0.41	2.45	0.32
	Pod	1.19	0.38	1.41	0.11	0.94	0.21	1.69	0.50	1.46	0.45
May 2000	Plant	3.15	0.43	3.29	1.16	2.80	0.41	2.10	0.35	4.93	1.31
	Leaf	4.19	1.15	3.91	1.46	3.64	0.96	3.18	0.59	7.13	1.71
	Pod	4.71	1.62	2.80	1.96	4.77	1.67	5.44	1.82	4.63	1.34
Dec. 2000	Plant					2.92	0.68			3.04	0.72
	Leaf					3.31	0.93			4.34	1.02
	Pod					0.94	0.33			1.56	0.34
June 2001	Plant					3.29	0.76			2.60	0.58
	Leaf					4.80	1.28			3.30	0.60
	Pod					3.58	1.13			2.44	0.60

Conclusions If tannin contents are used as selection criteria there is no need to separate plant parts. Samples of the whole plant can be used, saving a considerable amount of labour and laboratory work.

References
Burns, R.E. (1963). Methods of tannin analysis for forage crop evaluation. Technical Bulletin N.S., 32. Georgia Agriculture Experimental Station, 11pp.
Godoy, R, L.A.R. Batista & G.F. Negreiros (1994). Avaliação agronômica e seleção de guandu forrageiro (*Cajanus cajan* (L.) Millsp). *Revista Brasileira de Zootecnia*, 23, 730-742.
Lourenço, A.J. & E. Matsui (1994). Composição botânica da forragem disponível e da dieta selecionada por bovinos em pastos de capim-colonião consorciado com centrosema e ou galactia com acesso ao banco de proteína de guandu. *Revista Brasileira de Zootecnia*, 23, 100-109.
SAS Institute. SAS/STAT 1999-2001. User's guide: statistics, version 8, v.2, SAS Cary, NC, USA.

Use of green sulla forage for feeding. 1. Effects on lamb growth and gastrointestinal nematode parasite infestation

D. Giambalvo[1], A. Di Grigoli[2], M.L. Alicata[2], B. Formoso[2], A.S. Frenda[1] and P. Trapani[1]
[1]Dipartimento di Agronomia Ambientale e Territoriale, [2]Dipartimento S.En.Fi.Mi.Zo., sezione di Produzioni Animali; Università di Palermo, Viale delle Scienze, 90128 Palermo, Italy, Email: giardo@unipa.it

Keywords: sulla, ryegrass, condensed tannins, lamb growth, nematode

Introduction Recent studies have shown that some forage legumes containing condensed tannins (CT), such as sulla (*Hedysarum coronarium* L.), can reduce the gastrointestinal nematode burden in sheep (Niezen *et al.*, 1998) and increase post-ruminal protein availability (Waghorn *et al.*, 1994). This study aimed to evaluate the anthelmintic and nutritional properties of sulla forage in relation to its CT content. Thus, the growth performance and the level of nematode infestation of lambs fed sulla were compared with those of lambs fed ryegrass (*Lolium multiflorum* Lam. subsp. *wersterwoldicum*), lacking in CT.

Materials and methods Thirty-two male Comisana lambs aged 100 d were divided into two groups, fed *ad libitum* green forage of sulla (S) or ryegrass (R), cut daily, and 200 g/head per d of a concentrate for 49 d until slaughter. Within each group, eight lambs were treated in order to keep them parasite-free (SPF and RPF groups). During the trial, the lambs' live weight and feed intake were recorded, and faeces were obtained for faecal egg count (FEC) by a modified McMaster technique. At slaughter, gastrointestinal tracts were removed for nematode counting and identification. The forage was analysed for chemical composition; the CT content was determined by the butanol-HCl method. The data was analysed using the GLM procedure of SAS.

Results On average, sulla was higher in crude protein (CP) (17.9 *vs.* 11.9% of dry matter: DM) and lower in NDF (39.8 *vs.* 55.5%) than ryegrass and had a CT level equal to 1.7% (ranging from 1.5 to 2.3%). Lambs fed sulla showed a superior growth rate (225 *vs.* 147 g/d; s.e. 6.7; P • •0.0001) and final live weight (29.9 *vs.* 26.1 kg; s.e. 1.0; P • •0.013) (Figure 1), due to both higher forage intake (769 *vs.* 568 g/d DM; s.e. 15.9; P • •0.001) and lower feed conversion ratio (5.1 *vs.* 5.9; s.e. 0.28; P • •0.04). The CT intake of lambs fed sulla was 15.6 g/d on average. Growth performance did not differ between non-parasitisd and parasitised lambs, the level of infestation being low. Lambs given anthelminthic had FEC of zero. Diet did not affect FEC in parasitised lambs; nevertheless it showed a marked decrease with both forages at the end of trial (Figure 2), probably due to the improved environment and nutritional status of the lambs. The presence of nematode worms in the gastrointestinal tracts was low. Nevertheless, sulla was associated with a fewer total male adult nematodes (108 *vs.* 131). Most of these were of the *Teladorsagia circumcincta* species and were found in the abomasum (98 *vs.* 123 in S and R; s.e. 33.9), but these differences were not significant.

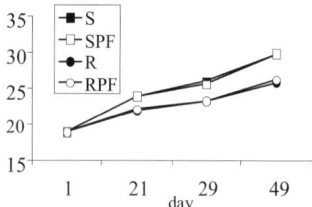

Figure 1 Lamb liveweight (kg) during experiment

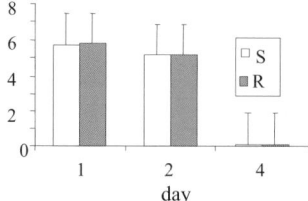

Figure 2 Faecal egg count (FEC) (100 eggs/g dry faeces) in parasitised lambs during experiment

Conclusions There was a positive effect of S on lamb growth in comparison with R. This was due to higher feed intake and better nutritional efficiency, linked to a high level of CP and, possibly, improved intestinal protein utilisation through CT in sulla inhibiting ruminal degradation. In spite of low levels of nematode infestation overall, sulla showed a tendency to reduced strongyle worm population in the abomasum.

Acknowledgements The research was funded by *Consorzio Bioevoluzione Sicilia* (BES)

References
Niezen, J.H., H.A. Robertson, G.C. Waghorn & W.A.G. Charleston (1998). Production, fecal egg count and worm burdens of ewe lambs which grazed six contrasting forages. *Veterinary Parasitology*, 80, 15-27.
Waghorn, G.C., I.D. Shelton, W.G. McNabb & S.N. McCutcheon (1994). The effect of condensed tannin in *Lotus pedunculatus* on nutritive value for sheep. Part 2. Nitrogenous aspects. *Journal of Agricultural Science*, 123, 109-119.

Use of green sulla forage for feeding. 2. Effects on lamb carcass and meat quality

A. Bonanno[1], M.L. Alicata[1], G. Tornambè[1], F. Mazza[1], G. Di Miceli[2] and D. Giambalvo[2]
[1]Dipartimento S.En.Fi.Mi.Zo., sezione di Produzioni Animali, Email: abonanno@unipa.it, [2]Dipartimento di Agronomia Ambientale e Territoriale; Università di Palermo, Viale delle Scienze, 90128 Palermo, Italy

Keywords: sulla, annual ryegrass, condensed tannins, lamb carcass, lamb meat

Introduction Diets with 2.5% of condensed tannins (CT) from carob pulp have been showed to reduce lamb carcass weight, yield and fatness, as a consequence of lower digestibility due to strong protein-tannin bonds. Moreover the diets lightened meat colour and negatively affected meat sensory properties (Priolo *et al.*, 2000). There is evidence that some tannin-rich legumes have weaker stability of the protein-tannin complex post-rumen (McSweeney *et al.*, 2001). Since the CT-containing sulla (*Hedysarum coronarium* L.) demonstrated improved lamb growth performance in comparison with CT-lacking annual ryegrass (*Lolium multiflorum* Lam. subsp. *wersterwoldicum*) (Giambalvo *et al.*, 2005), this study examined the impact of CT from sulla on carcass characteristics and meat quality.

Materials and methods Thirty-two male Comisana lambs, fed green forage of sulla (S=16) or ryegrass (R=16) and 200 g /head per d of a concentrate, were slaughtered at 150d of age. Weights of carcass, perirenal and pelvic fat and gastrointestinal content were recorded. Tissue components in the hind leg were determined. Meat quality was evaluated by pH, colour (L* a* b*, hue, chroma), thawing and cooking loss, Warner-Bratzel (WB) shear force determination and chemical analysis on *Longissimus dorsi* (LD) muscle. A panel of 28 members assessed in two sessions the meat in triangle tests. Data were analysed by the GLM procedure of SAS.

Results Lambs fed sulla forage had higher empty body weight than R lambs, so that they gave heavier carcass with higher dressing percentage and fatness (Table 1). As a consequence, S lambs showed higher incidence of fat tissue in hind leg and more lipids in LD muscle. Diet did not modify physical characteristics of LD meat (Table 2). Particularly, meat colour parameters (Table 2) were not affected by forage, in contrast to Priolo *et al.* (2000), although the level of CT found in the sulla (1.7% on average, ranging form 1.5 to 2.3%) was lower than that of carob pulp studied by these authors. In the triangle test, sensory panellists were unable to distinguish meat from the different forages at a significant level (21/56 correct indications).

Table 1 Performance of lambs at slaughter (a, b: P• 0.05; A, B: P• 0.01)

		Sulla	Ryegrass
Empty body weight	kg	24.3 A	19.7 B
Carcass at 24 h (CR)	kg	12.5 A	9.6 B
Net dressing at 24 h	%	51.3 A	48.7 B
Hind leg	kg	2.0 A	1.6 B
Hind leg	% CR	15.9	16.2
Periral and pelvic fat	g	133.8 A	82.8 B
Periral and pelvic fat	% CR	1.1 a	0.8 b

Table 2 Physical quality of *Longissimus dorsi* meat (no significant differences)

		Sulla	Ryegrass
pH at 24 h		5.9	6.0
Thawing loss	%	4.6	3.7
Cooking loss	%	25.5	26.3
Total loss	%	28.9	29.0
WB shear force	kg/cm^2	3.6	3.4
Lightness (L*)		36.8	38
Redness (a*)		14.8	14.2
Yellowness (b*)		7.8	7.5
Chroma		16.8	16.2
Hue angle		27.7	27.8

Conclusions A legume containing CT improved lamb carcass characteristics and did not negatively influence the rheological and sensoral properties of lamb meat. The beneficial effects of sulla could depend on the low content and type of tannins.

Acknowledgements The research was funded by *Consorzio Bioevoluzione Sicilia* (BES)

References

Giambalvo D., A. Di Grigoli, M.L. Alicata, B. Formoso, A.S. Frenda & P. Trapani (2005). The feeding green forage of sulla. 1. Effects on lamb growth and gastrointestinal nematode parasite infestation. *XX International Grassland Congress – offered papers* (in press).

McSweeney, C.S., B. Palmer, D.M. McNeill & D.O. Krause (2001). Microbial interactions with tannins: nutritional consequences for ruminants. *Animal Feed Science and Technology*, 91, 83-93.

Priolo, A., G.C. Waghorn, M. Lanza, L. Biondi & P. Pennisi (2000). Polyethylene glycol as a means for reducing the impact of condensed tannins in carob pulp: effects on lamb growth performance and meat quality. *Journal of Animal Science*, 78, 810-816.

Distribution of trace elements in plant parts of red clover (*Trifolium pratense* L.)

S. Ignjatovic, Z. Lugic, D. Sokolovic and J. Radovic
Agricultural Institute Serbia, Center for Forage Crops, Trg Rasinskih partizana 50, Krusevac, Serbia & Montenegro, Email: krmnobilje@ptt.yu

Keywords: trace elements, red clover, plant parts

Introduction There is little information on the distribution of minerals in plant parts although factors affecting mineral content in forages have been well investigated (Fleming, 1973; Whitehead *et al.*, 1985). The aim of this investigation was to determine the trace mineral content in plant parts of different cultivars of red clover (foreign and domestic) and to assess differences between cultivars. The existence of significant differences between cultivars would indicate the possibility of selecting cultivars to satisfy particular animal requirements for minerals.

Material and methods Six cultivars of red clover (*Trifolium pratense* L.): 2 foreign, Nike and Viola, and 4 domestic, K-9, K-17, K-27, K-39, were sown in 2002 as pure stands in three replicates. Clovers were harvested on 27 May 2003 at the beginning of flowering. Plants were divided into stems, leaves, petioles and flowers. Samples were analysed for trace element content (Fe, Mn, Zn and Cu) by AAS and data subjected to statistical analysis. Because of insufficient amount of sample, the mineral contents of flowers are not reported.

Results Trace element contents in plant parts of 6 red clover cultivars are shown in Table 1. In all cultivars examined leaves were higher in element contents than stems and petioles. The largest differences between plant parts were in Mn. Contenst of Mn in leaves were more than twice those in petioles and even 5 times more than in stems, as reported earlier for Lucerne (Ignjatovic *et al.*, 1998). Differences between cultivars in Fe content in leaves were not significant. There were significant differences in the content of Mn and Zn in leaves and petioles between cultivars and no differences were found for these elements in stems. Cultivar K-27 was lowest in Mn in leaves and petioles and highest in stems. Highest Zn was in K-17 in all plant parts. There were significant and very significant differences between cultivars in Cu content in leaves, stems and petioles. Highest Cu content was in K-39, both in leaves and petioles. This cultivar is very rich in Cu, because it also has high proportions of leaves and petioles in the dry matter.

Table 1 Mineral content in plant parts of red clover cultivars (mg/kgDM)

Cultivars	Fe			Mn			Zn			Cu		
	Leaf	Petiole	Stem	Leaf	Petiole	Stem	Leaf	Petiole	Stem	Leaf	Petiole	Stem
Nike	162	141	90	72	34	16	54	28	20	21	14	12
Viola	182	110	96	89	28	16	60	28	19	21	13	10
K-9	172	104	108	80	24	16	56	26	20	18	11	11
K-17	175	126	119	82	26	16	62	29	20	22	16	13
K-27	159	98	78	71	21	17	56	26	19	22	14	13
K-39	180	105	78	78	30	15	59	29	18	24	17	12
Mean	172	114	95	79	27	16	58	28	19	21	14	12
LSD 0.05	25.3	22.5	11.0	2.0	2.2	2.1	2.4	1.5	1.5	1.1	1.0	1.5
0.01	35.5	31.5	15.4	2.8	3.0	3.0	3.4	2.2	2.1	1.5	1.4	2.1

Conclusion There is appreciable variation in mineral content of the cultivars examined. It should be possible to select and develop new cultivars with high and low levels of particular minerals, depending on different needs and animal requirements.

References

Fleming, G.A. (1973). Mineral composition of herbage. In G.W. Butler and R.W. Bailey (ed.) Chemistry and Biochemistry of Herbage. Academic Press, New York, 529-566.
Ignjatovic S., B. Dinic, D.Kolarski & B.Urosevic (1998). Chemical composition of first and second cut of lucerne *(Medicago sativa)* of different stages of maturity. *Grassland Science in Europe*, 2, 729-732.
Whitehead, D.C., K.M. Goulden & R.D. Hartley. (1985). The distribution of nutrient elements in cell wall and other fractions of the herbage of some grasses and legumes. *Journal of the Science of Food and Agriculture*, 36, 311-318.

Potassium content and the balance between potassium and other minerals and crude protein in forage can have a big impact on dairy cow health

H. Eriksson
Department of Agricultural Research for Northern Sweden, SLU-Robacksdalen, Box 4097, 90403 Umea, Sweden, Email: Hary.eriksson@njv.slu.se

Keywords: mineral balance, dairy cow, health

Introduction Preliminary studies indicated that the mineral balance in forage may affect dairy cow health. A bigger study covering the four most northern counties in northern Sweden was conducted. The objective of this study was to compare data on harvested forage with farm data on milk yield and animal health.

Materials and methods Analysis from one year's forage cuts were investigated and used if potassium, calcium, magnesium and phosphorous had been tested. These data from different farms were combined with their production and health records from the official milk recording scheme for the following feeding year. If more than one analysis was available, the herd records were combined with the analysis with the highest potassium content. It was possible to combine forage and animal data from 487 farms with 15601 cows.

Results Statistical analysis (SAS) showed significant relationships between potassium content in forage and mastitis ($r^2 = 0.07$) and significant relationships between mastitis and milk yield ($r^2=0.05$). The data set was split into two parts - above and below the mean value 6.2 g Ca/ kg DM (min=2.3, max=14.7). This was done because inspection of the data indicated that in one region treatments for milk fever showed a declining pattern with increasing potassium, while the opposite tendency applied in the other regions. The major differences between the regions were in forage calcium content. On average, farms with less than 6.2 g Ca/kg DM had more treatments for lactation-related health disorders than farms with more than 6.2 g. (Table 1) Increasing milk yield with increasing potassium content in forage can also be seen in Table 1. Separate regression tests of K, P, Ca/P, K/Mg, K% x CP% or K/(Ca+Mg) indicated that the levels noted in Table 2 were associated with reduced need of treatments for health disorders The farms were grouped according to the number of analyses that were in the optimal range indicated in Table 2. The highest score was five, with no farms' analysis being in the optimal range for all six assessments. Farms with no forage analysis in the optimal range had the highest number of treatments and the treatment need was reduced with increasing number of analyses in the optimum range (Table 3).

Table 1 Total treatments (as % of number of cows) for all noted health disorders and mastitis and milk yield in relation to potassium and calcium content in forage

Potassium, g/ kg DM	<20		20-25		25-30		>30	
Calcium, g/ kg DM	<6.2	>6.2	<6.2	>6.2	<6.2	>6.2	<6.2	>6.2
Total treatments	52.8	48.7	52.5	53.9	60.8	57.2	65.1	54.9
Mastitis treatments	26.2	24.6	26.3	28.2	34.1	28.0	35.0	34.7
Milk yield, ECM kg/ cow	8239	7971	8459	8442	8643	8552	9006	8882

Table 2 Forage analysis intervals associated with reduced number of treatments for health disorders

Ca, g /kg DM	K, g/kg DM	P, g/kg DM	K/Mg	Ca/P	K/(Ca+Mg)	K% x CP% (3%K x 15% crude protein=45)
< 6.2	14-24	2.8-3.0	17-22	1.2-2.0	2.0-3.0	22-32
> 6.2	24-27	2.8-3.2	16-22	2.9-3.2	2.5-3.0	33-41

Table 3 Treatments (as % of number of cows) for health disorders in relation to analyses in optimum range

No. analyses in optimal range	0	1	2	3	4	5	6
<6.2 g Ca = % treatments	59.2	48.4	43.6	39.6	38.0	32.8	-
Std	45	26	26	26	21	19	-
>6.2 g Ca = % treatments	48.4	45.4	39.9	34.5	29.1	26.4	-
Std	34	29	19	16	20	3	-

Conclusions This study provides evidence of an association between forage mineral analysis and cow health. Forage mineral analysis will help in formulating supplementary feeds.

The effect of two magnesium fertilisers, kieserite and MgO, on herbage Mg content

M.B. O'Connor[1], A.H.C. Roberts[2] and R. Haerdter[3]
[1]AgResearch, Ruakura Research Centre, PB 3123, Hamilton, New Zealand, Email: mike.oconnor@agresearch.co.nz, [2]Ravensdown Fertiliser, P.O.Box 608, Pukekohe, New Zealand, [3]K+S Kali, Bertha-von-Suttner-Str.7, 34131 Kassel, Germany

Keywords: magnesium fertilisers, kieserite, MgO, pasture Mg

Introduction Supplementing Mg to dairy cows is widely practised in New Zealand. Various methods are used including drenching, pasture dusting, water trough treatment and adding to hay, silage and other feedstuffs (Young et al., 1979). Fertiliser Mg (calcined magnesite, MgO) is widely used to maintain soil Mg status but research has shown that using fertiliser Mg to achieve good soil, pasture and animal Mg status requires large inputs of Mg (120 kg/ha) and maintaining blood serum Mg status in dairy cows tends to be short-lived without further animal supplementation (O'Connor et al., 1987). The objective of these experiments was to test whether a more soluble Mg product like kieserite ($MgSO_4$), when applied to pastures, could achieve an immediate but short-term boost in Mg status when applied at critical times of the year.

Materials and methods Field trials were established in Northland in 2002 and Rotorua in 2003 in the North Island of New Zealand.. Each trial consisted of 2 products, MgO and kieserite, 3 rates of application (25, 50 and 100 kg Mg/ha) and 2 times of application (spring and autumn). Herbage samples were taken 5 to 6 times at 3-4 weekly intervals following either spring or autumn application and analysed for % Mg content.

Results Mg content in pasture was significantly higher from kieserite than MgO in the first sampling after application at both the Northland and Rotorua sites (Table 1). Both sites showed a marked rate effect to kieserite relative to MgO indicating that kieserite is a more soluble, quicker-acting material (Table 2). At subsequent samplings there was no difference between products at the Northland site but still differences at the third sampling at the Rotorua site before differences disappeared (Table 1). Other research has also indicated that long-term there will be very little difference between Mg fertilisers (Hogg & Karlovsky, 1968). Results suggest kieserite could be applied at critical times of the year to boost pasture Mg content and animal Mg requirements.

Table 1 Percentage Mg in pasture for control (no Mg) kieserite and MgO (mean of rates and times of application, samples taken at approximately monthly intervals post application)

Sample *	Northland				Rotorua			
	Control	Kierserite	MgO	SED	Control	Kiererite	MgO	SED
1	0.22	0.24	0.22	0.006	0.23	0.29	0.26	0.007
2	0.20	0.23	0.21	0.008	0.23	0.26	0.25	0.011
3	0.20	0.22	0.21	0.007	0.25	0.29	0.27	0.009
4	0.20	0.21	0.20	0.007	0.24	0.29	0.28	0.013
5	0.21	0.23	0.23	0.007	0.24	0.27	0.27	0.009
6					0.23	0.26	0.26	0.009

Table 2 Percentage Mg in pasture one month after application (mean of two sites)

Rate of Mg (kg/ha)	0	25	50	100
Kieserite	0.23	0.25	0.26	0.29
MgO		0.23	0.25	0.25

Conclusions The results from these trials demonstrate that kieserite is a much quicker acting fertiliser than MgO and could be used to provide a significant lift in Mg content of pasture when applied at critical times of the year. Although short-lived this effect could be important in terms of animal requirements.

References

Hogg, D.E., & J. Karlovsky (1968). The relative effectiveness of various magnesium fertilisers on a magnesium deficient pasture. New Zealand Journal of Agricultural research, 11,171-183.

O'Connor, M.B., M.G. Pearce,I.M. Gravett & N.R. Towers (1987). Fertilising with magnesium to prevent hypomagnesaemia (grass staggers) in dairy cows. Proceedings of the Ruakura Farmers' Conference, 39, 47-49.

Young, P.W., M.B. O'Connor & C. Feyter (1979). The importance of magnesium in dairy production. Proceedings of the Ruakura Farmers' Conference, 31,110-120.

Mineral status of some permanent range plants for grazing sheep in semi-arid areas of the Chaharmahal and Bakhtiari Province in Iran. 1. Some macro minerals

G.H.R. Shadnoush

Agriculture and Natural Resources Research Center, Shahrekord , Iran, Email: ghshadnoush@yahoo.com

Keywords: macro mineral, herbage, Chaharmahal & Bakhtiari , Iran

Introduction A deficiency of one or more of the essential minerals in animals can lead not only to a decrease in productivity but also to metabolic diseases in case of severe deficiencies (McDowell, 1985). In general, mineral, especially phosphorous, deficiencies and imbalances are reported from almost all tropical regions of the world (Matijovic & Durackova, 1994). This study was designed to measure the calcium, phosphorous, magnesium, potassium, sodium and chlorine contents of some plant species in the semi-arid rangeland of Chaharmahal & Bakhtiari Province in Iran, because little information is available at present.

Materials and methods This research investigated mineral status of dominant range plants *(Gramineae, Leguminosae, Compositae, Rosaceae* and *Umbeliferae)* in semi-arid areas of Chaharmahal and Bakhtiari Province. Samples were taken from five main non-grazed areas of the province during three stages of development - growing, blooming and seedling - between 1997 and 2000. All samples were analysed for Mg, Na, K, P, Ca, and Cl. The data were analysed using GLM of SAS (1996). The mineral status of each range plant was assessed by comparing the concentration of each mineral with established critical levels (Church, 1988; McDowell, 1985; NRC, 1985).

Results and discussion Means of mineral concentrations in the different species are given in Table 1.The content of magnesium for *Agropyron intermedium, Bromus tomentellus* and *Hordeum bulbosua* in the three growth stages was lower than the critical value required for ruminants (P<0.05). Sodium content in all species was lower than the critical value (P<0.05). Phosphorous content of *Agropyron intermedium, Bromus tomentellus, Cosina bakhtiarika, Scariolla orientalis* and *Astragalus spp* was lower than the critical value (P<0.05). Data obtained from these whole-plant samples suggested that semi-arid range land of Chaharmahal & Bakhtiari Province require mineral supplementation for grazing sheep specially for magnesium, sodium and phosphorous.

Table 1 Total means and standard error of some mineral concentrations of range plants in Iran (g/kg)

Species	Number	Ca	P	Mg	Na	K	Cl
Agropyron intermedium	56	5.1	1.55	1.06	0.34	15.14	4.8
Bromus tomentellus	53	5.7	1.45	1.07	0.39	15	5.2
Hordeum bulbosua	42	4.9	2.03	1.22	0.31	22.4	5.4
Cosina bakhtiarika	59	2.2	1.18	1.68	0.35	14.26	4.3
Scariolla orientalis	58	11.6	1.31	2.82	0.34	15.06	4.5
Vicia variabilis	30	18	2.03	2.43	0.33	17.23	6.7
Astragalus spp	39	18.5	1.5	3.03	0.37	12.27	3.9
Medicago sativa	44	22.2	1.75	2.47	0.36	17.24	5.2
Prangus ferulacea	53	23.1	1.89	2.9	0.39	20.95	3.3
Sanguisorba minor	44	17.5	2.1	4.94	0.29	15.13	5.1
SE	--	± 0.77	± 0.08	± 0.09	± 0.01	± 0.7	± 0.36

References

Church, D.C. & W.G. Pond (1988).Basic Animal Nutrition and Feeding. 3rd ed. John Willey and Sons, New York, U.S.A., 472 pp.

Matijovic, I. & A. Durackova (1994). Comparison of microwave digestion, wet and dry mineralization and solubilization of plant samples for determination of Ca, Mg, K, P, Na, Fe, Zn, Cu and Mn. Cornell. *Soil Science Plant Aria,* 25, 277-1288.

McDowell, L.R. (1985). Nutrition of Grazing Ruminants in Warm Climates 1st ed. Academic Press Inc., California, U.S.A., 443 pp.

National Research Council (1985). Nutrient Requirements, of Sheep. 6th rev.ed. National Academy Press. Washington D.C., U.S.A. 99pp.

SAS (1996). Release 6. 11, SAS Institute Inc., Cary, North Carolina, U.S.A.

Mineral status of some permanent range plants for grazing sheep in semi-arid areas of the Chaharmahal and Bakhtiari Province in Iran. 2. Trace minerals

G.H.R. Shadnoush
Agriculture and Natural Resources Research Center, Shahrekord, Iran, Email: ghshadnoush@yahoo.com

Keywords: trace minerals, herbage, Chaharmahal and Bakhtiari, Iran

Introduction The principal factors limiting the performance of grazing animals are low protein content of grasses, low energy intake due to high fibre content and mineral deficiencies or imbalances (McDowell, 1985). At least 26 mineral elements are required by at least one animal species (Church *et al.*, 1988). These elements are generally required in small amounts. They have a great deal of importance for body tissue growth as well as for physiological functions. Mineral deficiencies and imbalances have been reported from almost all tropical regions of the world. This study was designed to measure the copper, iron, manganese and zinc contents of some plant species in the semi-arid rangeland of Chaharmahal & Bakhtiari Province in Iran.

Materials and methods This research investigated the trace mineral status of dominant range plants *(Gramineae, Leguminosae, Compositae, Rosaceae* and *Umbeliferae)* in semi-arid areas of Chaharmahal & Bakhtiari Province. Samples were taken from five main non-grazed areas of the province during three stages of development - growing, blooming and seedling - between 1997 and 2000. All samples were analysed for Cu, Fe, Mn and Zn. The data were statistically analysed by GLM of SAS (1996). The mineral status of each range plant was assessed by comparing the concentration of each mineral with established critical levels (Church, 1988; McDowell, 1985; NRC, 1985).

Results and discussion Analysis of the whole plants revealed (Table 1) that the concentration of copper in most growth stages was lower than the critical value required for ruminants ($P<0.05$). The iron content in all species was sufficient for ruminants. The manganese content for all species except the *Compositae* was in the normal range. Zinc was deficient, except for *Vicia variabilis* in vegetative and flowering stages. Data obtain from whole plant samples suggested that semi-arid range of Chaharmahal & Bakhtiari Province require mineral supplementation for grazing sheep, specially for copper and zinc. It appeared that the effects of plant species and growth stage on plant mineral contents was more important than other factors, such as year of sampling, soil and climate.

Table 1 Total means and standard error of some trace mineral concentration of range plant in Iran (mg/kg)

Species	Family	Number	Cu	Fe	Mn	Zn
Agropyron intermedium	*Gramineae*	56	3.9	204	39.3	14.8
Bromus tomentellus	*Gramineae*	53	4.4	284	37.9	15.4
Hordeum bulbosua	*Gramineae*	42	5.2	162	32.3	17.1
Cosina bakhtiarika	*Compositae*	59	9.5	348	23.2	16.4
Scariolla orientalis	*Compositae*	58	7.9	368	29.5	18.6
Vicia variabilis	*Leguminosae*	30	8.4	402	37.4	28.4
Astragalus spp	*Leguminosae*	39	8.1	334	52.1	19.4
Medicago sativa	*Leguminosae*	44	9.6	345	34.1	22.3
Prangus ferulacea	*Umbeliferae*	53	7.6	496	60.6	24.1
Sanguisorba minor	*Rosaceae*	44	8.1	327	61.1	24.2
SE	--	--	± 0.32	± 57	± 2	± 0.96

References
Church, D.C. & W.G. Pond. (1988).Basic Animal Nutrition and Feeding. 3rd ed. John Willey and Sons, New York, U.S.A., 472 pp.
McDowell, L.R. (1985). Nutrition of Grazing Ruminants in Warm Climates 1st ed. Academic Press Inc., California, U.S.A., 443 pp.
National Research Council (1985). Nutrient Requirements, of Sheep. 6th rev. ed. National Academy Press. Washington D.C., U.S.A., 99pp.
SAS (1996). Release 6. 11, SAS Institute Inc., Cary, North Carolina, U.S.A.

Forage quality as related to mineral concentrations in tropical regions

L.R. McDowell
University of Florida, 125 Animal Science Bldg., P O Box 110910, Gainesville, FL 32611, USA, Email: McDowell@animal.ufl.edu

Keywords: tropical forages, minerals, deficiency, ruminants

Introduction Often tropical forages contain deficient or toxic concentrations of minerals for grazing livestock (McDowell, 2003). Tropical forages generally contain less of the more essential minerals than species grown in temperate regions (McDowell & Valle, 2000). Mineral elements in forages are dependent upon the interaction of a number of factors, including soil, plant species, stage of maturity, yield, pasture management and climate. The objective of this study was to determine the adequacy of minerals for ruminants in forages collected from seven tropical countries.

Materials and methods Forage samples were collected from six Latin American countries and one African country. The predominant forage genera, with sample numbers in parenthesis, were *Cynodon* (134), *Trachipogon* (118), *Paspalum* (104), *Andropogon* (94), *Pennisetum* (91), *Hyparrhenia* (63), and *Brachiaria* (46). Forage samples were collected, prepared and analysed for minerals by standardised procedures (Miles *et al.*, 2001).

Results Eleven mineral elements as related to critical concentrations are given in Table 1. Differences (p<0.05) were found among forage species and time of collection (wet vs dry season) within countries. The majority of forages for all countries were deficient in P, Na, Cu and Zn. Most countries also had forages deficient in Ca, Mg and Se, while most forages were adequate in K, Fe and Mn.

Table 1 Percentage forage minerals below ruminant requirements

Mineral level	Critical level[a]	Dominican Republic (69)[b]	Bolivia (84)	Colombia (36)	Guatemala (168)	Malawi (21)	Nicaragua (304)	Venezuela (198)
Ca, %	0.30	24	57	100	71	13	80	97
P, %	0.25	83	100	92	57	75	67	98
K, %	0.60-0.80	0	1	15	13	57	20	84
Na, %	0.06	78	100	100	88	97	80	84
Mg, %	0.20	33	64	56	76	31	96	94
Fe, ppm	30	0	0	0	0	3	13	0
Zn, ppm	30	86	81	74	49	94	89	96
Cu, ppm	10	64	100	100	92	91	100	99
Mn, ppm	30-40	10	0	0	24	3	41	5
Co, ppm	0.10	26	48	31	1	13	96	40
Se, ppm	0.10	48	47	74	49	96	18	96

[a]Critical levels (ruminants requirements) and complete reference citations for the seven countries are given in McDowell (2003); [b]Sample numbers in parenthesis

Conclusions Tropical forages are often deficient in essential minerals. Minerals most deficient were P, Na, Cu, Zn, Ca, Mg and Se.

References
McDowell, L.R. (2003). Minerals in Animal and Human Nutrition, 2nd ed., Elsevier, Amsterdam.
McDowell, L.R. & G. Valle (2000). Major minerals in forages. In D.I Givens & H.M. Omed (eds.). Forage Evaluation in Ruminant Nutrition, CABI Publishing, Wallingford, 373-397.
Miles, P.H., N.S. Wilkinson & L.R. McDowell (2001). Analysis of Minerals for Animal Nutrition Research; 3rd ed., Gainesville, FL, USA.

Towards truly "global" near infrared calibrations for protein and neutral detergent fibre in dried ground forages

D.J. Undersander[1], P. Berzaghi, P. Dardenne, P. Flinn, N.P. Martin, C. Paul, B.N.B. Büchmann, F. Mazeris, M. Lagerholm and I.A. Cowe
[1]University of Wisconsin, Dept. of Agronomy, 1575 Linden Drive, Madison WI 53706, USA, Email: djunders@facstaff.wisc.edu

Keywords: near infrared, protein, neutral detergent fibre

Introduction Over the past five years, Foss and DeLaval have sponsored the activities of a group of forage analysts with the aim of developing "global" Near Infrared (NIR) calibrations for parameters that are important in ruminant nutrition. The approach adopted has been based on the amalgamation of historical databases from centres worldwide and calibrations for protein and neutral detergent fibre (NDF) in dried ground forages have been developed based on databases that currently comprise approximately 30,000 records. Protein and NDF, while not the most important parameters in ruminant nutrition, were chosen for the initial calibration development exercise because of the amount of data available and because the methodologies adopted by different laboratories worldwide were relatively uniform. The aim was to create calibrations that would work for any forage type in any area of the world. Over the past two years, several trials have been carried out worldwide comparing the performance of "global" calibrations with the performance of locally developed calibrations for indigenous forages and based on reference values from local laboratories.

Materials and methods Standard methods were used to generate reference values for protein and NDF. Equations were developed by customer using PLS (partial least squares analysis) were compared to the global models based on artificial neural network (ANN) technology using a database that included spectra from all parts of the world and representing harvests for the last twenty years. Some tropical forages and all the major types of temperate forages, including legume and maize silages, were contained in the database.

Results In Table 1 we can see that the global model for protein gave similar statistics (standard error of prediction corrected for slope effects (SEP(C))) to those for the customer model while the global model for NDF was slightly worse than the customer model. It should be noted that the number of samples where the customer had suitable models for particular forages was 64 and 65 respectively while the global models predicted all forage samples available.

Table 1 Comparison of prediction statistics for Swedish data. Reference is customer wet chemical values

	Protein		NDF	
	PLS	ANN	PLS	ANN
n	63	111	64	111
RMSEP	0.56	0.95	3.09	3.62
SEP	0.54	0.58	2.55	3.08
SEP(C)	0.50	0.58	2.57	3.03
R	0.99	0.99	0.93	0.96

Where: PLS = model from partial least squares analysis; ANN = model from artificial neural network; RMSEP = root mean squared error of prediction (without bias correction); SEP = standard error of prediction (corrected for bias); SEP(C) = standard error of prediction corrected for slope effects

Conclusions The global calibrations would require slope and/or bias corrections before being used. This was because Scandinavian forages were not represented in the calibration database and local laboratories generated the reference values. The global models had poorer performance than the locally developed regional calibrations, but were able to handle a much wider range of samples with acceptable performance than could the regional calibrations. The benefits of having to maintain a single universal calibration rather than many species-specific calibrations are important to the economics of managing forage networks.

A rapid estimation of nitrogen bound to neutral detergent fibre in forages by near infrared reflectance spectroscopy

J. Bindelle[1], G. Sinnaeve[2], P. Dardenne[2], P. Leterme[3] and A. Buldgen[1]
[1]Faculté universitaire des Sciences agronomiques de Gembloux, Unité de Zootechnie, 2 Passage des Déportés, 5030, Gembloux, Belgium Email address: bindelle.j@fsagx.ac.be, [2]Centre wallon de Recherches agronomiques, Département Qualité des Productions agricoles, 24 Chaussée de Namur, 5030, Gembloux, Belgium, [3]École nationale vétérinaire de Lyon, Unité de Zootechnie, 1 avenue Bourgelat, 69280 Marcy l'Etoile, France

Keywords: forages, near infrared reflectance spectroscopy, neutral detergent fibre, nitrogen

Introduction Near infrared reflectance spectroscopy (NIRS) is widely used as a rapid method for the evaluation of the chemical composition or the nutritive value of foodstuffs (Givens et al., 1997). The determination of the neutral detergent fibre (NDF) bound N (NDF-N), which is highly variable in forages (Shayo & Udén, 1999), is expensive. The purpose of this study was to test the use of NIRS in the prediction of NDF-N in various forages.

Materials and methods The study used various fresh forages and hays from temperate, tropical and Mediterranean regions (n=288), consisting of grasses (n=131), herbaceous dicotyledons (n=38) and shrubs or trees (n=119). All samples were oven dried at 60°C and ground to pass a 1 mm mesh screen. After NIRS-spectra of the samples were recorded, a subset of 118 samples underwent a NDF extraction and N content of the residues were determined. Calibrations using the partial least squares procedure were developed for NDF-N within the sub-set and tested by cross validation. They were finally used to estimate the NDF-N of the 288 forages.

Results Table 1 indicates the accuracy of the prediction of NDF-N for the 3 groups of forages. SD/SEC and R^2 values show that the quality of the calibration is satisfactory for grasses and shrubs. The predictions for the dicotyledons are less accurate (SD/SEC < 3) due to a lower number of samples, but the predicting equation can however be considered as satisfactory ($R^2 > 0.85$).

Table 1 Range of NDF-N (g/kgDM) in the prediction subsets and quality of the predicting equations

	n	Minimum NDF-N	Maximum NDF-N	Standard error of calibration (SEC)	R^2	Standard error of cross validation	SD/ SEC
Grasses	49	1.02	16.11	0.185	0.920	0.198	3.54
Herbaceous dicotyledons	23	1.93	25.63	0.510	0.864	0.673	2.71
Shrubs and trees	46	0.14	21.17	0.353	0.931	0.387	3.79

Predictions on the 288 samples (Figure 1) show an important variability in the absolute NDF-N content and in the percentage of N bound to the NDF within each group of plants, and also between them. The dicotyledons and the shrubs have a significantly higher proportion of N bound to the NDF compared to the grasses (31 % vs 22 %, respectively). However, dicotyledons have a significantly higher non NDF-N content than the grasses, followed by the shrubs. For each group, significant regression equations linking NDF-N to total-N and NDF content were also found (R^2 from 0.80 to 0.89, p < 0.001).

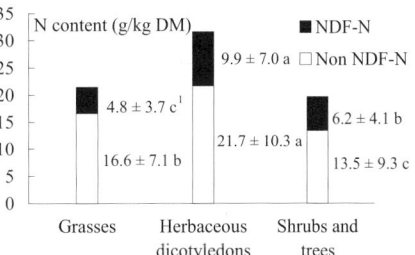

Figure 1 Predicted N fractions of the forages
[1]For a same fraction, values followed by different letters differ significantly (p<0.05)

Conclusions The present study showed that NIRS is a valuable method to offer a rapid estimation of the NDF-N content in forage resources. A variable and sometimes important part of the N is bound to the NDF, depending on the forage. It seemed also possible to obtain a reasonable estimate of the NDF-N fraction of a forage by considering the forage group and determining NDF and total-N contents.

References

Givens, D.I., J.L. De Boever & E.R. Deaville (1997). The principles, practices and some future applications of near infrared spectroscopy for predicting the nutritive value of foods for animals and humans. *Nutrition Research Reviews*, 10, 83-114.

Shayo C.M. & P. Udén (1999). Nutritional uniformity of crude protein fractions in some tropical browse plants estimated by two in vitro methods. *Animal Feed Science and Technology*, 78, 141-151.

Prediction of N fractions of warm-season grasses with near-infrared reflectance spectroscopy

S.W. Coleman[1], C.E. Johnson[2,3], B.A. Reiling[2,4] and P. Mislevy[2]
[1]USDA ARS Subtropical Agricultural Research Station, Brooksville, Florida 34601, Email:swcol@ifas.ufl.edu,
[2]University of Florida, Gainesville, Florida 32611, USA, [3]currently Colorado State University, Ft. Collins, Colorado, [4]currently University of Nebraska, Lincoln, Nebraska 68583

Keywords: N-fractionation, warm-season grasses, protein solubility, NIRS

Introduction Warm-season (C_4), the most common forage for beef production in much of the U.S.A., although having a higher productivity than temperate forages are of a lower quality. Current feeding standards (NRC, 1996) have adapted the Cornell Net Carbohydrate and Protein System (Sniffen et al., 1992) to more accurately characterize forage quality. As these procedures are tedious, data is limited on genetic and management factors influencing quality parameters in C_4 species. The objective of this research was to determine if near-infrared reflectance spectroscopy (NIRS) could be used to predict the various N fractions in three C_4 grasses.

Materials and methods The samples and data from two replicates of the experiment published by Johnson et al. (2001) were used to calibrate and validate the NIRS procedure. The design included five N fertilizer levels (0, 40, 70, 110, 147 kg ha^{-1}) and five harvest dates per year on bahiagrass (*Paspalum notatum*), bermudagrass (*Cynadon dactylon*), and stargrass (*Cynodon nlemfuensis*) over a two–year period (n=100 samples per species). Spectral data were collected on each dried, ground sample from 400-2500 nm with a NIRSystems 6500 spectrophotometer using Infrasoft International software. Calibration equations were developed using partial least squares regression. A robust method was used to evaluate the equations by deleting one harvest date from the calibration data and using samples from that date for validation, and repeating for each harvest date. Analysis of variance was conducted to determine existence of systematic bias due to species, harvest, or fertilizer level.

Results Calibration statistics (Table 1) indicate a good relationship with laboratory determined parameters, whereas fractions calculated by difference were more difficult to predict. Especially when calculated as a fraction of N. When the robust evaluation was used, small but significant bias existed among the interactions of species, harvest date and occasionally fertilizer level. Part of the bias was due to bahiagrass having a higher proportion of insoluble N fractions (e.g., ADIN). Though average bias was small, it increased with increasing ADIN levels indicating non-linearity of predicted values when compared to laboratory values.

Table 1 Calibration statistics for predicting N fractions with NIRS (n=~290)

N Fraction[a]	Mean	Sd	SEC[b]	R^2	SECV[b]	1-VR[b]	Bias[b]
Total	2.19	0.62	0.10	0.98	0.12	0.97	-0.020
TCAIN	1.51	0.34	0.07	0.96	0.08	0.95	0.002
SOLPN	1.32	0.29	0.07	0.95	0.08	0.93	0.003
NDIN	0.89	0.22	0.10	0.81	0.10	0.79	0.005
ADIN	0.18	0.06	0.02	0.90	0.02	0.85	0.001
A[c]	0.72	0.35	0.10	0.92	0.11	0.90	-0.027
B1[c]	0.20	0.12	0.09	0.47	0.09	0.43	0.005
B2[c]	0.30	0.20	0.10	0.71	0.12	0.65	0.004
B3[c]	0.76	0.20	0.10	0.73	0.11	0.71	0.015
C[c]	0.19	0.06	0.02	0.91	0.02	0.86	-0.001

[a]Percent of total DM
[b]n=number of samples for calibration after outlier elimination; SEC=Standard error of calibration; SECV = Standard error of cross validation; 1-VR = coefficient of determination (R^2) for cross validation; bias is overall bias from robust evaluation.
[c]A = NPN; B1 = soluble protein; B2 = moderate degradability rate; B3 = slowly degradable; c = undegradable N

Conclusions The results of this study demonstrate that NIRS can be used to predict fractions of N related to the solubility and degradability in the rumen. However, prediction of calculated fractions were rather poor, but could be calculated from predicted primary fractions such as total, NDIN and ADIN.

References
Johnson, C.R., B.A. Reiling, P. Mislevy, & M.B. Hall (2001). Effects of N fertilization and harvest date on yield, digestibility, fibre, and protein fractions of tropical grasses. *Journal Animal Science, 79*, 2439-2448.
NRC, (1996). Nutrient Requirements of Beef Cattle (7th Ed.). National Academy Press, Washington DC.
Sniffen, C.J., J.D. O'Connor, P.J. Van Soest, D.G. Fox & J.B. Russell (1992). A net carbohydrate and protein system for evaluating cattle diets: II. Carbohydrate and protein availability. *Journal Animal Science*, 70, 3562-3577.

Nutritive quality of silages by conventional laboratory methods and near infrared reflectance spectroscopy

P. Castro, G. Flores, A. González-Arráez, J. Castro and B. Fernández-Lorenzo
Centro de Investigaciones Agrarias de Mabegondo (CIAM), Apartado 10, 15080 La Coruña, Spain, Email: pilar.castro.garcia@xunta.es

Keywords: digestibility, near infra-red spectroscopy, detergent fibre, crude protein

Introduction Preservation of forages as silages is needed by dairy farmers in NW Spain to feed their cows during the dry (summer) and cold (winter) seasons. The objective of this work was to compare the prediction of *in vivo* digestibility values by conventional laboratory methods and by near infra-red reflectance spectroscopy (NIRS) for herbage and maize silages.

Material and methods The *in vivo* digestibility of 197 herbage and 93 maize silages from experiments and from farms was determined with sheep in the feeding unit of CIAM and their analysis by conventional laboratory methods was carried out to determine organic matter (OM), crude protein (CP), acid and neutral detergent fibres (ADF and NDF) and *in vitro* OM digestibility with rumen fluid (IVOMD) and pepsin-cellulase (PCOMD). Regression equations of *in vivo* OM digestibility (OMD) on laboratory parameters were obtained for both herbage and maize silages by Flores (2004). NIRS calibration equations were developed to determine OMD with the same sets of samples (Castro *et al.*, 2002, 2004) on a 6500 NIRSystem Spectrophotometer (Foss NIRSystem, Silver Spring, Washington, USA). Cross-validation values of OMD predicted by conventional methods were compared to those obtained by NIRS with WinISI 1.5 software (InfraSoft International, Port Matilda, PA, USA).

Results Best updated equations to predict *in vivo* OMD from laboratory parameters were those based on IVOMD and CP for herbage, $R^2_{cv} = 0.77$, SECV= 3.24 (for equation 1), and maize, $R^2_{cv} = 0.50$, SECV= 2.33 (2), silages

(1) OMD = 12.63 + 0.716 IVOMD + 0.521 CP (2) OMD = 12.77 + 0.701 IVOMD + 0.516 CP

Regression of cross-validation results from equations (1) and (2) and NIRS analysis ($R^2_{cv} = 0.84$, SECV= 2.76 for herbage and $R^2_{cv} = 0.55$, SECV= 2.13 for maize) on *in vivo* OMD values of herbages and maize silages are shown in Figures 1 and 2, respectively.

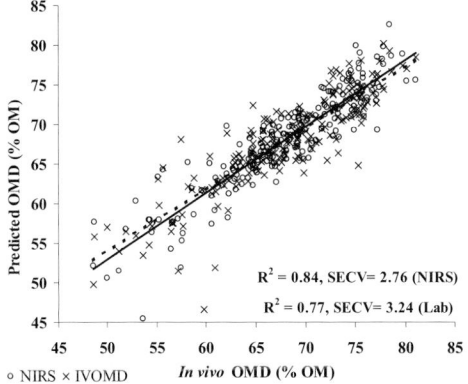

Figure 1 Organic Matter Digestibility (OMD) of herbage silages by conventional methods (Lab) and NIRS

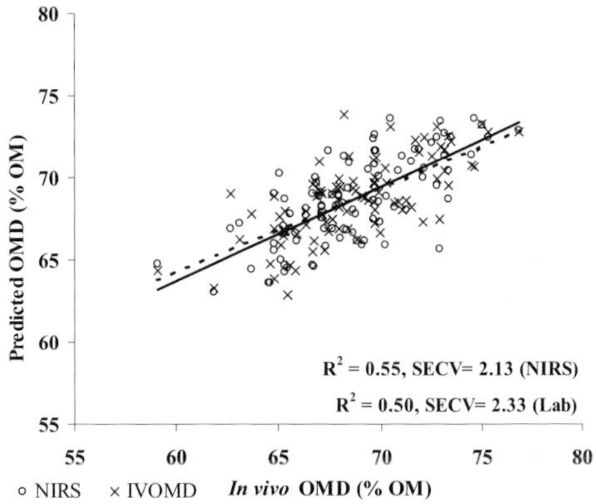

Figure 2 Organic Matter Digestibility (OMD) of maize silages by conventional methods (Lab) and NIRS

Conclusions The results from the present study demonstrate that NIRS analysis is the best method to predict *in vivo* OMD, not only because of practical reasons (faster, cheaper, easier) but also because of its higher precision.

References
Castro, P., G. Flores, A. González-Arráez & J. Castro (2003). Nutritive quality of herbages silages by NIRS: dried or undried samples?. *Grassland Science in Europe*, 7, 190-191.
Castro, P., G. Flores A. González-Arráez, J. Castro & B. Fernández-Lorenzo B. (2004). Análisis de ensilados de maíz mediante NIRS. *LIV Reunión Científica de la Sociedad Española para el Estudio de los Pastos*. Salamanca, Spain, 11-14 May 2004, 279-283
Flores, G. (2004). Evaluación de métodos de laboratorio para la predición de la digestibilidad *in vivo* de ensilajes de hierba y de maíz. *Tesis doctoral*. Universidad Politécnica de Madrid, 2004, 318 pp.

Sample preparation method for Near Infrared Reflectance Spectroscopy to predict fermentation quality of maize silage

H.S. Park[1], J.K. Lee[1] and M.C. Kim[2]
[1]Hanwoo Experiment Station, National Livestock Research Institute, RDA, Chahang-ri, Doam-myeon, Pyeongchang-gun, Kangwon-do, Korea, 232-952, Email: anpark@korea.com, [2]Department Animal Biotechnology, Cheju National University, 66 Jejudaehakno, Jeju-si, Jeju-do, Korea, 690-756

Keywords: near infrared reflectance spectroscopy, maize silage, fermentation quality

Introduction Using NIRS directly on undried silage can increase error due to variability in sample particle size, temperature and water content (Givens *et al.*, 1997). These problems can be overcome by grinding silages in frozen state with dry ice or liquid nitrogen, but such procedures are time-consuming and inconvenient due to cleanup required between samples and the need to thaw the sample for subsequent use. The objective of this experiment was to assess the effect of sample preparation methods on prediction of fermentation quality of maize silage, and to select an acceptable sample-preparation method for wet silage.

Materials and methods Maize silage samples (n=112) were collected from dairy farms in Korea. Each sample was subdivided into three treatments: i) Oven drying (65°C for 48h) then grinding (ODG); ii) Intact fresh (IF); iii) Liquid nitrogen grinding (LNG). For LNG, samples were immersed in liquid nitrogen (-196°C) for 30 min. followed by grinding. Concentrations of volatile fatty acids and lactic acid were determined based on methods of Fussell & McCalley (1987). Predictive equations were developed using modified partial least squares (MPLS) regression with internal cross-validation after scatter correction using SNV and Detrend.

Results Comparisons of NIR predicted parameters for the three sample preparation methods are shown in Table 1. Predictions of acetic acid concentrations were achieved with best accuracy in their intact fresh condition. The best predictions of propionic acid concentrations were obtained with LNG treatment (SEP=0.08, R^2=0.65), but ODG and IF preparation methods produced poor validation R^2. None of the NIRS calibration equations were satisfactory for the estimation of butyric acid. This is possibly due to the concentration of butyric acid in well fermented maize silage samples being low or not detectable. The best predictions of lactic acid concentrations were obtained using ODG and IF preparation methods.

Table 1 Accuracy of NIRS in sample preparation methods for volatile fatty acids and lactic acid

Parameters (%, DM)	Sample preparation	Calibration R^2	SEC	SECV	1-VR	Validation SEP	R^2
Acetic acid	ODG	0.91	0.16	0.23	0.82	0.21	0.73
	IF	0.95	0.10	0.19	0.85	0.13	0.89
	LNG	0.94	0.11	0.20	0.84	0.17	0.88
Propionic acid	ODG	0.66	0.12	0.14	0.50	0.11	0.32
	IF	0.63	0.09	0.11	0.44	0.12	0.27
	LNG	0.83	0.06	0.09	0.67	0.08	0.65
Butyric acid	ODG	0.27	0.01	0.15	0.19	0.004	0.00
	IF	0.19	0.003	0.004	0.04	0.007	0.00
	LNG	0.26	0.003	0.004	0.17	0.003	0.00
Lactic acid	ODG	0.79	0.70	0.81	0.73	0.58	0.80
	IF	0.90	0.43	0.64	0.78	0.75	0.77
	LNG	0.89	0.43	0.73	0.77	0.76	0.70

R^2, coefficient of determination; SEC, standard error of calibration; SECV, standard error of cross-validation; 1-VR, coefficient of determination for cross-validation; SEP, standard error of prediction.

Conclusions Analysis of undried silages by NIRS can provide accurate prediction of a wide range of fermentation products in maize silages.

References

Fussel, R.J. & D.V. McCalley (1987). Determination of volatile fatty acids (C2-C5) and lactic acid in silage by gas chromatography. *Analyst*, 112, 1213-1216.

Givens, D.I., J.L. De Boever & E.R. Deaville (1997). The principles, practices and some future applications of near infrared spectroscopy for predicting the nutritive value of foods for animals and humans. *Nutrition Research Reviews*, 10, 83-114.

Prediction of the feed values of maize silage by Near Infrared Reflectance Spectroscopy

H.S. Park[1], J.K. Lee[1], H.J. Ko[2], H.Y. Lee[2] and D.Y. Kil[2]
[1]Hanwoo Experiment Station, National Livestock Research Institute, RDA, Chahang-ri, Doam-myeon, Pyeongchang-gun, Kangwon-do, South Korea, 232-95, Email: anpark@korea.com, [2]Department of Agricultural Science, Korea National Open University, 169 Dongsung-dong Chongro-ku, Seoul, Korea

Keywords: near infrared reflectance spectroscopy, maize silage, digestibility

Introduction Until recently, feed evaluation of silages in official laboratories and feed factories was based on cutting date, chemical composition and the ammonia fraction. However, *in vitro* techniques have been developed based on rumen fluid or commercial enzymes to replace laborious, time-consuming and expensive digestibility experiments with animals. In this study the possibility of using near infrared reflectance spectroscopy (NIRS) to predict the chemical composition and digestibility of maize silage was examined.

Materials and methods Maize silage samples (n=112) were collected from dairy farms in Korea. Samples were dried in a forced-air drier at 60°C for 48h. All samples were analysed in duplicate. Ash content was determined by ashing at 550°C for 6h; CP, NDF, ADF and ADL were carried out according to AOAC methods. *In vitro* dry matter digestibility (IVDMD) was determined following the classical two-stage technique of Tilley & Terry (1963). Cellulase digestibility (COMD) was determined following the procedure of De Boever *et al.* (1986). Predictive equations were developed using modified partial least squares (MPLS) regression with internal cross-validation after scatter correction using SNV and Detrend.

Results The calibration and validation statistics for the prediction of feed values of corn silages were as shown in Table 1. The R^2 coefficients and standard errors of cross-validation were 0.93 (SECV 2.14), 0.96 (SECV 1.05), 0.90 (SECV 0.45), 0.88 (SECV 0.29), 0.92 (SECV 1.73), and 0.93 (SECV 0.29) for NDF, ADF, Ash, CP, IVDMD and COMD, respectively. Calibration statistics were good for all parameters. The results obtained on the independent validation set for the squared simple correlation coefficients (RSQ) are 0.83, 0.92, 0.91, 0.72 and 0.79 for NDF, ADF, CP, IVDMD and COMD, respectively.

Table 1 The calibration and validation statistics for the prediction of feeding values of maize silages

Parameters (%, DM)	Calibration				Validation	
	R^2	SEC	SECV	1-VR	SEP	R^2
Chemical composition						
NDF	0.93	1.73	2.14	0.89	1.98	0.83
ADF	0.96	0.79	1.05	0.94	1.10	0.92
ADL	0.80	0.46	0.54	0.70	0.51	0.73
Ash	0.90	0.34	0.45	0.83	0.41	0.86
CP	0.88	0.24	0.29	0.84	0.29	0.91
Digestibility						
IVDMD	0.92	1.22	1.73	0.83	1.57	0.72
COMD	0.93	1.20	1.74	0.91	1.48	0.79
IVTD	0.91	0.97	1.13	0.88	1.07	0.76

R^2, coefficient of determination; SEC, standard error of calibration; SECV, standard error of cross-validation; 1-VR, coefficient of determination for cross-validation; SEP, standard error of prediction.
IVDMD, *in vitro* dry matter digestibility; COMD, Cellulase digestibility; IVTD, *in vitro* true digestibility

Conclusions It was concluded that good prediction of quality parameters was obtained. The results showed that IVDMD as well as COMD, ADF and CP could be accurately predicted in maize silage using NIRS. Because of the higher costs and labour involved in the laboratory analysis of IVDMD and fibre analysis, NIRS will become the main analytical technique in feeding programmes and for farmer advice.

References

De Boever. J.L., B.G. Cottyn, F.X. Buysse, F.W. Wainman & J.M. Vanacker (1986). The use of an enzymatic technique to predict digestibility, metabolizable and net energy of compound feedstuffs for ruminants. *Animal Feed Science and Technology*, 14, 203-214.
Tilley, J.M.A. & R.A. Terry (1963). A two-stage technique for the *in vitro* digestion of forage crops. *Journal of the British Grassland Society*, 18, 104-111.

Near Infrared Spectroscopy to assess feeding value and antinutritional compounds in Legume species

M. Odoardi, S. Colombini, G. Piluzza and M. Confalonieri
Istituto Sperimentale per le Colture Foraggere, 29 V.le Piacenza, 26900 Lodi, Italy Email: m.odoardi@stanca.it

Keywords: NIRS, feeding value, antinutritional compounds, legumes

Introduction There is an increasing demand for information on the quality characteristics and chemical composition of forages in order to meet the demands of dietary specifications for feeding animals. Near Infrared (NIR) spectroscopy provides a tool for rapid and non-destructive analysis in agronomic and breeding programs of a number of chemical components of forages and grains. NIR spectroscopy in particular has the advantage of being able to simultaneously evaluate the samples for a number of qualitative traits of whole plants and seeds. In two experiments here presented, NIR Spectroscopy was used to predict: i) qualitative characteristics of field pea seeds and, as regards secondary metabolites responsible of detrimental or beneficial effects on animal nutrition ii) condensed tannins in legume forages, based on calibration sets of samples previously chemically analysed.

Materials and methods Up to 300 seed samples of 50 field pea cultivars of different geographical origin have been used in the first study. Standard methods were applied to determine crude protein (CP) and NDF content (% DM). In the latter case, the concentration of condensed tannins (CT) was evaluated on 320 sulla (*Hedysarum coronarium* L.) plant samples, determined according to Terrill *et al.* (1992). In both cases, the samples were scanned as dry ground powder in reflectance mode using a NIRSystems 5000 monochromator, and WIN-ISI version 2 software was used for spectral data collection, spectral processing and calibration development.

Results The chemical composition of a number of field pea seeds, determined on samples derived from comparative trials, allowed the characterization of the best performing materials to start specific breeding programs for selection of new high yielding genotypes with increased levels of crude proteins. In Table 1 the range and mean values of crude protein (CP) and NDF content in the two sets of samples randomly selected used for deriving the NIRS calibration equations are reported, together with the statistics of the calibration and validation procedures applied. The narrow degree of variation found with such materials could explain the good but not excellent coefficients of determination in validation ($r^2 = 0.88$ and 0.69 for CP and NDF, respectively) resulting from NIRS analyses, particularly for the prediction of CP, as usually found with different plant materials (Berardo *et al.* 1997). A wider variability on the contrary was found for the condensed tannin concentration in sulla (from 1.0 to 6.8% DM), which permitted a very good calibration ($r^2 = 0.91$, $SEP=0.42$), suitable to accurately predict CT percentage in sulla, as shown in Figure 1.

Table 1 Range of the Crude Protein and NDF content (%DM) and statistics of the calibration and validation sets of randomly selected pea meal samples used for developing NIRS equations

	CALIBRATION				
Trait	N	Mean ± SD	Range	R^2	SECV
CP	88	20.6 ± 1.0	17.6 – 23.7	0.86	0.42
NDF	95	13.2 ± 2.0	7.2 - 19.3	0.67	1.16

	VALIDATION				
Trait	N	Mean ± SD	Range	r^2	SEP
CP	58	20.7 ± 1.0	18.5 - 23.0	0.88	0.43
NDF	58	13.4 ± 1.4	10.7 – 18.5	0.69	1.47

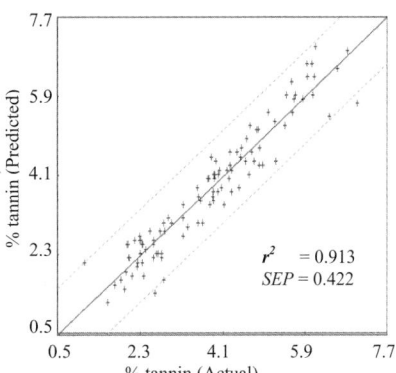

Figure 1 Relationship between actual and predicted % tannin in sulla

Conclusions The calibration and validation statistics presented in this work showed the potential of NIRS to predict both primary and secondary metabolites with good accuracy, confirming NIRS as an alternative approach for the rapid and reliable estimation of forage quality.

References

Berardo, N., F. Boccardi, E. Piccinini, A. Ursino & M. Odoardi, (1997). La spettroscopia NIR per l'analisi qualitativa dei foraggi. *Rivista di Agronomia*, 31(1), 208-211.

Terrill, T.H., A.M. Rowan, G.B. Douglas & T.N. Barry (1992). Determination of extractable and bound condensed tannin concentrations in forage plants, protein concentrate meals and cereal grains. *Journal of the Science of Food and Agriculture*, 58, 199-202.

Validation of faecal NIRS for monitoring the diet of confined and grazing goats

S. Landau, T. Glasser, L. Dvash and A. Perevolotsky
Department of Natural Resources and Agronomy, Institute of Field and Garden Crops, Agricultural Research Organization, the Volcani Center, P.O. Box 6, Bet Dagan 50250, Israel, Email: vclandau@agri.gov.il

Keywords: pasture, nutrition, animal-plant interaction, decision making

Introduction Goats are used for brush control and ecological management of Mediterranean grazing lands. Farmers are willing to cooperate with communities but they need an easy method to evaluate the daily intake of nutrients. A calibration of the chemical attributes of goats' diets was set-up, based on faecal near infrared (NIR) spectra (Landau *et al.*, 2004; Table 1). The accuracy of this methodology was estimated by using the standard error of cross-validation (SECV), which represents the variability in the difference between predicted and reference values when the equation is applied sequentially to subsets of data from the calibration data set. This procedure is justified in situations with calibration samples that are randomly selected from a natural population, but may give over-optimistic results, in particular if data are replicated. The standard error of prediction (SEP) represents the variability in the difference between predicted and reference values when the equation is applied to an external (i.e., not used in any step of the calibration) validation data set. (Naes *et al.*, 2002). The aim of the present study was to test the robustness of predicting dietary CP, *in vitro* dry matter digestibility (IVDMD), and NDF percentages in goats' diets, using faecal samples totally external to calibrations.

Materials and methods A validation data set was constructed by merging data including confined Saanen (n=36) and grazing Damascus and Sarda (n=8) goats. Predictions were made on faeces from experiments published prior to the establishment of calibrations and compared with actual dietary values. Accuracy was assessed from the slope and bias of regressions and from paired t-tests between actual and predicted values.

Table 1 Faecal NIRS calibrations (Landau *et al.*, 2004) of chemical dietary attributes (on DM basis) in Israeli Damascus goats. Diets consisted of hay and concentrates (n=60) and browse from up to three different species (n=83)

Constituent	Mean	Range	R^2	SECV
CP	12.2	7.7-16.9	0.98	0.53
NDF	37.9	28.5-50.1	0.94	1.53
IVDMD	60.0	41.3-80.0	0.98	1.98

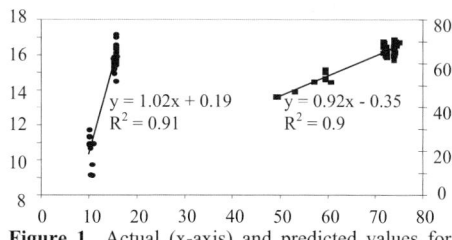

Figure 1 Actual (x-axis) and predicted values for dietary CP (• ; left y-axis) and IVDMD (• ; right y-axis) in the external validation set

Results Dietary CP was predicted with high linearity (R^2=0.91) and accuracy (slope 1.02, bias 0.2%, Figure 1). Predicted values of dietary CP differed from actual values for confined, but not grazing goats. The validation of dietary IVDMD was linear (R^2=0.90), but the slope differed from 1, and dietary IVDMD was significantly underestimated (Table 2). Dietary NDF was underestimated to a greater extent in grazing than in confined goats.

Table 2 Means of actual and predicted values for dietary percentages (on DM basis) of CP, NDF and IVDMD

Mean	CP (range 9.4-15.7%)		NDF (range 41.3-47.7%)		IVDMD (range 31.8-74.2%)	
	Actual	Predicted	Actual	Predicted	Actual	Predicted
Grazing	10.1	10.7	42.4[a]	38.4[b]	57.0[a]	52.0[b]
Confined	15.3[a]	15.8[b]	41.0[a]	40.0[b]	73.0[a]	66.6[b]

Separately for confined and grazing goats, means with different superscripts letters differ (paired t-test, P<0.05)

Conclusions Even though systemic bias needs correction, the present validation suggests that faecal NIRS predictions, in particular of dietary CP percentage, are sufficiently accurate for decision-making under farm conditions.

References

Landau, S., T. Glasser., L. Dvash & A. Perevolotsky (2004). Fecal NIRS to monitor the diet of Mediterranean goats. *South African Journal of Animal Science*, 34, 76-80.
Naes, T., T. Isakson, T. Fearn & T. Davies (2002). Validation. In: T. Naes, T. Isakson, T. Fearn & T. Davies (eds.) A user-friendly guide to multivariate calibration and classification. NIR Publications, Chichester, 155-177.

Are leaf traits suitable for assessing the feeding value of native grass species?

R. Al Haj Khaled, M. Duru and P. Cruz
*Research unit UMR 1248 Managing grassland farming systems, BP 27, F - 31326 Castanet -Tolosan - France,
Email: alhaj@toulouse.inra.fr*

Keywords: leaf dry matter content, leaf life span, specific leaf area, forage quality, native grass

Introduction Research on forage feeding value other than *in vivo* assessment can be roughly divided into three kinds of approach. The first aims to predict feeding value using a set of enzymatic or physical methods. A second approach is based on phenological stages of species. These approaches are mainly used for pure stands of improved grasses or legumes. However, for native grassland, a complex type of vegetation, a third approach, based on botanical records, has been proposed to rank grassland communities for their feeding value. The aim of this work concerns the third approach. We tested whether leaf traits (e.g. specific leaf area (SLA), leaf dry matter content (LDMC) and leaf life span (LLS)), assessed under non-limiting plant growth conditions, ranked the species in the same order as did chemical components and digestibility.

Materials and methods Seventeen grasses collected from their native habitats were sown separately according to a randomised block design with three replicates in Auzeville, France. The list of species is: *Agrostis capillaris, Anthoxanthum odoratum, Arrhenatherum elatius, Avenula pubescens, Brachypodium pinnatum, B. sylvaticum, Briza media, Cynosurus cristatus, Dactylis glomerata, Festuca ovina, Festuca rubra, Festuca arundinacea, Phleum pratense, Holcus lanatus, Lolium perenne, L. perenne* (cv) clerpin, and *Trisetum flavescens*. Plants were grown with non-limiting nutrients and water. Measurements were made during two growth cycles, corresponding to summer 2001 (I) and spring 2002 (II), of LDMC (leaf dry mass to saturated fresh mass ratio in mg/g) and SLA (ratio of surface area / dry weight of the blade in m^2/kg), following the protocol of Garnier *et al.* (2001). Four tillers were sampled per replicate in 2001 and five per replicate in 2002. The LLS (expressed in degree-days; 0°C basis) was determined from the beginning of each growth period. Blades were sampled for chemical composition and OMD. The sampling was done on the youngest fully expanded leaf in summer 2001, and on all the green blades in spring 2002. The 2001 samples of blades were analysed for fibre, cellulose, hemicellulose and lignin, with an enzymatic method used to estimate *in vitro* OMD. In 2002, chemical components of all green leaves and their OMD were estimated by NIRS (NIRS-OMD), calibrated and validated for a wide range of grasses and legumes.

Results Species ranked for LDMC and LLS in the same order (P• 0.05) as for fibre, hemicellulose and OMD of leaf blades, whereas ranking by SLA agrees with fibre, cellulose, lignin and OMD (Table 1). These correlations were significant, even though the data were obtained in different years, on different organs (youngest adult blades in 2001 and all the green blades in 2002) and by different analytical methods.

Table 1 Spearman's coefficients of correlation between leaf plant traits (average of two measurement periods for SLA, LDMC and LLS) against leaf composition (fibre content and its components) and OMD

	Design-1						Design-2			
	In vitro OMD	Fibre	Cellulose	Hemicell ulose	Lignin	NIRS-OMD	Fibre	Cellulose	Hemicell ulose	Lignin
SLA	0.46**	-0.32*	-0.64***	-0.13 n.s	-0.50***	0.65***	-0.58***	-0.55***	-0.41**	-0.60***
LDMC	-0.48***	0.62***	0.27 (□)	0.65***	0.28 (□)	-0.62***	0.60***	0.13 n.s	0.69***	0.60***
LLS	-0.61***	0.42**	0.28 (□)	0.35*	0.54***	-0.72***	0.67***	0.40**	0.62***	0.67***

(□), P• 0.10; *, P• 0.05; **, P• 0.01; ***, P• 0.001; NS, not significant. Design-1: in summer 2001, chemical and enzymatic methods were conducted on youngest adult blades for determine the fibre content, its components and IVOMD of blades, Design-2: in spring 2002, NIRS method was conducted on all the green blades of tillers

Conclusions We conclude that leaf plant traits are good indicators for ranking the grasses studied according to their feeding value. Furthermore, LDMC seems to be the most suitable trait to assess the specific feeding value because it ranks the species at least as well as other leaf traits and it is the easiest to measure.

Reference

Garnier, E., B. Shipley, C. Roumet & G. Laurent (2001). A standardized protocol for the determination of specific leaf area and leaf dry matter content. *Functional Ecology*, 15, 688-695.

Differences of energy density from plant species found in permanent grassland using the Cellulase Method in comparison to the Crude Nutrient Method

R. Bockholt, K. Friedel and F. Buske
The Faculty for Agricultural and Environmental Science at the Rostock University, Federal Republic of Germany, 18051 Rostock, Email: Renate.bockholt@uni-rostock.de

Keywords: forage value test, energy density, Cellulase Method, grasses, herbs, permanent grassland

Introduction If no special instructions have been given, the metabolic energy (MJ/kg DM) and the energy density (MJ NEL/kg DM) of grass samples are calculated by the agricultural test institute using a standard valuation formula as regards crude nutrients, crude fiber content, crude protein contents and crude ash. On the other hand, it is well known from feeding tests that permanent grassland grasses and herbs, which may be dominant under semi-intensive or extensive management, can have low digestibility and low energy density.

Material and methods A calculation of the difference has been made using 1500 data sets of 43 plant species from the peat soil grassland (1st - 11th week after 1st May). These data sets were tested by the laboratory of the institute using the Weender Forage Analyses for crude fiber content, crude protein content and crude ash content, in addition were tested using the Cellulase Method (Friedel, 1990) for digestibility. The energy density were calculated using 2 different methods and afterwards were calculated the differences of all the plants, samples and dates.

Results The energy density varies if all results between 8 and 2 MJ lactation per kg dry matter are taken into consideration. On average, the value valid for 43 plants will be below the limit of 6 MJ NEL, which applies to high-yield cows, between the 4th and 5th week after 1st May. It is noteworthy that both the positive and negative limits have been detected on herbs. The autochthonous plant species, on average, get a better assessment by 0,55 MJ NEL/ kg DM than they should have got when being tested with the unified crude nutrient formula, which applies for high-quality grassland grasses. The differences vary from species to species and increase with all species during the vegetation period. In some extreme cases individual plants investigated by means of the crude nutrient formula, have been given assessments that proved out to be by far too good by 2 MJ NEL.

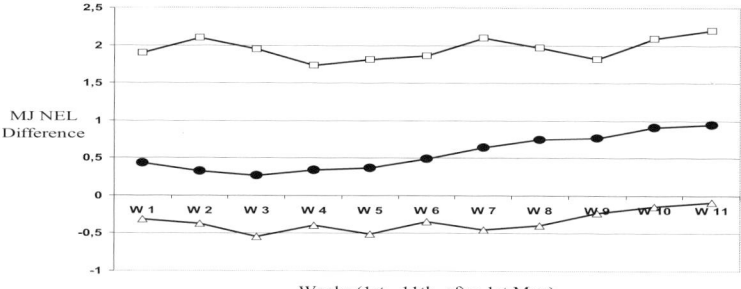

Figure 1 Differences of energy density in dependence on plant species and time in comparison of 2 methods, average (43 plant species), maximum = *Filipendula ulmaria*, minimum = *Glyceria maxima*)

Conclusions The crude nutrient method may lead to erroneous calculations of cow rations. By the additional inclusion of a digestibility examination, which can be done with the enzyme Cellulase or directly with the rumen fluid (Hohenheim forage value test), the special energetic forage value of autochthonous plants found in extensively managed permanent grassland can be investigated more precisely. Additional examinations of plant samples taken at the best cut-off date at the end of May demonstrated that the Hohenheim forage value test and the Cellulase Method coincided well provided that there was no infestation with fungi or putrefaction on dead leaves. However, it is known from examinations on forage that, in the opposite case when foliar fungi and putrefactive agents are found on necrotic cells, there are even considerable differences between the results obtained by the enzymatic Cellulase Method or the Hohenheim forage value test. Whilst the enzyme Cellulase is not capable of reacting to fungal toxins, the rumen fluid with its living micro organisms is sensitive to quality losses due to fungal toxins or substances produced by putrefaction.

References
Friedel, K. (1990). Die Schätzung des energetischen Futterwertes von Grobfutter mit Hilfe einer Cellulase - Methode. Wiss. Z. Uni Rostock, Nat. Reihe, 78 bis 86

Predicting intake from indigestible fibre

K.J. Moore and J.R. Russell

Iowa State University, Ames, Iowa, 50011, USA, Email: kjmoore@iastate.edu

Keywords: intake, fibre

Introduction Dry matter intake (DMI) of forages is often estimated as a reciprocal function of fibre concentration: DMI = fibre intake capacity / dietary fibre concentration (Mertens, 1987). This theoretical relationship is based on the concept that consumption of forage diets is limited by fill and that fibre represents the bulk of forage diets. This model, however, does not account for differences in DMI which should occur among forages with similar fibre concentrations but differing fibre digestibility. To account for these differences, we proposed an intake model where DMI is a reciprocal function of indigestible fibre concentration: DMI = c / C_I, where c = intake capacity for indigestible fibre and C_I is the concentration of indigestible fibre. This model assumes that livestock consuming forage diets of similar physical form but differing digestibility will consume a constant level of indigestible fibre. It applies only when DMI intake is regulated by fill and is not affected by digestible protein or energy concentrations. Since true DM digestibility is the inverse of indigestible fibre concentration, true digestible DM intake can be estimated as DDMI = c(1 - C_I)/C_I.

Materials and methods A feeding trial was conducted to determine the fill capacity of yearling heifers for C_I and to test the validity of the proposed DMI model. Mature and immature alfalfa, smooth bromegrass, and big bluestem were harvested, field dried, and baled. Immature and mature hays of each species were blended in ratios of 1:0, 2:1, 1:2, and 0:1 to produce diets with varying levels of digestible fibre. Each diet was individually fed *ad libitum* to four different heifers over four feeding periods and feed intake was recorded. Faecal output was determined using chromic sesquioxide as a marker. Diet and faecal samples were collected and analysed to determine fibre concentration using the neutral detergent method (Van Soest *et al.*, 1991).

Results Intake of fibre across the twelve diets ranged from 0.86 to 1.40% body weight (BW) with a mean of 1.11 and standard deviation of 0.163 (Table 1). Fibre intake was affected by species and maturity. Conversely, intake of indigestible fibre was relatively constant across diets with a mean intake of 0.48% BW and standard deviation of only 0.031. Therefore, based on the standard deviation, intake of total fibre was five times more variable than that for indigestible fibre. These data suggest that indigestible fibre intake is more constant across forages and should be a better indicator of fill than total fibre intake. Therefore, a fill model based on indigestible fibre should be more reliable and robust for predicting DMI than one based on total fibre. Intakes of DM and DDM were predicted using 0.48% BW as a fill constant (Fig. 1). The relationships between actual and predicted values were linear (R^2 = 0.81 and 0.89). In both cases, the slope was not significantly different from 1 and the intercept was different from 0. Therefore, both intake models resulted in reasonable intake estimates when using known undigested fibre concentrations for the calculations.

Table 1 Intake (% BW) of dry matter (DM), fiber (NDF), and undigested fiber (uNDF) of three forages fed at varying ratios of mature (M) to immature (I) hay

Forage	M:I	DM	NDF	uNDF
Alfalfa	1:00	2.22	1.01	0.50
	2:01	2.23	1.01	0.46
	1:02	2.40	1.08	0.45
	0:01	2.54	1.16	0.49
Bromegrass	1:00	1.66	0.86	0.45
	2:01	1.95	1.00	0.54
	1:02	2.16	1.09	0.46
	0:01	2.06	0.99	0.42
Bluestem	1:00	1.85	1.08	0.49
	2:01	2.10	1.25	0.5
	1:02	2.31	1.38	0.48
	0:01	2.37	1.40	0.49
Mean		2.15	1.11	0.48
SD		0.24	0.16	0.03

Conclusions The use of indigestible fibre for predicting potential intake of forages appears to be feasible. Several issues must be resolved in order to make these models useful for predicting intake of forages. First, a reliable *in vitro* assay for estimating *in vivo* indigestible NDF must be developed. Second, fill constants must be determined for different classes of livestock. Finally, because these models estimate potential intake, other factors that attenuate intake must be considered.

References

Mertens, D.R. (1987). Predicting intake and digestibility using mathematical models of ruminal function. *Journal of Animal Science,* 64, 1548-1558.

Van Soest, P.J., J.B. Robertson & B.A. Lewis (1991). Methods for dietary fiber, neutral detergent fiber, and nonstarch polysaccharides in relation to animal nutrition. *Journal of Dairy Science,* 74, 3583-3597.

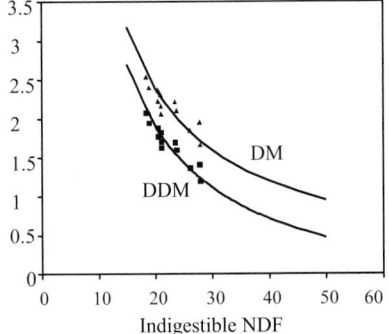

Figure 1 Actual and predicted intake(%BW) of dry matter (DM) and digestible dry mater (DDM) based on c = 0.48

Effect of cutting date on quality of red clover forage

S. Vasiljevic[1], S. Katic[1], V. Mihailovic[1], B. Cupina[2], D. Milic[1], A. Mikic[1], Dj. Karagic[1] and I. Pataki[1]
[1]Institute of Field and Vegetable Crops, Maksima Gorkog 30, 21000 Novi Sad, Serbia and Montenegro, Email: sanjava@ifvcns.ns.ac.yu, [2]Faculty of Agriculture, Trg Dositeja Obradovica 8, Novi Sad, Serbia and Montenegro

Keywords: red clover, *Trifolium pratense* L., cutting date, phenological phase, forage quality

Introduction Development stage or plant age is an important factor determining the chemical composition and quality of red clover forage (Ignjatovic *et al.*, 2001). In early spring, young red clover plants have large leaf mass, high contents of moisture, protein and minerals and a low fibre content. In the course of the growing season, under the effects of long days and high temperatures, the plant undergoes morphological changes: leaves grow more slowly, the stem elongates, dry matter yield increases and quality drops, especially digestibility and the contents of protein and minerals.

Materials and methods A trial with five red clover varieties (Junior, Diana, Milvus, K-17, Kolubara) was established in 1998 in the experiment field of the Institute of Field and Vegetable Crops. The size of the basic experimental unit was 5 m^2, and the seeding rate was 15 kg/ha. In the plant's second year samples were taken at three phenological phases (budding, beginning of flowering and full flowering) in each of three growth cycles (cuts) to study effects on quality as assessed by proximate analysis.

Results Table 1 shows that the red clover genotypes did not differ significantly from one another in the measured variables. Significant and expected differences were found between individual phenological phases (Table 1) and between cuts (Table 2). In each period, from budding to the beginning of flowering and from the beginning of flowering to full flowering, crude protein content fell by 2.2%. According to Džamic *et al.* (1970), who studied lucerne, the crude protein content is highest at the time of the second cutting (Table 2). In this study, the crude fibre content increased considerably, by about 5%, from budding to the beginning of flowering, which was in agreement with Popov (1971) (cited by Vuckovic, 1999). The ash content tended to go down with age.

Table 1 Chemical composition of red clover cut at different development stages

Variety	Phenological phase											
	Budding				Beginning of flowering				Full flowering			
	CP	CF	EE	Ash	CP	CF	EE	Ash	CP	CF	EE	Ash
Junior	21.1	21.3	1.9	9.3	18.4	25.8	1.5	9.1	16.4	27.6	1.9	8.0
Diana	20.8	21.7	1.8	9.3	19.0	25.9	1.6	8.7	16.7	27.6	1.7	7.7
Milvus	21.0	21.3	1.7	7.9	18.0	26.8	1.8	8.2	16.3	28.1	1.6	7.8
K-17	21.1	21.1	1.7	9.3	19.7	28.0	1.7	9.9	16.4	29.1	1.6	8.1
Kolubara	20.6	20.5	2.1	9.0	18.5	25.3	1.8	8.5	16.4	27.2	2.0	7.2
Average	20.9	21.2	1.9	8.9	18.7	26.3	1.7	8.9	16.4	27.9	1.8	7.8
LSD 0.05	1.5	2.4	0.7	0.6	2.5	3.6	0.7	1.4	1.7	3.0	0.5	1.7
0.01	2.1	3.5	1.0	0.9	3.6	5.3	1.0	2.1	2.5	4.4	0.7	2.5

Table 2 Contents of crude protein and crude fibre in red clover as affected by cut and by development stage

Development stage	Crude protein (%)			LSD		Crude fibre (%)			LSD	
	Cut 1	Cut 2	Cut 3	0.05	0.01	Cut 1	Cut 2	Cut 3.	0.05	0.01
Budding	20.1	21.5	21.2	1.1	1.6	20.8	20.9	21.9	1.9	2.7
Begin flowering	17.9	18.8	19.4	1.9	2.8	24.8	26.9	27.3	2.8	4.1
Full flowering	16.2	17.7	15.4	1.3	1.9	23.1	26.1	34.6	2.3	3.4

Conclusions These results demonstrated that the main quality parameters of red clover (crude protein and crude fibre) depended more on the phenological phase and the time of cutting than on genotype.

References
Džamic M., T. Stojisavljevic, I. Delic. & R. Curcic (1970). Prilog proucavanju kolicine i bioloske vrednosti aminokiselina i proteina u pojedinim otkosima lucerke. *Arhiv za poljoprivredne nauke*, 80, 3-16.
Ignjatovic S., J. Vucetic, Z. Lugic & B.Dinic (2001). Effect of growth stage on macro and trace elements content in red and white clover. *Journal of Scientific Agricultural Research*, 62, 309-316.
Vuckovic S. (1999). Forage Crops. Agricultural Research Institute SERBIA, Belgrade, p. 197.

Changes in crude protein content with advancing maturity in lucerne

S. Katic, D. Milic, V. Mihailovic, A. Mikic and S. Vasiljevic
Institute of Field and Vegetable Crops M. Gorkog 30, 21000 Novi Sad, Serbia and Montenegro, Email: katis@ifvcns.ns.ac.yu

Keywords: lucerne, crude protein content, forage quality

Introduction The main determinants of the quality of lucerne forage are digestibility and protein content (Julier *et al.*, 2001) as well as crude fibre content. In the early vegetative phases, the crude protein content of the leaves and stems is the highest and crude fibre content the lowest (Katic *et al.*, 2003). The aim of this study was to determine the rate of change in crude protein levels at different stages of growth and development.

Materials and methods Five lucerne genotypes in five phenological stages were studied in the second and third years (1997-1998) of the crop. Samples for analysis were taken at the five phenological phases according to Kalu & Fick (1981):

0. - Early vegetative: Stems • •15 long; no buds, flowers or seed pods.
1. - Mid-vegetative: Stems 16 - 30 cm long; no buds, flowers or seed pods.
2. - Late vegetative: Stems > 30 cm long; no buds, flowers or seed pods.
3. - Early bud: one or two nodes with visible buds, no flowers or seed pods.
4. - Late bud: • •three nodes with buds, no flowers or seed pods.

The crude protein (Kjeldahl) content in the five genotypes was determined in the five phenological phases during the second growth cycle in 1998.

Results The crude protein contents of the leaves and stems decreased from the early vegetative phase until budding by 3.81 g/kg per day in the leaves and 5.55 g/kg per day) in the stems. The crude protein content of the leaves decreased the most in the mid-vegetative phase and at early budding. In the stem, the levels of crude protein decreased steadily from the mid-vegetative stage until full budding (Table 1).

Table 1 Daily decrease g/kg per day of crude protein during different phenological stages in lucerne

Decrease in CP content (g/kg per day)	to Mid -veg. (1)	to Late veg. (2)	to Early bud (3)	to Late bud (4)	Mean
Leaf	6.02	2.42	4.86	1.94	3.81
Stem	8.8	5.7	4.98	2.74	5.55

At full budding, the crude protein content of the leaves decreased by 1.94 g/kg per day), while that of the stem dropped by 2.74 g/kg per day). Anderson *et al.* (1973) reported a daily decrease of crude protein content of 2.0 g/kg per day) in spring growth.

Conclusions The daily decline of protein content in the leaves was most rapid in the mid-vegetative phase. Crude protein in the stem decreased steadily from the mid-vegetative phase until full budding.

References

Anderson, M.J., G. F. Fries, D.V. Koplmand, & D.R. Waldo (1973). Effect of cutting date on digestibility and intake of irrigated first - crop alfalfa hay. *Agronomy Journal*, 65, 357-360.

Juiler, B., F. Guines, C. Ecalle & C. Huyghe (2001). From description to explanation of variations in alfalfa digestibility. Options méditerranéennes, Quality in lucerne and medies for animal production. *Proceedings of the XIV Eucarpia Medicago sp. Group Meeting*, Zaragoza, 45, 19-23.

Kalu, B.A & G.W. Fick (1981). Quantifying morphological development of alfalfa for studies of herbage quality. *Crop* Science, 21, 267-271.

Katic, S., V. Mihailovic, Dj Karagic, D. Milic & I. Pataki (2003). Yield, morphology and chemical composition of five lucerne genotypes as affected by growth stage and the enviroment, *Grassland Science in Europe*, 8, 376-379.

Proanthocyanidins from *Hedysarum, Lotus* and *Onobrychis* spp. growing in Sardinia and Sicily and their antioxidant activity

A. Tava[1], M.G. De Benedetto[1], D. Tedesco[2], G. Di Miceli[3] and G. Piluzza[4]

[1]*Istituto Sperimentale per le Colture Foraggere, v.le Piacenza 29, 26900 Lodi, Italy, Email: aldotava@katamail.com, [2]Dipartimento di Scienze e Tecnologie Veterinarie per la Sicurezza Alimentare, Università di Milano, v. Celoria 10, 20133 Milano, Italy, [3]Dipartimento di Agronomia Ambientale e Territoriale, Università di Palermo, v.le delle Scienze, 90128 Palermo, Italy, [4]CNR, Istituto per il Sistema Produzione Animale in Ambiente Mediterraneo Sez. Sassari, v. E. de Nicola, 07100 Sassari, Italy*

Keywords: proanthocyanidins, condensed tannins, quantification, antioxidant activity, quality

Introduction Proanthocyanidins (PA), or condensed tannins, are a class of natural polyphenolic compounds, occurring in numerous plant species, including a number of economically significant forage legumes. These compounds are polymers of flavan-3-ols, and typically contain from 2 to 20 units. Their biological significance is still being debated and, in recent years, a great deal of attention has been focused on their role in ruminant nutrition. Evidence has indicated that PA, in a moderate concentration (0.5-5% DM), may have considerable importance in protecting dietary proteins against microbial degradation in the rumen, and in preventing bloat. The antioxidant activity (AA) is also an important feature for animal well-being (Barry & McNabb, 1999). In order to study the PA content related to the antioxidant activity, samples of *Hedysarum, Lotus* and *Onobrychis* spp. from Mediterranean environments have been considered and investigated.

Materials and methods Forage plants analysed in this study were collected in hilly semi-arid areas of Sardinia and Sicily. The PA were extracted from the lyophilised plant samples with 30% aqueous acetone, and quantified by a spectrophotometric method following the butanol-HCl-Fe^{3+} assay (Porter *et al.,*1986). Delphinidin was used as standard, absorbance was read at 550 nm. After hydrolyses, a bidimensional TLC method was used to evaluate the monomers. The AA was performed by both FRAP (Ferric Reducing Antioxidant Power) and TEAC (Trolox Equivalent Antioxidant Capacity), as reported by Luximon-Ramma *et al.* (2002).

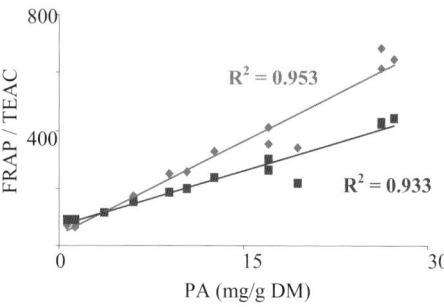

Figure 1 Correlation between PA and • •FRAP (• mol Fe(II) g^{-1} DM) and • •TEAC (• mol Trolox g^{-1} DM)

Results The species under investigation, both from Sardinia and Sicily, showed a wide range of PA content. In *Lotus* spp. (*L.cytisoides, L. corniculatus, L. edulis, L.ornithopodioides*) PA content was lower (from 0.6 to 6.0 mg g^{-1} DM), compared to the different genotypes of *H. coronarium* (PA quantified from 8.9 and 17.0) and *H. spinosissimum* (19.4). Higher values were detected in *H. glomeratum* (26.2) and *O. viciifolia* (26.2-27.2). From the TLC investigation delphinidin and cyanidin were found as the major detected monomers, together with other compounds not further investigated. The AA values were 65-174 FRAP and 87-156 TEAC for *Lotus* spp., 252-406 FRAP and 183-262 TEAC for *H. coronarium* and *H. spinosissimum*, and 649-686 FRAP and 422-443 TEAC for *H. glomeratum* and *O. viciifolia*. The correlation between PA content and AA is shown in Figure 1.

Conclusions These results demonstrate that the PA content in the plant material under study was usually within the range considered beneficial in ruminant diet. The AA of the extracts, expressed as FRAP and TEAC, was strongly correlated with the PA content.

Acknowledgements The research was financed by MiPAF project ANFIT

References

Barry, T.N. & W.C. McNabb (1999). The implication of condensed tannins on nutritive value of temperate forage fed to ruminants. *British Journal of Nutrition*, 81, 263-273.

Luximon-Ramma, A., T. Bahorun, M.A. Soobratee & O.I. Aruoma (2002). Antioxidant activity of phenolic, proanthocyanidin and flavonoid components in extracts of *Cassia ferula. Journal of Agricultural and Food Chemistry*, 50, 5042-5047.

Porter, L.J., L.N. Hrstich & B.G. Chan (1986). The conversion of procyanidins and prodelphinidins to cyanidin and delphinidin. *Phytochemistry*, 25, 223-230.

Dryland clovers: a phytochemical resource

L.P. Meagher[1], G.A. Lane[1], W. Rumball[1], M.A. Tavendale[1] and R.J. Lucas[2]
[1]AgResearch Ltd., Grasslands Research Centre, Palmerston North, New Zealand, Email: lucy.meagher@agresearch.co.nz, [2]Agriculture and Life Sciences Division, Lincoln University, Canterbury, New Zealand

Keywords: *Trifolium*, secondary metabolites, floral proanthocyanidins, flavanoids, mass spectrometry

Introduction Recent developments in the utilisation of phytoestrogens of red clover (Wuttke *et al.*, 2002) have encouraged us to investigate a wider range of *Trifolium* species for metabolites which could provide new product opportunities. The phytochemistry of the agronomically-important *Trifolium* species white (*Trifolium repens)* and red (*T. pratense*) clovers has been investigated in some detail (Foo *et al.*, 2000; Sivakumaran *et al.*, 2004). However numerous other clover species have been neglected in agriculture due to agronomic issues or the fact they are annuals and require more intensive management. While some of these clover species have been studied for their genetic diversity (Marshall *et al.*, 2002), investigations of the chemical composition of these specific species has not been reported.

Materials and methods Alsike (*T. hybridium*), ball (*T. nigrescens*), balansa (*T. michelianum*), Caucasian (*T. ambiguum*), strawberry (*T. fragiferum*) and subterranean (*T. subterraneum*) clover plants were grown at Grasslands Research Centre or Lincoln University. Aqueous acetone (3:7 v/v) extraction of leaves and flowers was performed on frozen plant material. The Sephadex LH-20 purified proanthocyanidin polymers of flowers were studied by acid catalyzed degradation with benzyl mercaptan and mass spectrometry. Flavanoid metabolites were identified from purified LH-20 fractions by analysis with liquid chromatography (LC)-ion trap mass spectrometry (MS).

Results and conclusions Analysis of proanthocyanidin polymers by thiolytic cleavage provides evidence of the identity of the terminal and extender units that make up the proanthocyanidin polymer, information on the mean degree of polymerisation (mDP), and the procyanidin to prodelphinidin unit ratio. Negative ion electrospray ionisation (ESI) and positive ion matrix-assisted laser desorption/ionisation time of flight (MALDI-TOF) mass spectrometry provides evidence of the compositional dispersion (range of DP) for oligomeric proanthocyanidin components. Methanolic fractions from LH-20 chromatography yielded a range of flavanoids and flavanoid glycosides, with some apparent contrasts between the clover species studied as determined by LC-MS.

The white clover floral prodelphinidins (Foo *et al.*, 2000) consist of terminal and extender units with nearly equal proportions of epigallocatechin and gallocatechin. The red clover floral procyanidins (Sivakumaran *et al.*, 2004) consist of extension units with epicatechin as the abundant flavan-3-ol and the terminating units dominated by catechin. The dramatic difference in the stereochemistry of the terminal and extender units observed for the red clover floral procyanidins contrasts with the mixture of *cis* and *trans* stereochemistry observed for white clover floral prodelphinidins. Other floral clover proanthocyanidin polymers were determined to be composed of either prodelphinidins or procyanidins exclusively, or possibly a mixture. The results of this study will present the biosynthetic trends for floral proanthocyandins from a range of clover species.

The LC-MS results indicate there is diversity amongst the quercetin, kaempferol and isorhamnetin derivatives extracted from the leaves of these dryland clovers. The targeted identification of clover metabolites such as the red clover isoflavones, could lead to future product opportunities.

References

Foo, L.Y., Y. Lu, A.L. Molan, D.R. Woodfield & W.C. McNabb (2000). The phenols and prodelphinidins of white clover flowers. *Phytochemistry*, *54*, 539-548.

Marshall, A.H., T.A. Williams, H.G. Powell, M.T. Abberton & T.P.T. Michaelson-Yeates (2002) Forage yield and persistency of *Trifolium repens* X *Trifolium nigrescens* hybrids when sown with a perennial ryegrass companion. *Grass and Forage Science*, 57, 232-238.

Sivakumaran, S., L.P. Meagher, L.Y. Foo, G.A. Lane, K. Fraser & W. Rumball (2004). Floral Procyanidins of the Forage Legume Red Clover (*Trifolium pratense* L.). *Journal of Agriculture and Food Chemistry*, 52, 1581 -1585.

Wuttke, W., H. Jarry, S. Westphalen, V. Christoffel & D. Seidlova-Wuttke (2002). Phytoestrogens for hormone replacement therapy? *Journal of Steroid Biochemistry and Molecular Biology*, 83.

Water-soluble carbohydrate (WSC) concentrations in Ireland and Norway of *Lolium perenne* differing in WSC genotype and receiving varying rates of N fertiliser

P. O'Kiely[1], L. Nesheim[2], P. Conaghan[1,3] and F.P. O'Mara[3]
[1]Teagasc, Grange Research Centre, Dunsany, Co.Meath, Ireland, Email: pokiely@grange.teagasc.ie, [2]The Norwegian Crop Research Institute, Kvithamar Research Centre, N-7500 Stjordal, Norway, [3]Dept. of Animal Science and Production, University College Dublin, Belfield, Dublin 4, Ireland

Keywords: water-soluble carbohydrate, grass, genotype, N fertiliser

Introduction Cultivars bred for elevated water-soluble carbohydrate (WSC) concentration may have improved grass ensilability and nutritive value. Increasing rates of application of N fertiliser generally reduce grass WSC concentration, although it is unknown if the response is similar for normal and elevated WSC genotypes or if these factors interact with growing conditions. This experiment evaluated the effects on grass WSC concentration of varying N fertiliser application rates to perennial ryegrass cultivars of high or normal WSC genotype grown in Ireland and Norway.

Materials and methods The experiment was conducted in Ireland (Grange) and Norway (Saerheim) in two years (Y). It was a split-plot randomised complete block design containing four replicates; replicates consisted of three or four main plots providing for successive harvests (early June (H1), early Aug. (H2) and late Sept.(H3) in Norway and late May (H1), early July (H2), mid-Aug. (H3) and early Oct. (H4) in Ireland). Within main plots, 2 cultivars x 4 (Norway) or 5 (Ireland) rates of inorganic N fertiliser (N_r) were fully randomised. The two diploid intermediate heading perennial ryegrass cultivars (Aberdart: selected for elevated WSC concentration and Fennema: control) were sown as monoculture plots. The rates of N fertiliser (calcium ammonium nitrate (CAN); 275 g N/kg) were equivalent to 0 (N_0), 40 (N_{40}), 80 (N_{80}), 120 (N_{120}) and 160 (N_{160} - Ireland only) kg N/ha. The remaining sub-plots received an application of CAN equivalent to 80 kg N/ha. All plots were harvested at each harvest period but only herbage from the main plots that recently received N_r were sampled.

Results Grass WSC concentrations are given in Table 1. Aberdart > Fennema in Ireland (169 vs. 158 g/kg DM) and Norway (263 vs. 235 g/kg DM) and Year 1 > Year 2. In Ireland, H1 had the highest values while in Norway values were highest for H1 and H2. Values decreased progressively from N_0 to N_{120} in Ireland and N_{40} to N_{120} in Norway. The differences between cultivars in response to N_r were maintained across sites, harvests and years - there were not significant CxN_r, $HxCxN_r$, CxN_rxY (except Y1 N_{80} in Norway) or $HxCxN_rxY$ interactions.

Table 1 Grass WSC concentration (g/kg DM) in Aberdart and Fennema swards at each harvest (H).

Cultivar (C)	Aberdart					Fennema					Statistical summary		
N rate (N_r)	N_0	N_{40}	N_{80}	N_{120}	N_{160}	N_0	N_{40}	N_{80}	N_{120}	N_{160}		*Sig.*	s.e.
Ireland - Y1													
H1	242	214	191	164	164	242	207	179	164	153	H	***	3.2
H2	240	201	171	147	128	233	168	174	136	132	N_r	***	3.0
H3	217	185	149	138	127	185	164	141	122	122	C	***	1.9
H4	202	162	148	137	135	181	154	134	116	123	Y	**	2.0
Ireland - Y2											HxN_r	***	6.2
H1	275	247	209	188	169	238	235	196	175	156	HxC	ns	4.2
H2	173	168	132	106	106	185	116	111	110	92	CxN_r	ns	4.2
H3	175	161	155	149	157	153	156	162	143	143	$HxCxN_r$	ns	8.7
H4	180	145	140	132	137	160	141	134	132	136			
Norway - Y1													
H1	350	364	308	241	-	319	331	300	221	-	H	***	3.8
H2	311	302	248	226	-	288	267	233	197	-	N_r	***	2.8
H3	291	295	230	206	-	246	266	223	204	-	C	***	2.0
											Y	***	2.0
Norway - Y2											HxN_r	***	5.6
H1	319	269	216	215	-	313	239	174	182	-	HxC	ns	4.5
H2	304	311	284	267	-	274	271	221	206	-	CxN_r	ns	3.9
H3	197	209	193	166	-	170	194	161	153	-	$HxCxN_r$	ns	7.4

Conclusions Grass WSC was higher and the difference between cultivars larger in Norway than Ireland. The negative effects of N fertiliser on WSC were similar for the two grasses across a range of conditions.

Characterisation of herbage from temperate organic pastures

E. Kuusela

University of Joensuu, Department of Biology, PO Box 111, FIN–80101 Joensuu, Finland, Email: eeva.kuusela@joensuu.fi

Keywords: organic dairy farming, pastures, herbage nutritive value

Introduction Grazing is an essential part of organic dairy farming systems. Although the nutritive value of herbage and herbage availability determine the intake and nutrient supply for grazing cows, the composition of typical herbage from organic pastures has been unclear.

Materials and methods Pre-grazing herbage mass and herbage nutritive value was assessed through entire grazing seasons (1996-1999) in six experiments at the Siikasalmi research farm (62°30'N, 29°30'E). Perennial sward mixtures contained white clover (*Trifolium repens*), alsike clover (*Trifolium hybridum*) red clover (*Trifolium pratense*) or birdsfoot trefoil (*Lotus corniculatus*) and complementary grasses were meadow fescue (*Festuca pratensis*), timothy (*Phleum pratense)* and smooth meadow-grass (*Poa pratensis*). Annual sward mixtures contained legumes (*Vicia villosa, Vicia sativa, Trifolium resupinatum* or *Trifolium repens*), barley (*Hordeum vulgare*) and Italian ryegrass *(Lolium multiflorum)*. No fertilisers were applied, except composted manure in Experiment 5. The average length of a rotation cycle was 21 days. After each grazing period the areas were topped with a mower to a height of 10 cm.

Results Organic pastures resulted in a moderate pre-grazing herbage mass with a good nutritive value (Table 1).

Table 1 Mean values of pre-grazing herbage mass and herbage nutritive value in six grazed experiments

Experiment	1. Clover-grass mixture	2. Clover/ Bidsfoot mixture	3. Clover-grass mixture	4. Clover-grass mixture	5. Annual legume-cereal-grass m.	6. Annual legume-cereal-grass m.
	1996-1998	1998	1996-1997	1998	1999	1999
Pre-grazing herbage mass (kg DM/ha)	1830	1664	1265	1945	1723	1827
Chemical content of DM						
- Ash (g/kg DM)	96.5	93.6	90.3	90.5	118	116
- Crude protein (g/kg DM)	184	165	177	170	206	217
- Neutral detergent fibre (g /kg DM)	510	447	521	453	433	391
In vitro digestibility of organic matter	0.754	0.758	0.753	0.748	0.751	0.760
Mineral content of DM						
- Calcium (g /kg DM)	7.1	8.4	Not measured			
- Magnesium (g /kg DM)	2.2	2.5	Not measured			
- Phosphorous (g /kg DM)	4.3	3.3	Not measured			
- Potassium (g /kg DM)	36.2	29.0	Not measured			

Discussion Herbage digestibility depends primarily on the state of maturity (length of grazing cycle) and also on plant species and growing conditions (Buxton, 1996). Crude protein content depends on soil N-availability and biological N-fixation (legumes). Despite different botanical compositions, mean digestibility was of a similar high level in all experiments, while crude protein values varied more. In contrast, Finnish N-fertilised grass pastures usually have a slightly higher digestibility and crude protein content. It has been suggested that extended rest periods between grazing episodes is one means of increasing the herbage mass of N-deficient pasture, even though this may be at the expense of herbage nutritive value (Delagarde *et al.*, 1997). Hence, in organic farming systems, the optimum length of rotation cycle is often a compromise between digestibility and the amount of regrowth, as measured by pre-grazing herbage mass. In this study, the mean Ca and Mg contents of legume- containing herbage were higher, but the P and K contents were similar to Finnish conventional pastures.

References

Buxton, D.R. (1996). Quality-related characteristics of forage as influenced by plant environment and agronomic factors. *Animal Feeding Science and Technology*, 59, 37-49.

Delagarde, R., J.L. Peyraud & L.Delaby (1997). The effect of nitrogen fertilization level and protein supplementation on herbage intake, feeding behaviour and digestion in grazing dairy cows. *Animal Feed Science and Technology*, 66, 165-180.

Evaluation of superoxide anion radical scavenging activities of plantains and pastures by electron spin resonance (ESR)

Y. Tamura[1] and T. Masumizu[2]

[1]National Agricultural Research Centre for Tohoku Region, Shimokuriyagawa Akahira 4, Morioka, Iwate 020-0198, Japan, Email:fumi@affrc.go.jp, [2]JEOL Ltd., 3-1-2 Musashino, Akishima, Tokyo 142-8501, Japan

Keywords: plantains, pastures, antioxidative activity, superoxide anion radical, electron spin resonance (ESR)

Introduction Producing animals without using feed-grade antibiotic growth promoters and chemical medicines is essential. In response, many scientists are now studying medicinal plants and herbs to identify and quantify those plants that may have a beneficial effect on animal production. Plantains have been used in herbal medicines and are being evaluated as a potential pasture species because of their medicinal values in animal health. In this study, antioxidant activities of plantains were compared to those of common pasture species to clarify the effects of plantains on animal health and production.

Materials and methods Two plantain (*Plantago lanceolata* L., *Plantago asiatica* L.) and three pasture species (*Lolium perenne* L., *Phleum pratense* L. and *Trifolium repens* L.) were used. Superoxide anion radical (O_2^-) scavenging activities of methanol-soluble extracts from freeze-dried leaves were determined by the spin-trapping method using an electron spin resonance spectrometer (ESR, JES-FA200 JEOL Ltd.), according to Sekine *et al.* (1998). The assays were made with and without ascorbate oxidase to distinguish the scavenging activities of antioxidant compositions from foliar ascorbate (Figure 1).

Results The scavenging activities of the extracts from the plantains and the pastures for O_2^- are shown in Figure 2. The extracts from the plantains exhibited much higher levels of O_2^- scavenging activity than those of the pasture species. The scavenging activities of *T. repens* were the lowest and those of *L. perenne* and *P. pratense* were between those of the plantains and *T. repens*. The results with/without the ascorbate oxidase treatment indicated that over 50% of the O_2^- scavenging activity in *T. repens* could be attributed to ascorbate, and that *P. lanceolata* and *P. asiatica* were due to bioactive compounds These species are known to contain compounds with high antioxidative activity, such as acteoside and plantamajoside.

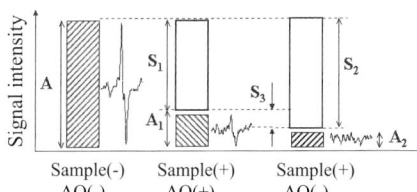

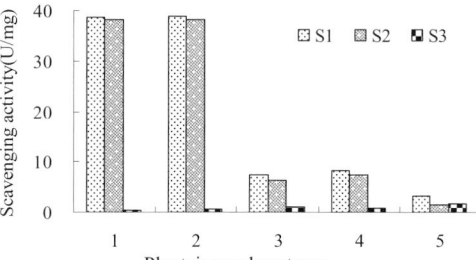

Figure 1 Estimation of scavenging activity
Note: sample(+) and (-) indicate with and without Samples. AO(+) and (-) indicate with and without ascorbate oxidase, respectively. A, A1 and A2; DMPO adduct signal intensity of control, with AO and without AO. S1, S2 and S3; Scavenging activity with AO, without AO and of ascorbate, respectively

Figure 2 Scavenging activities of the plantains and pastures.
Note: 1;*P. lanceolata* L, 2;*P. asiatica* L, 3; *L. perenne* L, 4;*P. pratense* L. and 5;*T repens* L.. S1, S2, S3; Scavenging activity without AO, with AO and of ascorbate, respectively

Conclusions *P. lanceolata* and *P. asiatica* exhibited significantly higher scavenging activity than the pasture species. The effects on animal health from *Plantago lanceolata* L. feeding, such as increased essential fatty acid in chickens (Yamamoto *et al.*, 2002) and decreased blood glucose in pigs (Fujii *et al.*, 2002), may therefore be related to the scavenging activity for O_2^-. These results indicate that effects of the plantains on animal health may be exerted through antioxidant composition and other bioactive compounds such as aucubin and catalpol. Further studies on the mechanisms underlying the effects of bioactive compounds on animals are needed.

References

Yamamoto, A., M. Nishiwaki & Y. Tamura (2002). [Effects of feeding *Plantago lanceolata* L. on the contents and compositions of fatty acid in broiler] (in Japanese). *Chikusann no Kennkyuu* 56, 685-687.

Takashi, S., M. Toshiki, M. Yoshie & N. Tsuji (1998). Evaluation of superoxide anion radical scavenging activity of shikonin by electron resonance. *International Journal of Phamaceutics*, 174, 133-139.

Fujii, Y., S. Kumagai & Y. Tamura (2002). [Meat quality of pig was increased by feeding *Plantago lanceolata* L.] (in Japanese), *Gendai nougyou* 2002, 5, 226-229.

Species and chlorine fertilisation affect dietary cation-anion difference of cool-season grasses

G. Bélanger[1], S. Pelletier[1,2], H. Brassard[1,2], G.F. Tremblay[1], P. Seguin[3], R. Drapeau[1], A. Brégard[2], R. Michaud[1] and G. Allard[2]

[1]Agriculture and Agri-Food Canada, 2560 Hochelaga Blvd., Sainte-Foy, QC G1V 2J3 Canada, Email: belangergf@agr.gc.ca, [2]Université Laval, Sainte-Foy, QC G1K 7P4 Canada, [3]Macdonald Campus, McGill University, Sainte-Anne-de-Bellevue, QC H9X 3V9 Canada

Keywords: mineral composition, timothy, milk fever, forage

Introduction The Dietary Cation-Anion Difference [DCAD = (Na + K) – (Cl + S); Ender *et al.*, 1971] is used in balancing rations for dry dairy cows. Low DCAD diets induce a mild, compensated metabolic acidosis that stimulates bone resorption, improves Ca homeostasis, and prevents milk fever. Dry cow rations contain a high proportion of forage and, therefore, forages fed two to four weeks prepartum should have a low or negative DCAD value. Our objectives were to evaluate the DCAD of five cool-season grass species grown in eastern Canada and to determine the effect of Cl fertilisation on the DCAD value of timothy (*Phleum pratense* L.).

Materials and methods In the first experiment, three cultivars of cocksfoot (*Dactylis glomerata* L.), two of tall fescue (*Festuca arundinacea* Schreb.), meadow bromegrass (*Bromus riparius* Rehm.), and smooth bromegrass (*Bromus inermis* Leyss.), and four of timothy were grown at three locations in Québec (Canada). Forage dry matter (DM) yield and mineral concentration (Na, K, Cl, and S) were measured in the spring growth and the summer regrowth of 2002; locations were considered a random effect. In the second experiment, we determined the effect of increasing rates of Cl fertilisation (0, 80, 160, and 240 kg Cl/ha) on timothy grown at four locations with contrasting soil K contents: two high (Ste-Anne-de-Bellevue, 289 kg K/ha; Normandin, 311 kg K/ha), one intermediate (St-Augustin, 199 kg K/ha), and one low (Ste-Perpétue, 123 kg K/ha). Forage DM yield and mineral concentration (Na, K, Cl, and S) were measured in the spring growth of 2003; locations were considered a fixed effect. All harvests were taken at the early-heading stage of development.

Results For both spring growth and summer regrowth, the DCAD of cocksfoot (CF) was the highest and that of timothy (T) the lowest; the DCAD of meadow bromegrass (MB), smooth bromegrass (SB), and tall fescue (TF) were intermediate (Figure 1). Cultivars within a species did not differ in DCAD. Chlorine fertilisation with CaCl$_2$ decreased timothy DCAD (Figure 2) at all four locations, and increased Cl concentration with no effect on DM yield. With no Cl fertilisation, the DCAD was lower (199 meq/kg DM) at the low soil K content location (Ste-Perpétue) than at the other three sites (365 - 459 meq/kg DM).

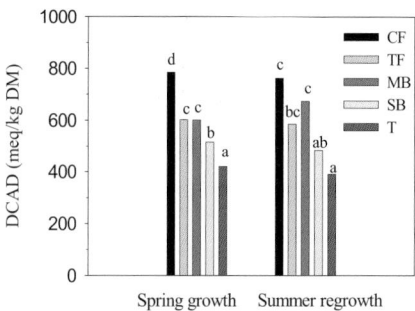

Figure 1 DCAD of five grass grass species in spring growth and summer regrowth (average of three locations)

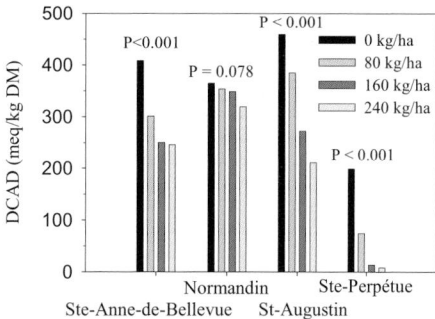

Figure 2 DCAD of timothy with increasing rates of Cl fertilisation at 4 locations. (Probabilities are for the linear effect of fertilisation)

Conclusions Timothy is the best cool-season grass for production of low DCAD forages for dry cows. Timothy with DCAD as low as 8 meq/kg DM can be produced using Cl fertilisation on a soil with a low K content.

Reference

Ender, F., I.W. Dishington & A. Helgebostad (1971). Calcium balance studies in dairy cows under experimental induction and prevention of hypocalcaemic *paresis puerperalis*. The solution of the aetiology and the prevention of milk fever by dietary means. *Zeitschrift f·r Tierphysiologie., Tierernährung und Futtermittelkunde*, 28, 233-256.

Forage quality of cool season pasture species under two rotational grazing height regimes

H.D. Karsten[1] and M. Carlassare[2]
[1]Crop & Soil Sciences Dept. 116 ASI Bldg, Pennsylvania State University,University Park, PA 16802 USA, Email: hdk3@psu.edu, [2]Dept. of Environmental Agronomy & Crop Sciences, Università degli Studi di Padova, Agripolis, Viale dell'Università 16, 35020 Legnaro Italia

Keywords: rotational-stocking grazing, forage quality, *Elytrigia repens* L., *Taraxacum officinale* Weber

Introduction To optimize animal and pasture performance in management intensive grazing systems, pasture production and quality often must be compromised. Rotationally-stocking pastures at slightly taller grazing heights can increase pasture productivity, but lower forage quality may limit animal performance. Our objective was to compare the forage quality of common cool season pasture species in the Northeastern U.S., under two rotational grazing regimes defined by slightly different grass heights.

Materials and methods A Pennsylvania mixed species pasture was divided into four blocks, and two rotational-stocking treatments ("tall" and "short") were randomly assigned to each block. Tall pastures were stocked with cattle when *Dactylis glomerata* L. extended height averaged 27 cm, cattle were removed when *Dactylis g.* residual height 7 cm; short pastures were stocked when *Dactylis g.* height was 20 cm and grazed to a residual height of 5 cm. Before each grazing event, we cut 12 (1998) or 18 (1999) forage samples from each of four treatment paddocks at the grazing regime residual height, and separated them into: *Dactylis glomerata* L., *Poa pratensis* L., *Elytrigia repens* L. and *Taraxacum officinale* Weber. Crude protein (CP), and neutral detergent fiber (NDF) were determined with near infrared reflectance spectroscopy analysis. Samples were grouped into seasons defined as: spring) grazed before 27 June; summer) grazed before September 22; and autumn) grazed after September 21. Data were analyzed using the MIXED model of SAS with grazing regime, seasons, species, and the interaction terms as main fixed effects, and year as random.

Results The species x season interaction was significant for CP and NDF. Crude protein increased from spring to autumn in all species, but with different trends (Fig.1a). In the three cool season grasses, NDF values tended to decrease from spring to autumn, while in *Taraxacum o.*, NDF tended to increase from spring to autumn (Fig 1b).

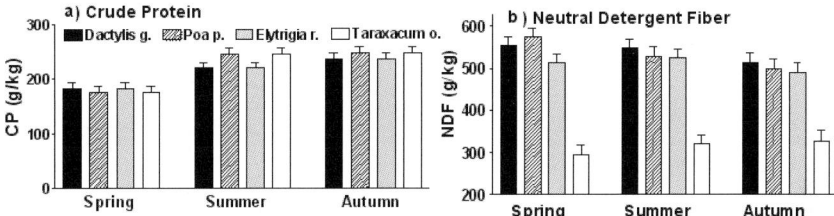

Figure 1 Crude protein (g/kg) and neutral detergent fiber (g/kg) in spring, summer, and autumn of *Dactylis glomerata*; solid bars, *Poa pratensis*: hatched bars, *Elytrigia repens*: grey bars, *Taraxacum officinale*: open bars

Grazing treatments significantly influenced species CP and NDF. On average, species grazed in short pastures had higher CP and lower NDF than species in tall pastures (243 vs. 218 g/kg, and 462 vs. 487 g/kg, respectively). Crude protein did not differ significantly among species. However, NDF of *Taraxacum o.* was significantly lower than the three cool season grasses (Table 1).

Table 1 Crude protein and neutral detergent fiber of the pasture species (** significantly different p < 0.05)

	Dactylis g.	s. e.	Poa. p.	s.e.	Elytrigia. r.	s.e.	Taraxacum o.	s.e.
				g/kg				
Crude Protein	213	11	223	11	243	11	243	11
NDF	539	19	534	19	510	19	315	**

Conclusions Forage quality of "weedy" *Elytrigia repens* was similar to *Dactylis glomerata.* and *Poa pratensis,* and sometimes better than *Poa pratensis*. Compared to the desirable cool season grasses, the NDF of *Taraxacum officinale was* significantly lower, and CP was similar. The slightly taller pasture height grazing regimes resulted in an average species CP reduction of 27g/kg and an average species NDF increase of 25g/kg. These differences may influence animal performance, but are within forage quality values reported for intensively grazed pastures in the Northeastern U.S.

Investigation into differences in palatability among Festulolium varieties as haylage

E. Touno[1], S. Kushibiki[2], H. Shingu[1] and A. Oshibe[1]
[1]*National Agricultural Research Center for Tohoku Region, 4 Akahira, Morioka, 020-0198 Japan, Email: etouno@affrc.go.jp,* [2]*National Institute of Livestock and Grassland Science, 2 Ikenodai, Tsukuba, 305-0901, Japan*

Keywords: Festulolium, silage, palatability differences, dairy cows

Introduction In Japan, paddy fields that are no longer used for cultivation of rice are being converted to cultivation of forage crops. Therefore, grass with greater wet resistance and higher quality is required. Festulolium is an interspecific hybrid between the *Lolium* and *Festuca* species and combines the characteristics of high-quality ryegrass and resistance to hostile environments from fescues (Thomas & Humphreys, 1991). Among festulolium varieties, there is wide variation in environmental resistance and feeding value. One festulolium variety, Paulita, shows superior wet resistance to Evergreen and the total digestible nutrients of Evergreen was similar to that of cocksfoot (cv. Kitamidori) (Touno *et al.*, 2004). In this study, we investigated palatability differences in festulolium varieties.

Materials and methods Two varieties of festulolium (Paulita (Pa) and Evergreen (Eg)) were studied. Timothy (*Phleum pratense* L. cv. Nosappu (Ty)) was also included as a reference species. These grasses were harvested at the heading stage and made into high dry matter silage (60%DM). Exp.1; Palatability of each silage was assessed by measuring dry matter intake (DMI) of mature castrated sheep for 1 hour after morning feeding with cafeteria feeding. Exp.2; Four lactating Holstein cows were assigned to a 3 x 3 Latin square design. The DMI of each silage was measured for 3 hours after morning feeding. Exp.3; A mature castrated bull fitted with ruminal cannulae was used for an *in situ* incubation trial. Nylon bags, each containing 5g dried sample milled through a 5mm mesh, were inserted in the rumen and removed at 3, 6, 12, 24, 48, 72, 96 h.

Results In Exp.1, the palatability of the grasses decreased in the order : Pa > Ty > Eg with Pa significantly superior (p < 0.05) to the other silages. The DMI of dairy cows increased linearly for 1 h after feeding with DMI for this period in the order : Pa > Ty >Eg, with Pa significantly higher (p < 0.05) than Eg. The DMI of silage after 3 h of feeding tended to decrease Pa > Ty > Eg, but there were no significant differences (Exp.2, Figure 1). The soluble fraction of Pa (15.52%) was twice that of Ty (6.65%) (p < 0.05), but there was no difference observed between those of Pa and Eg (13.41%). Rate of degradation of Pa (23.52%) was higher (p < 0.05) than that of the other silages (Ty = 18.56%, Eg = 18.25%) (Exp.3).

Conclusion The results of Exp.1 and Exp.2 indicated that the palatability of Pa silage was higher than that of the other silages. The higher degradability of Pa could have accounted for higher DMI after 3 h feeding but the significant difference between Pa and Eg in intake over 1 h could not be explained by only by the degradability of Pa. The olfactory and taste senses play an important role in the selection of feed (Morrison *et al.*, 1986). The odour of Pa silage was similar to caramel and we considered that this is probably one factor related to its high palatability.

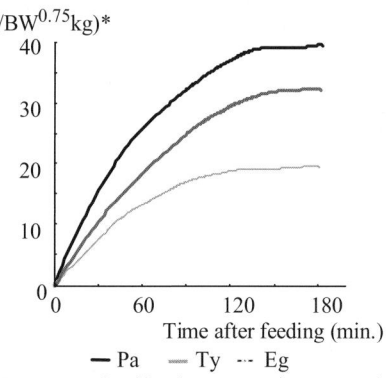

$(g/BW^{0.75}kg)^*$

Time after feeding (min.)

— Pa — Ty -- Eg

* Dry matter food intake measured every 5 min.[**]

Figure 1 Changes in food intake of silage for 3 hours after morning feeding for cows in midlactation

References
Morrison, W. H., R. J. Horvat & J. C. Burns (1986). GLC-MS analysis of the volatile constituents of *Panicum sp. Journal of Agricultural and Food Chemistry*, 34, 788-791.
Thomas, H. & M. O. Humphreys (1991). Progress and potential of interspecific hybrids of *Lolium* and *Festuca*. *Journal of Agricultural Science, Cambridge*, 117, 1-8.
Touno, E., T. Kondo & M. Murai (2004). [Feeding characteristics of Festulolium (cv. Evergreen) in the northeastern area of Japan] (in Japanese). *Grassland Science*, 50, 355-359.

The feeding value of silage made from peas grown alone or in mixture with cereals

A. Kirilov
Institute of Forage Crops, 5800 Pleven, Bulgaria, Email: kirilov_atanas@hotmail.com

Keywords: pea, pea-cereal mixture, silage, intake, digestibility

Introduction The interest in pea as a forage crop rich in protein does not decrease. In areas frequent summer drought pea (*Pisum sativum*) as a whole plant for forage gives assured yield and may be used for zero grazing, hay or silage. The winter varieties use winter-spring soil moisture better and give higher yield than the spring varieties, but they lodge, so it is necessary to sow them with supporting cereal crops to increase lodging resistance. The objective of the study was to compare the intake, digestibility, energy value and quality index of two wilted silage made from winter pea, variety Pleven 10, and from pea-cereal crop mixture.

Materials and methods The trials were carried out at the Institute of Forage Crops, Pleven. Silages were made from peas grown alone and from a pea-cereal (wheat and rye) mixture in a 25:75 ratio The crops were harvested at the stage of early wax ripeness of the first pods of pea and cereal components and they were wilted before ensiling. The silages both had very good fermentation characteristics: dry matter (DM) content 50.3 and 53.9%, lactic acid 9.6 and 8.4% of DM, acetic acid 0.91 and 0.71, respectively. Both silages had pH of 4.7 and no butyric acid. Six fine-wool wethers of average live weight of 58 kg were placed in cages for balance trials at two feeding levels, *ad libitum* (allowing 10% refusal) and restricted feeding to determine intake and digestibility. The forage quantity ingested, the refusal at *ad libitum* feeding and the faeces were recorded every day during the 8-day trial period., The net energy content was calculated as feed units for growth (FUG) (Todorov, 1995) (1 FUG = 6 MJ net energy for growth) from nutrients ingested and excreted in faeces. Using the results of DM intake during *ad libitum* feeding and FUG content, the forage quality index (QI) was calculated according to Kirilov (2000) (QI = DM intake*FUG/FUG for maintenance) with the maintenance requirement taken as 0.039 FUG/kg $W^{0.75}$ (Todorov & Dardjonov, 1995).

Results The intake and digestibility of pea silage were higher than those of the silage made from pea-cereal mixture (Table 1). This is probably due not only to the better palatability of the wilted pea silage, but also to the advanced developmental stage of cereals and their great presence in the mixture which caused an additional decrease in palatability. The lower digestibility resulted in a lower energy value of the silage of pea-cereal mixture 0.76 (4.56 MJ net energy for growth) *vs.* 0.83 FUG (4.98 MJ). During *ad libitum* feeding, the digestibility and energy values decreased a little, as compared to the limited feeding.

The lower intake and digestibility combined to give a lower forage quality index for the pea-cereal mixture compared to pea - with net energy intake 1.21 *vs.* 1.62 times maintenance requirement for the two silages.

Table 1 Intake, digestibility, energy value and quality index of silage from pea and pea-cereal mixture

Silage	Feeding	CP, g / kg DM	DM intake, g / kg $W^{0.75}$ *	OM digestibility, % **	FUG, per kg DM	Quality Index
	limited	154.8	42.3	$64.30^{a} \pm 1.45$	0.86	
Pea	*ad libitum*	154.8	$76.0^{a} \pm 0.6$	$63.30^{a} \pm 1.65$	0.83	1.62^{a}
Pea:cereals	limited	113.0	42.2	$58.14^{b} \pm 2.01$	0.81	
	ad libitum	113.0	$62.3^{b} \pm 4.8$	$57.15^{b} \pm 2.88$	0.76	1.21^{b}

*The means followed by different letters differ statistically significant; ** Mean ± Standard deviation

Conclusion The results of this study demonstrate the higher digestibility and intake, higher net energy content and higher forage quality index of pea silage as compared to the silage made of pea-cereal crop mixture at the stage of wax ripeness and with a cereal content of 75%.

References
Kirilov, A. (2000). Estimation of yield according to forage quality. *Grassland Science in Europe*, 5, 163-165.
Todorov, N. (1995). [Nutrient Requirements of Cattle and Buffalo] (In Bulgarian with English Summary). Publishing House NIS-UZVM, Stara Zagora, 217 pp.
Todorov, N. & T. Dardjonov (1995) [Nutrient Requirements of Sheep and Goats] (In Bulgarian with English Summary). Publishing House NIS-UZVM, Stara Zagora, 216 pp.

Annual legumes as an alternative for animal feeding in Cuba

M.F. Díaz, C. Padilla, E. Lon Wo, M. Castro, R. Herrera and R.O. Martínez
Instituto de Ciencia Animal. Apartado 24, San José de las Lajas, La Habana, Cuba, Email: mdiaz@ica.co.cu

Keywords: grain, forages, chemical analysis

Introduction Studies conducted in Cuba have demonstrated the importance of the agronomic and nutritional performance of the species *Vigna unguiculata* (cowpea), *Canavalia ensiformis* (jackbean), *Stizolobium niveum* (mucuna), *Lablab purpureus* (dolicho) and *Glycine max* (soybean) as feed sources for non-ruminant species. Under Cuban tropical conditions, and with minimum agricultural inputs, jackbean, dolicho and mucuna have attained forage yields between 4 and 6 t dry matter (DM)/ha and grain yields between 2.57 and 3.41 t/ha and cowpea and soybean have given yields of between 1 and 2 t/ha (Díaz 2000). This study was carried out to determine the chemical composition of grains and forages of these annual legumes in relation to their use in animal feeding.

Materials and methods Ten experiments were conducted to determine the chemical composition of forages (harvested when 100% of the plants were flowering), whole-crop forages (harvested when 100% of the plants had pods in the milky state) and grains in five species and fourteen varieties.

Results The results showed that the grains and forages had high content of nutrients needed for animal feeding. The forages and whole-crop forages were outstanding due to their higher content of fibrous components and soybean and cowpea were outstanding because of their high protein content. The grains had high content of protein and low fibre percentage. Soybean was also relevant due to its mineral content and cowpea, due to its low fibre proportion (Table 1). A good amino-acid balance was found with concentrations of essential amino acids similar or superior to the pattern of reference of FAO (D' Mello, 1995), for most of the varieties studied. Glutamic acid, aspartic acid, threonine, lysine and leucine were the most abundant amino acids in the species evaluated.

Table 1 Chemical composition of the legumes harvested as forage, whole crops or as grain (% of dry matter)

Species	CP	NDF	Ca	Mg	P	K
Forage						
S. niveum	14.16	56.71	0.99	0.33	0.21	0.64
C. ensiformis	15.57	57.91	1.85	0.51	0.20	0.38
L. purpureus	18.57	61.00	1.24	0.39	0.31	1.07
V. unguiculata	20.05	43.23	1.40	0.70	0.32	0.61
Glycine max	21.74	56.82	1.29	0.86	0.35	1.06
Whole crop	38					
S.. niveum		57.33	1.22	0.33	0.21	1.06
C. ensiformis	14.00	60.82	1.43	0.45	0.19	0.59
L. purpureus	14.52	62.21	1.35	0.46	0.27	0.75
V. unguiculata	15.06	43.67	1.29	0.78	0.30	1.75
Glycine max	17.36	39.98	1.05	0.12	0.33	1.44
	22.18					
Grain						
S. niveum	28.24	29.55	0.43	0.21	0.52	0.81
C. ensiformis	31.40	36.47	0.45	0.12	0.40	0.70
L. purpureus	25.16	39.06	0.51	0.21	0.60	1.23
V. unguiculata	25.18	10.86	0.42	1.39	0.28	0.34
Glycine max	43.11	13.33	0.47	0.38	1.17	0.84

Conclusions The chemical analyses supported the possibility of using these legumes as non-conventional feed sources in Cuba. Complementary nutritional studies made by this research group have demonstrated production technologies and the use of the grain of *Vigna unguiculata* for the feeding of broilers and growing pigs.

References

Díaz M.F. (2000). Producción y caracterización de forrajes y granos de leguminosas temporales para la alimentación animal. *Tesis de Doctor en Ciencias Agrícolas*, Instituto de Ciencia Animal, La Habana, Cuba. 91pp.

D' Mello, J.P.F. (1995). Under-utilized legume grains in non-ruminant nutrition. In: J.P.F D'Mello & C. Devendra (eds.) Tropical Legumes in Animal Nutrition. CAB International, p. 283.

Relative yields and nutritive value of whole crop rice harvested on four successive dates for forage in Korea

J.G. Kim, Y.C. Lim, E.S. Chung, S.H. Yoon, S. Seo and M.J. Kim
National Livestock Research Institute, Suwon 441-706, South Korea, Email: jonggk@rda.go.kr

Keywords: whole-crop rice, forage quality, productivity

Introduction About four million tons of forages are fed to ruminants in Korea, but half of them rely on rice straw as roughage and 0.6 million tons of forage was imported. The lack of forage results in increased imports of concentrate feeds and increased production cost. Now, Korea has about 1.1 million ha of rice fields, but as a consequence of world trade negotiation, Korea will open the rice market from next year. It is expected that due to aging farmers and lower rice price, about 0.2 million ha of paddy field will not be cultivated for grain rice. Therefore, we suggest that whole-crop rice cultivation for feeding beef and dairy cattle. The purpose of this study was to investigate relative yield and nutritive value of whole-crop rice grown in paddy fields in Korea.

Material and methods Two varieties of whole-crop rice (Suwon 468 and Dongjin) were harvested at four different growth stages (heading, milk, dough and ripe stage). Dry matter (DM) content of rice was determined by forced air drying at 65°C for 72h. The dried samples were ground and kept for the analysis. Crude protein (CP) was determined by Kjeldahl (AOAC, 1995), acid detergent fiber (ADF) and neutral detergent fiber (NDF) were measured by the method of Goering & Van Soest (1970), and *in vitro* dry matter digestibility (IVDMD) was determined by the method of Moore (1970).

Results Crude protein content decreased with progressed maturity at harvest. As harvest stage delayed, TDN concentration of whole-crop rice increased. The contents of ADF and NDF decreased with harvest maturity. This was in line with the results of Brundage *et al.* (1979), that the highest ADF and NDF content in oats were found at flowering stage. The highest DM yield was at the ripe stage and the early-maturity rice cultivar (Suwon 468) gave higher yields than the late cultivar (Dongjin). The DM yield of all varieties was low. This was associated with a late sowing date. Total digestible nutrient (TDN) yield followed a similar trend to DM yield.

Table 1 The content of crude protein (CP), ADF, NDF, TDN and IVDMD and yield of forage rice harvested at four different growth stages

Varieties	Harvest stage	CP (%)	ADF (%)	NDF (%)	TDN (%)	IVDMD (%)	DM (%)	Yield (kg/ha) DM	Yield (kg/ha) TDN
Dongjin	Heading	10.6	33.6	68.7	62.2	45.6	22.2	5,761	3,584
	Milk	7.6	32.9	65.8	62.9	51.5	31.0	7,811	4,914
	Dough	6.2	31.9	63.9	63.7	48.8	33.2	9,644	6,141
	Ripe	5.8	30.3	63.2	65.0	59.3	42.0	10,687	6,947
Suwon 468	Heading	13.5	33.5	67.9	62.5	55.9	20.0	6,038	3,772
	Milk	10.2	32.9	64.6	62.9	53.8	32.2	7,513	4,728
	Dough	8.6	30.7	63.1	64.6	57.5	41.2	9,798	6,332
	Ripe	8.6	31.1	63.2	64.3	53.7	43.2	12,137	7,807

Conclusions The experiments presented here show that whole-crop rice may provide a reasonably high quality feed for ruminants. It may cover the shortage of forage for ruminants in Korea and substitute for imported forage.

References
Association of Official Analytical Chemists (1995). Official Methods of Analysis. (16th ed.). AOAC, Arlington, Virginia.
Goering, H K., & P.J. Van Soest (1970). Forage fiber analysis. Agricultural Handbook 379, U.S. Government Print Office, Washington, DC.••
Moore, J E. (1970). Procedure for the two-stage *in vitro* digestion of forage. University of Florida, Department of Animal Science.
Brundage, A L., R L. Taylor & V.L. Burton (1979). Relative yields and nutritive value of barely, oat, and peas harvested four successive dates for forage. *Journal of Dairy Science*, 62, 740-745.

Nutritional evaluation of banana peelings from the various banana varieties in different regions of Uganda

J. Nambi-Kasozi[1], E.N. Sabiiti[1], F.B. Bareeba[1] and E. Sporndly[2]
[1]Faculty of Agriculture, Makerere University PO Box 7062 Kampala, Uganda, Email: jnambik@agric.mak.ac.ug, [2]Department of Animal Nutrition and Management, Swedish University of Agricultural Sciences, P.O. Box 7070, 750 07 Uppsala, Sweden

Keywords: banana peelings, composition, digestibility, varieties

Introduction In Uganda, peri-urban agriculture is very important in sustaining livelihoods of the increasing population in urban and peri-urban (U & PU) areas. However, this form of agriculture has a number of problems, feed shortage being the most important. The agricultural produce brought in from the rural areas is marketed in raw form hence increasing the crop wastes in the market areas. The crop wastes, if properly sorted, can serve as alternative feeds for the animals kept in the U & PU areas. Banana (*Musa spp*) peelings (BP) constitute the largest proportion of all the crop wastes in most markets in areas where they are the staple food crop and have become a popular feedstuff. Unfortunately, there is scanty literature on their nutritive and feeding value. Different varieties are grown in the different banana growing areas. The purpose of this study was to evaluate the nutritive value of BP of the different banana varieties from central, eastern and western Uganda. *In vitro* dry matter digestibility (INDMD) of the banana peelings was also determined.

Materials and methods Four samples of BP of each variety were collected from the market places over a period of four months from the central, eastern and western regions. The BP were analysed for crude protein (CP), acid detergent fibre (ADF), neutral detergent fibre (NDF), dry matter (DM), calcium and phosphorus. For *in vitro* fermentation, rumen liquor was collected from a fistulated animal into pre-warmed CO_2 filled thermos flasks and strained through a cheese cloth. Samples were weighed into 125ml flasks and 40ml of prepared medium was added to each flask. *In vitro* fermentation was then conducted using 10 ml inoculum for 48 hours. The DM digestibility was then determined. Dry matter, CP, Ca and P in the feeds were determined according to AOAC (1990). Acid detergent fibre and NDF were analysed according to Van Soest & Robertson (1985).

Results and discussion Dry matter was higher in varieties (Var) from the eastern region but lower and similar for varieties from other regions (Table 1). Phosphorus and calcium was higher in varieties from the western and central than in those from the eastern region. Crude protein, ADF, NDF and INDMD values did not differ among the varieties. The different soil types in the different areas are mostly likely responsible for the observed differences. The central and western regions have similar soils and gave varieties with similar composition.

Table 1 Chemical composition (%DM) and IVDMD of banana peelings from different regions

Component (%)	Eastern region			Central region			Western region		
	Var 1	Var 2	Var 3	Var 1	Var 2	Var 3	Var 1	Var 2	Var 3
Dry matter	23.7	27.8	24.9	17.3	17.1	18.6	18.0	18.6	17.3
CP	5.0	4.9	5.1	5.2	5.9	6.0	6.1	5.6	4.4
ADF	15.5	25.8	18.9	11.8	23.9	17.9	14.4	14.3	12.2
NDF	37.0	42.0	36.0	24.0	40.0	31.7	28.0	44.0	40.0
Phosphorus	0.23	0.21	0.27	1.00	1.80	0.34	1.00	1.00	0.80
Calcium	0.09	0.11	0.21	1.09	1.97	0.45	1.09	1.10	0.87
INDMD	77.6	79.5	82.0	78.4	84.5	77.4	70.6	87.7	79.5

Conclusion Banana peelings differ in nutritional composition depending on the region from which the bananas are grown. This should be taken into account when recommending feeding packages for animal production.

References
AOAC (Association of Analytical Chemists) (1990). Official Methods of Analysis. 15th edition. Inc. Arlington, Virginia, USA.
Van Soest, P.J. & J.B. Robertson (1985). Analysis of Forage and Fibrous Foods, a Laboratory Manual for Animal Science 613. Cornell University, Ithaca, New York, USA.

Effect of intensity of grassland management on chemical composition and content of structural saccharides in forage

J. Pozdíšek[1], P. Mičová[1], M. Svozilová[1] and A. Kohoutek[2]

[1]Research Institute for Cattle Breeding, Ltd. Rapotin, Czech Republic, Email: jan.pozdisek@vuchs.cz, [2]Research Institute of Crop Production, Prague 6 Ruzyně; Research Station of Grassland Ecosystems in Jevíčko, Czech Republic

Keywords: forage quality, nitrogen substances, crude fibre, neutral detergent fibre, acid detergent fibre

Introduction Forage quality has a crucial effect on animal performance and on grassland management. This paper contributed information on the effects of different methods of grassland utilisation in the Czech Republic.

Material and methods The experiment examined four methods of utilisation: intensive (4 cuts/ year – first on 15 May, followed by cuts at 45d intervals), middle intensive (3 cuts/ year – first on 30 May followed by cuts at 60d intervals), low intensive (2 cuts/ year – first on 15 June with a further cut after 90d), extensive (2 cuts/ year– first on 30 June with a further cut after 90 days). Each method of utilisation was divided with four levels of fertiliser (without fertiliser, $P_{30}K_{60}$, $N_{90}P_{30}K_{60}$, $N_{180}P_{30}K_{60}$). There were four replicates of each treatment. The FIBERTEC 2023 FIBERCAP FOSS TECATOR was use to analyse a structural fibre. The dominant species in the permanent sward were *Dactylis glomerata, Poa pratensis, Lolium perenne, Trifolium repens* and *Taraxacum*.

Results Average values for contents of crude protein and fibre are given in Table 1. The content of crude protein fell with reduced intensity of utilisation and cutting frequency in line with the results of Gaisler & Fiala (2003). Averaged across fertiliser treatments the content of crude fibre increased from 223 to 306 g/kg DM as intensity of utilisation fell in line with results of Pozdíšek *et al.*(2003) for different species of grass and red clover. With intensive utilisation the neutral detergent fibre (NDF) content averaged 486 g/kg DM compared with 598 g/kg DM for extensive utilisation, whilst the respective figures for ADF were 304 and 368 g/kg DM. The signifance of differences between treatments is indicated in Table 2. This stresses the large impact of method of utilisation and the substantial effect of fertiliser on crude protein content.

Table 1 Mean values of crude protein and fibre for the different treatments (g/kg DM)

Var.	1A	1B	1C	1D	2A	2B	2C	2D	3A	3B	3C	3D	4A	4B	4C	4D
CP	168	164	183	199	150	153	161	182	116	113	131	127	102	113	122	126
CF	221	226	224	221	241	245	243	241	277	305	292	308	308	292	305	318
NDF	485	476	488	495	509	516	524	513	566	572	558	602	594	566	606	627
ADF	297	290	321	307	322	317	323	314	356	363	349	355	377	365	356	376

Legend: 1 - Intensive   3 - Low intensive   A – zero fertiliser C – N_{90}PK fertiliser
    2 - Middle intensive   4 - Extensive   B – PK fertiliser   D – N_{180}PK fertiliser

Table 2 Dual-factorial variance of treatments in relation to crude protein and fibre ($\bullet\bullet=0,01$; F-values)

	d.f.	CP	CF	NDF	ADF
Utilisation	3	221.30^{++}	190.23^{++}	207.28^{++}	102.18^{++}
Fertiliser	3	32.16^{++}	2.08^{-}	10.15^{-}	0.46^{-}
Residuum		67.21	135.96	207.09	144.74

Conclusions These results confirm the large impact of method of utilisation (cutting frequency) on the contents of crude protein and fibre in grassland forage. Fertiliser application increased crude protein content.

Acknowledgement Partly supported by Ministry of Education CR, Prague No. MSM 2678846201

References

Gaisler, J. & J. Fiala (2003). Vliv hnojení a počtu sečí na botanické složení, výnos a kvalitu píce trvalého travního porostu. In: Ekologicky šetrné a ekonomicky přijatelné obhospodařování travních porostů Prague, 99-105.

Pozdíšek, J., A. Kohoutek, H. Jakešová & P. Divišová (2003). Kvalita píce jako významný faktor exploatace travních porostů In: Ekologicky šetrné a ekonomicky přijatelné obhospodařování travních porostů Prague, 224-237.

Changes in grass quality of coastal meadows in Estonia

T. Köster, K. Kauer, R. Viiralt and A. Selge
Estonian Agricultural University, Institute of Soil Science and Agrochemistry, Kreutzwaldi 64, 51014, Tartu, Estonia, Email: tints@eau.ee

Keywords: coastal grasslands, grass quality, crude protein

Introduction In Estonia the reason for reduction in coastal meadows and expansion of the stands of the common reed (*Phragmites australis*) is the discontinuation of traditional use of grasslands which were previously grazed and cut. *Phragmites australis* usually produces dense and monospecific stands at the waterline, where species richness is low. It can survive in ungrazed shore meadows, but it suffers from grazing (Tyler, 1969). The investigated *Phragmitetum australis* association had been influenced by grazing activities, but it remained rather sparse and due to that had relatively low productivity, reaching 3.11 t DM/ha. Traditionally coastal meadows have been used for grazing and have given stable quality of feed and good animal performance. The quality of fodder is an important factor for farms using the coastal areas for grazing. The aim of the study was to determine the changes in quality of the different plant associations in the coastal area.

Materials and methods In 2001-2003 the research area on the island Hiiumaa was established to investigate the plant associations in coastal grasslands. The investigated farm (total area 544 ha) in South-Eastern Hiiumaa embraced *c.* 400 ha of coastal meadows and it is partly grazed by horses and beef cows in addition to substantial grazing by birds. Six investigation sites were chosen to analyse the quality of plant associations at different times during the vegetation period. The height of cut was 2-5 cm for *Juncus gerardii* and 10-30 cm for *Phragmitetum* association. Plant material was analysed for crude protein (CP), acid detergent fibre (ADF), neutral detergent fibre (NDF). Digestible DM (DDM= 88.90-(0.779*ADF), %), DM daily intake (DMI=120/NDF, % of cattle weight), relative feed value (RFV= (DMI*DDM)/1.29, points) and metabolizable energy (ME) content were calculated on the basis of ADF and NDF (Nutrient Requirements of Dairy Cattle, 2001). Statistical analysis was by ANOVA and standard deviations (SD) are presented.

Results Crude protein is very sensitive to the stage of grass maturity. With *Phragmites australis* the protein content had very high value in the end of May but then dramatically dropped in June (Fig 1).

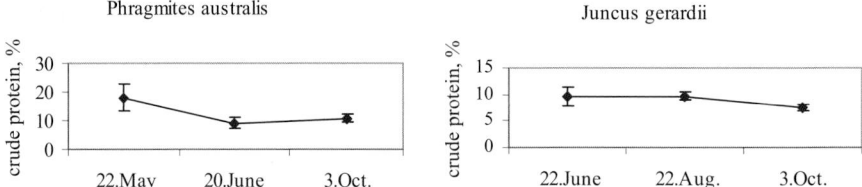

Figure 1 Changes in content of crude protein during the vegetation period in plant associations dominated by *Phragmites australis* and *Juncus gerardii* (n=5; M±SD)

The other components did not change significantly during the vegetation period. The content of NDF ranged 65-71%, ADF 34-40%, DDM 58-62%, DMI 1.7-1.8%, RFV 75-78 points and ME 8.9-9.4 MJ/kg. In comparison, the nutritive value was higher for the plant association dominated by *Juncus gerardii*: NDF ranged 59-64%, ADF 29-31%, DDM 61-66%, DMI 1.9-2.1%, RFV 93-105 points and ME 9.6-10.5 MJ/kg.

Conclusions Large areas of coastal meadows in Estonia have grown into reed and one possibility to stop that process is grazing. The *Phragmitetum australis* association was characterised by high content of protein at the early development stage, but this decreased rapidly during the vegetation period. Contents of ADF and NDF were higher in the *Phragmitetum* association than with *Juncus gerardii*, but did not change significantly between sampling dates. Consequently the calculated values for digestible DM and other aspects of feeding value were lower for *Phragmitetum* plant association.

References

Nutrient Requirements of Dairy Cattle (2001). Seventh Edition. National Academy Press. Washington, D.C.
Tyler, G. (1969). Studies in the ecology of Baltic sea-shore meadows. II Flora and vegetation. *Opera Botanica* 25, 101 pp.

Accumulation of zearalenone in herbage of winter pasture situated in West Poland

P. Goli•ski[1], M. Kostecki[2], B. Goli•ska[1], B.T. Goli•ska[3] and P.K. Goli•ski[2]
[1]Department of Grassland Sciences, [2]Department of Chemistry, [3]Department of Biochemistry and Biotechnology, Agricultural University of Pozna•; Wojska Polskiego 38/42, 60-627 Pozna•; Poland, Email: pgolinsk@au.poznan.pl

Keywords: forage quality, winter pasture, zearalenone

Introduction The importance of winter pastures in beef production in Europe has been growing steadily. In Poland, especially in its western part, there are already farms which utilise pasture swards during late autumn and winter. The major problem, however, is the quality of forage ingested by animals as it tends to deteriorate with the passage of the vegetation season with danger of accumulation of various mycotoxins (Laser *et al.,* 2003) of which the most important is zearalenone (ZEA).

Materials and methods During 2001-2003, an experiment was set up in Brody (52° 26' N, 16° 18' E)with a low-input pasture system in a paddock with a *Lolio-Cynosuretum* community dominated by *Poa pratensis.* There was a randomised Latin square design with three replicates (plot 20 m²). The factors were (a) time of pre-utilisation in summer (US) – last use at the beginning of June (Jun.), beginning of July (Jul.), beginning of August (Aug.) and (b) date of harvest in winter (HW) – at the beginning of November (Nov.), middle of December (Dec.), and end of January (Jan.). In each of the two years, the pasture was fertilised with 50 kg/ha N in the second half of August, in order to simulate the return of nutrients left with the faeces of grazing animals,. To estimate the yield of winter pasture plots and collect samples for chemical analyses an area of 10 m² of the sward was cut. Herbage from each plot (27 samples per year) was analysed for concentration of ZEA using ZaeralaTest[TM] column (Vicam USA) and HPLC method.

Results Accumulation of ZEA in herbage of winter pasture differed depending on US, HW and investigation year, being significantly higher in 2002/03. Higher concentrations of ZEA (19.4 ng/g DM) were found in the forage harvested in Jan. when compared to samples collected in Nov. and Dec. (6.6 and 4.5 ng/g DM, respectively). The highest concentration of ZEA for a single sample for each month are given in Table 1. According to literature data from Germany, dietary concentrations of ZEA in fodder higher than 500 ng/g DM may have adverse effects on heifers, dairy and suckler cows. The quantities of ZEA in the DM of winter pasture are shown in Figure 1. The highest number of samples in which ZEA exceeded the level of 3 ng/g DM (detection limit) occurred when the sward was harvested in Jan. (77.8%), indicating that the shorter the period of unfavourable weather conditions the lower the percentage of ZEA positive samples – 72.3% in Nov. and 61.2% in Dec.

Table 1 Highest concentration of ZEA in herbage of winter pasture (ng/g DM)

HW	US	2001/02	2002/03
Nov.	Jun.	14.20	9.52
	Jul.	8.84	23.58
	Aug.	12.50	7.92
Dec.	Jun.	5.87	10.93
	Jul.	4.88	3.05
	Aug.	6.48	17.77
Jan.	Jun.	6.44	36.42
	Jul.	7.03	98.93
	Aug.	47.89	79.10

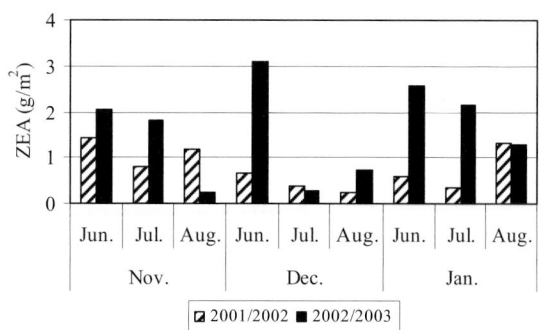

Figure 1 Quantity of ZEA in DM yield of winter pasture

Conclusions The results confirm that ZEA may decrease the quality of forage from winter pasture in West Poland. With increasing duration of sward regrowth, the accumulation of ZEA in herbage increased. The highest concentration of ZEA (on average, 19.4 ng/g DM) was recorded in forage harvested in end of January. Concentrations were though lower than those that have been reported to have adverse effects on cattle.

References
Laser, H., W. Opitz von Boberfeld, K. Wöhler & D. Wolf (2003). Effect of the botanical composition and weather conditions on mycotoxins in winter forage from grassland. *Mycotoxin Research*, 19, 87-90.

Diversity and variation in nutritive value of plants growing on 2 saline sites in south-western Australia

H.C. Norman, R.A. Dynes and D.G. Masters
CSIRO Livestock Industries, Private Bag 5, Wembley, WA 6913, Australia, Email: Hayley.Norman@csiro.au

Keywords: dryland salinity, biodiversity, nutritive value, saltbush, feeding value

Introduction In south-western Australia 10% or 1.8 million ha of the farmed area is affected by dryland salinity and a further 6 million ha are at risk of salinity (NLWRA, 2001). Animal production from saltbush (*Atriplex* spp.)-based pasture systems represents the most likely large-scale opportunity for productive use of saline land in the short to medium term. Feeding saltbush-based pastures as a maintenance feed during the prolonged autumn feed gap typical in Mediterranean-type climates maximises their economic value. The aim of this study was to explore the diversity and nutritive value of plants that typically persist in saltbush-based saltland pastures.

Materials and methods Two highly saline pastures were chosen at Meckering and Tammin in Western Australia. The Meckering site (19 ha) was situated 130 km east of Perth in the 400 mm annual rainfall zone. The Tammin site (12 ha) was situated 180 km east of Perth in the 325 mm annual rainfall zone. Saltbush was established at both sites more than 10 years before this study and no other plants species had been deliberately introduced. Plant diversity and quality were determined along 3 transects within each plot in autumn 2001. Diversity was measured through the Botanal technique (Mannetje & Haydock, 1963), ranking only the herbaceous component. All species were sampled for laboratory analysis of digestibility (pepsin-cellulase digestibility of the organic matter in the dry matter (P-CDOMD, Norman *et al.,* 2004), pepsin-cellulase organic matter digestibility (P-COMD), ash, acid-detergent fibre (ADF %OM) and crude protein (CP) concentrations.

Results The Meckering site contained 31 herbaceous plant species, of which 26 were volunteers and 7 were native to Australia. The Tammin site contained 24 plant species of which 19 were volunteers and 8 were native to Australia. Both sites contained greater botanical diversity than is found in adjacent non-saline areas. Of the feed on offer at the Tammin site, 60% was derived from halophytic shrubs, 38% grasses and 2% from forbs and legumes. Feed on offer at the Meckering site consisted of 43% halophytes, 52% grasses, 4% forbs and 1% legumes. The legumes provided the best quality biomass with both high high digestibility and CP content. There was considerable variation in P-CDOMD within the grasses. Many had sufficient energy for maintenance of an adult sheep however CP content was low. Some forbs and all halophytes accumulated excess salts. Although many of these plants have high P-COMD values, intake is likely to be restricted by salt and, therefore, sheep are unlikely to maintain themselves on these species alone. The halophytes provided a good source of CP.

Table 1 Mean nutritive values of plants collected from 2 saline sites in autumn

	#	P-CDOMD %	OMD %	CP %	ADF %OM	Ash %
Legumes	4	61.8 (0.6)	66.8 (1.7)	14.9 (1.2)	27.3 (2.9)	8.9 (2.0)
Grasses	16	56.8 (1.1)	62.2 (1.1)	4.4 (0.3)	27.8 (1.4)	9.1 (1.1)
Halophytes	16	48.0 (0.9)	64.2 (1.2)	9.2 (0.6)	14.4 (0.9)	25.5 (1.0)
Forbs	5	58.2 (2.1)	72.9 (7.1)	6.1 (0.5)	22.5 (4.4)	17.8 (6.2)

Numbers in parentheses are the s.e. of the means

Conclusions Mature saltbush pastures in Western Australia are rich in diversity. The variation in nutritive value within and between these species is significant. Legumes have the highest feeding value but are least salt-tolerant with a low biomass contribution. Halophytes and grasses were the largest contributors to biomass but the nutritive values of grasses (low digestibility and CP content) and halophytes (high ash content) suggest that sheep need a combination of both to maintain live weight. Given the opportunity to select, mature sheep should maintain live weight during autumn while grazing. The variation in nutritive value implies opportunities to improve the feeding value of saltbush-based pastures through manipulation of species composition and agronomic selection of high value species. A mixed sward of shrubs, grasses, legumes and forbs appears to be the most productive option, since species of different salt tolerance can occupy niches in these very heterogeneous areas. This mix in turn provides the best opportunity for sheep to select for energy, CP and salt.

References

Mannetje L.'t, & K.P. Haydock (1963). The dry-weight-rank method for the botanical analysis of pasture. *Journal of the British Grasslands Society*, 18, 268-275.
NLWRA (2001). Australian Dryland Salinity Assessment 2000: extent, processes, monitoring and management options, National Land and Water Resources Audit. Canberra.

Nutritional characteristics of four species of native pastures from the highlands of central Mexico

O.A. Castelán[1], A.A. Rayas[1], A. Espinoza[1], J. Estrada[1], C. Arriaga[1] and F. Mould[2]
[1]CICA-UAEMex, Instituto Literario No. 100, col. Centro CP 50000 Toluca, México, Email: oaco@uaemex.mx,
[2]Department of Agriculture, The University of Reading RG6 6AR Berkshire, UK

Keywords: native species, digestibility, crude protein

Introduction Research in Mexico has been devoted to nutritional characterisation of improved pastures, but native species of grasses have not been studied. The aim of this experiment was to evaluate the nutritional value of four species of native pastures and composite samples of them.

Material and methods *Pennisetum clandestinum* (PC), *Sporobolus indicus* (SI), *Trifolium amabile* (TA), *Eleocharis dombeyana* (ED) and pooled samples (PS), which included all the species, were collected. Sample selection was based on the preference showed by livestock that graze the four communities in Toluca valley. Four plots of native pastures were selected by community and sixteen plots were sampled in total. There were three harvest periods in 2003: July (P1), September (P2) and November (P3). All samples were analysed for CP, ME, NDF and *in vitro* gas production (GP). Digestibility of the NDF (dNDF) and DOM were estimated from the GP analyses and the ME content was calculated from DOMD according to AFRC (1993). A split-plot design was used, where the main plot was the species and growing periods were split plots with communities as blocks. Data was analysed by analysis of variance, Minitab v13.

Results and discussion Results for GP are presented in a separate paper (Rayas *et al.*, 2005). Significant differences (P<0.001) between species were observed for CP, EM, dNDF. Crude protein content in individual species ranged from 39 to 204 g/kg DM, with TA having the highest values in all harvest periods and SI the lowest. Significant differences (P<0.001) were observed for TA between periods, P1=204, P2=175 and P3=175 g CP/kg DM, and between this species and the other species (P<0.001). It was observed that the CP content in all species and the PS, decreased as the plants got older (P<0.001) (Table 1). Significant differences (P<0.001) were observed between species and growing periods in DOM and ME content; they ranged from 542 to 833 g/kg DM and 8.1 to 12.5 MJ/kg DM respectively.

Table 1 Crude protein, ME content, degradable organic matter and digestibility of the NDF of forage species by harvest period

Species	Period	CP g/kg DM	ME MJ/kg DM	DOM %	NDF g/kg DM	dNDF %
PS	1	124	12	80.5	591	82.1
	2	123	10.1	67.7	673	60.6
	3	72	10	66.9	659	51.6
PC	1	157	12.5	83.3	632	86.2
	2	147	10.9	73.3	654	68.5
	3	88	10.6	71.1	643	62.5
SI	1	94	11.9	79.4	722	84.2
	2	63	8.7	58.2	754	60.2
	3	39	9.4	62.7	774	57.4
TA	1	204	11.8	78.7	451	75.4
	2	175	10.8	72.1	487	56.9
	3	175	10.8	72	486	56.8
ED	1	149	11.3	75.5	595	80
	2	128	9.3	62.4	665	55.1
	3	97	8.1	54.2	635	36.3
S.E.M.		28.7	1.3	9.2	24.5	15.7
Species		***	***	***	***	***
Periods		***	***	***	***	***

CP=crude protein; ME=metabolizable energy; DOM digestibility organic matter; dNDF=digestibility of the NDF. ***Significance level for columns P< 0.001

The pastures with the highest values of DOM and ME (833, 805 g/kg DM and 12.5, 12 MJ/kg DM respectively) were PC and the PS. The lowest ME content and DOM in P1 and P3 was in ED. Significant differences (P<0.01) among species, periods and their interaction were identified for NDF content, which ranged from 451.2 in TA to 774.1 g/kg DM in SI. The NDF content was affected by harvest period and a highly significant (P<0.001) increase was observed in all the species from P1 to P3. Significant differences (P<0.05) were identified in the dNDF content between species, periods and their interaction. The dNDF ranged from 86.2% in PC to 36.3% in ED in P3.

Conclusions These results suggest that the species with highest nutritive value was TA, followed by PC and the poorest was ED.

References
AFRC (1993). Energy and Protein Requirements of Ruminants. CAB International, Wallingford, U.K.
Rayas, A.A., A. Espinoza, J. Estrada, C. Arriagal, F. Mould & O.A. Castelan (2005). Rumen degradation characteristics of four species of native pastures from central Mexico in three growing periods. *XX International Grassland Congress – offered papers* (in press).

Seasonal variation of forage productivity and quality of communally managed grassland in the N'komati river basin

D.W. Nguluve, C. Menezes, A. Buluveze and A.P. Laita

Instituto de Produção Animal, C.P: 1410, Maputo, Mozambique, Email: ngudamiao@yahoo.com

Keywords: forage, productivity, quality, grassland

Introduction Livestock production is increasing in Mozambique. This trend, however, is facing such challenges as land tenure, erratic and not well-distributed rainfall (resulting in floods or droughts), overgrazing, wildfires, and the unsustainable resource management practices of communities. The study objectives were to evaluate forage species occurrence and seasonal variation and to estimate grassland productivity, nutritive value and savanna carrying capacity.

Materials and methods Ten 8m x 8m exclosures were randomly established in low, transitory and high veld areas of the N'Komati river basin in southern Mozambique (semi-arid topical grassland). Each exclosure was divided into three sub-plots which were respectively cut every six weeks (6-W) throughout the year, cut once at the end of the rainy season (ERS) or cut once at the end of the dry season (DS) during 2002. Forage yield and crude protein (CP) were determined on samples from which inedible stalks were excluded and carrying capacity for the complete year (CC) was determined according to Handzel (1981).

Results The most frequently occurring species during the DS and ERS were *Panicum maximum*, *Digitaria ciliaris*, *Setaria sphacelata* and *Imperata cylindrica*. There were 50 % more species harvested and identified in the rainfall season compared to subsequent seasons. Native legumes did not occur frequently (except *Tephrosia* spp.), suggesting that wildfires, wildlife, and competing grass species may have suppressed them at the early growth stage. There were significant differences in yield between treatments (P=0.05, $LSD_{0.05}$=0.83. The total forage yield of 6W was 7.6 t/ha compared with only 4.0 t/ha and 2.0 t/ha for ERS and DS respectively. Lowveld (LV) produced on average twice the yield of the upland (HV) and transitory zones (TR). Significant differences (P=0.05; $LSD_{0.05}$ =2.0) between cutting treatments were also observed for CP content of the harvested herbage with a range from 4.3 % to 7.20 % and there were also significant differences between locations (Table 1). Forage CP declined as the harvested biomass advanced in phenological and physiological stages, especially at DS. Pasture carrying capacity (CC) increased with clipping frequency by 95 %, with the 6-W frequency and the lowveld pastures giving the most promising results (4.6 and 8.8 ha/LWU).

Table 1 Forage production, crude protein (CP) and carrying capacity for different clipping frequencies and locations

	6W	ERS	DS	LV	TR	HV
Forage production, t/ha	5.0[ab]	4.0[b]	2.0[c]	2.9[d]	3.0[d]	4.0[b]
CP, %	7.3 [a]	5.1[ac]	4.3[c]	5.0 [ac]	5.5 [ac]	6.2 [a]
Carying capacity, ha/LWU	4.6[a]	10.9 [b]	13.3[c]	8.8[d]	9.9[d]	18.3[e]

Values followed by transcripts (letters) in the same row are significantly different

Conclusions There is much variation of species composition, grassland productivity, nutritive value, and CC through the year. Forage yield increased at 6-W cutting intervals, while at the ERS and DS it was low. Crude protein was also highest with 6-W harvesting. The highest CC was at 6-W cutting (4.6 ha/LWU), in line with the higher forage productivity for this treatment. Range management strategies for communal pastures should strongly encourage approaches such as pasture deferment, rotational grazing and forage legume overseeding. The lowveld should be used more intensively during the dry season, since this area retains more moisture and thus has the potential for year-round grazing.

References

Forwood, J.R. & M.M. Magai (1992). Clipping frequency and intensity effects on big bluestem yield, quality, and persistence. *Journal of Range Management*, 45, 554-559.

Handzel, L. (1981). Range Management Handbook for Botswana. Ministry of Agriculture, Gaberone, 49-50.

Forage composition and quality of tankbed grassland ecosystems in Sri Lanka; a preliminary study of "Tabbowa" Tank

G.G.C. Premalal[1] and S. Premaratne[2]
[1]Pasture Division, Veterinary Research Institute, Peradeniya, Sri Lanka, Email: premalal_ggc@yahoo.com,
[2]Department of Animal Science, Faculty of Agriculture, University of Peradeniya, Peradeniya Sri Lanka

Keywords: grassland ecosystems, forage composition, free-grazing livestock, water catchments

Introduction Although, Sri Lanka does not have large natural lakes, the early settlers devoted their energy to build water bodies (tanks or reservoirs) in association with forest catchments to provide water for irrigation. The tank bed is a grassland area of the tank, which lies between the catchment forest and the present water level. In general, the tankbed has three major zones - lower, middle and upper - and in some seasons the middle and upper parts may be underwater. This tankbed area is dominated by grassland vegetation and is a valuable feeding ground for free-grazing livestock. The main objective of this study was to identify the common forage species and to investigate the nutrient composition of species most relevant to ruminant livestock in one of the larger tankbed grasslands in the country.

Materials and methods This study was carried out in Jan. to July 2003 at the "Tabbowa" tank, which is mostly grazed in common by free-grazing livestock. The mean annual rainfall and temperature of the area were about 1750 mm and 30° C respectively. Botanical composition of the tankbed was measured according to the predetermined contours in the lower, middle and upper zones using 1 x 1 m quadrats with three replicates. The edible part of the most common forage species were analysed in duplicate for crude protein (CP), neutral detergent fibre (NDF), lignin and minerals, viz; sodium, magnesium, calcium and phosphorus.

Results and discussion The floristic composition of the tankbed was very diverse. The plant species identified, belonged to 14 families, 14 genera and 29 species. Out of 29 species, *Cynodon dactylon* (67.94%) and *Cyperus rotundus* (22.84) were the dominant species. Density of species changed on the moisture gradient towards the slope. The density of *C. rotundus* was higher in the lower region while that of *C. dactylon* was higher in the middle. High densities of herb species such as *Alternanthera paronchoides*, *Coldenia procumbens*, *Aeschynomene indica* and *Mollugo pentaphyla* were found in upper region of the tankbed. Out of 29 species, 7 species were commonly consumed by the animals with their selective grazing. These species and their nutrient composition are shown in Table 1. The nutrient composition differed among the species. However, most of the species were rich in CP, NDF, and minerals and comparable with other improved forage species (Ibrahim, 1989).

Table 1 Nutrient composition of common forage species in "Tabbowa" tankbed grassland (%)

Species	DM	Ash	CP	NDF	Na	Mg	P	Ca
Cynodon dactylon	25.1	9.63	13.62	81.76	0.80	1.20	0.25	0.03
Cyperus rotundus	19.6	4.36	14.87	75.33	1.30	0.46	0.31	0.03
Lippia nodiplora	17.6	8.73	14.15	46.85	1.83	1.30	0.46	0.43
Alternanthera paronchoides	17.0	9.45	17.62	49.29	1.02	1.67	0.13	0.06
Panicum psilopodium	23.0	8.12	23.18	79.07	1.55	2.01	0.51	0.24
Securinega leucophyrus	42.0	5.75	18.09	34.06	0.87	1.79	0.27	0.25
Accacia leucoplea	33.5	3.93	19.18	31.17	0.62	0.91	0.19	0.03

Conclusions The vegetation of the "Tabbowa" tankbed grassland ecosystem is diverse and includes valuable forage species. This grassland type plays a major role for free-grazing livestock in the dry zone of the country. A detailed study of soil, plant and animal relationship and the socio-economic background of the farmers is required.

Reference

Ibrahim M.N.M. (1989). Feeding tables for ruminants in Sri Lanka. Fibrous Feed Utilization Project, Sri Lanka-Netherlands Livestock Development Programme.

Variations in nutritive values of two different desert forage plants growing in the United Arab Emirate environment

T. Ksiksi, A. Bamakhrama and M. Satri
Biology Department, UAE University P.O Box 17551 Al-Ain, UAE, Email: tksiksi@uaeu.ac.ae

Keywords: acacia, hamada, prosopis

Introduction Forage plants in desert environments have to withstand both shortages in resources and over-grazing during most years. This variation in resource availability is associated with variation in the nutritive values of the plant species. Moreover, these differences are not only seasonal but also vary between plant parts. Variations in nutritive values in key species, therefore, need to be better understood in order to sustainably feed livestock (Abdurazak *et al.*, 2000), because effective management requires adequate knowledge of the interaction between the animal, the pasture and the environment (Kassilly, 2002). Forage toxicity could, however, cause irreparable damage to production. In the Gulf region, and especially in the United Arab Emirates (UAE), *Acacia tortilis* and *Prosopis cineraria* are considered important sources of feed for livestock. The aims of the present study were (1) to improve our understanding of variations in the nutritive values of *Acacia tortilis* and *Prosopis cineraria* grown in the UAE and (2) to quantify differences between parts of the two species.

Materials and methods Three composite samples of the above-ground plant parts of each of the two species were collected from several stands around Al-Ain, UAE (25°N 56°E) between October 2003 and May 2004. Samples were kept in paper bags and brought to the laboratory shortly after collection. Analysis for P, Ni, Cr, Na, Ca, K, Mg, N and total carbohydrates was carried out following the procedure of Allen *et al.* (1974). Total nitrogen was estimated using the Kjeldahl method and ash was estimated by ignition at 500 °C for about 24 hours. Nitrogen concentration was multiplied by 6.25 to estimate crude protein (CP). A 2-way (species and plant part) analysis of variance was used to compare means.

Results Mean concentration of each element analysed and carbohydrates, ash and CP for leaves and stems for *A. tortilis* and *P. cineraria* are given in Table 1. *A. tortilis* leaves had higher Al, Ni, Cr, Na, Ca, Mg and N contents than those of *P. cineraria* (P<0.05). The high levels of aluminium in *A. tortilis* leaves may be a cause of toxicity in livestock. Crude protein was lowest for *P. cineraria* stems and was between 11.9 and 14.6% for *P. cineraria* stems and *A. tortilis* leaves, respectively. These results are higher than the average reported by Vercoe (1986). The pronounced variations in chemical content among these two species is an indication of how unpredictable nutrient supply can be in arid environments.

Table 1 Mean chemical composition of stems and leaves of *A. tortilis* and *P. cineraria* growing in the UAE

Species	Part	Al	P	Ni	Cr	Na	Ca	K	Mg	N	CHO	Ash	CP
Acacia	Leaf	2816	858	26.2	9.5	0.17	3.5	0.45	0.73	2.3	34.1	20.3	14.6
Acacia	Stem	257	1096	6.6	2.2	0.05	1.6	0.37	0.12	2.1	33.7	13.0	13.4
Prosop.	Leaf	584	856	7.4	3.6	0.04	2.7	0.47	0.50	1.9	37.0	17.9	11.9
Prosop.	Stem	156	761	3.0	1.4	0.11	1.1	0.59	0.20	1.1	37.0	8.5	6.9

Conclusions The results of the present study show that *A. tortilis* leaves have very high levels of aluminium which may lead to toxicity in livestock. Stems of *P. cineraria* had the lowest level of CP, which suggests that supplements may be needed if the ration contains a high proportion of stems.

References
Abdurrazak, S.A., T. Fujihara, J.K. Ondiek & E.R. Qrskov (2000). Nutritive evaluation of some acacia tree leaves from Kenya. *Animal Feed Science and Technology*, 85, 89-98.
Allen, S.E. H.M. Grimshaw, J.A. Parkinson & C. Quarmby (1974). Chemical Analysis of Ecological Materials. Blackwell Scientific Publications, Oxford.
Kassilly, F.N. (2002). Forage quality and camel feeding patterns in central Baringo, Kenya. *Livestock Production Science*, 78, 175-182.
Vercoe, T.K. (1986). Fodder potential of selected Australian tree species. In: J.W.Turnbull (ed.) Australian Acacias in Developing Countries. *ACIAR Proceedings no. 16*, Australian Centre for International Agricultural Research, Canberra. 95-100.

Factors affecting forage quality of native species in Iranian rangelands

H. Arzani, J. Torkan, H. Kaboli and M. Zohdi
College of Natural Resources, University of Tehran, Karaj- Iran, Email: harzani@ut.ac.ir

Keywords: forage quality, plant parts, phenological stage, quality index

Introduction Animal performance is closely correlated with the nutrient value of the forage available and this is affected by different factors (Arzani *et al.,* 2001). Crude protein content (CP), digestible dry matter (DDM) and metabolisable energy (ME) were considered particularly appropriate for evaluation of range forage quality. This paper reports on factors affecting the forage quality of range species grown in Iran.

Material and methods Four main types of experiments were conducted. The first experiment investigated the effects of climate, soil, and phenological stage on quality of five grass species - *Agropyron tauri* (Agta), *Agropyron trichophorum* (Agtr), *Bromus tomentellus* (Brto), *Festuca ovina* (Feov) and *Hordeum bulbosum* (Hobu). In the second experiment, the proportions and quality characteristics of plant parts in different phenological stages were measured. In the third experiments key factors for forage quality assessment were determined and in the fourth experiment the effects of location on forage quality at the same time of year were investigated.

Results Analysis of variance showed that environmental conditions affect forage quality. The magnitude of the effects ranked in the order phenological stage (greatest effect), climate and soil characteristics (least effect). Nutritive values differed significantly (P<0.05) both within and among plant parts and phenological stages. For each species, the leaves had the highest nutritive value. Based on correlation between factors, measurements of crude protein and acid detergent fiber are more important than others. Forage components (%) varied between species, within and among phenological stages (eg. Table 1).

Table 1 Interaction between species (sp.), phenological stage (St.) and climate (Cl.) and chemical composition

Sp.*St.*Cl.	St.*Cl.	Sp.*Cl.	Sp.*St.	Climate	Stage	Species	Variation
ns	ns	ns	ns	***	**	***	Ash%
ns	ns	***	***	***	***	**	Nitrogen%
ns	ns	***	***	***	***	***	CP%
ns	ns	ns	*	ns	ns	**	E%
**	ns	**	***	ns	***	***	CF%
**	ns	ns	***	*	***	**	ADF%
ns	ns	ns	**	ns	***	***	NDF%
ns	ns	ns	**	ns	ns	**	ME(Mj/kgDM)
ns	*	ns	**	ns	***	**	Cu(ppm)
ns	ns	*	**	ns	***	ns	Fe(ppm)
ns	ns	ns	ns	ns	***	***	Zn(ppm)
ns	ns	**	***	**	ns	***	Mn(ppm)
**	ns	***	***	***	***	***	Mg%
***	ns	*	**	**	***	***	K%
ns	ns	ns	ns	ns	ns	ns	Na%
ns	ns	*	**	*	***	ns	P%
*	ns	*	*	*	***	***	Ca%

N ns non significant; *(p< 0.05); **(p<0.01); ***(p<0.001)

Discussion The nutrient value of available forage depends on the plants present because different species had different nutrient values. Seasonal condition is important because this influenced the species present and their composition (Orr & Holmes, 1984). Arzani *et al.,* (2001) also reported that with progress of plant growth, the ratios of tissues giving protection and rigidity increased. Therefore, structural carbohydrates and fibre contents increased with increased plant maturity late in the growing season. Forage with higher leaf to stem ratio would result better animal performance. This is an important factor for selecting the correct time of grazing.

References
Arzani, H., J. Torkan, M. Jafari, & A. Nikkhah (2001). Investigation on effects of phenological stages and environmental factors on forage quality. *Journal of Agricultural Science,* 32, 385-397.
Orr, D.M. & W.E. Holmes (1984). Mitchell grasslands. In: G.N. Harrington, A.D. Wilson & M.D. Young (eds.) Management of Australia's Rangelands. CSIRO, 241-254.

Determination and comparison of forage quality of five species in different phenological stages in Alborz rangelands (Iran)

A. Ahmadi[1], H. Arzani[2] and A.A. Jaafari[3]
[1]Islamic Azad University, Email: bsahmadi@yahoo.com, [2]Natural Resources Faculty, University of Tehran, [3]Range and Forest Research Institute

Keywords: phenological stages, forage quality, near infrared reflectance spectroscopy, Alborz rangelands

Introduction In order to evaluate grazing capacity and integrated management of rangelands and grasslands , it is necessary to be aware of the quality of range plants. Phenological (growth) stage has the greatest effect on forage quality with most of the qualitative indices decreasing with the progress of the phenological stage (Holecheck *et al*., 2001). This research studied the changes in forage quality for species in the Alborz rangelands.

Material and methods Sampling was performed using a randomised - systematic pattern for three grasses and two forbs - *Agropyron tauri, Dactylis glomerata, Bromus tomentellus, Ferula ovina* and *Coronilla varia.* Samples were collected at three phenological stages as initial growth , full flowering and seed ripening in estival ranges of central Alborz (vardavard and gachsar). Plant samples were dried and milled in the laboratory and analysed by wet chemistry for important qualitative factors such as: DM, NDF, ADF, CP, ME, DDM, DMI, WSC, relative feeding value (RFV) (AOAC, 1990). In a further step, the plant samples were scanned by near infrared spectroscopy. The instrument was calibrated against reference data for chemical composition. Then, the best regression equations between two methods (NIR and Labaratory) were fitted based on statistical methods. An example of the relationship is given in Figure 1. The standard error of calibration (SEC), standard error of prediction (SEP) and coefficient of determination (R^2) were determined as precision criteria for NIR.

Results As shown in Figure 2, there were significant differences ($p<0.01$) between different species and phenological stages (except for DM). Also, interaction effects of plant species and growth stages were significant. Almost in all species CP, DDM and ME decreased with progress of growth stages, while ADF and NDF contents increased. Generally, *Coronilla varia* had highest forage quality based on estimated factors (RFV=257.9) and in this aspect, *Agropyron tauri* had lowest quality (RFV=92.1). The first phenological stage was higher than the other stages for nutritive factors (except WSC)

Conclusions In view of relatively small differences between the first and second pheonological stages in quality characteristics and the higher production and range readiness for later stages of growth, the second stage (flowering) was determined as the most suitable period for animal grazing in the framework of grazing systems. The high correlation between two methods supported the potential use of NIRS as a fast, precise and efficient technique for prediction forage quality of range plants (Givens *et al*., 2000).

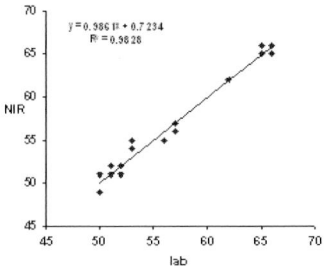

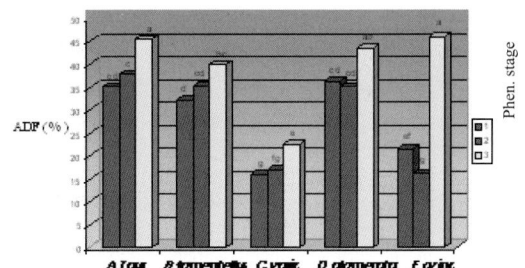

Figure 1 Regression relationship between two methods of NDF% measuring (NIR & Laboratory method.) in *Ferula ovina*

Figure 2 Interaction effect of species and phenological stages for ADF%

References

AOAC (1990). Official Methods of Analysis of the Association of Official Analytical Chemists. 15th ed. Washington D.C., USA.

Givens, D.L., E. Owen, R.F.E. Axford & H.M. Omed (2000). Forage Evaluation in Ruminant Nutrition. CABI Publishing,UK, 480 pp.

Holecheck, J.L., C.H. Herbel & R.D. Pieper (2001). Range Management Principles and Practices. 4th ed. Prentice Hall Publications, USA, 587 pp.

Variation in the quality of forage of six rangeland species in different phenological stages

Z. Ahmadi, H. Arzani and H. Azarnivand
Islamic AZAD University, Science & Research Campus, Iran, Email: aliabadi2004@yahoo.com

Keywords: forage quality, life forms, phenological stages, acid detergent fibre, crude protein, metabolisable energy

Introduction An understanding of forage quality is fundamental to the measurement of grazing capacity. In addition, knowledge of forage quality is necessary for planning grazing and developing range improvement and development programmes, such as planting and seeding of rangelands. Among the different factors that affect forage quality, phenological stage is particularly important (Cook, 1972; Caballero *et al.*,2001). The objective of this study was to investigate the effects of plant species, phenological stages and life form on forage quality indices.

Materials and methods Samples were collected from two highland ranges in the Vard Avard and Gachsar regions of Iran. There were six range species including *Bromus tomentellus* and *Dactylis glomerata* as grasses, *Frula ovina* and *Coronila varia* as forbs and *Salsola rigida* and *Artemisia aucheri* as shrubs. They were dried, ground and analysed according to standard methods.

Results There were significant differences between amounts of crude protein (CP), acid detergent fibre (ADF), digestible dry matter (DDM) and metabolisable energy between different species, phenological stages and life forms (P<0.01). Amongst the six species *C. varia* had the highest CP, ME and DDM with maximum values of 25.4%, 9.37 MJ/kg DM and 76.2% respectively, whilst the maximum value for ADF was 45.8% in *F. ovina*. In all species CP, ME, DDM decreased and ADF increased with advance in phenological stage. Among the three life forms and phenological stages, quality was highest in forbs at the vegetative stage and *C. varia* was the species with highest quality.

Conclusion These results suggest that the best time for grazing these rangelands was flowering stage because at this stage, there is a good compromise between yield (not reported here) and forage quality.

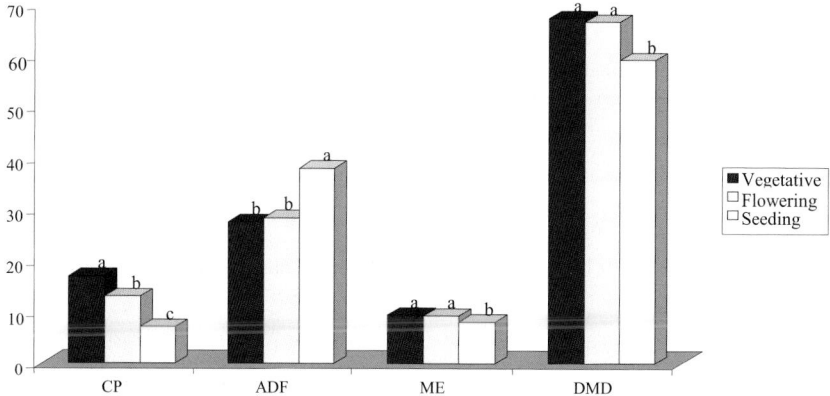

Figure 1 Values of crude protein, acid detergent fibre, metabolisable energy and dry matter digestibility at three phenological stages

References
Caballero, R., C. Alzueta, L. Tortiz, M.L. Rodriguez, C. Barro & A. Rebole (2001). Carbohydrate and protein fraction of fresh and dried common vetch at three maturity stages. *Agronomy Journal*, 93, 1006-1013.
Cook, C.W. (1972). Comparative Nutritive Value of Forbs, Grasses and Shrubs. USDA,General.Technical Report, INT– 1, 303-310.

Chemical composition and gas production of pasture dominated by *Artemisia frigida* and *Carex duriscula* species

G. Udval
The Research Institute of Animal Husbandry, Ulaanbaatar 210153, Mongolia,
Email: udval@gobi.initiative.org.mn

Keywords: pasture grass, gas production, organic matter digestibility, dry matter degradation

Introduction *In vitro* gas production from a feed sample incubated with a rumen fluid inoculum has been successfully used by Menke *et al.* (1979) and Steingass & Menke (1986) to predict the nutritive value of the substrate fermented. Recently, the gas test has been adapted so that the increase in gas production at a series of chosen time intervals is read off. The aim of this study was to describe the kinetics of fermentation of the feed incubated with reference to OMD, DMD, metabolisable energy (ME), neutral and acid detergent fiber (NDF, ADF), cellulose and lignin contents. Samples were harvested at different stages of maturity from *Artemisia frigida-Carex duriscula*-dominated pasture grasses and fed to goats.

Materials and methods Samples of *Artemisia frigida - Carex duriscula*-dominated pasture from the steppe zone of Mongolia were harvested at three stages: branching, flowering and fading. The samples (n=27) were dried and milled through a 1 mm screen. Dry matter was determined by drying to constant weight at 100°C. Ash was determined by ignition in a muffle furnace at 550°C for 8 h. Crude protein (CP) was determined as Kjeldahl nitrogen x 6.25. Neutral and acid detergent fibre (NDF and ADF) and lignin were determined according to Van Soest & Wine (1967). *In vitro* digestibility was determined by the procedure of Menke *et.al.* (1979) and Steingass & Menke (1986). Gas volumes were recorded after 3, 6, 12, 24, 48, 72 and 96 h of incubation.

Results The concentrations of NDF, ADF, lignin generally increased with maturity, whereas those of CP, ME and OMD decreased (Table 1). The degradation kinetics of the feed samples are shown in Table 2. Stage of maturity of the grasses was negatively correlated to *in vitro* DMD. Protein contents were significantly related (P<0.05) to ME (r = 0.980) and OMD (r = 0.987).

Table 1 Concentrations of cell wall constituents, lignin, CP, ash (all as % DM), OMD (%) and ME (MJ/kgDM)

Samples Name	Maturity	NDF	ADF	Lignin	CP	Ash	OMD	ME
Artemisia frigida-Carex duriscula	branching	53.9±0.01	37.5±0.05	16.3±0.1	13.6±0.5	8.47±0.3	63.7±0.6	9.2±0.3
	flowering	56.1±0.01	38.05±0.02	18.7±0.4	12.8±0.6	10.7±0.3	56.0±0.2	8.4±0.5
	fading	69.5±0.02	43.9±0.03	20.8±0.9	8.9±0.8	8.47±0.6	46.2±0.3	6.5±0.2

Table 2 *In vitro* DMD of the tested grasses at the different stages of maturity (gas volume, ml)

Incubation time, h	Stage maturity branching	flowering	fading
0	29.9±0.4	29.8±0.2	30.0±0.1
3	40.0±0.9	39.6±0.9	35.0±0.3
6	47.8±1.5	46.4±1.5	38.7±0.6
12	57.9±1.1	54.8±1.4	44.1±0.8
24	70.0±1.9	65.5±1.0	57.4±1.8
48	80.9±1.5	76.9±2.1	64.0±3.6
72	86.1±1.4	82.3±1.1	70.0±3.7
96	88.2±4.1	84.3±0.1	76.1±4.7

Cumulative gas production *in vitro* (ml/200 mg DM) of the tested samples was 44.9, 41.2, 37.8 for the branching, flowering and fading stage of maturity respectively.

Conclusions These trials demonstrate that the stage of maturity of the grass is negatively related to *in vitro* DMD, OMD and ME. The information on nutritive value will provide guidance on ways to utilise this important association.

References
Khazaal, K. & Ørskov, E. R. (1995). Prediction of apparent digestibility and voluntary intake of hays fed to sheep. *Animal Science,* 61, 527-538.
Menke, K.H., L. Raab, A. Salewski, H. Steingass, D. Fritz & W. Schneider (1979). The estimation of the digestibility and metabolizable energy content of ruminant feedstuffs from the gas production when they are incubated with rumen liquor *in vitro. Journal of Agricultural Science,* 93, 217-222.
Van Soest P.J. and Wine R.H. (1967). Use of detergents in the analysis of fibrous feeds. Determination of plant cell-wall constituents. *Journal of the Association of Official Analytical Chemists,* 50: 50-55.

Seasonal variation of crude protein content of different herbaceous, shrub and tree species

M.R. Mosquera-Losada, E. Fernández-Núñez and A. Rigueiro-Rodríguez
Departamento de Producción Vegetal. Escuela Politécnica Superior de Lugo, Spain, Email: romos@lugo.es

Keywords: dicotyledonous plant, monocotyledonous plant, twig, silvopastoral

Introduction Silvopastoralism is a sustainable way of land management that reduces fire risk due to the reduction of fuel under trees when plants are used as animal food. This is particularly important in areas like Galicia that have 16% of the fired area of Europe. Silvopastoral systems can contribute to environment conservation and provide feed for autochthonous breeds more adapted to mountain conditions, enhancing biodiversity conservation. In formation on seasonal changes in crude protein content of spontaneous species will indicate better management of pasture resources in mountain areas.

Materials and methods The experiment was conducted in A Fonsagrada (Galicia, NW Spain) at an altitude of 800m above sea level. Mean temperature during winter and spring was below 7°C and mean annual precipitation was around 1700mm. The chosen species grew spontaneously in acid soils (water pH of 5.01) with low nutrient availability and high percentage of saturated aluminium. Samples from each area were taken monthly using hand scissors and cutting young buds (including leaves and twigs) with diameter less than 0.5mm, sampling shrubs and trees up to a height of 1m, because this is the height that can be reached by animals. The analysed species were herbaceous: (*Agrostis duriaei* Boiss et Reuter ex Willk, *Agrostis stolonifera* L., *Holcus lanatus* L., *Lolium multiflorum* Lam, *Mentha suaveolens* Ehrh, *Plantago lanceolata* L., *Pteridium aquilinum* (L) Kuhn , *Rumex acetosa* L., *Trifolium pratense* L., *Trifolium repens* L.), shrubs (*Cytisus striatus* (Hill) Rothm, *Chamaespartium tridentatum* (L) P. Gibbs., *Daboecia cantabrica* (Hudson) C. Koch, *Erica arborea* L., *Rubus ulmifolius* Schott, *Ulex europaeus* L., *Ulex minor* Roth.) and trees (*Alnus glutinosa* gaertn , *Betula* sp L., *Corylus avellana* L., *Fagus sylvatica* L., *Fraxinus excelsior* L., *Quercus pyrenaica* Willd., *Quercus robur* L.) Samples were dried and milled and N was analysed after microkjeldahl digestion. Statistical analysis was by ANOVA.

Results Crude protein content ranged between 6.7-10%, 11-15%, 8-13% and 9-15% for herbaceous monocots, herbaceous dicots, shrubs and tree twigs, respectively. These values are normal for pasture species in the area and are sufficient to cover the requirements of cows and horses for maintenance (Mosquera *et al.*, 2000). Protein concentration was higher during the spring and lower during the winter for all the groups with herbaceous dicots having higher crude protein than herbaceous monocots in most of the periods. Tree species had higher protein content than shrubs, with the exception of autumn; most of the evaluated species were broadleaved and therefore lost leaves during the autumn. There were no significant differences between species groups during the winter as low temperature reduced plant development. Traditional management in the area consisted of tree branches being harvested for feeding animals during periods of pasture shortage like summer or winter.

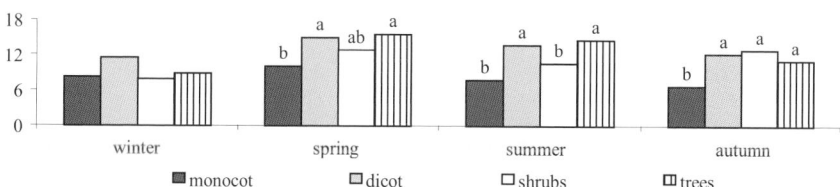

Figure 1 Seasonal concentration of crude protein for different classes of feed. Different letters indicate significant differences (P<0.05) between means

Conclusions The measured protein contents indicate that tree and shrub species can be used as a resource for animal feeding during the autumn and summer in the Galician mountain area at a time when spontaneous monocots had very low protein concentration. This allows a sustainable use of fodder resources in fragile areas.

References
Mosquera-Losada, R., A. González-Rodríguez & A. Rigueiro-Rodríguez. (2000). Ecología y manejo de praderas. Ed. Ministerio de Agricultura, Spain.
N.R.C. (1978). Nutrient Requirements of Domestic Animals. Ed. National Academy Press, USA.

Nutrient accumulation in leaves and soft twigs of *Moringa oleifera* Lam. at different growth stages in Western Highland of Cameroon

T.E. Pamo[1], B. Boukila[2], J.R. Kana[1], F. Tendonkeng[1], L.B. Tonfack[1] and M.C. Solefack Momo[1]
[1]*Animal Nutrition Laboratory, Department of Animal Production, FASA, University of Dschang. P.O.Box. 222 Dschang, Cameroon, E-mail: pamo_te@yahoo.fr / pamo-te@excite.com,*[2]*I.N.S.A.B. Université des Sciences et Techniques de Masuku, BP. 941 Gabon*

Keywords: *Moringa oleifera*, nutrients, growth stages, leaves, soft twigs

Introduction Moringa oleifera belongs to the Moringaceae family and is considered to have its origin in the south of the Himalayan mountains. The species is being introduced into the highland zone of Cameroon. It is a tree which has many valuable properties and it is of great nutritional and scientific interest. The objective of this experiment was to evaluate nutrient composition in leaves and soft twigs of M. oleifera at different growth heights when grown in the Western Highland of Cameroon.

Material and methods Seeds of M. oleifera from Nicaragua were directly planted on plots (1m x 1m), at 30cm x 30cm density on the clay soil at the Research and Teaching Farm of the University of Dschang located at an altitude of 1400 m in December 2003. Leaf samples and soft twigs of the plants were collected at different growth heights, i.e.: at 0.5, 1, 1.5 and 2 m for leaves and 1, 1.5 and 2 m for soft twigs respectively. Organic matter (OM), ash, crude protein (CP), lipids and phosphorous were analysed according to the methods described by AOAC (1990), while cellulose (Cell) and hemicellulose (H-cell) were determined as described by Goering & Van Soest (1970). For the non-fibrous sugar, the method described by Wattiaux (2002) was used. The results were submitted to ANOVA and the differences between means were tested by the Duncan multiple range test (Steel & Torrie, 1980).

Results Stage of growth had relatively less effect on OM (88.9-90.5%) and ash (9.5-10.7%) contents of Moringa leaves and soft twigs (Table 1). Lipids were significantly ($p<0.05$) higher in leaves than in soft twigs. While hemicellulose was significantly ($p<0.05$) higher in leaves than in soft twigs, cellulose was found to be greater in soft twigs than in leaves. Both constituents were relatively high in samples at 2 m growth. Crude protein contents were threefold higher in leaves than in soft twigs. At 0.5 m height CP of leaves (25.8%) and its phosphorous content (0.38%) was significantly ($p<0.05$) higher than in all other growth stages. Non-fibrous sugar was significantly ($p<0.05$) greater in 1 m height soft twig than in all other parts of the plant and growth stages (Table 1).

Table 1 Nutrient contents in leaves and soft twigs of M. oleifera at different growth heights

Plant height (m)	Plant parts	Chemical composition (%DM)							
		OM	Ash	P	CP	Lipids	H-cell	Cell	Non-fibrous sugar
0.5	Leaves	89.3	10.7	0.38	25.8	5.5	12.7	20.2	15.6
1.0	Leaves	88.9	10.1	0.20	22.2	6.5	12.4	21.1	17.5
	Twigs	90.5	9.5	0.12	8.2	2.4	7.6	40.2	22.7
1.5	Leaves	89.3	10.7	0.32	22.4	5.8	13.6	21.5	14.9
	Twigs	89.6	10.5	0.24	9.4	2.7	7.8	40.6	17.9
2.0	Leaves	89.8	10.1	0.27	20.8	5.7	13.8	26.6	16.4
	Twigs	90.2	9.8	0.18	8.5	2.6	7.9	42.2	18.1

Conclusion Results of the present study indicates that in the Western Highland of Cameroon, *Moringa oleifera* grows relatively well and crude proteins and phosphorous contents of leaves are highest in plants of 0.5 in height, whereas the fibrous contents increases with the age in leaves and soft twigs.

References
A.O.A.C. (1990). Official Methods of Analysis. 15th edition. Washington DC.
Goering, H.K. & P.J. Van Soest (1970). Forage fiber analysis. Agricultural Reseach Service. Agricultural Handbook no. 379, 1-11.
Steel, R.G. & J.H. Torrie (1980). Principles and Procedures of Statistics, McGraw Hill, New York, 635pp.
Wattiaux, M. A. (2002). Composition et analyse des aliments. In: J. Nysenholc and Y. Berger (eds) Recherche et Développement International du Secteur Laitier. Université de Wisconsin, Madison, USA. 4 pp.

Effect of type of tree leaves on intake, nutrient utilisation and rumen fermentation pattern in goat fed with *Cenchrus ciliaris* grass

S. Singh, S.S. Kundu and L.K. Karnani
Indian Grassland and Fodder Research Institute, Jhansi-284003 (UP) India, Email: sultan@igfri.up.nic.in

Keywords: goat, nutrient utilisation, tree foliage, fermentation pattern

Introduction Grasses and tree leaves/shrubs are the major feeds available from grasslands and grazing fields. There is though a dearth of information on the nutritional value of tree foliage although available information was collated by Shelton *et al.* (1995). This study evaluated the relative nutritive value of two foliages for goats.

Materials and methods Four adult goats (local breed) with average body weight of 29 kg were fed *Cenchrus ciliaris-Leucaena leucocephala* (CL) and *C. ciliaris-Gliricidia optiva* (CG) in 50:50 proportions in a switch-over experiment. Goats were maintained on each dietary regimen for 3-4 months. After 45 d, a digestion trial was conducted to evaluate relative dry matter intake (DMI) and nutrient utilisation. Rumen liquor was collected (0 & 4 h post feeding) to estimate volatile fatty acid (VFA) and N-metabolite production. Data was analysed according to Snedecor & Cochran (1967).

Results and discussion Crude protein and lignin contents were 20.10 and 8.10 % in *L leucocephala* and 16.87 and 2.76% in *G. optiva*, respectively. Dry matter intake (g/d) was comparable on the two diets. There was significantly (P<0.05) higher CP digestibility on CG than CL. This may be due to lower tannin and lignin contents in *G. optiva* than *L. leucocephala*. The DMD was 10.17 units higher on CG than CL. Digestibility coefficients of cell wall polysaccharides (NDF, ADF and cellulose) were 10.5, 11.0 and 12.8 units higher on CG compared with CL. Mean pH and TVFA values (0 and 4 h post feeding) in rumen liquor were 7.01 and 85.27 on CL and 6.6 and 89.10 meq/l on CG based diets. Total N (mg/100ml) tended to be higher in the rumen liquor of goats fed CL than CG. Concentrations of NH_3-N and TCA soluble-N were comparable on both the diets. Similar results for intake, nutrient digestibility and metabolite production for goats fed varying proportions of roughage and tree leaves have been reported by Adeloye (2000) and Bamikole *et al.* (2003).

Conclusions The goats utilised nutrients more efficiently from CG than CL based diets and rumen metabolite production was similar. *Gliricidia optiva* can be used as a quality fodder tree in grassland and silvipasture systems for improved goat production.

Table 1 Dry matter intake and nutrient digestibility in goats fed grass: tree leaf based diets

Diets			Nutrient digestibility (%)					
	g/d	g/kg w0$^{.75}$	DM	OM	CP*	NDF	ADF	Cellulose*
C. ciliaris-L. leucocephala	733.2	61.14	47.06	49.81	46.76	46.56	40.08	50.20
C. ciliaris-G. optiva	771.7	60.9	57.33	58.63	71.18	57.00	51.00	63.00

    * Differ significantly at P<0.05 level

Table 2 Rumen metabolite concentrations in goats fed grass: tree leaf based diets

Diets	Rumen metabolites (mg/100ml)				
	pH	TVFA	Total-N	NH_3-N	TCA-soluble-N
C. ciliaris-L. leucocephala	7.01	85.27	104.35	17.36	68.41
C. ciliaris-G. optiva	6.60	89.15	98.25	16.81	70.43

References

Adeloye, A.A. (2000). *Albizia lebbeck* in N supplementation of sorghum glume: nutrient digestion and N utilization of sheep. *Seventh International Goat Conference*, Tours, France, 84-85.

Bamikole, M. A., O.J. Babayemi, O.M. Arigbede and V. J.(2003). Nutritive value of *Ficus religiosa* in West African dwarf goats. *Animal Feed Science and Technology*, 105, 71-79.

Shelton, H. M., C.N. Piggin & J.L. Brewbaker (1995). Leucaena - opportunities and lmitations. *ACIAR Proceedings*, 57, Canberra, Australia.

Snedecor, G.W. & W.G. Cochran (1967). Statistical Methods, Sixth ed. IBH Publishing Co., Oxford and New Delhi, 258-298.

Ensiling characteristics and nutritive value of browse/maize forage mixtures

F.B. Bareeba, H. Kato and E.N. Sabiiti
Faculty of Agriculture, Makerere University, P.O.Box 7062, Kampala, Uganda, Email: fbareeba@agric.mak.ac.ug

Keywords: browse, maize, silage, dairy cows, intake

Introduction The practice of growing fodder tree and shrubs is being advocated for and adopted in smallholder dairy production systems. In Uganda, *Calliandra calothyrsus*, *Gliricidia sepium*, and *Leucaena leucocephala* have been identified and recommended as the most suitable species (Sabiiti, 2001). However tree foliage contains toxic compounds (Lowry, 1990), which may be alleviated by ensiling. The objective of the experiment was to study the ensiling characteristics of browse/ maize forage mixtures and their nutritive value when fed to lactating dairy cows.

Materials and methods Calliandra, Gliricidia and Leucaena tree foliages were ensiled with maize forage in a ratio of 1: 5 (dry matter (DM) basis). Silage was also made from maize forage alone. The silages were fed to Friesian dairy cows in mid-lactation in a 4x4 Latin square design with 28-d periods to determine DM intake, milk yield and composition. In addition, cows were supplemented with 4 kg/d of a commercial dairy meal.

Results All silages fermented well, but lactic acid content was higher (P<0.05) for maize silage compared to the browse/maize silages. Dry matter losses were higher (P<0.05) for the maize silage. Silage DM and total DM consumption were higher (P<0.05) for Calliandra and Gliricidia silages compared to Leucaena or maize silages. Milk yield followed the same trend. While milk fat was similar among treatments, milk protein was higher (P<0.05) for the maize silage compare to the other silages.

Table 1 Fermentation characteristics (% DM) of the silages

Silages	Calliandra	Gliricidia	Leucaena	Maize	s.e.m.
DM %	31.8[a]	28.5[b]	26.1[b]	23.3[bc]	1.0
CP %	16.7[a]	10.7[b]	11.9[b]	11.7[b]	1.1
Acetic acid	2.11	1.95	2.16	2.16	0.13
Butyric acid	0.12[b]	0.79[a]	0.65[a]	0.29[b]	0.09
Lactic acid	4.75[b]	3.26[c]	3.73[bc]	6.20[a]	0.40
pH	4.33	4.18	3.81	3.67	0.29
NH$_3$-N (% total N)	5.49	5.97	6.24	5.73	0.28
DM losses %	5.01[b]	0.28[c]	7.41[b]	13.50[a]	1.83

Table 2 Dry matter intake, milk yield and composition for cows fed the four silages

Silages	Calliandra	Gliricidia	Leucaena	Maize	s.e.m.
DMI (kg/d)					
Silage	10.50[a]	9.68[ab]	9.04[b]	9.23[b]	0.40
Total	14.16[a]	13.34[ab]	12.70[b]	12.89[b]	0.40
g/kgW$^{3/4}$	163.33[a]	154.74[ab]	145.85[b]	148.61[b]	3.65
Milk yield (kg/d)	9.72[a]	9.86[a]	9.46[b]	9.04[c]	0.07
BF %	3.86	3.81	3.81	3.81	0.04
4% FCM (kg/d)	9.56[a]	9.57[a]	9.19[b]	8.78[c]	0.09
Protein %	2.62[b]	2.65[ab]	2.62[b]	2.69[a]	0.02

Conclusion Inclusion of the browses did not affect silage fermentation. Silages made from a mixture of browse and maize resulted in higher DM intakes and milk yields of dairy cows than with silage made from maize alone.

References

Lowry, J.B. (1990). Toxic factors and problems: methods of alleviating them in animals. In: C. Devendra (ed.). Shrubs and Tree Fodders for Farm Animals. IDRC, Ottawa, Canada, 276pp.

Sabiiti, E.N. (2001). Pastures and range management. In: J.K. Mukiibi (ed.). Agriculture in Uganda. NARO, 237-297

The effect of fermentation of *Calliandra calothyrsus, Gliricidia sepium, Leucaena leucocephala* and maize forage on rumen degradation and microbial protein synthesis

H. Kato, F.B. Bareeba, E.N. Sabiiti and C. Ebong
Department of Animal Science, Makerere University, P.O. Box 7062, Kampala, Uganda, Email: fbareeba@agric.mak.ac.ug

Keywords: fermentation, degradation, microbial protein

Introduction Ensiling as a means of storing green fodder by acidification has a profound effect on the chemical composition of the resultant silage. Ensiling could therefore, ameliorate the effects of anti-nutritional factors associated with browses. The objective of the experiment was to determine fermentation characteristics and rumen degradation (D) of ensiled *Calliandra calothyrsus (C), Gliricidia sepium (G), Leucaena leucocephala (L)* and maize forage.

Materials and methods Chopped materials (<5cm) of the browses (leaf and petiole) and maize forage at milk stage were ensiled in triplicate 2kg lots in polythene bags, which were tightly packed and tied and were kept in the laboratory and allowed to ferment for 30 days. Their fermentation characteristics and resulting organic matter were determined. Degradation constants of fermented and unfermented browses and maize forage were studied using two fistulated *Bos indicus* steers. The exponential equation of McDonald (1981) was used to determine the constants. Total rumen microbial protein (TRMP) yield (g/kg DM) was estimated as: TRMP = (FOM/1000) x 150, where DM is dry matter and FOM = (ED x OM/1000) x OM, where ED is effective degradability and FOM is fermentable organic matter (Muia *et. al.,* 2001).

Results Maize forage had higher (P<0.05) lactic acid and lower (P<0.05) pH compared to the browses. Of the browses, *Gliricidia* fermented better with higher (P<0.05) lactic acid. Fermentation increased degradation of OM and RMP synthesis with *Gliricidia and Leucaena* but not with *Calliandra.* The poor degradation and RMP synthesis with *Calliandra* could be attributed to its high content of lignin and tannins, which have a binding effect (Fahey *et al.,*1980).

Table 1 Fermentation characteristics and degradation of the browse and maize forage silages OM

	Maize	C	G	L	SE
DM (%)	25.9[b]	35.6[a]	23.5[c]	24.4[bc]	0.73
Lactic acid (% DM)	5.0[a]	0.9[d]	2.8[b]	1.9[c]	0.21
PH	3.9[b]	5.4[a]	5.1[a]	5.3[a]	0.24
NH$_3$-N (% Total N)	10.3[a]	1.3[c]	5.6[b]	7.3[b]	0.03
Potential D(u)	566.7[b]	250.0[d]	554.1[c]	573.2[a]	t=1.968
Effective D(u)	371.1[b]	206.5[d]	432.6[a]	331.8[c]	t= 1.968
M P g/kg OM(u)	52.1	29.3	59.5	46.8	
Potential D(f)	843.9[a]	233.8[d]	810.3[b]	687.8[c]	t= 1.968
Effective D(f)	507.67[b]	207.6[d]	595.0[a]	374.0	t= 1.968
M P g/kg OM(f)	71.3	29.5	81.8	52.3	

[abc] Values having different superscripts in a row are significantly (P≤0.05) different
Unfermented (u), Fermented (f)

Conclusion *Calliandra* has poor fermentation with low levels of lactic acid. Fermentation increased rumen degradation of OM and RMP synthesis with *Gliricidia, Leucaena* and maize forage but not with *Calliandra.*

References
Fahey, Jr., G.C., S.Y. At-Haydari, F.C. Hindis & D.E. Short (1980). Phenolic compounds in roughages and their fate in the digestive system of sheep. *Journal of Animal Science,* 50, 1165-1172.
McDonald, I. (1981). A revised model for the estimation of protein degradability in the rumen. *Journal of Agricultural Science, Cambridge,* 96,251-252.
Muia, J. M. K., S. Tamminga,, P.N. Mbugua & J.N. Kariuki (2001). Rumen degradation and estimation of microbial protein yield and intestinal digestion of Napier grass (*Pennisetum purpureum*) and various concentrates. *Animal Feed Science and Technology,* 93,177-192.

Theme A: Efficient production from grassland

Section 6

Impacts of endophytic fungi and other biotic interactions on grassland production

Persistence of tall fescue and cattle grazing preference as affected by endophyte status

D.J. Lang[1], S.P. Wang[2], A. Tokilita[1], R. Given[1], M. Salem[1] and R. Elmore[1]
[1]Dept. of Plant and Soil Sciences, Mississippi State University, Mississippi, MS 39762-9555, USA, Email: dlang@pss.MsState.edu, [2]Laboratory of Quantitative Vegetation Ecology, Institute of Botany, Chinese Academy of Sciences, Beijing, 100093, China

Keywords: *Lolium arundinaceum*, *Neotyphodium coenophialum*, toxicity

Introduction Endophyte-infected (E+) grasses often exhibit increased survival, growth and resistance to herbivory compared to uninfected counterparts. Latch (1997) proposed a strategy for cultivar improvement of infecting elite cultivars with strains of *Neotyphodium coenophialum* that are non-toxic to livestock, but still able to convey the persistence advantage shown with wild-type, toxic endophyte. The strategy of re-infecting tall fescue (*Lolium arundinaceum* (Schreb.) Darbysh.) cultivars with naturally occurring, non-ergot-producing endophytes appears promising for removing animal toxicity symptoms and retaining agronomic performance (Bouton *et al.*, 2002; Hill *et al.*, 2002). The objective of this study was to compare the effect of endophyte status on grazing preference and persistence of tall fescue stand.

Materials and methods Tall fescue with three levels of endophyte [E-, toxic E+, and non-toxic E+ MAXQ® (NE+)] was no-till planted with a Tye® drill on 27 October 1999. Plots 6x20 m replicated 4 times within a 0.5 ha pasture on a Marietta soil (fine-loamy, siliceous, thermic Fluvaquentic Eutrochrepts) containing bermudagrass (*Cynodon dactylon*). Fertiliser was applied to supply 50 kg N/ha in the fall and spring of each year with lime, P and K applied according to soil test. Steers and/or cows (*Bos taurus*) were stocked continuously at 3000 kg live weight/ha from January through June each year and limit grazed during the summer, autumn, and early winter. Animals were observed for 1 h periods several times and their location within the plots was recorded every 2 minutes. Tall fescue stand coverage was determined visually from 2000 to 2003.

Results All groups of animals preferred to graze E- tall fescue (Table 1). Preference for novel E+ tall fescue was generally intermediate between toxic E+ and E- tall fescue with the exception of the three cows introduced to the plots in May 2002. These animals avoided both toxic E+ and novel E+ tall fescue. There was little preference for any endophyte status during January 2002, perhaps because of lower ergot alkaloid levels. Alkaloid levels were not measured in this study, but other workers have reported lower levels in mid-winter compared with spring or early summer in tall fescue (Belesky et al., 1987). Stand of E- tall fescue declined to 25% while stand of E+ and NE+ tall fescue remained greater than 75% and were similar to each other.

Table 1 Grazing preference of steers or cows for tall fescue as affected by endophyte status 2000 to 2002 and stand persistence of tall fescue 2000 2003

Endophyte Status	2000 April	2001 March	April	2002 January	April	May	2000 April	2001 March	2002 April	2003 May
	----------- Grazing Time min/animal/hr -------------						--------% Stand Persistence ---------			
E-	2.1	1.7	1.3	0.8	2.0	1.4	78	59	75	25
Toxic E+	0.5	1.0	1.0	0.7	0.5	0.6	90	70	87	85
Novel E+	1.0	1.2	1.2	0.9	1.6	0.4	88	71	85	78
LSD P<0.05	1.4	0.6	0.6	0.2	0.7	0.2	13	20	11	19

Conclusions Stand persistence of novel non-toxic endophyte-infected tall fescue was similar to toxic E+ tall fescue. Cattle preferred to graze E- tall fescue plots indicating that they could detect and avoid grazing tall fescue that was toxic E+ infected while their preference for NE+ tall fescue was generally intermediate.

References
Belesky, D.P., J.D. Robbins, J.A. Stuedemann, S.R. Wilkinson & D.J. Devine (1987). Fungal endophyte infection-loline derivative alkaloid concentration of grazed tall fescue. *Agronomy Journal,* 79, 217-220.
Bouton, J.H., G.C.M. Latch, N.S. Hill, C.S. Hoveland, M.A. McCann, R.H. Watson, J.A. Parish, L.L. Hawkins, & F.N. Thompson (2002). Reinfection of tall fescue cultivars with non-ergot alkaloid-producing endophytes. *Agronomy Journal,* 94, 567-574.
Hill, N.S., J.H. Bouton, F.N. Thompson, L. Hawkins, C.S. Hoveland & M.A. McCann (2002). Performance of tall fescue germplasms bred for high and low-ergot alkaloids. *Crop Science,* 42, 518-523.
Latch, G.C.M. (1997). An overview of *Neotyphodium*-grass interactions. In: C.W. Bacon & N.S. Hill (eds). *Neotyphodium*/Grass Interactions. Plenum Press, New York, 1-11.

Performance of animals grazing various tall fescue / endophyte combinations

A.A. Hopkins
Noble Foundation, Inc. 2510 Sam Noble Parkway, Ardmore, Oklahoma 73401 USA Email: aahopkins@noble.org

Keywords: alkaloids, health, *Neotyphodium coenophialum,* persistence, productivity

Introduction Endophyte (*Neotyphodium coenophialum*) infection can improve tall fescue persistence (Read & Camp, 1986), but animals grazing wild type tall fescue often suffer from reduced performance and poor health (Hoveland *et al.*, 1983). Novel endophytes do not produce the alkaloids implicated in animal toxicity and can help maintain stand longevity while minimizing animal health problems (Bouton *et al.*, 2002). The objectives of this research were to compare productivity and health of animals grazing various tall fescue / endophyte combinations, and the persistence of these combinations under grazing in the Southern Plains of the USA.

Materials and methods Three replications of tall fescue paddocks with novel (GA-5 MaxQ), wild type (GA-5 E+, KY-31 E+), or no endophyte (Dovey, GA-5 E-), were planted in 1999 in southern Oklahoma. Beef cattle grazed paddocks for 70 to 85 days in both autumn and spring, at stocking rates of about 1000 to 1600 kg liveweight / ha, from 2000 to 2004. Before and after grazing periods, cattle were weighed and body temperatures taken. Stand data were collected using a grid method. Endophyte infection level and type were monitored using commercial immunoblot and ELISA kits.

Results Average daily gain (ADG) did not differ among cattle in fall (Table 1). In spring, ADG of cattle was greatest from paddocks of Dovey and GA-5 MaxQ, intermediate for GA-5 E- and GA-5 E+, and least for KY-31 E+ (Table 1). Gains per acre followed the same pattern. Persistence of entries did not differ, with full stands being maintained after four years of grazing (Table 1).

Table 1 Animal gains averaged over three years, stands, and endophyte status (May, 2004) of 5 tall fescues

	g/day		kg / ha		Stand %		Endophyte %	
	Autumn	Spring*	Autumn	Spring *	2001*	2004	Wild type	Novel
Dovey	748	635 a	237	218 a	97b	80	3.0	0.0
GA5 E-	703	563 ab	221	190 ab	99a	95	8.8	0.0
GA5 MaxQ	771	667 a	244	227 a	99a	99	0.0	75.4
GA5 E+	708	467 b	226	162 b	97b	92	62.0	0.0
KY-31 E+	612	290 c	194	100 c	100a	100	75.8	0.0

*Means followed by a different letter within a column are significant (P < 0.10).

Animals on KY-31 E+ were roughly 0.8°C warmer than all other animals following spring grazing, otherwise body temperatures did not differ. Infection levels for wild type and novel endophyte pastures ranged from 62 to 76%; contamination levels for all pastures were less than 10% throughout the experiment.

Conclusions In the Southern Plains of the USA, where tall fescue is not widely used, stands on soils with good moisture holding capacity can persist well with moderate grazing pressure. Novel endophyes are a viable option for minimizing animal health and performance problems associated with tall fescue endophytes.

References
Bouton, J.H., G.C.M. Latch, N.S. Hill, C.S. Hoveland, M.A. McCann, R. H. Watson, J.A. Parish, L.L. Hawkins & F.N. Thompson (2002). Reinfection of tall fescue cultivars with non-ergon alkaloid-producing endophytes. *Agronomy Journal*, 94, 567-574.
Hoveland, C.S., S.P. Schmidt, C.C. King, J.W. Odom, E.M. Clark, J.A. McGuire, L.A. Smith, H.W. Grimes & J.L. Holliman (1983). Steer performance and association of *Acremonium coenophialum* fungal endophyte on tall fescue pasture. *Agronomy Journal*, 75, 821-824.
Read, J.C. & B.J. Camp (1986). The effect of fungal endophyte *Acremonium coenophialum* in tall fescue on animal performance, toxicity, and stand maintenance. *Agronomy Journal*, 78, 848-850.

Recovery of yearling calves from Fescue Toxicosis

G.E. Aiken[1], M.L. Looper[2], S.F. Tabler[2] and J.R. Strickland[1]
[1]*Forage-Animal Production Research Unit, Lexington, KY, U.S.A.* ,[2]*Dale Bumpers Small Farms Research Center, Booneville, AR, U.S.A., Email: geaiken@ars.usda.gov.*

Keywords: *Lolium arundinaceum,* tall fescue, ergot alkaloids, fescue toxicosis

Introduction Tall fescue (*Lolium arundinaceum* (Schreb.) S.J. Darbyshire) is widely utilized for grazing in the transition zone between the temperate and subtropical regions of the eastern U.S.A. Cattle grazing tall fescue frequently exhibit fescue toxicosis, a malady caused by consumption of toxins produced by the endophyte, *Neotyphodium coenophialum.* Symptoms of fescue toxicosis include retention of rough hair coat, increased body temperature and laboured respiration. Heat stress may be severe at onset of high ambient temperature and humidity. Transporting cattle exhibiting toxicosis can therefore be difficult because combined stresses of the toxicosis and transporting often result in high mortality. An experiment was conducted to measure trends in rectal temperatures for yearling steers following removal from tall fescue and placement on a fescue-free diet.

Materials and methods A grazing experiment was conducted at the USDA-ARS Dale Bumpers Small Farms Research Center in Booneville, AR to evaluate interactions between implantation with anabolic agents and stocking rate on steer weight gain. At the conclusion of the experiment on 22 June 2004, the steers (n = 36) were placed in a common pasture of bermudagrass (*Cynodon dactylon* (L.) Pers.), and rectal temperatures were collected at 0, 24, 48, 72, 144, 192, and 216 h after removal from tall fescue pastures. Upon removal from pastures, hair coats were rated as being rough, transitional, or sleek. Rectal temperatures were statistically analyzed with a PROC MIXED model of SAS evaluating previous treatment effects as a discrete variable and temporal effects as a continuous variable.

Results Previous treatments did not influence (P>0.10) trends in rectal temperatures. Over 95% of the calves had either rough or transitional hair coats, which clearly indicated that nearly all of the cattle were exhibiting toxicosis and heat stress (Figure 1). There was a quadratic (P < 0.001) change in rectal temperatures with time on the fescue-free diet (Figure 2). Rectal temperature increased from 0 to 24 h. This was likely related to an increase in mean daily temperatures from 26.3 to 29.7 °C between the first and second days of the experimental period. Rectal temperatures were high until approximately 120 h, but then showed a substantial decline in temperature by 196 h and a greater decline at 216 h. Rectal temperatures measured at 196 h (8 d) and 216 (10 d) indicated an improvement in health status.

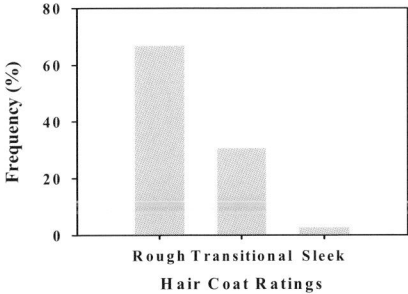

Figure 1 Hair coat ratings

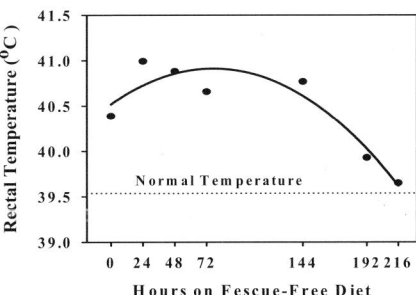

Figure 2 Relationship between hours on fescue-free diet and rectal temperature

Conclusion Results indicated that heat stress can be reduced in cattle exhibiting fescue toxicosis if they are provided diets free of endophyte-infected tall fescue for 8 to 10 days. Although complete alleviation of fescue toxicosis may not be claimed, an improvement in health status was apparent.

Reduce tall fescue toxicosis in *Festuca arundinacea* forage systems with legumes

J.C. Waller, A.E. Fisher, H.A. Fribourg and C.J. Richards
The University of Tennessee, 2505 River Drive, Knoxville, Tennessee, 37996-4574 USA, Email: jwaller@utk.edu

Keywords: *in vitro*, forage quality, *Neotyphodium coenophialum*, *Trifolium pratense*, *Trifolium repens*

Introduction Tall fescue (*Lolium arundinaceum = Festuca arundinacea*) is an important cool-season grass because of persistence and spring and autumn productivity. Most existing tall fescue pastures in Tennessee are Kentucky-31 (KY-31) and are infested with the endophytic fungus *Neotyphodium coenophialum* that causes tall fescue toxicosis. Symptoms in beef cattle include reduced rates of gain, poor conception rates, decreased dry matter intake, a long rough haircoat in summer, and very low serum prolactin. Earlier studies indicated that the performance of cattle grazing endophyte-free (EF) tall fescue is superior to that of cattle grazing endophyte-infected (EI) tall fescue (Fribourg *et al.*, 1995). However, EF tall fescue is not as persistent as EI. Addition of clovers to tall fescue pastures is a recommended practice. Our objective was to compare forage quality, *in vitro* dry matter disappearance, dry matter intake, and animal performance of EI and EF KY-31 pastures, and inclusion of legumes.

Materials and methods Experimental pastures were well established and located at the Blount Unit (35°49'N, 83°13'W) of the Knoxville Experiment Station. Forage treatments were EF KY-31 tall fescue, EI KY-31 tall fescue, EI Ky-31 with ladino white (*Trifolium repens*) and red clovers (*T. pratense*) (EI+Cl), and alternating groups of four 20-cm drill rows of EI and EF tall fescues (EI/EF). The strips of EI and EF in the same pasture evaluated a management strategy for improving EI pastures by adding EF areas to reduce the effects of the endophytes. Animal performance data were collected during spring and summer for two consecutive years. Masticate samples were obtained using ruminally cannulated steers (Olsen, 1991) from each pasture treatment for forage quality and *in vitro* digestibility estimates. Samples were placed on ice, transported to the laboratory, and frozen. Samples were air-dried (60°C forced air oven), ground to pass a 1-mm screen in a Wiley mill, and analyzed for crude protein (CP), neutral detergent fiber (NDF), and acid detergent fiber (ADF) via near-infrared technology (FOSS NIR Systems, Model 5000, Silver Spring, MD 20904). *In vitro* dry matter disappearance (IVDMD) was determined using modification of the Tilley and Terry (1963) two-stage *in vitro* technique. Weaned beef (*Bos taurus*) steers with an average initial weight of 329 kg grazed the pastures. Dry matter intakes (DMI) were determined using dosed chromic oxide fecal concentrations and IVDMD. Average daily gain (ADG) was determined by dividing the weight gained during the grazing season by the number of grazing days.

Results All experimental pastures contained excellent stands of forage and the EI+Cl contained about 25 to 35% clover each spring and decreased to 15 to 20% in summer.

Table 1 Forage quality *from masticate analyses, in vitro* dry matter disappearance and animal performance of tall fescue pastures

Parameter	Forage Treatments[1]			
	EI	EF	EI/EF	EI+Cl
CP, % d.m.	18.45[a]	18.04[a]	17.52[a]	24.11[b]
NDF, % d.m.	68.16[a]	68.66[a]	68.49[a]	49.89[b]
ADF, % d.m.	37.74[a]	38.28[a]	38.35[a]	31.13[b]
IVDMD, % d.m.	58.16[a]	58.42[a]	59.84[a]	65.12[b]
DMI, g/d	4696[a]	5654[b]	6177[b]	5520[b]
ADG, g/d	436[a]	717[b]	576[c]	553[c]

[1]Parameters not sharing superscripts are significantly different at $P < 0.05$

Conclusions Addition of clover to EI tall fescue increased forage intakes to that of EF or EI/EF fescue and also increased the quality and digestibility of diet consumed. This resulted in greater gains than those for steers grazing EI fescue without clover, but performance was reduced compared to that of steers grazing EF tall fescue.

References
Fribourg, H.A., J.C. Waller, J.H. Reynolds, M.A. Mueller & K.D. Gwinn (1995). Stand persistence of tall fescue cultivars free of or infested with *Acremonium coenophialum*. *Annales de Zootechnie*, 44 (Supp.), 124.
Olsen, K.C. (1991). Diet sample collection by esophageal fistula and rumen evacuation techniques. *Journal of Range Management*, 44, 515.
Tilley, J.M.A. & R.A. Terry (1963). A two-stage technique for the *in vitro* digestion of forage crops. *Journal of the British Grassland Society*, 18, 104-111.

Milk production from cows grazing perennial ryegrass pastures infected with wild or AR1 endophyte in New Zealand

S.J. Bluett, E.R. Thom and D.A. Clark
Dexcel, Private Bag 3221, Hamilton, New Zealand, Email: stephanie.bluett@dexcel.co.nz

Keywords: novel endophyte, ryegrass staggers

Introduction Most perennial ryegrass (*Lolium perenne*) cultivars in New Zealand are available with either the natural wild endophyte (*Neotyphodium lolii*); the AR1 novel endophyte (no lolitrem B or ergovaline production); or endophyte-free. Although wild endophyte protects ryegrass against insect attack, improving pasture persistence, it can also cause ryegrass staggers and reduced animal performance. Endophyte AR1 does not cause ryegrass staggers but still protects against insect pests such as Argentine stem weevil (*Listronotus bonariensis*). A 3-year farmlet experiment was carried out to evaluate the effects of AR1 and wild endophyte-infected ryegrass on pasture performance, milk production and cow health.

Materials and methods The experiment was conducted at Dexcel, Hamilton, New Zealand (37°47′S, 175°19′E, 40 m a.s.l.). Two farmlets (7 ha each) were sown with perennial ryegrass infected with either wild endophyte or AR1 endophyte, and with white clover (*Trifolium repens*). Farmlets were managed as self-contained systems using decision rules developed from other Dexcel farm systems experiments. Treatments were rotationally grazed by spring-calving Holstein-Friesian cows from September 2000 to May 2003, over 3 lactations (Bluett *et al.*, 2003). The mean stocking rate was 20 cows/farmlet or 2.9 cows/ha.

Results The AR1-infected ryegrass pastures were free of contamination from wild endophyte-infected ryegrass for at least 3 years after establishment, as confirmed by low concentrations of lolitrem B in ryegrass samples. Mean annual pasture production (15.0 vs 14.6 t DM/ha, s.e.d.=0.71, p=0.542) and ryegrass tiller density (3243 vs 3203 tillers/m^2, s.e.d.=242.2, p=0.874) were similar across AR1 and wild endophyte-infected ryegrass farmlets. A combined analysis over the 3 lactations, showed a significant 9% advantage in total milk production to cows grazing AR1-infected ryegrass pasture (Table 1). Cows grazing AR1-infected ryegrass produced more milk in summer (16.0 vs 14.7 L/cow/day, s.e.d.=0.48, p=0.009) and autumn (11.5 vs 10.3 L/cow/day, s.e.d.=0.47, p=0.026), and showed a similar trend in spring (19.8 vs 18.9 L/cow/day, s.e.d.=0.67, p=0.176). Milk composition was similar in all lactations. Ryegrass staggers occurred in cows grazing wild endophyte-infected pastures in January 2001, coinciding with the highest concentrations of lolitrem B over the 3 lactations (>3.5 mg/kg DM). Cow temperatures, respiration rates and plasma prolactin concentrations measured during periods of heat stress (>25°C) were seldom affected by endophyte treatment.

Table 1 Milk production and composition from cows grazing perennial ryegrass pastures infected with AR1 or wild endophyte (mean of 3 consecutive lactations)

	AR1	Wild	s.e.d.	p
Milk production				
Total (L/cow)	4016	3690	129.7	0.015
Daily (L/cow per day)	16.6	15.5	0.52	0.038
Milk composition				
Protein (%)	3.40	3.38	0.055	0.741
Fat (%)	4.28	4.25	0.100	0.799
Lactose (%)	4.92	4.89	0.033	0.412

Conclusions Results showed that renovating pastures with AR1-infected ryegrass can improve milk production without incidence of ryegrass staggers. On average over 3 lactations, cows grazing AR1-infected ryegrass pasture produced 326 L/cow more than those grazing wild endophyte-infected pasture. The magnitude of treatment differences varied from week to week, highlighting the benefits of measuring endophyte effects over the entire lactation and for consecutive years to allow for seasonal variations in alkaloid production. Although initially only a small portion of the farm would be regrassed with AR1-infected ryegrass, it would provide safe feed for animals affected by ryegrass staggers.

References

Bluett, S J., E.R. Thom, D.A. Clark, K.A. Macdonald & E.M.K. Minneé (2003). Milksolids production from cows grazing perennial ryegrass containing AR1 or wild endophyte. *Proceedings of the New Zealand Grassland Association, 65, 83-90.*

The effects of exposure to endophyte-infected tall fescue seed on faecal and urine concentrations of ergovaline and lysergic acid in mature gelding horses

C.L. Schultz, S.L. Lodge-Ivey, A.M. Craig, J.R. Strickland and L.P. Bush
University of Kentucky and USDA/ARS/FAPRU, Lexington, Kentucky 40546, USA,
Email: schultz@ffsru.tamu.edu

Keywords: feeding trial, metabolism, toxicity, short term, long term

Introduction Despite the good nutritive value of endophyte-infected tall fescue, consumption by livestock results in a decrease in both reproductive and growth performance due to ergot alkaloids produced by an endophytic fungus (Cross *et al.*, 1995). Little research has investigated the metabolic fate of ergot alkaloids and/or their metabolites in grazing horses. Thus, the objectives of this experiment were: a) to determine concentrations of ergovaline (EV) and lysergic acid (LA) in the faeces and urine of geldings exposed to tall fescue seed over a time course experiment and b) to measure the effects of alkaloid-containing tall fescue on nutrient digestibility and serum clinical enzyme profiles.

Materials and methods Mature geldings (394.2 ± 7.1 kg; n = 10) of mixed breeding were randomly assigned to one of two treatments: 1) control diet with endophyte-free tall fescue seed (EF) or 2) a diet containing endophyte-infected (EI) tall fescue seed at 0.5 mg ergovaline/kg of total diet. Lysergic acid in the EI diet was 0.84 mg/kg diet and 0.0 mg/kg in the control diet. Three distinct phases were established: no exposure (Control phase - ContP; 14 d); short-term exposure (Acute phase - AcuteP; 4 d), and longer-term exposure to ergovaline (Sub-acute phase - SAP; 21 d). Ergovaline and LA were quantified in the urine and faeces. Blood and rectal temperatures were collected daily during each phase. Serum was analyzed for creatine kinase (CK), alkaline phosphatase (AP) and aspartate aminotransferase (SGOT).

Results Differences due to treatment were undetected for rectal temperature (P = 0.97). Serum AP and CK were similar between treatment groups (Table 1). Serum SGOT was greater for the EF treated group than those in the EI group during the AcuteP (P = 0.09; 402.6 vs 269.6 U/L). Within the EI treatment group, SGOT was greater during the SAP than the ContP (P = 0.03). Geldings consuming the EI diet had faecal ergovaline concentrations of 0.0, 0.3, and 0.4 mg/kg for the ContP, AcuteP, and SAF, respectively. Within the EI group, ergovaline concentrations differed between the phases (P < 0.01) with the greatest amount excreted during the SAP.

Table 1 Rectal temperature (RT), serum enzyme, and ergovaline data for geldings fed an endophyte-free (EF) and endophyte- infected diet (EI)

| | Phase | | | | | |
| | Control | | Acute | | Subacute | |
Treatment	EF	EI	EF	EI	EF	EI
RT, °C[a]	38.5	38.6	38.6	38.7[b]	38.4	38.3[c]
SGOT, U/L[a]	329.8	228.9	402.6[d]	269.6[b, e]	407.9	326.6[c]
CK, U/L	239.2	228.2	311.9	231.8	293.9	231.3
AP, U/L[a]	222.1	213.6	247.2	233.4	207.4	210.4

[a] Phase effect (P < 0.10)
[b, c] Within treatment, means differed (P < 0.03)
[d, e] Treatments differed between EF and EI groups (P = 0.09)

Conclusions Concentrations of 0.5 mg ergovaline per kg diet or less had little effect on serum AP, CK, and SGOT, or rectal temperature. A majority of the ergovaline consumed by geldings was excreted in the faeces. Because no signs of decreased animal performance were observed, results suggest geldings grazing endophyte-infected tall fescue containing less than 0.5 mg ergovaline/kg DM may not experience fescue toxicosis.

References
Cross, D.L., L.M. Redmond & J.R. Strickland (1995). Equine fescue toxicosis: signs and solutions. *Journal of Animal Science*, 73, 899-908.

Ergovaline and ergovalinine and tall fescue content of pastures in central Kentucky

P.W. Long, J.C. Henning and L.P. Bush
University of Kentucky, Department of Agronomy, Lexington, KY 40546, USA, Email: pwlong@uky.edu

Keywords: tall fescue, endophyte, alkaloids, herbage

Introduction Kentucky has >2Mha of tall fescue (*Festuca arundianacea* Schreb.) grown mainly for livestock consumption. Many alkaloids in tall fescue are produced in a mutualistic association between tall fescue and an endophytic fungus *(Neotyphodium coenophialum)* (Long *et al.*, 2002). Ingestion of tall fescue by livestock may depress reproduction and growth (Schultz & Bush, 2002). Not all Kentucky fields of tall fescue are thought to be infected with endophyte, but forage samples from all surveyed pastures had measurable ergopeptine alkaloids. Therefore, it is reasonable to assume that some of the tall fescue plants in these fields were infected. We estimated tall fescue content and evaluated ergopeptine alkaloids in tall fescue monocultures and composite pastures of several central Kentucky horse farms.

Materials and methods In spring 2002 and 2003, tall fescue and composite grab herbage samples were selected randomly from 13 (2002) and 12 (2003) central Kentucky farm pastures every 2 weeks from March to June to determine the level of ergopeptine alkaloids present. Tall fescue frequency in the sward was estimated by a trained observer. Tall fescue and composite herbage samples, clipped manually from 6-10 sites/field, were dried and powdered for alkaloid analysis. Ergovaline and ergovalinine (E+E) analyses were done by HPLC with fluorescence detection.

Results E+E alkaloids were assayed in 631 samples (260 of tall fescue monocultures and 371 of composite herbage) on 75 fields. E+E levels in tall fescue were low (<0.50ppm) before 11 April and peaked between 23-30 May (Figure 1). E+E in composite herbage samples peaked in late June and were higher than tall fescue alone for the previous reporting period. Percent tall fescue in the pastures increased relative to other forage species for 2003 compared to 2002 (Figure 2). Environmental conditions (greater rainfall and cooler temperatures than normal) may have caused the increased percentage of tall fescue in the swards in 2003.

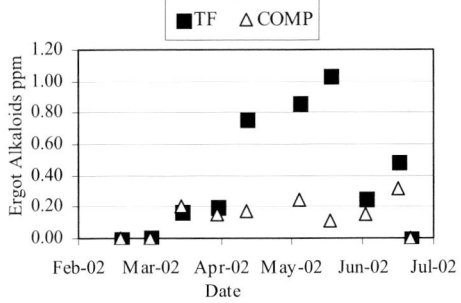 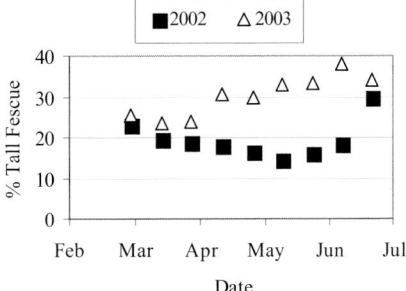

Figure 1 Seasonal levels of ergovaline + ergovalinine in tall fescue and composite herbage samples (ppm)

Figure 2 Percent tall fescue in central Kentucky pastures (spring 2002-03)

Conclusions E+E herbage samples from tall fescue plants and in multispecies pastures increased from near zero in late April, peaked in mid May, followed by secondary peak in late June. E+E exceeded 0.3ppm, a threshold for animal response, in herbage from some pastures with <20% tall fescue. Environment (rainfall and temperature) and farm management practices play a role in E+E levels and tall fescue content of pastures.

References
Long, W., J. C. Henning, B. Coleman, L. Lawrence, C. Peterson & A. Reinowski (2002). Overview of the mare reproductive loss syndrome monitoring program for 2002. *Proceedings of the First Workshop on Mare Reproductive Loss Syndrome*, Univ Kentucky, AES, SR-2003-1, 102-112.
Schultz, C. & L. P. Bush (2002). The potential role of ergot alkaloids in mare reproductive loss syndrome. *Proceedings of the First Workshop on Mare Reproductive Loss Syndrome*, Univ Kentucky, AES, SR-2003-1, 60-63.

Studies of seed characteristics of ecotypes of lucerne, *Bromus* and *Agropyron* in response to *Fusarium oxysporum* and *F. solani*

M.A. Alizadeh
Research Institute of Forest and Rangeland, P.O. Box 13185-116Tehran, Iran, Email: Alizadeh@rifr-ac.ir

Keywords: *Bromus*, lucerne, *Agropyron*, *Fusarium oxysporum*, *F. solani*, germination

Introduction Vigorous seeds and seedlings are more resistant to pathogens than non-vigorous seeds and seedlings (Kim, 1994). Therefore, it is necessary to assess seed and seedling performance in response to seed borne fungi.

Material and methods Seed samples were disinfected with detergent and placed in Petri dishes and inoculated with two levels of spores of two species of *Fusarim*. The samples were germinated in a germinator at 20°C with 1000 lux light under laboratory and greenhouse conditions. The percentage and speed of germination were recorded at days 3, 6, 9, 12 and 15 according to Maguire (1962). On day 15 the shoot : root ratios of randomly selected seedlings were measured according to Lekh & Khairwal (1993). Vigour index was measured according to Abdul-baki & Anderson (1973).

Results Vigour index in the greenhouse was reduced by both *Fusaruium* spp.. Level of infection gave contradictory results, because the ecotypes responded differently. The root/shoot ratio was not affected in the greenhouse, but in the laboratory *F. oxysporum* infection significantly reduced this ratio. Speed and percentage of germination were reduced by *Fusarium* infection. (Table 1).

Table 1 Mean of the main characteristics of seeds of 13 ecotypes of t *Agropyron, Bromus* and lucerne in response to two species of *Fusarium*

Vigour index		Root length /shoot (mm)		Speed of germination.		Germination (%)		
GRH	Lab.	GRH.	Lab.	GRH.	Lab.	GRH.	Lab.	
53.11a	64.7 a	0.42 a	1.02a	13.63a	15.07a	81.33a	95.69a	Control
40.55b	56.01 b	0.45 a	1.06a	10.67b	13.61b	64.33a	84.62b	SO1
42.27b	63.07 a	0.47 a	1.03a	11.79b	14.66a	67.62a	91.28a	SO2
38.48b	58.41 b	0.42 a	0.88b	10.06b	13.89b	58.67a	88.56a	OX1
37.51b	54.54b	0.49a	0.93 b	9.76b	12.51c	58.56a	80.62b	OX2
42.38	59.34	0.45	0.97	11.8	13.95	66.1	88.15	Mean
4.8	3.76	0.044	0.05	1.28	0.77	7.09	4.44	LSD

GRH=greenhouse, Lab=laboratory, SO1, SO2 = levels 1 and 2 of spore inoculation for *Fusarium solani*. OX1 and OX2= levels 1 and 2 of spore inoculation for *Fusarium oxysporum*. Data in columns with the same are not significantly different (P≤5%)

Table 2 Compound analysis of variance for seed characteristics of 13 ecotypes of species of *Agropyron, Bromus* and lucerne in response to two species of *Fusarium* under laboratory and greenhouse conditions

Vigour index	Root length /shoot (mm)	Speed of germination	Germination %	Df	
2801.33^{**}	8465.26^{**}	746.0^{**}	4741.26^{**}	1	Condition
7134.22^{**}	375.07^{**}	121.43^{**}	2680.58^{**}	12	Ecotype
2024.02^{**}	8.6^{**}	121.10^{**}	4118.50^{**}	4	Fungi
6229.86^{**}	354.87^{**}	120.28^{**}	2689.74^{**}	12	Condition x Ecotype
267.20^{**}	8.35^{**}	14.38^{ns}	661.55^{*}	4	Condition x Fungi
246.25^{**}	5.29^{**}	13.90^{**}	367.33^{*}	48	Fungi x Ecotype
201.04^{*}	5.30^{**}	15.10^{**}	389.28^{*}	48	Condition x Ecotype x Fungi
22.62	18.61	21.66	20.46		CV

*, ** and ns: Significant at the 5%, 1% levels and non-significant respectively

References
Abdul-baki, A. A. & J. D. Anderson, (1975). Vigour determination in soybean seed by multiple criteria. *Crop Science,* 13, 630-633.
Lekh, R. & I. S Khairwal,. 1993. Evaluation of pearl millet hybrids and their parents for germinability and field emergence. Indian *Journal of Plant Pathology,* 2, 125-127.
Kim, S.H., Z.R Choe, J.H., Kang, L.O., Copeland & S.G. Elias (1994). Multiple seed vigour indices to predict field emergence and performance of barley. *Journal of Seed Science and Technology,* 22, 59-68.
Maguire, J.D. (1962): Speed of germination: aid in selection and evaluation for seedling vigour. *Crop Science,* 2, 176-177.

The effects of symbiotic mycorrhizal fungi on drought tolerance and forage production of lucerne (*Medicago sativa*)

H. Panahpour

Gene Bank, Research Institute of Forests and Rangelands, P.O.Box 31585-343Tehran, Iran, Email: h.panahpor@ rifr-ac.ir

Keywords: lucerne, symbiosis, mycorrhizal fungi, drought tolerance, forage production

Introduction   Arbuscular mycorrhizal fungi (AMF) form beneficial symbioses with the roots of many plants. This association allows them to maintain themselves and grow well under relatively harsh conditions (Sieverding, 1986). They also improve the ability of plants to withstand or have enhanced water acquisition capability. AMF symbioses assist to extend crop and forage plants into arid and semi arid zones. In this research effects of AMF symbiosis were studied on drought tolerance and forage production of lucerne (*Medicago sativa*).

Materials and methods   Effects of identified mycorrhizal fungi in soil of an abandoned lucerne farm (Panahpour, 2003) was studied on drought tolerance and forage production of lucerne. This experiment was carried out in a greenhouse at the Alborz research centre, Karaj, during 2002. Five cultivars (Krisary, Australia, Bami, FAO, and Ghareh ionjeh), 2 types of soil (heated and unheated field soil) and five levels of irrigation (20%, 40%, 60%, 80% and 100% of field capacity) were arranged in factorial combinations in a completely randomised block with three replications. Plant height, tiller number, fresh and dry weight of stem, leaf and forage, leaf to stem ratio and some root morphological attributes (dry weight, length, branches number,) were recorded for 3 cuts.

Results   The performance of many traits in heated soil was significant and greater than unheated soil for both varieties and irrigation levels. The effects of heated soil also varied with cultivars   (Figure 1).Forage yield in heated soil was greater than field soil at all but the second irrigation levels (Table1).

Table 1 Interaction between (soil types× irrigation levels) on mean of dried forage in 3 cuts.

Irr. Levels	Mean of f.soil ± std.	Mean of hf.soil ± std.
10- 20%Fc	8.28 ± 3.48	7.11 ± 2.25
20-40%Fc.	12.48 ± 4.08	14.1 ± 5.28
40-60%Fc.	20.55 ± 4.59	24.06 ±11.94
60-80%Fc.	27.9 ± 6.9	33.96 ± 8.4
80-100%Fc.	24.06 ± 8.43	28.86 ± 10.5

f= field, hf = heated field, std= standard deviation , Fc.= field capacity, Irr=Irrigation, s=soil.

Figure 1   Interaction between (heated × field) soil on dried forage yield of varieties in 3 cuts

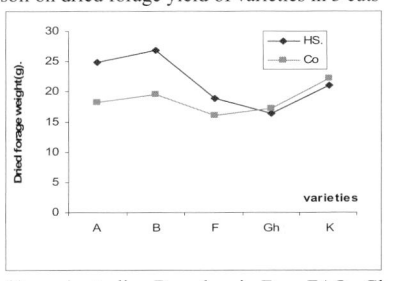

(A = Australia, B = bami, F = FAO, Gh = Ghareh ionjeh, K = krisary)

Conclusions   Germination of spores and hyphal growth of AMF occurred after the first and second heating of field soil respectively.The preference of heated soil in comparison to field soil may be stimulation and prolonged dormancy of mycorrhizal fungi spores, respectively (Tomerup, 1983).High performance of all   irrigations excluding 10- 20% levels of field capacity in heated soil vs. field soil indicate positive effects of AMF on drought tolerance of Lucerne. Low yield of Ghareh ionjeh and Krisary in heated soil was probably due to fungal disease infection (*Fusarium solani*) and high environmental temperature in the third cut (Figure1).

References

Sieverding, E. (1986). Influence of water regimes on VA mycorrhiza.IV. Effect on plant growth, water utilization and development of mycorrhiza. *Journal of Agronomy and Crop Science,* 150, 400-412.

Panahpour, H. (2003), Identification and verifying of symbiotic fungi (*Mycorhiza spp.*) on alfalfa (Medicago sativa), *Czech Journal of Genetic Plant Breeding,* 39, 173-177.

Tomerup,I.C. (1983),spore dormancy in vesicular-arbuscular mycorrhizal fungi. *Transactions of the British Mycological Society,* 81, 37- 45.

Selection of Australian root nodule bacteria for broad-scale inoculation of native legumes

R.G. Bennett[1], R.J. Yates[2], E.L.J. Watkin[2], G.W. O'Hara[2] and M.J. Dilworth[2]
[1]Cooperative Research Centre for Plant-based Management of Dryland Salinity, University of Western Australia, Australia, Email: rbennett@agric.uwa.edu.au [2]Centre for Rhizobium Studies, Murdoch University, Western Australia, Australia

Keywords: nitrogen fixation, promiscuity

Introduction The unique and diverse native Australian perennial legumes are under current investigation for use as pastures in Australian agriculture. Identification of root nodule bacteria (RNB) that can fix nitrogen effectively for the plant is a critical factor for the success of a legume species in agriculture (Howieson et al., 2000). Some legumes under investigation are relatively promiscuous (Lange, 1961). This trait may allow the development of a single, broad-scale inoculant that could allow inoculation of multiple species of agricultural importance, whilst more effective, specific RNB are developed in time. Aimed to identify strains that can form effective symbioses with several native legume species of potential interest to agriculture, this experiment screened putative indigenous RNB on 5 native legumes.

Materials and methods *Acacia acuminata* Benth. (Aa), *Jacksonia sericia* Benth. (Js), *Kennedia coccinea* Vent. (Kc), *Nemcia capitata* (Benth.)Domin (Nc) and *Swainsona formosa* (G.Don) Joy Thomps. (Sf), 5 native legume species, were inoculated with 20 indigenous RNB, 7 fast and 13 slow-growing strains (visible colony growth within 3 and 7 days, respectively). The plants were grown for 14wks under sterile conditions (Howieson et al. 1995). To estimate the effectiveness of inoculants, nodulation was scored and the dry weights of plant shoots were measured and compared with similar data from uninoculated controls.

Results Two fast-growing isolates showed differing cross-inoculation patterns, each nodulating 3 species (Js, Kc and Nc; and Kc, Nc and Sf). All 13 slow-growing strains, and one fast-growing strain, displayed a third cross nodulation pattern, where symbioses formed on four species (Aa, Js, Kc and Nc). Four fast-growing strains of RNB failed to form nodules with any tested host. The effectiveness of symbioses was highly variable between strains and hosts. No strain was the most effective inoculant on all hosts. The most effective RNB strain overall was slow-growing and was the most effective strain on Aa, the most effective slow-growing strain on Nc and Js, and the third most effective slow-growing strain on Kc.

Conclusions The nodulation of slow-growing RNB showed uniformly wide host ranges and a promiscuous habit. In contrast, the fast-growing strains were more specific in the hosts they nodulated. The effectiveness of the RNB tested was variable between hosts and RNB strains, and ranking of these strains is possible to select useful inoculants. It is evident that particular native legumes, such as the *Swainsona* species tested here, could require specific inoculum and may not nodulate with broad-scale inoculum, a finding supported by Yates (2004). That several RNB were moderately effective on several species is encouraging. That result suggests that there is an opportunity for further selection on a wider range of RNB to develop broad-scale inoculum for native pasture species in Australia.

References

Howieson, J. G., A. Loi & S. J. Carr (1995). *Biserrula pelicinus* L.--a legume pasture species with potential for acid, duplex soils which is nodulated by unique root-nodule bacteria. *Australian Journal of Agricultural Research*, 46, 997-1009.
Howieson, J. G., G. W. O'Hara & S. J. Carr (2000). Changing roles for legumes in Mediterranean agriculture: developments from an Australian perspective. *Field Crops Research*, 65, 107-122.
Lange, R. T. (1961). Nodule bacteria associated with the indigenous leguminosae of South-Western Australia. *Journal of General Microbiology*, 26, 351-359.
Yates, R. J., J. G. Howieson, K .G. Nandasena & G. W. O'Hara (2004). Root-nodule bacteria from indigenous legumes in the north-west of Western Australia and their interaction with exotic legumes. *Soil Biology and Biochemistry*, 36, 1319-1329.

A new napier grass stunting disease in Kenya associated with phytoplasma

A.B. Orodho[1], S.I. Ajanga[2], P. Jones[3] and P.O. Mudavadi[2]

[1]P.O. Box 1667 Kitale, Kenya, Email: aborodho@yahoo.com, [2]Kenya Agricultural Research Institute RRC Box 169 Kakamega, Kenya, [3]Plant Pathogens Interactions Division, Rthamstead Research, Harpenden, Herts. AL5 2JQ, UK

Keywords: blast, dairy farmers, mycoplasma

Introduction Napier grass (*Pennisetum purpureum* Schum) is a cultivated elephant grass native to Eastern and Central Africa forming the major livestock feed on East African smallholder dairy farms (Valk, 1990) as it is suitable for cut and carry for zero-grazing management systems. Although several plant pathogens have been described historically they were seldom severe. However, in 1970s there was an outbreak of snow mould fungal disease caused by *Beniowskia spheroidea* that attacked most varieties of napier grass. A napier grass variety clone 13 was bred which is resistant to the disease. In the 1990s two major outbreaks of napier grass diseases occurred in Kenya. In Central Kenya a napier grass head smut caused by *Ustilago kamerunensis* H Sydow and Sydow in 1992 and in Western Kenya a napier grass stunting disease was first reported in Bungoma in 1997. A similar stunting disease had been reported in Uganda (Tilley, 1969), which was suspected to be a virus transmitted by insects. This new outbreak of napier grass stunting disease is of major concern as it attacks all varieties of napier grass. The main objective of this study was to survey the extent of the disease and to identify the organism causing this disease.

Materials and methods A survey was carried in 2001 of 100 farmers' fields in five districts of Western Kenya by a multidisciplinary team of scientists to assess the occurrence of napier grass stunting disease, describe its symptoms, assess its spread and collect samples for laboratory analysis. Samples collected for laboratory analysis included leaves, roots, stems and insects found feeding on the whorls of diseased plants. Individual sections of the leaves, roots and stems from healthy and diseased samples were tested for possible fungal pathogens using a method developed by Lloyd and Pillay (1980). Further samples were used to test the presence of viruses. A third set of samples was taken to Rthamsted, U.K. to test for both virus and mycoplasma pathogens. At Rthamsted, yellowed and apparently healthy napier grass were grown under quarantine. A total DNA extraction was done from each sample for use as a template in a nested Polymerase Chain Reaction using phytoplasma 16S ribosomal DNA primers P1/P7 and R16F2n/R16R2. A band of 1250-bp rDNA product was amplified from all yellow leaves and in two of three apparently healthy leaves.

Results and discussion The survey and subsequent farm visits indicated that the disease had spread from the original district to five adjacent districts by 2001. The disease had spread to four more districts by 2003. Disease incidence ranged from 10-100%. The highest mean incidence recorded was in Bungoma and the least was in districts further away. No resistant napier grass variety was found. Cutting frequency, low soil fertility and water intensity stress intensify incidence and severity of the disease. The characteristic symptoms of the disease are yellowing of napier grass foliage, reduced leaf size, proliferation of tillers and shortening of internodes resulting in stunted growth. Laboratory analysis of external and internal morphology showed that the disease was not caused by either nematodes or fungus or virus. The DNA analysis showed that the sample grown had yellowed leaves and stunted growth and were phytoplasma positive. Refracted fragment length polymorphism (RFLP) analysis of amplimers showed similar patterns for all samples. While BLAST analysis showed the phytoplasmas to be the member of 16SrX1 (Rice yellow dwarf group). The higher homology 96% was with the 16SrX1 Bermuda grass white leaf group phytoplasma. Thrips, aphids and leaf hoppers were the main insects found feeding in the whorls of diseased plants.

Conclusion A phytoplasma similar to the rice yellow dwarf group is the casual organism of napier grass stunting disease in Western Kenya. Sugarcane and upland rice are possible hosts of this group of mycoplasma. A leaf hopper is the most probable vector. Movements of vegetatively propagated napier grass planting material provides a means for the rapid spread of the pytoplasma. There is a need to develop a napier grass variety resistant to this disease.

References
Llyod, H.L. & M. Pillay (1980). Development of an improved method for evaluating Sugarcane for resistance to smut. *Proceedings of the South African Technologists' Association*, 168-172.
Tilley, G.E.D. (1969). Elephant grass. Kawanda Agricultural Station Report, Kawanda, Uganda.
Valk, Y.S. (1990). Review report of the DEAF Survey during 1989. NDDP (M41/200) Ministry of Livestock Development, Nairobi, Kenya.

A study in Cuba of the biology, ecology and agroecological management of *Heteropsylla cubana* Crawford in *Leucaena leucocephala* (Lam.) de Wit

N. Valenciaga[1], M. Felicia Díaz[1], T.E. Ruíz[1], M. Fernández[2] and C. Mora[1]
[1]*Instituto de Ciencia Animal, Carretera Central Km 47 ½, San José de las Lajas, La Habana, Cuba, Email: nvalenciaga@ica.co.cu, *[2]*Centro Nacional de Sanidad Agropecuaria, San José de las Lajas, La Habana, Cuba*

Keywords: cattle, psyllid, silvopastoral systems

Introduction As a consequence of the increase in *Leucaena leucocephala* areas to counter the shortage of feed in Cuban cattle production, there is a risk of the development of *Heteropsylla cubana* Crawford (Hemiptera: Psyllidae) as a pest in silvopastoral systems with this legume. This psyllid is known to be the main phytophagous pest (Valenciaga, 2003), which produces damage in 95% of the apical region of branches. Since information on the identification and biology of a pest species is a necessary prerequisite for its management, a taxonomic, biological and ecological study was conducted to define *Heteropsylla* behaviour in Cuban conditions and elaborate the theoretical basis to propose management alternatives.

Materials and methods To achieve these objectives, we developed six main aspects of this work : taxonomy, biology, plant-insect relationships, effects of the feeding activity of *H. cubana* in *L. leucocephala*, ecological traits, biotic and climatic factors affecting the damage scale of *H. cubana,* theoretical basis for establishing management alternatives. Laboratory, semi-control and field experiments were carried out.

Results All the psyllids collected in the adult stage corresponded to *H. cubana;* the genitalia of the males and females were similar to those described by Crawford in 1914, whose results are consistent with those of Muddiman *et al.* (1992). The duration in days of the periods of pre-oviposition, oviposition, post-oviposition, ranged from 2 to 3 days and the sexual index was 1:2, i.e. there were 2 females for each male.

Spatial analyses according to the dispersion indices or average variance rate indicated that *H. cubana* tends to aggregation (b>1) independently of the diversification of the agricultural ecosystem. Oscillations of natural movement of the populations of eggs, nymphs and adults of *H. cubana* (figure 1) are similar for each development stage, with an increase in egg numbers/m2, in the first years of exploitation of the system evidence of an annual population peak in the period when temperatures start to decrease (October to December), i.e., end of the rainy season, beginning of the dry season.

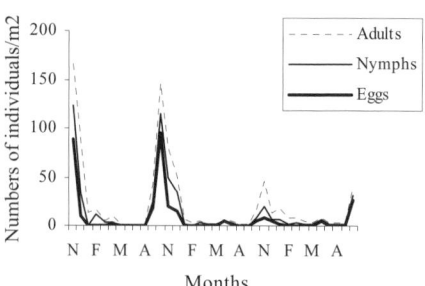

Figure 1 Oscillations in eggs, nymphs and adults of *H. cubana* with time

The results indicate that there is an active permanence of natural enemies in *L. leucocephala* sowings studied in Cuban conditions, whose levels are favored by the increase in plant biodiversity by providing feed to this species in its different growth stages. Therefore, we could state that the performance of the climatic factors determine better the population dynamics of this psyllid, compared to the biotic factors.

Conclusions These results constitute the first studies made in Cuba on the reaffirmation of *H. cubana* as the predominant species in *L. leucocephala* in the different regions sampled in Cuba. The life cycle of *H. cubana,* its space and time was elucidated. The natural enemies associated and performance of the climatic factors explain the regulation of its populations. This enables us to propose an agroecological management of the insect.

References

Crawford, D.L. (1914). A monograph of the jumping plant-lice of psyllidae of the New World. Bulletin of the United States National Museum. 85, 1-186.

Muddiman, S.B., I.D. Hodkinson & D. Hollis (1992). Legume-feeding psyllids of the genus *Heteropsylla* (Homoptera: Psylloidea). *Bulletin of Entomological Research*, 82, 73-117.

Valenciaga, N. (2003). Biología, ecología y base teórica para establecer las alternativas de manejo de *Heteropsylla cubana* Crawford en *Leucaena leucocephala* (Lam.) de Wit. *Tesis de Doctor en Ciencias Agrícolas*, Instituto de Ciencia Animal, La Habana, Cuba, 97.

Theme A: Efficient production from grassland

Section 7

Advances in sown tropcial legumes

Dual purpose cowpea for West Africa

S.A. Tarawali[1], I. Okike[2], P.K. Kristjanson[3], B.B. Singh[4] and P. Thornton[3]
[1]*International Livestock Research Institute (ILRI), PO Box 5689, Addis Ababa, Ethiopia, Email: s.tarawali@cgiar.org, [2]ILRI, Ibadan, Nigeria, [3]ILRI, Nairobi, Kenya, [4]International Institute of Tropical Agriculture (IITA), Kano, Nigeria*

Keywords: legume, grain, fodder

Background Cowpea (*Vigna unguiculata* (L.) Walp.) is grown as an intercrop with cereals in some 9M ha of West Africa, mostly in the dry savanna. Though grain yields are low (circa 500 kg/ha), it is a nutritious food and dry season fodder. The haulms (leaves and stems) are cut and stored after grain harvest. It aids soil fertility by fixing soil N and returning N via manure from ruminants fed with haulms. Up to the early 1990s, research had focused on developing high grain yielding varieties. Recognition of farmers' appreciation of multiple uses, in particular the fodder value and the increasing importance of crop residues as feed resources in much of West Africa where expansion of agricultural land and intensification mean reduced availability of land for planted forages, led to joint research by ILRI and IITA from 1994 onwards, resulting in identification of "dual purpose" varieties – with the potential to provide both good grain yields and quality fodder under farmer conditions.

Potential impact *Ex ante* impact assessment combining information from community discussion groups, village and household level surveys with crop models and GIS database layers has estimated the potential adoption and value of dual purpose cowpea in West Africa from 2000 to 2020 (Kristjanson *et al.*, 2001). Taking account of the heterogeneity in terms of market access and population density, two factors likely to influence adoption of dual purpose cowpea, this study estimates that of the 9M ha of cowpea, dual purpose varieties could be adopted on a consolidated area 1.4M ha of West Africa and potentially benefit 9.3M people (assuming proportions of land and human population are equal), with an internal rate of return to research investment of 50 to 103% (71% being the baseline figure) and a benefit:cost ratio 63 (subsequent sensitivity analysis gave a variation from 32 up to 127). Net present value (NPV), including a 5% discount, was estimated as US $606M. Whilst it is not yet possible to assess the accuracy of these 20-year horizon estimations, information from current research and development efforts especially those taking a holistic and farmer-focused approach (see Sanginga *et al.*, 2003; Tarawali *et al.*, 2003) suggests that such optimistic scenarios may not be unfounded.

Reasons for success Dual purpose cowpea varieties help farmers who have little land to obtain food and feed from the same area. Cowpea has other economic, ecological and social benefits. Farmers familiar with its management find it easy to adopt. The extension and research services and established networks promote the probability and intensity of cowpea adoption. Whilst cowpea varieties developed by ILRI and IITA were among the first to be promoted, the concept of including dual purpose features in national research has expanded, as evidenced by the increased inclusion of fodder parameters in cowpea research (Singh & Rachie, 1987; Singh *et al.*, 1997; Fatokun *et al.*, 2003). This gives cause for optimism for widespread adoption of such varieties.

References

Fatokun, C., S. A. Tarawali, B. B. Singh, P. M. Kormawa & M. Tamò Eds. (2003). *Challenges and Opportunities for Enhancing Sustainable Cowpea Production.* IITA, Nigeria. 433 pp.

Kristjanson, P., S. A. Tarawali, I. Okike, B. B. Singh, P. K. Thornton, V. M. Manyong, R. L. Kruska & G. Hoogenboom (2001). Genetically improved dual purpose cowpea: assessment of adoption and impact in the dry savannah region of West Africa. ILRI Impact Assessment Series 9. ILRI (International Livestock Research Institute), Nairobi, Kenya. 68 pp.

Sanginga, N., K. E. Dashiell, J. Diels, B. Vanlauwe, O. Lyasse, R. J. Carsky, S. A. Tarawali, B. Asafo-Adjei, A. Menkir, S. Schulz, B. B. Singh, D. Chikoye, D. Keatinge & R. Ortiz (2003). Sustainable resource management coupled to resilient germplasm to provide new intensive cereal-grain-legume-livestock systems in the dry savanna. *Agriculture Ecosystems and Environment,* 100, 305-314.

Singh, B. B., D. R. Mohan Raj, K. E. Dashiell & L. E. N. Jackai Eds. (1997). *Advances in Cowpea Research.* IITA, Nigeria and JIRCAS, Japan. 375 pp.

Singh, S. R. & K. O. Rachie Eds. (1985). *Cowpea Research, Production and Utilization.* John Wiley and Sons, UK. 460 pp.

Tarawali, S. A., B. B. Singh, S. C. Gupta, R. Tabo, F. Harris, S. Nokoe, S. Fernandez-Rivera, A. Bationo, V. M. Manyong, K. Makinde, & E. C. Odion, (2003). Cowpea as a key factor for a new approach to integrated crop-livestock systems research in the dry savannas of West Africa. In: *Challenges and opportunities for enhancing sustainable cowpea production.* Eds. C. A. Fatokun, S. A. Tarawali, B. B. Singh, P. M. Kormawa and M. Tamò. IITA, Ibadan, Nigeria. pp. 233-251.

Planted forage legumes in West Africa

S.A. Tarawali[1], P. Thornton[2] and N. de Haan[3]

[1]International Livestock Research Institute (ILRI), PO Box 5689, Addis Ababa, Ethiopia, Email: s.tarawali@cgiar.org, [2]ILRI, Nairobi, Kenya, [3]International Institute of Tropical Agriculture (IITA), Ibadan, Nigeria

Keywords: legumes, fodder banks, adopters

Background Planted forage legumes were introduced into West Africa circa 1950. Since then, a range of species and strategies for their introduction into farming have been evaluated. Approaches to both evaluation and use have changed considerably, especially in the past 15 years. Much of the early research was on-station and focused on using mainly *Stylosanthes* species as introduced pastures. The *Stylosanthes* "fodder bank" concept followed this, but with a fairly stringent "recipe" for farmers to manage and use the legume "bank" for strategic ruminant supplementation in the dry season. Later evaluation included more legume species, more participatory approaches, and identification of potential domains in relation to farmers' social, economic and biophysical situations. It also included a holistic view of the potential multiple roles of legumes in West Africa, especially in relation to the major mixed crop-livestock system, on which up to 80% of the population may depend. In the context of intensifying mixed crop livestock systems, dual purpose varieties of annual (food-feed) legumes (mainly cowpea and groundnut) are gaining popularity, especially in areas where farmers have good market access and pressure on land is high. [Dual purpose legumes are not considered in the assessment of adoption and potential in this manuscript]. Farmers in areas with poor market access and lower land pressure have adopted non-food legumes to a limited extent, especially *Centrosema pascuorum* (used as cut-and-carry from a 1-year planted sole plot)*,* and to a lesser extent, *Aeschynomene histrix* (grazed *in situ* from a 1-2 year planted fallow) planted for forage and soil fertility restoration.

Adoption and benefits An *ex-post* impact assessment of planted forages in West Africa from 1977-1997 focused mainly on impacts attributable to ILRI (or its predecessor, ILCA) interventions, especially fodder banks. It reported positive returns to the research investments (Elbasha *et al.*, 1999). The study reported that 27,000 farmers had planted about 19,000 ha of herbaceous legume forages in West Africa, a relatively small proportion considering the potential numbers of adopters and area in this region. Nevertheless, the estimated benefit by 1997 was $16.5 million from a research investment of $7 million, indicating a positive pay-off. Estimating to 2014 and using conservative estimates suggested a potential doubling of the benefits, with 40,000 adopters and an area of 32,000 ha. Benefits in this study included estimates for increased milk yield, weight gain, calving rate, calf and cow survival, crop grain and residue yields (from rotational effects).

Because adoption was limited, recent studies addressed the holistic role of legumes in the system, including social and economic factors in various agroecological zones. Because adoption is unlikely to be homogeneous over a wide area, but rather niche-specific, these studies included an estimate of potential legume adopters based on known adoption rates in villages in different agroecologies and over 4 different resource-use domains (related partly to population density and market access) over a 3-year period (2000-2002). Using some assumptions and extrapolations to the respective zones in West Africa, we estimate 33,000 potential adopters for *C. pascuorum* in the northern Guinea savanna (growing period, GP=151-180 days), and 22,000 for *A.histrix* in the derived savanna (GP=211-270 days).

The future The perception of fodder legumes for West Africa had changed over the past few decades, from a view that such species provided the answer to many feed related constraints, with the lack of adoption blamed on poor extension, seed availability, land tenure, fencing requirements (Tarawali *et al.*, 1999), to an acceptance that legumes can address some fodder needs in certain niches at specific times. Current research addresses today's challenge, which is to understand the institutions and processes necessary to scale up and out a broad range of fodder innovations. It works closely with farmers, takes cognisance of social and economic circumstances, provides a variety of options, of which forage legumes are just one, and will contribute to ensuring that appropriate legumes are available for a variety of niches.

References

Elbasha, E., P. K. Thornton & G. Tarawali (1999). *An* ex post *economic impact assessment of planted forages in West Africa.* ILRI Impact Assessment Series 2. ILRI (International Livestock Research Institute), Nairobi, Kenya. 68 pp.

Tarawali, G., V. M. Manyong, R. J. Carsky, P. V. Vissoh, P. Osei-Bonsu & M. Galiba (1999). Adoption of improved fallows in West Africa: lessons from mucuna and stylo case studies. *Agroforestry Systems,* 47, 93-122.

Fodder shrubs for improving incomes of dairy farmers in the East African highlands

S. Franzel[1], C. Wambugu[1], J. Stewart[2], J. Cordero[2] and B.D. Sande[3]
[1]World Agroforestry Centre, P.O. Box 30677, Nairobi, Kenya, s.franzel@cgiar.org, [2]Oxford Forestry Institute, Oxford OX1 3RB, UK [3]Forestry Resources Research Institute, P.O. Box 311, Kabale, Uganda

Keywords: *Calliandra calothyrsus, Leucaena trichandra*, adoption, fodder shrubs, agroforestry

Introduction Smallholder dairying is an important enterprise in the highlands of E Africa. Farm sizes average 1-2ha and zero-grazing, cut-and-carry systems predominate. Inadequate protein reduces milk production and forces many farmers to spend scarce cash on commercial dairy meal supplements. In 1991, on-farm trials on fodder shrubs were started in Embu District as a collaborative venture of the Kenya Agricultural Research Institute, the Kenya Forestry Research Institute, and the World Agroforestry Centre. *Calliandra calothyrsus* was released to farmers in 1995 and was followed by *Leucaena trichandra*, mulberry (*Morus alba*), and *Sesbania sesban*. Farmers produce seedlings of calliandra and trichandra in nurseries; mulberry is planted using cuttings. Farmers plant the shrubs in hedges along field and farm boundaries, on contour bunds, and intercropped with Napier grass. Within 1 year after planting, shrubs are ready to be pruned for feeding livestock. Most farmers cut them at a height of about 1m to ensure that they do not shade the adjacent crops (Franzel *et al.*, 2003).

Extent and benefits of fodder shrubs About 30K farmers have planted in Kenya, 23K in Uganda, and several thousand in Tanzania and Rwanda. These are minimum numbers, as they include only areas where we have reliable estimates. Most of these farmers feed shrub leaves to stall-fed, improved-breed dairy cows, others feed to improved-breed dairy or meat goats, or to local-breed cows and goats. From sample surveys, depending on the area surveyed, numbers of trees/farmer vary widely; averages range from 30-924. Seed quantities produced and distributed are not known as most seed flows through the informal sector. Beginning in the second year after planting, a farmer with 500 calliandra shrubs can provide about 6kg of fresh fodder leaves/dairy cow/day, earning an additional 62-122$US/year. The 53K farmers with fodder shrubs in Kenya and Uganda have circa 250 shrubs each; total net benefits are thus about 2.438M $US/year. In addition to milk production, other benefits from fodder shrubs include improved animal health, wood for fuel, seed sales, improved manure, bee forage for honey production, and stakes for vegetable production. Environmental benefits also are significant because many trees are planted along contour bunds, reducing soil erosion and fixing nitrogen.

Factors in achieving successful adoption (1) Demand for fodder shrubs by farmers is high, mainly because the shrubs save cash and need only small amounts of land and labour. (2) Market access is relatively good in the areas where adoption has occurred. (3) Participatory methods were used to design fodder shrub technology. Most on-farm trials were designed and managed by farmers, encouraging other farmers to innovate. (4) Partnerships between researchers and extension build local organisational skills and knowledge and help reach large numbers of farmers. (5) Dissemination through farmer groups, instead of to individual farmers, economises on extension resources and ensures greater farmer-to-farmer information exchange and dissemination. (6) Institutionalising Ugandan project activities into local government development plans helps to mobilize communities and create a sense of ownership among beneficiaries. (7) The species promoted are fast growing and easy to establish and manage. (8) Partners giving livestock to farmers require them to plant fodder shrubs as a precondition.

Factors slowing adoption (1) Extension services and NGOs are often unfamiliar with agroforestry practices and lack planting material. (2) Fodder shrub practices are relatively knowledge intensive and training is required intermittently over long periods, e.g., at planting, managing, and at harvesting/utilization time.

Future fodder shrub technology Because intensive livestock production is increasing rapidly in the E African highlands, the future is promising. Facilitating projects are needed to promote sustainable community-based seed production and distribution and to train extension staff in fodder shrub management. Research is needed also to diversify tree species and identify suitable species for semi-arid areas and high-altitude areas. There is also potential to include fodder shrub leaf meal in commercial feeds.

Reference

Franzel, S., C. Wambugu, P. Tuwei & Karanja, G. (2003) The adoption and scaling up of fodder shrubs in central Kenya. *Tropical Grasslands,* 37 (4), 239-250.

Stylo in India: much more than a plant for the revegetation of wasteland

C.R. Ramesh[1], S. Chakraborty[2], P.S. Pathak[3], N. Biradar[1] and P. Bhat[4]
[1]IGFRI Dharwad, India, Email: igfrirsd@sancharnet.in, [2]CSIRO Plant Industry, Australia, [3]IGFRI Jhansi, India, [4]BAIF, Dharwad, India

Keywords: *Stylosanthes seabrana*, leaf meal, poultry feeding, plantation horticulture, forestry

Introduction Since the 1950s introductions of *Stylosanthes scabra*, *S. hamata* and *S. guianensis* from Australia, South America, the USA and Africa have continued in India. Although no cultivar has been released, selections of *S. scabra*, *S guianensis* and *S. hamata* are used in a range of environmental and commercial production systems. A large seed industry spanning >400 ha and run by >600 smallholder farmers producing 800t seeds/ year supports this usage (Rao *et al.*, 2004). Stylo is mainly used in India for revegetation of wastelands where it reduces soil erosion and offers fodder for livestock (Pathak *et al.*, 2004). Relatively small use is made as supplementary feed for dairy and breeding farms, as pastures in sheep and goat farms, in urban forestry and as a cover crop in horticulture and agroforestry. There are probably some 20,000 ha under silvipasture and horticulture. In mixed crop-livestock farming system stylo has been a saviour for smallholder farmers in some arid areas. The recent success of *S. seabrana* as a multipurpose legume and the suitability of stylo leaf meal as a replacement for expensive constituents in commercial poultry feed formulations may further accelerate uptake.

Major reasons for success:
1.Government policy on wasteland development. Government-backed wasteland and watershed development programmes have created a sustainable demand for stylo seeds over the past three decades. This long running revegetation programme for village commons and problem lands has supported animal production by the rural poor and nomadic tribes. The 'Rajiv Gandhi Watershed Mission' in Madhya Pradesh is a good example, which covers a network of over 3,800 watersheds over 147,000 ha and developed in partnership with >100 NGOs. Others watersheds are developed by private-public partnerships through NGOs such as the Watershed Organization Trust, which has improved 160,000 ha in several states including Maharashtra.
2. The technology met a need. Stylo has a ready-made market for the degraded land covering well over 100 million ha. Smallholder farmers have used simple, effective and appropriate seed production technology to meet the strong demand for stylo seeds generated by the wasteland development programmes of the federal and state governments. The seed industry has expanded naturally and steadily over the last 25 years, from a handful of growers to several hundreds covering 40 villages.
3. The technology is simple and profitable. With excellent adaptation to infertile acid soils in arid zones, stylo is well suited to large tracts of problem soils to produce high quality forage and add 100-150 kg/ha N. Establishment and management is simple and does not require any specialised equipment. Stylo seed production in the major production region around Andhra Pradesh and Karnataka is more profitable than growing other commercial crops and it offers much-needed employment to rural women.
4. Strong network and partnerships between stakeholders. Stylo has featured prominently in deliberations of the fledgling sheep unions to the more formalised co-operative and village clusters. In recent years, strategic partnerships have been forged between poultry feed manufacturers and stylo leaf meal producer/co-operatives to exploit commercial opportunities. Some NGO's like the Bharatiya Agro Industries Federation (BAIF) and Nimbkar Agricultural Research Institute (NARI), have played a key role in forging these alliances.
5. Dedicated champions from private and public sectors have promoted stylo. Personnel from the Indian Grassland and Fodder Research Institute through the Dharwad regional station have run effective extension and training programmes for farmers. The NGOs, various farmers' co-operatives and several visionary farmers have persisted with stylo and promoted its use among end users. Departments of Animal Husbandry and Forestry in Kerala, Tamil Nadu, Karnataka and others have consistently supported and promoted the use of stylo.

Conclusion The use of stylo for wasteland and watershed development helped develop a large seed industry. Following success in soil stabilisation, water availability and forage production, watersheds are now commonplace on many privately owned village lands. Commercial success has come from its use in large-scale plantation and forestry projects. Use will accelerate with the widespread use of stylo leaf meal in poultry feed.

References
Rao, P. P., C.R. Ramesh, P.S. Pathak, Y. M. Rao & N. Biradar (2004). Recent trends in *Stylosanthes* seed production by small holders in India. In: S. Chakraborty (ed.) High-Yielding Anthracnose-Resistant *Stylosanthes* for Agricultural Systems. ACIAR, Canberra, Australia, 235-242.
Pathak, P.S., C.R. Ramesh & R.K. Bhatt (2004). *Stylosanthes* in the reclamation and development of degraded soils in India. In: S. Chakraborty (ed.) High-Yielding Anthracnose-Resistant *Stylosanthes* for Agricultural Systems. ACIAR, Canberra, Australia, 86-95.

Forage Arachis in Nepal: a simple success

A.D. Robertson

"Oaky Creek" Wilson's Downfall, MS 1983, Stanthorpe 4380, Australia, Email: halfmoon@halenet.com.au

Keywords: adoption, participatory, farming systems, Nepal

Introduction Nepali farming systems are remarkably diverse. Livestock play a central role in livelihoods and sustainable farming on most farms. There is a need for productive forage legumes that can fit existing farming patterns and that can be multiplied easily. A wide array of genetic material has been introduced recently into the cropping, cut-and-carry, grazing, and forestry systems, mainly in the Terai (Ganges Plain) and in the "mid-hills" to about 2km ASL. In 1999/2000, 8 lines of *Arachis pintoi* were introduced from CIAT, and additional *A. pintoi* and *A. glabrata* lines from Queensland. The introduced arachis was established on a small number of permanent sites to enable close observation, crossing and continual selection, and a reliable long-term supply of planting material. Concurrently, small samples were provided to a large number of smallholder farmers (>1000 in the first season alone) over a very diverse agro-ecological range, for evaluation, local demonstration, and the supply of planting material within the community. The programme has been based entirely on vegetative material since 2000. Most arachis establishment has been in intensive smallholder systems, involving cover cropping, mixed planting with productive grasses in back-yard areas, and establishment on terrace risers. There also have been trial plantings in ley systems and on communal land, including land-slips/slides and roadside cuttings.

Scale and benefits Forage arachis has become popularised in many communities; it is likely that 15-20K households are already involved. Farmers continue to expand on-farm areas and refine systems of management and utilisation. Some households have bulked up from a few slips to 1000-2000m^2 within 3 years. The arachis is used commonly in cut-and-carry systems for supplementary feeding of milking buffalo and cattle, goats, pigs, and poultry. Due to positive feeding responses, rapid expansion in plantings is occurring within individual farms. Establishment on communal areas, such as roadside cuttings, has been on a smaller scale and is much less significant.

Major reasons for success:

- **Farming systems and bulking-up** Most Nepali mid-hill farming systems are very intensive and offer many niches for forage arachis. Small farm sizes, and the small unit areas, initially targeted for on-farm forage development, are well suited to vegetative propagation. Nepali farmers generally prefer this to the use of seed, partly because of the quick and conspicuous results. Some bulking-up has been undertaken on a contract basis.

- **Farmer attitudes** New forage interventions in Nepal typically have very high uptake rates. This may be attributed partly to the long tradition of back-yard dairying that most households practise. Participating farmers commonly undertake their own screening, feeding new material to all classes of livestock and poultry. Forage arachis has stimulated more interest than most other forage introductions.

- **Groups and networks** Nepali farming communities are characterised by a high degree of organisation into enterprise groups, including livestock, milk marketing, and forest user groups. These groups facilitate rapid farmer-to-farmer exchange of planting material. Providing the forage arachis to selected farmers, who act as "farmer resource centres" for screening, demonstration, bulking up and local supply of cuttings to other group members, has been successful. This approach is important in the context of reaching remote communities, particularly in a conflict situation that precludes regular follow-up.

- **Participatory approaches** The immediate and direct involvement of smallholder farmers have streamlined screening, demonstration and multiplication. Development agencies have coordinated initial inter- and intra-district exposure visits and distribution of planting material to newly participating areas.

- **Mobilisation of field staff** Field development workers from diverse disciplines, including forestry, soil conservation, livestock, and women's empowerment, have been mobilised to assist in the delivery of the programme. This multi-disciplinary approach has enabled quicker and more widespread adoption.

The future Further establishment of scattered nucleus sites in new areas in the Terai and lower mid hills could result in secondary adoption of forage arachis by vast numbers of farmers at very low cost. The current range of genetic material should be expanded, particularly in terms of introducing more cold tolerance and more erect lines for cut-and-carry.

Stylo in China: a tropical forage legume success story

L. Guodao[1] and S. Chakraborty[2]
[1]Tropical Crops Genetic Resources Institute, CATAS, Danzhou, Hainan, China 57173, [2]CSIRO Plant Industry, Australia, Email: sukumar.chakroborty@csiro.au

Keywords: *Stylosanthes guianensis*, leaf meal, plantation horticulture, forestry

Introduction Although *Stylosanthes gracilis* was the first stylo to be introduced as a green manure cover crop for young rubber plantations in 1961, the *S. guianensis* cultivars Cook and Graham introduced in the 1970s and 1980s were largely responsible for the stylo revolution in China. Before serious anthracnose outbreaks, these cultivars covered over 13,000 ha in southern China. Anthracnose had shifted emphasis to *S. guianensis* CIAT184 and successful cultivars originate from this introduction (Guodao *et al.*, 2004). Well-adapted varieties are now available for much of southern China and in 2003, the total area of stylo exceeded 200,000 ha. Stylo development has been greatly aided by the release of commercial cultivars. Initially these were from Australia, but more recent cultivars have been selected from introduced accessions. Reyan No 2, Reyan No 5 Reyan No 7 and Reyan No 13 are selections from CIAT 184; Reyan No 10 originated from CIAT1283 and cultivar 907 was developed through Cr^{64}-γ radiation technology. These are suitable for utilisation for leaf meal production, pasture improvement, green manure and soil conservation and the variety Reyan No.5 is cold-tolerant.

Major reasons for success:

1. **A number of production systems generating strong market demand** China is deficient in high protein forage by about 15 million t and there is ongoing R&D to meet market demand for leguminous forage. Southern China supports an estimated 35 million head of ruminants, 74 million pigs and 830 million poultry on some 32 million ha of grasslands to meet the staggering 27% increase in the annual consumption of livestock products. Forest covers over 44% of the land in south China and agroforestry has become an ecologically sustainable option to supplement short-term income from annual crops. But it takes a long time for forest plantations to generate income. With privatisation offering incentives, smallholder farmers increasingly raise and sell animals such as pigs, ducks, chickens and goats. This has sharply increased the demand for both fresh and dried fodder.

2. **Stylo is well suited to many production systems** Stylo is well adapted to south China, it produces 15-22 t/ha per yr dry matter containing 15%-16% crude protein. Production of stylo leaf meal, pioneered by farmers in south China, has further enhanced the value of stylo as a nutritious fodder crop. Stylo intercropped in young plantation forests provides an early income stream from the fresh cut and carry forage, well before the 8-10 years it takes to produce the timber. When grown in hilly terraces with young rubber or horticultural plantations, stylo helps to conserve soil and water, control weeds and improve soil fertility and the growth of trees. Smallholders grow stylo to feed cattle, goat and pigs as freshly cut or cooked feed. They also and make crude meal as a feed supplement for livestock and poultry. High quality stylo meal and hay is produced and marketed by large farms using commercial drying and processing equipment.

3. **Simple and profitable technology** The labour-intensive forage, leaf meal and seed production technology in south China is appropriate for the large and relatively inexpensive labour force. Stylo has brought high economic benefits to farmers and returns of 1,500-22,500 Yuan/ha for stylo compare favourably with those from paddy (9,000-12,000 Yuan/ha) or sugarcane (9,000-10,800 Yuan/ha). Stylo provides employment, especially for local women. Improved technology, including the transplantation of vegetative cuttings, has increased commercial seed production to 225-300kg/ha in the otherwise poor yielding *S. guianensis*. This has helped to establish southwest Hainan as a commercial centre for the production and export of stylo seed.

4. **Effective international collaboration with strong national support** The Tropical Pasture Research Centre established in the late 1980s has a strong network of highly motivated research and extension personnel operating in Hainan, Guangdong and Guangxi. Access to improved germplasm and technology has come through international collaboration and the improvement and adoption of stylo technology comes from a successful application of the participatory research approach. Private and public sector partnership has helped to establish two leaf meal factories to process large quantities of stylo.

Conclusion Despite its relative short history, stylo in China has made a significant impact to a number of production systems, most notably in pioneering the use of dry leaf meal as animal feed. Commercial success has been underpinned by superior economic returns compared to other cash crops and through its use in plantation forest and horticulture, where stylo offers a quick return on capital in an otherwise long-term investment.

Reference

Guodao, L., B. Changjun, W. Dongjin & H. Huaxuan (2004). Currently used *Stylosanthes* cultivars in China: their development and performance. In: S. Chakraborty (ed.). High-Yielding Anthracnose-Resistant *Stylosanthes* for Agricultural Systems, ACIAR, Canberra, Australia, 153-157.

Stylo Adoption in Thailand: three decades of progress

C. Phaikaew[1] and M.D. Hare[2]
[1]Division of Animal Nutrition, Dept. of Livestock Dev., Bangkok 10400, Thailand, Email: chaisangp@dld.go.th,
[2]Faculty of Agriculture, Ubon Ratchathani University, Ubon Ratchathani 34190, Thailand

Keywords: *Stylosanthes guianensis, S. hamata*, stylo, Thailand

Background *Stylosanthes* forage legume was very popular in Thailand for >30 years. *S. humilis* (Townsville) was the first popular species (late 1960s). It tolerated heavy grazing and grew very well along roads on free-draining upland soils, but anthracnose destroyed it in 1976. *S. hamata* (Verano; more resistant) replaced it and stimulated of large-scale pasture development. From 1976-84, the Dept of Livestock Development (DLD) launched a project to improve 32Kha of communal grazing land by oversowing Verano (circa 250t seed/year). This project has been sustained until the present time. Stylo now is used mainly for private grazing and cut-and-carry feeding for cattle. However, farmers generally prefer to plant grass for higher forage yields. Perennial stylo (*S. guianensis*) was also used for >20 years for high quality, cut-and-carry backyard forage. Graham stylo was planted until 1996 but anthracnose damaged it and production ceased. Due to its good resistance to anthracnose and its high dry matter production, Tha Phra stylo (CIAT 184) replaced Graham stylo immediately. Due to its resistance to anthracnose and grazing tolerance, hybrid stylo seed (*Stylosanthes guianensis* var. *vulgaris* x var. *pauciflora* ATF 3308) is produced for export to South America,.

Seed Production For nearly 30 years, the Division of Animal Nutrition, DLD, has implemented a government-supported seed enterprise successfully (Phaikaew & Hare, 1998; Hare & Phaikaew, 1999). Since 1975, Thai village farmers and DLD stations have produced >4.5Kt of Verano, Graham and Tha Phra stylo seed.

Area planted Since 1975, stylo has been sown in >300Kha of grazing land (private land, communal areas and along roads). Verano is now naturalised, especially along roads in NE Thailand. If establishment failed, some areas were oversown many times. However, due to the decrease in communal grazing areas and the lack of persistence of perennial stylo with heavy grazing and frequent cutting, the stylo area has decreased in size.

Major reasons for adoption

- Sandy, acid soils and medium seasonal rainfall (1250mm), in which stylo species grow very well. All species survive over the 6-7 month dry season and prolific seeding contributes to their survival.
- Easy establishment, high germination and palatable forage makes stylo the most popular forage legume used by Thai farmers.
- Stylo has multiple uses as forage in cut-and-carry systems, grazing, hay, silage, cover crops, and as leaf meal protein in concentrate feeds for dairy, beef cattle, swine and poultry. Stylo planting improves soil fertility.
- Cheap stylo seed is available and produced by village farmers under the DLD programme.
- Farmers accept stylo as a good quality forage legume that can increase milk yield and reduce the cost of concentrate feed. Stylo is used in total mixed rations for feeding dairy cows and beef fattening.
- Regular extensive preparatory research, colourful brochures, publications and technical advice from DLD staff.

Future progress Thai farmers must develop better skills to manage perennial stylo (Tha Phra and Hybrid 3308) as legume protein banks for the dry season. To develop a seed export market, we must promote Thailand's reputation as a producer of top quality seed. Widespread use of perennial stylo has been limited due to lack of persistence and poor regrowth when it is cut as a mature stage. Hybrid stylo 3308 may be the ideal replacement, as it has good regrowth after cutting, no anthracnose disease and good grazing persistence.

References

Hare, M. D. & C. Phaikaew (1999). Forage seed production in Northeast Thailand: A case history. In: Loch, D.S., Ferguson, J.E. (eds). *Forage Seed Production Volume 2: Tropical and Subtropical Species* pp. 435-443. (CAB International, Oxon., UK).
Phaikaew, C. & M. D. Hare (1998). Thailand's experiences with forage seed supply. In: Horne, P.M., C. Phaikaew & W.W. Stur (eds) Forage Seed Supply Systems - *Proceedings of a workshop held at Tha Pra, Thailand,* 7-14. (CIAT Working Document No. 175, Los Banos, Philippines).

Sesbania grandiflora: a successful tree legume in Lombok, Indonesia

D. Hasniati[1] and M. Shelton[2]
[1]Faculty of Animal Science University of Mataram, Jl. Majapahit 62 Mataram 83125 Lombok, Indonesia
[2]Faculty of Natural resources, Agriculture and Veterinary Science, The University of Queensland, Brisbane Australia 4072, Email: m.shelton@uq.edu.au

Keywords: *Sesbania grandiflora*, Lombok, cut-and-carry

Introduction Sesbania (*Sesbania grandiflora*) is a multi-function tree. Its main use is as a livestock feed in southern Lombok, Indonesia, the major region for goat and cattle production. It is the main (and sometimes the only) component of ruminant diets (Dahlanuddin, 2001). A national program, aimed initially at improving soil fertility and replanting barren areas, formally introduced it to Lombok in the 1970s (Suseno, 1990).

Sesbania has the highest nutritive value and is the most widely used of all tree legumes available for livestock feeding in Lombok. It is planted in single rows along the bunds of rice paddies. The leaves are cut and fed fresh in a cut-and carry system; the branches are dried for firewood. Farmers harvest only the side branches of the tree to avoid tree mortality, and to make the trunk straight for pole timber when cut at around 3 years of age. It is also used as a nutritious vegetable, especially for nursing mothers as the local community believes it stimulates milk production. Southern Lombok has limited forest resources, making sesbania the main source of firewood and timber for both housing and animal pens.

Currently, sesbania is planted on approximately 25% of rice field bunds on Lombok, mostly on the southern part of the island (which is the main rice region in Lombok). Mature seeds that drop naturally during the dry season provide sufficient seedlings in the early wet season for transplanting onto the bunds. Each farmer plants an average of 520 plants, 40-60cm apart. From an estimate of the total length of the bunds over Lombok island, circa 65K small farmers plant sesbania.

Major reasons for success The Department of Agriculture introduced the sesbania concept, initially through demonstration and seed provision. At local level, several NGO's also were involved in encouraging farmers to plant it. In the early stages of the program, the local community was reluctant to adopt the technology because they believed that dead and decaying sesbania roots would make holes in the bunds, causing them to leak water from one field to another. Intensive extension activities by government and NGOs successfully convinced farmers that planting sesbania was worthwhile. By arrangement, landless livestock growers may plant sesbania on other farmer's rice field bunds; the livestock growers harvest the leaves leaving the trunk to the landowner, or the two parties may share the leaves and trunks. Landowners also may sell or barter the leaves to livestock growers.

Future potential and limitations to wider adoption Unfortunately, despite some attempts by the government to extend the system to other areas, sesbania plantings are concentrated mainly in southern parts of Lombok island. This may be related to lack of suitable soil types and conditions in other regions, or perhaps to lack of transfer of indigenous technology for planting to other farmers. The total area of rice field on Lombok Island is 167Kha, with an estimated total length of bunds of approximately 6,860 km. If all bunds were planted successfully with sesbania, the number and productivity of ruminant livestock in Lombok would increase significantly. The system has potential for expansion to other areas of Indonesia with similar agroecological conditions.

Conclusion Sesbania planting is a valuable and sustainable technology that fits well into smallholder farms. It allows farmers to produce high quality forage. Farmers are convinced that sesbania does not provide too much shade for rice crop production, which the addition of organic fertilizer from fallen sesbania leaves may improve. For these reasons, the agronomy of the area and methods of planting should be investigated further to expand the planting of sesbania to other areas of Indonesia.

References

Soeseno, S. (1990). *Turi Sebagai Antimurus Darah. TRUBUS*, 253, 266-267.
Dahlanuddin (2001). Forages commonly available for goats under farm conditions on Lombok, Indonesia. *Livestock Research for Rural Development*, 13, 1.

Leucaena: sustainable crop and livestock production systems in Nusa Tenggara Timur Province, Indonesia

C. Piggin[1] and J. Nulik[2]

[1]Australian Centre for International Agricultural Research, PO Box 1571, Canberra, Australian Capital Territory 2600, Australia, Email: piggin@aciar.gov.au, [2]Agricultural Technology Assessment Institute, Naibonat, Kupang, NTT Indonesia

Keywords: Leucaena leucocephala, cropping, livestock production

Introduction In the late 1800s/early 1900s, population increases, slash and burn cropping, wildfires, livestock, and weeds led to extensive losses of natural vegetation and land degradation in the semi-arid islands of Nusa Tenggara Timur province, E Indonesia. In the 1930s-60s, villagers, government institutions and NGOs recognized the need to reduce degradation and increase production. They developed and promoted more sustainable fallow systems, based on the use of leucaena (Leucaena leucocephala), introduced several centuries earlier from central America.

Major reasons for success (Piggin, 2003).
1. A recognized need for better systems In the 1930s, low farm productivity and poverty caused by serious land degradation led to the realization that serious efforts were needed to develop sustainable farming systems.
2. Failure of alternatives Attempts in the 1960s and 70s to control erosion and land degradation with physical structures and traditional terraces were unsuccessful because of labour requirements, costs and ineffectiveness.
3. Adaptation of leucaena to the local environment Leucaena is deep-rooted, drought-resistant, well adapted to semi-arid climates and low nutrient alkaline/neutral soils, relatively easy to establish and very persistent.
4. Compatibility of leucaena with local farming systems Leucaena is a robust plant that can persist and regenerate in traditional swidden cropping systems that involve regular and severe cutting and burning.
5. Capacity of leucaena to supply local needs Leucaena is a multi-purpose plant that contributes to a multitude of village needs, including fence timber, firewood, building timber, forage, mulch and seeds for ornaments and food.
6. Commitment of local leaders and groups Local village heads, NGOs, church groups, and government departments were committed to develop and demonstrate more sustainable leucaena systems to local villagers.
7. Creation of a favorable policy environment Regulations were instituted to tether/confine livestock in cropping areas, cropping credit for farmers planting leucaena, development of erosion prevention programs, obligatory planting of leucaena, and encouragement of cattle husbandry by livestock distribution schemes.
8. Effectiveness of leucaena Leucaena has been effective in reducing erosion, increasing infiltration and stream flows, suppressing crop weeds, improving soil N, and providing shade for mangoes, cocoa, pepper and cloves.
9. Contribution of leucaena to development of more commercial farming systems Leucaena has helped village farmers move from subsistence to commercial farming through the development of cattle and goat fattening and orchards of bananas, papaya, mangoes, coconuts, cloves, pepper, and cocoa.

Conclusions Contrasting systems have emerged and endured in the Amarasi and Sikka areas. They are excellent examples of the adoption and use of shrub legumes in village farming systems. The Amarasi system was developed in the 1930s and is based around the use of leucaena as a forest fallow rotation for corn and a forage for tethered cattle and goats. The Sikka system was developed in the 1960s and involves contour rows of leucaena to prevent erosion and create indirect terraces where corn, peanuts, and mungbeans are grown and mulched with leucaena clippings from the hedgerows. Weight gains of 1.3-1.7 kg/head per day have been recorded for tethered cattle fed with leucaena whilst maize yields can be doubled by including 2-3 years of leucaena in crop–fallow rotations. These systems each now cover about 50Kha, or 70% and 30% of the Amarasi and Sikka areas, respectively. They contribute substantially to farm production, wood supply, and stabilization of the resource base. Villager families farm about 2ha, suggesting that some 25K farm families may be growing leucaena in both Amarasi and Sikka.

References

Piggin, C.M. (2003). The role of Leucaena in swidden cropping and livestock production in Nusa Tenggara Timur Province, Indonesia. In: Agriculture: New Directions for a New Nation East Timor (Timor-Leste), ACIAR Proceedings, 113, 115 - 129.

Centrosema pascuorum in Australia's Northern Territory: a tropical forage legume success story

A.G. Cameron

NTDBIRD, Berrimah Agricultural Research Centre, Darwin, Northern Terrirory, Australia 0810, Email: arthur.cameron@nt.gov.au

Keywords: *Centrosema pascourum*, hay, feed cubes

Introduction Centurion (*Centrosema pascuorum*), an annual legume, was first sown for evaluation in the Northern Territory in the late 1970s (Cameron & McCosker, 1986; Clements *et al.*,1984). The bred cultivar Cavalcade (Line 2/2, Cameron, 2003b) was released in 1984, and Bundey (CPI 75115, Cameron, 2003a) in 1986. Seed availability limited the use of the legume for 10 years. In the Northern Territory, it is now used mainly as a hay crop, but is used also as a component of grazed mixed pastures. Most of the hay is made into feed cubes and pellets to feed cattle in the live export trade to southeast Asia. Up to 40% of the 50,000t of feed cubes and pellets used each year to feed 300,000 head of live export cattle, before, during and after shipping, are made from *Centrosema pascuorum*. Approximately 5,000ha of Cavalcade and Bundey are sown each year, mostly for hay. In 1998 annual production of hay and seed reached 20,000 and 177t, respectively. Over the last 5 years, on average 17,326t of hay valued at $2.61m has been produced. In the same period, on average 35.6t of seed valued at $0.234m has been produced and sold in the Northern Territory, interstate and overseas. While 2 farmers produced most of the seed, and 25 farmers produced most of the hay, >100 farmers make some use of the legume.

Factors in achieving successful adoption The use of *Centrosema pascuorum* met a need of farmers for a good quality hay crop. The hay was of better quality than most other species being used at the time (grasses, forage sorghums, Pearl millets and Verano (*Stylosanthes hamata*)), and also less demanding in time of harvest than some (Verano, cowpeas). Positive feedback and demand from end users, initially from larger pastoral properties, and later from cubing plants, coupled with an increase in price for hay, encouraged the use of this legume. Local Departmental Extension Officers strongly supported the use of Centurion. Departmental Seeds Officers also encouraged and supported farmers in the production of Certified Seed. The uptake of the legume, which was slow initially, was accelerated by the distribution of a large quantity of good quality seed in 1990 by the Government Department for on property testing.

Factors which constrained adoption These factors were the availability of seed, and the farmers' lack of experience with the legume that was a new species to agriculture in the Northern Territory, and to the world. This delayed largescale use of the legume for up to 10 years. After release, a Seed Increase Committee was responsible for distribution of Basic Seed for growing of seed increase crops. This committee distributed the limited amount of Basic seed available to a range of interested farmers. Unfortunately, these farmers had limited experience in growing seed crops and mostly, the crops were complete failures. After receiving a quantity of Prebasic Seed in 1986, the Department's Seed Section was able to produce over 600kg of Basic Seed in 1986 and 1987, and over 3t in 1989. This seed combined with 10 tonnes of seed produced by 2 farmers in 1989 and 1990 allowed hay producers to sow larger areas of Centurion from the 1990/91 wet season. The distribution of 2.5t of seed, grown by the Department and distributed free to farmers for the 1990/91 wet season, allowed them to become familiar with the species.

Conclusion While there will be the normal fluctuations in production and demand caused by variations in the weather and changes in live cattle export numbers, the use of Centurion is a permanent innovation that is likely to continue into the foreseeable future. Farming systems will change. Pastures and hay areas will need to be renovated and rotated to a grass to remove the build up of nitrogen fixed by the legume and to control the invasion of broadleaf weeds. Large quantities of good quality seed of this annual species need to be produced each year.

References

Cameron, A. G. (2003a). Bundey. NTDBIRD Agnote 427.

Cameron, A. G. (2003b). Cavalcade. NTDBIRD Agnote 415.

Cameron, A. G. & T. H. McCosker (1986). Introduced Pasture Species Screening on Mount Bundey Station. NTDPIF Technical Bulletin 97.

Clements, R. J., W.H. Winter & R. Reid (1984). Evaluation of some *Centrosema* species in small plots in northern Australia. *Tropical Grasslands*, 18, 83-91.

Arachis pintoi in the humid tropics of Colombia: a forage legume success story

E. Lascano Carlos[1], M. Peters[1] and F. Holmann[1,2]
[1]CIAT, AA 67-13 Cali, Colombia, Email: c.lascano@cgiar.org, [2]ILRI, AA 67-13, Cali, Colombia

Keywords: Brachiaria, legume pasture, milk production, on- farm research

Introduction Cattle liveweight gain and milk yield can be depressed significantly on grass alone pastures that degrade over time on the margins of tropical forest. Use of legumes in pastures is an alternative to minimize declines in quality and quantity of forage biomass and thus increase livestock production. From 1987-90, forage researchers in CIAT collaborated with several institutions in the Piedmont region of the Amazon basin in Caqueta, Colombia on selection of forage germplasm adapted to acid soils and with potential to reclaim large areas of degraded pastures in cattle farms of the region. The most successful pasture was the legume/ grass association of *Arachis pintoi* grown with several *Brachiaria* species. However, livestock producers in the region were not adopting the Arachis technology mainly because of lack of promotion, little knowledge on benefits and high seed cost. Thus, an inter-institutional on-farm project involving public and private institutions was carried out to document the on-farm benefits of Arachis-based pastures, train personnel of different institutions on establishment and utilization of Arachis pastures using participatory methods and initiate and catalyse an active extension transfer mechanism of the Arachis technology in the region.

Major reasons for success At the start of the project it was felt that the success in promoting Arachis would depend on establishing legume-based pastures in 10-15 pilot farms of the region. The strategy also considered that the10-15 key farmers initially selected to participate in the project would act as promoters of the *Arachis* technology to surrounding farmers. This in turn would ensure that a minimum of 100 farmers would be exposed and become adopters of the new pasture technology in the period covered by the project. However, given the high cost of the *Arachis* seed available in the market and the prevalence of absentee owners participating directly in the project, that strategy failed. Therefore, the extension phase of the project required an alternative diffusion approach of the *Arachis* technology. The main elements of the strategy were: (a) creation of a Technology Transfer Fund managed by NESTLE, who bought the milk produced by farmers participating in the project, (b) conducting a survey among all milk producers that sold milk to NESTLE to define interest in recuperating degraded pastures using Arachis, (c) contracting the multiplication and purchase of commercial seed of Arachis to fulfil demand among interested producers, (d) contracting tractors for timely land preparation and allowing farmers to pay for the partial cost of pasture establishment and (e) pay back the loan with milk delivered to NESTLE with no interest on the money. With this strategy >100 farmers established 3Kha of Arachis-based pastures in a 2-year period.

Future potential and limitations to wider adoption Lessons learnt during the course of the project included the need to have flexible research methods for on-farm pasture evaluation and to avoid absentee owners because they do not provide the necessary feedback to researchers and do not promote the technology. On-farm research using participatory methods cannot accomplish the ultimate goal of diffusion/ adoption of improved pasture technology by itself. Thus, alternative strategies for diffusion of new legume-based technology should be part of the overall objective of a pasture/livestock project. Future pasture development projects should define the R&D plans of relevant institutions present in the region in order to accomplish a multiplier effect through training.

Conclusions By identifying alternative methods to establish grass/ legume pastures, by in situ demonstrations of proper grazing management and by identifying "bottlenecks" for the adoption of Arachis, the project helped to identify ways to facilitate the diffusion of Arachis-based technology in the region. Undoubtedly a large, reliable market for fresh milk in the region contributed to the desire of most particepating and non-participating farmers in the project to consider investments on reclamation of degraded pastures. The promotion of legumes by the project also generated interest among livestock farmers to learn more about *Arachis* and its role in increasing milk yield and contributing to soil fertility.

References
Grof, B. (1985). Forage attributes of the perennial groundnut *Arachis pintoi* in a tropical savanna environment in Colombia. *Proceedings of the XVI International Grassland Congress,* Kyoto, Japan, 168-170
Lascano, C., G. Ruiz, J. Velasquez, J. Rozo & M. Jervis (1999). Developing improved pasture systems for forest margins. In Fujisaca Sam (ed) Systems and Farmer Participatory Research: Developments in Research on Natural Resource Management CIAT, Cali, Colombia. p 50-60.

Tropical kudzu (*Pueraria phaseoloides*): successful adoption in sustainable cattle production systems in the Western Brazilian Amazon

J.F. Valentim and C.M.S. Andrade
Agroforestry Research Centre of Acre – Embrapa Acre, Km 14 da BR-364, Caixa Postal 321, CEP 69908-970, Rio Branco, Acre, Brazil, E-mail: judson@cpafac.embrapa.br

Keywords: economic impacts, environmental benefits, legume, pasture reclamation

Introduction In 1976 the Program for Reclamation, Improvement and Management of Pastures in the Brazilian Amazon (PROPASTO), conducted by Embrapa, established on-farm experiments in the State of Acre. These experiments consisted of introducing and evaluating grass and grass-legume stands, both under cutting and grazing. Similar experiments were established in all states in the region. Since then, research has recommended new species of grasses, legumes and grass-legume associations for establishment of improved pastures in the Brazilian Amazon. The grass cultivars recommended were an instant success with farmers, with *Brachiaria brizantha* cv. Marandu becoming the predominant species, occupying approximately 80% of the total pasture area established in Acre until 1998. However, the legumes recommended had limited adoption and most failed to persist in the pastures after a few years under grazing, mainly due to the incidence of diseases and poor management. The exception occurred in Acre where the legume *Pueraria phaseoloides* (Tropical kudzu) became one of the most important forage resources and has maintained that position.

Technology adoption process Since the early 1980's Embrapa Acre has established grass-Tropical kudzu pastures in strategically selected areas owned by farmers that had been identified as leaders and innovators among their peers. Grass-legume associations were persistent and gave strong financial benefits due to increased carrying capacity from 0.5 to 1.5 animal units/ha and increased live weight gains from 90 to 360 kg/ha per year. Tropical kudzu showed excellent adaptation and high seed production in the environmental conditions of Acre. It soon became a cash crop for small farmers that started to use it also as an improved fallow in the reclamation of degraded agricultural land. This resulted in ready availability of low cost seeds in the market, making Acre a seed exporter to other regions of Brazil. Later in the 1980's, farmers began to be pressured by governmental and non-governmental organisations to reduce deforestation for establishment of new pastures and to reclaim degraded pasture areas. With the great distances from markets, poor roads and low profitability of extensive beef cattle production systems, the use of modern inputs such as fertilisers were not economically viable. At this point Embrapa Acre presented farmers once again with successful on-farm experiences with the availability of grass-Tropical kudzu pastures that, at that time, were 7 to 10 years old. After initial talks and field days conducted by researchers and extension agents, the owners of the demonstration farms became the true agents in the process of technology transfer. In 1989, there was a rush of farmers to establish grass-Tropical kudzu pastures and prices of the legume seeds increased from US$ 4.00 to US$ 11.00 per kg from July (harvest period) to November (the end of the planting season). Researchers had recommended a seed rate of 1.0 kg/ha of Tropical kudzu seeds in mixture with the grass seeds. However, in the next fifteen years it became common for farmers in Acre to use a seed rate of 1.0 kg of Tropical kudzu seeds for each 2.42 ha (one alqueire). It is estimated that Tropical kudzu is present in over 30% (480,000 hectares) of the total pasture area in Acre, in percentages ranging from less than 10%, in pastures under more intensive grazing, to over 90%, in areas where death of *B. brizantha* cv. Marandu is occurring, due to the lack of adaptation of this grass species to low permeability soils. In Marandu-Tropical kudzu pastures where the grass started to die, the legume acted as a buffer and saved farmers from complete bankrupcy. Tropical kudzu prevented weeds from completely taking over the pastures, supplied good quality forage for the animals and increased the nitrogen status of the soil. This allowed the farmers time to gradually re-establish their pastures.

Key factors for success (in order of importance): 1) the availability of appropriate technology and long term commitment of key players, particularly researchers from Embrapa Acre; 2) the socioeconomic situation of farmers and farming systems that were conducive to technological changes due to the increasing environmental restrictions on pasture area expansion; 3) farmer centred research and extension and capacity of local institutions to support the programme; 4) market access and strong financial benefits of the technology; and, 5) strategic partnership among stakeholders.

Conclusion Tropical kudzu is now considered a naturalised plant in the state of Acre, because of its widespread use in mixed pastures and in agricultural land reclamation for more than 20 years. Despite its low adaptation to high grazing pressures under rotational stocking, this legume continues to be an important forage resource in the cattle production systems in Acre. It is still the only legume used by farmers in the establishment of new pastures or in the reclamation of degraded pastures in the State. For more intensive production systems, *Arachis pintoi* cv. Belmonte has become the most important forage legume.

Forage peanut (*Arachis pintoi*): a high yielding and high quality tropical legume for sustainable cattle production systems in the Western Brazilian Amazon

J.F. Valentim and C.M.S. Andrade
Agroforestry Research Centre of Acre – Embrapa Acre, Km 14 da BR-364, Caixa Postal 321, CEP 69908-970, Rio Branco, Acre, Brazil, E-mail: judson@cpafac.embrapa.br

Keywords: economic impacts, environmental benefits, pasture reclamation, technology adoption

Introduction The State of Acre had 1.45 million ha of pastures and a cattle herd of 1.95 million heads in 2003. Since 1998, the increasing area affected by the death of Marandu grass (*Brachiaria brizantha*) led farmers in Acre to search for alternatives to maintain productivity and profitability of their production systems. However, the traditional strategy of converting primary forest areas into pastures has been severely restricted by strong enforcement of environmental legislation by state and federal agencies. This forced farmers to search for alternative technologies to reclaim degraded pastures and to intensify their production systems. Tropical kudzu (*Pueraria phaseoloides*), the major forage legume used in mixed pastures in Acre (480,000 ha), did not show good compatibility with some of the new grass species being established by farmers, such as African stargrass (*Cynodon nlemfuensis*), and also failed to persist when managed under rotational stocking with stocking rates above 1.5 animal units per hectare.

Technology adoption process In the beginning of 2000, farmers that traditionally collaborated with Embrapa Acre for on-farm validation of technologies demanded new legumes adapted for use in more intensive cattle production systems, which included rotational stocking management. At that time, *Arachis pintoi* (forage peanut) was in pre-recommendation phase for the environmental conditions of Acre. This had arisen from research initiated in 1990 that had led to the release of the cultivar Belmonte in 1999 in Bahia, Brazil,. In March 2000, one farmer started to establish *A. pintoi* cv. Belmonte in association with African stargrass in the process of reclaiming degraded pastures in low permeability soils where Marandu grass had died. Both the legume and the grass were manually planted using vegetative material (stolons). The initial success of this experience soon caught the attention of other farmers that were facing similar problems. In April 2001, about 20 farmers had established this legume in association with African stargrass, Marandu grass (both in well drained and low permeability soils), *B. decumbens* cv. Basilisk and *B. humidicola*. In December 2001, *A. pintoi* cv. Belmonte was officially recommended by Embrapa Acre for diversification of pasture ecosystems and also as a cover crop for soil protection in Acre. The news of the success of this legume in the reclamation of degraded pastures, and in the improvement of other still productive grass pastures, rapidly spread among farmers. By March 2004, close to 1,000 small, medium and big farmers of Acre had already introduced forage peanut into their pastures, some in almost 100% of their farms, with areas of up to 2,000 ha. It is estimated that forage peanut has been planted in association with grasses in approximately 65,000 ha in Acre. In some farms, these pastures have been successfully managed with 2.5 animal units/ha, with Nelore x Angus crossbreed steers ready for slaughter (255 kg of carcass weight) within 24 months and primiparous calving at 22-24 months of age. Recent features on national television networks, newspapers and rural magazines reporting the successful use of forage peanut in the Western Amazon (Acre), in the South (Rio Grande do Sul) and Northeast (Bahia) regions of Brazil have led to a strong demand for information and vegetative material of this legume by farmers from most parts of the country. In Acre, it was noted that many initially reluctant farmers became interested in planting this legume soon after the news on TV.

Key factors for success (in order of importance): 1) availability of appropriate technology; 2) socioeconomic situation of farmers and farming systems were conducive to technological changes due to the death of *B. brizantha* cv. Marandu and the increasing environmental restrictions on pasture expansion into new forest areas; 3) long term commitment of researchers of Embrapa Acre, who were champions in promoting adoption of grass-legume pastures; 4) farmer centred research and extension, market access and strong financial and environmental benefits of the technology; and, 5) strategic partnership among stakeholders and capacity of local institutions to support the programme. The use of farmers that were early adopters as instructors and their farms as demonstration sites of successful use of grass-forage peanut pastures were important factors in obtaining credibility and increasing adoption of this technology.

Conclusion The prospects for the use of grass-forage peanut pastures in tropical regions are very encouraging, especially in the humid climates. In the coastal region of Bahia, Brazil, there are reports of *B. humidicola*-forage peanut pastures more than 10 years old. In Acre, there are mixed pastures of Massai grass (*Panicum maximum* x *P. infestum* cv. Massai) and forage peanut still being productive nine years after planting, and African stargrass-forage peanut pastures established in 2000 present no evidence of legume decline.

Estilosantes Campo Grande in Brazil: a tropical forage legume success story

C.D. Fernandes[1], B. Grof[2], S. Chakraborty[3] and J.R. Verzignassi[4]
[1]Embrapa Gado de Corte/Uniderp, Caixa Postal 154, Campo Grande-MS, Brazil [2]Amiga Court, Palmwoods, Queensland, Australia, Email: bgrof@bigpond.com, [3]CSIRO Plant Industry, Austrália [4]Universidade Estadual de Maringá-PR, Brazil

Keywords: Estilosantes Campo Grande, legume, savannas

Introduction Estilosantes Campo Grande (ECG) is a mixture (80/20 by weight) of *Stylosanthes capitata* and *S. macrocephala*, derived through genetic combination of selected accessions by open crossing over 6 generations (Grof *et al.*, 2001). The cultivar is the result of >10 years of research on a collection of >1000 ecotypes of *Stylosanthes* species collected and maintained by Embrapa. The main use of ECG in Brazil is to improve grass-dominant pastures. It has good persistence with *Brachiaria* spp., *Panicum maximum* (Tanzania-1 and Mombaça) and *Andropogon gayanus*. It has excellent adaptation to low fertility sandy soils and is persistent under grazing. In regional trials, ECG was the best performer in S Brazilian Cerrados, where the dry season is less severe. Although it does not retain foliage after the rainy season in areas with long and intense dry season, it still contributes to nitrogen fixation and animal nutrition during the rainy season. It is a prolific seed producer and seedling recruitment is its main mechanism of persistence. More than 500t of ECG seed has been produced since its commercial release and the estimated area planted to this cultivar has increased from <500ha in 2000 to nearly 150Kha by 2004, mainly in the Brazilian Savannas.

Major reasons for success:
1. An excellent performance drives strong market demand Interest in this variety is increasing because of: (a) good adaptation to low fertility soils, mainly sand soils; (b) production of 12-15t DM/ha/year; (c) prolific seed production (250-500kg/ha); (d) seed is cheaper than that of *S. guianensis* cv. Mineirão and there are prospects of further reduction in seed production cost from mechanical seed harvesting; (e) anthracnose resistance; (f) >5 years persistence under grazing in well managed mixed *Brachiaria decumbens* pastures; (g) high natural reseeding and strong seedling recruitment in the pasture; (h) 180 kg/ha nitrogen fixation in protein bank; (i) 18-27% increase of cattle liveweight gain in mixed pastures with *B. decumbens* compared to grass alone and (j) increased biological activity on soils, making the system more sustainable.
2. Strong private-public partnership has led technology development The R&D leading to the development and release of ECG was a partnership between Embrapa Gado de Corte and Ribeirão Agropecuária and other farmers in Brazil. Practical on-farm experience of farmers was combined with the scientific and technical expertise of a multidisciplinary team of researchers to better define and identify problems and gaps in technology and develop solutions. Consequently, problems often were resolved rapidly and the 'best practice' solution was adopted quickly.
3. Simple and profitable technology Prolific seed production, high nutritional quality and persistence under grazing have been very important factors for its adoption by farmers. Seed is produced separately for each species and mixed at the point of sale. The seed price varies from 4-6U$/kg, which is cheaper than Mineirão. Recently introduced commercial mixtures of 1 part Mineirão and 3 parts ECG, sown at 2 kg/ha, have further reduced cost. A new technique of row sowing of EGC in established grass pastures, using no-tillage farming system that does not involve the removal of animals at sowing, has reduced cost further and increased flexibility and adoption.
4. Effective international collaboration with strong national support A team approach was used to develop agronomic principles and to extend knowledge on how best to establish and manage ECG for different farming/grazing systems. Scientists, seed producers and farmers worked together to develop an 'agronomy package' for use by farmers. The technology was well publicized by Embrapa and its collaborators through field days, tours, publications, one-on-one extension, and TV and radio interviews. National and international agencies, including Embrapa, seed growers and ACIAR, all supported, funded and promoted the 'quick uptake of the best technology'.

Conclusion ECG is the main legume currently used under grazing in the savannas of Brazil. Since its release in 2000, its use has increased steadily due its superior performance and development of the technology in close association with seed producers and farmers. By increasing productivity and sustainability while delivering significant economic benefits to the farmer, ECG offers high hopes of restoring the international competitiveness to the Brazilian Beef industry.

Reference
Grof, B., C. D. Fernandes & A. T. F. Fernandes (2001). A novel technique to produce polygenic resistance to anthracnose in *Stylosanthes capitata. Proceedings Nineteenth International Grassland Congress*, 525-526.

Butterfly pea in Queensland: a tropical forage legume success story

M.J. Conway

Department of Primary Industries & Fisheries, LMB 6, Emerald, Queensland, Australia 4720, Email: maurice.conway@dpi.qld.gov.au

Keywords: *Clitoria ternatea,* cv. Milgarra, butterfly pea, reasons for successful adoption

Background to success Butterfly pea (*Clitoria ternatea)* is a productive and persistent legume adapted to a range of soils and climates in northern (tropical & sub tropical) Australia. Cultivar Milgarra, a composite line of introduced and naturalised accessions, was released in Queensland in 1991. The area planted in Queensland is a measure of its success, reaching <500ha in 1996; 5Kha by 1998; 30Kha by 2000; 50Kha by 2002; and 100Kha by 2004. Butterfly pea is used by cattleman mostly to grow and finish cattle. It is not grown extensively as a ley legume yet. Farmers who crop and have cattle, plant it but usually on cropping soils with limitations e.g. low plant available water capacity, shallow, low N or duplex soils. Farmers without cattle generally choose to use inorganic nitrogen as an N-source for crops.

Many research trials were conducted throughout N Australia from the 1940's to 1990's. Data on establishment, production and adaptation was collected from small plot trials planted across a wide geographical area. Grazing sites produced data on grazing management and animal weight gain. Trials were conducted to determine row spacing; seeding rate and weed control. Appropriate techniques were developed to control butterfly pea in a subsequent crop at the end of a ley phase. Data on organic matter and total N accumulation were collected.

Major reasons for success

1. The technology worked Butterfly pea establishes quickly to produce a relatively cheap but high quality, productive pasture on soils previously considered 'difficult to establish'. It is an effective conduit for farmers who wish to change from farming to grazing for many reasons, including lifestyle choice, cost/price pressures of farming, or to avoid re-investment in expensive new farm machinery.

2. The technology met a need for a persistent legume that produced quality pastures to finish cattle at a younger age to meet market specifications. It also restores soil fertility. Nitrogen rundown is causing declining production from pastures and crops. Alternative legumes in this region have serious limitations.

3. A team approach was used to develop and extend agronomic principles for butterfly pea for varying farming/grazing systems by using the individual skills and experience of farmers coupled with good science and understanding of researchers and extension staff. Importantly, butterfly pea technology was incorporated successfully into an 'agronomy package' developed by farmers and scientists working together.

4. Champions from the farming and scientific communities constantly and enthusiastically promoted butterfly pea and its management. Field days, bus tours, publications and one-on-one extension were used. A video was produced and distributed to interested producers. A booklet (*'The Butterfly Pea Book - a guide to establishing and managing butterfly pea pastures in Central Queensland'*) was published. Extensive use by farmers occurred only after an 'agronomy' package was developed and extended.

5. The R&D was done in partnership, drawing on the practical experience of farmers and technical skill of scientists to develop solutions. Funding bodies funded R&D after commercial release. Problems and gaps in technology were identified and resolved and 'best practice' was developed and adopted. This allowed the new cultivar to achieve its potential.

6. High quality, cheap seed available. The decision not to apply Plant Breeders Rights to butterfly pea ensured that large quantities of high quality but cheap seed were available. This encouraged additional seed growers to grow increased seed quantities, which reduced the seed price (initially from $12 to $3/kg, currently $4-5/kg) and increased the numbers of farmers planting, area planted and planting rate.

Conclusions Growers did <u>not</u> adopt Butterfly pea when first released. The area planted in Queensland, especially in Central Queensland' is increasing each year but this occurred only after R&D by researchers, extension staff and farmers with financial assistance from funding bodies. Developing an agronomic package, demonstrating best practice and providing information on managing the pasture, animal performance, controlling weeds and producing quality seed ensured producers quickly adopted the new pasture legume. Proven establishment and management technology and a plentiful but cheap seed supply ensures that butterfly pea will continue to play an important role in providing quality pastures to finish cattle at a younger age. Butterfly pea will be used as a ley more often and on larger areas as cropping soils age and N declines or when the return from cattle decreases and crops increases.

Stylos: the broad acre legumes of N Australian grazing systems

J.P. Rains

Southedge Seeds Pty Ltd, PO Box 1502 Mareeba, Queensland, Australia 4880,
Email: johnr@southedgeseeds.com.au

Keywords: Stylo, adoption, legume

Background Early accidental introduction of *Stylosanthes humilis* into the N Australian savannas in the early 20th century and its contrived spread by the late 1960s led to the release of more perennial types. Townsville Stylo, as it was known, colonised large areas and had reached its climatic and agronomic limits by the 1970s. It had a major impact on beef production in areas where natural carrying capacities were relatively low. The precedent of Townsville Stylo directed attention to the genus Stylosanthes. This impact continued until 1974 when anthracnose virtually wiped it out. Fortunately 2 other *Stylosanthes* species, *S. hamata* cv. Verano and *S. scabra* cv Seca, were released around that time. These extended the ecological range of adaptation of the genus and were adopted widely to improve dry season cattle nutrition and broaden the opportunities for economic beef production on naturally infertile seasonally dry landscapes. Verano and Seca remain the most widely used. During that period *S. guianensis* came and went for various reasons in the high rainfall areas. *S. hippocampoides* has found a niche on sandy subtropical soils. Two cultivars, Primar and Unica, of a new species, *S. seabrana,* have been released for cold tolerance and suitability to fertile soils. New anthracnose tolerant *S. guianansis* varieties Nina and Temprano are creating interest in the hay industry. Most stylo sowings have been into native vegetation. Up to 1995 it was estimated that 1.15Mha were planted to stylos. Since then another 350Kha (estimated from seed sales) have been planted, making a total of 1.5Mha. With natural spread, an area of up to 3Mha would have some stylo presence. Prolonged drought during this recent period has reduced the annual area planted. The improvements were estimated in 1996 to have added $20M/year to returns from cattle across N Australia, a figure that might be closer to $30M now.

Major reasons for success

- Economic need There was an immense need for improved dry season nutrition in standing feed for cattle, which only a legume could provide. Carrying capacities and production on native grass pastures had serious limitations. Four-fold increases in stocking rates and 2-fold weight gains were obtained from fertilized stylo pastures at various sites in the mid 1980s.
- Ease of sowing and natural spread Aerial sowing, even over trees, was reliable because of ease of distribution, hard seed, and the ability of stylo to establish on undisturbed surfaces. There was an ongoing distribution mechanism via the digestive systems of ruminants because of the high, hard-seed factor of stylos. Up to 40% of mature seed survives this passage. Most properties across north Australia now have some stylos introduced by this method.
- Seed Production The pioneering of tropical pasture seed production in N Australia, along with purposeful cooperation between private and public sectors, resulted in reliable supplies of affordable, high-quality seed that has been available throughout the whole period.
- Extension Services A sustained, cooperative effort between industry and public research and extension across 4 states and 2 levels of government over 40 years provided development initiatives and financial support for the stylo program. Other opportunities for technical change, such as use of superphosphate, mineral supplementation feeding of cattle, and helicopter mustering, together with a healthy export beef market, facilitated progress in stylo pasture improvement.
- Sustainability As they attract pests, weeds and disease, and in some circumstances cause soil acidification and erosion, pure legume swards are unstable ecologically. Solutions lie with combining stylos with perennial grasses. Difficulties in the management of extensive swards still present challenges. Genetic variability in these 'wild species' has enabled them to adapt to a wide range of environments and to withstand ongoing pest and disease attacks.

On-going constraints and future potential The size and location of N Australian cattle properties, the isolation of graziers, and their conservatism and unwillingness to incur debt in periods of low profitability delayed adoption of stylo. Technology must be self-sustaining. To survive it must keep pace with biology, competition and market influences. There is a requirement for continuing access to new germplasm. This must take precedent over plant introduction barriers being applied through the ill-informed influence of the environmental movement. We need to accommodate sound environmental principles but reject unsound ones.

Leucaena in northern Australia: a forage tree legume success story

B.F. Mullen[1], H.M. Shelton[2] and S.A. Dalzell[3]
[1]School of Animal Studies, The University of Queensland, 4072 Australia, Email: b.mullen@uq.edu.au, [2]Faculty of Natural resources, Agriculture and Veterinary Science, [3]The University of Queensland, Brisbane Australia 4072, School of Land and Food Sciences, The University of Queensland, Queensland 4072, Australia

Keywords: leucaena, forage legume, adoption, grass-legume system

Introduction *Leucaena leucocephala* (leucaena) is a long-lived, perennial forage tree legume of very high nutritive value for ruminant production. In northern Australia, leucaena is direct seeded into hedgerows 5-10m apart, with grass species such as buffel grass (*Cenchrus ciliaris*) planted in the inter-row to form a highly productive and sustainable grass-legume pasture that cattle graze directly. It generally is grown on deep, fertile soils in sub-humid environments with average rainfall of 600-800mm/year. Steer gains of 275-300kg/head per year are achieved, with short-term daily gains over the main growing season >1kg/head. Being very deep-rooted, leucaena exploits moisture beyond the reach of grasses and remains productive well into the dry season. Once established, leucaena-grass pastures remain productive for >40 years.

Extent and Benefits Large-scale adoption has occurred in the past 5-10 years, with the area under production increasing from an estimated 35Kha in 1994, to 100Kha in 2004. Over 400 graziers have successfully established leucaena, with some planting as much as 2Kha. Leucaena-grass systems provide 3 bottom-line benefits contributing: (1) economic benefits from estimated annual gross earnings of 20M AU$; (2) social benefits stemming from the stable income and ease of management afforded by established leucaena pastures, and (3) environmental benefits through reduced soil erosion and dryland salinity, and increased soil fertility arising from the deep-rooted, permanent, leguminous nature of the system.

Major reasons for success

1. Leucaena technology meets farmers' needs Leucaena gives a profitable and highly sustainable option for beef producers targeting high-value export and domestic markets. It also offers an alternative to annual crops that are unreliable in this drought-prone environment, and to improved grass pastures that become N-deficient and unproductive within several years of planting and fail to support high liveweight gains.

2. Technological constraints have been addressed Many technological constraints had to be addressed over the past 40 years to facilitate successful integration of leucaena into Australian grazing systems. Working both individually and collaboratively, scientists and graziers have overcome the constraints. The discovery of *Synergistes jonesii*, a ruminal bacterium that can counter mimosine/DHP toxicity, was largely the work of a researcher, whereas advances in seed scarification technology were largely the work of a grazier. Partnerships between researchers and graziers have been important in addressing establishment issues of weed control and soil insect control, whilst teams of researchers have worked to broaden the agronomic potential of leucaena through germplasm evaluation and plant nutrition studies.

3. Dedicated champions Long-term commitment of dedicated and enthusiastic graziers, extension officers and scientists, working both collaboratively and individually, have developed the support services needed to overcome technological barriers to adoption.

4. Support services The industry is well supported now. The supports include reliable supplies of high quality seed (20-40t produced/year), contract planting services, effective herbicides and insecticides, mimosine-degrading bacteria, and high quality technical advice on establishment and management. Strong collaborations between graziers, extension workers and scientists have culminated in the formation of The Leucaena Network, a community organisation formed to address new research and development issues and to promote the responsible use of leucaena as a grazing resource. The Leucaena Network conducts grazier-oriented training courses and collaborative research projects with universities and government agencies.

The future Leucaena-grass systems are being rapidly adopted now by graziers in northern Australia, with plantings of 16Kha planned for the coming season and >1Mha of land in northern Australia being well-suited to leucaena production. The benefits of leucaena-grass systems now are considered sufficient to justify the high costs of establishment (250AU$/ha). However, leucaena has the potential to become an environmental weed that could limit future development of leucaena as a grazing resource. This is being debated currently. The Leucaena Network has engaged proactively with relevant government agencies and environmental groups, and has developed a Code of Practice to limit the spread of leucaena from grazing properties. Assuming a favourable resolution of this issue, and continuing strong beef cattle prices, leucaena will be adopted increasingly as a sustainable grazing resource by northern Australian graziers.

Aeschynomene and carpon desmodium: legumes for bahiagrass pasture in Florida

L.E. Sollenberger[1] and R.S. Kalmbacher[2]
[1]P.O. Box 110500, University of Florida, Gainesville, Florida, USA 32611-0500, Email: les@ifas.ufl.edu,
[2]Range Cattle Research and Education Centre, 3401Experiment Station, Ona, Florida, USA 33865-9706

Keywords: pasture legumes, legume persistence, legume establishment, adoption of legumes

Introduction Soils and climate are very diverse across Florida, and no single legume has state-wide adaptation. However, aeschynomene (*Aeschynomene americana*), an annual, and carpon desmodium (*Desmodium heterocarpon*) cv. Florida, a perennial, are the most commonly used legumes for grazing on the central and southern peninsula, which produces 65% of Florida's beef calves. Both grow well with bahiagrass (*Paspalum notatum*), which is the main pasture grass, with ~1M ha state-wide. Circa 65K ha of bahiagrass contain at least limited quantities of aeschynomene and 14K ha contain carpon desmodium.

The Florida Agricultural Experiment Station (FAES) did not release Aeschynomene. The first experimental planting was in 1952, and since then there have been many research trials and considerable extension effort, leading to adoption by cattlemen. Seed production reached a peak in the mid-1970s, when ~200 t/year were produced, but it has declined to 60 t/year currently. Circa 18% of cattlemen in the region currently grow aeschynomene in bahiagrass pasture, which is down from 57% in 1990.

Carpon desmodium was introduced into Florida in 1964 and the FAES released it in 1979. There have been many research trials and much extension effort, but carpon never achieved the popularity of aeschynomene. Seed production of carpon reached a peak in the late 1980s with ~20 t/year and currently is 4 t/year. Circa 5% of cattlemen now grow carpon desmodium.

Major reasons for adoption Aeschynomene was the first palatable, highly nutritious legume adapted to seasonally wet, relatively infertile soils of the region. Carpon desmodium was the first perennial, and although lacking the palatability and nutritional qualities of aeschynomene, it was persistent under close grazing. Many cattlemen sowed these legumes in bahiagrass to improve nutritive value, especially in late summer when livestock gain on bahiagrass is often poor. Seed was produced locally and was relatively inexpensive. Both legumes reseed themselves with proper management.

Why have these legumes not been more successful? Although partnership between cattlemen, seedsmen, research, and extension was strong, the difficulty of consistently establishing and maintaining these legumes in pasture was underestimated. Autumn is the season when aeschynomene produces seed and when the legume has the greatest potential impact on animal performance. Regular grazing in autumn reduces aeschynomene seed yield significantly and precludes development of the large seed bank needed for re-establishment year after year. Either seed yield or use as a fodder must be compromised, and most producers will use a currently available forage resource instead of limiting grazing to favour seed yield. Aeschynomene seedlings emerge in spring, but late-spring drought can be devastating to re-establishment in many years. Carpon desmodium use has been limited due to unreliable and/or slow establishment. This problem probably could be solved by research, but it has not received sufficient attention. Overall, when vagaries of weather are considered, neither legume is dependable. The probability of successfully establishing and maintaining these legumes is not great enough to offset the relatively high cost of seed and management. Even when successfully grown, economic benefits are not realized easily or readily in the extensive cow-calf production system in Florida.

Why has legume usage declined? There is less demand for seed of all forages in Florida. Little native land is being converted to planted pasture, and much planted pasture land has been converted to other uses such as citrus and urban development. Although considerable aeschynomene seed is still produced, about half is sold for wildlife plantings. Aeschynomene is currently US$6.50/kg and carpon is US$9.25/kg, so many cattlemen feel it is too expensive. Seed supply varies widely from year to year, and seedsmen are unwilling to maintain a large inventory because seed shelf-life is limited. Seedsmen are very selective on fields harvested due to tropical soda apple (*Solanum viarum*), a prohibited, noxious weed seed under the Florida Seed Law and the Federal Seed Act.

The future Cattlemen will continue to purchase and sow these legumes, but there is not likely to be resurgence in popularity. Some cattlemen have had many failures with legumes and are not receptive to their use. Cattlemen believe that grass-based systems with nitrogen fertiliser are more dependable and produce more forage. Less support for research and extension and increased focus on issues other than production agriculture will limit attention given to pasture legumes.

Rhizoma peanut: more than a 'lucerne' for subtropical USA

M.J. Williams[1], K.H. Quesenberry[2], G.M. Prine[2] and C.B. Olson[2]

[1]USDA, ARS, Subtropical Agricultural Research Station, Brooksville, Florida, USA, Email: mjwi@mail.ifas.ufl.edu, [2]University of Florida, Institute of Food and Agricultural Sciences, Gainesville, Florida, USA

Keywords: *Arachis glabrata*, hay, ornamental

Introduction Rhizoma peanut (*Arachis glabrata*) was introduced to Florida from South America in the 1930s. Selections 'Arb' (PI 118457) and 'Arblick' (PI 262839) were released in the 1960s, but their use was very limited due to slow establishment and low productivity. The University of Florida released 'Florigraze' (PI 421707) in 1978 and 'Arbrook' (PI 262817) in 1986. These cultivars produced much higher dry matter yields. Thereafter, rhizoma peanut began to gain commercial acceptance. These cultivars are used throughout the Gulf Coast region of the USA for commercial hay production, pasture, creep grazing, silage, balage, and living mulch (French *et al.*, 1994). It is estimated that circa 8 Kha of rhizoma peanut have been planted (Quesenberry, 1999). 'Ecoturf' (PI 262840), an *A. glabrata* introduction that is gaining wide spread acceptance as a low maintenance turf or ornamental, is the latest development with rhizoma peanut breeding at the University of Florida. In 2002, perennial peanut was selected as the "Plant of the Year" by the Florida Nurserymen and Growers Association. Current estimates are that rhizoma peanut sales (mainly hay, but also includes planting material and ornamental production) exceed $7M USD.

Major reasons for success:
1. The technology met a need: The US Gulf Coast region has a deficit in quality hay production. The state of Florida alone imports >$100M USD of hay/year, mainly for the horse and dairy industry. Studies had shown that rhizoma peanut has a nutritive value similar to lucerne (*Medicago sativa*). As it costs less and has high palatability, it is easy to see why rhizoma peanut hay is displacing western USA-produced lucerne in the horse, goat, and dairy industries in this region.
2. The technology worked: As a hay crop, rhizoma peanut production readily fits into existing regional farming systems where the use of vegetative material for stand establishment, a necessity with rhizoma peanut, was understood. Equipment developed for the vegetative establishment of hybrid bermudagrass (*Cynodon dactlyon*) was adapted for rhizoma peanut. Long persistence (>20 years), no requirement for nitrogen fertiliser, and relative freedom from pests are other factors that enhanced its adoption.
3. Rhizoma peanut production is profitable: Much land recently planted with rhizoma peanut was formerly in row crop production that had very marginal returns. Annual net profit from established rhizoma peanut hay can be >$1000 USD/ha and current demand for hay exceeds production. Also, the lower growing, non-forage cultivars are being used increasingly for landscaping and production of material for ornamental planting is a growing source of income.
4. There was a critical partnership between stakeholders: Support of the research, extension and producer sectors all contributed to the commercialization of rhizoma peanut. The Florida Department of Agriculture and Consumer Affairs maintains a web site on rhizoma peanut (http://www.fl-ag.com/peanuthay).
5. Dedicated champions were crucial: The components of demand and production technology were present. However, thanks to enthusiastic support over the years from a select group of producers, researchers, and extension personnel, particularly the late EC "Tito" French III and Chuck Paarlberg, rhizoma peanut avoided the fate of many other new crops that never emerged into commercial production.

Conclusion Rhizoma peanut was not adopted initially because research efforts targeted cattle producers and its production costs were too expensive for this use. Commercial success depended on the realization that rhizoma peanut could be used as a cash crop. Future research aims to develop ornamental cultivars and forage types with a wider range of adaptation.

References
French, E. C., G. M. Prine, W. R. Ocumpaugh, & R. W. Rice (1994). Regional experience with forage Arachis in the United States. In: P. C. Kerridge & B. Hardy (eds.). Biology and Agronomy of Forage Arachis. Centro Internacional de Agricultura Tropical (CIAT), Cali, Columbia, 169-186.

Quesenberry, K. H. (1999). Value of UF/IFAS forage legume cultivars to Florida livestock production. *Soil Crop Science Society Florida Proceedings*, 58, 23-27.

Leucaena production in Arid Botswana

F.P. Wandera[1], M. Karachi[2], S. Mangope[1] and B.M. Lefofe[1]
[1]*Department of Agricultural Research, Private Bag 0033, Gaborone, Botswana, Email: wpeter@gov.bw*
[2]*Egerton University, P.O Box 536, Njoro, Kenya*

Keywords: Leucaena, cold tolerance, productivity, quality, arid areas

Introduction The value of browse species as a source of nitrogen for grazing animals is restricted to wet seasons, with protein deficiencies being experienced by September (in dry winter season) in Southern Africa (Moleele 1998). This is when highly productive planted browse species would become useful to supplement the protein diet requirement of grazing animals (Morris & Du Toit 1998). Further, browse species can provide partly for protein requirement of intensive production systems, such as in feedlots and dairies. This paper reports work on the introduction and screening of *Leucaena* for Botswana conditions.

Materials and methods *Leucaena* species and a hybrid were evaluated for cold tolerance, productivity and quality at Morale for 3 seasons. The site (circa 23°S and 27°E) receives erratic rainfall that averages 450mm between October and May. Its temperature range is 30-22°C mean max and 17-3°C mean min, with absolute max and min of 40 and -6°C, respectively. The soils are haplic acrisols with a pH (H_2O) of 6.7, organic carbon 0.3%, available P 3 ppm, total N, 0% and Ca 1.7, Mg 0.6 and K 0.7 meq/100g. Rainfall during the experimental period amounted to 554.3, 284.3, 322.7mm for 1996/97, 1997/98 and 1998/99 growing seasons. Cold tolerance was rated on a scale of 0-5 (0=no leaf death and 5=maximum leaf death), 1 month after freezing temperatures (0-5°C) were recorded. Dry matter of edible component was determined by drying in forced draught oven at 65°C to a constant weight. The dry material was hammer-milled to pass through a 1mm sieve for nitrogen, calcium, phosphorus, fibre and digestibility analyses.

Results There was no correlation between cold tolerance and productivity in this genus (Table 1). Chemical composition shows that the high yielding accessions also had high N value (>3.2%) and equally high dry matter digestibility (55.6-57.3%), showing characteristics of high feed value. *L. leucocephala* dominated in production of high edible dry matter and feed value. Termites and wild animals were observed to be a problem during establishment.

Table 1 Cold tolerance, dry matter yield and feed value of some of *Leucaena* accessions in Botswana

Species	Acc. No.	Cold tolerance rating	Mean yield (kg/ha)	N	Ca	P	NDF	DMD
L. pallida		2.5	1,749	3.46	1.28	0.17	29.0	50.8
L. leucocephala	K 88A	3.0	1,979	3.46	1.68	0.22	36.0	47.4
L. esculenta subsp. Paniculata	79/92	3.0	1,795	3.84	0.94	0.19	30.5	50.1
L. leucocephala cv. Cunningham	K500	3.5	2,229	3.25	1.67	0.16	34.4	57.3
L. leucocephala	95/20	3.5	1,847	-	-	-	-	-
L. leucocephala cv. Taramba	K636	5.0	1,728	-	-	-	-	-
L. diversifolia x *L. Leuco.* (Kx3, F4)	4/95	5.0	2,099	3.63	0.95	0.16	29.1	55.6
LSD (P<0.05)			135	0.80	0.67	0.06	12.1	8.7

Conclusion *L. leucocephala* was productive under arid conditions prevailing in Botswana and can therefore be used to offset protein deficiency in grazing animals in such environments. Its medium to low cold tolerance indicates that the edible dry matter should be conserved for feeding in dry winters.

References
Moleele N.M. (1998). Enchroacher woody plant browse as feed for cattle. Cattle diet composition for three seasons at Olifants Drift, south-east Botswana. *Journal of Arid Environments,* 40, 255-268
Morris C.D. & L.P. Du Toit (1998). The performance of Boer goats browsing *Leucaena leucocephala* in Kwa Zulu-Natal, South Africa. *Tropical Grasslands,* 32, 188-194.

Agronomic evaluation of twenty ecotypes of *Leucaena* spp. for acid soil conditions in México

J.F. Enríquez-Quiroz, A. Hernández-Garay and A.R. Quero-Carrillo
INIFAP-*Veracruz*- CP *Iturbide 73, Salinas, San Luis Potosi. 78600, México, Email: queroadrian@colpos.mx*

Keywords: forage legumes, *Leucaena*, acid soils, forage production

Introduction *Leucaena leucocephala* Lam. (de Witt) has been shown to be a good forage producer and to posses good persistence under grazing conditions in México tolerating well the management of local cattlemen (Quero *et al.*, 2004). The *Leucaena* genus is native to Central America and Mexico (Hughes, 1998), but *L. leucocephala* is a low producer under acid soil conditions. The natural diversity is a good source of resistance to acid soil conditions resistance and to other adverse factors. Several *Leucaena* accessions were evaluated for production under acid soil conditions in tropical Mexico.

Materials and methods In order to compare 17 *Leucaena* species for forage production potential under acid soils conditions in the dry tropics, 20 ecotypes: local, from Oxford Forestry Institute (UK), and from the International Center for Tropical Agriculture (CIAT, Colombia) were established and evaluated at   CAEPAP (Campo Agricola Experimental del Papaloapan) in Isla, Veracruz (Table 1), under rainfed conditions in acrisol soils (4.7 pH). Plants were transplanted, under a randomised complete block design with three replicates. Plots consisted of a 9 m row with 7 m for evaluation in each replicate. Plants were allowed an eight-month establishment period. No fertilisation or soil amendment was used. Three and six months after transplantation, plants were measured for 1) plant size and 2) stem width. During the second year forage production was measured in cuts every two months.

Results *Leucaena lempirana* (1510mm height/13 mm stem width), *L. esculenta paniculata (142/15mm)*, and *L. leucocephala* (131/17mm; native) had fast development (P<0.05). *L. salvadorensis* did not grow well in the acid soil conditions. Six ecotypes did not provide enough plant material to be evaluated during the second year, although the plants survived: *Leucaena esculenta paniculata, L. diversifolia diversifolia, L. multicapitulata, L. lempirana*, and *L. pulverulenta*. After one year of evaluation, total forage production was superior for the three *L. leucocephala* accessions (P<0.05; Table 1); the ecotypes 9904 and Isla gave 1043 and 1128 g/plant per year, respectively. Most of the forage was produced during the rain season. The most productive material for the windy season was *L. leucocephala* 9904 (P<0.05) and for the dry season *L. leucocephala* cv. Isla and cv. 9904 with a production of 97 and 99 g/per plant, respectively (Table 1).

Table 1 Forage production (g/ plant) of *Leucaena* spp. at each cut and the total annual yield

Species	Accession	Harvest 1	Harvest 2	Harvest 3	Harvest 4	Harvest 5	Harvest 6	Total yield
Leucaena:			Rainy season		Windy		Dry season	
leucocephala	Isla	219	191	138	100	76	99	823
leucocephala Glabrata	34/92	213	204	180	108	73	97	875
leucocephala	17263	214	196	157	94	74	97	832
leucocephala	9904	206	191	162	104	82	96	841
leucocephala	774	221	193	165	101	76	87	844
collinsii	52/88	180	152	124	116	69	79	720
trichodes	61/88	136	118	76	55	50	50	485
Cratylia spp.	18957	106	117	94	56	72	82	527

Conclusions There is much variation for acid soil resistance within the evaluated species. Because of a higher production potential, it is suggested that studies on within-species acid soil resistance should concentrate on the genetic resources in *L. leucocephala.*

References
Hughes, E. C. (1998). Leucaena. Manual de recursos fitogenéticos. Tropical Forestry Papers. Oxford Institute. Plant Science Department, 280pp.
Quero, C. A. R., J. F. Enriquez Q. & A. Hernandez G. (2005). La pradera ideal y el ganadero ideal en los trópicos. In: INIFAP-CIFAP Veracruz (eds) XXV aniversario del CAEPAP, Veracruz, México, pp 21-33 (in press).

Desmodium velutinum: a high-quality shrub legume for acid soils in the tropics

R. Schultze-Kraft[1], M. Peters[2], N. Vivas[3], F. Parra[4] and L.H. Franco[2]
[1]University of Hohenheim (380), D-70593 Stuttgart, Germany, Email: rsk@uni-hohenheim.de, [2]Centro Internacional de Agricultura Tropical (CIAT), A.A. 6713, Cali, Colombia, [3]Universidad del Cauca, Popayán, Colombia, [4]Corporación Colombiana de Investigación Agropecuaria (Corpoica), Popayán, Colombia

Keywords: *Desmodium velutinum*, tropics, shrub legumes, dry matter production, forage quality

Introduction Drought tolerant legume shrubs can enhance the sustainability of smallholder production systems in the tropics through the provision, year-round, of high-quality feed and through their positive effect on soil. *Desmodium velutinum* (Willd.) DC. is one of the few shrub species that have been identified as (1) well adapted to acid tropical soils and (2) of good nutritive value (Schultze-Kraft, 1996). It is a perennial native to SE Asia and tropical Africa growing up to 3 m high, the velutinous (velvety) surfaces of its 1-foliolate leaves being a characteristic feature. It grows well on soils ranging from pH 4.0 to alkaline, prefers high rainfall (1000 to >3000 mm/year) but tolerates up to five dry months. Though previous work in West Africa (e.g., Mzamane & Agishi, 1986) and South America (e.g., Thomas & Schultze-Kraft, 1990) has indicated the potential of the species, there are only few studies and these are restricted to only one or a few accessions. It is important to assess the genetic diversity and agronomic variability in the germplasm collection of about 140 accessions that is now available in order to identify a core collection and accessions with promising agronomic performance and nutritive value. Possible relationships between accession origins, morphological-agronomic characteristics, and genetic diversity need to be assessed. The first-year results from a field experiment on forage yield and quality are presented here. The project is financially supported by the Volkswagen Stiftung, Hannover, Germany.

Materials and methods The experiment was carried out at the CIAT-Quilichao Experiment Station near Cali, Colombia (03°06' N, 76°31' E; altitude 990 m asl; average rainfall 1800 mm/year; Ultisol with pH 4.9 and high OM content). Accessions (137) of *D. velutinum* were sown into single-row plots with 5 plants each (1 m between plants, 1.5 m between rows). The experimental design was a randomised complete block with three replicates.

Results Table 1 shows that there is a large range in dry matter (DM) production and nutritive value. It is noteworthy that under the experimental conditions there were no major seasonal yield differences.

Table 1 Herbage (edible = <5 mm stem diameter) yield of 8-week regrowth in the wet (mean of 2 cuts) and dry season (1 cut), CP content and IVDMD (wet season) in the 137-accession world collection of *D. velutinum*

Accession		DM (g/plant)		CP (% N x 6.25)	IVDMD (%)
		Wet	Dry		
Whole collection	Range	14-325	20-346	17.2-26.1	58.9-76.2
	Mean (SD)	137 (82.6)	142 (81.9)	21.3 (2.1)	67.1 (4.2)
Promising accessions	CIAT 33443 (erect)	300	340	19.5	68.8
	CIAT 23981 (semi-erect)	281	283	22.1	68.2
	CIAT 33352 (erect)	195	276	23.7	70.2
	CIAT 13953 (erect)	302	274	21.1	70.5

Conclusions These initial results confirm the potential of *D. velutinum* as a high-quality shrub legume adapted to acid soils. Yields and CP protein contents are comparable to other shrubs (e.g., *Cratylia argentea*, *Flemingia macrophylla*) under similar conditions (CIAT, 2002). Of particular interest is the outstandingly high IVDMD. Consequently, *D. velutinum* appears to be a promising option as a protein and energy supplement to the diet of ruminants in tropical regions. Studies are required to assess the persistence potential of selected accessions under frequent cutting and under grazing.

References
CIAT (2002). Annual Report. Project IP-5: Tropical Grasses and Legumes, 131-140.
Mzamane, N. & E.C. Agishi (1986). *Desmodium velutinum*: A promising leguminous browse shrub of Nigeria's savannas. *PGRC/E ILCA Germplasm Newsletter,* 12, 24-26.
Schultze-Kraft, R. (1996). Leguminous forage shrubs for acid soils in the tropics. *Wageningen Agricultural University Papers*, 96-4, 67-81.
Thomas, D. & R. Schultze-Kraft (1990). Evaluation of five shrubby legumes in comparison with *Centrosema acutifolium*, Carimagua, Colombia. *Tropical Grasslands,* 24, 87-92.

Recent advances in *Stylosanthes* research in tropical America

B. Grof,[1] C.D. Fernandes[2] and J.R. Verzignassi[2]

[1]11Aminga Court, Palmwoods, 4555, Qld, Australia, E-mail: bgrof@bigpond.com, [2]Laboratório de Fitopatologia, EMBRAPA, Gado de Corte, C.P.154, Campo Grande MS, Brazil

Keywords: improved pastures, Cerrados, Brazil

Introduction The potential of neotropical savannas is very large for pasture-based livestock systems. There are 250 million ha of well-drained lowland savannas in the American tropics. Over 200 million ha are situated in the Cerrados region of Brazil. The Cerrados support 42% of the national herd. Although these ranges support large populations of livestock, productivity is generally low. Poor nutritive value of native pastures and monospecific swards of *Brachiaria* spp. is the principal cause of this low productivity, especially in the dry season when these grasslands often provide no more than 60% of the animal's maintenance requirements. The best option to increase pasture/livestock productivity is the use of improved pastures, particularly those based on adapted tropical legume-grass associations. Research in tropical America was concentrated on the evaluation and selection of *Stylosanthes* species adapted to low fertility acid-soils and resistant to anthracnose (*Colletotrichum gloeosporioides*). Anthracnose is considered the major limitation to the commercial use of *Stylosanthes* on a world-wide basis.

Materials and methods Significant advances in cultivar development of three species *S. capitata, S. macrocephala* and *S. guianensis* were made during the 1990s at the National Beef Cattle Research Center (EMBRAPA Gado de Corte/CNPGC). There are about 45 good species of *Stylosanthes,* of which 25 are native to Brazil, mainly to the Cerrados agroecosystem. *Stylosanthes* spp. are adapted to acid-soil savannas, tolerate Mn toxicity and high Al saturation of the bases. In general, stylos have a low requirement for soil phosphorus. *Stylosanthes* cv. Campo Grande was officially released by EMBRAPA in Brazil in 2000. This variety is the hybrid-derived progeny of 11 accessions of Venezuelan *S. capitata* and six accession of Brazilian origin. In order to maximise genetic diversity and obtain protection against anthracnose, seed of *S. macrocephala* was mixed with that of the composite hybrid population at the rate of 20% by weight of the total.

Results The "stacked" resistance genes of Brazilian and Venezuelan accessions compounded in cv. Campo Grande resulted in quantitative, multigene resistance to anthracnose. Studies conducted in the Colombian Llanos indicated 91.6% anthracnose susceptibility in native *Stylosanthes* var.vulgaris and 39.5% in var. pauciflora accessions (Miles & Lapointe, 1992). Significant genetic progress has been achieved by the *Stylosanthes* selection programme. Populations of *S. guianensis* var. vulgaris x var. pauciflora selected for resistance in Colombia succumbed to the disease in Brazil. Five cycles of recurrent selection and progeny testing of these selections in Brazil, SE Asia and Australia gave material with durable, quantitative, multigene resistance. Verzignassi (2001) recorded the response of 60 of these intervarietal hybrids of *S. guianensis* to 11 monosporic isolates of anthracnose. The 60 hybrids displayed 98.3% resistance to the disease. A highly significant aspect of the selection process for anthracnose-resistant *S. guianensis* is that it has been carried out in Brazil, the native habitat and probable centre of origin and diversity of the species, where extensive variability in virulence of the pathogen and specialised forms of the disease have been identified.

Conclusion Disease resistance of selected lines of *Stylosanthes capitata, S. macrocephala* and *S. guianensis* was confirmed in vastly different agroecosystems, in tropical South and Central America, SE Asia and Australia. Selections have been released in Thailand (M.Hare, personal communication, 2003), Central and South America (R. S. Bradley and G.Sauma, personal communication, 2003) and Australia (B. Grof, unpublished).

References

Miles J.W. & S.L.Lapointe (1992). Regional germplasm evaluation. In: Pastures for the Tropical Lowlands. CIAT publication no.211, 9-28.

Verzignassi J.R. (2001). Determinação da variabilidade genética de *Colletotrichum gloeosporioides* e avaliação agronômica de espécies de *Stylosanthes* spp. em Mato Grosso do Sul. Embrapa Relatório final.

Desmanthus: a new forage legume to improve wool growth in tropical Australia

J.H. de A. Rangel and C.A.M. Gomide
Embrapa Tabuleiros Costeiros, Av. Beira Mar, 3250, Praia Treze de Julho, 49025-040, Aracaju, Se., Brazil, Email: Rangel@cpatc.embrapa.br

Keywords: Desmanthus, animal nutrition, clean wool production, heavy clay soils

Introduction In tropical Australia, very short and erratic wet seasons are the critical factors in determining forage growth and animal production (Wheeler & Freer, 1986). Grasses are highly susceptible to low rainfall and animal production in such conditions become strictly seasonal. Improvements in meat and wool production by the introduction of stylo species (*Stylosanthes* sp.) into natural grasslands have been intensively reported (Gillard & Winter, 1984). However, there are currently no suitable introduced legumes for the *c.*28 million ha of Mitchell grass (*Astrebla* spp.) plains in heavy clay soils of western North Queensland, grazed predominately by wool-producing Merino sheep (Phelps, 1999). Members of the genus *Desmanthus* appear to offer the possibility for filling this role (Gardiner *et al.*, 2004). This work aimed to evaluate the potential of four *Desmanthus* accessions, in comparison with Verano stylo (*Stylosanthes hamata* cv. Verano), as alternative supplements for diets of Mitchell grass hay fed to Merino wethers in western North Queensland.

Material and methods Thirty-six Merino wethers (average liveweight 33.96 kg ∀1.82), were individually housed in metabolism cages and daily fed with 600 g/head/day of Mitchell grass (*Astrebla* spp.) hay alone or supplemented with 200 g/day of one of five legume hays (6 wethers per treatment). The hays were made from four *Desmanthus* accessions and from Verano stylo. The effects on intake and wool growth were measured.

Results Supplementary diets in general increased significantly nitrogen intake, metabolisable energy intake, clean wool production and wool yield (clean wool/grease wool * 100) (Table 1).

Table 1 Fibre content, digestibility, intake, and wool yield parameters in response to supplementation with legume hays

Supplement	Diets: Mitchell grass hay plus indicated supplement					
	Nil	Verano	*D.leptophyllus* CPI 38351	*D. virgatus* CPI 92803	*D. virgatus* CPI 78382	*D. virgatus* CPI 79653
Neutral detergent fibre %	55.3 a	55.0 a	51.2 ab	48.7 b	52.8 ab	51.6 ab
Acid detergent fibre %	52.7 a	50.3 a	45.0 b	43.4 b	48.2 ab	43.8 b
DM digestibility %	42.5 bc	46.5 a	43.8 abc	39.8 c	42.2 bc	44.9 ab
OM digestibility %	49.0 a	51.4 a	48.4 ab	45.3 b	48.6 ab	48.3 ab
Total Nitrogen intake (g/day)	7.6 d	9.4 bc	10.2 b	8.7 c	10.2 b	12.4 a
ME intake	110 c	139 ab	131 b	117 c	126 bc	144 a
Clean wool (mg/cm^2/day)	0.42 d	0.56 ab	0.55 abc	0.62 a	0.20 bcd	0.45 cd
Wool yield (%)	59.2 c	67.9 b	67.9 b	77.4 a	67.5 b	64.6 bc
Fibre diameter (microns)	18.9 a	20.4 a	19.8 a	19.1 a	19.8 a	20.5 a

Means followed by the same lower case letters in rows are not significantly different by Turkey (p< 0.05)

Conclusions These results, associated with evidence of the agronomic adaptation of *Desmanthus* genotypes (Gardiner *et al.* 2004) to the black clay soils of the Mitchell grass plains of northern Queensland, show the high potential of the legumes to improve wool growth in that region.

References
Gardiner, C.P., L. Bielig, A. Schlink, R. Coventry & M. Waycott (2004). *Desmanthus* – a new pasture legume for the dry tropics. *Fourth International Crop Science Congress*, Brisbane, (in press).
Gillard, P. & W.H. Winter (1984). Animal production from *Stylosanthes* based pastures. In: H. M. Stace & L.A. Edye (eds.) The Biology and Agronomy of Stylosanthes**,** 405-432.
Phelps, D. (1999). Mitchell grass: Long-term wool production and grazing pressure. DPI Note, File No: SW0025, July 1999. Http://www.dpi.qld.gov.au/sheep/4993.html.
Wheeler, J.L. & M. Freer (1986). Pasture and forage: The feed base for pastoral industries. In: G, Alexander & O.B. Williams (eds.) The Pastoral Industries of Australia, 5th Edition, Sydney University Press, Sydney, 165-182.

Enhancement of grassland production through integration of forage legumes in semi-arid rangelands of Kenya

P.N. Macharia[1], J.I. Kinyamario[2], W.N. Ekaya[2] and C.K.K. Gachene[2]
[1]KARI-Kenya Soil Survey, PO Box 14733, 00800 Westlands, Nairobi, Kenya, Email: pnmacharia@hotmail.com,
[2]University of Nairobi, PO Box 30197, 00100 GPO, Nairobi, Kenya

Keywords: forage legumes, grassland production, rangelands

Introduction Livestock production in semi-arid rangelands of Kenya is limited by the seasonal quantity and quality of fodder. Kirkman & Carvalho (2003) stated that these inter- and intra-seasonal quality and quantity fluctuations result in nutrient deficits that severely limit livestock production potential. The objective of this experiment was to study the effect of three forage legumes on the production of natural pastures in semi-arid rangelands of Kenya.

Materials and methods The study was conducted in Kajiado District in semi-arid south-eastern Kenya which has a low and erratic annual rainfall of about 500 mm which occurs in two distinct seasons. The long rains (LR) season occurs between March and May while the short rains (SR) season occurs between Oct. and Dec. each year. A legume and grass integration experiment was set up in 2002 with the treatments *Neonotonia wightii* (Glycine), *Macroptilium atropurpureum* (Siratro) and *Stylosanthes scabra* (Stylo) planted as pure stands as well as mixed stands with the natural grass. The dominant grasses in the pasture were *Dichanthium insculpta, Chloris roxburghiana, Themeda triandra* and *Eragrostis superba*. The dry matter (DM) yield was measured by cutting at 0.15 m height on a bi-monthly basis. Excavation of whole plants to determine their root depths was done after five months of growth, while stem counts and length measurements were done after two seasons of growth.

Results and discussion By the third season, Glycine in monoculture produced more (P=0.0021) DM than the other treatments (Table 1). It produced more than three times the DM of Siratro, even though both species have the same prostrate growth habit. This can be explained by the fact that after two seasons of growth Glycine produced 14 stems/plant which were 2.55 m long, while Siratro produced 8 stems/plant which were 1.85 m long.

Table 1 Seasonal dry matter production (t/ha)

Treatments	Seasons			Mean
	LR 2003	SR 2003/4	LR 2004	
Grass	3.3	3.2	3.0	3.2[b]
Glycine	4.5	11.2	9.5	8.4[a]
Siratro	0.8	3.8	2.7	2.4[b]
Stylo	2.7	2.8	4.3	3.3[b]
Grass+Glycine	3.2	4.9	3.3	3.8[b]
Grass+Siratro	4.2	4.1	4.0	4.1[b]
Grass+Stylo	2.3	3.4	2.2	2.6[b]
Mean	3.0	4.8	4.1	
SE (Treatments) = 0.997: SE (Seasons) = 0.653				

significantly different (P=0.05)

Further, the three forage legumes were self seeding with new seedlings spontaneously germinating at the onset of the rainfall season in March and Oct. each year. The tap roots of Glycine, Siratro and Stylo were to 0.80, 0.95 and 0.85 m depths respectively, well beyond the rooting depth of the grasses, which had about 80% of their roots located between 0-0.3 m depth.

Means followed by the same superscript are not

Conclusions The results demonstrate that Glycine, Siratro and Stylo contributed to the DM production of the natural pasture. The DM yield of Glycine (8.4 t/ha) in monoculture indicates its suitability for hay production and also as fodder in a 'cut and carry' feeding system especially for lactating and sick cows left around the homesteads during the dry seasons when the bigger herds move away in search of pasture and water. This is the time when the natural pasture is at its lowest level in terms of quality and quantity. The three forage legumes are also difficult to uproot and are deep rooted. These are desirable attributes with respect to propagation and naturalisation within pastures. For the legumes to germinate and establish into the natural pasture, there is a need to accord them a competitive advantage over the grasses at least during the first season, so as to allow the roots of the legumes to penetrate the 0.3 m depth reached by most fibrous roots of the grasses.

Reference

Kirkman, K.P. & P.C. de Faccio Carvalho (2003). Management interventions to overcome seasonal quantity and quality deficits of natural rangeland forages. *Proceedings of the Seventh International Rangelands Congress*, 1289-1297.

The dry matter yield and nutritive value of wet tolerant tropical forage legumes in single cropping or mixed cropping with gramineous forage crops in drained paddy field

M. Tobisa[1], Y. Nakano[2], K. Okano[2], M. Shimojo[2] and Y. Masuda[2]
[1]Faculty of Agriculture, University of Miyazaki, Miyazaki, 889-2192, Japan, Email: mtobisa@cc.miyazaki-u.ac.jp, [2]Faculty of Agriculture, Kyushu University, Fukuoka, 812-8581, Japan

Keywords: drained paddy field, dry matter production, forage legume, mixed cropping

Introduction In Japan the production of rice has been controlled since the 1970's and some parts of the paddy fields have been laid off for forage production. However, in poorly-drained fields or fields with high ground water table, forage species with high tolerance of wet conditions are required. The tropical forage legumes *Aeschynomene americana* cv. Glenn (Glenn) and *Macroptilium lathyroides* (L.) Urb. cv. Murray (phasey bean) have a high wet endurance (Bishop *et al.*, 1985; Tobisa *et al.*, 1999) and show high dry matter productivity (Skerman *et al.*, 1988; Tobisa *et al.*, 1999). The objective of this experiment was to evaluate the dry matter yield and nutritive value of Glenn and phasey bean in single cropping or mixed cropping with gramineous forage crops in drained paddy fields.

Materials and methods The experiment was conducted in a drained paddy field adjoining a rice paddy field at the Kyushu University Farm. Tropical forage legumes (Glenn and phasey bean), gramineous forage crops (Japanese barnyard millet (cv. White panic and Aoba millet), maize (cv. Snow dent 123) and sorghum (cv. Ultra sorgo)) were used. On 23 June, the paddy field was sown at 2kg/ha in rows with a distance of 50 cm between rows with alternate row sowings of the legume and the gramineous forage crop. Plants were harvested on 2 and 22 September and 18 November (data not shown), and measurements made for dry matter yield (DMY), *in vitro* dry matter digestibility (IVDMD) (Goto & Minson 1977) and crude protein (CP). Digestible dry matter yield (DDMY) and CP yield (CPY) were calculated.

Results Dry matter yield of the single-cropped Glenn was similar to that of the single-cropped sorghum when cut on 22 September and on 18 November. The DDMY of Glenn-sorghum mixed crop was higher than that of the single-cropped sorghum when cut on 22 September. The mixed crops of forage legumes and gramineous forages showed higher total CPY compared with the single crop of gramineous forage when cut on 22 September.

Conclusions The results of the present study demonstrated that Glenn has a high DMY, DDMY and CPY. The mixed crop of forage legume with a gramineous forage with high wet tolerance provided a good forage production system for the drained paddy field.

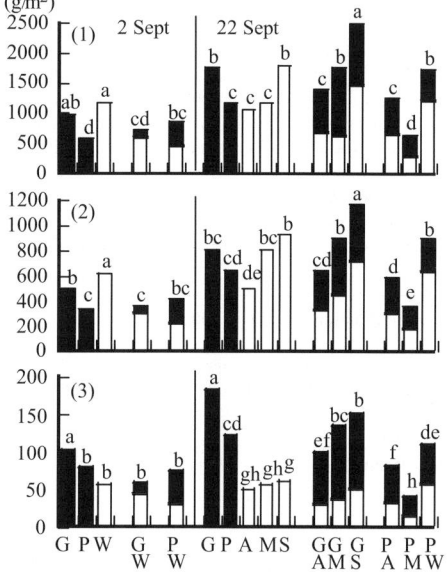

Figure 1 Dry matter yield (DMY,1), digestible dry matter yield (DDMY,2) and crude protein yield (CPY,3) of single and mixed crops in the drained paddy field
G: Glenn, P: Phasey bean, W: White panic, A: Aoba millet, M: maize, S: Sorghum.
The values followed by different letters are significantly different at P<0.05 at each sampling time.
■ : Legume, □ : Gramineous forage.

References

Bishop H.G., D. H. Ludke & M.T. Rutherford (1985). Glenn jointvetch: A new pasture legume for Queensland coastal areas. *Queensland Agricultural Journal*, 111, 241-245.

Goto I., & D. J. Minson (1977). Prediction of the dry matter digestibility of tropical grasses using a pepsin-cellulase assay. *Animal Feed Science and Technology*, 2, 247-253.

Skerman P. J., D. G. Cameron & F. Riveros (1988). In: Tropical Forage Legumes. FAO. Rome. 205-211.

Tobisa M., M. Shimojo, K. Okano & Y. Masuda (1999). Growth habit of tropical forage legume genus *Aeschynomene* in the drained paddy field and upland field. *Grassland Science*, 45, 248-256. In Japanese.

Experiences with establishing legumes as part of a ley pasture in a low-input farming system of the Eastern Amazon, Brazil

S. Hohnwald[1], B. Rischkowsky[1], A.P. Camarão[2], J.A. Rodrigues-Filho[2], R. Schultze-Kraft[3] and J.M. King[1]

[1]*Institute of Agronomy and Animal Production in the Tropics and Subtropics, Georg-August University of Göttingen, Kellnerweg 6, 37077 Göttingen, Germany, Email: shohnwa@gwdg.de,* [2]*Embrapa Amazônia Oriental, Trav. Dr. Enéas Pinheiro S/N-Marco, CEP-66095-100, Belém - Pará, Brazil,* [3]*University of Hohenheim (380), Garbenstr. 13, 70599 Stuttgart, Germany*

Keywords: *Arachis pintoi, Cratylia argentea, Chamaecrista rotundifolia,* grass-legume pasture, humid tropics

Introduction In the Eastern Amazon extensive pasture management by smallholders can result in degradation of grassland, leading to unproductive and abandoned agricultural areas (Dias-Filho, 2003). To avoid long and costly restoration of those areas, ley-systems with alternating cropping and pasture phases might offer a promising solution. The inclusion of N-fixing legumes is seen as a suitable method to improve soil fertility during the pasture phase. As mono-cultures seem not to be appropriate, given the phytodiverse climax vegetation in the humid tropics, a combination of various shrub and herbaceous legume species is proposed. Thus, this paper tests a grass-legume mixture to replace the fallow phase in the slash-and-burn agriculture practised by smallholders.

Materials and methods A researcher-managed on-farm experiment was set up at Igarapé-Açu (47°30'W/1°2'S). Alternating 5 m broad strips of herbaceous *Arachis pintoi* Krapov. & W.C. Gregory cv. Amarillo and the grass *Brachiaria humidicola* (Rendle) Schweick. were planted on three 0.3 ha plots. Rows of the shrub *Cratylia argentea* (Desv.) Kuntze cv. Veraniega were planted in the centre of the *A. pintoi* strips and rows of the shrub *Chamaecrista rotundifolia* (Pers.) Greene var. *grandiflora* in the respective grass strips. After an establishment phase of one year the plots were grazed in a rotational system with average stocking rates of 1.48 LU/ha per year[1] in the 1st year and 1.23 LU/ha per year in the 2nd year. Legume nodulation, population dynamics, and growth were evaluated during the first two years of the grazing phase.

Results The results indicate unsatisfactory performance of the grass-legume pasture. All legume species had poor nodule production. In the first three grazing months, *A. pintoi* developed well but nearly disappeared during the dry season (08/00-12/00) and stabilised at a low level in the second year (figure 1). *C. argentea* was heavily grazed at a much earlier stage than expected. Neither legume provided much forage biomass and the strips were heavily invaded by the grass in 2000. *C. rotundifolia* established well and the bushes grew rapidly, as they were hardly consumed (data not shown). However, the chosen accession died off after two years, but due to prolific seeding seedlings invaded all over the pasture plots, entering the *A. pintoi/C. argentea* strips and even suppressing the grass (Figure 1), which therefore dropped to a low level in 2001.

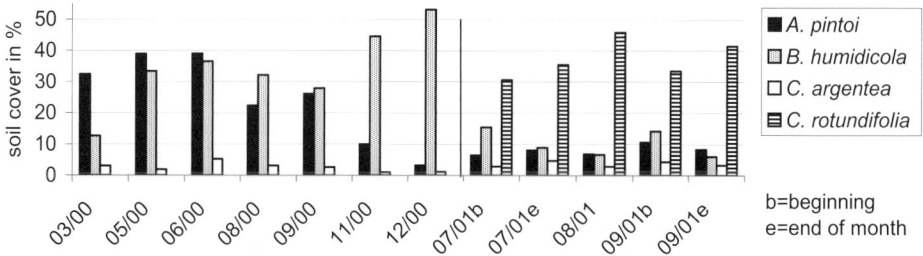

Figure 1 Soil cover percentages in the *Arachis pintoi/Cratylia argentea strips* during the two grazing phases, showing the invasion by *B. humidicola* (2000/2001) and *C. rotundifolia* (2001), respectively (n=1336)

Conclusions The tested grass-legume mixtures cannot successfully fulfil their function in a ley system and have yet to be improved regarding species choice and speed and length of establishment phase (Hohnwald, 2002).

References
Dias-Filho, M.B. (2003). Degradação de pastagens: processos, causas e estratégias de recuperação. Embrapa Amazônia Oriental, Belém.
Hohnwald, S. (2002). A Grass-Capoeira Pasture Fits Better Than a Grass-Legume Pasture in the Agricultural System of Smallholdings in the Humid Brazilian Tropics. Cuvillier Verlag, Göttingen.

Variation within the species *Macroptilium atropurpureum* regarding adaptation to grazing

C.K. McDonald[1] and R.J. Clements[2]
[1]CSIRO Sustainable Ecosystems, 306 Carmody Road, St Lucia, Queensland, 4067, Australia, Email: cam.mcdonald@csiro.au, [2]Crawford Fund, 29 Holmes Crescent, Campbell, ACT, 2612, Australia

Keywords: legume, grazing, characterisation

Introduction The twining legume *Macroptilium atropurpureum* cv. Siratro was released around 1960 (Hutton 1962) and the rust resistant cultivar Aztec was released in 1994 (Bray & Woodroffe 1995). The species showed great potential for pastures in northern Australia and was planted over some 220 Kha in the 1960's and 70's. The species was high yielding and readily eaten by cattle. However, by the early 1980's the species had declined dramatically in grazed pastures. Clements (1989) showed that a major problem with Siratro was the frequency of removal of growing points. This leads to less regrowth after grazing, less seed set and hence less regenerative capacity. Accessions with greater branching characteristics may overcome this problem.

Materials and methods Seeds of 175 elite accessions from the CSIRO collection were scarified and germinated at room temperature in Petri dishes. When the radicle had just emerged (1-5mm long), germinated seeds were planted 1.5cm deep in 18cm diameter pots. There were 3 plants/pot and 4 pots of each accession, arranged in a randomised block design. Measurements included: days to emergence, cotyledon node height, number of nodes on days 15 and 30 and branches on day 30, days to flowering and seed yield. The data were used to determine the age of branching (in plastochron units) and probability of branching. Plants were subsequently transplanted to the field and plant yield and rust resistance measured.

Results Branching characteristics among the accessions had a wide range (Table 1). As branching was a key desirable characteristic, the results are summarised in 3 groups (see Table 1). Group 2 included cv. Siratro. Group 1 had a lower age at branching, higher probability of branching, earlier flowering date and higher early seed yield than the other 2 groups. Plant yields were similar across all 3 groups but group 1 had a higher proportion of plants with cotyledon nodes below ground and a higher proportion of accessions with a zero rust rating. Eight accessions from group 1 had zero rust ratings and cotyledon nodes below ground level. Of these, 3 (CPI 91352, CPI 84578 and CQ 1392) yielded >2000 kg/ha.

Table 1 The number of accessions, median age at branching (in plastochron units), probability of branching, days to first flower, early seed yield, total seed yield and plant yield, and percentage of accessions in group with zero rust rating and cotyledon nodes below ground, for the 3 groups based on branching (0-2, 3-5 and 6-8) (standard errors are given in brackets)

Group	No. of Branches at day 30	No. of Access-ions	Age at branch-ing	Prob. of branch-ing	% Cot. nodes below ground	Days to 1st flower	Early seed yield (g)	Plant yield (kg/ha)	Zero rust rating (%)
1	6 to 8	45	3.7 (0.7)	0.8 (.12)	49	77 (15)	2.9 (3.0)	1760 (401)	47
2	3 to 5	111	4.0 (0.7)	0.6 (.16)	43	85 (19)	2.8 (2.7)	1540 (399)	37
3	0 to 2	19	4.7 (1.1)	0.3 (.17)	26	100 (23)	1.1 (1.2)	1640 (302)	26

Conclusions There are several accessions with more desirable characteristics for grazing and survival than those of cv. Siratro. As well as having more branches, those with cotyledon nodes below ground will be more frost tolerant. Also, these flower earlier than Siratro and some resist rust although the resistance will be from a single gene rather than multiple genes as for the bred line cv. Aztec. Cultivar Aztec was not available at the time of this experiment but has similar characteristics to cv. Siratro. Further development with this species should be made with accessions from group 1.

References
Bray R. A. & T. D. Woodroffe (1995). *Macroptilium atropurpureum* (DC.) Urban (atro) cv. Aztec. *Australian Journal of Experimental Agriculture*, 35, 121.
Clements R. J. (1989). Rates of destruction of growing points of pasture legumes by grazing cattle. *Proceedings XVI International Grasslands Congress, Nice, France*, 2, 1027-1028.
Hutton E. M. (1962). Siratro – a tropical pasture legume bred from Phaseolus atropurpureus. *Australian Journal of Experimental Agriculture and Animal Husbandry*, 2, 117-125.

Production and persistence of a native pasture-*Arachis pintoi* association in the humid tropics of Mexico

E. Castillo-Gallegos[1], L. 't Mannetje[2] and A. Aluja-Schunemann[3]
[1]CEIEGT, FMVZ-UNAM, Martínez de la Torre, Ver. 93600 México, Email: pime11302002@yahoo.com.mx,
[2]Wageningen University, The Netherlands, [3]UADY, Mérida, México

Keywords: native pasture, *Arachis pintoi*, persistence

Introduction *Arachis pintoi* (Ap) is a highly persistent legume. This study aimed at: (i) determining if introducing Ap CIAT 17434 into a NG pasture affected its standing dry matter (SDM) and botanical composition (BC, %) and (ii) verifying if Ap would be as persistent as it is with introduced grasses (IG).

Materials and methods The experiment was carried out from Jan. 1998 to Dec. 2000 in the State of Veracruz at 20° 02' N, 97° 06' W, altitude 112 m. a. s. l. The hot and humid climate has mean yearly rainfall of 1930 mm and average maximum and minimum monthly temperatures of 29°C and 19°C. The soils are clay-loam, acid Ultisols (pH 4.1-5.2) of low fertility. The treatments were native grass pasture (NG) and NG +Ap grazed for 1 day and rested for 20. Stocking rate was 2 cows/ha in the wet season and 3.2 cows/ha the remaining time. Botanical composition (BC) distinguished Ap, native grasses (NG), introduced grasses (IG), broad-leafed weeds (BW), narrow-leafed weeds (NW), and native legumes (NL). Flower number, stolon length, rooted nodes and soil seedpod reserves were recorded at regular intervals.

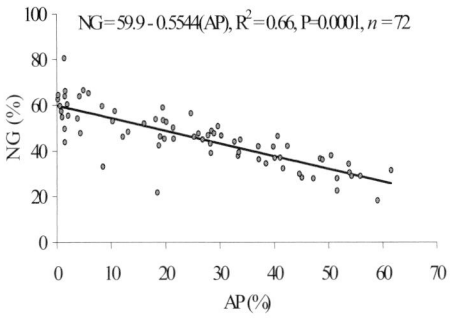

$$NG = 59.9 - 0.5544(AP), R^2 = 0.66, P = 0.0001, n = 72$$

Figure 1 Relationship between Ap and Ng contents

Results SDM of NG+Ap was consistently higher (P<0.05) than that of NG: 3758 vs. 3233 in 1998, 3893 vs 2999 in 1999, and 3301 vs. 2889 in 2000. The treatment effect was significant (P<0.05) on NG and NW, but it did not affect (P>0.05) other components. The effect of year was significant (P<0.05) on all components, except on IG (P>0.05). Ap and IG increased with the years, while NG; BW, NW and NL decreased, but the tendency was not as strong (Table 1). Sixty six percent of the variation in NG content was explained by AP content (Figure 1). Flowering was zero from Dec. to Feb, but stolon length, rooted nodes and soil seed reserves increased from year to year.

Table 1 Botanical composition and reproductive traits of *Arachis pintoi*

Trait	1998		1999		2000	
	NG	NG+Ap	NG	NG+Ap	NG	NG+Ap
Bot. composition (%)						
AP	-----	14.7 [A]	-----	25.2 [B]	-----	36.6 [C]
NG	67.3 [a]	52.4 [b]	69.4 [a]	44.6 [b]	68.9 [a]	40.4 [b]
IG	7.6 [a]	4.4 [a]	6.0 [a]	7.1 [a]	10.4 [a]	3.8 [b]
BW	15.5 [a]	16.0 [a]	12.9 [a]	9.9 [a]	12.8 [a]	11.6 [a]
NW	4.3 [a]	7.8 [b]	5.9 [a]	9.1 [b]	3.7 [a]	5.9 [a]
NL	5.3 [a]	4.7 [a]	5.8 [a]	4.1 [a]	4.2 [a]	1.7 [b]
Reproduction of A. pintoi						
flowers/m^2	-----	41 [A]	-----	98 [B]	-----	346 [C]
m/m^2 of stolons	-----	10 [A]	-----	19 [B]	-----	48 [C]
rooted nodes/m^2	-----	205 [A]	-----	397 [B]	-----	516 [C]
seed pods/ha (10^6)	-----	0.9 [A]	-----	1.6 [B]	-----	8.6 [C]
kg/ha of seed pods	-----	123 [A]	-----	207 [B]	-----	765 [C]

Year means with different capital letter are significantly different (P<0.05); treatment means with different lower case letter are significantly different (P>0.05)

Conclusions AP content increased with the years mostly due to its strong stoloniferous habit that led to profuse flowering and high level of seed pod reserves, which confirm its great persistence (Ibrahim & Mannetje, 1998).

Management strategies to keep legume content around 30% are needed to maintain a high potential for DM production.

Reference

Ibrahim, M. A. & Mannetje, L.'t (1998). Compatibility, persistence and productivity of grass-legume mixtures in the humid tropics of Costa Rica. 1. Dry matter yield, nitrogen yield and botanical composition. *Tropical Grasslands* 32, 96-104.

Performance of dual-purpose cows on a native pasture-Arachis pintoi association in the humid tropics of Mexico

E. Castillo-Gallegos[1], J. Jarillo-Rodríguez[1], E. Ocaña-Zavaleta[1], B. Marín-Mejía[1], L. 't Mannetje[2] and A. Aluja-Schunemann[3]
[1]CEIEGT, FMVZ-UNAM, Martínez de la Torre, Ver. 93600 México, Email: pime11302002@yahoo.com.mx, [2]Wageningen University, The Netherlands, [3]UADY, Mérida, México

Keywords: milk yield, dual-purpose systems, *Arachis pintoi*, native pastures

Introduction Native grasslands (NG) are the main feed supply of dual-purpose (DP) cows of the Mexican humid tropics. NG comprise about 85% of *Paspalum*, *Axonopus* and *Cynodon* species, about 5% of native legumes, mainly of *Desmodium*, and the remaining 10% are narrow and broad leafed weeds. *Arachis pintoi* (AP) is a persistent grazing tolerant tropical legume. In association with sown grasses, it has improved dry matter (DM) yield, nutritive quality of forage, and milk yield up to 9 kg/cow/day (González *et al.*, 1996). The objective was to determine if productive performance of DP cows could be improved by the introduction of AP into NG grassland.

Materials and methods The experiment was carried out from 1998-2001 in the State of Veracruz in a hot (23.5 °C mean temperature) and humid (annual rainfall 1980 mm) climate with acid soils (pH 4.5-5.2) of low fertility (<2 ppm of avail. P). Treatments were NG and NG+AP (CIAT 17434) sown in 1996; no fertiliser was used during the experiment. A 1-day grazing/20-day rest system was used. Stocking rate was 2 cows/ha from Feb. to Oct., and 3.2 cows/ha the remaining time. F1 (Holstein x Zebu) DP cows were used that calved from Mar.-July each year. The cows were milked once a day (8:00 AM). Lactation length was on average 200 days and drying-out occurred when liquid saleable milk yield (SMY) fell to < 3 kg/day, or till the last week of Jan. to keep a 1-year production cycle. One kg of DM/head/day of molasses was given during milking. The calves suckled ½-hour after milking and for ½-hour at 2:00 PM from1998 to 2000; there was no afternoon suckling in 2001. Calves grazed separately from their dams and consumed 0.9 kg of DM/calf/day of concentrate (13% CP, 11.2 MJ EM/kg DM) up to weaning (4 months). The liveweight (LW) of cows (LWC) was recorded monthly, and that of calves (LWc) every week before and after suckling, to calculate by difference the daily milk intake (DMI). The daily gains (ADG) and losses (DWL) of the cows before and after peak LWC and the calves daily gain (ADGc) were estimated by regressing the LW (Y) against days (X), the regression coefficient being an estimate of daily LW change. Data were analysed separately for each year.

Results The treatment effect was not consistent from year to year either for ADG or DWL (Table 1), and while the ADGc was significantly (P>0.05) higher in NG+Ap than in NG in 3 out of the 4 years, the differences were too small to be of agronomic significance (Table 2). The DMI of NG+Ap was significantly lower than that of NG in 1998, but there was no difference (P>0.05) between 1999 and 2001. There was a difference (P<0.05) between treatments in SMY only in 1999.

Conclusion The introduction of AP into the NG grassland could not improve productive performance of the DP cows.

Table 1 Average daily gains (ADG, kg/cow) before peak LWC and daily weight loss (DWL, kg/cow) after peak LWC

Year	ADG		DWL	
	NG	NG+Ap	NG	NG+Ap
1998	0.772 [a]	0.759 [a]	0.426 [a]	0.137 [b]
1999	0.635 [a]	0.140 [b]	0.038 [a]	0.020 [a]
2000	0.191 [a]	0.807 [b]	0.303 [a]	0.513 [a]
2001	0.869 [a]	0.601 [a]	0.580 [a]	0.321 [b]

ADG and DWL values within a year, with different letter are statistically different (P<0.05).

Table 2 Calf daily gains (ADGc, kg/calf/day) and milk intake (DMI, kg/calf/day), and cow saleable milk yield per lactation (SMY, kg/cow)

Year	ADGc		DMI		SMY	
	NG	NG+Ap	NG	NG+Ap	NG	NG+Ap
1998	0.59 [a]	0.51 [b]	4.2 [a]	3.6 [b]	1212 [a]	1299 [a]
1999	0.70 [a]	0.74 [b]	4.7 [a]	4.6 [a]	1175 [a]	1465 [b]
2000	0.56 [a]	0.57 [a]	4.0 [a]	3.9 [a]	1229 [a]	1214 [a]
2001	0.57 [a]	0.54 [b]	2.1 [a]	2.1 [a]	1356 [a]	1336 [a]

ADGc values and DMI and SMY m eans within a year, with different letter are statistically different (P<0.05)

References
González, M. S., L. M Van Heurk, Pezo, D. A. & P. J. Argel, (1996). Producción de leche en pasturas de estrella Africana (*Cynodon nlemfuensis*) solo y asociado con *Arachis pintoi* o *Desmodium ovalifolium*. *Pasturas Tropicales* 18, 2-12.

Response of *Arachis pintoi* to grazing intensity when associated with different grasses

C.M.S. Andrade[1], R. Garcia[2], J.F. Valentim[1] and O.G. Pereira[2]

[1]*Embrapa Acre, Caixa Postal 321, CEP 69908-970, Rio Branco, AC, Brazil, Email: mauricio@cpafac.embrapa.br,* [2]*Universidade Federal de Viçosa, Departamento de Zootecnia, CEP 36571-000, Viçosa, MG, Brazil*

Keywords: *Brachiaria brizantha*, forage peanut, *Panicum maximum*, persistence

Introduction Lack of legume persistence is one of the main reasons for poor utilisation of grass-legume pastures in the tropics. *Arachis pintoi* (forage peanut) is currently the most promising forage legume for the humid tropics, mainly because of good persistence under grazing (Grof, 1985; Fisher & Cruz, 1995). The objective of this work was to show how two accessions of *A. pintoi* react to increasing herbage allowance levels when associated with two different grasses.

Materials and methods Two grazing experiments were carried out at the Experimental Station of Embrapa Acre (10°01'59"S and 67°42'13"W), in Rio Branco, AC, Brazil. In experiment 1, a nine-year-old *Panicum maximum* cv. Massai and *A. pintoi* Ac 01 pasture was grazed at three herbage allowance (HA) levels (9.0, 14.5 and 18.4% body weight (BW)), from October 2002 to December 2003. In experiment 2, a three-year-old *Brachiaria brizantha* cv. Marandu and *A. pintoi* Ap 65 pasture was grazed at four HA levels (6.6, 10.3, 14.3 and 17.9%BW), from January to December 2003. Pastures were rotationally stocked and botanical composition was measured pre-grazing in each grazing cycle. Only results for the final grazing cycle are presented.

Results Initial average forage peanut percentages (FP%) were 5% and 4% when associated with Massai grass and Marandu grass, respectively. These low FP% can be explained since both pastures were under-utilised at the onset of the experimental periods. Final FP% increased linearly as HA was reduced, in both experiments (Figure 1), confirming that *A. pintoi* is favoured when managed under higher grazing intensities, as demonstrated previously for cultivars Amarillo (Fisher & Cruz, 1995; Ibrahim & Mannetje, 1998) and Belmonte (Santana *et al.*, 1998). In contrast to prostrate legumes like *Desmodium ovalifolium*, whose high grazing resistance is due to avoidance because of low palatability, *A. pintoi* is a palatable legume (Lascano, 2000). Its prostrate and stoloniferous growth habit, with growing points protected from grazing, explains the high grazing resistance. Therefore, the increase of FP% with low HA was not related to selectivity, but primarily to sward structure modification.

Conclusions The results of the present study demonstrate that *A. pintoi* can be successfully associated with Massai grass or Marandu grass under rotational stocking in the Western Brazilian Amazon, but to obtain a significant legume content swards should be grazed sufficiently hard to avoid excessive shading of the legume.

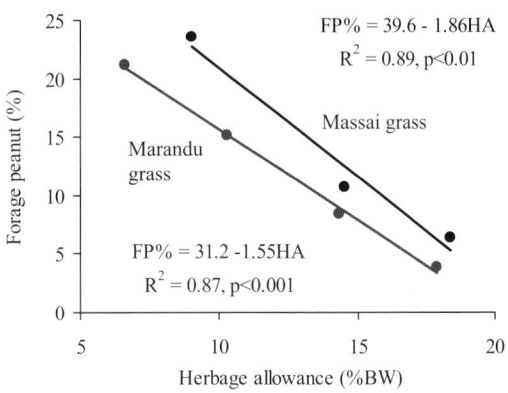

FP% = 39.6 - 1.86HA
$R^2 = 0.89$, p<0.01

Massai grass

Marandu grass

FP% = 31.2 -1.55HA
$R^2 = 0.87$, p<0.001

Figure 1 Effect of herbage allowance on forage peanut percentage

References

Fisher, M.J. & P. Cruz (1995). Algunos aspectos de la ecofisiología de *Arachis pintoi*. In: P.C. Kerridge (ed.) Biología y Agronomia de Especies Forrajeras de Arachis. CIAT, Cali, 56-75.

Grof, B. (1985). Forage attributes of the perennial groundnut *Arachis pintoi* in a tropical savanna environment in Colombia. *Proceedings of the Fifteenth International Grassland Congress*, 168-170.

Ibrahim, M.A. & L.'t. Mannetje (1998). Compatibility, persistence and productivity of grass-legume mixtures in the humid tropics of Costa Rica. 1. Dry matter yield, nitrogen yield and botanical composition. *Tropical Grasslands*, 32, 96-104.

Lascano, C.E. (2000). Selective grazing on grass-legume mixtures in tropical pastures. In: G. Lemaire, J. Hodgson, A. Moraes, C. Nabinger & P.C.F Carvalho (Eds.) Grassland Ecophysiology and Grazing Ecology. CAB International, Wallingford, 249-263.

Santana, J.R., J.M. Pereira & C.P. Resende (1998). Avaliação da consorciação de *Brachiaria dictyoneura* Stapf com *Arachis pintoi* Krapov & Gregory sob pastejo. In: Reunião Anual da Sociedade Brasileira de Zootecnia, 35., 1998, Botucatu: SBZ, 1 CD-ROM.

Selection of Forages for the Tropics (SoFT) – a database and selection tool for identifying forages adapted to local conditions in the tropics and subtropics

B.C. Pengelly[1], B.G. Cook[2], I.J. Partridge[3], D.A. Eagles[1], M. Peters[4], J. Hanson[5], S.D. Brown[1], J.L. Donnelly[1], B.F. Mullen[1], R. Schultze-Kraft[6], A. Franco[4] and R. O'Brien[4]

[1]CSIRO Sustainable Ecosystems, St Lucia, Queensland Australia 4067, bruce.pengelly@csiro.au, [2]Department of Primary Industries and Fisheries, Gympie, Queensland Australia 4570, [3]Department of Primary Industries and Fisheries, Toowoomba, Queensland Australia 4350, [4]CIAT, A.A. 6713, Cali Colombia, [5]ILRI, P.O. Box 5689, Addis Ababa Ethiopia, [6]University of Hohenheim (380), 70593 Stuttgart Germany

Keywords: tropical forages, selection, information sheets

Introduction Rising populations and incomes in developing countries are likely to double demand for livestock products by 2020 (Delgado *et al.* 1999). This strong demand has potential to improve profitability for farmers but will require improved animal feeding in both semi-intensive crop-livestock and more extensive livestock systems. Forages usually are the most cost-effective option to supply feed demands, particularly for ruminant-, but also for pig- and poultry- production. It is critical to select the most suitable forages for the local system and conditions. Small- and even larger-scale farmers depend heavily on advice from extension and development agencies, and from seed companies, but this advice often is limited by inexperience and the difficulty in accessing reliable information. Expert information on an extensive range of tropical forages is now readily available through the SoFT database.

Database development and structure Forage research over the last 50 years has identified many useful tropical grasses and legumes. Information on their adaptation and use has resided in peer-reviewed literature, research reports with limited distribution and, often most importantly, in the memories of forage agronomists with decades of experience. The SoFT database has accessed these information sources to define the adaptation and use of >200 forages, and has integrated this knowledge into a user-friendly database. The database has 4 main features: (i) information in fact sheets on the adaptation, uses and management of forage species, cultivars and elite accessions; (ii) a selection tool built on LUCID™ that enables easy identification of best-bets based on 19 criteria (Table 1); (iii) a bibliography of >6000 references and abstracts on forage diversity, management and use; (iv) a collection of photographs and images of species to help in their identification and use. The database selection tool is an expert system based on the experiences of >50 forage specialists who have worked for many years in tropical and subtropical regions of Africa, Tropical USA, Central and South America, South and South-east Asia and Australia.

Table 1 Selection criteria available in the SoFT database to select the most suitable forages for environments and uses

Climate/farming system attributes	Soil environment attributes	Plant attributes
Latitude x altitude	Soil pH	Plant family (legume or grass)
Rainfall (average annual)	Level of available soil Al/Mn	Life cycle
Length of dry season	Level of soil salinity	Growth form
Inundation	Soil drainage	Stem habit
Intended forage use	Soil texture	Cool season growth
Grazing pressure	Soil fertility	Frost tolerance (foliage damage)
Shade environment		

Conclusions The SoFT project has summarised information on tropical forage adaptation and use from available literature and experiential sources. The SoFT database on CD and the Internet will allow researchers and advisors to select those forages most suitable for local conditions. It is also a valuable teaching tool for colleges and universities. CIAT (International Centre for Tropical Agriculture) will undertake updates of SoFT.

Acknowledgments We acknowledge gratefully the financial support of ACIAR, BMZ and DFID for this project, and thank the many forage experts from around the world who freely gave their time and experiences to make development of the SoFT database possible.

Reference

Delgado, C., M. Rosegrant, H. Steinfeld, S. Ehui & C. Courbois (1999). Livestock to 2020: The Next Food Revolution. Food, Agriculture and the Environment Discussion Paper No. 28. International Food Policy Research Institute, Washington, D.C.

Theme A: Efficient production from grassland

Section 8

Grassland management

Yield and nutritive value of heading and headless sorghum×sudangrass hybrids in response to cutting frequency

J.K. Lee[1], H.S. Park[1], Y.C. Lim[1], S. Seo[1], B.H. Paek[1] and J.H. Fike[2]
[1]Hanwoo Experiment Station, National Livestock Research Institute, RDA, Chahang-ri, Doam-myeon, Pyeongchang-gun, Kangwon-do, South Korea, 232-952, Email: leejk58@rda.go.kr [2]Crop and Soil Environmental Sciences Department, Virginia Tech. Blacksburg, VA 24061, USA

Keywords: cutting frequency, heading and headless variety, sorghum × sudangrass hybrid

Introduction Summer annual forages contribute greatly toward solving the problem of roughage supply for cattle in Korea. These forages support high levels of dairy and beef production during hot summer months when the quality and production of perennial herbage decreases due to unfavourable climatic conditions (Olson, 1971). This study investigated the effects of cutting frequency on dry matter (DM) yield and nutritive value of heading versus headless varieties of sorghum×sudangrass hybrid.

Materials and methods Heading (cv. 'T. E. Haygrazer') and headless (cv. 'Jumbo') sorghum×sudangrasses were established on 27 May, 1999. Fertiliser applications (N 210 and K 150 kg/ha) were distributed equally across establishment and all but final harvest dates; P (150 kg ha) was applied at establishment only. The forages were cut 1, 2, or 3 times/year. Estimates of yield were determined by harvesting all forage in each plot. Fresh forage was oven-dried for 72 h at 75°C, weighed and converted to DM yield. Forage nutritive value was evaluated by ADF and NDF (Goering & Van Soest, 1970), and *in vitro* dry matter disappearance (IVDMD; Moore, 1970). Heading type was the main plot and cutting frequencies were sub plots.

Results DM yields of heading vs. headless types of sorghum×sudangrass hybrids (7,687 vs. 9,857 kg/ha, respectively) did not differ significantly (Table 1). For both types, DM yields for multiple-cut treatments were highest at Cut 1. The low yields at Cut 3 of the 3× cutting treatment were likely due to the short growing period. Concentrations of ADF and NDF were slightly greater in the headless type sorghum×sudangrass. Levels of ADF were highest at Cut 1 for both varieties. Levels of NDF were highest at Cut 1 in the headless type, but highest at Cut 2 in the heading type. IVDMD was numerically greater in the headless variety than in the heading type and was numerically greatest in Cut 2 for both varieties but differences among harvest treatments were not significant.

Table 1 Dry matter (DM) yield of sorghum×sudangrass hybrid

Treatments		DM yield (kg/ha) per cut			
Type	Cutting frequency	1st	2nd	3rd	Total
Heading	1×	9,130			9,130
	2×	6,670	1,010		7,680
	3×	4,890	1,000	360	6,250
	Mean				7,687
Headless	1×	13,420			13,420
	2×	7,340	1,330		8,670
	3×	4,800	1,850	830	7,480
	Mean				9,857
LSD (0.05)	Heading Type (A)				NS
	Cutting frequency (B)				NS
	A×B				NS

NS: not significant

Conclusions Although there were no significant yield differences, cultivation of headless types with one annual harvest is satisfactory. Headless sorghum × sudangrass hybrids can produce high levels of dry matter with very good nutritive value for summer forage in Korea.

References
Goering, H. L. & P. J. Van Soest (1970). Forage fiber analysis. Agricultural Handbook, No. 379. USDA.
Moore, J.E. 1970. Procedures for the two-stage *in vitro* digestion of forages. p. 5001.1–5001.3. *In* L.E. Harris (ed.) Nutrition research techniques for domestic and wild animals. Vol. 1. Utah State Univ., Logan.
Olson, T. C. (1971). Yield and water use by different populations of dryland corn, grain sorghum, and forage sorghum in the western corn belt. *Agronomy Journal*, 63(1): 104-107.

The effect of defoliation interval on regrowth of tall fescue

K.A. Adamczewski and D.J. Donaghy
The Tasmanian Institute of Agricultural Research, PO Box 3523, Burnie, Tasmania 7320, Australia, Email: Danny.Donaghy@utas.edu.au

Keywords: tall fescue, defoliation, regrowth, dry matter

Introduction   Herbage yield, persistence and quality optimise when defoliation interval is based on physiological indicators, such as leaf regrowth stage. Examples include ryegrass (Fulkerson & Donaghy, 2001), cocksfoot (Rawnsley *et al.*, 2002), prairie grass (Fulkerson *et al.*, 2000) and kikuyu (Reeves *et al.*, 1996). Yield, persistence and quality optimise because leaf regrowth stage relates closely to plant energy reserves, which generally peak as the number of live leaves/tiller maximise. More frequent defoliation than the optimum reduces energy reserves and leads to a smaller root system, fewer tillers and retarded growth rate (Fulkerson & Donaghy, 2001). Based on plant physiological development, the optimum defoliation interval for tall fescue has not been defined.

Methods   Each leaf regrowth stage ('leaf stage') was defined as the time taken to produce 1 fully expanded leaf/tiller. Treatments of tall fescue in the glasshouse were based on defoliation intervals of 1-, 2- and 4-leaf stages of regrowth, with treatments ending when the 1-leaf defoliation had been completed 4 times, the 2-leaf defoliation 2 times, and the 4-leaf defoliation once. Half of the plants was harvested destructively just after cessation of defoliation treatments (H_1) and the other half after regrowth to the next 4-leaf stage (H_2), which took 70 days. The dry matter (DM) yield was determined at each harvest event for leaves (plant material >50mm), stubble (plant material <50mm) and roots (plant material below ground).

Results   Root-, leaf- and stubble- DM yield was closely related to defoliation interval. Less frequent defoliation gave a higher DM yield, not only immediately after the treatment period (H_1), but also at H_2. At H_1, plants defoliated at the 4-leaf stage had a significantly higher ($P<0.001$) root-, leaf- and stubble- DM yield than the other two treatments (Table 1). Plants defoliated at the 2-leaf stage also had a significantly higher ($P<0.001$) DM yield of all plant components than plants defoliated at the 1-leaf stage. The pattern was identical at H_2, except for root DM yield, which showed no significant difference ($P>0.05$) between defoliation treatments at the 1- or 2-leaf stage.

Conclusions   Compared to more frequent defoliation at the 1- or 2-leaf stages, defoliating tall fescue plants at the 4-leaf stage maximised regrowth of all plant components. The deleterious effect on regrowth of frequent defoliation was still present 70 days after the end of treatments, most likely due to depletion of energy reserves. Roots were the only plant component that could recover in the short-term after repeated defoliation at the 1-leaf stage.

Table 1 Tall fescue leaf, root and stubble dry matter (mg/plant) immediately following defoliation treatments (H_1) or after 70 days (H_2)

	Leaf stage	H_1	H_2
Leaf (mg/plant)	1L	1512	1852
	2L	2654	2991
	4L	3199	3877
LSD (*P*=0.05)		351	352
Root (mg/plant)	1L	1582	2228
	2L	3537	4193
	4L	4810	8260
LSD (*P*=0.05)		299	2725
Stubble (mg/plant)	1L	390	581
	2L	752	1002
	4L	1080	1530
LSD (*P*=0.05)		41	111

References

Fulkerson, W. J. & D. J. Donaghy (2001) Plant soluble carbohydrate reserves and senescence – key criteria for developing an effective grazing management system for ryegrass-based pastures: a review. *Australian Journal of Experimental Agriculture*, 41, 261-275.

Fulkerson, W. J., J. F. M. Fennell & K. Slack (2000) Production and forage quality of prairie grass in comparison to perennial ryegrass and tall fescue in subtropical dairy pastures. *Australian Journal of Experimental Agriculture*, 40, 1059-1067.

Rawnsley, R. P., D. J. Donaghy, W. J. Fulkerson & P. A. Lane (2002) Changes in the physiology and feed quality of cocksfoot during regrowth. *Grass and Forage Science*, 57, 203-211.

Reeves, M., W. J. Fulkerson & R. C. Kellaway (1996) Forage quality of kikuyu (*Pennisetum clandestinum*): the effect of defoliation and nitrogen fertiliser application and in comparison with perennial ryegrass (*Lolium perenne*). *Australian Journal of Agricultural Research*, 47, 1349-1359.

Italian ryegrass and whole crop cereal mixture: effect of sowing rate and maturity on variety on yield and botanical composition in northern latitudes

O. Niemeläinen, O. Nissinen and M. Kontturi
MTT Agrifood Research Finland, FIN-31600 Jokioinen, Finland, Email: oiva.niemelainen@mtt.fi

Keywords: cropping systems, Italian ryegrass, whole crop cereals

Introduction The profitability of a dairy farm using whole crop cereals (WCC) is better in Finland than that of a farm producing combine harvested cereals and grass silage (Turunen, 2000). The main reason for that is the decrease in the machinery costs. However, quality of the WCC yield can vary considerably depending on the ear- straw ration. This may lead to problems in feeding. In this experiment we studied the effect of sowing rate and growing time of the cereal cultivar on the botanical composition, quality and yield of the WCC harvest. The objective was to study if it would be possible to increase the grass component in the WCC harvest by cultivation management to such a level that the WCC yield would be feasible to be used in feeding with pure grass silage. In this paper we present results of the botanical composition of the WCC harvest.

Materials and methods Early (E) Artturi cv. and late (L) Inari cv. of barley were sown at three densities: 500, 350 and 200 viable seeds/m² in Jokioinen (60°49 N, 23°30 E), Finland. Italian ryegrass (IRG) cv. Turgo was sown on the whole experimental area right after sowing the cereal treatments at 850 viable seeds/m². Nitrogen fertiliser application was 110 kg N/ha at establishment and 60 kg N/ha after the harvest of WCC. Results of 2002 and 2003 are presented. Plot size was 2,0 x 10 m and the harvested area was 1,5 x 10 m. Experimental design was a completely randomised block with four replicates. WCC was harvested at the dough stage. Botanical composition was estimated on a 500 g sample taken from the harvested yield from each plot, and the yields were corrected to 100% DM based on DM percentage of each yield component. Autumn was exceptionally dry in both years. In 2003 the experiment was irrigated after the harvest of WCC.

Results In 2002 treatments of the E cv. were harvested on July, 25. and of the L cv. on August,1. The second cut was delayed by early snow and was taken on October, 29. In 2003 the E cv. was harvested on August 4. and L cv. on August, 12. The second cut was taken on September 24. Dry conditions affected negatively the yield of 2nd cut of IRG in 2002 (Figure 1). The contribution of IRG at WCC harvest ranged from 9 to 38 percent in 2002, and from 14 to 37 percent in 2003.

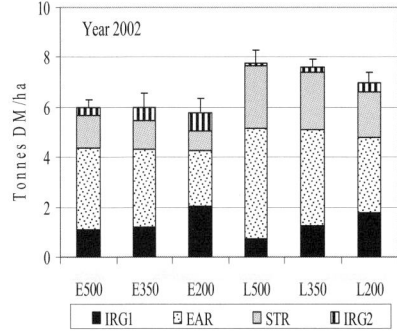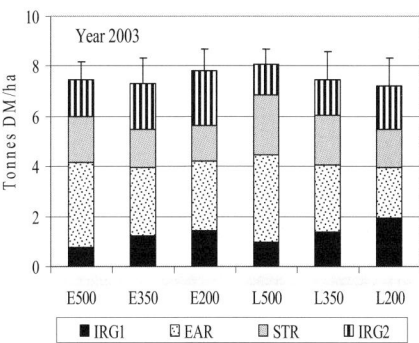

Figure 1 DM yield of ears, straw and Italian ryegrass at WCC harvest and in the 2nd cut in 2002 and 2003 in Jokioinen, Finland. Bar indicates standard deviation in the total DM yield of the growing season

Conclusions The ressults indicate that by using IRG in WCC cultivation the grass component in the WCC harvest can be raised so that the use of WCC would be easier than that of pure cereal WCC.

Reference

Turunen, H. (2000). Kokoviljasäilörehun viljelyn tuotantokustannukset ja kannattavuus maidontuotannossa. (In Finnish) Summary: Production costs and profitability in the cultivation of whole crop silage in milk production. Agricultural Economics Research Institute, Finland. Working papers 6/2000. 54 pp.

Italian ryegrass and barley mixture for forage production: effect of harvesting time on yield and quality in northern latitudes

O. Nissinen, O. Niemeläinen and M. Kontturi

Lapland Research Station, MTT Agrifood Research Finland, FIN-96900 Saarenkylä, Finland, Email: oiva.nissinen@mtt.fi

Keywords: cropping systems, Italian ryegrass, quality, whole crop cereals

Introduction In northern latitudes forage cereal and Italian ryegrass (IRG) mixtures provide a good source of forage in situations where perennial swards have suffered winter damages (Nissinen, 1994). In this experiment harvesting time of the first cut of an IRG-barley mixture was studied to optimise the yield and quality in the growing season. The objective of the study was to assess if harvesting an IRG-barley mixture later than two weeks after heading, as currently recommended, offers benefits.

Materials and methods An early barley cultivar was sown at a density 200 viable seeds/m^2 with IRG at 850 seeds/m^2 in Rovaniemi (66°35 N, 26°10 E) in Finland. The first harvest was taken: a) at heading of barley (H), b) two weeks after heading (H2), c) at early dough stage (ED), and d) at late dough stage (LD). Treatment H was harvested three times in the season in both years and treatment H2 three times in 2002 and two times in 2003. Treatments ED and LD were harvested twice in the season. N fertiliser application at establishment was 80, 90, 100 and 100 kg/ha for H, H2, ED and LD, respectively. Total N application in the season was 200, 200, 160 and 160 kg/ha, respectively. The plot size was 1,5 x 8 m. Experimental design was a completely randomised block with four replicates. Botanical composition was estimated on an approximately 1000 g sample taken from the harvested yield from each plot, and the yields were corrected to 100% DM based on the DM percentage of each component.

Results In 2002 first cut was taken 10.7., 23.7.,1.8., and 9.8. in H, H2, ED, and LD treatments. The last cut (2nd or 3rd cut) was taken 11.9. In 2003 the respective harvest dates were: 16.7., 30.7., 5.8., and 7.8., and the last cut was taken 11.9. The DM yield of IRG and ear and straw component of barley in the first cut and the total yields are shown in Figure 1. Share of IRG at the DM yield of the first cut ranged from 17 to 22 percent in 2002, and from 15 to 22 percent in 2003. Digestibility of organic matter (DOM) in the first harvest was 735, 699, 707, and 729 g/kg in 2002 in H, H2, ED and LD. In 2003 the DOM of ED was clearly lower (632 g/kg) than that of H2 (701 g/kg) and LD (697 g/kg) (Data for H is not available). OMD of regrowth of IRG was above 800 g/kg.

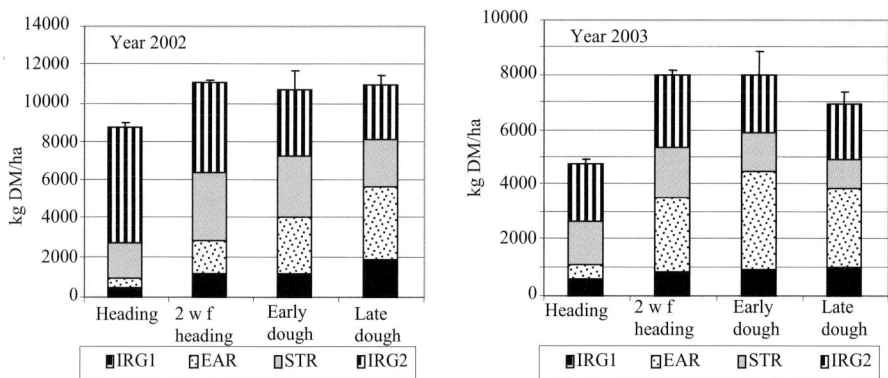

Figure 1 DM yield of IRG, and ears and straw of barley in the 1st cut, and of regrowth of IRG2 in 2002 and 2003. The bars indicate standard deviation in the total DM yield

Conclusions The data indicate that harvesting an IRG-barley mixture later than currently recommended (within 2 weeks from heading) may provide benefits in production costs (fewer cuts). However, the quality of the first harvest requires more investigation. High yields can be obtained of IRG-barley mixtures.

Reference

Nissinen, O. (1994). The utilization of green forage plants in crop farming in northern Finland. In: C.A. Scott Smith (Ed). *Proceedings of the 1st Circumpolar Agricultural Conference* Whitehorse, Yukon, Canada September 1992. p. 115-116.

Use of novel spatial presentations of plant species to improve legume abundance

J.M. Sharp, M.J. Jeger, R.W. Fraser and G.R. Edwards
Department of Agricultural Sciences, Imperial College London, Wye Campus, High Street, Wye, Ashford, Kent TN25 5AH, UK, Email: joanna.sharp@imperial.ac.uk

Keywords: white clover, sheep grazing, spatial arrangement, pasture composition, plant population dynamics

Introduction The benefits of using white clover (Trifolium repens) in pasture grazed by sheep have been widely recognised. However, clover is considered inadequate and risky as the main source of nitrogen input, since its abundance in the pasture is patchy, low (typically less than 20%) and shows great year-to-year variation. This is thought to be due to the costs of nitrogen fixation, competition with grass, the preference for clover by sheep and patchy dung and urine deposition (Schwinning & Parsons, 1996). One possible solution may be the spatial separation of clover from grass, which would remove inter-specific competition, allowing clover to grow unimpeded in a greater abundance than previously observed. Spatial separation can occur over a range of spatial scales, from narrow strips of alternating clover and grass to complete separation, where half a pasture is clover while the other half is grass. This in turn may have a significant impact on the processes occurring within the pasture, such as plant growth and spread, nitrogen cycling and animal behaviour.

Materials and methods The following 4 treatments were used in a field site established in 2001, in 0.15 ha plots, grazed continuously by a Suffolk flock: (1) mixed pasture, and 50:50, by ground area, (2) alternating 1.5m strips of grass and clover monocultures (partial separation), (3) 3m strips (partial separation) and (4) adjacent monocultures (full separation). As part of a longer study, in spring 2004, herbage production using exclosure cages and sward surface height were measured as described by Frame (1993). Grazing behaviour was observed on 3 occasions to estimate intake.

Results Table 1 shows the mean herbage production. Analysis of variance showed that production of white clover in spatially separated plots was significantly greater ($F_{1,6}$=19.91, P<0.01), although overall herbage production was significantly greater in mixed plots than spatially separated plots ($F_{1,6}$=20.24, P<0.01). Combining observed grazing behaviour with sward surface height (SSH), as described by Rook *et al.*, (2002), a selection coefficient was calculated (Figure 1).

Table 1 Mean herbage production (kg DM/ha/day)

Treatment	Total	Clover
Mixed	37.67	1.52
1.5m strips	22.71	4.75
3m strips	23.04	3.48
Adjacent monocultures	25.79	4.63

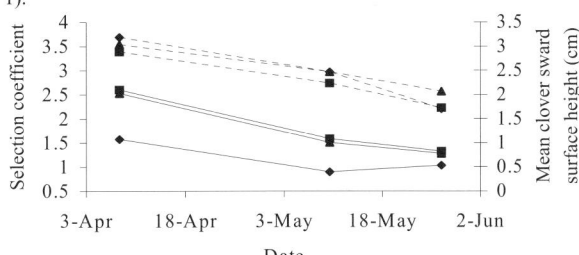

Figure 1 Mean clover SSH (- -) and selection coefficient (—) for clover by grazing sheep on 1.5m (▲), 3m (♦) strips and adjacent monocultures (■)

Conclusions The initial results of this study showed that spatially separating grass from clover produced higher dry matter content of clover in the sward, compared to mixed pastures. A selection coefficient of spatially separated treatments >1 (Figure 1) indicates that the sheep were actively selecting clover, despite the declining SSH. Further work with this system to estimate pasture composition through time and an economic evaluation of the method of spatial separation is expected to incorporate additional experimental work and a cellular automata-based model (Schwinning & Parsons, 1996).

References

Frame, J. (1993). Herbage Mass. Sward Measurement Handbook (2nd Edition). Davies, A., R. D. Baker, S. A. Grant & A. S. Laidlaw, Reading, The British Grassland Society. pp 39-68.

Rook, A. J., A. Harvey, A. J. Parsons, P. D. Penning & R. J. Orr (2002). Effect of Long-Term Changes in Relative Resource Availability on Dietary Preference of Grazing Sheep for Perennial Ryegrass and White Clover. *Grass and Forage Science*, 57(1): 54-60.

Schwinning, S. & A. J. Parsons (1996). A Spatially Explicit Population Model of Stoloniferous N-Fixing Legumes in Mixed Pasture with Grass. *Journal of Ecology*, 84(6): 815-826.

Effects of grass suppression on legume abundance in a naturalised pasture

C. Hepp[1], I Valentine[2] and P.D. Kemp[2]

[1]Instituto de Investigaciones Agropecuarias (INIA), Chile, Email: chepp@tamelaike.inia.cl, [2]INR, Massey University, Palmerston North, New Zealand

Keywords: grass suppression, haloxyfop herbicide, hill country, legume abundance, soil moisture, white clover

Introduction Low abundance and poor persistence of legumes is a generalised problem in hill country pastures in New Zealand, even at adequate soil phosphorus levels (Woodfield & Caradus, 1996). Likely causes of low legume contents in swards include lack of soil moisture, high temperatures (Barker *et al.*, 1993), frequency and intensity of defoliation (Suckling, 1975; (Lambert *et al.,* 1982) and increased grass competition due to increasing fixed nitrogen in the soil (Lambert *et al.*, 1982). On this later point there is anecdotal evidence that suppression of the grass component with herbicide will boost clover content, colloquially called 'chemical topping'.

Materials and methods Plots of resident pasture (H0) were maintained or treated with the herbicide haloxyfop at a rate of 9.0 g ai/10 *l* water in autumn 2000 (H1) to suppress existing grass species. In year 2, grass suppression treatments were H0, H1 (from the previous year) and H2 (resident pasture herbicide treated in autumn 2001), so that short-term and residual effects on swards could be compared. Live standing biomass, botanical composition and white clover growing point density (GPD) were determined on 24 sward cores per replicate at each harvest. Swards were sampled five times in 2000 (July-Dec.) and seven times in 2001 (July-March), at approximately monthly intervals. After sampling, plots were cut using a rotary mower (25–30 mm height). Phosphorus was applied to a target of 30 Olsen P using triple superphosphate.

Results In 2000/01 suppressing grass with herbicide more than doubled the average legume content of the standing biomass (7.1% *v* 27.6%, Figure 1). In the following season the average white clover content increased from 24.5% in H0 to 86.2% in H2, with the residual herbicide effect maintaining 37.9% in H1. The grass suppression did not affect yield in the year of spraying but provided a significant response the following year (Table 1). The general increase in legume content in the sward with grass suppression was probably connected to an early removal of grass competition. This generated open spaces that could in turn be colonised by the spread of white clover stolons reflected in the increase in white clover growing points (Table1).

Table 1 Total yield and growing point density in 2001/02

	Total annual yield (kg DM/ha)	Growing point density white clover (/m²)
H0	8404	580
H1	9818	3065
H2	7715	473
SEM	520	157

Conclusions Legume abundance in naturalised pasture in New Zealand is strongly affected by competition from the resident grass component of the sward. 'Chemical topping' is an effective, if expensive, way of increasing the density of the white clover component.

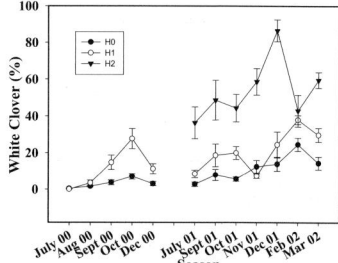

Figure 1 The response of white clover to the herbicide suppression of grass (% DM)

References

Barker, D.J., J.A. Lancashire, S.C. Moloney, N. Dymock, D.R. Stevens, J.D. Turner, D. Scott & W.J. Archie (1993). Introduction, production, and persistence of five grass species in dry hill country. 8. Summary and conclusions. *New Zealand Journal of Agricultural Research*, 36, 61-66.

Lambert, M.G., P.C. Luscombe & D.A. Clark (1982). Soil fertility and hill country production. *Proceedings of the New Zealand Grassland Association*, 43, 153-160.

Suckling, F.E.T. (1975). Pasture management trials on unploughable hill country at Te Awa. III. Results for 1959-69. *New Zealand Journal of Experimental Agriculture*, 3, 351-436.

Woodfield, D.R. & M.C. Caradus (1996). Factors affecting white clover persistence in New Zealand pastures. *Proceedings of the New Zealand Grassland Association*, 58, 229-235.

The interaction of management with botanical composition of irrigated grass-legume pasture mixtures in the Intermountain West USA

J.W. MacAdam, T.C. Griggs and G.J. Mileski
Utah State University, Logan, Utah 84322-4820 U.S.A, Email: jenmac@cc.usu.edu

Keywords: irrigated pasture, *Lotus corniculatus*, *Trifolium repens*, rotational stocking management

Introduction Beef produced on semi-arid range and milk produced in confinement are the main agricultural commodities in the semi-arid western USA. The studies reported here were undertaken to determine the suitability of irrigated pasture as an alternative to traditional beef and dairy production systems. The clipping and grazing studies were not run concurrently or within the same field, but were successive steps in selecting mixtures best-suited for rotational stocking of irrigated pastures in the Intermountain West. Summaries of productivity data have been reported elsewhere (MacAdam, 2002; MacAdam et al., 2004).

Materials and methods Treatments consisted of binary mixtures of introduced grasses (*Bromus riparius*, *Dactylis glomerata*, *Festuca arundinacea*, or *Lolium perenne*) and legumes (*Lotus corniculatus* or *Trifolium repens*) under clipping or grazing. Clipped plots were 1.5 x 6 m, and were fertilised with 15-5-10 NPK at each harvest . Grazed paddocks were 15 x 22 m and were fertilised with 56 kg/ha N in spring. Treatments were either managed as for season-long grazing, or a hay crop was taken followed by clipping or grazing. Paddocks were grazed or harvested according to regrowth rate of each mixture, so harvest number per year varied among treatments. *L. perenne* plots were harvested at 15 cm down to 4 cm. *B. riparius D. glomerata*, and *F. arundinacea* plots were harvested at 25 cm down to 8 cm. Treatments in the clipping study were replicated five times, and treatments in the grazing study were replicated four times.

Results Percentage *T. repens* was always higher than that of *L. corniculatus* in the clipping study. In the grazing study, percentage *L. corniculatus* was always higher than that of *T. repens*, regardless of management, and highest when a hay crop was taken prior to grazing. Management had no effect on percentage *T. repens* under grazing. *L. perenne* mixtures always had the highest legume percentage, and under grazing, *D. glomerata* always had the lowest. Under grazing, *F. arundinacea* and *B. riparius* had similar percentages of legumes.

Table 1 Percentage legume of clipped (1997-1999) or grazed (2001-2003) mixtures determined by NIRS

Clipped	L.c.	s.e.m.	T.r.	s.e.m.	Clipped	L.c.	s.e.m.	T.r.	s.e.m.
	Season-long clipping					Hay crop followed by clipping			
B. r.	12.8	3.05	18.8	3.29	*B. r.*	16.8	2.60	25.0	7.88
D. g.	8.2	3.03	14.1	1.99	*D. g.*	9.5	2.15	12.6	2.11
F. a.	6.7	1.91	17.6	3.27	*F. a.*	9.3	1.57	19.0	3.86
L. p.	35.5	3.99	55.2	6.03	*L. p.*	43.7	4.00	57.1	4.56

Grazed	L.c.	s.e.m.	T.r.	s.e.m.	Grazed	L.c.	s.e.m.	T.r.	s.e.m.
	Season-long rotational stocking					Hay crop followed by rotational stocking			
B. r.	17.3	2.77	8.2	2.20	*B. r.*	36.8	4.04	7.5	1.52
D. g.	4.1	1.20	0.8	0.56	*D. g.*	12.7	2.06	0.0	0.31
F. a.	17.3	1.48	8.2	1.67	*F. a.*	37.4	2.35	5.4	1.39
L. p.	44.3	2.33	32.3	5.27	*L. p.*	59.9	1.68	28.6	5.34

Conclusions In the western U.S.A., mixtures of *L. perenne* and *T. repens* are short-lived and have resulted in pasture bloat even when bloat preventatives were used. *L. corniculatus* is non-bloating, established readily in our high pH soils, and was retained well under rotational stocking management. Utilisation was comparable to mixtures containing *T. repens*. *L. corniculatus* is tolerant of the periodically dry soils of irrigated pastures, and is recommended as the legume component of irrigated pasture mixtures for the semi-arid western USA.

References

MacAdam, J.W. 2002. Grass and forage legume mixes – what's hot and what's not! pp. 79-88 *In* J.E. Brummer and C.H. Pearson (ed.) Proceedings of the Intermountain Forage Symposium Colorado State Univ., Tech. Bull. LTB 02-1.

MacAdam, J.W., T.C. Griggs, & G.J. Mileski. 2004. Productivity of grass and legume components in irrigated pastures under rotational stocking in the Intermountain West. p. 450 *In* K. Cassida (Ed.) *Proceedings of the American Forage Grassland Council*, June 12-16, 2004, Roanoke, VA.

The input of forage legumes in sustainable grassland systems

Z. Kadziuliene and L. Sarunaite
The Lithuanian Institute of Agriculture, LT-58344 Akademija, Kedainiai distr., Lithuania, Email: zkadziul@lzi.lt

Keywords: grassland, legumes, ley, nitrogen

Introduction There is increased interest in sustainable grassland systems. One step towards sustainability is expansion of legume use, because of their potential to fix and transfer nitrogen (N) to subsequent crops. However, legumes can also have negative aspects, such as difficulties in establishment (Porqueddu *et al.*, 2003), lack of persistence, N loss (Scholefield *et al.*, 2002) and accumulation of soil borne disease agents (Kadziulis, 2001). The large variability within legume swards and between years in pastures and leys has encouraged us to search for possibilities to achieve stability of their inputs in sustainable grassland systems.

Materials and methods Between 1997-2001 field trials were carried out on a loam (Cambisol) using red clover (*Trifolium pratense*), white clover (*T. repens*) and lucerne (*Medicago sativa*) in mixture with grasses in various combinations for frequent (F) and less frequent (LF) grazing (with 5-6 and 4-5 grazings per season) and for 3 cuts per season in a ley. The grasses were also sown in pure stands and either fertilised with 0 or 240 kg N/ha /season.

Results The lucerne/grass and lucerne/white clover/grass swards produced high dry matter (DM) yield in all years under grazing, whilst white clover/grass and pure grass swards were less productive, especially in the 4th year (Table 1). Under cutting the lucerne based swards were high yielding in all three years of ley use. Lucerne/grass and lucerne/white clover/grass swards were most suitable for three years' use in pastures or leys, red clover/grass and white clover/red clover/grass swards only for one-two years' leys. DM yield was influenced by composition and grazing frequency. Lucerne competed well even under frequent grazing. All legume/grass swards were superior to grasses fertilised with 0 or 240 kg N/ha. Apparent N_2 fixation highest also in lucerne based swards.

Table 1 Annual yield of different swards on pasture and in ley

Swards	DM t/ha/use year				Apparent N_2 fixation kg/ ha/ year			
	1st		4th		1st		4th	
Pasture	F	LF	F	LF	F	LF	F	LF
Trifolium repens/grasses	5.62	6.56	1.08	2.47	133	138	11	0
Medicago sativa/grasses	6.51	7.55	2.30	3.03	176	208	51	11
T.repens/*M.sativa*/grasses	5.96	6.96	1.62	3.02	150	162	30	19
Grasses/ N_{240}	5.74	7.54	1.53	3.04	-	-	-	-
$LSD_{.05}$ (A factor/B factor)	0.333/0.124		0.320/0.130		7.11/3.55		9.49/4.74	
Ley	1st	2nd	3rd		1st	2nd	3rd	
Trifolium pratense	7.97	6.46	2.01		191	128	27	
Medicago sativa	9.23	8.85	6.74		266	208	188	
T.pratense/*T.repens*/grasses	7.94	6.03	2.41		154	100	18	
M.sativa/*T.repens*/ grasses	8.84	9.21	7.13		224	183	188	
Grasses/ N_{240}	9.13	5.63	3.67		-	-	-	
$LSD_{.05}$	0.356	0.429	0.518		9.81	12.6	16.5	

LSD, least significant difference (P< 0.05)

Conclusions The inputs of the legumes in sustainable grassland systems greatly depends on legume species and sward management. Lucerne-based swards showed the highest potential DM yield and apparent N_2 fixation. Forage legumes, especially lucerne, can be an essential factor for sustainability of grassland systems.

References

Kadziulis, L. (2001). Increasing the share of legumes in a crop rotation by alternated growing of clover species and lucerne. *Grassland Science in Europe*, 6, 51- 54.

Porqueddu, C., G. Parente, & M. Elsaesser (2003). Potential of grasslands. *Grassland Science in Europe*, 8, 11-20.

Scholefield, D., M., Halling, M., Tuori, M Isolahti,., U. Soelter, & A.C. Stone, (2002) Assessment of nitrate leaching from beneath forage legumes. *Landbauforschung Voelkenrode*, SH 234, 17-25.

The effect of cutting regime and cultivar on longevity of pure lucerne stands in Latvia

Z. Gaile
Research and Study farm "Vecauce" of Latvia University of Agriculture, Akademijas street 11.a, Auce, Latvia, LV-3708 Email: zinta@apollo.lv

Keywords: *Medicago sativa*, American cultivars, yield, longevity

Introduction Lucerne growing is important for high and excellent quality yields of hay or silage without application of N fertilisers, as well as for increasing crop diversity in crop rotation systems. It is comparatively expensive to establish lucerne stands in Latvia. Different aspects can affect longevity of lucerne stands: suitability of cultivar to specific conditions, soil characteristics, different stress conditions and cutting management, including frequency and height of cutting, and critical rest period in autumn (Sheaffer *et al.*, 1988). The aim of our study was to evaluate the persistence of lucerne stands without renovation while obtaining reasonable yields.

Materials and methods A series of five field trials was carried out at the research and study farm "Vecauce" of Latvia University of Agriculture (56° 28' N, 22° 53' E). Four trials were performed using a three-cut regime in the harvest years following an establishment year – 1993-2001 (8 varieties), 1994-2001 (the same 8 varieties), 1995-2001 (8 other varieties), 1996-2001 (24 varieties). The fifth trial, 1999-2004, used 10 varieties and three different cutting regimes – traditional three-cut schedule as for previous series, three-cut schedule by calendar date (May 25, July 10, August 20), four-cut schedule (three-cut schedule by calendar plus a 4th cut on October 10). The traditional three-cut schedule is based on known stand persistence: 1st cut in bud stage or in the 1st ten day period of June, 2nd cut in 10-25% full bloom, 3rd cut in early October. In every trial two or more regional (Latvian, Lithuanian, Estonian) cultivars were included together with others from North America. Stand longevity was measured by yield in specific years against yield of the first full harvest year. Stand density was evaluated visually in percentage from that in the autumn of the establishment year.

Results In each trial yields differed significantly with well performing local regional cultivars. North American cultivars also showed highly productive persistency. The trials have shown which specific cultivars can be recommended for Latvian conditions (Table 1). The benefit of the local traditional three-cut regime with respect to production persistency was confirmed. However, some modern cultivars can be persistent and high yielding under a four-cut regime (trial 5) and other work (Gaile & Kopmanis, 2004) showed that forage quality was then improved. In agreement with other trials (e.g. Sheaffer *et al.*, 1988) high yields were obtained over a range of stand densities. In trial 5 a significant decrease of 25% on average was observed in 2004, mainly explained by sharp varying conditions in winter/spring 2003/2004 and the spread of diseases; cultivar differences were also substantial (p<0.01).

Table 1 Yields from the five trials (kg DM per single-row plot for 1st and 2nd trial; t DM/ha 3rd, 4th, 5th trial)

Trial No	Duration (years)	1st full harvest year		Last harvest year		Stand density[1], %
		Average	Range	Average	Range	
1	8	2.33	1.65 – 2.65	2.29	0.63 – 3.11	-
2	7	3.03	2.70 – 3.65	2.63	0.90 – 3.48	-
3	6	18.89	14.66 – 21.89	15.20	6.78 – 18.78	61.3
4	5	22.90	16.87 – 28.68	15.41	10.14 – 21.15	53.5
5	5	18.38	14.63 – 20.73	13.93	10.20 – 16.71	53.6

[1]Average stand density in last year relative to establishment year = 100 %

Conclusions The results of these studies identified lucerne cultivars most suitable for Latvian conditions. The traditional three-cut regime proved advantageous to long term stand production and persistence but a four-cut regime could improve forage quality in some cultivars. High yields were obtained in spite of variable stand densities.

References

Gaile Z. & J. Kopmanis (2004) Harvest management effect on productivity and forage quality of ten lucerne varieties. *Grassland Science in Europe*, 9, 924-926.
Scheaffer C.C., G.D. Lacefield & V.L. Marble (1988) Cutting Schedules and Stands. In: A.A. Hanson (ed) *Alfalfa and Alfalfa Improvement*. ASA, CSSA, SSSA, Madison, Wisconsin, USA, 411-437.

Improvement of lucerne cutting management: the relative impact of initial organic reserves, cutting height and residual leaf area on forage yield

J.C. Avice[1], F. Meuriot[1], M.L. Decau[1], A. Morvan-Bertrand[1], M.P. Prud'homme[1], F. Gastal[2], J-C. Simon[1], J.J. Volenec[3] and A. Ourry[1]

[1]UMR INRA / UCBN 950, EVA, Nutritions NCS, Université, 14032 Caen Cedex, France, Email: avice@ibfa.unicaen.fr, [2]UEPF INRA, Domaine du Chêne, 86600 Lusignan, France, [3]Department of Agronomy, 915 W. State St., Purdue University, West Lafayette, Indiana 47907-2054, USA

Keywords: forage production, cutting intensity, N reserves

Introduction Less lucerne (*Medicago sativa* L.) is now grown because of difficulties arising from 2 interacting characteristics: productivity and stand persistence. Optimisation of these two parameters depends highly of the cutting management (cutting height and/or frequency) and of the taproot N reserves. For example, Avice *et al.* (1997) showed that lucerne shoot regrowth is relates closely to taproot soluble protein concentrations (especially vegetative storage protein: VSP). However, it is not known how stubble C-N reserves and/or residual leaf area (both depending of the cutting management) influence the contribution of taproot reserve-derived C-N supply to regrowing lucerne shoots after defoliation. This study aimed to estimate the role of stubble C/N reserves or residual leaf area (RLA) on the contribution of taproot N reserves to shoot regrowth of lucerne after cutting.

Materials and methods After 75d of greenhouse culture, lucerne plants were cut and transferred in a growth cabinet and submitted to the following treatments: 2 cutting heights (6 or 15cm), 2 RLAs (completely defoliated: 0% or not defoliated: 100%), and 2 initial N reserve levels (High N: HN or Low N: LN). Meuriot *et al.* (2003) gave details for general conditions of culture and for obtaining plants with initial High N or Low N reserves. The effects of these treatments were followed on forage production and N reserve dynamics within stubble and taproots during 27d of post-cutting regrowth.

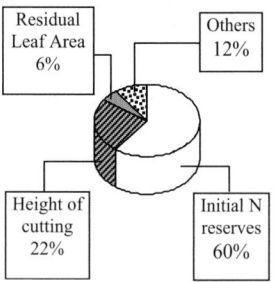

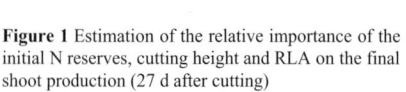

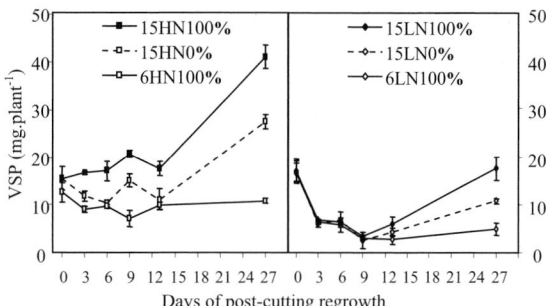

Figure 1 Estimation of the relative importance of the initial N reserves, cutting height and RLA on the final shoot production (27 d after cutting)

Figure 2 Changes in VSP contents of alfalfa taproot after harvest following two cutting heights (6, 15cm), two initial N reserves status (HN, LN) and two residual leaf area (RLA0%, RLA100%).Vertical bars indicate ± SEM for *n*=3 when larger than the symbol

Results On stepwise regression (Figure 1), initial taproot N status explained >60% of the variance of final shoot DM production. Cutting height and RLA explained about 22 and 6%, respectively, of the variance of final forage production (Figure 1). Increase of cutting height or RLA increased stubble C/N supply to regrowing shoots, which partly reduced the contribution of taproot C/N reserve to new shoots. Irrespective of initial N reserve, taproot VSP level at the end of regrowth was 4-fold more important in plants cut at 15cm than in plants cut at 6cm (Figure 2). Whatever the initial N reserve, the final amount of taproot VSP (Figure 2) decreased 33% in plants cut at 15cm and completely defoliated (RLA0%) compared to non-defoliated plants (RLA100%).

Conclusion Initial taproot N reserve affected lucerne forage production more than cutting height or RLA levels did but cutting height and RLA strongly affected the contribution of taproot N reserves to regrowing shoots and the level of re-accumulation of taproot N reserves. Therefore cutting height and RLA levels may have more impact on final shoot production at the following cutting / regrowth cycle. To improve lucerne winter survival and persistence and spring herbage regrowth, we should consider lucerne management strategies that increase cutting height (and RLA) in the penultimate harvest in autumn in cold regions with significant winter stress.

References
Avice, J. C., A. Ourry, G. Lemaire, J. J. Volenec & J. Boucaud (1997). Root protein and vegetative storage proteins are key organic nutrients for lucerne shoot regrowth. *Crop Science*, 37, 1187-1193.
Meuriot F., J. C. Avice, M. L. Decau, J. C. Simon & A. Ourry (2003). Accumulation of N reserves and vegetative storage protein (VSP) in taproots of non-nodulated lucerne (Medicago sativa L.) are affected by mineral N availability. *Plant Science*, 165, 709-718.

The effect of Sowing Date and Autumn Management on Sainfoin *(Onobrychis viciifolia)* Regrowth and Yield

Z.G. (Zhigang) Liu and G.P.F. Lane
The Royal Agricultural College, Cirencester, GL7 6JS, UK, Email: zhigang.liu@rac.ac.uk

Keywords: sowing date, autumn management, regrowth and yield

Introduction   Due to its characteristics (palatability, non-bloating, high protein, high voluntary intake etc.: Frame, 1998), sainfoin was a traditional forage legume in the UK, grown widely during the 17-19th century (Bland, 1971). It has almost disappeared in recent years. The rise of organic farming and the need for home-grown protein may encourage the return of sainfoin. This experiment aimed to explore the impact of sowing date and autumn management on the growth and yield of sainfoin in UK.

Materials and methods   A trial was established in 2003 at the Royal Agricultural College, Cirencester: 6 main treatments (sowing dates: Apr, May, Jun, Jul, Aug & Sep) and 2 sub treatments (autumn cut v non-cut) in a randomised block design with 3 replicates on plots of 2m× 4m. The variety was a landrace sainfoin, Cotswold Common. The seed rate was 90 kg hulled seed/ha. Seed was broadcast and raked into soil to circa 0.5-1.0cm deep and rolled. One harvest was obtained in Aug 2003 (data not shown) and a 2 more in May and Jul 2004; 30kg P_2O_5/ha was applied after harvest in 2003 and 30kg P_2O_5 + 20 kg K_2O/ha after harvest 1 and 40 kg K_2O/ha after harvest 2 in 2004. MCPA/MCPB was used to control broadleaved weed at first compound leaf at trifoliate stage in 2003 and Carbetamax was used to kill grass weeds and chickweed in Jan 2004. Autumn cuts were taken in Sep 2003.

Results   Table 1 shows the yields (t DM/ha) of harvests 1 and 2 in 2004. Sowing dates had significant effects on yield; sowing in Aug and Sep significantly reduced yields (P<0.001). Table 2 shows the yields (t DM/ha) of autumn cut v non cut treatments. There were no significant differences.

Table1 Mean yield of harvests 1 and 2 in 2004

Treatment	Harvest 1	Harvest 2	Total*
Apr	9.15	2.62	11.76 a
May	9.68	2.82	12.50 a
Jun	8.53	2.67	11.20 a
Jul	8.77	2.72	11.49 a
Aug	3.42	2.21	5.50 b
Sep	2.23	1.78	4.01 b

Table 2 Mean yield of Autumn cut v non-cut

Treatment	Non-cut*	Cut*
April	11.76a	12.46a
May	12.50a	12.10a
June	11.20a	11.09a
July	11.49a	11.76a

* Means with the same letter in the same row or column are not significantly different

Conclusions   One year's results show that sowing at any time between April and July gave similar yields in the following season. Sowing August and September greatly reduced forage yields in the following season. Autumn (September) cutting in the establishment year had little effect on yields in the following season.

References
Bland, B.F. (1971). Crop Production: Cereals and Legumes. Academic Press: London & New York
Frame, J. (1998). Temperate Forage Legumes. CAB International.

Rangelands in the Mediterranean zone of Croatia

J. Rogosic

University of Split, Department of Natural Resources, Split, Livanjska5, 21000 Split, Croatia, Email jozo@oss.unist.hr

Keywords: Croatian rangelands, vegetation, range condition, productivity, grazing

Introduction Rangelands dominate the landscape of the Mediterranean part of Croatia along the Adriatic coastline, occupying 83% of the agricultural land (1.7 million ha) and 40% of the entire country. The proportion of rangelands is considerably higher in the Mediterranean littoral than in other ecological regions of Croatia. Sheep and goats are widespread in the region, comprising 76% of total sheep and goat numbers in Croatia. However, in comparison to other regions of Croatia, livestock production in the Mediterranean zone is not well developed. Continuous grazing begun too early in the growing season has caused substantial rangeland degradation and a decline from potential productivity. The improvement strategy is to rely on controlled grazing systems, in which priority is given to restricting grazing pressure in the early spring, improving animal distribution and introducing rotational grazing practices. The more difficult task is to regulate animal numbers. Although the general climate of southern Croatia is Mediterranean, there is a gradient inland from the coast. Dry summer stress, combined with a long history of man's influence on the natural vegetation, has resulted in the formation of several contrasting rangeland types. The main types are pasture vegetation, shrublands (maquis and garrigues) and forested ranges (Horvatic, 1975).

Pasture vegetation All kinds of environments are found on the Croatian littoral from Adriatic islands to the mountain ranges running parallel to the coast. The vegetation is in the form of dry Mediterranean grasslands, which are used for grazing and cutting, and in herbaceous vegetation in the interstices of rocky ground covering extensive areas. The pasture vegetation includes two vegetation classes - *Thero-Brachypodietea* and *Festuco-Brometea*. Mediterranean rocky ground pasture and grasslands represent about 45% of the total rangelands in the region, excluding arable land (Table 1).

Mediterranean shrublands The shrublands of the Croatian littoral are at low elevations. They contribute 31% of the rangelands and include two types of vegetation. The first is Mediterranean evergreen maquis and deciduous thickets, which grow several meters high in dense almost impenetrable stands. The other type is Mediterranean evergreen garrigues (chaparral), which is relatively low vegetation of mostly thinned thickets dominated by dwarf shrubs. Mediterranean forest and woodland ranges occupy about 24% of the total area of rangelands (Table 1).

Table 1 Rangelands in Mediterranean Croatia

Class of natural resources	Area (ha)	% of total area
1. Pastures and grasslands	775,000	38.37
2. Maquis and garrigues	534,000	26.44
3. Mediterranean forest	422,000	20.89
4. Arable lands	289,000	14.30
Total	2,020,000	100.00

Range condition and productivity Because of centuries of poor grazing management rangelands have seriously deteriorated in the Mediterranean Croatia (Rogosic, 1995). This degradation has reduced forage production from rangelands and increased the variability from one vegetation type to another and from one environment to the next. On average, the grazing capacity is estimated at about 2 animal unit months (AUM)/ha in the grasslands, 1 AUM/ha in the shrublands, and 0.5 AUM/ha in the forested ranges. (An animal-unit month is the amount of dry forage required by one animal unit (one cow, five sheep or five goats) for one month based on a forage allowance of 11.8 kg/d). The grazing capacity can be increased considerably, especially in grasslands and shrublands, through improvements such as fertilisation, reseeding of improved range species, or conversion of shrublands to grasslands by prescribed burning and seeding.

Priorities in research The political and social climate is ripe for changes in rangelands and livestock management to enhance the productivity of the precious natural resources of Croatia.

References

Horvatic, S. (1975). Neuer Beitrag zur Kenntnis der Syntaxonomie der Trocken-Rasen und Steintriften-Gesellschaften der ostadriatischen Karstegebuetes. In: Problems of Balkan Flora and Vegetation, *Acta Botanica Croatica,* 2,300-310.

Radinovic, S. & D. Supe (1994). Family estate of Dalmatia users of credits for the development of cattle Breeding. *Stocarstvo,* 48, 9-17.

Rogosic J. (1995). Determination of the forage values of the Mediterranean garrigues and maquis communities and their utilization. Doctoral Dissertation, Faculty of Agronomy, University of Zagreb.

Plant arrangement effects on dry matter production and nitrogen fixation of berseem clover: annual ryegrass mixture

C. Attardo, G. Di Miceli, A.S. Frenda, D. Giambalvo and C. Scarpello
Dipartimento di Agronomia Ambientale e Territoriale, Università di Palermo, 90128 Palermo, Italy, Email: giardo@unipa.it

Keywords: berseem - ryegrass mixture, plant arrangement, nitrogen fixation, nitrogen content

Introduction Agronomic factors affect the productivity and efficiency of cereal-legume intercropping systems (Ofori & Stern, 1987). This research aimed to determine the effects of different plant arrangement on hay yield, nitrogen (N) content and N fixation of berseem clover-annual ryegrass mixture in a Mediterranean semi-arid environment.

Material and methods The site was a hilly area of Sicily (37°30'N; 13°31'E; 178m a.s.l.) in a deep, clayey, well structured soil, previously cropped for wheat. The experiment in 2002/03 was a complete randomized block design with 4 replicate. Treatments were: berseem clover (*Trifolium alexandrinum* L. cv Lilibeo) in pure stand (B); annual ryegrass (*Lolium multiflorum* Lam subsp. *wersterwoldicum* cv Elunaria) in pure stand (R); and their mixture arranged in alternate rows (BRa), or in the same row (BRs). Pure stands were seeded at 50kg seeds/ha. The intercrop design was based on replacement principle sown with in 50:50 berseem:annual ryegrass. The ^{15}N isotope dilution technique was used to estimate N fixation by berseem clover (10 kg N/ha as ammonium sulphate at 10 at.%^{15}N excess). All plots were cut at the start of berseem flowering; total above-ground dry matter yield (DM) and N concentration (as % DM) were determined for each species.

Results and Conclusions The pure berseem clover stand and its mixtures had similar DM yields. DM yields of both mixtures were significantly higher than annual ryegrass in pure stand (Figure 1). The arrangement of plants in the mixture (BRs vs. BRa) had no significant effect on DM yield but the contribution of berseem was significantly higher in BRs (75.4%) than in BRa (63.5%).

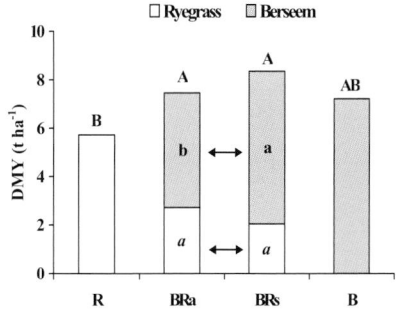

Figure 1 DM yield of berseem and ryegrass in pure stands (B and R) and in mixture with different spatial arrangement (BRa alternate rows; BRs same row)

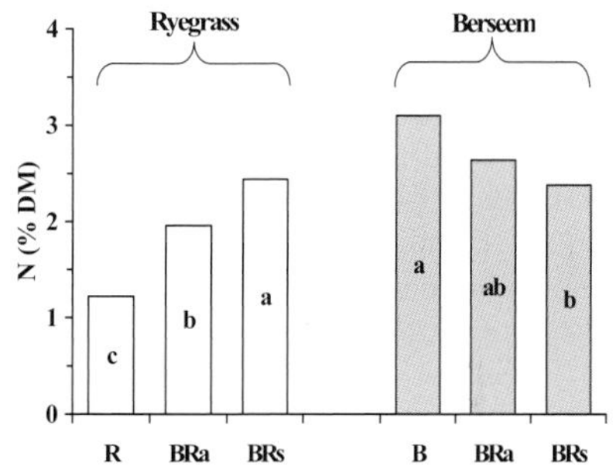

Figure 2 N concentration (%DM) of the harvested berseem and ryegrass in pure stand and in mixture (BRa alternate rows; BRs same row)

Intercropped berseem always had a significantly higher percentage of N derived from N_2 fixation (Ndfa), than monocropped berseem clover (90.1, 87.4 and 77.4% Ndfa respectively in BRs, BRa and B). Annual ryegrass intercropped with berseem had a consistently higher N% than ryegrass grown in pure stand (Figure 2). Moreover, mixing the components within the row, compared to sowing in alternate rows, significantly increased the N% in annual ryegrass. In contrast, the N% of berseem was much higher in pure stand than in mixtures, particularly when it was intercropped in the same row with annual ryegrass. Plant arrangement in the same row can improve the efficiency of berseem clover-annual ryegrass intercrop system.

Acknowledgements The research was funded by Patti Territoriali "Magazzolo-Platani"

References
Ofori, F., W. R. Stern (1987). Cereal-legume intercropping systems. *Advances in Agronomy,* 41, 41-90

Long-term (9-year) response of two semiarid grasslands to prescribed fire in the southwestern USA

R.L. Pendleton[1], B.K. Pendleton[1] and C.S. White[2]
[1]USDA Forest Service, Rocky Mountain Research Station, 333 Broadway SE, Albuquerque New Mexico 87102, USA, Email: rpendleton@fs.fed.us, [2]Department of Biology, University of New Mexico, Albuquerque, New Mexico 87131, USA

Keywords: arid grasslands, fire, soil fertility, drought

Introduction Historically, arid grasslands of SW USA experienced fire return intervals of 5-10 years. During the last 100 years, however, fire has been a rare event. Recent expansion of woody plants in arid grasslands has prompted managers to re-introduce fire as a tool to reduce abundance of woody plants and maintain perennial grass cover. The use of fire in desert grasslands poses unique challenges, however, due to extreme variability in rainfall patterns. Our research examines vegetation response to repeat fire in 2 desert grassland ecotones near Albuquerque, New Mexico (35.05°N 106.60°W).

Materials and methods Pre- and post-fire vegetation cover and soil fertility data were taken annually on 3 permanent transects within each of 8 X 1-ha plots located at the Bernalillo and West Mesa watersheds. In Nov 1995 and again in Jan 1998, 4 plots at the Bernalillo site were burned, 4 plots at West Mesa were burned once in February 1996. A wildfire burned all West Mesa plots in July 2001. Potentially mineralisable nitrogen (PMN) of surface soils collected under shrubs and grasses was determined by the sum of soil inorganic N following moist incubation (White & Loftin, 2000). Plots were inventoried for *Juniperus monosperma* (whole plots) and *Opuntia* spp. (belt transects) in February 2002.

Results Compared with unburned plots, burned plots at Bernalillo had fewer numbers of juniper (p<0.0005) and *Opuntia imbricata* (p<0.001), and smaller size patches of *O. phaeacantha* (p<0.01). Surviving junipers on burned plots were larger (p=0.0463), indicating that fire killed smaller trees. Burning reduced cover of woody shrubs and subshrubs for 2-3 years after fire at both Bernalillo and West Mesa sites (Figure 1). Perennial grasses recovered within 2 years given above-average precipitation (1996-98), but required 5-6 years to recover during a period of drought (1999-2003). After the wildfire of 2001, perennial grass cover was higher on previously-burned plots (p=0.0286), indicating that subsequent fires may have less detrimental effects than the initial reintroduction. Cover of fire-susceptible grasses was also higher on previously-burned plots. Soil fertility (PMN) under grass on twice-burned plots at both locations was higher than under shrubs for a brief window of 1-2 years after the second fire (significant cover type x collection date interaction; p≤0.0245).

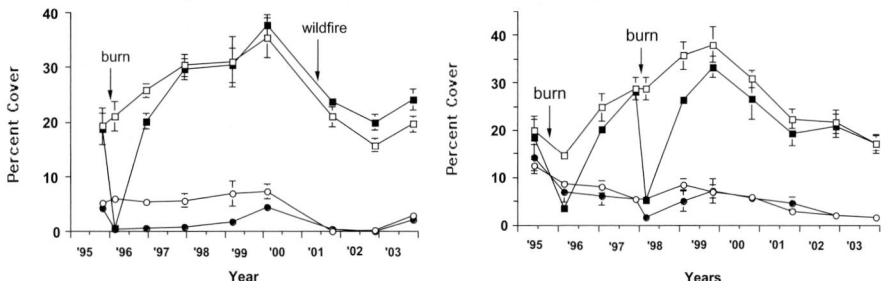

Figure 1 Perennial grass (•) and shrub (•) cover for West Mesa (left) and Bernalillo (right) sites. Filled symbols indicate prescribed burn treatment

Conclusions Fire successfully reduces woody plant and succulent cover and maintains healthy stands of native grasses. Recovery time depends on pre- and post-fire rainfall patterns that are very variable in desert grasslands. Repeat fire events may do less harm to grass regeneration than the initial fire reintroduction. This is especially important for fire-susceptible grasses such as *Bouteloua eriopoda*, a signature species of Chihuahuan desert. The well-documented pattern of greater soil fertility under shrubs may represent communities removed from the influence of fire. Repeat fire may increase soil fertility and stimulate below- and above-ground grass production.

References

White, C. S. & S. R. Loftin (2000). Response of 2 semiarid grasslands to cool-season prescribed fire. *Journal of Range Management*, 53, 52-61.

Studies on rehabilitation and recovery of degraded shrub rangeland in central China

G.. Xiao

P.O Box57, Zheng Zhou College of Animal Husbandry, P. R China , 450011, Email: xiao_g99@hotmail.com

Keywords: rehabilitation, recovery, rangeland

Introduction Central China used to be the most important region for animal production, but over-grazing, over-cultivation and deforestation all resulted in a decline in shrub rangeland productivity as well as increasing soil degradation (JiZhou, 1989). Soil erosion, salinisation and desertification have become serious problems. The productivity of shrub rangeland decreased from 3,000~4,000kg/ha to 1,500~2,000kg/ha from the 1960s to the 1990s (Wei, 1991). Since 1990, measures have been taken to solve this problem with promising results.

Approaches *1. The combination of indigenous forage with exotic forage, based on ecological principles.* Different indigenous and exotic forages were chosen according to different climate and soil conditions, especially rainfall, evaporation capacity, light pattern and soil texture (Table 1). *2. The effective combination of agricultural cropping system with rangeland conservation.* Farming systems with combinations of cereal crop-economic plant-forage plants were established. These can increase the multiple crop index and through rational rotation produce more and better forage for animal production and promote soil rehabilitation. *3. The development of save-grain agriculture to avoid rangeland reclamation.* Swine and poultry used to be the dominant domestic animals, but now herbivorous animals are developing steadily. They not only consume less grain than swine and poultry, but also can utilise agricultural by-products such as stalks and straw.

Table 1 The ecological combination of indigenous forage with exotic forage

Sites	Indigenous forage			Exotic forage	
1	*Festuca arundinacea*	*Medicago sativa*	*Artemisia argyi*	*Coronilla varia*	*Poa annua*
	Lemperata cylindrica	*Lespedeza florribunda*		*Puccinellia chinampoensis*	
	Amorpha fruticosa				
2	*Albizia julibrissin*	*Bothriochloa ischaemum*		*Trifolium repens*	*Lolium perenne*
	Lespedeza bicolor	*Cleistogenes aquarrosa*		*Sorghum sudanense*	
3	*Pennisetum flaccidum*	*Astragalus adsurgens*		*Festuca arundiance*	*Onobrychis viciaefolia*
	Lemperata cylindrica	*Themeda triandra*		*Medicago sativa*	
4	*Bothriochloa ischaemum*	*Carex lanceolata*		*Trifolium pratense*	*Dactylis glomerate*
	Albizia ulibrissin	*Vicia kioshanica*		*Trifolium hybridum*	
5	*Medicago falcata*	*Lemperata cylindrica*		*Astragalus*	*Agropyroh cristatum*
	Calamagrostis epigejos	*Setaria viridis*		*Coronilla varia*	

Sites 1 to 5 are different habitats with different indigenous species

Results The production of crop combination systems has increased compared with traditional farming systems. Grain output and by-product yield (including forage plants) increased by 12.5 % and 46 % respectively. The level of protein from straw and forage plants increased by 25-30 %. Crop combination systems reduced the utilisation of rangelands. Central China produces about 1.3×10^8 t agricultural by-products (including straw and vine etc.).That is more than 20 times the quantity of forage produced on local rangeland. When ammoniated straw and stalk silage were substituted for roughage, and processed peanut or cottonseed cake and bran were fed instead of grain, production cost was much reduced. Since 1992, over 3×10^6 t of grain was saved for human consumption each year. This relieved pressure on arable land and also avoided rangeland degradation. Since 1990, with the development of herbivorous animal production, manure production increased by 50-65% in Henan Province. Many feedlots adopted the gravity drain pit system to collect manure. Application of organic fertiliser in agriculture has increased 15% and degraded land area has decreased by 5.6 %.

Disscussion The development of grain -save husbandry, along with ecological combinations of forages, recycling of manures and good husbandry, will reduce pressure on shrubland and let degraded land lie fallow and recover..

References

Ji Zhou (1989). The recovery and construction of natural grassland. *Journal Henan Animal Husbandry*, 16, 18-21.

Wei, M. (1991). Studies on wild forage resources in Henan Province. *China Grassland*, 69, 7-11.

Reasons for the premature decline in *Astragalus adsurgens* stands in Kerqin sandy land

Q.Zh. Sun[1], Z.L. Wang[1], J.G. Han[2], Y.W. Wang[2] and G.R. Liu[3]

[1]*Grassland Research Institute, Chinese Academy of Agricultural Sciences, Huhhot, 010010, China, Email: Sunqz@126.com,* [2]*Department of Grassland Science, College of Animal Science and Technology, China Agricultural University, Beijing, 100094,* [3]*Grassland Service Station of Chifeng City, Chifeng, 024000, China*

Keywords: *Astragalus adsurgens* cv. Shadawang, degradation, phosphorus, root rot disease

Introduction Diseases partly account for reductions in *Astragalus adsurgens* stand longevity. The effect of some cultural practices on the control of pests and diseases have been reported (Hou, 1986; Nan, 1996), but few reports have detailed the relationship among soil fertiliser status, diseases and premature stand decline. This study was conducted to investigate these relationships in order to extend the longevity of *Astragalus adsurgens* stands.

Materials and methods *Astragalus adsurgens* for this research was established in 1993 at 6-7kg/ha. No irrigation or fertilisers were applied after establishment. Disease incidence was calculated as

$$\text{Disease incidence} = \frac{\text{The number of infected plants} \times 100\%}{\text{The total number of investigated plants}}$$

Plants infected index was divided into 5 classes based on symptoms according to Flood & Isaac (1978) and Latunde-dada & Lucas (1982). Disease index of the stands was calculated according Liu Ruo (1998) as

$$\text{Disease index} = \frac{\sum(\text{Plant infected index} \times \text{plant number}) \times 100\%}{5 \times \text{total number of plants}}$$

Results In the third and fifth year after establishment, phosphorus content in soil, plant leave and stem significantly decreased compared to the first year (Table 1); however, root rot diseases and plant mortality increased. The pathogens accounting for the root rot disease were *Fusarium solani, Fusarium oxysporum* and *Fusarium moniliforme*. The stand showed premature degradation in the 5th to 6th year after establishment.

Table 1 The changes of phosphorus in soil and in *Astragalus adsurgens* plants

Years after establishment	Phosphorus content (P_2O_5 mg/ 100g)			Stand condition (%)	
	Soil	Leaf	Stem	Plant mortality	Disease index
0	1.86 a	—	—	—	—
1	1.54 a	0.5401 a	0.0710 a	66.3	54.8
2	—	—	—	78.9	55.6
3	0.54 b	0.1407 b	0.0291 b	100.0	77.7
4	—	—	—	100.0	86.2
5	0.11 c	0.1066 c	0.0233 c	100.0	92.6
6	—	—	—	100.0	95.0

Note: data in a column followed by different letters are significantly different ($p < 0.05$)

Conclusion Premature degradation of *Astragalus adsurgens* were attributed to phosphorus deficiency in soil and root rot diseases caused by the fungi *Fusarium solani, F. oxysporum*, and *F. moniliforme*.

References

Flood J.& I. Issac (1978). Reaction of some cultivars of lucerne to various isolates of *Verticillium albo-atrum. Plant Pholtology,* 27,166-169.

Hou, T. J. (1986). Diseases of *Astragalus adsurgens* in western Liauning province and Inner Mongolia, *Grasslands of China,* 3, 40-43.

Latunde-dada, A.O. & J.A. Lacas (1982). Variation in resistance to *Verticillium* wilt within seeding populations of some varieties of lucerne. *Plant Patholology,* 31,179-186.

Liu, R. (1998). Science of Grassland Protection. Agriculture Press of China. 259-263.

Nan, Zh. B. (1997) *Astragalus adsurgens* disease and their distribution charactors. *Acta Prataculturae Sinica,* 14, 30-34.

Improvement of the Mediterranean agro-silvopastoral system Montado

J. Potes, H. Babo and D. Navas

Estação Nacional de Melhoramento de Plantas (INIAP), Apartado 6, 7350-951 Elvas, Portugal, Email: jmirapotes@hotmail.com

Keywords: botanical composition, grazing system, pasture improvement, fertilisation, reseeding

Introduction "Montado" is an old man-made agro-silvopastoral system, with three vegetation components (trees, shrubs and herbs) used for multiple species animal husbandry in extensive production systems (Potes & Babo, 2003). Several authors have reported on the improvement of Mediterranean annual self-reseeding pastures (Espejo-Diaz, 1996) and increased stocking rates.

Methodology A randomised block experiment carried out with four replicates and four treatments: T1-maintenance of the existing ecosystem; T2-liming (2000kg/ha) and P fertilisation (46 kg of P_2O_5/ha); T3-T2 plus introduction of annual self reseeding legumes without soil disturbance (direct seeding); T4-T2 plus introduction of annual self reseeding legumes with soil disturbance. Grazing was carried out in autumn (O), winter (I) and spring (P): only by cattle (PB), only by goats (PC), only by sheep (PO) or by the three animal species together (PM) at the same grazing pressure of 1780 kg/ha liveweight. The trial lasted three years. The frequency of grasses, legumes and 'other species', organic matter (OM), crude protein (CP) and digestibility of OM (DOM) were determined.

Results The pastures consisted for 70% to 80% of grasses. The frequency of legumes could be increased up to 50% with liming and P fertilisation (Figure 1). However, the introduction of legume seeds could also increase legumes to 37%.

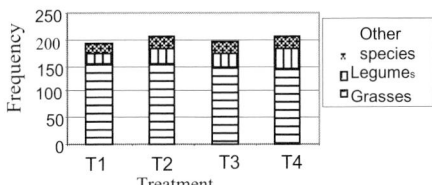

Figure 1 Frequency of grasses legumes other species for treatment

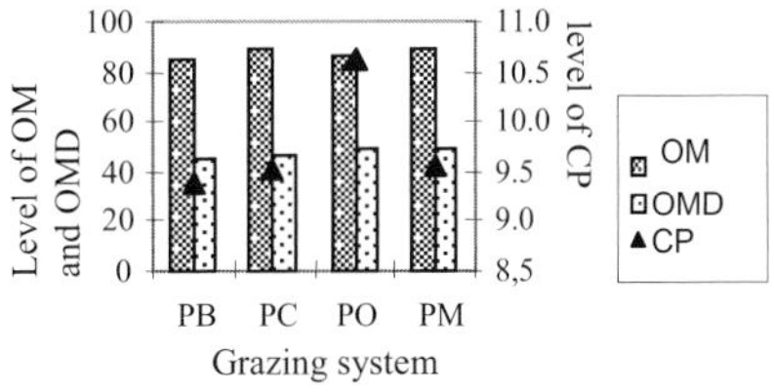

Figure 2 The influence of grazing system on the organic matter (OM), organic matter digestibility (OMD) and crude protein (CP)

The most effective control of other plant species was obtained by goat grazing (PB 22, PO 30 and PC 5%). Sheep grazing resulted in significantly higher crude protein levels (Figure 2).

Conclusions This experiment confirmed the value of fertilisers (liming and P), plant introduction and grazing management for production system improvement (Lazenby & Tow, 2001).

References

Espejo-Diaz, M. (1996): "Evaluation and improvement of efficacy of animal production systems using natural resources by grazing ruminants in Mediterranean areas: 1st results". In: "The optimal exploitation of marginal Mediterranean areas by extensive ruminant production systems" EAAP Publication No. 83, pp. 105-112.

Lazenby, A. & P. G. Tow (2001): "Some Concluding Comments". In: "Competition and Succession in Pastures" Edited by PJ Tow and A. Lazenby. CABI Publishing Wallingford: pp. 305-314

Potes, J. M & H. Babo, (2003) "Montado, an old system in the new millennium". Proceedings of the VII International Rangeland Congress, *African Journal of Range & Forage Science,* 20, 141.

Tiller dynamic of guineagrass (*Panicum maximum*) under defoliation

T. Clavero and R. Razz
Centro de Transferencia de Tecnología en Pastos y Forrajes. Facultad de Agronomía. Universidad del Zulia. Apdo 15098. Maracaibo 4005. Venezuela, Email: tclavero@hotmail.com

Keywords: defoliation, tiller, guineagrass

Introduction Guineagrass (*Panicum maximum*) is a perennial warm season bunchgrass native to Africa that has been introduced to many tropical areas. Its management is important because of its important role in animal production. Total production of herbage and the persistence of tufted grasses can be markedly reduced when the grass is defoliated too frequently and too intensively (Clavero, 1997). In order to determine the proper management for optimum production and long term persistence, it is important to study the effect of cutting practices on tiller dynamics of guineagrass.

Materials and methods Plants were grown from seed in a sandy clay soil mixture in 50 kg plastic pots. Irrigation and weed control were provided in order to maintain active growth. After establishment, a uniformity clipping was performed and treatments were initiated using the factorial combinations of four clipping frequencies (14, 28, 42 and 56 days) and three heights (20, 40 and 60 cm). Treatments were arranged in a split-plot in a randomised block with three replications. Measurements included basal, aerial and dead tillers. Number and type of tillers per plant were recorded at the time of harvest. The initial number of main tillers of each plant was recorded and used as a covariate in subsequent statistical analysis.

Results and discussion The number of basal tillers increased as cutting interval was reduced (Table 1). The number of basal tillers varied from 48 (14 days) to 26 (56 days) which resulted in a decrease of approximately 45.8% of basal tillers. However, tillers were visibly shorter and less vigorous with more frequent cutting.

Table 1 Basal, aerial and dead tillers of guineagrass

Tillers	Frequency of harvest (days)											
	14			28			42			56		
	Cutting height (cm)											
	20	40	60	20	40	60	20	40	60	20	40	60
Basal	43[a]	48[a]	46[a]	32[b]	33[b]	29[c]	36[b]	35[b]	38[b]	29[c]	28[c]	26[c]
Aerial	12[d]	19[c]	21[b]	14[d]	24[b]	23[b]	19[c]	29[a]	28[a]	17[c]	27[a]	23[b]
Dead	38[a]	34[a]	27[b]	33[a]	27[b]	26[b]	29[b]	17[c]	16[c]	20[c]	18[c]	16[c]

Values in the same row with different superscripts are different (P<0.05)

Frequency of harvest had less influence on aerial tiller numbers than cutting height. Number of tillers per plant declined as defoliation height decreased. When defoliation increased from 20 to 40 cm aerial tiller numbers increased by approximately 60%. The number of dead tillers per plant increased as the frequency of harvest was shortened and intensity of defoliation was increased. The number of dead tillers approximately halved as the cutting height was increased from 20 cm to 40 and further decreased at the cutting height of 60 cm. At low cutting heights the apical meristem of a tiller is removed, so that it is no longer capable of initiating new leaves unless on axillary bud at or above the bottom node initiates growth (Carvalho *et al.*, 2001). There were more, taller and vigorous tillers of plants defoliated at longer interval and at greater cutting height, which must be due to the depletion of carbohydrate reserves. Also under these conditions, the plants failed to achieve high levels of light interception before subsequent defoliation. In this trial, the more severe treatments probably caused morphological changes at plant level, which affected the tiller population density and herbage biomass. Frequent and intensive defoliation must be avoided in order for guinea grass to persist. However, this has negative consequences for herbage quality.

References

Clavero, T. (1997). Tiller dynamics of dwarf elephantgrass (*Pennisetum purpureum* cv Mott) under defoliation. *Proceedings of the XVII International Grassland Congress*, Winninpeg and Saskatoon, Canada, June 1997, paper 345.
Carvalho, D., C. Matthew & J. Hodgson (2001). Effect of aging in tillers of *Panicum maximum* on leaf elongation rate. *Proceedings of the XIX International Grassland Congress*, S•o Pedro, S•o Paulo, Brazil, pp. 41-42.

Effect of plant population and phosphorus fertilizer application on dry matter and seed yield of two lablab (*Lablab purpureus*) varieties in Botswana

K.C. Kawonga[1], F.P. Wandera[2], S. Mangope[1], P. Mutshewa[3] and B.M. Lefofe[1]
[1]DAR, Box 10 Mahalapye, Botswana, Email: mkawonga@yahoo.com, [2]DAR, P/B 0033, Gaborone, Botswana
[3]DAPH, Box 50, Mahalapye, Botswana

Keywords: lablab, dry matter, stem size

Introduction Lablab has high potential as a protein source to grazing livestock especially during the dry season in arid Botswana. It produced 8.5t dry matter (DM)/ha and had 14% crude protein with 60% digestibility (APRU, 1988; Aganga, 2003). Lack of agronomic data on lablab production was probably the reason why some farmers in Botswana got yields as low as 300 kg/ha (APRU, 1987). Therefore, a trial of two lablab varieties was conducted to determine the effect of plant population and phosphorus (P) on DM yield.

Method Over two seasons from 2002-2004, a trial was conducted at Morale (23° 34'S and 26° 50'E; Haplic Acrisol), Maun (19° 56'S 25⁰ 40'E; Gleyic Luvisol) and Pandamatenga (18° 32'S 23⁰ 29'E; Utric Vertisol). A 2x3x3 factorial experiment, with treatments arranged in a randomised complete block design with 2 replicates each, tested 2 lablab varieties (Highworth and Rongai) x 3 plant populations (53333[S1], 26667[S2] and 17778[S3] plants/ha) x 3 P fertiliser rates (0, 40 and 80 kg P/ha). Fertiliser was placed in pre-marked planting stations and was mixed with soil before placing the seed. Two seeds of lablab were planted at 4cm deep. The seedlings were thinned to 1 plant/station at 14 days after emergence. DM yield was determined gravimetrically at 50% flowering stage after samples were dried at 80°C for 4 days. Using a Vernier calipers, stem diameter of 5 representative plants/plot was measured at a 20 mm height above ground. A SAS statistical package analysed the data after log transformation.

Results Rainfall at Panda, Maun and Morale was 512.4, 332.1 and 326.7mm, respectively, which gave yields (1.9-4.8 t DM/ha) that were within the range observed in semi-arid areas (Murthy & Colucci, 1999). High plant population (S1) tended to produce more DM and plants with significantly (P• 0.05) smaller stem size than S2 and S3 (Table 1). When plant population was reduced, plants produced more shoot branches (unpublished data) and developed bigger stems, thereby compensating for yield losses. Rongai produced bigger plants and ultimately more DM than Highworth (Table 1). Phosphorus fertilizer did not (P• 0.05) influence lablab DM yield but slightly increased stem size. Only Highworth set seed and produced 561 and 192 kg/ha at Maun and Morale, respectively.

Table 1 Effect of plant population and fertiliser P application on lablab performance in Botswana

Treatment	Yield (t DM/ha)				Stem size (mm Ø)			
Population (Plants/ha)	Maun	Morale	Panda	Mean	Maun	Morale	Panda	Mean
53333	1.49a	1.02a	1.72a	1.41a	2.55b	2.33b	2.36b	2.41b
26667	1.35a	0.93a	1.63ab	1.30b	2.68a	2.49a	2.51a	2.56a
17777	1.39a	0.86a	1.52b	1.25b	2.71a	2.54a	2.52a	2.59a
P (kg/ha)								
0	1.45a	0.97a	1.59a	1.34a	2.57b	2.46a	2.44a	2.49b
40	1.35a	0.95a	1.68a	1.33a	2.63ba	2.44a	2.46a	2.51ab
80	1.43a	0.89a	1.61a	1.31a	2.73a	2.47a	2.48a	2.55a
Variety								
Rongai	1.53a	0.97a	1.64a	1.38a	2.69a	2.53a	2.53a	2.58a
Highworth	1.28b	0.90a	1.61a	1.27b	2.59a	2.38b	2.40b	2.46b

Conclusion Plant populations of 26667-53333 plants/ha maximized lablab DM yield. A Higher plant population (S1) produced plants of smaller stem thickness. Rongai tended to be more productive than Highworth, however only the latter that set seed under these environment.

References

Aganga, A. A. (2003). Chemical Composition of *Lablab purpureus* at different stages of growth and its use in feeding growing Tswana goats. *A.A. Bull. Animal Health Prod. Afr.* **51**, 23-29.

Animal Production Research Unit (APRU, 1987). Livestock and Range Research in Botswana. APRU Annual Report. MoA. Gaborone. p16-22.

Animal Production Research Unit (APRU) 1988. Livestock and Range Research in Botswana. APRU Annual Report. MoA. Gaborone. p19.

Murthy, A. M. & P.E. Colucci (1999). A tropical forage solution to poor quality ruminant diets: A review of *Lablab purpureus. http://www.cipav.org.co/lrrd/lrrd11/2/colu.htm.*

Forage yield and structural traits of Tanzaniagrass (*Panicum maximum*) at four canopy heights

C.A.M. Gomide, A.C. Ruggieri, R.A. Reis, J.A. Gomide, J.H.A. Rangel and E.O. Almeida
EMBRAPA-CNPGL, Av. Beira Mar 3250, Aracaju-SE 49025-040 Brazil, Email: cagomide@cpatc.embrapa.br

Keywords: yield, herbage mass, leaf-stem ratio, tiller density

Introduction Pasture forage production is based on the growth of tillers (Hodgson, 1990). Although the effect of canopy height and structural traits on productivity of temperate grasses are well known e.g. (Bircham & Hodgson, 1983; Binnie & Chestnut, 1994) tiller studies on tropical pasture species are scarce.

Material and methods A trial was carried out in 3x3m plots to evaluate the effects of height on total forage biomass, forage yield and structural traits of the canopy of *Panicum maximum* cv. Tanzania. Canopy heights of 20, 35, 50 and 65cm were studied in a completely randomised design, with three replications. Evaluations were performed in summer from Dec. 2003 to March 2004. Weekly cuts were taken to maintain experimental heights, and 0.5x0.5m samples were collected for yield estimates. Harvests were also taken fortnightly at ground level to assess total herbage biomass, its fraction components and tiller number.

Results Vegetation structure was affected by canopy height (Table 1). Dead material increased with plant height. Forage yield in Dec-Jan was 50% more than in Feb-March (Table 2), with significant differences for the 20 and 35 cm heights only. Surprisingly, the leaf/stem ratio was not affected by canopy height, but it was by summer period (P<0.05), with lower values being observed toward the end of summer. Tiller population density declined and tiller weight increased with canopy height, according to the tiller size/weight compensation law. Heavier tillers were observed in the first period as compared to the second (Table 2). Forage biomass increased with canopy height quadratically and linearly ($r^2 = 0.97$) in the first and second period, respectively.

Table 1 Effect of canopy height and summer period on structural traits of cv. Tanzania

	Height (cm)				Period	
	20	35	50	65	Dec-Jan	Feb-Mar
Dead Material (kg/ha)	867 b	1029 ab	1139 a	1085 ab	933	1127
Leaf/Stem ratio	2.51	2.20	2.18	2.35	2.68 a	1.94 b
Tiller / m^2	441 a	412 ab	408 ab	332 b	400	396
Tiller weight (g)	0.72 d	1.05 c	1.35 b	1.66 a	1.35 a	1.09 b

Means followed by different letters differ by Tukey test (p<0.05)

Table 2 Mean and total herbage mass in cv. Tanzania according to canopy height and summer period

	Height (cm)				
	20	35	50	65	
Period	Mean fortnightly herbage mass (kg DM / ha/week)				Average
Dec-Jan	1,164 Aa	1,003 Aa	940 Aa	576 Ab	920.7
Feb-Mar	553 Ba	564 Ba	713 Aa	652 Aa	620.5
Average	858.5	783.5	826.5	614	
	Total Herbage Mass (kg DM/ha)				
Dec-Jan	3,579 Ac	4,595 Abc	7,106 Aa	5,546 Ab	5,206
Feb-Mar	2,634 Bc	3,993 Ab	4,455 Ba	5,091 Aa	4,043
Average	3,107	4,294	5,780	5,318	

Means followed by different capital letters in columns and lower case letters in rows, differ by Tukey (p<0.05)

Conclusions Cv. Tanzania should be managed to maintain a canopy height range from 20 to 50cm for maximum DM yields

References
Binnie, R.C.& D.M.B. Chestnutt, (1994). Effect of continuous stocking by sheep at four heights on herbage mass, herbage quality abd tissue turnover on grass/clover and N- fertilized grass sward. *Grass and Forage Science*, 49, 192-202.
Birchan J.S. & J. Hodgson, (1983). The influence of sward condition on rates of herbage growth and senescence in mixed sward under continuous stocking management. *Grass and Forage Science*, 38, 323-331.
Hodgson, J. (1990). Grazing Management: Science to Practice. England: Longman Group Ltd. 203p.

Forage yield in two tropical grasses at different cutting intervals and N levels

E. Cortes-Díaz[1], H. Diaz-Solis[2], A. Saldívar-Fitzmaurice[2], W. Grant[2], A. Martínez-Garza[2] and A. Martínez-Hernandez[1]

[1]C.U.R. Anahuac, Centros Regionales, Universidad A. Chapingo, km 38.5 carr. México-Texcoco, México 56230, Email:ecodia@yahoo.com.mx, [2]Universidad A. de Tamaulipas, Ciudad Victoria, Tamaulipas 34871, Mexico

Keywords: tifton68, Coastcross-1, nitrogen fertilisation, cutting

Introduction Cutting interval and N level determine forage yield in grasses (Whitehead, 1995). Coastcross-1 (*Cynodon dactylon* and Tifton 68 (*Cynodon spp*) are tropical grasses of high forage yield potential (Burton, 1972; Burton *et al.*, 1993).

Materials and methods Forty treatments were evaluated in a split-split-plot design, two grasses, five cutting intervals and four N levels. Experimental design was a completely randomised block with four replications. Cutting height was 3 cm, N was applied after each cut at a rate to reach the target annual level of N. The experiment lasted a full year.

Results Grass X cutting interval X N level interaction was not significant (P>0.05). Tifton68 showed higher (P<0.05) mean yield than Coastcross-1 across all cutting and N levels. Tifton 68 yielded 21% more forage than Coastcross-1 (Table 1). Both grasses had the highest (P<0.05) yield when cut every 4 or 6 weeks across all N levels (Table 1) and increased (P<0.05) their forage yield as N level increased up to the highest rate of 1200 kg N/ha/year (Table 2).

Table 1 Annual forage yield (t/ha) in two tropical grasses under five cutting intervals. Mean of four N levels

Grass	Cutting interval (weeks)					Mean
	2	4	6	8	10	
Tifton 68	40	41	42	35	40	40 a
Coastcross-1	29	39	34	31	30	33 b
Mean	35 b	40 a	38 a	33 b	35 b	

Table 2 Annual forage yield (t/ha) in two tropical grasses at different N level. Mean of five cutting intervals

Grass	N level (kg/ha/year)			
	0	400	800	1200
Tifton 68	26	36	44	53
Coastcross-1	21	28	36	46
Mean	23 d	32 c	40 b	49 a

Conclusions Annual forage yield of Tifton68 and Coastcross-1 were responsive to cutting interval and N level and Tifton68 gave higher yield than Coastcross-1.

References

Burton, G.W. (1972). Registration of Coastcross-1 Bermudagrass. *Crop Science,* 12,125.

Burton, G.W., R.W. Gates & G.M. Hill. (1993). Registration of Tifton 68 Bermudagrass. *Crop Science,* 33, 644-645

Whitehead, D.C. (1995). Grassland Nitrogen. CAB International, Wallingford, UK.397 pp.

Root systems in tropical pasture restoration treatments in Rondônia, Brazil

M.C. Piccolo[1], K.C. Augusti[1], C. Neill[2], L. Fante Junior[3], M. Bernoux[1] and C.C. Cerri[1]
[1]Centro de Energia Nuclear na Agricultura, C.P.96, 13400.970, Piracicaba, SP, Brazil, Email: mpiccolo@cena.usp.br, [2]The Ecosystems Center, Woods Hole, MA, 02543, USA, [3]UNIMEP, C.P.68, 13400.901, Piracicaba, SP, Brazil

Keywords: Amazon, Brazil, Rondônia, pasture, roots

Introduction   Soil management can influence physical and chemical soil properties, with fundamental differences in root system development. Our objectives were to quantify carbon (C) and nitrogen (N) concentrations, dry mass and length of roots subjected to different restoration treatments of degraded pasture.

Materials and methods   This research was conducted in an existing 63 ha pasture area established in 1983 in the process of degradation at Fazenda Nova Vida, in central Rondônia (10°30'S, 62°30'W), Brazil. The experiment consisted of four random blocks (replicates) of three pasture reformation treatments: (1) Control *Brachiaria brizantha* + weeds; (2) Conventional tillage followed by planting of forage grass and fertiliser (NPK + micronutrients) and (3) Herbicide (2.4 D) followed by fertiliser (NK + micronutrients). Root systems were evaluated using the profile wall method 5 months after the start of soil management (March, 2002). Soil was collected from four pits (100 x 50 x 30 cm) of each treatment, from 0-5, 5-10 and 10-20 cm depths. Root samples were separated by sieving into different diameter classes (very fine, fine and coarse roots). The roots of different diameters were analysed with "SIARCS" (Integrated system for root and soil vegetation covering evaluation) to determine root length and root dry mass per unit soil volume (Crestana *et al.*, 1994). Root samples were oven-dried at 60°C and ground for analysis of total C and N in composite samples per diameter class and treatment. Contents of C and N were determined by dry combustion with a LECO CN-2000 Elemental Analyser.

Results   Contents of C and N were higher in coarse roots for all treatments. The C: N ratios ranged from 33 to 50; 36 to 48 and 24 to 37, in control, ploughing and herbicide treatments, respectively. Root C concentrations were higher in 10-20 cm layer (Table 1). Root dry mass and length decreased with depth in the soil profile, and both were higher in herbicide treatment compared with control (Figure 1). Ploughing showed negative influenced to root mass and length (Figure 1).

Table 1 Root carbon and nitrogen contents under restoration pasture treatments in Rondônia, Brazil

Depth (cm)	Control			Plowing			Herbicide		
	VF	F	C	VF	F	C	VF	F	C
	Carbon (%)								
0-5	21	20	35	23	24	33	20	19	41
5-10	18	22	36	19	31	35	22	22	42
10-20	22	27	46	26	27	48	19	25	45
	Nitrogen (%)								
0-5	0.6	0.6	0.7	0.5	0.7	0.8	0.7	0.8	1.4
5-10	0.5	0.5	0.7	0.5	0.7	0.7	0.6	0.7	1.2
10-20	0.6	0.8	1.0	0.5	0.7	1.1	0.5	0.9	1.2
	C:N ratio								
0-5	33	34	47	43	37	41	29	24	30
5-10	34	41	50	36	45	48	36	30	34
10-20	37	36	44	52	37	44	36	27	37

VF = very fine roots (diameter< 0.3mm); F = fine roots (diam. = 0.3mm) and ; C = coarse roots (diam. = 0.5mm)

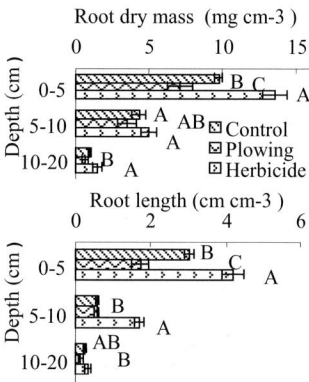

Figure 1 Roots dry mass and roots length. Mean values (n=4)•• standard error. Tukey test (p<0.05)

Conclusions   Herbicide treatment had advantages in root mass and length and roots C and N contents, 5 months after the soil restoration management. This treatment, did not have competition with weed plants on the root system development, and had an extra source of nutrients from the weed roots to forage grass roots.

References
Crestana S., Guimarães M.F., Jorge L.A. C., Ralish R., Tozzi C.L., Torre-Neto A., Vaz C.M.P. (1994) Avaliação de distribuição de raízes no solo auxiliada por processamento de imagens digitais. *Revista Brasileira de Ciência do Solo*, 18, 365-371.

Economic methodology for pasture grass and legume seed production

D.M. Cino , G. Febles, M.F. Díaz and F. Funes
Instituto de Ciencia Animal. Carr. Central Km 47 ½ S. José de las Lajas, Habana, Cuba, dcino@ica.co.cu

Keywords: economic methodology, seed production

Introduction The importance and complexity of the industrial process of seed production is known. Thus, conditions should be established for achieving and efficiently controlling the activity in order to know production costs, selling prices and to guarantee economic efficiency. The objective of this paper was to evaluate from an economic point of view, seed production of species of tropical grasses and legumes based on a preliminary methodology facilitating the control of the whole activity.

Materials and methods Figure 1 shows the steps in the process of seed production. Seed yields obtained at the Institute of Animal Science in Cuba for nine legumes and two grasses, were used for the economic analysis. Cost records were prepared considering the costs of the activity based on cost from establishment to storage. The method involved the collection of data sets recording different operations for collecting seeds in Cuba. It was considered that a profit margin between 15 and 20% was needed to achieve economic feasibility.

Results The economic analysis (Table 1) showed that costs/kg varied between 2.2 and 10.8 USD. To increase the quality of the analysis it is necessary to achieve better assessment of the methodological activities involved in the production process and strict control should be established to reduce costs. For commercialisation, each producer can then approach the market with precise knowledge of his production costs. The market will be determined by the seed volume obtained and the quality of the product offered.

Conclusions The methodology is a step ahead for the seed pasture industry in Cuba where, in the past, the economic control of the whole process was very poor.

Figure 1 Stages in evaluating costs and economic feasibility of production

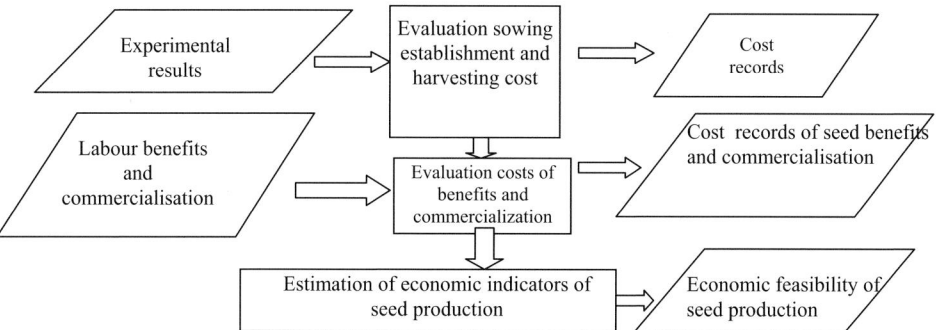

Table 1 Production costs and commercialisation prices for different grasses and legumes

Pasture species	Production, t/ha	Costs, USD/kg	Commercialisation prices, USD/kg
Glycine	0.30	6.39	10.90
Vigna	0.60	2.92	3.50
Canavalia	0.85	2.20	2.60
Centrosema	0.20	8.62	10.00
Siratro	0.16	10.87	13.00
Mucuna	0.80	2.25	2.70
Stylosanthes	0.20	8.68	10.40
Leucaena	0.55	3.22	3.80
Guinea	0.10	5.44	6.50
Brachiaria	0.80	6.79	7.80

Theme A: Efficient production from grassland

Section 9

Integrated production systems

Integration of forage production in rice-based cropping systems for mitigating forage crisis of ruminant livestock – studies in Bangladesh

M.A. Akbar, M.S.U. Bhuiyan[1] and M.S. Islam[2]

Dept. of Animal Nutrition, [1]Dept. of Agronomy and [2]Dept. of Agricultural Economics, Bangladesh Agricultural University, Mymensingh 2202, Bangladesh; Email: maakbar@bttb.net.bd

Keywords: rice/forage, integration, ruminants, milk, Bangladesh

Introduction Intensive rice cropping in Bangladesh is causing alarming shortages of forage and low ruminant productivity. This is also causing degeneration of soil fertility. Therefore it is imperative to identify some approach to integrate fodder production into rice cropping systems on rural farms. The integration of legume forage may improve soil fertility and soil structure, thus enhancing crop yield and may provide high quality feed for livestock (Haque 1992). Studies were done to investigate the effects of rice/forage integration on forage yield, soil fertility and also on milk yield of cows fed on the grown forages.

Materials and methods Several legume forages were tested for yield, nutrient composition and nutritive value. The best two of these, *Sesbania rostrata* and *Lathyrus sativus*, were selected for two seasons (wet and dry, respectively) for application under farm conditions. Studies were carried out for 3 consecutive years, from 1997 to 2000. Yield of forages and effects of feeding them to cows on milk production were studied. Soil nutrient status (nitrogen & OM) and rice yield before and after forage cultivation were monitored simultaneously.

Results Both on-station and on-farm production and nutritive quality of the forages proved promising. On-farm yield of *Sesbania* and *Lathyrus* were 25 and 15 t/ha. Crude protein contents were high, 34.9 and 24.5%, respectively. Soil nitrogen and organic matter were significantly (P<0.05) improved due to both of the forage cultivation. Rice yield in both seasons (wet and dry) was also increased (P<0.01) due to forage cultivation (Fig. 1). Supplementation of both forages (*Sesbania* and *Lathyrus*) with rice straw diets fed to lactating cows under on-farm conditions significantly (P<0.01) increased mean milk yield by 21% and 19%, respectively, over that of control with no supplementation. The results are shown in Table 1.

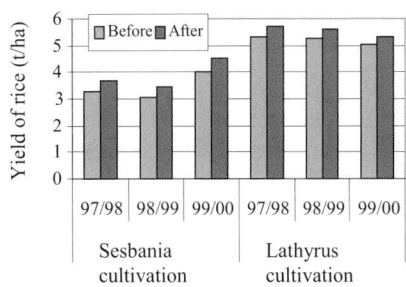

Figure 1 Effect of forage cultivation on rice

Table 1 Feeding forages to cows on milk yield

Year	Milk yield (kg/d) Supplement		SEM	Significance
	None	With		
Sesbania rostrata				
97/98	1.77[a]	2.00[b]	0.061	*
98/99	1.58[a]	1.98[b]	0.069	**
99/00	1.54[a]	1.97[b]	0.058	**
Lathyrus sativus				
97/98	2.10[a]	2.52[b]	0.033	**
98/99	1.34[a]	1.63[b]	0.052	**
99/00	1.12[a]	1.28[b]	0.030	*

Conclusion *Sesbania* and *Lathyrus* are high yielding and high grade forages in terms of nutritional quality. When cultivated under farm conditions they increased soil fertility and rice yield and also improved milk production of dairy cows when fed with straw-based diets. These two forages are promising for cultivation in integration with rice in two seasons to mitigate forage shortage for ruminants in Bangladesh.

References

Haque, I. (1992). Use of legume biological nitrogen fixation in crop/livestock production systems. In: K Mulongoy, I Mueye & DSC Spencer (eds.) *Biological Nitrogen Fixation and Sustainability of Tropical Agriculture*, 423-437.

Evaluating the economic and environmental sustainability of integrated farming systems

C.A. Rotz[1], M.A. Sanderson[1], M. Wachendorf[2] and F. Taube[2]
[1]*Pasture Systems and Watershed Management Research Unit, USDA/Agricultural Research Service, Building 3702, Curtin Road, University Park, Pennsylvania, 16802, USA, Email: alrotz@psu.edu, [2]Crop Science and Plant Breeding Institute, Christian-Albrechts-University, Kiel, Germany*

Keywords: nutrient management, farm, economics, environment, simulation

Introduction Economic and environmental sustainability has become a major concern for forage-based animal production in Europe, North America and other parts of the world. Development of more sustainable farming systems requires an assimilation of experimental and modelling research. Field research is critical for supporting the development and evaluation of models, and modelling is needed to integrate farm components for predicting the long-term effects and interactions resulting from farm management changes. Experimentally supported simulation provides a tool for evaluating and comparing farming strategies and predicting their effect on the watershed, region and beyond.

Materials and methods Field experiments on sandy soils in northern Germany have determined nitrogen fluxes in grass and maize silage production systems (Wachendorf *et al.*, 2004). Grass production included defoliation strategies of all silage harvests, silage harvest followed by grazing later in the season and all grazed. Nitrate leaching and denitrification losses of nitrogen were monitored on plots receiving various applications of fertilizer and manure. These production systems were simulated using a whole farm model (Rotz & Coiner 2004), and simulated nitrogen fluxes were verified by field data. The model is now being used as a research and teaching tool to evaluate cropping, feeding and manure management strategies in integrated crop and animal farming systems and to predict their environmental impacts.

Results To illustrate an application of the model, all grass and grass and maize silage systems were simulated for a typical dairy farm in northern Germany with and without the use of rotational grazing. Farm net return or profit increased with the use of maize silage or grazing (Figure 1). Nitrogen losses increased with the use of grazing but decreased with the use of maize silage. An appropriate combination of grass, maize silage and grazing was used to maximize farm profit while maintaining nitrogen losses. These results are not directly transferable to other sites such as the soil and climate conditions of the northern USA, but through the use of the model these and other management practices are being evaluated in other regions of the world. Farm and watershed scale models are also being linked to predict the effects of farm management on the surrounding region and beyond. A version of the farm model is available at http://pswmru.arsup.psu.edu

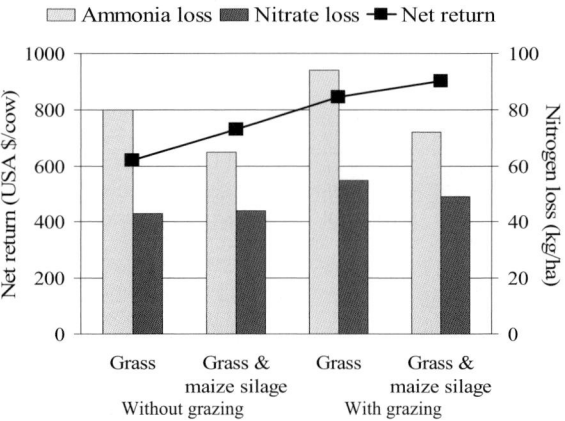

Figure 1 Farm net return and nitrogen loss for production systems using grass and maize silage with or without grazing

Conclusions When properly supported with field observations, whole-farm simulation provides an effective research and teaching tool to integrate farm processes and predict farm performance, profitability and environmental impact. This tool is very useful to develop, evaluate and transfer more sustainable grassland systems to commercial livestock production.

References

Rotz C.A. & C.U. Coiner (2004). Integrated Farm System Model, Reference Manual. Available: http://pswmru.arsup.psu.edu (Accessed August 2004).
Wachendorf M., M. Buchter, H. Trott & F. Taube (2004). Performance and environmental effects of forage production on sandy soils. II. Impact of defoliation system and nitrogen input on nitrate leaching losses. *Grass Forage Science*, 59, 56-68.

Simulation of pasture phase options for mixed livestock and cropping enterprises

L. Salmon, A.D. Moore and J.F. Angus
CSIRO Plant Industry, GPO Box 1600, Canberra, ACT, Australia, 2601; Email: libby.salmon@csiro.au

Keywords: decision support tool, lucerne, subterranean clover, wheat, mixed farming

Introduction In southern Australia, 50% of grain-producing farms also run beef and/or sheep enterprises. Legume pasture leys are used to replace soil nitrogen and manage crop disease risks. Deep-rooted perennials, predominantly lucerne (*Medicago sativa*), are replacing annual *Trifolium subterraneum*-based leys to increase pasture production. They also have the environmental benefits of limiting soil acidity, rising water tables and dryland salinity. After recent droughts depletion of soil water by lucerne has penalised wheat yields. Decision support tools can help farmers evaluate the long-term effects of grazed annual and perennial leys on animal and crop production at the whole farm level.

Materials and methods FarmWi$e, a modular decision support tool (Moore 2001) was used to simulate wheat yield, pasture production and animal production in rotations of 3 years of wheat and 3 years of either lucerne or self-regenerating sub clover at Lockhart, NSW (mean annual rainfall 500mm). The water balance model used in the simulations was an updated version of that described by Moore *et al.,* (1997). Continuous simulations from 1954-2003 used daily weather records and a description of a Red Dermosol soil from the region. The management module of FarmWi$e was used to simulate the recommended practices of undersowing lucerne in the 3rd year of wheat, lucerne removal in mid-October of the 3rd year of the pasture phase and residue cultivation prior to sowing wheat. Simulations assumed maximum root depths in this soil of 0.7m for sub clover, 1.3m for wheat, 2.4m for lucerne and that soil nitrogen and disease did not limit pasture or crop growth. The effects of each pasture type on water balance were modelled but any differences in the effect of lucerne and sub clover on soil structure or soil nutrient dynamics were not. In each system the pasture phase was grazed by weaned 30 kg 4 month-old Border Leicester × Merino sheep to achieve a live weight of 45 kg. Six simulations of each of the rotations were run, each commencing in a different year of the rotation. The difference in wheat yield for the two pasture rotations was calculated for the each of the 50 years simulated (1954-2003).

Results Average wheat yields over 50 years of 3.90 and 3.75 t/ha for the sub clover and lucerne wheat rotations, respectively, agree with district performance. However yield differences between rotation types depended on the year of the crop phase (Figure 1). In Year 1 of the wheat phase, yields from crops sown after lucerne were, on average, 0.14 t/ha greater than crops sown after sub clover but were similar in Year 2 and 0.56 t/ha less in Year 3. Lucerne removed more water from the soil profile than sub clover but summer rainfall during the long fallow period before the first wheat crop often replenished the soil profile. In Year 3, wheat yields in lucerne rotations were penalised by competition from the undersown lucerne. Distributions of yield differences show that in 2% of years the penalty in Year 1 of the wheat phase after lucerne exceeded 1.2 t/ha. These years were the severe droughts of 2002 and 2003; the recent bad experiences of local farmers are therefore exceptional and are not necessarily a sound basis on which to choose pasture species.

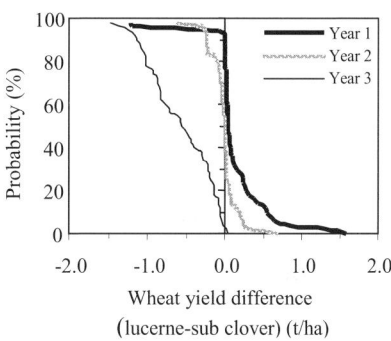

Figure 1 Distribution of differences between simulated lucerne and sub clover rotations in wheat yields for each year of the wheat phase

Conclusions Decision support tools enable a whole farm approach to assess the relative benefit of lucerne to both crop and livestock enterprises on farms and environmental sustainability.

References
Moore, A.D. (2001). FarmWi$e: a flexible decision support tool for grazing systems management. In: Gomide, J.A., W.R.S. Mattos & S.C. da Silva (eds.), *Proceedings of the Nineteenth International Grassland Congress*, Sao Pedro, Brazil. FEALQ, Sao Pedro, Brazil, pp. 1045-1046.
Moore A.D., J.R. Donnelly & M. Freer (1997). GRAZPLAN: decision support systems for Australian grazing enterprises. III. Pasture growth and soil moisture submodels, and the GrassGro DSS. *Agricultural Systems* 55, 535-582.

Overseeding cereal rye and annual ryegrass into soyabean for forage as part of a multifunctional cropping system

L.B. Smith and R.L. Kallenbach
Plant Sciences Unit, University of Missouri, Columbia, Missouri 65211, USA, Email: lbswb7@mizzou.edu

Keywords: overseeding, multifunctional, annual ryegrass, cereal rye

Introduction In the lower Midwest, the longest period of inadequate forage supply from pasture is from mid-December through mid-March (Matches & Burns, 1995). Livestock producers in this region are looking for high quality forage for winter grazing (Kallenbach *et al.*, 2003). Annual ryegrass (*Lolium multiflorum* Lam.) and cereal rye (*Secale cereale* L.) are two forages that Missouri's beef producers are interested in to extend the grazing season. The objective of this research was to determine how seeding date impacts the establishment, growth, and forage production of annual ryegrass and cereal rye when planted into soyabean fields as part of a multifunctional cropping system.

Materials and methods The experiment was conducted at Bradford Research and Extension Centre, near Columbia (38°57'N 92°20'W). 'Saddle Pro' annual ryegrass and 'Winter grazer 70' cereal rye were overseeded into standing soyabean (*Glycine max* L.) at 3 different stages of crop growth (Table 1). An unseeded control was included. The annual ryegrass was overseeded into the soyabean crop at 39 kg/ha of pure live seed, and cereal rye was overseeded at 140 kg/ha of pure live seed. Forage yield for each treatment was measured when the average height of the 5 replications was 20-25 cm. All forage was harvested using a sickle mower set to leave a 7.6 cm stubble. In the following late April, the annual ryegrass and cereal rye were terminated to allow for row crop planting.

Table 1 Annual ryegrass and cereal rye overseeding treatments based on soybean developmental stage

Species to be overseeded	Developmental stage of soyabean	Description of treatment
Annual ryegrass	R 5.5	2 weeks before leaf drop
Cereal rye	R 5.5	2 weeks before leaf drop
Annual ryegrass	R 6.5	At leaf drop
Cereal rye	R 6.5	At leaf drop
Annual ryegrass	R 8	At harvest
Cereal rye	R 8	At harvest
-	-	Control

Results Forage treatments did not alter soyabean yield. Figure 1 shows the cumulative forage yields for the treatments. Cereal rye seeded at R 5.5 and R 6.5 yielded equally (average of 4000 kg/ha), but the R 5.5 treatment produced 25% more than the cereal rye seeded at R 8. Annual ryegrass showed a similar trend with the R 5.5 and R 6.5 yielding equal amounts of forage (mean 3200 kg/ha), but the R 6.5 treatment yielded 20% more than the treatment seeded at R 8. When seeded at R 5.5, cereal rye yielded 30% more than the annual ryegrass.

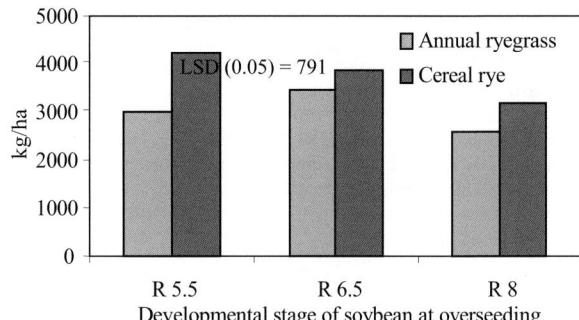

Figure 1 Total forage yield

Conclusion Livestock operations in the lower Midwest can use cereal rye and annual ryegrass overseeded into soyabean for winter grazing. Overseeding at R 5.5 cereal rye produced more forage than annual ryegrass and overseeding at R 6.5 or R 8 produced similar yields for both cereal rye and annual ryegrass. All treatments yielded >2500 kg/ha for the season, which would be an adequate source for winter forage.

References

Kallenbach R.L., G.J. Bishop-Hurley, M.D. Massie, M.S. Kerley & C.A. Roberts (2003). Stockpiled annual ryegrass for winter forage in the lower Midwestern USA. *Crop Science,* 43, 1414-1419.

Matches A.G. & J.C. Burns (1995). Systems of grazing management, pp179-192. In R.F. Barnes *et al.* (eds.) Forages. Volume II. The science of grassland agriculture. 5th edition. Iowa State University Press, Ames, IA.

Previous grass-lucerne mixtures affect barley yield and quality in a semiarid location of the Canadian prairie region

P.G. Jefferson, F. Selles, R.P. Zentner and R. Lemke
Agriculture and Agri-Food Canada, Semiarid Prairie Agricultural Research Centre, P.O. Box 1030, Swift Current, Saskatchewan, S9H 3X2, Canada, Email: JeffersonP@agr.gc.ca

Keywords: crop rotation, N uptake, tillage, greenhouse gases

Introduction In the semiarid region of the Canadian prairies perennial forages are not rotated with annual crops because previous experiments reported negative impacts (Kilcher and Anderson 1963; Campbell *et al.* 1990). However, previous research used persistent species while short-lived species could have less adverse effect. Our objective was to compare three grass species in three lucerne mixtures terminated with tillage or herbicide for effects on barley grain, N concentration, and N uptake.

Materials and methods Intermediate wheatgrass (*Elytrigia intermedia*) cv. Chief (IWG), Dahurian wildrye (*Elymus dahuricus*) cv. Arthur (DWR), and slender wheatgrass (*Elymus trachycaulus*) cv. Revenue (SWG) were seeded in monoculture, or mixture with lucerne (*Medicago sativa*) cv. Beaver, a persistent winterhardy cultivar or Nitro, an annual non-winterhardy cultivar in 1998 and 1999. Half of each trial was tilled or sprayed with glyphosate herbicide to terminate the forage stands in 2002 and 2003. Barley (*Hordeum vulgare*) cv. Harrington was seeded without N fertiliser. Grain and straw yield, and grain and straw N concentration were determined and N uptake was calculated. N_2O emission was monitored on selected treatments.

Results Nitrogen availability to barley, as shown by N uptake, was improved by including lucerne in the previous short-lived grass (DWR and SWG) stand (Table 1). In 2002, a year with above-average precipitation, the inclusion of lucerne increased barley grain yields but did not affect grain N concentration. Conversely, in 2003, a dry year, inclusion of lucerne did not affect grain yield but increased grain N concentration. Tillage increased grain yield compared to no-tillage in 2002 but not in 2003 (data not shown). Emissions of N_2O during the barley phase was not affected by previous forage mixture (data not shown).

Table 1 Effect of grass species and lucerne mixture on lucerne composition of forage in the year before barley, barley grain yield, barley grain N concentration and N uptake for two years

	Lucerne (g/g)		Barley (kg/ha)		Grain N (g/kg)		N uptake (kg/ha)	
Grass species	2001	2002	2002	2003	2002	2003	2002	2003
DWR	0.23	0.33	1265	1224	14.8	23.0	31	48
IWG	0.10	0.12	681	1045	14.6	19.4	23	35
SWG	0.23	0.27	1352	1231	14.8	22.8	35	48
SE	0.02	0.01	89	70	0.3	1.6	2	2
Lucerne cultivar								
Beaver	0.57	0.72	1450	1099	14.9	24.9	37	50
Nitro	0.01	0.01	986	1222	14.6	20.7	27	43
None	0.01	0.01	862	1180	14.7	19.6	24	37
SE	0.02	0.01	78	63	0.2	1.6	1	2

Conclusions Barley after short-lived perennial grasses such as DWR and SWG yielded more in 2002 and had higher grain protein in 2003 than barley after IWG. The inclusion of lucerne improved N availability to the barley in both years. These forages have potential for inclusion in annual crop rotations in semiarid regions of the Canadian prairies.

References
Campbell CA, RP Zentner, HH Janzen & KE Bowren (1990) Crop rotation studies on the Canadian Prairies. Agriculture Canada, Research Branch, Publication 1841/E. Ottawa, Ont. 133 pp.
Kilcher MR & LJ Anderson (1963) Wheat yields and soil aggregation after perennial grasses in a semi-arid environment. *Canadian Journal of Plant Science,* 43, 289-294.

Canavalia brasiliensis: a multipurpose legume for the sub-humid tropics

A. Schmidt[1], M. Peters[1], L.H. Franco[1] and R. Schultze-Kraft[2]
[1]Centro Internacional de Agricultura Tropical (CIAT), A.A. 6713, Cali, Colombia, Email: a.schmidt@cgiar.org,
[2]University of Hohenheim (380), D-70593 Stuttgart, Germany

Keywords: *Canavalia brasiliensis*, dry season feed, green manure, drought resistance, nutritive value

Introduction *Canavalia brasiliensis* Mart. ex Benth. ("Brazilian jackbean") is a weakly perennial, prostrate to twining herbaceous legume with a wide natural distribution in the New World tropics and subtropics. In comparison with *C. ensiformis* ("jackbean"), research reports on *C. brasiliensis* are scattered and restricted to studies done in Latin America. The species develops a dense and extensive, deep-reaching root system and subsequently tolerates a 5-6 month dry period. Based on studies that generally were done with only one genotype, it is adapted to a wide range of soils, including very acid, low-fertility soils. Its main use is as green manure, for fallow improvement and erosion control. Due to medium biomass decomposition, nutrient release of *C. brasiliensis* green manure has the potential to synchronise well with the nutrient demand of the succeeding crop and may lead to high N recovery rates. Whereas the high concentration, in *Canavalia* seeds, of antinutritive substances such as toxic amino acids (e.g., canavanin), lectins (e.g., concanavalin Br) and trypsin inhibitors, there is little information on the nutritive value of the herbage of this species (Schloen *et al.*, 2004). In order to develop multipurpose legume germplasm for smallholder systems in the sub-humid tropics, we initiated a *C. brasiliensis* germplasm screening experiment and engaged with farmers in Central America to integrate this legume into local maize-bean production systems. First promising results are reported.

Materials and methods *Germplasm characterisation*: The available collection of 53 accessions of *Canavalia brasiliensis* was sown at the onset of the rainy season 2004 at CIAT's research station in Santander de Quilichao near Cali, Colombia. A Randomized Complete Block design with 3 replications was employed. Plot size was 3 m x 2 m. Parameters of evaluation included ease of establishment (soil cover, vigour: rating 1-5 with 5 being the most vigorous), DM yield and forage quality (CP; IVDMD, tannins, fibre and lignin) across seasons; selected accessions were also analysed for canavanin content.

On-farm work: *C. brasiliensis* accession CIAT 17009 was sown in 5 m x 5 m plots in 3 replicates at the end of each rainy season (October) since 2001 in San Dionisio, Matagalpa, Nicaragua, to allow for total plot cover before the onset of the 6-month dry season. In the subsequent rainy season (May), plots were slashed and maize planted into the *C. brasiliensis* mulch. Maize yields and other crop parameters were compared with traditionally fertilised and fallowed maize plots.

Results *Germplasm characterisation*: Three months after transplanting, *C. brasiliensis* accessions were well established with a mean of almost 65% (± 24.7%, range 55-92%) soil cover, and a mean vigour rating of 2.6 (± 1.1, range 2-5). Accessions CIAT 808, 7319, 7648, 7970, 8557, 17008, 17009, 18515, 20095, and 20096 had soil covers of 85% or above and a vigour rating of 4 to 5.

On-farm work: *C. brasiliensis* plots remained green during the dry season producing 3.5-4 t/ha of biomass, and reducing wind/soil erosion and weed pressure. Maize yields after *C. brasiliensis* (5.59 kg/plot) were significantly higher than those after traditional fallow (2.18 kg/plot), and slightly higher than those obtained with traditional fertiliser (5.16 kg/plot). Farmer confirmed good plant establishment, and fast growth and cob development.

Conclusions *C. brasiliensis* is a promising species for use as green manure. Smallholder farmers in Central America appreciate it because of its robustness over a wide range of soils and climates, its biomass production and its green manure effect. Through its dry season tolerance, the legume opens a significant time window for soil improvement without affecting grain production during the rainy season. There might be also a high potential as a dry season feed, but feed values for animals must be determined. Initial observations suggest considerable diversity in the collection.

References

Schloen M, M Peters & R Schultze-Kraft (2004) *Canavalia brasiliensis* Mart. ex Benth. http://www.fao.org/ag/AGP/AGPC/doc/GBASE/Default.htm (in press).

Relative forage yield of intercropped lucerne (*Medicago sativa* L.) and winter forage cereals

T.W. Pereyra, H.R. Pagliaricci and A.E. Ohanian
Departamento de Producción Animal, Universidad Nacional de Río Cuarto, Córdoba. Argentina, Email: tpereyra@ayv.unrc.edu.ar

Keywords: intercropping, lucerne, winter forage cereals

Introduction In tropical regions of the world intercropping is mainly associated with the production of maize for grain, while in temperate areas it is associated with the efficient production of forage. There is increasing interest in the development of cereal - legume intercropping in some temperate regions (reviewed by Mason & Pritchard, 1987). The aim of this study was to compare the relative forage yield of intercropped lucerne and cereal.

Materials and methods Lucerne pasture and forage cereals were grown as monocultures and as eight intercropped treatments comprising three sowing dates - Early (Ea), intermediate (In) and late (La) – with different species and cultivars of cereal sown with lucerne - : oat short cycle (O shc), oat long cycle (O lc), triticale (Tr) and barley (Ba).). Relative Forage Yield (RFY) of the mixtures was determined, according to the method of Trenbath (1976), which calculates RFY = ½ (Ai/Ap + Bi/Bp), where Ai and Bi are the yields of each component A and B in the intercrop and Ap and Bp are yields of the components in monoculture., respectively. Values for RFY greater than 0.5 indicate that the intercropped system is a better use of the area than growing the two components in monoculture. . A split-plot experimental design in randomised blocks was used. The results were analysed by ANOVA.

Results The RFYs differed significantly (p<0.05) between intercropped treatments (Table 1). Those involving early and intermediate sowing did not differ significantly (p≥0.05), while RFYs of late-sown treatments were significantly higher than those with intermediate sowing date, having the highest relative yield values of 0.82 and 0.83.

Table 1 Relative forage yield of eight intercropped lucerne-winter cereal treatments

Intercrop	RFY
1. Lucerne + O shc-Ea	0.78 [b a]
2. Lucerne + Ba-Ea	0.78 [b a]
3. Lucerne + O shc-In	0.73 [b]
4. Lucerne + Ba-In	0.74 [b]
5. Lucerne + Tr-In	0.71 [b]
6. Lucerne + O lc-In	0.72 [b]
7. Lucerne + Tr-La	0.83 [a]
8. Lucerne + O lc-La	0.82 [a]
C.V. (%)	5.57
Probability	0.04

Means with no common superscripts are statistically significant (p<0.05)

Conclusions Our preliminary results suggest that intercropping lucerne with forage cereals is a feasible alternative to monocultures for forage production, allowing efficient use of the area for animal production. The values obtained show that it is possible to obtain between 70% and 82% of the forage produced in a given area compared with the combined yields from the two components grown on twice the area.

References
Mason, W. K & K.E. Pritchard (1987). Intercropping in a temperate environment for irrigated fodder production. *Field Crops Research*, 16, 243-253.
Trenbath, B.R. (1976). Plant interactions in mixed crop communities. In: M. Stelly (ed.). Multiple Cropping. American Society of Agronomy, Madison, 129-169.

The effectiveness of nitrogen rates on winter wheat and white clover bi-cropping grown for silage

J. Sowiński
Crop Production Department, Agricultural University of Wrocław, 25 Norwida , 50-375 Wrocław, Poland,
E-mail: sowinski@ekonom.ar.wroc.pl

Keywords: direct sowing, bi-cropping, winter wheat, white clover, nitrogen rates

Introduction Whole-crop cereals harvested for silage cover *c.* 500,000 ha in Europe (Wilkins & Kirilov, 2003). Generally dry matter (DM) yield of small-grain cereals is lower than that of maize. In some investigations, DM yield (of high nutritional value) reached 15 t/ha (Balsdon *et al.* 1997; Clements *et al.* 1997). Whole-crop silage produced in a bi-cropping system offers more balanced forage compared to pure cereals and legumes. Nitrogen rates can be decreased with bi-cropping. The aim of this investigation was to compare nitrogen effectiveness using two methods of winter wheat cultivation: direct drilling into stubble and bi-cropping with white clover.

Methods In the years 1998-2001 field experiments, with four replicates were carried out in the south-west of Poland. Two sowing methods of winter wheat were compared: direct drilling into stubble and bi-cropping. In both systems five application rates of N fertiliser were applied. For comparison, the control field was not treated with nitrogen. The measurements included DM, crude protein (CP) and energy yields and the effectiveness of N use. The results were analysed statistically using STATISTICA software.

Results Nitrogen fertilisation in the system of direct sowing into stubble increased DM, CP and energy yields to a greater extent than in the bi-cropping system (Table 1). Winter wheat and white clover bi-cropping resulted in 26% higher DM, 95% higher CP and 25% higher energy yields compared with direct sowing into stubble. Increases occurred at all rates of N application.

Table 1 Dry matter, crude protein and energy yields and effectiveness of nitrogen use with the two systems

Sowing method	Nitrogen rates kg N/ha	Dry matter yield		Crude protein yield		Energy yield	
		t/ ha	kg DM/ kg N	kg/ha	kg CP/ kg N	UFL	UFL/ kg N
Bi-cropping	0	6.07	-	676	-	4286	-
	30	6.20	4.4	686	0.33	4418	4.4
	60	6.65	9.7	715	0.65	4697	6.9
	90	6.72	7.2	781	1.17	5153	9.6
	120	6.65	4.8	790	0.95	5631	11.2
Direct sowing	0	3.20	-	209	-	2636	-
	30	4.26	35.2	279	2.30	3262	20.9
	60	5.63	40.5	356	2.44	4155	25.3
	90	5.92	30.2	476	2.97	4411	19.7
	120	6.60	28.3	544	2.79	4859	18.5
Significance		n.s.	-	n.s.	-	n.s.	-
Bi-cropping		6.46	-	729	-	4837	-
Direct sowing		5.12	-	373	-	3865	-
Significance		***	-	***	-	***	-

Conclusions The impact of N rates on DM, CP and energy yields was lower in the bi-cropping system, because of N fixing capabilities of legumes. The highest effectiveness with direct sowing into stubble was observed with 30 and 60 kg N/ha. Nitrogen applied before ear emergence had less influence on DM and energy yield, but a bigger impact on CP yield. Winter wheat and white clover bi-cropping system can be recommended for ecological farming.

References
Balsdon, S.L., T.M. Martyn, R.O. Clements & S. George (1997). The potential of a clover: cereal bi-cropping system to produce high quality whole-crop silage. *Proceedings British Grassland Society Fifth Research Conference*, 85-86.
Clements, R.O., G. Donaldson, G. Purvis & J. Burke J. (1997). Clover: cereal bi-cropping. *Aspects of Applied Biology*, 50, 467-469.
Wilkins, R.J. & A.P. Kirilov (2003). Role of forage crops in animal production systems. *Grassland Science in Europe*, 8, 283-291.

The influence of winter wheat and white clover bi-cropping system on white clover sward parameters

J. Sowiński
Crop Production Department, Agricultural University of Wrocław, 25 Norwida street, 50-375 Wrocław, Poland,
E-mail: sowinski@ekonom.ar.wroc.pl

Keywords: white clover, sowing method, growing point, stolon, length

Introduction Whole-crop silage produced in a bi-cropping system represents a low-input forage production system (Clements *et al.*, 1997). Depressing competition of white clover (by mowing or spraying with herbicides) when winter wheat starts its growth is necessary in this system. Winter wheat also competes with white clover during crop growth. The number of growing points as well as the length and weight of stolon are the main parameters that characterise the persistence of white clover (Jorgensen & Ledgard, 1997; Marriott & Haystead 1990). The purpose of the present investigation was to study the effects of a bi-cropping system on some parameters of white clover.

Methods Field experiments were conducted in 1998-2001. A sward of white clover (cv Grassland Huia) was established in April 1998. Winter wheat (cv Kobra) was directly drilled into white clover in autumn 1998, 1999 and 2000. Before winter wheat was sown, the white clover sward was sprayed with a low rate of glyphosate (2 l/ha). The data were compared with those obtained with a pure white clover sward without herbicide (control). There were four replicates of each treatment. The sward samples were taken from 0.02 m^2 areas six times during the entire experiment. Directly after sampling, soil was separated and white clover stolons were washed. The number of growing points, length and weight of stolons were measured. . The results were analysed statistically by analysis of variance (STATISTICA software).

Results Spraying with glyphosate and winter wheat competition in the bi-cropping system decreased the growth of white clover. The number of growing points was lower for almost every measurement. On average, the number of growing points of the control was 37% higher than that in the bi-cropping system and the lengths of stolon in the control were 133% higher. The weight of stolon in the bi-cropping system was higher than that in the control only in autumn 2000, when the value for bi-cropping was 62% higher. On other dates, the stolon mass was higher in the control. On average, the weight of stolon in the control was 62% higher than that in the bi-cropping system.

Table 1 The effects of sowing method on some parameters of white clover

Measurement date	Growing points (x 1000/m^2)		Length of stolon (m/m^2)		Weight of stolon (g/m^2)	
	Bi-cropping	Control	Bi-cropping	Control	Bi-cropping	Control
23-10-98	2.5	4.6	36	169	15	150
23-04-99	5.9	7.8	75	187	57	134
30-11-99	4.2	8.1	78	220	99	351
18-04-00	9.7	8.2	99	166	111	167
22-11-00	4.8	7.4	113	207	344	210
25-04-01	5.6	8.5	79	171	111	193
Significance	*	***	*	n.s.	***	**
Average	5.4	7.4	80	186	123	200
Significance	**		***		*	

Conclusions The combination of herbicide treatment and sowing winter wheat significantly decreased the growth of white clover and influenced the persistence of white clover. After two years of bi-cropping, white clover parameters decreased. In the south-west of Poland, the bi-cropping system can be recommended for a maximum of three years in the same field.

References
Clements, R.O., N. Koefoed, G. Donaldson, J. Burke & G. Purvis (1997). Exploitation of a sustainable low-input and reduced-output system for arable crops. Report for final year to EU, 468 pp.
Jorgensen, F.V. & S.F. Ledgard (1997). Contribution from stolons and roots to estimates of the total amount of N_2 fixed by white clover (*Trifolium repens* L.). *Annals of Botany*, 80, 641-648.
Marriott, C.A. & A. Haystead (1990). The effect of defoliation on the nitrogen economy of white clover: re-growth and the remobilization of plant organic nitrogen. *Annals of Botany*, 66, 465-474.

Options for improved biomass production in feeding systems for dairying in high rainfall environments in New Zealand

J.M. de Ruiter[1], D.R. Wilson, S. Maley and S.M. Henton
New Zealand Institute for Crop & Food Research, Private Bag 4704, Christchurch, New Zealand,
www.crop.cri.nz, Email: deruiterj@crop.cri.nz

Keywords: supplementary feeds, nitrogen balance, maize, brassicas, climatic risk

Introduction New Zealand dairy production has expanded into marginal climates and soil types on the premise of excellent profitability and efficient utilisation of forage. Annual pasture production in the cool West Coast of the South Island (rainfall 2042-2933 mm) is <9000 kg compared with the national mean of 15,000 kg. Increased farm production and feeding of high quality biomass, from imported feed or supplementary feed crops grown on-farm, are needed to improve milk solid output. Small plot trials with spring and early summer-sown brassicas, cereals and maize were the focus for development of systems to maximise and manage the seasonal feed supply. The effect of sowing time, fertiliser timing and rate of N and K fertiliser application were studied to quantify the risks of crop failure in the high rainfall and low radiation environment. The aim was to increase forage supply/ha in a predominantly grass-based system and reduce associated risks to environmental sustainability.

Materials and methods Trials of cereals (triticale cv. Rocket and barley cv. *Boss*), maize (hybrid *Elita*) and brassicas (leafy turnip cv. *Pasja*, rape cv. *Maxima* Plus, and turnip cv *Barkant*) were sown at coastal Barrytown (BT) and inland Ikamatua, (IK) locations. The cereal experiment was a split-split plot comprising sowing dates (27 Oct., 20 Nov., 17 Dec. at BT; 27 Nov., 17 Dec. at IK) and three N/K applications comprising basal N and K, and factorial N and K (medium and high rates for each nutrient). The maize experiment was a split plot design with dual sowing dates (27 Oct. and 27 Nov. at BT; 20 Oct. and 19 Nov. at IK) and three fertiliser levels (control, high inputs and demand). Brassica trials were factorial designs comprising location (x2), sowing date (x2) and cultivar. N and K balances were calculated for respective treatments in cereal and maize experiments allowing for plant uptake, pre- and post-season soils levels, and fertiliser application. Crops were managed for optimal biomass production within the constraints of the experimental treatments. Climatic risk analysis was performed for maize using a simulation model (Wilson *et al.* 1995) to predict forage yield, maturation time and harvest index (HI) using daily maximum/minimum temperature and solar radiation for Reefton (inland), Westport (northern coastal) and Hokitika (southern coastal).

Results Cereal yields were 7.6-13.0 t/ha. Fertiliser treatment and site effects accounted for most variation. Monthly delay in sowing caused yield losses of 1 t/ha. Medium fertility treatments achieved the best cereal yields. Maize yields were comparable with model predictions. Predicted crop failure was higher with later sowing, more northerly latitudes and distance from coast. (Table 1). By late March, turnips yielded up to 14 t/ha (8 t/ha in bulb) from 4 months growth. Total leaf yields of rape and cv. *Pasja* did not exceed 8 and 7 t/ha, respectively.

Table 1 Model predictions of risk of maize crop failure (percent of years tested). Crops failed if killed by frost before maturity, low yield (<14 t/ha) or incomplete grain fill (HI <0.4)

Sowing date	Reefton (23 years)	Westport (12 years)	Hokitika (21 years)
1 Oct.	9	8	5
15 Oct.	4	8	5
1 Nov.	17	17	10
15 Nov.	39	17	38

Conclusions Sowing time and choice of crop type ensured forage supply as follows: standing crops of brassicas for grazing (February-April); cereal harvest from January-March for whole-crop silage; and maize harvest from March-April for silage. High rates of fertiliser caused excessive leaching losses of N and K in cereals and maize. Best yields required moderate rates of N and K in frequent applications. Factors that influenced productivity and quality most were: *Fusarium* head blight in triticale; incomplete grain filling in late-sown maize from lower than average temperature and radiation during January-March; and chewing insects on brassicas during establishment. Overuse of N and K fertiliser, with leaching into ground water, occurred only with high fertiliser rates. Risk of crop failure induced by low temperature and radiation shortfall was apparent only in maize. Adverse effects of high rainfall, causing crop lodging losses, occurred only in heavily fertilised cereals.

Reference

Wilson D.R., R.C. Muchow & C.J. Murgatroyd (1995). Model analysis of temperature and solar radiation limitations to maize potential production in a cool climate. *Field Crops Research,* 43, 1-18.

Implications of the use of grazing sheep on kiwi fruit orchard

C.H.E.C. Poli[1], R.C. Gomes[2], P. Cinel Filho[2], M.F. Gomes[2], A. Zborowki[2], G. Pires[2] and J.L. Rigon[2]
[1]Embrapa Pecuária Sul/Paraná, P.O. box 242, CEP 83411-000, Colombo, PR, Brazil, Email: cpoli@cnpf.embrapa.br, [2]Fepagro, Rua Gonçalves Dias, 570, CEP 90130-060, Porto Alegre, RS, Brazil

Keywords: silvipastoral system, ewes, agrosystems

Introduction In the southern part of Brazil there is an important area of kiwi fruit, mainly cultivated by small farmers. The use of sheep under trees of kiwi fruits could be an interesting alternative for small farmers to reduce their mowing costs, to improve their income and to provide meat for the farmers' family. However there is a lack of information about the damage that the animals could cause to kiwi fruit plants. The objective of this study was to monitor the effect of the use of sheep on a kiwi fruit orchard.

Material and methods The study was carried out from 12 July to 18 Sept. 2001 in a 1.0 ha kiwi fruit orchard at the Experimental Station of Caxias do Sul, FEPAGRO, RS, Brazil. Ten corriedale pregnant ewes with average liveweight of 46 kg were grazed on this area. Grass availability and the damaged caused by the sheep were assessed three times (12 July, 24 Aug. and 18 Sept.) during this period. The grassland was sampled by cutting ten quadrats (0.4 x 0.6 m) to ground level. The damage caused on the plant trunks was visually assessed through the number of trees that were gnawed by sheep. Two areas of approximately 15 m^2 (5 x 3 m) were isolated from animal grazing to verify the development of the grassland without the access of sheep.

Results and discussion The dry matter (DM) yield of the grassland fell from almost 5,000 kg/ha to about 728 kg/ha in 68 d (Figure 1). The number of plants damaged did not increase until the DM herbage mass fell below 1,000 kg/ha. Therefore, the increase in the number of damaged plants seemed to be related to the low amount of herbage mass available in the last assessment period. In this period, the animals might have been looking for new sources of food, gnawing the trunks of the trees. According to Rattray *et al.* (1987), pasture cover below 1,000 kg/ha limits pasture intake by sheep.

However even when the herbage mass was reasonably high, the animals had still gnawed some plants. This suggests that pasture cover is not the only variable that influences the decision by sheep to gnaw trunks of plants. There may be similarities to studies when more than one pasture species is offered (Poli *et al.*, 1997). The ruminants are always sampling what is available and thus consume a mixed diet, but when availability is reduced, the animals change their strategy and graze randomly.

Conclusions The results of the present study demonstrates that the amount of grass offered was not enough to avoid the animals damaging kiwi fruit plants. Although the sheep were good grass mowers and a potential source of income for small farmers, it is important to ensure high pasture cover and to protect plant trunks when using sheep in kiwi fruit orchards.

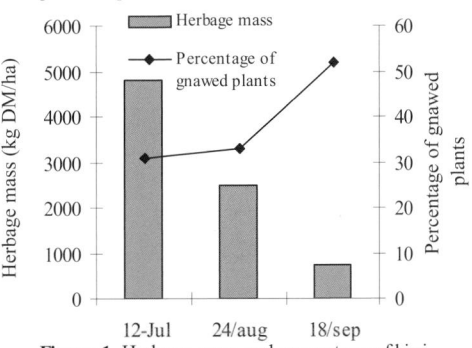

Figure 1 Herbage mass and percentage of kiwi fruit plants gnawed by sheep on an orchard, Caxias do Sul, RS, Brazil

References
Poli, C.H.E.C., J. Hodgson, G.P. Cosgrove & G.C. Arnold (1997). Partial preference of grazing cattle for contrasting legume swards. *Proceedings XVIII International Grassland Congress*, I: section 5-5.
Rattray, P.V., K.F. Thompson, H. Hawker & R.M.W. Summer (1987). Pastures for sheep production. In: A.M. Nicol (ed.) Livestock Feeding on Pasture. *New Zealand Society of Animal Production. Occasional Publication*, 10, 89-104.

Production systems to integrate livestock grazing and grain production in southern Brazil and Midwestern USA

R.M. Sulc[1], A. Moraes[2], S.J. Alves[3], A. Pelissari[2], P.C.F. Carvalho[4] and C.R. Lang[2]
[1]*The Ohio State University, Columbus, OH, USA 43210, Email: sulc.2@osu.edu,* [2]*Universidade Federal do Paraná, Curitiba – PR 80035-050, Brazil,* [3]*Instituto Agronomico do Paraná, Londrina – PR 86060-220, Brazil,* [4]*Universidade Federal do Rio Grande do Sul, Porto Alegre – RS 91501-970, Brazil*

Keywords: integrated systems, grazing, no-tillage, crop rotation

Introduction Agriculture in the USA and Brazil has undergone similar and dramatic changes in the past 20 years. In both countries, production systems have become increasingly specialized. Large farms are characterized by single enterprises, simple crop rotations, and livestock production is segregated from grain production. The lack of diversification and high production costs expose producers to risk from economic swings of single enterprises and greater reliance on pesticides and synthetic fertilizers to maintain profitability, along with greater risk of soil erosion from continuous row crop production. Scientists in southern Brazil and Ohio are collaborating to develop no-tillage systems that integrate livestock grazing with cash grain production. The goal is diversified production systems that are profitable as well as biologically and environmentally sound.

Strategies of integration Two primary strategies are being followed for integrating livestock grazing and grain production: rotation of annual grain crops with perennial pastures and livestock grazing of winter cover crops or annual pastures in rotation with summer grain crops. The first strategy seems to be readily adopted by livestock producers, especially when pasture renovation is desirable. The second strategy often receives resistance from grain producers, because they believe grazing livestock will significantly compact the soil and adversely affect soil properties such that subsequent grain yield will be compromised.

Systems research To address producer concerns, a systems approach to studying soil-plant-animal interactions was carried out in southern Brazil since 1994 and is now being developed at several institutions in the USA. The research is providing a basis for best management practices for integrating livestock grazing and cash grain production on the same land base in a sustainable and ecologically sound manner. Evaluations include species for autumn and winter grazing, stocking density and grazing intensity, timing of grazing events, and animal performance on winter annual pastures while considering the impact on nutrient cycling, soil physical and chemical properties, pest cycles, and subsequent grain production. Brazilian studies have shown that winter grazing does not compromise grain yield of soybean (Glycine max L.) and corn (Zea mays L.); winter grazing may even increase yield provided animal stocking and grazing are managed appropriately (Mello and Assmann, 2002; Moraes et al., 2003). Soil physical properties are affected by animal traffic, but biologically amelioration of soil compaction from grazing appears possible via pasture and grain crop root system growth under strategically managed grazing (Moraes and Lustosa, 1997). In Brazil, managed grazing of high quality forages by steers with improved genetic potential has yielded average daily gains exceeding 1.2 kg/day on winter annual pasture and 0.8 kg/day on productive perennial summer pasture, resulting in live weight production of 650 kg/ha during the winter season (annual pastures rotated with summer grain crops) and 1600 kg/ha on permanent summer pasture (210 days). Steers have reached market weight in 18 months on grass pastures. In Brazil, the no-tillage integrated systems can improve profits 8-fold over the average extensive stocker grazing systems and 1.5-fold over soybean grain production systems. In Ohio, evaluations are focusing on extending the grazing season with cereal forages established within corn, soybean, and winter wheat grain rotations to reduce use of expensive stored feed during the winter months. Studies are in progress to evaluate winter grazing strategies to minimize adverse impacts on soils and subsequent grain production. The results have motivated grain and livestock producers to experiment and learn first-hand of the financial, biological, and environmental benefits of integrated systems. Integrated livestock grazing – grain systems show tremendous promise to increase profitability and sustainability of farms in Brazil and the USA.

References

Mello N.A. & T.S. Assmann (ed.) (2002) I Encontro de Integração Lavoura-Pecuária no Sul do Brasil, Pato Branco, Brazil. 14-16 August 2002. CEFET-PR, Pato Branco – PR, Brazil.

Moraes A. & S.B.C. Lustosa (1997) Efeito do animal sobre as caraterísticas do solo e produção da pastagem. In Jobim et al. (ed.) *Simpósio Sobre Avaliação de Pastagens com Animais*, Maringá, Brazil. Univ. Estadual de Maringá, Maringá – PR, Brazil, 129-149.

Moraes A., R.M. Sulc, C.R. Lang, D.J. Barker, A. Pelissari, L.A. Lucchesi & M.B. Magalhaes (2003) Corn productivity following winter annual pasture with different grazing intensities and nitrogen rates. *Agronomy Abstracts* [CD-ROM computer file]. ASA, Madison, WI.

Effect of dairy effluent on turnip yields

J.L. Jacobs, G.N. Ward and F.R. McKenzie
Department of Primary Industries, 78 Henna Street, Warrnambool, Vic 3280, Australia, Email: joe.jacobs@dpi.vic.gov.au

Keywords: leaf, root, nitrogen, potassium, dry matter

Introduction Dairy effluent is a significant point source in the pollution of waterways. Only 50% of dairy farms in the dryland regions of Victoria, Australia, have suitable dairy effluent systems of which only 25% are managed effectively (IRIS Research 2000). Despite many farmers viewing effluent as an undesirable waste, it contains relatively large amounts of agronomically valuable nutrients especially nitrogen (N) and potassium (K). Results are reported from the first two years of a three year study comparing turnip leaf and root dry matter (DM) responses to a range of dairy effluent rates.

Materials and methods This study was conducted on a commercial dairy farm (38°14'S, 142°55'E) in western Victoria on a Mottled-Sodic, Eutrophic, Brown Chromosol (Isbell 1996) soil. In both years (2002, 2003) following silage harvesting, the experimental area was grazed and ploughed. Within a week, the area was power harrowed and sown to turnips (*Brassica rapa* cv Barkant) at a rate of 2 kg/ha. From 6–8 weeks after sowing, effluent was applied at 15 mm/ha/d, providing six treatment levels of 0, 15, 30, 45, 60 and 75 mm to random plots (12 m x 12 m) replicated six times in a randomised block design. DM yield was estimated 14 weeks after sowing, (6 quadrats [1.0 m^2]/plot were weighed individually and sub sampled for DM determination). An analysis of variance (ANOVA) (GenStat Committee 2000) with significance declared if $P<0.05$ was conducted.

Results Dairy effluent composition (Table 1) shows that whilst total N content remained relatively constant, the proportion as ammonia-N altered markedly with 89% being in this form in year 1 and 62% in year 2. Other nutrients, (P, Na, K) increased in year 2. In Year 1, effluent at 45 mm and higher increased ($P<0.05$) turnip leaf DM yield compared to the control, whilst at 60 mm and above it also increased ($P<0.05$) leaf DM yield compared to 15 and 30 mm (Figure 1). At 30 mm or higher there was an increase ($P<0.05$) in root DM yield compared to the control. In year 2, effluent increased ($P<0.05$) leaf and root DM yields compared to the control.

Table 1 Effluent composition

	Year 1		Year 2	
	Mean	s.d	Mean	s.d
pH	8.0	0.0	8.1	0.16
P (mg/l)	23.3	1.	34.3	3.14
N (mg/l)	155	10.	157	5.16
NH$_3$-N (mg N/l)	137.5	5.	97.8	11.57
Na (mg/l)	507.5	17.	581.7	40.21
K (mg/l)	445	12.	480	42.4

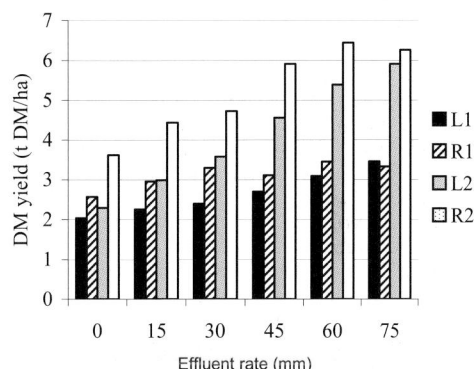

Figure 1 Effect of dairy effluent on turnip leaf (L)and root (R) DM yields (t DM/ha) in year 1 and 2

Conclusions Results indicate that application of dairy effluent can have a marked effect on turnip leaf and root DM yields. Responses in year 2 are likely to have been higher due to extensive damage caused by cabbage moth (*Plutella xylostella* (L.) in year 1. This work will continue for a further year.

References
Genstat 5 Committee (1997). 'Genstat 5.41 Reference Manual'. Oxford Science Publications, Oxford, UK.
IRIS Research, (2000). A survey of Natural Resource Management on Australian Dairy Farms. Technical report.
Isbell R.F. (1996). Australian Soil and Land Survey Handbook - The Australian Soil Classification. CSIRO

Effect of dairy effluent on turnip nutritive characteristics

J.L. Jacobs, G.N. Ward and F.R. McKenzie
Department of Primary Industries, 78 Henna Street, Warrnambool, Vic 3280, Australia; Email: joe.jacobs@dpi.vic.gov.au

Keywords: crude protein, metabolisable energy, nitrogen, potassium

Introduction In southern Victoria, high summer temperatures and low rainfall lead to low pasture growth and a decline in nutritional value until rainfall commences in autumn. Annual forage crops such as turnips often are used to fill the summer feed gap. Jacobs & Ward (2003) observed that dairy effluent applied at low rates could improve turnip DM yields and crude protein content. Results from the first two years of a 3-year study comparing a range of effluent application rates on turnip leaf and root nutritive characteristics are reported.

Materials and methods This study was conducted on a commercial dairy farm (38°14'S, 142°55'E) in western Victoria on a Mottled-Sodic, Eutrophic, Brown Chromosol (Isbell 1996) soil. In both years (2002 ,2003) following silage harvesting, the experimental area was grazed and ploughed. Within a week, the area was power harrowed and sown to turnips (*Brassica rapa* cv Barkant) at a rate of 2 kg/ha. From 6–8 weeks after sowing, effluent was applied at 15 mm/ha/d, providing six treatment levels of 0, 15, 30, 45, 60 and 75 mm to random plots (12 m x 12 m) replicated 6 times in a randomised block design. To determine nutritive characteristics, 6 quadrats (1.0 m^2)/plot were collected and sub-sampled for leaf and root 14 weeks after sowing. An analysis of variance (ANOVA) (GenStat Committee 2000) with significance declared if $P<0.05$ was conducted.

Results In both years, dairy effluent contained approximately 29 kg P, 156 kg N, 545 kg Na and 460 kg K/ML. In year 1, leaf (L1) and root (R1) CP was increased ($P<0.05$) at 60 mm compared with the control (Figure 1). In year 2, leaf (L2) CP was increased ($P<0.05$) at 45 mm and higher compared with the control. Leaf water soluble carbohydrate (LWSC) content was reduced ($P<0.05$) at the highest rate of effluent application, compared with all other treatments (Table 1). Root (R2) CP was increased ($P<0.05$) at 45 mm and higher compared with the control, whilst NDF content (RNDF) was higher ($P<0.05$) in the control compared with all effluent applications, except the 15 mm.

Table 1 Effect of dairy effluent on turnip leaf water soluble carbohydrate (WSC) content and root neutral detergent fibre (NDF)content (%DM) in year 2

	0	15	30	45	60	75	l.s.d
L WSC	24.1	23.0	23.5	22.4	22.3	18.9	2.05
R NDF	21.2	21.0	18.9	19.4	19.7	18.9	1.02

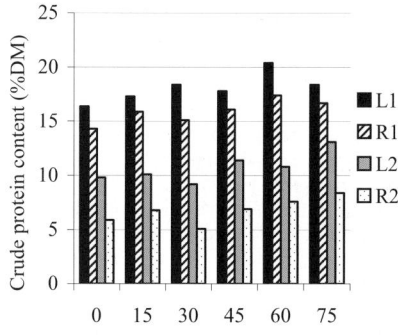

Figure 1 Effect of dairy effluent on turnip leaf and root crude protein content (%DM)

Conclusions The application of dairy effluent had a marked effect on turnip leaf and root CP content. Pasture CP content during summer is often low (<10%) and cost of purchasing additional CP is high. The reduction in NDF content is unlikely to be an issue given the fibrous nature of available pasture and conserved feeds used at this time. The use of dairy effluent on turnips may provide greater flexibility in feeding options to maintain milk production during summer.

References

Genstat 5 Committee (1997). 'Genstat 5.41 Reference Manual'. Oxford Science Publications, Oxford, UK.
Isbell R.F. (1996). Australian Soil and Land Survey Handbook - The Australian Soil Classification. CSIRO.
Jacobs J.L. & G.N. Ward (2003). Effect of different rates of dairy effluent on turnip DM yields and nutritive characteristics. *Proceedings of the 11th Australian Agronomy Conference*, Geelong, Victoria.

Implications of land use changes on the yields in dry matter, energy and protein of range and crop fields in Zamfara Reserve, northwestern Nigeria

B.S. Malami[1], P.H.Y. Hiernaux[2], H.M. Tukur[1] and B. Rischkowsky[3]
[1]Department of Animal Science, Usmanu Danfodiyo University, Sokoto, Nigeria, E-mail: bmalami@yahoo.com, [2]Centre for Agriculture in the Tropics and Subtropics, University of Hohenheim, Stuttgart, Germany, [3]Institute of Livestock Ecology, Justus Liebig University, Giessen, Germany

Keywords: forage productivity, land use, rangeland, crop-livestock interaction

Introduction The Zamfara reserve is a 235,500 ha grazing land within the Sudan savannah zone (12° 10' - 13° 05 N; 6° 30' - 7°15' E) of north western Nigeria. Rainfall varies from 500mm in the north to 800mm in the south of the reserve, and is restricted within the months of May - September. The reserve is an important grazing site for the herds of sedentary, transhumant and agro pastoralists. Population growth within the farming communities in the reserve has led to the conversion of more grazing land to croplands. This work was carried out to evaluate the quantity and quality of herbage on the natural range and the croplands, in order to have an insight on the effect of increased cropping activity on biomass availability in the reserve.

Materials and methods Total biomass in the range was estimated at intervals of five weeks for one year. Density and leaf mass of trees and shrubs was estimated at the end of the rainy season. These measurements were carried out along 1000m long transects in the north, centre and south of the reserve. Biomass productivity in the crop fields was measured at the end of the rainy season in low, medium and high productivity farms selected in the north, centre and south of the reserve. A diagonal transect of 100m was used in each field and measurements taken at every 10m. Samples of biomass from the range and crop fields were analysed for CP and gas production. The latter was used to calculate ME (Menke et al., 1979).

Results Total forage biomass (excluding grains), ME and DCP were higher in the crop lands compared to the range (Table 1). Density of woody plants was however higher in the range. Thus, leaf mass from trees and shrubs was higher in the range compared to the crop fields.

Conclusion These results suggest better forage productivity under crop cultivation. However, intensive crop cultivation could have a negative impact on resource conservation and biodiversity. This is clearly indicated by the very low density of the ligneous strata in crop fields, which could have a negative effect on nutrient recycling and soil erosion. Therefore, among others, integration of crop and livestock production should aim to conserve natural woodland to enhance sustainable resource utilisation.

Table 1 Mean aerial utilizable plant biomass, metabolisable energy and digestible crude protein on natural range and crop fields of Zamfara reserve

	Crop fields	Natural range
Density of woody plants (No./ ha)		
Trees	0.19	18.3
Shrubs	1.41	120.1
Biomass (kg DM /ha)		
Herbaceous mass	-	563.0
Farm weeds	135.1	-
Crop residues from:		
Sorghum	586.7	-
Millet	1721.3	-
Cowpea	260.5	-
Leaf mass (kg DM /ha)		
Trees and shrubs	21.4	105.6
Biomass		
Total forage mass (Kg DM /ha)	2725.0	668.6
Energy (MJ /ha)		
Sorghum	4165.6	-
Millet	13426.1	-
Cowpea	2318.5	-
Farm weeds	1310.5	-
Tree and shrub leaves	128.4	633.6
Herbaceous layer	-	5517.4
Digestible CP (Kg /ha)		
Sorghum	6.5	
Millet	37.9	
Cowpea	20.8	
Farm weeds	7.0	
Tree and shrub leaves	1.7	8.2
Herbaceous layer	-	45.0

References

Menke K.H., L. Raab, A. Salewski, H. Steingass, D. Fritz & W. Schneider (1979). The estimation of the digestibility and the metabolizable energy content of ruminant feeding stuffs from gas production when they are incubated with rumen liquor *in vitro. Journal of Agricultural Science, Cambridge,* 93, 217-222.

The suitability for organic cattle beef production of mixed farming systems in the highlands of north east Portugal

J.M. Pires[1], M. Rodrigues[1], F. Sousa[1], A. Bernardo[2], J.C. Pires[1], J. Cabanas[1], H. Resendes[1], M.J. Ferreira[1], M.I. Silva[1] and N. Moreira[3]

[1]Mountain Research Center, Escola Superior Agrária, 5301-855 Bragança, Portugal, Email: jaime@ipb.pt, [2]Direcção Regional de Agricultura de Trás-os-Montes, Montalegre, 5470 Montalegre, Portugal, [3]Crop Science Department, Universidade de Trás-os-Montes e Alto Douro 5001-911 Vila Real, Portugal

Keywords: agricultural systems, land use, ranching, meadow, organic cattle feeding stuffs

Introduction The EC Reg. 1804/99 takes account of animal production in organic farming. However, these specifications may limit implementation and expansion of organic animal production, due to environmental and system constraints. Mixed farming, as defined by Spedding (1988) and Grigg (1996), is commonly practiced in the NE highlands of Portugal. Two farms were studied in this region in order to evaluate their suitability for organic cattle beef production, taking account of the technical specifications of the EC regulation.

Materials and methods Two farms were monitored for a year (autumn 2002-03); one located near Montalegre (F1) (41° 36' N, 7° 55' W and 950 m asl) and the other near Vinhais (F2) (41° 53' N, 6° 58' W and 700 m asl). The long-term annual rainfall is 1531 and 741 mm and the annual mean temperature 9.6 and 12 °C, respectively. The farms, F1 and F2, produce beef cattle from the "Barrosã" and "Mirandesa" breeds, respectively. Farm activities and components such as inputs, outputs, yield and flows between state variables were recorded.

Results Legumes are present in meadows, although with low mean values, i.e. 7 % (F1) and 15 % (F2). Farm F1 has a larger area of grassland and other forage crops than F2 (96.4 % compared to 40.6 %), less cropland based on cereals and crucifers (11.5 % compared to 40.6 %), 7-8 times less off-farm nitrogen (inorganic-N fertilisers) and grazing accounts for a higher proportion of the cattle diets (67.3 % compared to 51.9 %). Considering these as criteria for suitability for organic farming, F1 is better suited for organic beef production.

Table 1 Characteristics of farms F1 and F2 (total annual values from autumn 2002 to autumn 2003)

| | Montalegre farm (F1) | | | | Vinhais farm (F2) | | | |
| | Area | | Feedingstuffs | | Area | | Feedingstuffs | |
Land use	(ha)	(%)	(DM, t/yr)	(%)	(ha)	(%)	(DM, t/yr)	(%)
Meadow	22.6	63.1	$120.1^G + 69.0^H$	68.6	16.2	29.1	$88.2^G + 52.9^H$	76.2
Turnip + beetroot					1.5^1	2.7	3.1^F	1.7
Maize (regional)	3.9^1		21.1^F	7.7	0.2^1		0.4^F	0.2
Fallow ^{2}					9.6	17.2		
Wheat (grain) ^{2}					0.3	0.5		
Rye (grain) ^{2}					6.3	11.3		
Rye (forage)	4.1^1	11.5	11.2^G	4.1	0.4^1		2.1^F	1.1
Oat (forage) ^{2}					1.4	2.5	34.2^H	18.5
Vegetable-garden	0.2^1				0.2^1			
Long-term fallow ^{2}					3.5	6.3	4.3^G	2.3
Shrubs and forest	7.8	21.8	54.0^G	19.6	7.2	12.9		
Chestnut	1.3	3.6			9.7	17.4		
Total	35.8	100.0	275.4	100.0	55.7	100.0	185.2	100.0

Other data per farm, F1 and F2, respectively: concentrates 5,120 and 1,280 kg; feedingstuffs area 34.5 ha and 22.6 ha; stocking rate (livestock units/ha of feedingstuffs area) 0.58 and 0.35; livestock weight 8,942 and 7,467 kg; deadweight 615 and 249 kg; off-farm N 10 and 75 kg.
[1]Double crop; [2]Cereal-fallow rotation; ^{G}Grazing; ^{H}Hay; ^{F}Green forage

Conclusions The simpler mixed farming system practised in F1, located in a more mountainous humid region, is approaching a ranching system, as defined by Grigg (1996). This farm, using only seven state variables, five of which are directly related to cattle, seems to better fulfil the specifications for organic animal production.

References
Grigg, D. B. (1996). The Agricultural Systems of the World. An Evolutionary Approach. Cambridge University Press, Cambridge, 358 pp.
Spedding, C. R. (1988). An Introduction to Agricultural Systems. Elsevier Applied Science Publishers, Essex, 189 pp.

Black medick – a beneficial companion crop for use in organic grass production

R. Machá••and B. Cagaš
OSEVA PRO Ltd., Grassland Research Station at Zubri, Hamerská 698, CZ-756 54 Zubri, Czech Republic, Email: machac@quick.cz

Keywords: perennial ryegrass, timothy, organic seed production

Introduction Organic farmers must use only organically produced seed for establishing new meadows and for, renovation and undersowing of old pastures, in accordance with EC regulations. Therefore an important and difficult goal is to obtain enough seed of grasses without the use of pesticides and inorganic fertilisers. The seed yield is closely related to the number of fertile tillers, which depends on adequate nitrogen in the soil. Growing grasses for seed with a legume, as a companion crop, is one possibility for providing a source of organic nitrogen. Aamlid (1999) claimed that growing timothy together with white clover or alsike clover can produce yields of timothy grass seed comparable to conventional production. The need for more information about growing grasses with leguminous crops was emphasised by Marshall & Humphreys (2002) and was the subject of this research.

Materials and methods Seed yields of perennial ryegrass (*Lolium perenne* L.) cv. Ba•a and timothy (*Phleum pratense* L.) cv. Sobol, grown together with companion legumes were compared in a 3-year field trial. A multifactorial trial was established in April (timothy) and August (ryegrass) 2000 at the Grassland Research Station at Zubri. Timothy was undersown into spring wheat as a cover crop, but perennial ryegrass was sown directly. The trial consisted of two grass species (factor 1); three leguminous companion crops (factor 2), i.e. diploid red clover (*Trifolium pratense* L.) cv. Start, white clover (*T. repens* L.) cv. Vyso•an and black medick (*Medicago lupulina* L.) cv. Ekola, and N-nutrition (factor 3), i.e. N-transfer from current legumes, the organic N from previously harvested legumes (mulching) or organic manuring with slurry. Each treatment combination had four replicates. The seed yield from all combinations of the factors was compared with that from conventional grass seed production and tested by ANOVA.

Results The seed yield of timothy ranged from 158 to 863 kg/ha. Timothy with black medick produced 676 kg/ha (15 % lower than conventional practice) in the 2nd harvest year (the seed of legumes was harvested in the 1st year) and 255 kg/ha (29 % higher than conventionally grown seed; highly significant difference) in the 3rd year. Total seed yield of timothy (for harvest years 2 and 3) is shown in Table 1. The combination of perennial ryegrass with black medick gave the best results, with overall weighted mean (organic treatments) of 529 kg/ha in the first harvest year, 472 kg/ha in the second harvest year and 167 kg/ha in the third harvest year. Compared with the conventional grass seed production, the yield in the 1st year was 8 % higher, in the 2nd year 32 % lower, and in the 3rd year it was 40 % lower. Total seed yield of three harvest years for perennial ryegrass is shown in Table 2.

Table 1 Seed yield of timothy (kg/ha)

Treatment Companion crop	Bacterial nodules	Mulching	Organic manure	Conventional
Red clover	533	582	858	730
White clover	680	581	797	700
Black medick	1012	716	1072	928
Without legume				912

Table 2 Seed yield of perennial ryegrass (kg/ha)

Treatment Companion crop	Bacterial nodules	Mulching	Organic manure	Conventional
Red clover	779	414	843	1173
White clover	944	558	1130	1431
Black medick	990	677	1307	1491
Without legume				1480

Conclusions The 3-years field trial showed that timothy and perennial ryegrass grown for seed together with black medick are able to produce seed yields comparable to those produced conventionally especially when the medick is combined with organic manuring.

Acknowledgements Supported by the Ministry of Agriculture of the Czech Republic, Project No. QD 0004

References

Aamlid, T.S. (1999). Organic seed production of timothy (Phleum pratense) in mixed crops with clovers (*Trifolium ssp.*). *Proceedings of 4th International Herbage Seed Conference*, Perugia, 28-32.
Marshall, A.H. & M.O. Humphreys (2002). Challenges in organic forage seed production. *Proceedings of the COR Conference*, 26-28 March 2002, Aberystwyth, 53-54.

Effect of temporary grasslands of different age, composition and management on winter wheat yields in a crop rotation

B. Deprez, R. Lambert and A. Peeters
Laboratory of Grassland Ecology, Catholic University of Louvain, Place Croix du Sud 5 bte 1, B-1348 Louvain-la-Neuve, Belgium, Email: deprez@ecop.ucl.ac.be

Keywords: grassland, red clover, winter wheat, fertility transfer

Introduction Organic nitrogen (N) accumulates in the soil in temporary grasslands. This accumulation is especially important when mineral N fertilisation is high. Legume-based temporary grasslands may also supply the soil with nitrogen through fixation of atmospheric N, for instance in organic farming. When ploughing temporary grassland, great amounts of mineral nitrogen can be released through the process of decomposition and mineralisation of soil organic matter. This mineral nitrogen can be taken up by succeeding crops. The aim of this experiment was to evaluate the effect of temporary grasslands of different ages, N fertilisation levels and legume contents on winter wheat yields in a crop rotation.

Materials and methods The experiment was established on a sandy loam soil. From 1999 to 2002, grassland was sown each year in four complete randomised blocks. Each block was divided to give the following treatments: perennial ryegrass (*Lolium perenne*, cv Merlinda) (Lp) with no N fertilisation (N_0), Lp with 200 kg N/ha (N_{200}), Lp with 400 kg N/ha (N_{400}) and a non-fertilised red clover (*Trifolium pratense*, cv Merviot)/perennial ryegrass (*Lolium perenne*) mixture (Tp+Lp). In Oct. 2003, all grassland plots at the four different ages (1, 2, 3 and 4 years) were ploughed and winter wheat (*Triticum aestivum*, cv Drifter) was sown. The previous Lp N_0 swards were used to establish a calibration curve by fertilising wheat with 0, 50, 100, 150 and 200 kg N/ha.year (in three dressings). On the previous Lp N_{200}, Lp N_{400} and Lp+Tp swards, no N fertiliser was applied to the wheat. In Aug. 2004, wheat was harvested and grain dry matter (DM) yields were measured.

Results Grain DM yields of wheat cultivated on the previous red clover based plots were similar for all years to that of the 50 kg N/ha fertilised reference wheat (Table 1). The DM yield of non-fertilised wheat sown after the 4 year old Lp N_{200} grassland was similar to that of wheat fertilised with 50 kg N/ha. For the younger grasslands (1, 2 and 3 years old) no differences were noted between the previous Lp N_{200} grassland and the non-fertilised reference wheat (Table 1). Grain yields of non-fertilised wheat cultivated after the previous Lp N_{400} grassland of 3 and 4 years old were similar to that obtained for the 100 kg N/ha fertilised reference wheat while wheat yields after 2 and 1 years old Lp N_{400} grasslands were, respectively, similar to the 50 kg N/ha and non-fertilised reference wheat (table1). Yields of wheat cultivated after previous Lp N_{200} and Lp N_{400} significantly increased with the age of the grass (Figure 1).

Table 1 Grain DM yield (t/ha) and statistical analysis (α=5%)

Fertilisation		Age of sward (years)			
Grassland (kg/ha)	Wheat (kg/ha)	4	3	2	1
0	0	3.9 nop	4.3 mno	3.3 pq	3.2 Q
0	50	5.6 ghij	5.7 ghi	4.9 jklm	4.6 klmn
0	100	6.7 ef	6.9 e	6.0 fg	6.0 fg
0	150	8.3 bc	8.5 abc	7.3 de	6.9 e
0	200	9.1 a	8.6 ab	7.8 cd	7.4 de
200	0	5.0 ijklm	4.5 lmno	3.8 opq	3.4 pq
400	0	7.2 de	6.8 e	5.2 hijk	3.9 opq
Tp+Lp (0)	0	5.8 gh	5.5 ghij	5.2 hijkl	5.2 hijk

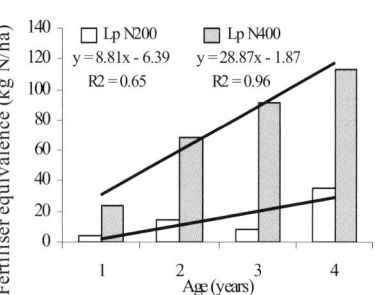

Figure 1 Fertiliser equivalence for Lp N_{200} and Lp N_{400} of different ages

Conclusions Introducing red clover based temporary grasslands in a crop rotation is of potential value in organic farming. Indeed, the present results show wheat grain DM yields of up to 5.8 t/ha are obtained and are not significantly influenced by the age of the sward; the mean production of wheat grain is around 5.0 t DM/ha in Belgian organic farms in the same region. In conventional farming, the N fertilisation of wheat can be reduced by 63 kg/ha (mean of the 4 years) after red clover mixtures (data not shown). Grasslands of 4 years old allow a saving of 36 (N_{200}) and 113 kg N/ha (N_{400}) for the fertilisation of a succeeding wheat depending on the N fertilisation level. This may be of great interest considering the price and the environmental impact of mineral fertilisers.

Riparian management in intensive grazing systems for improved biodiversity and environmental quality: productive grazing, healthy rivers

S.R. Aarons[1], M. Jones-Lennon[1], P. Papas, N. Ainsworth, F. Ede and J. Davies
[1]Ellinbank Research Institute, PIRVic Ellinbank, RMB 2460 Hazeldean Rd, Ellinbank, Victoria, 3821 Australia, Email: sharon.aarons@dpi.vic.gov.au & www.dpi.vic.gov.au/vro/biodiversity/riparian

Keywords: biodiversity, weeds, water quality, woody debris

Introduction Within high rainfall intensive grazing systems of southern Victoria, riparian zones are often degraded due to vegetation clearing, stock access and inappropriate farm management. Streams in these landscapes often have poor water quality and reduced biodiversity due to degraded terrestrial and aquatic ecosystems. Improved management of riparian zones depends on developing tools and practices for integration into productive grazing systems. This paper describes the approaches used and the tools developed in the 'Productive Grazing, Healthy Rivers: Improving riparian and in-stream biodiversity' project.

Materials and methods The project sites are on intensive beef and dairy properties that occur within the 4 high rainfall (>750mm) bioregions (NRE 1997) of southern Victoria (Figure 1). During the project development phase (September 2001-April 2002) an extensive literature and data review was undertaken followed by consultation with key stakeholders involved in riparian zone management in the grazing industries. Research gaps identified in this process were used to develop the 6 activities currently underway in the research and development phase (June 2002-July 2005) of the project.

Results The review of riparian biodiversity assets (NRE 2002) identified the need to survey riparian biodiversity actively on grazing properties in the study area. Current riparian management actions often differed from recommended guidelines. Eight key riparian management issues were identified, which were used to develop 6 research and development modules. These are: biodiversity surveys of 40 paired, fenced and unfenced, riparian sites on dairy and beef properties in southern Victoria; investigation of the impact of introducing small woody debris on biodiversity of fenced replanted riparian sites; surveys of 36 sites to identify factors affecting native tree recruitment in fenced riparian sites; development of a weed decision support tool to assist landholders manage riparian weeds; assessment of riparian condition of 107 dairy farm sites; water and soil monitoring along the riparian zone of 2 commercial dairy farms to identify farm management impacts and to monitor changes after fencing and revegetation (Aarons et al., 2005).

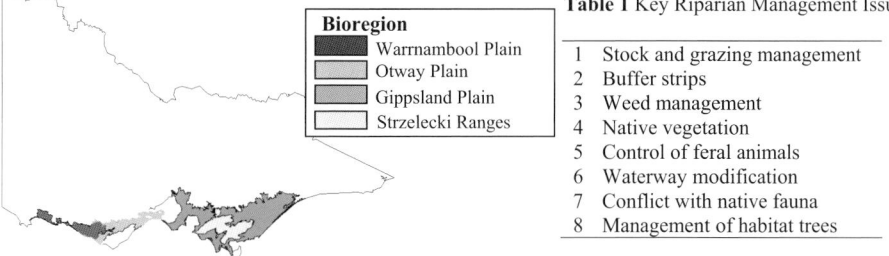

Bioregion
- Warrnambool Plain
- Otway Plain
- Gippsland Plain
- Strzelecki Ranges

Table 1 Key Riparian Management Issues

1	Stock and grazing management
2	Buffer strips
3	Weed management
4	Native vegetation
5	Control of feral animals
6	Waterway modification
7	Conflict with native fauna
8	Management of habitat trees

Figure 1 Project location showing the 4 study bioregions

Conclusions Key riparian management issues identified by the team and stakeholders were the basis of the research activities developed. Biodiversity survey data indicates that fenced sites on beef and dairy farms have greater native biodiversity. Unfenced riparian areas on dairy farms in SE Victoria are generally in poor condition, with on-farm activities contributing to degraded riparian zones. On the other hand, weed species often are associated with fenced riparian areas. A riparian weed decision support tool was developed to assist farmers to manage fenced riparian land. Information from this project is disseminated regularly to landholders and natural resource managers in the project study area.

References

Aarons, S.R., A. Melland & C.J.P. Gourley (2005). Grazing management impacts on the riparian zone and on water quality. XX International Grasslands Congress – offered papers (in press).

NRE (1997). Victoria's biodiversity: Directions in management. Department of Natural Resources and Environment, East Melbourne, Victoria.

NRE (2002). Biodiversity conservation in intensive grazing systems: Riparian and in-stream management. Department of Natural Resources and Environment, Ellinbank, Victoria.

Response of guinea grass (*Panicum maximum* Jacq) to application of cow dung in South West Nigeria

O.S. Onifade, J.A. Olanite, A.O. Jolaosho, M.O. Arigbede and N.K. Tijani
Dept. of Pasture and Range Management, College of Animal Science and Livestock Production, University of Agriculture, Abeokuta, Ogun State, Nigeria, E-mail: femionifade@yahoo.com

Keywords: Guinea grass, cow dung, forage yield, crude protein content

Introduction The yield of forage species from the world's grazing land is limited by poor soil and unproductive species (Jones & Wild, 1975; Cooke, 1982). The use of manure on pasture land not only represents a low cost disposal method but also a means of recycling nutrients for plant growth and counteracting the decreasing organic matter content in most agricultural soils. In agro-pastoral production systems, the interaction between crops and livestock is important. Manures are used mainly to complement inorganic fertiliser in the production of food crops. There is a dearth of information on the response of pasture grass to application of cow dung and so the response of two *Panicum maximum* ecotypes to cow dung was evaluated.

Materials and methods The experiment was carried out at the Teaching and Research Farm of the University of Agriculture, Abeokuta, Nigeria. There were four treatments consisting of two ecotypes of *Panicum maximum* – Ntchisi and local. These received either 30 kg N/ha as cow dung or no dung (control). A randomised complete block design was used with three replicates. Crown splits (about 20 cm in length) of the two grasses were transplanted 50 cm between and within rows into a cultivated seedbed in May, 2002. Cow dung was broadcast one week after transplanting (WAP). The heights (to the tip of the youngest leaf) and dry matter (DM) yields were estimated every four weeks. Crude protein was determined according to the methods of AOAC (1990).

Results The height of the local ecotype of *Panicum maximum* (119 cm) was consistently higher (P• 0.05) than the improved Ntchisi ecotype (99 cm). The mean heights of both ecotypes at 4, 8, 12, 16 and 20 WAP were 47, 81,137,155 and 152 cm, respectively. The values from week 12 to 20 were similar (P=0.05). The differences in height between Ntchisi + Cow dung (109 cm) and Local – Cow dung (102 cm) were not significant. The greatest height (136.5cm) was for Local + cow dung and the least (88.2cm) for Ntchisi – cow dung. There was no stem component when measured four weeks after planting (Table 1). Application of cow dung to Ntchisi enhanced the leaf and stem DM yields, except in week 8 for the leaf component. Total DM yields at 8, 16 and 12 weeks were 430, 750 and 810% higher than the initial cut, respectively. Throughout the growing season, the two control treatments had the lowest total DM yields. The leaf, stem and total DM yields reached maximum values at week 12 but were similar to those at week 16. The CP contents for the stem components ranged from 9.2 to 12.1% and 7.3 to 10.3% at weeks 8 and 12 respectively. The controls had lower CP contents compared to the treatments which received cow dung.

Table 1 Effects of cow dung application on dry matter yields (t/ha) of two ecotypes of *P. maximum*

Treat.	4WAP Leaf	4WAP Total	8WAP Stem	8WAP Leaf	8WAP Total	12WAP Stem	12WAP Leaf	12WAP Total	16WAP Stem	16WAP Leaf	16WAP Total
N + CD	0.43	0.43	1.11	0.74	1.85	2.44	1.07	3.51	2.22	1.05	3.27
L + CD	0.35	0.35	0.48	0.99	1.47	1.64	0.87	2.51	1.54	0.79	2.33
N - CD	0.21	0.21	0.82	0.44	1.26	1.07	0.76	1.83	1.03	0.56	1.59
L - CD	0.81	0.81	0.22	0.95	1.17	0.83	0.64	1.47	0.95	0.53	1.48
Mean	0.29	0.29	0.66	0.78	1.44	1.49	0.84	2.33	1.44	0.73	2.17
SEM	0.05	0.05	0.13	0.07	0.11	0.25	0.06	0.30	0.02	0.07	0.27

WAP = Weeks after transplanting, N = Ntchisi, L = Local, CD = Cow dung

Conclusion Application of cow dung increased growth, DM production and CP content of *Panicum maximum* (Ntchisi) compared to the local ecotype over a period of twenty weeks.

References
AOAC (1990). Association of Official Agricultural Chemists, Official Methods of Analysis Washington DC.
Cooke, G.W. (1982). Fertilizing for Maximum Yield. 2nd ed. Grenada Publishing Coy, London.
Hodges, R.D. (1991). Soil organic matter: Its central position in organic farming. In: Wilson, W.S. (ed.) Advances in Soils Organic Research: The impact on Agriculture and the Environment. The Royal Society of Chemistry, Redwood Press, Wiltshire, U.K.
Jones, M.J. & A. Wild (1975). Soils of the West African Savanna. CAB Technical Communication No. 55, Harpenden, UK.

Nitrogen use efficiency of specialized dairy farms in Flanders: evolution and future goals

F. Nevens, I. Verbruggen, M. Meul and D. Reheul
*Flemish Policy Research Center for Sustainable Agriculture, Potaardestraat 20, 9090 Gontrode, Belgium,
Email: frank.nevens@ugent.be*

Keywords: nitrogen use, eco-efficiency, dairy farms, Flanders

Introduction Efficient use of nutrients is one of the major aims of eco-efficient and sustainable agricultural production systems. We determined the nitrogen use efficiency of a representative set of specialised dairy farms in Flanders, between 1989-1990 and 2000-2001 and set achievable eco-efficiency targets for sustainability

Materials and methods Based on data of the local Farm Accountancy Data Network, we established farm-gate or whole-farm N balances of specialised dairy farms, for 1989-1990 (n=334) and for 2000-2001 (n=148). Nitrogen inputs included purchased concentrates, forages and by-products, straw (or sawdust), animals, mineral fertiliser, manure, biological fixation and deposition. The N output included exported milk, animals, manure and crops. The farm-gate N surplus was calculated as total N input – total N output. The farm N use efficiency was defined as 100 * N output / N input. These results were compared to those of Dutch experimental farms or farm-groups (references available from the authors). Finally, we proposed achievable targets to reach a given low N surplus and eco-efficiency (N-surplus per litre milk).

Results The average N surplus of the farms in the study decreased during the study period from 378 kg N/ ha per year in 1989-1990 to 238 kg N/ha per year in 2000-2001 (Table 1). The corresponding N use efficiencies were 15.1 and 22.0 %, respectively. This significant progress was mainly due to a significant decrease in the use of mineral fertiliser and, to a lesser extent, reduced concentrate use. The N output (in milk production) remained unchanged (Table 1). The farms moved from a 1989-1990 eco-efficiency level of 15 to 40 l milk/kg N surplus (Figure 1, zone A) to 20 to 60 l milk/kg N surplus in 2000-2001 (Figure 1, zone B), while a level of 60 to 100 l milk/kg N surplus seems achievable (Figure 1, zone C). Further, an absolute maximum level of 150 kg/ha for N surplus is necessary in order to comply with the European Nitrates Directive (Verbruggen *et al.*, 2004). Hence, an optimum zone for sustainable and eco-efficient dairying in Flanders can be delimited (Figure 1, zone C').

Table 1 Flemish dairy farms:
N balance (kg N/ha per year)

		1989	2001
N input			
Mineral fertiliser		238	128
Concentrates		104	76
Manure		25	29
Straw		1	1
Forages		26	17
Deposition		50	48
Fixation		2	6
	Total	446	305
N output			
Milk		47	49
Animals		19	16
Crops		2	2
	Total	68	67
N surplus		378	238
N use efficiency (%)		15.1	22.0

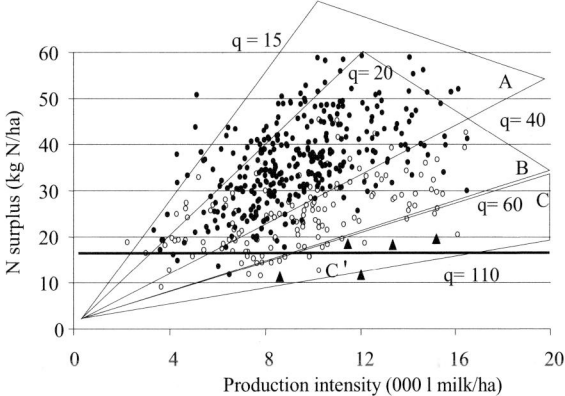

Figure 1 Farm-gate N surpluses in relation to production intensity: Flemish dairy farms in 1989-1990 (•) and in 2000-2001(•). Dutch experimental farms or farm groups (•). Isoquants of eco-efficiency (q, l milk/kg N surplus)

Conclusions During the past 15 years, Flemish dairy farms have successfully made considerable efforts to increase their N use efficiency with the average farm N surplus showing a significant decrease from 378 kg N/ha per year in 1989-1990 to 238 kg N/ha per year in 2000-2001. Nevertheless, further progress can still be made: an eco-efficiency of 60 to 110 l milk/kg N surplus and a maximum farm gate N surplus of 150 kg N/ha per year seem relevant and achievable targets for a sustainable future for Flemish dairy farming.

Reference
Verbruggen, I., F. Nevens, D. Reheul & G. Hofman (2004). Nitrogen use and nitrogen use efficiency on Flemish dairy farms. Flemish Policy Research Centre for Sustainable Agriculture, Publication 6, 58pp.

Evaluation with simulation of lucerne-based cropping systems to combat dryland salinity in Australia

W. Chen[1], M.J. Robertson[2] and W.D. Bellotti[1]
[1]School of Agriculture & Wine, The University of Adelaide, Roseworthy Campus SA 5371 Australia, Email: wen.chen@adelaide.edu.au, [2]CSIRO Sustainable Ecosystems, St Lucia Qld 4067 Australia

Keywords: APSIM simulation, phase cropping, companion cropping, dryland salinity, drainage, lucerne

Introduction Dryland salinity is one of the most significant forms of land degradation that farmers face in Australia. There are currently 2.5 million ha affected by dryland salinity in Australia, and this may rise to 15 million ha over the next 30 to 100 years if no action is taken. National field experiments suggest that adoption of cropping systems that integrate deep-rooted perennials, such as lucerne, are important to reduce dryland salinity. This paper reports simulation results with APSIM (The Agricultural Production Systems Simulator), that have been used to explore climate, soil and agronomic factors affecting effectiveness of lucerne-based phase and companion cropping systems in sustaining crop yield and reducing deep-water drainage in South Australia.

Materials and methods The APSIM farming system simulator has been developed in Australia to facilitate analysis of complex production and sustainability issues of agricultural systems (Keating *et al.*, 2003). Simulations were conducted for Roseworthy, South Australia, with annual rainfall of 436 mm, 65% of which falls in the winter. The soil type is a red-brown earth with a soil texture of sandy loam in the top over clay to heavy clay in subsoil. The maximum effective rooting depth and plant-available water capacity were estimated respectively to be 1.8m, and 173mm for lucerne, and 1.6m and 143mm for wheat. The simulations were run from 1950 to 2000 with APSIM version 2.1. The simulated cropping systems were continuous wheat (CW), lucerne-based phase (LW-phase, terminated early E or late L) and companion (LW-companion) cropping. See footnote to table for further details of treatments.

Results Simulated drainage in LW-phase and LW-companion systems was much lower than in CW (continuous wheat) (Table 1). Compared with lucerne-based cropping systems, drainage in CW system occurred more frequently over the simulated period (Figure1). Although annual drainage in LW-companion was reduced significantly (below national target of 10mm or 1% of annual rainfall), reduction in wheat yield compared with CW system was high. The LW-phase systems seemed to provide the best balance of sustaining wheat yield and effectively reducing annual drainage in this environment, although there was some yield loss when lucerne stands were terminated late (LW-phase-L).

Table 1 Summary of simulated average wheat yield and annual deep drainage (% of annual rainfall in parentheses) in different cropping systems

Cropping systems[1]	Yield (kg/ha)	Annual drainage (mm)
CW	3,206	13 (3%)
LW-phase-E	3,350	8 (2%)
LW-phase-L	2,679	4 (0.9%)
LW-companion	1,706	2 (0.4)

[1]Continuous wheat (CW). Wheat was sown following three years lucerne, when lucerne stands being terminated in Nov. (LW-phase-E) and April (LW-phase-L). Wheat was sown into established lucerne stands (LW-companion).

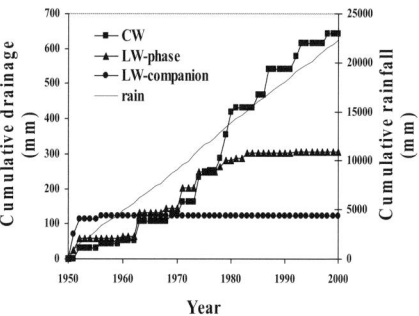

Figure 1 Simulated drainage in different systems

Conclusions Lucerne-based phase cropping systems appear promising in the Roseworthy environment, combining reduced deep drainage with acceptable grain yield. For LW-companion system, further experimentation is required on agronomic options for suppressing lucerne growth, enhancing crop competitiveness and thus reducing crop yield loss. This location-specific analysis needs to be extended to additional locations and the whole farm economics of changing from annual crop and pasture systems to a perennial forage based farming system needs to be considered.

Reference
Keating, B.A. *et al.* (2003). An overview of APSIM, a model designed for farming systems simulation. *European Journal of Agronomy*, 18, 267-288.

Australian pasture systems: the perennial compromise

L.W. Bell and M.A. Ewing
CRC for Plant-based Management of Dryland Salinity & School of Plant Biology, The University of Western Australia, 35 Stirling Highway, Crawley WA 6009, Australia, Email: lbell@agric.uwa.edu.au

Keywords: integration, annual pastures, perennial pastures, sustainability, agricultural systems

Introduction Dryland salinity, soil acidification and weed herbicide resistance challenge traditional agricultural production systems in south Australia. The pasture component of such systems rely on annuals like *Trifolium subterraneum* and *Medicago* spp. Replacing annual with perennial pastures allows some redress of the sustainability challenges, but few well-adapted species are available (Ewing & Dolling 2003). A range of perennial species are under evaluation to supplement current options. Some of these new perennial pastures may need modified production systems that allow full expression of their productive potential, especially when integrated with annual crops including cereals, pulses and oil seeds. Integrated systems rely on spatial or temporal segregation of pastures from crops. The necessary characteristics of plants for likely systems are discussed.

Systems Very important factors to design systems into which new species might be embedded are: (1) There is a trade-off between persistence and productivity. In very arid, low productivity environments, new pasture species need to have specific advantages to warrant adoption. (2) New species may vary in plant form (i.e. woody shrubs or trees to herbaceous species). Woody species, though less productive, have deep roots, maintain green leaf area and differ in growth pattern. Their purpose may be only to boost water use and fill gaps in feed supply. (3) Production systems must be flexible enough to respond to temporal changes in profitability between component enterprises. (4) The system must reflect feasible investment and input needs and have enough low risk associated with the technology to encourage wide-scale and rapid uptake.

Phase farming rotates pasture with crops. Pasture phases are flexible and allow farmers to change to crop production when desired. Farmers can integrate perennials into the pasture phase. However, the need to re-establish pastures at the start of each phase means that establishment cost must be low. Species well suited to this system must have a cheap source of seed, high early vigour, compete with weeds and withstand grazing in the year of sowing. Lucerne (*Medicago sativa*) is successful. It can be used to manage dryland salinity by establishing a dry soil buffer below the root depth of annual species to reduce recharge of groundwater tables (Latta *et al.*, 2001). It also aids weed management and fixes nitrogen for subsequent crops.

Alley systems are used for woody species when spatial separation of plants is more suitable. This enables annual pastures or crops to be grown between rows of fodder shrubs or trees. Plants used in this system generally are slow to establish but are long-lived. Tagasaste (*Chamaecytisus proliferus*), Leucaena (*Leucaena leucocephala*) and Saltbush (*Atriplex spp.*) are successful. These have deep roots and can access the water table to increase water use of agricultural systems (Lefroy *et al.*, 2001). They also improve the continuity of feed supply.

Intercropping (or companion cropping) is where another species is grown amongst a crop. This has been explored little with perennial pasture species. Suitable species would be leguminous, have low winter activity to reduce the competition on the accompanying crop, prostrate habit to avoid contamination of crop products and responsiveness to opportunities outside the crop-growing season. This system may enable sustainability objectives such as high water use to be achieved whilst continuing to produce crops.

Conclusion Integration of perennial pastures can improve sustainability of agricultural systems. Development of new species should address essential considerations. Production systems that integrate perennial pastures may need to evolve as new species are developed.

References
Ewing M.A. & P.J. Dolling (2003). Herbaceous perennial pasture legumes: their role and development in southern Australian farming systems to enhance system stability and profitability. In S. J Bennett. (ed.) New perennial legumes for sustainable agriculture, Univ of Western Australia Press: Perth, 3-14.
Latta R.A., L.J. Blacklow & P.S. Cocks (2001). Comparative soil water, pasture production, and crop yields in phase farming systems with lucerne and annual pasture in Western Australia. *Australian Journal of Agricultural Research,* 52, 295-303.
Lefroy E.C., R.J. Stirzaker & J.S. Pate (2001) The influence of Tagasaste (Chamaecytisus proliferus Link.) trees on the water balance of an alley cropping system on deep sand in south-western Australia. *Australian Journal of Agricultural Research,* 52, 235-246.

Theme A: Efficient production from grassland

Section 10

Industrial products from grassland

Grass pellet bioenergy in the Northeastern USA

J.H. Cherney and D.J.R. Cherney
Cornell University, Ithaca, NY 14853, Email: jhc5@cornell.edu

Keywords: grass, biofuel, pelleting

Introduction Grass pellets are a renewable energy supply that combines low technology/small-scale with local production/consumption for a cost effective energy system. There have been significant recent advances in pellet furnace technology and some pellet stove manufacturers now claim their stoves are capable of burning biomass with 5-6% ash content. Cool-season grasses have not been considered acceptable for pelleting and direct combustion in the past due to high ash content. Rain after harvest, however, has been shown to leach significant amounts of potassium and chlorine from grass (Sander, 1997). High yields are possible under lax harvest management (Cherney et al., 2003), producing grass biomass with potassium content as low as 0.5%. Ash content of cool-season grasses is dependent on species, plant maturity, soil type, leaching before and after cutting, loss of high-ash plant parts, and soil contamination. Our objective was to develop a strategy for practical commercial production of relatively low ash cool-season grass biomass.

Materials and methods A series of experiments were conducted. 1) Reed canarygrass (*Phalaris arundinacea* L.) primary growth (3 replicates, 6 m x 30 m plots) was overwintered in 2003 and evaluated the following spring, 2) Reed canarygrass (3 replicates, 6 m x 30 m plots) was evaluated for effects of delayed harvest following cutting on ash contents in 2003, 3) Twenty-one tall fescue (*Festuca arundinacea* L.) varieties (3 replicates, 1.8 m x 6.1 m) were evaluated for ash content during the spring growing season of 2003, and 4) Four fields of mixed cool-season grasses were cut on Aug. 3, 2004, tedded as needed, and harvested Aug. 17, 2004 using commercial hay equipment in 2004 to evaluate ash reduction strategies. All four soils were silt loams. Primary species were timothy (*Phleum pratense* L.), reed canarygrass, tall fescue, alfalfa (*Medicago sativa* L.), tall oatgrass (*Arrhenatherum elatius* L.), and bedstraw (*Galium aparine* L.).

Results 1) Reed canarygrass overwintered in the field in 2004 in New York State became lodged flat with 100% loss of harvestable yield. Ash content of hand-harvested samples in April, 2004 averaged 3.78% ash (0.58 SD). 2) Reed canarygrass (7.15% ash) was cut and left in the field up to 29 days during the summer of 2003, and ash content of the forage declined to 4.19%. 3) There were no significant differences in ash content among 21 tall fescue varieties during the spring of 2003, and ash content declined during spring growth (Figure 1). 4) Three bales (320 kg each) were sampled from each field and forage from all four fields was less than 4% ash (Table 1). This ash value includes any soil contamination due to the use of field-scale harvest equipment and is therefore a practical value. Gross energy values were somewhat higher than barley (Hordeum vulgare L.) straw (4.59% ash, 17.7 MJ/kg) from adjacent fields.

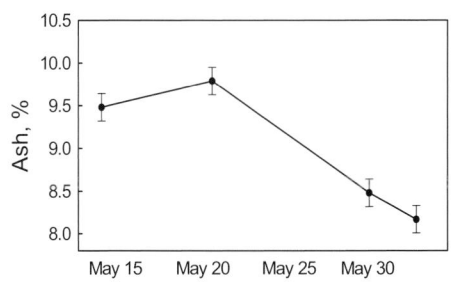

Figure 1 Change in ash content over time for 21 varieties of tall fescue (SED = vertical lines)

Table 1 Ash and gross energy content (DM basis) of mixed grass meadows harvested for pellet fuel

Field	Primary species (% of stand)	Total forage ash content, %	Energy, MJ/kg
A	Timothy (90%)	3.77	21.1
B	Timothy (56%) Alfalfa (32%)	3.84	19.9
C	Bedstraw (42%) Mixed grasses (58%)	3.85	19.6
D	Tall oatgrass (41%) Mixed grasses (59%)	3.78	18.7
	SED	0.30	0.76

Conclusions Mixed species meadows in the Northeastern USA can be managed to produce biomass with less than 4% ash content.

References

Cherney, J.H., D.J.R. Cherney & M.D. Casler (2003). Low intensity harvest management of reed canarygrass. *Agronomy Journal,* 95, 627-634.

Sander, B. (1997). Properties of Danish biofuels and the requirements for power production. *Biomass and Bioenergy,* 12(3), 177-183.

Theme A: Efficient production from grassland

Section 11

Grass and forage agronomy

Annual forage legume response to herbicides labelled for lucerne establishment

T.J. Butler, R. Bow and J.P. Muir
The Noble Foundation, 2510 Sam Noble Parkway, Ardmore OK 73401, USA, Email: tjbutler@noble.org

Keywords: herbicide tolerance, forage legume yield

Introduction Weed competition reduces stand establishment, thus lowering forage production and quality. However, there are no herbicides labelled for annual legume establishment, despite several labelled for the establishment of lucerne (*Medicago sativa*). Some of these may be useful in the establishment and production of annual forage legumes. Lucerne herbicides have greater potential for use on other legumes, since they have grazing and feeding clearance. The objective of this paper is to summarize annual legume yield response trials to herbicides labelled for lucerne establishment.

Materials and methods Herbicides were applied either pre-emergent (benefin and EPTC) or early post-emergent (imazethapyr, imazamox, imazapic, bromoxynil and 2,4-DB) to 18 annual legumes (Table 1) with a CO_2 backpack sprayer delivering 140 L/ha over the 2002-04 growing seasons. Forage dry matter (DM) yields were estimated by hand clipping 3 randomly placed quadrats (30.5 x 61 cm) from each treatment.

Results *Lathyrus* and *Pisum* spp. were most tolerant of imazethapyr, imazamox and 2,4-DB; annual *Medicago* spp. were most tolerant of EPTC, imazethapyr and 2,4-DB; *Strophostyles* spp. were most tolerant of benefin, imazethapyr and imazamox; *Trifolium* spp. were most tolerant of 2,4-DB (Conrad & Stritzke, 1980; Grichar et al., 1993); and *Vicia* spp. were most tolerant of imazethapyr, producing the same or more forage than the untreated controls.

Table 1 Forage yield of annual legumes in response to herbicides labelled for lucerne

Herbicide Treatment	Rate (kg ai//ha)	*Lathyrus/ Pisum-* Peas[1]	*Medicago* - Annual medic[2]	*Strophostyles-* Wild Bean[3]	*Trifolium-* Clovers[4]	*Vicia-* Vetch[5]
		----------------------------kg DM/ha----------------------------				
No Herbicide	-	4480 b	2087 b	1336 a	1950 bc	4884 b
Pre-emergent						
Balan (benefin)	1.34	4155 b	2578 b	1580 a	1490 c	4888 b
Eptam (EPTC)	3.90	4548 b	3619 a	701 bc	2266 b	5320 b
Post-emergent						
Pursuit (imazethapyr)	0.052	5941 a	3637 a	1172 a	2095 b	7305 a
Raptor (imazamox)	0.026	5910 a	3753 a	1206 a	2536 b	5426 b
Plateau (imazapic)	0.052	247 d	0 c	893 b	0 d	451 d
Buctril (bromoxynil)	0.84	3441 c	2413 b	264 c	2346 b	3998 c
Butyrac (2,4-DB)	1.12	5069 ab	4126 a	397 bc	3255 a	4587 b

Means of : [1] *P. sativum* and *L. hirsutus* VNS and AU groundcover; [2] *M. lupulina* BEBLK, *M. orbicularis* Estes, *M. polymorpha* Armadillo, and *M. minima* Devine; [3] *S. helvula* and *S. leiosperma*; [4] *T. vesiculosum* Yuchi and Apache, *T. hirsutum* Overton R-18, *T. incarnatum* AU robin, and *T. nigrescens VNS*; [5] *V. villosa* VNS and AU early cover, and *V. sativa VNS*.

Conclusions Herbicides labelled for lucerne establishment have potential for use in establishing annual legumes. However, based on growth stage and environment, legume species vary in their susceptibility. Therefore more trials are needed to determine the level of safety of herbicides to forage legumes at varying growth stages and different environments.

References

Conrad, J. D. & J. F Stritzke (1980). Response of arrowleaf clover to post emergent herbicides. *Agronomy Journal,* 72, 670-672.
Grichar, W. J., A. J. Jaks, G. W. Evers & A. M. Schubert (1993). Clover response to selected post emergent herbicides. Forage Research in Texas http://forageresearch.tamu.edu/1993/cloverpostemergency.pdf Texas A&M University Agricultural Research and Extension Centre.

Comparative seed strike of temperate, sub-tropical and native grasses and herb species under contrasting environments in southern Australia

Z.N. Nie, K.F.M. Reed, F. Cameron and B. Clark
CRC for Plant-based Management of Dryland Salinity; Department of Primary Industries, Primary Industries Research Victoria, Private Bag 105, Hamilton, Vic. 3300, Australia, Email: zhongnan.nie@dpi.vic.gov.au

Keywords: seedling establishment, viable seed, pastures evaluation

Introduction The role of deep-rooted perennials in reducing recharge to mitigate dryland salinity has been recognised widely in Australia recently. Poor seedling establishment is a key limiting factor for the expression of genetic merit for some perennial pasture species. Nie *et al.* (2004) investigated seedling establishment, and its relationship with rainfall and temperature, of a range of perennial grass and herb species in southern Australia. This paper reports seed strike of a range of perennial pasture species in 2 contrasting environments. There was significant interaction between species and site on seed strike. Environmental conditions caused different establishment outcomes within a diverse set of perennial forage species.

Materials and methods Seed of 24 grass and herb varieties was sown at Yatchaw (142°01'E, 37°50'S) and Warrak (143°07'E, 37°14'S) in Victoria, Australia in Aug-Sep 2002. The soil was a duplex clay loam (Olsen P = 29 mg/kg) at Yatchaw, and a sedimentary sandy loam (Olsen P = 17 mg/kg) at Warrak. The rainfall from sowing to seedling counts was 124 and 90mm for Yatchaw and Warrak, respectively. The sown species included 6 temperate grasses, 5 sub-tropical grasses, 3 native grasses and 2 herbs. Seedlings were counted 6-9 weeks after sowing. Seed strike was calculated as the proportion of established seedlings from the number of viable seed sown. The analysis was performed using the method of residual maximum likelihood in Genstat.

Results There was an interaction ($P<0.01$) in seed strike between species and site (Table 1). Most temperate grasses had a higher seed strike at Warrak than at Yatchaw whereas the seed strike of sub-tropical and native grasses and the 2 herb species was higher at Yatchaw than at Warrak. This was attributed to quicker responses of temperate grasses to 17mm of rain immediately after sowing at Warrak. Overall, most temperate grasses and the 2 herb species at Yatchaw appeared to have higher seed strike than the herbs at Warrak, which were higher than the native and sub-tropical grasses. The number of viable seed, based on recommended rates, varied with species from 75-884 seeds/m^2. Mean seedling density varied from 13-242 seedlings/m^2.

Table 1 Viable seed sown (VS, seeds/m^2), mean seedling density (MSD, seedlings/m^2), and seed strike (%) of some grass and herb species 6-9 weeks after sowing in spring 2002. Data for seed strike were transformed (T) and back transformed data are presented as percentages

Species	Variety	VS	MSD	Seed strike (T)		Seed strike (%)	
				Yatchaw	Warrak	Yatchaw	Warrak
Phalaris aquatica	Australian	278	124	3.8	3.8	44	44
Dactylis glomerata	Porto	405	72	2.3	2.2	16	15
Festuca arundinacea	Fraydo	884	242	2.7	3.1	22	30
Lolium perenne	Avalon	422	197	3.6	3.8	39	45
Lothopyron ponticum	Dundas	304	123	3.4	3.8	36	45
Bromus stamineus	Gala	240	141	3.8	4.5	44	62
Microlaena stipoides	Wakefield	381	70	2.3	1.9	16	11
Pennisetum clandestinum	Whittet	75	13	1.7	1.6	8	8
Plantago lanceolata	Tonic	247	75	3.5	2.8	36	24
Chichorium intybus	Grouse	370	99	3.2	2.6	32	20
s.e.d.				0.52**			

Conclusions Because of the strong interaction in seed strike between species and site, special attention should be given to variations in environmental conditions while establishing the perennial species. Sub-tropical grass, native grass and herbs generally had a poorer seed strike under the more stressful environment at Warrak.

References
Nie, Z. N., K. F. M. Reed, G. M. Lodge, G. Moore, T. Albertsen, S. M. Miller, A. D. Craig, S. P. Boschma, B. Hackney, B. Dear & M. L. Mitchell (2004). Evaluation of pasture perennial grass and herb species for use in recharge areas in temperate Australia. *Proceedings of the Salinity Solutions Conference,* Bendigo, Victoria (CD ROM).

Perennial grass emergence and establishment using a micro-nutrient seed treatment

C.D. Clements and J.A. Young
*US Department of Agriculture, Agricultural Research Service, 920 Valley Road, Reno, Nevada, 89512, USA,
Email: charlie@scs.unr.edu*

Keywords: restoration seeding, GERM-N-8, nitrogen

Introduction Resource managers have become increasingly frustrated with restoration seeding failures in semi-arid and arid environments. In response to this frustration, some resource managers have attempted restoration seedings using non-conventional methodologies such as propriety seed treatments. The exact nature of these propriety treatments is often confidential, but they generally consist of either nutrient or micro-nutrient enrichment or inoculation with unspecified micro-organisms. One of the more popular propriety seed treatment used in Nevada, USA, is GERM-N-8®. This product is a suspension of nutrients (N 2%, P 14%, and K 3%) applied to dry seed. Resource managers often report excellent success using these propriety treatments, but lack of experimental design make it impossible to assign cause and effect.

Materials and methods Dry seed of 8 native, *Poa secunda, Festuca idahoensis, Elymus lanceolatus, Elymus elymoides, Pascopyrom smithii, Hesperostipa comata,, Achnatherum hymenoides, Psuedoroegneria spicata* and 1 introduced grass, *Agropyron desertorum* were treated with the propriety seed treatment GERM-N-8® at a rate of 182 g per 45 kg of each seed species. Treated and untreated seed was sown by hand in October 2001 at a rate of 12 seeds per 30 cm row, or 120 seeds per 300 cm plot and replicated 3 times at 2 separate locations in north-western Nevada, USA. The first site, Beddell Flat is at an elevation of 1581m and received an average of 21.25 cm of precipitation over the 2 years of this study. The site is dominated by *Artemisia tridentata* ssp. *wyomingensis* with a *Achnatherum thurberianum* understorey. The second site, Granite Peak is at an elevation of 1780 m and received an average of 26.5 cm of precipitation over the 2 years of this study. The habitat is dominated by *Artemisia tridentata* ssp. *vaseyana* with an understorey of *Achnatherum thurberianum, Festuca idahoensis, Elymus elymoides*, and *Poa secunda*. Treatments were checked monthly from October 2001 through August 2003 as initial sprouting, mortality, and persistent establishment were recorded.

Results The initial sprouting of *Elymus elymoides, Hesperostipa comata*, and *Psuedoroegneria spicata* seedlings were significantly (P • •0.05) greater when treated with GERM-N-8 at the Beddell Flat site. This did not hold true at the Granite Peak site nor was this the case for the other seed species tested. The application of treating these perennial grass seeds with GERM-N-8 did not significantly (P • • 0.05) enhance seedling establishment. In fact, *Elymus lanceolatus* at the Beddell Flat site and the introduced *Agropyron desertorum* at the Granite Peak site had significantly (P • •0.05) more establishment than their treated counter parts. Untreated *Agropyron desertorum* established as a rate of 1 per 60 cm compared to 1 per 120 cm when treated with GERM-N-8.

Conclusion Resource managers may very well be visually experiencing success using some of these propriety products as favourable climate conditions, site potentials, and other factors can play an important role in any seeding success or failure. Nitrate-N is the ingredient most likely to enhance seed germination. The application rate of 3 mg of 0.3% nitrate-N in this particular propriety product, GERM-N-8®, is far below the rate of enrichment that has been shown to enhance grass seed germination in our laboratory. The establishment of 1 perennial grass per 30 cm is desired on arid and semi-arid rangelands when attempting to suppress exotic invasive weeds such as *Bromus tectorum*, the introduced *Agropyron desertorum* was the perennial grass species that came closest to achieving this density of all the species we experimented with.

Density and germination characteristics of seeds of *Bromus tectorum* in field seedbanks

J.A. Young and C.D. Clements
US Department of Agriculture, Agricultural Research Service, 920 Valley Road, Reno, NV 89512, USA, Email: jayoung@scs.unr.edu

Keywords: bioassay, invasive weeds, nitrate, gibberellin

Introduction *Bromus tectorum* is a highly invasive exotic weed that has spread over millions of hectares of grazing land in the semi-arid regions of far western North America. The annual grass is an important grazing resource, but herbage production is highly variable among years, depending on the amount and periodicity of precipitation. When production is abundant, the accumulations of fine-textured, early-maturing herbage increase the chance of ignition and the rate of spread of wildfires. On certain years the area burned in such fires may be several million hectares. Such fires destroy forage resources and degrade watershed quality on extensive areas as well as threaten human property and lives. *Bromus tectorum* plants can produce a very large number of caryopses. Caryopses that fail to find a safe site for germination acquire a dormancy in the field that leads to the building of large seedbanks (Young *et al.*, 1968). It is critical for managers to have estimates of the size and extent of *Bromus tectorum* seedbanks during the planning and implementation of pasture restoration treatments. Our objective was to obtain an estimate of *Bromus tectorum* seedbank size and germinability through bioassay of samples obtained from a variety of plant communities and to relate this information to site characteristics easily ascertained by pasture managers.

Materials and methods In the autumn of 2003, before germination of the current year's crop of *Bromus tectorum* caryopses had initiated, we collected 1,000 samples from the seedbed surface (5 by 10 by 2.5 cm) in 100 different pasture communities infested with the annual grass. The samples were placed in individual cups in the glasshouse, moistened with tap water and emerging seedlings identified and removed weekly for 4 weeks (wk). After the initial 4 wk, the process was repeated after enriching the samples with a solution containing 0.1 M potassium nitrate. After 4 wk the process was repeated after enrichment with 0.014 M gibberellin.

Results The emergence subsequent to the initial wetting with water of the bioassay samples provides an estimate of the readily germinable seedbank. The average *Bromus tectorum* seedling emergence was 3,900 per m^2. The range was 0 (6% of samples) to 8,200 per m^2. Subsequent enrichment of the germination substrate with a nitrate source increased emergence by an average of 15%. Subsequent enrichment with gibberellin increased seedling emergence by an additional 10%. Nitrate and gibberellin enrichment of the samples reduced the number of samples without any *Bromus tectorum* seedling emergence to less than 1% of the total. Enrichment provides an estimate of the viable, but dormant seeds in the seedbank (Evans & Young, 1975). The site characteristic most highly correlated with the density of the *Bromus tectorum* seedbank was the recent wildfire history of the collection site. Sites burned the previous year had much lower seedbanks, while sites burned two to five years before sampling had markedly higher seedbanks of both initially germinable and initially dormant *Bromus tectorum* caryopses compared to sites with no recent history of wildfires. This reflects the extreme dynamics of *Bromus tectorum* populations following wildfires (Young *et al.*, 1976). The presence of litter on the soil surface and the visible presence of *Bromus tectorum* caryopses on the soil surface were also highly correlated with increased seedbank density.

Conclusions Recent wildfire history and visible characteristics on the surface of seedbeds can be used to estimate the density of *Bromus tectorum* seedbanks.

References
Evans, R. A. & J. A. Young (1975). Enhancing germination of dormant seeds of downy brome. *Weed Science*, 23, 354-357.
Young, J. A., R. A. Evans, & R. E. Eckert, Jr. (1968). Population dynamics of downy brome. *Weed Science*, 18, 41-48.
Young, J. A., R. A. Evans, & R. R. Weaver (1976). Estimating potential downy brome competition after wildfires. *Journal of Range Management*, 29,322-325.

Potential biological control agents for *Nassella neesiana* (Poaceae) invading Australian native grasslands

F.E. Anderson[1], M.L. Díaz[1] and D.A. McLaren[2]
[1]*CERZOS - UNS, Camino La Carrindanga Km 7, 8000, Bahia Blanca, Argentina, Email: anderson@criba.edu.ar,* [2]*DPI Frankston, CRC for Australian Weed Management, PO Box 48, Frankston 3199, Australia*

Keywords: biological control, fungi, *Puccinia nassellae, Uromyces* cf. *pencanus*

Introduction   The introduction and proliferation of exotic stipoid grasses over the past 100 years seriously threatens agricultural productivity and the integrity of Australia's indigenous flora and fauna, particularly its grasslands (McLaren *et al.*, 1998). The full effect on biodiversity by the spread of these grasses is unknown but likely to be major (Hocking, 1998). Conventional control techniques have not stopped the invasion adequately, so it is a priority to find control options to achieve an effective management strategy. A biological control project against *Nassella trichotoma* and *N. neesiana* was initiated in 1999 in Argentina. We report on the most recent findings on two pathogens, *Puccinia nassellae* and *Uromyces* cf *pencanus*, selected on the basis of previous results (Anderson *et al.*, 2004), as potential biological control agents against *N. neesiana*, a South American species that can dominate both pasture and native grasslands in Australia.

Materials and methods   *Puccinia nassellae,* Specificity tests: Six Australian accessions of *N. neesiana* (4 plants from each) were tested for their susceptibility to 3 different rust isolates. Inoculations were performed by spraying a suspension of urediniospores in water. Inoculated plants were kept in an environment cabinet (19-21°C, 100% RH, 12 h photoperiod) for 2-3 weeks when plants were assessed for the presence of uredinia. *Uromyces* cf *pencanus,* Life cycle: Trials to germinate teliospores were performed following Evans (1987). Specificity tests: Inoculations used aqueous spore suspensions (for *N. neesiana*) or direct deposition of dry spores on leaves of the other species followed by incubation as described above. The species tested were: *N. neesiana* (7 Australian accessions*), N. trichotoma, N. tenuissima, Stipa brachychaeta* and *Poa ligularis* (one Argentinian accession of each).

Results   Three Australian accessions of the weed were susceptible to one of the *P. nassellae* isolates. No infection was achieved with the others. This rust, studied since early stages of this project, has proved to be highly specific, as infection was achieved in most inoculation trials only when isolates and plants originated from the same site. This posed the problem of finding isolates that could infect the Australian accessions which need to be controlled. One such isolate has been found. Morphological characteristics of *U.* cf. *pencanus* uredinia and telia found in the field appear to coincide with those of *U. pencanus* (Greene & Cummins, 1958), but this rust is known to produce aecia on its grass host, and these have not been found to date. Teliospores did not produce basidiospores under any of the treatments, so it was not possible to study whether or not these infected *N. neesiana*, completing the life cycle on this host. This information is vital to assess the convenience of the rust as a biocontrol agent and to confirm its identity. All but one of the Australian accessions of *N. neesiana* were susceptible to this rust. None of the other species were infected, indicating an acceptable level of specificity at this stage.

Conclusion   These findings are encouraging. They justify the continuation of investigations, which should now include further specificity testing against Australian native grasses.

References

Anderson F., W. Pettit, L. Morin, D. Briese & D. McLaren (2004). Pathogens for the biological control of weedy stipoid grasses in Australia: completion of investigations in Argentina. In: Proceedings of the XI International Symposium on Biological Control of Weeds (eds. Cullen, J.M., Briese, D.T., Kriticos, D.J., Lonsdale, W.M., Morin, L. and Scott, J.K.) CSIRO Entomology, Canberra, Australia.

Evans, H. C. (1987). Life-cycle of *Puccinia abrupta* var. *partheniicola*, a potential biological control agent of *Parthenium hysterophorus. Transactions of the British Mycological Society,* 88, 105-111.

Greene, H. C. & G. B. Cummins (1958). A synopsis of the uredinales which parasitize grasses of the genera *Stipa* and *Nassella. Mycologia,* 50, 6-35

Hocking, C. (1998). Land management of *Nassella* areas: implications for conservation. *Plant Protection Quarterly,* 13, 86-90

McLaren, D. A., V. Stajsic & M. R. Gardener (1998). The distribution and impact of South/North American stipoid grasses (Poaceae: Stipeae) in Australia. *Plant Protection Quarterly,* 13, 62-70

A new herbicide (GF-839) for long-term control of annual and perennial broad-leaved weeds in grassland

S.A. Egerton, A.D. Bailey and L.A. Brinkworth
Dow AgroSciences, Latchmore Court, Brand Street, Hitchin, Hertfordshire, SG5 1HZ, UK, Email: segerton@dow.com

Keywords: aminopyralid, fluroxypyr

Introduction The new herbicide GF-839 is a combination of a new active substance aminopyralid and the fully approved active substance fluroxypyr in the quantities 30 g ae/l aminopyralid + 100 g ae/l fluroxypyr. It is an emulsion, water in oil formulation (EO), and will be sold as a foliar acting herbicide for the long-term control of annual and perennial broad-leaved weeds in grassland. Globally aminopyralid can be used for weed control in range and pasture situations and plantations; in addition, uses in oilseed rape and cereals are also being explored. Aminopyralid is the most active halopyridine yet discovered and as a synthetic hormone it poses a low risk of resistance. *Rumex obtusifolius* (broad-leaved dock), *R. crispus* (curled leaf dock), *Cirsium arvense* (creeping thistle), *C. vulgare* (spear thistle), *Urtica dioica* (common nettle), *Ranunculus repens* (creeping buttercup), *Taraxacum officinale* (dandelion) and *Stellaria media* (common chickweed) are all pernicious, persistent weeds of grassland in Europe. If left unchecked they can lead to significant reductions in sward quality and quantity as well as spreading to neighbouring areas. In the UK 1.1M ha of grassland are infested with thistles, and 400,000 ha with more than $1/m^2$, equating to a potential loss of 1Mt DM / year. Docks at an infestation level of 10% cause potential silage losses of 10%. There are currently various products on the market for control of these weeds but GF-839 differs in that it is the first new compound to be developed primarily for the grassland market for over 30 years, and offers reliable long-term control of all of these weeds, in combination with good grassland management practice, whilst also offering a high degree of selectivity to grass.

Results During 2002 and 2003, 125 trials were carried out in established grassland (grass more than 1 year old) to evaluate the spectrum of activity and dose rate of GF-839. All field trials were carried out in accordance with EPPO guidelines. In season control of all target weeds from 2 l/ha GF-839 was over 95%. Figure 1 shows that long-term control (12-18 months after application) of perennial weeds was also excellent compared to market standards.

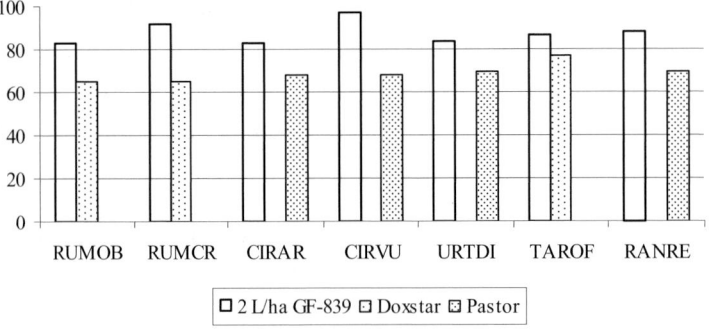

Figure 1 Long term percent control of Perennial weeds in grassland

Doxstar contains 100 gae/l fluroxypyr + 100 gae/l triclopyr. Pastor contains 50 gae/l clopyralid, 75 gae/l fluroxypyr + 100 gae/l triclopyr

Yield and quality data from 20 trials in established grassland and new leys (which included the label rate of 2 l/ha and the double rate of 4 l/ha of GF-839) demonstrated that GF-839 may be used on new or established grass from the three true leaf stage. Phytotoxicity data from 145 trials showed no serious long-term injury in any trial and data from 6 species screens on 14 of the most commonly sown and invasive grass species in the UK showed that GF-839 at the label rate and 2n rate is safe to apply to new and old grass pastures. Data from 9 cutting interval trials show that to allow maximum translocation of GF-839 to the roots the grass may be cut 7 days after application.

Conclusion Together with good husbandry and management techniques, GF-839 is a novel, useful and effective tool to be used in an integrated approach to improving the quality of grassland.

Field performance of an annual medic tolerant of sulfonylurea herbicide residues

J.H. Howie and C.A. Bell
South Australian Research and Development Institute, Waite Campus, Urrbrae, SA 5064, Australia, Email: howie.jake@saugov.sa.gov.au

Keywords: *Medicago littoralis,* triasulfuron, residual effect

Introduction Sulfonylurea (SU) herbicides such as triasulfuron, chlorsulfuron and metsulfuron-methyl are used extensively in the cereal-livestock zones of temperate Australia. They are regarded by farmers as effective, cheap and safe-to-apply herbicides with useful levels of residual activity in the year of application. However these residues can persist into following years, particularly in areas with alkaline soils and low rainfall, where their breakdown by microbial action and chemical hydrolysis is significantly reduced. Regenerating pasture legumes typically used in Australian ley farming systems are highly intolerant of even very low residues of SU herbicides (e.g. < 1ppb; Heap, 2000) resulting in severe stunting, reduced dry matter production, lower seed yields, poor persistence and decreased N fixation. In this study we compare the field performance of an artificially induced mutant cultivar (FEH-1) of annual strand medic (*Medicago littoralis*) with putative tolerance to sulfonylurea herbicide residues (Heap, 2000) with the cultivar Herald, its intolerant strand medic parent.

Materials and methods Triasulfuron was applied as a pre-emergent herbicide at four rates (0, 7.5, 13 and 26 g.a.i./ha) to wheat in a low rainfall site at Waikerie, South Australia (sandy loam, pH 8.3_w) in May 2002. Herald and FEH-1 were subsequently sown (22/5/2003) into the wheat stubble containing the herbicide residues. Shoot dry weight was assessed at 15 weeks post sowing, seed yield at plant senescence and seedling regeneration on June 26 2004. The trial design was a randomised block with three replicates; plots were 8 x 1.2m.

Results Increasing rates of triasulfuron residues reduced Herald shoot dry weight, seed yield and seedling regeneration by > 50% (see Table 1). FEH-1 however demonstrated good tolerance to the triasulfuron residues from all application rates for all parameters measured.

Table 1 Shoot dry weight, seed yield (2003) and regeneration (2004) of Herald and FEH-1 in soil treated with triasulfuron at 0, 7.5, 13 and 26 g.a.i./ha (2002)

Rate	Shoot dry weight (kg/ha)		Seed yield (kg/ha)		Regeneration (plants/m^2)	
	Herald	FEH-1	Herald	FEH-1	Herald	FEH-1
0	3422	3022	652	678	1532	1472
7.5	2314	2868	564	690	1323	1488
13	2442	3192	511	690	945	1543
26	1618	2863	290	790	728	1549
LSD 5%	799		177		439	

Conclusions These field results agree with the findings of Heap (2000) from in-vitro enzyme assays, soil and foliar dose response pot experiments and pilot field studies (Howie *et al.*, 2002) that FEH-1 has increased tolerance to a range of ALS inhibitors including SU herbicides. In the presence of SU residues, FEH-1 has greater dry matter production and seed yields. Improved seed yields (and thus seed reserves) result in improved regeneration, greater legume dominance and competitiveness with weeds. Where SU residues exist, the more vigorous root systems of SU tolerant medics are likely to have better nodulation and N fixation, increased tolerance of root diseases, increased ability to extract soil moisture and nutrients and an enhanced ability to take advantage of the residual weed control of SU herbicides. This increased pasture productivity and pasture legume dominance will benefit the livestock component of a cereal/pasture rotation as well as the cereal phase, which will benefit from improved organic N levels and reduced carryover of cereal root diseases. FEH-1 is scheduled to be released as Australia's first SU residue tolerant pasture legume in 2006.

Acknowledgments Funding by GRDC is gratefully acknowledged, as is the valuable technical assistance provided by Peter Schutz, SARDI.

References
Heap, J. (2000). Increasing *Medicago* resistance to soil residues of ALS-inhibiting herbicides. PhD thesis, University of Adelaide.
Howie, J.H., J. Heap, C. Preston & R.M. Nair (2002). Development of an annual medic tolerant of sulfonylurea herbicide residues. In: J.A. McComb (Ed.) Plant Breeding for the 11th Millennium. Proceedings of the 12th Australasian Plant Breeding Conference, Perth, WA, 15–20 September 2002, 767-769.

Seed size and its relationship with crop establishment, productivity and nutritive value in genotypes of maize for silage

C. Pérez M.[1], M.R. Tovar G.[2*], G. García[1], A. Carballo[1], G. Mendoza[1], T. Vásquez[1], F. González[1] and M. Crosby[1]
[1]Colegio de Postgraduados, [2]INIFAP-CEVAMEX, México, Email: claudiatlaxcala@yahoo.com.mx

Keywords: physical quality, maize genotypes, yield, nutritional value

Introduction For a high dry matter (DM) yield of forage maize an adequate population of plants is required, which is related to germination, vigour, and seed size (Ellis, 1992). The objective of the present study was to investigate seed size and its relationship to establishment, productive and nutritional potential of silage maize, which has not before been reported in the literature.

Materials and methods The study was done in 2002 with nine genotypes at the Colegio de Postgraduados and the Experimental Station Valle de Mexico (INIFAP) (PECSL and PECT. Research consisted of two phases 1) the physical seed quality and 2) field establishment, production and nutritional value of maize silage. The design was a randomised block with four replications and a factorial array of treatments. Physical quality of the seed was determined in terms of shape, size, flat large and medium, length (LS), width (AS) and thickness (ES). The percentage establishment was determined whilst the emergency remained constant at both sites. The harvest was done at the silage stage (30-35 % DM of the whole plant). The variables recorded were: female flowering (FF), protein yield (CPY) and digestible dry matter yield (DDMY), crude protein (CP) and in vitro DM digestibility (IVDMD).

Results and discussion Significant ($P \leq 0.001$) differences among genotypes were observed in all the variables, but for seed size only the physical and nutritional quality and PCY were significant (Table 1).

Table 1 Comparison of means for physical quality variables, productivity and nutritional value in genotypes of maize silage

Genotype	LS (cm)	WS (cm)	TS (cm)	PECT (%)	PESL (%)	CPY (t/ha)	DDMY (t/ha)	CP (%)	IVDMD (%)
H157	1.1	0.8	0.3	89.6	89.1	1.8	15.6	7.6	64.4
H135	1.2	0.8	0.4	89.5	77.9	2.0	16.4	8.5	71.7
A791	1.1	0.8	0.4	91.1	84.0	1.8	12.7	8.0	56.7
VS2000	1.4	0.7	0.4	87.1	83.6	1.6	13.4	8.1	65.8
Campeón	1.7	0.8	0.4	94.0	92.1	2.1	16.8	8.8	71.5
Promesa	1.2	0.8	0.4	95.9	92.4	1.8	13.9	8.7	68.3
HS2	1.2	0.7	0.4	94.8	94.0	1.8	16.6	7.9	70.7
VS22	1.3	0.7	0.4	90.9	92.0	1.7	14.1	8.4	70.2
H358	1.0	0.7	0.4	85.6	52.1	1.7	14.6	8.1	70.9
Average	1.3	0.7	0.4	90.9	84.1	1.6	13.4	7.4	61.0
Significance	**	**	**	**	**	**	**	**	**
DMSH	0.07	0.07	0.04	9.7	10.2	0.23	2.09	0.08	1.08

** $P < 0.001$

Campeón, VS22 and VS2000 had the best physical seed quality. The greater percentage of establishment at both sites were obtained with Promesa, HS2, Campeón and VS22. Cultivars H157, H135, Campeón and HS2 had the highest CPY and DDMY. Campeón, Promesa, H135 and VS22 showed the highest CP concentrations. H135, Campeón, H358, HS2 and VS22 had the highest IVDMD On the other hand, LS and FF were negatively correlated, indicating that genotypes with greater seed size are maturing earlier (FF) as was the case with cv. VS22. Likewise, the LS had the highest correlation with CP (r = 0.78) and IVDMD (r = 0.67). This result agrees partially with Ries & Everson (1973) who reported a positive association between CP and seed size in wheat.

Conclusions Genotype affects establishment, productivity and nutritional value of maize silage. The seed size influences the physical and nutritional quality. Genotypes with the best physical seed quality were Campeón, VS22 and VS2000, whilst in field establishment Campeón, HS2 and Promesa were superior. Campeón, H135, HS2, H157, VS22 and H358 were the best in productivity and nutritional value.

References
Ellis, R.H. (1992). Seed and seedling vigour in relation to crop growth and yield. *Plant Growth Regulation*, 11, 249-255.
Ries, S.K., and E.H. Everson (1973). Protein content and seed size relationships with seedling vigour of wheat cultivars. *Agronomy Journal*, 65, 884-181.
S.A.S. (2000). Statistical Analysis System Institute. Inc. Cary, NC. USA.

Comparison of technologic and economic parameters of drills for grassland oversowing

L. Gonda and M. Kunský

Grassland and Mountain Agriculture Research Institute (GMARI), SK-97421 Banská Bystrica, Slovakia, Email: gonda@vutphp.sk

Keywords: grassland, oversowing technologies, shallow rocky soil

Introduction Direct grassland oversowing is a very important technology in Slovakia as some 350,000 ha (42 %) of grassland are on unploughable sites that dominate mountain regions. Slope inclinations >16° and shallow soil layers (150-200mm) limits ploughing on these sites (Tiley & Frame, 1988).

Materials and methods After 2000, 2 types of drills for direct oversowing were developed: SPP 6 and SP 16, respectively, for band- and strip- oversowing (see Table 1). Both prototypes were tested in mountain conditions (Gonda, 2003) together with series-produced machinery: drill VREDO 125.07.05 (Netherlands) and HORSCH SE 3 (Germany), used for row and total (wide) oversowing, respectively. Soil was shallow rocky clay (<200 mm), altitude 400-600 m and slope inclination 12°; seed rate 40kg grass/clover mixture /ha (20/80 clover/grass).

Results Table 1 compares technological and economic parameters of 4 main oversowing systems of grassland renovation in Slovakia.

Table1 Technical and economical evaluation of oversowing technology in upland grasslands in Slovakia

Technology	Machinery Tractor Drill	Engine power (kW) Working width (mm)	Productive efficiency[#] (ha/h)	Price (€)	Technological costs		Technology system	Sward profile
					(€/h)	(€/ha)		
Row oversowing	ZETOR - 7540	57.0	-	31010.0	23.74	19.79	W s	passive
	VREDO 125.07.05	2 500.0	1.20	15972.6				
Strip oversowing	ZTS 123 45	90.5	-	39345.7	32.68	38.91	W s̄	active discs destroyed sward 22%
	SP 16	2 400.0	0.84	15444.4				
Band oversowing	ZTS 163 45	121.0	-	48892.2	41.50	44.63	M t	active heads destroyed sward 38%
	SPP 6	2 700.0	0.93	26649.2				
Total (wide) oversowing	ZTS 183 45	136.0	-	80504.0	50.51	50.51	W s	active heads destroyed sward 100%
	SE 3 HORSCH	3 000.0	1.00	26164.7				

[#] Full production: no idle time   w... width of the row; r... row distance; d ... row depth

Row oversowing needs little energy and is cheap. It suits less-demanding soil conditions but not highly degraded grassland because it only partially destroys the turf (5%) and the microrelief stays undestroyed. **Strip oversowing** best suits shallow rocky soils with considerably degraded grassland. It does not destroy the microrelief and does not cause soil erosion or herbage contamination. **Band oversowing** needs much energy and is expensive. It mainly suits deep plateau-land soils without stones and considerably degraded grassland. It destroys turf (38%) and creates excellent conditions for root development, but can cause erosion at slope inclinations >12°. Strongly devastated microrelief needs rolling to prevent soil contamination of herbage. **Total (wide) oversowing** has the highest energy demands and costs of the 4 methods. It suits severely degraded grassland where total destruction of the root system of the original sward is needed and excellent conditions for the development of a dense sward are created. It cannot be used on shallow and rocky soils.

Conclusions The 4 oversowing technologies do not compete with each other. Instead, they offer a choice after consideration of all soil, climatic and energetic conditions and the degradation degree of the grassland.

References

Gonda L. (2003). Výskum a modelovanie aplikácie techniky a technológií pre obhospodarovanie trávnych porastov v horských a podhorských oblastiach (výskumná správa), VÚTPHP Banská Bystrica, Sk, 94 pp.

Tiley, G. E. D & J. Frame (1988). Sward establishment and renovation without ploughing. *Proceedings of the 12th General Meeting of the European Federation*, Dublin, Ireland pp. 199-203.

Repeated strip-seeding of a legume-grass mixture into permanent grassland in the Czech Republic

A. Kohoutek, P. Komárek, V. Odstr•ilová and P. Nerušil
Research Institute of Crop Production Prague, Research Station of Grassland Ecosystems Jevicko, K. H. Borovskeho 461, Czech Republic 569 43, Email: vste@seznam.cz

Keywords: grassland, botanical compounds, strip-seeding

Introduction Although ruminants do not belong to the main consumers of protein supplements, there is a possibility to decrease the use of feed grain by increasing the protein content in the forage from permanent grassland. This paper describes strip-sowing of a legume-grass mixture. Saved protein supplements can then be used for feeding pigs and poultry.

Materials and methods Strip-seeding trials were established on a fluvisoil at Jevicko, Czech Republic in a mild climatic region (average annual temperature 7.5 °C, annual rainfall 629 mm, altitude 330 m). Strip-sown grassland (SG) was compared with permanent grassland (PG). Strip-seeding in 1991 (seeding machine SE 2-024), 1996, 2000 and 2003 (seeding machine for strip-seeding - prototype) was done with the same mixture and seed quantity (29 kg/ha). The mixture had this composition: Festulolium hybrid (*Lolium multiflorum.* x *Festuca arundinacea*) cv. Felina (12 kg/ha), Perennial ryegrass (*Lolium perenne*) cv. Sport (8 kg/ha), Cocksfoot (*Dactylis glomerata*) cv. Niva (4 kg/ha), Red clover (*Trifolium pratense*) cv. Kvarta (3 kg/ha), White clover (*T. repens*), cv. Huia (2 kg/ha). The original grassland and the strip-sown alternative were fertilised with 30 kg/ha P as super phosphate and 60 kg/ha K as potash salt. This paper reports dry matter (DM) production and corrected DM production of strip-sown legumes and grasses (corrected DM production = DM production x percentage of botanical group / 100).

Results Average DM productions of SG of all years was 29% higher than PG (Table 1). DM production of strip-sown legumes plus grasses (SL+G) amounted to 59.1 % of total DM yield. Substantial modification of the botanical composition in favour of strip-sown species was acquired by annually repeated strip-seeding. DM production of SL (mostly red clover) was highest in the first year after strip-seeding, but in following years only traces of SL were recorded, which is in accordance with the general persistence of red clover in central European for two to three years. SGr have higher and more regular DM production than SL, because the most productive and persistent grass species were used in the mixture (mainly *Dactylis glomerata* and Festulolium hybrid).

Table 1 DM production (t/ha)

Var./year	1991	1992	1993	1994	1995	1996	1997	1998	1999	2000	2001	2002	2003	AVG
PG	3.83	3.78	3.30	3.66	8.32	7.79	5.81	4.03	7.26	6.15	6.45	8.02	5.69	5.40
SG	3.54	8.49	5.56	5.91	10.41	8.44	9.31	6.29	8.59	5.90	6.91	8.28	6.54	6.97
LSD$_{0,05}$	0.78	1.69	1.34	1.45	2.98	2.71	2.03	2.13	2.79	1.38	1.65	1.69	1.43	.
LSD$_{0,01}$	0.91	1.95	1.56	1.71	3.58	3.21	2.40	2.52	3.30	1.63	2.06	2.00	1.75	.
SL	0.38	2.38	1.20	0.52	0.00	0.00	4.71	1.79	0.02	0.26	3.32	1.28	0.38	1.22
SGr	0.56	3.60	2.17	3.74	3.30	3.72	3.66	3.00	3.17	3.01	2.36	3.61	2.41	2.90
SL+Gr	0.94	5.98	3.37	4.26	3.30	3.72	8.37	4.79	3.19	3.27	5.68	4.89	2.79	4.20

SL: strip-sown legumes; SGr: strip-sown grasses; SL+G: total strip-sown legumes + grasses

Conclusions Strip-seeding of a legume-grass mixture repeated every 3 to 5 years would contribute substantially to increased productivity of PG. The proportion of strip-sown species and especially legumes is highest in the first yield year after strip-seeding.

Acknowledgements Support of the research project No. VZ01 'Creation, calibration and validation of sustainable and productive cropping systems' is gratefully acknowledged.

Strip-seeding of red clover, lucerne, alsike clover, white clover and sainfoin into grassland in central Europe

P. Komárek, A. Kohoutek, V. Odstrčilová and P. Nerušil
Research Institute of Crop Production Prague, Research Station of Grassland Ecosystems Jevicko, K. H. Borovskeho 461, Czech Republic 569 43, Email: vste@seznam.cz

Keywords: grassland, legumes, strip-seeding

Introduction Strip-seeding of legumes into grassland improves forage quality and adds fixed nitrogen (N) to grassland, which decreases the need for mineral N.

Materials and methods Strip-sown legumes were established on a fluvisoil at Jevicko, Czech Republic (altitude 330 m, mean annual temperature 7.5 °C, annual rainfall 629 mm, of which 397 mm during the growing period) in temporary grassland of *Dactylis glomerata* (60 %), *Poa pratensis* (5 %), *Alopecurus pratensis* (20 %) and *Arrhenatherum elatius* (15 %) in 2000 (Table 1). The plots were fertilised with 35 kg/ha P 100 kg/ha K. Rainfall in 2002 was normal but 2003 was an extremely dry year.

Results Red clover was the most successful of the strip-sown species with 30 – 47 % of the dry matter (DM) production in 2002 and 12 – 33 % in 2003 (Table 1). Tetraploid cultivars had a higher proportion in the sward, especially cv. Beskyd and the newly bred cultivar (nbc) RH, than diploid cv. Tábor. Alsike clover (17 and 7 % in 2002 and 2003, respectively), lucerne (12 and 6 %), white clover (15 and 4 %) performed reasonably and sainfoin (8 and 4 %) was the least successful. Two nbc's of tetraploid red clover were very successful too: (a) nbc RH, which was bred to increase N fixation ability (Šimon & Jakešová 2004) and (b) nbc V, which was bred for higher resistance against viruses. Red clover maintained high DM production in the summer of the extremely dry year 2003.

Table 1 Treatments with legumes strip-seeding, DM production, and strip-sown cultivars

Treatment no.	Strip-sown species	Cultivar	Seed quantity (MVS/ha)[1]	DM production (t/ha)			
				2002		2003	
				Total DM	Strip-sown species	Total DM	Strip-sown species
1	red clover (4n)	Vesna	8.0	7.91	2.93	4.46	1.01
2	red clover (4n)	Radegast	4.0	7.70	2.29	4.42	0.62
3	red clover (4n)	Beskyd	8.0	8.20	3.69	4.60	1.24
4	red clover (4n)	RH (nbc.)	8.0	8.17	3.69	5.13	1.71
5	red clover (4n)	V (nbc.)	8.0	6.32	2.34	4.27	0.87
6	red clover (2n)	Tábor	8.0	7.30	2.21	4.35	0.54
7	lucerne	Zuzana	8.0	7.71	0.97	4.49	0.28
8	alsike clover	Táborský	15.0	8.15	1.40	4.00	0.26
9	white clover	Jordán	15.0	6.59	0.99	4.42	0.18
10	sainfoin	Višňovský	5.0	6.31	0.50	4.42	0.18
average	.	.	.	7.44	2.10	4.46	0.69
$LSD_{0.05}$	.	.	.	2.69	0.66	1.45	0.29
$LSD_{0.01}$	.	.	.	3.23	0.79	1.74	0.36

[1] MVS/ha: Seed quantity in millions of viable seeds per ha; nbc = newly bred cultivar

Conclusions The selected legumes showed to be very suitable for strip-seeding in central Europe, particularly red clover because of its it has wide tolerance Tetraploid cultivars were more productive than diploid ones.

Acknowledgements Funds for this study were provided by the research project No. VZ01 'Creation, calibration and validation of sustainable and productive cropping systems'.

Reference
Šimon, T. & Jakešová, H. (2005) Relationships among nitrogenase activity, dry forage and water soluble carbohydrates content of selected red clover genotypes. In: Grassland – a Global Resources, XX International IGC 2005, Dublin, 2005 (*in press*).

Persistence of timothy in mixture with smooth meadow grass

H. Björnsson
Agricultural Research Institute, Keldnaholti, 112 Reykjavík, Iceland, Email: holmgeir@rala.is

Keywords: timothy, smooth meadow grass, persistence, cutting date, competition

Introduction Timothy (*Phleum pratense)* is the most common grass species in Iceland for the production of high quality herbage for dairy cows. For this purpose timothy is cut early, often around heading. The high quality, however, is at the cost of limited persistence of timothy. If cut late, three to four weeks after heading, timothy cultivars of northern origin can sometimes dominate for a long time, especially if the aftermath is not cut or grazed. Earlier experiments have shown that, relative to a late cutting treatment, the percentage timothy in the first harvest is reduced by one unit for each week that the harvest is moved forward (Helgadóttir & Hermannsson, 1991). This effect is cumulative over the years and the result applies to swards where all of the fertiliser is applied in spring and smooth meadow grass (*Poa pratensis*) is present in the sward, either sown or invaded. Experiments were run to further study the persistence of timothy under different cutting treatments and N applications and the effect of cultivars of both timothy and smooth meadow grass.

Materials and methods Timothy (cvs Adda, Vega and Saga) was sown in a 2:1 mixture with the smooth meadow grass cv. Lavang in 1995 in a $2\times2\times2\times3$ factorial experiment with three replicates. Treatments were applied in 1996–1998, the plots harvested and the harvest sampled for botanical analysis between 1996 and 1999. Harvest dates of 1st and 2nd cut were a 2×2 factorial, arranged on main plots. Dates of 1st cut were June, 27–30 and July, 15–17 and dates of 2nd cut were August, 23–25 and September, 6–9. N (180 kg/ha) was all applied in spring or split with 60 kg/ha applied after 1st cut. In 1999 the experiment was cut once, measuring the accumulated effects of treatments. N applied was 100 kg/ha and the plots were harvested on July, 5–6. In a series of three experiments between 1999 and 2002 twelve cultivars of smooth meadow grass were tested in pure stand and in mixture with timothy cv. Adda. The N level was 150 kg/ha in split application.

Results Timothy was dominant at the first harvest, 92% and 95% on June, 27 and July, 15, 1996, respectively. The timothy cultivars are all of northern origin with little regrowth potential so that the split application of N favours the smooth meadow grass. In 1996 timothy was 85% and 78% in 2nd cut for all N applied in spring and as split application, respectively. Timothy declined gradually with time and the results presented are for the final harvest in 1999 only. Date of 2nd cut had little if any effect on the persistence of timothy and results are not shown. In Table 1 results are presented for two way combinations of timothy cultivars with each of the factors date of 1st cut and split nitrogen application. The standard error of difference is not valid for direct comparison of cutting dates. Cv. Adda is the most northern of the three cultivars and has persisted the competition better than the others, especially when cut late. On plots with late 1st cut of cv. Adda and all fertiliser applied in spring timothy amounted to 79.5% of the harvest, while on nearby plots with pure cv. Adda and similar treatment it was 85% of the yield, indicating that smooth meadow grass was competing at this cutting date. In the experiments with smooth meadow grass cultivars the over all average of timothy in the herbage was 71.6–73.4% with four cultivars and with the other eight cultivars timothy was in the range 61.1–68.3%. Timothy was in most cases relatively competitive against cultivars that are used as turf grass. The cultivar used in the first experiment, Lavang, was among those most competitive against timothy. Yield differences among cultivars were small. The two cultivars yielding highest in pure stand left a sward with low timothy content in the mixture, whereas none of the cultivars with high timothy content in the mixture were low yielding.

Table 1 Percentage timothy in 1999, SED=2.23

| | Mean date of 1st cut | | N application | | |
	29 June	16 July	Spring	Split	Mean
Adda	56.8	73.1	72.5	57.5	65.0
Vega	49.3	59.6	61.3	47.6	54.4
Saga	51.6	54.7	61.1	45.1	53.1
Mean	52.6	62.5	65.0	50.0	

Conclusions Smooth meadow grass is often sown with timothy in order to secure a dense sward, thus reducing the risk of soil contamination in the harvest. This shortens the lifetime of the timothy, especially if cut early and fertilised after the 1st cut in order to get high quality forage. The highest yielding cultivars of smooth meadow grass are not necessarily the best ones for use in mixture with timothy.

Reference

Helgadóttir, Á. & J. Hermannsson (1991). The effect of management on botanical changes in a newly established grass sward and the need for reseeding. In: Varig grasmark – til slått og beite. NJF seminar nr. 196, Kolbotn, Norge, 11.-13. March 1991, 75-81.

Performance of timothy cultivars in monoculture or in mixtures with meadow fescue in Finland

M. Kari[1] and M. Rinne[2]
[1]Experiment farm of Kesko Agro, Hahkialantie 57, FI-14700 Hauho, Finland, Email: maarit.kari@kesko.fi,
[2]MTT Agrifood Research Finland, FI-31600 Jokioinen, Finland

Keywords: dry matter accumulation, cultivar, yield, primary growth, regrowth

Introduction Timothy (*Phleum pratense*) and meadow fescue (*Festuca pratensis*) are dominating grass species in Nordic countries with severe winters. Dry matter (DM) yield potential, forage quality and winter hardiness are of main interest for commercial cultivars. Further, it is advantageous to take into account the relation between primary growth and regrowth (P/R) DM yield when planning mixtures for different purposes and areas. In the current experiment, three different types of timothy were evaluated: a northern type cv. Jonatan, a southern type cv. Ragnar and an intermediate type cv. Grindstad. In addition, meadow fescue cv. Kasper and mixtures with all timothy cultivars and cv. Kasper were studied.

Materials and methods The trial was carried out in Hauho, Finland (61°N), at the experimental farm of Kesko Agro and sown in May 2000 in monoculture with three replicates for each treatment. N fertilisation was set at 90 kg/ha for primary growth and 80 kg/ha for regrowth. Harvesting dates for primary growth and regrowth, respectively, were June, 18 and August,29, 2001, June, 10 and August, 1, 2002, and June, 17 and August, 12, 2003. All cultivars were harvested simultaneously at the time of heading in cv. Grindstad. Samples were dried at 60°C for 20 hours and DM yield recorded. Combined samples for the replicates were analysed for digestible organic matter (D-value) and crude protein (CP) by NIRS in the laboratory.

Results The total DM yield of meadow fescue cv. Kasper was lower than that of all timothy cultivars (Table 1). Monoculture timothy gave higher DM yield in both harvests compared to mixtures with meadow fescue. For individual cultivars, cv. Ragnar performed better in monoculture in primary growth, and both cvs. Ragnar and Grindstad in regrowth, than mixed with meadow fescue. In primary growth, the D-value of cv. Kasper was numerically higher than in the timothy cultivars. In regrowth, the variation in D-value was small. The average D-value in primary growth was higher than in regrowth (674 vs 658 g/kg DM) with the current harvesting strategy applied. All the D-values were slightly lower than recommended for silage based milk production in Finland (680-700 g/kg DM). The differences in growth type between the cultivars were estimated as DM produced in primary growth per DM produced in regrowth (P/R). The low value for cv. Kasper describes the typically good regrowth ability of meadow fescue, though total DM yield remained lower than in timothy. Northern type cv. Jonatan and intermediate type cv. Grindstad produced more DM in first harvest than cvs. Ragnar and Kasper. The DM yield decreased significantly with progressing age of the leys. For example, total DM yield was 13398, 10224 and 7545 kg/ha for years 2001, 2002 and 2003, respectively. This demonstrates the decreasing yield potential as the age of the ley increases, but in the present experiment, the amount of decline is confounded with varying annual weather conditions.

Conclusions Different types of timothy differ in yield potential. Southern and intermediate types of timothy had positive effects on regrowth DM yield under climatic conditions of Southern Finland. Mixtures with meadow fescue are considerable options for practical farming despite poorer performance compared to monocultures of timothy in the present trial. Mixed leys with components with different potential for yield, regrowth, winter hardiness and resistance to stress are likely to stabilise annual variation in forage production.

Table 1 Performance of timothy cultivars Grindstad (G), Jonatan (J) and Ragnar (R) as monoculture or as mixtures with meadow fescue cv. Kasper (K) in primary growth and in regrowth in a two cut system over three years in Finland

	G	J	R	K	G+K	J+K	R+K	SEM	C1	C2	C3	C4	C5
Primary growth (P)													
DM yield (kg/ha)	6188	5715	5595	4309	5934	5554	5132	144.5	***	*	*	**	
CP yield (kg/ha)	549	519	456	408	501	503	452	11.8	***	*	o	***	***
D-value (g/kg DM)	672	663	671	693	666	675	675	-					
Regrowth (R)													
DM yield (kg/ha)	5683	4459	5492	4556	4928	4390	4790	139.6	***	***	***		***
CP yield (kg/ha)	301	302	341	364	312	294	365	11.3	***			*	*
D-value (g/kg DM)	658	672	654	646	656	668	654	-					
DM yield (P/R)	1.122	1.333	1.031	0.914	1.205	1.281	1.072	0.0373	***		***	o	***

C1=timothy vs meadow fescue, C2=Monoculture timothy vs mixtures, C3=G vs J, C4=G vs R, C5=J vs R

Yield components in annual ryegrass and oats grown in association and monoculture

V. Pablo, A. Martínez-Hernández, E. Cortés-Díaz and E. Ojeda
Departamento de Zootecnia, Universidad Autónoma Chapingo Km. 38.5 Carr. México-Texcoco, Chapingo Texcoco, CP 56230 Mexico, Email: pedroarturo@correo.chapingo.mx

Keywords: leaf yield, nitrogen level

Introduction Earliness of oats and higher growth rate of annual ryegrass later in the season explain the higher forage yield of annual ryegrass+oats association over monocultures (Améndola & Morales, 1997). However, changes in yield components of the species grown in association compared to monoculture have not been explored. This study aimed to determine leaf, stem and dead matter yield in annual ryegrass and oats when grown in association and monoculture at different nitrogen (N) levels.

Materials and methods Annual ryegrass (AR) and oats (O) in monoculture and associated (AR+O) were grown with 4 different levels of N: 0, 50, 100 or 150kg/ha. Experimental design was completely random with 5 replicates; the experimental unit was a pot with 2 plants of each species in monoculture and 1 of each species in the association. Three cuts were done at 5cm high. Statistical analyses were on leaf, stem and dead matter yields (dry matter basis) from the three cuts, comparing yield components in annual ryegrass grown in monoculture versus in association and in oats grown alone versus in association.

Results AR+O showed 4 and 17% higher forage yield, respectively, across N levels than O and AR (data not shown). Annual ryegrass grown in AR+O compared to AR showed 27, 56 and 68% higher (P<0.05) leaf, stem and dead matter yields, respectively. In both AR+O and AR, annual ryegrass reached the highest leaf and dead matter yields with 100kg N/ha (Table 1). Stem yield of oats was 30% higher when grown in monoculture than in association, and in both O and AR+O the highest stem yield of oats was reached with 100kg N/ha.

Table 1 Yield (g/m^2; sum of 3 cuts) by components in annual ryegrass and oats in monoculture and association at different levels of added N

	N level (kg/ha)	AR	AR+O	Mean	O	AR+O	Mean
(a) Leaf	0	133	325	279 b	114 b	151 ab	132
	50	298	390	344 b	184 ab	222 a	203
	100	390	455	422 a	238 a	157 ab	197
	150	422	536	479 a	238 a	162 ab	200
	Mean	336 b	427 a		193	173	
(b) Stem	0	49	81	65	106	124	114 b
	50	65	124	94	168	151	159 ab
	100	81	103	92	222	130	176 a
	150	87	130	108	220	146	183 a
	Mean	70 b	109 a		179 a	138 b	
(c) Dead matter	0	5	5	5 b			
	50	11	22	16 b			
	100	22	43	32 a			
	150	27	38	32 a			
	Mean	16 b	27 a				

Conclusions Annual Ryegrass yielded higher leaf, stem and dead matter when grown in association with Oats than in monoculture. This response was associated with higher forage yield of Annual Ryegrass+Oats mixture than monocultures. The higher yield of the association came from changes in yield components and not only from different growing cycle.

References

Amendola Massiotti, R. & B. Morales Méndez (1997). Competition between oats and annual ryegrass under grazing. *Proceedings of 18th International Grassland Congress,* Saskatoon, Saskatchewan, Canada, 119- 120

Agronomic value of mixture of perennial rye-grass cultivars: preliminary results

W.G. do Nascimento[1], F. Surault, J.C. Emile and C. Huyghe
Institut National de la Recherche Agronomique, 86600 Lusignan, France. Email: huyghe@lusignan.inra.fr,
[1]*granted by CAPES, Brazil*

Keywords: perennial ryegrass, forage yield, nitrogen, exploitation regime

Introduction Mixtures of grass and legume species are commonly used in sown grasslands. Mixtures have been shown to be favourable for stable production over cycles and years due to a succession of species over time (Mosimann & Charles, 1996 ; Nie et al, 2004). However, little is known whether the genetic variation in pure stands has an influence on the agronomic value and its variation over seasons.

Materials and methods Different initial genetic compositions were sown in spring 2003 at 25 kg/ha. Nine cultivars of perennial ryegrass from three maturity groups (early (E) (cv. Hamilton, Belramo and Vital), intermediate (I) (cv. Herbie, Brest and Milca) and late (L) (cv. Ohio, Kerval and Barlatan) were sown as pure stands or in mixtures of two or three cultivars. In mixtures, cultivars were either from the same or from different maturity groups. A total of 19 genetic compositions was studied. Two N levels (N+ : 250 and N- : 60 kg N/ha/yr) were applied. Swards were cut frequently, every 21 days, or infrequently, every 40 days. The experimental design was a split-plot with three replicates. In spring and summer 2004, these cutting regimes resulted in three and two cuts respectively. Statistical analysis was performed on the cumulated DM of the cuts in spring and summer 2004.

Results All three main factors (genetic composition, N fertilisation, cutting regime) had a significant effect on DM yield. As expected, the high fertilisation level and the infrequent cutting regime yielded more. Among the different genetic compositions, when grown in pure stands, the early heading cultivars yielded more than the late heading ones (Fig. 1) thanks to the production of stemmier forage. In every maturity group, mixtures of cultivars of similar maturity had a forage yield similar to the pure stands. Thus, in the first growing season, increasing the within-sward genetic variation did not influence the forage production. Binary mixtures of cultivars of different maturity groups tended to behave as the earliest heading one (figure 1). The mixture of three cultivars, each of different maturity, behaved as the intermediate group.

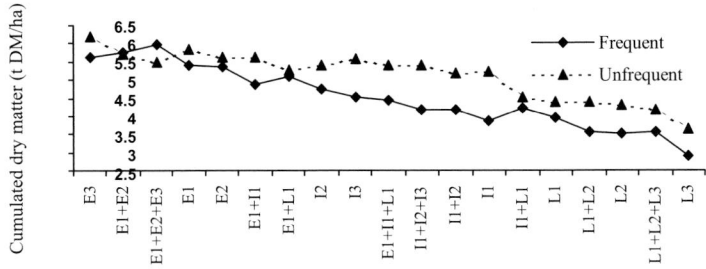

Figure 1 Dry matter yield of the swards with different initial genetic composition under two cutting regimes

Conclusion The genetic composition x exploitation regime interaction was significant mainly because of the highest production of the intermediate group with the infrequent cutting regime. This is due to high biomass production of this group only as a consequence of a high proportion of stems. The N fertilisation x exploitation regime interaction was also significant. A large difference in DM production among cutting regimes was observed with high N fertilisation only.

References

Mosimann, E.& J.P. Charles(1996). Conception des mélanges fourragers en Suisse. *Fourrages*, 145, 17-31
Nie, N.Z., D.F. Chapman, J. Tharmaraj, & R. Clements (2004). Effects of pasture species mixture, management, and environment on the productivity and persistence of dairy pastures in south-west Victoria. 2. Plant population density and persistence. *Australian Journal of Agricultural Research*, 55, 637-643.

Developments in the use of plantain (*Plantago lanceolata*) cultivars in New Zealand pastures

A.V. Stewart and H.G. Judson
PGG Seeds, PO Box 3100, Christchurch 8015, New Zealand, Email: alan.stewart@pggseeds.com

Keywords: plantain. *Plantago lanceolata,* New Zealand, herbs

Introduction The use of pasture herbs, such as chicory, is commonplace in New Zealand in recent years. This has stimulated interest in other herb species such as plantain (*Plantago lanceolata*) that often occurs as a ubiquitous weed in temperate pastures throughout the world. In the last decade 2 improved commercial cultivars, Grasslands Lancelot (Rumball *et al.*, 1997) and the erect, winter active Ceres Tonic (Stewart, 1996), have been bred in New Zealand for use in pastures. These cultivars have useful agronomic features that make them valuable for grazing. They are productive in mixtures, palatable to grazing animals, and tolerate a wide range of soils and dryland conditions (Stewart, 1996; Stewart & Charlton, 2003).

Plantain in New Zealand pastures Plantain, often with added chicory and red clover, is used widely with perennial ryegrass and white clover in pasture mixtures. Autumn establishment and winter activity of Tonic Plantain compare well with those of perennial ryegrass. Thus, plantain has few limitations to establishment and use in mixed pastures. It usually persists longer than red clover or chicory in pastures under common grazing practices and many 5-year-old pastures still contain up to 5-10% plantain in the sward as a tertiary component. Plantain's superior heat tolerance often allows it to exceed 10% of the sward in the north of New Zealand.

Plantain and minerals Plantain herbage contains a higher concentration of Ca, Mg, Na, Cl, S, P, Zn, Cu and Co than pastures based on perennial ryegrass-white clover. Lambs grazing pure plantain swards have higher liver concentration of Cu (2250 vs 716 µmol/kg FW) and Se (671 vs 380 nmol/kg FW) than those grazing perennial ryegrass (Moorhead *et al.*, 2002). Animals grazing plantain also retain more Ca, Mg and Na than those grazing perennial ryegrass (Wilman & Derrick, 1994).

Plantain and enhanced animal performance With appropriate management, plantain supported summer growth rates in young lambs of 222 g/day. Increased intake at a common allowance (Moorhead *et al.* 2002) and a faster rumen degradation rate of plantain than perennial ryegrass (24.6 vs. 11.7%; Burke *et al.* 2000) may explain enhanced animal performance. Older and stemmier herbage has a reduced animal performance (51g/day, Robertson et al., 1995, 84-141g/day, Fraser & Rowarth, 1996) due to a much reduced digestibility. As herbage quality depends on the stage of plant development, grazing management should aim to keep plantain at a relatively leafy stage and utilise seed-head before lignification.

Conclusion The useful features of these productive bred plantain cultivars make them a valued component of pasture mixtures for sheep, cattle and deer in New Zealand. However, like chicory, the lack of suitable herbicidal controls for serious broadleaved weeds can limit plantain use in some situations.

References

Burke, J.L; G.C, Waghorn, I.M, Brookes, G.T, Attwood, E.S Kolver (2000). Formulating total mixed rations from forage – defining the digestive kinetics of contrasting species. *New Zealand Society of Animal Production* 60: 9-14.

Fraser, T.L. and J.S. Rowarth, (1996). Legumes, herbs or grass for animal performance. *Proceedings of the New Zealand Grassland Conference* 58: 49-52

Moorhead A.J.E, H.G. Judson and A. Stewart (2002). Liveweight gain of lambs grazing 'Ceres Tonic' plantain (*Plantago lanceolata*) or perennial ryegrass (*Lolium perenne*). *NZ Soc. of Animal Production* 62: 171-173

Robertson, H.A., J.H. Niezen, G.C., Waghorn, W.A.G., Charleston, and M. Jinlong (1995). The effect of six herbages on liveweight gain, wool growth and faecal egg count of parasitised ewe lambs. *New Zealand Society of Animal Production 55: 199-201*

Rumball W, R.G. Keogh, G.E. Lane, J.E. Miller, and R.B. Claydon (1997). `Grasslands Lancelot' plantain (*Plantago lanceolata* L.) *New Zealand Journal of Agricultural Research* 40: 373-377

Stewart A V (1996). Plantain (*Plantago lanceolata*) - a potential pasture species. *Proceedings of the New Zealand Grassland Association* 58: 77-86

Stewart A V, J.F.L. Charlton (2003). Pasture and forage plants for New Zealand. *Grassland Research and Practice Series No 8, Second Edition. October 2003* 96 pp

Wilman, D. and R.W. Derrick (1994). Concentration and availability to sheep of N, P, K, Ca, Mg and Na in chickweed, dandelion, dock, ribwort and spurrey, compared with perennial ryegrass. *Journal of Agricultural Science* 122: 217-223

Yield and quality of annual ryegrass grown in pure stand and in mixtures with squarrosum clover

M.E.V. Lourenço and P.M.M. Palma
Universidade de Évora, Departamento de Fitotecnia, Apartado 94, 7002-554 Évora, Portugal, Email: melouren@uevora.pt

Keywords: productivity, nutritive value, *Lolium multiflorum* Lam., *Trifolium squarrosum* L

Introduction The objective of this study was to evaluate the importance of growing annual ryegrass in mixtures instead of ryegrass alone in order to reduce nitrogen application and thereby lowering production costs, and environmental pollution.

Materials and methods The treatments consisted of annual ryegrass as pure stand (AA), 75% of annual ryegrass + 25% clover squarrosum (25% TS), 50% of annual ryegrass + 50% clover squarrosum (50%TS) , 25% of annual ryegrass + 75% clover squarrosum (75%TS). In each treatment there were four sub-treatments of nitrogen levels applied after seeding and each harvest: 0, 50, 100 and 150 kg/ha. The experiment was conducted under rainfed conditions for two years (2001/02 and 2002/03). The first year was a normal year (598 mm) with a very rainy autumn, and the second year was dry (478 mm), as compared to the average for 1941/70 (574 mm). The field trial was conducted as a split-plot design at "Revilheira" Farm, 60 km from Évora. The soil type was a vertic luvisol with 16 and 16 mg/kg of available phosphorus, 70 and 104 mg/kg available potassium, 1.16 and 1.26% organic matter, and pH (H_2O) values 6.2 and 5.6, respectively in 2001 and 2002. Dry matter (DM) yield, crude protein, and DM digestibility were determined according to Lourenço & Palma (2001).

Results Dry matter yields (Figure 1) tended to increase with nitrogen application especially for ryegrass alone and the mixture with 25% TS, at the two highest nitrogen levels, probably because when nitrogen is applied the grass benefits more than the legume (Vallis *et al.*, 1977). The greatest benefit of increasing *Trifolium squarrosum* in the mixture occurred with respect to crude protein yield (Figure 2) when nitrogen applied was lower than 50 kg/ha. The trends for digestible DM yield (Figure 3) were similar to those for DM yield.

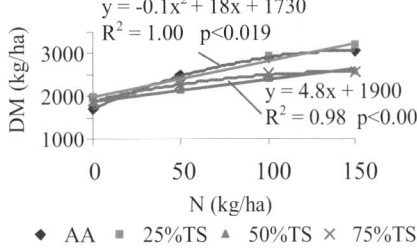

Figure 1 Dry matter yield over the two-year period

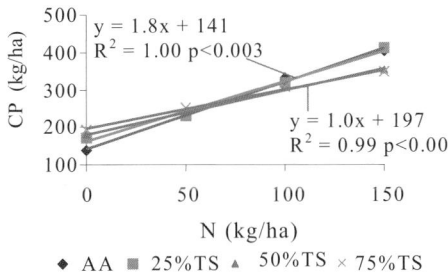

Figure 2 Crude protein yield over the two-year period

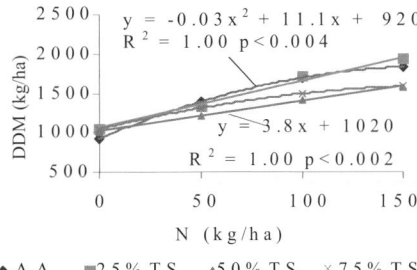

Figure 3 Digestible dry matter yield over the two-year period

Conclusions This study shows the importance of increasing the percentage of the clover in the mixtures, at the lowest nitrogen levels, especially with respect to crude protein yield. As the nitrogen rates were increased, there was a trend to get higher yields with ryegrass alone and with the mixture with 25% *Trifolium squarrosum*.

References

Lourenço, M.E.V. & P.M.M. Palma (2001). The effect of plant population on the yield and quality of annual rye-grass. *Proceedings of the XIX International Grassland Congress*, 416-417.
Vallis I., E.F. Henzell & T.R. Evans (1977). Uptake of soil nitrogen by legumes in mixed swards. *Australian Journal of Agricultural Research*, 28, 412-425.

Quality legume-based forage systems for contrasting environments: COST Action 852

A. Helgadottir[1], J. Connolly[2], M Fothergill[3], R.P. Collins[3], C. Porqueddu[4], A. Lüscher[5], M.T. Sebastia[6], M. Kreuzer[7] and M. Wachendorf[8]

[1]Agricultural Research Institute, Keldnaholt, 112 Reykjavík, Iceland, Email: aslaug@rala.is, [2]University College Dublin, Belfield, Dublin 4, Ireland, [3]IGER, Plas Gogerddan, Aberystwyth, Ceredigion, SY23 3EB, UK, [4]CNR-ISPAAM, Via De Nicola, 07100 Sassari, [5]Swiss Federal Research Station for Agroecology and Agriculture, 8046 Zurich, Switzerland, [6]Technology and Forestry Centre of Catalonia, 25280 Solsona, Spain, [7]ETH Zurich, ETH centre, 8092 Zurich, Switzerland, [8]Christian-Albrechts-University, 24098 Kiel, Germany

Keywords: forage legumes, grass-legume mixtures, genetic resources, legume utilisation, sward management

Introduction Agricultural systems that reduce environmental degradation, sustain agricultural productivity and economic viability, maintain stable rural communities, enhance the quality of life and respond to increasing demand for livestock products are promoted in developed countries. Though major challenges exist, forage legumes, adapted to a wide range of soil types, climatic conditions and management systems, will become increasingly important components of sustainable agricultural production systems in Europe. Temporal and spatial variation in legume performance often occurs. Compared to pure grass systems, legume-based systems may lead to increased N losses. To what extent ruminants can use the protein from forage legumes, and whether there are differences among species and cultivars are unknown. To improve reliability and the range of forage legumes, we must understand the constraints of environment, the reasons for divergence between species potential and actual performance and the most efficient way to use the herbage. The COST Action 852 - Quality legume-based forage systems for contrasting environments was set up to help resolve some of these questions.

Objectives and benefits The overall objective of Action 852 is to increase the quantity and quality of home-grown proteins from regionally adapted legume-based forage systems. This will benefit: society by producing high quality animal products (e.g. low in residues, high in fatty acids) coupled with reduced environmental impact, the farming industry by developing more reliable systems (species selection and management), and the scientific community by stimulating active communication between scientists by a multidisciplinary approach. To achieve this, Action 852 has 3 working groups, each with its specific objectives and common experimental protocols. The countries in Europe with sites in the common protocols range from Iceland to Greece and Finland to Spain. There are sites in North America and Australia also.

Working Group 1: Legume genetic resources The main areas of study are: mechanisms of adaptation of the legume plant and its associated micro-organisms, the genetic basis of adaptation, and breeding of plants/micro-organisms (including both traditional and molecular techniques). A common experiment started in 15 sites in 12 countries to determine whether populations of forage legumes with a wide genetic base are more productive and stable over time and location than those with a narrower genetic base.

Working Group 2: Sward management The main areas of study are: mechanisms resulting in successful sward establishment, mechanisms affecting competition and complementarity in mixed swards, sward dynamics, and nitrogen flows in legume-based systems. A common experiment has started in 38 sites in 20 countries to examine whether there are benefits of species mixtures over monocultures in terms of productivity, persistence, resistance to invaders and increased nutrient retention over broad environmental gradients.

Working Group 3: Forage utilisation The main areas of study are: animal intake and grazing behaviour, quality aspects of legume-based fresh and ensiled forage, and the mechanisms of N-flows within the ruminant. A common experiment has been developed at 9 sites in 7 countries to examine the quality of legume-based fresh and ensiled forage and investigate how herbage composition relates to N use efficiency and milk and meat quality.

Conclusions COST 852 is still in its early days, yet already we have seen the power of internationally integrated experimentation. Such interaction has the potential to extend beyond the confines of the action and sets up a framework for true international exchange of research findings.

Higher yield and fewer weeds in four-species grass/legume mixtures than in monocultures: results from the first year at 20 sites of COST action 852

L. Kirwan (Ireland)[1], G. Bélanger (Canada), J. Finn (Ireland), M. Fothergill (UK), B. Frankow-Lindberg (Sweden), R. Garcia-Sarrion (Spain), A. Ghesquiere (Belgium), P. Golinski (Poland), A. Helgadóttir (Iceland), M. Jørgensen (Norway), Z. Kadžiuliene (Lithuania), D. Nyfeler (Switzerland), P. Nykänen-Kurki (Finland), G. Parente (Italy), V. Vasileva (Bulgaria), R. Collins (UK), J. Connolly (Ireland), A. Lüscher (Switzerland), C. Porqueddu (Italy) and M.T. Sebastià (Spain)

See the authors' affiliations at http://www.COST852.com [1]Email: Laura.Kirwan@ucd.ie

Keywords: grass / legume mixtures, forage yield, invasion, simplex design

Introduction Utilisation of grass/legume mixtures instead of grass monocultures is a sensible alternative for low input, efficient agricultural systems that reduce production costs, promote environmental policy and maintain a living countryside. Consequently, widely adapted forage legumes will become increasingly important. Instability of simple grass / legume mixtures with only one grass and one legume species is a major problem (Wachendorf et al., 2001). An experiment was established in 39 sites in Europe, Australia and Canada within COST Action 852 to: (1) assess the benefits of grass / legume mixtures in terms of forage production, (2) test whether the combination of fast and slow-growing species improves the stability of the mixtures and (3) assess response patterns over a large environmental gradient.

Materials and methods The experiment consisted of 30 plots, each sown with a mixture of varying proportions of 2 grass and 2 legume species, or a monoculture of one of the 4 species after a simplex design (Cornell, 1990). The design was duplicated at 2 sowing densities (100 and 60 %). One species of each functional group was fast-and the other was slow-growing. Results from year 1 were analysed from 20 sites covering a North-South gradient from Iceland to Spain. Data were available only for the first harvest in 5 of the sites. Three mixture types were tested: North European (NE), Central European (CE) and Mediterranean (Med). The swards were managed according to local farming practices.

Results and Discussion The mean yield of mixed plots was compared with the mean for monoculture plots for the total yield for year 1 in 20 sites (Table 1). The averaged monocultures never performed better than the averaged mixtures and the average yield of the mixed plots was significantly greater than the average yield of the monoculture plots in 12/20 sites. The average percentage unsown species for year 1 in mixed plots was lower than in monoculture plots for 19/20 sites, significantly so in 15 sites. This suggests the existence of synergistic yield effects among the sown species and greater resistance to invasion by unsown species in the mixtures.

Conclusions: The results of year 1 suggest that mixing forage species from different functional types, grasses and legumes, enhances productivity and decreases the proportion of unsown species. The observed advantages of the mixtures were consistent for most sites, and the results are assumed to be reliable due to the broad gradient considered in this study.

Table 1 Ratio of Mixed / Mono for total yield and % unsown at 20 sites for the first year after establishment

Country	Total		% Unsown			
	Mix/Mon	Sig	Monoculture	Mix/Mon	Sig	
CE	1.4	<0.001	8.8	0.26	<0.001	
CE	1.4	<0.001	10.4	0.28	<0.001	
CE	1.3	0.011	15.9	0.12	<0.001	
CE	1.4	<0.001	33.1	0.50	<0.001	
CE	1.3	0.002	33.7	0.59	0.004	
CE	1.65	0.002	38.7	0.07	<0.001	*
CE	1.0	NS	45.2	0.72	0.006	
CE	1.1	NS	67.1	0.72	0.076	
CE	1.2	<0.001	2.5	0.50	NS	*
CE	1.0	NS	64.5	1.05	NS	
CE	1.0	NS	80.5	0.85	NS	
Med	1.72	<0.001	33	0.92	NS	
Med	1.97	<0.001	29.6	0.03	<0.001	
Med	1.0	NS	71.3	0.66	0.023	
Med	1.4	NS	68.4	0.58	0.008	
Med	1.2	NS	54.9	0.54	0.007	*
NE	1.3	<0.001	40.6	0.15	0.001	
NE	1.3	<0.001	38.5	0.15	<0.001	
NE	1.45	<0.001	30.4	0.33	0.001	*
NE	1.08	0.065	93.1	0.91	0.004	*

Results for sites marked * were for the first harvest

References

Cornell, J. A. (1990). Experiments with Mixtures: designs, models, and the analysis of mixture data, 2nd ed. John Wiley & Sons, New York, 632 pp.

Wachendorf, M., R. P. Collins, A. Elgersma, M. Fothergill, B. E. Frankow-lindberg, A. Ghesquiere, A. Guckert, M. P. Guinchard, A. Helgadottir, A. Lüscher, T. Nolan, P. Nykänen-kurki, J. Nösberger, G. Parente, S. Puzio, I. Rhodes, C. Robin, A. Ryan, B. Stäheli, S. Stoffel, F. Taube & J. Connolly (2001). Overwintering and growing season dynamics of Trifolium repens L. in mixture with Lolium perenne L.: A Model Approach to Plant-Environment Interactions. Annals of Botany, 88: 683-702

Adaptation, compatibility and acceptability of grass-legume pastures in the Andean region of Colombia

E. Cárdenas and E. Castro
Universidad Nacional de Colombia 30th Avenue with 45th Street, Bogotá, Colombia, Email: eacardenasr@unal.edu.co

Keywords: *Pennisetum clandestinum, Lotus corniculatus*, adaptation, acceptability

Introduction In Colombia, the specialised dairy production system is located in the high altitude Andean region. Its main feed resources are pure stands of *Pennisetum clandestinum* or/and *Lolium* spp. Nevertheless, the present market conditions require highly competitive and quality forages year round. These forages should be produced with low inputs (irrigation, fertilisers and agrochemicals) and be resistant to pests and diseases. The objective of this research was to evaluate the adaptation, compatibility and acceptability of introduced forage species for sustainable pasture management.

Materials and methods Ten grass species were planted mixed with *Lotus corniculatus*. *P. clandestinum* fertilised with 50 kg of N/ha after each cut was used as a control. All species were planted in 2.5x 5.0 m plots interspersing rows of legume and grass species. Treatments (grass-legume combinations) were arranged in a split plot design with three replicates. During establishment (6 months), vigour, ground cover, pests and diseases were measured. After establishment, plots were cut and the botanical composition, biomass production (kgDM/m^2) and nutritional quality (CP, NDF, ADF and IVDDM) were determined for 45 and 70 d regrowths in the dry and rainy season. At the end of the trial, the relative acceptability index (RAI) was measured with heifers of 300 kg liveweight in 45 days old regrowth (Maass *et al*, 1999).

Results and conclusions During establishment, the legume and the grasses showed good development and compatibility. The aerial biomass production was greater for the 70 d regrowth (Table 1) particularly in the rainy season. The highest production was obtained in mixtures of *L. corniculatus* with *Festuca arundinacea, Festuca rubra, Bromus catharticus* and *P. clandestinum*. The proportion of legume was similar between cuts. The legume took over in *Phleum pratense* and *Festuca pratense* plots (> 80% legume). The 45 d cut gave higher nutritional quality. The higher CP concentration in the grasses was observed for naturalised *P. clandestinum* (14%), *D. glomerata (15.5%)* and *F. arundinacea (15.1%)*. Values for *L. corniculatus at 45 d were* CP 26.5 % and DIVMS of 68.6%. The mixture with the highest RAI was *F. arundinacea* and *P. clandestinum* (naturalised). It was concluded that the most promising mixtures were *F. arundinacea* and *P. clandestinum* (naturalised) with *L. corniculatus*.

Table 1 Aerial biomass production, legume proportion and relative acceptability index

| *Lotus corniculatus* + | Regrowth | | | | | | RAI |
| | 45 days | | | 70 days | | | 45 days |
	(gDM/m^2)	Leg (%)		(gDM/m^2)	Leg (%)		
P. clandestinum (without *Lotus*) + Fert	62.4	abcd	-	-	183.4	abcd	-	1.46
Bromus catharticus	62.5	abcd	57.3	abc	186.1	abcd	50.9	cd	0.74
Festuca rubra	100.3	A	60.6	abc	257.6	a	50.1	cd	0.47
Dactylis glomerata	81.2	abc	22.1	d	215.6	abc	27.8	d	0.94
Festuca arundinacea	80.5	abc	64.1	abc	236.5	ab	48.9	cd	1.82
Pheum. Pratense	26.1	D	90.8	a	35.0	cd	83.7	ab	0.46
Pennisetu clandestinum (introduced)	60.5	abcd	62.8	abc	231.8	ab	57.3	bcd	1.42
Anthoxanthum odoratum	83.2	ab	62.2	abc	86.8	abcd	58.4	bcd	1.35
Holcus lanatus	82.2	abcd	48.9	bcd	199.2	abcd	45.8	cd	0.45
Dactylis glomerata (cv. Knaulgrass)	43.1	bcd	79.9	ab	47.5	bcd	67.8	abc	0.71
Festuca pratense	32.8	cd	81.2	ab	16.8	d	94.4	a	0.43
Pennisetum clandestinum (naturalised)	95.2	A	35.4	cd	75.3	abcd	53.0	bcd	1.76
Average	66.7***		60.8***		147.6***		58***		1.00

References
Maass, L. B, E. C Lascano & E. A. Cárdenas (1999). La leguminosa arbustiva *Codariocalyx gyroides*. 2. Valor nutritivo y aceptabilidad en el piedemonte amazónico, Caquetá, Colombia. *Pasturas Tropicales*, 3, 12-18

New cultivars for high quality, persistent legume-grass pastures in the southern USA

J.H. Bouton
University of Georgia, Athens, Georgia 30602 USA (now Noble Foundation, Ardmore, Oklahoma 73401 USA), Email: jhbouton@noble.org

Keywords: MaxQ tall fescue, Durana white clover

Introduction In the southern USA one cannot depend on perennial legume-grass pasture systems to have persistently high nutritive quality. 'Jesup' tall fescue, a cultivar with better persistence in the region than 'Kentucky 31', was re-infected with a non-ergot alkaloid producing strain of the *Neotyphodium coenophialum* fungal endophtye (MaxQ™) and found to give persistence equal to Jesup with its endemic strain (E+), but without animal toxins (Bouton *et al.*, 2002). 'Durana' white clover (Bouton *et al.*, 2004) was developed from regional ecotypes of *Trifolium repens f. hollandicum* germplasm and found to improve animal performance in both E+ and endophyte-free (E-) versions of Jesup. This experiment aimed to assess the ability of Jesup MaxQ when inter-planted with Durana white clover to provide persistent, high quality pasture in the southern USA.

Materials and methods Treatments were Jesup MaxQ paddocks (0.91ha) either inter-planted with Durana white clover (WC), or fertilised with 120kg N/ha annually (N), with 2 replicates/treatment. Forage available yield was determined initially and every 4 wk during the spring-summer grazing season by sampling ten, 0.09 m^2 quadrate random samples throughout each paddock. Botanical composition was determined by hand separation of these samples into their components (e.g., tall fescue, clover, other). A put and take system was used to adjust stocking rate of grazing animals (beef steers; initial weight circa 223kg). Two steers were designated initially as testers and the others as grazers. The testers stayed on the paddock all the time. Animals were weighed every 28 days for two seasons (2003, 2004).

Results Moister conditions in 2003 allowed a longer grazing period (126 days) than in 2004 (77 days). Due to these conditions, available forage supply and total animal gain in 2003 were higher than in 2004 (Figure 1). Higher forage yields were found for the grass alone treatment with N-fertiliser than when the grass was grown with white clover. However, in each year, the addition of Durana white clover increased both daily and total animal gains over the N-fertilised treatment. Durana also composed 29% of the available forage supply when averaged for both years (35% in 2003 and 23% in 2004).

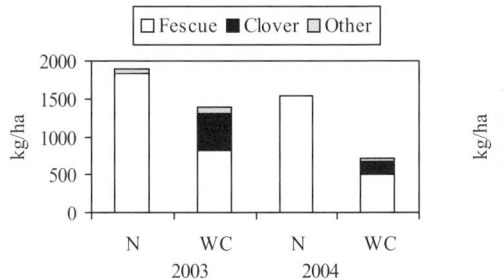
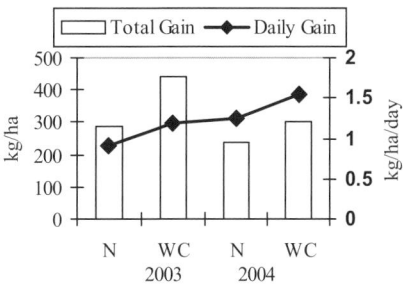

Figure 1 Available forage supply (left panel) and animal performance (right panel) of Jesup MaxQ tall fescue paddocks inter-planted with Durana white clover (WC) or grown alone with nitrogen fertiliser (N)

Conclusions In both years, both cultivars performed well when inter-planted. Additions of Durana white clover to the available forage supply increased animal performance even when grown with a high quality grass like Jesup MaxQ. These paddocks will be monitored in future years to assess whether the 2 cultivars will continue to provide a high quality, persistent legume-grass pasture for the region.

References
Bouton, J. H., G. C. M. Latch, N. S. Hill, C .S. Hoveland, M. A. McCann, R. H. Watson, J. A. Parish, L. L. Hawkins, & F. N. Thompson (2002). Re-infection of tall fescue cultivars with non-ergot alkaloid producing endophytes. *Agronomy Journal*, 94, 567-574.
Bouton, J. H., D. R. Woodfield, C. S. Hoveland, M. A. McCann & J. R. Caradus (2004). New white clover cultivars for the southeastern USA. In: K. Cassida (ed.) Proceedings of 2004 Conference of American Forage and Grassland Council (AFGC), 13, 338-342.

Seed productivity of *Festulolium* and *Lolium x boucheanum* varieties

I.J. Gutmane and A.M. Adamovich
Latvia University of Agriculture, Liela iela 2, Jelgava, LV 3001, Latvia, Email: alexadam@cs.llu.lv

Keywords: *Festulolium, Lolium x boucheanum*, seed production

Introduction Hybrid ryegrasses are intermediate lines between perennial ryegrass and Italian ryegrass as regards growth parameters, productivity and resistance. Good quality and digestibility of forage and higher yield are the main advantages of hybrid ryegrass over perennial ryegrass. Climatic conditions in the Baltic States and North Europe are not favourable enough for ryegrass cultivation but *Festulolium* is a prospective forage crop for this area (Nesheim, 2000, Adamovich, 2003). Ryegrasses, including hybrid ryegrasses, surpass *Festulolium* in forage quality. However, *Festulolium* persistance (resistance, winter-hardiness and, thus, high productivity) are noteworthy. Due to yield quality and competitive productivity, *Festulolium* may rank equally with timothy and fescue, the main grasses of this climatic zone.

Materials and methods Field experiments (2001-2003) were conducted on sod-podzolic soil to determine the productivity of *Festulolium* (*Lolium ssp. x Festuca ssp.*) and hybrid ryegrass (*Lolium x boucheanum*) seeds. The seed fields were developed using both local and foreign varieties of *Festulolium* of different origin: 'Ape' (control), 'Lofa', 'Hykor', 'Perun', 'Punia', and 'Tapirus' and 'Ligunda', 2 foreign hybrid ryegrasses. The seeding rate was 600 germinating seeds/m². Two rates of mineral fertilisers were applied in the seed production year: P104, K150, N90 kg/ha, or P104, K150, N120 kg/ha.

Results Lodging occurred on both fertiliser backgrounds in all varieties except 'Hykor'. The lodging resistance was only 2.3-3.3 points. Table 1 shows mean seed yield and yield structure data on both fertiliser backgrounds.

Table 1 Seed productivity and yield structure of *Festulolium* and *Lolium x boucheanum*

Varieties (FB)	Winter hardiness (points)	Seed yield (kg/ha)	Seed yield (%)	TKW (g)	Generative tillers (p/m)	Flowerhead Length (cm)	Flowerhead Weight (g)
Ape (LV)	8.5	1397	100	3.8	1915	23.2	0.67
Lofa (DLF)	6.8	1547	111	4.0	1950	23.7	0.69
Hykor (DLF)	7.5	963	68	2.8	1600	17.1	0.57
Perun (DLF)	7.5	866	62	4.2	1550	25.4	0.83
Tapirus (DSV)	6.5	665	95	4.2	1390	21.4	0.72
Ligunda (DSV)	2.5	640	46	2.9	892	18.7	0.46
Punia (LT)	8.0	800	57	3.9	1830	23.4	0.70
LSD 0.05		105					

Mean seed yields were relatively high (0.61-1.58 t/ha). 'Tapirus', a late maturing hybrid ryegrass, produced similar seed yield (0.63 and 0.70 t/ha) on both fertilizer treatments. *Festulolium* gave higher seed yields (0.76-1.58 t/ha). The variety 'Lofa' (*L. multiflorum x F. arundinacea*) produced the highest seed yield (1.58 t/ha) at the fertiliser rate N120. The 1000 seed mass for the varieties ranged from 2.89-4.26 g. 'Tapirus' and 'Perun', respectively, produced the coarsest seed (4.23 and 4.19 g). Flowerhead length ranged from 16.5-25.9 cm. Flowerhead mass was 0.44-0.88 g. *Festulolium* 'Perun' had the longest flowerhead with the greatest mass (25.4 cm and 0.84 g, respectively).

Conclusions *Festulolium* and hybrid ryegrasses are prospective forage grasses for the Baltic States and North Europe. Due to quality and competitive productivity, these forage grasses equally rank with other grasses grown in Latvia. Some foreign *Festulolium* and hybrid ryegrass varieties are suitable for seed production under agroclimatic conditions of Latvia.

References
Adamovich A. & O. Adamovicha (2003). Productivity and forage quality of *Festulolim*/legume mixed swards in response to cutting frequency. *EGF, Grasland Science in Europe*, 8, 453-456.
Nesheim L. & I. Bronstad (2000) Yield and winter hardiness of *Festulolium (Festuca x Lolium)* in Norway. *EGF, Grassland Science in Europe*, 5, 238-240.

Productive longevity and yield quality of galega-grass swards

A.M. Adamovich
Latvia University of Agriculture, Liela iela 2, Jelgava, Latvia, LV 3001, Email: alexadam@cs.llu.lv

Keywords: fodder galega, grasses, mixtures, productivity, yield quality

Introduction Fodder galega (*Galega orientalis* Lam.) is an early maturing, very productive perennial forage legume that fixes atmospheric nitrogen. Unlike other legumes, pure stands of fodder galega provide stable yields of green feed and seeds and do not thin out between the years. The symbiotic potential of fodder galega to grow in mixtures with grasses can be exploited to produce ecologically safe forage and animal products.

Materials and methods Field trials (1986-2003) aimed to study continuous green forage production from fodder galega-grass swards in the intensive growth stage. Mixed swards (n=35; 13 binary- and 22 multi- species) were developed on stagnic luvisol and sod-podzolic soils. Binary- and multi-species seed mixtures contained 40% fodder galega (*Galega orientalis* Lam.) 'Gale' and 60% grasses from 13 grass species. The plots were fertilised: P 40 and K 150 kg/ha, and N 0, N $90_{(45+45)}$. Swards were cut 2-4 times during the growing season. Plant samples were analysed for dry matter (DM); crude protein (CP; by modified Kjeldahl); crude fibre (CF), neutral detergent fibre (NDF) and acid detergent fibre (ADF; van Soest, 1980).

Results In 23 production years of pure galega, the following mean yields (t/ha) of DM and CP were attained in early flower: DM 9.68 and CP 1.84 on stagnic luvisol; DM 8.44 and CP 1.70 on sod-podzolic soils. Fodder galega significantly surpassed other forage legumes as regards productive longevity; fluctuations in its DM yields were insignificant between years. Inclusion of a grass species in a mixture increased yield by 26-32 % already in production year 1. Split application of the fertiliser N 90 reduced the proportion of galega in a sward and decreased DM yields by 1.49 t/ha at 3-fold cutting treatments, compared to unfertilized plots. Frequent, 4-fold cutting reduced the productivity of galega-grass mixtures; total DM yield decreased by 2.82 t/ha or 30.8% in all experimental plots at 4-cut treatments.

Table 1 Mean yield quality of fodder galega/grass swards (1998-2002, cut 1 and cut 2)

Cutting regime	Index of quality	Composition of swards						Mean in mixtures
		fodder galega	number of species in mixtures					
			2	3	4	5	6	
3-fold	galega in DM yield, g/kg	924.0	532.6	579.4	552.2	560.3	527.2	550.3
	ME, MJ/kg DM	10.4	10.5	11.2	11.7	10.9	11.3	11.1
	NDF, g/kg DM	522.5	504.6	469.4	475.3	480.4	455.3	477.0
4-fold	galega in DM yield, g/kg	860.3	530.2	490.1	502.6	475.4	439.2	487.5
	ME, MJ/kg DM	11.4	11.7	12.2	12.1	11.0	12.0	11.8
	NDF, g/kg DM	4235	410.8	430.4	437.6	459.3	440.2	435.7

Fodder galega in pure stands at the branching and bud stages excelled. Its crude protein content (g/kg DM) was high (297±21), particularly in plant leaves (342); in fodder galega in mixed swards it fell to 257±16 in the bud stage and 203±24 in the early flower stage. Mean metabolizable energy (ME) content (MJ/kg DM) of fodder galega and mixed galega-grass stands was 10.9 and 11.4±1.1, respectively (Table 1). Mean NDF content (g/kg DM) in galega-grass mixtures was </=456±27, compared with 522±32 in pure galega stands in early flower. Different proportions of plant leaves and their position in canopy structure in galega-grass mixtures, compared to pure galega swards could explain this. Slower growth of associated grasses in a mixed sward slowed maturation and ageing of grass leaves, compared to galega, thus contributing to production of quality forage.

Conclusions Fodder galega in pure stands or in mixtures with grasses of various growth patterns is productive, of high quality and persists for long periods but 3 species mixtures were the most productive. Competitive grasses in the mixtures reduce productive longevity of swards compared to pure galega stands.

References
Raig H., Nommsalu H., Meripold H., Metlitskaja J. (2001) Fodder galega. Estonian Research Institute of Agriculture, Saku, 141 pp.

Red clover in monoculture or in association with grasses?

A. De Vliegher and L. Carlier
*Department of Crop Husbandry and Ecophysiology, B. van Gansberghelaan 109, B-9820 Merelbeke, Belgium,
Email: a.devliegher@clo.fgov.be*

Keywords: seed mixtures, clover, cover crop

Introduction Sowing grass or clover under a cereal cover crop or after the harvest of the cereal is common practice, but there still are some questions about seed mixtures regarding monocultures or mixtures, especially when the fodder crop has to be productive for more than 2 years. In this experiment seed mixtures with clovers and grasses in a cereal cover situation were compared in terms of DM yield, energy and protein content and proportions of grass, clover and weeds.

Materials and methods The treatments were *Trifolium pratense* in monoculture or with Lolium *multiflorum, L. perenne* or *Phleum pratense* and *T. repens with L. perenne.* Botanical composition, DM yield and quality parameters were determined. The experiment was carried out in 2002-22003 at 2 locations in Belgium.

Results Sowing seed mixtures under a cereal cover crop resulted in a significantly higher DM yield in the year of sowing in comparison with sowing after harvest in both locations. In the second year, an equal DM yield was noted for both sowing periods in Lovendegem, but in Merelbeke sowing under a cover crop gave significantly lower DM yield because the establishment of this sowing was inferior. Red clover (*T. pratense*) in monoculture or in association with grasses yielded more than white clover (*T. repens*) with perennial ryegrass (*L. perenne*) in the first and second year. The differences in DM yield were significant in the sowing under cover crop in Lovendegem and in the sowing after harvest in Merelbeke. In Lovendegem, with a low N input (about 115 kg N/ha in spring 2002 and 2003) there were no differences in DM yield between a monoculture of red clover and mixtures with grasses. In Merelbeke, with a medium N input (27 kg N/ha in the first and 215 kg N/ha in the second year) the association *T. pratense* + Italian ryegrass (*L. multiflorum*) yielded significantly more than the monoculture red clover; the other associations didn't. The red clover content was the lowest in the association with *L. multiflorum* (average 44%) and the highest in the association with *Phleum pratense* (average 76%). *L. perenne* was in between with a clover content of 60%. The white clover content was very high in Lovendegem (low N-fertilisation) and substantial in Merelbeke (215 kg N/ha). Red clover in monoculture contained higher quantities of weed. Weed development decreased when red clover was sown in association with *Lolium* spp. (De Vliegher *et al.* 2004). The companion grass also affected the nutritive value. There was a tendency that the *T. pratense* in monoculture contained less energy (795VEM/kg DM) than associations with *L. multiflorum* (810VEM/kg DM) or *L. perenne* (815VEM/kg DM). The association with *Phleum pratense* had about the same energy content as the red clover monoculture (799VEM/kg DM) and this could be explained by the small amount of grass in the mixture. There was no influence on the protein content apart from the mixture with *L. multiflorum* where a lower DVE (-16 g/kg DM) and OEB content (-39 g/kg DM) were measured. The association *T. repens - L. perenne* had a higher energy level (872VEM/kg DM) than *T. pratense – L. perenne* (815VEM/kg DM) but there was no difference in protein content, measured by DVE and OEB.

Conclusions In general *T. repens* + *L. perenne* gave a lower DM yield in comparison with *T. pratense* in monoculture or in association with grass species. The effect of *Phleum pratense* on DM yield and energy content in the first and second year was negligible. The addition of *L. multiflorum* reduced the clover content significantly and showed a tendency to a higher DM yield, a higher energy content and a lower protein content. In general, intermediate varieties of *L. perenne* seemed to be the best partner for *T. pratense,* because it gave a good balance between clover and grass in the sward and cutting was tolerated very well.

References
De Vliegher, A., Carlier L. & Haesaert G. (2004). Behaviour of seed mixtures with clover, sown in a cereal
    cover crop or after the harvest of the cereals. COST Action 852 Workshop in Ystad (Sweden) 4p. (In press).

Forage yield and seed bank production with new annual legumes for the dryland conditions in the Araucanía Region in the south of Chile

O. Romero, A. Catrileo, C. Rojas and A. Loi
Instituto de Investigaciones Agropecuarias, Centro Regional de Investigación Carillanca. Casilla 58-D, Temuco, Chile, Email: oromero@carillanca.inia.cl

Keywords: *Ornithophus compressus, B. pelecinus,* forage legumes, seed production

Introduction In dryland areas of the IX Region of Chile, with a dry period of 150 days with a negative hydrologic balance beef production is based on natural pasture, with low P soils, which produces a low DM yield. The sown pasture for this area is fescue in mixture with subclover. The subclover, has persistence problems and erratic production (Romero & Rojas, 2001). Hard seeded serradella (*Ornithophus compressus),* soft seeded serradella *(O.sativus)* and biserrula *(Biserrula pelecinus*) are annual forage legumes and well adapted to low P levels in the soil (Oram, 1990). The objectives of the present study were to evaluate the 3 legumes species and different cultivars in terms of dry matter (DM) yield and the ability to form seed banks to improve the persistence of the legumes in a ley farming system for beef production.

Materials and methods During the 2003-2004 growing season different cultivars of the annual *O. compressus*, *O.sativa, B. pelecinus* were studied in dryland conditions in the Regional Center-Carillanca, Temuco, Chile (38° 41˝S -72° 25˝ W, Alt. 200 m.s.l.). The sowing was in autumn at a rate of 40 kg/ha, for serradella and 20 kg/ha for biserrulla. The DM yield, and seed production were recorded. The experimental design was a randomised block with three replicates.

Results The highest DM yields were obtained with *O. compressus* cv. Avila and *O. sativus* cv. French and Cadiz (P≤0.05) (Figure 1). Biserrulla had low DM yields during the first year of establishment. The highest seed pod production was obtained with *O. compressus* cv. Avila (Table 1).

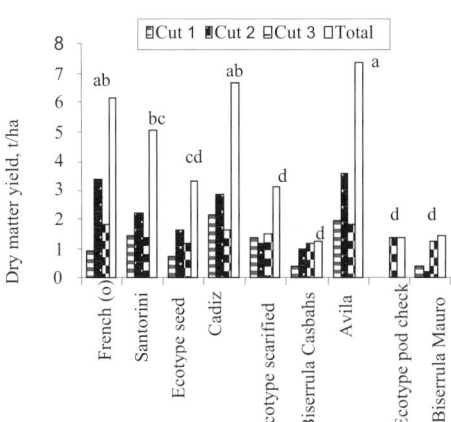

Figure 1 Dry matter yield per cut and total kg/ha

Table 1 Seed production (pods/ha). IX Region. Chile

Legumes species	Distribution of harvested pods kg/ha		
	In the stem	Soil surface	Total pods
O.sativus French	2.993	1.161	4.155 ab
O.sativus cv. Cadiz	2.618	384.0	3.002 ab
O.compressus cv. Santorini	3.252	393.1	3.645 ab
O.compressus ecotype seed without pods	1.789	949.3	2.739 ab
O.compressus ecotype scarified	1.937	384.0	2.322 ab
O.compressus cv. Avila	3.888	850.7	4.739 a
O.compressus ecotype with pods	259	145.1	404 b
B.pelecinus cv. Casbah	501	154.1	655 b
B.pelecinus cv. Mauro	2.674	557.9	3.232 ab

Letters in column indicate significant differences (P≤0.05)

Conclusions The highest DM yield was obtained with "Avila" and "French and Cadiz serradella" cultivars. Seed production in serradella was high for all the cultivars. Both species of serradella are well adapted to the dryland conditions in the Mediterranean climate of IX Region in the southern of Chile.

References
Oram, N.R. (1990). Register of Australian herbage plant cultivars. Third edition. CSIRO. Australia. p. 111-168.
Romero, Y. O. y Rojas, G. C. (2001). Producción de materia seca de nuevas leguminosas forrajeras como alternativas para el secano de la IX Región. *In* Resúmenes de la XXVI Reunión Anual SOCHIPA. Santiago. p. 434-435.

An appraisal of the potential for soybeans in the United Kingdom

C.A. Sawyer, G.P.F. Lane and W.P. Davies
Royal Agricultural College, Cirencester, Gloucestershire, GL7 6JS, UK, Email: charlotte.sawyer@rac.ac.uk

Keywords: soybeans, protein, forage

Introduction Soybean is a most important crop worldwide, accounting for 56% of world oilseed production and 69% of world protein meal consumption in 2003 (Soystats, 2004). Since their introduction in the early 1800's, forage soybeans have been grown widely in USA. Used originally as a forage crop, this use largely had been forgotten until Dr T. E. Devine (United States Department of Agriculture, Agricultural Research Service; USDA-ARS) released 4 new forage cultivars. Soybeans potentially offer UK farmers a high quality protein source in a short season and also meet the requirements of supermarket chains to remain GM free. This paper aims to indicate whether soybeans can be grown successfully in the UK.

Materials and methods Field trials at the Royal Agricultural College in 2000-01 and 2003-04 investigated the potential of soybeans as a forage source for on-farm feeding. The site was in the Cirencester area on a calcareous, stony, reddish brown clay (Sherborne series). The trials focused on varietal selection, fertiliser applications, inoculant use and harvest date. All forage varieties were sourced from the USDA-ARS. Grain varieties originated mainly from eastern Europe, with one entry from Canada. Samples were assessed for DM Yield, Crude Protein and Fibre Content (NDF). Scores of resistance to lodging were noted also.

Results Forage yields varied between seasons and between harvest dates; forage varieties yielded more than grain types at later harvests. Largely due to predation at emergence, plant populations were variable. Covering the crop with fleece immediately after drilling in 2004 increased plant populations (data not shown). A pre-emergent herbicide, used in all seasons, controlled most grass weeds but was ineffective against *Solanum nigrum* and *Bilderdykia convolvulus*. Mean CP values ranged from 11-14% DM; values up to 20% DM were recorded but not consistently. The best adapted forage types yielded >10t DM/ha in a poor season (Table 1).

Table 1 Mean yield and nutrient content of whole crop soybeans evaluated in Cirencester

Variety	Type	Year	Source	Origin	DM Yield t/ha	CP % DM	NDF % DM
Tara	Forage	2003	USDA-ARS	USA	11.5	11.31	37.4
Donegal	Forage	2001	USDA-ARS	USA	12.2	12.46	40.2
7P116	Forage	2001	USDA-ARS	USA	8.9	13.68	43.4
8GH-85-2	Forage	2001	USDA-ARS	USA	7.1	13.26	44.8
Altesse	Grain	2003	Premium Crops	Canada	7.04	13.92	45.3

Discussion In order to assess the suitability of whole crop soybean for use as a forage crop in the UK, a SWOT analysis shows its benefits and limitations:

Strengths	Weaknesses
Crude Protein content: 11-14 (max 20) %DM	Cost and effectiveness of weed control
Yield: 7-12t (max 18) t DM/ha	Varietal choice
Short growing season	Few herbicides approved
Low fertiliser input	Lack of literature and guidelines for successful cropping.
No disease problems to date	High seed cost
Opportunities	**Threats**
Non GM protein	Predators (birds, rabbits)
Export	Weed control
Nitrogen fixation- reduced fertiliser costs	Frost
	Risk of lodging in wet seasons

Conclusion These initial trials suggest that soybeans may be a viable option as a high yielding forage crop for UK conditions. However, substantial difficulties must be overcome.

References

Soystats (2004). Soybean Production 2003. [online]. Soystats. Available from: http://www.soystats.com/2004 [Accessed 15/08/04]

Production and non-production functions of grassland in an upland region of Slovakia

J. Cunderlik[1], L. Gonda[1] and J. Tomaskin [2]

[1] Grassland and Mountain Agriculture Research Institute, Mladeznicka 36, 974 21 Banska Bystrica, Slovak Republic, Email: cunderli@vutphp.sk,[2] Faculty of Natural Sciences, Matej Bel University, Akademicka 13, Banska Stiavnica, Slovak Republic

Keywords: grassland biomass, non-production function, grassland roots, ecosystems, nutrients

Introduction The importance of grassland lies chiefly in its production of good quality forage that is utilised by cattle and sheep. In upland and mountain regions the non-production functions of grassland such as landscape enhancement or water catchment are especially important. These functions are performed mainly through the sward tillering zone (boundary area between above-ground vegetation and roots) and root characteristics (Jancovic, l985). The objective of this research was to study the biomass above, at, and below ground level of three sward types at different fertilisation levels.

Materials and methods The production of total biomass (CB) was evaluated in three grassland ecosystems: seminatural (SG), overdrilled (OG) and sown (TG) at Radvan (Banska Bystrica,460m a.s.l.) over six years (1993-98). Soil type was cambisol (pH 4.3). Seminatural grassland (Poa- Trisetum; Arrhenatherium) was oversown with a grass/legume mixture. All three grassland types were fertilised in the same pattern: 1- no fertilisers, 2- $P_{30}K_{60}$, 3- $P_{30}K_{60}+N_{90}$ and 4- $P_{30}K_{60}+N_{180}$ (kg/ha). The following measurements were made on a dry matter (DM) basis: production of above-ground biomass (NB) by a 3-cut regime; production of the tillering zone (ZO), which was defined as the sheep on the soil surface boundary and comprised two parts- underground biomass to the depth 1.5-2.0 cm and above-ground biomass to the height of 2.0-3.0 cm; production of root biomass (Ko) to the depth of 10 cm. The sum of these components (CB) represents the total biomass. Nutrient uptake (N, P, K, Ca and Mg) levels in the components were also calculated from standard analysis of the elements.

Results Selected mean data on DM and N uptake are presented in Table 1. The biomass DM of the components differed among the sward types with SG having the highest ZO, Ko and CB values. The lowest NB, Ko and CB values were for TG. The order of biomass DM was consistently Ko>ZO>NB for all sward types. The trend for the fertiliser treatments was for increased component biomass DM as fertilisation increased from nil to $N_{180} P_{30} K_{60}$. In terms of N production, the order of the sward types was SG>OG>TG and for fertiliser treatments, in line with level of applied N. For the other parameters of P, K, Ca and Mg (data not shown) CB biomass was similar for swards SG and OG with TG being lowest.

Table 1 Biomass dry matter (t/ha) and nitrogen (kg/ha) of sward components (sward type meaned over fertiliser treatment and fertiliser treatment meaned over sward type)

Sward	Sward type			Fertiliser treatment			
	SG	OG	TG	1	2	3	4
DM production							
NB	2.10	2.20	1.95	1.57	1.70	2.37	2.70
ZO	5.73	5.10	5.43	4.77	5.07	5.73	6.10
Ko	8.52	7.78	7.28	7.20	7.90	7.90	8.43
CB	16.36	15.08	14.66	13.54	14.67	16.00	17.23
N production							
NB	212	160	141	109	121	170	214
ZO	105	98	91	89	93	100	107
Ko	119	115	103	103	114	111	122
CB	436	370	335	301	328	381	443

Conclusions Treatment SG with its non disturbance had greater ZO and Ko than the other sward types. This treatment maintained a similar NB production to OG and TG. All swards responded positively to increased fertilisation. At upland sites, sward renovation may not be a better option than managing the existing seminatural sward whether for above-ground production or for maintaining a sward for a non-production function.

References

Jancovic, J. (1985). Vplyv hnojenia na korene travnych porastov. [Effects of fertiliser application on grassland roots.] Agrochemia, 25, 2, 43-45.

Row spacing and productivity of Russian wild rye pastures in semiarid environments

J.D. Berdahl, S.L. Kronberg, J.R. Hendrickson and J.F. Karn
*USDA-ARS, Northern Great Plains Research Lab., P.O. Box 459, Mandan, North Dakota 58554-0459, USA,
Email: berdahlj@mandan.ars.usda.gov*

Keywords: semiarid climate, seeded pasture, Russian wildrye, row spacing

Introduction   To sustain forage yields in dry years in semiarid climates, row spacings >59cm have been recommended for Russian wild rye [*Psathyrostachys juncea* (Fisch.) Nevski] (Lawrence & Heinrichs, 1968). However, wide row spacings promote weed invasion, soil erosion, and elevated plant crowns resulting in a rough, "washboard" ground surface (Kilcher, 1961). Jefferson and Kielly (1998) suggested a 30-cm row spacing for optimum sustainable forage yields in Russian wild rye in the semiarid prairie region of Canada. This study aimed to evaluate the relationship between row spacing and productivity of Russian wild rye at two semiarid sites near Mandan, in the northern Great Plains region of the USA (46° 48' N latitude, 100° 55' W longitude).

Materials and methods   Four Russian wild rye cultivars were seeded in rows spaced from 15-90 cm apart in 15-cm increments at two field sites. Data are presented for the 15, 45, and 90-cm spacings in Tables 1 and 2. Soils were a sandy-loam at site 1 and a silt-loam at site 2. Dry matter (DM) yields were measured from a single annual harvest for 3 years at each site when plants were at the hard-dough stage of maturity. Crude protein (CP) content was measured at site 2 for 2 years when plants were at the hard-dough stage and again after forage regrowth.

Results   DM yields varied significantly (p<0.01) among years at both sites (Table 1), primarily in response to differences in precipitation among years. Yields tended to be greater for the 15-cm spacing than the other row spacings when the tests were newly established, but differences among row spacings diminished in succeeding years. A significant (p<0.01) row spacing x year interaction for DM yield was present at site 1 but not at site 2. Severe drought in 2002 resulted in few reproductive tillers being produced and little regrowth after harvest. In 2002, forage regrowth had greater CP concentration (p<0.05) than forage at the hard-dough stage of development in July, but differences among row spacings were not significant at either sampling date (Table 2). In 2003, forage regrowth at the 90-cm row spacing had greater CP concentrations than the 15-cm spacing, but DM content of the regrowth, a measure of succulence, was equal for the two extreme row spacings.

Table 1 Mean DM yields of 4 Russian wild rye cultivars at 3 row spacings in different years at two sites (Mg/ha)[#]

Row spacing (cm)	Site 1			Site 2		
	1999	2000	2001	2002	2003	2004
15	6.58	4.22	3.56	2.08	5.82	3.25
45	6.57	3.90	3.09	1.89	5.31	3.19
90	5.69	3.56	3.42	1.67	5.13	3.16

[#] Differences among row spacing means within years were not significant (p ≤ 0.05)

Table 2 Mean CP concentrations of 4 Russian wild rye cultivars at three row spacings at different sampling dates (g/kg)[#]

Row spacing (cm)	2002		2003	
	9 July	28 Oct	8 July	22 Oct
15	132a	138a	92a	161b
45	123a	135a	71b	186ab
90	129a	144a	83ab	204a

[#] Means within a column followed by a different letter were significantly different (p ≤ 0.05)

Conclusions   DM yields among row spacings at the July harvest dates were essentially equal by the third production year at both field sites. Even though forage regrowth from the wide row spacing treatments may have higher CP concentrations in some years, potential weed invasion and other problems would negate any advantage for row spacings >45 cm for Russian wild rye pasture.

References
Jefferson, P. G. & G. A. Kielly (1998). Reevaluation of row spacing/plant density of seeded pasture grasses for the semiarid prairie. *Canadian Journal of Plant Science,* 78, 257-264.
Kilcher, M. R. (1961). Row spacing affects yields of forage grasses in the Brown soil zone of Saskatchewan. Publ. 1100, Research Branch, Canada Dept. of Agriculture, Ottawa, ON, Canada.
Lawrence, T. & D. H. Heinrichs (1968). Long-term effects of row spacing and fertiliser on the productivity of Russian wild ryegrass. *Canadian Journal of Plant Science,* 48, 75-84.

Brachiaria and *Panicum* productivity at different sites within the Brazilian Amazon

F.F.C. Mello[1], C.E.P. Cerri[2], C.C. Cerri[2] and C.A.C. Crusciol[1]

[1]Faculdade de Ciências Agronômicas, UNESP, P.O. Box 237, Botucatu, SP, Brazil, Email: ffcmello@fca.unesp.br, [2]CENA, Universidade de São Paulo, P.O. Box 96, 13400-970, Piracicaba, SP, Brazil

Keywords: grass productivity, Brazilian Amazon, *Panicum*, *Brachiaria*

Introduction Over the last 25 years more than 70 M ha of native vegetation in Brazil have been replaced by pastures for beef production. The substitution of native vegetation on such a large scale with African grasses (mainly of the genera *Brachiaria* and *Panicum*) is likely to have an impact on nutrients and organic matter composition, as well as a regional impact on hydrology and water quality.

Materials and methods The literature was searched to gather data on grass productivity for the genera *Brachiaria* and *Panicum*, found in Amazonian pastures. Only data from unfertilised pastures and grasses on Oxisol or Ultisol (which together represent about 60% of the soils in the Brazilian Amazon) were selected for this study. Therefore, 17 cases for *Brachiaria* and 15 for *Panicum* were obtained from the following publications: Valentim & Moreira (1994); Souza-Filho *et al*. (1990); Costa *et al*. (1989); Costa (1989); Gonçalves *et al*. (1982); and Azevedo *et al*. (1982).

Results and discussion Mean productivity of *Brachiaria* was 8.22 t DM/ha/year and for *Panicum* was 9.55 t DM/ha/year (Table 1). In both cases, the differences in productivity can be explained by the variation in the genetic cultivars more adapted or more susceptible to warm and humid tropic weather. No correlation was found between grass productivity, soil properties (texture, pH, carbon, bulk density, etc) and geographic location. However, *Brachiaria* and *Panicum* showed higher productivity compared to native pastures (about 3 t DM/ha/year, Camarão & Souza Filho, 1999) in the Brazilian Amazon.

Table 1 Grass productivity in different sites within the Brazilian Amazon

County	Coordinates		Weather[1]		Soil	Productivity (t DM/ha/year)	
	Lat (S)	Long (W)	Prec (mm)	T (°C)	type	*Brachiaria*	*Panicum*
Rio Branco	09° 58'	67° 29'	1989	24.9	Oxisol	-	18.00 (1.41)
Ariquemes	09° 55'	63° 03'	2270	24.5	Oxisol	7.00	-
Vilhena	12° 44'	63° 08'	1941	22.7	Oxisol	7.96	-
Porto Velho	08° 46'	63° 05'	2200	25.6	Oxisol	9.90 (5.06)	9.90 (8.22)
Presidente Médici	11° 71'	62° 15'	1825	25.0	Ultisol	4.33 (0.54)	5.35 (1.64)
Ji-Paraná	11° 17'	61° 55'	2270	24.5	Ultisol	11.52 (5.41)	5.20
São João do Araguaia	04° 50'	48° 55'	2081	26.2	Ultisol	8.62 (0.96)	9.30 (1.33)

Values in brackets refer to standard deviation. [1]Mean annual precipitation and annual temperature

Conclusion *Brachiaria* and *Panicum* pastures present high productivity levels when well managed (mainly controlling weeds and maintaining an adequate animal stocking rate) even when they were not fertilised. Moreover, cultivated pastures are on average 35% more productive than Amazonian native pastures.

References

Azevedo, G.P.C., A.P. Camarão & E.A.S. Serrão (1982). Introdução e avaliação de forrageiras no município de São João do Araguaia, Estado do Pará.Boletim de Pesquisa Embrapa – CPATU – Pará, 47, 23pp.

Camarão, A.P.& A.P.S.Souza Filho (1999).Pastagens nativas da Amazônia. Embrapa Amazônia Oriental, 150pp.

Costa, N.L. (1989). Avaliação agronômica de gramíneas forrageiras sob três níveis de adubação fosfatada Comunicado Técnico Embrapa – UEPAE – Rondônia, 80, 1-4.

Costa, N.L., C.A. Gonçalves, S.N. Botelho & J.R.C. Oliveira (1989). Níveis de calagem e fósforo na formação de pastagens de *Brachiaria humidicola* em Rondônia. Comunicado Técnico Embrapa, RO 82, 1-5.

Gonçalves, C.A., J.C. Medeiros & J.R.C. Oliveira (1982). Introdução e avaliação de gramíneas e leguminosas forrageiras em Rondônia. Botetim de Pesquisa Embrapa – UEPAE – Rondônia, 01, 35pp.

Souza Filho, A.P.S., P.R.L. Meirelles & D.M. Pimentel (1990). Introdução e avaliação de gramíneas forrageiras em área de várzea do Amapá. Boletim de Pesquisa Embrapa – UEPAE – Amapá, 07, 16pp.

Valentim, J.F.& P. Moreira (1994). Vantagens e limitações dos capins tanzânia-1 e mombaça para a formação de pastagens no Acre. Comunicado Técnico Embrapa – CPAF – Acre, 60, 1-3.

The effect of nitrogen fertilisation on the morphological development and growth rate of star grass (*Cynodon nlemfuensis*)

P.P. del Pozo[1], R.S. Herrera[2] and A. Hameleers[3]
[1]*Faculty of Veterinary Medicine, Agricultural University of Havana, Cuba, Email: delpozo@isch.edu.cu,* [2]*Institute of Animal Sciences, Havana, Cuba,* [3]*The agricultural Research Institute of Northern Ireland, Hillsborough, Co. Down, BT26 6DR, UK*

Keywords: *Cynodon nlemfuensis*, nitrogen fertilisation, growth rate

Introduction Nitrogen is one of the main inputs used in forage production systems to increase productivity. However, in Cuba, the availability of fertiliser N is limited and therefore if applied, needs to be used with high efficiency. Previous studies (Johnson, 2001, Del Pozo, 2003) investigated the effects of N on growth, carbohydrate and protein content but did not consider morphological changes in *Cynodon nlemfuensis*. A study was therefore undertaken investigating how N application influences morphological development of the plant and how these changes might affect the efficiency of use of applied N.

Materials and methods The experiment was carried out at the research facility of the faculty of Veterinary Medicine in Cuba in the year 2001, using a 10-year old pasture of *Cynodon nlemfuensis*. A total of four growth cycles (two in the rainy season and two in the dry season) of each 12 weeks, was studied over one year. Average temperature ranged from 24 to 31 ^{0}C, with a maximum of 30.7 ^{0}C and a minimum of 19 ^{0}C with 1308mm rainfall. The experiment consisted of two treatments (0 or 50 kg N per growth cycle) and a split plot design was used with three replicates. For every growth cycle the herbage was cut back to a height of 5 cm. Subplots were harvested on a weekly basis (week 2-12) and DM yield for the different components was estimated.

Results No N X age interactions were found within season for DM yield but the effect of season was significant (p >0.05). Maximum growth rates were achieved for the N treatments in week 7 of the dry season (0.36 tDM/ha) and week 5 (0.663 tDM/ha) of the rainy season. N response was 34 and 52 kgDM/kg N for the dry and wet season, respectively. Proportion of leaf in the DM was unaffected by either N application or season but tended to decrease for al treatments from 0.6 in week 2 to 0.45 in week 12.

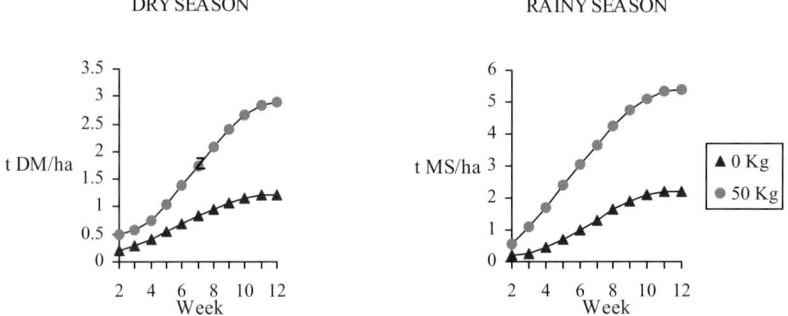

Figure 1 DM yields of *C, nlemfuensis* in the dry and wet seasons

Conclusions The response to N was substantially higher during the wet season compared to the dry season, while leaf proportion in the DM was unaffected by N application. Lower levels of N application should be used in the dry period.

References
Del Pozo, P.P., R.S Herrera,.& M. García (2003). Dynamics of carbohydrates and crude protein contents in star grass (*Cynodon nlemfuensis*) with and without nitrogen application. *Cuban Journal of Agricultural Science*, 36, 265-275.
Johnson, C.R., B.A., Reiling, P., Mislevy, & M.B. Hall, (2001). Effect nitrogen fertilization and harvest date on yield, digestibility, fiber and protein fractions of tropical grasses. *Journal of Animal Science*, 79, 2439-2448.

Comparison of a biometric method with clipping and weighing method for estimating the yield of *Artemisia sieberi* (case study Zarand-e-Save region)

H. Dianati[1], M. Abedi[1], E. Shahriary[2] and H. Arzani[2]
[1]Tarbiat Modaress University, College of Natural Resources, Range Management Department, P. O. Box: 46414-356, Noor, Mazandaran, I. R. Iran, Email: dianatitilaki@yahoo.com, [2]Tehran University, College of Natural Resources, Range Management Department, P. O. Box: 31585-4314, Karaj, Tehran, I. R. Iran

Keyword: biometric method, yield, *Artemisia sieberi*, Iran

Introduction Forage production is a most important vegetation attribute in rangeland analysis and evaluation and use in management practices. *Artemisia sieberi* covered about 47% of the rangeland area in Iran and scientists need to know the best method for analysis and evaluation of this species. Clipping and weighing has high precision, but this method is time-consuming and expensive and alternative methods are required. Russian scientists developed a biometric method based on plant dimensions and suggest this method for rangelands in the Middle East (discussed by Dianati, 2003). In America, Muray (1982) used plant dimensions for yield estimation and regression models for estimating production. This method required the selection of a lot of individual plants for each species to draw curves and estimate forage production. The critical stage of this method is the determination of the relationship between plant dimensions and yield. The research showed that in *A. sieberi* the best relationship with yield is from height (H) and the sum of diameters ($D_1 + D_2$). This relationship should be calculated separately for each species. In this study this method was tested with the Artemisia type in Iran.

Materials and methods The Artemisia type in one key area of the Zarand-e-Save region was selected. For testing the application of this method plants were divided into two groups - middle and small size. For each group, 20 individual plants were selected randomly and after measuring height and the sum of diameters, the plants were clipped and weighed. Forage production curves were then estimated. Measured and estimated data were compared by two paired T tests with Minitab software.

Results Mean and SE of means for measured and estimated groups are given in Table 1. There were significant differences in the middle-sized plants ($p < 0.01$) but not for the small-sized plants ($p > 0.05$).

Table 1 Measured and estimated values for the weight of plants of *Artemisia sieberi* (g)

Group		Mean	SE mean
Middle-sized Artemisia	Measured	10.4**	1.21
(30< H< 60 cm)	Estimated	14.4**	1.26
Small-sized Artemisia	Measured	5.7^{ns}	0.624
(H< 30 cm)	Estimated	6.25^{ns}	0.598

Conclusion This study illustrates that plant size affects the applicability of this method, because there was a good relationship for small plants, but not for middle-sized plants. This method is cheap, easy and not destructive. The method should be tested with other *Artemisia sieberi* types with the possibility of developing pooled equation modeling across the Middle East.

Reference

Muray, R.B. & M. Q. Jacobson (1982). An evaluation of dimension analysis for predicting shrub biomass. *Journal of Range Management,* 35, 451-454.
Dianati,H. (2003). Investigation of a biotic ecological factor on rangeland production and condition in desert ecosystem. PhD thesis. N. C. X. A

Root density in *Panicum maximum* cv. Tanzania monoculture and in a mixture with *Leucaena leucocephala* with different densities in Mexico

H.J. Delgado Gòmez[1], L. Ramirez Avilés[2], J. Ku Vera[2], J. Escamilla Bencomo[3] and P.A. Velázquez Madrazo[2]
[1]*Universidad del Zulia, Facultad de Ciencias Veterinarias, Núcleo Agropecuario, Ciudad Universitaria, Apartado 15252, Maracaibo, 4500-A, Estado Zulia, Venezuela, Email: delgado7255@hotmail.com,* [2]*Universidad Autónoma de Yucatán, Facultad de Medicina Veterinaria y Zootecnia, km 15.5, Carretera Mérida – Xmatkuil. Apartado 4-116 Itzimná, Mérida, 97100 Yucatán, México,* [3]*Centro de Investigación Científica de Yucatán, Calle 43 N° 140 Col. Chuburná, Mérida, 97200 Yucatán, México*

Keywords: *Panicum maximum* cv. Tanzania, *Leucaena leucocephala*, plant density, root dynamics

Introduction In Yucatan cattle production is limited by forage availability during the dry season. L. leucocephala has good nutritive value (24 – 30% CP) and can stand drought and grazing, therefore its use in mixture with grasses is recommended. However, in association both species could compete for light, water and nutrients. The aim of this study was to assess the effect of introduction of L. leucocephala with different densities on root density of P. maximum.

Materials and methods The experiment was located in Yucatan, Mexico, with mean annual rainfall below 1200 mm, soils were Luvisoles and Leptosoles fertilised with 50N 80P 30K. Treatments were a mixture combining P. maximum cv. Tanzania and L. leucocephala, with two densities (5,000 and 10,000 plants/ha) and monoculture of P. maximum. The experiment was carried out between November 2001-2002. A randomised block design with four replicates of each treatment was used. Root density (RD) was estimated by the soil core-break method (Escamilla et al., 1991).

Results RD of *P. maximum* declined (P<0.01) as depth increased. *P. maximum* monoculture had the highest RD (4.45 mg/cm³). In the mixture RD diminished (P<0.01) as distance from the trees increased, with averages 3.59 – 2.92 mg/cm³ in the mixture with 5,000 plants/ha. In the mixture with 10,000 plants/ha it was the opposite with averages 2.32 - 2.86 mg/cm³ during the dry season (figure 1). In the rainy season RD in *P. maximum* monoculture was 3.17 mg/cm³, and it was lower nearby the stems of *L. leucocephala* in both plant densities (3.15 - 3.58 and 2.90 - 3.49 mg/cm³) for 5,000 and 10,000 plants/ha respectively. RD of *P. maximum* in the mixture with 5,000 plants/ha was similar to that in *P. maximum* cv. Tanzania monoculture. As plant density increased RD diminished, probably due to competence for growth resources (Odhiambo *et al.*, 2001).

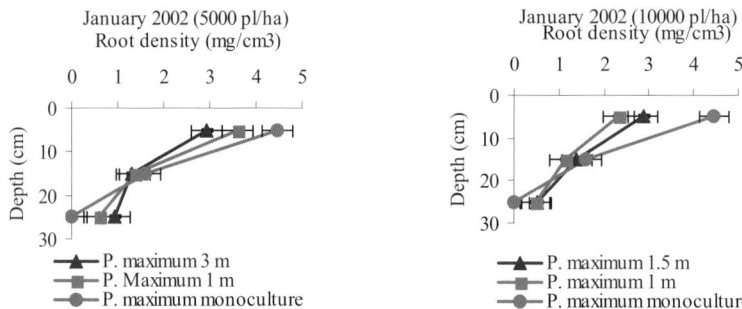

Figure 1 Root density in mixture and monoculture of *P. maximum* in the dry season

Conclusions Introduction of *L. leucocephala* with density of 10,000 plants/ha affected RD of *P. maximum* cv. Tanzania negatively. Association of both species also resulted in decreased root density in pasture next to *L. leucocephala* stems. Root density was higher in the dry than in the rainy season. *P. maximum* cv. Tanzania monoculture had the highest RD. It is important to assess interactions of pastures mixed with woody perennials to develop silvopastoral systems.

References
Escamilla, J.A., N.B. Comerford & D.G. Neary (1991). Soil-core break method to estimate pine root distribution. *Soil Science Society American Journal*, 55, 1722-1726.
Odhiambo, H.O., C.K. Ong, J.D. Deans, A. Wilson, A.H. Khan & J.I. Sprent (2001).Roots, soil water and crop yield: tree crop interactions in a semi-arid agroforestry systems in Kenya. *Plant and Soil*. 235, 221-233.

Leaf appearance and elongation in Panicum maximum cv. Tanzania tillers of varying ages

D. Nascimento Jr.[1], R.A. Barbosa[2], V.P.B. Euclides[2], S.C. da Silva[3] and R.A. Torres[2]
[1]Universidade Federal de Viçosa, Viçosa, MG, Brasil, Email: domicio@ufv.br, [2]Embrapa – Gado de corte, Caixa Postal 154, Campo Grande, MS, 79002-970, Brasil, [3]Universidade de São Paulo, Piracicaba-SP, Brasil

Keywords: grazing frequency, grazing intensity, light interception, morphogenesis

Introduction A sward may be considered as a tiller population of varying ages and sizes, and these different age groups are likely to present distinct behaviour in terms of growth and herbage production. However, there is very little information on how tiller age, in association with grazing management practices (e.g. frequency and intensity of grazing), alter morphogenetic characteristics and, therefore, herbage production. Against this background, the present experiment had the objective to evaluate leaf appearance and elongation in *Panicum maximum* cv. Tanzania tillers of different age groups when submitted to intermittent grazing regimes.

Material and methods Treatments consisted of combinations between three grazing intervals and two post-grazing residues. Grazing intervals corresponded to the time interval necessary to reach 90, 95 and 100% sward light interception (LI) during regrowth, and post-grazing residues were 25 and 50 cm. During the grazing period (July 2003 - May 2004), tagging and counting of tillers were performed every grazing cycle, generating at the end of the experiment the different tiller age groups used for measurements. Tillers were classified as young (less than 2 months old), mature (between 2 and 4 months old) and old (more than 4 months old) (Carvalho, 2002). Twelve tillers from each age category were randomly chosen and submitted to measurements of leaf appearance (leaves/tiller/day) and elongation (cm/tiller/day) rate.

Results Leaf appearance rate (LAR) was higher for young than for mature and old tillers, regardless of treatment (Figure 1). Shorter grazing intervals (90 and 95% LI) were responsible for larger differences in LAR from young and mature tillers. On the other hand, the long grazing interval (100% LI) associated with 25 cm residue did not affect LAR in either young or mature tillers. Leaf elongation rate (LER) varied considerably with tiller ageing, with young tiller presenting higher values of LER in all treatments, except for the 100% LI/25 cm treatment (Figure 2). Further, differences in LER from young and old tillers increased as grazing interval decreased, and did not exist when grazing interval was long (100% LI).

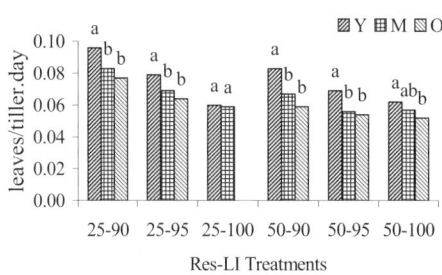

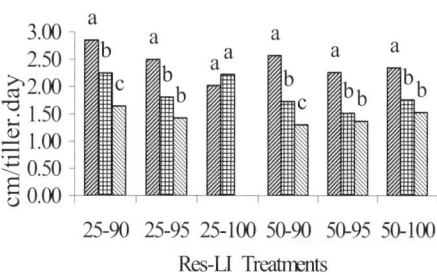

Figure 1 Leaf appearance rate.
Means followed by the same letter within treatment bars are not different (P>0.10)

Figure 2 Leaf elongation rate.
Means followed by the same letter within treatment bars are not different (P>0.10)

Conclusions The ageing process of tillers can result in progressive reduction in tiller vigour and growth. Grazing management practices that allow for a high turnover in tiller population (younger tillers) may revert this trend and ensure conditions to sustain and/or increase herbage production.

References

Carvalho, D.D. (2002). Leaf morphogenesis and tillering behaviour in single plants and simulated swards of Guinea grass (*Panicum maximum* Jacq.) cultivars.. Ph.D. Thesis – Massey University, Palmerston North,, New Zealand,155p

An eco-morphological examination of tiller and stolon dynamics in a *Zoysia japonica* sward

M. Ito, Y. Ueda, M. Kodama and T. Okajima
Niigata University, 2-8050, Igarashi, Niigata, 950-2181, Japan, Email: shoot@agr.niigata-u.ac.jp

Keywords: multiple-node, population density, stolon, tiller, *Zoysia japonica*

Introduction Japanese lawn grass (*Zoysia japonica*), which dominates in grazed semi-natural grasslands in Japan, is being reconsidered recently, because of its high adaptability to poorer conditions and its aggressive creeping habit in open fields. *Zoysia* has a unique potential for indeterminate multiple-node generation in the stolon tip and differential tiller formation at two tillering sites of every multiple-node (Ta in the bottom node and Tb in the mid-part), so that it displays contrasting behaviour in stolon extension and aerial tiller production in various situations (Ito *et al.*, 2003). In this study, we examined the population structure of various tillering modules of *Zoysia* clones in a dense sward.

Materials and methods During the growing season in 2000, changes of the erect tiller population were recorded at a ca. 4-week-interval in 4 quadrats (40cm×40cm) in an experimental *Zoysia* sward (100m^2), which was fertilised with 24:24:24 g/m^2/year of N:P$_2$O$_5$:K$_2$O and defoliated every two weeks from May 19 to October 20. In the same sward, *Zoysia* sods (10cm×20cm) with 3 replicates were collected as well on June 20, August 10, and September 27, and washed out to remove soil and litter for dissection. Number of stolon apices in sods, multiple-node number on stolon segments, stolon length, and developmental state (foot note Table 1) of primary tillers on each tillering site of multiple-nodes were examined.

Results and Discussion The mean population density of total erect tillers was about 15,000 shoot/m^2 during the season examined (Figure 1). Soon after the onset of growth in early May, 34 % of total existing tillers bore ears, and the tiller population declined sharply after the first defoliation in mid May. Decreased tiller density was recovered instantly in late May and early June. New tiller emergence was fairly active during the seasons (ca. 3 % against the total number of existing tillers as counted one week after each defoliation), and it resulted in gradual but steady rise in tiller density. Active generation of erect Ta tillers from stolons coincided with increasing tendency of tiller density, whereas lesser appearance of Tb tillers was always observed (Table 1). Thus, the majority of existing erect tillers seemed to be composed of Ta tillers (53 to 65 % against whole erect tillers) and was supplemented with smaller numbers of Tb tillers (ca. 10 %). The second major component was secondary and tertiary tillers, which attained 27 to 36 % of total erect tillers in the sward. Creeping tillers were rare, i. e., the most erect primary tillers emerged from stolons generally kept upward extension for top growth. The total number of multiple-nodes on stolons bearing various primary tillers averaged ca. 13,000/m^2, and the total stolon length kept a level of ca. 200m/m^2 throughout the seasons, while the stolon apex density stayed < 500/m^2. There is a general tendency of preferential development of existing stolons in *Zoysia* plants growing in a dense sward condition.

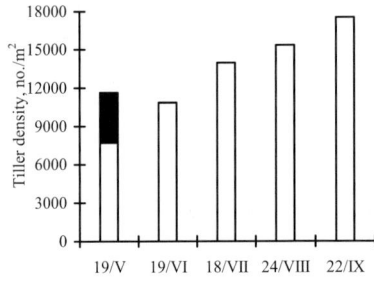

Figure 1 Changes in erect tiller population in *Zoysia* sward. LSD (P<0.05); 3275
■ Reproductive tillers

Table 1 Population density (no/m^2) of primary tillers with different developmental state, as measured on each tillering site

	Date	Et	Sp	Dm	St	Dd*
	20/VI	6167	1067	767	117	10250
(Ta)	10/VIII	7067	533	367	83	7033
	27/IX	9400	350	667	0	5050
	20/VI	833	167	14900	0	2467
(Tb)	10/VIII	1417	500	11417	100	1650
	27/IX	2167	133	11867	17	1283

*Primary tillers were classified as; Et=erect tiller, Sp=sprouting bud; Dm=dormant bud, St= stoloniferous, and Dd=dead tiller.

Reference
Ito, M., M. Kodama, Y. Ueda & T. Okajima (2003). Regularity in developmental patterns of stolons and tillers of *Zoysia japonica* Steud. plants growing under a spaced-plant condition. *Grassland Science*, 49, 438-443.

An investigation on ecological aspects of crested and intermediate wheat grasses in semi-steppe vegetation of Iran

M. Amirkhani and M. Mesdaghi
Department of Range Management, College of Natural Resources, University of Tarbiyat Moudarres, Nour, Iran, Email: maasoome_amirkhani@yahoo.com

Keywords: *Agropyron cristatum, Thinopyrum intermedium*, chemical composition, phenological stages, Golestan National Park

Introduction   Crested and intermediate wheat grasses (*Agropyron cristatum* (L.)Gaertn., *Thinopyrum intermedium* (H.)Beauv.) are adapted to relatively dry conditions in Iran and have a significant role in providing good forage quality for domestic sheep and wild ungulates in summer rangelands. These grasses occur at altitudes of 1,200 to 1,800 m. *A. cristatum* is a bunch grass with diverse spikes and medium height (40 cm) and is common on open and exposed knolls, whereas *T. intermedium*, with height of 115 cm and with long rhizomes, is found in more moist niches in gully bottoms. The objectives of this study were to determine forage values, canopy coverage, production, local distribution, and phenological stages of these species in Golestan National Park, which is representative of the semi-steppe zone in Iran.

Material and methods   Phenological stages, canopy coverage and production of two species of *A. cristatum* and *T. intermedium* were recorded at representative sites. The foliage of ten plants of each species were harvested at three stages of vegetative growth (VG), full flower (FF) and seed ripening (SR) and analysed for nitrogen (N) and acid detergent fibre (ADF). Crude protein (CP), dry matter digestibility (DMD) and metabolisable energy (MED) were then calculated using the following prediction equations:
%N=0.16CP, %DMD=83.58-0.824ADF+2.626%N, and MED=0.17%DMD-2 (Standing Committee on Agriculture, 1990). Statistical analysis was by ANOVA.

Results   The phenological stages, cover, production and chemical composition of the species are shown in Figure 1 and Table 1.

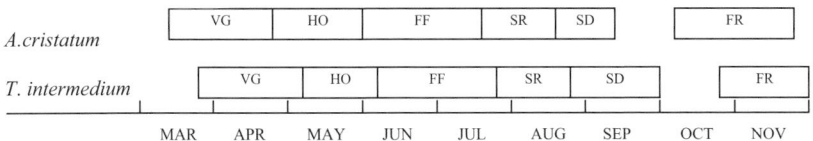

Figure 1 Phenological stages of *A. cristatum* and *T. intermedium*. VG=Vegetative growth, HO=Heads out, FF=Full flower, SR=Seed ripening, SD=Seed dissemination, FR=Full regrowth

Table 1 Canopy cover (%), production and chemical composition of *A. cristatum* and *T. intermdium*

Species	Cover (%)	Production (kg/ha)	Chemical composition (%)			
			Crude protein		Acid detergent fibre	
			Mean	StDev	Mean	StDev
A. cristatum	5	180	9.8	2.6	42.8	3.9
T. intermedium	8	200	9.5	2.6	45.1	5.9

There were significant differences between stages for chemical composition of both species (p<0.05), but there were not any differences between the species (p>0.05). With advance in maturity, CP decreased and ADF increased. Additional information was obtained on the number of flowering stalks, seeds per spike, seed germination and distribution of these species, but this is not presented here.

Conclusion   Although the contribution of these two species was relatively small in relation to production and forage quality of rangelands in the study area, there is need for further work on a larger scale.

References
Dewey, D.R. & K H. Asay (1975). The crested wheatgrass of Iran. *Crop Science*, 15, 844-849.
Dewey, D.R. (1978). Intermediate wheat grasses of Iran. *Crop Science*, 18, 43-48.
Standing Committee on Agriculture (1990). Feeding Standards for Australian Livestock. Ruminants. CSIRO, Australia.

The effect of topographic factors on the productivity of mountain grasslands in northwestern Benin

T.H. Avohou and B. Sinsin
Laboratoire d'Ecologie Appliquée, Université d'Abomey Calavi (LEA/UAC), 01 BP 526 Cotonou, Benin, Email: bsinsin@bj.refer.org

Keywords: grasslands, mountains, topography, exposure, biomass

Introduction The Atacora mountains range in northern Benin (660 m altitude) is a special ecosystem in the sudanian zone because of the overriding importance of topographic factors and shallow soils. The vegetation over this mountain range consists of shrub and tree savannas, woodlands and fallows. More and more cattle herds graze on this range. But the functioning of this ecosystem is still unknown like many others in the tropical zone (Sene & Zingari, 2001). This study aims to determine the impact of topography and mountain side exposure on the productivity of the Atacora mountains grasslands in the Atacora mountains.

Materials and methods Five grassland types respectively dominated by *Loudetia flavida* Stapf, *Andropogon schirensis* Hochst, *Andropogon tectorum* Schumach & Thonn, *Andropogon gayanus* Kunth and *Hyparrhenia involucrata* Stapf, were identified for the study. Three protected plots of 10 m × 10 m were set up on the mountain range per grassland type, per topographic level (hilltop, middle side, lower steep and valley) and per type of mountain side exposure (east or west). Seven sub-plots of 1 m^2 were sampled randomly inside each protected plot. Maximum standing biomass was cut from these sub-plots and weighed at the end of the rainy season in order to assess herbaceous dry matter production. The average productivity of grasslands was determined for each topographic level and for each type of exposure based on the biomass production of the five plant communities. Data were analysed through analysis of variance with significance assessed at the 0.05 level.

Results and discussion The results showed that topography did not influence productivity (F = 1.94; df = 3; p = 0.1053) (Table 1). These results, which differed from those of Fournier *et al.* (1982), are explained by the soil properties in relationship to the mountains' gradient. Soil depth and texture did not vary significantly according to topography, because water run-off and soil erosion are very intense due to the steepness of the slopes (average gradient of 40%). Because of this hydrological erosion, the accumulation of soil particles is limited down the slopes. However, the type of exposure influenced productivity (F = 4.47; df = 2; p = 0.0048), with herbaceous biomass being lower on eastern than western sides although soil properties did not differ according to the side (Table 2). This difference in biomass production is explained by direct exposure of the eastern sides of the range to a cold and dry wind called "Harmattan", making the eastern sides drier than western sides. This wind is normally experienced from Nov. to March and impacts on the vegetation structure of this relatively high mountain range, as mentioned by Tchamié & Bouraïma (1997).

Table 1 Productivity (Pd (t DM/ha)) according to topography (NP: number of plots; Alt.: altitude (m); Depth: soil depth (cm); %H: grass water content (%); M. steep: middle side; L. steep: lower steep)

	NP	Alt.	Depth	Pd.	%H
Hilltop	4	450	22.0a	6.51	67.1ab
M. steep	8	400	13.5a	5.29	68.9b
L. steep	7	350	14.3a	5.84	70.7ab
Valley	3	350	33.9b	5.88	76.6a

Depth: p=0.0000; Pd: p=0.1053; %H: p=0.0345
Values followed by the same letter are not significantly different at 0.05 level based on Newman Keuls test

Table 2 Productivity [Pd (t DM/ha)] according to side exposure (E. side: east side; W. side: west side; NP: number of plots; Alt.: altitude (m); Depth: soil depth (cm); %H: Grass water content (%))

	NP	Alt.	Depth	Pd.	%H
Hilltop and valley	7	450	24.9a	6.24a	73.6a
E. side	7	400	12.7b	4.97b	66.3c
W. side	8	370	15.9b	6.10a	69.3b

Depth: p=0.0001; Pd: p=0.0048; %H: p=0.0291
Values followed by the same letter are not significantly different at 0.05 level based on Newman Keuls test

References

Fournier, A., O. Hofmann & J-L. Devineau (1982). Variation de la phytomasse herbacée le long d'une toposéquence en zone soudano-guinéenne, Ouango-Fitini (Côte d'Ivoire). *Bulletin de l'IFAN*, T.44, Series A, No. 1-2, 70-77.

Sene E.H. & P.C. Zingari (2001). Montagnes dans les tropiques: Le défi des forêts. *Bois et forêts des tropiques*, 270 (4), 63-72.

Tchamié T. & M. Bouraïma (1997). Les formations végétales du plateau de Soudou-Dako dans la chaîne de l'Atakora et leur évolution récente (Togo). *Journal of the Botanical Society of France*, 3, 89-94.

Dry matter production and nutritive quality of wild Guinea grass (Panicum maximum) grown along roadsides in Sri Lanka

S. Premaratne[1] and G.G.C. Premalal[2]

[1]Department of Animal Science, Faculty of Agriculture, University of Peradeniya, Peradeniya, Sri Lanka, Email: suep@pdn.ac.lk, [2]Pasture Division, Veterinary Research Institute, Peradeniya, Sri Lanka

Keywords: Guinea grass, roadside, productivity, quality

Introduction Wild Guinea grass (*Panicum maximum*) was introduced into Sri Lanka in the 1820s for forage purposes and has now naturalised in most ecological zones, ecosystems and habitats including roadsides with the exception of hilly and semi-arid parts of the country. The enormous distribution of the grass throughout the country has contributed much to supplying livestock feeds, soil erosion control, and improvement of soil fertility. The objective of this study was to investigate the growth, dry matter (DM) production and nutritive quality of wild Guinea grass along roadsides in different ecological zones.

Materials and methods The study was conducted in five ecological zones in which the Guinea grass is naturally abundant: Low Country Dry zone (LCD), Low Country Wet zone (LCW), Low Country Intermediate zone (LCI), Mid Country Wet zone (MCW) and Mid Country Intermediate zone (MCI) during the North East Monsoonal rainy period (Oct. 2002 to March 2003). Grass patches, which were flowering or near to panicle initiation stage, were selected for sampling. Plant density (culms/m^2), tiller production (active tillers/m^2) and fresh yield/m^2 were recorded randomly in five replicates from each zone at different times during this period. Sub samples were taken from the same locations and analysed for DM, crude protein (CP) and, neutral detergent fibre (NDF) and lignin.

Results There were significant difference (p< 0.05) in plant density, tiller production and DM production between the climatic zones, while no significant differences (P< 0.05) were observed in CP, NDF and lignin (Table 1). The highest DM production observed in MCW could be due to high and well-distributed rainfall and nutrient accumulation along roadsides due to the sloping topography towards the road in this zone. Seasonal and low annual precipitation seemed to have resulted in low DM production in LCD. Mean values recorded for CP, NDF and lignin contents were comparable with other studies (Peiris & Ibrahim, 1985; Gutmanis *et. al.*, 2001)

Table 1 Mean growth, production and nutritive quality of roadside Guinea grass

Climatic Zone	Plant density (culms/m^2)	Tiller Production (tillers/ m^2)	DM Production (kg/m^2)	CP%	NDF%	Lignin%
LCD	6.00 [bc]	103.72 [b]	0.28 [d]	10.01	71.89	8.30
LCW	5.22 [c]	93.39 [c]	0.41 [c]	9.79	72.08	8.77
LCI	8.55 [a]	125.00 [a]	0.50 [b]	9.63	73.06	8.05
MCW	5.77 [bc]	96.17 [c]	0.58 [a]	9.82	74.17	8.34
MCI	4.33 [cd]	80.06 [d]	0.50 [b]	10.27	72.90	8.21
SEM.	0.21	2.09	0.01	0.34	0.50	0.23

SEM= Standard error of the mean. Means within a column having same superscript is not different (P< 0.05)

Conclusion Annual DM production of 20,000 - 25,000 kg/ha of satisfactory quality from roadside Guinea grass contributes substantially to livestock feed supplies in all parts of the country.

References

Gutmanis, D., V.B.G. Lourenco, & M.T Colozza,. (2001) Nutritive quality of tropical grass sown under a pine plantation. *Proceedings of the XIX Grassland Congress*, San Pedro, Brazil, 11-21 February 2001, pp. 663-664.

Peiris, H. & M.N.M. Ibrahim, (1985). Effect of intensity and frequency of defoliation on the yield and nutritive value of unfertilized Guinea grass (*Panicum maximum* ecotype A). *Sri Lanka Veterinary Journal*, 33, 11-18.

The use of multivariate analysis in tropical grass and legume seed production in Cuban regions

G. Febles[1], R. Baños[2], S. Yáñes[2], V. Torres[1], T. Ruiz[1] and F. Funes[2]
[1]*Instituto de Ciencia Animal, km 47½ Carretera Central, La Habana, Cuba, Email: feblesgjva@yahoo.es,*
[2]*Instituto de Investigaciones de Pastos y forrajes, Ministerio de la Agricultura, La Habana, Cuba*

Keywords: multivariate analysis, tropical seed production

Introduction Seed production is an important activity in developing countries where pastures are the main source for animal feeding (Febles *et al.* 2003). Another outstanding aspect is the mathematical analysis used when a large number of species, varieties and ecotypes are used in the same study. The objective of this paper was to examine the use of multivariate analysis in studies on the effects of edaphoclimatic factors on seed production from tropical grasses and legumes.

Materials and methods Six grasses and legumes were sown in six Cuban provinces without fertilisation and irrigation. Seed yields were measured together with the range of climatic and edaphic factors listed in Table 1 and the data analysed by the principal component method

Results Table 1 shows the effects of the edaphoclimatic factors on seed yields. Amongst the species effects were greatest for *Brachiaria decumbens* and least for *Neonotonia wightii* (data not shown). The complete data set demonstrates that the lowest yield for legumes was *Pueraria phaseoloides* in Camagüey province (40.6 kg/ha) and the highest for *Leucaena. leucocephala* in Guantánamo province (170 kg/ha). For the herbaceous legumes, the highest value was for *Macroptilium atropurpureum* in Guantánamo and the lowest for *Teramnus labialis* in S. Spíritus .

Table 1 Most preponderant values in edaphoclimatic variables with coefficients above 0.70 and their main components

| | Species | | | |
Climatic	*P. phaseoloides*	*L. leucocephala*	*B. brizantha*	*P. maximum*
Mean temperature	-	1 (.74)	1 (-.71)	1 (.78)
Minimum temperature	-	-	1 (-.76)	1 (-.82)
Maximum temperature	-	-	1 (-.77)	-
Rainy season	-	-	1 (.88)	3 (.71)
Dry season	3 (-.73)	-	3 (.90)	3 (.71)
Relative humidity	1 (-.72)	-	1 (.73)	1 (.81)
Total rainfall	-	1 (-.79)	1 (.81)	1 (-.77)
Light hours	2 (-.69)	-	4 (.82)	2 (.84)
Soil physics				
Dryness	1 (-.92)	1 (-.86)	2 (.80)	2 (.84)
Effective depth	2 (-.69)	2 (-.91)	1 (.96)	-
Pedregosity	-	-	-	-
Soil chemistry				
K	1 (.76)	-	1 (.96)	1 (.98)
Organic matter	1 (.83)	1 (.88)	1 (.83)	1 (.72)
pH	1 (.79)	1 (.73)	1 (.97)	1 (.88)
P	1 (.88)	2 (.85)	2 (-.86)	1 (.98)

() values between parenthesis represent most preponderant values

Conclusions The relationships obtained demonstrate the large effects of edaphoclimatic factors on seed production and the value of multivariate analysis in both understanding factors determining seed yield and the identification of appropriate sites for production of seed of particular species. Most of the principal components had positions 1 and 2 in the analysis.

Reference

Febles, G., V. Torres, T.E. Ruiz, L. Martínez, H. Díaz and A. Noda (2003). The use of multivariate analysis to evaluate the production of seeds in accessions of Leucaena leucocephala in Cuba. *Cuban Journal of Agricultural Science*, 37, 299-304.

Theme A: Efficient production from grassland

Section 12

Overcoming seasonality of production

Forage production and nitrogen status in mixed fodder crops

G. Convertini, M. Maiorana and D. Ferri
Istituto Sperimentale Agronomico, Bari, 70125, Italy, Email: convertini.isaba@tuttopmi.it

Keywords: fodder crops, yields, nitrogen status, SPAD, Nitracheck

Introduction In Southern Italy, the lack of rain during the summer period is one of the main factors limiting fodder crop production. Another very important parameter, linked to drought, is N fertilisation. In these conditions, it is necessary to find mixtures of legumes and grasses able to ensure good production and quality in the driest months, and to rationalise N fertilisation through control of the nutritional status of the crops at the beginning of spring. By this approach, it is possible to adjust N application during the cropping cycle. The aim of this research was to evaluate in a hilly area of Apulia Region the production of several mixtures of annual grasses and legumes, cropped in temporary grassland, and to investigate their N nutritional status.

Materials and methods Ten mixtures of different species and cultivars of grasses: Oat (O), Italian ryegrass (IRG) and legumes: crimson clover (CC), sea clover (SC), Egyptian clover (EC), burr medic (BM), common vetch (CV) were compared in plots of 8 m^2. In each trial year, all crops received 50 kg N/ha in March. Nitrogen status of plants was determined by measuring the green index of leaves (SPAD method) (this is strongly correlated with chlorophyll content (Schepers *et al.*, 1992)), and analysing the nitrate content of the basal parts (stalks) of the plants (Nitracheck method). Total N uptake was also calculated (dry matter x N content). Data obtained were analysed by ANOVA (SAS Institute, 2001).

Table 1 Dry matter production and plant nitrogen status

Mixtures	Dry matter (t/ha)	SPAD	Nitrate content (mg/kg)	N uptake (kg/ha)
1 = O, CC, V	10.6[a]	46.6[a]	142.9[d]	302.0[ab]
2 = IRG, CC, V	10.0[a]	44.6[ab]	163.6[d]	240.1[b]
3 = O, CC, V	9.9[a]	47.2[a]	179.0[d]	285.0[ab]
4 = IRG, CC, V	9.4[a]	42.6[bc]	258.4[c]	229.1[b]
5 = O, SC, V	8.7[a]	47.7[a]	198.0[cd]	273.1[ab]
6 = IRG, SC, V	9.2[a]	40.9[c]	488.2[a]	234.8[b]
7 = O, EC, V	10.8[a]	47.3[a]	404.9[b]	316.1[a]
8 = IRG, EC, V	10.3[a]	42.4[bc]	207.8[c]	264.6[ab]
9 = O, BM, V	9.9[a]	46.1[a]	316.1[b]	317.5[a]
10 = IRG, BM, V	9.7[a]	41.5[bc]	440.8[a]	273.7[ab]

Values with different letters in each column are significantly different at P≤0.05 (SNK test)

Results The best production of dry matter (mean values of 2000 and 2001) was obtained with mixture 7 (Table 1), but differences were not statistically significant. SPAD values of the mixtures with oat were significantly better than those of mixtures with Italian ryegrass, while nitrate contents showed this trend: 6, 10 > 7, 9 > 4, 5, 8 > 1, 2, 3. There were significant positive correlations between SPAD and both N content and N uptake (Table 2). Dry matter was negatively correlated with nitrate and positively correlated with N uptake. On the whole, the contents of chlorophyll in leaves and of nitrate in stalks seem to be good N indicators for the mixtures. In particular, SPAD values indicate N status and would allow adjustment to N fertilisation. On the other hand, the nitrate content of crop stalks, with its negative correlation with dry matter, seem to indicate that when nitrates accumulate in plants, nitrogen nutrition efficiency is very low. This probably occurs because in Southern Italy temperature and water availability affect nitrogen status, mainly affecting the transformation of mineral N to organic N. This step is very important for dry matter accumulation.

Table 2 Correlation coefficients among bio-agronomic parameters and nitrogen indicators

Parameters	SPAD	Nitrate content	N content	N uptake
Dry matter	-0.02042 n.s.	-0.60712 **	0.02012 n.s.	0.56338 ***
SPAD		-0.23403 n.s.	0.44795 *	0.44007 *
Nitrate content			-0.13722 n.s.	-0.51222 *
N content				0.69153 ***

*, **, *** = significant at the P<0.05, 0.01 and 0.001 levels respectively; n.s. = not significant

Conclusions The results obtained so far have shown that: (1) all of the mixtures provided good production of dry matter; (2) direct measurements of leaf chlorophyll contents and stalk nitrate concentrations give an effective indication of the nitrogen status of plants and have good correlations with yields and N uptake.

References
SAS Institute Inc. (2001). User's Guide, Version 8.02. SAS/STAT, Cary, USA.
Schepers, J.S., D.D. Francis, N. Vigil & F.W. Below (1992). Comparison of corn leaf N concentration and chlorophyll meter reading. *Communications in Soil Science and Plant Analysis*, 23 (17 & 30), 2173-2187.

The effects of strategic nitrogen fertiliser application during the cool season on the composition of a perennial ryegrass-white clover pasture in the Western Cape Province of South Africa

J. Labuschagne[2], M.B. Hardy[2] and G.A. Agenbag[1]
[1]University of Stellenbosch, Private Bag X1, Matieland 7602, South Africa, Email: johanl@elsenburg.com,
[2]Department of Agriculture Western Cape, Private Bag X1, Elsenburg 7607, South Africa

Keywords: clover content, perennial ryegrass-white clover, strategic nitrogen

Introduction Application of fertiliser N to stimulate DM production of perennial ryegrass-white clover pastures during the cool season can be an important management tool. Application of fertiliser N should however maintain clover contents between 30 and 50 percent (Martin, 1960; Harris, 1994). The aim of the study was to develop a better understanding of the effect of a strategic N fertiliser application during the cool season on the grass-clover balance and to identify possible management guidelines that would maximise dry matter production without suppressing clover content to values lower than required to maintain the benefit of clover in the pasture.

Materials and methods A perennial ryegrass-white clover pasture was established under irrigation in autumn 1999 and treatments commenced in autumn 2000. The trial was laid out as a randomised complete block with a 4 x 5 factorial arranged in a split-plot design (Snedecor & Cochran, 1967). Four N levels (0, 50, 100, and 150 kg N ha^{-1}) were applied during five seasons (late-April/early-May [autumn], early June [early winter], mid-July [late winter], late-August [early spring] or late-September/early October [late spring]) with four replicates. Each plot received a single annual application of fertiliser nitrogen after cutting, the timing of application depending on season. To facilitate soil N studies the pasture was grazed at four weekly intervals in summer and mowed five weekly from late autumn to late spring. Clover content was determined 5 and 10 weeks after N application.

Results Increased fertiliser N rates resulted in increasingly lower clover percentages (Table 1). No differences in clover percentage were recorded between the 100 and 150 kg N ha^{-1} application rates five weeks after fertiliser N application. The effect of season of application was inconsistent due mainly to different initial clover percentages. Partial recovery of clover content was noted at the end second re-growth cycle (10 weeks) but did not reached the same levels as at the 0 kg ha^{-1} treatment combinations. Clover content (%) of treatments that received 50 kg N ha^{-1} normally recovered to the same levels as measured at the 0 kg N ha^{-1} treatments within 4-5 regrowth cycles. The predicted clover content (%) after fertiliser N application during a specific season can be described by a linear function: $y = a + (b_1x_1) + (b_2x_2)$ where x_1 is the initial clover percentage and x_2 the fertilizer N rate.

Table 1 Clover percentage of a perennial ryegrass-white clover pasture five and ten weeks after fertiliser N application during year 2000 at Elsenburg, South Africa

Season	Five weeks Fertiliser N rate (kg ha^{-1})				Mean
	0	50	100	150	
Autumn	66.47	49.79	47.06	39.65	50.75 a*
Early winter	55.48	46.89	41.07	39.12	45.64 ab
Late winter	52.91	36.24	32.66	26.31	37.03 bc
Early spring	47.4	29.06	21.18	19.04	29.17 c
Late spring	50.42	32.24	22.02	20.00	31.17 c
Mean	54.53 a	38.84 b	32.80 c	28.82 c	38.75
	Ten weeks				
Autumn	55.48	47.21	46.95	43.82	48.37 a*
Early winter	52.91	49.06	43.39	36.11	45.37 ab
Late winter	47.4	38.93	35.58	22.41	36.08 bc
Early spring	50.42	35.79	23.44	19.99	32.41 c
Late spring	57.42	41.44	30.11	27.05	39.00 abc
Mean	52.73 a	42.48 b	35.89 c	29.88 d	40.25

*Means of a specified sampling time in the same column or row followed by the same letter are not significantly different (P<0.05). Nitrogen rate x Season of application = NS

Conclusions This study showed that the application of 50 kg N ha^{-1} did not reduce the clover content to less than 30% during any season if an initial clover content of *ca* 47% was present before fertiliser N application. The application of 150 kg N ha^{-1} in the current study resulted in numerous treatment combinations to have clover percentages of less than 30%. Initial clover content (%) of the pasture, the season when N application is planned and pasture productivity will dictate the fertiliser N rate to be applied.

References
Harris, S.L. (1994). Nitrogen and white clover. *Dairy Research Corporation*, Hamilton, p 22-27.
Martin, T.W. (1960). The role of white clover in grassland. *Herbage Abstracts,* 30. 159-164.
Snedecor, G.W. & W.G. Cochran (1967). *Statistical Methods. Sixth Ed.* The Iowa State University Press, Ames, Iowa, USA.

The effects of strategic nitrogen fertiliser application during the cool season on perennial ryegrass-white clover pasture production in the Western Cape Province of South Africa

J. Labuschagne[2], M.B. Hardy[2] and G.A. Agenbag[1]
[1]University of Stellenbosch, Private Bag X1, Matieland 7602, South Africa, Email: johanl@elsenburg.com,
[2]Department of Agriculture Western Cape, Private Bag X1, Elsenburg 7607, South Africa

Keywords: dry matter production, perennial ryegrass/white clover, strategic nitrogen fertilisation

Introduction Low dry matter (DM) production of perennial ryegrass-white clover pastures during the cool season in the Western Cape Province of South Africa is of major concern and difficult to address successfully. Perennial ryegrass is able to respond to fertiliser N at temperatures where white clover plants are almost dormant (Frame, 1994; Hatch & Macduff, 1991). The objective of this study was to determine DM response of a perennial ryegrass-white clover pasture to fertiliser N applied during different seasons.

Materials and methods The experimental detail is given in Labuschagne et al. (2005). Dry matter production measured 5 weeks after fertiliser N application is discussed here.

Results On average the highest DM yields were obtained with spring applications of fertiliser N and lowest yields when N was applied in early winter (Table 1). Fertiliser N application could be used to stimulate DM production during a predetermined period. Although 150 kg N/ha applied in late spring generally gave the best results in terms of DM produced, the 50 kg N/ha treatments resulted in more efficient conversion of N applied to additional DM produced (Table 2). The predicted DM production after fertiliser N application during a specific season can be described by a linear function: $y = a + (b_1x_1) + (b_2x_2)$, where x_1 is the initial clover percentage and x_2 the fertiliser N rate. Annual DM production at the different treatment combinations was at comparable levels. The increased DM production during the first regrowth cycle after N application, was nullified by increasingly lower DM production as N rate increased from regrowth cycle three, resulting in annual DM production at comparable levels over all N rates.

Table 1 Dry matter production (t/ha) of perennial ryegrass-white clover pasture in response to fertiliser N rate and season of application

Season	Nitrogen (kg/ha) 0	50	100	150	Mean (S)
Autumn	2.01 fgh*	2.27 e	2.45 de	2.66 cd	2.35**
E. winter	0.95 m	0.99 lm	1.22 jkl	1.31 ijk	1.11 c
L. winter	1.13 klm	1.50 i	1.84 h	2.00 gh	1.62 b
E. spring	1.42 ij	2.25 ef	2.69 bcd	2.82 bc	2.30 a
L. spring	1.50 i	2.23 efg	2.91 ab	3.07 a	2.43 a
Mean (N)	1.40 d	1.85 c	2.22 b	2.37 a	1.96

Table 2 Nitrogen use efficiency (kg additional DM/kg N applied) of perennial ryegrass-white clover as a result of different fertiliser N rates and season of application

Season	Nitrogen (kg/ha) 50	100	150	Mean (S)
Autumn	9.7	7.2	6.4	7.8 cd*
E. winter	7.6	6.6	5.0	6.4 d
L. winter	10.3	9.0	7.5	9.0 bc
E. spring	14.8	12.1	9.0	12.0 a
L. spring	12.7	9.9	9.0	10.5 ab
Mean (N)	11.0 a	9.0 b	7.4 c	

*Means followed by the same letter are not significantly different (P<0.05). **Means (bold) in the same column or row followed by the same letter are not significantly different (P<0.05).

*Means in the same column or row followed by the same letter are not significantly different (P<0.05). Nitrogen rate x Season of application = NS

Conclusions Autumn, early and late winter, as well as early spring, applications of fertiliser N were shown to be valuable in reducing the negative effect of the winter gap. Late spring application was however too late to be of any significance in managing the winter gap as pasture productivity (without N application) was at the same - or higher - levels compared to DM production at the onset of the winter gap.

References
Frame, J. (1994). Soil fertility and grass production; nitrogen. In: J. Frame (ed.). Improved Grassland Management. Farming Press Books, Redwood Press, Melksham, Wiltshire, UK.
Hatch, D.J. & J.H. Macduff (1991). Concurrent rates of N_2 fixation, nitrate and ammonium uptake by white clover in response to different root temperatures. Annals of Botany, 67, 265-274.
Labuschagne, J., M.B. Hardy & G.A. Agenbag (2005). The effects of strategic nitrogen fertilizer application during the cool season on the composition of a perennial ryegrass-white clover pasture in the Western Cape province of South Africa. XX International Grassland Congress – offered papers (in press).

Soil compaction in cropland pastures used for winter grazing

B.F. Tracy
Department of Crop Sciences, University of Illinois, 1102 S. Goodwin Ave. Urbana, IL 61820, Email: bftracy@uiuc.edu

Keywords: cropland pastures, cover crops, winter grazing, soil compaction

Introduction In the northern United States, forage availability on cool season pastures declines rapidly late in the growing season (Moser & Hoveland 1996). To supplement low forage availability in the fall and winter, producers can graze livestock on cropland pastures containing annual cover crops and crop residues. Managed properly, cropland pastures can provide livestock with abundant, high quality forage that lasts long into winter. A potential issue with cropland pastures is that presence of cattle on moist, non sod-bearing soils could lead to severe soil compaction. The objective of this particular study was to determine how winter grazing on cropland pastures would effect soil compaction and subsequent crop yield.

Materials and methods The cropland pastures evaluated in this study are part of an integrated pasture and row crop experiment located on the Dudley Smith research farm in central Illinois, USA. Soil compaction was compared among 4 crop/forage systems used on the farm: 1) ungrazed corn fields, 2) summer pastures, 3) winter cropland pastures with corn residues, and 4) winter cropland pastures with cover crops. Soils across the 90 ha farm ranged from silt loams and silty clay loams. Beef cattle (1.1 cows/ha) were released into the cropland pastures in November 2003 and they grazed pastures until March 2004. Soil compaction, expressed as penetration resistance, was measured in late March, when soils were moist. Twenty electronic penetrometer measurements (1 cm cone diameter) were taken to a depth of 45 cm in each system. Soil moisture was measured gravimetrically from 6 locations on each field. Corn yields were recorded in September.

Results Penetration resistance was greater at 46 cm depth compared with 15 cm (Figure 1). Penetration resistance differed significantly at 46 cm (One Way ANOVA, P = 0.07, df = 3, 8). Apparent soil compaction was greatest in the winter cropland pastures and highest under the cover crops. The adjacent corn fields and summer pastures exhibited significantly less compaction. Soil moisture at the time of sampling did not differ significantly among the 4 locations (P > 0.10). Corn yields on the cropland pastures ranged from 12.3 to 14.9 Mg/ha in 2004 and were virtually identical to yields from ungrazed corn fields.

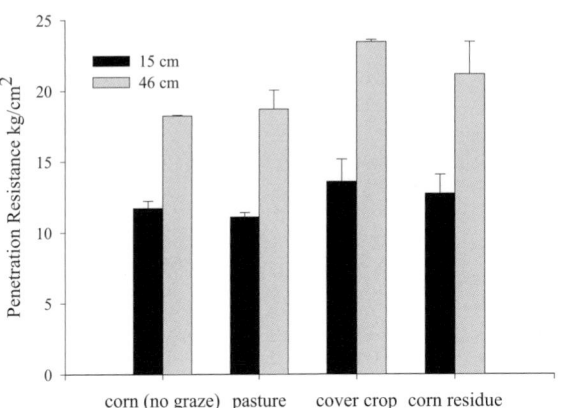

Figure 1 Soil compaction measured at 15 and 46 cm depths in the 4 crop and forage systems (March 2004). Error bars are 1 SE

Conclusions Although cropland pastures provide badly needed forage in the late autumn and winter, preliminary conclusions suggest cattle trampling can cause significant soil compaction at depth (45cm). Compaction was worse under cover crops where cattle tend to spend more time grazing during the winter. Subsequent corn yields were unaffected by the apparent soil compaction. Compaction probably did not affect yields in 2004 because growing season conditions were relatively wet. In drier years, compaction may affect yields more significantly since it could restrict root growth in deeper soil.

References Moser, L.E., and C.E. Hoveland. 1996. Cool-season grass overview, *In:* L. E. Moser, et al. (eds) Cool-season forage grasses. *Agronomy Monograph* 34. ASA, CSA and SSSA, Madison, Wisconsin.

Tissue damage and nutritive value of warm-season grasses following a freeze

S.W. Coleman[1], J. Breman[2], A.S. Blount[2] and K.H. Quesenberry[3]
[1]USDA ARS STARS, Brooksville, FL 34601 USA, Email: swcol@ifas.ufl.edu; [2]North Florida Research and Education Center, Marianna, FL 32446 USA; [3]Univ. Florida Agronomy Department, Gainesville, FL 32611 USA

Keywords: C4 grass, freeze tolerance, neutral-detergent fiber

Introduction Bahiagrass (*Paspalum notatum* Flugge) is a major forage for livestock in the subtropics of the U.S.A. However, it is subject to freeze damage with minimal winter regrowth, and is generally considered a poor grass for stockpiling due to poor quality of the residue. Bahiagrass genotypes have been found showing a range of leaf freezing tolerance in the in the field (-3^0 C) (Blount et al., 2001). Other C_4 grasses have been reported to have genotype-specific tolerances to below-freezing temperatures ranging from -3 to -10^0 C (Sakai & Larcher, 1987). This research was begun to try to understand the processes that take place following freeze injury to bahiagrass. Three grasses native to the Midwest, big bluestem (*Andropogon gerardii* Vitman cv Alamo), Indiangrass (*Sorghastrum nutans* (L.) Nash cv Lometa), and switchgrass (*Panicum virgatum* L. cv Kaw) and commonly used for stockpiling were used for comparison.

Materials and methods Randomized complete block design of 6 genotypes was established in north Florida in rows of eight single plants. Three bahiagrass cultivars ('Argentine', 'Pensacola' and 'Tifton 9'), with Argentine being the commercial cold-sensitive cultivar, were compared to several native grasses. Plots were fertilized twice, staged on 1 July 2003 to 6.08 cm for bahiagrass and 15.24 cm for native grasses, and allowed to grow thereafter to stockpile dry matter until a killing frost would occur. Harvests included a pre-freeze event baseline (16 Nov 2003), freeze event -3.3C (30 Nov 2003) followed by harvest at 3, 7, 14, and 28 d post-freeze. Harvested material was dried at 80°C and separated into leaf and stem. Leaf tissue cold tolerance (LTCT) was rated after 28 d using the USDA G.R.I.N. scale (1= no damage or 100% green leaf, 9= total top growth damaged or 0% green leaf). Crude protein (CP) and neutral detergent fiber (NDF) were determined on all samples using NIRS calibrated with 20% of the samples.

Results Single plant yield and percent leaf were different among genotypes (Table 1). Switchgrass had the highest yield and bahiagrass had the highest leaf percent. Leaf yield was more similar, but switchgrass produced more than twice as much leaf of any other cultivar. Tifton 9 showed the most resistance to freeze damage whereas the native grass cultivars and Argentine bahiagrass were completely damaged. Crude protein of native grasses and Argentine declined dramatically following the freeze (Fig. 1). Tifton 9 increased slightly in CP concentration and remained above 8%. At this level, Tifton 9 would provide maintenance protein for dry beef cows during the winter. Pensacola would be marginal. Neutral detergent fiber increased following the freeze for all grasses except Tifton 9, but more dramatically for switchgrass and Argentine.

Conclusions Midwest native species can accumulate tremendous amounts of dry forage in the subtropical U.S.A. for stockpiling but requires protein supplement. While bahiagrass has the reputation of being very poor in quality after a freeze, this study suggests at least one variety (Tifton 9) maintains quality quite well following a freeze and may not require additional supplement for non-lactating cows.

Table 1 Yield, leaf percent, and cold damage for six cultivars of C4 grasses

Cultivar	Yield[a], g DM plant[-1]	Leaf[a], %	LTCT[b]
Big bluestem	46.3[a]	21.0[a]	9.00[c]
Indiangrass	99.7[b]	22.7[a]	9.00[c]
Switchgrass	222.0[c]	21.0[a]	9.00[c]
Argentine	16.5[a]	80.2[b]	9.00[c]
Pensacola	21.9[a]	81.3[b]	8.54[b]
Tifton 9	17.4[a]	88.7c	6.71[a]

[a]Averaged across harvest dates.
[b]At day 28 post-freeze.

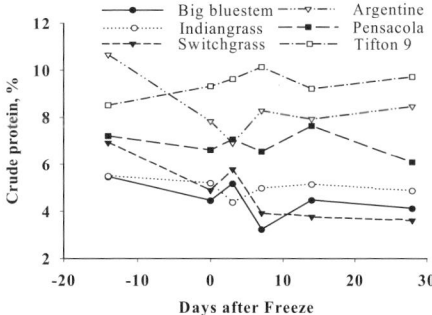

Figure 1 Changes in crude protein content of C4 grasses following a freeze (-3 °C)

References
Blount, A.R., K.H. Quesenberry, P. Mislevy, R.N. Gates & T.R. Sinclair (2001). Bahiagrass and other Paspalum species: An overview of the plant breeding efforts in the Southern Coastal Plain. In: D. Lang (Ed*.) Proceedings of 56th Southern Pasture and Forage Crop Improvement Conference*, pp. 52-55.
Sakai, A. & W. Larcher (1987). Frost survival of plants: Responses and adaptation to freezing stress. Springer-Verlag, Heidelberg, Germany.

Effect of seed rate of Trifolium repens in pasture overdrilling

P. Goliński, S. Kozłowski and B. Golińska
Department of Grassland Sciences, Agricultural University of Poznań, Wojska Polskiego 38/42, 60-627 Poznań, Poland, Email: pgolinsk@au.poznan.pl

Keywords: overdrilling, sward improvement, white clover

Introduction In the region of Wielkopolska, unfavourable climatic conditions, particularly periodical shortage of precipitation, have contributed to a rapid degradation of pastures in dairy farms. In grass-clover mixtures *Trifolium repens* (*Tr*) is found to disappear very quickly from the sward. In consequence the DM yield and herbage quality in summer is low. One of the methods of improving of pasture sward and reducing the seasonality of forage production is overdrilling (OD). Many factors affect the success of this undertaking (Sheldrick 2000). This research investigated the response to one easily adjustable factor, that of seed rate (SR).

Materials and methods During 2001-2002 an experiment was set up, in Brody (52° 26' N, 16° 18' E) to evaluate the effect of different SR (3, 4, 5, 6 kg/ha) of Tr cv. Barbian on the success of OD. A block design was used with four replicates (plot size 8 m × 2.5 m). The pastures were situated on poorly mineralised Histosols soils (pH_{KCl} – 6.5, N_t – 0.62%, P_2O_5 – 67.9 mg/100 g, K_2O – 30.0 mg/100 g, Mg – 7.1 mg/100 g). At the time of OD the proportion of Tr in the sward yield was 2.3%. The area was prepared by using a rototiller followed by rolling. The seeds were sown in spring (early April 2001) using a Vredo drill. Fertiliser was applied each year at a rate of: N – 80 kg/ha, P_2O_5 - 80 kg/ha, K_2O - 140 kg/ha and 4-5 regrowths were harvested. The effectiveness of OD was evaluated on the basis of the number of seedlings per m^2 25 and 50 days from sowing as well as the proportion of Tr in the sward (samples were separated and the fractions were weighed). Dry matter yield was measured on an area of 15 m^2. The herbage was dried in a forced-draught oven at 60 °C. Crude protein (CP) and acid detergent fibre (ADF) were also measured on selected plots using commonly accepted methods. F-tests were applied to main effects. Means were separated by the LSD and were declared different at the p<0.05 level.

Results Increasing the SR of *Tr* resulted in a greater number of seedlings per unit area (Table 1). There was a positive correlation between the proportion of *Tr* in the yield and SR. Increasing the SR from 3 to 4 kg/ha, also increased the proportion of *Tr* in the sward during the first year of utilisation from 14.0% to 20.3%. Increasing SR of *Tr* did not have a significant effect on sward yields. The improvement in sward botanical composition obtained following the introduction of *Tr* had a significant impact on herbage quality. Over the two-year period it was found that increasing SR from 3 to 4 kg/ha resulted in a significant increase in CP concentrations (from 184 to 215 g/kg DM) and a decline in ADF content (from 252 to 217 g/kg DM). The impact on herbage quality of increase in SR from 4 to 6 kg/ha was negligible.

Table 1 Effect of seeding rate (kg/ha) of *Trifolium repens* (*Tr*) in pasture overdrilling

Seeding rate	3	4	5	6	$LSD_{0.05}$
No. of seedlings per m^2 after 25 days from sowing	168	277	336	529	30.4
No. of seedlings per m^2 after 50 days from sowing	91	150	213	264	24.3
Proportion of *Tr* in sward in sowing year (%)	16.6	21.2	22.9	25.2	3.67
Proportion of *Tr* in sward in year of utilisation (%)	14.0	20.3	21.0	20.5	2.23
Total yield of DM in sowing year (t/ha)	6.84	6.81	6.87	6.84	ns
Total yield of DM in year of utilisation (t/ha)	9.75	10.02	9.89	10.04	ns
Herbage quality – means from two years of pasture utilisation					
CP (g/kg DM)	184	215	206	218	14.3
ADF (g/kg DM)	252	217	233	230	6.8

Conclusions Increasing the SR of *Tr* from 3 to 4 kg/ha resulted in a significant increase in the proportion of this species in the sward and an improvement of herbage quality. Increasing the SR of *Tr* from 3 to 6 kg/ha for pasture OD did not have a significant effect on total DM yield.

Reference

Sheldrick, R.D. (2000). Sward establishment and renovation. In: A. Hopkins (ed.) Grass. Its Production and Utilization. Blackwell Science, Oxford, 13-30.

Breeding and evaluation of forage soyabeans

T.E. Devine
Sustainable Agricultural Systems Laboratory, Building 001, BARC-West, Beltsville Agricultural Research Center, 10300 Baltimore Ave., Beltsville, MD 20705, USA, Email: DevineT@ba.ars.usda.gov

Keywords: forage soybeans, plant protein

Introduction The principal use of soyabean in the US in the early 1900s was as livestock forage. Soyabeans are less expensive to establish than small seeded perennial legume forages and can provide legume protein after winter killing of perennial legumes. Soyabean can improve production distribution by vigorous growth during the hot summer season when traditional perennial legumes are less productive.

Materials and methods Three soyabean cultivars were bred for forage production: Donegal, Derry and Tyrone. At Orange, VA, Tyrone and the grain-type cultivar Hutcheson were interplanted with pearl millet and sorghum and tested for height and forage yield. At Ames, IA, Derry and Donegal were tested with Hutcheson in four replications for yield and forage quality (Darmosarkoro, 2001).

Results At Orange the forage soyabean cultivar grew taller than Hutcheson and was more competitive with pearl millet than Hutcheson (Table 1). All soyabeans were overgrown by the sorghum. At Ames the forage lines showed a yield advantage over Hutcheson (Table 2). For the earlier-maturing cultivars Donegal and Hutcheson crude protein declined in mid-season and then increased. For all cultivars, IVDMD decreased over the growing season.

Table 1 Height and yield at Orange, VA, 1996

Soyabean lines and associated species	Height (cm)		Yield t/ha
	Grass	Soybean	
Sorgo 10	231	0	9.34
P. Millet	137	0	6.94
Hutcheson	0	91	6.25
Hutcheson & Sorgo 10	236	106	8.74
Hutcheson & P. Millet	142	99	8.56
Tyrone	0	180	8.00
Tyrone & Sorgo 10	239	147	9.68
Tyrone & P. Millet	150	168	9.79

Table 2 Yield, *in vitro* dry matter digestibility (IVDMD) and crude protein (CP), g /kg, Ames, IA, 1994

Cultivar	Forage yield t/ha	Days after planting								
		46	60	74	88	102	116	130	144	
Derry	11.5	702	644	590	558	566	550	537	564	IVDMD
		260	219	180	154	158	157	155	150	CP
Donegal	10.2	698	634	595	581	574	618	597	626	IVDMD
		243	198	168	142	160	184	184	197	CP
Hutcheson	8.1	703	666	625	608	596	595	598	609	IVDMD
		260	229	195	177	171	169	190	183	CP
LSD (0.05)	1.7	13.2	17.2	20.0	17.5	19.9	19.2	43.5	27.0	IVDMD
		12.4	14.3	11.6	12.5	9.9	11.4	20.2	12.9	CP

Conclusions The forage cultivars grew significantly taller than conventional grain-type soyabean cultivars, suggesting their ability to better compete in the sward with other tall growing species such as pearl millet. Yields of the forage soyabeans were higher than yields of conventional grain cultivars.

Reference

Darmosarkoro, M., M.M. Harbur, D.R. Buxton, K.J. Moore, T.E. Devine & I.C. Anderson (2001). Growth, development, and yield of soybean lines developed for forage. *Agronomy Journal*, 93, 1028-1034.

Effect of utilisation date on the yield and quality of a semi-natural grass stand in winter

J. Skládanka

MendelUniversity of Agriculture and Forestry, Department of Fodder Production and Grassland Management, Zemědělská 1, 613 00, Brno, Czech Republic, Email: sklady@mendelu.cz

Keywords: yield, quality, winter grazing

Introduction Prolongation of the grazing period into the winter season reduces the costs of feedstuffs for dry cows. To enable the use of pastures in winter it is necessary to carry out the last cut or grazing cycle in June or July and to postpone the last grazing until the first week of August. (Opitz von Boberfeld, 1997). The time of the last utilisation in summer influences the yield and quality of herbage during the winter season (Bartholomew *et al.*, 1997). This research studied changes in yield and quality of a semi-natural grass stand in autumn and in winter following preparatory cuts taken at different times in summer.

Material and methods The experiment was carried out over two years in the Českomoravská vrchovina highland (Czech Republic) at the altitude of 553 m. The average annual temperature and precipitation were 7.4 $^{\circ}$C and 736 mm respectively. Fertiliser at the rate of 50 kg/ha N was applied in the first half of August. The main harvesting cuts (H) were taken in November, December and January following last preparatory cuts (P) for each of H taken in June, July and August. Assessments were made of dry matter (DM) yield and the digestibility of organic matter (DOM) estimated by the Hohenheim test (Menke & Steingass, 1987). Data were analysed using ANOVA and the statistical programme Statistica 6.0.

Results In both years, yields of DM and DOM decreased from November to January (Table 1). A colder autumn in 2001/2002 was reflected in lower yields of DM and quality of the grass stand in December and January in that year than in the following year. As the time of the last preparatory cut was delayed from June to August, it resulted in a significant decrease in forage DM in the November to January period had little effect on DOM

Table 1 Effects of preparatory cuttings and the period of main stand use on DM yields (t/ha) and DOM (%)

		Preparatory cutting (P)					
		2001/2002			2002/2003		
	Main period of harvesting (H)	June	July	August	June	July	August
DM	November	2.35	0.73	0.37	3.13	2.07	1.61
	December	0.68	0.32	0.19	2.30	0.89	0.65
	January	1.06	0.36	0.15	1.36	0.82	0.75
		$LSD_{(p<0.05)} = 0.37$			$LSD_{(p<0.05)} = 0.53$		
		P = *** H = *** P x H = ***			P = *** H = *** P x H = NS		
DOM	November	44.8	48.6	52.4	56.9	58.6	60.6
	December	35.9	36.5	36.2	51.6	55.8	51.7
	January	35.2	35.8	35.8	52.6	54.3	57.9
		$LSD_{(p<0.05)} = 2.87$			$LSD_{(p<0.05)} = 4.82$		
		P = ** H = *** P x H = *			P = NS H = ** P x H = NS		

Conclusions The data show wide variation between years in the quantity and quality of herbage produced in the November – January period. This was most likely due to variation in weather conditions in these years. Under conditions of the Czech Republic, the best time in the autumn to harvest the grass is in November and exceptionally also in December. The most suitable dates of the last preparatory cuts were June and July. The benefit of increased DOM following the preparatory cut in August did not compensate for the relatively low production of DM.

References

Bartholomew, H.M., S.L. Boyles, B. Carter, E. Vollborn, D. Miller, D & R.M. Sulc (1997). Experiences of eight Ohio beef and sheep producers with year-round grazing. *Proceedings of the Eighteenth International Grasland Congress,* 127-128.

Menke, K.H. & H. Steingass (1987). Schätzung des energetischen Futterwertes aus der in vitro mit Pansensaft bestimmten Gasbildung und der chemischen Analyse. *Übersicht zur Tierernährung*, 15, 59-94.

Opitz v. Boberfeld, W. (1997). Winteraußenhaltung von Mutterkühen in Abhängigkeit vom Standort unter pflanzenbaulichen Aspekt. *Ber. Ldw.*, 75, 604-618.

Study on re-growth and nutritional potentials of *Eleusine indica L* in Chitwan, Nepal

P.R. Regmi[1], N.R. Devkota[2], M. Sapkota[2] and M.P. Sharma[2]
[1]*Nepal Agricultural Research Council, Nepal, Email: prajwalsquare@yahoo.com,* [2]*Institute of Agriculture and Animal Science, Rampur Campus Tribhuvan University, Nepal*

Keywords: re-growth, digestibility, fertiliser, cutting height, tiller

Introduction The efforts to use exotic fodder species to solve the problem of green roughage scarcity in Nepal, have had only limited success because the species are not persistent. Use of local forage species, such as *Eleusine indica*, which is widely adapted and tolerant of repeated cutting (Lowry *et al.*, 1992), could be possible solution to the problem. The objective of this study was to understand the re-growth potential of *E. indica* with respect to nitrogen fertiliser application and cutting management and to determine its feeding value in the dry season.

Materials and methods The field trial was carried out in Chitwan. It had a split plot design with three replicates. The main plot treatments were nitrogen fertiliser (0, 50, 100 kg N/ha), and the sub-plot treatments were cutting height (2, 4, and 6 cm from the base). The plot size was 4.5 x 2.7 m^2 with row-row distance of 30 cm and plant-plant distance of 15 cm. Tiller numbers/plant, forage re-growth and proximate constituents were recorded. The digestibility trial was carried out with local *Khari* goats of around 1 year age and digestibility of the test feed was compared with that of seasonally available grass (*Melia azedarach* and *Imperata cylindrica* in equal parts).

Results Increased number of tillers per plant (46[b,], 57[ab] and 66[a] at 0, 50 and 100 kg N/ha respectively for the April count; P<0.05) and higher re-growth (Table 1) was obtained with 100 kg N/ha. However, cutting height had no effect (P<0.05) on forage re-growth. The cumulative forage re-growth for four harvests during the dry season (Jan-May) was about 11 t/ha with 100 kg N/ha and 2-cm cutting height. This is comparable with the productivity of other cultivated and indigenous grasses found in Nepal (Pande, 1997; Relwani, 1979). Crude protein % increased with nitrogen level (11.9[b], 13.0[a] and 13.3[a] at 0, 50 and 100 kg N/ha respectively; P<0.05). The CP content of *E indica* in this study was higher than that of the majority of grasses grown in the dry season in Nepal (Pande, 1997). Crude fibre (19.98 % to 22.27%) and ash (16.3 % to 21.3%) were higher than reported by Serra *et al.* (1997). The grass was fairly digestible (Table 2), palatable and had no untoward effect on goats.

Table 1 Effect of different levels of N on re-growth (g/ha) of *E. indica* at different harvesting dates in Chitwan

Date of harvest/ Rate of fertiliser	Jan.1 (n=9)	Feb15 (n=9)	Apr. 1 (n=9)	May 15 (n=9)
0 kg N/ha.	77.1 (2.94)	45.4 (3.15)	330 (32.5)	181 (20.28)
50 kg N/ha.	68.33 (7.79)	53.25 (4.45)	376 (13.0)	243 (11.7)
100 kg N/ha.	94.6 (6.28)	91.1 (4.69)	465 (37.6)	304 (26.9)

Figures in parentheses indicate standard error of mean.

Table 2 Digestibility coefficients of *E. indica* and local seasonal grass fed to goats in Chitwan

Chemical constituents	Digestibility coeff. of fodder / forage	
	E. indica (n=4)	(*I. cylindrica* + *M. azedarach*) (n=4)
DM	53.5 (1.23)	52.9 (2.11)
CP	66.6 (1.54)	57.6 (1.47)
CF	69.6 (2.05)	47.2 (2.41)
EE	53.7 (0.87)	41.6 (1.30)

Figures in parentheses indicate standard error of mean; and n= number of observations.

Conclusions The study indicates that *E. indica* is a nutritious, fairly palatable, and easily digestible grass. It has no untoward effect on goats. It appears that it can be cut or grazed as low as 2 cm from the ground without jeopardising its persistency. The species has a potential in overcoming the problem of green roughage scarcity during the dry season in Chitwan, Nepal, and other similar areas in different parts of the world.

References
Lowry, J.B., R.J. Petheram & B. Tangendjaja (1992). Plants fed to village ruminants in Indonesia. ACIAR Technical Report No. 22, 3-29, Australian Centre for International Agricultural Research, Canberra.
Pande, R.S. (1997). Fodder and Pasture Development in Nepal, Udaya Research and Development Services (P.) Ltd., 1-12.
Serra, A.B., S.D. Serra, E.A. Orden, L.C. Cruz, K. Nakamura & T. Fujihara (1997). Variability in ash, crude protein, detergent fiber and mineral content of some minor plant species collected from pastures grazed by goats. *Asian-Australasian Journal of Animal Science*, 10, 28-34.
Relwani, L.L. (1979). Fodder Crops and Grasses. Indian Council of Agricultural Research, 59-85.

Seasonality of forage production of coastcross-1 with different sources and applications of phosphorus

J.C. Pinto, Í.P.A. Santos, A.E. Furtini Neto, A.R. Morais, E.E. Mesquita, D.J.G. Faria and I.F. Andrade
Animal Science Department, Universidade Federal de Lavras, Lavras, Minas Gerais, Brazil, Email: josecard@ufla.br

Keywords: *Cynodon dactylon*, phosphate of Arad, phosphate of Araxa, triple superphosphate

Introduction Brazil presents high potential for meat production from pastures. However, the feeding of ruminants depends on the conditions and the climate. Approximately 80% of the annual production of dry matter (DM) occurs in the period October to March (spring - summer). In the autumn and winter production is low associated with high humidity and low temperatures in the south and low rainfall in the tropical north. The situation is exacerbated by inadequate management practices and low soil fertility, particularly low levels of phosphorus (P). The objective of this experiment was to evaluate the seasonality of production of DM of coastcross-1 (*Cynodon dactylon*) with different sources and applications of P for two consecutive years.

Materials and methods The field experiment was conducted from August 2000 to October 2003 in an area of the Universidade Federal de Lavras, Lavras, Minas Gerais, Brazil at 21°14 ' S and 40°00 ' W, with an average altitude of 919 m, characterised by a climate Cwb, with two very defined seasons; a wet one, from October to March, and a dry one, from April to September. Mean temperature and precipitation are 19.4°C and 1,526.7 mm, respectively (Brasil, 1992). The experimental design was a complete randomised block with split plots. The main plots received different sources of P [Triple superphosphate (TS); Reactive phosphate (RP), Arad, and Natural phosphate (NP), Araxa] and the subplots received different applications of P (0, 40, 80 and 120 kg/ha of P_2O_5). Dry matter production was measured for two consecutive years with cuts every 35 days in the wet season and one cut at the end of the dry season, after about 200 days of poor growth.

Results The DM production in the wet season increased linearly with increase in P application rate ($Y = 12.1 + 0.010X$; $R^2 = 0.74$; $p<0.01$). However, DM production in the dry season and total for the year were affected by the interaction between source and application rate of P ($p<0.01$) (Figure 1 a and b). There was a linear increase in DM production in the dry season for the plants fertilised with NP ($Y = 4.3 + 0.0067X$; $R^2 = 0.53$; $p<0.05$) whereas there were quadratic relationships with RP ($Y = 3.9 + 0.0346X - 0.00015X^2$; $R^2 = 0.71$; $p<0.05$) and TS ($Y = 4.1 - 0.020X + 0.0003X^2$; $R^2 = 0.99$; $p<0.001$). Annual DM production increased by 30.8 and 19.2 kg/ha for each kg of P_2O_5 supplied by RP ($Y = 16.2 + 0.031X$; $R^2 = 0.54$; $p<0.05$) and TS ($Y = 15.4 + 0.019X$; $R^2 = 0.92$; $p<0.001$), respectively. There was a quadratic relationship for NP ($Y = 16.0 + 0.062X - 0.00042X^2$; $R^2 = 0.99$; $p<0.001$). The maximum production of 18.25 t/ha was obtained with 73.9 kg/ha of P_2O_5 supplied by NP (Figure 1b). Averaged across treatments in the wet and dry seasons were 12.7 and 4.7 t/ha, representing 73% of the annual production in the wet season and 27% in the dry season.

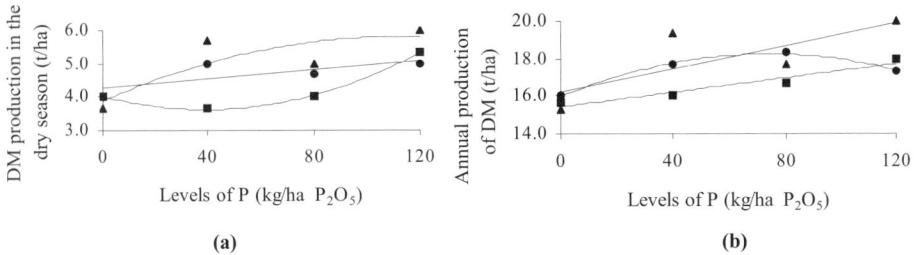

(a)   (b)

Figure 1 Dry matter production in the dry season (a) and the complete year (b) as a function of the sources [TS (■), RP (▲) and NP (●)] and application rates of P

Conclusions Phosphorus fertilisation increased DM production by coastcross-1 independent of the season and the reactive phosphate (Arad) gave higher production. Some 70-80% of the production occurred in the wet season.

Acknowledgement Research supported by FAPEMIG

Reference
Brasil, Ministério da Agricultura e Reforma Agrária (1992). Normais climatológicas 1961-1990. MARA, Brasília, 84pp.

Levels of concentrate for grazing Nelore crossbred steers in the dry period of the year in Brazil

I.F. Andrade, A.A.F. Baião, E.A.M. Baião and J.C. Pinto
Universidade Federal de Lavras, UFLA, Caixa Postal, 37- 37200-000 Lavras, MG- Brazil, Email: iandrade@ufla.br

Keywords: pasture supplementation, *Brachiaria brizantha*, dry season, protein supplementation, daily gain

Introduction In Brazil, the fattening of supplemented grazing steers accounts for many finished animals in the dry season. This work had the objective of evaluating the effect of levels of concentrate on the fattening of crossbred Nelore steers grazing *Brachiaria brizantha* pasture in the dry period of the year.

Materials and methods Twenty four crossbred uncastrated Nelore male cattle, averaging 30 months of age and weight of 295 kg were used. During the experimental period, the animals grazed a *Brachiaria brizantha* pasture of 11.50 ha with average forage dry matter (DM) availability of 7, 635 and 3,495 kg/ha at the beginning and the end of the experiment, respectively. The treatments were composed of increasing levels of concentrate fed as a percentage of cattle body live weight (LW) to each steer: T1 - 0; T2 - 0.4; T3 - 0.8; T4 - 1.2%. The experimental design was a randomised complete block with six replicates. The initial LW was the blocking factor. The concentrate was made of coffee husk 44%, soybean meal 10%, cotton meal 3%, wheat meal 15%, and sorghum panicles 28%. The experiment was conducted at the Animal Science Department, UFLA, from June to September, 2001, lasting 114 d. The animals were weighed every week without fasting and *Brachiaria* pasture was sampled every 30 d. Pasture samples were analysed for DM, crude protein, ash, NDF, and ADF. The software SISVAR was used for statistical analysis (Variance Analysis System of Balanced Data).

Results and discussion The steers rejected some of the concentrate and did not consume the intended amounts. The real levels of intake were, T1-0; T2-0.37; T3-0.76 and T4-0.87%. The average final live weight was 319 kg/steer and the daily gain was 0.277 kg/steer. There was a significant effect of the concentrate level on daily live weight gain (DLWG) with a linear relationship ($P < 0.05$) (Figure 1).

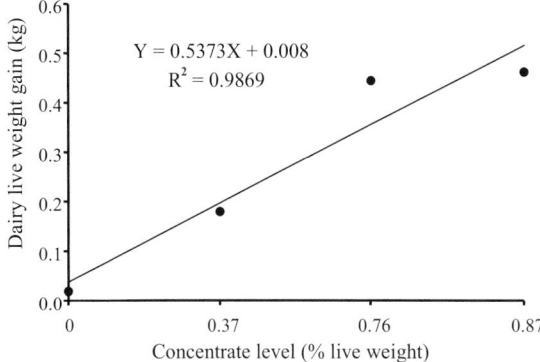

$$Y = 0.5373X + 0.008$$
$$R^2 = 0.9869$$

Figure 1 The effect of concentrate levels on daily live weight gain

Conclusions It was concluded that finishing Nelore steers grazing *Brachiária brizantha* and fed concentrate during the dry season, is technically and economically viable. The highest LWG was at the level of 0.87%, but the better bio-economic performance was with the intake of 0.76% of concentrate.

Acknowledgement Research Project sponsored by: FAPEMIG – Fundação de Amparo à Pesquisa do Estado de Minas Gerais

Reference
Bomfim, M.A.D, C.A.P. Rezende, P.C.A. Paiva, I.F. Andrade, J.A. Muniz & A.R.P. Silva (2001). Níveis de concentrado na terminação de novilhos holandês x zebu suplementados a pasto na estação seca. *Ciência e Agrotecnologia*, 25, 1457-1466.

Comparing yield and quality of milk from dairy cows fed stockpiled annual ryegrass (*Lolium multiflorum* L.) and cereal rye (*Secale cereale* L.)

L.E. Meinhardt and R.L. Kallenbach
Plant Sciences Unit, University of Missouri, 210 Waters Hall, Columbia, Missouri 65211, USA, Email: lemadf@mizzou.edu

Keywords: annual ryegrass, cereal rye, conjugated linoleic acid, dairy cow

Introduction Stockpiling annual ryegrass and cereal rye provides a low cost substitute to hay and creates an excellent source of feed during winter (Kallenbach *et al.*, 2003). In addition to lowering feed costs, grazing increases the conjugated linoleic acid (CLA) content of milk compared to feeding hay. Previous research suggested that forage species might differ in their ability to alter milk CLA content during the growing season (Wu *et al.*, 1997). However, research is needed to determine if different forage species used for winter and early spring grazing impacts the CLA content of milk. The objective of this experiment was to compare yield and quality of milk when cows graze annual ryegrass or cereal rye in late winter and early spring.

Materials and methods Thirty dairy cows in early lactation grazed for 14 weeks from 3 March to 3 June 2004 on two forage treatments (annual ryegrass and cereal rye). Treatments were replicated three times in six, 1.6 ha pastures that were subdivided into sixteen paddocks of equal size. Animals were rotated between paddocks on a 12-h grazing schedule. Additional cows were added as needed to maintain equal forage availability between treatments. In addition to forage, cows received 6.35 kg of grain supplement per day. Available forage was estimated weekly from each paddock using a rising-plate meter, which was calibrated every 21 d by clipping ten, 0.81 x 4.57 m strips from pre- and post-grazed paddocks. The mass from the harvested strips was used in a multiple regression equation to estimate available forage (Sanderson *et al.*, 2001). Forage and grain supplements were sampled weekly for quality analyses. Milk yield and quality was monitored on a weekly basis. Milk quality was measured from composite milk samples collected at consecutive a.m. and p.m. milkings. The composite samples were weighted based on milk yield of individual cows.

Results Cereal rye produced forage earlier in the season and sustained nearly 33% more animals than annual ryegrass from March to May (Figure 1). Annual ryegrass growth in early spring was slower and thus a two-week rest period in early April was needed to allow it to recover. However, over the entire season, annual ryegrass provided 76 days of grazing, which was 14 days more than cereal rye. There were no significant differences (*P* >0.05) among treatments for pasture intake or yield of milk, milk fat, or milk protein. Milk yield, fat, and protein for annual ryegrass and cereal rye was 30.0 and 30.8 kg, 4.04 and 4.04 g/kg, and 3.17 and 3.26 g/kg for treatments, respectively. However, averaged over the 14-week experiment, annual ryegrass contained 10.4 mg CLA/g of fat, 20% more CLA than milk from cows grazing cereal rye (*P* <0.02) (Figure 2).

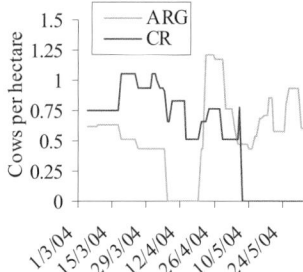

Figure 1 Stocking rate (454 kg cow basis) for annual ryegrass (ARG) and cereal rye (CR) pastures

Conclusions Our results show that cereal rye and annual ryegrass are practical forages for late winter/early spring on pasture-based dairies. While cereal rye was better for late winter, annual ryegrass persisted longer into spring. This study also stresses the importance of further research on CLA content of milk when different forage species are used.

References

Kallenbach, R.L., G.J. Bishop-Hurley, M.D. Massie & C.A. Roberts (2003). Stockpiled annual ryegrass for winter forage in the lower Midwestern USA. *Crop Science*, 43,1414-1419.

Sanderson, M.A, C.A. Rotz, S.W. Fultz & E.B. Rayburn (2001). Estimating forage mass with a commercial capacitance meter, rising plate meter, and pasture ruler. *Agronomy Journal*, 93, 1281-1286.

Wu, Z., L.D. Satter, V.R. Kanneganti & M.W. Pariza (1997). Paddock containing red clover compared to all grass paddocks support high CLA levels in milk. US Dairy Forage Research Center Research Summary. Univ. of Wisconsin, Madison, WI, 94 pp.

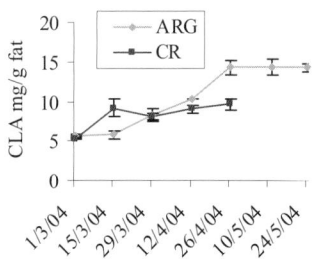

Figure 2 CLA content of milk from cows grazing annual ryegrass (ARG) and cereal rye (CR) pastures

A grazing method to solve the lack of pastures in the dry season of tropical areas with long periods of drought

R.O. Martinez, R. Tuero and M.F. Díaz
Instituto de Ciencia Animal. Apartado 24 San José de las Lajas La Habana Cuba, Email: romartinez@ica.co.cu

Keywords: grazing, dry season, milk yield, *Pennisetum purpureum*

Introduction   Rotational grazing systems used in tropical areas in Latin America do not solve the great difference in pasture availability between the dry and the rainy season. The main studies on rational grazing (Voisin, 1963) were performed in temperate areas where the deficit of feeds in winter may only be solved with external feeds such as forages and silages produced out of the grazing system. The objective of this work was to demonstrate that it is possible to maintain pasture availability throughout the year with the use of a *Pennisetum purpureum* clone (Cuba CT-115) adapted to grazing (Martínez *et al.*, 1995), in spite of the dry season.

Materials and methods   A dairy unit with 130 cows of the Siboney de Cuba breed (5/8 Holstein, 3/8 Zebu) was used for eight years. The cows grazed 60 ha divided into 80 plots. The grazing system was rational, according to Voisin (1963), with two days of occupation on average. The milking cows grazed first and the dry cows afterwards (leaders and followers). The system was organised with the pastures Cuba CT-115 (*P. purpureum*) and star grass (*Cynodon nlemfuensis*). The principal variable was the progressive increase in the area of CT-115, up to 30 % of the total area. In the rainy season, the area with CT-115 was separated to store feed for the dry season. The system was not replicated, but was of sufficient scale to be representative of farm conditions. Measurements of yield and quality of the pastures and animal production were made. Neither irrigation nor chemical fertilisers were used. The cows were milked twice and grazed 18 h each day.

Results   It was found that the gradual increase of the grazing area with CT-115 up to 30 % of the total area (18 ha) allowed the production and storage of the necessary biomass to balance the needs in the dry season. There were three rotations on the pasture Cuba CT-115 in the dry season (15 Nov.- 15 May) and two in the rainy season (15 May- 15 Nov.). With star grass, there were two rotations in the dry season and four in the rainy season. One hectare of Cuba CT-115 fed 700 cow days in the 180 d of the dry season (total of 12600 for the 18 ha), while star grass, on the rest of the area, produced 285 cow days/ha (total of 12000 for the 42 ha). In the rainy season, the biomass provided by star grass was higher (550 cow days/ha), compared to Cuba CT-115 (350 cow days/ha). The daily intake of pasture dry matter (calculated using the chromic oxide technique) was 10 to 12 kg /cow. Table 1 presents the increase in milk yield in the dairy unit over time. Figure 1 illustrates the stability in monthly milk yield (average for three years) and shows little influence of the rainy season.

Table1 Increase in performance over time (milk in thousands of litres)

Per year	1995	1997 (1)	1999	2001 (2)	2003
Total milk	78.9	165.5	155.4	191.9	232.9
L/cow per day	3.27	6.7	6.40	7.55	8.18
Birth rate (%)	52.1	90.2	94.5	87.8	94.8
Milk/ha	1314	2747	2590	3198	3882
CT-115, ha	5.7	12.0	18	20	20

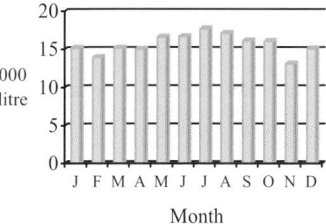

Figure 1 Monthly milk production

(1) Start artificial rearing, (2) start 30% CT -115

Conclusions   It is concluded that with adequate management of each pasture in every season and the combination of species of long and short growth cycles, standing feed could be stored and enough regrowth produced to keep 2.3 cows/ha in grazing throughout the year, without bringing feed from outside the grazing system, despite the 180 d of low rainfall (200 mm from 15 Nov. to 15 May).

References

Voisin, A. (1963). Productividad de la Hierba. Editorial Tecnos. S.A. Madrid.
Martinez, R.O., R.S. Herrera, R. Cruz & V. Torres (1995). Tissue culture and mutation breeding in tropical pastures. *P. purpureum*:another example for obtaining new clones. *Cuban Journal of Agricultural Science*, 30, 1-11.

Seasonal herbage accumulation of different dairy pasture types in southern Australia

J. Tharmaraj[1], D.F. Chapman[1] and Z.N. Nie[2]
[1]The University of Melbourne, Parkville, Victoria, Australia 3010, Email: jthar@unimelb.edu.au, [2]Department of Primary Industries, Private Bag 105, Hamilton, Victoria, Australia 3300

Keywords: pasture types, herbage accumulation, seasonal feed supply, dairy systems

Introduction Perennial ryegrass pastures, which are the mainstay of dairy feeding systems in southern Australia, are characterised by strong spring growth, little summer/autumn growth, and poor persistence. These limitations impose costs to farm businesses through the purchase of additional fodder to fill feed gaps, and regular re-sowing of pastures. The objective of the research reported here was to investigate the potential for alternative pasture types with different seasonal growth characteristics to improve the seasonal distribution of feed supply and overcome some of the limitations associated with perennial ryegrass.

Materials and methods The experiment was conducted over 3 years at 3 sites in southwest Victoria, Australia: DemoDairy, near Terang (38°14′S, 142°54′E, mean annual rainfall 740 mm), Heytesbury (38°57′S, 142°92′E, 1000 mm), and Naringal (38°40′S, 142°72′E, 840 mm). Five pasture types were established in 3 replicate blocks of 0.1 ha plots in April 2001. The pasture types were: 1) short-term winter active (STW), based on Italian ryegrass; 2) long-term winter active (LTW), based on Mediterranean tall fescue types; 3) long-term summer active (LTS), based on Continental tall fescue types plus chicory and red clover; 4) perennial ryegrass pastures with high N inputs (210 kg N / ha / year, compared to 90 kg in all other treatments) (RHN); and 5) perennial ryegrass (Control). Plots were grazed 10 – 12 times per year depending on site and season, and herbage accumulation (HA) was estimated by difference between post-grazing pasture mass and the subsequent pre-grazing pasture mass measured using a calibrated rising plate meter.

Results HA was greater in spring than in any other season for all pasture types (Table 1). However, there was a significant season x pasture type interaction in HA, due to greater HA in the LTS treatment compared to all other treatments in summer and lower HA from the LTS treatment compared to all other treatments in winter. The LTS pasture produced 25 – 30% of total annual HA in the summer months across all 3 sites, whereas the control treatment consistently produced only 15 – 20% of total annual HA in summer (Figure 1). Chapman & Kenny (2005) estimated that each additional kg DM / ha grown and grazed *in situ* in summer is worth an extra $0.24 operating profit (before the cost of growing the extra feed is included), while the equivalent value for extra feed in winter is $0.14. Thus, the additional 1310 kg DM / ha grown in summer in LTS compared to control (Table 1) is potentially worth an additional $314 / ha in operating profit, while the loss of 840 kg DM / ha in winter for LTS means $118 / ha less profit. On balance, LTS pastures could increase operating profit by nearly $200 / ha compared to perennial ryegrass assuming equal efficiency of growing and using both pastures. RHN grew extra feed in spring (Table 1) which has relatively low economic value (Chapman & Kenny 2005).

Table 1 Effects of pasture type on seasonal HA (mean of all years across the three sites)

	Pasture type				
	STW	LTW	LTS	RHN	Con
			(t DM / ha)		
Autumn	2.51c	2.46b	2.67c	2.66c	2.45c
Winter	3.69b	3.68a	2.80c	3.75b	3.64b
Spring	4.14a	3.92a	4.43a	5.37a	4.53a
Summer	2.11d	2.11c	3.86b	2.62c	2.55c
Total	12.45	12.17	13.76	14.40	13.17

s.e.m. season = 0.054 (P<0.01)
s.e.m. season x pasture type = 0.131 (P<0.01)
Letters apply to means within columns

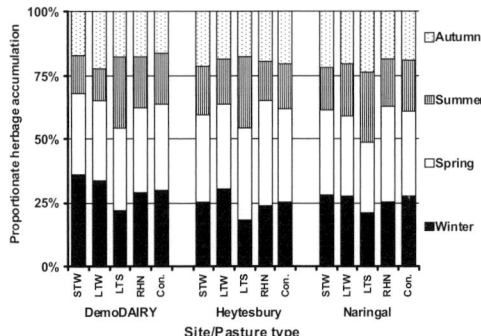

Figure 1 Proportion of total annual HA grown per season

Conclusions A more-even seasonal pattern of feed supply can be achieved using alternatives to perennial ryegrass. Gains appear possible in summer through use of tall fescue-based pastures, and these should translate into worthwhile economic returns for farm businesses.

References

Chapman, D.F. & S. Kenny (2005). Alternative feedbase systems for southern Australia dairy farms. 3. Economic returns from extra dry matter consumption. *XX International Grassland Congress – offered papers* (in press).

Alternative feedbase systems for southern Australia dairy farms: 1. Predicted pasture/crop consumption and farm financial performance

D.F. Chapman[1], S. Kenny[1] and D. Beca[2]
[1]*The University of Melbourne, Parkville, Victoria, Australia 3010, Email: d.chapman@unimelb.edu.au,*
[2]*BecaZuur Consulting Pty, 16 Grange Road, Warrnambool, Victoria, Australia 3280*

Keywords: dry matter consumption, pasture types, fodder crops, dairy systems, modelling

Introduction Traditional perennial ryegrass-based pastures have significant limitations for efficient feeding of dairy cattle in dryland dairy regions of southern Australia. These include strong seasonality of growth, with 50 – 60% of total annual dry matter arriving in spring and little or no growth during summer. There is clear potential for improving total forage production and the seasonality of forage supply in these regions through the use of alternative pastures (Nie *et al.* 2004) and fodder crops. This series of papers applies a modelling approach to investigate the potential improvements in farm productivity and profitability resulting from their use.

Materials and methods Three analytical tools were used 'in series' to simulate non-irrigated, seasonal calving dairy farms 'typical' of the Terang district in southwest Victoria, Australia (38°14´S, 142°54´E). The pasture growth model DairyMod (Johnson *et al.* 2003) was parameterised using appropriate soil physical properties and plant physiology, and used to calculate long-term (1961 – 2000) mean monthly pasture harvest rates. These were then entered into the dairy farm system model UDDER which was parameterised to represent management policies typical of a farm in either the top 40% or top 10% of farms in the region using benchmark data. The farms were 125 ha, calving in autumn, and stocked at 1.8 or 2.25 cows / ha for the respective farms. Predicted total milk production (95,780 or 125,920 kg milksolids respectively) and the inputs required to produce this milk were analysed in 'Red Sky' to estimate the resulting farm operating profit and return on assets. 'Base' models (100% farm area in perennial ryegrass) were compared with: 1) a winter cereal crop grown for silage on 10% of farm area, 2) a summer brassica crop (turnips) grown on 10% of farm area, 3) combination of 2 with annual ryegrass sown on 10% of farm area, 4) combination of 1 and 2 (double cropping) on 15% of farm area, 5) oversowing annual ryegrass into the perennial ryegrass base, 6) combination of 4 and 5, and 7) summer shoulder pasture (tall fescue) sown on 100% of farm area. Harvest rate data were from DairyMod in all cases except for the summer crop where yields were estimated from local survey information. Feeding and other policies were adjusted in UDDER as necessary to efficiently use the additional feed available in each system.

Results Moving from the top 40% farm to the top 10% farm yielded the largest improvements in DM consumption (Table 1), reflecting higher base soil fertility and pasture growth of the latter which was utilised effectively by a higher stocking rate. Alternative feedbase systems resulted in 0.9 – 1.6 t DM/ha extra consumption for both farm types, and improved financial performance. Regression analysis using all simulations predicted that, for each additional tonne of DM consumed / ha, an additional $71 / ha or $102 / ha operating profit is available for the top 40% and top 10% farms respectively. Double cropping and summer shoulder pasture sat above this general relationship (Table 1) due to better seasonal distribution of forage supply.

Table 1 Model predictions for selected feedbase systems for 2 farm types

	DM consumed (t/ha)		Profit ($/ha)		Return on assets	
	Top 40%	*Top 10%*	*Top 40%*	*Top 10%*	*Top 40%*	*Top 10%*
Base ryegrass 10.0	6.7	8.6	750	1314	6.3	10.0
Double crop 11.4	8.0	10.2	930	1500	7.8	11.4
Ryegrass oversowing 10.4	7.7	9.5	836	1380	6.8	10.4
Summer shoulder pasture	7.6	9.6	985	1488	8.2	11.3

Conclusions The models successfully predicted system-level outcomes of changing the feedbase on southern Australian dairy farms. Alternative feedbase systems can improve forage consumption and financial return, but the biggest improvements will come from improved management and utilisation of the base pasture.

References

Johnson, I.R., D.F. Chapman, A.J. Parsons, R.J. Eckard, & W.J. Fulkerson (2003). DairyMod: A biophysical model of the Australian dairy system. In: *Proceedings of the Australian Farming Systems Conference,* Toowoomba, September 2003 (http://afsa.asn.au/)
Nie, Z.N., D.F. Chapman, J. Tharmaraj & R. Clements (2004). Effects of pasture species mixture, management and environment on the productivity and persistence of dairy pastures in south west Victoria. 1. Herbage accumulation and seasonal growth pattern. *Australian Journal of Agricultural Research,* 55, 625 - 636.

Alternative feedbase systems for southern Australia dairy farms. 2. Seasonal variability

S. Kenny[1], D.F. Chapman[1] and D. Beca[2]

[1]The University of Melbourne, Parkville, Victoria, Australia 3010, Email: d.chapman@unimelb.edu.au,
[2]BecaZuur Consulting Pty, 16 Grange Road, Warrnambool, Victoria, Australia 3280

Keywords: dry matter consumption, seasonal variability, dairy systems, modelling, risk

Introduction The standard feedbase on non-irrigated dairy farms in southern Australia is perennial ryegrass-dominant pasture supplemented by concentrate feeds, silage and hay to fill seasonal feed gaps. Using models, Chapman *et al.* (2005) concluded that dairy producers in this region can increase forage consumption and operating profit through the use of summer-active pastures and double-cropping (winter cereal grown for silage, followed by a summer grazing crop). However, these results were based on long-term average pasture and crop growth rates and therefore do not account for seasonal variability associated with climatic variation, which is important in southern Australia. This paper investigates the interaction between seasonal conditions and feedbase system to determine the potential risk associated with changing to alternative pastures or crops.

Materials and methods The general modelling approach is described by Chapman *et al.* (2005). The least 'reliable' seasons in southwest Victoria are autumn (timing of rain after the summer) and spring (timing of onset of moisture stress). Pasture growth outcomes were simulated in DairyMod for each year 1961 – 2001. Growth rates from years with poor autumns or springs were used to synthesise monthly harvest rates for years which included these seasonal outcomes alone, or together. The converse situations (early autumn break and long spring) were also modelled. Harvest rates were entered into UDDER and Red Sky (Chapman *et al.* 2005) to estimate milk production and return on assets. Analyses were conducted for top 40% and top 10% farms (Chapman *et al.* 2005), but only results for the first of these are presented here. Feedbase systems modelled were: Base (perennial ryegrass), double cropping (winter cereal followed by summer turnip crop), oversowing of annual ryegrass into the perennial ryegrass base, and summer shoulder pasture (based on tall fescue).

Results The growth of perennial ryegrass pastures was highly variable in time (Figure 1). Poor autumns (late break of season) occurred in about 20% of years, and poor springs (early end to season) in about 10% of years. 'Average' annual pasture growth curves were seldom seen and it was rare for two consecutive years to be alike.

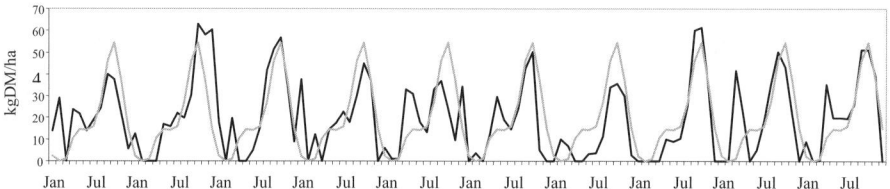

Figure 1 An example of predicted daily growth rates of perennial ryegrass pastures for 1991 – 2000 inclusive (dark, variable line), compared to the long-term mean 1961 - 2000 (lighter, regular line)

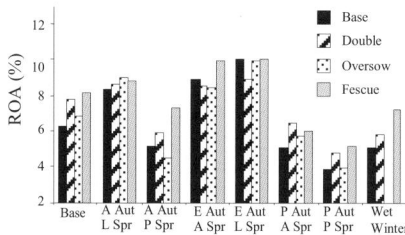

A = average; E = early; L = long; P = poor
Aut = autumn; Spr = spring

Figure 2 Predicted return on assets for different seasonal scenarios and feedbase systems

Reference

Chapman, D.F., S. Kenny & D. Beca (2005). Alternative feedbase systems for southern Australia dairy farms. 1. Predicted pasture/crop consumption and farm financial performance. *XX International Grassland Congress - offered papers* (in press).

Return on assets was adversely affected by poor autumns and springs for all feedbase systems due to increased purchased feed costs (Figure 2). However, fescue-based pasture and double cropping showed more consistent financial returns in the face of inter-annual climate variability (CV of return on assets 22%) compared to ryegrass-based systems (CV 31%). This is due to the deeper root systems of tall fescue compared to ryegrass, and security of yield of the winter crop (growing season April – October) insulating the double cropping system against the effects of dry seasons.

Conclusions All feedbase systems are subject to seasonal risk, but staying with ryegrass-based systems appears more risky than adopting alternatives such as tall fescue or crops

Alternative feedbase systems for southern Australia dairy farms. 3. Economic returns from extra dry matter consumption

D.F. Chapman and S. Kenny
The University of Melbourne, Parkville, Victoria, Australia 3010, Email: d.chapman@unimelb.edu.au

Keywords: dry matter consumption, feed gaps, dairy systems, modelling

Introduction Growth rates of the 'traditional' perennial ryegrass pasture frequently fail to meet the seasonal feed requirements of herds in non-irrigated dairy systems in southern Australia, leading to a dependence upon additional feed at these times of the year. Farmers commonly purchase this feed off-farm, which can be costly. Growing extra feed on-farm may be more cost effective but will require additional inputs such as N fertiliser and alternative pastures/crops. The gross return to dairy farms of growing extra feed at certain times of the year can be estimated by connecting biophysical models of pasture growth to farm systems models and financial analysis tools (e.g. Chapman *et al.* 2005). Farmers can then measure additional costs incurred in growing this feed against the margin available to help decide on cost-effective feeding strategies. This paper reports such an analysis for non-irrigated dairy farms in southwest Victoria using different calving policies.

Materials and methods The general modelling approach is described by Chapman *et al.* (2005). In this analysis, only the top 40% farm type with 100% perennial ryegrass pasture was used. Estimated pasture dry matter consumption through *in situ* grazing on this farm was 6.0 t / ha / year, with an additional 0.6 t DM / ha / year conserved as silage. We then asked the question: if an additional 10% of the total DM consumed by grazing on this 'base' farm (i.e. an additional 0.6 t DM / ha / year) was available for *in situ* grazing, what would be the gross economic return to the farm, and how does this return differ if the 0.6 t DM is spread evenly throughout the year versus confined to each of the 4 seasons of the year? Adjusted harvest rates were entered into the base UDDER simulation, and farm management policies such as supplementary feeding, N fertiliser use, and silage and hay conservation adjusted to utilise all of the extra feed available, either through direct grazing or via fodder conservation. This allowed us to determine the time of year when additional home-grown feed is most valuable in an economic sense. Since this may depend on calving pattern and the associated seasonal pattern of feed demand, we compared a seasonal calving herd (calving over 60 days from 10th May) with a split calving herd (60% of herd start calving on 1st March and 40% on 1st August) at the same stocking rate (1.8 cows / ha).

Results When expressed in $ per extra kg DM / ha grown, additional feed consumed in summer and autumn was worth more than additional feed consumed in any other season for both seasonal and split calving herds (Table 1). This reflects low pasture growth rates in these seasons, and heavy dependence on purchased feed. The seasonal conditions. For example, an

Table 1 Impact of 10% extra feed consumed on gross farm return

	Total DM consumed	RoA (%)	Seasonal calving Change in op profit $	$/extra kgDM/ha	Split calving $/extra kgDM/ha
Base	6014	6.3			
All year	6615	6.9	9,585	0.13	0.21
Autumn	6616	7.4	16,325	0.22	0.36
Winter	6616	7.0	10,440	0.14	0.21
Spring	6615	6.9	8,800	0.12	0.15
Summer	6616	7.5	18,290	0.24	0.30

additional kg DM grown with N fertiliser at current prices and assuming a response efficiency of 10 kg DM per 1 kg N applied would cost an estimated 8.7 cents. This would be highly profitable after an early autumn break, but marginally profitable in spring for a seasonal calving dairy herd.

Conclusion The timing of feed supply is critical to dairy farm profitability. Growing extra feed when it is easiest to produce (i.e. spring) may yield only limited improvements in farm financial return.

References

Chapman, D.F., S. Kenny & D. Beca (2005). Alternative feedbase systems for southern Australia dairy farms. 1. Predicted pasture/crop consumption and farm financial performance. *XX International Grassland Congress- offered papers* (in press).

Year-round forage systems for beef cows and calves

J.P. Fontenot[1], W.M. Clapham[2], W.S. Swecker[1], Jr., D. Fiske[1], J.B. Hall[1], J. Fike[1] and G. Scaglia[1]
[1]*Virginia Polytechnic Institute and State University, Blacksburg, VA 24061, Email: cajunjoe@vt.edu,* [2]*USDA-Agricultural Research Service, Beaver, WV 25813, USA,*

Keywords: beef cows, forages, systems

Introduction Beef cow systems in the USA are based on forages with little or no concentrates fed. Tall fescue *(Festuca arundinacea* Schreb. L.*)* is one of the important pasture forages in the lower Northeast and upper South (Allen *et al.*, 2001). Limited research has been conducted on year-round all forage systems based on cool season forages. Stockpiling tall fescue in late summer-early fall provides good quality forage that is usually grazed rather than harvested. Forage systems including tall fescue and clover *(Trifolium repens* L.*)* produced excellent performance in beef cows and calves, with minimum inputs (Allen *et al.*, 2001). The present experiment is a component of a larger initiative, Pasture-based Forage Systems for Appalachia. The specific objective of this experiment is to evaluate different forage systems for beef cows and calves.

Materials and methods Initially, 108 Angus crossbred cows were allotted to six forage systems with three replicates. The cows remained in the same systems thereafter, and received only the forage produced in the systems. Tall fescue was a component (45%) of each system. The systems were: 1) fescue-fescue/clover, three paddock system, 0.9 ha/cow; 2) same as system 1 except 0.7 ha/cow; 3) same as system 2 except rotationally grazed; 4) fescue, fescue/clover, alfalfa *(Medicago sativa* L.*)*-orchardgrass *(Dactylis glomerata* L.*),* rotationally grazed, 0.7 ha/cow; 5) fescue, fescue/clover, switchgrass *(Panicum virgatum* L.*)*, three paddock system, 0.7 ha/cow; 6) fescue, fescue/birdsfoot trefoil *(Lotus corniculatus* L.*)*, fescue/lespedeza *(Michx. cuneata* (Dumont) G. Don*)*, three paddock system, 0.7 ha/cow. The three-paddock system was as described by Blaser *et al.* (1977). For all systems fescue was stockpiled in late summer-early fall. In the spring cows were oestrous synchronised, bred artificially, then exposed to bulls. Calves were creep grazed in all systems. Forages were sampled during the growing season to determine chemical composition and estimate forage mass.

Results Performance of cows and calves for the first two years is given in Table 1. Pregnancy rate was lower for year 1 (2001-02) than year 2. This is attributable to a serious drought and shortage of high quality forage. Ample forage was available in year 2, and is reflected in pregnancy rates of 89-100%. In year 1 pregnancy rate was lowest (P < .05) for system 1, and highest (P < .05) for system 3. There is no explanation for the low conception rate for system 1 in year 1. In year 2, pregnancy rates were not significantly different. Weaning weights of the cattle were satisfactory for all systems and differences in years 1 and 2 were not significant (P > .05). Crude protein content of the forages usually met or exceeded the levels required for beef cows (NRC, 1996). Fibre components (ADF, NDF) indicated that the available energy content of the forages was sufficient for the cows and calves.

Table 1 Performance of cows and calves

	Forage system					
	1	2	3	4	5	6
Pregnancy rate, %						
2001-02	56	78	89	67	78	67
2002-03	94	100	94	100	89	94
Calf weaning weight, 222 days, kg						
2001-02	238	247	235	237	230	236
2002-03	250	268	246	232	231	252

Conclusions The results indicate that the fescue based systems were satisfactory for beef cows and calves. The choice of system by beef producers will depend on resources available on the farm, and cost of establishing and maintaining the forages.

References

Allen, V.G., J.P. Fontenot, D.R. Notter & R.C. Hammes, Jr. (1992). Forage systems for beef production from conception to slaughter: I. Cow-calf production. *Journal of Animal Science,* 70, 576-587.
Blaser, R.E., R.C. Hammes, Jr., J.P. Fontenot & H.T. Bryant (1977). Forage-animal systems for economic calf production. *Proceedings of the XIII International Grasslands Congress.* Leipzig, FRG, 2, 1541.
NRC (1996). Nutrient Requirements of Beef Cattle. National Academic Press, Washington, D.C.

The use of forage supplements to overcome seasonal shortages of grazed herbage in dairy production systems

A. Hameleers and D.J. Roberts
*SAC, Sustainable Livestock Systems Group, Crichton Royal Farm, Dumfries, UK, *Present address: The Agricultural Research Institute of Northern Ireland, Hillsborough, Co. Down BT26 6DR, UK, Email: ahameleers@yahoo.com*

Keywords: buffer feeding, forage supplementation, dairy cows

Introduction In most dairy production systems, grazed herbage is potentially the cheapest forage resource. However, while the availability is affected by seasonal changes in herbage growth, and/or between year variations in climatic conditions, the requirements of dairy production systems tend to remain constant throughout the season. This paper summarises five experiments that examined the effect of the characteristics of the forage supplement (dry matter, DM; metabolisable energy, ME, content; Type) and the effect of stage of lactation and access method to the supplement on animal performance.

Materials and methods Five experiments were carried at Crichton Royal Farm, Dumfries, UK, between 1992-1997 and involved between 20 to 48 Holstein Friesian dairy cows per experiment, using a continuous design and experimental periods of 4-5 weeks (see Table 1 for treatments). Animals grazed predominantly perennial ryegrass swards (*Lolium perenne*), using a continuous grazing system, with a sward surface heights of between 6 and 12 cm. Animals were milked twice daily. In four studies, the animals had access to the forage supplements for one hour after each milking in a feed passage in separate treatment groups. In the fifth experiment the animals had access to the forage supplement either twice daily after milking, continuously in the grazing area, or were housed overnight with access to the forage supplement. Animal performance was recorded and herbage and forage supplement intake were estimated using the *n-alkane* technique.

Results As shown in Table 1 increasing DM-content of the forage supplement from 30% to 80 % did not affect supplement or total forage intake, while increasing the ME content of the forage supplement from 8.4 to 10.4 MJ/kg DM increased supplement intake, although total forage DM intake and corresponding production responses were unaffected. Both early- and late-lactation animals consumed the same amount of forage supplement but total DM intake was different. Comparing precision-chopped grass silage with big-bale grass silage and maize silage showed no significant differences in supplement intake, total forage intake and production. When comparing access method, it was found that twice-daily access resulted in the lowest supplement intake and that overnight access resulted in the highest forage supplement intake. However, no significant differences in terms of overall animal performance were observed as a result of access method.

Table 1 Effects of different supplement treatments on intake and milk yield

	Treatments			SED
DM of straw mixture	303 g/kg	541 g/kg	798 g/kg	
Milk yield (kg/d)	21.7	22.3	23.2	1.23
Total intake (kgDM/d) *	12.5 (4.5)	15.0 (4.7)	14.4 (5.5)	1.15 (0.679)
Stage of lactation (days after calving)	Early (71.8 ± 3.95)	Late (218.5 ± 17.1)		
Milk yield (kg/d)	26	18.2		0.58**
Total intake (kgDM/d) *	15.5 (3.7)	12.9 (3.6)		0.67** (0.533)
ME content supplement	10.4 MJ/kgDM	8.4 MJ/kgDM		
Milk yield (kg/d	25.0	23.3		1.17
Total intake (kgDM/d) *	16.9 (5.3)	16.7 (2.3)		1.01 (0.511)
Supplement type	Maize silage	Big bale silage	Precision-chopped silage	
Milk yield (kg/d)	33.3	32.8	34.1	2.37
Total intake (kgDM/d) *	19.7 (1.3)	19.7 (1.8)	22.2 (2.3)	1.31 (0.57)
Access Method	Twice daily	Overnight	In field	
Milk yield (kg/d)	36.1	37.4	36.6	1.03
Total intake (kgDM/d) *	21.4 (1.6)	22.4 (5.4)	22.5 (2.4)	1.54 (0.671)

* Value between brackets represents forage supplement intake ** p<0.01

Conclusions Forage supplementation can be used to overcome seasonal shortages of forage supply and the characteristics of the forage supplement or access method can be used to manipulate forage supplement intake, but this does not result in differences in milk yield.

An agronomic evaluation of grazing maize combined with companion crops for sheep in northwestern KwaZulu-Natal, South Africa

C.S. Dannhauser[1] and E.A. van Zyl[2]
[1]School of Agriculture & Environmental Science, University of the North, Private Bag X1106, Sovenga, 0727, South Africa, Email: chrisd@unorth.ac.za, [2]Dundee Research Station, PO Box 626, Dundee, 3000 South Africa

Keywords: grazing maize, companion crops, winter feed

Introduction Northwestern KwaZulu-Natal (KZN), in South Africa, is well known for its sheep production from natural rangeland in summer (October to May). During winter however, the nutritional value of the rangeland cannot maintain young growing sheep or pregnant and lactating ewes. With this in mind Lyle (1991) suggested the use of planted pastures for the winter. Crichton, Gertenbach & Henning (1998) and Esterhuizen & Niemand (1989) suggested maize crop residues for both cattle and sheep during winter, whereas Moore (1997) evaluated grazing maize (not harvested) for this purpose. He found that the protein content of the crop was inadequate and for this reason, protein rich companion crops were evaluated in this study.

Materials and methods Maize was intercropped with 14 different crops for three consecutive seasons, on the Dundee Research Station, Dundee, KZN, South Africa. Ten of the crops are given in Table 1 (see the conclusions for the others). Yellow maize was planted in blocks of 60m x 40m in a tramline layout. Rows alternated with a spacing of 90cm and 270cm. The companion crops were planted by hand between the maize rows on two different planting dates: PD1 during late January and PD2 during late February. A randomised plot design with two replications was used. The first season was used for technique evaluation and in the next two seasons maize grain yield, total dry matter (DM) production of companion crops and nutritional values were measured.

Results The rainfall for the three different seasons (July–June) was 575, 620 and 964 mm respectively (long-term average is 782.9mm). In the second season the maize grain production amounted to 3.9t/ha and in the third season to 4.6t/ha. The DM production of the companion crops and their nutritional value are given in Table 1.

Table 1 The average DM production (t/ha) and nutritional value of the companion crops

Companion crop	DM Production (t/ha)				Nutritional value
	Season 2		Season 3		
	PD 1	PD2	PD 1	PD2	CP (%)
Raphanus sativus	4.40a	1.74 bc	2.18a	1.20 b	20.16
Avena sativa	3.03 b	3.27a	0.75 b	1.14 b	4.23
Pennisetum glaucum	2.35 bc	2.49ab	0.41 b	0.63 bc	10.00
Ornithopus sativus	2.39 bc	1.71 bc	1.94a	2.04a	12.23
Vicia dasycarpa	1.65 cde	2.04 bc	1.92a	1.21 b	18.05.
Secale cereale	1.95 c	1.11 bc	0.18 b	1.18 b	13.43
Triticale hexaploide	1.48 cde	0.65 bc	0.00 b	0.21 c	15.04
Lablab purpureus	0.72 ef	1.46 bc	0.66 b	0.14 c	14.65
Glycine max	0.17 f	0 c	0.13 b	0.00 c	15.06

Figures with the same roman letters do not differ significantly (P≤0.05) for DM production

Conclusions From a DM production point of view *Ornithopus sativus* and *Raphanus sativus* can be recommended for their high production and high nutritional value (CP). *Vicia dasycarpa* had a high nutritional value. The DM production of *Pennisetum glaucum* and *Avena sativa* was relatively high, but the nutritional value was marginal. *Glycine max* and *Lablab purpureus* showed a high nutritional value, but DM production was low. [*Lolium multiflorum, Sorghum, Trifolium vesiculosum, Eragrostis teff* and *Bromus wildenowii* (not mentioned in Table 1) cannot be recommended].

References

Crichton, J. S., W. D. Gertenbach & P. W. van H. Henning (1998). The utilization of maize-crop residues for over wintering livestock (1). Performance of pregnant beef cows as affected by stocking rate. *South African Journal of Animal Science,* 28 (1), 9-15.
Esterhuizen, C.D. & S.D. Niemand (1989). Oesreste van ses kontantgewasse vir die oorwintering van skape. Inligtingsdag, Nooitgedacht-Navorsingstasie, Ermelo. Bl.85-107.
Lyle, A.D. (1991). The use of supplementary licks for sheep on summer and winter Sourveld. Sheep in Natal. Co-ordinated Extension Committee of Natal (5.2.1991). KwaZulu-Natal Department of Agricultural, Private Bag X9059.Pitermaritzburg. 3200.
Moore, A. (1997). Proewe wys lam-afronding op mielies werk. *Landbouweekblad,* 7 Maart 1997:10 -12.

An evaluation of grazing value of maize and companion crops for wintering lactating ewes

E.A. van Zyl[1] and C.S. Dannhauser[2]
[1]*Dundee Research Station, PO Box 626, Dundee, 3000 South Africa, Email: vanzyle@dunrs.kzntl.gov.za,*
[2]*School of Agriculture & Environmental Science, University of the North, Private Bag X1106, Sovenga, 0727, South Africa*

Keywords: grazing maize, companion crops, winter-feed, Merino sheep

Introduction Northwestern KwaZulu-Natal (KZN), in South Africa, is well known for its sheep production from natural rangeland in summer (October to May). During winter, however, the nutritional value of the rangeland cannot maintain young growing sheep or pregnant and lactating ewes. With this in mind, Lyle (1991) suggested the use of planted pastures for the winter. Crichton *et al* (1998) and Esterhuizen & Niemand (1989) suggested the use of maize crop residues for both cattle and sheep during winter, whereas Moore (1997) evaluated grazing (not harvested) maize for this purpose. He found that the crude protein content of the crop was inadequate and for this protein-rich companion crops were evaluated in this study.

Materials and methods Maize, with eight different companion crops, was evaluated as winter feed for lactating Merino ewes and lambs on the Dundee Research Station, KZN, South Africa. One ha plots were planted with 0.5 ha maize and 0.5 ha companion crop adjacent to each other. The following companion crops were used: *Ornithopus sativus (Os), Vicia dasycarpa (Vd), Raphanus sativus (Rs), Ladlab purpureus (Lp), Glycine max (Gm), Vigna unguiculata (Vu), Avena sativa (As), Secale cereale (Sc)* and maize (M) alone. The maize and summer companion crops were planted during late November, each year, and the winter crops during February. Ten ewes and their lambs were allocated to each treatment and they were allowed to select between the maize and the companion crops. Whenever the experimental animals did not ingest sufficient of the available biomass, dry ewes were added. Grazing potential and growth of ewes and lambs were measured. The experiment was conducted over three consecutive years.

Results The rainfall data for the three years (from July-June) were 442.3mm, 792.7mm and 862.6 respectively, with the long-term average being 782.9mm. During year 1 the low rainfall resulted in poor performance of some companion crops and only M+Gm, M+Lp and M+Sc were grazed. During years 2 and 3 all treatments were evaluated. Although the rainfall in the first year was below average, M+Gm managed to carry 24.8 small stock units (SSU)/ha, with a total liveweight gain of 334 kg/ha. In a normal rainfall situation (year 2) M+Lp and M+Rs carried more than 20 SSU/ha and liveweight gains were 318 and 346 kg/ha respectively. During year 3, with a high rainfall, M+Rs and maize alone carried more than 32.5 SSU/ha, with liveweight gains of 376 and 341 kg/ha respectively.

Table 1 Grazing capacity per treatment

Treat-ment	SSU/ha/100days*			
	Season1	Season2	Season3	Mean
M+Gm	24.8	18.3	27.6	23.6
M+Vu	-	16.5	25.7	21.1
M+Lp	19.2	22.2	40.1	27.2
M+As	-	18.8	20.8	19.8
M+Os	-	16.9	27.6	22.2
M+Vd	-	19.6	25.3	22.4
M+Rs	-	20.0	32.5	26.2
M+Sc	20.4	-	-	-
Maize	-	20.0	38.7	29.3

Table 2 Total liveweight gain per treatment (kg/ha)

Treat-ment	Liveweight gain (kg/ha)			
	Season1	Season2	Season3	Mean
M+Gm	334	272	293	300
M+Vu	-	249	361	306
M+Lp	230	318	294	281
M+As	-	247	267	257
M+Os	-	270	238	254
M+Vd	-	233	291	262
M+Rs	-	346	376	360
M+Sc	285	-	-	-
Maize	-	274	341	307

*1 ewe = 1 small stock unit (SSU) and 1 lamb (average 20 kg) = 0.5 SSU

Conclusions When below average rainfall seasons are experienced, maize + *Glycine max* (soybeans) can be expected to be an appropriate winter feed. During normal rainfall seasons maize + *Lablab purpureus* (dolichos), maize + *Raphanus sativus* (Japanese radish) and maize alone are appropriate winter feeds. Maize + *Vigna unguiculata* (cowpeas) may be an appropriate winter feed during a higher rainfall season.

References

Esterhuizen, C.D. & S.D. Niemand (1989). Oesreste van ses kontantgewasse vir die oorwintering van skape. Inligtingsdag, Nooitgedacht-Navorsingstasie, Ermelo, Bl.85-107.

Lyle, A.D. (1991). The use of supplementary licks for sheep on summer and winter Sourveld. Sheep in Natal. Co-ordinated Extension Committee of Natal (5.2.1991). KwaZulu-Natal Department of Agricultural, Private Bag X9059.Pitermaritzburg. 3200.

Graze-out plus: filling forage gaps in the Southern Great Plains, USA

B.K. Northup, W.A. Phillips and H.S. Mayeux
USDA-ARS Grazinglands Research Laboratory, El Reno, OK 73036, USA, Email: bnorthup@grl.ars.usda.gov

Keywords: forage systems, intensive grazing, sward costs, tall fescue

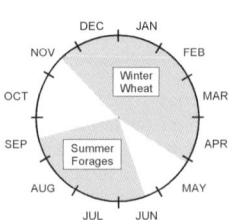

Figure 1 Primary forage system used to graze yearling stocker cattle in the Southern Great Plains

Introduction Putting low-cost gain on yearling cattle with forages is a significant agricultural activity in the Southern Great Plains. The primary forage system within the area has two components: winter wheat (*Triticum aesitivum*) grazed from fall through spring (Redmon et al., 1995), and warm-season perennial grasses for summer grazing. This system has significant gaps (Fig. 1) when high-quality forage is not readily available (September-November and May-June). Introduced cool-season perennial grasses have longer growing seasons than wheat, and could help fill these gaps. This experiment tested the function of an introduced cool-season perennial grass, new to the southern Great Plains, in a stocker production system involving intensive grazing of paddocks.

Materials and methods Studies were conducted during 2003-2004 on 1.8 ha paddocks (n=3) planted to non-toxic endophyte-infected tall fescue (*Festuca arundinacea* var. 'Jessup Max-Q'). In October through mid-November (35 d), cattle (200 kg body weight (BW)) were assigned to fescue paddocks. Thereafter, they were moved to paddocks of annual winter wheat, and grazed through April. Cattle (300 kg BW) were then moved back to the fescue paddocks from late-April through May (32 d). Cattle grazed fescue paddocks at high densities (7.4 hd/ha, ~ 3 times normal densities) during the two grazing periods. Standing crop (±1 s.e.) was measured at the start of grazing periods, and livestock gains were determined at different stages of the study. Establishment and maintenance (fertilizer, weed control, application costs) costs were recorded and used to compare cost of gain of forages related to annual costs of sward management.

Results Fescue establishment costs were 2.7 times that of wheat (Table 1); 58% of this cost was seed, interest, labour, and equipment. However, when amortized over the planned stand life (7 y), establishment costs were only 48% of total annual costs, which were less than total annual costs for wheat. Fescue paddocks produced 2030(±305) kg/ha by mid-October and 3630(±870) kg/ha by mid-April. These paddocks produced 259 (in fall) and 238 (in spring) stocker-grazing days during these periods. Fescue paddocks generated 511 kg/ha of gain at times when wheat or warm-season pasture could not readily supply forage, for an additional 69 kg gain per head. Based on the historic potential market value of gain for the region (US$ 1.10/kg), an additional US$ 50/hd return above sward costs could be produced by incorporating fescue paddocks into a wheat-based system.

Table 1 Mean (±s.e.) sward costs, grazing management and livestock responses on wheat and fescue paddocks

	Wheat	Fescue
Sward Costs, US$/ha		
Establishment [1]	234 (5)	635 (7)
Annual maintenance	--	100 (5)
Total annual	234 (5)	191 (4) [2]
Grazing Management		
Grazing season, d	171 (6)	67 (2)
Stocker grazing days, d/ha	589 (32)	497 (13)
Land, ha/hd	0.77 (0.3)	0.14 (0.1)
Livestock Response, kg		
ADG [3]	1.15 (0.1)	1.03 (0.1)
Gain / ha	678 (45)	511 (13)
Gain / hd	188 (12)	69 (5)
Forage Effectiveness		
Cost of gain, US$/kg [4]	0.34 (0.06)	0.37 (0.04)

1. Costs included seed, fertilizer, labour, equipment, and interest. Fescue costs also included first year deferment.
2. Annual maintenance plus 1/7th of establishment cost.
3. Average daily gain.
4. Based on total annual costs related to swards.

Conclusion Despite high establishment costs, paddocks of non-toxic endophyte-infected tall fescue were effective at generating gains by yearling cattle, and did so during time periods when traditional forages were not available. This capacity would allow graziers in the Southern Great Plains to partially fill fall and spring forage gaps with high quality forage, and extend the grazing season. Such an extended grazing season would allow graziers to change marketing strategies for their cattle or possibly develop grass-fed beef for niche markets if wheat, cool-season perennials and warm-season forages were combined in integrated systems.

References

Redmon, L., G.W. Horn, E.G. Krenzer & D.J. Bernardo (1995). A review of livestock grazing and wheat grain yield: boom or bust? *Agronomy Journal*, 87, 137-147.

A modified forage system for stocker production in the Southern Great Plains, USA

B.K Northup, W.A. Phillips and H.S. Mayeux
USDA-ARS Grazinglands Research Laboratory, El Reno, OK 73036, USA, Email: bnorthup@grl.ars.usda.gov

Keywords: forage systems, intensive grazing, sward costs, wheatgrass

Introduction Putting low-cost gain on yearling cattle with forages is an important agricultural activity in the Southern Great Plains. The primary forage system within the area incorporates two forages: winter wheat (*Triticum aestivum*) for grazing in fall through spring, and warm-season grasses in the summer (Fig. 1). These systems have significant gaps in time when high-quality forage is not available. This study tested the function of introduced cool-season perennial grasses in filling the spring gap, and their capacity as large-scale replacements for winter wheat.

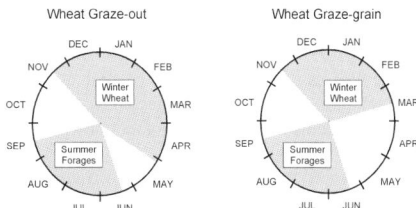

Figure 1 Traditional forage systems used to graze yearling stocker cattle in the Southern Great Plains

Materials and methods Studies were conducted during 2002-2004 on 2.0 ha paddocks (n=27) planted to wheat (n=9) or one of two introduced varieties (n=9 per variety) of wheatgrass (*Thinopyrum intermedium* and *T. ponticum*). The fall grazing period began in mid-November, with cattle (200 kg body weight (BW)) assigned to paddocks (wheat, 2.5 hd/ha; perennials, 4.1 hd/ha) and allowed to graze until perennial grasses terminated growth in early winter. All the cattle were moved to ungrazed paddocks of wheat for winter grazing (January-February), and then returned to their original perennial or wheat paddocks (plus put-and-take animals) for spring grazing. Spring period cattle (310 kg BW) were allowed to graze until May 1 (wheat, 7.1 hd/ha) or June 1 (perennials, 4.5 hd/ha). Carrying capacity was determined at the start of grazing periods and livestock weight gains were measured for each period. Establishment and maintenance (fertilizer, weed control, application costs) costs were used to compare performance of wheat and wheatgrass swards during fall and spring grazing periods.

Results Establishment costs of perennials were 2.3 times the cost of wheat (Table 1); 49% of this cost was seed, interest, labor and equipment. When amortized over the planned stand life (7 y), establishment costs were 47% of total annual costs (maintenance + establishment), which were lower than winter wheat. The perennial paddocks produced enough forage in the fall and spring to support a 23% increase in stocker grazing days. However, daily gains (ADG) were 75-81% of levels recorded for wheat Perennial paddocks generated slightly lower total gains than wheat (538 vs. 559 kg/ha) during the two grazing periods, despite longer grazing periods and increases in stocker grazing days. However, costs of gain related to total annual sward costs were 26% lower for the wheatgrasses. Based on the historic potential market value of gain for the region (US$ 1.10/kg), the perennial swards produced 43 US$/ha in additional gross (not including animal production costs) potential value above sward costs.

Conclusion Despite high establishment costs, wheatgrasses could be effective components of forage production systems in the Southern Great Plains. While daily gains were below those of winter wheat, their lower cost and extended spring grazing season can make such forages viable options for stocker production systems. Their value to graziers will depend on establishment and maintenance costs (both wheat and perennials), risk of stand failure, potential reductions in fixed costs, and the amount of land that must be planted to perennials to augment wheat and warm-season species in a three-forage system.

Table 1 Mean (±s.e.) sward costs, management, and livestock responses on wheat and wheatgrass paddocks

	Wheat	Perennials
Sward Costs, US$/ha		
Establishment [1]	234 (5)	549 (21)
Annual maintenance	--	89 (8)
Total annual	234 (5)	167 (4) [2]
Management and		
Fall Period		
Length, d	69 (1)	60 (8)
Stocker grazing days, d/ha	171 (30)	245 (47)
ADG, kg [3]	1.26 (0.2)	0.94 (0.2)
Gain, kg/ha	215 (19)	228 (64)
Spring Period		
Length, d	36 (5)	63 (6)
Stocker grazing days, d/ha	255 (35)	282 (23)
ADG, kg [3]	1.35 (0.1)	1.09 (0.1)
Gain, kg/ha	344 (64)	310 (25)
Total Gain, kg/ha	559 (48)	538 (96)
Cost of Gain, US$/kg [4]	0.42 (0.05)	0.31 (0.06)

1. Costs included seed, fertilizer, labor, equipment, and interest. Perennials also included first year deferment.
2. Annual maintenance plus 1/7th of establishment cost.
3. Average daily gain
4. Based on total annual costs related to swards.

Nutrient composition of Napier grass (Pennisetum purpureum) and Napier grass silages made with different additives

A.A. Aganga, U.J. Omphile and J.C. Baitshotlhi
Botswana College of Agriculture, P/Bag 0027, Gaborone. Botswana, Email: aaganga@bca.bw

Keywords: nutrient composition, napier grass, silages, additives

Introduction Forage contributes about 73% and 95% of the diets of ruminants fattened on grain supplements and on natural ranges respectively (Sarwar & Nisa, 1998). Napier grass is one of the highest yielding tropical forage grasses and was shown to provide a good quality silage when it was supplemented with molasses, as the fermentation quality was not affected by the high storage temperature (40ºC). The objectives of this study were to determine the chemical composition of silages when ensiled alone or mixed with additives.

Materials and methods This study was conducted at the Botswana College of Agriculture (BCA) farm. Napier grass regrowth was harvested manually with a sickle starting with 50cm growth height at the first harvest then monthly at every 25cm of additional growth for five months until the grass attained 1.5m height. Results are presented here only for the first stage of growth. After harvesting, the grass was cut into 2-4cm length and 500g of the cut forage were placed in plastic bags in 4 replicates of each of 6 treatments. The treatments for silage making involved the following additions to 500g of chopped Napier grass (a) no additive, (b) 0.25% $CaCO_3$, (c) 10% sorghum meal, (d) 5% molasses, (e) 1% Urea + 5% molasses and (f) all the additives combined. The sealed plastic bags were stored in the dark at room temperature for 21 d. The nutrient components of Napier grass and its silage were determined according to the procedures of AOAC (1996). The neutral detergent fibre (NDF), acid detergent fibre (ADF), acid detergent lignin (ADL) and *in vitro* true digestibility (IVTD) with rumen liquor were determined according to Van Soest (1994). The data were subjected to analysis of variance, mean values were separated by the Duncan's new multiple range test using SAS (2000).

Results Chemical composition and digestibility of Napier grass and silages are shown in Table 1 harvested at 0.5m cutting height. Inclusion of additives to Napier grass harvested this growth stage improved silage quality significantly compared to fresh Napier grass harvested at the same height.

Table 1 Nutrient and mineral composition of Napier grass and Napier grass silages harvested at 0.50m height of growth (chemical component as % of dry matter, unless otherwise specified)

Type	IVTD	pH	ADF	ADL	CP	ASH)	Ca	P	Fe ppm
FG	64		37.0	3.5	13.3	5.0	0.8	0.7	294
GS	64	4.4	35.5	3.5	13.8	5.0	1.1	0.8	331
GS+$CaCO_3$	64	4.2	33.5	3.5	13.7	5.5	2.2	0.9	319
GS+ so	70	4.1	22.5	3.3	13.9	5.3	1.6	1.0	308
GS+M	68	4.0	28.5	3.0	13.8	5.1	1.9	0.9	341
GS+UM	71	4.1	28.5	2.5	16.5	5.2	2.0	0.9	350
GS+all additives	73	4.2	20.5	2.0	16.6	7.3	2.7	1.0	361
s.e.m	4.2	0.2	1.0	0.8	0.8	0.5	0.1	0.04	5.9

FG= fresh grass, GS= grass silage, so =sorghum, M = molasses, UM=urea and molasses

Conclusions Inclusion of nutrient additives improved nutritive quality of Napier grass silage. The results (not presented here) show that the nutritive quality of Napier grass declines with maturity.

References
AOAC (1996). In: P. Cunniff (ed.) Official Methods of Analysis. 16th Edition. Association of Official Analytical Chemists Publishers, Arlington, Virginia.
Sarwar, M & M.U Nisa (1998). Effect of nitrogen fertilization and storage of naturity of *Pennisetum purpureum* on its chemical composition, dry matter intake ruminal characteristics and digestibility in buffalo bulls. *Asian Australasian Journal of Animal Sciences*, 12, 1035-1039.
SAS (2000). Statistical Application Systems. SAS Institute Inc, Carey, NC.
Van Soest, P.J. (1994). Nutritional Ecology of the Ruminant (2nd ed.). Comstock Publishing Association, Cornell University Press, Ithaca.

Effect of agronomic management on feeding value of Festulolium hybrids for winter pasture

W. Opitz v. Boberfeld and K. Banzhaf
Justus-Liebig-University Giessen, Department of Grassland Management and Forage Growing, Ludwigstr. 23, D-35390 Giessen, Germany, Email: Wilhelm.Opitz-von-Boberfeld@agrar.uni-giessen.de

Keywords: winter-grazed herbage, *Festulolium, Festuca arundinacea*, metabolisable energy

Introduction *Festulolium spp. are* considered to combine the distinctive winterhardiness of *Festuca* species with the high forage quality of *Lolium* species. Such cultivars may be particularly appropriate for winter pastures, but knowledge about quality aspects during winter under central European conditions is not available. The objective of this research was to determine forage quality of four *Festulolium* cultivars (*festucoid* type or *loloid* type*)* during winter under varying agronomic treatments.

Materials and methods The experiment was established on the Research Station near Giessen (160 m above sea level), central Germany. Pure stands of the cultivars Felina *(festucoid)*, Lofa *(loloid)* and Hycor *(festucoid)*, of *Festulolium pabulare*, Perun *(loloid)*, of *Festulolium braunii*, and Kora, of *Festuca arundinacea* (as "standard"), were observed over two years. Treatments examined the influence of pre-utilisation (accumulation since June or July) and date of winter harvest (Dec. or Jan.). Energy concentration, estimated as metabolisable energy (ME) using formula 16e of Menke & Steingass (1984) is presented. Dry matter (DM) yield and ergosterol concentration (Schwadorf & Müller, 1989) were also determined. All results were examined by analysis of variance with P < 0.05 as the level of significance and least-significance differences (= LSD) were calculated.

Results

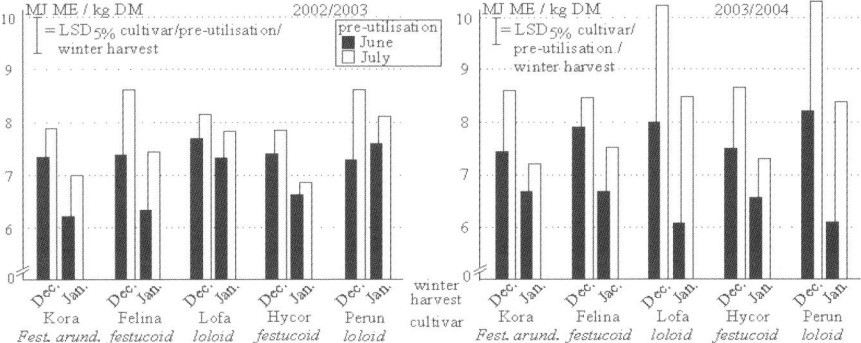

Figure 1 Effect of cultivar, pre-utilisation and winter harvest date on ME observed over two years

The *Festulolium* cultivars with *festucoid* character, Felina and Hycor and the *Festuca arundinacea* cultivar Kora had the highest yields during both winters. Yields of the *loloid* cultivars Lofa and Perun were clearly lower. Energy concentrations during winter were mainly influenced by the period of accumulation since summer (Figure 1) with the growths pre-utilised in July, frequently having higher values than those saved since June. This difference was particularly marked for the *loloid* cultivars. In the milder winter of 2003/2004, with generally higher growth rates before winter, energy values decreased from December to January. The effect of species is evident in both years, with values being higher for the *loloid* varieties. The higher yields of the *festucoid* cultivars (and *Festuca arundinacea)* are associated with more advanced maturity resulting in lower energy concentrations. Furthermore, the ergosterol-concentrations of the *loloid* cultivars indicate a lower durability to fungal infections.

Conclusions These results demonstrate that the *festucoid* cultivars of *Festulolium* are better adapted to utilisation as saved herbage during winter grazing. However, regarding yield productivity and feeding value the hybrids did not surpass the *Festuca arundinacea*-cultivar.

References

Menke, K.H. & H. Steingass (1987) Schätzung des energetischen Futterwertes aus der *in vitro* mit Pansensaft bestimmten Gasbildung und der chemischen Analyse. 2. Mitteilung: Regressionsgleichungen. *Übers. Tierern.* 15, 59-94.
Schwadorf, K. & H.M. Müller (1989). Determination of ergosterol in cereals, feed components and mixed feed by liquid chromatography. *Journal of the Association of Official Analytical Chemists*, 72, 457-462.

Ensilability and silage quality of different *Festulolium* hybrids in comparison to *Festuca arundinacea*

K. Banzhaf and W. Opitz v. Boberfeld
Justus-Liebig-University Giessen, Department of Grassland Management and Forage Growing, Ludwigstr. 23, D-35390 Giessen, Germany, Email: Wilhelm.Opitz-von-Boberfeld@agrar.uni-giessen.de

Keywords: *Festulolium,* water-soluble carbohydrate, pH, silage quality

Introduction *Festulolium* hybrids as cool-season grasses may be used as dominant species for winter pastures in year-round outdoor livestock systems. The utilisation of these species during summer is limited due to low intake as a fresh pasture grass by grazing ruminants. Therefore, ensiling the primary growths of these hybrids may be an alternative approach to using these species during the growing season. However, information on the quality of *Festulolium* silages under central European conditions is not available. The objective of this research was to determine ensilability and silage quality of four *Festulolium* cultivars (of *festucoid* or *loloid* type*)* compared to one *Festuca arundinacea* cultivar.

Materials and methods The experiment was established on the Research Station near Giessen (160 m above sea level), central Germany. The primary growths of the cultivars Felina *(festucoid)*, Lofa (*loloid*) and Hycor *(festucoid)* of the species *Festulolium pabulare*, Perun (*loloid*), of *Festulolium braunii*, and Kora, *Festuca arundinacea* (as "standard"), were harvested in the beginning of June, pre-wilted (= 32 % dry matter, DM) and ensiled with a storage period of 90 d in a laboratory ensiling experiment. To characterise the ensilability, the concentration of water-soluble carbohydrate (WSC) (Yemm & Willis, 1954) and the buffering capacity (BC) (Weissbach, 1967) were determined and the WSC/BC-ratio was calculated. Silage quality was assessed by pH (potentiometric determination), lactic acid concentration (colorimetric determination) and concentrations of volatile fatty acids and ethanol (gas chromatorgraphy).

Results In both years the cultivars Lofa and Perun had the highest WSC concentrations due to their loloid attributes (Table 1). As the concentrations of the festucoid cultivars, including *Festuca arundinacea,* are at a lower level, the factor cultivar is important. This influence is also reflected in the WSC/BC-ratio. The ratios of Lofa and Perun are always above the required value of 2 (for DM of 30 %). Related to their higher buffering capacity and their lower WSC concentrations, the WSC/BC-ratios of Kora, Felina and Hycor were below that required ratio in 2000. This result suggests comparatively better ensilability of the loloid cultivars Lofa and Perun, but there was little evidence of this in the quality aspects

Table 1 Effect of cultivar on forage and silage (DM 32 %) composition

Cultivar	Kora	Felina	Lofa	Hycor	Perun	LSD$_{5\%}$
Year 2000						
WSC (g /kg DM)	55.9	48.7	131.4	55.5	100.1	12.1
WSC/BC-ratio	1.4	1.2	3.1	1.4	2.3	0.31
pH	4.3	4.3	4.2	4.2	4.1	0.69
Lactic acid (g/kg DM)	41.7	43.7	49.9	41.3	52.2	17.3
Acetic acid (g/kg DM)	9.0	9.9	10.3	8.1	9.7	3.34
Year 2003						
WSC (g/kg DM)	104	112	156	105	160	16.5
WSC/BC-ratio	2.7	3.7	5.0	3.5	4.3	0.67
pH	4.0	4.0	4.0	4.0	4.1	0.69
Lactic acid (g/kg DM)	37.4	53.0	46.7	37.7	34.3	8.04
Acetic acid (g/kg DM)	10.4	11.6	12.4	10.8	13.6	1.41

determined in the silages. None of the silages exceeded the critical pH of 4.5. Furthermore, concentrations of lactic and acetic acid were adequate and concentrations of other volatile acids, including butyric acid, were negligible.

Conclusions The loloid *Festulolium* hybrids show a higher concentration of water-soluble carbohydrates and an adequate WSC/BC-ratio compared to the other varieties. This suggests that their ensilability might be better. However, the determined aspects of silage in all the *Festulolium* hybrids were comparable to those for *Festuca arundinacea.*

References
Weissbach, F. (1967). Die Bestimmung der Pufferkapazität der Futterpflanzen und ihre Bedeutung für die Beurteilung der Vergärbarkeit. *Tagungsber. Deut. Akad. Landwirtschaftswiss*, Berlin, 92, 211-220.
Yemm, E.M. & A.J. Willis (1954). The estimation of carbohydrates in plant extracts by anthrone. *Biochemistry Journal*, 57, 85-97.

Ensiling characteristics and ruminal degradation of Italian ryegrass with or without wilting and added cell wall degrading enzymes

Y. Zhu[1], H. Jianguo[1], Z. He[1], X. Qingfang[1], B. Chunsheng[1] and N. Nishino[2]
[1]Institute of Grassland Science, China Agricultural University, No 2 Yuan mingyuan Xilu,Haidian District Beijing 100094,China, Email:yuzhu3@sohu.com; [2]Department of Anima and Technology, Faculty of Agriculture, Okayama University, Japan

Keywords: wilting, enzymes, ryegrass, silage, degradation

Introduction The previous experiment (Yu zhu et al.,1999) has shown that the efficacy of added enzymes varied greatly according to the DM content of the material crop. The silage DM did not alter the effects of enzymes on the in vitro digestion of NDF (Yu zhu et al.,1999, Yu zhu et al.,2000). The aim of this experiment was to study the effect of wilting and enzymes on fermentation quality, chemical composition and *in situ* digestion of Italian ryegrass (*Lolium multiflorum* Lam.) silage.

Materials and methods Primary growth of Italian ryegrass was harvested at the late heading stage. They were chopped into approximately 25 mm length and ensiled in laboratory silos (1 L) directly or after being wilted for 2 h with or without added cell wall degrading enzymes. The enzymes (1:2 mixture of Acremonium and Trichoderma cellulase based on avicelase activity) were added at 0.1 g /kg just before ensiling. Triplicate silos for each treatment were stored for 45 d at room temperature, then sampling for the analysis of fermentation quality (Yu zhu et al.,1999). Three castrated mature goats about 19 kg body weight were used. Nylon bag incubation was conducted using the silages samples and the disappearance of DM and NDF after 0,3,6,12,24,48 and 72 h was determined. The parameters explaining the degradation were estimated by non-linear regression analysis using Syatat (Ver 5.2 for Macintosh) and subjected to two way analysis of variance with wilting and addition of enzymes as main factors.

Results Wilting increased the contents of DM, CP, NDF, ADF, ADL, and the buffering capacity, decreased the WSC content, and had little effects on fructose, glucose and sucrose contents. This difference suggested a significant amount of fructosan. The addition of enzymes increased ($P<0.01$)the DM, and decreased the NDF and ADF contents, but did not affect the CP and ADL contents of silage. Higher contents($P<0.01$) of NDF, ADF and ADL were recorded in wilted rather than direct-cut silage. There remained more WSC ($P<0.01$) in enzyme-treated silage, while the difference appeared less when treated with wilted crops. The main fermentation quality and degradation characteristic of Italian ryegrass silage were show in Table 1.

The *in situ* degradation of DM was reduced ($P<0.01$) by wilting. The addition enzymes increased ($P<0.01$) the degradation at 3, 6, and 12h of incubation, while the effect on the DM degradation was diminished with wilted crops. The degradation of NDF was also enhanced by enzymes ($P<0.05$) at the initial incubation time. However, at 6 and 12h of incubation, the effect of enzymes was found to be opposite in the silage which was wilted prior to the treatment.

Table 1 Fermentation quality and degradation characteristics of silage

Item	Direct cut -E	+E	Wilted -E	+E	Pooled SE	ANOVA E	W	E×W
pH	4.11	3.98	4.88	4.62	0.08	*	**	NS
Lactic acid (g/kg DM)	59.7	67.3	21.3	27.7	7.61	NS	**	NS
Butyric acid (g/kg DM)	17.9	6.43	35.8	32.2	6.05	NS	**	NS
NH3-N (g/kg N)	90.2	93.5	108	97.3	5.43	NS	NS	NS
Dry matter								
Potential degradation (g/kg)	791	801	750	790	17.7	NS	NS	NS
Rate of degradation	0.047	0.045	0.047	0.034	0.006	NS	NS	NS
Neutral detergent fibre								
Potential degradation (g/kg)	702	676	673	693	36.8	NS	NS	NS
Rate of degradation	0.045	0.043	0.043	0.034	0.006	NS	NS	NS

NS; not significant,*;$P<0.05$,**;$P<0.01$, –E; no enzyme, +E; with enzyme, W; wilting

Conclusions Added enzymes may have a potential of enhancing the digestibility of silage, although the benefits would be hindered by wilting. It appeared necessary to consider, in addition to the species, the DM content of the forage, when cell wall degrading enzymes were used to improve the silage utilization.

References

Yu Zhu Nishino N, (1999). Ensiling characteristics and ruminal degradation of Italian ryegrass and lucerne silages treated with cell-wall degrading enzymes. *Journal of the Science of Food and Agriculture,* 11:111-117.
Yu Zhu Nishino N, (2000). Fermentation of rhodesgrass and guineagrass silages with or without wilting and added cell wall degrading enzymes. *Grassland Science,* 4:235-239.

Influence of grass species and sample preparation on ensiling characteristics

D.J.R. Cherney[1], M.A. Alessi[2] and J.H. Cherney[1]
Cornell Univeristy, Ithaca, NY, USA 14853[1] and Universitá Degli Studi Di Palermo, Italy[2]. Email: djc6@cornell.edu

Keywords: laboratory silo, forage, fermentation

Introduction Laboratory silos are considered a practical method of comparing a number of treatments (O'Kiely, 1993). Cherney et al. (2004) reported that vacuum-sealed polyethylene bags effectively ensiled corn silage samples in the laboratory. Grasses, with their inherently higher buffering capacities and lower sugar levels, generally are more difficult to ensile. Objectives were to evaluate the influence of species and chopping (whole vs. shredded) on pH and volatile fatty acid profile of grasses ensiled in vacuum-sealed polyethylene bags and to assess the suitability of this method as a laboratory ensiling method.

Materials and methods Four replicates of three grass species, orchardgrass (*Dactylis glomerata*, L.), reed canarygrass (*Phalaris arundinacea* L.) and tall fescue (*Festuca arundinaceae* Schreb), first- and second cutting, were ensiled whole or shredded with a chipper-shredder. Forages (500g) were ensiled in polyethylene bags as previously described (Cherney et al., 2004). Lactic acid, acetic acid, propionic, butyric acid were determined on forages at 0, 2, 4, 8, 16, 24, and 30 d post ensiling.

Results There was little or no butyric or propionic acids in the silages, indicating that the silages did not undergo clostridial fermentation. Lactic acid and acetic acids, accounting for most of the total acids, varied by species and harvest date (Table 1). Ensiled forages dropped rapidly in pH and were relatively stable beyond 4 days (Figure 1). Kung and Shaver (2001) indicated that grass silages typically range in pH from 4.3 to 4.7. There were species differences, with orchardgrass pH tending to be lower than reed canarygrass. Shredded orchardgrass and reed canarygrass silages had lower pH than whole silages. Despite species and processing differences, pH of silages tended to be within the typical range, suggesting that vacuum-sealed polyethylene bags are an acceptable method of laboratory ensiling.

Table 1 Mean silage lactic acid (% of total acids), acetic acid (% of total acids), and latic:acetic acid as influenced by forage species and harvest (average of forages ensiled >8 days).

VFA	Lactic	Acetic	Lactic:Acetic
OG1[1]	58.3[b2]	37.3[b]	1.53[ab]
OG2	51.2[c]	45.6[a]	1.21[a]
RC1	70.0[a]	25.8[c]	2.61[c]
RC2	56.9[b]	38.4[b]	1.56[b]
TF1	57.7[bc]	37.0[b]	1.52[ab]
TF2	61.3[b]	34.0[b]	1.79[b]

[1]OG=orchardgrass, RC=reed canarygrass, TF=tall fescue, 1= first harvest, 2 = second harvest.
[2]Means within a column with different superscripts differ ($P < 0.05$).

Figure 1 Effect of species and chopping on pH (SEM=standard error of the mean).

Legend:
- Chopped orchardgrass, SEM=0.033
- Chopped reed canarygrass, SEM=0.032
- Whole reed canarygrass, SEM=0.032
- Chopped tall fescue, SEM=0.038
- Whole tall fescue, SEM=0.38

Conclusions Despite inherent problems in all small scale silo systems, laboratory silos are an accurate and reliable experimentation unit. Fermentation characteristics suggested that all samples in this study were well ensiled within 8 days of ensiling. We conclude that it is possible to use vacuum-sealed plastic bags to ensile temperate grasses to assess treatment differences.

References

Cherney, D.J.R., J.H. Cherney & W.J. Cox (2004) Fermentation characteristics of corn forage ensiled in mini-silos. *Journal of Dairy Science* (In press).

Kung, L. and R. Shaver (2001) Interpretation and use of silage fermentation analysis reports. *Focus on Forage* 3(13):1-5.

O'Kiely, P. (1993) Influence of a partially neutralised blend of aliphatic organic acids on fermentation, effluent production and aerobic stability of autumn grass silage. *Irish Journal of Agricultural and Food Research* 32:13-26.

Calculation of forage value and suitability for silage of autochthonous plant mixtures found in peat soil grassland in relation to the cut-off date

L. Dittmann and R. Bockholt

Faculty for Agricultural and Environmental Science at the Rostock University, 18059 Rostock , Germany, Email: lisa.dittmann@auf.uni-rostock.de

Keywords: grassland, forage value, autochthonous plants, calculation

Introduction The semi-extensive management of permanent grassland results in autochthonous plant mixtures whose productivity and forage quality dynamics are relatively unknown, but important for its utilisation. There are enormous differences in forage value and the suitability for silage depending on botanical composition, cutting date or grazing date,. The time-based changes in the forage value are of economic interest for the farmers and important for determining a utilisation strategy

Material, methods and results The base for the calculation scheme was the classification data for the vegetation stages of 42 autochthonous grasses and herbs found in peat soil grassland (Bockholt & Buske, 1997, 2001) with measured data on growth height and analysis for crude protein, crude fibre, ash, digestibility, energy density, water-soluble carbohydrates, buffer capacity and nitrate content). Information on the cultivated grasses *Lolium perenne* and *Lolium multiflorum* were included. The digestibility and energy density were estimated using the cellulase method described by Friedel (1990). An instrument was created with MS-Excel for the easy assessment of the time-dependent parameters. After alteration of any selection or input data, there is an automatic re-calculation and output in the form of a table and a graph (see Figure 1).

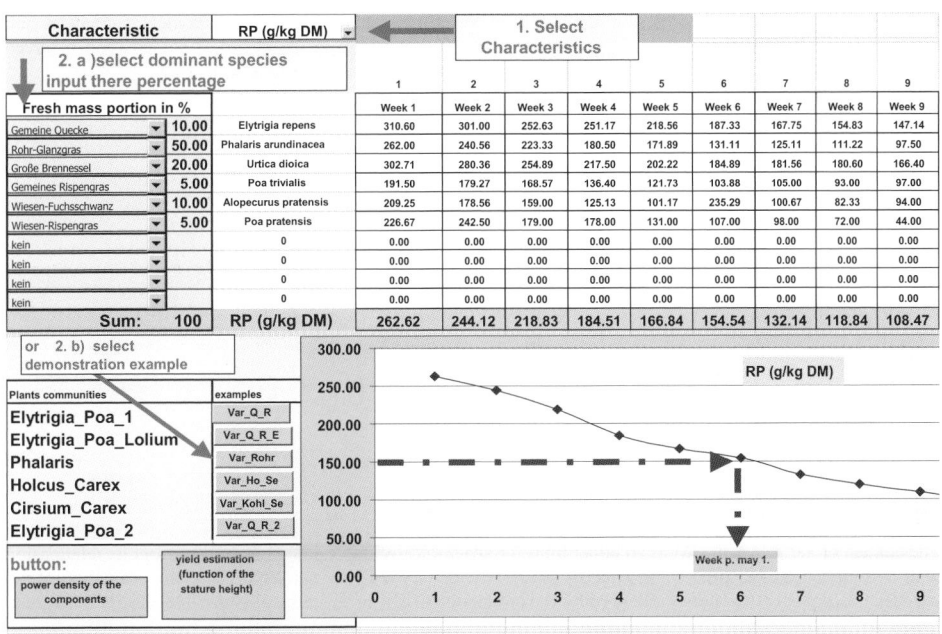

Figure 1   MS-EXCEL-spreadsheet with demonstration data (raw protein)

Conclusions The demonstrated calculation plan provides valuable information to help decision making. It can be extended and updated with data of additional species.

References

Bockholt, R. (2001). Futterwert und Siliereignung der häufigsten autochthonen Pflanzenarten des Niedermoorgrünlandes. *Arch. Acker - Pfl. Boden*, 47, 183-199.

Bockholt,R. & Buske, F. (1997). Variationsbreite des Futterwertes von Niedermoorgrünland unter Berücksichtigung der häufigsten autochthonen Pflanzen.*Das wirtschaftseigene Futter*, 43, 5-20.

Friedel, K. (1990). Die Schätzung des energetischen Futterwertes von Grobfutter mit Hilfe einer Cellulasemethode. *Wiss. Z. Univ. Rostock, Nat.Reihe*, 39, 78-86.

A new system for the evaluation of the fermentation quality of silages

E. Kaiser and K. Weiß
*The Institute of Animal Science, Humboldt-University of Berlin, Invalidenstraße 42, D-10115 Berlin, Germany,
Email: ehrengard.kaiser@agrar.hu-berlin.de*

Keywords: silage quality, fermentation quality, fermentation process

Introduction Depending on the content of nitrate in green forage, the pattern of fermentation products in silages differ significantly (Weiß & Kaiser, 2001). The systems, which are now common in practice for evaluating the quality of silage fermentation, characterise fermentation quality incorrectly because the evaluation is influenced by the chemical composition of green forage. The aim of this work was to derive an evaluation system for fermentation quality, which is independent from the chemical composition of green forage.

Materials and methods Under laboratory conditions, 570 silages were produced from different green forages of known chemical composition. Fermentation quality parameters were selected which were suitable to characterise all stages of fermentation quality independent of the chemical composition of the green forage. An evaluation system was developed on the basis of the relations between the parameters of undesirable decomposition (butyric acid (BA), acetic acid (AA), ammonia (NH_3)) considering recent information of metabolism in silages during the fermentation process. It was applied to 3503 silages from green forages with unknown chemical composition obtained from farms of different regions in Germany.

Results and discussion The results confirmed that all stages of fermentation quality (anaerobic stability, "turn over" of fermentation products and increased spoilage) can be evaluated by BA and AA concentration exclusively (Kaiser *et al.*, 1999, 2000). The parameters pH-value and ammonia content in silages are inappropriate for evaluation, because they are influenced by variation in the chemical composition of green forage (see also Kaiser *et al.*, 2000). The suggested new estimation system is presented in Table 1.

Table 1 Evaluation system for the fermentation quality of silages from contents of butyric acid and acetic acid

Butyric acid (% DM)	Points	Acetic acid (% DM)	Points (Discount)	Evaluation Score	Evaluation Mark
0 - 0.3	100	≤ 3.0	0	90 to 100	1
> 0.3 - 0.4	90	> 3.0 - 3.5	-10	72 to 89	2
> 0.4 - 0.7	80	> 3.5 - 4.5	-20	52 to 71	3
> 0.7 - 1.0	70	> 4.5 - 5.5	-30	30 to 51	4
> 1.0 - 1.3	60	> 5.5 - 6.5	-40	<30	5
> 1.3 - 1.6	50	> 6.5 - 7.5	-50		
> 1.6 - 1.9	40	> 7.5 - 8.5	-60		
> 1.9 - 2.6	30	> 8.5	-70		
> 2.6 - 3.6	20				
> 3.6 - 5.0	10				
> 5.0	0				

The content of 3.0 % AA in DM as an upper limit for anaerobically stable silages is derived from its relationship with BA and ammonia. If the content of BA is low, the classes are very narrow because the evaluation of the fermentation quality is strongly influenced by the production of BA in anaerobically unstable silages from green forage low in nitrate.

Conclusions From the evaluation of 3503 silages made under practical conditions, this new system, based only on the content of BA and AA, was able to characterise the fermentation quality of all green forage silages, including maize, more correctly than previous systems.

References
Kaiser, E., K. Weiß & R. Krause (1999). Vorschlag zur Beurteilung der Gärqualität von Grassilagen. *Proceedings 111. VDLUFA-Kongreß*, Halle, 385-388.
Kaiser, E., K. Weiß & R. Krause (2000). Beurteilungskriterien für die Gärqualität von Grassilagen. *Proceedings of the Society for Nutritional Physiology*, 9, 94.
Weiß, K & E. Kaiser (2001). Fermentation patterns in silage depending on chemical composition of herbage. *Grassland Science in Europe*, 6, 150-153.

The potential of different forage combinations for green-chop silage

T.L. Knight[1], T.J. Fraser[1], T.A. White[1] and M.G. Hyslop[2]
[1]AgResearch Limited, PO Box 60, Lincoln 8152, New Zealand, Email: trevor.knight@agresearch.co.nz, [2]Heinz-Wattie's Ltd, PO Box 16083, Christchurch 8004, New Zealand

Keywords: cereals, legumes, silage, Canterbury, New Zealand

Introduction On the Canterbury Plains of New Zealand (NZ) there is an opportunity on cropping farms, between summer harvest and autumn/winter sowing, to grow forage crops to make high quality silage. Recently, cereal cultivars have been specifically bred for forage production and suitability for whole-crop silage (de Ruiter *et al.* 2002), and also high legume (e.g. sulla) content forage mixes have resulted in high quality silages with high lactic acid and soluble carbohydrate content (Niezen *et al.* 1998). This trial aimed to determine the yield potential of various cereal/legume forage mixtures summer sown and harvested for silage in autumn.

Materials and methods The trial was sown into a cultivated seedbed on 15 Jan 03 at Lincoln, Canterbury, NZ. Binary mixtures of legumes and cereals/grass were created by sowing 7 cereal/grass (15 x 3m) split plots (6 different species plus no cereal) within 4 different legume main plots (replicated 4 times) (Table). Plots were irrigated as required. Three 1 m row lengths were cut from each plot on 25 Mar 03 and dissected into species, weighed, dried and reweighed. Wilted and chopped herbage from each plot was used to make silage in 20 l plastic bag lined buckets. Buckets were opened in May 03 and assessed for colour, mould and odour.

Results Legume and total dry matter (DM) yield from the pea treatments was significantly higher than any of the bean treatments (Table 1; $P<0.05$). Bean mixtures tended to be dominated by cereals whereas pea mixtures achieved the intended 50:50 split. The yield of triticale, barley and oats were significantly higher than wheat and ryecorn. Individual and total yield for Italian ryegrass was significantly lower than the cereals. Visual and olfactory assessment indicated all species combinations resulted in well preserved silage.

Table 1 Cultivar, sowing rate and main effect mean yields of legume, cereal/grass and total

Common name Legumes	Botanical name	Cultivar	Sow rate kg/ha	Legume kg DM/ha	Cereal/grass kg DM/ha	Total kg DM/ha
Peas	*Pisum sativum* L.	Magnus	220	3735	2744	6112
Dwarf beans	*Phaseolus vulgaris* L.	Labrador	150	1210	4435	5006
Runner beans	*Phaseolus coccineus* L.	Scarlet	190	951	5083	5308
Haricot beans	*Phaseolus vulgaris* L.	Navy bean	170	1225	4900	5425
LSD (P<0.05)[a]				378	492	560
Cereals/grass		No cereal	0	3378	0	3378
Barley	*Hordeum vulgare* L.	Boss	28	1068	5371	6439
Wheat	*Triticum aestivum* L.	Sapphire	24	1665	4108	5772
Oats	*Avena sativa* L.	Stampede	27	1289	5695	6984
Triticale	*Triticum (x Triticosecale)*	DoubleTake	28	1291	5488	6779
Italian ryegrass	*Lolium multiflorum* L.	Tabu	7	2271	1660	3931
Ryecorn	*Secale cereale* L.	Rahu	13	1499	3466	4965
LSD (P<0.05)[a]				363	534	541

[a]LSD (P<0.05) means "Least significant difference" at 5% level of significance.

Conclusions In spite of low cereal sowing rates, the bean species did not yield well and therefore are considered unsuitable for the production of high silage yields. Similarly Italian ryegrass produced low yields, compared to the cereals, and has limited suitability. Peas in combination with barley, oats or triticale produced greater than 6 t DM/ha over a 69 day period and therefore offered the greatest potential to make substantial yields of well preserved legume/cereal silage. These silages would be ideally suited to supplement intensive livestock systems such as pastoral dairy farming.

Acknowledgements The authors thank Dave Saville for statistical analysis.

References
de Ruiter, J.M., R. Hanson, A.S. Hay, K.W. Armstrong & R.D. Harrison-Kirk (2002). Whole-crop cereals for grazing and silage: balancing quality and quantity. *Proceedings of the NZ Grassland Association*, 64, 181-189.
Niezen, J.H., G.C. Waghorn, T.B. Lyons & D.C. Corson (1998). The potential benefits of ensiling the forage legume sulla compared with pasture. *Proceedings of the NZ Grassland Association*, 60, 105-109.

Effect of a new microbial strain as an inoculant on the quality of maize silage

J.G. Kim, J.S. Ham, E.S. Chung, S. Seo and J.K. Lee
National Livestock Research Institute, Suwon 441-706, South Korea, Email: jonggk@rda.go.kr

Keywords: inoculant, silage, maize, organic acid

Introduction Lactic acid bacteria play a key role in making silage from forage, and lactic acid bacteria selected from good silage could be expected to be suitable inocula for making good silage. Thus, the purpose of this study was to examine the such novel lactic acid bacteria for making high quality maize silage.

Materials and methods Maize was harvested at the ripe stage. It was ensiled in experimental silos (20 l capacity) with or without microbial additives (C3-2, C11-4, B13-1, B14-1, C9-1) and stored at room temperature for 60 d. Crude protein (CP) was determined by the Kjeldahl method (AOAC, 1995), and acid detergent fiber (ADF) and neutral detergent fiber (NDF) by the method of Goering & Van Soest (1970). A pH meter was used to measure pH (HI 9024; Hanna Instrument Inc. UK; Kim *et al.*, 2000). Volatile fatty acids were analysed by gas chromatography Model 3400; Varian Co., USA) and lactic acid by HPLC(HP-1100 ; Hewlett-Packard Co., USA).

Results Chemical composition of the silage is shown in Table 1. All silages were well preserved. The pH value and acetic acid contents of additive-treated silages were lower and lactic acid content was higher than those of the control. There was a trend for acetic acid contents to be lowest and lactic acid to be highest with B13-1 and C9-1. Generally, additives decrease the butyric acid content of silage, but the butyric acid content of this experiment control silage was lower than that of treatments. Crude protein contents of the silages were increased, but ADF and NDF contents of the silages did not differ between treatments. The most common change in the chemical composition of forage during ensiling is reduction of CP and increase of structural carbohydrates (ADF and NDF). In this experiment, CP content showed the general trend, but ADF and NDF content were decreased in comparing with fresh material. Dry matter content was increased in comparison with the fresh material.

Table 1 Acidity (pH), organic acid, , dry matter (DM), crude protein (CP), acid detergent fiber (ADF) and neutral detergent fiber (NDF) content of fresh maize and maize silage with or without inoculant treatment

Microbes	pH	Organic acid (% of DM)			DM (%)	CP (%)	ADF (%)	NDF (%)
		Acetic	Butyric	Lactic				
Fresh material	-	-	-	-	30.4	8.3	32.4	50.3
Control	3.69	1.16	0.15	3.94	34.5	6.8	29.3	45.3
C3-2	3.67	0.93	0.32	5.79	31.5	7.9	30.2	47.7
C11-4	3.74	1.06	0.14	4.72	35.4	7.7	26.9	44.7
B13-1	3.66	0.80	0.30	8.64	33.0	7.8	31.2	49.2
B14-1	3.62	1.06	0.30	8.39	33.1	7.2	29.9	49.1
C9-1	3.69	0.87	0.29	8.61	34.0	7.3	31.1	50.1
Mean	3.68	0.98	0.25	6.68	33.6	7.5	29.8	47.7
LSD (0.05)	0.81	0.22	0.16	2.58	2.5	0.4	NS	NS

* "C" originated from maize silage. "B" is originated from barley silage. The number is area code.

Conclusions These results indicate that *L. plantarum* C9-1 was effective as an inoculant for maize silage. This culture was named NLRI 201 and registered in the Korea Agricultural Culture Collection (KACC-91067).

References

Association of Official Analytical Chemists (1995). Official Methods of Analysis.16th ed. AOAC, Arlington, Virginia.
Goering, H. K. & P. J. Van Soest (1970). Forage fiber analysis. Agricultural Handbook 379, U.S. Government. Printing Office, Washington, DC.

Moisture control, inoculant and particle size in tropical grass silages

S.F. Paziani, L.G. Nussio, D.R.S. Loures, L.J. Mari, J.L. Ribeiro, P. Schmidt, M. Zopollatto, M.C. Junqueira and A.F. Pedroso
Department of Animal Science, University of São Paulo, ESALQ, Av. Pádua Dias 11, Piracicaba, SP, 13418-900, Brazil, Email: nussio@esalq.usp.br

Keywords: grass silage, *Panicum maximum*, silo losses, moisture, wilting

Introduction Decreased fermentation and spoilage losses with improved aerobic stability during feed out can be accomplished by several strategies, such as wilting, addition of microbial additives and moisture absorbents. Particle size reduction may increase bulk density and improve the fermentation. The objective of this trial was to evaluate the effects of particle size, moisture content and a microbial additive on chemical-physical parameters and losses in silages made from Tanzania grass.

Material and methods The trial was carried out during the summer on a 90 d vegetative regrowth cut of Tanzania grass (*Panicum maximum*) which was harvested and ensiled with the following treatments: T1 - fresh forage, large particle size, no microbial additive; T2 - fresh forage, small particle size, no microbial additive; T3 - wilted forage, large particle size, no microbial additive; T4 - fresh forage, large particle size, no microbial additive + ground pearl millet grain (GM); T5 - fresh forage, small particle size, microbial additive (Ecosyl®, UK). Pressed bag silos (40t each) with 2.7m diameter were packed under pressure (80pounds/inches^2) and opened after 90 d storage. A core sample (30x30x30cm) was taken weekly for analysis. Spoilage losses were measured daily as a % of the silage unloading rate. Chemical analyses were carried out according to AOAC (1980), mean particle size following Lammers *et al.* (1996) and porosity according to Williams (1994).Repeated measurements were taken in a complete randomised design during eight weeks and analysed using a mixed procedure (SAS, 1996).

Results Wilting and pearl millet grain addition increased the dry matter (DM) content (Table 1). The small particle size in the forage did not increase wet or DM silage bulk densities (Table 2), even though the addition of pearl millet grain showed a trend for higher DM density-DMD (156 kg/m^3) compared to the other treatments. Forage wilting tended to lower the wet density of the silage (460 kg/m^3), but DM density was not affected due to the compensatory effect of the higher DM content. Reducing the particle size in the forage (T2 and T5) did not reduced the porosity, in contrast to the expected results. This may have arisen because fewer and larger pores with longer forage were compensated by many smaller pores. The wilted forage (T3) showed higher losses when compared to the addition of pearl millet (29% vs 18%). Particle size reduction did not change the spoilage losses (P=0.60) but the addition of bacterial inoculant showed a trend (P=0.09) for increased losses.

Table 1 Chemical parameters of tropical grass silages

Parameters	T1	T2	T3	T4	T5
DM, %	24.8	24.0	27.7	28.5	24.0
CP, % DM	9.2	10.2	9.6	11.0	8.5
NDF, % DM	67.8	69.4	69.0	49.8	69.3
ADF, % DM	45.0	45.4	46.4	33.7	45.4
ASH, % DM	10.9	10.5	11.2	8.3	10.8
WSC, % DM	1.8	1.8	2.4	1.4	1.2
N-NH$_3$, % total N	8.2	5.8	4.6	2.4	10.1
pH	4.9	4.9	4.8	4.8	4.7

Table 2 Physical parameters of tropical grass silages

Parameters	T1	T2	T3	T4	T5
Mean particle size, cm	2.4	2.2	3.4	2.2	2.0
Sieve retention, %	47.4	53.1	67.4	54.0	36.9
Bulk density, kg/m^3	535[a]	523[a]	460[b]	505[ab]	487[ab]
DMD, kg/m^3	142[ab]	131[b]	135[ab]	156[a]	122[b]
Porosity, %	45[b]	52[a]	50[ab]	48[ab]	55[a]
Spoilage losses, %	17[ab]	14[b]	29[a]	18[ab]	23[ab]

[a,b](P<0.05)

Conclusions High spoilage losses suggested that wilting may not be a suitable strategy for ensiling tropical grasses when harvested with larger particles and stored in pressed bag silos. The bacterial inoculant also increased spoilage losses during feed out.

Acknowledgement Financial support provided by FAPESP, SP (Brazil)

References
AOAC (1980). Official Methods of Analysis.13th .ed. Washington, 1015 pp.
Lammers, B.P., D.R. buckmaster & A.j. Heinrichs. (1996). A simple method for the analysis of particle sizes of forage and total mixed rations. *Journal of Dairy Science*, 79, 922-928
Williams, A.G.(1994). The permeability and porosity of grass silage as affected by dry matter. *Journal of Agricultural Engineering Science*, 59, 133-140.

Factors affecting bag silo densities and losses

R.E. Muck[1] and B.J. Holmes[2]

[1]USDA, Agricultural Research Service, US Dairy Forage Research Center, Madison, Wisconsin 53706 USA, Email: remuck@wisc.edu, [2]Biological Systems Engineering Dept., University of Wisconsin-Madison, Madison, Wisconsin 53706 USA

Keywords: bag silo, density, loss, lucerne, maize

Introduction Bag silos (polyethylene tubes, 30 to 90 m length, 2.4 to 3.7 m diameter, 0.22 mm thick) are used on approximately one-third of the dairy farms in the U.S.A. for making silage, and the level of adoption is increasing rapidly. Unfortunately, almost no research data have been published on these types of silos. Our objective was to measure densities and losses in bag silos at three farms, looking for causes of variation in both.

Materials and methods Bag silos made on three research farms over the course of two years were monitored at filling and emptying. These consisted largely of lucerne and whole-crop maize silages. All loads of forage entering the bags were weighed and sampled. Average density was calculated based on bag length and nominal bag diameter. At emptying, the weight of all silage removed from a bag was recorded. Any spoiled silage not fed was weighed, sampled and specifically identified on the emptying log. A grab sample from the face of each silo was taken periodically, one per filling load. Factors influencing density and losses were determined through data analysis using a combination of the CORR, GLM, and STEPWISE procedures in SAS®.

Results Over two years, 47 bag silos were made at the three farms, 23 of lucerne, 23 of whole-crop maize and 1 of red clover. Density ranged from 160 to 280 kg dry matter (DM)/m³. Density increased as DM content increased (Figure 1). The operator and how the bagging machine was set were important factors affecting density. The same bagging machine (Kelly-Ryan, KR) was used at the Arlington (Arl) and West Madison (WM) farms, and Arl consistently got higher densities. The Prairie du Sac (PDS) farm had higher densities the second year after training from a manufacturer's representative. Density declined with longer particle size (Fig. 2). Kernel processing in maize reduced density at PDS where there was a planned comparison.

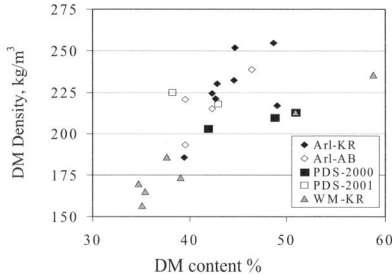
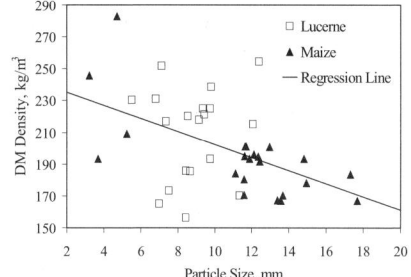

Figure 1 Dry matter density in hay crop bag silos as affected by dry matter content, farm (Arl, PDS, WM), and bagging machine (Ag-Bag, AB; KR)

Figure 2 Dry matter density in bag silos as affected by average particle size at ensiling

Dry matter losses were measured on 39 of the bag silos and ranged from 0 to 40%. Average DM losses were 9.2% invisible plus uncollected losses and 5.4% spoilage losses for a total loss of 14.6%. Six silos had excessive spoilage losses (>15%) due to damaged plastic or overly dry silage (>40% DM) being fed out in warm weather. In contrast, 11 silos had no spoiled silage, and 15 bags had less than 5% spoilage loss, representing bags with spoilage largely at the ends. Invisible losses were reduced in high porosity silages (where spoilage losses were exacerbated), greater in warm weather, and affected by emptying procedures (reduced at WM where bag silos were emptied in 2 to 3 one-day periods as opposed to daily removal for cattle at PDS and Arl). Spoilage losses in bags without damaged plastic were greater in dry, porous silages, from emptying silos in warm weather, and at lower feed out rates. Both invisible and spoilage losses were not affected by crop or bagging machine.

Conclusions These results indicate that low DM losses (<10%) are regularly achievable in bag silos. However, deviations from good management (harvesting between 30 and 40% DM, operating the bagging machine to get a smooth bag of high density, monitoring routinely for and patching holes, and feeding out at a minimum of 300 mm/d) can result in substantial (>25%) losses. Because higher losses occur during warm weather, silage from the best preserved bags should be reserved for summer use.

Round-bale silage preparation of rice straw

Y. Cai, C. Xu, N. Yoshida and M. Ogawa
National Institute of Livestock and Grassland Science, Nishinasuno, Tochigi 329-2793, Japan, Email: cai@affrc.go.jp

Keywords: fermentation quality, lactic acid bacteria, rice straw, round bale silage

Introduction Rice straw is an important feed resource for ruminants. In Japan, rice straw cannot be fully dried due to the usually humid autumn season, which leads to about 70% of the production being ploughed back or incinerated. Therefore, the development of techniques to enhance the long-term preservation and quality of rice straw is of great importance. In this work, a new lactic acid bacterium was used as a silage inoculant, and its effect on round-bale silage preparation from fresh rice straw was examined.

Materials and methods Fresh rice straw of Koshihikari cultivar was obtained from a field in Saitama, Japan, on October 2002. Silage was prepared using a round-bale system. Chikuso-1 (*Lactobacillus plantarum*, Brand seed Ltd., Sapporo, Japan; Cai *et al.*, 2003) was used as an inoculant.

Table 1 Fermentation quality of rice straw silage

	Silage ensiled for 65 days		Silage ensiled for 300 days	
	Control	Chikuso-1	Control	Chikuso-1
pH	5.67^b	3.77^a	5.75^b	3.85^a
Dry matter (%)	65.73	65.97	64.56	63.24
Lactic acid (% FM)	0.17^a	2.06^b	0.22^a	1.86^b
Acetic acid (% FM)	0.16	0.18	0.35	0.27
Propionic acid (% FM)	nd	nd	nd	nd
Butyric acid (% FM)	0.14	nd	0.35	nd
Ammonia N (g/kg FM)	0.28^b	0.09^a	0.45^b	0.10^a

FM, fresh matter; nd, not detected. Chikuso-1: *Lactobacillus plantarum* ; a,b Values are means of three silage sample Means in the same silage row with different superscripts are significantly different (P < 0.05)

Results The moisture content of the fresh rice straw after harvest was 65%. Its content of water-soluble carbohydrates and crude protein were 5% and 4% of dry matter, respectively. The inoculant strain Chikuso-1 was a Gram-positive and catalase-negative rod that did not produce gas from glucose, formed L(+) and D(-) lactic acid and grew under a low-pH condition. After storage for 65 and 300 d, silages inoculated with Chikuso-1 were well preserved and exhibited significantly (P<0.05) lower pH, butyric acid and ammonia-nitrogen, and significantly (P<0.05) higher lactic acid content, as compared to control silages (Table 1). During silage fermentation, the control silages displayed mould growth, whereas in Chikuso-1-inoculated silages, moulds were at or below the detectable level.

Conclusions These results showed the growth potential of *Lactobacillus plantarum* Chikuso-1 and its beneficial effects on rice-straw silage, suggesting that this strain could help achieve higher quality and longer preservation of this type of silage.

Reference
Cai Y., M. Fujita, M. Murai, M. Ogawa & N. Yoshida (2003). Application of Lactic acid bacteria (*Lactobacillus plantarum* Chikuso-1) for silage preparation. *Grassland Science*, 49, 477-485.

The effects of offering grass or maize silages to in-lamb ewes on body weight and condition changes, colostrum yield and quality

T.F. Crosby, P.J. Quinn, J.J. Callan, P. Reilly, B. Flynn, D. Cunningham and T. Massey
University College Dublin, Faculty of Agriculture, Belfield, Dublin 4, Ireland, Email: frank.crosby@ucd.ie

Keywords: grass, maize, ewe, colostrum

Introduction Hay and more recently grass silage (GS) have been the traditional feeds for sheep in Ireland over the winter period. Alternatives such as maize silage (MS) are becoming increasingly important as winter forage sources especially for cattle. This study sought to evaluate grass silage and maize silage when offered to pregnant ewes.

Materials and methods Sixty four oestrus synchronised, Suffolk-cross, twin-bearing ewes were individually penned and offered either grass silage or maize silage at the rate of 1.1 times the previous day's intake. The forage diet was supplemented from d 98 of gestation with 400g/d of a barley (35.2%), molassed beet pulp (35.1%), soyabean meal (22.2%) based concentrate. Ewes were weighed and body condition scored at the beginning and end of the experiment. Following lambing, the ewe's udder was covered for 24-h, to prevent suckling by the lambs, and the ewes were hand milked at 1-h, 10-h and 18-h. At each milking measured quantities of colostrum were fed to each lamb. The lambs were blood sampled at 24-h and analysis carried out for immunoglobulins level in the colostrum (Fahey & McKelvey, 1965) and blood (McEwan, *et al.*, 1970).

Results The feed analysis and ewe performance data are presented in Tables 1 and 2 respectively. The forages offered were high quality and well preserved. The higher dry matter (DM) maize silage gave a 16.8% increase in silage DMI compared with grass silage. The lower CP in the maize was partially offset by the higher DMI to give similar levels of crude protein intake. Reflecting the higher DMI with the maize silage, these ewes gained more weight and lost least condition. The higher DMI with maize silage was not reflected in higher colostrum or IgG yields or in the efficiency of IgG absorption.

Table 1 Analysis of silages and concentrate

	Grass Silage	Maize Silage	Conce-ntrate
Dry Matter (%)	24.4	27.1	86.5
pH	3.6	3.7	-
Composition (g/kg DM)			
Crude protein	112	81	208
Crude fibre	356	261	85
NDF	587	572	-
Ash	74	40	59
Ether extract	26	29	12
Starch	-	280	215
Buffering Capacity (mEq/kg DM)	1023	695	-
%DMD (in vitro)	69.7	66.2	94.3
ME (MJ/kg DM)	10.1	9.7	13.8

Table 2 Effect of diet on ewe and lamb performance

	Grass Silage	Maize Silage	SEM	sig
Silage DMI (kg/d)	0.95	1.11	0.041	**
Total DMI (kg/d)	1.29	1.47	0.037	**
Total MEI (MJ/d)	14.3	15.6	0.40	*
CPI (g/d)	178	169	3.2	*
Weight change (kg)	10.9	12.9	0.66	*
Body score change	-0.22	-0.08	0.043	*
Gestation length (d)	148.4	146.4	0.34	***
Litter weight (kg)	9.87	9.53	0.237	ns
Colostrum: (ml)				
1-h	525	510	60.9	ns
10-h	618	654	40.6	ns
18-h	596	566	31.8	ns
IgG yield to18-h (g)	79	75	3.9	ns
IgG absorption (%)	14.7	13.3	1.92	ns

Conclusions The ewes on maize silage were closer to their energy requirements as reflected in their lower losses in body condition score and higher liveweight gain in late pregnancy. For the commercial producer, supplementing maize silage with 400g/hd per d of concentrates at a flat rate over the last seven weeks of pregnancy may result in the body reserves of well-fleshed ewes not being fully utilised. Consequently, it may be possible to reduce the level of concentrate supplementation given to the ewes in late pregnancy when maize silage rather than grass silage is used, provided the protein content of the supplement is increased.

References

McEwan, A.D., E.W. Fisher, I.E. Selman & W.J. Penhale (1970). A turbidity test for the estimation of immune globulin levels in neonatal calf serum. *Clinica Chimica Acta*, 27, 155-163.

Fahey, J.L. & E.M. McKelvey (1965). Quantitative determination of serum immunoglobulins in antibody agar plates. *Journal of Immunology*, 94, 84-90.

The effects of offering hay, pit-stored grass silage or big-bale silage to pregnant ewes on ewe and lamb performance

T.F. Crosby, P.J. Quinn, J.J. Callan and T. McGrane
University College Dublin, Faculty of Agriculture, Belfield, Dublin 4, Ireland, Email: frank.crosby@ucd.ie

Keywords: grass, big-bale silage, ewe, colostrum

Introduction Hay and pit silage have been used extensively for feeding sheep over the winter period, but in recent years, especially on smaller farms, big-bale silage has become increasingly popular. However, there is limited comparative information on the use of big-bale silage for sheep, especially in relation to the effects of chopping and growth stage at harvest. This study compared the performance of ewes offered hay, pit silage, or big-bale grass silage made from either chopped or unchopped grass and cut from the same field at the same time.

Materials and methods Sixty four twin bearing ewes were offered either unchopped or chopped big-bale silage, double-chopped pit silage or hay, supplemented with 400g daily of either a molassed sugar beet pulp or a barley-based concentrate in a 4 x 2 factorial design experiment for the last eight weeks of pregnancy in order to investigate feed intake, ewe weight and body score changes, colostrum yield and immunoglobulin levels. All forages were cut on the same day and wilted for 1 (pit silage), 2 (big bale) or 7 days (hay). The ewes were individually penned and offered the forages daily at proportionately 1.1 times each ewe's previous days intake. They were weighed and body condition scored at the beginning and end of the experiment and hand milked at 1-h, 10-h and 18-h post partum. The colostrum was fed back to the lambs. Colostrum and blood taken from the lambs at 24-h were analysed for immunoglogulin levels.

Results There were no forage x concentrate interactions and the data for forage composition and the main forage effects are presented in Table 1. There were no performance differences between the two concentrate types offered. The pit silage had a lower digestibility value. Ewes offered the pit silage had a lower intake of dry matter (DM), metabolisable energy (ME) and crude protein (CP) than any other treatment and these ewes lost more weight and tended to produce less colostrum. Intake and performance were very similar for the remaining three treatments with no significant differences in DM intake, ME intake and body condition score. Crude protein intake was however significantly higher for the two big-bale silages than for hay, whilst the loss in liveweight was greater for unchopped big-bale silage than for hay.

Table 1 Forage analysis and ewe performance data in relation to forage type

Forage type	Hay	Pit silage	Big-bale silages		SEM
			Unchopped	Chopped	
Dry matter (%)	86.5	19.3	27.6	27.5	---
pH	---	3.92	4.28	4.18	---
Crude protein (g/kg DM)	73.7	90.9	92.5	91.4	---
Crude fibre (g/kg DM)	371	378	355	343	---
In-vitro DMD (g/kg DM)	64.2	60.2	65.1	64.0	---
Forage DM intake (kg/d)	1.26[a]	0.86[b]	1.15[a]	1.17[a]	0.06
Total ME intake (MJ/d)	16.2[a]	12.3[b]	15.2[a]	15.4[a]	0.42
Crude protein intake (g/d)	161[a]	149[c]	175[b]	176[b]	4.0
Weight loss d98-24h post-partum (kg)	-0.4[b]	2.5[a]	1.7[a]	0.49[ab]	0.88
Body score change (range 0-5)	-0.40[a]	-0.30[ab]	-0.19[a]	-0.32[ab]	0.090
Gestation length (d)	146.3	146.8	146.2	147.1	0.08
Litter weight (kg)	10.3	9.4	9.9	9.7	0.45
Colostrum yield to 18-h (ml)	2039	1753	1917	1974	152.2
IgG absorption (%)	21.4	21.4	21.8	24.4	2.46

[a,b,c] Means with different superscripts are significantly different (P<0.05)

Conclusions The results demonstrate the higher intake of hay and big-bale silage over pit silage, even though the pit silage had been double chopped in the field and was basically produced from the same raw material. In relation to animal performance, chopping before making the big-bale silage did not appear to be justified. The lower protein content of the hay was partially compensated for by the higher intake. In this experiment, all forages were individually offered daily. This would have made access to the long forages easier and because they were wilted for a longer period and had a higher DM, could possibly have enhanced intakes in these treatments relative to intakes that would occur in many farm conditions.

The effect of harvesting strategy of grass silage on milk production

K. Kuoppala[1], M. Rinne[1], J. Nousiainen[2] and P. Huhtanen[1]
[1]MTT Agrifood Research Finland, Animal Production Research, FI-31600 Jokioinen, Finland, Email: kaisa.kuoppala@mtt.fi, [2]Valio Ltd, Farm Services, P.O. Box 10, FI-00039 Valio, Finland

Keywords: primary growth, regrowth, digestibility, maturity, milk production response

Introduction Timing of harvest in primary growth of grass is a major factor affecting D-value (digestible organic matter, g/kg DM) of silage and dry matter (DM) consumption and milk production of dairy cows (Rinne, 2000). The objective of this research was to investigate whether there is a similar pattern in regrowths of grass.

Materials and methods Six grass silages, two from primary growth (PG) and four from regrowth (RG), were prepared from a mixed timothy (*Phleum pratense*) meadow fescue (*Festuca pratensis*) sward in 2002 at Jokioinen (61°N). The silages were slightly wilted (on average 4 h), precision chopped and ensiled in bunker silos with a formic acid based additive (4.1 l formic acid/t). Silages from PG were cut on 5 June at early (E) and on 17 June at late (L) stage of growth. Regrowths from both PG cuts were harvested on 29 July at early (EE and LE) and on 12 Aug. at late (EL, LL) stage of growth. These six silages were fed to 24 Finnish Ayrshire cows in a cyclic change-over design supplemented with 12 or 8 kg concentrate daily. The results presented are pooled over concentrate treatments. Silages were fed *ad libitum* and concentrates in three equal meals per day. Indigestible NDF (INDF) content of the silages was measured with a 12-day rumen incubation in a nylon bag. The *in vivo* D-value of the silages was determined with sheep fed at maintenance level by total collection of faeces.

Results D-value declined by 5.0 g/d with advancing growth stage in PG while the decline in RG was 3.6 (EE• EL) and 2.5 (LE• LL) g/d (Table 1). Silage and total DM intake of PG silages was higher (P<0.001) than that of RG silages (Table 2), as could be expected based on higher average D-value. Progressing maturity in PG decreased silage intake by 0.48 kg and energy corrected milk (ECM) production by 0.60 kg per 10 g decline in D-value. In RG, the decline in D-value did not decrease silage intake but ECM production decreased by 0.14 (EE• EL) or 0.46 (LE• LL) kg per 10 g decline in D-value. The results suggest, that milk production potential of RG silages was lower than PG silages as silage intake and milk production of EE and LE were not higher than L although D-value was. The confounding effect of silage DM content, which was clearly lower in RG silages, may have contributed to this.

Table 1 Description of the silages

	E	L	EE	EL	LE	LL
Date of harvest	5Jun	17Jun	29Jul	12Aug	29Jul	12Aug
Leaves in timothy, g/kg DM	489	340	503	390	639	529
DM yield, t/ha	3.3	5.1	4.2	4.9	3.1	3.8
DM content, g/kg	270	278	224	319	209	308
Chemical composition, g/kg DM						
Ash	74	62	87	88	91	84
Crude protein	151	115	129	118	153	126
NDF	513	598	566	562	549	549
INDF	50.3	97.2	70.5	93.3	59.8	79.1
D-value, g/kg DM	704	644	659	609	664	629

Table 2 Dry matter (DM) intake and milk production (kg/day) of dairy cows

	E	L	EE	EL	LE	LL	SEM	C_1	C_2	C_3	C_4	C_5
Silage DM intake	16.2	13.3	12.2	12.2	12.8	12.9	0.13	***	***	***		
Total DM intake	24.8	22.0	20.8	20.8	21.5	21.5	0.13	***	***	***		
Milk	33.3	30.7	29.5	29.2	31.6	29.9	0.38	***	***	**		*
ECM#	36.2	32.6	30.6	30.0	32.6	31.0	0.39	***	***	*		*

Energy corrected milk. Orthogonal contrasts: C_1= E vs. L; C_2= PG vs. RG; C_3= EE+EL vs. LE+LL; C_4=EE vs. EL; C_5=LE vs. LL. Statistical significance: *** P<0.001, ** P<0.01, * P<0.05, o P<0.10.

Conclusions Harvesting at a more advanced growth stage decreased D-value and milk production potential in PG of grass. In general, all RG silages were of moderate quality and variation of quality was smaller than in PG. The production responses to changes in D-value were greater in PG silages than in RG silages. It seems that at comparable D-value, the milk production potential of PG silages is slightly higher than that of RG silages.

References

Rinne, M. (2000). Influence of the timing of the harvest of primary grass growth on herbage quality and subsequent digestion and performance in the ruminant animal. Academic Dissertation, University of Helsinki, Department of Animal Science, Publications 54: 42 p. + 5 encl.,
http://ethesis.helsinki.fi/julkaisut/maa/kotie/vk/rinne/.

Theme A: Efficient production from grassland

Section 13

Animal-plant relations

Chew-bites, jaw movement allocation and bite rate in grazing cattle as identified by acoustic monitoring

E.D. Ungar, N. Ravid, T. Zada, E. Ben-Moshe, R. Yonatan, S. Brenner, H. Baram and A. Genizi
Department of Agronomy and Natural Resources, Institute of Field and Garden Crops, The Volcani Center, POB 6 Bet Dagan, Israel 50250 Email: eugene@volcani.agri.gov.il

Keywords: ingestion, behaviour, mastication, efficiency

Introduction Bite rate derives from the time budget of the biting and chewing processes of intake, which are both performed by jaw movements. A new type of jaw movement was revealed by acoustic monitoring in cattle - the "chew-bite" -which chews herbage already in the mouth and harvests fresh herbage with the same jaw movement (Laca *et al.*, 1992). Chew-biting should enable the animal to reduce the total number of jaw movements performed per bite without reducing the number of chews per bite. We examined the variation among individuals in the allocation of jaw movements between the three types, and its relation to bite rate.

Materials and methods Nine Israeli-Holstein dairy heifers grazed pristine, continuous and homogeneous expanses of oats six weeks after sowing. Grazing sessions were recorded on video with acoustic monitoring, using a microphone on the forehead of the animal. Sward height was measured before grazing. A 5-minute session of uninterrupted grazing was extracted from these video recordings. The acoustic signal was sequenced aurally; each sound burst produced by a jaw movement was classified as a pure chew, pure bite or chew-bite. Chews per bite = the number of pure chews and chew-bites divided by the number of pure bites and chew-bites. Jaw movements per bite = total jaw movements divided by the number of pure bites and chew-bites.

Results Jaw movements maintained a virtually uninterrupted, regular rhythm of sounds (bite, chew or chew-bite). Mean rate of jaw movement was 78.9 min^{-1}, with a coefficient of variation (CV) of only 6%. There was high variability among animals in the allocation of jaw movements (CV for proportion chew-bites = 50%). On average, chew-biting accounted for 39% of jaw movements, the same proportion was allocated to pure chews, and the remaining 22% of jaw movements were pure bites. The proportions of pure chews and pure bites traded off directly against chew-biting. The mean biting rate was 48.2 min^{-1} (CV = 13%), and the animals invested 1.66 (CV = 13%) jaw movements per biting action and performed 1.27 (CV = 11%) chewing actions per biting action. As the proportion of chew-biting increased, bite rate increased and jaw movements per bite declined. Chews per bite showed no clear response to any of the variables (Figure 1). Linear regression of the number of chewing actions (pure chews or chew-bites) on the number of biting actions (pure bites or chew-bites) and pre-grazing sward height explained 87% of the variation.

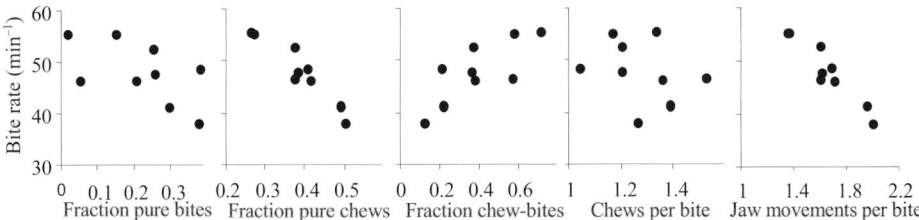

Figure 1 Relationships between bite rate and the proportion of jaw movements allocated to pure bites, pure chews and chew-bites, chews per bite and jaw movements per bite. Each point is a different animal

Conclusions Differences in bite rate among individuals derive primarily from differences in jaw movement allocation between the three types rather than differences in the rate of jaw movements. This allocation would appear to be constrained by chewing requirements. Animals that invest fewer jaw movements per bite by employing a high proportion of chew-bites may be more efficient grazers by virtue of a higher bite rate. Chew-bites can conceivably enable the animal to regulate bite weight to bring loading and processing rates into balance, thereby weakening the rationale to maximize bite weight. In most foraging environments this would allow the animal to improve diet quality.

References

Laca, E.A., E.D. Ungar, N.G. Seligman, M.R. Ramey & M.W. Demment (1992). An integrated methodology for studying short-term grazing behaviour of cattle. *Grass and Forage Science*, 47, 81-90.

Spatial heterogeneity of seasonal grazing pressure created by herd movement patterns on hilly rangelands using GPS and GIS

A.I. Arnon[1], E.D. Ungar[2], T. Svoray[1], A. Perevolotsky[2], M. Shachak[1], H. Baram[2], R. Yonatan[2], E. Ben-Moshe[2], S. Brenner[2] and D. Barkai[2]

[1]Ben-Gurion University of the Negev, POB 653 Beer-Sheva, 84105 Israel, Email: amirisra@bgumail.bgu.ac.il,
[2]Department of Agronomy and Natural Resources, the Volcani Center, POB 6 Bet Dagan 50250, Israel

Keywords: Bedouin, rangeland, heterogeneity, GPS, GIS

Introduction   The spatial heterogeneity of grazing pressure on extensive rangelands has management implications (Adler et al., 2001) but it has traditionally been difficult to quantify. Combination of technologies based on GPS (Global Positioning System) and GIS (Geographic Information Systems) is a quantum leap in our ability to address this issue. These tools were used to estimate the spatial heterogeneity of grazing pressure at a farm scale, and examine the relation between local landscape features and local grazing pressure.

Materials and methods   The study site is in the hilly, semi-arid region of Israel (31°20' N 34°45' E), populated by a mixed herd of sheep and goats (400 animals) which is shepherded as a group across the landscape, with a fixed night corral and watering point. The herding route was tracked on 78 days in the green season of 2003 (Feb to June), using a tagged goat, harnessed with a GPS rover unit (Trimble GEII Explorer) that recorded a position every 0.5min. The routes were overlaid on GIS raster layers containing data on abiotic factors, at a resolution of 25x25m/cell. For each GPS location, 25min of animal presence was accrued to each of the 8 closest raster grid cells, based on animal number and the estimated area occupied by the stationary herd.

Results   The area available to the herd was 9648 raster grid cells (627ha), of which 7312 (457ha) had non-zero animal presence. Average velocity based on adjacent GPS locations was 0.28m/s. The total animal presence time accrued was 65736h, yielding an average of 9h/cell, visited at least once, or 144hr/ha. The frequency distribution of grazing pressure for the area visited (Figure 1a) was highly skewed to the right, with a long tail (not all shown) reaching a maximum of 2000hr/ha. Of the area grazed, 67% was frequented less than the expected mean. Presence was greater on the shallower slopes (<9°) and lower on the steeper slopes (>13°) than expected randomly (Figure 1b). Presence according to distance from the night corral deviated strongly from random, with a strong preference for the 800-1000m category (Figure 1c). Presence according to aspect showed a small increase for North and decrease for East (Figure 1d).

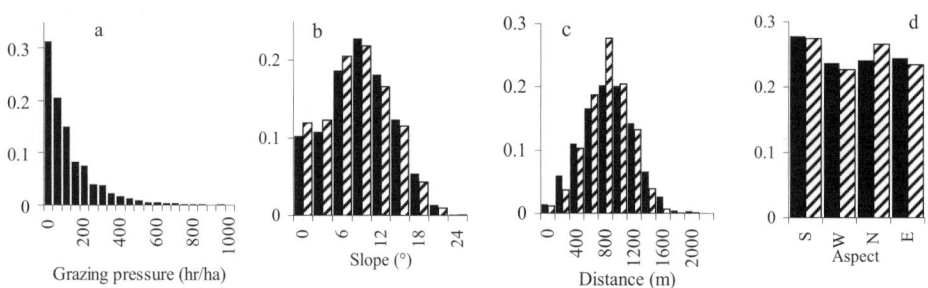

Figure 1 Frequency distributions of (a) grazing pressure, (b) slope, (c) distance from night corral, (d) aspect. Y axis is relative proportion of observations. In (b), (c) and (d), black = landscape, striped = animal presence

Conclusions   Using GPS and GIS, it is feasible to map and analyse the cumulative seasonal grazing pressure imposed by a herd over an entire grazing season. These tools also enable expected spatial distributions for null hypotheses to be computed that are site-specific. The observed spatial use deviated strongly from random or highly systematic patterns. Statistical analyses of observed versus expected distributions are planned using filtered data sets to reduce autocorrelation.

References

Adler, P. B., D. A. Raff & W. K. Lauenroth (2001) The effect of grazing on the spatial heterogeneity of vegetation. Oecologia, 128, 465-479.

Development of a 2-dimensional video-acoustic tool for monitoring bite placement

W.M. Griffiths[1,3], V. Alchanatis[2], R. Nitzan[1], V. Ostrovsky[2], E. Ben-Moshe[1], R. Yonatan[1], S. Brener[1], H. Baram[1] and E.D. Ungar[1]

[1]Department of Agronomy and Natural Resources, Institute of Field Crops, Agricultural Research Organisation, the Volcani Center, P.O.B. 6, Bet Dagan 50250, Israel, Email: wendy.griffiths@agresearch.co.nz, [2]Department of Testing and Advanced Technology, Agricultural Research Organisation, the Volcani Center, P.O.B. 6, Bet Dagan 50250, Israel, [3]Present address: AgResearch Ltd, Invermay Agricultural Centre, Puddle Alley, Private Bag 50034, Mosgiel, New Zealand

Keywords: bite area, bite placement, grazing, inter-bite distance

Introduction Studies of grazing behaviour conducted at the spatial scale of a feeding station demonstrate that intake rate declines with increasing depletion, a response attributed to an increase in bite overlap (Ginnett *et al.*, 1999; Ungar *et al.*, 2001). In order to understand the rules that govern bite placement, a methodology is required that can map the sequential placement of bites on the sward surface. We developed a video-acoustic tool to achieve this and report the findings of using the tool on small uniform patches of herbage.

Materials and methods Four Israeli Holstein heifers grazed patches (0.34 m^2) of Lucerne (*Medicago sativa* L.). Treatments were three depletion levels of 6, 18 and 30 bites. Grazing sessions were recorded on video with acoustic monitoring, using a microphone on the forehead of the animal. After grazing, a grid was placed over the patch and filmed as a calibration image. Each grid cell was mapped as grazed or un-grazed. From the video record, a single frame representing bite location (the position of the mouth immediately prior to severance) was extracted for each bite or chew-bite jaw movement identified from the sound track. The screen coordinate of the most forward position of the muzzle was extracted for each bite (Figure 1 (a)). Using the calibration image and appropriate geometric procedures, the screen coordinates were converted to coordinates on the sward surface. Assumptions were made regarding bite depth and the displacement between the marked point and the bite centre. The locations of predicted bite coordinates were compared to the grazed/un-grazed status of the grid cells.

Results The bite location pathways obtained were broadly S-shaped (Figure 1 (b-d)). Bite placement was neither random nor highly systematic. Inter-bite distance was consistent across depletion levels at approximately 13 cm. There was broad correspondence between the predicted bite locations and the mapped status of the grid cells (Figure 1 (b-d)). The observed pattern of grazed/un-grazed grid cells could not be predicted precisely by simply assuming a maximum distance of impact from bite centre to grid cell centre. The best balance in the number of correct predictions of grazed and un-grazed cells was achieved with a 9-cm distance of impact. Discrepancies were probably due to the fact that mouth position immediately prior to severance is not synonymous with the surface area of the vegetation grasped within the bite, primarily a reflection of tongue sweeping movements.

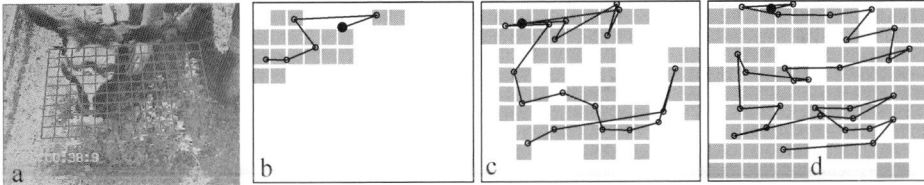

Figure 1 (a) Bite image frame with overlaid calibration grid, **(b-d)** Mapped grid cell status (shaded = grazed) and predicted centre of bite coordinate (•) for depletion levels of 6, 18 and 30 bites, respectively, for one animal. Line indicates bite sequence beginning from the solid black dot.

Conclusions The sequence and the approximate location of bites removed from a patch can be successfully determined using a single camera and acoustic monitoring. When bite depth is close to the average, the error in the estimation of the bite coordinate on the sward surface is not expected to exceed a few centimetres. The results suggest that this tool could be useful to test hypotheses regarding bite placement and inter-bite distance.

References

Ginnett, T.F., J.A. Dankosky, G. Deo, & M.W. Demment (1999). Patch depression in grazers: The roles of biomass distribution and residual stems. *Functional Ecology*, 13, 37-44.

Ungar, E.D., N. Ravid, & I. Bruckental (2001). Bite dimensions for cattle grazing herbage at low levels of depletion. *Grass and Forage Science*, 56, 35-45.

The sound of chewing

J.R. Galli[1], C.A. Cangiano[2], E.A. Laca[3] and M.W. Demment[3]
[1]Facultad de Ciencias Agrarias, Universidad Nacional de Rosario, C.C. 14. (2123) Zavalla, Santa Fe, Argentina, Email: jgalli@agatha.unr.edu.ar, [2]E.E.A. INTA Balcarce, Departamento de Producción Animal, C.C. 276 (7620) Balcarce, Buenos Aires, Argentina, [3]Department of Plant Sciences, The University of California, Davis, California, USA 95616

Keywords: acoustic analysis, grazing, ingestion, rumination, intake

Introduction Acoustic biotelemetry has been proposed as a way to count ingestive bites and chews of grazing animals. Recent work has indicated the possibility that detailed analysis of 'sounds of chewing' contains information about other characteristics of the ingestive process that can be used to study grazing behaviour of free ranging animals (Laca & Wallis DeVries, 2000), or to monitor stall-fed animals in more detail.

Materials and methods Steers (n=3; 284-316kg) were offered 4 levels (75, 150, 225, and 300 g DM) of 4 forages (fresh lucerne, dry lucerne, dry oats, fresh cocksfoot), at 2 particle lengths, in a factorial design with 32 treatments. They had previous experience with all forages and were offered each treatment once, for enough time to allow them to consume most of the forage. Until the animal chewed and swallowed all food, which took between 100-700s, wireless microphones (Nady 155 VR, Nady Systems, Inc., Oakland, California) transmitted the sounds to the sound track of a VHS video recorder. Rubber foam, placed on the animal's forehead and fastened to a halter, protected the microphones. The recorded sounds were digitised and analysed using Cool Edit Pro (Syntrillium Software 2002). After removing the "silent" intervals between chews, total energy flux density (EFD, pJ/m^2) was determined for the chewing sounds of each session. Average intensity (amplitude or loudness, AI) of chews, EFD/unit time eating, number of chews, and time of chewing were measured also.

Results EFD and AI (Table 1) were good predictors of the dry matter intake measured in each session for all forages (R^2=0.80), grouped in fresh and hay (R^2=0.86), or each type of forage (R^2=0.88). The best regression models combined total EFD and AI. However, treatment did not affect AI, and AI alone was not a good predictor of intake (R^2=0.08). The prediction equations tended to differ between dry and fresh forages. This is consistent with the documented "crunchiness" measured for human foods.

Table 1 Relationship between dry matter intake and chewing sound

Forage type	No. predictors	Coefficients in the model			R^2
		EFD (•$_f$)	AI (•$_g$)	•$_d$	
All	1	103.1	(NA)	48.8	0.59
	2	142.0	-37.4	136.5	0.80
Fresh	2	125.9	-28.2	103.3	0.89
Hay	2	158.1	-42.5	153.3	0.83
Lucerne	2	143.3	-41.9	157.9	0.80
Grass	2	146.6	-34.8	123.7	0.81

N=90

Model: Dry Matter Intake (g DM)=•$_d$+•$_f$Total EFD+•$_g$AI

Conclusions These data and previous work (Laca & Wallis DeVries, 2000), show that the energy of chewing sounds related strongly with the amount of forage ingested. While the sounds of feeding hold considerable potential for more accurate assessments of ingestive behaviours, sound characteristics contain considerable information related to the intake and the nature of the ingested forage. Grazing sounds of free-ranging animals may be monitored telemetrically and recorded automatically to make inferences about intake.

References

Laca, E. A. & M. F. Wallis DeVries (2000). Acoustic measurement of intake and grazing behaviour of cattle. *Grass and Forage Science*, 55:97-104.

Rich information in the acoustic signals from feeding and grazing in ruminants

M.W. Demment[1], J.R. Galli[2], C.A. Cangiano[3] and E.A. Laca[1]

[1]Department of Plant Sciences, The University of California, Davis, California, USA 95616, Email: mwdemment@ucdavis.edu, [2]Facultad de Ciencias Agrarias, Universidad Nacional de Rosario, C.C. 14. (2123) Zavalla, Santa Fe, Argentina, [3]E.E.A. INTA Balcarce. Departamento de Producción Animal. C.C. 276 (7620) Balcarce, Buenos Aires, Argentina.

Keywords: acoustics, grazing, intake, monitoring, rumination

Introduction Because of their impact on productivity and the environment, feeding behaviour, ingestion and rumination are critical to understand intake in grazing ruminants. Many systems, mainly mechanical, have been developed to measure ingestive behaviour. However, these systems have problems, including mechanical failure and the inability to distinguish between the complex jaw movements of prehension and ingestion (Laca *et al.*, 1994). The sounds generated by these behaviours are rich in information that holds potential not only to distinguish and count behaviours, but also identify aspects of the nature of the foods ingested.

Materials and methods Steers and sheep were fed dry or fresh leaves or stems of lucerne or 'grasses'. All sounds were recorded using wireless microphones (Nady 155 VR, Nady Systems, Inc., Oakland, California) protected by rubber foam, placed on the animal's forehead and fastened to a halter where the transmitter was attached. Sound was recorded on the sound track of a VHS video recorder. Comparison of sound with recorded visual observations allowed objective matching of behaviours and sounds.

Results Prehensile manipulative movements, bites and bite chews were distinguished easily in both cattle and sheep and were similar in waveform between animal species. Sound intensity, silence/chew and time/chew differed between hay and fresh lucerne. Regardless of plant part consumed, chewing sounds were louder, shorter and more frequent for fresh than dried lucerne. The amount of energy flux density in the chewing sound/unit of NDF consumed had a significant interaction between food type and plant part (P• 0.036). The value for hay leaf was significantly lower than for fresh stem, or for leaf and dried stem, confirming the conclusion of Galli *et al.* (2003) that fresh tissues produce sounds that are related to free water present in cells.

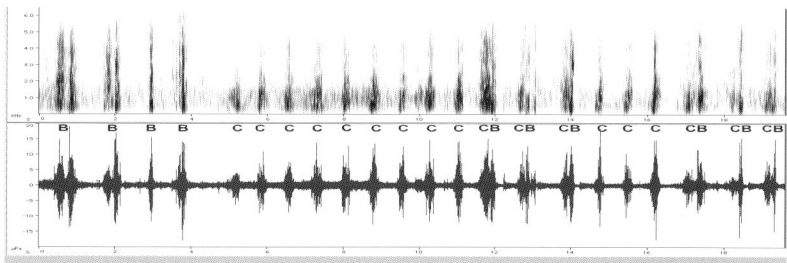

Figure 1 Spectrograph and waveform of grazing behaviour by an animal exhibiting bites (B), chews (C), and compound jaw movements (CB)

Conclusions Sounds produced in the ingestive process can be recorded accurately and simply to distinguish between ingestive bites and other types of jaw actions that other systems cannot distinguish. These sounds be used as the basis for a more accurate way to monitor feeding behaviour, particularly in free-ranging animals. The sounds generated in feeding, ingestion and rumination have rich patterns that may have potential for telemetric monitoring of dietary aspects of intake in grazing animals.

References

Galli, J. R., C. A. Cangiano & E. A. Laca (2003) Analisis acustico de la masticacion y del consume de forrajes frescos y secos en novillos. Revista *Argentina de Produccion Animal Supl.* 1 (in press).

Laca, E. A., E. D. Ungar & M. W. Demment (1994) Mechanisms of handling time and intake rate of large mammalian grazers. *Applied Animal Behaviour Science*, 39, 3-19.

Using the n-alkane technique to estimate the herbage intake and diet composition of cattle grazing a *Miscanthus sinensis* grassland

Y. Zhang[1], Y. Togamura[2] and K. Otsuki[2]
[1]Animal Science and Technology College, China Agricultural University, West Road 2 Yuan Ming Yuan, Beijing, P.R. China 100094, Email: zhangyj@cau.edu.cn, [2]National Institute of Livestock and Grassland Science, Senbenmatsu, Nishinasuno, Tochigi, Japan 329-2793

Keywords: n-alkane technique, intake, diet composition, *Miscanthus sinensis*

Introduction Plant wax alkanes are now widely used as marker substances (Dove & Mayes 1991) for the estimation of forage intake and diet composition of grazing herbivores. The objective of this study was to evaluate this method with cattle grazing a *M. sinensis* grassland in Japan.

Materials and methods Four cattle were continuously stocked on the grassland from 28 July to 24 August. The sward consisted of 85% *M. sinensis*, 5% *Pleiablastus chino* and 10% *Aralia elata* and *Lespedeza bicolour.* Animals were dosed with a controlled release device capsule (Captec[TM], New Zealand).. Faecal samples of individual animals were taken once daily by hand, immediately after faecal excretion on the ground. Meanwhile, herbage samples of each species were hand plucked. Alkane concentrations were determined in the samples as described by Zhang *et al.* (2002). The proportion of each species consumed were calculated using the software package 'Eatwhat' (Dove & Moore, 1995). The alkane concentrations in the diet were corrected using the proportions of the pasture species and the alkane concentrations in each species. Herbage intake was calculated on the basis of the C_{33}/C_{32} ratio using the equation described by Dove & Mayes (1991).

Results The animals had consumed 10 - 40% *P. chino* and 3 - 12% *L. bicolour* each day. Average DM intakes ranged from 1.6 - 2.3% of live weight (MBW) and were significantly. ($P < 0.05$) related to the change in body weight (Table 1).

Table 1 Daily herbage intake on *M. sinensis* grassland

Animal	MBW (kg)	Daily herbage intake (kg) Mean ±SD	DMI as % of body weight	Change in body weight (kg/day)
1	388.5	6.17±1.02	1.59	0.04
2	523.5	11.74±0.55	2.24	0.78
3	385	8.69±0.71	2.26	0.22
4	418.5	6.98±1.04	1.67	-0.19

MBW: mean body weight over a 27 days period; SD: standard deviation.

Conclusions The relative proportions of the different species in the diet was successfully estimated using the alkane technique. Using the diet composition correction, herbage intake of *M. sinensis* grassland can be successfully determined.

References

Dove, H. & R.W. Mayes (1991). The use of plant wax alkanes as marker substances in studies of the nutrition of herbivores: a review. *Australian Journal of Agricultural Research* 42, 913-925.

Dove, H. & A.D. Moore (1995). Using a least-squares optimization procedure to estimate diet composition based on the alkanes of plant cuticular wax. *Australian Journal of Agricultural Research* 46, 1535-1544.

Zhang, Y. J., Y. Togamura & K. Otsuki (2002). Differences in the n-alkane concentration of four wild plants species in Japan. *Grassland Science* 48, 50-52.

Estimating pasture intake by cattle using alkanes and a known amount of supplement

E. Charmley[1], J.L. Duynisveld[1] and H. Dove[2]
[1]Crops and Livestock Research Centre, Nappan, Nova Scotia, B0L 1C0, Canada, Email: Charmleye@agr.gc.ca,
[2]CSIRO Plant Industry, Canberra, 2601, Australia

Keywords: intake, pasture, alkane

Introduction The alkane ratio method for estimating pasture intake involves calculating the fecal ratio of plant (endogenous) and exogenous alkanes. This method is effective for sheep, although the delivery mechanism for the exogenous alkanes has presented challenges in cattle (Charmley *et al.* 2003). Dove *et al.* (2003) have shown that the relative concentration of components in a mixed diet can be estimated from fecal alkane concentrations using least squares methods. Further, if the amount of one dietary component is known, then the amount of all components, and hence intake, can be determined. In this trial beeswax was added to barley (BWB) giving the mixture a unique alkane composition. Known amounts of this mixture were then fed to cattle grazing three sward types.

Materials and methods In a balance study with 6 steers, intake of silage and BWB was measured, and used to estimate alkane recovery by collecting total faeces. Least squares methods were used to estimate the proportion of BWB consumed from concentrations of alkanes in faeces, BWB and silage. Knowledge of the proportion and amount of BWB fed was then used to estimate forage intake. In the grazing study, cattle were given 4 kg BWB/d while rotationally grazing one of three pasture types; native (a mixture of *Poa pratensis*, *Phleum pretense* L., *Festuca pratensis* Huds. and *Trifolium repens* L.); timothy/red clover (*Trifolium pratense*) and tall fescue (*Festuca arundinacea* Shreb.). Each sward was grazed by 5 heifers (450 kg LW) and replicated twice. Pasture intake was measured over 4 d on two occasions. Intakes based on the difference in DM yield at the beginning and end of the 4 d grazing period (sward method). This method was compared with DM intake estimated from alkane concentration in faecal grab samples taken daily over the 4 d period (alkane method) using least squares as described above.

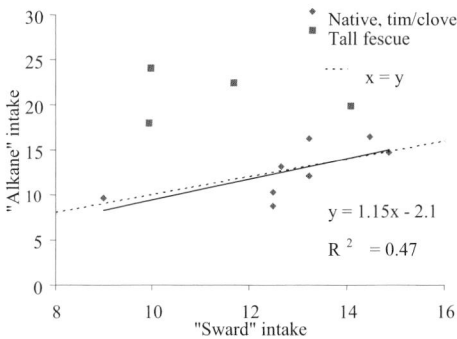

Figure 1 Relationship between predicted and pasture DM intake. Equation does not include tall fescue

Results Fecal alkane recovery ranged from 64 to 100%. Estimated silage intake in the metabolism study was 4.0 kg/d (SD±0.43), similar to the observed value of 3.83 kg/d (SD±0.008). The variation in discrepancy ranged from 0.07 to 0.96 kg. On pasture, the alkane method was unable to predict intake of the tall fescue sward. However, within native and timothy/red clover swards the relationship between predicted and observed methods was close, although the R^2 was only 0.47 (Figure 1).

Conclusions Least squares methods successfully estimated supplement and forage proportions in silage-based diets, and complex swards with at least 3 forage types and was thus a useful method for predicting intake from the known amount of supplement fed. Surprisingly, in the simplest sward, comprising over 95% of tall fescue, the method did not work. This may be due to the inability to measure accurately pasture DM yield in tall fescue and selection for minority species due to low palatability of tall fescue.

References
Charmley, E., D.R. Ouellet, D.M. Veira, R. Michaud, J.L. Duynisveld and H.V. Petit (2003). Estimation of intake and digestibility of silage by beef steers using a controlled release capsule of n-alkanes. *Canadian Journal of Animal Science*, 83, 761-768.
Dove, H., E. Charmley, & K.V. Kleven (2003) Using n-alkanes to estimate intakes of mixed forages by feeding a known amount of an alkane-labelled supplement. *Canadian Journal of Animal Science*, 83, 641.

An evaluation of the *n*-alkane technique for determining diet composition in animals grazing complex swards

M.D. Fraser, J.M. Moorby, V.J. Theobald and R. Jones
Institute of Grassland and Environmental Research, Plas Gogerddan, Aberystwyth, SY23 3EB, UK, Email: mariecia.fraser@bbsrc.ac.uk

Keywords: diet composition, grazing, *n*-alkanes, alcohols

Introduction The *n*-alkane profiles of epicuticular waxes derived from different plant species are sufficiently distinct to allow assessment of the proportions of different herbages in two-component mixtures including perennial ryegrass/white clover, heather/hill grass and rush/perennial ryegrass (Dove & Mayes, 1996). Evidence suggests the reliability of such estimates declines as the number of different dietary components increases. However, recent studies have shown analysis of additional compounds, including long-chain fatty alcohols, may improve discrimination between different dietary components. The aim of this experiment was to quantify the accuracy of such methods for determining the diet composition of animals grazing complex swards.

Material and methods Twelve mature, barren Welsh Mountain ewes were zero-grazed on defined diets of heathland plant species. After a 22-day adaptation period, the dry matter weight and botanical composition of feed offered and refused were determined over a 7-day measurement period. A representative sub-sample of each dietary component was then analysed for *n*-alkane (C21-C35) and long-chain fatty alcohol (C20-C30) concentrations. The Solver function of Microsoft Excel was used to estimate diet composition based on the relative proportional profile of *n*-alkanes and fatty alcohols in the faeces of each animal. Solver was set to alter the proportions of diet components to minimise a function consisting of the sum of squares of differences between diet and faeces profiles and the r^2 value of a linear regression of the two. Lin's concordance correlation coefficients (r_c) between estimated proportions of the major components of the diets and the actual values for each animal were calculated, as described by of Dhanoa *et al.* (1999), for a range of models (n=14 in total) using different alkanes and fatty alcohols, with measured faecal recoveries or recoveries assumed to equal to 1. Results are presented for five of the best models: C21-C35 alkanes, no alcohols, measured recoveries (P1); C23-C35 alkanes, plus alcohols, recoveries = 1 (P2); C25-C35 alkanes, no alcohols, measured recoveries (P3); C27-C35 alkanes, no alcohols, measured recoveries (P4); and C27-C35 alkanes, no alcohols, recoveries = 1 (P5).

Results No single model could predict all dietary components accurately (Table 1). One of the main components, *Calluna vulgaris*, was very well predicted in most cases (r_c close to 1), whilst *Vaccinium myrtillus* was not well estimated by any model (r_c close to 0) but accounted for a small proportion of the diet. Including alcohols in the models did not always improve the r_c, and the accuracy was not necessarily reduced by assuming recovery rates of 1.

Table 1 Mean actual and best predicted diet component percentages from a range of models, with the concordance correlation coefficient (r_c) of the best fit model for each component (all components live growth except [†])

| Dietary component | Actual | Prediction models (best 5 of 14) | | | | | r_c |
		P1	P2	P3	P4	P5	
Molinia caerulea	41.4	35.8	42.6	40.3	42.2	35.3	0.435
Calluna vulgaris	18.2	16.9	9.0	17.9	17.7	17.4	0.970
Vaccinium myrtillus	0.21	0.25	0.15	0.24	0.25	0.24	0.081
Erica tetralix	0.17	0.08	0.15	0.11	0.11	0.08	0.647
Juncus effusus	6.96	6.55	4.74	7.04	7.40	0.55	0.547
Festuca spp.	1.72	1.60	1.95	1.52	1.49	5.74	0.118
Carex spp.	3.95	2.73	17.83	4.46	4.90	0.71	0.153
[†]Dead grass	21.1	17.8	12.8	21.8	19.5	25.5	0.241
Moss	6.4	18.3	10.8	6.7	6.4	14.5	0.141

Conclusions These results demonstrate that *n*-alkanes can be used to estimate several components within the diet of animals grazing complex swards. The most appropriate model for predicting diet composition will depend on which plant types are likely to be the main components within the diet.

Acknowledgements This work was funded by Defra, the Countryside Council for Wales, and English Nature.

References
Dhanoa, M. S., S. J. Lister, J. France & R. J. Barnes. (1999). Use of mean square prediction error analysis and reproducibility measures to study near infrared calibration equation performance. *Journal of Near Infrared Spectroscopy*, 7, 133-143.
Dove, H. & R. W. Mayes. (1996). Plant wax components: a new approach to estimating intake and diet composition in herbivores. *Journal of Nutrition*, 126, 13-26.

Estimating pasture intake by dairy cows

D.R. Cosgrove and D.P. Cooper
College of Agriculture, Food and Environmental Science University of Wisconsin, River Falls, Wisconsin 54022,
Email: dennis.r.cosgrove@uwrf.edu

Keywords: pasture yield, intake, dairy cows

Introduction   Proper nutrient management planning minimizes the environmental impact of manure from dairy farms. Manure output from dairy cows can be predicted from feed intake (Wilkerson *et al.*, 1997). Weighing feed and refusals each day can determine accurately the feed intake of dairy cows in confinement. Intake determination is more difficult for dairy cows on pasture (Vasquez & Smith, 2000). As part of a larger study aimed at estimating manure production of dairy cows on pasture, this study compares 3 methods for estimating pasture yield and feed intake.

Materials and methods   Six rotationally grazed dairy farms were selected to represent varying geographical areas of Wisconsin. Before each grazing event, $1m^2$ samples were clipped to a height of 5 cm to determine yield. Yield was also estimated using a self-made acrylic pasture plate, where 1 cm = 194 kg/ha, and by measuring canopy height, where each 1 cm = 134 kg/ha (Cosgrove, 1999). These measurements were repeated after grazing in order to calculate animal intakes. Milk yields were obtained for the 24- hour period during the grazing event. Other feed inputs, such as corn grain, silage, dry hay, etc., were recorded. Predicted milk yield based on intake measurements were compared with actual milk yields to test the veracity of the different intake measurements in the pasture. Data was analysed as a completely randomised design using farms as replicates.

Results   Table 1 shows pasture intake, calculated using yield estimates from the three different methods. Actual clipping of pasture samples and using a plate meter gave similar intake estimates. Estimating pasture intake using yields based on height alone gave significantly higher intake figures. These intake estimates were then used to predict milk production. These predictions were compared to actual bulk tank milk yield measurements (Table 2). Pasture yield estimated by the clipping and plate meter methods gave milk yield predictions that were similar to the actual milk yields. Using height alone predicted a significantly higher milk yield than was actually measured.

Table 1 Pasture intake as estimated by three different methods[1]

Method	Intake (kg/cow/day)
Clipping	12.0a
Plate meter	15.6a
Height	34.3b
SE	4.5

[1]Means followed by the same letter are not significantly different (P<0.05)

Table 2 Milk production of grazing cows as estimated by three different methods[1]

Method	Milk production (kg/day)
Clipping	24.8a
Plate meter	27.0a
Height	36.9b
Actual milk weight	25.1a
SE	2.4

[1]Means followed by the same letter are not significantly different (P<0.05)

Conclusions   Using clipping and plate meters to estimate pasture yield and animal intake on pasture are more reliable methods that using pasture height alone.

References

Cosgrove, D. R. (1999). Evaluation of a simple method for measuring pasture yields. *Proceedings of the 23rd Forage Production and Use Symposium,* Appleton, WI.23:123-125.

Vasquez, O. P. & T. R Smith (2000). Factors affecting pasture intake and total dry matter intake in grazing dairy cows. *Journal of Dairy Science*, 83:2301.

Wilkerson, V. A., D. R. Mertens & D. P. Casper. 1997. Prediction of excretion of manure and nitrogen by Holstein dairy cattle. *Journal of Dairy Science*, 80:3193.

Theoretical considerations on a one-parameter approach to compare actual and estimated compositions of multi-component diets

C. Elwert and M. Rodehutscord
Institute for Nutritional Sciences, Martin-Luther-Universität Halle-Wittenberg, 06099 Halle /S., Germany, Email: Christian_Elwert@web.de

Keywords: diet composition estimation, accuracy

Introduction The composition of ingested herbage mixtures can be estimated using the alkane technique (Dove & Moore, 1995), with the accuracy of the estimate assessed by linear regression of estimated and actual proportions of the dietary components (Dove, 1992). However, although the linear regression might not differ from the line of equality, large discrepancies may occur within individual components (Hoebee *et al.*, 1998). This paper presents an approach to compare actual and estimated diet compositions using only one parameter.

Describing the problem The x^2-test ($\sum (o_i-e_i)^2/e_i$), with o and e being observed and expected frequency of component i, is commonly used to statistically compare expected and observed distributions. In grazing situations animals may completely avoid some species, although this may not be detected by visual observation. Consequently, those species will be included as potential components in the estimate of diet composition. In balance trials simulating such selective intake, the accuracy of the resulting estimate of diet composition cannot be analysed by the x^2-test, because the denominator for the component selected against becomes zero.

Solution Estimates of diet composition are not independent for each dietary component, since the proportions of all components have to add up to 1. Thus, let individual dietary components be the dimensions of a multi-dimensional space. The constraint of diet composition (all proportions sum up to 1) results in a (n-1)-dimensional object, which represents all combinations of dietary components possible (including single-component diets). Therefore, this object contains both estimated as well as actual diet. The 'distance' (D) between estimated and actual diet can be calculated according to Pythagora's theorem as the square root of the sum of the squared differences between the estimated and known proportion of each component:

$$D = \sqrt{\sum_{i=1}^{n} (a_i - e_i)^2}$$

with a and e representing actual and estimated proportions respectively, of dietary component i. It should be noted that D is a theoretical parameter of the 'similarity' of actual and estimated diet composition, thus conclusions about individual components cannot be drawn directly.

Discussion Although Distance D indicates the degree of similarity between actual and estimated diet composition, it is not a statistical test of differences. In contrast to the Mean Discrepancy ($MD=D \cdot (1/n)$), which yields the (absolute) difference between the actual and estimated proportion averaged across all dietary components, D is independent of the number of components (Table 1). A value of D of 70 g/kg is suggested as a limit for assuming diet similarity. This corresponds to a maximum difference in a single dietary component of 70 g/kg (or in diets of up to four species, 50-60 g/kg), if this difference is compensated equally by all other components (Table 1). Note, that D accounts only for absolute differences between actual and estimated proportions, a possible inclusion of relative differences requiring further research.

Table 1 Differences between actual and estimated proportion of dietary components (g/kg) and resulting *Distance (D)* and *Mean Discrepancy (MD)*[#]

Case	Dietary component				D	MD
	1	2	3	4		
(1)	+49.5	-49.5	/	/	70	49.5
(2)	+49.5	0	-49.5	/	70	40.4
(3)	+57.2	-28.6	-28.6	/	70	40.4
(4)	+60.6	-20.2	-20.2	-20.2	70	35.0

[#] Due to the mathematics of calculating D and MD, the unit for both parameters are g/kg

References
Dove, H. (1992). Using the n-alkanes of plant cuticular wax to estimate the species composition of herbage mixtures. *Australian Journal of Agricultural Research*, 43, 1711-1724.
Dove, H. & A.D. Moore (1995) Using a least-squares optimization procedure to estimate botanical composition based on the alkanes of plant cuticular wax. *Australian Journal of Agricultural Research*, 46, 1535-1544.
Hoebee, S.E., H. Dove & D.I. Officer (1998) Using plant wax alkanes to estimate the species composition of sub-tropical grass mixtures. *Animal Production in Australia*, 22, 364.

Elasticity of ingestive behaviour and intake in sheep associated with food diversity on plurispecific swards

C. Cortes[1,2], J.C. Damasceno[2], G. Bechet[1], J. Jamot[1] and S. Prache[1]
[1]Unité de Recherches sur les Herbivores, INRA Clermont-Ferrand/Theix, 63122 St-Genès-Champanelle, France, Email: prache@clermont.inra.fr, [2]Departamento de Zootecnia, Universidade Estadual de Maringá (UEM), Avenida Colombo, 5790, Maringá, Paraná, Brazil

Keywords: grazing behaviour, dietary choices, grazing, intake

Introduction Animals on heterogeneous swards generally opt for a varied diet. This may stimulate their intake, unless searching constraints limit intake rate (Champion *et al.*, 1998). However the management of plurispecific swards presents a risk of overgrazing the preferred species and undergrazing the less-preferred species. This study aimed to test the effect of type of diversity and type of management on the elasticity of ingestive behaviour and intake in sheep.

Material and methods Five treatments were compared with 5 groups of 5 dry INRA 401 ewes during 5 periods, using a latin-square design: L=grazing a monoculture of *Lolium perenne cv.* Herbie; F=grazing *Festuca arundinacea cv.* Florine; FLF=L+F=grazing conterminal monocultures, 0.50:0.50 by ground area, the animals having a free choice between both species; DLF=grazing L from 1600-0900h and F from 0900-1600h, the choice being made by the researcher; ILF=grazing a mixture of both species that were finely imbricated. Sward height was maintained at 9 cm by regular mowing. Each period comprised a 5-day adaptation period and a 5-day measurement sub-period. Dietary choices and intake were measured using the n-alkanes technique. Grazing time was assessed using the automatic Ethosys system (Scheibe *et al.*, 1998). Data were analysed using the SAS-GLM procedure and the Duncan contrast procedure for pair-wise comparisons.

Results The *Lolium perenne/Festuca arundinacea* association is a good model to study foraging behaviour and diet selection. Preference for ryegrass was marked (Table 1); n-alkane profiles between both species were steady and contrasting. The proportion of *Lolium* in the diet was 76.9, 75.0 and 78.2% for FLF, DLF and ILF respectively; the proportion of *Lolium* in the corresponding swards was 50, 52 and 65% respectively. Each treatment, period and animal had a significant effect on intake and ingestive behaviour. Intake and intake rate was higher in L than in F. Intake was higher in DLF and ILF than in L (+13% and +17%, respectively). Grazing time was higher in FLF and ILF than in L (+15% and +17%, respectively). There was no difference in diet selection, intake and grazing behaviour between FLF, DLF and ILF. Mediated via an increase in grazing time rather than an increase in intake rate, intake was higher on mixed ryegrass-tall fescue swards than on ryegrass (the preferred monoculture).

Table 1 Ingestive behaviour and intake of ewes offered monocultures of *Lolium perenne*, monocultures of *Festuca arundinacea,* or swards associating both plant species, maintained at 9 cm-high

Treatment	F	L	FLF	DLF	ILF	s.e.
Lolium in total biomass (%)			50	52	65	
Lolium in the diet (%)			76.9 a	75.0 a	78.2 a	7.7
Intake (g OM/day)	1557 c	1822 b	2028 ab	2055 a	2130 a	358
Grazing time (min/day)	383 bc	366 c	420 ab	397 abc	430 a	53
Intake rate (g OM/min)	3.996 b	5.400 a	4.900 a	5.006 a	4.867 a	0.902

Within a row, values with different letters differ significantly (P<0.05)

Conclusions Offering a choice of herbage species to ewes increased intake, mediated via an increase in grazing time rather than an increase in intake rate. There was no interaction with the type of diversity (conterminal monocultures vs mixture) or the type of management of the diversity (free choice vs directed choice) on ingestive behaviour and intake.

References

Champion, R. A., S. M. Rutter, R. J. Orr & P.D. Penning (1998). Costs of locomotive and ingestive behaviour by sheep grazing grass or clover monocultures or mixtures of the two species. 32. International Congress of the ISAE. Clermont-Ferrand, France. 213.

Scheibe, K. M., Th. Schleusner, A. Berger, K. Eichhorn, J. Langbein, L. Dal Zotto & W. J. Streich (1998). ETHOSYS® - New system for recording and analysis of behaviour of free-ranging domestic animals and wildlife. *Applied Animal Behaviour Science*, 55, 195-211.

Diversity of diet composition decreases with conjoint grazing of cattle with sheep and goats

A.M. Nicol[1], M.B. Soper[1] and A. Stewart[2]

[1]Agriculture and Life Sciences Division, PO Box 84, Lincoln University, New Zealand, Email: nicol@lincoln.ac.nz, [2] PGG Seeds, PO Box 3100, Christchurch, New Zealand

Keywords: diet selection, cattle, sheep, goats

Introduction Conjoint or mixed grazing can affect the diet selected by each species (Nicol & Collins, 1990). Diet similarity coefficients are often used to compare *pairs* of diets (Krebs, 1999). However this approach is awkward when a number of contrasts are required in a multifactorial comparison. Species diversity is a descriptor of a *particular* environment. Many models provide an estimate of *species* diversity, the most common of these being a log-normal distribution (Tokeshi, 1996). We tested whether this model could be applied to dietary *components* selected from a pasture, and thus provide a coefficient of dietary diversity for the *individual* diets of cattle, sheep and goats when grazed alone or in mixtures, which could then be statistically compared.

Methods Groups of cattle, sheep and goats grazed alone (CA; SA; GA) or as cattle plus sheep (CS; SC) and cattle plus goats (CG; GC), during a progressive defoliation (4400 to 1550 kg DM/ha) of a ryegrass/white clover pasture during 20 days in summer. Oesphageal extrusa (OE) samples were collected from two of each animal species on most days (n = 116 OE samples). The samples were dissected into six botanical components (grass leaf, grass stem, grass seed head, clover leaf and petiole, clover flower and dead material). The dietary diversity coefficient (k) was estimated for each OE by an iterative minimisation of the sum of the squared deviations of the observed proportion of each component and the predicted proportion (P) of each component from $P_R = 100$ $(1-k) k^{R-1}$, where R = the rank of each component in the observed OE composition. Dietary diversity coefficients were compared by analysis of variance using species (cattle, sheep and goats) and grazing environment (alone or mixed), and their interaction, as fixed effects.

esults and discussion Simple correlations between predicted and observed OE composition ranged from 0.87 to 0.99. The mean k for cattle was significantly greater than that for sheep and goats (Table 1) and the effect of conjoint grazing was to significantly reduce the diversity of the diet.

Table 1 The dietary diversity coefficient (k) and *in vitro* dry matter digestibility of oesophageal extrusa of cattle, sheep and goats grazed alone or conjointly

	Dietary diversity coefficient (k)				*In vitro* DMD (g DMD/kg DM)			
	Alone	Conjoint	Average	sem	Alone	Conjoint	Average	sem
Cattle	0.582	0.522	0.552 c		613	595	604 c	
Sheep	0.505	0.431	0.468 d	0.025	619	641	630 d	6
Goats	0.504	0.401	0.452 d		617	632	624 d	
Average	0.53 a	0.451 b		0.021	616 a	624 b		5

Values followed by a different letter (a or b in rows, c,d in colomns) are significantly different (P<0.01)

Decreased diet diversity was associated with a higher *in vitro* digestibility of OE for sheep and goats, especially when they were conjointly grazed with cattle. This probably reflects their ability (smaller incisor arcade breadth), and opportunity when mixed with cattle (SC and GC), to exploit their dietary preferences. In contrast the reduced diet diversity of CS compared with CA (significant interaction), was associated with a reduction in *in vitro* digestibility, suggesting that the quality of the diet of cattle suffered when they were in competition with sheep.

References

Krebs, C.J. (1999). Ecological methodology. Benjamin/Cummings, Addison Wesley Longman, California, USA, 701 pp.

Nicol A.M. & H.A. Collins (1990). Estimation of the horizons grazed by cattle, sheep and goats during single and mixed grazing. *Proceedings of the New Zealand Society of Animal Production* 50, 49-53.

Scott, D. (1993). Constancy in pasture composition? *Proceedings Seventeenth International Grassland Congress,* 1604-1606.

Tokeshi, M. (1996). Species co-existence and abundance: patterns and processes. In: A. Takuya, S.A. Levin, M. Higashi (eds.) Biodiversity – An ecological perspective. Springer-Verlag New York, 35-55.

Cattle and sheep mixed grazing: 1. species equivalence

R.D. Améndola-Massiotti, S.J.C. González-Montagna and P.A. Martínez-Hernández
Universidad Autónoma Chapingo, Posgrado en Producción Animal, Programa de Investigación en Forrajes, Carretera México-Texcoco.Km 38.5, Chapingo, Edo. Méx., CP56230, México, Email: r_amendola@yahoo.com
Keywords: heifers, ewes, stocking rate, herbage intake

Introduction The effects of mixed grazing of cattle and sheep depend on stocking rate (SR) and species ratio (Nicol, 1997). Calculations of SR and species ratio require the use of species equivalence. Equivalents are often estimated in terms of intake requirements related to live weight (LW), while maintenance energy requirements are calculated on the basis of $LW^{0.75}$. Freer (1981) stated that $LW^{0.9}$ would be more appropriate for comparisons of intake requirements for maintenance of sheep and cattle. Nonetheless, Nolan & Connolly (1977) stated that the equivalent is system-specific and depends on the species being considered. The objective of this experiment was to estimate species equivalence for a dairy system based on grazing in temperate Mexico.

Materials and methods The experiment was carried out at Chapingo, Mexico (19°29' N, 98°54' W,2240 m a.s.l.), between 15 March and 25 May 2000. Holstein heifers (initial LW 336±7 kg) and pregnant Suffolk ewes (initial LW 75.8±0.8 kg) were used. There we 8 treatments, 3 heifers were used per mixture treatment and numbers of ewes varied to achieve the proportions of species shown in Table 1. Nine paddocks of *Medicago sativa* and *Dactylis glomerata* of 0.46 ha were strip-grazed for 7 days each. Three blocks of 3 paddocks were formed and treatments were randomly allotted to paddocks; only one paddock was grazed per week, the paddocks of the first and second week of each block were grazed by three treatments (each on separate areas), while the third-week paddock was grazed by the remaining two treatments. Fresh areas of pasture were allotted when already grazed areas reached 10 cm height (falling disc); no back fencing was used. The areas effectively grazed after 7 days were measured. Herbage dry matter (DM) intake was measured using chromium oxide as external marker and *in situ* digestibility of hand plucked herbage samples.

Results Excluding the treatment of grazing by ewes only, a linear equation was calculated of area allotted (m^2/d) on number of ewes: y = 97.9 + 4.29 x, R^2 = 0.94, (P < 0.05) (Table 1); the intercept and the regression coefficient represent the area allotted to 3 heifers and the additional area allotted per ewe, respectively. The equivalent based on area allotted resulted in 7.6 ewes per heifer [(97.9/3)/4.29]. The equivalent based on average herbage DM intake in Table 1 (8.76 and 2.31 kg DM/animal/d) for heifers and ewes, respectively) resulted in 3.8 ewes per heifer (8.76/2.31), corresponding with the estimate of intake requirements using $LW^{0.9}$ ($336^{0.9}/75.8^{0.9}$), as recommended by Freer (1981). This difference between estimates of equivalents, expresses the higher efficiency of herbage utilisation under mixed grazing than under single species grazing by heifers, which was due to grazing by sheep below 10 cm height (data not shown here).

Table 1 Area allotted and herbage intake of heifers and ewes under single species grazing and mixed grazing with different ewes to heifer's ratios

Nr. of heifers	Nr. of ewes	Area allotted (m^2/d)	Intake	
			(kg DM/heifer/d)	(kg DM/ewe/d)
3	0	93.4	8.17 (0.49)	
3	6	131.6	10.50 (0.33)	2.08 (0.15)
3	9	148.2	7.68 (0.49)	2.07 (0.15)
3	12	116.0	9.54 (0.33)	1.98 (0.15)
3	24	218.6	8.98 (0.33)	2.52 (0.15)
3	36	269.5	6.49 (0.49)	2.49 (0.15)
3	48	287.6	9.93 (0.33)	2.49 (0.15)
0	15	91.5		2.54 (0.15)

Conclusions The use of an equivalent based on DM intake (3.8 ewes/heifer), which concurs with intake requirements based on $LW^{0.9}$, would neglect benefits of more efficient pasture utilisation under mixed grazing; the use of the higher equivalent based on area allotted (7.6 ewes/heifer) should lead to better performance of the system.

References

Freer, M. (1981). The control of food intake by grazing animals. In: F. H. W. Morley (Ed.). Grazing Animals. Elseviers Scientific Publishing Company, Amsterdam, 105-124.
Nicol A. M. (1997). The application of mixed grazing. In: J. G. Buchanan-Smith, L. D. Bailey and P. Mc Caughey (Eds.) *Proceedings of the XVIII International Grassland Congress*, Winnipeg and Saskatoon, Canada, 525-534.
Nolan, T. & J. Connolly. (1977). Mixed stocking by sheep and steers. A review *Herbage Abstracts*, 47, 367-364.

Cattle and sheep mixed grazing: 2. competition

S.J.C. González-Montagna, P.A. Martínez-Hernández and R.D. Améndola-Massiotti
Universidad Autónoma Chapingo, Posgrado en Producción Animal, Programa de Investigación en Forrajes, Carretera México-Texcoco.Km 38.5, Chapingo, Edo. Méx., CP56230, México, Email: r_amendola@yahoo.com

Keywords: heifers, ewes, pasture utilisation, replacement series

Introduction The outcome of mixed grazing depends on the degrees of complementarity and competition between animal species. Complementarity increases the utilisation of herbage resource but competition may be desirable when one grazing species has a higher priority ranking in the farming system. A species wins in the competition by harvesting a higher proportion of the available herbage than the other (Nicol, 1997). De Wit (1960) used the replacement series based on degrees of substitution of species, for the quantification of the outcome of mixtures experiments. The use of species equivalence is required in order to apply this approach to the analysis of the outcome of a mixed grazing experiment. This paper aims to discuss the competition between heifers and ewes in a dairy system, where heifers have a higher priority ranking. Data on species equivalence, reported in Paper 1, were used for this purpose.

Materials and methods The species equivalence based on average herbage dry matter intake (DMI) estimated in Paper 1 corresponded with the estimate of DMI requirements based on live weight (LW), using $LW^{0.9}$. Analysis using the replacement series approach was carried out *a posteriori*. Species proportions (SP) were calculated as SP= • •$LW^{0.9}$ of the species/• •$LW^{0.9}$ of both species (average LW 361 and 75.7 kg for heifers and ewes, respectively). DMI/ha of each species (DMI_a) was calculated as $II_a \times n_a/AA \times 10000$, where II_a is the DMI/animal of species a, n_a is the number of animals of species a and AA is the daily area allotted (m^2). The proportions DMI/ha of each species in total DMI/ha were calculated thereafter and linear regression equations of those proportions on the proportions in total $LW^{0.9}$ were developed.

Results Total DMI increased with the proportion of ewes, i.e., the inclusion of ewes increased the efficiency of herbage utilisation (Figure 1a). There was a narrow relationship (R^2=0.97, p<0.0003) between proportions of DMI and proportions of $LW^{0.9}$, with small deviations from the 1:1 relationship (Figure 1b). Both species appeared to be more competitive to some extent when in high proportions of total $LW^{0.9}$. Therefore, heifers were slightly more competitive than ewes if their proportion in total LW^{09} was >0.47 (Figure 1b).

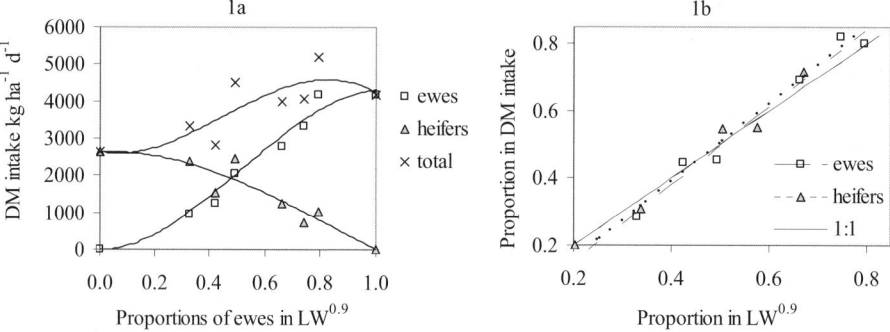

Figure 1 Herbage DMI of heifers and ewes related to proportions of ewes in total $LW^{0.9}$ (1a) and DMI of each species as proportion of total DMI related to proportions of the species in total $LW^{0.9}$ (1b)

Conclusions Ewe inclusion increased the efficiency of utilisation of herbage. One might recommend ewe inclusion at <0.53 of total $LW^{0.9}$ because that gave a slight competitive advantages to heifers (the species with higher priority ranking in the farming system).

References
de Wit, C. T. W. (1960). On competition. Verslag Landbouwkundig Onderzoek No. 66, Pudoc, Wageningen, The Netherlands, 1-82.
Nicol, A. M. (1997). The application of mixed grazing. In: Buchanan-Smith, J. G., L. D. Bailey & P. Mc Caughey (Eds.) *Proceedings of the XVIII International Grassland Congress*, Winnipeg and Saskatoon, Canada, 525-534.

Biomass vertical distribution in a grazed grassland under monoespecific and mixed grazing

C. Saroff, S. González, A. Ohanian, H. Pagliaricci and Y.R. Chiaramello
Facultad de Agronomía y Veterinaria de la Universidad Nacional de Río Cuarto, Ruta 36 km 601, X5804BYA, Río Cuarto, Córdoba, Argentina, Email: csaroff@ayv.unrc.edu.ar

Keywords: pasture, mixed pasturing, biomass

Introduction Mixed grazing is defined as the use of the same forage resource for more than one herbivore species. It has been shown that different herbivore species have specific grazing modalities (Black and Kenney, 1984), which may differentially modify the structure of the pasture. The aim of this study was to evaluate the biomass vertical distribution in a sward with mixed grazing.

Materials and methods Two treatments with two random repetitions were applied in a pasture of *Medicago sativa* L., *Trifolium repens* L., *Festuca arundinacea* Schreb., *Dactylis glomerata* L. y *Bromus unioloides* H.B.K. One treatment was grazed by Aberdeen Angus steers of 204±26 kg LW, while the other, was grazed by Aberdeen Angus steers, 204±26 kg LW, and Corriedale ewes of 48±12.7 kg LW. The stocking rate was 715 kg LW T ha^{-1} for both treatments. In the mixed treatment, the stocking rate was composed by 60% of steers (429 kg LW) and the 40% of ewes (286 kg LW). A rotational stocking of 7 days of grazing and 35 days of rest was implemented. To determine the vertical distribution of the biomass, it was divided into three strata: lower (a basal part of 7 cm), and the rest were divided into two equal part (medium and upper strata). During two cycle, four plots (0.44 ha each) were sampled before and after of grazing. A factorial experimental design was used in randomly selected blocks. The values obtained were subjected to analysis of variance.

Results The biomass availability showed significant differences only in the distribution in strata.

Table 1 Accumulated biomass (kg DM.ha^{-1}) per stratum affected by grazing method. Cycle I (12/02/2003)-Cycle II (26/03/2003)

STRATA	CYCLE I	CYCLE II	TOTAL
Upper	386.1 c	123.3 b	1169.9 c
Medium	1353.9 a	797.2 a	4738.3 a
Lower	865.0 b	714.4 a	2806.7 b
GRAZING METHOD			
Steers	786.3 ns	555.5 ns	2912.9 ns
Mixed	950.4 ns	534.4 ns	2897.3 ns
R^2	0.67	0.76	0.84
C.V. (%)	37.5	35.2	25.0
Probability			
Stratum	0.0016	0.0004	0.0008
Grazing	0.1647	0.7277	0.9600
Str.x Graz.	0.6973	0.6974	0.7400

Table 2 Biomass remnant (kg DM.ha^{-1}) per stratum affected by grazing method. Cycle I (19/02/2003)-Cycle II (02/04/03)

STRATA	CYCLE I	CYCLE II	TOTAL
Upper	47.3 b	52.2 c	178.9 b
Medium	647.8 a	587.7 b	2716.6 a
Lower	878.9 a	1165.5 a	3349.4 a
GRAZING METHOD			
Steers	523.7 ns	485.5 b	2016.6 ns
Mixed	525.5 ns	718.1 a	2146.6 ns
R^2	0.74	0.86	0.93
C.V. (%)	45.4	36.9	20.1
Probability			
Stratum	0.0043	0.0002	0.0002
Grazing	0.9834	0.004	0.5800
Str.x Graz.	0.9913	0.0961	0.9500

The data of availability and residue show the greatest decrease in the higher and medium strata. The lower stratum increased, which demonstrates that the animal species used in the experiment preferred to consume the higher plant strata, which concentrate the highest quality. Hodgson (1994) observes that pastures in a reproductive state offer a heterogeneous mixture of leaves and shoots at all levels. According to him, both height as well as leaf and shoot proportions have different effects on the selective pasturing behavior of the animal species used. In this experiment there were no significant differences in the pasturing habits of the animal species used, which may be due to the short duration of the evaluation period.

Conclusions
The treatments used in the first year of the experiment did not modify the vertical structure of the pasture. This may be due to the short duration of the evaluation period. It may also be due to a difference in the stocking rates observed at the end of the experiment, as the monoespecific treatment was had a higher stocking rate than the mixed treatment.

References
Black, J.L. and P.A. Kenney (1984). Factors affecting diet selection by sheep. II Height and density of pasture. *Australian Journal Agricultural Research*, 35, 565 – 678.
Hodgson, J. 1994. Manejo de pastos: teoría y práctica. Ed. Diana. México. 252 p.

A sward based method to estimate herbage selection of grazing dairy cows

F. Taube, M. Wachendorf and J. Baade[1]
[1]University of Kiel, Department of Grass and Forage Science/Organic Agriculture; Hermann Rodewald Strasse
9, D-24118 Kiel, Germany, Email: ftaube@email.uni-kiel.de

Keywords: pastures, forage mixtures, selection index, biodiversity

Introduction Diet selection of grazing animals is influenced by sward composition and vertical sward structure. Grazing studies were established in northern Germany (Kiel, Schleswig – Holstein state) to determine if selective grazing behaviour in a mixed sward can be measured by a sward based method. The hypothesis that active selection of different functional groups of forages can be documented by using the selection index (Figure 1, Hodgson, 1990) and regressive approaches vs time was tested.

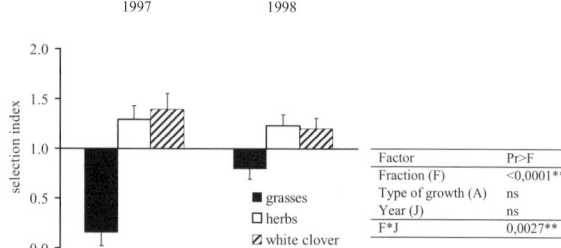

Figure 1 Selection index of dry matter intake (according to Hodgson, 1990) for grasses, herbs and clover (means of growths, 1997 and 1998)

Materials and methods Pastures were a multi-species permanent grassland with grasses (*Lolium perenne* exceeded 80% of the grass fraction), white clover (*Trifolium repens*) and herbs (*Taraxacum officinalis* exclusively). The measurements were conducted in a rotational grazing system during 1997 and 1998. In both years we used three grazing periods each of 5 days of grazing. Clip boards were used to divide the samples into layers of 5 cm each to separate the herbage mass in different plant groups. This procedure was conducted daily during each grazing period (four replicates), starting the day before grazing began. Four non-grazed replicates were used as control. Analysis of variance was performed with the procedure MIXED of SAS software taking into account the factors fraction (grass, clover, herbs), type of growth (spring, summer, autumn) and year (1997, 1998).

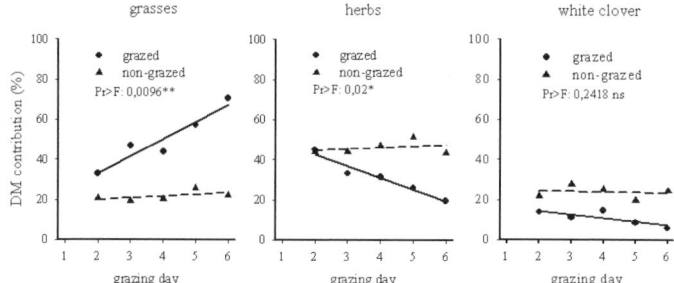

Results Selection index of dry matter intake indicates that cows selected herbs and white clover in preference to grasses (Fig.1). However, the selection index gives no indication as to whether a species in a specific layer has been selected actively or not. Depending on the vertical architecture of a

Figure 2 Effects of defoliation (R, grazed, non-grazed) and time (T, grazing day within growth period) on the contribution of grass, herbs and white clover (% of DM) of the grazed and non-grazed swards (1997, mean of growth periods). Shown is the interaction RxT

species in a mixed sward, stratification effects (passive selection) can overlap with active selection processes. Data from the different layers of grazed and non- grazed swards indicate, that selection of herbs was an active selection process, while selection of white clover was a passive selection process due to the stratification of the sward by the grazing animal. Selection of herbs and white clover obviously depends on the amount of these species in the offered forage. When the percentage of herbs exceeded 40% in offered forage (Fig. 2 ;1997) a significant selection could be documented, while a smaller percentage of herbs (less than 25%) in the sward did not enhance dry matter disappearance of herbs (1998).

Conclusions Herbage disappearance and crude protein content of disappeared herbage indicated that by using a sward based method in an accurate way, quantity and quality of consumed forage can be estimated in its dynamic within the grazing period.

Reference

Hodgson, J. (1990) Grazing Management – Science into Practise. Longman Scientific Technical, New York

Influence of giving a choice of grazing or maize silage offered in the field simultaneously on diet selection of lactating dairy cows

O. Hernandez-Mendo[1] and J.D. Leaver[2]
Imperial College, University of London. Wye, Near Ashford, Kent, UK
Present address: [1]Ganaderia-CP. Km. 36.5 Carr Mexico-Texcoco. Montecillos, Edo. de Mexico 56230 Mexico,
Email: ohmendo@colpos.mx, [2]Royal Agricultural College. Cirencester. Gloucestershire GL7 6JS UK

Keywords: diet selection, grazing, dairy cows

Introduction Changing the times of access to grazing and to maize silage (MS) offered indoors affects the relative intake of each feed, but has little effect on dairy cow performance (Hernandez-Mendo & Leaver, 2000). Offering the MS in the grazing paddock as an instantaneous alternative to grazing should give an insight into the factors influencing feed intake, and may provide a means of alleviating the high rate of decline in milk yield of grazing dairy cows. The objective of this study was to examine the response in diet preferences and performance of lactating dairy cows when grazing a perennial ryegrass sward and having access to maize silage *ad libitum* simultaneously in spring, at two concentrate levels (CL) and two sward heights (SH).

Materials and methods This 35 day study was conducted in spring under continuous stocking and involved 24 multiparous Holstein Friesian cows. Treatments were arranged in a 2x2 factorial design, with two replicates, and CL (0 and 6 kg/cow: 5.5kg barley + 0.5kg soybean, fresh basis) and SH (4-6 and 8-10cm) as the main variables. Fresh MS was offered daily in bins in the field to each group of 3 cows, according to the previous day's intake, and daily intake recorded as the difference between maize silage offered and refused. Herbage dry matter intake (DMI) and total DMI were estimated indirectly from individual animal energy requirements (AFRC, 1993) and the ME concentration of the diets. Animal behaviour was recorded during two 48h observation periods.

Results Cows strongly preferred grazing to maize silage, as illustrated by the high proportion of grazing time (Table 1). For example, mean herbage DMI was 9.9 kg/d, compared with 5.1 kg DM/d for maize silage. Offering concentrates decreased grazing time (GT) and intake rate (IR) of herbage, but had no significant effect on maize silage eating behaviour. Intake of grazed herbage increased with SH even though GT was similar with both sward heights. This was due to IR of herbage being significantly greater at the high sward height.

Table 1 Main means on animal behaviour, diet preference and animal performance of dairy milking cows when grazing and having access to maize silage *ad libitum* simultaneously in the field

		GT	TMS	P-GT	IR, g DM/min		DMI, kg DM/d		LWCh	MY
		(min/d)			HB	MS	MS	HB	kg/d	kg/d
	0	407	69	0.83	30.1	77.4	5.3	12.3	+1.23	25.1
CL	6	363	66	0.82	21.0	74.0	4.8	7.5	+1.35	25.8
	s.e.d.	9.9	7.2	0.035	0.67	7.47	0.26	0.57	0.103	0.41
	Significance	0.05	NS	NS	0.001	NS	NS	0.001	NS	NS
	4-6cm	389	77	0.80	22.1	74.1	5.7	8.7	+1.20	24.7
SH	8-10cm	380	58	0.85	29.0	77.3	4.4	11.1	+1.38	26.1
	s.e.d.	9.9	7.2	0.035	0.67	7.47	0.26	0.57	0.103	0.41
	Significance	NS	NS	NS	0.01	NS	NS	0.01	NS	0.05

GT, grazing time; TMS, time spent eating maize silage; P-GT, GT/GT+TMS; IR, intake rate; DMI, dry matter intake; HB, herbage; MS, maize silage; LWCh, liveweight change; MY, milk yield; NS, Not significant

Conclusions When grazing cows were offered maize silage in the field, cows preferred grazed grass to maize silage. Increasing sward height and increasing concentrate feed level had a greater impact on herbage intake than offering maize silage. The preference for grazing could be due to the higher digestibility of the herbage, and to improved palatability factors of herbage compared to maize silage. The high rate of intake of maize silage may be more beneficial at times when herbage is in short supply or when the feed value of herbage is poor. While it may be possible to exploit preferences for grazing vs maize silage in practice, further research is needed to provide a better understanding of diet preferences under grazing conditions.

References

Hernandez-Mendo O. & J. D. Leaver (2000). Combining grazing with different periods of access to an indoor diet to alleviate high rates of decline in milk yield of dairy cows. *Proceeding of the British Society of Animal Science*. p.76.
AFRC (1993). Energy and protein requirements of ruminants. CAB International. Wallingford, UK.

A simple vegetation criterion (NDF content) may account for diet choices of cattle between forages varying in maturity stage and physical accessibility

C. Ginane and R. Baumont

INRA, Unité de Recherches sur les Herbivores, Centre de Clermont-Theix-Lyon, F-63122 Saint-Genès Champanelle, France, Email: ginane@clermont.inra.fr

Keywords: diet choice, prediction, cattle, intake rate, neutral detergent fibre

Introduction The management of extensively grazed pastures requires an understanding and prediction of the diet choices of herbivores grazing on vegetation that is qualitatively (maturity stage) and quantitatively (biomass, sward height) heterogeneous. The Optimal Foraging Theory (OFT, Stephens & Krebs, 1986), bases its predictions on the relative energy intake rate (EIR) of forages. However, as EIRs are difficult to assess at pasture and are subject to wide intra- and inter-individual variations, another vegetation criterion was sought (accessibility, quality), by-passing the animal's influence, to predict cattle diet choices quantitatively.

Materials and methods The results of two grazing and two complementary indoor experiments (Ginane *et al.*, 2002; Ginane *et al.*, 2003) were pooled. Eighteen-month old heifers were able to choose, throughout the day for approximately 7 days, between two forages (standing swards or hays), varying in relative maturity stage (vegetative *vs.* reproductive) and physical accessibility (sward height). Forages were characterized by their protein and fibrous chemical composition, their digestibility (measured *in vitro*), and their intake rates (measured *in situ* on the animals used in the choice experiments). These measurements yielded EIR values.

Results Diet choices were significantly and positively linked to forage EIR ratio (EIR of the vegetative forage/EIR of the reproductive forage, Figure 1A), consistent with OFT. Among the different criteria tested the difference in NDF content between forages (reproductive-vegetative) was the one most closely related to diet choices (Figure 1B). The close relation between diet choices and the neutral detergent fibre (NDF) criterion may arise because the NDF content is linked to (i) forage prehensibility, as it takes into account sward resistance to defoliation and mastication (Sauvant *et al.*, 1996), (ii) forage ingestibility, as it partly expresses forage fill effect in the rumen (Mertens, 1994), and (iii) forage digestibility, which varies inversely with NDF content.

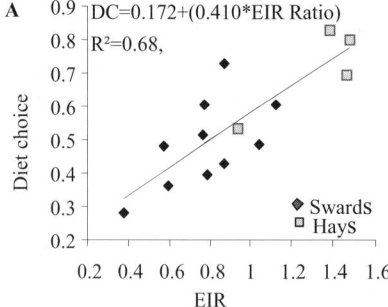 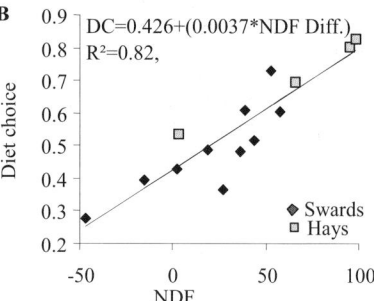

Figure 1 Linear regressions equations of observed diet choices (DC) with (**A**) EIR ratio of forages (vegetative/reproductive) and (**B**) difference in NDF content of forages (reproductive-vegetative)

Conclusions This study indicates that forage NDF content, a common and easy-to-measure criterion, may be useful for predicting cattle diet choices in heterogeneous pastures.

References

Ginane, C., R. Baumont, J. Lassalas & M. Petit (2002). Feeding behaviour and intake of heifers fed on hays of various quality, offered alone or in a choice situation. *Animal Research*, 51, 177-188.

Ginane, C., M. Petit & P. D'Hour (2003). How do grazing heifers choose between maturing reproductive and tall or short vegetative swards? *Applied Animal Behaviour Science*, 83, 15-27.

Mertens, D.R. (1994). Regulation of forage intake. In: G.C. Fahey Jr (eds.) Forage quality, evaluation and utilisation. American Society of Agronomy, Madison, Winsconsin, 450-493.

Sauvant, D., R. Baumont & P.Faverdin (1996). Development of a mechanistic model of intake and chewing activitites of sheep. *Journal of Animal Science*, 74, 2785-2802.

Stephens, D.W. & J.R. Krebs (1986). Foraging theory. Princeton University Press, Princeton, NJ, USA.

Diet selection in a pasture of Panicum maximum cv. Mombaça with different sward heights

T.C.M. Genro[1], Ê.R. Prates[2], F.F. Cardoso[1] and L.R.L. San'Thiago[3]
[1]Embrapa Pecuária Sul, Postal box 242, 96401-970, Bagé, RS, Brazil, Email: cristina@cppsul.embrapa.br, [2]UFRGS, Av. Bento Gonçalves, 7712 Postal box 15.100, 91501-970, Porto Alegre, RS – Brazil, [3]Embrapa Gado de Corte, Postal box 154, 79002-970, Campo Grande, MS, Brazil

Keywords: leaf blade, n-alkanes, non-negative least squares, stem and sheath

Introduction The n-alkane method allows to measure dry matter (DM) intake and to determine the proportion of grass and legume ingested by ruminants grazing mixed pastures. Dove & Mayes (1996) pointed out that using this method, beyond estimating intake of individual plant species, it may also be possible to determine intake of anatomical components of the plant. The aim of this work was to use the non-negative least squares procedure to estimate diet selection based on plant alkanes and animal faeces.

Materials and methods The experiment was conducted at the Embrapa Beef Cattle Centre, Campo Grande, MS to study intensive beef production based on tropical grasses under rotational grazing throughout the year. The grazing system included 16 paddocks, a 2-day grazing period followed by a 30-day resting period. In this study, samples of stem+sheath and leaf blade were collected in 20 cm stratified layers of a pasture of Panicum maximum cv. Mombaça during three periods of the year: in the middle of the dry season (Period 1), at the beginning (Period 2) and at the end of the of the wet season (Period 3). Before grazing, three representative areas of the paddock were selected for sampling. Six Nelore steers dosed during 12 days, twice a day, with alkane C_{32} were used to estimate DM intake (DMI). The determination of n-alkanes within the range of C-chain between 27 and 35 was performed according to Mayes et al. (1986). N-alkane profiles were corrected for faecal recoveries using mean values found in literature (Dove & Mayes, 1996). The diet selection was calculated by the "Eatwhat" program described by Dove & Moore (1995), using the non-negative least squares algorithm.

Results The percentage of diet selected within the strata of Mombaça grass pasture and the DMI expressed as percent of live weight for the three periods studied are given in Table 1. In the first period, the program estimated that DMI was divided among three strata: stem+sheath 0-20 cm, leaf blade 0-20 cm above ground level, and leaf blade 40-60 cm. In the other two periods, the stratum preferred was stem+sheath > 20 cm. According to these results, the steers ingested over 60 % from this layer, in stead of the leaves from higher strata. These data, however, are not supported by the literature, where it was established that bovines prefer eating leaves from the top of the pasture. This may be due to problems in the alkanes determination, or perhaps, another possibility is that the programme was designed to work with diet selection between different plant species, but not within parts of the same plant, particularly of tall tropical plants such as Panicum spp..

Table 1 Diet selected (%) within the strata of Mombaça pasture and the dry matter intake (DMI, % LW)

Period	Stratum height (cm)							
	Stem 0-20	Stem >20	Leaf 0-20	Leaf 20-40	Leaf 40-60	Leaf 60-80	Leaf >80	DMI (%)
			Diet selected, % of intake					
1	35.6	0	30.6	0	33.8	NE*	NE	2.26
2	0	60.5	NE	0	39.5	0	NE	2.72
3	0	84.0	0	0	.5	4.3	11.2	1.93

*NE= Stratum not existent

Conclusions The use of non-negative least squares procedure to determine diet composition based on the plant alkanes does not seem to adequately estimate plant part selection within the plant profile.

References
Dove, H. & R.W. Mayes, (1996). Plant wax components: a new approach to estimating intake and diet composition in herbivores. American Journal of Nutrition, 126, 13-26.
Dove, H & A.D. Moore, (1995). Using a least-squares optimisation procedure to estimate diet composition based on the alkanes of plant cuticular wax. Australian Journal of Agricultural Research, 46, 1535-1544.
Mayes, R.W., C.S Lamb,. & P.M Colgrove,. (1986). The use of dosed and herbage n-alkanes as markers for the determination of herbage intake. Journal of Agricultural Science, 107, 161-170.

Effect of time spent on pasture on grazing behaviour in Latxa dairy sheep

A. Garcia-Rodriguez, N. Mandaluniz and L.M. Oregui
NEIKER, A.B. - Granja Modelo de Arkaute, Apartado 46, 01080 Vitoria-Gasteiz, Spain, Email: loregi@neiker.net

Keywords: dairy ewes, grazing behaviour

Introduction In the Basque country the dairy ewe production system is based on a high pasture utilisation by means of partial time grazing where ewes spend a limited number of hours outdoors (Oregui *et al.*, 1997). This production system, which is also typical of other Mediterranean areas, is different from that in northern Europe where continuous grazing is the usual management. There is a lack of knowledge concerning grazing behaviour under these conditions. This research investigated the effect of time spent on pasture on grazing behaviour.

Materials and methods The experiment was conducted over four weeks during the spring of four consecutive grazing seasons (2000-2003). In each year 48 multiparous Latxa dairy ewes were blocked into homogeneous groups of 12 on the basis of lactation number, days in lactation, milk yield and body weight. Initial mean values for milk yield, body weight and days in milk for the different years were respectively 1333, 1307, 1583 and 1818 ml/d, 62.5, 68.4, 62.2 and 58.3 kg, and 28, 40, 42 and 44 d. Four resulting groups were randomly assigned to one of the following experimental treatments: i) 4 hour-access to pasture (4H), two groups, or ii) 7 hour-access to pasture (7H), two groups. Each group of ewes was allocated to a different paddock. Sward heights were measured twice a week and were kept between 6 and 8 cm. As indoor supplements each ewe received 260 g dry matter (DM) of a concentrate, at milking time, and 250 g DM of lucerne hay, after the evening milking. Animal behaviour was recorded once a week with assessments of ewe activity (grazing, resting or moving) made by two observers every five minutes during the time on pasture.

Results The total time spent grazing was longer in 7H than 4H (246 *vs.* 196 min d^{-1}, P<0.001). Although these differences are statistically significant, from a quantitative point of view the utilisation of the 180 supplementary minutes on pasture of 7H was limited as only 50 min/d (28%) of this time was spent grazing. As a consequence less efficient utilisation of the time spent on pasture is observed in 7H compared with 4H groups (36 *vs.* 49 min/h, P<0.001 respectively). In terms of grazing distribution throughout the period on pasture (Figure 1), the grazing activity was similar in both groups, with the maximum being in the first hour (56 *vs.* 52 min/h, P>0.05, for 7H and 4H respectively). However, during the rest of the time, 4H maintained higher grazing times compared with 7H, that showed rapid decreased in the time spent grazing after the first hour to reach a constant grazing time at the third hour.

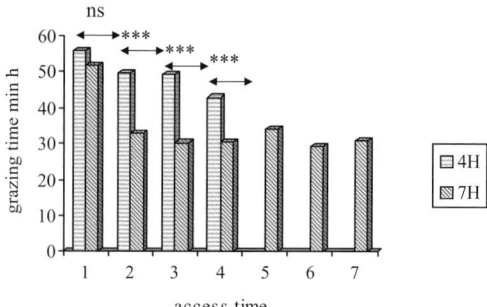

Figure 2 Grazing behaviour of ewes with limited access time (4H or 7H) to pasture.
*: P<0.05; **: P<0.01; ***:P<0.001

It appears that animals learn that they will stay a limited number of hours on pasture and adjust their grazing behaviour. As a consequence, whereas 7H took longer resting times resulting in a more constant grazing pattern along the day, 4H grazed more intensively and efficiently, but not enough as to compensate for all the difference on access time. However, this grazing behaviour can only be achieved if ewes are managed in a very constant way.

Conclusion Under the dairy Latxa production system, time spent on pasture causes a significant modification in grazing behaviour resulting in a higher efficiency of time use by those ewes with less time at pasture, but this does not completely compensate for the differences in access time.

Reference
Oregui, L.M., J. Garro, M.S. Vicente & M.V. Bravo (1997). Estudio del sistema de alimentación en las razas ovinas Latxa y Carranzana: utilización de los pastos comunales y suplementación en pesebre. *Información Técnica Económica Agraria*, 93, 173-182.

Time budget on major activities of livestock grazing heterogeneous natural range and crop fields in semi-arid Nigeria

B.S. Malami[1], P.H.Y. Hiernaux[2], H.M. Tukur[1] and J. Steinbach[3]

[1]*Department of Animal Science, Usmanu Danfodiyo University, Sokoto, Nigeria, bmalami@yahoo.com,* [2]*Centre for Agriculture in the Tropics and Subtropics, University of Hohenheim, Stuttgart, Germany,* [3]*Institute of Livestock Ecology, Justus Liebig University, Giessen, Germany*

Keywords: grazing activity, rangeland

Introduction Semi-arid rangelands of West Africa provide herbs, trees and shrubs, which together with crop residues form the main sources of feed for the livestock population. Feed supply in this region is characterised by a progressive decline in quantity and quality with advancing dry season. It was reported that walking ability as well as watering frequencies affect the productivity of grazing livestock (Dicko and Sangare, 1984). This study tests the hypothesis that advancing season increases both time spent walking as well as feeding, with a switch from grazing to browsing.

Materials and methods The study was conducted in the 235,500ha Zamfara reserve, northwestern Nigeria. The reserve falls within the Sudan-savannah zone (12° 10' – 13° 05'N; 6° 30' – 7° 15'E). Annual rainfall ranges from 500 mm in the north of the reserve to 800mm in the southern part, with an interannual variation of 22 – 32%. The rainy season is restricted to the months of May to September. Thirty-six indigenous ruminants (12 bulls, 12 rams and 12 bucks) were used in the study, which lasted from July 2002 to June 2003. The study period was divided into ten five-week periods. During each period, three bulls, three rams and three bucks were randomly selected and their grazing behaviour was observed and recorded after every five minutes for ten hours (from 08.00 to 18.00 hours) for two consecutive days. The experiment was designed as complete randomized block design, analysis of frequencies was performed on the data using SPSS 11.5 version.

Results Over the year, grazing and browsing constituted the major activities of all the livestock species, with a peak in May, June and July (Figure 1). Peak period of browsing was in May for all the species, indicating deterioration of the herbaceous layer. Time spent on grazing and browsing was higher compared to other activities. Time spent on grazing was higher (P<0.05) for cattle (45%) compared to sheep and goats (39 and 36%). The reverse was observed for browsing. For the three species, time spent on grazing was higher (P<0.05) during the rainy season compared to the dry season. Time spent walking and resting did not differ (P > 0.05) between the species and the periods.

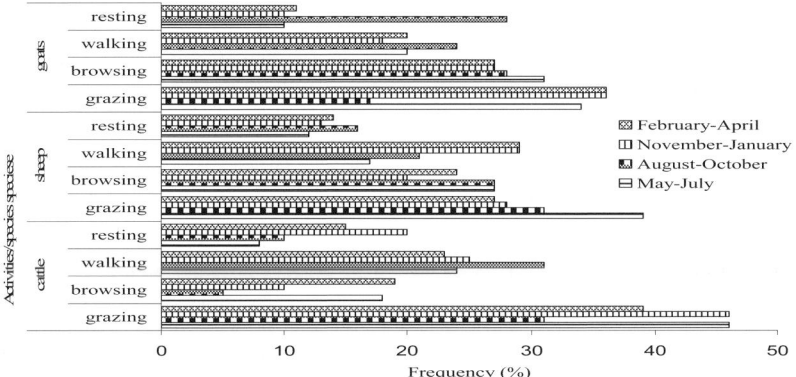

Figure 1 Mean time allocated to main activities by cattle, sheep and goats grazing heterogeneous semi-arid range and crop fields

Conclusion The increase in browsing activity during the dry season confirms the hypothesis that decline in feed quality and quantity increases the feeding time, with a switch from grazing to browsing. However, walking and resting periods were not affected by the seasonal availability of feed.

Reference

Dicko, M.S. and M. Sangare (1984) Feeding behaviour of domestic ruminants in sahelian zone. *Proceeding of the 2nd International Congress*, Adelaide, 388- 390.

The diet of free-ranging beef cattle in a semi-arid savanna of eastern Namibia

A. Rothauge[1], A.L. Abate[2] and G.N. Smit[3]
[1]Neudamm Agricultural College, P/Bag 13188, Windhoek, Namibia, Email: arothauge@unam.na, [2]Dept of Animal Science, University of Namibia, P/Bag 13301, Windhoek, Namibia, [3]Dept of Animal, Wildlife and Grassland Sciences, University of the Free State, P.O. Box 339, 9301 Bloemfontein, South Africa

Keywords: diet selection, beef cattle, semi-arid savanna, principal forages

Introduction Beef ranching is the most important agricultural enterprise in arid and semi-arid SW African countries. It earns foreign exchange via beef exports to the EU and very many rural people depend on cattle pastoralism for their livelihood. However there is no published information on what cattle eat in such extensive systems. Therefore, it is difficult to optimize grazing strategies and to prevent degradation of rangeland.

Materials and methods The diet selected by free-ranging beef cattle in a semi-arid Namibian savanna was observed from 2001 to 2003. A 2 x 4 factorial design investigated treatment effects on cattle and rangeland: 2 cattle types (large-frame AfrikanerXSimmental versus small-frame purebred Sanga) X 4 systematically increasing stocking rates (low, targeting 30 ha/LSU, to high, targeting 10 ha/LSU). Diet selection by 6 randomly chosen cows/treatment was determined by bite-counting for 10 min/cow, repeated on 4 consecutive days early in the grazing cycle (Narjisse, 1991; Ortega *et al.*, 1995). Treatment plots (mean 142 ± 28.9ha) were not replicated. Bites were converted into dietary abundance (% frequency) and transformed by arcsine before ANOVA-GLM analysis. Abundance data were pooled for the 3 main treatments cattle type, stocking rate and season of the year.

Results and discussion The perennial grasses, *Schmidtia pappophoroides*, *Anthephora pubescens*, *Eragrostis lehmanniana* and *Stipagrostis uniplumis*, were the principal forage species (those used most often), contributing 33.7 ± 18.23, 14.5 ± 19.65, 11.0 ± 10.50 and 7.5 ± 9.54%, respectively ($P<0.01$), to cattle diet. The 3 main treatments affected utilization ($P<0.01$, Figure 1). *S. pappophoroides* was the only grass that was both a principal forage species and maintained a sizeable presence in the sward, making it a valuable ecological indicator. Woody forage species (e.g. *Acacia mellifera*, a major invasive species in these parts) and non-graminiferous herbaceous forages (e.g. *Nidorella resedifolia*, an annual, indigenous weed characteristic of disturbed sites) became important to cattle only during the drier seasons and at the higher stocking rates (Figure 1).

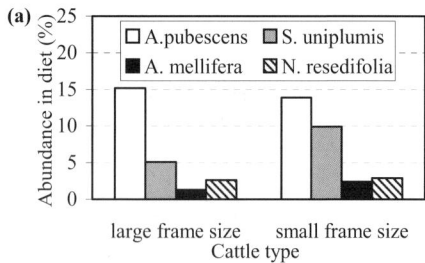

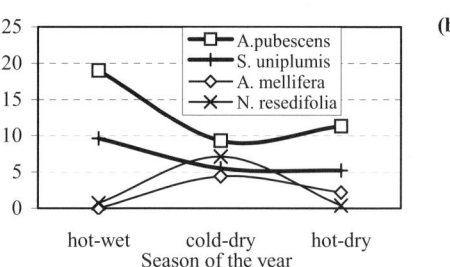

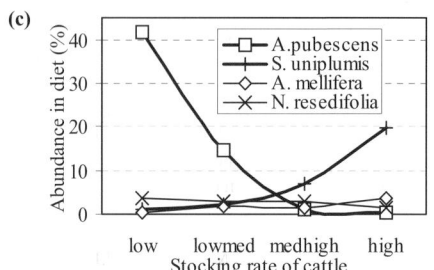

Figure 1 Abundance (%) of four forages in cattle diets, as affected by **(a)** type, **(b)** season and **(c)** stocking rate

Conclusions The principal forages of free-ranging beef cattle in a semi-arid Namibian savanna consisted of 4 perennial grasses that, together, constituted 67% of the diet. Their persistence in the grass sward should be encouraged through appropriate grazing strategies. Utilization shifted towards woody plants and herbs at high stocking rates and in the drier seasons.

Acknowledgements The support of the IFS (grant B/3183-1) is acknowledged gratefully.

References
Narjisse, H. (1991). Feeding behaviour of goats on rangeland. In: P. Morand-Fehr (ed.) Goat Nutrition, ch. 2. Pudoc, Wageningen, The Netherlands.
Ortega, I.M., F.C. Bryant & D.L. Drawe (1995). Contrasts of oesophageal fistula versus bite-count techniques to determine cattle diets. *Journal of Range Management*, 48, 498-502.

Effects of simulated high-sugar forages on grazing behaviour of sheep

G.P. Cosgrove and A.J. Parsons
AgResearch Grasslands, PB 11008, Palmerston North, New Zealand, Email: gerald.cosgrove@agresearch.co.nz

Keywords: grazing behaviour, ryegrass, water soluble carbohydrate, white clover

Introduction Sustaining an optimum composition in mixed-species pasture and in the diet of grazing animals is constrained by many factors. Altering the concentration of water soluble carbohydrate (WSC) of one species to improve its nutritional value for animals may not deliver the benefits in the assumed manner if it also affects preference and the balance of species in the pasture. Furthermore, associated changes in other constituents (e.g. fibre, protein) make it difficult to attribute animal responses solely to the manipulated trait (e.g. Lee et al. 2000). The objective of this study was to simulate changes in the concentration of water soluble carbohydrates in ryegrass or white clover, independently of changes in other plant constituents, and determine the effect of this trait alone on grazing behaviour.

Materials and methods Twelve, 2-year old rumen-fistulated Romney sheep were stocked on either perennial ryegrass (*Lolium perenne*, cv. Yatsyn) or white clover (*Trifolium repens* cv. Pitau) pastures maintained at 7 cm sward surface height. Each sheep received an infusion directly into the rumen of either water (control) or sugar solution (250 g/l sucrose) at the rate of 1 ml/min grazing ryegrass or 1.6 ml/min grazing clover. The rates of delivery and concentration were calculated to simulate ingestion of a 'high' sugar ryegrass or white clover nominally containing an additional 100 g WSC/kg DM. A remote control switch, operated by an observer, synchronised the infusion with each grazing bout between dawn (05:00 hrs) and dusk (21:00 hrs) for 3 consecutive days. The total time spent grazing each day and the number and duration of bouts was recorded to describe the animal response. The experimental unit was a group of 3 sheep treated for 3 days, and treatments were replicated 3 times by repeating the procedure with sheep re-randomised among treatments.

Results There were no significant interactions between forage species and infusion treatment and only the infusion treatment main effect is shown. Sugar infused sheep received 90 g sucrose/day. The total WSC concentration in the diet was 130 g/kg DM and 215 g/kg DM for the control and infused sheep, respectively. Sheep eating the simulated high-sugar forages grazed for less time in total per day, and they grazed in shorter but more frequent bouts (Table 1).

Table 1 Effect of high sugar forage on some aspects of grazing behaviour

	Total grazing time (mins/day)	Duration of bouts (mins/bout)	Number of bouts (per day)
Control	325	48	6.8
'High' sugar	295	39	7.6
Signif.	P<0.01	P<0.05	P<0.05

Conclusions Synchronising the infusion of sugar with eating allowed the sheep to regulate the total intake of WSC (dietary and infusion sources) from the simulated high sugar forages (the infusion stopped whenever sheep ceased grazing). They responded to the higher concentration of WSC in ryegrass and white clover by altering the duration and frequency of grazing bouts. These changes reflect the animals attempt to balance the inflow and metabolism of the higher dietary WSC concentration. These behavioural attempts to control or even limit their intake to maximise fitness (Newman et al. 1995) rather than maximising daily intake create a challenge for delivering nutritional (or other) traits to grazing animals via manipulating the forage. The reduction in the total duration of eating each day provides an early indication that manipulating WSC in one species might affect the proportion in the diet of the species containing that trait and over time, the composition of the sward.

References

Lee, M.R.F., E.L. Jones, J.M. Moorby, M.O. Humphreys, M.K. Theodorou, J.C. MacRae, & N.D. Scollan (2000). Production responses from lambs grazing on *Lolium perenne* selected for high water soluble carbohydrate. In: A.J. Rook & P.D. Penning (eds) Grazing Management. British Grassland Society Occasional Symposium No. 34, 45-50.

Newman, J.A., A.J. Parsons, J.H.M. Thornley, P.D. Penning, & J.A. Krebs (1995). Optimal diet selection by a generalist grazing herbivore. *Functional Ecology*, 9, 255-268.

Grazing preference, herbage production and quality of diploid and tetraploid *Lolium perenne* cultivars in Southern Chile

O.A. Balocchi and I.F. López
*Faculty of Agriculture, Institute of Animal Production, University Austral of Chile, PO Box 567, Valdivia, Chile,
Email: obalocch@uach.cl*

Keywords: *Lolium perenne*, dry matter production, ploidy

Introduction *Lolium perenne* is the most important plant species used in sown pasture in Southern Chile. Cultivars used are mainly diploid. However, in recent years some tetraploid cultivars have been introduced into the country. The objective of the study was to determine the effect of the ploidy of the cultivars on herbage production, nutritive value, grazing preference and utilisation of the herbage.

Materials and methods The study was conducted in Southern Chile between 39° 47` 46`` and 39° 48`` 54` S and 73° 13` 13`` and 73° 12` 24`` W from April 1999 to May 2002. Cultivars used were Quartet (4n), Gwendal (4n), Pastoral (4n), Napoleon (4n), Anita (2n), Jumbo (2n), Aries (2n) and Yatsyn 1 (2n). When the average sward height reached 20 cm, all plots (50m^2 each) were grazed simultaneously by 8 dairy cows for 24 hrs. Before and after grazing, sward height (sward stick), dry matter availability (strip of 9 m^2 per plot, 4 cm from ground level) and nutritive value (CP and D value) were determined. Grazing preference was assessed visually every 5 min. during 2.5 hrs after the afternoon milking. Utilisation efficiency was calculated as a relationship between apparently consumed forage and pre-grazing DM availability. During the three years 22 grazing events were evaluated in a randomised block with eight cultivars and three replicates.

Results Overall, diploid cultivars showed greater herbage mass accumulation than tetraploid cultivars (Table 1). In environments with higher levels of stress (mainly, low soil fertility and summer drought), as occur in the soil and climatic conditions of Southern Chile, diploid cultivars, due to their stress tolerance=, were able to show advantages in herbage production, in relation to tetraploid cultivars. No significant differences were obtained in the annual average CP content. Nevertheless, tetraploid cultivars had a greater D value than diploid cultivars, except during the third year when the difference was not statistically significant. These results are in agreement with the morphological features of tetraploid cultivars, which have a higher cell content/cell wall ratio that confers a greater digestibility to the plant (O'Donovan, 2001).

Table 1 Herbage mass accumulation, D value and CP content, grazing preference, residual dry matter availability and pasture utilisation percentage of diploid and tetraploid *L. perenne* cultivars in Southern Chile

Year	Herbage production (ton DM/ha/year)		Crude protein (g/kg DM)		D value (% of the DM)		Grazing preference (Minutes/plot)		Residual Dry matter (kg DM/ha)		Utilisation efficiency (%)	
	D	T	D	T	D	T	D	T	D	T	D	T
1	12.9a	11.9 b	162 a	157 a	75.1 b	76.4 a	19,9 b	24,7 a	518 a	382 b	75,9 b	80,8 a
2	10.1a	8.9 b	164 a	168 a	72.9 b	74.2 a	20.4 a	26.1 b	382 a	285 b	68.9 a	74.0 b
3	9.9 a	8.8 b	200 a	202 a	76.9 a	77.9 a	27.0 a	29.0 a	224 a	168 b	89,2 b	91,0 a

D: diploid cultivars; T: tetraploid cultivars. Values in rows with different letter are different (P< 0.05)

Dairy cows grazed more time on tetraploid cultivars. Considering, additionally, the availability of residual dry matter after grazing (measured over 4 cm), and the percentage of pasture utilisation (over 4 cm), the diploid cultivars were less intensively grazed, suggesting a lower consumption by the cows.

Conclusions In the soil and climatic conditions of Southern Chile diploid cultivars of *Lolium perenne* showed greater herbage mass accumulation than tetraploid cultivars. Nevertheless, tetraploid cultivars showed higher D value and were more intensively grazed by dairy cows.

References

O'Donovan, M. (2001). The Influence of Grass Cultivars on Milk Production.
  http://www.teagasc.ie/research/reports/dairyproduction/4572/eopr-4572.htm.

The timing of daily grazing on annual ryegrass or sulla forage: the effects on milk yield and composition of Comisana ewes

G. Di Miceli[1], L. Stringi[1], C. Scarpello[1], P. Iudicello[1], A. Bonanno[2] and A. Di Grigoli[2]
[1]Dipartimento di Agronomia Ambientale e Territoriale, [2]Dipartimento S.En.Fi.Mi.Zo., sezione di Produzioni Animali; Università di Palermo, Viale delle Scienze, 90128 Palermo, Italy, Email: giardo@unipa.it.

Keywords: daily grazing time, sulla, annual ryegrass, milk yield, milk composition

Introduction The timing and duration of grazing greatly affect the response of animals. Night grazing, in addition to grazing during the day, seems to be one of the most important practices for improving animal performance (Bayer et al., 1987). Many authors have already underlined the major benefits of night grazing, such as improved body condition, reduced heat stress, increased forage intake and milk production. However, labour constraints, insecurity, damage to crops by animals are considered as the main reasons for not practising day and night grazing. The aim of this research was to improve knowledge about the effects of daily grazing time in ryegrass or sulla forage on milk yield and composition of Comisana ewes.

Materials and methods The research was carried out in a semi-arid, hilly area of Sicily (37°30'N; 13°31'E; 178 m a.s.l.). For 5 weeks from 3 May 2004, two plots of 2.500 m^2 area, sown with *Lolium multiflorum* subsp. *wersterwoldicum* (R) and *Hedysarum coronarium* (S), respectively, were grazed by Comisana ewes. There were two daily grazing times for each forage resource: 8h (from 08.00 h to 16.00 h) and 24h. Thus, 40 ewes were homogeneously divided into 4 groups, on the basis of milk yield, lactation phase (102±30 d) and liveweight, and assigned to the treatments: R8, R24, S8, S24. During the night, R8 and S8 ewes were housed in an open shelter, without feeding supplements. Once a week individual and bulk milk production in each group was recorded and sampled. Individual milk samples were analysed for fat, protein and lactose content. Casein, whey protein and non-protein N (NPN) were determined in bulk milk samples

Results and conclusion Herbage mass on offer was relatively high in all plots throughout the trial (> 3 t DM/ha); post grazing herbage mass never fell below 2.0 t DM/ha. Grazing time and forage species significantly increased milk yield and energy when ewes grazed for 24h compared to 8h and on S than on R, probably due to higher forage intake in both cases (Table 1). The content of fat and protein fractions in milk of ewes grazing on R decreased significantly when the grazing time was extended from 8h to 24h, as a consequence of the increase in milk yield. On the contrary, significant increases in fat, protein and casein content were observed in the milk of ewes in S24 compared to S8, although the milk yield increased. Ewes on S8 had higher milk yield than on R8, but the quality was lower, whereas the S24 group had higher milk yield and quality than R24. These results can be related to the content in condensed tannins (CT) of sulla.

Table 1 Effects of grazing time and forage on milk yield and composition

	R8	R24	S8	S24	Significance		
					Grazing time (GT)	Grazed forage (GF)	GT*GF
Milk yield (g/d)	599a	705b	786b	907c	**	***	
Milk energy (kcal/d)	730 a	821 a	830 a	1007 b	***	***	
Fat (%)	8.2a	7.6b	6.7c	7.1d		***	***
Lactose (%)	4.5a	4.7b	4.6ab	4.7b	*		
Protein (%)	6.3a	6.1b	5.9b	6.4a			***
Casein (%)	5.3a	5.0b	4.8c	5.2a		*	***
Whey protein (%)	1.2a	1.0b	1.2a	1.2a	*	***	**
NPN mg/ml	36.8ab	35.6a	40.6b	40.6b		*	

*= P • •0.05; **= P • •0.01; ***= P • •0.001; a,b,c,d: P • •0.05

In fact, CT decreased the time spent eating the main meals and increased the number of small meals (Landau et al., 2000). In this way, CT favoured a longer eating time, then a slowdown of the ruminal transit, improving the degradation of the tannin-protein complex and the protein utilization. In conclusion, the daily grazing time of 24h can be a suitable management practice to improve ewes milk yield, in particular on S meadow.

Acknowledgements The research was financed by MiPAF project ANFIT.

References

Bayer, W., H. Suleiman, R. Kaufmann & A. Waters-Bayer, (1987). Resource use and strategies for development of pastoral systems in sub-umid West Africa. The case of Nigeria. *Quarterly. Journal of International Agriculture* 26, 58-71.

Landau, S., N. Silanikove, Z. Nitsan, D. Barkai, H. Baram, P.D. Provenza & A. Perevolotsky (2000). Short-term changes in eating patterns explain the effects of condensed tannins of feed intake in heifers. *Applied Animal Behaviour Sci*ence, 69, 199-213.

The effects of proximity to watering points on vegetation parameters in winter rangelands of Chahe-Nou, Damghan, Iran

R. Khalifeh-Zadeh and M. Mesdaghi
Department of Range Management, University of Agricultural and Natural Resources, Gorgan, Iran,Email: khalifehzadeh_r@yahoo.com

Keywords: watering points, Chahe-Nou, vegetation parameters, palatability

Introduction The Chahe-Nou with 115 mm rainfall and high temperatures (30 °C) is located 95 km south of Damghan. It is grazed intensively by nomadic and sedentary sheep and goats during winter. Proximity to watering points affects the rangelands vegetation. With increasing distance from the watering points more palatable species increase, whilst grazing decreases. The areas nearest the watering points contain more unpalatable undesirable invasive species (Department of Natural Resources, 1999). The objective of this research was to evaluate the effects of proximity to watering point on vegetation parameters at two watering points.

Material and methods The effects of proximity to watering point on vegetation parameters, soil moisture, and palatability of plant species were analysed by MANOVA, ANOVA, and CCA. In each of two watering points, 8 transects of 2 km were established randomly, and each 100 m of transect was considered as a treatment (Fusco, *et al.*, 1995). In each treatment, four plots of 4-m^2 were sampled systematically for canopy coverage, species richness, and density and two soil samples were taken randomly by using auger for moisture measurement. The associations of pairs of plant species based on presence/absence data were analysed by using chi-squares. The most important species were *Artemisia siberi, Salsola rigida, Zygophyllum euripterum, Haloxylon persicum,* and *Peganum harmala.* Palatability was rated as I (the most palatable), II (moderately palatable), and III (least palatable).

Results Multiple vegetation parameters of canopy coverage, richness, and density were significant (p<0.05) for all treatments. Density and species richness were significant (p<0.05), and canopy coverage and soil moisture were not significant (p>0.05). By increasing distances from watering points, the palatability of plant species improved. Although vegetation parameters were significantly different according to MANOVA discriminant analysis was not consistent with this. This was due to sparse vegetation around watering points which caused a high value for Wilks' Lambda index. The Relative Cover+ Relative Density+ Relative Frequency of *Artemisia siberi* as winter forage decreased while the other unpalatable species invaded near watering points (Figure 1).

Conclusion There were negative associations between the two species of *Artemisia siberi* and *Peganum harmala,* but there were random associations among other species. It can be concluded that grazing intensity was concentrated on 400-500 meters from watering points which can be considered as critical zone.

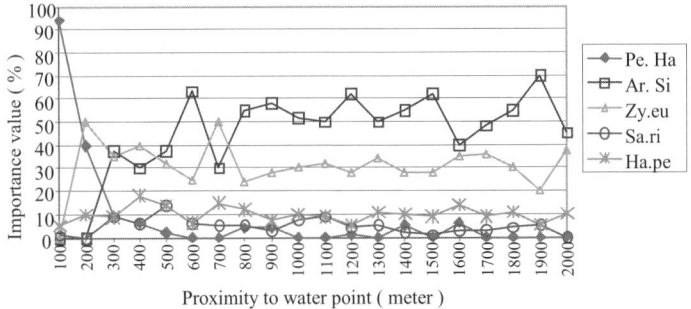

Figure 1 The effects of proximity of watering point on importance value of different species

References

Department of Natural resources. (1999). Range management plan for winter rangelands of Chahe-Nou. The Damghan Jahhad Organization. Iran.

Fusco, M. Holechek, J. Tembo, A. Daniel, & M. Cardenas. (1995). Grazing influence on watering point vegetation in the Chihuhuan desert. *Journal of Range Management*, 48(1), 32-38.

How are distances between grazing cows determined?

M. Shiyomi

Faculty of Science, Ibaraki University, Bunkyo 2-1-1, Mito 310-8521, Japan, Email: shiyomi@ mx.ibaraki.ac.jp

Keywords: cattle herd, distances between individuals, pattern formation

Introduction Although domestic cows are protected from their natural enemies, they form herds when grazing like their wild forbears did. The herding instinct is used to manage cattle because it is easier to control a herd than separate individuals. This study examined how distances between individual grazing cows are determined.

Materials and methods The coordinates of six Holstein cows in a one-dimensional, fenced grassland of 88 m × 6 m were recorded every 5 minutes (Figure 1a). The cow individuals were numbered A, B, …, F, and distance between two individuals A and B was expressed by A-B or B-A. For each pair of cows, the contribution of each of the five effects was calculated using the sum of squares of data, based on the temporal changes in the distances between individuals.

Model In a grazing herd, complicated interrelationships operate between individuals. We assumed that the distance between any two individuals was based on the following five factors (Figure 1b): the behaviour of the entire herd (measured as the distance between the leftmost and rightmost cows: the herd length effect); repulsive and attractive forces operating directly between two individuals within the herd (direct effect); the effect of third individuals on the distance between two individuals (half-indirect effect); the effect of unconnected pairs on the distance between two individuals (indirect effect); and residual effects that cannot be explained by the other four factors (random/involuntary movement effect). These five factors were analysed using regression analysis.

Results Overall, the random movement effect made the largest contribution (38.8%) to the temporal variation in the distance between two cows, followed by the herd length (25.6%), direct (21.0%), and half-indirect (12.5%) effects. The contribution of the indirect effect was negligible. Contributions of the herd length, direct effects, half-indirect effect and random movement effect between two individuals, measured using sum of squares of temporal variation of distances are shown in Figure 2. The results revealed that one (F) of the six cows was persecuted by the other cows.

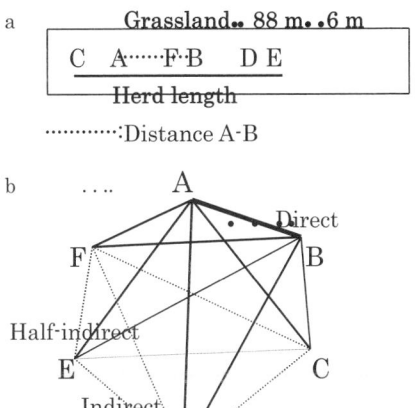

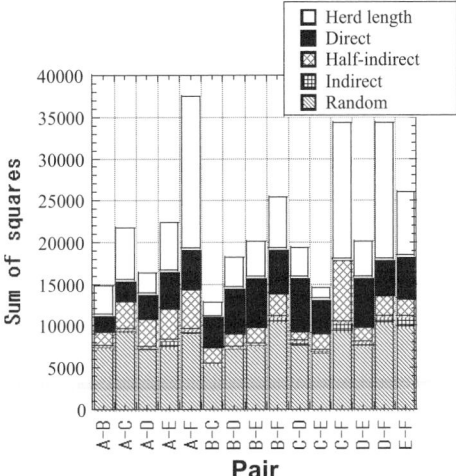

Figure 1 Six cows in a grassland (a), and direct, half-indirect and indirect effects to distance A-B (b)

Figure 2 Contributions of the herd length, direct effect, half-indirect effect, indirect effect and random movement effect, measured using sum of squares of temporal variation of distances between two individuals

Conclusions Many mathematical models in crowd formation studies operated on the premise that individuals act like physical particles (*e.g.*, Gueron & Levin, 1993). By contrast, in this study, each individual was supposed to have its own characteristics, although they were not complete.

References

Gueron, S. & Levin S.A. (1993) Self-organization of front patterns in large wildebeest herds. *Journal of Theoretical Biology*, 165, 541-552.

Grazing behaviour of lambs in different production systems

C.H.E.C. Poli[1], A.L.G. Monteiro[2], C.S. Gomes[2], G.J. Bosquetto[2], J.R. Dittrich[2], S. Gilaverte[2] and H.V.L. Piazetta[2]
CNPq, [1]Embrapa Pecuária Sul/Paraná, P.O. box 242, CEP 83411-000, Colombo, PR, Brazil, Email: cpoli@cnpf.embrapa.br, [2]Universidade Federal do Paraná, Curitiba, PR, Brazil

Keywords: grazing time, ruminating time, idling time, biting rate

Introduction Ingestive behaviour is an important component of a grazing system (Fryxell *et al.*, 2001). The understanding of lamb behaviour patterns in different production systems is crucial to management and to assess the impact of the production system on the use of resources by animals. The objective of this study was to determine the grazing behaviour of lambs in different production systems.

Materials and methods The experiment was carried out at the Experimental Research Station of UFPR, Pinhais, PR, Brazil, in a randomised block design with three replications. The grazing behaviour of four Suffolk lambs/plot grazing bermuda grass (*Cynodon dactylon* hybrid Tifton 85) were compared in three production systems: (1) lambs weaned at 60 days of live; (2) lambs with their mothers until sent for slaughter; (3) the same as treatment 2, but the lambs were supplemented daily (1% of live weight) with a concentrate (18% CP and 80% TDN) in creep feeding. The stocking rate was adjusted every 14 days to maintain 1.000 kg/ha of green DM. The experiment was carried out from weaning to slaughter. During this period the animals were observed for three days, using the method of Jamieson & Hodgson (1979). Individual animal activity: eating, grazing, suckling and eating concentrate, ruminating and idling were recorded from dawn to dusk at intervals of 10 minutes. Between each 10 minutes of recording, rates of biting were measured using the 20 bite method of Forbes & Hodgson (1985). The grazing behaviour assessment was carried out in three different periods 6, 13 and 17 December.

Results The fact that biting rate was not significantly (P<0.05) different between treatments indicates that the sward structure was not having a strong effect on grazing activity. The weaned lambs spent significantly (P<0.05) more time eating and less time idling than in the two other treatments (figure1). In contrast, the animals that had access to a concentrate spent significantly less time eating and more time idling. This indicates, that the sward allowance did not restrict intake and the animals were substituting grass for concentrate, and the presence of the ewe affected the grazing behaviour of the animals. Although the time spent suckling was relatively low (6.7 ± 1.054 vs. 5.8 ± 1.863 min/period), demonstrating that the ewe's milk was not important for the lamb's diet, the presence of the ewe might have had an important effect on a social ewe-lamb interaction. The animals spent more time eating in the morning than in the afternoon. Therefore, the most important meal for growing animals is in the morning. The time spent ruminating was not different

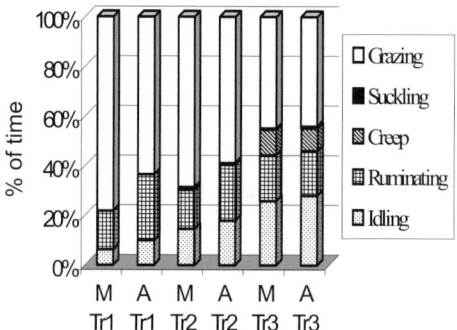

Figure 1 Effect of different lamb production systems (T1, T2 and T3) on the proportion of time spent grazing, suckling, eating in the creep feeding system, ruminating and idling in the morning (M) and afternoon (A)

(P<0.05) between treatments, but in treatment 1 and 2 the ruminating time was greater in the afternoon. It means that the amount of concentrate offered to the animals was not enough to reduce the ruminating time, but altered the way the ruminating time was distributed during the day. The production system altered mainly the grazing and idling times of lambs. Ruminating time was not different between production systems.

Conclusions The presence of the ewe and the supplementation in different production system has an important effect on lamb grazing behaviour.

References

Fryxell, J.M., C.B.D Fortin .& Wilmshurst, J. (2001) On the scale dependence of foraging in terrestrial herbivores. *Proceedings of the 19th. International Grassland Congress*, Saõ Pedro, Brazil, 271-275.

Forbes, T.D.A. & Hodgson, (1985) Comparative studies of the influence of sward conditions on the ingestive behaviour of J. cows and sheep. *Grass and Forage Science*, 40, 69-77.

Jamieson, W.S. & J. Hodgson, (1979). The effect of daily herbage allowance and sward characteristics upon the ingestive behaviour and herbage intake of calves under strip-grazing for grazing dairy cows. *Grass and Forage Science,* 34, 69-77.

Intake, digestibility and rate of passage of grass in grazing by light breed horses on different pastures

M. Kawai[1], N. Yabu[1], T. Asa[1], K. Deguchi[2] and S. Matsuoka[1]
[1]Obihiro University of Agriculture and Veterinary Medicine, Obihiro, Hokkaido 080-8555, Japan, Email: kawaim@obihiro.ac.jp, [2]Hokkaido Animal Husbandry Experiment Station, Shintoku, Hokkaido 081-0038, Japan

Keywords: intake, digestibility, passage rate, horses, grazing

Introduction In a previous study, grazing light breed horses could ingest CP and DE requirements for maintenance without supplements. However, their grazing behaviour, such as biting and chewing efficiency, which related to the passage rate of forage in the digestive tract and fibre digestibility, was affected by the pasture conditions (Kawai *et al.*, 2004). In this study, the DM intake, digestibility and mean retention time (MRT) of grass in light breed horses were determined and compared in spring, summer and autumn on an improved pasture.

Materials and methods Three mature light breed geldings were grazed on a pasture (2.7 ha), which consisted mainly of Kentucky bluegrass, from May to October by set-stocking. Experiments were carried out in two consecutive years with different sward conditions (H: high and L: low sward height) each in June, Aug. and Oct. for 13 days. Sward heights (cm) were 36.4 and 22.7, 21.1 and 14.8, 14.2 and 9.7, and herbage mass (g DM/m^2) were 163.3 and 95.6, 77.3 and 48.8, 44.0 and 31.7 in June, Aug. and Oct., respectively. The horses were offered 100 g of pellets each, which consisted of 10% chromic oxide, at 8:00 and 17:00 every day. Ytterbium (Yb)-marked grass was prepared by the immersion method and was fed to the each horse (200g) on day 9 in each experiment. Faecal samples of each horse were collected for daily from day 9 to day 13 at about 4-h intervals. Voluntary intake and digestibility were determined by a double-indicator method using chromic oxide and acid insoluble ash, and MRT was calculated with the following expression using excretion pattern of Yb in faeces (Blaxter *et al.*, 1956); MRT=$\sum$mt/$\sum$m (m: amount of Yb in faeces, t: time after Yb dosing)

Results Dry matter intake decreased linearly from 15.2 to 12.2 kg/d as the herbage mass was reduced (Table 1, r=0.61, P<0.01). There was a positive correlation between passage rate and dry matter intake (Figure 1), NDF (r=0.63, P<0.01) and ADF (r=0.70, P<0.01) digestibility, respectively.

Table 1 Dry matter intake (DMI), digestibility, mean retention time (MRT) and passage rate (PR) of grass in light breed horses[#]

	Jun.		Aug.		Oct.	
	H	L	H	L	H	L
DMI (kg/d)	15.2[a]	14.5[a]	13.5[ab]	13.9[ab]	13.5[ab]	12.2[b]
Digestibility (%)						
DM	67.3[ab]	64.5[abc]	68.6[a]	66.2[ab]	59.9[c]	63.9[bc]
NDF	65.6[a]	54.8[bc]	66.0[a]	57.0[b]	51.5[bc]	50.2[c]
ADF	61.4[a]	53.7[b]	62.4[a]	57.0[ab]	41.9[c]	41.3[c]
MRT (h)	18.8[ab]	18.9[ab]	19.9[a]	16.3[c]	20.0[a]	17.4[bc]
PR (%/h)	23.2[a]	20.4[a]	22.6[a]	24.8[a]	20.2[a]	14.6[b]

[#]Effects of herbage conditions on all variables were significant (p<0.05)

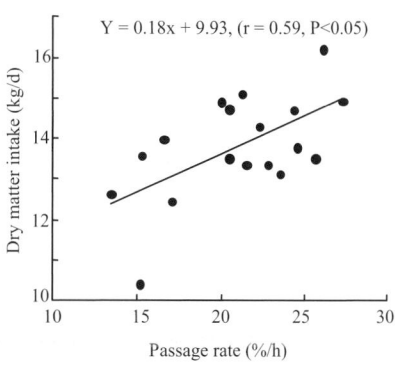

Figure 1 Relationship between passage and dry matter intake in horses

Y = 0.18x + 9.93, (r = 0.59, P<0.05)

Conclusions The results of this study demonstrate that pasture condition affected the dry matter intake and digestibility of grass in light breed horses. The relationship between MRT and DM intake or digestibility was not clear, but the passage rate was affected by the intake and digestive availability of fibre fraction.

References

Blaxter, K.L., N.M. Graham, & F.W. Wainman (1956). Some observations on the digestibility of food by sheep, and on related problems. *The British Journal of Nutrition*, 10, 69-91.
Kawai, M., N. Yabu, T. Asa, K. Deguchi & S. Matsuoka (2004). Biting and chewing behavior of grazing light breed horses on different pasture-conditions. *Proceedings of the 38th International Congress of the International Society for Applied Ethology*, 167 pp.

Sustainable semi-arid grazing management based on indigenous Shona practices prior to introduction of western ideas in Zimbabwe

O. Mugweni and R. Mugweni
Njeremoto Biodiversity Institute, P.O. Box 135 Masvingo, Zimbabwe, Email:muweni@zol.co.zw

Keywords: semi-arid areas, grazing management, Southern Africa, grazing systems

Introduction In the Shona culture the land, i.e. the plants, animals, soil, water, air and others, evolved with herding animals. Hence, the absence of one results in the destruction of the other. It is argued that the conventional grazing management belief that too many animals cause overgrazing is a misconception of the semi-arid savanna environments of Southern Africa where these environments evolved with thousands of herding grazers and mega-faunas such as elephants, wildebeests and buffalo. The objective of the research is to establish that grazing with an adequate recovery period for grazed plants, as a result of domesticated animals being managed effectively rather than staying on the same piece of land too long (continuous grazing) or returning too soon to the grazed area (rapid rotational grazing systems), can reverse the process of land degradation and the low water table of semi-arid rangelands, and can improve biodiversity by engaging in communal herding of livestock.

Materials and methods The research is a four-year study which started in October 2002 at Njeremoto Biodiversity Institute near Chatsworth ICA in Zimbabwe. The results of the first two years of the study are reported. The systems study involves intervention strategies which are being implemented in two areas, which are grazing and arable areas. The grazing area, which is 200 hectares in extent, is divided into three zones, A, B and C, and is grazed as shown in Table 1 below:

Table 1 Period of controlled grazing

Period	A	B	C	Arable
Early summer grazing (Nov. to Jan.)	X			
Late summer grazing (Feb. to May)		X		
Full summer recovery (Nov. to May)			X	
Dry season grazing in arable area (June to Oct.)				X

Communal herding of livestock is practiced in summer once in each zone. Monitoring of vegetation is done through transect and fixed-point methods. As well as this data, information is also recorded by means of fixed-point photography, video recordings and field work notes.

Intervention strategies in the arable area, which is 50 ha in extent, include organic farming, permaculture, water-harvesting, fodder production, and planting multi-purpose and fruit trees. The area is grazed in the dry season (June to October) after harvest.

Results The following are the results of the study obtained to date. Controlled grazing with high animal impact causes many plants to grow with tight plant spacing and increased vigour. The recovery times used has produced a multi-species pasture of healthy, tight plant communities with a good age distribution. There is increased grass cover, as well as reduced mature capping, bare ground, gully formation and sheet erosion. There is an increase in productivity of stover, feed, fruits and fodder on the arable area. The social structure through the community herding of cattle is changed and is beneficial to the young children who will be the environmental managers of the future.

Conclusion The research findings to date reveal that indigenous knowledge systems complemented with modern methods of investigation may result in effective and productive, and potentially sustainable land management practices for semi-arid rangelands.

Managing the reproductive development of grasses by grazing practices

J.P. Theau[1], L. Chazelas[2], P. Ansquer[1], J. Viegas[1], O. Stefanini[2], M.L. Petit[2], P. Cruz[1] and M. Duru[1]
*[1]Research unit UMR 1248 Managing grassland farming systems, BP 27, F-31326 Castanet-Tolosan, France,
Email: jtheau@toulouse.inra.fr, [2]FRCICAM Limousin, Lycée Agricole de Naves, F-19460 Naves, France*

Keywords: native vegetation, pluri-specific grasslands, growth, heading control

Introduction Grazing natural grassland communities is necessary for both productive (feeding herbivores) and environmental (maintenance of open landscapes) objectives. Management guidelines should take into account the functional diversity of plant species between and within grassland communities. The management of the heading stage of grasses by grazing is an important tool to maintain acceptable forage quality and to avoid the seeding of low-valued species. The heading stage should be managed even in extensive systems and this needs a good knowledge of the phenological development of dominant species. In this work we illustrate the approach through the study of the development of four contrasting grasses and analysing the consequences for grazing management. This work concerned only diversity of the plant components of the grassland ecosystem.

Materials and methods Three native populations of *Lolium perenne* (*Lp*), *Dactylis glomerata* (*Dg*) and *Bromus hordeaceus* (*Bh*), components of a natural grassland and an improved variety of *L. perenne* (Ohio) growing in a sown grassland were studied in grassland of the Limousin area (centre-west of the Massif Central in France, 900 m a.s.l.). From 21 April to 28 May, 20 reproductive tillers (10 tillers x 2 transects) of each species were harvested every 7 days. Both, plant and apex height from the soil level were recorded. The position of apices within the tillers was observed by dissection. Heading development was expressed on a thermal time basis (base temperature 0°C). After apex ablation these species do not have a further reproductive cycle (Gillet, 1980).

Results Stem elongation was very regular within the grass populations but highly variable between them. This would preclude the use of the sward height as an indicator to manage these complex communities (Figures 1a and 1b). Stem elongation was similar in *Dg* and *Lp* up to an apical height of 8 cm (a threshold height for animal intake), whereas *Bh* reached this height 12 days earlier. On the other hand, *Lp* Ohio, which was selected for its late development, attained it 16 days later. The relationship between plant and apex height was specific to grass populations. Nevertheless, whatever the studied material, for an apex height of about 8 cm the plant height was always too high (from 25 to 45 cm) to allow efficient grazing in the first cycle of a rotational system and the proportion of apices removed by grazing will be very different between species and varieties. Thus, 80% of apices reached the 8 cm threshold at 453, 533, 513, 666 degree days for *Bh*, *Lp*, *Dg* and Ohio (Figure 1c).

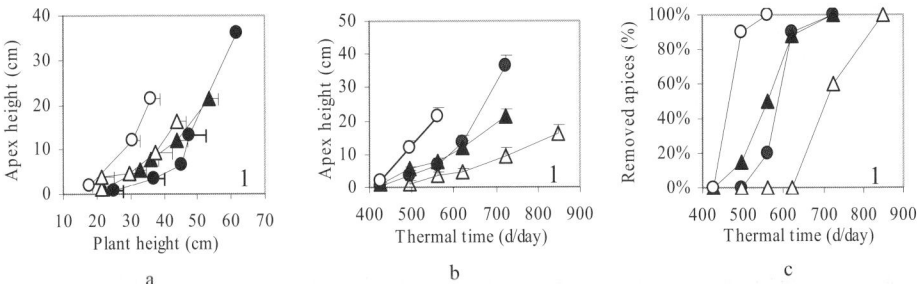

a   b   c

Figure 1 Height of reproductive apices of three native grasses (• : *Bh*; • : *Lp*; • : *Dg*) and a commercial variety of *L. perenne* (• •: Ohio) related to the plant height (1a) and the thermal time (1b) and differences in values of thermal time at which 80% of the apices of each material could be removed by grazing (1 c)

Conclusions These results indicate that the control of heading would not be feasible at the time of the first grazing cycle, because many apices will not be removed. In this case, an early first grazing could be recommended in order to reduce plant height and thus allows a maximal removal of apices during the following grazing cycle. We also showed that good knowledge of the functional diversity existing within and between grasslands could be used to decide the best method of managing the reproductive cycle of grasses by grazing.

References
Gillet M. (1980). Les graminées fourragères. Description, fonctionnement, application à la culture de l'herbe. Ed. Gauthier-Villars.

Impacts of strategic grazing on density and ground cover of naturalised hill pasture

Z.N. Nie, P.E. Quigley and R. Zollinger
Department of Primary Industries, Primary Industries Research Victoria, Private Bag 105, Hamilton, Vic. 3300, Australia, Email: zhongnan.nie@dpi.vic.gov.au

Keywords: deferred grazing, plant density, *Austrodanthonia, Austrostipa, Microlaena*

Introduction Low ground cover by perennial species is a major problem in naturalised pasture on steep hill country in southern Australia. This leads to water and nutrient runoff, recharge to groundwater, and soil erosion, all of which impact on the environmental sustainability and profitability of grazing enterprises. Restoration of perennial components, particularly the native grasses for these marginal land classes, is of great importance for improving water balance, halting land degradation (Ridley *et al.* 1997), extending growing season, and increasing pasture production. The objective of this study was to use strategic grazing management to increase the ground cover and plant population density of perennial species in steep hill country.

Materials and methods The site was on a naturalised hill pasture near Ararat, Victoria, Australia (143°08'E, 37°25'S). The soil was a sedimentary clay loam; mean rainfall was 600mm/year. From Oct 2002 to May 2003, 4 treatments were imposed in a randomised complete block design with 3 replicates. Treatments were: (1) short-term deferred grazing (SD): pastures not defoliated between Oct 2002 and Jan 2003; (2) long-term deferred grazing (LD): from Oct 2002 to May 2003; (3) late-start deferred grazing (LSD): from Nov 2002 to May 2003; and (4) set-stocking treatment (ST). The pasture initially was dominated by annual grasses (>65%, mainly *Vulpia bromoides*) with about 30% perennial grasses. Most perennial grasses were native grass such as *Austrodanthonia* spp., *Austrostipa* spp. and *Microlaena stipoides*. Using an optical point quadrat, ground cover was measured 3 times between Jan and May 2003 and plant population density was measured by collecting 10 X 80mm diameter soil cores/plot in Oct 2003. Data were analysed using a General ANOVA model (Genstat Committee, 2000).

Results Grazing initially had no effect on ground cover. ST reduced (*P*<0.01) ground cover by Mar 2003 (Figure 1). This was due mainly to the removal of senescent annual grass by grazing from Jan to Mar, a period of feed shortage. ST ground cover increased from Mar to May, but was still lower (*P*<0.1) than other treatments. Treatments SD, LD and LSD significantly (*P*<0.05) increased tiller density of perennial grass, but did not affect the densities of annual grass, *Romulea rosea* and annual clovers, compared with ST (Table 1). There were no significant differences in the densities of all plant types between SD, LD and LSD.

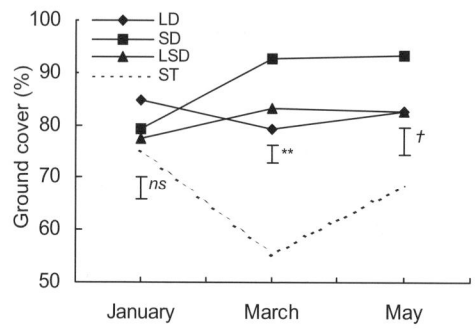

Table 1 Plant population density (tillers of grasses, plants of other species/m^2) of perennial grass (PG), annual grass (AG), *Romulea rosea* (RR), and annual clover (AC) in October 2003

Treatment	PG	AG	RR	AC
SD	8800	4040	2230	670
LD	9450	3890	3300	570
LSD	8380	2310	2820	480
ST	6430	7310	3540	1160
s.e.m.	*515**	*1955*	*759*	*462*

*P<0.05

Figure 1 Ground cover under various grazing treatments (Vertical bars=s.e.; *ns*=not significant; **P<0.01; $^\dagger P$<0.1)

Conclusions Set stocking in this environment is likely to decrease ground cover due to removal of dead annual grasses by grazing in summer/autumn. Deferred grazing has the potential to reduce the risk of a high proportion of bare ground on steep slopes, and to increase the density of perennial species. The long-term effects of various deferred grazing regimes on plant population density and ground cover may become cumulative and need further investigation.

References

Genstat Committee (2000) Genstat for Windows. Release 4.2, 5th Edition. VSN International Ltd., Oxford, UK.
Ridley AM, White RE, Simpson RJ, Callinan L (1997) Water-use and drainage under phalaris, cocksfoot and annual ryegrass pastures. *Australian Journal of Agricultural Research*, 48, 1011-1023.

The impact of continuous grazing by free ranging sheep on the structure and botanical composition of grassland as determined by multivariate analysis

G.A. Heshmati
Gorgan University of Agricultural and Natural Resources Sciences, P. O. Box 386, Gorgan, Iran, Email: heshmati@gau.ac.ir

Keywords: grassland, disturbed, sensitive, stable zones

Introduction Grazing shapes the botanical composition of vegetation at the landscape level (Oksanen *et al.*, 1995). Sheep seek spatially scattered plants of nearly constant and high nutritional value. There is strong interaction between the grazing behaviour of the sheep and the structure of the plant community that it grazes (Squires, 1981). This paper examines the situation in a grassland vegetation grazed by free-ranging sheep in a semi arid region of North Eastern Iran.

Materials and methods Twelve indices of soil surface condition (Tongway, 1994) were assessed in each of five quadrats within each transect. The data were collected from five transects in five contiguous 1 m^2 quadrats along a 50 m transect located about 100 m from each other in excloser and exposer areas at each site. Vegetation cover for each species was measured in each quadrat. Detailed analyses have been made of plant and soil characteristics and the data sets analysed using multivariate techniques.

Results By plotting plant and soil features against distance from water, the grazing gradient method distinguished three common grazing gradient zones (Figure 1). These three zones were the disturbed (degraded) zone, the sensitive zone (the boundary region between them) and the "outer" zone located from closest to the water point to furthest from the water point, respectively. Based on the dendrogram derived from UPGMA cluster analysis of these features, overgrazing around the trough has apparently resulted in an increased density of unpalatable species, decrease of palatable species, and decrease the percentage of crustose soil lichen cover.

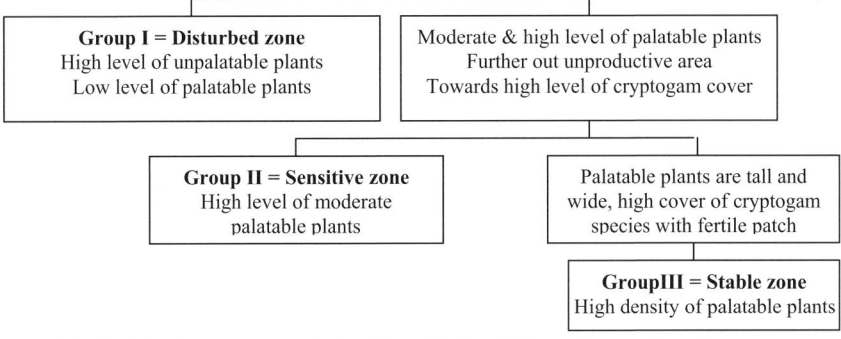

Figure 1 Stylised dendrogram group relationships with plant indicator features for each group

Conclusions In other studies, Crisp (1978) considered that under continuous grazing an increase in unpalatable species and a decrease in palatable species occurs, although these palatable plant species may be lost by excessive browsing. There was an area between the outer and degraded zone where the rangeland condition was different from the other two. This area was called the sensitive zone. The rate of increase of less palatable plants to palatable and start of reducing cryptogam cover are the best indicators for changing the direction of sensitive zone to disturbed zone. It seems that large and abundant palatable plants are the best indicators of stable zone.'

References

Crisp, M. D. (1978). Demography and survival under grazing of three Australian semi-desert shrubs. *Oikos* 30, 520-528.
Oksanen, L., J. Moen, & T. Helle, (1995). Timberline patterns in northernmost Fennoscandia relative importance of climate and grazing. *Acta Botanica Fennica* 153, 93-105.
Squires, V. R. (1981). Livestock management in the arid zone. Melbourne, Australia, Inkata Press.
Tongway, D. J. (1994). *Rangeland soil condition assessment manual*. CSIRO, Division of Wildlife and Ecology,Canberra, Australia

Herbage intake and animal performance of cattle grazing *Brachiaria brizantha* cv. Marandu under continuous stocking

S.C. Da Silva, D.O.L. Sarmento, L.K. Molan, F.M.E. Andrade, A.F. Sbrissia, A.V. Lupinacci, A.C. Gonçalves and D.E. Oliveira
University of São Paulo, Piracicaba, Avenida Pádua Dias, 11, Brazil, Email: scdsilva@esalq.usp.br

Keywords: herbage intake, sward structure, grazing management, animal performance

Introduction Grazing management affects sward structure, which in turn influences plant and animal responses. With the objective of understanding causal relationship between sward structure and animal responses, the present experiment evaluated the daily herbage intake and live weight gain of growing cattle on *Brachiaria brizantha* cv. Marandu pasture during summer (Dec. 2001 to Mar. 2002).

Material and methods Four sward surface heights (10, 20, 30 and 40 cm) were maintained by continuous stocking and variable stocking rates for 12-months. Treatments were assigned to experimental units according to a complete randomised block design with four replications (1200 m^2 per plot). Grazing was performed initially by 280 kg LW Nelore and Canchim heifers from Dec.2001-Jun. 2002 and replaced by 230 kg heifers during Jul.-Dec.2002. Two testers per plot were used for estimates of intake and weight gain with variable stocking rates to control target sward condition. Spatial distribution of sward components was evaluated using an inclined point quadrat (Warren Wilson, 1960) and herbage mass and its morphological composition at the time of evaluations were determined from four 0.30x0.37 m quadrats/plot on a monthly basis. Cattle were weighed at four-weekly intervals. Daily herbage intake and its morphological composition were estimated using controlled release capsules of n-alkane (C_{32} and C_{36}) (Dove *et al.*, 1988) in Jan./Feb. 2002. Animals (32) were dosed with alkanes on January 14th (day 0) and faecal samples collected on a daily basis from day 8 till day 12.

Results Herbage mass increased and herbage as well as leaf bulk density decreased with increasing sward height (Table 1). There was no difference in sward leaf-to-stem ratio. Daily herbage intake was higher on taller (30 and 40 cm) than on shorter (10 and 20 cm) swards (Table 2), with the lowest value recorded on 10 cm swards. Intake of leaf material, however, was not so clear-cut. Stocking rate decreased (8.3, 5.3, 4.1 and 2.9 stock units/ha for the 10, 20, 30 and 40 cm swards, respectively (1 stock unit = 450 kg LW) and live weight gain increased with increasing sward height, with the highest value recorded on the 40 cm swards.

Table 1 Herbage mass (HM – kg DM/ha), leaf-to-stem ratio (L-S), herbage (HBD) and leaf bulk (LBD) densities (mg/cm^3)

	Sward surface height (cm)			
	10	20	30	40
HM	4630[d]	8210[c]	11920[b]	14420[a]
	(512)	(512)	(512)	(512)
L-S	0.72[a]	0.84[a]	0.78[a]	0.69[a]
	(0,080)	(0,091)	(0,080)	(0,081)
HBD	4.7[a]	4.3[a]	4.1[a]	3.7[a]
	(0,31)	(0,33)	(0,31)	(0,31)
LBD	1.5[a]	1.3[ab]	1.2[b]	0.9[c]
	(0.10)	(0.11)	(0.10)	(0.10)

Means followed by the same lower case letters in lines are not significantly different (P>0.10). Values between parentheses correspond to the standard error of the mean.

Table 2 Daily herbage (HI) and leaf (LI) intake (kg DM/100 kg LW.day) and live weight gain (LWG) of beef cattle heifers (kg LW/animal per day)

	Sward surface height (cm)			
	10	20	30	40
HI	1.25[b]	1.57[ab]	1.63[a]	1.77[a]
	(0.06)	(0.07)	(0.08)	(0.07)
LI	0.23[b]	0.44[a]	0.50[a]	0.40[a]
	(0.05)	(0.05)	(0.05)	(0.04)
LWG	0.15[c]	0.48[bc]	0.58[ab]	1.00[a]
	(0.11)	(0.11)	(0.11)	(0.11)

Means followed by the same lower case letters in lines are not significantly different (P>0.10). Values between parentheses correspond to the standard error of the mean.

Conclusions Herbage intake and animal performance responded to variations in sward structure, indicating that sward targets like height can be used to plan, control and monitor grazing management practices.

References
J. Warren Wilson. (1960). Inclined point quadrat. *New Phytologist*, 58, 92-101.
H. Dove, M. Freer & J.Z. Foot. (1988). Alkane capsules for measuring pasture intake. *Proceedings of the Australian Nutrition Society 13*, Melbourne. Australian Academy of Science, p 131.

Tiller population density and sward stability of Brachiaria brizantha continuously stocked by cattle

A.F. Sbrissia, S.C. Da Silva, L.K. Molan, D.O.L. Sarmento, F.M.E. Andrade, A.V. Lupinacci and A.C. Gonçalves
University of São Paulo, Piracicaba, Avenida Pádua Dias, 11, Brazil, Email:scdsilva@esalq.usp.br

Keywords: tillering dynamics, sward surface height, tiller population

Introduction   Tiller population density is one the most important parameters of sward structure and its evaluation is normally included in studies of sward dynamics. Moreover, a greater level of understanding is achieved when the survival of successive tiller generations is monitored. (Matthew *et al.*, 2000). This would help to explain seasonal variation in tiller populations based on tiller appearance and death rates. While *Brachiaria brizantha c.v.* Marandu occupies up to 70 million hectares of cultivated grassland in Brazil, little is known of its ecophysiology. The objective of this work was to calculate survival probability of *B. brizantha* tillers and identify seasonal variation on sward stability.

Materials and methods   Treatments corresponded to four sward surface heights (SSH) (10, 20, 30 and 40 cm), maintained by continuous stocking and variable stocking rate with cattle, and were assigned to experimental units (1200 m²) according to a complete randomised block design, with four replications. Tillering dynamics were evaluated on 4 circular plastic frames (30 cm diam.) in each paddock. Initially all tillers inside the frames were counted and tagged with plastic rings. New counts and taggings were performed on a monthly basis using different colours. Tiller population density was evaluated separately in 0.25 m² quadrats. Survival probability was calculated as proposed by N.R. Sackville Hamilton (non published manuscript) (Bahmani *et al.*, 2003).

Results   Tall swards had consistently lower tiller populations (TP) than short ones, with highest and lowest values registered during summer (rainy season) and winter/early spring (dry season), respectively (Table 1). This decrease in TP with time of the year was relatively larger for the 10 cm swards, probably due to their greater susceptibility to shortage of rainfall at that time of the year. Despite the highest tiller population density during summer, the 10 cm swards had the lowest survival probability (Dec, Jan and Feb) (Figure 1), indicating that the high tiller appearance rates of those swards were not enough to compensate for the reduced survival of tillers. On the other hand, tall swards were relatively more stable than short swards throughout the experiment.

Table 1 Seasonal tiller population density (tillers/m²) of *B. brizantha* swards continuously stocked by cattle

| Season | Sward surface height (cm) | | | | Mean | s.e.m. |
	10	20	30	40		
Summer	1301	1178	1059	914	1113 [A]	22.3
Autumn	1081	1009	969	746	951 [B]	22.3
Winter	958	877	656	523	753 [D]	22.3
Early Spring	949	831	665	486	732 [D]	22.3
Late Spring	934	881	830	658	826 [C]	22.3
Mean	1069 [a]	978 [b]	865 [c]	692 [d]		
SEM	18.1	18.1	18.1	18.1		

Means followed by the same lower case letters in lines and upper case letters in columns are not different (P>0.10)

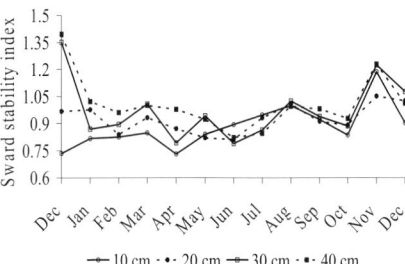

Figure 1 Tiller survival probability diagram

— 10 cm · • · 20 cm — 30 cm · • · 40 cm

Conclusions   Results of tiller population density need to be interpreted carefully, since stability is an important feature of population dynamics and resistance to grazing. *B. brizantha* should not be grazed shorter than 10 cm during summer.

References

C. Matthew, S. G. Assuero, C. K. Black & N.R. Sackville Hamilton. (2000). Tiller dynamics of grazed swards. In: G. Lemaire, J. Hodgson, A. Moraes, P.C.F. Carvalho & C. Nabinger (Eds.) Grassland Ecophysiology and Grazindg Ecology. CABI Publishing, Wallingford, 127-150.

I. Bahmani, E.R. Thom, C. Matthew, R.J. Hooper & G. Lemaire. (2003). Tiller dynamics of perennial ryegrass cultivars derived from different New Zealand ecotypes: effects of cultivar, season, nitrogen fertiliser, and irrigation. *Australian Journal of Agricultural Research*, 54, 803-817.

Effects of vegetation structure and plant height when grazed on persistency of meadow fescue pasture

K. Sudo, Y. Ogawa and K. Umemura
National Agricultural Research Center for Hokkaido Region, Shinsei, Memuro, Hokkaido 082-0071, Japan, Email: ksudo@affrc.go.jp

Keywords: meadow fescue, vegetation structure, plant height, number of tillers, persistency

Introduction An intensive grazing technique using meadow fescue (Mf) pasture has been developed in northern Japan, where soil freezes in winter. It has been shown that the appropriate plant height of Mf pasture when grazed for persistency is about 27 cm. When Mf and perennial ryegrass (Pr) pastures were grazed at the same plant height of 20 cm, vegetation of Mf pasture declined and plant length of Mf pasture was longer than that of Pr pasture (Sudo *et al*., 2002). These phenomena might be due to the effects of differences in grass species and plant height when grazed on vegetation structure, but the mechanisms are not clear. This study was conducted to elucidate the mechanisms of these phenomena. Data on plant height and length obtained in previous studies were reviewed, and pot tests were carried out to reproduce the phenomena.

Materials and methods Measurements of plant height (natural canopy height) and plant length (length of straightened stem and leaf) before grazing were carried out using a ruler over a period of 7 years in Mf and Pr pastures, and relationships between them were analysed. In addition, Mf and Pr pastures were established in September, and dry weights of Mf and Pr were measured every 5 cm from the ground to 30 cm by the stratified clip method in October of the following year. Pot tests for Mf and Pr were carried out between April and August; seedlings were transplanted in a pot (113 mm in diameter and 140 mm in height) and they were cut when plant length reached a fixed length (24 or 32 cm) to a fixed cutting height (5 or 13 cm). Plant lengths of 24 and 32 cm simulated plant heights of 20 and 27 cm, respectively, and cutting heights of 5 and 13 cm simulated high and low stocking intensities, respectively. The effects of grass species and cutting management on cutting frequencies and above ground production during the test and on root weight and numbers of tillers at the end of the test were analysed as a two-factor factorial design.

Results A positive correlation was found between plant height and length (Figure 1). Although regression coefficients to estimate plant length from plant height could be pooled as 1.13 between Mf and Pr pastures, constants differed 1.8, so plant length of Mf was 1.8 cm greater than that of Pr at the same plant height. Distribution of mass between 0 and 5 cm above ground in the Mf pasture was significantly lower than that in the Pr pasture (Figure 2). In the pot tests, significant differences in cutting frequencies, above ground production and root weight were not found between the species, and interaction between species and cutting management was found in number of tillers. Cutting frequency was higher in the order of plots 24-13, 24-5 and 32-13 (plant length - cutting height cm), and above ground production and root weight were decreased in plot 24-5. The number of tillers in plot 24-5 was less than the numbers in other plots for Mf but not for Pr.

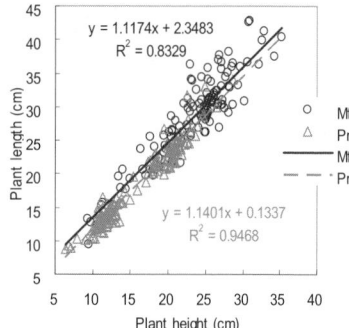

Figure 1 Relationships between plant height and length of Mf and Pr

Conclusions Compared to Pr, Mf was a less erect plant type and distribution of mass in the basal part of community was low. Thus herbage mass of Mf pasture after grazing becomes less if plant height after grazing is the same as that of Pr pasture. Moreover root weight and number of tillers of Mf decreased when Mf pasture of 20 cm in plant height was intensely grazed. It is concluded that these Mf properties cause deterioration in conditions of regrowth, resulting in a decline in vegetation.

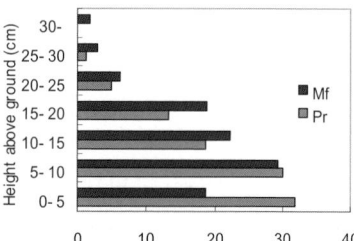

Figure 2 Distribution of mass in every 5 cm above ground in Mf and Pr pastures

Reference

Sudo, K., K. Ochiai, T. Ikeda & K.Umemura (2002). Effects of management on yield, nutritive value and persistency of intensively grazed meadow fescue pasture. *Grassland Science*, 48, 421-427.

Effect of grazing management on herbage accumulation of lucerne-orchard grass sward

J.E. Zaragoza, A. Hernández-Garay*, J. Pérez-Pérez, S. González-Muñoz, J.G. Herrera-Haro, G.F. Osnaya, P.A. Martínez-Hernández, A.R. Quero-Carrillo and J.F. Enriquez-Quiroz

Colegio de Postgraduados, Km 35.5 Carretera México-Texcoco, Montecillo, Texcoco, Edo. de México. CP. 56230. México, Email: hernan@colpos.mx

Keywords: herbage accumulation, grazing frequency and intensity

Introduction Throughout most of México, lucerne (*Medicago sativa*) is the primary forage legume used in the dairy industry. Unfortunately, lucerne does not grow in late autumn and winter due to adverse weather. Recent studies with lucerne have suggested that the inclusion of a companion grass will invariably increase the seasonal distribution and total annual yield of swards (Laidlaw & Teuber, 2001). However the management of mixed swards containing lucerne is difficult as a grazing frequency or intensity which suits one species may be detrimental to the other. Changes in balance between grass and legume, especially in grazed swards, have been observed. In México mixtures of lucerne-orchard grass have a good persistence and productivity. However the explanation for this is unclear. This study examined the effects of different grazing management practices on lucerne-orchard grass production and seasonal distribution

Materials and methods The experiment was conducted from July 2000 to June 2001 at the research unit FES-Cuautitlan, UNAM, State of México, on a well-established mixture of lucerne (cv. CUF101) - orchard grass (Dactylis glomerata cv. Potomac). Swards were rotationally grazed by sheep every 28 and 35 days to residual heights of 3-6 cm (hard), 7-10 (medium) and 11-14 cm (lax). Treatments were arranged in a 2 x 3 factorial design with four replicates. Plot size was 100 m^2. Herbage mass was determined from two 0.25 m^2 quadrats per plot, harvested to ground level before and after each grazing. In the middle of each season, from the two ground-level quadrat cuts, one pooled sub-sample of herbage from each plot was used to determine pre-grazing botanical composition.

Results There were effects of grazing frequency and intensity on seasonal and total herbage accumulation, but there was no frequency x intensity interaction (Table 1).Total herbage accumulation was 24 and 7% more for hard than for lax and medium grazing, respectively (P< 0.05). During spring, summer and winter herbage accumulation decreased as grazing severity decreased from hard to lax (P<0.05). Total herbage accumulation was 4.5 % higher in 28-day than in the 35-day grazing interval. Herbage accumulation was 29 and 30 % greater with 28 than 35 day grazing frequency in summer and winter. In contrast, during autumn and spring 35 day grazing frequency produced 27 and 14 % more than 28 day intervals. Longer photoperiod and warmer temperatures explain summer results, however it is hard to explain the results in winter.

Table 1 Herbage accumulation (kg DM/ha) of lucerne-orchard grass swards under different grazing management

	Summer	Autumn	Winter	Spring	Total
Grazing intensity					
HG	6790	3720	5970	8890	25370
MG	6100	3880	5820	7920	23720
LG	5050	3320	4900	7260	20520
Significant	**	NS	*	**	***
Grazing frequency					
28	6730	3210	6280	7490	23710
35	5230	4070	4840	8560	22700
SEM[a]	430	270	390	450	560
Significant	***	***	***	**	*

* < 0.05; ** < 0.01; *** <0.001; NS no significant differences. [a] Standard error of the least square means.
HG = hard grazing; MG = medium grazing; LG = lax grazing.

Conclusions The results of this study show that to increase herbage accumulation grazing interval must be 35 days during spring and autumn and 28 days during summer and winter. The response was greater under hard grazing than medium and lax grazing.

References

Laidlaw A. S & N. Tueber (2001). Temperate forage grass-legume mixtures: advances and perspectives. p. 85-92. In J.A. Gomide et al. (Ed.) *Proceedings of the IXX International Grassland Congress*, Saõ Pedro (SP) Brazil, 11-21.

Root and vigor response of big bluestem to summer grazing strategies

E.M. Mousel[1], W.H. Schacht[1], L.E. Moser[1] and C.W. Zanner[2]
[1]University of Nebraska, Department of Agronomy and Horticulture, Lincoln, Nebraska, USA, Email: emousel2@unl.edu, [2]University of Nebraska, School of Natural Resource Sciences, Lincoln, Nebraska, USA

Keywords: Animal Unit Month (AUM), root mass-density, organic reserves

Introduction Warm-season grasses e.g., big bluestem (*Andropogon gerardii* Vitman) are great potential sources of summer forage in eastern Nebraska. Frequent, intensive defoliation can reduce root mass and limit root distribution. Quantifying root structure response to multiple defoliation events in a grazing situation is critical to develop management plans for these types of grasses. This experiment aimed to quantify the cumulative effects of timing and frequency of grazing on root structure and organic reserve estimates in big bluestem pastures.

Materials and methods A pasture experiment was conducted in 1999, 2000, and 2001 using a 2 x 2 x 2 factorial with 4 replications, and the following factors and levels: (i) May grazing (M) or May deferred (NM), (ii) June grazing at a late vegetative stage (Jv) or June grazing at an early elongation stage (Je), and (iii) late-summer grazing in early August and early September (AS) or late summer grazing in September only (S). These factors and levels resulted in 8 grazing-date treatments applied to 0.05-ha paddocks using yearling steers (*Bos taurus*) (227 kg) at a cumulative recommended stocking rate of 9.9 AUM/ha (1 AUM=310 kg forage dry matter/month). Following three years of grazing, organic reserves were estimated by measuring etiolated tiller growth within 1.0 m^2 quadrats (Reece *et al*. 1997). Five soil cores (6.6 x 120 cm) were extracted from each paddock and divided into 30 cm segments. Root material was manually washed and separated from soil material, and then dried and weighed. The proportion of root mass to core segment volume was calculated to estimate root-mass density.

Results Timing of grazing and rest-period interval seemed to have a much larger effect than did frequency of grazing on organic reserve estimates and root-mass density in the upper 30 cm of the soil profile. However grazing in May and early June, did not appear to affect root structure or level of organic reserves. Etiolated tiller weight in paddocks grazed at either level in May, the elongation stage in June (Je) and in early August and early September (AS) was 48% lower (P<0.1) than all other grazing-date combinations (Figure 1). Similarly, this grazing-date combination resulted in a 30% reduction (P<0.1) in root-mass density compared to the other grazing strategies (Figure 2). It appears that having less than 40 days rest between the late June and early August grazing periods was inadequate for optimum recovery of big bluestem plants.

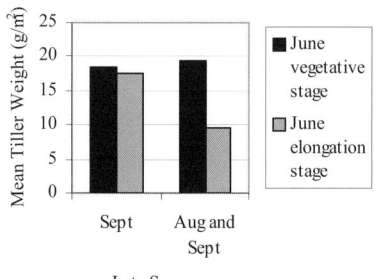
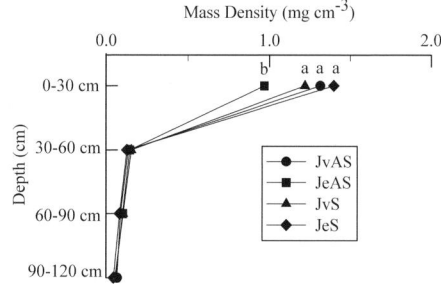

Figure 1 Mean tiller weight of etiolated big bluestem tillers for June and late summer levels of grazing following 3 years of grazing

Figure 2 Mean root-mass density of big bluestem for June and late summer levels of grazing averaged over both levels of May following 3 years of grazing

Conclusions Grazing big bluestem pasture at the elongation stage in June and early August reduces root density and organic reserves of big bluestem. Management strategies for big bluestem pasture should include early season rest periods of 20-30 days and >40 days of rest between defoliation events in late summer to avoid reductions in root-density and plant persistence.

References

Reece, P. E., J. T. Nichols, J. E. Brummer & R. K. Engel. (1997). Technical Note: Field measurement of etiolated tiller growth of rhizomatous grasses. *Journal of Range Management*, 50, 175-177.

Sheep grazing during drought collapses the perennial grass resource in Australian semi-arid wooded grasslands

K.C. Hodgkinson[1], S. Marsden[1] and W.J. Muller[2]
[1]CSIRO Sustainable Ecosystems, Canberra, Act (Australian Capital Territory) 2911, Australia, Email: ken.hodgkinson@csiro.au, [2]CSIRO Mathematics and Information Sciences, Canberra ACT 2600 Australia

Keywords: grasslands, grazing, rainfall

Introduction Grazing of sheep in arid grasslands is risky; sudden shifts to lower functional states may occur when the ecosystem is stressed (Scheffer *et al.*, 2001). To avoid the stresses that shift states, easy-to-recognise critical thresholds need to be identified (Westoby *et al.*, 1989). Preliminary analysis of perennial grass survival in a drought indicated a critical threshold based on co-occurrence of drought and grazing. Crossing this threshold collapses grass populations (Hodgkinson, 1994). Here we examine the relationships between basal area change and rainfall and grazing levels based on a 10-year period and propose a management guideline.

Materials and methods From 1986 to 1996, 6 Merino wethers continuously grazed each of 7 paddocks, each 4-15 ha. The landscapes within each paddock were similar and typical of banded mulga (*Acacia aneura*) woodland. Exclosures (3 x 3m) in each zone (Tongway & Ludwig, 1990) in each paddock were controls. In each zone of each paddock 15 m square quadrats were located randomly. At intervals of 6 months, grass basal area was estimated by adding the basal areas of individual grass plants.

Results Basal area expanded when rainfall was >100 mm for *E. eriopoda* and >150 mm for *T. mitchelliana*. Species (C3 versus C4) determined expansion more than rainfall amount, grazing level or zone location. Below these rainfall thresholds, the less palatable *E. eriopoda* contracted only a little but plants of the highly palatable *T. mitchelliana* were nearly eliminated when grazed during the prolonged drought from early 1991-1995. Contraction of *T. mitchelliana* during this drought was generally greater in the woodland than upslope grassland zone for both the heavily grazed and ungrazed treatments.

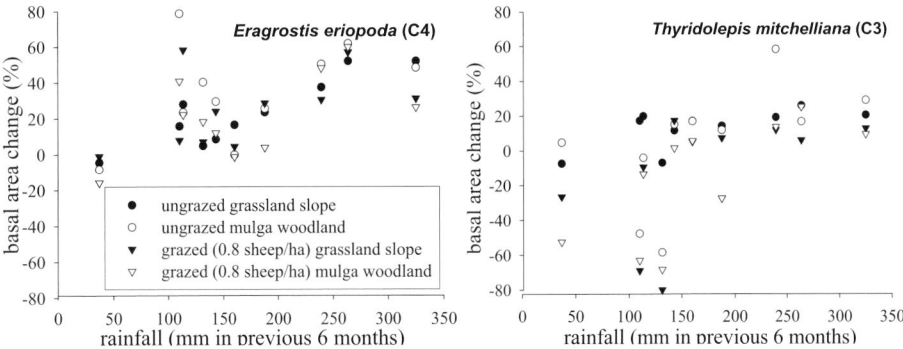

Conclusions When sheep graze heavily in this banded woodland, drought contracts the basal area of some palatable grasses, largely by plant death but shrinkage may be due also to death of some tillers. These plant death times are termed death traps because the mass deaths are set by grazing and sprung by drought. It should be possible to minimise the deaths by reducing sheep density to very low or zero levels as drought conditions begin (3 month rainfall < 75 mm). Such grazing management is termed "Tactical Grazing" because the decision to reduce numbers is made on the basis of an approaching drought.

References

Hodgkinson, K. C. (1994). Tactical grazing can help maintain stability of semi-arid wooded grasslands. In: Proceedings XVII International Grassland Congress, Palmerston North, New Zealand, 75-76.

Scheffer M.S., S. Carpenter, J.A. Foley, C. Folkes & B. Walker (2001). Catastrophic shifts in ecosystems. *Nature*, 413, 593-596.

Tongway, D. J. & J. A. Ludwig (1990). Vegetation and soil patterning in semi-arid mulga lands of Eastern Australia. *Australian Journal of Ecology*, 15, 23-34.

Westoby, M., B. Walker & I. Noy-Meir (1989). Opportunistic management for rangelands not at equilibrium. *Journal of Range Management*, 42, 266-274.

Rainfall and grazing impacts on the population dynamics of *Bothriochloa ewartiana* in tropical Australia

D.M. Orr and P.J. O'Reagain
Dept. Primary Industries and Fisheries, PO Box 6014, Rockhampton Mail Centre, Queensland 4702, Australia, Email: david.orr@dpi.qld.gov.au

Keywords: plant survival, plant size, *Bothriochloa ewartiana*

Introduction *Bothriochloa ewartiana* (desert bluegrass) is a palatable, native perennial (C4) grass of considerable importance to the northern Australian grazing industry. However, little is known of the interaction between grazing pressure and the highly variable rainfall found in this area, on its population dynamics. This paper reports interim results (1998-2004) from a long-term study, in which its population dynamics were examined under 3 grazing strategies.

Materials and methods An extensive grazing study was established in December 1997, at "Wambiana" (20^{0}34' S 146^{0}07'E) near Charters Towers, to assess the relative ability of five grazing strategies, replicated twice, to cope with rainfall variability in terms of their effects on animal production, economics and resource condition (O'Reagain and Bushell 1999). Paddock sizes are 93-117 ha and are arranged so that each paddock contains similar portions of each of 3 soil types. Long-term mean annual rainfall is 630 (range 109-1653) mm, with most falling between November and March. The vegetation is open *Eucalyptus* savanna overlying an herbaceous layer of C$_4$ grasses. Permanent quadrats (n=20; each 50 x 50 cm) delineating 40 *B. ewartiana* plants, were established in each replicate of 3 grazing treatments to examine the persistence of *B. ewartiana* under grazing. These treatments were light stocking (8 ha/steer), heavy stocking (4 ha/steer) and rotational rest (6 ha/steer with 33% of the pasture rested annually during the wet season). The dynamics of these *B. ewartiana* plants were charted annually between 1998 and 2004 using the methodology of Orr *et al.* (2004).

Results Plant size increased between 1999 and 2001 in response to above average summer rainfall but fell dramatically after 2002 in response to severe drought. There were no differences (P>0.05) in plant size between the 3 treatments although there was a clear trend for plant size to be greater under light grazing (Figure 1a). Plant survival did not differ (P>0.05) between treatments with high survival between 1998 and 2002 but there was evidence of accelerated plant death between 2002 and 2004 in response to the severe drought (Figure 1b).

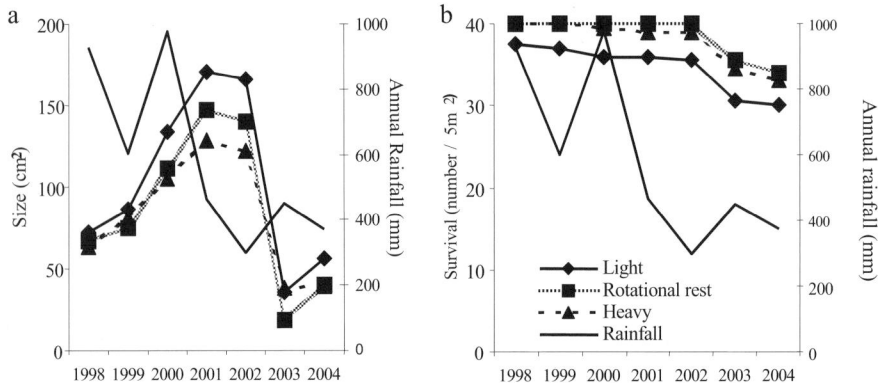

Figure 1 Changes in (a) plant size (b) plant survival of *B. ewartiana* between 1998 and 2004 at Wambiana

Conclusions Rainfall variability rather than grazing pressure appeared to have the greater impact on the dynamics of *B. ewartiana* between 1998 and 2004. This result indicates some resilience under grazing by this grass but we expect a more pronounced grazing impact in future years of this study.

References
O'Reagain, P. J. & J. J. Bushell (1999). Testing grazing strategies for the seasonally variable tropical savannas. *Proceedings 6th International Rangelands Congress*, Durban, 485-486.
Orr, D.M., Paton, C.J. and Reid, D.J. (2004). Dynamics of plant populations in *Heteropogon contortus* (black speargrass) pastures on a granite landscape in southern Queensland. 1. Dynamics of *H. contortus* populations. *Tropical Grasslands*, 38, 17-30.

Herbage production of Tanzania grass (*Panicum maximum* cv. Tanzania) submitted to combinations of frequencies and intensities of grazing by cattle

D. Nascimento Jr.[1], R.A. Barbosa[2], V.P.B. Euclides[2], S.C. Da Silva[3] and R.A. Torres[2]
[1]*Federal University of Viçosa, Viçosa, MG, Brazil, Email: domicio@ufv.br,* [2]*Embrapa – Gado de Corte, P.O.Box 154, Campo Grande, MS, 79002-970, Brazil,* [3]*University of São Paulo, Piracicaba, SP,13418-900, Brazil*

Keywords: light interception, grazing management, growth and senescence

Introduction Animal production from pastures is a complex process comprising three main stages: herbage growth, consumption by grazing animals and conversion into animal products (Hodgson, 1990). Utilisation is the stage where the grazier finds greater flexibility for management, probably because most processes related to harvest of the produced herbage by the grazing animals are very responsive to manipulation and control of defoliation practices. The objective of this study was to evaluate herbage production of a *Panicum maximum* cv. Tanzania pasture submitted to combinations of frequencies and intensities of grazing.

Material and methods Treatments comprised six rotational grazing systems characterised by combinations of three grazing intervals (time interval after grazing for the grass canopy to reach 90, 95 or 100% of incident light - LI) and two post-grazing residues (25 and 50 cm). These were allocated to experimental units (2500 m^2) according to a complete randomised block design with three replications. Sward light interception was monitored during regrowth with a canopy analyser (AccuPAR Linear PAR / LAI ceptometer) and grazing was carried out by a variable number of 200 kg Nelore steers depending on the herbage mass available. Grazings lasted no longer than three days. Herbage accumulation was determined by cutting all herbage at ground level within six 1 m^2 quadrats, randomly located in each experimental unit. Cuts were taken pre and post-grazing. Herbage samples were hand-sorted into leaf, stem and dead material. The samples were then dried in an oven at 65 °C and weighed. Total and leaf herbage accumulation were calculated as the difference between successive post and pre-grazing herbage masses. The experimental period lasted 309 days, from 11 July, 2003 until 15 May, 2004.

Results Highest herbage accumulation was recorded for the 95% LI and 25 cm residue treatment. At 90% LI production was lower than at 95 and 100% LI, probably due to the smaller leaf area available for capturing all the incident radiation, limiting the growth process. Treatments with 100% LI had a higher proportion of dead material at grazing than the 90 and 95% LI (24.8 and 11.9%, respectively). Leaf lamina accumulation did not differ between the 90 and 95% LI, suggesting that the herbage accumulated in 90% LI was practically all leaf lamina. Treatments with 100% LI resulted in the lowest values of leaf lamina accumulation, with a higher proportion of stems compared to 90 and 95% LI (17.6 and 15.3%, respectively), particularly when associated with the 50 cm post-grazing residue (17%).

Table 1 Total herbage accumulation (kg DM/ha)

Residue	Light interception (%)			
cm	90	95	100	Mean
25	11740^{Ab} (770)	15120^{Aa} (770)	11620^{Ab} (770)	12830^{A} (444)
50	9440^{Bb} (770)	11940^{Ba} (770)	12710^{Aa} (770)	11360^{B} (444)
Mean	10590^{b} (544)	13530^{a} (544)	12170^{a} (544)	12100 (314)

Numbers within parentheses correspond to standard error of mean
Means followed by the same upper letter in columns are not different (P>0.10). Means followed by the same lower letter in lines are not different (P>0.10)

Table 2 Total leaf accumulation (kg DM/ha)

Residue	Light interception (%)			
cm	90	95	100	Mean
25	9000^{Ab} (392)	10600^{Aa} (392)	8030^{Ab} (392)	9210^{A} (226)
50	8360^{Aa} (352)	8060^{Ba} (392)	6750^{Bb} (392)	7720^{B} (226)
Média	8680^{a} (277)	9330^{a} (277)	7390^{b} (277)	8470 (160)

Numbers within parentheses correspond to standard error of mean
Means followed by the same upper letter in columns are not different (P>0.00). Means followed by the same lower letter in lines are not different (P>0.10)

Conclusions Herbage as well as leaf lamina accumulation in Tanzania grass under rotational grazing can be controlled by means of adjustments in grazing interval and intensity. Optimum regrowth interval was associated with the pre-grazing condition of 95% LI and grazing intensity with the post-grazing residue of 25 cm.

References
Hodgson, J. (1990). Grazing Management: Science into practice. New York: John Wiley & Sons, 203p

Herbage accumulation and vegetation structure in Tanzâniagrass (*Panicum maximum*, Jacq. cv Tanzania) pasture submitted to regime of intermittent stocking

A.C. Ruggieri, C.A.M. Gomide, E.R. Janusckiewicz, E.D. Contato and R.A. Reis
Instituto de Zootecnia, C.P. 63, Sertãozinho, SP, Brazil CEP 14160-900, Email: ruggieri@iz.sp.gov.br

Keywords: yield, herbage mass, sward persistence, leaf/stem ratio, tussock perimeter

Introduction Knowledge of pasture structure characteristics such as height, tillering density, leaf area index (LAI) and herbage mass (HM), together with growth rates and forage accumulation, can become important tools to indicate improvements to the production system. This experiment evaluated the structural and productive characteristics and persistence of a Tanzaniagrass pasture.

Material and methods The pastures were grazed (cows with medium weight of 450 kg) from Dec. 2002 to March, 2004 in a completely randomised design with 2x2 factorial scheme with two rest periods (25 and 35 d) and two residue heights post grazing (30 and 50 cm) and three replicates. Before each grazing, measurements of the medium height, light interception and the LAI of each paddock were made to verify the effect of the residue pos- grazing and rest period on the growth cycle for the following grazing. The total HM was assessed by cutting 1m^2 samples at ground level. Harvested samples were separated into the fractions leaf, stem and dead material. Plant cover was evaluated by sampling a straight transect line (Carvalho *et al.*, 2003) after the exit of the animals from the pasture. The data were analysed by the procedure GLM of SAS.

Results There were differences in HM between the first and second year, but these arose only in the treatment with residue 30 cm and 25 d rest. The higher HM in the first year had a higher dead material %, while leaf:stem ratio (L/S) differed between the grazing cycles in the first year. There were differences in plant cover (Table 2) among the treatments with values being higher with 25 d compared with 35 d of rest. The average of plant cover was 48%, in line with this being a bunch grass with rapid growth, high soil cover and low plant density. The perimeter of the tussocks was higher for the longer rest period, associated with these treatments having longer available time for the plants to develop. However the most severe grazing intensity with the shortest rest stimulated tillering and resulted in higher plant population.

Table 1 Effect of treatments on herbage mass, dead material and leaf:stem ratio during two years

	residue 30, rest 25	residue 50, rest 25	residue 30, rest 35	residue 50, rest 35	average
			Total HM (t/ha)		
Year 1	4.7a	5.8a	5.7a	6.4a	5.7A
Year 2	5.2a	5.1a	4.2b	5.1a	4.9B
			Dead material (%)		
Year 1	12.76a	14.24a	14.06a	15.02a	14.02A
Year 2	9.49a	10.77a	8.57a	11.15a	9.99B
			L/S		
Year 1	3.59 a	3.00 ab	2.48 bc	2.26 c	2.83A
Year 2	2.65a	2.62a	2.85a	2.60a	2.68A

Table 2 Occurrence of plants, perimeter of tussocks and height of tussocks in relation to treatments

Treatments	Plant cover (%)	Perimeter of tussocks (cm)	Height of tussocks (cm)	LAI
Residue 30, rest 25	56.67A	118.34BC	55.33C	6.3
Residue 50, rest 25	50.00AB	90.63C	60.57B	6.5
Residue 30, rest 35	46.67B	123.96B	57.10C	6.0
Residue 50 cm, rest 35	40.00C	159.33A	89.76A	5.6

Values with different letters differ significantly at P<0.05

Conclusion The management system with shorter rest gave more plants, while longer rest periods gave larger tussocks. Higher residues and longer rest periods resulted in taller plants.

Reference
Carvalho, D.D., A.A.G. Pagano & J.F. Figueiras (2003). Cobertura de solo e tamanho de touceiras em pastagens de capim Aruana e Tanzânia In: Reunião anual da SBZ,40, 2003, Santa Maria. Annals, CD rom.

Seasonal variation of taproot biomass and N content of lucerne crops under contrasting grazing frequencies

E.I. Teixeira, D.J. Moot, H.E. Brown and M. Mickelbart
Agriculture and Life Sciences Division, Lincoln University, Canterbury, New Zealand, Email: teixeie2@lincoln.ac.nz

Keywords: lucerne, modelling, root reserves

Introduction Taproot nitrogen reserves (TN, kg N/ha) a function of N concentration within taproots (N%) and taproot biomass (TBM) are a major determinant of lucerne (*Medicago sativa* L.) growth rates after defoliation and in early-spring (Avice *et al.*, 1997b). Several studies have shown that N% changes with seasons (Cunninghan & Volenec, 1998) and defoliation frequencies (Avice *et al.*, 1997a). However the seasonal pattern of TBM deserves further investigation as the dynamics of root reserves is a weak point in lucerne simulation models (Confalonieri & Bechini, 2004). The objective of this experiment was to assess the seasonal variation in TN through the measurement of TBM and N% of lucerne crops subjected to two defoliations frequencies.

Materials and methods A two year old fully irrigated 'Kaituna' lucerne crop was subjected to 28 or 42-day grazing rotations to induce different levels of N and biomass in taproots. The experiment was conducted at Lincoln University, NZ (43°S and 172°28'E) from 12 June 2002 to 10 June 2003 in a randomized complete block design with four replicates. Taproot biomass (300 mm depth) was excavated at the end of each regrowth cycle. Samples were analysed for total N.

Results From winter to summer TBM decreased at a rate of ~8 kg DM/ha/day for both defoliation regimes (Figure 1a), while from summer to autumn TBM increased and was fully restored to ~3.0 t/ha in the 42-day crop but was 1.0 t/ha less (P<0.05) in the 28-day crop. The seasonal pattern for N% was similar and decreased from 2.1% in the winter of 2002 to a minimum of 1.4% in the 42-day crop and 1.0% in the 28-day crop (Figure 1b). Both crops increased the N% in autumn, but the 28-day crops remained 20% lower (P<0.05) than 42-day. Increases in TN (P<0.05) were greater at ~25 kg/ha in summer to 50 kg/ha in autumn for the 42-day crop than the 28 day crop where the increase was always less than 25 kg/ha.

Conclusions Seasonal variations and two fold increases in TN were identified between summer to autumn in 42-day crops. This was caused by an increase in both the N concentration in taproots (i.e. N%) and the storage capacity of the root system (i.e. TBM). The similarity in the direction of response of TBM and N% to environmental signals and defoliation managements suggests the inclusion of both variables is required to accurately access shoot growth rates in lucerne simulation models.

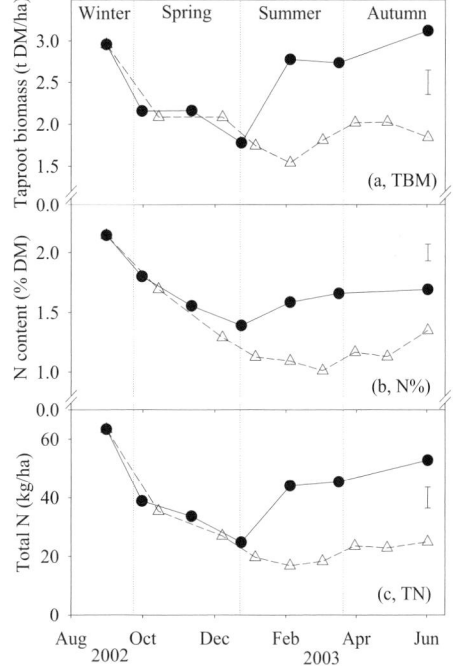

Figure 1 Seasonal variation of TBM (a), N% (b), and TN (c) in lucerne crops subjected to 28-day (●) and 42-day (●) grazing rotation. Bars indicate one SEM

References

Avice, J., G. Lemaire, A. Ourry, & J. Boucaud (1997a). Effects of the previous shoot removal frequency on subsequent shoot regrowth in two *Medicago sativa* L. cultivars. *Plant and Soil,* 188, 189-198.

Avice, J., A. Ourry, G. Lemaire, J. Volenec, & J. Boucaud (1997b). Root protein and vegetative storage protein are key organic nutrients for alfalfa shoot regrowth. *Crop Science,* 37, 1187-1193.

Confalonieri, R. & L. Bechini (2004). A preliminary evaluation of the simulation model CropSyst for alfalfa. *European Journal of Agronomy,* 21, 223-227.

Cunninghan, S. & J. Volenec (1998). Seasonal carbohydrate and nitrogen metabolism in roots of contrasting alfalfa (*Medicago sativa* L.) cultivars. *Journal of Plant Physiology* 153, 220-225.

Sward composition, forage yield, and grazing effects in kura clover and grass mixtures

P. Jeranyama[1], R. Leep[2], T. Dietz[2] and D. Min[3]
[1]Plant Science Department, South Dakota State University, Box 2207A, Brookings, South Dakota 57007 USA, Email: peter.jeranyama@sdstate.edu, [2]Department of Crop and Soil Sciences, Michigan State University, East Lansing, Michigan 48824 USA, [3]Upper Peninsula Experimental Station, Chatham, Michigan 49816, USA

Keywords: forage yield, sward composition, grazing

Introduction Rotational stocking on mixed pastures of cool season grasses (C3) and kura clover (*Trifolium ambiguum* Bieb.) can be a sustainable way to reduce cattle feed costs. Kura clover is a long-lived rhizomatous perennial legume of good forage quality under grazing. However, the suitability of diverse grass species in binary mixtures with kura clover has not been reported extensively. This experiment aimed to evaluate sward composition, forage dry matter yield, and post-grazing residue in mixed kura clover and C3 grass pastures.

Materials and methods C3 grasses (n=7) were seeded into kura clover on a fine sandy loam soil at Lake City, Michigan. Both P and K were added according to soil test; no additional N was added to the pastures during the experiment. Kura clover seeds were inoculated with *Rhizobium meliloti* before planting in a replicated study. Because kura clover is slow to establish, plots in this experiment were not grazed in the seeding year. A conditioned Simmental beef herd mob-grazed (Bittman & McCartney, 1994) pasture mixtures in years 2 and 3 at a stocking rate of 6-10 animals/ha/day. Grazing started before emergence of inflorescence on grasses and continued until an average stubble height of 5-8cm was left on the most preferred grass. On average, grazing lasted 2 days depending on the availability of forage mass. Pastures were stocked rotationally 6 times in 2 years. Sward composition and forage yield were assessed by clipping a 0.25m^2 quadrat before grazing. To determine residual forage mass, each plot was harvested to 5cm height post-grazing using a flail type harvester.

Results The high forage dry matter yield (DM) in the cocksfoot (*Dactylis glomerata* L) mixture (up to 4t/ha), was due largely to high grass content (85-88%) in the sward. Inversely, the low DM yield with the Kentucky bluegrass (*Poa pratensis* L.) mixture was due to a low grass content (38-47%) in the sward (Table 1). Low DM yields at Grazing Event 6 was due to cumulative grazing effects resulting in slow plant recovery. Self recruiting white clover (*T. repens* L.) contributed more than kura clover (3-21%) to the swards. Total clover contribution in swards ranged from 12-41%. On average, weeds comprised <12% of sward DM yield. The high residue at Grazing Event I compared with Event 6, was due mainly to the high forage mass available, resulting in spoilage.

Table 1 Pre-grazing dry matter yield (DM), grass and clover content, and post-grazing residues in kura clover and C3 grass mixtures

Species	Grazing Event 1				Grazing Event 6			
	DM (t/ha)	% grass	% clover	[#]% post	DM (t/ha)	% grass	% clover	% post
Kentucky bluegrass	2.7	47	36	85	0.7	38	41	4
Cocksfoot	4.0	85	12	74	1.3	88	6	7
Perennial ryegrass	2.7	69	22	69	0.7	59	23	1
Reed canarygrass	3.1	79	13	55	1.0	60	16	3
Smooth bromegrass	3.6	72	21	61	0.7	64	12	11
Tall fescue	2.5	66	21	85	0.8	75	12	5
Timothy	2.6	61	25	75	0.5	47	16	3
LSD (0.05)	0.8	18	23	NS	0.5	20	22	4

[#]post-grazing residues

Conclusions It was difficult to maintain kura clover in swards. Self recruiting white clover replaced kura clover as the dominant clover in the sward; white clover, ladino clover and red clover contributed greatly to the total clover content in the range of 12-41%. Cocksfoot was too aggressive to allow other species to grow. The grass composition in sward largely dictated the forage yield.

Reference

Bittman, S. & D. H. McCartney (1994). Evaluating alfalfa cultivars and germplasms for pastures using the mob-grazing technique. *Canadian Journal of Plant Science*, 74, 109-114.

Production and plant density of Sulla grazed by sheep at three growth stages

H. Krishna and P.D. Kemp
Institute of Natural Resources, Massey University, Palmerston North, New Zealand, Email: p.kemp@massey.ac.nz

Keywords: *Hedysarum coronarium*, persistence, grazing frequency, herbage mass

Introduction Sulla is one of the few temperate forage legumes that contain enough condensed tannins to improve the efficiency with which livestock use protein (Marshall *et al*. 1979). However, it usually is productive only for approximately 14 months in New Zealand, and little is known of its response to grazing. This paper reports on the production and persistence of Sulla cv. Necton, when using growth stage as the criterion for time of grazing by sheep in a maritime, temperate environment.

Materials and methods Sulla cv. Necton was sown (15 kg/ha of inoculated seed) in October in medium fertility, Typic Fragiaqualf soil (pH 6, Olsen P 12 µg/g) near Palmerston North, New Zealand. A randomised complete block design with 3 treatments (LV (late vegetative), MSE (mid-stem elongation), and EF (early flowering) growth stages) and 4 blocks was used. Individual treatment plots were 29 X 6.6 m. Each time it reached the set growth stages, Sulla was grazed with mature Romney ewes for 365 days after sowing (DAS). The first grazing of LV, MSE, and EF was 83, 90, and 111 DAS, respectively. Grazing intensity was set at approximately 70% of herbage removed, including most leaves. Herbage mass was measured pre- and post-grazing using ground level cutting in three 0.3 m^2 quadrats/plot. Leaf and stem were dissected in the pre-grazing samples before drying.

Results Sulla was highly productive in all treatments and its EF stage had the highest 365-day herbage mass accumulated for three grazings (Table 1). The mean post-grazing herbage masses for the EF, MSE, and EF treatments were 1,616, 1,465, and 1,972 kg DM/ha, respectively. The residual herbage consisted almost entirely of stem. The ratio of leaf mass : stem mass was lower at EF than at LV or MSE stages (Table 1). In February, the grazed herbage in the EF treatment had 2.3% N and a DM digestibility of 72%. Plant density was greater in the EF treatments after 365 days (Table 1), but all treatments failed to persist >14 months. The grazing of LV and MSE treatments in winter was the main cause of their lower plant density than the EF treatments (Table 2).

Table 1 Net herbage mass accumulation, leaf: stem ratio and plant density of Sulla cv. Necton over 365 days from sowing under infrequent, hard grazing with sheep

Growth stage	No. of grazings	Herbage mass kg DM/ha	Leaf: stem ratio	Plants/m^2 0 DAS	365 DAS
LV	4	21,780	2.0	67	15
MSE	4	22,020	2.1	62	16
EF	3	24,700	1.3	45	32
LSD 5%		1990	0.2	NS	6

Table 2 Plant density of Sulla cv. Necton in spring (September) after being grazed or not grazed in winter (June/July) at the late vegetative (LV) and the mid-stem elongation (MSE) growth stages, by sheep

Treatment	Plant/m^2 LV	MSE
Grazed in winter	14	16
Ungrazed in winter	34	39
LSD 5%	4	4

Conclusions The productivity of Sulla cv. Necton was confirmed with >20 t DM/ha in all treatments when grazed hard (3-4 times) during the 365 day period post sowing. Winter grazing damaged plant density and it was difficult to use winter growth. Grazing at early flowering avoided winter grazing and increased plant survival, but decreased feed quality through increased stem. Necton Sulla is highly productive under infrequent grazing but its intolerance of winter grazing and short lifespan limits its usefulness.

References
Marshall, D.R., P. Broue, J. Munday (1979). Tannins in pasture legumes. *Australian Journal of Experimental Agriculture and Animal Husbandry*, 19, 192-197.

Productivity and grazing capacity of five typical natural rangelands for Yaks in the Alpine region of China

Y. Xuebing [1,2], G. Yuxia [3], Z. He [1] and W. Kun [1]
[1]China Agricultural University, Beijing 100094, P.R.C., Email: yxbbjzz@163.com, [2]Zhengzhou College of Animal Husbandry, zhengzhou 450002, P.R.C.3.Henan Agricultural University, zhengzhou 450002Chin,

Keywords: Alpine rangeland, productivity, feed value, grazing

Introduction The Qinghai-Tibet plateau has greatly aroused the interest of scientists as an uncommon rangeland resource of great agro-ecological importance. Yak (*Bos grunniens*) is a unique, vulnerable ungulate. The objective of this experiment was to evaluate the productivity and feed value of five natural rangelands.

Materials and methods The experiment was conducted in Tianzhu county of Gansu province during 1998 and 1999. Dynamic changes in yield and feed value (content and degradability in sacco of DM,OM,ADF,CP) of five groups of alpine rangelands, comprising *Elymus nutans, Kobresia capillifolia, Polygonum viviparum, Rhododendron spp.+ Carex spp.* and rangeland elevations (m) are: A 2880, B 2930, C 2980, D 3030 and E 3080. Grazing capacity was based on the grazing intake of a mature Yak cow (Yan *et al.*,2003) and the Chinese Law on the grazing rate for protecting rangeland from deterioration. Degradability of mixed samples from each rangeland was calculated according to Orskov & Mcdonald (1979).

Results Content and degradability of each nutrient showed had the highest values in July and August except for ADF content (Figure 1). The maximum grazing capacities were 338,272,290,245,184 heads/day/ha for each alpine rangeland, respectively (Table 1).

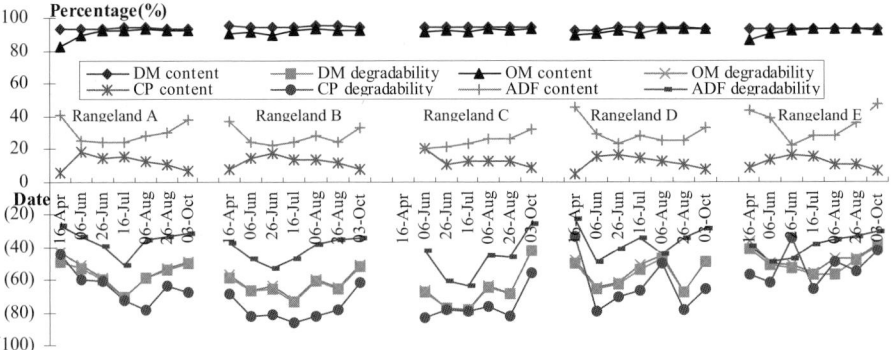

Figure 1 Dynamics in content and degradability of each nutrient in five rangelands

Table 1 Grazing capacity(heads/day/ha) of five natural rangelands at different time

Date	16th April	6th June	26th June	16th July	6th August	26th August	3th October
Grassland A	8	58	125	265	338	261	163
Grassland B	13	67	117	190	272	232	207
Grassland C	\	45	87	150	210	290	193
Grassland D	33	37	80	192	245	165	100
Grassland E	38	63	80	106	184	130	100

Conclusions Grazing capacity of each rangeland was highest between July and August, but feed value decreased with growth. Based on intake and nutrition, grazing in summer and supplementary feeding in winter reasonable optimal for Yaks. On the plateau, the proposed ratio of natural to cultivated grasslands is 10:15.

References

Yan Xuebing, Wang Xi & Guo Yuxia (2003).Determination of several physiological parameters of intake and digestion of grazed Yaks. *Chinese Journal of Herbivores*, 23, 5-7.
Orskov, E. R. & I. Mcdonald (1979). The estimate of protein degradability in the rumen from incubation measure must be weighted according to rate of passage. *Journal of Agricultural Science Cambridge*, 29, 499-503.

Effect of stocking rates on plant morphology in the inner Mongolia steppe of China

S.P. Wang[1], Y.F. Wang[1], Z.Z. Chen[1], B. Patton[2] and P. Nyren[2]
[1]Institute of Botany, Chinese Academy of Sciences, Beijing 100093, China; Email: wangship@yahoo.com,
[2]Central Grasslands Research Extension Center, North Dakota State University, ND 58483, USA

Keywords: stocking rate, evolution, bunchgrass, rhizomes, stolons

Introduction During the long period of co-evolution with herbivores, range plants have adapted and developed resistant mechanisms in response to grazing (Briske, 1991). The objective of this experiment was to determine the morphological response of a number of the dominant plant species in the Inner Mongolia steppe of China to stocking rate.

Materials and methods There were four stocking rates (no-grazing, 1.33, 4.00 and 6.67 sheep/ha) on plots that had been subjected to these grazing pressures for six years. Of *Stipa krylovii*, *Cleistogenes squarrosa* and *Artemisia frigida*, 20 bunches were chosen per treatment, while of *Leymus chinensis* (rhizomatous plant), 20 plots (30 cm x 30 cm) were selected per treatment. Plant height (H, cm), biomass (g DM per shoot or bunchgrass) (B), number of the shoots on the two adjacent reproductive nodes or tillers in a bunchgrass (S) (P and V were the reproductive and vegetative shoots for *Stipa krylovii*, respectively), length between two nodes along rhizome or stolon (D, cm) and maximum diameter of the *Cleistogenes squarros* canopy (R) were measured.

Results Heavy grazing decreased H and B of individual range plants (Table 1). The distance between nodes along rhizome or stolon was reduced by grazing for *L. chinensis* and *A. frigida*. Light grazing increased shoot density for *L. chinensis* and reproductive tillers for *S. krylovii*. Heavy grazing decreased the range of canopy for *C. squarrosa*.

Table 1 Morphologies of individual range plants under different stocking rates

Plant species	Item	Stocking rate (sheep/ha)				Plant species	Item	Stocking rate (sheep/ha)			
		0	1.3	4.0	6.7			0	1.3	4.0	6.7
Leymus	H	18.1a	16.8a	9.7b	8.6b	*Cleistogenes*	H	11.8a	8.9bc	6.5c	3.5d
chinensis	B	3.3a	4.9a	0.8b	0.5b	*squarrosa*	B	31.9a	6.1b	5.9b	3.9c
	S	1.8b	2.6a	1.3c	1.4bc		R	14.5a	9.9b	8.2b	7.3d
	D	2.6a	2.1b	2.0bc	1.6c		S	64.2a	67.4a	35.3b	44.3b
Stipa	H	51.6a	42.9b	41.2b	33.8d	*Artemisia*	H	8.2a	9.7a	4.9b	2.7b
krylovii	B	33.6a	40.1a	15.5b	5.0c	*frigida*	B	8.8a	/	2.4b	1.6b
	P	3.2b	5.7a	2.7c	1.2d		S	4.9a	/	4.0a	5.8a
	V	29.3a	26.5a	20.9b	12.4c		D	3.0a	/	2.0ab	1.2b

Conclusions Plants with rhizomes or stolons may have adapted to grazing by increasing the reproductive node density. Fragmentation of bunchgrasses may be their strategy under heavy grazing. Therefore, 1.3 sheep/ha seems to be optimum in terms of plant height and biomass of individual plant.

Acknowledgement This paper was supported by project of KSCX2-SW-107

Reference
Briske, D.D. (1991). Developmental morphology and physiology of grasses. In R.K. Heitschmidt & J. W. Stuth (eds.) Grazing management: an ecological perspective. Timber Press, Portland, OR., 85-108.

Effect of stocking rate on a *Stipa breviflora* Desert Steppe community of Inner Mongolia

G. Han[1], W.D. Willms[2], M. Zhao[1], A. Gao[1], S. Jiao[1] and D. Kemp[3]
[1]College of Ecol. and Env. Sci., Inner Mongolia Agric. Univ., Hohhot, Inner Mongolia 010018 P.R. of China, Email: nmghand@hotmail.com, [2]Agric. and AgriFood Canada, PO Box 3000, Lethbridge, Alberta, Canada T1J 4B1, [3]University of Sydney, PO Box 883, ORANGE, New South Wales 2800, Australia

Keywords: plant cover, above ground annual production

Introduction Stocking rate is an important factor in grazing management. The stocking rate defines utilization and ultimately grazing pressure, which in turn affects grassland sustainability. Grassland sustainability is partly defined by its species composition and ultimately by its productivity. These attributes are unique for specific plant communities and the effect of stocking rate must be established for each in order to understand the community response to grazing and to determine its carrying capacity. While some information exists on the effects of stocking rate on livestock production in the *Stipa breviflora* Griseb. Desert Steppe (Wei *et al.*, 2000), the effects on the plant community are not understood well. This study aimed to determine the effects of stocking rate on the species composition and productivity of that community.

Materials and methods The study was conducted on the Inner Mongolian Plateau (41° 47' N, 111° 54' E, average annual precipitation, 280 mm: elevation, 1450 m asl: soil, Light Chestnut) in May, 2004. The dominant species were *S. breviflora*, *Artemisia frigida* Willd. and *Cleistogenes songorica* (Roshev.) Ohwi. The site had been moderately degraded. A 4 (stocking rates) x 3 (replicates) study commenced in 2004. Each paddock was 4.4 ha and stocked with 0 (control), 0.46 (light), 0.91 (moderate) or 1.35 (heavy) sheep units/ha per yr in a randomized complete block design. Standing crop was estimated monthly from May to October using 10 moveable cages (1.5 x 1.5 m) that were randomly located within each paddock. Total above ground annual production (ANPP), and ANPP of individual species, were estimated by the cumulative difference method of paired grazed and ungrazed plots (1 x 1 m). The cages were moved to a new location after each harvest.

Results Total ANPP was greater (P<0.05) with the control and light stocking rate, than with the moderate and heavy stocking rate (Table 1). The ground cover of the plant community decreased linearly with increasing stocking rate. Similar trends were observed for individual species, with the greatest difference occurring between the control and light grazing (Table 1).

Table 1 Effect of stocking rate on the ANPP (kg/ha per yr) and ground cover (%) of the whole plant community and of selected species in the Desert Steppe of Inner Mongolia

Stocking rate	Plant community	*Stipa breviflora*	*Artemisia frigida*	*Cleistogenes songorica*	*Convolvulus ammannii*	*Heteropappus altaicus*
			Above ground net primary production (Mean±SD)			
Control	108.8±5.3a	12.9±2.0a	71.9±5.3a	5.0±0.8a	6.7±1.1a	1.5±0.3a
Light	81.4±4.8b	6.2±2.0b	53.2±5.7b	3.5±0.8ab	1.3±0.9b	0.2±0.1b
Moderate	62.3±6.8c	6.3±1.5b	56.0±3.6b	3.1±1.2ab	0.1±0.0b	0.1±0.1b
Heavy	51.1±3.7c	6.9±2.0b	34.9±4.8c	1.0±0.6b	0.8±0.5b	0.1±0.1b
			Ground cover (Mean±SD)			
Control	23.3±0.9a	3.8±0.8a	15.0±1.1b	2.6±4.2a	1.4±0.2a	0.6±0.1a
Light	22.2±0.9a	3.3±0.6ab	18.2±1.3a	1.8±2.4b	0.9±0.2ab	0.6±0.1a
Moderate	16.6±0.8b	2.6±0.3ab	11.8±0.7c	2.3±2.0ab	1.1±0.2ab	0.2±0.1b
Heavy	10.6±0.4c	2.3±0.1b	7.8±0.3d	1.3±0.8b	0.8±0.1b	0.1±0.0b

Means within a subset of a column having a common letter are not different (*P* > 0.05)

Conclusion A sustainable stocking rate on the *Stipa breviflora* Desert Steppe community appears to be <0.91 (moderate) sheep/ha per yr. This stocking rate is below 1.1 sheep/ha per yr, which was recommended by Han *et al.* (2000), based on liveweight gain. These results are not unexpected because the optimal stocking rate for livestock often exceeds the optimal to maintain a desirable plant community.

Acknowledgements This work was funded by the National Natural Science Foundation of China (30060056,30360022)

References

Wei, Z., G. Han, J. Yang & Xiong Lu (2000). The response of *S. beviflora* community to stocking rate. *Grassland of China*, 6, 1-5.

Han, G., B. Li, Z. Wei & H. Li (2000). Liveweight change of sheep under 5 stocking rates in *Stipa breviflora* Desert Steppe. *Grassland of China*, 6, 4-6.

Effect of stocking rate and grazing system on fine and superfine Merino wool production and quality on native swards of Uruguay

I. De Barbieri, F. Montossi, E.J. Berretta, A. Dighiero and A. Mederos
*Instituto Nacional de Investigación Agropecuaria (INIA), Ruta 5, km 386, PC: 45000, Tacuarembó, Uruguay,
Email: idebarbieri@tb.inia.org.uy*

Keywords: fine wool, native swards, stocking rate, grazing system

Introduction Modern textile tendencies show that consumers prefer light, soft, resistant, natural, and comfortable clothes, for which fine and superfine wools are in great demand, particularly at the high value markets (Whiteley, 2003). The main objective of the present study was to define sustainable stocking rates and grazing systems on native swards for fine and superfine wool production in the Basaltic region of Uruguay.

Materials and methods The trial was carried out for two years (October 2001 to October 2003), using the same 72 Australian Merino whethers in both years. The evaluated factors were stocking rate (SR, 5.3 and 8.0 animals/ha) and grazing system (GS, continuous -CG- and 21 days strip grazing -21G-), and their combination. The experimental area was 6.0 ha, (1.5 ha per treatment). The variables measured each 21 days were: a) on pre grazing swards: herbage mass (DM, kg DM/ha), sward height (H, cm), botanical composition (BC) and nutritive value; b) on all animals: liveweight (LW, kg) and condition score (CS, units); and c) on wool (annually): fleece weight (FW, kg), fibre diameter (FD, μ) and other quality characteristics. The design was a complete randomised block, arranged in a factorial structure, the main factors being: SR and GS at two levels each.

Results The SR of 5.3 animals/ha gave higher DM yield and H than the SR of 8.0 animals/ha (Table 1). The overall sward botanical composition and nutritive value was not affected by SR and GS. The BC (annual average) was: green grass leaf 48.0%, green grass stem 5.5%, dead material 40.0% and other species 6.5%; nutritive values were: CP 8.0%, NDF 68.5% and ADF .8%. BC and nutritive value were affected only by SR, essentially during spring and summer. A SR increase from 5.3 to 8.0 animals/ha resulted in higher sward quality. SR affected LW, CS and FW, associated with its effect on sward quantity and quality. Wool quality was not affected by SR (Table 2). GS did not significantly affect the variables related to animals and wool (Table 2).

Table 1 Effect of SR and GS on sward quantity characteristics

Variable	SR			GS			SR*GS				
	5.3	8.0	P	21G	CG	P	5.3*21G	5.3*CG	8.0*21G	8.0*CG	P
DM (kg MS/ha)	3043a	1745b	**	2734a	2054b	**	3593a	2493b	1875c	1615d	**
H (cm)	10.8a	6.1b	**	9.4a	7.5b	**	12.3a	9.3b	6.5c	5.7c	**

**= P<0.01. a, b, c, d = means with different letters between columns differ significantly (P<0.05).

Table 2 Effect of SR and GS on animal traits

Variable	SR			GS			SR*GS
	5.3	8.0	P	21G	CG	P	P
LW (kg)	51.2a	46.5b	**	48.5	49.2	ns	ns
CS (units)	3.5a	3.1b	**	3.3	3.3	ns	ns
FW (kg)	3.53a	3.29b	*	3.34	3.48	ns	ns
FD (μ)	18.5	18.4	ns	18.3	18.6	ns	ns

ns = P>0.05, *= P<0.05 and **= P<0.01

Conclusions Stocking rate had the major impact on forage production and quality, liveweight and fleece weight production and wool quality in fine and superfine Merinos. Grazing system had a minor effect on production and quality). These results suggest that the advantage of using controlled grazing systems is limited for quantity and quality of fine and superfine wool production on native swards. The information generated in this study in the Basaltic region, highlights the possible implementation of high quality wool production systems with an interesting economical return when controlled grazing system, suitable stocking rate and known animal genetics merits are used.

References

Whiteley, K. (2003). Características de importancia en lanas finas y superfinas. In: Seminario Internacional de Lanas Merino finas y superfinas. Roberto Cardellino (ed.) SUL, INIA, CLU& SCMAU. p. 17-22.

Effect of forage legume species and stocking rate of lambs on sward characteristics in Uruguay

F. Montossi, D.F. Risso, R. San Julián, M. Iglesias, N. Ramos, I. De Barbieri, R. Cuadro and A. Zarza
Instituto Nacional de Investigación Agropecuaria (INIA), Ruta 5, km 386, PC: 45000, Tacuarembó, Uruguay,E-mail: fmontossi@tb.inia.org.uy

Keywords: stocking rate, lambs, legume species, sward height

Introduction The sheep industry is a mayor component of the pastoral industries, given its importance for the Uruguayan economy. In the last decade, sheep farmers have been more interested in low cost technologies to enhance productivity and profit. The objective of this study, conducted in the Basaltic region of Uruguay, was to evaluate the effect of legume species and stocking rate of lambs on sward structure, production, composition and nutritive value.

Materials and methods The trial was carried out from May 30 to September 18, 2001, using 128 Corriedale whether lambs 8-9 months of age and 24 kg initial fasted liveweight. Factors evaluated were forage legume species (Spp; *Lotus corniculatus* cv. INIA Draco -D-, *Lotus pedunculatus* cv. Maku -M-, *Lotus subbiflorus* cv. El Rincón -R- and *Trifolium repens* cv. LE Zapicán -TB-) and stocking rates (SR; 8 and 12 lambs/ha). The experimental area was 13.36 ha, divided into two blocks, each divided into eight plots (one per treatment; 8 animals per plot). A 14 days strip grazing system was used. The improved sward was two years old. The evaluated variables, each 14 days, were: a) on sward (pre and post grazing); herbage mass (kg DM/ha -DM-) and sward height (cm -H-), botanical composition (BC), nutritive value (NV); c) fasted liveweight (LW) and LW gain (LWG). The experimental design was a randomised block with a sub-divided plots arrangement.

Results Spp affected significantly (pre grazing) DM, H and BC and NV. Spp did not significantly affect either post grazing herbage mass or NV, but affected H, and BC (Table 1). Low SR compared with high SR gave higher DM production and H, without altering BC. Spp had an important effect on sward characteristics and production. TB had higher forage production, because the plots were dominated by the clover, which resulted in a higher NV and better sward structure. The differences found between D and M in comparison with R can be ascribed to the higher contribution and better vertical distribution of the former species in the sward, which in turn determined differences in the NV forage consumed and lamb accessibility to them. SR had a smaller impact than Spp on sward characteristics. SR caused differences in sward production and tongue accessibility, but it was not sufficiently important to cause important differences in NV. These high levels of production of the Basaltic production systems can be explained by the favourable climatic conditions for pasture growth. Spp affected LWG (176, 182, 150 and 221 g/lamb/day; *P*<0.01) and LW (43.2, 43.9, 40.4 and 48.1 kg; *P*<0.01), for D, M, R and TB, respectively. The higher productivity obtained on TB swards compared with the intermediate position of D and M and the poorer performance of R, are associated with the differences found in DM, H, BC and NV between species. SR significantly (*P*<0.01) affected LWG (193 and 171 g/l/day;) and LW (45.1 and 42.7 kg), The low SR gave higher LWG, but the high SR superior LW, These differences were mainly linked with DM and H variables.

Table 1 Effects of Spp and SR on sward characteristics and nutritive value

	Variable	Spp					SR			Spp*SR
		D	M	R	TB	P	12	8	P	P
Pre graz.	DM (kg DM/ha)	2583ab	2456b	1982b	3125a	*	2449b	2624a	*	ns
	H (cm)	10.8b	7.8c	7.1c	12.9a	**	9.2b	10.1a	*	ns
	Legume leaf (%)	22.5b	25.3ab	13.2c	28.1a	**	22.9	21.7	ns	ns
	Legume stem (%)	26.2a	28.8a	10.4b	30.1a	**	23.2	24.5	ns	ns
	Crude Protein (%)	19.0b	20.5a	18.3c	21.0a	**	19.4	20.0	ns	ns
Post graz.	DM (kg DM/ha)	2398	2376	2153	2335	ns	2187b	2445a	**	ns
	H (cm)	7.6b	5.9c	6.2c	10.2a	**	6.8b	8.1a	**	ns
	Legume leaf (%)	13.7c	19.5b	6.1d	21.5a	**	15.5	14.8	ns	ns
	Legume stem (%)	24.1b	28.6a	8.8c	30.3a	**	23.8	22.2	ns	ns
	Crude Protein (%)	16.0	18.8	14.8	18.1	ns	16.9	16.9	ns	ns

D = *Lotus corniculatus*; M = *Lotus pedunculatus*; R = *Lotus subbiflorus*; TB = *Trifolium* repens; ns = *P*>0.05; *= *P*<0.05; **= *P*<0.01. Means with different letters between columns differ significantly (*P*<0.05).

Conclusions The experimental results indicate the high potential of all the legumes evaluated for lamb production and the high fertility soils of the Basaltic region. *Trifolium repens* cv. LE Zapicán gave the best performance and *Lotus subbiflorus* cv. El Rincón the worst.

The effect of stocking rate and lamb grazing system on sward performance of *Trifolium repens* and *Lotus corniculatus* in Uruguay

F. Montossi, R. San Julián, M. Nolla, M. Camesasca and F. Preve
Instituto Nacional de Investigación Agropecuaria (INIA), Ruta 5, km 386, PC: 45000, Tacuarembó, Uruguay, Email: fmontossi@tb.inia.org.uy

Keywords: stocking rate, grazing system, lambs, sward height

Introduction Lambs have a great potential to diversify and stimulate meat and wool production and economical returns within the industry. The main objective of this study was to evaluate different feeding and management alternatives for the production of high quality wool and meat as well as their effects on sward characteristics in the Basaltic region of Uruguay.

Materials and methods The experiment was carried out from May 22 till September 10, 2001, using a two year old mixed sward of *Trifolium repens* (cv. LE Zapicán) and *Lotus corniculatus* (cv. San Gabriel) grazed by 60 Corriedale lambs (8 months of age; 27 kg initial liveweight). The effects studied were stocking rate (SR; 12 and 24 lambs/ha) and grazing system (GS; continuous, CG; strip, SG; and 7 days rotational grazing, 7G). The experimental area was 3.68 ha, divided into 6 plots. The variables measured were (pre and post grazing): a) on sward (each 14 days): herbage mass (ton DM/ha -DM-, ton green DM/ha -GM- and ton green leaf DM/ha -GL-), sward height (cm, H), botanical composition (BC) and nutritive value (NV); and b) on animals (each 14 days): liveweight (LW) and LW gain (LWG); and c) on carcasses (at slaughter): cold weight (CCW) and fat cover (GR). The design was a complete randomised block arranged in a factorial structure.

Results Before and after grazing, the increased SR reduced DM and GM and H (Table 1; *P*<0.01) without affecting NV and BC. GS had a significant effect (*P*<0.01) on sward variables (DM and GM and GL and H) having in general, higher values for CG, and lower and similar for SG and 7G, particularly after grazing. BC and NV of post grazing forage were not affected by GS. Overall, SR had higher effect than GS on the sward characteristics. As the experiment progressed DM and H values remained very stable, evolving into a sward dominated by *Trifolium repens* with a substantial increase in NV. SR affected LWG (210 vs. 168 g/d, *P*<0.01), final LW (50.0 vs. 45.3 kg, *P*<0.01), CCW (23.4 vs. 20.9 kg, *P*<0.01) and GR (16.7 vs. 11.9 mm, *P*<0.01) for 12 and 24 lambs/ha, respectively. GS influenced LWG (197, 191 and 180 g/d, *P*<0.01) and final LW (48.5, 47.8 and 46.6 kg, P<0.05), for CG, 7G and SG, respectively. CG showed higher CCW (23.2 vs. 22.3 and 21.0 kg, *P*<0.01) and GR (17.1 vs. 14.2 and 11.6 mm, *P*<0.01) compared with 7G and SG. The increase in SR did not alter wool production and quality, but reduced CCW and GR and boneless leg. Implementation of more controlled grazing systems produced a progressive reduction in LWG and LW, without effecting wool production and quality. Rotational grazing systems produced lighter carcasses with low fat, without modifying the other evaluated carcass characteristics. Within the range of SR used and the sward maintained during this short fattening period, the implementation of a more controlled grazing system would not be justified biologically and economically.

Table 1 Effects of SR and GS on sward characteristics (pre and post grazing)

	Variables	SR			GS				SR*GS						
		12	24	P	C	S	7	P	12-CG	12-SG	12-7G	24-CG	24-SG	24-7G	P
Pre	DM (t DM/ha)	2.42a	2.13b	**	2.15b	2.40a	2.28ab	*	2.37ab	2.61a	2.28b	1.93b	2.20b	2.27b	*
	GM (t DM/ha)	2.12a	1.85b	**	1.83b	2.16a	1.97b	**	2.06b	2.32a	1.99b	1.60b	2.01b	1.95b	*
	GL (t DM/ha)	1.00a	0.92b	*	0.86c	1.07a	0.95b	**	0.96bc	1.09a	0.94c	0.75d	1.05ab	0.97bc	*
	H (cm)	15.8a	13.2b	**	11.6c	18.3a	17.0a	**	14.9a	18.3a	14.9a	8.3c	15.6b	15.6b	**
Post	DM (t DM/ha)	2.17a	1.72b	**	2.07a	1.97ab	1.80b	**	2.34a	2.31a	1.86b	1.81b	1.62b	1.74b	**
	GM (t DM/ha)	1.83a	1.39b	**	1.78a	1.60b	1.45b	**	2.07a	1.84b	1.58c	1.49cd	1.36d	1.32d	ns
	GL (t DM/ha)	0.76a	0.57b	**	0.77a	0.64b	0.58b	**	0.88a	0.71b	0.68b	0.66bc	0.57c	0.47d	ns
	H (cm)	10.2a	6.2b	**	11.0a	6.1c	7.4b	**	14.0a	7.6c	9.0b	8.1bc	4.7e	5.8d	**

ns = *P*>0.05; *= *P*<0.05 and **= *P*<0.01. Means with different letters between columns differ significantly (*P*<0.05)

Conclusions This study shows the high productive potential of mixed swards of high nutritive value and stocking rate, where grazing system played a minor role in affecting animal productivity during this short autumn-spring lamb fattening system. All the lambs coming from the different treatments achieved the requirements of the Heavy Lamb Market of Uruguay, producing between 280 and 440 kg of animal liveweight/ha, in a 110 days period, demonstrating the great potential of white clover dominated swards for lamb production in the Basaltic production systems.

Effects of livestock grazing on the shrub vegetation biomass in the 'Sierra de Guara' Natural Park (Spain)

J.L. Riedel, I. Casasús, A. García, A. Sanz, M. Blanco, R. Revilla and A. Bernués
CITA - Gobierno de Aragón. Apdo. 727, 50080 Zaragoza, Spain, Email: jlriedel@aragon.es

Keywords: grazing, shrub invasion, biomass, landscape, protected areas

Introduction The 'Sierra de Guara' Natural Park (80.7 Kha) is a Mediterranean mountain area in Huesca, south of the Spanish Pyrenees. Shrub and forest pastures dominate the Park. They are grazed mainly by sheep, but also by suckler cattle and goats. Average stocking rate is 0.15 LU/ha. As in other European mountain areas, agricultural activities have declined during the last few decades. This has caused a process of secondary vegetation succession towards shrub invasion, with consequent landscape changes. This study aimed to quantify the effect of grazing on shrub vegetation biomass.

Materials and methods Six locations, representative of different areas of the Park and different sheep grazing management regimes, were selected. At each location, 2 plots (10x10 m) were fenced to avoid grazing. All individual shrubs were identified along a fixed transect located inside and outside the plots (1x10 m). The volume of each individual shrub (height, length and width) was measured before the grazing season in 3 consecutive years (2001-2003) to estimate shrub biomass. Biomass was then related to volume for the 7 predominant species (77.5% of total number of shrubs), using prediction equations (Torrano, 2001) for *Genista scorpius* (L.) DC.; *Buxus sempervirens* L.; *Prunus spinosa* L.; *Thymus sp.*; and *Dorycnium pentaphyllum* (L.); and in-situ developed equations for *Santolina chamaecyparissus* L. (Biomass (g DM)= 3551.1 x Volume (m^3) (R^2= 0.9258; P<0.001)) and *Echinospartum horridum* (Vahl.) Rothm. (*Genista horrida* (Vahl) DC.) (Biomass (g DM)= 7252.8 x Volume (m^3) (R^2= 0.9753; P< 0.001). The annual increment rate of shrub biomass was calculated.

Results Biomass accumulation occurred in both Grazed and Non-grazed areas (Figure 1). The increment rates were respectively 31.9 and 14.9% (NS) in Non-grazed and Grazed areas in 2001-02, 46.6 and 29.1%, (NS) in 2002-03 and 80.0 and 42.2% (P<0.01) for the entire period (2001-03). In common with results obtained by Bartolomé *et al.* (2000) in a similar area, grazing reduced but did not stop the increment of shrub biomass. Nevertheless, Casasús *et al.* (2003) observed no increment of shrub biomass in grazed areas with higher stocking rate during a 6-year study in the Pyrenees. In the current study, differences were found between species. Increment of biomass was null in *Thymus sp.* in Grazed areas (P<0.01) and small in *Genista scorpius* (NS). The effect of grazing was minor with the other species. This phenomenon was related to animal preference towards different species.

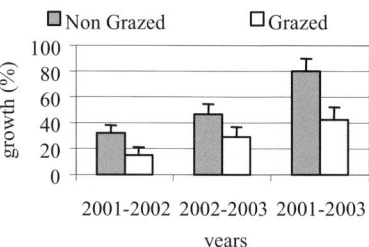

Figure 1 Biomass growth rate in Grazed and Non-Grazed areas (% of initial values)

Conclusions There was a strong trend towards shrub invasion in the Park at current stocking rates. Although the effects of grazing were not strong enough to prevent shrub invasion, Grazed areas had less invasion. Adequately managed grazing livestock systems could be effective to modulate vegetation dynamics, and thus could preserve the natural resources and landscape in protected mountain areas.

Acknowledgements Research work funded by INIA and Gobierno de Aragón. Students in receipt of grants from Gobierno de Aragón - Fondo Social Europeo, INIA and Consejo Nacional de Ciencia y Tecnología (CONACYT, México).

References
Bartolomé, J., J. Franch, J. Plaixats, & N. G. Seligman. (2000). Grazing alone is not enough to maintain landscape diversity in the Montseny Biosphere Reserve. *Agriculture, Ecosystems and Environment,* 77, 267-273.
Casasús, I., A. Bernués, A. Sanz, J. Riedel, & R. Revilla. (2003). Utilization of Mediterranean forest pastures by suckler cows: animal performance and impact on vegetation dynamics. In: Animal production and natural resources utilisation in the Mediterranean mountain areas. XII symp. HSAP-FAO. 5-7 June, Ioannina, Greece.
Torrano, L. (2001). Utilización por el ganado caprino de espacios forestales invadidos por el matorral y su impacto sobre la vegetación del sotobosque. Doctoral Thesis. Zaragoza University.

Grazing behaviour and selection of browse species by cattle, sheep and goats on natural pasture in the Sahelian zone of Burkina Faso

H.O. Sanon[1], C. Zoungrana-Kabore[2] and I. Ledin[3]

[1]INERA, Station de Farako-ba, BP. 910 Bobo-Dioulasso, Burkina Faso, Email: oumou_sanon@hotmail.com, [2]IDR, Université Polytechnique de Bobo-Dioulasso, Burkina Faso, [3]Department of Animal Nutrition and Management, SLU, P.O. Box 7024, SE-750 07 Uppsala, Sweden

Keywords: feeding behaviour, browse species, cattle, sheep, goats

Introduction Pastures in semi-arid countries, are subject to seasonal variability. Browse species that are less dependent on rainfall, are highly valued. Goats are browsers, while cattle and sheep are grazers. However, faced with a scarcity of feed resources, especially in the late dry season, all animal species fall back on browse species. The objective of this study was to estimate feeding behaviour and browse species utilisation by cattle, sheep and goats on natural pasture in different seasons, and concurrently appreciate the indigenous knowledge on browse species in the study area.

Materials and methods The study was undertaken in the Sahelian zone of Burkina Faso. A herd of cattle (22), a flock of sheep (25) and one of goats (34) were followed during three consecutive days each month from June 2003 to April 2004: Rainy season - RS (June to Sept.), post rainy season – PRS (Oct. to Dec.) and dry season – DS (Jan. to April). Two mature females of each animal species were randomly selected each day and their activities recorded every 15 minutes. While browsing, the species and plant parts eaten were identified. The woody flora inventory was made in the same area to determine the contribution of different species to the available browse. Four types of pasture were identified and three random locations with a surface area of 1 ha were selected to count individual trees and shrubs in each type.. Concurrently a formal survey was conducted with farmers exploiting these pastures to investigate indigenous knowledge on browse species. The species names were given in Fulani and the corresponding scientific name was identified using a flora.

Results Overall, 42 species of 17 families and 28 genera were found on the rangeland. Woody plants density, flora diversity and contribution of browse species are given in Table 1. The activities of cattle, sheep and goats (mean animal per day and per season) on pasture in different seasons are shown in Figure 1. The feeding activities declined for all animal species from the rainy to the dry season, while resting and ruminating activities were increasing. This decline was more important with cattle that relied on herbage for feeding. They browsed (leaves and litter) about 4,5% of time spent on pasture during the study period. Sheep made a shift in their feeding activities from grazing to browsing in the dry season (27%). Browsing was the main activity of goats in all seasons (52% of the time). Cattle browsed on 10 species with *Guiera senegalensis* frequently selected. *Combretum micranthum, G.senegalensis* and *Balanites aegyptiaca* were the most important among the 18 browse species selected by sheep. Goats avoided only 12 species in the range. *Acacia senegal, Balanites aegyptiaca* and *Pterocarpus lucens* were the most preferred.

Table 1 Plants density, flora diversity and contribution of browse species on pastures

Pasture Type	Plants density /ha	No. of species	Browse Sp (%) contribution
Shrubby Steppe	222 ± 67	15	96.0
Sparse Woody Steppe	517 ± 186	27	87.3
Hollow Formation	1051 ± 66	35	83.0
Tiger bush	1590 ± 283	25	86.3

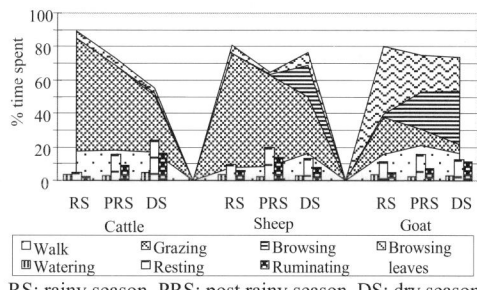

RS: rainy season, PRS: post rainy season, DS: dry season

Figure 1 Grazing behaviour of cattle, sheep and goats

The farmers identified a total of 56 browse species during the survey. Compared to the results of the flora inventory, 11 species were cited but not found during the inventory, while 5 species not cited by farmers were found.

Conclusions Farmers had good knowledge of the browse species present in the area and their classification depended on the availability of the species, their nutritive value and also on other uses. Most of the woody plants in the study area were browse species and played an important role in animal feeding, especially in the dry season. Cattle and sheep grazing in the same area in the rainy season could have great pressure on herbaceous cover that could result in degradation of this resource; while cattle and goats grazing could be advantageous.

Intake by lactating goats browsing on Mediterranean shrubland

M. Decandia, G. Pinna, A. Cabiddu and G. Molle
Istituto Zootecnico e Caseario per la Sardegna 07040 Olmedo, Italy, Email: mdecandia@tiscali.it

Keywords: intake, dairy goats, shrubland

Introduction In Mediterranean regions goat feeding systems are mainly based on shrubland that contain a wide variety of species. There are only a few equations for predicting feed intake of stall-fed goats (Luo *et al.*, 2004). The objective of this study was to develop a model for predicting the intake of lactating goats browsing on Mediterranean shrubland.

Materials and methods A database of mean treatment observations (N=44) from goat feeding studies was analysed. The studies were conducted with dairy goats that browsed for 5-7 hrs/day on 5 ha shrubland (Decandia *et al.*, 2000a,b). The goats received 200 g of concentrate and 200 g of ryegrass hay/day. The shrubland contained many tanniferous species and a low proportion of herbaceous species. The variables analysed were: body weight and body condition score (BCS); supplement intake; pasture intake, and botanical composition of the diet (Kababya *et al.*,1998); chemical composition of the diet (Meuret *et al.*, 1995); botanical composition of pasture (Daget & Poissonet, 1969); fat corrected milk (FCM) and milk urea.

Results The variables most strongly related to the intake were FCM, CP and polyphenolic tannin level in the diet. The effect of supplementation was not significant. Two prediction equations for pasture intake were (1)DMI = -18.63 + 6.75 CP + 0.02 FCM; N=38; R^2=0.77; P<0.001; (2) DMI = 52.54 +0.037 FCM – 16.44 PT/CP; N=40; R^2=0.59; P<0.001, where: DMI=DM intake (g /Kg $BW^{0.75}$); FCM=fat (4 %) corrected milk yield (kg);

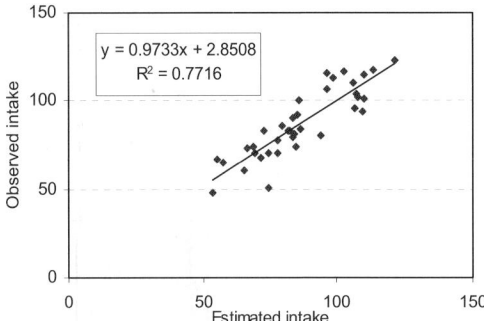

Figure 1 Relationship between observed and estimated pasture DM intake (g/kg $BW^{0.75}$)

(% DM); PT/CP= ratio between polyphenolic tannins and CP in the diet. Equation 1 has a higher R^2 and is easier to implementation at farm scale than Equation 2. Comparing predicted with observed values, a strong relationship was found (R^2=0.77; P<0.001; a, b not statistically different from 0 and 1, Figure 1). CP in the diet was related to the percentage of grass (GRAP) in the pasture and milk urea level (mg 100 ml[-1]; MU): (3)CP = 5.10 + 0.20 GRAP + 0.119 MU; N=31; R^2=0.82; P<0.001.Using this relationship, a two-step prediction model of DM intake of browsing goats is proposed. Step 1: on the basis of GRAP and MU, CP content in the diet can be estimated (Equation 3). Step 2: knowing FCM and dietary CP level, the DMI at pasture can be predicted (Equation 1).

Conclusions This model provides a useful tool for estimating the intake of goats browsing Mediterranean shrubland rich in tanniferous species, with a low percentage of herbaceous species.

Acknowledgement This study has been funded by Italian Ministry of Agriculture and Forestry.

References
Decandia M., Sitzia M., Cabiddu A., Kababya D., Molle G., (2000a). The use of polyethylene glycol to reduce the anti-nutritional effects of tannins in goats fed woody species. *Small Ruminant Research* 38, 157-164.
Decandia M., Molle G., Sitzia M., Cabiddu A., Ruiu P.A., Pampiro F., Pintus A., (2000b). Responses to an antitannic supplementation by browsing goats. *Proc of VII° International Conference on goats* 15-18 May 2000 Tours, France. pp 71-73.
Luo J., Goetsch A. L., Nsahlai, I. V., Moore, J. E., Galyean, M. L., Johnson, Z. B., Sahlu, T., Ferrell C. L., Owens F. N. (2004) Voluntary feed intake by lactating, Angora, growing and mature goats. *Small Ruminant Research* 53, 357-378.
Kababya D., Perevolotsky A., Bruckental I., Landau S., (1998). Selection of diets by dual-purpose mamber goats in Mediterranean woodland. *Journal of Agricultural Science Cambridge* 131, 221-228.
Meuret M., Bartiaux-Hill N. and Bourbouze A., (1985). Evaluation de la consommation d'un troupeau de chèvres laitières sur parcours forestier: - Méthode d'observation directe des coups de dents, - Méthode du marquer oxyde de chrome. *Annales de Zootechnie*, 34: 159-180.

Effect of forest grazing in summer on grazing behaviour, heart beat and heat production of beef cows

H. Tobioka, M. Fukumoto and S. Takeda
School of Agriculture, Kyushu Tokai University, 869-1404 Choyo-son, Aso-gun, Kumamoto, Japan, E-mail: htobioka@ktmail.ktokai-u.ac.jp

Keywords: cattle, forest, shade, heart-beat, heat production

Introduction In Japan summer is very hot and humid, particularly in daytime, therefore the supply of shade to animals is important. At the same time, a lot of forest and partial forest area is under-utilised in most of the mountain areas. Introduction of animals to forest areas might result in not only less labour demanding animal management, but also in the efficient weeding in the forest (Sugimoto *et al.*, 1999). We compared the grazing behaviour, heart beat and heat production of beef cows in partial forest with those in the normal grazing place outside the forest.

Materials and Methods The experiment was undertaken in summer. One study site was normal hill side (reference place (RP), mean temperature 24°C) with artificial shading of a feeding shed and some trees and the other was a partial forest range (forest place (FP), mean temperature 25°C). The animals were 2 mature/2 old and 2 mature/1 old cow in RP and FP, respectively. The grazing behaviour was observed and the heart beat measured. Expired gas samples were taken with a change of heart beat and analysed by mass-spectrometry and wet gas meter. Heat production was estimated by Brower's equation (1963). The data were analysed by 4 ways of analysis of variance.

Results There were 3 peaks of eating behaviour in RP. However it was distributed throughout the day in FP. The animal spent ca. 4 hr in the shade in RP and on and of in FP. The grazing behaviour under trees was distributed throughout the day (Fig.1 and 2). The ratio of rumination time to eating time tended to be lower in FP than in RP. The mean heart beat of animals in FP was much lower than that in RP (P<0.05). The regression equation of heat production *vs.* heart beat was also different between the RP and FP groups (P<0.05). Heat production per metabolic body weight tended to be lower in FP than in RP, reflecting the lower heart beat of FP. Heat production was also affected by place, grazing behaviour and age of the animal (P<0.05-0.01).

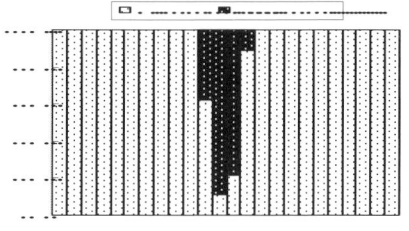 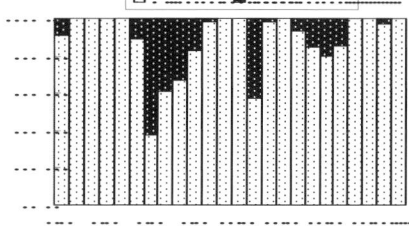

Figure 1 Behaviour in reference place   **Figure 2** Behaviour in forest place

Conclusions The grazing behaviour, heart beat and heat production of animals in Forest Place was greatly different from those of normal grazing (in Reference Place). This suggests the favourable effect of forest on animal grazing in a hot summer**.**

References
Sugimoto, Y., Y, Nakanishi K., Hirata & H. Tobioka, (1999). Forest-pastoral system in the mountainous area of Kyushu, Japan. *Proceeding of the International Workshop on Conservation and Utilization of Land Resources in Less Favored Areas with Special Emphasis on the Roles of Livestock and Technology*, Matsue City, Japan, 92-93.

Brower, E. (1965). Reports on sub-committee on constants and factors, In: K.L Blaxter, (Ed.) *Energy Metabolism.* Academic Press, London, 441-443

Cattle grazing management effects on pasture composition in semi-arid woodlands

T.J. Hall[1] and J.R. Douglas[2]

[1]DPI&F, PO Box 308, Roma, Queensland 4455, Australia, Email: trevor.hall@dpi.qld.gov.au, [2]"Verniew", Mitchell, Queensland 4465, Australia,

Keywords: grazing management, pasture composition, burning, tree competition, woodland

Introduction Manipulating grazing pressure, controlling tree competition and burning are the main options for cattle farmers to manage land in subtropical Australian *Eucalypt* woodlands. These can contain >175 herbaceous and 60 woody species, but only 5 are desirable perennial and productive grass (Silcock *et al.,* 1996). Here we describe the responses of some perennial grasses to cattle grazing pressure, tree competition and spring burning.

Methods Two experiments in *Aristida/Bothriochloa* (wiregrass/bluegrass) native pasture in a poplar box woodland (*Eucalyptus populnea*) in inland Queensland (25^0 45'S; 148^0 25'E) measured the effects of cattle grazing pressure, tree competition and spring burning on pasture composition between 1994 and 2002. Experiment 1 had 3 grazing pressures; low (25% utilisation of end of summer pasture), medium (50%) and high (75%), and 2 tree competition levels; trees killed (stem injected herbicide) and live trees, by 2 replications, in 12 paddocks of 4-30 ha; while experiment 2 had 2 burning regimes; annual spring burn after 25 mm of rain and no burning, with the same 2 tree competition levels, by 3 replications in 12 plots each of 1 ha. Species composition (as a % contribution to total pasture yield) was recorded by visual dry weight ranking of the 6 highest yielding species in 50-354 quadrats ($0.25m^2$)/paddock (experiment 1) and in 50 quadrats (experiment 2) at the end of each summer. Transformed data were analysed by 2-way ANOVA with randomised blocks in the Genstat program.

Results and Discussion In 2002 after 8 years, perennial grasses showed variable responses within the desirable, intermediate and undesirable species groups to the 3 management treatments (Table 1).

Table 1 Effect of grazing pressure, tree competition and burning on composition (%) of desirable[d], intermediate[i] and undesirable[u] grasses in Eucalypt woodland after 8 years (* indicates significant difference P<0.05)

Grass species	Grazing pressure[1]			Tree competition[1]		Spring burning[2]	
	Low	Medium	High	Cleared	Treed	Burn	No burn
Aristida ramosa [u]	29.0	30.9	11.2*	22.1	25.3	5.5*	17.1
Bothriochloa bladhii [d]	2.3	0.2	0.1	1.4*	0.4	7.1*	0.7
Bothriochloa decipiens [i]	9.1*	23.5	18.1	9.6*	24.2	22.0	23.8
Chloris divaricata [i]	2.9	3.2	6.5	5.0	3.4	2.1	0.5
Chrysopogon fallax [d]	1.9	5.0*	2.1	1.2*	4.8	13.0*	4.4
Cymbopogon spp. [i]	4.3*	0.2	0	2.4	0.6	1.0	8.7*
Dichanthium sericeum [d]	6.2	6.9	1.6*	7.2*	2.6	9.6*	2.6
Heteropogon contortus [d]	3.3*	0.6	0	1.5	1.1	8.8*	2.0

[1] experiment 1; [2] experiment 2

Of the desirable species, clearing and spring burning increased *Bothriochloa bladhii* (Forest bluegrass); low grazing pressure and burning increased *Heteropogon contortus* (black speargrass); killing trees and burning increased *Dichanthium sericeum* (Queensland bluegrass) but high grazing reduced it; killing trees reduced *Chrysopogon fallax* (golden beard grass) but medium grazing pressure and burning increased it. High grazing pressure maximised intermediate grasses, like *Chloris divaricata* (windmill grass). Low grazing pressure and killing trees reduced *Bothriochloa decipiens* (pitted bluegrass) but burning did not. Low grazing pressure and no burning increased *Cymbopogon* species (barbwire grasses). High grazing pressure and spring burning decreased the undesirable *Aristida ramosa* (purple wiregrass), but tree competition did not affect it.

Conclusion To improve landscape stability and cattle productivity in this semi-arid Eucalypt community, one can use strategic management to manipulate plant composition, encourage desirable grasses, maintain intermediate species and discourage undesirable components.

References

Silcock, R. G., P. G. Filet, T. J. Hall, E. T. Thomas, K. A. Day, A. M. Kelly, P. K. Knights, B. A. Robertson & D. Osten (1996). Enhancing pasture stability and profitability for producers in Aristida/Bothriochloa woodlands. Queensland Department of Primary Industries, Final Report, Oct. 1992-Jun. 1996. pp. 157.

Simulations of woodland grassland transitions caused by elephant

K.J. Duffy and S. Moyo
Centre for Systems Research, Durban Institute of Technology, Box 1334, South Africa, Email: kevind@dit.ac.za

Keywords: models, elephant reintroduction, elephant vegetation dynamics

Introduction In South Africa, reintroduction of wildlife on small to medium sized farms is common. A primary concern for the landowners who introduce elephant is the effects they will have on tree and grass densities. It is possible that elephant impact can exacerbate a shift from woodland to grassland. In this paper it is shown how simulations can possibly assist in understanding the possible dynamics involved.

Simulation methods Global characteristics of the dynamics between elephant and trees are simulated using ordinary differential equations (given in Stretch & Duffy, 2003). A range of parameters from literature data applicable for African Savannah's were used (Stretch & Duffy, 2003). These differential equations ignore spatial heterogeneity that can be simulated using a grid-based simulation. A grid is placed on a vegetation map of a region and elephant and tree dynamics are simulated as elephant move from square to square.

Results Data from the Pongola Game Reserve (PGR) in South Africa, a relatively small game ranch (75 km^2), where elephant were reintroduced gave initial tree reduction by elephant as 0.035 /year at a density of 0.3 elephant/ km^2 (Duffy *et al.*, 2002). Using equations alone no combinations of parameters could mimic this trend. In the model, one parameter *teq* represents the overall trees needed per elephant in their range Varying *teq* revealed that it is possible to recreate the first few years of elephant growth together with tree removal in the PGR using a value of *teq* = 8000 trees/elephant. Simulations using this value for the next 60 years result in a constant decrease in trees (Figure 1). Although grass is not directly part of the model it is evident that this would represent a switch from wooded Savannah to grassland.

Different areas of the region have similar elevation, species composition and distance from water, but have different elephant impact (Duffy *et al.*, 2002). This difference is found to be as large as a factor of four and is quite surprising because it is assumed that food preference is based on environmental conditions and food species composition. Using the grid-based model it is possible to recreate these differences for individual runs. This result is dependent on the stochastic nature of elephant movement over a short time frame. This indicates that while the data in Duffy *et al.* (2002) is useful for understanding initial elephant impacts it cannot be used conclusively for global model parameters. It is interesting to reconsider the density of trees required for elephant equilibrium *teq* (the only one based on PGR data). By varying this parameter it is possible to show at what value the density of trees can persist.

Conclusions The differential equation model used here is simple and these types of model have been considered theoretically in detail before. They can be used to understand the dynamic trends of a system if parameterisation is done. For a specific situation the parameterisation can be difficult. Certain parameters depend on the characteristics of the exact environmental conditions, especially in cases where reserves are small. However, combining these with grid methods, global trends can be considered.

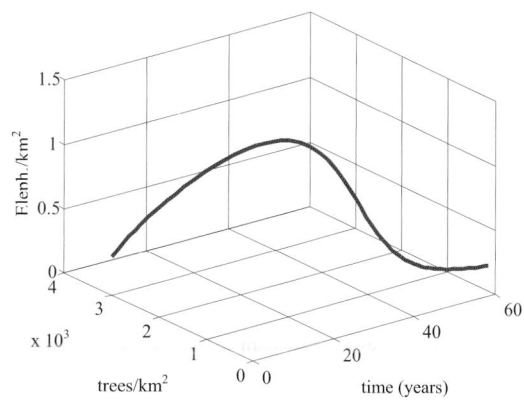

Figure 1 Model trajectory for tree density against elephant density over time for PGR. Initial conditions are: trees =30000 trees/km^2 and elephant =0.3 elephant/km^2 ; the trajectory tends to an equilibrium (E_0=0.2 elephant/km^2 and T_0=1000 trees/km^2) representing low tree density

References

Duffy, K., R. van Os, S. Vos, J. van Aarde, G. Ellish & A. Stretch (2002). Estimating the impact of reintroduced elephant on the trees of a small reserve. *South African Journal of Wildlife Research,* 32, 23-29.

Stretch, A-M. & K. Duffy (2003). Understanding global elephant and tree dynamics with ode's. *Mathematics and Computer Education,* 37, 184-192.

Toxicity in beef cattle grazing *Leucaena leucocephala* in Queensland, Australia

S.A. Dalzell, D.J. Burnett, J.E. Dowsett, V.E. Forbes and H.M. Shelton
School of Land and Food Sciences, The University of Queensland, Queensland 4072, Australia, Email: s.dalzell@uq.edu.au

Keywords: DHP, mimosine, *Synergistes jonesii*

Introduction Improved pastures based on the leguminous shrub *Leucaena leucocephala* (leucaena) are the most productive, profitable and sustainable for beef cattle production in northern Australia. Leucaena forage contains the toxic, non-protein amino acid mimosine, which is rapidly converted to 3-hydroxy-4(1H)-pyridone (DHP) upon ingestion by grazing cattle. This is a potent goitrogen and appetite suppressant. Animals suffering severe DHP toxicity exhibit distinctive symptoms (e.g. hair loss, excessive salivation, goitre and weight loss), while subclinical DHP toxicity can suppress live weight gain by 30-50% without producing any obvious symptoms. Prior to the discovery and introduction of the DHP-degrading rumen bacteria *Synergistes jonesii* into Australia in 1982, DHP toxicity severely limited animal performance from leucaena pastures and was a major impediment to adoption. Initial rumen inoculation of cattle in Australia with *S. jonesii* successfully protected them against DHP toxicity and the bacterium appeared to be easily and rapidly transmitted between grazing animals. Consequently many scientists and graziers believed that a single inoculation of a herd with *S. jonesii*, combined with simple ongoing herd management, was sufficient to overcome the problem of DHP toxicity. However, during the 2003 drought there were several reports of severe leucaena toxicity (including animal deaths) in cattle grazing leucaena in Queensland. Toxicity was evident even in herds that had followed recommended control measures. Preliminary results are presented of a study, designed to ascertain the prevalence and possible causes of leucaena toxicity in Queensland cattle herds. Meat and Livestock Australia Limited funded this research (NBP.340).

Materials and methods Forty-four (44) cattle herds grazing leucaena were randomly tested for toxicity. Paired urine and rectal grab faecal samples were collected from 385 animals. Urinary concentrations of mimosine, 3,4-DHP and 2,3-DHP were measured using HPLC analysis (Tangendjaja & Wills, 1980). Faecal delta carbon ($\bullet^{13}$C) radioisotope analysis (Jones *et al.*, 1979) determined the proportion of leucaena in each animal's diet.

Results Urine of all the cattle tested contained trace amounts of mimosine, indicating none was suffering acute mimosine poisoning. However, there was considerable variation in urinary concentrations of 3,4-DHP (1-2000 ppm) and 2,3-DHP (1-1800 ppm). Herds generally fell into 3 categories of toxicity status: i) 21 herds (48%) were completely protected as animals had low urinary concentrations (<200 ppm) of both 3,4-DHP and 2,3-DHP; ii) 9 herds (20%) were unprotected as animals had high (>200 ppm) urinary concentrations of both 3,4-DHP and 2,3-DHP; and iii) 14 herds (32%) where animals had low concentrations (<200 ppm) of 3,4-DHP but high concentrations (>200 ppm) 2,3-DHP.

Detoxification processes were working efficiently in the protected herds, as these animals were consuming diets containing high proportions (mean 35•3% DM intake) of leucaena. In the remaining herds, where leucaena intake ranged from 7 to 59% of the diet, subclinical DHP toxicity occurred even though all but one grazier had attempted to introduce the bacterium to their cattle. This might reflect poor grazier understanding of effective rumen inoculation procedures and herd management strategies. The frequent observation of significant amounts of 2,3-DHP in the urine samples was most unexpected. Our current understanding of the microbial detoxification of leucaena is that *S. jonesii* is the only bacterium capable of degrading 3,4-DHP to 2,3-DHP. Once this step has occurred, *S. jonesii* and a suite of other rumen bacteria rapidly breakdown 2,3-DHP into harmless by-products. In other work, 2,3-DHP has only been observed as a transitory degradation product present in low concentrations in the urine of protected animals. We are unsure why this process was not working efficiently. Herd management practices will be correlated with herd 2,3-DHP status to determine possible causes of this phenomenon. The concentrations of 2,3-DHP measured could limit animal performance by appetite suppression.

Conclusions Leucaena toxicity is still a significant issue in Australia, as *S. jonesii* and other rumen bacteria did not adequately protect 52% of the herds tested. In addition to further research, an extension program is required to inform graziers of effective inoculation and herd management strategies that prevent mimosine/DHP toxicity.

References

Jones, R.J., M.M. Ludlow, J.H. Troughton & C.G. Blunt (1979). Estimation of the proportion of C3 and C4 plant species in the diet of animals from the ratio of natural ^{12}C and ^{13}C isotopes in the faeces. *Journal of Agricultural Science Cambridge*, 92, 91-100.
Tangendjaja, B. & R.B.H. Wills (1980). Analysis of mimosine and 3-hydroxy-4(1H)-pyridone by high-performance liquid chromatography. *Journal of Chromatography*, 202, 317-318.

Metabolic profiling of heathland plants in the diet of sheep

G.G. Allison[1], M.D. Fraser[1], J.M. Moorby[1], J. Kopka[2], A. Erban[2] and V.J. Theobald[1]
[1]Institute of Grassland and Environmental Research, Plas Gogerddan, Aberystwyth SY23 3EB, UK, Email: gordon.allison@bbsrc.ac, [2]Max Planck Institute of Molecular Plant Physiology, Am Mühlenberg 1, D-14467 Golm, Germany

Keywords: diet composition, GC/TOF-MS, metabolite profiling, ruminant nutrition

Introduction Little is known about how plant biochemistry influences the grazing behaviour of animals grazing heterogeneous vegetation communities. Furthermore, most biochemical profiles of grassland species are restricted to major nutritional characteristics. Recent developments in analytical techniques have made possible the detailed analysis of minor components, which can potentially affect animal feeding preferences, performance and health. Gas chromatography/time of flight mass spectroscopy (GC/TOF-MS) coupled with automated library annotation is ideally suited to the acquisition of detailed metabolite profiles of plant extracts (Wagner *et al.*, 2003) and can be applied to other matrices such as blood and faeces. In this study GC/TOF-MS was used to identify metabolites within heathland plants, and to investigate which of these metabolites were present and absent within plasma and faeces from sheep consuming mixtures of these plants.

Material and methods Twelve mature Welsh Mountain ewes were zero-grazed on defined diets of heathland plant species for 29 days. The average diet composition was (as a percentage of total dry matter): *Molinia caerulea*, 41.4%; *Calluna vulgaris*, 18.2%; *Vaccinium myrtillus*, 0.21%; *Erica tetralix*, 0.17%; *Juncus effuses*, 6.96%; *Festuca* spp., 1.72%; *Carex* spp., 3.95%; dead grass, 21.1%; and moss, 6.4%. Methanolic extracts were prepared from freeze-dried and milled samples of each diet mixture, each major plant group within the diet mixtures and faecal output. These extracts, together with blood plasma samples were analysed by GC/TOF-MS (Wagner *et al.*, 2003). Individual compounds that were predominantly found in one of the nine plant groups (i.e. more than 75% of the quantity of a compound summed across all plant groups was attributable to a single plant group) were investigated further to determine their presence in plasma and faeces of the sheep fed the diets.

Results Metabolite profiling by GC/TOF-MS resolved several hundred compounds in the plant, faeces and plasma samples. Library searching allowed over 100 of these to be tentatively identified. All of the plant groups, with the exception of dead grass, contained several metabolites that were largely specific to that group. These group specific metabolites represent many metabolic processes and include plant sugars and phosphorylated sugars, e.g. mannitol (*Molinia*) and inositol-1-phosphate (*Erica*); amino compounds, e.g. allantoin and spermidine (*Vaccinium*); and one phenolic compound, 1-caffeoylquinic acid (*Erica*). Some of these plant group-specific metabolites were also detected in the sheep plasma and faeces (Table 1). Although many of the metabolites found in the samples remain unidentified, it may become possible to identify some of these unknown metabolites as new additions are made to the mass spectral library or by use of alternative analytical approaches.

Table 1 Numbers of identified (ID) and unidentified (UID) metabolites that predominated in each plant group, together with the numbers of these also found in plasma and faeces (all categories were live growth except dead grass)

Plant group category	Plant group ID	Plant group UID	Faeces ID	Faeces UID	Plasma ID	Plasma UID
Molinia caerulea	2	3	1	0	1	1
Calluna vulgaris	0	11	0	4	0	3
Vaccinium myrtillus	2	17	1	10	2	9
Erica tetralix	2	9	0	6	0	4
Juncus effusus	1	5	1	3	1	2
Festuca spp.	0	6	0	4	0	1
Carex spp.	0	2	0	0	0	2
Dead grass	0	0	0	0	0	0
Moss	0	1	0	0	0	0

Conclusions Using GC/TOF-MS it is possible to quickly characterise plant species beyond basic nutritional composition. Further work is required to determine the nutritional significance of the presence or absence of indicator plant compounds in plasma or faeces, in order to interpret their role in influencing diet selection.

Acknowledgements This work was supported by BBSRC, and carried out on samples collected as part of research funded by DEFRA, CCW and English Nature.

References

Wagner, C., M. Sefkow, J. Kopka (2003). Construction and application of a mass spectral and retention time index database generated from plant GC/EI-TOF-MS metabolite profiles. *Phytochemistry*, 62, 887-900.

Lactating ewes were strongly attracted to salt when spread on sodium-deficient undergrazed hill pastures

N. Mandaluniz[1], J.C. Ruiz and R.J. Lucas[2]
[1]NEIKER, A.B., Apdo. 46, E-01080, Vitoria-Gasteiz, Spain, Email: nmandaluniz@neiker.net, [2]Plant Sciences Group, PO Box 84, Lincoln University, Canterbury (New Zealand)

Keywords: grazing behaviour, hill pasture, management tools, sodium deficiency

Introduction New Zealand hill and high country pastures for sheep and cattle cover very variable topography. Shady south face hill country pastures on steeper slopes are grazed inefficiently because of their mature state and low nutritive value. Flatter more fertile areas reflect more frequent defoliation and stock camping behaviour. These inland pastures are deficient in sodium (Na) (Aspinall et al., 2004) because they are distant from coastal sea spray. Management alternatives are sought to improve pasture quality on steep, shady aspects. The provision of different nutrients such as Na or nitrogen (N), which are deficient in these under-grazed areas of pasture, is expected to improve herbage nutritive value and its attractiveness to animals. Consequentially this could be expected to improve pasture utilization.

Materials and methods A south-facing 80 ha paddock at Mt. Grand Station, Central Otago, New Zealand (500-800 m altitude, mean slope 20°) was studied. In December 2003 and in the presence of sheep, different nutrients, urea (46%N, 100kg N/ha), sulphur superphosphate (19% S and 8% phosphate, 500kg product/ha) and salt (100kg NaCl/ha), were spread on 30m x 8m plots using 3 replicates of a 2^3 factorial design. Untreated strips between plots were at least 10m wide. Merino ewes were set stocked at 5/ha on the paddock from September to December and had access to all 80 ha at all times. Grazing pressure was measured as bare ground cover (%). Cocksfoot was sampled for macronutrient % in April 2004.

Results and discussion Lactating ewes were strongly attracted only to the plots where salt was applied and their intensive trampling caused a high proportion of turf damage and a significant ($P<0.001$) increase in bare ground plus litter cover (40-60% in salt plots vs. 21-23% in non-salt and control plots). The salt craving could have been due to the Na requirement of lactating ewes (Edmeades & O'Connor, 2003) and/or Na deficiency in these hill pastures (Aspinall et al., 2004). Cocksfoot (20% cover in the studied area) grown on N deficient hill pastures is likely to have low grazing preference for sheep (Edwards et al., 1993). Salt-treated cocksfoot had (p=0.08) higher content of Na (0.061±0.007 vs. 0.041±0.007); S+P treated plots had similar (p=0.10) P (0.37±0.02 vs. 0.31±0.02) and significantly (p=0.03) more S (0.46±0.03 vs. 0.37±0.03). Urea-treated samples had significantly (p=0.02) higher N content (2.63±0.11 vs. 2.21±0.12) but more intensive grazing of N treated plots was not observed.

Conclusions Salt application has potential to improve grazing management in hill pastures. The strong attraction of sheep for salt suggests that its application with seed in the presence of sheep may enhance pasture establishment when broadcast onto non-arable hill pastures. Salt application without sheep present may provide Na supplementation through plant uptake but it is not known if sheep can detect grass with increased Na%. Further studies are needed to determine the best application time and rate of salt. The long-term effect of salt on soils and pastures will indicate the economic value of these suggestions.

Acknowledgements Struthers Trust and the Dominion Salt Ltd., for financial assistance.

References
Aspinall, A., N. Mandaluniz, L. Hight & R. J. Lucas (2004). Sodium deficiency in Canterbury and Central Otago sheep pastures. *Proceedings of New Zealand Grassland Association*, 66, 227-232.
Edmeades, D.C. & M.B. O'Connor (2003). Sodium requirements for temperate pastures in New Zealand: a review. *New Zealand Journal of Agricultural Research*, 46, 37-47.
Edwards, G.R., R. J. Lucas & M. R. Johnson (1993). Grazing preference for pasture species by sheep is affected by endophyte and nitrogen fertility. *Proceedings of New Zealand Grassland Association*, 55, 137-141.

Theme B: Grassland and the environment

Section 14

Climate change

Long-term responses of a mesic grassland to manipulation of rainfall quantity and pattern

A.K. Knapp, J.M. Blair, P.A. Fay, M.D. Smith, S.L. Collins and J.M. Briggs
Department of Biology, Colorado State University, Ft. Collins, Colorado, 80523, USA, Email: aknapp@lamar.colostate.edu

Keywords: climate change, productivity, tallgrass prairie

Introduction Climatic variability is an inherent feature of grassland biomes, with large fluctuations in temperatures combined with precipitation regimes characterised by floods and severe drought occurring on both an interannual and seasonal scale. Global climate models and emerging data indicate that extremes in precipitation regimes are increasing worldwide coupled with increases in temperature. Thus, variability in spatial and temporal patterns of water availability in grasslands, as directly influenced by altered precipitation patterns and indirectly by increased temperatures, will likely increase in the future. The objectives of our experiments were to experimentally manipulate rainfall amount and temporal patterns (amount and timing of individual rainfall events) to assess soil, plant, community and ecosystem responses to this projected climate change.

Materials and methods Two long-term experiments in undisturbed mesic grassland in eastern Kansas are ongoing: a 10-yr water supplementation study (Knapp *et al.,* 2001), and a 7-yr experiment in which growing season rainfall patterns (variability) have been manipulated without altering total amounts (Fay *et al.,* 2000). Both experiments are fully replicated with the irrigation study designed to alleviate seasonal water limitation, and the rainfall variability experiment designed to increase rainfall event size concurrent with lengthening periods between events by 50%, but with no change in total growing season rainfall amount.

Results Water availability limited aboveground net primary production (ANPP) in 8 of 10 the years and long-term records of interannual variability in rainfall were strongly correlated with temporal variability in ANPP (r^2 = 0.62). Within-season increases in precipitation extremes characterised by fewer, larger rainfall events with longer intervening dry periods, led to a suite of responses including reduced ANPP and soil respiration (Figure 1) and shifts in plant community structure. Increased temporal variability in soil water content was strongly related to responses in these key C responses, more so than mean soil water quantity.

Conclusions Results of these two long-term studies demonstrate that even relatively mesic grasslands (annual precipitation ca. 830 mm) are quite sensitive to alterations in precipitation regime. The greater range in treatments (rainfall amount and pattern) that can be experimentally

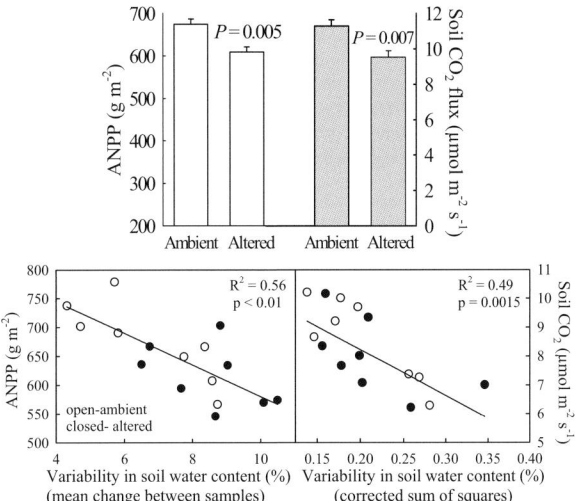

Figure 1 Top panel: Mean ANPP and mean soil CO_2 flux in grassland plots exposed to either ambient or altered rainfall patterns. Bottom panels: Negative relationship between ANPP, soil CO_2 flux, and variability in soil water content, suggesting that climate change-induced increases in the temporal variability of soil water content will have significant effects on ecosystem processes.

imposed permits predictions of grassland responses to climate change that extend beyond those based on past ecosystem responses to climate change constrained by the historic range of climatic divers in an ecosystem.

References

Fay, P.A., J.D. Carlisle, A.K. Knapp, J.M. Blair & S.L. Collins. (2000). Altered rainfall timing and quantity in a mesic grassland ecosystem: design and performance of rainfall manipulation shelters. *Ecosystems,* 3, 308-319.
Knapp, A.K., J.M. Briggs & J.K. Koelliker. (2001). Frequency and extent of water limitation to primary production in a mesic temperate grassland. *Ecosystems* 4, 19-28.

Forage grass phenology in relation to climate change

G. Żurek

Plant Breeding and Acclimatisation Institute, Botanical Garden, 85-687 Bydgoszcz, Poland, Email: gzurek@interia.pl

Keywords: climate change, ecotypes, phenology

Introduction Phenological phases of plants, such as heading or flowering are mainly driven by environmental factors such as pests, diseases, competition, soil properties, genetics, age and, most importantly, weather conditions (Menzel & Fabian, 1999; Menzel 2000). At the end of the last century there was an emerging recognition that phenological records can be especially useful in environmental monitoring and it has gained the UK government approval as an indicator of climate change (Sparks *et al.*, 2000). The aim of this work was to analyse long-term trends in the mean heading date for Polish ecotypes of three forage grass species.

Material and methods Phenological records for 11,921 Polish ecotypes of three forage grass species (*Poa pratensis*, smooth-stalked meadow grass; *Dactylis glomerata*, cocksfoot and *Festuca pratense*, meadow fescue) were used. Each year from 1973-2003 new sets of ecotypes were collected from natural habitats and planted in a three-replicate design with 30 plants per ecotype. Ecotypes were grown in one place (Bydgoszcz, Poland 53°10'28"N, 018°02'49"E). Heading was recorded as the number of days from 1 April to the moment when 'emerged' tillers were visible on 30% of spaced plants of a particular ecotype (not less than 3 tillers per plant), The Jones data set (Jones *et al.*, 1999) was used as reference temperature data. It contains annual data of temperature time series at a grid-box resolution of 5° latitude by 5° longitude over the period 1880-2000. Annual temperatures are expressed as *departures from normal temperature* i.e. means of the 1961-1990 reference period. Data from the grid-box covering most area of Poland (between 50° - 55° North and 15° - 20° East) was selected.

Results Trends in heading were all negative and significant for smooth-stalked meadow grass and cocksfoot (Table 1). There was also a significant positive trend for the temperature data. Mean heading was negatively and significantly correlated with temperature data (for cocksfoot: $r = -0.82$, $P = 0.00$; for smooth-stalked meadow grass: $r = -0.68$, $P = 0.01$). It is therefore clear that along with increasing air temperature (ca. 1°C / 31 years), mean heading of the above grass species was significantly advanced from 10 to 11 days for cocksfoot and smooth-stalked meadow grass respectively.

Table 1 Changes in mean heading of three forage grass species

Species	No. of ecotypes	Heading (days) [1]		Linear trend (days/year)
		1973	2003	
Smooth-stalked meadow grass	2488	48.5	38.2	-0.34 **
Cocksfoot	6018	51.7	42.1	-0.32 **
Meadow fescue	3415	57.6	51.1	-0.22 ns

** - significance of trend at $p < 0.05$, [1] – calculated from linear regression equations (y=ax+b) where x were years, y – observed days to heading of given species in years

Conclusions These results support the statement that changes in plant behaviour will not only happen in the near future but they have already happened. Over the last three decades the growing season in Europe has been lengthened by 10.8 days on average (Menzel & Fabian, 1999). Along with extending growing season and faster heading of forages, the total pasture period for animals will also be extended. Yields could be increased most frequently in areas where temperature approximates the optimum for crop growth. However, although many factors of changing climate will have direct effect on plants, the duration of these effects and their impacts at the levels of population and ecosystem are still relatively unknown.

References

Jones, P.D., D.E. Parker, T.J. Osborn & K.R. Briffa (1999). Global and hemispheric temperature anomalies - land and marine instrument records. In: Trends: A Compendium of Data on Global Change. Carbon Dioxide Information Analysis Center, Oak Ridge National Laboratory, U.S. Dept. of Energy, Oak Ridge, TN.

Menzel, A. (2000). Trends in phenological phases in Europe between 1951 and 1996. *International Journal of Biometeorology*, 44, 76-81.

Menzel, A. & P. Fabian (1999). Growing seasons extended in Europe. *Nature*, 397, 659.

Sparks, T.H., E.P. Jeffree & C.E. Jeffree (2000). An examination of the relationship between flowering times and temperature at the national scale using long-term phenological records from the UK. *International Journal of Biometeorology*, 44, 82-87.

The effect of extremes in soil moisture content on perennial ryegrass growth

A.S. Laidlaw

Applied Plant Science Division, Department of Agriculture and Rural Development, Plant Testing Station, 50 Houston Road, Crossnacreevy, Belfast BT6 9SH, UK, Email: scott.laidlaw@dardni.gov.uk

Keywords: soil moisture, climate change, perennial ryegrass, growth model

Introduction Seasonal distribution of rainfall in the UK and Ireland is predicted to become more variable (Sweeney, 2003). The problems of excessive soil moisture on grass utilisation and the effect of deficit in soil moisture on grass growth are well known. However the effect of excess rainfall on the growth of sown grass is less clear and is usually not taken into account in grass growth models. This study was carried out to investigate the potential impact of excess moisture on perennial ryegrass growth in the field and to investigate the relative effect of soil moisture varying from deficit to excess levels on perennial ryegrass tiller development.

Materials and methods A grass growth model was run against DM yields from plots in the intermediate heading Recommended List trials in Northern Ireland, over the years 1994-2003. This was done to identify any years which had lower yields than would have been predicted from known responses to light temperature and moisture deficit and which could be ascribed to excess soil moisture. The model was based on that of Johnson & Thornley (1985) adapted to take account of soil moisture deficit. The effect of varying levels of soil moisture (brought up to 0.5, 0.75, 1.0 and 1.25 field capacity (FC) from a mean of 0.43, 0.64, 0.86 and 1.11FC by applying distiller water 3 times per week) on grass leaf growth was studied on perennial ryegrass cv Mamout. The grass was grown in pots (20 cm in diameter) filled with a medium loam in 6 replicates in a glasshouse with supplementary heat and light during winter and nutrients applied to simulate intensive management. Water was prevented from draining from the 1.25 FC treatment by lining the pots with polythene. Grass was cut to 4 cm when the experiment commenced and every 4 weeks thereafter for 4 regrowth periods. Leaf extension rate (LER) was measured on 16 marked tillers per pot weekly. Canopy net photosynthesis was measured on week 8 by the closed method (PAR 700 μmol/m^2/s, mean temperature 19.6°C, air flow rate 35 litres/min) and LAI was determined at the harvest at the end of week 8.

Results The model predicted actual yields in 9 out of the 10 years reasonably well. However for 2002 the mean rainfall from January to October was 915 mm compared with a mean for the other 9 years of 749 (range 612-848 mm) and the model prediction was very poor, over-predicting by about 5 t/ha (Figure 1). Near steady state soil moisture levels resulted in the excessively high level treatment having significantly lower long term LER and lower LAI at week 8 than the 0.64-0.75FC treatment but higher efficiency of photosynthesis per unit LAI (Table 1). The driest soil treatment produced the expected adverse effects.

Table 1 Mean effect of prolonged near steady state soil moisture content on tiller and canopy development

		Week 8		
	LER (mm/d)	LAI	Net photos. g CO$_2$/m^2/h	Net photos. per LAI unit
0.50FC	11.7	4.8	1.4	0.29
0.75FC	15.8	6.4	1.9	0.30
1.00FC	14.7	5.5	2.4	0.44
1.25FC	13.6	4.4	1.6	0.36
LSD	2.04	0.71	0.41	0.106
Prob.	0.005	<0.001	<0.001	0.032

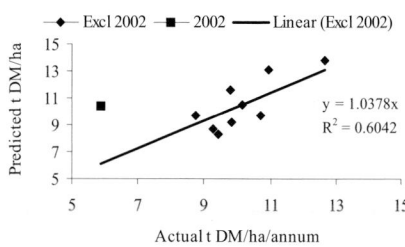

Figure 1 Actual and predicted yields of perennial ryegrasses in cultivar evaluation trials 1994 to 2003

Conclusion These results show that prolonged periods of high soil moisture content may have a significant impact on grass growth in intensively managed systems. If rainfall patterns change as predicted in the grass growing regions of the UK and Ireland, models of grass growth, used in grassland management decision support systems, should take the effect of excess soil moisture into account.

References

Johnson, I.R. & J.H.M. Thornley (1985). Dynamic model of the response of a vegetative grass crop to light, temperature and nitrogen. *Plant, Cell and Environment*, 8, 485-499.

Sweeney, J. (2003). *Climate Change Scenarios and Impacts for Ireland (2000-LS-5.2.1-M1) Final Report.* Environment Protection Agency, Johnstown Castle, Wexford, Republic of Ireland, 229pp.

Shifts in N-efficiency of different farm types in response to climate change

S. Dueri, P.L. Calanca and J. Fuhrer
Agroscope FAL Reckenholz, Swiss Federal Research Station for Agroecology and Agriculture, Air Pollution/Climate Group, CH-8046 Zurich, Switzerland, Email: juerg.fuhrer@fal.admin.ch

Keywords: climate change, N-efficiency, farm

Introduction Climate change may affect European farms, but – in contrast to individual crops - the sensitivity of whole farming systems has not been the subject of much research. At the farm level, where different farm units are linked through the availability and flow of nitrogen (N), effects on individual crops are interlinked, and through shifts in grasslands and related animal production with altered nutrient flows. Ideally, N flows into the system and N-export with products should be equal, and thus N-use-efficiency (NUE), expressed as the ratio of N export to N loss, would be maximal. The objective of this study was to test the effect of gradually changing temperature (T) and precipitation (P) on NUE of two farm types under Swiss conditions.

Model description A Stella© model (CH-Farm) with a monthly time step was developed for a mixed dairy/arable farm. CH-Farm is a statistical model, which uses relations inferred from functional models to represent the effect of climate change on arable crops (CropSyst) and grasslands (Pasture Simulation Model, PaSim). CH-Farm is implemented with a limited number of input variables and relations describing the effect of T and P on N fluxes between 5 subsystems: soil, grass, crops, animals (milking cows) and manure (Figure 1). The model was applied to 2 farm types and driven by a set of climate scenarios with 20 years of historical data (1980-2000), followed by a progressive shift in T and P until 2100 (Table 1), with (+) or without (-) consideration of CO_2 fertilization effects.

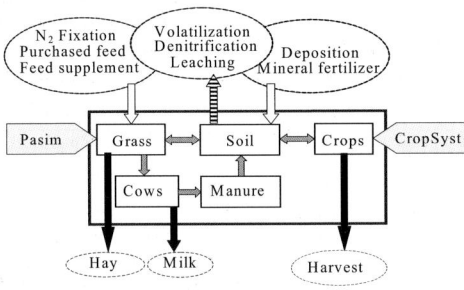

Figure 1 Structure of farm model CH-Farm

Table 1 Climate scenarios and farm types

Scenario	ΔT (2100)	ΔP (2100)
'Warm-wet' (WW)	+2 °C	+5%
'Hot-dry' (HD)	+5 °C	-20%

Farm types:
Farm D : 30 cows, 8 ha cropland (mainly dairy)
Farm C : 15 cows, 15 ha cropland (mainly arable)

Results NUE of the farm with mainly dairy production increased or remained steady over time (Figure 2); this was due to an extension of the period of grass growth and accelerated plant development leading to higher grassland and pasture production. Conversely, the farm with more arable land showed decreasing NUE, which was related a loss of crop productivity. For each farm type, NUE was generally higher with the WW scenario compared to the corresponding HD scenario because of stronger limitations of productivity by T and water of the latter. Scenarios considering effects of elevated CO_2 (+) increased NUE due increased plant N uptake from the soil, and correspondingly less N loss.

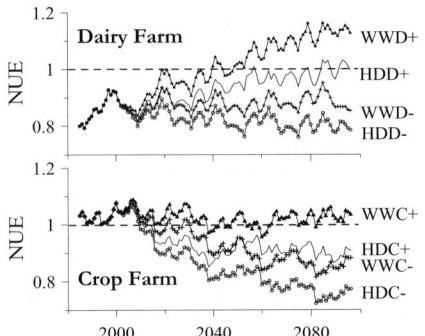

Figure 2 Trend in NUE for different farming systems and climate scenarios

Conclusions

1. Grassland and pasture productivity benefit from the longer growing season, while crop production decreases due to acceleration plant growth.
2. NUE of arable farms is more likely to decline with climate change, i.e. more N is lost per unit of production, while NUE of dairy farms remains steady or increases.
3. Consequently, the distribution of land between different production units has an important effect on the trend of NUE.
4. Drier (HD) scenarios cause lower NUE compared to wetter (WW) scenarios. With HD scenarios, limitation of production by T is more important than effects of changed precipitation.

Catch-up in response to elevated CO₂ – a study of genotypes of 12 grassland species

E. Wright[1], J. Connolly[1] and A. Luescher[2]
[1]University College Dublin, Dublin 4, Ireland, Email: Elaine.wright@ucd.ie, [2]Agroscope FAL Reckenholz, Reckenholzstrasse 191CH-8046 Zürich Switzerland

Keywords: catch-up, CO₂, global change

Introduction Differential growth enhancements for species or genotypes under elevated CO₂ can lead to changes in the composition of plant communities. Under a Rich-get-richer hypothesis, species that constitute a large proportion of a community (the dominants) will increase their dominance at elevated CO₂ (Bazzaz and Garbutt, 1988). Under the alternative Catch-up hypothesis the smaller components of communities will benefit proportionately more than dominants from elevated CO₂ conditions (tested at the level of individual plants in a monoculture in Wayne and Bazzaz, 1997). A recent review (Poorter & Navas, 2003) provided no evidence for differential growth enhancements by dominant or subordinate species. We examine this question at the genotype level for genotypes of 12 grassland species.

Materials and methods The data are from a study on the effects of CO₂ enrichment on genotypes within plant species (for details see Luescher *et al.*, 1998). Twelve native perennial species from 3 functional groups Grass (*Lolium perenne, Lolium multiflorum, Arrhenathecult elatius, Dactylis glomerata, Festuca pratensis, Holcus lanatus, Trisetum flavescens*), Non-Legume Dicotyledon (*Rumex obtusifolius, Rumex acetosa, Ranunculus friesianus*) and Legume (*Trifolium repens, Trifolium pratense*)) were grown in a FACE system. A maximum of 14 genotypes were selected for each species. Three replicates of each genotype were grown at both ambient (350 p.p.m.) and elevated (600 p.p.m.) CO₂ levels. The genotypes were grown in competition with a background matrix of the grass species *Lolium Perenne*. Harvests were taken 3 times a year for 3 years with yields for the first harvest not included in the analysis. A Structural Relationship model (Kendall & Stuart, 1979), fitted by Maximum Likelihood, was used to discriminate between the hypotheses. The model for each species was

$$E(z_{aij}) = \mu_{ai} \text{ and } E(z_{eij}) = \mu_{ei} = \alpha + \beta\mu_{ai},$$

where z_{aij} and z_{eij} are the logarithm of the yield of the jth replicate of the ith genotype at ambient (a) and elevated (e) CO₂ level respectively and β is the slope of the structural relationship. A value of $\beta < 1$ supports the Catch-up hypothesis and a value > 1 supports the Rich-get-richer hypothesis.

Results The majority of species-harvest combinations support the Catch-up hypothesis ($\beta < 1$), in most cases significantly so (Table 1). Nine species have a majority of species-harvest combinations supporting this hypothesis with 3 species showing complete support. Two species (*Festuca pratensis* and *Rumex obtusifolius*) have a majority of species-harvest combinations supporting the Rich-get-richer hypothesis. There was a general decrease in the average slope (β) over the 3-year period.

Table 1 Number of species-harvest combinations supporting the two hypotheses

Supporting	Significant	Non-Significant	Total*
Catch-up ($\beta < 1$)	48	17	65
Rich-get-richer ($\beta > 1$)	2	25	27

*Four combinations were excluded due to missing values

Conclusions For several species there is support for the Catch-up hypothesis. If this occurred within a community consisting solely of genotypes of a species, the composition would shift in favour of the subordinate genotypes under elevated CO₂ conditions.

References
Bazzaz, F.A. & K. Garbutt (1988). The response of annuals in competitive neighbourhoods: effects of elevated CO₂. *Ecology*, 69, 937-946.
Kendall, M. & A. Stuart (1979). The advanced theory of statistics, Vol. 2 Inference and relationship, 4th ed., Charles Griffin & Company Limited, London & High Wycombe. pp. 399-443.
Lüscher, A., G.R. Hendrey & J. Nösberger (1998). Long-term responsiveness to free air CO₂ enrichment of functional types, species and genotypes of plants from fertile permanent grassland. *Oecologia*, 113, 37-45.
Poorter, H. & M.L. Navas (2003). Plant growth and competition at elevated CO₂: on winners, losers and functional groups. *New Phytologist*, 157, 175-198.
Wayne, P.M. & F.A. Bazzaz (1997). Light acquisition and growth by competing individuals in CO₂-enriched atmospheres: consequences for size structure in regenerating birch stands. *Journal of Ecology*, 85, 29-42.

Yield progress of perennial ryegrass and silage maize – genetic gain or climate change?

A. Herrmann, A. Kornher and F. Taube
Institute of Crop Science and Plant Breeding, Grass and Forage Science/Organic Agriculture, Christian-Albrechts University, Kiel, Germany, Email: aherrmann@email.uni-kiel.de

Keywords: climate change, modelling, yield, perennial ryegrass, silage maize

Introduction Gains in annual dry matter yield (DMY) from breeding achieved during the last decades are reported to range between 2.5 and 6% per decade for perennial ryegrass (Wilkins & Humphreys, 2003). In contrast, accelerated progress in improving DMY has been achieved for silage maize, varying between 8 and 13% per decade (Lauer *et al.*, 2001). These gains are mainly attributed to (i) genetic yield potential increase, (ii) improved crop management and (iii) increased stress tolerance. The potential impact of climate change on yield progress, however, is disregarded in most studies. The objective of this study therefore was to quantify the contributions of climate change and breeding on yield progress of perennial ryegrass and silage maize by comparing results of long-term simulation studies with data from official variety tests.

Materials and methods Twenty-five year long-term simulations (1979-2003) using the weather-driven *FOPROQ* model (Kornher *et al.*, 1991) were run for an early (Gremie) and a late (Vigor) perennial ryegrass grown at a site in northern Germany. A 4-cut system for silage production was assumed together with N fertiliser application of 360 kgN/ha (120/80/80 /80). Corresponding *FOPROQ* simulations for silage maize comprised three early cultivars at two sites (Kiel, Schuby) in northern Germany and two mid-early varieties at five sites (Augsburg, Freising, Nürnberg, Regensburg, Ulm) in southern Germany. Model calibrations are based on multi-year, multi-site field trials. Official variety tests comprise 25 and 18 years of data for perennial ryegrass and silage maize, respectively.

Results and discussion Over the last 25 years mean annual temperature increase ranged between 0.04 °C (Augsburg) and 0.07 °C (Kiel), which was mainly caused by higher temperatures during the vegetation peiod and was accompanied by slightly reduced precipitation. Due to the lower temperature optimum for growth of ryegrass compared to silage maize, temperature increase resulted in less positive effects for ryegrass (Figure 1). Climate-driven yield gains of 0.02 t/ha per year are in the range reported, but variety tests on the other hand show a yield decrease. However, comparability was somewhat limited since the N fertiliser regime was not constant over years. Variety tests for silage maize production in northern Germany indicate

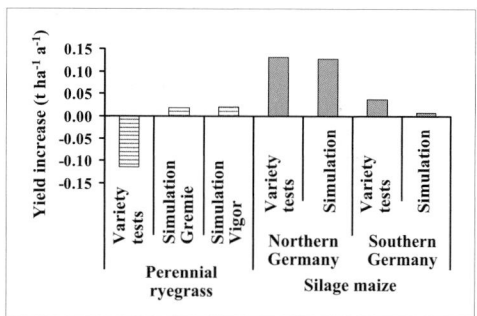

Figure 1 Trends in simulated and observed yield of perennial ryegrass (1979-2003) and silage maize (1986-2003)

yield gains of 0.13 t/ha per year over the last two decades, which, however, can completely be attributed to climate change. Yield gains in southern Germany were substantially lower, caused by a lower temperature increase and a consistently higher temperature. Overall, the study indicates a low contribution of breeding to yield progress of maize and perennial ryegrass. Our results are confirmed by Duvick & Cassman (1999), who found little evidence for an increase of maize grain yield in north-central United States.

Conclusions The yield gains observed for silage over the last two decades can be attributed mainly to climate change, i.e. temperature increase. Simulations for perennial ryegrass showed a slight, climate-driven yield gain, which, however, was not reflected in the variety tests.

References

Duvick, D.N. & K.G. Cassman (1999). Post-green revolution trends in yield potential of temperate maize in the North-central United States. *Crop Science*, 39, 1622-1630.
Kornher, A., P. Nyman & F. Taube (1991). Ein Computermodell zur Berechnung der Qualität und Qualitätsveränderung von gräserdominierten Grünlandaufwüchsen aus Witterungsdaten. *Das Wirtschaftseigene Futter*, 37, 232-248.
Lauer, J.G., J.G. Coors, & P.J. Flannery (2001). Forage yield and quality of corn cultivars developed in different eras. *Crop Science*, 41, 1449-1455.
Wilkins, P. & M.O. Humphreys (2003). Progress in breeding perennial forage grasses for temperate agriculture. *Journal of Agricultural Science*, 140, 129-150.

Reproductive allocation in *Brassica kaber*

C. Brophy[1], J. Connolly[1], P.M. Wayne[2] and D.J. Gibson[3]

[1]*University College Dublin, Dublin 4, Ireland, Email: caroline,brophy@ucd.ie,* [2]*Harvard University, Cambridge, Massachusetts, 02138 USA,* [3]*Southern Illinois University, Carbondale, Illinois 62901-6509 USA*

Keywords: reproductive allocation, *Brassica kaber*, truncated regression

Introduction Understanding the reproductive biology of weeds is essential to understanding their role in semi-natural plant communities. The annual forb *Brassica kaber* is an important agricultural weed in the mid-western regions of the U.S.A. Linear regression is often used to establish the relationship between the reproductive mass (R) of a plant (as measured by number of flowers, seeds or mass of reproductive organs) and its dry mass (W) (Sletvold, 2002). Some plants do not have any reproductive output (R=0). These zero values cause a problem for linear regression. Schmid *et al.* (1994) suggested using truncated regression as a solution. In this presentation we use an extended version of this method to model reproductive allocation in *Brassica kaber.*

Materials and methods In an experiment on *Brassica kaber* (described in detail in Wayne *et al.,* 1999) seeds of *B Kaber* were sown in eighty four 25 cm diameter pots at 6 densities, 1, 2, 4, 8, 16 and 32 plants per pot and at a CO_2 level of 350 µL/L (ambient) or 700 µL/L (elevated). The experiment was structured as three blocks, each with two CO_2 concentrations and 6 densities. Density x CO_2 combinations were replicated twice within each block except for those at the lowest density, which were replicated four times. The dry weight of each plant and the dry weight of its reproductive organs were measured at harvest. Data from two pots was missing. In all 704 plants were measured.

Assuming that R is normally distributed given W, truncated regression is a likelihood based technique in which values for which there is reproductive output contribute as in ordinary linear regression to the estimation of the regression coefficients. An adjustment is made in the method for plants with zero reproduction.

We extend the truncated regression approach of Schmid *et al.* (1994) in two ways. The plants in a pot are not independent of each other and we include a random term in the model to allow for pot-to-pot variation. We also explore whether the allocation to reproduction is related not just to the mass of the plant but also to the mean mass of its neighbours in its own pot (Ratio = W/Mean). In the model fitted the reproductive mass is related to the mass of the plant, the mean mass of its neighbours in the pot (both on the log scale) and to the level of CO_2 at which it is grown. The coefficients of the model were estimated using SAS NLMIXED.

Results and conclusions Estimates of the regression coefficients of the fitted model are displayed in Table 1. The interaction between plant mass and CO_2 indicates support for the catch-up effect, *i.e.* elevating CO_2 increases the reproductive allocation of smaller plants proportionately more than larger ones. The significant Ratio term shows that reproductive allocation is increased for plants that are larger relative to their neighbours. The random pot effect is highly significant.

Table 1 The model parameter estimates and significance levels (L indicates the logarithm scale was used)

Variable	Coefficient estimate	Standard error	Significance level
Intercept	-4.539	0.213	***
LW	0.203	0.126	ns
LRatio	0.951	0.131	***
CO_2	0.774	0.265	**
LW*CO_2	-0.233	0.111	*

Significance levels: ns = non significant, * = P < 0.05, ** = P < 0.01, *** = P< 0.001

References

Wayne P.M., A.L. Carnelli, J. Connolly & F.A. Bazzaz (1999). The Density-Dependence of Plant Responses to Elevated CO_2: I. Stand-Level Productivity. *Journal of Ecology,* 87, 183-192.

Schmid B., W. Polasek, J. Weiner, A. Krause & P. Stoll P (1994). Modelling of Discontinuous Relationships in Biology with Censored Regression. *American Naturalist,* 143, 494-507.

Sletvold N. (2002). Effects of plant size on reproductive output and offspring performance in the facultative biennial *Digitalis purpurea. Journal of Ecology,* 90, 958–966.

Targeted seasonal climate forecasts offer more to pastoralists

D.H. Cobon[1], J.N. Park, K.L. Bell[1], I.W. Watson[2], W. Fletcher[2] and M. Young[2]
[1]Emerging Technologies, DPI&F, 203 Tor Street, Toowoomba, Q 4350, Australia, Email: david.cobon@dpi.qld.gov.au, [2]Department of Agriculture, PO Box 483, Northam, WA 6401, Australia

Keywords: targeted forecasts, lead-time, forecast period, pastoralism, decision making

Introduction The existing forecast systems such as the Southern Oscillation Index (SOI) phase system (Stone et al., 1996) and the Sea Surface Temperature (SST) phase system (Drosdowsky 2002) produce rolling three monthly forecasts with lead-times of either zero (SOI phase) or 1 month (SST phase). Both forecasts are re-issued monthly. This approach leaves little time for pastoralists to consider the forecast and then make changes to management decisions. In addition the forecast period can often be of little interest because of the seasonal pattern of rainfall.

Targeted forecasts have a forecast period that is important for the region of interest and they are issued for that period counting down from long lead-times of 3 to 6 months, to zero months. If management decisions are normally made during the count down period the forecast can be useful provided there is adequate skill. Skill should be evident both spatially and temporally. Statistical forecast systems have lower skill with increasing lead-time. As such, multiple forecast systems may be used to deliver the skill needed at longer lead-times provided they have a published mechanism describing their physical action and that temporal coherence of forecast skill is maintained counting down to zero lead-time.

Materials and methods Two pastoral regions in Australia were used to demonstrate the targeted forecast approach, western Queensland (101 rainfall stations) and the Gascoyne Murchison region of Western Australia (82 rainfall stations). Preferred forecast periods identified by pastoralists were November to March in western Queensland and January to March in Gascoyne Murchison. The skill of the both SOI and SST phase systems was determined using the Kruskal Wallis (KW) non-parametric statistical test. The proportion of stations that had a significant KW (probability KW≥0.90) was used as a measure of both temporal and spatial coherence in forecast skill. To assess whether the percentage of significant stations was real and not due to chance, we used the confidence intervals of a proportion and found that the minimum number of stations needed for the lower 95% confidence interval to equal or exceed 10% was 17% and 18% respectively for the two regions. The following classification was used to describe the level of forecast skill in each region; 0-18% of locations (no skill), 19-30% of locations (useful skill), 31-40% (moderate skill), 41-50 (high skill) and >50% (very high skill).

Results and discussion Targeted forecasts of useful skill or better can be issued with lead-times of 5 months in western Queensland and 4 months in Gascoyne Murchison (Table 1). Forecasts targeted to an important forecast period in the region can be issued with lead-times that provide pastoralists with sufficient time to change management in light of the impending seasonal conditions. Limited confidence in long lead forecasts, low incidence of significance of individual phases in some regions, difficultly in defining meaningful spatial and temporal boundaries, limited consensus on assessing adequate levels of skill and problems with climate change influencing some statistical forecast systems are challenges that scientists working in climate applications face in promoting the useful application of targeted forecasts.

Table 1 Percentage of stations with a significant KW (prob KW≥0.90) for western Queensland (WQld) and Gascoyne Murchison (GM), forecast period (FP) and length of record of the rainfall data (LOR)

Region	FP	LOR	Lead-time (months)					
			0	1	2	3	4	5
W Qld	Nov-Mar	1891-2001	59[A]	66[A]	36[A]	73[B]	64[B]	87[B]
GM	Jan-Mar	1915-1998	32[A]	44[B]	44[B]	27[B]	22[B]	15[B]

[A] SOI phase, [B] SST phase

Acknowledgements This work was funded by the Land Water & Wool climate sub-program and the Gascoyne Murchison Strategy

References
Drosdowsky, W. (2002). SST Phases and Australian Rainfall. *Australian Meteorological Magazine, 51* (1), 1-12.
Stone, R.C., G.L. Hammer, & T. Marcussen (1996). Prediction of global rainfall probabilities using phases of the Southern Oscillation Index. *Nature, 384,* 252-255.

Potential climate change impacts on beef production systems in Australia

D.H. Cobon[1], K.L. Bell[1], G.M. McKeon[2], J.F. Clewett[1] and S. Crimp[2]
[1]Emerging Technologies, DPI&F, Tor St., Toowoomba, Q 4350, Australia, Email: david.cobon@dpi.qld.gov.au,
[2]Climate Impacts & Natural Resource Sciences, DNRM, 80 Meiers Road, Indooroopilly, Q 4068, Australia

Keywords: climate change, native pasture, beef production, modelling

Introduction There is increasing evidence suggesting that Australia's climate is changing due to enhanced levels of greenhouse gases and that it will continue to change (Pittock 2003). Climate changes are partly established, however the impact on systems, industries and process are unclear. Industry distribution reflects climatically imposed boundaries and the relative profitability of alternative land use. Climate change may negatively impact some existing industries but create opportunities for others. This study provides an assessment of the likely impacts of plausible climate change on the beef industry in central Queensland.

Materials and methods Climate change scenarios based on output from a range of General Circulation Models (GCM) were generated for Emerald (23°31'S, 148°10'E), Brigalow Research Station (near Banana, 24°28'S, 150°08'E) and Gayndah (25°37'S, 151°37'E) in central Queensland. Changes in pasture growth, stocking rate, beef production, runoff, deep drainage, basal area and utilisation of pasture, and pasture growth days were assessed for 2030 and 2070. Each projection was for 30 years and combinations of higher (H) and lower (L) levels of predicted temperature and rainfall were generated to form four combinations that consisted of higher temperature/higher rainfall (HH), HL, LH and LL (Table 1). The most likely climate change scenario for north-east Australia is either HL or LL. A perennial native pasture model called GRASP was used to simulate the range of variables under climate change and CO_2 enrichment circumstances. An average CO_2 enrichment scenario was applied where the base CO_2 level was 355ppm in 1990, 452ppm in 2030 and 603ppm in 2070.

Results and discussion The results are summarised in Table 1. Climate change impact on beef production is likely to be considerable in the absence of adaptation or mitigation strategies. Changes in rainfall will have larger impacts on pasture growth and production of beef cattle than changes in temperature. For 2030, a reduction in pasture growth of between 9 and 14% is most likely, reducing stocking rates and beef production. For 2070, these reductions will increase most likely to around 55 to 64%. Increases in the growing season length and enhanced pasture growth efficiency (higher CO_2) failed to compensate for lower moisture levels. Higher levels of pasture utilisation and lower basal area place added pressure on the natural resource, lower runoff will reduce surface water supplies and large reductions in deep drainage of water may reduce the threat of salinity. The challenge is for scientists and industry to anticipate the impacts of climate change and develop regional adaptation strategies that will allow the continuation of profitable beef production.

Table 1 Average control values across the three locations and percentage change from control for different climate change scenarios for 2030 and 2070. Temperature change is a rise in absolute values

	1990 Control	2030 HH	2030 HL	2030 LH	2030 LL	2070 HH	2070 HL	2070 LH	2070 LL
Inputs to GRASP from GCM									
Rain	671 mm	+7%	-13%	+7%	-13%	+20%	-40%	+20%	-40%
Max temp	28.7 °C	0.8-1.1	0.8-1.1	0.1-0.2	0.1-0.2	2.3-3.3	2.3-3.3	0.4-0.5	0.4-0.5
Min temp	14.7 °C	0.9-1.2	0.9-1.2	0.1-0.2	0.1-0.2	2.7-3.7	2.7-3.7	0.4-0.6	0.4-0.6
Outputs from GRASP (% difference from control)									
Growth	4094kg/ha	7	-14	9	-9	15	-64	19	-55
Stock rate	23 hd/km^2	6	-15	10	-9	13	-64	20	-53
Beef prod	49 kg/ha	11	-16	13	-11	25	-70	28	-61
Runoff	31 mm	9	-22	10	-23	29	-62	30	-62
Drainage	31 mm	21	-57	40	-49	54	-96	125	-93
Basal area	3 %	5	-12	5	-11	13	-39	14	-39
Utilisation	23 %	-2	7	-3	6	-3	35	-6	43
Beef prod	185 kg/hd	4	0	2	-2	10	-9	5	-12
Growth days	79 %	5	1	2	-1	12	-3	6	-7

References
Pittock, B. (2003). Climate Change: An Australian Guide to the Science and Potential Impacts. Australian Greenhouse Office.

Impact of climate change on potential distribution and relative abundance of the migratory grasshopper (Orthoptera: Acrididae) in the prairie ecosystem of Canada

O. Olfert and R. Weiss
Agriculture and Agri-Food Canada, Saskatoon, SK S7N 0X2, CANADA, Email: olferto@agr.gc.ca

Keywords: grasshoppers, climate change, bio-climatic modelling

Introduction Climate is the dominant force determining the distribution and abundance of most insect pest species. There has been considerable concern in recent years about climatic warming caused by human activities and the affects of these changes on agriculture in North America. Warming conditions may affect insect populations by altering timing of emergence, increased growth and development rates, shorter generation times and reduced overwintering mortality (McCarthy *et al.* 2001). Given that the magnitude of predicted temperature change associated with global warming is beyond the historical experience of modern agriculture computer models are one method by which researchers can study the possible impacts of climate change.

Materials and methods Model output was generated with CLIMEX[TM] 2.0, based on climate data derived from splined data. A 0.5^0 world grid dataset was used for input into the models. Models were run with the North American dataset (n=12452 grids). Incremental scenarios were created for all possible combinations for temperature $(0, +1, +2, +3, +4, +5, +6, +7^0$ of climate normal temperature for each grid) and precipitation (-20%, -10%, 0%, +10%, +20% of climate normal precipitation for each grid). Contour maps were generated by importing Ecoclimatic Indices into ArcView [TM] 8.1 (ESRI Inc 2001). Ecoclimatic Index values weree used to identify locations where climatic conditions are conducive to the development of pest populations.

Results Prior to utilizing the bioclimatic model for *M. sangunipes* (Gage *et al.* 1976) to predict the potential impacts of climate change in the grasslands of Canada, the model was validated by comparing predictions to known distributions of grasshoppers using approximately 50 years of population data. The results of the analysis indicated that potential grasshopper abundance and distribution varied with different combinations of predicted precipitation and temperature changes. Figure 1a, representing current Canadian conditions, provides a baseline for comparison against each of the climate change scenarios. For example, even a two degree increase in mean temperature and a 20% decrease in precipitation can potentially have a significant increase in risk to grasslands productivity due to the increase in grasshopper population density and distribution (Figure 1b).

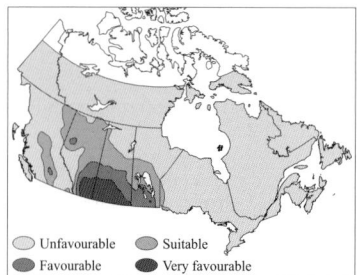 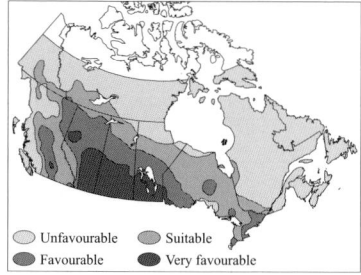

Figure 1 (a) Map on the left represents the grasshopper density and distribution under current conditions; (b) Map on the right predicts the density and distribution of grasshoppers with a two degree increase in mean temperature and a 20% decrease in precipitation

Conclusions The results of the present study showed that grasshopper range and distribution are positively and negatively affected based on specific combinations of temperature and precipitation. The model predicted that increased temperatures would result in the northward extension of *M. sanguinipes*. The study demonstrates that grasshopper populations, associated with observed meteorological conditions, can be used as analogues to estimate risk to grasslands of North America from climate change.

References

Gage, S.H., M.K. Mukerji & R.L. Randell. 1976. A predictive model for the seasonal occurrence of three grasshopper species (Orthoptera: Acrididae) in Saskatchewan. *The Canadian Entomologist*, 108, 245-253.
McCarthy, J., O.F. Canziani, N.A. Leary, D.J. Dokken, & K.S. White (eds.). 2001. *Climate Change 2001: Impacts, Adaptation, and Vulnerability. Contribution of Working Group II to the Third Assessment Report of the Intergovernmental Panel on Climate Change,* Cambridge University Press, Cambridge, UK, pp. 235-342.

Nitrogen fluxes in grassland in response to inter-annual climate variability

P. Calanca and J. Fuhrer
Agroscope FAL Reckenholz, Swiss Federal Research Station for Agroecology and Agriculture, Air Pollution/Climate Group, CH-8046 Zürich, Switzerland, Email: pierluigi.calanca@fal.admin.ch

Keywords: inter-annual variability, N-pools, N-fluxes

Introduction Most model studies dealing with the response of grassland dynamics to climate change (e.g. Thornley & Cannell, 1997; Riedo *et al.*, 2000) consider either step changes or a smooth transition in the climatic drivers, therefore neglecting short-term variations. However, the inter-annual variability of the climatic and edaphic elements (Calanca, 2004) can be substantial, with important consequences for the fluxes of nitrogen (N). The objective of this study is to investigate the response of nitrogen fluxes to this type of variability with the help of the Pasture Simulation Model (PaSim), looking at years considerably differing in terms of temperature, radiation and precipitation conditions.

Model and model experiment PaSim is a process-oriented model that simulates dry matter production and fluxes on C, N, water and energy in permanent grasslands. For the present experiment, we considered an extensively managed grassland located at Oensingen, on the Swiss Plateau. The model was driven with hourly meteorological data for 2002 and 2003. While the summer of 2002 was close to the norm, the summer of 2003 was characterized by drought conditions, with temperature, precipitation and radiation anomalies significantly above, below and above the respective long-term mean. The model was started with the soil organic pools in equilibrium. The clover fraction of the sward was set at 0.25. The management consisted of 3 cuts each year.

Results Table 1 presents selected averages of N-pools and N-fluxes for June-July-August. The differences between 2002 and 2003 are striking. Drought conditions during the summer of 2003 completely shut down leaching (Leach), reduce grass growth, thus limiting the biological N fixation (BNF), and foster biomass turnover. As a consequence of an increased flux of residue (F_{res}) to the soil organic pools and a sustained mineralisation, mineral N (N_{min}) availability in 2003 is four times as large as in 2002. Higher nitrification rates are explained by soil water contents at or below field capacity, while higher denitrification rates and N_2O emissions follow from a larger availability of nitrate.

Table 1 Simulated N-pools and N-fluxes of an extensively managed grassland at Oensingen (pools in kg-N/ha, fluxes in kg-N/ha/d), and difference 2003-2002 relative to 2002. Averages for June-July-August

year	N_{min}	F_{res}	Nitrif	Denitrif	Leach	N_2O-Em	BNF
2002	1.1	0.72	0.10	0.001	0.002	0.001	1.60
2003	5.3	1.08	0.31	0.011	0.000	0.005	0.90
rel. diff	4	0.5	2	10	---	4	-0.5

Conclusions The present results demonstrate that the inter-annual variability of climatic conditions can affect the fluxes of N in grassland by orders of magnitude. This has to be borne in mind in studying the consequences of anthropogenic climate change, given a probable increase in this type of variability, at least in mid-latitudes (Schär *et al.*, 2004). Weaknesses of the present investigation are: (i) a constant fraction of clover; (ii) the lack of a N_2O uptake by the soil in PaSim; (iii) the sensitivity of the results to the initial conditions of the organic pool.

References

Calanca, P. (2004). Interannual variability of summer mean soil moisture conditions in Switzerland during the 20th century: A look using a stochastic soil moisture model. *Water Resources Research*, 40, W12502, doi:10.1029/2004WR003254.

Riedo, M., D. Gyalistras & J. Fuhrer (2000). Net primary production and carbon stocks in differently managed grasslands: simulation of site-specific sensitivity to an increase in atmospheric CO2 and to climate change. *Ecological Modelling*, 134, 207-227.

Schär, C., P.L. Vidale, D. Lüthi, C. Frei, C. Häberli, M.A. Liniger, & C. Appenzeller (2004). The role of increasing temperature variability in European summer heatwaves. *Nature*, 427, 332-336.

Thornley, J.H.M. & M.G.R. Cannell (1997). Temperate grassland responses to climate change: an analysis using the Hurley Pasture Model. *Annals of Botany*, 80, 205-221.

Theme B: Grassland and the environment

Section 15

Greenhouse gases

Effect of red and white clover added to a rye grass-based diet on intake, fibre digestion and methane release of dairy cows

H.A. van Dorland, M. Kreuzer, H. Leuenberger and H.R. Wettstein
*Institut of Animal Science, Animal Nutrition, ETH-Zürich, ETH-Zentrum/LFH A1, CH-8092 Zürich, Switzerland,
Email: anette.van-dorland@inw.agrl.ethz.ch*

Keywords: methane, clover, fibre digestion, dairy cows

Introduction Forage legumes like white and red clover are widely grown in association with grass, with the intention to improve the quality of grass-based diets. However little is known about the effect of either white or red clover added to a grass-based diet on methane release, and existing studies are not conclusive. The objective of this study, applying the respiratory chamber technique, was to determine the effect of red and white clover added to a rye grass-based diet on intake, fibre digestion and methane release of dairy cows.

Material and methods Thirty-six Holstein Friesian and Brown Swiss dairy cows, weighing on average 637±55 kg and yielding 25.7±3.7 kg milk/day, were allocated to one of six experimental diets. These consisted of rye grass (60%, DM basis; fresh or ensiled) (*Lolium perenne* L., cultivar 'Fennema') either mixed with red clover (40%) (*Trifolium repens*, cultivar 'Pirat') in fresh (RCF) and ensiled form (RCS) or with white clover (40%) (*Trifolium pratense*, cultivar 'Klondike') in fresh (WCF) and ensiled form (WCS). Control diets consisted of rye grass-silage diets either unsupplemented (RGS) or supplemented with maize gluten (RGS+) as a control for the potential protein supply by the clovers. All diets were supplemented with barley to meet each cow's calculated extra energy requirement for milk production. Feed and faeces were collected individually. Open-circuit respiratory chambers were used to quantify methane emission from the individual cows.

Results Total feed intake was similar in all groups with 18.8± 0.9 kg DM/d as was milk yield. Fibre content of the white clover and red clover silage diets were higher compared to the other diets (ADF, g/kg: 266 and 261 vs. 235, 232, 243 and 246 for diets WCS vs. RCF, WCF, RGS+, RGS and RCS, respectively). The corresponding ADF intakes were 5.4, vs. 4.5, 4.2, 4.6, and 4.3, and 4.8 kg/d (P<0.05). Digestibility of organic matter (0.72 ±0.01) was quite similar for all diets. Digestibility of ADF was slightly higher for WCF and WCS (0.65 and 0.64, respectively) compared to groups RCF, RGS+, RGS, and RCS (0.60, 0.61, 0.63, and 0.61). Methane emission relative to fibre intake (Figure 1) was higher with the ensiled rye grass diets compared to the legume silage diets (P<0.05, contrast analysis). Methane (l) per kg of DM intake was higher with the pure rye grass diets compared to all legume diets with 34.5 and 33.5 for RGS+ and RGS vs. 30.6, 31.7, 31.4, and 32.2 for RCF, WCF, RCS and WCS.

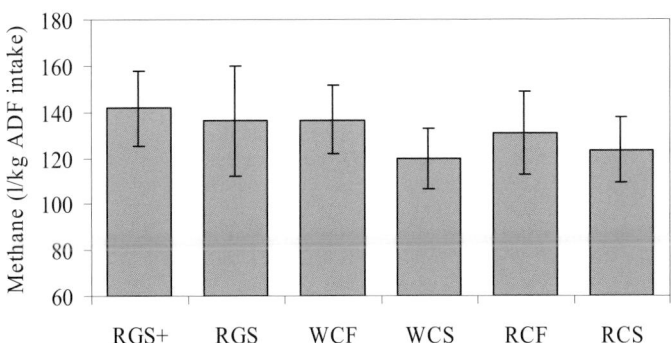

Figure 1 Effect of diet on methane release per kg ADF ingested

Conclusion These results suggest methane emissions from dairy cows fed ensiled herbage containing white or red clover may be lower than that of dairy cows fed a pure ensiled rye grass-based diet. This may be related to the chemical composition of the diets (Hindrichsen *et al.*, 2004).

Acknowledgement This study was supported by the Swiss Federal Office of Education and Science, Berne.

References
Hindrichsen, I.K., H.-R. Wettstein, A. Machmüller, C.R. Soliva, K.E. Bach Knudsen, J. Madsen & M. Kreuzer (2004). Effects of feed carbohydrates with contrasting properties on rumen fermentation and methane release in vitro. *Canadian Journal of Animal Science*, 84, 265-276.

Sources and sinks of greenhouse gases from European grasslands and mitigation options: the 'GreenGrass' project

J.F. Soussana[1], V. Allard, P. Ambus, C. Amman, P. Berbigier, C. Campbell, P. Cellier, E. Ceschia, P. Ciais, J. Clifton-Brown, S. Czóbel, R. Domingues, T. de Groot, R. Falcimagne, C. Flechard, J. Fuhrer, G. Gaborit, L. Horváth, A. Hensen, M.B. Jones, S. Jones, G. Kasper, K. Klumpp, P. Laville, C. Martin, C. Milford, Z. Nagy, A. Neftel, E. Nemitz, J.E. Olesen, A. Patterson, K. Pilegaard, A. Raschi, R. Rees, U. Skiba, P. Stefani, S. Salètes, P. Smith, M.A. Sutton, Z. Tuba, A. van Amstel, A. van den Pol-van Dasselaar, N. Viovy, N. Vuichard, M. Wattenbach, T. Weidinger

[1]*INRA, Grassland Ecosystem Research, Agronomy Unit, 234 Av. du Brézet, F-63100 Clermont-Ferrand, France, Email: soussana@clermont.inra.fr*

Keywords: carbon sequestration, temperate grasslands, climate change, greenhouse effect

Introduction Adapting the management of grasslands may be used to enhance carbon sequestration into soil, but could also increase N_2O and CH_4 emissions. In support of the European post-Kyoto policy, the European 'GreenGrass' project (EC FP5, EVK2-CT2001-00105) has three main objectives: i) to reduce the large uncertainties concerning the estimates of CO_2, N_2O and CH_4 fluxes to and from grassland plots under different climatic conditions and assess their global warming potential, ii) to measure net greenhouse gas (GHG) fluxes for different management which reflect potential mitigation options, iii) to construct a model of the controlling processes to quantify the net fluxes and to evaluate mitigation scenarios by up-scaling to a European level.

Materials and methods Net exchange of greenhouse gases (CO_2, N_2O and CH_4) was measured at nine European grassland sites, using eddy correlation techniques and soil respiration chambers for CO_2, static chambers, soil diffusion tubes and eddy correlation for N_2O and an *in-situ* SF_6 tracer technique for the emission of CH_4 by herbivores at grazing. First estimates of the annual net global warming potential from these grassland sites have been calculated. A quality analysis/quality check procedure has been used for the flux data. To gain further understanding, detailed experiments on micro-plots and pasture monoliths have been established at four sites with a range of management and mitigation options. The components of the net ecosystem exchange of CO_2 and the mean residence time of C in above and below-ground compartments were studied in two of these experiments by measuring continuously above and below-ground CO_2 exchanges and by steady-state or pulse labeling with $^{13}CO_2$. A detailed and mechanistic grassland ecosystem model (PASIM) has been improved to better simulate N_2O and CH_4 emissions from grasslands. After being parameterized, this model was tested against the flux data of the nine sites and has been used to assess the mitigation potential of field-level grassland management options. A simulation tool, FARMSIM, was designed in order to describe in a consistent way the C and N fluxes in cattle breeding farms and calculate the net balance of greenhouse gas emissions from 8 representative farms. Continental upscaling of net CO_2, N_2O and CH_4 exchanges with European grasslands were simulated applying a more detailed (PASIM) and a simpler (DNDC model) approach and preliminary maps of greenhouse gas fluxes have been prepared. Finally, a sensitivity analysis of the grassland carbon stocks to climate change has been run using the RothC model.

Results Net ecosystem exchange in the first year resulted in an average sink activity of 0.27 kg C m^{-2} yr^{-1} for atmospheric CO_2 with a coefficient of variation of 57 % among grassland sites. Assuming that 70% of the C that is harvested as hay or silage is respired within one year, the average net annual biome productivity was estimated at -0.13 kg C m^{-2} yr^{-1} (range -0.40 to +0.11). A further offset of the grassland sink activity occurred through the emissions of N_2O and CH_4. When converted to C equivalents, N_2O emissions reached on average, 6 % (0.015 kg equivalent CO_2-C m^{-2} yr^{-1}) of the NEE. Moreover, at the three sites where estimates are currently available, the CH_4 emissions from cattle reached 0.08 kg equivalent CO_2-C m^{-2} yr^{-1}, that is 32% the average NEE for these sites. The N_2O emission factor (EF1, according to IPCC Tier 2 methodology) reached 0.69%, on average, but with a large coefficient of variation (107 %) between sites and years. Pasture monolith experiments showed that both past and current extensive grassland management favored C sequestration, partly by increasing the mean residence time of C in soil organic matter fractions. This was also consistent with results from the PASIM model. First simulation results with FARMSIM for a case study dairy and beef farm in France showed that, on an annual average, the farm was a net source (1.7 tC-CO_2 equivalent ha^{-1} yr^{-1}) of GHG. European upscaling and sensitivity analyses were conducted for PASIM and DNDC focusing on management parameters and with RothC focusing on climate scenarios. First simulation results have shown that potentially productive grasslands are more likely to become larger sources of GHG when they are managed close to their optimum than less productive grasslands and that grasslands are likely to behave more frequently as sources of CO_2 under a changed climate.

Conclusion The results from this project will be used to refine the emissions factors used in national inventories, and address the mitigation potential of management scenarios in support of the EU response to Kyoto.

Modelling effects of agricultural policies on regional greenhouse gas emissions from cattle raising production systems in Baden-Württemberg (southwest Germany)

H. Neufeldt[1], M. Schäfer[2], E. Angendt[2], M. Kaltschmitt[1] and J. Zeddies[2]
[1]Institute for Energy and Environment, Torgauer Str. 116, 04347 Leipzig, Germany, Email: henry.neufeldt@ie-leipzig.de, [2]Department of Farm Management, University of Hohenheim, 70593 Stuttgart, Germany

Keywords: coupled geoecological economic model, agricultural policies, greenhouse gas emissions

Introduction In the light of the anthropogenic climate change and the resulting need to mitigate greenhouse gas (GHG) emissions, policies are needed which efficiently abate GHG emissions in the agricultural sector. However, reliable estimates of regional GHG abatement potentials in the agricultural sector are rare because the models do not integrate the economic and environmental effects of different agricultural policies and are generally restricted to a single-gas approach. Coupling an economic sector model with a process-oriented ecosystem model can overcome this gap and thus provide realistic *exante* information of socioeconomically and environmentally sustainable agricultural policies.

Materials and methods The process-oriented ecosystem model DNDC (Li, 2000) was coupled with the economic farm model EFEM (Angenendt *et al.*, 2002) via a database interface, and georeferenced soil, land use, management, and climatic data were used to simulate regionally disaggregated GHG emissions of grassland soils and cattle raising (dairy and meat) production systems in the state of Baden-Württemberg (southwest Germany). EFEM simulates production systems on farm level using Linear Programming to maximise return rates. EFEM was run for a series of realistic and feasible policy measures, including a 30€ tax per ton CO_2 on all emitted GHG, and a tripling of N fertilizer costs. Simulated environmental and economic effects of reduced livestock densities are being prepared.

Results Table 1 shows that both the 30€ tax on GHG and the 3-fold N fertilizer price reduced gross margin by about 10% and abated GHG by around 8% when compared to the reference scenario. Raising the fertilizer price was slightly more efficient with respect to reducing soil borne emissions whereas the tax on GHG led to a stronger reduction of ruminant CH_4 emissions due to a slightly altered feeding scheme. The soil emissions of the simulated policy measures were lower than the reference because less mineral fertilizer-N was applied whereas the application of organic N as slurry or farmyard manure remained nearly the same. Organic fertilizer application rates were only slightly reduced because livestock density was unaffected from management responses to the policy measures. Regional livestock density, however, explains nearly 90% of GHG emission variation in cattle raising production systems (Figure 1).

Table 1 Economic and environmental effects of mitigation strategies on dairy production systems

	Reference	GHG tax	N price
Gross margin (€ ha^{-1})	1260	1120	1140
GHG$_{total}$ (kg CO$_2$-eq ha^{-1})	5110	4690	4710
GHG$_{soil}$ (kg CO$_2$-eq ha^{-1})	1120	980	940
N$_2$O$_{soil}$ (kg N ha^{-1})	3.71	3.28	3.15
Fertilizer N$_{min}$ (kg ha^{-1})	108	75	60
Fertilizer N$_{org}$ (kg ha^{-1})	50	43	45
Livestock density (LS ha^{-1})	1.03	1.03	1.03

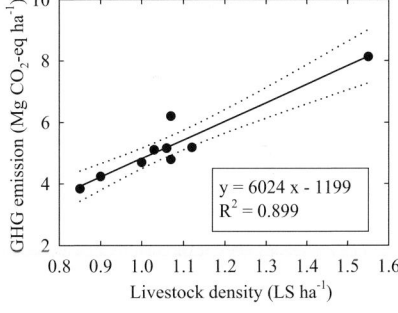

$$y = 6024 x - 1199$$
$$R^2 = 0.899$$

Figure 1 GHG emission vs. livestock density

Conclusions The results of the study suggest that taxes on either GHG or N fertilizers are efficient measures to reduce GHG emissions in cattle raising production systems. Since livestock density is so closely correlated to the overall GHG emissions, extensification is probably the most effective mitigation measure. Further simulations will show how a reduction of livestock density will affect environmental and economic parameters.

References

Angenendt, E. M. Schäfer, J. Zeddies, H. Neufeldt & M. Kaltschmitt (2002). Modelling of spatially disaggregated sectoral greenhouse gas balances and evaluation of regional mitigation strategies. Globale Klimaerwärmung und Ernährungssicherung, *Hohenheimer Umwelttagung*, 34, 117-122.

Li, C.S. (2000). Modeling trace gas emissions from agricultural ecosystems. *Nutrient Cycling in Agroecosystems*, 58, 259-276.

Modelling nitrous oxide emissions from grazed grasslands in New Zealand

S. Saggar[1], D.L. Giltrap[1], C. Li[2], C.B. Hedley[1], K.R. Tate[1] and S. Lambie[1]
[1]Landcare Research, Private Bag 11052, Palmerston North, New Zealand, Email: SaggarS@landcareresearch.co.nz, [2]Institute for the Study of Earth, Oceans, and Space, University of New Hampshire, Durham, NH 03824, USA

Keywords: dairy-grazed, excretal inputs, IPCC-approach, NZ-DNDC model, sheep-grazed

Introduction Spatial and temporal variability are major difficulties when quantifying annual N_2O fluxes at the field scale. New Zealand currently relies on the IPCC default methodology (National Inventory Report, 2004). This methodology is too simplistic and generalised as it ignores all site-specific controls, but is also not sufficiently flexible to allow mitigation options to be assessed. Therefore, a more robust, process-based approach is required to quantify N_2O emissions more accurately at the field level. Denitrification-decomposition (DNDC) is a process-based model originally developed (Li *et al.*, 1992) to quantify agricultural nitrous oxide (N_2O) emissions across climatic zones, soil types, and management regimes. This has been modified to represent New Zealand grazed grassland systems (Saggar *et al.*, 2004). More recent modifications include measured biomass C and N parameters in perennial pasture and compaction impacts on the soil water dynamics. Further validation tests have been conducted against observed soil moisture and gas fluxes. Here we i) assess the ability of a modified DNDC model "NZ-DNDC" to simulate N_2O emissions; ii) compare the measured, modelled and IPCC-estimated N_2O emissions from dairy- and sheep-grazed pastures; and iii) give preliminary results for upscaling the model to provide preliminary regional emissions estimates.

Materials and methods Nitrous oxide measurements were made periodically between April 2001 and June 2003 from sheep- and dairy-grazed and ungrazed areas at three sites. To account for the spatial variability, 18 chambers were randomly located *c.* 20 m apart along a Z-shaped transect to measure the fluxes of N_2O from the grazed area (~1 ha); two chambers were located in the ungrazed area (~0.005 ha). The modified DNDC model, hereafter named "NZ-DNDC", was then used to simulate N_2O emissions from the pastures grazed by dairy cattle. Full descriptions of the experimental methodology used and the model modifications are reported in Saggar *et al.* (2004).

Results Large spatial and temporal variations were observed in the N_2O fluxes measured from the grazed area over all sites. Large fluxes were generally observed after each grazing and rainfall event, and were followed by a decline. The modified NZ-DNDC model simulated well the average daily N_2O fluxes from the control and grazed grassland sites. Significant differences in emission rates between the sheep-grazed and dairy pastures

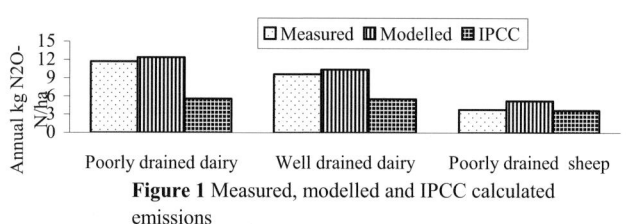

Figure 1 Measured, modelled and IPCC calculated emissions

exist. The NZ-DNDC was able to predict the annual measured emissions from all the sheep- and dairy-grazed and ungrazed grassland sites very well. The NZ-DNDC model was also able to pick up the differences in emissions resulting from differences in soil texture and grazing regime (Figure 1). The IPCC methodology cannot account for such influences. The model will be used to provide regional estimates of N_2O emissions.

Conclusions The overall comparisons of predicted and measured annual emissions indicate NZ-DNDC should be applicable to the simulation of N_2O emissions from a range of grazed grasslands. NZ-DNDC offers a solid beginning to develop regional- and national-scale inventories with known levels of uncertainties.

References
Li C, S. Frolking & T.A. Frolking (1992) A model of nitrous oxide evolution from soil driven by rainfall events: 1. Model structure and sensitivity. *Journal of Geophysical Research,* 97, 9759–9776.
National Inventory Report New Zealand 2004: Greenhouse Gas Inventory 1990-2002 (including common reporting format (CRF) for 2002), April 2004. New Zealand Climate Change Office, Wellington, New Zealand. Also online: http://www.climatechange.govt.nz
Saggar S, R.M. Andrew, K.R. Tate, C.B. Hedley, N.J. Rodda, & J.A. Townsend (2004). Modelling nitrous oxide emissions from New Zealand dairy grazed pastures. *Nutrient Cycling in Agroecosystems.* 68, 243–255.

Cattle overwintering areas in middle-European conditions – important "point" sources of nitrous oxide emissions

M. Šimek[1,2], J. Hynšt[1,2], P. Brůček[1,2], J. Čuhel[1,2], D. Elhottová[1], H. Šantrůčková[1,2] and V. Kamír[3]
[1]Institute of Soil Biology, Academy of Sciences of the Czech Republic, Na Sádkách 7, 370 05 České Budějovice, Czech Republic, Email: misim@upb.cas.cz, [2]University of South Bohemia, 370 05 České Budějovice, Czech Republic, [3]Farm Borová, 382 08 Chvalšiny, Český Krumlov, Czech Republic

Keywords: nitrous oxide, nitrogen, grassland, cattle, soil

Introduction Nitrous oxide (N_2O) emissions in grazed grasslands are strongly influenced by animal excreta (Fowler *et al.*, 1997). In addition, soil compaction caused by animal traffic significantly influences soil physical conditions and thus directly or indirectly impacts on the microbial processes producing N_2O. In the Czech Republic pastures are mostly located in hilly and mountain areas. During the growing season, cattle are typically grazing, while during the winter the animals are concentrated near the animal house on a relatively small plot called an "overwintering area". The objective of this study was to estimate the fluxes of N_2O from a typical overwintering area, where the combined effects of soil compaction and deposition of urine and dung occur.

Materials and methods The overwintering area was at Borová Farm in South Bohemia. The area was 4 ha and it was used by 90 cows. Three microsites were identified, differing in extent of animal impact: 1 - severe impact (totally destroyed surface soil), 3 - moderate impact (visible changes in soil surface and vegetation), 5 - light to no impact (mostly unaffected soil and vegetation). Gas fluxes were determined 40 times between Oct. 2002 and May 2004 using portable chambers and samples were analysed using gas chromatography (Šimek *et al.*, 2000).

Results The time course of N_2O emissions is given in Figure 1. In autumn 2002, the emissions of N_2O remained at a low "background level" for several weeks (around 5 g N/ha per d) in all microsites. They increased slightly in the middle of Nov. and again in Jan. 2003. However, substantial increases in N_2O fluxes were recorded from microsites 1 and 3 much later, in April and the beginning of May 2003, respectively. In contrast, gaseous losses from site 5 were relatively low. In late autumn 2003, peaks of N_2O emissions were found in microsites 1 and 3 indicating favourable conditions for microbial processes affecting N transformations. Large fluxes of N_2O from microsites 1 and 3 were then recorded in late spring 2004, similarly to a previous year's emission pattern. Total N_2O fluxes were estimated to be about 10, 8, and 1.5 kg N_2O-N/ha per y in microsites 1, 3, and 5, respectively.

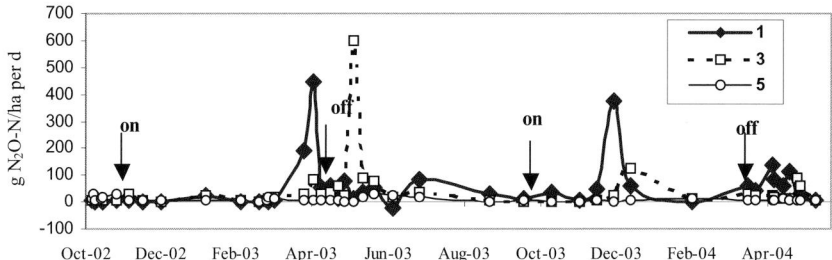

Figure 1 Fluxes of N_2O from three differently impacted microsites in a cattle overwintering area. Note: **on** and **off** inserted indicate when cattle were moved on and off the overwintering area

Conclusions These results indicate that the cattle overwintering plots, which are common in hilly areas in the Czech Republic, represent very important "point" sources of N_2O emissions. Although these specific pasture sites are relatively small, they can contribute substantially to overall N_2O fluxes.

Acknowledgements The research was supported by the EU project MIDAIR (No EVK2-CT-2000-00096) and by the Grant Agency of the Czech Republic (No 526/04/0325).

References
Fowler, D., U. Skiba & K.J. Hargreaves (1997). Emissions of nitrous oxide from grasslands. In: S.C. Jarvis and B.F. Pain (eds.) Gaseous Nitrogen Emissions from Grasslands. CAB International, Wallingford, 147-164.
Šimek, M., J.E. Cooper, T. Picek & H. Šantrůčková (2000). Denitrification in arable soils in relation to their physico-chemical properties and fertilization practice. *Soil Biology and Biochemistry*, 32, 101-110.

The use of long-term modelling in analysing N_2O abatement strategies in dairy pastures

R.J. Eckard[1], I.M. Johnson[2] and D.F. Chapman[1]

[1]ILFR, The University of Melbourne, Parkville, Victoria 3010, Australia, Email: rjeckard@unimelb.edu.au, [2]IMJ Consultants, PO Box 1590, Armidale, NSW 2350, Australia

Keywords: Best Management Practices, greenhouse gas emissions, grazing, nitrous oxide

Introduction Concerns about the environmental impact of nitrogen (N) losses in Australia, in particular, nitrous oxide emissions are related to the rapid increase in N application on dairy pastures. Computer modelling is the most suitable method available to assess the potential of best management practices (BMP) to reduce field losses, as direct field measurements are frequently limited by the short term nature of many field trials.

Materials and methods DairyMod is a biophysical model (Johnson *et al., 2003)* focussing on pasture and cropping systems, and associated greenhouse gas emissions. The model incorporates the principal rotational grazing management strategies that are used in Australia, including modules for pasture growth and utilisation by grazing animals, animal metabolism including milk production, water and nutrient dynamics, and management strategies for fertiliser, irrigation and grazing. The output includes daily and annual estimates for all the principal fluxes in the system, including methane and nitrous oxide. DairyMod was parameterised using data from an experimental site where N losses were measured for 3 years from grass/clover pastures grazed by dairy cows, receiving either 0 (0N) or 200 kg N/ha (200N) as urea (Eckard *et al., 2003*). A 42 year historical simulation was conducted using long-term climate data for the location.

Results Mean measured N_2O emissions for the 3 years of the field experiment were 4.6 and 7.2 kg N_2O/ha for the 0N and 200N rates, respectively, close to the 42-year modelled means of 2.9 and 7.5 kg N_2O/ha. However, over the 42-year simulation, predicted N_2O emissions ranged between 0.1 and 7.4 kg N_2O/ha for the 0N rate, and 1.5 to 14.4 kg N_2O/ha for the 200N rate (Figure 1). If the field experiments were run from 1959 to 1961 mean N_2O emissions could have been 0.2 (0N) or 1.9 (200N) kg N_2O/ha.y; perhaps insufficient to justify a targeted abatement strategy. However, mean N_2O emissions between 1978 and 1980 were 3.5 (0N) or 9.6 (200N) kg N_2O/ha, justifying the development of BMPs aimed at reducing annual N inputs.

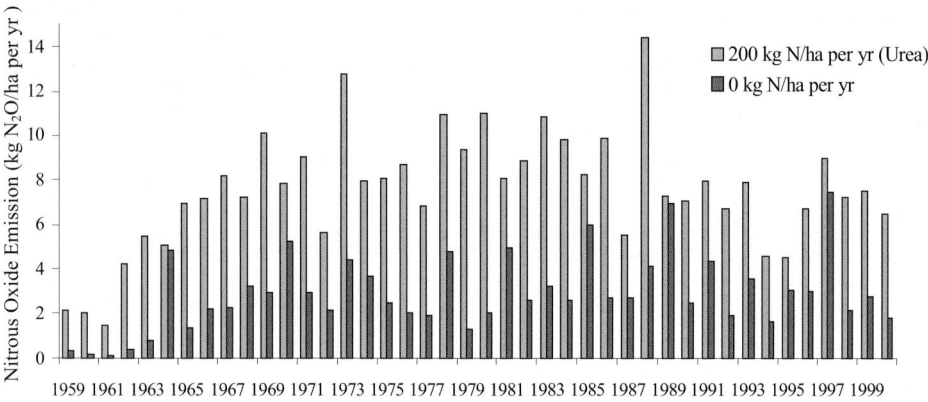

Figure 3 Annual total of N_2O emissions from a grazed dairy pasture with or without 200 kg N/ha

Conclusions Simulation modelling is an essential tool in evaluating the longer-term impacts of BMPs developed from short-term field experiments, particularly for highly variable processes like N_2O emissions.

References

Eckard R.J., D. Chen, R.E. White & D.F. Chapman (2003). Gaseous nitrogen loss from temperate grass and clover dairy pastures in south eastern Australia. *Australian Journal of Agricultural Research,* 54, 561-570.

Johnson I.R., D.F. Chapman, A.J. Parsons, R.J. Eckard & W.J. Fulkerson (2003) DairyMod: a biophysical simulation model of the Australian dairy system. *Australian Farming Systems Conference*, Toowoomba, Queensland. http://afsa.asn.au

A comparison of N_2O emissions after application of dairy slurry on perennial grass or bare soil prior to planting an annual crop in coastal British Columbia, Canada

S. Bittman, C.G. Kowalenko, A. Friesen and D.E. Hunt
Pacific Agri-Food Research Centre, Box 1000, Agassiz, BC, Canada, V0M 1A0

Keywords: nitrous oxide, dairy slurry, perennial grass

Introduction Because of restrictions on land application of manure in autumn and winter, dairy farmers in coastal British Columbia (BC) must apply half of their annual manure supply from mid-Feb. to mid-April. Although two thirds of their land is in perennial grass, most of this manure is applied to bare soil, usually maize land, prior to planting. This is done for convenience and to avoid damaging grass stands with equipment traffic. Farmers are encouraged to allocate more manure to grass to minimise soil NO_3 concentrations after maize harvest, because maize takes up less N than grass, and the bare fields after harvest are subject to wintertime leaching. However, the effect of this practice on emissions of N_2O is not known. Our objective was to compare the effect of spring application of manure on bare land and on grass with respect to emissions of N_2O. A second objective was to compare early, late and split applications of manure.

Materials and methods The trial (randomised complete block design with 6 replications) was conducted in 2001 and 2002 on a silty loam soil at Agassiz, BC. This coastal location has moderate summer and winter temperatures and about 1500 mm of annual precipitation that falls mainly in autumn and winter. Liquid dairy manure (dry matter of 4.0 and 6.9, total ammoniacal N (TAN) of 1.6 and 2.0 g/kg, total N of 4.0 and 6.9 g/kg, and pH of 7.8 and 7.4, in 2001 and 2002, respectively) was surface-broadcast by hand on a tall fescue sward or on bare soil prior to planting an annual crop. The manure was applied as a single dose of 100 kg TAN per ha in late Feb (TSUM 285 in 2001 and 235 in 2002) or early April (TSUM 533 in 2001 and 457 in 2002), or the dose was split evenly between the two dates (grass only). N_2O and soil NO_3-N (0-15 cm depth) were measured regularly on all plots from the first manure application until early to mid-June, prior to crop planting. At this time, emission rates were low, with relatively little difference among treatments. Emissions of N_2O were measured by drawing air from vented static chambers into a continuous N_2O analyser (Thermo Electron Corp. 46C, Franklin, MA, USA).

Results and Discussion Cumulative N_2O emissions were 6-10 times ($P<0.05$) greater from bare soil than from grass plots treated with manure (Table1). Emissions were similar or greater from bare soil without manure than from grass with manure. On bare soil, early application produced more N_2O in 2001 but less in 2002, while the effect of date of manure application on grass was comparatively small. Herbage production and N-uptake (May harvest) were generally greater from early than from late manure applications while split applications were intermediate (not shown), which is consistent with previous studies and local recommendations. The grass plots averaged very low soil NO_3 levels (Table 1) even soon after manure application (not shown). The bare soil plots, including the un-manured plots, had much higher average soil NO_3 levels than the grass plots (Table 1). Cumulative N_2O emission was linearly related to average concentration of soil NO_3 (r^2=0.92). Low concentrations of soil NO_3 are typical in grassland and reflect rapid uptake of ammonia and NO_3 by the grass crop and by soil microbes. The results suggest that farmers should apply manure on grass fields in early spring to minimise both emissions and residual soil NO_3 in autumn. Methods for reducing damage to swards will need to be implemented. The effect of harvested or incorporated winter cover crops on emissions needs to be examined.

Conclusions This study shows that N_2O emission is greatly reduced when spring manure is applied on perennial grassland rather than on bare soil prior to planting annual crops. The reason appears to be characteristically low soil NO_3 levels under grass. Date of manure application has little effect on grass but can either increase or decrease emissions on bare soil. For annual crops, manure should be applied just before or after planting to minimise soil NO_3 concentrations.

Table 1 Cumulative N_2O emissions and mean soil NO_3-N (Feb.-June) as affected by dairy slurry applied at 100 kg TAN ha^{-1} to bare soil and perennial grass

Treatment		N_2O Emissions kg N_2O-N ha^{-1}		Soil NO_3 kg NO_3-N ha^{-1}	
		2001	2002	2001	2002
Bare soil	No manure	0.64c[#]	0.72c	6.9c	8.1c
Bare soil	Manure Feb 27/26	2.24a	2.19b	19.2a	19.1b
Bare soil	Manure April 2/8	1.77b	2.99a	10.5b	22.2a
Grass	No manure	0.10d	0.30c	0.2d	0 d
Grass	Manure Feb 27/26	0.12d	0.45c	0.4d	0 d
Grass	Manure April 2/8	0.21d	0.48c	0.7d	0 d
Grass	Manure Feb/April	0.16d	0.47c	0.4d	0.1d

[#] values in column not followed by the same letter are significantly different at P=0.05

A simulation study of nitrous oxide emissions from a fertilised, grazed grassland site in Ireland

M.J. Hawkins[1], J. Connolly[1], B. Hyde[2], M. Ryan[2] and R.P.O. Schulte[2]
[1]University College Dublin, Dublin 4, Ireland, Email: michael.hawkins@ucd.ie, [2]Teagasc Environmental Research Centre, Johnstown Castle, Wexford, Ireland

Keywords: greenhouse gases, nitrous oxide, simulation model

Introduction Nitrous oxide (N_2O) emissions from grazed grassland are understood to be strongly influenced by the availability of a mineral N source, soil temperature and soil water content (Skiba & Smith, 2000). We derive an empirical model of emissions based on these factors and use it to simulate daily emissions from a fertilised, grazed grassland site in Wexford, Ireland, under different application schedules and climatic conditions for the period 1994 to 2001, inclusive (Table 1).

Table 1 Simulated scenarios

Sim. 1	Fertiliser (N) application on the first day of each month between February and September, inclusive.
Sim. 2	N application on the first day of February, April, June and August.
Sim. 3	As Sim. 1, with N application on the days that soil moisture is at a weekly minimum.
Sim. 4	As Sim. 1, with N application on the days that soil moisture is at a weekly maximum.

Materials and methods A two-year rotational grazing experiment was conducted recently in Teagasc's Environmental Research Centre, Johnstown Castle, Wexford, Ireland, to study N_2O emissions from fertilised grazed pasture. A generalised linear mixed model was fitted to this data and the fixed effects regression component was then used in the simulation exercise.

Table 2 Total annual simulated emissions (kg N_2O-N/ha per yr)

Year	Mean Soil Temp. (°C)	Mean Vol. Soil Moisture (%)*	Sim. 1 (210)**	Sim. 2 (210)	Sim. 3 (210)	Sim. 4 (210)	Sim. 1 (390)	Sim. 2 (390)	Sim. 3 (390)	Sim. 4 (390)
1994	11.5	55	13	12	13	13	16	21	16	18
1995	10.3	55	13	13	14	14	19	25	20	22
1996	12.1	51	10	8	10	10	10	8	10	10
1997	11.5	55	12	10	12	12	14	13	13	15
1998	11.3	56	13	11	12	13	16	18	15	17
1999	11.2	53	11	9	11	12	13	12	12	14
2000	10.9	53	11	9	11	11	11	12	11	12
2001	10.8	54	10	8	10	10	11	9	11	12
Mean	11.2	54	12	10	12	12	14	15	14	15
CV (%)	3	5	12	17	12	13	24	40	25	27

*Based on estimates of soil moisture deficits. **Figures in brackets represent N application rate (kg/year)

Results and Conclusions Initial results (Table 2) suggest that inter-annual variability due to climatic factors increases with the amount of applied fertiliser and decreases with frequency of application. Increasing application frequency is more effective at reducing total emissions than restricting applications to days when the soil is relatively dry.

Acknowledgements This work was part funded under the Irish Government's ERTDI programme 2000 – 2006.

References
Skiba, U., and Smith, K.A., 2000. The control of nitrous oxide emissions from agricultural and natural soils. Chemosphere – Global Change Science, Vol. 2, 379 – 386.

Nitrogen leaching from a timothy sward following application of anaerobically digested cattle slurry in Japan

T. Matsunaka, T. Sawamoto, H. Ishimura, K. Takakura and A. Takekawa
Rakuno Gakuen University, Ebetsu, Hokkaido, 069-8501, Japan, Email: matsunak@rakuno.ac.jp

Keywords: NH_3 volatilisation, N_2O emission, NO_3 leaching

Introduction The objective of this study was to quantitatively evaluate nitrogen (N) leaching from a *Phleum pratense* pasture following application of anaerobically digested cattle slurry by monitoring N uptake by the grass, NH_3 volatilisation, N_2O emission, and NO_3 leaching in a lysimeter trial.

Materials and methods The experiment was conducted with 12 lysimeters (3×3 m^2 in area and 1.7 m in depth) sown with timothy in Hokkaido. About 50% of N in the slurry was NH_4-N and the remainder was organic N. NO_3-N was not detected in the slurry. Treatments (with two replications) were as follows, 1) 8N-S-OCT, 2) 16N-S-OCT, 3) 8N-S-APR, 4) 16N-S-APR, 5), 16N-C-APR-JUN, 6) 0N, where 8N and 16N indicate the application of 8 (standard level) and 16 (double level) g NH_4-N/m^2, respectively. S and C indicate the application of the slurry and chemical fertiliser ((NH_4)$_2$SO$_4$), respectively. OCT, APR, and JUN refer to the months of application. JUN corresponds to the first post cutting. N leaching derived from the slurry and chemical fertiliser was estimated as the difference from those of the 0N treatment. The trial was conducted from Oct 2000 to Oct 2003.

Results The major results were as follows: (1) N uptake by the grass corresponded to 18–30 % of the applied N. Standard level application of the slurry and application in APR increased the ratio. (2) The amount of NH_3 volatilisation to the atmosphere corresponded to about 13% of the applied N with no relation to the rate and the time of the slurry application. (3) The amount of NO_3-N leaching corresponded to 4–10 % of the applied N, which was higher in the application in OCT than in APR. (4) The amount of N_2O-N emission corresponded to less than 0.1 % of the applied N. (5) The amount of (1)–(4) was slightly lower than the applied NH_4-N in the slurry.

Conclusions N leaching derived from the slurry (the total amount of (2), (3), and (4)) corresponded to 17–22 % of N in the slurry, in which the NH_3 volatilisation was the major component. NO_3-N leaching could be controlled by the standard level application in spring.

Fluxes of CO$_2$ and N$_2$O from soils of a grazed pasture in Ireland

S. Kumar[1], M. Abdalla[1], P. Ambus[2], A. McCourt[1], J. Clifton-Brown[3], J. Burke[4], M. Jones[1] and M.L. Williams[1]
[1]Department of Botany, Trinity College, University of Dublin, Dublin 2, Ireland, Email: willimsm@tcd.ie, [2]Plant Research Department, Riso National Laboratory, Roskilde DK4000, Denmark, [3]IGER, Aberystwyth, Wales, [4] Oak Park Research Centre, Teagasc, Carlow, Ireland

Keywords: soil respiration, nitrous oxide, carbon dioxide, grassland, Ireland

Introduction As part of a European research program on greenhouse gas flux measurements from agricultural systems, a three year study of both CO$_2$ and N$_2$O fluxes from grassland soils has been established in Ireland. Results presented in this paper represent a preliminary exercise in the modelling of soil respiration.

Materials and methods Weekly measurements of soil respiration and N$_2$O flux were made on a grazed *Lolium/Trifolium* grassland situated at the Teagasc Research Centre in Co. Carlow, Ireland from July 2002 and October 2003. Measurements of soil respiration were made using a portable IRGA system (PP systems, UK), whilst measurements of N$_2$O flux consisted of the collection of gas samples from static chambers, and their subsequent analysis by GC.

Results The pattern of CO$_2$ and N$_2$O release from the soil over a two year period is shown in Figure 1. The results suggest correlation between N$_2$O release and soil respiration from early October 2003 (day 650) through May 2004 (day 850). The range of flux values obtained for both gases were of the same scale as other published data on European grasslands. The similarity

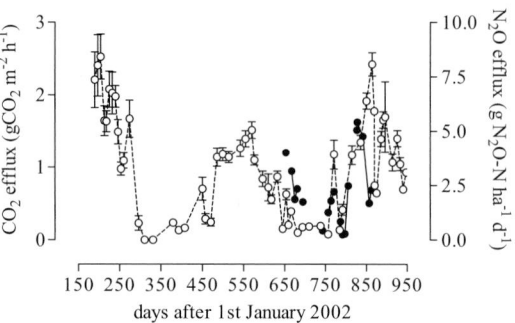

Figure 1 Soil fluxes of CO$_2$ (open circles) and N$_2$O from a grazed *Lolium/Trifolium* grassland

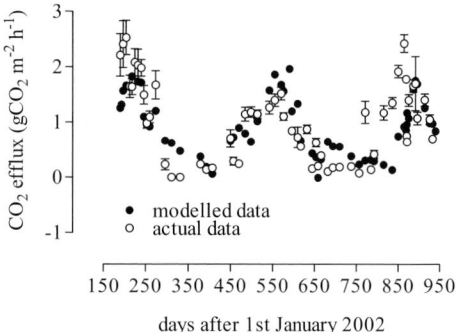

Figure 2 Comparison of modelled and field data for soil respiration

in pattern of N$_2$O release with soil respiration may be expected given that soil respiration is largely the result of microbial activity. However, addition of 155 kg/ha N-fertiliser to selected plots on the 7th April 2004 (day 98) resulted in a peak of N$_2$O release that was not reflected by any change in soil respiration (data not shown). Using soil temperature and volumetric soil water content data, an empirical model for soil respiration was established that could account for approximately 64% of the variation of the CO$_2$ flux. A comparison of modelled and field data for soil respiration is illustrated in Figure 2. Using this model with soil temperature and water content data collected every 30 minutes over the two year period, an annual release of 2.6 x 10^{-3}tC/ha from the soil due to respiration was calculated for this site.

Conclusions A measurement of the soil CO$_2$ and N$_2$O released from grazed *Lolium/Trifolium* grassland in Ireland has shown close similarities in their temporal pattern of flux. Peaks in N$_2$O release were strongly associated with fertiliser N application. Correlation of soil CO$_2$ with soil temperature, soil water content and aboveground biomass has shown temperature to be the major limiting factor on soil respiration.

Acknowledgements We acknowledge the financial support of the EU through their Framework programme in funding this research.

Effects of tropical legumes with contrasting tannin contents and mixtures of them on *in vitro* ruminal fermentation and methanogenesis

H.D. Hess[1], F.L. Valencia[2], P. Avila[2], C.E. Lascano[2] and M. Kreuzer[1]

[1]*Swiss Federal Institute of Technology, Institute of Animal Science, ETH-Centre/LFW, CH-8092 Zurich, Switzerland, Email: dieter.hess@inw.agrl.ethz.ch, [2]Tropical Grass and Legume Project, Centro Internacional de Agricultura Tropical (CIAT), Cali, Colombia*

Keywords: legume mixtures, tannins, ruminal fermentation

Introduction   Previous *in vitro* experiments have showed that the supplementation of a low-quality grass diet with *Arachis pintoi* or *Cratylia argentea* (legumes low in tannins) enhanced ruminal fermentation and methane release, whereas the supplementation with *Calliandra calothyrsus* (rich in tannins) decreased methane release and suppressed organic matter (OM) degradation and N turnover (Hess *et al.*, 2003). It was hypothesised that a mixture of tanniniferous legumes with legumes free of or low in tannins would result in a methane-suppressing effect but without impairing ruminal nutrient degradation.

Materials and methods   Three *in vitro* experiments were performed with an 8-fermenter RUSITEC-apparatus (Rumen simulation technique) to evaluate the fermentation characteristics of legumes with contrasting contents of condensed tannins and their mixtures. In every experiment, four basal diets consisting only of legumes were evaluated in four replicates. The daily dry matter (DM) supply to the fermenters was maintained constant at 15 g. In experiment 1, the 4 legumes *A. pintoi, C. argentea, C. calothyrsus* and *Flemingia macrophylla* were compared. In experiment 2, the effects of a partial or complete replacement (0, 0.25, 0.5 and 1 of total of diet DM) of *C. calothyrsus* by *C. argentea* were investigated. In experiment 3, the influences of a partial replacement (0.5 of diet DM) of *F. macrophylla* by *C. argentea, C. calothyrsus* or by a mixture of both were evaluated.

Results   In experiment 1, apparent degradation of OM and crude protein (CP) was highest for *A. pintoi*, intermediate for *C. argentea* and lowest for *F. macrophylla* and *C. calothyrsus* (p<0.05). Methane release relative to OM degraded was highest for *A. pintoi* and *C. argentea*, intermediate for *F. macrophylla* and lowest for *C. calothyrsus* (p<0.05). In experiment 2, apparent degradation of OM and methane release relative to OM degraded were lowest with *C. calothyrsus* alone and increased linearly (p<0.001) with increasing proportion of *C. argentea*. When 0.25 or 0.5 of *C. calothyrsus* was replaced by *C. argentea*, only minor changes were observed in N turnover. However, the complete replacement of *C. calothyrsus* enhanced N turnover. With the exception of methane release, which was suppressed (p<0.05), only minor changes in fermentation were observed when 0.5 of *F. macrophylla* was replaced by *C. calothyrsus* in experiment 3. In contrast, the inclusion of *C. argentea* increased (p<0.05) apparent organic matter degradation and methane release per gram of OM degraded. When 0.5 of *F. macrophylla* was replaced by *C. calothyrsus* no changes in apparent CP degradation and ruminal N turnover occurred (p>0.05). However, the inclusion of *C. argentea* resulted in an improvement of CP degradation and N turnover.

Conclusions   The results from the present study confirmed the high nutritional value of *A. pintoi* and *C. argentea* and the potential of tannin-rich legumes to suppress methanogenesis. The shrub legume *F. macrophylla* seems to have a slightly higher feeding value than *C. calothyrsus* and was less effective in suppressing methanogenesis. The results also suggest that the use of mixtures of *C. calothyrsus* and *C. argentea* is a way to mitigate methane emission but simultaneously affects the nutritional value of the diet. This is in contrast with the observations made in a previous experiment (Hess *et al.*, 2004) where the supplementation of a low-quality grass with a mixture of these legumes increased OM degradation without enhancing methane emission relative to OM degraded. Therefore we postulate that the effects of legume mixtures on rumen fermentation not only depend on the quality and proportion of the different legumes, but also on the remaining components of the diet. If this is so, the evaluation of legume mixtures has to be done in combination with grasses which represent the most important diet ingredient for ruminants in tropical smallholder livestock systems.

Acknowledgement   This study was supported by the Swiss Agency for Development and Cooperation (SDC).

References
Hess, H.D., L.M. Monsalve, C.E. Lascano, J.E. Carulla, T.E. Díaz & M. Kreuzer (2003). Supplementation of a tropical grass diet with forage legumes and *Sapindus saponaria* fruits: effects on *in vitro* ruminal nitrogen turnover and methanogenesis. *Australian Journal of Agricultural Research*, 54, 703-713.
Hess, H.D., F.L. Valencia, L.M. Monsalve, C.E. Lascano & M. Kreuzer (2004). Effects of tannins in *Calliandra calothyrsus* and supplemental molasses on ruminal fermentation *in vitro*. *Journal of Animal and Feed Science*, in press.

Methane production by cattle grazed at two stocking rates on a semi-natural grassland

C.S. Pinares-Patiño[1,2], Y. Rochette[1], M. Fabre[1], J.-P. Jouany[1] and C. Martin[1]
[1]URH, INRA Clermont-Ferrand/Theix, 63122 Saint-Genès-Champanelle, France, Email: cesar.pinares@agresearch.co.nz, [2]AgResearch Ltd., Grasslands Research Centre, Private Bag 11008, Palmerston North, New Zealand

Keywords: methane, cattle, stocking rate, grazing

Introduction Global warming induced by the human-enhanced concentrations of greenhouse gases (GHG) in the atmosphere is a major environmental concern of our day. Enteric methane (CH_4) is the most important GHG associated with grazing livestock. The emissions of methane may be influenced by environmental conditions and grazing management. Stocking rate (SR), the number of animals per unit of land area, remains the simplest management tool in pastoral farming. However, little is known about the influence of SR on CH_4 emission. The objective of this study was to compare the CH_4 emissions from cattle managed under low and high SR.

Materials and methods The study was conducted during the 2002 and 2003 grazing seasons on a 6.7 ha semi-natural grassland (Massif Central, France; 1250 m altitude). The area was subdivided into two paddocks and continuously grazed under low (LSR: 1.1 livestock units (LU)/ha) or high (HSR: 2.2 LU/ha) SR, each involving seven 20-months-old Holstein-Friesian heifers (initial body weight: 442±28.5 kg in 2002 and 451±44.5 kg in 2003). The study comprised four experimental periods: late spring (P1), mid summer (P2), late summer (P3) and early autumn (P4), each comprising a 6-d acclimatisation period (d 1 to 6), followed by an 8-d (d 7 to 14) sample collection period. Feed organic matter intake (OMI, kg/d) by individual cows was calculated using their mean faecal OM output (kg/d) and OM digestibility (OMD) estimations. Daily CH_4 productions were determined using the sulphur hexafluoride (SF_6) tracer technique.

Results In both grazing seasons SR treatments did not differ in daily CH_4 production (g/d) (Table 1). However, the rate of CH_4 emission per unit of digestible OMI (g/kg DOMI) was lower in the HSR system than in the LSR system (Table 1). The latter was probably due to the differences between systems in feed OMI and OMD (both higher in HSR), suggesting that the high herbage mass observed in the LSR system had adverse effects upon OMI and OMD. There was no clear pattern of period (P) effects upon CH_4 emission in either grazing season.

Table 1 Effects of stocking rate treatments (T; LSR and HSR), periods (P) and T×P interaction on CH_4 emission by heifers, expressed as g/d or g/kg digestible OM intake (DOMI)

Periods		P1	P2	P3	P4	s.e.m.	T	P	T × P
				Grazing season 2002					
g/d		224.7ac	237.6a	204.4c	263.2b	8.0	ns	***	ns
g/kg DOMI									
	LSR	32.0aA	41.3bcA	47.7cA	38.3abA	2.6	ns	ns	***
	HSR	40.7aB	31.0bB	34.1abB	35.3abA				
				Grazing season 2003					
g/d									
	LSR	162.7aA	208.5bA	213.1bcA	229.2cB	6.6	ns	***	**
	HSR	176.6aA	215.5bA	207.8bA	199.4bA				
g/kg DOMI									
	LSR	36.5a	40.3a	37.0a	36.9a	2.9	*	**	ns
	HSR	33.6a	39.2a	29.3ab	22.4b				

a–c, means within a row which do not share a common letter are significantly ($P < 0.05$) different
A,B, means within the same column with different letters are significantly ($P < 0.05$) different
*, **, ***, ns indicates statistical significance (*, $P < 0.05$; **, $P < 0.01$; ***, $P < 0.001$; ns, not significant)

Conclusion The results of this study are in general agreement with those found by McCaughey et al. (1997) and suggest that SR effects upon CH_4 emission should be evaluated on the basis of animal production efficiency.

Acknowledgement Research part of the *Greengrass* Project (European Contract N° EVK2-CT2001-00105).

Reference
McCaughey, W.P., K. Wittenberg & D. Corrigan (1997). Methane production by steers on pasture. *Canadian Journal of Animal Science*, 77, 519-524.

Cattle slurry amended with nitrification inhibitors: effects on nitrous oxide, dinitrogen and methane emissions

J.P. Carneiro[1], L. Cardenas[2], D. Hatch[2], H. Trindade[3], J. Hawkins[2], D. Scholefield[2] and D. Chadwick[2]

[1]*Escola Superior Agrária de Castelo Branco, Qta. Sra. de Mércules, Ap.119, 6001-909 Castelo Branco, Portugal, Email: jpc@esa.ipcb.pt, [2]Institute of Grassland and Environmental Research, North Wyke, Okehampton EX20 2SB, UK, [3]Universidade de Trás os Montes e Alto Douro, Ap.1013, 5001-911 Vila Real, Portugal*

Keywords: nitrous oxide, slurry, methane, inhibitor, nitrification, denitrification

Introduction In recent decades, a very intensive dairy farming system has been developed in northern Portugal. Considering the appreciable amounts of slurry generated and the small farm areas for spreading, this activity involves large annual inputs of N with the risk of undesirable consequences for the environment. Emission of greenhouse gases from agricultural sources, such as N_2O (originating from nitrification and denitrification) and CH_4, therefore need to be reduced. The aim of this laboratory experiment was to evaluate the effectiveness of two nitrification inhibitors (dicyandiamide (DCD) and 3,4-dimethylpyrazole phosphate (DMPP)) in reducing nitrification and subsequent denitrification after being added to cattle slurry.

Materials and methods Nitrification inhibitors were added along with sieved (4 mm) cattle slurry to a sieved (2 mm) arable soil (Cambisol) collected from the 0-20cm layer and incubated under simulated autumn conditions for Portugal of temperature (12h day/12h night cycle of 20/10°C) and water content (90% field capacity). Soil mineral N analyses gave: 0.86 mg NH_4^+-N and 4.62 mg NO_3^--N (/kg dry soil). Total N of the slurry was 0.13% (fresh basis). Treatments, with three replicates, consisted of a control (without slurry), slurry treated with DCD (at 5% slurry-N), slurry treated with DMPP (at 1% slurry-N) and slurry only. Slurry was applied at the equivalent of 70 m^3/ha (~90 kg N/ha). An automated laboratory incubation system permitted direct, independent measurements of N_2O, N_2 and CH_4 fluxes, which were measured approximately every 2 h over a period of 36 d. After 32 days incubation, (phase 1, nitrifying conditions), the O_2 supply to the incubation system was removed to induce anaerobic conditions in the soils (phase 2, denitrifying conditions).

Results During phase 1, N_2O emissions were only evident in soil receiving slurry without inhibitors, with peak emissions on days 15-17 (~ 0.22 kg N/ha per d). There were clear diurnal patterns in emissions that were highly correlated with temperature. There were no discernible emissions of N_2, and CH_4 was only emitted during the first 4.5 d from the slurry-treated soils. Significantly greater amounts of CH_4 were emitted from soil receiving slurry +DCD (0.31 kg C) compared with slurry only treatment (0.18 kg C). In phase 2, there were much greater amounts of N_2O from soil receiving slurry only and control treatments (~1.0 kg N/ha per d) compared with soils receiving slurry plus inhibitors (~0.1 kg N/ha per d). During this phase, emissions of N_2 were greatest with the two inhibitor treatments, reaching 1.0-1.5 kg N/ha per d with no emissions of CH_4. Considering the results of both phases, the amounts of gaseous N emitted were 4.8, 5.8, 3.9 and 4.5 kg N/ha from the control, slurry-only, slurry +DCD and slurry +DMPP treatments, respectively.

Conclusions Results show that, relative to the slurry only treatment, the N_2O emissions were 62%, 4% and 10% from soils receiving the control, slurry +DCD and slurry +DMPP treatments, respectively. Considering the total N flux emitted, the differences between treatments were less evident (83%, 68% and 78%, respectively). Nevertheless, nitrification inhibitors provided important reductions in N emissions, particularly when considering effects on global warming. However, the amounts of CH_4-C emitted from the slurry plus inhibitor treatments were greater than those from the slurry only treatment (with DCD exceeding DMPP emissions by 29%) and this needs to be set against the apparent benefits of inhibitors.

Methane emissions of dairy cows in Irish spring-calving, grass based production systems

F.P. O'Mara and D.K. Lovett
Department of Animal Science, Faculty of Agri-Food and the Environment, University College Dublin, Belfield, Dublin 4, Ireland, Email: frank.omara@ucd.ie

Keywords: methane, enteric fermentation, dairy cows

Introduction Countries which sign up to the Kyoto Protocol on greenhouse gas emissions are required to return annual inventories of greenhouse gas emissions, including methane emissions from farm livestock. Countries with large agricultural emissions are encouraged to derive specific emission factors (annual production of methane per animal in a specified animal class) for their livestock populations, rather than relying on default emission factors provided by IPCC (1996). In this paper, we describe how an emission rate of methane from enteric fermentation for spring-calving dairy cows in Irish production systems has been derived.

Materials and methods The process had four main steps. Firstly, the country was divided into three regions: 1) south, 2) east, and 3) north and west. The number of cows in each region was obtained from CMMS (2003). The production system in each region was defined (data source in parentheses) in terms of calving date (CMMS, 2003), dates of winter housing and spring turn-out to grass (expert opinion), milk yield and composition (national statistics), concentrate feeding level (a farm survey conducted in 2003 involving 215 farms with spring-calving dairy cows), forages fed (farm survey) and cow live weight (expert opinion). Secondly, the daily energy requirement of cows in each region was calculated by month based on maintenance requirements, milk yield and composition, requirements for foetal growth, and gain or loss of bodyweight (Jarrige, 1989). Thirdly, the composition of the diet of cows in each region was described by month and daily intake was calculated by reference to the daily energy requirement. The concentrate allowance was fixed while forage intake varied according to energy requirements. Fourthly, daily methane emissions were calculated from metabolisable energy intake using the Mitscherlich two equation of Mills et al. (2003), however a constant of 0.065 of gross energy intake was applied when the diet was grazed grass plus 3 kg or less of concentrate supplement/day. The daily emissions were summed to give annual emissions per region, a weighted national average was then calculated.

Results and discussion The predominant feeding system in all regions was grass silage only prior to calving, grass silage supplemented with concentrates (5.3, 5.9 and 6.3 kg/d in regions 1, 2 and 3, respectively) post-calving but before turnout, and minimal concentrates during the grazing season. Annual milk yield was 4700 kg/cow (assumed to be the same between regions). Feed intake and methane emissions are shown in Table 1. There was little difference between regions in methane emissions, but the average national value of 106.8 kg/yr is higher than the IPCC (1996) default value of 100 kg. This has implications for Ireland's greenhouse gas inventory, but to determine the full implication, an equivalent figure for 1990 needs to be derived.

Table 1 Regional feed intake, calving date and methane emissions of spring calving dairy cows in Ireland

	No of cows (000)	Mean calving date	Annual DM intake (kg/hd) Grass	Silage	Concentrate	Annual CH_4 emissions (kg/hd)
South	629	12 March	3005	1279	463	106.6
East	350	15 March	2594	1644	556	106.8
North & west	96	18 March	2147	1911	767	107.6

Acknowledgement The authors acknowledge the support of the Environmental RTDI Programme 2000-2006, financed by the Irish Government under the National Development Plan and administered on behalf of the Department of the Environment and Local Government by the Environmental Protection Agency.

References
CMMS (2003). Cattle Movement and Monitoring System Statistics Report. The Department of Agriculture and Food, Dublin.
IPCC (1996). Greenhouse Gas Inventory Revised Methodology. Guidelines for National Greenhouse Gas Inventories. Vol. 3. Bracknell, UK.
Jarrige, R. (1989). Ruminant Nutrition. Recommended Allowances & Feed Tables. John Libbey Eurotext, London-Paris.
Mills, J.A.N., E. Kebreabs, C.M. Yates, L.A. Crompton, S.B. Cammell, M.S. Dhanoa, R.E. Agnew & J. France 2003. Alternative approaches to predicting methane emissions from dairy cows. *Journal of Animal Science*, 81, 3141-3150.

An examination of the diurnal variability in nitrous oxide emissions

B.P. Hyde, A.F. Fanning, M. Ryan and O.T. Carton
Teagasc Environmental Research Centre, Johnstown Castle, Wexford, Ireland, Email: bhyde@johnstown.teagasc.ie

Keywords: diurnal variability, nitrous oxide, temperature

Introduction It is generally assumed in field experiments, that the measurement of nitrous oxide (N_2O) using enclosed chambers for a period of 1 hour can be used to provide an estimate of daily emission rates. In the majority of studies, emission measurements are conducted between 0900 and 1300 h. However, clearly defined diurnal cycles in N_2O emission rates have been observed from both agricultural and forest soils in temperate regions as a consequence of diurnal fluctuations in temperature (Blackmer *et al.*, 1982; Ball *et al.*, 1999; Baggs *et al.*, 2002). The objective of this study was to quantify the diurnal variation in N_2O emissions from a grassland soil receiving two rates of nitrogen (N) fertiliser inputs under ambient spring and summer conditions.

Materials and methods A system of six mini-lysimeters was constructed using soil cores (150 mm in diameter and 100 mm deep) taken from a perennial ryegrass sward (*L. perenne* cv Tyrone). Two N treatments were imposed at rates equivalent to 100 and 150 kg N/ha 2 days before the start of the measurement period. The study was undertaken during the spring and summer of 2003. Emission measurements were undertaken on a 2-hourly basis for 2 full days (i.e., 48 hours) in each experimental period. Nitrous oxide concentrations were measured using gas chromatography. Soil surface and air temperature were monitored throughout. The effect of N treatment on diurnal emission rates in both spring and summer was explored using residual maximum likelihood variance components analysis (REML). Pearson correlation coefficients were used to identify relationships between N_2O emissions and temperature.

Results Clearly defined diurnal cycles were identified in both experimental periods which were not affected by N treatment. There were significant differences in emission rates between N treatments within each period. Diurnal variations in N_2O emission rates in spring were markedly out of phase with diurnal variations in soil temperature with minimum and maximum emission rates occurring at midday and night-time, respectively. Minimum and maximum N_2O emission rates as measured on day 1 were 67 and 117 µg N_2O-N /m^2/h, respectively, at 100 kg N/ha and 49 and 61 µg N_2O-N /m^2/h, respectively at 150 kg N/ha on day 1. Minimum and maximum emission rates on day 2 were 70 and 353 µg N_2O-N/m^2/h, respectively, at 100kg N/ha and 48 and 220 µg N_2O-N/m^2/h, respectively, at 150 kg N/ha. Correlations were weak and negative and non-significant at the 95% confidence interval. Soil moisture content and soil mineral N status would appear to have been the overriding contributing factors. In contrast, N_2O emissions followed diurnal variations in soil temperature with a time-lag of four hours between maximum soil temperature and maximum emission rates during the summer measurement period. Maximum emission rates were found in early afternoon with minimum emission rates occurring in early morning. Minimum and maximum N_2O emission rates, as measured on day 1, were 1361 and 5952 µg N_2O-N/m^2/h, respectively, at 100 kg N/ha and 3171 and 6582 µg N_2O-N/m^2/h, respectively, at 150 kg N/ha. Minimum and maximum emission rates on day 2 were 318 and 2230 µg N_2O-N/m^2/h, respectively, at 100kg N/ha and 1123 and 5040 µg N_2O-N/m^2/h, respectively, at 150 kg N/ha. Correlations with temperature were positive and significant at the 95% confidence interval.

Conclusions The diurnal variations in N_2O emission rates observed were due to temperature, soil moisture content and mineral N status. The variations observed confirm the difficulties in assessing daily emission values based on a single measurement for a short period during the 24-hour day. However, this is often the only logistically feasible approach in large scale-field experiments. Low cost methodologies need to be developed which enable the continuous measurement of N_2O emission rates using enclosed chambers.

Acknowledgements This work was part funded under the Irish Government's ERTDI programme 2000 – 2006.

References

Baggs, E.M., R.M. Rees, K. Castle, A. Scott, K.A. Smith & A.J.A. Vinten (2002). Nitrous oxide release from soil receiving N-rich crop residues and paper mill sludge in eastern Scotland. *Agriculture, Ecosystems and Environment*, 90, 109-123.

Ball, B.C., A. Scott & J.P. Parker (1999). Field N_2O, CO_2 and CH_4 fluxes in relation to tillage, compaction and soil quality in Scotland. *Soil and Tillage Research*, 53, 29-39.

Blackmer, A.M., S.G. Robbins & J.M. Bremner (1982). Diurnal variability in the rate of emission of nitrous oxide from soils. *Soil Science Society of America Journal*, 46, 937-942.

The effect of soil type and climate on modelled greenhouse gas emissions derived from pasture based milk production systems

D.K. Lovett[1], L. Shalloo[2], P. Dillon[2] and F.P. O'Mara[1]
The Department of Animal Science, University College Dublin, Lyons Research Farm, Co. Dublin, Ireland, Email: dan_lovett@talk21.com, [2]Dairy Production Department, Teagasc, Moorepark Production Research Centre, Fermoy, Co. Cork, Ireland

Keywords: pasture, soil type, climate, greenhouse gas emissions

Introduction The ability of spring calving dairy farmers to exploit herbage production can be limited by soil type and climatic conditions. Previous work, using the Moorepark Dairy System Model (MDSM) (Shalloo *et al.*, 2004) demonstrated differences in terms of biological and production efficiency for two contrasting sites. This study models whole farm greenhouse gas (GHG) emissions (Lovett et al., in press) from two dairy systems, the Moorepark Standard System (MSS, Co. Cork) and the Kilmaley Standard System (KSS,Co. Clare), classified as lower and high rainfall(1025 and 1614 mm yr) and free draining versus poor draining respectively.

Materials and methods Farm size, herd size and N application required to fill 468,000 kg of milk quota of 36.0 g kg^{-1} fat differed between sites (36.4 and 50.2 ha^{-1}, 74.49 and 85.85 cows and 330 and 238 kg N ha^{-1} yr^{-1} for the MSS and KSS systems respectively). The grazing period of MSS was 250 days while that of KSS was 150 days and consequently the proportion of grazed grass within the diet differed (0.693 versus 0.440). Using this production data, methane (CH_4), nitrous oxide (N_2O) and carbon dioxide (CO_2) emissions were predicted using emission factors relevant to Northern Europe (Lovett et al., submitted). Both direct on-farm emissions (e.g. CH_4 from enteric fermentation) and indirect emissions (those associated with inputs used to maintain farm productivity, for example CO_2 equivalents associated with the production of concentrates) were predicted.

Results Annual, direct and indirect greenhouse gas emissions were 373.8 and 455.9 and 534.8 and 626.3 Mg CO_2 equivalents for the MSS and KSS systems respectively. The source of these emissions can be seen below in Table 1. Methane, N_2O and CO_2 emissions at MSS were 0.794, 0.957 and 0.918 respectively of those for the KSS. Relating animal productivity to GHG emissions revealed that emissions were lower at MSS (0.794 and 1.143 kg CO_2 equivalents per kg milk for direct and total emissions) than at KSS (0.969 and 1.338 kg CO_2 equivalents for direct and total emissions). Farm profitability before tax, as predicted by the MDSM (Shalloo *et al.*, 2004), was lower for the KSS (€18,270) than the MSS (€46,765).

Table 1 Predicted GHG emissions (CO_2 equivalents) for the Kilmaley and Moorepark standard systems

System	KSS	MSS
Direct sources of GHG emission		
Enteric fermentation	262.31	233.05
Silage effluent	53.27	25.90
Slurry/FYM storage and spreading	67.47	42.26
Excreta at pasture	8.56	12.70
Fertiliser/lime application	45.26	46.57
Diesel use	19.04	13.29
Soil sink	-2.64	-1.91
Indirect sources of GHG emission		
GHG via N leaching	41.25	39.69
GHG via NH_3	9.52	6.71
GHG associated with electricity	18.99	20.74
GHG from fertiliser production	79.65	82.50
GHG from concentrate feeds	16.72	8.53
GHG associated with diesel/lime	6.92	4.80

Conclusion Emissions of GHG's per unit of saleable product were greater at the high rainfall poor-draining Kilmaley site than the lower rainfall free-draining Moorepark site. This is principally due to the longer housing period and lower milk yields per animal for the KSS combining with a greater dependence on conserved silage and purchased concentrate feeds relative to the MSS. Increased geographic concentration of dairy herds in the South West of Ireland can be expected due to enhanced financial returns in the future. This could be expected to reduce GHG emissions derived from the national dairy herd.

References

Lovett, D.K., L. Shalloo, P. Dillon, F.P. O'Mara. A systems approach to quantify greenhouse gas fluxes from pastoral dairy production as affected by management regime. *Agricultural Systems*, (in press).

Shalloo, L., P. Dillon, J. O'Loughlin, M. Rath & M. Wallace (2004). Comparison of a pasture-based system of milk production on a high rainfall, heavy-clay soil with that of a lower rainfall, free-draining soil. *Grass and Forage Science,* 59, 157-168.

The effect of dairy cow genotype on modelled greenhouse gas emissions derived from pasture based milk production systems

D.K. Lovett[1], L. Shalloo[2], P. Dillon[2] and F.P. O'Mara[1]
The Department of Animal Science, University College Dublin, Lyons Research Farm, Co. Dublin, Ireland
Email: dan_lovett@talk21.com, [2]Dairy Production Department, Teagasc, Moorepark Production Research Centre, Fermoy, Co. Cork, Ireland

Keywords: greenhouse gas emissions, pastoral milk production, genotype, systems

Introduction A three-year systems comparison study was undertaken to see if progressively increasing the genetic potential for milk production of the dairy cow is desirable within pastoral based systems of spring milk production (Kennedy *et al.*, 2002). The production data was inputted into the Moorepark Dairy System Model (MDSM) (Shalloo *et al.*, 2004) to describe the economic, biological and production efficiency of each system. Output was then used to model whole farm greenhouse gas (GHG) emissions (Lovett et al., in press) from the nine systems studied (three concentrate levels by three genotype levels). Only the genotype effects are reported.

Materials and methods Based on their pedigree index (LP, < 100 kg, MP, 100 to 200 kg and HP, 200 to 300 kg), dairy cows of three different genetic potentials were fed three different levels of concentrate (376, 810 and 1540 kg cow yr^{-1}). Consequently, farm and herd size required to fill 468,000 kg of milk quota (at 36.0 g kg^{-1} fat) differed (29.7 to 39.10 ha^{-1} and 57.43 to 74.49 cows respectively). Annual N application rate and grazing days were fixed at 250 days and 330 kg N/ha. Using this production data, methane, nitrous oxide and carbon dioxide emissions were predicted using emission factors relevant to Northern Europe (Lovett et al., submitted). Both direct on-farm emissions (e.g. CH_4 from enteric fermentation) and indirect emissions (those associated with inputs used to maintain farm productivity such those associated with concentrate production) were simulated.

Results Annual, direct and total (direct and indirect) GHG were 351.6, 356.4 and 368.3 and 503.6, 510.1 and 527.1 Mg CO_2 equivalents for the LP, MP and HP systems respectively. The distribution of some of the GHG emissions by livestock category can be seen below in Table 1. Emissions of GHG's in relation to animal productivity increased with increasing genotype from 0.748 to 0.783 and 1.076 to 1.126 kg CO_2 equivalents per kg milk for direct and total emissions respectively. Farm profitability before tax, as predicted by the MDSM (Shalloo *et al.*, 2004), was €44,446, €43,935 and €43,344 for the LP, MP and HP systems respectively.

Table 1 Predicted GHG emissions (Mg CO_2 equivalents) for spring based milk production as affected by genotype

System	LP	MP	HP
Direct GHG emissions	351.6	356.0	368.3
Indirect GHG emission	503.6	510.1	527.1
Direct sources of GHG emissions by herd category			
Lactating cows			
Enteric fermentation	185.2	180.1	176.2
Slurry/FYM storage and spreading	32.1	30.7	29.9
Excreta at pasture	9.8	9.6	9.4
Total	227.1	220.4	215.5
Non-lactating cows			
Enteric fermentation	35.6	44.1	56.1
Slurry/FYM storage and spreading	3.9	4.8	6.1
Excreta at pasture	2.3	2.9	3.7
Total	41.8	51.8	65.9

Conclusion Improving cow genotype for milk production reduces GHG emissions from the lactating herd. However, reduced herd fertility means that to fulfil quota allowances increased numbers of non-productive cattle must be maintained. Consequently, GHG emissions associated with the non-productive herd increase, with this increase being in excess of the reduction achieved within the lactating herd. Changes in breeding policy may have implications for national GHG emissions. In particular the recent introduction of a balanced breeding index to Ireland to simultaneous improve milk yield and fertility could reducing whole farm dairy sector GHG emissions.

References

Kennedy, J., P. Dillon, P. Faverdin, L. Delaby, F. Buckley, & M. Rath (2002). The influence of cow genetic merit for milk production on response to level of concentrate supplementation in a grass-based system. *Animal Science*, 75, 433-445.

Lovett, D.K., L. Shalloo, P. Dillon, F.P. O'Mara. A systems approach to quantify greenhouse gas fluxes from pastoral dairy production as affected by management regime. *Agricultural Systems,* (in press).

Shalloo, L., P. Dillon, J. O'Loughlin, M. Rath & M. Wallace (2004). Description and validation of the Moorepark Dairy System Model. *Journal of Dairy Science*, 87, 1945-1959.

Methane of animal origin in cattle fed high or low tannin sorghum silage

T.T. Berchielli[2], S.G. Oliveira[2], M.P. Pedreira[2], O. Primavesi[3], M.A. Lima[3] and R. Frighetto[3]
[2]*Faculdade de Ciências Agrárias e Veterinárias – Unesp, Via de Acesso Prof. Paulo Donato Castellane, Mail Code 14884-900, Jaboticabal – SP, Brazil, Email: ttberchi@fcav.unesp.br, [3]Empresa Brasileira de Pesquisa Agropecuária, Rod. Washington Luiz, Km 234, Caixa Postal 339, CEP 13560-970, São Carlos – SP, Brazil*

Keywords: beef cattle, intake, methane, sorghum silage, tannin

Introduction The harmful or beneficial effects associated with the presence of tannin in plants depends on its concentration and form. One of these beneficial effects is the reduction of the production of methane in the rumen (Woodward et al., 2001). This study financed by FAPESP aimed to evaluat the effect of diets containing low or high tannin sorghum silage supplements on dry matter intake (DMI) and methane production in cattle.

Material and methods Eight Nelore steers were fed low or high tannin sorghum silage plus additional supplements of urea or concentrate. The methane eructed by the animals was collected by means of a capillary tube connected to a PVC tube adapted to the neck of the animals according to the methodology of Johnson & Johnson (1995).

Results Dry matter intake was higher for animals receiving diets supplemented with concentrate (Table 1). Roughage source had no effect on DMI and methane production suggesting that there was no effect of tannin concentration in the evaluated silages (Table 2). Diets containing 60% concentrate yielded the highest methane production per animal and per kg live weight, this is in agreement with the study of Berchielli *et al.* (2003). Relating methane production to DMI revealed no difference between forage or supplement sources, suggesting that methane production per animal should not be considered as an isolated factor.

Table 1 Dry matter intake and methane production in beef cattle

Treatments		DMI (kg/d)	Methane (g animal/d)	Methane (g/kg LW)	Methane (g/kg DMI)
LTSU		3.50	49.52	0.22	14.02
LTSC		5.82	66.63	0.31	11.66
HTSU		3.72	49.27	0.22	12.73
HTSC		5.83	70.44	0.32	11.62
Coefficient of variation (%)		35.60	41.84	27.84	21.65
			Main Effects		
Roughage	SBT	4.66	58.07	0.26	12.84
	SAT	4.78	59.85	0.27	12.18
Supplement	Concentrate	5.83 b	68.53 b	0.31 b	11.64
	Urea	3.61 a	49.40 a	0.22 a	13.38

LTSU – Low tannin silage + urea, LTSC – Low tannin silage + concentrate, HTSU – High tannin silage + urea, HTSC – High tannin silage + concentrate.
Means followed by different letters in the same column are significantly different at 5 %, according to the test of Tukey.

Conclusions Increasing the concentration of tannins within the silage did not promote any reduction in the production of methane. The production of methane as a function of DMI indicates that the use of dietary energy must be maximized so that methane production during the production cycle can be reduced in relation to production per unit of animal product.

References

Berchielli, T.T., M.P. Pedreira, & S.G Oliveira (2003). Determinação da produção de metano e pH ruminal em bovinos de corte alimentados com diferentes relações volumoso:concentrado. In: *Reunião Anual da Sociedade Brasileira de Zootecnia*, Anais. Santa Maria: SBZ, CD-ROM.

Johnson, K.A., & D.E. Johnson (1995). Methane emissions from cattle. *Journal of Animal Science*, 73, 2483-2492.

Woodward, S.L., G.C. Waghorn, & M.J. Ulyatt (2001). Early indications that feeding Lotus will reduce methane emissions from ruminants. *Proceedings of the New Zealand Society of Animal Production*, 61, 23-26.

Field measurements of ruminal methane of cattle grazing tropical grasses

M.S. Pedreira[2], O. Primavesi[3], M.A. Lima[3], R. Frighetto[3], T.T. Berchielli[2]
[1]Project granted by FINEP, [2]Faculdade de Ciências Agrárias e Veterinárias – Unesp, Via de Acesso Prof. Paulo Donato Castellane, mail code 14884-900, Jaboticabal – SP, Brazil, Email: ttberchi@fcav.unesp.br, [3]Brazilian Agricultural Research Corporation - Embrapa, Rod. Washington Luiz, Km 234, Caixa Postal 339, mail code 13560-970, Sao Carlos – SP, Brazil

Keywords: dairy cattle intake, pasture, sulphur hexafluoride

Introduction Ruminal methane production represents energy losses from ingested feed that should be utilised to maintain body weight or to generate products. Quantitatively, daily methane production varies according to amount and quality of ingested dry matter (DM), as well as physiological status of the animal.

Material and methods Holstein and Zebu cross-bred heifers, and dry and lactating cows, grazing fertilised *Panicum maximum* cv. Tanzania and *Brachiaria decumbens*, as well as unfertilised *B. decumbens* during summer and autumn were used in this study. Measurements of methane emissions were taken directly, using the sulphur hexafluoride tracer gas method, as described by Johnson & Johnson (1995).

Results Methane production, in g/animal/day (Table 1), was greater from cows than heifers, from lactating cows than dry cows, and greater from Holstein heifers and lactating cows than Zebu-bred cattle of the same categories. Data match findings of Holter & Young (1992), who reported that different methane emission rates occurred among breeds and animal categories, mainly as function of the size of the gastric compartments, and of animal nutritional requirements. Heifers did not present variations in methane production as a function of the forages used. Methane production (g/kg of $LW^{0.75}$) was different for contrasts VC vs. NV, VL vs. VS and NH vs. NM, but not for the contrasts VLH vs. VLM, VSH vs. VSM and heifers of both breeds and on both pastures. The same behaviour was observed with methane when expressed as g/kg of DM intake.

Table 1 Contrasts among categories, breeds and pastures for mean methane production by dairy cattle

Contrasts	Methane production		
	g/d	g/kg of $LW^{0.75}$	g/kg of DM intake
VC vs. NV	311.3 vs. 200.9*	2.83 vs. 2.1*	21.3 vs. 17.5*
VL vs. VS	353.8 vs. 268.8*	3.3 vs. 2.3*	23.2 vs. 19.3*
VLH vs. VLM	393.2 vs. 314.5*	3.36 vs. 3.2	21.5 vs. 24.9
VSH vs. VSM	271.1 vs. 266.4	2.17 vs. 2.5	17.8 vs. 20.8
NH vs. NM	205.7 vs. 196.1*	2.0 vs. 2.2*	16.4 vs. 18.6*
Nhi vs. NHe	233.6 vs. 177.8	2.18 vs. 1.8	18.0 vs. 14.8
Nmi vs. NMe	211.6 vs. 180.6	2.44 vs. 2.1	20.1 vs. 17.2

* Significant by Student test (P<0,05). Contrasts – cows and heifers (VC vs. NV), lactating cows and dry cows (VL vs. VS), Holstein and Zebu-bred lactating cows (VLH vs. VLM), Holstein and Zebu-bred dry cows (VSH vs. VSM), Holstein and Zebu-bred heifers (NH vs. NM), Holstein heifers on Panicum + concentrate and Brachiaria (NHi vs. NHe) and Zebu-bred heifers on Panicum + concentrate and Brachiaria (NMi vs. NMe).

Conclusions Methane production varied as a function of physiologic stage of animals and breed. Methane production by heifers grazing forages with different qualities supplemented or not with grain concentrate did differ.

References

Johnson, K.A. & D.E. Johnson (1995). Methane emissions from cattle. *Journal of Animal Science*, 73, 2483-2492.
Holter, J. B. & A.J. Young (1992). Nutrition, feeding and calves: methane prediction in dry and lactating Holstein cows. *Journal of Dairy Science*, 75, 2165-2175.

A systems approach to managing greenhouse gases on New Zealand sheep and beef farms

M.G. Lambert and H. Clark
AgResearch Grasslands, PB 11008, Palmerston North, New Zealand, Email: greg.lambert@agresearch.co.nz

Keywords: greenhouse gases, sheep, beef, systems

Introduction Agriculture contributes more than 50% of New Zealand's greenhouse gas (GHG) emissions, mainly through release of methane (CH_4) and nitrous oxide (N_2O) from pastoral farms. Decisions on implementation of mitigation strategies will be made by individual farmers, who seek also to maintain financial performance in the face of declining terms of trade for commodities, hence leading to pressure to further intensify production. New Zealand (NZ) sheep & beef farms are typically hill country properties with a mixture of steep and easier topography, and year-round grazing of mainly permanent pastures. Specific GHG mitigation technologies will be difficult to incorporate into these systems and most of them are a considerable distance from commercial application. Farming system change is currently the only effective mitigation approach likely to be readily available to those farmers in the short term.

Materials and methods Methane and N_2O emissions were calculated for a theoretical 500 ha hill country sheep & beef farm operating under 3 different farming scenarios involving increasing levels of fertiliser (phosphorus, sulphur and nitrogen) inputs, lambing percentage, forage crop use, a change in enterprise design from all breeding and sale of offspring to finishing of lambs for slaughter and the purchase of young cattle for finishing in preference to maintenance of a breeding cow herd to generate young cattle. Low, Medium and High intensity scenarios were simulated using the whole-farm decision-support model Stockpol to predict production and economic outcomes, and the nutrient budgeting decision-support model Overseer to predict GHG outcomes. Effects of replacement of areas of pasture with plantation forestry on the steepest, least productive areas of the farm, to mitigate increases in GHG emissions as a result of intensification through carbon (C) sequestration, were also estimated.

Results Production (feed intake, meat & wool production) and financial returns increased as level of inputs increased across the Low to High scenarios (Table 1). Similarly, system efficiencies (meat & wool produced per unit feed and per unit GHG emission, and meat & wool and $ earned per unit GHG emission) increased. However, CH_4 and N_2O emissions also increased. The markedly increased N_2O emission for the High scenario was a consequence of increased fertiliser N use.

Table 1 Results of farming scenario simulations for Low, Medium and High intensification scenarios

	Low	Medium	High
Feed intake (t DM/ha per yr)	6210	7719	10501
Net meat & wool production (kg/ha per yr)	163	258	366
Financial returns ($ gross margin/ha per yr)	411	601	934
Feed to product conversion efficiency (kg product/t feed intake)	26.3	33.4	34.8
Production per unit GHG emission (kg product/t CO_2 equivalent)	47.3	59.8	60.9
Financial return per unit GHG emission ($/t CO_2 equivalent)	119	139	156
Methane emissions (t CO_2 equivalent/ha)	2.76	3.45	3.78
Nitrous oxide emissions (t CO_2 equivalent/ha)	0.70	0.86	2.22
Total GHG emissions (t CO_2 equivalent/ha)	3.46	4.31	6.00

Conversion of 30 ha and 81 ha of pasture on Medium and High intensity scenarios respectively to plantation forestry, maintained total CH_4 and N_2O emissions at the same level as for Low intensity but resulted in a reduction in livestock returns of $13,000 and $61,000 for Medium and High respectively; however balanced against this were average (over the production cycle) annual returns from sale of timber of $30,000 and $81,000 respectively. Cash flow for costs and returns of forestry enterprises differ for annual livestock and 30-year forestry production cycles, and this would be an issue in transition to the combination of livestock and forestry.

Conclusions There are existing feasible system solutions involving changed land use that would enable New Zealand's Kyoto commitments to be met whilst meeting farmer requirements for ongoing financial viability. Design and implementation of such systems would require that whole-farm GHG budgets pay regard not just to CH_4 and N_2O emissions but also to C sequestration. Although CH_4 is currently the most important agricultural GHG in New Zealand, increased N fertiliser use will markedly increase the relative importance of N_2O.

Hedgerow systems and livestock in Philippine grasslands: GHG emissions

D.B. Magcale-Macandog, E. Abucay, R.G. Visco, R.N. Miole, E.L. Abas, G.M. Comajig and A.D. Calub
Institute of Biological Science, University of the Philippines Los Baños, College, Laguna, Philippines, Email: macandog@pacific.net.ph

Keywords: hedgerow system, livestock, N_2O emission, methane emission, GHG emissions

Introduction Hedgerow systems are widely adopted in the smallholder farms in the sloping grassland areas of Claveria, Mindanao, Philippines. The system is effective in addressing soil erosion problems and in conserving the topsoil. *Gmelina arborea* and *Eucalyptus deglupta* are two fast-growing timber species that are planted in hedgerow systems while maize is planted in the alley areas in between the hedgerows. Livestock holdings are widespread in Claveria, with 74% of the households having livestock. Cattle and carabao are the most common livestock in smallholder farms providing draught power for land preparation and transportation. In hedgerow systems, fodder tree leaves and crop residues are fed to livestock, while animal manure is added to the soil. Thus, these systems may serve as both a source and sink of methane and nitrogen oxides, depending on the management practices and component trees and crops of the system. This study aims to estimate methane emissions from livestock holdings and nitrogen oxide emissions through fertilization, tree litterfall and decomposition, maize residue incorporation and livestock manure from *G. arborea* and *E. deglupta* hedgerow systems.

Materials and methods Experimental plots were established in 1 and 7-yr old *E. deglupta*- and *G. arborea*-hedgerow systems with maize planted in the alley areas. The treatments are different combinations of tree species, tree age, and tree spacing. Inorganic N and P fertilizer, and maize crop residues were applied in the maize crop. Maize biomass, grain yield, tree litterfall and leaf litter decomposition were measured. A survey of 300 households in Claveria was conducted to gather information on livestock holding and management.

Results The major sources of N inputs in the different hedgerow systems are the maize crop residues (FCR) and synthetic nitrogen fertilizer (FSN) (Fig.1). Other sources include animal manure (FAW) and tree leaf litter. Since the average animal holding is quite small, nitrogen input from animal waste is small. Direct soil N_2O emissions from the plots range from 2.11 to 5.17 kg N ha^{-1}yr^{-1}. Direct soil N_2O emissions from 1-year old hedgerow systems are significantly higher than emissions from 7-year old hedgerow system. Local values for N excretion from cattle and carabao were 12.3 kg and 14.2 kg, respectively; much lower than the default values of 40 kg for both non-dairy cattle and carabao given by IPCC (1997). Enteric fermentation of cattle and carabao (11,352 kg and 3,410 kg, respectively) and swine manure management (2,786 kg) were the main sources of CH_4 emissions from livestock holdings in 300 Claveria households (Table 1).

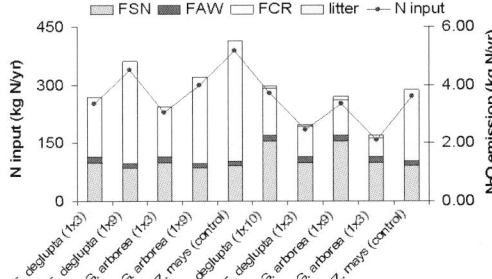

Figure 1 Nitrogen inputs and N_2O emissions in *E. deglupta* and *G. arborea* hedgerow systems

Table 1 Methane (CH_4) emissions from enteric fermentation and manure management per animal type

Livestock Type	Number of animals	EF	Enteric Fermentation	EF	Manure Management
Non-dairy cattle	258	44	11,352	2	516
Carabao	62	55	3,410	3	186
Goat	46	5	230	0.22	10.12
Swine	398	1.5	597	7	2,786
Poultry	1252	-	-	0.023	28.8
Total			15,589		3,526.92

Conclusions In tree-based hedgerow systems, crop residue incorporation and fertilizer application are the major sources of nitrogen inputs. Direct soil N_2O emissions from the plots range from 2.11 to 5.17 (kg N yr^{-1}) with significant N_2O emissions in 1-year old hedgerow systems than 7-year old hedgerow system. Use of local values for N excretion factors will reduce uncertainties in the estimates of N excretion from animal manure. Enteric fermentation of cattle and carabao and swine manure management were the main sources of CH_4 emissions from livestock holdings in 300 Claveria households.

Reference
IPCC (1997). The Revised 1996 Guidelines for National Greenhouse Gas Inventories (Workbook and Reference Manuals). Intergovernmental Panel on climate Change. OECD. Paris, France.

Theme B: Grassland and the environment

Section 16

Carbon sequestration

Quantification of CO_2 exchange in grassland ecosystems of the world using tower measurements, modeling and remote sensing

T.G. Gilmanov[1], M.W. Demment, B.K. Wylie, K. Akshalov, D.D. Baldocchi, L. Belelli, J.A. Bradford, G.G. Burba, R.L. Coulter, W.A. Dugas, W.E. Emmerich, L.B. Flanagan, A.B. Frank, J. Fuhrer, M.R. Haferkamp, M.B. Jones, D.A. Johnson, T. Laurila, A. Lohila, T.P. Meyers, P.C. Mielnick, J.A. Morgan, M. Nasyrov, C.E. Owensby, M.S. Pekour, K. Pilegaard, A. Raschi, N.Z. Saliendra, M.J. Sanz, P.L. Sims, R.H. Skinner, J.F. Soussana, A.E. Suyker, L.L. Tieszen, Z. Tuba, R. Valentini, S.B. Verma and E.A. Laca
[1]*Department of Biology and Microbiology, South Dakota State University, Ag Hall 304, Box 2207B, Brookings, South Dakota, 57007, USA, Email: tagir.gilmanov@sdstate.edu*

Keywords: nontropical grasslands, gross primary production, ecosystem respiration, net CO_2 flux partitioning, light-use efficiency

Introduction Grasslands cover significant areas in nontropical regions, perform essential biogeochemical functions and represent important natural and agricultural resource. Nevertheless, in contrast to forests and agroecosystems, no flux measurement-based global summary of their CO_2 exchange, sequestration potential, and role in mitigation of the greenhouse effect were available.

Materials and methods Data sets of continuous CO_2 flux measurements at 32 towers from nontropical grasslands in North America, Europe and Asia were analysed to estimate the major parameters for CO_2 exchange and light-use efficiency. Tower-derived net CO_2 exchange (F_c) was partitioned into gross primary productivity (P_g) and ecosystem respiration (R_e) components ($F_c = P_g - R_e$) by using physiologically-based analysis of ecosystem-scale light-response functions (Gilmanov *et al.*, 2003; Gilmanov *et al.*, 2004). Methods of nonlinear multivariate analysis were used to identify relationships between P_g, R_e and the normalised difference vegetation index (NDVI) and other factors.

Results Maximum daily rates of P_g and R_e achieved values of 63 g $CO_2/m^2/d$ and 54 g $CO_2/m^2/d$, respectively and were approached not only in highly productive tallgrass prairies, but also in southern arid grasslands in years with high precipitation. Maximum values of annual P_g (5200 g $CO_2/m^2/a$) and R_e (4700 g $CO_2/m^2/a$) were observed in tallgrass prairies of Oklahoma and Texas, respectively. The lowest P_g (< 400 g $CO_2/m^2/a$) was recorded on the grazed mixed prairie in Montana during a drought year. The lowest R_e (< 500 g $CO_2/m^2/a$) was estimated for the semidesert grassland in Uzbekistan. Depending on weather conditions and management (grazing, fire), net ecosystem CO_2 exchange of grasslands varies from +2800 g $CO_2/m^2/a$ (early post-fire succession in tallgrass prairie, strong sink) to less than –1500 g $CO_2/m^2/a$ (grazed mixed/tallgrass prairie during a drought year, a strong source). Maximum values of daily light use efficiency $\varepsilon = P_g$/(incident PAR) were achieved in warm temperate grasslands of Europe and eastern U.S. (ε_{max} = 34 – 40 mmol CO_2/(mol incident PAR)), followed by tallgrass and mixed prairies (ε_{max} = 32), while the lowest light use efficiency values were recorded in the semidesert of Uzbekistan (ε_{max} = 9). Grassland CO_2-flux data showed significant correlation (r) between P_g and R_e and NDVI, with r in the range 0.55 to 0.92 for P_g and NDVI, and 0.45 to 0.94 for R_e and NDVI. Maximum r-values were achieved in ecosystems with highest phytomass (tallgrass prairies, warm temperate grasslands). For some sites statistically significant relationships P_g = f(NDVI, X_{i1}, ..., X_{in}), R_e = g(NDVI, X_{j1}, ..., X_{jm}) were established, where X_i represents some aspect of climate, soils or vegetation. The R^2 for these relationships in many cases achieved values higher than 90%.

Conclusions Superposition of these functions on GIS data layers of NDVI and X_i for areas ecologically similar to the tower sites represents a defensible method for scaling-up tower CO_2 flux measurements allowing regional quantification of CO_2 balance on grasslands with implications for continental and global carbon budgets (Gilmanov *et al.*, 2004; Wylie *et al.*, 2004).

References

Gilmanov, T.G., D.A. Johnson & N.Z. Saliendra (2003). Growing season CO_2 fluxes in a sagebrush-steppe ecosystem in Idaho: Bowen ratio/energy balance measurements and modelling. *Basic and Applied Ecology*, 4, 167-183.

Gilmanov, T.G., D.A. Johnson, N.Z. Saliendra, K. Akshalov & B.K. Wylie (2004). Gross primary productivity of the true steppe in Central Asia in relation to NDVI: Scaling-up CO_2 flux measurements. *Environmental Management*, DOI: 10.1007/s00267-003-9157-7.

Wylie, B.K., T.G. Gilmanov, D.A. Johnson, N.Z. Saliendra, K. Akshalov, L.L. Tieszen, B.C. Reed & E. Laca (2004). Intra-seasonal mapping of CO_2 flux in rangelands of northern Kazakhstan at one-kilometer resolution. *Environmental Management*, DOI: 10.1007/s00267-003-9156-8.

Carbon accumulation under *Brachiaria* pastures in the Brazilian Cerrado in relation to their productivity

R.M. Boddey, S.P. Braz, R.S.M. dos Santos, B.J.R. Alves and S. Urquiaga.
Embrapa-Agrobiologia, C. P. 74.505, Seropédica, 23851-970, RJ, Brazil, Email: bob@cnpab.embrapa.br

Keywords: [13]C abundance, *Brachiaria* pastures, carbon sequestration, productivity

Introduction   Vast areas of land within Brazil has been converted from either the native vegetation of the Amazon and Atlantic forests or the native vegetation (NV) of the central savanna region (the Cerrado) to *Brachiaria* spp pastures. Productive pastures (PPs) in these regions eventually accumulate soil carbon (C) in excess of levels under native vegetation.   However the effect of pasture degradation on soil C stocks has not been reported. We compared soil C stocks under both NV and *Brachiaria* pastures varying in productivity according to the indicators of Oliveira *et al.* (2004) over 5 sites in the Cerrado region.

Materials and methods   The 5 sites (cronosequences) in the Cerrado region were situated in the districts of Luz (Minas Gerais - MG), Chapadão do Sul and Dourados (both in the State of Mato Grosso do Sul - MS), Goiânia (Goiás - GO) and Penápolis (São Paulo - SP). Pasture productivity was ranked according to rate of regrowth and deposition of plant litter, soil microbial biomass and the light fraction of the soil organic matter. Soil organic C was estimated from samples taken for bulk density measurements, texture (sand/silt/clay), total N (Kjeldahl) and total C and ^{13}C abundance. Soil C stocks were corrected for a soil mass equal to that under the NV (0-100 cm depth) at each site (Neill *et al.*, 1997).

Results   The dynamic/biological indicators of pasture productivity identified by Oliveira et al. (2004) were successful in ranking pasture productivity. Productive pastures showed significantly higher C stocks than those under native Cerrado vegetation (NV) at all sites (Table 1). Highest C stocks under NV and PPs were associated with high clay content. Degraded pastures (DPs) had lower C stocks relative to PPs and at sites with coarser textured soils, sometimes lower than under the original NV. In PPs C derived from the Brachiaria (C_{BR}) was found present to the maximum sampling depth (100 cm), but under DPs there was no sign of C_{BR} below 40 cm depth (Figure 1).

Table 1 Indicators of pasture productivity (regrowth and litter deposition – g m^{-2}/14days) and C stocks (to 100cm depth – Mg ha^{-1}) at 2 sites in the Brazilian Cerrado

Site/ parameter	Forest	Productive pasture	Degraded pasture
Luz (MG)			
Regrowth	nd	264**	119
Litter	nd	37*	27
C stock (1m)	57.1	62.6	53.1
Chap[#] (MS)			
Regrowth	nd	235	206
Litter	nd	114**	35
C stock (1m)	117.0	164.6	138.0

[#] Chapadão do Sul (MS), nd (not determined)
* and **, difference between PP and DP not significant at p<0.05 and p<0.01, respectively

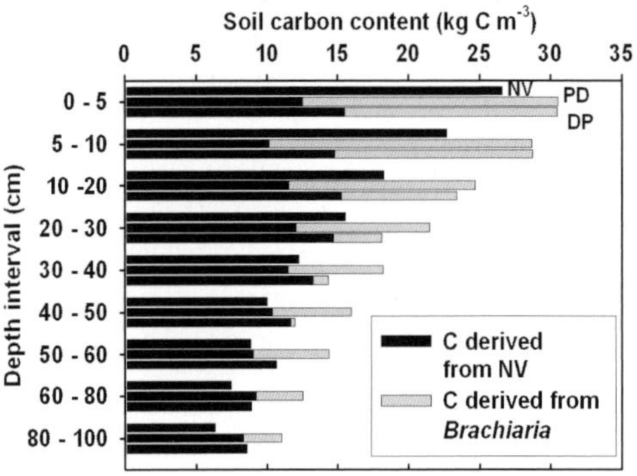

Figure 1 Carbon derived from NV and *Brachiaria* in soil under native Cerrado vegetation and productive (PP) and degraded (DP) pastures. Site: Luz (MG).

Conclusions   Under PPs soil C stocks are high and occur at depths of up to 100 cm. However, when pasture productivity falls due to lack of maintenance and fertiliser application or excessive grazing, C derived from the grass is lost and appears to be almost entirely eliminated from depths greater than 40 cm.

References

Neill,C., Melillo,J.M., Steudler,P.A., Cerri,C., Moraes,J.F.L.d., Piccolo,M.C. & Brito,M. (1997). Soil carbon and nitrogen stocks following forest clearing for pasture in the southwestern Brazilian Amazon. *Ecological Applications*, 7, 1216-1225.

Oliveira, O.C.de, Oliveira, I.P.de, Urquiaga, S., Alves, B.J.R. & Boddey, R.M. (2004). Chemical and biological indicators of decline/degradation of *Brachiaria* pastures in the Brazilian Cerrado. *Agriculture, Ecosystems and Environment*, 103, 289-300.

Comparison of methodological tools in tropical soil Carbon sequestration field research

B. van Putten[1] and M.C. Amézquita[2]

[1]*Department of Mathematical and Statistical Methods, Biometris, Wageningen University and Research Centre, Bornsesteeg 47, 6708 PD Wageningen, The Netherlands, Email: Bram.vanPutten@wur.nl,* [2]*Carbon Sequestration Project - The Netherlands Cooperation CO-010402. CIPAV-U.Amazonia-CIAT-CATIE-Wageningen University. CIAT's Science Park, A. A. # 67-13, Cali, Colombia*

Keywords: model, carbon sequestration, extrapolation, land management system, tropical

Introduction Models play a crucial role in studying complex systems like soil Carbon sequestration processes (Hanson *et al.*, 2001). A Carbon sequestration research project is currently under way in pasture and silvo-pastoral systems of four Tropical American ecosystems. A main research question is the identification of Land Management Systems (LMS) that exhibit optimal soil Carbon sequestration capacity (*identification question*). Another main issue is the extrapolation of data in space and time (*extrapolation question*).

Material and methods From each of the ecosystems, farms were selected, based on having long-established LMS of abovementioned type. Soil Carbon stocks (SCS) were measured in permanent plots equidistantly along transects, at four depths, using two space replications per system and 12 sampling points per system/replication. Observations in the corresponding Native Forest (NF) were made following the same lines. Here we compare statistical techniques with techniques based on chemical/biological/physical process knowledge, represented by process-based simulation (PBS) models like CENTURY, RothC, CANDY, DAISY and DNDC.

Results Data on optimal allocation of observations within plots will be reported elsewhere. An appropriate statistical technique with respect to *identification* is to consider plots of each farm as a random block. As plots are the 'experimental units', the statistically dependent SCS measurements of each plot should be summarized using spatial statistics theory. Statistical inference, based on multiple comparisons hypothesis testing, is possible under the model assumption that, at the farm level, the various LMS were randomly assigned to the plots at the time the NF partially was cleared. Comparison with 'baseline' NF leads to the required solution. In order to obtain sufficient power of the resulting test, one should take care to have sufficient 'experimental units'. As PBS models do not deal with experimental error, and consequently cannot discriminate between systematic effects and random variation, they are not appropriate for solving the identification question. Statistical modelling in principle could be useful for *extrapolation* purposes as well, by using a regression model $y=\beta_0+ \beta_1x_1+ \beta_2x_2+ \beta_3x_3+...+\varepsilon$, where y is SCS, $x_1, x_2, x_3,...$, are outcomes of regressors (among them: 'driving variables'), $\beta_0, \beta_1, \beta_2, \beta_3,...$, are regression coefficients, and ε is the experimental error. Prediction of SCS in an unvisited place P is: $(y^\wedge)_P = b_0+b_1(x_1)_P+b_2(x_2)_P+ b_3(x_3)_P +...$, where b_is are least squares estimates, and outcomes of regressors could be found e.g. using GIS. If time is among regressors, prediction in time is possible as well. Predictions can be given including confidence bounds. A disadvantage of the method is that it requires quite a lot of experimental observations in a wide range of experimental conditions (the regressors). Extrapolation using PBS models requires considerably less observations but a lot of input data instead. Most of PBS models have been developed in temperate zones. It is dangerous to apply them (unmodified) in tropical situation, as turnover rates are higher, and small inaccuracies in the determinants of these rates may have very strong effects on the results (van Keulen, 2001). PBS model predictions are given without any accuracy. In the evaluation of PBS models, existing long-term experiments are usually the only source of data (Powlson, 1996). Such long-term experiments have difficulties that mainly arise from improper experimental design, improper sampling methods, and poor record keeping (Glendining & Poulton, 1996). In short, application of PBS models for extrapolation purposes have a large number of problems and uncertainties.

Conclusions If sufficient data is available, statistical methodology is preferable to using PBS models.

References

Glendining, M.J. & P.R. Poulton (1996). Interpretation Difficulties with Long-Term Experiments. In: D.S. Powlson, P. Smith & J.U. Smith (eds.) Evaluation of Soil Organic Matter Models, using existing long-term datasets. Springer, Berlin, 99-109.

Hanson, J.D., M.J. Schaffer & L.R. Ahuja (2001). Simulating Rangeland Production and Carbon Sequestration. In: R.F. Follett, J.M. Kimble & R. Lal (eds.) The potential of U.S. Grazing Lands to Sequester Carbon and Mitigate the Greenhouse Effect. CRC Press LLC, Boca Raton, 345-370.

Powlson, D.S. (1996). Why Evaluate Soil Organic Matter Models? In: D.S. Powlson, P. Smith & J.U. Smith (eds.) Evaluation of Soil Organic Matter Models, using existing long-term datasets. Springer, Berlin, 3-11.

van Keulen, H. (2001). (Tropical) soil organic matter modelling: problems and prospects. *Nutrient Cycling in Agroecosystems,* 61, 33-39.

Dynamics of long-term carbon sequestration on rangelands in the western USA

G.E. Schuman[1], L.J. Ingram[2], P.D. Stahl[2] and G.F. Vance[2]
[1]High Plains Grasslands Research Station, USDA, ARS, Cheyenne, Wyoming 82009 USA, Email: Jerry.Schuman@ars.usda.gov, [2]Department of Renewable Resources, University of Wyoming, Laramie, WY 82071 USA

Keywords: carbon sequestration, grazing management, semi-arid environment, drought, exclosure

Introduction Rangelands in the USA occupy 161 million hectares of land. Worldwide, rangelands occupy about half of the land area and account for more than 1/3 of the world's terrestrial carbon (C) reserves. Because of their large land area, rangelands have the potential to sequester a significant amount of additional atmospheric C. Schuman et al. (2001) estimate that rangelands and marginal croplands restored to grasslands in the USA can sequester 64 million metric tonnes C/ha/yr if properly managed. The objective of this research was to evaluate the long-term effects of grazing on soil C storage in a northern mixed-grass prairie (NMP).

Materials and methods Organic C dynamics were assessed on NMP paddocks that were grazed season-long at light (CL) and heavy stocking rates (CH) and on non-grazed exclosures (EX) after 10 and 20 years following initiation of grazing in 1983. Soil organic C and aboveground C components were assessed on one permanent 50 m transects established in each of two replicate paddocks. Soil were sampled in 1993 and 2003 at 10 m intervals on each transect for C measurements and samples were also collected at the 10 and 30 m locations for bulk density determination to enable calculation of C mass.

Results Grazing increased soil organic C storage after 10 years, Table 1. Grazing enhanced plant residue decomposition and early season photosynthesis resulting in increased soil C storage compared to the EX. No differences were evident between CL and CH grazing after 10 years. However, significant soil C was lost from the soil profile between 1993 and 2003 under CH grazing compared to CL and EX. The nearly 30% loss in soil organic C indicates that CH grazing can have significant negative impacts on C cycling during extended periods (2000-2003) of drought that occurred during the second 10 year period. This loss of C is supported by Morgan *et al.* (2004) who reported lower net system CO_2 assimilation, using CO_2 flux measurement technology, in a short-grass prairie ecosystem with CH grazing during the drought years of 2002 and 2003 compared to moderately grazed systems and EX. Continuous, heavy season-long grazing has resulted in a major shift in plant community composition and decreased above-ground biomass. Whereas, C_3 perennial grasses predominate in the CL treatment, C_4 perennial grasses dominate in CH, and C_3 perennial grasses and forbs co-dominate in the EX.

Table 1 Soil organic C mass (kg C/ha x 10^2) under various grazing treatments on northern mixed-grass prairie in Wyoming USA

Soil Depth (cm)	1993			2003			Change (%) From 1993 to 2003		
	EX	CL	CH	EX	CL	CH	EX	CL	CH
0-15	282b	351a	359a	273a	320a	260a	-3	-9	-28*
0-30	479b	580a	583a	473b	542a	425b	-1	-7	-27*
0-60	881b	919ab	1013a	805b	925a	705b	-9	+1	-30*

Different lower case letters indicate significant differences between grazing treatments (within a year and soil depth), P≤0.10. Asterisks indicate a significant difference in C mass between years, P≤ 0.10.

Conclusions This study demonstrates that proper grazing management under normal precipitation can enhance C sequestration in NMP rangelands compared to no grazing (or exclusion of grazing) over a 20 year period. In addition, heavy season-long grazing resulted in plant community shifts, lowered production, and loss of SOC from the soil profile during extended drought periods. Therefore, proper grazing management is paramount to ensure sustainable long-term production and C sequestration by these ecosystems.

References
Morgan, J.A., D.R. LeCain, J.D. Reeder, G.E. Schuman, J.D. Derner, W.K. Lauenroth, W.J. Parton & I.C. Burke (2004). Drought and grazing impacts on CO2 fluxes in the Colorado shortgrass steppe. *Abstracts Eighty-ninth Annual Meeting of the Ecological Society of America*, Portland, OR, USA, (In press).
Schuman, G.E., J.E. Herrick & H.H. Janzen (2001). The dynamics of soil carbon in rangelands. In: Follett, R.F., Kimble, J.M. & Lal, R. (eds.) The Potential of US Grazing Lands to Sequester Carbon and Mitigate the Greenhouse Effect, Chapter 11, Lewis Publishers, Boca Raton, Florida, USA, 267-290.

Net ecosystem productivity of a grassland in comparison with an arable and a forest ecosystem

P.A. Davis[1], K. Black[1], J. Clifton-Brown[2], G. Lanigan[2], J. Burke[3], A. Fortune[3], M.B. Jones[2] and B. Osborne[1]
[1]Botany Department, University College Dublin, Belfield, Dublin 4, Ireland, Email: Phillipdavis2002@hotmail.com, [2]Botany Department, Trinity College, University of Dublin, Dublin 2, Ireland, [3]Teagasc Oak Park Research Centre, Carlow, Ireland

Keywords: eddy covariance, net ecosystem productivity; carbon sequestration, land use

Introduction Grassland, arable farming and forests are the major land use categories in Ireland and it is, therefore, important to know the carbon-source/sink strengths of these land use types. Forest ecosystems are also an important and fast growing land use category in Ireland. Here we present a comparison of the net ecosystem productivity (NEP) of these three land use categories (grass land, arable and forest) from three sites.

Methods The grassland (*Lolium perenne* dominated) and arable (spring barley) sites are situated at the Oak Park Research Station, Teagasc, Co. Carlow in SE Ireland and the forest site (Sitka spruce plantation) is located 30 km West in Co. Laois. The grassland site was cut for silage in June and grazed by cattle until the autumn. Eddy covariance measurements of CO_2 and H_2O fluxes were made with the EdiSol system, described by Moncrieff *et al.* (1997). The sonic anemometer (soluent R3, Gill Instruments Ltd., Lymington, England) and gas analyser intake were situated at a height of 1.5 – 1.9 m on the arable site 2 m at the grass site and 18 m at the forest site. The CO_2 and H_2O concentrations were measured using a closed path analyser (Li-7000, Li-Cor Inc., Lincon, NE). Carbon export was determined from biomass samples collected at silage cut and during grazing for the grassland and at harvest for the spring barley. Samples were oven dried (80ºC) before mass was determined.

Results All three sites showed the expected seasonal variations in NEP with the major CO_2 uptake occurring in the spring and summer months and less uptake or losses occurring during winter months (Figure 1). Uptake at the arable site stopped in mid July when the Barley ripened and senesced. Fluctuations in summer NEP at the grassland were associated with changes in leaf area index caused by silage cuts and grazing. The arable site showed the greatest losses of carbon during winter months due to the lack of plant cover. Small losses were also observed between November and February at both the grassland and forest sites: this was associated with low temperatures and short day lengths. Differences in the carbon uptake of the three systems were mainly associated with differences in the duration of plant cover (Table 1). Management, carbon export and duration of plant cover also influenced the carbon losses and, therefore, total sequestered carbon.

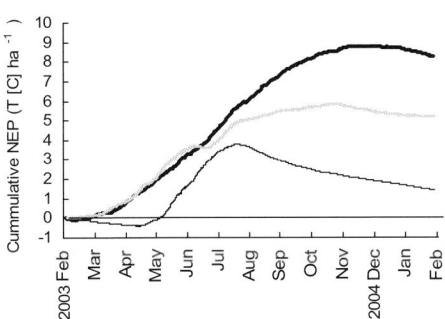

Figure 1 Comparison of cumulative NEP for the three ecosystems. Black = arable; grey = grassland; bold = forest

Table 1 Annual NEP (tC ha⁻¹), carbon export (tC/ha) sequestered carbon (tC ha⁻¹) and period of plant cover (months) for the three ecosystems form Feb 1 2003 to Jan 31 2004. Negative values indicate carbon loss

	Grass land	Arable	Forest
NEP	5.2	1.4	8.2
Carbon export	4.5	3	0
Sequestered carbon*	0.7	-1.6	8.2
Period of plant cover	12	4	12

Calculated as, NEP – export. This excludes losses due to run off and leaching.

Conclusions Forest systems act as the strongest sinks for carbon dioxide and arable systems act as the strongest sources of carbon dioxide. The Grassland was a small sink for carbon dioxide. Carbon export in biomass and other management practices caused the greatest loss of carbon for the arable and grassland. Duration of plant cover and leaf area index appear to have the largest effect on carbon uptake.

References

Moncrieff, J.B., J.M. Massheder, H. de Bruin, J. Elbers, T. Friborg, B. Heusinkveld, P. Kabat, S. Scott, H. Soegaard & A. Verhoef. (1997). A system to measure surface fluxes of momentum, sensible heat, water vapour and carbon dioxide. *Journal of Hydrology,* 188-189, 589-611.

Dormant-season carbon fluxes in humid-temperate pastures

R.H. Skinner

USDA-ARS, Pasture Systems and Watershed Management Research Unit, Building 3702 Curtin Road, University Park, Pennsylvania 16802, USA, Email: howard.skinner@ars.usda.gov

Keywords: carbon sequestration, pastures, winter, photosynthesis

Introduction Human induced increases in atmospheric CO_2 through the burning of fossil fuels and deforestation are considered to be a primary cause of rising global temperatures. However, carbon sequestration by terrestrial ecosystems has reduced the rate of CO_2 accumulation in the atmosphere. Because of their vast size, grazing lands have the potential to sequester significant quantities of carbon, slowing the increase in atmospheric CO_2 and reducing the risk of global warming. Although C uptake during the growing season can be substantial, losses following defoliation and drought, and during the winter months can significantly reduce annual sequestration, frequently turning grazing lands into net carbon sources. The objective of this research was to quantify the magnitude of carbon sources and sinks for humid-temperate grazing lands under a range of management practices, and, in particular, examine the impact of dormant-season fluxes on annual carbon storage.

Materials and methods The study was conducted on two pastures near State College, Pennsylvania, USA. The grass-based pasture was dominated by a mixture of cool-season grasses including, cocksfoot, tall fescue, smooth brome, and Kentucky bluegrass. The lucerne-based pasture was intermixed with patches of cocksfoot, smooth brome, dandelion, Kentucky bluegrass, and tall fescue. Pasture-scale CO_2 fluxes were quantified using an eddy covariance CO_2 flux system. This report focuses on CO_2 fluxes during the winters of 2002-03 and 2003-04. During the winter of 2003-2004, transects were clipped to the soil surface on 4 December, 6 January, and 11 March, and live leaves were scanned to determine green leaf area.

Results The dormant season was defined as the period beginning when net daily C uptake became negative in the autumn and ending when daily uptake became positive again in the spring. The beginning of the dormant season varied by as much as a month, depending on pasture type and year (Table 1). In contrast, the end of the dormant season varied by only a week. Even though the lucerne dormant season was 20 d longer than that of the grass pasture in 2003-04, total C loss from the lucerne pasture was reduced by 17% due to a 28% decrease in the daily rate of C efflux compared with the grass pasture. Some green leaf area was observed in both pastures throughout the winter of 2003-04 (Leaf Area Index (LAI) = 0.03 to 0.50). The lucerne pasture had greater LAI at the beginning of December, but lower LAI than the grass pasture in both January and March. In the absence of snow cover, photosynthetic uptake occurred at daytime air temperatures well below freezing, beginning at -3° C in the lucerne and -8° C in the grass pasture. However, night time efflux from the system was greater than daytime uptake so that the pastures remained C sources throughout the winter. Daytime uptake was greater in the grass compared with the lucerne pasture. Night time losses, however, were also greater in the grass pasture, making the lucerne pasture less of a C source for the entire day despite having reduced photosynthetic uptake.

Table 1 Characterization of dormant season for two pastures in central Pennsylvania

	2002-2003		2003-2004	
	Grass	Lucerne	Grass	Lucerne
Starting date	27th November	-	11th November	26th October
Ending date	31st March	31st March	2nd April	6th April
Duration (d)	125	-	144	164
Total Flux (g C m^{-2})	28.3 (2003 only)	12.8 (2003 only)	87.2	72.5
Daily Flux (g C m^{-2}/d)	0.31 (2003 only)	0.14 (2003 only)	0.61	0.44

Conclusions Even though pastures are generally C sources during the winter, there were days at both sites and in both years when light and temperature conditions were conducive to C uptake. Because of lower net daily effluxes, total dormant-season C loss from the lucerne pasture was less than that from the grass pasture, despite the fact that the lucerne pasture had lower photosynthetic rates and a dormant season that was 20 d longer than that of the grass pasture. At the beginning of the growing season, 25 to 39 d were required to recapture the C that was lost during the previous dormant season.

The effect of legume/grass pasture on soil organic carbon

Z. Kadziuliene and A. Slepetiene
The Lithuanian Institute of Agriculture, LT-58344 Akademija, Kedainiai distr., Lithuania, Email: zkadziul@lzi.lt

Keywords: soil organic carbon, pasture, C:N ratio

Introduction Soil organic carbon content is an important integral indicator of soil fertility. The extent at which the soil can sequestrate carbon depends on the nature of agricultural production, land use and soil type (Follet, 2001). More than two thirds of the annual grassland biomass production is allocated to below ground structures and deep humus layers are common in grassland (Körner, 2002). Grasslands differ markedly in species composition, utilisation purpose, fertilisation and this can significantly change their effect on soil organic content (Hassink & Neeteson, 1991). The objective of study was to estimate changes in the soil organic carbon content and C:N ratio after pure grass and mixed swards of legumes and grass at different grazing frequencies.

Materials and methods The experiment was conducted in Dotnuva (55°24'N) on a loamy *Cambisol*. Soil pH varied from 6.5 to 7.0, humus content was 2.5-3.2 %, C content 14.5 g/kg, available P 50-80 mg and K 100-150 mg/kg. The experiment involved swards consisting of different legumes and grasses with frequent (F) and less frequent (LF) grazing. Grazing intensity was 2-2.5 cows/ha and the grazing season was 150 d. The *Lolium perenne* swards were with and without nitrogen fertilisation. Soil organic carbon content and C:N ratio were studied in 2001-2002 in the third and fourth year of grassland use. Soil N was determined by the Kjeldahl method and the Ponomariova-Plotnikova-modified Tyurin method was used for the organic carbon content.

Results In the soil under different swards and grazing frequency, the content of organic carbon ranged from 15.53-17.83 g/kg (Table 1). The soil organic carbon content varied little during the experimental period in this soil which was rich in organic carbon. The content of organic carbon increased slightly in the soil under legume/grass swards. Grazing frequency of swards did not have any significant effect on the organic soil carbon content. The C:N ratio varied slightly depending on the composition of swards. The C:N ratio was lower with frequent grazing, because the N content of the soil had increased.

Table 1 The effect of different swards on soil organic carbon content and C:N ratio

Swards (A factor)	Carbon content, g/kg							
	2001 spring		2001 autumn		2002 spring		2002 autumn	
	F[1]	LF[1]	F	LF	F	LF	F	LF
Trifolium repens/L. perenne	16.87	16.23	17.43	16.63	17.83	16.63	17.00	16.13
Medicago sativa/ L . perenne/Poa pratensis	16.16	16.07	16.7	16.93	16.93	16.77	15.87	16.53
T .repens/M .sativa/L. perenne	16.63	16.40	16.80	17.46	17.60	17.57	16.30	17.10
L .perenne/ N_0	17.20	15.9	17.60	17.30	17.33	16.93	16.90	16.97
L .perenne/ N_{240}	16.57	16.07	17.00	17.30	16.60	16.30	17.47	17.07
LSD.05 A factor/ B factor	0.796/ 0.325		0.850/ 0.347		0.783/ 0.320		0.821/0.335	
	C:N ratio							
T. repens/L. perenne	10.00	9.54	9.43	9.89	9.72	9.62	9.39	9.85
M. sativa/ L. perenne/P. pratensis	10.50	9.30	9.50	10.07	9.75	9.89	9.14	10.26
T. repens/M.sativa/L. perenne	9.74	9.63	9.57	9.85	9.66	9.96	9.60	9.90
L.perenne/ N_0	9.98	9.59	9.69	10.06	10.11	10.01	9.78	10.07
L.perenne/ N_{240}	9.89	9.55	9.61	9.91	9.07	9.49	9.59	9.96
LSD.05 A factor/ B factor	0.284/ 0.116		0.182/ 0.074		0.214/ 0.087		0.259/ 0.106	

B factor: F[1] frequent grazing, 5-6 grazings; LF[1] less frequent - 4-5 grazings/season
LSD, least significant difference (P< 0.05)

Conclusions The soil under grazed swards was rich in organic carbon. A wider C:N ratio was identified in the soil under less frequent grazing.

References

Follet R.F. (2001). Soil management concepts and carbon sequestration in cropland soils. *Soil and Tillage Research*, 61, 77-92.
Hassink J. & J.J. Neeteson (1991). Effect of grassland management on the amounts of soil organic N and C. *Netherlands Journal of Agricultural Science*, 39, 225-236.
Körner C. (2002.) Grassland in a CO₂-enriched world. *Grassland Science in Europe*, 7, 611-624.

Carbon sequestration in irrigated pastures

G.E. Shewmaker[1], J.A. Entry[2] and R.E. Sojka[2]
[1]The University of Idaho, Twin Falls Research and Extension Center, Twin Falls, Idaho 83303-1827, USA, Email: gshew@uidaho.edu, [2]USDA Agricultural Research Service, Northwest Irrigation and Soils Research Laboratory, 3793 North 3600 East, Kimberly, Idaho 83341, USA

Keywords: carbon sequestration, irrigation, pasture

Introduction Carbon sequestration potential for irrigated grazing lands is significant. We measured organic and inorganic C stored in southern Idaho soils having long-term land use histories that supported native sagebrush vegetation (NSB), irrigated pasture systems (IP), irrigated conservation tillage sites (ICT), and irrigated moldboard plowing systems (IMP). This study estimates the amount of possible organic, inorganic and total C sequestration if irrigated pasture land was expanded by 10%.

Materials and methods The study area is the Snake River Plain between 42°30' and 43°30' N and 114°20' and 116°30' W, 860 to 1300 m elevation. The climate is temperate semi-desert with cool, moist winters and hot, dry summers. Annual precipitation is 175 to 305 mm. Average annual temperature is 9 to 10°C and soils are typically well-drained loams and silt loams (Aridisols) derived from loess deposits overlying basalt. Soil samples were taken from 3 NSB, 3 IP, 3 ICT (past 8 years), and 3 IMP sites. Ten random 2.4-cm diameter replicate soil cores were taken from each site and partitioned into 0-5, 5-15, 15-30 and 30-100 cm depths. Roots greater than 1.0-cm diameter were measured separately. Inorganic C in each sample of mineral soil was determined by titration (Loeppert & Suarez 1996). Concentration of organic C in each sample of mineral soil was determined by the Walkley-Black procedure (Nelson & Sommers 1996). Total C ha^{-1} to a 100-cm depth of mineral soil was calculated assuming 0.44 g C g^{-1} organic matter with correction for soil bulk density.

Results Inorganic C and total C (inorganic + organic C) in soil were higher in IP than NSB (Table 1). If irrigated pasture land was expanded by 10%--meaning NSB land was converted to IP--a possible gain of 9.6 x 10^8 Mg total C (1.7 % of the total C emitted in the next 30 year) could be sequestered in soils worldwide. If irrigated agricultural land was expanded worldwide and NSB was converted to IP, while an equal amount of less-productive rain fed agricultural land was returned to native grassland, a possible gain of 9.3 x 10^9 Mg total C (11.9 % of the total C emitted in the next 30 yr) could be sequestered in soils. The following figures formed the basis of the calculations: Land area in irrigated crop land in the Western United States is 2.43 x 10^7 ha and worldwide is 2.60 x 10^8 ha. % C_s/C_{EW} = C sequestered (C_s) divided by the amount of C projected to be emitted worldwide during the next 30 years, which is 5.7 x 10^{10} Mg C (C_{EW}) multiplied by 100.

Table 1 Total C sequestered in soils during a 30-year period by a 10% conversion from NSB to irrigated conservation tillage and irrigated pasture lands; and 10% conversion from irrigated moldboard plow to IP

Vegetation conversion		Western United States		Worldwide	
	Mg C ha^{-1}	Mg C	% C_s/C_{EW}	Mg C	% C_s/C_{EW}
NSB to irrigated conservation tillage	8.0	1.9 x 10^7	0.03	2.1 x 10^8	0.37
NSB to IP	35.6	8.7 x 10^8	1.53	9.3 x 10^9	16.32
10% of irrigated moldboard plow to IP	37.1	9.0 x 10^7	0.16	9.6 x 10^8	1.68

Conclusions The expansion of irrigated pasture land would significantly increase C sequestration in soils world-wide. Land use shift from relatively low productivity rain fed agricultural land to temperate forest or native grassland could also cause meaningful reductions in atmospheric CO_2.

References

Nelson, D.W., & L.E. Sommers (1996). Total Carbon, Organic Carbon and Organic Matter. In J.M. Bigham (ed). Methods of Soil Analysis. Part 3, Chemical and Microbiological Properties. American Society of Agronomy, Madison, Wisconsin, 961-1010

Loeppert, R.H. & D.L. Suarez (1996). Carbonate and Gypsum. In: D.L. Sparks, A.L. Page, P.A. Helmke, R.H. Loeppert, P.N. Soltanpour, M.A. Tabatabai, C.T. Johnson & M.E. Sumner (eds.) Methods of Soil Analysis. Part 3, Chemical Methods. American Society of Agronomy, Madison, Wisconsin, 437-474.

The carbon balance of long term and newly established temperate grasslands

J. Clifton-Brown[1]*, G. Lanigan[1], R.B. Taylor[1], J.I. Burke[2] and M.B. Jones[1]
[1]Botany Department, Trinity College, University of Dublin, Dublin 2, Ireland. [2]Teagasc, Oak Park Research Centre, Carlow, Ireland. * present address: IGER, Aberystwyth, SY23 3EB, UK, Email: jcbrown@bbsrc.ac.uk

Keywords: carbon cycle, carbon sequestration, Lolium perenne

Introduction Carbon (C) sequestration is the process of removing CO_2 from the atmosphere and storing it in C pools of varying lifetime. Storage can be in the form of above-ground biomass, below-ground biomass or recalcitrant organic and inorganic C in the soil. While the processes of C sequestration are ultimately regulated at the molecular level, management practices and climate can greatly affect the way in which terrestrial ecosystems sequester C (Soussana et al., 2004). Because temperate grasslands account for a significant portion of the agricultural and semi-natural land-cover in N-W Europe any increase in the potential of temperate grasslands to store or sequester C could help remove CO_2 from the atmosphere. Currently, this has particular significance because under Article 3.4 of the Kyoto Protocol of the United Nations Framework Convention on Climate Change countries can count this sequestration as a contribution to reducing greenhouse gas emissions. The objective of this experiment was to quantify the effects of severe disturbance on the carbon balance of long-term grasslands. This was achieved by measuring the carbon balance during ploughing and re-sowing of grassland mesocosms.

Materials and methods We used mesocosms containing grassland monoliths to compare the continuous carbon balance of mature and newly-sown grasslands. Eight monoliths were extracted from an old pasture dominated by Lolium perenne which had not been reseeded for ~30 years. Each 40 x 40 x 120 cm monolith was placed in a metal box and covered by a polythene enclosure. Soil water content was maintained at a constant volumetric water content of ~30% by an automated irrigation system. Above and below ground diurnal CO_2 fluxes were measured for each mesocosm. We simulated ploughing in four of the mesocosms by applying a general herbicide (27 April) 23 d prior to inverting the top ~20cm of soil by hand (14 May) and sowing with Lolium perenne (var. Cashel and Greengold). The remaining four mesocosms were retained as controls. We report the carbon balance of resown and undisturbed monoliths over a period of five months.

Results The net carbon balance for undisturbed and ploughed mesocosms are presented in Table 1. Between 27 April and 14 May, following the herbicide treatment (sprayed phase), the sink strength of the treated mesocosms was severely weakened. Following ploughing and reseeding on the 14 May all C uptake ceased (not shown). From May until August, during sward establishment (ploughed & reseeded and recovery phases), the treated mesocosms were a weaker C sink than the controls. However, from mid August onwards, reseeded mesocosms were a stronger C sink than the controls (full canopy phase). Overall, controls sequestered 4.4 t ha^{-1} more C than resown mesocosms, but this difference was not significant (P=0.09) due to heterogeneity in new grass establishment in the resown mesocosms.

Table 1 Carbon uptake for re-seeded (treated) and undisturbed (control) monoliths in four periods. SE = standard error. P < 0.01=***, P < 0.1=**, ns = not significant.

	Control		Treated			
Period in 2004	t C ha^{-1}	SE	Phase	t C ha^{-1}	SE	P
20 April - 27 April	0.7	0.16	Pre-treatment	0.7	0.13	ns
27 April - 14 May	2.1	0.21	Sprayed	0.7	0.16	***
15 May - 15 June	2.5	0.24	Ploughed & Reseeded	0.5	0.07	***
16 June - 15 Aug	6.2	0.33	Recovery	3.7	1.24	**
16 Aug - 10 Sept	0.8	0.39	Full canopy	2.4	0.36	***
Total	12.4	1.15		8.0	3.69	ns

Conclusions Reseeding retarded C sink activity for a period of 3 months. Over the five-month period of measurements reported here undisturbed controls fixed more carbon than re-sown grassland. Longer term measurements are needed to determine if the vigorous growth of newly sown grassland from 3 months after establishment can recover C uptake lost at re-sowing.

Reference

Soussana J-F, P. Loiseau, N. Vuichard, E. Ceschia, J. Balesdent, T. Chevallier & D. Arrouays (2004). Carbon cycling and sequestration opportunities in temperate grasslands. Soil Use and Management, 20, 219-230.

Carbon sequestration in pasture and silvo-pastoral systems in ecosystems of the Latin American tropics

M.C. Amézquita and M. Ibrahim
Carbon Sequestration Project - The Netherlands Cooperation CO-010402. CIPAV-U.Amazonia-CIAT-CATIE-Wageningen University. CIAT's Science Park, A. A. # 67-13, Cali, Colombia, Email: m.amezquita@cgiar.org

Keywords: tropics, tropical pastures, agroforestry systems, soil carbon stocks

Introduction Conversion of forests to pastures has been the most important land use change in tropical America (TA) in the last fifty years. After deforestation and pasture establishment many areas have been abandoned due to productivity declines arising through mismanagement. Over 60% of the TA's pasture area is degraded. Recent interest in carbon sequestration and environmental considerations might suggest partial reforestation of current pastoral areas but this has implications for the socio-economic welfare of farmers and food availability. However, combining agricultural production with environmental objectives (particularly carbon sequestration) could provide a sustainable alternative. Here we present 3-years of research on the evaluation of soil carbon stocks (SCS) in long-established pasture and silvo-pastoral systems (10-16 years of commercial production) relative to native forest (positive control) and degraded land (reference control) under four ecosystems of TA: Andean hillsides (Colombia), sub-humid and humid tropical forest (Costa Rica) and humid tropical forest, Amazonia (Colombia).

Materials and methods A soil sampling design controlling factors affecting SCS (site conditions, slope or main gradient, land use system, and soil depth) was used. Field research was conducted at farm level, in farmer networks within the project ecosystems. SCS were evaluated at four soil depths (0-10, 10-20, 20-40 and 40-100 cm) using 2 space replications per system and 12 sampling points per system/rep. Total C, oxidisable C, total N, P, CEC, pH, soil texture and bulk density were evaluated at each soil pit and depth. Total, oxidisable and stable C (expressed as the difference between total C and oxidisable C) were corrected for bulk density and expressed as C/ha for each soil depth. Statistical comparisons of SCS between systems were based on fixed soil mass but without subdivision in soil horizons as modified by Buurman *et al.* (2004).

Results Data from Andean hillsides suggests that although native forest possesses the highest SCS in this ecosystem (234 and 186 Mg/ha/1m-equivalent for sites 1 and 2, respectively), improved pasture systems increase SCS while B. decumbens with trees (162 and 152 Mg/ha/1m-equivalent for sites 1 and 2) show higher SCS than natural regeneration systems (fallow land and secondary forest, with 156 and 142 Mg/ha/1m-equivalent for sites 1 and 2) and degraded pasture (156 and 97 Mg/ha/1m-equivalent for sites 1 and 2). On the contrary, SCS estimates from the humid tropical forest (Atlantic coast, Costa Rica) show that pasture systems such as I. ciliare, B .brizantha + A. pintoi, A. mangium + A. pintoi and B. brizantha in monoculture (208, 194, 168 and 134 Mg/ha/1m-equivalent, respectively) had statistically higher stocks than native forest (128 Mg/ha/1m-equivalent) and this in turn had statistically higher stocks than degraded pasture (94 Mg/ha/1m-equivalent). Similar rankings were obtained in the humid tropical forest of Amazonia, Colombia, where B. humidicola and B. decumbens pastures (monoculture and legume-associated) showed higher SCS than native forest. SCS data suggest that in hot and humid environments improved pasture systems show SCS comparable or higher than native forest, therefore representing attractive solutions for C storage.

Conclusions Results suggest that in the tropical ecosystems of Latin America studied, improved pasture and silvo-pastoral systems show SCS levels comparable or even higher than those from native forest, depending on climatic and environmental conditions (altitude, temperature, precipitation, topography and soil). Our research indicates that these systems should be considered as attractive and viable C-improved systems.

References

Buurman, P., M. Ibrahim & M.C. Amézquita (2004). Mitigation of greenhouse gas emissions by silvopastoral systems: optimism and facts. *Proceedings of the Second International Congress in Agroforestry*, Mérida, Mexico.

The nature of sequestered carbon in different Irish mineral soils

C.M. Byrne[1], D. Fay[2], J.A. Ferreira[1] and M.H.B Hayes[1]
[1]*Department of Chemical and Environmental Sciences, University of Limerick, Ireland, Email: corinna.byrne@ul.ie,* [2]*Teagasc, Environmental Research Centre, Johnstown Castle, Wexford, Ireland*

Keywords: carbon sequestration, sand incubation, humic acid, fulvic acid, humification process

Introduction Humic substances (HS) provide the major sinks for carbon (C) in soils. Although HS have a degree of resistance to microbial degradation, they are degraded in time. Humin, the HS component in association with the soil mineral colloids, has greatest resistance to degradation. To understand the extent to which soil can be a sink for C it is important to know the soil mineralogy, and to be aware of aspects of the structures of the humic components. Enhanced biological oxidation occurs in soils in long term cultivation. Its effects can be observed by comparing the amounts and compositions of the humic components in cultivated soils with those in the same soil types in long term grassland. Three such paired soils were included in the study as well as three grassland soils that are in new lysimeter studies at the Environmental Research Centre, Johnstown Castle. In a study of the humification process, maize (*Zea mays L.*, a C4 plant, with a $\delta^{13}C$ value of the order of –12) was incubated in calcareous organic C-free sand, and the products were studied in the same way as those from the soils

Materials and methods Soil samples were H^+-exchanged and exhaustively extracted (until the extracts had negligible colour) with 0.1 M NaOH adjusted to pH 7, then with the NaOH solution adjusted to pH 10.6, then at pH 12.6, and finally with 6 M urea in the 0.1 M NaOH solution. Sand (20 Kg, 38% $CaCO_3$) was incubated with 3.6 Kg of maize. Prior to extraction (on a yearly basis) the $CaCO_3$ was removed (1 M HCl), and the system was then treated as for the soils. Humic acids (HAs) were precipitated at pH 1, dialysed to remove the salt, then freeze dried. Fulvic acids (FAs) were subjected to the standard XAD-8 treatment, recovered in 0.1 M NaOH, H^+-exchanged (IR-120 resin), and freeze dried. Each sample was titrated (potentiometric), subjected to $\delta^{13}C$, neutral sugar, and amino acids analyses, and to infrared, solid state NMR, and ESR spectroscopies. X-ray diffraction was used to classify the soil clays.

Results Organic matter (OM) contents of soils in long-term cultivation were significantly less (about 50%) than those of the same soils in long-term grassland. Approximately 30% of the soil OM content comprised of HAs and FAs, and humin was invariably the most abundant component. The compositions of the HAs and FAs in the different fractions highlighted the greater extents of oxidation/humification in the cultivated soils. This was illustrated by the titration data, and by the carboxyl (160-180 ppm) and O-aromatic (140-160 ppm, includes phenols) resonances in the NMR spectra. Figure 1 is one illustration of how CP/TOSS, Dipolar Dephasing, and Chemical Shift Anisotropy (CSA) NMR spectra show differences between HAs isolated from the same soil type in cultivation and in grassland management, and the data show distinct differences between these and the HA products from plants undergoing humification. The extensive data we have for a variety of soil types clearly show distinctive differences between the humic components in the different soil types. The components can be related to the types and contents of the clays and to the drainage regimes.

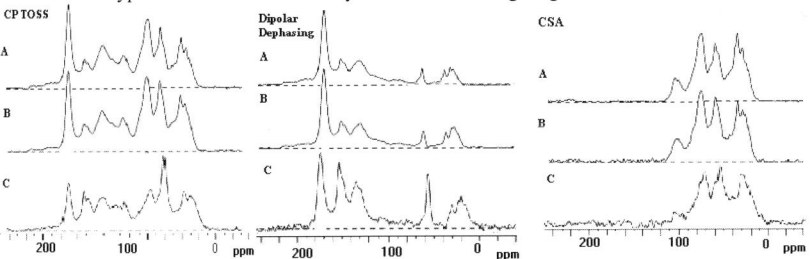

Figure 1 CP/TOSS, Dipolar Dephasing, and Chemical Shift Anisotropy (CSA), spectra of humic acids isolated at pH 12.6 from a coarse sandy loam cultivated (A) and uncultivated (B) soil, and from the sand/maize incubation system (C).

Conclusions The variety of data that will be presented, from a range of analytical and spectroscopic procedures, will show clearly that mineral grassland soils are good sinks for C. These vary, however, in their abilities to sequester C. The variations can be related to the nature of the vegetation, to the soil aeration, to the water holding capacity, and to the amounts and compositions of the inorganic as well as the organic colloids, and to the management. Research is needed to understand how humin, the major OM component, is protected.

A basis for designing policies to optimize soil carbon sequestration in Southeastern US Grasslands

J.R. Brown and D.L. Faulkner
USDA NRCS, Las Cruces, New Mexico and Richmond, Virginia USA, Email: joelbrow@nmsu.edu

Keywords: increasing soil carbon, improved grassland management, policy and program options

Introduction Increasing the amount of carbon (C) stored in terrestrial ecosystems is an important part of most national greenhouse gas (GHG) management strategies. Among the policy and program options available to achieve increased C sequestration, improved management of grasslands offers an attractive option to both reduce atmospheric concentrations of C and enhance environmental co-benefits (soil quality, water quality, food and fibre production, and wildlife habitat). In the United States, incentives for applying improved land management practices come primarily via federal government conservation programs administered by the United States Department of Agriculture (USDA). While private sector markets offer opportunities to increase the range and amount of incentives, low rates of C accumulation per ha and small land holdings per producer will likely limit the impact of the private sector on producer actions. This is especially true in the humid grasslands of the Southeastern U.S.

Discussion Conant *et al.* (2002) have shown that Southeastern U.S. pastures have been the source of relatively high amounts of C sequestration over the past two decades. However, most of that increased C storage can be attributed to changes in land use from cropland to pastureland, primarily in response to land retirement programs. Because most of the land use conversions occurred between 1985 and 1995, when interest in the Conservation Reserve Program was most active, it is likely the rapid soil C increase phase has passed (Follett *et al.,* 2001). Thus, strategies to reduce emissions from agricultural systems and increase soil C should shift from land use change to land and livestock management.

Improved grazing management increased soil C 0.41 Mg C ha^{-1} per year across a range of soil types (Conant *et al.,* 2003). The soil, climate and management combinations in the study were representative of the Southeastern U.S. Enhancing fertility also increased soil carbon, although use of synthetic N fertilizers or legumes will contribute to atmospheric nitrous oxide emissions and in part offset some of the benefits accruing from increased soil C.

In addition, improved livestock management in humid pastures of the Southeastern U.S. has been shown to reduce methane emissions by improving diet quality and forage conversion efficiency in ruminants. Improving forage quality has the potential to reduce methane emissions/kg of product from enteric fermentation by individual animals. Improving calf crop %, weaning weights and herd genetics also have the potential to reduce methane emissions per unit of beef harvested. All of these practices have been shown to have positive economic benefits regardless of the associated GHG emissions. The initial capital cost requirements (cash cost hurdle) can be a significant impediment to adoption of improved grazing management technology and management knowledge, but in many situations this is more a perceived obstacle than a real one. The necessary infrastructure for improved management can be incrementally adopted which can allow investors to minimize negative cash flow impacts during the transition phase.

Conclusions All of the practices identified as decreasing GHG emissions and increasing soil C sequestration in Southeastern U.S. pastures are based on improved management. While land use change may have a greater impact in the 15-20 years following conversion, it is unlikely that rates of change over the past 20 years will extend into the future. Thus, we propose that policies to facilitate improved GHG management in SE pastures should enhance programs to deliver technical assistance, rather than payments, to landowners and managers.

References

Conant, R.T., K. Paustian & E.T. Elliott (2002). Pastureland use in the Southeastern U.S.: Implications for carbon sequestration. In: J.M. Kimble, R. Lal & R.F. Follett (eds) Agricultural Practices and Policies for Carbon Sequestration in Soil. CRC Press, Boca Raton Florida, USA, 423-432.

Conant, R.T., J. Six & K. Pautian (2003). Land use effects on soil carbon fractions in the southeastern United States. I. Management intensive versus extensive grazing. *Biology and Fertility of Soils,* 38, 386-392.

Follett, R.F., J.M. Kimble & R. Lal (2001). The potential of U.S. grazinglands to sequester soil carbon. In: R.F. Follett, J..M. Kimble & R. Lal (eds.) The Potential of U.S. Grazing Lands to Sequester Carbon and Mitigate the Greenhouse Effect. CRC Press, Boca Raton, Florida, USA, 401-430.

Scaling up of site carbon dynamics to predict the carbon dynamics in Kazakhstan, Central Asia

B.K. Wylie[1], T.G. Gilmanov[2], S.L. Stensaas[1], N.Z. Saliendra[3], D.A. Johnson[4], K. Akshalov[5], A.B. Frank[6], L. Zhang[1], R.F. Doyle[1], E.A. Laca[7] and M.W. Demment[7]
[1]USGS EROS Data Center, SAIC, Sioux Falls, South Dakota 57198, USA, Email: wylie@usgs.gov, [2]South Dakota State University, Brookings, South Dakota 57007, USA, [3]USDA Forest Service, North Central Research Station Forest Sciences Lab, Rhinelander, Wisconsin 54501 USA, [4]USDA-ARS Forage and Range Research Lab, Logan, Utah 84322 USA, [5]Barakev Kazakh Research Institute for Grain Farming, Shortandy, Kazakhstan, [6]Northern Great Plains Res. Lab, USDA-ARS, Mandan, North Dakota 58554 USA [7]Agronomy and Range Science, University of California, Davis, California 95616 USA

Keywords: carbon dynamics, mapping, grassland productivity, ecosystem respiration

Introduction Climate and management determine whether rangelands are net carbon sources or carbon sinks. Regional carbon dynamics of the Kazakh Steppe has not previously been documented. The objective of this study is to quantify the regional carbon flux dynamics of these extensive steppes.

Materials and methods Carbon flux towers measure the net flux of carbon dioxide (CO_2) between the ecosystem and the atmosphere. Detailed light curve analysis partitions these net fluxes into ecosystem respiration (R_e) and photosynthesis (gross primary production, P_g). A positive flux of carbon represents movement from the atmosphere to the land surface. We combined values at flux tower locations from the spatial and temporal data sets of photosynthetically active radiation (PAR), temperature, precipitation, Normalized Difference Vegetation Index (NDVI), and NDVI derived metrics with flux tower derived gross primary production (P_g) and ecosystem respiration (R_e) into a 10-day time step training database. Regression tree models were parameterized using training databases from Northern Great Plains and Kazakh Steppe flux towers. These regression tree models were then applied across the Kazakh Steppe to map growing season 10-day P_g and R_e. Annual fluxes were estimated using gap filled winter fluxes from the winter of 2001-2002 based on methods used by Gilmanov et al. (2004). Refinement in regional estimation of winter fluxes in the Kazakh Steppe would improve the annual NEE maps. Inter-annual and spatial variability of annual net ecosystem exchange (NEE) were investigated using inter-annual statistics and spatial moving window statistics.

Results Regression trees used for mapping of growing season 10-day P_g and R_e were robust and accurate (Table 1). The Kazakhstan winter flux estimate (1.25 g CO_2/m² per day) was used with growing season NEE to estimate annual NEE for each year from1998 to 2001. The area associated with inter-annual carbon sinks were fifty five times larger than inter-annual carbon source area. Carbon sources are seen surrounding wetland areas within Kazakhstan and may be attributable to wetland dynamics not being adequately captured by the MODIS land cover. The annual average NEE for the Kazakh Steppe was relatively stable with the greatest value occurring in 2000. This resulted in a increase in CO_2 across the four years as seen in Figure 1. The local variance images identified areas with lower NEE surrounding some urban areas. Although this was not common, it may identify overgrazed areas with reduced carbon sequestration and potential for improvement.

Table 1 Kazakh Steppe Regression Tree Accuracies

	R^2	Standard Error
P_g (g CO_2 m^{-2} day^{-1})		
Jackknife Years	0,81	3,24
All Training	0,83	1,77
R_e (g CO_2 m^{-2} day^{-1})		
Jackknife Years	0,63	2,70
All Training	0,77	1,33

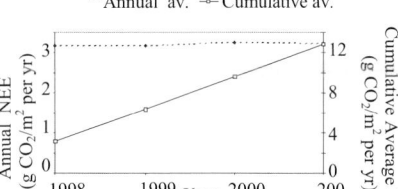

····Annual av. ─○─Cumulative av.

Figure 1 Annual and Cumulative NEE average

Conclusion The results of this study show that regression tree algorithms were effective when applied regionally to the Kazakh Steppe. This approach maximized the utility of intercontinental flux towers for regional mapping. The Kazakh Steppe has more average cumulative sinks than sources. Land use changes from soil carbon depleting spring wheat to grassland represent carbon sequestration opportunities.

References

Gilmanov, T.G., D.A. Johnson, N.Z. Saliendra, T.J. Svejcar, R.F. Angell, & K.L. Clawson (2004). Winter CO_2 fluxes above sagebrush-steppe ecosystems in Idaho and Oregon. Agricultural and Forest Meteorology, (in press).

Soil carbon sequestration under three years of no-till forage cropping systems

D.H. Min, J.D. DeYoung and R.H. Leep
Michigan State University, Upper Peninsula Experiment Station, Chatham, Michigan, 49816, USA, Email: mind@msu.edu

Keywords: soil carbon sequestration, forage, alfalfa, orchardgrass, no-till

Introduction The effects of reduced tillage on soil organic carbon (SOC) are generally well established (Chan *et al.*, 2002; Cabardella & Elliott, 1992). The effects of different crops are also somewhat understood (Drinkwater *et al.*, 1998). However, many of these studies are done in the laboratory to study the effects of crop residues on SOC. Many forage-based systems have very little crop residue returned to the field. What residue does return is often in the form of manure or compost, which is usually broken down much more rapidly than most crop residues. The objective of this study is to assess the several no-till forage cropping systems used extensively in Michigan, USA for their potential to sequester SOC levels.

Materials and methods For this study, three different no-till forage cropping systems were used at two different locations with two very different soils and climates. The first was at the Upper Peninsula Experiment Station (UPES), in Chatham, Michigan, USA (Spodosol, short-growing season, cold winters). The second was at the Kellogg Biological Station (KBS), in Hickory Corners, Michigan, USA (Alfisol, temperate climate, mild winters). The three cropping systems included 3 years of continuous silage corn (*Zea mays*) (C), 3 years of continuous lucerne (*Medicago sativa*) (A), and a mixture of lucerne and cocksfoot (*Dactylis glomerata*) (A+O). Two organic matter treatments (OM+, organic matter added; OM-, no organic matter added) were applied using dairy slurry manure at UPES, and using a manure and sawdust compost at KBS (applied at a rate of 3362 kg Carbon/ha). Three soil cores (3.25 cm diameter) were taken from each plot and composited by depth (0-5 cm). The soil was analyzed for total carbon and nitrogen using a Costech ECS 4010.

Results At the KBS location, in the 0-5 cm soil horizon, the A+O treatment resulted in significantly higher levels of soil carbon than lucerne grown alone (Table 1). The additions of compost also resulted in significantly higher levels of soil carbon than the no compost treatment. At UPES, there was no difference between cropping and organic matter treatments in the 0-5 cm depth. However, UPES had much higher levels of soil carbon than KBS. There was no interaction effect between cropping systems and OM treatments in either location.

Table 1 Soil carbon (kg ha[-1]) under different no-till cropping systems and organic matter treatments at two different sites (KBS and UPES)

Treatment	KBS	UPES
	0-5 cm	
C	9035 [ab#]	15 552
A	8083 [b]	15 470
A+O	9448 [a]	15 091
OM +	9403 [a]	15 339
OM -	7943 [b]	15 482

[#] means within same column followed by the same letter are not statistically different (*P<0.05*)

Conclusions Diversification of cropping systems with fine-rooted crops (i.e., orchardgrass) seemed to pay off with higher soil carbon levels at KBS. At UPES, however, the soil already held an average of 43% more carbon than KBS. This likely resulted from the climate and the previous history of the region being more conducive for building soil carbon stores. But it appears that they may have reached threshold. The soils at UPES are therefore good candidates for the maintenance of soil carbon, while the soil at KBS is a good candidate for additional soil carbon sequestration. By incorporating a fine rooted crop into existing forage production practices; it may be possible to continue building soil carbon levels and thereby sequester more carbon in these soils.

References

Cambardella, C.A. and E.T. Elliott (1992). Particulate soil organic-matter changes across a grassland cultivation sequence. *Soil Science Society of America Journal,* 56, 777-783.

Chan, K.Y., D.P. Heenan & A. Oates (2002). Soil carbon fractions and relationship to soil quality under different tillage and stubble management. *Soil and Tillage Management,* 62, 133-139.

Drinkwater, L.E. P. Wagoner & M. Sarrantonio (1998). Legume-based cropping systems have reduced carbon and nitrogen losses. *Nature,* 396, 262-265.

Theme B: Grassland and the environment

Section 17

Biodiversity in grassland

A new system for plant experiments on biodiversity or multi-species competition

J. Connolly and L. Kirwan
University College Dublin, Dublin 4, Ireland, Email: john.connolly@ucd.ie

Keywords: multiple species, competition, biodiversity, experiment design, statistical models

Introduction Considerable discussion in recent years has focused on the design of competition and biodiversity experiments (Connolly *et al.*, 2001a; Allison, 1999). Few agronomic experiments with >2 plant species have been conducted in greenhouse conditions (Gibson *et al*, 1999) or in the field (Connolly *et al.*, 2001b). In many experiments the effects of density and initial species size have been confounded. The effects of species richness and evenness also are confounded frequently. The proposed system provides a framework of design and analysis, in which to address questions of function at community level and of structure and competition at the level of species. It provides a set of statistical models that allow the separate assessment of initial overall abundance, species richness, species evenness and environment.

Materials and methods *Experimental designs:* The proposed experimental designs are based on the simplex (Cornell, 1990) at each of at least 2 overall densities. For *s* species, each experimental community will consist of up to s species and can be represented as a point in an *s-1* dimensional simplex. For 3 species (Figure 1), the simplex vertices represent monocultures of each species and the central point a mixture in which the initial abundance of all species is equal. A design consists of stands defined by 2 simplexes, each at a different levels of total initial abundance The simplex methodology provides a simple framework for selecting communities. The design selected will depend on *s*, the number of species, on whether one is primarily interested in questions of function or structure, the complexity of the models to be fitted and considerations of design power. Designs for large number of species can be constructed which are not excessively large in number of experimental stands required.

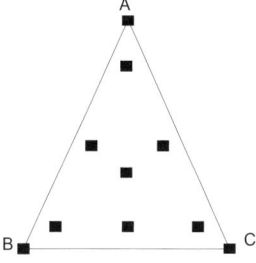

Figure 1 Simplex design for 3 species

Statistical Models: Some statistical models are proposed for the analysis of data from these experiments. For functional responses, such as yield, the models contain terms that assess the effects of species identity, overall initial abundance, environment, species richness and evenness. The richness terms are interpretable as synergistic and antagonistic interspecific relationships whose impact on the response depends on stand evenness. For structural responses, the RGRD (relative growth rate difference) models, proposed by Connolly and Wayne (2005), allow assessment of the effects of species identity, initial species abundance and environment as determinants of change in community biomass composition.

Examples Data from a 4-species experiment, using 2 grass and 2 legume species and an experiment with 5 weed species common in pastures in Switzerland, illustrate these models.

Results and conclusions The analyses of stand yield and resistance to unsown species in the 4-species experiment show strong identity effects and synergistic interaction between species, but a weak effect of overall initial abundance. The analysis of the 5-species experiment shows strong effects of species identity and environment but weak effects of initial species abundance on change in community composition. The design and models proved to be a flexible framework to address questions in multispecies experiments.

References
Allison G.W. (1999). The implications of experimental design for biodiversity manipulations. *American Naturalist*, 153, 26-45.
Connolly, J., P. Wayne & F.A. Bazzaz (2001a). Interspecific Competition in Plants: How Well Do Current Methods Answer Fundamental Questions? *The American Naturalist,* 157, 107-125
Connolly, J., H. C. Goma & K. Rahim (2001b.) The information content of indicators in intercropping research. *Agriculture, Ecosystems and Environment*, 87, 191-207.
Connolly, J. & P. Wayne (2005) Assessing determinants of community biomass composition in two-species plant competition studies. *Oecologia*, 142: 450-457.
Gibson, D.J., J. Connolly, D. C. Hartnett & J. D. Weidenhamer (1999). Essay review: Designs for greenhouse studies of interactions between plants. *Journal of Ecology,* 87, 1-16
Cornell, J.A. (1990) Experiments with mixtures. John Wiley and Sons Inc. New York pp 632.

Native grasses seeded into a cool-season pasture encouraged by low resource availability

J.E. Doll[1], R.L. Cates[2] and R.D. Jackson[1]
[1]Agronomy Department, University of Wisconsin-Madison, 1575 Linden Drive, Madison, Wisconsin, USA; Email: jedoll@wisc.edu, [2]Department of Soil Sciences, University of Wisconsin-Madison, 1525 Observatory Drive, Madison, Wisconsin, USA

Keywords: grazing, prairie, warm-season grass, cool-season grass, regression tree

Introduction   Native prairie of the Upper Midwest, which was dominated by warm-season (C_4) grasses, now exists as relatively small relict and restored patches (Curtis 1959). Re-introduction of natives into grazed agroecosystems would promote genetic, species, and landscape diversity. Extensive re-introduction of C_4 grasses will require a shift away from the paradigm of maximizing production because C_4 grasses have higher C:N ratios than C_3 grasses rendering them inferior forage species. Nonetheless, there is great interest amongst the grazing community of the Upper Midwest in establishing native grasses as a means of improving wildlife habitat

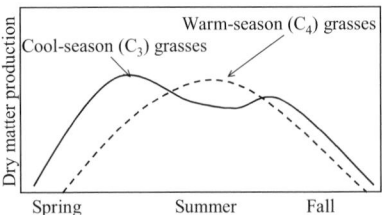

Figure 1 Phenological differences between warm-season (C_4) and cool-season (C_3) grasses.

and increasing belowground carbon storage. That said, given the different phenologies of these 2 functional groups (Figure 1), a relatively even distribution of C_3 and C_4 species in grazed pastures theoretically could provide a more even distribution of forage production.

Materials and methods   In November 2001 we drill-seeded native grass (C_4) species *Andropogon gerardii* (big bluestem), *Panicum virgatum* (switchgrass), and *Sorghastrum nutans* (Indiangrass) into a 0.5-ha pasture dominated by introduced (C_3) species *Festuca elatior* (meadow fescue), *Festuca arundinacea* (tall fescue), and *Dactylis glomerata* (orchardgrass).

Seed application rate was ~1 g/m². Native grass seed was collected locally by members of the Aldo Leopold Foundation, Baraboo. Before this seeding (October 2001), cattle were grazed at high intensity in this pasture to reduce existing sod cover. In August 2003 we counted the number of C_4 grass tillers in 32 quadrats, each 3×1m², and visually estimated cover of C_4 and C_3 grasses, forbs, legumes, and bare ground. At each quadrat, 4 small soil cores were collected (2 cm dia × 15 cm deep) and composited for net N

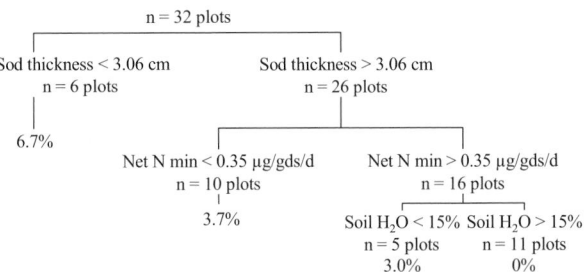

Figure 2 Regression tree showing values of the most important variables (ranked top to bottom) for predicting C4 grass cover (%). Data are from 32 arbitrarily selected plots within a C3 pasture where C4 grasses were interseeded 2 yr earlier.

mineralization assays and gravimetric water content. A shovel was used to pry up sod in each quadrat such that its overall thickness could be quantified. The response variable, C_4 cover, was subjected to regression tree analysis, which sequentially selects the predictor variable and a value of that predictor variable that results in the greatest amount of explained variability in the response variable (De'ath & Fabricius 2000).

Results   While the C_4 grasses were clearly present throughout the pasture of the 32 plots, they represented a very small fraction of the total forage, i.e., usually <5 tillers/m². Sod thickness was the strongest predictor of C_4 cover (Figure 2). Where sod thickness was <3.06 cm, C_4 cover averaged 6.7%. Where sod thickness was >3.06 cm (26 plots), 10 plots where net N min was <0.35 µg/g dry soil/d maintained average C_4 cover of 3.7%. Soil H_2O content was the final variable selected indicating that virtually no C_4 cover was found where soil H_2O was >15%.

Conclusions   The results indicate that successful establishment of C4 grasses in C3 pastures requires manipulation of below ground resource availability. Experiments are currently underway to determine combinations of grazing, burning, and soil amendments to confer a short-term competitive advantage to C4 grasses in resource-rich pastures.

References

Curtis J.T. (1959). The vegetation of Wisconsin: an ordination of plant communities. University of Wisconsin Press, Madison.

De'ath G., & K.E. Fabricius (2000). Classification and regression trees: a powerful yet simple technique for ecological data analysis. *Ecology,* 81, 3178-3192.

The Ag-Biota project: a preliminary assessment of potential indicators of biodiversity in agricultural grasslands

G. Purvis, A. Anderson, A.J. Helden and L. Kirwan
Department of Environmental Resource Management, Faculty of Agri-Food and Environment, University College Dublin, Belfield, Dublin 4, Ireland, Email: gordon.purvis@ucd.ie

Keywords: biodiversity, grassland, bio-indicators, arthropods, Hymenoptera

Introduction   In compliance with European committments to the Convention on Biological Diversity, the conservation and enhancement of biodiversity within agricultural land is a primary objective of current agri-environmental measures. However, there is a widespread lack of information concerning the effectiveness of agri-environmental schemes in Europe, (Kleijn & Sutherland, 2003). In large part, this is due to uncertainties about selection of appropriate biological indicators of biodiversity (Buchs, 2003; Duelli & Obrist, 2003).

Materials and methods   A range of sampling methods was used to measure the diversity of sward flora and animal taxa present in moderately to intensively managed grassland systems at ten locations in S.E. Ireland. We restrict ourselves here to an initial analysis of arthropod catches made in August 2003 with a Vortis suction sampler from individual fields at these locations. A pooled sample comprising 30 random $0.02m^2$ areas each sampled for 10 seconds, was collected from each site. Six arthropod groups were identified to various taxonomic levels (see Table 1). Correlations between the taxon richness of these groups and the total diversity of all *other* arthropods were calculated by bootstrapping (Efron and Tibshirani, 1993) and the significance of calculated mean correlation coefficients was determined over 1,000 randomisations (Manly, 1997).

Results   In total, 15 Araneae; 38 Staphylinidae; 32 other Coleoptera; 15 Hemiptera; 15 Families and 73 Genera of parasitoid Hymenoptera; 25 Diptera taxa were collected. The number of Families of parasitoid Hymenoptera was significantly correlated with the total diversity of all other arthropods (Table 1). However, an increase in the level of taxonomic resolution for Hymenoptera to generic level did not improve this correlation.

Table 1 Summary of mean correllation coeficients between the taxon richness of individual groups (at indicated taxonomic levels) and total arthropod diversity in agricultural grasslands

Arthropod Group	Correlation coefficient	95% confidence intervals		Significance (p = 0.05)
Araneae (species)	0.123	-0.623	0.685	NS
Staphylinidae (species)	0.556	-0.230	1.000	NS
Other Coleoptera	0.033	-0.685	0.717	NS
Diptera (family)	-0.258	-0.877	0.568	NS
Hemiptera (species[1])	0.189	-0.604	0.881	NS
Hymenoptera (family)	0.729	0.138	0.997	*
Hymenoptera (genus)	0.630	-0.010	0.979	NS

[1]Except Aphididae, which were identified to morpho-species

Conclusions   In this preliminary analysis, we have an early indication that the diversity of parasitoid Hymenoptera Families may be a suitable surrogate indicator for wider arthropod diversity in agricultural grasslands. The relatively broad range of feeding habit within parasitoid Families on other arthropods may underlie this relationship. Failure to improve this correlation at an enhanced level of taxonomic resolution could be the consequence of small sample size (number of study sites) and much narrower host range of parasitoids at generic level. These factors may have resulted in a considerable degree of randomness in the numbers of wasp Genera caught. Increasing our sample size (number of sites and/or area sampled per site) could improve the utility of wasp Genera as a surrogate biodiversity indicator, but only at the cost of additional taxonomic effort.

Acknowledgement   The Ag-Biota project is funded under the National Development Plan by the ERTDI Programme 2000-2006 (Project no. 2001-CD/B1-M1).

References

Buchs, W. (2003) Biodiversity and agri-environmental indicators – general scopes and skills with special reference to the habitat level. *Agriculture, Ecosystems and Environment*, 98, 35-78.
Duelli, P. & Obrist, M.K. (2003) Biodiversity indicators: the choice of values and measures. *Agriculture, Ecosystems and Environment*, 98, 87-98.
Efron, B. & Tibshirani, R. (1993) *An introduction to the Bootstrap*. Chapman & Hall, New York.
Kleijn, D. & Sutherland, W.J. (2003) How effective are European agri-environmental schemes in conserving and promoting biodiversity? *Annals of Applied Ecology*, 40, 947-969.
Manly, B.F.J. (1997) *Randomisation, bootstrap and Monte Carlo methods in biology*. Chapman & Hall, London.

Impact of grazing regimes on mean sward height: implications for the management of bird habitats in agricultural landscapes

M. Tichit[1], D. Durant[2] and E. Kernéïs[2]
[1]UMR INRA SAD APT, INAPG, 16 rue Claude Bernard, 75231 Paris cedex 05, France, Email: tichit@inapg.fr,
[2]Domaine INRA SAD,17450 Saint-Laurent de la Prée, France

Keywords: grazing, sward structure, agricultural grasslands, bird habitats

Introduction Grazing in wet grasslands is a key process to manage foraging and nesting habitats for waders. Grazing has positive and negative effects related to the importance of sward conditions for these species and to nest-trampling by cattle. For settlement and nesting, lapwings need a short sward ($\leq$10cm; see Durant *et al.,* this congress). However, when lapwings settle in early spring, grasslands seldom are grazed yet, due to low soil carrying capacity. We studied the effect of autumn and winter grazing regimes on sward structure in early spring, and the effect of grazing regime in early spring on sward structure at hatching. We tested the hypothesis that the delayed effects of grazing on sward height could promote lapwing breeding habitats.

Material and methods The study was of permanent wet grasslands in Rochefort marsh (French Atlantic coast), grazed by cattle. The impact of grazing on mean sward height was assessed on 8 fields from 2002-04. Using a swardstick, and walking representative transects (60-80 points/ha), sward height was measured in early and mid-spring. Herbage N index was measured as proposed by Lemaire & Salette (1984). Stocking rates (LU days)/period were calculated for: autumn, winter [year (n-1)], and early spring and mid-spring [year (n)].

Results Stocking rate and initial mean sward height were the two main variables that influenced early spring sward height significantly (Table 1).

Autumn and winter stocking rates mainly determined mean sward height in early spring but autumn stocking rate had a more significant effect (Model 1, Figure 1). For model 2, in mid-spring (hatching), early spring stocking rate and initial sward height had both significant effects. Interaction between stocking rate and initial sward height was significant also, indicating that stocking rate effect was greater for short swards (Figure 1). In both analyses, N index had no significant effect.

Table 1 Results of regression models (with year as a random effect)

Dependant	Variables	df	Estimate	SE	p-value
Model 1	Intercept	2	1.251	0.028	$\leq$0.01
	Autumn stocking rate*	25	-0.085	0.025	$\leq$0.01
Mean sward height* before spring grazing (settlement)	Winter stocking rate*	25	-0.086	0.041	0.05
Model 2	Intercept	2	0.526	0.157	$\leq$0.1
	Early spring stocking rate* (SR$_{es}$)	24	-0.575	0.154	$\leq$0.001
Mean sward height* at hatching	Initial sward height* (SH$_i$)	24	0.746	0.135	$\leq$0.001
	SR$_{es}$ x SH$_i$	24	0.385	0.142	0.012

*Log10 transformed

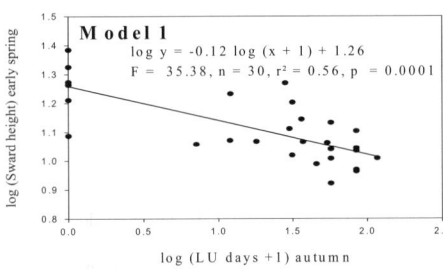

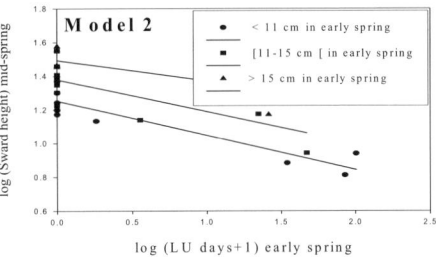

Figure 1 Representations of model 1 and model 2

Conclusions Heavy grazing in autumn and early spring, which slows re-growth in the next period, is needed to create a short sward at critical stages for breeding lapwing. However heavy grazing in spring should be assessed carefully in relation to potential negative effects related to nest-trampling (in spring).

References

Lemaire, G. & J. Salette (1984). Relationship between herbage growth and N uptake in a pure stand. I. - Environmental effects. *Agronomie*, 4: 423-430.

Effects of plant species diversity in multifunctional grasslands on avian communities

K.K. Bakker[1] and K.F. Higgins[2]
[1]College of Arts and Sciences, Dakota State University, 820 N. Washington, Madison, South Dakota, USA 5704, Email: kristel.bakker@dsu.edu, [2]South Dakota Cooperative Fish and Wildlife Research Unit, USGS, South Dakota State University, Brookings, South Dakota, USA 57007

Keywords: grassland birds, plant diversity, North America, sown grasslands, native sod prairies

Introduction Grasslands, planted or native sod, can provide multiple simultaneous functions (e.g., hay-forage production, biomass fuel products, bird nesting cover, soil and water conservation). Bird use of and abundance in grasslands is dependent on vegetation structure, size, and surrounding landscape (Bakker et al., 2002) but little is known about avian relationships to plant species diversity. We evaluated the relationship between plant species diversity and avian community structure in 5 grassland types. The results have application to grazing, haying, burning, and conservation practices applied to native and sown grassland stands and to conservation strategies for declining grassland bird populations.

Methods Birds and vegetation were surveyed on 86 grassland sites in eastern South Dakota and western Minnesota, USA, during the 1999-2004 breeding seasons. Bird species occurrence and density were calculated from surveys of 2-ha fixed-width belt transects. Plant species diversity, height-density readings, litter depth, and grass, forb and woody vegetation heights were recorded to evaluate vegetative structure. Grassland types consisted of switchgrass (n=13) and intermediate wheatgrass (n=16) monotypes, cool-season (n=18) and warm season (n=21) mixes containing 3-6 plant species, and native sod prairies (n=18) containing ≤119 species. We used Analysis of Variance (ANOVA) to evaluate vegetation structural differences between fields and the effects of plant species diversity on grassland bird communities.

Results Mean litter depths were significantly greater in cool and warm season mixes. Warm season mixes and native prairie had significantly lower height-density readings than other grassland types. Clay-colored Sparrows (CCSP), and Grasshopper Sparrows (GRSP) had their highest and Savannah Sparrows (SASP), Bobolinks (BOBO), and Western Meadowlarks (WEME) had their 2nd highest frequency of occurrence in native prairie (Table 1). Sedge Wrens (SEWR) occurred more often in warm season mixes. All 5 grassland types supported the highest density of at least one grassland bird species (Table 1). The density of 3 species was highest in warm season mixes. Dickcissel (DICK) occurrence and density were similar in all grasslands. Total avian richness increased directly with plant diversity, except in cool season mixtures.

Table 1 Mean grassland bird species occurrence (Occ) (%) and density (Dens) (birds/100 ha) by grassland type

Avian Species	Switchgrass		Wheatgrass		Warm Season		Cool Season		Native Sod	
	Occ	Dens	Occ	Dens	Occ	Dens	Occ	Dens	Occ	Dens
SEWR	7.1	28.5	33.3	29.0	75.0*	85.0*	50.0*	64.0*	12.5	9.5
DICK	21.4	39.5	33.3	37.5	10.0	12.5	27.8	39.0	18.7	12.5
CCSP	35.7	3.5	0.0	0.0	15.0	10.0	5.6	2.8	18.7	15.5
SASP	28.6	18.0	58.3	70.0*	45.0	30.0	33.3	36.0	43.8	31.5
GRSP	21.4	13.5	0.0	0.0	45.0*	57.5*	22.2	22.0	56.2*	47.0*
BOBO	7.1*	3.5*	75.0	140.0	55.0	135.0	50.0	61.0	68.7	65.5
WEME	7.1	7.0	25.0	25.0	55.0	60.0*	33.3	22.0	56.2	37.5*
Richness	2.39 (0.37)		1.56 (0.24)		2.95 (0.85)		2.17 (0.31)		3.29 (.025)*	

*Significantly (p≤0.05) different from other values for occurrence or density for that species

Conclusions Grasslands with greater plant diversity had higher avian richness, occurrence and density. We recommend that high priority be placed on conserving remaining native prairie. Our results indicate that a high diversity of plant species should be incorporated into sown grassland mixtures to benefit the majority of grassland birds.

References

Bakker K.K., D.E. Naugle & K.F. Higgins (2002). Incorporating landscape attributes into models for migratory grassland bird conservation. Conservation Biology, 16, 1638-1646.

Diversity and adaptation of perennial plants from North Africa: legumes and grasses

A. Abdelguerfi[1], M. Laouar[2], K. Abbas[3] and M. M'Hammedi Bouzina[4]
[1]Lab-RGB, INA Belfort, El Harrach 16200 Alger, Algérie, Email: aabdelguerfi@yahoo.fr, [2]INRAA Belfort, El Harrach 16200 Alger, Algérie, [3]INRAA, Unité de Sétif, Algérie, [4]Université de Chlef, Chlef, Algérie

Keywords: species, diversity, forage, pastoral, adaptation

Introduction A variety of climates, soil types and reliefs characterise North Africa (Algeria, Morocco, and Tunisia). These natural conditions have produced a large diversity of environments, landscapes, plant formations and flora. Several factors have contributed to the spread of a wide floristic variety: the influence of the Mediterranean in the north and in the north-east, the Atlantic in the west, and the Sahara in the south, as well as the presence of mountain ranges, particularly the Rif, the Tell Atlas, the Sahara Atlas, the Middle Atlas, the High Atlas, and the Anti-Atlas. This diversity of landscapes, environments and ecosystems has generated many different agricultural and breeding practices. Ranges as well as ecosystems favouring animal breeding constitute the most important formations and occupy the largest areas. Climatic factors, especially rainfall and temperature in the growth period, are the most important limitations to the production of high quality forage and/or pasture. This paper describes the floristic diversity of perennial herbaceous legumes and grasses, and their particularities, place and importance in agriculture in the region.

North African floristic diversity Morocco, Algeria and Tunisia have a very high number of plant species. Morocco has 3,700 species (4,200 including sub-species ; M'Hirit & Maghnouj, 1997). Algeria and Tunisia have 3,139 and 2,162 species, respectively. **Morocco** has circa 400 fabaceous species; the fabaceae family is among the richest endemic species, with circa 63 species of the 550 endemic species. Morocco also has 300 poaceous species. *Ononis, Astragalus,* and *Trifolium* genera include between 40 and 50 species; *Lotus* and *Vicia* genera include between 30 and 35 species. The endemism of the North African flora is very high. **Tunisia** has 100 genera and 197 species of grasses, but only 36 genera and 216 species of papilionaceous plants (Nabli, 1989). There are 12, 14, 15, 16, 20, 20, and 28 species, respectively, of *Lotus, Bromus, Lathyrus, Vicia, Ononis, Medicago, Astragalus,* and *Trifolium* genera (Nabli 1989). The absence of high mountains in Tunisia prevents the growth of important endemic flora, as is the case in Algeria and Morocco; there are 34 endemic Tunisian taxa, about 6 of which are fabaceous plants and 1 is a poaceae plant. **Algeria** has important endemism in fabaceae and poaceae plants. The Algerian Sahara has 74 grass genera with 204 species, 19 of which are endemic and 30 legume genera with 154 species, 22 of which are endemic. On the other hand, the Maghreb countries are considered as the centre of a genetic diversity for several genera.

Particularities of some species North African perennial grasses have a winter dormancy in the very cold areas and a marked summer dormancy because of the dry climate and the intense heat. Also, contrary to a large number of European species, most North African species can grow again (alternative production of sucker-ear) after each cutting. These characteristics permit the local populations to resist often unfavourable climatic conditions. Perennial legumes also cease vegetative growth in summer because of the drought and in winter because of the cold in some mountainous regions. The vegetative rest is more or less marked depending on the conditions of the environment (extent of the drought or cold). Some legumes, and Rhizobium, have high resistance to salinity.

Place of herbaceous perennial plants Most North African legumes and grasses are utilised in pastures and very few species are cultivated. They rarely are found in artificial environments but they are widespread in natural environments (meadows, ranges, etc.).

Conclusion North Africa has a huge pool of valuable perennial grasses and legumes for forage and/or pasture. These species have particular adaptation characteristics: many species resist cold and especially drought. North African farmers use very few species at present. To characterise and, especially, to evaluate the main perennial herbaceous plants, grasses and legumes of the area, are urgent priorities. Creation and widespread use of adapted cultivars could increase and improve the production of forage and pasture in the region.

References
M'Hirit, O. & M. Maghnouj (1997). Stratégie de conservation des ressources génétiques forestières au Maroc. Actes Editions, Rabat, Maroc, 123-138.
Nabli, M.A. (1989). Essai de synthèse sur la végétation et la phyto-écologie tunisiennes. I. Eléments de botanique et de phyto-écologie. Faculté des Sciences de Tunis-MAB, Tunisie, 247 pp.

Comparison of compositional changes in multi-species grass/legume mixture experiments across three Nordic countries (Iceland, Sweden and Finland) over two years

C. Brophy[1], A.M. Gustavsson[2], A. Helgadottir[3], O. Nissinen[4] and J. Connolly[1]
[1]University College Dublin, Dublin 4, Ireland, Email: caroline.brophy@ucd.ie, [2]Swedish University of Agricultural Sciences, SLU, Department of Agricultural Research for Northern Sweden, Norrländsk jordbruksvetenskap, Crop Science, Box 4097, S-904 03 UMEÅ, Sweden, [3]Agricultural Research Institute, Keldnaholti, 112 Reykjavik, Iceland, [4]MTT Agrifood Research Finland, Lapland Research Station, Tutkijantie 28, FIN-96900 Saarenkylä, Finland

Keywords: compositional change, mixtures, legumes, grasses

Introduction A multi-site experiment was established at 39 sites across Europe, Australia and Canada within COST Action 852 to: (1) assess the benefits of grass/legume mixtures over monocultures, (2) test the stability of mixtures and (3) evaluate the consistency of the observed patterns over broad environmental gradients. This paper compares the results from 3 Nordic sites using the same plant species: Korpa Experimental Station in Iceland, Lapland Research Station in Finland and Piteå in Sweden to investigate if compositional changes in mixtures are consistent across the sites.

Materials and methods A common experiment is ongoing at each of the 3 sites. See: http://www.cost852.com/ for full details of the experimental design. The experiment comprises 22 plots containing 11 mixtures of 4 species (*Phleum pratense* (G_1, fast growing grass), *Poa pratensis* (G_2, slow growing grass), *Trifolium pratense* (L_1, fast growing legume) and *Trifolium repens* (L_2, slow growing legume)) from 2 plant functional groups, legumes and grasses, sown at 2 densities. The 22 plots were sown with different proportions of the 4 species, varying from domination by one species to equal proportions of all 4. We analysed the average percentage contribution of each of the 4 species to stand biomass for the initial harvest (IH) and the final harvest (FH) for the first 2 years of the experiment for the 3 sites, where available. The analysis for each site fitted all main effects and interactions among the factors mixture, density and species and the main effect of harvest and its 2-factor interactions with the 3 other factors. We summarise the results of the species x harvest interaction here. We also analysed the average percentage of unsown species over the same time period. These data were analysed similarly to exclude the factor species. Because the 3 sites did not begin the experiment at the same time, the available data varied from site to site.

Results and Conclusions The fast growing grass, G_1, (Sweden and Iceland) or the fast growing legume, L_1, (Finland) was the dominant sown species in the initial harvest (Yr 1) (Figure 1). The change between the initial and final harvest for Yr 1 was consistent across the 3 sites: the proportion of G_1 decreased while the proportion of L_1 increased (see Table 1 for significance of tests). This pattern between initial and final harvest changed in Yr 2 for Iceland, as the slow growing species (G_2 and L_2) began to establish and increase their contribution. The proportion of unsown species was higher in the initial harvest than in the final harvest for 2 of the 3 sites in Yr 1. This trend also was observed in Iceland in Yr 2 (Figure 2 / Table 1).

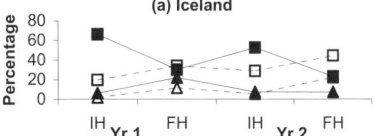

Figure 1 Average percentage contribution of sown species at four harvests for each country

(a) Iceland

(b) Finland

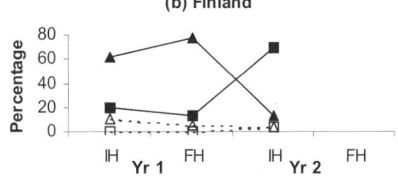

(c) Sweden

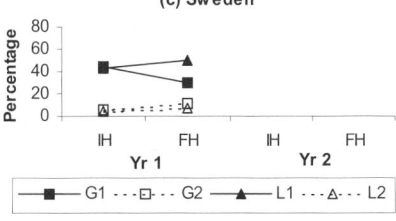

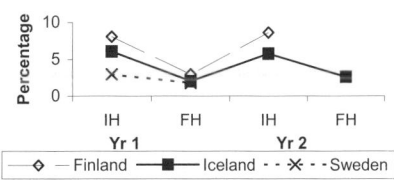

Figure 2 Average percentage of unsown species for four harvests at each country

Table 1 Significance of changes in species percentage from initial to final harvest in year 1 and year 2

	Year 1			Year 2		
	G1 ↓	L1 ↑	Unsown ↓	G2 ↑	L2 ↑	Unsown ↓
Iceland	***	***	***	***	***	**
Finland	*	***	**	-	-	-
Sweden	***	**	ns	-	-	-

With significance levels: *** p<0.0001; ** p < .01; * p<0.05; ns=non-significant

Site effects on the composition of multi-species grass/legume mixtures during sward establishment

R.P. Collins[1], M. Fothergill[1] and J. Connolly[2]

[1]IGER, Plas Gogerddan, Aberystwyth, SY23 3EB, UK, Email: rosemary.collins@bbsrc.ac.uk, [2]University College Dublin, Dublin 4, Ireland

Keywords: establishment, forage yield, multi-species mixtures, species dynamics

Introduction Under the auspices of COST Action 852, a multi-site experiment investigated issues of community structure, function and their interrelationships in multi-species grass/legume mixtures (Collins et al., 2004; Sebastià et al., 2004). We describe the effects of 2 contrasting sites on forage yield and species composition during sward establishment.

Materials and methods Identical experiments were established at 2 sites in Wales, site 1 (AB) at Aberystwyth (30m ASL) and site 2 (BM) at Bronydd Mawr (400m ASL). The experiment used 4 species. Legume (L) and grass (G) functional groups were subdivided further into fast growing, competitive species (subscript '1'), and relatively slow growing, persistent species (subscript '2'). The species used were perennial ryegrass cv. Fennema (G_1); cocksfoot cv. Cambria (G_2); red clover cv. Merviot (L_1), and white clover cv. Alice (L_2). Mixtures were constructed using a range of species proportions based on monocultures sown at 2 densities. High density (HD) monoculture plots were sown at rates of 40, 30, 15 and 5 kg/ha for G_1, G_2, L_1 and L_2 respectively; low density (LD) monocultures were sown at 60% of these rates. This paper describes the development of mixtures sown with equal proportions of the 4 species (i.e. each sown at 25% of its monoculture rate). The response was derived from the mean of yields in the HD and LD plots. The experiment was sown in site AB in 2002 and in site BM in 2003. Results from the first 2 harvests of the following year (2003 for AB and 2004 for BM) were analysed separately. The contributions of individual species to total sward yield were analysed using ANOVA.

Results and Discussion There were no differences between the 2 sites in total sward yield in either harvest (Table 1). However, the yields of individual species differed significantly from each other at both sites in both harvests. Averaged over sites, perennial ryegrass (G_1) was the highest yielding and cocksfoot (G_2) the lowest yielding species in the mixtures in harvest 1. In harvest 2, however, whilst cocksfoot remained the lowest yielding species overall, red clover (L_1) greatly increased its yield to become the highest yielding mixture component. There were significant species x site interactions at both harvests, brought about by the contrasting performance of red clover at the 2 sites. Thus, in the lowland AB site, red clover was consistently high yielding and by harvest 2 had become the dominant component of the mixture. In the upland BM site, red clover was low yielding in harvest 1 and, although its relative performance had improved by harvest 2, it did not become dominant. The mixture in the BM site in harvest 2 was relatively evenly balanced between the 4 species. The results give an indication of the strong influence of environment on the processes occurring during the establishment phase of sward development in multi-species mixtures.

Table 1 Individual species yield and total sward yield (kg/ha) of multi-species mixtures from harvests 1 and 2 at 2 sites in Wales

	Site	G_1	G_2	L_1	L_2	Total Yield
Harvest 1	AB	961	70	995	227	2253
	BM	1204	422	118	165	1909
Harvest 2	AB	426	95	3898	544	4963
	BM	858	468	668	734	2728

Sig. level and lsd for Harvest 1: Species (P=0.006) 190.7; Site (ns); Species x Site (P=0.045) 621.9
 Total Yield: Site (ns)
Sig. level and lsd for Harvest 2: Species (P<0.001) 469.5; Site (P=0.005) 332.0; Species x Site (P<0.001) 664.0
 Total Yield: Site (ns)

References

Collins R. P., J. Connolly and C. Porqueddu (2004). Effects of legume genetic diversity on the productivity of legume/grass mixtures – COST Action 852. In: (Lüscher A. et al., eds) Land use systems in grassland dominated regions. Proceedings of 20th Meeting of European Grassland Federation, Luzern, Switzerland. pp. 486-488.

Sebastià M. T., A. Lüscher, J. Connolly, R. P. Collins, I. Delgado, A. De Vliegher, P. Evans, M. Fothergill, B. Frankow-Lindberg, A. Helgadóttir, C. Iliadis, M. Jørgensen, Z. Kadžiuliene, O. Nissinen, D. Nyfeler and C. Porqueddu (2004). Higher yield and fewer weeds in grass/legume mixtures than in monocultures – 12 sites of COST action 852. Ibid. pp. 483-485.

Unplanned fires and sustainability of a semi-arid rangeland in South Africa

H.A. Snyman
Department of Animal, Wildlife and Grassland Sciences, University of the Free State, PO Box 339, Bloemfontein 9300 South Africa, Email: snymanha.sci@mail.uovs.ac.za

Keywords: aboveground phytomass production, basal cover, root weight

Introduction Whether due to lightning or human negligence, large-scale unplanned fires affect large areas of the semi-arid southern African rangelands during the dormant winter (June-August) period (Everson 1999). These fires cause enormous fodder flow problems, and also influence ecosystem functioning (Snyman 2003). The short-term impact (2 years) of an unplanned fire on the sensitivity of above- and belowground productivity of a semi-arid rangeland was therefore investigated.

Materials and methods The research was conducted in a semi-arid summer rainfall region of South Africa (30^0 15^1S, 27^010^1E, altitude 1652 m). Annual rainfall is 623 mm of which 65% occurs from November to March. The study area is situated in the moist, cool Highveld grassland and consisted of a dense sward of perennial grasses such as *Cymbopogon plurinodes*, *Themeda triandra*, *Digitaria eriantha* and *Elionurus muticus*. Soils in the study area are mostly fine sandy loams. The burning treatment was a single accidental wind-driven head fire. The research was conducted on six plots of 10x10 m^2 each, half of them set out on the burnt and half on unburnt patches. All plots were excluded from livestock grazing for the 2-year (2001/02 to 2002/03) trial period. Basal cover was determined with a bridge-point apparatus every 4 months. Regrowth was determined at the end of each growing season (April) by clipping all grasses to a height of 30 mm in eight randomly selected quadrats of 1m^2 each. Root weight was obtained by extracting root cores with a soil drill (70 mm diameter), at 50 mm depth intervals to a total depth of 600 mm. Roots were removed by washing through a 2.0 mm sieve each of 20 cores randomly selected/plot. One-way analyses of variance at 95% confidence level was computed for basal cover, above- and belowground production.

Results and discussion At 4 and 2 months after the fire, basal cover for all species in the burnt area averaged 40 and 18% less, respectively, than that in unburnt rangeland (P<0.01). *C. plurinodes* and *E. muticus* tufts had the greatest fire damage, while *E. chloromelas*, *D. eriantha* and *T. triandra* were less badly affected. In the first season after the fire, total seasonal production from the burnt rangeland was 1268 kg/ha less than from the unburnt area, 1452 versus 2720 kg/ha, respectively (P<0.01). Two years after the fire, production from burnt and unburnt rangeland differed by only 109 kg/ha (P>0.05). Root weight declined most, by as much as 57%, over the first 100 mm depth (P<0.01). For the two growing seasons, root weight decreased with depth for both the burnt and unburnt rangeland (P<0.01). Root weight increased in the burnt rangeland over all depths for the second season following the fire and was lower (P<0.01) than that of unburnt rangeland only in the first 50 mm depth class. The top 0-300 mm soil layer, which is mostly responsible for production, had 89% of the root weight averaged for the two seasons for unburnt rangeland. The root/shoot ratio decreased (P<0.01) with burning over the first season after the fire from 1.46 to 1.20. The decrease in aboveground phytomass due to burning for the first and second growing seasons was 1268 and 109 kg/ha, respectively, compared to the 1757 and 402 kg/ha decrease respectively of root weight.

Conclusions Accidental or unplanned fires contribute towards anthropogenic droughts in semi-arid areas, and therefore influence the short-term sustainability of rangeland ecosystems. Belowground growth is more sensitive to burning than aboveground growth. These results can serve as scientific guidelines in claims for damages and short-term risk management of semi-arid rangelands after accidental fires.

References

Everson C.S. (1999). Veld burning in different vegetation types. In: N.M. Tainton (ed.) Veld management in South Africa. University of Natal Press, Pietermaritzburg, South Africa, 228-235.
Snyman H.A. (2003). Fire and the dynamics of semi-arid grassland; influence on plant survival, productivity and water-use efficiency. *African Journal of Range and Forage Science*, 20, 29-39.

Effects of burning on grassland vegetation cover on the northeastern side of the Alborz ranges in Iran

F. Amiri, M.R. Chaichi and A. Atrakchali
Department of Rangeland Sciences, Science and Research Branch, The Islamic Azad University, Poonak, Tehran, Iran, Email: rchaichi@ut.ac.ir

Keywords: fire, species diversity, succession

Introduction   Golestan National Park is located on the northeastern side of the Alborz ranges, Golestan province, Iran. Because of the special vegetation cover and being located close to two wet and dry weather areas, this park is vulnerable to fire hazards. Between 1957 and 2004 more than 67 fires have been reported in the park. The international importance of the park requires a careful study on fire effects on vegetation cover, phytomass production, grass diversity and successional process after fires.

Materials and methods   The measurements were recorded in the Takhti Yeylagh grassland site in Golestan National Park. The site had experienced an accidental fire in 1995, which burned more than 720 ha. One area each of burned and unburned (control) were identified and separated on 1/50,000-scaled topography maps and 1/20,000-scaled aireal photos. Vegetation types in both control and burned areas were identified on similarity in soil and topography. In each site 36 samples were taken by stratified randomisation. A double sampling method of transects and quadrats was employed to identify the vegetation types and species frequency in each site (control and burned). The Shannon Index was used to evaluate species diversity in control and burned sites.

Results   The percent vegetation cover of annual and perennial grasses in both control and burned sites significantly ($P<0.05$) increased after fire (Table 1). The results correspond to reports by Engle, *et al* (1998) and Decastro & Kauffman (1998). However, the vegetation cover of forbs significantly decreased in burned areas. The total vegetation cover in burned sites significantly increased compared to the control (Table 1). No species diversity differences were observed within five years after the fire between control and burned sites (Shannon index 0.95 and 1.02). These results are in contrast with Fensham's (1999), who reported a significant difference in tropical Eucalyptus forest ten years after fire in Australia.

Table 1 Vegetation cover of plants in control and burned sites in Golestan National Park five years after fire

Vegetation type	Vegetation cover percent (Burned site)	Vegetation cover percent (Control site)
Annual an biennial grasses	56.9a	21.2c
Perennial grasses	4.7e	12d
Forbs	3.4e	30.6b

Means with the same letter are not significantly different at 5% level

Conclusions   The reduction in perennial grasses and significant increase in annual grasses (secondary species) in burned site compared to control indicates a preventing effect of fire on positive succession towards climax. The increment of grasses after each fire makes the range ecosystem in Golestan National Park more vulnerable to fires because of huge leaf senescence, especially in annuals.

References
Decastro, E. A. & J. B. Kauffman (1998). Vegetation gradient of above ground biomass, root and consumption by fire. *Journal of Tropical Ecology,* 14, 263-283.
Engle, D. M., Mitchell, R. L. & R. L, Stevens 1998. Late growing season fire effect in mid successional tall-grass prairies. *Journal of Range Management*, 51, 115-121.
Fensham, R. J. 1999. Interactive effects of fire frequency and site factor in tropical Eucalyptus forest. *Australian Journal of Ecology*, 15, 65-72.

The influence of tree thinning and subhabitat differentiation on *Panicum maximum* and *Urochloa mosambicensis* of a bush encroached semi-arid savanna in South Africa

G.N. Smit

Department of Animal, Wildlife and Grassland Sciences, University of the Free State, P.O. Box 339, Bloemfontein 9300, Republic of South Africa, Email: smitgn.sci@mail.uovs.ac.za

Keywords: bush encroachment, grass-tree competition, grass species composition

Introduction The productivity of grasses is most important for herbivores in extensive semi-arid savannas. In these areas an increase in woody plant abundance, commonly referred to as bush encroachment, suppresses grasses. This is the main reason why landowners often consider thinning or even total clearing of all woody plants. The aims of this study were to investigate the influence of different intensities of tree thinning and subhabitat differentiation on the grass layer of a semi-arid South African savanna.

Materials and methods The study was conducted in the Limpopo Province of South Africa on a site dominated by the tree species *Acacia erubescens* and *Combretum apiculatum* on sandy soil. The mean long-term seasonal rainfall is 416 mm and usually occurs from October to March. The treatments consisted of 6 plots, each 1 ha, of which the control plot was left undisturbed (100% plot), and the others thinned to 70, 50, 30, 10 and 0 % of the tree density of that of the 100% plot (3679 tree equivalents (TE)/ha, 1 TE = a tree of 1.5 m). Three subhabitats were distinguished: between trees, under live tree canopies and areas previously under the canopies of the removed trees. The grass species composition, based on frequency of occurrence, was determined using the wheel point apparatus. In all, 400 point-observations/plot were recorded, with proportional sampling of the various subhabitats. Data were recorded over a period of 5 seasons after the tree thinning treatments.

Results and discussion Total seasonal rainfall was 481, 277, 335, 495 and 335 mm, respectively, for the 5 seasons after the tree thinning treatments. Grass species (n=17) were recorded in the study area. Only 2 species, *Panicum maximum* and *Urochloa mosambicensis,* made up the bulk of the grass layer. Subhabitat differentiation proved very important, with *P. maximum* preferring the canopied subhabitats, notably under leguminous trees, while *U. mosambicensis* preferred the uncanopied subhabitat. In response to the tree thinning treatments these 2 grass species displayed a distinct negative relationship (P<0.01) (Figures 1 & 2). *Panicum maximum* initially responded favourably to the high intensity tree thinning operation, but quickly declined in these plots, while *U. mosambicensis* displayed a corresponding increase in those plots (Figures 1 & 2).

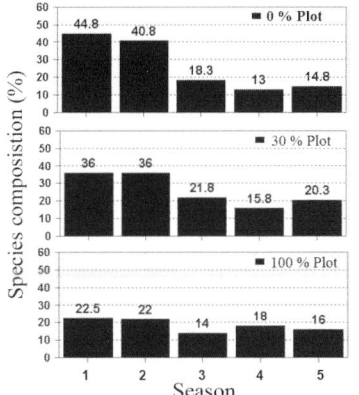

Figure 1 Contribution of *Panicum maximum* to the total grass species composition during the five seasons after tree thinning

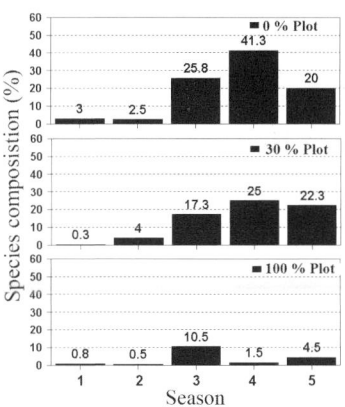

Figure 2 Contribution of *Urochloa mosambicensis* to the total grass species composition during the five seasons after tree thinning

Conclusions The data confirmed the suppressive effect that trees at high densities have on the grass layer and that tree thinning/clearing is essential to restore the production potential of the herbaceous layer. However, the potential loss of *P. maximum* (a highly desirable perennial grass), in favour of *U. mosambicensis* (a less desirable annual grass) at high intensities of tree thinning, confirmed the value of trees in this savanna ecosystem. It is concluded that tree thinning to circa 30% of the initial tree density is the most desirable and that total tree clearing should be avoided.

Maintaining grassland plant diversity while controlling woody plant encroachment

J. Stubbendieck, K.L. Kottas, S.J. Tunnell and S.J. Palazzolo
Department of Agronomy and Horticulture, University of Nebraska, P.O. Box 83095, Lincoln, Nebraska 68583-0915, USA, Email: jstubbendieck@unl.edu

Keywords: herbicides, plant diversity, restoration *Rhus glabra*, tallgrass prairie

Introduction The tallgrass prairie region of the United States is a fragmented grassland ecosystem. Much of the 1 to 2% of the remaining prairie is being degraded by invading woody plants, which frequently results in a shift from grassland to woodland. Smooth sumac (*Rhus glabra* L.), a shrub native to the region, can rapidly increase in density and become dominant in the plant community. Prescribed fire alone is not a constraint to this resprouting species (Stubbendieck *et al.*, 2003). Herbicides are useful tools for managing woody plants in grasslands, but the negative response of desirable plants to herbicides is a concern. Our objective was to control smooth sumac with selective herbicides while minimizing their negative influence on the forb community.

Materials and methods Two experiments, replicated in time, were conducted on a remnant tallgrass prairie in east Nebraska to evaluate smooth sumac control and forb response to spring-applied herbicides, application rates, and broadcast spray and wick application techniques. The study was designed as a randomized complete block with 3 replicates and a control and 13 treatments (Table 1) applied to plots (7 x 10 m) in burned and unburned prairie. Smooth sumac stem density was determined before treatment, and 1 and 4 months after treatment. Forb frequency was recorded in the autumn after spring treatment and in the following autumn. Forb species richness was determined by tabulating the total number of forbs in each plot. A mixed model analysis of variance was used to assess treatment differences (Littell *et al.*, 1996).

Table1 Herbicide treatments and rates applied as broadcast spray or with a wick

Broadcast Spray Treatments	Wick Treatments
1.06 and 2.13 kg ae 2,4-D ester/ha	1.40 kg ae 2,4-D amine/ha
0.15 kg ae picloram + 0.56 kg ae 2,4-D ester/ha	0.20 kg ae picloram + 0.74 kg ae 2,4-D amine/ha
0.20 kg ae picloram + 0.84 kg ae 2,4-D ester/ha	1.48 kg ae triclopyr/ha
1.26 kg ae triclopyr + 0.42 kg ae clopyralid/ha	1.11 kg ae glyphosate/ha
1.26 kg ae triclopyr/ha and 2.24 kg ae triclopyr/ha	0.74 kg ae picloram/ha
0.56 kg ae picloram/ha	Control

Results All herbicide treatments reduced smooth sumac stem density significantly (an average of >90%), but no distinct advantage was detected as regards type of herbicide, application rate, or method of application. Prescribed burning did not increase smooth sumac mortality. Burning was expected to make smooth sumac more susceptible to herbicides, but it resulted in increased stem density. With few exceptions, herbicide type, application rate, or application method did not influence forb frequency. Leadplant (*Amorpha canescens* Push), classified as a shrub, is an important indicator species and a desired component of tallgrass prairie (Stubbendieck & Conard, 1989). Frequency of leadplant was greatest where 2,4-D was applied selectively with a wick, while western ragweed (*Ambrosia psilostachya* DC.) and Missouri goldenrod (*Solidago missouriensis* Nutt.) were greatest where glyphosate was applied with a wick. In experiment 2, annual sunflower (*Helianthus annuus* L.) increased and was the only forb that herbicide affected significantly. We believe that the seed bank of this early successional annual was sufficient to increase plant recruitment after reduction in smooth sumac. Species richness varied by treatment in experiment 1, with the greatest species richness in plots where picloram and 2,4-D were applied with a wick. Forb species richness did not differ among treatments in experiment 2.

Conclusions Herbicide use was a viable method to reduce woody plant encroachment to restore grasslands under our experimental conditions. All herbicides effectively controlled smooth sumac, without detected adverse effect on the native forb community. Therefore, economics may influence selection of the herbicides and application equipment to be used.

References

Littell, R. C., W.W. Stroup & R. D. Wolfinger (1996). SAS system for mixed models. SAS Institute, Cary, North Carolina, 633 pp.

Stubbendieck, J., M. J. Coffin & L. M. Landholt (2003). Weeds of the Great Plains. Nebraska Department of Agriculture, Lincoln, Nebraska, 605 pp.

Stubbendieck, J. & E. C. Conard (1989). Common legumes of the Great Plains. University of Nebraska Press, Lincoln, Nebraska, 330 pp.

The effect of different utilisation methods on composition of semi-natural grassland

J. Jančovič, Ľ. Vozár and Ľ. Jančovičová
Slovak Agricultural University, Tr. Andreja Hlinku 2, SK-949 76 Nitra, Slovakia, Email: Jan.Jancovic@uniag.sk

Keywords: semi-natural grassland, botanical composition, cessation of fertilising

Introduction There has been much research on application of mineral fertilisers to permanent grasslands to maximise production of above-ground phytomass (Folkman & Jančovič 1990), but the changes arising after exclusion of fertilisers have not been studied in Slovak Republic up to now despite current interest in low input ststems. This paper reports on the effects of cessation of fertiliser inputs for an eight-year period.

Materials and methods The changes in botanical composition were investigated on permanent grassland (association *Lolio-Cynosuretum cristati*) on a site in the Strážov Hills (central part of the Slovak Republic). A small-plot experiment was established with four replicates (area of the harvest plot was $10m^2$). In the years 1986-1989, grassland was harvested at the grazing stage four times a year, in the years 1990-1993 at the hay-making stage (2-3 cuts). During the years 1994-2001 fertilisation was omitted and only one cut was realised at the time of maximum biomass production according to the method of Rychnovská *et al.* (1987). Before each cut, botanical analysis was assessed on particular treatments by the method of projective dominance for the purpose of determining changes in botanical composition of grassland. Fertiliser treatments in the years 1986-1993 were as follows: 1. K – non-fertilised control, 2. PK – constant rate of P and K (35 kg P/ha and 70 kg K/ha), 3. N_{60} – 60 kg N/ha was applied in spring (+PK), 4. N_{120} – 80 kg N/ha was applied in early spring, 40 kg N/ha after the 1st cut (+PK), 5. N_{240} – 100 kg/ha was applied in early spring, 80 kg N/ha after the 1st cut and 60 kg N/ha after the 2nd cut (+PK).

Results In the initial year (1986) of the experiment grasses dominated in all treatments. Cover was compact without any blank places. By 2001, after eight years of intense fertilising and utilisation and after eight years exclusion of fertilisation and minimum harvest (1cut) changes of a degraded character had happened in botanical composition (Table 1). The covers had markedly thinned and blank places represented 16% (treatment 1) to 40% (treatment 5) of the total area. After 16-years observation (1986-2001) only seven species dominated in the respective treatments (Figure 1).

Table 1 Composition of grassland in 1986 and 2001 (%,±SE)

Year	Treatments	Botanical groups			
		Grasses	Legumes	Other herbs	Blank places
1986	1	70.75±2.2	0.75±0.3	29.5±2.2	–
	2	69.0±1.7	5.25±0.6	25.75±1.6	–
	3	64.0±2.7	1.75±0.3	34.25±2.8	–
	4	57.0±2.9	3.0±0.4	40.0±3.1	–
	5	66.5±1.9	1.0±0.2	32.5±2.2	–
2001	1	42.0±3.0	12.0±5.0	30.0±2.0	16.0±6.0
	2	40.0±6.0	5.0±0	37.5±3.5	17.5±2.5
	3	28.0±0	2.0±1.0	40.0±6.0	30.0±5.0
	4	48.5±5.5	1.0±1.0	31.5±5.5	19.0±1.0
	5	37.5±4.5	–	22.5±5.5	40.0±10.0

Conclusions The results of the present study demonstrate that intense utilisation without fertilising reduced the number of species and the dominance of grasses and increased the portion of dicotyledonous plants, mainly those forming ground leaf rosette (*Plantago lanceolata and Taraxacum officinale*). The highest stability of dominance during 16-year investigating period was shown by the following seven species – *Agrostis tenuis* Sibth., *Anthoxanthum odoratum* L., *Festuca rubra* L., *Festuca pratensis* L., *Achillea millefolium* L., *Alchemilla vulgaris* L and *Taraxacum officinale* auct. non Weber.

This paper was supported by the Slovak Grant agency VEGA No. 1/2425/05.

Acknowledgement Grant agency Slovak university of Agriculture No. 703/02190

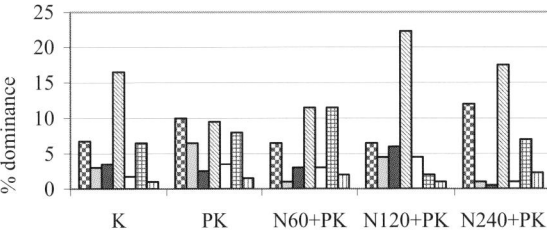

Agrostis tenuis Sibth.
Anthoxantum odoratum L.
Festuca pratensis Huds.
Festuca rubra L.
Achillea millefolium L.
Alchemilla vulgaris L.
Taraxacum officinale auct non. Weber

Figure 1 Dominance of the most widespread species (2001)

References
Folkman, I. & J. Jančovič (1990). Uplatnenie stupňovaných a striedavých dávok dusíka na TTP. Záverečná správa. VŠP Nitra, 71 pp.
Rychnovská, M. *et al.* (1987). Metody Studia Travinných Ekosystémů. Academia, Praha.

Impact of the agricultural use on the biodiversity of a *Festuca rubra* meadow

I. Rotar, F. Păcurar, R. Vidican and N. Sima
University of Agricultural Sciences and Medicine, Faculty of Agriculture, 3-5 Mănăştur Street, 3400, Cluj-Napoca, Romania, Email: roxanavidican@yahoo.com

Keywords: meadow, biodiversity, fertilisation

Introduction Technological inputs into meadow ecosystems trigger significant changes in the sward. In this paper we present the effect of organic and mineral fertilisation on the biodiversity of a *Festuca rubra* meadow.

Materials and methods The experimental field was located in the Apuseni Mountains, Romania, at an altitude of 1150 m, an annual rainfall of 1200 mm/year and an annual average temperature of 4° C. Two experiments were carried out where different quantities of organic and mineral fertilisers were applied.

Results Important changes in the pasture vegetation were recorded after a four year period. After the organic and mineral fertilisers are applied, The plant groups in the swards after this period are given in Table 1. Some species increase their cover following fertiliser application (eg. *Trisetum flavescens, Trifolium repens, Trifolium pratense, Vicia cracca, Centaurea pseudophrygia, Pimpinella major, Stellaria graminea*), others decrease (eg. *Festuca rubra, Agrostis capillaris, Luzula campestris, Alchemilla vulgaris, Plantago lanceolata, Plantago media, Potentila erecta, Hypochoeris radicata*), while some disappear when certain quantities of manure are applied (Table 1). Some species increase their proportion by cover (*T. flavescens, A. capillaris, T. pratense, V. cracca, C. biennis, P. major, Taraxacum officinale, S. graminea*), some partially lose their share (*F. rubra, Lotus corniculatus, A. vulgaris, C. pseudophrygia, Hypericum maculatum, P. lanceolata, P. media etc*), while other disappear at certain rates (Table 1).

Table 1 Plant cover following four years of application of organic or inorganic fertilisers at different rates

	Control	10 t/ha manure	20 t/ha manure	30 t/ha manure
Poacee (%)	42.9	30.2	21.3	30.7
Fabacee (%)	9.4	29.2	35.2	28.5
OBF (%)	44.3	53.6	51.8	45.1
Species that disapear		*Arabis hirsuta, Leontodon autumnale, Prunella vulgaris, Scabiosa columbaria, Thymus dacicus, Carex pallescens, Gentiana praecox*	*Campanula patula Polygala vulgaris*	*Luzula campestris, Centaurea pseudophrygia, Cerastium glomeratum, Ranunculus bulbosus, Trollius europaeus, Viola declinata*
	control	50N 25P_2O_5 25K_2O	100N 50 P_2O_5 50K_2O	150N 75P_2O_5 75K_2O
Poacee (%)	31.6	19.4	33.0	53.3
Fabacee (%)	6.7	24.9	22.7	9.6
OBF (%)	53.0	51.9	46.8	39.2
Species that disapear		*Carex pallescens, Arnica montana, Campanula patula, Gymnadenia conopsea, Leontodon autumnale, Potentilla erecta, Polygala vulgaris, Prunella vulgaris, Thymus dacicus*	*Arabis hirsuta Centaurea pseudophrygia Leontodon autumnale Scabiosa columbaria Trollius europaeus Carlina acaulis*	*Luzula campestris Plantago media Rhinanthus minor Viola declinata*

OBF-plants from other botanical families (forbs)

Conclusions Agricultural use significantly influences the biodiversity of *Festuca rubra* meadows, determining major changes in the sward by increasing proportion of cover of some species, by reducing the cover of others and by causing the disappearance of some species from the sward.

References
Păcurar F. & I. Rotar, (2004). Maintaining biodiversity and increasing the production of dry matter on mountain meadows. *Grassland Science in Europe*, 9, 216-218.
Rotar, I., F. Păcurar, R. Vidican & N. Sima (2003). Effects of manure/sawdust fertilisation on *Festuca rubra* type meadows at Ghetari (Apuseni Mountains). *Grassland Science in Europe*, 8, 192-194.

Long-term effect of levels of N-, P-, K-supply on the Shannon-Index for two pastures located in Central Germany

J.F. Oerlemans and W. Opitz von Boberfeld
Justus-Liebig-University Giessen, Department of Grassland Management and Forage Growing, Ludwigstr. 23 D-35390 Giessen, Germany, Email: Judith.Oerlemans@agrar.uni-giessen.de

Keywords: Shannon-Index, biodiversity, fertilisation, mowing pasture, *Lolio-Cynosuretum*

Introduction Intensive grassland production, including the use of mineral fertilisers, has degraded the botanical diversity of grassland communities seriously (Chapman, 2001). There is little information on optimal amounts of soil nutrient availability to maintain/regenerate species-rich grassland communities. This study in Central Germany aimed to quantify the long-term effect of different N-, P-, K-supply combinations on biodiversity, expressed in terms of the Shannon-Index (Sh-Id), of 2 pastures classified as *Lolio-Cynosuretum*.

Materials and methods In a Latin square design with 3 replicates, 2 pastures on mineral soils at 260 and 360m above sea-level, respectively, were fertilised with 27 different N-, P-, K-supply combinations yearly from 1986-2002. Unfertilised soil exchangeable P and K levels were 1.3 and 13.0 mg/100 g, respectively, for the 260m site and 3.0 and 9.4 mg/100 g for the 360m site. Dry matter proportions of the species were estimated in spring 2002 using the Klapp/Stählin method (Klapp, 1929) and the Sh-Id (Magurran, 1988) was calculated for each plot.

Results N-supply had most influence on Sh-Id at both sites (Figure 1). K-supply had a significant effect at the 360m site, and P x K and P x N interactions had minor influences. Sh-Id for fertiliser combinations without N-supply was higher at the 360m than the 260m site. This was due mainly to more species at this site without N-supply. Increasing N-supply on the 360m site encouraged *Elymus repens* and *Alopecurus pratensis* dominance and exclusion of less aggressive species; overall species and Sh-Id values decreased. Increasing the P- or K-supply on this site within a specific supply combination had an ambiguous effect on the Sh-Id. Although K-supply reduced the number of herb species, higher P- or K-supply sometimes increased the evenness of the species abundance distribution, resulting in higher Sh-Id values. N-supply encouraged *Holcus lanatus* and *Alopecurus pratensis* dominance and reduced the species number on the 260m site but species reduction was less than at the 360m site. N-fertilisation lowered Sh-Id only moderately at the 260m site because N-supply did not influence the evenness of species abundance distribution. Despite low original soil P levels, P-fertilisation had no effect on the Sh-Id at this site.

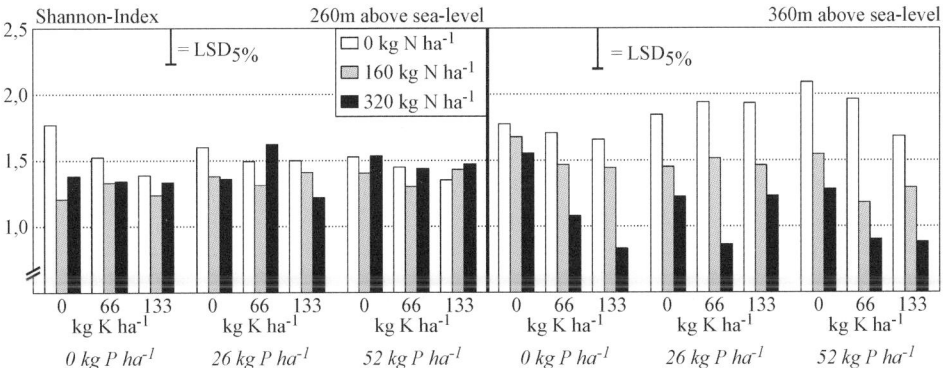

Figure 1 Shannon Index in relation to long-term N-, P-, K-supply for 2 pastures in Germany

Conclusion Nil-fertilisation does not necessarily result in highest biodiversity. Instead, the maintenance of long-term combinations of P- and K-supply, at levels specific to different sites, can increase biodiversity significantly.

References
Chapman, R. (2001). Recreated botanical diverse grassland. In: P.G. Tow & A. Lazenby (eds.) Competition and Succession in Pastures. CAB International, Wallingford, 261-282.
Klapp, E. (1929). Thüringische Rhönhutungen. *Wissenschaftliches Archiv der Landwirtschaftlichen Abteilung A, Archiv Pflanzenbau, 2,* 704-786.
Magurran, A. E. (1988). Ecological Diversity and Its Measurement. Princeton University Press, New Jersey.

The effect of NPK fertilisation on structure and species composition of grasslands

F. Hrabě and J. Skládanka
Mendel University of Agriculture and Forestry, Department of Fodder Production and Grassland Management, Zemědělská 1, 613 00, Brno, Czechoslovakia, Email: sklady@mendelu.cz

Keywords: fertilisation, grassland sward, grasses, legumes, herbs

Introduction Long-term research into the effect of N+PK nutrition enables an objective evaluation of the trend and rate of succession and a prediction of changes including production development in permanent grassland (Hrabě *et al.*, 1991). Former research (Hrabě & Halva, 1993) and evaluation within an eco-system concept (Rychnovská *et al.*, 1994) demonstrated marked changes in the species composition of grass communities with the application of N+PK fertilisation that did not correspond to the sward type and site conditions.

Material and methods The experimental site is situated in the Českomoravská vrchovina highland at an altitude of 650 m a.s.l., total annual rainfall is 786 mm, mean annual temperature is 6.3°C. The period of study was 1992-2002. Fertilisation treatments were H_0 (no NPK application), H_1 (30 kg P/ha, 60 kg K/ha); H_2 (90 kg N/ha, 30 kg P/ha, 60 kg K/ha), H_3 (180 kg N/ha, 30 kg P/ha, 60 kg K/ha). The grassland swards used were a seminatural sward (PG) of the *Sanguisorba-Festuceum comutatae* type and a newly-sown sward (SG) established in 1991 (sown with a grassland mixture: *Trifolium pratense* 3.0 kg/ha, *Trifolium repens* 2.0 kg/ha, *Festulolium* 12.0 kg/ha, *Dactylis glomerata* 8.0 kg/ha, *Lolium perenne* 4.0 kg/ha). The measurements made were the changes in proportion of grasses, legumes and herbs, the effect of dominant plants and the effect of newly-sown legumes and grasses.

Results With intensive N+PK nutrition, the grass component (66-80%) dominated farm fodder from both sward types. In the absence of legumes, the remaining proportion was herb biomass. As compared with PG, the contribution of herbs in SG was about 1/3 lower. The extensive PG nutrient treatment, i.e. the variant with zero fertilisation and with only PK fertilisation, resulted in a stand type with a slight predominance of herbs (<50 %) and a proportion of legumes ranging from 3.5-9.7%. In the extensive SG nutrient treatment, the grass component was dominant, and the proportion of legumes was similar to that in PG (4.1-8.0%). Dominant species (Table 1) in both sward types and in all variants of fertilisation were *Alopecurus pratensis, Poa pratensis, Sanguisorba officinalis* and *Polygonum bistorta.*

Table 1 Proportion (%) of some dominant grass and forbs species in meadow herbage

Species	Sward type	Fertilisation variant			
		H0	H1	H2	H3
Poa pratensis	PG	8.4	11.9	13.6	22.2
	SG	5.7	8.9	9.8	12.8
Alopecurus pratensis	PG	12.8	14.6	25.0	31.8
	SG	7.6	14.3	12.3	17.2
Sanquisorba officinalis	PG	10.4	8.3	6.7	4.2
	SG	4.0	2.8	0.3	0.7
Polygonum bistorta	PG	4.4	21.5	15.2	10.8
	SG	6.6	3.7	7.7	5.1

Conclusions The NPK fertilisation increased the proportion of grasses in both sward types and in SG decreased the herbs. The legume contribution was low, only 4-8% in SG. There was speedy regeneration of species composition in SG, from 8 species to 29 species.

Acknowledgements The work originated from the grant support MSM 432100001.

References
Hrabě F., P. Blížkovský, B. Kubíková & Z. Pospíšil (1991). Variability in the development of plant species in natural and resown permanent meadows. Report EGF Symposium, Grassland renovation and weed control in Europe. Graz, 227-230.
Hrabě F. & E. Halva (1993). Limits of forage production and the efficiency of grassland management. In: J. Rychnovská (ed) Structure and functioning of seminatural meadows. Elsevier, Amsterdam, 165-192.
Rychnovská M., D. Blažková & F. Hrabě (1994). Conservation and development of floristically diverse grasslands in central Europe. In: J. Frame (ed) *Grassland and Society Proceedings of the 15ᵗʰ Meeting of the EGF*, 267-277.

Effect of aspect and animal movement on a temperate mountain grassland structure

S. Mendarte[1], I. Amezaga[2], I. Albizu[1], A. Ibarra[1], I. Mijangos[1] and M. Onaindia[2]
[1]NEIKER, Nekazal Ikerketa eta Garapenerako Euskal Erakundea, Basque Institute for Agricultural Research and Development, Berreaga, 1. 48160 Derio, Spain, Email: smendarte@neiker.net, [2]Department of Plant Biology and Ecology, The University of the Basque Country, POB 644, 48080 Bilbao, Spain

Keywords: extensive pastures, cover, functional groups, graminae, legumes

Introduction Mountain pasture systems are maintained by the combination of physical and environmental factors and human activity leading to highly complex ecosystems (Watkinson & Ormerod 2000). Differences between pastures are usually due to physical conditions and animal pressure (Fynn & O´Connor 2000). In extensive livestock production systems in the Basque Country (northern Spain), livestock graze in mountain pastures from May until October. This study examined the effect of aspect and animal movement on grassland structure.

Materials and methods The study was performed in the Spring of 2000 in 3 grasslands (each 100 ha) (North, South and Southwest aspects) and in 4 zones (each 10 ha) related to the animal movement (Water point, Hut, Extensive and Resting zone) in the Natural Park of Aralar (11000 ha) (Basque Country, NE-Spain), where more than 3000 ha of original beech forest have been maintained as grassland by grazing since Neolithic times. At each zone, 3 sites (1 ha each) were selected and 10 random quadrats (0.5 X 0.5 m) were used in each site to determine the percentage cover of the plant species present, which were then were grouped into functional groups (graminae, legumes, herbs, woody plants and non productive soil) for the analysis.

Results In general, graminae were the dominant functional group, the most common species being *Agrostis capillaris* and *Festuca* gr. *rubra*. Legumes, mainly *Trifolium repens,* had the highest percentage cover in the zones with highest herbivore pressure (Hut zone and Water point). Cover of herbs (e.g. *Bellis perennis, Carex caryophillea, Potentilla erecta, P. montana*) was the highest in the lower grazing pressure zones, namely the Extensive and Resting zones, while woody plants (e.g. *Thymus praecox, Daphne laureola*) had higher cover in the zone with the lowest animal and herbivore pressure (Resting zone) (Figure 1). The Extensive zone, with an intermediate herbivore pressure, had intermediate values of most of the considered functional groups. The Resting zone, mainly in the South and Southwest aspects, was the most different one. The relative cover of bare (non productive) soil, possibly due to the rough conditions of the area (steep and windy), and woody species were higher than in the other zones (Figure 1). Although graminae were dominant in those grasslands, in the Southwest aspect their importance was lower than in the other two aspects (Figure 1). This is mainly seen in the Resting zone where species such as *Carex caryophillea, Gallium saxatile, Hieracium pilosella, Luzula campestris and P. erecta* have a considerable presence.

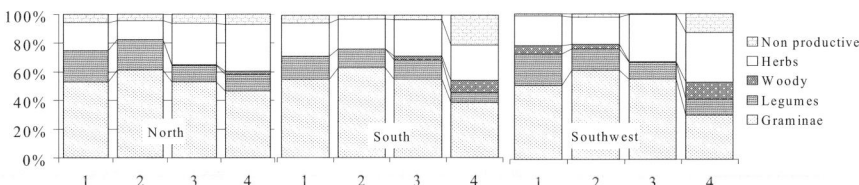

Figure 1 Percentage cover of the studied functional groups of species. 1: Hut zone; 2: Water point; 3: Extensive zone; 4: Resting zone.

Conclusion The cover of graminae and legumes was highest in zones with highest herbivore pressure. The Resting zone, with the lowest herbivore pressure, had the lowest percentage of graminae and the highest of non graminoids, i.e. legumes and herbs. When the pressure was intermediate, the groups of species were more evenly distributed, especially in the Southwest aspect.

References

Watkinson A.R. & S.J. Ormerod (2001). Grasslands, grazing and biodiversity: editor´s introduction. *Journal of Applied Ecology*, 38, 233-237.

Fynn R.W.S. & T.G. O´Connor (2000). Effect of stocking rate and rainfall on rangeland dynamics and cattle performance in a semi-arid savanna, South Africa. *Journal of Applied Ecology,* 37, 491-507.

Do species and functional diversity indices reflect changes in grazing regimes and climatic conditions in northeastern Spain?

F. de Bello[1], J. Leps[2] and M.T. Sebastià[1]

[1]*Laboratory of Plant Ecology and Forest Botany, Forestry and Technology Centre of Catalonia, pujada seminari s/n. E-25280 Solsona, Lleida, Spain, Email: fradebello@ctfc.es, [2]Department of Botany, Faculty of Biological Sciences, University of South Bohemia. Na Zlate stoce 1 CZ-370 05 Ceske Budejovice, Czech Republic*

Keywords: land use, disturbance, Mediterranean rangelands, sheep

Introduction Understanding the mechanisms that maintain biodiversity in various ecosystems enables the development of management practices that prevent degradation (Canals & Sebastia, 2000). Each diversity index reflects some compositional properties and could be influenced differently by stress and disturbance factors (Magurran, 2004). In this study, we aim to reveal 1) which management practices and environmental factors affect biodiversity in rangelands of northeastern Spain and 2) the relationship between species diversity and functional diversity (SD and FD).

Methods Species frequencies were measured in natural slope vegetation areas, along a gradient of sheep grazing pressure (high, low, abandonment). Five locations were selected across a steep altitudinal and corresponding climatic gradient from Mediterranean rangelands to natural sub-alpine grasslands. The factorial design was 3 sheep grazing intensities x 5 locations x 2 aspects x 2 replicates = 60 plots. We calculated: 1) species richness (number of species); 2) species diversity according to Shannon (H^1) and Simpson (1-D); 3) species evenness according to Pielou and Camargo (see Canals & Sebastia, 2000); 4) species rarity (number of species classified "rrr") and 5) functional diversity (FD). The average of the pair-wise species differences weighted for their relative frequencies was used as a measure of the functional diversity in such a way that, if difference in all species pairs equals 1, the Simpson index is obtained (Shimatani, 2004). Species differences were calculated on the basis of 8 life history traits for the 467 species found. Three-way ANOVA was used and slope inclination (°) was covariate in the model. Duncan post-hoc test was performed to detect mean differences.

Results The effect of the studied factors on the diversity indices are given in Table 1. Climatic variables were the most important factor affecting species and functional diversity. Generally, diversity indices were lowest in a water-stressed environment (in the driest sites and southern aspects) and increased toward moist areas. FD reached its peak at intermediate elevations but the values in the wettest site resembled those in dry sites. Grazing enhanced species diversity, but no effect was found on functional diversity; species rarity was higher in abandoned areas. There was no clear relationship between species and functional diversity indices.

Table 1 Results of 3-way ANOVA for various diversity indices. Letters indicate different means. (*=P<0.05;**=P<0.01;***=P< 0.001)

	Richness	Shannon	Simpson	E Pielou	E Camargo	Rarity	FD
Slope (covariable)	NS	NS	NS	NS	NS	NS	NS
Aspect (south-north)	*	**	*	**	0.055	NS	***
Location (along climatic gradient)	***	***	***	***	***	***	**
Grazing intensity	0.060	**	*	***	***	*	NS
Aspect x Location	*	0.091	*	NS	NS	0.072	***
Aspect x Grazing	NS	NS	NS	NS	NS	**	NS
Location x Grazing	NS	*	*	*	**	***	***
Aspect x Location x Grazing	NS	NS	NS	*	**	NS	***
R^2 adj	0.60	0.74	0.76	0.73	0.71	0.67	0.86
GRAZING (abban/low/high)	a/b/b (+)	a/b/b (+)	a/b/b (+)	a/a/b (+)	a/a/b (+)	b/a/a (-)	NS

Conclusion Water stress decreases species diversity. Species rich grasslands are not functionally different. Grazing increased species diversity only, having no clear effect on functional diversity. Adequate grazing pressure on natural slope vegetation maintained the diversity of species.

References

Canals, R.M. & M.T. Sebastià (2000). Analyzing mechanisms regulating diversity in rangelands through comparative studies: a case in the south-western Pyrennees. *Biodiversity and Conservation*, 9, 965–984.

Magurran, A.E. (2004). *Measuring Biological Diversity*. Blackwell Publishing, Oxford. 260 pp.

Shimatani, K (2004). On the measurement of species diversity incorporating species differences. *Oikos* 93, 135-147

Plant functional types and grazing management in Mediterranean grassland: an 11-year synthesis

M. Sternberg[1], Z. Henkin[2], A. Perevolotsky[2], M. Gutman[2] and E.D. Ungar[2]
[1]Department of Plant Sciences, Tel Aviv University, Tel Aviv 69978, Israel, Email: MarceloS@tauex.tau.ac.il,
[2]Department of Natural Resources, ARO, The Volcani Centre, PO Box 6, Bet Dagan 50250, Israel

Keywords: cattle, diversity, long-term, plant cover, response groups

Introduction Mediterranean ecosystems have high seasonality in resource availability, high year-to-year rainfall variability, many annual plants in the forage and a long history of grazing and disturbance. These facts, and results of previous studies (Sternberg *et al.*, 2000) suggested that stocking rates could be increased above those of traditional grazing. However, the long-term effects of intensive grazing on the vegetation were not fully known. This study aimed to evaluate the responses of plant functional types (PFT) and community structure to different regimes of cattle grazing in Mediterranean grassland, particularly at high stocking rates.

Materials and methods The experiment was conducted at the Karei Deshe Experimental Farm in north-eastern Israel (lat. 32° 55'N, long. 35° 35'E; altitude 150 m a.s.l; mean annual rainfall 570 mm). Treatments (6) were defined by the combination of grazing intensity (**M**oderate: 0.55 cows/ha per yr; **H**eavy 1.1 cows/ha per yr and Very Heavy: 2.2 cow/ha per yr), duration of grazing (**C**ontinuous vs. **S**easonal) and season of grazing (**E**arly: Jan.-March vs. **L**ate: April-October). Statistical analyses included ANCOVA with treatments, year and interactions as main effects, while grazing pressure (GP) of the present and former year, rainfall of present and previous year and plant biomass were used as covariates. Contrasts analyses included grazing regime (**C** vs. **S**), timing of grazing (**E** vs. **L**), stocking rates, and wet vs. dry years and springs, respectively.

Results Grazing treatment did not affect species diversity; year and rainfall distribution did, particularly in years with dry springs (Figure 1). Inter- and intra-seasonal rainfall variation was a dominant factor in the expression of different grazing treatments on plant community structure. Reduction in cover of tall grasses was correlated with an increase in cover of less palatable PFT's, such as annual thistles and crucifers.

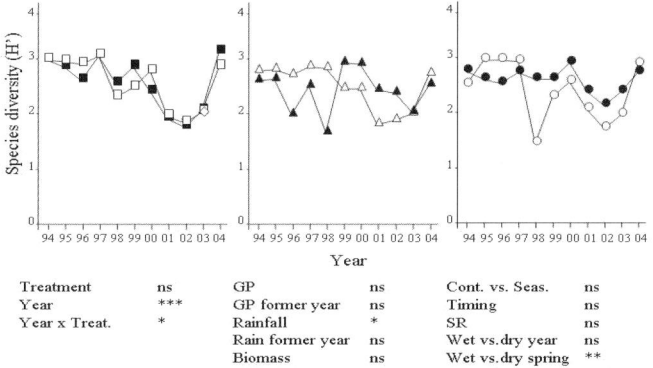

Treatment	ns	GP	ns	Cont. vs. Seas.	ns
Year	***	GP former year	ns	Timing	ns
Year x Treat.	*	Rainfall	*	SR	ns
		Rain former year	ns	Wet vs.dry year	ns
		Biomass	ns	Wet vs.dry spring	**

Figure 1 The effects of grazing on species diversity (Shannon-Weaver index - H'). Treatments: (a) CM (continuous moderate, black squares) vs. continuous heavy, CH (white squares); (b) S-HE (seasonal heavy early, white triangles) vs. S-HL (seasonal heavy late, black triangles); (c) S-VHE (seasonal very heavy early, white circles) vs. S-VHL (seasonal very heavy late, black circles).

Conclusions Persistence of dominant species and the relatively small change in plant cover of the plant functional types suggest that the community was stable in spite of wide variation in grazing regimes and climate. Grazing pressure could be increased to higher stocking rates without comprising the sustainability of the system. Eastern-Mediterranean grasslands seem to be adapted to grazing due to their long history of human association.

References
Sternberg M., M. Gutman, A. Perevolotsky, E.D. Ungar & J. Kigel (2000). Vegetation response to grazing management in a Mediterranean herbaceous community: a functional group approach. *Journal of Applied Ecology*, 37, 224-237.

Floristic composition and species richness model in winter rangelands of northeastern Iran

A. Rashtian and M. Mesdaghi
Department of Range Ecology, University of Agricultural Sciences, Gorgan, Iran, Email: arashtian@yahoo.com

Keywords: flora, vegetation type, species richness, life form, grazing

Introduction Due to the variable annual rainfall, the plant communities of rangelands in northeastern of Iran are fluctuating and the productivity is highly variable. This ecosystem covers about 500,000 ha and is one of the important winter ranges in Iran. In this fluctuating ecosystem, species diversity and floristic composition are changing annually (Pabot, 1967). Under heavy grazing most of the range species were extinct, but under light and moderate grazing, species richness was improved, fitting with the humped-back model (Wilkinson, 1999). The objectives of this research were 1) to determine floristic composition in three vegetation types and 2) to show the effects of rainfall, slope, aspect, and elevation on species diversity.

Material and methods The flora of range species were collected from representative stands of three vegetation types of woodland, grassland, and shrubland and classified according to Raunkauer's life forms. Species richness was measured using a modified Whittaker's plot. The simple regression model: $S = \beta_\circ + \beta_1 \log X + \varepsilon$, was used, where S is species richness, $\beta_\circ$ is the intercept, β_1 is the slope of regression line, ε is random errors, and X's are the plot sizes of 0.1, 1, 10, 100, and 1000 m^2 (Stohlgren *et al.*, 1995). A group regression model was used (Zar, 1999) to compare changes in species richness per unit area (β_1) and the number of species in the smallest plot ($\beta_\circ$).

Results The study area includes 198 plant species belong to 133 genus and 44 families. There were significant differences in richness based on the longevity of species in the three vegetation types (p<0.05). The richness of annuals is higher than perennials in all vegetation types. The biological types of therophyte were dominated in all vegetation types. There were significant differences among biological types and life forms (p<0.05) and the trends of life form from shrubland to woodland are shown in Figure 1. Plant richness was usually higher in northern compare to southern aspect .

Discussion and conclusions In general, annuals and ephemerals are dominant, which are the characteristic of areas with highly variable rainfall. When there is enough rain, especially in spring, the seeds of many annuals germinate and species richness much increases, so the diversity of this ecosystem is highly variable (Mesdaghi, 1993) and the results contradict the humped-back model. In a dry year, the plant diversity decreases and the pastures are in poor condition (Figure 2).

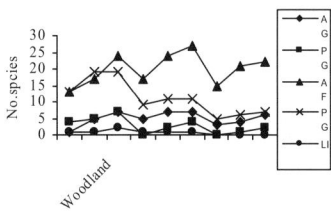

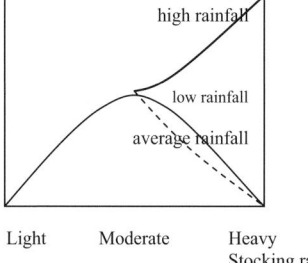

Figure 1 The trends of growth form in different fluctuating annual grassland vegetation types

Figure 2 The plant diversity model for communities under three levels of rainfall

References
Mesdaghi, M. (1993). Vegetation analysis of semi-arid regions in northeastern Iran. *Proceedings of the Seventeenth International Grassland Congress*, 56-57.
Pabot, H. (1967). Pasture development and improvement through botanical and ecological studies. FAO publication,. No.2311, FAO, Rome.
Stohlgren, T.J., M.B. Falkner & L.D. Schell (1995). A modified-whittaker nested vegetation sampling method. *Vegetatio*, 117, 113-121.
Wilkinson, D.M. (1999). The disturbing history of intermediate disturbance. *Oikos*, 84, 146-147.
Zar, H.J. (1999). Biostatistical Analysis. Fourth ed. Prentice Hall Inc.

Population changes of invasive annuals in California annual and perennial grasslands

R.J. King

USDA Natural Resources Conservation Service, 1301 Redwood Way, Suite 215, Petaluma, California, 94954 USA, Email: Richard.King@ca.usda.gov

Keywords: invasive, annual grassland, perennial grassland

Introduction and methods Some non-native annual forbs and grasses can be invasive in California's annual grassland region. *Taeniatherum caput*-medusae (L.) Nevski and *Centaurea solstitialis* L. can spread quickly, become nearly monospecific stands, and reduce carrying capacity for livestock and wildlife. *Centaurea* also can be toxic to horses. *Lathyrus hirsutus* L. is less invasive, but the seed is toxic to various livestock. Since 1991, the author has raised beef cattle on 16 ha of predominantly annual grassland where these 3 species occur, and a family member has raised cattle on approximately 40 ha of adjacent property. Surface soils in this hilly land are loamy (1/3) or high shrink-swell clay (2/3); *Centaurea* does not grow on the clay. This paper reports observed population changes of these 3 invasive annual species on both farms from 1991-2004 on the following 6 sites:

1. (12 ha)—Domestic oat hay dry farmed for decades until 1991; trial seed mix of perennial species broadcast,
2. (40 ha)—Oat hay dry farmed for decades until about 1980 when *Dactylis glomerata* L. and/or *Phalaris aquatica* L. were planted as pasture; only 50-60% of this area were successful plantings; then sheep and cattle raised here until 1991; *Centaurea* populations were suppressed by sheep utilisation,
3a. (4 ha)—Dry farmed prior to 1941; pastured with cattle since 1941 along with Sites 3b and 3c below,
3b. Asphalt & gravel roadbed abandoned circa 1935-40 within a corner of Site 3a; litter/forbs/grass cover 95%,
3c. Relict area strip (3 x 15m) within Site 3a and also *pastured* since 1941; likely never cropped due to limited size and access; predominantly native perennial species; part of an elevated island between 2 roads cut into a hill,
4. Roadside relict area strip (3 x 15m) adjacent to Site 3c but *not pastured* since 1941; likely never cropped.

Cattle did not overgraze the 3 invasive species. All sites except Site 3b had areas of perennial grassland; *Dactylis* stands developed in Site 1 after 1991. Sites 1, 2, & 3a also had substantial areas of annual grassland.

Results Table 1 shows population responses under cattle use since 1991 for each site. Each species typically formed distinct patches within paddocks. The patch density of all 3 species became greater in Site 1 than in any other site, and plant density of invasive species was often much higher too. *Taeniatherum* in Site 3a was minor in extent, expanded relatively little, and patches had much lower plant density in comparison with Site 1. The 3 invasive species were absent or their cover was minor in all well-developed perennial grass stands.

Table 1 Population changes of three invasive annual species from 1991–2004

Sites Observed	Spp.* Present Onsite in 1991			Spreading in Early Years			Currently Spreading			Currently Static			Currently Decreasing		
1	(T)	(C)	L	T	C	L	T	--	L	--	C	--	--	C#	--
2	--	(C)	L	--	C	--	--	--	--	--	C	L	--	--	--
3a	T	--	--	T	--	--	T	--	--	--	--	--	--	--	--
3b	--	C	--	--	--	--	--	--	--	--	C	--	--	--	--
3c	--	(C)	--	--	--	--	--	--	--	--	(C)	--	--	--	--
4	--	(C)	--	--	--	--	--	--	--	--	(C)	--	--	--	--

*Species: T (*Taeniatherum*), C (*Centaurea*), L (*Lathyrus*), ( ) very few plants, # (perhaps in some patches)

Conclusions Only in Site 1 were all 3 species invasive, and they showed higher patch and plant densities. The probable loss of soil organic matter and changes in soil microbes from farming (Steenwerth *et al.*, 2003) better explains this invasiveness than a possible lack of pathogenic microbes from native soils where these 3 Eurasian species originate (Callaway *et al.*, 2004). Because cropping ceased much earlier on Sites 2 & 3a, soil organic matter and soil microbes may have increased more than on Site 1. These observations suggest that management of annual grasslands to increase perennials and/or soil organic matter may help suppress these 3 invasive species.

References
Callaway, R. M., G. C. Thelen, A. R. Rodriguez & W. E. Holben (2004). Soil biota and exotic plant invasion. *Nature*, 427, 731-733.
Steenwerth, K. L., L. E. Jackson, F. J. Calderon, M. R. Stromberg & K. M. Scow (2003). Soil microbial community composition and land use history in cultivated and grassland ecosystems of coastal California. *Soil Biology and Biochemistry*, 35, 489-500.

Number and viability of seeds recovered from faeces of ruminant animals

A.O. Jolaosho, O.S. Onifade, O.M. Arigbede, J.A. Olanite and T.O. Akinola
Dept. of Pasture and Range Management, College of Animal Science and Livestock Production, University of Agriculture, Abeokuta, Ogun State, Nigeria, Email: ajolaosho@yahoo.com

Keywords: cattle, sheep, goats, seeds, faeces, seed viability

Introduction Ruminants play a significant role in the dissemination of plant seeds as a result of ingested seeds during grazing on pasture escaping digestion that are voided with the faeces and returned to the seed bank (Russi *et al.*, 1992).

Materials and methods This study was carried out at the Teaching and Research Farm of the University of Agriculture, Abeokuta, Nigeria for four months from October 2003 to January 2004. Two breeds each of the three types of ruminants were selected for the study. They were White Fulani and N'dama breeds of cattle, West African Dwarf (WAD) and Yankassa breeds of sheep and West African Dwarf (WAD) and Red Sokoto breeds of goat. Three samples of faeces were collected early in the morning weekly. The faecal samples were weighed and dissolved in water and passed through sieves to remove the seeds. The recovered seeds were then air-dried at room temperature and counted under a magnifying lens to determine the total number of seeds in the faeces. The seeds were also put into Petri dishes to determine their viability. The trial was a 3x2x4 factorial with three replicates.

Results Although, the number of seeds recovered per unit weight of the dry faeces was highest in cattle (P<0.05), cattle had the least percentage viable seeds and goats the highest (Table 1). This may be because goats are browsers; they ingest feed with little or no mastication unlike cattle and sheep according to Simao Neto *et al.* (1987). The number of seeds recovered from the faeces reached the highest value (P<0.05) in November, in the early dry season similar to the result of Jones and Simao Neto (1987). The highest percentage viability (P<0.05) of seeds recovered was in January (Table 2).

Table 1 Effects of Animal types on population and viability of seeds recovered from the faeces of ruminant animals

Animal types	Dwts	TNS	No/g Dwts	TNVS	%via
Cattle	132.8[c]	92.0[a]	0.69[a]	5.1[b]	4.7[c]
Sheep	211.8[a]	87.3[b]	0.41[b]	44.2[a]	8.1[b]
Goat	138.9[b]	3.4[c]	0.21[c]	0.9[c]	16.9[a]

Table 2 Population and viability of seeds at different months of the dry season

Months	Dwts	TNS	No/g Dwts	TNVS	%via
Oct	150.1[d]	36.5[c]	0.25[c]	1.57[d]	6.97[c]
Nov	167.5[a]	102.2[a]	0.60[a]	6.15[b]	7.58[b]
Dec	165.7[b]	68.9[b]	0.43[b]	54.99[a]	7.70[b]
Jan	161.4[c]	36.1[d]	0.22[d]	4.28[c]	17.43[a]

#Means in the same column with different superscripts are significantly different at $P \leq 0.05$
**Dwts – Dry weights of faeces; No – Number; TNS – Total Number of seeds; %via - % viability; TNVS – Total Number of viable seeds.

Conclusions The results of this study showed that small ruminants, especially goats, disseminate more viable seeds than cattle even though the number of seeds recovered from the faeces of cattle was higher. Also, seed dissemination through ruminants is better practised during the late dry season when the seeds are highly viable and close to the onset of rains.

References
Russi, L., P.S. Cocks, & E.H. Roberts (1992). Seed bank dynamics in a Mediterranean grassland. *Journal of Applied Ecology, 29, 763 – 771.*
Jones, R. M. & M. Simao Neto (1987). Recovery of pasture seeds ingested by ruminants. The effect of the seed in the diet and quality of seed recovery from sheep. *Australian Journal of Experimental Agriculture, 27, 253-256.*
Simao Neto, M., R. M. Jones & D. Ratcliff (1987). Recovery of Pasture seeds ingested by ruminant: Seeds of six Tropical pasture species fed to cattle, sheep and goat. *Australian Journal of Experimental Agriculture, 27, 239-246.*

The effect of sheep grazing at two stocking rates on the seedling recruitment of grassland forbs

J. Isselstein, N. Kowarsch, S. Bonn and M. Hofmann
Institute of Agronomy and Plant Breeding, University of Goettingen, Von-Siebold-Str. 8, 37075 Goettingen, Germany, Email: jissels@gwdg.de

Keywords: oversowing wildflower seeds, seedling establishment, sward management

Introduction Limitations for seedling recruitment are major constraints to maintain and enhance plant species diversity in productive grasslands (Bakker & Berendse 1999). Grass sward condition plus species-specific requirements for germination and survival determine the recruitment success. Therefore, a field experiment investigated the establishment of oversown seeds from wildflower forbs in relation to grass sward management.

Materials and methods A blend of freshly ripened seeds from wildflower species (100 seeds/species per m^2 in 9m^2 subplots) was broadcast in late summer over a species-poor permanent grassland dominated by *Lolium perenne* and *Trifolium repens*. The sward management treatments were (three replications): (1) Sheep grazing at a low stocking rate, target compressed sward height (CSH) 12 cm, 1.2 standard livestock units (SLU)/ha; (2) sheep grazing at a moderate stocking rate (target CSH 6 cm, 2.8 SLU/ha; (3) no grazing, control treatment, two cuts/yr. Sheep were stocked continuously from April to October and animal numbers were adjusted weekly according to CSH. No fertiliser was applied. Established seedlings were counted one year after sowing. The yield share of sown species was determined in the standing crop 3 years after sowing.

Results Seedling establishment from oversown seed was generally low (Table 1) but sward management had significant effects. Averaging over of all species, seedling establishment was much higher at the moderate stocking rate compared to both the low stocking rate and the cutting treatment (control). At moderate stocking, percentage of short grass patches (<6 cm sward height) varied from 21-59 % during the season while it was only 2-18 % for the low stocking rate. Reduction of the competitive strength of the existing grass sward by frequent defoliation has been shown to be of overriding importance for the survival of emerging seedlings (Hofmann & Isselstein 2004). Surprisingly, this holds true irrespective of the species because species like *Daucus carota*, *Tragopogon pratensis* or *Centaurea jacea*, which are susceptible to frequent defoliation at later stages of their life, also were dependent on frequent defoliation at the seedling stage. The long-term establishment of the oversown species was evaluated according to their contribution to the harvestable yield 3 years after sowing. The yield percentage of oversown species summarized over all species was 7, 165, and 62 g/kg standing crop dry matter for the low stocking, the moderate stocking, and the control treatment, respectively. There was a tendency that in the cutting treatment, species that are common on infrequently defoliated grasslands produced higher yields compared to the number of established seedlings than in the grazed treatments.

Table 1 Established seedlings (% of sown seed) of grassland forbs one year after oversowing of the seeds

Treatment[#]	*Centaurea jacea*	*Daucus carota*	*Lathyrus pratensis*	*Leontodon autumnalis*	*Lotus corniculat.*	*Plantago lanceolata*	*Tragopog. pratensis*	*Trifolium pratense*	mean
Low	0	0.7	6.7	0	0.4	0	0	0.4	1.0
Moderate	3.7	18.9	33.3	52.6	10.4	53.0	34.1	27.0	29.1
Control	0	4.1	16.7	1.1	0	3.7	2.2	2.6	3.8
p[$]	*0.022*	*0.057*	*0.365*	*0.021*	*0.035*	*0.023*	*0.035*	*0.030*	*0.038*

[#]*Treatment: low/moderate stocking rate, control=no grazing, two cuttings/yr*
[$]*Level of significance, Kruskal-Wallis-Test*

Conclusions To establish wildflower species from seed by oversowing, grass swards should be adapted to a frequent defoliation for up to one year after the oversowing. Management after that should consider the requirements of the adult plants, i.e., frequent defoliation for pasture species, infrequent for hay meadow species.

References
Bakker J.P. & F. Berendse (1999). Constraints in the restoration of ecological diversity in grassland and heathland communities. *Trends in Ecology and Evolution*, 14, 63-68.
Hofmann M. & J. Isselstein (2004). Seedling recruitment on an agriculturally improved mesic grassland: The influence of disturbance and management schemes. *Applied Vegetation Science*, in press.

The influence of local immigration and extinction of species on spatial heterogeneity of vegetation in semi-natural grasslands in Japan

T. Yasuda[1], M. Shiyomi[2], T. Egawa[2], K. Sei[2], R. Ishikawa[2] and S. Takahashi[3]
[1]Yamanashi Institute of Environmental Sciences, 5597-1, Kenmarubi, Kamiyoshida, Fujiyoshida, Yamanashi 403-0005, Japan, Email: yasuda@yies.pref.yamanashi.jp, [2]Laboratory of Ecology, Faculty of Science, Ibaraki University, 2-1-1, Bunkyo, Mito 310-8512, Japan, [3]National Institute of Livestock and Grassland Science

Keywords: extinction, grazing, immigration, spatial heterogeneity

Introduction Spatiotemporal variation in the local immigration and extinction of species in a community may form and change the spatial heterogeneity (*SH*) of vegetation but few studies have evaluated the influences of these processes on *SH*. *SH* often occurs in grassland grazed by cattle and sheep. Understanding the formation and dynamics of *SH* is important because *SH* profoundly affects local and regional ecological processes. We propose a new way to quantify the effect of the local immigration and extinction of species on *SH*, and try to clarify the influence of the processes on *SH*.

Materials and methods **(1) Field survey:** The National Institute of Livestock and Grassland Science site (36º55'N, 139º58'E, about 300m ASL) was in a flat semi-natural grassland grazed by cattle. Quadrats (n=100, each 50x50cm) were located by stratified random sampling. In an 8-year survey (1995 to 2002), we recorded all species that occurred in each quadrat. **(2) Data analysis:** *SH* was measured as the mean dissimilarity in species composition among quadrats at a site within a year. *SH* was defined as $(S_a+S_b)/_{100}C_2$, where S_a is number of species found only in quadrat A, S_b is number of species found only in quadrat B and $_{100}C_2$ is the number of all two combinations of quadrats, that is $_{100}C_2 = 4950$. To classify species as either inferior or superior, we also calculated the occurrence (*p*) of each species. Inferior species have a low occurrence ($0<p\leq0.5$), and the superior species have a high occurrence ($0.5<p\leq1$). Also, we divided each group (inferior and superior species) into two groups as regards change in *p* from time *t* to time *t*+1. As a result, all species were classified into one of the following four groups: *Ii*, inferior species that increased in *p* from *t* to *t*+1; *Ie*, inferior species that decreased in *p*; *Si*, superior species that increased in *p*; *Se*, superior species that decreased in *p*. We also developed a new method (not described here) to calculate the degree of contribution of change in *p* of each species to increase or decrease in *SH*. Using this approach, we quantified the degree of contribution of the above 4 groups to *SH* change. This equation describes the change of *SH*: $SH_{t+1} = SH_t + f(Ii) + f(Ie) + f(Si) + f(Se)$, where $f(...)$ is the degree of contribution of each groups to change of *SH*. For example, $f(Ii)$ is the increment of *SH* from *t* to *t*+1, due to the increase of *p* for inferior species. We calculated the degree of contribution of each group, and compared the effect of each group on *SH* change.

Results *SH* increased from 1995-98 (Figure 1). Between these periods, $f(Ii)$ was >0 and therefore increased *SH* (Figure 2). However, $f(Ie)$ was <0 and decreased *SH*. For the superior species, $f(Si)$ and $f(Se)$ were near zero and had almost no effect on *SH* change ($f(Si) + f(Se)$ was close to 0). Thus, the increase in *SH* from 1995-98 was because the immigration rate of inferior species exceeded their extinction rate ($f(Ii) + f(Ie) >0$). *SH* declined from 1998-2002 (Figure 1) because the extinction rate of inferior species was less than their immigration rate ($f(Ii) + f(Ie) <0$, $f(Si) + f(Se)$ was close to 0).

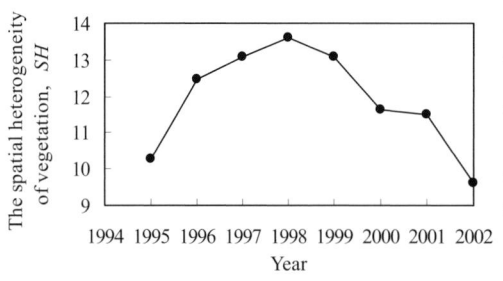

Figure 1 Dynamics of the spatial heterogeneity of vegetation (*SH*) from 1995 to 2002

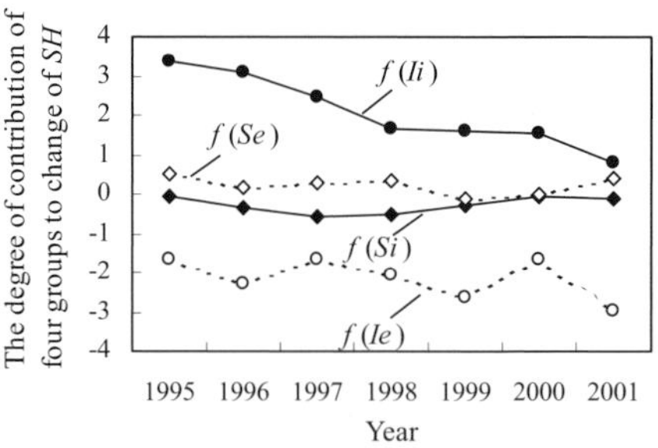

Figure 2 The degree of contribution of species to *SH* change

Discussion The balance between local immigration and extinction of inferior species was the main influence on the dynamics of *SH*. By contrast, superior species had little direct effect on the dynamics of *SH*. However, local plant community structure and the response of superior species to grazing may influence the dynamics of inferior species. The data suggest that local immigration and extinction of inferior species have a direct influence on the formation of spatial heterogeneity in grazed grasslands.

Long-term trends of community structures in tall grassland vegetation under three treatments in northeastern Japan

M. Tsutsumi[1], S. Sakanoue[2], S. Takahashi[2], Y. Yamamoto[1], Y. Saito[1] and S. Itano[1]
[1]*National Institute of Livestock and Grassland Science, Nishinasuno, Tochigi 329-2793, Japan, Email: mcot@affrc.go.jp,* [2]*National Agricultural Research Centre for Hokkaido Region, Sapporo 062-8555, Japan*

Keywords: cutting, *Miscanthus sinensis*, grazing, species composition, species diversity

Introduction In order to acquire fundamental knowledge on sustainable use and conservation of grasslands in an Asian monsoon climate, we have investigated since 1982 the vegetation of typical Japanese tall-grass, *Miscanthus sinensis* (Japanese plume-grass) dominant grassland. Three treatments, cattle-grazing, cutting and abandoning were carried out at the study site. We report the results of this research here, in particular focusing on the analysis of the changes in community structure and the influence of different utilization.

Materials and methods Our study site was an *M. sinensis* grassland located in northeastern Japan. The grassland was divided into four paddocks, where grazing (two paddocks, Paddocks Ga and Gb, hereafter), cutting (Paddock C) and abandoning (Paddock A) treatments were conducted. One transect was set at each of Paddocks Ga and Gb, and two transects at each of Paddocks C and A. In 1994, as investigation at Paddock Gb was discontinued, an additional transect was set at Paddock Ga. Along each transect, ten 2 × 2 m permanent quadrats were set. Vegetation survey was conducted from late August to early September in every year. Within each quadrat, coverage, community height, and cover and the longest length of respective occurring species were measured. Data analysis was performed on each transect. Species diversity of each transect was estimated as an average of the number of species occurring within respective quadrats. Changes in species composition from 1982, the year we started this research, were calculated as an average of a beta diversity index (β_T, Wilson & Shimida 1984) between data from 1982 and any given year on respective quadrats.

Results Figure 1a shows the changes in species diversity. Species diversity increased with grazing treatment, while it remained stable with cutting and abandoning treatments in the long term. Figure 1b shows the changes in species composition from 1982. The species composition in Paddocks Ga and A started to change from the middle 1980's, and kept on changing gradually after that time, while in Paddock C it did not change largely in the long term.

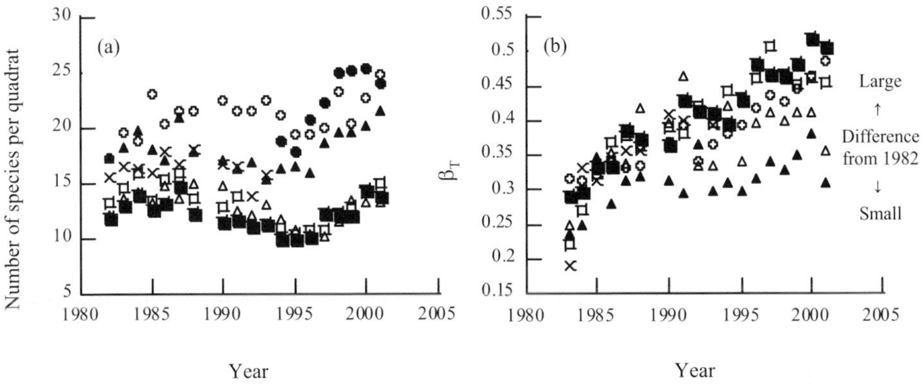

Figure 1 (a) The changes in species diversity and (b) the changes in species composition from 1982.

○ Ga-1 ● Ga-2 □ A-1 ■ A-2
× Gb △ C-1 ▲ C-2

Conclusions In *M. sinensis* grassland in an Asian monsoon climate, grazing management promoted species diversity in the community, and cutting management made the community structurally stable.

References

Wilson M.V. & A. Shimida (1984). Measuring beta diversity with presence-absence data. *Journal of Ecology*, 72, 1055-1064.

Is biodiversity declining in the traditional haymeadows of Skye and Lochalsh, Scotland?

G.E.D. Tiley[1] and D.G.L. Jones[2]
[1]Scottish Agricultural College (SAC), Auchincruive, Ayr, KA6 5HW,UK, Email: Karen.Crighton@sac.co.uk,
[2]SAC Advisory Office, Somerled Square, Portree, Isle of Skye, IV51 9EH, UK

Keywords: biodiversity, haymeadows, management

Introduction Species-rich haymeadows have developed on crofts in the Isle of Skye and Lochalsh Districts of north-west Scotland as a result of a century or more of traditional land use. This has involved long rotations of late cutting for hay with aftermath grazing by cattle and short breaks for cropping. The traditional haymeadows are increasingly coming under threat from changes taking place in the countryside. A survey of the main haymeadows still remaining in Skye and Lochalsh was carried out during 2003 to assess the current botanical composition, management and conservation value, and to compare with earlier surveys.

Materials and methods Grassland sites (31 in Skye; 18 in Lochalsh) were recorded, including several surveyed earlier (Orange, 1987; Hutcheon, 1997). The presence and estimated abundance of plant species were recorded as Dafor values for communities and Domin scores in 2m x 2m quadrats. Croft owners or managers provided data on present, past and intended future managements. Based on floristic composition, uniformity, stability and area of the constituent communities, each site was assigned a subjective conservation value.

Results Conservation values of 5/13 sites on Skye, recorded in the Orange (1987) survey, have deteriorated radically due to neglect, invasion by *Juncus effusus* and heavy grazing by sheep. The other 8/13 sites showed a sharp reduction in the occurrence of orchid species and of globeflower (*Trollius europaeus* L.). As compared with the survey by Hutcheon (1997), deterioration of conservation values was similar on sites in Lochalsh. Table 1 shows the frequencies of current managements observed during the 2003 survey.

Table 1 Haymeadow management observed

Management	No of sites (%)
Not cut or grazed	18
Cut only	8
Cut and grazed	34
Grazed only	40
Of which % grazed	
By cattle	47
By sheep	28
By cattle and sheep	25

Discussion and conclusions The widespread traditional croft management that led to the development of haymeadows has only a tenuous presence on a few scattered crofts in Skye and Lochalsh. Older crofters still make hay but big bale silage is the norm now, mainly because it allows for flexibile management to cope with the prevailing weather. Other threats to the maintenance of biodiversity are replacement of cattle grazing by intensive sheep, invasion of rushes and changes in social structure leading to decrofting. Croft managers may require incentives to encourage them to continue to use traditional management systems.

Acknowledgement In conjunction with Scottish Natural Heritage, the Highland Regional Council Biodiversity Project commissioned the 2003 survey of haymeadows in Skye and Lochalsh.

References
Hutcheon,K.(1997). Vegetation survey of Drumbuie, Kyle of Lochalsh. National Trust for Scotland.
Orange,A.(1987). A survey of haymeadows and associated grasslands in Skye, Ardnamuchan, Sunart and Lochaber. Nature Conservancy Council, Edinburgh.

Development of saline vegetation on embanked grasslands at the Baltic Sea coast after 10 years of extensive pasture use

R. Bockholt, S. Schmitz and S. Noel
The Faculty for Agricultural and Environmental Science at the Rostock University, Federal Republic of Germany, 18051 Rostock, Email: Renate.bockholt@.uni-rostock.de

Keywords: saline grassland, plant community, permanent grassland, extensive pasture, Juncus gerardii

Introduction Since the establishment of the National Park "Vorpommersche Boddenlandschaft" in 1992, the objective has been the restoration the grasslands, which was used intensively before. The soil type is a fine- sand gley. Management of embanked grasslands has been changed into extensive management. Fertilisation, renewal of the sward and water control by drainage have been stopped. Instead of the previous use of cutting for silage, mother cows (1 cattle/ ha) were grazing the areas extensively.

Methods The grazing areas in the region were mapped by GPS (Global Positioning System) and then compared with the original state recorded 10 years before. The different plant communities were evaluated from the agricultural, ecological and nature-preservation points of view. Separate soil samples were taken of individual plant communities and analysed. Furthermore, the yields, the nutrient contents of forage and the botanical composition of the vegetation were determined in test areas depending on the frequency of use (0 up to 3 harvests per year).

Results Instead of the original grass sward consisting of uniform sown (1992) *Lolium perenne and Festulolium*, there is now a small-area mosaic as result of 10 year's development (2001). It comprises 4 plant communities, with different dominant plant species: 1. *L.perenne, Elytrigia repens* 53%, 2. *Juncus conglomeratus, J. effusus* 27%, 3. *Alopecurus geniculatus* 11%, 4. *Eleocharis uniglumis, Juncus gerardii* – saline grassland 9% of the area. After 10 years, these plant communities show a similar percentage of organic soil matter and only slight differences in soil nutrient contents. However, these differences of ecological parameters (soil moisture, soil salinity) according to Ellenberg and of forage value according to Klapp were significant. The differences in salinity of the soil, of soil moisture and of ground water levels were significant. Yield and forage value in test areas were also significantly different. Diversity of plant species was highest in the *Lolium/Elytrigia* plant community, whilst the *Eleocharis uniglumis/Juncus gerardii* saline-grassland contained the highest number of protected plants included in the Red List of Mecklenburg, Western Pomerania. All plant communities include some typical plant species of saline grasslands. On the whole, after 10 years, there are 11 protected plant species contained on the Red List. These protected plants can also be found in all 4 plant communities (*Aster tripolium, Centaurium pulchellum, Eleocharis uniglumis, Lotus tenuis, Juncus conglomeratus, Juncus gerardii, Plantago media, Ranunculus flammula, Triglochin maritimum, Triglochin palustre, Tripleurospermum maritimum*).

Table 1 Biodiversity in plant communities of regions after 10 years extensive use in 2001

Dominant species of plant communities	Number of species 2001 A / B	Number of species from Red List
Lolium perenne / Elytrigia repens	58 /48	3
Juncus conglomeratus / Juncus effusus	50 /45	5
Alopecurus geniculatus	34 / 24	2
Eleocharis uniglumis/ Juncus gerardii	40 / 25	7

Table 2 DM yield and forage value on the test areas (3 harvests, 1999 – 2001)

Dominant species of plant communities	Klapp value number	Yield (t DM* ha⁻¹)
Lolium perenne / Elytrigia repens	6.7	6,9 a
Juncus conglomeratus / Juncus effusus	4.4	5,2 ab
Alopecurus geniculatus	5.5	3,0 b
Eleocharis uniglumis/ Juncus gerardii	4.9	3,4 b

Conclusions From ecological and nature-preservation points of view all the different plant communities are very valuable. Wetting these areas will result into a further increase of plant communities of the types 2, 3, 4 as well as a further reduction of the yield and of the forage value. In comparison to the starting point (12,2 t DM/ha) the level of yield is just only 57 to 25 percent.

Extensive management of sheep grazing in upland sown grassland: long-term effects on plant species composition

C.A. Marriott, G.T. Barthram, T.G. Common, J.H. Griffiths, J.M. Fisher and K. Hood
The Macaulay Institute, Craigiebuckler, Aberdeen AB15 8QH, UK, Email: c.marriott@macaulay.ac.uk

Keywords: botanical diversity, *Lolium perenne*, abandonment, *Trifolium repens*

Introduction Changes in the Common Agricultural Policy have led to the development of agri-environment schemes to deliver environmental goods from grasslands. These schemes encourage more extensive grazing systems, and change the emphasis from animal output to issues such as increasing biodiversity. Lower stocking densities are expected to promote the development of a heterogeneous habitat and associated compositional changes in plant species. The long-term effect of more extensive sheep management, combining cessation of fertiliser and lower grazing intensity, on botanical composition and animal output in upland sown grassland has been studied at 3 sites (Marriott *et al.*, 2002) since 1990. We describe changes in vegetation at one site between 1990 and 2004.

Materials and methods The site was on a freely draining brown forest soil at Sourhope Research Station (2°14' W 55°29' N, 367m ASL) in SE Scotland. An intensive, control treatment (4F) was maintained from Apr until mid-Nov at a sward height of 4cm and was fertilised (150kg/ha N and 20kg/ha each of P_2O_5 and K_2O/year). Three extensively managed treatments received no fertiliser and were grazed to maintain two sward heights, 4cm (4U) or 8cm (8U), or were ungrazed (UN). Swards were grazed by Scottish blackface ewes, with their single lambs from May until weaning in mid-Aug. Height treatments were maintained by adjusting ewe numbers in response to weekly measurements of sward surface height. The plots were circa 0.45 ha, and each treatment was replicated twice in a randomised block design. Vegetation composition was measured at 18 permanent locations/plot, using an inclined point quadrat to measure percentage specific frequency and a 0.5 x 0.5m grid quadrat to count the number of species present. Percentage specific frequency of unsown species was analysed by REML, a generalised analysis of variance suitable for unbalanced data (treatment UN was not measured in years 11 and 13). The number of species (grasses, forbs, legume, rushes and sedges, total) was analysed by analysis of variance.

Results There was a large and rapid increase in specific frequency of unsown species in treatment UN (Table 1) and the sown species virtually disappeared within 5 years. The differences between grazed treatments were smaller and varied between years, but the specific frequency of unsown species was generally higher in treatment 8U than 4F from 6 years after the experiment began. Increases in the number of species present did not necessarily accompany changes in specific frequency. The total number of species and the number of forbs increased (p<0.01) in treatment UN, exceeding the numbers in all other treatments. Some forb species were found only in treatment UN, e.g. *Cirsium arvense*, *Conopodium majus*, *Endymion non-scriptus*, *Galium aparine* and *Viccia cracca*. In the grazed treatments, there was a small increase in the number of species in treatment 4U, due mainly to an increase in the number of grass species (p=0.067), but no difference in treatments 8U or 4F.

Table 1 Species composition

Year	4F	4U	8U	UN	s.e.d.[1]
% Specific frequency of unsown species					
1990	29.3	26.3	36.6	33.8	4.94
1994	46.3	48.8	51.9	100	(4.38)
2003	47.5	58.5	68.1	-	
Total number of species					
1991	16.0	15.5	18.0	15.0	2.74
1995	15.0	16.0	19.0	13.0	(1.47)
2004	17.5	19.0	17.5	23.5	
Number of forb species					
1991	4.0	5.0	4.0	2.5	2.10
1995	4.0	4.5	3.5	2.5	(0.91)
2004	5.5	5.5	4.0	10.0	
Number of grass species					
1991	10.5	9.0	12.0	11.5	1.22
1995	9.0	9.5	13.5	10.5	(0.91)
2004	10.5	11.5	11.5	13.0	

[1] S.e.d. in parenthesis to compare within treatment

Conclusions Species composition changed rapidly when grazing was removed and swards were abandoned. However, it appears to be difficult to increase botanical diversity in upland sheep systems simply by removing fertiliser and reducing grazing intensity. This is most likely due to a lack of seed sources of new plant species in sown grasslands. The presence of a diverse range of vegetation types within the local area could improve on the colonisation by new plant species. Otherwise, reseeding with desired species may need to be considered.

Reference

Marriott, C.A., G.R. Bolton, G.T. Barthram, J.M. Fisher & K. Hood (2002). Early changes in species composition of upland sown grassland under extensive grazing management. *Applied Vegetation Science*, 5, 87-98.

Grazing, biodiversity and pastoral vegetation in the South Sudanien area of Burkina Faso

E. Botoni-Liehoun[1] and P. Daget[2]

[1]INERA Farako-Bâ, BP 910 Bobo Dioulasso, Burkina Faso, Email: edwigebot@hotmail.com, [2]Cirad-emvt, TA 30/E, campus international de Baillarguet 34398 Montpellier cedex5, France

Keywords: biodiversity, grazing pressure, *Isoberlinia doka* forest

Introduction Grazing impact on plant diversity is dominated by two contradictory views. In some studies, it has been found to lead to an increase in diversity and in other studies to a decrease associated with dominance of a few species (Nösberger *et al*, 1998, Hiernaux, 1998). In an *Isoberlinia doka* forest ecosystem, considered as the climax vegetation in the South Sudanien area of Burkina Faso, a study was carried out to assess the impact of grazing on the diversity of herbaceous species. The *Isoberlinia doka* forest is one type of South Sudanien savanna. The woody stratum is open and allowed development of a continuous stratum of graminae dominated by Andropogonea such *Andropogon ascinodis and Hyparrhenia spp. .*

Materials and methods Seven sites (4 m X4 m) had been protected in three areas which had been submitted to three levels of grazing pressure according to the duration and the season of grazing:
- Level 0: No grazing pressure. Two sites had been surveyed in a protected forest
- Level 1: Low grazing pressure. Two sites had been also surveyed in a pastoral area. This unoccupied area had been managed only for pastoral use since 2001. It received cattle from May to February.
- Level 2: High grazing pressure. Three sites had been surveyed in the village of Torokoro which is submitted to silvopastoral pressure.

Individual animals pressure cannot be identified because of common use of the pastureland. Measurements of floristic richness, forage production and forage quality (pastoral value) were made according to Daget & Poisssonet (1972).

Results The higher the grazing pressure the greater was the floristic richness of herbaceous plants. However, the added species were unpalatable (e.g. *Spermacoce* spp. and *Indifofera* spp.). Species diversity, measured by the Shannon index, was higher in grazed than in the ungrazed vegetation. Grazing allowed other species to alter the balance of the native grasses such as *Andropogon ascinodis, Andropogon shirensis, Schyzachyrium sanguineum* and *Hyparrhenia* spp. Forage production and its quality was lower when plant biodiversity (floristic richness, specific diversity) increased.

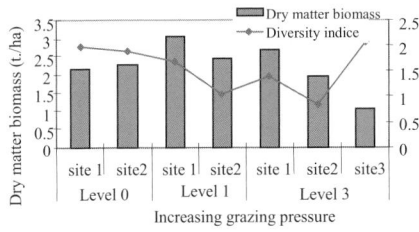
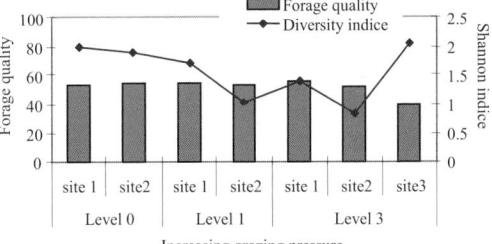

Figure 4 Forage production evolution according to grazing pressure

Figure 5 Forage quality according to grazing pressure

Conclusion In the South Sudanien savannah of Burkina Faso, grazing pressure led to increased plant diversity. But this is not favourable to livestock sustainability because of a reduction in forage productivity and its quality. These results show that a high biodiversity is not a good indicator for high productivity of pastoral vegetation.

References

Hiernaux P. (1998). Effects of grazing on plant species composition and spatial distribution in rangelands of the sahel. *Plant Ecology*, 33, 387-399.

Nösberger, J., M. Messerli & C. Carlen (1998). Biodiversity in grassland. *Annales de Zootechnie*, 47, 383-393.

Daget P. & Poissonet J., (1972). Un procédé d'estimation de la valeur pastorale des fourrages. *Fourrages*, 46, 31-39

The effect of grazing on rare and common grasses and forbs in the Mediterranean coastal desert of Egypt

A. El-Keblawy[1] and A. Ramadan[2]
[1]Dept of Biology, Faculty of Science, UAE University, P.O. Box 17551, Al-Ain, UAE, Email: a.keblawy@uaeu.ac.ae, [2]Dept of Botany, Faculty of Science, Suez Canal University, Ismailia, Egypt

Keywords: grasses, forbs, diversity, grazing, species abundance

Introduction Large parts of the rangelands in the Arab countries are either overgrazed or gradually deteriorating due to large numbers of livestock and unrestricted grazing (Assaeed, 1997). Protection of vegetation against grazing in desert environments has been suggested as a feasible approach to halting land degradation and rehabilitating rangelands (El-Keblawy, 2003). However, little is known about the response of different plant life forms and life cycles to grazing. The objectives of this study were to address this point and to test the response of rare and common forbs and grasses to protection against grazing.

Materials and methods Thirty seven stands, each of 10 x 10 m (this area approximate the minimal area according to species area curve), were selected by stratified sampling to cover the prevailing habitat and community variations inside and outside Al-Arish airport in the desert Mediterranean coastal region of Egypt, which was fenced against grazing for about 35 years. In each stand, the present species were recorded and their density was estimated in 10 quadrats of 1 m^2. Species richness of the vegetation inside and outside the exclosure was calculated as the average number of species per stand. Species diversity was calculated using Shannon-Weaver Index, which evaluates the relative evenness of species, on the basis of absolute density.

Results Protection against grazing increased significantly the total number of species, species richness, species diversity, density and frequency (P<0.05). The response of forbs to protection was greater than that of grasses. The number of grasses, their density, frequency, richness and diversity were greater in the protected than in grazed areas by 67%, 237%, 28.9%, 29.1% and 8.2%, respectively, but those of forbs were greater by about 140%, 760%, 179%, 178% and 56.6%, respectively. The differences in the values between grazed and protected areas were greater for rare grasses and forbs than for common ones. The number of rare grasses was significantly lower than that of rare forbs in both grazed and protected areas, so rare grasses attained greater density than rare forbs in the protected areas (Table 1). It is interesting to note that all rare grasses were annuals. Also, common perennial grasses were more represented in grazed (4 species) than in protected area (2 species). The density (individuals /100m^2) of rare annual grasses increased from 0.6 in grazed to 212 in protected areas, indicating that protection would help in conserving rare grasses in desert rangelands.

Table 1 Effect of grazing on some community attributes of common and rare grasses and forbs. N: number of species

Abundance	Grazing Status	Grasses					Forbs				
		N	Den.	Freq.	richness	Div.	N	Den.	Freq.	richness	Div.
Common	Grazed	10	101.0	16.9	2.9	0.52	24	23.6	11.2	5.8	0.98
	Protected	15	510.0	22.4	3.8	0.66	48	191.0	28.4	14.7	1.52
Rare	Grazed	1	0.6	2.9	0.1	0.0	3	0.7	2.2	0.3	0.0
	Protected	2	211.8	12.5	0.2	0.0	16	40.6	15.3	2.4	0.52
Community	Grazed	12	139.0	16.6	3.6	0.73	28	17.2	8.7	6.3	1.06
	Protected	20	468.4	21.4	4.7	0.79	67	147.8	24.3	17.5	1.66

Conclusions The significant increase in community attributes, especially for rare species, in the protected area indicates that protection could conserve these species in desert rangelands. Stimulation of ramification through animal disturbances and/or the greater ability of the individual plants to overcompensate for consumed parts after grazing (El-Keblawy, 2003) would explain the greater representation of common than rare grasses in grazed area. This would also indicate that common grasses are more resistant to overgrazing.

References

Assaeed, A.M. (1997). Estimation of biomass and utilization of three perennial grasses in Saudi Arabia. *Journal of arid Environments*, 36, 103-111.

El-Keblawy, A. (2003). Effect of protection from grazing on species diversity, abundance and productivity in two regions of Abu-Dhabi Emirate, UAE. In: Desertification in the Third Millennium, A.S. Alsharhan, W.W. Wood, A.S. Goudie, A. Fowler & E. Abdellatif (Eds.), Swets & Zeitlinger Publisher, Lisse, The Netherlands, 217-226.

Plant species richness at three levels of range potential in semi-steppe vegetation of Iran

M. Mesdaghi
Department of Range Ecology, University of Agricultural Sciences, Gorgan, Iran, Email: mesdagh@yahoo.com

Keywords: grazing intensity, protection from grazing, species numbers

Introduction Semi-steppe vegetation with a mean annual precipitation of 250 mm covers 18.5% of Iran (Pabot, 1967). Under full protection, bunch grasses are dominant in this grassland ecosystem. Golestan National Park, (100,000 ha), is representative of this semi-steppe vegetation zone. It has been protected for 40 years and can be considered as a reference area (RA) of semi-steppe vegetation. Under moderate grazing the The Gorokhoud Protected Area acts as a buffer zone to the National Park, with grazing based on permits issued by the Range and Forest Organisation. Parts of the protected area can be considered as a key area (KA), the palatable forbs and grasses were weakened and the dominant species of the steppe region invaded the semi-steppe ecosystems. Under severe grazing in critical areas (CA), most of the perennial grasses and legumes were eliminated and unpalatable spiny and cushion shrubs and annuals invaded. Under heavy grazing the composition of forage species shifts towards undesirable species, especially ephemeral species, and ultimately the plant richness declines. In contrast, under light to moderate grazing the plant diversity improves. However, improving the plant diversity under full protection has been questioned by many authors (West, 1993, Naveh & Whittaker, 1979).

Material and methods The Protected Area (KA) is located in vicinity of Golestan National Park (RA). There are no controls on the other rangelands in the vicinity of this protected area, and parts of them, especially near the villages, can be considered as CA. The species richness of three representative stands were measured at each level of range potential using the Whittaker's plot (Shamida, 1984). The linear model of $S = \beta_{\circ} + \beta_1 \log X + \varepsilon$, was used where S is number of species and X is the plot sizes of 0.1, 1, 10, 100, 1000 m^2. To compare the changes of species richness per unit of area (β_1) and number of species at the lowest plot size ($\beta_{\circ}$) at each level of range potential, group regression model was used (Zar, 1999).

Results The results of group regression between pairs of range potential are shown in Figure 1. The slope and intercept of regression lines at RA and KA were not significantly different (P>0.05), so it could be assumed that the data of these two range potentials belong to the same parent population. CA and RA in Figure 1b had only different intercepts and CA and KA in figure 1c had different slopes and different intercepts so that it was impossible to combine pairs of data to one regression, therefore species richness of these areas were different. A list of desirable and undesirable species was provided.

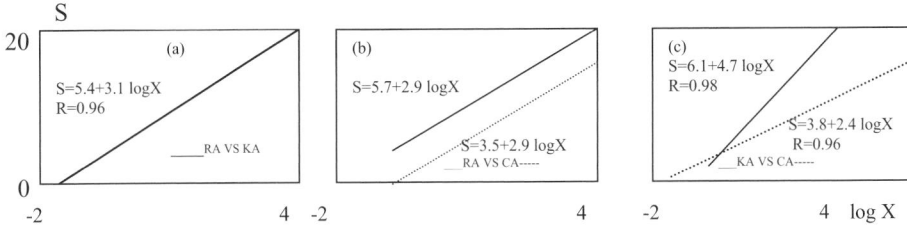

Figure 1 Comparison of plant richness at three level of range potential. (a) with the same plant richness for RA and KA, (b) with same richness in unit of plot, but different intercept at RA and CA (c) with different plant richness at KA and CA

Conclusions Species richness under moderate grazing in KA is the same as in the ungrazed area in RA, but under severe grazing in CA species richness is less and undesirable annuals and sod species hare invaded. It is important to note that species richness should be weighed for life duration and desirability of plants.

References

Naveh, Z.& R. H.Whittaker (1979). Structural and floristic diversity of shrublands and woodlands in northern Israel and other Mediterranean Areas. *Vegetatio*, 41, 171-190.

Pabot, H. (1967). Pasture development and improvement through botanical and ecological studies. *FAO*. No.2311.

Shamida, A. (1984). Whittaker's plot. *Israeli Journal of Botany*, 33, 41-46.

West, N. E. (1993) Biodiversity of rangelands. *Journal of Range Management*, 46, 2-13.

Zar, H. J. (1999). Biostatistical analysis. 4th.Edition, Prentice Hall, Inc.

Plant community structure of midland grassland of the Flooding Pampa in relation to grazing management

A.M. Rodríguez, E.J. Jacobo and V.A. Deregibus
Departamento de Producción Animal, Facultad de Agronomía, Universidad de Buenos Aires. Av. San Martín 4453, Buenos Aires C1417DSE, Argentina, Email: adrianar@coneau.gov.ar

Keywords: plant community structure, species diversity, temperate grasslands

Introduction The impact of grazing on plant community structure and ecosystem functioning is a key issue for range management. Although excessive grazing may often lead to land degradation and loss of biodiversity, maximisation of livestock production requires high stocking rates. The aim of this study was to evaluate the adequacy of intermittent grazing to improve the condition of midland grassland in Flooding Pampa. We compared the responses of functional groups and the changes of species diversity of the plant community over three years under intermittent and continuous grazing regimes.

Materials and methods A 3-year experiment was conducted, from 1993 to 1996. Total annual rainfall decreased from 1465 mm in 1993 to 845 mm in 1996. From March 1989, traditional continuous grazing was replaced by intermittent grazing on four commercial farms located in different sites of the Flooding Pampa. In each site, an adjacent farm with similar traits (total surface, soil type, vegetation communities, land use) managed under continuous grazing was assessed as a control. Average stocking rate in all farms was 1 ha/breeding cow. The pair of farms at each site constituted the replications of the experiment. The proportion of basal cover of different functional groups, litter and bare soil were monitored. Plant species diversity was calculated using Shannon-Wiener diversity index (H'). Analysis of variance (ANOVA) techniques were used to evaluate the effect of grazing method and year on each response variable and to test the significance of differences between H'.

Results Grazing method strongly affected the presence of most functional groups and other structural variables. Basal cover of C_3 annual grasses (Figure1a) and C_3 perennial grasses (Figure1b) were higher, while cover of C_4 prostrate grasses (Figure 1c) was much lower under intermittent grazing. Intermittent grazing increased cover of litter (Figure 1f) and reduced the amount of bare soil (Figure1e). As the years became drier, C_4 tussock grasses tended to decrease (Figure1d) and C_3 annual and perennial grasses increased (Figures 1a and b) during the experimental period. However, total plant basal cover (avg. 71%) and plant species diversity (H'=2.61 for continuous and H'=2.45 for intermittent grazing, p=0.25) were not affected by grazing method across the years.

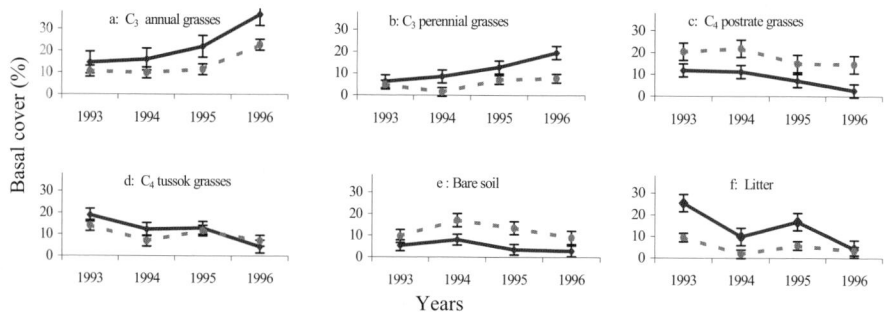

Figure 1 Basal cover of functional groups, bare soil and litter of midland grassland of Flooding Pampa under intermittent grazing (solid line) or continuous grazing (interrupted line). Bars indicate standard errors

Conclusions Intermittent grazing increased the cover of functional groups of high forage value, like C_3 annual and perennial grasses, while total basal cover and community biodiversity were not affected. These changes in community structure imply an improvement in rangeland condition and secondary productivity (Jacobo *et al.*, 2000). These results suggest that productivity and conservation goals may be compatible using this grazing method in Flooding Pampa midlands.

Reference

Jacobo, E.J, A.M. Rodriguez, J.L. Rossi, L.P. Salgado & V.A. Deregibus (2000). Intermittent grazing and production of Italian ryegrass on Argentinian rangelands. *Journal of Range Management*, 53, 483- 488.

Overgrazing influence on the presence of legumes in a natural pasture of Sardinia

L. Salis[1], M. Vargiu[1,2], E. Spanu[1] and F. Loche[1]
[1]Centro Regionale Agrario Sperimentale, Viale Trieste 111, Cagliari, Italy, Email: foraggisassari@tiscali.it, [1,2] foraggicoltura@cras.sardegna.it

Keywords: natural pasture, botanical composition, legumes, forage production

Introduction The knowledge of forage production and botanical composition in natural pasture is essential to plan forage crop systems. Floristic balance often changes due to overgrazing, which affects forage quality and causes the disappearance of less competitive and more palatable species. This trial aimed to evaluate the forage yield in a natural pasture and to verify the effect on botanical composition of overgrazing by dairy ewes.

Materials and methods In a typical pastoral hill located in central Sardinia, a native marginal pasture of 10 ha was studied regarding its herbage production in 2002 and from the floristic point of view in 2003. According to long-term traditions and usage in this area, the field was grazed with a high stocking rate. In the first year, 15 exclusion cages of 18 m^2 were placed to evaluate the offered herbage yield; the grazed forage was determined by the difference technique (Frame 1981). All forage samples collected in 0.5 m^2 were partitioned in floristic groups to determine their botanical composition and oven-dried to evaluate the content of dry matter. In the second year the field was divided in two parts; one part was grazed in winter and spring and the other was ungrazed. In the spring period, the floristic composition in both fields was determined by linear analysis (Daget & Poissonet 1969; Pignatti 1982) with 15 lines each.

Results Forage dry matter yield in the first year was 3.2 t/ha; only half was utilized due to its low quality. Overgrazing influenced botanical composition; in comparison with the herbage collected in the cages, the percentage of legumes in forage production decreased from 17 to 1%, whereas the percentage of grasses increased from 67 to 94% (Figure 1).

Table 1 Predominant species and specific contribution (%) in the two swards

Predominant species	Ungrazed area SC%	Grazed area SC%
Hordeum marinum Huds.	20.6	6.3
Phalaris caerulescens Defs.	14.1	4.3
Bromus hordeacus L.	12.5	7.3
Vulpia ligustica Lk.	8.7	8.6
Trifolium subterraneum L.	5.8	0.2
Gliceria glabra L.	5.1	0.7
Vulpia sicula Lk.	5.0	2.8
Lotus ornithopodioides L.	2.7	0.1
Elymus caput medusae L.	2.0	6.6
Avena sterilis L.	1.4	7.6

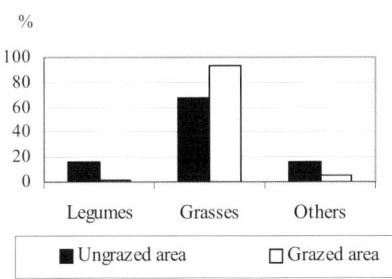

Figure 1 Forage yield percentage of floristic groups in the preserved and pastured areas after grazing *(All data differ significantly at P<0.01, t test)*

Floristic data confirmed this trend in the second year. In the pastured area, *Trifolium subterraneum* L. and *Lotus ornithopodioides* L., the most represented legume species, showed a minor presence in comparison with the preserved area, 0.2% versus 5.8% and 0.1% versus 2.7%, respectively (Table 1).

Conclusions Overgrazing influenced the quality of pasture production, reducing the presence of legume species. In each of the examined years and adopting two different methods to estimate botanical composition, the preservation from overgrazing increased the presence of legumes in the sward. The low frequencies of legumes assessed in the overgrazed sward show the risk of their disappearance, with possible loss of local biodiversity.

References
Daget P. & J. Poissonet (1969). Analyse phytologique des praries; application agronomique. CNRS, CEPE Montpellier, Doc 48, 67 pp.
Frame J. (1981). Sward measurement handbook. Herbage mass 39-69 pp. Ed. Hodgson J., R.D. Barker, A. Davies, A.S. Laidlaw, J.D. Leaver. Hurley, Great Britain.
Pignatti S. (1982). Flora d'Italia. Edizioni Agricole.

Changes in floristic diversity associated with sheep grazing management on a karst pasture

M. Vidrih, F. Batič and K. Eler
Department of Agronomy, Biotechnical faculty, University of Ljubljana, Jamnikarjeva 101, 1000 Ljubljana, Slovenia, Email: matej.vidrih@bf.uni-lj.si

Keywords: karst pasture, sheep grazing, Shannon diversity index

Introduction Greater plant diversity, richness and lower primary production are more characteristic for karst (calcareous) pastures than for lowland grassland. In relation to the level of animal grazing required, light and moderate levels are usually most appropriate (Hart, 2001). From the conservation point of view grazing intensity should be variable between sites, and between parts of large sites, and timed to provide for the requirements of different species in different seasons (Dolek & Geyer, 2002).

Materials and methods Five sites with different grazing intensity (heavy (2x), moderate, light and zero) were located on the slope of the karst mountain Vremščica (820 m a.s.l.; $45^0 41$' N; $14^0 12$' E), where controlled sheep grazing management was initiated a decade previously. On the first heavily grazed site the shrub vegetation was also cut prior to the start of sheep grazing but the other sites had a herbaceous sward from the beginning. At each site, the vegetation on four replicate plots of 100 m^2 was mapped from 1999 to 2001 using the method of Braun-Blanquet. Shannon diversity index and evenness (Magurran, 2004) were calculated from data obtained in spring and summer mapping each year.

Results The pasture community at each site consisted of more than 32 species. With the removal of shrubs, vegetation secondary succession of the sward had started on the first heavily grazed site, and that is why the number of plants (121) was so high. But in the following two years a decrease in species was observed because grazing led to a more stable and dense sward. From 32 to 58 species were found under zero, moderate and heavy grazing. *Brachypodium rupestre* and *Calamagrostis varia* were prevailing grasses on the first heavily grazed site, but on the zero-grazed site *Bromus erectus* and *Carex humilis* dominated. Only a small proportion of the species found were legumes. The Shannon diversity index was lower (3.69 to 3.70) in the zero-grazed site and the contributions of the different species to total cover were more unequal (lower evenness - not shown here). Shannon indices indicated that the second heavily grazed site had fewer diverse plant communities than the lightly or moderately grazed sites.

Table 1 The Shannon diversity index and species number on five sites from 1999 to 2001 with respect to grazing intensity

	Shannon diversity index					Species richness				
Year/Intensity	heavy	heavy	moderate	light	zero	heavy	heavy	moderate	light	zero
1999	4.52c	3.80a	4.13b	3.92a	3.70a	121d	32a	40b	51c	48b
2000	3.81a	3.79a	3.94b	4.10b	3.71a	95c	41a	43a	55b	45a
2001	3.75a	3.85b	4.27c	4.31c	3.69a	91c	39a	39a	58b	41a

Means within a row followed by the same letter are not significantly different at P=0.05 according to Duncan's multiple range test.

Conclusions These results show that controlled sheep grazing management can preserve sward development in karst pastures. In the long-term a higher diversity can be expected when grazing swards on karst pastures with sheep at moderate or light intensity. At the same time, a heavy grazing intensity in those areas can act as a strong modifying factor to create a more productive grassland community (fewer species, but more productive sward) or as a tool for re-establishing the sward on abandoned land.

References
Dolek, M. & A. Geyer (2002). Conserving biodiversity on calcareous grasslands in the Franconian Jura by grazing: a comprehensive approach. *Biological Conservation*, 104, 351-360.
Hart, R.H. (2001). Plant biodiversity on shortgrass steppe after 55 years of zero, light, moderate, or heavy cattle grazing. *Plant Ecology*, 155, 111-118.
Magurran, A.E. (2004). Measuring biological diversity. Blackwell Publishing, Oxford, 178 pp.

Forage legume persistence in mixtures with native and introduced grasses at a semiarid location on the Canadian prairies

P.G. Jefferson, A.D. Iwaasa and M.P. Schellenberg
Agriculture and Agri-Food Canada, Semiarid Prairie Agricultural Research Centre, P.O. Box 1030, Swift Current, Saskatchewan, S9H 3X2, Canada, Email: JeffersonP@agr.gc.ca

Keywords: mixtures, legume persistence, native grasses

Introduction Cultivars of native grass species with adequate nutritive value for summer and early fall grazing by beef cattle are becoming available for seeding in the Canadian prairie region (Jefferson *et al.* 2004). Mixing native grass species with introduced legumes could improve forage quality but little information is available on legume persistence with these species. This experiment aimed to determine the persistence of 3 forage legumes when seeded with 3 native grasses compared to 3 introduced grasses.

Materials and methods The grasses were 3 native species: northern wheatgrass (*Elymus lanceolatus*) cv. Elbee, green needlegrass (*Stipa viridula*) cv. Lodorm, and western wheatgrass (*Pascopyrum smithii*) cv. Walsh; and three introduced species: meadow bromegrass (*Bromus riparius*) cv. Fleet, crested wheatgrass (*Agropyron cristatum*) cv. Kirk, and Russian wildrye (*Psathyrostachys juncea*) cv. Swift. They were seeded in monoculture or in mixture with lucerne (*Medicago sativa*) cv. Rangelander, sainfoin *(Onobrychis viciaefolia)* cv. Nova or 'Oxley' cicer milkvetch (*Astragalus cicer*) cv. Oxley. The experiment was divided into 2 harvest dates, 01 Aug and 01 Sep. Forage yield and botanical composition of mixtures were determined for both dates. Legume content, averaged across 6 grasses, 2 harvest dates and 4 replicates, are presented for each year, 1999 to 2003, inclusive.

Results Lucerne content in the forage was higher than cicer milkvetch or sainfoin content (Figure 1). Due to severe drought, all 3 legume species declined in 2001. Lucerne content rebounded to nearly 50% by 2003 but sainfoin and cicer milkvetch contents did not recover to pre-drought levels. Lucerne dominated mixtures with green needlegrass (0.68 content in 2000) while it was more compatible with meadow bromegrass (0.45 content in 2000) Cicer milkvetch appeared to be more compatible with native grasses than with introduced grasses (data not shown).

Conclusions Native grasses can be seeded in mixture with introduced legumes but lucerne will dominate the stands. Cicer milkvetch appears to be more compatible with native species than lucerne for seeding mixtures but suffers from a lack of persistence during drought.

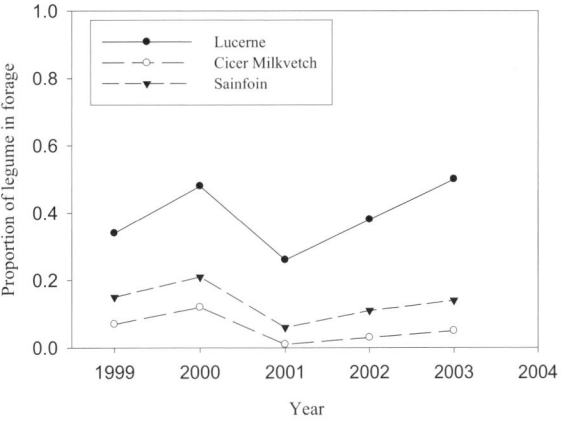

Figure 1 Legume content of forage by year as affected by species averaged over six grass species

References

Jefferson, P.G., W. P. McCaughey, K. May, J. Woosaree & L. McFarlane (2004). Forage quality of seeded native grasses in the fall season on the Canadian prairie provinces. *Canadian Journal of Plant Science*, 84, 503-509.

Floristic composition as a parameter of the quality of the grassland type *Festucetum vallesacae* in theStara Planina hilly-mountainous region of Serbia

Z. Nesic, Z. Tomic, S. Mrfat-Vukelic, M. Zujovic, I. Djalovic and S. Djordjevic
Institute for Animal Husbandry, Autoput 16, Beograd, SCG, Serbia, Email: zonen@eunet.yu

Keywords: floristic composition, quality, natural meadows, Stara Planina Mountain

Introduction With the increase of sea level and changes of climatic conditions, the possibilities for growing cultivated forages decreases. Therefore natural grasslands will become more important in relation to livestock nutrition. This will increase the importance of the nutritive value of these natural grasslands that have very diverse and dynamic floristic composition. The quality of the grassland depends on species categorised as grasses, legumes and other species. Other species were often regarded as harmful in regard to quality, however, they often contain many medicinal and stimulating substances that may have beneficial effects on animals and on the quality of animal products (Djordjevic-Milosevic, 1997). Conversely, there are also weed and harmful species among grasses and legumes. however, among species from other families there are also useful species. To provide nutrition of livestock from quality grasslands melioration measures are necessary. This paper reports on the composition of grasslands in Stara Planina and provides information on the proportion of useful species.

Materials and methods The grassland type *Festucetum vallesiacae* was sampled at four locations on Stara Planina at 750-800m above sea level. Samples were taken from an area of $1m^2$. Floristic composition was determined on a weight basis and qualitative categorisation of species was carried out according to Kojic (1990, 2001).

Results The *Festucetum vallesiacae* grassland had 77 species. By weight the grassland comprised 65.2% grasses, 24.5% legumes and 10.2% other plants. Tomić *et al.* (2003) reported that the natural grasslands of Stara Planina at 800m above sea level comprised 42.4% grasses, 44.0% legumes and 13.6% other species. The most present species was *Festuca vallesiacae,* followed by: *Agrostis capillaris, Cynosurus cristatus, Poa violacae, Trifolium campestre, Medicago lupulina, Lotus corniculatus, Galium verum, Achillea millefolium*, as characteristic species of association.

Table 1 The composition of the association *Festucetum vallesiacae* categorised in qualitative groups (% by weight)

Association	Useful grasses	Useful legumes	Other useful sp.	Weeds	Total useful
Festucetum vallesiacae	14.47	26.32	3.95	55.26	44.74

When categorised as in Table 1 the highest percentage was for weeds 55.26%, with useful plants totalling 44.74%. The weed group included worthless, harmful and poisonous plants among which were a great number of grasses and legumes. This highlights the point that high weight presence of grasses in grassland does not give a good indication of the quality of the grassland. The weed species with significant presence were *Festuca ovina, Dorychnium herbaceum,Thymus serpyllum, Teucrium chamaedrys,Linum, catharticum, Holcus lanatus* and *Hypericum perforatum.* These indicated the low quality of the grassland.

Conclusion The association *Festucetum vallesiacae* had a total of 77 plant species, of which 42 are categorised as worthless and harmful, including grasses, legumes and other plants. These species comprised 55.26% of the grassland by weight. Because of low quality and problems with growing cultivated forages in this region, measures for revitalisation of natural grasslands are necessary in order to obtain forage of better quality and safety.

References
Đorđević-Milošević, S. (1997). Produkcija prirodnih travnjaka različitih visinskih pojaseva Stare Planine. *Biotehnologija u stočarstvu*, 13, 51-69.
Kojić, M. (1990). Livadske biljke. Naučna knjiga, Belgrade.
Kojić, M., S. Mrfat-Vukelić, S.Vrbničanin, Z. Dajić & S. Stojanović (2001). Korovi livada i pašnjaka Srbije. Institut za istraživanja u poljoprivredi SRBIJA, Belgrade.
Tomić, Z., S. Mrfat-Vukelić, M. Petrović, M. Žujović, Z. Nešić & V. Krnjaja (2003). Useful leguminous plant of natural pastures of Stara Planina Mountain as quality components. Biotechnology in Animal Husbandry, *7th International Symposium, Modern Trends in Livestock Production*, 19, 5-6, 421-426.

Floristic composition as parameter of quality of ass. *Agrostietum vulgaris*

Z. Tomic, S. Mrfat-Vukelic, Z. Nesic, M. Zujovic and S. Djordjevic-Milosevic
Institute for Animal Husbandry, Autoput 16, Beograd, SCG, Serbia, Email: zotom@mail.com

Keywords: floristic composition, useful grasses, useful legumes, other species

Introduction In hilly-mountainous region of Serbia, meadows and pastures are the main sourcees of roughage feeds and grazing for ruminants. On Stara Planina mountain, in SE Serbia, meadows and pastures with different plant associations and a wide range of species of differing nutritional value predominate. Until the 1930s, and even until now, only species of the grass family and leguminous plants were desired and all other plants were considered worthless or harmful Klapp (1986). The main criteria to evaluate whether a meadow plant species is a weed are: is it poisonous?; is it suitable for consumption by domestic animals as regards its morphology?; do animals eat it or not?; is it nutritious and digestible? This study aimed (1) to quantify the proportion of useful plants, other species and weeds based on the presence of certain plant species at 3 locations in ass. A*grostetum vulgaris* and, (2) based on this data, to establish the quality potential for ruminant nutrition. Measures to improve the quality of grassland arise from the results obtained.

Materials and methods By the end of July 2004, grassland yield on Stara Planina mountain at 650-800m a.s.l. was determined at 3 locations. In 2 replicates, samples were taken from an area of $1m^2$ to determine the floristic composition. The proportion by weight of different species was determined and a categorization based on quality according to Kojic (1990, 2001) was carried out.

Results *Agrostietum vulgaris*-type grassland occupied circa 30% of grass areas in Serbia. Productivity often was very low and biomass of medium to poor quality. Unfavourable floristic composition was the main cause. It reflected low soil nutrient levels and poor management, since *Agrostietum vulgaris*-type meadows are man-made (of secondary origin), created through the effects of cutting forests and mowing and fertilisation. They also grew on areas used to grow agricultural crops. They rely on human intervention for their preservation and they are under constant change as regards quality and quantity. In ass. *Agrostietum vulgaris* in the Rudanj highlands, 75 species were determined, although the total number could be up to 100. Table 1 shows the number of quality species of grass, legumes, other species and weed species. The total number of species in ass. *Agrostietum vulgaris* at all 3 locations was relatively small and varied from 23 on 1, and 26 on the remaining 2 higher locations. *Agrostis vulgaris* represented circa 20% of biomass at location I, 27% at location 2 and 10% at location 3. At all locations, *Arrhenatherum elatius, Dactylis glomerata, Festuca arundinacea, Lolium perenne, Festuca rubra, Poa bulbosa* and *Poa pratensis* were useful grasses.*Lotus corniculatus Trifolium repens Trifolium incarnatum, Medicago lupulina, Trifolium campestre* were useful legumes. *Achillea millefolium* was another useful species, and *Rhinanthus minor* was included amongst the harmful and poisonous species.

Table 1 Numbers of species and their % in biomass in quality groups in ass. *Agrostietum velgaris* at 3 locations

Association Location	Ass. *Agrostietum vulgaris*									
	Useful grass		Useful legumes		Useful others		Weeds		Useful sp.total	
I Gulenovci 650 m	8	34.78	9	39.13	1	4.35	5	21.74	18	78.26
II V.Odorovci 730 m	8	30.77	5	19.23	1	3.85	12	46.15	14	53.85
III Mojinci 800 m	7	26.92	9	34.61	-	-	10	38.46	16	61.54

Conclusion The number of species in ass. *Agrostietum vulgaris* was very low at 3 locations on Stara Planina mountain (650-800m a.s.l.); 23 were at the lowest location and 26 at the other 2 locations. Useful grass species represented 26.9-34.7% of biomasss, useful legumes were 19.2-39.1%, and useful other species were 21.7-46.1%. Considering that the association is man-made, optimal NPK fertilisation will improve the quality of grassland with higher participation of weeds, which will contribute to higher production of animal feed.

References
Klapp, E. (1986). Visen und Weiden. Verlag Paul Parey. Berlin-Hamburg
Kojić, M. (1990). Livadske biljke. Naučna knjiga, Beograd
Kojić, M., S. Mrfat-Vukelić, S. Vrbničanin, Z. Dajić & S. Stojanović (2001). Korovi livada i pašnjaka Srbije. Institut za istraživanjea u poljoprivredi "Srbija", Beograd

Forage and livestock productivity on pastures of differing plant diversity

B.F. Tracy

Department of Crop Sciences, University of Illinois, 1102 S. Goodwin Ave. Urbana, IL 61820, Email: bftracy@uiuc.edu

Keywords: pastures, plant diversity, forage yield, cattle performance, rotational grazing

Introduction Plant diversity and its function in grassland ecosystems has been the focus of many recent studies, and debate, in the ecological literature (Wardle 1999, Loreau and Hector 2001). We know less about the role of plant diversity in pastures used solely for agricultural production (Sanderson et al. 2004). The objective of this study was to learn how pastures planted with increasing levels of plant diversity would affect forage yields and beef cattle performance.

Materials and methods In August 2001, 9 pastures (3-5 ha) were sown with 3, 5 and 8 forage species at the Orr Beef Research Center in western Illinois, USA. Pastures were rotationally grazed by beef cattle from 2002 to 2004. Cow-calf pairs (black purebred Simmental) were assigned to pastures to achieve an initial socking rate of ~2.5 animals/ha. Initial and final calf weights were measured on two consecutive days to assess performance. Forage yields were estimated using a rising plate meter technique.

Results Figure 1 shows mean standing crop from rotationally grazed paddocks in 2003. Forage yields were similar most of the season until August when the 3 species mixtures were more productive. Forage yields in 2002 showed a similar trend. Calf gain per hectare was higher in 2003 but showed no differences among the different pasture mixtures (Figure 2).

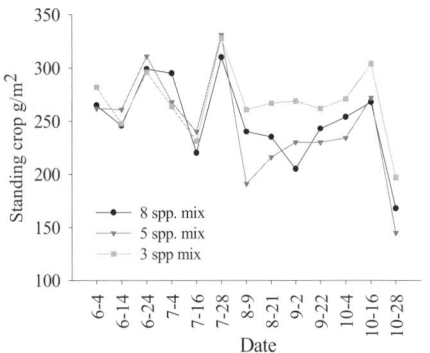

Figure 1 Forage standing crop from June to Nov. 2003

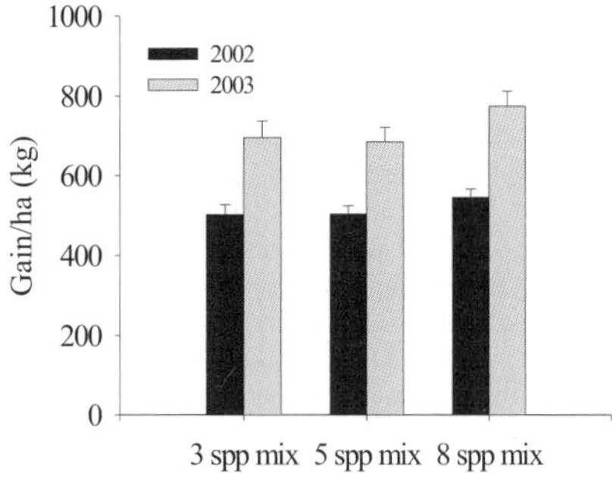

Figure 2 Calf weight gain per hectare in 2002 and 2003

Conclusions Results from this study support previous research. Good forage yields and stability can be achieved by planting 2 or 3 forage species that are well adapted to local conditions. Yield differences among the mixtures were not great enough to affect cattle performance. Overall, it appears that sowing additional plant species beyond a simple mixture of 2 perennial grasses and 1 legume does little to improve agricultural production.

References

Loreau, M., & A. Hector (2001). Partitioning selection and complementarily in biodiversity experiments. *Nature* 412:72-76.

Sanderson, M.A., R.H. Skinner, D.J. Barker, G.R. Edwards, B.F. Tracy, & D.A. Wedin. (2004). Plant species diversity and management of temperate forage and grazing land ecosystems. *Crop Science*. 44:1132-1144.

Wardle, D.W. (1999). Is sampling effect a problem for experiments investigating biodiversity -ecosystem function relationships? *Oikos* 87:403-408.

Effect of *Acacia caven* on the productivity and botanical composition of low-rainfall Mediterranean grassland in central Chile

D. Troncoso, N. Farías and R. Franco
The Catholic University of Maule, Carmen 684, Curicó, Chile, Email: dtroncos@hualo.ucm.cl

Keywords: Mediterranean grassland, *Acacia caven*, distance from trees

Introduction The dryland of central Chile (33° to 38° S) presents a Mediterranean-type climate. Typical local grassland is characterised by multiple annual herbaceous species, which coexist with the small tree *Acacia caven*, covering up to 3 m hectares. Over the past 100 years, the main use of this land has been for charcoal production from trees and uncontrolled grazing by sheep and cattle in rotation with wheat. Previous research has reported that *Acacia caven* can improve grassland condition (Olivares *et al*, 2000). The objective of this study was to analyse the influence of *Acacia caven* on productivity and botanical composition of grassland.

Material and methods The experiment was located in the dryland of central Chile (36° 06' latitude south and 71° 49' longitude west), on a site with clay loam soil and poor fertility; available N, P_2O_5 and K_2O, was 8 ppm, 1 ppm and 117 ppm, respectively. Total grassland biomass and botanical composition was measured identifying legumes, grasses and other species groups ('t Mannetje 2000), at 30 cm, 150 cm and 300 cm from 7 randomised *Acacia caven* trees. Grassland was sampled on August 30, October 05 and November 14, 2001. Total rainfall was 1,084 mm, 85% concentrated between May and August, a normal year. Sampling was with a 0.1 m^2 square quadrat, cutting manually at ground level, weighing in field and sorted into different groups of species. Subsamples were taken for DM, which was determined at 70°C for 72 hours oven–dried.

Results Distance from tree trunk affected total biomass of grassland on the first sample date (Table 1) and botanical composition, benefiting grasses (Table 2). At 150 cm and 300 cm, contribution of legumes increased while grasses decreased.

Table 1 Mean total biomass of grassland vegetation (g m^{-2}) according to distance from trunk (cm) and measurement date

Distance to trunk	Measurement date		
	Aug. 30	Oct. 05	Nov. 14
30	212 a	533	743
150	95 b	533	815
300	109 b	499	731
s.e.d.	20.9	82.0	103.7
	**	n.s.	n.s.

**<0.01; n.s., not significant. Different letters in each column indicate statistical difference between distance to trunk.

Lolium multiflorum, Briza sp., Avena barbata, Hordeum murinum, and *Bromus beterianus,* were the main grasses. Main legumes were *Trifolium subterraneum* and *Trifolium dubium*. Other species found were *Erodium sp., Plantago lanceolata, Juncus sp.* and *Oxalis articulata*.

Table 2 Botanical composition of grassland vegetation (% DM contribution of groups) according to distance from trunk and measurement date

Measurement date		Distance to trunk (cm)		
		30	150	300
August 30	Grasses	41 a	24 b	12 b
	Legumes	0 b	22 b	20 b
	Other	59 a	54 a	68 a
	s.e.d.	6.5	9.9	8.5
October 05	Grasses	59 a	21 a	14 b
	Legumes	0 b	46 a	32 ab
	Other	41 a	33 a	54 a
	s.e.d.	6.2	11.5	10.4
November 14	Grasses	65 a	33 ab	32 ab
	Legumes	0 c	11 b	15 b
	Other	35 b	56 a	53 a
	s.e.d.	8.3	11.5	12.1

Within measurement date, different letters in each column indicate statistical differences (P<0.01) between botanical group.

Conclusions *Acacia caven* had a positive effect on both dry matter availability and botanical composition. Grass growth was promoted directly under the tree, while legume content was higher under the tree canopy. Therefore, *Acacia caven* grassland ecosystem are productive and should be conserved.

References

Mannetje, L. 't (2000). Measuring biomass of grassland vegetation. In: L. 't Mannetje & R.M. Jones (Eds.) Field and laboratory methods for grassland and animal production research. CABI Publishing, Wallingford, 151-178.
Olivares, A., M.T. Serra & F. Venegas (2000). Relación entre la altura de *Acacia caven* (Mol.) Mol. y su área de influencia en la pradera anual con exposición norte y sur. *Avances en Producción Animal* 25(1-2), 77-84.

The relationship between species diversity and productivity of cool-season grassland

S.E. Florine, K.J. Moore, S.L. Fales and R.L. Hintz
Department of Agronomy, Iowa State University, Ames, Iowa, 50011, USA, Email: kjmoore@iastate.edu

Keywords: diversity, species richness, biomass productivity

Introduction Iowa grasslands consist mainly of introduced cool-season grasses and forbs. Many of these species are well adapted and have become naturalised. Most of these grasslands are located on marginal sites with heterogeneous soils and topography. Consequently, there is significant variation in the botanical composition and biomass productivity within and across grassland sites. This experiment aimed to evaluate the botanical composition and biomass productivity of representative grasslands and to determine if there was a relationship between grassland species diversity and biomass production across sites.

Materials and methods In south central Iowa, 10 grassland sites were selected as 'random' survey locations. Within each site, depending on the area of the site, 6 or 10 sampling areas were selected along transects. Sites 3, 6, and 9 each had 6 sampling areas, whereas sites 1, 2, 4, 5, 7, 8, and 10 had 10. In total, there were 88 sampling areas. Using a sampling frame, botanical composition of the sward was determined within each of these areas in late June. A frame ($1m^2$) was placed over the plant canopy at 2 locations/sampling area. Every species in the frame was determined and ranked in order from most to least predominant and a percentage cover was estimated for the respective sampling areas. Species richness, the Shannon-Weaver diversity (H'), and evenness (J') were determined for each site (Peet, 1974). Biomass yield was determined by clipping herbage within the frames to a height of 2.5cm, drying at $60^{\circ}C$ for 48h, and weighing. Univariate statistics were calculated for each site and the relationship between biomass production and species richness was evaluated using multiple linear regression.

Table 1 Species richness, diversity, and evenness at ten grassland sites

Site	Species Richness	H'	J'
1	10	0.93	0.93
2	12	0.92	0.86
3	11	0.96	0.92
4	10	0.90	0.90
5	9	0.82	0.86
6	7	0.73	0.86
7	6	0.68	0.87
8	14	1.06	0.92
9	5	0.57	0.81
10	5	0.58	0.83
Overall	26	1.09	0.77

Results Across all sites 26 plant species were identified. Species richness ranged from 5-14 species among the 10 sites. Smooth bromegrass (*Bromus inermis* Leyss.), Kentucky bluegrass (*Poa pratensis* L.), tall fescue (*Festuca arundinacea* Schreb.), and birdsfoot trefoil (*Lotus corniculatus* L.) were the most abundant species with frequencies of 82, 40, 38, and 34% respectively. Shannon-Weaver diversity (H') over all sampling areas was 1.09 with an evenness of 0.77. Diversity ranged from 0.57-1.06 and evenness from 0.81-0.93. There was a positive relationship between diversity and evenness with higher diversity sites exhibiting less dominance (1 – J').

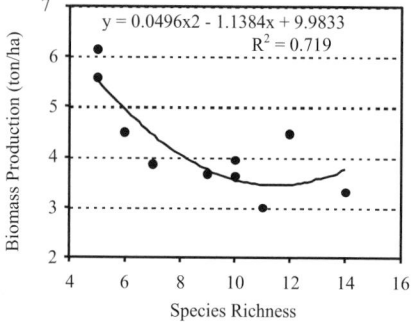

$y = 0.0496x2 - 1.1384x + 9.9833$
$R^2 = 0.719$

Figure 1 Relationship between species richness and biomass production

Biomass production varied from approximately 3-6t/ha across sampling sites. Mean biomass production over all sites was 4.2t/ha. There was a negative relationship between species richness and biomass production ($R^2=0.72$). Sites with the lowest species richness had the highest yield; sites with the highest species richness had the lowest yield. This relationship is contrary to that often postulated (Sanderson *et al.*, 2004), but agrees with that reported White *et al.*, (2004) for New Zealand grassland sites.

Conclusions Species diversity may be a good indicator of the biomass productivity of grassland sites in south central Iowa.

References
Peet, R. K. (1974). The measurement of diversity. *Annual Revue of Ecology and Systematics,* 5, 285-307.
Sanderson, M. A., R. H. Skinner, D. J. Barker, G. R. Edwards, B. F. Tracy & D. A. Wedin (2004). Plant species diversity and management of temperate forage and grazing land ecosystems. *Crop Science,* 44, 1132-1144.
White, T. A., D. J. Barker & K. J. Moore (2004). Vegetation diversity, growth, quality and decomposition in managed grasslands. *Agriculture Ecosystems and Environment,* 101, 73-84.

Plant functional diversity increases biomass production in the establishment of perennial herbaceous polycultures

V.D. Picasso and E.C. Brummer
Department of Agronomy and Graduate Program in Sustainable Agriculture, Iowa State University, 1207 Agronomy Hall, Ames, IA, 50011,USA, Email: vpicasso@iastate.edu

Keywords: functional diversity, perennials, polycultures, productivity

Introduction Natural grasslands are functionally diverse mixtures of perennial species and provide a model for sustainable agriculture systems. There is strong evidence for positive relationships between species and functional diversity and ecosystem processes like productivity and stability (Loreau *et al.*, 2001). This research aimed mainly to study the effect on biomass production (BM) of increasing plant functional diversity in agriculturally relevant perennial herbaceous polycultures during their establishment years.

Materials and methods Perennial species (n=8, from 4 functional groups) were assembled in 52 combinations including no crop, all the monocultures and some polycultures of 2, 3 and 4 functional groups, in a balanced design that did not confound species and functional diversity. Species were legumes (*Medicago sativa, Trifolium repens,* and *Desmanthus illinoensis*), cool-season grasses (*Dactylis glomerata* and *Agropyron intermedium*), warm season grasses (*Panicum virgatum* and *Tripsacum dactyloides*), and one composite *Helianthus maximiliani*. In April 2003, rectangular plots 4m x 3m, separated by 1.5m turf walkways, were sown in 2 sites in Iowa. The experimental design was a randomised complete block with 3 replicates/location. Plots were tilled; no fertilisers, lime, or herbicides were applied. Plots were mowed twice/year. In May and July 2004, BM was sampled with two $0.09m^2$ quadrats/plot. Cumulative BM data are presented for 2004.

Results Total BM production and BM of the planted species increased while weed BM reduced significantly with functional groups richness (Figure 1). Functional group composition had a major effect on those variables. Plots with legume species and cool season grasses produced 44% more on average than other plots. Since the design included a balanced combination of species, these results can be understood as an example of complementarity between functional traits, rather than individual species effects (e.g. the inclusion of legumes). The presence of warm season grasses in the mix did not increase productivity at this time of the season.

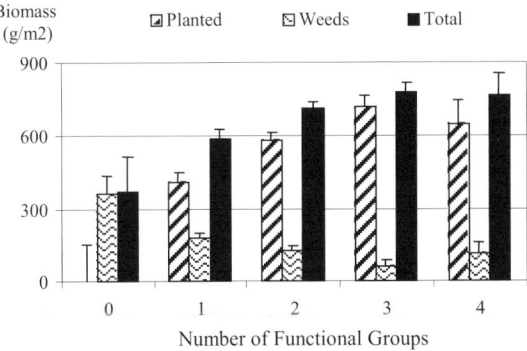

Figure 1 Biomass production (g/m^2) for planted species, weeds and total vs. number of functional groups Standard error bars are shown.

Conclusions This project is part of a long-term study on the agricultural relevance of functional diversity for various ecosystem processes. These initial results confirm the importance of functional diversity in increasing biomass productivity and resistance to weed invasion, even in species selected for agriculture in highly productive soils. Research is ongoing and more data are needed to verify the long-term effects of such diversity.

References
Loreau, M., S. Naeem, P. Inchausti, J. Begtsson, J. P. Grime, A. Hector, D. Hooper, M. Huston, D. Raffaelli, B. Schmid, D. Tilman & D. Wardle (2001). Biodiversity and Ecosystem functioning: current knowledge and future challenges. *Science,* 294, 804-808.

Relationships between productivity, quality and traits in seven co-occurring grass species

L.S. Pontes[1], J.F. Soussana[1], P. Carrère[1], F. Louault[1], J. Aufrère[2] and D. Andueza[2]
[1]INRA, Unité d'Agronomie, Clermont-Ferrand, France. Email: lpontes@clermont.inra.fr, [2]INRA, Unité de Recherche sur les Herbivores, Theix, France.

Keywords: plant traits, nutritive value, yield, cutting frequency, nitrogen

Introduction The impacts of management options sometimes have dramatic effects on botanical composition which in turn affect yield and nutritive value. A functional analysis of vegetation response may help to understand and predict the impact of changes in grassland management in a more general way screening for plant traits that may control productivity and nutritive value.

Materials and methods Monocultures of *Anthoxanthum odoratum (Ao), Dactylis glomerata (Dg), Elymus repens (Er), Festuca rubra (Fr), Holcus lanatus (Hl), Lolium perenne (Lp)* and *Poa pratensis (Pp)* were grown at two cutting frequencies (3 and 6 cuts per year; C- and C+) and at two levels of inorganic N supply (120 and 360 kg N $ha^{-1}yr^{-1}$; N- and N+) in a complete block design with 3 replicates at Theix (900 m a.s.l., Massif-Central, France). In spring and summer, plant traits were measured after 3 weeks of regrowth on 10 tillers selected at random: elongated plant height (PHe), sheath length (SL), number of growing (NG) and of mature (NM) leaves. After rehydration, the length (LL), area (LA) and fresh mass (LFM) of the youngest fully expanded leaf were measured. Leaf dry matter content (LDMC) and specific leaf area in dry matter (SLA) were then calculated. Leaf lifespan (LLS) and phyllochron (Ph) were determined in labelled tillers. Aboveground DM yield (Y) and the tiller density per unit of ground area (TD) were measured at the cutting date, and the nutritive value of the cut herbage was estimated. Forage samples were analysed for *in vitro* dry matter digestibility (IVDMD) (pepsin-cellulase, Aufrère & Demarquilly, 1989) and fibre fractions (NDF and ADF) (Goering & Van Soest, 1970) via near-infrared reflectance spectroscopy (NIRS). Soluble cellular content (1-NDF) (C1), hemicellulose (NDF-ADF) (C2) and cellulose plus lignin (ADF) (C3) were calculated from the fibre fractions. Relationships between production, quality and traits were assessed by simple correlations and through principal components analysis.

Results Traits describing tiller and leaf size were most strongly correlated to DM yield (Figure 1). Large grasses (e.g. *Dg*) with a low tiller density and a short leaf lifespan were most productive. Digestibility (DMD) was positively correlated with SLA (r=0.49) and negatively correlated with LDM (-0.54). As expected, DMD was positively correlated with C1 and negatively correlated with C2 and C3. Grasses with high tiller density (*Fr* and *Pp*) displayed a low nutritive value (r=-0.44). The data also show that both DM yield and digestibility were positively related to with SL (r=0.40, r=0.46). All correlations were highly significant (P<0.001).

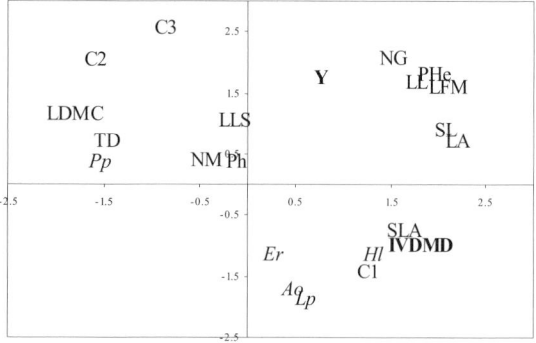

Figure 1 Projection of traits, agronomic characteristics (in bold) and species (italic) in the plan formed by the first two components of a PCA (axis 1 = 42%; axis 2 = 27%). Each value is the mean of 12 plots (four treatments).

Conclusions The first axis of the PCA appears to contrast competitive (large size, productive, high SLA, low LDM, low TD) and conservative grasses. The second axis indicates a trade-off between productivity and quality, by opposition between yield and related traits (such as tiller, sheath and leaf size) on the one hand, and DMD and SLA on the other. For the species studied, leaf and stem size were good predictors of aboveground productivity, while SLA and LDM of leaves were good predictors of DMD. These preliminary conclusions will be further tested on more species by analysing the coupled changes in traits, productivity and quality in relation to grassland management (cutting frequency and N supply) and species diversity (comparing monocultures and mixtures).

References
Aufrere J. & Demarquilly C., (1989). Predicting organic matter digestibility of forage by two pepsin-cellulase methods. *Proceedings of the XVI International Grassland Congress*, Nice, France, Vol. 2, 877-878.
Goering, H.K. & Van Soest, P.J., (1970). Forage fiber analysis, ARS Agric. Handbook, 379, 1-12. (USDA).

Specific leaf area on fresh matter basis: a soft trait for leaf thickness?

J. Debril[1], E. Kerneïs[1], S. Carré[3], P. Cruz[2] and F. Gastal[3]
[1]*INRA, F-17450 St. Laurent de la Prèe, Email: debril@stlaurent.lusignan.inra.fr*, [2]*INRA, BP27, F-31326 Castanet-Tolosan*, [3]*INRA-UEPF, F-86600 Lusignan.*

Keywords: native species, leaf traits, species ranking

Introduction Leaf dry matter content (LDMC) and the specific leaf area (SLA), are leaf traits frequently measured in field studies. Unfortunately, leaf thickness (LT) rarely is measured because that is time consuming and needs special equipment. Specific leaf mass (SLM), or leaf dry matter content/unit of leaf area, has been correlated positively to leaf density (Van Arendonk & Poorter 1994) but very few studies have examined its correlation with leaf thickness (Witkowsky & Lamont 1991). Since leaf volume is basically related to its water content, specific leaf area on a fresh matter basis (SLA$_W$) could be consistently correlated to leaf thickness. Hard traits are linked to plant functions but are difficult to measure. Therefore, it is necessary to find traits easier to measure and correlated to hard traits in order to integrate the latter in field studies on natural vegetation. The aim of this study was to find out to what extent SLA$_W$ is a reliable estimator of LT.

Materials and methods Eleven wild species (7 grasses, *Agrostis capillaris (Ac)*, *Agrostis stolonifera (As)*, *Alopecurus bulbosus (Ab)*, *Dactylis glomerata (Dg)*, *Elytrigia repens (Er)*, *Hordeum secalinum (Hs)*, *Lolium perenne (Lp)*; 3 sedges, *Carex divisa (Cd)*, *Carex riparia (Cr)*, *Eleocharis uniglumis (Eu)* and one rush, *Juncus gerardi (Jg))* were colleted in the fresh marshland of West France (March 2004). After immersion in water (6h at 4°C), the fresh mass of the youngest fully expanded and hardened leaf blade, and its area (*Delta-T area meter*) were measured, and then SLA$_W$ computed. Leaf sections were taken on fresh materials at 1/3 of the distance from ligule to leaf tip (Figure 1), and the area of transverse blade sections and lamina width were measured by image analysis (Aphelion software). Leaf thickness was estimated as the ratio between area and width of blade sections. Linear regression and Spearman's rank test were performed to analyse the correlation between traits and the similarity of SLA$_W$ and LT to rank species.

Results SLA$_W$ and leaf thickness correlated linearly and negatively after transformation to their natural logarithms (Figure 2). Spearman's rank test shows that the two variables inversely rank the 11 species (correlation coefficient of -0.88) with a high level of signification (P-value of 0.0007***).

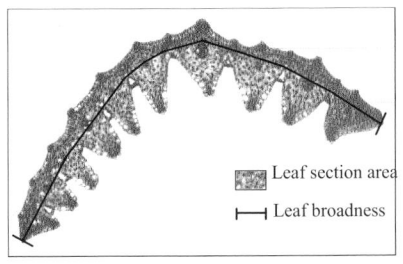
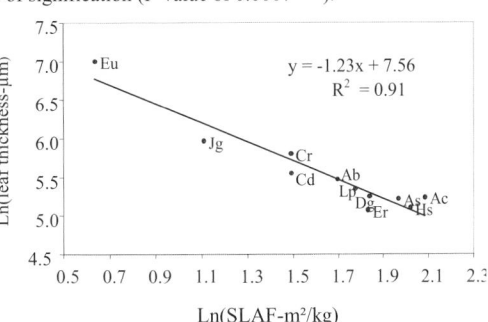

Figure 1 Blade transverse section of *Ab*   **Figure 2** Relation between SLA$_W$ (m^2.kg^{-1}) and LT (µm)

Conclusions Results illustrate the strong relationship between LT and SLA$_W$ as indicated by the relation previously shown by Garnier & Laurent (1994) between LT and the leaf specific mass expressed on a fresh mass basis (which is the reverse of SLA$_W$). Furthermore our data show that SLA$_W$ is highly reliable for ranking grass and other native species along a gradient of leaf thickness.

References

Garnier, E. & G. Laurent (1994). Leaf anatomy, specific mass and water content in congeneric annual and perennial grass species. *New Phytologist*, 128, 725-736.
Van Arendonk, J.J.C.M. & H. Poorter (1994). The chemical composition and anatomical structure of leaves of grass species differing in relative growth rate. Plant, Cell and Environment, 17, 963-970.
Witkowski, E.T.F. & B.B. Lamont (1991). Leaf specific mass confounds leaf density and thickness. Oecologia, 88, 486-493.

The potential for using the alkanes and long-chain alcohols of plant cuticular wax to distinguish the contribution of different plant species to a mixed root mass

H. Dove and T.P. Bolger
CSIRO Plant Industry, GPO Box 1600, Canberra, ACT 2601, Australia, Email: hugh.dove@csiro.au

Keywords: plant competition, hydrocarbons, aliphatic alcohols

Introduction In mixed pastures, plants compete below ground for soil water and nutrients, just as they compete above ground for light. Quantifying below-ground competition is difficult, partly because of the difficulty of measuring the contribution of different plant species to a mixed root mass. For some years, the hydrocarbons (alkanes) of plant cuticular wax have been used to quantify the species composition of the diet of herbivores (see Mayes & Dove, 2000). More recently, the long-chain aliphatic alcohols (LCOH) of plant wax have also proved useful markers (Bugalho *et al.*, 2004). Plant roots also contain cuticular alkanes and these may be used to discriminate between roots coming from different species (Dawson *et al.*, 2000). We report an extension of this concept, using a combination of cuticular alkanes and LCOH to discriminate between root tissues from plant species commonly found in or sown as pastures in southeastern Australia.

Materials and methods Nine plant species (Figure 1) were grown in pots in a glasshouse, harvested while still vegetative and separated into root and shoot material. Cuticular alkanes and LCOH were extracted from freeze-dried, ground root material (12 samples/species) and assayed by gas chromatography (Bugalho *et al.*, 2004). Differences in alkane or LCOH profiles between plant species were examined using canonical variates analysis (CVA). Canonical variates scores from alkanes and LCOH were compared by orthogonal Procrustes rotation to determine if the discriminatory information provided by LCOH was additional to that provided by alkanes.

Results Alkanes with carbon-chain lengths C21-C35 were detected and, in confirmation of earlier work (Dawson *et al.*, 2000), their concentrations were <20 mg/kg OM, an order of magnitude lower than those typical of shoot tissue (Bugalho *et al.*, 2004). Root LCOHs with chain lengths C20-C32 were detected; concentrations (5-400 mg/kg OM) were also lower than those reported in shoots (Bugalho *et al.*, 2004), but were in general much higher than root alkane concentrations. The CVA showed that the concentration profiles of both alkanes and LCOH in root tissue from the nine pasture species differed, and that they significantly discriminated between roots from the different species (P<0.001). Orthogonal Procrustes rotation indicated that the discrimination based on LCOH was additional to that based on alkanes. CVA based on the combined use of alkanes and LCOH achieved complete discrimination between roots from the 9 species (Figure 1: P<0.001).

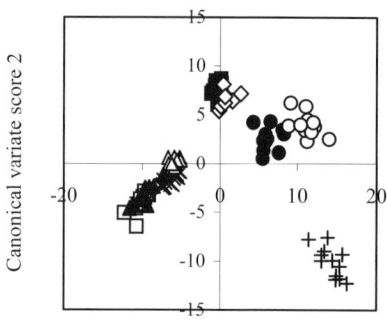

Canonical variate score 1

Figure 1 Canonical variates analysis of combined data for alkanes and LCOH in root tissue. The species listed were significantly discriminated by their canonical scores 1, 2 and (in parenthesis) score 3 (all P<0.001). X *Lolium perenne* (1.76); • •*Vulpia bromoides* (6.69); • •*L. rigidum* (-3.99); • • *Dactylis glomerata* (-2.33); •• *Austrodanthonia richardsonii* (-1.72); + *Trifolium repens* (-2.97); • •*Medicago sativa* (-2.73); • •*Phalaris aquatica* (-4.74); • •*T.subterraneum* (8.46)

Conclusions Between-species differences in alkanes and LCOH concentrations in root tissue from pasture plants, separately and especially in combination, are sufficient to provide a chemical approach for assessing the contribution of different plant species to a mixed root mass. In further work, we are validating this technique using known mixtures of roots from different species.

References
Bugalho M.N., H. Dove, W.M. Kelman, J.T. Wood & R.W. Mayes (2004). Plant wax alkanes and alcohols as herbivore diet composition markers. *Journal of Range Management* 57, 259-268.
Dawson L.A., R.W. Mayes, D.A. Elston & T.S. Smart (2000). Root hydrocarbons as potential markers for determining species composition. *Plant, Cell and Environment* 23, 743-750.
Mayes, R.W. & H. Dove (2000). Measurement of dietary nutrient intake in free-ranging mammalian herbivores. *Nutrition Research Reviews* 13, 107-138.

Does niche complementarity explain the relationship between biodiversity and ecosystem functioning in managed grasslands?

N. Buchmann and A. Kahmen

Institute for Plant Sciences, ETH Zuerich, ETH Zentrum LFW C56, Universitaetsstr. 2, CH-8092 Zuerich, Switzerland, Email: nina.buchmann@ipw.agrl.ethz.ch

Keywords: biodiversity, productivity, resource use, grasslands, nitrogen

Introduction Niche complementarity was suggested to largely explain the positive relationship noted between plant diversity and productivity in some recent studies. This suggests that an increasing number of species exploits resources more efficiently and thus enhance ecosystem functions. This hypothesis, however, implies that niches occupied by different plant species are rather distinct so that niches from extinct or missing species stay unoccupied by the remaining species of an ecosystem. This experiment tested if plant species occupy different and distinct niches with respect to soil N uptake, being a possible functional explanation for the biodiversity ecosystem functioning relationship.

Materials and methods We tested the niche complementary hypothesis using 3 semi-natural grasslands with different plant diversities. We characterized ecological niches with respect to nitrogen acquisition chemically, spatially and temporally by injecting ^{15}N labeled nitrate, ammonium and glycine in the soil of separate 1 m^2 plots at 2 different depths twice/year. After labeling, plant species in each plot were harvested and analysed for uptake of ^{15}N. We performed multivariate ordination techniques to estimate shifts in niche occupancy.

Results Grassland plant species did not differ in their chemical, temporal and spatial N uptake patterns (Fig. 1, arrows), but in their respective uptake rates. N uptake rate was independent of treatment and of a plant's abundance. N uptake rates, however, correlated negatively with a plants N concentration (Fig. 2). Therefore, plant functional groups seem to rely on N pools other that soil N (N fixation, internal N recycling) and can thus be classified according to nitrogen use strategies into: grasses, tall herbs, small herbs and legumes. All groups showed highly distinct N use strategies across different communities, so that so that a positive effect of functional group diversity and ecosystem functioning is to be expected.

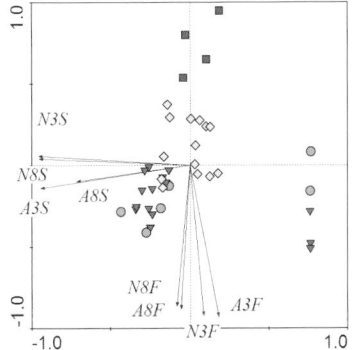
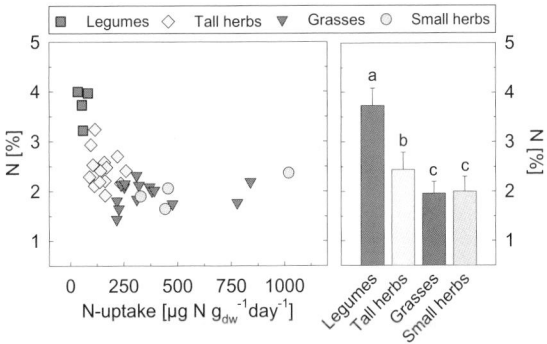

Figure 1 Principal component analysis (PCA) of N uptake by species (symbols) and treatments (N = NO$_3^-$, A = NH$_4^+$, 3 = 3 cm treatment, 8 = 8 cm treatment, F = spring, S = summer)

Figure 2 Average daily N uptake of plants in four functional groups in relation to above-ground N concentration (left panel). Average N concentration per functional group (right panel; with SD)

Conclusions No differences in spatial, temporal or chemical N uptake were observed among the investigated grassland species. However, functional groups showed disdtinct N use strategies. Our data suggest that species loss will not impair ecosystem functions that relate to N use strategies as long as functional diversity, i.e. functional group diversity in maintained. However, if biodiversity drops below a critical level where functional diversity declines, severe effects on the functioning of the ecosystem are to be expected.

Linking community and ecosystem ecology by developing a grassland ecosystem model (GEMINI) with interactions between plant, herbivore and soil microbial populations

J.F. Soussana, S. Witzmann, P. Loiseau, S. Fontaine, P. Carrère, C. Picon-Cochard and F. Louault
*INRA, Grassland Ecosystem Research, Agronomy Unit, 234 Av. du Brézet, F-63100 Clermont-Ferrand, France,
Email: soussana@clermont.inra.fr*

Keywords: simulation, plant functional traits, vegetation dynamics, C cycle, N cycle

Introduction Simulation models may help to understand the functional role of plant and soil biodiversity for C and N cycles and for intake by herbivores in semi-natural grassland ecosystems. Detailed models of grassland ecosystems calculate C, N, water and energy fluxes without accounting for the species dynamics in the plant and soil communities. Schwinning & Parsons (1996) proposed a simple pasture growth model that includes mixed grass and clover components. This model was, however, restricted to 2 plant functional groups and it excluded the dynamics of the soil organic matter. The role of competitive interactions between at least 2 functionally distinct soil microbial communities that would mediate a priming effect and account for the limitation by energy of soil organic matter decomposition has generated renewed interest recently (Fontaine *et al.*, 2004). Moreover, recent studies (Personeni & Loiseau, 2004) show that root traits partly control decomposition. To make further progress, we have developed a modelling approach of the interactions between plant populations and microbial groups within a grassland patch. This model aims to link the dynamics of plant and microbial groups with biogeochemical cycles. It allows the testing within a single framework of some of the main hypotheses proposed to account for the functional role of biodiversity in grasslands.

Model description GEMINI (Grassland Ecosystem Model with INdividual Based Interactions) is a deterministic object-oriented simulation model. It runs in a robust, flexible and portable platform developed in C++ (C Builder 6.0, Borland™ under Windows™) and based on a unified numerical integrator. This platform allows the coupling of numerical models. A model is seen as a tree of modules, each having its own numerical integrator variables. These objects are referenced in runtime modifiable lists. A flexible graphical front-end using the VCL library of C++ Builder is generated from the tree of modules. GEMINI consists of vegetation- and soil- sub-models, coupled with environment- and management- modules. The vegetation model, named CANOPT (Soussana & Oliveira-Machado, 2000), is an individual-based model of the growth of mixed pasture species, which describes explicitly the shoot and root morphogenesis of plants and the competition for light and for inorganic N within a multi-layer canopy and soil. It consists of six modules: (1) management options (grazing and/or cutting mode, N fertilizer supply); (2) environment module, which calculates the microclimate and the inorganic N balance; (3) plant growth and allocation module, which simulates the C and N balance and the allocation of growth to the shoot structures, leaf proteins and roots; (4) shoot morphogenesis module, which computes the demography and the size of shoots and roots; (5) competition module, which calculates radiation (PAR) and N partitioning among plant populations; (6) animal module, which calculates the defoliation of each species at grazing and animal returns. The grazing time/bite has 3 components: a fixed prehension time, a mastication time and a time to sort between mixed plant populations. The C-N soil model (named SOILOPT) describes the dynamics of 4 soil organic compartments, each with a fixed C:N ratio. Two functional microbial groups (e.g. bacteria and fungi) degrade respectively fresh litter and soil organic matter (FOM and SOM decomposers). The 2 microbial groups differ in their potential growth rate, type of substrate used for growth and mineral N requirements. This allows simulation of the energy limitation of SOM decomposers.

Conclusions The model captures some of the feedbacks between plant species dynamics, soil microbial groups and soil C and N dynamics. It shows the role of differences between plant functional types for ecosystem processes such as above-ground productivity and litter decomposition. It also allows analysis of complementarities between plants, which contribute to the functional role of biodiversity in grasslands.

References

Fontaine, S., A. Mariotti & L. Abbadie (2004). Carbon input to soil may decrease soil carbon content. *Ecology Letters*, 7, 314-320.
Personeni E. & P. Loiseau (2004). How does the nature of living and dead roots affect the residence time of carbon in the root litter continuum? *Plant and Soil* (in press).
Schwinning, S. & A. J. Parsons (1996). Analysis of the coexistence mechanisms for grasses and legumes in grazing systems. *Journal of Ecology*, 84, 799-813.
Soussana, J. F. & A. Oliveira-Machado (2000). Modelling the dynamics of grasses and legumes in cut mixtures. In Grassland ecophysiology and grazing ecology. CAB International Pub., Oxon UK, pp. 169-190.

Study of characteristics of soil animals in halophilous plant communities of *Leymus chinensis* grasslands of northeast in China

X. Yin, Y. Zhang and W. Dong

College of Urban and Environmental Science, Northeast Normal University, Changchun, Jilin Province. P.R. China 130024, Email: yinxq773@nenu.edu.cn

Keywords: *Leymus chinensis* grassland, halophilous plant community, soil animals

Introduction We have researched soil animals in 8 types of halophilous plant communities of *Leymus chinensis* grasslands of Northeast China to characterise soil animal groups and explain the role and function of soil animals in grassland ecosystems (Richard & Roger, 1998) and provide a scientific basis for research to improve alkaline lands in these grasslands.

Methods We investigated within *Leymus chinensis* grasslands the plant communities *Aeluropus litoralis*(A), *Puccinellia tenuifolia* (B), *Suaeda hetroptera* (C), *Suaeda glauea* (D), *Suaeda corniculate* (E), *Kochia sieversiana* (F), *Artemisia anethifolia* (G)and *Puccinellia chinampoesis* (H). Each plant community was sampled randomly at 4 sites. Sample sizes were 50cm × 50cm (for large-scale soil animals) and 10cm ×10cm (for middle-small-scale soil animals). We sampled at depths of 0-5cm, 5-10cm, 10-15cm, 15-20cm and 20-30cm. Animals were separated from soil by handpicking, Tullgren funnels and Baremann funnels (Jun-ichi AOKI, 1973).

Results A total of 784 soil animals belonging to 50 groups of 3 phyla, 4 classes, 14 orders and 36 families were found. There were 25 groups of large-scale soil animals. There were 3 dominant and 7 frequent groups. The individual numbers of both dominant and frequent groups accounted for 93.8% of the total. There were 35 groups of middle-small-scale soil animals. There were 3 dominant and 12 frequent groups. The individual totals of dominant and frequent groups accounted for 90.9% of the total. All of these groups were the basic components of soil animal populations and they were distributed widely in *Leymus chinensis* (Yin Xiuqin, 2003). Trends in group numbers and individuals for both large-scale and middle-small-scale soil animals in different halophilous plant communities were different (Figure 1). In order to analyse the relationship between soil animals and different communities, we calculated the diversity index of large-scale and middle-small-scale soil animals in different halophilous plant communities. The results are shown in Table 1. Vertical changes of large-scale and middle-small-scale soil animals in different halophilous communities were different.

Table 1 The diversity index of soil animals in different halophilous plant communities

Community No.	A	B	C	D	E	F	G	H
Large-scale	1.82	1.74	0.90	2.30	1.61	1.98	1.82	1.50
Middle-small-Scale	1.40	2.71	2.43	2.29	2.41	2.46	1.16	2.07

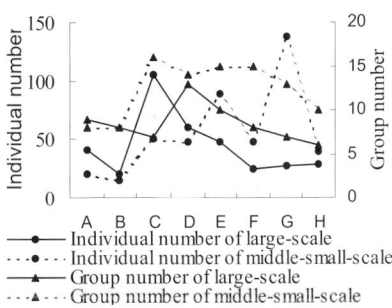

Figure 1 Number of individuals and groups of large-scale and middle-small-scale soil animals in different halophilous communities

Conclusions In the horizontal structure, for large-scale soil animals, the individual numbers of C was highest and B was lowest. However, the group numbers of D was highest and H was lowest. For middle-small-scale soil animals, the individual numbers of G was highest, B was lowest. While the group numbers of C was highest, A and B were the lowest. The group number and diversity index of soil animals were positively correlated. In 8 types of halophilous plant communities the numbers of groups and individuals of soil animals decreased with increasing soil depth.

References

Jun-ichi AOKI, (1973). The zoology of soil, Hokuryukan, Tokyo, 5-25.

Bardgett, R. & R. Cook (1998). Functional aspects of soil animal diversity in agricultural grassland. *Applied Soil Ecology,* 10, 263-276

Yin Xiuqin, Wang Haixia & Zhou Daowei (2003). Characteristics of soil animals communities in different agricultural ecosystem in the Songnen Grassland of China. *Acta Ecologica Sinica,* 23, 1071-1078

Spatial distribution of soil macroinvertebrates in a dry steppe (South-Eastern Siberia, Russia)

K.B. Gongalsky
A.N. Severtsov Institute of Ecology and Evolution, Russian Academy of Sciences, Leninsky pr., 33, Moscow, 119071, Russia, Email: kocio@mail.ru

Keywords: soil, macrofauna, spatial distribution

Introduction   Soil macroinvertebrates are important components of ecosystems. They play a key role in decomposition processes and turnover of the most of elements. Adequate estimation of abundance and biomass of these animals is fundamental for understanding their input in steppe ecosystems. Asian steppes of Russia are poorly studied. Therefore, baseline soil invertebrate composition, abundance and rules of distribution were estimated.

Materials and methods   The sampled site was in dry steppes on Kashtanozem soils, 30 km east of Krasnokamensk (50°05•N, 118°15•E), in the Chita region, South-East Siberia. The map is in: Gongalsky 2003. The relief is represented by small hills with a maximal elevation of 800-1300 m above sea level. The climate is sharply continental. Soils freeze up to 3 m below the surface. Dominance of bushes of *Caragana stenophylla* with *Agropyron cristatum, Stipa baicalensis, Leymus chinensis, Artemisia frigida, A. mongolica, Adenophora gmelinii, Bupleurum scorzonerifolium* is the general feature of the steppe vegetation. In August 2000, 144 intact soil cores to the depth of 8-12 cm were collected at the site. Each core was from an area of 76 cm^2. Samples collected formed a grid of 24x6 units. Samples were placed into separate marked plastic bags and then hand-sorted in the laboratory. Litter, soil and pebble (fraction >2.7 mm) mass, water holding capacity (WHC), pH and loss on ignition (LOI) were measured in every soil sample using standard methods in the laboratory.

Results   Macrofauna numbered 139.8• 4.3/m^2. Phytophagous animals (54.2%) dominated the community; the rest were saprophagous (23.9%) and predatory animals (21.9%). Larvae of Tenebrionidae (36.1/m^2) and Curculionidae (6.3/m^2) were the most abundant. Dipterans larvae (mostly Therevidae, 27.1/m^2) dominated also. Predatory insects included carabids, both imagines (6.3/m^2) and larvae (7.2/m^2) were less abundant. In general, the macrofauna consisted of beetles at different stages of development. Although stones and pebble correlated negatively with soil faunal abundance (r=-0.569 and -0.673, respectively) p values were >0.05. WHC was the only significant correlation with animals (r=0.810). Soil quality (both physical and chemical parameters) defined the abundance of the soil macrofauna. Spatial distribution was heterogeneous (Figure 1).

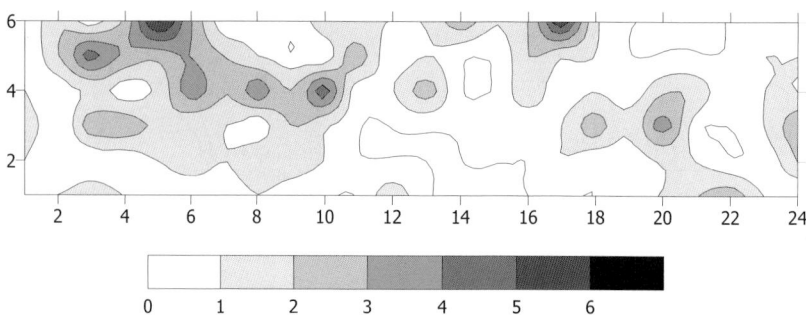

Figure 1   Spatial distribution of total animal number in a dry steppe. Axes are marked by the samples' numbers, the size of each plot is 2.4x0.6 m.

Conclusions   Although the population is similar to those in European steppes (Ghilarov 1965), soil macroinvertebrates of South-Eastern Siberian steppes are characterized by a unique complex of traits. The spatial distribution of macroinvertebrates in the studied site was heterogeneous, which has to be taken into account while extrapolating abundance and biomass data to large areas.

Acknowledgements   The study was supported by Russian Foundation for Basic Research (grant 03-05-64127).

References
Ghilarov M.S. (1965). Zoological methods in soil diagnostics. Nauka Publ., Moscow, pp. 278. (in Russian).
Gongalsky K.B. (2003). Impact of pollution caused by uranium production on soil macrofauna. *Environmental Monitoring and Assessment,* 89, 197-219.

Grazing effects on spatial microdistribution of soil macroinvertebrates in a steppe of European Russia

F.A. Savin, K.B. Gongalsky and A.D. Pokarzhevskii
A. N. Severtsov Institute of Ecology and Evolution, Russian Academy of Sciences, Leninsky pr., 33, Moscow, 119071, Russia, Email: savfe@mail.ru

Keywords: soil, macrofauna, spatial distribution, pasture

Introduction   Measures of ecosystems and their populations include biomass, production and trophic composition and animal spatial distribution. Grazing of grasslands influences spatial distribution of vegetation seriously. The same effects on soil macroinvertebrates are less well studied. Spatial distribution is usually studied at a coenosis level, and not at a studied point level, although the scale of sampling has a giant importance in estimation of organism's distribution. The aim of our study was to estimate grazing impact on large soil invertebrates in steppe ecosystems in Chernozem Nature Reserve in Russia.

Materials and methods   The sampling site was in Kursk Region, 500km south of Moscow. Material was collected from 2 plots in June 2001. Plot 1 was a mixed-herbaceous meadow-steppe (ungrazed steppe, US) kept untouched since 1947. Plot 2 (grazed steppe, GS) was the same type of steppe with a pasture pressure of 2 cows/ha. At each plot, 144 intact soil cores were collected to a depth of 8-12cm. Each one had an area of 76 cm^2. Samples collected formed a grid of 24 x 6 units. Samples were placed into separate marked plastic bags and then hand-sorted in the laboratory. Standard laboratory methods were used to measure litter, soil and pebble (fraction >2.7 mm) mass, water holding capacity (WHC), pH and loss on ignition (LOI) in every soil sample. Data analysis included descriptive statistics, estimation of sample normality and by kriging in Surfer 6.0 to create surface diagrams to show the spatial distribution of the parameters.

Results   Root- and litter- mass, respectively, were 2 and circa 20 times higher at Plot 1 (US) than at Plot 2 (GS). Root- and litter- mass dictated invertebrate presence and distribution. Macroinvertebrate Abundance in US ($339/m^2$) was higher, than in GS ($246/m^2$). Staphylinidae imagoes ($72/m^2$), Scarabaeidae larvae ($63/m^2$) and Julidae diplopods ($50/m^2$) dominated in US. Larvae of Curculionidae ($40/m^2$), Elateridae ($39/m^2$) and Scarabaeidae ($31/m^2$) dominated in GS. Myriapods were almost absent at GS, where their numbers/m^2 were circa 60-times less than in US. Figure 1a shows aggregated US macrofaunal distribution: areas with 12 fauna/sample bordered on empty areas. Total macrofaunal abundance in US and GS had CVs of 73.7 and 72%, respectively. The difference between GS and US was mainly due to lack of litter in GS; this tended to exclude herpetobiontic animals. The proportion of geobionts at GS and US differed markedly: 67 vs. 36%, respectively.

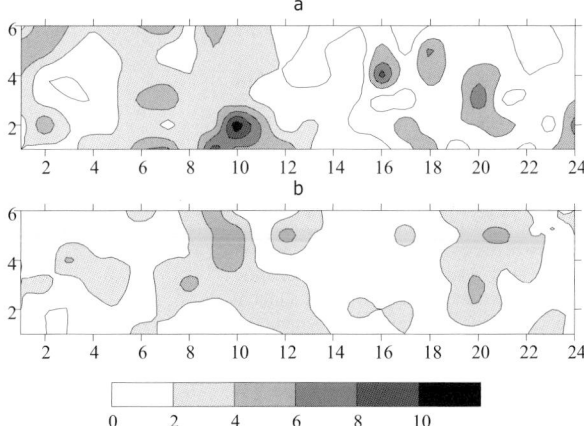

Figure 1 Spatial distribution of numbers of total invertebrates in ungrazed (a) vs. grazed (b) steppe. Axes are marked by the samples' number, following the 24 x 6 grid used for taking core samples; each plot was 2.4x0.6m.

Conclusions   Grazing of a European Russian steppe decreased heterogeneity of distribution of soil-living invertebrates in the ecosystems. The effect was due mostly to destruction of abundant litter mass (a suitable habitat) to support herpenobiont animals.

Acknowledgement   The Russian Foundation for Basic Research supported this study (grant 03-05-64127).

Grassland arthropod species richness in a conventional suckler beef production system and one compatible with the Irish agri-environment scheme (REPS)

A.J. Helden, A. Anderson and G. Purvis
Department of Environmental Resource Management, Faculty of Agri-Food and the Environment, University College Dublin, Belfield, Dublin 4, Ireland, Email: alvin.helden@ucd.ie

Keywords: biodiversity, grassland management, Araneae, Hemiptera

Introduction Grassland management practices, such as grazing, strongly affects the biodiversity of grassland arthropods; increasing grazing intensity causes a general decline in species richness (Morris, 2000). One of the aims of the Rural Environment Protection Scheme (REPS) is to conserve and enhance biodiversity within Irish agricultural land (Feehan *et al.*, 2002). In order to determine the effectiveness of this aspect of REPS, one must compare the relative biodiversity of grassland under REPS with that of conventionally managed grassland. Aiming to determine whether species richness was higher in REPS-compatible compared with a standard system of management, we measured the species richness of grassland arthropods within two contrasting grassland treatments within an experimental study of suckler beef production.

Materials and methods Grassland arthropods were sampled from the Systems of Suckler Beef Production experiment at Teagasc Grange, County Meath. The experiment compared 2 treatments: standard system (0.65ha/cow unit, 225kg N/ha); and REPS-compatible (0.82ha/cow unit, 88kg N/ha). The experiment involved 4 blocks, each containing one replicate of both treatments. The individual replicates were sub-divided into 3 grazing paddocks, grazed in a fixed sequence within each treatment and between blocks, with 2 blocks being grazed concurrently. Insects were sampled in Aug 2003 using a Vortis suction sampler. One sample, consisting of 10 randomly placed sub-samples of 10s duration, was taken per paddock, giving 3 nested samples per replicate. Depending on the taxon, 5 groups of arthropods were identified to species, morphospecies, genus or family. The 5 groups were Araneae (species); Coleoptera (species); Diptera (family); Hemiptera (species and morphospecies); parasitic Hymenoptera (genus).

Results The number of species, or equivalent, recorded for the 5 arthropod groups were: Araneae 7; Coleoptera 43; Diptera 23; Hemiptera 17; parasitic Hymenoptera 43. Treatment comparisons were carried out using the log transformed (ln+1) number of species per grazing paddock, which was incorporated into a nested analysis of variance. There were no significant differences between blocks or treatments for Coleoptera, Diptera or Hymenoptera. Block was not significant for Araneae and Hemiptera, respectively ($F_{3,4} = 0.42$ & $F_{3,4} = 0.85$), but the REPS-compatible treatment had significantly more species than the standard system ($F_{4,16} = 3.17$ p < 0.05 & $F_{4,16} = 4.84$; p<0.01, respectively). Figure 1 shows the mean number of species per grazing paddock for each treatment.

Conclusions Although these grasslands had relatively low species richness of arthropods, significant treatment effects were found. The species richness of both Araneae and Hemiptera, but not the three other arthropod groups, was significantly higher in REPS-compatible than in conventionally managed grassland. This provides evidence that REPS-can fulfil, at least partially, its aim of maintaining and enhancing biodiversity. The differences between the arthropod groups may reflect contrasting mobility and relationships with vegetation structure. Araneae and Hemiptera would appear to be suitable groups for studying the effect of grassland management on arthropod biodiversity.

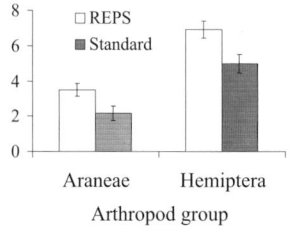

Figure 1 Species richness of Araneae and Hemiptera

References

Feehan, J., D. A. Gillmor, & N. E. Culleton (2002). The impact of the Rural Environmental Protection Scheme (REPS) on plant and insect diversity. *Tearmann: Irish Journal of Agr-environmental Research*, 2, 15-28.

Morris, M.G. (2000). The effects of structure and its dynamics on the ecology and conservation of arthropods in British grasslands. *Biological Conservation*, 95, 129-142.

Comparison of epigeic spider (Arachnida : Araneae) assemblages in winter wheat agroecosystems of the European part of Russia

R.R. Seyfulina

Department of Invertebrate Zoology, Faculty of Biology, Lomonosov Moscow State University, Vorobyevy Gory, Moscow 119992 Russia, Email: r-seyfulina@yandex.ru

Keywords: Araneae, spiders, agroecosystem, agrocoenosis, agrolandscape

Introduction Spiders comprise 20-80% of predatory fauna (Ferguson *et al.*, 1984) and are important in controlling dangerous pests (Horner, 1972). Although different from most natural associations, cropland spider complexes resemble meadow communities and their diversity varies from tens to more than 300 species. About 20 species, referred to as agrobionts (Luczak, 1979), are common to all European agroecosystems.

Materials and method Species composition, dominance structure, spatial distribution and seasonal activity dynamics of epigeic spider assemblage were studied in winter wheat fields and surrounding habitats of 2 distant regions, which differ in agricultural landscapes and climate conditions. The sampling sites were situated in the centre (Moscow Area (M), 55°59′N, 37°24′E) and in the south of the European part of Russia (Krasnodar Province (K), 45°03′N, 39°18′E). In M the agricultural fields are of medium size (10-15ha) and surrounded by mixed forests. In K large local fields (up to 100ha) are separated by shelter belts. Sampling with pitfall traps was performed for 2 crop seasons (from spring till harvest) on 2 fields different per region per year (M 1998-99; K 1999-2000). Traps were set in the cropland, viz. in field edges (FE) (10m from the field border) and in the field centre (FC) (150m in M; 200 and 400 m in K), as well as in grassy field margins (FM) at 3-5m from the field, and bordering strips of adjacent habitats.

Results In M, 132 spider species from 15 families were recorded, 112 and 68 species in agroecosystems (with and without taking the FM into account, respectively). In K, 100 species from 18 families were recorded, 89 and 72 species in agroecosystems (with and without FM). In K and M, 35 species were common to both regions. Jackard and Sørenssen indices gave similarity values of 18 and 30%. In M, FM had a significantly higher species diversity according to Margalef, Shannon, Berger-Parker indices. In K, shelter belts, FM and FE, had no significant difference among the species diversity. Species diversity declined from FM toward FC, where it was minimal in both regions. However, species diversity did not change considerably across the central zone of large fields (sites 200- and 400 m apart).

Family composition and dominance structure changed from across FM and adjacent habitats toward FC. Spider assemblage of M croplands typically had higher contents of Linyphiidae and Tetragnathidae compared to FM, with the opposite pattern for Lycosidae; Lycosidae were a superdominant group in K. These families constituted 95% of the total numbers. *Oedothorax apicatus* (M, K), *Araeoncus humilis* (M)*, Erigone dentipalpis* (M), *Pachygnatha degeeri* (M), *Pardosa agrestis* (M, K)*, P. palustris* (M), *Trochosa robusta* (K) were species with preference for cropland. In general, intrazonal agrobiontous species dominated in FC. With the exception of singletons, all spider species in cropland were reported from FM also. Most FM dominant species were almost absent from FC. On the other hand, FC dominant species also dominated in FE, while species prevailing in FE did not necessarily remain so (or were even abundant) in FC. Peak activity of nearly all dominant species was observed in the first half of the vegetation season. Maximum activities of both epigeic spiders and wheat pests (*Oscinella* flies, *Phyllotreta* beetles, *Lemma* beetles and shield-backed bug larvae) contemporized, suggesting spider involvement in pest control.

Conclusions Spider species diversity in agroecosystems tended to be similar to that in adjacent natural habitats, with a marked decline in FC. High species diversity in FM and FE does not imply high diversity in the rest of the field, because not all spiders abundant in FM penetrated into croplands beyond FE. Agrobiont species occured in the field regardless of its size. They rapidly colonized winter crops at the beginning of vegetation season. With time, some species left FC with senescent vegetation. The spider species composition in agrocoenoses was similar among different regions and the list of dominant species was rather constant. Crop characteristics, local agrolandscape, particularly, proximity of grassy margins to the crops, and weather conditions were the factors affecting the activity of epigeic spiders.

References

Ferguson, H. J., R.M. McPherson & W.A. Allen (1984). Ground- and foliage-dwelling spiders in four soybean cropping systems. *Environmental Entomology,* 13, 975-980.

Horner, N. V. (1972). *Metaphidippus galathea* as a possible biological control agent. *Journal of Kansas Entolomological Society*, 45, 324-327.

Luczak, J. (1979). Spiders in agrocoenozes. *Polish Ecological Studies,* 5, 151-200.

Grassland and avian biodiversity within Irish agriculture

B.J. McMahon and J. Whelan
Ag-Biota Project, Dept. Environmental Resource Management, Faculty of Agriculture, UCD, Dublin 4, Ireland,
Email: barry.mcmahon@ucd.ie

Keywords: intensive grassland, farmland birds

Introduction In the last quarter of the 20th century, populations of farmland birds have declined markedly, representing a severe threat to biodiversity (Donald et al., 2001). Because the vast majority of Irish farmland is devoted to intensive grassland, it is important to establish what quality of habitat this provides for biodiversity, especially avian. This study aimed to establish the avian species on the selected sites, compared with the total number of species that have been recorded on Irish farmland in recent years, as documented by the Complete Guide to Irish Birds (Dempsey & O'Clery, 2002).

Methods Five intensive grassland sites were selected as part of the Ag-Biota project. One site was in Co. Meath (Grange (GR)); 2 were in Co. Wexford (Johnstown Castle (JC) and an associated commercial site (JCC)) and 2 were in Co. Tipperary (Solohead (SH) and an associated commercial site (SHC)). The total species richness was recorded using line transects (Bibby et al., 2000) across agricultural grassland during 4 resamples in summer. The cumulative species richness was calculated for the 5 sites and this was compared to the total species richness found on Irish farmland as recorded in the Complete guide to Irish Birds (Dempsey & O'Clery, 2002).

Results The total number of species recorded on Irish grassland is 56. The number of species recorded in this study varied from 24 (SH, SHC) to 28 (JC). Figure 1 shows (as percentages) the total possible species richness compared to the recorded species richness over the 5 sites.

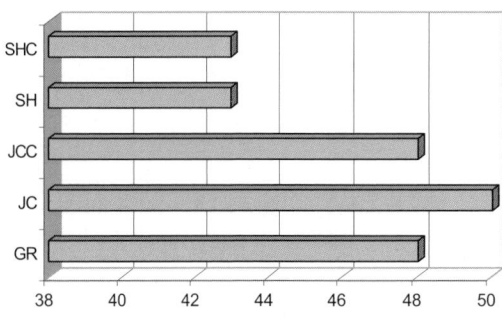

% of grassland species

Figure 1 Total species richness (%) compared to recorded species richness

Conclusions This study highlights that intensive grassland is a sub-optimal habitat for many farmland bird species. Further research is needed on how to optimise bird species richness and diversity on Irish grassland in a way that does not significantly impinge on agricultural production.

References
Bibby, C.J., N. D. Burgess, D. A. Hill & S. H. Mustoe (2000). Bird Census Techniques. Academic Press: London.
Dempsey, E. & M. O'Clery (2002). The complete guide to the birds of Ireland. Gill and McCillian.
Donald, P. F., R. E. Green & M. F. Heath (2001). Agricultural intensification and the collapse of Europe's farmland bird population. *The Royal Society London*, 298, 25-29.

The effect of burning abandoned reclaimed land in early spring on the distribution of an endangered grassland bird species – the Japanese Marsh Warbler

T. Sugiura, E. Ochiai, M. Baba and H. Kobayashi
Kitasato University, Toawada Aomori 034-8628 Japan, Email: sugiura@vmas.kitasato-u.ac.jp

Keywords: burning, common reed, reclaimed land, courtship display, *Locustella pryeri*

Introduction The Japanese Red Data Book of Birds lists the Japanese Marsh Warbler (*Locustella pryeri pryeri*) as an endangered species. It feeds and breeds in-reed dominated (*Phragmites australis* (Cav.) Trin.ex.Steud.) grassland in N Japan. Expanses of reclaimed land (Hotokenuma) are its largest breeding grounds. Each April, fire management is used to rid the area of the dead reed material amassed from the previous year. Circa 1 month after burning, the species returns to the area to inhabit and breed in the unburned areas. Fire use to manage grasslands is an important tool for the conservation and management of bird habitat (Pon *et al.*, 2003, Kirkpatrick *et al.*, 2002). We investigated the size and distribution of the unburned areas and the monthly changes in the number of the courtship displays observed in breeding season. This study aimed to draft recommendations for use to conserve the microhabitat of this species from the results.

Materials and methods The site was 24ha of the densest area of the species in Hotokenuma. We divided the site into 4 x 6ha sites and the size and distribution of the burned and unburned area were recorded immediately after the fire in 2002 and 2003, respectively. The locations of courtship displays in the Japanese Marsh Warbler were documented every 2 weeks over a 20-week period from early May to early September in 2002 and 2003.

Results The unburned areas in early April were 2.98±1.86ha in 2002 and 0.74±0.37ha in 2003, respectively. The number of individual courtship displays by Japanese Marsh Warblers in the 24ha from June to July was 54.2±2.8 in 2002 and 29.2±2.2 in 2003, respectively. The number of birds began to decrease in August and the warblers had disappeared by September. Upon arrival at Hotokenuma, the birds congregated in the burned area for a few days. However, they moved toward the remaining mature stands of common reed. Correlation between the area of unburned reeds and the number of birds in June in 2002 and 2003 was significant (p<0.05).

Conclusion Japanese Marsh Warbler displays were seen in the areas of common reed that remained unburned in the early breeding season. They then moved to the burned areas where regrowth of vegetation had begun. The size of the unburned area in early spring affected the number of the birds at the site. Considerable care is needed when undertaking controlled burns because these activities impact markedly on bird numbers and breeding activities at a site. This in turn has important implications for bird conservation.

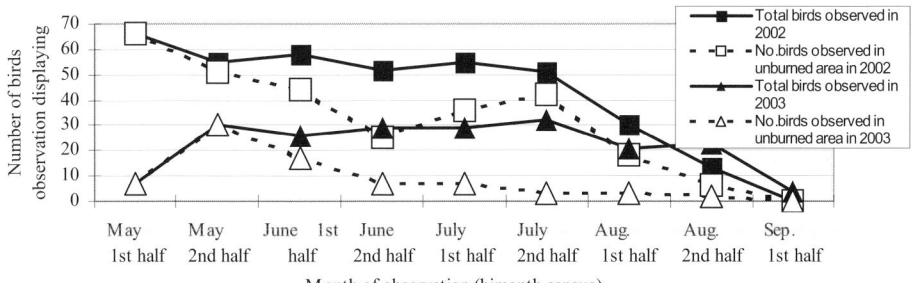

Figure 1 Number of birds observed displaying in burned and remained areas

References

Pons P., B. Lambert, E. Rigolot & R. Prodon (2003). The effects of grassland management using fire on habitat occupancy and conservation of birds in a mosaic landscape. *Biodiversity and Conservation,* 12(9), 1843-1860.

Kirkpatrick C., S. DeStefano, R. W. Mannan & J. Lloyd (2002). Trends in abundance of grassland birds following a spring prescribed burn in southern Arizona. *Southwestern Naturalist,* 47(2), 282-292.

Lapwing and redshank nesting sites on coastal marshes: does sward structure matter?

D. Durant[1], M. Tichit[2] and E. Kernéïs[1]
[1]Domaine INRA SAD, 545 route du bois maché, 17450 Saint-Laurent de la Prée, France, [2]UMR INRA SAD APT-INAPG, 16 rue Claude Bernard, 75231 Paris cedex 05, France.

Keywords: breeding waders, grazing, sward structure, agricultural grasslands

Introduction Grazing is central to the debate on wildlife conservation. Agricultural grasslands are the main breeding areas for many waders and grazing is very important in grassland use by these species. Waders, ground-nesting birds, are very sensitive to sward structure for nesting (Milsom *et al.*, 2000). As a marsh has different grazing regimes, all fields are not equally suitable for waders. This study evaluated the factors affecting lapwing and redshank selection of nesting sites. Also, the hypothesis was tested that sward structure (mean grass height, frequency of tussocks) is an important factor affecting this selection.

Material and methods Angles-Longeville marsh, 4700 ha of grassland in the Marais Poitevin (France), was studied in 2004. Most of the 250 fields chosen were grazed by cattle or mowed. Birds were counted every 7-10d (February-July) using the "field by field method" (Bibby *et al.*, 1992). Each field was visited during the hatching period of lapwings and redshanks (late March and early May, respectively). Habitat variables measured were: (1) <u>Ground habitat</u>: mean grass height (cm), measured with a "sward stick" at 4 m intervals along representative transects (30 measurements/ha); tussock frequency (%); rill density (m/ha); field wetness (%), (2) <u>Landscape</u>: field area (ha); boundary index (presence of trees and/or fence), (3) <u>Disturbance</u>: distance to nearest road (m).

Results Mean grass height and field wetness were the main significant variables from the original list of habitat variables (Table 1).

Probability for a field with at least one pair of lapwing or redshank at hatching decreased with grass height (Figure 1). In both species, the surface wetness also was a crucial factor, positively affecting field selection.

Table 1 Lapwing & redshank binary logistic regressions. *Log10 transformed

Species	Variables	df	Chi-	Estimate	SE	p-value
Lapwing	intercept	1	12.07	- 4.83	1.39	≤ 0.001
	grass height*	1	17.13	6.03	1.46	≤ 0.0001
	field wetness	4	25.52	- 0.94	0.38	≤ 0.0001
	likelihood ratio	223	233.11			0.31
Redshank	intercept	1	8.52	- 4.18	1.43	≤ 0.01
	grass height*	1	15.83	4.90	1.23	≤ 0.0001
	field wetness	4	25.26	- 2.63	0.72	≤ 0.0001
	grass height × field	1	4.63	- 1.93	0.90	≤ 0.05
	likelihood ratio	231	155.65			1.0

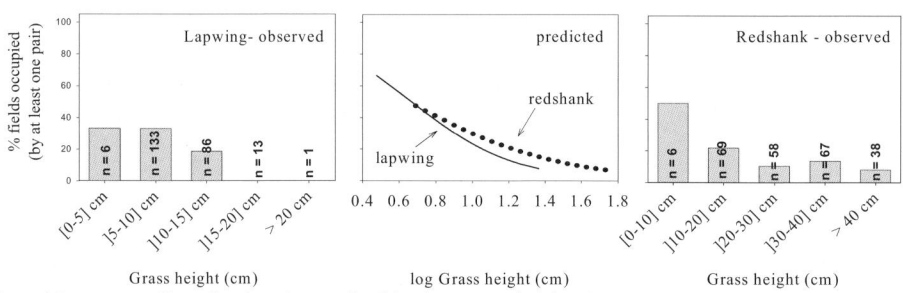

Figure1 Responses of breeding lapwings and redshanks to grass height

Conclusions Lapwings preferred grass mainly ≤10 cm (Figure 1). Redshanks also used short swards but exploited a larger range of heights (up to > 40 cm). Grass height was an important factor affecting the selection of nesting sites in waders. The 2 species preferred different grass heights (short/longer swards). To maintain or enhance biodiversity of breeding waders in agricultural landscapes, one must maintain diversity of grazing regimes to produce various grass heights that favour various wader species.

References

Bibby C.J., N.D. Burgess & D.A. Hill (1992). *Bird Census Techniques*. Academic Press, London, UK
Milsom T.P., S.D. Langton, W.K. Parkin, S. Peel, J.D. Bishop, J.D. Hart & N.P. Moore (2000). Habitat models of bird species' distribution: an aid to the management of coastal grazing marshes. *Journal of Applied Ecology*, 37, 706-727.

Management of grasslands used by waders: integrating time and key spatial scales of grazing processes

M. Tichit[1], D. Durant[2], O. Renault[3] and E. Kernéïs[2]
[1]UMR INRA SAD APT-INAPG, 16 rue Claude Bernard, 75231 Paris cedex 05, France, Email: tichit@inapg.fr,
[2]Domaine INRA SAD, 17450 Saint-Laurent de la Prée, France, [3]Chaire d'Ecologie des Populations et
Communautés, INAPG, 16 rue Claude Bernard, 75231 Paris, France

Keywords: breeding waders, grazing, spatial scales, timing

Introduction Agriculture has many functions. Mainly through agri-environment schemes, farmers are asked to manage grasslands of special value to conserve biodiversity. Assessment of grazing as an ecological factor of variation of grassland characteristics is needed to understand how grazing contributes to grassland management for species conservation. Several wader species use wet grasslands preferentially for nesting and foraging. Like many ground-nesting birds, they are very sensitive to the sward structure (see Durant et al., this congress).

Dynamic interactions We must understand the nature of dynamic interactions between grazing processes and wader preference of for certain sward structures. Different species select different sward structure and their timing of breeding varies. Therefore different grazing regimes are likely to be needed at different critical periods. On the basis of previous research, we propose some thoughts to guide future work on this subject, according to two main questions: (1) Does grassland suitability to the species depend on timing and intensity of grazing? (2) What influence has grassland distribution on its suitability as a bird habitat?

Integrating timing and intensity of grazing and key spatial scales *Timing and intensity of grazing:* Niche differentiation between species exists as regards grazing intensity, which is a critical influence on habitat suitability (Tichit et al., 2004). For example, intense autumn grazing improves suitability as a lapwing habitat. The needs of curlew and redshank contrast sharply to that. Curlews and redshanks select fields with a spring grazing intensity above and below average, respectively. Thus, conflicts of interest may emerge from particular grazing regimes being adopted for the conservation of certain species. Also, nest-trampling influences breeding success. Mitigation of those negative effects related to livestock density may require a subtle trade-off. There is need to investigate a 'threshold date', before which grazing may have detrimental effects. Also, the delayed effects of autumn grazing (Tichit et al., this congress) should considered as a way to increase the attractiveness of fields for lapwings in the following spring. *Key spatial scales:* Because species-environment relationships operate at field and coarser scales, such as the landscape scale, we contend that grassland management should take into account several spatial scales of observation. The definition of these scales should be made from the spatial resolution of bird behaviour and should be based on relevant ecological traits of breeding waders (for example, home range size). Modelling habitat suitability for 5 wader species, we showed that habitat preferences are not built on the same ecogeographical variables for all species (Renault et al., 2004). Two groups of birds need consideration. Group 1 (migratory species, like lapwing and black-tailed godwit) is very sensitive to landscape variables (distance to water, distance to mowed pastures). Group 2 (nesting species, like lapwing and redshank), is attracted more by land use at field scale instead. At the landscape scale, we showed that birds are distributed non-randomly and that some sectors of the marsh are more attractive than others. Therefore, studies should integrate at least 3 spatial scales: •*Field scale:* Mean grass height and heterogeneity, which depend on grass growth and grazing intensity, are crucial in the choice of nest site in breeding waders (Durant et al., this congress). •*Scale of a block of a few adjacent fields:* This scale is defined by the spatial resolution of the distribution of birds in the breeding season: breeding birds usually are not restricted to a single field, but move from one field to another. This supports the idea that it is important to take field surroundings into account. •*Landscape scale:* Landscape may provide clues used by birds for habitat selection. For example, the proportion of grazed pastures in the landscape probably explains its degree of use by waders.

Conclusions Increasing evidence advocates the importance of maintaining a some heterogeneity in agricultural landscapes in time and at various spatial scales. The management of heterogeneity at those multiple scales results from several farmers acting according to their own purpose. We conjecture that coordination between farmers may be crucial to improve bird habitat conservation through grazing practices.

References

Renault, O., T. Potter & M. Tichit (2004). Variability of suitable habitats for waders: does grazing management help? 55th Meeting of European Association for Animal Production, Bled, 5-8 September 2004. http://www.eaap.org/bled/LNCS2_06.htm

Tichit, M., T. Potter & O. Renault (2004). Grazing intensity as a tool to assess positive side effects of livestock farming systems on wading birds. 55th Meeting of European Association for Animal Production, Bled, 5-8 September 2004. http://www.eaap.org/bled/LPM3_13.htm

A comparison of restored native grasslands and exotic grass pastures as wintering habitat for declining grassland bird species in the Southeastern United States

A.B. McMellen and S.H. Schweitzer
Warnell School of Forest Resources, University of Georgia, Athens, Georgia 30602, USA, Email: mcmellen@uga.edu

Keywords: native grass restoration, grassland birds, Georgia

Introduction Southeastern grasslands were not pristine when the first Europeans arrived in the 15th century. American Indians had modified the landscape through centuries of fire use, cultivation, and other activities (Denevan, 1992). However, native southeastern grasslands did not evolve with disturbance from intensive grazing. Livestock and intense grazing pressure arrived with the Europeans. Modifications to southeastern grasslands by the early 1900s included exclusion of fire, intensive grazing, and introduction of cultivated, sod-forming grasses, which resulted in an increase in hardwood trees and shrubs, changes in herbaceous species composition, and the near extirpation of native warm-season species such as switch grass (*Panicum virgatum*), big bluestem (*Andropogon gerardi*), little bluestem (*Schizachyrium scoparium*), Indian grass (*Sorghastrum nutans*), and eastern gamma grass (*Tripsacum dactyloides*) (Rasnake, 1992). A mosaic of cultivated pastureland, cropland, pine (*Pinus* spp.) plantations, and mixed pine-hardwood forests has replaced the grassland and grassland savanna habitats that were present in the southeast. Most pastures now are planted in introduced cool and warm-season grass species, such as fescue (*Festuca arundinacea*), bermuda (*Cynodon dactylon*), and bahia grass (*Paspalum notatum*). Several songbird species are associated closely with the structure of the native bunchgrass-forb community. Species including the Henslow's sparrow (*Ammodramus henslowii*), Bachman's sparrow (*Aimophila aestivalis*), loggerhead shrike (*Lanius ludovicianus*), northern bobwhite (*Colinus virginianus*), eastern meadowlark (*Sturnella magna*), and Savannah sparrow (*Passerculus sandwichensis*) have experienced precipitous declines, likely due to landscape-level changes in habitat. This project aimed to identify the most effective method of creating and maintaining grassland habitat for declining grassland birds in the southeastern United States within an open agricultural landscape and forest openings.

Methods Twelve plots were established in central Georgia, 6 in an open agricultural landscape and 6 within a forested area. Three plots within each landscape were prepared in autumn 2001 and planted in spring 2002 with a mixture of little bluestem, big bluestem, and Indian grass at a rate of 7.84kg PLS/ha. The remaining sites were kept under current management of annual mowing and periodic burning. Fescue and bahia grasses dominated the control plots. The bird community was identified with mist netting during January-March 2003 and 2004. Each bird was banded, measured, and released. Capture rate (number of birds captured/mist net hour) was used as an index of avian abundance.

Results Data were analysed within year and within landscape. In 2003, there was no significant difference between capture rates in open planted (0.3863±0.154 birds/net hour) and open control (0.2199±0.1819 birds/net hour, p=0.14) plots, but forest planted plots had a significantly higher capture rate (0.306±0.206 birds/net hour) than the forest control plots (0.0341±0.017 birds/net hour). When capture rates for only grassland obligate sparrows were considered, the planted treatments within both landscapes yielded significantly more sparrows than the control treatment (open planted=0.298±0.049 sparrows/net hour, control=0.089±0.083 sparrows/net hour, p=0.01; forest planted=0.275±0.228, control=0.011±0.009, p=0.058). In 2004, capture rates were much lower in all plots. Capture rates did not vary between planted and control plots in either landscape (open plant v. control p=0.23; forest plant v. control p=0.089), although the data within the forested landscape tended towards a significant difference. High variance associated with low capture rates may have precluded the detection of a statistical difference. Sparrow capture rates were not significantly different between treatments within either landscape.

Conclusions Grassland bird winter abundances vary dramatically from year-to-year within the same landscape making year-to-year comparisons difficult. Native grass restoration may provide an important management tool for wintering grassland birds, especially within a forested landscape.

References
Denevan, W.M. (1992). The pristine myth: the landscape of the Americans in 1492. *Association of American Geographers,* 83, 369-385.
Rasnake, M. (1992). Management of warm season bunch grasses. *Proceedings of the 48th Southern Pasture and Forage Crop Improvement Conference*, Auburn, AL, 38-39.

Agricultural intensification: have sown pastures damaged the environment?

J.G. McIvor, C.K. McDonald, N.D. MacLeod and J.J. Hodgkinson
CSIRO, 306 Carmody Road, St Lucia, Queensland 4067,Australia, Email: john.mcivor@csiro.au

Keywords: ecological condition, catchments, land management, south-east Queensland

Introduction Concerns are growing about the impact of agriculture on the environment. Particular concerns have been expressed about the impacts of intensive agriculture (e.g. cropping involving fertiliser, pesticides, limited germplasm, fuels, etc) on biodiversity, and both on-site (e.g. soil health and fertility) and off-site resources (e.g. pesticide contamination). Less intensive agriculture (e.g. sown pastures) can also have undesirable impacts. In the woodlands of eastern Australia sown pasture development has been associated with loss of native vegetation and wildlife habitat, accelerated soil acidification, salinisation and poor tree health (McIntyre *et al.*, 2002). Based on this, there have been calls to restrict such development to less than 30% of the area (McIntyre *et al.*, 2002). However, there have been no direct measurements of environmental impacts in relation to the proportion of sown pastures in an area.

Materials and methods Thirty small (*c.* 500 ha) catchments were selected near Crows Nest (27.3°S; 152.1°E) in south-east Queensland. As part of a larger study on the impacts of land use and management (MacLeod *et al.*, 2004), detailed field surveys were conducted to determine the area of sown pastures in each catchment, and to estimate values of a number of indicators of ecological condition within those areas.

Results Figure 1 shows relationships between the proportion of a catchment that was developed to sown pastures and selected measures of ecological condition. None of the relationships were strong (see r^2 values), and some were positive with ecological condition improving as the proportion of sown pasture increased.

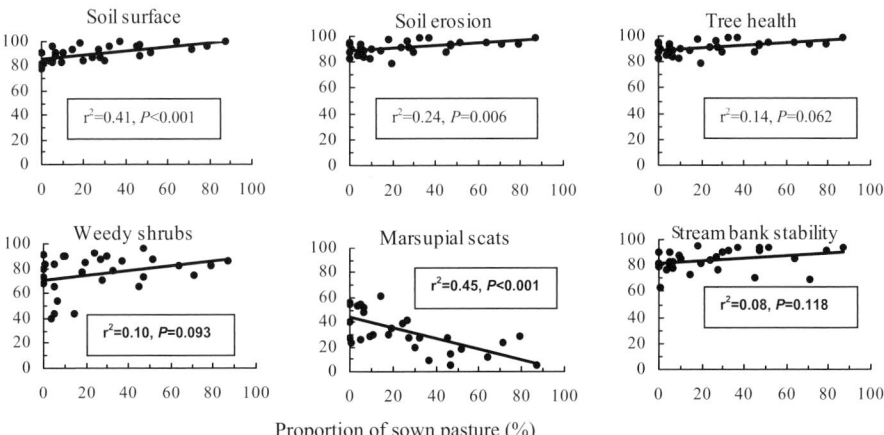

Figure 1 Relationships between the proportion of a catchment developed to sown pastures and some indices measuring ecological condition. Index values (y=axis) range from 0 to 100 where 100 = best condition.

Discussion Since declines in some measures of ecological condition have been associated with increases in sown pastures, why were the results of this study different? There are possible reasons. It may take time for many changes to occur – this is possible, but many of the pastures were 50 years old. Condition indicators are influenced by many factors in addition to sown pasture development and these factors may have been more important determinants of condition. Ground cover was high in most pastures and this was reflected in good soil surface condition (friable, no crusting or sealing, litter cover) and little erosion. Tree health was also poor in some native pastures. Activity of marsupials declined when there was more sown pasture but probably reflects the associated decline in woodland cover offering less protection and shelter to the animals.

References
MacLeod, N. D., J. G. McIvor, C. K. McDonald & J. J. Hodgkinson (2004). Report on Project CSE7 to LWA.
McIntyre, S., J. G. McIvor, & K. M. Heard (eds) (2002). Managing and Conserving Grassy Woodlands. CSIRO Publishing, Melbourne, 253 pp.

The use of ecological indicators in studies of ecological recovery for sustainable management of grazed grassland ecosystems

G.A. Heshmati
Gorgan University of Agricultural and Natural Resources Sciences, P. O. Box 386, Gorgan, Iran, Email: heshmati@gau.ac.ir

Keywords: state and transition model, indicators, grasslands

Introduction Early warning systems that depend on the selection of suitable indicators of thresholds are useful (Friedel, 1991). An ideal indicator should be unbiased, sensitive to changes, predictive, referenced to threshold values, data transformable, integrative and easy to collect and communicate (Liverman *et al.*, 1988). Methods for selection of indicators for assessing ecosystem health are being developed (Pyke *et al.*, 2002). This paper examines the situation in grassland vegetation grazed by sheep in a semi arid region of northeastern Iran.

Materials and methods A systematic transect sampling method was chosen at two grassland sites. The data were collected from five transects in five contiguous 1 m^2 quadrats along a 50 m transect located about 100 m from each other in exclosure and exposure areas at each site. Vegetation cover for each species was measured in each quadrat. Twelve indices (soil cover-interception of raindrops, soil cover-obstruction to overland flow, crust integrity, cryptogam cover, erosion features, eroded material, litter cover, litter incorporation, soil microtopography, surface nature, crust slake test and soil texture) were assessed in each of the five quadrats within each transect. The empirical factors to evaluate a set of nominal factors by comparing ranks of each on their implicit scales of function (low scores equivalent to degraded zone, high scores to stable zone) were used.

Results Five zones were recognised, of which zone (A) had light disturbance and played no role in development and maintenance of the status of its vegetation (Fig. 1). There was a transitional state between zones (A) and (B), the vegetation cover changed. Under certain disturbance intensities the process of change proceeded to the next zone (B). By further increasing the disturbance intensity, there was a certain level beyond which the vegetation did not sustain its stability at this zone (B) and was beginning to decline. This was the beginning of the third ecological transition zone which was between second (B) and the third zones (C) and it led into the critical threshold. The zones (D) and (E) were the further degraded zones with no, or minimum, recovery potential.

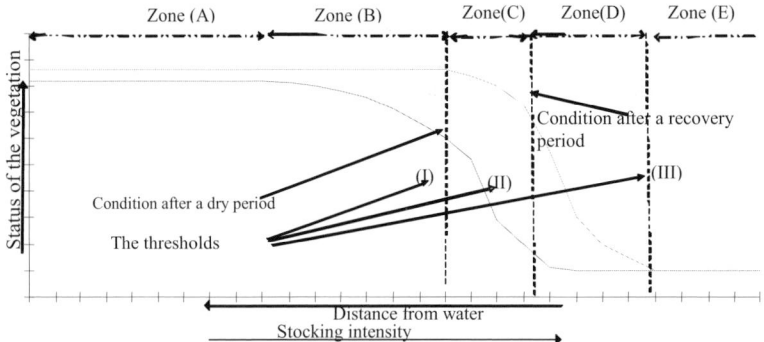

Figure 1 A simplified model of the dynamics of perennial forage plants under differing degrees of grazing pressure

Conclusions A conceptual model was developed that presents possible hypothetical mechanisms for transitions between states of a grassland community. Each state represents a combined series of indicative plant physiognomic features and soil surface characters that are readily identifiable in the field. The plant and soil characteristics at each state define the states in the model. Transitions between these states occur following changes in management practices such a grazing regimes, location of watering points and season of use.

Reference
Friedel, M.H. (1991). Range condition assessment and the concept of threshold: A viewpoint. *J. Range Management,* 44, 422-426.
Liverman, D. M., M. E. Hanson, Brown, B.J. and Merideth, R.W. Jr. (1988). "Global sustainability: toward measurement." *Environmental Management,* 12(2), 133-143.
Pyke, D.A., Herrick, J.E., Shaver, P. A. and Pellant, M. (2002). Rangeland health attributes and indicators for qualitative assessment. *Journal of Range Management* 55, 584-597.

Investigation on the temporal variation of vegetation cover in Karaj river basin (1973-1993)

M. Mohseni Saravi[1], A. Malekian[1] and B. Mohammadi Golrang[2]
[1]Faculty of Natural Resources, University of Tehran, Iran, Email: saravi@nrf.ut.ac.ir, [2]Agricultrure and Natural Resources Research Center of Khorasan Province, Mashhad, Iran.

Keywords: vegetation, temporal variation, Karaj basin

Introduction Area changes in vegetation cover depend on several climatic and edaphic factors as well as direct and indirect human activities. Vegetation maps in a region are mosaics of various associations and types, which clearly show their characteristics and provide a useful tool for classification purposes. Kochler (1967) suggested that vegetative forms and species are 2 most important factors for descriptive purposes. This research aimed mainly to determine the vegetation types based on floristic-physiognomic forms and comparison of vegetation cover maps of 2 different dates to clarify the temporal variations of vegetation cover over this period. This detection of change would be useful to suggest range improvement practices for the area.

Materials and methods Vegetation cover of Karaj river basin was studied on a scale of 1:50000. Floristic samplings were conducted from aerial photos. The percentage of crown cover, forage, litter, fine stones and bare soil were estimated in each plot established along the transects. Cain's method (Cain, 1959) was used to analyse the vegetation units. Then, Kochler's method (Kochler, 1967) was used to classify the vegetation types based on the floristic comparisons and site condition. Vegetation maps of the region were prepared for 1973 and 1993 and compared to detect changes.

Results From the interpretation of aerial photos of 1973, a large part of the study area (26.9%) had no vegetation cover but 7.7% was covered with alpine vegetation. *Astragalus-Acantholimon-Onobrychis* covered the largest area of the region (15.13%). In 1993, about 17 vegetation types were determined, of which *Astragalus-Psathyrostachys* was the most dominant. Also *Alopecurus textiles- Agropyron cristatum* covered the lowest area over this period (Table 1).

Table 1 Variation of vegetation cover over the period of 1973-1993

Vegetation type	Area percentage (1973)	Area percentage (1993)	Variation
Hor.fragilis-Br. persicus	5.91	5.08	-0.83
Hor.violaceum-Fes. ovina	1.53	1.53	0
Oryzopsi- Bromus- Stipa	0.51	0	-0.51
Alopecurus-Ag. cristatum	0.07	0.07	0
Hor.fragilis-Ag. taurii	8.16	7.16	-1
Fes. Spectabilis- Ag. taurii	0.99	0.69	-0.3
Oryz. molinioides- Br.tomentelus	1.1	0	-1.1
Catabrosa-Catamagrstis-Phragmites	0.3	0	-0.3
Hor.violaceum-Tri.repens	0.49	0.52	0.03
Berberis-Rosa	0.78	0	-0.78
Astragalus- Psathyrostachys	13.52	18	4.48
Astragalus-Acantholimon-Onobrychis	15.13	15.1	-0.03
Others	51.51	51.85	

Conclusion Comparison of vegetation maps for 1973-1993 show that some types have appeared recently (*Bromus persicus-Agropyron tauri*) but the density of more palatable species (*Oryzopsis, Bromus, Stipa*) has decreased. Although some vegetation types have not disappeared the density of species and their palatable associates have reduced seriously. In some areas poisonous plants have covered large areas and the area of agricultural lands has increased. Range improvement activities, such as grazing management and planning, enclosure, reducing the number of livestock and seeding, are recommended for the area.

References

Cain, S. A. (1959). Manual of vegetation analysis, New York Harper, 325pp.
Hansen, D., & F. Churchill (1961). The plant community, Ronald Press Company, New York.
Kuchler, A. W. (1967). Vegetation mapping, The Yonald Press Company, New York.

Operating systems of the meadows in semi-arid region of Algeria

K. Abbas[1], M. Abdelguerfi-Laouar[2], A. Abdelguerfi[3], T. Madani[4] and A. Mebarkia[5]
[1]INRA of Algeria, Unit of Sétif, Algeria, Email: abbaskhal@yahoo.fr, [2]INRA of Algeria, Laboratories of Vegetable Physiology, Algiers, Algeria, [3]INA, Laboratory-RGB, El-Harrach 16200 Algiers, Algeria, [4]University of Sétif, Laboratory of Biology, Algeria, University Center of El Taref, Department of Agronom, Algeria

Keywords: meadow, agricultural production system, livestock production, diversity, semi arid region

Introduction Permanent meadows are the base of the fodder resources and also provide environmental services (MAP France, 2002). The function of fodder production is no longer solely to ensure bulk agricultural production. It must also allow the development of livestock products of good quality, contribute to environmental protection, the quality of the landscape and ensure a viable economic activity in the rural areas. These areas have decreased greatly (from 1 million ha at the beginning of the century (Lapeyronie, 1982) to less than 300,000 ha in 2000), in particular in the semi-arid zones at high altitude; however they contribute very effectively to natural diversity and the fight against desertification (Faye & Alary, 2001). The lack of a global solution to agricultural development and the intensification of production in certain sectors (cultivation of cereals) have induced a continued loss of meadow areas. This paper is the first study of the diversity of production systems comprising natural meadows in a small area of Algeria on a randomly selected sample of farms.

Results Meadow are exploited by production systems combining dryland cultivation of cereals with sheep production and of mixed farming – livestock production systems generally having irrigation resources. It constitutes a significant support for dairy production only in two situations:

•• small-scale mixed-farming – dairy production
•• cereal production that includes sheep and dairy production, but particularly when meadows play a significant role in the fodder system.

In these two types of production system, the quality of the meadows appears suitable for extensive pasture systems – mowing supported by irrigation and organic manure. Unfortunately these situations account for only 25% of the studied sample, showing overall that the meadows depend on dairy production and the future of state farms.

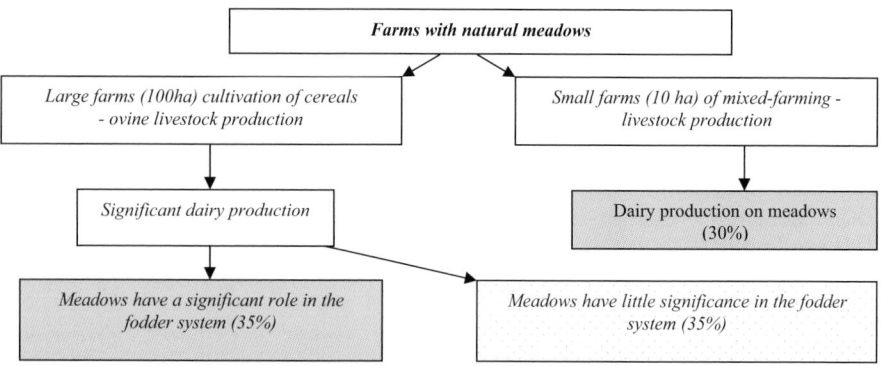

Figure 1 Diagrammatic representation of the diversity of the exploitations having meadows

References
Faye B. & V. Alary (2001). Les enjeux des productions animales dans les pays du Sud, INRA *Production Animale*, 14, 3-13.
Lapeyronie A. (1982). Les productions fourragères méditerranéennes. Tome-I- Généralités, caractères botaniques et biologiques. Techniques agricoles et productions méditerranéennes. G.P. Maisonneuve et Larose, Paris, France.
Ministère de l'Agriculture et de la Pêche (MAP), France (2002). Vers une intégration de la diversité biologique et paysagère pour une agriculture durable. Conférence paneuropéenne à haut niveau sur l'agriculture et la biodiversité. Strasbourg, 4 mars 2002.

Protection of agrobiodiversity: model calculations in Rhineland-Palatia: costs and implications for farmers

H. Bergmann

Department of Agricultural Economics, University of Göttingen, Platz der Göttinger Sieben 5, D-37073 Göttingen, Germany, Email: hbergma1@gwdg.de

Keywords: agri-environmental schemes, cost efficiency, grassland extensification, agrobiodiversity

Introduction Biological conservation and production use the same areas of land in less favoured areas. Grassland in these areas makes an important contribution to the protection of agro-biodiversity. However, under existing market conditions and production needs, the use of low yielding grasslands is not economically efficient. The objective of this study was to analyze the economic consequences of different mowing strategies in a small region in Rhineland-Palatia (Germany) that served the protection of two butterfly species.

Materials and methods The impacts of extensification measures for nature protection on herbage quality and yield have been described from a literature analysis. The effects of different cutting dates on quality and yield were calculated, based on a function by Opitz von Boberfeld (1994). Before these calculations, the calculated yields/ha were qualified in relation to nutritional requirements of cattle. The method of standard gross margin calculations was used to determine the compensation payments for the profit foregone by farmers that adopt alternative mowing regimes. It was assumed that farmers purchased concentrates as an additional fodder to compensate for the loss of energy yields and the compensation payments were calculated accordingly. Because most farmers in the region used meadows for silage, the calculations were based on silage production with a base energy yield of 52 GJ NEL/ha.

Results The literature on cattle and horse nutrition suggests that fodder with an energy content >6 MJ NEL/kg DM is usable in intensive cattle production, while fodder of 4-6 MJ NEL/kg DM is mostly usable in horse and heifer nutrition. Due to its low quality, silage harvested with a first cut in August is not recommended for use in horse nutrition or for cattle (except for yearlings between 12-18 months old; Figure 1). Therefore, for mowing regimes with a first cut after the beginning of August, farmers must be compensated for the complete loss of use of the meadow. The curve of calculated compensations (Figure 2) shows three stages, (1) until 15 June, with compensation costs of about 200 €/ha, (2) from 15 June to 1 August, with compensation costs increasing to 1000 €/ha and (3) after 1 August, when a total loss was presumed, with compensation costs of 1156 €/ha.

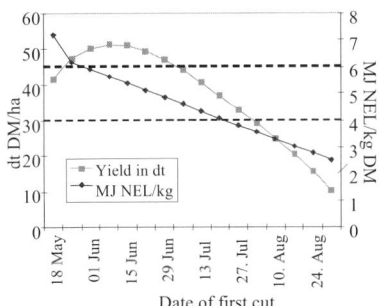

Figure 1 Effect of different mowing dates on MJ NEL/kg, DM/ha and their usability Source: Own calculations based on Opitz von Boberfeld (1994)

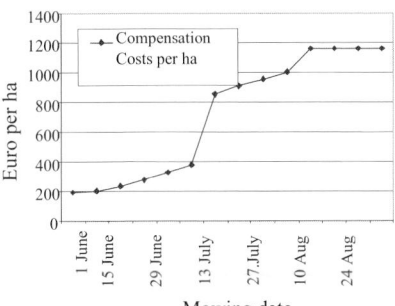

Figure 2 Compensation Costs in dependence of mowing date (1 cut a year) Source: Own calculations

Conclusions Linearity in compensation calculations is not a realistic way to reimburse farmers for extensification requirements. Justifiable compensation amounts can be calculated only in combination with specific grassland science, cattle nutrition and ecological knowledge.

References

Opitz von Boberfeld (1994). Grünlandlehre, UTB 1770, Stuttgart.

The Victorian Volcanic Plains grassland: past, present and future

S.G. Clark[1], J. Crosthwaite[2], J. Dorrough[3], J.R. Hirth[4], Y. Ingeme[5], J. Mavromihalis[3] and V. Turner[3]
[1]Department of Primary Industries, Hamilton, Victoria 3300 Australia, Email: steve.clark@dpi.vic.gov.au,
[2]Department of Sustainability and Environment, East Melbourne, 3002 Australia, [3]DSE, Heidelberg, 3084
Australia, [4]DPI, Rutherglen, 3685 Australia, [5]DSE, Hamilton, 3300 Australia

Summary The Grasslands of the Victorian Volcanic Plains have been replaced with introduced pastures and crops since European settlement with the loss of many plant and animal species. What remains, on public and private land, has high conservation value and needs urgent protection.

Keywords: grassland, native, Australia, conservation

Introduction Prior to European settlement in the 1830s the grassland of the Victorian Volcanic Plains covered 21,000 km^2 of the plains west of Melbourne. European agricultural practices have severely impacted upon this grassland and only a few remnants remain of what was described in 1836 as "*... enchantingly beautiful – extensive rich plains all around with gently sloping hills in the distance, all thinly wooded and having the appearance of an immense park. The grasses, flowers and herbs that cover the plains are of every variety that can be imagined...*" (Lunt *et al.,* 1998). Although the grassland remnants are poorly understood and threats remain, there is a growing recognition of their intrinsic value and the need to protect them.

Past These basalt plains were formed by five million years of volcanic activity. The grassland flora contained over 550 species of grasses, lilies, daisies, orchids and other forbs (Conley & Dennis, 1984). It was lightly grazed by kangaroos and emus and was habitat for specialised mammals, birds and reptiles. It was also home to Aborigines for over 20,000 years. One of the first European settlers reported "*a region more extensive than Great Britain, equally rich in point of soil, and which now lies ready for the plough in many parts, as if specially prepared by the creator for the industrious hands of Englishmen*" (Lunt *et al.,* 1998). Within 20 years, livestock had spread across the plains. Immediate impacts included displacement of Aborigines, loss of bird and mammal species, destruction of lichen layer, soil erosion, weed invasion and the first signs of salinity. Cultivation, cropping, introduction of the rabbit, swamp drainage, reduction in fire frequency, fertiliser use and closer settlement fragmented the grasslands and led to the extinction of many plant species.

Present High quality grassland remnants now total less than 3,000 ha, mostly in small road, railway and cemetery reserves, and still under threat. Larger areas, invaded by introduced species, can still be found on many private properties, particularly on non-arable, stony areas (Table 1). Some are valued by their owners, others are at risk as financial pressures increase and technology allows access to areas previously considered too rocky to plough. All are likely to suffer ongoing degradation.

Table 1 The flora diversity of three privately owned grasslands on the Victorian Volcanic Plain

Site	Native species	Introduced species
Hamilton	47	28
Darlington	55	23
Birregurra	21	15

Management options to maintain biodiversity are poorly understood and seldom studied. Government and private organisations and individuals are now working to save grassland remnants through grazing management research, flora and fauna surveys, revegetation programmes, seed collection and establishment research and farmer education.

Future Although adapted to the climate and low fertility soils, the perceived economic value of the grassland is low compared with well-fertilised European pastures. Farmers will need financial incentives and management skills to retain their grassland remnants. Research contributing to the understanding of financial and management issues confronting grassland farmers has begun and will need to be ongoing.

Conclusions This once vast, species-rich grassland has been reduced to fragments and is in great danger of disappearing altogether. Much of the damage occurred so soon after European settlement that it is hard to know exactly what was lost. Only through coordinated research and extension activities and financial assistance to grassland farmers can the remnants be preserved. There are promising indications that this is beginning to occur.

References
Lunt, I., T. Barlow & J. Ross (1998). Plains Wandering – Exploring the Grassy Plains of South-eastern Australia. Victorian National Parks Association and the Trust for Nature (Victoria), Melbourne, 152pp.
Conley, D. & C. Dennis (1984). The Western Plains – A Natural and Social History. Australian Institute of Agricultural Science, Colac, 116pp.

Contributions of the United States Department of Agriculture Natural Resources Conservation Service to conserving grasslands on private lands in the United States

L.P. Heard

USDA/NRCS, Wildlife Habitat Management Institute, 100 Webster Circle, Suite 2, Madison, Mississippi 39110 USA, Email pete.heard@ms.usda.gov

Keywords: grassland, private lands, biodiversity

Introduction The future of biodiversity in the USA is tied inseparably to activities taking place on private lands. Agriculture is by far the most important user of these lands, with about 50% or 900M acres managed as private cropland, grassland or rangeland. Decisions made by America's farmers and ranchers directly affect grasslands and their impact on food supply, biodiversity, soil protection and water quality. Agricultural programs and policies in the USA have had a large influence on the choices available to farmers and ranchers in land management. Since the 1930s, USDA's Natural Resource Conservation Service (NRCS) has been working with farmers, ranchers, and other land managers to promote conservation of natural resource through the nation's 3000 soil and water conservation districts. The Conservation Title of the 1985 Farm Bill, amended in 1996, raised the importance of biodiversity /wildlife in the delivery of conservation programs to the nation's privately owned lands. NRCS is charged with developing and delivering the proper grassland establishment techniques to landowners and evaluating the results. Recognising the opportunities and challenges related to conserving and enhancing fish and wildlife habitat, NRCS created the Wildlife Habitat Management Institute (WHMI) in 1997 as part of the NRCS National Science and Technology Consortium. WHMI was to interact with academic institutions, partner agencies, non-government organisations and others to develop and disseminate scientifically based technical materials to NRCS field staffs and others to enhance delivery of sound habitat management principles and practices, including grasslands to America's land users.

Materials and methods NRCS/WHMI works with scientists from various institutions to develop tools useable in the field and to evaluate the response of the utilised tools to develop native grasslands. Work on tool development was started first, recognising results could be 5-6 years away. As tool development began, an evaluation of wildlife response to previous Farm Bill programs was undertaken. The effort was intended to get a starting point to guide future planning for WHMI. Planning by WHMI staff is critical, as only 7 staff guide the effort to develop technical tools, such as jobsheets and technical notes used in management recommendations on grassland and other private land habitats across the USA.

Results A Comprehensive Review of Farm Bill Contributions to Wildlife Conservation (Heard et al., 2000) included reviews of impacts on grasslands. The Conservation Reserve Program (CRP) added 18M acres of grassland, mostly native, in the Midwest,. These grasslands were beneficial to grassland birds and contributed to a 30% (10.5M) increase in ducks from the northern Great Plains. The new grasslands did not produce an immediate increase in population of many grassland birds but decreased the precipitous decline that was being experienced. Conversion to native grassland in the southeast has been limited by the use of exotic forage grasses, rapid succession, and mowing. Most of the southeast habitat improvement from native grasses has come from the development of agricultural field edges. Although the above report reviewed wildlife responses, the response of native grass establishment on soil protection, water quality and biodiversity has been positive also.

In addition to the early report above, NRCS/WHMI continues to work with partners to develop techniques and to evaluate the impacts of native grassland establishment on wildlife. The following are current projects designed to move NRCS forward in its ability assist agricultural producers to conserve and increase native grasslands within their operations: (1) Bird response to grassland management in the northeast, (2) Lesser prairie chicken habitat in the southwest plains, (3) Evaluating wetland restorations in the Gueydan Prairie in coastal Louisiana. WHMI will continue to work to enhance the habitat of species of concern such as northern bobwhite and sage grouse.

References

Heard, L. P., A. W. Allen, L. B. Best, S. J. Brady, W. Burger, A. J. Esser, E. Hackett, D. H. Johnson, R. L. Pederson, R. E. Reynolds, C. Rewa, M. R. Ryan, R. T. Molleur, P. Buck (2000). A comprehensive review of Farm Bill contributions to wildlife conservation, 1985-2000. I n W. L. Hohman and D. J. Halloum (eds). U.S. Department of Agriculture, Natural Resources Conservation Service, Wildlife Habitat Management Institute, Technical Report, USDA/NRCS/WHMI-2000.

Biodiversity in grassland: Bangladesh perspective

B. Hossain
Institute for Environment and Development Studies, 5/12-15, Eastern View (5th floor), 50, D.I.T Extension Road, GPO Box-3691, Dhaka-1000, Bangladesh, Email: iendesbk@accesstel.net

Keywords: Bangladesh, grassland, biodiversity, policy

Grassland in Bangladesh   The rapidly increasing human population in Bangladesh has caused widespread damage to and disturbance of natural habitats and a loss of indigenous wildlife. There are now very few, if any, extensive patches of grassland in Bangladesh and any that might remain are inundated for two-thirds of the year with no alternative refugia available. Most remaining grassland areas are fragmented, heavily used and harvested up to three times a year. Furthermore, the reed lands of northeast Bangladesh were leased out for paper production and are reported to have been entirely destroyed and settled by encroachers. Tall grasslands around rivers and lakes were also utilised in Bangladesh, and were dominated by ekra (ikora) *Erianthus ravaneae*, nal "*Orundo karka*" (presumably either *Phragmites karka* or *Arundo donax*), ullu *Saccharum cylindricum* (possibly *Saccharum* or *Imperata cylindrica*), hogla *Typha elephantina* and hargoza *Acanthus ilicifolius*.

Unfortunately, there has been no extensive study or analysis of the ecological need and importance of grassland in Bangladesh and there is no current research focusing on grassland flora and fauna and its conservation.

Problems in relation to biodiversity   Population growth, overuse of resources, unplanned building projects and expansion of agriculture on to less productive lands have destroyed critical habitats, decreased biodiversity and created erosion and run-off. There has been massive deforestation and most surviving tropical forests and freshwater floodplains have been negatively affected by human activities. These developments have had adverse effects in the income of the poor. Fish catches have been drastically reduced and loss of biodiversity in the forests has meant less food, fodder, medicine and shelter for the poor.

Biodiversity policies   After the UNCED event in 1992, and in the face of growing campaign pressure by environmental organisations, the government took actions to ensure the conservation and sustainable use of biological and genetic resources and related knowledge, culture and practice. Actions were required in order to maintain and improve diversity as a means of sustaining the life support and healthcare system of the people of Bangladesh and to protect biological and genetic resources and the related knowledge from pollution, destruction and erosion. It was also necessary to protect and maintain indigenous knowledge in these areas.

Initial small efforts to conserve biodiversity was started after the independence of Bangladesh. In 1973 the Government declared the Bangladesh Wildlife (preservation Amendment) Order by repealing the previously enacted Wild Animal Laws to protect flora and the Government declared the Brick burning (control) Act. in 1990 and amended the Forest Act (1927) by imposing provisions for heavier punishment in 1990. In 1998 the government enacted a law entitled 'Biodiversity and Community Knowledge Protection Act of Bangladesh'.

As a part of the policy actions, to conserve, breed and develop flora and fauna under the laws the Government established three national parks, three wildlife sanctuaries, one game reserve (for elephants) and two botanical gardens in different parts of Bangladesh. The primary objective of the national park is to protect and preserve scenery, flora and fauna in the natural state. The objectives of the wildlife sanctuaries are to create undisturbed breeding grounds for wildlife, vegetation and soil and water. By the end of 2015 the Government will establish 10% of the total reserved forest as national parks, wildlife sanctuaries, game reserves and natural reserved areas in the Chittagong Hill Tracts.

An action plan has been developed for biodiversity conservation. This will include the preparation of an inventory and a management plan for three areas of tropical forest in south-east Chittagong, Jinjira Island and Tanguar Haor of Sunamganj. The main objectives of the projects are (i) promotion of sustainable development through institutions at the national level and the national planning process; (ii) improvement of environmental management of biodiversity and (iii) raising the level of environmental awareness of various sections of society.To protect and support the rights, knowledge, innovations and practices of local and indigenous communities, it is now the practice to involve the local communities in a number of biodiversity conservation steps and projects. The participation of local communities in the Sundarban Bio-diversity Conservation Project (SBCP) is notable in this regard.

The full beneficial effects of these measures to protect biodiversity have, however, still to be realised.

The Global Environment Programme (GEF) and United Nations Development Programme (UNDP) "Supporting the Conservation of Grassland Systems in Africa"

W.A. Rodgers and M. Niamir-Fuller
UNDP GEF, Box 30552 Nairobi, Kenya, Email: alan.Rodgers@undp.org

Keywords: livelihoods, biodiversity, international programmes, African grasslands, partnerships

Introduction The Global Environment Facility (GEF) was created after the World's Environmental Summit in Rio in 1992. The GEF provides funding for developing countries to meet their responsibilities and commitments under global conventions. The GEF is the financing mechanism for the Convention on Biological Diversity (CBD) and the United Nations Framework Convention for Combating Desertification (CCD). The conservation and wise use of grasslands can be supported through a variety of funding opportunities. The GEF channels support through Implementing Agencies, of which UNDP specialises in technical assistance and capacity building. The HQ of UNDP is in New York. There are Regional Offices in all continents and offices across all developing nations. The CBD focuses on ecosystems and grasslands form an important part of GEF's mountain ecosystem and dryland ecosystem programmes. The CCD concerns are accessed by GEF through a new programme for sustainable management to overcome land degradation.

GEF projects The GEF works with countries through its Secretariat, based in Washington DC, and its main Implementing Agency Partners (UNDP, UNEP and the World Bank). Synopses of four GEF projects via UNDP that are underway or in planning stages in Africa are presented in accompanying papers.

•• SABONET: (Southern Africa Botanical Network) which has supported herbaria across southern Africa to upgrade plan taxonomic products (Steenkamp & Rodgers, 2005). The main outputs are detailed grass (Poaceae) checklists for ten countries; plus a computerised mapping output showing distribution of all grass species in southern Africa by Degree grid square. A detailed flora and description of the montane "nyika" grassland of Malawi is a more detailed output

•• Montane Grassland Conservation Project in southern Lesotho (Rodgers, 2005). Lesotho, on the Maloti-Drakensberg watershed, has an extensive area of montane grassland and afro-alpine health, with cliffs and bogs of considerable biodiversity interest. The project has worked with national and local governments and communities to set up small-scale conservation programmes with regulated grazing areas to maintain grassland production and biodiversity values.

•• Southern Africa Grasslands Project takes a broad look at the landscape level planning pressures that threaten the highveld grasslands of South Africa (Maze & Rodgers, 2005). These largely natural grasslands are under intense development threat for cultivation, forestry and urban expansion.

•• The Southern Highlands Montane Grasslands of Tanzania Project is in preparation (Davenport & Rodgers, 2005). These grasslands, at elevation over 2,800m, harbour high levels of plant biodiversity with several localised and rare endemics. The grasslands are not included within Tanzania's extensive but mammal-oriented protected area system, and are under threat from cultivation, exploitation, invasions and fire.

The GEF and grasslands more widely Many broader GEF projects address grassland issues amongst other ecosystems and other objectives. The land degradation portfolio focuses on arid and semi-arid areas across Central and South America, Africa and Asia. There are projects in the Patagonian grasslands of Argentina, high grazing areas of Morocco, and Pakistan, the arid lands of Kenya and the Mongolian Plains. The World Initiative for Sustainable pastoralism has just been initiated and will promote sustainable rangeland management around the world. Under the biodiversity focal area, the operational programme emphasises dryland systems with several grassland / wooded grassland conservation programmes in Africa (eg Indigenous Vegetation Support in Mali, Kenya and Botswana), and other continents. The multiple agency indigenous vegetation project is highlighted as it shows linkages from on-ground conservation (UNDP), to applied research on arid grasslands through UNEP and the University of Oslo.

References

Davenport, T. & W.A. Rodgers (2005). UNDP-GEF grasslands projects: the Tanzania montane grasslands project. *XX International Grassland Congress - offered papers* (in press).

Maze, C. & W.A. Rodgers (2005). UNDP-GEF grasslands project: mainstreaming biodiversity into productive landscapes: the southern African grasslands programme. *XX International Grassland Congress - offered papers* (in press).

Rodgers, W.A. (2005). UNDP-GEF grasslands project: conserving mountain biodiversity in southern Lesotho. *XX International Grassland Congress - offered papers* (in press).

Steenkamp, Y. & W.A. Rodgers (2005). UNDP-GEF grasslands project: the Southern Africa Botanical Network "Sabonet". *XX International Grassland Congress - offered papers* (in press)

UNDP-GEF grasslands project: conserving mountain biodiversity in southern Lesotho

W.A. Rodgers
UNDP GEF, Box 30552 Nairobi, Kenya and UNDP Country Office, UN House, Maseru. Lesotho, Email: alan.Rodgers@undp.org

Keywords: biodiversity, protection, endemic species, grasslands, Lesotho

Background The Kingdom of Lesotho contains some 70% of the Drakensberg-Maloti Mountains, recognised as the Eastern Mountains "Centre of Biodiversity and Endemism" of southern Africa. The Mountains have globally significant plant diversity, with unique habitats and high endemism. These resources have been increasingly degraded by a grazing regime based on communal access, with reduced regulatory capability. Lack of ownership has restricted investment in conservation. Lesotho has the lowest Protected Area coverage of any nation in Africa (<0.4%). Biodiversity is thus at risk.

Project approach This United Nations Development Programme (UNDP) Global Environment Programme (GEF) project (1999-2005) provided two distinct but complementary interventions. The first was to work with government and communities to create a network of small protected sites, targeting specific biodiversity values. The second objective addressed conservation more broadly, by seeking to incorporate biodiversity values in rangeland management systems. This required inputs to policy review as well as developing incentive and regulatory systems within central, district and community organisations.

The project focused on the rangelands of Quthing District in south Lesotho, but linked with other biodiversity initiatives in the country. The project worked through national and district institutions including Range and Conservation Divisions in the Ministry of Agriculture, the National University of Lesotho and NGOs. The National Environment Secretariat (NES) provides oversight and coordination.

The grassland flora There are several classifications of vegetation and floristic communities for southern Africa as a whole. The vegetation types of Lesotho have been assessed within two broad categories -the lower veld grasslands and the higher mountain grasslands. The main grasslands, with their approximate areas are:

Moist lower grassland	6,689 km²	Below 1,700m asl
Afro-Mountain grassland	15,484 km²	1,700m to 2,500m asl
Alti-Mountain grassland	7,118 km²	2,500m to 3,480m asl

None, or tiny areas only, of these vegetation types are represented in the protected area systems of southern Africa.

The Alti-Mountain biome has two subdivisions: (a) the temperate alpine belt with *Erica / Helichrysum* heathland and *Merxmuellera / Festuca* temperate grassland (b) the temperate- sub-tropical alpine mixed grassland with *Merxmuellera / Themeda / Harpochloa*.

The Afro-Mountain biome also has two subdivisions: (a) the sub-tropical / sub-alpine belt with *Themeda / Eragrostis* sub-tropical grassland (b) the sub-tropical montane belt with *Catalepis* and *Cymbopogon* sub-tropical grasslands.

In addition to these `Zonal' vegetation types, three `Azonal' vegetation categories can be recognised in the higher altitudes, wetlands (largely bogs and mires), riverine gorges, cliffs and talus etc.

Endemism There are an estimated 30% endemics out of a total 1,750 taxa on the Eastern Mountains. The most important family for endemics is Asteraceae with 118 endemics out of a total of 167 species. Strict Lesotho endemics number about 50 higher plant species and many more lower plants. A large proportion of the 30 % endemics are found in the heathlands and the bogs of the upper alpine belt. It is these two categories that form the globally significant biodiversity value (the entry point for GEF biodiversity project eligibility). Endemic plant taxa include: *Helichrysum palustre, Helichrysum qathlambanam, Kniphofia hirsuta* (red hot poker), *Crassula qoatihambensis, Dianthus basuticus* (orchid), *Brownleea spp* (4) (orchids), *Dierama jucundum* (harebell), *Saniella verna* (an endemic genus). At least two endemics are recognised to be endangered: *Aloe polyphylla*, the spiral aloe threatened by illegal trade, and *Aponogeton ranunculiformis*, a submerged water plant confined to a few small pools.

Conclusion The project is now winding down. Principal outputs have been much greater awareness of biodiversity with government and communities, the integration of biodiversity concepts in livestock policies and a network of community managed biodiversity areas – within larger format grazing land-use plans.

UNDP-GEF grasslands project: the Southern Africa Botanical Network "SABONET"

Y. Steenkamp[1] and W.A. Rodgers[2]
[1]SANBI, Private Bag X101, Pretoria, South Africa, [2]UNDP GEF, Box 30552, Nairobi, Kenya, Email: alan.Rodgers@undp.org,

Keywords: biodiversity, herbaria, geographical information system, taxonomic services, southern Africa, Poaceae

The Project This six-year biodiversity capacity building project closes at the end of June 2005. The principal outcomes have been greatly strengthened capacities within, and interaction between, the national herbaria of southern Africa, in order to improve plant taxonomic outputs to better serve the needs of conservation end-users. Grassland plants have in many ways been the main regional focus. The project was based in the National Botanical Institute (NBI) of Southern Africa (now the South Africa Biodiversity Institute – SANBI) and had components in Angola, Botswana, Lesotho, Malawi, Mozambique, Namibia, Swaziland, Zambia and Zimbabwe. The NBI provide regional coordination and technical support.

Major project outputs have been:
- Training: with 33 degree level (BSc, MSc) awards, and a variety of in-service training programmes.
- Greater infrastructural support for herbaria, including cupboards, computers, vehicles.
- Computerisation of the region's plant collections into an interactive database.
- Increased collaboration between herbaria and botanical gardens in the region and globally.
- Greater collaboration between herbaria and end users of taxonomic services (conservationists, EIA expertise, genetic resource specialists and bio-prospectors medicinal plant expertise etc).
- Over 40 major publications on southern African plants with a considerable emphasis on grasses.

SABONET and grasses and grasslands The project entry point has been species based, but using plant collections to inform conservation planners at the site and ecosystem level. Major contributions have been:
- Formal published grass checklists (Poaceae) for all countries; with varying degrees of ecological and bio-geographical content.
- Computerised database on all grass collections for all of southern Africa, with details on site, habit, habitat etc.
- SABONET is also in the process of producing a "Poaceae GIS" electronic information product. This GIS system contains the Poaceae datasets from all the participating herbaria, and can produce simple and composite species, genus, and family distribution maps (based on one degree squares) for the entire SABONET region (as a whole, or for countries separately). Each "dot" on the map produced by this system can be linked back to herbarium specimen(s). It can also produce species richness maps, endemism maps, collector effort and information on useful species (such as wild rice or potential fodder grasses), With other overlays, such as Protected Area systems, rainfall and topography data etc, the maps will enable many other ecological functions. This product needs no separate GIS software to function. It can also be expanded to include data from other plant families.
- Computer-based map distributions of all grass species, across southern Africa; based on one degree grid squares. This allows the mapping of endemic species, mapping of centres of diversity, mapping localities of potentially useful species such as wild rice, etc.
- Field-based exploration of critical sites, with a major expedition to the Nyika mountain plateau in Malawi, with new species and locality records, and a major publication.
- A full edition of the widely distributed SABNET Newsletter devoted to Southern Africa Grasses and Grasslands, with information on statistics, uses, economics and educational materials.
- Site-based descriptions and lists; with an emphasis on:-
  - The Nyika Montane grasslands of Malawi, a major publication.
  - Arid sites in Namibia and Botswana.
  - Montane sites in Lesotho.
- The project has provided an entry point for the new "Important Plant Area" programmes, including South Africa, Namibia and Mozambique.
- Provided detailed botanical support to the major global hotspot programmes in southern Africa.

The SABONET website is:- www.sabonet.org and publications may be ordered from <sabonetpub@sanbi.org>

UNDP-GEF grasslands project: mainstreaming biodiversity into productive landscapes: the southern African grasslands programme

C. Maze[1] and W.A. Rodgers[2]
[1]South Africa Biodiversity Institute. Private Bag X101, Pretoria. South Africa, Email: alan.Rodgers@undp.org,
[2]UNDP GEF Box 30552, Nairobi, Kenya

Keywords: biodiversity, ecological services, protection, grasslands, integrated land use, South Africa

Background This is an exciting new initiative under the second strategic priority for the Global Environment Programme (GEF)'s Biodiversity Focal Area: "Mainstreaming biodiversity into productive landscapes and sector". The rationale is that whilst grassland biomes cover some 30% of South Africa, (within montane, coastal and high-veld systems) less than 3% is formally protected at national, provincial or private land-owner levels. Over 40% of the grasslands have been totally converted to other land usages and 30% is degraded. Forces of degradation and conversion (cultivation, forests, urban spread) continue. The issues of conservation are of land use and putting in place incentives to encourage beneficial use patterns (open grazing land and set aside areas) and using a set of disincentives to discourage negative land usages in sensitive areas.

The grasslands are significant for biodiversity in southern Africa with:
- 52 important bird areas, and at least ten globally threatened bird species.
- 50% of South Africa's endemic mammal species.
- 42% endemic reptile species.
- At least 3,370 plant species with an estimated endemism rate of 20%.

The project The project is in the final stages of development, and has a broad-based livelihood and environment goal: *"Ecological services provided by grasslands are sustained and contributing to economic development and poverty alleviation"*. The approach to meeting this goal is to persuade the main production sectors (e.g. agriculture) to contribute to biodiversity conservation priorities. This will be done by focusing on three project outcomes:
- The integration of biodiversity principles into agricultural, forestry and urban sectors through institutional capacity building.
- Increased knowledge and partnership within and between stakeholders.
- Developing market based management instruments as incentives for land use.

The key sectors involved in grassland land-use are:
- Agriculture; with more land being cultivated. Cultivation leads to habitat fragmentation and species loss. Cultivation brings with it chemical control of pests and increasing levels of fertiliser application. However, croplands provide 10% of employment and contribute 4% to national GDP. Intensive cultivation is essential to feed the nation. Key issues are for biodiversity proponents to identify "biodiversity hotspots" within the land use matrix, and advocate for agricultural planning to accept environmental principles - including maintaining biodiversity.
- Afforestation Sector; with 1.35 million ha of plantations, produces jobs and product benefits, but has considerable environmental impact, especially on sensitive watersheds.
- Urbanisation: the Greater Gauteng region will soon be Africa's largest urban development. Urban expansion needs to be planned on environmental principles.
- Rangeland: rangeland is the least damaging sector when stocking densities are correct, but overgrazing, over-fencing, fragmentation; invasive weed infestation and erosion are problems.

The project is expected to start in 2006; with multiple national, local and civil society partnerships. Partnership is expected with South Africa's strong research and training institutes to provide the scientific underpinning to planning. This project takes advantage of South Africa's strong institutional support for biodiversity, including recent biodiversity policy and a new biodiversity legislation.

UNDP-GEF grasslands project: the Tanzania montane grasslands project

T. Davenport[1] and W.A. Rodgers[2]
[1]Wildlife Conservation Society, Mbeya, Tanzania,[2] UNDP GEF Box 30552, Nairobi, Kenya, Email: alan.Rodgers@undp.org

Keywords: biodiversity, protected areas, grasslands, Tanzania, stakeholders, plants, birds

Background The Tanzania Southern Highlands and adjacent Nyika montane grasslands in Malawi form a distinct centre of plant diversity and endemism. The area is characterised by complex geology with old basement mountains and much more recent volcanoes (Mount Rungwe) adjacent to the rift valet faulting with Lakes Malawi (Nyasa) and Rukwa. Lake proximity generates rainfall up to 3,000 mm per annum. The maximum altitude is 3,000 m asl.

Whilst the biodiversity values of these montane grasslands (and some adjacent montane forests) have long been known to biologists, they have fallen through the cracks in national Protected Area system planning. Tanzania's conservation Protected Areas have traditionally focused on large mammals of the savannah systems and more recently the primate values of forests have led to forested national parks. But grasslands, with no trees or large mammals, were not the mandated concerns of Forest or Wildlife Departments. There was a major ecosystem gap in the protected area system. The realisation of this gap, coincided with the realisation that high-altitude grasslands which were once wilderness were now under threat for cultivation of potatoes and the growing importance of the wild orchid bulb trade as a dry-season food stuff in neighbouring Zambia ("chikanda").

The project This is a Global Environment Programme (GEF) Priority One project for Biodiversity – aimed at ensuring sustainability of Protected Area Systems. There are two entry points, one aiming to fix the "gap" by designing sustainable Protected Areas in this new biome (Parks, Reserves and buffer-zone community sustainable use areas). The other is working with the different protected area institutions (Parks Authorities, Wildlife and Forestry Divisions at national and local levels, and civil society) to develop capacity to plan, implement and monitor grassland protected areas.

Biodiversity values are high: the Kitulo Plateau has been described as the Serengeti of Flowers. At least 40 species of vascular plants are known to be unique to the Southern Highlands, and many more restricted to the Highlands and the Nyika Plateau in Malawi. The block-faulted Uporoto, the Rungwe volcanics and the older metamorphic Kipengere Range all give rise to distinct botanical assemblages. For example, *Protea praticola* is found only between 2300 and 2400 m on Mt Mbogo in Umalila, the balsam *Impatiens leedalii* is restricted to the edge of Numbi in Kipengere and the bell flower, *Cyphia rupestris* can be found only on Kitulo and Mbeya. Kitulo.The plateau itself has long been recognised as an area of outstanding botanical importance, with 31 plant species as national endemics, and the presence of as many as 42 species of terrestrial orchid.

The avian importance of Southern Highland grasslands is reflected in the fact that six areas have been designated as 'Important Bird Areas', the Tanzania / Malawi mountains are an 'Endemic Bird Area' (No. 105). Six species have been designated by IUCN / BirdLife International as 'category one' species (globally threatened). Three of these are listed as 'vulnerable' and three as 'near-threatened'.

The project will work with government and civil society stakeholders to develop capacity for the management of grassland protected areas. Outputs will include a strategic plan for a system of protected areas in the southern highlands, model management plans, sustainable use strategies and partnership agreements between different management agencies. A detailed M and E framework will allow monitoring of selected biodiversity values to show conservation impact.

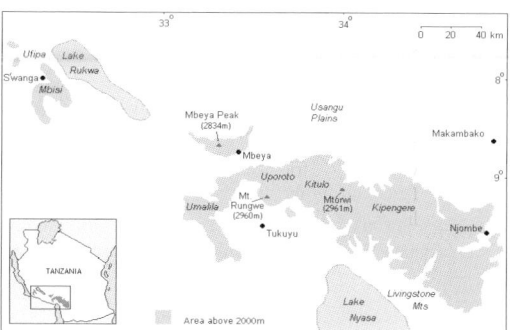

Long term results for the naturalisation of river valley grassland in the lower floodplains of the river Oder

G. Schalitz and A. Behrendt
Centre for Agricultural Landscapes and Land Use Research Müncheberg, Research Station, Gutshof7, 14641 Paulinenaue, Germany, Email: fspaul@zalf.de

Keywords: flood plains, naturalisation, flooded grassland, fodder quality

Introduction The flooded grassland area in the lower reaches of the River Oder covers about 10,000 ha, of which 4,000 ha are situated on the German side and the remainder in Poland. In spite of extreme flooding conditions (annual winter flooding from 15 Dec. to 15 April and occasional summer flooding) the area was used very intensively in GDR times. After the political change in 1989, the management of the cross-border German-Polish National Park, which was situated in this area, was changed abruptly to extensive grassland utilisation. This paper reports on changes in the composition and productivity of the grassland.

Material and methods The development and composition of grassland in the experimental area has been scientifically assessed since the end of the 1960s with plant community evaluations after Klapp and investigation of yield and quality in several plot and field experiments (Schalitz, 2003).

Results and discussion Before intensification there were diverse plant communities on the heavy clay soil. Due to the extreme flooding conditions they were less species-rich than normal grassland swards. With the increase in intensification since 1970 (involving N fertiliser application up to 360 kg N/ha, three cuts and intensive pasture) only two plant communities remained: *Alopecuretum pratensis* (Meadow foxtail meadow) and *Phalaridetum arundinacea* (Reed canary community).

The foxtail meadow lost its original abundance of species, whereas *Agropyron repens* in some places increased its part of the yield up to nearly 90 %. In spite of high fertilisation, yield did not increase any more and fodder quality stagnated or decreased. The reed canary community, formerly rich in *Carex species* and herbs, became a high yielding monoculture due to nitrogen. Annual yield from three cuts reached 10 t DM/ha with the production of good quality conserved silage and hay. After the change to complete extensiszation in 1990 (no fertiliszation, one or two cuts, two grazing rotations) about 15 years were needed to recover the original combination of plant species. In the dryer areas of *Alopecuretum pratensis* (50-90 cm groundwater level in summer) there has been a very positive development of the sward with increases in *Alopecurus pratensis, Poa pratensis* et. *trivialis, Vicia cracca, Trifolium repens, Leontodon autumnale* etc. Yields on the fertile soils scarcely decreased. Table 1 indicates enrgy contents and yields with extensive management

Table 1 Guideline for energy content and energy yield of naturalised *Alopecuretum pratensis* with extensive utilisation on the fixed date of 1 June

Sites in flooding area	Dry matter yield (t/ha)	Net energy content, NEL (MJ/kg DM)	Net energy yield (MJ/ha)
Dry to fresh	4.4	6.6	29,040
Fresh to moist	4.7	6.2	29,140

In wet zones of *Phalaridetum arundinacea, Carex species* spread massively, as well as herbs including *Senecio aquaticus, Rorippa amphibia, Ranunculus repens, Polygonum amphibium, Lythrum salicaria, Lysimachia vulgaris, Eleocharis palustris, Barbarea stricta, Caltha palustris, Cardamine pratensis, Thalictrum flavum* and many others. Because of these changes there was a drastic decrease in fodder quality. The establishment of the National Park was mainly carried out to achieve total reservation, particularly to provide a reservoir for the birds *Acrocephalus schoenobaenus* and *paludicola*. Consequently these areas are no longer in agricultural utilisation.

Reference
Schalitz, G. (2003). Vorschlag für einen Bewertungsrahmen Grünland in den Überflutungspoldern des Deutsch-Polnischen Nationalparks. Amt für Flurneuordnung und ländliche Entwicklung, Prenzlau.

Theme B: Grassland and the environment

Section 18

Grassland and water resources

The effects of water availability on plant growth in *Sesleria albicans* – dominated grasslands in the Burren, Co. Clare

P. Moran, S. Ryan and B. Osborne
Botany Department, University College Dublin, Belfield, Dublin 4, Ireland, Email: patrick.moran@ucd.ie

Keywords: limestone pavement, water availability, gas-exchange, fluorescence

Introduction The Burren is a karstic region in the west of Ireland characterised by large areas of exposed limestone pavement with sparse vegetation. Despite the prevailing oceanic climate and high rainfall, substrate volumetric water content values are similar to those of semi-arid habitats due to high run-off. As a consequence, plants growing on the pavement regularly experience water deficit during the summer months. *S. albicans*, a species reported to be tolerant of water deficits, is one of the most abundant species growing on the limestone pavement. The objective of this study was to determine the impact of water availability on the plant performance of a number of species commonly occurring on the limestone pavement.

Materials and methods Substrate volumetric water content, rainfall and substrate/air temperature were recorded at half hour intervals and logged. The physiological responses of a number of species grown in a glasshouse (under ambient conditions of light, heat and relative humidity), including *S. albicans*, T*eucrium scorodonia* and *Corylus avellana*, exposed to differing substrate volumetric water content were examined using fluorescence and gas-exchange techniques. In addition, measurements were carried on plants growing in the field during the period May 2004 – December 2004 (concentrating on *T.scorodonia*) to examine the impact of variations in water availability on plant performance.

Results Substrate volumetric water content (%) was observed to follow a cyclical pattern associated with rainfall events and subsequent run off (Figure 1). All of the plants examined showed a marked physiological response to changes in substrate volumetric water content, with *T. scorodonia* having a three-fold increase in maximum photosynthetic rate as a result of a 6% increase in substrate VWC (Figure 2). *S. albicans*, *T. scorodonia* and *C. avellana* showed similar responses to water deficit with significant reductions in maximum photosynthetic rate and smaller changes in fluorescence parameters, associated with decreased substrate volumetric water content.

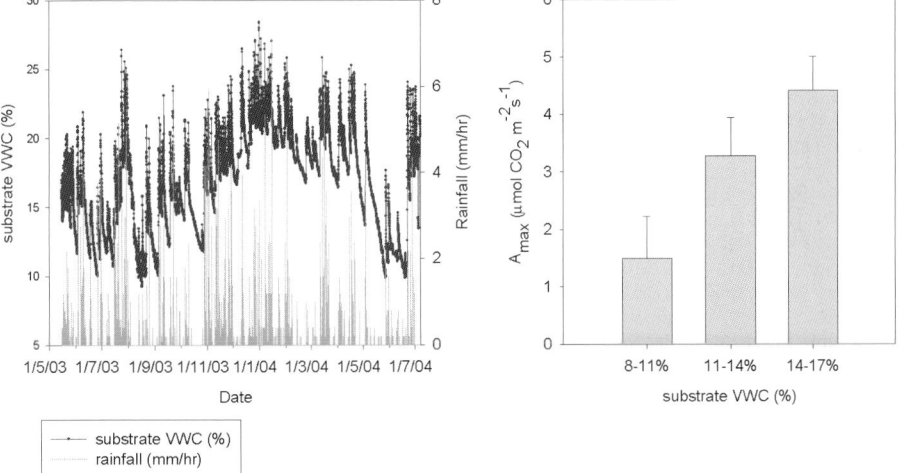

Figure 1 Variation in substrate volumetric water content with rainfall

Figure 2 Variation in maximum photosyntetic rate (A_{max}) of T. scorodonia (μmol CO_2 m^{-2}s^{-1}) with variation in substrate volumetric water content (%)

Conclusions Cyclical variations in water availability result in corresponding alterations in photosynthetic performance. Of the species examined, *S. albicans* showed the smallest response to variations in water availability indicating that its success under these conditions is related to internal or morphological adjustments that result in the maintenance of plant water balance.

Forage yield and soil moisture content in *Panicum maximum* cv. Tanzania monoculture and in a mixture with *Leucaena leucocephala* with different densities in Mexico

H.J. Delgado Gòmez[1], L. Ramirez Avilés[2], J. Ku Vera[2], J. Escamilla Bencomo[3] and P.A. Velázquez Madrazo[2]
[1]*Universidad del Zulia, Facultad de Ciencias Veterinarias, Núcleo Agropecuario, Ciudad Universitaria, Apartado 15252, Maracaibo, 4500-A, Estado Zulia, Venezuela, Email: delgado7255@hotmail.com,* [2]*Universidad Autónoma de Yucatán, Facultad de Medicina Veterinaria y Zootecnia, km 15.5, Carretera Mérida – Xmatkuil. Apartado 4-116 Itzimná, Mérida, 97100 Yucatán, México,* [3]*Centro de Investigación Científica de Yucatán, Calle 43 Nº 140 Col. Chuburná, Mérida, 97200 Yucatán, México*

Keywords: plant density, forage yield, soil moisture

Introduction Cattle production is limited by forage availability during the dry season since water and soil fertility are the main factors limiting production. Leucaena leucocephala has good nutritive value (24-30% CP). It can stand drought and grazing and so its introduction into pastures is recommended as an alternative to forage production during the dry season. The aim of this study was to assess the effect of the introduction of L. leucocephala with different densities on biomass production of P. maximum and soil water content.

Materials and methods The experiment was in Yucatan, Mexico, with mean annual rainfall below 1200 mm with Luvisole and Leptosole soils fertilised with 50N 80P 30K. Treatments were a mixture combining Panicum maximum cv. Tanzania and L. leucocephala, with two densities (5,000 and 10,000 pl/ha) and monoculture of P. maximum. The experiment was carried out between Nov. 2001 and Nov. 2002. A randomised block design with four replicate plots of each treatment was used. Dry matter yield (DM) was calculated including tree and pasture biomass on a surface basis (Hodgson et al., 1981). Yield was estimated every 42 d during rains and every 62 d during the drought. Soil gravimetric moisture content (SWC) (Aguilera & Martínez, 1996) was estimated weekly at 30, 60 and 90 cm depth in Luvisoles soil and 20 and 40 cm depth in Leptosoles soil.

Results Dry matter yield differed between treatments harvested during the dry and rainy seasons (P<0.05). The monoculture of *P.maximum* cv. Tanzania had a yield of 22.5 t/ha per year compared with the yield of grass in the mixture of 21.3 and 19.3 t/ha per year plus yields of *L. leucocephala,* at 2.4 and 4.9 t/ha per year at 5,000 and 10,000 pl/ha, respectively) (Figure 1). In both Luvisoles and Leptosoles SWC diminished (P<0.01) as soil depth increased. Water was limited in Leptosoles due to soil depth (0-40 cm) and competition between trees and grass for water and nutrients. This resulted in a low SWC (P<0.05) of 23.9% and 24.4% in the *P. maximum-L. leucocephala* mixture with densities of 5,000 and 10,000 pl/ha, respectively, and 26.7% in the *P. maximum* monocultures. This may explain why yields in mixtures are lower than in monocultures (Lehmann *et al.*, 1998). In Luvisoles, SWC was similar in monoculture and mixture.

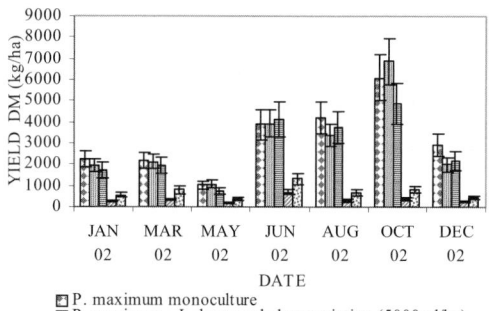

P. maximum monoculture
P. maximum - L. leucocephala association (5000 pl/ha)
P. maximum - L. leucocephala association (10000 pl/ha)
L. leucocephala (5000 pl/ha)
L. leucocephala (10000 pl/ha)

Figure 1 *P. maximum* DM yield of mixture and monoculture

Conclusions Introduction of L. leucocephala with density of 10,000 pl/ha reduced DM yield, while 5,000 pl/ha had no effect on DM yield. The SWC was higher in monocultures. It is important to assess ecological interactions of pastures mixed with woody perennials in developing silvopastoral systems.

References
Aguilera, C.M. & E.R. Martínez (1996). Relaciones Agua-Suelo-Planta-Atmósfera, 4th ed. UNCH, México, 35-152.
Hodgson, J., R.D. Baker, A. Davies, A.S. Laidlaw & J.D. Leaver (1981). Sward Measurement Handbook. British Grassland Society, 277 pp.
Lehmann, J., I. Peter, C. Steglich, G. Gebauer, B. Huwe & W. Zech (1998). Below-ground interactions in dryland agroforestry. *Forest Ecology and Management*, 11, 157-169.

Effects of supplemental irrigation on berseem seed crop in a semi-arid Mediterranean environment

G. Amato, D. Giambalvo, C. Scarpello and P. Trapani

Dipartimento di Agronomia ambientale e territoriale - Università di Palermo - Viale delle Scienze 90128, Palermo, Italy, Email: amato@unipa.it

Keywords: *Trifolium alexandrinum*, seed production, supplemental irrigation

Introduction Berseem seed production in Mediterranean environments is strongly influenced by soil water availability, particularly during spring growth. A long-term study (11 years) in Sicily recorded seed yields of between 0 and 1600 kg/ha, for an annual rainfall range of 289 to 867 mm (Stringi *et al.*, 2001). It was proposed that water irrigation during sensitive growth stages could increase and stabilize seed yield. This research investigated the response of berseem seed crop to low levels of irrigation applied at different growth stages.

Materials and methods Field trials were conducted for 2 years (1999/00 and 2000/01) in a typical, semi-arid Mediterranean environment of Sicily, Italy (37°33'N, 13°30'E, 150 m a.s.l) on a vertic xerochrept. The experiments were arranged in a randomised block design with 4 replications. Recommended cultural practices for a berseem seed crop were applied. The seed of cv Lilibeo was sown on 8 January 2000 and 12 December 2000 at the rate of 30 kg/ha. The crop was cut on 13 April 2000 and 3 April 2001 and herbage was removed. During the first trial year 5 irrigation treatments were applied: RC, rainfed control (not irrigated); AC, 1 irrigation (irr.) soon after the cut; EB, 1 irr. at early bud; 2I, 2 irr. soon after cut and at early bud; IC, irr. control, 3 irr. at vegetative stage (VS, 30 March), AC and EB. During the second trial year a sixth treatment was included: EF, 1 irr. at early flower; in IC 4 irr. were applied including VS (29 March) AC, EB and EF. Water was applied to the plots using overhead sprinklers at the rate of 40 mm during each irrigation session. Compared to average conditions, 1999/00 was very dry, particularly in autumn and spring, with a total rainfall of 393 mm (26% less than normal) and temperatures were lower in winter and higher in autumn and spring, whereas 2000/01 was quite wet with a total rainfall of 620 mm (16% above normal) and higher temperatures throughout the cropping season, particularly in spring.

Table 1 Seed yield (SY, kg/ha) and total above-ground dry biomass at seed harvest (BY, t/ha) of berseem for the various irrigation treatments in the two cropping seasons

Irrigation treatment	SY		BY	
	99/00	00/01	99/00	00/01
Rainfed control (RC)	398 b[+]	1240 d	2.6 c	6.5 c
1 irr. after cut (AC)	392 b	1240 d	3.1 b	7.0 b
1 irr. at early bud (EB)	515 a	1550 b	3.7 b	7.7 b
1 irr. at early flower (EF)	n.a.[§]	1690 a	n.a.	7.1 b
2 irr. AC and EB	512 a	1390 c	4.8 a	7.8 b
3-4 irr. (VS, AC, EB, EF[#])	571 a	1420 c	5,6 a	8,7a

[+] Means followed by the same letter are not significantly different (Fisher's protected LSD test at P ≤ 0.05). [§] n.a., treatment not applied. [#] EF only in 00/01

Results and conclusions In rainfed conditions, the seed yield of the wettest year (2000/01) was about three times that of the driest year (1999/00), confirming how greatly the amount of rainfall during the crop cycle can influence crop productivity. Compared to rainfed crop, in both trial years, water addition of soon after herbage removal resulted in an increase in vegetative growth, but did not influence the reproductive phase (Table 1). So, significant increases in total above-ground biomass, but not in seed production, were observed. However, the addition of the same limited amount of water at a later stage of growth, i.e. the early bud stage, prompted significant increases in both seed yield (+29 and +25%, respectively, over two years) and total biomass (+42 and +18%). Delaying irrigation further, i.e. at the first open flower, (applied only in 2000/01 season) led to the highest seed yield, and a significant increase of about 36 and 9%, compared to the non-irrigated control and irrigation at early bud, respectively. Multiple irrigations (2I and IC), compared to the single irrigation treatment at early bud, in 1999/00 had no effect on seed yield but resulted in greater vegetative growth only, whereas in 2000/01 caused a significant decrease of seed yield. Seed quality parameters (germination, MTG, hardness) were not affected by irrigation treatment. Overall, the research showed the possibility of increased seed production of berseem clover in semi-arid Mediterranean environments by merely applying a very limited amount of water (40 mm). Further research to investigate supplemental irrigation at even later growth stages and/or greater quantities of water is needed.

Acknowledgements The research was funded by P.O.M., Misura 2, Project B26.

References

Stringi, L., D. Giambalvo, G. Amato & A.S. Frenda (2001). Influenza di tecniche di lavorazione del suolo sulla produzione di seme di trifoglio alessandrino. *Proceedings of the Thirty-fourth Congress of the Italian Society of Agronomy,* 228-229.

Primary productivity and water use of the perennial grass, *Cenchrus ciliaris*, in arid environments

L. Mnif and M. Chaieb
Faculty of Sciences, Sfax 3018 B P802, Tunisia, Email: elobna@yahoo.fr

Keywords: *Cenchrus ciliaris*, accession, rain-use-efficiency, biomass, spikes

Introduction *Cenchrus ciliaris* is a perennial grass that may be suitable for the restoration of *Rhanterium* steppes (Chaieb *et al*., 1991). In this study, four *Cenchrus ciliaris* accessions from Tunisia from a range of climate and soil conditions, likely to vary in their adaptation to drought, were evaluated for productivity, rain-use-efficiency and reproductive output at Sfax in southern Tunisia. The suitability of these accessions for the restoration of *Rhanterium* steppes is considered.

Material and methods Four *Cenchrus ciliaris* accessions were collected from different regions in southern Tunisia : Bou Hedma (A1), Tozeur (A2), Raâs Jedir (A3) and Sidi Toui (A4). In October 2002 twenty four 6x3 m plots were established and assigned randomly to each of the four accessions under study (i.e. six replicates). Sixteen seeds of each accession were sown in each plot. The experiment was carried out throughout the 2002-2003 growing season. Plots received supplementry irrigation. The number of spikes in each plant were counted and the above-ground parts harvested to estimate the standing biomass, after drying at 105°C for 24 h. Rain-use-efficiency (RUE) was calculated by dividing the above ground biomass by the volume of rainfall plus irrigation water supplied to each plot. Differences between accessions in numbers of spikes per plant and biomass accumulation were tested for significance by one way ANOVA.

Results Aboveground dry matter accumulation by the end of growing period differed significantly between accessions. Accessions from Sidi Toui (A4) and Bou Hedma (A1) were the most productive. This variability emphasises the importance of polymorphism in phytomass production in this species. The differences in average biomass between accession (Table 1) were not significant (p>0.05). Spike production was highly variable. It was greater for accession A3 and A4 than for A1 and A2. The number of spikes per plot was not correlated with biomass accumulation ($r^2 = 0.075$). Accession A3 showed the highest biomass allocation to spikes. Differences in rain-use-efficiency between *Cenchrus ciliaris* accessions were substantial (Table 1). Accessions A4 and A1 were the most efficient. Similar results were found by Mseddi *et al.* (2003) in a three years study: dry matter production ranging from 0.91± 0.26 g dry matter/m^2 per mm to 1.63 ± 0.55 g dry matter/m^2 per mm.

Table 1 Mean biomass and spike production per plant and rain-use-efficiency of *C ciliaris* accessions

	A1	A2	A3	A4
Biomass (g dry matter/m^2)	126.9± 37	65.7±25	78.4±16	139±42
RUE (g dry matter/m^2 per mm)	0.39± 0.014	0.21±0.02	0.22±0.05	0.45±0.02
Number of spike per plant	84.3±9	94.5±6	160.0±17	162.0±18

Conclusions Accessions A1 (Bou Hedma) and A4 (Sidi Toui), which were the most productive and utilise rainwater most efficiently, are worth developing for reintroduction into degraded *Rhanterium* steppes. They are very promising for reseeding into depleted grazing lands and may help to meet the grazing needs of livestock in south Tunisia. This study supports previous studies carried out in the arid bioclimate of Tunisia, and confirms the relationship among the variability, the genetic polymorphism and the biological performance of *Cenchrus ciliaris*, a species threatened with extinction in Tunisian ecosystems.

References

Chaieb M, C. Floret, E. Le Floc'h & R. Pontanier (1991). Life history strategies and water resource allocation in five pasture species of Tunisian arid zone. *Arid Soil and Research and Rehabilitation*, 6, 1-10.
M'seddi K, L. Mnif, M. Chaieb & M. Neffati (2003). *Cenchrus ciliaris* productivity in relation to rainfall : correlations between fodder yield components, *Proceedings of the Seventh International Rangelands Congress*, 1414.

Evaluation of sweet grain sorghum silage for dairy cows as an alternative to irrigated maize silage

W.G. do Nascimento[1], Y. Barrière, X. Charrier, C. Huyghe and J.C. Emile
Institut National de la Recherche Agronomique, 86600 Lusignan, France, Email: emile@lusignan.inra.fr,
[1]*granted by CAPES, Brazil*

Keywords: sweet grain sorghum, water resource, dairy cows, maize, silage

Introduction Under European dairy cattle rearing conditions, whole plant maize silage is the main part of the dairy cow's diet especially during the winter season. Nevertheless maize production can be limited in some areas because summer rainfall is insufficient and so irrigation is necessary. Grain sorghum hybrids, and especially sweet sorghum types, are potentially of great interest to avoid this water consumption (Lemaire *et al.*, 1996, Legarto, 2000). For this reason we evaluated in 2003 the benefits and limits of a sweet grain sorghum silage for dairy milk production, compared to an irrigated maize silage. We paid particularl attention to forage quality and yield, environmental effects and animal performance.

Materials and methods Two ha of each of the 2 hybrids (*Sorghum bicolor* (L) Moench, cv Topsilo and *Zea mays* L., cv Cocagne) were grown in Lusignan under standard farming practice. Irrigation was provided to the corn crop only to prevent summer drought stress. Harvest, storage and feeding of both silages were strictly comparable. Sixteen Holstein-Friesian dairy cows (30.6 kg milk, 635.5 kg BW, 65 d DIM) were paired and assigned to one of the 2 experimental diets i.e. maize silage (MAS) or sweet sorghum silage (SSS) for 15 weeks. Cows were fed the silages individually *ad libitum*. The diets were complemented with concentrates (both diets), cereals (SSS only, 1 kg wheat) or urea (MAS only, 0.05 kg). Forage yield and composition, intake and animal performance were measured.

Results Treatment means are given in Table 1 (crops and forages) and Table 2 (animal performances). Summer rainfall were lower than usual (93 mm between 15 June and harvest date) and maize was irrigated five times to provide 152 mm. Temperature was exceptionally high in August (19 days above 30 °C). Dry matter yield and grain content were lower for sorghum than maize. The SSS was of lower quality than MAS based on fibre content and digestibility. Cows fed SSS had a significantly lower intake than cows fed MAS. However this could be partly related to the lower DM content. SSS cows were also fed grain and so probably lowered silage intake. Milk production and milk protein content were significantly higher with the MAS diet. Neither milk fat content nor body weight changes differed between diet treatments.

Table 1 Crops and forages characteristics

	SSS	MAS
Seeding date	14 May	24 April
Harvest date	25 August	28 August
S. rainfall, mm	93	93
Irrigation, mm	0	152
DM Yield, t/ha	11.7	18.0
Grain content, %	10.3	44.3
NDF content, %	49.2	40.6
Digestibility, %	66.3	76.4

Table 2 Animal performance

	SSS	MAS
Silage % of DM	26.2[#]	35.2
Silage DM intake	13.6[#]	17.0
Total DM intake	**18.9[#]	**22.2
Milk yield	24.1[#]	28.8
Fat content	4.56	4.39
Protein content	3.05[#]	3.25
BW variation, kg	+15.7	+ 9.0

[#] The effect of the diet was significant (p<0.001)

Conclusions The results of the present study confirm that non-irrigated sweet sorghum is able to provide adequate biomass. It also has a high efficiency for milk production provided its lower feeding value is taken account of and corrected within the total diet (Barriere *et al.*, 2003). This allows water to be either saved or to be more efficiently managed in ruminant rearing areas which experience summer drought. This type of sorghum could become a suitable alternative to maize if its agro-environmental and economic benefits are confirmed.

References

Barriere, Y., C. Guillet, D. Goffner & M. Pichon (2003). Genetic variation and breeding strategies for improved cell wall digestibility in annual forage crops. A review. *Animal Research*, 52,193-228.

Legarto, J. (2000). L'utilisation en ensilage plante entière des sorghos grains et sucriers : intérêts et limites pour les régions sèches. *Fourrages*, 163, 323-338.

Lemaire, G, X. Charrier & Y. Hebert (1996). Nitrogen uptake capacities of maize and sorghum crops in different nitrogen and water supply conditions. *Agronomie*, 16, 231-246.

Enhancing water use efficiency on irrigated dairy pastures with nitrogen fertiliser

F.R. McKenzie, J.L. Jacobs and G.N. Ward
Department of Primary Industries, 78 Henna Street, Warrnambool, Vic 3280, Australia, Email: frank.mckenzie@dpi.vic.gov.au

Keywords: dry matter, perennial ryegrass, urea

Introduction Low summer rainfall in southwest Victoria, Australia, restricts pasture growth and limits milk production. One fifth of dairy farmers in the region have some capacity to irrigate during summer. Irrigated dairy pastures are relatively poor utilisers of water with water use efficiencies (WUE) of about 1 t DM/ML water (Ward *et al.* 1998). Using nitrogen (N) fertiliser may increase dry matter (DM) yields for a given amount of water. Data on N response efficiencies from irrigated pasture in southwest Victoria are lacking. Two experiments determined the potential of N fertiliser to maximise the conversion of irrigated water to pasture DM.

Materials and methods Experiments ran from mid-October 2003 to the end of April 2004 on commercial farms. Farm 1 (38°28'S, 142°45'E) was under fixed sprinkler irrigation and Farm 2 (38°27'S, 142°42'E) under centre pivot irrigation. Perennial ryegrass dominated (84 to 88%) both experimental sites. Soil tests (0 to 10 cm) in October 2003 revealed: 26 and 28 mg/kg P (Olsen method), 120 and 230 mg/kg K (Skene method), 19 and 24 mg/kg S (CPC method) and pH$_{(water)}$ 6.7 and 6.0 for Farms 1 (dark-grey sandy loam) and 2 (dark brown-grey loam) respectively. Following each grazing 0, 25, 50, 75, and 100 kg N/ha (urea, 46%N) was applied to 12m x 24m plots, replicated three times in a randomised block design. Weekly neutron probe measurements determined irrigation scheduling and quantity. During the experiment there were nine and seven grazings for Farms 1 and 2 respectively. Three applications of a P, K and S blend (each equivalent to 8 kg P/ha, 27 kg K/ha and 8 kg S/ha) were applied at six-week intervals to both sites. Pasture DM consumed was measured as the difference between pasture DM before and after grazing using a calibrated rising plate meter (Earle and McGowan 1979). An analysis of variance (GenStat Committee 2000) with significance declared if P<0.05 was conducted.

Results Multiple N applications of 25 to 100 kg/ha increased (P<0.05) DM consumed at both sites by a total of between 1.7 to 5.3 t/ha for the summer irrigation period (Table 1). While the improvements in WUE were greatest (68 to 69% and 33 to 40% for Farms 1 and 2 respectively) with N at 75 to 100 kg/ha, the corresponding N response efficiencies were lower than when 25 kg N/ha was applied.

Table 1 Effect of treatment (kg N/ha/grazing), total pasture consumed (t DM/ha), extra pasture produced (t DM/ha), nitrogen response efficiency (kg DM/kgN), water use efficiency (t DM/ML) and improvement in water use efficiency (%) for Farms 1 and 2

Treatment	Total N applied		Total consumed		Extra produced		N response efficiency		WUE		Improvement in WUE	
Farm	1	2	1	2	1	2	1	2	1	2	1	2
0	0	0	7.7	9.1	-	-	-	-	0.85	1.34	-	-
25	225	175	10.3	10.8	2.6	1.7	11.6	9.7	1.14	1.59	34	19
50	450	350	11.5	11.5	3.8	2.4	8.4	6.9	1.27	1.69	49	26
75	675	525	13.0	12.8	5.3	3.7	7.9	7.0	1.44	1.88	69	40
100	900	700	12.9	12.1	5.2	3.0	5.8	4.3	1.43	1.78	68	33
l.s.d. (P<0.05)	-	-	1.34	1.27	-	-	-	-	-	-	-	-

Total water (irrigation plus rain) from mid-October 2003 to 30 April 2004: 905 and 681mm/ha for Farms 1 and 2

Conclusions Pasture DM consumed increased with N fertiliser. While 75 to 100 kg N/ha resulted in the highest WUE, applications of 25 kg N/ha following each grazing gave best N responses (10 to 12 kg DM/kgN). Nitrogen fertiliser improved the conversion of a fixed amount of water (irrigation plus rain) into pasture DM.

References
Earle D.F., & A.A. McGowan (1979). Evaluation and calibration of an automated rising plate meter for estimating dry matter yield of pasture. *Australian Journal of Experimental Agriculture and Animal Husbandry*, 19, 337-43.
Genstat 5 Committee (1997). 'Genstat 5.41 Reference Manual'. Oxford Science Publications, Oxford, UK.
Ward G, S. Burch, J. Jacobs, M. Ryan, F. McKenzie & S. Rigby (1998). Effects of sub-optimal irrigation practices on dairy pasture production in south west Victoria. *Proceedings of the Ninth Australian Agronomy Conference*, 254-257.

Responses of irrigated pasture nutritive characteristics to summer nitrogen fertiliser

F.R. McKenzie, J.L. Jacobs and G.N. Ward

Department of Primary Industries, 78 Henna Street, Warrnambool, Victoria 3280, Australia, Email: frank.mckenzie@dpi.vic.gov.au

Keywords: crude protein, metabolisable energy, dairy pasture, perennial ryegrass, urea

Introduction Low summer rainfall in southwest Victoria, Australia, restricts pasture growth and reduces pasture nutritive value thereby limiting potential milk production. One fifth of dairy farmers in the region have some capacity to irrigate during summer and nitrogen (N) fertiliser is used to enhance pasture dry matter (DM) yield. Data on the effects of N fertiliser on irrigated pasture nutritive characteristics during summer in southwest Victoria are lacking. Two experiments determined the potential of N fertiliser to improve pasture nutritive (crude protein, CP and metabolisable energy, ME) value during summer.

Materials and methods Experiments ran from mid-October 2003 to the end of April 2004 on commercial farms. Farm 1 (38°28'S, 142°45'E) was under fixed sprinkler irrigation and Farm 2 (38°27'S, 142°42'E) under centre pivot irrigation. Perennial ryegrass dominated (84 to 88%) both experimental sites. Soil tests (0 to 10 cm) in October 2003 revealed: 26 and 28 mg/kg P (Olsen method), 120 and 230 mg/kg K (Skene method), 19 and 24 mg/kg S (CPC method) and $pH_{(water)}$ 6.7 and 6.0 for Farms 1 (dark-grey sandy loam) and 2 (dark brown-grey loam) respectively. Following each grazing 0, 25, 50, 75, and 100 kg N/ha (urea, 46%N) was applied to 12m x 24m plots, replicated three times in a randomised block design. Weekly neutron probe measurements determined irrigation scheduling and quantity. Fifteen pasture samples per plot (each approximately 100 cm^2), to ground level, were taken before each grazing from the 0, 50 and 100 kg N/ha treatments for CP and ME content analysis. While there were nine and seven grazings with N applications for Farms 1 and 2 respectively (McKenzie *et al.* 2005), sampling for ME and CP content was not conducted on the final grazing at either farm. Analysis of variance (GenStat Committee 2000) with significance declared if P<0.05 was conducted.

Results For most observations, N applications increased (P<0.05) both pasture CP and ME content relative to no N (Table 1). There were generally no additional gains in either CP or ME above those obtained at 50 kg N ha/grazing when 100 kg N/ha/grazing was applied. For both sites, 50 kg N/ha per grazing increased pasture CP and ME content by an average of 5.5 % DM and 0.8 MJ/kg DM, respectively.

Table 1 Effect of nitrogen treatment (kg N/ha per grazing) on pasture crude protein (CP, %DM) and metabolisable energy (ME, MJ/kg DM) content of irrigated dairy pasture during summer for Farms 1 and 2

Farm 1	12/11/03		2/12/03		16/12/03		5/01/04		20/01/04		3/02/04		18/02/04		15/03/04	
Treatment	CP	ME	CP	ME	CP	ME	CP	ME	CP	ME	CP	ME	CP	ME	CP	ME
0	13.5	11.8	11.7	11.0	11.6	10.4	9.1	10.2	16.8	10.6	15.5	10.8	14.3	10.5	14.1	10.4
50	16.3	12.2	18.3	11.7	16.7	11.1	20.6	11.6	22.2	11.3	24.0	11.8	20.7	12.0	20.1	11.4
100	18.9	12.2	21.5	12.2	19.2	11.5	25.1	11.7	22.8	11.1	23.8	12.0	23.0	12.1	24.9	11.8
l.s.d. (P<0.05)	2.89	0.48	4.8	0.67	4.03	0.76	4.76	0.80	3.79	0.73	3.94	0.63	4.60	0.65	5.53	0.65

Farm 2	30/10/03		21/11/03		15/12/03		9/01/04		2/02/04		1/03/04	
Treatment	CP	ME	CP	ME	CP	ME	CP	ME	CP	ME	CP	ME
0	18.3	11.9	21.2	11.6	14.9	10.6	21.4	11.4	18.3	11.5	14.3	10.7
50	26.9	12.2	22.6	11.5	21.1	11.7	23.4	11.9	21.7	12.1	18.9	11.8
100	24.7	12.1	25.0	11.9	20.5	11.6	24.2	11.7	24.0	12.2	25.0	11.9
l.s.d. (P<0.05)	3.34	0.40	3.59	0.52	1.78	0.23	1.70	0.48	4.30	0.65	6.16	0.61

Conclusions Summer pasture CP and ME content can be increased by N fertiliser up to 50 kg/ha per grazing. At 50 kg N/ha/grazing, pasture CP and ME content was increased by an average of 5.5 %DM and 0.8 MJ/kgDM. Together with improved dry matter yield from N use (McKenzie *et al.* 2005), these results have important implications for the feeding of high producing dairy cows during summer when pasture nutritive values are generally low and the purchase of supplements high in CP may not be economical.

References

Genstat 5 Committee (1997) 'Genstat 5.41 Reference Manual'. Oxford Science Publications, Oxford, UK.

McKenzie FR, Jacobs JL and Ward G (2005) Enhancing water use efficiency on irrigated dairy pastures with nitrogen fertiliser. *XX International Grasslands Congress – offered papers* (in press).

Balancing water use efficiency and milk production in the sub-tropics

M.N. Callow and S.J. Kenman
Department of Primary Industries and Fisheries, Queensland, Mutdapilly Research Station, Peak Crossing, Queensland, Australia 4306, Email: mark.callow@dpi.qld.gov.au

Keywords: water use efficiency, pastures, forage crops, dairy cows

Introduction Queensland dairy farmers have had to confront in the last 5 years deregulation of the milk pricing system, resulting in a 25% reduction in farm gate price for milk in the year 2000, and drought. Many storage dams are significantly below capacity and regulatory authorities have imposed restrictions on irrigation water allocations. Major changes in farm business strategies were needed to overcome the shortfall in milk income. Production systems had to change to deliver more milk more efficiently and become more profitable. A farmlet study was developed in the sub-tropical dairy region of Queensland to evaluate 5 very different farm systems identified by a group of experts as capable of tripling production whilst achieving a 10% return on assets and 600,000 L/labour unit. This paper compares the water use efficiencies and milk production of these systems.

Materials and methods The 5 farmlets included M1 - raingrown tropical grass pastures and oats, M2 - 20% of the farm area planted to irrigated annual ryegrass and the remaining area to raingrown tropical grass pastures, M3 - 10% of the farm irrigated and planted to annual ryegrass and the remainder to forage crops, M4 - 90% of the farm planted to irrigated temperate pastures and summer forages, and M5 - a feedlot whose feedbase consists of irrigated temperate and tropical crops. All farmlets received equivalent to 3 t dry matter (DM)/head/year of purchased concentrate. Each farmlet consisted of 20 cows whose calving pattern reflected the forage production system. Defoliation practice, fertiliser and irrigation management were similar across the farmlets. Paddocks were managed to an agreed best practice, and forage growth rates were measured at each rotation. Milk production from forage was calculated using reverse feeding standards. Water use efficiency (WUE) was determined by either dividing milk from home grown forage or forage utilisation by irrigation plus rainfall.

Results The most water-efficient system for milk production was the feedlot system M5 using maize silage and barley hay as conservation, which avoided herbage wastage associated with grazing (Table 1). The high efficiency was in part at least associated with growing a crop of barley that recorded the highest WUE of all C3 species monitored with 2.1 t DM/ML. The herd in this system was milked 3 times/day and was fed as a feedlot hence the metabolic efficiency of this herd was higher compared to the remaining grazing systems, which also contributed to the high WUE of milk production. When comparing the grazing systems M1 to M4, the high irrigation, high quality temperate pasture system M4 recorded the highest WUE of milk production (Table 1). However, this system contained perennial temperate species whose WUE was comparatively low (less than 1.5 t DM/ML) compared to the short-lived annual species and summer forages. So although this system had the highest milk production it recorded the lowest WUE of forage production (1.1 t DM/ML). In contrast, the M3 cropping system recorded the highest WUE of forage with 1.9 t DM/ML. This system had the largest land area and contained the highest proportion of C4 forage crops that maximised total herbage utilisation and WUE of forage. But, the forage quality of C4 species was comparatively low compared to the C3 species so the WUE of milk production for this farmlet was low.

Table 1 Water use efficiency for milk production ('000 L milk/ML) and forage utilisation ('000 kg DM/ML), milk from forage ('000 L), and rainfall and irrigation (ML) from April 2003 to March 2004

Farmlet system	Milk from forage (ML)	Water ML/farmlet			Water use efficiency	
		Rainfall*	Irrigation	Total	Milk	Forage
M1	78.6	66.7	-	66.7	1.18	1.4
M2	70.4	65.3	4.1	69.4	1.01	1.3
M3	83.0	73.7	5.0	78.6	1.06	1.9
M4	91.3	54.3	17.0	71.3	1.28	1.1
M5 (Feedlot)	114.3	41.4	9.5	50.9	2.25	1.2

* Derived from total rain received multiplied by winter and summer farmlet areas 2003/2004

Conclusions Optimising WUE by selecting highly water-efficient species had a greater effect on milk efficiency than the amount of water received through irrigation and rainfall. For instance, although the M5 system received the lowest volume of water, this system recorded the highest volume of milk from home grown forage. The WUE of forage production was increased with the selection of a feedbase that contained higher water use efficient species. However, exclusively selecting for water use efficient species will not maximise milk production. Farmers will need to find a balance between optimising WUE and milk yield for their farm system.

Rooting pattern distribution and spatial variability of Italian ryegrass (Lolium multiflorum Lam) in a Mediterranean region

M.R.G. Oliveira[1], F.C. Brasil[1,2,3], Q.I. Monteiro[1] and R.O.P. Rossiello[2]
[1]ICAM -University of Évora, Apartado 94, 7002-554, Évora, Portugal, Email: mrol@uevora.pt, [2]Soil Science Department, Federal Rural University of Rio de Janeiro, Brazil, BR 465, km 7, 23850-000 Seropédica – RJ; [3]CAPES-Brazil

Keywords: root morphology, minirhizotron, digital image

Introduction It is estimated that less than 10% of the studies on pastures and forages have evaluated the subterranean biomass production. The objective of this study was to evaluate for a Mediterranean region the rooting characteristics and spatial variability of Italian ryegrass *(Lolium multiflorum Lam)* under two different soil water status conditions.

Materials and methods The study was carried out in two experimental plots with irrigated and rainfed ryegrass in a vertic Luvissol soil in a Mediterranean climate. Root length density (RLD) and root dry mass (RDM) in relation to depth were determined from auger root samples (500 cm^3). Also three minirhizotron tubes (5 cm i.d.) were installed per plot at an angle of 30º to the vertical and to a depth 0.5m for phenological studies. During the growing season digital images (1.8 x 1.35 cm) were recorded using a BTC system (Bartz Technology, Santa Barbara, CA), recorded at 1.34 cm depth intervals. Images were analyzed using RooTracker v2.0 software (Duke University) for morphological parameters quantification (Root length, number, diameter, and color).

Results The pattern of root distribution and spatial variability shows in general that more than 80% of RLD and 75% of RDM is concentrated in the first 10cm layer and maximum rooting depth was 60 cm (Table 1). The response was similar for the other sampling dates. The relationship is closely described by a negative exponential equation (van Noordwijk, 1993) for both root parameters. In terms of specific root length the data show a decrease (207.8 m/g at 10cm depth to 143.2 m/g at 20 cm depth) probably as a result of high soil bulk density. For both parameters the coefficient of variation increased with depth but the variability observed for the top layers was lower than that from other grassland studies. Minirhizotron data obtained throughout the growing season show that for all situations a significant amount of fine roots (• •0.5 mm) and root branching occurred in the top 10cm of the soil profile. In rainfed ryegrass, after the second harvest (May) there was a strong decrease in fine root production and root mortality increased which contributed to cessation in shoot regrowth. Meanwhile irrigated ryegrass gave two more cuts although new root production rate declined by 30 – 40 %.

Table 1 Ryegrass root distribution and spatial variability (data from the first sampling date)

Depth (x) (cm)	Root Length Density (y) (m/500 cm^3)	CV (%)	Roots (%)	Root Dry Mass (y) (g/ 500 cm^3)	CV (%)	Roots (%)
0 -10	157.4 • •48.64	31	80.1	0.76 • •0.16	21	75.25
10 -20	25.3 • •6.32	65	12.88	0.18 • •0.10	58	17.82
20 – 30	8.5 • •7.43	87	4.33	0.04 • •0.02	55	3.96
30 – 40	3.3 • •3.69	112	1.68	0.02 • •0.02	97	1.98
40 – 50	1.8 • •2.10	120	0.92	0.01 • •0.01	99	0.99
50 - 60	0.3 • •0.82	245	0.15	0.00	245	0
Equation	$y(RLD)$ • •191.29 $e^{•11.355x}$ R^2 = 0.98			$y(RDM)$ • •6.8347 $e^{•5.8873 x}$ R^2= 0.95		

Conclusions The ryegrass rooting pattern followed an exponential decrease with depth with a strong concentration, about 75-80%, of root density on the top 10cm, mostly fine roots. This is a disadvantage for ryegrass production in Mediterranean conditions caused by soil water conditions in those top layers.

References

Oliveira, M.R.G.; Van Noordwijk, M.; Gaze, S.R.; Brower, G.; Bona, S.; Mosca, G.; Hairiah, K. (2000). Auger sampling, in-growth cores and pinboard methods. In: A.L. Smit, A.G. Bengough, C. Engels, M. van Noordwijk, S. Pellerin & S.C. van de Geijn (eds.) Root Methods : A Handbook. Springer-Verlag Heidelberg, 175-210

Van Noordwijk, M.(1993). Roots:length, biomass, production and mortality. Methods for root research: In: Anderson, J.M. and Ingram, J.S.I.(eds). Tropical soil biology and fertility, a Handbook of Methods. CABI Publishing, Wallingford, 132-144.

Effect of timing and intensity of drought on perennial ryegrass seed yield

R.J. Martin, R.N. Gillespie and S. Maley
*New Zealand Institute for Crop & Food Research Limited, Private Bag 4704, Christchurch, New Zealand,
Email: martind@crop.cri.nz*

Keywords: *Lolium perenne* L., ryegrass, seed, drought, evapotranspiration

Introduction Perennial ryegrass seed worth about $50 million is produced annually in Canterbury, New Zealand (Rowarth 1998). Ryegrass seed production in New Zealand is often affected by drought, reducing both seed number and seed size (Rolston *et al.*, 1994). Irrigation management recommendations are not currently available for farmers growing ryegrass seed crops. To quantify the effect of water stress on perennial ryegrass seed yield, we carried out an experiment in a rainshelter where rainfall was excluded from experimental plots otherwise exposed to normal weather (Martin *et al.*, 1990).

Materials and methods 'Bronsyn' perennial ryegrass was sown on 11 April 2002 in 15 cm rows at a seeding rate of 8 kg/ha. Treatments applied were: (1) full irrigation weekly from late winter to harvest, adding the weekly actual soil moisture deficit each time; (2) no irrigation from early winter to harvest, (3) irrigation to field capacity in late winter then no irrigation until harvest; (4-8) same amount of water as (1) except (4) drought at head emergence; (5) drought from head emergence to harvest; (6) drought at peak flower; (7) drought from peak flower to harvest; (8) drought from early seed fill to harvest. The experiment was a randomised complete block design with three replicates. Each 5 m x 3 m plot had its own trickle irrigation supply and was harvested on 6 January 2003 by cutting two 0.5 m^2 quadrats to ground level.

Results Total seed yields and % first grade seed were markedly affected by intensity of water deficit as measured by maximum potential soil moisture deficit (MPSMD) (Table 1). Seed yield was closely related to head numbers, but % 1st grade seed was reduced in season-long drought and later drought treatments. First grade seed yields were closely related to maximum potential soil water deficit with a small additional reduction in yield in late drought treatments (Figure 1).

Table 1 MPSMD (mm), total seed yield (kg/ha), % 1st grade seed and head numbers/m^2

Tmt	MPSMD (mm)	Total seed yield	% 1st grade seed	Head no/m^2
1	184	2380	79	1980
2	548	990	48	1010
3	478	1530	49	1370
4	368	2090	72	1850
5	500	1730	56	1640
6	372	2090	69	1710
7	410	1530	55	1470
8	332	1870	60	1760
LSD (5%)		525	8.2	302

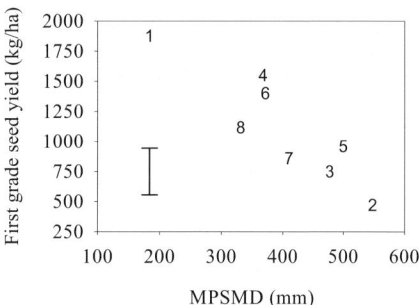

Figure 1 First grade seed yield (at 11% m.c.) v. MPSMD. Bar is LSD (5%). Numbers are treatments

Conclusions Water stress at any time reduced ryegrass seed yields, mainly through lower head numbers, but the effect was accentuated in crops with high biomass, where late drought reduced seed size. Perennial ryegrass seed crops should therefore be irrigated on the basis of actual or potential soil moisture deficits rather than at certain stages of development, with special attention late in the season when deficits occur quickly and seed size, as well as head number, can be reduced.

References
Martin, R.J., P.D. Jamieson, D.R.Wilson & G.S. Francis (1990). The use of a rainshelter to determine yield responses of Russet Burbank potatoes to soil water deficit. *Proceedings Agronomy Society of New Zealand*, 20, 99-101.
Rowarth, J.S. (1998). Practical herbage seedcrop management. Lincoln University Press, New Zealand. 243pp.
Rolston, M.P., J.S. Rowarth, J.M. DeFilippi & W.J. Archie (1994). Effects of water and nitrogen on lodging, head numbers and seed yield of high and nil endophyte perennial ryegrasses. *Proceedings Agronomy Society of New Zealand*, 24, 91-94.

Study on transpiration rates of Vicia villocea and Bromus inermis species

S.H.R. Sadeghi[1] and N. Rahimzadeh[2]
[1]Department of Watershed Management Engineering, [2]Department of Rangeland Management Engineering, College of Natural Resources and Marine Sciences, Tarbiat Modares University, Noor, Mazandaran, Iran, Email: sadeghi@modares.ac.ir

Keywords: water cycle, transpiration, *Bromus inermis, Vicia villocea*, range management, Iran

Introduction Ecohydrology is concerned with the interaction between the hydrological and plant processes. Some aspects of the hydrologic cycle, such as transpiration and interception have received little attention owing to difficulties in field measurements. Quantifying the components of water balance for a watershed is crucial for understanding the dominant hydrologic processes occurring in a basin (Flerchinger & Cooley, 2000). Water use by vegetation is controlled by the water uptake by roots, the transfer of liquid water through plants and vapour loss from the leaf surfaces by the opening and closure of the stomata (Roberts, 2000) i.e. transpiration. Comparison of transpiration of rangelands species is a prerequisite for improving range management. The present study is a preliminary comparison in transpiration between two important Iranian rangeland species, viz. the legume, *Vicia villocea* and the grass, *Bromus inermis*.

Materials and methods Four or five shoots of *Vicia villocea* and *Bromus inermis* were planted in small polyethylene pots of surface area 50.27 cm^2 in five replicates. The amount of daily evapotranspiration was measured by determining the weight lost during each day weighing with scales of 0.0001 g accuracy. Un-planted pots were also weighed to determine evaporation from the soil surface. Transpiration was calculated by subtracting evaporation from evapotranspiration taking account of the amount of water lost by drainage or applied by irrigation. The study was carried out from 29 May to 3 July 2004 under relatively natural conditions. Mean transpiration rates of the species were determined and statistically compared.

Results The average daily transpiration of *Bromus inermis* and *Vicia villocea* is shown in Figure 1. The average daily transpiration during the study period was 5.166 and 5.358 ml per pot (standard deviation of 2.392 and 2.360 ml) for *Bromus inermis* and *Vicia villocea*, respectively. The minimum and maximum values of transpiration for *Bromus inermis* varied from almost zero to 10.06 ml per pot whereas the corresponding values for *Vicia villocea* were 1.11 and 9.94. The differences between the means for the two species were not significantly different at the 95% significance level.

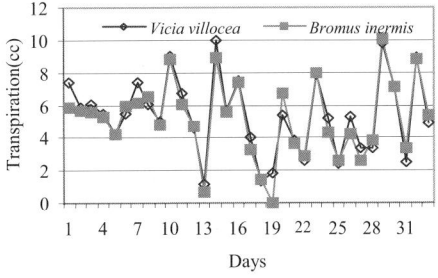

Figure 1 Daily transpiration of *Bromus inermis* and *Vicia villocea*

Conclusions These results demonstrate that daily transpiration rate of the studied species are similar. The variation was probably due to climatological factors, such as temperature and relative humidity as discussed by Roberts (2000), as well as to physiological characteristics of the plants (Anderson 1981). More studies on other species for longer study periods are recommended.

References

Anderson, M.C. (1981). The geometry of leaf distribution in some south-eastern Australian forests, *Agricultural Meteorology*, 25, 195-205.

Flerchinger G.N. & K.R. Cooley (2000). A ten year water balance of a mountainous semi-arid watershed, *Journal of Hydrology*, 237, 86-99.

Roberts, J., (2000). The influence of physical and physiological characteristics of vegetation on their hydrological response, *Hydrological Processes*, 14, 2885-2901.

Influence of water stress on root/cladode ratio and water-use efficiency of *Opuntia ficus-indica* and *O. robusta*

H.A. Snyman

Department of Animal, Wildlife and Grassland Sciences, University of the Free State, P.O. Box 339, Bloemfontein 9300 South Africa, Email: snymanha.sci@mail.uovs.ac.za

Keywords: cactus pear, root length, root and cladode mass, root thickness, water-use efficiency

Introduction Due to the regular occurrence of drought in southern Africa, there is a need for more research on drought tolerant fodder plants such as *Opuntia* (cactus pear) species. In contrast with the cladode system, the roots of cactus pear have received little attention, however, they certainly differ from other plants, as they develop xeromorphic characteristics, enabling the plant to survive prolonged periods of drought (Nobel, 1988). The influence of different water applications was evaluated in terms of root and cladode mass, water-use efficiency (WUE) and root length for one-year-old plants of *Opuntia ficus-indica* (cultivar Morado) and *O. robusta* (cultivar Monterey).

Materials and methods The cladodes were planted in pots (210 mm diameter and 550 mm deep soil) and grown in the greenhouse at day/night temperatures of 25-30/15-18 ^{0}C. The water treatments applied were 0-25%, 25-50%, 50-75% and 75-100% depletion of total plant available water. After four weeks of keeping the soil at field water capacity, the water stress treatments started. The planted pots were washed out when reaching the different water levels, after which root lengths, root and cladode mass and WUE (dry matter production per mm water used) were determined. The length of the washed roots was measured by using a modified infrared root length counter (Rowse & Phillips, 1974). The thickness at the end of the root where die-back took place was measured by a vanier calliper. The number of side roots per tap root was also determined by measuring twenty roots, randomly selected, in each pot. A 2x4 (*Opuntia* species and water treatment) factorial experiment with two replications for each water treatment was conducted.

Results and discussion The root mass for *O. ficus-indica* and *O. robusta* decreased (P<0.01) from 27.5 to 10.6 g/plant and 18.6 to 9.9 g/plant respectively with increased water stress. Due to the finer root system of *O. robusta* the root mass was lower (P<0.01) than that of *O. ficus-indica*. In contrast, the root length increased (P<0.01) with water stress for *O. ficus-indica* and *O. robusta* from 55.3 to 68.2 m/plant and 51.3 to 59.7 m/plant respectively. The root/cladode ratios for *O. ficus-indica* and *O. robusta* decreased (P<0.01) with water stress from 0.136 to 0.092 and 0.144 to 0.075, respectively. On average for all water treatments the roots of *O. ficus-indica* and *O. robusta* composed only 12% and 10% respectively of the total plant biomass. These low root/cladode ratios are supported by the 0.14 found by Nobel (1988) for *Opuntia* species. Water-use efficiency decreased (0.091 to 0.072 g/mm) (P<0.01) with an increase in water stress for *O ficus-indica*, while in contrast increased (0.029 to 0.081 g/mm) (P<0.01) for *O. robusta*. The side roots per tap root increased (P<0.01) with water stress with as much as 35 and 70 for *O. ficus-indica* and *O. robusta* respectively. The influence of water stress on root die back was clearly observed in *O. ficus-indica* but less noticeable in *O. robusta*. The average thickness at the end of the roots for *O. ficus-indica* and *O. robusta* where die back took place was 0.9 and 0.3 mm respectively for the highest water stress treatment. The water percentage in the cladodes decreased with water stress from 90.3 to 86.9% and 91.3 to 86.4% for *O. ficus-indica* and *O .robusta* respectively.

Conclusions The unique and efficient root system of the cactus pear can not be over emphasized if taking into account the small amount of roots in the total plant biomass. It was clear that *O. ficus-indica* is more sensitive to water stress than *O. robusta*. *Opuntia* species is a multifunctional crop (fodder and fruit) which can be of great value in both developed and undeveloped countries, because of its ability to utilise arid areas to its full potential.

References

Nobel, P.S. (1988). Environmental Biology of agaves and cacti, Cambridge University Press, Cambridge New York, USA, 260 pp.

Rowse, H.R. & D.A. Phillips (1974). An instrument for estimating the total length of root in sample. *Journal of Applied Ecology*, 11, 309-314.

Variability in tolerance to water stress by *Holcus lanatus* L., *Bromus valdivianus* Phil. and *Agrostis capillaris* L. accessions

I.F. López and O. Balocchi
Instituto de Producción Animal, Universidad Austral de Chile, Valdivia, Chile, Email: ilopez@uach.cl

Keywords: water shortage, water deficit, ecotype, phenotypic variability, phenotypic plasticity

Introduction *Holcus lanatus* L. (Hl), *Bromus valdivianus* Phil. (Bv) and *Agrostis capillaris* L. (Ac) are frequently present in the naturalised pasture of the Chilean humid region, which has a summer drought with two distinguishable areas according to average summer rainfall: a Northern area (Long summer drought, LSuD: 136-186 mm;) and a Southern area (Short summer drought, SSuD: 186-338 mm). It was hypothesised that plant species have colonised areas with different water deficits during summer through differentiated drought tolerance, which would imply ecotype generation.

Materials and methods Three accessions of *H. lanatus* (Hl), *B. valdivianus* (Bv) and *A. capillaris* (Ac) were collected from the LSuD area and three from the SSuD area. From each of these, three individual tillers were established in pots in a glasshouse. The water stress levels applied were: 100% field capacity (FC), 50% FC, and 25% FC. The water stress period was equivalent to that of Valdivia (1,196 accumulated growing degree days). Leaf appearance rate (LAR), leaf extension rate, leaf length (LL), number of leaves (LN), tiller number (TN), extended height (DH), tiller weight (TW), and sheath (SW) and laminae (LW) weights were evaluated. Following this period, the plants were trimmed to a height of 3 cm and accumulated herbage mass (HM1) was measured. All pots were irrigated to 100% FC for 30 d (resilience period), DH was measured, and accumulated herbage mass (HM2) evaluated to ground level. Total accumulated herbage mass (TAHM) was calculated. The design was a complete block with factorial arrangement of the treatments, with four blocks. Results were analysed using ANOVA and Canonical Variate Analysis (CVA).

Results The results from ANOVA and CVA were in agreement. For Hl, the first two canonical variates of the CVA explained 75% of the differences between accessions, such that TW, LW, LL, DH, TN significantly diminished due to water stress (CAN1: 53%). The differences due to accession features were explained by CAN2 (22%), with LSuD2 behaving differently ($P<0.001$) from the other accessions, with higher LW, TW, HM1 during the water stress period, but with lower TN. The differences between LSuD2 and the other accessions were enhanced with increasing water stress. The results showed that only LSuD2 had a high water stress tolerance. The variability in behaviour due to water stress amongst the other Hl accessions was most likely due to phenotypic plasticity. The differences amongst accessions measured at the end of the resilience period were due to the differences shown during the stress period. The other variables that were measured did not explain differences amongst accessions.

The Bv accessions performed differently under all water treatments ($P<0.001$), but these differences were enhanced by water stress. The first three canonical variates explained 79.4% of the accession variation: CAN1 (40%) explained differences between the accessions in TN and DH in contrasting directions; CAN2 (23%) accounted for differences between the accessions in TW and HM1. The effects of the water stress period on the performance of accessions were shown by a regular pattern along CAN2. The differences amongst accessions were expressed as the combination between CAN1 and CAN2. Since all of the Bv accessions performed differently from each other, no groups were formed. The results indicated that the drought survival strategy of Bv would be through the presence of ecotypes. The other variables that were measured did not explain differences amongst Bv accessions.

The Ac accessions did not show differences and their performance was not affected by water stress.

These results show that plant species have different survival strategies for water stress, which are based on a combination of the presence of ecotypes and phenotypic plasticity. The extent that both strategies are present would be expected to affect the stability of grasslands.

Conclusions In the Chilean humid region, *B. valdivianus* showed a high intra-species diversity, through the presence of ecotypes, which allow it to survive through the summer drought. *Agrostis capillaris* survived this environmental constraint through its high phenotypic plasticity. *Holcus lanatus* showed a mixed survival strategy with both phenotypic plasticity and ecotype presence.

Development of a breeders' toolkit for drought resistance in a *Lolium/Festuca* hybrid

J. Humphreys, I.P. Armstead and M.W. Humphreys
*Institute of Grassland and Environmental Research, Plas Gogerddan, Aberystwyth, Ceredigion SY233EB, UK,
Email: jan.humphreys@bbsrc.ac.uk*

Keywords: introgression-mapping, drought resistance, *Festuca glaucescens, Lolium multiflorum*

Introduction *Lolium multiflorum* (Lm) is considered an ideal grass for European agriculture. However, existing high-quality forage Lm cultivars have been bred for intensive systems in benign environments, and have proved to be insufficiently robust to meet many of the environmental challenges that face extensive agriculture in more extreme conditions. Genes for persistency, tolerance of cold, drought and poor soils, can be found in currently under-exploited native *Festuca* ecotypes. These *Festuca* ecotypes cannot however compare with Lm cultivars for productivity or quality of forage under favourable conditions. *Festuca glaucescens* (Fg) is of Mediterranean origin and as such is adapted to drought and heat stress. The object of this work was to introgress a single chromosome segment of Fg containing genes for drought resistance into a diploid Lm background. Subsequent to the introgression of a Fg chromosome segment, Fg markers were mapped and a prototype toolkit developed to follow the genes for drought resistance through a breeding programme.

Materials and methods A Fg ($2n = 4x = 28$) genotype was hybridised onto a synthetic autotetraploid Lm cultivar Roberta ($2n = 4x = 28$). The F_1 hybrid as a male was backcrossed onto a diploid Lm cultivar ($2n = 2x = 14$) to generate triploid BC_1 plants which were further backcrossed onto diploid Lm cultivars to produce diploid BC_2 plants. Nine BC_2 populations each of 30 plants were used in a drought test carried out in polythene lined brick bins in a glasshouse. Water was withheld for 14 weeks and recovery assessed by dry matter production 4 weeks after re-irrigation. Genomic *in situ* hybridisation (GISH) analysis was used to detect the presence of introgressed Fg chromosome segments. Amplified fragment length polymorphism (AFLP) (Vos *et al.*1995) using 100 primer pair combinations and sequence tagged site (STS) markers were used to target the introgressed Fg segment. Mapping of the markers was carried out with 96 genotypes of a BC_3 population using JoinMap® 3.0 (Van Ooijen and Voorrips, 2001). The BC_3 population was then subjected to a drought test as described above.

Results BC_2 genotype P194/208/19 was selected as having the best recovery growth with a mean of 6gm dry weight. The Lm parent under the same conditions performed very poorly with a yield of 0.075gm while the Fg parent produced 1.2 gm of dry matter. GISH analysis of P194/208/19 showed 14 Lm chromosomes with a single terminal Fg segment on the satellite of chromosome 3. Nine AFLP markers and an STS marker Fg71673 discriminated clearly between the Lm and Fg derived sequences. The genetic distance of the markers within the Fg segment was estimated to be 32cM with Fg71673 at one end of the map. During the time that the BC_3 mapping population was assessed for drought resistance the stress in the glasshouse was particularly severe with maximum temperatures of >40°C during 20 days and > 50°C over 7 other days. Ten of the BC_3 plants survived the combination of drought and heat and were able to re-grow during the 4 week recovery period. Each of these plants contained the entire set of 10 Fg markers ($p < 0.001$) which was evidence that an intact Fg chromosome segment was necessary to enhance the drought resistance of the Lm.

Conclusion Good drought and heat tolerance but poor establishment is a characteristic of Fg, its growth rate is slow and it regularly enters quiescence during the summer months. However the Fg translocation onto Lm chromosome 3 had enhanced the drought resistance of the Lm with no compromise in its forage yield. The STS marker Fg71673 co-segregated at one end of the genetic map with the genes for drought resistance and has now been fluorescently labelled for high throughput use on the ABI 3100 Genetic Analyser. This will be used as prototype breeders' toolkit for cultivar development. Another such suitable marker will be sought at the other end of the map to bracket the Fg genes for drought resistance enabling the whole Fg segment to be followed through a breeding programme.

References
Van Ooijen, J., Voorrips, R. (2001) JoinMap® 3.0, Software for the calculation of genetic linkage maps. Plant Research International, Wageningen, the Netherlands.
Vos, P. Hogers, R., Blecker, M., Rijans, M., Van der Lee, T., Hornes, M., Frijters, A. Pot, J., Poleman, J., Kuiper, M. & Zabeau, M. (1995) AFLP: a new technique for DNA fingerprinting. *Nucleic Acid Research,* 23, 4407-4414.

Impact of land use on water quality in River Njoro Watershed, Kenya

W.A. Shivoga, M. Muchiri, S. Kibichi, J. Odanga, S.N. Miller, T.J. Baldyga and C.M. Gichaba
Egerton University, P.O. Box 536, Njoro, Kenya, Email: shivogawa@yahoo.co.uk, [2]Department of Renewable Resources, University of Wyoming, Box 3354, 1000 E. University Dr., Laramie, WY 8207, USA

Keywords: upland land use, subwatershed, downstream water quality, riparian zone

Introduction Water resources within the River Njoro watershed have become degraded due to high population growth rate and change in land use upsetting environmental stability. Land cover classification using Landsat images (Baldyga *et al.*, 2004) shows loss of about 20% of forested areas between 1986 and 2003 in the watershed. The forested and large-scale farm areas have been converted mainly into small-scale mixed agriculture and human settlements. These changes have impacted negatively on the ecological integrity and hydrologic processes in the watershed (Shivoga, 2001) but little is known about the influence of specific land uses on water quality of the river.

Materials and methods Data recorded from ten sampling sites along River Njoro were used to examine the contribution of nutrients from subwatersheds upstream draining each of the sites. Standard Digital Elevation Model GIS analysis was used to determine the spatial distribution of land cover types and subwatershed contributing runoff to the sites in the river. Water and sediment samples were collected for chemical analysis related to upstream land use types and size of subwatersheds.

Results The mid-stream portion of the river near Egerton University, with industrial, human settlement and agricultural land uses, accounted for the highest cover, the lower the P loss from the subwatershed There was, however, significant decrease in nutrient levels downstream indicating natural purification as the river flows through an area of large- scale farming with dense riparian vegetation. Small-scale farms and bare lands contribute over 55% of the phosphorus (P) load to the River Njoro. The size of the subwatershed accounts for about 53% variability in soluble P in the river. Grassland cover had a negative relationship with P loss (R^2 = 0.4171) indicating that the more the grass cover, the lower the P loss from the subwatershed.

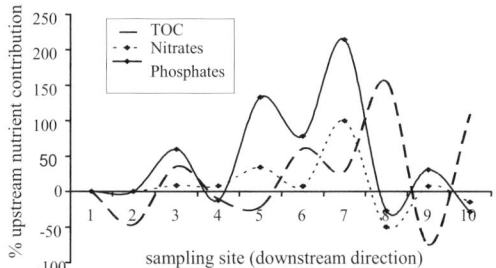

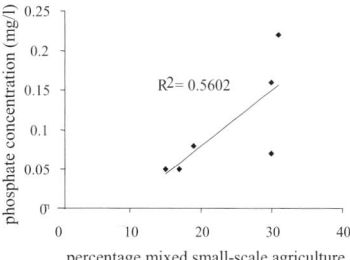

Figure 1 Downstream variation in contribution , of phosphate nitrate and organic matter

Figure 2 Relationship between proportion of land comprising small scale mixed farming and phosphate concentration in the river contribution of nutrients

Conclusions Quantification of land use in subwatersheds is important for characterising water quality in the River Njoro watershed. Upland land uses are as important as near-stream land uses. Intact riparian corridors along the river provide natural purification and recovery of the ecological integrity of the River Njoro. Grassland cover reduces the P loss from the watershed. The intact riparian zone retains nutrients from the larg-scale farms surrounding the river, thereby significantly reducing the contribution of nutrients from the arable land to the river.

References
Baldyga, T.J., S.N. Miller, W.A. Shivoga & C. Maina-Gichaba (2004). Assessing the impact of land cover change in Kenya using remote sensing and hydrologic modeling. *Proceedings of the 2003 American Society for Photogrammetry & Remote Sensing Annual Conference*, Denver, CO, May 23-28, 2004.
Shivoga, W.A. (2001). Influence of hydrology on the structure of invertebrate communities in two streams flowing into Lake Nakuru, Kenya. *Hydrobiologia*, 458,121-130.

Phosphorus transfer to river water from grassland catchments in Ireland

H. Tunney[1], P. Jordan[2], G. Kiely[3], R. Moles[4], G. Morgan[3], P. Byrne[4], W. Menary[2] and K. Daly[1]

[1]Teagasc, Johnstown Castle, Wexford, Ireland, Email: htunney@johnstown.teagasc.ie, [2]University of Ulster, Coleraine, BT52 ISA, UK, [3]University College, Cork, Ireland, [4]University of Limerick, Ireland

Keywords: phosphorus, transfer, water, grassland

Introduction In Ireland it is estimated that at least half of phosphorus (P) loss to water is from agricultural sources and National and European Union policy and legislation aim at reducing phosphorus (P) loss to water in order to reduce eutrophication. In Ireland, the average soil test P (STP) levels increased ten-fold, from less than 1 to over 8 mg Morgan P per 1 soil over the past 50 years, reflecting increased P inputs in fertiliser and animal feed. One of the main objectives of this three-year research programme, started in 2001, was to investigate P loss to water in grassland catchments.

Materials and methods Phosphorus loss to water was studied in three catchments (with nested sub-catchments), one in the north (Oona, Co. Tyrone; shale soil), centre (Clarianna, Co. Tipperary; limestone soil) and south (Dripsey, Co Cork; old red sandstone soil) of the island. This involved setting up field stations for the collection of hydrological and water chemistry data in the nested catchments at different scales (Table 1), investigating the loss of different P fractions and suspended solids in the river water (runoff) under various seasonal, meteorological, hydrological and soil conditions. The mean intensity of grassland farming and STP were broadly similar in the smaller subcatchments of the three catchments.

Results The differences in hydrology, rainfall and soil types between the three catchments were reflected in runoff and P export to water.The Dripsey and Oona catchments had broadly similar total P (TP) exports, of the order of 2 kg P/ha per year (Table 1). This level of loss is higher than the level of about 0.5 kg TP/ha that is considered compatible with good water quality. In general Oona had higher SS than the other two catchments, probably reflecting soil type and more intensive runoff. In contrast, Clarianna had a several fold lower P export per unit land area despite having broadly similar STP levels and agricultural intensity as Dripsey and Oona. The Clarianna had less runoff and has mainly thick calcareous Quaternary deposits which retain P more effectively than the other two catchments. Losses of P per unit area were influenced by catchment size and STP.

Table 1 Rainfall, evapotranspiration (ET), catchment area, runoff, and mean values in the river water for P fractions and corresponding loss per ha (load) for total P (TP), particulate P (PP) and dissolved reactive P (DRP) for the three catchments from 1 Jan. to 31 Dec. 2002. The error statistics (%E) were calculated for loads from the 95% confidence limit least squares regression equations used to gap fill time series water chemistry data

Basin	Rain	ET	Area	Runoff	TP			PP			DRP		
	mm	mm	km²	mm	mg/l	kg/ha	%E	mg/l	kg/ha	%E	mg/l	kg/ha	%E
Dripsey	1833	362	0.17	1206	0.22	2.66		0.049	0.60		0.15	1.85	
			2.11	1080	0.23	2.48		0.099	1.07		0.11	1.14	
			14	1037	0.15	1.60		0.057	0.59		0.08	0.81	
Oona	1366	352	0.15	611	0.39	2.40	2.89	0.239	1.46	3.57	0.08	0.51	3.45
			0.62	894	0.27	2.41	2.07	0.157	1.40	3.57	0.06	0.52	3.57
			88.5	817	0.38	3.13	6.40	0.203	1.66	5.77	0.11	0.90	4.62
Clari-anna	1091	493	0.8	603	0.11	0.69	3.77	0.078	0.47	3.62	0.02	0.15	4.00
			7.3	435	0.07	0.30	1.67	0.049	0.21	1.43	0.01	0.06	5.00
			13.6	416	0.04	0.17	5.88	0.021	0.09	6.67	0.01	0.06	8.33
			29.8	434	0.05	0.23	8.26	0.021	0.09	10.0	0.03	0.11	6.36

Conclusions Hydrology and soils in some catchments (e.g. Clarianna) can minimise the loss of P compared with other catchments (Oona and Dripsey) with broadly similar mean STP and grassland farming practices and these factors are important determinants in P transfer from catchments. The relative importance of factors influencing P transfer from grassland to water will help in agreeing the most appropriate management practices to help reduce loss to water.

Grazing management impacts on the riparian zone and water quality

S.R. Aarons, A. Melland and C.J.P. Gourley

Ellinbank Research Institute, PIRVic Ellinbank, RMB 2460 Hazeldean Rd, Ellinbank, Victoria, 3821 Australia,
Email: sharon.aarons@dpi.vic.gov.au

Keywords: soil nutrients, piezometers, phosphorus, sediment

Introduction Inappropriate farm management activities such as stock access to creeks, and poor fertiliser and effluent management can negatively impact riparian zones and waterways, contributing to increased in-stream nutrient, sediment and microbiological loads and loss of riparian biodiversity, amongst other impacts. Nutrient budgets for dairy systems indicate that on-farm nutrient accumulation and redistribution is common (Gourley 2004), which in large part is due to the uneven distribution of dairy cow dung and the nutrients they contain (Aarons *et al.*, 2004). The '*Gippsland Dairy Riparian Project – Environmental Monitoring module*' was established in Jan. 2003 to monitor the impact of dairy farm management and changed riparian zone management on the riparian zone and water quality.

Materials and methods The research site is located on two adjacent commercial dairy farms which incorporate 1.7 km of the Sandy Creek, a dairy-dominated catchment, in southeastern Victoria. Six weirs and 11 nests of piezometers were installed at the site to assist in determining the impact on soil and water of tracks, paddock management, dairy shed and stand-off areas, effluent ponds, and effluent-irrigation of pasture. Monitoring of soil (nutrients), stream (height, nutrients, solids, microbiology), groundwater (levels, nutrients), habitat condition (stream temperature, canopy cover) and rainfall, commenced in late autumn (May) 2003.

Results Intensive soil sampling identified nutrient accumulation zones near the creek that could contribute to degraded water quality (Figure 1). Soil P and EC concentrations, as well as total N, total K and EC concentrations in the shallow groundwater were higher in these zones compared with nearby areas. Weekly creek grab samples (Figure 2) indicated that suspended solid concentrations were elevated downstream of the track crossing the creek (at Weir 2), while increased total N concentrations occurred downstream of areas where run-off from pasture was observed (Weir 3).

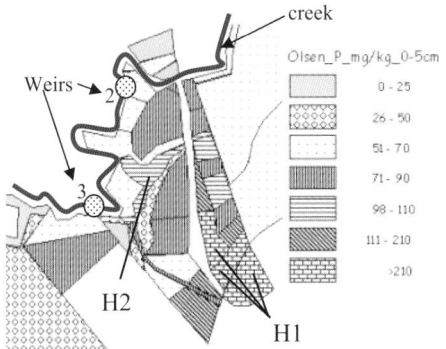

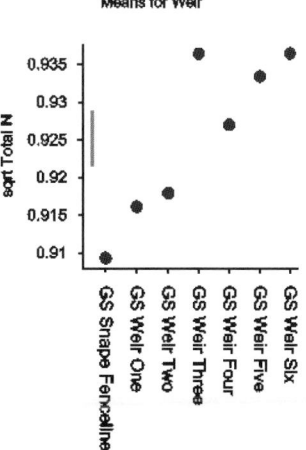

Figure 1 Soil Olsen P (mg/kg) in the 0-5cm layer in each of the sample areas. Those areas with the highest (H1) P levels are indicated. Movement of water from H2 into the creek is observed upstream of weir 3

Figure 2 Mean total nitrogen (square root) of grab samples collected at seven locations along the creek. Fenceline – where the creek enters the property; Weirs 1 to 6 – progressively located downstream. See Figure 1 for locations of Weirs 2 and 3

Conclusions The riparian and in-stream data indicate that dairy farm management activities are having a measurable impact at the research site. Nutrient accumulation within the landscape in areas that can have a potential negative impact on water quality, appears to be associated with increases in stream nutrient concentrations. On-farm activities that can improve riparian zone condition and the waterway have been identified and used to develop practical on-farm management actions.

References

Aarons, S.R., C.R. O'Connor & C.J.P. Gourley (2004). Dung decomposition in temperate dairy pastures: I Changes in soil chemical properties *Australian Journal of Soil Research*, 42, 107-114.

Gourley, C. (2004). Improved nutrient management on commercial dairy farms in Australia. *Australian Journal of Dairy Technology*, 59, 152-156.

Denitrification under pastures on permeable soils helps protect ground water quality

M.P. Russelle[1], B.A. Browne[2], N.B. Turyk[2] and B. Pearson[2]
[1]USDA-Agricultural Research Service-Plant Science Research Unit, 1991 Upper Buford Circle, Room 439, University of Minnesota, St. Paul, Minnesota 55108-6028, USA, Email: russelle@umn.edu, [2]College of Natural Resources, University of Wisconsin, 800 Reserve Street, Stevens Point, Wisconsin 54481, USA

Keywords: soil nitrate, ground water, nitrate leaching, dissolved organic carbon, nitrous oxide

Introduction   Pastures have been implicated in ground water contamination by nitrate, especially in humid regions with thin or sandy soils (Stout *et al.*, 2000). Significant losses can occur even under low N input, because available N from excreta patches often exceeds plant uptake capacity. Lack of evidence that appreciable nitrate leaching was occurring in established Midwestern USA pastures led us to test the hypothesis that denitrification was preventing or remediating nitrate loading. Higher denitrification rates have been found in the relatively limited number of trials since Ball & Ryden (1984) first reported the significance of this process in pastures.

Materials and methods   At three grazing dairy farms located on soils with high hydraulic conductivity in central Wisconsin, USA, multiport ground water wells were established on the up-gradient and down-gradient edges of at least one paddock and in a field under corn-soybean management on a nearby confinement dairy farm. Inorganic N was determined in the upper 1.2m of soil. In one paddock and in the corn-soybean field, an intensive grid of mini-piezometers was established to determine the range of variation in *in situ* dissolved solids and gases in ground water, sampled by pumping-induced ebullition (Browne, 2004). Two independent experiments were conducted in a growth chamber on intact soil cores (6-cm diam. by ~60-cm long) from one paddock, with or without fresh dairy cow excreta applied at the start of each 28- to 31-d incubation period.

Results   Ground water samples from the multiport wells indicated that nitrate was leaching at substantially smaller rates than under other agricultural practices in the area. Although differences in soil nitrate concentration were evident between excreta spots and background areas on several sampling dates, no differences in dissolved organic carbon were detected. The intact soil core experiments provided convincing evidence that urine increased denitrification. Soil pH, ammonium, and nitrate concentrations followed patterns reported by others, the first two increasing rapidly after urine application, and nitrate increasing after about 7 days. Smaller changes occurred under fresh dung than fresh urine. Nitrous oxide emission over 4 weeks was 3-fold higher with dung and 9-fold higher with urine than the control soil. Methane emission was 20-fold higher with dung than either urine or no treatment, whereas CO_2 emission quadrupled with either excreta. In the field, there was tremendous spatial variability in ground water chemistry (dissolved nutrients and gases). Figure 1 shows that the dissolved denitrified N (measured as dissolved N_2 gas in excess of atmospheric N_2) was higher as a percentage of total nitrate (nitrate + denitrified N) in groundwater beneath the pasture (n >60 sites per sampling time) than beneath the arable field (n >20 per sampling time), and ancillary measurements (e.g., dissolved organic C and dissolved O_2) supported this result. In contrast, dissolved N_2O was lower under the pasture than corn. Further research is underway to determine the variation in dissolved gas concentrations in ground water under other grazed paddocks and arable fields, the sources of the dissolved organic carbon, the proportion of N lost as N_2 in these systems, and the potential of denitrification to reduce nitrate loading of ground water in the region.

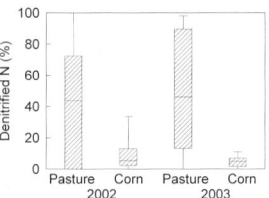

Figure 1 Denitrified N (excess N_2 as a percent of nitrate+denitrified N) in ground water [percentiles of data are shown by whiskers (10th and 90th), boxes (25th and 75th), and horizontal line (50th)]

Conclusions   This is the first report of the wide variation in dissolved gas composition in ground water under pastures. The field evidence and results from intact soil cores lend support to the hypothesis that denitrification may remove substantial amounts of N from pastures. Enhanced denitrification may benefit water quality more generally as ground water moves from or toward adjacent arable cropland.

References

Ball, P. R. & J. C. Ryden (1984). Nitrogen relationships in intensively managed temperate grasslands. *Plant and Soil*, 76, 23-33.

Browne, B. A. (2004). Pumping-induced ebullition: A unified and simplified method for measuring multiple dissolved gases. *Environmental Science and Technology*, 38, 5729-5736.

Stout, W. L., S. L. Fales, L. D. Muller, R. R. Schnabel, & S. R. Weaver (2000). Water quality implications of nitrate leaching from intensively grazed pasture swards in the northeast USA. *Agriculture, Ecosystems, and Environment*, 77, 203-210.

Implementation, participation and evaluation of a voluntary water quality protection program for grazingland owners and managers

M.R. George, J.M. Harper, S.R. Larson, R.E. Larsen, N.K. McDougald and D.J. Lewis
Rangeland Watershed Program, Agronomy and Range Science Department, University of California, Davis, California 95616, USA, Email: mrgeorge@ucdavis.edu

Keywords: water quality, water quality protection practices

Introduction In 1990, California's range livestock industry began working with the state's water quality regulatory agency to develop a voluntary producer participation programme to protect water quality on privately owned grazinglands. In 1995 they implemented a voluntary programme of surface water protection supported by extension education and technical assistance conducted by University of California and USDA Natural Resources Conservation Service. Past studies have shown that education programmes are crucial to voluntary pollution control programmes in agriculture (EPA 1990) and that ranchers will change grazing management practices in response to extension education programmes (Richards and George 1996). The objective of this project was to conduct an extension education programme that facilitated water quality planning and implementation of water quality protection practices by range livestock producers.

Materials and methods To facilitate completion of water quality protection plans a Ranch Water Quality Planning Short Course was developed, tested and improved from 1994-96. Beginning in 1997 this short course was conducted for private grazingland owners throughout California. During the short course ranchers learned about (1) non-point source pollution associated with ranching, (2) state and federal water quality regulations and regulatory agencies, (3) basin water quality assessments, (4) self-assessment of pollution sources on their own property, (5) practices that protect water quality, (6) cost-share programmes, and (7) how to monitor pollution sources and practice effectiveness. Plans developed during the short course included: ranch descriptions, ranch goals, ranch maps, basin water quality status, pollution source self-assessments, existing and planned water quality protection practices and monitoring procedures. The short course curricula can be reviewed and downloaded from http://californiarangeland.ucdavis.edu.   In 2002-2003 producers who participated in the short course were surveyed to determine their water quality protection activities following the short course.

Results From 1997 to 2004 more than 1000 producers attended 60 short courses in 31 counties. Two-thirds of those attending short courses completed water quality plans for their ranches totalling more than 500,000 ha. Two-thirds of the short course participants who completed the survey implemented water quality protection practices. There was a significant relationship ($p<0.001$) between plan completion and practice implementation. Ninety percent of those who implemented water quality protection practices had completed ranch water quality plans. The majority of the respondents managed ranches less than 2000 ha in size. Half of the respondents raised beef cattle. We found that personal funds invested in water quality protection practices exceeded cost-share funds until costs exceeded $2000. The survey revealed that initially respondents took the course to avoid regulation, but upon short course completion, became more proactively involved in controlling non-point source pollution because they found pollution sources during their self-assessments. Fewer respondents implemented a monitoring programme than implemented water quality protection practices.

Conclusions These results indicate that the range livestock industry initiated voluntary program supported by education, was an effective means for helping grazingland owners and managers to voluntarily address non-point source pollution on their properties.

References

EPA (1990). Rural clean water program, Lessons learned for a voluntary non-point source control experiment. EPA 440/4-90-012. U.S. Environmental Protection Agency. Washington, District of Columbia, 29pp.

Richards, R. & M.R. George (1996). Evaluating changes in range management practices though extension education. *Journal of Range Management*, 49, 76-80.

Profitable and sustainable grazing systems for livestock producers with saline land in southern Australia

N.J. Edwards[1], D. Masters[2], E. Barrett-Lennard[3], M. Hebart[1], M. McCaskill[4], W. King[5] and W. Mason[6]
[1]South Australian Research and Development Institute, Struan Agricultural Centre, PO Box 618, Naracoorte, SA, 5271, Australia, Email: edwards.nick@saugov.sa.gov.au, [2]CSIRO Livestock Industries, Private Bag No. 5,Wembley, WA, 6913, Australia, [3]Department of Agriculture Western Australia, Locked Bag No. 4, Bentley Delivery Centre, WA, 6983, Australia, [4]Department of Primary Industries, Private Bag 105, Hamilton, Victoria, 3300, Australia, [5]NSW Dept of Primary Industries, Orange Agricultural Institute, Forest Road, Orange, NSW, 2800, Australia, [6]SGSL Coordinator, PO Box 2157 Orange, NSW, 2800, Australia

Keywords: dryland salinity, saltbush, puccinellia, tall wheat grass

Introduction Dryland salinity affects over 2.5 M ha in Australia, mostly in southern states and is expanding at 3-5% per year (NLWRA, 2001). The prognosis is for considerable expansion of the area affected by salinity and waterlogging (12–17 M ha at equilibrium), because groundwater levels continue to rise and only small-scale land management programmes have been implemented. In addition, many waterways are increasingly saline, especially in the Murray Darling Basin and in Western Australia (WA). Sustainable Grazing on Saline Land (SGSL) addresses the need to make productive use of saline land and water resources. Its research component operates at 12 sites across WA, South Australia (SA), Victoria and New South Wales (NSW) and consists of coordinated activities that have regional relevance and contribute nationally. The programme seeks to develop and demonstrate profitable and sustainable grazing systems on saline land that have positive environmental and social impacts. Whilst there are different priority research issues at each site, data collection is governed by common measurement protocols for salt and water movement, biodiversity, and pasture and animal performance in order to make comparisons and data sharing across sites practical.

The research programme In WA research is spread across seven sites, representing about 4.3M ha of salt affected land. These include two large (about 50 ha) sites (near Tammin and Yealering) that allow comparisons between unimproved land and land improved to current best practice, using a saltbush (*Atriplex* spp.)-based system with and without improved understorey species. Other sites have been established at Yealering, Lake Grace, Wubin, Meckering and Grong Grong (NSW) to examine factors affecting the composition, growth, grazing management, utilisation and value of saltbush-based pastures to sheep. Saline areas in the upper south east of SA are subject to both severe waterlogging and inundation in winter, when rising groundwater brings salt to the root zone and soil surface, inhibiting plant growth, seed set and survival in spring and early summer. Research here is focused on a puccinellia (*Puccinellia ciliata*)-based pasture where the impacts of fertiliser and addition of balansa clover (*Trifolium michelianum*) into existing puccinellia stands are being assessed under continuous and strategic grazing. Maintaining the persistence of balansa clover is a key challenge. In Victoria the targeted areas are characterised by shallow water tables, which are often saline but where winter waterlogging and inundation are an added challenge. The research here is focusing on use of tall wheat grass (*Lothopyron ponticum*) and annual legumes to provide quality out of season grazing compared to unimproved pastures. The targeted areas in NSW, the Lachlan and Macquarie catchments in the central west of the state, are characterised by high and rising salt load and electrical conductivity levels and generally small discharge areas close to waterways. Research is assessing the impacts of a salt-tolerant, perennial grass-based pasture (tall wheat grass dominant), compared to volunteer/naturalised pasture, on pasture and animal production, and water, soil and salt movement off-site. All projects will be assessed for their impact on whole farm economics.

Conclusions This ambitious project is testing current best-bet options for animal production from saline land. Outputs will include clarifying the environmental impacts and quantification of the production and economic benefits of grazing saline land. Extension products to assist farmers to make better decisions about managing these land types will boost their confidence to incorporate more saline land into their whole-farm management plans for environmental, economic and social outcomes. A significant component of this national network of projects and sites is its links with, and the participation of, farmers through the research being located on commercial farms, the involvement of local advisory groups and formal and informal links with a national network of over 125 farmer initiated small-scale projects testing locally relevant options for managing saline land.

Acknowledgements SGSL is an initiative of Australian Wool Innovation Pty Ltd, with Land and Water Australia and support from the Cooperative Research Centre for Plant-based Management of Dryland Salinity and Meat and Livestock Australia.

References
NLWRA. (2001). Australian dryland salinity assessment 2000: extent, impacts, processes, monitoring and management options. National Land & Water Resources Audit, Canberra.

A new perennial legume to combat dryland salinity in south-western Australia

L.W. Bell, M.A. Ewing, M. Ryan, S.J. Bennett and G.A. Moore
CRC for Plant-based Management of Dryland Salinity & The University of Western Australia, Nedlands Western Australia A 6100, Australia, Email: lbell@agric.uwa.edu.au

Keywords: perennial pastures, canary clover, alfalfa, recharge control, water use

Introduction Dryland salinity has devastated large tracts of productive land in Australia. This has resulted from the clearing of native perennial vegetation and its replacement with annual crops and pastures. As annual plants are shallow rooted and only use water during their winter-spring growing season, unutilised rainwater leaks into groundwater tables which rise and bring stored salt to the soil surface. The adoption of deep rooted perennial pasture plants that increase the water use can help to manage dryland salinity whilst maintaining productivity. However, new plants are needed as few perennial pasture options currently exist. Preliminary research into the potential of hairy canary clover (*Dorycnium hirsutum* (L.) Ser.) to increase water use is presented.

Materials and methods Six replicate plots (2.5 m x 4 m) of *Dorycnium hirsutum*, lucerne cv. Sceptre (*Medicago sativa* L.), annual burr medic cv. Santiago (*Medicago polymorpha*) and a bare ground control, were established in 2002 at Merredin in the Western Australian wheat-belt, a low rainfall (315 mm annual mean) site with a mediterranean climate. Soil moisture content was monitored under these pastures using a neutron moisture meter from September 2002 to April 2004 at approximately 3 week intervals.

Results Establishment was slow in 2002 with an extremely dry winter growing season. During the first summer lucerne dried the soil more than the annual medic, but little difference was observed in soil water between *D. hirsutum* and the annual medic (Figure 1). Both annual and perennial species dried the soil during spring 2003, but rainfall during summer increased soil water under the annual medic while both perennials maintained a drier soil profile. *D. hirsutum* and lucerne dried the soil to a depth of 180 cm in the 2003/04 summer increasing the soil moisture deficit to 80 mm greater than that recorded under the annual medic.

Conclusions Both lucerne and *D. hirsutum* dried the soil profile more than annual medic, thereby creating an additional buffer of 80 mm that would need to be exceeded before drainage could occur. Similar studies of lucerne have shown it to be

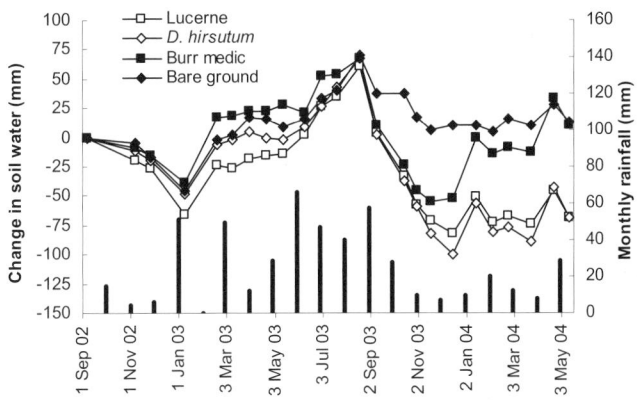

Figure 1 Monthly rainfall (bars) and • soil water (< 3 m) under bare ground, annual medic, lucerne and *D. hirsutum* pastures from September 2002 to April 2004 at Merredin, Western Australia

effective at reducing recharge (Latta *et al.*, 2001 & 2002; Ward *et al.*, 2001). Water use of *D. hirsutum* was equal to lucerne in the second year. This was surprising given the different breeding history and adaptation of these species to a semi-arid mediterranean environment. We conclude that *D. hirsutum* shows some promise for reducing groundwater recharge and helping to manage dryland salinity in farming systems of the Western Australian wheat-belt.

References

Latta R.A., L.J. Blacklow & P.S.Cocks (2001). Comparative soil water, pasture production, and crop yields in phase farming systems with lucerne and annual pasture in Western Australia. *Australian Journal of Agricultural Research*, 52, 295-303.

Latta R.A., P.S. Cocks & C. Matthews (2002). Lucerne pastures to sustain agricultural production in south-western Australia. *Agricultural Water Management*, 53, 99-109.

Ward P.R., F.X. Dunin & S.F. Micin (2001). Water balance of annual and perennial pastures on a duplex soil in a Mediterranean environment. *Australian Journal of Agricultural Research*, 52, 203-209.

Selecting grassland species for saline environments

M.E. Rogers, A.D. Craig, T.D. Colmer, R. Munns, S.J. Hughes, P.M. Evans, P.G.H. Nichols, R. Snowball, D. Henry, J. Deretic, B. Dear and M. Ewing
Cooperative Research Centre for Plant-Based Management of Dryland Salinity, Perth, Western Australia 6009, Australia, Email: MaryJane.Rogers@dpi.vic.gov.au

Keywords: genetic diversity, plant salt tolerance, soil salinity

Introduction In Australia, around 5.7 million hectares of agricultural land are currently affected by dryland salinity or at risk from shallow water tables and this figure is expected to increase over the next 50 years (LWRA, 2001). Most improved grassland species cannot tolerate the combined effects of salt and waterlogging and, therefore, the productivity of sown grasslands in salt-affected areas is low. However, there is potential to overcome the lack of suitably adapted fodder species by introducing new, salt and waterlogging-tolerant species and by diversifying the gene pool of proven species. Potential species include exotic, naturalised and native Australian grass, legumes, herb and shrub species that are halophytes and non-halophytes. A collaborative national project in southern Australia commenced in 2004 with the objective of evaluating a range of forage species for saline environments.

Materials and methods The project involves glasshouse and field research. Forage germplasm is being acquired from Australian and International Genetic Resource Centres and by direct collection from centres of natural diversity for salt and waterlogging tolerance. Plant material will be evaluated for salt and waterlogging tolerance under glasshouse conditions before promising species are assessed in the field and validation phases undertaken in saline environments. It is envisaged that some priority plant material will be available and recommended to primary producers by year 6 of the project.

Results and discussion Table 1 lists the priority plant genera that have been identified with potential for salt and waterlogging tolerance and that will be evaluated in this research project (Rogers *et al.*, 2004). Initially, priority will be given to the development of superior legume cultivars, since generally these are less tolerant than their companion grasses, yet are considered "drivers" of the system, being the providers of nitrogen and forage of high nutritive value. Within this project, species will be ranked according to forage nutritive value, biomass, ground cover potential, seasonality, ease of establishment, seeding potential, persistence (eg. perenniality, palatability, drought tolerance, crown exposure etc) and for their potential weediness. Species will also be recognised for their role in areas where there is lateral and/or mosaic variation in the salt/waterlogging profile, and where a mixture of species with a range of salt tolerance levels may give the most productive option.

Table 1 High priority legumes, grasses, herbs and shrubs with salt or waterlogging tolerance

Plant category	Genera
Legumes	*Astragalus, Ceratoides, Glycyrrhiza, Hedysarum, Lotus, Medicago, Melilotus, Swainsona, Trifolium, Trigonella, Viminaria*
Grasses	*Aeluropus, Chloris, Cynodon, Dactyloctenium, Distichlis, Enteropogon, Eragrostis, Festuca, Lachnogrostis, Leptochloa, Paspalum, Pennisetum, Porteresia, Puccinellia, Saccharum, Sporobolus, Stenotaphrum, Thinopyrum, Zoysia,*
Herbs	*Cichorium, Plantago, Ptilotus*
Shrubs	*Acanthus, Atriplex, Chenopodium, Maireana, Minuria, Rhagodia*

Conclusion Finding new forage species that are adapted to saline and periodically waterlogged land will provide new options to manage dryland salinity within Australia and internationally.

References

National Land and Water Resources Audit (2001) Australian dryland salinity assessment 2000. Extent, impacts, processes, monitoring and management options. National Land and Water Resources Audit, Canberra. 129pp.

Rogers, M.E., A.D.Craig, S.J. Bennet, C.V. Malcolm, A.J. Brown, W.S. Semple, T.D. Colmer, P.M. Evans, S.J. Hughes, R. Munns, P.G.H. Nichols, G. Sweeney, B.S. Dear & M. Ewing (2004). Fodder plants for the salt-affected areas of southern Australia. Scoping Document 2 - CRC Plant Based Management of Dryland Salinity. ISBN 1740521021. 36pp.

Early spring surface runoff from grassland and arable land

S. Hejduk and K. Kasprzak
Mendel University of Agriculture and Forestry, Department of Fodder Production and Grassland Management, Zemedelská 1, 613 00 Brno, The Czech Republic, Email: hejduk@mendelu.cz

Keywords: snow thawing, grassland, runoff, floods

Introduction Surface runoff is regarded as an undesirable phenomenon because it deprives plants and soil of precipitation water and reduces its penetration underground. It is also the cause erosion and flooding. The occurrence and depth of a frozen soil layer is the main factor which determines the amount of surface runoff in winter. A well-developed surface and/or sub-surface layer of frozen soil is practically impenetrable for water. This layer results from ice-forming processes, which are influenced by snow melting due to diurnal fluctuations in temperature in early spring, partial thaws, winter rainfalls, and thermocapillary processes taking place in frozen soil.

Materials and methods Experiments were carried out at a research station in Brno–Kníničky, Czech Republic. Within the period of 1965 – 2002, selected crops were cultivated on six south facing sloping experimental plots with loam soil (20 m^2 each). Water infiltrating into the soil and increasing the soil humidity is expressed by the equation $\varphi_r = H_r / (H_s - H_e)$ where φ_r = runoff coefficient, H_r – sum of surface runoff, H_s – total precipitation, H_e – total evaporation (average daily evaporation from the snow cover being 0.21 mm/day). The surface runoff coefficient from precipitation is calculated from the onset of snow cover to its melting during the main period of thawing.

Results In the period 1965-2002, the average proportion of runoff water flowing from the soil surface during the early spring was 35 %, 53 %, 58 %, 60 % and 68 % on plots with ploughed up stubble of winter cereals, shallowly loosened fallow, cereal and maize stubble, grassland and lucerne, respectively. Measured values of runoff in individual winters are presented in Table 1.

Table 1 Coefficients of surface runoff on plots with individual crops during winter

Winter season of (Year)	Date of onset and end of snow cover	Number of days with frozen soil	Max. depth of frozen soil (cm)	Characteristics of individual plots			
				Ploughed up winter cereals	Maize stubble	Grassland	Lucerne
				Coefficient of surface runoff (%)			
1996/97	20 Dec.- 20. Feb.	54	15	29	-	-	66
1998/99	16 Dec.-27 Feb.	86	15	54	83	78	-
1999/00	19 Dec.-1 Feb.	41	17	57	77	71	-
1984/85	26 Dec.-8 Feb.	86	40	66	-	96	-

Differences among individual plots resulted from different conditions of formation of the frozen soil layer. In loosened soils (ploughed up winter cereals) the ice-forming processes were less intensive (due to a lower number of capillary pores) and the ice crust disappeared more quickly during spring thawing. On the other hand, on plots with compacted soil (stubble, grassland), the ice layer was thicker, and more compact, and took longer to disappear and surface runoff was more intensive.

Conclusions In early spring, the intensity of surface runoff originating from melting snow and/or rainfall depends on the existence and depth of the frozen soil layer. Its quality is dependent above all on physical condition of soil, type and amount of winter precipitation and nature of the thawing process itself. The colder the winter and the longer the period of subzero temperatures, the higher are the losses resulting from surface runoff and the higher the risk of floods occurring in the early spring. Grassland stands, stubbles and compacted soil constitute a higher risk of the occurrence of surface runoffs (floods) in early spring than soil after ploughing. Extensively managed grass (cut only once a year) is the only exception to this rule as minimal use of machinery and a favourable microclimate promote an intensive proliferation of organisms in the soil (edaphon) which result in soil loosening.

Acknowledgement Result achieved were obtained and processed under the financial support of the Czech Science Foundation (GACR), Research Plan No. 526/02/P061.

Theme B: Grassland and the environment

Section 19

Soil quality and nutrients

Effect of mineral nutrition on red clover leaf area index

B. Cupina[1], P. Erić[1], S. Vasiljević[2], V. Mihailovic[2] and D. Milic[2]
[1]Faculty of Agriculture, D. Obradovića 8, 21 000 Novi Sad, Serbia and Montenegro, Email: cupinab@polj.ns.ac.yu, [2]Institute of Field and Vegetable Crops, M. Gorkog 30, 21 000 Novi Sad, Serbia and Montenegro

Keywords: *Trifolium pratense*, mineral nutrition, leaf area index

Introduction The legume red clover (*Trifolium pratense*) fixes its own nitrogen (N), but requires P and K fertilisation. There are no recent reliable data in the domestic literature, on the amounts of P and K recommended to farmers; present recommendations are often either inadequate or excessive. Red clover mineral nutrition is significantly affected by soil and weather conditions (Taylor & Quesenberry, 1996). The objective of our two-year study was to enable rational fertiliser application in accordance with soil type and agro ecological conditions.

Material and methods The trial was carried out in the field conditions during 2000 and 2001. The red clover cv. Kolubara, was grown in monoculture plots, with 20 cm row distance. The experimental design was a randomised block with four replicates. Soil analyses showed slight acidity (pH 6.64), mean N content 0.20 % and P and K contents 12.10 and 14.0 mg/100g of soil, respectively. Fertiliser rates were based on the soil supply with these elements and their removal by yield. The soil N content was 45 kg/ha (determined by the N-min method), and there were three different P and K rates (P_1-40, P_2-80 and P_3-120 kg/ha; K_1-60, K_2-120 and K_3-180 kg/ha, and control, 12 treatments in total. Treatments were applied in one application during the winter. At maturity, dry matter (DM) yield (t/ha) and leaf area index-LAI (m^2/m^2) were determined. LAI was determined on the basis of leaf DM weight and leaf area (Sarić *et al.*, 1986).

Results and discussion Taking the two study years together, the highest LAI was obtained in the P_1K_2 treatment, followed by P_2K_1 and P_3K_1. The lowest LAI was recorded in the control (Table 1). In the first year of study, three cuts produced an average DM yield of 14.5 t/ha, while in the second year, also from three cuts the average yield was 13.8 t/ha. The correlation coefficient between LAI and DM yield was (r=0.25) was not significant in contrast to Vasiljevic *et al.* (2003), who reported a large effect of LAI on DM yield.

Table 1 Effect of mineral nutrition on red clover leaf area index-LAI (average for two years)

Cut	Control	P_2	K_2	P_1K_1	P_2K_1	P_3K_1	P_1K_2	P_2K_2	P_3K_2	P_1K_3	P_2K_3	P_3K_3	Average	LSD (5%)
First	4.2	4.3	4.1	4.2	4.8	4.2	4.7	4.4	3.5	4.2	4.3	4.3	4.2	0.51
Second	2.3	2.6	2.4	2.4	2.4	2.6	2.5	2.3	2.2	2.4	2.1	2.4	2.4	0.35
Third	0.7	0.6	0.6	1.2	1.3	1.3	1.5	0.8	1.1	1.0	0.7	1.0	0.8	0.21
Total	6.7	7.5	7.1	7.8	8.5	8.1	8.7	7.5	6.8	7.6	7.1	7.7	7.4	0.85

Conclusion Red clover cv Kolubara achieved high DM yields, confirming that the species has a high genetic yield potential. Results were influenced by precipitation.

References
Sarić, M. (1986). Praktikum iz fiziologije biljaka. Naučna knjiga, Beograd, 245 pp.
Taylor, N.L. & K.N. Quesenberry (1996). Red clover science. Kluwer Academic Publisher, Dordrecht, 226 pp.
Vasiljevic, S., G. Surlan, B. Cupina, P. Eric, A. Mikic & Dj. Karagic (2003). Effect of yield and leaf area indicators (LAI, Lad) on green forage yields in red clover (*Trifolium pretense* L.). *Proceedings of the 20th General Meeting of the European Grassland Federation*, Luzern, Switzerland, Book of Abstracts p. 85

Sources of N used for growth following defoliation in *Panicum maximum*[1]

P.M. Santos[2,4], B. Thornton[3] and M. Corsi[2]
[1]*Financial support: FAPESP and SEERAD,* [2]*University of Sao Paulo, Av. Pádua Dias, 11, CEP 13400-000, Piracicaba-SP, Brazil, Email: patricia@cppse.embrapa.br,* [3]*The Macaulay Institute, Craigiebuckler, Aberdeen AB15 8QH, Scotland, UK,* [4]*Present address: Embrapa Southeast – Cattle Research Centre, Rod. Washington Luiz, km 234, Caixa Postal 339, CEP 13568-800, São Carlos – SP, Brazi,*

Keywords: C_4, defoliation, N-mobilisation, N-uptake, regrowth

Introduction The nitrogen (N) supplied to growing leaves from root uptake and mobilisation from senescing tissues may be reduced following defoliation. However, morphological adaptation of the shoot to prior defoliation occurs (Matthew *et al.*, 2002), which may affect the potential N supply from remaining leaves. This study determined the degree to which plants of *Panicum maximum* utilised current root uptake and mobilisation to supply N to growing leaves and side tillers following defoliation.

Materials and methods Plants were grown in sand culture supplied with a complete nutrient solution containing 1.5 mol.m^{-3} NH_4NO_3 and cut weekly for seven weeks to a height of either 15 or 30 cm. This established plants differing in shoot morphology. Eight weeks after the first cut, plants were defoliated for a final time to remove either 0, 25, 50, 75 or 100 % of the area of each individual leaf blade on the main tiller. Both immediately and one week after the final defoliation plants were harvested and separated into roots, stem, old leaves, young leaves and side tillers. ^{15}N tracer techniques were used to determine the degree to which defoliated *P. maximum* utilised current root uptake and mobilisation to supply N to growing leaves and side tillers

Results In most instances N was supplied to young leaves and side tillers from both mobilisation (unlabelled N) and current root uptake (labelled N). In all treatments for both young leaves and side tillers current root uptake supplied more N than mobilisation. Generally, the relative contribution of root uptake to the total N increase was greater in young leaves than in side tillers, indicating that the allocation pattern of mobilised N differed from that of N derived from current root uptake.

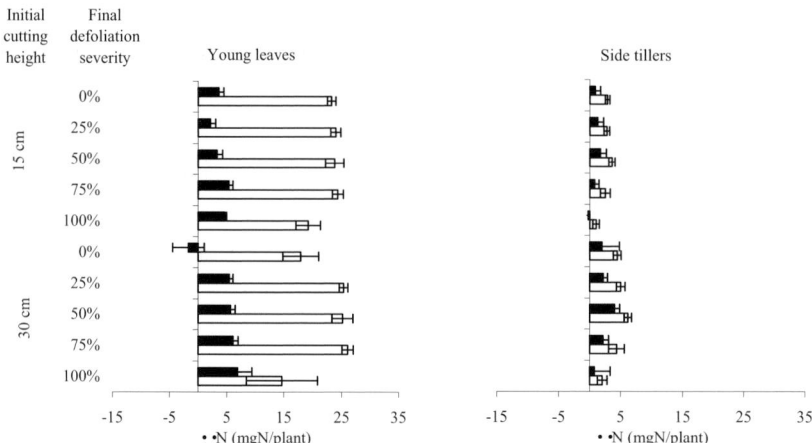

Figure 1 Change in labelled (•) and unlabelled (■) (•) N (mg/plant) of the young leaves and side tillers of *P. maximum* over a 7 day period following the final defoliation. Values are means (• •SE) of 4

Conclusions In defoliated plants of *P. maximum*, the young leaves and side tillers obtained N from both root uptake and mobilisation. In all instances current root uptake was the major source of N. The allocation pattern of mobilised N differed from that of N derived from current root uptake.

Reference

Matthew C., S.G. Assuero, C.K. Black & N.R. Sackville Hamilton (2000). Tiller dynamics of grazed swards. In: G. Lemaire,. J. Hodgson, A. Moraes, P.C.F. Carvalho & C Nabinger (Eds.) *Grassland Ecophysiology and Grazing Ecology*. CAB International, Wallingford, 27- 150.

Growth characteristics of kikuyu grass with different sources and doses of phosphorus

J.C. Pinto, Í.P.A. Santos, A.E. Furtini Neto, A.R. Morais, E.E. Mesquita, D.J.G. Faria and I.F. Andrade
The Animal Science Department of the Universidade Federal de Lavras, Lavras, Minas Gerais, Brazil, Email: josecard@ufla.br

Keywords: leaf area index, leaf area ratio, *Pennisetum clandestinum*, specific leaf area

Introduction Growth is defined as the increase in size, volume and mass as a function of time. Growth analysis allows evaluating the final growth of the plant as a whole and the contribution of the different organs in total growth (Benincasa, 1988). The experiment had as objective to evaluate specific leaf area (SLA), leaf area per unit of leaf DM, leaf area ratio (LAR), leaf area per unit of whole plant DM, leaf weight ratio (LWR), leaf weight per unit of plant weight, leaf area index (LAI), leaf area per unit of soil area, leaf/stem ratio (LSR), leaf weight per unit stem weight, of 35 days old kikuyu grass with different sources and doses of P.

Materials and methods The field experiment was conducted from August, 2000 till October, 2003 at the Universidade Federal de Lavras, southern Minas Gerais, Brazil, (21°14' S ; 40°00' W, altitude 918.84 m). The climate is Cwb, with a very defined wet season (October to March) and dry season (April to September) and average precipitation of 1,526.7 mm The experimental design was a split plot within a completely randomised block, with source of P [Triple superphosphate (TS), Phosphate of Arad (PAd) and Phosphate of Araxa (PAx)] as plots and doses of P (0, 40, 80 and 120 kg/ha of P_2O_5) as subplots. Cuts were made every 35 days in the wet season.

Results SLA, LAR and LWR were affected by the sources and doses of P, however LSR only varied with the doses of P (Table 1). LAI was not affected ($p > 0.05$) (mean 3.27). There was a linear increase of SLA when fertilised with PAd and a quadratic one for TS and PAx. LAR increased linearly to increased doses of P as PAd. LWR was quadratically inversely related to SLA. Fertilisation with PAx gave a maximum value of LWR of 0.6237 g/g at 72.3 kg/ha and with TS it was of 0.5213 g/g at 23.2 kg/ha of P_2O_5. Irrespective of the source of P, the LSR was increased to a maximum of 1.97 at the rate of 52.5 kg/ha of P_2O_5. In the range of 40-80 kg/ha of P_2O_5 the plants had higher LSR and consequently higher LWR.

Table 1 Adjusted regression equations for SLA, LAR, LWR and LSR of kikuyu grass, as a function of the doses of P (X)

Growth characteristics	Source of P	Regression equations	R^2	Prob.
	Triple superphosphate	Y= 0.0421 - 0.000142X + 0.000003X^2	0.94	<0.001
SLA (m^2/g)	Phosphate of Arad	Y= 0.04098 + 0.00011X	0.90	<0.001
	Phosphate of Araxa	Y= 0.07457 - 0.00107X + 0.000006X^2	0.99	<0.001
	Triple superphosphate	Y= 0.0213	-	-
LAR (m^2/g)	Phosphate of Arad	Y= 0.0203 + 0.000053X	0.73	<0.001
	Phosphate of Araxa	Y= 0.0209	-	-
	Triple superphosphate	Y= 0.5105 + 0.00093X - 0.00002X^2	0.99	<0.05
LWR (g/g)	Phosphate of Arad	Y= 0.5083	-	-
	Phosphate of Araxa	Y= 0.3992 + 0.00636X - 0.000044X^2	0.98	<0.001
LSR		Y= 1.6917 + 0.01056X - 0.00011X^2	0.86	<0.05

Conclusions Phosphorus fertilisation particularly from more reactive sources, stimulated growth of the kikuyu grass, leading to increments of SLA, LAR, LWR and LSR.

Acknowledgement Research supported by FAPEMIG.

References
Benincasa, M.M.P (1988). Análise de plantas: noções básicas. FUNEP, Jaboticabal, 1988. 41pp.

Sustainable pastures for the high altitude Andean tropics of Colombia

E. Cárdenas and L. Panizzo
Universidad Nacional de Colombia 30th Avenue 45th Street Bogotá, Colombia, Email: eacardenasr@unal.edu.co

Keywords: nitrogen, pastures, grass, *Lotus*, sustainable grassland farming

Introduction Dairy production systems in the high altitude Andean region of Colombia (>2.600 m.a.s.l.) use large amounts of nitrogen (N) fertilisation. Due to the inefficient use of N by the grass, it contaminates surface and ground water resulting in the eutrophication of lakes and rivers. It contributes to increased atmospheric NOx, greenhouse gas and acid rain. Therefore, the effect of different species of grasses mixed with *Lotus corniculatus* on N soil balance was evaluated.

Materials and methods Ten grass species were sown with *L. corniculatus* in a complete randomised block design with three replicates. An additional pasture of naturalised *Pennisetum clandestinum* fertilised with 400 kg of N/ha/year as urea was used as a control. Botanical composition, canopy biomass production and N forage concentration were measured during the dry and the rainy season on 45 days-old regrowth . The model proposed by Thomas *et al.* (1992) was used to estimate the amount of N cycled in the soil, fixed by the legume, N retained by dairy cattle and total soil N balance.

Results and conclusions The pastures with a good legume proportion and a positive N balance were the newly introduced *P. clandestinum, Bromus catharticus, Festuca rubra, F. arundinacea* and *Dactylis glomerata* cv. Knaulgrass. *P. clandestinum* fertilised (control) and that grown with the legume showed a negative N balance (Table 1).

Table 1 Nitrogen balance in pastures with *Lotus corniculatus*

Specie	Biomass production (grass+leg) (kgDM/ha/year)	Legume proportion (%)	N percent/MS (%)	Amount of N (kg/ha/year) Up take by plants	Recycled*	Balance
P.clandestinum (naturalised) + 400 kg N/ha	3.830 de	-	2.2	84	44	- 40 +/- 7.9
L. corniculatus + *P. clandestinum* (nat.)	4.933 d	22	3.0	148	119	- 29 +/- 3.8
P. clandestinum (newly introduced)	7.713 bc	51	3.3	254	286	32 +/- 9.7
Bromus catharticus	8.006 bc	34	2.8	224	223	- 1 +/- 8.6
Festuca rubra	10.135 a	35	2.6	263	275	12 +/- 8.2
Dactylis glomerata	6.871 c	15	2.8	192	138	- 54 +/- 2.8
Festuca arundinacea	10.440 a	42	3.0	313	334	21 +/- 9.7
Phleum pratense	8.198 bc	72	3.5	287	380	93 +/- 9.6
Anthoxantum odoratum	8.844 ab	42	3.2	283	292	9 +/- 5.1
Holcus lanatus	8.064 bc	38	2.3	185	216	31 +/- 9.7
Dactylis glomerata var. Knaulgrass	2.197 e	57	3.2	70	85	15 +/- 2.4
Festuca pratense	8.358 bc	66	2.9	242	342	100 +/- 10
Average	7.299 ***	43	2.9	212	228	

Conclusion Some introduced grasses mixed with *Lotus* have a favourable N balance in the high altitude Andean region of Colombia and the currently used *P. clandestinum* is inefficient in the uptake of applied N.

Reference

Thomas, R., C. Lascano, J. Sanz, M. Ara, J. Spain, R.Vera & M. Fisher (1992). The role of pastures in production systems. In: International Center for Tropical Agriculture (CIAT). Pastures for the tropical lowlands. Cali, Colombia, 121-144.

Seasonality of growth in grass-clover swards under repetitive nitrogen application

M. Nassiri and A. Elgersma
Department of Agronomy, Faculty of Agriculture, Ferdowsi University of Mashhad, Iran, Email: mnassiri@ferdowsi.um.ac.ir

Keywords: seasonality, nitrogen, grass-clover sward

Introduction The cohabitation of grass and clover is possibly due to asynchrony in their growth patterns, and to the beneficial effects of fixed nitrogen (N) on grass. Incompatibility of clover persistence with N fertilisation has been frequently reported (Nassiri and Elgersma, 2002). However, limited information is available regarding the effect of repetitive application of N in mixed swards. This research aims to study the balance between species in response to application of increasing rates of N throughout the growing season.

Materials and methods Two white clover (cvs. Alice and Gwenda) and two perennial ryegrass (cvs. Barlet and Heraut) were used to make four different mixtures. The mixtures were grown under two N levels, 0 (-N) and 150 kg N ha-1 (+N) in split doses. In +N mixtures, 30 kg ha-1 N was applied five times during the growing season. Monoculture of the clovers (without N) and ryegrasses (with same levels of N as mixtures) were also established. Experimental design was a split plot repeated in two complete randomized blocks (Nassiri and Elgersma, 1998 for details). All plots were cut at an approximate average target yield of 2000 kg dry matter (DM) ha-1, leading to 5 cuts during the growing season. The cut material was separated into grass and clover. DM of species was measured after drying for 48 hours at 70 °C. In this paper average DM of grass and clover cultivars are presented.

Results Relative distribution of annual DM of species over the regrowth periods (Fig. 1) showed maximal grass growth in the first period, declining towards the end of season. The opposite pattern was observed in clover, however. In grass monocultures the seasonal variation in DM (expressed by coefficient of variation, CV) decreased significantly (P < 0.05) with increasing N levels. In the +N mixtures the CV of grass and clover DM was lower than in the -N mixtures (P < 0.05), but a significantly higher CV of total DM was obtained in the +N mixtures (Fig. 1). This led to a more even distribution of total DM in the -N compared to the +N mixtures (Fig. 1). The distribution of seasonal yield of grass and clover showed similar patterns in mixtures and monocultures. Grass had the highest proportion of its annual DM in the first cut (corresponding to its reproductive growth). This declined remarkably during summer. However, for clover the opposite pattern was observed.

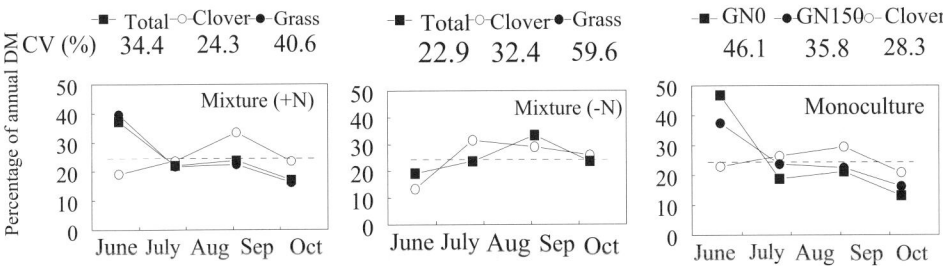

Figure 1 Mean seasonal distribution of grass, clover and total DM expressed as percentage of the total yield of the 5 regrowth periods. Coefficient of variation (CV) is also shown for all regrowth periods.

Conclusions Reliance of grass on N fixed by clover resulted in reduced interspecific competition, increased possibility of coexistence and higher total yield through synchronized growth pattern of species. However, this harmony was disturbed in +N swards, where grass still had a high seasonal variation, while the clover content was not sufficient to overcome this variation. Therefore, the seasonal distribution of the total yield in the +N swards followed the variable pattern of the grass component.

References
Nassiri, M. & A. Elgersma (1998). Competition in grass-clover mixtures under cutting. 2. Leaf characteristics, light interception and dry-matter production during regrowth. *Grass and Forage Science*, 53, 367-379.
Nassiri, M. & A. Elgersma (2002). Effects of nitrogen on leaves, dry matter allocation and regrowth dynamics in *Trifolium repense* and *Lolium preenne* in pure and mixed swards. *Plant and Soil*, 246, 107-121.

Effects of soil compaction by animal trampling on growth of *Agropyron repens*

M.R. Chaichi, B. Ataeian and M. Mohseni Saravi
Department of Rangeland Sciences, Science and Research Branch, The Islamic Azad University, Poonak, Tehran, Iran, Email: rchaichi@ut.ac.ir

Keywords: animal trampling, soil compaction, growth characteristics, *Agropyron repens*

Introduction Animal trampling is the most important factor that increases soil compaction beyond the soil elastic property in rangelands under heavy grazing intensities (Lull, 1959). The complicated ecological interactions in a rangeland ecosystem make it difficult to predict the impact of trampling under different conditions. However, soil properties and vegetation composition are sensitive to different grazing management practices. This research, which was carried out in simulated conditions, sought to establish the contribution of soil compaction resulting from animal trampling over a long period to the decline in rangeland condition

Materials and methods The Lar watershed is located 84 km northeast of Tehran, Iran. The experimental area (73,000 ha) was sub-divided into three sub-sample areas: the Reference (Control) area which had not been grazed for the last 30 years; the Key area grazed moderately (1 sheep/ha); and the heavily-grazed Critical area (more than 3 sheep/ha). At the end of the grazing period (late May to late Sept.), 12 intact soil samples were collected from each area using metal pots of 15cm diameter and 30cm height. To simulate different soil compactions from animal trampling, mechanical (pressing machine) and physical (freezing) techniques were employed. The pressing machine was used to increase the bulk density of soil from the Reference site to that of the Key (R• K) and Critical (R• C) sites and from the Key site to that of the Critical site (K• C) while freezing was used to reduce the bulk densities of soil samples from the Critical site to those of the Key (C• K) and Reference sites (C• R) and from Key site to Reference site (K• R). The indicator plant of the reference area (*Agropyron repens*) was then sown in the pots to assess growth under the different treatments. The sampling was repeated for two years (2000 and 2001). The data was analysed as a completely randomised block design with four replicates. A compound analysis was used to compare the results from the two years of experiments.

Results Plant height, tiller number and plant biomass were significantly affected by the interaction between compaction treatments and year. The highest plants were obtained in K in 2000 (26.5cm) while the shortest plants were in C• K in 2001 (Table 1). The artificial increase in soil bulk density increased tiller number per plant and whole plant biomass (R• K and R• C treatments). Although soil compaction has been suggested to reduce grass germination (Valentin, 1989), the germination percentage, whole plant biomass and other growth characteristics of *Agropyron repens* were not significantly reduced by artificial soil compaction. For the unamended soils, biomass was highest for the reference treatment (R) which had received the least compaction.

Table 1 Plant height (cm), tiller number per plant and whole plant biomass (kg/plant) for *Agropyron repens*

Plant characteristics	Soil compaction treatments								
	R	K	C	R• K	R• C	K• R	K• C	C• R	C• K
Plant height									
2000	24a	26.5a	24a	22.9a	14bc	22ab	22.9ab	23a	25.5a
2001	12cd	22.5ab	5.5c	7.5cd	6cd	12cd	7cd	6cd	4.9d
Tiller number									
2000	20.1bc	15cd	12e	20.9bc	20bc	18c	16cd	12.5de	14d
2001	18.8bc	42a	13d	30.5b	53a	41a	27b	14d	17c
Plant biomass									
2000	8.8a	5.2bc	2.1ef	5.8b	8.3a	3.0cdef	4.8bcd	2.5def	3.6cdef
2001	2.0ef	3.3cdef	1.0c	5.4bc	3.8bcde	2.9cdef	2.5def	2.2f	1.9ef

Conclusions With the soils used here, compaction due to trampling was not the factor responsible for less biomass and poorer growth of high quality plant species in heavily grazed sites. It appears that climatic factors as well as some soil edaphic conditions (e.g. organic matter, depth) other than soil compaction play a significant role in vegetation establishment and production in these rangelands.

References
Lull, H.W. (1959). Soil compaction on forest and rangelands. Forestry Service U.S. Department of Agriculture, Miscellaneous Publication, 768.
Valentine, J.F. (1989). Range Development and Improvement. Academic Press, San Diego, California.

Effect of fertilisation with macro and micronutrients on the mineral content of *Leucaena leucocephala*

S. Camacaro
Facultad de Agronomía. Universidad Central de Venezuela. Apartado 4579. Maracay Z.P. 2101. Venezuela, Email: camacaros@agr.ucv.ve

Keywords: *Leucaena leucocephala*, fertilisation, mineral content

Introduction   The Venezuelan cattle industry has been developed in areas with soil limitations, causing a negative impact on animal production, since the pastures are of medium to low quality and quick maturation (Chacón, 1996). On the other hand, forage legumes have been introduced to improve the nutritive value of the biomass on offer. Growing interest has been shown in *Leucaena leucocephala* in recent years due to its multiple benefits (Duguma *et al.*. 1988). However the use of varieties not adapted to acid soils can be a problem, largely due their low tolerance to the toxicity of the aluminium and to phosphorus deficiency (Shelton & Jones. 1994). The objective of this experiment was to evaluate the effect of the fertilisation on the mineral content of *L. leucocephala*.

Materials and methods   The legume was sown in single rows in plots of 1x2 m a pasture of *Cynodon sp.*. There were sixteen treatments in a randomised block design, including two controls (negative and positive), consisting of combinations of Mg, K, Cu and Zn with a missing element, applied at doses of 20, 70, 3 and 3 kg/ha, respectively. Samples of potentially edible material (leaves and stems Ø < 6 mm) were taken every three months, to determine N, K, Mg, Cu and Zn concentrations.

Results   There was no treatment effect (P>0. 05), however a seasonal effect was detected (P < 0. 01) (Table 1). The variations in the reported values could be due to soil humidity (García *et al.*, 1998), cutting frequency and parts of the plant that were sampled (Razz *et al.*, 1995). A tendency to increase mineral concentration of the plants with the application of the elements applied was observed, although the differences did not reach significance.

Table 1 Effect of fertilisation on the minerals content of *Leucaena leucocephala*

Period	N	Mg	K	Zn	Cu
		%		ppm	
Dry	3.70 [b]	0.51 [a]	1.33 [b]	26.90 [a]	5.25 [b]
Rainy	4.60 [a]	0.29 [b]	1.54 [a]	31.35 [a]	18.30 [a]

Conclusions   It was observed that in most of the cases where the higher values were obtained, the treatment included the element analysed. This seems to indicate that the effect could be more significant if combinations of different levels of those elements had been evaluated. There is some experience in this respect in legume fertilisation trials, using appropriate statistical designs that permit the exploration of multiple combinations of elements at different levels (Chacón, 1989).

References
Chacón, E. (1996). Manejo de recursos alimenticios para la ganadería de doble propósito y lechería tropical, con énfasis en pastoreo. I Seminario Internacional de Montería de Ganado de Doble Propósito Gyr lechero y Búfalo. Memorias. Montería, Colombia, 1-34.
Chacón, E. (1989). Efecto de la fertilización con azufre y micronutrimentos sobre las características cuantitativas y cualitativas y aceptabilidad de la asociación *Brachiaria mutica* x *Teramnus uncinatus* por bovinos a pastoreo. Trabajo de Grado de Maestría en Producción Animal. Facultades de Agronomía y Ciencias Veterinarias. Universidad Central de Venezuela. Maracay, Venezuela. 206pp.
Duguma, B.; Kang. B. & D. Okali. (1988). Effect of liming and phosphorus application on performance of *Leucaena leucocephala* in acid soil. *Plant and Soil,* 110, 57-61.
García, L.; X. Rincón.; G. Pirela.; T. Clavero. & O. Ferrer (1998). Efecto de diferentes láminas de riego sobre el perfil mineral de la *Leucaena leucocephala* (Lam) de Wit. *Revista de la Facultad de Agronomía* (LUZ), 15 (2), 199-207.
Shelton. H. & Jones, R. (1994). Opprotunities and limitations in *Leucaena*. In: Opportunities and limitations (Eds. H. Shelton. C. Piggin y J. Brewbaker). Proceedings. Bogor, Indonesia, 16-22.

Effects of applied quantity of phosphorus fertiliser on phosphorus content in plant tissues of lucerne (*Medicago sativa*) and seed yield in North-western China

Y.W. Wang[1], J.G. Han[1], S.M. Fu[2] and Y. Zhong[2]
[1]*Department of Grassland Science, China Agricultural University, Beijing, China 100094, Email: wangyunwen120@sohu.com,* [2]*Chengdu Daye International Investment Co. Ltd. Chengdu, China 610016*

Keywords: lucerne, phosphorus fertiliser, tissue phosphorus concentration, seed yield

Introduction Phosphorus concentration in plant tissue can be a useful index of P deficiency in lucerne, P fertiliser recommendations and monitoring of effectiveness of current P fertiliser practices (Jacobsen & Surber, 1995). The objective of this study was to measure P concentration in different lucerne plant parts and seed yield in relation to P fertiliser application rates in order to improve recommendations for lucerne seed production.

Materials and methods The experiment was located in north-western China ($39°37'$N, $98°30'$E, altitude 1480 m). The soil was an irrigated desert earth (Chinese soil classification), classified as silty clay soil with pH 8.5. Total soil P and available P content was 0.607 g/kg and 12.47 mg/kg, respectively. The lucerne was established with 60 cm row space and a seeding rate of 4.0 kg/ha. The experimental design was a randomised block with 3 replications of 7 P fertiliser application rates laid out in spring of 2001. Ten to fifteen plants of 15-20 cm high were taken per replicate in the regrowth period in spring of 2002; roots were 30-40 cm long. A sample of each plant part was taken by combining the plant parts of three replicates. P in each plant tissues and seeds were determined spectrophotometrically.

Table 1 The total P content of plant parts response to applied P fertiliser rates*

P_2O_5 kg/ha	Total P content (%DM)			
	Roots	Stems	Leaves	Seeds
0	0.09	0.23	0.34	0.45
60	0.20	0.27	0.42	0.56
120	0.19	0.30	0.40	0.63
180	0.21	0.31	0.41	0.64
240	0.23	0.28	0.40	0.62
300	0.20	0.29	0.40	0.65
360	0.23	0.30	0.42	0.67

Table 2 The equations of the relationship of total P content of plant parts to applied P fertiliser rates

Plant parts	Equation	Pr>F	R^2	S.E.
Roots	$Y=0.1138+0.000869X-1.65\times10^{-6}X^2$	0.05	0.77	0.028
Stems	$Y=0.2386+0.000530X-1.09\times10^{-6}X^2$	0.07	0.73	0.017
Leaves	$Y=0.3629+0.000405X-7.94\times10^{-7}X^2$	0.28	0.47	0.024
Seeds	$Y=0.4717+0.00134X-2.35\times10^{-6}X^2$	0.01	0.89	0.031

* Effects of P fertiliser application on all alfalfa plant parts were significant ($p<0.001$)

Results Addition of 60 to 360 kg P_2O_5/ha resulted in a relatively greater increase in P concentration for roots than other plant parts (Table 1). The responses of lucerne seed yield and seed yield components to P application treatments in two seasons were not significant ($P>0.05$), addition of 120 kg P_2O_5/ha treatment showed highest seed yield of 794.1 kg/ha and 757.1 kg/ha in both years (data not listed). The P contents in seeds, roots and stems were highly correlated to P application rate (r^2 = 0.89, 0.77, and 0.73, respectively) and to a lesser extent to P content in leaves (r^2 = 0.47). Quadratic regression equations for roots and seeds were significant ($P\leq0.05$).

Conclusions The concentration of P in roots can be a good indicator of soil P fertility. The critical concentration of P in roots to determine P fertiliser recommendations for lucerne plant growth or seed production deserves further evaluation.

References

Jacobsen, J. S. & G. W. Surber (1995) Alfalfa/grass response to nitrogen and phosphorus application. *Communications in Soil Science and Plant Analysis,* 26, 1273-1282.

Influence of P fertility and grazing on plant species in a temperate Australian pasture

J.O. Hill[1], R.J. Simpson[1], A.D. Moore[1], J.T. Wood[2] and D.F. Chapman[3]
[1]*CSIRO Plant Industry, PO Box 1600, Canberra, ACT, Australia 2601, Email: Jacqueline.Hill@csiro.au,*
[2]*Statistical Consulting Unit, Graduate School, Australian National University, Acton, ACT, Australia 2601,*
[3]Institute of Land and Food Resources, University of Melbourne, Victoria, Australia 3010

Keywords: grazing, basal cover, plant competition, *Phalaris aquatica*, phosphorus

Introduction Graziers in temperate Australia are increasing their use of P fertiliser so they can run more stock and maintain profitability. However, intensification changes grassland botanical composition and perennial grass cover can be reduced. Perennial grasses are important because they improve production stability, reduce deep drainage and slow the rate of soil acidification. This study examined how P fertility and grazing affected the botanical composition of pasture based on *Phalaris aquatica*, a key perennial grass in south-eastern Australia.

Materials and methods *Field experiment* Continuously-grazed *P. aquatica* and *Trifolium subterraneum*-based pastures growing on a yellow chromosol soil at Hall (ACT, Australia), were left unfertilised (extractable P (Colwell 1963) 0-10 cm depth 8-12 mg P/kg soil), or were fertilised optimally (Colwell P 20-25 mg P/kg) for six years. Basal cover of plants was determined across three replicates by the point quadrat method in Sept. 1999, 2000 and 2001. *Glasshouse experiment* Pasture species present at the field site (Table 1) were grown in steam pasteurised topsoil (pH(CaCl$_2$) 4.8; Colwell P 10 mg P/kg) from an unfertilised paddock. Pots contained 1.5 kg of soil and were supplied a basal dressing of all nutrients except P, which was supplied as KH$_2$PO$_4$ at rates from 0 to 84 mg P/pot. Each species was harvested when the shoots of its highest P treatment reached ~6 g dry weight.

Results The most abundant species in the unfertilised system (Figure 1) were either effective at extracting P from unfertilised soil (e.g. *P. aquatica*, *Vulpia* spp.), had a low critical P requirement (e.g. *Vulpia* spp.) (Table 1), or were able to fix N (*T. subterraneum*). *P. aquatica* and *Bromus* spp. were more abundant in fertilised pasture at the low stocking rate (which gave underutilised pasture). Abundance of species in the intensive grazing system was inversely related to maximum relative growth rate (RGR, determined at high soil fertility).

Table 1 Relative P uptake rate (RPUR) and critical P (amount of P required for 90% maximum shoot growth) of *P. aquatica* (Pa), *Holcus lanatus* (Hlan), *Bromus molliformis* (Bm), *Vulpia* spp. (Vspp), *Arctotheca calendula* (Ac), *Lolium rigidum* (Lr), *T. subterraneum* (Ts) and *Hordeum leporinum* (Hlep)

Pasture species present at the site	Pa	Hlan	Bm	Vspp	Ac	Lr	Ts	Hlep
RPUR unfertilised soil (mg/g per d)	91.9	66.1	54.9	79.9	n/a	81.1	26.3	36.1
Critical P (mg P/pot)	22.0	8.0	25.1	13.0	14.7	13.9	34.0	28.5

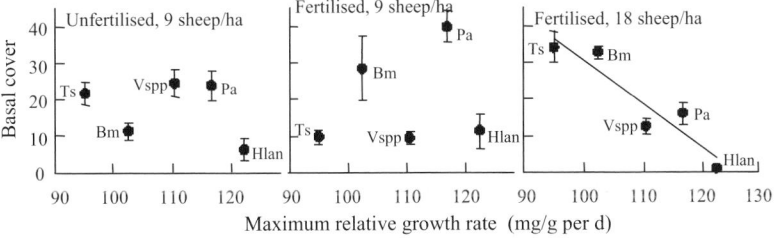

Figure 1 Maximum RGR (glasshouse) and mean basal cover (1999-2001) of main species (>5% cover) in the field. Bars = 2x sem.

Conclusions The C-S-R model (Grime 2001) adequately explains species abundance in the unfertilised pasture (dominated by stress-tolerant species) and the fertilised low stocking rate pasture (tall competitor species). It was expected that plants with high RGR would also dominate the fertilised high stocking rate pasture (Lambers & Poorter, 1992), but abundance and RGR were negatively correlated. We propose that this is because grazing pressure was a major factor controlling the success of plants in the intensive system and that slow growth either assisted a species to avoid being grazed or was associated with lower grazing preference. These propositions require further investigation in order to improve our ability to predict changes in pasture composition.

References
Colwell, J.D. (1963). The estimation of the phosphorus fertiliser requirements of wheat in southern New South Wales by soil analysis. *Australian Journal of Experimental Agriculture*, 3, 190-198.
Grime, J.P. (2001) Plant Strategies, Vegetation Processes and Ecosystem Properties. Wiley, New York.
Lambers, H. & H. Poorter (1992). Inherent variation in growth rate between higher plants: A search for physiological causes and ecological consequences. *Advances in Ecological Research*, 23, 187-261.

Forage quality and yield of berseem clover and annual ryegrass grown in pure and mixed stands in relation to different N application rates

C. Scarpello, L. Stringi, G. Amato, D. Giambalvo, P. Trapani and C. Attardo
Dipartimento di Agronomia Ambientale e Territoriale, Università di Palermo, Viale delle Scienze - 90128 Palermo, Italy, Email: giardo@unipa.it

Keywords: *Trifolium alexandrinum*, annual ryegrass, nitrogen fertilisation, nitrogen concentration

Introduction Grass-legume mixtures offer several advantages over monocultures in forage-animal production systems (Haynes, 1980). In fact, in grass-legume mixtures forage yield and quality are generally higher compared to grass monocultures also due to more efficient soil N-utilisation (Ta & Faris, 1987); furthermore grasses often utilise some of the N fixed by legumes (Malhi *et al.*, 2002). Legumes do not generally require the addition of N fertiliser due to symbiotically fixed N, but the yield of the grass component in a mixture may be further improved with N application. The objective of this study was to investigate forage yield and quality of berseem and annual ryegrass grown in pure stands, and in mixture, at different N fertiliser rates.

Material and methods The research was carried out during two successive seasons (2001/02 and 2002/03) in a semi-arid, hilly area of Sicily (37°30'N; 13°31'E; 178 m a.s.l.) in a deep, clayey, well structured soil and with wheat as the previous crop. The experiment was set up as a split-plot design with six replications. Main plots were N application rates (0, 50 and 100 kg/ha). Sub-plot treatments consisted of berseem (*Trifolium alexandrinum* cv Lilibeo) in pure stand (B); annual ryegrass (*Lolium multiflorum* subsp. *wersterwoldicum* cv Elunaria) in pure stand (R), and their mixture (BR). Pure stands were seeded at 50 kg/ha. The intercrop design was based on the replacement principle, with B and R sown in alternate rows in a 0.5:0.5 ratio. All plots were first cut 100 days after sowing; a defoliation frequency of 28 days was used (3 cuts in 2001/02 and 4 cuts in 2002/03). The total amount of N fertiliser for each treatment was applied in split doses: at crop emergence and soon after the 1st and the 2nd utilisation. All plots were harvested at 5 cm stubble height and dry matter yield (DMY) and N concentration, for each species, were determined.

Results and conclusions Berseem in pure stand gave higher mean DMY over the 2 years than the mixture and than annual ryegrass in pure stand (Table 1). Berseem in pure stand did not show any significant DMY advantage from the application of N; on the contrary, a significant yield increase was recorded in annual ryegrass in pure and mixed stands, due to N. Application of N reduced the clover content in the mixture from 63 to 50%. Annual ryegrass grown in mixture with berseem consistently had a higher N concentration in the DM than annual ryegrass in pure stand (Table 1). Application of N significantly increased the N concentration of annual ryegrass in pure and mixed stand, particularly at the highest N rate. The N concentration in berseem varied

Table 1 Dry matter yield and N concentration of annual ryegrass (R) and berseem (B) in pure and in mixed stand (BR) and of their components in mixture (R_{BR}; B_{BR}) (Mean values over two years)

N treatment	Dry matter yield (kg/ha)					N concentration (% on DM)			
	R	B	BR	R_{BR}	B_{BR}	R	R_{BR}	B	B_{BR}
0 N	3401	6525	5719	2111	3608	2.14	2.51	3.70	3.51
50 N	4189	6544	6129	2745	3384	2.12	2.64	3.69	3.51
100 N	5344	6648	6376	3176	3200	2.39	2.99	3.70	3.59
LSD (P<0.05)		205			459		0.084		ns
LSD (P<0.01)		272			618		0.113		

between 3.51 and 3.70% DM, and was similar for berseem in pure stand and in mixture, and for N treatments. The results indicate that in a Mediterranean environment the annual ryegrass-berseem clover intercrop might be a suitable alternative to conventional pure stands for improving forage quality and also better exploiting environmental resources, bearing in mind the different competitiveness between species with regard to N availability.

References

Haynes, R.J. (1980). Competitive aspects of the grass-legume association. *Advances in Agronomy* 33, 227-261.
Malhi, S.S., R.P Zentner .& K. Heier (2002). Effectiveness of alfalfa in reducing fertilizer N input for optimum forage yield, protein concentration, returns and energy performance of bromegrass-alfalfa mixtures. *Nutrient Cycling in Agroecosystems*, 62, 219-227.
Ta, T.C.& M.A. Faris (1987). Effects of alfalfa proportions and clipping frequencies on timothy-alfalfa mixture. II. Nitrogen fixation and transfer. *Agronomy Journal*, 79, 820-824.

Effect of different phosphorous sources and levels on the productive behaviour of a *Lotus pedunculatus* cv. Grasslands Maku oversown pasture

R.E. Bermúdez and W. Ayala
*INIA National Institute of Agricultural Research, Uruguay, R8 Km 281, Treinta y Tres, Uruguay, PC 33000,
Email: rbermudez@tyt.inia.org.uy*

Keywords: Lotus, phosphorous sources, phosphorous response, oversown pastures, rhizomes

Introduction The organic meat production protocol of Uruguay (INAC, 2003) requires that the animals graze pastures that receive no chemical fertilisers. The oversown legume pastures in Uruguay used to be fertilised with soluble phosphorous (P) sources that are not accepted by the protocol. The relative efficiency of different P sources would be useful data for farmers. This information is not available for the acid soils of the eastern region of Uruguay. *Lotus pedunculatus* cv. Grasslands Maku is one of the most adapted legumes to be included in this type of pasture, as the sown area has increased in the last few years. The objective of this experiment was to evaluate the P levels response and the relative efficiency of different P fertilisers for forage production.

Material and methods The experiment was established on a farm (32° 20′S) with the following chemical soil characteristics pH5.3, OM 4.0%, P Ac. Citric4.5 • g/g, K 0.29 meq/100g and Al 0.41 meq/100g. The treatments were four levels (0, 13, 26 and 39 kg of P/ha) of natural ground phosphate rock (NPR) with a relation P soluble/P total (Pt/Ps) of 4.4/12.2, one level (26 kg of P/ha) of granulated partially acidulate reactive phosphate rock (H) (Ps/Pt) of 6.1/11.8 and one level (26 kg of P/ha) of granulated superphosphate (S) (Ps/Pt) of 9.2/10.0.
The design was a randomised block with two replicates and three blocks; plot size was 2*5 m. Three cuts were taken on 24/9/02, 29/10/02 and 10/4/03 at 4 cm height. The cutting area per plot was 6.27 m^2, a sub sample of 1 kg was taken to estimate dry mater (DM) content and lotus content by manual separation. The measurements were total dry matter (TDM), lotus DM (LDM) (kg/ha). Rhizome length (m/m^2), diameter (mm) and dry weight mass (g/m^2) were estimated in autumn 2003, taking four cores of8 cm of diameter and 5 cm depth per treatment.

Results The TDM production was not affected (P>0.05) by P levels but the LDM yield response was 65.4 kg/ha/kg P (P<0.01) applied as NRF, which added between 0 and 39 kg of P/ha (Table 1). There were no differences (P>0.05) between P sources in TDM production but NRF increased LDM production 24 % (P<0.05), with the H source giving an intermediate response. Rhizome lengths were not affected (P>0.05) either by P level or by P sources.

Table 1 Effect on total DM (TDM), lotus DM (LDM) production and rhizome characteristics

	P (kg /ha)	TDM (kg/ha)	LDM (kg/ha)	Rhizome Length (m/m^2)	Diameter (mm)	Dry weight (g/ m^2)
NRF	0	7745	1647	56	2.2	4,0
	13	8308	2498	85	2.6	6,1
	26	7443	3348	68	1.8	3,9
	39	7288	4198	109	2.1	5,4
	SEM	463 (6)	1222 (6)	22.4 (4)	0.94 (4)	1.6 (4)
NRF	26	7443	3348	68	1.8	3,9
H	26	4199	3682	67	2.0	3,0
S	26	4719	2980	129	2.1	7,1
	SEM	220 (6)	350 (6)	21.9 (4)	0.73 (4)	1,3 (4)

Conclusions TDM and rhizomes were not significantly affected (P>0.05) by P levels of NRF and by the different P sources evaluated. There were a significant LDM P response (P<0.01) of 65.4 kg/ha per P applied as NRF. NRF was significantly more efficient (P<0.05) than S producing LDM. NRF could be efficiently used to produce organic meat on this type of soil.

Reference
INAC, 2003 Protocolo de producción de carne natural. http://www.naturalmeaturuguay.com/sis-prod.shtml

Effects of sowing date and phosphorus fertiliser application on winter survival of lucerne cv. Aohan in the northern semi-arid region of China

Z.L. Wang[1], Q.Zh. Sun[1], Y.W. Wang[2], Zh.Y. Li[1] and Sh.F. Zhao[3]

[1]Grassland Research Institute, Chinese Academy of Agricultural Sciences, Huhhot, 010010, Email: wangzongli@sina.com, [2]Department of Grassland Science, College of Animal Science and Technology, China Agricultural University, Beijing, 100094, [3]Grassland Service Station of Linxi County, Linxi County, 024550

Keywords: *Medicago sativa*, fertilisation, semi-arid region, winter survival

Introduction In the northern semi-arid region of China, winter survival is always a limiting factor for lucerne production, because low temperatures and a dry climate in winter (Zhou *et al.*, 1993; Ma, 2000; Sun & Gui, 2001; Sun *et al.*, 2003). An experiment was conducted to find an appropriate sowing date and P application rate in order to improve lucerne winter survival.

Materials and methods The study was conducted in Linxi county, Inner-Mongolia. The experimental design waas a randomised block with 3 replications, which included sowing time, and fertiliser application rates. Each plot measured 5.0 m ×2.0 m. Sowing of lucerne cv. Aohan occurred on 28 May,1997 and at eight dates in 1998. Two levels of P fertiliser were applied with 15 kg/ha N on 30 May, in 1997 and on 25 June, in 1998, respectively.

Results Sowing date influenced winter survival significantly (Table 1). Shoot number per plant, buds per plant before winter and winter survival decreased at dates later than early June. P and N fertilisers applied at sowing increased winter survival in both years and winter survival increased with P application rate (Table 2). Compared to the control treatment, 75.0 kg P_2O_5/ha + 15 N kg/ha increased lucerne winter survival by 15.1% and 45.9% in 1997 and 1998, respectively.

Table 1 Responses of seeding dates in 1998 on winter survival of alfalfa

Sowing date	Shoot numbers per plant	Buds per plant	Winter survival rate (%) *
18 May	3.5-4.0	5.0-5.7	98.6
30 May	3.5-4.0	5.0-6.3	95.5
5 June	3.0-3.5	4.5-6.0	91.8
12 June	1.0-2.0	2.0-2.5	63.7
25 June	1.0-1.5	1.5-2.2	47.2
4 July	1.0	1.0-2.3	38.2
16 July	1.0	0.0-1.5	18.4
27 July	1.0	0.0-1.0	6.5

Table 2 Effects of fertiliser on lucerne winter survival

Sowing date	Fertiliser rates (P_2O_5+N) kg/ha)	Plant crown diameter (m)	Winter survival rate (%)
30 May, 1997	0	0.47	81.5
	37.5□15.0	0.53	89.6
	75.0□15.0	0.72	96.6
25 June, 1998	0	0.26	47.2
	37.5□15.0	0.51	78.3
	75.0□15.0	0.67	93.1

* Effects of seeding date on winter survival rate were significant (p<0.01)

Conclusion In the northern semi-arid region of China, seeding date of lucerne cv. Aohan should not be later than early June. The application of 75.0kg P_2O_5/ha + 15 kg N /ha at the time of sowing is recommended in order to increase winter survival.

References

Ma, Zhiguang (2000). The technology of alfalfa industrialization in northern semiarid region of China. ICET2000-Session 6: Technology innovation and sustainable agriculture, 531-534.

Sun, Qizhong □X. Y. Hao &Y. Q. Wang (2003). The study of alfalfa on winter survival. The secondary convention of alfalfa development in China, 34, 37.

Sun, Qizhong & R. Gui (2001). The freezing injury and preventing methods of Aohan alfalfa in the Chifeng Region. *Acta Agri-Culturae Boreali-Sinica*, 16,136-142.

Zhou, Xingmin, Y. L. Feng & D. J. Cai. (1993). Study on the introduction of cold-resistant alfalfa cultivars in the northern cold region of china. *Animal Husbandry and Veterinarian*, 11, 4-7.

Organic and mineral fertilisation of temporary grassland

M. Maiorana, F. Montemurro and F. Fornaro
Istituto Sperimentale Agronomico, Bari, 70125, Italy, Email: michele.maiorana@tin.it

Keywords: cocksfoot, lucerne, municipal solid waste compost

Introduction In Italy, the need to reduce the application of chemical fertilisers and to dispose of different kinds of bio-wastes (municipal solid waste (MSW), olive mill waste, distiller's grains) has led to more compost being made from these materials. Since 1995 our Institute has carried out several studies on different crops. The results obtained so far for grain and industrial crops (Montemurro *et al.*, 2003; Maiorana *et al.*, 2004) appear very interesting. But rarely have the composts been applied to fodder crops. Therefore, this research is aimed to evaluate the effect of MSW-compost on temporary grassland of cocksfoot (*Dactylis glomerata*) and of lucerne, (*Medicago sativa*) in Southern Italy (Apulia Region).

Materials and methods The experiment was established in the autumn of 2001 at Rutigliano (41° 01' N, 4° 39' E, 90 m. a. s. l.) on the experimental farm of the Institute. Two types of fertilisation were compared on plots of 7 m² each: organic (with MSW-compost) and mineral (with 50 kg N per ha for cocksfoot and 75 kg P_2O_5 per ha for lucerne). The experimental design was a split-plot with 3 replications, assigning the split to crops and the plot to fertilisation treatments. In the field, MSW compost was applied yearly in the same amount of element as chemical fertilisers. Crops were grown under dry conditions, except for a post-sowing irrigation of 300 m³ of water per ha. At the time of harvest, green forage production was determined for each centre plot. Samples of 1 kg of forage were oven-dried at 105 °C, to measure dry matter content and crude protein concentration was determined on samples of 100 g after oven-drying at 80 °C for 24-36 hours. The experimental data were submitted to a separate analysis of variance for each crop.

Results The results obtained during the first two years (Table 1) show that cocksfoot with mineral nitrogen fertilisation produced significantly higher yields with a higher protein concentration than cocksfoot with compost. This is probably because the mineral nitrogen present in the compost becomes available over a longer time and not only during the growing period. In contrast, there were no significant differences on production or crude protein concentration of lucerne.

Table 1 Effects of treatments on production and protein concentration

Treatments	Green forage (t ha⁻¹)	Dry matter (t ha⁻¹)	Plant height (cm)	Protein content (%)
COCKSFOOT				
MSW-compost	15. 6[b]	4.3[b]	35.5	20.8[b]
N mineral	20.6[a]	5.9[a]	39.6	25.3[a]
LUCERNE				
MSW-compost	43.6	10.8	43.75	23.0
P_2O_5 mineral	47.7	11.8	45.22	23.4

Values with different letters in columns are significantly different at P≤0.05

Conclusions On the whole, the results obtained till now have shown that MSW compost was less effective than N fertiliser on cocksfoot. However compost had the same effect as P_2O_5 on lucerne. The research will continue to assess the possible long-term effects of MSW compost, not only on cocksfoot performance, but also on the possible accumulation of heavy metals in the soil.

References

Maiorana, M., F. Montemurro, G. Convertini & D. Ferri (2004). Concimazione minerale ed organica del girasole: effetti su produzione e qualità. *Agroindustria*, 3 (1), 35-38.

Montemurro, F., D. Ferri, G. Convertini & M. Maiorana (2003). MSW-compost application in Mediterranean soils: evaluation of agronomic effects and environmental impact. *Proceedings of the 4ᵗʰ International Conference of ORBIT Association on Biological Processing of Organics: Advances for a Sustainable Society*, Perth (Australia), 295-303.

Effect of animal manure on forage yield and quality of pangolagrass and soil fertility

F.H. Hsu, K.Y. Hong and C.H. Hsieh
Livestock Research Institute, Council of Agriculture, Hsinhua, Tainan, Taiwan, Email: fhhsu@mail.tlri.gov.tw

Keywords: *Digitaria decumbens,* manure, forage yield, quality

Introduction Animal wastes may cause environmental pollution. Lu & Hsu (2004) reported that N utilisation in the manure by pangolagrass was 10-28%. Objectives of this study were to determine the effect of animal manure on forage yield and quality of pangolagrass and soil fertility.

Materials and methods Pangolagrass pasture received 9 treatments (annual application per ha): CK: no fertiliser, CF_1: chemical fertiliser N : P_2O_5 : K_2O=200 : 72 : 75 kg, CF_2: twice of CF_1, CA_1: cattle manure with N 600 kg, CA_2: twice of CA_1, HO_1: hog manure with N 600 kg, HO_2: twice of HO_1, CH_1: chicken manure with N 600 kg, CH_2: twice of CH_1. Dry matter yield (DMY) was measured. The contents of crude protein (CP), Cu and Zn in plant and pH, electric conductivity (EC), organic matter content (OM), Cu and Zn in the soil were determined. The experiment was carried out from April 9, 1997 till October 16, 2001.

Results Pangolagrass applied with CH_1, CH_2 and CA_2 produced the highest DMY (Table 1). The highest CP contents were observed for HO_2, CA_2, CH_2 and CF_2. The treatments applied with more manure had higher CP contents. The Cu contents of pangolagrass applied with manure were higher than the other treatments. The Zn contents of pangolagrass were highest for HO_1, HO_2 and CA_1. The pH values, EC, the contents of OM, Cu and Zn in soil increased with manure application except Cu contents for CH_1 and CH_2. Both contents of Cu and Zn in the soil were highest in HO_2. Similar results were also observed with pangolagrass (Hsu *et al.,* 1999) and napiergrass (*Pennisetum purpureum*) (Hong *et al.,* 2000).

Table 1 Effect of manure on DMY and chemical compositions in plant and soil of pangolagrass pasture

Treatment	Plant					Soil			
	DMY	CP	Cu	Zn	pH	EC	OM	Cu	Zn
	Mg/ha	%	----------mg/kg------			dS/m	%	--------- mg/kg------	
CK	12.6^e	5.6^f	7.9^e	53.9^{bc}	4.99^d	0.077^d	2.26^e	2.05^{de}	8.7^d
CF$_1$	16.9^{cde}	6.6^e	8.8^{de}	59.7^{abc}	4.65^e	0.113^{bcd}	2.47^{de}	1.32^{efg}	7.6^d
CF$_2$	15.6^{de}	8.2^{abc}	9.4^{cd}	48.0^c	4.55^e	0.103^{cd}	2.54^{de}	0.97^{fg}	7.7^d
CA$_1$	19.4^{bcd}	7.4^{cde}	10.5^{abc}	64.5^{ab}	6.10^b	0.153^{bc}	3.42^{bc}	2.43^{cd}	23.0^c
CA$_2$	24.1^{ab}	8.5^{ab}	10.1^{abc}	54.3^{bc}	6.13^b	0.293^a	5.57^a	3.74^{ab}	47.4^a
HO$_1$	18.0^{cd}	7.6^{bcd}	10.8^{ab}	71.3^a	5.66^c	0.167^b	3.13^{cd}	3.22^{bc}	31.4^b
HO$_2$	21.3^{bc}	8.8^a	9.7^{bcd}	70.6^a	5.70^c	0.260^a	3.75^{bc}	4.48^a	49.2^a
CH$_1$	26.3^a	6.9^{de}	10.3^{abc}	48.0^c	7.49^a	0.257^a	3.42^{bc}	1.56^{def}	33.3^b
CH$_2$	24.1^{ab}	8.2^{abc}	11.3^a	57.4^{bc}	7.58^a	0.293^a	4.05^b	0.47^g	46.0^a

[a, b, c]Means with the same letters in the same column are not significantly different at 5% level by multiple range test

Conclusion The results showed that forage yield and quality of pangolagrass increased with manure application. The manure could prevent soil acidification. The OM content in the soil increased and soil fertility improved after manure application. However, higher contents of Cu were observed in soil with pig manure applied and higher contents of Zn in soil with all types of animal manure.

References
Hong, K. Y., F. H. Hsu & C. H. Lu (2000). Effects of cattle and hog manure on forage yield and quality of napiergrass and soil fertility. *Taiwan Livestock Research,* 33, 84-94.
Hsu, F. H., K. Y. Hong & C. H. Lu (1999). Effects of cattle and hog manure on forage yield and quality of pangolagrass and soil fertility. *Journal of the Agricultural Association of China,* 187, 101-107.
Lu, C. H. & F. H. Hsu (2004). Effects of manure application on the forage yields of napiergrass and pangolagrass and N uptake efficiency. *Taiwan Livestock Research,* 37, 351-358.

Sustainability of permanent grassland on a low moor soil with different N and K nutrient management

K. Orlovius[1] and J. Pickert[2]
[1]Agricultural Service of K+S KALI GmbH, Bertha-von-Suttner Straße 7, D-34131 Kassel, Email: kristian.orlovius@kali-gmbh.com, [2]Landesamt für Verbraucherschutz und Landwirtschaft, Gutshof 7, D-14641 Paulinenaue Germany

Keywords: permanent grassland, low moor soil, N and K fertilisation, dry matter yields, K contents in the plants

Introduction After the reunification of Germany fertilisation with P and K was strongly reduced in the eastern states of Germany due to the poor financial situation of the farms. Particularly on sites with a low nutrient delivery capacity, such a nutrient management strategy implies the risk of decreasing soil fertility. On a low moor soil with permanent grassland a 4 - year trial with different N and K fertilisation was set up to study the development of dry matter production and K concentrations in the grass.

Materials and methods The fertiliser trial was conducted in Paulinenaue about 50 km Northwest of Berlin on a low moor soil with permanent grassland with an average water table at 80 cm during the growing season. The trial was carried out with four randomised replicates. In each year three cuts were harvested and analysed. The climatic conditions, with an average precipitation of 513 mm and 9 °C mean temperature, can be classified as continental. Before the beginning of the experiment the nutrient status of the soil was analysed showing: pH_{CaCl2}: 6.0, 84 mg/kg P_{DL}, 110 mg/kg K_{DL} and 183 mg/kg Mg_{CaCl2}. The fertiliser treatments comprised increasing amounts of K rates with 0-77-160-240-320 kg/ha K_2O at a constant N rate of 160 kg/ha. With two additional treatments the effect of 0 and 160 kg/ha K_2O without N fertilisation was tested. All K rates were given in one application at the start of growth; the N fertilisation was divided in 80 kg/ha N at the start of growth and 80 kg/ha after the first cut.

Results Omission of K application led to a steady decline in dry matter yields and K content in the plants. Without N application the average yield difference between the control and 160 kg/ha K_2O amounted to 1 t/ha (Table 1). The negative effect of K omission was aggravated through N application. At the N rate of 160 kg/ha the yield losses between the highest K rate and the control plot were 3.5 t/ha in the first year and the gap between 0 and 160 kg/ha K_2O widened tremendously to 8.1 t/ha in the fourth year. From the beginning of the trial the very K deficient situation on both controls (without and with N fertilisation) was confirmed. At the low rates of K and N fertilisation during the duration of the trial the K concentration in the plant DM decreased further to a level below 0.7% K. The increasing K rates led to increased K leaf concentrations in all years. A K application of 240 kg/ha K_2O at 160 kg/ha N increased the K concentration to exceed the desired 2% K. Without N application the same level was already attained with 160 kg/ha K_2O due to the significantly lower yield level.

Table 1 Effect of different N and K fertilisation on the development of DM yields and K concentration in a 4-year experiment

N	K_2O	year					year				
		1	2	3	4	mean	1	2	3	4	mean
kg/ha		DM (t/ha)					K concentration (% K in DM)				
0	0	3.59	4.00	4.63	1.96	3.55	0.74	0.73	0.55	0.65	0.67
0	160	4.32	4.59	4.85	4.22	4.50	2.06	2.26	2.16	2.10	2.15
160	0	5.90	4.32	4.29	1.32	3.98	0.87	0.67	0.58	0.50	0.65
160	77	8.37	8.51	9.83	7.31	8.50	1.31	1.16	0.99	0.96	1.11
160	160	8.74	9.34	10.66	8.26	9.25	1.64	1.76	1.34	1.35	1.52
160	240	8.81	9.67	10.98	9.18	9.66	2.06	2.32	2.16	2.14	2.17
160	320	9.35	10.12	10.71	9.43	9.90	2.45	2.74	2.37	2.05	2.40

Conclusions The results of the present study show very strong effects of N and K fertilisation on DM yield and K concentration in the grass. On a low moor soil an NK reduction led within a few years to a very fast decline in DM yields to less than 2 t/ha, which cannot sustain a sound economic system. An extensive grassland management (without N fertilisation but 160 kg/ha K_2O) led to a low but stable yield level of 4.5 t/ha. An intensive grassland management with 160 kg/ha N and K fertilisation based on the K uptake could sustain the yields at a high level of 9-10 t/ha DM without exceeding the K content in the growth.

Effect of mineral fertiliser levels on the yield quality of perennial ryegrass

S.R. Bumane and A.M. Adamovich
Latvia University of Agriculture, Liela iela 2, Jelgava, Latvia, LV 3001, Email: alexadam@cs.llu.lv

Keywords: *Lolium perenne*, fertilisation, yield quality

Introduction Perennial ryegrass is a short-lived bunch-grass with shallow root system. The plant is nutritious and palatable and stands up to hard grazing. It will not do well under poor conditions, where fertility or rainfall is low. Perennial ryegrass is a major component in different seed mixtures that are used for grassland management and forage production. This plant is the predominant component in nearly all pasture mixtures, with perennial ryegrass and white or red clover forming the basis for permanent pasture for dairy production and cattle. This grass species plays an important role in grassland productivity and forage quality. Ryegrass requires high fertility levels for good production. The significance of ryegrass to agriculture is reflected by the huge investments in research. The aim of work is to determine the balanced rate of mineral fertilisers in ryegrass seed field.

Materials and methods Field experiments (2000 – 2002) were carried out on sod podzolic sandy loam soil (Luvic Phaeozem, WRB 1998), pHKCl-6.5, plant available, P-48 and K-169 mg/kg (Egner-Riehm), soil organic carbon-21 g/kg (Tyurins' method). Perennial ryegrass 12 kg/ha was planted using a Nordsten seed drill in May 1999, 2000 and 2001 after field preparation. The following mineral fertilisers rates were used: N – 0, 30, 60, 90, 120 kg/ha, P applied at 0, 13, 26, 39 and 52 kg/ha, K at 0, 33, 66, 100 and 133 kg/ha. The biomass, DM content and plant material chemical composition were determined by standard method. Mean were separated by LSD and were declared different at the p<0.05 level.

Results Perennial ryegrass tetraploid *cv.* 'Spidola' was developed at the Skriveri Research Center of the Latvia University of Agriculture. Perennial ryegrass on the first cut produced the highest DM yield (5.5 – 5.8 t/ha) using increased mineral fertiliser rates. The ranges of data obtained are presented in Table 1.

Table 1 Dry matter yield and chemical composition of perennial ryegrass 'Spidola' (first cut, mean 2000–2002)

Fertiliser rate, kg/ha			DM yield, t/ha	Content in DM, g/kg				
N	P	K		CF	CP	Ash	Fats	Digestibility
0	0	0	2.1	266.7	77.4	43.8	22.5	596
0	26	66	2.4	273.6	80.3	44.7	22.6	586
30	13	33	3.3	263.2	83.3	45.3	24.4	608
30	39	100	3.8	272.8	83.8	47.4	26.8	594
60	26	66	4.4	264.4	82.1	46.3	26.5	625
60	26	133	4.6	272.8	91.3	50.5	27.0	602
90	39	33	4.7	274.9	95.9	53.1	30.9	607
90	39	100	5.5	272.8	95.9	53.1	28.5	608
120	26	66	5.8	289.2	107.1	56.4	33.4	588
120	52	133	5.4	272.2	118.1	60.9	36.2	594
$LSD_{0.05}$			0.5	17.3	9.4	3.7	4.5	34

At the end of heading stage, perennial ryegrass 'Spidola' gave comparatively high crude fibre content (267 to 289 g/kg DM) at all investigated mineral fertiliser rates that significantly affected the chemical composition and digestibility of forage. Crude protein content in perennial ryegrass at this developmental stage is 80 – 118 g/kg DM and mineral content is 43.8 – 60.9 g/kg DM. Optimal values for ruminants nutrition for K (20 g/kg DM), Ca (3-7 g/kg DM) and P (2 g/ha DM) are realized through forage obtained. Mg concentration exceeding 2.0 g/kg DM, is given as a critical value for hypomagnesaemia in farm animals. In our investigations P, K and Ca changed within the range of these parameters, but Mg content accounted only for 1.4 – 1.7 g/kg DM.

Conclusion Application of balanced quantities of N, P and K fertilizers provides comparatively high DM yield of perennial ryegrass with good herbage quality.

References

Whitehead, D.C. (1972) Chemical composition, In: Speding, C.R.W. and Diekmahns, E.C. (eds), Grasses and legumes in British agriculture, 99 - 132.

Effects of cattle slurry, their solid and liquid fractions and mineral N fertilizers on Italian ryegrass and maize forage yield

H. Trindade, J. Coutinho and N. Moreira
CECEA Universidade de Trás os Montes e Alto Douro, Ap.1013, 5001-911 Vila Real, Portugal, Email: htrindad@utad.pt

Keywords: nitrogen, cattle-slurry, solid-liquid separation, maize, Italian ryegrass

Introduction Solid-liquid slurry separation techniques expand possibilities to improve slurry use efficiency and to reduce its negative environmental impact. These possibilities arise from the different behaviour of the two fractions concerning the release of nutrients, namely nitrogen (N), due to different C:N ratios ($\approx$30 for the solid fraction and $\approx$7 for the liquid fraction).

The aim of this work was to evaluate the effect of cattle-slurry and their solid and liquid fractions applied at sowing time of Italian ryegrass (*Lolium multiflorum* cv. Andrea) and forage maize on crop dry matter (DM) yield. In the NW region of Portugal these two crops are the bases of an intensive double forage cropping system with silage maize grown from May to October and Italian ryegrass during the winter season.

Materials and methods The experiment was carried out in the NW region of Portugal between October 2002 and October 2003. The soil was a deep sandy loam derived from granite and classified as dystric cambisol. The trial was laid out as a 2 factor factorial design with three replications and a strip-plot layout. Five fertiliser treatments (main plots) consisting of the application at sowing of Italian ryegrass + maize, were 48 + 56 t/ha of dairy cattle slurry (S), 40 + 56 t/ha of slurry-liquid fraction (LF), 30 + 30 t/ha of slurry-solid fraction (SF), 70 + 140 kg N/ha of N mineral fertiliser (M) and a non-fertilised control (C). These were factorially combined with the top-dressing of mineral N fertiliser (strip plots) at the rate of 0 and 50 kg N/ha on the Italian ryegrass and 0 and 100 kg N/ha on the maize crop. Ammonium nitrate was used as mineral fertiliser and both crops were harvested in a single cut.

Results Crop DM yields were significantly (p<0.01) affected by both factors under study (Table 1). There was no interaction between the factors. On the Italian ryegrass crop, treatments S and LF with 50 kg N/ha as mineral fertiliser top-dressed gave the highest DM yields. Lowest DM yields were obtained from treatments SF, M and C when no fertiliser N was applied as top-dressing. On Italian ryegrass top-dressing of N increased DM yield, on average, by 2.2 t DM/ha. On the maize crop, the highest DM yield was obtained with the treatment LF. Top-dressing of 100 kg N/ha allowed only a further increase in forage yield of less than 1 t DM/ha of the LF treatment.

Table 1 Effect of fertiliser treatment at sowing and rate of mineral N top-dressed on forage DM yield (t DM/ha) of Italian ryegrass and maize

	Italian ryegrasss			Maize		
	N top-dressed (kg N/ha)		Average	N top-dressed (kg N/ha)		Average
	0	50	yield	0	100	yield
Fertiliser treatments at sowing:						
Control not fertilised (C)	4.1	6.2	5.1 b	12.6	14.2	13.4 b
Cattle-slurry (S)	5.9	9.4	7.6 a	19.8	18.6	19.2 a
Slurry liquid Fraction (LF)	7.0	8.4	7.7 a	21.3	22.5	21.9 a
Slurry solid fraction (SF)	4.1	5.3	4.7 b	16.5	20.5	18.5 ab
Mineral N fertiliser (M)	4.0	6.7	5.4 b	19.5	18.4	18.9 ab

Data within a column followed by the same letters do not differ at P<0.05 level, Tukey test

Conclusions As expected, the two slurry fractions showed different agronomic effects. Slurry-liquid fraction showed a high fertiliser value possibly due to its N content and fast release of N. If slurry-solid fraction promotes the temporary N immobilisation, as was found in a laboratory experiment for 2-3 months (results not shown here), its application at sowing of the Italian ryegrass crop may contribute to the reduction of N losses by leaching and denitrification during the winter period. Further work studying these aspects is in progress.

Combinations of nitrogen and sulphur for signal grass yield

F.A. Monteiro and C.P. Silveira
Soil and Plant Nutrition Departament, ESALQ/USP, Av. Pádua Dias, 11, Zip Code 13418-900, Piracicaba-SP, Email: famontei@esalq.usp.br

Keywords: *Brachiaria decumbens*, macronutrients, plant tops, roots

Introduction Signal grass (*Brachiaria decumbens*) is grown in Brazilian pastures, and the increase in forage yield of such pastures is achieved by fertilization. Nitrogen (N) is the nutrient mostly demanded for increasing grass productivity, and its utilization creates a demand for other nutrients, such as sulphur (S). These two nutrients are well related in plant metabolism, but the S nutrition of signal grass must be better understood. The use of a fractional factorial makes possible the study of several rates of these two nutrients, that combined with the response surface methodology allows anyone to find out the responses to these rates combinations. The objective of this research was to obtain the responses in dry matter yield of plant tops and roots of signal grass grown under N and S combinations.

Materials and methods An experiment with *Brachiaria decumbens* was carried out in greenhouse conditions, with ground quartz as substrate, during Summer-Fall seasons, at Piracicaba, São Paulo State, Brazil. A 5^2 fractional factorial experiment, based on Littell & Mott (1975), was set in randomized blocks, with four replications. Nitrogen rates (14; 126; 210; 336 and 462 mg L^{-1}) were combined with S rates (3.2; 12.8; 32; 64 and 80 mg L^{-1}) and the 13 nutrient solutions were prepared according to Sarruge (1975) nutrient solution, modified for N and S rates. Plants had two growth periods, the first one with 36 days after the seedling transplanting into the pots, and the second 34 days following the first harvest. Plant tops and roots were dried (65°C) and weighed. Data was analyzed by using response surface methodology through the use of the Statistical Analysis System (SAS, 1996).

Results Nitrogen x sulphur interaction was significant (P<0.05) for plant tops dry matter yield in the two harvests (Figures 1 and 2), and also for roots dry matter. Polynomial models fitted well for the studied variables. The use of N at 210 mg L^{-1} (usual rate as in the Sarruge's solution) is far below the rate that results in the maximum signal grass yield. Increasing yield by supplying higher amount of N demanded the supply of higher rate of S, and maximum forage production was obtained with high rates of both N and S.

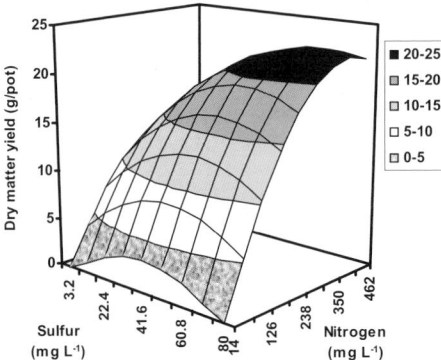 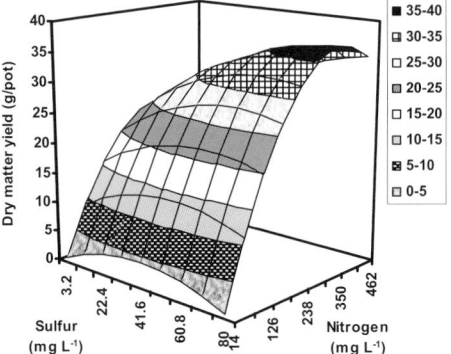

Figure 1 Above ground dry matter yield in the first harvest of *Brachiaria decumbens*, as related to the combinations of nitrogen and sulphur rates

Figure 2 Above ground dry matter yield in the second harvest of *Brachiaria decumbens*, as related to the combinations of nitrogen and sulphur rates

Conclusion Sulphur must be combined with nitrogen in the fertilization of signal grass to attain high forage yield.

References

Littell, R.C. and G.O. Mott (1975). Computer assisted design and analysis of response surface experiments in agronomy. *Soil and Crop Society of Florida Proceedings*, 34, 94-97.

Sarruge, J. R. (1975). Soluções nutritivas. *Summa Phytopathologica*, 1, 231-233.

SAS Institute Corporation (1996). Propriety software release 6.08. Cary, 1996.

Supplementation of cattle with rock phosphate and urea treated straw to improve manure quality and crop yields in the Sahel zone of Senegal

M. Cissé, M. N'Diaye and C.M. N'Dione

Senegalese Institute of Agricultural Research (ISRA), LNERV, BP 2057, Dakar, Sénégal, Email: maicisse@refer.sn

Keywords: supplementation, minerals, manure, crop yield

Introduction Mineral deficiencies are a major constraint in improving animal production and crop yield in the Sahel zone (Cissé *et al.*, 1996). Millet (*Pennisetum glaucum*) and groundnut (*Arachis hypogaea*) are two major food and cash crops in this zone. The purpose of this study was to assess effects of supplementing grazing cattle with rock phosphate and nitrogen enriched diets on animal performances, and the effects of the application of their manure on crop yield in a pearl millet-groundnut rotational system located in N Senegal.

Material and methods The study was conducted with 12 farmers. Sixty Gobra cattle, 52 females and 8 males, were equally allotted to a control (group 1) and to 3 other groups which received, during the dry season, concentrates based on phosphorus or/and nitrogen supply after pasture grazing. Cattle received 75 g/animal per day of Thiès rock phosphate in 30 l of water in group 2, 500 g of 4% urea-treated millet stover and 1 kg of peanut cake and 800 g of millet bran/animal per day in group 3, and combining the diet offered in the groups 3 and 2 for group 4. Cattle body condition was scored monthly (Cissé *et al.*, 2003) and manure produced during nights was recorded daily, collected and sun dried. The farm experiment was a millet (*var.* souna 3) groundnut (*var.* Fleur 11) rotational cropping system with 5 treatments: control (no manure), manure from unsupplemented animals (group 1), and manure from cattle group 2, 3, and 4, respectively. During the rainy season, manure was applied at 4 t/ha to millet. Groundnut was planted in the following year without renewing manure application. Plant growth and yield were measured at 24, 52 days and at harvest.

Results and discussion There were important changes in body condition score (BCS) according to the supplement given to cattle. Controls lost ($p<0.05$) 0.9 points in BCS (3.6 *vs* 2.7) while cattle supplemented with rock phosphate mixed in water maintained their BCS at 3.5 points. Animals from groups 3 and 4 gained ($p<0.01$), respectively, 0.7 (2.8 *vs* 3.5) and 0.9 points (3.1 *vs* 4) of BCS; this being in part due to the high energy content of their diet. After 28 days' growth and at harvest, millet and groundnut plant populations were not significantly influenced by manure application. At 52 days, manured plants were slightly taller than the controls. Enriching manure resulted in positive response in groundnut plant leaf number and height (Table 1). Millet grain yield increased by 24 to 68% depending on the diet offered to animals. The control without manure provided the lowest yield and the highest production was obtained with additional supply of P and N by manure. However, compared with the production from plots manured by control animals, the gain in millet grain yield due to manure enriched in P and N (i.e., 264 kg/ha) was higher than the sum of the gains due to supplementation either in P (73 kg/ha) or in N only (92 kg/ha). The residual effect of manure on groundnut yield resulted in an increase of 11 to 25% over the yield from the unmanured plots.

Table 1 Effect of manure on number of leaves/plant 52 days after planting, on plant height and on grain yield

Treatment	Direct effect on millet		Residual effect on groundnut			
	Grain yield, kg/ha	% increase	Number of leaves	Plant height, cm	Grain yield, kg / ha	% increase
Control	599c	-	51.93a	19.09a	683b	-
Manure from group 1	744b	24	51.80a	20.79c	742ab	9
Manure from group 2	817b	36	52.56a	20.16b	756ab	11
Manure from group 3	836b	39	57.55b	20.27b	842ab	23
Manure from group 4	1008a	68	58.75b	21.24d	857a	25

Means followed by different letters in the same column are different at $p< 0.05$.

Conclusions This trial demonstrated several advantages of supplementation. However, a better response to crop yields could be expected if the animals were stabled in fields, because of the increase in nutrients cycling both from faecal and urinary excretions.

References

Cissé M., H. Guérin & E. Prince (1996). Les carences minérales existent au Sénégal. Comment corriger ce déficit nutritionnel en élevage? Etudes et documents, *ISRA (ed.), vol.7, no. 1, 33 p.*

Cissé M., A. Korréa, I. Ly & D. Richard (2003). Change in body condition of zebu cattle under different level of feeding. Relationship with body lipids and energy. *Journal of Animal Feed Science, 12, 485-495.*

Nitrogen response of spring and winter wheat to biosolids compared to chemical fertiliser

W. Kato[1,2], O.T. Carton[1], D. McGrath[1], H. Tunney[1], W.E. Murphy[1] and P. O'Toole[2]
[1]Teagasc, Johnstown Castle Research Centre, Wexford, Ireland, Email: kato@vmas.kitasato-u.ac.jp, [2]University College Dublin, Dublin 4, Ireland,

Keywords: biosolids, nitrogen response, spring wheat, winter wheat

Introduction Irish sewage sludge production was over 30,000 t/year in the 1990s (EPA, Ireland, 2003). Application to agricultural land is a management option for this organic material as it results in the recycling of the nutrients they contain for crop production. The EU Directive (91/271/EEC) encourages the recycling of sewage sludge as biosolids to agriculture. However, up to 1999, only about 5 % of biosolids produced was applied to agricultural land. In this study, several biosolids and a chemical fertiliser were used to assess N availability for spring and winter wheat (*Triticum aestivum*) production in a pot experiment.

Materials and methods The experiments were carried out in a solarium from May to July 2001 for spring wheat, and from December 2001 to June 2002 for winter wheat. Three types of biosolids [anaerobic biosolid (AB), dried biosolid (DB) and lime biosolid (LB)], cattle slurry (CS) and chemical fertiliser (CF) were used as N sources. The materials were applied at rates of 90 and 180 mg N/pot for spring wheat and 180 and 360 mg N/pot for winter wheat. There was also a control treatment with zero-N in each case. P and K were applied at sowing time to meet crop requirements. In each pot (area 227 cm^2) 2 kg of loam shale soil (87.5 % dry matter, pH 5.9) was placed over 1 kg of sand. The wheat was sown at a rate of 1.0 g/pot and harvested at the vegetative stage. The dry weight and N content were measured.

Results and discussion The N uptake (mg/pot) for all the treatments and the equations for the linear responses to CF in spring and winter wheat are shown in Figure 1. The N uptakes from AB, LB and CS (90 and 180 mg N/pot) in spring wheat were not much different from each other, however, CF was the highest and DB was significantly the lowest. For example, in the 180 mg N/pot treatment were CF 198.3 (a), AB 115.4 (b), LB 102.0 (b,c), CS 88.1 (c) and DB 31.6 (d) mg/pot (means with a letter in common are not significantly different). In contrast, there were some different trends between the treatments in N uptake for winter wheat (180 and 360 mg N/pot); the CF treatment gave a lower slope than in spring wheat, in addition LB and DB gave a higher yield than CF. For example, in the 180 mg N/pot treatment N uptakes were LB 49.3 (a), DB 33.2 (b), CF 24.8 (c), AB 22.2 (c) and CS 17.2 (c) mg/pot. From the above, the efficiency (E) of N uptake in the biosolids relative to CF (Table 1) was calculated by the following formula (Pommel, 1995): E (%) = (A$_1$/A$_2$)×100, where A$_1$ = slope for biosolids and A$_2$ = slope for CF.

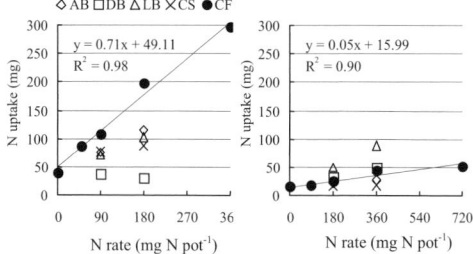

◇ AB □ DB △ LB × CS ● CF

spring: y = 0.71x + 49.11, R^2 = 0.98

winter: y = 0.05x + 15.99, R^2 = 0.90

Figure 1 N uptake (mg per pot) for spring (left) and winter wheat (right)

The relative efficiency of each biosolid in spring wheat was 48, 39 and 30 % for AB, LB and CS, respectively, while DB was negative. In contrast, LB and DB showed higher responses (over 100 %) in winter wheat. This may indicate that calcium or organic bound N in LB and DB was converted to more available forms before or during winter wheat growth.

Table 1 Relative efficiency (%) of each biosolid as an N source compared to chemical fertiliser for N uptake

	AB	DB	LB	CS	CF
Spring wheat	48	<0	39	30	100
Winter wheat	45	116	249	15	100

Conclusions Biosolids can be used to replace part of the N requirements of wheat. The relative efficiency of LB and DB for winter wheat is higher than CF, however, biosolid N for spring wheat is generally lower than in CF. Biosolids cannot be relied upon to supply sufficient N to produce full crops of spring wheat unless applied the previous year. Therefore, supplementary fertiliser N should be used with this crop.

References
EPA, Ireland (2003) Urban Waste Water Discharges in Ireland, A Report for the Years 2000/2001. Environmental Protection Agency, Ireland. 99pp.
Pommel, B. (1995) Value of a heat-treated sludge in the phosphorus fertilization. *European Journal of Agronomy* 4, 395-400.

Nitrogen balance and soil nitrates in suckler cow pastures fertilised with mineral fertiliser, pig slurry or cattle compost

I. Dufrasne[1], J.F. Cabaraux[2], L. Istasse[2] and J.L. Hornick[2]
[1]Experimental Farm, University of Liège, chemin de la ferme, 6, B 39, Email: Isabelle.Dufrasne@ulg.ac.be,
[2]Nutrition Unit, University of Liège, Boulevard de Colonster, 4000 Liège Belgium,

Keywords: fertilisation, suckler cows, nitrogen, slurry, compost

Introduction A code of good practice was established by each European member state according to the EU Nitrate Directive. In Belgium, the nitrogen (N) inputs on pastures from slurry or compost are limited to 210 kg N/ha. Bigger quantities can be applied if the farmer follows a programme of additional measurements, including soil nitrate (NO_3) analysis. This investigation aimed to measure animal performance, N balance and soil NO_3 in pastures fertilised with mineral N, pig slurry or cattle compost, the pastures being grazed by Belgian Blue cows and their calves.

Materials and methods During two consecutive years, two blocks of pasture were divided into three plots each. One plot of each block was fertilised with cattle compost (C), the second one with pig slurry (S) and the third one with mineral N (min N). Grass was analysed for chemical and botanical composition. The animals were weighed at regularly intervals and a blood sample was taken in order to measure plasma urea. Soil cores were taken up to 60 cm depth (0-30 and 30-60 cm) in November and March in order to determine the soil NO_3 content. N inputs consisted of biological fixation, atmospheric N (35 kg/ha) and N from fertilisation. N in live weight gains and in grass silage represented N outputs.

Results N inputs by fertilisation were 166, 161 and 80 kg N/ha in C, S and min N plots, respectively. In terms of N efficiency (30 % in C and 50 % in S), N inputs were 54 kg in C and 76 kg in S. The types of fertilisation did not significantly affect animal performance. Plasma urea did not differ significantly, so N excretions by the cows were probably similar in the three groups (Table 1; Ciszuk & Gebrekziabher, 1994). N surplus was lower in min N. By calculating the inputs on the basis of N efficiency in compost and slurry, the surpluses were slightly lower in C and L than in min N. The apparent N efficiency (N output/N input) was higher in C. NO_3 contents in soil were generally lower than 20 kg/ha except in S and min N at the autumn of the first year (Figure 1). There were no significant differences between methods of fertilisation in NO_3 content

Table 1 Plasma urea concentration and N Balance

	Compost	Slurry	Mineral nitrogen
Plasma urea (mg N/l)	195	200	188
Nitrogen balance (kg N/ha)			
Atmospheric	35	35	35
Legume fixation	78	71	73
Fertiliser input	165	161	80
Total input	279	266	188
Output	37	38	38
Surplus Balance	242	228	150
Surplus (efficient N)	131	144	150
Apparent N efficiency (%)	13	14	21

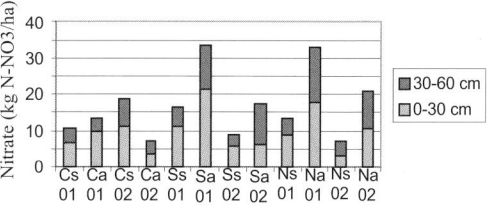

Figure 1 NO_3 content in soil of grazed pastures fertilised with cattle compost, pig slurry or mineral N fertiliser. (C: compost, S: slurry, N: mineral N fertiliser, S: spring; a: autumn, 01: first year; 02: second year)

Conclusion The use of pig slurry and cattle compost as compared with mineral N allowed identical animal performance but increased N balance and reduced apparent N efficiency. However the NO_3 content in soil and N excretion by animals were not increased by the use of slurry or compost. The low NO_3 contents suggested low NO_3 leaching with the three types of fertilisation.

Reference

Ciszuk P. & T. Gebrekziabher (1994). Milk urea as an estimate of urine nitrogen of dairy cows and goats. *Acta Agriculturae Scandinavica*, 44 , 87-95.

Reducing soil phosphorus buildup from animal manure application

G.W. Evers
Texas A&M University Agricultural Research and Extension Center, P. O. Box 200, Overton, Texas 75684 USA,. Email: g-evers@tamu.edu

Keywords: soil phosphorus, animal manure

Introduction Animal manure is an excellent plant nutrient source for pastures but increasing soil P level over time is a major environmental problem (Sims *et al.*, 1998). The increase in residual soil P is due to the difference in N-P ratio in the manure and forage crop requirements (Robinson, 1996). All the N in animal manure is normally utilised but only from 20 to 40% of the P is taken up. With moderate rates of manure application, nitrogen becomes the limiting nutrient for grasses. The objective of this study was to see if combining commercial N fertiliser with broiler litter would enhance forage yield and P uptake sufficiently to reduce residual soil P levels.

Materials and methods The study site was a Coastal bermudagrass (*Cynodon dactylon* (L.) Pers.) hay meadow on a Keithville very fine sandy loam (fine silty, siliceous, thermic Glossaquic Paleudalfs) in northeast Texas. Each October from 1998 to 2002 the Coastal bermudagrass was over-seeded with annual ryegrass (*Lolium multiflorum* L.) at 33.6 kg/ha or crimson clover (*Trifolium incarnatum* L.) at 22.4 kg/ha. Broiler litter was applied at 9 t/ha the first two years and 4.5 t/ha the last two years each Oct. to the annual ryegrass-bermudagrass (RB) system and in April to the crimson clover-bermudagrass system (CB). Fifty-six kg N/ha was applied from one to four times/year in Dec., March, May and/or July to the RB system and one to three times/year in April, June, and/or July to the CB system. Yield and P uptake were determined annually and residual soil P was determined at the end of the four-year study. Soil P was determined by the Vanadomolybdic acid color method (Jackson, 1958).

Results Maximum annual dry matter yield of both systems was about 13.5 t/ha. In the RB system, maximum annual yield occurred with 2 to 3 N applications/year. In the CB system, there was no yield response to commercial N fertiliser which implies that between the broiler litter and the clover providing N through symbiotic N_2-fixation, they were meeting the N requirements of the bermudagrass. In the RB system, maximum annual P uptake averaged 46.4 kg/ha that occurred with one to three N applications, depending on the year. The maximum annual P uptake in the CB system averaged 41.3 kg/ha, with no significant difference between treatments in three of the four years. The slightly higher P uptake by the RB system may be due to more efficient nutrient uptake by the fibrous root system of annual ryegrass than the tap root morphology of crimson clover in the CB system. After four years, residual soil P at the 0-15 and 15-30 cm depths were lower in treatments receiving N fertiliser than the broiler litter only treatment in the RB system. In the CB system, there were no differences in residual soil P between the no N treatment and the N fertiliser treatments at the 0-15 depth, which is in agreement with the yield and P uptake data. Several of the N fertiliser treatments did have lower residual P levels than the no N treatment at the 15-30 cm depth. At the 0-15 cm depth, residual soil P in the no N treatment in the CB system (21.7 kg P/ha) was equal to, or less than, the N fertiliser treatments in the RB system (19.0-34.1 kg P/ha).

Conclusions Combining N fertiliser with broiler litter enhanced P uptake and reduced residual soil P. Substituting crimson clover for annual ryegrass was as effective as applying N fertiliser in the RB system for reducing residual soil P.

References
Jackson, M.L. (1958). Phosphorus determinations for soils. In: M.L Jackson (ed.). Soil Chemical Analysis. Prentice-Hall Inc., Englewood Cliffs, N.J., 134-182.
Robinson, D.L. (1996). Fertilization and nutrient utilization in harvested forage systems - southern forage crops. In: R.E. Joost & C.A. Roberts (eds.). Nutrient Cycling in Forage Systems. Foundation for Agronomic Research, Manhattan, Kansas 66502, 65-92.
Sims, J.T., R.R. Simar, & B.C. Joern. (1998). Phosphorus loss in agricultural drainage: historical perspective and current research. *Journal of Environmental Quality*, 27, 277-293.

Effects of nitrogen fertiliser on nitrate leaching and production of autumn-sown Italian ryegrass in a double-cropping system on a New Zealand dairy farm

E.R. Thom[1], A.A. Judge[2], R.N. Jensen[1], M.S. Sprosen[2], S.F. Ledgard[2] and W.D. Catto[3]

[1]Dexcel, Private Bag 3221, Hamilton, New Zealand, Email: errol.thom@dexcel.co.nz, [2]AgResearch, Ruakura Research Centre, Private Bag 3123, Hamilton, New Zealand, [3]Ballance Agri-Nutrients, Private Bag 12503, Tauranga, New Zealand

Keywords: forage crops, annual ryegrass, *Lolium multiflorum*, nitrogen losses, maize, *Zea mays*

Introduction On intensive dairy farms in New Zealand, winter Italian ryegrass crops are combined with summer maize silage crops in double-cropping systems. Limited data (Davies & Neilson, 1975) showed variable ryegrass yield responses to nitrogen (N) fertiliser when grown after maize. Nitrogen leaching losses were not measured in this experiment but Ledgard *et al.* (1988) showed that late autumn/early winter N applications are vulnerable to leaching. Different rates of N fertiliser were applied to Italian ryegrass grown after maize to assess yield responses and levels of nitrate leaching.

Materials and methods Two experiments were conducted at Dexcel (37°47'S, 175°19'E, 40 m a.s.l.) on a free-draining Horotiu silt loam soil (Umbric Vitrandept). Italian ryegrass was drilled (20 kg/ha) into maize stubble in mid April of Years 1 and 2 in areas double-cropped for three years out of pasture. Treatment plots (5 x 2m) were arranged in a four replicate randomised block design. Plots received totals of 0, 40, 100, 160 and 220 (Year 2 only) kg N/ha as urea (46% N) in equal split applications at 24, 40, 90 and 109 d after drilling. Plot herbage was cut 80, 135 and 170 d from drilling and a sub sample was analysed for dry matter (DM) and N contents. Porous ceramic leachate collectors (3/plot) were inserted 60 cm below the soil surface and leachate was sampled 6-8 times throughout the winter drainage period. This was analysed for nitrate and ammonium-N by high-pressure liquid chromatography. Drainage was estimated using lysimeters containing intact soil cores.

Results Strong linear yield responses to N fertiliser occurred at all harvests in Year 1 up to 160 kg N/ha and in Year 2 up to 100 kg N/ha. Cumulative yields were 2.7, 3.5, 4.2 and 4.8 t DM/ha (SED=0.3) for 0, 40, 100 and 160 kg N/ha in Year 1; comparable Year 2 data were 3.9, 4.5, 5.9 and 5.9 t DM/ha, with plots receiving 220 kg N/ha yielding 6.2 t DM/ha (SED=0.3). Most (>94 %) of the total inorganic N leached was as nitrate-N (Table1).

Table 1 Nitrate-N leached over the total drainage period

kg N/ha	Year 1		Year 2	
	kg/ha	mg/l	kg/ha	mg/l
0	20.6	6.5	14.0	6.6
40	30.9	9.8	22.2	11.3
100	15.7	5.0	34.5	17.5
160	31.7	10.1	35.3	18.0
220	-	-	41.5	21.0
Average SED	6.2	2.1	12.1	6.1

In both years there were high inorganic-N levels in soil (85 and 93 kg N/ha, respectively) before the first N application. Year 1 June drainage was >2 times higher than that in Year 2, with reduced leaching losses for the 0N treatment in Year 2. A significant (p<0.05) increasing trend in nitrate leaching occurred as N fertiliser increased in Year 2, but not in Year 1. Average nitrate-N concentrations in the leachates were above the recommended drinking water threshold (11.3 mg N/l) at the three highest N rates in Year 2 but, possibly because of lower annual (310 vs 198 mm/yr, respectively) drainage, exceeded it only in the 160 kg N/ha treatment during Year 1. In Year 1, herbage nitrate-N concentrations in the 100 and 160 kg N/ha treatments were near the critical animal health threshold (0.21 % of DM) at the first and second harvests. In Year 2, this was the case for all N treatments at harvest 1, the three highest N rates at harvest 2, but only the highest rate at harvest 3.

Conclusions Large improvements in winter yield of Italian ryegrass occurred in response to N fertiliser, although in Year 2 it was near maximum with •100 kg N/ha. Responses ranged from 13-20 kg DM/kg N and were higher than in typical local perennial ryegrass-based pastures. However, the high rates coincided with high nitrate-N concentrations in grass and drainage and highlight the conflict between production and environment.

References

Davies, D.J.G. & B.A. Neilson (1975). An economic appraisal of double cereal cropping. *Proceedings of the Agronomy Society of New Zealand*, 5, 21-26.
Ledgard, S F., K.W. Steele & C. Feyter (1988). Influence of time of application on the fate of ^{15}N-labelled urea applied to pasture. *New Zealand Journal of Agricultural Research* 31, 87-91.

Nitrogen and phosphorus losses in runoff on beef production systems

M.A. Alfaro, F.J. Salazar, N.G. Teuber, S.P. Iraira and L.A. Ramirez
National Institute for Agricultural Research (INIA), Remehue Research Centre, P.O. Box 24-O, Osorno, Chile,
Email: malfaro@remehue.inia.cl

Keywords: water pollution, eutrophication, nutrient losses

Introduction In West Europe countries between 37 and 82% of the nitrogen (N) and between 27 and 38% of the phosphorus (P) reaching water sources come from agriculture and a strong correlation between the number of animals per area unit and N and P contribution to waters has been shown (Issermann, 1990). There are few data about the environmental impact of beef production systems in Southern Chile. The objective of this experiment was to quantify N and P losses in runoff (surface and subsurface) with two different stocking rates in Southern Chile.

Materials and methods The experiment was carried out in winter 2004. Two stocking rates were tested (3.5 and 5 steers/ha) in closed systems (2 ha each). Grazing was carried out with Holstein-Friesian steers (200 kg initial live weight) under rotational grazing, in a permanent pasture on an Andisol (40°35'S, 73°12'W). In each system, three surface lisymeters (5x5m) were established (Scholefield & Stone, 1995), and runoff collected periodically (ground level and at 50 cm depth). Water samples were kept at 4°C until analysis for N ($N-NO_3^-$ and $N-NH_4^+$) and reactive P. Total N and P losses were calculated as the product of drainage and nutrients concentration in the runoff samples.

Results Total drainage during the experimental period was 978 and 1002 mm for the 3.5 and 5 steers/ha treatments, respectively. Most of water moved down the soil profile (>60 cm depth). Subsurface runoff represented only a small proportion of the water flows, confirming that in this soil type surface runoff and leaching are the most important pathways for nutrients transfer and losses. Results showed that during the experimental period no differences were found between the two stocking rates tested for N leaching losses (on average 16 and 17 kg N/ha for the 3.5 and 5 steers/ha treatments, respectively; data not shown). Total losses in runoff varied between treatments.

Table 1 Drainage (% of the total drainage), average nutrient concentrations in surface runoff samples (mg/L) and N and P losses (g/ha) in paddocks grazed by 3.5 and 5 steers/ha, in Southern Chile

Stocking rate	3.5 steers/ha	5 steers/ha
Drainage		
Surface runoff	34%	37%
Subsurf. runoff (0-50 cm)	3%	2%
Leaching (60 cm)	63%	61%
Average surface runoff concentrations (Range)		
NH_4^+	52 ± 19.9	34 ± 11.4
	(1 – 116)	(1 – 82)
NO_3^-	37 ± 10.7	17 ± 5.0
	(1 – 64)	(1 – 33)
Reactive P	1 ± 0.3	2 ± 0.5
	(0 –3)	(0 – 7)
Total losses in runoff (surface plus subsurface)		
Nitrogen	36 ± 0.9	15 ± 0.5
P	3.5 ± 0.29	1.4 ± 0.111

Total N and P losses in runoff were 2.4 and 2.5 times greater in the treatment with the lower stocking rate.

This could be explained by differences in animal behaviour of both treatments. Where the higher stocking rate was used, there was lower grass availability, so that animals spent more time walking in the paddock looking for feed and there was a greater distribution of the faeces and a reduction in the nutrients concentration in a specific area. Because of the small amount of water measured as subsurface runoff, total nutrient losses in this pathway was small (on average they represented only 5% of the total losses in runoff). Ammonium losses from surface runoff were on average 200% greater than those measured for $N-NO_3^-$, probably due to the direct effect of urine patches. The soil bulk density did not vary between treatments (0.49 g cm^{-3}).

Conclusions The stocking rates studied differed in the magnitude of nutrient losses in runoff. In these grazing systems, on average 95% of the N and P lost in runoff was lost in the surface pathway. Most of available N lost in runoff was lost as ammonium. Further studies need to be carried out to fully understand the mechanisms controlling N and P losses in Southern Chile.

Acknowledgments This research was funded by the International Fundation for Science (IFS), grant W/3550-1.

References
Isermann, K. (1990). Share of agriculture in nitrogen and phosphorus emissions into the surface waters of Western Europe against the background of their eutrophication. *Fertilizer Research*, 26, 253-269.
Scholefield, D. & A.C. Stone (1995). Nutrient losses in runoff water following application of different fertilisers to grassland cut for silage. *Agriculture Ecosystems and Environment*, 55, 181-191.

Effect of long-term nutrient management strategies for pastures on phosphorus in surface runoff and soil quality

G.L. Mullins, J.P. Fontenot, G.A. Alloush, G. Johnson, D.G. Boyer, V.G. Allen and G. Scaglia
Virginia Polytechnic Institute and State University, Blacksburg, Virginia 24061 and USDA-Agricultural Research Service Beaver, WV 25813, USA, Email: gmullins@vt.edu

Keywords: water quality, nutrient management, poultry litter, soil quality

Introduction Manure, whether mechanically applied or deposited by grazing animals, has been associated with increased risk of non-point source pollution, especially phosphorus. This is especially true in areas where the industry, especially poultry, has been concentrated in geographical areas that are grain deficient, resulting in a reliance on imported grain for poultry feed. Intensification has resulted in the production of large quantities of poultry manure, within relatively small geographical areas. Surplus litter is typically land applied as a nutrient source or used as an animal feed. The objective of this project was to evaluate the effects of long-term nutrient management strategies using poultry litter as a feed and fertiliser for grazed pasture systems in the Shenandoah Valley of Virginia on soil quality, selected soil chemical characteristics and P losses in surface runoff.

Materials and methods The original experiment was established in 1994 and consisted of stocker cattle grazing endophyte-free tall fescue (*Festuca arundinacea* schrub. KY-31). Treatments included: (1) no fertiliser control (i.e., no feeding of broiler litter or soil application of fertiliser or litter) (C); (2) surface application of inorganic fertiliser (AIF); (3) surface application of broiler litter (ABL); and (4) feeding broiler litter (FBL) to the grazing cattle (all they would eat). Applied broiler litter and inorganic fertiliser supplied the same amount of total N, P, and K as was fed as broiler litter to the steers (FBL) in the previous year. The amount of broiler litter fed to the cattle and land applied averaged 2600 kg/ha per yr (29 kg P/ha per yr). The experimental area was divided into 12 x 1.03-ha paddocks (three paddocks per treatment), with four steers grazing from Dec. through Sept./Oct. Surface applications of broiler litter and inorganic fertilisers were made in late spring (June) each year. Soil samples from the surface 0-10 cm were collected in the autumn of each year for routine chemical analysis

Phosphorus losses in surface runoff were measured in 2001 (7th-year of the trial) using protocols established for the National P Runoff Project (Sharpley *et al.*, 1999). Portable rainfall simulators were used to evaluate the relationships between P concentrations in surface runoff and soil test P (STP). Two runoff plots were located in each paddock on a slope of 5-8% and within each plot, paired 0.75 x 2-m subplots (one 1.5 x 2-m plot split along the long axis) were used. Soil was sampled adjacent to the runoff plots prior to the simulated rainfall events. Runoff was collected for a 30-min period and analysed for dissolved inorganic P (DPi), molybdate reactive P (MRP), total dissolved P (TDP), particulate P (PP) and total P (TP) (Pierzynski, 2000). The paddocks were also sampled in 2001 and the autumn of 2004 for selected soil chemical and soil quality characteristics.

Results Feeding broiler litter (FBL), ABL and AIF resulted in a significant increase in the levels of Mehlich-1 extractable soil P in the surface 0-5 cm layer as compared to the no fertiliser control (C). Concentrations of DPi, MRP and TDP in simulated surface runoff were higher in AIF, ABL, and FBL as compared to the C treatment. These concentrations in runoff were higher in AIF and ABL as compared to FBL, indicating that feeding broiler litter to cattle decreased the potential mobility of P. This may be due in part to the absorption of soluble P from the litter by the animals. Particulate P losses were lower than the MRP fraction, constituting 27, 42, and 27% for AIF, ABL and FBL, respectively, compared to 50% in the C treatment. Regression analysis showed that there was a significant relationship between soluble P concentrations (DPi, MRP and TDP) and selected soil test P methods. Soil analysis data suggests that fertilisation improved soil chemical and physical characteristics.

Conclusions Results of this long-term study demonstrate that the use of broiler litter as a fertiliser or as a feed for cattle increased the levels of extractable soil nutrients and improved soil physical characteristics of grazed pastures. The results also demonstrated that soluble P losses in surface runoff using simulated rainfall increased with increasing soil P levels, but the risk of P losses in surface runoff was no greater with using broiler litter as a nutrient source as compared to inorganic fertiliser.

References

Sharpley, A.N., T. Daniel, B. Wright, P. Kleinman, T. Sobecki, R. Parry & B. Joern. (1999). National research project to identify sources of agricultural phosphorus loss. Better Crops with Plant Feed. No. 4. (http://www.sera17.ext.vt.edu/publications/National_P/National_P_protocol%20.pdf).
Pierzynski, G.M. (2000). Methods of phosphorus analysis for soils, sediments, residuals, and water. Southern Cooperative Series Bulleting No. 396. SERA-IEG 17 Regional Publication. (http://www.sera17.ext.vt.edu/publications/sera17-2/pm_cover.htm).

Nitrate leaching in the meadows of Western Lithuania with different liming and fertilisation levels

R. Butkute and N. Daugeliene
Lithuanian Institute of Agriculture, Vezaiciai Branch, Gargzdu 29, LT-96216, Vezaiciai, Klaipedos District, Lithuania, Email: ruta@vezaiciai.lzi.lt

Keywords: nitrate leaching, lysimeter, permanent meadow, liming, nitrogen fertilising

Introduction Investigations of recent decades in the Baltic region including Lithuania showed that intensive liming and mineral fertilisation are the main factors achieving maximum crop yield. However, this farming method had a negative effect on soil quality and nutrient leaching became more intensive. Grassland fertilisation with nitrogen (N) helps plants to assimilate P, K, Ca, Na, but it may increase leaching of nutrients (Niczyporuk & Jankowska, 1995). Nitrate leaching is usually low from grasslands, primarily due to their long growing season compared to arable crops (Korsaeth *et.al.*, 2003). The objective of this research was to determine the effect of liming and N fertilisation on lysimeter water contamination by nitrate (NO_3).

Materials and methods The investigation was carried out in Western Lithuania, on the eastern part of sea-coast lowland with a moderately warm climate. The prevailing soils are haplic-luvisols (LVh) (FAO-Unesco, 1997), which are low acidity loam ($pH_{KCl} < 5.5$), with low to medium amounts of mobile P_2O_5 (72.5-127.0 mg/kg) and mobile K_2O (74.0-149.5 mg/kg). Four different pH levels (5.0-5.5, 5.6-6.0, 6.1-6.5, 6.6-7.0) were established by liming. Two sward types were established: white clover/grass, fertilised with $N_0P_{60}K_{60}$ and grass only fertilised with $N_{120}P_{60}K_{60}$.

Results The lowest NO_3 concentrations in the lysimeter water were at soil pH level 1 in both fertiliser treatments. At the near neutral pH level 4 NO_3 concentrations were 3 to 3.5 times higher than at the most acid pH level 1. NO_3^- concentrations were 1.1-2.4 times higher in the N_{120} than the N_0 plots (Figure 1).

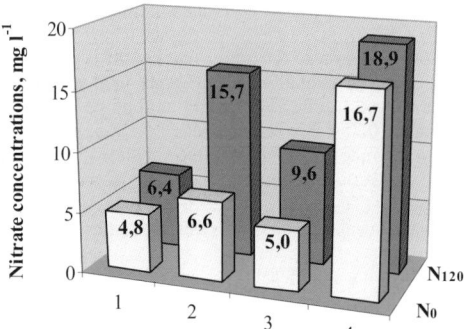

Conclusions Fertilising with N_{120} and liming to alkaline pH levels increased NO_3 concentrations in the leached water. However, NO_3^- concentrations in all treatments met the EU Council Nitrates Directive **91/676/EEC** limit (< 50 mg l^{-1}).

Figure 1 Effect of different N fertilization and liming on nitrate concentrations in lysimeter water, mg l^{-1}.
pH levels: **1** - 5.0-5.5, **2** - 5.6-6.0, **3** - 6.1-6.5, **4** - 6.6-7.0

References
FAO-UNESCO (1997). Soil map of the world revised legend with corrections and updates. Technical Paper 20. - ISRIC Wageningen, 140pp.
Korsaeth A., Bakken L.R. & Riley H. (2003). Nitrogen dynamics of grass affected by N input regimes, soil texture and climate: lysimeter measurements and simulations. In: *Nutrient Cycling in Agroecosystems*, 66 (2), 181-199.
Niczyporuk A. & Jankowska-Huflejt H. (1995). Influence of nitrogen, phosphorus and potassium (NPK) fertilization on the yielding and feeding value of the meadow sward. Ann. Warsaw Agricultural University - SGGW. Agr.- N.29, 45-50.

Mechanical aeration and liquid dairy manure: application impacts on grassland runoff water quality and yield

T.J. Basden, S.B. Shah and J.L. Miller
West Virginia University Extension Service, P.O. Box 6108 Morgantown WV, 26506 USA, Email: tom.basden@mail.wvu.edu

Keywords: aerator, simulated rainfall, nitrogen, phosphorus, total suspended solids

Introduction Wet weather on heavy soils reduces oxygen availability in the root zone and reduces forage yields. Mechanical aeration can improve forage yield in these soil conditions. Research has shown that under certain conditions, mechanical aeration can increase yield by improving drainage and aeration (Davies *et al.*, 1989); aeration can also increase depression, storage and infiltration thus reducing surface runoff and improving nutrient distribution in the root zone. Aeration on sloping, fertilised grassland can provide environmental (Douglas *et al.*, 1995) and agronomic benefits. The objectives of this study were to evaluate the runoff water quality and agronomic impacts of mechanical aeration and liquid dairy manure (LDM) applied to hillside grasslands.

Materials and methods A mechanical Aerator (Model: AerWay®) was evaluated on pasture consisting of cocksfoot with 10-20% lucerne, which had not received LDM for 6 years. The soil is a well-drained silt loam. The experiment was RCB with four treatments and three replicates. Each treatment block (2.5mx 2.5m) received 67 mm of simulated rainfall (SR) to generate runoff. Each block was surrounded on three sides with metal borders and on the fourth (down-slope) side with a runoff collector. Treatments included: 1) control, no aeration and no liquid dairy manure (CTL), 2) aeration only (AER), 3) manure only (MAN), 4) aeration and manure (AER+MAN). Runoff water quality analysis included: nitrate, ammoniacal-N, Total N (TKN), Dissolved Reactive Phosphorus (DRP), total P, Total Suspended Solids (TSS), runoff depths and rainfall leaving plots.

Results Water quality impacts included 1) Runoff Depth; Aeration during spring did not improve infiltration of water. 2) Nutrients; Nutrient concentrations in the simulated runoff events were higher with LDM and were unaffected by aeration. Aeration reduced losses of three or more nutrient species (N and P) in two of six SR events only in manured plots. Concentrations and loadings indicated that aeration of manured plots was more effective in reducing DRP losses than other species (Table 1). Total mean loadings of individual nutrient species in SR were reduced by •26% by AER+MAN vs MAN. 3) Suspended Solids; TSS concentrations were significantly higher with aeration but not with LDM application. Loadings of TSS from AER were >30% higher than the other three treatments. Forage yield impacts 1) In two of three harvests, MAN increased forage yield significantly vs CTL and AER while AER+MAN reduced yield vs MAN in one harvest. 2) Compared with MAN, total forage yields with CTL, AER, and AER+MAN were 78%, 67%, and 81%, respectively.

Table 1 Pollutant loadings in the first simulated runoff event

Treatment	Mean (SD)[a] nutrient loading (g/ha)					Mean (SD)[a] TSS loading (kg/ha)
	NO$_3$-N	Ammoniacal-N	TKN	DRP	Total P	
CTL, control	27 (17)	4c[b] (4)	269 (101)	23(9)	32b (10)	2.9 (1.1)
AER, aeration	19 (3)	1c (2)	254 (92)	45 (55)	27b (4)	1.8 (0.6)
MAN, manure	40 (18)	112a (18)	531 (24)	210 (20)	186a (21)	2.4 (2.0)
AER+MAN Aeration+Manure	40 (17)	72b (14)	644 (347)	151 (160)	181a (102)	4.8 (2.4)
LSD[d]	(NS)	17	(NS)	(NS)	79	(NS)

[a]Mean and standard deviation based on three replicates;[b]Treatment means, followed by the same letter are not significantly different at • ₌0.05;[c]ANOVA; [d]Fisher's least significant difference (p=.05)

Conclusions Aeration partially improved runoff water quality from manured grassland but adversely affected crop yield and nutrient uptake. Further studies should be performed on pastures with livestock traffic. Also, there is need for aerators that minimise surface soil disturbance to reduce TSS losses.

References

Davies, A., W. A. Adams & D. Wilman (1989). Soil compaction in permanent pasture and its amelioration by slitting. *Journal of Agricultural Science,* 113, 189-197.

Douglas, J. T., C. E. Crawford & D. J. Campbell (1995). Traffic systems and soil aerator effects on grasslands for silage production. *Journal of Agricultural Engineering Research,* 60, 261-270.

The effect of agricultural practices on a dairy farm on nitrate leaching to 1m

M. Ryan[1], K. McNamara[2], D. Noonan[1], O.T. Carton[1], C. Brophy[3], J. Connolly[3] and E. Houtsma[3]
[1]Teagasc, Johnstown Castle, Wexford, Ireland, [2]Teagasc, Moorepark, Fermoy, Co.Cork, Ireland, [3]Dept. Statistics and Actuarial Science, University College, Dublin, Ireland, Email: mryan@johnstown.teagasc.ie

Keywords: leaching, nitrate, dairying

Introduction Dairy farms, in Ireland, carry the highest stock densities and use the highest rates of fertiliser nitrogen (N). They constitute the highest risk of nitrate leaching, especially where soils are thin or free-draining. The effect of 4 grass managements on leaching was studied on a dairy farm having free-draining soils overlying Karst limestone. This was a new, farm-comprehensive approach to nitrate leaching which had not been carried out previously.

Materials and methods Leaching was measured over three years (2001-2004) under 4 managements: grazing (dirty water + N fertiliser); 2-cut and 1 cut silage & grazing (slurry (S) + fertiliser N). The mean N inputs were as shown in Table 1. Eight ceramic cups per plot, in 3 replicate plots of each management were used to collect water from1 m deep using 50 k Pa suction over the winter drainage seasons.

Table 1 N inputs (kg/ha) to 4 management systems on dairy farm, 2001-2003

	Fertiliser	S/DW	Total	*Organic
DW	244	54	298	227
2 cut	337	25	362	79
Grazed	266	--	266	230
1 cut	296	22	318	146

*Organic = Recycled N in grazing @ 108 kg/LU/yr

Results Mean, drainage water, nitrate-N concentrations for 2 of the three years are shown in Table 2.

Table 2 Effect of management on mean, drainage water, nitrate-N concentrations (mg/l)

	2001-02			2002-03		
	Mean	Maximum	Drainage (mm)	Mean	Maximum	Drainage (mm)
Dirty water	12.1	24.7		8.1	20.3	
2 cut silage	11.8	22.0		5.3	26.9	
Grazed	4.9	11.1		2.1	5.0	
1 cut silage	1.9	4.8		0.9	3.5	
Overall mean	7.9	15.7	440	4.2	13.9	539

A repeated measures ANOVA of the data showed statistically significant effects ($p<0.05$) of management in both years. Grazed and 1 cut had lower mean concentrations than DW or 2 cut;1 cut had lowest while DW had highest concentrations.

The results relate to the extreme soil conditions that might be experienced on a dairy farm, i.e., very free-draining soil and mean annual total overall N inputs (including recycled N) of 482 kg/ha. Based on 2001-02, the nitrate-N concentrations in the DW and 2 cut treatments are of concern. All mean and maximum values for 2001-02 were higher than those for 2002-03. This shows the importance of collecting data over a number of years to gauge the range of nitrate-N concentrations in drainage water.

Conclusions: Overall mean nitrate-N concentrations recorded were < MAC (11.3 mg/l) for drinking water but may exceed environmental water quality standards. Dirty water management had the highest nitrate-N concentrations in both years, but improvements, i.e., reduced load and less fertiliser N can reduce its impact. High nitrate-N in the 2-cut silage management in year 1 reflected a pre-experimental history of high N inputs. The results relate to drainage water at the 1 m depth and are indicative only as to what may occur in the groundwater. Bartley (2003) studied nitrate-N concentrations in the groundwater on the same farm and confirmed the designation of the area as having an important aquifer of extreme vulnerability.

References
Bartley, P. 2003. Nitrate responses in groundwater under grassland dairy agriculture. Ph.D, Thesis, Trinity College, Dublin.

Agro-ecological effects of sewage sludge application to orchard grass (Dactylis glomerata) fields

G.E. Merzlaya[1], R.A. Afanasiev[1], M.E. Shibaeva[2] and I.A. Arkhipchenko[2]

[1]All-Russia Research Institute of Agrochemistry named after D.N. Pryanishnikov, Moscow, Russia, Email: bolsh@mail.cnt.ru, [2]All-Russia Research Institute of Agricultural Microbiology, Saint-Petersburg-Pushkin, Russia

Keywords: dry matter production, chemical composition, sewage sludge, soil microbiology

Introduction Municipal sewage sludge can have a great fertilisation effect. However, its application is limited by high amounts of heavy metals, which have adverse effects on soil biota and plants (Ladonin et al., 2002). Our work was aimed at investigating the sludge effect on a soil-plant system.

Materials and methods The investigation was carried out as a micro-field experiment with two kinds of heat-digested sewage sludge of Moscow: 1 from filter-presses and 2 from sludge fields with a storage life exceeding 10 years. Two sludge doses were studied (10 and 35 t/ha of dry matter DM). The experiment was carried out in the field in small plastic cylinders without bases (size 0,5x0,5 m depth 0,4 m) filled with soil. In 1999 orchard grass was sown under the cover of barley for green forage. Peat-podzol clay loam was used in the experiment. The top layer contained 0.6% of organic carbon, 3.5 mg/100 g P_2O_5, 7.8 mg/100 g K_2O, pH $_{KCl}$ 4.0. The soil micro-flora was studied using the methods described by Alef & Nannipieri (1998).

Results Sludge application had a great agro-chemical effect (Table 1). Average DM yield in comparison with the control increased during the experimental period (1999 - 2003) with the increase of the sludge dose. Sludge from filter-presses in the amount of 35 t/ha gave the maximum increase.

Table 1 Average production of perennial grasses over 5 years

Experiment versions	DM g/m²	Increase	
		g/m²	%
Without fertiliser (control)	277	-	-
Sewage sludge 1, 10 t/ha	365	88	32
Sewage sludge 1, 35 t/ha	569	292	105
Sewage sludge 2, 10 t/ha	372	95	34
Sewage sludge 2, 35 t/ha	460	183	66
$N_{180}P_{60}K_{100}$ fertiliser	661	384	139

For all doses of both kinds of sludge there were no differences between the sludge treated and control plots in DM content of cadmium (0,40-0.42 mg/kg), nickel (0,4-0,6 mg/kg) and copper (3-5 mg/kg). At the same time the content of phosphorus in grass forage increased up to 0,43% in comparison with the control (0,33%) as well as with mineral fertiliser application (0,35%). With both types of sewage sludge the total amount of micro-organisms was slightly higher (10%) in comparison with the control. The same applied for ammonifying, ammilolythic micro-flora and oligonitrophyls. The amount of fungi was slightly lower (7%), which is a positive factor.

Conclusions Sewage sludge application of 10 and 35 t/ha DM increased production of orchard grass by 32-105%, with increasing phosphorus content in the plants. Content of Cd, Ni, Cu in plants with sludge application was the same as in the control plants. Sludge application increased microbiological soil activity stimulating growth of aerobic bacterial micro-flora.

References
Alef K.& P. Nannipieri (1998). Methods in applied soil microbiology and biochemistry. Academic press, U.K., 576pp.
Ladonin V.F., G.E. Merzlaya & R.A. Afanasiev (2002). [Strategies of utilization of sewage sludge and its compost in agriculture] (in Russian) Moscow, 140pp.

The effect of steel industrial residue-enriched soil on the initial growth and heavy metal profiles of elephantgrass

L.P. Passos, M.F. Saldanha, M.C. Vidigal, J.C. De Crignis, S. Sozzi, F.J.S. Lédo and R.C. Oliveira
Embrapa Dairy Cattle, The National Dairy Cattle Research Centre, Juiz de Fora, 36016-210 MG, Brazil, Email: lpassos@cnpgl.embrapa.br

Keywords: contamination, pollution , *Pennisetum purpureum*, heavy metals, steel industrial residue

Introduction Heavy metal contamination of industrial sites are becoming a matter of growing concern. In spite of the substantial progress in the assessment of the influence of steel industrial plant waste on soil and water (Adamo *et al.*, 2002), studies on the immediate responses of cultivated plants are still scarce. The objective of this experiment was to verify the short-term effects of soil added phosphate mud (P mud) or metallurgical scale (M scale), which are trace element-rich steel industry residues, on the initial uptake and heavy metal profiles of elephantgrass (*Pennisetum purpureum*).

Materials and methods Seedlings of elephantgrass cv. Napier were grown in soil containing 0, 1, 2, 3 or 4 g/kg of either P mud (with a Fe level of 107,000 mg/kg) or M scale (with a Fe level of 344,000 mg/kg), harvested after a 60-day growing period in a greenhouse. Biomass production was determined as leaf area (L Area, in cm^2), number of basal tillers, and dry weight (DW, in g) of leaves (Lf DW), stems (St DW), stubs (Sb DW) and roots (Rt DW). Heavy metal content in the stem base was measured by the ICP-AES technique (Jung *et al.*, 2002). The experiment was conducted as completely randomised block with four replicates.

Results Biomass production was not influenced by either P mud or M scale (Table 1), although a consistent tendency of decreasing Rt DW was observed with enhanced levels of M scale. The number of basal tillers was linearly increased by the addition of P mud. Regarding the heavy-metal profiles in the stem base, the sole significant alteration was a linear augmentation in Zn content in response to P mud application (Figure 1) - approximately, from 30 to 82 mg/kg DW. The contents (g/kg DW) of the other elements ranged as follows: Cd, Co and Mo (0.12 to 1.10), B, Ni and Cu (5 to 70), Al, Cr, Mn, Na and Pb (42 to 215), Ca, Fe and Mg (820 to 1,800), and K (14,000 to 16,200).

Table 1 Effect of residue level (g) on plant variables [#]

g	Residue	L Area	Lf DW	St DW	Sb DW	Rt DW
0		1,882.7	18.1	30.10	6.78	12.54
1	P mud	2,073.9	16.36	25.64	8.93	13.67
	M scale	1,940.5	16.42	21.82	5.21	10.01
2	P mud	2,072.9	12.66	17.36	6.75	13.21
	M scale	1,931.8	18.32	25.98	4.54	8.43
3	P mud	2,062.4	16.57	22.56	7.45	12.13
	M scale	2,028.6	19.05	26.00	5.10	6.66
4	P mud	2,215.7	16.96	23.29	6.94	9.85
	M scale	1,971.6	15.16	18.33	7.07	6.60

[#] All mean differences were not significant (p>0.005)

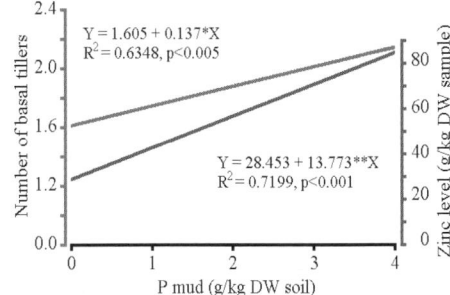

$Y = 1.605 + 0.137*X$
$R^2 = 0.6348, p<0.005$

$Y = 28.453 + 13.773**X$
$R^2 = 0.7199, p<0.001$

Figure 1 Effect of P mud on the number of basal tillers and on the zinc level

Conclusions Short-term effects on elephantgrass were caused by P mud, but not by M scale. The main responses were an increase in the number of basal tillers and in the stem base Zn content, which reached toxic levels. Considering the available data with other plant species exposed to heavy metal-contaminated soil, further long-term evaluations of the residues examined in the present study could lead to a better assessment of alternatives for soil rehabilitation.

References
Adamo, P., M. Arienzo, M.R. Bianco, F. Terribile & P. Violante (2002). Heavy metal contamination of the soils used for stocking raw materials in the former ILVA iron-steel industrial plant of Bagnoli (southern Italy). *Science of the Total Environment*, 295, 17-34.
Jung, M.C., I. Thomson & H.T Chon (2002). Arsenic, Sb and Bi contamination of soils, plants, waters and sediments in the vicinity of the Dalsung Cu-W mine in Korea. *Science of the Total Environment*, 295, 81-89.

Effect of adding moraine soil or shell sand into peat soil on soil properties and grass yields in western Norway

L.S. Sognnes[1], G. Fystro[2] and S. Øpstad[1]
[1]The Norwegian Crop Research Institute, Fureneset Rural Development Centre, NO-6967 Hellevik I Fjaler Norway, Email: livper.sognnes@c2i.net, [2]The Norwegian Crop Research Institute, Løken Research Centre, NO-2940 Heggenes, Norway

Keywords: leys, mineral materials, peat soil, yield, water infiltration

Introduction Cultivation and utilisation of peat soil leads to problems related to its high water content. The problems have become more pronounced with the increasing weight of agricultural machinery and more frequent harvesting. Increased particle density, reduced porosity and decreased potential plant-available water was found after incorporation of sand to peat in experiments conducted in the north of Norway (Sveistrup & Haraldsen, 1995). Peat soil has a weak soil skeleton, low bearing capacity, poor thermal properties and insufficient soil aeration. The objective of this study was to investigate the impact of added mineral material to peat soil to improve characteristics important for more optimal plant growth and management practices in the future.

Materials and methods A field experiment was conducted from 1978 to 1995 on peat soil (H5-H6 von Posts scale) at Fureneset Rural Development centre (61•48`N, 5•03`E) in Norway. In 1977 either moraine soil or coarse shell sand was added as evenly as possible in different amounts to experimental plots using bulldozers and tractors. The treatments were: no addition, 200, 400 and 800 m^3/ha moraine soil or shell sand respectively. All plots were fertilised with 750 kg/ha compound NPK fertilizer (16-3-15/18-3-5) in spring and 500 kg/ha after first harvest. The grass was harvested two times each year.

Results Adding mineral materials to the peat led to an increase in grass production both in the first (1978-1987) and second period (1989-1995); the year 1988 was not included due to re-sowing. In the first 10-year period, the increase was most pronounced after shell sand compared to moraine soil addition. After re-sowing, the moraine soil treatments demonstrated an additional increase in dry matter yield (Table 1).

Table 1 Dry matter (DM) yields and herbage timothy contents in the periods 1978-87 and 1988-95 after the incorporation of moraine soil and shell sand in different amounts to a peat soil in 1977

	DM yield, kg/ha		Timothy, %	
	1978-87	1989-95	1978-87	1989-95
Control	1040	816	33.4	38.8
Mineral type:				
Moraine soil	1080	887	33.9	43.6
Shell sand	1153	865	42.7	49.0
Applied amounts:				
200 m^3/ha	1123	858	38.1	41.5
400 m^3/ha	1118	898	38.6	47.1
800 m^3/ha	1108	873	38.2	50.3
Sign. of effects[a]:				
Control vs. material treatments	***	**	**	**
Material type	*	ns	**	*
Amount	ns	ns	ns	**
Material type* Amount	ns	***	ns	*
Year	***	*	***	***
Material type*Year	**	***	***[c]	ns
Amount*Year	ns	ns	ns	ns

[a]*, **, *** and ns; significant at P<0.05, P<0.01, P<0.001 and not significant, respectively

Conclusion Mineral materials added to a fuel peat soil improved conditions for plant growth. The observed increase in infiltration rates combined with reduced total porosity and soil water content after additions of mineral materials are considered to be important contributing factors to the improved growth conditions. For economical reasons, addition of more than 400 m^3/ha mineral materials cannot be recommended.

References
Sveistrup, T. & T.K. Haraldsen (1995). Effects of sand application and soil compaction on yields of leys and soil properties in peat soils in northern Norway. *Norwegian Agricultural Research*, 9, 133-146.

Pasture production after sewage sludge and liming application on highlands in North West Spain

M.L. Rigueiro-Rodríguez[2], A. López-Díaz[1] and M.R. Mosquera-Losada[2]
[1]Dpto Biología y Producción de los Vegetales, Centro Universitario de Plasencia, Univ. Extremadura, Avda. Virgen del Puerto, nº2, 10600 Plasencia-Cáceres, Spain, Email: lurdesld@unex.es, [2]Dpto. Producción Vegetal, EPS Lugo, Univ. Santiago de Compostela, Campus Universitario s/n, Lugo (Spain)

Keywords: sewage sludge, lime, pasture production, soil acidity

Introduction In recent years, a sewage sludge surplus has been created in the EU countries, due to the Urban Waste Water Treatment Directive 91/271/CEE. Therefore, it is necessary to find adequate disposal for these residues in accordance with EU policy. Organic matter and nutrient sewage sludge contain principally N, indicating that it could be used as fertiliser. The main risk of this residue is its heavy metal content, whose solubility is usually increased as soil pH declines. The objective of this experiment was to determine the effect of liming and sewage sludge application on pasture production in a silvopastoral system located on acid soil

Materials and methods The experiment was located in Lugo (NW Spain) at 510 m.a.s.l. on very acid soil. In autumn 1997, a pasture mixture of *Lolium perenne*, *Dactylis glomerata* and *Trifolium repens* was sown under a 5-yr old *Pinus radiata* stand. Treatments consisted of no fertilisation (NF), three sewage sludge doses (L1: 160 kg total N/ha; L2: 320 kg total N/ha; L3: 480 kg total N/ha, the same treatments with lime (2.5 t CO_3 Ca/ha), which was applied in autumn 1997, and mineral fertilisation (MIN: 500 kg/ha 8:24:16). Fertilisation treatments were applied in spring 1998 and 1999. In the second year of the experiment, three cuts took place in May, July and November.

Results In the first two years, pasture production increased with sewage sludge doses in May and July (figure 1). Mineral fertilisation had no effect on production because this treatment reduced the pH significantly from 5.3 with sewage sludge to 4.3. No statistically significant differences were detected in pasture production in July due to liming. In the last cut, there were no differences between treatments.

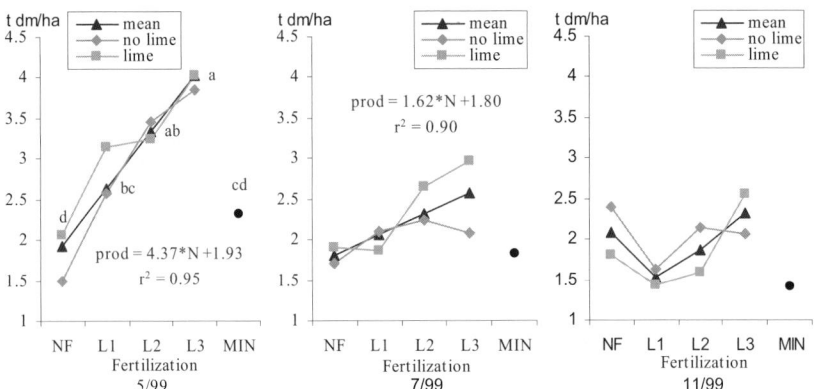

Figure 1 Pasture production (prod) (t DM/ha) after fertilisation treatments in May, July and November 1999. N: N doses (t/ha); NF: no fertilisation; L1: low sewage sludge doses (0,16 t total N/ha); L2: medium sewage sludge doses (0,32 t total N/ha); L3: high sewage sludge doses (0,48 t total N/ha); MIN: 500 kg/ha 8:24:16; lime: 2,5 t CO_3Ca/ha. Different letters indicate significantly differences between fertilisation treatments.

Conclusions In the first two cuts, sewage sludge application increased pasture production proportionally to doses applied. In the third cut, no response was detected among treatments due to the lack of sewage sludge residual effect as a result of low temperatures. Mineral fertilisation had no effect on pasture production due to pH reduction from sewage sludge application

Agronomic characterisation of sewage sludge: residual effects

M.R. Mosquera-Losada and A. Rigueiro-Rodríguez
Crop Production Department. High Politechnic School, University of Santiago de Compostela, Email: romos@lugo.usc.es

Keywords: compost, pellets, anaerobic

Introduction The progressive implementation of the EU Urban Waste Water Treatment Directive 91/271/EEC is increasing the quantities of sewage sludge requiring disposal. Agronomic use of sewage sludge should be based on crop fertiliser requirements and heavy metals content. Sewage sludge fertiliser potential depends on the stabilisation treatments of sewage sludge: composting, anaerobic digestion, aerobic digestion or pelleting (EPA, 1994). The objective of this experiment was to evaluate the fertiliser effectiveness and residual effects of three types of sewage sludge treated in three different ways.

Materials and methods The experiment was carried out in Lugo (NW Spain). The design of the experiment was a randomised block with three replicates. Treatments consisted of the application of three types of waste: compost, anaerobically digested sewage sludge and pelleted sludge on plots of 8 m^2 at doses of 160 or 320 kg total N/ha equivalent. Each treatment was applied in autumn in 6 plots in the first year of the experiment, but in the second year, only half of them were fertilised with the same treatments in order to determine the residual effect of sewage sludge. Two additional treatments were also established: No fertilisation and two inorganic fertiliser rates of 500 and 1000 kg/ha of 8:24:16 (NPK). The two first harvests of spring of the second year of study are presented in this paper.

Results Pasture production of the two first cuts of spring is shown in Figure 1. Pasture production in the first year was increased by anaerobically digested and pelletised sewage sludge application; however, urban compost did not produce any effect, in spite of taking into account the high proportion of inert material (30% plastic), as doses were based on N content. There were not residual effects of organic fertilisers. EPA (1994) indicated that a residual effect of 10% and 5% for these two sewage sludge types could be expected, but this was not the case in the present experiment. This was probably due to the high precipitation of the area and adequate temperatures for mineralisation. Fertiliser effectiveness was higher with anaerobically digested and pelleted treatments than with compost, but only at higher rates. Similar results were found in the first year experiment (Mosquera *et al.*, 2004).

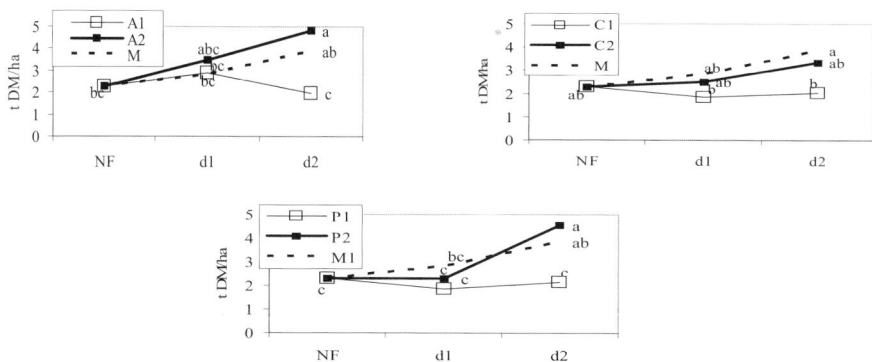

Figure 1 Two first cut spring pasture production. M:mineral; NF: no fertilisation; C:compost; P:pellet; A:anaerobic. Number identifies the application. Different letters indicate significant differences

Conclusions The results of the present study demonstrated that in areas with precipitation above 1200 mm residual effects of compost, anaerobically digested or pelletised sewage sludge cannot be expected. Effectiveness was higher with anaerobically digested and pelletised urban waste than with compost.

References

EPA, (1994). Land application of sewage sludge. A guide for land appliers on the requirements of the federal standards for the use of disposal of sewage sludge, 40 CFR. Part 503.

Mosquera-Losada M.R., E Fernández-Núñez,. M.L. López-Díaz, M. Rois-Díaz, & A. Rigueiro-Rodríguez (2004). Pasture establishment response to fertilisation with different types of sewage sludge. *Proceedings of the VIII Congress European Agriculture*, 771-772.

Increase in forage maize production by bacterial fertilisers

D. Egamberdiyeva[1] and G. Hoflich[2]
[1]University of Agriculture, Centre of Agroecology, Tashkent 700140, Uzbekistan, Email: dilfuza_egamberdiyeva@yahoo.com, [2]Institute for Primary Production and Microbial Ecology, ZALF, Muencheberg, Germany

Keywords: bacterial fertilisers, maize, nutrient uptake

Introduction Farmers in many countries value green material generated from maize as a high quality forage. Its inclusion in dairy cow diets can improve forage intake, increase animal performance and has the potential to reduce production costs (Phipps, 1994). Restrictions on the use of chemical fertilisers and a renewed interest in organic sustainable farming systems in general, has restored attention to crop rotations The increase of maize production using bacterial fertilisers also gives alternative ways to reduce chemical fertilisers in forage production. Increased uptake of nutrients such as N, P, and K and crop yield was reported with rhizobacteria (Lazarovits & Nowak, 1997). This paper presents studies carried out to evaluate the effects of bacterial fertilisers on maize production.

Material and methods An experimental field site was established on Salmtieflehm-Fahlerde (Arbeitsgruppe Boden, 1994), Germany, in a randomised block design with six replicates (plot size: 15 m^2, harvest was carried out after 3 months). The preceding crops were yellow lupin (*Lupinus luteus* L.) with under-sown cocksfoot (*Dactylis glomerata* L.). Farmyard manure (30 t/ha fresh weight) was mixed into the soil by a milling machine before the sowing of the maize. The seeds were inoculated with the bacterial preparation (10^8 cfu / g peat). The criteria for growth promotion were shoot dry matter and the N, P, K and Mg content of the plants. *Rahnella aguatilis* 6 and *Pantoea agglomerans* 050309 were used as bacterial preparations. The data were analysed with two-way ANOVA and Student-Newman-Keuls test for testing significant differences (P = 0.05) of main effects.

Results Increases in plant growth and nutrient uptake were recorded for treated plants (12-leaf stage) in the field experiment (Figure1). Strain *Rahnella aguatilis* 6 gave the best performance and resulted in a 23% increase in plant growth over the control. The various bacterial inoculants differentially influenced the N, P, K, and Mg contents of plant components. The content of K was increased significantly in both treatments. Only strain *Rahnella aguatilis* 6 resulted in a significant increase in N uptake.

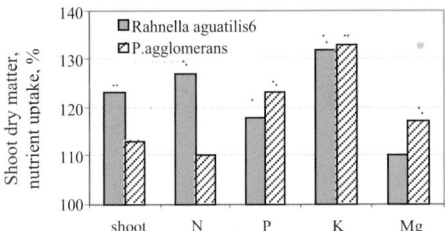

Figure 1 Effect of inoculation with *Rahnella aguatilis* 6, and *Pantoea agglmerans* 050309 on shoot dry matter and nutrient uptake of maize in field experiments with loamy sand (control=100)

Conclusion The experiments showed that maize dry weight and nutrient content such N, P, K, Mg uptake may be increased using bacterial fertilisers. This biological approach will decrease the use of chemical fertilisers in forage production.

References
Lazarovits, G. & J. Nowak (1997). Rhizobacteria for improvement of plant growth and establishment. *Horticultural Science*, 32, 188-192.
Phipps, R.H. (1994). Complementary forages for milk production. In: P.C. Garnsworthy & D.J.A. Cole (eds.) Recent Advances in Animal Nutrition. Butterworths, London, 215-230.

Soil organic matter along a degradation gradient in a semi-arid rangeland of South Africa

C.C. du Preez[1] and H.A. Snyman[2]

[1]*Department of Soil, Crop and Climate Sciences, Email: dpreezcc.sci@mail.uovs.ac.za,* [2]*Department of Animal, Wildlife and Grassland Sciences, University of the Free State, PO Box 339, Bloemfontein 9300, South Africa*

Keywords: organic carbon, rangeland degradation, sandy loam soil

Introduction In a semi-arid South Africa rangeland on a sandy loam soil, artificial manipulations to create various rangeland conditions typical of different levels of livestock impact were shown to result in considerable losses of organic matter after 15 years (Du Preez & Snyman, 1993). In this follow-up study on a similar soil, we regularly measured the organic matter content following the conversion of rangeland in a good to a poor ecological condition, an undisturbed bare soil and a cultivated soil respectively. The changes in soil organic matter content after five years are reported.

Materials and methods The study was conducted in a semi-arid, summer rainfall (560 mm annually) region. Mean maximum monthly temperatures range from 17^0C in July to 33^0C in Jan.. Rangeland typical of the Dry Sandy Highveld Grassland was selected on a Rhodic Epigleyic Luvisol with three distinct horizons (A: 0-300 mm, B1: 300-600 mm and B2: 600-1200 mm) containing 10, 24 and 42% clay respectively. The experimental layout was a fully randomised design consisting of four treatments (rangeland in good and poor condition, undisturbed bare soil and cultivated soil) replicated thrice on plots, each measuring 2x15 m. The botanical composition and basal cover of the rangeland in the good and poor conditions was typical of that on commercial farms nearby. The rangeland was artificially kept in the above-mentioned conditions by removing, with minimum disturbance, all undesirable species. On the disturbed bare soil any plants that had emerged, were removed monthly by hoeing, while the cultivated soil was ploughed twice annually to a depth of 200 mm. Just before introduction of the treatments (April 1993), and five years later (April 1998), soil samples were collected from each plot for organic C analysis as an indicator of organic matter. The samples were taken at 0-25, 25-50, 50-75, 75-100, 100-150, 150-200, 200-250 and 250-300 mm depth intervals at five randomly selected points in each plot. Composite samples for each depth interval were prepared by combining and thoroughly mixing the five cores of 20 mm diameter. Two-way analyses of variance at the 95% confidence level were computed for organic C to identify significant differences.

Results and discussion The organic matter content of all 12 plots as indicated by organic C was similar (P>0.05) in April 1993 before the treatments were introduced. During the experimental period this measure of organic matter content remained the same (P>0.05) for the rangeland in good condition. Therefore the organic matter content of the good rangeland in April 1998 is representative of that of the poor rangeland, undisturbed bare soil and cultivated soil before the introduction of these three treatments. The three treatments resulted in a decline (P<0.05) in organic C within five years. In the 0-25 mm layer the poor rangeland, undisturbed bare soil and cultivated soil contained 14, 26 and 46% less organic C, respectively than the good rangeland. These differences declined with depth and phased out below 100 mm. These results are in accordance with previous research on a similar soil, viz. where conversion of rangeland in a good to poor ecological condition resulted in a considerable loss of soil organic matter (Du Preez & Snyman, 1993). Loss of organic matter from the upper 50 mm of soil of the rangeland in poor condition in this study averaged 18% within five years, while in the preceding study it averaged only 28% after 15 years. It seems therefore that the rate of soil organic matter loss on account of the conversion of rangeland in good to poor condition is initially high, but the rate then decreases until a new equilibrium is reached. A major reason for the loss of soil organic matter from the rangeland in poor condition compared to that in good condition was 7,703 kg/ha less (P<0.05) aboveground phytomass production over the five year period due to differences in botanical composition and a decrease (P<0.05) in basal cover from 9.0 to 3.4 %. The decline of organic matter in the undisturbed bare soil can be assumed to be representative of what happens in bare patches found in degraded rangeland.

Conclusion The results illustrate the importance of sound rangeland management for maintaining organic matter at acceptable levels to ensure sustainable productivity. Once organic matter is lost, restoration is very slow.

Reference

Du Preez, C. C. & H. A. Snyman. (1993). Organic matter content of a soil in a semi-arid climate with three long-standing veld conditions. *African Journal of Range and Forage Science*, 10, 108-110.

Influence of grazing on soil microbial communities on a mixed grass prairie ecosystem

L.J. Ingram[1], G.E. Schuman[2], P.D. Stahl[1] and J. Buyer[3]
[1]Dept. Renewable Resources, University of Wyoming, Laramie, WY USA 82071, Email: lachy@uwyo.edu, [2]USDA-ARS High Plains Grasslands Research Station, Cheyenne, WY, USA 82009, [3]USDA-ARS Sustainable Agricultural Systems Laboratory, Beltsville, MD, USA 20705

Keywords: microbial communities, grazing management, semi-arid environment, phospholipid fatty acids

Introduction The grazing of ungulates is the predominant use for much of the world's semiarid rangelands. Grazing these lands can result in significant changes not only in the vegetation community but also in the soil physical, chemical and biological properties. Changes in soil physical and chemical properties and the plant community can potentially lead to changes in soil microbial communities which may have long-term ramifications for nutrient cycling and carbon (C) sequestration. The objective of this research was to ascertain the influence of three long-term grazing treatments on soil microbial communities.

Materials and methods In May 2003, two replicates of three different grazing treatments were sampled; an ungrazed exclosure (EX); grazed season-long continuous light (CL) and; a grazed season-long continuous heavy (CH) treatment. These three grazing treatments had been initiated in 1983. In each replicated paddock, sampling was undertaken along a permanent 50 m transect and soils sampled at 0-5, 5-15, 15-30 and 30-60 cm depth increments. After sieving (< 2 mm) and air-drying, soils were rewetted and incubated and microbial respiration measured, using a base-trap method, at three (3d-MR) and 21 d (21d-MR). Nitrogen (N)-mineralisation was also determined by calculating the amount of inorganic N (Ni: NO_3 + NH_4) produced over the course of the 21 d incubation. Microbial biomass carbon (MBC) was determined using the chloroform fumigation-incubation method. A second set of soil samples were collected at the 0-5 and 5-15 cm depth increments along the transects for phospholipid fatty acids (PLFA) biomarker analyses.

Results There were distinct treatment differences in the 3d-MR, 21d-MR, and N-mineralisation in the order CL > EX >> CH. There was 8% greater nitrate in the CL treatment compared to the EX and CH treatments suggesting possible differences in the nitrifying populations between the CL treatment compared to the EX and CH treatments. Microbial biomass C was higher in the CL than the EX treatment and both were much greater than the CH treatment. Multiple analysis of variance using PLFA biomarkers indicated a significant grazing treatment effect at both depths examined. Results show the microbial community structure in the CL treatment was distinct from that in the CH grazed treatment whereas the EX treatment was intermediate.

Discussion and conclusion The results of this work suggests that in these prairie ecosystems, which have co-evolved with native ungulates over many millennia, light grazing (the CL treatment was grazed at ~35% lower stocking rate than that recommended by the Natural Resources Conservation Service) is beneficial to the soil community. Previous work by Hamilton & Evans (2001) found that clipping of a C_3 plant stimulated the release of root exudate C compounds which in turn increased microbial biomass as well as NO_3, NH_4 and potential N-mineralisation, compared to unclipped plants. Many of the changes observed in the microbial communities probably also reflect the influence that grazing has had on the plant community composition. Cool season (C_3) perennial grasses are the predominant group under the CL treatment, whereas C_4 (warm-season) perennial grasses dominate CH, and C_3 perennial grasses and forbs co-dominate the EX. It would appear that under very heavy grazing, much of the C that would normally be available for root exudation is instead put into the production and replacement of grazed plant parts. The changes in N-mineralisation were probably due to the C_4 dominated plant communities tieing up more N in plant material as compared to the C_3 dominated plant communities with this reducing the availability of inorganic N in the soil (Wedin, 1999). This change, together with potential variation in root exudation, appears to have given rise to different microbial communities (as evidenced by the PLFA results and the observed differences between other measures of microbial activity) evolving in each of the three grazing treatments.

References
Hamilton, E.E., & D.A. Frank. (2001). Can plants stimulate soil microbes and their own nutrient supply? Evidence from a grazing tolerant grass. *Ecology*, 82, 2397-2402.
Wedin, D.A. (1999). Nitrogen availability, plant-soil feedbacks and grassland stability. *Proceedings of the Sixth International Rangeland Congress*, Vol. 1, 193-197.

Soil microbial diversity of an artificial *Caragana korshinskii* plantation on the loess plateau of China

W. Zhang[1,2], H.L. Wei[3], H.W. Gao[1] and Y.G. Hu[2]

[1]*Institute of Animal Science, Chinese Academy of Agricultural Sciences, Beijing 100094, China, E-mail: cauzhangwei@163.com,* [2]*College of Agronomy and Biotechnology, China Agricultural University, Beijing 100094, China,* [3]*Institute of Microbiology, Chinese Academy of Sciences, Beijing 100080, China*

Keywords: 16S rDNA gene, biodiversity, phylogenetic analysis, loess plateau

Introduction Peashrub (*Caragana korshinskii*) is an important dune-fixation plant on the loess plateau of China and is valuable in ecological environment construction of North-western China. For determining relationships between peashrub and soil microbes, three clone libraries of 16S rDNA from rhizoplane, rhizosphere and bulk soil communities of peashrub were constructed with a culture-independent approach. The data obtained from three clone libraries were used to investigate the magnitude of vegetative changes in the microbial community, and to search for general ecological relationships.

Materials and methods Sampling from peashrub rhizoplane, rhizosphere, and bulk soil was done in October 2003 at Wuzhai, Shanxi Province (39°00'22.9"N, 111°45'39.7"E, with a ground water depth of 1362 m). Metagenomic DNA was prepared by the CTAB-SDS method. Three 16S rDNA clone libraries were constructed with pMD18-T, and every clone was sequenced through reverse and forward M13 primers. Phylogenetic tree construction was carried out by the nearest neighbour method. And the Shannon-Weaver index (H'), Simpson index (D) and the relative index were chosen to characterise the microbial communities.

Results One hundred seven positive clones were randomly selected and fully sequenced (about 900 bp) from 3 clone libraries of 16S rDNA genes. Based on sequence comparison, 3 phylogenetic trees were obtained and the closest relatives of most clones gave 13 distinct clusters except 7 sequences were closely associated with environmental clone sequences, i.e., uncultivated bacteria (Table 1). Bacteria were the dominant group in each of the 3 libraries, though their percentage decreased with the distance from peashrub roots. If rhzioplane, rhizosphere and bulk soil are figured as one line, the number of α- *Proteobacteria* and low-G+C gram positive bacteria decreased but high-G+C bacteria and archaea increased along the line. Otherwise, the *Acidobacterium* division was a major group in rhizosphere and bulk soil samples but was not detected in the peashrub rhizoplane.

Table 1 Summary of the phylogenetic assignments of 107 SSU rDNA clones made from a Loess Plateau *C. korshinskii* plantation soil within the major taxa

Major taxon and group	No. (%) of organisms identified			
	Rhizoplane	Rhizosphere	Bulk soil	Total
Archaea	2(5.0)	6(13.9)	6(25.0)	14(13.1)
Crenarchaeota	2(5.0)	6(13.9)	6(25.0)	14(13.1)
Eucarya	4(10.0)	0	0	4(3.7)
Plants	4(10.0)	0	0	4(3.7)
Bacteria	33(82.5)	34(79.1)	15(62.5)	82(76.6)
Green nonsulfur	0	1(2.3)	1(4.2)	2(1.9)
Low-G+C gram positive	2(5.0)	0	0	1(0.9)-
High-G+C gram positive	2(5.0)	3(6.9)	3(12.5)	8(7.5)
Cytophaga-Flexibacter-Bacteroides	2(5.0)	1(2.3)	0	3(2.8)
Planctomyces	1(2.5)	2(4.7)	1(4.2)	4(3.7)
Verrucomicrobium	1(2.5)	2(4.7)	1(4.2)	4(3.7)
Proteobacteria	26(65.0)	12(27.9)	5(20.8)	43(40.2)
α	6(15.0)	5(11.6)	2(8.3)	13(12.1)
β	2(5.0)	2(4.7)	0	4(3.7)
γ	18(45.0)	4(9.3)	3(12.5)	25(23.4)
δ	0	1(2.3)	0	1(0.9)-
Acidobacteria	0	13(30.2)	4(16.7)	17(15.9)
Unclassified	1(2.5)	3(6.9)	3(12.5)	7(6.5)
Total clones sequenced	40	43	24	107

Conclusions The microbial richness and diversity index was lower in bulk soil, which reflects the soil character and environmental condition at the loess plateau. But as a long-term planting of peashrub, many plant-associated bacteria appeared in the rhizoplane and rhizosphere of peashrub, which probably play a great role in peashrub growth and survival in an abominable environment.

Soil enzyme activities as bio indicators of soil pH and fertility in temperate grassland

I. Mijangos, A. Ibarra, I. Albizu, S. Mendarte and C. Garbisu
*NEIKER, Basque Institute of Agricultural Research and Development, Berreaga 1, E-48160 Derio, Spain,
Email: imijangos@neiker.net*

Keywords: bioindicator, enzyme, fertility, grassland, soil

Introduction In recent years, biological indicators are being used to estimate the continued capacity of a given soil to function (*i.e.*, soil health). After all, biological processes are intimately linked with the maintenance of soil structure and fertility, being more sensitive to changes in the soil than conventional physicochemical parameters. Soil enzymes, as mediators and catalysts of vital soil functions, offer great potential as integrative indicators of soil health (Dick *et al.*, 1996). The main aim of the current work was to study the potential of soil enzyme activities as biological (more precisely, biochemical) indicators of soil physicochemical properties as well as of soil fertility in different temperate grasslands.

Materials and methods Two different types of forage production systems were studied, *i.e.*, an intensive crop rotation (cereals in winter/fodder maize in summer) and a temporary meadow for hay, during three consecutive years. The following soil physicochemical parameters were determined (MAPA, 1994): pH, moisture (weight/weight), organic matter content, total N, P Olsen, exchangeable Ca^{2+} and Mg^{2+}, extractable K^+, % Al saturation, CEC, and C/N ratio. Finally, plant biomass production (PBP) and soil enzyme activities [*i.e.*, arylsulfatase (S), β-glucosidase (G), acid phosphatase (P), urease (U), and dehydrogenase (DH)] were determined as described by Dick *et al.* (1996).

Results A range of randomly selected plots in the meadow, that represented the whole interval of pH values characteristic of our region, showed a range of DH values that were significantly correlated, following an exponential pattern, with soil pH (Fig. 1). DH has been reported to be related to intracellular processes that occur in viable microbial cells and is usually determined to estimate overall microbiological activity of soil (Dick *et al.*, 1996). A similar but weaker correlation between DH and pH was also observed in soils belonging to the intensive crop rotation plots (y= 0.24 $e^{0.94 x}$, R2= 0.77). Most interestingly, DH appeared a good indicator of PBP in the meadow (Fig. 2), again showing a stronger relationship here than in the intensive crop rotation (data not shown). This is an expected result since, according to Skujinš (1978), in unmanaged ecosystems or low-input agricultural systems, such as our meadow, a stronger relationship between soil enzyme activity and PBP might be expected. Finally, S, G, and U showed positive correlations (p>0.001) with soil pH and organic matter content. All enzyme activities showed strong correlations among themselves.

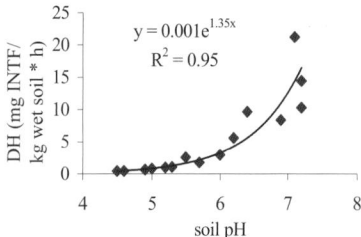

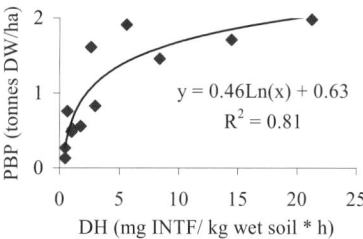

Figure 1 Relationship between pH and DH in meadows

Figure 2 Relationship between DH and PBP in meadows

Conclusions Dehydrogenase activity appears a good indicator of soil pH and plant biomass production, especially in the less perturbed agricultural system studied (*i.e.*, a temporary meadow). Arylsulfatase, β-glucosidase and urease also showed potential as bio-indicators of soil pH and organic matter content.

References:
Dick, R.P., D.P. Breakwell & R.F. Turco (1996). Soil enzyme activities and biodiversity measurements as integrative microbiological indicators. In: J.W. Doran & A.J. Jones (eds.) Methods for Assessing Soil Quality. SSSA, Madison, 247-271.
MAPA (1994). Métodos Oficiales de Análisis III. Mº de Agricultura, Pesca y Alimentación, Madrid, 662pp.
Skujinš, J. (1978). History of abiontic soil enzyme research. In: R. G. Burns (Ed.) Soil Enzymes. Academic Press, New York, 1-49.

The effect of nitrogen fixation and plant species on ammonium and amino acids soil contents

F. Lesuffleur, S. Diquélou, A. Bré and J.B. Cliquet
UMR INRA-UCBN 950 Ecophysiologie Végétale, Agronomie et Nutrition N,C,S. IRBA, Université de Caen, Esplanade de la Paix, 14032 Caen Cedex, France, Email: lesuffleur@ibfa.unicaen.fr

Keywords: N fixation, ammonium, amino acids, soil contents

Introduction Grass and legume intercropping systems, particularly ryegrass-clover mixed cultures, increase grasslands productivity, due to the clover nitrogen (N) fixation and to the transfer of part it to the companion grass. The objective of this experiment was to determine if N fixation and plant species could modify ammonium and amino acids soil contents.

Materials and methods Two weeks after germination, plants of white clover (*Trifolium repens*) cv. Huia and perennial ryegrass (*Lolium perenne*) cv. Bravo were transferred into pots filled with soil collected from a natural grassland sward in Haras du Pin, Normandy, France. Plants were grown in a controlled environment room and supplied with nutrient solution to achieve nitrate concentrations equivalent to 0 (N0), 50 (N50) or 180 (N180) kg/ha. Control pots were left without any plants in the same conditions to study plant effects on soluble N profiles of the soil. Eighty days after transplanting, 4 replicate pots for each treatment were harvested. Samples of 10 g homogenised soil were extracted using KCl 0.5M to determine ammonium and amino acids contents. Only significant differences ($P < 0.05$, Anova or Kruskal-Wallis tests) are discussed.

Results Ammonium concentration was high in the clover rhizosphere and decreased with the nitrate fertilisation rate (Table 1). This can be related to an inhibition of N fixation by high N levels as observed by numerous authors. In case of ryegrass, ammonium rhizospheric concentration was low and increased with N fertilisation. As for ammonium, amino acids concentration in the clover rhizosphere decreased with the nitrate fertilisation rate. Glycine (Gly), serine (Ser) and glutamate (Glu) were the major amino acids found in the soil solution, while aspartate (Asp), asparagine (Asn) and arginine (Arg) where also recovered in significant amounts (Figure 1). Comparison of the free amino acid profiles of non-rhizospheric soil and soil from clover or ryegrass rhizosphere suggests that the soil was not influenced by the presence of the roots or by the plant species.

Table 1 Ammonium and amino acids soil contents

	Ammonium (mM)			Amino acids (µM)		
	N0	N50	N180	N0	N50	N180
Clover	1.79	0.41	0.26	69	37	28
Ryegrass	0.10	0.17	0.74	38	19	40
Mixture	0.27	0.10	0.17	36	32	31

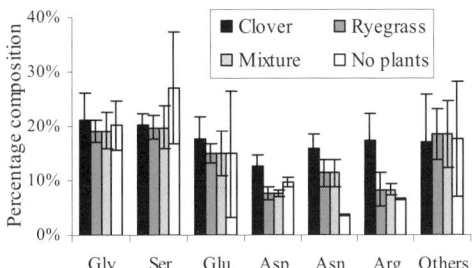

Figure 1 Effect of plant species on amino acids soils profiles

Conclusions There was a significant effect of N fixation and plant species on ammonium and amino acids (specially for clover) soil pools. However, no effect was observed on soil amino acids profiles. The low proportion of amino acids compared with the amount of ammonium in the soil solution confirms recent data by Jones *et al.* (2004) showing that these compounds do not accumulate in grassland soils because they are turned over very rapidly by the microbial activity. As found by Lipson *et al.* (1999), glycine, serine and glutamate are among the most abundant free amino acids found in soil. In contrast to ammonium, which has a nutrient role for plants, these amino acids could be implied in the selection of rhizospheric bacteria by plants.

References
Jones, D.L., D. Shannon, D.V. Murphy & J. Farrar (2004). Role of dissolved organic nitrogen (DON) in soil N cycling in grassland soils. *Soil Biology and Biochemistry*, 36, 749-756.
Lipson, D.A., T.K. Raab, S.K. Schmidt & R.K. Monson (1999). Variation in competitive abilities of plants and microbes for specific amino acids. *Biology and Fertility of Soils*, 29, 257-261.

Soil constraints (pH and aluminium) for legume performance in hill country of Uruguay

W. Ayala and R. Bermúdez
National Institute of Agricultural Research of Uruguay, Casilla de Correo 42, Treinta y Tres, Uruguay CP 33000, Email: wayala@tyt.inia.org.uy

Keywords: lotus spp., white clover, pH, aluminium tolerance

Introduction Pastoral areas in eastern Uruguay have soils with pH 5.5 or lower (Mas, 1978), which is frequently associated with the acid soil syndrome (Cregan, 1980). As pH drops below 5.5, aluminium (Al) concentration can increase to toxic levels. These conditions may adversely affect growth of introduced legumes. A way to overcome these constraints is by the use of tolerant species. The objective of this study was to evaluate the adaptation and productivity of different legumes under environments with restrictions in pH and Al concentrations.

Materials and methods Four legumes (*Lotus subbiflorus* "El Rincón" (LR), *Lotus corniculatus* "INIA Draco" (LD), *Lotus pedunculatus* "Grasslands Maku" (LM) and *Trifolium repens* "Zapicán"(WC)) were oversown in March 2002 in a hill country acid soil (pH 5.3) of eastern Uruguay (32° 20′ S, 54° 10′ W). Chemical soil properties were P4.5 μg P/g, K0.29 and Al0.41 me/100 g soil, respectively. Plots were top dressed annually with natural phosphate rock at 26 kg P/ha. During two years, two cutting heights were applied at monthly intervals (4 or 10 cm height) in the periods October-February, November-February and December-February. A split plot design with three replicates was used. Measurements included herbage dry matter (DM) production and species contribution (native grasses, sown legume) during year 1 and 2 and soil seed reserves at the end of year 1.

Results The two-year average of total herbage DM production showed differences between legumes, with the total yield of the annual lotus (LR) being 23% higher than the average of the perennial legumes (Table 1). Perennial legumes did not differ between themselves. Differences were explained by the higher native grasses' contribution during summer for the LR pasture. There was a significant interaction between species and cutting height (Table 1), with the legume yield of LM defoliated at 4 cm being 23% higher than at 10 cm. The other legumes (LR, LD and WC) did not show differences in yield as a consequence of cutting height. The relative herbage production of oversown legumes was LM 100, LR 59, LD 35 and WC 18(%). For soil seed reserves, a significant interaction of species x closing period was detected (Table 2). The annual legume (LR) produced significantly more seeds than the perennial legumes (LD, LM and WC) in October and November. There were no significant differences between perennial legumes in soil seed reserves.

Table 1 Total dry matter and legume dry matter (kg/ha) of 4 oversown legumes (2- year averages)

Species	Total DM	Legume DM	
		4 cm	10 cm
LR	8072	2198	2567
LD	6474	1311	1436
LM	6869	4315	3520
WC	6333	609	758
Significant factors			
Species (S)	*		**
Cutting height (I)	ns		ns
Closing dates (R)	**		ns
SxI	ns		*
SEM (n)	325 (36)		225 (18)

ns= not significant; *= p<0.05; **=p<0.01; SEM= standard error of the mean; n=observations for each mean

Table 2 Soil seed reserves (seeds/m^2) in March 2003 of 4 legumes for different rest dates in spring

Closing dates	LR	LD	LM	WC
October	50300	4050	3583	3883
November	12683	4216	2250	1383
December	7233	3367	833	2450
Significance (species x dates)		**		
Standard error (n)		2823(12)		

**=p<0.01; SEM=standard error of the mean; n=observations for each mean

Conclusions A high degree of adaptation and tolerance to soil constraints (pH and Al) in the hill country was observed in Grasslands Maku, in comparison with other lotus species and white clover. The annual lotus (El Rincón) invested more resources in reproductive processes than perennial legumes.

References

Cregan, P.D. (1980). Soil acidity and associated problems – guidelines for farmer recommendations, AG bulletin No. 7. New South Wales Department of Agriculture, Australia.
Mas, C. (1978). Región Este. In Pasturas I V. CIABB. *Miscelánea,* 18, 37-64.

Grazing effects on some soil characteristics in Lar rangelands

S.A. Javadi, M. Jafari and G.H. Zahedi
Department of Range Management, Islamic Azad University of Iran, Tehran, Iran, Email: sadynan@yahoo.com

Keywords: soil degradation, potassium, phosphorus, nitrogen, pH, organic matter

Introduction Heavy grazing pressures jeopardise the sustainability of the ecosystem by reducing soil fertility (Dormaar *et al.*, 1998). Different results have been reported of grazing intensities on soil chemical properties, which stem from climate, soil, vegetation, management and kind of animal (Dormaar *et al.*, 1998; Javadi, 2003; Sanadgool, 2002).

Materials and Methods The site was in northern Iran, 84 km east of Tehran with 410 mm/year mean precipitation, which falls mostly as snow. The main type of livestock is sheep. The dominant species were *Agropyron spp., Bromus tomentellus, Poa bulbosa, Astragalus spp and Thymus kotschyanus.* Three adjacent grazing areas of ca. 4 ha each were selected: reference area (ungrazed & closed), key area (moderately grazed) and critical area (heavily grazed). Twenty soil samples were taken from each layer (0-10cm, 10-30cm) in each sampling period (early, mid and late grazing). The samples were air-dried and passed through a 2 mm sieve prior to chemical analysis.

Results Organic matter, total C and N were higher in the reference area, however, P, K and pH were higher in the critical area. Values were higher in the surface layer except for pH.

Table 1 Nutrient concentrations in three areas at each depth (A: 0-10cm, B: 10-30cm)

Areas	K(ppm)		P(ppm)		%N		%C		pH	
	A	B	A	B	A	B	A	B	A	B
Reference	760b	565d	42.8b	23.8e	.34a	.13d	3.4a	1.1c	6.1e	6.2d
Key	707c	484e	45.6a	17.7d	.27c	.13d	2.8b	1.1c	6.3c	6.2d
Critical	1000a	736bc	47.0a	26c	.3b	.14d	2.8b	1.2c	6.6b	6.9a
SEM	14.48		0.8366		0.006090		0.05502		0.02578	

Means with the same letter are not significantly different. (P =0.05)

Conclusions Reduction in soil C and organic matter in critical areas could be caused by grazing, which reduces vegetation cover, root biomass and nutrient recycling in the rangeland. Our results are similar to those of Willms *et al.* (1990) and Sanadgool (2002). The differences observed in soil N among the three areas maybe related to changes in vegetation type. Percentage of legumes was higher in the reference area compared to the other areas (Javadi, 2003). Increment in K and P under heavy grazing was probably due to more urine and faeces and fewer plants being present to use the available P and K. Higher P, K, N and C content in the surface layer could be due to higher organic matter, litter and vegetation cover. The pH results were similar to those reported by Dormaar (1998), who found that the increase in pH was an indicator of soil loss. With increased grazing pressure the depth of the soil profile decreases, resulting in organic matter being closer to the surface. Results of this study seem to indicate that heavy grazing reduces soil quality by changing plant growth form, species composition and reduction in vegetation cover as well as soil trampling.

References
Dormaar, J.F. & W D. Willms (1998) Effect of forty-four years of grazing on fescue grassland soils. *Journal of Range Management,* 51, 122-126.
Javadi, S.A. (2003) Investigation of grazing effects on some vegetation and soil characteristics. M.Sc thesis, Tehran University.
Sanadgool, A. (2002) Effects of short-term grazing on animal production, vegetation and soil in *Bromus tomentellus* pasture. PhD thesis, Tehran University.
Willms W.D., S. Smoliak & J.F. Dormaar (1990) Vegetation response to time – controlled grazing on mixed and fescue prairie. *Journal of Range Management* 43, 513-517.

Accumulation and decomposition rates and N, P and K returned to the soil by the litter of tropical legumes and grasses

G. Crespo, I. Rodríguez, M.F. Días and S. Lok
Instituto de Ciencia Animal, Carretera Central Km 47 ½, San José de las Lajas. La Habana, Cuba, Email: gcrespolopez@yahoo.es

Keywords: litter, grasses, legumes

Introduction The return of plant nutrients through litter decomposition of legumes and grasses is important for the sustainability of grassland ecosystems (Sánchez *et al.*, 1989). Therefore, the selection of species that produce high amounts of decomposable litter is important . The objective of the present work was to study the production and decomposition rates of litter from various legumes and grasses in Cuba.

Materials and methods The accumulation and decomposition rates and N, P and K returned to the soil by the litter of the legumes *Desmodium ovalifolium, Pueraria phaseoloides, Stylosanthes guianensis* and *Neonotonia wightii/Macroptilium atropurpureum* mixture, and the grasses *Panicum maximum, Brachiaria decumbens, Cynodon nlemfuensis* and a mixture of native grasses was investigated. Legumes and grasses were sown in plots of 400 and 225 m², respectively. The monthly quantity of litter accumulation was determined by the method of Crespo & Fraga (2002). The monthly disappearance rate of litter was determined by the method of Thomas & Asakawa (1993). Samples of litter collected monthly were analysed to determine N P, K (spectrophotometer) and lignin (Van Soest & Wine 1968).

Results Among legumes, the mixture *N. wightii/M. atrorpurpureum* produced the highest accumulation rate (Table 1), whereas *B. decumbens* and *C. nlemfuensis* produced the highest rates among grasses. Litter of *D. ovalifolium* and *S. guianensis* disappeared 180 days after deposition whereas total disappearance of litter of the other legumes occurred 210 days after deposition. The rate of litter disappearance was low in grasses, especially in *C. nlemfuensis, P. maximum* and *B. decumbens*. Of these grasses one year after deposition 30, 15 and 10 % of the original litter deposition remained, respectively. Figure 1 shows the behaviour of litter decomposition in *D. ovalifolium* and *C. nlemfuensis*

Table 1 Litter produced by legumes and grasses (kg DM/m²/year)

Legumes	kg DM/ m²/ year	Grasses	kg DM/ m²/ year
M.atropurpureum G.wightii	1.86[a]	P. maximum	0.16[d]
P. phaseoloides	0.90[b]	B. decumbens	0.28[a]
D. ovalifolium	0.86[b]	C. nlemfuensis	0.22[b]
S. guianensis	0.70[b]	Native grasses	0.20[bc]
SE±	0.09**	SE±	0.07**

** (P<0.001)

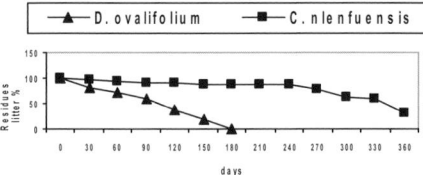

Figure 1 Decomposition rate of litter

Conclusions Legumes accumulated more litter than grasses. Total decomposition of legume litter occurred between 180-210 days, while in grasses it took more than 360 days. The mixture of *M. atropurpureum / G. wightii* produced a higher quantity of litter than the remainder single legumes.

References
Crespo, G., J Ortiz, A. Pérez. & S. Fraga (2001). Accumulation and decomposition rates and N, P, K released by the litter of perennial legumes. *Cuban Journal of Agricultural Science,* 35,39.
Crespo, G. & S. Fraga (2002). Production of litter and nutrients returned to the soil by *Cajanus cajan* (L.) Millsp and Albizia lebbeck (L.) Benth species in silvopastoral systems. Technical note. *Cuban Journal of Agricultural Science,* 36,397.

Effect of earthworm manure fertilisation on biomass production and mineral content of *Digitaria swazilandensis*

A. Paredes, S. Camacaro and W. Machado

Facultad de Agronomía, Universidad Central de Venezuela, Apartado Postal 4579, Maracay Z.P. 2101, Venezuela, Email: camacaros@agr.ucv.ve

Keywords: *Digitaria swazilandensis*, earthworm manure, mineral fertilisers, biomass production, mineral concentration

Introduction Chemical fertilisers are expensive and can cause environmental contamination, affecting the biodiversity; it is therefore desirable to use organic manure that is cheaper and can be prepared on the same farm. Earthworm manure is the product of worms that transform organic matter (Salinas & Rojas, 1995), which improves the physical, chemical and biological properties of the soil (Pérez, 1993; Ravera & De Sanso, 1999). The importance of these improvements justifies the use of the worm humus in fertilisation programs in grasslands. Therefore, the objective of this experiment was to evaluate the effect of the earthworm manure on the biomass production and minerals concentration of *Digitaria swazilandensis*.

Materials and methods Plots of 2 x 5 m of *D. swazilandensis* sown three years before the experiment were used. The soil is sandy, low clay, low pH and medium fertility. A randomised design was used, with three repetitions and seven treatments (1 to 7): 0, 125, 250, 500, 750 kg/ha earthworm manure, respectively, 125 kg/ha earthworm manure + 700 kg/ha of 15-15-15 (N,P,K)and 750 earthworm manure kg/ha + 350 kg/ha of 15-15-15, respectively. Total, leaf, stem and weed biomass, as well as concentration of N, P, K, S and Ca were measured. Four samplings were carried out at 6 weekly intervals.

Results Biomass data are shown in Table 1 and mineral content in Table 2. There was no treatment effect (P>0.05) for any of the biomass variables, except stems (P < 0.01). Neither was there any effect (P>0.05) on mineral concentration, except for S (P>0.01). There were no significant differences (P>0.05) among treatments for weed biomass.

Table 1 Effect of the earthworm manure on the biomass of D. *swazilandensis*

Treatments	Biomass (kg DM/ha)			
	Total	Leaves	Stems*	Weeds
1	2004	945.75	879.42	61.67
2	2328	1149.58	910.83	33.33
3	2548	1254.42	1062.67	7.67
4	2536.3	1113.08	1244.42	43.33
5	2367.2	1194.33	881.50	47.00
6	3006.3	1325.33	1545.08	6467
7	1863.8	1062.58	1400.25	45.67

* (P<0.001)

Table 2 Effect of the earthworm manure on the concentration of minerals of D. *swazilandensis*

Minerals (%)			
N	P	K	S*
0.99	0.26	1.51	0.63[a]
1.17	0.25	1.96	0.48[c]
1.04	0.25	1.74	0.53[bc]
1.08	0.28	1.79	0.57[ab]
1.07	0.28	1.84	0.54[bc]
1.08	0.24	1.84	0.55[b]
1.14	0.26	1.92	0.55[b]

* (P<0.01)

Conclusions Although there was no treatment effect and the experiment was very short, a time effect was observed on the biomass production and on the mineral concentration (K and S). Also a time x treatment interaction was detected for S. It is recommended that experiments with earthworm manure should be carried out for longer periods of time.

References

Pérez. I. 1993. La lombriz de tierra, potencial y perspectivas de su producción. En: E. Cardozo (Ed.) III Simposium de Especies Animales Subutilizadas. Universidad "Ezequiel Zamora". Guanare, Venezuela, 14-33.

Salinas,Y, & J. Rojas. 1995. Efecto de las lombrices en el mejoramiento de las características físico-químicas del suelo. Tesis de Ingeniero agrónomo. Facultad de Agronomía, Universidad Central de Venezuela. Maracay, Venezuela. 25pp.

Ravera, R. & De Sanzo, C. Como criar lombrices rojas californianas. Programa de Autosuficiencia Regional (Alta Vista)http://www.geocities.com/hotsprings/spa/9449/lombriz/libro/index.html. Buenos Aires, Argentina

Fire and nutrient cycling in shortgrass steppe of the southern Great Plains, USA

P.L. Ford[1] and C.S. White[2]
[1]USDA Forest Service, Rocky Mountain Research Station, Albuquerque, New Mexico, Email: plford@fs.fed.us,
[2]University of New Mexico, Albuquerque, New Mexico, USA

Keywords: nutrient cycling, fire, shortgrass steppe, potentially mineralisable nitrogen, plant cover

Introduction Fire in semi-arid grasslands releases nutrients bound up in organic matter and accelerates the rate of decomposition in the soil. This research experimentally tested effects of season and frequency of fire on nutrient cycling dynamics in shortgrass steppe. The objective was to identify if fire treatments have the ability to increase potential grassland productivity relative to untreated 'reference condition' grassland. Many such studies focus on short-term, direct effects of fire. However, this study is part of a long-term, 18-year study examining both direct, and indirect effects of fire in the growing *vs.* dormant season at return intervals of 3, 6 and 9 years.

Materials and methods The study is located in semi-arid shortgrass steppe in the southern Great Plains of northeastern New Mexico, USA (36° 31' 20" N, 103° 3' 30" W). The never-ploughed, ungrazed, 160-ha site has mostly native vegetation with the sod-forming *Buchloë dactyloides* and the bunchgrass *Bouteloua gracilis* being the dominant plant cover. The experimental design was completely randomised with 5 treatments and 5 replicate 2-ha plots per treatment. Treatments were 3-year dormant- (3D) and growing-season (3G) burn cycles (twice burned), 6-year dormant- (6D) and growing-season (6G) burn cycles (burned once), and unburned, reference condition (RC) plots. The first two rounds of treatments were applied in 1997 and 2000 with measurements taken during the drought year of 2003. Response variables included % bare ground, litter and live perennial grass cover, soil organic matter content, pH, sodium adsorption ration, field available nitrogen (N) as nitrate and ammonium, potentially mineralisable N (PMN), and other soil and plant nutrients.

Results The only suggested difference in cover variables among treatments (p = 0.07) was between the unburned (RC) plots with a mean of 60% litter cover and the 3-year dormant-season (3D) fire treatment with a mean of 46% litter cover (Figure 1, Table 1). There were significant differences in the levels of boron, calcium, and sodium in vegetation among treatments, but no generalities could be made regarding fire effects. Depending on the frequency, fire can either increase or deplete soil nutrients. The main differences among soil variables occurred between the 6D and 3D fire treatments (Table 1). When differences were suggested, the 6D treatment always had a significantly higher mean than 3D (p = 0.06), but did not significantly differ from the other treatments (Table 1 and Figure 1). In addition, ammonium was significantly higher in 6D than in RC.

Table 1 Tukey's Studentised Range Test means (values
the same letter are not significantly different (alpha = 0.10))

Grouping		Mean (SD)	n	Treatment
Litter %	A	60 (1)	5	RC
	AB	54 (5)	5	6G
	AB	54 (8)	5	6D
	AB	49 (9)	5	3G
	B	46 (10)	5	3D
PMN:	A	41 (8)	5	6D
Nitrate	AB	36 (6)	5	RC
μg/kg	AB	35 (6)	5	6G
	AB	32 (4)	5	3G
	B	30 (2)	5	3D
PMN:	A	0.70 (.09)	5	6D
Ammonium	AB	0.52 (.13)	5	3G
μg/kg	AB	0.51 (.07)	5	6G
	B	0.49 (.19)	5	RC
	B	0.47 (.10)	5	3D

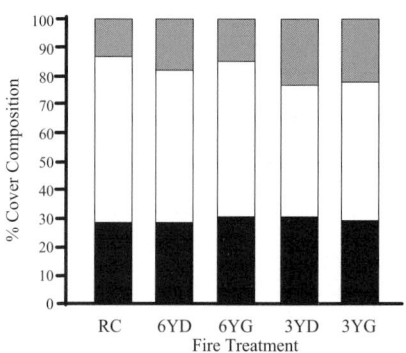

Live grass cover (black), Litter (white), Bare ground (grey)

Figure 1 Ground cover by treatment

Conclusions The current results of this long-term study suggest that in semi-arid grasslands a 3-year fire frequency (burned twice in 6 years), regardless of season, may be too short, and may cause a greater loss of litter and limiting N resources, than other frequencies. The 6-year dormant-season fire (i.e. burned once in 6 years), is the only fire treatment that shows the potential for increased site production relative to 'reference condition' unburned grassland.

Soil quality under permanent and annual pastures: its implications for soil microbial activity and nutrient turnover

R.J. Haynes[1], R.M. Milne[1] and N. Miles[2]
[1]School of Applied Environmental Sciences, University of KwaZulu-Natal, Private Bag X01, Scottsville 3209, South Africa, Email: haynesd@ukzn.ac.za, [2]Cedara Research Station, KwaZulu-Natal Department of the Agriculture, Environmental Affairs, Private Bag X9059, Pietermaritzburg 3200, South Africa

Keywords: microbial biomass, soil organic matter, tillage

Introduction Dairy farming in humid, subtropical parts of South Africa is based on permanent kikuyu grass (*Pennisetum clandestinum*) swards. However, there is a shortage of feed during the winter because low temperatures limit kikuyu growth. As a result, annual pastures incorporating temperate grasses, are grown for winter feed production. The grasses used are typically annual ryegrass (*Lolium multiflorum*) and sometimes perennial ryegrass (*Lolium perenne*). Kikuyu is so invasive in the locality that it becomes dominant within a few years, even if the field is sown to perennial ryegrass. For that reason, the swards are usually incorporated using a rotary cultivator, each summer and resown. Concern has arisen in recent years about soil degradation that is possibly occurring under annually-tilled pastures. Observations have suggested that with time, production has progressively declined and this is suspected to be related to soil degradation, particularly a decrease in organic matter content and a loss of related soil microbial and physical properties. The purpose of this study was to compare the effects of permanent pasture and annually-tilled ryegrass pasture on soil organic matter content, size and activity of the microbial biomass and aggregate stability.

Materials and methods Four commercial dairy farms were selected in the Tsitsikamma region of the Eastern Cape, South Africa. Three fields were chosen. They were under: (1) annually tilled ryegrass, (2) permanent kikuyu pasture and (3) undisturbed native vegetation. Study fields were about 20ha in size. The fields were divided into 100m^2 plots and three separate plots were randomly chosen and sampled to a depth of 10cm. Twenty five samples were taken from each plot and these were bulked and sieved (<2mm). Soils were analysed for organic C, soluble C, microbial biomass C, basal respiration, arginine ammonification rate, the activities of protease, phosphatase and sulphatase and aggregate stability according to the methods described by Dominy & Haynes (2002) and Nsabimana *et al.* (2004).

Results and discussion In comparison with soils under sparse, native grassy vegetation, those under both annual ryegrass and permanent pasture had a higher soil organic C content on very sandy soils of the eastern end of the region. By contrast, in the higher rainfall western side, where native vegetation was coastal forest, there was a loss of organic matter under both types of pasture. Nevertheless, soil organic C, soluble C, microbial biomass C, basal respiration, arginine ammonification rate, protease, phosphatase and sulphatase activities and aggregate stability were substantially less under annual than permanent pasture at all sites. These results reflect the degrading effect of annual tillage on soil organic matter and the positive effect of grazed permanent pasture on soil organic matter, nutrient turnover and aggregation. The lower protease, phosphatase and sulphatase activities under annual than permanent pasture suggest a lower rate of turnover of N, P and S in the annually-tilled soil. This reflects a much lower organic matter (organic N, P and S) content in the tilled soil and the substantial nutrient turnover rate under grazed permanent pasture. There was also a difference in measured soil properties with soil depth. For example, organic C and related soil microbial properties decreased markedly with increasing soil depth under native vegetation and permanent pasture. This is because the input of fresh organic material decreases with depth. By contrast, under annual pasture there was no significant stratification of organic matter. This is due to regular downward redistribution of organic matter during annual tillage. Thus, while microbial activity and nutrient turnover concentrated in the surface few cm under permanent pasture, they are more evenly distributed in the surface 15cm under annual pastures. It was concluded that annual pasture involving conventional tillage results in a substantial loss of soil organic matter, soil microbial activity and soil physical condition under dairy pastures and that systems that avoid tillage and/or increase the longevity of temperate pastures need to be developed.

References
Dominy, C.S. & R.J. Haynes (2002). Influence of agricultural land management on organic mater content, microbial activity and aggregate stability in the soil profiles of two Oxisols. *Biology and Fertility of Soils,* 36, 298-305.
Nsabimana, D. R.J. Haynes & F.M. Wallis (2004). Size, activity and catabolic diversity of the soil microbial biomass as affected by land use. *Applied Soil Ecology*, 26, 81-92.

Micro-field assessment of soil erosion and surface runoff using mini rainfall simulator in upper River Njoro watershed in Kenya

J.O. Onyando[1], M.O. Okelo[1], C.M. Gichaba,[1] W.A. Shivoga[1] and S.N. Miller[2]
[1]Egerton University P.O. Box 536, Njoro, Kenya, Email: jonyando@yahoo.com, [2]Department of Renewable Resources, University of Wyoming, Box 3354, 1000 E. University Dr., Laramie, WY 8207, USA

Keywords: rainfall simulator, erosion, run-off, agricultural land use

Introduction Soil erosion and surface runoff are consequences of integration of several factors and processes within a catchment. The use of a rainfall simulator and run off plots provides a valuable research tool and are often used in soil erosion and surface runoff studies. Cheruiyot (1984) used this approach to study infiltration rates and sediment yield in Kiboko, Kenya. The present study used the same method but with a mini-rainfall simulator (Kamphorst, 1987) to study the effects of different land use treatments on soil loss and surface runoff.

Materials and method The study was carried out on run-off plots, which were used to assess soil erosion and surface runoff in the River Njoro watershed in Kenya. The watershed is currently suffering from severe degradation due to increased settlement and subsequent human activities. There was a randomised block design with five land use treatments and three replicates (sites) per treatment. The site plots were mapped using GPS and plotted in a GIS environment. On every study site, rainfall was applied at an average rate of 10 mm/h on the three plots using the rainfall simulator. This had dimensions of 0.4 m x 0.25 m to give a plot size of 0.1 m². Soil erosion and surface runoff generated from the five plots of different land uses were measured. The soil characteristics bulk density, texture, organic matter content and pH were measured at each experimental site.

Results and discussion Table 1 shows a summary of soil loss from every land use type and other soil properties. Soil erosion decreased in the order agricultural land, deforested areas, grazing land, exotic trees and least in indigenous forest. The values of mean soil loss from each area were 86, 31, 18, 2 and zero g/0.1 m² respectively. There were no significant differences (P<0.05) in soil loss between the following land use areas: deforested, grazing, exotic and indigenous forest. However, there were significant differences between agriculture land use and all the other land use areas. The highest surface runoff was on grazing and the values decreased in the order agriculture, deforested areas, exotic trees and indigenous forest land use. The respective values of mean surface runoff from these areas in ml were 1200, 920, 860, 380 and 20. Statistical analysis at (p<0.05) revealed that there were significant differences between surface runoff from all land use areas except between agriculture and deforested areas and agriculture and grazing lands.

Table 1 Mean soil loss and soil properties for the five land use types in Upper River Njoro watershed

	Bulk density (g/cm³)	Organic matter (%)	Soil pH	Soil texture	Mean soil loss (g/0.1 m²)	Mean surface runoff (ml)
Agriculture	0.85	5.7	6.2	Clay loam	86	920
Grazing	1.05	5.0	5.9	Clay loam	18	1200
Exotic	0.95	6.2	6.4	Clay loam	2	380
Deforest	0.78	10.1	5.8	Sandy clay loam	31	860
Indigenous forest	0.74	9.4	6.2	Sandy clay loam	0	20

Conclusion and recommendation Soil erosion increased in the order of indigenous forest, exotic trees, grazing land, deforested areas and agricultural land whilst surface runoff increased in the order of indigenous forest lands, exotic trees, deforested areas, agricultural land and grazing lands. Since soil erosion and surface runoff depend on rainfall, several watershed characteristics and management practices, many of which could not be investigated with the mini simulator, the results obtained are considered quite preliminary and give only a general impression of relative soil loss and surface runoff. It is recommended that more detailed studies be carried out with a simulator that can generate different rainfall intensities.

References

Cheruiyot, S.K. (1984). Infiltration rates and sediment production of a Kenya bushed grassland as influenced by vegetation and prescribed burning. MSc thesis, Range Science Department, Texas A & M University College, Station Texas.
Kamphorst, A. (1987). A small rainfall simulator for the determination of soil erodability. *Netherlands Journal of Agricultural Science.* 35:407-415.

Validation of the software "Recycling of Nutrients" in dairy-farms of western Cuba

I. Rodríguez, G. Crespo, M.F. Días and S. Fraga
Instituto de Ciencia Animal, Carretera Central Km 47 ½, San José de las Lajas. La Habana, Cuba, Email: idrodri@yahoo.es, irodriguez@ica.co.cu

Keywords: models, recycling, software, grassland, dung

Introduction To maintain a good productivity in the soil-plant-animal system it is necessary to achieve an equilibrium between input and output of nutrients and energy. Models and simulation software have been proposed for this purpose. For dairy farms, models have been utilised to show the behaviour of different aspects, such as: feed utilisation, energy flow, digestibility values and daily weight gains (Freer *et al.,* 1970; Assis & France, 1983, Bruce *et al.,* 1984). However, there are few models showing the interaction between soil-plant-animal components of these systems. The objective of the present study was to validate the software "Recycling of Nutrients" proposed by Ortiz (2000) for dairy farms of western Cuba.

Materials and methods The validation was conducted in 9 dairy farms of western Cuba. Data necessary to apply the software corresponding to the year 2002 were: area of the sward, botanical composition, stocking rate, cattle categories, milk production, daily weight gain, births and sales, of animals, forages and supplements consumed, fertilisation (organic or mineral), type of soil, annual rainfall, extraction of nutrients and production by the sward, dung and urine deposition, biologically N fixation, and others.

Results There was a negative balance of P in all dairy farms (Table 1), and the application of this nutrient is necessary in order to achieve the stability of these systems. Also, K had a negative balance in the majority of the dairy farms. The N balance was satisfactory in the soils and systems where *Leucaena leucocephala* and N fertiliser were present (Figure 1).

Table 1 Annual balance of nutrients in the different dairy farms

Farms	Soil			System		
	N	P	K	N	P	K
1	-17	-8	-76	22	-1	-34
2	-66	-17	-134	18	-1	-50
3	-22	-10	-64	11	-4	-28
4	-11	-10	-54	19	-3	-24
5	120	-3	-23	129	-0.1	-12
6	427	-8	-21	454	-2	5
7	99	-19	108	161	-8	-46
8	49	-8	-17	117	26	-4
9	-14	-7	-52	13	-1.0	-23

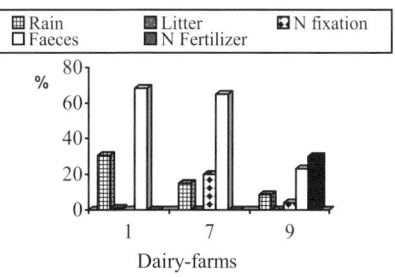

Figure 1 Different N input sources in the dairy farms 1, 7 and 9

Conclusions Software showed that faeces constitute the main source of nutrients in the systems, nevertheless the non-uniformity of deposition in the sward and the high accumulation in the cowshed constitute serious problems to obtain an efficient recycling of nutrients. Therefore, the software showed that is necessary to return to the sward the faeces accumulated in the cowshed and introduce legumes in these systems. The validation trials showed that the software could be a useful tool to manage the nutrient balances in dairy farms.

References
Assis A. G. & J. France (1983). Modelling dairy cattle feeding in the South East Region of Brazil. *Agricultural Systems,* 12, 129.
Bruce J. M., P. J Broadbent. & J. H. Topps (1984). A model of the energy System of lactating and pregnant cows. *British Society of Animal Production,* 30, 2.
Freer M., J. L. Davidson, J. S Armstrong. & J. R. Donnelly (1970). Simulation of summer grazing. Proce*edings XI International Grassland Congress*, Surfers Paradise, 913-917.
Ortiz. R. J. (2000). Modelación y simulación matemática del reciclaje de N, P y K en sistemas de pastoreo vacuno en Cuba. PhD Thesis. Instituto de Ciencia Animal. La Habana Cuba, 112 pp.

Modelling urine nitrogen production and leaching losses for pasture-based dairying systems

I.M. Brookes and D.J. Horne
College of Sciences, Massey University, Private Bag 11 222, Palmerston North, New Zealand, Email: I.Brookes@massey.ac.nz

Keywords: dairy production, nitrogen losses, pasture feeding, supplementary feeds

Introduction Urine from dairy cattle grazing pastures with high crude protein (CP) concentrations is a major source of N lost in drainage water from New Zealand farms. This paper provides predictions of urinary N leaching losses for a range of stocking rates and levels of supplementation.

Method Urinary outputs (kg N/ha) were estimated for dairy herds on notional 100 ha farms stocked at 2.35 (L), 3.00 (M) or 5.00 (H) cows/ha. Farm L was assumed to grow 13,000 kg pasture dry matter (DM)/ha annually. For farms M and H, 100 and 200 kg fertiliser N/ha were applied, to give annual pasture yields of 14,080 and 15,160 kg DM/ha, respectively. A feed budget identified periods when pasture (P; 20% CP) required supplements of either maize silage (MS; 7% CP) or pasture silage (PS; 16% CP) to maintain at least 1,700 kg grass DM/ha. Supplements were fed from Feb. to Sept. on farm M and in all months on farm H, accounting for 16% and 47% of the total DM intakes. Urinary N outputs were estimated by the Cornell Net Carbohydrate and Protein System model (Fox *et al.*, 2004), and leaching losses by the Nitrogen Leaching Estimation (NLE) model of Di & Cameron (2000).

Table 1 Annual N intake, urine N output and N leached

Stocking rate/Supplement	L	MMS	MPS	HMS	HPS
Dietary CP (% DM)	20.0	17.9	19.1	13.9	17.4
Dietary N intake (kg/cow)	150	136	145	107	135
Urine N output (kg/cow)	72	58	67	39	59
(kg/ha)	169	174	200	196	296
N in urine patches* (kg/ha)	889	725	833	490	740
N leached (kg/ha)	7	16	21	23	44

* Assumes urine from a cow covers ~ 0.08 of the area

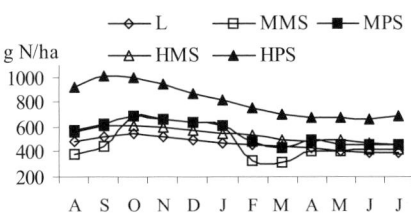

Figure 1 Monthly urine N outputs per ha

Results Dietary CP %, urinary N outputs and N leaching losses are given in Table 1 and Figure 1. Although the CNCPS model is reported to under-predict the ratio of urinary to faecal N, the results described here are supported by Mulligan *et al.* (2004), who showed a positive linear relationship between urinary N output and N intake, in cows fed supplements with high or low CP%. They also observed a decrease in the proportion of urinary N in the total excreta N when the supplement CP% concentration was reduced. Replacing pasture with a supplement of lower CP% can reduce the concentration of N under urine patches. On farm M, this occurs mainly in early spring and autumn, as no supplements were required in late spring and summer. The extent of N leaching was related to N fertiliser use and N concentration in the urine patches. Despite a two-fold increase in stocking rate and the higher N fertiliser usage, N concentration under urine patches was 50% lower on farm H with MS than on farm L. The reduction in urinary N production per cow and the increased area of urine deposition on the HMS farm results in N leaching losses that are not much greater than those for the M systems. However, when pasture silage was fed on farm H, N leaching losses were much higher.

Conclusions This analysis suggests that where increases in stocking rate are accompanied by the use of maize silage, intensification may not pose an increased risk of nitrogen leaching from dairy farms. There is a need for quantification of urine N output on farms as affected by stocking rate and diet, and for measurement of the exact area occupied by urine patches.

References

Di, H.J. & K.C. Cameron (2000). Calculating nitrogen leaching losses and critical nitrogen application rates in dairy pasture systems using a semi-empirical model. *N Z Journal of Agricultural Research*, 43, 139-147.

Fox, D.G, L.O.Tedeschi, T.P.Tylutki, J.B.Russell, M.E.Van Amburgh, L.E.Chase, A.N.Pell & T.R.Overton (2004). The Cornell Net Carbohydrate and Protein System model for evaluating herd nutrition and nutrient excretion. *Animal Feed Science and Technology*, 112, 29-78.

Mulligan, F. J., P. Dillon, J.J. Callan, M. Rath & F.P. O'Mara (2004). Supplementary concentrate type affects nitrogen excretion of grazing dairy cows. *Journal of Dairy Science*, 87. 3451-3460.

Management options to reduce N-losses from ploughed grass-clover

J. de Wit, G.J. van der Burgt and N. van Eekeren
Louis Bolk Institute, Hoofdstraat 24, 3972 LA Driebergen, The Netherlands, Email: j.dewit@louisbolk.nl

Keywords: nitrate leaching, N-losses, grass-clover, silage maize

Introduction Nitrate (NO_3^-) leaching from grassland can be kept at acceptable levels, but is often high after ploughing for grassland renewal or for silage maize/grain production. In on-farm research with several organic farmers, management options are being explored to save scarce organic manure and to reduce N-losses.

Materials and methods A trial was conducted on a loamy löss (2.6% OM in 0-30 cm) with four treatments: A) 'standard' farmer practice, i.e. application of 20m³ of slurry in early spring (21 April), mowing grass (26 May), soil ripping, application of 18m³ slurry, ploughing and sowing of maize (1 June); B) and C) were similar to A, but with no slurry application before ploughing, or no slurry at all, respectively. D) as C), but with early soil ripping (27 April) to enhance N-availability. Field history included three years of arable crops followed by one-year grass-clover (> 50% clover). Mineral N-availability was measured 9 times at regular intervals, maize production was assessed by harvesting three rows of 3 m/plot. Mineral N-availability in the layer of 0-30 cm was modelled by the soil-N flow model NDICEA (Koopmans & Bokhorst, 2002).

Results The recorded mineral N-availability was lower than standard Dutch advice (185 kg N/ha), while recorded residual N was higher than the maximum Dutch advice to attain NO_3 leaching <50 mg/l, i.e. 90 kg N (Table 1). The recorded production follows mineral N-availability reasonably closely except for the early soil ripping treatment. Figure 1 shows the NDICEA-results, matching the recorded mineral N-levels closely (Table 2) except for the late growing period, possibly due to underestimation of the capillary capacity resulting in a predicted decomposition rate being lower than reality (2003 was an extremely dry year). Figure 1 also shows NDICEA-results of "treatment E", being similar to D but with silage maize following three-year-old grass-clover. Mineral N-availability is predicted to be much higher due to the decomposition of a much higher amount of easily available soil OM (roots and living soil organisms).

Table 1 Maize production and mineral availability for different management options

Treatment	Maize production (ton DM/ha)	Mineral N (kg/ha) at: 21-6 (0-60 cm)	Mineral N (kg/ha) at: 25-11 (0-90 cm)
A	16.8	136	135
B	15.8	133	100
C	15.3	113	58
D	16.1	175	103

Table 2 Recorded mineral N-availabity (0-30 cm, kg/ha) for different treatments at different dates

	23-4	7-5	6-6	21-6	3-9	25-11
A	25	23	83	107	65	58
B	20	18	89	110	26	46
C	13	15	71	88	24	37
D	10	66	121	143	77	49

Conclusion Silage maize production following grass-clover was increased by manure application but N-losses, both residual- N and denitrification losses (not shown), increased much more. This is particularly true for grass-clover leys older than 2 years. Early soil ripping enhances mineral N-availability and silage maize production without applying scarce organic manure, but also reduces grass production and increases potential N-losses. Therefore, future experiments will be concentrated on maize production without destroying the grass-clover to facilitate a more direct transfer of fixated N from clover to maize and possibly lower inherent N-losses.

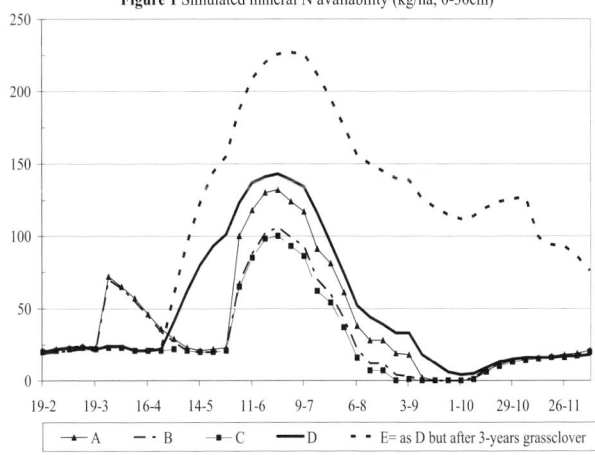

Figure 1 Simulated mineral N availability (kg/ha, 0-30cm)

References

Koopmans, C. J. & J. Bokhorst, 2002. Nitrogen mineralisation in organic farming systems: a test of the NDICEA model. *Agronomie* 22, 855-862

Nutrient cycling in a tropical grazing land ecosystem of southern India

K.S.T.K. Karunaichamy[1] and K. Paliwal[2]
[1]*Rubber Research Institute of India, Kottayam-686 009, India, Email: Karunai@rubberboard.org.in,*
[2]*Department of Plant Sciences, Madurai Kamaraj University, Madurai – 625 021, India*

Keywords: accumulation, annual uptake, live shoot, nutrient cycling

Introduction The nutrient component of any ecosystem operates in a dynamic state through a series of inputs and outputs of the essential elements. Nutrients from plants are continuously transferred to soil via litter formation, which act as a reservoir for the plants in an ecosystem. Most of the studies on nutrient budgets and flux rates have been reported in forest communities and to a lesser degree in grassland. The present study was, therefore, aims to understand the variation of calcium (Ca) and magnesium (Mg) in the vegetation compartments and to quantify annual budget by estimating the annual uptake from soil and its release to soil.

Materials and methods The study was conducted at Maramalai, Kanniyakumari District, Tamil Nadu ($8^0 10'$ N; $77^0 27'$ E). *Cympogon flexuosus* Watson is the major dominant species of the grazing land ecosystem. Aboveground vegetation was harvested at monthly intervals laying ten quadrats (50X50 cm) randomly. Litter was collected separately. In order to sample the root, three soil monoliths (25X25X30 cm) were excavated from the centre of each harvested quadrats. Soil samples were collected monthly by digging to 30 cm depths. Ca and Mg concentration in plant components and soil were estimated using an atomic absorption spectrophotometer. The transfer of nutrients between various compartments and the release of nutrients through root and litter disappearance were calculated following Singh & Yadava (1974).

Results Nutrient content in plant components were maximum in the live shoot followed by litter, root and dead shoots (Table 1). Generally live shoots contain a greater amount of nutrients than the root. The decline of nutrient concentration from live shoot to dead shoot is a common phenomenon in temperate (Agrawal, 1988) as well as in tropical grazing lands (Karunaichamy, 2003). Nutrient levels may be attributed to their relative requirement in the metabolic process and to availability of the nutrients in the grazing land ecosystem. Mg concentration in root was positively correlated with soil moisture in different months as evident from the regression, Y=0.034+0.013X (r=0.88, P<0.05). Annual uptake and release of Mg was maximum followed by Ca (Table 2). From the total uptake, large amount of nutrients was channelled to live shoots than root. The amount of nutrients returned to the soil through root was more than litter decomposition. It is interesting to note that major portion of nutrients was retained by the soils. Less than 1% of Mg and 6% of Ca was channelled through biological circulation.

Table 1 Nutrient concentration (%) in various components (± SE; n = 5)

Nutrients	Live shoot	Dead shoot	Litter	Root
Ca	0.25 ± 0.03	0.17 ± 0.03	0.29 ± 0.04	0.22 ± 0.03
Mg	0.29 ± 0.02	0.18 ± 0.02	0.33 ± 0.06	0.28 ± 0.04

Table 2 Net uptake, release and retention of nutrients (kg/ha/yr)

Nutrients	Soil	Uptake	Retention	Release
Ca	166	102 (100)	88 (85.4)	15 (14.6)
Mg	1256	122 (100)	103 (85.0)	18 (15.0)

Values parenthesis is percentage relative to uptake

Conclusions The results of the present study demonstrate that cycling of Ca and Mg in a tropical grazing land at Maramalai is regulated by their greater accumulation in live shoots and slow recycling through decomposition. The present study clearly indicates that grazing land of this study area not only reduces the nutrient economy of the system but also slow down its circulation within the plant biomass. The lowest rate of transfer of nutrients from live shoots to dead shoots and the lower amount of nutrient releases through litter and root in the grazing land becomes clear.

References

Agrawal, A.K. (1988). Nutrient structure and dynamics in a temperate grassland community of Western Himalaya (Garhwal), India. *Tropical Grasslands*, 22(1), 33-39.
Karunaichamy, K.S.T.K. & K. Paliwal (2003). Nutrient dynamics and inventory in a tropical grazing land ecosystem at maramalai, southern India. In:N. Allsopp, A.R. Palmer, S.J. Milton, K.P.Kirkman, G.I.H. Kerley, C.R. Hurt & C.J. Brown (eds.). *Proceedings of the VII International Rangeland Congress*, 84-86
Singh, J.S. & P.S. Yadava (1974). Seasonal variation in composition, plant biomass and net primary productivity of a tropical grassland at Kurukshetra, India. *Ecological Monographs*, 44, 351-376.

Integrated nutrient management for natural grasslands of mid-hills of Himalayas

I. Dev, S. Narayana, R. Singh, B. Misri, S. Sareen and S. Radotra
Regional Research Centre, Indian Grassland and Fodder Research Institute, CSK HPKV Campus, Palampur (H.P) 176 062 India, Email: inderdev@lycos.com

Keywords: integrated nutrient management, natural grassland, Himalayas

Introduction Livestock rearing is an important pursuit in mountain farming in India and plays a crucial role throughout the country. The preponderance of marginal and small landholdings (about 82%) in hilly regions does not allow the farmers to allocate even a small part of their land exclusively for forage production. In Himachal Pradesh state of India about 1.16 m ha (20% of the total area) is under permanent pastures and other grazing lands and none of the natural grasslands are fertilised in any form. Existing grasslands have deteriorated to such an extent that their carrying capacity is only 1.05 ACU (Adult Cattle Unit, with an average body weight of 350 kg)/ha (Vashist *et al.*, 2000). Biofertiliser-based technologies could be appropriate and cost effective approaches that are easy to adopt and eco-friendly. Response may arise from increased populations of phosphate solubilisers in the rhizosphere in P- deficient soils resulting in mobilisation of insoluble phosphorus (Raghu & Mac Rac, 1967). The study was undertaken with the main objectives of assessing the effects of biofertilisers on productivity and quality of natural grassland and the level of N and P substitution by biofertilisers.

Materials and methods The study was undertaken at a community natural grassland in the mid-hills of Himachal Pradesh state (1300 m altitude, 32^0 6' N-latitude and 76 3' E- longitude in north western Himalaya) in India during 2001and 2002. Thirteen treatments comprised a control (no application); recommended application of N and P (60 and 40 kg /ha); sole application of Azotobacter, Phosphobacteria; Azotobacter + Phosphobacteria; nine combinations of 50, 75 and 100% of the full N rate with 50, 75 and 100% P along with Azotobacter + Phosphobacteria. The trial had a randomised block design with three replicates.

Results *Heteropogon contortus* (62.89 %), *Eulalia* sp (15.34%), *Chrysopogon fulvus* (11.96), *Arundinella nepalensis* (3.37%) and *Cyperus deformis* (1.53%), were the dominant species in the grassland. Sole application of Azotobacter produced 0.27 t/ha of herbage dry matter (DM) compared to the control and the combined use of the two bacteria produced a further response of 0.36 t DM/ha (Table 1). Application of 50, 75 and 100% N and P along with Azotobacter and Phosphobacteria as well as the treatment with the recommended application of N and P produced more herbage biomass compared to no application of fertilizers or the sole application of Azotobacter and Phosphobacteria. Application of 75 % of N and P with Azotobacter + Phosphobacteria gave significantly higher herbage production compared to the sole bacterial applications, control and 50% N and P application. The herbage yield from the treatments with the bacterial additions together with 75% N and 100% P or 100% N and 75% P were similar to that with full application of 100 % N and P. Chemical analysis did not indicate any significant differences in crude protein content.

Table 1 Impact of biofertilsers on herbage production (t DM/ha) (average of two years)

Treatments	
T1- Control	2.07
T2- Azotobacter	2.34
T3- Phosphobacteria	2.04
T4- Azotobacter + Phosphobacteria	2.70
T5- 50% N + 50% P + Azotobacter + Phosphobacteria	2.90
T6- 50% N + 75% P + Azotobacter + Phosphobacteria	2.96
T7- 50% N + 100% P + Azotobacter + Phosphobacteria	2.98
T8- 75% N + 50% P + Azotobacter + Phosphobacteria	3.05
T9- 75% N + 75% P + Azotobacter + Phosphobacteria	3.08
T10- 75% N + 100% P + Azotobacter + Phosphobacteria	3.35
T11- 100% N + 50% P + Azotobacter + Phosphobacteria	3.02
T12- 100% N + 75% P + Azotobacter + Phosphobacteria	3.33
T13- 100% N + 100 % P + Azotobacter + Phosphobacteria	3.45
T14- Recommended N and P	3.29
SE of mean values	0.06
CD	0.17

Conclusions Azotobacter and Phosphobacteria could substitute about 25 % of the requirement for N and P, presumably through N fixed by Azotobacter and P solubilised by Phosphobacteria.

References
Raghu ,K & J.C Mac Rac (1967). Effect of nitrogen mineralization and fixation and selected bacteria. *Canadian Journal of Microbiology*, 13, 621.
Vashist, G.D, P. Mehta, A. Kumar, S.K. Sharma & D.C. Katoch (2000). A study of socio-economic aspects of forage and fodder crops. A case study of availability and requirement in Himachal Pradesh. Dept of Agricultural Economics, CSK HPKV, Palampur, 121 pp.

Theme B: Grassland and the environment

Section 20

Multifunctional grassland

Abiotic resource efficiency of grassland production systems in north-west Europe

M. Wachendorf and F. Taube
Institute of Crop Science and Plant Breeding, Christian-Albrechts-University, 24098 Kiel, Germany, Email: mwach@email.uni-kiel.de

Keywords: nitrate leaching, energy efficiency, grazing, white clover, carbon dioxide emission

Introduction Nitrate leaching and energy efficiency are key criteria of a resource efficient grassland production system. A four-year field experiment was conducted to evaluate the effects of N input and defoliation system on both criteria. The objective of this experiment was to develop strategies which help to facilitate future grassland production with maximum environmental friendliness.

Materials and methods The field experiment consisted of all combinations of four defoliation systems (GO=grazing only; MS I/II=mixed systems with one cut and two cuts in spring, respectively, and subsequent grazing; CO=cutting only), four mineral N application rates (0 to 300 kg N/ha per yr), and two slurry levels (0 and 20 m³ slurry/ha per yr) (Trott *et al.*, 2004). An old sward on a sandy soil in northern Germany was oversown with seeds of white clover/grass in the year before the start of the experiment. Grazing was carried out with heifers. Leaching losses were assessed by ceramic cups installed at 60 cm below ground level (Wachendorf *et al.*, 2004). Fossil energy input comprised all field and farm activities, including both direct (diesel use for field operations) and indirect (fossil energy input in the manufacture and distribution of fertilisers, pesticides, *etc.*) contributions (Kelm *et al.*, 2004). Energy output was the net energy yield in grassland production. Energy data were converted to CO_2 emissions with similar emission factors for diesel and mineral N fertiliser (82.6 and 81 kg CO_2/GJ, respectively).

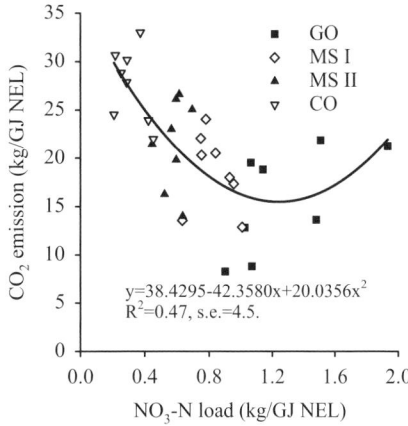

Figure 1 Relationship between CO_2 emission and nitrate-N load in the drainage water from grassland (GO=grazing only; MS I/II=mixed systems with one cut and two cuts in spring, respectively, and subsequent grazing; CO=cutting only; NEL=net energy lactation). Average of 1997-2002

Results Nitrate leaching losses were strongly affected by defoliation system, with lowest values in CO (NO_3 concentrations of the leachate were generally below the EU threshold of 50 mg NO_3 /l) and highest values in GO (up to 250 mg NO_3 /l, corresponding to 114 kg N/ha per yr). Intermediate NO_3 losses occurred in both mixed systems. In all systems, NO_3 losses increased with increasing N input. Without mineral N fertiliser application, the total input of fossil energy ranged from 4.2 GJ/ha (pasture) to 11.2 GJ/ha (cutting-only), of which the greatest proportion was due to diesel use. With increasing mineral N fertiliser rates, the total energy input increased linearly by 38.7 MJ/kg N (pasture) to 44.7 MJ/kg N (cutting-only). Cutting systems produced the highest CO_2 emission per GJ NEL and the lowest NO_3-N loads (Figure 1). With increasing grazing intensity, CO_2 emission was reduced whereas N loads increased continuously. The CO_2 emissions were lowest in unfertilised pastures, but increased again with increased input of mineral N fertiliser.

Conclusions The environmental benefits from cutting systems with regard to nitrate leaching losses should be regarded in the context of significantly lower energy efficiency and higher CO_2 emissions. Unfertilised, white clover-based pastures or mixed systems combine low CO_2 emissions with moderate N leaching losses. Hence, such systems may provide promising strategies for future grassland production.

References

Kelm, M., M. Wachendorf, H. Trott, K. Volkers & F. Taube (2004). Performance and environmental effects of forage production on sandy soils. III. Energy efficiency in forage production from grassland and maize for silage. *Grass and Forage Science*, 59, 69-79.

Trott, H., M. Wachendorf, B. Ingwersen & F. Taube (2004). Performance and environmental effects of forage production on sandy soils. I. Impact of defoliation system and nitrogen input on performance and N balance of grassland. *Grass and Forage Science*, 59, 41-55

Wachendorf, M., M. Büchter, H. Trott, and F. Taube (2004). Performance and environmental effects of forage production on sandy soils. II. Impact of defoliation system and nitrogen input on nitrate leaching losses. *Grass and Forage Science 59*, 56-68.

CAP reform and nitrate restrictions: implications for Irish grass based dairy production systems

A.M. Butler[1], M. Wallace[1] and P. Dillon[2]

[1]Department of Agribusiness, Extension and Rural Development, University College Dublin, Dublin 4, Ireland, Email: annemarie.butler@ucd.ie, [2]Teagasc, Moorepark, Fermoy, Co. Cork, Ireland

Keywords: linear programming, dairy cows, decoupling, nitrates directive

Introduction The benefit of a systems approach to analysing production situations has long been recognised in agricultural research. The development and application of production-oriented dairy models offer tremendous capabilities for both encompassing the realities faced by producers while also considering the adaptation possibilities available to them in light of internal and external forces of change. This farm level dairy model represents one such approach. The objectives of the study were: (1) to develop a comprehensive farm-level model of Irish milk production systems and (2) to apply the model to identify optimal adaptation strategies of dairy farmers within the context of European policy reform. This paper examines the implications of both the Luxembourg Agreement and the imposition of the Nitrates Directive on Irish dairy systems.

Materials and methods The farm model employs a linear programming framework which was constructed in Excel and solved using GAMS software. The model consisted of 388 activities and 218 resource constraints with the objective being profit maximisation. The comprehensiveness of the model is recognised by the detailed specification of alternative configurations of activities and constraints. Activities include forage production, dairy cow activities, subsidiary activities, purchase of inputs and sale of outputs, land, housing and quota, nutrition, labour and capital. The technical coefficients pertaining to production were based on a series of research experiments at Moorepark. An initial model application was specified assuming an area farmed of 50 ha and a milk quota of 468,000 kg (fixed for the purposes of this example). The farm model was initially solved to identify the optimum system of milk production under a baseline situation (2003). The model was then resolved to reflect the policy environment under a decoupled system as is forecast for 2007. The model was finally solved to reflect the imposition of both the decoupled system and the introduction of the EU Nitrates Directive. This directive details maximum organic N and total N limits of 170kg/ha and 260kg/ha respectively.

Results Table 1 presents a comparison of key results from model optimisations. There was a considerable reduction in N application on the grazing ground with the imposition of the Nitrates Directive. This resulted in a substantial increase in the level of concentrate supplementation while grass and silage intakes fell. While the inclusion of a beef enterprise was profitable under both the baseline and decoupled scenarios, it subsequently fell from the optimal plan. Farm net profit (FNP) over the period reduced substantially. From the baseline to the decoupled and nitrates scenario, FNP fell by €17,377, which represented a 26% drop in income.

Table 1 Comparison of key model results

	Baseline	Decoupled	Decoupled + Nitrates Directive
Grazing level of N	330 kg/ha	330 kg/ha	131 kg/ha
Number of cows	75	75	71
Month of calving	Feb./Mar	Feb./Mar	Feb./Mar
Avg. milk/cow (kg)	6577	6577	6587
Feed budget (kg DM/cow)			
Grass	3738	3738	3575
Silage	1465	1465	967
Concentrate	291	291	861
Beef animals	10	10	0
S.R. (lu/ha)	2.46	2.46	2.05
Farm Net Profit	€67,896	€57,064	€50,519

Conclusions The results indicate that dairy producers face considerable income pressures from both imposed and pending EU policy amendments. Minimisation of this negative income effect may be achieved through subsequent adaptation and configuration of system resources as shown. The dairy model allows such adaptation strategies to be identified and examined.

Species-rich grassland as an ecological good in an outcome-based payment scheme

E. Bertke, R. Marggraf and J. Isselstein
Research Centre Agriculture and the Environment, Georg-August University of Goettingen, Am Vogelsang 6, 37075 Goettingen, Germany, Email: ebertke@gwdg.de

Keywords: biodiversity, outcome-based payment scheme, species-rich grassland

Introduction Agriculture plays an important role in protecting the biodiversity of the rural environment. Since the reform of the EU's common agricultural policy (CAP) in 1992, agri-environment schemes have been supported by the EU within the framework of the second pillar of CAP. In these programmes, farmers were rewarded for environmental services. The predominantly action-oriented programmes imply particular disadvantages; they tend to lack economic efficiency and to fail with regard to the conservation and improvement of biodiversity (Kleijn & Sutherland, 2003; Wilhelm, 1999). This situation was the starting point for the development of an outcome-based payment scheme. We focus on the process of defining ecological goods - in particular with regard to grassland - as the results of ecological services, which are to be remunerated by means of this innovative payment system.

Payment system The payment scheme differs from current agri-environment programmes in four main aspects. First, it is an outcome-based scheme. Farmers will not be rewarded for particular actions, but for the results of ecological services, i.e. the ecological 'goods' of plant species diversity. Secondly, the design of the programme is based on fundamental components of market economies, such as supply and demand. Farmers can offer ecological goods voluntarily, for example, in a bidding procedure. Thirdly, the payment scheme is organised regionally in accord with the EU principle of subsidiarity. Lastly, the preferences of the local population will be taken into account in a participatory approach, with a regional advisory board making decisions according to local demand for ecological goods. This board consists of local stakeholders from nature conservation, agriculture and local government (Gerowitt *et al.*, 2003). These four aspects will improve the cost-efficiency of schemes aimed at the conservation of biodiversity in agricultural landscapes and will increase the social acceptance of environmental programmes. The described system of reward was developed for the administrative district Northeim in the south of Lower Saxony (Germany), but is transferable to other regions.

The definition of ecological goods connected to grassland Ecological goods have to be defined by transparent floristic criteria that meet specific requirements: (i) the ecological benefits should exceed those achievable merely by good farming practice - a direct connection between the criteria and grassland management is therefore necessary; (ii) they have to be clearly defined - farmers should easily be able to demonstrate fulfilment; (iii) floristic criteria should imply additional benefits to fauna and abiotic resources; and (iv), criteria have to be adapted to regional conditions and therefore have to be assessed on the basis of regional investigations.

The aims of the production of "ecological goods grassland" are the preservation of grassland on marginal sites in particular, the promotion of species-rich grassland and the protection of regionally endangered plant communities. In relation to these requirements, two criteria are suitable for the definition of the "ecological goods grassland": (a) the number of species per area unit, and (b) a catalogue of grassland species that are adapted to extensive grassland management and that are characteristic of regionally endangered plant communities. The number of plant species is strongly dependent on grassland management, e.g. N-fertilisation and the frequency of defoliation. This relationship allows the setting of a specific number of species as a minimum level for reward. Moreover, the total number of species is correlated with the number of forbs in the grassland, so that we can use the number of forb species per area unit as an indicator for the total plant species diversity, saving time and effort compared to counting all species. Three quality levels of "ecological goods grassland" have been defined. The lowest level is defined by the number of forbs ($>=8$ species of herbs/12.6 m^2), whilst higher levels (grassland II and III) are defined by this same number of forbs and additionally by the occurrence of species from the above-mentioned catalogue. The ecological goods are achieved when the required criteria are met on the whole field. Standardised methods for the determination and the control of such goods have been developed.

References

Gerowitt, B., J. Isselstein & R. Marggraf (2003). Rewards for ecological goods – requirements and perspectives for agricultural land use. *Agriculture, Ecosystems and Environment*, 98, 541-547.
Kleijn, D. & W.J. Sutherland (2003). How effective are European agri-environment schemes in conserving and promoting biodiversity? *Journal of Applied Ecology*, 40, 947-969.
Wilhelm, J. (1999). Ökologische und ökonomische Bewertung von Agrarumweltprogrammen: Delphi-Studie, Kosten-Wirksamkeits-Analyse und Nutzen-Kosten-Betrachtung. Peter Lang, Frankfurt am Main, Berlin.

Silvopastoral systems: analyses of an alternative to open swards

J. McAdam[1], M.R. Mosquera-Losada[2], V. Papanastasis[4], A. Pardini[3] and A. Rigueiro-Rodríguez[2]
[1]Dept. of Applied Plant Science, Queen's University Belfast and Applied Plant Science Division, Dept. of Agriculture and Rural Development, Newforge Lane, Belfast, BT9 5PX, UK, jim.mcadam@dardni.gov.uk,[2] Crop Production Department. University of Santiago de Compostela. Campus de Lugo. 27002 Lugo, Spain, [3]Department of Agronomy and Land Management, University of Florence, Italy, [4]Laboratory of Rangeland Ecology, Aristotle University, 54124 Thessaloniki, Greece

Keywords: silvopasture, multiple land use, ecosystems, environment, production, western Europe

Introduction Silvopasture is a sustainable land use management practiced in most continents in the world including parts of southern Europe, but is not broadly used in northern and western Europe. The importance of this practice has been recognised and the last draft of the EU regulation by the European Agricultural Fund for Rural Development (http://europa.eu.int/comm/agriculture/capreform/rurdevprop_en.pdf) includes specifically funding for establishment of agroforestry practices in Europe. This paper discusses the advantages of managing semi intensive grassland within a silvopastoral system from an ecological, productive and social point of view in the south, south-central and western countries of Europe.

Materials and methods A comparison between intensively managed open swards (with high nitrogen inputs and concentrate use) and silvopastoral land use was made based on a review of research and statistical data from Greece, Italy, Spain, France and United Kingdom grouped in three areas: Mediterranean, Atlantic with dry summers and Atlantic with humid summers.

Results The main advantages of silvopastoral systems from a productive, ecological and social point of view and across the different climatic regions of Europe are shown in Table 1. Generally, productivity is higher in silvopastoral systems, as they combine short and long term returns from land. Other relevant aspects are that silvopastoral systems will enhance production of other products, such as medicinal plants or mushrooms. Silvopastoral systems will also allow extension to the grazing season because pastures growing under trees usually have more favourable temperature (lower in summer and higher in winter) and humidity profiles than open swards. This will reduce concentrate use per year. Stocking rates will also be lower than those in intensive systems. From an ecological point of view, silvopastoral systems present a good tool for carbon sequestration and reduced nitrogen contamination with N efficiency being promoted through uptake by tree roots. Biodiversity is enhanced through the creation of spatial diversity in the habitat, which increases the number of species and the continuity between forests and cropped land. Silvopastoral systems are needed in those areas where fire risk is important. Overstocking promotes erosion, compared with forestry land, as well as reducing regeneration. This can be solved when regeneration is an objective through use of tree protectors or by fencing areas. From a social point of view these systems enhance rural population stabilisation and tourism is favoured by multiple land use.

Table 1 Qualitative characterisation of advantages of silvopastoral systems (G: Greece; I: Italy; F: France; Sp: Spain; I: intensive grassland; S: Silvopastoral; pop: population; stab: stabilisation; **P**: productive; **E**: Ecological; **S**: Social; Med: Mediterranean, DS: Dry summer; HS: Humid Summer).1: highest value, 2: lowest value

		Med	Atlantic DS		Atlantic HS				Med	Atlantic DS		Atlantic HS	
		G,I,Sp	Sp,F		UK				G,I,Sp	Sp,F		UK	
		I	S	I	S	I	S			I	S	I	S	I	S

		I	S	I	S	I	S			I	S	I	S	I	S
P	Global productivity/year	2	1	2	1	2	2	E	Fire risk reduction	1	1	1	1	-	-
	Animal production	1	2	1	2	1	2		Erosion reduction	2	2	2	2	2	2
	Animal product quality	2	1	2	1	2	1		Carbon sink	2	1	2	1	2	1
	Short term return	1	2	1	2	1	2		Animal welfare	2	1	2	1	2	1
	Long term return	2	1	2	1	2	1		Biodiversity	2	1	2	1	2	1
	Other products	2	1	2	1	2	1		N leaching reduction	1	2	1	2	1	2
	Extended grazing	2	1	2	1	2	1	S	Tourism	2	1	2	1	2	1
	Stocking rate	1	2	1	2	1	2		Rural pop. stab.	1	1	1	1	1	1
	Concentrate use/year	1	2	1	2	1	2		Landscape amenity	2	1	2	1	2	1

Conclusions This study demonstrates that silvopastoral systems are a means of managing grassland in a sustainable way as they increase the output of animal and other products, increase biodiversity, extend the grazing season, reduce concentrate use, stocking rate, fire risk, erosion, habitat fragmentation and N leaching whilst enhancing landscape amenity and carbon storage compared with high input farms.

Sustainable grazing on saline land in Western Australia – multidisciplinary research linking producers and scientists

H.C. Norman, D.G. Masters, M.G. Wilmot, A.J. Rintoul, R. Silberstein, E. Lefroy and T. York
CSIRO Livestock Industries, Private Bag 5, Wembley, WA 6913, Australia, Email: Hayley.Norman@csiro.au

Keywords: dryland salinity, animal production, participatory research, biodiversity, hydrology

Background Dryland salinity is one of the most critical environmental issues challenging Western Australian farmers. Currently 10% of the cropping zone (1.8 million ha) is salt-affected and this is predicted to increase dramatically in the next 50 years (NLWRA, 2001). Animals grazing saline pasture systems represent the most likely large-scale opportunity for economic return from saline land in the short to medium term. To date, few farmers have invested in large-scale revegetation of saline land as the economic return from grazing has not been perceived to cover costs. Furthermore other benefits of saltland pasture systems, such as biodiversity, water use and improved quality of animal products have not been quantified.

Profitable and sustainable grazing on saline lands in Western Australia – site 1 research project This project brings together several leading farmers with scientists from across a large range of disciplines to assess the impacts of saline pasture systems on water, salinity and biodiversity in addition to animal production and profitability. Producers, animal scientists, hydrologists, ecologists, agronomists and agricultural economists have worked together from project initiation through to experimental design and implementation. The primary research site is situated on a 10,000 ha sheep (5,000 Merino ewes) and cropping farm in Tammin, one of the most salt-affected parts of southwestern Australia. The producer is a leading advocate of saltland pastures and has revegetated nearly half of the 2,000 ha of saltland on his farm. The research is spread across 100 ha and divided into five treatments encompassing; unimproved volunteer saltland pasture, saltbush (*Atriplex* spp.) and volunteer pasture (typical of most saltland pastures in this area) and saltbush with a sown 'high quality' (legume) understorey (thought to be the best system for animal production). There are several areas where plant agronomists are testing new pastures species and demonstrating alternative options.

Returns from animal production are expected to be key drivers in the adoption of saltland pastures. The animal research aims to quantify the benefits of grazing saltbush-based pastures during the autumn when conventional annual pastures are of poor quality in mediterranean-type climates. Plant diversity, biomass production, nutritive value and selection by animals are key components of this research. Results to date suggest that substantial decreases in the cost of establishment of the pastures will be required to justify returns from animal production.

We aim to quantify the environmental impact of saline pasture systems on water, salinity and biodiversity. Two 25 ha plots of unimproved saltland were isolated to allow comparison of water run-off, salt loss, sediment and nutrient fluxes. After one year of measurement, one plot was planted with the saltbush and sown understorey system to allow measurement of any hydrological benefits of the saltland pasture system from establishment through to maturity. Data demonstrates that the saltbush pastures are using more rainfall than the unimproved pastures. This could have a large impact on the adoption of saltbush for recharge control on valley floors.

The biodiversity value of the saltland pasture treatments is being compared with adjacent remnant vegetation and unimproved saltland. The biodiversity research encompasses elements of landscape function (measured using a Landscape Functional Index), microbial respiration, invertebrate and plant diversity (species and functional groups). Preliminary data suggests that saltland pastures are botanically diverse. The modified saltbush plots have the highest levels of microbial biomass in the soil (a sign of soil health).

The research plots have been characterised extensively for cross-site comparison and to aid in the extrapolation of data to other farms. This project is one of four research projects operating over 12 sites across Australia. Synthesis of knowledge is being fostered through common protocols for site characterisation and data collection, a customised database and the development of national themes. Research outcomes are being communicated through field days, scientific publications and formal communication products such as brochures and books.

Acknowledgement This project is part of the national Sustainable Grazing on Saline Lands project supported by Land, Water and Wool, the CRC for Plant-based Management of Dryland Salinity, CSIRO and the Department of Agriculture, Western Australia.

Reference
NLWRA (2001). Australian Dryland Salinity Assessment 2000: Extent, Processes, Monitoring and Management Options. National Land and Water Resources Audit, Canberra.

Energy use and energy use efficiency of specialised dairy farms in Flanders

M. Meul, F. Nevens, I. Verbruggen and D. Reheul
*Flemish Policy Research Center for Sustainable Agriculture, Potaardestraat 20, 9090 Gontrode, Belgium,
Email: marijke.meul@ugent.be*

Keywords: energy use, eco-efficiency, dairy farms, Flanders

Introduction Our highly mechanised agriculture largely depends on ever declining stocks of fossil energy and hence contributes to global warming through the emission of greenhouse gases. Therefore, energy use (efficiency) is an important aspect of eco-efficient and sustainable agricultural production systems. In this study, we estimated direct and indirect energy use on a representative set of specialised dairy farms in Flanders (334 farm datasets in 1989-1990 and 147 farm datasets in 2000-2001) and we calculated their energy use efficiency. We studied the observed evolution between 1989-1990 and 2000-2001.

Materials and methods Direct energy used during farm operations was calculated based on the amount of diesel, lubricants, electricity and other energy carriers directly used on the farms. Calculated indirect energy (i.e. energy consumed during the production of farm inputs) included the energy used for the production of mineral fertilisers, seeds, pesticides, concentrates, forages and machines.

Results The results in Table 1 show that the major part of the energy used on dairy farms can be attributed to production of concentrates and mineral fertiliser. In fact, indirect energy use is much higher than direct energy use, where diesel use is the most important factor. Compared to 1989-1990, energy use per ha has decreased significantly in 2000-2001, mainly caused by a large decrease in the use of mineral fertilisers. In Figure 1, we consider energy input vs. milk production for each farm. Some 95% of all farms have an energy productivity between 14 and 40 l milk/ 100 MJ. In 2000-2001, the farms are generally more efficient than in 1989-1990: they have a higher average milk production per unit of energy used (27 l/100MJ and 22 l/100MJ respectively). The most energy-efficient farms have an energy productivity of about 40 l milk/100 MJ. They combine a high milk production (achieved by good livestock management, high milk production from forage or a high livestock density) with a low energy use (mainly achieved by low use of mineral fertiliser and diesel). The least efficient farms have an energy productivity < 14 l milk/100 MJ. They are often extensive farms (± 5000 l milk/ha) with a low livestock density and a relatively high use of concentrates, mineral fertiliser and diesel.

Table 1 Flemish dairy farms: energy use per ha

	1989-1990		2000-2001	
	MJ	%	MJ	%
Diesel	5522	12.9	5771	16.8
Contract work diesel	1503	3.5	2015	5.9
Lubricants	524	1.2	534	1.6
Electricity	2768	6.5	2185	6.4
Other energy sources	0	0.0	24	0.1
Direct energy	**10317**	**24.2**	**10529**	**30.8**
Mineral fertiliser	15245	35.7	8798	25.7
Seeds	150	0.4	150	0.4
Pesticides	189	0.4	219	0.6
Concentrates	13642	31.9	11639	34.0
Forages	950	2.2	506	1.5
Machines	2228	5.2	2426	7.1
Indirect energy	**32404**	**75.8**	**23738**	**69.2**
Total energy use	**42721**		**34267**	

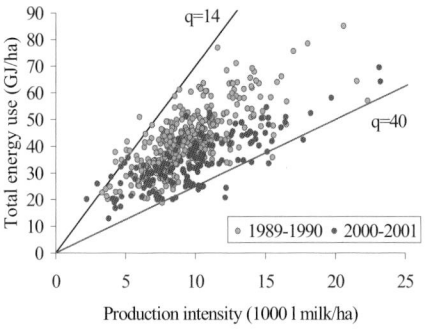

Figure 1 Energy use vs. milk production, q = energy productivity (l milk/100 MJ)

Conclusions Flemish diary farms can save most energy by decreasing the use of mineral fertiliser and diesel. Using less concentrates would also lower the energy use, but this might have a negative impact on milk production (Verbruggen *et al.*, 2004). Our results show that an energy productivity of 40 l milk/100 MJ is an achievable eco-efficiency goal. The most efficient farms can reach average to high milk production with a lower than average energy use. The least efficient farms are often extensive farms (low milk production per ha).

Reference

Verbruggen, I., F. Nevens, D. Reheul & G. Hofman (2004). [Nitrogen use and nitrogen use efficiency on Flemish dairy farms] (in Dutch). Flemish Policy Research Center for Sustainable Agriculture, Publication 6, 58 pp.

Agroecosystem performances of livestock farms in a mountain area of Sicily

L. Stringi, G. Alfieri, G. Amato, A.S. Frenda, D. Giambalvo and P. Iudicello
Dipartimento di Agronomia ambientale e territoriale - Università di Palermo - Viale delle Scienze 90128, Palermo, Italy, Email: lstringi@unipa.it

Keywords: energetic efficiency, agroecosystem performance indicators

Introduction Agroecosystem performance indicators (APIs) represent instruments for studying agroecosystem performance via an input/output approach and a knowledge base, with the aim of improving the sustainability level of the farm's activity (Tellarini & Caporali, 2000). This research used APIs to evaluate the influence of stocking rate on the performance (in terms of energy) of farms in a mountainous area of northern Sicily, Italy.

Materials and methods Forty six livestock farms in the Madonie and Nebrodi mountains (850-1660 m asl) were examined. The farms were classified on the basis of stocking rate: L <0.75 LU/ha; M 0.75-1.50 LU/ha; H >1.50 LU/ha. The farm parameters recorded (structure, crop and livestock management, production, external resources) were used to calculate input and output energy values from which APIs were derived according to Tellarini & Caporali (2000). The results were tested for significance using PROC GLM procedure (SAS, 1996).

Table 1 Some structural and livestock management parameters for the three classes of stocking rate

	L[#]	M[#]	H[#]
Mean LU/ha[+]	0.56	1.00	2.36
Total farm area (ha)	125.3	35.1	34.4
Wood (%)	13.5	5.3	0.0
Permanent pastures (%)	72.9	56.1	29.4
Forage crops (%)	10.4	23.6	48.2
Cereals (%)	3.2	15.0	22.4
Only grazing (d/year)	247	180	45
Housing (d/year)	56	104	292
Hay (kg/LU per d)	1.7	3.4	5.2
Concentrate (kg/LU per d)	0.9	2.0	5.5

[#] Number of farms: L=25; M=14; H=7; [+] LU: livestock unit (500 kg liveweight)

Table 2 Some performance and structural indicators for the three classes of stocking rate

Indicators (Gj/Gj)	Stocking rate			P level of significance		
	L	M	H	L *vs* M	L *vs* H	M *vs* H
Total farm input	0.78	0.76	0.57	ns	0.006	0.019
Total internal input	1.37	1.60	2.14	ns	0.007	ns
Total external input	2.37	1.90	0.82	ns	0.046	ns
DoNES*	0.05	0.08	0.05	ns	ns	ns
Farm autonomy	0.58	0.50	0.32	ns	0.001	0.009
Immediate removal	0.26	0.34	0.42	0.041	0.003	ns

*Dependence on non-renewable energy sources

Results and conclusions The land use was markedly different at the different farm stocking rates: with increased stocking rate, permanent pasture decreased, whilst that of forage crops increased (Table 1). Consequently a shorter period of exclusive grazing and a greater feed supply was recorded. The total farm input indicator (Gj obtained from the production process per Gj from any source introduced into the system) was, on average, 0.72, similar to the findings of other studies (e.g. Risoud & Chopinet, 1999). The farms with the highest stocking rate showed a significantly lower efficiency than the other two groups (Table 2).

The indicator of dependence on non-renewable energy sources was, on average, very low and the differences among the farm groups were not significant. With the increase in stocking rate the farm autonomy indicator (ratio of input produced on the farm to total input) fell and the immediate removal indicator (ratio of output destined for final consumption to total output) increased. Thus energy efficiency can be increased by increasing the level of internal transfer (re-use). Furthermore, the autonomy indicator was positively related to the efficiency of performance of external inputs (the ratio of total output to external input), showing that external inputs decrease as the efficiency with which they are used increases. On the whole, the best agroecosystem performance, in terms of energy, was found on farms with a lower stocking rate, a higher proportion of permanent pasture and a longer period of exclusive grazing.

References

Risoud B. & B. Chopinet (1999). Efficacité énergétique et diversité des systèmes de production agricole. Application a des exploitations bourguignonnes. *Ingénieries EAT*, 20, 17-25.

SAS (1996). User's Guide: Statistics. Version 6. SAS Institute Inc., Cary, NC.

Tellarini V. & F. Caporali (2000). An input/output methodology to evaluate farms as sustainable agroecosystems: an application of indicators to farm in central Italy. *Agricultural Ecosystems and Environment*, 77, 111-123.

Offered papers

Low input techniques for firebreak covering: agronomic aspects

S. Caredda[1], A. Franca[2], F. Sanna[1] and G. Seddaiu[3]
[1]Dipartimento di Scienze Agronomiche e Genetica Vegetale Agraria, Sassari, Italy, [2]CNR - ISPAAM, Sassari, Italy, Email: a.franca@cspm.ss.cnr.it, [3]Dipartimento di Scienze Ambientali e delle Produzioni Vegetali, Ancona, Italy

Keywords: fire prevention, firebreaks, green cover

Introduction In Sardinia, wild fire prevention is traditionally "passive", based on mechanical removal of vegetation and upper soil layers. This has a dramatic negative effect on plant diversity and soil erosion. This experiment concerns "active" prevention of wildfires, based on green covering and grazing of firebreaks. The objective was the establishment of a persistent green cover, to be grazed by animals to reduce fuel accumulation.

Materials and methods In the protected woodland of Pabarile, S.Lussurgiu (OR), Sardinia, a 3-year experiment was carried out on a firebreak at 840 m a.s.l., with sub-acid soil on basaltic rock. This was a bare ground cover firebreak that had been degraded after repeated removal of cover. Two sowing techniques (minimum tillage vs. no-tillage) and four 'cover crops' - ANGLONA (*Medicago polymorpha* 'Anglona'), CAMPEDA (*Trifolium subterraneum* 'Campeda'), MIXTURE (mixture of 2/3 'Anglona' + 1/3 'Campeda') and NATURAL (natural canopy as a control) were compared, using a completely randomised block design with three replicates. The seed rate was 20 kg/ha and all plots were fertilised with 40 kg N and 80 kg P_2O_5/ha. The floristic composition and soil cover rate dynamics were observed for three years using transects.

Results All the cover crop treatments reduced the bare ground percentage (20÷40%) compared with traditionally managed firebreaks (Caredda *et al.*, 2002). Minimum tillage didn't improve significantly the establishment of introduced species (Figure 1). Low input green covering led to a light but significant increase in subclover (P≤0.05, LSD = 8%). There was an interaction between ground cover and tillage method; lack of tillage increased significantly the bare ground in the MIXTURE treatment. Green covering, combined with grazing, led to a strong reduction of unpalatable species, generally the highly combustible species (*Cistus spp.*, *Inula viscosa*, *Rubus spp.*, etc.), and to a soil cover close to 80-90% (Figure 2).

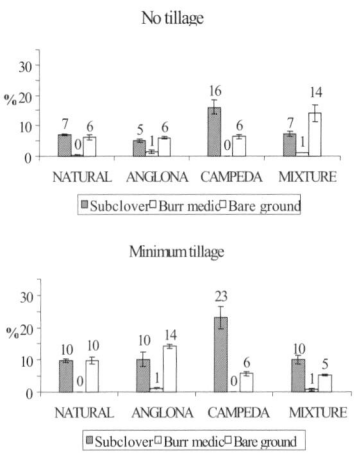

Figure 1 Botanical composition and bare ground in spring

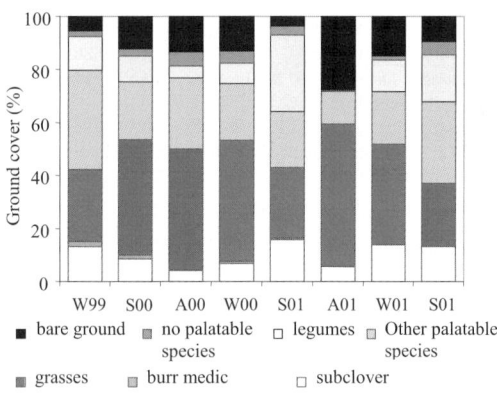

Figure 2 Dynamics of vegetation in the over-sown firebreak (W, winter; S, spring; A, autumn)

Conclusions A low-input technique involving no-tillage and a low seed rate allowed the establishment of an adequate herbage cover and lowered environmental risks through wildfires (reducing fuel accumulation by grazing) and soil erosion (reducing the naked soil) whilst providing green biomass for grazing.

Reference

Caredda S., A. Franca & G. Seddaiu, (2002). Firebreaks over-sowing: an alternative tool for the wildfire risk reduction in Sardinia. *Grassland Science in Europe*, 7, 908-909.

Cost efficiency of agri-environmental schemes – Model calculations in Central Europe for grassland extensification

H. Bergmann
Department of Agricultural Economics, University of Göttingen, Platz der Göttinger Sieben 5, D-37073 Göttingen, Germany, E-mail: hbergmann@gwdg.de

Keywords: agri-environment schemes, cost efficiency, grassland extensification, agro-biodiversity

Introduction The use of grassland in less favoured areas is important for the protection of agro-biodiversity. However, under existing market conditions the use of low yielding grasslands is not efficient economically. For this reason the European Union is paying specific support to make extensive grasslands economically more attractive. The objective of this study was to analyse whether the existing agri-environmental scheme was the cheapest way to protect extensive grasslands.

Materials and methods Four simulated dairy farms were used in this study (Bergmann, 2004). Using Linear Programming, the sum of standard gross margins per farm was maximised. All farms had a milk quota of 300.000 kg/year and incorporated the possible alternative use of grassland with suckler cows. The land use of the farms was: Farm I: 45 ha arable farm land, 5 ha grassland, Farm II: 25 ha arable farm land, 25 ha grassland, Farm III: 10 ha arable farm land, 40 ha grassland, Farm IV: 5 ha arable farm land, 45 ha grassland. The compensation necessary was calculated to test if the existing agri-environmental scheme would be capable of compensating farmers for the losses as a result of the extensification measure.

Results The existing scheme in Rhineland-Palatia pays €204/ha for one cut per year after June 14th. The existing payment was high enough to compensate dairy farms with a high proportion of grasslands, but not dairy farms with a high proportion of arable land. While farms I and II needed a fixed compensation amount (€232 /ha), farms III and IV required amounts depending on their acceptance of the program (from 0 to €/354ha). Of the 200 ha all 4 farms are using 42,5 ha for wheat 10,91 ha for other cereals (Table 1, column 1) An increase in scheme participation increases maize production and a decrease suckler cows.

Table 1 Production effects of extensification premiums

Crop	Premium supposed						Tendency
	0	100	150	204	225	235	
Wheat	42.50	40.00	36.80	35.00	35.00	34.29	↓
Rye	0.00	0.00	0.00	0.00	0.00	0.00	↔
Barley	10.91	8.00	7.52	7.52	7.52	2.69	↓
Triticale	0.00	0.00	0.00	0.00	0.00	0.00	↑
Rape	10.91	8.00	7.52	7.52	7.52	2.69	↓
Maize (Silage)	16.62	24.96	29.11	30.91	30.91	41.29	↑
Fallow	4.05	4.05	4.05	4.05	4.05	4.05	↑
Intensive Grassland	115.00	60.38	46.13	36.75	34.50	4.50	↓
Extensive Grassland	0.00	54.63	68.88	78.25	80.50	110.50	↑
Suckler cows	35.16	0.00	0.00	0.00	0.00	0.00	↓
Total specific farm cost	0	€5.663	€10.332	€17.215	€18.112	€25.967	
Costs paid with premium	0	€11.144	1€4.051	€17.215	€16.422	€22.542	

Conclusions Existing compensation is too low for farms with a high proportion of arable land. Consequently, only farms with a high proportion of grassland (in this case 62%) will participate. Extensification of grassland results in an increase in maize production with negative effects on groundwater quality (Scheringer, 2002). It is cheaper to implement specific farm schemes than general mean compensation schemes if only a part of the area in a boundary is needed to fulfil specific ecological goals.

References
Bergmann, H. (2004). Berechnung von Kosten für Maßnahmen zum Schutz von gefährdeten Maculinea-Arten, UFZ Berichte 2/2004; Leipzig
Scheringer, J. (2002). Nitrogen on dairy farms: balances and efficiency, Excelsior, Hohengandern.

Phytosociological and economical properties of some water-meadows of Nemunas, Lithuania

L. Baležentienė and E. Venskutonienė
The Lithuania University of Agriculture, Studentų 11, LT - 53347 Akademija, Kaunas distr., Lithuania, Email: ligita@nora.lzuu.lt

Keywords: phytosociology, water-meadow, feed value, management

Introduction The water meadows of Nemunas in Lithuania have a high biological diversity as well as economic value. This paper describes the syntaxonomic type of meadow communities, plant species diversity of some water meadows of Nemunas, estimating the productivity of these grasslands in the summers of 2002 and 2003 and the optimal method of management.

Material and methods Four water meadows under various ecological conditions were investigated. The phytosociological analyses and identification of plant species were carried out in relation to soil type and water regime (Balevičienė *et al.*, 1998; Whittaker, 1980). Available P and K in soil chemical composition (ChC) were determined by Egner *et al.* (1960) (A-L method based on extraction with Ammonium (A) Lactate (L)).

Results The main four plant communities in the water meadows of Nemunas were formed on alluvial soils with different chemical composition, mechanical structure and groundwater level (Table 1). These features of ecotopes determine the management of meadows.

Table 1 Properties of investigated water meadow communities

Plant community	Soil type	Chem.comp. at 25 cm depth			Hay yield
		pH $_{KCl}$	pH $_{KCl}$	pH $_{KCl}$	t/ ha
Arrhenatheretum elatioris Br.-Bl. Ex Scherrer 1925	Medium soggy *Hapli-Calcaric Fluvisoil*; light/ medium loam on loamy sand	8.47	15.5	14.1	3.43
Arrhenatheretum elatioris v. *Calamagrostis epigejos*	Light soggy *Hapli-Calcaric Fluvisoil*; light loam on loamy sand	7.4	15.4	14.1	2.41
Phalaridetum arundinacea (Koch 26) Libbert 1931,	Light soggy *Hypogleyi-Calcaric Fluvisoil*;	7.0	14.4	13.9	3.17
Alopecuretum pratensis Regel 1925	''	7.0	14.2	13.8	4.19
Alopecuretum pratensis v. *Bromus inermis*	''	7.0	14.4	14.0	4.21
Deschampsietum caespitosae Horvatić 1930	Strong soggy *Hypergleyi-Calcaric Fluvisoil*; loam / loamy sand or turf	6.5-7.0	15.9	14.1	2.15
LSD $_{05}$					0.11

Conclusions *Alopecuretum pratensis* v. *Bromus inermis* communities are the most productive, but their management is aggravated by frequent flooding. The harvesting conditions are better in drier meadows, where productive *Arrhenatheretum elatioris* communities with good feed value grow. The hay of *Phalaridetum arundinacea* and *Deschampsietum caespitosae* is of low digestibility. The management of these communities is complicated because of frequent flooding or hummocks.

References
Balevičienė, J., B. Kizienė, Ž.Lazdauskaitė &D. Patalauskaitė (1998). Vegetation of Lithuania. Meadows / BI. Kaunas-Vilnius, 269 pp. (in Lithuanian)
Egner H., Riehm H. & Domingo W.R. (1960). Untersudrungen über die chemische für die Beurteilung des Nährstoffzustandes der Boden. Kungl. *Lontbrunkshögsholans Annaler*, 26, 199-215.
Whittaker R.H. (1980). Classification of plant communities. Hague-Boston-London, 287-399.

Role of grasslands and grassland management for biogeochemical cycles and biodiversity. Setting up long-term manipulation experiments in France

G. Lemaire[1], J.F. Soussana[2], J.C. Emile[4], A. Chabbi[4], F. Louault[2], P. Loiseau[2], B. Dumont[3] and X. Charrier[4]
[1]Unité d'Ecophysiologie, INRA 86600 Lusignan, France, Email: lemaire@lusignan.inra.fr, [2]Unité d'Agronomie INRA, Site de Crouël, 234 avenue du Brézet, 63039 Clermont-Ferrand, France, [3]Unité de Recherches sur les Herbivores, INRA Theix, 63 St-Genès Champanelle, France, [4]Domaine Expérimental INRA, 86600 Lusignan, France

Keywords: N cycle, C cycle, soil organic matter, vegetation dynamics, greenhouse gas

Introduction Land use for grassland is recognised to have some beneficial effects for biodiversity and the environment: (i) regulation of the water cycle and protection of soils against erosion, (ii) accumulation of organic matter in soil and sequestration of atmospheric C, (iii) regulation of the N cycle and attenuation of the risk for N leaching, (iv) recycling of nutrients and improvement of soil quality, (v) improvement of biodiversity of vegetation, soil microbes and micro- and meso-fauna. All these effects depend upon the management of the grassland: cutting vs. grazing, stocking density, level of N inputs. Management decisions often result from short-term objectives, whereas the soil-vegetation interactions are long-term processes. Therefore, a steady state is usually not reached, which makes it difficult to determine the overall environmental effects of changes in land use and in grassland management.

Research questions and need for long-term experiments Long-term agro-ecosystem manipulation experiments are needed to relate changes in land use and management at the landscape level with their environmental consequences, which are partly determined by the fluxes, residence times and the balance of major elements, such as C, N and P. There is a need (i) to identify and characterise the compartments of the soil organic matter that play key roles, (ii) to quantify some of the key internal fluxes and to monitor at the boundaries of the system fluxes to the atmosphere and hydrosphere, (iii) to investigate the functional role of plant, microbial and soil fauna diversity with the aim of characterising the response of the whole system to the disturbance induced in the long term by contrasted management systems.

Long-term experimental design Two long-term experimental sites have been set up and will be starting in 2005. The first is located at Theix (Massif Central, 900 m a.s.l.) and comprises perennial semi-natural grasslands which have previously received intensive inputs. Three contrasting levels of herbage use will be compared without N fertiliser by manipulating stocking density (sheep and cattle grazing) in a randomised block design with four (2000 m^2) replicate paddocks. Further comparisons between grazing and cutting will be obtained by including mown plots at three contrasting levels of inorganic N supply. The second, at INRA Lusignan, will study ley farming systems. Sown grasslands of a mixture of perennial ryegrass, cocksfoot and tall fescue, with or without the addition of white clover, will be included for three or six years within an arable crop rotation (maize, wheat, barley) receiving different levels of N application. Two control treatments are included: the arable crop rotation alone, in order to analyse the effects of sown grasslands on soil dynamics and on environmental fluxes, and a long-term sown pasture.

On both sites, and for each treatment, herbage and crop production and exports of nutrients will be measured. The main soil state variables will be monitored at regular time intervals, as well as the diversity of plant and soil (microbes, micro- and meso-fauna) communities. Water balance will be calculated through profiles of soil water content using TDR probes and the soil solution will be regularly sampled by ceramic cups and plate lysimeters to determine its composition and to calculate fluxes. On nearby plots of about 3 ha and for some of the treatments, CO_2 and H_2O fluxes will be monitored using the eddy covariance technique. Collections of soil, plant and water samples will be kept and made available for further determination. A database system will be constructed for information exchange. Given the large size of the experimental plots, sub-plots can easily be accommodated to further analyse some of the key processes (e.g. by using isotopic tracers) and relate them to the state variables of the system.

Conclusion These new experimental sites for studying the long-term development of the soil-vegetation systems of grasslands under contrasting management regimes should give the opportunity to strongly integrate disciplinary research on vegetation dynamics, soil biology, soil physics and chemistry and sward-herbivore interactions. Previously such studies have too often been carried out separately in different experiments and at different sites.

SAFE: a framework for assessing sustainability levels in agricultural systems

X. Sauvenier[1], C. Bielders[2], V. Brouckaert[1], V. Garcia[1], M. Hermy[3], E. Mathijs[4], B. Muys[3], J. Valckx[3], N. Van Cauwenbergh[2], M. Vanclooster[2], E. Wauters[4] and A. Peeters[1]
[1]*Laboratoire d'Ecologie des Prairies, UCL, Croix du Sud, 5 bte 1, 1348 Louvain-la-Neuve, Belgium, Email: sauvenier@ecop.ucl.ac.be,* [2]*Unité de Génie Rural, UCL, Croix du Sud, 2 bte 2, 1348 Louvain-la-Neuve, Belgium,* [3]*Laboratorium voor Bos, Natuur en Landschap, Vital Decosterstraat, 102, 3000 Leuven, Belgium,* [4]*Afdeling Landbouw- en Milieueconomie, KUL, Willem de Croylaan, 42, 3001 Leuven, Belgium*

Keywords: sustainability, principles, criteria, indicators, integration

Introduction Evaluating the sustainability of agricultural systems is a major challenge for scientists, policy makers and farmers. Numerous sets of indicators have recently been designed, both at national and international levels. However, most of these initiatives focus only on environmental aspects of sustainability, indicators are often selected arbitrarily and usually do not fit in a consistent, comprehensive and universally applicable framework. This paper presents an original framework for integrating the information contained by indicators into a single quantitative measure of agricultural sustainability in order to facilitate comparison and diagnosis.

Methodology For each of the three sustainability pillars (environmental, economic and social), SAFE defines hierarchical levels - principles, criteria, indicators and reference values - reflecting the multiple functions that an agricultural system should maintain or enhance in order to be sustainable (Table 1). An exhaustive list of indicators was built and submitted to experts (scientists, policy makers and farmers) for evaluation by the Delphi method (Okoli & Pawlowski, 2004) and against a specific set of criteria (Table 2). Multivariate analysis determined a core set of indicators per criterion. Selected indicators were calculated at different scales (parcel, farm and landscape) and converted to a reference value. Fuzzy evaluation (Cornelissen *et al.*, 2001) allowed the rescaling of indicator values on a continuous scale of sustainability values: S_i [0-1]. Finally, S_i was integrated at the criterion level (weighted average of S_i), at the principle level (average of S_c), at the pillar level (average of S_{pi}) and at the sustainability level (average of S_{pr}). The SAFE framework was tested on four experimental farms.

Table 1 Hierarchical framework

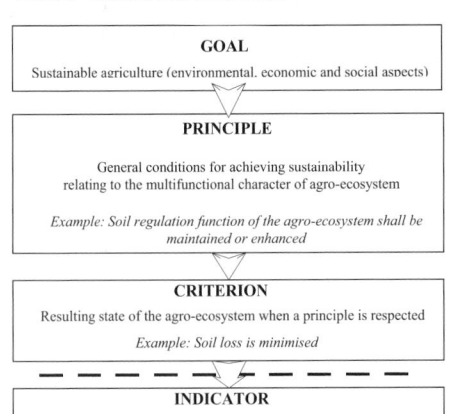

Table 2 Criteria for indicator selection

1. **Discriminating power in time and space**
→ Ability to discriminate between changes due to external factors and management, in space and time
2. **Analytical soundness**
→ "Is the indicator scientifically valid?"
3. **Measurability / Cost and time consumption**
→ "Is the use of the indicator justified in terms of cost and time consumption?"
4. **Transparency**
→ "How understandable is the indicator?"
5. **Policy relevance**
→ "Does the indicator help in monitoring policy measures effects and in identifying areas where policy action is needed?"
6. **Transferability**
→ "Does the indicator relate to general practices of major farm types?"
7. **Relevance to sustainability criteria**
→ "Is the indicator a relevant measuring tool for the sustainability criterion/criteria it is related to?"

Results and conclusions The SAFE framework ("P/C/I" hierarchy) provided a consistent approach for evaluating the sustainability of agricultural systems as a whole (holistic approach) and at different scales. The framework is filled with indicators selected on a scientifically sound basis and SAFE integrates progressively the information of selected indicators in a single quantitative measure of sustainability (S).

References
Cornelissen, A.M.G., J. van den Berg, W.J. Koops, M. Grossman & H.M.J. Udo (2001). Assessment of the contribution of sustainability indicators to sustainable development: a novel approach using fuzzy set theory. *Agriculture, Ecosystems and Environment*, 86, 173-185.
Okoli C. & S.D. Pawlowski (2004). The Delphi method as a research tool: an example, design considerations and applications. *Information & Management*, (in press).

Ecosystem management in pasture communities: tools from restoration ecology

S.C. Goslee and M.A. Sanderson
USDA-ARS Pasture Systems and Watershed Management Research Unit, Building 3702, Curtin Road, University Park, PA 16802 USA, Email: Sarah.Goslee@ars.usda.gov

Keywords: ecosystem management, pasture, plant community

Introduction Pasture systems have the potential to improve both economic and environmental sustainability in agricultural communities worldwide. To achieve maximum benefits, pasture plant communities must be tailored to the climate, the site type and the goals of the producer. Little is known about how to accomplish this, beyond very broad recommendations. We approached the problem by adapting a three-step conceptual framework from restoration ecology for use in managed pasture communities of the northeastern United States. The three steps, inventory, assessment and remediation, were designed for managers interested in restoring degraded native ecosystems, but can be applied equally well in managed ecosystems.

The first step, inventory, involves a detailed examination of what is present in pasture plant communities, and how community composition relates to soil factors, climate and management. Assessment is the comparison of the existing ecosystem to a target system. For restoring degraded systems, the appropriate target is usually obvious, but in managed ecosystems, the choice of target condition is much more problematic. The target must be appropriate for the climate, topography and soils, be sustainable both economically and ecologically, and satisfy the requirements of the producer. Potential requirements for pasture systems include total production, growing season length, invasion resistance and drought tolerance. Remediation is the process of moving from the existing state to the target by altering management to produce and maintain the desired state. We present here the results from regional inventory and initial assessment phases.

Methods The inventory phase was begun in 1998 with detailed sampling of plant community composition and soil properties on farms across the northeastern United States. Plant sampling was conducted using the multiscale modified Whittaker plot, which has subplots of 1 m^2, 10 m^2, 100 m^2 and 1000 m^2 (Stohlgren et al., 1995). Soil properties were measured using a standard soil test pooled across the pasture. As part of developing appropriate reference conditions for various objectives, we have been using the Pasture Condition Score, a ten-element semi-quantitative index, as a general assessment tool (Cosgrove et al., 2001). Additional management information has been collected from producers whenever possible. Relationships between plant community composition and environmental factors were assessed using the Mantel test (Mantel, 1967).

Results We have sampled 126 pastures on 42 farms over the past seven years. Many of these pastures have been sampled repeatedly, for a total of 227 modified Whittaker plots. We have found 286 plant species. Most of these were rare: 181 were found in less than 5% of the plots, and only 14 were found in more than half. Two species, white clover and dandelion, were nearly ubiquitous, found in 98% of the plots. Plant community composition was strongly related to latitude and related to the topographic factors elevation and aspect. Phosphorus, pH, organic matter and texture were the soil properties most closely related to vegetation. Although the overall Pasture Condition Score was not strongly related to plant community composition, the relationships with certain individual elements - percentage cover, uniformity of use, standing dead and plant vigour – were highly significant.

Discussion We found far more species in pastures than were expected. Although most were rare, a few were nearly ubiquitous. Any management plan for pasture vegetation should assume that some of these species will be present, whether planted or not. The strong significant relationships between location, topographic factors, soil properties and plant community composition illustrate clearly the importance of including environmental information when choosing the management goals for a pasture system. While the survey results presented here are a valuable first step, additional information on plant species function alone and in mixtures is required to develop appropriate target communities. This is being addressed by ongoing greenhouse and small-plot studies. The results of these experiments will allow us to evaluate the ecosystem function of existing pastures based on inventory data, and to design pasture plant communities to meet economic and environmental goals. Restoration ecology, which deals with creating and maintaining particular ecosystem types, can provide valuable lessons for the management of pasture ecosystems.

References
Cosgrove, D., D. Undersander & J.B. Cropper (2001). Guide to pasture condition scoring. USDA-NRCS Grazing Lands Technical Institute.
Mantel, N. (1967). The detection of disease clustering and a generalized regression approach. *Cancer Research*, 27, 209-220.
Stohlgren, T.J., M.B. Falkner & L.D. Schell (1995). A Modified-Whittaker nested vegetation sampling method. *Vegetatio*, 117, 113-121.

Effects of tree and tillage systems on the productivity of the herbaceous stratum in silvopastoral systems in the southwest of Córdoba, Argentina

O. Plevich, C. Saroff, C. Cholaky, T. Pereyra, O. Barotto and H. Pagliaricci
Facultad de Agronomía y Vetrinaria de la Universidad Nacional de Río Cuarto. Ruta 36, Km 601, X5804BYA. Río Cuarto, Córdoba, Argentina, Email: oplevich@ayv.unrc.edu.ar

Keywords: silvopastoral system, forage biomass, tillage

Introduction   In the southwest of Córdoba, Argentina, there are lands with severe water erosion, due to the interaction of rolling pampas, high intensity precipitation, loam soil, and farming systems based on annual crops (Cantero *et al.*, 1998). In an attempt to mitigate the erosive processes, a silvopastoral system was established in which winter forage was combined with trees. To improve the physical condition of the soil, two tillage systems were implemented. The objective of this paper was to determine the effect of trees and tillage systems on the production of forage.

Materials and methods   A silvopastoral system with two different tree species was established in a split plot experiment with two replicates in 1998. The system included a double row of trees planted at each side of an "alley" of 21 m.   The species of trees were *Pinus elliottii* and *Eucalyptus viminalis*. The forage was *Avena sativa* in 2003. Two systems of reduced tillage, one superficial and the other deep, were used. To estimate the forage biomass, plots were established at two distances from the trees (2 and 10.5 m) and one plot without trees, adapting the method of Acuña *et al.* (1984).   A sample of four observations of forage biomass was taken from each replicate at the end of the cropping cycle in 2003. Data were analysed by ANOVA.

Results   The forage biomass associated with the arboreal species is shown in Table 1 and the biomass associated with tillage treatments in Table 2. The forage biomass production was significantly superior (p<0.05) when forage grew in the "alley" associated with *Pinus elliottii*, than when it grew in the plot either associated with *Eucalyptus viminalis* or without trees.   The forage production with deep tillage was significantly higher (p<0.05) than that with superficial tillage. These findings suggest that the water evaporation rate in the "alley" was lower than that in the plot without trees. However, in the "alley" with *Eucalyptus* the forage biomass close to the tree was significantly lower, possibly due to the effect of allelopathy. A greater level of root exploration was observed in the deep tillage treatment, which made it possible for *Avena sativa* to have a greater quantity of water for biomass production.

Table 1 Forage biomass in different planting treatments

Treatments	Forage biomass (kg/ha)
Pinus elliottii	1912 a
Eucalyptus viminalis	1079 b
Field without trees	1262 b

Table 2 Forage biomass in different tillage treatments

Treatments	Forage biomass (kg/ha)
Reduced deep tillage	1578 a
Reduced superficial tillage	1258 b

Values followed by the same letter do not differ significantly by the Fisher (LSD) test at p<0.05

Conclusions   The forage biomass production was superior in the plots with deep tillage and with *Pinus elliottii* in the fifth year of the experiment. However, it should be noted that the wood biomass has not been assessed, but this is part of the future research agenda

References
Acuña, H. P., P. Soto & P. Melin (1984). Método para estimar el crecimiento de las praderas de secano, por medio de cortes en ausencia de pastoreo. *Agricultura Técnica (Chile)* ,44, 325-333.
Cantero, A., M. Cantú, J.M. Cisneros, J.J. Cantero,M. Blarasin, A. Degioanni, J. Gonzalez, V. Becerra, H. Gil, J. De Prada, S. Degiovanni, C. Cholaky, M. Villegas, A. Cabrera & Y E. Carlos (1998). Las Tierras y Aguas del Sur de Córdoba. Propuesta para un Manejo Sustentable. Editorial UNRC, 119 pp.

Effect of tree species and density on pasture production in Galicia, Spain

A. Rigueiro-Rodríguez, E. Fernández-Núñez and M.R. Mosquera-Losada
Departamento de Producción Vegetal. Escuela Politécnica Superior Lugo. España, Email: romos@lugo.usc.es

Keywords: *Pinus radiata, Betula alba,* silvopasture, nitrogen efficiency

Introduction Galicia produces 50% of the forest products of Spain. Livestock production earns 62% of the income of the agrarian sector in Galicia. Afforestation has been very important in the last decade to such an extent that the area of forest and woodlands now covers 62% of Galicia. It is necessary to increase the rate of return on investments in planted forests in order to avoid rural depopulation through improvement of rural development and welfare. This paper reports on the effect of combining pastures with trees.

Materials and methods The study was conducted in an acid (pH$_{water}$ 5.5) sandy and soil in Castro Riberas de Lea, Lugo ,Galicia, NW Spain, with a mean annual precipitation of 800 mm and mean monthly temperature of 11.6 °C over the last 20 years. *Pinus radiata* and *Betula alba* were planted in 1995. Each species was planted at two densities (870 and 2500 trees/ha) and each plot consisted of 25 trees. The trees of the 2 species had a mean height of 5 and 4 m in 2001, respectively. *P. radiata* was pruned to 2 m at the start of 2001. Experimental pasture plots were established within the tree plots after ploughing and fertilisation with 154 m^3 of dairy sewage sludge (160 kg N, 85.9 kg P$_2$O$_5$ and 23.4 kg K$_2$O per ha) and sowing with 25 kg/ha of *Lolium perenne* cv. Brigantia, 4 kg/ha of *Trifolium repens* cv. Artabro and 4 kg/ha of *T. pratense*. Inorganic fertiliser at the rate of 500 kg/ha of 8:24:16 (N,P,K) and 40 kg/ha of ammonium nitrate were applied between 1998 and 2001at the end of March and after the second harvest. Herbage was sampled in May, June, July and December by cutting all the herbage between 4 inner trees.

Results Riguerio *et al.* (2001) reported the first results of this experiment. Pasture production was significantly higher in plots with a low tree density, but higher than yields obtained from the open sward. The year of study (2001) had abnormal precipitation, with 955 mm in May, only 12mm in June and 70mm in July, equivalent to 40%, 48% and 26% of the 20-year mean rainfall. Rain may cause leaching of nutrients and therefore reduce initial pasture production in sandy soils. The presence of tree cover can reduce N leaching as possibly occurred in the plots of low tree density compared with the open sward (Fig. 1). However, pasture production was reduced at the high tree density because of reduced light penetration to the sward. Tree species also affected initial pasture production, being higher with *B. alba* as the canopy of this tree allows greater light transmission to the sward. Pasture production at low tree density was twice that in the open sward at the July harvest, as it reduced the effect of drought.

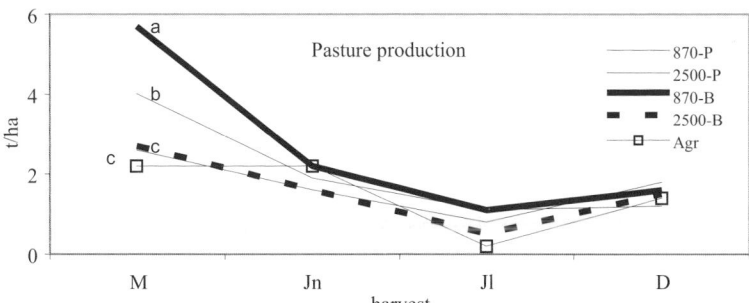

Figure 1 Mean herbage production in combination with *Pinus radiata* (P) and *Betula alba* (B) and open sward (Agr.) Letters indicates significant differences between means at each harvest.

Conclusion Silvopastoral systems can extend the grazing season with at the proper tree density, improving nitrogen efficiency.

Reference

Rigueiro-Rodríguez, A., R. M. Mosquera-Losada, & E. Gatica-Trabanini (2000). Pasture production and tree growth in a young pine plantation fertilized with inorganic fertilisers and milk sewage sludge in northwestern Spain. *Agroforestry Systems,* 48, 245-256.

Evolution and ageing of *Brachiaria brizantha* pasture component in a silvopastoral system

M. Rakocevic[1], F.C. de Oliveira, J. Ribaski and O.J. Lavoranti
[1]*Embrapa Florestas, Estrada da Ribeira Km 111, P. O. Box 319, 83411-000 Colombo, Paraná, Brazil, Email: mima@cnpf.embrapa.br*

Keywords: grass regeneration, leaf area index, morphology

Introduction The main causes of the decline of *Brachiaria* pasture in the tropics are lack of maintenance, fertilisation and excessively high animal stocking rates (Boddey *et al.*, 2004). *Brachiaria brizantha* has two predominant morphological forms: 1) relatively high stems with grouped tillers at a high position constructed from short leaves (bunch type); 2) low stems with long leaves, which in time evolve into the first type (Kanno *et al.*, 1999). The concept of *Brachiaria* management involves animals entering to commence grazing when the pasture is 50-60cm high and being removed at 25-30cm, leaving a regrowth period of at least 30 d (Alves *et al.*, 1996). The aim of this study was to determine the impact of two distinct regrowth periods (short versus very long) on the morphology of *Brachiaria brizantha* Hochst. ex A. Rich. (Bb), in a silvopastoral system (SPS) with *Corymbia citriodora* Hook. (Cc).

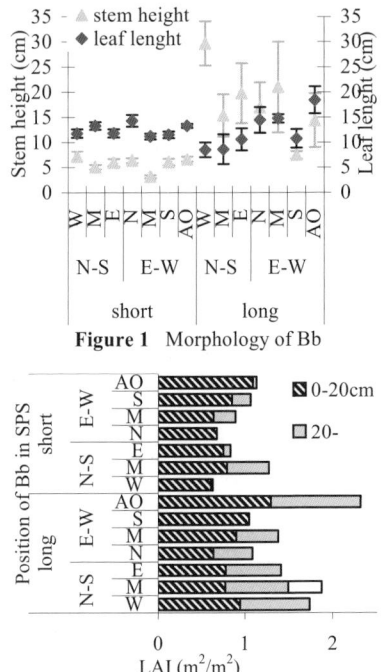

Figure 1 Morphology of Bb

Figure 2 Leaf area index per layer

Material and methods The experiment was conducted during 2001-2002 on twelve-year-old SPS established on 70 ha, with Cc and Bb components planted on the contours (30 m between tree strips) in Tamboara, north-western of State of Paraná, southern Brazil (23° 10´ latitude, 52° 27´ longitude), average altitude of 480m. The climate is Cfa (Köppen´s). *Brachiaria* in SPS is used for rotational pasture by adult zebu cattle. A completely randomised experimental design was used, with seven treatments hierarchically classified in distinct positions of Bb in SPS defined by Cc orientations: North-South (N-S) and East-West (E-W). An open area (AO) of pasture was used as control. The other treatments were distributed in the middle (M) between tree strips - 15 m, and 3 m distant from tree lines, corresponded to W(est) and E(ast) side of N-S tree orientation, and N(orth) and S(outh) side on E-W tree orientation. The orthogonal comparison within and between short (Oct. 2001) and long (Oct. 2002) periods of Bb regrowth was evaluated by F test. Leaf area index (LAI) per layer of 20cm was calculated from three replicates, while stem height and leaf length were calculated for ten replicates in 2001, and from three replicates in 2002.

Results At the end of the long regrowth period, Bb stems were significantly higher (Figure 1) than with short regrowth, and there was a greater leaf area (Figure 2) distributed in upper layers of the grass canopy. The leaf length showed a tendency to be shorter in long regrowth, especially on the N-S orientation, where there was an increased LAI in upper layers (20-40cm and even in 40-60cm). The highest LAI was developed in AO and in the middle of transects (M), with an important part distributed in the upper horizons of the canopy. At these sample points, the inter-specific competition for the light is less important, but intra-specific competition partially blocks tillering close to the soil surface.

Conclusions Long grass regrowth induced morphological changes in Bb (appearance of bunch type tillers) which irreversibly led to ageing of the Bb canopy. These results question the concept of allowing *Brachiaria* regrowth to a canopy height of 60cm.

References

Alves, S.J. & C.V. Soares Filho (1996). Braquiária. Comissão paranaense de avaliação de forrageiras. *Forragicultura no Paraná*, Londrina, 183.

Boddey, R.M., R. Macedo, R.M. Tarré, E. Ferreira, O.C. de Oliveira, C. P. de Rezende, R.B. Cantarutti, J.M.Pereira, B.J.R. Alves & S. Urquiaga (2004). Nitrogen cycling in *Brachiaria* pastures: the key to understanding the process of pasture decline. *Agriculture, Ecosystems and Environment*, 103, 389-403.

Kanno, T., C. M. Macedo, V.P.B.Euclides, A.E. Bono, J. Santos, C.M. Rocha & L.G. Beretta (1999). Root biomass of five tropical grass pastures under continuous grazing in Brazilian savannas. *Grassland Science*, 45, 9-14

The influence of tree thinning and tree species on the dry matter yield of grasses of a bush encroached semi-arid savanna in South Africa

G.N. Smit

Department of Animal, Wildlife and Grassland Sciences, University of the Free State, P.O. Box 339, Bloemfontein 9300, Republic of South Africa, Email: smitgn.sci@mail.uovs.ac.za

Keywords: *Acacia erubescens*, bush encroachment, *Combretum apiculatum*, grass-tree competition

Introduction The *Acacia erubescens-Combetum apiculatum* dominated savanna of South Africa is water-limited and an increase in woody plant abundance suppresses the grasses. This is the major reason why thinning or total clearing of all woody plants is often considered by landowners. The objectives of this study were to investigate the influence of intensity of tree thinning and tree species on grass yields in a semi-arid South African savanna.

Materials and methods The study was conducted in the Limpopo Province of South Africa on sandy soil. The mean long-term seasonal rainfall is 416 mm. The treatments consisted of five, 1 ha plots (200 m x 50 m) of which the control plot was left undisturbed (100 % plot) and the others thinned to the equivalents of 70 %, 50 %, 30 % and 10 % of the tree density of the 100 % plot (3 679 tree equivalents (TE)/ha, 1 TE = a tree of 1.5 m). The mean height of the trees was 2.57 m (SE ± 0.067), with a mean canopy diameter of 2.05 m (SE ± 0.074). The dry matter (DM) yield of grasses within the five tree-density plots was determined during five seasons following tree thinning. Grasses were harvested in 60 quadrats (0.5 m^2) per treatment, randomly placed under leguminous and non-leguminous trees, respectively. The leguminous tree group comprised of *A. erubescens, A. tortilis, A. nilotica, Dichrostachys cinerea* and *Peltophorum africanum*. The non-leguminous tree group comprised of *C. apiculatum, Euclea undulata, Terminalia sericea* and *Grewia* species.

Results and discussion The total seasonal rainfall for the five seasons following tree thinning were 481, 277, 335, 495 and 335 mm, respectively. The annual grass DM yields (Figures 1 and 2) followed the rainfall pattern, but within any season yields were generally higher in the thinned plots compared to the control plot, confirming the suppressive effect of the woody plants on the grass layer. However, the highest yields were not always recorded in plots with the highest intensity of tree thinning. Grass yields were generally higher under leguminous trees compared to non-leguminous trees. This can mainly be ascribed to one grass species, *Panicum maximum*. A higher soil nutrient status under leguminous trees is considered the main reason for these differences.

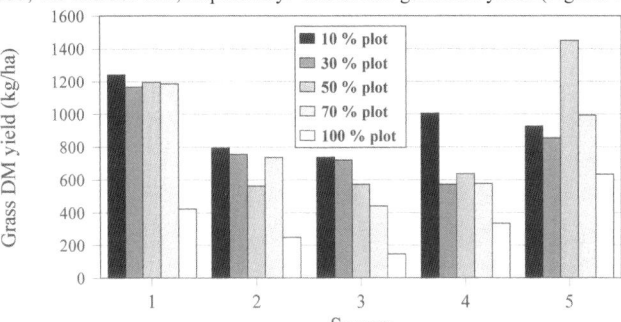

Figure 1 Grass DM yields under the canopies of leguminous tree species

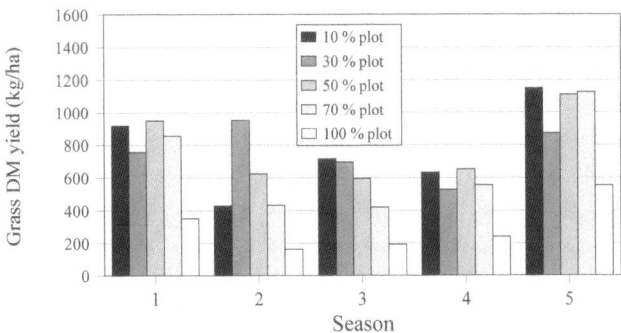

Figure 2 Grass DM yields under the canopies of non-leguminous tree species

Conclusions Considerable benefits in terms of grass DM yield, can be derived from tree thinning, but some trees need to be retained. In this regard the leguminous trees proved more advantageous to grass yields and these trees should preferably be retained.

Productivity of a *Leucaena leucocephala*-*Cynodon nlemfuensis* silvopastoral system with sheep in Yucatan, Mexico

J.G. Escobedo-Mex[1] and L. Ramirez-Aviles[2]
[1]*Instituto Tecnologico Agropecuario 2, Conkal, Yucatan, Mexico;* [2]*Universidad Autonoma de Yucatan, FMVZ, Email: escobmex@yahoo.com.mx*

Keywords: *Cynodon nlemfuensis*, *Leucaena leucocephala*, sheep, grazing

Introduction Animal production in the tropics of Mexico is based on grazed grasslands of low productivity; this type of production system has reduced the areas of natural vegetation and damaged the ecology (erosion of flora, fauna and soil). Silvopastoral technologies may improve the welfare and economic conditions of the rural population and, consequently, preserve their natural resources. The current work was designed to assess the introduction of *Leucaena leucocephala in* a silvopastoral system with *Cynodon nlemfuensis* (star grass) grazed by sheep.

Materials and methods The experiment was carried at the ITA 2, located in Conkal (20° 59' N, 89° 39' W), Yucatan, Mexico. The climate of the area is Aw_o (García, 1998) with a mean annual rainfall of 850 mm and mean temperature of 26.5 °C. Soils are Lithosols with pH 7-8 (Duch-Gary, 1988). The study was conducted from September 1999 till February 2000. The design was a completely randomised block with three repetitions and a 2 x 2 factorial arrangement. The experimental factors were 1) Systems (a. star grass alone; b. star grass and leucaena; 2) Seasons of grazing (a. late-rainy season; b. dry season). Availability of green dry matter (GDM), percentages of leaf, stem and dead material of the forage grass, and daily liveweight gain (DLG) of Pelibuey sheep was recorded. Experimental plots were 525 m^2 in size, which were rotationally grazed, with 7 days grazing and 28 days rest during the late-rainy season, and 5 days grazing with 35 days rest during the dry season. There were 16 grazing cycles with 24 young male sheep (Pelibuey).

Results The GDM yield in the late-rainy season was greater (p< 0.0001), with 2,678 kg GDM/ha, than, in the dry season with 2,272. The GDM was lower (p<0.01) with star grass alone than with star grass and leucaena, (2,350 and 2,600 kg/ha, respectively). The leaf:stem index was significantly (p<0.001) lower during the late-rainy season (0.28) than in the dry season, (0.39), but was similar (p>0.05) in the 2 systems. At the beginning of the trial, leucaena was at the flowering phase, resulting in many flowers and pods before pruning. The many branches per plant of leucaena (6.5 during the rainy and 4.9 in the dry season), resulted in a high leaf availability. The stem diameter was similar in both seasons. The DLG of Pelibuey sheep was 29 and 46 g for star grass alone and star grass and leucaena, respectively (p<0.01). However, DLG was similar (p>0.05) in the 2 seasons. The low GDM yield contribution of leucaena to the system could be associated with the climatic and soil conditions of the experimental site; in addition, the pruning may have reduced the leucaena yield. DLG was higher in the leucaena- star grass treatment. Alayon *et al.* (1998) found a positive effect of *Gliricidia sepium* as a supplement to low quality *C. nlemfuensis* hay on DM production, organic matter and crude protein intake with Pelibuey sheep. It should be possible then to expect a similar effect with leucaena.

Conclusions The use of leucaena in a silvopastoral system with *C. nlemfuensis* grasslands increased the GDM of forage biomass during the late-rainy season and, consequently, the DLG of Pelibuey sheep in Yucatan.

Acknowledgements The first author is grateful to CONACYT, Mexico for a scholarship The authors are grateful to IFS (Sweden) and CoSNET (Mexico) funds to carry out the present study.

References
Alayon, J.A., L. Ramirez-Aviles & J.C. Ku-Vera (1998). Intake, rumen digestion, digestibility and microbial nitrogen supply in sheep fed *Cynodon nlemfuensis* supplemented with *Gliricidia sepium*. *Agroforestry Systems*, 4, 115-126
Duch-Gary, J. (1988). La conformacion territorial del Estado de Yucatan. Los componentes del medio fisico. Centro Regional de la Peninsula de Yucatan. Universidad Autonoma Chapingo. Mexico, D.F. pp. 295-395
Garcia E. 1998. Modificaciones al sistema de clasificacion climatica Köeppen (para adaptarlo a las condiciones de la Republica mexicana). 3ª Edición. México, D.F. p. 120

Agroforestry systems in Cuba: some aspects of animal production

J.M. Iglesias, L. Simón, L. Lamela, I. Hernández, M. Milera and T. Sánchez
Estación Experimental de Pastos y Forrajes "Indio Hatuey ", CP 44280, Matanzas, Cuba Email: iglesias@indio.atenas.inf.cu

Keywords: agroforestry systems, animal production

Introduction The silvopastoral systems, that nowadays constitute scientific achievements of the Grasses and Forages Research Station "Indio Hatuey ", have been developed from the results of investigations that were carried out since the 1980s, to improve the productivity of natural pastures through the introduction of valuable herbaceous species and tree legumes. Those investigations also determined the essential elements of pasture management such as the optimal stocking rates for low input systems and suitable methods of grazing to obtain sustainability of grasslands.

Material and methods Among the diverse types of Silvopastoral systems under study, the protein banks and multiple associations of legumes and grasses have contributed much to the development of sustainable dairy and meat production, and could be considered as systems that can be extended and to the farmers and that integrate well with the production objectives of Cuban cattle production.

Results *Leucaena leucocephala* has been the most frequently used tree in Cuban silvopastoral systems and it has also contributed much to experimental data that demonstrate the real advantages of agroforestry (Table 1). However, it is not the only species used. Others such as *Albizia lebbeck, Erythrina berteroana, E. poeppigiana, Gliricidia sepium, Bauhinia purpurea* and *Morus alba*, have been tested with success and appear to be important elements of diversification of plant communities in silvopastoral systems in Cuba.

Table 1 Effect of different silvopastoral systems with low external inputs on performance of young fattening bulls

Production System	Genotype	Accumulated gain (g/day)	Live weight at slaughter (kg)	Age (months)
Association of P. maximum with L. leucocephala	Zebu	621,8	413,7	24
Association of P. maximum with L. leucocephala	1/2 H x 1/2 C	525,6	376,3	26
Association of P. maximum with L. leucocephala	5/8 H x 3/8 C	491,6	357,1	28
Protein Bank of L. leucocephala (25 % of the total area) + Natural pasture	Zebu	394,0	355,0	24
Protein Bank of L. leucocephala (25 %) + P. maximum (80 kg of N)	Zebu	555,0	372,5	25
Protein Bank of L. leucocephala (25 % of the total area) + A. gayanus	Zebu	487,0	449,0	29
Association of P. maximum with Albizia lebbeck	Zebu	729,0	409.2	24
Association of P. maximum with L. leucocephala	Zebu	788,0	424,0	24

Conclusion The main results obtained on the use of agroforestry for animal production in Cuba are 1. Daily live weight gains of between 500 and 600 g in young bulls for fattening, with an average production of around 800 kg of meat per ha annually. 2. Daily milk production of 7-10 kg/cow (14-25 kg/ha), without supplements. 3. Daily live weight gains of between 400-525 g in growing replacement heifers, which allows a live weight for reproduction of 290-300 kg at 20-27 months of age. 4. Minimal use of external inputs to the system

Optimising forage production on degraded lands in the dry tropics through silvopastoral systems

P.S. Pathak
Indian Grassland and Fodder Research Institute, Jhansi 284003, India, Email: pathak36@yahoo.com

Keywords: silvopastoral system, degraded land, livestock production, ecological restoration

Introduction In India, 187 M ha out of a total area of 328 M ha face the problem of land degradation, mostly due to water and wind erosion. The problems are aggravated by poor land cover and increasing pressure of human and livestock populations. There is over-exploitation of the scarce resources of forage and firewood. Several techniques, including watershed based silvopastoral land use have been proposed (Patil & Pathak, 1977). Tree, grass and legume based systems have been tried after land treatment to reduce runoff and soil loss while meeting the forage needs of the livestock and firewood for cooking in many studies (Debroy & Pathak, 1983). Results of an operational research project on silvopastoral systems are reported in this paper.

Materials and methods Watershed-based land treatments such as staggered contour trenches, bunds, diversion drains, gabion structures were coupled with an appropriate silvopastoral model to rehabilitate four different types of degraded lands: degraded revenue lands and forests, salt-affected ravines and undulating terrains around Jhansi (central India). The experiment involved the tree species *Acacia tortilis, Albizia amara, A. lebbeck, Hardwickia binata* and *Leucaena leucocephala* planted at three spacings in associations with *Cenchrus ciliaris, Chrysopogon fulvus, C. setigerus, Dichanthium annulatum* and *Stylosanthes hamata, Macroptilium atropurpureum* on the degraded lands. Control plots were not given any soil conservation treatments, except protection from grazing. Initial application of fertiliser was 40 kg N, 40 kg P and 20 kg K/ha followed by 20 kg N/ha every year. Use of leguminous nurse crops seeded on the mounds on the side of trenches to assist better tree growth and edaphic enrichments, controlled grazing by livestock for testing the value of biomass produced and bio-economic modelling were attempted to evaluate the fitness of different interventions (Pathak *et al.*, 1995).

Results The treatment with 600 contour trenches (3x0.4x0.5 m) per ha followed by seeding of grasses and legumes and tree planting (4x4 m) reduced soil loss from 17.8 to 1.26 t/ha/yr in 4 years. It also increased species richness and biodiversity of the natural vegetation. There were also improvements in the physical, chemical and biological properties of the soil. The production of herbaceous and woody biomass of less than 1 t/ha/yr was increased to 10 t/ha/yr with a 10-year rotation. There was a 7-fold improvement in crude protein. The system was evaluated by a mixed herd of cattle, sheep and goats grazing year-round in a deferred rotational system at the equivalent of 1 adult cattle unit per ha stocking rate after 7 years of establishment of the system. During the monsoon period the heifers grew 500 g/head/day without any supplemental feeding. During the remaining part of the year the animals gained 250-350 g/head/day after supplementation of concentrates and tree leaves. The models were able to predict the harvest cycle along with the yield to assist land managers. The benefit/cost ratio and the internal rate of return (IRR) from this project was 1.42 and 18 % respectively, ensuring the possibility of getting support from banks for the rehabilitation of dry degraded lands. Based on these studies, the species have been identified for scaling up forage and firewood production from different types of degraded lands in India (Pathak *et al.*, 1995). Currently the Government is allocating >20 M ha degraded forest to the landless under Joint Forest Management, where this technology has a great scope to assure environmental conservation together with forage and firewood supply. It is assumed that if 50 % of the degraded lands are allocated to this technology, it will be possible to meet the deficits of fodder in the country (Pathak & Roy, 1994).

Conclusions Silvopastoral systems of degraded land management assured conservation of natural resources along with supply of fodder, grazing and firewood in addition to the environmental amelioration. Grazing livestock within the carrying capacity produced high levels of individual livestock production. Such projects have been found economically viable and environmentally sound. It has a great applicability as a technology.

References

Deb Roy , R. & P.S. Pathak. (1983). Silvipastoral research and development in India. *Indian Review of Life Sciences.* 3, 247-264.

Pathak, P.S. & M.M. Roy (1994). Silvipastoral system of production, *Bulletin*, IGFRI, Jhansi. pp. 55.

Pathak, P.S., S.K. Gupta & P. Singh (1995). IGFRI approaches: Rehabilitation of degraded lands : *Bulletin*, IDRC - IGFRI, Jhansi. pp. 23.

Patil,B.D. & P.S.Pathak (1977) Energy plantation and silvipastoral systems for rural areas. *Invention Intelligence,* 12(1-2), 79-87.

The valuation of service of recreation function of Dalinor National Natural Reserve

G. Qiao[1], H. Wang[1], G. Han[2] and M. Zhao[2]
[1]College of Economics and Management, [2]College of Ecological Environmental Science. Inner Mongolia Agricultural University, Hohhot 010019 China, Email: qgh@sohu.com

Keywords Dalinor National Natural Reserve, recreation value, travel cost valuation method, valuation

Introduction Dalinor National Natural Reserve is located in the west of Keshketeng Banner of Chifeng City, Inner Mongolia Autonomous Region (IMAR). It covers a total area of 119,413.35 ha, and is in drought or semi-drought continental climatic zone. DNNR is a comprehensive natural reserve made up of different ecosystems including lakes, prairies, wetlands, and forests. Dalinor Lake in DNNR is regarded as the paradise of birds, covering an area of 238 km^2 between sandy lands and grasslands. DNNR was established in 1987 as a provincial natural reserve and was upgraded to a National Natural Reserve by the State Council of China in 1997. DNNR not only serves as an ideal resort for eco-tourism but also serves as an important scientific research base. The value of service of recreation function of DNNR depends heavily on the environmental quality that offers tourists various ecosystem amenities and opportunities for viewing wild plants and animals. So it is meaningful to value the recreation function of DNNR

Method and model This paper values the service of recreation function of the DNNR using travel cost valuation method (TCM) approach which is widely applied to the valuation of all kinds of public goods and recreation activities (Xuedayuan, 1998). This method is specially appropriated for the evaluation of recreation function of ecosystems with low entrance fee or fee free and limited demand for the recreation activities. Based on the principles of TCM (Xuedayuan, 1998), the value of service of recreation function is the sum of total travel cost, consumer surplus, travel time value and other expenses. A regression model is used to determine the correlation between the visiting rate and the variables such as the population, annual wages, and travel cost and travel duration (Chenfu, Zhangjie, 2001). In light of correlation analysis, travel duration and travel cost are the most remarkable factors that affect visiting rate. However, travel duration can be transformed into time cost, i.e. opportunity cost of time, therefore the visiting rate is directly correlated with the travel cost. A questionnaire is developed and random sampling approach was used. More than 450 questionnaires were answered, of which 376 were reasonable. It is supposed that total population; annual average wage and travel duration keep stable, and travel cost 101 yuan, which is the expenses from the nearest place Keshketeng banner of Chifeng City to DNNR, is taken as basic expenses. With the increasing of travel cost, the amount of tourists decreases to zero. At this time, the highest cost is 2447 yuan (1 euro=10.7 yuan) according to our investigation, With the help of SAS software, a regression model of person-time and travel cost was established and tested based on the data collected.

$Y=16998-17.165996+0.004271 \ x^2$ Y: travel person-time in DNNR; X: increment of expenses According to Clawson-Knetsh curve

Results Total travel cost $=\sum$travel cost/person·time×travel person-time in DNNR from every region = 58,883,580 yuan ;

$$CS= \int_p^{pm} Y(X)dx = \int_{101}^{2447} (16998-17.165996x + 0.004271x^2)dx =9,429,800 \text{ yuan}$$

CS: consumer surplus; x: total travel cost; Y(x): Clawson-knetch curve Pm: travel cost at which there is no tourist in departing places; p: travel cost to and fro between the nearest departing region and DNNR.
The sum of total travel cost, consumer surplus, and travel time value and other expenses is 100,228,140 yuan. Not all of the travel cost of tourists should be included in the value of recreation function of DNNR; because 75% of 376 visitors visited DNNR for viewing and admiring prairies and Dalinor Lake. So the total value of service of recreation function of DNNR is 100,228,140×75%=75,171,100 yuan.

Conclusions and discussion It is estimated that the total value of the service of recreation function of DNNR is RMB 75,170,100, namely 629.5yuan/ha. Comparing with the value of recreation of Zhangjiajie National Forest Park 6051.07 yuan/ha and the value of recreation of Changbaishan Mountain Biosphere Reserve 2134 yuan/ha, the value of recreation of DNNR is on the low side. The main reason for this is possibly the difference among the actual value of them or the different methods adopted, which needs to be identified further. The result can be of some interest to both policy-making and compensating of natural resources of DNNR.

Acknowledgements This research was funded by the National Natural Science Foundation of China (30360022). Many thanks to Professor Carl. E. Olson, University of Wyoming, USA.

References
Xue Dayuan. (1998) *Economic Valuation of Biodiversity* (M). China Environment and Science Press: 69-76
Chen Fu, Zhang Jie (2001) Analysis on Capitalization Accounting Of Travel Value —A Case Study of Jiu Zhai Gou Scenic Spot (J). *Journal of Nanjing University (Natural Science)*. 37(3): 296-302

Significance of grasslands in protected forest areas

C.M. Mishra, A. Kumar, B.C. Tiwari and S.L. Singh
Forest Research Institute, U.P., Kanpur – 208024, India, Email: chandramm@rediffmail.com

Keywords: protected area, sanctuary, wildlife

Introduction This paper describes case studies of grassland formation in the protected forest areas of the Indian state of Uttar Pradesh. This state has a total forest/tree cover of 8.84%, whereas the protected forest areas comprise only 2.54%.The protected areas constitute, one national park, eleven wildlife sanctuaries and thirteen bird sanctuaries.

Ecological status of grassland at Dudhwa National Park The park is situated on the Indo-Nepal border in Lakhimpur Kheri district of Uttar Pradesh state and was established in 1977. The park is famous for the presence of swamp deer. According to a recent census the estimated population of herbivorous wildlife was about 20,000.The forest throughout the park is interrupted by wide stretches of mesophyllous grassland. Recent surveys have shown that these grasslands were organised into a number of recognisable assemblages, called communities. Each community in turn was characterised by one dominant species imparting a characteristic physiognomy to the vegetation. The area of grasslands was found using satellite remote sensing technique to be about $85.71 km^2$, or 12.67% of the total. Table 1 indicates the value of different parameters, recorded during the study. The productivity was estimated through clipping quadrats of size $1m^2$, using the dry weight method (Milner & Hughes, 1968).

Table 1 Name, annual productivity (t/ha) and composition (% ground cover of dominant species) of the dominant grassland communities in the Dudwa National Park

Community name	Productivity	Composition (%)
Narenga porphyrocoma	17.90	76
Bothriochloa pertusa	9.92	60
Themeda arundinacea	13.89	82
Arundo donax	20.84	69
Desmostachya bipinnata	15.86	56

Ecological status of Samaspur Bird Sanctuary at Raebareli The Samaspur Bird Sanctuary, covering an area of 8 km2, is situated about 40 km from Raebareli city. It is an ideal habitat for migratory and native birds. An initial survey of the lake revealed a rich vegetational diversity of aquatic flora with vegetative production of approximately 4 t/ha per yr. However, there is ample scope to enrich the habitat by planting suitable tree, grass and legume species and to increase the carrying capacity of the sanctuary. Hence, to increase the carrying capacity of the sanctuary, afforestation with a suitable tree grass legume model with certain aquatic flora has been suggested in Table 2. The choice of species will be an important feature in seeking to achieve sustained productivity.

Table 2 Suitable tree, grass, legume species with a list of certain aquatic plants appropriate for Samaspur Bird Sanctuary

Tree spp.	Grasses and legumes	Aquatic spp.
Acacia nilotica	*Brachiaria mutica*	*Spirodela*
Acacia leucophloea	*Chloris* spp.	*Lemna*
Acacia auriculiformis	*Panicum maximum*	*Wolffia*
Albizzia lebbek	*Panicum notatum*	*Wolfiella*
Albizzia procera	*Stylosanthes* spp.	*Hydrilla*

Conclusions This investigation of one national park and one bird sanctuary revealed huge biological diversity with high productivity but at the same time all the sites require careful scientific management for sustained productivity for maintenance of a proper food-chain in between herbivorous and carnivorous wildlife.

Reference

Milner, C. & R.E. Hughes (1968). Methods for the measurement of primary production of grasslands. IBP Handbook No. 6, Blackwell Scientific Publication, Oxford, 70 pp.

Sustaining the multi-functionality of the Zamfara reserve in semi-arid Nigeria: what is the role of co-management?

B.F. Umar

Department of Agricultural Economics and Extension, Usman Danfodiyo University, PMB 2346, Sokoto, Nigeria, Email: bfumar@yahoo.com

Introduction Unhealthy competition and conflicts among the diverse users (herders, farmers, fishermen, etc.) of the natural resources (pasture, water, land, etc.) in the Zamfara reserve, Nigeria have undermined the reserve's capacity to serve its intended multi-functional roles. The reserve (3, 650 km^2), which was established in 1919 with 4 enclave villages where farmers live and cultivate crops, was meant also to provide pasture and water for Fulani herdsmen. Vast numbers of people are, however, becoming landless or near landless in the reserve. Powerful non-local actors are forcing their way into the reserve area and are extracting resources with no respect to traditional customs and rules. Umar (2004), for example, has shown how politicians and traditional rulers, with the support of the Zamfara state government, have converted over 100 ha of common grazing land in the reserve to private farms. This paper offers a model for co-management of the shrinking natural resources in the reserve, which the paper assumes, may help in maintaining the reserve's multi-functional nature. Co-management is a pluralist approach to managing natural resources, incorporating a variety of partners in a variety of roles, generally to the end goals of environmental conservation, sustainable use of natural resources and the equitable sharing of resource-related benefits and responsibilities (Viswanathan, undated).

Materials and methods Experience gained through participation in the EU-sponsored INCO-DC project, which has been conducting research on how to revitalise the Zamfara reserve, was used as the basis for designing the co-management model. This paper assumes that if co-management is adopted it will provide a situation in which all social actors in the reserve will negotiate, define and guarantee amongst themselves a fair sharing of the reserve's management functions, entitlements and responsibilities.

Results The paper suggests mechanisms for involving all major stakeholders in establishing co-management institutions in the reserve. Some of the mechanisms include 1) a participatory, bottom-up approach, 2) building on existing popular local institutions such as the Miyetti Allah Cattle Breeders Associations, farmers' clubs, etc., 3) making the process flexible and adaptable to fit local contexts, complexities and needs, 4) allowing co-management groups to determine their own boundaries and membership, their management structures and procedures, and their constitutions, bye-laws, rules, sanctions, and natural resources management plans and 5) providing training and sensitisation for the stakeholders. The paper has further identified 1) roles the stakeholders should play in co-management of the reserve, 2) ways for strengthening and sustaining the co-management institutions if established, and 3) the need for incorporating participatory monitoring and evaluation in the co-management institutions as a tool for ensuring and sustaining success.

Conclusions The paper concludes that if co-management is adopted in running the affairs of the Zamfara reserve it will provide a situation in which all stakeholders will negotiate, define and guarantee amongst themselves a fair sharing of the reserve's management functions, entitlements and responsibilities. Using the approach in the reserve may thus not only ensure sustaining its multi-functionality but may also help in ensuring peaceful coexistence among its multiple users.

References

Umar, B.F. (2004). Management of pastoral-agricultural conflicts in Zamfara State, Nigeria. Unpublished PhD Dissertation, Department of Sociology, Bayero University, Kano, Nigeria.Viswanathan, K.K. (undated). Co-management of natural resources – implications for government. International Center for Living Aquatic Resources Management (ICLARM).

Theme C: Delivering the benefits from grassland

Section 21

Adoption of new technology

Integrity of indigenous knowledge systems in natural resource management: the case of the arid and semi-arid Baringo herders of Kenya

P.M. Makenzi and A.A. Aboud
Egerton University, P.O. Box 422, Egerton, Kenya, Email: pmakenzi@yahoo.com

Keywords: educational levels, indigenous and exogenous knowledge systems, natural resource management

Introduction Ineffective management of natural resources in arid and semi-arid lands (ASALs), resulting in resource depletion, rapid loss of biodiversity and environmental degradation, is of great concern globally. The Baringo herders in the ASALs of Kenya have been branded as perpetrators of this vice, with the blame placed particularly on their traditional livestock management, utilising indigenous knowledge systems (IKS). These IKS involve livestock mobility and maximisation, and have been regarded as being outdated and inefficient in meeting the challenges and demands for environmental conservation and sustainable management of the natural resources. A common reaction of the government has been to advocate modern interventions that are based on exogenous knowledge systems (EKS), involving sedentary livestock raising and destocking. These EKS are, however, not performing as well as expected, since they are not adapted to the ASALs ecological conditions and the herders' socio-economic and cultural situations (Aboud *et al* 1997; Makenzi, 2003). This study empirically explored the above propositions, in order to test the integrity of the IKS, in relation to EKS and the herders' levels of education.

Materials and methods An *ex post facto* study in the form of a sociological survey was used to interview 300 randomly sampled herders in six administrative divisions of Baringo District. A structured questionnaire was completed with household heads, soliciting empirical information to assess the influences of the two knowledge systems (individually and in combination), on the levels of natural resource management (NRM) efforts and relationships with the educational levels of the herders,. Multiple regression analysis was used to determine the influences, as indicated by the beta regression coefficient (β) values.

Results As shown in the relationship path model (Figure 1), the strongest influence on the levels of NRM was from the use of both IKS and EKS in combination ($\beta=0.43$), while the weakest was from the use ITK only ($\beta=0.09$). This latter effect was not statistically significant. Educational levels seemed to exert a strong direct influence on the levels of NRM ($\beta=0.33$), which increased significantly when the influence was indirect, through the use of ETK only ($\beta=0.41$), and through use of both ITK and ETK ($\beta=0.43$).

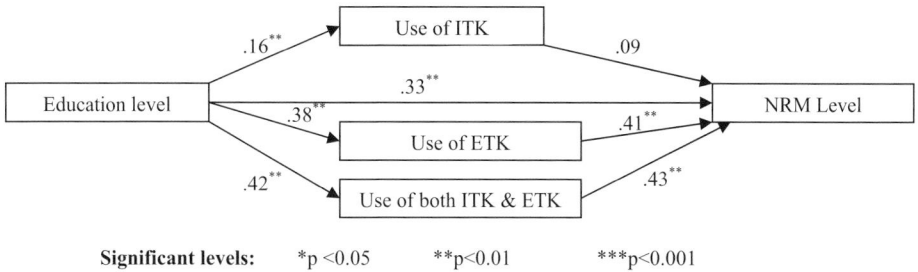

Significant levels: *p <0.05   **p<0.01   ***p<0.001

Figure 1 Relationship path model showing regression coefficients, beta (β)

Conclusions These results support the proposition that use of modern, scientific ETK enhances the level of NRM and production, but the influence will be even greater if the ETK is used in combination with the IKS. However, to ensure high levels of NRM levels among the herders, the educational levels need to be high. Hence, educating the operators and managers of the natural resources, in this case, the herders and extension agents, is basic to effective NRM and improved livestock production and conservation.

References
Makenzi, P.M. (2003). Indigenous Knowledge in Natural Resources Management in the ASALs: A Case of Biodiversity Conservation in Baringo, Kenya. Published PhD thesis. Moi University, Eldoret, Kenya, 98-118.
Aboud, A., F. Abdikadir & M. Hileman (1997). Pastoralist Life-styles and Development Paradigms: The Contradictions and Conflicts. Research Document. Kenya Pastoralist Forum, Nairobi, Kenya, 14-18.

The role of the PHARE Programme and the Danish-Lithuanian Project on the development of the grassland production and of dairy-beef cattle husbandry

J. Jatkauskas and V. Vrotniakiene
Lithuanian Institute of Animal Science, R. Žebenkos 12, LT-5125 Baisogala, Radviliškis distr., Lithuania. Email: lgi_pts@siauliai.omnitel.net

Keywords: grassland, dairy-beef cattle, silage, Programme, Project

Introduction Animal production is one of the priorities for economic development in Lithuania. Dairy-beef cattle husbandry has remained an important branch of animal production under the conditions of agricultural reform. Favorable agroclimatic conditions for grass cultivation, old traditions of agricultural production and the ability to train farmers in new techniques creates good conditions for the development of dairy-beef cattle husbandry. It is most appropriate to use cattle types suitable for grass feeding with low inputs of concentrates. This paper reports on two development projects that sought to increase animal production through grassland improvement and the production of high-quality silage.

PHARE Programme "Human Resource Development and Support to Grassland Management" In 1992-1993 farmers in Lithuania produced practically no silage because they had neither appropriate machinery nor silos and no equipment for mechanisation of feeding. Lithuanian farmers basically had no knowledge on new technologies of grass production, renovation and reseeding of pastures and leys. The main goal of the programme was to increase the number of farmers utilising new grassland management technologies and the area of land used in this way. This would lead to improved grassland and good quality silage production. Two model demonstration plots were established for grassland husbandry and grass production on the grounds of the Lithuanian Institute of Agriculture and the Lithuanian Institute of Animal Science. In addition, several farmers, willing to demonstrate new technologies were selected in collaboration with Lithuanian Agricultural Advisory Service. Furthermore advanced machinery was introduced in collaboration with machinery manufacturers. The programme progressed grassland management development in Lithuanian farms and led to increased milk and beef production per hectare.

Danish – Lithuanian Project "Lithuanian Dairy Farms Demonstration Project" The dairy sector was identified as an agricultural sector with export potential and with good natural conditions for production. Many farmers lack entrepreneurial skills, as well as knowledge on how to integrate forage production of high quality into the total milk production system and knowledge on how to manage stable milk production throughout the year. Earlier projects had mainly focused on specific parts of a dairy farm enterprise, such as grassland management, rather than on the farm as a whole. A demonstration project was established as a result of cooperation between The Lithuanian Institute of Agriculture, Lithuanian Agricultural Advisory Service, Lithuanian Institute of Animal Science and Danish Institute of Agricultural Sciences. The project includes case studies on three private farms. Grass was the most important part of the feed on each farm. It was clear that there had been a need to improve the pastures and leys and the grazing regimes in order to improve the amount and quality of silage production. The farmers have started to improve the grassland by reseeding new grass swards and introducing silage making. The quality of the grass markedly increased and milk yield per herd ranged form 5035 – 6849 kg per cow during the period of the project. It was found that interdisciplinary effort was required for the programme to be effective.

Conclusions The PHARE Programme and the Danish-Lithuanian Project pushed forward grassland management in Lithuanian farms and led to increased high quality milk and beef production.

Reference

Christensen J., J. Hermansen, I. Kristensen, T. Kristensen, J. Jatkauskas, J. Gutauskas & D. Sakickas (2003). A case study on improving sustainability in dairy farming systems in Lithuania. Livestock Farming Systems in Central and Eastern Europe. *EAAP Technical Series Number 3*, Netherlands, 141-154.

Farmer Field Schools in expanding cultivation to newly reclaimed land in Ismailia in Egypt

O. Niemeläinen[1], M. Komulainen[2], Y. Ahmed[3], M. El-Sayes[3] and A. El-Adawy[3]

[1]MTT Agrifood Research Finland, Plant Production, FIN-31600 Jokioinen, Finland, Email: oiva.niemelainen@mtt.fi, [2]ProAgria MKL, Association of Rural Advisory Centres, PL251, FIN-01301 Vantaa, Finland, [3]EFARP Project, Ismailia Agricultural Research Station, P.O. Box 320, Ismailia, Egypt

Keywords: development cooperation, farmer field schools, interdisciplinary research, participatory research

Introduction The use of irrigation systems is expanding in Egypt to facilitate cultivation on sandy areas that were previously desert. Many of the farmers starting on the new areas are undergraduates and others with no background in farming. Much support is required when they start farming. In addition, farmers moving to the new lands from the Nile Delta meet completely new challenges because the poor and infertile sandy soils require different management from the clay soils at the Delta. Ismailia Agricultural Research Station (IARS) of the Agricultural Research Center of Egypt focuses on research on how to cultivate the newly reclaimed sandy soils.

The Ministry for Foreign Affairs of Finland has supported the research and advisory work of IARS for the last fifteen years through the Egypt – Finland Agricultural Research Project (EFARP). This has been implemented as a twinning project between the Agricultural Research Center of Egypt and the equivalent organisation in Finland: MTT Agrifood Research Finland. In this paper we describe the Farmer Field School (FFS) approach applied to forage and animal production on smallholder farms from 2000-2004.

Background The EFARP has had three main phases: I) Support to research infrastructure and human capacity building (1990-94), II) Interdisciplinary research (1994-98), and III) Extension activities (1998-2004). During the years of the EFARP research activity (involving the disciplines of soil science, forage production, plant protection, animal husbandry and veterinary science), research results were obtained to facilitate the production of recommendation packages for extension agents and for farmers. Close collaboration between the research and advisory services was encouraged in EFARP. The project also tried to involve farmers in the process of choosing research priorities. In the Extension phase, 64 facilitators (advisory persons running the FFSs) were trained both in technical matters and in training skills to run the Farmer Field Schools in the villages. Most of the training programmes were carried out at IARS and training on technical matters was provided by the research staff. When facilitators were working on the field, refreshment training was provided twice a month on new subjects. These training sessions proved to be a fruitful opportunity to exchange experience amongst the facilitators themselves and with research personnel.

Application of Farmer Field Schools to forage production and utilisation Each facilitator was responsible for four FFS groups and 6-14 farmers in each of her/his group. The FFS groups were formed twice a year: in September for winter season and in May for summer season. Each group was formed to operate over the coming growing season but the majority of the groups continued to meet over the following season, thereby covering the whole year. In winter season 2003/2004 of the 196 FFS groups 108 were men's groups, 10 were women's groups, and 78 were mixed groups. The total number of members in these groups was 1893. Fifteen of the 64 facilitators were women. These groups met every second week at the same point (field) where the facilitator gave a special introduction to topics related to the activities on the field or animal husbandry and they studied the growth of the demonstration plots. The facilitator participated in the follow-up discussion. Exchange of ideas and experience between the group members was important and the introduction of personal decision-making ideas by group members was encouraged. The project supported activities to be demonstrated at the FFS meeting sites, e.g. testing of new forage species or cultivars, and arranged materials for demonstrations, such as silage making, urea treatment of straw, Rhizobium inoculation. A 'Farming Guide' book (Anon, 2004) was produced in collaboration with research and advisory personnel as one tool to help facilitators to carry out their work in the field.

Results and conclusions The results obtained through the FFS activity were excellent. The facilitators were very motivated and the participants gained valuable information for their profession and established valuable information channels and networks for future needs. The FFS approach proved to be a valuable way to increase interaction between research, advisory and farming activities and to help all of them.

Reference

Anon. (2004). Farming Guide. Practical Help for Smallholders in Ismailia. (In Arabic) Eds M. Komulainen, O. Niemelainen, R. Ojala, Y. Ahmed, Y. Shawky & A. El-Adawy. Egypt – Finland Agricultural Research Project. University Printing Office, Ismailia. 192 pp. (CD in English of the text without pictures is available from the Ismailia Agricultural Research Station, Ismailia, Egypt).

Forage development in the Nepal mid-hills: new perspectives

A.D. Robertson

"Oaky Creek" Wilson's Downfall, MS 1983, Stanthorpe 4380, Australia, Email: halfmoon@halenet.com.au

Keywords: groups, participatory, quick start, Nepal

Introduction Nepali hill farming communities are typically poor and remote, and are currently severely affected by conflict. The challenge is to define simple approaches which can generate results within this context. Livestock are central to livelihoods and to the sustainability of farming, with rain-fed agriculture dependent on inputs of manure-based compost. Stall feeding has increased dramatically with the adoption of community forestry and general preclusion of grazing. A broad landscape approach to forage development is increasingly being adopted, with concurrent on-farm interventions, such as intercropping and back-yard forage, and off-farm interventions, such as landslide stabilisation with forages, development of forest understory, and reinforcement of degraded grazing areas with forage. Considerable work has been undertaken in the mid-hills (below 1800m ASL) with very limited higher altitude programs to 4,000m ASL. Until the late 1990s forage development was restricted to use of a very narrow array of genetic material. To accommodate the agro-ecological diversity, broad mixtures are now commonly used, encompassing species with known potential locally, and some peripheral commercial and pre-release material for testing to refine recommendations. Productive erect cut-and-carry grasses including Mott Napier (*Pennisetum purpureum* cv Mott) are popular. A suite of legumes including *Stylosanthes.guianensis*, *Chamaecrista rotundifolia* cv Wynn, *Aeschynomene americana* cv Glenn, *Aeschynomene villosa*, *Neonotonia wightii*, forage arachis (*Arachis pintoi*), and *Leucaena leucocephala* have been successful in various niches.

Scale and benefits Since the late 1990s, it is estimated that at least 40,000 households have been involved in the adoption of the newly introduced genetic material. There are now more than 20 new species actively promoted. Additionally, more than 35,000 school students became involved in only two districts during 2004. Adoption rates of the forage packages are as high as 80%, with preliminary "social mobilization" unnecessary. Farmer-farmer exchanges have been encouraging. Most forage from on-farm and cultivated communal plantings is used in cut-and-carry systems for feeding goats and dairy and draft bovines, with increasing use also for poultry and swine. On-farm forage development has dramatically reduced labour requirements for forage collection. Farmers report benefits from forage legume introduction to crop areas, in terms of stabilizing crop production. Landslide stabilisation from direct seeding has been successful on many sites.

Major reasons for success:

Farmer attitudes and farmer groups Farmers perceive the lack of good quality forage to be a major constraint on livestock productivity. Women commonly spend more than four hours per day on fodder collection; they welcome any intervention to reduce this burden. Traditional involvement of the majority of households in milk production for home consumption or sales is considered to be a major factor in achieving higher adoption rates than could be expected, for example, in South-east Asia or most of Africa. Nepali hill farmers are commonly coordinated into focus groups, such as livestock groups, community forest user groups (of which there are now more than 10,000) and various women's empowerment groups; this presents exceptional opportunities for intervention and efficient delivery of technology, with high rates of farmer-farmer adoption locally.

Farmers and participatory research The Nepali Government capacity for conventional forage research is negligible in the context of the vast agro-ecological diversity. Hence large numbers of widely scattered farmers and farmer groups are now directly involved in screening new development strategies and genetic material. Experience has shown the necessity of including simple and reliable strategies (such as back-yard or terrace-riser forage) and some conspicuous and reliable genetic material including Mott napier and forage arachis. Such species have had a high rate of spontaneous lateral adoption locally, although technology transfer over larger distances has typically been slow; this reinforces the importance of initiating work at many widely scattered sites.

The future Remoteness of communities and the current conflict preclude regular visitation. Simple and flexible technical packages and technology delivery mechanisms, with the capacity to provide quick and conspicuous results for poor and remote communities, are central to success; recent programmes have demonstrated the potential for reaching large numbers of poor hill farmers. It is now necessary to maintain access to improved genetic material, to improve the supply of good quality seed, to involve more development agencies and community based organisations in delivery of the technology, to trial technologies in new environments, and to streamline adoption by facilitating exchange visits for farmers from new areas.

Factors influencing the adoption of fodder production techniques by milk producers in Dhankuta District, Eastern Nepal

C.P. Ferris and R. Nelson
The Agricultural Research Institute of Northern Ireland, Hillsborough, Co. Down BT26 6DR, UK, Email: conrad.ferris@dardni.gov.uk

Keywords: Nepal, fodder production, adoption, milk producers

Introduction Many districts are food-deficient in Nepal. Increased milk consumption could improve human health and nutrition, while milk sales can provide families with a valuable source of income. However, milk production in Nepal is low, mainly because of the poor nutritional status of livestock, which can be attributed in part to a fodder deficit. However there is potential to increase fodder production, and some innovative fodder production techniques have been tried and proven locally. These include planting of fodder trees, fodder grasses and fodder crops. As the level of adoption of these techniques is variable, a survey was undertaken to quantify adoption levels by milk producers in Eastern Nepal, together with factors influencing adoption levels.

Materials and methods This survey, in which 68 farmers were questioned, was conducted at the Milk Collection Centres of Chungbang, Hattikarka and Hille (Dhankuta District, Eastern Nepal). For each of 3 forage production technologies identified, farmers were classified as: (1) not aware of the technology, (2) aware of but not having tried the technology, (3) Having tried, but subsequently dropped the technology, (4) Having tried, but uncertain about adoption, and (5) Having definitely adopted the technology. Farmers' situational characteristics (farm size, number of milk-producing livestock and family size), personal characteristics (age of head of household, education level and economic status) and sociological characteristics (local-cosmopolitan, frequency of listening to agricultural news on the radio, degree of extension contact) were identified also. SPSS statistical package was used to test relationships between level of adoption of techniques and 'farmer characteristics'.

Results The degree of adoption of fodder tree and fodder grass production was high (85.3 and 79.4%, respectively). Most of the remaining respondents were aware of, but had not tried the technology (Table 1). Only 35.3% of farmers had adopted the practice of fodder crop planting, with 41.2% aware of the technology, but not having tried it. Fodder crops differ from fodder trees and fodder grasses in that their planting often competes directly with the production of human food resources. For example, while fodder grasses and fodder trees will normally be planted on terrace risers and areas of ground unsuitable for cultivation, fodder crops are grown on land that could otherwise be used for producing food for direct human consumption.

Table 1 Percentage adoption of forage production techniques (n=68)

	1) Not aware of technology	2) Aware of, but not tried	3) Tried but dropped	4) Tried but uncertain	5) Adopted
Fodder tree planting	1.5	13.2	0	0	85.3
Fodder grass planting	1.5	10.3	4.4	4.4	79.4
Fodder crop planting	16.2	41.2	4.4	2.9	35.3

A number of significant relationships were identified between the level of adoption of these innovative techniques, and situational personal and sociological characteristics, as follows: the degree of adoption of fodder tree planting was positively related to farm size (P<0.05), family size (P<0.01) and frequency of listening to agricultural news on the radio (P<0.01), while the degree of adoption of fodder grass planting was positively related to frequency of listening to agricultural news on the radio (P<0.05). As regards the latter, the survey did not reveal if this was a direct 'cause and effect' relationship, or if farmers who listen to the agricultural news more frequently are more innovative, and as such, more likely to adopt fodder production techniques. There was also a positive correlation between the adoption score for fodder crop planting and the number of milk-producing livestock on a farm (P<0.01). This may reflect the fact that farmers with large numbers of livestock are likely to have sufficient land for the production of their direct food needs, and as such, may have excess land available for fodder crop production.

Conclusions The level of farmer adoption of fodder tree and fodder grass planting was high, while the level of adoption of fodder crop planting was considerably less. The latter is likely due to fodder crop production as a livestock feed competing directly with the production of human food resources.

Acknowledgements Emily Sarah Montgomery Travel Scholarship (Queen's University Belfast)

A situation analysis of ley pasture utilisation in the Western Downs and Maranoa regions of S Queensland, Australia

K.J. Sibson[1], L. Bahnisch[1], R. Routley[2] and K. Taylor[3]
[1]University of Queensland, Gatton, Queensland 4343, Australia, Email: k.sibson@uq.edu.au, [2]Department of Primary Industries and Fisheries, Roma, Queensland 4455, Australia [3]Department of Primary Industries and Fisheries, Miles, Queensland 4415, Australia

Keywords: ley pasture, livestock production systems

Introduction Previous studies have shown that the uptake of ley pasture systems in S Queensland's grain growing region has been slow when compared with mixed farming systems in S Australia. This is despite their demonstrated benefits to subsequent crops, livestock production and the environment. A survey was conducted that aimed to determine the level of utilisation of ley pastures in the Western Downs and Maranoa regions of S Queensland, and the possible constraints to their adoption, and benefits arising from their use. The survey also aimed to determine the commonly used pasture species, the reasons for their use and their contribution to the livestock component of mixed farms.

Materials and methods The survey was conducted in spring 2004 using a closed answer questionnaire. The target population was mixed farmers in 8 shires from the Western Downs and Maranoa regions of S Queensland. A stratified random sampling strategy was used. The target population was stratified by shires and the sample size in each shire was determined by proportional allocation. Initial telephone contact was made with the selected farmers to gain their consent to participate and the questionnaire was then posted to each participant. The survey had a 48% response rate, with 65 surveys completed correctly and returned.

Results Of mixed farmers surveyed, 66% integrate ley pastures into their farming systems, with 46% of these using 2 or more different ley pasture types. This is much higher than the level of adoption in 2002, estimated at 38% (Lawrence, 2002). On average 63, 48 and 38% of respondents who use ley pastures indicated that the unreliability of establishment, the cost of establishment and the lag time with no productivity from either crops or livestock when using ley pastures are problems. For non-adopters, these establishment factors are a major deterrent for not integrating ley pastures into their systems with 82 and 68% respectively seeing unreliability and cost as problems. The positive aspects of ley pastures such as their ability to improve livestock growth rates and soil structure and increase soil nitrogen, grain yield and protein were readily agreed to by the majority of mixed farmers surveyed, with less then 5% disagreeing with these statements. In general, the age, gender or size of property owned did not significantly influence (P<0.05) farmers' opinions on the benefits and problems associated with ley pastures. However, whether or not the mixed farmer had adopted the practice influenced some opinions (Table 1). A number of different ley pasture species are used in the region with lucerne (*Medicago sativa*) being the most common (41% of cases). In half of these cases, lucerne was grown with annual medics and/or tropical grasses. The ley pastures are used mainly for growing and fattening cattle. On most types of ley pastures grown cattle growth rates generally were between 0.5-1.0kg/hd/day, but sheep growth rates often were unknown.

Table 1 Mean ratings of statements regarding ley pastures by adopters and non-adopters. Ratings are strongly agree (1), neutral (3) through to strongly disagree (5)

Statement	Adopters	Non-adopters
Helps meet target market specifications better*	2.35	2.75
The cost of establishment is too high	2.81	2.41
Leys do not increase livestock production*	4.33	3.50
The lag time with no production from crop or livestock is a problem	2.88	2.50
Allows spreading of risk between crop & grazing*	2.07	2.62
Leys are uneconomical*	3.76	3.11
Establishment of ley's is too unreliable*	2.62	2.10

*statements in which ratings differed significantly (P<0.05) between adopters and non-adopters

Conclusions Ley pastures are used widely in the Western Downs and Maranoa regions of S Queensland. While adopters indicated problems with their use, they were very positive about the benefits. In contrast, non-adopters tended to be more negative about both the problems and benefits, indicating a need for further investigation.

References
Lawrence, D. (2002). Perceptions of Western Farming Systems project, how it has helped and how have farming practices changed since 1995? In: Western Farming Systems Results Booklet, Department of Primary Industries & Fisheries, Brisbane.

Rotational grazing demonstration with beef cattle on conservation reserve land in Adams County, Iowa, USA

S.K. Barnhart[1], B. Peterson[2], C.O. Nelson[3], R. BreDahl[4], J. Klein[5] and R. Sprague[5]
[1]Agronomy Department, Iowa State University, Ames, Iowa 50011USA, Email: sbarnhar@iastate.edu, [2]Iowa NRCS, Des Moines Iowa 50319 USA, [3]Iowa State University Extension, Corning, Iowa, USA, [4]Iowa State University Extension, Creston, Iowa, USA, [5]NRCS, Corning, Iowa, USA.

Keywords: technology adoption, grazing, beef cattle

The United States Dept. of Agriculture's Conservation Reserve Program (CRP) is a voluntary program available to agricultural producers who will enroll erosive, marginally productive cropland for a 10 to 15 year period. In return, participants are provided annual rental payments and cost-share assistance to establish and maintain long-term, resource-conserving vegetative cover to improve the quality of water, control soil erosion, and enhance wildlife habitat. Since the inception of the CRP, policymakers, conservationists, farmers, and rural residents have been concerned about the likely fate of program land after the contracts expire. Most of the existing research, whether it relies on farm surveys or computer models, suggests that a significant proportion, perhaps more than 50 percent, will move back into row-crop production. Many rural residents in areas in which the CRP has significantly affected agricultural production would prefer to see the land returned to some form of agricultural activity, competitive with intensive row-crop production but with management and technologies that lead to acceptable environmental consequences.

The Adams County CRP Grazing Demonstration was initiated to demonstrate the production potential of well managed livestock grazing systems on highly erodible, marginally productive CRP land similar to 2.5 million hectares in the surrounding southern Midwest U.S corn/soybean 'belt'. The project is an interagency, cooperative effort sponsored by the Southern Iowa Forage and Livestock Committee.

The demonstration consists of three rotationally stocked grazing systems. A 4-paddock and a 13-paddock system were established in 1991. An 18-paddock system was added in 1992. Pasture vegetation is primarily perennial, cool-season grass-dominant. Two of the systems are stocked with crossbred beef cow/calf pairs, and the other with crossbred steers averaging 281 kg. Pasture management technologies demonstrated include: rotational stocking; pond water access using electric, solar and 'nose' pumps; dispersed paddock water stations; numerous types of temporary and semi-permanent electric fencing materials; improved lane design; fertilization and legume oversowing; the conversion of some paddocks to perennial, warm-season grasses; species composition assessment techniques; weed control alternatives; harvest and conservation of excess forage; sampling and testing for the tall fescue (Schedonorus phoenix (Scop.) Holub [= Festuca arundinacea Schreb.; also = *Lolium arundinaceum* (Schreb.) S.J. Darbyshire] endophyte (*Acremonium coenophialum*), and early calf weaning. The unreplicated demonstrations have continued for 13 years. This longevity has provided a visual performance of new pasture and grazing technologies over the range of environmental conditions occurring in this part of the U.S. The mean animal and pasture performance data for the demonstrations is contained in Tables 1 and 2.

Table 1 Performance summary of the 18-paddock steer grazing demonstration (1997-2003)

	Range	Mean
Area grazed (ha)	26-30	28
Steers at start	75-98	87
Mean initial weight (kg)	270-304	281
Initial stocking (steers/ha)	2.9-3.3	3.1
Steers sold mid-season	0-65	38
Steers grazing to season's end	25-76	48
Days on pasture	108-159	130
Steer gain (kg/ha)	251-355	288
Live weight gain (kg)	0.82-1.03	0.89

Table 2 Performance summary of cow/calf pairs demonstration (1991-2003)

	4-paddock	13-Paddock
Area grazed, ha	9.1	14
Ha/pair	0.68	0.64
Grazing Days	144	145
Calf ADG, kg	1.07	1.05
Mean calf gain, kg/ha	223	236
Cow wt. change, kg	35	28
Cow condition[1]change	+0.3	+0.3

[1]Body Condition Score System, 0-9 point scale

The outreach effort has reached the residents of the area. There have been 36 field days and tours conducted for producers, specifically highlighting the pasture demonstration area and its technologies. It has been used as a 'field classroom' for 15 grazing schools for producer and agri-business professionals conducted in the area and over 450 K-12 school students have used the demonstration site as a field classroom.

Sustaining grass-legume pastures for cow-calf herds: a case study

J.L. Caddel[1], D.D. Redfearn[1] and R.L. Woods[2]

[1]*Plant and Soil Sciences Department, Oklahoma State University, Stillwater, OK 74078,USA, Email: john.caddel@okstate.edu, [2]Oklahoma Cooperative Extension Service, Muskogee, OK 74401, USA*

Keywords: grasslands, mixed pastures, beef cattle

Introduction An on-going pasture demonstration study has been used since 1988 to demonstrate methods to improve pasture production for small beef cow-calf herds. Many cattle enterprises are not economically viable because poor management decisions lead to excessive stocking rate, ineffective fertilisation programmes etc., leading to a dependence on hay purchases. The initial objective, continued until 2001, was to maintain one cow-calf pair per ha without purchasing forage or grain produced off farm. More recently, reducing the dependence on harvested forage has been added as an objective.

Materials and methods This study used 52 ha from 1988 to 1995 and 73 ha since 1995. The farm is located at 35.7^O latitude and 95.6^O longitude and represents approximately 2.6 million ha in the Cherokee Prairie Resource Area of eastern Oklahoma in the USA. Soils are 65 to 120cm deep, gently sloping, somewhat poorly drained to moderately well-drained with a loamy surface layer and a clayey subsoil. The elevation is 180 m above sea level. Average annual precipitation is 1166 mm with average monthly temperature ranging from 27.8°C in July to 2.3°C in January. Bermudagrass (*Cynodon dactylon* (L.)) and tall fescue (*Festuca arundinacea* Schreb.) are the most common perennial grasses in this area and exist in these pastures because of natural encroachment and intentional introduction. Other grasses became established in some pastures with hairy crabgrass *(Digitaria sanguinalis* (L.) Scop.), Old World bluestem (*Bothriochloa* spp.), dallisgrass (*Paspalum dilatatum* Poir.), and annual ryegrass (*Lolium multiflorum* Lam.) making important contributions. Major forage legumes were red clover (T*rifolium pratense* L.), white clover (*T. repens* L.), arrowleaf clover (*T. vesiculosum* Savi), and subterranean clover (*T. subterraneum* L.). Other legumes used with, variable contributions, included lucerne (*Medicago sativa* L.) and annual lespedezas (*Kummerowia* spp.). Crimson clover (*T. incarantum* L.), birdsfoot trefoil (*Lotus corniculatus* L.), and rose clover (*T. hirtum* All.) were established in the pastures, but they did not persist or contribute significantly to the overall system. Limited N fertiliser was applied, which allowed legumes to fix nitrogen. Lime, phosphorous, and potassium were surface-applied according to soil analysis. Cows and calves were rotated among seven pastures by allowing the cattle to move to a new pasture when available forage was consumed. Hay was harvested from pastures when excess forage was produced. This hay was then fed when standing forage was sufficiently deficient to potentially cause cattle to lose an excessive amount of body condition. During the first seven years, hay was fed for up to 180 days. To minimise the number of hay-feeding days, N-fertilised pastures were added to the system. These pastures were used for stockpiling forage to be grazed during late autumn and winter when little forage was produced because of cold temperatures and again during the summer months. Nitrogen fertiliser (50 kg/ha) was applied in early September to maintain yield and quality.

Results and conclusions The original objective of maintaining a cow-calf pair per ha between 1989 and 1995 was achieved, but this relied heavily on mechanically-harvested forage. The reason for the excessive number of hay-feeding days, was that legumes were re-established during late summer. Thus, it was necessary to minimise standing forage during September and October. Legumes themselves produced too little autumn growth to contribute to winter grazing. A combination of legume/grass mixtures and N-fertilised stockpiled grasses, reduced the hay-feeding days to as few as 30. The combination of proper stocking rate, a type of rotational stocking, correct fertilisation, interseeding forage legumes into well-adapted grasses and using stockpiled grass during winter resulted in a sustainable pasture programme requiring less labour and mechanical inputs. Small adjustments are needed annually to improve the system and account for irregular weather patterns, but small cow-calf herds can be economically positive and contribute to the small farm.

References

Woods, R.L. & J.L Caddel (1994). Managing clover pastures in eastern Oklahoma. *Proceedings of the American Forage and Grassland Council,* 3, 167-171.

Redfearn, D.D., J.L. Caddel, R.L. Woods, & K. Barnes (2003). Integrating pasture management and herd size strategies. *Proceedings of the American Forage and Grassland Council,* 12, 74.

Redfearn, D.D. & R.L Wood (2002). Pasture options, short and long-term. Eastern Research Station Field Tour Proceedings http://oaes.pss.okstate.edu/agronstations/1haskell/haskelltour2002/redfearn.htm.

Woods, R.L (2004). Cow-calf cost estimates for forage and fertilizer. Eastern Research Station Field Tour Proceedings http://oaes.pss.okstate.edu/agronstations/1haskell/haskelltour2004/woods.htm.

A forage area of expertise team: the Michigan approach to applied research and extension

R.H. Leep[1] and D.H. Min[2]

[1]Department of Crop and Soil Sciences, A464 PSSB, East Lansing, Michigan 48824, USA, Email: leep@msu.edu, [2]Upper Peninsula Experiment Station, E3774 University Drive, PO Box 168, Chatham, Michigan 49816, USA

Keywords: self-directed, team, extension, research

Introduction   Agricultural Experiment Stations and the Cooperative Extension Service have traditionally contributed to the economic, social, human, and environmental capital of the United States. Despite this, both institutions have experienced declining federal budget support and increasing competition for resources (Hamm 1997; Hood & Schutjer 1990; Knutson & Outlaw 1994; Paarlberg 1992). Michigan State University Extension, in partnership with the Michigan Agricultural Experiment Station, implemented self-directed area of expertise (AOE) teams as its major educational development and delivery model. AOE teams grew out of experiences with previous temporary research/Extension teams and quick response professional groups operating within traditional line responsibilities for research and Extension units. An extension forage area of expertise team was organized in 1999 to deliver educational programmes in forage management and conduct applied forage research. Funds were made available directly to the forage team from central extension administration.

Materials and methods   A forage area of expertise team was organized at Michigan State University in 1997 as a highly trained group of Extension and Experiment Station employees fully responsible for planning, implementing, and evaluating educational programmes in a self-directed manner.   The team has two co-chairs; one from the campus and another from off-campus. The on-campus co-chair has a joint research-Extension faculty appointment or responsibility. The team co-chairs provide leadership on a yearly rotational basis. Co-chairs are selected by the team and serve as facilitators. The team develops its own micro-vision, mission and operating procedures and develops a plan for programme delivery and curricula for staff development. Involvement of stakeholders is used for information input for programme/project selection, direction. The team consists of 15 active members and includes an additional 20 people on the list-serve who participate less frequently in team efforts.   The team is expected to be entrepreneurial and develop self-supporting educational programmes.

Results   Since its inception, the forage team has developed four major educational programme modules. Each program module consisted of a series of 10-12 MS Powerpoint presentations of 30-45 minutes duration. An accompanying notebook is also used to complement the Powerpoint presentations. Educational programme modules include: Building skills in grazing management, mastering the art of grazing management, advanced forage management, and utilization of dairy forages. Evaluations were conducted for all programmes. The team has developed numerous grant proposals and received grants from non-profit organizations, state and federal agencies, and private companies. The forage team has been recognized for its excellence by receiving a significant number of state and national awards for its educational programmes.

Conclusions   The forage AOE team, which connects field, campus, and stakeholders, and ties research to Extension with an interdisciplinary, problem-solving focus, has produced results that improve peoples' lives. Feedback from both campus-based and field staff members has been very positive. A trend of enhanced motivation among field staff members and stronger credibility with agricultural stakeholders has emerged as a result of the AOE approach. Improved credibility has translated into renewed pride among many stakeholders for "their" land-grant university, and this helps assure continued public support into the 21st century.

References

Hamm L.G. (1997). Assessing the role of public research and Extension policies in promoting improved performance of the agro-food marketing system." *American Journal of Agricultural Economics*, 79, 646-650.

Hood L. & W. Schutjer (1990). Cooperative Extension's role in shaping the future of land grant universities. *Choices*, 5, 22.

Knutson R.D. & J. Outlaw (1994). Extension's decline. *Review of Agricultural Economics*, 16, 465-475.

Paarlberg D. (1992). The land grant college system in transition. *Choices*, 7, 45.

Technology transfer through a network of producer groups

D.J. Thomson and R. Smith

Russell House, South Stoke, Bath, BA2 7DW, England, Email: tact@btinternet.com, Business Link Gloucestershire, Chargrove House, Shurdington, Cheltenham, Glos. GL51 4GA, England

Keywords: groups, information structure, R & D hub, distilled information, integrated delivery

Objective and approach The objective was to enhance the rate of uptake of research and development (R & D) in the context of the whole farm business. This paper reports on two pilot projects - SWARD (South West Agricultural and Rural Development – Devon and Cornwall) and GARD (Gloucestershire Agricultural and Rural Development). The projects were focused on improving the quality of decision making at a time of change, through the provision of R & D results and information within an integrated group support service. Key aspects of the approach are outlined below.

Information and communication structure A network of self-selected, common interest producer groups was linked to a research and information 'hub'. Results from R & D were distilled from publications and reports and disseminated to the groups through the group leaders. Group leaders, who were producers, distributed and discussed the results and information within their groups.

Services Distilled R & D results and information, the provision of research reports, signposting and audio conferencing within groups were all services provided within the projects.

Training A training bursary was available to be used on a group or individual basis. A flexible approach was adopted in relation to training. Groups could choose the form of training which included engaging the services of leading UK and overseas consultants, study visits to centres of research and leading farmer exponents and acquiring technical and business skills. The type, mode of delivery and training provider were requested by the groups.

Flexibility and minimum bureaucracy The projects were flexible to enable support services to respond to participants' needs and requirements and to enable groups to evolve and change. Bureaucracy was minimised.

Professional support The professional support staff facilitated the establishment and initial operation of the groups. They (a) set up and managed the information and communication structure, based on the hub, (b) distilled the research information, (c) cultivated ownership of the problems, solutions and the project and (d) helped participants to achieve their individual and group goals.

Participation and group structure The project Chairmen were farmers. All group leaders were farmers. All group members were farmers. Groups were self-selected. Incentives to participate were provided. Participants owned the projects.

Results Some 146 groups were established, including grazing groups with participation of 1,400 agricultural and rural businesses. The hub provided 880 technical digests and notes. Over 5,000 training, study visit and learning days were taken up. Signposting services directed participants to funds, grants, conferences and appropriate workshops.

Conclusions Self-selected, common interest groups are an effective mode of technology transfer. Ease of access to distilled R & D information is vital to the process of technology transfer. Farmer to farmer exchange within groups is an essential component in the process of change. Appropriate training can take many forms. Effective technology transfer can be achieved within integrated projects which encompass technical, training and business components.

Acknowledgements Business Link Devon & Cornwall and Gloucestershire; Seale Hayne (University of Plymouth), DEFRA, EU, Learning and Skills Council, Cornwall County Council, Devon County Council, Gloucestershire Farmers' Club, Stroud and Cotswold District Councils and South West Regional Development Agency, participants and professional support staff who contributed to making the projects successful.

Using abandoned paddy fields for grazing in Northern Japan

M. Nashiki, H. Narita and Y. Higashiyama
The National Agricultural Research Center for Tohoku region, Akahira 4 Morioka, 020-0198 Japan, Email: na493@yahoo.co.jp

Keywords: grazing, abandoned paddy field, global positioning system, ryegrass, millet

Introduction The number of abandoned paddy fields is increasing in Japan, because the government has been regulating rice production. It has been recommended that the abandoned paddy fields be used for stock raising. However, there is sometimes a lack of information about the land on the abandoned paddy fields and farmers want to know the cost of fencing and the best grass species to use. This paper seeks to provide farmers with information on how to begin to use the abandoned paddy fields as pastures in hilly rural areas in Japan.

Materials and methods Three aspects were considered. 1) Measurement of abandoned paddy fields: A handheld GPS (Garmin eMap) with an external antenna (GA27C) and tape measure (tape) were used to survey a flat land area with slender shape of about 0.38 ha and consisting of 11 small paddy fields in Morioka, Iwate. Results from the two methods were compared. 2) Construction of an electric fence including fence energiser, posts, galvanised wire, etc. The cost was compared with a conventional barbed-wire fence. 3) Grass species for abandoned paddy fields: Annual ryegrass (two commercial varieties), perennial ryegrass (one commercial variety) and millets for feed (two commercial varieties) were examined for water tolerance and productivity in a former paddy field where soil moisture was high.

Results Table 1 shows that there was no significant difference between the GPS and tape measurements of land area. The accuracy of the handheld GPS is around 4.8m, but this inexpensive GPS with its high performance and ease of use is ideal for measuring land for grazing. A barbed-wire fence cost (1,060Eur) is 1.4 times as expensive as an electric fence (754Eur) even in a small field of 0.38 ha with 344 m perimeter. Furthermore, electric fences require less labour to construct and cost less. Herbage yields and intakes are shown in Table 2. All species examined grew well in the high soil moisture conditions of the former paddy fields, but both yield and intake of annual ryegrass and millet species tended to be higher than perennial ryegrass.

Table 1 Comparison between measurements by handheld GPS with external antenna and tape measure about area and perimeter of a certain land (0.38 ha)

	GPS (A)	Tape (B)	Ratio (A/B*100)	t-test (P<0.01)
Area (m^2)	3,789	3,890	97.4	ns
Perimeter (m	351	344	102.0	ns

Table 2 Hebage yields and intake in former paddy field

Species (CV)	Total yield	Total intake
	—(DMkg/ha)—	
Annual ryegrass (Nagahahikari)	5,450ab	4,925a
Annual ryegrass (Akiaoba)	5,127ab	4,377a
Millet (White Panic)	8,912a	7,654a
Millet (Green Millet)	7,378ab	6,300a
Perennial ryegrass (Friend)	4,055b	3,593a

The means followed by the same letter are not significantly different at the 5% level.

Conclusions Most abandoned paddy fields are located in hilly rural areas and usually are small with complex shapes. It is important to know the land areas for managing grazing on these fields. A handheld GPS with an external antenna is easy to handle and useful for surveying these fields easily and with reasonable accuracy. The electric fence is easy to use in these fields and the construction cost and labour are less than for a conventional barbed-wire fence. Annual ryegrass and millet species are suitable for former paddy fields.

Utilisation of whole-crop rice silage as a feed for ruminants

N. Yoshida[1], M. Ogawa[1], Y. Cai[1], H. Nemoto[2] and M. Ishida[3]
[1]National Institute of Livestock and Grassland Science (NILGS), 329-2793,Japan, Email: norio55@affrc.go.jp,
[2]National Institute of Crop Science (NICS), 305-8518, Japan, [3]National Agricultural Research Center (NARC),
305-8666, Japan

Keywords: rice breeding, harvester, silage, feeding

Introduction A national project on the utilisation of whole-plant rice (_Oryza sativa_ L.) as a feed has developed in Japan since 2000. The use of a home-produced forage may be useful to reassure consumers concerned with risks of BSE and foot and mouth disease. Furthermore, Japanese farmers have been obliged to convert one million ha of paddy field from rice to other crops.

Materials and methods We started to develop four technical methods for the production and utilisation of whole-crop rice: 1) breeding forage rice cultivars with resistance to disease and lodging, 2) the development of machinery suitable for paddy fields, 3) the techniques of making silage and 4) technology for feeding ruminants.

Results Six forage cultivars, Kusayutaka, Hoshiaoba, Kusanohoshi, Kusahonami, Yumeaoba and Nishiaoba, have been bred for feed by National Institutes by 2004. These cultivars have improved resistance to disease and lodging and the grains do not readily shed. The dry matter (DM) yield/ha of these cultivars is 15-20 t with good cultivation and management. The most important development has been the development of harvesting machines for wet paddy fields. The machine combining the harvester, equipped with crawler wheels and a roll-baler, has been developed by the cooperation of National Institutes and machinery companies. One is a combine type and the other is a flail type (Figure 1). Both are equipped with additional sprayers for silage additives. The third point was to attain the long-term storability of whole-crop rice silage. It was difficult to get high quality fermentation because of the hollow structure of the culm, scarce lactic acid bacteria and high moisture content of forage rice. New additives, e.g. Chikuso-1 (Kai _et al._, 2003), of lactic acid bacteria have been developed by NILGS for rice silage. They have enabled the production of silages with as high fermentation quality as grass silage even with whole crop rice of high moisture content. Fourthly, the TDN of new cultivars has become shown by _in_ vivo trials to be 52.9-57.7% and crude protein content is influenced by level of fertiliser use, but averages 6.7% of the DM.

We have clearly shown that a lot of whole-crop rice silage will be consumed provided silage quality is adequate. It can be fed at maximum rates of 6 kg DM/day in dairy cows (Ishida _et al._, 2000), and 3kg DM in Wagyu and 5-6kg DM in reproductive cattle .

Figure 1 Harvest of forage rice by developed machines

Table 1 The changes in planted area of forage rice (ha)

Region	1999	2000	2001	2002	2003
Hokkaido	0	0	40	54	109
Tohoku	19	29	372	551	1053
Kanto-Tokai	34	85	347	616	852
Chugoku-Shikoku	5	15	140	258	390
Kyusyu-Okinawa	74	373	1334	2029	2725
Total	1318	502	2233	3509	5214

Discussion and conclusions The area of forage rice was 5,214 ha in 2003 with ruminants consuming it as silage (Table 1). Japanese agriculture is now obliged to move to more sustainable and natural conditions with improved self-sufficiency of food and feed. Traditional scenes of paddy fields in Japan will be maintained. However, there are still many technical problems to solve in increasing feed production in the temperate monsoon zone, including Asian countries.

References

Cai Y., Y. Fujita, M. Murai, M. Ogawa, N. Yoshida, A. Kitamura & T. Miura (2003). [Application of lactic acid bacteria (_Lactobacillus plantarum_ Chikuso-1) for silage preparation of forage paddy rice] (in Japanese). _Grassland Science_ ,49, 477-485.

Ishida M, M.R. Islam, S. Ando, M. Sakai & N. Yoshida (2000). [Preliminary observation on milk yield and nutrients utilization by holstein cows fed the round baled silage of the newly developed variety of whole crop rice,"Kanto-shi-206"] (in Japanese). _Kanto Journal of Animal Science_, 50, 14-21.

Development of a grazing land management education program for northern Australia's grasslands and grassy woodlands

C.R. Chilcott, M.F. Quirk, C.J. Paton, B.S. Nelson and T. Oxley
Queensland Department of Primary Industries and Fisheries, Animal Research Institute, Yeerongpilly, Queensland 4105, Australia, Email: Chris.Chilcott@dpi.qld.gov.au

Keywords: grazing land management, carrying capacity, decision tools

Introduction Recognition of the potential to enhance grazing land management to meet the goal of sustainable beef production has been increasing over the past decade. Recognition of the relationship between poor land management and negative off-site environmental impacts, such as soil erosion and a decline in the condition of rivers and adjacent near shore coastal areas from sediment transport, has increased also. This concern has matured somewhat to include the critical link between land condition and production, and the threat to sustainable carrying capacity that comes from declining land condition. Concurrently, interest has increased in optimising the use of pasture, e.g. through the development of infrastructure (watering points, fencing), through more pro-active management e.g. alternative grazing systems, spelling of pastures, and through pasture development. In fact, it can be argued that achieving production goals while improving and maintaining the health of the land has become the major on-property issue for northern Australian graziers.

The increasing demand for better information and decision tools to support grazing land management has been accompanied by investment into relevant research and development (R&D). However, despite this investment there has been limited uptake of new management information. Consequently, this has limited further investment into strategic, long-term R&D. An education product was developed to address the lack of adoption of management information. The product was targeted specifically to address the needs of grazing land management as identified by graziers in northern Australia.

Materials and methods The Grazing Land Management (GLM) package was developed in response to identification by industry (Meat and Livestock Australia) of the need for a 'product' that would enhance management of grazing lands in northern Australia by transfer of information to graziers. This 'product' includes: (1) description/presentation of the principles, concepts and relationships underlying sustainable grazing land management; (2) the technical process or framework that supports planning, decision-making, and implementation; and (3) the design and delivery that would both interest and genuinely assist producers. Decision support and educational tools were developed to aid the adoption of principles, for example, determination of the (sustainable) carrying capacity of land types within a property's paddocks according to land condition. The financial implications of management decisions also were assessed using a case study property with representative land types and a grazing business structure appropriate for the local region. The GLM education package is based on a workshop format and includes a technical manual, participant's workbook and technical manual.

Results Development of the GLM package highlighted the need to provide regionally-specific information on which graziers could make informed decisions. Accordingly, education packages have been developed for specific regional ecosystems, based on (water) catchment boundaries. Four regional packages - the Burdekin Catchment, Burnett Catchment, Victoria River District and Mitchell Grasslands of Queensland have been completed and delivery to graziers continues. GLM packages for another five regions (Queensland Murray-Darling Basin, Fitzroy Catchment, Northern Gulf Catchment, Mulga woodlands, and Central Australia rangelands) are being developed now.

Conclusions The development and delivery of GLM packages has allowed past R&D efforts to be communicated to graziers in a manner that facilitates adoption of sustainable land management practices. A number of unique tools were developed including a land condition framework, a carrying capacity calculator, and land-type information sheets. The success of the package (identified by graziers and R&D organisations) has attracted additional investment for further development and development in other regions. Importantly, workshop participants have indicated overwhelmingly a willingness to adopt land management practices or adjust stocking rates according to the principles promoted in the education program. This highlights the success of the program as a trigger to adopt improved land management, and the need for on-going extension support to ensure sustained improvements in land condition over the longer term.

Forage-Animal Production Research Unit (FAPRU): establishment of a new USDA-ARS research location

J.R. Strickland, G.E. Aiken, I.A. Kagan and R.D. Dinkins
USDA/ARS/Forage-Animal Production Research Unit, Lexington, Kentucky 40546, USA, Email: jstrickland@ars.usda.gov

Keywords: animal-plant interface, animal performance, forage utilisation, horse, cattle

Introduction Forages are vital to the success of grazing livestock production systems. Forages provide a low cost source of nutrients for animal production (Barnes & Nelson 2003; Ball *et al.*, 1996). Limited fundamental (i.e., genomic, proteomic, metabolomic) research on the effects of environment and management on plant quality and production and the effects of plant metabolites (i.e., nutrients, anti-quality factors, nutraceuticals) on animal performance has hindered our ability to improve the productivity of forage-based enterprises. There is insufficient information for reliable prediction of animal performance in response to plant metabolites. To address these issues, USDA-ARS established **FAPRU** (Forage-Animal Production Research Unit) in 2003 at U Kentucky, Lexington. Its mission is to improve the productivity, profitability, competitiveness and sustainability of forage-based enterprises through improved understanding of the fundamental biological processes that occur at the animal-plant interface.

Materials and methods To accomplish the FAPRU mission, multidisciplinary teams work to identify, evaluate, and manipulate genetic and physiological factors to enhance animal performance and forage plant quality and production. Transfer of technologies and management systems to forage-based enterprises ultimately realises the FAPRU mission. To aid the success of the mission, FAPRU has recruited scientists with expertise in genomics, metabolomics, and grazing systems. Expertise in other areas, such as proteomics, plant breeding, pasture ecology, and nutrient intake and metabolism are being recruited or tapped through FAPRU's partnership with U Kentucky and other research locations. Also, FAPRU has funded development of laboratory competency in genomics, proteomics, metabolomics and real-time ultrasonic imaging. To validate its mission, focus its research on producer concerns and identify research priorities, FAPRU held a Focus Group Meeting with stakeholders (August 19-20, 2004). Stakeholders were selected from the transition zone of the Eastern and Midwestern States. This transition zone is characterised by a transition from warm-season forages in the south to cool-season forages in the north.

Results Drs. James Strickland (Research Leader) and Glen Aiken (Research Animal Scientist) were hired in 2003. Drs. Isabelle Kagan (Research Plant Physiologist) and Randy Dinkins (Plant Molecular Geneticist) were hired in 2004. This staffing provided FAPRU with expertise in the areas of genomics, metabolomics, and grazing systems. One scientist position still remains to be filled with expertise in rumen ecology. Large laboratory equipment purchases by FAPRU include a genetic sequencer, a RT-PCR, a LC/MS/MS, 2 GC/MS systems (ion traps; one with pyrolysis), an accelerated solvent extractor, an ultrasound with vascular and 3D tissue imaging capabilities, and a spot cutter and picker for 2D gels as well as an automated station for MALDI plate preparation (in partnership with U Kentucky). The Focus Group Meeting identified 4 research priorities: (a) tall fescue toxicosis and best management practices, (b) forage utilisation, (c) expert systems, and (d) environment-plant-animal interactions. These research priorities are currently being formulated into the FAPRU 5 year research plan.

Conclusions The complexity of the biological processes and interactions associated with the animal-plant interface and sustainability of forage-based enterprises necessitates the use of multidisciplinary interinstitutional research teams to address adequately the issues facing forage-based enterprises in the 21st century. FAPRU is excited about potential collaborations and current and future opportunities to improve the sustainability of forage-animal production systems world-wide.

References
Ball D.M., C.S. Hoveland & G.D. Lacefield (1996). Southern Forages, 2nd Edition. Potash & Phosphate Institute and the Foundation for Agronomic Research. Georgia: Norcross.
Barnes R.F. & C.J. Nelson (2003). Forages and grasslands in a changing world. In Barnes, Nelson, Collins, Moore (eds.) Forages: An Introduction to Grassland Agriculture. 6th Edition. Iowa State Press. Iowa: Ames.

A systems approach to assessing the viability of grazing legume systems across Europe

C.F.E. Topp[1], C.J. Smith[1], L. Wu[2] and G. Molle[3]
[1]Land Economy Research, SAC, Edinburgh, EH9 3JG, UK, Email: Kairsty.Topp@sac.ac.uk, [2]Crop and Soil Research, SAC Aberdeen, AB21 9YA, UK, [3]Istituto Zootecnico e Caseario per la Sardegna, Olmedo, Italy

Keywords: legumes, Europe, model

Introduction Forage legumes are important in grassland farming throughout much of the world because of their ability to fix atmospheric nitrogen, and hence they are expected to play an increasingly prominent role in low-input grazing farming systems in Europe. Nevertheless, the socio-economic impact of the adoption of the "new" legume based grazing technologies have been poorly researched (Rochon *et al.*, 2004). Thus a methodology has been developed to:

- Assess the on-farm costs and benefits of including different legume crops for animal production;
- Determine the types of management systems and environmental conditions under which forage legumes may play a major part as grazing crops in production systems in Europe; and
- Determine the wider social, economic and environmental implication of widespread adoption of the "new" technologies.

Modelling methodology The methodological framework is outlined in the schematic diagram. Mathematical programming (MP) provides the ability to examine optimal resource deployment to competing and complementary enterprise activities. In order to supplement the historical records, a biophysical models system of grass–legume growth and animal production has been developed. The response to inputs and yields can be used to infer the normal expectations and variability for a specific grass-legume-animal system on a particular soil with particular environmental characteristics and outside inputs.

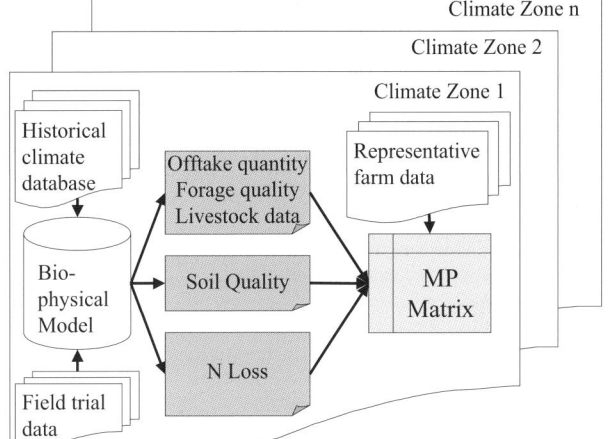

In this project, representative farms are defined in each partner country and parameterised at steady state, average input/output data for resource use and endowment. The field trial data has been processed through the biophysical modelling system to allow for suitable farm level information to be generated given modelled historical climate databases for a range of climatic zones across Europe (Topp & Doyle, 2004). This modelled production database is used to generate the representative data required for the MP models. In this way, the novel grazing technology systems can be plugged into a single year representative farm model and a comparative assessment made in terms of their financial competitiveness. Additionally, an exploration of the environmental trade off delivered can be made through applying multiple objectives to the model in more than the financial dimension (e.g. N Loss and soil structure).

Conclusions The modelling framework will thus assess the financial competitiveness and the environmental consequences of the "new" legume technologies. The model system will provide the opportunity to explore adoption scenarios as encouraged by differing policy options whilst at the same time exploring the environmental consequences.

Acknowlegements This work was funded by the EU Commission within the LEGGRAZE project (QL K5 CT2001—2328).

References

Rochon, J.J., C.J. Doyle, J.M. Greer, A. Hopkins, G. Molle, M. Sitzia, D. Scholefield & C.J. Smith (2004). Grazing legumes in Europe: a review of their status, management, benefits, research needs and future prospects. *Grass and Forage Science*, 59, 197-214.

Topp, C.F.E & C.J. Doyle (2004). Modelling the comparative productivity and profitability of grass and legume systems of silage production in Northern Europe. *Grass and Forage Science*, 59, 274-292.

Land Stewardship for the 21st Century: pasture and livestock management workshop for novices

L.A. Redmon, G.M. Clary, J.J. Cleere, G.W. Evers, V.A. Haby, C.R. Long, L.R. Nelson, F.M. Rouquette, Jr. and G.R. Smith
Texas A&M University Agricultural Research and Extension Centre, PO Box 38, Overton, Texas, USA, 75684; Email: l-redmon@tamu.edu

Keywords: pasture management, livestock management, land grant university, land stewardship

Introduction Land ownership patterns in Texas and the southern USA are changing. Since 1994 (Wilkins *et al.*, 2000) consumers interested primarily in recreational purposes have become the predominant owners of rural land. This land ownership change has created potential land stewardship problems associated with natural resource management. Few, if any, new landowners have any training related to the soil-plant-animal interface. New land owners need linkage with subject matter experts from land grant universities in a relaxed instructional setting while providing the opportunity for question and answer sessions. Thus, a programme was developed by a multi-disciplinary, multi-agency team at the Texas A&M University (TAMU) Agricultural Research & Extension Centre at Overton targeting novice landowners. The main goals in developing the programme were to a) provide basic information regarding management of soil-plant-animal resources that leads to sound, economic decisions and good land stewardship; and b) introduce the programme participants to the educational resources available to them through the land grant university system.

Materials and methods Faculty at TAMU-Overton, representing Texas Cooperative Extension and the Texas Agricultural Experiment Station as well as TAMU Departments of Soil and Crop Sciences, Animal Science, and Agricultural Economics developed the Pasture & Livestock Management Workshop for Novices. The programme is a fee-based, intensive 3-day event that targets novice or inexperienced landowners. Topics covered include:

- Soil resources and soil fertility
- Plant growth and development
- Adapted forage species
- Weed management
- Using forage legumes in the pasture system
- Developing forage systems
- Stocking rate & grazing systems
- Nutrient requirements of livestock
- Animal selection and management
- Record keeping

Frequent discussion sessions enable participants to begin to more fully understand the complexity of resource management. In addition to classroom lectures, Workshop attendees also spend approximately 50% of their Workshop experience in field laboratory exercises related to management practices. To determine programme effectiveness, pre-tests and post-tests, which evaluate attendees' knowledge about various soil, plant, and livestock topics, are administered to Workshop participants. Exit surveys are also conducted to obtain feedback from Workshop participants regarding the overall quality of their experience.

Results The Workshop has been a success due to the unique collaborative work between faculty from agencies and departments at TAMU-Overton. Results of the Workshop include:

- Participants were introduced to the soil-plant-animal interface.
- Effects on soil and water conservation and improved air and water quality associated with appropriate stocking rates was highlighted.
- Fertiliser application based on soil test to reduce negative effects on the environment while improving forage production.
- Introduction to IPM strategies for managing weeds rather than the traditional "mow" or "spray" only approach.
- Participants were made aware of the need for quality animal health care.
- Equipment for planting, fertilizing, herbicide application, etc., was demonstrated.
- Alternative production systems compared to livestock production were illustrated.To date, pre-test scores versus post-test scores indicate a significant improvement in knowledge of the Workshop participants during the short 3-day event.

Conclusions The Workshops have improved significantly the participants' abilities to be good land stewards, while increasing the profit potential of their production systems. Workshop participants also were introduced to the functions of the land grant university system.

Reference
Wilkins N., R.D. Brown, R.J. Conner, J. Engle, C. Gilliland, A. Hays, R. Douglas Slack & D.W. Steinbach (2000). Fragmented Lands: Changing Land Ownership in Texas. MKT-3443.

New Zealand pastoral systems: a current perspective

F.R. Duder[1], R.B. Green[2], W.D. Catto[3], D.R. Woodfield[4] and B.R. Guy[5]
[1]3/19 Selwyn Ave Mission Bay Auckland 1005, email, rduder@xtra.co.nz,[2]Agricom NZ Ltd   P O Box 539 Ashburton 8300, [3]Ballance AgriNutrients Ltd Private Bag 12503 Tauranga 3020,[4]AgResearch Grasslands Private Bag 11008 Palmerston North 5301, [5]Byreburn Ltd., RD 5, Aorangi Rd, Feilding 5600, New Zealand

Keywords grazed pasture, plant breeding, pasture utilisation, fertiliser, environment, human resources

Introduction New Zealand's diverse grassland resource of 13.5 M ha of permanent pasture, tussock or alpine grasslands underpin an intensive pastoral industry worth NZ$13 billion in 2004. The pastoral industry involves 37,000 farmers and a service industry of about 215,000 persons. It produces 27 M prime export lambs and 180,000 t of predominantly crossbred wool from 40 M ewes; 13 B litres of milk from 5.2 M dairy cattle; 800,000 t of beef from 4.7 M beef cattle and 33,000 t of venison from 2 M deer. Pastures and forage crops, mostly non-irrigated, are grazed 'in situ' by animals through controlled grazing management. The emphasis is on optimising utilisation at each grazing without penalising feed intake, pasture regrowth or persistance. Stocking rate and feed budgeting, along with a flexible stock trading policy are used to match animal demands with seasonal fluctuations in forage supply. The timing and duration of mating are varied to match anticipated seasonal growth. In dairy farming more condensed calving patterns have assisted in extending lactation length.

Improved pastures Improved "permanent" pastures are based mainly on perennial ryegrass (*Lolium perenne*) and white clover (*Trifolium repens*) with varying levels of minor species such as red clover (*Trifolium pratense*), browntop (*Agrostis spp*), tall fescue (*Festuca arundinacea*), cocksfoot (*Dactylis glomerata*), *Paspalum dilatatum* and Kikuyu (*Pennisetum clandestinum*). All of NZs improved pasture plants originated from overseas germplasm and increasingly from adapted ecotypes such as Mangere ryegrass and Huia white clover (Stewart & Charlton, 2003). Pasture plant breeding has delivered genetic gains (> 1% per year) for target traits such as forage yield and quality, seasonal yield and improved animal health (Woodfield & Easton, 2004). The development of non-toxic endophyte strains that alleviate ryegrass staggers and fescue toxicosis has improved animal productivity of sheep and cattle in NZ. Forages with increased feed intake and nutritional value (e.g. high non-structural carbohydrate grasses, and herbs and legumes containing condensed tannins) have also improved animal performance.

Fertiliser and environment New Zealand farmers apply 3 M t of fertiliser annually, as superphosphate and nitrogen (N). Fertiliser N use has increased twofold in the past eight years but average use is still below 150kg N/ha per annum.   Major changes have occurred in assessing plant nutrient demands of grazed pastures. Nutrient budgeting on an individual farm basis is a widespread tool for balancing environmental issues with economic considerations. Increased emphasis is placed on designing fertiliser programmes and farm systems that match economic goals with environmental sensitivity. In a deregulated environment, industry has developed Fertilizer Quality Assurance (Fertmark), and application (Spreadmark) schemes, a fertiliser Code of Practice (a world's first effects-based code for both fertiliser formulation and its application) and accreditation of Consultants.

Human resources   The deteriorating age structure of NZ farmers, research, extension and Agribusiness personnel remains a pivotal issue facing NZ agriculture.   Recent initiatives by the dairy industry and Agribusiness sector are attempting to reverse this critical shortage of younger people. A dairy industry campaign "Lets Talk Dairying" has increased public awareness while the Meat and Wool NZ monitor farms have increased community awareness of key issues and have improved uptake of new technologies, and decision making. Farm Environmental Awards (e.g. Ballance Farm Environmental Award), environmental protection initiatives by dairy industry and regional councils (e.g. Clean Streams Accord) and voluntary farmer land retirement and protection of tussock, grassland, bush and wetlands, are evidence of a renewed awareness of the guardianship role of farmers in the rural environment.

Summary and conclusions Grasslands play a major role in NZ's economic well-being. They are based on improved germplasm, low fertiliser and labour inputs and continuous *in situ* grazing, emphasising pasture utilisation. The ongoing evolution of these grasslands is driven by research and innovation but matched with environmental protection and ecological awareness of NZ's unique and scarce natural fauna and flora.

References
Stewart, A.V. & D.C.L. Charlton (2003). Pasture and Forage Plants for New Zealand. Grassland research and practice series. No8, 96pp
Woodfield, D.R. & H.S. Easton (2004). Advances in plant breeding for animal productivity and health. *NZ Veterinary Journal*, 52, 300-310.

A university course on management intensive grazing

D.L. Zartman

*Department of Animal Sciences, The Ohio State University, 2027 Coffey Road, Columbus, Ohio 43210, USA,
Email: zartman.3@osu.edu*

Keywords: education, university instruction, teaching

Introduction Education of new talent in good grassland management is important. Farmer-oriented extension- and outreach- programs do this, but USA university curricula rarely include it. Some livestock production courses contain a segment on grazing management, but there seem to be no credit-bearing, formal courses in USA universities except in the Ohio State University (OSU). Our course teaches management of intensive grazing (MIG) for three credit hours in a quarter system. It has been a popular elective course for the past 6 years. Every year that they have entered, OSU students have won all prizes in the 18–22 age division in the American Forage and Grassland Council and the Ohio Forage and Grassland Council essay competitions. This proves the students' knowledge. On graduation, students often move to professional positions or production agriculture, where they can apply the skills learned. Members of Ohio organizations and government agencies, who are concerned about the environment and family farm survival, enthusiastically support the course.

Materials and methods The course - 3 credit hours - is offered every spring quarter as an elective. Students must learn the principles of forage and livestock management to optimise the utilization of sunlight, soil, water and air to produce goods useful to humankind without diminishing the resource base while generating a profit. They develop plans for managing two different grazing units, one simple and one complex, which they then establish and operate for the entire quarter. Panels of 2-3 students meet the instructor at the grazing units every evening to rotate the pastures, assess the conditions of the pastures and the cattle, and learn animal management strategies. To learn scientific principles applicable to MIG, the students learn how to measure weight change in the cattle, forage density, analyse plant tissue, and test soil. They also develop a business plan for a grazing operation of their choice, write an essay on the topic of forages and grazing (mentioned above), and participate in 3 field trips to commercial farms with alternative livestock species.

Results To gauge student acceptance, the course was offered first in 1999 as a group studies exercise. Since then, the course has been taught officially with a mean of 35.2 (range 31-41) students/class. Before joining the course, students will have majored in different areas, such as animal sciences, agricultural education, agricultural communications, agricultural systems management, agricultural business and crop science.

Several students have entered graduate programs, become vocational agriculture teachers, taken positions in government agencies or entered into grazing-based farming. They are now leaders in the grazing network in Ohio, along with other farmers who have adopted MIG methods. Several students were not members of farming families. They have now become advocates for forage-based agriculture. Two of the students returned to the Professor to do undergraduate Honours research in forage management using livestock. Another student became a high-school teacher in New York City where he is able to impart the utility of forages in ecological concerns. Two students spent 1-2 years in New Zealand to further their knowledge. On returning to USA, one established his 130-cow dairy farm. The other is in graduate work to earn her Teacher Education certification.

These students are encouraged to bring their parents or other family members to the classes, field laboratories or the field trips. In this manner, families get involved in the adoption of MIG or the transition from extensive set stocking to MIG methodology. The course thus affects many more people than the members of the class. Local businessman, Mr. Bob Evans, has supported the costs of the field trips each year; and one host, the Earl McKarns family, treats the class to a wonderful lunch on the patio when we visit their progressive beef cattle operation. These experiences teach the students they can enter the farming business with modest financial conditions and soon advance to very comfortable farming incomes. They do not have to inherit or marry to get enough resources to farm. At one field trip location, they witness the utilization of goats to reclaim farmland that has gone wild with neglect over a long time. They see how to develop natural springs into watering systems for livestock. Members of the Ohio State University Extension Forages Team, the Natural Resources Conservation Service of the United States Department of Agriculture, and grazing equipment suppliers are frequent guest lecturers in the class to acquaint the students with the many people available to help them with their plans.

Foragebeef.ca web site: a model for technology transfer

D. McCartney, R. Weisenburger, K. Ziegler, G. Vaillancourt and G. Hutton
Agriculture and Agri Food Canada, & Alberta Agriculture Food and Rural Dev. Lacombe Alberta Canada,
Email: mccartneyd@agr.gc.ca

Keywords: scientific reviews, web based learning

Introduction   We live in an age of information overload. As budgets for technology transfer of scientific information and extension education continue to fall, new ways to disseminate agricultural knowledge are needed. Research findings, published in many scientific journals and reports, are seldom readily available to extension agents and farmers. Over time some of this material is lost. This program aimed to locate the best information on various forage and beef topics relative to Canada and the Northern USA and to summarise them in condensed form and in scientific review papers for a web-based site.

Materials and methods   North American, European and Australian web sites were trawled in a massive initial computer search for information on a broad range of forage and beef cattle topics. The search concentrated on locating research information and agriculture fact sheets summarising the information developed by government extension and research agencies. This information was copied unto a master CD in PDF or text format and organised into different topics within forage or beef subject matter. Copies of the CD were then distributed to extension personal across the region for their assessment in selecting the best 5-10 fact sheets or research summaries for each topic. The extension agents then developed the 10 most important information points or knowledge nuggets for each topic, based on the information coming from the research summaries or fact sheets. The research community then located or developed scientific review papers that summarised the scientific research on the various topics. These scientific review papers were submitted for publishing in scientific journals and later posting on the web site. The project took 4 years to develop and involved various Canadian Agriculture funding agencies, researchers and extension workers from across Canada.

Results   With massive advertising at universities and in the agriculture press across Canada and northern USA, Foragebeef.ca web site became operational in Dec 2003. The home page consists of the latest forage and beef news stories, weather and cattle market reports and an index of all the forage and beef topics. Upon selecting a forage or beef topic the knowledge nuggets, the best fact sheets and, if available, the scientific review papers appear. The web links takes the reader to the original web source for the fact sheet or research summary report. An editorial committee has been established to monitor the site and add new research summaries to the site in the future.

Conclusions   The Foragebeef.ca web site has attracted >3000 visits/month from 40 different countries to date. New topics and information are added continually to the site. Other countries and agencies can use this model to consolidate and transfer agriculture information to the farming community, researchers, teachers, extension agent and the general public in the future.

Forages for horses programmes

G.W. Wilson[1], R.L. Hendershot[2] and J. Hoorman[3]
[1]Ohio State University Extension, Hancock County, 7868 CR 140 Suite B, Findlay, OH 45840 United States, Email: wilson26@ag.osu.edu, [2]NRCS, 831 College Avenue, Suite B, Lancaster, OH 43130 United States, [3]Ohio State University Extension, Hardin County, 1 Courthouse Square, Suite 40, Kenton, OH 43326 United States

Keywords: horses, forages, hay, pasture

Introduction A survey by the American Horse Council in 1996 showed there were 6.9 million horses in the USA with 1.9 million horse owners and 7.1 million people involved in allied industries. The value of the USA horse industry to the gross national product is $25.3 billion. Nationally, 2.2 % of households own a horse but 4.9 % of households want to own a horse. Ohio's horse industry has 192,000 horses (7th nationally) and generates $776 million per year. Most of the 48,500 homes with horses in Ohio have 2-5 horses with 1-2 ha of land. Many exercise lots and high-use areas are little more than mud lots. Since each horse needs a minimum of 0.8 ha for feed, many pastures and hay fields are over-grazed and poorly managed leading to soil erosion, nutrient management problems with excess manure, and water quality problems. Most horse owners have a need for basic education to help them make good decisions on pasture and horse management. The educational resources directed at the 263,500 Ohioans involved in the horse industry are minor compared to its size. The objectives of the Ohio Horse Program are: 1) To increase awareness, knowledge and skills for horse owners on managing hay fields and pastures to produce quality horse forages, 2) To change management practices of horse owners to produce higher quality forages by learning how to better evaluate, produce, store and manage quality forages, 3) To develop a curriculum and provide a notebook of indexed referenced material to all programme participants, 4) To establish a grass plot programme to compare forage varieties for yield and quality, and 5) To provide field day and pasture walk experiences.

Materials and methods A forages for horses curriculum was developed to create a notebook and PowerPoint presentations on forage species selection and establishment; soil fertility; hay quality, evaluation and storage; pasture management and renovation; poisonous plants; digestive physiology, horse nutrition; manure management; and economics and marketing. The curriculum was taught in two three-hour sessions (six hours total) at 21 seminars around the state of Ohio in the last four years. Attendance ranged from 15 to 100 with 630 total participants. Over 1100 people have attended seminars presented to Equine Affair (a large trade show in Columbus, Ohio), Ohio Forage and Grassland Conference, and Farm Bureau events. A 400-page notebook was compiled with 10 sections as a reference guide, and 500 of these notebooks sold in two years. A forage plot and grazing research programme was established with the University of Findlay Equestrian Center located on a 30 ha site with 350 horses. The research goal is to study and collect data on mainly grass varieties and mixes. Most horse owners forage preference is for a grass-based diet coming from either hay or pasture. The following research plots have been established: 1) 32 replicated grass variety plots (three replicates) to collect yield, quality, and persistence forage data. 2) A horse preference grazing study to evaluate four different grasses replicated three times, and 3) Evaluating 20 new native cool-season grass varieties of Virginia Wild Rye (three replicates). Two annual Forages for Horses Field Days have been conducted (175 people) along with pasture walks at participants' farms.

Results Over 1700 people attended 21 two-night seminars or seven statewide events in the last four years. Interest in the programme continues to grow. In-depth evaluations were conducted for nine Forages for Horses seminars (374 total participants) in two years. Participants had an average of three horses and a range of 2 to 50 horses on an average of 3.3 ha. Eight different topics were presented at each session with a rating of usefulness from "not useful" = 1 and "very useful" = 4. The topics and respective scores were: Nutrition of Horses on Pasture (3.5), Fertility & Soils (3.3), Pasture Establishment & Renovation (3.7), Understanding Plant Growth (3.3), Forage Species Selection (3.4), Poisonous Plants (3.8), Management of Tall Fescue (3.2), and Hay Quality and Storage (3.7). The most useful topics were Poisonous Plants – followed closely by Hay Quality and Storage, and Pasture Establishment & Renovation. The impact of these seminars was rated by asking if the seminar met expectations. Results indicated 62% exceeded expectations, 38% meeting expectations, with no one indicating it fell short of expectations. Participants attending the two Forages for Horses Field Day Programs indicated 83% exceeded expectations and a rating of 3.53 on a 4-point scale for overall usefulness of the information.

Conclusions The Forages for Horses Program and the forage research plots helped horse owners maximise economic value from their forage resources and manage their pasture and hay fields to minimise environmental problems.

Forages for horses

J.C. Fisher, D.H. Samples and R.A. Sherman
The Ohio State University, Pike Co. 120 S. Market St. Waverly, Ohio 45690, USA; Email: fisher.7@osu.edu

Keywords: horses, forages, education

Introduction According to the 1997 Census of Agriculture, Ohio had 11,668 horse farms. An OSU survey estimated nearly 250,000 horses in Ohio. The large number of horses in the state has prompted many questions from producers to Extension personnel on the topic of improving forage resources. While many horse producers are proficient at equine management, they have not had formal training in forage production and management. Often, traditional agronomic programs don't target this audience. Many horse producers do not have the land mass, and/or experience to utilise large mechanical equipment for renovation and management

Materials and methods The Ohio State Integrated Forage Management Team is comprised of Extension Agents, Forage Extension Specialists and an NRCS Grazing Specialist. This team considered identified equine producer requests and recognised needs to develop a "Forages for Horses" notebook. Partial funding was provided by the *Ohio Forage and Grasslands Council*. This notebook is used as a curriculum for a short course that covers various forage issues. References can be resourced to develop educational programs based on local needs.

Notebook and Programs topics include: (1) Horse Nutrition; (2) Soil and Forage Fertility; (3) Species Selection and Establishment; (4) Plant Growth Physiology; (5) Hay Quality and Storage; (6) Pasture Management; (7) Poisonous Plants; (8) Manure Management; (9) Economics and Marketing. The authors designed a curriculum that entails 2-3 evenings of lecture and a day for a pasture walk to observe and apply what was learned. The agents have conducted 6 such programs in their Extension district with >110 participants. Evaluation methods were used to measure program expectations, teaching effectiveness, and producer demographics.

Results Of the respondents, 72% own <5 horses and 52% have <4 ha of pasture. Survey information and individual consultations determined that most horse owners have no forage resources to adequately support their enterprise. Also, 74% of respondents indicated that they had <4 ha of hay. These results indicate that educational programs on forage management, grazing practices, and hay production are greatly needed. Overall program content was rated by participant level of satisfaction. Of the attendees >97% felt that the "Forages for Horses" program either met or exceeded their expectations. Likert scale evaluations of the authors' presentations ranked 4.29-4.60 (1=not useful; 5=very useful). Participants responded unanimously that they would recommend this program to someone else.

Discussion The producers participating in the "Forages for Horses" programs were asked to self-identify their own on-farm challenges. They further substantiated the above results by prioritising their educational needs in Overstocking, Pasture Management, Renovation, and Hay Quality/Quantity. Results from this survey and clientele requests will be used to help determine the emphasis of future programs. After attending the entire program, participants were asked to indicate which practices they would most likely adopt immediately on their farm. "Managing Forage Maturity" was identified as the greatest need. Many producers would like to have an in-depth educational program on the benefits of rotational grass management. Other topics to be adopted included Establishment/Renovation of Forages, Soil Fertility, Poisonous Plant Identification, and Hay Quality Improvement. Most respondents did not have access to tillage, seeding, or spray equipment needed for pasture establishment or renovation. Extension educators have an opportunity to work with these producers to demonstrate these management methods and provide resources to accomplish these tasks.

Theme C: Delivering the benefits from grassland

Section 22

The contribution of participatory and on-farm research

The paired-paddock model as an agent for change on grazing properties across southeast Australia

J.P. Trompf[1], P.W.G Sale[1] and G.R. Saul[2]

[1]Agricultural Sciences, La Trobe University Melbourne 3086, Australia 3086, Email: J.Trompf@latrobe.edu.au,
[2]Department of Primary Industry, PB105, Hamilton, 3300 Australia

Keywords: practice change, adoption, pasture, fertiliser, stocking rate

Introduction From the mid 1970s to the mid 1990s the low productivity of wool and beef producing farms in the high rainfall zone (>550 mm annual rainfall) in south east Australia has been a major contributing factor to the difficulties faced by farmers in this region. This was despite research from the Long-term Phosphate Experiment at Hamilton in south west Victoria indicating that there is considerable potential to increase the productivity and profitability of wool production (Saul, 1994). By implementing the productive pasture technology (PPT) that involves increased rates of fertiliser on pastures containing productive species and increased stocking rates to utilise the extra pasture grown (Trompf & Sale, 2000), gains in excess of $A200/ha on a gross margin basis can be regularly achieved. However wool and beef producers were reluctant to adopt the technology. In 1993 the Grassland's Productivity Program (GPP) was initiated to assist producers to develop skills and gain confidence in their ability to manage more productive pastures on their farms. In brief, groups of 4-6 farmers were assisted by an experienced facilitator to compare current management practice in one paddock with PPT in an adjacent paddock. Over 500 wool and beef producers in south east Australia have been exposed to the paired-paddock model, firstly in the GPP from 1993 to 1997 and more recently in the Triple P Program. This paper reports on the effectiveness of the paired-paddock model in assisting pastoral producers to adopt PPT.

Methods A longitudinal study of the impacts of the paired-paddock model was undertaken using a series of surveys of participants at the beginning and end of the programs. In addition, there were in-depth interviews undertaken and a comparative financial analysis of farm profitability before and after involvement.

Results and discussion Marked changes in pasture productivity settings across the whole farm were measured among participants during the 1990s (Table 1). Fertiliser use (kg P/ha) more than doubled, stocking rate (dse/ha) increased by 54% and participants had more than half of their farms under the PPT by 1999. As a result the participating farms reduced their cost of production of wool by 15% and increased net farm income by 64%.

Table 1 Changes in pasture productivity settings from 1993 to 1999[#]

	1993	1995	1997	1999
Fertiliser use (kg P/ha)	5.6	9.2	11.9	12.8
Stocking rate (dse/ha)	9.1	10.3	11.7	14
Farm area under PPT (%)	0	10	29.8	51.5

[#]Productivity settings differed significantly between years (p<0.05)

Other fundamental changes among participants included a 4-fold increase in the assessment of pasture and livestock, a doubling in the number of flocks spring lambing and a doubling in the number of producers focusing on production per hectare as a key driver of farm profitability.

The in-depth interviews indicated that GPP participants gained knowledge and new skills in managing the PPT. In addition there were changes in attitude such as an increased awareness of the productive capacity of their land, which lead to an increased in their confidence regarding the future viability of their farm.

Conclusion These results demonstrate that the paired-paddock model is an effective agent for increasing the productivity and profitability of grazing farms in the high rainfall zone of south east Australia. This is attributed to the additive and interactive effect of the paired-paddock comparison, the guidance provided by the facilitator, the group interaction and the skills training. These components enabled producers to compare and manage the new and old technology side-by-side on his/her own farm, in a supportive group environment. The repeated witnessing of the fence-line comparison highlighted the performance of the PPT which lead to profound change in attitude and practice among participating producers.

References

Saul, G.R. (1994). Productive and profitable grazing systems in western Victoria. In 'Proceedings of the Ninth Annual Conference of the Grasslands Society of New South Wales', 66-72.

Trompf, J.P. & P.W.G. Sale (2000). The paired-paddock model as an agent for change on grazing properties across south-east Australia. Australian Journal of Experimental Agriculture, 40, 547-556.

Development of a toolkit for participatory management of rural watersheds in Kenya

L.W. Chiuri[1], F.K. Lelo[1], M.W. Jenkins[2] and S.N. Miller[3]
Faculty of Environment and Resources Development, Egerton University, P.O. Box 536, 20107 Njoro, Kenya, Email: drchiuri@yahoo.com, [*] [2]*Civil and Environ. Engineering Department, University of California, One Shields Ave., Davis CA 95616, USA,* [3]*Department of Renewable Resources, University of Wyoming, Box 3354, 1000 E. University Dr., Laramie, WY 8207, USA*

Keywords: community participatory, toolkit, watershed management, Kenya

Introduction Effective public participation is a foundation for sustainable watershed management, yet there are no demonstrated methods for or examples of its achievement in tropical semi-arid rural grassland watersheds of Kenya which support critical downstream water services. Within the Sustainable Management of Watersheds (SUMAWA) multidisciplinary international research project, a set of tools has been developed and tested to engage local communities and stakeholders in a dialogue and decision-making process to improve the development and management of the River Njoro Watershed in Kenya and reverse declining water quality and quantity problems. A toolkit manual based on the experience is under preparation for general distribution.

Materials and methods Participatory rural appraisal (PRA) data gathering (community map; benefits analysis; resource flow chart; seasonal calendar; institutional analysis), problem analysis (problem list; causes and coping strategies; trends; pair wise ranking matrix), and opportunity assessment tools (opportunity listing, assessment, and ranking; action planning) used in Kenya in development work were modified and adapted to focus on river problems in the River Njoro Watershed. Professional facilitators, accompanied by project scientists, led a series of two-hour discussions with a group of 25-40 community representatives over a two-week period in 2002-2003 in five communities residing along the river. Outcomes were documented and synthesised.

Results The discussions revealed 19 different common pool resource uses of in-river water, river bank/bed materials, and riparian buffer vegetation by local community members and other actors in the watershed (Jenkins *et al.*, 2004). Livestock-related uses were high on the list and widespread. These uses were shown to be critical for meeting very basic domestic human and livestock needs, and supplementing income for economic survival. The discussion tools also influenced stakeholders' perceptions and attitudes as they went through the process of learning and enriching each other's knowledge base. A unifying theme for action emerging from the process was the protection and rehabilitation of the riparian buffer zone which has been badly undermined by population pressure and lack of management (Table 1). The next challenge in the public participation process will be development of local rules and enforcement mechanisms to protect the riparian corridor when water, fodder, and other riparian materials are considered open-access resources.

Table 1 Common themes for community action in the River Njoro Watershed

1. Restoring and protecting the riparian buffer reserve (zoning riparian areas)
2. Local enforcement of laws on river pollution (e.g., community patrols)
3. Education, awareness raising, and training on need for and rules to protect riparian buffer, trees and river water
4. Infrastructure rehabilitation and new construction for water supply (livestock and human) and sanitation
5. Agroforesty and riparian tree planting programs
6. Developing a riparian management plan for the watershed
7. Clarification and enforcement of laws governing water abstractions from the river

Conclusion A toolkit has been developed and preliminarily tested that demonstrates achievement of community participation in the sustainable management of watersheds in Kenya. The tools provide a process for identifying local communities' perceptions and priorities for watershed management and common objectives and options on which a watershed-wide action plan can be built. The study has also demonstrated how partnerships between communities, scientists, policy-makers and other stakeholders can be built which are likely to lead to opportunities for locally-based cooperative action. The results will be used in the next stage of the SUMAWA public participation process to blend communities' indigenous knowledge with modern science and technology to identify and select specific rehabilitation and management actions for the riparian zone along the Njoro Watershed in a series of tiered watershed-wide workshops.

References

Jenkins, M.W., F.K. Lelo, L.W. Chiuri, W. Shivoga, & S.N. Miller (2004). Community perceptions & priorities for managing water & environmental resources in the River Njoro Watershed Kenya. *Proceedings 2004 World Water & Environmental Resources Congress*, Jun 27-Jul 1, 2004, Salt Lake City, UT. ASCE, Reston, VA.

Spatio-temporal scales of animal grazing in herding systems

H. Barani
Department of Range Management, Gorgan University of Agricultural Sciences and Natural Resources, Gorgan, Goletan Prov. Iran, Email: barani@gau.ac.ir

Keywords: grazing, shepherd, herder, grazing distribution, grazing pattern

Introduction A holistic approach is needed for natural resources management and this demands understanding of the role of all the components in a system. Animal grazing is one of the most important processes in rangeland ecosystems. In rangeland utilisation in Iran, the human herder is crucial in achieving sound management. This study focused on the role of the human as a grazier (herder).

Materials and methods The study was focused on pastoralism culture in the eastern Alborz, in the north of Iran. This was considered as an anthropologically unknown region. Transhumant groups utilise the rangeland ecosystems of this region. The pastoralism culture and pastoralists' knowledge that is associated with grazing processes was investigated through participatory research during a one-year cycle. Informants were selected carefully and the data were collected through in-depth interviews with them and the researcher occasionally role playing as an amateur shepherd. The data were analysed by concept mapping (Daly, 2004). This paper is based on a part of the research findings.

Results The findings show four spatial and temporal levels: pasture, paddock, foraging site and foraging station. Human decisions affect the size and number of spatial units and grazing duration in the first, second and the third levels. The pasture (*charagah* in Persian) is an area allocated to feed a flock during a part of the year. Each stockholder owns two to four pastures in upland or lowland ranges. In some cases, pastures are held in common. Both singular and joint owners must consider governmental acts and professional limitation to end the grazing season in particular pastures. The paddock (*roogah* in Persian) can be considered as a given part of a pasture. It may be allocated to graze animals for one to several days. In summer pastures, a flock may be divided into two or three goups and a good paddock must be allocated to lambs. Herders' knowledge and experiences may lead to rotational grazing (i.e. paddock by paddock) of a pasture. Foraging site (*towgah* in Persian) is a part of the paddock (for example, a distinct slope) that is considered as an area to graze a flock for a few minutes to a few hours. The shepherds were familiar with the extent of these sites and their characteristics, according to their habits, interests and experiences. For example, some shepherds are able to monitor and mnage a large flock, so they locate a large area as a foraging site for such a flock. Foraging station (*kalafgah* in Persian) refers to a small area at which the grazing animal stops and defoliates desirable plants. The grazing time at a foraging station depends highly on non-human factors, such as plant and animal characteristics. At this level shepherds may have an indirect influence on grazing duration by increasing flock movement.

Conclusions The results highlight different spatio-temporal grazing scales in herding systems. In these systems, the role of the grazier (herder) is more important than the role of the grazing animal in determining spatial or temporal patterns and scales. The shepherd decides where to graze animals specially at the three major levels. Subsequently the models or explanations that are needed to describe spatial and temporal grazing scales in herding systems are basically different from those which are formulated for non-herding systems in developed countries (e.g., Friedel, 1994; Baily *et al.*, 1996).

Herding systems have evolved through the centuries. These systems are mainly adapted to the local environment and people's culture. Range management theories should be re-thought to consider the role of the human as a grazier and his influence on grazing management. This is particularly important for countries in which herders play an integral role in exploiting range ecosystems.

References

Baily, D.W., J.E. Gross, A. Laca, L R. Rittenhouse, M.B. Coughnour, D.M. Swift & P.H.L. Sims (1996). Mechanisms that result in large herbivore grazing distribution patterns. *Journal of Range Management,* 49, 380-400.

Dally, B. J. (2004). Using concept maps in qualitative research. In: A.J. Canas, J.D. Novak & F.M. Gonzalez (eds.). Concept Maps: Theory, Methodology and Technology. Proceedings First International Conference on Concept Mapping, Pamplona, Spain, //:httpCMC.ihmc.us/papers/cm2004-060.pdf.

Friedel, M.H. (1994). How spatial and temporal scale affect the perception of change in rangelands. *Rangeland Journal,* 16, 16-25.

Building decision tools for sustainable grassland management: a case study of participatory research in La Réunion

S. Gerbaud, V. Blanfort, P. Thomas, Ph. Lecomte and J.P. Choisis
CIRAD Pôle Elevage, Ligne Paradis, 7, chemin de l'Irat 97410 Saint Pierre,France,
Email: gerbaud_sophie@yahoo.fr

Keywords: innovation process, decision tools, pasture management, sociology

Introduction During the last 30 years, cattle breeding has developed in La Reunion Island (France, Indian Ocean) with strong support from local authorities and extension structures. The Union des Associations Foncières Pastorales (UAFP) initiated and still sustain a large effort on grassland improvement. During the same period, CIRAD managed different research programmes on agronomic aspects of pasture management). Scarcity of space, the volcanic soil and tropical climatic conditions contributed to repeated forage shortages during the dry season, making forage production a major issue for cattle breeders. In 1991, CIRAD launched a research programme to elaborate decision tools for assessing and improving pasture management. A partnership was developed with UAFP, with the local institution being involved in the elaboration of the tools and now being completely in charge of their use to advise cattle breeders on pasture management.

Material and methods A joint workshop (INRA, CIRAD, national and overseas French Agricultural Research Institutions) on decision support tools gave the opportunity to study this particular programme as a completed R&D innovation process, from its start until the transfer of the tools to extension. The aim was to provide an *ex post* evaluation of the innovation process with a sociologically orientated approach, in order to improve management of research programmes with professional partnerships and the ways to produce operational tools.

Results A chronological review of the process enabled identification of the main stages of the innovation process. A grassland ecology PhD candidate started the programme. Regular exchanges between scientists, breeders and professionals partners and the local research funding authorities enabled the research team to identify the needs of the breeders and technical advisors and pointed out a problem in grazing management and fertilisation. In contrast, the first plan for the research programme had focused on floristic aspects of pasture management. As a result, decision tools using sward height for rotational grazing management and use of soil and plant mineral nutrient levels for fertilisation advice were chosen and adapted to the local situation. This stage involved experimentation on six farms that were partners of the research team for other actions. Technical advisers, who were already following these farms, were involved in every intervention made by the research team. Judging by the first results, and successful enlargement of technical advice for fertilisation and rotational grazing to others farms, the tools proved to be successful. Transferring the tools to UAFP occurred naturally after a five-year research period. The tools have so far been used for advice on more than 100 farms and have given significant results in improving grassland management and reducing forage shortage during the dry season.

The study of this programme also enabled interesting conclusions to be drawn about managing a participative innovation process. In this case the transfer of the tools to the professional structure (UAFP) was all the easier because it occurred simultaneously with the transfer to UAFP of the main technician who was a former trainee working in the CIRAD research programme. His hiring by UAFP was agreed by both structures at the time of the transfer. In the first stages mainly scientists of CIRAD were involved, then quickly, their importance decreased and technicians became more and more important. There was a concentration of knowledge and advice about pasture in only one technician of UAFP, despite many farmers were, and are still, demanding advice. This ensured consistent and efficient advice, because this person had obtained ten years experience, but it raises the issue of technical support on pasture management in the future. To improve the dissemination of information, UAFP has just published a compendium of technical notes (Barbet-Massin *et al.*, 2004) which summarises all the scientific and practical knowledge of CIRAD and UAFP about grasslands on the island.

We also focused on several other functions of the decision tools. The elaboration process enables people to define more precisely their demand for information, so that the tools represent a medium for dialogue and exchanges between the adviser, the scientists and the farmer and implicitly promote self questioning by the farmers and changes of their opinion about pastures. Lastly, the use of tools adds to the power of the adviser and may enable him to better assess the quality of farmers' practices. Further conclusions are still in development about ways to manage innovation processes and to build decision tools in order to improve research efficiency and the implementation of results.

Reference

Barbet-Massin, V., P. Grimaud , A. Michon. & P. Thomas (2004). Guide Technique pour la Création, la Gestion et la Valorisation des Prairies à La Réunion. UAFP. CIRAD-Pôle Elevage.

Potential and constraints for animal feed as an objective of poor farmers in participatory research with multipurpose forage crops in Central-America

R. van der Hoek[1], M. Peters[2] and V. Hoffmann[1]
[1]University of Hohenheim (430a), 70593 Stuttgart, Germany, Email: reinvdhoek@xs4all.nl, [2]International Centre for Agricultural Research (CIAT), Multipurpose Tropical Grasses and Legumes, AA 6713 Cali, Colombia

Keywords: participatory research, multipurpose forages, animal feed, Central-America

Introduction Multipurpose forage crops can play an important role in improving the environmental and socio-economic sustainability of smallholder production systems in fragile environments. However, since the forage technology development framework has not been sufficiently applicable for poor farmers, adoption of especially legumes has been generally low (Peters et al., 2001). In a participatory research effort with smallholder farmers in Honduras focused at forage based technologies, food security turned out to be the main selection criterion whereas animal feed was secondary. Since animal feed related activities (farmer-led forage seed systems, production of dry season feed) have been identified as promising income generating options for poor farmers in the hillsides of Central-America, a further analysis was carried out to identify the (mainly household related) factors inducing or inhibiting farmers to opt for production of animal feed.

Materials and methods A group of 150 farmers with different levels of resource endowment representing the typical maize and beans based agricultural system of central Honduras conducted over 200 experiments in their own fields with several grasses (e.g. *Brachiaria brizantha*), leguminous crops (mainly several varieties of *Vigna unguiculata*) and shrubs (e.g. *Cratylia argentea*). The choice of research methods and parameters was determined simultaneously by both farmers and researchers. A dichotomous logistic regression model was used to examine the variables influencing the inclusion of animal feed as an objective (Table 1). The independent variables were identified by a Principal Component Analysis.

Table 1 Definition of variables used in animal feed regression model

$$\ln(\text{ObjectiveFeed}) = \beta_0 + \beta_1\text{Altitude} + \beta_2\text{LandTenure} + \beta_3\text{BuyMaize} + \beta_4\text{Ureamaize} + \beta_5\text{MaizeYield} + \beta_6\text{CattleNr} + e_i$$

Variable	Definition
ObjectiveFeed	1: yes, 0: no
Altitude	1: low (< 800 masl), 0: other (≥ 800 masl)
LandTenure	Land tenure: 1: full or semi land ownership, 0: other
BuyMaize	Maize bought for consumption: 1: yes, 2: no
Ureamaize	Level of urea application on maize (kg/ha)
MaizeYield	Maize yield (kg/ha)
CattleNr	Number of cattle

Results Altitude had no significant influence on the inclusion of animal feed production as an objective. Full or semi landownership increased the chance of feed being an objective by 24%, controlling for other variables in the model (p=.005). Farmers who depend on purchased maize from outside are 17% more willing to include feed production as an objective than those who are self sufficient in maize production (p=.025). Every extra 100 kg/ha urea application on maize increases the chance of feed being an objective by 22% (p=.025). A yield increase of 100 kg/ha maize augments the chance of feed being an objective by 1% (p=.033). An increase of one unit of cattle increases the chance of feed being an objective by 2% (p=.001).

Conclusions Results indicate that farmers owning land, applying fertiliser and owning cattle are more likely to include animal feed as a research and production objective than the poorer farmers, except for those who are not self-sufficient in maize. Farmers without full decisive power over their land are reluctant to engage in animal feed production. Whereas research and development work can continue to be directed at all farmer categories in Central-American hillsides, special attention is justified for farmers without full land ownership and those who depend on outside acquired basic grains for their food security.

References

Peters, M., P. Horne, A. Schmidt, F. Holmann, P.C. Kerridge, S.A. Tarawali, R. Schultze-Kraft, C.E. Lascano, P. Argel, W. Stür, S. Fujisaka, K. Müller-Sämann, & C. Wortmann (2001). The role of forages in reducing poverty and degradation of natural resources in tropical production systems. *Agricultural Research and Extension Network Paper* No. 117. ODI, London.

Improvement of grassland through community participation in the Middle Awash Valley of Ethiopia

E. Abule[1], G. Getachew[2], A. Gezahegn[3], A. Nigatu[4] and B. Shimeles[5]
[1]Adami Tulu Research Center, P.O.Box 35, Ethiopia, Email: abule_ebro@yahoo.com, [2]GL-CRSP PARIMA, c/o ILRI, P.O. Box 5689, Ethiopia, [3]Debre Zeit Research Center, P.O. Box 32, Ethiopia, [4]ILRI,P.O.Box 5689, Ethiopia, [5]CARE-Ethiopia, P.O. Box 4710, Addis Ababa

Keywords: grassland improvement; community participation

Introduction The natural resources of the grassland in the middle Awash valley of Ethiopia are subjected to competing claims: development to generate revenue for the state, conservation of wildlife and wilderness areas, as well as use for local production. The combination of climatic conditions causing drought and the over use of the natural resource can be cited as the primary cause of grassland deterioration in the area. Since the problems of the grasslands are complex and multi-dimensional, they are not amenable to quick and easy fixes. Hence, if sustainable progress is to be achieved, the responsibility for change must be in the hands of the communities and household themselves. Pastoral communities, in collaboration with CARE-Ethiopia, local government and other partner NGOs embarked on grassland improvement activities that were based on traditional activities. The objective of this study was to assess the condition of the traditionally-improved grazing lands.

Materials and methods Six traditionally-improved (IM) grazing lands locally known as 'kalo' and three nearby unimproved (UNIM) communal grazing lands were assessed at the peak of the growing season in 2003. Species composition and % bare ground were determined using the wheel point (Tidmarsh & Havenga, 1955). The frequency of occurrence of each species and bare ground was expressed as a percentage of the total number of points and grass dry matter (DM) yield was analysed using GLM.

Results The dominant grass species in the traditionally-improved and unimproved sites was *Chrsysopgon plumolosus*. The percentage of bare ground and non-grass species was higher in the unimproved grassland sites (bare, 11.9%; non-grass, 6.10%) than in the traditionally improved (bare, 5.9%; non-grass, 4.31%) ones (Figure 1). The non-grass species in both areas were *Cyperus bulbosus*, *Tribulis terristris*, *Tephrosia* species, *Indigofera* species, *Sida ovata* and *Edostemon terticaulis*. The grass DM yield of the traditionally-improved grassland sites (mean for improved, 590 kg/ha) was higher (p<0.001) than that of the un-improved ones (mean for unimproved, 268 kg/ha) and there was an improvement in grass DM yield by 119% (Table 1).

Table 1 Grass dry matter yield (kg/ha) at each site

Sites	Yield
IM1	830
IM2	600
IM3	830
IM4	470
IM5	455
IM6	355
UNIM1	230
UNIM2	340
UNIM3	235

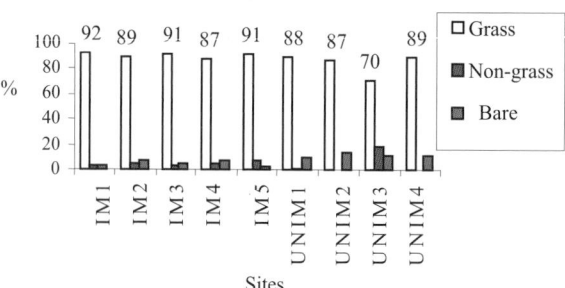

Figure 1 The percentage of grass, non-grass and bare ground in improved (IM) and un-improved (UNIM) grassland sites

Conclusion The establishment of communal kalo is proving a plausible measure towards recovering the degraded grasslands. Optimal and sustainable use of the grassland requires an adequate rest period for the grass species to grow, to seed and be able to accumulate reserves for the next growing season. The purpose of kalo is to reserve an increased standing crop of forage for the dry season. In view of the need to match existing feed resources to livestock needs increased focus is required on the use of non-grazing feed resources as alternative feed resources and on improving markets and linking pastoral communities to the export market.

Reference

Tidmarsh, C.E.M. & C.M. Havenga (1955). The wheel point method of survey and measurement of semi-open grasslands and Karoo vegetation in South Africa. *Memoirs of the Botanical Survey of South Africa*, No. 29.

Linking research to development in pastoral communities of northern Kenya: recent experiences and key findings in participatory research approaches

G.A. Keya, M. Ngutu, A. Adongo and I. Tura
Kenya Agricultural Research Institute (KARI), Marsabit Research Centre, P.O Box 147-6500 Marsabit Kenya.
Email: karimar@africaonline.co.ke, gakeya@kari.org

Keywords: pastoralists, participatory research, technology, Kenya

Introduction   Over 50% of Kenya's land mass is arid. The mainstay of the local economies of these areas is nomadic pastoralism. Arable agriculture is limited to the few oases and mountain zones. Due to the harsh climate, there are few alternative livelihood options available to the local communities of mostly herders. Efforts to improve agricultural productivity have often been hampered by low adoption of available technologies. This low adoption is linked to many factors, notably the non-involvement of the stakeholders in the research process and the weak linkage between research and extension. To address these shortcomings, new approaches were tested with the aim of improving acceptability and use of agricultural technologies and knowledge in a pastoral community of northern Kenya.

Materials and methods   Several approaches were used in the course of the research. For each of the key phases of the research cycle, care was taken to employ methods that were most appropriate in enhancing community participation and ensuring gender sensitivity. Collaboration between KARI and relevant stakeholders was also emphasised. The diagnostic (constraint identification) phase of the research cycle included participatory and rapid rural appraisals, detailed exploratory and diagnostic surveys in which pastoralists from selected parts of Marsabit district were involved in rigorous self-evaluation exercises. These revealed the major impediments to improved living standards in the district. These exercises involved over 5,000 pastoralists/agro-pastoralists and were done in 38 target areas. Identification, development, testing, and dissemination of technologies aimed at solving the identified constraints were carried out on farm/on site using the farming systems and participatory learning and action research (PLAR) approaches. The aim was to ensure that pastoralists participated in the whole process of technology development and testing and so owned the research findings. Some of the tools used included community based planning workshops, farmer study tours, feedback workshops, training/demonstration workshops, participatory monitoring, and field days. Pastoralists/farmers were not left out of the institutional research planning process and were invited to participate in institutional planning workshops and research advisory committee meetings both at centre and national level. The process of catalysing adoption involved empowering farmers to demand technologies through the Agricultural Technology and Information Response Initiative (ATIRI). In this approach farmers demanded technologies from KARI by submitting competitive proposals. To ensure a sustained process, partnerships were developed with relevant and willing stakeholders with clear memoranda of understanding (MoU). The major intervention areas included natural resource management, improved crop production through water harvesting technologies, livestock husbandry and health, and development and marketing of livestock products.

Results   The main benefits of the use of participatory methods were the increase in the number of pastoralists/farmers adopting beneficial technologies and an increased awareness by pastoralists of the need for research in different fields touching on their daily lives. This was demonstrated by the relatively high adoption rates reported from the different projects being implemented; for example a project on protection of selected woodland recruitment sites using red paint recorded 100% adoption. The adoption rates of water harvesting and water saving technologies was equally high, with at least 125 non-test farmers adopting within three years. Adoption of improved movable young stock housing reached over 100% in just 33 months. Three mini-dairies were established and are managed by community groups. The groups were trained on micro-enterprise management and group dynamics. A constant and reliable market was established for about 400 farmers in the processing areas resulting in improved household income from milk sales. In rainy seasons prices have stabilised at a higher level than previously offered by brokers. The outlets of the mini-dairies substantially reduce the rate of milk spoilage previously experienced due to lack of markets. The dairies have created direct employment for about 10 non-group members with a total monthly income of over $200. This income supports more than 60 family members at household level. Partnership with a collaborating local non-governmental organisation resulted in the building of three additional small-scale milk processing units with support from Farm-Africa.

Conclusion   Participatory approaches contributed to increased adoption of technologies by pastoral and agro-pastoral groups of northern Kenya. Farmers were appreciative of the research benefits accruing to them through their active involvement. It is recommended that participatory approaches be institutionalised as a way of bridging the gap between research and extension for accelerated development in pastoral areas.

Priority tree fodder species in the Maasai silvopastoral system of Kajiado district, Kenya

E.C. Kiptot
The Kenya Forestry Research Institute, P.O Box 20412, Nairobi, Kenya, Email: ekiptot@yahoo.com

Keywords: key informants, silvopastoral system, tree fodder resources, fodder evaluation

Introduction In an effort to improve pastoral land management systems, donor agencies have spent millions of dollars over the last two decades on research and development programmes. However, most of these programmes have been unsuccessful. The reasons for the failure are increasingly clear. Current thinking argues that more attention needs to be given to local knowledge systems and rural people's participation in development planning and implementation. The involvement of local communities in the research process is critical and has been shown by Ego (2001) to lead to the design of appropriate interventions. The main objective of this study was to elicit pastoralists indigenous knowledge on priority fodder species and the criteria used in evaluation.

Materials and methods A formal household survey, semi-structured interviews using key informants, tree inventory and group consensus method were used to elicit indigenous knowledge from informants.

Results Table 1 shows priority fodder species and pastoralists ratings based on selected criteria. Species evaluation was based on fodder quality (linked to nutritive factors) and physiological/physical/agronomic characteristics of the various species. From the pastoralists perspective, fodder quality is determined by: i) the ability of fodder to increase milk production and butter fat content, ii) the ability of fodder to fatten livestock and iii) the palatability of fodder. Physiological/physical/agronomic attributes which were also considered are: i) the ability of fodder to satisfy, ii) drought resistance, iii) ability to withstand multiple browsing, iv) forage biomass, v) fodder availability during the dry season, and vi) presence of both edible fruit/pods and leaves.

Table 1 Key informants' ratings of 10 priority fodder species for cattle across selected criteria considered important to them

Fodder species in order of priority	Effect on milk prod.	Fattening	Drought resistant	Resistant to browsing	Palatability
	mean scores[a] and standard deviations in parentheses				
Acacia tortilis	3.6[a] (0.54)	3.7 (0.65)	3.8 (0.44)	3.2 (0.47)	4.0 (0.00)
Grewia tembensis	2.0 (0.60)	2.7 (0.45)	1.8 (0.54)	2.1 (0.58)	3.1 (0.56)
Sericomopsis hildebrandtii	3.3 (0.78)	3.2 (0.56)	1.4 (0.48)	2.6 (0.94)	3.6 (0.96)
Phyllanthus sepialis	3.0 (0.57)	2.6 (0.75)	1.3 (0.44)	1.7 (0.78)	3.6 (0.64)
Lonchocarpus eriocalyx	1.2 (0.44)	1.2 (0.40)	3.1 (0.58)	2.1 (0.88)	1.5 (0.50)
Acacia mellifera	2.4 (0.56)	1.7 (0.57)	3.4 (0.49)	3.6 (0.43)	2.8 (0.52)
Grewia bicolor	2.6 (0.44)	2.7 (0.45)	3.0 (0.40)	2.5 (0.50)	2.4 (0.56)
Salvadora persica	1.3 (0.64)	1.4 (0.56)	3.8 (0.34)	3.5 (0.50)	2.6 (0.63)
Cordia ovalis	2.0 (0.82)	1.9 (0.94)	3.6 (0.99)	3.4 (0.66)	1.8 (0.33)
Acacia brevispica	2.9 (0.78)	2.6 (0.64)	3.8 (0.78)	2.5 (0.74)	2.9 (0.34)

[a] Pastoralists were asked to rate species based on selected criteria, a rating of 4 =very good, 3 = good, 2=average and 1=poor.

Conclusions This research has shown that Maasai pastoralists use various skillful criteria to evaluate fodder species. One of the immediate implications of this, is for the scientific community to consider farmers'/pastoralists criteria when screening trees for fodder value so as to maximize on the potential of these species. This can be done by linking up scientific knowledge of animal nutrition with pastoralists' knowledge and objectives. For instance, if a pastoralist's objective is fattening of livestock for sale, then fodder species that are known to increase animal body weight should be promoted so that the pastoralist can realise his objectives.

Reference

Ego, W.K. (2001). "Evaluation of some suitable grass species for the Rangelands of Kenya" *Our Environment our life* Series 3 No 1. Agroforestry Research for Intergrated Development in Arid and Semi arid lands of Kenya. KEFRI/BADC Nairobi , Kenya

Participatory plant breeding in Uruguay

D. Real[1,2], G.A. Ferreira[1], D.F. Risso[1] and C. Mas[1]

[1]National Institute of Agricultural Research, INIA Tacuarembó, Ruta 5 km 386,CP 45000, Tacuarembó, Uruguay, Email: dreal@cyllene.uwa.edu.au, [2]Cooperative Research Center for Plant-Based management of Dryland Salinity, The University of Western Australia, University Field Station, 1 Underwood Avenue, Shenton Park, WA 6009, Australia

Keywords: participatory plant breeding, forage, legumes

Introduction   The introduction of forage legumes into the native pastures of Uruguay is considered an environmentally safe method with which to increase the level of production through the addition of biologically fixed nitrogen. An integrated plant breeding programme was initiated at INIA (National Agricultural Research Institute), Uruguay, with the aim of developing forage legumes able to persist and produce in co-existence with the native (grass dominant) vegetation under cattle and sheep grazing. The programme was conducted in parallel with rhizobial strain selection. During 1998 to 2001 at Glencoe Research Station (32° 01'32''S lat; 57° 00° 39''W long), 326 temperate and subtropical forage legume species were evaluated in the basaltic region of Uruguay. This study provided a set of data with which the legume species could be ranked according to their performance (Real *et al*., 2004) so that the best 10% could be selected for further evaluation and breeding. The best species according to objectively measured characters (i.e. forage and seed yield) could be very valuable to certain farming systems and farmers, but might be of little value for other farmers and their systems. A significant relationship between the farm decision making unit and the farming systems has been reported for this region (Ferreira, 1997). Participatory research has proved to be an effective way to include farmers' local knowledge in plant breeding programmes (Dusseldorp & Box, 1993).Therefore, to aid the conventional plant breeding approach, a farmer participatory breeding group was formed.

Materials and methods   The 13.6 million ha. of native grasslands could be divided into several regions according to soils and climate. There are also different farming systems and farmers within regions, with different approaches to farming and decision-making schemes. One of the aims was to select a group, to include as many different farmers to represent the existing variability. The size of the group was restricted to 35 farmers, all of them being leaders in their communities. Three meetings per year (spring, autumn and winter) were conducted from 2001 to the present time, in which the farmers were able to score (1 to 5 and halves) the coded species in the field according to their perception of the value of the species. After the field work, a meeting was held to present the results of the previous meetings and to discuss the impressions from the field work.

Results   The outcome of this participatory group has exceeded all the expectations that were considered at the start in 2001. The group has not only helped in the selection of the best 10% of the species, but also has remained as a working group for the next phase of the breeding programme in which the best 40 species were evaluated from 2002 to 2004. Moreover, the group, together with INIA scientists, has prioritised in autumn 2004 the best four species in which INIA will start their breeding programmes. In 1999, INIA initiated the breeding of the subtropical forage legume *Lotononis bainesii* Bak (*L. bainesii*), strongly prioritised by the group. The cultivar 'INIA Glencoe' was released in 2003 (Real & Altier, 2005). The first seed multiplication areas for this species were at INIA research stations and also on farms belonging to group members.

Conclusions   The participatory plant breeding approach has helped to prioritise efforts and resources towards species that would be welcomed and needed by the farming community. Also, farmers feel part of the breeding process, having made a real input whilst understanding the long timeframe that usually is required. Therefore, the demand for the new species or cultivars will already exist even before the release. Participation will also improve the adoption process as appears to be the case with 'INIA Glencoe'.

References

Dusseldorp, D.V. & L. Box (1993). Local and scientific knowledge: developing a dialogue. In: W. de Boef, K. Amanor, K. Wellard & A. Bebbington (eds.) Cultivating Knowledge. Genetic Diversity, Farmer Experimentation and Crop Research. Intermediate Technology Publications, London, 20-26.

Ferreira, G. (1997). An evolutionary approach to farming decision making on extensive rangelands. University of Edinburgh. Faculty of Science and Engineering, I.E.R.M. PhD thesis. 372 pp.

Real, D., C.A. Labandera & J.G. Howieson (2004). Performance of temperate and subtropical forage legumes for over- seeding native pastures of the basaltic region of Uruguay. *Australian Journal of Experimental Agriculture* (in press).

Real, D. & N. Altier (2005). Breeding for disease resistance, forage and seed production in Lotononis bainesii Baker. *New Zealand Journal of Agricultural Research* (in press).

Statistical precision of a replicated farm grazing trial versus replicated paddock trials

K.P. Vogel[1], D.E. Bauer[2] and L.E. Moser[3]

[1]USDA-ARS, 344 Keim Hall, Univ. of Nebraska, PO Box 830937, Lincoln, NE 68583-0937, USA, Email: kpv@unlserve.unl.edu, [2]Coop. Ext., Ainsworth, NE 69210-1696, USA, [3]Dept. Agronomy & Horticulture, Univ. of Nebraska, Lincoln, NE -68583-0915, USA.

Keywords: statistics, grazing, gain, standard error

Introduction The experimental unit for animal average daily gain (ADG) and gain/ha in grazing trials is the paddock. Grazing trials on research stations often are conducted using small paddocks because animal and land costs restrict the number of treatments, replicates, and animals per paddock. Land and animal restrictions can be reduced by conducting trials on farms using animals provided by cooperating farmers. Farmers typically want only a single replicate on their farms and as result, virtually all on-farm trials in the USA and elsewhere have been un-replicated demonstration trials from which estimates of experimental error cannot be obtained. Farms can be used as replicates but concerns about statistical precision and the ability to detect treatment differences to date has limited the use of this design in the USA to a single study which we conducted in the Central Great Plains in the 1990's. Our objective is to compare the statistical precision of this on-farm grazing trial with replicated paddock trials on research stations in the same geographical region.

Materials and methods An on-farm grazing trial in which farms were replicates was conducted in north central Nebraska using three farms in different counties (Bauer *et al.*, 1995). Two cultivars each of smooth bromegrass (*Bromus inermis* Leyss.), intermediate wheatgrass, western wheatgrass (*Pascopyron smithii* (Rydb.) Löve) and one cultivar of crested wheatgrass were evaluated. Strains of switchgrass (*Panicum virgatum* L.), crested wheatgrass (*Agropyron* spp.), and intermediate wheatgrass (*Thinopyrum intermedium* (Host) Barkworth & D.R. Dewey) were evaluated in replicated, paddock trials by Anderson *et al.* (1988), Vogel *et al.* (1993) and Moore *et al.* (1995), respectively, in eastern Nebraska (Table 1). In all trials, one-year-old (yearling) beef cattle were used as the tester animal. Standard errors were used to compare statistical precision among trials.

Results Standard error of the mean for both average daily gain and gain per ha were lower in the on-farm trial than in any of the replicated small paddock trials (Table 1). Because of their larger size, paddocks in the on-farm trial were stocked by more animals which reduced the experimental variation for animal gain and gain per hectare. Paddock size and number of animals also had an effect on SE's in the research station trials.

Table 1 Standard errors (SE) for average daily gain (ADG) and gain/ha for beef yearlings in replicated paddock trials versus an on-farm trial in which farms were replicates. All trials were conducted in Nebraska, USA

Trial	Duration (years)	Replicates (n)	Treatments (n)	Paddock size (ha)	Cattle/ paddock	SE (mean) ADG (kg)	SE (mean) gain/ha (kg)
Res. station trials							
Anderson *et al.*	3	4	3	0.4	3 to 4	0.07	13
Vogel *et al.*	3	4	2	0.8	4	0.04	8
Moore *et al.*	2	3	4	0.4	3	0.07	20
On-farm trial							
Bauer *et al.*	2	3	7	2.5 to 3.2	12 to 24	0.02	6.5

Conclusions Well-managed grazing trials in which farms are replicates can provide greater statistical precision than small paddock trials on research stations.To be successful, trials should have clear objectives, test a small number of treatments and farmers must participate in the management of the trials.

References

Anderson, B., J.K. Ward, K.P. Vogel, M.G. Ward, H.J. Gorz & F.A. Haskins. (1988). Forage quality and performance of yearlings grazing switchgrass strains selected for differing digestibility. *Journal of Animal Science,* 66, 2239-2244.

Bauer, D., L.E. Moser & K.P. Vogel (1995). Grazing evaluation of cool-season grasses in the Nebraska Sandhills in replicated ranch trials. *Agronomy Abstracts*, p. 165.

Moore, K.J., K.P. Vogel, T.J. Klopfenstein, R.A. Masters & B.E. Anderson. (1995). Evaluation of four intermediate wheatgrass populations under grazing. *Agronomy Journal*, 87, 744-747.

Vogel, K.P., B.C. Gabrielsen, J.K. Ward, B.E. Anderson, H.F. Mayland & R.A. Masters. (1993). Forage quality, mineral constituents, and performance of beef yearlings grazing two crested wheatgrasses. *Agronomy Journal*, 85, 584-590.

Using adult learning principles to plan participatory action research for incorporating ley pastures in sustainable mixed farming systems

C.A. Hall

CSIRO, PO Box 102, Toowoomba, Q 4350 Australia, Email: Cristine.hall@csiro.au

Keywords: adult learning principles, ley pastures, participatory action research

Introduction In the northern grain growing region of Australia, the use of ley pastures (tropical legumes and grasses) can increase soil fertility and restore soil structure on degraded cropping land. This is seen by some researchers as a viable component of a sustainable mixed farming system incorporating crops, pastures and livestock. It has been recognized that the expertise of multiple sources can contribute to successful participatory action research (PAR) (Kemmis & McTaggart 1988) so a workshop was organized as the 'Plan' phase of a research project action learning cycle (ALC) to determine best management principles of incorporating ley pastures in sustainable mixed farming systems.

Methods The workshop was designed using 7 adult learning principles (Table 1) (Fell 1986). Four researchers from three agencies and 8 local producers attended a half-day workshop (ALP 3) in a comfortable environment (ALP 2). Each participant introduced themselves (ALP 1) and stated their expectation of the workshop (ALP 3). Current research in tropical ley legumes was presented (ALP 1, 3 & 6) complemented by contributions from producers on practical farm management (ALP 1). Refreshers and breaks were used throughout (ALP 2 & 6). An 'on-farm' field trial was designed from sub-group discussion (ALP 1, 4 & 5), the associated action plan was developed by the whole group (ALP 1,4 &5) and the workshop was evaluated at the completion of the workshop against the participant's original expectations (ALP 3 & 7).

Table 1 Description of 7 adult learning principles

ALP	Description
1	Build on local experience; use the knowledge within the group/individual.
2	Make the learning environment comfortable and encouraging.
3	Ensure the learning activity meets the needs of the clients and relates to the problem of the group.
4	Involve the audience in planning their own learning experience.
5	Have activities that involve, are stimulating and are participatory
6	Allow time for people to reflect on their learning, take difficult subjects slowly and be open to questioning.
7	Build group and individual confidence that they are making progress towards their goals.

Results Input from all worksh op participants produced an 'action plan' as the output from this workshop. The action plan included the design of an 'on-farm' replicated field trialwith 7 treatments of pure legumes, mixed grasses or alternate strips of legumes and mixed grasses to research options of incorporating ley pastures in sustainable mixed farming systems. The designing of the field plan was complemented by a matrix of required action within a timeframe and individual participants taking responsibility for its implementation so that the 'Act' phase of the research project ALC could be realized.

Conclusions Using ALP as a workshop framework was successful in completing the 'group-owned' 'Plan' phase of the research project ALC. Involving both farmers and researchers was important in determining not just the desired 'on-farm' trial to explore relevant research but the 'action plan' required for the ALC to continue. With a PAR approach, farmers and researchers can collaborate to explore farm management options for sustainable farming systems by developing more than just a demonstration trial of the possible solutions recognized as relevant within the farming community, but also a scientifically rigorous research trial with set parameters that can be analysed and acknowledged in the scientific community.

References

Fell (1986) Extension and Communication Skills Workshop Course Manual. QDPI, Brisbane.
Kemmis, S. & R. Mc Taggart (Eds.) (1988). The Action Research Planner. 3rd edition, Deakin University, Victoria 3217, Australia.

Evaluation of narrow-row forage maize in field-scale studies

W.J. Cox, J.H. Cherney and D.J.R. Cherney
Cornell Univ., Dep. of Crop and Soil Sci., Ithaca, NY, USA 14850, Email: wjc3@cornell.edu

Keywords: forage maize, nitrogen management, participatory research, forage quality

Introduction   Some dairy producers in the northeastern USA adopted narrow row (0.38 m) maize forage production in the mid-1990s because of its 5% dry matter (DM) yield advantage (Cox & Cherney, 1998). These dairy producers, however, continued to plant forage maize at high plant densities (125,000 plants/ha) under high N fertility (225 kg N/ha), despite research that indicated that forage maize had optimum DM yields and forage quality when planted at the recommended 100,000 plants/ha under 175 kg/ha of N fertility (Cox & Cherney, 2001). We evaluated forage maize at 0.38 and 0.76 m (conventional) row spacing under recommended vs. high plant densities and N fertility on a large dairy farm in New York. The objective of the study was to demonstrate to dairy producers that narrow-row forage maize does not require high plant densities and N fertility for optimum DM yield and forage quality.

Materials and methods   We formed a farmer-researcher partnership to conduct field-scale studies (5-10 ha) on a large dairy farm with field-scale narrow-row equipment. We evaluated first, second, and third-year forage maize at recommended vs. high plant densities and N fertility for three years for a total of nine comparisons. The work crew on the farm performed all field operations, including applications of dairy manure, tillage, planting, spraying and harvesting. We sampled for soil NO_3-N and plant N concentrations at the 6th leaf stage (V6), silking and at harvest. We also measured neutral detergent fiber (NDF), NDF digestibility and *in vitro* true digestibility (IVTD) at harvest. Years were considered random and year in rotation and row spacing were fixed in a combined analysis of variance (ANOVA). A mixed model was used to analyse the data using PROC MIXED (SAS Inst., 1999). Mean separations were conducted using Fisher's Protected LSD (P= 0.05).

Results   When averaged across years and rotations, narrow-row maize at high vs. recommended plant densities and N fertility had greater soil NO_3-N concentrations at planting (Table 1). All treatments, however, had similar soil NO_3-N and whole plant NO_3-N concentrations at the V6 stage, ear-leaf N concentrations at silking, plant N concentrations at harvest and DM yields at harvest (Table 1). Also, NDF, NDF digestibility, and IVTD did not differ significantly between narrow-row maize at high vs. recommended plant densities and N fertility (data not shown). Narrow-row maize at high vs. recommended N fertility, however, had more than twice the residual soil NO_3-N concentrations at harvest (Table 1). The doubling of residual soil NO_3-N concentrations and the non-significant 3.25% DM yield advantage of narrow-row maize at high N fertility demonstrated to dairy producers that narrow-row forage maize did not benefit from high vs. recommended plant densities and N fertility.

Table 1 Soil NO_3-N, plant N, and DM yields of forage maize when averaged across rotations and years

Row Spacing	Soil NO_3-N			Plant N			DM yield
	Planting	V6	Harvest	V6	Silking	Harvest	
	----------------mg/kg----------------			--------------------g/kg----------------------			t/ha
0.76 m	21	54	11	42.1	26.0	10.5	17.6
0.38 m	27	49	10	41.7	25.8	10.6	18.3
0.38 m High	37	49	21	41.2	26.2	10.6	18.9
LSD 0.05	10	NS	9	NS	NS	NS	0.7

Conclusions Dairy producers in New York with more than 700 cows are classified as a Concentrated Animal Feeding Operation (CAFO) and must have a Nutrient Management Plan that follows Cornell University guidelines. Based on the results of this study, Cornell maintained guidelines of a 175 kg N/ha limit for forage maize production, regardless of row spacing. The results of this study helped dairy producers in New York with more than 700 cows understand why there is a 175 kg N/ha recommended limit for narrow-row maize production in New York.

References
Cox, W.J, & D.J.R. Cherney (2001). Row spacing, plant density, and nitrogen effects on corn forage. *Agronomy Journal*, 93, 597-602.
Cox, W.J., D.J.R. Cherney & J.J. Hanchar (1998). Row spacing, hybrid, and plant density effects on corn silage yield and quality. *Journal of Production Agriculture*, 11, 128-134.
SAS Institute (1999). SAS User's Guide. Statistics. SAS Institute., Cary, NC.

Assessment and improvement of the efficiency of nitrogen use on commercial dairy farms

D.J. Roberts, K.A. Leach and J. Goldie
SAC Dairy Research Centre, Dumfries, Scotland, DG1 4SZ, UK, Email: d.roberts@au.sac.ac.uk

Keywords: nitrogen efficiency, dairy systems

Introduction Dairy farming systems have a low efficiency of converting nitrogen (N) into milk protein, due to the many transfers which occur in the production process. Losses of N from the system can be detrimental to the environment and represent wasted inputs. At SAC, in a systems research project, management changes achieved increases in nitrogen efficiency (milk N output/ N inputs) (NE) from 23 to 34% in a grass-clover based system (GC), and 13 to 21% in a purchased fertiliser based system (GN) (Leach & Roberts, 2002). Nitrogen surplus (NS = N inputs - N output in sold produce) was reduced from 184 to 90 kg N/ha in GC and from 369 to 258 kg N/ha in GN. This work was then incorporated into a participatory research project, to obtain data on N balances in commercial dairy systems and investigate the effects of suggesting management changes to improve NE.

Materials and methods Nine dairy farmers agreed to participate in this project from 1997 to 2001. All were subscribers to a detailed enterprise recording system, which included milk sales, feed and fertiliser inputs and stocking rate. Farmers were provided with information from the SAC dairy systems project and possible strategies to improve NE were discussed with them. Nitrogen efficiency and NS were evaluated each year for each dairy unit, excluding youngstock and any other farm enterprises. Nitrogen output/ha was estimated from annual milk sales, milk protein content and stocking rate. Nitrogen input was calculated from nitrogen fertiliser used, amount and protein content of purchased animal feed and estimates of deposition in rain and fixation (where there were clover based swards receiving less than 250 kg N/ha per year).

Results The largest changes in nitrogen output/ha were due to changes in stocking rate rather than changes in milk yield/cow or milk protein content. Reductions in nitrogen input/ha were mainly due to reductions in fertiliser use. Farm 1 - was unable to make changes to the system due to other farming constraints. Farm 2 - increased forage area by 20% and reduced nitrogen fertiliser use over whole unit by 24%. Farm 3 - reduced fertiliser nitrogen use (kg/ha) by 28% in year 2, but then increased forage area, cow numbers and fertiliser use in subsequent years. Farm 4 - converted to organic status milk output remained the same with forage area increasing by 36% and nitrogen fertiliser use falling to zero by year 4. Farm 5 - had the lowest fertiliser use at the start of the project and this was further reduced in year 2 but in year 4 there was a fall in output and a decision made to increase fertiliser use to grow more home-grown forage. Farm 6 - also moved to organic status from year 1; the decrease in efficiency in year 4 was due to increased nitrogen fixation, more purchased feed and a decline in milk sales. Farm 7 - improvements in nitrogen efficiency were due to reductions in nitrogen fertiliser use in years 2 and 4. Farm 8 - increased purchased feed use by 33%, although from a low level of 725 kg DM/cow per year but increased forage area and reduced fertiliser use by 77%. Farm 9 - increased cow numbers at start of project and in year 4 reduced nitrogen fertiliser use from 351 to 207 kg/ha.

Table 1 Nitrogen efficiency on farms (NE, %) over four years

Farm	1	2	3	4	5	6	7	8	9
Year 1	17	13	15	15	23	10	13	17	17
Year 2	16	14	24	18	28	21	18	20	21
Year 3	17	15	16	22	22	27	15	21	19
Year 4	Na	18	19	31	13	23	18	23	25

Conclusion The results of the study demonstrate the low N efficiencies of UK commercial dairy farms, which were closely linked to N fertiliser application rate. Improvements were made in N efficiency on most farms but not up to the levels from the research systems. The main effects were due to increases in stocking rate and reduction in fertiliser use rather than increases in output per cow. There were year to year variations that arose from effects of season on forage yield and quality and farmers deciding to make other management decisions based on short and long term economic considerations at the expense of nitrogen use efficiency.

Reference

Leach K.A. & D.J. Roberts (2002). Assessment and improvement of the efficiency of nitrogen use in clover based and fertilizer based dairy systems. 1. Benchmarking using farm gate balances. *Biological Agriculture and Horticulture*, 20, 143-155.

On-farm participatory research is an essential step towards achieving successful adoption of innovation: 'Lifetime Wool' a case study

C. Oldham[1], P. Barber[2], M. Curnow[1], S. Giles[1], D. Gordon[2] and A. Thompson[2]
[1]Department of Agriculture, Locked Bag No. 4, Bentley Delivery Centre, Western Australia, Australia 6983, Email: coldham@agric.wa.gov.au, [2]Department of Primary Industries, Hamilton, Private Bag 105, Hamilton, Victoria 3300, Australia

Keywords: on-farm, participatory research, lifetime wool, innovation

Introduction 'Lifetime Wool' project (LTW) is a national project that is developing new nutritional guidelines for the management of ewe flocks across Australia funded by farmers through Australian Wool Innovation (AWI – EC298; 2001-2008). A large replicated plot-scale experiment was used to define the dose-response of current production (wool and reproduction from the ewe) and future production (survival, growth and wool from progeny over their lifetime) to a range of levels of ewe nutrition (Thompson & Oldham, 2004). However, "farmers and research workers have long realised that the difference between the results obtained on experimental plots and those obtained by farmers is of crucial importance" if farmers are to be convinced to adopt new technology (Davidson & Martin, 1968). Hence, the LTW was designed from the start to include four distinct phases: (i) plot-scale research (2001 – 2003; see Oldham *et al.* 2006); (ii) on-farm paddock-scale research (2003 - 2005); (iii) whole-farm systems modelling (see Young *et al.* 2004); and (iv) on-farm demonstration or 'road-testing' of the draft guidelines (2005-2007).

Results and Discussion The results of the first 2 years of plot-scale experiments strongly suggested that all of the key ewe and progeny characteristic responded to increasing levels of nutrition during late pregnancy and lactation with an asymptotic dose response (Oldham *et al.* 2005). Similarly, the initial whole-farm modelling suggested that it was feasible to describe an annual liveweight/condition score profile for ewe flocks (joining to joining) that would 'optimised' whole-farm profit. Hence, the next challenge was to compare the performance of commercial flocks managed to follow either the maximum performance liveweight/condition score profile or a 'best-bet' optimum profile to estimate the erosion in the difference between nutritional treatments when transferred to the farm scale. Project staff closely supervised this phase. Flocks were weighed and condition scored monthly by project staff on 15 farms across southern Australia and changes in feeding levels/stocking rate were budgeted in close consultation with the co-operating farmers. These experiments were primarily aimed at confirming that results at the paddock-scale would fall on the same nutritional response surface as those from the plot-scale experiments, exploring the performance of twins verses single progeny but most importantly developing the management process for the new guidelines in conjunction with 'potential champions'. These experiments were conducted on flocks joined in January to April 2003, thus only 3/15 results for adult wool production of progeny have been collected. However, the overall results to date reflect the differences seen at the plot-scale and all the co-operating farmers have shown their potential as future 'Lifetime Wool Champions' by committing to the final Demonstration Phase of the project. The objectives of the demonstration phase are to test that the key activities and tools in the draft 'LTW guidelines' are; **feasible** (practical to implement) and **effective** (lead to an increase in profit) **in the hands of our target audience**. The key message underlying these draft guidelines is "You must measure to manage – let the animals do the talking". This activity started in January 2005 and it is seen as the key to the success of the overall project; it involves participatory research in its ultimate form. The 15 co-operators in the experimental phase will be observed/monitored as they attempt to manage a flock to the new guidelines; managing a flock to follow as closely as possible a predetermined annual liveweight or condition score profile using the tools developed by the project. On these farms project staff will also measure the performance of a control flock managed, as the farmer would have done before they were exposed to LTW. In addition, a further 85 farmers have volunteered to 'road-test' the guidelines by attempting to manage a flock to the LTW guidelines in consultation with LTW staff. The additional farms are associated in groups of 5 or 6 around the original 15 farms and are supported by LTW staff during 5 group meetings per year.

References
Davidson. B.R. & B.R. Martin (1968). Experimental research and farm production. University of Western Australia Press, 4pp.
Oldham. C., M. Ferguson, B. Paganoni, A. Thompson, G. Kearney and M. A. van Burgel (2005). The impact of the level of feed-on-offer available to Merino ewes during winter-spring on the wool production of their progeny as adults. Proceedings XX International Grasslands Congress (this volume)
Thompson, A.N. & C.M. Oldham (2004). Lifetime Wool 1. Project overview. *Animal Production Australia,* 25, 326. www.publish.csiro.au
Young. J.M., A.N. Thompson and C.M. Oldham (2004). Lifetime Wool 15. Whole-farm benefits from optimising lifetime wool production. *Animal Production Australia,* 25, 338. www.publish.csiro.au

The contribution of participation to the grassland research of the Louis Bolk Institute

J. de Wit, T. Baars and N. van Eekeren
Louis Bolk Institute, Hoofdstraat 24, 3972 LA Driebergen, Netherlands, Email: j.dewit@louisbolk.nl

Keywords: participation, experiential science, mutual learning, grass-clover

Introduction Participatory research methods are well described, but the contribution of participation to the R&D process often remains unreported. In this paper some benefits of participatory on-farm research (OFR) carried out by the Louis Bolk Institute (LBI) are highlighted.

Methodology Most research at the LBI on grass-clover management is executed together with organic farmers, mainly focussing on the introduction of clover, management options to affect the clover content of the sward and grass-clover production with limited manure application (de Wit *et al.*, 2004). The OFR is mutually planned and executed by farmers and scientists, focussing on the facilitation of mutual learning and integration of experiential and experimental knowledge (Baars *et al.*, 2004). Besides OFR, LBI is also involved in more traditional research, demonstration and extension activities.

Results On-farm research offers opportunities to include specific relevant agro-ecological conditions. Thus, it was possible for the LBI to asses the effect of P-fertilisation on organically managed grass-clover on soils with a very low P-status, conditions which are hard to find on a research station. More importantly, however, was the possibility of transforming rather general R&D topics (originating from discussions with organic farmers organisations, funding agencies, etc.) into specific conditions through the mutual planning of the OFR by the scientist and the farmer. In this example, it became clear that the farmer's objective was in optimising feed production given the low P-status and pH of his soil, but also given the limited amount of organic manure available at the farm and the objective to avoid slurry application in spring on part of his land due to nature objectives (birds). Thus, the trial took note of these constraints. Later, during field visits it became clear that results of such a trial with a specific set of conditions and objectives are easier to communicate to other farmers than results from formal research. Group discussions revealed that this might be because the interpretation of the results is easier for farmers if all relevant conditions are transparent and because the set of conditions and objectives is a clearly recognisable ideotype relevant for the development of Dutch organic agriculture.

Participatory OFR also offers opportunities to incorporate farmer's experiences into the formal scientific system. Due to continuous observations of phenomena under variable conditions (years, soils, management, etc.) farmers may be better capable to include relevant conditions in their experience than can be expected of formal research in a relatively new research area. Experience of the LBI, however, shows that this requires regular contact between scientists and farmers, and observational qualities of the farmer related to the particular subject (a pioneer farmer in grassland management is often not a pioneer in cattle breeding). For example, grassland scientists of the LBI are often contacted by farmers and extension workers to advise on problems with (too high or low) clover content of their grassland. However, it is impossible to review all known relevant (management and agro-ecological) conditions in hindsight, and, thus, truly adequate advice or specific hypotheses to understand the events are hardly ever formulated as a result of these irregular contacts. During a regular contact the influence of these known factors can be filtered. For example, a distinct negative effect of hybrid ryegrass on clover content was established in OFR with a farmer interested in clover-rich, short-term leys, sown in autumn. Clear understanding of the relevant context proved essential some years later, when too high clover contents in leys became a common problem for organic farms with favourable soil conditions using modern, highly competitive, clover varieties in leys used mainly for cutting. It was possible to use the very competitive hybrid ryegrass to control clover content below 50-60% under these conditions.

Conclusions Research and development programmes can benefit from participatory OFR through the transformation of general into specific problems during the mutual planning by scientists and farmes of OFR, and through the incorporation of farmer's experiences into the formal scientific system. Thereby, OFR will speed up the search for and dissemination of adequate innovations. Major perquisites for successful participatory OFR are direct and regular contact between scientist and farmer, keen interest of the farmer in the specific problem and of the scientist in farmer's conditions and objectives.

References

Baars, T., L. Veltman & N. van Eekeren (2004). Farmer's experiences and scientific on-farm experimentation integrated in an experiential science approach. *Grassland Science in Europe*, 9, 1181-1183.

Wit, J. de, M. van Dongen, N. van Eekeren & E. Heeres (2004). Handboek Grasklaver. Report LV54. Louis Bolk Instituut, Driebergen, NL, 109pp.

Farmer-driven research for developing models of successful low input dairy farms of small to medium size in the American midwest

D.G. Johnson[1], M.V. Rudstrom[1], R. Imdieke[2], E. Ballinger[3] and G.J. Cuomo[1]
[1]West Central Research and Outreach Center, University of Minnesota, Morris, MN, USA 56267, Email: dairydgj@mrs.umn.edu, [2]19560 68th Street North East, New London, MN, USA 56273, [3]Animal Science, Room 156, 6118, University of Minnesota, 1364 Eckles Avenue, Saint Paul, MN, USA 55108

Keywords: dairy herd, on-farm research

Introduction The dairy industry in the upper Midwest continues to evolve with a drastic reduction in the number of dairy farms with less than 100 cows that utilise tie-stall housing. Many of the farms that do remain are at a critical point where facilities require renovation or replacement that is not economic. Rural communities have fewer residents engaged in agriculture to participate in the local economy and the rural landscape includes many farm sites that are abandoned. Low input dairies are an alternative system developed by farmers as a grass-roots movement. Low input farms may include grazing, outdoor housing throughout the year, crossbreeding, group housing of calves, etc. Barns formerly used to stable cows may be renovated to provide a milking centre. Low input dairy farmers are eager to participate in on-farm research when they determine the direction of research. We will describe one trial designed to answer questions determined by farmers and outline our research approach to designing effective reduced input dairy farms. The objective is to identify the essential components of management and organisation for low input dairy farms in the American Midwest.

Methods Sets of 6-8 low input dairy farmers in southern and central Minnesota were convened and asked to identify researchable issues on their farms for which they would invest time and effort to conduct research in accordance with a structured protocol. Three questions emerged: 1. Can grazed heifers achieve a growth rate of 0.9 kg/day? 2. Does grazing cost less than confinement housing? 3. Will grazing generate as much profit as field crop production? One farmer agreed to provide two replicates of 30 Holstein heifers (300kg) for grazing and feedlot comparisons for three consecutive years. A research protocol was negotiated with the farmer. A fifth year stand of lucerne was fenced for pasture and heifers were rotated twice weekly.

Results There was no significant difference in the growth rate which achieved the target of 0.9 kg/day. Grazing was a lower cost method of growing heifers than a feedlot each year. The average daily cost over three years was $1.29 in the feedlot and $0.94 on pasture. Returns from pasture were contrasted with maize grain, soybeans and lucerne grown for hay. Returns per hectare were maize grain, -$36.53; soybeans, $74.45; lucerne hay, $225.19; and grazing, $328.51. The farmer participated in the analysis and has presented the results at several public gatherings, including the annual meeting of the American Forage and Grassland Council.

Further steps There is a need for an evaluation of the components of successful low input dairy production systems. Traditional research utilises a series of comparative trials in a controlled setting. Our approach obtains dairy systems information for a diverse set of reduced input, moderate size dairy farms and a more intensively managed prototype farm. This is a system for analysing production and management information, consolidating the results and testing the results by modelling. The method provides results that have authenticity prized by farmers and a methodology that is accepted by scientists.

- Data collection from 10 low input farms
    - Pasture, feed purchase, milk sales, other inputs
    - Document range of performance in the field
- Intensive data from the prototype
    - Create a base herd to initiate the model
    - Include results from discipline studies
- Simulation with Simherd (DK) and other models
    - Identify characteristics of successful systems
    - Create a map for further research

Summary Participatory research yields authentic information which is very credible with farmers. When farmers have a strong voice in identifying researchable methods they take ownership of the process and provide high quality data. There is a relatively narrow window of time to solve complex systems relationships. On-farm survey data for modelling systems can be obtained in greater abundance and at lower cost than exclusive reliance on research stations.

Management of gastro-intestinal parasite pressure, under grazing in organic farms: development of a Decision Support System through the mobilisation of a participative research process

D. Jamar[1], V. Decruyenaere[1], Y. Seutin[1], D. Stilmant[1], L. Perriaux[1] and P. Stassart[2]
[1]Farming Systems Section, CRA-W, 100 rue du Serpont, B-6800 Libramont, Belgium, daniel.jamar@skynet.be,
[2]Society, Economy, Environment and Development, ULg, av Longwy, B-6700 Arlon, Belgium

Keywords: participative research, organic meat production, consumption food chain

Introduction Under grazing, gastro-intestinal parasite management remains a major problem in ruminant production systems, more especially in systems respecting organic farming rules following their obligations (1) to perform grazing as soon as pedo-climatic conditions are adapted and (2) to use anti-parasitic products only in a curative way. Surprisingly, this problem has not been highlighted by the different stakeholders in the food chain from cattle meat production to consumption. This could arise from the lack of a clear and pertinent norm or infestation threshold that would allow differentiation between preventive and curative treatments. Such a norm would indicate to the breeder whether he was permitted to treat his herd or not. The question is how to involve the stakeholders in a participative research process in order to develop a decision support system (DSS) adapted to their needs, when there is, initially, no clear demand for such a system. We present the steps followed to develop such a DSS: (1) stakeholder sensitisation to the question, (2) data recording with farmers and developing the DSS principle, (3) data processing and DSS calibration and validation.

Stakeholder sensitisation A multidisciplinary team comprising an agronomist, a sociologist and an economist was formed to help analyse, develop and increase awareness of the organic cattle meat food chain 'producer – retailer – consumer'. Their work highlighted the need to increase the sustainability of this chain through the development of a new meat product. Organic cattle meat must be different from the conventional 'lean and tender' young Belgian Blue bull meat, which represents the majority of the Belgian cattle meat market. One way to induce product differentiation, from its colour, flavour and nutritional quality aspects, lies in increasing the importance of grazing in the feeding scheme of these bulls. So the obligation to include grazing in organic farming, could become a resource to achieve product differentiation and to improve the sustainability of the market. This was the starting point for a collaborative project dealing with the food chain and including a research consortium and several organic producers. The objectives of this project are to differentiate a new food chain product (organic beef meat) and through true cooperation with the research consortium to develop a new shared approach to management, where each producer has his own conception of what is a good grassland, a good stock, a good breeder etc. This evolution of approach significantly modified the researcher question from the identification of a pertinent threshold to perform only curative anti-parasitic treatment to the co-definition of the grazing management scheme to promote in order to optimise stock performance.

Data recorded and DSS development principle To reach such a target, parameters characterising grassland quality, animal intake and parasitic pressure have been recorded on 14 farms, together with the recording of animal health status and performance (average daily gain, ADG). Grassland quality was assessed by measurement of sward surface height, sward quality and floristic composition, whilst behaviour assessment involved calculation of expected intake and observed intake, through NIRS analysis of, respectively, grass and faeces. Coproscopy and plasmatic pepsinogene were used to assess parasitic pressure. All parameters were recorded and discussed in association with farmers, in order to be in complete interaction with them and to be able to consider their points of view. The key steps in the DSS are (1) estimation of potential intake from sward quality, animal type and weight, (2) model potential ADG, (3) compare predicted ADG with observed ADG, (4) seek reasons for the difference between observed and predicted levels of intake and performance, considering the level of parasitic pressure and/or the general health status and/or the sward availability.

First results and conclusions The results from the 2003 grazing season, which was characterised by dryness and low parasitic infestation, demonstrated the good performance of the relation used to predict ADG (r=0.724; N=13; p<0.01). Data from 2004 will allow improvement in the calibration of the DSS (i.e. the link between parasitic pressure and animal performance below the potential) and give further experience on its usefulness in co-defining, with the farmer, the causes of levels of animal performance observed under grazing. This DSS will be used more as a 'diagnostic' tool than a 'predictive' one. It will provide a comprehensive quantitative framework for exchanges between the farmer and his advisory service.

Acknowledgement This research is funded by the Second Program of Sustainable Development Support of the Belgian Science Policy.

Organic herbage seed production in Wales – working with farmers to develop the technology

H. McCalman and A.M. Marshall

Institute of Grassland and Environmental Research, Plas Gogerddan, Aberystwyth, Ceredigion, SY23 3EB, UK, Email: heather.mccalman@bbsrc.ac.uk

Keywords: farmer participation, organic, herbage seed

Introduction The National Assembly Government of Wales is providing incentives for organic farming through its agri -environment schemes and has set a target of 10% by 2010. The organic systems in Wales are grassland based, including some with crop rotations and grass-clover leys. Reseeding currently relies on 60% conventional seed, but this derogation to the organic standard of 100% will be removed by August 2005. Conventional methods of seed production are not acceptable under organic standards. Following farmer discussion group meetings highlighting the difficulty of sourcing and the cost of organic seed, a feasibility project to tackle some of the practical challenges of organic seed production was set up with the IGER Grassland Development Centre and local farmers. The key challenges for Welsh and UK organic seed producers range from weed control to harvesting methods (Marshall & McCalman, 2003) as well as creating links between farmers and seed companies to build a local organic forage seed industry. Building on on-going plot work at IGER, this project is developing techniques for field-scale seed production working with a group of farmers and seed companies.

Materials, methods and results Four farmers with a range of farm types and systems were recruited from within organic discussion groups. Field plots were designed with the farmers, each looking at different aspects and approaches to organic forage seed production and providing a range of demonstration points. The topics explored included the (a) use of white clover (sown as a companion crop) as a nitrogen source for a hybrid ryegrass seed crop; potential of different fertility-building legumes (white clover/red clover/vetch/lupins/crimson clover) sown in the year before the grass seed crop (b) response of different grass species (perennial ryegrass/hybrid ryegrass/timothy) to a red or white clover fertility- building phase and (c) Integration of herbage seed crops into a whole organic farm system.

Importantly, the decision making process for the management of plots was guided by the participating farmers and other group members. On one site weed control is an important issue. On another capitalising on fertility build stimulated a good discussion particularly when taking into account the practicalities of integrating herbage seed production into a crop rotation that fits the farm and meets seed crop regulations. Initial results showed that producing seed crops that meet official seed standards is not a problem. Using white clover as a companion to a hybrid ryegrass seed crop produced a seed yield of 650kg/ha. However there was insufficient nitrogen within the systems to enable a silage cut in the spring of the harvest year, a common practice amongst conventional seed growers. Where a fertility-building crop was used prior to the grass seed crop, the number of fertile tillers was comparable with a conventional seed crop. Where perennial ryegrass was grown after lupins, there was insufficient nitrogen in the system for the crop to lodge, an important characteristic for attaining high seed yields. However this was achieved by sowing the perennial ryegrass with a white clover companion. To optimise input of interested parties ('stakeholders'), the participating farmers met with Organic Seed Certification and NIAB seed certification personnel to explore the issues in organic forage seed production and to develop a better understanding of the challenges involved for all. Input from these other interested parties is invaluable so that the problems and challenges are tackled together.

Conclusions This project has confirmed that organic seed production may be feasible in Wales. Working with farmers on commercial farms has enabled rapid adoption of techniques into farm practice, and, supported by IGER, given the farmers confidence to develop methods on their own farms. Inclusion of other stakeholders at the outset has improved the understanding of the key issues for all parties. There are limitations to this approach - key challenges are to engender the 'ownership' of the project to the farmer and to ensure timeliness of procedures, e.g. accuracy of recording at harvest. The work is on-going and the interest and enthusiasm of the farmers has continued to increase following on-farm meetings and discussions.

Acknowledgements The financial support of the Farming Connect Programme of the National Assembly Government of Wales is gratefully acknowledged.

Reference

Marshall, A.H. & H. McCalman (1993). The use of white clover as a source of nitrogen for organic grass seed crops. *Proceedings of the Fifth International Herbage Seed Conference*, Gatton, Australia, 59-63.

Farmer adoption; ten years of productive pasture systems in southern Australia

G.R. Saul[1], H.L. Quinn[1] and J.T Trompf[2]
[1]Department of Primary Industry, PB 105, Hamilton, Victoria 3300, Australia, Email: geoff.saul@dpi.vic.gov.au, [2]J.T. Agri-source, 2A Bradley Drive Mill Park, Victoria 3082, Australia

Keywords: fertiliser, adoption, farmers, stocking rate, sheep

Introduction Southern Australian sheep and beef farmers have been slow to adopt technology related to grazing management and pasture utilisation despite clear evidence of a strong link between utilisation (stock per ha) and profitability. Between 1971-95, the average stocking rate on farms was 10-12 dry sheep equivalent per hectare (dse/ha) (Anon 2004). Results from the Hamilton Long-term Phosphate Experiment (Cayley et al., 2002) show higher pasture production, herbage digestibility, stocking rates and profitability as phosphorus fertiliser applications increase. In 1993, the Grassland Productivity Program (GPP) started in the winter rainfall areas of southern Australia (Trompf & Sale 2000), initiated by the Grassland Society of Southern Australia, funded by the wool industry. In brief, groups of 4-6 farmers were assisted by experienced advisors to compare current management practice in one paddock with productive pasture technology (PPT) in an adjacent paddock. PPT consisted of appropriate fertiliser application; pasture manipulation to balance grass and legume content and higher stocking rates to ensure utilisation of the herbage grown. Over 300 farmers participated in GPP between 1993-2003. This paper reports the impact on the grazing industry 10 years after PPT was introduced.

Results Since 1971, the annual physical and financial performance of 50-75 farms in the 450-800 mm rainfall regions of south-western Victoria has been monitored by the Department of Primary Industries (Anon 2004). Farms voluntarily participate in the Farm Monitor Program (FMP) and stay in the program for around 5-10 years. Farms in the FMP are 40-4600 ha (mean 900 ha) and on average run 5500 sheep and 260 cattle.

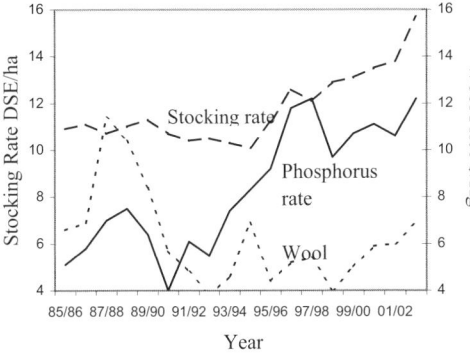

Figure 1 Stocking rate (dse/ha), wool price ($A/kg) and fertiliser use (kg phosphorus/ha) for farms in the FMP 1985-2003

During the 1980's, fertiliser use (mainly phosphorus) was driven largely by wool prices with a peak application rate in 1990 when wool prices reached $A11/kg, followed by a rapid decrease when wool prices fell. However, from around 1994, phosphorus use increased steadily, from 4 kg/ha to 12 kg/ha in 2003, coinciding with GPP and the promotion of PPT. This increase occurred despite wool prices remaining relatively low. In response to higher fertiliser applications, farmers increased stocking rates by 50% from 10 dse/ha in 1994-95 to about16 dse/ha in 2002-03. Farms in the FMP with the highest return to equity continue to use more fertiliser/ha, run higher stocking rates and have higher gross margins per ha than average farms (Anon 2004).

Conclusions GPP fundamentally changed farmer views on fertiliser use, stocking rate and profitability (Trompf & Sale 2000). The Long-term Phosphate Experiment provided the scientific understanding of the grazing systems in southern Australia while GPP allowed farmers to experience and understand the principles of pasture production and utilisation on their own farms. Farmers now understand that fertiliser is a non-discretionary input and that stocking rates (utilisation) must be increased to use the extra higher quality pasture grown. These principles have now moved beyond the farms involved in the targeted extension program (GPP) and into the wider farming community. Based on conservative estimates of $A85/ha higher gross margin ($14/dse and an additional 6 dse/ha), and implementation over 1M ha of southern Australia high rainfall regions, the additional productivity is worth around $A85M/year.

References

Anon (2004). Farm Monitor Project; Summary of results 2002-03. Pub. Department of Primary Industries, Hamilton, Victoria, Australia.

Cayley, J.W.D., G.R. Saul & M.R. McCaskill (2002). High fertility pastures in south-west Victoria can be economically and environmentally sustainable. *Wool Technology and Sheep Breeding,* 50, 724-729.

Trompf, J.P. & P.W.G. Sale (2000). The paired paddock model as an agent for change on grazing properties across south-eastern Australia. *Australian Journal of Experimental Agriculture,* 40, 547-556.

Participatory development of a forage grass cultivar

M.D. Casler[1], P.G. Pitts[2], P.C. Bilkey[3] and C.A. Rose-Fricker[4]
[1]USDA-ARS, Dairy Forage Research Center, 1925 Linden Dr. West, Madison, WI 53706-1108, USA, Email: mdcasler@wisc.edu, [2] 3709 Hwy C, Spring Green, WI 53588 USA, [3]AgResearch Intl., 7841 E. Oakbrook Cir., Madison, WI 53719 USA; [4]Pure Seed Testing, 3057 G St., Hubbard, OR 97032 USA

Keywords: Spring Green festulolium, *Festulolium braunii*, forage breeding, selection

Introduction Perennial forage grasses exist in both nature and agriculture as a highly heterogeneous mixture of genotypes. Extreme environments, fluctuating environments, and severe managements can impose selection pressures that will result in loss of unadapted genotypes. Mortality of unadapted genotypes leads to dominance of fewer highly adapted genotypes which may be useful as superior germplasm in other similar environments.

Materials and methods Parents of 'Spring Green' festulolium were selected as survivors from old research plots in mixture with lucerne (*Medicago sativa* L.) and from an old pasture near Spring Green, WI. Plants were sent to Oregon where they were selected for resistance to rust (*Puccinia graminis* Pers.) and seed was produced for testing and cultivar multiplication. Individual plants of Spring Green, Tandem, and Kemal festulolium were tested for freezing tolerance in a growth chamber for 3 d at -11°C after a 35-d hardening period. Survival of individual plants and individual tillers within surviving plants was counted after 30 d of recovery in the glasshouse. The three cultivars were planted in field trials in 1997 at 14 locations ranging from the northcentral to the eastern USA. Forage yield and survival were determined in 1998 and 1999.

Results Spring Green had higher plant and tiller survival compared to both of its parents in the freezing test (Table 1). This improved cold tolerance led to greater survival under field conditions with the improvement being more pronounced for locations that had the harshest winter conditions. Forage yield Spring Green over two years was also higher than Kemal festulolium, which made up 50% of the pedigree of Spring Green.

Table 1 Mean performance of Spring Green festulolium relative to its two commercial parent cultivars. Data adapted from Casler *et al.* (2002)

Cultivar	Plant survival at -11°C %	Tiller survival at -11°C %	Forage yield t/ha	Survival at all locations# %	Survival at six locations# %
Spring Green	56	48	3.91	65	52
Tandem	33	41	3.98	56	37
Kemal	3	25	3.74	61	43
LSD (0.01)	19	15	0.10	3	4

Survival was measured at 14 field locations. Six of these locations were classified in USDA hardiness zones 2 to 5, with the most severe winter conditions

Discussion and conclusions Natural selection of surviving festulolium plants under harsh field conditions resulted in progeny populations with improved freezing tolerance compared to their parents. This improved freezing tolerance translated into improved survival under field conditions, most noticeably at the locations with the coldest winter temperatures. Thus, there is considerable natural variability for freezing tolerance within festulolium germplasm and this variation can be used to make genetic improvements that can be captured under realistic field conditions. Spring Green festulolium represents a relatively new paradigm in the development of forage grass cultivars for the USA, the direct involvement of a forage producer throughout the development and marketing phases of the process. Peter Pitts was largely responsible for identification of one of the more important selection plots, for obtaining grant funding to conduct much of this research, and for identifying commercial markets for the new cultivar. Over 500,000 kg of seed have been sold during the past five years, making this a highly successful cultivar. Over 100,000 kg of certified organic seed was produced in 2004.

Reference

Casler, M.D., P.R. Peterson, L.D. Hoffman, N.J. Ehlke, E. C. Brummer, J.L. Hansen, M.J. Mlynarek, M.R. Sulc, J.C. Henning, D.J. Undersander, P.G. Pitts, P.C. Bilkey & C.A. Rose-Fricker (2002). Natural selection for survival improves freezing tolerance, forage yield, and persistence of festulolium. *Crop Science*, 42, 1421-1426.

Livestock producers and researchers – a case study of an effective partnership

R.D.B. Whalley[1], I.H. Simpson[2] and W.K. Mason[2]
[1]Centre for Ecology, Evolution and Systematics, Botany, University of New England Armidale, NSW 2351 Australia Email: rwhalley@pobox.une.edu.au, [2]PO Box E9009 East Orange NSW 2800 Australia

Keywords: sustainable grazing, research, impact, applicability

Introduction The Sustainable Grazing Systems Programme (SGS) ran from 1996-2002 with 11 regional producer committees (Regional Producer Network - Simpson *et al.*, 2003) and 6 research sites (National Experiment – Andrew & Lodge, 2003) distributed throughout the high rainfall zone of temperate Australia. Each regional committee had a core of producers, with invited scientists and extension practitioners and a paid facilitator. The chair was always a producer and rotated annually. Each regional committee conducted a number of paddock-sized demonstrations of improved grazing management practices on a number of farms within the themes; grazing management and weeds, high input systems, innovative grazing methods for production and the SGS adoption process. Each demonstration was on a producer's property assisted by a local management committee with objective data collected by the regional facilitator. The National research sites investigated the effects of grazing management on sustainable livestock production through five themes; water, nutrients, pastures, animals and biodiversity. Producers had a significant input into the design of the National Experiment and each site had a local advisory committee.

The overall aim of the SGS programme was to foster the development and adoption of practices that improved grazing management in the higher rainfall zone of temperate Australia. The programme was spectacularly successful and over 8,000 producers made changes to their management as a result of this programme which they anticipated would yield financial (78%) and sustainability (81%) benefits (Allan *et al.*, 2003).

National Forum 2000 Though local interactions between producers and researchers were occurring, a National Forum in 2000 brought together about 100 producers and regional facilitators from all regions, researchers from all research sites and members of the national steering group, to review progress and plan for the final two years of the programme. The Forum was highly successful in initiating the process of amalgamating information from producer-led farm demonstrations with research results from formal experiments to produce a major impact on farm practices. The Forum was significant for being the first formal attempt at national integration.

Format of the Forum The National Forum was held over three days with a social gathering on the evening preceding the first day. Delegates had previously been assigned to one of 12 mixed (producers, researchers, facilitators) discussion groups each of about eight people. The groups were encouraged to meet and get to know each other on the first evening. Two of the days of the Forum comprised plenary sessions; one led by the producers and the other led by the researchers. To demonstrate and promote equality of input, the format was the same in each case. Each theme was introduced by either a producer or researcher and the main outcomes of either the Regional Producer Network sites or the National Experiment with respect to each theme, were briefly summarised. The mixed groups then discussed the relevance of these outcomes for producers, researchers and more sustainable grazing. The main points were recorded for a Summary Team and also verbally presented to the whole session. At the end of the final session, the Chair of the Summary Team presented the outcomes from the Forum and participants were given a two page summary.

Outcomes The outstanding feature of this forum was the close association of producers and researchers in small groups discussing the outcomes from farm-based demonstrations and formal experiments. Having each group represented at each table meant that every producer and researcher was exposed to the other perspective. It was agreed that producer involvement in research planning is critical and that the whole farm system must be encompassed including its social and personal dimensions. Involvement of producers in assessing the outcomes of each theme of the National Experiment was also deemed essential. This forum resulted in strong and positive interactions and in the words of one producer, "My image of scientists as dull and boring has been destroyed", and one researcher "Real producer involvement with researchers was perhaps the greatest innovation in SGS".

References
Allan, C.J., W.K. Mason, I. Reeve & S. Hooper (2003). Evaluation of the impact of SGS on livestock producers and their practices. *Australian Journal of Experimental Agriculture*, 43, 1031-1040.
Andrew, M.H. & G.M. Lodge (2003). The sustainable grazing systems national experiment. 1. Introduction and methods. *Australian Journal of Experimental Agriculture*, 43, 695-709.
Simpson, I.H., G. Kay & W.K. Mason (2003). The SGS regional producer network: a successful application of interactive participation. *Australian Journal of Experimental Agriculture*, 43, 673-684.

Adoption of Participatory Rural Appraisal: a case study from China

A. Chu[1], A. Meister[1], P. Guo[2], J. Reid[1], B. Nowak[1], S. Morris[1], J. Hodgson[1], P. Matthews[1], P. Gregg[1], K. Cai[3], J. Xie[4] and X.Y. Li[2]

[1]Massey University, Palmerston North, New Zealand, Email: a.chu@massey.ac.nz, [2]China Agricultural University, Beijing, China, [3]Yunnan Institute of Geography, Kuming, China, [4]Department of Agriculture, Guizhou, China

Keywords: technology transfer, Participatory Rural Appraisal

Introduction There are many models of technology transfer. They vary from the linear "scientist-extension worker-farmers" model to the integrative "natural resource management" model (Jiggins, 1993). International experience has shown that for small holding farmers in developing countries a farmer driven model based on participatory approaches (the Participatory Rural Appraisal (PRA) Model) is more effective and efficient.

Methods Three PRA workshops were conducted in China: at Chengdu, Zhangye and Pingliang. At the Chengdu workshop approximately 37% of the participants were senior government officers. Traditionally, these people were considered as "sabboteurs" (Pretty *et al.* 1995) by development practitioners because they tended to dominate the proceedings and could result in non-participation by lower ranking officers and farmers. At the two subsequent workshops (Zhangye and Pingliang) more lower ranking extension officers were selected for training. In order to assess the degree of adoption of PRA methodology by the participants, two post-workshop surveys were conducted at 9 months (March 2002) and 3 years (Jan. 2004) after the workshops respectively. Table 1 summarises the key findings.

Table 1 Adoption of PRA by participants

Workshops	Chengdu (Oct. 2000)	Zhangye (June 2001)	Pingliang (June 2001)
No .of participants	46	30	34
No. of organisations represented	22	20	21
Senior ranks (%)	37	26	15
Survey March 02, No. responded (%)	*24 (52%)*	*24 (80%)*	*30 (88%)*
Survey Jan. 04, No. responded (%)	*22 (48%)*	*13 (43%)*	*6 (17%)*
Was PRA useful? Yes (%)	68%	38%	33%
No. organisations using PRA in Planning	12 (55%)	5 (25%)	1 (5%)

Results and conclusion There was a greater degree of adoption by participants from the Chengdu workshop and PRA application was more successful even though there were more potential "sabboteurs" than in the other two workshops. Amongst the success stories were (1) a rabbit factory moved from a loss situation to profitability within two years of the manager adopting PRA when dealing with farmers (2) a research centre had 150% improvement in the number of projects approved by head office after the scientists had involved farmers in identifying research priorities, and (3) Tibetan herdsmen accepted modern veterinary practices after the veterinarian had used a PRA approach to combine traditional Tibetan herbal medicine with modern medicine. These results were reported separately by Chu (2003). One of the lessons learned is that the PRA method by itself is not enough to cause any significant changes in technology adoption by farmers. To be effective, at least in the Chinese "top down" context, the institutional framework behind the development projects needs to be built around the participatory philosophy and has to be fully supported by the Government authorities. Otherwise PRA simply goes through the motions, but does not result in any sustainable changes. It is concluded that one has to be flexible in applying the principles of PRA in developing countries as each could have its own unique political and social environment through which adoption of the PRA methodology would occur.

References

Chu, A.C.P. (2003). Agricultural education for rangeland rehabilitation : a case study from China, In: Global Perspective in Range Rehabilitation and Prevention of Desertification. *Proceedings of the 2003 Obihiro Asia and the Pacific Seminar on Education for Rural Development* (OASERD) UNESCO-APEID, 15-22.

Jiggins, J. (1993) From technology transfer to resource management. In: M.J. Baker (ed.). Grasslands for Our World, SIR Publishing, Wellington, New Zealand, 184-191

Pretty, J.N.,I. Guijt, J. Thompson & I. Scoones (1995). Participatory Learning and Action: A Trainers Guide. IIED Participatory Methodology Series. IIED, London.

On-farm research to improve the measure of variability of forage production across the landscape for evaluating economic risk in forage-based enterprises

E.B. Rayburn

P.O. Box 6108, West Virginia University, Morgantown, West Virginia, 26506, U.S.A., Email: ed.rayburn@mail.wvu.edu

Keywords: stochastic budget, deterministic budget, yield risk, economic risk

Introduction Experiment station trials provide forage crop mean yields and standard deviations (SD) useful in evaluating risk in forage production (Rayburn, 2003). However, in the northeast USA many of these sites are on valley soils atypical of hill-farms in the Appalachian Mountains. This study used on-farm and experiment station research to evaluate the variability of forage yield across a range of soils. This information was used in stochastic budgets to evaluate the economic risks in forage production on soils differing in yield potential.

Materials and methods On-farm and experiment station trials of nitrogen fertilised grass and grass-clover mixtures were summarised into 27 yield classes based on soil and treatment yield potential. Classes contained 3 to 67 site years (348 site years in all). The class yield SD was regressed against mean yield. Deterministic and stochastic budgets were used to evaluate economic risk based on the calculated break even price (BEP) for hay. Stochastic budgets were run using Excel add-in @Risk (Palisade, 2002).

Results Yield SD was a linear function of mean yield (Figure 1), expressed by the regression:

$$\text{yield SD} = 0.371 + 0.234 \text{ mean yield}$$
$$R^2 = 0.86, \text{SD}_{\text{Reg}} = 0.266$$

When yield mean and SD were used in hay budgets to determine economic risks, stochastic budgets gave median BEP 18% greater than the deterministic budget at moderate yields and 6% greater at high yields (Table 1). Stochastic budget BEP at the 5% and 95% probability levels for moderate yields were 79% and 157% of the deterministic budget BEP. At high yield levels stochastic budget BEP at the 5% and 95% probability levels were 84% and 136% of the deterministic budget BEP.

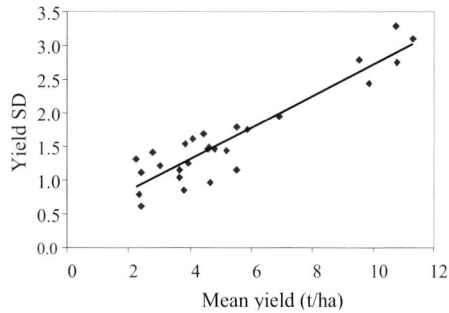

Figure 1 Yield standard deviation versus mean yield

Conclusions On-farm research provided improved estimates of forage yield and SD across soils and treatments differing in yield potential and improved the estimate of economic risk in forage production across the range of conditions occurring on farms. Stochastic budgets gave median costs higher than deterministic budget values and provided an estimate of the range of BEP that the manager is likely to experience. Differences between stochastic and deterministic budgets were greater at moderate production levels typical of farms in the study area. There was less difference between BEP estimated by these budgets at high production levels achieved on valley soils typical of experiment stations.

Table 1 Comparison of deterministic and stochastic budgets on calculated break even price of hay produced at moderate and high yields

Budget type	Moderate yield			High yield		
	Probability level			Probability level		
	Mean	p 0.05	p 0.95	Mean	p 0.05	p 0.95
			Yield (t/ha)			
Deterministic	6.74			11.22		
Stochastic	6.74	3.50	10.06	11.25	6.40	16.34
			Break even price ($US/t)			
Deterministic	53.38			40.08		
Stochastic	62.82	42.02	83.79	42.46	33.80	54.34

References

Palisade Inc. (2002). @Risk- Risk analysis and simulation add-in for Microsoft Excel. Palisade Inc. 31 Decker Rd. Newfield, NY USA 14867.

Rayburn, E.B. (2003). Production risk of cool-season grasses in the Northeast United States. In: K. Cassida (ed.). *Proceedings of the American Forage and Grassland Council,* 12, 197-201.

LeyGrain: a participatory action-learning model for ley pastures in cropping systems

D.L. Lloyd, B. Johnson and S.M. O'Brien
PO Box 102, Toowoomba, Queensland 4350, Australia, Email: david.lloyd@dpi.qld.gov.au

Keywords: ley pastures, action learning, profitability, Australia

Introduction   Since the 1930s, crop/pasture rotation systems have been used in the wheat-sheep belt of temperate southern Australia to maintain the productivity and environmental sustainability of farming systems (Puckridge & French, 1983). Yet, in the northern grain belt of Australia, there is limited adoption of ley pastures, owing to inherently fertile and well-structured vertisol soils. However, soil fertility decline now costs the grain industry about $450 m per year. Legume-based leys are an option for improving soil OM and N and providing other benefits to cropping and livestock production systems (Lloyd *et al.*, 1991). Despite strong one-on-one extension processes since the 1950s, the adoption of crop/pasture rotation systems is less than one-tenth of that in southern Australia.

A participatory learning decision support model   The LeyGrain action learning process has been developed and is being implemented to focus groups of farmers on the need for ley pastures in their cropping systems and to develop their decision making skills in managing them. It is based on the participatory learning and research technique of continuous improvement and innovation. Its objective is to improve the knowledge and skills of farmers in understanding the profitability, benefits and management of these pastures and to promote an attitudinal change that results in their greater implementation. The process is based on four workshops, each of which is followed by on-farm action learning.

Workshop 1: "Benefits and profitability" Farmers believe that crop/pasture systems are less profitable than cropping systems. Thus, they fail to apply precision and timeliness in operations at the crop and pasture interfaces, predisposing to economic failure of the system. The LeyGrain farm profitability model, PRECaPS, based on MSExcel with risk analysis included, enables farmers to compare the profitability (as taxable income) of a range of crop and crop/pasture systems, using their own data or data derived in any other way. It has demonstrated the crop/pasture rotation is more profitable and a better hedge against risk than cropping alone, in all but the best 25% of years when grain prices are high. Knowledge of the bio-physical benefits of pasture phases in cropping lands is also provided for farmers in this workshop.

Workshop 2: "Getting started" - developing knowledge and skills of the benefits of leys, paddock selection, species selection, crop/pasture sequences and establishment. Well-adapted annual and perennial grasses and legumes are available to include in flexible crop/pasture sequences that are based on the commodity prices of grain crops, beef and lamb. Precision and timeliness in establishment is critical to profitability.

Workshop 3: "Making it work" – grazing and pasture management principles, providing high quality feed for targeted livestock markets, animal health. High quality pasture options can be integrated and managed to produce continuous high live weight gains to meet high value markets.

Workshop 4: "Back to crop" – removing the pasture, timing the removal to replenish soil mineral nitrogen and water, reconsidering the crop and pasture sequences. Managing the interface between pasture and crop to minimise the time out of production and replenishing soil water and mineral nitrogen is vital to profitability.

LeyGrain has been road tested with six groups of farmers ranging in number from 5 to 25, with whom the action learning process is on going and continuous. The process utilises the expertise of many disciplines. It ensures that the formal workshops are followed by action on farm - pasture sowings and so on. Thus, it is embedding on-farm comparisons and research within an extension framework. Thereby, the focus on pastures is maintained. Farmers have been surprised by economic analyses that have consistently demonstrated the profitability of ley pasture systems and the economic hedge value of pastures in the system. The process is in its early stages, but farmers are now committing to increased pasture establishment.

References

Lloyd, D.L., K.P. Smith, N.M. Clarkson, E.J. Weston & B. Johnson, (1991). Sustaining multiple production systems 3. Ley pastures in the subtropics. *Tropical Grasslands*, 25, 181-188.
Puckridge, D.W. & R.J. French (1983). The annual legume pasture in cereal ley farming systems in southern Australia. *Agriculture, Ecosystems and Environment*, 9, 229-267.

Farmer-directed on-farm experimentation examining the impact of companion planting barley and oats on timothy-lucerne forage establishment in central Newfoundland

D. Spaner[1] and A.G. Todd[2]
[1]Department of Agricultural, Food and Nutritional Science, 4-16D Agriculture/Forestry Centre, University of Alberta, Edmonton, Alberta, Canada T6G 2P5, Email: Dean.Spaner@ualberta.ca, [2]Agriculture and Agri-Food Canada, P.O. Box 39088, St. John's, Newfoundland, Canada, A1E 5Y7

Keywords: timothy, lucerne, barley, oat, underseeding

Introduction Growing barley or oats in the year of forage establishment is a common agronomic practice in marginal growing regions, but is often not recommended to growers in Newfoundland. Spaner & Todd (2003) reported that barley seeded at rates of 100-150 kg seed/ha and undersown with a timothy-clover mixture (harvested at mid-milk) resulted in the planting year in greater forage yield of poorer quality than pure-stand timothy-clover. A barley seeding rate of 100 kg seed/ha did not impede forage production in the subsequent year.

Materials and methods Three experimental trials were planted on a dairy farm in central Newfoundland in 1998, 1999 and 2000. Treatments consisted of 1) Chapais barley planted at each of three seeding rates (22.5, 45 and 67.5 kg seed/ha) together with a forage companion crop and 2) Nova oat planted at the same three seeding rates together with a forage companion crop. The forage companion crop consisted of Champ timothy planted at a rate of 5.4 kg certified seed/ha together with Caribou lucerne planted at a rate of 10 kg certified seed/ha. Planting-year harvests were between 26 July and 5 August in each year, when the cereals were in the soft dough stage. Harvests in the production year were conducted between 21 June and 7 July, when lucerne had reached the late-bud stage.

Results The lucerne-timothy mixture companion planted with Chapais barley yielded 11% more forage dry matter than with Nova oats in the year of planting, but the two species did not differ for any forage quality trait. Increasing companion cereal crop seeding rates resulted in more tillers, but the forage contained less lucerne and had lower protein and higher neutral detergent fibre contents. Forage establishment was not impeded when companion planted with barley or oat at any of the seeding rates studied. There were no differences between treatments for forage yield or quality in the production year.

Technical conclusions In the present study, barley companion planting resulted in higher planting-year yields than companion planted oats. Nevertheless, there was a greater percentage of lucerne in the lowest seeding rate of the barley companion crop than with oats. Percent timothy in the forage harvested in the planting year was not altered by any treatment. Similarly, forage establishment was not impeded when companion planted with barley or oat at any of the seeding rates studied. Indeed, increasing companion cereal planting from 22.5 to 67.5 kg/ha in the planting year resulted in a 9% increase in forage dry-weight yield in the year following planting, with no difference in quality at any treatment level. Results from the present experiment indicate that a seeding rate of 67.5 kg/ha may be employed in central Newfoundland when companion planting barley or oats with a forage mixture of timothy-lucerne.

Farmer-directed on-farm experimentation This experiment was undertaken as the result of three imperatives: to foster closer relationships with farmers, to demonstrate varieties and technologies directly to the farming community and to maintain research activity with shrinking research dollars. There are limitations to on-farm experimentation such as this: 1) Most management decisions were dictated by practical considerations of the farm; 2) Farmers use large machinery, plant large areas and can not easily alter seeding rates to conform to exact numbers of plants/ha; 3) This large scale of experimentation implies that there will be a great deal of within-plot variation; and 4) This experiment was conducted on one farm and therefore the statistical inference space for these data is, strictly speaking, this farm. Nevertheless, local farmers are presently using these results in their day-to-day farming operations.

Reference
Spaner, D. & A.G. Todd. (2003). *Journal of Agronomy and Crop Science*, 189, 273-279.

Theme C: Delivering the benefits from grassland

Section 23

Improved livelihoods from grassland

Use of forage legumes to restore overgrazed natural grasslands in Uganda

E.N. Sabiiti, S. Mugasi and F.B. Bareeba
Makerere University, Faculty of Agriculture, P.O. Box 7062 Kampala, Uganda, Email: Esabiiti@agric.mak.ac.ug

Keywords: grasslands, forages, legumes, degraded, dry matter

Introduction The natural grasslands of Uganda support over 95 % of the country's livestock. They are also used by wildlife and protect soil resources from heat and erosion. Unfortunately, the pastoral/agro-pastoral communities which derive their livelihoods from these grasslands have in turn degraded them through overgrazing and uncontrolled burning, thus lowering their productivity (pasture and animal production) and biodiversity. The natural grasses (e.g. *Panicum maximum, Brachiaria brizantha, Setaria anceps, Themeda triandra*) mature rapidly and lose quality. Furthermore, the indigenous legumes (e.g. *Neonotonia wightii, Desmodium adcsendens, Indigofera errecta*) are less persistent and productive to maintain feed quality and hence animal production. There is a need to introduce into the grassland ecosystem alternative forage legumes that combine both persistence and productivity.

Materials and methods A total of eight exotic forage legume species (Table 1) and one indigenous species (*Neonotonia wightii*) were planted into natural grass (*Brachiaria brizantha, Setaria anceps* and *Themeda triandra*) plots on four farms using over-sowing techniques (Sabiiti, 2003). Observations on emergence, seedling vigour, leafiness and reseeding ability were carried out after emergence through flowering to seed production in one year. Subsequently, herbage was clipped to determine dry matter (DM) yield and % crude protein (CP).

Results All the introduced forage species performed better than the indigenous species (*Neonotonia wightii*) (Table 1). The exotic legumes all performed better in terms of seedling emergence, seedling vigour, and leafiness and reseeding ability than the native legume. The inclusion of forage legumes into the natural grasslands more than doubled the DM yield (Table 2), and increased the percent CP above the minimum animal intake requirement of 7% (Humphreys, 1978). The most productive species in terms of DM production were *M. atroperpureum, C. pubescens, D. uncinatum, D. intortum* and *C. rotundifolia.*

Table 1 Performance of the nine forage legume species in natural grass plots in Mbarara, Uganda

Species	% Emergence*	Vigour	Leafiness	Reseeding ability
Macroptilium atropurpureum	80	10	8	6
Centrosema pubescens	75	8	7	6
Cassia rotundifolia	80	9	7	10
Desmodium intortum	75	8	8	5
Desmodium uncinatum	75	8	8	5
Stylosanthes guianensis	60	7	5	5
Stylosanthes scabra	60	6	5	5
Neonotonia wightii	50	5	5	4
Desmanthus virgatus	70	7	6	8

Score scale: 1-3 = Poor adaptability, 4-6 = Moderate adaptability, 7-10 = High adaptability (scores were based on %emergence, vigour, leafiness and reseeding ability); *Based on viable seed sown

Table 2 Productivity of natural grasslands planted with exotic forage legumes species in Mbarara, Uganda

Farm	Grass plots	Mean DM (kg/ha)	Mean % CP
1	No legume	1875	6.2
	Legume	5142	8.1
2	No legume	1724	5.9
	Legume	4462	8.4
3	No legume	1460	5.7
	Legume	3870	8.9
4	No legume	1054	6.2
	Legume	4275	9.5

Conclusion This study shows that it is possible to increase grassland production in terms of herbage yield and quality by introducing exotic forage legumes into pastoral/agro-pastoral production systems.

References
Humphreys, L.R. (1978). Tropical Pastures and Fodder Crops. Longman Group Ltd. 135 pp.
Sabiiti, E.N. (2003). Eco-bio-techniques in the rehabilitation of overgrazed/bushed rangelands in East Africa. Paper presented in workshop at Seventh International Rangeland Congress, Durban.

Carbon sequestration in desertified rangelands of *Hossein-abad*, Iran: a participatory approach

F. Amiraslani
Forest, Rangeland and Watershed Management Organization, #131,West Zartosht Avenue, Valiasr Street, Tehran, Iran, Email: f_amiraslani@yahoo.com

Keywords: carbon sequestration, Iran, rangelands

Introduction The Hossein-abad (H.A) area is located in Southern Khorasan Province of Iran and covers some 148,000 ha.. This is one of the poorest regions in the country and has a large area of degraded rangelands. .Following a request from local people and in line with national and global goals, a carbon sequestration initiative has been funded by I.R.Iran and GEF(Global Environmental Facility) from April 2003. The objective is to promote and model carbon sequestration through developing range species in cooperation with local people and using a participatory approach. The immediate target beneficiaries are the people living in the project area and communities living in the watershed basin.

Approach The project addresses three areas: capacity building, social communication and carbon sequestration.

Capacity building While the main development objective is to sequester carbon through the rehabilitation of desertified areas, the project will involve capacity building for communities who will carry out the activities to sequester carbon. The project will empower communities, especially the poor and other marginalised groups in H.A. and within the larger sub-watershed to manage their own resources, as well as to have the capacity and therefore, the confidence to influence policies and to access support from outside. Following the results of questionnaires, capacity building workshops have been held and local people have been completely familiarised with the project. An inception team composed of local men and women have been trained and become knowledgeable and skilled in related subjects including the aim of project.

Social communication Social communication is a necessary ingredient for the successful launching and replication of the project and will receive considerable attention.

Carbon sequestration Carbon sequestration through rehabilitation of the area will involve the planting and reseeding of at least 9,000 ha of degraded land with various woody and non-woody species, with the woody component providing about 80% of the cover. The proposed species are mainly indigenous, have been verified by the local population. They include *Haloxylon persicum, H. aphylum, Atriplex canescens, A. leucoclada, Calligonum spp., Zygophyllum spp., Amygdalus spp.* (Almond)*, Berberis spp*, and wild pistachio. These are going to be planted by the local population in designated areas based on the results of comprehensive study . According to the results of the baseline studies results and experiences of local people, the selection of appropriate species will be based on climatic and topographic conditions.

Conclusions Although the project is still at an early stage, it has demonstrated that desertified lands can be cost-effectively reclaimed by and for the benefit of the local people and that there is significant potential to sequester carbon in plants and soil in these areas for overall global benefit. To the present the most quantifiable improvements in livelihoods resulting from the project are (a) most of the local communities have been trained in various aspects of livelihood support activities and (b) according to the baseline studies, about 10% improvement in the Human Development Index (HDI) and in overall productivity and income generating activities of the H.A area have been achieved

The lessons drawn here from true collaborative rehabilitation and management of natural resources could be applied in similar places in other countries with similar topography.

Herders and wetland degradation in northern Cameroon

E.T. Pamo, F. Tendonkeng and J.R. Kana
University of Dschang, FASA, Dept. of Animal Sci. P.B. Box 222 Dschang, Cameroon, Email: pamo_te@yahoo.fr

Keywords: Cameroon, wetland, degradation, pastoralists, adaptation

Introduction Livestock rearing in Northern Cameroon is carried out under two majors systems: the nomadic and the transhumance production systems (Pamo & Pamo, 1991). Nomadism is the practice of wandering from place to place, while transhumance involves seasonal displacement of flocks from one area to another by herders. These production systems involved large grazing areas, which may encompass different ecosystems. The Yaére, the only wetland of the northern Cameroon, is the major dry season grazing lands for livestock and wildlife. The main characteristic of this wetland is that the whole area is excluded from grazing during the growing season as a result of large scale flooding. Thus the major forage species (i.e. *Echinochloa pyramidalis, Oryza longistaminata, Hyparrhenia rufa, Echinochloa stagnina*) can set seed thereby ensuring their continued dispersal, establishment, and survival during the subsequent rainy season. In 1979, an upstream dam of 28 km with an additional 20 km embankment along the Logone river was build to store water for a rice irrigation project. This suppressed flooding over some 60 000 ha, and seriously affected the hydrological regime over another 200 000 ha. Major perennial forage species were gradually replaced by less palatable annual species such as *Sorghum arundinaceum*. This paper investigates how herders coped with the induced degradation of this dry season grazing land.

Wetland use before the dam construction The Yaére was a multifunctional human use area. Exploitation varied by site and by season, corresponding to the dynamic character of the flood plain and the cultural background of the floodplain users. During the dry season the area played a fundamental role in sustaining the rural economy on a regional scale. Fish and muskwari were exported from the area and herds from many parts of the North Cameroon, Chad, and Nigeria were provided with fresh pasture and water. During this period, the crude protein content of rain-fed pasture declines and the surrounding savanna pasture becomes desiccated. The accessibility of the flood plain and its primary products are of vital importance to livestock and wildlife. Pastoralists with cattle, sheep and goats moved into the area as the dry season progressed.

Adaptation to wetland degradation Pastoralists used flexible strategies to mitigate the effect of a high-risk environment. This required a formal or mainly informal set of rules allowing herders access to different ecological areas of the region to use different resources at different seasons of the year. Their survival is attributable to a wide spectrum of adaptive strategies. Some were ecologically based, while others depended on socio-economic and cultural mechanisms. The ecologically, as well as economically, based strategies rely on herd maximization which is achieved by herd diversification. The use of different livestock species has ecological and economic implications. Wetland degradation creates forage scarcity and leads to the poor spatial herd distribution. Different species then fill different ecological niches which may be more efficient as each species prefers to graze certain plant species. Increased mobility was also widely used and involved resource exploitation mobility (Oba & Lusigi, 1987), carried out in response to unpredictable forage and water availability, and escape mobility involving long distance migration to escape the combined effects of range degradation and decreased rainfall. Resource exploitation mobility allowed utilization of a widely dispersed forage resource at the times when it was rare. The distance covered, the routes followed, the length of stay in an area and the degree of flexibility built into the system varied from year to year. The number of displacements during the dry season depended on the state of available resources and of livestock. Such patterns of land use allowed for a high degree of fluidity and variation in the pastoralist system and provided an opportunity to individual herd owners to respond independently to seasonal fluctuations. Escape mobility, involving long-distance migration, was implemented to escape the combined effects of range degradation and reduced rainfall.

Conclusion Degradation of rangeland creates long-term economic and ecological disasters and diverts scarce resources to relief programmes. These observations from Northern Cameroon demonstrate that policy options for economic development require the assessment of economic, social and environmental function of any ecosystem before deciding to initiate and implement any project. In fragile environments, such as North Cameroon, measures should be taken to reinforce the ability of pastoralists to move between different ecological areas. Inevitably such a process requires limits to be placed on the numbers allowed to use a particular area in any season. Social and cultural background of the local population suggests that this will never be an easy strategy to implement in the short term.

References
Oba, G. & W..J. Lusigi (1987). An overview of drought strategies and land use in African Pastoral system. Pastoral Network Paper 23a, ODI, London, 1987.
Pamo, T.E. & C.T. Pamo (1991). An evaluation of the problems of open range use system in northern Cameroon. *Tropicultura*, 9.3, 125-128. Belgique.

Grazing prohibition programme and sustainable development of grassland in China

X.Y. Hou and L. Yang
Department of Research Management, Chinese Academy of Agricultural Sciences. Zhongguancun Nandajie 12, Beijing, 100081,China, Email: houxy16@caas.net.cn

Keywords: grazing prohibition programme, grassland sustainable development

Introduction Prohibition of grazing is now the main grassland management measure in China. From 1999, prohibition of grazing has been implemented on a trial basis in some areas. From 2001, the grazing prohibition programme (GPP) has been carried out in five provinces (Shaanxi, Gansu, Hebei, Jilin and Yunnan) and two autonomous regions (Inner Mongolia and Ningxia), with the objective of protecting and restoring grassland by seasonal or yearly banning of grazing with subsidiary assistances. The area within which grazing was prohibited of 2.93×10^7 ha in 2001 was increased to over 3.33×10^7 ha in 2004. With a view to improving the GPP and ecological reconstruction, we conducted a survey in some counties to review the relationship between GPP and the sustainable development of grassland.

Methods Six sampled counties (Chinba'erhuzuo county, Ewenke county, Kerqin youyizhong county, Hangjin county, Etuoke county, Wulatehou county) were in the Inner Mongolia autonomous region, two sampled counties (Gangcha county and Haiyan county) were in Qinghai province, and one sampled county (Songpan county) was in Sichuan province. One hundred and sixty-three households, 42 officers in the rural and pasture area and 11 officers in counties were interviewed individually with a survey questionnaire used during the interviews.

Results From 1998, the methods of feeding livestock from grassland in China began to change. The ratio of households which herded livestock dropped from 73.8% in 1998 to 65.9% in 2002, and the percentage of households which fed livestock indoors rose from 4.3% in 1998 to 15.2% in 2002. Some 62.4% of surveyed herders agreed with GPP, while only 22.4% disagreed (Table 1). Economic concerns were the main barrier in the course of promoting GPP. The survey showed that just 8.5% of herders had benefited from GPP, while 27.9% saw losses. Meanwhile, 72.5% of interviewed officers believed GPP will decrease the benefit to herders in the long term and 94.0% of interviewed herders believed that they should receive subsidies because of GPP. More than half of the surveyed herders proposed that subsidies of 200 Yuan/ ha per year would be acceptable. The survey of herders showed that several factors presented difficulties in promoting GPP. These included shortage of starting capital (85.7%), lack of forage (66.7%), absence of technical guidance (42.9%) and traditional concepts (42.9%).

Table 1 Some survey data relating to the grazing prohibition programme

Attitude toward GPP	Percent	Critical factors in GPP	Ranking	Subsidies required (Yuan/ ha per yr)	Percent
Agree	62.4%	Overpopulation	1	50	31.0%
Disagree	22.4%	Risk resist intention	2	100	18.0%
Neutral	9.7%	Lack of forage	3	200	51.0%

Conclusions An increase in costs, a decrease of benefits and a shortage of starting capital should affect implementation of GPP. Traditional grazing systems, which lack modern scientific approaches and techniques cannot support sustainable development of grassland in China any more. So it is suggested that: firstly, government should provide practical financial aids for herders to solve the current difficulties in this special transitional period; secondly, technical and scientific assistances should be provided in order to transform the traditional grazing system; for example, construction of basic facilities, subsidised loans, scientific guidance on grazing and suitable livestock species for indoor feeding.

Degraded rangeland: can the balance be restored in the absence of satisfactory range management practices?

F.J. Mitchell, R.G. Bennett, B.D. Forbes and R.N. Reynolds

Dept of Agriculture and Environmental Affairs, Private Bag X9059, Pietermaritzburg, 3200, South Africa, Email: mitchellf@dae.kzntl.gov.za

Keywords: degraded rangeland, overstocking, food security

Introduction The rangelands of KwaZulu-Natal play a fundamental role in the wealth and security of communal populations who are dependent on these forage-producing lands for their livelihoods. In most communal areas of the Province, there is an absence of satisfactory range management practices and the utilization of resources is generally non-sustainable. A major threat to the productivity of rangeland is inappropriate land use, such as overgrazing and incorrect burning practices, leading to extensive degradation of both the vegetative and soil components. Range vegetation and soil reserves show vastly reduced productivity. Degradation also results in increased susceptibility to erosion, loss of vegetative cover and palatable species, loss of biodiversity and reduced productivity, directly threatening food security of vast numbers of people in the rural areas. Social issues such as weakened and marginalized traditional authorities and reduced control of resource utilization is partly responsible. In addition, the value placed on livestock for draught power, meat, milk and other products, and for financial security against calamity, entrenches a reluctance to diminish stock numbers. Alternative strategies to reduce pressure on stressed range systems need to be formulated in participation with affected communities to, among other benefits, increase the contribution from animals to household security.

Materials and methods Three case studies are assessed to determine the impact of a lack of satisfactory range management on rangeland productivity, species change and soil erosion. Field studies and remote sensing data were utilized to identify the nature and extent of degradation, current land use practices and opportunities to close the gap between potential production and current land use. A predictive tool to pre-empt degradation in agro-ecological zones of the Province was developed.

Results Degradation patterns assessed from long term conservation and communally managed research areas identified ecosystems which are so fragile that rehabilitation is uneconomical, particularly when the pressure on the system cannot be reduced. A fifty year research study has provided valuable information on the ability of various soil types to recover after severe mismanagement. The study indicates an increase in tree density from 477 trees/ha to 6266 trees/ha over 23 years on shales with erosion still active, while palatable grasses have increased to 120% of benchmark on dolerite sites. A technological package has been developed that can be used on farms of all sizes, with a participatory approach ensuring the transfer of technology, mentorship and integrated natural resource management systems ensuring success. Alternative strategies encompassing cover cropping, improved maize production for residue availability, organic manures, improved animal health, planted pastures for soil conservation and grazing, alley cropping and water harvesting have been identified within the community-driven approach. One communal study area was severely overgrazed, with stock losses every winter due to lack of fodder. Subsequent to intervention strategies, maize yields improved from <1 ton per hectare (t/ha) to 5.8 t/ha, planted area increased from 8 ha to 130 ha, providing greater household food security and providing additional residue for over-wintering cattle. Farmers' resource management skills and the adoption of sustainable agronomic practices was evident. Marketing and on-farm value adding opportunities were instituted. Animal deaths were reduced by more than 50%, livestock sales were encouraged and 54 jobs were created within the community directly attributed to improved agricultural productivity.

Conclusions To protect soil, water and vegetation and to achieve sustainable utilization of resources, traditional carrying capacity and grazing pressure ideologies need to be replaced by alternative, integrated strategies which are community driven to ensure adoption of sustainable resource management techniques. A food-feed system which increases the yield of food crops, sustains or increases soil fertility and provides improved fodder from residues is essential for communal systems, particularly in marginal areas where degradation erodes food security and where reduction of stock numbers is culturally sensitive.

References

Camp K.G.T., F.J. Mitchell, R.G. Bennett & P.P. Whitwell (2001). Unlocking Agricultural Potential: The KwaZulu-Natal Bioresource Programme. *Proceeding of the Thirty-fifth South African Society for Agricultural Extension Congress,* 194-200.

Liengme, D.P., B.D. Forbes & K.G.T. Camp (2003). Reclamation of degraded land in the Weenen Valley Bushveld. *African Journal of Range and Forage Science* (incorporating the Proceedings of Seventh Rangeland Congress. Durban. South Africa) 20, 157-174.

Contribution of grasses to soil fertility and improved livelihoods

G.P. Ojha[1] and B.K. Dhital[2]
[1]Capital Research Center, PO Box 9737 Kathmandu, Nepal, Email: gpojha2002@yahoo.com, [2]Sustainable Soil Management Programme, Bakhundole, Lalitpur, Nepal

Keywords: vegetable, grasses, soil fertility, gender, empowerment

Background and approach Vegetable farming is increasing in Nepal as it provides better economic returns than growing other crops, especially in areas that have easy access to markets. Vegetable farming demands intensive care and balanced supplies of nutrients. Therefore, farmers cultivate vegetables near their residence and because vegetable growing is more profitable, farmers allocate more resources, including organic manure, for its cultivation. In general, using more organic manure on vegetables means that less organic manure is available for non-vegetable crops and farms, unless alternative arrangements are made for producing more organic manure or manure of higher quality.

There is a risk that the soil fertility of non-vegetable farms belonging to households that have been growing vegetables for a long period of time may have deteriorated due to low use of organic manure. It is possible though that farmers may have produced some mechanisms to cope with this problem. A study was conducted in Paang village, Parbat district, Nepal, where the District Agriculture Development Office, with support from Sustainable Soil Management Programme, has carried out an action research programme to improve the livelihoods of the farming community through the integration of socio-economic and environmental activities. The general objective of the study was to assess the impact of vegetable farming on soil fertility and socio-economic aspects including the changes in gender role. The methodology used was (1) participatory rural appraisal and (2) soil sample analysis. Levels of significance were assessed by the t test.

Results The findings showed that soil fertility on non-vegetable areas had not been significantly affected by growing vegetables nearby, The lands growing and not growing vegetables were receiving 18,500 and 1,300 kg/ha more manure respectively compared to the amount applied before vegetable growing started. This arose because the households were producing more manure by using different strategies. A significant contribution to this increase had been made by the increased quantity of grasses in the community-managed forest and the use of crop weeds for animal bedding. Stall-feeding, improved housing of animals and use of urine were other contributing factors. Significant differences in soil fertility between vegetable growers and non-growers were not yet taking place, as the differences were only 0.005% for N, 12.7 kg/ha for P_2O_5, 4.6 kg/ha for K_2O and 0.10 for organic manure. Together with vegetables, yields of cereals were gradually increasing in both types of farms and this had contributed to the additional average annual household income of 5651 Nepalese rupees.

Appreciable changes had taken place in gender roles by households participating in vegetable cultivation and marketing, with women no longer limited to work inside households. They proved to be capable in the market place as well as on the land. The special ability of women to keep patience contributed to obtaining better market prices for their products. With their participation in marketing, women were found to have developed skills in salesmanship, gathering and using market information, developing seasonal crop calendars to fetch higher prices and some of them have built-up leadership capabilities. The households also experienced increased food security and enhanced family harmony.

Policy recommendations Integrated soil management has brought both economic and social benefits to vegetable growers without affecting the soil fertility on lands being used for vegetables and non-vegetables. Development agencies should look through the eyes of farmers and plan and implement activities holistically to enhance the sustainable livelihoods of farmers.

Transhumance in protected areas in Benin

E.A. Sogbohossou, M. Houinato, C. Tamou, K. Sounkere and B. Sinsin
Laboratory of Applied Ecology, Faculty of Agronomy, University of Abomey-Calavi. 01 BP 526 Cotonou, Benin, Email: mrhouinat@yahoo.fr

Keywords: transhumance, protected areas, impacts, Benin

Introduction Every year, protected areas and regions in West Africa receive transhumant herds. This movement of herds from the dry zone (the Sahelian region) to more humid costal zones is a tradition for the Fulani people. In general, protected areas in West Africa are located at the border of the Sahelian zones through which most transhumants must pass. This periodic movement has an impact on natural resources and the people in the reception zones, especially around and in the protected areas. The objectives of the study were to define and describe the type of transhumant cattle breeding systems around these protected areas, to deduce impacts on population and environment of this system and to provide suggestions for better management of cattle breeding and transhumance around protected areas

Materials and methods Data were mainly collected by interviewing all stakeholders (farmers, herders, breeding and forestry administrators, rangers) and conducting a survey of the cattle. A spatial study of rangelands (Boudet, 1991) was conducted in order to estimate their carrying capacity and to assess the impact of cattle, especially transhumant cattle, on rangelands in the study area in two national parks (Pendjari and W). Data analyses were performed by Arcview®, Excel ®and Minitab®.

Results and discussion Fulani specialise in cattle breeding around the northern Benin protected areas. Farmers rarely breed cattle. For Fulani, cattle play a social, cultural and economic role, but for farmers avoiding breeding their cattle is mainly a way to save money. Compared with Pendjari national park, the number of cattle received in the W national park increases each year. Transhumants come from drier countries of Africa. Reasons for mobility are diverse, according to foreigners or national herders. During the rainy season, cattle move essentially to avoid conflicts with farmers and the high humidity of the rangelands. In the dry season, herders move to find water and forage for their cattle. Table 1 shows the carrying capacity of rangelands in the two protected areas and the number of cattle that can be supported.

Transhumance has both negative and positive impacts on the natural resources and on livelihoods (Sournia, 1998). Transhumance leads to the degradation of grasslands (tree and herbaceous strata) and creates conflicts between cattle and wildlife. It also creates conflicts between the different socio-professional groups living around protected areas.

Table 1 Carrying capacity of grasslands and supported capacity in protected areas, Benin

Protected area	Carrying capacity (ha/TLU per year)	Corresponding need in land (TLU/ha per year)	Number of supportable cattle	Number of cattle present
Pendjari	0.6	1.7	435	> 4,335
W	0.82	1.43	149,923	> 286,347

Conclusion This study shows that the protected areas in West Africa receive cattle herds from many other countries each year. These herds impact on local communities and natural resources. To solve the problem or to improve management, a participatory approach should be adopted involving every stakeholder.

References

Boudet, G. (1991). Manuel sur les Pâturages tropicaux et les cultures fourragères. Coll. Manuel et Précis d'Elevage, Fourth ed., Ministère de la Coopération et du Développement,. Paris, France, 266pp.
Sournia, G. (1998). Transhumance et pastoralisme. In: Les aires protégées d'Afrique francophone. ACCT, 26-31.

Land subdivision, heterogeneity, and declining food security for African Pastoralists

R.B. Boone

Natural Resource Ecology Laboratory, 1499 Campus Delivery, Colorado State University, Fort Collins, Colorado, USA, Email: rboone@nrel.colostate.edu

Keywords: patch isolation, forage choice, livestock mobility, semi-arid, Pastoralists

Introduction Pastoral livestock inhabit landscapes that are spatially heterogeneous and have forage patches that pulse in their value to animals. Mobile pastoralists have evolved movement patterns to maximize use of these ephemeral food sources. In pastoral communities across Africa, changes in land tenure policy and socioeconomic pressures have caused pastoralists to decrease their mobility. Pastoralists recognize that shrinking access to land reduces their options to find forage, and theory suggests that the capacity of land to support herbivores decreases as a power of the square root of area accessible. We used ecosystem modelling in South Africa and Kenya to quantify declines in the number of livestock that can be supported under subdivision.

Study areas and methods A generic but flexible application of the SAVANNA ecosystem model was adapted from an application from the semi-arid Vryburg area of the North-West Province of South Africa. A 300 km^2 area was subdivided into progressively smaller parcels until it was composed of thirty 10 km^2 parcels, simulations run for each parcel, and changes in the number of cattle across the entire 300 km^2 summed (Boone and Hobbs 2004). A more representative application of SAVANNA was created for southern Kajiado District, Kenya. In Kajiado, lands were divided into group ranches beginning in the 1960s, and subdivided into individual parcels owned by Maasai pastoralists and agro-pastoralists, a process still ongoing. Simulations were run for three intact group ranches, and for each ranch, 20 randomly placed parcels for each area of 10, 5, 3, and 1 km^2.

Results When a 300 km^2 area in South Africa was subdivided into smaller parcels, a linear decline in the number of cattle the area could support occurred (Figure 1). Animals in an isolated patch with poor forage could not move to better forage patches, leading to a decline in the population. At the smallest parcel area, 19% fewer animals could be supported. In Eselenkei Group Ranch, Kajiado District, the number of livestock that could be supported declined by 25% when subdivided to 1 km^2 parcels (Figure 2) (Boone et al., In press), associated with reduced access to heterogeneous forage patches, and greater travel costs associated with accessing water. Declines in group ranches were not equal. In Olgulului Group Ranch,

the livestock population on 10 km^2 parcels was 20% lower than for the intact ranch, but did not decline as subdivision continued. The ranch had low heterogeneity, so that as land was subdivided, the variety of forage patches did not change. The most productive group ranch, Osilalei, livestock did not decline under subdivision.

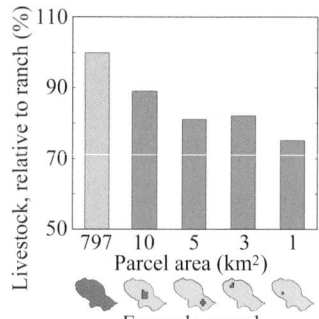

Figure 1 Cattle on a subdivided area South Africa

Figure 2 Livestock supported of on Eselenkei under subdivision

Conclusions Declines in livestock populations were dramatic in modelling results from South Africa, and for some group ranches in southern Kajiado, Kenya. The Maasai of Kajiado are facing diminishing food security due to large increases in human population without concurrent increases in livestock. Based on our research, declines in food security will be exacerbated by subdivision. National and international governments may expect to offset losses of 25% of livestock populations if pastoralists are sedentarized. There are some benefits to subdividing lands, and additional costs; we encourage stakeholders to retain open access, so that benefits may be enjoyed without the concurrent declines in livestock populations.

References

Boone, R.B., S.B. BurnSilver, P.K. Thornton, J.S. Worden, & K.A. Galvin. In press. Quantifying declines in livestock due to land subdivision in Kajiado District, Kenya. *Rangeland Ecology & Management*.

Boone, R.B., & N.T. Hobbs. 2004. Lines around fragments: effects of fencing on large herbivores. *African Journal of Range and Forage Sciences* 21:147-158.

Factors related to marketing successes for fibre producers in Middle Asia

R.B. Boone[1] and K.A. Galvin[1,2]

[1]*Natural Resource Ecology Laboratory, Colorado State University, Fort Collins, Colorado, USA, Email: rboone@nrel.colostate.edu,* [2]*Department of Anthropology, Colorado State University, Fort Collins, Colorado, USA*

Keywords: cashmere, wool, goats, camels, Kazakhstan

Introduction Following the collapse of the Soviet Union in the early 1990s, the economic well-being of livestock producers of Kazakhstan and Kyrgyzstan declined dramatically (see Kerven 2003; Kerven *et al.*, 2003). Like the economies in general, the livestock economies are slowly recovering and restructuring. Livestock producers have been encouraged by international market prices to raise sheep, goats, camels, and animals producing specialty fibre. Fine-fibre sheep and goats remain in Kazakhstan and Kyrgyzstan, but marketing of fibres from the region is not ideal. As examples, sheep pelts are not sorted and graded, which is expected by international buyers, and cashmere is shorn and sold in bulk. Lastly, marketing opportunities are limited, technology, transportation infrastructure, and market information is lacking, and the bargaining power of individual fibre producers is weak. Under support from the U.S. AID Global Livestock-Collaborative Research Support Program (GL-CRSP) project, "Developing institutions and capacity for sheep and fibre marketing in Central Asia," we seek to understand the spatial relationships that can help determine success in fibre marketing.

Methods We have gathered data on marketing success and marketing opportunities for livestock producers in the Zhanakurgan region of Kylz Orda Oblast, Kazakhstan. In cooperation with the Kyrgyz Sheep Breeders Association, similar data have been gathered within Chui, Talas, and Naryn Regions of Kyrgyzstan. Interviews were carried-out that quantified activities, costs, and incomes for producers. Interviews continue, with data from 25-30 livestock producers in each study site our goal, citing, for instance, transportation costs, decision-making, and knowledge of markets. Spatial data are being extracted from moderate resolution topographic maps and satellite images, combined with data from informants and the literature. Modelling will be used to measure the benefit of transporting fibre to larger markets, versus the costs associated with transport.

Results and Discussion Some interviews have been conducted in Zhanakurgan, quantifying marketing access and prices. Interviews are ongoing in the Talas, Bishkek, and Naryn regions of Kyrgyzstan. Spatial data that will be used to calculate distances travelled have been gathered for Zhanakurgan, Kazakhstan (e.g., Figure 1). The topographic maps were scanned, georectified, and merged to create spatial imagery from which distances can be measured accurately. Regional and national maps have been collected. From the spatial data, the distances to livestock markets, for instance, will be measured. Our analyses are ongoing, but spatial data appear adequate. Data from the Soviet era are ample, and our collaborators inform us that much of the data remain up-to-date.

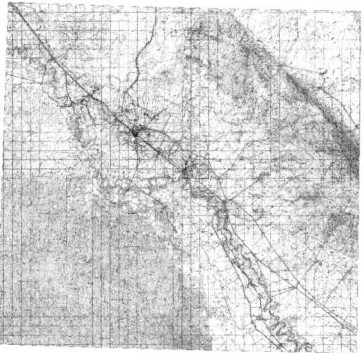

Figure 1 Four 1:200,000 scale topographic maps incorporating Zhanakurgan were merged and positioned geographically, representing an area 160 km wide

References:

Kerven, C. (2003). Prospects for pastoralism in Kazakstan and Turkmenistan: from state farms to private flocks. Routlege and Kegan Paul, London, UK

Kerven, C., I.I. Alimaev, R. Behnke, G. Davidson, L. Franchois, N. Malmakov, E. Mathijs, A. Smailov, S. Temirbekov & I. Wright (2003). Retration and expansion of flock mobility in Central Asia: costs and consequences. *Proceedings of the Seventh International Rangelands Congress,* Durban, South Africa. 543-556.

Task force development to provide education and leadership to the meat goat industry

J.C. Fisher, L.A. Nye and D.A. Mangione

The Ohio State University Extension, Pike Co. 120 S. Market St. Waverly, Ohio 45690, USA, Email: fisher.7@osu.edu

Keywords: meat goats, leadership, education

Introduction Chevon is the most frequently consumed meat in the world. Meat goat production is increasing because of the economic value of goats as efficient converters of low-quality forages into quality meat, milk, and hide products for specialty markets of health conscious, ethnic, and faith based consumers. Estimates of national marketing indicate that U.S. meat goat production is nearly 500,000 head less than demand. Where resources are limited, meat goats can be raised efficiently and profitably on small farms, so the country could become self-sufficient in meat goats.

Engaging resources Meat goats, as an enterprise, did not have supporting infrastructure such as a commodity based organisation, university sponsored education and research, or well known marketing channels. To address these needs, a task force was formed and directed by personnel of The Ohio State University Extension and consists of producers, multi-disciplinary faculty, ethnic and faith based community leaders, other state universities and colleges, allied industry, and other interested persons. The mission of the *Ohio Meat Goat Industry Task Force* is to enhance the production and marketing of meat goats through education and practical experience. The objectives of the *Ohio Meat Goat Industry Task Force* are to: 1) identify and access emerging ethnic markets having a preference for goat meat in their diet, 2) develop producer networks, alliances and/or cooperatives to meet the demands of emerging markets, and 3) provide leadership for education and research.

Education Extension members of the task force have developed the *Ohio Meat Goat Production and Budgeting Fact Sheet* as a guide for establishing this value added enterprise. Agents have designed and conducted regional workshops, seminars, and on-farm tours to transfer knowledge to 1200 participants. Education, production, and marketing topics are discussed in the *Buckeye Meat Goat Newsletter*. The website http://south.osu.edu/cle/news.htm has been developed to enhance the exchange of production and marketing information.

Building leadership capacity Leadership development has been a primary objective of the Ohio Meat Goat Task Force. Producer members have been instrumental in the formation of the *Buckeye Meat Goat Association*. This group has developed by-laws and articles of incorporation to promote and market commercial meat goat in Ohio. Three producer-driven marketing networks have been established. Recent group marketing efforts have increased average revenue by $10.00/100kg and reduced marketing cost by $3.00/head. Task force members are developing leadership among emerging ethnic and faith-based consumers as a social approach to building infrastructure of the meat goat industry for market development of fresh chevon. This foundation infrastructure will create value-added opportunities for refugees in our urban centres and small farms in Ohio. Additionally, economic development in the creation of agricultural jobs will do much for community development in the rural/urban interface.

Developing an industry The task force has received $63,000 in Research and Extension grants to conduct on going feasibility studies of ethnic markets, Ohio's processing infrastructure, and development of farmer/consumer cooperatives. On-farm meat goat research encompasses breed comparisons, forage utilisation, and developing benchmark data. Progress continues in the ability to market a fresh and safe product directly to emerging ethnic and faith-based consumer populations to capture the most value. Behavioural changes include an increase in farmers producing for emerging markets, an increase in communication between producers and markets, and coordination for consumers, retailers, and producers through functional marketing partnerships that fit the social and ecological paradigm.

Conclusion The Ohio Meat Goat Task Force is a model for engaging resources and building leadership capacity to generate income and enhance sustainability of farm businesses. The collaboration of faculty, producers, allied industry, and ethnic cultures combines expertise and leadership with applied experience to foster entrepreneurship. Research has identified ethnic market preferences, processing infrastructure and capacity, and economically viable production systems. Education provides farm businesses capacity to build leadership, share knowledge, and network resources to capture value-added marketing opportunities.

Cashmere marketing is a new income source for Central Asian livestock farmers

C. Kerven[1], S. Aryngaziev[2], N. Malmakov[2], H. Redden[1], A. Smailov[3] and K.A. Galvin[4,5]
[1]*Macaulay Institute, Aberdeen AB15 8QH, UK, Email: Kerven_Behnke@compuserve.com, [2]Mynbaevo Sheep Breeding Institute, Kazakstan Scientific Centre for Livestock and Veterinary Research, Kazakstan, [3]Institute of Pasture and Fodder, Kazakstan Scientific Centre for Livestock and Veterinary Research, Almaty, Kazakstan, [4]Natural Resource Ecology Laboratory, Colorado State University, Fort Collins, CO 80523-1499 USA, [5]Department of Anthropology, Colorado State University, Fort Collins, CO 80523-1787 USA*

Keywords: Kazakstan, farmers, goats, cashmere, marketing

Introduction Some indigenous goats in the Central Asian republics of Kazakstan, Kyrgyzstan and Tajikistan produce good quality cashmere (Millar 1986). International processors have recently been buying this cashmere. (Kerven *et al.,* 2005), but Central Asian producers are not equipped to take full advantage of these new marketing opportunities. The U.S. AID Global Livestock-Collaborative Research Support Program project, "Developing Institutions and capacity for sheep and fiber marketing in Central Asia" is working to increase the income of small-scale livestock farmers through improved cashmere marketing.

Methods Project activities are targeted at the Kyzl Orda region of southwest Kazakstan, semi-arid rangelands (100-300 mm annual precipitation) in which high quality cashmere goats have been identified. Information is collected on the marketing chain from farmers to local traders and to international processors. Farmers are shown demonstrations of cashmere harvesting by combing and fibre quality assessment, and receive global market information. The project is also assessing the quality of cashmere according to international standards and training national livestock scientists.

Results Central Asian producers are generally unable to distinguish good from poor quality cashmere. Raw cashmere requires particular processing techniques by industrial processors who have tight specifications for quality and reward quality with higher prices. Producers and local traders lack global market information on demand and prices. Producers sell individually to traders rather than pooling their fibre to gain higher prices. Mean prices in 2004 in the project area were $2-3/kg for whole fleeces. Mongolian farmers received on average $22/kg and Chinese farmers got $31/kg for raw combed and sorted cashmere (ACDI/VOCA 2004; Schneiders 2003/4). Strong international demand continues for cashmere. Central Asian countries could learn from Mongolia's experience, where herders now gain their main income from cashmere sales (World Bank 2003). Poorer farmers in remote mountainous and desert regions tend to have more goats than sheep, partly because goats reproduce faster than sheep, often producing twins and kidding twice a year. They also cost less to feed over winter than fine wool sheep. In Kazakstan, goat populations have been rising since independence, from 700,000 in 1992 to 1.4 million in 2003. Goats are preferred by poorer farmers trying to restock since the reduction of sheep numbers from 34 million in 1992 to 10 million in 2003.

Conclusions Kazak livestock farmers have the potential to increase their incomes through improved marketing of high value cashmere, especially where alternative income sources are extremely scarce. Currently, such farmers rely on selling live animals, which is not sustainable for very small flocks. Local goat breeds thrive in the semi-arid shrub ecology of Kazakstan's rangelands. In 2004 traders were offering Kazak farmers from $11-19/kg for combed and sorted cashmere, which very few farmers were able to supply. This is a sharp contrast to $0.20/kg offered by traders for coarse sheep wool produced by local breeds. Based on the respective amounts of cashmere and wool produced, one Kazak goat could have yielded a gross income of approximately $4.75, while income from the wool of a coarse-woolled sheep would be $0.50. Enabling Central Asian farmers to realize the full value of their cashmere requires farmer training and dissemination of market information, improved goat breeding, cooperative marketing, capacity-building of national scientists and better connections to international markets.

References
ACDI/VOCA Gobi Regional Economic Growth Initiative (2004). www.pactworld.org/programs/country/mongolia/mongolia_gobi.htm
Kerven, C., S. Aryngaziev, N. Malmakov, H. Redden & A. Smailov (2005). Cashmere marketing: A new income source for Central Asian livestock farmers. GL CRSP Research Brief 05-01-WOOL. www.glcrsp@ucdavis.edu
Millar, P. (1986). The performance of cashmere goats. *Animal Breeding Abstracts*, 54, 181-199
Schneiders (2003/2004). www.gschneider.com/marketreports
World Bank (2003). Impact of institutional and trade policy reforms: Analysis of Mongolia's cashmere sector. http://poverty.worldbank.org/files/14794_Mon_Csm_Mixed_Rep_Mar03_BBL.pdf

Production strategies of livestock herders in the grasslands of Kazakhstan: implications for the marketing of fine fibres

K.A. Galvin[1,2], C. Kerven[3], R.B. Boone[2] and A. Smailov[4]
[1]Department of Anthropology, Colorado State University, Fort Collins, Colorado, USA, Email: Kathleen.Galvin@colostate.edu, [2]Natural Resource Ecology Laboratory, Colorado State University, Fort Collins, Colorado, USA, [3]Macaulay Institute, Aberdeen, Scotland, [4]Institute of Pasture and Fodder, Kazakstan Scientific Center for Livestock and Veterinary Research, Almaty, Kazakstan

Keywords: Central Asia, pastoralism, household production strategies

Introduction Goat populations have been rising in Kazakhstan over the past ten years since independence and goats are preferred by farmers trying to restock. Quality of cashmere production is the key to profitable and sustainable sales to world markets for this luxury good. However, Kazakhstan did not develop a cashmere industry in the Soviet period so today goats are sheared rather than combed and little profit is made from cashmere. Goats, as well as sheep and camels are currently multi-purpose animals providing income from sales of animals, cashmere, milk and meat. This will change as the terms of trade change for high quality cashmere and households comb for fine down.

Methods Research on village household surveys in three villages (60 households) of Zhanekurgan District among livestock herders was conducted in 2004. The research was carried out under the auspices of the U.S. AID Global Livestock-Collaborative Research Support Program (GL-CRSP) project, "Developing institutions and capacity for sheep and fibre marketing in Central Asia".

Results In Zhanekurgan ownership of flocks are dominated by goats. Table 1 shows the household ownership by sheep and goats by villages. Sheep command a higher market live weight price than goats, as mutton is the preferred meat for Kazaks. Higher prices for live sheep are obtained in the regional market city of Turkestan, where consumers have more income and can afford to buy more sheep. In Turkestan prices per live weight were $1.25/kg for sheep and $0.65/kg for goats. But in the local district town market, live weight prices for sheep were only $0.10/kg higher than for goats. In the 2003 marketing season, coarse sheep wool sold for only US 0.20 cents per kg while goat fibre containing cashmere was sold for US $2.0 per kg. The fibre returns per goat are approximately four times higher than per sheep: - coarse sheep produces 2 kg wool worth $0.50 - goat produces 1 kg fibre worth $2.0.

Table 1 Household ownership of sheep and goats

Village	Mean private sheep	Range.	Mean private goats (no. h/holds)	Range
Kruash	56	0-400	34	12-180
Kosenka	71	7-650	31	0-250
Jailma	134	0-700	89	5-400

Conclusions The GL-CRSP project activity on livestock product marketing is making Central Asian livestock researchers, government officials, local traders and producers appreciate the international market value of cashmere and camel hair. However, households have yet to realize the increased value associated with fibers as their production strategies are geared towards meat. The long-term goal of the research is to increase income of small-scale livestock farmers by annually harvesting high value cashmere and camel hair.

Adding value to grasslands through certified organic beef production

G.A. Ferreira, O. Pittaluga, C. Mas, S. Revello and R. Tellería
Instituto Nacional de Investigación Agropecuaria (INIA), Ruta 5 km 386 Tacuarembó, Uruguay, Email: gferre@inia.org.uy

Keywords: organic meat, production systems, adding value

Introduction New demands are being made for safe beef from low input production systems (low input of energy, pesticides, other chemically synthesised products and hormones and GM free) by the main retailers and consumers (Howard, 2004). These present an opportunity for adding value to grazing production systems and to reinforce relationships among farmers, agro-industrialists, exporters and consumers and to show advantages of positive externalities of grasslands (Meister, 2001).

Materials and methods A project with two slaughterhouses located in the North of Uruguay was developed in order to produce certified organic beef. The main objective was to certify extensive grassland production systems where the animals grazed mainly native grasslands, i.e. the most common system to produce meat in Uruguay (Ferreira & Pittaluga 2002). The organic programme started with INIA support working with PUL and Tacuarembó slaughterhouses in 2000 and the number of certified farms to August 2004 is shown in Table 1.

Table 1 Certified organic beef production programs

Slaughterhouses	PUL	Tacuarembó	Total
Farms	130	140	270
Hectares	323903	419813	743716
Beef cattle	244641	258461	503102
Sheep	284920	326166	611086
AU/ha	0.74	0.62	0.67

Source: PUL & Tacuarembó Data bases of farms certified by SKAL International

Results Currently, the two plants are exporting organic certified meat mainly to Europe and USA and the average premium price is around 12% for the farmers. The market niche is growing and the market forecast for the next few years is that demand will continue to increase (Regmi & Dick 2001). The programme has been approved by SKAL International, KRAV Sweden and USDA Organics to certify that the meat produced is organic beef. Most of the foreigner brokers and buyers are amazed at the quality of the natural environment of Uruguay to produce quality grassland beef. This reinforces the results obtained by Environmental Sustainability Index (ESI) "that is a measure of overall progress towards environmental sustainability, developed for 142 countries, that permits cross-national comparisons of environmental progress in a systematic and quantitative fashion. It represents a first step towards a more analytically driven approach to environmental decision-making" (Columbia University 2002). Uruguay has been ranked as number 6 because of its clearly natural resources. Therefore, this programmes adds value to grasslands in different ways, i.e. by i) increasing the economic value of the products, ii) improving cooperation between farmers, slaughterhouses and exporters iii) applying sustainable technology and methods approved by the organic protocol iv) supporting decision making to enhance positive externalities of grasslands to the environment

Conclusions The organic beef programme implemented in Uruguay by two slaughterhouses working with INIA technical assistance, shows that is possible to develop an innovative programme supported by a strategic alliance among agricultural researchers, farmers and slaughterhouses.

References
Columbia University (2002) Environmental Sustainability Index http://www.ciesin.columbia.edu/indicators/esi/
Ferreira, G. & O. Pittaluga (2003). Evaluación Económica de Sistemas de Engorde Bovino y Ovino para la
Howard, M., (2004). Developing the Uruguayan Food Chain. Draft Report. Project N°: GRP-P89 Issue: 1.0
    Producción de carnes Diferenciadas. En: Seminario de Actualización Técnica: Producción de Carne Vacuna
    y Ovina de Calidad. Serie Actividades de Difusión. N° 317
Meister, A (2001) Dilemma: increase in human food production or use of grasslands for environmental and / or
    social purposes. *Proceedings of the Nineteenth International Grassland Congress*, Brazil, 1013-1014.
Regmi, A. & J. Dick (2001): Effects of Urbanization and Global Food Demand, in Changing Structure of Global
    Food Consumption and Trade. In Regmi, A. (ed), Agriculture and Trade Report. WR 501-1, USDA, ERS.

Alternative land use options for Philippine grasslands: a bioeconomic modeling approach using the WaNuLCAS model

D.B. Magcale-Macandog, E. Abucay and P.A.B. Ani
Institute of Biological Science, University of the Philippines Los Baños, College, Laguna, Philippines 4031,
Email: macandog@pacific.net.ph

Keywords: *Imperata cylindrica*, *Eucalyptus deglupta*, hedgerow, WaNuLCAS model, land-use change, bioeconomic modelling

Introduction In the Philippines, pure grasslands occupy 1.8 million ha and another 10.8 million ha (33% of the country's total land area) is under extensive cultivation mixed with grasslands and scrub. Most of these grasslands are under-utilised and dominated by *Imperata cylindrica*. *Imperata* grasslands generally represent areas of degraded soils that are acidic, low in organic matter and susceptible to erosion. However, conversion of these grassland areas into upland farms planted to annual crops and perennial trees is proliferating at a fast rate. This is triggered by the interacting factors of rapidly increasing population, the system of landholding, scarcity of jobs and the declining arable area in the lowlands.

Materials and methods The biophysical and economic consequences of land-use change from *Imperata* grasslands to continuous maize and agroforestry (*Eucalyptus deglupta* + maize hedgerow) systems were assessed using bioeconomic modeling. The Water, Nutrient and Light Capture in Agroforestry Systems (WaNuLCAS) model (van Noordwijk, Lusiana & Khasanah, 2004) was used to examine tree and crop growth and productivity, soil fertility changes, soil erosion and water balance. The different land-uses were modeled in the sloping upland areas of Southern Philippines characterised by rugged topography, clayey soils and annual rainfall of about 2500 mm.

Results Simulation showed that the dynamics of nutrients (N and P) in the systems differ. More than half of the total nitrogen in the three systems is tied up in the soil organic matter (SOM). Leaching and lateral flow are the main avenues of nitrogen losses in the three systems. Much of the P (90%) is tied up in SOM and immobilised in the *Imperata* grasslands.

Results of modeling the water balance of the three systems showed that *Eucalyptus*-maize hedgerow system had the highest subsurface flow and surface run-off (Table 1) compared with the other two systems. Maize cropping and *Imperata* grassland had significantly more drainage compared with the agroforestry system.

Simulation results also showed significant competition for light between trees and crops under the *Eucalyptus*-maize hedgerow system. Maize yield was initially higher in the continuous annual cropping system (2.4 t/ha) than under the *Eucalyptus*-maize hedgerow system (1.8 t/ha).

The benefits obtained from the maize cropping system is the grain yield, from the *Eucalyptus*-maize hedgerow system the benefits are maize grain yield and *Eucalyptus* timber, while biomass from the *Imperata* grassland is the harvested and sold as roofing material. Cost benefit analysis showed that the *Eucalyptus*-maize hedgerow system had the highest NPV after 9 years of simulation (P 304,323), compared with the *Imperata* grassland (P 10,722) and continuous maize (P 20,872).

Table 1 Water balance (li /m^2) in the three land use systems

Component	Agroforestry	Continuous crop	*Imperata* grassland
Surface	18,311	18,284	18,243
Subsurface flow	214,206	204,186	205,121
Drainage	4,151	153,844	156,255
Soil	274	9,555	7,331
Canopy	1,962	342	342
Crop	4,753	21,514	21,514
Tree	9,237		
Total	249,150	407,720	408,800

Conclusion This study has shown that land-use change from *Imperata* grasslands or continuous maize cropping system to *Eucalyptus*-maize hedgerow systems provide significant improvements to a range of biophysical and economic measures of productivity and sustainability.

Reference

Van Noordwijk, M., B. Lusiana & N. Khasanah (2004). WaNuLCAS version 3.1, Background on a model of water nutrient and light capture in agroforestry systems. International Centre for Research in Agroforestry (ICRAF), Bogor, Indonesia.

Are new farming systems based on perennial pastures in south west Australia more profitable?

P. Sanford[1] and J. Young[2]
[1]West Australian Department of Agriculture and CRC for Plant-based Management of Dryland Salinity, 444 Albany Highway, Albany, Australia 6330, Email: psanford@agric.wa.gov.au, [2]Farm Systems Analysis, RMB 309, Kojonup, Australia 6395.

Keywords: perennial plants, whole farm profit, salinity, south west Australia

Introduction Traditional farming systems in south west Australia based on annual plants have been shown to use insufficient water leading to excess leakage below the root zone, groundwater rise and eventually salinisation of the landscape. Introduction of deep-rooted summer active perennial plants can significantly increase water-use thus reducing the risk of salinisation. However the adoption of perennials by farmers is also dependent on their effect on economic factors. This paper reports an analysis of the impact of perennials on whole farm profit.

Materials and methods A model of the farming system known as MIDAS was used to undertake this analysis (Morrison *et al.*, 1986). The model was paramatised with a representative 2000 ha farm in the Albany Eastern Hinterland catchment receiving 600 mm annual rainfall and comprising of three soil types, deep sand, shallow sand over clay and deep sand over clay. Livestock enterprise was a self replacing Merino flock utilising surplus ewes for crossbred lamb production. Three different systems where analysed. 1. Traditional farming system - 30% crop, 70% annual pasture subterranean clover (*Trifolium subterraneum*) based, stocked at 8.5 dse/ha. 2. Current best practice - 30% crop, 23% annual pasture subterranean clover based, 47% perennial pasture either lucerne (*Medicago sativa*) alone or kikuyu (*Pennisetum clandestinum*)-subterranean clover, stocked at 10 dse/ha. 3. Future farming system - 30% crop, 70% perennial pasture either lucerne alone or kikuyu-subterranean clover or tall fescue (*Festuca arundinacea*)-subterranean clover, stocked at 12 dse (dry sheep equivalents)/ha. Leakage values were estimated using a farm scale hydrologic model (Beverly *et al.* 2003).

Results Farm profit and leakage results are presented in Table 1. As expected there is a substantial decrease in the leakage of water below the root zone (69 to 23 mm) as the area of the farm under perennials increased from 0 to 70%. Encouragingly farm profit increased ($32 to $104 per ha per yr) as leakage decreased. Increased profit was driven by higher pasture yield, superior feed quality in summer and autumn, reduced supplementary feed and higher stocking rates. The optimum area of the farm under varying perennial options is presented in Figure 1. With only one perennial option in this case, lucerne, the optimum proportion of the farm is around 25% in terms of profit. If another perennial option with a different growth pattern, such as kikuyu, is added the optimum increases to about 50% and with a third the optimum reaches to 75%. This result is quite promising as large areas of the landscape need to be planted to perennials to minimise the impact of salinisation.

Table 1 Estimated farm profit ($/ha per yr) and leakage (mm) beneath contrasting farm systems

	Traditional farming system	Current best practice	Future farming system
Farm profit ($/ha/yr)	$32	$69	$104
Leakage below root zone (mm)	69 mm	46 mm	23 mm

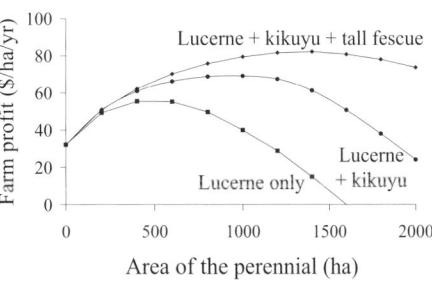

Figure 1 The effect of varying the area of perennials on farm profit

Conclusions This analysis suggests that it is possible for farmers to substantially increase farm profit while reducing the risk of salinisation through planting a high proportion of their farm to perennial pastures.

References
Beverly, C., A. Avery, A. Ridley, & M. Littleboy (2003). Linking farm management with catchment response in a modelling framework. *Proceedings of 11th Australian Agronomy Conference*, Victoria, Australia, 1-4.
Morrison, D., R. Kingwell, D. Pannell, & M. Ewing (1986). A mathematical programming model of a crop-livestock farm system. *Agricultural Systems*, 20, 243-268.

Lucerne production and economics on stakeholder ranches on Coastal Plain acid soils

V.A. Haby[1], A.T. Leonard[1], G.M. Clary[2], F.M. Rouquette, Jr.[1] and L.A. Redmon[2]
[1]Texas A&M University System, Texas Agricultural Experiment Station, P.O. Box 200, Overton, Texas, 78684, USA, Email: v-haby@tamu.edu, [2]Texas Cooperative Extension, P.O. Box 38, Overton, Texas, 78684, USA

Keywords: *Medicago sativa*, Alfisols, Ultisols, soil acidity

Introduction Research using small plots determined factors restricting lucerne (*Medicago sativa*) production on Coastal Plain soils of the southern USA. These acid soils, primarily Ultisols and Alfisols, become deficient in B when limed to pH 6.8 or 7.0 for lucerne. Other production-limiting soil problems include poor drainage and aeration, subsoil acidity, and low natural fertility. The objective of this study was to evaluate stand longevity on hectare-size stakeholder fields when growth-limiting factors were eliminated by site selection and soil treatment.

Materials and methods Selection criteria developed from small-plot studies were used to locate sites suitable for lucerne. Soil samples to 1.2 m were evaluated for adequate drainage and aeration. Each 0.3 m depth was tested for pH of 5.5 or greater to avoid potential phytotoxic levels of subsoil Al. The surface 0.15 m depth was analyzed for limestone and fertiliser requirements including B. Selected sites were treated according to guidelines developed for lucerne production on Coastal Plain soils. Half of each site was seeded to 'Amerigraze 702' and half to 'GrazeKing', both at 28 kg/ha. GrazeKing, a fall dormancy 5 variety, and Amerigraze 702, a fall dormancy 7 variety, were selected based on superior performance in an earlier grazing-tolerant variety evaluation. At low soil-test P, phosphorus was applied at 134 kg P_2O_5/ha for the seedling year. Potash was applied preplant and after the second and fourth cuttings with the annual application exceeding 335 kg/ha. Sulphur and magnesium were applied after the fourth cutting with total rates approximating 80 and 40 kg/ha, respectively. Yield-estimate samples from each regrowth were clipped from four random, one-metre quadrats for each variety.

Results Yield and forage density indicated the production potential and sustainability of lucerne on Coastal Plain soils (Table 1). Hay yields ranged from about 32 to 43 t/ha among the four locations and primarily were attributable to soil selected and stakeholder management. Amerigraze 702 was more sustainable compared to GrazeKing at all locations. Net returns in dollars (US) after deduction of production costs served as a significant aspect of technology transfer and educational enhancement for all stakeholders and ranged from about $ 1616 to $ 3437/ha.

Table 1 Yield, stand density, establishment costs, hay value, production expenses, and net returns from four years of lucerne production on cooperating stakeholder ranches in the USDA-SARE funded research program

Yields, inputs/returns 4-yr totals	Griffin Ranch	Prud'homme 7-P Ranch	Taylor Ranch	Riley Ranch
Yield, t/ha @ 88% D.M.[†]	39	43	37	32
Amerigraze 702, stand after 4 years, %	47	57	76	54
GrazeKing, stand after 4 years, %	13[‡]	53	64	41
	------------------------------$/ha------------------------------			
Establishment costs	573.96	624.25	805.87	846.69
Hay value (@ $148.81/t)	6211.35	6911.88	5904.45	5063.82
Production expenses[§]	3357.35	3474.50	3302.12	3447.14
Net returns	2854.00	3437.38	2602.33	1616.68

[†] Average for Amerigraze 702 and GrazeKing varieties
[‡] GrazeKing on wetter soil with extensive invasion of common bermudagrass (*Cynodon dactylon*)
[§] Includes production costs, custom hay harvesting and hauling, interest, and overhead (machinery and equipment, land, and 4-year prorated establishment costs)

Conclusions Lucerne production on these Coastal Plain soils was profitable and sustainable for at least four years where, in previous attempts to grow lucerne, stands would not last more than two years. Keys to sustainable lucerne production include selection of well-drained sites with subsoil acidity above pH 5.5 to 1.2-m deep to avoid phytotoxic Al, fertilisation of these soils with B after liming to pH 7, and adequate fertilisation with P, K, and other plant-essential elements determined to be deficient by soil analysis. With critical site selection and proper management, lucerne hay production can offer substantial economic returns on Coastal Plain soils.

Economic efficiency of the production in dairy sheep breeding

M.V. Stoykova

Institute of Forage Crops, 89 General Vladimir Vazov St, 5800 Pleven, Bulgaria, Email: mayast6@yahoo.com

Keywords: dairy sheep farms, economic efficiency

Introduction Bulgaria is a country of rich grassland and forage resources and has a long tradition in sheep breeding. In spite of these conditions, the economic efficiency of milk production by sheep is low (Panayotov & Kostadinova, 1999). The objective of the study was to determine the real economic efficiency of production at twelve Bulgarian dairy farms.

Materials and methods During the period 1999-2003 twelve dairy farms were studied. In the farms, the sheep ('Blackface Pleven' breed) were housed in stables for 160-180 d and at pasture for 200-210 d. Economic analysis was carried out following the methodolgy of Koteva (2004). The data used were from primary bookkeeping accountancy of the farms, as well as information from the Regional Centres for Sheep Breeding and Reproduction in the towns of Pleven and Lovech.

Results The average values for the experimental period of characters to describe the economic efficiency of production at the farms are presented in Table 1. The average milk productivity and gross output (the milk produced multiplied by its market price) are low. There is a background of high costs so that the net income per ewe and profitability are low. The results demonstrate a low economic efficiency (as expressed by profitability) per ewe at the farms studied. The causes may differ by farms and relate to factors internal to the farm and external factors. Technical factors including poor feeding resulted in reduced milk productivity and contributed to poor economic results. However, the main reason is the high prime cost of milk, which almost equals the market price of milk. It is noteable that Farm 1 had by far the highest milk yield, net income and profitability with the lowest prime cost of milk. Further analysis for the reasons for success at that farm would be worthwhile.

Table 1 Productivity, production costs (in lev) and economic efficiency for twelve Bulgarian dairy sheep farms (1999-2003)

Farm	Milk yield (kg/ew)	Gross output (lv/ewe)	Production costs (lv/ewe)	Net income (lv/ewe)	Prime cost of milk (lv/ewe)	Profitability (%)
1	127.9	233.52	135.74	97.78	0.62	72.03
2	55.4	108.43	116.28	-7.85	0.87	-6.75
3	69.3	109.44	120.00	-10.56	0.89	-8.80
4	77.1	121.88	112.86	9.02	0.78	7.99
5	66.6	114.09	109.05	5.04	0.80	4.62
6	66.7	109.67	114.13	-4.46	0.87	-3.91
7	65.9	133.90	121.58	12.32	0.76	10.13
8	48.9	97.69	105.28	-7.59	0.91	-7.21
9	50.6	95.61	105.92	-10.31	0.93	-9.73
10	93.8	130.94	107.32	23.62	0.71	22.01
11	66.2	106.04	105.84	0.20	0.84	0.19
12	68.4	98.25	107.42	-9.17	0.92	-8.54
Mean	66.6	121.62	113.45	8.17	0.83	6.00

Conclusions This study suggests that if either average milk yield, or its market price increase, then better economic results and higher efficiency of production will be obtained. The farmer can not directly influence the present low price of milk in Bulgaria, so that the only way to improve profitability is to increase technical efficiency. In this context the efficient use of forage resources is particularly important.

Acknowledgments The author wishes to thank cordially Mr. Ivan Katerov for the critical reading, helpful suggestions and anglicising of the manuscript.

References

Panayotov, D. & N. Kostadinova (1999). Comparative economic analysis of incomes in the breeding of sheep from different productive direction, *Agricultural Economics and Management* (Sofia), 44, 31-33.

Stoykova, M. (2004). Economic efficiency of production in dairy sheep breeding. *PhD thesis*, (Sofia), 34.

Koteva, I., (2004) Analysis of the efficiency of some organizations in agriculture, *Agricultural economics and management* (Sofia), vol.1, 40.

A study of labour use on Irish grassland farms specialising in suckler production

H. Leahy[1,2], E.G. O' Riordan[1] and D.J. Ruane[2]
[1]Teagasc, Grange Research Centre, Dunsany, County Meath, Ireland, Email: southwestcavancdp@eircom.net,
[2]Dept. of Agribusiness, Extension & Rural Development, University College Dublin, Belfield, Dublin, Ireland

Keywords: labour study, suckler cows, grassland farms

Introduction Important structural changes have taken place within the agricultural workforce in recent years. There has been a persistent decline in the proportion of the total workforce engaged in agriculture. Demand for labour by farmers has become a major issue, and if this shortage continues it will have a significant impact on the development of Irish farms in future (Ruane et al., 2001).

Materials and methods Data were collected from 115 spring calving suckler farmers distributed evenly across the east and west of the country. Each farmer was randomly assigned to 1 of 4 groups for data collection. Each group was allocated a week in the month during which they recorded time spent undertaking predefined tasks using the timesheet method. Starting and finishing time for each farm task was recorded by the farmer daily over 3 consecutive (incl. weekends) days. Task duration, length of working day and discretionary time was recorded.

Results Figure 1 outlines the labour input for the 115 farms, (mean size of 54 cows and 72 hectares. As the grazing season progressed the time allocated to feeding stock declined from 2.24 hours per day in March to 0.23 hours in July, and began to increase in August as suckler stock were supplemented with concentrates at grazing. Time allocated to feeding stock continued to increase through the autumn reaching 1.28 hours in November as stock were housed for the winter. Cleaning tasks declined initially only to increase slightly in June as silage pits were made ready to receive silage, it declined again and began to increase again in October and November as stock were housed for the winter. Animal husbandry

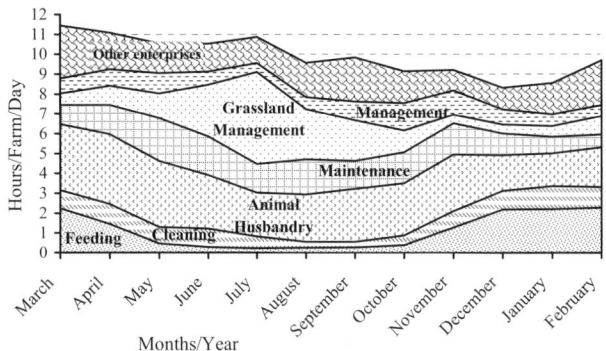

Figure 1 Total labour input (hours per farm per day) required to carry out predefined tasks on a sample of 115 suckler farms

tasks were at their highest in March, April and May, averaging 3.37 hours per day over the 3 months. Animal husbandry tasks reached a peak in April as the calving season progressed and calf numbers increased on farm. Animal husbandry tasks increased again in September as autumn calving and weaning began. Grassland management increased in May and June, peaked in July at 4.63 hours/farm/day, with the majority of time being taken up by silage making, topping and slurry spreading tasks, grassland management remained high in August and September, and decreased over the winter period as stock were taken off the land. Farm management tasks changed minimally throughout the farm year, increasing slightly in September, October and November (as farmers traded stock). Other enterprises were high in February and peaked in March at 2.25 and 2.66 hours per farm per day respectively. This was mainly as a result of lambing and sowing cereals (50% of the sample had sheep as another enterprise, and 27% of the sample had cereals as another enterprise).

Conclusions Total labour input on the farm peaked in the spring-time. Animal husbandry emerged as the most time consuming task over spring. Labour input was elevated in July also; this may be attributable to the unseasonably wet summer of 2002, where farmers contended with inclement weather to harvest silage and spread slurry. Labour-use decreased after July. The goal for the typical beef farmer is to improve living standards by reducing stress or by affording the opportunity for off farm income.

References
Ruane, D.J. & J.F. Phelan (2001). Making labour more attractive on Irish Farms: A case study of South Tipperary in the Republic of Ireland. *Proceedings of the Fifteenth European seminar of extension education, Congress*, Wageningen, Netherlands, August 2001, 1-15.

Economic efficiency of milk production for different sizes of modern dairy farms in Hamadan Province, Iran

Y. Rostami[1], M. Koopahi[2] and S.A. Mohaddes[3]
[1]Animal Science and Forest & Rangeland Research Institutes, Iran, Email: yousef_rostami@yahoo.com, [2]Agricultural Economics Department, Faculty of Agriculture, University of Tehran, Iran, [3]Khorasan Natural Resources and Livestock Research Center, Mashhad, Iran

Keywords: efficiency, farm size, production function

Introduction Iran has limited water resources and attention has been focused on selecting agricultural sub-sectors, such as dairy farming, that maximise economic returns to this limited resource. The study measures the levels of production efficiency of a sample of modern farms located in the Hamadan province of north-west Iran. Though the production levels of modern farms are higher than those of traditional farms, levels of production efficiency are in many cases well below potential resulting in lower profit margins and less investment in dairy farming. This study was concerned with quantifying efficiency in a sample of modern dairy farms.

Materials and methods A sample of 148 modern dairy farms was included in the study. Based on the number of cows, the farms were divided into three groups: a) <50 heads b) 50-100 heads and c) >100 heads. The study used a classic Cobb-Douglas production function. As indicated in Figure 1, efficiency levels were decomposed into technical and allocative components according to the method of Farrell (1957).

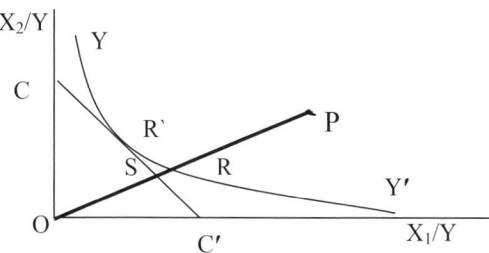

Figure 1 Technical and allocative efficiency

The curve YY' is the isoquant representing the efficient combination of inputs x_1 and x_2 needed to produce a given level of output, whilst CC' is the iso-cost-line with slope equal to the ratio of unit input costs of x_1 and x_2. The tangency point between the isocost line and the isoquant is the least-cost combination of inputs x_1 and x_2 to produce the level of output represented by isoquant YY'. The rate of technical efficiency of an individual producer can be determined by measuring their position relative to the 'best practice' frontier defined by the isoquant. If we consider the producer represented by point P, the technical efficiency is equal to OR/OP. If this ratio equals 1 then the producer is operating at a point on the isoquant and is said to be technically efficient. Conversely, a ratio of less than one indicates that the producer is technically inefficient as the same level of output could be produced using less input. Allocative efficiency measures whether or not the producer is operating at the point of least cost combination of inputs. In the case of producer P this equals OS/OR. The producer is said to be allocatively inefficient if the level of output represented by the isoquant could be produced at lower cost. An overall measure of economic efficiency comprises the composition of the allocative and technical efficiency components and is represented as OS/OP.

.

Results The production functions and rates of economic efficiency estimated in the study differed according to farm size classification. All three models provided satisfactory explanation of the production data with R^2 higher than 0.5 in each case. The variables feed cost, herd size and labor cost were the most important factors that affected milk production in different farm size. Based on the estimated models, economic efficiency was 0.62, 0.78, and 0.72 for the farm sizes below 50, 50-100 and more than 100 cows respectively.

Discussion and conclusions This study indicates that the three groups of dairy farms lacked adequate managerial knowledge on production methods and this was the primary cause of low efficiency. In order to improve efficiency, the manpower of the farms require adequate training. Furthermore, there is variation among farms in the use of technology such as artificial insemination. Consequently, there are opportunities for many farmers to improve performance through the use of improved technology and infrastructural facilities. In addition, appropriate extension advice is required to assist farmers in improving their economic efficiency.

Reference

Farrell, M.J. (1957). The measurement of productive efficiency. *Journal of the Royal Statistical Society*, 120 (Series A (General), Part III), 253-281.

Agro-pastoralists concerns over the *Prosopis* tree: the case of the IlChamus of Baringo District, Kenya

A.A. Aboud, F.W. Lusenaka, C.I. Lenachuru and P.K. Kisoyan
Department of Natural Resources, Egerton University, P.O. Box 536, Njoro, Kenya, Email: eu-crsp@africaonline.co.ke

Keywords: *Prosopis*, agro-pastoralists, attitudes, environmental degradation

Introduction The *Prosopis* tree was introduced to the arid and semi-arid lands of eastern Africa in the 1970s, through governmental forest development agencies to curb environmental degradation and provide fodder for small stock. A number of other benefits were also then attributed to the tree. However, the *Prosopis* tree has turned out to be a cause of serious concerns, as it has invaded, dominated and almost totally removed all grass and short vegetation species from pasturelands. In Baringo District of Kenya, *Prosopis* has been the worst enemy of the local IlChamus agro-pastoralists (Lenachuru, 2003), who have now raised much concern over the species, calling for its complete eradication, and threatening to sue the government for damages caused by the tree. This paper examines the case against *Prosopis* species, based on the agro-pastoralists' perspectives, and the numerous benefits of the tree that the agro-pastoralists apparently unaware of, or simply refuse to attach value to.

Materials and methods An exploratory survey solicited knowledge, opinions and attitudes on the merits and demerits associated with the *Prosopis* tree. This used 73 key informants and resource persons among the IlChamus. The findings of the survey were presented at a workshop held to discuss and generate recommendations on integrated management of the *Prosopis* tree in Kenya. A total of 65 participants, including agro-pastoralists, researchers and scientists, and other stakeholders, attended the two-day workshop to share knowledge and experiences, and to develop approaches for effective management of the invasion, spread and control of the tree, and its sustainable utilisation by the pastoralists. This paper is concerned with the demerits, as argued by the agro-pastoralists, relative to the merits as presented by the experts.

Results While the IlChamus' perception of the demerits of the species is clear and detailed, their perception of the merits are less so. The numerous merits of the species, as advocated by the experts, include the fact that the tree provides nutritious fodder for small stock, flowers vigorously and continuously making it ideal for bee pasturage, (since the nectar makes good white honey), and provides good wood, timber and numerous products that offer commercialisation opportunities for socioeconomic improvement of households. Ecological benefits and various beneficial domestic uses were also listed. On the other hand, the agro-pastoralists strongly argued against the species, as evidenced by their opinions, listed in Table 1.

Table 1 Agro-pastoralists' opinion about *Prosopis*

Opinion about Prosopis	Strongly Agree	Agree	Not Sure	Disagree	Strongly Disagree
It is fast spreading and invasive	72%	26%	1%	1%	-
It eliminates other plants	81%	19%	-	-	-
Its creates bare ground	83%	16%	1%	-	-
It is injurious to livestock	64%	28%	5%	2%	1%
It is injurious to humans	35%	13%	26%	17%	9%
It should be totally eradicated	87%	11%	-	2%	-

Conclusion Despite the array of merits in favour of the *Proposis* tree presented by the experts, the IlChamus agro-pastoralists seemed adamant and were not persuaded. This is probably because of their bad experiences with the tree and the realisation that the associated economic and the ecological benefits are unachievable in the short term. Hence their call for complete eradication of the species from their pastures, farmlands and homesteads. In essence, they seem to concur with Cable (1977) who concluded that *Prosopis* roots exert a stronger "pull" on the soil water than do the grasses. This, presumably is the fatal characteristic that makes *Prosopis* a killer plant and a serious threat to the existence of the IlChamus community.

References
Cable, D.R. (1977). Seasonal use of soil water by mature velvet mesquite. *Journal of Range Management,* 30, Jan. 1977.
Lenachuru, C. I. (2003). Impacts of *Prosopis* Species in Baringo District. In: S.K. Choge and B.N. Chikamai (eds.) *Proceedings of Workshop on Integrated Management of Prosopis Species in Kenya.* Global Environmental Facility (GEF), Kenya Forestry Research Institute (KEFRI) and Forest Department (FD). Republic of Kenya.

Of grasslands and guns: natural-resource based conflict among the Waso Borana pastoralists of northern Kenya

A.D. Jillo[1], A.A. Aboud[1] and D.L. Coppock[2]
[1]Department of Natural Resources, Egerton University, P.O. Box 536, Njoro, Kenya, Email: eu-crsp@africaonline.co.ke, [2]Department of Environment & Society, Utah State University, Logan, Utah, USA 84322-5215

Keywords: pastoralism, participatory research, rangeland development

Introduction The once productive, arid rangelands of northern Kenya, traditionally dominated by a mix of woody species (*Acacia, Commiphora, Cordia* spp.) and graminoids (*Tetrapogon, Aristida, Chrysopogon* and *Sporobolus* spp.) have gradually deteriorated in ecological condition over recent decades (Herlocker, 1999). A major factor considered to be responsible for this trend is the disintegration of traditional systems of land stewardship. Traditional authority has waned in northern Kenya and has often been replaced by open-access tenure, overseen by ineffectual government administrators. Couple this with frequent droughts that typify this zone, as well as expanding populations of people and livestock, and the net result is increased competition for diminishing quantity and quality of grazing and water resources. Local people throughout northern Kenya have reportedly entered a survival mode of existence where the incidence of armed conflict has increased because resource-based disputes have intensified (Smith *et al.*, 2000). The objective of this research was to investigate and quantify the views of the Waso Boran people, one of many ethnic groups in the northern Kenyan rangelands, concerning the causes, and possible solutions, for their conflicts that revolve around natural resources.

Materials and methods Social survey research, founded on structured questionnaires, was conducted during two years to quantify attitudes and opinions of 540 household heads from among the Waso Boran community concerning natural-resource based conflict.

Results Survey respondents most commonly mentioned the following factors as the greatest contributors to resource-use problems in their area: (1) influx of modern weapons (noted by 96% of respondents); (2) shortage of water (74%); (3) wildlife predation on livestock (70%); and (4) shortage of grazing (66%). Accordingly, most respondents (94%) felt that curbing the flow of weapons could reduce conflict. Another 93% felt that efforts to rehabilitate traditional systems of resource tenure would help the situation. A reduction in conflict could allow use of what are now "no-man's lands" and allow implementation of technical improvements to grazing capacity and water resources. Survey respondents felt that other outcomes from conflict include: (1) a gradual aggregation of people towards urban centres (93% of respondents) as well as complete land alienation (46%). These factors in turn have contributed to changes in the diet of pastoral people (from livestock to more grain-based foods) and have exacerbated the incidence of human and livestock diseases as settlements have become more concentrated. Mixed effects of conflict on food security have been noted, with some people having increased access to grain-based diets as they migrate closer to urban centres. Conflict is also believed to hinder livestock marketing by 44% of respondents. Conflict reportedly reduces the mobility of pastoralists and so they are less able to cope with drought shocks (83% of respondents) or routinely access water points and forage reserves (79 to 82%).

Conclusions Fear of violent conflict is reportedly pervasive in the Waso Borana region of northern Kenya. Although increased populations of people and livestock are likely to be the root causes of these problems, the respondents tended to identify symptoms that revolve around enhanced competition for natural resources. There has been a systemic failure of government to provide a secure environment for pastoral production systems. Until commitment is made by government to reduce conflict and begin to restore the confidence of local people in resource access and governance, technical intervention to enhance forage or water supplies will be irrelevant. Relief, rather than development, will also continue to dominate the social agenda. First and foremost, control of weapons proliferation appears paramount. Associated efforts by government policy makers to protect rights of local people to land are also vital. Some of the problem has international roots, however. Unrest within neighbouring countries such as Ethiopia and Somalia occasionally spills over into northern Kenya, and arms may originate from these sources. International coordination is therefore required in any long-term, viable solution.

References

Herlocker, D. (1999). Vegetation dynamics. In: D. Herlocker (ed.) Rangeland Resources in Eastern Africa: Their Ecology and Development. German Technical Cooperation, Nairobi, 17-29.

Smith, K., C. Barrett & P. Box (2000). Participatory risk mapping for targeting research and assistance: with an example from east African pastoralists. *World Development*, 28, 1945-1959.

Community perceptions of vulnerable key ecological resources in Baringo, Kenya

M.N. Mutinda[1], A.A. Aboud[1] and D.L. Coppock[2]
[1]Department of Natural Resources, Egerton University, P.O. Box 536, Njoro, Kenya, Email: eu-crsp@africaonline.co.ke, [2]Department of Environment & Society, Utah State University, Logan, Utah, USA 84322-5215

Keywords: pastoral risks, participatory research, landscape

Introduction Key resources in arid lands are often relatively small patches of seasonal grazing or water access that critically support entire livestock production systems (Scoones, 1993). When these are lost, production systems may be destroyed. An early-warning system is needed whereby key resources at risk can be identified and protected. The Baringo District of north-central Kenya has endured decades of resource abuse and high rates of population growth—breakdowns of traditional systems have occurred and food relief is common (Little, 1992). Despite this situation, most production system research in the past has been conducted at local scales of resolution. The advent of Geographic Information Systems (GIS) technology, however, allows investigations to scale-up. Precise mapping of resource problems is now possible, and such maps can provide useful communication tools to better address issues. We have undertaken a hierarchical approach that focuses on the district, divisions, localities and communities. At the largest spatial scales we rely on social science methods to assess perceived key resources at risk according to community leaders, while at smaller scales we use ecological methods to verify and quantify resource vulnerability. Here we report on the first phase of research involving surveys of community leaders.

Materials and methods In this first research phase we interviewed 136 regional leaders as key informants, widely selected from across seven administrative divisions. Four divisions were comprised of arid pastoral zones while three were comprised of semi-arid agro-pastoral zones. Respondents were asked to rank (1 being highest) the most important key resources they perceived to be at risk in their respective divisions, and postulate sources of risk. Friedman's non-parametric analysis of variance was used to assess repeatability of rankings.

Results Preliminary data indicates that the respondents felt, overall, that herbaceous forage, water, land, and livestock were the main resources that were most vulnerable in Baringo District (Table 1). Hundreds of specific problem sites were identified and mapped as a result of these interviews. Differences in ranking ($P<0.05$) between arid and semi-arid zones were observed from these data. Water and land in general were more highly ranked as vulnerable in the agro-pastoral zones, while forage and water were more highly ranked as vulnerable in the arid zones; water problems were pre-eminent for the district overall ($P<0.05$). Most causes of resource vulnerability were related to population pressure and inadequate management of natural resources (Table 1).

Table 1 Identification and ranking of vulnerable key ecological resources by key informants (n=78 interviews for the arid divisions and n=58 interviews for the semi-arid divisions)

Resource	Causes of vulnerability and loss of the resource	Mean ranking of resource vulnerability	
		Agro-pastoral zones	Pastoral zones
Grazing	Sedentarisation; climate; uncontrolled grazing *Prosopis* invasion; crop farming; insecurity	3	1
Water	Poor catchments; excessive abstraction; pollution; silting; inadequate sources	1	2
Land	Increased populations; degradation	2	4
Livestock	Diseases; lack of grazing and water	4	3

Conclusions Interviews of community leaders have proven useful in a first-cut assessment of vulnerable key resources at a district scale of resolution. Patterns will be verified from future site-based observations. Overall, water was deemed as the most vulnerable resource across the arid and semi-arid spectrum, and population pressure and poor management appeared to be the root causes of most resource-related problems.

References
Little, P. D. (1992). The Elusive Granary: Herder, Farmer, and State in Kenya. Cambridge University Press, Cambridge, 212 pp.
Scoones, I. (1993). Wetlands in drylands: Key resources for agricultural and pastoral production in Africa. *Ambio*, 20, 366-371.

High elevation grasslands as a crucial resource to ranchers of northern New Mexico

A.M. McSweeney and C. Raish

USDA Forest Service, Rocky Mountain Research Station, Albuquerque, New Mexico, USA, Email: amcsweeney@fs.fed.us

Keywords: high-elevation grasslands, grazing allotments, permittee ranchers

Introduction High-elevation grasslands of northern New Mexico (NM), located at the southern tip of the Rocky Mountains in the western United States, are a crucial resource for small-scale, family-owned ranches. Due to evolution of land acquisition in northern New Mexico, many of these lands are in public ownership, and ranchers must now rely upon government-managed grazing allotments for pasturing their livestock. Regulations and management decisions governing these lands, along with competition for use (e.g. recreation), can significantly affect the viability and survival of ranching throughout the area (Raish & McSweeney, 2003).

The place The mountain grasslands of northern New Mexico are coniferous forest clearings at altitudes exceeding 2400 m. Small meadows and grassy slopes are often dominated by cool-season, tall bunchgrasses such as *Danthonia, Deschampsia, Festuca, Koeleria, Muhlenbergia,* and *Poa* (Allred, in press). Owing to elevation, they receive greater precipitation than lower lying regions of the state, resulting in more dependable pasturage for livestock. This area has provided an historic resource for generations of ranching families.

The problem Many families of this region are of Hispanic origin, continuing a ranching heritage that began with Spanish colonization in 1598. Land use and ownership were confirmed by grants from the Spanish Crown and later by the Mexican government. Changes began to occur after U.S. conquest in 1848, resulting in the loss of over 80% of the grant lands (Westphall, 1965). The forest and grassland portions were acquired by government land agencies (US Forest Service and Bureau of Land Management), and are now used only under federal grazing permits. Approximately 34% of the six counties of northern New Mexico is now federally controlled (Eastman*, et al.*, 2000). Loss of grazing land has diminished the ranchers' ability to sustain their livelihoods from agriculture and, in turn, affects the viability of rural communities (Raish & McSweeney, 2003).

The people While income from livestock may not fully support a majority of these families, it is used in a variety of ways. In our study, we found that 93% of the permittee ranchers use part of the income from livestock to improve the ranching operation, 58% for basic living expenses, and 48% for family emergencies. In northern New Mexico, retaining family land and traditional values is regarded more highly than acquisition of material possessions or monetary gain. When asked to prioritize family goals and values (Figure 1), 95% of the participants ranked better quality of life and continuance of traditional values as more important than an increase in family income (Raish & McSweeney, 2003).

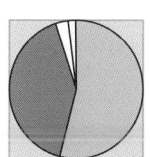

☑ Quality of life

◼ Maintain traditional values

☐ Increase farm income

☐ Respect in community

Figure 1 Family Goal Priorities

Conclusion Despite ongoing problems with land loss, encroaching urbanization, conflicts over land use, and government regulations, ranch families retain a strong attachment to the land. These mountain grasslands provide more than forage for cattle and sheep. The high-elevation grasslands of northern New Mexico play a role in maintaining the tradition, heritage, and culture of a land-based people.

References

Allred, K.W. (in press). A field guide to the grasses of New Mexico, 3rd ed. Department of Agricultural Communications, New Mexico State University, Las Cruces, New Mexico, U.S.A.

Eastman, C., C. Raish & A.M. McSweeney (2000). Small livestock operations in northern New Mexico. In: R. Jemison & C. Raish (eds.) Livestock Management in the Southwest: Ecology, Society, and Economics. Elsevier Science, Amsterdam, 523-554.

Raish, C. & A. M. McSweeney (2003). Economic, social, and cultural aspects of livestock ranching on the Española and Canjilon Ranger Districts of the Santa Fe and Carson National Forests: a pilot study. RMRS-GTR-113. U. S. Department of Agriculture, Forest Service, Rocky Mountain Research Station, Fort Collins, Colorado, U.S.A.

Westphall, V. (1965). The public domain in New Mexico 1854-1891. The University of New Mexico Press. Albuquerque, New Mexico, U.S.A.

Avenues for enhancing traditional livelihoods from grasslands: income diversification among pastoral women's groups in southern Ethiopia

S. Desta[1], D.L. Coppock[2], S. Tezera[3] and G. Gebru[3]
[1]PARIMA Project, International Livestock Research Institute, P.O. Box 30709, Nairobi, Kenya, Email: s.desta@cgiar.org, [2]Dept. Environment & Society, Utah State University, Logan, Utah, USA 84322-5215, [3]PARIMA Project, International Livestock Research Institute, P.O. Box 5689, Addis Ababa, Ethiopia

Keywords: risk management, participatory research, Boran

Introduction The rangelands of Africa remain home to millions of people who try to make a living by raising livestock on natural forage. Recent increase in human and livestock populations, however, along with a lack of economic development, has relegated many people to poverty and vulnerability. The semi-arid Borana Plateau of southern Ethiopia is a case in point. About 250,000 people herd one million head of livestock there. Thousands of animals die in periodic droughts and people are food insecure. It has been proposed that one way to better manage risk in this system is through economic diversification to reduce vulnerability (Desta & Coppock, 2002). The need to better address problems requires that local human capacity be built and solutions carefully targeted. To this end some members of the USAID-funded Pastoral Risk Management (PARIMA) project have adopted participatory research methods where scientists, communities, and development agents share power in a process of problem solving.

Materials and methods Work reported here has been conducted with five, semi-settled pastoral communities in southern Ethiopia since 2000. The approach involves community problem diagnosis and formulation of solutions referred to as Participatory Rural Appraisal (PRA; Lelo *et al.,* 2000). This involves an initial, week-long assessment of a community, with researchers as facilitators. This is followed by preparation of a community action plan (CAP) by the community in partnership with a development agent wherein intervention priorities are set based on relevance to priority problems and the ability of the community to make a sustainable contribution to solutions. The role of the PARIMA project has been to help select communities and development agents, conduct the PRA, solicit funding for CAPs and engage in monitoring and evaluation. Constraints and the means to overcome them involves another process called "action research" (Brown & Tandon, 1983). The PARIMA project has also become involved in linking pastoralists across the Ethio-Kenya border as well as networking various players in livestock marketing chains.

Results Community priority problems from several PRAs conducted in peri-urban settings indicate that shortages of food, water, health care, and lack of education are prominent. The CAPs have focused on the problem of lack of livelihood diversification that underlies these issues, and hence creation of non-formal education (NFE) centres, grass-roots micro-finance (savings and loan) operations, and training in small-scale business skills have been most in demand. Females and males are active participants, but newly formed women's groups tend to take leading roles. Several hundred people have received loans after sufficient savings were generated. Loan repayment has been 100% over the past two years. Most loan recipients have made profits, and the common approach has been to engage in livestock trade, although successes have also been observed in areas of petty trade. Provision of NFE permits people to achieve a base level of literacy and numeracy that accelerates their performance in entrepreneurial endeavours. Efforts by PARIMA to promote livestock marketing have involved networking among pastoral producers, development agents, and exporters. Recent successes in having pastoral groups supply small ruminants for export have been achieved.

Conclusions Use of participatory action research has yielded relatively quick benefits to pastoral communities, but requires researchers to change their values and approaches. Mentoring communities to help them achieve a novel vision of future possibilities is a major ingredient of success. Attempts to strengthen livelihoods helps create enhanced opportunities to better manage forage and other natural resources.

References
Brown, L.D. & R. Tandon (1983). Ideology and political economy in inquiry: action research and participatory research. *Journal of Applied Behavioral Sciences*, 19, 277-294.
Desta, S. & D.L. Coppock (2002). Cattle population dynamics in the southern Ethiopian rangelands, 1980-97. *Journal of Range Management,* 55, 439-451.
Lelo, F., J.O. Ayieko, R.N. Muhia, S.K. Muthoka, H.K. Muiruri, P.M. Makenzi, D.M. Njeremani & J.W. Omollo (2000). Egerton PRA Field Handbook for Participatory Rural Appraisal Practitioners. Third Edition. The PRA Programme, Egerton University, Njoro. 89.

Theme C: Delivering the benefits from grassland

Section 24

Tools for grassland management

Pastures from Space – Application of satellite-derived pasture predictions improve the profitability of Australian sheep producers

S.G. Gherardi[1], L. Anderton[1], J. Sneddon[2], C. Oldham[1] and G. Mata[3]

[1]*Department of Agriculture, Locked Bag No. 4, Bentley Delivery Centre, Western Australia, Australia 6983, Email: sgherardi@agric.wa.gov.au, [2]Graduate School of Management, University of Western Australia, Crawley, Western Australia, Australia 6009, [3]CSIRO Livestock Industries, Private Bag 5, Wembley, Western Australia, Australia 6913*

Keywords: pastures from space, satellite, sheep, pasture growth rate

Introduction Pastures from Space, a collaborative program between CSIRO Livestock Industries and the Western Australian state Departments of Agriculture and Land Information, has developed the capacity to measure both the biomass and growth rate of annual pasture in the winter rainfall regions of southern Australia using satellite images (Edirisinghe *et al.,* 2002). Producer groups were set up to pilot test the delivery of satellite-derived pasture growth rate (PGR, kg dry matter/hectare.day) and biomass (feed on offer or FOO, kg dry matter/hectare) predictions for paddocks on individual farms in Western Australia. This paper reports on the value to Australian sheep producers of satellite-derived PGR information on pastures.

Methodology Producer groups were established at each of five localities situated in the major sheep producing areas in Western Australia. For each of the 51 co-operating producers detailed maps of their farms showing all external and paddock boundaries were entered into a geographic information system computer database for the delivery of PGR. Near-real time predictions and 7-day forecast PGR were delivered weekly for each of the paddocks on their farms via Pasture Watch™, a farmer friendly tool for downloading and analysing PGR data (Wiese *et al.*, 2004). The producer groups met on 4-5 occasions throughout the pasture growing season. At the meetings, each of the producers provided feedback on the reliability, timeliness and accuracy of the PGR predictions being delivered and shared how they were using the information to make management decisions on their farms. Beginning and end of season surveys were used to provide a qualitative evaluation of the usefulness and benefits of PGR, and six case studies provided detailed analysis of the economic value to the farm business.

Results All producers surveyed found the PGR information easy to access. Of the forty-three producers who completed both surveys, thirty three (77%) used the PGR information. Over 75% used the PGR information, at a frequency from weekly to monthly. The majority (91%) of these producers reported that they had used the PGR information to make stock management decisions. Decisions about feed budgeting (70%), planning (64%) and stocking rates (58%) were the next most frequent management decisions. Around one quarter (27%) employed PGR information to make decisions about liveweight, 21% to make land use decisions and 12% to manage the fibre diameter profile of the wool grown. Overall there were highly positive responses to the compatibility, usefulness and benefits of the technology.

In all six case studies the use of the PGR information improved the profitability of the producers' sheep enterprise. The increase in profit ranged from a gross margin of AUD\$23 to AUD\$332/winter grazed hectare. The increased profit resulted from better utilisation of pasture through more effective feed budgeting and the introduction of new management techniques into the farming system. The producers recognised PGR as a valuable tool when applied to decisions about the use of a range of management techniques such as increasing stocking rates, feedlotting of wethers, whether or not to agist livestock, application of fertilisers and conservation of fodder during spring. The information on PGR was also found to improve producers' confidence in decision-making and helped reduce their levels of stress.

Conclusions Both the surveys and case studies reported in this paper have clearly demonstrated a significant increase in the profitability of those farm businesses that have used satellite-derived pasture information. The vision of the Pastures from Space program is to provide cost effective, reliable, timely and accurate satellite-derived PGR and FOO predictions that will enable producers to substantially increase the productivity and profitability of their farming enterprise.

Acknowledgements Funding by the Australian Wool Innovation and contribution of the co-operating producers.

References
Edirisinghe, A., G.E. Donald, M.J. Hill & D.A. Henry (2002). Precision management of feed supply through the timely delivery of biomass and growth rate estimates of Western Australian annual pastures. *Twenty-ninth International Symposium on Remote Sensing of Environment*, Buenos Aires, Argentina.
Wiese, R., S. Gherardi, G. Mata & C. Oldham (2004). Pasture Watch – a farmer friendly tool for downloading and analysing Pastures from Space data. Agribusiness Sheep Updates, Perth, Western Australia, 79-80.

Diagnosing nitrogen, phosphorous and potassium status of natural grassland in the presence of legumes

C. Jouany, P. Cruz, J.P. Theau and M. Duru.
UMR 1248 ARCHE - INRA, BP 27, 31326 Castanet -Tolosan Cedex, France, Email: cjouany@toulouse.inra.fr

Keywords: diagnosis, grassland, legume, nutrition index.

Introduction In most temperate areas, sustainable management of grassland ecosystems has to deal with evaluation and management of N, P and K resources. For this purpose, appropriate diagnostic systems are needed in order to manage fertilisation accordingly. The nutrient index method based on nutrient concentrations in plant tissues relative to the degree of growth has been developed; it relies on critical curves which serve for diagnostic: for N, the critical curve gives the optimum N concentration for different levels of biomass accumulation in swards, for P and K optimum concentrations are a linear function of sward N concentration (Duru & Thélier-Huché, 1997). However limitations in the use of P nutrient index (PNI) were reported when the herbage contained a large proportion of white clover (Jouany *et al.*, 2004). Our objectives were to verify whether similar behaviour were observed with other legumes and for K and N nutrition indices (KNI, NNI) as well.

Materials and methods This work was conducted on meadows and native grasslands of the Ercé valley in the Central French Pyrenees (0°30E, 42.48N). These grasslands are located between 600 and 1000 m asl. In spring 2001, 10 grasslands have been characterized for N, P, and K Nutrition Indices. Biomass samples were collected for dry matter yield measurement, and sorted into 2 fractions: non-legumes and legumes fraction where trifolium pratense, lotus corniculatus, medicago sativa, represented more than 75 % of the total legume fraction. Total N, P and K concentrations were measured on oven dried, (80°C) ground milled herbage (0.5 mm). Nutrition indices were calculated according to Duru & Ducrocq (1997). They were 3 repetitions plots sampled per grassland.

Results The average nutrition indices calculated on the non-legume fraction were always higher than for the mixed one. For a given sward, the difference between indices calculated for mixed sward and non legume fraction increased with the legume content of the sward (Figure 1), the relationship is similar to the one obtained for clover in previous work. Calculated PNI and KNI from mixed sward concentrations resulted in an underestimation of the sward nutrition while difference between NNI for mixed sward and NNI for non legume fraction increases with the legume content and leads to an overestimation of sward N nutrition status (Figure 2).

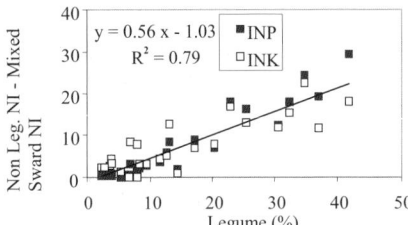

Figure 1 Difference between non legume fraction PNI and (KNI) and mixed sward PNI (KNI) as a function of legume content

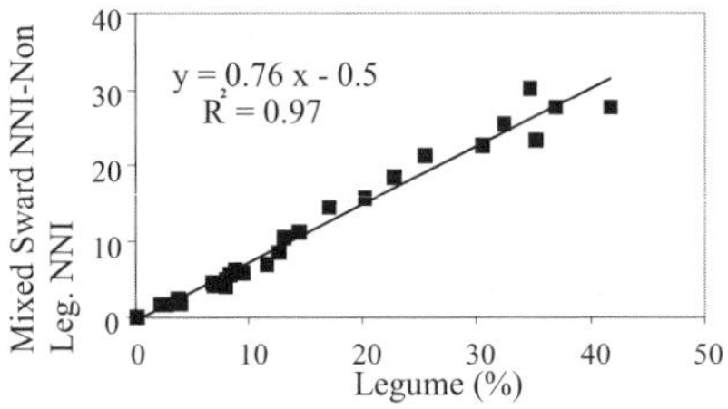

Figure 2 Difference between mixed sward NNI and legume fraction NNI as a function of legume content

Conclusions High legume proportions limit the use of nutrient diagnosis systems. Whatever the diagnostic index used, measurements on mixed swards could substantially underestimate sward PK status and overestimate N status. We suggest basing diagnosis on the non-legume fraction when nutrition indices are used for fertiliser recommendations.

References

Duru, M., & H. Ducrocq (1997). A nitrogen and phosphorus herbage nutrient index as a tool for assessing the effect of N and P supply on the dry matter yield for permanent pastures. *Nutrient Cycling in Agroecosystems*, 47, 59-69.

Duru, M., & L. Thélier-Huché (1997). N and P-K status of herbage : use for diagnosis of grasslands. In: G. Lemaire & I.G. Burnes (eds.) Diagnostic procedures for crop N management and decision making. INRA, pp 125-138.

Jouany, C., P. Cruz, P. Petibon & M. Duru (2004). Diagnosing phosphorus status of natural grassland in the presence of white clover. *European Journal of Agronomy*, 21, 273-285.

Fertiliser responses and soil test calibrations for grazed pastures in Australia

C.J.P. Gourley[1], A.R. Melland[1], K.I. Peverill[2], P. Strickland[1], I. Awty[1] and J.M. Scott[3]

[1]*Primary Industries Research Victoria, Ellinbank, Victoria, Australia, Email: Cameron.Gourley@dpi.vic.gov.au,*
[2]*KIP Consultancy Services Pty Ltd, Wheelers Hill, Victoria, Australia,* [3]*University of New England, Armidale, NSW, Australia*

Keywords: fertiliser, nutrients, pasture growth response

Introduction   On-farm management of fertiliser is of major economic significance to the Australian grazing industries, based on expenditure on fertiliser and higher farm productivity that fertiliser use supports. However the application of fertiliser has traditionally been an inexact and inefficient process (Peverill *et al.* 1999) and there is increasing pressure for nutrient losses from agriculture to be minimised. The improved adoption and application of tools like soil testing can make substantial improvements in nutrient use efficiency but interpretation needs to be based on the best available information. This paper reports on the collation of current and historical experimental data relating to pasture production - fertiliser response relationships (nitrogen, phosphorus, potassium and sulphur) for various pasture types, climatic zones and soils across Australia.

Materials and methods   A national team of scientists and fertiliser agronomists from all states of Australia has contributed to the collation of a comprehensive set of pasture production-fertiliser response data from field studies. This has involved identifying and collating previous reviews, published papers, departmental reports and where available, unpublished material. These data sets have been integrated using a relational database to derive the most appropriate response relationships available for the grazing industries in Australia. The national database has been used to provide regionally specific and scientifically validated soil test calibrations for improved pastures.

Results   More than 350 experimental data sets have been collated consisting of circa 2600 sites and >3800 experimental trial years. The number of sites is made up of around 479 N, 662 P, 692 K and 810 S trials, with a total of 615 N, 1313 P, 933 K and 974 S experimental trial year. Not surprisingly, experimental data sets ranged in quality, scope and complexity. Less than 33% had enough statistical rigour to enable nutrient response curves to be generated. Only a few studies involved a number of sites in various regions with different soil types and climatic zones. Most studies provided simpler data sets, mostly at a single field site with a single nutrient applied at only 2 or 3 rates. Such data sets could not be used to establish nutrient response curves. There was a limited capacity to combine experimental data sets as methodologies often differed markedly and site data was inadequate. In some single site studies, soil test levels were strongly related to pasture response to fertiliser applications (variance accounted for (VAF) > 0.9), but when applied to a range of soils and environments, they invariably lack precision (VAF ranging from 0.0-0.5). The addition of other variables such as soil type and climatic zones only marginally improved these relationships.

Conclusions   More than 50 years of experimental data relating to the response of pasture to soil nutrient availability has been compiled and analysed across the pasture-based grazing industries of Australia. This extensive exercise has highlighted the lack of precision in many response relationships for N, P, K and S and the difficulty in combining historical data sets to assist in extrapolating across soil types and regions. It is proposed that appropriate soil tests may still require regional calibration and further research should adopt standard experimental methodologies.

References
Peverill K.I., L.A. Sparrow & D.J. Reuter eds. (1999) "Soil Analysis; an interpretation manual" CSIRO Publishing, Australia.

Potential for forecasting UK summer grass growth from the North Atlantic Oscillation

P.S. Kettlewell[1], J. Easey[1], P.D. Hollins[1], T. Martyn[2] and D.B. Stephenson[3]
[1]*Crop and Environment Research Centre, Harper Adams University College, Newport, Shropshire TF10 8NB, UK, Email: pskettlewell@harper-adams.ac.uk* [2]*Institute of Grassland and Environmental Research, North Wyke Research Station, Okehampton, Devon EX20 2SB, UK* [3]*Department of Meteorology, University of Reading, Earley Gate, PO Box 243, Reading RG6 6BB, UK*

Keywords: NAO, climate, irrigation, rainfall

Introduction The North Atlantic Oscillation (NAO) is a large-scale atmospheric circulation pattern which is well-known to influence the UK winter climate (Wilby *et al.*, 1997). Recently, it has been shown that the winter NAO also affects summer rainfall in the UK (Kettlewell *et al.*, 2003). Since water supply is an important limitation to summer grass growth in many parts of the UK, the winter NAO may influence summer growth. The objective of this study was to test the hypothesis that there is a relationship between the winter NAO and summer grass growth using data from reference plots at North Wyke in Devon.

Materials and methods Plots of perennial ryegrass (cv. Cropper) were established each year from 1982 to 1992 at North Wyke. In the year following establishment four series of plots were cut every four weeks in rotation and dry matter yield recorded. Growth rates were calculated for each week according to the method of Corrall and Fenlon (1978) and the mean growth rate calculated for the conventional climatological summer (June, July, August) corresponding to weeks 22 to 35 inclusive. The plots were duplicated with one set of plots unirrigated and the other set irrigated to field capacity each week. The mean summer growth rate each year was regressed against the preceding winter (December, January, February) NAO index of Hurrell (taken from http://www.cgd.ucar.edu/~jhurrell).

Results The summer growth rate of the unirrigated plots showed a clear relationship with the preceding winter NAO index. Years with very high winter NAO indices had summer growth rates only about half those of years preceded by intermediate or low winter NAO indices (Figure 1a). In contrast, the growth rate of irrigated plots was high in almost all years irrespective of the preceding winter NAO index (Figure 1b).

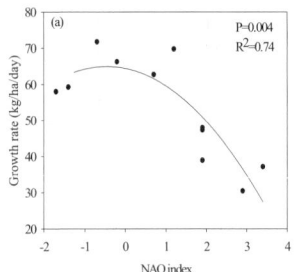 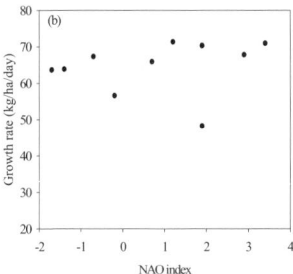

Figure 1 Effect of the winter North Atlantic Oscillation on summer grass growth rate in (a) unirrigated and (b) irrigated plots at North Wyke 1983-1993

Conclusion These results indicate that summer grass growth at North Wyke is dependent on the preceding winter NAO and that the relationship may be mediated through soil water supply. The results are consistent with previous work showing that a high winter NAO index tends to be followed by a dry summer in England and Wales (Kettlewell *et al.*, 2003). Analysis of local weather and soil moisture data is in progress to confirm the mechanism. The relationship is based on a very small dataset at one site only and needs to be confirmed at other sites. If the effect proves to be widespread throughout the UK, a general forecast of summer grass growth may be feasible with a lead time of at least three months. This may be of benefit to farmers in assisting with planning e.g. deciding the area to be sown with maize or spring cereals for wholecrop silage.

References
Corrall, A.J. & Fenlon, J.S. (1978). A comparative method for describing the seasonal distribution of production from grasses. *Journal of Agricultural Science, Cambridge*, 91, 61-67.
Kettlewell, P.S., D.B. Stephenson, M.D. Atkinson & P.D. Hollins (2003). Summer rainfall and wheat grain quality: relationships with the North Atlantic Oscillation. *Weather*, 58, 155-164.
Wilby, R.L., G. O'Hare & N. Barnsley (1997). The North Atlantic Oscillation and British Isles Climate variability, 1865 – 1996. *Weather*, 52, 266-276.

Modelling grass productivity in the Brazilian Amazon

C.E.P. Cerri[1], K. Paustian[2], C.C. Cerri[1], F.F.C. Mello[1], M. Bernoux[1], K. Coleman[3] and E. Milne[4]

[1]*Centro de Energia Nuclear na Agricultura (CENA), Universidade de São Paulo, P.O. Box 96, 13400-970, Piracicaba-SP, Brazil, Email: cepcerri@cena.usp.br,* [2]*Colorado State University, Ft Collins, CO 80523, USA,* [3]*IACR-Rothamsted, Harpenden, Hertfordshire AL5 2JQ, UK,* [4]*University of Reading, P.O. BOX 233, Reading, UK*

Keywords: modeling grass productivity, Brazilian Amazon, Century model, GEF-SOC project

Introduction The Amazon Basin covers an area of 7 million km^2, and the central part is almost entirely located within Brazilian territory. This region has the highest rates of deforestation in the world, and the total area deforested now exceeds 600,000 km^2. Cattle pasture represents the largest single use (about 70%) of this once-forested land in most of the Brazilian Basin, with an estimated area of 20 million hectares. Our main objective was to simulate grass productivity in different forest to pasture chronosequences within the Brazilian Amazon.

Materials and methods We used data collected from nine forest to pasture chronosequences within the Brazilian Amazon (Figure 1) to simulate grass productivity using the Century Ecosystem Model (Parton *et al.*, 1987). Century is a model of soil organic matter and nutrient dynamics that emphasises the decomposition of soil organic matter and the flux of C and N within and between different components. The grassland/crop and forest systems have different plant production submodels that are linked to a common soil organic matter and nutrient cycling submodel, described by Paustian *et al.* (1997).

Results and discussion Our results suggested that modeling techniques can be successfully used for simulating grass productivity in Amazonian pastures (Table 1). Statistical tests between modeled and measured grass productivity data for the nine studied sites gave a correlation coefficient (r) of 0.78, Root Mean Square Error (RMSE, showing total error) of 10.63 %, Coefficient of Determination of -0.004 and a mean difference (M, showing bias) of -0.05 t DM/ha per yr (data not shown). The effects of the conversion of tropical forest to pasture on total soil C and N, were analysed by Cerri *et al.* (2004). Results were used not only to evaluate soil C dynamics but also to indicate soil C sequestration opportunities for the Brazilian Amazon. Scenarios of different pasture management practices and climate change were also explored at these chronosequences.

Figure 1 Location of the forest to pasture chronosequences in the Brazilian Amazon

Table 1 Simulated and measured grass productivity values for forest-pasture chronosequences within the Brazilian Amazon

Chronosequence	Weather		Grass Productivity (t DM/ha/yr)		
	T (^{o}C)	P (mm)	Measured	Simulated	Difference
Água Parada	27.2	1750	11.58	10.89	0.69
Bosque	27.2	1750	11.58	11.22	0.36
Piquiá	27.2	1750	11.58	10.94	0.64
Entre Rios	27.9	2591	10.20	12.06	-1.86
Sao Joao	27.2	1750	10.20	12.04	-1.84
Teixeira	26.7	2075	15.30	15.72	-0.42
Benjamim	25.6	2200	11.58	11.39	0.19
Lenk	25.6	2200	11.58	12.03	-0.45
Rancharia	25.0	1280	9.80	7.52	2.28

Conclusions Simulations using the Century model showed similar grass productivity results to those measured in previous studies. Moreover, modeling studies in these forest to pasture systems have important implications, for example, on the calculated CO_2 emissions from land-use change in national greenhouse gas inventories.

References

Cerri,C.E.P., K.Paustian, M.Bernoux, R.Victoria, J.Melillo & C.C.Cerri (2004). Modeling changes in soil organic matter in Amazon forest-pasture conversion with the Century model. *Global Change Biology*, 10, 815-832.

Parton,W.J., D.S.Schimel, C.V.Cole & D.S.Ojima (1987). Analysis of factors controlling soil organic matter levels in Great Plains grasslands. *Soil Science Society of America Journal*, 51,1173-1179.

Paustian,K., E.Levine, W.M.Post & I.M.Ryzhova (1997). The use of models to integrate information and understanding of soil C at the regional scale. *Geoderma*, 79, 227-260.

Simulation of lablab pastures

J.O. Hill, M.J. Robertson, A.M. Whitbread and B.C. Pengelly
CSIRO Sustainable Ecosystems, Queensland Bioscience Precinct, 306 Carmody Road, St. Lucia, Queensland, Australia 4067, Email: Jacqueline.Hill@csiro.au

Keywords: APSIM, GrazFeed, tropical legume

Introduction The potential of legume-based pastures to address declining soil nitrogen on marginal cropping soils is increasingly recognised in northern Australia, as such there is a need for cost benefit analysis of pastures and crops in a mixed farming system. In highly variable rainfall environments, biophysical modelling may be the best way of identifying and quantifying interactions with mixed crop-livestock systems on a seasonal basis. This paper describes a case study where both animal productivity and lablab pasture production is simulated. Lablab (*Lablab purpureus*) is an annual tropical legume widely used as a short-term legume phase in crop-pasture rotations, providing high quality forage for animal production and a low risk nitrogen input for crop production.

Materials and methods A simulation capability for lablab growth was developed using the APSIM-Legume model framework (Agricultural Production Systems simulator; Robertson et al., 2002). The lablab model was tested for its ability to predict leaf biomass (the portion of lablab that is consumed by animals) against measured data from northern Australia. Using daily weather records from 1957 to 2004 the production of leaf biomass was simulated each season for annual lablab on a 150 cm deep Vertosol at Gayndah, Qld, Australia (25°39'S 151°45'E). A relationship between lablab leaf biomass and liveweight gain (LWG) of *Bos indicus* steers was derived by running simulations of the GrazFeed animal biology model (Freer et al., 1997) with increasing leaf weights. This relationship was used to predict a range of LWG over the seasons.

Results An example of the model's ability to simulate leaf biomass at Gatton, Qld, is shown in Figure 1. The long-term simulation of lablab production under dryland conditions at Gayndah indicates that at least 1000 kg/ha of leaf biomass will be produced in 50% of years, and at least 500 kg/ha will be produced in 80% of years (Fig. 2). This corresponds to a LWG of at least 0.75 kg/head/day in 80% of years assuming a stocking rate of 1 steer/ha (Fig. 3). At a stocking rate of 2 steers/ha, and assuming that steers were consuming approximately 10 kg lablab/day, this rate of LWG could be maintained for 25-100 days depending on biomass produced.

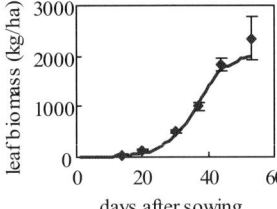
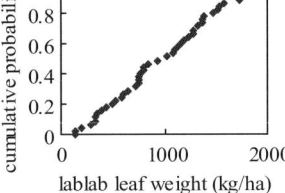
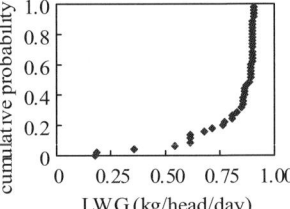

Figure 1 Measured values of biomass at 14, 20, 30, 37, 44 and 53 days after sowing (± s.e.m.) and the simulated leaf biomass of lablab

Figure 2 Cumulative distribution function for seasonal lablab leaf production at Gayndah (1957-2004)

Figure 3 Cumulative distribution function for a range of LWG (1957-2004)

Conclusions The APSIM lablab model was able to simulate leaf biomass production with a high degree of precision. The long term simulation of lablab production under dryland conditions indicated a low risk of crop failure, and the potential for relatively high rates of LWG of *Bos indicus* steers. The persistency of animal productivity will depend on subsequent rainfall, plant regrowth and stocking rates. Work is currently in progress to link the GrazFeed and APSIM models in order to simulate lablab production and animal production dynamically with a daily time step.

References

Freer, M., A.D. Moore & J.R. Donnelly (1997). GRAZPLAN: Decision support systems for Australian grazing enterprises II. The animal biology model for feed intake, production and reproduction and the GrazFeed DSS *Agricultural Systems*, 24, 77-122.

Robertson, M.J., P.S. Carberry, N.I. Huth, J.E. Turpin, M.E. Probert, P.L. Poulton, M. Bell, G.C. Wright, S.J. Yeates & R.B. Brinsmead (2002). Simulation of growth and development of diverse legume species in APSIM. *Australian Journal of Agricultural Research*, 53, 429-446.

A herbage growth model for different types of natural grassland

J. Viégas[1], M. Duru, P. Cruz, J.P. Theau, P. Ansquer and C. Ducourtieux
Research unit UMR, 1248 Managing grassland farming systems, BP 27, F-31326 Castanet, Tolosan, France,
[1]*UFSM, Santa Maria - RS - Bolsista CAPES, Brazil; Email: jviegas@smail.ufsm.br*

Keywords: modelling, senescence, reproductive phase, functional traits

Introduction The aim of this work was to extend existing growth models established for pure stands to a wide range of grassland communities. For this purpose we built a simple growth model, including sub-models for radiation interception and use. Parameters for the effect of nutrient rates (N, P) and defoliation regimes were based on a plant trait database. Senescence and reproductive processes were particularly considered because of their importance in late spring growth. The model makes it possible to simulate the daily biomass production as a function of both environmental factors and the functional type of the dominant species in the community.

Materials and methods Grasslands were characterised by the functional type of dominant plants as defined by Ansquer *et al.* (2004), according to their leaf dry matter content (LDMC), i.e. for example, type **A**: *Lolium perenne*; type **B**: *Dactylis glomerata*; type **C**: *Festuca rubra* and type **D**: *Brachypodium pinnatum*. Vegetation types also were associated with differences in leaf life span (LLS), beginning of stem elongation (BSS from February 1st), and beginning of flowering (BF from February 1st), which all were expressed on a thermal time (ST) basis. The model structure derives from Duru *et al.* (2002). Plant traits were used to parameterise the growth model over the reproductive phase (BSS and BF), and the senescence sub-model (LLS). The values of LLS, BSS and BF characterising the different types vegetation were 500, 600, 800 and 1000; 500, 700, 900 and 1100; 1200, 1400, 1600 and 1800 for the types A, B, C and D respectively. The model was evaluated by comparing the simulations to the growth recorded during spring 2002 on 8 natural grasslands located in the Pyrenees region, France. Four grasslands were close to type A, and 4 were intermediate between types B and C (type BC). The nutrient index (Ni) was assessed using plant analysis, and was an input variable of the growth model, being calculated accordingly Duru *et al., 2002*. The Ni affects the radiation utilisation efficiency, and the growing differences of the LAI. It also modifies the intensity of the reproductive process.

Results The model estimated adequately the daily biomass production (Figure 1). In the case of type A, the model's predictions differed from the observations with an average root mean square deviation (RMSD) of 43 g/m^2 (Ni=77%). The model is less capable of simulating the growth of grasslands with lower nutrient levels. For type BC the model results deviated from the observations with an average RMSD of 36 g/m^2 (Ni=65%).

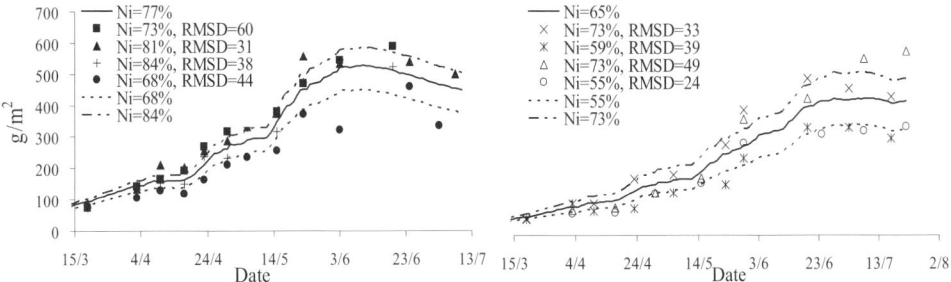

Figure 1 Estimated and measured herbage biomass accumulation on grassland of type A with an initial biomass (Wo) of 40 g/m^2 and a leaf area index (LAI) of 1.0 on February 1st (1a), and type BC with Wo= 40 g/m^2 and LAI=0.2 (1b). The Ni for all situations and the RMSD for the simulation in relation to the measured data are given in the legends. Full lines correspond to the simulated herbage mass for the average Ni of the four grasslands, and dotted lines correspond to the simulated herbage mass for the lowest and highest Ni.

Conclusions The results show that the proposed models are well suited for the purpose; however more validation analyses are needed, specifically for type D vegetation.

References

Ansquer P., J.P. Theau, P. Cruz, J. Viegas, R. Al Haj Khaled & M. Duru (2004). Caractérisation de la diversité fonctionnelle des prairies naturelles. Une étape vers la construction d'outils pour gérer les milieux à flore complexe. *Fourrages*, In press.

Duru M., H. Ducrocq, C. Fabre & E. Feuillerac (2002). Modelling net herbage accumulation of an orchardgrass sward. *Agronomy Journal*, 94, 1244-1256.

Variation of LDMC and SLA relationship between growth forms in natural grasslands

J. Viegas[1], P. Cruz[2], J.P. Theau[2], C. Jouany[2], P. Ansquer[2], R. Al Haj Khaled[2], O. Therond[3] and M. Duru[2]
[1]UFSM, Santa Maria-RS, Brazil (bolsista CAPES), Email: jviegas@smail.ufsm.br, [2]UMR 1248 ARCHE - INRA, BP 27, 31326 Castanet -Tolosan Cedex, France, [3]Chambre d'Agriculture de la Haute Garonne, 28 rue d'Eaunes 31605, Muret, France

Keywords: specific leaf area, leaf dry matter content, grasses, rosettes, fertility

Introduction In agro-ecological studies, there is a growing interest in measuring both leaf dry matter content (LDMC) and specific leaf area (SLA). This interest lies on the fact that leaf traits are linked to gradients of environmental factors and ecosystem functions. Working with three contrasting wild species, Garnier *et al.* (2001) proposed a model linking these two traits. The model shows a relatively simple non linear and negative correlation between LDMC and SLA. Nevertheless, none of the species used to build the model were *grasses* (GRA) or *forb rosettes* (ROS = i.e. dicotyledonous with large entire leaves and absence of stem at the vegetative stage); the species which make the largest contribution to the standing biomass of most natural grasslands. Furthermore, due to the divergent range of LDMC (and not SLA) values between these growth forms, Cruz *et al.* (2002) proposed that grass records alone could be used as an indicator of fertility gradients. The aim of this paper was to analyse discrepancies in the LDMC – SLA correlation with respect to model predictions in order to consider them in any development of LDMC-based tools for the management of natural vegetation.

Material and methods Data for LDMC and SLA were recorded in Pyrenean grasslands (France) growing in rich (site A = SA) and low (site B = SB) fertility levels and were compared to the model proposed by Garnier *et al.* (2001). Leaf traits were measured following the protocol proposed by these authors. The fertility level was assessed by the nitrogen nutrition index (Ni) (Duru *et al.*, 1997). Three GRA and two ROS, representing 60 and 10% of the biomass were measured in site A (Ni=88%), and six GRA and four ROS (45 and 30 % of the biomass) in site B (Ni=49%). Forty individuals were measured for each species at each site (Table 1).

Results The relationships between LDMC and SLA for grasses and rosettes diverge in opposite ways with regard to the model (Figure 1). For GRA, the model underestimates the SLA with an average root mean square deviation (RMSD) of 8.2 m²/kg. The opposite is observed for ROS, with a RMSD of 15.6 m²/kg. The average SLA is the same for the two groups; however the corresponding LDMCs are higher for GRA than ROS (more than 100 mg/g). Fertility had an effect on trait values but not on their relationship (Table 1).

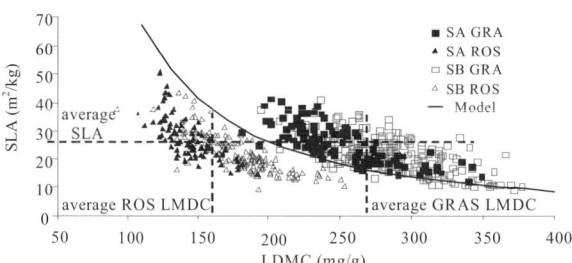

Figure 1 Relationships between LDMC and SLA for *grasses* and *forb rosettes* compared with the model curve (SLA=11.3 10^4 xLDMC^{-1.58}, Garnier et al., 2001)

Conclusions We conclude that there is not a single relationship between LDMC and SLA. Growth forms which are abundant in natural grasslands show very different LDMC values for the same SLA value. In order to use LDMC as an indicator of fertility gradients, we have to avoid mixing data belonging to different growth forms.

Table 1 LDMC and SLA for different growth forms in two Pyrenean sites that differ in nitrogen index (Ni)

Site/ Ni%	Growth Form	LDMC (mg/g)	SLA (m²/kg)
A/88	Grass	252	25.70
	Rosette	143	27.44
B/49	Grass	281	22.25
	Rosette	178	22.48

References

Cruz, P., T. Duru, O. Therond, J.P. Theau, C. Ducourtieux, C. Jouany, R. Al Haj Khaled & A. Ansquer (2002). Une nouvelle approche pour caractériser les prairies naturelles et leur valeur d'usage. *Fourrages*, 172, 335-354.

Garnier E., B. Shipley, C. Roumet & G. Laurent (2001). A standardized protocol for the determination of specific leaf area and leaf dry matter content. *Functional Ecology,* 15, 688-695.

Duru M., G. Lemaire & P. Cruz (1997). The nitrogen requirement of major agricultural crops. Grasslands. In: G. Lemaire (ed.) Diagnosis of N Status in Crops, Springer-Verlag, 59-72.

Adapting the CROPGRO model to predict growth and perennial nature of bahiagrass

S.J. Rymph[1], K.J. Boote[1] and J.W. Jones[2]
[1]*University of Florida Agronomy Department, P.O. Box 110500 Gainesville, Florida 32611 USA, Email: kjb@ifas.ufl.edu, *[2]*University of Florida Department of Agricultural and Biological Engineering, P.O. Box 110570, Gainesville, Florida 32611 USA.*

Keywords: crop model, forage, tissue N concentration, tropical perennial grass, *Paspalum notatum*

Introduction The objective of this research was to modify an existing crop growth model for ability to predict growth and composition of bahiagrass (*Paspalm notatum* Flügge) in response to daily weather and management inputs. The CROPGRO–CSM cropping systems model has a generic, process-oriented structure that allows inclusion of new species and simulating cropping sequences and crop rotations. An early adaptation of CROPGRO-CSM "species files" for bahiagrass over-predicted growth during late fall through early spring, and totally failed in re-growth if all foliage was lost from freeze damage. Revised species parameters and use of "pest damage" offered only a partial solution. Three processes, absent from the annual CROPGRO-CSM model, contributed to prediction of excessive cool-season growth: (1) no provision for storage (reserve) structures, (2) lack of winter dormancy, and 3) freeze damage killed all leaves at once and resulted in crop death. In addition, the model lacked the CO_2-concentrating effect of C_4 photosynthesis in the leaf photosynthesis routines. Therefore, we modified the source code of CROPGRO to include these processes to improve biological accuracy of re-growth patterns and prediction of seasonal patterns of growth (Rymph *et al.*, 2004).

Materials and methods A new plant organ (STOR) was added to simulate stolons, thereby serving as a perenniating sink and source for storing carbohydrate and N. Partitioning of new growth among leaves, stems, roots, and STOR is assumed to shift toward STOR with increasing vegetative maturity. New functions were added to promote re-growth after harvest and in the spring. These include increasing the mobilisation of N and carbohydrate (CH_2O) from STOR progressively more rapidly as LAI falls below 3.0 and more rapidly as whole-plant N status increases from 30 to 70% of potential N-status. Functions for dormancy were added to regulate the degree of partitioning to STOR versus leaf and stem, and to regulate rate of mobilisation of CH_2O and N from STOR. Dormancy is initiated when day-length is <12.5 h, becoming progressively stronger (maximum) as day-length reaches 10.5 h. The strength of this dormancy signal acts to increase partitioning to STOR, decrease mobilisation of N and CH_2O from STOR and roots, and reduces herbage and root growth (because of partitioning shift). As day-length increases above 10.5 h in spring, the process is reversed.
The freeze damage process was modified to use a "death constant" that reduces leaf and stem mass 5% for each degree of minimum daily temperature <-5°C. A lethal freezing temperature threshold was defined as the low temperature required to kill the STOR organ.
A "CO_2-concentrating factor" was added to the leaf-level photosynthesis code to simulate C_4 photosynthesis, accounting for effect of high CO_2 concentrations in bundle sheath chloroplasts of C_4 plants. This factor reduces sensitivities of quantum efficiency and leaf photosynthetic rate to atmospheric CO_2 concentration.

Results Five experiments from Texas and Florida were simulated with the modified version of CROPGRO. Predicted herbage mass, herbage N concentration, and herbage N mass were compared to measured values (303 observations). Seasonal patterns of growth were more accurately predicted to mimic low leaf and stem growth during winter, and increase in STOR mass during fall through early spring. STOR mass is used (depleted) to support re-growth in spring and after each harvest. The crop now survives winter freeze events that kill all green foliage. Performance of the new model version was improved with better index of agreement (d-index) values for predicted herbage mass of 0.81 and 0.82 (out of a possible 1.0) using leaf-level and daily canopy photosynthesis options, respectively, compared to 0.71 and 0.73 for the older CSM version (slope, intercept, and r^2 were also improved). Prediction of herbage N concentration was similarly improved. Predicted leaf-level photosynthetic response to elevated CO_2 was of the same magnitude as observed in phytotron measurements. Predicted N-stress in the model is excessive, but the cause has not been identified. Additional work is needed on N-related parameters of the crop and soil to reduce the predicted N stress and refine model response to N.

Conclusions The modified forage version of CROPGRO marks a significant step in adapting this model to more accurately reflect the perennial and seasonal patterns of organ growth of bahiagrass. Summer vs. winter patterns of herbage production, herbage N concentration, leaf, stem, root, and stolon growth are more accurately predicted. Sensitivity analyses show reasonable leaf growth and stolon storage dynamics after cutting harvest.

References
Rymph S.J. (2004). Modeling growth and composition of tropical perennial forage grasses. Ph.D. diss. University of Florida, Gainesville.

State and transition model of lowland grassland in Flooding Pampa

E.J. Jacobo, A.M. Rodríguez and V.A. Deregibus
Departamento de Producción Animal, Facultad de Agronomía, Universidad de Buenos Aires. Av. San Martín 4453, Buenos Aires 1417, Argentina, Email ejacobo@agro.uba.ar

Keywords: temperate grasslands, plant community dynamics, functional groups, grazing methods

Introduction Rainfall conditions are considered to be a major factor in determining vegetation structure in temperate grasslands with grazing playing a secondary role (Biondini *et al.*, 1998; Sternberg *et al.*, 2000). In order to analyse the relative importance of both factors on the lowland community of the Flooding Pampa we compared the responses of functional groups under both intermittent and continuous grazing regimes over a 3-year period of important inter seasonal rainfall variation. The results are presented in a state and transition model.

Materials and methods The experiment was conducted from 1993 to 1996. Total annual rainfall was 1465 mm in 1993, 1156 mm in 1994, 576mm in 1995 and 845 mm in 1996. Since March 1989, traditional continuous grazing was replaced by intermittent grazing in four commercial farms located in different sites of the Flooding Pampa. In each site, a near farm managed under continuous grazing was assessed as a control. Two main classifying effects were examined: grazing method and year and sites were treated as blocks. Average stocking rate in all farms was 1 breeding cow/ha. Proportion of basal cover of different functional groups was monitored. Analysis of variance (ANOVA) was used to evaluate the effect of grazing method and year on each response variable. Specific states and transitions were proposed based upon measurements of the vegetation made over the analysed period.

Results Hydrophytic grasses and sedges dominated this community in the extremely wet year (1465 mm), when water saturation of soils occurred during several month, irrespective of the grazing method. As annual rainfall fell (575 mm), sedges decreased and plant community followed different pathways depending on grazing management: under continuous grazing, C$_4$ grasses and forbs replaced sedges, whereas under intermittent grazing hydrophytic grasses maintained an important cover and legumes tended to increase (Figure 1).

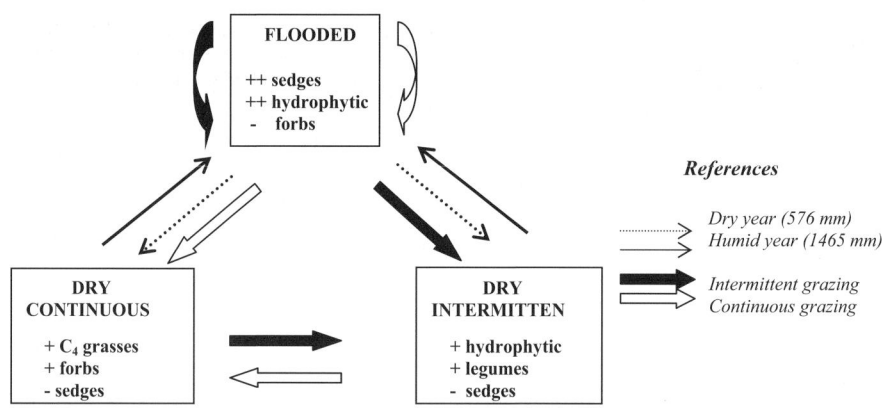

Figure 1 States and transitions of lowland community in the Flooding Pampa

Conclusions Opposing with semi arid grasslands, grazing effect on plant composition in lowland community of Flooding Pampa grasslands was inversely related to annual rainfall. In humid years, grazing method did not change plant composition. In dry years, intermittent grazing promotes functional groups composed of high forage value species, suggesting an improvement in rangeland condition.

References

Biondini, M.E., B.D.Patton & P.E. Nyren (1998) Grazing intensity and ecosystem processes in a northern mixed-grass prairie, USA. *Ecological Applications*, 8, 469-479.

Sternberg, M., M. Gutman, A. Perevolotsky, E.D. Ungar & J. Rigel (2000) Mediterranean herbaceous community: a functional group approach. *Journal of Applied Ecology*, 37, 224-237.

The influence of growing degree days on *Robinia pseudoacacia* browse quality and productivity in the southeastern USA

L.J. Unruh Snyder[1], J.P. Mueller[1], J-M. Luginbuhl[1], K.E. Turner[2] and C. Brownie[1]
North Carolina State University, Raleigh, North Carolina 27695, USA, Email: Paul_Mueller@ncsu.edu, [2]USDA ARS, Beaver, West Virginia 25813 2409 Williams Hall, Box 7620, North Carolina State University, Raleigh, North Carolina 27695 USA

Keywords: goat, heat units, herbage mass, prediction equations, *Robinia pseducacacia*, tannins

Introduction The possibility of estimating browse quality and productivity of black locust (BL; *Robinia pseudoacacia* L.) herbage from accumulated air temperature heat units (growing degree days, GDD) could be a valuable tool for researchers and graziers in efficient allocation of feed resources. Accumulated air temperature heat units (GDD) above a 10° C base have been used to predict several forage quality constituents (Onstad & Fick ,1983).

Materials and methods Data were collected for two years to determine the relationship between GDD and herbage mass (HM), estimates of herbage quality N, IVTDMD , NDF, ADF, CELL, and ADL), and estimates of anti-quality agents (Folin-reactive phenolics [FR-phenol], condensed tannins [CT], and hydrolysable tannins [HT]) of a five-year old stand of BL in Raleigh, NC. Herbage mass samples were hand plucked and used to evaluate browse quality variables.

Results With the exception of IVTDMD (58%), 2-yr means of BL herbage quality estimates were high (4.2% N, 37% NDF, and 26% ADF). The 2-yr means for CELL and ADL were 8 and 16%, respectively. Concentrations of FR-phenol, CT, and HT averaged over years were 7.9, 7.6, and 8.0% DM, respectively. Condensed tannins have been reported to depress voluntary intake and ruminal fibre digestion (Terrill *et al.*, 1992). Regression analyses were performed by year using GDD as the independent variable. In 1999, there was a significant relationship between GDD and HM, H, NDF, ADF, CELL, ADL, N, and IVTDMD (Table 1). In 2000, the only significant relationships with GDD were N, ADF, and CELL. High rainfall in 2000 caused a flush of new herbage growth that most likely masked the maturity effects demonstrated in the previous year. Horner (1988) reported that non-linear trends might exist when dilution effects caused by leaf expansion occur.

Conclusions In years with moderate to low rainfall, without multiple growth flushes, GDD appear to be closely related to HM, NDF, ADF, CELL, ADL, N and IVTDMD. Under these conditions GDD could serve as a useful predictor of herbage quality and productivity. In both years GDD was a poor predictor of BL tannin concentrations

References

Horner, J.D (1988). Astringency in douglas fir foliage in relation to phenology and xylem pressure potential. *Journal of Chemical Ecology*, 14, 1227-1237.

Onstad, D.W., & G.W. Fick (1983). Predicting crude protein, in vitro true digestibility, and leaf proportion in alfalfa herbage. *Crop Science*, 23, 961-964.

Terrill, T.H., G.B. Douglas, A.G. Foote, R.W. Purchas, G.F. Wilson, & T. N. Barry (1992). Effect of condensed tannins upon body growth, wool growth and rumen metabolism in sheep grazing sulla (*Hedysarum coronarium*) and perennial pasture. *Journal of Agricultural Science*, 119, 265-273.

Table 1 Linear regression equations for NDF, ADF, CELL, ADL, N, IVTDMD, FA-phenol, CT, HT, H, and HM as a function of GDD from 01 March 1999 and 2000, Wake County, North Carolina, USA[†]

Year Regression Equations	SE[‡]	r^{2}[§]
1999 (g/kg except for HM = kg /ha)		
NDF=134.4 +0.34(GDD)	23.7	0.93***
ADF=76.6 +0.24(GDD)	24.9	0.85***
CELL=52.5+0.04(GDD)	3.3	0.92***
ADL=23.8+0.19(GDD)	26.1	0.77**
N=58.2-0.02(GDD)	2.9	0.65***
IVTDMD=885.7 -0.38(GDD)	45.4	0.82***
CT=9.3 -0.003(GDD)	1.1	0.34*
HM=-434.7 +2.2(GDD)	293.7	0.77***
2000 (g/kg)		
ADF=373.2 -0.07(GDD)	28.3	0.32*
CELL=104.9 -0.01(GDD)	6.8	0.26*
N=46.8 -0.008(GDD)	1.9	0.58***

[†] NDF, neutral detergent fibre; ADF, acid detergent fibre; CELL, cellulose; ADL, acid detergent lignin; N, nitrogen; IVTDMD; in vitro true dry matter disappearance; FR-phenol, Folin-reactive phenolics; CT, condensed tannins; HT, hydrolysable tannins; HM, herbage mass; GDD, growing degree days $=\sum\{[(T\ air_{max} + T\ air_{min}) / 2] - 10\ °C\}$ (from 01 March); [‡] SE= Standard error of mean; [§] r^{2} = coefficient of simple determination; *, **, *** Significant at P= 0.05, 0.01, and 0.001 levels, respectively.

Modelling tiller density dynamics in a grass sward

M. Hirata

*Division of Grassland Science, Faculty of Agriculture, University of Miyazaki, Miyazaki 889-2192, Japan,
Email: m.hirata@cc.miyazaki-u.ac.jp*

Keywords: model, tiller population density, grass sward, bahia grass

Introduction Simulation models are useful tools for grassland management. Among many quantitative and qualitative attributes describing vegetation of grasslands, density of plant population is important because of its close relationship with persistence of grasslands (Hirata, 2004), which in turn is crucial for sustainable agricultural production and/or conservation of the environment, wildlife and recreational resource. Although various models have described grassland vegetation, relatively few models have dealt with plant population density. This paper presents a model describing dynamics in tiller population density in a grass sward.

The model The framework of the model is shown in Figure 1. Since the present model is intended to be a submodel of an integrated model for canopy dynamics in a grass sward, it requires daily herbage mass as an input. The model also needs daily mean air temperature, annual nitrogen fertiliser rate and month of the year as inputs, and initial tiller density at the commencement of a simulation.

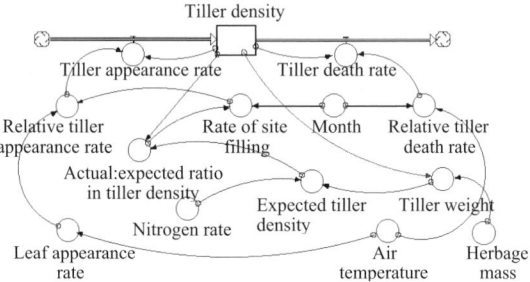

Figure 1 The framework of the model.

Results The model was parameterized or calibrated to data from bahia grass swards under different nitrogen fertiliser rates and cutting heights (Hirata & Pakiding, 2004). The simulated tiller densities mostly showed good agreement with observed densities, i.e. tiller densities increased as the cutting height decreased and as the nitrogen rate increased under the lowest cutting height (Figure 2). The parameterisation results were thus successfully tested. The model was then validated against data from bahia grass swards under cattle grazing (4 years data) and under different cutting heights (3 years data). As a whole, the validation results were acceptable (not shown as figures).

Conclusions The results show potential value of the present model as a submodel of an integrated model of a grass sward canopy.

References

Hirata, M. (2004). Canopy dynamics in bahia grass (*Paspalum notatum*) swards. In: S.G. Pandalai (ed.) Recent Research Developments in Crop Science, Vol. 1. Research Signpost, Kerala, India, 117-145.

Hirata, M. & W. Pakiding (2004). Tiller dynamics in bahia grass (*Paspalum notatum*): an analysis of responses to nitrogen fertiliser rate, defoliation intensity and season. *Tropical Grasslands*, 38, 100-111.

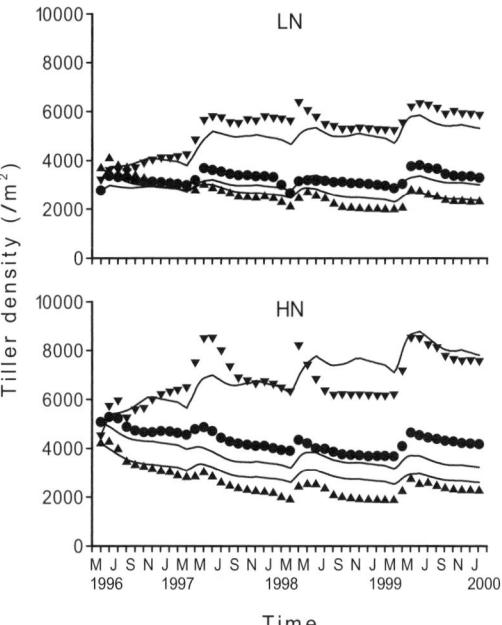

Figure 2 Observed (symbols) and simulated (lines) tiller population densities in bahia grass swards under different nitrogen rates and defoliation intensities (parameterisation results). LN=low nitrogen (5 g/m²/year), HN=high nitrogen (20 g/m²/year), Low (2 cm, ▼), medium (12 cm, ●) and high (22 cm, ▲) heights of cutting above ground.

Leaf and tiller dynamics in centipede grass and bahia grass

M.A. Islam and M. Hirata
Division of Grassland Science, Faculty of Agriculture, University of Miyazaki, Miyazaki 889-2192, Japan,
Email: a.islam@cc.miyazaki-u.ac.jp

Keywords: leaf appearance, death and detachment, tillering, centipede grass, bahia grass

Introduction Centipede grass (*Eremochloa ophiuroides* (Munro) Hack.) is a warm-season perennial which has received the attention of farmers and researchers as a new forage resource for sown pastures in the low-altitude regions of south-western Japan where bahia grass (*Paspalum notatum* Flügge) has been widely used (Islam & Hirata, 2005). Leaf and tiller dynamics provide the basis to explain variation in production and canopy structure of a grass sward (Rhodes & Collins, 1993), and knowledge of the dynamics can be used as a tool for sward management. However, this information is lacking for centipede grass (Islam & Hirata, 2005). The aim of this study was to obtain information on leaf and tiller dynamics of centipede grass in comparison with bahia grass.

Materials and methods There were eight plots (1×1 m) each of which consisted of a turf (0.3×0.3 m) of centipede grass (cv Common) at the centre with the remainder grown with bahia grass (cv Pensacola). The plots received two levels of N (low=5 g/m^2 per yr, high=20 g/m^2 per yr; 4 plots for each N level). Five vegetative tillers were randomly tagged in each of the grasses in each plot, and rates of leaf appearance (LAR), death (LDR) and detachment as litter fall were measured at monthly intervals by recording number and state of leaves on the tillers. The number of daughter tillers initiating from the axils of the youngest six leaves were also counted.

Results The effects of N on all measures were minimal. LAR was lower in centipede grass than in bahia grass (0–0.13 vs 0–0.23 leaves/tiller per d) while LDR (0.004–0.195 leaves/tiller per d) and leaf detachment rate (0.004–0.184 leaves/tiller per d) were similar for both grasses, resulting in lower number of live leaves per tiller in centipede grass than in bahia grass (1.9–7.1 vs 3.8–10.9). For both grasses LAR was positively related to air temperature with maximal LAR lower in centipede grass than in bahia grass (Figure 1). The number of daughter tillers per tiller in centipede grass was four times higher than in bahia grass (1.45 vs 0.36) due to the higher probability of tillering at axils of leaves. The probability of having daughter tillers at the axil of the sixth leaf in centipede grass and bahia grass averaged 41 and 6% respectively over the growing season (May–Oct.), and 28 and 0% respectively over the remaining time (Nov.–March) (Figure 2).

Conclusions Centipede grass had lower LAR and fewer live leaves per tiller than bahia grass. However, centipede grass showed much higher tillering ability and number of daughter tillers per tiller than bahia grass. This is a key mechanism by which centipede grass forms a denser, leafier sward and attains similar production to bahia grass (Islam & Hirata, 2005).

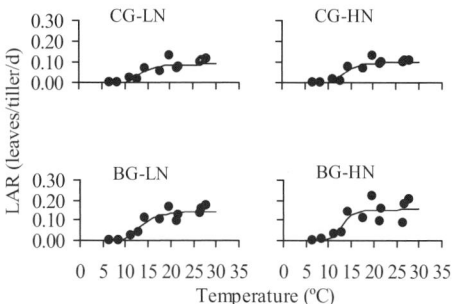

Figure 1 Relationship between LAR and mean daily air temperature in centipede grass (CG) and bahia grass (BG) with low (LN) and high (HN) N. Equations: CG-LN, y=0.089((x-6.6)/7.0)$^{4.5}$/(1+((x-6.6)/7.0)$^{4.5}$), r=0.89, p=0.004; CG-HN, y=0.099((x-6.6)/7.0)$^{4.5}$/(1+((x-6.6)/7.0)$^{4.5}$), r=0.94, p=0.003; BG-LN, y=0.145((x-6.6)/7.1)$^{3.6}$/(1+((x-6.6)/7.1)$^{3.6}$), r=0.93, p=0.006; BG-HN, y=0.155((x-6.6)/6.5)$^{5.3}$/(1+((x-6.6)/6.5)$^{5.3}$), r=0.84, p=0.019

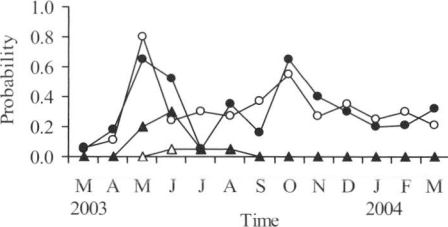

Figure 2 Probability of having daughter tillers at the axil of the sixth youngest leaf for low (○) and high (●) N in centipede grass, and low (△) and high (▲) N in bahia grass. Symbols are not visible when overlaping

References

Islam, M.A. & M. Hirata (2005). Centipede grass (*Eremochloa ophiuroides* (Munro) Hack.) – its growth behaviour and multipurpose usages: a review. *Grassland Science*, 51, (in press).

Rhodes, I. & R.P. Collins (1993). Canopy structure. In: A. Davies, R.D. Baker, S.A. Grant & A.S. Laidlaw (eds.) Sward Measurement Handbook. British Grassland Society, 139-156.

Modelling winter grass growth and senescence

D. Hennessy[1,2,3], S. Laidlaw[3], M. O'Donovan[2] and P. French[2]
[1]Teagasc Beef Research Centre, Grange, Dunsany, Co. Meath, Ireland, dhennessy@grange.teagasc.ie [2]Teagasc Dairy Research Centre, Moorepark, Fermoy, Co. Cork, Ireland [3]Queens University Belfast, Crossnacreevy, Belfast, BT6 9SH, Northern Ireland, UK

Keywords: tissue turnover, winter, model, modelling

Introduction In temperate climates, because net grass growth in winter is low, most grass growth models deal with the main growing season (Mar-Oct in the N Hemisphere), with little emphasis on grass growth in winter (Nov-Feb). However, grass tissue turns over continuously (Hennessy et al., 2004) and the fate of herbage entering the winter is important in extended grazing season systems. This study aimed to model winter grass growth for the period 15 Oct 2001 to 28 Jan 2002 for a range of autumn closing dates (1 Sep, 20 Sep and 10 Oct) by modifying an existing model, so that the amount of green leaf could be predicted at intervals over the winter.

Materials and methods The model of Johnson and Thornley (1983) was selected for modification. This is a vegetative grass growth model incorporating leaf area expansion and leaf senescence. It was run in Excel. The model suited the requirements of this study as it characterises leaves according to age (in line with tissue turnover concepts) and it was designed for an established vegetative grass crop supplied with unlimited nutrients and water. Tillers are mainly vegetative in Ireland during the winter, and water and nutrients are seldom limiting. As tissue turnover data were available from two sites, data from site 1 (Grange) were used to develop coefficients to modify the model, while data from site 2 (Moorepark) were used to validate the model. Mean daily air temperature and radiation, initial leaf area index (LAI) and amount of leaf in each age category were inputs to the model. The output predicted LAI at intervals over the winter. Daily meteorological data for the experimental period and latitude for the site, and initial lamina and sheath weight/unit area and LAI on 15 Oct were the inputs to the modified model. Leaf appearance rate (LAR) was modified for the winter period based on a simple regression equation between measured LAR and temperature. Coefficients for leaf senescence rate (LSR) of the 2nd and 3rd youngest leaves were derived from Grange data, based on the flux of material between leaf age categories. These coefficients were varied and the model run until the output (predicted LAI) was similar to the measured LAI for each of the closing date treatments at Grange. The output of the modified model was tested against actual LAI from Moorepark. Measured and predicted LAI were compared using mean squared prediction error (MSPE). Mean prediction error (MPE) was calculated also.

Results The model predicted the rapid decline in LAI on the 1 Sept and 20 Sept closing date treatments (Table 1.) However, the model predicted an increase in LAI on the 10 Oct treatment, which did not occur. Overall MSPE was 0.47. Most of the variation in the model was random (0.728) and LAI was not consistently over or under predicted (low mean bias value of 0.085). Where differences between measured and predicted LAI occurred they were short lived e.g. on 5 Nov on the two earlier closing treatments. This may be explained by an over-prediction of LAI as autumn moves into winter at the end of Oct/start of Nov.

Table 1 Measured and predicted LAI over the winter for 3 closing dates for Moorepark (the validation site)

Closing date	1 September		20 September		10 October	
	Measured LAI	Predicted LAI	Measured LAI	Predicted LAI	Measured LAI	Predicted LAI
5 Nov	5.47	4.14	4.02	2.76	2.82	2.11
26 Nov	2.98	3.37	3.61	2.72	2.53	2.61
17 Dec	2.35	2.67	2.61	2.46	2.34	2.84
7 Jan	2.23	2.22	2.32	2.20	2.28	2.85
28 Jan	1.36	1.11	1.59	1.26	2.17	2.42

MSPE = 0.47; MPE = 0.25; R^2 = 0.65; Mean bias = 0.085; Line bias = 0.187; Random variation = 0.728

Conclusions This study suggests that it is possible to model winter leaf growth and senescence, and hence LAI, when swards are closed in autumn at a range of dates and at more than one sward state or site. This has potential for use as the basis of a winter grass growth model.

References

Hennessy, D., P. French, M. O'Donovan & A. S. Laidlaw (2004). Tissue turnover during the winter in a perennial ryegrass sward. *Grassland Science in Europe*, 9, 766-768.

Johnson, I. R. & J. H. M. Thornley (1983). Vegetative crop growth model incorporating leaf area expansion and senescence, and applied to grass. *Plant, Cell and Environment*, 6, 721-729.

Leaf dry matter content of native grassland species under contrasting N and P supply

P. Cruz, C. Jouany, J.M. Enjalbal and M. Duru
Research unit UMR 1248 Managing grassland farming systems, BP 27, F - 31326 Castanet -Tolosan, France,
Email: cruz@toulouse.inra.fr

Keywords: species ranking, leaf functional traits, nitrogen, phosphorus

Introduction The management of native grasslands - herbaceous vegetation with a broad diversity of flora and a large range of uses - must meet the requirements of environmental conservation and improvement of the quality of agricultural production. For this purpose we need tools for diagnosing the state of the vegetation in order to design, evaluate, and apply management practices to attain these objectives. These tools must be simple and quick to use and should not require botanical skills. Leaf dry matter content (LDMC) has been proposed as a good indicator of both fertility gradients and species preference for habitats (Cruz *et al.*, 2002). The aim of this work was to test the robustness of this leaf trait to rank species for differences in their growth strategies and nutrient acquisition.

Material and methods Seeds of native populations of species having a strategy of resource conservation, i.e. *Brachypodium pinnatum* (Bp) and *Festuca rubra* (Fr) or resource capture, i.e. *Lolium perenne* (Lp), *Dactylis glomerata* (Dg) and *Rumex acetosella (Ra)* were harvested in natural meadows of the Ercé Valley (French Central Pyrenees) and cultivated under glasshouse conditions (60% of incoming solar radiation) in pots filled with an inert sand. The experimental design was a factorial 2x2 with six replicates combining two levels of nitrogen (1.9 and 0.05 meq/l for N+ and N- respectively) and two levels of phosphorus (1.85 and 0.7 meq/l for P+ and P- respectively). Dry matter (DM) of roots, leaves and stems were measured 12 weeks after sowing. Leaf DM content was determined on all of the green leaves after the last harvest as the ratio leaf oven-dry mass / leaf fresh mass (mg/g). The measured leaves were considered to be turgid because the plants had been thoroughly watered. Spearman's rank correlation between nutrition levels was performed on Statistix for Windows.

Results Nutrition levels for P and N had a strong effect on mass allocation among plant organs (Table 1). However, species with a conservative growth strategy (*Bp* and *Fr*) were less affected than others, principally under N shortage. In spite of large differences observed in mass allocation among species and nutrition levels, LDMC values were stable and the species ranking was unaffected. The Spearman's coefficient of correlation was equal to 1 whatever the nutritional levels compared.

Table 1 Relative (%) dry matter of leaf (l-DM) stems (s-DM) and root (r-DM) respective to 100% in N+P+ treatment (n=6)

		l-DM	s-DM	r-DM
	N+P-	0.8	0.5	6.6
Lp	N-P+	23.4	18.7	135.7
	N-P-	0.6	0.3	5.2
	N+P-	0.5	0.4	1.4
Dg	N-P+	15.5	20.9	37.3
	N-P-	0.5	0.4	1.6
	N+P-	2.0	1.4	7.7
Fr	N-P+	40.2	43.7	126.0
	N-P-	1.7	1.3	6.3
	N+P-	5.2	2.2	9.3
Bp	N-P+	103.2	87.6	115.0
	N-P-	4.5	2.3	8.1
	N+P-	0.2	0.2	0.9
Ra	N-P+	11.7	7.8	33.1
	N-P-	0.3	0.2	0.93

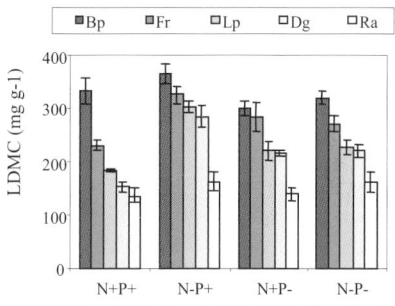

Figure 1 Species ranked by LDMC values under contrasting levels of N and P supply (see M&M for species and nutritive solution codes). Bars are the standard deviation of means (n=6)

Conclusions The LDMC of water-saturated leaves can be considered as a biological marker of native species, largely independent of both growth conditions (mineral nutrition in this case) and plant growth strategies. This attribute makes of this leaf trait a useful indicator to be included in management tools of natural grasslands based on a functional approach to the vegetation (Lavorel and Garnier, 2002).

References

Cruz P, P. Sire, Al Haj Kalhed, J.P. Theau, O. Therond & M. Duru (2002). Plant functional traits related to growth strategies and habitat preference of native grass populations. *Grassland Science in Europe*, 7, 776-777.
Lavorel S. & E. Garnier (2002). Predicting changes in community composition and ecosystem functioning from plant traits: revisiting the Holy Grail. *Functional Ecology,* 16, 545-556.

The use of near infrared reflectance spectroscopy (NIRS) to follow the leaf/stem ratio of legumes during drying

D. Stilmant[1], V. Decruyenaere[1], C. Clément[1], P. Dardenne[1] and N. Grogna[1]
[1]Walloon Agricultural Research Centre - Farming Systems Section, 100 rue du Serpont, B-6800 Libramont, Belgium, Email: stilmant@cra.wallonie.be, [2]Agricultural Product Quality Department

Keywords: drying losses, sward morphological composition, sward analytical tool

Introduction Legume-rich mixed swards allow the production of a high quantity protein-rich forage with low nitrogen input. Nevertheless, during hay or silage making, dry matter losses as high as, 40 and 25 % have been recorded (Ciotti & Cavallero, 1979; Stilmant et al., 2004). These losses have mainly been linked to the high sensitivity to physical loss of legume leaves during drying. The development of a tool to characterise leaf losses or leaf/stem ratio during drying will help us to define the technical approach to reach the best compromise between quality loss reduction and good pre-wilting of legum-rich mixed swards. The aim of the present work was to test the potentialities of near infrared reflectance spectroscopy (NIRS) to quantify legume leaf/stem ratio in mixed grass-legume swards. The mixtures tested were perennial ryegrass-white clover (PR-WC), perennial ryegrass-red clover (PR-RC), timothy-red clover (T-RC) and cocksfoot-lucerne (C-L) swards. This technique has been successfully used to quantify leaf/stem ratio in pure perennial ryegrass swards (Leconte et al. 1999).

Materials and methods Material used in this study came from 64 sward samples harvested at two stages of development (flowering and vegetative stages) on the mixed swards listed above. These swards were also used to study drying losses (Stilmant et al., 2004). The samples were sorted, by hand separation, into four fractions : grass, legume leaf, legume stem and rest (dead material, weeds, …). With these different fractions, 883 samples were created : 140 to 270 samples per grass-legume association. Legume stem varied between 18 and 95 %, legume leaf varied between 0 and 50 % while grass and rest materials were in the 0 to 50 % range. All samples were submitted to NIRS analysis (NIRS system monochromator 5000). Spectral data, in the range of 1100 – 2500 nm by 2 nm steps were correlated to legume leaf or stem, grass and rest fractions. Calibrations were developed according to the Partial Least Square procedure with cross validation of the ISI software.

Results and Conclusions With R^2 always higher than 0.95 and SD/SECV ratio higher than 4 (Williams, 2004), the performances of the different calibrations (Table 1) allow their use to quantify the evolution of the different fractions of a grass-legume mixture during drying as illustrated, for legume leaf fraction in Figure 1 (modified from Stilmant et al., 2004). According to these results, NIRS appears a promising tool to define the best management rules to follow during the drying of legume-rich mixed swards.

Table 1 Calibration performances for the different swards fractions (SD : Standard Deviation, SECV : Standard Error in Cross Validation)

	N	Mean	SD	R^2	SECV	SD/SECV
Legume leaf	868	20.2	12.77	0.98	2.24	5.7
Legume stem	869	50.9	15.93	0.97	2.80	5.7
Grass	869	15.7	12.30	0.97	2.25	5.5
Rest	872	12.7	8.49	0.96	1.96	4.3

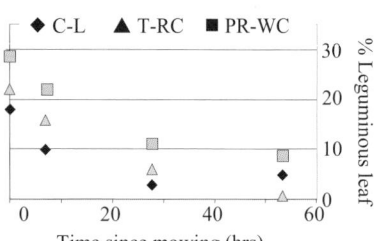

Figure 1 Leaf ratio evolution of legumes during hay making (Stilmant et al. 2004)

References
Ciotti A. & A. Cavallero (1979). Haymaking losses in cocksfoot, lucerne and a cooksfoot-lucerne mixture in relation to conditioning and degree of drying at harvest .Occasional Symposium 11, British Grassland Society, 214-220.
Leconte D., P. Dardenne, C. Clément & Ph. Lecomte (1999). Near infrared determination of the morphological structure of rye grass swards. In: A.M.C. Davies & R. Giangiacomo (eds.) Near Infrared Spectrometry : Proceedings of the 9th International Conference, NIR publications, 41-44.
Stilmant D.,V. Decruyenaere, J. Herman & N. Grogna (2004). Hay and silage making losses in legume-rich swards in relation to conditioning. Grassland Science in Europe, 9, 939-941.
Williams, P. (2004). Near-Infrared Technology – Getting the Best Out of Light. A Short Course in the Practical Implementation of Near-infrared Spectrometry for the User. PDK Grain, Manitoba, Canada.

Near infrared spectroscopy of faeces to predict diet quality in grazing animals: development of a portable system

D.R. Tolleson and J.W. Stuth
Texas A&M University Department of Rangeland Ecology and Management, 2126 TAMU College Station, Texas, USA 77843-2126, Email: tolleson@cnrit.tamu.edu

Keywords: diet quality, faeces, grazing, near infrared spectroscopy, portable

Introduction Faecal near infrared spectroscopy (FNIRS) has been used to predict dietary crude protein (CP) and digestible organic matter (DOM) in grazing animals (Stuth *et al.*, 2003, Coates 2000). Development of robust FNIRS calibrations can be time consuming and costly, thus hindering the application of FNIRS in developing countries. Delivery of samples to central laboratory facilities is dependant upon adequate transportation infrastructure. A "take the laboratory to the samples" approach is being tested in Mongolia using a portable FNIRS laboratory (Stuth *et al*,. 2004). The initial step in this process is to duplicate the performance of a static laboratory procedure with portable equipment. The objective of this study was to determine the effectiveness of re-creating existing FNIRS diet quality calibration models on a portable spectrometer.

Materials and methods Near infrared absorbance spectra (800-1700 nm) were collected on an Ocean Optics® NIRS512 portable spectrometer from 42 faecal samples which had previously been used to develop sheep (*Ovis aries*) diet quality predictive equations (Li *et al.*, 2004) on a bench top spectrometer (400-2500 nm). Diet reference chemistry:faecal spectrum calibration pairs in the current study were thus, the values for CP and DOM as reported by Li *et al.* (2004) matched with new spectra obtained on the portable spectrometer. To test biological relevance of the spectra, calibration was also attempted for random numbers between 1 and 100. Partial least squares regression (n=6 factors) was used to develop predictive equations.

Results Absorbance spectra from the portable spectrometer are visibly similar to those from the bench-top spectrometer. To illustrate, 3 samples are represented (Figure 1). Predictive equation performance statistics are listed in Table 1.

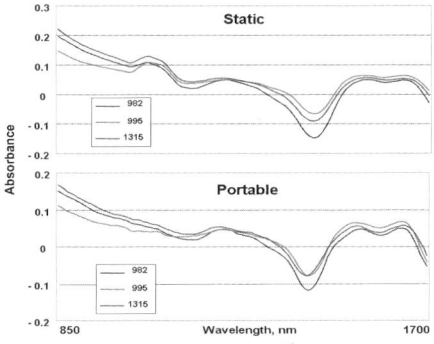

Figure 1 Comparison of 1st derivative spectra, 3 faecal samples, static vs. portable NIRS

Table 1 Portable FNIRS calibration statistics: CP, DOM, and random numbers

Constituent	RSQ	SEC	SEP
CP	0.96	0.97	1.20
DOM	0.92	0.84	1.25
Random #'s	0.32	12.3	17.2

RSQ and SEC represent the full calibration model. (n = 42)
SEP represents a 10 sample validation subset predicted with a smaller (n = 32) calibration model.

Conclusions Preliminary efforts with portable FNIRS indicate that it is possible to re-create calibration equations developed on a bench top, static laboratory spectrometer.

References

Coates D.B. (2000). Faecal NIRS - What does it offer today's grazier? *Tropical Grasslands,* 34, 230-239.
Li H., J.W. Stuth, D.R. Tolleson & S.L. Kronberg (2004). Near infrared reflectance spectroscopy (NIRS) of feces to determine dietary crude protein and digestible organic matter in sheep. *Proceedings: Society for Range Management meetings*, Salt Lake City, Utah.
Stuth J.W., A.A Jama & D.R. Tolleson (2003). Direct and indirect means of predicting forage quality through near infrared reflectance spectroscopy. *Field Crops Research*, 84, 45-56.
Stuth J.W. (2004). GOBI FORAGE: Forage Monitoring Technology to Improve Risk Management by Herders in the Gobi Region of Mongolia. Global Livestock CRSP Annual Accomplishments Report. UC Davis, California.

Potentialities of near infrared spectroscopy to assess nitrogen, phosphorus and potassium nutrient status of grasslands in the Reunion Island

Ph. Lecomte[1], V. Blanfort[1], M. Duru[2], P. Thomas[3] and P. Grimaud[1]
[1]*Centre International en Recherche Agronomique pour le Développement (CIRAD), Pole Elevage; 97410, St Pierre, La Réunion Isld. France, Email: philippe.lecomte@cirad.fr,* [2]*Institut National de la Recherche Agronomique (INRA), Toulouse, France,* [3]*Union des Associations Foncières et Pastorales (UAFP), La Réunion, France*

Keywords: near infrared spectroscopy, nitrogen, potassium, phosphorus, grasslands

Introduction   Controlled mineral fertilisation practices are an important component for   sustainable management of grasslands. The assessment of available nutrients for plants and the general recommendations on the level of phosphorus and potassium to apply to grasslands are classically based on classical soil analysis and average regional levels. For nitrogen, mid or long term recommendations cannot easily be derived solely from soil composition, because it may be rapidly leached from the soil. Recent approaches tended to show that herbage plant N (Lemaire & Gastal, 1997), P, K (Duru & Huché, 1997) mineral analyses associated with actual biomass measurement could be useful for the calculation of combined nutrient indices (IN, IP, IK). Expressing these indices along references curves with a standard optimum value of 100, indicates the limiting factors or excess in the mineral feeding of the plants. It provides a diagnosis of the main nutrient status at a specific local plot situation. The step has been successfully implemented to provide local advice in the management of grasslands on Reunion Island (Blanfort, 1998). Nitrogen content can be predicted from NIRs, but this technique is less used for the other elements. However, the concern is here more related to the development of a combined index, it appeared interesting to test the potential of NIRs to predict these or to rank grasslands according to nutrition levels.

Material and methods A large set of 900 milled herbage samples referenced for N, P, K % and DM content were scanned using Nirsystem 6500 (400-2500 nm). Index IN, is calculated according to nitrogen reference content dilution in the measured sward dry biomass DMHA, t/ha: (IN = 100 × N% / 4.8 × DMHA -.32). The IP and IK indexes are adjusted ratios of P and K to N content (IP = 100 × P% / (0.15 + 0.065 × N%); IK = 100 × K% / (1.6 + 0.525 ×N%)). Spectral and calculated indexes were put into calibration. (SNVD scatter correction, 2.5.5.1 pretreatment using the Modified Partial Least Squares procedure, WinisiII 1.5 software).

Results Within the ranges 19 - 106 (IN), 29 -199 (IK), 38 -143 (IP), the calibration cross validated standard error (Secv) and R squared (R²cv) values were respectively : 7.5-0.81 for IN; 9.2-0.90 for IK and 11.9-0.62, for IP. The narrowness and precision of the relationships for N and P indices were lower than generally observed with organic components, the NIRs predictions appear highly acceptable when compared to a standard optimum of 100 (Figure 1).

Conclusions This research supports the interest in using NIRs for rapid diagnosis of grass mineral status, particularly when one considers the low cost of scanning and the delays that occur with classical analysis. The same scanning can be used to predict the feeding value of the forage samples. The data obtained in this study will be used to test PLS2 discriminant models for predicting classes of mineral excess/deficiency.

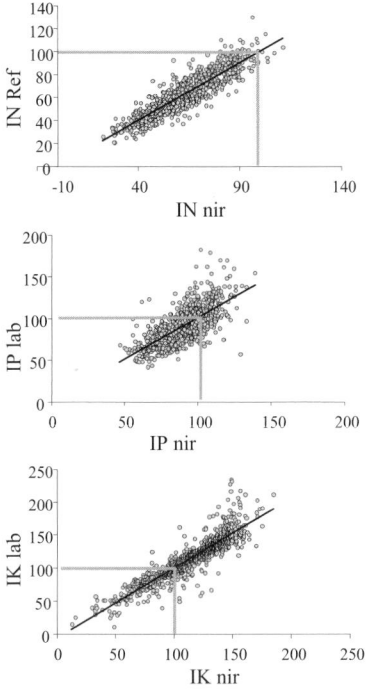

Figure 1 NIR predicted and reference calculated values for IN IP. IK

References
Lemaire, G. & F. Gastal. (1997). N uptake and distribution in plant canopies. In: G. Lemaire (ed.) Diagnosis of the Nitrogen Status in the Crops. Springer Verlag, 3-44.
Duru, M. & L.Thélier (1997). N and P-K status of herbage : use for diagnosis of grasslands. In: I.N.R.A. (ed.) Diagnostic Procedures for Crop N Management and Decision Making, 125-128.
Blanfort, V. (1998). Agro-écologie des pâturages d'altitude à La Réunion. Cirad - Université de Paris-Sud.. 299 pp + annexes.

The prediction of biological nitrogen fixation

C.F.E. Topp[1], C.A. Watson[2], R.M. Rees[3] and I. Sanders[2]

[1]Land Economy Research, SAC, Edinburgh, EH9 3JG, UK, Email: Kairsty.Topp@sac.ac.uk, [2]Crop and Soil Research, SAC Aberdeen, AB21 9YA, UK, [3]Crop and Soil Research, SAC, Edinburgh, EH26 OPH, UK

Keywords: N fixation, models

Introduction In organic farming systems, biological nitrogen (N) fixation is crucial for short-term productivity and long-term sustainability. However, the estimation of biological N fixation is fraught with difficulties, and many equations attempt to estimate the process. As part of an organic research programme, biological N fixation was measured by the ^{15}N dilution technique in the ley phases of 2 experimental organic ley-arable rotations at 2 sites, between 1997 and 2000. Hence, N fixation has been determined on N partitioned to above-ground biomass. The measured values have been compared with N fixation estimates calculated from the equations proposed by Korsaeth & Eltun (2000) and Hogh-Jensen *et al.* (2004).

Materials and methods Equations of Korsaeth & Eltun (2000) and Hogh-Jensen *et al.* (2004), both consider the yield of white clover and the N content of the legumes. The equation of Korsaeth & Eltun (2000) includes the age of the ley and the maximum fraction of fixed N in the legume, which is adjusted for the quantity of added N fertiliser/manure. Also, the N fixation is corrected for the net accumulation of N fixed below stubble height. By contrast, in the equation of Hogh-Jensen *et al.* (2004), N fixation includes below-ground N in the root and stubble, the below-ground transfer to grass, the above-ground transfer to the grass by grazing animals and the immobilisation of fixed N into the organic soil pools, with the age of the ley impacting on the values of these factors. In order to compare the model with the measurements of above-ground fixation (Sanders *et al.* 2001), only factors relating to the roots and stubble were included in the calculation of N fixation. Theil's inequality coefficient (Theil, 1970), which has a value of between 0 and 1, with 0 indicating a perfect fit, was used to compare the performance of the models with the measured values.

Results The performance of the models differs between the age of the ley and year (Figure 1). The poorest predictions were observed in 2000, with Theil's inequality coefficients of 0.11 and 0.08 for Korsaeth's and Hogh-Jensen's models respectively. In 1997-1999, Theils' inequality coefficient ranged from 0.02-0.03 for Korsaeth's model and 0.003-0.01 for Hogh-Jensen's. Predictions for the 1st year ley across all years were worst, and for the 4th year ley were best.

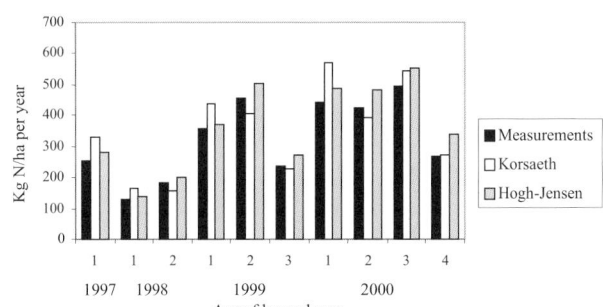

Figure 1 N fixation for age of ley and year

Conclusions The performance of the models varies between the age of the ley and year. However, Theil's inequality coefficient illustrates that the performance of both models is relatively good. Based on the assumptions of the model described by Hogh-Jensen *et al.* (2004), the total N fixation for the second year ley is 182 kg/ha/yr, which is 62% higher than the estimate excluding transfers and immobilisation. The unavailability of reliable field estimates of N fixation is hampering an adequate evaluation of these models currently.

Acknowledgments The Scottish Executive Environment and Rural Affairs Department funded this work.

References

Hogh-Jensen, H., R. Loges, F.V. Jorgensen, F.P. Vinther & E.S. Jensen (2004). An empirical model for quantification of symbiotic nitrogen fixation in grass-clover mixtures. *Agricultural Systems*, 82, 181-194.

Korsaeth, A. & R. Eltun (2000). Nitrogen mass balances in conventional, integrated and ecological cropping systems and the relationship between balance calculations and nitrogen runoff in an 8-year field experiment in Norway. *Agriculture, Ecosystems and Environment*, 79, 199-214.

Sanders, I., C.A. Watson, R.M. Rees & D. Atkinson (2001). Effects of ley arable rotation on soil nitrogen fixation. *Aspects of Applied Biology*, 63, 61-66.

Theil, H. (1970). Economic Forecasts and Policy, 2nd edn. North Holland Publishing Company, Amsterdam.

Assessment of the Nitrogen Nutrition Index (NNI) by the nitrogen concentration of the upper part of the sward

B. Deprez, R. Lambert and A. Peeters
Laboratory of Grassland Ecology, Catholic University of Louvain, Place Croix du Sud 5 bte 1, B-1348 Louvain-la-Neuve, Belgium, Email: deprez@ecop.ucl.ac.be

Keywords: grassland, nitrogen status, nitrogen index

Introduction The Nitrogen Nutrition Index (NNI) is based on the concept of a critical nitrogen concentration, defined as the minimum N concentration in the aerial biomass necessary to obtain maximum yield. Values of N_{crit} are high at the start of the growing period and decline during growth, in relation to dry matter accumulation (t DM/ha) according to the equation of Lemaire & Salette (1984): $N_{crit}(\%) = 4.8(DM)^{-0.32}$. The NNI is calculated as the ratio of the actual N concentration of the sward to the N concentration required at a similar biomass to sustain non-limiting growth and biomass accumulation (Lemaire & Gastal, 1997). However, the evaluation of NNI requires the determination of aerial biomass per unit area, in addition to the determination of N concentration. This is time-consuming and is a practical limit to using NNI on farms. The objective of this experiment was to assess NNI by a more practical and easier method based on the determination of the N concentration in the upper leaves, as proposed by Gastal *et al.* (2001).

Materials and methods The experiment was conducted in 2001 and 2002 on eight contrasting sites throughout Belgium. Four sites were common to both years. Each site had one plot of perennial ryegrass (*Lolium perenne*) with no N fertilisation (N_o) and one plot receiving 50 kg N/ha (N_{50}) in four replicates. In each plot, aerial biomass (cutting height = 7 cm) and total N concentration were measured in May 2001 and 2002. Samples of herbage were also selected and cut with scissors at 10 cm from the tip of the longest leaf and total N concentration (N_{up}) was determined.

Results The linear regression between NNI and N_{up} in 2001 and 2002 (Figure 1) showed a significant correlation ($r^2 = 0.78$). The slopes and intercepts of the regression did not differ statistically between the two treatments. The slope of the regression was slightly lower than that obtained by Gastal *et al.* (2001). According to Duru *et al.* (1997), NNI values above 1 indicate that N nutrition is excessive, values between 1 and 0.8 indicate that N nutrition is satisfactory, and values below 0.8 indicate that N nutrition is significantly limiting. In our experiment, NNI values between 0.2 and 0.7 were found indicating a limiting N nutrition of the sward.

In Gastal's experiment, the NNI values ranged between 0.25 and 1.25.

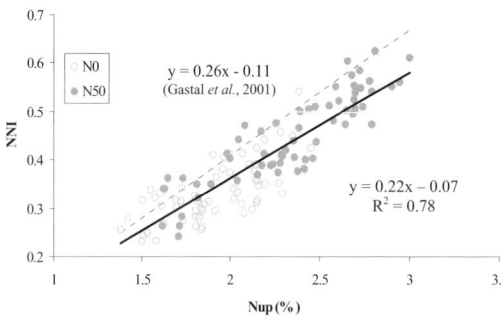

Conclusions The present results confirm the possibility of using N_{up} to evaluate sward N status (NNI). For most studies, the precision of N_{up} (4.6%) is sufficient and thus the determination of herbage biomass per unit area is no longer necessary, eliminating the most time-consuming step of the conventional NNI sampling procedure. Moreover, N_{up} determination allows the comparison of swards from different sites, different times of growth and different years of harvest with an acceptable approximation.

Figure 1 Relation between NNI of the sward and N concentration of the lamina at the 10 cm top of the canopy (N_{up}). Gastal's reference equation is in dashed line

References

Duru, M., G. Lemaire & P. Cruz (1997). The nitrogen requirement of major agricultural crops. Grassland. In: G. Lemaire (ed.) Diagnosis of the Nitrogen Status in Crops, 59-72.

Gastal ,F., A. Farruggia A. & J. Hacquet (2001). The Nitrogen Nutrition Index of grass can be evaluated through determination of N concentration of upper leaves. *Proceedings of the 11th Nitrogen Workshop*, Reims, France, 449-450.

Lemaire, G. & F. Gastal (1997). N uptake and distribution in plant canopies. In: G. Lemaire (ed.) Diagnosis of the Nitrogen Status in Crops, 3-43.

Lemaire, G. & J. Salette (1984). Relation entre dynamique de croissance et dynamique de prélèvement d'azote pour un peuplement de graminées fourragères. I. Etude de l'effet du milieu. *Agronomie*, 4, 423-430.

Compilation of a database of research information on legume based grazing systems; a part of the Leggraze research project

J.J. Rochon[1], G. Molle[2], A. Hopkins[3], J.M. Greef[4]
[1]Institut Universitaire de Technologie, 66962 Perpignan, France, Email: rochon@univ-perp.fr, [2]Istituto Zootecnico e Caseario per la Sardegna, 07040 Olmedo, SS, Italy, [3]Institute of Grassland and Environmental Research, North Wyke, EX20 2SB Okehampton, UK, [4]Federal Agricultural Research Centre, 38116 Braunschweig, Germany

Keywords: legumes, grazing, database, website

Introduction The establishment of a publicly accessible web-resident database of published and current European research on agronomy, animal production and environmental impact of legume based grazing systems is reported. This database facilitates the sharing of information among the partners of the "Low input animal production based on forage legumes for grazing systems" (Leggraze), a research project funded by the UE (QL K5 CT-2001-02328). It also forms an important tool for transferring the results of the project to the wider research community and to end users in the agricultural sector and to policy makers at national and community level.

Materials and methods The database uses the database server MySQL. It allows, starting from a form, the search for a document according to several criteria (authors, year of publication, language and or keywords (from 1 to 8). The database is hosted on the server of the University of Perpignan (France). The address is: http://www.univ-perp.fr/leggraze/bibliography.htm.

Articles are sourced from over 40 journals and conference proceedings. The update of the database is carried out by the indexing of each new issue of The European Journal of Agronomy, Grass and forage Science, Small Ruminant Research and Animal Research.

Results The database contains 310 articles written by 501 authors form 20 countries, provided by the partners of the project. The number of key words is 149. The articles can be gathered around 19 topics (Table1).

Table 1 Main topics of the articles included in the database

Mixed swards	102
Forage quality	67
Persistence of plants	57
Animal production	57
Harvest management	47
Tannins	45
Fertilisation	41
Pasture production	37
Plant morphology	29
Environmental influence	26
Establishment of sward	26
Annual legumes	25
N2-fixation/Leaching	19
Soil fertility	18
Grass/Legume interaction	16
Dietary preference	16
Grazing behaviour	16
Phytopathology	8
Genetic resources	7
Monography	7

Conclusion This web-resident database visited more than 2000 times is an important way of diffusion of the results of the Leggraze project. We hope that it will be maintained and updated after the end of this research project to provide for the research community a useful tool for the study of grazed legume pastures.

Forage monitoring technology to improve risk management decision making by herders in the Gobi region of Mongolia

J.P. Angerer[1], J.W. Stuth[1], D. Tsogoo[2], D. Tolleson[1], D. Sheehy[1], U. Gombosuren[2] and S. Granville-Ross[2]

[1]Department of Rangeland Ecology and Management, Texas A&M University, 2126 TAMU, College Station, Texas 77843-2126, USA; Email: jangerer@cnrit.tamu.edu, [2]Mercy Corps, P.O. Box 76,1 Ulaanbaatar 49, Mongolia

Keywords: drought, Mongolia, forage monitoring

Introduction In the period from 1999 to 2002, Mongolia experienced a series of droughts and severe winters that lowered livestock numbers by approximately 30% countrywide. In the Gobi region, livestock mortality reached 50% with many households losing entire herds (Siurua & Swift 2002). In March 2004, a program was initiated by the United States Agency for International Development (USAID) through the Global Livestock Collaborative Research and Support Program (GLCRSP). The goal of this program is to develop forage monitoring technologies that provide early warning of drought and winter disaster to improve livestock herder decision making in the Gobi region. The program has two major objectives: (1) to develop a regional forage monitoring system that provides near-real time spatial and temporal assessment of current and forecasted forage conditions, and (2) to develop a communication infrastructure that provides herders with data on forage conditions to assist them in making timely and specific management decisions.

Materials and methods The protocol for the forage monitoring was similar to that used in East Africa and the USA (Stuth *et al.,* 2003). In May 2004, 120 monitoring sites in the Gobi Region (Figure 1) were selected in 3 aimags (Gobi-Altai, Bayankhongor, and Ovorkhangai). At each site, vegetation, soil, and grazing data were collected for input into the PHYGROW forage production model (Stuth et al. 2003). PHYGROW is driven by near real-time climate data acquired from the National Oceanic & Atmospheric Administration's (NOAA) CMORPH system (Joyce *et al.,* 2004). The forage model outputs for the monitor sites are then coupled with satellite data for Normalized Difference Vegetation Index (NDVI) using geostatistics to create surface maps of forage yield and deviations from long-term average. Statistical forecasting (Autoregressive Integrated Moving Averages (Stuth *et al.,* 2003)) is used to project forage conditions for 90 days into the future.

Results Vegetation parameter collection has been completed at each monitor site. Grassland communities being monitored range from mountain steppe to desert grasslands. Livestock being monitored include cattle, sheep, goats, yaks, camels and horses. NOAA CMORPH climate data is being archived and data extraction tools have been developed to provide data visualization and inputs into the PHYGROW model. Communication protocols are being developed to provide herders in the region with current forage conditions and 90-day forecasts on a 14-day cycle. The Rural Business News unit of Mercy Corps in Mongolia will publish the data in their weekly newspaper and radio broadcasts to herders. The data will be available on WWW also.

Conclusions A forage monitoring system is being developed to provide early warning for below normal forage or catastrophic winter conditions on grasslands in the Gobi region of Mongolia. Information from this system will allow herders to have near real-time information to reduce risk of livestock mortality and protect the ecological stability of the grassland resources. This information will also help herders in the region to better cope with risk and market access.

Figure 1 Location of monitoring sites within the three aimags in the Gobi region of Mongolia

References

Joyce R.J., J.E. Janowiak, P.A. Arkin & P. Xie (2004). CMORPH: A method that produces global precipitation estimates from passive microwave and infrared data at high spatial and temporal resolution. *Journal of Hydrometeorology,* 5, 487-503.

Siurua H. & J. Swift (2002). Drought and zud but no famine (yet) in the Mongolian herding economy. *IDS Bulletin, Institute of Development Studies,* 33, 88-97.

Stuth J., J. Angerer, R. Kaitho, K. Zander, A. Jama, C. Heath, J. Bucher, W. Hamilton, R. Conner & D. Inbody (2003). The Livestock Early Warning System (LEWS): Blending technology and the human dimension to support grazing decisions. *Arid Lands Newsletter* No. 53, May/June.

Modelling the effect of breakeven date in spring rotation planner on production and profit of a pasture-based dairy system

P.C. Beukes, B.S. Thorrold, M.E. Wastney, C.C. Palliser, G. Levy and X. Chardon
Dexcel Ltd, Private Bag 3221, Hamilton 2001, New Zealand, Email: pierre.beukes@dexcel.co.nz

Keywords: grass silage, farm cover, management

Introduction The breakeven date is the expected date when pasture supply exceeds cow demand. This date is used to plan the rotation rates, slow during the winter, when pasture growth is low and cows are dry, to a fast rotation in spring, when growth is accelerating and most cows lactating. This date is influenced by regional climate, mainly rainfall and soil temperature, which affects timing and rate of growth acceleration. The objective of this modeling exercise was to explore the effect of the breakeven date on milksolids (MS), grass silage, farm cover and economic farm surplus (EFS) over different climate years for the Canterbury region of New Zealand.

Materials and methods Observed starting farm covers, herd data, silage stacks, irrigation (585 mm) and fertilizer (200 kg N/ha) schedules from Lincoln University Dairy Farm (LUDF) near Christchurch, Canterbury for the 2002/03 season were used to initialize Dexcel's Whole Farm Model (WFM, Wastney *et al.*, 2002), first with the observed breakeven date of 25 September (Strategy 1) and then with a breakeven date of 20 October (Strategy 2). Both strategies were simulated for five seasons (1994/95 to 1998/99) using observed climate data from LUDF and a stocking rate of 3.65 cows/ha. Seasons were simulated individually (no carry-over) starting 1 June for 365 days. Economic results are presented assuming a payout of NZ$ 3.90/kg MS.

Results Averages of model predictions over five climate years are given in Table 1, and model predictions of seasonal changes in farm covers for the 1996 season are shown in Figure 1. An early breakeven date (Strategy 1) resulted in faster rotations earlier in the season with the consequence that lactating cows obtained a larger proportion of their demand from pasture and could be fed less grass silage compared to Strategy 2 (Table 1). The faster rotations earlier resulted in slower recovery of farm covers for Strategy 1, however, covers had recovered to Strategy 2 level before Christmas (Figure 1). With higher covers early in the season more silage could be made with Strategy 2 (Table 1). Generally grass silage has lower nutritional value than spring pasture (Holmes *et al.*, 2002), which explains why MS production per cow and per hectare was lower for Strategy 2 (Table 1). The effect of silage feeding on production was most pronounced in the early part of the season when the breakeven date determines rotation length and the proportion of the daily intake from pasture and from silage. The significantly higher EFS for Strategy 1 (Table 1) confirmed the benefit of feeding quality pasture during peak lactation, although it might be at the expense of farm covers and silage making.

Table 1 Predicted averages (±SD) with results of paired t-tests

Parameter	Strategy 1	Strategy 2	Significance (P)
Silage fed (kg DM/ha)	1882±702	2050±726	0.19
Silage made (kg DM/ha)	2557±535	2816±540	0.12
MS (kg/cow)	399±5.5	394±4.9	0.02
MS (kg/ha)	1455±20	1439±19	0.01
EFS (NZ$/ha)	2261±265	2180±256	0.02

Figure 1 Effect of breakeven date on average farm cover for the 1996 season

Conclusions The results of this modeling exercise show that for the Lincoln farm a breakeven date of 25 September for the spring rotation planner is more profitable than a date of 20 October. Optimum breakeven dates for different farms will depend on stocking rate and local climate.

References
Wastney, M.E., C.C. Palliser, J.A. Lile, K.A. Macdonald, J.W. Penno & K.P. Bright (2002). A whole-farm model applied to a dairy system. *Proceedings of the New Zealand Society of Animal Production*, 62, 120-123.
Holmes, C.W., I.M. Brookes, D.J. Garrick, D.D.S. Mackenzie, T.J. Parkinson & G.F. Wilson (2002). Milk production from pasture. Principles and practices. Massey University, Palmerston North, New Zealand, 601.

Development of a model simulating the impact of management strategies on production from beef cattle farming systems based on permanent pasture

M. Jouven and R. Baumont
INRA, Unité de Recherches sur les Herbivores, Centre de Clermont-Theix-Lyon, F-63122 Saint Genès Champanelle, France, Email: mjouven@clermont.inra.fr

Keywords: simulation, grazing systems, animal intake, animal performance, vegetation dynamics

Introduction Grazing systems in Europe increasingly have to meet environmental objectives, which influence management strategies. A deterministic model describing farming system dynamics is being developed in order to elucidate interactions between nature-friendly management practices, as for example late (after flowering) hay harvest or moderate stocking rate, and agricultural output.

Model presentation The model predicts, with a daily time step, farm system operation and agricultural output from a given farm structure and management strategy. It is built and calibrated for non-intensive French beef cattle farming systems based on permanent pasture. The model is made up of four interacting sub-models. A grassland resource sub-model adapted from Carrère et al. (2002) predicts grass growth and quality at the paddock level, from soil quality, vegetation functional traits and climatic data. A herd performance sub-model based on INRA (1989) calculates weight gain and milk production from energy intake, for an average cow and calf. A feeding sub-model predicts selective intake at pasture for an average cow and calf, and herd feeding indoors. A management sub-model decides on herd movements, concentrate supplementation to achieve intermediate objectives (calf weight at sale, cow body condition score at calving) and day of hay harvest. The whole model and individual sub-models are undergoing sensitivity analysis and validation.

Simulation example The effects of stocking rate and proportion of late hay harvest were examined in a 2x2 factorial design. The model was run for 6 climatic years, with stocking rate (SR) 1.4 or 1.2 LSU/ha, half the surface planned for hay harvest (half for 1 cut, half for 2 cuts), with 50% or 100% late 1st cut, and calves sold at 9 months, 320 kg. SR change was performed by altering the number of animals. Simulation results are given in Table 1; the model being deterministic, standard deviations refer to discrepancies between climatic years.

Table 1 Simulated annual agricultural output (mean ± SD); DOM=digestible organic matter

Management strategy	Days at pasture	Grazed DOM (t/calf)	Grazed DOM (t/cow)	Harvested DOM(t/cow)	Concentrate consumption at pasture (kg/calf)	Concentrate consumption housed (kg/cow)
SR 1.4 – 50% late	172 ± 7	0.35 ± 0.08	1.69 ± 0.07	1.27 ± 0.12	180 ± 13	36 ± 33
SR 1.2 – 50% late	178 ± 10	0.34 ± 0.03	1.75 ± 0.10	1.47 ± 0.13	184 ± 13	20 ± 23
SR 1.4 – 100% late	170 ± 7	0.32 ± 0.02	1.68 ± 0.07	1.15 ± 0.16	204 ± 57	64 ± 40
SR 1.2 – 100% late	177 ± 10	0.34 ± 0.03	1.74 ± 0.09	1.30 ± 0.19	204 ± 50	53 ± 49

Lowering SR increases grass availability at pasture, and thus lengthens grazing season (+7 days); consequently, grazed DOM per cow is higher (1.75 vs 1.69 t/cow), even though grazed DOM per day is unchanged (9.86 kg/cow). Lowering SR also increases harvested DOM per cow (1.39 vs 1.21 t/cow); therefore, winter concentrate consumption per cow tends to be reduced (37 vs 50 kg/cow). 100% late 1st cut has little effect on grazed DOM at pasture, but reduces harvested DOM per cow (1.23 vs 1.37 t/cow) via a decrease in mean hay digestibility (0.59 vs 0.63) that is not compensated by an increase in dry matter yields (2.01 vs 2.02 t/cow). Thus, concentrate consumption for cows is higher for 100% late 1st cut (59 vs 28 kg/cow). Overall, whatever the management strategy, calves performance remains unchanged. Simulated agricultural output varies widely with climatic years, as in a real farm.

Conclusions This model may offer a useful tool to support discussion between research, advisory and environmental services. It can be enriched by new management rules and a biodiversity score sub-model.

References

Carrère P., C. Force, J.F. Soussana, F. Louault, B. Dumont & R. Baumont (2002). Design of a spatial model of perennial grassland grazed by a herd of ruminants: the vegetation sub-model. *Grassland Science in Europe*, 7, 282-283.
INRA (1989). Ruminant Nutrition: recommended allowances and feed tables. Jarrige, R. (ed.), John Libbey Eurotext, Paris, 389 pp.

GrassCheck: monitoring and predicting grass production in Northern Ireland

P.D. Barrett[1] and A.S. Laidlaw[2]
[1]Agricultural Research Institute of Northern Ireland, Large Park, Hillsborough, Co. Down, BT26 6DR, UK,
[2]Department of Agriculture and Rural Development for Northern Ireland, Plant Testing Station, Crossnacreevy,
Belfast, BT6 9SH, UK, Email: scott.laidlaw@dardni.gov.uk

Keywords: herbage, growth, model, dairy, management

Introduction Grass budgeting is a key management practice on dairy farms to balance grass supply on paddocks with grass demand by the grazing herd. Grass budgets must be pre-emptive to be effective. The uncertainty of grass production and the difficulty in quantifying both current and forecasted rates of growth hamper effective budgeting and paddock management. Grass growth rates are highly variable both in time and space. Therefore, they vary greatly between locations at any given time and also across the season at any given location. Figure 1 shows the pattern of growth rates recorded at the Agricultural Research Institute of Northern Ireland (ARINI) in the two seasons before this project. The GrassCheck project was established in Northern Ireland to quantify current rates of grass growth and grass quality and to predict growth rates for up to 2 weeks in advance. The project will run from 2004 until 2006. This paper outlines the project and reports on its findings after one year.

Methodology A total of 6 sets of perennial ryegrass plots were established at 3 Department of Agriculture and Rural Development for Northern Ireland (DARDNI) sites in Northern Ireland: at ARINI, Hillsborough; Greenmount Campus, Antrim and The Plant Testing Station, Crossnacreevy. Plots, circa 1.5 x 5.0m, were cut 4cm above ground with a motor scythe. A total of 365kg N/ha was applied over the growing season. Each set of plots consisted of 9 plots comprising 3 series of 3 replicates. Only one series was cut/week under a sequential weekly cutting regime. Therefore, to simulate rotational grazing, all plots were given 21 days regrowth. Also, growth rates were predicted for the next 2 weeks using the ARINI GrazeGro growth model (Barrett *et al.*, 2004). Growth rate and grass quality were determined rapidly from the plots. They, plus the 2-week predicted growth, were reported to the farmers in weekly bulletins in the local farming press and on DARDNI websites.

Results Growth rates for 2004 varied considerably across the season and deviated consistently from the 5-year mean growth rate line determined over the 5 years preceding the project. Figure 2 shows the pattern of growth rates for 2004. Accumulated grass production from Mar to Sep 2004 was 15.2% above the mean for the previous 5 years. The GrazeGro growth model provided a reliable indication of future rates of grass growth. Figure 2 also shows predicted growth rates. $R^2=0.78$ for predicted output regressed against observed growth rates until the middle of Sep 2004.

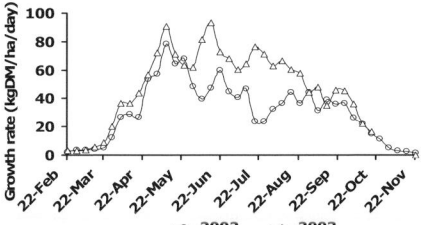

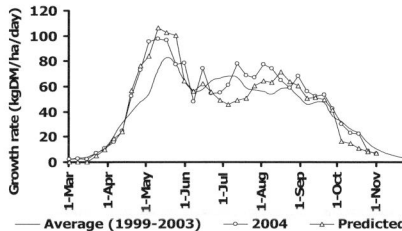

Figure 1 Variation in growth rate observed at ARINI 2002-03

Figure 2 GrassCheck 5-year mean, actual and predicted growth rate for 2004

Conclusion Given the variability of growth, the need for a monitoring programme for grass growth and quality was demonstrated well. Weekly GrassCheck bulletins throughout the season provided accurate and timely indication of growing conditions to farmers and advisors in Northern Ireland. Given the good precision of the GrazeGro model, accurate estimates of predicted growth rates were given to aid in decision-making for grassland management and grass budgeting procedures.

References

Barrett P. D., A.S. Laidlaw & C. S. Mayne (2005). GrazeGro: A European herbage growth model to predict pasture production in perennial ryegrass swards for decision support. *European Journal of Agronomy*, (In press).

A farmer friendly feed budget calculator for grazing management decisions in winter and spring

M. Curnow and M. Hyder

Department of Agriculture Western Australia, 444 Albany Hwy, Albany 6330 Western Australia, Email: mcurnow@agric.wa.gov.au

Keywords: feed budget, electronic calculator, grazing, sheep

Introduction The Western Australian (WA) environment is Mediterranean with annual legume/grass pastures and a 6 month growing season. In autumn where over grazing can impact pasture establishment and in spring, prior to senescence, when under grazing can mean significant losses of efficiency are crucial times for grazing management. Pasture utilisation is typically low (25-35%) due to conservative stocking regimes; key to increasing productivity is increasing pasture utilisation (Grimm, 1998). Increased level of productivity require farmer sophistication in the way they feed budget. To this end, satellite technology is being used to provide farmers in southern Australia with weekly estimates of pasture growth rate (PGR; kg DM/ha/d) and monthly estimates of Feed on Offer (FOO; kg DM/ha) (Kelly *et al*. 2003). In addition the Green Feed Budget Paddock Calculator (GFBC) was developed to provide a simple and accessible electronic calculator which utilises this new information to assist farmers to feed budget and to make more accurate and timely stocking rate decisions.

Materials and methods The GFBC is a computer-based tool that reflects the key tactical decisions that graziers have to make throughout the growing season. It uses feed intake data generated from Grazfeed® where the inputs have been modified to fit clover-dominant annual pastures. It has six scenarios, outlined in the main menu (Figure 1). The scenarios are divided into two phases - establishment (autumn- winter) and vegetative (late winter-spring) - to take account of the differences in pasture morphology (e.g. height, % dry matter). Each scenario provides a single point calculation and has an archive that allows paddock level recording. With the aim of developing a tool specifically for WA legume/grass pastures it was 'product tested' by farmer groups participating in the 'Pastures from Space' project (Kelly *et al*. 2003), who are keenly involved in improving pasture assessment accuracy and lifting pasture utilisation. Feed back on the design was incorporated and the original calculation screens expanded to include feed mix calculators and other improvements.

Accessibility is a key design factor and due to slow rural internet line speeds the calculator is available as a runtime CD. The calculator is also free on the Department's website where additional information can also be sourced. Further research into feed intake, energy requirements and production targets of the sheep are being regularly incorporated into the calculator via new versions or updated screens on the web site.

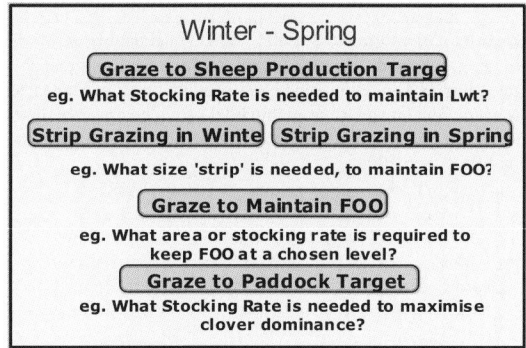

Figure 1 Choice of Winter-Spring scenarios in the calculator

Discussion The challenge was to build a simple, yet accurate product that allows the best tactical decisions to be made, according to the grazing system. Frequently simple decision tools suffer from not adequately reflecting the complexity of the biological system and therefore suffer inaccuracy. However, this potential source of error is small in comparison with the error inherent in pasture assessment. The benefits of this product are that it encourages frequent use, can be used in the paddock, and the skill and confidence gained by regular feed budgeting encourage farmers to explore whole farm budgeting (which is the ultimate end point). Other products available on the market tend to be complex and ask the grazier to enter a lot of data prior to calculation, which often contributes little to the output, leaving many farmers with a time consuming and intimidating product.

References

Grimm, M., (1998). Tactical grazing strategies for annual pastures, Proceedings Grassland Society of Victoria, Victoria, Australia, 67.

Kelly, R., A.Edirisinghe, G.Donald, C.Oldham & D.Henry, (2003). Satellite based spatial information on pastures improves Australian sheep production. *1st European Conference on Precision Livestock Farming*, June 15-18: 93-98.

Feed planning – methods used by "expert" farmers

D.I. Gray, W.J. Parker, E.A. Kemp, P.D. Kemp, I.M. Brookes, D. Horne, P.R. Kenyon, C. Matthew, S.T. Morris, J.I. Reid and I. Valentine
College of Sciences, Massey University, Private Bag 11 222, Palmerston North, New Zealand, Email: D.I.Gray@massey.ac.nz

Keywords: decision-making, farmer knowledge, feed budgeting, planning, tactical management

Introduction Although formal feed planning has been heavily promoted in New Zealand, relatively few farmers have adopted this approach (Nuthall & Bishop-Hurley, 1999). Reasons for non-adoption have been identified, but little is known about how farmers manage their pastoral farms in the absence of formal feed planning. To this end, the feed management processes used by three successful (expert) farmers were investigated.

Materials and methods A case study design was adopted and the criteria for case selection were farm productivity and expertise in feed management. One sheep and beef (7,770 s.u.) and two dairy (220 and 330 cows respectively) farmers were selected. Monthly semi-structured interviews and field observations were used to collect data over two years. Interview data were transcribed verbatim and analysed to develop models of the farmers' feed planning processes.

Results and discussion The farmers separated the year into four planning horizons and alternated between formal and informal feed planning. Formal feed planning was used when critical decisions had to be made, accurate pasture measurement could be undertaken and the level of environmental uncertainty was perceived to be low (late autumn and winter). Informal feed planning was used at other times of the year. At its simplest level, the informal process used by the farmers was to recall a successful plan from the past. This largely sub-conscious approach required much less cognitive effort than more formal planning. These predefined or *"typical"* plans were used except when circumstances forced the farmers out of "*plan mode*" and into "*decision mode*" to choose between plans. A broadly defined set of farm state conditions was required at the start of the planning period for the *"typical"* plan to be feasible. If conditions were outside this range, the plan was modified. Plans were also modified in response to prior learning, previously made strategic and tactical decisions and significant changes in the market. To modify a plan, the farmers postulated the nature of the change required and then tested its feasibility. At its simplest, this required the modification of some simple heuristics (rules of thumb) in the plan, e.g., the substitution of one type of supplement (grass silage) to make up for the loss of another (forage crop). More complex changes (e.g., the introduction of maize silage) usually took several iterations of adjustments and the associated use of mental feed budgeting to quantify the impact of the change.

Once a plan for a period had been developed, what we term *"micro-budgets"* was used to control its implementation in the face of uncertainty. These micro-budgets operated at a paddock rather than the whole-farm level and were used for time frames of 2 – 4 weeks. For example, during summer, each time the dairy farmers shifted the herd, they estimated the post-grazing residual. They then estimated, given current climatic conditions, how much pasture would grow between now and when the herd returned to the paddock in 3 – 4 weeks time. From this they estimated the proportion of the herd's diet that the paddock was likely to supply at the next grazing. This gave them 3 – 4 weeks warning of an impending feed deficit and allowed them time to evaluate alternative options. As this process was repeated every day, the farm's future feed position was being continually updated. The sheep farmer used a similar process over spring when his sheep were set-stocked. Pasture cover in each paddock was recorded fortnightly in the sheep block. Feed demand for the number of sheep and cattle in each paddock for the next two weeks was estimated and compared with expected pasture growth for that paddock. The size of the feed deficit or surplus over the next two weeks was derived. Stock per paddock was then adjusted to ensure animals were fed to appetite and pasture quality was maintained.

Conclusions Pasture management expertise is integral to successful livestock farming in New Zealand. Despite this, understanding of how farmers plan and control grazing decisions remains limited. Here we have provided some insight into how some farmers manage effectively without formal feed budgeting. Extrapolation of these results from "experts" to the wider farming community must be taken with some care. Nevertheless, the results suggest that rather than continue to unquestioningly promote the adoption of formal feed planning, researchers, extension agents and consultants need to look at how farmers currently manage their feed, and at their requirements for simple, low cost (time and capital) management tools.

References
Nuthall, P.L. & G.J. Bishop-Hurley (1999). Feed planning on New Zealand farms. *Journal of International Farm Management*, 2, 100-112.

Grassland monitoring system for sustainable utilisation in Inner Mongolia, China. 1. Concept of real-time monitoring system and estimation of biomass using NOAA/NDVI

T. Akiyama[1], K. Kawamura[1], H. Yokota[2] and Z.Z. Chen[3]

[1]River Basin Research Center, Gifu University, 1-1, Yanagido, Gifu 501-1193, Japan, Email: akiyama@green.gifu-u.ac.jp, [2]Nagoya University, Togo, Aichi 470-0151, Japan, [3]Institute of Botany, Chinese Academy of Sciences, Beijing 100093, China

Keywords: climate factor, desertification, NOAA/NDVI, real-time monitoring, steppe

Introduction Steppe grassland in Inner Mongolia, China, is threatened by desertification mainly because of overgrazing by animals. Grassland production varies from year to year and from place to place and is much affected by climate conditions and grazing intensity. The questions addressed in these three papers are the development of a real-time monitoring system to conserve the ecosystem whilst at the same time achieving sustainable livestock farming. The development of satellite sensors with remarkable resolution made it possible to provide precise, timely and site specific information on grassland. Geographic Information Systems (GIS) and enhancement of accuracy of Global Positioning Systems (GPS) became powerful tools for assessing grassland ecosystems.

Experiment site The Bayinxile Livestock Farm (3,730 km^2) is situated in the Xilingol steppe in northern Inner Mongolia, approximately 400 km north of Beijing. The average altitude is 1,300 m above sea level. Climatically, this area belongs to the continental middle temperate semi-arid zone. The mean annual temperature is -0.4ºC; while, the coldest month (January) is -19.5ºC, and the warmest month (July) is 20.8ºC. Annual precipitation is about 300 mm, concentrated in summer. Total livestock numbers increased to 252,700 SU (sheep unit) in December 2001. Thus the average grazing intensity for the total grassland area is 0.67 SU/ha, including land cut for winter feed.

General scheme of the monitoring system Grassland conditions will be determined by the balance between grass production (GP) and herbage intake by animal (HI). When HI exceeds GP, grassland will be degraded, and when GP exceeds HI, it will be conserved and will recover (Figure 1). The rate of GP is determined by climate, soil etc, while HI is determined by grazing intensity, including stocking rate, animal species, palatability etc. Our system gives real-time monitoring of GP and HI using satellite data, GPS, GIS and mathematical models. We will (1) describe the general concept and estimate steppe biomass using NOAA data, in this paper, (2) show real-time monitoring of grass biomass and quality, and animal behaviour using satellite data and GPS, in the second paper, and (3) estimate herbage intake and animal production under free grazing conditions in the Xilingol steppe in the third one.

Results Clipped quadrats (178) from 20 sites were correlated with Normalised Difference Vegetation Index (NDVI) derived from NOAA data. There was a positive correlation between NOAA/NDVI and aboveground biomass (AB, gDM/m^2) (r = 0.62, P<0.01) for all 20 sites (Kawamura *et al*., 2003). The equation was, AB= 450.91 NDVI – 58.99. Seasonal NDVI changes were modelled by multiple regression analysis using NOAA/NDVI observed data as the dependent variables and air temperature and precipitation as the independent variables. Here, the seasons were divided into three phases of growth. The results are shown in Figure 2 (Kawamura *et al*., 2004). The seasonal NDVI changes were predicted well, except for 1994 and 2000.

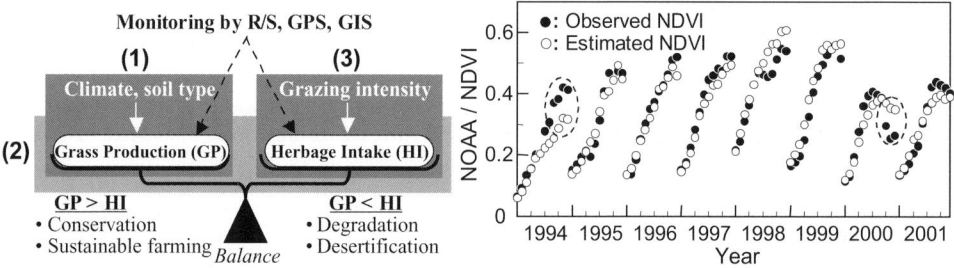

Figure 1 Concept of real-time monitoring system

Figure 2 Seasonal NDVI changes estimated by climate factors

References

Kawamura, K., T. Akiyama, O. Watanabe, H. Hasegawa, F.P. Zhang, H. Yokota, & S. Wang (2003). Estimation of aboveground biomass in Xilingol steppe, Inner Mongolia using NOAA/NDVI, *Grassland Science*, 49, 1-9.

Kawamura, K., T. Akiyama, H. Yokota, M. Tsutsumi, O. Watanabe & S. Wang (2004). Estimation of model for NOAA/NDVI changes of meadow steppe in Inner Mongolia using meteorological data, *Grassland Science*, 49, 547-554.

Grassland monitoring system for sustainable utilisation in Inner Mongolia, China. 2. Real-time monitoring of grass and animal interaction using satellite data and GPS

K. Kawamura[1], T. Akiyama[1], H. Yokota[2], M. Inoue[2], T. Yasuda[3], O. Watanabe[4] and Y. Wang[5]
[1]River Basin Research Center, Gifu University, 1-1 Yanagido, Gifu 501-1193, Japan. E-mail: kawamura@green.gifu-u.ac.jp, [2]Graduate School of Bioagricultural Sciences, Nagoya University, Japan, [3]Division of Natural Sciences, Yamanashi Institute of Environmental Sciences, Japan, [4]Laboratory of Levee Vegetation Management, National Agricultural Research Center for Western Regions, Japan, [5]Chinese Academy of Sciences, China

Keywords: Global Positioning Systems, grazing behaviour, Terra MODIS, sheep, Xilingol steppe

Introduction Overgrazing is one of the primary causes of desertification in Inner Mongolia grassland. A previous paper estimated herbage quantity and quality (Kawamura *et al.*, 2005), and quantified the grazing intensity on grass biomass using Terra MODIS satellite, Global Positioning Systems (GPS) and GIS (Kawamura *et al.*, 2003). The aim of this study is real-time monitoring of both grass biomass and animal behaviour to evaluate the effect of grazing intensity (GI) on grass growth rate during the growing season using Terra MODIS satellite and GPS.

Materials and methods Seasonal changes of grass biomass were estimated in grassland near the Xilin river (12 km^2) using time series of Terra MODIS imageries. Sixteen-day MODIS/EVI composite data (MOD13GQK), from the EROS Data Center, USA, for 6 April to 11 August 2004, was used in this study. The biomass measurements were carried out on 18-19 May and 8-10 July 2004 at the same ten sites. The experimental farm, with total area of 5 km^2, was selected for measuring the spatial distribution of sheep. Here, 720 sheep (360 female, 60 castrated and 300 under 1-year-old) were grazing during June to September. Five sheep were fitted with GPS (four with handy type GPS (Sony HGR3S) and one with GPS collar (Televilt, Posrec Collar 600UD)). To quantify grazing intensity, the grazing distribution map for the flock of sheep was created using a grid cell method from the tracking data recorded by GPS. Grazing intensity in each cell was calculated by the numbers of sheep visits from 1 June to 19 July, 2004.

Results There was a significant relationship between green biomass (GBM) and MODIS/EVI (GBM = 675.836 EVI - 52.414, R^2 = 0.769, P<0.01). This equation was used to make the distribution map of GBM between early April and early August. The spatial distribution of the flock of sheep between June 1 and July 19 is shown in Figure 1. Grass growth rate between late April and late July (ΔG = (GBM$_{July 26}$ - GBM$_{April 22}$) / 180 days) decreased with increasing grazing intensity (Figure 2).

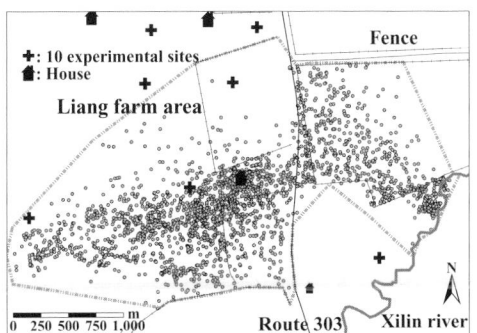

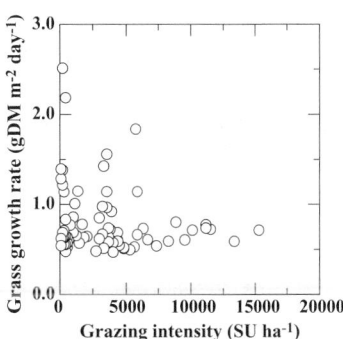

Figure 2 Relationship between GI and grass growth rate between June 1 and July 19

Figure 1 Spatial distribution of herd of sheep from June 1 to July 19

Conclusions GBM and grazing behaviour were monitored with integrating field experiment, satellite remote sensing, GPS and GIS tools. Grass growth rate and GBM were affected by grazing intensity. These results suggested that grazing intensity could be used as a primary parameter when estimating grass production with climate factors.

References

Kawamura, K., T. Akiyama, H. Yokota, M. Tsutsumi, O. Watanabe & S. Wang (2003). Quantification of grazing intensities on plant biomass in Xilingol steppe, China using Terra MODIS image. *International Workshop organized by Working Group VII/6 ISPRS*. 21 October 2003, Kyoto, Japan.

Kawamura, K., T. Akiyama, H. Yokota, M. Tsutsumi, T. Yasuda, O. Watanabe & S. Wang (2005). Comparing MODIS vegetation indices (VIs) with AVHRR NDVI for monitoring the forage quantity and quality in Inner Mongolia grassland, China, *Grassland Science,* (In press).

Grassland monitoring system for sustainable utilisation in Inner Mongolia, China. 3. The estimation of herbage intake of sheep during grazing the natural grassland

H. Yokota[1], K. Kawamura[2], T. Akiyama[2], M. Inoue[1], M. Kondo[1], K. Kita[1] and Y. Wang[3]
[1]Graduate School of Bioagricultural Sciences, Nagoya University, Japan, Email: y941210m@agr.nagoya-u.ac.jp, [2]River Basin Research Center, Gifu University, Japan, [3]Chinese Academy of Sciences, China

Keywords: herbage intake, sheep, grazing, digestibility, steppe

Introduction Grassland condition depends on a balance between growth rates of grasses and herbage intake by animals. In the previous two reports the general concept was described of the monitoring system using satellite data, GPS and GIS and real-time monitoring of grass biomass and quality and animal behaviour. This paper reports the estimation of herbage intake by sheep which had been raised by a farmer in the Inner Mongolia steppe under a typical grazing system with no supplement feeds except salt and also estimation of the growth rate of young sheep.

Material and methods The farmer raised a flock of 720 sheep (360 female, 60 castrated and 300 under 1 year old) on about 5.01 km² of typical grazing grassland near the Xilin River Basin in Inner Mongolia. The sheep grazed from about 6 a.m. to 8 p.m., and walked 10 to 12 km a day. Of the herd, five sheep (one under 1 year old, one castrated male of 2 years old and three females of 2 to 4 years old) were selected for the estimation of herbage intake and five sheep under 1 year old for the estimation of growth rate. The experiment for herbage intake was done from 8 to 19 July, 2004, and the experiment for growth rate was from 1 June to 1 July, 2004. Herbage intake was estimated from faecal output and *in vitro* digestibility of the herbage (Tilley & Terry, 1963), determined on samples from seven areas of grazed grassland. Total faeces was estimated by the dilution method, with chromic oxide administered every morning and evening.

Results Dominant species of herbage in the grazed grassland are shown in Table 1. Dry matter (DM) digestibility of the mixed grasses from areas 1 to 7 ranged from 58.8 to 65.0 %. Herbage DM intake of the sheep was 2.18 to 4.46% of body weight, but the value for sheep 47 was low, because the sheep was born in early spring this year. She grazed with her mother all day, and sometimes suckled. Herbage intake of the other four sheep was from 2.9 to 4.5% of body weight. Body weight gain of the sheep under 1 year old was 0.203 kg/day.

Table 1 Dominant grass of the area and its *in vitro* digestibility

area	dominant grasses [1]	digestibility (%) dry matter	crude protein
1	1, 2, 3	64.5	79.9
2	2, 3, 7, 8	65.0	80.0
3	1, 2, 7	62.0	80.4
4	2, 5	63.5	80.9
5	2, 3, 4, 7	61.6	81.4
6	5, 6, 9	63.5	74.5
7	5, 9	58.8	78.0
mean		62.7	79.3

1) Figures show dominant grasses in the area
1 *Stipa* spp.
2 *Cleistogenes squarrosa* (Trin.) Keng
3 *Agropyron cristatum* (L.) Gaertn
4 *Carex* spp.
5 *Artemisia frigida* Willd.
6 *Leymus chinensis* (Trin.) Tzvel.
7 *Potentilla acaulis* L.
8 *Potentilla tanacetifolia* Willd. ex Schlecht.
9 *Salsola collina* Pall.

Table 2 Daily herbage dry matter intake of the sheep

sheep No.	age, sex	BW. kg	DM herbage intake g/day	% of BW
35	3F	42.33	1793	4.24
42	4F	48.33	2155	4.46
44	4F	56.28	1834	3.26
47	1F	28.88	629	2.18
56	2C	45.44	1318	2.90

Conclusion A combination of above-ground biomass and animal behaviour monitored using satellite data, GPS and GIS (described in the two previous reports) and herbage intake, digestibility and animal growth rate as estimated here gives a mathematical tool for the real-time monitoring of this precious ecosystem .

References
Tilley, J..M..A. & R.A. Terry (1963). A two-stage technique for the *in vitro* digestion of forage crops. *Journal of the British Grassland Society*, 18, 104-111.
Kawamura, K., T. Akiyama, O. Watanabe, H. Hasegawa, F.P Zhang, H. Yokota, & S. Wang (2003). Estimation of aboveground biomass in Xilingol steppe, Inner Mongolia using NOAA/NDVI, *Grassland Science*, 49, 1-9.
Kawamura, K., T. Akiyama, H. Yokota, M. Tsutsumi, O. Watanabe & S. Wang (2004). Estimation of model for NOAA/NDVI changes of meadow steppe in Inner Mongolia using meteorological data, *Grassland Science*, 49, 547-554.

Using geospatial information technologies to identify factors affecting grazing distribution on grasslands

W.H. Schacht[1], A. Guru[2], P.E. Reece[1], J.D. Volesky[1] and D.C. Cotton[2]
[1]Department of Agronomy and Horticulture, and [2]Communications and Information Technology, University of Nebraska-Lincoln, P.O. Box 830915, Lincoln, NE, 68583-0915 USA, Email: wschacht1@unl.edu

Keywords: global positioning systems, geographic information systems, grazing distribution

Introduction   The relationship between environmental and management factors and grazing livestock distribution is fundamental to understanding and improving grazing systems. With the advent of geospatial information technologies, global positioning systems (GPS) and geographic information systems (GIS) have been used to improve the efficiency and effectiveness of quantifying the distribution of livestock grazing in response to various independent variables (Bailey et al., 2001). The specific objective of this project was to develop a tool that enables managers and students to identify and study the effect of management and environmental factors on grazing livestock distribution.

Development of the tool   Data for development of the tool were collected at the University of Nebraska's Barta Brothers Ranch, a 2,350 ha ranch in the eastern Nebraska Sandhills. Six cows (*Bos taurus*) were fitted with GPS collars and grazed freely with a herd of cow-calf pairs. Their locations were recorded at 5- to 10-minute intervals during two summer grazing periods in 2003. Following each period, the collars were removed and the data were transferred to a personal computer. A GIS software (GRASS) was the infrastructure used for data processing and analyses. A standard digital elevation model file of the ranch was imported into GRASS as the base topographical map. Software tools were used to create animations and present analyzed data in tables and graphs.

The tool   The tool is a web-based, user-friendly, online system that enables the user to freely investigate the relationship between several independent variables (e.g., topographical position or livestock water location) and cattle distribution. The first feature of the tool is an animation of cow movement in a selected pasture and time period. When viewed on the screen, the cows appear as dots, fast forwarding their way over a three-dimensional image of the pasture (Figure 1). In analyzing the GPS data, users have numerous options and select the pasture, date(s), hours of the day, and the independent variable to be included in the analyses (Figure 2). Results of analyses are presented in tabular or graphical form allowing the user to view the relationship between grazing distribution and a selected environmental or management variable.

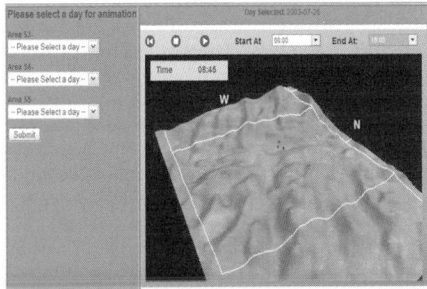

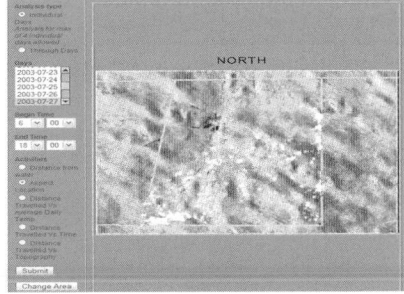

Figure 1 Screen capture of the three-dimensional animation representing the movement of the cows in a pasture for a selected time period

Figure 2 Screen from which users select date(s), hours of the day, and independent variable for analysis

Conclusions   The tool demonstrates to users the environmental and management factors that affect grazing distribution. Livestock producers are intrigued by the technology and recognize the potential of GPS/GIS systems to enhance the management of grazing-based enterprises by aiding in selection of water sites and planning fencing.

References
Bailey, D.W., G.R. Welling, and E.T. Miller. 2001. Cattle use of foothills rangeland near dehydrated molasses supplement. *Journal of Range Management* 54, 338-347.
Schacht, W.H., J.D. Volesky, and S.S. Waller. 1996. Proper livestock grazing distribution on rangeland. NebGuide G80-504-A, Cooperative Extension, Institute of Agriculture and Natural Resources, University of Nebraska.

Heterogeneous nutrient distribution across dairy grazing systems in southeastern Australia

C.J.P. Gourley, I. Awty, P. Durling, J. Collins, A. Melland and S.R. Aarons
Ellinbank Research Institute, PIRVic Ellinbank, RMB 2460 Hazeldean Rd, Ellinbank, Victoria, 3821 Australia, Email: Cameron.Gourley@dpi.vic.gov.au

Keywords: dairy farms, soil nutrients, phosphorus, potassium

Introduction The Australian dairy industry is largely based on a grazed pasture system, although most cows also consume substantial amounts of imported feed (Fulkerson & Doyle 2001). This trend is expected to increase as the Australian dairy industry continues to intensify. Fertiliser inputs of nitrogen (N), phosphorus (P), potassium (K) and sulphur (S) are still viewed as necessary to maintain adequate pasture and milk production despite the fact that most dairy farms are in net positive balance for all of these nutrients (Reuter 2001). Nutrient losses from dairy farming regions and eutrophication of waterways has gained strong public and political attention and intensive pasture systems are no longer seen as 'clean and green'. An important aspect of a viable dairy industry in the future will be more refined nutrient management planning.

Materials and methods Soil samples (0-10 cm) were collected from paddocks on 30 commercial dairy farms (Victoria, Australia) and tested for P, K, S, and pH (water). Farm sizes and intensity of inputs varied, with between 14 and 54 paddocks tested per farm. Soil test information was spatial presented and simple nutrient budgets calculated for each paddock or set of paddocks on each dairy farm, using mostly readily available information such as milk production, purchased feed, harvested hay and silage, and feeding strategies.

Results Soil P, K and S levels were unevenly distributed across the farms. For example a soil phosphorus distribution is provided (Figure 1). In general higher nutrient levels were associated with night paddocks, calving paddocks, sacrifice paddocks and dairy effluent application areas. Low nutrient levels were associated with remote farm locations and hay and silage areas. The most spatially variable measure was K, particularly on farms where the dairy unit was at one end; least variable was soil pH. Farmer response to the spatial presentation of their soil test information and the description of nutrient flows within their farm was highly positive. In most instances, farmers recognised that imported nutrients in purchased fodder could offset fertiliser costs. Changes in fertiliser management included deciding to apply no fertiliser to high fertility areas, changing fertiliser blends to better balance nutrient requirements, or increasing rates to identified areas of nutrient deficiency.

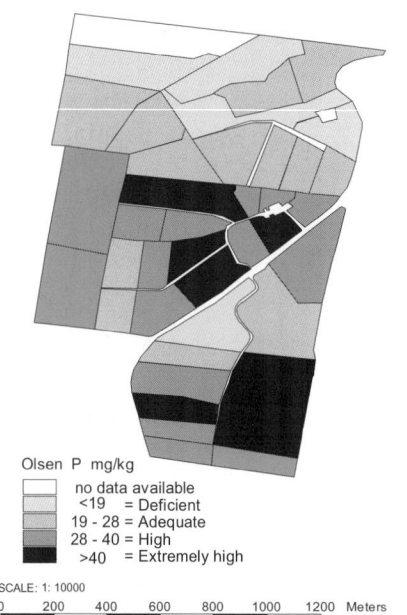

Olsen P mg/kg
- no data available
- <19 = Deficient
- 19 - 28 = Adequate
- 28 - 40 = High
- >40 = Extremely high

SCALE: 1: 10000

0   200   400   600   800   1000   1200 Meters

Figure 1 Soil phosphorus distribution across a dairy farm in Victoria

Conclusions The improved adoption and application of tools, such as soil testing and nutrient budgeting can make substantial improvements in fertiliser decisions, increasing productivity and profitability, while reducing adverse environmental impacts. Advances in analytical methods and procedures are continuing to refine fertiliser recommendations and reduce costs, while GPS mapping can provide a greater capacity for 'whole-farm' nutrient planning. Nutrient budgets are gaining acceptance as an indicator of sustainable nutrient practices, and provide a useful educational tool to assist dairy farmers to more effectively account for nutrient inputs, redistribution and losses within the farm in Australia.

References

Fulkerson, B. & P. Doyle (2001). 'The Australian dairy industry.' (Dept. Natural Resources and Environment: Tatura, Victoria).

Reuter. D.J. (2001). Nutrients – farm gate nutrient balances. Australian national land and water audit 2001.Http://audit.ea.gov.au/ANRA/land/farmgate/Nutrient_Balance.pfd

Evaluation and determination of the appropriate method for assessing optimum utilisation rate of *Eurotia ceratoides* in upland grasslands of Iran

F. Amiri and M.R. Chaichi
Department of Rangeland Sciences, Science and Research Branch, The Islamic Azad University, Poonak, Tehran, Iran, Email: fazel16760@yahoo.com, rchaichi@ut.ac.ir

Keywords: measurement methods, percent utilisation, Analytical Hierarchy Process

Introduction Determination of optimum utilisation rate for different range species is an important factor in assessing range grazing capacity. The vast rangelands in Iran with diverse vegetation types require an accurate, economic and quick method to determine the optimum utilisation rate for different range species. This experiment was conducted to determine the most appropriate method to determine the optimum utilisation rate for *Eurotia ceratoides*, which is one of the important grass species in upland grasslands in Iran.

Materials and methods The measurement methods were classified in three categories according to sampling size and then for each method the number of samples was determined using the appropriate statistical procedures (Bonham, 1989). The time spent in field and laboratory as well as all costs (equipment and labour) were measured for each sample and each method. To determine the accuracy of the applied methods, the statistical method of Estimating Sampling Sizes was employed (Cook *et al*, 1986). The collected data were analyzed by the AHP statistical method using "Expert Choice" software.

Results Stem counting was the quickest and most economic method, while the paired caging (control) method appeared to be the most expensive and time consuming method (Table 1). There was a significant difference ($P<0.05$) in the mean measured utilisation rate of *Eurotia ceratoides* between paired caging (control) and other methods, except for height-weight before and after grazing and ocular (double sample) estimation methods. The accuracy test showed that height-weight before and after grazing, ocular (double sample) estimation and paired caging (control) methods with 4.5, 8.3 and 8.6 % estimation faults (k) respectively were the most reliable methods. These results in respect of high costs and time requirement for the paired caging (control) method agree with results reported by Klingman *et al*. (1943). The stem counting method appeared to be economic and quick, but it was not accurate and reliable, supporting the results reported by Pechanec *et al*. (1937).

Table 1 Mean values for utilisation rate of *Eurotia ceratoides* assessed by different measurement methods

Methods	Percentage utilisation	Time minutes	Costs Rials	% estimation faults (k)	AHP
Paired cage (control)	56.33a	1866	170075	-------	0.0866b
Before and after grazing	55.40a	1546	155215	8.3	0.007b
Ocular estimate (double sample)	60.00a	150	39643	8.6	0.128b
Height-weight	61.93a	374	35073	4.5	0.134b
Stem count	43.68b	136	13631	16	0.233a
Reference units	40.43bc	359	37515	10.5	0.1162c
Production index	33.64dc	429	43900	18.5	0.0981dc
Plant count	30.44d	273	27988	23.4	0.0481d
Twig length	25.35d	666	67194	25	0.0684d

Means with the same letter are not significantly different at 5% level

Conclusion The results obtained by Analytical Hierarchy Process in this experiment showed that the appropriate method (quickest, most economic and most accurate) to measure the optimum utilisation rate of *Eurotia ceratoides* in upland grasslands of Iran is the height-weight method.

References
Bonham, C. D. (1989). Measurements for Terrestrial Vegetation. John Wiley and Sons. Inc., New York.
Cook, C. W. & W. Stubbendieek (1986). Range Research: Basic Problems and Techniques. Society of Range Management. USA.
Pechanec, J.F. & G.D. Pickford (1937). A comparison of some methods used in determining percentage utilization of range grasses. *Journal of Agricultural Research*, 54, 753-765.
Klingman, D., S.R. Miles & G.O. Mott. (1943). The cage method for determining consumption and yield of pasture herbage. *Journal of the American Society of Agronomy*, 35, 739-746.

The use of digital imagery for the assessment of green biomass in native pastures

A.F. Southwell[1], G. McKenzie[1], J.M. Virgona[1], A.M. Ridley[2], P. Eberbach[1]
[1]School of Agriculture, Charles Sturt University, Wagga Wagga 2678, Australia, Email: asouthwell@csu.edu.au
[2]Department of Primary Industries, Rutherglen, Victoria, Australia

Keywords: biomass, cover, percentage green

Introduction A practice common to pasture research is the assessment of green leaf. In Australia, where the water use of plants is becoming an increasingly important issue due largely to its implications for dryland salinity, it is imperative that accurate and repeatable methods for characterising the amount of green leaf in pastures be used. The assessment of green leaf has been approached in many ways in the past with varying degrees of success and accuracy. The most accurate way is to physically harvest an area of pasture and separate the green component to make the relevant measurements. For many situations, this may not be suitable particularly due to the destructive, laborious nature of the activity. Many techniques have been tried but they vary in such areas as accuracy, the quantitative nature of the output, repeatability, destructiveness, complexity and labour and equipment expenses ('t Mannetje 2000). The project aim was to determine if digitally derived green cover measurements could act as a remote substitute for percentage green biomass in pastures.

Methods From summer 2003 to winter 2004, a series of photographs were taken of three native pastures all varying in composition, growth patterns and water use abilities. A digital camera was mounted on a box constructed to maintain constant illumination on an area of pasture 0.1m² in size. The image produced was analysed using a software program designed to calculate the number of pixels between selected colour thresholds and convert them to a measurement of percentage green cover. This was then compared to the green biomass percentage calculated by harvesting the area of pasture, sorting green biomass from dead and determining the dry-weight of each component. Data was collated and a regression analysis carried out to determine the relationship between the two measurements.

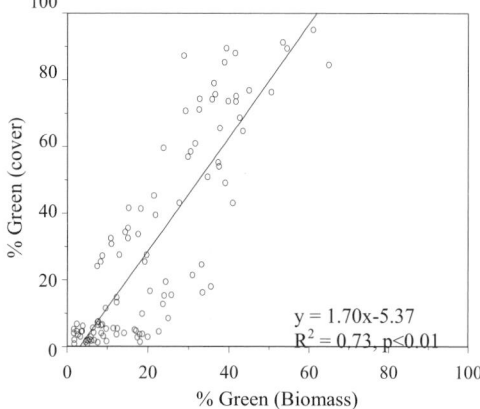

Figure 1 A comparison of digitally derived % green cover and % green biomass measurements for three types of native pastures (< 1000kg DM/ha removed)

Results A linear relationship was found and 73% of the changes in percentage green biomass explained by the digitally derived green cover measurements (Figure 1) when samples with less than 1000kg DM/ha were omitted to improve this relationship.

Discussion The use of digital image technology was an inadequate descriptor of pasture percentage green biomass. Though the r^2 values were comparable to other studies using different techniques for similar traits (Künnemeyer *et al.*, 2001 = 0.7, Denison & Russotti 1997 = 0.76 to 0.98), the level of error indicates this is not a suitable alternative. This can be attributed to: a two dimensional image being compared to the three dimensional measure of biomass and when manually collecting the data the fact that the human eye can isolate shades of green outside the colour thresholds set by computer software when green biomass is low. Other factors influencing the success of the digital analysis technique may include total biomass, composition and the phenological stage of the pasture. Interestingly, the middle section of the greenness scale where visual estimation techniques are least accurate (Hatton *et al.*, 1986) is where the digital analysis technique was most accurate.

References

Denison, R.F., & R. Russotti (1997). Field estimates of green leaf area index using laser-induced chlorophyll fluorescence. *Field Crops Research,* 52, 143-149.

Hatton, T.J., N.E. West & P.S.T.G. Johnson (1986). Relationships of the error associated with ocular estimation and actual total cover. *Journal of Range Management,* 39, 91-92.

Künnemeyer, R., P.N. Schaare & M.M. Hanna (2001). A simple reflectometer for on farm pasture assessment. *Computers and Electronics in Agriculture*, 31, 125-136.

't Mannetje L., (2000). Measuring biomass of grassland vegetation. In: L. t'Mannetje (ed.) Field and laboratory methods for grassland and animal production research. CABI Publishing, Wallingford, 168-169.

A simple theoretical model for calculating agricultural value of grasslands

G. Nagy

Debrecen University Agricultural Center, Debrecen, Böszörményi út 138. H-4032, Hungary, Email: nagyg@helios.date.hu

Keywords: agricultural value, model, ground cover, yield potential, forage quality

Introduction In spite of the emerging new social demands for non-material grassland products and services (nature reservation, environmental protection, landscape and amenity), the productivity of grasslands is going to remain in the mainstream of overall grassland use in many regions and countries of the world. Strategic planning of future grassland use in a region needs reliable information on the agricultural value of grasslands. The purpose of our research programme was to find a model for calculating the agricultural value of grasslands, in order to assist the classification of grassland productivity for a given area.

Agricultural value – definition and content The agricultural value of grassland (AVG) refers to the production potential for agricultural products (milk, meat and fibre) of a given grassland. In practical terms, it reflects the productivity of grasslands. The productivity of grasslands is first of all determined by the genetics of grassland components (species and cultivars). The sward composition is influenced by ecological conditions and farming practices. The AVG therefore gives reliable information on the productivity of permanent grasslands.

Theoretical bases for AVG The AVG can be derived from the agronomic characteristics of the grass plants. Yield potential and the overall quality of a species (or cultivar) have been selected as the two most important features, as these have major effects on agricultural value. A positive correlation is found between these two features and agricultural value. As AVG is related to grassland fields and areas, it is necessary to examine the efficiency with which grass swards can utilise the productive space of the ground. The productive space of the ground has two dimensions, vertical and horizontal ones. The efficiency of the vertical use of the productive space of the ground is determined by the height of plants, which is evaluated in the yield potential of a plant. The efficiency of horizontal use of the productive space can be evaluated by the ground cover of the sward components. Theoretically, the ground cover also has a positive correlation with AVG, provided weeds and poisonous plants are excluded.

The mathematical model of AVG The AVG is built up by the individual agricultural values of grass components. The contribution of grass plants to AVG is described by the following equation:

$$AVG_o = \frac{1}{100} \times \sum_{i=1}^{n} GC_i \times YP_i \times FQ_i, \text{ where}$$

AVG_o = overall agricultural value of a grassland; GC_i = individual ground cover of grass plants; YP_i = individual yield potential category of grass plants; FQ_i = individual forage quality category of grass plants.

Numerating the constituents of AVG For the practical use of the model, components have be numerated. Ground cover is expressed as a percentage, visually estimated during field visits. For the yield potential and forage quality, categories are set up for the individual grass plants. Categories for yield potential: 1 very low; 2 low; 3 medium; 4 good; 5 outstanding. Categories for forage quality: 1 very poor; 2 poor; 3 medium; 4 good; 5 excellent. The assessment of grasses to categories may be done according to specific local experimental results, as well as by using the scientific literature.

Classification of grasslands based on AVG Using the ground cover % and category numbers for YP and FQ, the AVG may range between 0 and 25. Using the AVG model calculation, a 5-rank grassland classification system has been set up:

AVG	Description of agricultural value of grassland categories
< 5.0	very poor grassland
5.1 - 10.0	poor grassland
10.1 - 15.0	medium grassland
15.1 - 20.0	good grassland
> 20.1	very good grassland

Discussion and conclusion In several examples, the use of the AVG model successfully supported the former empirical classification of Hungarian grasslands. The model is a promising tool for planning regional grassland use of different intensities and for different purposes.

The role of Proper Use Factor model for the prediction of available forage in rangeland in the south of Iran

G.R. Badjian[1], I. Dahlan[1], M. Shahwahid H.O.[1] and A.A. Mehrabi[2]
[1]University Putra Malaysia, 43400 Serdang, Selangor, Malaysia, E-mail: badjian@yahoo.com, [2]University of Tehran, Karaj, Iran

Keywords: proper use factor, available forage, rangeland, Iran

Introduction In Iran, 80 to 90% of the livestock production is associated with rangeland but 48% of the rangeland has been classified as in poor condition (Farahpour, 2002). Available Forage (AF) is that portion of the forage production accessible for use by a specified kind or class of grazing animal. Estimates of AF by plant species, consumption by the animal, and the contribution of the forage to the animal's diet must be synchronised with each other in the same time frame (Valentine, 2001). The Proper Use Factor (PUF) is the degree of utilisation of current year's growth of the vegetation that, if continued, will achieve management objectives and maintain or improve the long-term productivity of the site. The objective of this study was to identify and determine the PUF of forage that cause effect the qualitative and quantitative aspects of AF in rangeland of Bakkan, south Iran.

Materials and methods The research involved the identification and determination of integrated PUFs that had impacts on AF production in rangeland with consideration of local traditional experience. Aerial photo interpretation, interpreted maps of soil types and rangeland evaluation were done to derive the slope, soil, range condition, and range trend factors of PUF respectively. Field survey and laboratory analyses were done to assess the palatability of the produced forage and determine a coefficient factor for analysing the PUF. Techniques include both formal and participatory research and GIS for analysis and integration.

Results and discussion Factors that determined the selection of key grazing areas and species were: topography or slope properties (SL); soil properties (SOP), range conditions (RC), range trend (RT) and palatability (PL) or vegetation class, as shown in the following equations:

$PUFsp = f$ (minimum rate of: RC, RT, SOP, SL and PL)

$AFsp = PUFsp * Pspi * Svt$ (kg DM/ha) (if PUF< PL) or $AFsp = PLsp * Psp * Svt$ (kg DM/ha) (if PL <PUF)

AFsp is the available forage of expected species, Psp is the production of expected species in vegetation type, and Svt is the area of vegetation type.

These equations indicated the important role of PUF as a forage index and PL as a coefficient factor for range grazing. Therefore, the amount of AF will largely be based on the coefficient factor of PUF or PL. If a pasture is continuously grazed for the grazing season, PUF would be approximately 50% (i.e., take half and leave half); if the pasture was in a planned grazing system, PUF might be 60% (Society for Range Management, 1991). The minimum coefficient ratios of PUF and PL were used as an index for measuring the amount of AF. Table 1 shows the factors and their coefficient rates used for construction of the AF model. Application of PUF for prediction of AF in Bakkan indicated that, depending on the situation, 15 - 50% of the produced forage in rangeland can be removed by grazing animals without damaging the forage plants.

Table 1 Proper Use Factors (PUF) for calculation of Available Forage

Factors	Classification	Definition	Coefficient rate %
Slope %	1,2,3,4	0 - >40	0 - 50
Soil property	1,2,3,4	1 - 17	20 - 50
Range condition and trend	1,2.3…,12	Good to very poor condition and up to down trend	15 - 50
Palatability (Vegetation class)	1,2,3,4,5,6	Class I,II,III	15 - 50

Conclusion Due to limitation to range production and sensitivity to degradation in dry or semi-dry regions like Bakkan, south Iran, this procedure can be applied to help decision makers prepare rangeland management plans.

References

Farahpour, M. (2002). A Planning Support System for Rangeland Allocation in Iran. PhD Thesis. Wageningen University. Netherlands, 186pp.
Society for Range Management (1991). A Glossary of Terms Used in Range Management. 3rd ed. USA.
Vallentine, J.F. (2001).Grazing Management.2nd ed. Academic Press, USA, 659pp.

Assessment and monitoring of grazing lands in the northeastern United States

M.A. Sanderson[1], S.C. Goslee[1], J.B. Cropper[2] and R.B. Bryant[1]

[1]*USDA-ARS Pasture Systems and Watershed Management Research Unit, Building 3702, Curtin Road, University Park, Pennsylvania 16802 USA, Email: mas44@psu.edu,* [2]*USDA-NRCS East National Technology Support Centre, Greensboro, North Carolina, USA*

Keywords: pasture health, pasture condition, environment

Introduction The Pasture Condition Score System (Cosgrove *et al.*, 2001) was developed as a monitoring and management tool on grazing lands The system considers 10 indicators of soils, plants, and animals including percent desirable plants, plant cover, plant diversity, plant residue, plant vigor, percent legume, uniformity of use, livestock concentration areas, soil compaction, and soil erosion. The indicators are assigned a score according to detailed criteria and the scores are summed to give an overall score for a pasture, or relevant grazing unit. The score is then interpreted, indicating if some type of management change or treatment is necessary. We tested the Pasture Condition Score system on farms across the northeast USA.

Materials and methods We applied the system to 138 pastures on 32 farms across the Northeast. Both beef and dairy farms that used either rotational or continuous stocking were included. On each farm, two to eight pastures in different landscape positions were rated according to the Pasture Condition Score System. Each indicator was scored on a 1 to 5 scale with 1 representing an unacceptable condition and 5 representing the optimum condition. The entire pasture was walked, examined, and the 10 indicators and six causative factors were scored according to the guidelines in Cosgrove *et al.* (2001).

Results Across all farms, 44% of the pastures fell into the category of needing only minor changes to management and another 40% fell into the category of needing some improvements (Table 1). At the extremes, we found only a few "perfect" pastures and none fell into the lowest category of "major problems." There were 14% where the pasture condition score indicated that immediate changes were needed. Examining the indicator scores for the group of pastures with the lowest overall scores, it is clear that plant diversity and percent legume have a large influence on scores along with "uniformity of use" (Table 1). The legume component of these pastures was almost nonexistent and the pastures were typically dominated by one grass species. Uniformity of use was low on these pastures because they were overgrown and there was a great deal of spot grazing. A score of 2 corresponds to 25 to 50% of the area either ungrazed or grazed very little. Other indicators of resource degradation (soil erosion and compaction, livestock concentration areas) were mid-range and were not driving the low scores. This indicates that many of the potential problems on the pastures would be relatively straightforward to resolve.

Table 1 Distribution of pasture condition scores for 138 pastures surveyed on 32 farms in the northeastern USA along with the scores for the 10 pasture indicators on the lowest scoring pastures in the survey (n=20)

Distribution of pasture condition scores		Scores for indicators on lowest scoring pastures	
Pasture condition score category	% of pastures	Indicator	Average score[#]
10-15 Major problems	0	% Desireable plants	2.4
16-25 Immediate changes needed	14	Plant Cover	1.9
26-35 Improvements needed	44	Forage diversity	1.6
36-45 Minor changes needed	40	Plant residue	2.0
46-50 No changes needed	2	Plant vigor	3.1
		% Legume	1.1
		Uniformity of use	1.7
		Livestock concentration areas	3.1
		Soil compaction	2.7
		Soil erosion	3.6

[#]An individual indicator score of 1 indicates that major management changes are necessary, whereas a score of 5 indicates that no changes are needed

Conclusions Land degradation was not the main cause of poor pasture condition. Data indicated that low forage diversity, lack of legumes, and low uniformity of use, were the driving factors. There is a need for rapid objective methods to quantify this information rather than relying on visual estimates.

Reference

Cosgrove, D., D. Undersander & J.B. Cropper (2001). Guide to pasture condition scoring. USDA-NRCS Grazing Lands Technology Institute. Washington, D.C., USA.

On-farm information: a valuable tool for the sustainable management of mountain pastures in protected natural areas

N. Mandaluniz[1], A. Bernués[2], A. Igarzabal[1], I.J.L. Riedel[2], R. Ruiz[1], A. Sanz[2], I. Casasús[2] and L.M. Oregui[1]
[1]NEIKER, Granja Modelo de Arkaute, PB 46, E-01080, Vitoria-Gasteiz, Spain, Email: nmandaluniz@neiker.net,
[2]CITA, Gobierno de Aragón, PB 727, E-50080, Zaragoza, Spain

Keywords: mountain pastures, commercial farms, management, land use

Introduction Mountain pastures have traditionally been maintained by livestock. The analysis of data concerning farms' characteristics, productive-reproductive management and land use of commercial farms can constitute a real approach to study these systems and the changes that are occurring. This information is necessary to develop new utilisation guidelines, making compatible livestock production and conservation of natural resources. This paper describes a methodological framework to study the issues described above through some examples taken out from a wider research project (Mandaluniz *et al.*, 2003).

Materials and methods The study was carried out in 3 Protected Natural Parks located in the north of Spain with different agro-climatic conditions and livestock utilisation regimes (more details in Mandaluniz *et al*, 2003). The methodology used implied a multidisciplinary approach; this paper is focused in the characterisation of farming systems and grazing management by means of information collected from commercial farms. Farms using pastures within these Natural Areas were selected to carry out the study. A structured questionnaire was designed to determine the main characteristics of farming systems and collect farmers objectives and concerns. Breeding and grazing calendars were distributed to farmers in order to describe the importance of the grazing period in the different production systems, and livestock performance throughout the grazing season was analysed.

Results and discussion One of the main characteristics of the farming systems was the large heterogeneity (Mandaluniz *et al.*, 2003). In general, farms are quite extensive in terms of communal pastures utilisation, which cover 40-54% of annual energetic requirements in beef cattle (Casasus *et al.*, 2002; Mandaluniz *et al.*, 2004). Some factors that can reduce rangelands utilisation were pointed out i.e. lack of water, communications, degradation of vegetation, reproductive intensification, etc. The farmers suggested that some of these aspects should be improved by the Administration in order to optimise pasture use. Parturition time occurred mainly indoors and late-pregnant and lactating animals are fed indoors during the winter. Finally, there were differences in animal performance depending on year, grazing area and productive state of animals (Casasus *et al.*, 2002; Mandaluniz *et al.*, 2004).

Conclusions Data collected from commercial farms allowed the determination of characteristics and constraints of these systems and the needs and concerns of the farmers. However, it was not always easy to collect reliable information or to find technical solutions to some problems using this methodology. In this sense, it is interesting to make a technical monitoring by case study research to study constraints and concerns identified in these commercial farms. An in-depth dynamic analysis of productive-reproductive results and grazing management will allow evaluation of different productive alternatives (including grazing management, feeding and reproduction) and discuss them with farmers according to their needs.

Acknowledgement This work has been financed by project INIA RTA 02-086-C2-2. Authors would like to acknowledge collaborator farmers, associations and the Direction of the respective Parks.

References
Casasus, I., A. Sanz, D. Villalba, R. Ferrer & R. Revilla (2002). Factors affecting animal performance during the grazing season in a mountain cattle production system. *Journal of Animal Science*, 80, 1638-1651.
Mandaluniz N., I. Casasús, A. Bernués, A. Aguirrezabal, J.L. Riedel, A. Sanz & L.M. Oregui (2003). Tools for sustainable management of mountain pastures: application to three Protected Natural Areas in northern Spain. First Joint Seminar of the Sub-Networks FAO-CIHEAM on Sheep and Goat Nutrition and on Mountain and Mediterranean Pastures. Granada.
Mandaluniz, N., A. Legarra & L.M. Oregui (2004). Mountain pasture utilisation by free-ranging beef-cattle in the Natural Park of Gorbeia (Basque Country). *Proceedings of the Fifty-fifth European Association of Animal Production*, Bled, Slovenia.

Using Landsat Imagery to Analyse Land Cover Change in the Njoro Watershed, Kenya

T.J. Baldyga[1], S.N. Miller[1], K.L. Driese[1] and C. Maina-Gichaba[2]
[1]Department of Renewable Resources, Department 3354, University of Wyoming, 1000 E. University Avenue, Laramie, Wyoming 82071, USA, Email: tbaldyga@uwyo.edu, [2]Department of Geography, Egerton University, PO BOX 536, Njoro, Kenya

Keywords: land cover, remote sensing, Landsat

Introduction In developing nations where resources are scarce and increased population pressures create stress on available resources, methods are needed to examine effects of human migration and resultant changes in land cover. Widespread availability and low cost of remotely sensed imagery and Geographic Information Systems (GIS) are making such methods a reality to develop quantitative resource mapping and land cover change detection in developing nations (Sheng et al., 1997). However, difficulties arise in tropical regions when trying to analyse traditional vegetation bands (Bands 3 and 4), or indices such as NDVI because saturated pixels limit spectral distinction.

Materials and methods Band separability for 9 informational classes was measured for a Landsat 7 image acquired in Kenya's Rift Valley (Path 169, Row 60) on 4 February 2003. Baldyga et al. (2004) showed that vegetation diversity and temporal variability resulted in large classification errors using bands 2, 3 and 4 in an unsupervised classification in 4 scenes captured for this region. Band separability analysis indicates that in this region the nine identified spectral classes are best distinguished using a four-dimensional image consisting of bands 4, 5 and 6 and the tasselled cap transformation for brightness (TC1). Nine informational classes were identified for this project and a combination of unsupervised and supervised classification methods were used to classify the 4-dimensional image.

Results Baldyga et al. (2004) achieved only 41% accuracy with unsupervised classification; errors were most frequent in distinguishing agricultural lands from grasslands. This has serious implications, as response to land cover change is not linear (Baldyga et al., 2004). The current classification (Table 1) was only 75% accurate; the greatest error was in classifying Barren areas. Barren areas in the region change seasonally and annually, so the error is not surprising given that ground truth data collection was impossible on the acquisition date of the Landsat image. Shrublands and Riparian area were classified as Agriculture and Forest respectively. In all cases of misclassification, at least one adjacent cell was classified as the accuracy assessment point. Several points were collected using a range finder and calculating the location, rather than collecting an actual GPS coordinate at the point due to inaccessibility. All misclassified Grasslands cells were classified as Agriculture or Forest and located near transitional areas.

Table 3 Error matrix resulting from accuracy assessment

Land Cover Class	Map Total	Number Correct	Producer's Accuracy	User's Accuracy
Open Water	5	5	100%	100%
Urban	3	2	67.00%	67.00%
Agriculture	33	21	64.00%	81.00%
Barren	10	1	10.00%	10.00%
Forest	25	15	60.00%	79.00%
Grasslands	95	86	91.00%	78.00%
Wetlands	—	—	—	—
Riparian	1	0	0.00%	0.00%
Shrublands	1	0	0.00%	0.00%
Total:	173	130		
Overall Accuracy:	75.14%			

Conclusions We believe the classification accuracy, using the bands and enhancements indicated above, was much higher than indicated. Refining the classification process by incorporating ancillary data will improve results in Riparian and Agricultural areas. Classified land cover scenes are input to GIS-based models as part of a systems approach to understanding watershed dynamics. Therefore, developing accurate classification methods in rapidly changing tropical landscapes is critical, as migration into these fertile areas puts pressure on scarce resources.

References
Baldyga, T. J., S. N. Miller, W. Shivoga & C. Maina-Gichaba (2004). Assessing the impact of land cover change in Kenya using remote sensing and hydrologic modelling. *Proceedings of the American Society for Photogrammetry and Remote Sensing Annual Meeting*, Denver, Colorado, May 23-28, 2004.
Sheng, T. C., R. E. Barrett & T.R. Mitchell (1997). Using geographic information systems for watershed classification and rating in developing counties. *Journal of Soil and Water Conservation*, 55(2), 84-89.

Forage suitability group report: a tool for grassland management

J.B. Cropper[1] and G.L. Peacock[2]
USDA, Natural Resources Conservation Service, [1]East National Technology Support Center, Northwood Building, Suite 410, 200 E. Northwood Street, Greensboro, North Carolina 27401, USA, Email: james.cropper@gnb.usda.gov, [2]Central National Technology Support Center, 501 West Felix Street, Building 23, P.O. Box 6567, Fort Worth, Texas 76115 USA

Keywords: forage production, soil map unit component, grassland, Microsoft Access[TM]

Introduction Forage suitability groups (FSG's) are USDA-Natural Resources Conservation Service (NRCS) interpretative reports used to develop conservation plans for forage-producing farms and ranches and provide grassland resource information to producers. These electronic reports use soil properties and climatic data to develop forage selection, management recommendations, seasonal distribution of growth, and yield potentials for groups of soil map unit components that have like agronomic characteristics. The information contained in a FSG report can help the user develop proper livestock-forage balances, grazing management plans, pasture and haycrop renovation options, and land treatment measures.

Materials and methods The FSG's are being developed by local NRCS grassland experts and soil scientists for each Major Land Resource Area (MLRA) in the US with significant forage production occurring in it. They use the NRCS National Soil Information System (NASIS) and other relevant soil databases to group similar response soil map unit components together. Once the groupings are completed, evaluated, and accepted, interpretative reports are written for each group. Also contained within FSG's are regional climatic data that affect yield and forage selection and management recommendations. If a MLRA contains very dissimilar climatic conditions due to its terrain or extent that would impact the contents of the FSG report, the dissimilar areas are separated into land resource units (LRU's). Guidance in producing FSG reports is contained in the NRCS *National Range and Pasture Handbook*, revision 1, Chapter 3, Section 2 (USDA-NRCS, 2003). Microsoft Access[TM] is the database used to store the information and generate a report for each FSG currently. Once the template for FSG's is completed for the NRCS Ecological Science Information System, all FSG reports will reside in that database.

Results Each FSG report contains the following sections: general information (FSG name, identifier number, & MLRA or LRU locale), physiographic features, climatic features, soil interpretations, soil map unit list, adapted species list, production estimates, growth curves, soil limitations, management interpretations, management dynamics, and documentation. The user has the option of selecting only those sections of most interest to them. Currently with Microsoft Access, this is done by selecting Tools, then Office Links, and then Publish It with MS Word[TM]. Once in MS Word, the user can extract what they want. An example section excerpt follows:

Forage Growth Curves

Growth Curve Number:	PA1205
Growth Curve Name:	Orchardgrass-Kentucky bluegrass-white clover, 120-140 day growing season
Growth Curve Description:	Orchardgrass pasture with K. bluegrass and white clover components 20-30% each by weight

Percent production by month

Jan	Feb	Mar	Apr	May	Jun	Jul	Aug	Sep	Oct	Nov	Dec
0	0	0	15	30	22	8	6	14	5	0	0

Conclusions The FSG's reduce the redundancy of providing the same agronomic interpretations for each soil map unit component when many may share much the same soil properties with several other named soils. FSG reports are stored in a convenient format for retrieval and editing. For FSG's sharing similar interpretations with other FSG's, report contents can be copied, then edited, and given a new identifier and name to reduce repetitive report inputs. Reports of FSG's can show besides what we know, what we do not know. They put together the storehouse of soil information in NASIS with forage research trials and on-farm yield trials observation on identified soil map unit components. Missing quantified data can be painfully obvious. They can, however, unleash a stream of useful known data for anyone querying the database. They can also point to needed data collection for action to strengthen the data bank.

References
USDA-NRCS (2003). National Range and Pasture Handbook, revision 1. Chapter 3, Section 2 and Exhibits. Washington, District of Columbia, 61pp.

Pasture condition scoring

J.B. Cropper

USDA, Natural Resources Conservation Service, East National Technology Support Center, Northwood Building, Suite 410, 200 E. Northwood Street, Greensboro, North Carolina 27401 USA, Email: james.cropper@gnb.usda.gov

Keywords: pasture condition, diversity, indicators, inventory, pasture management

Introduction A pasture condition score sheet has been developed for use in the United States. It has rating criteria for key indicators that are used to ascertain if some areas of pasture management could be improved. It can also help evaluate what is causing less than desirable pasture conditions. Pasture condition scoring involves the visual and tactual evaluation of ten indicators that rate a pasture's overall condition. The ten indicators are: percent desirable plants, plant cover, plant diversity, plant residue, plant vigor, percent legume, uniformity of use, livestock concentration areas, soil compaction, and erosion (sheet and rill, gully, streambank and shoreline, and wind). Six causative factors that impact pasture plant growth and vigor are also rated: soil fertility status, soil pH status, severity of use, forage species suitability, episodic climatic conditions, and insect and disease pressure. Regionally, levels of salinity, sodicity, and toxic elements (e.g. aluminum) can also be measured and rated where they commonly affect pasture productivity, stability, and forage species selection. Indicators and causative factors receiving the lowest scores can be focused upon and corrective actions taken as warranted.

Materials and methods Two USDA-NRCS publications, "Guide to Pasture Condition Scoring" (Cosgrove *et al.*, 2001) and the "Pasture Condition Score Sheet" (Cosgrove *et al.*, 2001), were distributed across the US in 2001. The Guide describes ten indicators and six causative factors vital for assessing pastures while maintaining productivity and environmental viability. The Score Sheet is used to record pasture conditions. Each pasture is rated using the score sheet criteria (Figure 1) and some basic inventory methods. Each indicator's or causative factor's condition is estimated and scored separately on a score sheet using a range of 1 (lowest) to 5 (highest). The indicator scores can be totalled or left as an individual score and compared with the other nine indicators. Causative factors are simply scored individually to see which ones are furthest from the ideal (score of 5). Many of the indicators/causes are simply rated visually. For estimating percent desirable plants, plant diversity, and percent legume in multiple species swards, the Dry-Weight-Rank method is recommended. Soil fertility test results for phosphorus (P), potassium (K), and pH are necessary to rate soil fertility and pH causative factors. Soil compaction is rated by using a soil penetrometer or simple probe to compare soil strength between fenceline and treaded areas. Some indicators and causative factors are rated highest at some moderate level considered optimal (e.g. 4 to 5 forage species from at least two major functional groups will score higher for plant diversity than a near monoculture or a very diverse species, but from one dominant functional group, pasture). The same holds true for percent legume, soil test P and K, plant nitrogen status, soil pH, severity of use, and plant residue. Too much or too little of any one of these may not good for the environment, livestock, or the pasture sward.

Farm or Ranch Site:					Date:			
					Pasture unit description			
Indicators					#1	#2	#3	#4
Percent desirable plants								
Percent plant cover by weight that is desirable forage:								
1	2	3	4	5	4	5	5	3
<20	20-40	40-60	60-80	>80				

Figure 1 First pasture condition indicator showing layout of Pasture Condition Score Sheet with entries

Results By using the pasture condition score sheet over time and keeping the records, trends in decline or improvement can be detected so if need be, any changes in treatment can be done. Consultants, farmers, and various farm service agencies' personnel all can find these two pasture condition documents useful. With a complete inventory of pasture condition, proper remedies to correct the defined problems are more easily found.

Conclusions More definitive work needs to be done on plant diversity's role in pasture productivity. Present plant diversity criteria attempt to define the grazing ideal, but it is based on limited, inconclusive research work.

References
Cosgrove, D., D. Undersander & J. Cropper (2001). Guide to pasture condition scoring. USDA-NRCS Grazing Lands Technology Institute. Washington, District of Columbia.
Cosgrove, D., D. Undersander, and J. Cropper (2001). Pasture condition score sheet. USDA-NRCS Grazing Lands Technology Institute. Washington, District of Columbia.

Disappearance of residual dry matter on annual grassland in the absence of grazing

W.E. Frost[1], J.W. Bartolome[2] and K.R. Churches[1]
[1]University of California Cooperative Extension, 311 Fair Lane, Placerville, California, USA, Email: wefrost@ucdavis.edu, [2]Environmental Sciences, Policy and Management, University of California, Berkeley, USA

Keywords: utilisation, mulch, management, annual grassland

Introduction Residual dry matter (RDM) is a standard used by grassland managers for assessing the level of grazing use on annual grasslands and associated savannas and woodlands. Residual dry matter is the old plant material left standing or on the ground at the beginning of a new growing season. It indicates the combined effects of the previous season's forage production and its consumption by grazing animals of all types. The standard assumes that the amount of RDM remaining in the fall, subject to site conditions and variations in weather, will influence subsequent species composition and forage production, in addition to providing soil protection and protect against nutrient losses (Bartolome, et al., 2002). While RDM is measured at the beginning of a new growing season, grazing does not always occur continuously up to this time. Managers do not have information to predict the disappearance of residual dry matter due to physical and chemical breakdown during a period of non-grazing. In this study the rate of RDM disappearance during the summer (non-growing) period on annual grasslands was investigated.

Materials and methods Nine 9.29 square meter (100 square foot) exclosures, protected from grazing by fencing, were located across 4 counties in the central Sierra Nevada mountains of California. At each location, within these exclosures, six .093 square meter (1 square foot) plots were randomly assigned to measure either peak standing crop at the end of the growing period (3 plots) or residual dry matter just prior to the beginning of the following growing season (3 plots). Exclosure were relocated, on the same ecological site, after measurement of residual dry matter. Data were collected from 1998 through 2003. Data was converted to reflect the percentage weight disappearance per 30 day period during the dry summer period between the time of peak standing crop and the beginning of the next growing season.

Results Residual dry matter disappearance varied by location and year. However, these differences were not consistent among locations, i.e. no location consistently had amounts of RDM disappearance that were consistently greater than or less than other locations. The RDM disappearance at individual locations ranged from a high of 13.3% per 30 day period to a low of no disappearance over the dry summer period. Yearly averages (per 30 day period) ranged from a high of 9.4% in 1998 to a low of 4.4% in 2003.

Conclusions The results of the study demonstrate that the amount of residual dry matter, by weight, will average a decrease of 7% per 30 day period from the time of peak standing crop of annual herbaceous species to occurrence of the germinating rain in the fall. The time of peak standing crop is generally accepted to be the time at which the vast majority of annual species cease growth, demonstrated by a change of colour from green to yellow/brown. With the information from this study, grassland managers, for the first time, will be able to determine the amount of herbaceous material that must be left at peak standing crop to insure adequate amounts of residual dry matter at the time of the first fall rains to provide for site protection. In situations where conservative use and a higher residual dry matter standard is appropriate or desired, grassland managers should plan utilizing the higher observed rate of residual dry matter disappearance, 13% per 30 day period. Thus, management of grazing animals can be altered to optimise the utilisation of annual herbaceous production while maintaining the residual dry matter to provide site protection and insure long term productivity.

References

Bartolome, J.W., W.E. Frost, N.K. McDougald & J.M. Connor (2002). California guidelines for residual dry matter (RDM) management on coastal and foothill annual rangelands. University of California Division of Agriculture and Natural Resources Publication 8092, 8.

The impact of concentrate price on the utilization of grazed and conserved grass

P. Crosson[1,2], P. O'Kiely[1], F.P. O'Mara[2], M.J. Drennan[1] and M. Wallace[2]
[1]Teagasc, Grange Research Centre, Dunsany, Co. Meath, Ireland, Email: pcrosson@grange.teagasc.ie, [2]Faculty of Agri-Food and the Environment, University College Dublin, Belfield, Dublin 4, Ireland

Keywords: beef production, systems, mathematical model, linear programming

Introduction A linear programming model was designed and constructed to facilitate the identification of optimal beef production systems under varying technical and policy scenarios. The model operates at a systems level and most activities that could occur in Irish spring-calving, suckler beef production systems are included. In this paper, the components of the model are described together with a simple application of the model involving changing concentrate prices.

Model description The model was developed using a mathematical programming methodology which encompasses the following characteristics: 1) a range of possible activities, 2) various constraints to prevent free selection from the range of activities, and 3) an objective which can be quantified (Dent *et al.*, 1986). It is a single year steady-state design. The fundamental unit on which the model is based is the cow unit. Due to the predominance of pasture-based systems in Ireland a detailed set of grazing options that are typical of those available to Irish cattle farmers is specified. Model details are specified on a monthly basis. This enables it to respond to monthly fluctuations in feed supply and animal requirements. Financial budgets (Teagasc, 2003) assign a cost or revenue to each activity and thus the program identifies the optimal net farm margin. Nutritional specifications are described in terms of net energy (NE) requirements subject to a maximum intake capacity.

Model application A scenario investigating the impact of a change in concentrate price on optimal systems is presented. Concentrates generally are the most costly feedstuffs and the cost of concentrates influences farm margin and system operated to a large extent. The influence of concentrate price on the optimal system is presented below (Table 1) together with the resulting impact on net margin (Figure 1). Above €140/tDM it was found that there was no response to further increases in concentrate price. Therefore, results are presented for the range €100/tDM to €140/tDM.

Table 1 Production results for concentrate price change scenario

Concentrate price (€/tDM)	100	105	110	115	120	125	130	135	140
Area for grazing (ha)	60.0	60.0	60.0	50.0	50.0	49.0	45.3	35.7	35.7
Area for grass silage (ha)	0.0	0.0	0.0	0.0	0.0	1.0	4.7	14.3	14.3
Land rented (ha)	10.0	10.0	10.0	0.0	0.0	0.0	0.0	0.0	0.0
Total N applied (kg/ha)	199.4	196.3	196.3	196.3	196.3	202.4	217.7	243.4	243.4
Concentrates fed (t)	94.5	84.5	84.5	70.4	70.4	58.8	39.1	6.0	6.0
Suckler cow numbers	61.3	61.2	61.2	51.0	51.0	53.7	56.4	56.2	56.2

Up to around €120/tDM all animals are finished on concentrate based diets but above this steer progeny are finished off grass and from about €130/tDM all progeny are finished off grass. Stock numbers initially decrease with an increase in concentrate price but recover somewhat after the change to grass-based finishing.

Conclusions Concentrate price impacts crucially on optimal systems driving both the operated finishing system and grass silage requirements. The model can be used also to analyse current or prospective scenarios. Future changes in agricultural policy can be investigated routinely. Whilst the production data are based mainly on performances obtained at Grange, the parameters can be modified to reflect other situations.

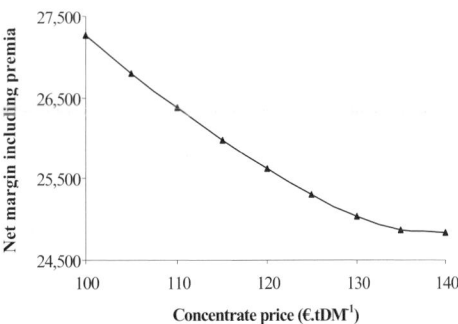

Figure 1 Change in net margin with increasing concentrate price

References
Dent J.B., S.R. Harrison & K.B. Woodford (1986). Farm Planning with Linear Programming: Concept and Practice. London: Butterworths, pp. 32-52.
Teagasc (2003). Management Data for Farm Planning, Teagasc, 19 Sandymount Avenue, Dublin 4, Ireland.

An interactive, web-based module to teach the principles of silage fermentation

M.H. Hall
Department of Crop and Soil Sciences, The Pennsylvania State Univ., University Park, PA 16802 USA, Email: mhh2@psu.edu

Keywords: teaching, silage fermentation

Introduction The forage-livestock industry represents a major agricultural enterprise throughout the world. In the United States, forages comprise 40-60% of the dairy cow's diet with many producers feeding 50-100% of this forage as silage (Kempisty, 1997). Harvesting forages as silage continues to increase as the size of farms increase and the mechanisation of silage harvesting allows timely harvest of a greater area. There are many factors that affect the quality of silage and producers need to better understand these factors to optimise their silage making process. Our objective was to develop a web-based learning module to help users better understand the processes involved in silage fermentation and how management practices can influence those processes.

Module development and description Introductory and background information about the various processes involved in silage fermentation (Pitt, 1990) are presented using JavaScript (Sun Microsystems, Inc., Santa Clara, CA). This portion of the module contains text and non-interactive graphics that visually illustrate the principal phases and reactions that occur during the silage making process. When the user has completed this section and feels comfortable with the basic concepts of fermentation they can move to the interactive portion of the module by clicking on the Interactive Module icon on the menu bar.

The interactive "what if" simulation segment of the module that follows forage through completion of the fermentation process was constructed using Flash (Macromedia, Inc., San Francisco, CA). This segment begins with a series of five questions about the silage making process that determine the results of the simulation model. Assistance is available while answering these questions by clicking on either the "Why we ask" or "Factors" icon associated with each question (Bolsen, 1997).

Clicking the "Run Simulation" button at anytime will graphically depict the fermentation process over time and highlight potential consequences of making silage too dry, too wet, with low carbohydrate or lactobacillus bacteria content, or not packing adequately. From the simulation window, the user can restart or pause the simulation or go back and change their responses to the questions.

This interactive learning module allows individuals to develop an understanding of the dynamic silage fermentation process. By providing a rich visual presentation along with the ability to create different management scenarios, it allows learners to explore the key management issues involved in making high-quality silage. Running the simulation model multiple times under various conditions designated by the user results in a more comprehensive understanding of the many dynamic and multifaceted processes involved in forage fermentation.

Conclusions The fermentation of forage into silage is a dynamic and multifaceted process that can be conceptually challenging to understand. Our objective was to develop a web-based learning module to help people better understand the processes involved in silage fermentation and how management practices could influence them. The module was developed for a broad audience including extension and government agency personnel, individual producers and consultants in forage management. It can be used effectively in presenting forage fermentation concepts during group training sessions as well as for individual training and consultations. The fermentation learning module can be accessed at
http://www.forages.psu.edu/topics/hay_silage/preservation/silage_preserv/index.html.

References
Bolsen, K.K. (1997). Issues of top spoilage losses in horizontal silos. In: Silage: Field to Feedbunk. *Proceedings from the National Silage Production Conference*, Northeast Regional Agricultural Engineering Service, Ithaca, NY, 137-150.
Kempisty, L.H. (1997). The importance of silage. In: Silage: Field to Feedbunk. *Proceedings from the National Silage Production Conference*, Northeast Regional Agricultural Engineering Service. Ithaca, NY, 1-2.
Pitt, R.E. (1990). The biology of silage fermentation. In: Silage and Hay Preservation. NRAES-5. Northeast Regional Agricultural Engineering Service. Ithaca, NY, 5-20.

Theme C: Delivering the benefits from grassland

Section 25

Decision support for grassland systems

Pâtur'IN: a user-friendly software tool to assist dairy cow grazing management

L. Delaby, J.L. Peyraud and P. Faverdin
INRA – UMR Production du Lait, 35590 Saint Gilles, France, Email: luc.delaby@rennes.inra.fr

Keywords: grazing management, decision-support system, dairy cows

Introduction The feeding of dairy cows at pasture presents many technical, economic and environmental advantages, while benefiting from a very favourable image. However, the management of grazed land is a complex game of strategy in which the farmer applies decisions in order to manage two unstable and uncertain fluxes of change: growth of grass and intake of the herd. Many tools (platemeter, etc.) and overall methods (local stocking rate references, farm cover, etc.) have been developed as aids to grazing management. Nevertheless, few decision-support systems are currently available that make it possible to anticipate and assess the consequences of a given decision in a dynamic way (Peyraud *et al.*, 2004). The objective of this article is to present the structure and functions of Pâtur' IN, a software tool designed to help the management of dairy cow grazing.

Structure and functions of Pâtur'IN After defining the structure of the farm, which comprises the plots, herds and available feed supplements, the user can, on the one hand, record all the events taking place on the farm during the grazing season and then, on the other hand, simulate various scenarios of sward use. A scenario consists of a succession of events (grazing, cutting, fertilisation, etc.) that act on the "Growth" and "Intake" functions according to rules of decision defined by the user (Figure 1). The "Growth" function makes it possible to calculate the biomass present on each day starting from a standard growth grid that can be modified by the user according to the local context. This grid is then adapted to each plot for each day using various models integrating the effects of climatic conditions, N fertilisation, amount of biomass present, etc. The "Intake" function calculates the quantities of grass ingested by the herd based on the intake capacity of the cows (INRA, 1988) and adapted to the characteristics of the pasture (sward height, pasture allowance, etc.). During a simulation, the user defines a set of rules of decision as well as the order of use of the plots. From this, the software program can determine, for each grazed plot, either the residence time in a rotational grazing system, or the land area to be offered each day in a strip grazing system. The user interface (Figure 1) allows a visualisation of the condition of each plot for each day, by means of various colours and the display of simulated events in the form of a grazing schedule. The whole set of results for a given simulation is available in the form of text file. Pâtur' IN can be run on a PC and is now available in an English version.

Conclusion The advantage of Pâtur'IN is that it makes use of data available at farm level, thus enabling farmers to carry out regular updating of the data according to the actual progression of the grazing season. In this way, users are able to develop simulations adapted to the particular context of each farm.

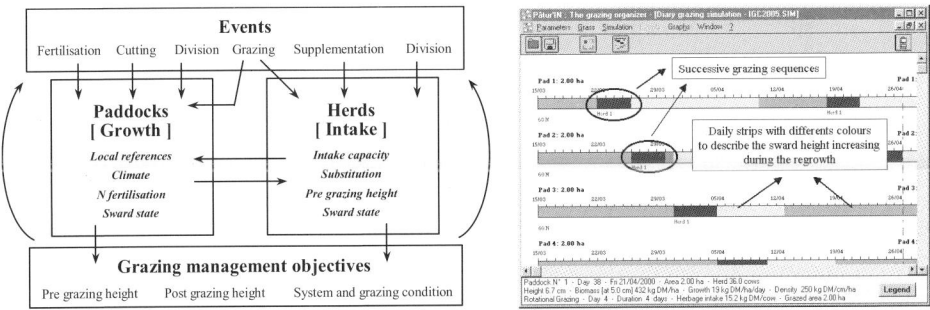

Figure 1 A graphical representation of the events and functions used in Pâtur'IN software tool and a screen copy of the user interface

References
INRA (1988). Ruminant Nutrition, R. Jarrige (ed.). INRA & John Libbey Ltd, Paris and London, 389 pp.
Peyraud, J.L., R. Mosquera-Losada & L. Delaby (2004). Challenges and tools to develop efficient dairy systems based on grazing: how to meet animal performance and grazing management. *Grassland Science in Europe*, 9, 373-384.

External validation in northwest Spain of a decision support system for grazing dairy cows (Grazemore)

A. González-Rodríguez, J. López Díaz and O.P. Vázquez Yáñez
Centro de Investigaciones Agrarias de Mabegondo, Xunta Galicia. Apartado 10 - 15080 A Coruñ,a, Spain, Email: antonio.gonzalez.rodriguez@xunta.es

Keywords: grass growth, herbage intake, grazing management, milk prediction, decision support

Introduction A model to predict intake and milk production of cows on grazed grass (Grazemore) was developed (Mayne *et al.*, 2004) for farmers to increase reliance on the grassland resources of the farm and reduce the tendency towards intensive dairy production in most countries of the Atlantic Arc of Europe (González, 2003). Climatic conditions and fertiliser use is the basis of the herbage growth model, but under practical conditions the estimation of the nutrient supply to cows is much affected by management, number of cows and the area of each paddock grazed. The intake capacity of the animal and the ingestibility of the feeds drive the herbage intake model, considering cow characteristics, pregnancy, body condition and stage of lactation, as well as supplementation (concentrates and forages) at grazing. The decision support system (DSS) fits both models to farm conditions and it is now at the farm validation phase in the six EU countries involved.

Materials and methods On an experimental farm in Galicia, three groups (A, B, C) of 30 Friesian spring-calving dairy cows, supplemented with zero, 4 and 8 kg/cow of concentrate, rotationally grazed separate areas of ryegrass-white clover pasture, from March to July (20 weeks). Herbage mass, grass utilisation (intake), pre- and post-grazing herbage samples, and milk production were measured and compared with the output of the model.

Results Predicted (y) and observed (x) data from all groups had linear relationships. For milk responses, y = $0.898x + 2.47$ ($R^2 = 0.625$; Pearson correlation (Pc), 0.791; $p < 0,001$) and for herbage mass at grazing, y = $0.807x + 646$ ($R^2 = 0.665$; Pc, 0.926; $p < 0,001$).

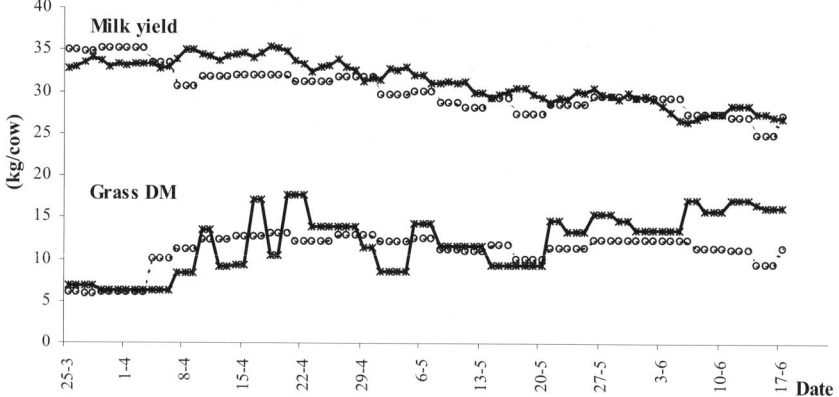

Figure 1 Predicted (dotted) and observed (line) milk yield (higher) and grass DM intake (lower) in group C

Conclusions This decision support model (Grazemore) will enable farmers to predict grass growth, intake, grass depletion per paddock, and milk yield of the herd on a daily basis under real and simulated weather conditions. It could become an important tool to increase the confidence on grazed grass as a main farm resource.

Acknowledgement We acknowledge European Union funding for this project under the Framework 5 Directive.

References

Mayne, C.S., A.J. Rook, J.L. Peyraud, J. Cone, K. Martinsson & A. Gonzalez (2004). Improving sustainability of milk production systems in Europe through increasing reliance on grazed pasture. *Grassland Science in Europe*, 9, 584-586.

González Rodríguez, A. 2003. Low input grazing system for dairy production in northwest Spain. *Grassland Science in Europe*, 8, 491-494.

The Grazemore decision support system for grazing management of dairy cows

M. Hetta[1], M. Norrsken-Eriksson[2], S. Persson[2], E. Larsson[1], L. Karlsson[1], N. Alvarez-Torre[2], H. Eriksson[1] and K. Martinsson[1]

[1]Dept. of Agricultural Research for Northern Sweden, Swedish University of Agricultural Sciences, 904 03 Umeå, Sweden, Email: marten.hetta@njv.slu.se, [2]Division of information technology, Swedish University of Agricultural Sciences, 901 83 Umeå, Sweden

Keywords: herbage growth, herbage intake, milk production, grazing calendar

Introduction Low prices of concentrates and the high demands for good management of grass growth associated with grazing, have led to lower utilisation of grazed grass in Europe. The use of decision support systems (DSS) with valid predictions of herbage growth (HG) and milk yield (MY) may improve grazing management for dairy farmers. The aim of this study was to explore the possibilities to improve grazing management of dairy cows in Europe, by developing a DSS within the European Union project, Grazemore.

Material and methods The Grazemore DSS was developed for rotational grazing systems with perennial ryegrass (*Lolium perenne* L.) and white clover (*Trifolium repens* L.) swards. The DSS is a large simulation platform displaying the effect of variance of management and environment on, MY (kg cow/day), herbage intake (HI) (kg DM/cow per d) and HG (kg DM/ha per d). The solution of the DSS can be described as a group of bank accounts. The "accounts" (paddocks) are replenished with grass with individual growth rates (interest rates) predicted by an HG model (Barrett *et al.*, 2004). The removal of grass from the paddocks acts as withdrawals, and is predicted by an HI model (Delagarde *et al.*, 2004). The DSS can also optimise a suggested grazing and cutting calendar for the farm depending on the management and feeding preferences of the user. Different grazing scenarios can be biologically and economically evaluated under different climatic conditions.

Simulations are available from 1 March to the end of October based on, N fertiliser input, daily measurements of average temperature (°C), precipitation (mm) and photosynthetic active radiation (MJ/m^2). The DSS performs daily predictions of herbage mass (HM) (kg DM/ha), HG, organic matter digestibility (OMD, %), crude protein (g/kg) and white clover contribution (% DM) for each paddock. Milk yield and HI are predicted as herd averages for the residence period in each individual paddock, depending on the grazing management, supplementary feeding and the status of the herd. The DSS predictions of HM on a paddock level and MY on herd level have been externally evaluated by Centro de Investigaciones Agrarias Mabegondo, Spain, on 27 farms in five countries in Western Europe during the 2004 season. The evaluation is based on weekly measurements of HM in individual paddocks and herd average MY, during the residence in the paddocks.

Results The preliminary results from the on-farm validation of the DSS are presented in Table 1.

Table 1 Mean values for milk yield (kg/head per d) for herds and herbage mass (kg DM/ha) in paddocks, observed and predicted by the Grazemore DSS 1.0 on 27 farms in Europe during 2004; regression analysis between observed and predicted values and mean predicted error (MPE) and relative predicted error (RPE)

Parameter	n	Observed		Predicted		Difference		Regression analysis			Statistical analysis	
		mean	sd	mean	sd	mean	sd	a	b	R^2	MPE	RPE (%)
Milk yield	1732	25.2	3.61	25.1	4.37	0.17	3.26	11.1	0.56	0.46	3.27	12.9
Herbage mass	2392	2904	985	2985	1365	-82	1233	1853	0.35	0.24	1236	42.6

Conclusions Preliminary results from the on farm validation indicate that the Grazemore DSS may be used as a management tool to improve the use of grazed grass in Europe.

References

Barrett, P.D., A.S. Laidlaw & C.S. Mayne (2004). Development of a European herbage growth model (The EU Grazemore project). *Grassland Science in Europe*, 9, 653-655.

Delagarde R., P. Faverdin, C. Baratte, M. Bailhache & J.L. Peyraud (2004). The herbage intake model for grazing dairy cows in the EU Grazemore project. *Grassland Science in Europe*, 9, 650-652.

Participative decision mechanisms for sustainable development in co-operative livestock systems in Europe

E. Ruoss[1], A. Boltshauser[1] and P. Hofstetter[2]

[1]*Entlebuch UNESCO Biosphere Reserve, Chlosterbüel 28, CH-6170 Schüpfheim, Switzerland, Email: e.ruoss@bluewin.ch,* [2]*Education and Advisory Centre for Agriculture, Chlosterbüel 28, CH-6170 Schüpfheim, Switzerland*

Keywords: participative process, sustainable development, alpine livestock systems, indicators, visualisation tools

Introduction   Alpine pastures have been used for centuries and have a specific economic, ecological and cultural history that gives local identity. Alpine pastures, used only in summer, are endangered due to modern farming methods and economic conditions. The consequences include loss of biodiversity, traditionally used landscapes and socio-cultural identity in marginal regions (Riseth *et al.*, 2003). As the Entlebuch UNESCO Biosphere Reserve was established by its inhabitants in a participative process, sustainable development in alpine pastures is also implemented by stakeholder participation. The methodology of participative decision mechanisms were used in two EU-projects: LACOPE: Landscape development, Biodiversity and Co-operative Livestock Systems in Europe, developing references for sustainable development in marginal regions and VisuLANDS: Visualisations Tools for Public Participation in the Management of Landscape Change. The main objective was to improve participative decision mechanisms using visualisation tools.

Materials and methods   Ten alpine pastures were selected. The farmers, NGO's, local trade organisations, local tourism providers and experts from national and local public authorities were invited to a series of workshops. The models of System Dynamics and 3-D visualisation tools were used - applied simultaneously in the workshops and adapted for the regional conditions. Local knowledge and more detailed information was gained from the workshops, interviews, questionnaires (n = 150) and context analysis and was used to develop indicators and measures for designing scenarios for the use of livestock in random regions in mountain areas.

Results   Stakeholders developed scenarios starting with a SWOT analysis, defining main targets and indicators. The indicators integrated in the model of System Dynamics show interdependences and connections between indicators and gives a better understanding of the entire system (Coyle, 1998). The model shows limitations and possibilities for controlling and changing land use management for a future development.

According to the results of the model, scenarios for changes in land use were developed and consequences for the landscape were presented with the 3-D visualisation tool, showing changes and providing support for the discussion about sustainable development of landscape management, considering ecological and economical conditions. System Dynamics and 3-D visualisation tools were important instruments, improving participative processes. Considering a balance of economic development and maintenance of socio-cultural factors improving biodiversity of alpine pastures, co-operative structures and the management of alpine pastures have to be maintained and adapted to changed conditions (Kalies *et al.*, 2003). Strategies developed in participative processes have a broader acceptance in the region and accelerated changes in management and increased the sensitivity of farmers for future challenges.

Conclusion   Applying Dynamic System models and 3D visualisation tools support participative decision mechanisms by allowing the present situation and possible scenarios for the future to be seen clearly. Both instruments are worth adopting for better understanding of complex systems and can be used for modelling scenarios (Vennix, 1999). Participative decisions have a broad acceptance, initiate early adaptations of structural changes and maintain sustainable development.

References

Coyle G. (1998). The practice of system dynamics: milestones. Lessons and ideas from 30 years experience. *System Dynamics Review*, 14, 343-365.

Kalies M., D. Scholle & G. Kaule (2003). Spatial analysis to implement extensive grazing systems in Germany. *Natur und Landschaft*, 78, Jahrgang, Heft 3, 100-108.

Riseth J., G.R. Karlsen & B. Ulvevadet (2003). Governance of Co-operative Livestock Systems (CLS) in Europe. NORUT, Social Research Ltd. Report No. 07/2003.

Vennix J.A.M. (1999). Group Model Building. Facilitating Team Learning Using System Dynamics. Chichester.

Allocating grazing resources with KansasGrazer® and making management decisions in a stocker operation

J.L. Moyer[1] and J.O. Fritz[2]
[1]KSU Southeast Agricultural Research Center, Parsons, KS 67357, USA, Email: jmoyer@oznet.ksu.edu,
[2]Department of Agronomy, Kansas State University, Manhattan, KS 66506, USA

Keywords: grazing allocation, decision aid, animal unit month, tall fescue, white clover

Introduction Management decisions for forage-beef cattle production systems are complex because of the many interrelated factors in the plant-animal complex. Evaluation of a system and effects of any changes to it are critical, however, because of the impact that any factor may have on the operation. A computer program to evaluate forage-beef cattle systems would enable producers to make more informed management decisions. Several such programs are available in Kansas, but they are not widely used because of their limited scope, user-friendliness and/or flexibility. Thus, an earlier program derived from the KYBEEF model by Bullock *et al.* (1983) was modified for Windows PC compatibility. Available information was used to develop additional grazing models to broaden applicability throughout the state, resulting in KansasGrazer®.

Materials and methods A typical operation for Labette County, Kansas, was developed from averages found in the 2002 US census (USDA, 2004). The average operation defined as a "farm" was slightly more than 162 ha. Approximately 66% of farmland was cropland, but about 29% was used only for pasture. Thus, we designated 40 ha of "cropland" in tall fescue for grazing as "high" in relative productivity for the model, since tall fescue is an important grazing resource in Labette County. Another 40 ha of tall fescue was categorised as "medium" productivity, and 8 ha each were designated as grazed woodland, one parcel in native warm-season perennial grass in "fair" condition, the other in tall fescue-lespedeza of "low" productivity. Stored forage, amounting to 49 AUM (animal-unit months) came from 8 ha of warm-season native meadow. For simplicity and because grazing (Coffey *et al.*, 1996) and economic (Burton *et al.*, 1994) data were developed at that location, we grazed stocker steers from March to November. Carrying capacity in AUM's was calculated for four available management alternatives with the 80 ha of tall fescue using KansasGrazer®, then costs and returns were projected using data from the sources listed previously.

Results Output of the model and economic projections of six management alternatives were calculated. Production with no nitrogen (N) fertilisation or legumes with the 80 ha of tall fescue supported 40 steers, with the stipulated "cushion" of 60-70 AUM's per season, and less than 49 AUM of spring requirement for stored feed. The variable costs were relatively low, resulting in a total net return of $1204. Adding spring N (90 kg/ha at $0.62/kg) increased costs, but produced a stocking level of 105 steers, resulting in a net return of $3725. Adding 56 kg/ha of autumn N increased carrying capacity to 139 steers, but higher costs resulted in a reduced net return of $3659. If large white clover with no N is grown in mixture with the 80 ha of high-producing tall fescue and only spring N is used on the medium-producing 80 ha pasture, 111 stockers can be carried, but net return is increased to $7984. If large white clover replaces the use of N on all of the pasture considered, 116 head are carried for a net return of $11836. Finally, the replacement of the 40 ha of tall fescue on cropland assumed to have toxic endophyte with non-toxic tall fescue resulted in a return of $11901.

Conclusions Production and returns can be projected for many options in this scenario, but six ways of managing tall fescue were demonstrated. A cow-calf enterprise could also be added, and cropland and/or crop residues used for forage. Flexibility of KansasGrazer® also allows the replacement of default values for forage production and animal gain when model assumptions do not apply.

References

Buller, O., R.R. Murry & G.L. Posler. (1983). Computer programs for forage management and utilization (FMUP I and FMUP II). Department of Economics Paper 82-191-D. Kansas State University, Manhattan, Kansas.

Burton, R. O., Jr., P. T. Berends, J. L. Moyer, K. P. Coffey & L. W. Lomas (1994). Economic analysis of grazing and subsequent feeding of steers from three fescue pasture alternatives. *Journal of Production Agriculture*, 7, 482-489.

Coffey, K. P., J. L. Moyer & F. K. Brazle (1996). Performance by steers grazing infected and noninfected fescue pastures with and without ladino clover interseeding. In: 1996 Agricultural Research. Kansas Agricultural Experimental Station Report of Proress, 761, 5-7. Available online at: http://www.oznet.ksu.edu/library/crpsl2/srp761.pdf.

USDA (2004). 2002 Census of Agriculture - County Data. National Agriculture Statistics Service. Available online at: http://www.nass.usda.gov/census/census02/volume1/index2.htm.

A farmer-based decision support system for managing pasture quality on hill country

I.M. Brookes and D.I. Gray

College of Sciences, Massey University, Private Bag 11 222, Palmerston North, New Zealand, Email: I.Brookes@massey.ac.nz

Keywords: feed budgeting, micro-budgeting, tactical management, sheep production, pasture quality

Introduction   Despite considerable effort to promote formal feed budgeting in New Zealand, survey data suggests it is only adopted by 20% of farmers (Nuthall & Bishop-Hurley, 1999). Recent work (Gray *et al.*, 2003) has identified that farmers may use a different approach - micro-budgeting - to manage feed. Rather than operate at a whole farm level, micro-budgeting focuses at the paddock level. This paper describes micro-budgeting as used by a high performing hill country sheep and cattle farmer to manage pasture quality over spring and a decision support model developed to help other farmers undertake this process.

Farmer practice   During spring, the farmer identifies separate sheep and cattle blocks. Different areas of the sheep block are allocated to mixed-age and two-tooth ewes, separated into triplet-, twin- and single-bearing ewe blocks, and a hogget block containing dry and lambing hogget areas. The cattle block is separated into rising one-year (R1) and rising two-year (R2) areas. The sheep are set-stocked from Sept. to early Jan. and the cattle are grazed on a 12-15 d rotation within each block. The farmer monitors pasture cover in each paddock fortnightly. Feed demand for the next two weeks is compared with forecasts of pasture growth for each paddock, and stock numbers adjusted to match feed demand with supply. The aim is for sward height on the sheep and cattle blocks to not exceed 1200 kg DM/ha and 1500 kg DM/ha respectively, thereby maintaining pasture quality and minimising within- and between-block variation. Options include shifting stock from areas with a feed deficit to those with a feed surplus; buying additional cattle for paddocks with a surplus; increasing stocking rate; and freeing up areas for additional stock, a forage crop or production of grass silage.

Decision support system   A decision support system has been developed on an Excel spreadsheet (Table 1). Paddock number and area are entered in columns 1-2. The stock class is identified in columns 3-5, by age, birth rank and lambing date, and stock numbers in column 6. Feed demand per head is calculated from notional live weight, docking percentage and mean lambing date for each stock class (column 7). Per hectare demand in each paddock (column 8) is calculated from the number of grazing stock multiplied by feed demand (kg DM/head) divided by the paddock area (ha). Estimated pasture growth rates (PGR) are entered in column 9, and pasture cover at the start of the period in column 10. Final pasture cover is calculated (column 11) and compared with the target covers entered in column 12. The paddock area or stock numbers required to meet target cover are calculated in columns 13-14. This information is used to decide if an area can be freed for other stock, or if additional stock can be shifted from paddocks that are short of feed into paddocks with a surplus. Columns 15-16 provide an estimate of the stock numbers to be added to or removed from each paddock to ensure target pasture cover levels are reached. In this example, additional single-bearing ewes or R1 cattle may be placed in paddocks in this block. The cattle may come from the cattle block, if it is short of feed, or they may be purchased.

Table 1 A decision support system for managing pasture quality on hill country

1	2	3	4	5	6	7	8	9	10	11	12	13	14	15	16
						Intake kg DM		PGR kg DM	Pasture cover kg DM/ha					Extra stock	
Paddock		Stock										Required		Ewes	Cattle
No.	ha	Class			No.	/head	/ha	/ha per d	Start	Final	Target	ha	No.	S	R1
1	5.0	MA	S	E	53	2.57	27.2	31.0	1170	1223	1200	4.72	56	3	1
2	4.5	MA	S	E	60	2.57	34.3	31.0	1230	1184	1200	4.65	58	-2	-1
3	5.2	MA	S	E	70	2.57	34.6	31.0	1250	1200	1200	5.20	70	0	0

Conclusions   This decision support system for feed management is modelled on the practice of an expert farmer. The approach is quite different from the methods normally advocated by extension agents, but may prove more attractive for use on farm. This work suggests that the development of effective decision support systems for farmers requires an in-depth understanding of how they currently manage their feed.

References

Gray, D.I., W.J. Parker, E.A. Kemp, P.D. Kemp, I.M. Brookes, D. Horne, P.R. Kenyon, C. Matthew, S.T. Morris, J.I. Reid, & I. Valentine (2003). Feed planning – alternative approaches. *Proceedings of the New Zealand Grassland Association, 65*, 211-218.

Nuthall, P.L & G.J. Bishop-Hurley 1999. Feed planning on New Zealand farms. *Journal of International Farm Management. 2*, 100-112.

Enhancing grasslands education with decision support tools

H.G. Daily, J.M Scott and J.M. Reid
University of New England, Armidale, New South Wales 2351, Australia, Email: Helen.Daily@une.edu.au

Keywords: Internet, simulation, wool industry, grazing systems, problem solving

Introduction We have successfully used Decision Support Tools (DST) relevant to the management of grazing enterprises to enhance problem solving skills of undergraduates in Australia. Tools such as GrassGro™ (Moore *et al.*, 1997) and GrazFeed™ (Freer *et al.*, 1997) are accessed from a central server by authorised users at many widely dispersed Universities across Australia using remote access to thin-client technology via an Internet portal. This has been supplemented with training for lecturers. Experience in developing appropriate teaching and learning materials and the reliable delivery of simulation software to many clients has enhanced learning outcomes at tertiary level. We are also trialling the use of DST to other learning sectors.

Materials and methods Support from the Australian government's Department of Education Training and Youth Affairs, and Australian woolgrowers through Australian Wool Innovation Ltd have been used to extend and consolidate the use of thin-client technology for distributing grazing models to undergraduate students at institutions around Australia on request from their lecturers on a fee-for-service basis. In collaboration with CSIRO, we have trained over 30 lecturers in the use of these DST and their application in tertiary education programs such as rural science, agricultural economics and natural resource management. A Decision Support Specialist provides assistance to lecturers in the development of training materials based on the grazing management DST, GrassGro™, and the grazing ruminant nutrition DST, GrazFeed™. Registered lecturers and students log on to the eD-Serve portal (http://ed-serve.une.edu.au) to access a range of DST and other customised relevant information (Daily *et al.*, 2003a). We are piloting approaches to create awareness of grazing management decision support tools amongst wool producers and secondary school agriculture students.

Results We have increased the awareness of key profit drivers, risk management and the role of DST in decision making in the grazing industries, particularly the wool industry in Australia over 5 years. Students from 11 campuses of 8 Australian universities have accessed grazing DST through eD-Serve, and a survey of lecturers using these DST in their teaching provided evidence of their support for the system, and the benefit it provided to their teaching. Student surveys have revealed that they recognise that the acquisition of skills with these commercially available DST is especially relevant training for their future employment. Climate datasets for other international localities are currently being assembled so that simulations related to a range of international sites can be conducted by students, thus broadening their global perspective. This thin-client delivery system is also being developed for accessing other legacy software related to applications across the agricultural industries.

Conclusions This e-learning project using DST has demonstrated that excellent grasslands science can be readily acquired by undergraduate students and others interested in learning about complex grassland ecosystems through the use of DST in education. The exercises are tailored to explore scientific principles, illustrate profit drivers and/or test risk management in grazing enterprises, and overall, to develop problem solving skills and "systems" thinking in students (Daily *et al.*, 2000). Internet access to DST widens the experience of students and will increasingly provide opportunities for collaboration across Australia and internationally.

References
Daily, H.G., G.N. Hinch, J.M. Scott & J.V. Nolan (2000). The use of a decision support program to facilitate the teaching of biological principles in the context of agricultural systems.
http://www.tedi.uq.edu.au/conferences/teach_conference00/abstractsA-H.html#Daily
Daily, H.G., J.M. Scott, & J.M. Reid (2003a). Internet delivery of Decision Support Tools for teaching nationally. In G. Crisp, D. Thiele, I. Scholten, S. Barker and J. Baron (Eds), *Interact, Integrate, Impact: Proceedings of the Twentieth Annual Conference of the Australian Society for Computers in Learning in Tertiary Education*. Adelaide, 7-10 December, 2003.
Freer, M., A.D. Moore, & J.R. Donnelly (1997). GRAZPLAN: Decision Support Systems for Australian Grazing Enterprises. II. The Animal Biology Model for Feed Intake, Production and Reproduction and the GrazFeed DSS. *Agricultural Systems,* 54, 77-126.
Moore, A.D., J.R. Donnelly & M. Freer (1997). GRAZPLAN:Decision Support Systems for Australian Grazing Enterprises. III. Growth and Soil Moisture Submodels, and the GrassGro DSS. *Agricultural Systems,* 55, 535-582.

Systems simulation assists land capability estimation in Australia's temperate grasslands

R.J. Simpson[1], L. Salmon[1], P. Graham[2], A.D. Moore[1], A. Stefanski[1], D.J. Marshall[1] and J.R. Donnelly[1]

[1]CSIRO Plant Industry, PO Box 1600, Canberra, ACT, Australia 2601, Email: Richard.Simpson@csiro.au, [2]NSW Department of Primary Industry, Yass, NSW, Australia 2582

Keywords: GrassGro, land capability, simulation, soil water-holding capacity, stocking rate

Introduction Intensification of production in the water-limited grasslands of temperate Australia has increased the need to quantify their sustainable carrying capacity. Empirical rainfall-based rules for estimating stocking rate fail when used in districts with differing weather patterns, or when soil and pasture resources limit the utilisation of rainfall. Grazing systems simulation should help to overcome these problems because local conditions can be taken into account. This study investigated the impact of soil resources on potential stocking rate, profitability and production risk in a local climatic area of the southern tablelands of NSW, Australia.

Materials and methods The GrassGro decision support tool (Moore et al., 1997) was used to simulate Merino wethers continuously grazing pasture at Bookham, NSW (rainfall 775±206 mm/year, mean±sd). Pasture mass was measured at a reference farm ("Kia-Ora"; 34°48.2'S, 148°34.9'E) at monthly intervals (1998-2003) and the data were used to test model predictions. Inputs to simulations included daily weather records, soil physical properties and a description of the sheep genotype. Sheep were fed for maintenance whenever liveweight declined below a threshold. Pasture was simulated as dominantly annual grass-*Trifolium subterraneum*. Subsequently, GrassGro was used to explore the impact of soil fertility at the farm and to simulate alternative scenarios at high soil fertility, in which the plant-available water holding capacity (PAW) was varied to reflect different soils in the area. All other inputs were kept constant. Relative profitability was determined from simulations (1965-2003) that used costs and prices suitable for 2004. Optimum stocking rate was defined as the lowest stocking rate that achieved the "best" combination of high median annual profit and low below-median variation in profits. Production risk was defined as the proportion of years with profit <$100/ha.

Results The test simulation explained 80% of the variation in available green pasture at the reference farm (regression analysis n=63; one outlying point excluded). Mean annual pasture yield was predicted to be 5.3 t DM /ha for unfertilised paddocks (6 wethers/ha) but increased to 10.5 t DM/ha for paddocks simulated to reflect field applications of P, S, K and Mo (18 wethers/ha). A survey of ten paddocks across the district (~1,800 km²) showed that soils varied independently in depth, PAW and root depth (root zone PAW range: 64–128 mm). Variations in soil properties of this magnitude may occur within a single farm. Skeletal soils also occur, but in relatively small areas and were not encountered in the survey. Carrying capacity of paddocks at high fertility was examined by assessing productivity relative to root zone PAW. For this climate, there was a critical PAW (~45 mm) below which production and profitability declined and production risk increased markedly (Figure 1).

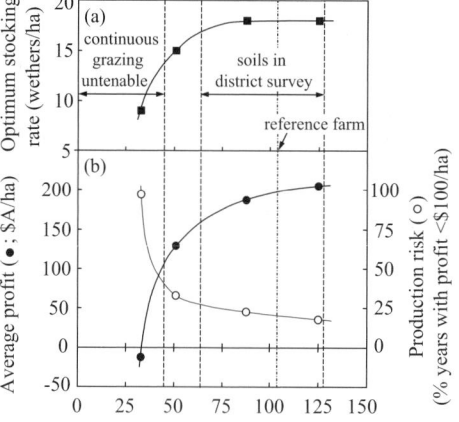

Figure 1 Stocking rate, profit and risk in relation to root zone PAW (from simulations: 1965-2003)

Conclusions Continuous grazing appeared to be unsustainable on soils with <45 mm PAW. The soils in the survey were predicted to carry 17-18 wethers/ha with similar, low production risk. However, substantial differences in relative profitability were predicted with the highest PAW paddock returning 28% more profit than the lowest paddock. The optimum stocking rate is not achieved on all farms experiencing similar climate. It was concluded that differences in fertiliser management, rather than soil characteristics were the likely cause.

Acknowledgement Work supported by Australian Wool Innovations Ltd and Bookham Agriculture Bureau.

Reference

Moore, A.D., J.R. Donnelly & M. Freer (1997). GRAZPLAN: Decision support systems for Australian grazing enterprises. III. Pasture growth and soil moisture submodels and the GrassGro DSS. *Agricultural Systems*, 55, 535-582.

Determination of optimal grazing management for dairy cows in Galicia (Spain) using a decision support system

O.P. Vázquez Yáñez, A. González Rodríguez and J. López Díaz
Centro de Investigacións Agrarias de Mabegondo (CIAM), Apartado 10, 15080 A Coruña (Spain), Email: orlando.vazquez.yanez@xunta.es

Keywords: dairy cows, decision support system, grazing management

Introduction GRAZEMORE is a decision support system (DSS) oriented to evaluate dairy grazing management decisions on pasture growth and milk production. The system integrates a herbage growth model (Barrett *et al.*, 2003) and a herbage intake model (Delagarde *et al.*, 2004) and has been validated in several EU countries. The objective of this work was to use this DSS to determine optimal grazing management of dairy cows at three supplementation levels and three grazing management strategies under Galician conditions.

Materials and methods The GRAZEMORE DSS was used to simulate eight grazing scenarios using the average weather data of the CIAM in Galicia (NW Spain). The scenarios considered a herd of 50 dairy cows with a continuous calving pattern, grazing 13.6 ha divided into 17 paddocks from 1 May to 30 October. The potential peak milk yield was estimated to be 36 kg/d. The scenarios compared where three concentrate supplementation levels (0, 4 and 8 kg DM concentrate/d) and three management options (a daily pasture area offered (DPAO) of 0.2, 0.3 and 0.5 ha/d). The scenario 8 kg DM/d vs. 0.5 ha/d was excluded due to DSS restrictions. The number of rotations during the grazing season was 7, 9 and 12 for a DPAO of 0.2, 0.3 and 0.5 ha/d respectively, but during the five initial rotations only half of paddocks where grazed with the remaining paddocks being assigned to silage production.

Results Average pasture allowance, pasture intake, total intake, and daily milk production between 15 May to 30 October are shown in Table 1. These results show that increasing DPAO resulted in higher pasture allowance (from 15.7 to 36.3 kg DM/d) and, as a consequence, pasture intake was raised from 9.9 to 13.2 kg DM/d. Increasing supplementation level reduced pasture intake and the magnitude of the reduction was enlarged with increased pasture allowances (0.4 kg DM in low DPAO vs. 0.6 kg DM in high DPAO for a total supplementation of 4 kg DM/d). Daily milk production ranged from 9.8 kg/d for 0.2 ha/d DPAO and no concentrate, to 27.0 kg/d for 0.3 ha/d DPAO and 8 kg DM concentrate. Total silage production in the low DPAO system (46 t DM) was almost twice the silage production in the high DPAO. Grass utilization (total pasture cut and consumed/pasture produced) increased with DPAO but decreased with concentrate supplementation level.

Table 1 Pasture and milk production for the scenarios simulated

Scenarios	Area offered (ha/d)							
	0.2			0.3			0.5	
Supplementation (kg Conc/cow)	0	4	8	0	4	8	0	4
Pasture allowance (kg DM/cow per d)	15.7	16.3	17.1	21.3	22.5	24.4	33.3	36.3
Pasture DM intake (kg DM/cow per d)	9.9	9.5	8.8	11.4	10.9	9.8	13.2	12.6
Total DM intake (kg DM/cow per d)	9.9	13.5	16.8	11.4	14.9	17.8	13.2	16.6
Milk (kg/cow per d)	9.8	18.6	25.2	13.1	21.5	27.0	17.2	24.8
Total pasture intake (t DM)	94.9	91,0	83.6	108.7	104.1	93.7	126.0	120.5
Total silage production (t DM)	46.4	46.4	46.5	33.2	33.2	33.3	23.9	23.9
Total grass produced (t DM)	165.9	166.1	166.3	161.4	161.7	162.0	160.4	160.8
Grass utilization (%)	85%	83%	78%	88%	85%	78%	93%	90%

Conclusions This work shows the sensitivity of the grazing system to different grazing and supplementation strategies and the ability of GRAZEMORE to assist in optimal management decisions. Future work must consider the impact of the grazing management system on the economics of the farm over the whole year.

References

Barrett P.D., A.S. Laidlaw & C.S. Mayne (2003). An Herbage Growth Model to Predict Pasture Production in Predominant Perennial Ryegrass Swards for use in a Decision Support System to Aid Pasture Management in Northwest Europe. Submitted: *The Journal of Agricultural Science.*

Delagarde R., P. Faverdin, C. Baratte, M. Bailhache & J.L. Peyraud (2004). The Herbage Intake Model for Grazing Dairy Cows in the EU Grazemore Project. *Proceedings of the Twentieth General Meeting of the European Grassland Federation*, Luzern, Switzerland.

Forecasting forage yields using the ARIMA model in pastoral areas of East Africa

R.J. Kaitho, J.W. Stuth, J. Angerer and A.A. Jama
*Texas A&M University, Dep. of Rangeland Ecology and Management, College Station, Texas 77843, USA,
Email: rkaitho@cnrit.tamu.edu*

Keywords: forage, forecasting, auto-aggressive integrated moving-average procedure, pastoral areas

Introduction Predicting forage supply is an age old quest for pastoralists, particularly in fragile and drought-prone areas of Africa. Traditional methods of forecasting forage used by many communities have become less effective due to climate change, frequent droughts and decline of grazing areas. Conflicts relating to available forage and water resources are increasing, because more marginal lands are put to crop production. A new forage forecasting technology has been developed that provides a comprehensive view of current forage condition (Stuth *et al.*, 2004). A multiple species grazing land plant growth hydrology based model (PHYGROW) was parameterised with site-specific soil, plant community, grazer data that was spatially linked with satellite weather and predicted daily available forage (Rowan, 1995). The objective of this study was to explore use of the Auto-Regressive Integrated Moving-Average (ARIMA) procedure in forecasting a 30, 60 and 90-day available forage.

Materials and methods The Livestock Early Warning Systems (LEWS) project has developed a monitoring system to assess emerging trends in forage supply and animal condition on pastoral rangelands of Ethiopia, Kenya, Uganda and Tanzania. The PHYGROW model was parameterised on 400 sites in the study region and decadal runs made from Jan. 2002 to June 2004. The LEWS country teams selected 81 sites for model verification. Fifty, $0.5\ m^2$ quadrats were sampled on each validation site representing the 8 x 8 km grid using a comparative yield method (Haydock & Shaw, 1975). Regression equations converted the rankings into actual forage values. After each dekad, 30, 60 and 90-day forage forecasts were estimated using the ARIMA model with Normalised Difference Vegetation Index (NDVI) as covariate according to Box & Jenkins (1994).

Results The results indicated a good relationship between forage yield estimations with PHYGROW and field observations (R^2= 0.96 and SEP= 161 kg/ha). The ARIMA time series forecasting methodology provided suitable projections well within normal sampling errors. The observed R^2 and SEP (kg/ha) values for the 30, 60 and 90 day forecast of grazeable standing crop were 0.93/139, 0.84/206, and 0.71/254 respectively (Figure 1).

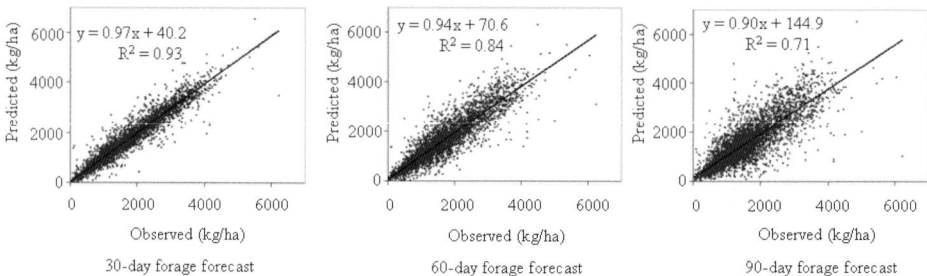

Figure 1 Relationship between observed and predicted 30, 60 and 90-day forage forecasts

Conclusions This methodology allows a new powerful mechanism for decision makers to visualise with a progressive 90-day analysis window emerging "hot spots" (spatial areas of forage scarcity/abundance) that are difficult to perceive and determine if they are going to recover or worsen.

References
Box, G P., G.M. Jenkins & G.C. Reinsel (1994). Time Series Analysis: Forecasting and Control. 3rd ed. Prentice Hall, Upper Saddle River, NJ.
Haydock, K.P. & N.H. Shaw (1975). The comparative yield method for estimating dry matter yield of pasture. *Australian Journal of Experimental Agriculture and Animal Husbandry*, 15, 663-670.
Rowan, R.C. (1995). PHYGROW Model Documentation Version 2. Ranching System Group, Department of Range land Ecology and Management. Texas A & M University, College Station.
Stuth, J.W., J. Angerer, R. Kaitho, A. Jama & R. Marambii (2004). Livestock Early Warning System for Africa Rangelands. In: V. Boken (ed.). Agricultural Drought Monitoring Strategies: A Global Study. Oxford Press. Oxford, UK (in press).

Sensitivity analysis of a growth simulation for finishing lambs

P.C.H. Morel, B. Wildbore, I.M. Brookes, P.R. Kenyon, R.W. Purchas and S. Ramaswami
College of Sciences, Massey University, Private Bag 11-222, Palmerston North, New Zealand, Email: P.C.Morel@massey.ac.nz

Keywords: lamb finishing, simulation modelling, pastoral systems

Introduction A stochastic lamb growth simulation model with a set of heuristic rules has been developed to evaluate management strategies for a solely pastoral grazing system in New Zealand (Morel *et al*., 2005). In the present paper the results of a sensitivity analysis for this model are presented.

Method In the sensitivity analysis, only one parameter was changed at a time and the others were kept at their default values. For each parameter combination, the farm gross margin ($/yr per ha) for a one-year period (FGM= returns from lamb sales minus the costs of lamb purchases and of pasture consumed) was calculated 1000 times for a 100 ha farm. The parameters investigated (default value) were: lamb buying price (220c/kg live weight); selling price (450 c/kg carcass weight); pasture cost (11c/kg dry matter (DM)); annual pasture production (10,956 kg DM/ha), initial pasture cover (1,500 kg DM/ha), minimal pasture cover (1,200 kg DM/ha) and initial stoking rate (15 lambs/ha).

Results The gross margin per ha (FGM) with default values was $856.7 ±$13.36 (mean ± SD). The relationship between initial stocking rate and FGM was curvilinear, with FGM increasing from $826.6 to $856.7 as stocking rate increased from 12 to 15 lambs/ha and then decreasing to $825.3 for 18 lambs/ha. The changes in FGM (±SD) with changes in each of six parameters from the defaults values are presented in Figure 1. Changes in financial parameters had a greater impact on FGM than changes in pasture parameters. A 1% change in lamb buying price, selling price, or pasture cost were equivalent to ±$51.4, ±$71.5 and ±$11.4 changes in FGM, respectively.

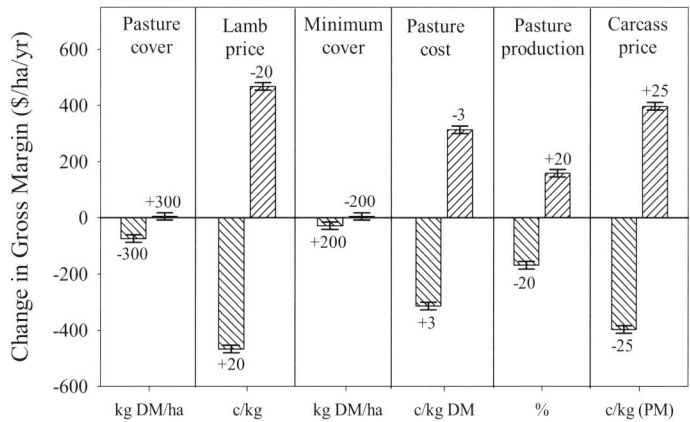

Figure 1 Changes in the gross margin per hectare with changes in each of six parameters with standard deviation bars for the default situation based on 1000 runs

A 1 % increase (decrease) in initial pasture cover, mimimum cover or total pasture production were equivalent to $0.22 (-$3.7), -$1.72 ($0.26) and +$8.4 (-$8.0), changes in FGM, respectively. The FGM decreased by $30.76 for each percentage point decrease in feed allowance from a default value of *ad lib* feeding.

Conclusions It is concluded that this model provides an efficient means of evaluating the relative importance of a number of changes to a system of lamb meat production on pasture.

Reference
Morel, P.C.H., B. Wildbore, I.M. Brookes, P.R. Kenyon, R.W. Purchas & S. Ramaswami (2005). A growth simulation model for finishing lambs in a pastoral system. *XX Grassland Congress – offered papers* (in press).

A model to evaluate buying and selling policies for growing lambs on pasture

P.C.H. Morel, B. Wildbore, I.M. Brookes, P.R. Kenyon, R.W. Purchas and S. Ramaswami
College of Sciences, Massey University, Private Bag 11-222, Palmerston North, New Zealand, Email: P.C.Morel@massey.ac.nz

Keywords: lamb finishing, simulation modelling, pastoral systems

Introduction In pastoral sheep finishing systems, farmers aim to maximize profitability by deciding on when and how many animals to buy and/or sell, while taking into account feed availability and current prices. This paper describes a stochastic lamb growth simulation model with a set of heuristic rules, which has been developed to financially evaluate different management strategies for growing lambs on pasture.

Model inputs Feed supply as pasture dry matter (DM; kg) is described in terms of daily DM growth rates, metabolisable energy concentration in pasture (ME; MJ/kg DM), minimum and maximum pasture DM covers and pasture utilisation (%). The financial parameters used are lamb buying price ($/kg live weight), a selling price ($/kg CW) based on carcass weight (CW) and fatness (GR; mm) and pasture cost (c/kg DM). The stocking rate at the start of the period is also an input parameter. Simulations have been run for a 100 ha farm in New Zealand, starting with 25 kg live weight weaned lambs in January.

Lamb growth The stochasticity in the program is in the form of normally-distributed multiplication factors (1±SD) for: live weight (LW; kg; 1±0.12) at the start, dry matter intake (DMI; kg DM/d; 1±0.01), metabolisable energy for maintenance (MEm; 1±0.0033) and net energy per kg gain (NEg; 1±0.0066). Each of these characteristics is unique for each lamb. The daily growth rate (ADG; g/d) for a lamb is calculated as follows:

$$DMI = C \times 0.04 \times 70 \times (LW/70) \times (1.7 - (LW/70))$$

where 70 kg represents the mature live weight and DMI is reduced by factor C to ensure that pasture cover does not fall below the desired minimal value of 1200 kg DM/ha.

ME intake (MEI) = DMI × ME $MEm = LW^{0.75} \times [0.39 / (((0.35 \times ME) / 18.4) + 0.503)]$

ME for growth (MEg) = MEI - MEm

NEg = 2.5 + 0.35 × LW   ADG = MEg × ((0.0424 × ME) + 0.006) / NEg

CW (kg) = -2.04 + 0.473 × LW;   GR (mm) = -10.8 + 1.2 × CW

Random normally-distributed deviations with means of zero are added to the CW (SD=±0.67kg) and GR (SD= ±2.6mm) values (Garrick *et al.*, 1986).

Heuristic rules The lambs are sold at weekly intervals, but only when 50 or more have reached 45 kg LW. At the same time as lambs are sold, new lambs may be purchased, with the number bought being calculated as a function of pasture cover and the predicted length of time to grow lambs to 45 kg.

Model outputs The model calculates the predicted farm gross margin for a one-year period (returns from lamb sales minus the costs of lamb purchases and of pasture consumed) and can be used to investigate the effects of different management options on profitability. Results of a sensitivity analysis conducted with this model are presented elsewhere (Morel *et al.*, 2005).

References

Garrick, D.J., R.W. Purchas & S.T. Morris (1986). Consideration of alternative lamb drafting strategies. *Proceedings of the New Zealand Society of Animal Production, 46,* 49-54.
& S. Ramaswami (2005). Sensitivity analysis of a growth simulation for finishing lambs. *XX International Grassland Congress – offered papers* (in press).

CaNaSTA – Crop Niche Selection for Tropical Agriculture, a spatial decision support system

R. O'Brien[1,2], M. Peters[1], R. Corner[2] and S. Cook[1]

[1]*International Center for Tropical Agriculture, AA 6713, Cali, Colombia, Email: r.obrien@cgiar.org,*
[2]*Department of Spatial Sciences, Curtin University of Technology, GPO BoxU1987 Perth WA 6845, Australia*

Keywords: spatial decision support system, tropical agriculture, tropical forages, CaNaSTA

Introduction Farmers in the developing world frequently find themselves in uncertain and risky environments, often having to make decisions based on very little information. Risks for smallholder farmers are often critical because of their poverty. In addition, in the tropics and subtropics, the natural environment is spatially and temporally variable and often harsh, thereby increasing the uncertainty faced by these farmers. This research aims to improve forage adoption decisions in the developing world, thereby increasing sustainable intensification and ultimately contributing to increased sustainable world food production and the alleviation of under-nutrition.

Spatial Decision Support System Decision support can facilitate the decision process by making available relevant data and knowledge. Spatial Decision Support Systems (SDSS) work with explicitly spatial data, and outputs usually include maps. An SDSS has been developed called CaNaSTA (Crop Niche Selection for Tropical Agriculture). The engine of the tool is Bayesian probability modelling. Six main criteria were identified for model selection. These are the ability to work with small datasets, the ability to work with expert knowledge and the ability to predict a range of species' responses. In addition, a low structural complexity is required as well as ease of communication and the ability to implement the DSS spatially.

Probability calculations CaNaSTA calculates probability distributions for each forage species under specific sets of environmental conditions. These are related to predictor variables using Bayesian probability modelling techniques, with data drawn from forage trials and expert knowledge, including the forage knowledge base SoFT (Pengelly *et al.*, 2005). Probability of adaptation is classified as 'excellent', 'good', 'adequate' or 'poor', based on data in an existing forage database (Barco *et al.*, 2002). The predictor variables used are elevation, annual rainfall, length of dry season, soil pH, soil texture and soil fertility. Model outputs include a score value based on the probability distribution and a certainty value associated with the distribution. Stability measures are derived from changes in distribution when variables change states. From these, a ranked list of recommended species is calculated, along with suitability maps.

Results Results from CaNaSTA were compared with results from three existing tropical forage knowledge bases and direct elicitation from forage experts, highlighting a number of strengths of CaNaSTA. Firstly, species are not automatically excluded when one variable is unsuitable, as all other variables may be highly suitable. Secondly, the score and ranking system allows more suitable species to be considered first, rather than the user being presented with an unranked list of all species which fit the criteria. Finally, CaNaSTA produces suitability maps dynamically; most other available knowledge bases do not have inherent spatial functionality and maps can only be produced on an ad-hoc basis.

Conclusions Incorporating spatial capabilities into an agricultural DSS, as in CaNaSTA, allows more informative output of results and allows spatial variability to be made explicit, both of results and of uncertainties related to the results. Even with limited data, results can be obtained which support the farmer's decision-making process. When uncertainties are made explicit, farmers can then make less-risky decisions by taking these uncertainties into account. Providing access to decision support through an SDSS, such as CaNaSTA, ensures that the information is delivered in a consistent and robust manner. Trial data and expert knowledge previously inaccessible to farmers are made available so that decisions taken are better informed.

References

Barco, F., M.A. Franco, L.H. Franco, B. Hincapie, C. Lascano, G. Ramírez & M. Peters (2002). Forrajes Tropicales: Base de Datos de Recursos Genéticos Multipropósito, Versión 1.0 (Tropical Forages Database) http://www.ciat.cgiar.org/catalogo/producto.jsp?codigo=P0219.

Pengelly, B.C., B.G. Cook, I.J. Partridge, D.A. Eagles, M. Peters, J. Hanson, S.D. Brown, J.L. Donnelly, B.F. Mullen, R., Schultze-Kraft, A. Franco & R. O'Brien (2005). Selection of Forages for the Tropics (SoFT) – a database and selection tool for identifying forages adapted to local conditions in the tropics and subtropics. *XX International Grassland Congress – offered papers* (in press).

An internet-based tool for use in assessing the likely effect of intensification on losses of nitrogen to the environment

N.J. Hutchings, B.M. Petersen, I.S. Kristensen, N. Detlefsen and M.S. Jørgensen
Dept. of Agroecology, Danish Institute of Agricultural Sciences, Research Centre Foulum, P.O. Box 50, 8830 Tjele, Denmark, Email: nick.hutchings@agrsci.dk

Keywords: nitrogen, environmental impact assessment, livestock farm, nitrate, ammonia

Introduction The EU Nitrates, Habitat and National Emissions Ceilings directives and the Kyoto Agreement mean that agricultural losses of NO_3, NH_3 and N_2O are under scrutiny by national and international environmental authorities. When farmers wish to intensify their operations, the authorities must then assess the likely environmental impact of the change in operation. The FARM-N internet tool was developed to help farmers and authorities agree how the farm will be structured and managed in the future, and to provide an objective assessment of the environmental losses that will result.

Methods The farmer must describe the current and proposed farm structures and management, in terms of the number and type of livestock to be kept, the animal housing and manure storage facilities to be used, the land that will be available to the farm and details of field management. The latter include the crop rotation and type of manure spreading equipment to be used. For ruminant livestock farms, the production of milk and livestock has to be consistent with the animal feeding practice, the choice and productivity of the crops chosen, the sale of crop products and the amount of additional animal feed imported. To assist in this process, the tool contains two models. The first is based on a series of decision rules and enable an agronomically-sensible crop rotation(s) to be constructed, based on an input combination of crop type and area planted. It then determines the applications of manure and mineral fertiliser that should be applied to each crop, dependent on the amount and quality of manure that is available. The amount and quality of manure is, at this stage, assumed to be equal to the standard values in the Danish nutrient management regulations. The second model predicts N excretion from livestock, N losses from animal housing, manure storage and fields and crop responses to N inputs. In situations where livestock are partly fed on home-grown products, there is a closed cycle between crop production and protein content, N excretion by the livestock, N losses in animal housing, storage and after manure spreading and the N available to the crop. This means that these models have to be used iteratively until a consistent N flow is achieved. References to the models used in the FARM-N tool can be found on www.farm-n.dk.

The tool calculates a farm N balance and uses relatively simple models to partition this balance between ammonia emissions from animal housing, manure storage and field application, nitrous oxide and dinitrogen losses, nitrate leaching and changes in the soil N. The results are presented in the form of a simple table. The tool can store a number of future farm structure and management scenarios, enabling the farmer and environmental case officer to explore different structural and management options.

Discussion Discussions with target users indicate that they are satisfied with the technical aspects of the tool but disagree amongst themselves concerning the role of soil C and N in the future scenarios. Under Danish conditions, both the models and monitoring (Heidmann *et. al.*, 2001) suggest that there is an accumulation of C and N in the soil of grassland farms. This is thought to be because most Danish farms were mixed livestock/arable until 30-40 years ago and the soils are still adjusting to the higher inputs and larger contribution of grassland to the crop rotation. The disagreements relate to whether this accumulated C and N should be considered semi-permanent storage or a large source of N that could be lost to the environment, should management practices change.

Reference

Heidmann, T., J. Nielsen, S.E. Olesen, B.T. Christensen & H.S. Østergaard (2001). [Changes in the C and N content of culitvated soils: Results from the soil sampling grid 1987-1998] (in Danish). Danish Institute of Agricultural Sciences Report no. 54, 73 pp.

Software PPBB_MX: potential productivity modelling of Brachiaria brizantha (cultivars Marandu and Xaraés)

E.R. Detomini and D. Dourado Neto

Agricultural Department - University of São Paulo, Piracicaba, SP. Mail Box 09 CEP 13418-900, Brazil, Email: detomini@esalq.usp.br

Keywords: modelling, C4 grasses, software, *Brachiaria brizantha*

Introduction Recent improvements in computer capacity and technology allow models to be built to simulate the attributes of many agricultural processes and systems. Although *Brachiaria brizantha* is the most cultivated tropical grass species in Brazil, there is no single tool to predict its production under optimal conditions. The objective of this paper is to present PPBB_MX software to calibrate and simulate (using a stochastical procedure) the shoot and total biomass potential productivity (output variables) of *Brachiaria brizantha* as a function of the following input variables: local latitude, season (from cutting date – Julian day), length of regrowth (time, days) and climate attributes (global solar radiation and air temperature).

Material and methods From use of an iterative method and field experimental data, obtained with high nitrogen supply and optimal water conditions (Detomini, 2004), a Visual Basic for Windows software was built – the PPBB_MX (Dourado Neto *et al.*, 2004). This was developed from physiological principles according to a model concept based on the energy relations between the plant and the atmosphere (Heemst, 1986). Throughout the regrowth period roots and shoots were sampled. Fifteen equations were needed to predict simulated shoot dry matter values (kg/ha). These related to solar declination, day length, available photosynthetic active radiation flux density, degree-days, relative plant growth, leaf area index, canopy extinction coefficient, fraction of light interception by canopy, CO_2 assimilation, gross photosynthesis rate, accumulated gross photosynthesis, respiration rate, total biomass accumulation, root partitioning and shoot biomass accumulation.

Results An iterative method was used to create the procedures F1 (calibration function) and F2 (calibration factor). These were used to find, respectively, optimal values for CO_2 assimilation and root partitioning (both from general empirical equations) to calibrate the model against the field data, and then to simulate total biomass (including roots, kg/ha) and shoot biomass (Fspa) (kg/ha). From historical data for air temperature and radiation, which give a distribution of probabilities, the program generates 1000 randomised numbers for each climatic attribute for each day following the cutting date. Finally, a probabilistic distribution is generated providing 1000 possible values of FSpa. The histogram in Figure 1, for example, resulted from a simulation considering a regrowth of 66 d following a cutting date of 22 Nov. in Piracicaba-Brazil (latitude: 22.73). This location has a good set of historical data for atmospheric attributes. In this case, the normal distribution suggested some value between 16156 and 17458 kg/ha, as the probable magnitude of synthesised shoot biomass.

Conclusions From a stochastical procedure, the software PPBB_MX simulates satisfactorily the shoot and total potential biomass productivity (output variable) of *Brachiaria brizantha* (cultivars Marandu and Xaraés) as a function of the input variables season (from cutting date – Julian day), time of regrowth (days), climate attributes (daily means of global solar radiation and air temperature) and the local latitude.

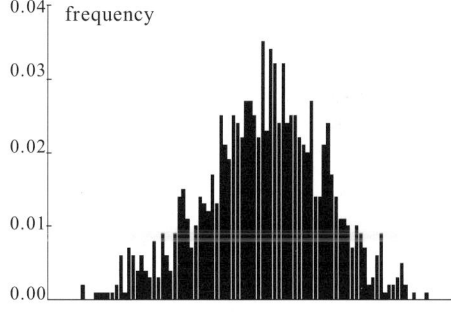

Figure 1 Simulation of potential shoot biomass production of *B. brizantha* cultivar Marandu - 1000 values

References

Detomini, E.R. (2004). Modelagem da produtividade potencial de *Brachiaria brizantha* (variedades cultivadas Marandu e Xaraés). Master Thesis, University of São Paulo – Escola Superior de Agricultura "Luiz de Queiroz"., Piracicaba, 112 pp.

Dourado Neto, D., E.R. Detomini & P.A. Manfron (2004). Programa computacional PPBB_MX: Produtividade potencial de *Brachiaria brizantha*, variedades cultivadas Marandu e Xaraés - v.1.00 [users manual]. USP/ESALQ-LPV, Piracicaba, 80pp

Heemst, H.D.J, van (1986). Physiological principles. In: H. van Keulen and J. Wolf (eds.). Modelling of Agricultural Production: Weather, Soils and Crops. Pudoc, Wageningen, 13-26.

N-mineralisation and phosphorous: important elements in decision support for grassland systems

A.L. Nielsen[1] and C.C. Hoffmann[2]

[1]Natlan, Agro Business Park, 8830 Tjele, Denmark, Email: lisbeth.nielsen@agropark.dk, [2]National Environmental Research Institute, 8600 Silkeborg, Denmark

Keywords: organic soils, N-leaching, P-leaching, grazing intensity, cutting

Introduction Leaching of N and P from extensively managed grasslands on organic soils varies considerably. In environmentally sensitive areas it is important to diminish leaching by appropriate agricultural management. In Denmark low grazing intensity and management without fertilisation have been given a high priority. The type of soil has not been equally in focus, and it seems that the effect of cutting, compared to grazing, resulting in a higher removal of nutrients (e.g. Benke *et al.*, 1992) can be used more strategically. The objective of this case study was to combine data from management strategies with data from leaching examinations on organic soils to elucidate the differences between type of management and type of soil for the potential leaching of N and P.

Materials and methods The study was carried out at two separate sites 2 km apart, referred to as 'West' and 'East'. At 'East' the effect of three management strategies on soil mineral N was examined in a block design with three replicates: a) continuous grazing with steers (compressed sward height (CSH): 6 cm), b) two cuts, c) two cuts with deep litter (20 t/ha) average 83 kg N of which 6 kg NH_4N, 17 kg P and 134 kg K. Leaching of N and P was recorded at high and low grazing intensity in 'West' and 'East'. Continuous high intensity grazing with steers and sheep in separate paddocks aimed at a CSH of 6 cm. At low grazing intensity the number of animals/ ha was 50% (steers) or 65% (sheep) of the number at high intensity. The lowest level of the water table varied between 30 and 55 cm below soil surface in the years of the experiment (Hald *et al.*, 2003a).

Results Initial soil analyses for respectively 'West' and 'East' were $pH(CaCl_2)$ 5.6 and 4.7, total N (%) 1.8 and 2.7, soil organic matter (%) 49 and 66, C:N-ratio 15.6 and 13.9, N-mineralisation measured by incubation at $20°C$ 2.0 and 4.6 kg N/ha per d. Soil mineral N increased through the growing season on the grazed plots compared to the cutting treatments and soil mineral N in spring was lower in plots with cutting and deep litter compared with the other treatments (Table 1). Leaching of N and P was considerably higher in 'East' compared with 'West' (Table 2). There was no difference in N and P leaching between high and low grazing intensity.

Table 1 Soil mineral N (kg N/ha) with three management strategies ('East'), average of 1998-2000

	Soil sampling	Continuous grazing	Two cuts	Two cuts, deep litter	LSD*	Average of 20 grasslands**
Change through the	0-20 cm	21.0	0.9	-1.0	20.8	-1
growing season	20-40 cm	2.1	3.6	1.9	(7.2)	9
Mineral N	0-20 cm	30.0	33.8	28.1	(9.6)	42
in spring	20-40 cm	30.6	30.1	22.5	7.5	14

*LSD: Management, least significant difference (p<0.05), **Different management intensity (Hald et al. 2003b)

Conclusions This case study demonstrates that the level of N-mineralisation is important when making decisions about management. Where the level of N-mineralisation is high it is possible to remove N from the soil when supplying the correct amounts of limiting nutrients. Where depletion of nutrients is required for the environment, cutting can be used for a number of years, but the management should be adjusted according to development in soil conditions and the intentions for the area.

Table 2 Concentration of N and P in field drains*

		Mean	Std. Error	N**
Total N	'West'	2.6	0.13	69
mg/l	'East'	11.1	0.41	69
Total P	'West'	0.4	0.03	58
mg/l	'East'	1.8	0.14	58

*Data from two similar grasslands with 1.3 mg N/l and 0.13 mg P/l (Grant, R., pers. communication)
**N, number of samples

References

Benke, M., A. Kornher & F. Taube (1992). Nitrate leaching from cut and grazed swards influenced by nitrogen fertilization. *Proceedings 14th General Meeting of the European Grassland Federation*, 184-188.

Hald, A.B., C.C. Hoffmann & L. Nielsen (2003a). Ekstensiv afgræsning af ferske enge. DIAS Report 91, 191 pp.

Hald, A.B., A.L. Nielsen, K. Debosz & J.H. Badsberg (2003b). Restoration of degraded low-lying grasslands: indicators of then environmental potential of botanical nature quality. *Ecological Engineering*, 21, 1-20.

Artturi assists Finnish advisers and farmers to succeed in grass-based dairy production

M. Rinne[1], P. Huhtanen[1], K. Kuoppala[1], H. Nikander[1], J. Nousiainen[2], M. Hellämäki[2], L. Nyholm[2], J. Helminen[2], K. Lampinen[3], M. Maisi[4] and M. Korhonen[4]

[1]MTT Agrifood Research Finland, FI-31600 Jokioinen, Finland, Email: marketta.rinne@mtt.fi, [2]Valio Ltd, P O Box 10, FI-00039 Valio, Finland, [3]ProAgria Association of Rural Advisory Centres, P O Box 251, FI-01301 Vantaa, Finland, [4]Kemira GrowHow Ltd., P O Box 900, FI-00181 Helsinki, Finland

Keywords: digestibility, forecast, internet, decision support, growth model

Introduction Artturi is a collective name for a wide range of services. It is a common tool for different bodies who share an interest in strengthening grass-based dairy production in Finland: research, advisory service and industries. The Service is named after A. I. (Artturi Ilmari) Virtanen, the Finnish scientist who was awarded the Nobel prize in 1945, partly based on his work in developing the ensiling process of grass. The Artturi web site is available in Internet at: http://www.agronet.fi/artturi. Access to Artturi Services is free and no registration is required. The language used is Finnish. During summer 2003, 15,000 visits were recorded at the web site.

Description Artturi Service combines several tools in one concept:
●Grass harvest time assistance (see detailed description below)
●Extensive selection of advisory material on grass production, harvesting and ensiling techniques, interpretation of feed analysis etc.
●Feed analysis provided by laboratories of Valio Ltd. Correct information of the nutritional quality of grass silage is a key factor for its succesful utilisation in dairy cow feeding.
●Local farmers' groups focusing on forage production, which are orginised by ProAgria.
●Artturi examination, which measures the level of knowledge of advisors. The names of Artturi experts, i.e. persons who have passed the examination, are published on the Artturi web site.

Grass harvest time assistance Finnish environmental conditions emphasise correct timing of harvest, because grass development and concomitant decline of its nutritional quality is extremely rapid during primary growth. The quality of grass is expressed as concentration of digestible organic matter in dry matter (D-value, %). The D-value of grass declines on average by 0.5 %-units per day in primary growth. The harvest time assistance consists of the following components:
●D-value estimates are presented separately for grasses and red clover during primary growth, based on growth models. The D-value is calculated from cumulative temperature and geographical location (Rinne et al., 2001). Cumulative temperature for the current day and a 5-day forecast is provided by the Finnish Meteorological Institute. The D-values are presented as maps (Figure 1) and in a numeric form for every municipality in Finland. D-value alerts are also available as SMS text messages into mobile phones.

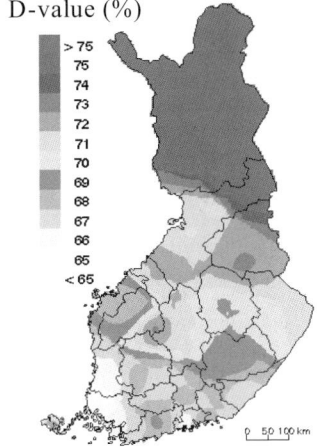

D-value (%)

> 75
75
74
73
72
71
70
69
68
67
66
65
< 65

0 50 100 km

Figure 1 The D-value map of grass based on cumulative temperature on 26 June 2004 shows great geographical variation in Finland

●Frequent samples from grass are collected from practical dairy farms around Finland, analysed by NIR and results presented on the web site the following day. The samples are used to develop further the grass growth models.
●Verbal description of progress in harvest is reported by advisory personnel from different parts of the country.
●D-values based on growth models can also be obtained subsequently for any date in primary growth and any municipality. These D-values may be used to simulate feed analysis.

Further development Artturi provides a wide range of non-commercial information on forage production and utilisation. The flexible structure and present financial support from Finnish Ministry of Agriculture and Forestry facilitate further development of the service to cover feed budgeting, economics and nutrient balances in different forage-based milk production systems. The new components will be based on biologically sound production responses both in plant and milk production derived from research conducted at MTT.

References

Rinne, M., J. Nousiainen, I. Mattila, H. Nikander & P. Huhtanen (2001). Digestibility estimates based on a grass growth model are distributed via Internet to Finnish farmers. *Proceedings of the Nineteenth International Grassland Congress*, 1072-1073.

Theme C: Delivering the benefits from grassland

Section 26

Participatory research and decision support systems

Involvement of Maasai pastoralists in participatory rangeland management planning and implementation

E. Kiptot[1], W. Ego[2], D. Ochieng[3] and A. Mohamed[3]
[1]Kenya Forestry Research Institute, P.O Box 20412 Nairobi, Kenya, Email: ekiptot@yahoo.com, [2]Kenya Agricultural Research Institute, Kiboko Regional Research Centre, P.O Box 12 Makindu, Kenya, [3]ARIDSAK Project, P.0 Box 87, Kibwezi, Kenya

Keywords: re-seeding, action plans, indigenous knowledge, rangeland management

Introduction The pastoral Maasai lifestyle was and still is traditionally based on subsistence dairy and meat production. But with population increase, the rangelands can no longer sustainably support livestock production systems. Most of the rangelands which are used for grazing have been subdivided and partially cleared to pave way for cultivation, because of increased population pressure (Ego et al., 1999). This has led to a tendency to overgraze, thus impacting negatively on secondary production from the range. In order to effectively reverse this trend, the users of the rangeland resources were brought together to analyse constraints and opportunities for sustainable use, so that they could develop action plans for the improvement of the rangelands.

Materials and methods With the help of local leaders, several community workshops were organised and held in six sub-locations of Mashuru division, Kajiado district. These workshops brought together all stakeholders who were willing to share ideas and learn from each other.

Results Priority grass and tree species for forage/fodder were identified through a pair-wise matrix ranking exercise. Five priority grasses identified were; *Digitaria melanjianus, Cynodon dactylon, Themada triachadra, Dactyloctenium aegyptiam* and *Cenchrus ciliaris*. Problems analysed by participants and opportunities are presented in Table 1. Each sub-location drew it's own community action plan. Tasks and responsibilities spelled out in the action plans were shared out amongst key stakeholder groups. Re-seeding of priority grasses, bush management, protection of valuable tree species and soil conservation are among technologies that were set up in on-farm demonstration plots in representative farms selected by the community so that others could learn from them. On -station trials were also undertaken concurrently to assess the performance of the prioritised grass species. Monitoring of the activities has been a joint venture between the communities, government extension staff and researchers. One way of monitoring and evaluation has been through field days and exchange tours.

Table 1 Problem analysis for rangelands in the Kajido District

Problems	Causes	Copping strategies	Opportunities
Shortage of forage/fodder	-Drought -Overgrazing -Trampling -Cutting of trees for charcoal and cultivation -Bush encroachment -Overstocking -Wildlife menace	-Nomadism -Destocking -Paddocking (Use of *olopololi/olale* -Bush management	-Planting and protecting useful fodder trees and shrubs, de-stocking -KWS to confine wildlife in game reserves/parks, paddocking/*olopololis* -Fencing to keep off wild animals -Awareness creation -Training (seminars, tours and field days) -Reseeding of important grass species
Bush encroachment	-Inadequate knowledge on bush management -Lack of labour	-Clear bush	- Bush Control -Training in bush management
Soil erosion	-Overstocking	-De-stocking	-Terracing, planting of grass, de-stocking

Conclusions Technologies developed through this approach stand a high chance of being successful because they address immediate needs and problems of the community as spelled out in the action plans. The fact that the community was involved in project planning and implementation using their own knowledge gave them confidence and motivation to succeed.

Reference

Ego, W.K, E.C. Kiptot, A.M. Mohammed & D. Ochieng (1999). Community Action Plan for the Management of the Natural Rangelands in Kajiado and Makueni Districts. The case of Mashuru, Kibwezi and Kanthonzweni Divisions. Research Report. Agroforestry Research for Intergrated Development in Arid and Semi arid lands of Kenya. KEFRI/BADC Nairobi , Kenya.

Management of pasture quality for sheep on New Zealand hill country

D.I. Gray, J.I. Reid, P.D. Kemp, I.M. Brookes, D. Horne, P.R. Kenyon, C. Matthew, S.T. Morris and I. Valentine
College of Sciences, Massey University, Private Bag 11 222, Palmerston North, New Zealand Email: D.I.Gray@massey.ac.nz

Keywords: decision-making, farmer knowledge, feed budgeting, planning, tactical management

Introduction The control of pasture quality over spring is central to the achievement of high levels of sheep performance on hill country. Despite this, with the exception of the work of Lambert *et al.* (2000), little is known about how farmers actually manage pasture quality. The purpose of this research was to describe how a high performing hill country farmer manages pasture quality on their sheep area over spring and from this develop a framework that will assist other farmers improve their pasture management.

Materials and methods The case study farmer (647 ha, 7,770 s.u.) was selected because of his high levels of performance for the district and expertise in tactical feed management. Data collection was primarily through monthly semi-structured interviews supported by field observations. Interview data were transcribed verbatim and analysed using qualitative techniques to develop a model of the farmer's decision-making processes.

Results and discussion The control of sheep pasture quality requires farmers to make important strategic and tactical decisions (Figure 1). Strategic decisions aim to match feed supply with pasture growth over the spring and maintains grazing pressure so that average pasture cover (APC) levels do not exceed 1200 kg DM/ha. Key decisions in this area include lambing date, stocking rate, sheep performance levels, pasture cover at set-stocking, stock purchase and sale dates, shearing policy and weaning date. Equally important are the tactical decisions to minimise within- and between-block variation in pasture cover levels ($\approx$ 1200 kg DM/ha) during mid- to late-spring. Key tactical decision areas include: (1) ensuring the correct distribution of pasture cover at set-stocking, (2) setting stocking rate and pasture cover levels at set-stocking for the different sheep mobs that best match feed demand to pasture growth, (3) integrating cattle to help control the steeper contour sheep paddocks and (4) using fortnightly monitoring and micro-budgeting to match feed demand with feed supply.

Figure 1 Methods used by the case farmer to manage pasture quality on his sheep area

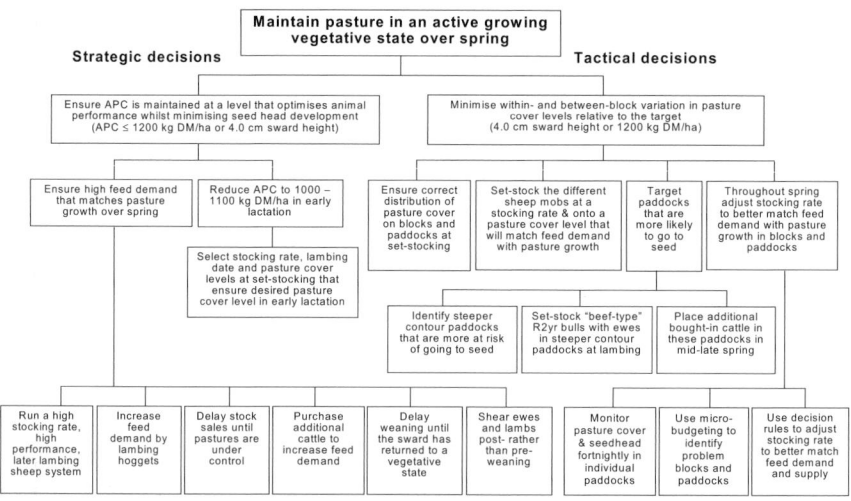

Conclusions This study highlights that the control of pasture quality on hill country is complex, requiring farmers to make a range of important strategic and tactical decisions. The model presented in this paper provides a framework that other farmers can use to improve their management of pasture quality on hill country.

References

Lambert, M.G., M.S. Paine, G.W. Sheath, R.W. Webby, A.J. Litherland T.J. Fraser, & D.R. Stevens (2000). How do sheep and beef farmers manage pasture quality. In: *Proceedings of the New Zealand Grassland Association*, 62, 117 - 121.

Decision support system for grassland-based sheep production in the Chilean Patagonia

R.R. Vera, S. Morales and C. Aguilar

Departamento de Zootecnia, Pontificia Universidad Católica de Chile, Casilla 306-22, Santiago, Chile, Email: rverai@puc.cl

Keywords: Patagonia, sheep, pasture growth, model, decision-making

Introduction Temporal decision-making in sheep production systems of the Chilean Patagonia is influenced by forage availability in existing paddocks, which in turn is determined by previous use, localisation, climate and soil. Also, production strategies are evolving, with a trend towards fat lamb production, introduction of new breeds and a reduction in the importance of wool production. The objective of this study was to develop an initial version of a spatial empirical simulation model based on extremely limited information, to predict forage growth and sheep production so that decision-makers can assess alternative production strategies.

Materials and methods The first version of the model was developed for a 5,077 ha ranch, subdivided into 21 paddocks, located in an above-average rainfall area, for which satellite and aerial photographs were available, together with limited data on pasture availability and some estimates of pasture growth in contrasting paddocks (Schiappacasse, 2000). Existing paddocks were composed of naturalized and native species, including *Agrostis sp., Holcus lanatus, Gunnera magallanica, Empetrum rubrum, Trifolium repens, Festuca gracillima* and others. Sites and their condition (Gastó *et al.*, 1999) were mapped (Schiapacasse, 2000) into a GIS, and were superimposed on a layout of existing paddocks from a LANDSAT image. Potential dry matter (DM) yields were calculated for each paddock weighted for the type and size of sites present in each one and these estimates were used to adjust current growth rates. The animal sub model simulated forage removal and utilization by grazing sheep. In addition, to simulate animal production, the model was used to estimate the potentially utilizable forage in each paddock based on existing proper use factors (SAG, 1998), and the respective potential carrying capacity.

Results Potential forage yields ranged between 1,900 and 6,900 kg DM/ha, and the estimated carrying capacity varied between 0.1 and 0.7 adult sheep equivalent/ha. The simulation of lamb growth in two consecutive years (1999 and 2000) was accurate; r^2 between observed and simulated lamb weights was 0.99 for both years, and the intercepts and slopes of the equations did not differ (P> 0.05) from zero and one respectively. Subsequently, production scenarios involving fattening of Corriedale vs. Corriedale x Texel lambs on different pasture types were analyzed. Crossbred lambs grazed on the best native pastures reached slaughter weight (35 kg) 30 days earlier (163 vs. 194 d) than purebreds with clear short-term economic benefits. When lambs were grazed on sown pastures, age at slaughter decreased further (116 and 132 d respectively). On the other hand, paddocks located in the drier stepparian ranges did not allow lambs to reach the target weight within the time window available without supplementary feeding, but the cost per kg was higher than in the previous case, although net economic benefits were still achieved. Removal of lambs prior to the onset of the winter would result in a reduction of grazing pressure, lower maintenance costs of the flock, and possibly enhanced pasture regrowth in the following spring, but these potential benefits remain to be elucidated.

Conclusions The model represents the first attempt to synthesize the very limited information available in the Chilean Patagonia on grazed pasture systems. Although it was able to accurately simulate animal production for one large, varied, ranch, the pasture sub model must be further elaborated so that water balance, temperature and possibly P status drive growth rates. These developments require field research to collect the relevant information, an effort that is ongoing. The scenarios simulated highlight the strong link between available pasture resources, the possibilities for system diversification and the resulting bio-economic performance.

References

Gastó, J., F. Cosio & I. Aránguiz (1999). Método holístico-empírico de cálculo de la capacidad sustentadora y de la productividad ganadera potencial de los sitios. Provincia Estaparia Muy Fría Secoestival o Patagonia Occidental. *Ciencia e Investigación Agraria*, 26, 125-138.

SAG (1998). Guías de condición para los pastizales de la ecoregión estaparia fría de Ayesén. Proyecto FNDR-SAG XI Región de Aysén: Levantamiento para el ordenamiento de los ecosistemas de Aysén. 95 p.

Schiappacasse, P. (2000). Caracterización de sitio y condición del pastizal en el sistema ovino de la estancia "Las Coles", Provincia Boreal Húmeda Fría, XII Región. Thesis. Facultad de Agronomía, Universidad Católica de Valparaíso, Valparaíso, Chile.

Accelerating the impacts of participatory research and extension: lessons from Laos

J.E. Millar[1], V. Photakoun[2] and J.G. Connell[3]

[1]Charles Sturt University, PO Box 789, Albury, New South Wales, Australia, 2640, Email: jmillar@csu.edu.au, [2]National Agriculture and Forestry Extension Service, PO Box 1888, Vientiane, Lao PDR, [3]CIAT International Centre for Tropical Agriculture, PO Box 6766, Vientiane, Lao PDR.

Keywords: participatory research, rural extension, scaling out, livelihood impacts

Introduction The role of participatory research and extension in stimulating farmer uptake and adaptation of beneficial technologies has been demonstrated over the last two decades in both developed and developing countries. The challenge is to move beyond simply trialling new technologies with farmers on a small scale to enabling significant livelihood impacts across larger numbers of households, villages and districts. This paper presents results of a project in Lao PDR exploring ways to accelerate and spread localised impacts in complex upland farming systems.

Methods The project uses action research techniques with district and provincial extension staff to investigate how changes in extension processes and organisational learning can lead to effective scaling out of impacts at the farming systems level. A variety of extension methods are being trialled to demonstrate potential farming system changes to new villages and farmers at critical times. These include cross visits (taking a group of farmers from one or more villages to visit other farmers), case studies, champion farmers and farmer group development. Case studies have been developed from farmer interviews to document how and why some farmers are gaining benefits from their use of forage and livestock technologies. Organisational learning is being explored using facilitated workshops, on the job mentoring, observations, semi-structured interviews and focus group interviews with staff.

Results Thirty case studies have been developed to document the range of impacts emerging from farmers using forages for livestock. These impacts cover a range of environmental, economic and social benefits to upland households (eg reduction in labour, increase in livestock weight gain, lower calf mortality, less damage to crops from wandering stock, increased income, healthier livestock). An adaptive approach to case study development has been used in that district staff learn by experience and from peer review. Case studies have been found to be effective when used as an extension tool to introduce new villages to potential livestock production systems. The use of system sketches along with photos was more effective than photos or text alone in terms of gaining farmer interest and willingness to trial new forages. Cross visits and the use of champion farmers are more popular for farmer learning and problem solving than case studies. However these methods require greater time commitment and cost for the benefit of fewer farmers. On the other hand, farmer awareness of the potential impacts of adopting new forage varieties or livestock systems was greatest amongst those that went on a cross visit, resulting in less demand for technical support from district staff. Barriers to scaling out impacts such as farmer need or capacity to use forages, livelihood constraints, information pathways (eg through kinship), lack of forage material, market influences, level and quality of technical information and support, village leadership and farmer group development will be explored.

Conclusions Preliminary results from this research show that to accelerate and spread impacts emerging from participatory research requires a shift from simply demonstrating use of a technology, to showing achievable and significant impacts from systems changes. A mix of extension tools are required to provide farmers with the information and support they need at critical times. Case studies are an effective method for demonstrating options for system changes across a wide range of farming households and different areas. Cross visits and discussion groups are more useful for practical applications, problem solving and village planning. Moving from participatory research with individual farmers to more villages and households requires a concerted effort to build the skills and understanding of researchers, extensionists and farmers in providing learning opportunities.

References

Connell, J., J. Millar, V. Photakoun & O. Pathammavong (2004). Strategies for enabling scaling up: technology innovation and agroenterprise development. Poverty Reduction and Shifting Cultivation Stabilisation in the Uplands of Lao PDR: Technologies, approaches and methods for improving rural livelihoods. January 27-30, 2004 at Luang Prabang, Lao PDR.

Horne, P.M. & W.W. Stür (2003). Developing agricultural solutions with smallholder farmers - how to get started with participatory approaches. *ACIAR Monograph No. 99*, ACIAR, Canberra. 119pp.

Millar, J., V. Photakoun & P. Horne (2003). Accelerating the impacts of participatory research and extension on shifting cultivation farming systems in Lao PDR. First Australian Farming Systems Conference, Toowoomba, September 2003.

A decision support system for monitoring livestock diet quality and performance: verification study on cattle, Adami Tulu, Ethiopia

A.A. Jama[1], D. Tolleson[1], J.W. Stuth[1], A. Ebro[2], K. Zander[1] and R. Kaitho[1]
[1]Texas A&M University, Department of Rangeland Ecology and Management, College Station, Texas 77843, USA, Email: jwstuth@cnrit.tamu.edu, [2]Adami Tulu Research Center, Ziway, Ethiopia

Keywords: near infrared reflectance spectroscopy, NUTBAL-PRO, crude protein, digestibility, Arsi cattle

Introduction Fecal profiling technology based on near infrared spectroscopy (NIRS) has been widely used in the U.S. to predict the diet quality and performance of free-ranging animals (e.g. Leite & Stuth 1995, Lyons & Stuth 1992, Lyons et al., 1993, Stuth et al., 1999, Tolleson et al., 2001). This technology is linked with the Nutritional Balance Analyzer (NUTBAL-PRO) model to form the core of a nutritional advisory system for livestock producers in the United States. This model predicts changes in body weight and condition for a broad range of livestock classes for cattle, sheep and goats. To test the system's transferability and usefulness to livestock producers in the developing countries a small trial was conducted in the Rift Valley of Ethiopia to evaluate the validity of the advisory system in East Africa using the NIRS equations developed in the United States.

Materials and methods The study was conducted at the Adami Tulu Research Center near Ziway, Ethiopia. Composite fecal samples were collected once a week for 12 weeks from fifteen Arsi steers (mean weight 235 kg) grazing in previously deferred paddocks. The fecal samples were dried at 60° C and sent to a lab in Ethiopia for NIRS scanning. Weight of the steers and weather variables (Min/Max temperatures, wind speed, relative humidity) were also recorded during sampling periods for use in the NUTBAL-PRO model. Dietary Crude Protein (CP) and Digestible Organic Matter (DOM) were predicted with the US functional near infrared imaging (FNIR) equation currently used for commercial samples.

Results Steers were able to select diets at a fairly constant level of DOM throughout the trial even though CP increased with time through the trial. The weekly average weight changes of the steers estimated by NUTBAL-PRO were compared to the corresponding measured weight (Figure 1). The maximum difference between actual and generated body weight was always less than 6 kilograms which translates to about 2.3% of the measured weight of the steers. Both actual and estimated weight changes followed similar trends during the trial.

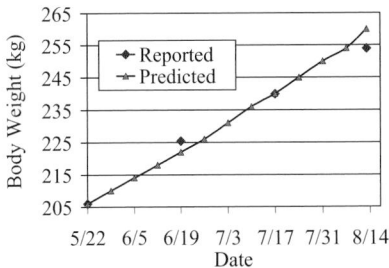

Figure 1 Comparison of average NIRS/NUTBAL predicted and actual body weight changes of 5-yr old Arsi steers grazing in a verification site at Adami Tulu, Ethiopia, 1998

Conclusions Arsi steers represented mature and castrated males typical of draft animals found throughout Ethiopia. Weight changes estimated by NUTBAL-PRO in combination with NIRS adequately tracked the observed weights, indicating a potential for the technology to enhance the nutritional management of cattle in Ethiopia.

References

Leite, E.R., & J.W. Stuth (1995). Fecal NIRS equations to assess diet quality of free-ranging goats. *Small Ruminant Research,* 15, 223-230.

Lyons, R.K. and J.W. Stuth. 1992. Fecal NIRS equations predict diet quality of free ranging cattle. Journal of Range Management. 45:614-618.

Lyons, R.K., J.W. Stuth, J.E. Huston & J.P. Angerer (1993). Predictions of the nutrient composition of the diets of supplemented versus unsupplemented grazing beef cows based on near-infrared reflectance spectroscopy of feces. *Journal of Animal Science*, 71, 530-538.

Stuth, J.W., M. Freer, H. Dove & R. K. Lyons (1999). Nutritional management for free-ranging livestock. In: H.G. JUNG & G.C. FAHEY JR., (eds.) Nutritional Ecology of Herbivores. Savoy, Illinois, USA: ASAS, 696-751.

Tolleson, D., J. Stuth, W. Vandervorste, D. Steffen, J. Hermann & D. Schmidt (2001). Prediction of weight gain in breeding heifers via the NIRS/NutbalPro system. Proceedings: Society for Range Management Meetings, Kona, Hawaii, USA.

Theme C: Delivering the benefits from grassland

Section 27

The role of the International Grassland Congress and Grassland Societies in technology interaction and influencing policy

Irish Grassland Association: delivering the benefits from grassland

S. Flanagan

Teagasc, Research Centre, Athenry, Co.Galway, Ireland, Email: sflanagan@athenry.teagasc.ie

Keywords: Ireland, grassland, leading forum, technology interaction

Introduction "Ireland was designed for grass production", Paddy O'Keeffe, Past-President

Description The Irish Grassland Association (IGA) is Ireland's leading forum for discussing the science of grass and animal production, the economics and finances of dairy, beef cattle and sheep farm enterprises. Membership is 800 and is a lively mix of progressive farmers (70%), research scientists, advisers and agri-business personnel. The IGA is a member of the European Grassland Federation and has close ties with the Ulster Grassland Society, Fermanagh Grassland Club and the British Grassland Society.

Technology interaction Research and Development have always been core components of IGA meetings. The IGA has been a significant driving force in the use of knowledge by dairy, beef cattle and sheep producers for exploiting grazed grass as the cheapest form of feed for ruminant production. In particular, the IGA has drawn heavily on the skills and knowledge of scientists at the research centres: Moorepark, Grange, Belclare (now Athenry), Johnstown Castle and Hillsborough. But it was not always so. The IGA was founded in 1946 and its formative years were characterised by major constraints to successful farming arising from lack of knowledge. Basic information on soil fertiliser requirements, grass varieties, reseeding and animal nutrition was non-existent. Founder members of IGA conducted trials on their own lands and showed that it was possible to improve greatly on a grazing season that was then limited to 5-6 months. Matters improved in the 1950's as a result of research at Johnstown Castle. Production targets for Irish grassland based on measurement were identified for the first time. But today's view of grassland being fully integrated in a farming system scarcely existed By 1960, however, IGA interests were centred on inter-relationships of animal and pasture and on livestock performance. Eminent speakers from abroad were invited and became part of the Association's platform, e.g. Prof. McG. Cooper from Newcastle. But it was the research programme of An Foras Taluntais (now Teagasc) under Director Tom Walsh in the 1960/70's that gave new impetus to Irish agriculture and resulted in the pursuit of knowledge on all sectors of grass farming and ruminant production. Production norms for Irish grassland and animal output in response to fertiliser N inputs were identified. For the first time grass-based livestock production systems were developed based on sound scientific principles and measures of efficiency and were widely adopted by IGA members especially on EU entry in 1973.

Ideas for low cost, profit-focused animal production have continued to flow and questions of major import discussed, e.g. the infertility problem in high yielding herds. Grass intake requirements by the high yielding cow and responses, or lack of, to concentrates have been defined. New Irish-bred grasses with higher spring yields are available as a result of research at Oakpark and Crossnacreevy. Extended grazing has developed as a two-way process from farmer to researcher and back again, exposing a need for more precision in sward management for both late and early season growth. The more grass that is grazed instead of silage and concentrates, the lower the costs. Outdoor pads for wintering cattle offer a capability for lifting cattle enterprise competitiveness. Their adoption in practice around the country represents present day technology uptake. Today, some stark messages are being delivered from the IGA platform in the context of the EU decoupled regime and future options are being analysed as never before.

These examples were selected to illustrate the strong links and interactions between member farmers and researchers that became embedded in the mindset of the IGA from day one. Each group thrived on each other's abilities to mutual advantage; a pattern of inter-dependence that continues to flourish today.

Activities (1) Members participate in: (a) dairy, beef cattle and sheep conferences each held annually, in grass farm study tours both at home and to temperate grassland zones overseas; (b) analyses of results and development options with other top farmers, scientists and advisers; (c) keeping up to date with developing technologies which can increase farm profits; (2) publication of annual proceedings of the IGA containing the papers presented by internationally eminent speakers; (3) submissions to Government on policies for agriculture, e.g. independent research as a basic need, Nitrates Directive.

Future plans Continue to measure progress in knowledge and to expand the horizons. Serve as a leading edge organisation for technology transfer where the best scientific information is analysed and applied in farm practice to maximise the productivity and profitability of Ireland's grasslands. Continue an instrumental role in the strategic direction of agriculture.

The role of the British Grassland Society in technology transfer

J.M. Crichton

British Grassland Society, PO Box 237, University of Reading, Reading, Berkshire, RG6 6AR, UK, Email: office@britishgrassland.com

Keywords: membership, technology transfer, communication

History and objectives Since its formation in 1945, the British Grassland Society (BGS) has been active in disseminating information to the grassland and associated sectors of agriculture and has adapted to a change from production-based agriculture to its current output mix of private and public goods. The BGS has as its remit: (a) improved production and utilisation of grass and forage crops for the promotion of agriculture and the public benefit, and (b) the advancement of education and research in grass and forage crop production and utilisation, and the publication of results.

Membership Currently standing at around 1,000, BGS members are mainly from the UK but with a growing proportion (around 20%) from overseas. Members come from the research community, consultancy and supply industries and those in education, together with many practising farmers. In addition BGS provides technology transfer support to a network of affiliated local grassland societies with some 4,500 members throughout the UK.

Technology transfer Technology transfer is achieved through a programme of conferences, meetings and grassland events, when BGS promotes the results of research findings to those within the grassland industry and the public at large. As a registered charity, it is an independent body with learned society status. All its meetings are open to members and non-members and its published material lies within the public domain.

Publications 'Grass and Forage Science' (GFS), edited by Professor J A Milne, is a highly regarded, peer-reviewed, publication and is both the scientific Journal of the British Grassland Society and the Official Journal of the European Grassland Federation. It publishes the highest quality academic papers within grass and forage crop production, management and utilisation as well as non-agricultural aspects of grassland management such as recreational and amenity use and the environmental implications of grassland systems. Sitting as a companion publication, the magazine 'Grass & Forage Farmer'(GFF), with a circulation of some 6,000, is aimed primarily at farmer members of BGS and the local grassland societies with the aim of supplying up-to-date technical information. Scientific findings from GFS are distilled into farmer-friendly messages in GFF.

Conferences and meetings The BGS engages with its members and the public through its regular programme of conferences and meetings including:

- A three-day **Summer Meeting,** usually held in July, including visits to farms and research stations.
- An annual **Winter Meeting** which, in 2004, became 'mobile' for the first time with a series of regional one-day meetings aimed at attracting local farmers who might find it impossible to leave the farm for longer periods.
- **BGS Research Conferences** aimed at giving younger research workers the opportunity to present papers in theatre or poster format and also provides a forum for work which may be at an early stage of development.
- **Occasional Conferences** on 'hot' subject areas. BGS publishes the proceedings as the Occasional Symposium Series and collections of conference papers - these being available for sale after a meeting so widening the availability of the information presented.

Research and development initiatives Whilst BGS does not directly fund R&D, it is pro-active in influencing direction of funding by Government and other agencies through the work of its R&D Committee. In particular, the Society maintains a list of priorities for R&D, which it distributes widely within the research community and to Government departments. The wealth of experience and expertise within its membership provides a wide knowledge base from which to source information and individuals are regularly co-opted to assist in such initiatives. The R&D Committee also provides the focus for BGS input into wider Government policy issues by responding to consultation documents directly and through its affiliation to the Institute of Biology. Furthermore, BGS is prepared to act to fill a 'knowledge gap' if one is identified. A recent project, 'Grass '99' was launched to initiate on-farm grazing groups throughout the UK to enable farmers to communicate with, and learn directly from, each other. This kick-started a whole new way of 'self-help' groups which continue to thrive.

Conclusion The BGS exists to facilitate two-way dialogue involving scientists and practitioners with communication being its key objective. It provides a vital link so that technology transfer can flourish within the UK and globally to the benefit of all actively involved in grassland and forage crop science and practice.

Portuguese Society of Pastures and Forages

J.M. Potes, E.V. Lourenço and T. Carita
Estação Nacional de Melhoramento de Plantas, P.O. Box 6, Elvas, Portugal, Email: jmirapotes@hotmail.com

Keywords: pastures, forages, conservation, animal production

Introduction The Portuguese Society of Pastures and Forages (SPPF) was born from a course on pastures and forages held in the National Plant Breeding Station of Elvas (where the headquarters of SPPF are located) in 1979. It was organised by the head of the Pastures and Forages Department, David Gomes Crespo, who was the first and founder member. According to Abreu *et al.* (1999), the structure established was that of a Scientific Society and was considered by the Government as a Service of Public Interest. It gathered inspiration from the British Grassland Society, the Association Francaise pour la Production Fourragére, the Sociedad Española para el Estúdio de los Pastos and the European Grassland Federation. Presently it has 1450 members including research workers, technicians, farmers and students (number not included). The number of members was initially 100 and now there are 450 active members.

Activity The SPPF activity falls mainly into the organisation of two annual meetings in spring and autumn. Spring meetings are held usually over three to five days with the vote of the General Assembly determining the location and topic, usually following the proposal from a group of members from a specific region of the country at the previous annual meeting. There have until 2004 been 25 Spring meetings. They have covered all the country with meeting been held twice at some places. Autumn meetings are usually held for one day, and concentrate on a specific topic. Because of that, they almost always have taken place at the Estação Nacional de Melhoramento de Plantas. Up to now, 26 Autumn meetings have been held.

The programme of Spring meetings comprises thematic sessions with the presentation of conferences, oral and poster presentations, and discussions. There are also visits to enterprises with activity related to the topics of the meeting and round tables coordinated by farmers. The conclusions are presented to the members of the government and regional authorities present in the closing session. The programme also includes the meeting of the General Assembly and the annual dinner of the Society. Every two years there are elections for the leader bodies of the SPPF. In the Autumn meetings, there is a presentation of the topic by one or more experts. When the session takes place in an auditorium it is followed by a discussion or a field visit.

Public service The most important public service of the SPPF is the publication of the periodical -Revista Pastagens e Forragens. Twenty two volumes have been published, usually one volume by year as a scientific publication. According to the Foundation for Science and Technology (FCT), which sponsors the periodical, it is one of the ten most used national periodicals in the domain of agricultural sciences. The SPPF also has another series of publication - Boletim da SPPF- for technical papers with more irregular periodicity.

Occasionally the SPPF has conducted studies or promoted the organisation of working groups upon institutional demands.

Plans for future We plan to pursue the actions developed up to know and to make a greater effort in technology transfer from the research centres to the farmers. We intend to produce more Technical Bulletins for technology description and promotion and to organise meetings directly with farmers. Also a more interventionist role will be taken as a professional association to support policymakers at the national level. Finally a more campaigning role will be taken with different grassland societies through the European Grassland Federation to promote the interests of grassland science.

References

Abreu, J. M. & N. Farinha (1999). Os vinte Anos da SPPF. *Revista Pastagens e Forragens*, 20, III – VII.

The activities and aspirations of the Estonian Grassland Society

A. Selge, R. Viiralt and T. Köster
The Estonian Agricultural University, 56 Kreutzwaldi St, 51014 Tartu, Estonia, Email: ares@eau.ee

Keywords: Estonian Grassland Society, activities, policies, aspirations

Introduction The aim of the Estonian Grassland Society (EGS) is to develop the flow of grassland-based knowledge between different stakeholders and multidisciplinary research and to create international contacts. The EGS is a non-profit organisation and more than 100 interdisciplinary researchers, plant breeders, advisors and farmers are actively involved in the work of the society. The interdisciplinary activities are becoming more and more attractive and important and bring together different target groups. At present seminars, which are focused on the adaptation producers to the EU rules and support system for farmers are attractive. Nowadays the essential goal for the EGS is to find solutions for integrating efficient grassland management and biodiversity, including socio-economic aspects. Novel approaches to combine the benefits for wildlife and the grass producer will be highlighted in the future EGS activities.

History of the Estonian Grassland Society The Estonian Grassland Society has a rather long history. The foundation meeting of the Estonian Meadow and Pasture Development Society was held on 31 Jan., 1930 in Tallinn. Although the activities of the Society were stopped in 1940 by the Soviet authorities, in ten years of its operation a strong basis had been laid for the development of grassland culture even in the situation of occupation that lasted for the next fifty years. The activities of the Estonian Meadow and Pasture Development Society were started again in Tartu on 16 April, 1993 under the name of the Estonian Grassland Society (Selge & Viiralt, 2004). In 1994, the Estonian Grassland Society was accepted as a full member of the European Grassland Federation.

Development of the Estonian Grassland Society More than 70 years ago the main tasks of the Society were to organise field days and seminars where farmers were introduced to modern techniques for the development of grassland and pastures as well as the study and improvement of the theoretical and practical bases of the grassland husbandry (Older, 2000). Some of these issues remained the same today and will also be important in the future. But at the same time the role of the EGS for the rural life has changed. The EGS activities are meant not only for dairy farmers, but also for the whole rural community and environment for developing the multi-functionality of grassland systems. The EGS works in close co-operation with different research institutions, as well as with advisory and farmers' organisations. The board of the EGS plans, leads and supports regular activities such as seminars, workshops, field-trial days, visits to advanced farms, etc. Feedback from the monitoring, organised by the EGS, is a useful tool for decision-makers in the Ministry of Agriculture. At the moment there are important and attractive seminars at the farm level, which introduce the EU rules and support systems to farmers. At a national level the issue of how to integrate efficient grassland management and biodiversity has to be solved. In Estonia two groups, thinking and acting differently, have emerged: a) environmentalists focus their research on maintaining the diversity of nature and wildlife, and agriculture is regarded by them as an environmentally hazardous factor, b) agricultural producers who have focused their operations on getting profit in order to stay and be competitive in the market. The interests of both groups should not be ignored and it is essential to find solutions to meet the goals of the two groups. The aim of the EGS is to find ways for introducing possibilities to realise common aims and objectives through common activities.

Conclusions

1. The EGS activities are meant not only for dairy farmers, but for the whole rural environment, for developing the multi-functionality of grassland systems.
2. The results of farmers' feedback, obtained by the EGS, serve as the bases of evaluating and planning the activities of the Estonian Grassland Society.
3. The essential goal for the EGS is to find solutions on how to integrate efficient grassland management and biodiversity, including socio-economic aspects.
4. The bonding of different researchers, advisors and farmers should increase the role of the Estonian Grassland Society, and through the Ministry of Agriculture, will have its impact on agricultural policy.

References

Older, H (2000). Conventional and ecological grassland management: comparative research and development. International Symposium, Tartu, Estonia, 1.

Selge, A, & R. Viiralt (2004). Network of extension service and Grassland Society toward an effective grassland management in Estonia. *Grassland Science in Europe*, 9, 1187-1189.

The role of the New Zealand Grassland Association in technology interaction and policy evolution

M.W. Calder

New Zealand Grassland Association, P O Box 8099, Gardens, Dunedin, 9030, New Zealand, Email: arfitty@xtra.co.nz

Keywords: grassland establishment, management, communication, farmers, agribusiness

Introduction The New Zealand Grassland Association (NZGA) aims for a two-way communication between grassland researchers and pastoral farmers on research findings and farming trends and opportunities. Annual conferences, special symposia, print and electronic publications as well as the web site are used to further these aims. Policy evolution is indirect, using the relationship with agribusiness sponsors and members to promote research objectives.

NZGA The NZGA was established in 1931 as a result of a meeting involving agricultural policy makers, scientists, academics and commercial representatives called to discuss "the betterment of research, investigation, demonstration or instruction in grassland management".

Membership The NZGA places strong emphasis on recruiting and retaining farmer members as they are considered important sources of information about the application of technology, areas for further investigation, problems and possible improvements. Agribusiness representatives are also encouraged for their feedback. Currently the membership is around 1400, balanced between farmers/producers (some 45%) and scientists (25%), consultants (5%), educators (10%) and agribusiness (15%).

Technology transfer The NZGA provides a forum for farmers, researchers, consultants, educators and commercial people to gain the latest information on the latest scientific progress and production advances in grassland establishment and management, and to exchange information farming trends, opportunities, problems and developments. The NZGA organises an annual conference where the venue moves around the country, and at least one session focuses on topics of practical interest or technology transfer and provides take home messages for producers. In addition there are field trips to selected farms to back up the messages delivered in the presentation sessions.

Special symposia have been introduced in recent years to cover research developments on particular topics in greater depth than might be achieved during a conference; again at least one session is devoted to practical issues. Examples have been Ryegrass Endophyte, Deer Nutrition and Management, and Legumes for Dryland Pastures. Further symposia are being developed.

Communication In addition to publishing the proceedings of each conference and symposium the NZGA produces a newsletter three/four times per year and regular "e-zines" are issued. The NZGA web site www.grassland.org.nz is also used to communicate and encourage membership. Visitors can search the database of all our publications and see abstracts of papers; however full versions of the papers are restricted to financial members. The association publishes a CD containing the papers from each annual conference, which is sold to members at a modest price.

Scholarships and awards High standards in the grassland farming industry are encouraged through annual awards for regional farming excellence and a special award for outstanding contributions to grassland industry and technology, promoted by the NZ Grassland Trust. The Trust also provides annual scholarships for agricultural science students.

Sponsorship and commercial relationships Sponsors are an important part of the operations of the Association, with emphasis on the agribusiness sector. Some government research funding agencies were committed, but as a result of the declining direct financial support considerable effort has been directed into developing and maintaining relations with the commercial sector, particularly farm supply, seed and fertiliser companies.

Policy evolution Until recently the NZGA employed no permanent staff, so policy development and promotion was undertaken informally via interested members, or *ad hoc* sub committees. However the close association the NZGA enjoys with government research funders and providers, as well as the connections with agribusiness enterprises ensures that the Association's views are registered.

Reference

Woodfield D. R. & J.F.L. Charlton (1989). NZ Grassland Association: Over 50 years of communication in grassland farming. *Proceedings of the New Zealand Grassland Association*, 50, 55-64.

Slovenian Grassland Society: science, profession and practice

B. Kramberger[1], T. Vidrih[2], J. Čop[2] and M. Vidrih[2]

[1]*Faculty of Agriculture, University of Maribor, Vrbanska 30, 2000 Maribor, Slovenia, Email: branko.kramberger@uni-mb.si*, [2]*Department of Agronomy, Biotehnical Faculty, University of Ljubljana, Jamnikarjeva 101, 1000 Ljubljana, Slovenia*

Keywords: grassland society, Slovenia, less favoured areas

Description of Society The Slovenian Grassland Society (SGS) was established in 1993. It has around 120 members. A half of members are active farmers, around 10% are scientists, the rest are employed in extension services or other agricultural enterprises (seed companies, administration bodies, etc.)

Highlights of actions Publishing professional and scientific papers represents one of the important activities of the society (e.g. Čop, 1998; Kramberger, 1998; Vidrih, 1999; Vidrih, 2004). In addition, the organisation of professional and scientific meetings and expert guidance and organisation of panel discussions of complex problems related to grassland management and forage production are important activities. The panel discussions, covering various topics (Table 1), are held at our annual meetings. Through oral presentations and discussions the participants get familiarised with the latest findings of scientists. Also field trips are organised for individual case studies. These discussions are covered by reporters from professional agricultural magazines. By doing this, topical issues are regularly reported to the wider professional audience as well as general public. In order to enssure wide involvement in the activities, the annual general meeting of the SGS and its panel discussions are held at a different place in Slovenia each year. Society also has a web page, where all interested in grassland management can find much information (http://www2arnes.si/surtvidr/index.htm).

Table 1 Foci of each year's annual meeting of the Slovenian Grassland Society

Year	Focus	Place	Region of Slovenia
1994	Grassland renovation by direct drilling	Vrhnika	Central
1995	Recultivation of hilly pastures by animals as a tool	Nova Gorica	Southeast
1996	Pasture management of hilly areas	Bohinj	Northwest
1997	Cereal sod seeding into pasture sward	Postojna	Southwest
1998	Effectiveness of subsidies for pasture cultivation	Zreče	Northeast
1999	Cattle wintering on pastures	Pivka	Southwest
2000	Experiences with seeding of grass clover mixtures	Ptuj	East
2001	Summer animal grazing on ski field terrains	Maribor	Northeast
2002	Recultivation of abandoned arable crop fields by grazing	Rakičan	East
2003	With animal grazing to food of good quality for people	Rodik	Southwest
2004	Grassland management on Ljubljana marsh and surrounded hills	Vrhnika	Central

Up to now, our panel discussions (Table 1) have focused primarily on grassland cultivation and pasture management. Consequently, in Slovenia there is a trend away from indoor animal production to production based on pastures. The spread through Slovenia of sward renovation by direct drilling owes much to the activities of SGS in the last decade. To most farmers who decide upon such an approach to sward management, machinery services are offered in order to facilitate reseeding with special drills.

Activities to influence policy or public perception Apart from carrying out the planned activities, the SGS members strive also to affect public perception and have an influence on the creation of agricultural policy in various governmental bodies and professional commissions.

Plans for future We intend to stress the importance of the multi-purpose grassland areas. We intend to strengthen the membership and start publishing a professional journal, which would, for the time being, be issued annually.

References
Čop, J. (1998). Vpliv pogostosti rabe na botanično sestavo travne ruše ter pridelek in kakovost zelinja. *Sodobno kmetijstvo*, 31, 195-198.

Kramberger, B. (1998). Sestavljanje mešanic za setev. *Sodobno kmetijstvo*, 31, 187-189.

Vidrih, M. (1999). Domače živali naj se pasejo. *Kmečki glas*, 56, (num. 17), 9.

Vidrih, T. (2004). Ukrepi Programa razvoja podeželja in pašna reja. *Sodobno kmetijstvo*, 37, 35-36.

Role of Hellenic Range and Pasture Society in technology interaction and policy evolution

T.G. Papachristou, P.D. Platis, V. Papanastasis and A. Ainalis
Hellenic Range and Pasture Society, Forest Research Institute, 570 06 Vassilika, Thessaloniki, Greece, Email: tpapachr@fri.gr

Description The Hellenic Range and Pasture Society (HRPS) was founded in 1992 and is a non-profit scientific association. It is composed of 74 members with a common interest in the study, management, and rational use of rangelands and related ecosystems (e.g. pasturelands). The objectives for which the corporation was established are: a) to develop an understanding of rangeland ecosystems and of the principles applicable to the management of rangeland resources (soil, plants, water, and animals), b) to assist all who work with rangelands and pasturelands to keep abreast of new findings and techniques in the science and art of rangeland and pasture management, c) to create a public appreciation of the economic and social benefits to be obtained from the rangeland environment, and d) to promote professional development of its members.

Activities Over the past twelve years, ten scientific events have been organised by the HRPS with a goal of developing among its members a sense of identity with the profession, to make positive and imaginative contributions to the field of rangeland resources management, and to provide information and guidance to the general public in this area. These events involved five meetings, four national congresses and one international congress, which dealt with the following themes: *Meetings*, 1) The role of rangelands in rural development, 2) Range science and alternative land uses, 3) Relation among livestock, wildfires, and environment, 4) The natural environment of Mygdonia area, 5) Strategy for rational management of rangeland resources; *National Congresses*, 1) Sustained utilisation of rangelands and pasturelands, 2) Range science at the threshold of the 21st Century, 3) Range science and development of mountainous regions, 4) Rangelands of lowland and semi-mountainous areas: means of rural development; *International Congress*, 1) Woody plants in Europe. The above activities established the HRPS in the national scientific society and with many rangeland users. It is acknowledged that the HRPS's activities help them to realise that rangelands produce a wide variety of goods and services desired by society, including wildlife habitat, livestock forage, water, mineral resources, wood products, wild land recreation, open space, and natural beauty. This increases the appreciation that the proper use and management of rangeland is vitally important to people everywhere. Also, members of HRPS, by participating at relevant TV programmes or by being interviewed by newspapers or radio stations, have promoted to their best the activities of the HRPS and informed public opinion in parallel.

Plans for future initiatives Doubtless the 12-year presence of HRPS has influenced the scientific society. However, we appear to play a minor role, if any, in the development of policies that affect rangelands. In fact, government and land management agencies in Greece lack policies directed specifically towards the conservation and management of rangeland resources or the well-being of pastoral populations that depend on them. Policy-making that affects rangelands, is driven largely by non-technical factors, because politics and perceived short-term economic and social needs prevail over science and resource management. The lack of policies that address rangeland conservation and management is an indication of the relatively low priority of resource management for many rangeland inhabitants and users, and consequently for their political representatives. We argue that the links between science, policy and management must be strengthened, through adaptive management, participatory research, and clear and proactive communication by rangeland professionals. We also believe that rangeland professionals and its scientific society can no longer afford to view policy-making as outside our domain of expertise and appropriate involvement. To be effective in the policy arena, our society and it's members must also improve their communication skills. We have to share our knowledge and concerns with policy-makers, and to do so we must develop expertise in communicating ecological complexity in simplified terms. Our job will be made easier if we involve the public more directly in our science, through participatory research and adaptive management. Such collaborative learning efforts may have the added benefit of increasing the scientific literacy of the public, so that people better appreciate the underlying complexities and uncertainties associated with rangeland science and management. Finally, we must take a more active role in influencing policies that directly and indirectly affect rangelands. At the same time, however, we have to apply the basic principles of our science that we have painstakingly built over the past decades. These principles, when applied, can make a difference in natural resource conditions and people profitability.

Technology interaction and policy evolution; Grassland Society of Southern Australia inc

L.L. Bennison

Grassland Society of Southern Australia Inc. PO Box 1349 Warragul Victoria Australia 3820, Email: office@grasslands.org.au

Background The Grassland Society of Southern Australia was formed in 1959 to provide a forum for the transfer of information, ideas and experiences in relation to all aspects of grassland establishment, maintenance, utilisation, persistence and research. The membership of over 2,000 is composed primarily of primary producers (80%) with extension and research scientists comprising 10%, agribusiness 8% and academia 2%. The major interests of members are wool, lamb and beef production, dairying and cropping. The society produces bi-monthly newsletters, holds an annual conference with attendances between 400 and 600 annually and has a branch structure that hosts local events. The society is focused on disseminating information from researchers to primary producers.

Highlights of actions to improve technological interaction

The Grasslands Productivity Programme (GPP) The GPP has been an important event in the history of the wool industry in south-east Australia. It empowered participants to seek information, gain skills and to challenge and intensify their pasture management. Furthermore it provided hope to some stakeholders in the wool industry that it was possible for wool producers to become more productive and thereby lower the cost of producing wool. Papers reporting on this programme are presented at IGC, Dublin, by Geoff Saul and Jason Trompf. The GPP provided farmers with a comparison of their current farm practices and a management package of increased fertiliser application, increased stocking rate and improved management skills. The novel approach to on-farm extension has since been widely adopted across a range of agricultural industries.

The Pasture Web Database A joint project by the University of Melbourne Institute of Land and Food Resources and the Grassland Society of Southern Australia (formerly the Grassland Society of Victoria) saw the launch of a web-based interactive database on Pasture Species and Cultivars in 2001. The database holds information on 37 species and 242 cultivars. A workshop was held in 2003 with Prof. David Hannaway (USA) and Prof. David Chapman (Australia) to explore whether or not the website could include GIS technology and spatial layering in the future.

Activities to influence policy

Best Practice for Phosphorus Fertiliser Application on Improved Pasture in Victoria In the late 90's several dry years exacerbated problems with excessive nutrient in waterways resulting in eutrophication and algal blooms. Farmers were often accused of being the source of the problem and reports in the media were often inaccurate and unfounded. In a proactive step the Society released Fertiliser Best Practice Guidelines to increase the understanding of farmers of the implications of fertiliser application. This was picked up and carried by all the major fertiliser industries and is still used as a resource today.

Leasing Kit Another reflection of our changing society is the increased age of our farmers and the large amounts of capital required by younger people to begin farming. The information kit was developed with guidelines for lessees and lessors to encourage the leasing of land. Many older farmers, whilst not wanting to continue working on the properties, were not ready to leave the farm. This presented an ideal entry point for younger farmers with energy and some capital but not sufficient to purchase land.

Plans for future initiatives

Pastures from Space At the time of writing the Society is organising a workshop with the Commonwealth Scientific and Industrial Research Organisation (CSIRO) and representatives from State Departments of Agriculture to evaluate whether or not the Pastures from Space program can be adapted for commercial use in the eastern states of Australia. The program uses satellites, modelling and software to predict and interpret pasture growth rate. It is currently calibrated for annual pasture and the society is keen to explore its use with perennial pasture in the south-eastern states of Australia for on-farm use.

Safe Off Site A second project under consideration is the Safe Off Site computer software program which is an online data back-up program. As well as providing a service to members, it has the advantage of providing a royalty stream to the Society. A pilot will be offered to members in early 2005.

Role of the Grassland Society of Bosnia and Herzegovina in grassland agriculture

S. Alibegovic-Gbic and M. Bezdrob
Faculty of Agriculture University of Sarajevo, Zmaja od Bosne 8, 71000 Sarajevo, Bosnia and Herzegovina, Email: senija@lsinter.net

Keywords: Grassland Society, grassland management, ecology

Introduction The biggest part of Bosnia and Herzegovina is covered with hilly and mountainous terrain, which, from the agricultural perspective, determines the manner of soil usage. Of all agricultural areas in Bosnia and Herzegovina, 55.49% are covered by grassland with that figure increasing to 79.14% in mountainous regions. Therefore, grassland agriculture and its cousin livestock production are the most significant part of the Bosnian economy. In the current situation natural grasslands are largely neglected and give a low yield of forage of low quality. Many areas are completely abandoned and have been gradually overgrown with weeds and bushes. Yet a special trait and advantage of Bosnian grasslands is the diversity of plant types and it is important that this is maintained. In that sense, all organisations which could contribute to the development of grassland farming and livestock production, as well as to the development of rural areas in general are of great importance, because of the significant contribution that they could play in the overall sustainable development of our country. One of the organisations which could and should play an important role in achieving more rapid development of this sector is a grassland society.

The Grassland Society The Grassland Society of Bosnia and Herzegovina was founded in 2001, but because of the political instability in our country, the society, even with all the effort invested, was not officially registered until June 2004, so its activities have been very modest. We were mostly working on how to find a way to register the Society and to start some activities at the state level. The society currently counts 52 members: ca. 10% are farmers, 60% agronomists and 30% are scientists and researchers. During the meting in Lucerne in 2004, our Grassland Society became a full member of the European Grassland Federation.

Future activities Our Grassland Society works on improving the communication network between agricultural specialists, farmers, environmentalists, and all others whose interest lies in improving grassland utilisation and its role in preserving the environment and tourism. One of the main goals of the society is to achieve professional management of production, conservation and utilsation of forage. In that sense, the society will organise professional and scientific seminars, symposiums and lectures about how to manage grasslands and the production, preservation and utilisation of forage. It is also our intention to organise discussion sessions about legal issues cocerning grasslands and agriculture, environment and ecology when that is necessary.

Further significant aims of the association are: promoting scientific studies on proper grassland management resulting in the improvement of forage yield and forage quality, while at the same time maintaining grassland biodiversity and a healthy environment; introducing and expanding organic production; researching the relationship between grassland management and ecology, and finding the optimal way to conserve and utilise forage in animal production.

The Grassland Society plans to start publishing scientific and professional publications and to stimulate publication of professional innovations within these regular publications.

Conclusion The activities of the Grassland Society of Bosnia and Herzegovina should contribute to sustainable livestock farming, rural development and the overall economic situation in our country.

Institutions, structures and topics of grassland research in Germany – From science to practice

J. Isselstein[1] and H. Hochberg[2]

[1]Institute of Agronomy and Plant Breeding, Grass and Forage Research, University of Goettingen, Von-Siebold-Str. 8, 37075 Goettingen, Germany, Email: jissels@uni-goettingen.de, [2]Thuringia State Institute of Agriculture, Department of Grass and Forage Production, Bahnhofstr. 1a, 99869 Wandersleben, Germany

Keywords: grassland, research, management, Germany, herbage production, multi-function grasslands

Introduction In Germany, grassland covers some 30% of the agriculturally used land. It forms the basis of forage production in dairy and beef cattle husbandry as well as sheep and horse feeding. The intensification of grassland farming during the last fifty years was made possible through extended basic and applied research in all fields of grassland management. This included the improvement of forage species and varieties by plant breeding, the adaptation of botanical knowledge for the control of the botanical composition of permanent grass swards, the application of regular fertilisation, the improvement of the grazing management, the increased frequency of utilisation or herbage conservation by ensiling. In addition to the general improvement of forage production, the refinement of production measures in animal husbandry has led to a marked increase of efficiency in dairy and beef cattle farming. Production-orientated research was well funded until the late 1980's. Since then, the awareness of adverse side effects of the intensification of grassland farming, such as the loss of biodiversity, the pollution of the environment mainly by excess nitrogen and phosphorus, and the emission of greenhouse gases has grown. In addition, with the reform of the European Union Agricultural Policy in 1992, the rental costs for grassland decreased as did the stocking rates. On marginal sites, grassland is now at risk of being abandoned from agricultural use. Thus, increasing forage production and refining production measures have lost priority in grassland research and multiple function grasslands have become the main target of research.

Organization of scientific and applied grassland research Grassland research in Germany is performed by Universities, State Institutes, and institutions of the countries within the federal state. Basic research is mainly located in the Universities and the state institutions, whereas applied science is performed by institutions of the federal countries. The latter are also responsible for the support of the extension services. The transfer of research results from science to practice is generally achieved by publication in scientific journals, publication in popular series for the extension services and the farming community, and by congresses with participants from science and practice.

Grassland societies The Arbeitsgemeinschaft Grünland und Futterbau (AGGF, 'German Grassland Society'), is the main scientific association for grassland research in Germany. It was founded in 1950. Later, the AGGF joined the Gesellschaft für Pflanzenbauwissenschaften (GPW, German Society of Agronomy) and became a working group within the GWP. It is the general aim of the AGGF to enhance the scientific progress in grassland management and to represent this field to neighbouring scientific disciplines, to national and international politics and to society as a whole. It also represents German grassland research in international scientific associations. At present, the AGGF has some 250 members, half of them being from science and the other from applied science and practice. There is a regular annual meeting every year with 100 to 150 participants. These meetings have a general topic which is chosen according for its relevance to actual grassland farming. In addition to this annual conference, there is a session for grassland research within the annual conference of the GPW. Here, mainly scientists from basic research meet. In 1991, the Deutsche Grünlandverband (DGV) was founded. It is the aim of the DGV to support farmers, nature conservation groups and policy makers in their efforts to maintain the agronomic and ecological performances of grassland. The focus is on applied research and on the application and transfer of recent results of scientific research into the practice of grassland management. The DGV has some 1200 members, mainly farmers, and it organizes a regular meeting every year.

Recent challenges for grassland research The change in priority from forage production to multi-function aspects of grassland management in recent years has been accompanied by a lowering of the position of grassland research among the scientific community. Basic disciplines in biology and ecology have successfully established strong research in the field of grassland ecology and nature conservation, e.g. on the relationship of biodiversity with productivity and ecosystem functioning or on mega-herbivore grazing for landscape management. Agriculturally-orientated grassland research is challenged to develop production systems that consider and utilise recent findings on the ecology of grassland for efficient animal husbandry. Unless basic findings are integrated into efficient and profitable farming systems, their adoption in practice will remain poor. Therefore, grassland research in Germany needs to be linked to neighbouring scientific disciplines more efficiently. It should further develop its particular strength to bridge science and practice by a systems approach.

The Spanish Society for the Study of Pastures: 45 years promoting better pasture knowledge and management

A. San Miguel and S. Roig

E.T.S. Ingenieros de Montes. Universidad Politécnica de Madrid, Ciudad Universitaria s/n E-28040 Madrid, Spain, Email: asanmiguel@montes.upm.es

Keywords: grassland, research, rangeland, pastoralism

Introduction The Spanish Society for the Study of Pastures (SEEP) (http://www.seepastos.es) was founded in 1960 through the initiative of Prof. G. González with the intention of combining efforts and initiatives on pasture studies from many points of view. The Society has the objective of promoting the knowledge and improvement of Spanish pastures, regarding every issue related with pastoral science: typology; ecology and functions of pastoral ecosystems; plant production; feeding value; animal production; economics; sociology and agricultural policy, etc. The term 'Pasture' is considered in a broad sense to include every plant part, individual or community capable of being used for livestock or wildlife feeding. Since the inception, the Society has enrolled as members university graduates, engineers, farmers, institutes and companies interested in the soil-vegetation-livestock complex.

Members The current number of members is 284. At present the Society includes members in all Spanish autonomous communities who, in accordance with the Society's initial aims, develop research and technical activity on multiple aspects and at many centres: Universities (Faculties of Biology, Veterinary Science, Pharmacy, Agronomy..., Forestry, Environment, and Agricultural Schools, etc), Scientific Research Council Institutes (CSIC), National Institute of Agricultural Research, Agriculture Research centres from autonomous communities, Agriculture Ministry, etc.

Activity Society activity is mainly developed through the publication of the research journal *Pastos*. It also organises an annual scientific meeting at a different location each year, trying to visit and cover all the diversity of pastoral and productive systems of Spain. Proceedings are published for each annual scientific meeting. The Society also promotes and organises some other scientific and technical or informative activities that are related to pastures, such as the recent Pasture Nomenclature (Dictionary). The most important such current activity is the research project Typology, Cartography and Characterisation of Spanish Pastures

The Spanish Pastures Project This project is described in http://www.carm.es/cagr/cida/pastos/index.html. After 40 years of Society activity, the Society Vice-President, Prof. C. Ferrer, highlighted the fact that there was much more knowledge about Spanish pastures than was actually being used or was even accessible to potential users. Therefore, through his initiative, a project to transfer research results was funded by the Spanish National Institute for Agricultural Research (INIA) and almost every Autonomous Community. The aim of that Project was to summarise, standardise and systematise all the outstanding information and research results on Spanish pastures dispersed in more than 2000 scientific publications. This project joins together more than 200 Spanish researchers, working in 19 regional working groups -one per autonomous community with two in the two larger communities - and with six different subjects (each one with its own coordinator): Cartography, Natural Pastures, Agricultural Pastures, Feeding Value, Animal Production and Socioeconomics. The project is coordinated by Prof. C. Ferrer and Prof. A. San Miguel.

Basic dictionary on Spanish pastures At the 2001 research meeeting, the S.E.E.P. general assembly approved a first version of a basic dictionary on Spanish pastures (Ferrer *et al.*, 2002) that standardises the main Spanish terms related with pastures. However, a more general glossary or dictionary of pasture terms will be prepared in the near future. A dictionary of latin and common English names of herbaceous and shrubby Iberian plants has also been published by Ferrer & Broca (2002).

Collaboration with the Centre for Animal Food Information (CIA) The Society maintains a collaboration project with the Centre for Animal Food Information (CIA) (Córdoba University) (http:/www.uco.es/organiza/servicios/apoyo/nirs/cia1.htm), with the objective of compiling and standardising all the outstanding information related to the characteristics of products for animal feeding.

References

Ferrer, C. & A. Broca (2002). Diccionario de nombres vulgares en inglés de las especies herbáceas y arbustivas de la flora Ibérica. *Pastos*, XXXI, 45-123.
Ferrer, C., A. San Miguel & L. Olea (2002). Nomenclátor básico de pastos en España. *Pastos*, XXXI, 7-44.

Activity of Japanese Society of Grassland Science

K. Sugawara[1], S. Saiga[2], K. Tateno[3], Y. Cai[3], N. Yoshida[3], T. Takamizo[3] and M. Goto[4]
[1] President of Japanese Society of Grassland Science, c/o National Institute of Livestock and Grassland Science, 768 Senbonmatsu, Nasushiobara, Tochigi 329-2747, Japan, [2] Vice President of Japanese Society of Grassland Science, Ueda, Morioka, Iwate 020-8550, Japan,[3] National Institute of Livestock and Grassland Science, 768 Senbonmatsu, Nasushiobara, Tochigi 329-2747, Japan, Email: takamizo@affrc.go.jp, [4]Mie University, TSU, Mie, 514-8507, Japan

Keywords: Japanese Society of Grassland Science, journals, Grassland Science, Japanese Journal of Grassland Science

Background The Japanese Society of Grassland Science (JSGS) was founded in 1954 for the purposes of progressing grassland and forage crop sciences and fostering grassland agriculture and better management of grassland for animal production in Japan. From the first, the members of JSGS have included interdisciplinary scientists from forage crop science, forestry, animal science, agribusiness and many related fields. In the 50 years since its foundation, JSGS has made large contributions to the progress of both science and industry in Japan. The number of JSGS members is now declining slightly, but there are still about 950 including 800 individual members and 150 organisations or private companies. The profile of the current members is mainly scientists working in university or governmental and private research institutes.

Main activities Major activities of the JSGS are (a) to hold the annual meeting (usually in spring), (b) to publish the journals of JSGS, "Grassland Science" in English and " Japanese Journal of Grassland Science" in Japanese and (c) to confer Research Awards

Recent developments The internationalisation of research activity and the evolution of grassland science to include environmental science have led to changes in the approach of the Society. It is important that information and experiences are shared with scientists all over the world. To respond to this situation a "Japan-Korea-China Symposium on Grassland Agriculture and Animal Production" was held successfully on the 50th anniversary of JSGS in Hiroshima, with the help of both the Chinese and Korean Societies of Grassland Science. It was agreed that this trilateral symposium should be held regularly and the next will be in Lanzhou, China, in 2006.

Because Asian scientists in grassland science still did not have an international journal for our special field, JSGS decided to publish a new international journal in English. The journal covers all the fields of grassland science; cultivation and management of grassland, physiology, morphology, pathology, breeding and genetics of forage plants, forage conservation and nutritive value, animal behaviour in grassland, grassland environment and landscape, ecology, soil and soil organisms, farm machinery, economics and others related to grassland. The journal will be published quarterly, March, June, September and December. We hope that this journal will contribute to the progress of grassland research and grassland agriculture towards the establishment of sustainable agriculture in Japan, Asian countries and the world.

The Tropical Grassland Society of Australia Incorporated

D.M. Orr[1], M.H Shelton[2] and C.A. Hall[3]

[1]*Department Primary Industries and Fisheries, PO Box 6014, Rockhampton Mail Centre, Queensland 4702, Australia, Email: david.orr@dpi.qld.gov.au,* [2]*School of Land and Food Science, University of Queensland, St Lucia, Queensland, 4072, Australia,* [3]*CSIRO Sustainable Ecosystems, PO Box 102, Toowoomba, Queensland, 4350, Australia*

Introduction The Tropical Grassland Society of Australia was formed in 1962 and became incorporated in 1987 and has the following aims:

- To publicise information of interest to primary producers and scientists
- To improve the relevance of research and adoption of technology through the flow of ideas between scientists and producers
- To publicise the findings of Australian pasture research and development to overseas workers, and to draw on their experience for application in Australia

Membership The Society currently provides two types of members – journal membership ($75 per annum) and Newsletter membership ($25). Journal membership is aimed primarily at research and extension staff who receive both the journal and the newsletter, whereas Newsletter membership is aimed principally at commercial producers and those other members who require less technical information. Currently, the Society has 270 members. Numbers were boosted in 2000 when 40 members of the newly formed Leucaena Network joined as newsletter members. A review of Society membership and activities between 1962 and 1995 (McDonald, 1996) has highlighted the changing nature of the Society. To address the issue of declining membership, the current executive is exploring closer association with other Grassland Societies in southern Australia.

Executive committee The Executive committee comprises a President, Vice President, Past President, Secretary, Treasurer together with the Journal Editor and Newsletter Editor and meets at 2-monthly intervals. Executive positions are usually held for one term, however, the Constitution allows for these positions to be held for more than one term, as is often the case for Secretary and Treasurer. The President usually serves for three years, initially as Vice President, moving to President in the second year and then as Past President in the third year. The Society attempts to rotate the President's position between Producer and Scientist/Extension members.

Journal and newsletter The Society publishes *Tropical Grasslands* quarterly comprising contributed papers, conference proceedings and book reviews. An average of 30 contributed papers have been published annually between 1994-2004: 40% from Australian authors and the remainder from 40 other countries – mainly Africa, South America and south east Asia. Manuscripts are managed by a Society funded, part-time Journal Editor who is assisted by a panel of Associate Editors and an Editorial Advisory Board. For a charge of $50, authors have been provided with hard copy reprints although, from 2005, reprints will be available only in Abode Portable Document File format. *Tropical Grasslands* is indexed in *Current Contents*. The quarterly newsletter provides articles of interest, highlights research findings and acts as a forum for exchanging views between members.

Conferences and field days The Society has traditionally held a Tropical Pasture Conference each five years. These conferences have been highly successful and have provided a major financial boost to the Society's funding. The next conference is due in 2005 but has been delayed due to overlap with the International Grassland Congress and the Society is considering a joint conference in 2006 with the New South Wales and Victorian Grasslands Societies. The Society conducts field days on commercial properties and on research stations around themes of topical interest and reports from these field days are published in the Newsletter.

Website The Society maintains a web site www.tropicalgrasslands.asn.au which lists information about the Society. A feature of this site is "Pasture Picker" which provides information about the useful characteristics of a range of grass and legume species and a selection tool, based on climate and soil criteria, that allows users to select species suitable for their particular conditions.

Future The Society faces an uncertain future with declining membership due primarily to the greatly reduced number of professionally-trained pasture scientists in northern Australia. From the 1970's peak, the number in government employment has declined substantially. This has occurred despite a buoyant pastoral industry and a strong demand for pasture technology. Survival may require collaboration or amalgamation with other societies.

Reference

McDonald, C. K. (1996). The Tropical Grassland Society of Australia Inc. – our current status. *Tropical Grasslands*, 30, 341-344.

Comments on the present activity of Czech specialists in grassland cultivation

B. Cagaš

OSEVA PRO s.r.o., Výzkumná stanice Rožnov-Zubří, Hamerská 698, 756 54 Zubří, Czech Republic, Email: cagas@iol.cz

Keywords: grassland cultivation, forage crops, lawns, research institutes, universities

Background Unlike some other countries, there is no officially established forage crop or grassland society in the Czech Republic. This, however, does not mean that activity in this field is lacking. Several institutions are involved in the development of grassland research. These are predominantly: Grassland Commission of the Section of Plant Production of the Czech Academy of Agricultural Science, Forage Crop Commission of the Bohemian-Moravian Union of Plant Breeders, Bohemian-Moravian Association of Plant Breeding and Seed Production and Association of Grass and Clover Seed Growers. Members of these institutions are researchers and breeders of grasses and legumes, university lecturers from departments which deal with these aspects, workers in forage crop seed companies and the general agricultural public.

Uses of grassland As in other European countries, grassland cultivation in the Czech Republic has undergone dramatic changes. Even though grasses and legumes still remain the basis of ruminant nutrition, especially grasses have become popular in non-agricultural uses, i.e. in cultural landscape formation and preservation and in environment protection and improvement. Grasses are widely used in landscape revitalisation and in restoration of localities damaged by anthropogenic activity to its original state. This is certainly reflected both in the composition of the species and their proportions and in professional discussions within the grassland public. Careful attention has recently been given to the impact of the Czech Republic joining the European Union, predominantly in relation to agricultural production, forage crop seed production, environment, and the position of applied agricultural research.

Research organisations Those who take part in this discussion and predominate in running research projects and putting them to use are research and teaching staff from private organisations such as OSEVA PRO s.r.o., the Grassland Research Station Rožnov – Zubří based in Zubří (curator of genetic resources of grasses in the Czech Republic, agronomy of grass growing for seed, utilisation of grasses for power production purposes, etc.), the Forage Crop Research Institute Troubsko Ltd. based in Troubsko (curator of the collection of genetic resources of legumes, legume agronomy and seed production, theoretical principles of breeding etc.), the Grass Ecosystem Research Station Jevíčko (re-seeding and improving the quality of permanent grasslands, forage quality, etc.), the Plant Breeding Station Větrov and Domoradice (subsidiary of OSEVA UNI a. s.), the Plant Breeding Station Hladké Životice and the Plant Breeding Station TAGRO Červený Dvůr s.r.o., specialised in breeding forage and lawn varieties of grasses and leguminous crops. The Forage Crop Departments of the Czech Agricultural University in Prague, the Mendel University of Agriculture and Forestry in Brno and the Faculty of Agriculture in České Budějovice coordinate or staff research projects concerning grassland cultivation. In both agricultural universities research into lawns, their establishment and maintenance has recently been of great importance.

Discussion and exchange of information Where do the interchange of ideas and discussions about new procedures in grassland cultivation take place? At annual sessions of scientific boards of the above-mentioned research institutes, workshops on particular topics organised by the Forage Crop Commission of the Czech Academy of Agricultural Science, annual workshops organised by plant breeders and the annual workshop on grass seed production organised jointly by the Grassland Research Station and the Association of Grass and Clover Seed Growers. The source of information is an almanac Pícninářské listy. An increasing role of grassland cultivation is shown by active participation in the annual seminars and seminar proceedings Trávníky (Lawns), published by Bonus agency. The importance of the role of grasslands can also be seen in books published this year on the utilisation of grasses for forage crop and lawn purposes which were the outcome of cooperation of the staff of research institutes and universities.

Czech grassland growers do not, however, have any umbrella organisation, but they actively present the results of mainly applied research into forage crop and grassland production, at home and abroad and disseminate the information among Czech users involved in agriculture and in other areas.

American Forage and Grassland Council technology interaction and policy development

E.K. Twidwell and W.A. Tucker

American Forage and Grassland Council, Georgetown, TX 78627, USA, Email: etwidwell@agctr.lsu.edu.

Keywords: forage, organisation, coalition

Brief description The American Forage and Grassland Council (AFGC) is a national organisation which has been in existence since 1968. Membership of AFGC is about 2,500. The membership of AFGC is divided into three main sectors: private, public and industry. The private sector has the largest membership (60%), and private members are usually producers that are engaged in some type of agricultural enterprise involving the use of forages. The public sector members (30%) are educators and other government agency personnel that work with the general public. The industry sector (10%) involves various companies that deal with the forage industry. The AFGC Board of Directors is composed of 18 members, 6 from each sector. Most of the AFGC membership belongs to an affiliate council. There are currently 25 affiliate councils in the United States, most of which are located in the eastern, southern and midwestern regions of the country. There is one affiliate council located in Canada (Ontario). One of the major strengths of AFGC lies in its diversity of membership among the three sectors. The primary core purpose of AFGC is to advance forage agriculture and grassland stewardship. This organisation has the vision to be recognised as the leader and voice of economically and environmentally sound forage agriculture.

Actions to improve technology interaction The development of an AFGC website (www.afgc.org) in 2000 has provided a good means of communication among AFGC members and headquarters. Information on this site includes forage production information, links to other forage societies and general AFGC information. In 2005 people were able to register on-line for the annual conference via the AFGC website. In 2003 AFGC launched a monthly electronic mailing to those who have their Email address on file with AFGC. This publication is called "Forage Progress" and is compiled from presentations made at annual AFGC conferences. A quarterly publication, "The Forage Leader" is mailed directly to members and contains information dealing with AFGC events and activities. This publication also contains articles written by producers and forage researchers. The AFGC executive committee uses conference-calling technology to meet on a monthly basis and discuss AFGC matters.

Activities to influence policy or public perception Through its affiliate council structure and the diversity of the board's membership, AFGC is engaged in multiple issues across a broad spectrum of topics. The Director of Legislative Affairs works with the Legislative Committee to prioritise those areas of focus most in need of resource allocation. The Director reports to the board and participates monthly with the Executive Committee. This organisation shares office space in Washington, DC, with the Tri Societies, the Weed Science Society of America and the Society for Range Management (SRM). Most of the work at the federal level involves securing and maintaining research funding for forage-related projects nationwide. As the pre-eminent pasture and harvested forage organisation working primarily in those areas of the country with adequate rainfall, AFGC often partners with SRM to represent the more arid areas of the country on those issues of national or global relevance. Most recently the National Forage Coalition was formed to focus on issues specific to harvested forages. Partnering with AFGC in this effort are the National Hay Association and the National Alfalfa Alliance. Each organisation has three individual representatives on the coalition. In the conservation arena AFGC is a founding member of the Grazing Lands Conservation Initiative and participates with them both in policy development and funding issues specific to grasslands conservation. Issues forums at the Annual Conference and participation in numerous industry and public policy workshops keeps AFGC engaged with a changing network of responders and influencers. Segment-specific or regional-specific trade groups also receive policy influence assistance through their membership in AFGC. Examples of the latter would include the United States Dairy Forage Research stakeholder group and the Northeast Pasture Consortium. In 2005 AFGC developed a certification programme for individuals working in the forages area. This programme was initiated to ensure that individuals have met the basic qualifications necessary to carry out professional work in the forages area.

Future initiatives This organisation plans to have some influence on how the forage component is written into the next Farm Bill. In the past several Farm Bills, forages have received little recognition. The AFGC is currently working with SRM in planning a joint conference for 2008 in Louisville, Kentucky. The last time a joint AFGC-SRM meeting was held (1999), it was a resounding success.

The Ulster Grassland Society

J. Morrison

Secretary, Ulster Grassland Society, College of Agriculture, Food and Rural Enterprise, Greenmount Campus, Antrim BT41 4PU, Northern Ireland, Email: james.morrison@dardni.gov.uk

Keywords: Northern Ireland, grassland, technology transfer

Background   Northern Ireland is a grass-growing region. Grassland accounts for 92% of the area farmed (excluding hill land). So the great majority of farm businesses in the Province are, at least partly, dependent on grass as a source of ruminant feed and grassland farming makes a major impact on the environment and landscape of the Province.

The Ulster Grassland Society was formed in response to a realisation that grass should be treated as a crop in its own right with potential to produce high output in higher rainfall areas which include Northern Ireland. Although there had been informal meetings held since the 1940s on grassland improvement, the inaugural meeting of the Society was held in 1960. Membership was drawn from farming, extension, industry and farmers' organisations, education, research and administration, and this continues to reflect the composition of the current membership. At its inception, the UGS became affiliated to the British Grassland Society (BGS) and has grown to become the largest grassland society affiliated to the BGS with about 600 members.

Aims   The aims of the UGS are, in summary, to encourage development and adoption of new ideas and practices related to efficient and effective use of grassland and to provide opportunities for information on these to be exchanged and discussed. These are met mainly by the Society's core activities i.e. the Annual Conference, visits (often incorporating a mini conference) and competitions. However the Society's other activities, outlined below, contribute to enhancing these aims.

Conference and Visits   The annual one-day Conference provides an opportunity for eminent national and international speakers to address the membership on up-to-date developments in grassland farming and is a forum for research workers to discuss their current work. The Summer meeting, in contrast, involves visits to farms, commercial units or research establishments to see at first hand specific examples of development or adoption of the latest technology into grassland systems. The Autumn meeting has evolved into a farm walk in the morning followed by a mini-conference after lunch. The subjects covered in the mini-conference, comprising short papers presented by specialists followed by a discussion panel, mainly embrace topical issues and are often relevant to the theme of the morning visit. The summation of these activities ensures that the members are exposed to the latest developments in grassland research and technology.

Competitions   The UGS runs a Grassland Farmer of the Year Competition which comprises three components i.e. Grazing, Silage and Environmental Practice and a Maize and Alternative Forages Competition. The UGS Grassland Farmer of the Year competes with the Fermanagh Grassland Club winner and the overall winner in the Province then enters the National BGS competition. Evaluation of Environmental Practice has been introduced in response to the increasing importance placed on grassland farming in environmental conservation and enhancement in the Province in recent years. Participation in the competitions and the constructive comments of judges, including their presentations to the membership at the Annual Conference, help raise standards of grassland farming and are an effective means of transferring technology.

Other activities   The UGS is effective in mustering help in times of crisis for grassland farmers. For example in years when winter feed has been scarce the Society has organised special emergency meetings or has publicised and supported those organised by Government and commercial firms. The Society has responded to government enquiries and proposals relevant to grassland farming e.g. to a recent governmental review of agricultural and food education and R & D in the Province. A student membership scheme encourages membership of younger grassland enthusiasts, offering free UGS membership during study, sponsored attendance at meetings and travel bursaries.

Conclusions   Ever tightening environmental constraints, unstable product prices, and declining government support (both economic and technical) are exerting unprecedented pressure for change on grassland farming. An organisation such as the UGS, which is free from commercial vested interest, has a unique role to play in helping the grassland farmer and the specialist interact to ensure that grassland in the Province continues to be the principal raw product for the agricultural industry and to increase its role as a resource to be managed for the protection of the environment.

The role of grassland societies in the west of Scotland

G.E.D. Tiley
Scottish Agricultural College (SAC), Auchincruive, Ayr KA6 5HW, UK, Email: Karen.crighton@sac.co.uk

Keywords: Grassland Society, livestock, technology transfer

History   Local grassland societies were first established in the United Kingdom in the 1950s, under the guiding hand of the national, British Grassland Society, itself founded in 1946. In the west of Scotland two local societies were formed: 1) the South West Scotland Grassland Society in 1962, covering the former counties of Ayrshire, Dumfries, Kirkcudbright and Wigtown; and 2) the Central Scotland Grassland Society for Lanark, Stirling, Renfrew, Dumbarton and Clackmannan. The declared aim of these two societies was to promote good grassland farming in all its aspects amongst members and to identify opportunities for improved grassland management, all to the benefit of agriculture and the public good.

Operation   Membership is drawn from farmers, advisers, researchers, staff in commercial firms and any others with an interest in grassland. Society affairs are run by an Executive Committee, mainly of farmers with a farmer as chairman. From the outset there has been a close working relationship with the Scottish Agricultural College (SAC) and Hannah Research Institute, both of which have provided organisational help.

Activities   In common with other local societies, the main activities are: organised visits to grassland farms, where good grass and livestock management can be seen, including new developments; winter evening meetings at which leading exponents are invited to speak on topical subjects; organisation of competitions for members, including silage, hay, sward, innovations and environmental competitions; circulation of grassland literature from the national society, newsletters and commercial leaflets. The Central and South West Scotland Societies publish their own journal, 'Greensward', which is currently at its 46th issue (Greensward, 1962-2004). The societies sponsor prizes for excellence to students of grassland and farming in the local SAC college at Auchincruive, Ayr, together with the provision of travel assistance for young family members within the Society.

Achievements   The most important achievement has always been the provision of a forum for informal discussion of common problems or new techniques and concepts with fellow enthusiasts for grassland and its products. The west of Scotland is one of the most favoured areas in Europe for the production of grass and its livestock derivatives. A long tradition in the west of Scotland has resulted in many farms with dairy, beef and sheep stock managed to very high standards. The grassland societies have sought to promote this enthusiasm and to foster a spirit of competition and improvement with the ultimate aim of sustained profitability. Liaison with and support from the agricultural colleges, research organisations and commercial firms have greatly strengthened these aims.

Awards   An example of popular activity by grassland societies is the annual silage competition. In South West Scotland this has been held annually since 1974. The competition involves on-farm judging of 8-12 silages with the best chemical analysis. The nominated Judge is usually a prominent dairy farmer who can accurately assess production and utilisation characteristics. The local Society winner then enters a Scottish national event along with the winners from other Scottish societies, and the winner of this competition finally competes at a UK national level competition. South West Scotland winners have won the Scottish event 12 times and the UK competition twice, achieving runner-up position on two further occasions. It also reached runner-up position in the recently introduced UK National Grassland Management competition. Judging of silages has become progressively more sophisticated with the advancing knowledge of animal nutrition and developments of modern analytical techniques.

Summary and conclusion   Grassland societies play an important role in technology transfer between Research and Development and practical commercial farming. Group meetings of enthusiastic farmers with common interests can demonstrate and promote the adoption of new ideas, equipment or systems. Lack of vested interest and the impartiality of the societies albeit under scrutiny of practical economic reality, have served to accelerate their intermediary, liaison role. With increasingly rapid changes in the agricultural scene and a decline in publicly supported R & D farms, local grassland societies will have an increasingly important role in future technology transfer.

References
Greensward (1962 –2004). *Journal of the South West and Central Scotland Grassland Societies*, Nos. 1-46.

The Leucaena Network: a grazier advocacy organisation ensuring the future of a valuable forage resource for northern Australia

K. McLaughlin[1], B.F. Mullen[2] and H.M. Shelton[3]

[1]The Leucaena Network, 9Palm Crt, Yeppoon, 4703 Australia, Email: maxtrax@cyberinternet.com.au, [2]School of Animal Studies, The University of Queensland, 4072 Australia, [3]Faculty of Natural resources, Agriculture and Veterinary Science, The University of Queensland, Brisbane Australia 4072

Keywords: leucaena, community organisation, participatory development

Introduction *Leucaena leucocephala* is a productive and sustainable forage tree legume for beef cattle production in northern Australia. Following a protracted period of research and development to overcome agronomic and social constraints, substantial adoption by graziers is now occurring (Mullen, these proceedings). However, a recent challenge has threatened future development, viz. the perception by some environmentalists that leucaena is an environmental weed. In addition, production-oriented support from public research and development organisations has diminished significantly over the past 20 years. The Leucaena Network (TLN) was formed in July 2000, primarily to counter the anti-leucaena movement, but has since developed as an advocacy organisation promoting the many beneficial aspects of leucaena forage systems. The structure, aims and achievements of this unique organisation are outlined below.

The Leucaena Network This is a community organisation comprised primarily of cattle graziers, but also of research and development staff from government agencies and universities. The charter of TLN is to, "*Promote the responsible development of leucaena for productive and sustainable grazing and agroforestry systems to build stronger rural communities*". The TLN aspires to being proactive, rather than reactive in response to opportunities and industry needs. The formal structure of TLN is typical, with an executive committee elected at an annual general meeting and committee meetings held every two months. Operational expenses are met from income derived from membership fees and from industry support. There are currently 100 members.

Addressing the weed issue The leucaena-grass system in northern Australia is based on *Leucaena leucocephala* subsp. *glabrata* (forage leucaena). Unfortunately, the closely related subspecies, *L. leucocephala* subsp. *leucocephala* (weedy leucaena), is a minor environmental weed in northern Australia. Both subspecies have weed potential in ungrazed, coastal environments. Consequently, environmental lobby groups have pressured government to curtail the planting of forage leucaena by graziers. The TLN has addressed this issue through both advocacy and action: 1) A Code of Practice for the responsible use of leucaena by graziers was developed. The Code lists measures to minimise the production and spread of leucaena seed in grazed systems, and outlines measures to eradicate any seedlings that recruit beyond the paddock boundary. 2) TLN engaged with the Environmental Protection Agency, the Department of Natural Resources and Mining, and the Department of Primary Industries and Fisheries (DPIF), to develop a policy proscribing the use of leucaena. Following substantial input from TLN, the policy supports the responsible use of leucaena as a grazing resource, but highlights the weed potential of ungrazed, or irresponsibly grazed leucaena. 3) TLN was successful in attracting funding for a weed eradication programme targeting highly visible infestations of weedy leucaena. In most instances these are infestations of the naturalised weedy leucaena and are not the result of grazier plantings. The proactive approach of TLN to the weed issue averted the unnecessary banning of leucaena for future planting.

Research collaboration The network has determined research priorities to address production limitations and environmental issues. It currently coordinates two Meat and Livestock Australia-funded projects with the University of Queensland worth over AU$600,000: 1) Development of a hybrid cultivar resistant to psyllid insects through a recurrent selection program (Dalzell *et al.*, these Proceedings a); and 2) An investigation of the causes and extent of mimosine toxicity in Queensland (Dalzell., these Proceedings b). Research proposals have been submitted to various funding bodies to: 1) Promote leucaena for salinity mitigation in Queensland; 2) Quantify reductions in greenhouse gas (methane) emissions from cattle grazing leucaena pastures; 3) Study the carcass quality of cattle grazing leucaena-grass pastures; and 4) Develop sterile forage and timber cultivars.

Grazier training courses In collaboration with The University of Queensland and DPIF, TLN has developed a highly acclaimed, grazier-oriented training programme entitled "Leucaena for Profit and Sustainability". This two-day course is conducted on-farm in leucaena production areas. Practical lectures on agronomy, plant nutrition, grazing management, mimosine management and economics are followed by farm walks and group discussion sessions. Five courses were conducted in 2004 and were attended by over 100 graziers in total.

The future To ensure the future use of leucaena as a valuable forage resource, TLN aims to pre-empt and address ongoing concerns of weed potential, expand its grazier membership and continue to prioritise research issues and attract appropriate funding. By successfully addressing these issues TLN will continue to prosper.

The *Lotus* Newsletter: an electronic *Lotus* research community

M. Rebuffo

INIA, National Institute of Agricultural Research, La Estanzuela, Colonia, Uruguay, Email: rebuffo@inia.org.uy

Keywords: Lotus Newsletter, Lotus, research

Introduction The *Lotus* Newsletter (*LN*) was created by Dr. W.F. Grant (former editor) in 1971, integrating scientists working on diverse aspects of research on *Lotus* spp. The *LN* received further impetus under the editorship of Dr. P.R. Beuselinck, who shaped the electronic version (http://www.psu.missouri.edu/lnl/). The newsletter aims to provide a vehicle for the exchange of information where opinions, in addition to established facts, can be presented. One of the strengths of the *LN* has been the lead article for each issue and the extensive bibliographic listing published every year, which facilitates access to recent literature. As a medium to let everyone know what research is being carried on in different parts of the world, it helped to prevent overlap of projects and to form research alliances.

Broadleaf birdsfoot trefoil (*Lotus corniculatus* L.), narrowleaf trefoil (*L. glaber* Mill.), and big trefoil (*L. uliginosus* Schkuhr.) are the most common commercial species of *Lotus*, but there are more than 150 other species. Within *LN* volumes the first indications of a biotechnology revolution are recorded. This eventually identified *L. japonicus* as a plant model for genetic research on symbiosis and physiological processes. The *LN* had pulled together information on all aspects of *Lotus* spp. research and provided a wide view of the *Lotus* community. The *Lotus* community through the Newsletter could play a key role in exploiting the close relationships between the model species and forage *Lotus* species, thus assisting the development of the latter.

Objectives The *LN* seeks to extend researchers information world wide. The consolidation of the information about research activities, the development of a readily available and easy to use comprehensive media for researcher profiles, the publication of annual issues and the development of a virtual library are among the main objectives. The target audience is the research community, to ensure the linkage between molecular and applied research.

The research community The highly diverse nature of the research community involved with *Lotus* shows both the strength of the Newsletter and the challenge to it. Some 166 researchers from 28 countries working on 31 species of the genus *Lotus* and related genera (eg. *Podolotus*, *Acmispon*) participate in *LN*. The distribution of researchers within areas and species reflects the structure of the *Lotus* community (Table 1). The number of researchers is distributed evenly between model *Lotus* and the main cultivated species (*L. corniculatus*). Whereas research on model *Lotus* is spread across 23 countries, research on cultivated species is concentrated in their main areas of utilisation. U.S.A. represents the main research group on *L. corniculatus* (26%), while Argentina leads the research group on *L. glaber* (61%) and almost half of *L.uliginosus* researchers come from Uruguay. At the other end of the spectrum, one or two scientists develop research on 24 miscellaneous species.

Table 1 Number of countries and researchers within each *Lotus* species

Species	Countries	Researchers
Model		
L.japonicus	23	73
Cultivated		
L.corniculatus	17	77
L.glaber	8	36
L.uliginosus	10	29
L.subbiflorus	2	6

Progress and perspective Researchers provided information on main activities, species, and subjects of research for the web page. Its development has progressed thought several interactions based upon members recommendations. The current main page is at the following URL: http://www.inia.org.uy/sitios/lnl/index.html Static information currently limits the web effectiveness in serving multiple users. An essential step towards a dynamic newsletter is the revised information. The thorough documentation of researchers' activities, together with the monthly updating of researchers' profiles, considerably enhanced their subsequent use, enabling the rapid location of potential information/partners. Links to various genetic resources for genetic and genomic research in *L. japonicus* developed in Europe and Japan, as well as forage and legume web pages, avoid useless duplications. The counter inclusion and the email relay list "Lotus Network" helped monitor the utilisation of *LN*. Assembling currently existing materials will be the future task, in parallel with developing the overall system design. A virtual library will pool together the information available at *LN*, complementing the traditional information available and minimising the time necessary to find information or partners.

Grassland technology interaction and policy evolution in Canada

D. McCartney[1] and P. Jefferson[2]
[1]Agriculture and Agri Food Canada Research Centre 6000 C and E Trail Lacombe Alberta T4L1W1, Canada, Email: mccartneyd@agr.gc.ca, [2]Agriculture and Agri Food Canada Research Centre , POB. Box 1030 Swift Current Sask. S9H 3X2, Canada

Keywords: www.Foragebeef.ca, grazing management, grazing schools

Introduction Canada is the world's second largest country covering approximately 10 million km^2 (McCartney & Horton, 1997). About 90% of Canada is uninhabited with 90 percent of Canadians living within 500 km of the American border. The forage resources used by livestock grazing and the production of forage crops covers over 36 million ha of Canada's land base (3.6%) and is divided into 72% native range (26 million ha), 11% cultivated pastures (4 million ha) and 17% forage crops (6 million ha) There are 25 million ha in grain and oilseed crops (McCartney & Horton, 1997).

Forage/ Grassland Organisations Canada has several organisations that promote research and extension education on environmentally sustainable grassland management. Many grassland managers/farmers participate in grassland management educational programmes, but there is no national producer-driven organisation that represents the interests of the grazing industry to government policy makers.

The Canadian grassland/ rangeland scientific community comprising of about fifty University Professors and Agriculture and Agri Food Canada Research Scientists are either active members of the Canadian Society of Animal Science, the Canadian Society of Agronomy, or the Society of Range Management. In some provinces, farmers and ranchers participate in the activities of Provincial Forage Councils or local Forage Associations. Their activities include sponsoring seminars, grazing schools and provincial grazing conferences usually one to two days in length, news letters, field days and demonstration projects of local interest to cattle grazers and forage producers. Across Canada, there would be approximately 1000 to 1500 producers involved in these activities. On a national level the Expert Committee on Forage Crops, which is a sub committee of Canadian Agri-Food Research Council, helps co ordinate and provide input into research and extension activities and has limited input into federal policy and research priorities of the grassland industry.

Technology interaction Canadian research and extension personal have developed www.Foragebeef.ca a web site that summarises forage and beef research applicable to Canada. A large emphasis is placed on grazing management. The Western Forage Beef Group, an amalgamation of federal research scientists and provincial forage and beef extension agronomists at Agriculture and Agri Food Research Centre Lacombe Alberta, concentrate on research and extension to extend the grazing season and promote grazing management on seeded pastures. Many provincial governments promote environmentally sustainable grazing management systems through producer meetings, extension publications and provincial web site and telephone information centres.

Influence on policy and public perception The Canadian Cattlemen's Association is the largest producer organisation that presents policy concerns to Provincial and Federal governments. The Expert Committee on Forage Crops and National Forage/ Beef Research Reviews of federal research programmes at Agriculture and Agri Food Canada Research Centres has input into future research needs relating to grassland management. The Prairie Farm Rehabilitation Administration is a federal organisation responsible for the delivery of federal programmes that promote environmentally sustainable grassland management practices. Ducks Unlimited and Nature Conservatory are North America donor-sponsored organisations that maintain grazing lands and promote grazing management strategies to the public as well as the cattle producer. Canadians have an appreciation for grasslands, but there is no united producer voice or funding agency that promotes grassland research and grassland education at the University level. This is critical for the future of grassland research and extension of these new technologies in the future.

Reference

McCartney, D. & R. Horton (1997). Canada's Forage Resources. XVIII International Grassland Congress, Winnipeg/Saskatoon, Canada.

The great mission of the Chinese Grassland Society: grassland strategic research

F. Hong

The Chinese Grassland Society, Yuanmingyuan Xilu 2, Haidian District, Beijing 100094,China, Email:
dengbo67@hotmail.com

Keywords: Chinese Grassland Society, strategic research, sustainable development, grassland

Introduction to the Society The Chinese Grassland Society (CGS) was founded in 1979 in Beijing. Since 1991, it has become a national science society authorised by the State Ministry of Civil Affairs. The basic task of CGS is to unite the national grassland scientists and technicians for academic exchanges at home and abroad, to popularise grassland science, to train qualified personnel for the further development of grassland science and to accelerate the development of grassland science. The Society is affiliated with ten professional committees and sponsors the academic journals: Acta Agrestia Sinica ,Acta Prataculturae Sinica and China Grassland.

The activities of the Society The CGS and its branches have been charged by their of annual meetings to improve technology interaction by involving some non-members (such as enterprisers and farmers) and achieving their participation. International conferences or forums are convened from time to time. The CGS played a part in the foundation last year in Lanzhou, northwest China, of the Sino-America Grassland Research Institute for technology interaction. Activities to influence policy and public perception are always being carried out, especially to stress the importance of protecting the grassland environment. The CGS also proposed that the Chinese government should carry out the following measures to protect the grassland : (1) adopt planting structure on the three-dimension; (2) reuse farmland for planting forages; (3) cease to graze on pastures that are to be renewed; (4) extend nature reserves in grassland regions; (5) set up professional departments in governments to enforce the grassland law.

Plans for future The main task of the CGS from now onwards is to increase strategic research on Chinese grassland sustainable development with the aim of confirming the function and position of grassland for achieving historic change in the development of the nation. The present status of Chinese grassland and animal husbandry are as follows: (1) there is an imbalance between investment and output; (2) management is outdated and the market is poor; (3) the environment is degrading, and the ability to resist calamities is weakening; (4) the ecological flexibility is declining and the frequency of natural calamities is increasing; (5) the grassland resource is in crisis; (6) there is severe desertification of grassland. Unfortunately research on grassland ranks very low in the popularity of science and technology (Yutang Li, 2001; Shouxiang Shi and Fengjuan Li, 2004).

Priorities for research We should do our best to build the frame for grassland development strategy and provide the scientific bases for macro-policy making. The research will involve: (1) the grassland industry's historic evolvement and modern civilisation; (2) theoretical research and its move to practical progression; (3) the strategic status and functions of grassland; (4) some important strategic problems in the grassland; (5) the study of the grassland systems and grassland development.

References

Shouxiang Shi & Fengjuan Li (2004).Implementation of a basic grassland protection system to promote sustainable development of the grassland industry. *Pratacultural Science*, 21, No. 3, 49-52.

Yutang Li (2001). The exploitation of grassland resources and strategies for China sustainable development in the future. *Grassland of China*, 23 No.3, 64-66.

Change in grassland science: implications for training, research and Grassland Societies

G. Lemaire[1], R.J. Wilkins[2] and J. Hodgson[3]

[1]INRA, Unite d'Ecophysiologie des Plantes Fourragères, 86600, Lusignan, France, Email: lemaire@lusignan.inra.fr, [2]Institute of Grassland and Environmental Research, North Wyke, Okehampton EX20 2SB, England, [3]Institute of Natural Resources, Massey University, Palmerston North, New Zealand

Keywords: grassland, multi-functional, research, training, organisations

Background In most of the world the priority for production-oriented research has been succeeded by the need for grassland research to focus on systems which satisfy requirements relating to the stability and protection of land, water and atmospheric resources and to biodiversity, in addition to production efficiency. This dictates not only a new approach to research, but also new approaches for the organisation of research, the training and development of research scientists and the activities of Grassland Societies and associated organisations.

Research approaches The concept of multi-functionality provides a new framework for all disciplines in agricultural research. Scientific objectives, methods of investigation and models have to be reconsidered to give an integrative approach at a range of scales where different functions can be evaluated. The multiple functions of grassland demand a genuinely inter-disciplinary approach to research. To achieve such objectives it is necessary to produce integrated knowledge, new concepts and new tools at the different levels of organisation of grassland agro-ecosystems: (i) the field plot, where the basic biogeochemical processes are acting, (ii) the farming system, where coherent management procedures are combined, (iii) the landscape, where multi-functionality, interaction between different land uses and overall impact can be evaluated and (iv) the region, where socio-economic and political factors become important. We see a requirement for networks of long-term experiments with different grassland ecosystems and contrasting managements. The evolution of vegetation, soil, populations of organisms, biogeochemical cycles and environmental fluxes need to be assessed and the information developed through process-based models. The research needs to be progressed together with associated socio-economic studies.

Implications for research organisation and funding Programmes on integrated land use often span the responsibilities of different Government Departments, Research Councils and Institutes, making it a major challenge to achieve the timely initiation of appropriate programmes. One would expect there to be benefits from a research structure involving a single body responsible for scientific research, rather than strong sectoral research councils, as in the UK, but we do not see evidence that single bodies have made more rapid progress in this area. Good communication between the different funding bodies and scientists in the different disciplines is of crucial importance. A good model may be the recently established Rural Economy and Land Use programme in the UK (www.escr.ac.uk/relu/). This has committed funding from three Research Councils and two Government Departments and its own Programme Director. Funds are available both for capacity building and for inter-disciplinary research projects. The funding is, however, for projects of up to four years duration, whereas some of the research required needs secure funding for at least ten years.

Implications for education and training The requirement for inter-disciplinary programmes comes at a time when most of the recent entrants into agricultural and land use research receive their initial training in one of the specialised physical or biological sciences and may have little appreciation of other disciplines. We need to consider how they can be developed to fully participate in and eventually lead wide inter-disciplinary projects. Should initial training emphasise 'content and awareness', rather than 'depth and detail'? There are requirements for higher degree students to carry out projects in inter-disciplinary teams and for young researchers to move out of their disciplinary comfort zones to participate in more broadly-based projects. Progress needs to be made in giving better recognition of the contributions of individual scientists to collaborative programmes.

Implications for Grassland Societies and associated organisations There are responsibilities in this changed situation not only for research funding bodies, but also for professional organisations. National Grassland Societies should develop activities that are appropriate to the new requirements for multi-functional grassland. They need to embrace scientists and practitioners from areas beyond their traditional base and form alliances and joint activities with bodies centrally involved in health, ecology and socio-economics. Most importantly they should provide opportunities for contact and discussion between scientists working at different levels and in different disciplines. Inter-continental and continental bodies, such as the International Grassland Congress, the International Rangeland Congress and the European Grassland Federation could play key roles through (i) the provision of a forum for the early exchange of views of scientists, (ii) the evolution of research networks and network experiments and (iii) assistance with their promotion and the subsequent dissemination of research results. These organisations have a particularly important role in fostering positive attitudes to the value and feasibility of inter-disciplinary research.

Citizen scientists: efforts by the Tri-Societies to inject science into US policymaking

E. Bergfeld and K. Glasener

American Society of Agronomy, Crop Science Society of America, Soil Science Society of America, 677 South Segoe Rd, Madison WI, 53711, USA, Email: ebergfeld@agronomy.org

Keywords: science policy, grassroots advocacy, political action committee (PAC)

Description of the societies The Tri-Societies (American Society of Agronomy, Crop Science Society of America and Soil Science Society of America), the largest life science professional societies in the United States (US) dedicated to the agronomic, crop and soil sciences, encompass approximately 18,000 members globally and include approximately 14,000 certified professionals who are a direct conduit to the farming communities across the US. The Tri-Societies' US-based federal science policy programmes have been active for decades. There is a Washington, DC-based Director of Science Policy, Congressional Science Fellows (CSF) and Science Policy Interns.

Highlights of actions to improve technology interaction Science policy activities encompass education and advocacy. Tools employed include media relations workshops and special sessions on how to communicate with policymakers, the development of Rapid Response Team electronic listserves and an internet-based Knowledge Database which together allow for rapid access to scientific expertise. Most recently, the Societies implemented an online grassroots advocacy programme that specifically utilises enhanced technology for rapid communication between targeted segments of our membership and their elected representatives. The Societies continue to survey the membership and certified professionals regarding issues of public importance. In addition, regular communications occur with all members via a monthly hard copy and online newsletter and a biweekly electronic newsletter.

Activities to influence policy/public perception In 1986, the Tri-Societies first became involved in the CSF programme. A total of 26 Fellows, supported by the Tri-Societies, have had the opportunity to work on science policy issues within the US Congress. Thirteen, or 50%, of the Tri-Societies' CSFs have remained very engaged in science policy in Washington, D.C., working in Congress as professional staff, with one of the federal agencies or as professional lobbyists. Semi-annually, society leadership and members travel to Washington, DC, to take part in congressional visits day (CVD) activities during which they are given a "hands on" orientation in how to interact with policy makers and meet with scientists from other disciplines. During CVD these scientists are afforded the opportunity to pay visits to their Congressional delegations at which they urge support for the agricultural sciences. In 2004, in order to tap into the more than 18,000 members and 14,000 certified professionals, the Societies developed the Science Policy Action Alert, an electronic grassroots advocacy resource that provides easy and quick online access for our members to contact their members of Congress about issues of critical importance to the societies. We continue to evaluate the effectiveness of this service.

Plans for future initiatives The Tri-Societies plan to identify and train a core group of citizen scientists, in the districts and states of key members of Congress, who can and will communicate rapidly and efficiently with policy makers and thus affect policy and funding for agricultural sciences. We are also considering development of a series of Congressional educational briefings at key districts/states and at the national level in order to ensure that legislation is based on sound science. The Tri-Societies and certified professionals continue to consider the potential of a Political Action Committee (PAC). PACs are utilised to raise funds, which are then used to support select campaigns and hence gain access to key members of Congress. Ideally, with time, the Tri-societies will develop a Congressional champion for the agricultural sciences who can promote their interests.

Ultimately, scientific societies need to highlight the priority of communicating science to the public as well as to policy makers. Scientists, as individuals, are not traditionally well-known for their ability to communicate and market the benefits of their research to lay audiences. It is critical for scientific societies to engage on issues of importance to the grassland community, as well as the broader scientific community and general public in order to be recognised and heard. Societies must train members to be effective communicators, as well as to understand the value of such communications, in order for members to continue to be active and effective participants at the grassroots level of advocacy.

Author index

Browne, B.A.	692	Carton, O.T.	577, 720, 728
Brownie, C.	869	Carvalho, D.D.	204
Brůček, P.	567	Carvalho, P.C.F.	388
Brummer, E.C.	63, 643	Casasús, I.	538, 896
Bryant, D.	245	Caselli, A.G.	156
Bryant, R.B.	895	Casler, M.D.	824
Büchmann, B.N.B.	258	Castelán, O.A.	236, 287
Buchmann, N.	647	Castillo-Gallegos, E.	345, 346
Buckley, K.E.	182	Castro, E.	426
Buldgen, A.	259	Castro, J.	261
Buluveze, A.	288	Castro, M.	280
Bumane, S.R.	716	Castro, P.	190, 261
Burba, G.G.	587	Cates, R.L.	604
Burggraaf, V.T.	242, 243	Catrileo, A.	431
Burke, J.	572, 591, 595	Catto, W.D.	723, 797
Burke, J.L.	229, 230	Cavalcante Jr., F.N.	91
Burke, J.M.	223	Cellier, P.	564
Burnett, D.J.	544	Cerri, C.C.	372, 435, 863
Bush, L.P.	308, 309	Cerri, C.E.P.	435, 863
Buske, F.	267	Cervantes-Martinez, T.	108
Busso, C.A.	214	Cervantes-Santana, T.	108
Butkute, R.	726	Ceschia, E.	564
Butler, A.M.	756	Chabbi, A.	765
Butler, T.J.	407	Chadwick, D.	575
Buyer, J.	736	Chaichi, M.R.	612, 706, 891
Byrne, C.M.	597	Chaieb, M.	678
Byrne, P.	690	Chakraborty, S.	320, 322, 330
Cabanas, J.	392	Chapman, D.F.	135, 460, 461, 462, 463, 568, 709
Cabaraux, J.F.	721	Chardon, X.	881
Cabiddu, A.	540	Charmley, E.	493
Caddel, J.L.	788	Charrier, X.	679, 765
Cagaš, B.	393, 946	Chazelas, L.	517
Cai, K.	826	Chen, W.	398
Cai, Y.	481, 792, 944	Chen, Z.Z.	533, 886
Calanca, P.	552, 559	Chennaoui, H.	93
Calder, M.W.	937	Cherney, D.J.R.	403, 474, 816
Calixto, S.	109	Cherney, J.H.	403, 474, 816
Callan, J.J.	482, 483	Chiaramello, Y.R.	501
Callow, M.N.	682	Chilcott, C.R.	793
Calub, A.D.	583	Chiuri, L.W.	806
Camacaro, S.	707, 743	Choisis, J.P.	808
Camarão, A.P.	343	Cholaky, C.	768
Cameron, A.G.	326	Chu, A.	826
Cameron, F.	408	Chudy, A.	172
Camesasca, M.	537	Chung, E.S.	97, 98, 281, 478
Campbell, C.	564	Chunsheng, B.	473
Cangiano, C.A.	490, 491	Churches, K.R.	900
Caplis, J.	174, 191, 192	Ciais, P.	564
Carballo, A.	414	Cinel Filho, P.	387
Cárdenas, E.	426, 704	Cino, D.M.	373
Cardenas, L.	575	Cissé, M.	719
Cardoso, F.F.	505	Clapham, W.M.	464
Caredda, S.	762	Clark, B.	408
Carita, T.	935	Clark, D.A.	307
Carlassare, M.	277	Clark, H.	582
Carlier, L.	430	Clark, S.G.	664
Carneiro, J.P.	575	Clary, G.M.	796, 848
Carneiro, L.M.T.A.	62	Clavero, T.	368
Carré, S.	645	Cleere, J.J.	796
Carrère, P.	644, 648	Clément, C.	874

Clements, C.D.	409, 410	Cruz, A.M.	95
Clements, R.J.	344	Cruz, L.O.	123
Clewett, J.F.	557	Cruz, P.	209, 266, 517, 645, 860, 865, 866, 873
Clifton-Brown, J.	564, 572, 591, 595	Cuadro, R.	170, 171, 536
Cliquet, J.B.	739	Čuhel, J.	567
Cobon, D.H.	556, 557	Cuitiño, M.J.	61
Cogan, N.O.I.	80	Culleton, N.	188
Coleman, K.	863	Cunderlik, J.	433
Coleman, S.W.	260, 451	Cunningham, D.	482
Colgan, S.L.	175	Cuomo, G.J.	820
Collins, J.	890	Ćupina, B.	89, 269, 701
Collins, R.P.	78, 245, 424, 425, 610	Curnow, M.	818, 884
Collins, S.L.	549	Czóbel, S.	564
Colmer, T.D.	696	Da Silva, S.C.	439, 520, 521, 527
Colombini, S.	264	Daget, P.	631
Colombo, C.A.	88	Dahlan, I.	894
Colozza, M.T.	91	Daily, H.G.	911
Comajig, G.M.	583	Daly, K.	690
Common, T.G.	630	Dalzell, S.A.	92, 333, 544
Conaghan, P.	273	Damasceno, J.C.	497
Concha, A.	247	Damiani, F.	244
Condon, F.	61	Dannhauser, C.S.	466, 467
Confalonieri, M.	264	Dardenne, P.	258, 259, 874
Connell, J.G.	928	das G. Morais, M.	233
Connolly, J.	424, 425, 553, 555, 570, 603, 609, 610, 728	Daugeliene, N.	726
		Davenport, T.	671
Contato, E.D.	528	Davies, J.	395
Convertini, G.	447	Davies, W.P.	432
Conway, M.J.	331	Davis, P.A.	591
Cook, B.G.	348	de A. Rangel, J.H.	340
Cook, R.	74	De Barbieri, I.	170, 171, 535, 536
Cook, S.	917	de Bello, F.	620
Cooper, D.P.	495	De Benedetto, M.G.	271
Čop, J.	938	de Camargo, L.G.H.	210
Coppock, D.L.	853, 854, 856	De Crignis, J.C.	730
Cordero, J.	319	de Groot, T.	564
Corner, R.	917	de Haan, N.	318
Corsi, M.	702	de Oliveira, F.C.	208, 770
Cortes, C.	497	de Resende, K.T.	156
Cortes-Díaz, E.	371, 420	de Riek, J.	127
Cosgrove, D.R.	495	de Ruiter, J.M.	386
Cosgrove, G.P.	509	de Sousa, F.B.	123
Cotton, D.C.	889	de Souza, N.A.	210
Coulman, B.E.	114	De Vliegher, A.	430
Coulter, R.L.	587	de Wit, J.	749, 819
Coutinho, J.	717	Dean, D.B.	225
Cowe, I.A.	258	Dear, B.	696
Cox, W.J.	816	Debril, J.	645
Craig, A.D.	149, 696	Decandia, M.	540
Craig, A.M.	308	Decau, M.L.	360
Crespo, G.	742, 747	Decruyenaere, V.	821, 874
Crichton, J.M.	934	Deguchi, K.	515
Crimp, S.	557	del Pozo, P.P.	436
Cropper, J.B.	895, 898, 899	Delaby, L.	164, 905
Crosby, M.	414	Delgado Gòmez, H.J.	438, 676
Crosby, T.F.	482, 483	Demment, M.W.	490, 491, 587, 599
Crosson, P.	901	Deprez, B.	394, 878
Crosthwaite, J.	664	Deregibus, V.A.	634, 868
Crusciol, C.A.C.	435	Deretic, J.	696
Crush, J.R.	124	Desta, S.	856

Faverdin, P.	905	Fraser, M.D.	494, 545
Faville, M.J.	124	Fraser, R.W.	355
Fay, D.	597	Fraser, T.J.	153, 477
Fay, P.A.	549	Fregadolli, F.L.	156
Fearon, A.M.	227	Freitas, D.	156
Febles, G.	373, 444	Freitas, J.C.T.	91
Felicia Díaz, M.	314	French, P.	872
Ferdinandez, Y.S.N.	114	Frenda, A.S.	250, 363, 761
Ferguson, M.	138	Fribourg, H.A.	306
Fernandes, C.D.	330, 339	Friedel, K.	267
Fernandes, F.D.	110	Friesen, A.	569
Fernandes, M.A.M.	187	Frighetto, R.	580, 581
Fernández, M.	314	Fritz, J.O.	909
Fernández-Lorenzo, B.	261	Frohlich, A.	126
Fernández-Núñez, E.	295, 769	Frost, W.E.	900
Ferreira, D.S.	156	Fu, S.M.	708
Ferreira, G.A.	813, 845	Fugui, M.i	119
Ferreira, J.A.	597	Fuhrer, J.	552, 559, 564, 587
Ferreira, M.J.	392	Fukagawa, S.	207
Ferri, D.	447	Fukumoto, M.	541
Ferris, C.P.	165, 785	Funes, F.	373, 444
Fievez, V.	183	Furtini Neto, A.E.	456, 703
Figueroa, M.	247	Fychan, R.	246
Fike, J.	194, 351, 464	Fystro, G.	731
Finn, J.	425	Gaborit, G.	564
Fisher, A.E.	306	Gachene, C.K.K.	341
Fisher, J.C.	801, 842	Gaile, Z.	359
Fisher, J.M.	630	Galli, J.R.	490, 491
Fiske, D.	464	Galvin, K.A.	841, 843, 844
Flanagan, L.B.	587	Gao, A.	534
Flanagan, S.	933	Gao, H.W.	737
Flechard, C.	564	Garbisu, C.	738
Fletcher, W.	556	García, A.	538
Flinn, P.	258	García, G.	414
Flores, G.	261	García, M.	95
Flores, R.	106	García, P.	190
Florine, S.E.	642	Garcia, R.	347
Flynn, B.	482	Garcia, V.	766
Fontaine, S.	648	Garcia-Moya, E.	103
Fontenot, J.P.	464, 725	Garcia-Rodriguez, A.	168, 506
Forbes, B.D.	837	Garcia-Sarrion, R.	425
Forbes, V.E.	544	Gasperin, C.	187
Ford, J.L.	81	Gastal, F.	202, 203, 212, 360, 645
Ford, P.L.	744	Gebru, G.	856
Formoso, B.	250	Genizi, A.	487
Fornaro, F.	713	Genro, T.C.M.	505
Forster, J.W.	80	George, M.R.	693
Fortune, A.	591	Gerbaud, S.	808
Fothergill, M	424, 425, 610	Getachew, G.	810
Fraga, N.	95	Gezahegn, A.	810
Fraga, S.	747	Ghamari Zare, A.	71
Franca, A.	762	Ghamkhar, K.	82
Francioso, M.	126	Gherardi, S.G.	859
Francisco, F.G.	62	Ghesquiere, A.	425
Franco, A.	348	Giacomini, A.A.	204
Franco, L.H.	338, 382	Giambalvo, D.	250, 251, 363, 677, 710, 761
Franco, R.	641	Gibson, D.J.	555
Frank, A.B.	587, 599	Gichaba, C.M.	689, 746
Frankow-Lindberg, B.	425	Gierus, M.	240
Franzel, S.	319	Gilaverte, S.	514

Giles, S.	818	Haby, V.A.	796, 848
Gillespie, R.N.	684	Haerdter, R.	254
Gilliland, T.J.	132, 227	Haferkamp, M.R.	587
Gilmanov, T.G.	587, 599	Hall, C.A.	815, 945
Giltrap, D.L.	566	Hall, J.B.	464
Ginane, C.	504	Hall, M.H.	902
Given, R.	303	Hall, T.J.	542
Glasener, K.	955	Halling, M.A.	131
Glasser, T.	265	Ham, J.S.	478
Glimp, H.A.	214	Hameleers, A.	436, 465
Godoy, R.	87, 249	Han, B.	112, 113
Goldie, J.	817	Han, G.	113, 534, 775
Golińska, B.	285, 452	Han, J.G.	161, 366, 708
Golińska, B.T.	285	Hanada, M.	155
Goliński, P.	285, 425, 452	Hansen, D.K.	158
Goliński, P.K.	285	Hanson, J.	348
Gombosuren, U.	880	Hardy, M.B.	448, 449
Gomes, C.S.	514	Hare, M.D.	323
Gomes, M.F.	387	Harper, J.M.	693
Gomes, R.C.	387	Harris, M.	88, 91
Gomide, C.A.M.	340, 370, 528	Hasniati, D.	324
Gomide, J.A.	370	Hassen, A.	215
Gonçalves, A.C.	520, 521	Hatch, D.	575
Gonda, L.	415, 433	Hatfield, R.D.	220
Gongalsky, K.B.	650, 651	Hattori, I.	207
Gontijo Neto, M.M.	109	Hawkins, J.	575
González, F.	414	Hawkins, M.J.	570
González, S.	501	Hayes, M.H.B	597
González-Arráez, A.	261	Haynes, R.J.	745
González-Montagna, S.J.C.	499, 500	He, Z.	117, 473, 532
González-Muñoz, S.	523	Heard, L.P.	665
González-Rodríguez, A.	906, 913	Hebart, M.	149, 694
Gordon, D.	818	Hedley, C.B.	566
Goslee, S.C.	767, 895	Hejduk, S.	697
Goto, M.	944	Helden, A.J.	605, 652
Gourley, C.J.P.	691, 861, 890	Helgadottir, A.	424, 425, 609
Graham, P.	912	Hellämäki, M.	921
Grant, W.	371	Hellman, P.	247
Granville-Ross, S.	880	Helminen, J.	921
Gray, D.I.	885, 910, 926	Hendershot, R.L.	800
Greef, J.M.	879	Hendrickson, J.R.	434
Green, R.B.	797	Henkin, Z.	621
Gregg, P.	826	Hennessy, D.	872
Griffiths, J.H.	630	Henning, J.C.	309
Griffiths, W.M.	489	Henry, D.	696
Griggs, T.C.	357	Hensen, A.	564
Grigsby, K.N.	158	Henton, S.M.	386
Grimaud, P.	876	Hepp, C.	356
Grof, B.	330, 339	Hermy, M.	766
Grogna, N.	874	Hernández, I.	773
Gunter, S.	148	Hernandez-Garay, A.	103, 337, 523
Guo, P.	826	Hernandez-Mendo, O.	503
Guo, Y.D.	122	Herrera, R.	95, 280, 436
Guodao, L.	322	Herrera-Haro, J.G.	523
Guru, A.	889	Herrmann, A.	554
Gustavsson, A.M.	609	Heshmati, G.A.	519, 660
Gutman, M.	621	Hess, H.D.	573
Gutmane, I.J.	428	Hetta, M.	907
Guy, B.R.	797	Heywood, S.	213
Gyüre, P.	151	Hiernaux, P.H.Y.	391, 507

Higashiyama, Y.	791	Ibarra, A.	619, 738
Higgins, K.F.	607	Ibrahim, M.	596
Hill, J.O.	709, 864	Igarzabal, A.	896
Hintz, R.L.	642	Iglesias, J.M.	773
Hirano, K.	146	Iglesias, M.	536
Hirata, M.	870, 871	Ignjatovic, S.	125, 252
Hirth, J.R.	664	Imdieke, R.	820
Hochberg, H.	942	Ingeme, Y.	664
Hocking Edwards, JE.	149	Ingram, L.J.	590, 736
Hodgkins, C.	179	Ingwersen, B.	127
Hodgkinson, J.J.	659	Inoue, M.	887, 888
Hodgkinson, K.C.	525	Iraira, S.P.	724
Hodgson, J.	826, 954	Ishida, M.	792
Hodkinson, T.R.	126	Ishii, Y.	207
Hoffmann, C.C.	920	Ishikawa, R.	626
Hoffmann, V.	809	Ishimura, H.	571
Hoflich, G.	734	Islam, M.A.	377, 871
Hofmann, M.	625	Isselstein, J.	625, 757, 942
Hofstetter, P.	908	Istasse, L.	721
Hohnwald, S.	343	Itano, S.	627
Hollins, P.D.	862	Ito, M.	440
Holloway, J.W.	137, 189	Iudicello, P.	511, 761
Holmann, F.	327	Iwaasa, A.D.	637
Holmes, B.J.	480	Jackson, R.D.	604
Hong, F.	953	Jacobo, E.J.	634, 868
Hong, K.Y.	714	Jacobs, J.L.	389, 390, 680, 681
Hood, K.	630	Jafari, A.	64, 71
Hoorman, J.	800	Jafari, M.	292, 741
Hopkins, A.	879	Jahufer, M.Z.Z.	81
Hopkins, A.A.	304	Jakesova, H.	73
Horan, B.	136	Jama, A.A.	914, 929
Horne, D.	748, 885, 926	Jamar, D.	821
Hornick, J.L.	721	Jamot, J.	497
Horváth, L.	564	Jančovič, J.	615
Hoskin, S.O.	139	Jančovičová, Ľ.	615
Hossain, B.	666	Jank, L.	109, 110
Hou, X.Y.	836	Janusckiewicz, E.R.	528
Houinato, M.	839	Jarillo-Rodríguez, J.	346
Houtsma, E.	728	Jatkauskas, J.	782
Howie, J.H.	413	Javadi, S.A.	741
Hrabě, F.	618	Jefferson, P.	381, 637, 952
Hsieh, C.H.	714	Jeger, M.J.	355
Hsu, F.H.	714	Jenkins, M.W.	806
Hu, Y.G.	737	Jensen, R.N.	723
Hubbell, D.	148	Jentsch, W.	172
Hughes, S.J.	696	Jeranyama, P.	530
Huhtanen, P.	484, 921	Jianguo, H.	473
Humphreys, J.	231, 688	Jiao, S.	534
Humphreys, L.R.	39	Jillo, A.D.	853
Humphreys, M.O	130	Jinfeng, Y.	118, 119
Humphreys, M.W.	688	Johnson, B.	828
Hunt, D.E.	569	Johnson, C.E.	260
Hussein, H.S.	232	Johnson, D.A.	587, 599
Hutchings, N.J.	918	Johnson, D.G.	820
Hutton, G.	799	Johnson, G.	725
Huyghe, C.	421, 679	Johnson, I.M.	568
Hyde, B.	570, 577	Jokś, W.	121
Hyder, M.	884	Jolaosho, A.O.	396, 624
Hynšt, J.	567	Jones, C.	78
Hyslop, M.G.	153, 477	Jones, D.G.L.	628

Macedo, M.C.M.	106, 109	Maze, C.	670
Macháč, R.	393	Mazeris, F.	258
Machado, W.	743	Mazza, F.	251
Macharia, P.N.	341	McAdam, J.	758
MacLeod, N.D.	659	McCalman, H.	822
Madani, T.	662	McCartney, D.	799, 951
Mader, T.L.	175	McCaskill, M.	694
Magalhães, J.R.	123	McCaughey, W.P.	182
Magcale-Macandog, D.B.	583, 846	McCourt, A.	572
Maina-Gichaba, C.	897	McDonald, C.K.	344, 659
Maiorana, M.	447, 713	McDougald, N.K.	693
Maisi, M.	921	McDowell, L.R.	176, 257
Makenzi, P.M.	781	McFarlane, JD.	149
Malami, B.S.	391, 507	McGee, M.	167
Malekian, A.	661	McGrane, T.	483
Maley, S.	386, 684	McGrath, D.	720
Malmakov, N.	843	McGrath, S.	126
Mandaluniz, N.	168, 506, 546, 896	McIvor, J.G.	659
Mangione, D.A.	842	McKenzie, F.R.	389, 390, 680, 681
Mangope, S.	336, 369	McKenzie, G.	892
Mannetje, L. 't	345	McKeon, G.M.	557
Marggraf, R.	757	McLaren, D.A.	411
Marghali, S.	93	McLaughlin, K.	950
Mari, L.J.	479	McMahon, B.J.	654
Marín-Mejía, B.	346	McMellen, A.B.	658
Marita, J.M.	220	McNabb, W.C.	195
Marley, C.L.	246	McNamara, K.	728
Marrakchi, M.	93	McNeil, D.L.	199
Marriott, C.A.	630	McSweeney, A.M.	855
Marsden, S.	525	McWilliam, E.L.	222
Marshall, A.H.	76, 77, 78, 79, 80, 245	Meagher, L.P.	272
Marshall, A.M.	822	Mebarkia, A.	662
Marshall, D.J.	912	Mederos, A.	535
Martens, S.	131	Mee, J.F.	136
Martha Jr., G.B.	110	Meeske, R.	141, 142, 143, 144
Martin, C.	564, 574	Mehrabi, A.A.	894
Martin, N.P.	258	Meinhardt, L.E.	458
Martin, R.J.	684	Meister, A.	826
Martínez, A.	190	Melland, A.	691, 861, 890
Martínez, M.J.	190	Mello, F.F.C.	435, 863
Martínez, R.O.	280, 459	Melo, G.M.P.	235
Martínez-Garza, A.	371	Menary, W.	690
Martínez-Hernandez, A.	371, 420, 499, 500, 523	Mendarte, S.	619, 738
Martinsson, K.	154, 907	Mendes-Bonato, A.B.	100, 101
Martyn, T.	862	Mendoza, G.	414
Mas, C.	813, 845	Mendoza, M.	108
Mason, W.	694, 825	Menezes, C.	288
Massey, T.	482	Mengli, Z.	118
Masters, D.	286, 694, 759	Merzlaya, G.E.	729
Masuda, Y.	342	Mesdaghi, M.	441, 512, 622, 633
Masumizu, T.	275	Mesquita, E.E.	456, 703
Mata, G.	859	Meul, M.	397, 760
Mathijs, E.	766	Meuriot, F.	212, 360
Matsunaka, T.	571	Meyers, T.P.	587
Matsuoka, S.	515	Michaelson-Yeates, T.P.T.	76, 77, 78, 80
Matthew, C.	203, 885, 926	Michaud, R.	276
Matthews, P.	826	Mickelbart, M.	529
Maudet, P.	181	Mičová, P.	283
Mavromihalis, J.	664	Mielnick, P.C.	587
Mayeux, H.S.	468, 469	Mihailović, V.	89, 90, 269, 270, 701

Mihók, S.	151	Morrison, J.	948
Mijangos, I.	619, 738	Morvan-Bertrand, A.	130, 360
Mikic, A.	89, 90, 269, 270	Moser, L.E.	524, 814
Mikić, V.	70	Mosjidis, J.A.	223
Milera, M.	773	Mosquera-Losada, M.R.	201, 295, 732, 733, 758, 769
Miles, N.	745		
Mileski, G.J.	357	Mould, F.	236, 287
Milford, C.	564	Mousel, E.M.	524
Milić, D.	89, 269, 270, 701	Moyer, J.L.	909
Millar, J.E.	928	Moyo, S.	543
Miller, J.E.	223	Mrfat-Vukelic, S.	638, 639
Miller, J.L.	727	Muchiri, M.	689
Miller, R.K.	189	Muck, R.E.	480
Miller, S.N.	689, 746, 806, 897	Mudavadi, P.O.	313
Milne, E.	863	Mueller, J.P.	869
Milne, R.M.	745	Mugasi, S.	833
Min, D.	530	Mugweni, O.	516
Min, D.H.	600, 789	Mugweni, R.	516
Minchin, F.R.	241	Muhuyi, W.B.	162
Miole, R.N.	583	Muir, J.P.	407
Mishra, C.M.	776	Mullen, B.F.	333, 348, 950
Mislevy, P.	107, 260	Muller, W.J.	525
Misri, B.	751	Mulligan, F.J.	231
Mitchell, F.J.	837	Mullins, G.L.	725
Mitchell, R.	99, 115, 116	Munns, R.	696
Mittelmann, A.	123	Murphy, J.J	231
Mizen, K.A.	74	Murphy, J.P.	164
Mizukami, Y.	122	Murphy, P.	188
Mnif, L.	678	Murphy, W.E.	720
Mocha, M.	190	Murray, B.E.	185
Mohaddes, S.A.	851	Mutinda, M.N.	854
Mohamed, A.	925	Mutshewa, P.	369
Mohammadi Golrang, B.	661	Muys, B.	766
Mohseni Saravi, M.	661, 706	N'Diaye, M.	719
Molan, L.K.	520, 521	N'Dione, C.M.	719
Moles, R.	690	Nagy, G.	151, 893
Molle, G.	540, 795, 879	Nagy, Z.	564
Moloney, A.P.	191, 193	Nájera, A.I.	185
Monahan, F.J.	193	Nakanishi, Y.	146
Monteiro, A.L.G.	187, 514	Nakano, M.	238
Monteiro, F.A.	718	Nakano, Y.	342
Monteiro, Q.I.	683	Nambi-Kasozi, J.	282
Montemurro, F.	713	Narasimhamoorthy, B.	69
Montossi, F.	170, 171, 535, 536, 537	Narayana, S.	751
Moorby, J.M.	494, 545	Narita, H.	791
Moore, A.D.	379, 709, 912	Nascimento Jr., D.	439, 527
Moore, G.A.	695	Nashiki, M.	791
Moore, K.J.	268, 642	Nasri, H.	64
Moot, D.J.	199, 200, 529	Nassiri, M.	705
Mora, C.	314	Nasyrov, M.	587
Moraes, A.	388	Navas, D.	367
Morais, A.R.	456, 703	Neftel, A.	564
Morales, S.	927	Neill, C.	372
Moran, P.	675	Nelson, B.S.	793
Moreira, N.	392, 717	Nelson, C.O.	787
Morel, P.C.H.	915, 916	Nelson, L.R.	129, 796
Morgan, G.	690	Nelson, R.	785
Morgan, J.A.	587	Nemitz, E.	564
Morris, P.	219, 241	Nemoto, H.	792
Morris, S.	826, 885, 926	Nerušil, P.	416, 417

Nesheim, L.	131, 273	
Nesic, Z.	638, 639	
Neufeldt, H.	565	
Nevens, F.	397, 760	
Nguluve, D.W.	288	
Ngutu, M.	811	
Niamir-Fuller, M.	667	
Nianogo, A.J.	152	
Nichol, W.W.	145	
Nichols, P.G.H.	696	
Nicol, A.M.	498	
Nie, Z.N.	135, 408, 460, 518	
Nielsen, A.L.	920	
Niemeläinen, O.	353, 354, 783	
Nigatu, A.	810	
Nikander, H.	921	
Nikkhah, A.	237	
Nishino, N.	473	
Nissinen, O.	353, 354, 609	
Nitzan, R.	489	
Noci, F.	193	
Noel, S.	629	
Nolla, M.	537	
Noonan, D.	728	
Norman, H.C.	286, 759	
Norriss, M.G.	145	
Norrsken-Eriksson, M.	907	
Northup, B.K	469, 468	
Nousiainen, J.	484, 921	
Nowak, B.	826	
Nulik, J.	325	
Nussio, C.M.B.	210	
Nussio, L.G.	479	
Nute, G.E.	193	
Nutt, L.	85	
Nye, L.A.	842	
Nyfeler, D.	425	
Nyholm, L.	921	
Nykänen-Kurki, P.	425	
Nyren, P.	533	
O'Brien, B.	188	
O'Brien, R.	348, 917	
O'Brien, S.M.	828	
O'Connell, K.	231	
O'Connor, G.A.	176	
O'Connor, M.B.	254	
O'Connor, P.	136, 188	
O'Donovan, M.	872	
O'Hara, G.W.	312	
O'Kiely, P.	131, 273, 901	
O'Loughlin, J.	163	
O'Mara, F.P.	164, 174, 191, 192, 231, 273, 576, 578, 579, 901	
O'Neill, B.	167	
O'Reagain, P.J.	526	
O'Toole, P.	720	
Ocaña-Zavaleta, E.	346	
Ochiai, E.	655	
Ochieng, D.	925	
Odanga, J.	689	

Odoardi, M.	264	
O'Donovan, M.	164	
Odstrčilová, V.	416, 417	
Oerlemans, J.F.	617	
Ogawa, M.	481, 792	
Ogawa, Y.	522	
Ohanian, A.	383, 501	
Ohtani, S.	238	
Ojeda, E.	420	
Ojha, G.P.	838	
Okajima, T.	440	
Okamoto, M.	155	
Okano, K.	342	
Okello, S.	140	
Okelo, M.O.	746	
Okike, I.	317	
Okumura, D.	155	
Olanite, J.A.	396, 624	
Oldham, C.	138, 818, 859	
ole Sinkeet, S.N.	162	
Olesen, J.E.	564	
Olfert, O.	558	
Oliván, M.	190	
Oliveira, A.P.	235	
Oliveira, D.E.	520	
Oliveira, M.P.	106	
Oliveira, M.R.G.	683	
Oliveira, R.C.	730	
Oliveira, S.G.	580	
Olson, C.B.	335	
Olsson, V.	231	
Olyott, P.	74, 75, 76, 77	
Omphile, U.J.	470	
Onaindia, M.	619	
Onifade, O.S.	396, 624	
Onyando, J.O.	746	
Opitz von Boberfeld, W.	471, 472, 617	
Øpstad, S.	731	
Oregui, L.M.	168, 506, 896	
O'Riordan, E.G.	850	
Orlovius, K.	715	
Orodho, A.B.	313	
Orr, D.M.	526, 945	
Osborne, B.	591, 675	
Oshibe, A.	278	
Osnaya, G.F.	523	
Osoro, K.	185, 190	
Ostrovsky, V.	489	
Otsuki, K.	492	
Ourry, A.	360	
Owens, M.K.	137	
Owensby, C.E.	587	
Oxley, T.	793	
Pablo, V.	420	
Păcurar, F.	616	
Padilla, C.	280	
Paek, B.H.	194, 351	
Paganoni, B.	138	
Pagán-Riestra, S.	234	
Pagliaricci, H.	383, 501, 768	

Pagliarini, M.S.	100, 101	Petersen, B.M.	918
Palazzolo, S.J.	614	Peterson, B.	787
Paliwal, K.	750	Petit, M.L.	517
Palliser, C.C.	881	Peverill, K.I.	861
Palma, P.M.M.	423	Peyraud, J.L.	905
Pamo, E.T.	835	Phaikaew, C.	323
Pamo, T.E.	296	Phillips, W.A.	468, 469
Panahpour, H.	311	Photakoun, V.	928
Panizzo, L.	704	Piazetta, H.V.L.	514
Paolocci, F.	244	Picasso, V.D.	643
Papachristou, T.G.	939	Piccolo, M.C.	372
Papanastasis, V.	758, 939	Pichard, G.	221, 239
Papas, P.	395	Pickert, J.	715
Pardini, A.	758	Picon-Cochard, C.	648
Paredes, A.	743	Piggin, C.	325
Parente, G.	425	Pilegaard, K.	564, 587
Park, H.S.	194, 262, 263, 351	Piluzza, G.	264, 271
Park, J.N.	556	Pinares-Patiño, C.S.	574
Parker, W.J.	885	Pinna, G.	540
Parra, F.	338	Pinto, J.C.	456, 457, 703
Parsons, A.J.	509	Pires, G.	387
Parsons, M.J.	129	Pires, J.C.	392
Partridge, I.J.	348	Pires, J.M.	184, 392
Passos, L.P.	123, 730	Pittaluga, O.	845
Pataki, I.	90, 269	Pitts, P.G.	824
Paterniani, R.S.	105	Platis, P.D.	939
Pathak, P.S.	320, 774	Plevich, O.	768
Paton, C.J.	793	Pokarzhevskii, A.D.	651
Patterson, A.	564	Poli, C.H.E.C.	187, 387, 514
Patterson, D.C.	165	Pomroy, W.E.	139
Patton, B.	533	Poncet, C.	248
Paul, C.	258	Pontes, L.S.	644
Paustian, K.	863	Poozesh, V.	209
Paziani, S.F.	479	Porqueddu, C.	424, 425
Pedreira, M.P.	580	Possenti, R.A.	96
Pedreira, M.S.	581	Potes, J.	367
Pedroso, A.F.	479	Potes, J.M.	935
Peeters, A.	394, 766, 878	Potter, G.D.	158
Pekour, M.S.	587	Powell, H.G.	76, 77
Pelissari, A.	388	Pozdíšek, J.	283
Pelletier, S.	276	Prache, S.	497
Pendleton, B.K.	364	Prado, C.H.B.A.	210
Pendleton, R.L.	364	Prado, O.R.	187
Pengelly, B.C.	348, 864	Prates, Ê.R.	505
Pereira, O.G.	347	Premalal, G.G.C.	289, 443
Peres, A.M.	184	Premaratne, S.	289, 443
Perevolotsky, A.	265, 488, 621	Preve, F.	537
Pereyra, T.	383, 768	Primavesi, O.	580, 581
Pérez, M.	414	Prine, G.M.	335
Pérez, H.P.	111	Prud'homme, M.P.	130, 360
Pérez-Pérez, J.	103, 523	Purchas, R.W.	915, 916
Peri, P.L.	199	Purvis, G.	605, 652
Perriaux, L.	821	Qiao, G.	775
Perry, I.G.	123	Qingfang, X.	473
Perryman, B.L.	214	Quero-Carrillo, A.R.	103, 337, 523
Persson, S.	907	Quesenberry, K.H.	335, 451
Peters, M.	327, 338, 348, 382, 809, 917	Quigley, P.E.	518
		Quinn, H.L.	823
		Quinn, P.J.	482, 483
		Quirk, M.F.	793

Radotra, S.	751
Radović, J.	70, 252
Rahimzadeh, N.	685
Rains, J.P.	332
Raish, C.	855
Rakocevic, M.	208, 770
Ramadan, A.	632
Ramaswami, S.	915, 916
Ramesh, C.R.	320
Ramirez Avilés, L.	438, 676, 772
Ramirez, L.A.	724
Ramirez-Restrepo, C.A.	222
Ramos, A.K.B.	110
Ramos, N.	536
Ramos-Santana, R.	234
Rancane, S.	72
Rangel, J.H.A.	370
Raschi, A.	564, 587
Rashtian, A.	622
Rath, M.	136, 163
Rathour, R.	84
Ravid, N.	487
Rayas, A.A.	236, 287
Rayburn, E.B.	827
Razz, R.	368
Real, D.	813, 85
Rebuffo, M.	61, 951
Redden, H.	843
Redfearn, D.D.	788
Redmon, L.A.	796, 848
Reece, P.E.	889
Reed, K.F.M.	408
Rees, R.	564
Rees, R.M.	877
Regmi, P.R.	455
Reheul, D.	397, 760
Reid, J.	826, 885, 926
Reid, J.M.	911
Reiling, B.A.	260
Reilly, P.	482
Reis, R.A.	156, 235, 370, 528
Renault, O.	657
Resende, R.M.S.	109
Resendes, H.	392
Rethman, N.F.G.	215
Revello, S.	845
Revilla, R.	538
Reynolds, R.N.	837
Ribaimont, F.	245
Ribaski, J.	208, 770
Ribeiro, J.L.	479
Ribeiro, T.M.D.	187
Richards, C.J.	306
Richardson, R.I.	193
Ridley, A.M.	892
Riedel, J.L.	538, 896
Rigon, J.L.	387
Rigueiro-Rodríguez, A.	201, 295, 733, 758, 769
Rigueiro-Rodríguez, M.L.	732
Rim, Y.W.	97, 98
Rinne, M.	419, 484, 921
Rintoul, A.J.	759
Rischkowsky, B.	343, 391
Risso, D.F.	170, 171, 536, 813
Risso-Pascotto, C.	100, 101
Roberts, A.H.C.	254
Roberts, D.J.	465, 817
Robertson, A.D.	321, 784
Robertson, M.J.	398, 864
Rochette, Y.	574
Rochon, J.J.	879
Rodehutscord, M.	496
Rodgers, W.A.	667, 668, 669, 670, 671
Rodrigues, M.	392
Rodrigues-Filho, J.A.	343
Rodríguez, A.M.	634, 868
Rodríguez, I.	742, 747
Rodríguez-Barreira, S.	201
Rodríguez-Carías, A.A.	234
Rogers, M.E.	696
Rogosic, J.	362
Roig, S.	943
Rojas, C.	431
Romero, A.	95
Romero, O.	431
Rose-Fricker, C.A.	824
Rossiello, R.O.P.	683
Rostami, Y.	851
Rotar, I.	616
Rothauge, A.	508
Rotz, C.A.	378
Rouquette Jr., F.M.	158, 796, 848
Routley, R.	786
Ruane, D.J.	850
Rudstrom, M.V.	820
Ruggieri, A.C.	370, 528
Ruiz, J.C.	546
Ruiz, R.	896
Ruiz, T.	314, 444
Rumball, W.	272
Ruoss, E.	908
Russell, J.R.	268
Russelle, M.P.	692
Ryan, M.	570, 577, 695, 728
Ryan, S.	675
Rymph, S.J.	867
Sá Morais, J.	184
Sabiiti, E.N.	140, 282, 298, 299, 833
Sadeghi, A.A.	237
Sadeghi, S.H.R.	685
Saggar, S.	566
Saiga, S.	944
Saito, Y.	627
Sakanoue, S.	627
Salazar, F.J.	724
Saldanha, M.F.	730
Saldívar-Fitzmaurice, A.	371
Sale, P.W.G	805
Salem, M.	303
Salètes, S.	564

Saliendra, N.Z.	587, 599
Salis, L.	635
Salmon, L.	379, 912
Samples, D.H.	801
San Julián, R.	170, 171, 536, 537
San Miguel, A.	943
San'Thiago, L.R.L.	505
Sánchez, C.	111
Sánchez, T.	773
Sande, B.D.	319
Sanders, I.	877
Sanderson, M.A.	378, 767, 895
Sandral, G.A.	85
Sanford, P.	847
Sanna, F.	762
Sanon, H.O.	539
Santoni, M.M.	210
Santos, Í.P.A.	456, 703
Santos, P.M.	87, 210, 249, 702
Šantrůčková, H.	567
Sanz, A.	538, 896
Sanz, M.J.	587
Sapkota, M.	455
Sarath, G.	99, 115, 116
Sareen, S.	751
Sarmento, D.O.L.	520, 521
Saroff, C.	501, 768
Sarunaite, L.	358
Sato, K.	207
Satri, M.	290
Saul, G.R.	805, 823
Sauvenier, X.	766
Savin, F.A.	651
Sawamoto, T.	571
Sawyer, C.A.	432
Sbrissia, A.F.	520, 521
Scaglia, G.	464, 725
Scarpello, C.	363, 511, 677, 710
Schacht, W.H.	524, 889
Schäfer, M.	565
Schellenberg, M.P.	637
Schmidt, A.	382
Schmidt, P.	479
Schmitz, S.	629
Scholefield, D.	575
Schreurs, N.M.	195
Schulte, R.P.O.	570
Schultz, C.L.	308
Schultze-Kraft, R.	338, 343, 348, 382
Schuman, G.E.	590, 736
Schweitzer, S.H.	658
Scollan, N.D.	179, 180
Scott, J.M	861, 911
Scott, S.L.	182
Sebastià, M.T.	424, 425, 620
Seddaiu, G.	762
Seguin, P.	276
Sei, K.	626
Selge, A.	284, 936
Selles, F.	381

Seo, S.	281, 351, 478
Seutin, Y.	821
Seyfulina, R.R.	653
Shachak, M.	488
Shadnoush, G.H.R.	255, 256
Shah, S.B.	727
Shahriary, E.	437
Shahwahid H.O., M.	894
Shaik, S.A.	223
Shalloo, L.	163, 578, 579
Sharma, M.P.	455
Sharma, S.K.	84
Sharma, T.R.	84
Sharp, J.M.	355
Shawrang, P.	237
Sheehy, D.	880
Shelton, H.M.	324, 333, 544, 945, 950
Sherman, R.A.	801
Shewmaker, G.E.	594
Shibaeva, M.E.	729
Shimeles, B.	810
Shimojo, M.	342
Shin, C.N.	86
Shingu, H.	278
Shivoga, W.A.	689, 746
Shiyomi, M.	513, 626
Shouji, A.	146
Sibson, K.J.	786
Silberstein, R.	759
Silva, M.I.	392
Silveira, C.P.	718
Sima, N.	616
Šimek, M.	567
Simoes, F.C.	91
Simon, J-C.	212, 360
Simón, L.	773
Simon, T.	73
Simpson, I.H.	825
Simpson, R.J.	709, 912
Sims, P.L.	587
Sinclair, T.R.	94, 107
Singh, B.B.	317
Singh, R.	751
Singh, S.	84, 297
Singh, S.L.	776
Sinnaeve, G.	259
Sinsin, B.	442, 839
Sizer, E.	79
Sizer-Coverdale, E.	78
Skiba, U.	564
Skinner, R.H.	587, 592
Skládanka, J.	454, 618
Sledge, M.K.	69
Slepetiene, A.	593
Smailov, A.	843, 844
Smit, G.N.	508, 613, 771
Smit, H.J.	128, 186
Smith, C.J.	795
Smith, G.R.	83, 796
Smith, K.F.	80

Tornambè, G.	251	van Putten, B.	589
Torres, F.E.	102	van Wijk, A.J.P.	132
Torres, R.A.	439, 527	van Zyl, E.A.	466, 467
Torres, V.	444	Vance, G.F.	590
Touno, E.	278	Vanclooster, M.	766
Tovar, G.	111, 414	Vargiu, M.	635
Tracy, B.F.	450, 640	Vasileva, V.	425
Trapani, P.	250, 677, 710	Vasiljevic, S.	89, 269, 270, 701
Tremblay, G.F.	276	Vásquez, T.	414
Trifi-Farah, N.	93	Vázquez Yáñez, O.P.	906, 913
Trindade, H.	575, 717	Vázquez, J.I.	111
Trompf, J.P.	805	Vecchies, A.C.	80
Trompf, J.T	823	Velázquez Madrazo, P.A.	438, 676
Troncoso, D.	641	Venskutoniené, E.	764
Tsogoo, D.	120, 880	Vera, E.	111
Tsutsumi, M.	627	Vera, R.R.	927
Tuba, Z.	564, 587	Vera, U.M	111
Tuck, E.	213	Verbruggen, I.	397, 760
Tucker, W.A.	947	Verdenal, A.	202
Tuero, R.	95, 459	Verma, S.B.	587
Tukur, H.M.	391, 507	Verzignassi, J.R.	330, 339
Tunnell, S.J.	614	Vidican, R.	616
Tunney, H.	690, 720	Vidigal, M.C.	123, 730
Tura, I.	811	Vidrih, M.	636, 938
Turner, K.E.	869	Vidrih, T.	938
Turner, L.R.	205, 226	Viegas, J.	517, 865, 866
Turner, V.	664	Viiralt, R.	284, 936
Turyk, N.B.	692	Vilela, L.	110
Tweed, J.K.S.	179, 180	Viovy, N.	564
Twidwell, E.K.	947	Virgona, J.M.	892
Udval, G.	294	Visco, R.G.	583
Ueda, Y.	440	Vivas, N.	338
Umar, B.F.	777	Vogel, K.P.	99, 115, 116, 814
Umemura, K.	522	Volenec, J.J.	360
Undersander, D.J.	258	Volesky, J.D.	889
Ungar, E.D.	487, 488, 489, 621	Volpe, P.L.O.	206
Unruh Snyder, L.J.	869	Vozár, Ľ.	615
Urquiaga., S.	588	Vrotniakiene, V.	782
Usberti Jr., J.A.	62, 88, 91, 96, 104, 105	Vuichard, N.	564
Usberti, R.	62, 88, 91, 96, 206	Wachendorf, M.	378, 424, 502, 755
Vaillancourt, G.	799	Waghorn, G.C.	229, 230, 242, 243
Valadez-Moctezuma, E.	108	Wallace, M.	163, 756, 901
Valckx, J.	766	Waller, J.C.	306
Valencia, F.L.	573	Wallsten, J.	154
Valencia-Chin, E.	234	Wambugu, C.	319
Valenciaga, N.	314	Wandera, F.P.	336, 369
Valentim, J.F.	328, 329, 347	Wang, H.	775
Valentine, I	356, 885, 926	Wang, S.P.	303, 533
Valentini, R.	587	Wang, Y.	533, 887, 888
Valle, C.B.	102	Wang, Y.W.	366, 708, 712
van Amstel, A.	564	Wang, Z.L.	366, 712
van Burgel, M.A.	138	Ward, G.N.	389, 390, 680, 681
Van Cauwenbergh, N.	766	Warden, J.	85
van den Pol-van Dasselaar, A.	564	Warrington, B.G.	137, 189
van der Burgt, G.J.	749	Wastney, M.E.	881
van der Hoek, R.	809	Watanabe, O.	887
van Dorland, H.A.	563	Watkin, E.L.J.	312
van Eekeren, N.	749, 819	Watson, C.A.	877
van Heerden, J.M.	68	Watson, I.W.	556
van Niekerk, W.A.	215	Wattenbach, M.	564

Wauters, E.	766	Xiuwen, H.	119
Wayne, P.M.	555	Xu, C.	481
Webb, K.J.	213	Xuebing, Y.	117, 532
Wei, H.L.	737	Yabu, N.	515
Wei, X.H.	160	Yamada, T.	122
Weidinger, T.	564	Yamamoto, Y.	627
Weisenburger, R.	799	Yáñes, S.	444
Weiß, K.	476	Yang, L.	836
Weiss, R.	558	Yang, Q.	66, 67
Werner, J.C.	105	Yasuda, T.	626, 887
Wettstein, H.R.	563	Yates, R.J.	312
Wever, A.C.	181	Yayota, M.	238
Whalley, R.D.B.	825	Yin, X.	649
Whelan, J.	654	Yokota, H.	886, 887, 888
Whitbread, A.M.	864	Yonatan, R.	487, 488, 489
White, C.S.	364, 744	Yoon, S.H.	281
White, T.A.	153, 477	York, T.	759
Widdup, K.H.	81	Yoshida, N.	481, 792, 944
Wildbore, B.	915, 916	Young, J.	409, 410, 847
Wilkins, P.W.	65	Young, M.	556
Wilkins, R.J.	954	Yuxia, G.	117, 532
Wilkinson, N.S.	176	Zada, T.	487
Williams, A.	79, 80	Zahedi, G.H.	741
Williams, B.A.	228	Zander, K.	929
Williams, M.J.	335	Zanner, C.W.	524
Williams, M.L.	572	Zaragoza, J.E.	523
Williams, T.A.	74, 75, 76, 77, 78	Zartman, D.L.	798
Willms, W.D.	112, 113, 534	Zarza, A.	170, 171, 536
Wilmot, M.G.	759	Zborowki, A.	387
Wilson, D.R.	386	Zeddies, J.	565
Wilson, F.E.A.	227	Zentner, R.P.	381
Wilson, G.W.	800	Zhang, L.	599
Wilson, PR.	139	Zhang, W.	737
Winters, A.L.	241	Zhang, Y.	492, 649
Witkowska, I.	181	Zhao, M.	112, 113, 534, 775
Witzmann, S.	648	Zhao, Sh.F.	712
Wolters, L.	127	Zhong, Y.	708
Wood, J.T.	709	Zhou, H.	66, 67
Woodfield, D.R.	242, 243, 797, 81	Zhu, Y.	473
Woods, R.L.	788	Ziegler, K.	799
Woodward, S.L.	229, 230, 242, 243	Zohdi, M.	291
Wright, E.	553	Zollinger, R.	518
Wu, J.P.	160	Zopollatto, M.	479
Wu, L.	795	Zoungrana-Kabore, C.	539
Wylie, B.K.	587, 599	Zujovic, M.	638, 639
Xiao, G..	365	Żurek, G.	550
Xie, J.	826	Zwierzykowski, Z.	121
Xinmin, X.	118		